German Technical Dictionary

Wörterbuch für Technik Englisch

Langenscheidt

Routledge

German Technical Dictionary
2nd Edition

Wörterbuch für Technik Englisch
2. Auflage

Volume One
German – English

Band 1
Deutsch – Englisch

London and New York

This edition first published 2004
by Routledge
11 New Fetter Lane, London EC4P 4EE

Simultaneously published in the USA and Canada
by Routledge
29 West 35th Street, New York, NY 10001

Routledge is an imprint of the Taylor & Francis Group

© 1996 Routledge and © 2004 Langenscheidt Fachverlag GmbH,
München, und Routledge Ltd./Taylor & Francis Books Ltd., London

All rights reserved. No part of this book may be reprinted or reproduced or utilised in any form or by any electronic, mechanical, or other means, now known or hereafter invented, including photocopying and recording, or in any information storage or retrieval system, without permission in writing from the publishers.

British Library Cataloguing in Publication Data
A catalogue record for this book is available from the British Library

Library of Congress Cataloging in Publication Data

ISBN 0-415-33586-8

Inhalt / Contents

Vorwort	VII	Preface	VIII
Aufbau und Anordnung der Einträge	IX	Features of the dictionary	IX
Hinweise zur Benutzung des Wörterbuches	XI	Using the dictionary	XIV
Im Wörterbuch verwendete Abkürzungen	XVII	Abbreviations used in this dictionary	XVII
Wörterbuch Deutsch-Englisch	1	German-English Dictionary	1
Anhänge		Appendices	
Abkürzungen	911	Abbreviations	911
Umrechnungstabellen	929	Conversion tables	929
Längenmaße	929	Length	929
Flächenmaße	929	Area	929
Raummaße	930	Volume	930
Winkelmaße	930	Angle	930
Zeit	931	Time	931
Masse	931	Mass	931
Kraft	932	Force	932
Leistung	932	Power	932
Energie, Arbeit, Wärme	933	Energy, work, heat	933
Druck	934	Pressure	934
Magnetfluss	935	Magnetic flux	935
Magnetische Flussdichte	935	Magnetic flux density	935
Magneto-EMK	935	Magnetomotive force	935
Magnetische Feldstärke	936	Magnetic field strength	936
Ausleuchtung	936	Illumination	936
Leuchtdichte	936	Luminance	936
Chemische Elemente	937	Chemical elements	937

Vorwort

Das vorliegende Werk basiert auf der 1996 erschienenen 1. Auflage des Routledge-Langenscheidt Fachwörterbuches *Technik, Deutsch-Englisch*. Die dynamisch sich fortentwickelnden Fachsprachen machten eine gründliche Überarbeitung des Werkes und eine substanzielle Erweiterung und Modernisierung der darin enthaltenen Termini erforderlich.

Der vorhandene Wortbestand wurde in den einzelnen Fachgebieten auf Genauigkeit, Richtigkeit und Vollständigkeit überprüft und um über 5.000 aktuelle Begriffe aus 32 Fachgebieten erweitert. Stark ausgeweitet wurden die Bereiche Telekommunikation, neue Medien und Internet, Elektrotechnik und Elektronik. Andererseits wurden veraltete Begriffe systematisch eliminiert.

Das daraus entstandene Werk deckt alle Bereiche der modernen Technik und des zugrundeliegenden naturwissenschaftlichen Wissens ab. Zusätzlich zu einer breiten Grundlage technischer Ausdrücke aus traditionellen Bereichen wie etwa Maschinenbau, Bauindustrie, Elektrotechnik und Elektronik enthält dieses Wörterbuch auch Vokabular aus neuen und hochaktuellen Fachgebieten wie etwa Sicherheitstechnik und Mobiltelefonie.

Die einzelnen Fachgebiete sind proportional zu ihrer Anwendungshäufigkeit vertreten, sodass also ein etabliertes und umfangreiches Gebiet wie der Maschinenbau etwa 8.000 Stichwörter aufweist, während ein Gebiet, in dem sich das Vokabular im Augenblick noch stark in der Entwicklung befindet, wie etwa nichtfossile Energiequellen, dagegen mit etwa 800 Stichwörtern vertreten ist.

Es wurde sorgfältig darauf geachtet, das wichtigste technische Grundvokabular vollständig aufzunehmen, auch wenn einige Einträge bereits in allgemeinsprachlichen Wörterbüchern zu finden sind. Obwohl manchmal in bestimmten Fachgebieten auch andere Übersetzungsvarianten möglich wären, haben wir stets den in der Fachwelt gängigsten Terminus angegeben.

Das neue Wortgut wurde von führenden Experten aus den wichtigsten Fachgebieten erarbeitet. Aufgrund seiner langjährigen lexikografischen Erfahrung als Herausgeber von verschiedenen technischen Fachwörterbüchern brachte Herr Dipl-Ök. Karl-Heinz Radde den nötigen fachlichen Überblick mit, um die Neubearbeitung zu koordinieren.

Mit diesem Wörterbuch wurde somit in erster Linie ein hochaktuelles Nachschlagewerk für Studenten aller Universitäten, Fachhochschulen und Fortbildungseinrichtungen mit technischer Ausrichtung, für professionelle Übersetzer und Dolmetscher geschaffen, aber auch für Fachjournalisten, Juristen, Geschäftsleute und alle weiteren Fachleute, die einen verlässlichen, praxistauglichen terminologischen Ratgeber benötigen.

Der Verlag bittet die Nutzer dieses Wörterbuchs um Nachsicht für etwaige verbliebene Fehler und ist für sachdienliche Hinweise und konstruktive Kritik, die einer weiteren Verbesserung dieses Werks dienen können, stets dankbar. Bitte senden Sie diese an den Langenscheidt Fachverlag GmbH, Postfach 40 11 20, D-80711 München.

Der Verlag

Preface

This dictionary is based on the 1st edition of Routledge-Langenscheidt's technical dictionary *Technik, Deutsch-Englisch*, which appeared in 1996. The continuing dynamic development of technical terminology made a thorough revision of the work essential. This has meant a substantial expansion and modernization of the terms it contained.

The existing lexical material in the individual technical fields was checked for accuracy, correctness and completeness, and over 5,000 current terms from 32 specialized areas have been added. The fields of telecommunications, new media/Internet, electrical engineering and electronics were greatly expanded, while obsolete terms were weeded out.

The resulting dictionary covers all fields of modern technology and the science that underlies them. In addition to a broad base of technical expressions from such traditional fields as engineering, construction, electrical engineering and electronics, this dictionary also contains vocabulary from new and topical fields such as security technology and mobile telephony.

The individual areas are represented in proportion to their frequency of application, so that an established and extensive field like engineering has some 8,000 headwords, whereas a field in which the vocabulary is still undergoing rapid development, such as non-fossil sources of energy, is represented by about 800 headwords.

Care has been taken to include the basic technical vocabulary in its entirety, even though some entries may be found in general dictionaries as well. Although alternative translations would have been possible in certain specialized areas, we have always included the term most commonly used in specialist circles.

The new material was processed by leading experts in the most important specialized areas. In view of his many years of lexicographical experience as editor of various technical dictionaries, Mr. Karl-Heinz Radde provided the expertise necessary to coordinate the work of revision.

This dictionary is thus mainly intended as an up-to-the-minute reference work for students of all universities, technical colleges and advanced training institutions with a technical slant, for professional translators and interpreters, technical journalists, lawyers, businesspeople and all other specialists who need a reliable and practical guide to terminology.

The Publishers crave the indulgence of users of this dictionary for any errors that may remain, and welcome any useful comments or constructive criticism which might help to improve the work. Please send your comments to Langenscheidt Fachverlag GmbH, Postfach 40 11 20, D-80711 München.

The Publishers

Aufbau und Anordnung der Einträge / Features of the dictionary

Das folgende Textbeispiel illustriert Aufbau und Anordnung der Einträge. Weitere Erläuterungen und Hinweise zur Benutzung befinden sich auf den Seiten XI–XIII.

The main features of the dictionary are highlighted in the text extracts on the opposite page. For a more detailed explanation of each of these features and information on how to get the most out of the dictionary, see pages XIV–XVI.

X

Sachgebietskürzel in alphabetischer Reihenfolge helfen beim Finden der korrekten Übersetzung

Britische und amerikanische Varianten werden voll ausgeschrieben und entsprechend gekennzeichnet

Die Angabe von Zusammenhängen ergänzt die gegebenen Informationen und unterstützt die Suche nach dem passenden Übersetzungsäquivalent

Deutsche Einträge sind streng alphabetisch geordnet

Sowohl für den deutschen Eintrag wie für die englische Übersetzung werden bei Abkürzungen Querverweise auf die entsprechende Vollform gegeben

Runge *f* 1. <Eisenbahn> stanchion, upright; 2. <Fertig> upright; 3. <Kfztech> post *(zwischen Wagenseite und Radachse)*
Rungenpalette *f* <Trans> post pallet
Run-up-Bereich *m* <Lufttrans> run-up area
Runzelbildung *f* <Kunststoff> silking
runzelige Oberfläche *f* <Ker & Glas> cockled surface
runzelige Oberfläche *f* **durch ungleichmäßige Kühlung** <Ker & Glas> *(BE)* chill mark, *(AE)* chill wrinkle
Runzelkorn *n* <Foto> reticulation *(einer Emulsion)*
Runzellack *m* <Kunststoff> wrinkle paint
Rupertstropfen *m* <Ker & Glas> Prince Rupert drop
rupfen *v* <Lebensmittel> pluck *(Geflügel)*
Rupfen *n* 1. <Papier> picking; 2. <Textil> hessian
rupfende Kupplung *f* <Kfztech> grabbing clutch
Rupffestigkeit *f* <Papier> picking resistance
Rüsche *f* <Textil> ruffle
Rüschung *f* <Textil> gathering
Rushhour *f* <Trans> peak hour, rush hour
Ruß *m* 1. <Ker & Glas> soot; 2. <Kunststoff> carbon black
Russell-Saunders-Kopplung *f (L-S-Kopplung)* <Kerntech> Russell-Saunders coupling *(l-s coupling)*
Rußpunkt *m* <Maschinen> smoke point
Rußschwarz *n* <Erdöl> carbon black *(Petrochemie)*
Rüsteisen *n* <Wassertrans> chain plate *(Segeln)*
rüsten *v* <Maschinen> set-up
Rüsten *n* <Maschinen> setting up *(einer Maschine)*
Rüster *m* <Wassertrans> elm *(Bauholz)*
Rüstzeit *f* 1. <Comp & DV> set-up time; 2. <Druck> makeready time *(einer Druckmaschine)*; 3. <Fertig> set-up time *(Maschine)*; 4. <Textil> set-up time
Ruthenium *n (Ru)* <Chemie> ruthenium *(Ru)*
Ruthenium... <Chemie> ruthenic
Rutherfordium *n (Rf)* <Chemie> rutherfordium *(Rf)*
Rutherford-Streuung *f* <Phys> Rutherford scattering
Rutsche *f* 1. <Bau, Ker & Glas> chute; 2. <Maschinen, Mechan> chute, slide
Rutschen *n* 1. <Bau> slippage; 2. <Elektrotech> slip; 3. <Lufttrans, Trans> skidding
rutschfest *adj* 1. <Bau, Sicherheit> nonslip, slip-resistant; 2. <Trans> nonskidding; 3. <Wassertrans> nonslip
rutschfester Boden *m* <Bau, Sicherheit> non-skid floor
rutschfestes Schuhwerk *n* <Sicherheit> nonslip footwear
rutschfestes Sicherheitsschuhwerk *n* <Sicherheit> nonslip safety footwear, slip-resistant safety footwear, slip-resistant safety shoes
rutschfrei *adj* <Trans> nonskidding
Rutschkupplung *f* <Maschinen> slip clutch
rutschsicherer Bodenbelag *m* <Sicherheit> antislip floor covering
Rutschsicherung *f (ASD)* <Kfztech> antiskid device, antislip device, ASD *(Bremsanlage)*
Rutschung *f* <Kohlen> slide
Rüttelbeton *m* <Bau> vibrated concrete
Rüttelformmaschine *f* <Fertig> *(BE)* bumper
rütteln *v* <Kfztech> vibrate
Rütteln *n* <Trans> jolting
Rüttelplatte *f* <Abfall> tumbling station
Rüttelrost *m* <Fertig> shake-out
Rüttelschaffußwalze *f* <Bau> vibrating sheepsfoot roller
Rüttelsieb *n* <Chemie> vibrating screen
Rüttelsiebrost *n* <Kohlen> vibrating grizzly
Rütteltisch *m* 1. <Maschinen> concussion table; 2. <Verpack> jarring table
rüttelverdichten *v* <Bau> vibrate *(Beton)*
Rüttler *m* <Bau> vibrator
Ruzahl *f* <Heiz & Kälte> soot number

Subject-area labels given in alphabetical order show appropriate translation

British-English and American-English variants are given in full and are labelled accordingly

Contexts give supplementary information to help locate the right translation

German terms are ordered in strict letter-by-letter order

Cross-references between abbreviations and full forms are shown for both the German term and the English translation

Hinweise für die Benutzung des Wörterbuches

Reihenfolge der Anordnung

Alle Einträge sind alphabetisch angeordnet. Zusammengesetzte Einträge erscheinen unter ihrem Basiswort. Bei Wortzusammensetzungen werden erst jeweils die einzelnen Wörter abgehandelt. Dabei erscheinen getrennt geschriebene Wörter in einem Wortnest. Daran anschließend folgen die zusammengeschriebenen Begriffe.
Zum Beispiel:

Destillation *f* 1. <Erdöl> distillation *(Raffinerietechnik)*; 2. <Thermod> distillation
Destillation *f* **im Vakuum** <Chemtech> vacuum distillation
Destillation *f* **mittels Sonnenenergie** <Nichtfoss Energ> solar distillation
Destillationsanlage *f* <Chemtech> still
Destillationsapparat *m* 1. <Chemtech> distiller, distilling apparatus, still; 2. <Labor> still

Stoppliste

Alle Stichwörter sind nach ihrem ersten Element, dem Basiswort, angeordnet, es sei denn, dieses Element ist ein Artikel, eine Präposition, eine Konjunktion, ein Pronomen oder ein anderes ausgegliedertes Wort.
In solchen Fällen erscheint das Stichwort als erstes gültiges Element; die eigentlich voranzustellenden Bestandteile (hinter einem Schrägstrich stehend) werden wie nachgestellte Wörter behandelt.
Zum Beispiel:

Abquetschfläche/mit <Fertig> landed
Austreiberlappen/mit <Fertig> tanged

In Einträgen mit zusammengesetzten Stichwörtern werden bestimmte und unbestimmte Artikel, Pronomen, Präpositionen und Konjunktionen bei der alphabetischen Sortierung berücksichtigt:
Zum Beispiel:

Übertragung *f* 1. <Comp & DV, Elektriz, Kontroll, Maschinen, Mechan, Phys> transfer, translation, transmission; 2. <Telekom> transfer *(Information)*; transmission
Übertragung *f* **aus dem Speicher** <Comp & DV> copy-out
Übertragung *f* **im Klartext** <Telekom> clear transmission, plain text transmission
Übertragung *f* **in den Speicher** <Comp & DV> copy-in
Übertragung *f* **innerhalb der Sichtweite** <Telekom> line-of-sight transmission, LOS transmission
Übertragung *f* **mit Geheimhaltung** <Telekom> secure transmission

Zusammensetzungen aus dekliniertem Adjektiv und Substantiv sind unter den jeweiligen Adjektiven entsprechend ihrer Endung aufgeführt.
Zum Beispiel:

segmentierte Abtastung *f* <Fernseh> segmented scanning
segmentierte Aufzeichnung *f* <Fernseh> segmented recording
segmentiertes Multiprozessorsystem *n* <Telekom> segmented multiprocessor system

Zusammensetzungen aus Substantiv und Verb (bzw. Partizip) sind unter der Grundform des Substantivs aufgeführt.
Zum Beispiel:

Gegenlicht beleuchten v/im <Foto> backlight
Handvorschub ausfräsen v/mit <Fertig> rout *(Blech)*
Segelstellung bringen v/auf <Lufttrans> feather *(Luftschraube)*

Reflexive Verben werden unter ihrer Grundform alphabetisiert:
Zum Beispiel:

abmelden v/sich 1. <Comp & DV> log off, log out; sign off *(vom System)*; 2. <Mobilkom> sign off
nähern v/sich 1. <Math> approximate to *(einer Lösung oder einem Grenzwert)*; 2. <Wassertrans> approach *(Navigation)*

Kleingeschriebene und großgeschriebene Einträge werden separat aufgelistet. Dabei werden kleingeschriebene Zeichen vor großgeschriebenen alphabetisiert.
Zum Beispiel:

da *(Deka...)* <Labor> da *(deca...)*
D/A *(Digital-Analog-...)* <Aufnahme, Comp & DV, Elektronik, Labor, Telekom> D/A *(digital-analog)*
DA *(direkter Zugriff)* <Comp & DV> DA *(direct access)*

Einträge, die mit Zahlen oder Sonderzeichen beginnen, stehen am Wörterbuchanfang. Zahlen oder Symbole im Wortinneren spielen dagegen bei der alphabetischen Reihenfolge keine Rolle.
Zum Beispiel:

β *(Phasenkonstante)* <Akustik, Elektriz> β *(phase constant)*
ϵ 1. <Hydraul> *(kinematische Wirbelzähigkeit)* ϵ *(kinematic eddy viscosity)*; 2. <Phys> *(durchschnittliche kinetische Molekularenergie)* ϵ *(average molecular kinetic energy)*
4-D-Verstärkung *f* <Raumfahrt> 4-D reinforcement
3-Stufen-Implementierung *f* <Comp & DV> three-tier implementation
90°-Verschiebung *f* <Elektronik> quadrature phase shifting *(Phasenverschiebung)*
a 1. <Akustik> *(akustische Absorption)* a *(total acoustic absorption)*; 2. <Metrol> *(Ar)* a *(are)*

Umlaute werden wie a, o, u eingeordnet. Berücksichtigt werden diese nur bei gleichlautenden Begriffen ohne Umlaut. In diesem Fall erscheint der Begriff mit dem Umlaut hinter dem Begriff ohne Umlaut.
Zum Beispiel:

loten *v* <Funktech> sound *(Navigation)*
Loten *n* <Bau> plumbing
löten *v* 1. <Fertig> solder, sweat; 2. <Funktech, Maschinen> solder
Löten *n* <Elektriz, Fertig, Maschinen> soldering

Homographen

Alle Einträge sind mit einem Label versehen, das die Wortklassenzugehörigkeit angibt. Eine vollständige Liste dieser Labels befindet sich auf Seite XVII.
Einträge mit gemeinsamem Basiswort, jedoch unterschiedlicher Wortklassenzugehörigkeit, erhalten für jede Wortklasse einen eigenen Eintrag. Die Einträge sind in der Reihenfolge Verb, Adjektiv, Adverb, Substantiv aufgeführt.
Zum Beispiel:

ziehen v 1. <Bau> drive *(Schrauben)*; 2. <Bau> draw; 3. <Fertig> cut *(Nuten)*; 4. <Papier> draw, pull; 5. <Trans> haul
ziehen lassen v <Lebensmittel> infuse
ziehen v **und übergeben** v <Comp & DV> drag and drop
Ziehen n 1. <Kunststoff> drawing; 2. <Maschinen> pulling; 3. <Papier> draw, pull; 4. <Trans> haulage
Ziehen n **eines Nagels** <Maschinen> pulling out of a nail

Reihenfolge der Übersetzungen

Auf jedes Stichwort der Ausgangssprache folgen ein oder mehrere Labels, die das technische Fachgebiet angeben, in dem das Wort benutzt wird. Eine vollständige Liste dieser Labels mit Erklärungen befindet sich auf den Seiten XVII–XVIII.
Wenn dasselbe Stichwort in mehr als einem technischen Fachgebiet benutzt wird, so werden entsprechend mehrere Labels angegeben. Sie werden stets in alphabetischer Reihenfolge aufgeführt.
Falls die angegebene Übersetzung in mehr als einem Fachgebiet benutzt wird, so folgt die Übersetzung jeweils im Anschluss an die entsprechenden Labels.
Zum Beispiel:

Netzmittel n <Foto, Kunststoff, Lebensmittel, Meerschmutz, Phys, Textil> wetting agent

Hat ein Stichwort verschiedene, von seinen jeweiligen Fachgebieten abhängige Übersetzungen, so ist die zutreffende Übersetzung jeweils nach dem entsprechenden Kürzel zu finden.
Zum Beispiel:

ablassen v 1. <Fertig> bleed *(Flüssigkeit)*; 2. <Maschinen> bleed, drain *(Flüssigkeit)*; deflate *(Luft aus Reifen)*; 3. <Raumfahrt> jettison *(Raumschiff)*; 4. <Sicherheit> lower; 5. <Umweltschmutz> discharge *(Wasser)*

Veranschaulichende Phrasen folgen unmittelbar auf das Stichwort und sind durch einen fett formatierten Punkt (•) gekennzeichnet.
Zum Beispiel:

Riemen m 1. <Kfztech> belt; 2. <Maschinen> band, belt; 3. <Mechan, Papier> belt; 4. <Wassertrans> oar *(Rudern)*
• **die Riemen einlegen** <Wassertrans> ship the oars *(Rudern)* • **einen Riemen ausrücken** <Maschinen> throw a belt off • **einen Riemen spannen** <Kfztech, Maschinen> tighten a belt

Zusatzinformationen

Bei Übersetzungen werden alle deutschen Substantive mit Geschlechtsangabe versehen. In vielen Fällen wird über das Stichwort noch zusätzliche Information gegeben, die über die Benutzung des Wortes Aufschluss gibt. Solche Informationen zum Benutzungszusammenhang können verschiedene Formen annehmen:

(a) bei einem Verb ein typisches Subjekt oder Objekt:

flachspritzen v <Fertig> quench *(Kunststoffe)*
flechten v <Bau> bend *(Bewehrungsstahl)*

(b) typische Substantive, die mit einem bestimmten Adjektiv verwendet werden:

maschinengestrichen adj <Verpack> machine-coated *(Papier)*
mehrfach ungesättigt adj <Chemie, Lebensmittel> polyunsaturated *(Fettsäure)*

(c) Wörter, die für Substantive einen typischen Bezug erläutern:

Fassungsvermögen n 1. <Bau, Elektriz> capacitance, capacity *(Behälter)*; 2. <Fertig> holding capacity *(von Zuführvorrichtungen)*; 3. <Heiz & Kälte> freezer capacity *(Tiefkühlgerät)*; 4. <Maschinen> capacity; 5. <Wassertrans> volume *(Maßeinheit)*

(d) Informationen, die das Fachgebiet noch weiter eingrenzen:

Niccolit m <Chemie> niccolite, nickel arsenide *(Mineralogie)*
Ölschiefer m <Erdöl> oil shale *(Erdölgeologie)*

(e) Umschreibungen oder ungefähre Äquivalente:

Fluttor n <Wassertrans> floodgate *(Schleuse)*

Wenn innerhalb eines Fachgebiets verschiedene Übersetzungen für ein Stichwort möglich sind, so sollen hier die Zusatzinformationen anzeigen, welche im jeweiligen Zusammenhang die zutreffendste ist. Die Zusatzinformationen folgen jeweils der/den Übersetzung/en, auf die sie sich beziehen. Beim Vorhandensein mehrerer Zusatzinformationen verdeutlichen Semikola den jeweiligen inhaltlichen Zusammenhang.
Zum Beispiel:

schleifen v 1. <Anstrich, Bau> grind; 2. <Fertig> cut *(Edelstein)*; sandpaper *(Holz)*; grind *(Werkzeug)*; 3. <Maschinen> grind, sharpen

Querverweise

Geographische Varianten der Übersetzungen, sowohl orthographischer als auch lexikalischer Art, werden immer angegeben und voll ausgeschrieben:
Zum Beispiel:

Luftbetankungssonde f <Lufttrans> *(AE)* flight refueling probe, *(BE)* flight refuelling probe, *(AE)* in-flight refueling probe, *(BE)* in-flight refuelling probe

Im Hauptteil des Wörterbuches sind auch Abkürzungen und ihre ausgeschriebenen Formen in streng alphabetischer Reihenfolge angeordnet. Bei jedem Eintrag wird die vollständige Information – inklusive Übersetzungen und Querverweisen zur Vollform respektive Abkürzung – aufgeführt.

KARS *(kohärente Antistokes-Raman-Streuung)* <Strahlphys, Wellphys> CARS *(coherent anti-Stokes Raman scattering)*

Hat eine Abkürzung je nach Fachgebiet unterschiedlich ausgeschriebene Formen, so erscheinen diese jeweils nach dem oder den Fachgebietskürzeln.

Zum Beispiel:

MP 1. <Comp & DV, Elektriz, Maschinen> *(Mikroprozessor)* MP *(microprocessor)*; 2. <Verpack> *(Metallpapier)* MP *(metallic paper)*

Außerdem sind Abkürzungen im Anhang des Bandes als separate Liste aufgeführt. Dies erleichtert die Suche, wenn die genaue Form einer Abkürzung nicht bekannt ist.

Using the dictionary

Placement of terms

All entries are in alphabetical order.
Compound entries appear under their base word. Words that also appear in word compounds are treated in their individual form first. Words written separately appear in a block of compounds. Terms written in one word are given as the next entry.
For example:

Destillation f 1. <Erdöl> distillation *(Raffinerietechnik)*; 2. <Thermod> distillation
Destillation f **im Vakuum** <Chemtech> vacuum distillation
Destillation f **mittels Sonnenenergie** <Nichtfoss Energ> solar distillation
Destillationsanlage f <Chemtech> still
Destillationsapparat m 1. <Chemtech> distiller, distilling apparatus, still; 2. <Labor> still

Stoplists

All headwords are arranged according to their first element, the base word, unless this element is an article, a preposition, a conjunction, a pronoun, or other purely functional word.
In such cases the headword appears as the first valid element; the element that actually precedes it is given after an oblique (/).
For example:

Abquetschfläche/mit <Fertig> landed
Austreiberlappen/mit <Fertig> tanged

In entries with compound headwords, definite and indefinite articles, pronouns, prepositions and conjunctions are taken into account in determining the alphabetical order:
For example:

Übertragung f 1. <Comp & DV, Elektriz, Kontroll, Maschinen, Mechan, Phys> transfer, translation, transmission; 2. <Telekom> transfer *(Information)*; transmission
Übertragung f **aus dem Speicher** <Comp & DV> copy-out
Übertragung f **im Klartext** <Telekom> clear transmission, plain text transmission
Übertragung f **in den Speicher** <Comp & DV> copy-in
Übertragung f **innerhalb der Sichtweite** <Telekom> line-of-sight transmission, LOS transmission
Übertragung f **mit Geheimhaltung** <Telekom> secure transmission

Compounds consisting of a declined adjective and a noun are listed under the adjective in accordance with ist ending. For example:

segmentierte Abtastung f <Fernseh> segmented scanning
segmentierte Aufzeichnung f <Fernseh> segmented recording
segmentiertes Multiprozessorsystem n <Telekom> segmented multiprocessor system

Compounds consisting of a noun and a verb (or a participle) are listed under the noun's base form.
For example:

Gegenlicht beleuchten v/im <Foto> backlight
Handvorschub ausfräsen v/mit <Fertig> rout *(Blech)*
Segelstellung bringen v/auf <Lufttrans> feather *(Luftschraube)*

Reflexive verbs are listed under their base form. For example:

abmelden v/sich 1. <Comp & DV> log off, log out; sign off *(vom System)*; 2. <Mobilkom> sign off
nähern v/sich 1. <Math> approximate to *(einer Lösung oder einem Grenzwert)*; 2. <Wassertrans> approach *(Navigation)*

Entries beginning with the same small and capital letters are listed separately, those beginning with a small letter being given first.
For example:

da (Deka...) <Labor> da *(deca...)*
D/A (Digital-Analog-...) <Aufnahme, Comp & DV, Elektronik, Labor, Telekom> D/A *(digital-analog)*
DA (direkter Zugriff) <Comp & DV> DA *(direct access)*

Entries beginning with numbers or special characters are given at the beginning of the dictionary. Numbers or symbols occurring inside a word do not affect the alphabetical order.
For example:

β *(Phasenkonstante)* <Akustik, Elektriz> β *(phase constant)*
ε 1. <Hydraul> *(kinematische Wirbelzähigkeit)* ε *(kinematic eddy viscosity)*; 2. <Phys> *(durchschnittliche kinetische Molekularenergie)* ε *(average molecular kinetic energy)*
4-D-Verstärkung f <Raumfahrt> 4-D reinforcement
3-Stufen-Implementierung f <Comp & DV> three-tier implementation
90°-Verschiebung f <Elektronik> quadrature phase shifting *(Phasenverschiebung)*
a 1. <Akustik> *(akustische Absorption)* a *(total acoustic absorption)*; 2. <Metrol> *(Ar)* a *(are)*

Umlaute are treated like a, o, u in the alphabetical order. They receive no special consideration. However, if an identical term exists with no umlaut, then the term featuring an umlaut appears behind the one without an umlaut. For example:

loten v <Funktech> sound *(Navigation)*
Loten n <Bau> plumbing
löten v 1. <Fertig> solder, sweat; 2. <Funktech, Maschinen> solder
Löten n <Elektriz, Fertig, Maschinen> soldering

Homographs

All entries have a label indicating what class of word they belong to. A complete list of these labels may be found on page XVII.
Entries with a common base word but belonging to different word classes receive a separate entry for each. The entries are given in the sequence: verb, adjective, adverb, noun.
For example:

ziehen v 1. <Bau> drive *(Schrauben)*; 2. <Bau> draw; 3. <Fertig> cut *(Nuten)*; 4. <Papier> draw, pull; 5. <Trans> haul
ziehen lassen v <Lebensmittel> infuse
ziehen v **und übergeben** v <Comp & DV> drag and drop
Ziehen n 1. <Kunststoff> drawing; 2. <Maschinen> pulling; 3. <Papier> draw, pull; 4. <Trans> haulage
Ziehen n **eines Nagels** <Maschinen> pulling out of a nail

Ordering of translations

Every term is accompanied by one or more labels indicating the technological area in which it is used. For a complete list of these labels and their expansions, please see pages XVII–XVIII.
Where the same term is used in more than one technological area, multiple labels are given as appropriate. These labels appear in alphabetical order. Where a term has the same translation in more than one technological area, this translation is given after the sequence of labels.
For example:

Netzmittel n <Foto, Kunststoff, Lebensmittel, Meerschmutz, Phys, Textil> wetting agent

When a term has different translations according to the technological area in which it is used, the appropriate translation is given after each label or set of labels.
For example:

ablassen v 1. <Fertig> bleed *(Flüssigkeit)*; 2. <Maschinen> bleed, drain *(Flüssigkeit)*; deflate *(Luft aus Reifen)*; 3. <Raumfahrt> jettison *(Raumschiff)*; 4. <Sicherheit> lower; 5. <Umweltschmutz> discharge *(Wasser)*

Illustrative phrases follow immediately after the headword and are introduced by a boldface bullet (•).
For example:

Riemen m 1. <Kfztech> belt; 2. <Maschinen> band, belt; 3. <Mechan, Papier> belt; 4. <Wassertrans> oar *(Rudern)*
• **die Riemen einlegen** <Wassertrans> ship the oars *(Rudern)* • **einen Riemen ausrücken** <Maschinen> throw a belt off • **einen Riemen spannen** <Kfztech, Maschinen> tighten a belt

Supplementary information

The gender of all the German nouns used in the translations is given. In many cases additional data is given about a term in order to show how it is used. Such contextual information can be:

(a) the typical subject or object of a verb, for example:

flachspritzen v <Fertig> quench *(Kunststoffe)*
flechten v <Bau> bend *(Bewehrungsstahl)*

(b) typical nouns used with an adjective, for example:

maschinengestrichen adj <Verpack> machine-coated *(Papier)*
mehrfach ungesättigt adj <Chemie, Lebensmittel> polyunsaturated *(Fettsäure)*

(c) words indicating the reference of a noun, for example:

Fassungsvermögen n 1. <Bau, Elektriz> capacitance, capacity *(Behälter)*; 2. <Fertig> holding capacity *(von Zuführvorrichtungen)*; 3. <Heiz & Kälte> freezer capacity *(Tiefkühlgerät)*; 4. <Maschinen> capacity; 5. <Wassertrans> volume *(Maßeinheit)*

(d) information which supplements the subject-area label, for example:

Niccolit m <Chemie> niccolite, nickel arsenide *(Mineralogie)*
Ölschiefer m <Erdöl> oil shale *(Erdölgeologie)*

(e) a paraphrase or broad equivalent, for example:

Fluttor n <Wassertrans> floodgate *(Schleuse)*

If a headword in a given specialist area can have different translations, the additional information will indicate which is the most suitable translation in the relevant context. The additional information follows the translation(s) to which it refers. If several items of additional information are given, the contexts are separated by semicolons.
For example:

schleifen v 1. <Anstrich, Bau> grind; 2. <Fertig> cut *(Edelstein)*; sandpaper *(Holz)*; grind *(Werkzeug)*; 3. <Maschinen> grind, sharpen

Cross-references

Geographical variants, both spelling and lexical, are given in full when they are translations.
For example:

Luftbetankungssonde f <Lufttrans> *(AE)* flight refueling probe, *(BE)* flight refuelling probe, *(AE)* in-flight refueling probe, *(BE)* in-flight refuelling probe

Both abbreviations and their full forms are entered in the main body of the dictionary in alphabetical sequence. Full information – including translations and cross-references to the full form or abbreviation as appropriate – is given at each entry. For example:

KARS *(kohärente Antistokes-Raman-Streuung)* <Strahlphys, Wellphys> CARS *(coherent anti-Stokes Raman scattering)*

If the same abbreviation has different meanings in different specialist areas, the relevant form is given after the abbreviation for the relevant specialist area(s).
For example:

MP 1. <Comp & DV, Elektriz, Maschinen> *(Mikroprozessor)* MP *(microprocessor)*; 2. <Verpack> *(Metallpapier)* MP *(metallic paper)*

Abbreviations are also listed in a separate alphabetical sequence at the back of this volume to allow browsing in cases where the exact form of the abbreviation is not known.

Im Wörterbuch verwendete Abkürzungen / Abbreviations used in this dictionary

Wortarten / Parts of speech

abbr	Abkürzung	abbreviation
adj	Adjektiv	adjective
adv	Adverb	adverb
f	Femininum	feminine
fpl	Femininum Plural	feminine plural
m	Maskulinum	masculine
mpl	Maskulinum Plural	masculine plural
n	Neutrum	neuter
npl	Neutrum Plural	neuter plural
v	Verb	verb

Geographische Kürzel / Geographic codes

AE	Amerikanisches Englisch	American English
BE	Britisches Englisch	British English

Fachgebietskürzel / Subject-area labels

Abfall	Abfallwirtschaft	Waste Management
Akustik	Akustik	Acoustics
Anstrich	Anstrichtechnik	Coatings Technology
Aufnahme	Aufnahmetechnik	Recording Engineering
Bau	Bauwesen	Construction
Chemie	Chemie	Chemistry
Chemtech	Chemotechnik	Chemical Engineering
Comp & DV	Computertechnik & Datenverarbeitung	Computer Technology & Data Processing
Druck	Druckereiwesen	Printing
Eisenbahn	Eisenbahnbau	Railway Engineering
Elektriz	Elektrizität	Electricity
Elektronik	Elektronik	Electronics
Elektrotech	Elektrotechnik	Electrical Engineering
Erdöl	Erdöltechnologie	Petroleum Technology
Ergon	Ergonomie	Ergonomics
Fernseh	Fernsehtechnik	Television
Fertig	Fertigungstechnik	Production Engineering
Foto	Fotografie	Photography
Funkort	Funkortung	Radio Location
Funktech	Funktechnik	Radio Technology
Geom	Geometrie	Geometry
Gerät	Geräte & Instrumente	Instrumentation
Heiz & Kälte	Heizungs- & Kältetechnik	Heating & Refrigeration
Hydraul	Hydraulische Anlagen	Hydraulic Equipment
Ker & Glas	Keramik & Glas	Ceramics & Glass
Kerntech	Kerntechnik	Nuclear Technology
Kfztech	Kraftfahrzeugtechnik	Automotive Engineering
Kohlen	Kohlentechnik	Coal Technology
Konstzeich	Konstruktionszeichnung	Engineering Drawing
Kontroll	Kontrolltechnik	Control Technology
Künstl Int	Künstliche Intelligenz	Artificial Intelligence
Kunststoff	Kunststoffindustrie	Plastics
Labor	Laboreinrichtungen	Laboratory Equipment

Lebensmittel	Lebensmitteltechnik	Food Technology
Lufttrans	Lufttransport	Air Transportation
Maschinen	Maschinenbau	Mechanical Engineering
Math	Mathematik	Mathematics
Mechan	Mechanik	Mechanics
Meerschmutz	Meeresverschmutzung	Marine Pollution
Metall	Metallurgie	Metallurgy
Metrol	Metrologie	Metrology
Mobilkom	Mobilkommunikation	Mobile Communications
Nichtfoss Energ	Nichtfossile Energiequellen	Fuelless Energy Sources
Optik	Optik	Optics
Papier	Papier & Pappe	Paper & Board
Patent	Patente & Warenzeichen	Patents & Trademarks
Phys	Physik	Physics
Qual	Qualitätssicherung	Quality Assurance
Raumfahrt	Raumfahrttechnik	Space Technology
Regelung	Regelungs- & Steuerungstechnik	Industrial Process Measurement & Control
Sicherheit	Sicherheitstechnik	Safety Engineering
Strahlphys	Strahlenphysik	Radiation Physics
Strömphys	Strömungsphysik	Fluid Physics
Teilphys	Teilchenphysik	Particle Physics
Telekom	Telekommunikation	Telecommunications
Textil	Textiltechnik	Textiles
Thermod	Thermodynamik	Thermodynamics
Trans	Transportwesen	Transportation
Umweltschmutz	Umweltverschmutzung	Pollution
Verpack	Verpackungstechnik	Packaging
Wassertrans	Wassertransport	Water Transportation
Wasserversorg	Wasserversorgungstechnik	Water Supply Engineering
Wellphys	Wellenphysik	Wave Physics
Werkprüf	Werkstoffprüfung	Testing of Materials

Warenzeichen / Trademarks

Bei der Kennzeichnung von Wörtern, die nach Kenntnis der Redaktion eingetragene Warenzeichen darstellen, wurde mit größter Sorgfalt verfahren. Diese Wörter sind mit ® gekennzeichnet. Weder das Vorhandensein noch das Fehlen solcher Kennzeichnungen berührt die Rechtslage hinsichtlich eingetragener Warenzeichen.

Every effort has been made to label terms which we believe constitute trademarks. These are designated by the symbol ®. The legal status of these, however, remains unchanged by the presence or absence of any such symbol.

α 1. <Akustik, Funktech, Phys, Strahlphys> *(Absorptionskoeffizient)* α *(absorption coefficient)*; 2. <Geom> *(Alpha)* α *(Alpha)*; 3. <Mechan> *(Winkelbeschleunigung)* α *(angular acceleration)*; 4. <Optik> *(Absorptionsfaktor)* α *(absorption factor)*; 5. <Optik> *(optischer Drehwinkel)* α *(angle of optical rotation)*
β *(Phasenkonstante)* <Akustik, Elektriz> β *(phase constant)*
ε 1. <Hydraul> *(kinematische Wirbelzähigkeit)* ε *(kinematic eddy viscosity)*; 2. <Phys> *(durchschnittliche kinetische Molekularenergie)* ε *(average molecular kinetic energy)*
μ 1. <Elektriz> *(Permeabilität)* μ *(permeability)*; 2. <Funktech> *(Verstärkung)* μ *(amplification factor)*; 3. <Thermod> μ
μ H *(Hall'sche Mobilität)* <Funktech> μ H *(Hall mobility)*
μ V *(Mikrovolt)* <Elektriz, Elektrotech> μ V *(microvolt)*
Ω *(Volumen in Phase)* <Phys> Ω *(volume in phase space)*
Φ 1. <Akustik> *(Winkelverdrängung)* Φ *(angular displacement)*; 2. <Akustik> *(Geschwindigkeitspotenzial)* Φ *(velocity potential)*
σ 1. <Bau> σ ; 2. <Kohlen, Kunststoff, Mechan> σ ; 3. <Phys, Thermod> σ
τ *(Relaxationszeit)* <Akustik> τ *(relaxation time)*
θ 1. <Hydraul> *(Wärmewiderstand, absolute Temperatur, thermodynamische Temperatur)* θ *(absolute temperature)*; 2. <Phys> *(Wärmewiderstand)* θ *(thermal resistance)*; 3. <Thermod> *(absolute Temperatur, thermodynamische Temperatur)* θ *(absolute temperature)*; 4. <Thermod> *(Wärmewiderstand)* θ *(thermal resistance)*
θ D *(Debye'sche Temperatur)* <Phys, Thermod> θ D *(Debye temperature)*
θ K *(Einstein'sche Temperatur)* <Thermod> θ K *(Einstein temperature)*
32-Bit-Busarchitektur f <Elektronik> MCA, microchannel architecture
3-D *(dreidimensional)* <Maschinen, Phys> 3-D *(three-dimensional)*
3-dB-Breite f <Funktech> half-power beamwidth
3-dB-Koppler m <Telekom> three-dB coupler *(Mikrowellen, Faseroptik)*
3-dB-Punkt m <Telekom> three-dB point
4-D-Verstärkung f <Raumfahrt> 4-D reinforcement
3-Stufen-Implementierung f <Comp & DV> three-tier implementation
90°-Verschiebung f <Elektronik> quadrature phase shifting *(Phasenverschiebung)*

A

a 1. <Akustik> *(akustische Absorption)* a *(total acoustic absorption)*; 2. <Metrol> *(Ar)* a *(are)*
A 1. <Akustik, Aufnahme, Comp & DV, Elektriz, Elektronik, Funktech, Phys, Wassertrans, Wellphys> *(Amplitude)* A *(amplitude)*; 2. <Chemie> *(Affinität)* A *(affinity)*; 3. <Elektriz, Elektrotech, Fertig, Funktech, Metrol, Phys> *(Ampere)* A *(ampere)*; 4. <Elektriz> *(lineare Stromdichte)* A *(linear current density)*; 5. <Elektrotech, Fernseh, Funktech, Phys, Wassertrans> A *(Anode)*; 6. <Kerntech> *(Aktivität)* A *(activity)*; 7. <Kerntech> *(Massenzahl)* A *(mass number)*; 8. <Phys> *(Aktivität, Schallstärke)* A *(activity)*; 9. <Phys> *(Isotopenmasse, Massenzahl)* A *(mass number)*; 10. <Teilphys> *(Massenzahl, Nukleonenzahl)* A *(mass number)*

Å *(Angström)* <Metrol> Å *(angstrom)*
a0 *(Bohr'scher Radius)* <Phys> a0 *(Bohr radius)*
AACS *(Luftfahrtfunkdienst)* <Raumfahrt> AACS *(airways and air communications service)*
AADT *(durchschnittliches Tagesverkehrsaufkommen pro Jahr)* <Trans> AADT *(annual average daily traffic)*
AAP *(akustischer Akzeptanzpegel)* <Akustik> ACI *(acoustic comfort index)*
AB *(akustischer Blindleitwert)* <Akustik> BA *(acoustic susceptance)*
abätzen v <Fertig> etch
Abbau m 1. <Chemie> breakdown, decomposition; 2. <Kunststoff> degradation, mastication; 3. <Textil> breakdown
Abbau m **einer Verbindung** <Telekom> clear down
abbaubar adj <Abfall> degradable
abbauen v 1. <Bau> dismantle, strip down; 2. <Maschinen> dismount; 3. <Meerschmutz> foul; 4. <Telekom> clear; take down, tear down
Abbauen n <Bau> striking *(Schalgerüst)*
Abbauförderstrecke f <Kohlen> gate road
Abbaumittel n <Kunststoff> peptizer
Abbauprozess m <Verpack> destruction process
Abbaurate f <Erdöl> depletion rate *(Lagerstättenkunde)*
Abbaustrecke f <Kohlen> coal drift, gate road
Abbaustufe f <Abfall> stage of decomposition
Abbeizbehälter m <Anstrich> strip tank
abbeizen v 1. <Anstrich> strip; 2. <Bau> scour, strip; 3. <Fertig> pickle
Abbeizen n <Bau, Wassertrans> pickling *(Schiffinstandhaltung)*
Abbeizer m <Bau> paint stripper; remover *(Farbe)*
Abbeizmittel n <Chemie> scouring agent
Abbe'sche Theorie f <Phys> Abbe theory
Abbe'sche Zahl f 1. <Ker & Glas> Abbe coefficient; 2. <Phys> Abbe number *(Dispersion)*
Abbe'sches Refraktometer n <Phys> Abbe refractometer
AB-Betrieb m <Elektronik> class AB mode
Abbiege... <Trans> turning
abbiegen v <Bau> bend down *(Bewehrungsstahl)*
abbiegender Verkehr m <Trans> turning traffic
Abbiegespur f <Trans> turning
Abbiegeströme mpl <Trans> turning movements
Abbild n 1. <Comp & DV> image, map; instance *(virtuelle Nachbildung eines Körpers)*; 2. <Ergon> image • **als Abbild wiedergeben** <Comp & DV> image
Abbild-Aktualisierung f <Comp & DV> image refreshing
Abbildbereich m <Comp & DV> image area
abbilden v 1. <Comp & DV> image; map *(Speicher)*; 2. <Math> map *(zuordnen)*
Abbilden n <Comp & DV> mapping
Abbildträger m <Comp & DV> image carrier
Abbildung f 1. <Bau> projection; 2. <Comp & DV> image, picture; map *(des Speichers)*; 3. <Druck> figure, illustration; 4. <Fernseh> image formation; 5. <Math> mapping; 6. <Patent> figure *(einer Zeichnung)*; 7. <Telekom> imaging
Abbildungsfehler m <Phys> aberration
Abbildungsmaßstab m <Konstzeich> reproduction scale
Abbildungsmechanismus m <Optik> imaging mechanism
Abbildungssystem n <Optik> imaging system
Abbildverarbeitung f <Comp & DV> image processing
Abbild-Wiederherstellung f <Comp & DV> image restoration
abbimsen v <Fertig> pumice
Abbindebeginn m <Bau> initial set
Abbindebeschleuniger m 1. <Bau> accelerator *(Beton)*; 2. <Chemtech> accelerator *(Zement)*

Abbindefaden

Abbindefaden m <Lufttrans> lacing cord
Abbindemittel n <Bau> curing compound
abbinden v 1. <Bau> break, condition; cure *(Kleber)*; 2. <Bau> set; 3. <Fertig> harden *(Kleber)*; 4. <Telekom> lace *(Kabel)*
Abbinden n 1. <Bau> cure, hydration, setting *(Zement)*; breakdown *(Emulsion)*; 2. <Chemtech> setting *(Zement)*; 3. <Fertig> hardening *(Kleber)*
Abbindetemperatur f <Bau> setting temperature
Abbindtemperatur f <Bau> cure temperature
Abbindzeit f 1. <Abfall> setting time; 2. <Bau> cure period, cure rate, setting time; 3. <Fertig> pot life *(Kleber)*
abblasen v <Kerntech> blow off
Abblasventil n 1. <Fertig> bleeder; 2. <Hydraul> blow valve
abblättern v 1. <Bau> chip, peel, spall; 2. <Fertig> scale
Abblättern n 1. <Fertig> scaling; 2. <Textil> peeling
Abblend... <Elektrotech, Kfztech> dimmed, dipped
abblenden v 1. <Elektrotech> dim; 2. <Foto> stop down
Abblendlicht n <Kfztech> *(BE)* dimmed headlight, dipped beam, *(AE)* dipped headlight, *(AE)* passing light
Abblendschalter m <Elektrotech> dimmer switch
abböschen v <Bau> batter, slant, slope
Abbrand m <Kerntech> burnout
abbrechen v 1. <Bau> demolish *(Gebäude)*; 2. <Comp & DV> abort, cancel; 3. <Ker & Glas> *(BE)* break off, cap; 4. <Kontroll, Raumfahrt> abort; 5. <Telekom> terminate *(Programm)*
Abbrechen n <Chemtech> breaking
Abbrechen n **des Flaschenbodens** <Ker & Glas> breaking-off of base
Abbrems... <Raumfahrt> retardation, retrograde
abbremsen v 1. <Chemie> retard *(Elektronen)*; 2. <Phys> decelerate; 3. <Raumfahrt> retard *(Raumschiff)*
abbremsend adj 1. <Phys> decelerated; 2. <Raumfahrt> retrograde
Abbremsorbit m <Raumfahrt> retrograde orbit
Abbremsrakete f <Raumfahrt> retardation rocket
Abbremsung f <Mechan, Phys> deceleration
Abbremszeit f <Comp & DV> stop time
abbrennen v <Chemie> deflagrate *(explosionsfrei)*
Abbrennschweißen n <Bau> flash welding
abbröckeln v <Bau> crumble away
Abbruch m 1. <Comp & DV> abort, cancel, hang-up, termination; 2. <Telekom> disconnection
Abbruchabfall m <Abfall, Bau> demolition waste, rubble
Abbruchbedingung f <Kontroll> truncation condition
Abbruchfehler m <Comp & DV> truncation error
Abbruchhöhe f <Lufttrans> critical altitude
Abbruchkolonne f <Bau> breakdown gang
Abbruchmaterial n 1. <Abfall> demolition waste, rubble; 2. <Bau> demolition waste
Abbruchzustand m <Kontroll> truncation condition
abbrühen v <Thermod> scald
ABC *(automatische Helligkeitsregelung)* <Fernseh> ABC *(automatic brightness control)*
ABC-Hubschrauber m <Lufttrans> ABC helicopter, advanced blade concept helicopter
abdachen v <Fertig> point *(Zahnräder)*
Abdachen n <Fertig> pointing *(Zahnräder)*
abdämmen v <Wasserversorg> block off *(gegen Wasser)*
Abdampf m <Hydraul, Lebensmittel, Maschinen> dead steam, exhaust steam, waste steam
Abdampfapparat m <Chemtech> evaporator
abdampfen v 1. <Chemtech> vaporize; evaporate *(Flüssigkeiten)*; 2. <Thermod> boil away
Abdampfen n <Chemtech> vaporization *(Lösemittel)*
Abdampfgefäß n <Chemtech> evaporating vessel
Abdampfkasserolle f <Chemtech> evaporating pan

Abdampfkessel m 1. <Chemtech> evaporating boiler; 2. <Heiz & Kälte> exhaust steam boiler
Abdampfschale f 1. <Chemtech> evaporating dish, evaporating pan, evaporating basin; 2. <Labor> evaporating basin, evaporating dish, evaporating pan
Abdampfturbine f <Heiz & Kälte> exhaust steam turbine
Abdampfventil n <Hydraul> exhaust valve
Abdeck... <Druck, Elektronik, Verpack> masking
Abdeckband n 1. <Druck> masking paper; 2. <Verpack> masking tape, surface protection tape
Abdeckbanddichtung f <Verpack> band sealing
Abdeckblech n 1. <Bau> flashing; 2. <Mechan> base plate
Abdeckblende f <Elektronik> mask
Abdeckblockstein m <Bau> coping block
abdecken v 1. <Bau> revet *(Böschung, Fundament)*; cope *(Träger)*; 2. <Kerntech> blanket
Abdecken n <Bau> coping
Abdecker m <Lebensmittel> knacker
Abdeckhaube f <Lufttrans> cowl *(Motor)*
Abdeckmaterial n 1. <Abfall> covering material; 2. <Druck> resist
Abdeckplatte f 1. <Bau> cap; 2. <Fertig> cover *(Bettbahn)*; apron *(Walzen)*; 3. <Ker & Glas> *(BE)* cover tile; 4. <Kerntech> cover gas discharge line, cover slab; 5. <Lufttrans> access panel
Abdeckscheibe f <Fertig> sealing plug *(Kunststoffinstallationen)*
Abdeckschicht f <Kunststoff> resist coating
Abdeckstein m <Bau> capstone, coping stone
Abdeckung f 1. <Bau> cover, covering; 2. <Erdöl> cap rock; 3. <Kerntech> cover cap; 4. <Kfztech> blind *(Kühler)*; 5. <Mechan> guard; 6. <Sicherheit> cover; 7. <Wasserversorg> coping
Abdeckung f **einer Hemisphäre** <Telekom> hemispherical coverage
Abdeckung f **für Schaftrippe** <Lufttrans> cover strip of root rib
abdestillieren v 1. <Thermod> distil; 2. <Chemtech, Thermod> distil off
Abdicht... <Bau, Chemie, Mechan> sealing
abdichten v 1. <Abfall> line; seal *(Deponie)*; 2. <Bau> caulk, make impermeable, proof, seal; 3. <Mechan> pack; 4. <Papier> seal; 5. <Wassertrans> seal *(Schiffbau)*
Abdichten n <Bau> sealing
Abdichtleiste f <Bau> window bar
Abdichtmittel n <Chemie> sealant
Abdichtung f 1. <Bau> packing; 2. <Erdöl> sealing; 3. <Fertig> sealing *(Kunststoffinstallationen)*; 4. <Maschinen> sealing; 5. <Mechan> seal
Abdichtungslage f <Bau> barrier
Abdichtungsmasse f 1. <Bau> sealing agent; 2. <Verpack> lining compound
Abdichtungsmittel n <Bau> sealant
Abdichtungspolster n <Sicherheit> protective padding
abdrehen v 1. <Fertig> dress *(Schleifkörper)*; true *(Schleifscheibe)*; 2. <Wasserversorg> turn off *(Hahn)*
Abdrehen n <Fertig> dressing, truing *(Schleifscheibe)*
Abdrehvorrichtung f <Maschinen> truing attachment, turning attachment
Abdrift f 1. <Lufttrans> drift *(Seitwärtsbewegung des Flugzeugs)*; 2. <Meerschmutz> drift; 3. <Raumfahrt> drift *(Raumschiff)*
Abdriftbereich m <Elektronik> drift region
Abdriftwinkel m 1. <Lufttrans> drift angle; 2. <Wassertrans> leeway angle
Abdrosselung f <Lufttrans> stall *(Kompressor, Turbomotor)*

Abdruck m 1. <Druck> impression; 2. <Fertig> (AE) mold, (BE) mould
abdruckbares Zeichen n <Druck> printing character
Abdruckbarkeit f <Druck> printability
Abdrücksignal n <Eisenbahn> backing signal
Abdrückversuch m <Maschinen> hydraulic test
Abduktion f <Ergon> abduction
Aberration f <Metall, Optik, Wellphys> aberration
Aberrationskreis m <Optik> circle of aberration
abfackeln v <Thermod> burn off
Abfackeln n <Erdöl> flaring (Verbrennung)
Abfall m 1. <Abfall> junk; 2. <Bau> tailings; 3. <Elektronik> fall-off (der Spannung); 4. <Elektrotech> voltage drop (der Spannung); 5. <Hydraul> fall (Niveau); 6. <Mechan> junk, scrap; 7. <Papier> waste; 8. <Qual> scrap; 9. <Telekom> roll-off (Kurve); 10. <Verpack> (AE) garbage, (BE) rubbish; 11. <Wasserversorg> refuse, spillage
Abfallablauf m <Kerntech> waste outlet
abfallarme Technologie f <Abfall> clean technology, low-waste technology
Abfallaufbereitung f 1. <Abfall> waste recovery, waste recycling; 2. <Umweltschutz> waste recovery
Abfallausgrabung f <Kerntech> disinternment of waste
Abfallbecken n **der Sortiermaschine** <Ker & Glas> grader waste pond
Abfallbehälter m 1. <Abfall> (AE) garbage can, (BE) rubbish bin, waste container; 2. <Kerntech> waste canister (für Endlagerung)
Abfallbehandlung f <Abfall> waste processing, waste treatment
Abfallbeseitigung f 1. <Abfall> refuse disposal, waste disposal; 2. <Wasserversorg> waste disposal
Abfallbörse f <Abfall> waste exchange market
Abfallbrennstoff m <Abfall> waste fuel
Abfallbunker m <Abfall> waste hopper
Abfallcontainer m <Abfall> caster-equipped container
Abfalldeponie f <Wasserversorg> refuse dump
Abfalldesinfektion f <Abfall> waste disinfection
abfallen v 1. <Bau> sink; slope (Gelände); 2. <Comp & DV> decay; 3. <Wassertrans> bear away; fall off (vom Wind beim Segeln)
Abfallen n 1. <Elektrotech> drop-out; 2. <Fernseh> decay
abfallende Flanke f <Elektronik> trailing edge (eines Impulses)
abfallende Schneide f <Fertig> (AE) leading tool edge
abfallender Bogen m <Bau> rising arch
Abfallentsorgung f <Abfall> garbage disposal, waste disposal
Abfallentsorgung f **im Meer** <Abfall> ocean disposal
Abfallentsorgung f **vor Ort** <Kerntech> on-site waste disposal
Abfallentsorgungsanlage f 1. <Abfall> mechanical waste disposal system, waste disposal plant; 2. <Sicherheit> mechanical waste disposal system
Abfallentsorgungseinrichtung f <Abfall, Sicherheit> waste disposal facility
Abfallerzeuger m <Abfall> waste generator, waste producer
Abfallerzeugung f <Abfall> waste formation, waste generation, waste production, waste stream
Abfallextraktionssystem n <Verpack> waste extraction system
abfallfreie Technologie f <Abfall> NWT, nonwaste technology
Abfallgärung f <Abfall> fermentation of refuse
Abfallgesetz n <Abfall> Waste Recycling and Disposal Act
Abfallkondensatpumpe f <Kerntech> waste condensate pump

Abfallkonzentration f <Kerntech> waste concentration
Abfalllagerung f <Abfall> waste storage
Abfallmaterial n <Verpack> scrap material
Abfallprodukt n 1. <Maschinen> by-product; 2. <Umweltschmutz> waste product (unverwertbar)
Abfallrate f <Fernseh> decay rate
Abfallrecycling f 1. <Abfall> recycling of refuse, waste recycling; 2. <Sicherheit> waste recycling
Abfallrelais n <Elektriz> release relay
Abfallrückholung f <Kerntech> disinternment of waste
Abfallsäure f 1. <Abfall> waste acid; 2. <Textil> spent acid
Abfallschicht f <Abfall> waste layer
Abfallsortieranlage f <Abfall> refuse separation plant, refuse sorting plant, sorting plant, waste sorting plant
Abfallsortierung f **am Anfallsort** <Abfall> source separation
Abfallspannung f <Elektrotech> drop-out voltage
Abfallstoff m <Abfall> junk, waste
Abfallstreifen m <Fertig> skeleton (Ausschneiden)
Abfallstrom m 1. <Abfall> flow of waste; 2. <Elektriz, Elektrotech> drop-out current
Abfallverarbeitungsanlage f <Abfall> refuse processing plant
Abfallverbrennungsanlage f <Abfall> refuse incinerator
Abfallverbrennungsofen m <Umweltschmutz> incinerator
Abfallvermeidung f <Abfall> waste avoidance
Abfallverursacher m <Abfall> waste producer
Abfallverwertung f <Abfall> recycling of refuse, recycling of waste, waste processing, waste recycling, waste treatment
Abfallverwertungsanlage f 1. <Abfall> waste treatment plant; 2. <Umweltschmutz> waste utilization plant
Abfallwirtschaft f <Abfall> waste management
Abfallzeit f 1. <Elektriz> drop-out time; 2. <Elektronik> decay time; fall time (Transistor); 3. <Elektrotech> decay time; 4. <Funktech> fall time; 5. <Lufttrans> release time (Verkehr); 6. <Phys> fall time (Impuls)
Abfallzellstoff m <Abfall> waste pulp
Abfallzerkleinerer m <Abfall> waste crusher, waste crushing plant, waste disintegrator
Abfangung f 1. <Raumfahrt> intercept; 2. <Sicherheit, Trans> safety stop
Abfangen n **eines Flugzeugs** <Lufttrans> aircraft interception
Abfangbogen m <Trans> flare-out
abfangen v **und Höhe halten** v <Lufttrans> level out
Abfangen n 1. <Chemie> scavenging; 2. <Raumfahrt> intercept; 3. <Wassertrans> recovery
Abfangkeil m <Erdöl> slip (Bohrtechnik)
Abfangpunkt m <Raumfahrt> intercept point
Abfangseil n <Sicherheit, Trans> safety stop cable
Abfärben n <Druck> set-off
abfasen v <Bau, Maschinen> chamfer
Abfasen n <Bau> (AE) beveling, (BE) bevelling
Abfasung f <Maschinen> bevel
Abfasungswinkel m <Maschinen> angle of bevel
Abfederung f <Kerntech> whipping
abfeilen v <Maschinen> file
abfeimen v <Ker & Glas> skim
Abfeimen n <Ker & Glas> skimming
Abfeimnische f <Ker & Glas> skimming pocket
Abfeimstange f <Ker & Glas> skimming rod
abfendern v <Wassertrans> fend off
Abfertigung f <Lufttrans> check-in, handling
Abfertigungsgebäude n <Lufttrans> passenger terminal
AbfG (Abfallgesetz) <Abfall> Waste Avoidance and Management Act, Waste Disposal Act

abfiltrieren v <Chemtech> filter out
Abflachen n <Elektronik> (AE) leveling, (BE) levelling (Kurvenverlauf)
abflammen v <Textil> gas, singe
Abflammen n <Textil> singeing
abflanschen v <Fertig> notch (Werkstofftrennen)
abflauen v <Wassertrans> calm down (Wind)
abfließen v <Bau> outflow
Abflug m <Lufttrans> departure
Abflugflughafen m <Lufttrans> airport of departure, departure airport
Abflugleitstrahl m <Lufttrans> outbound beam (Navigation)
Abflugstation f <Lufttrans> departure terminal
Abflugsteuerkurs m <Lufttrans> outbound heading (Navigation)
Abfluss m 1. <Abfall> effluent; 2. <Elektrotech> leakage; 3. <Hydraul> sink; 4. <Maschinen> drain; 5. <Meerschmutz, Nichtfoss Energ, Umweltschmutz> runoff; 6. <Wasserversorg> discharge, runoff
Abfluss m im Oberflächenbereich <Abfall> surface runoff
Abflussbecken n <Hydraul, Labor> sink
Abflussgraben m <Hydraul> gutter
Abflusskanal m <Wasserversorg> distributing canal, effluent channel, tailrace, tailrace tunnel
Abflusskoeffizient m <Wasserversorg> runoff coefficient
abflussloses Einzugsgebiet n <Wasserversorg> blind drainage area
Abflussmenge f <Wasserversorg> discharge, rate of flow
Abflussregler m <Nichtfoss Energ> discharge regulator
Abflussrinne f 1. <Lufttrans> guttering; 2. <Mechan> gutter; 3. <Nichtfoss Energ> sluice box; 4. <Wasserversorg> flume, sluice box
Abflussrohr n 1. <Bau> drainpipe; 2. <Hydraul> flow pipe; 3. <Wasserversorg> downpipe; discharge lift, discharge pipe (einer Pumpe)
Abflussschleuse f <Hydraul> sink
Abflussventil n <Hydraul> escape valve
Abflussvorrichtung f <Wassertrans> drain
Abfolge f <Kontroll> sequence
abfördern v <Heiz & Kälte> discharge
Abfrage f 1. <Comp & DV> inquiry, query, request; 2. <Gerät, Regelung> scan; 3. <Patent> inquiry; 4. <Telekom> inquiry, polling, scan
Abfrage f auf Absatzbasis <Comp & DV> phrase query
Abfrage f auf Satzbasis <Comp & DV> sentence query
Abfrage f mit Platzhaltern <Comp & DV> wild card query
Abfragebearbeitung f <Comp & DV> query processing
Abfragecode m <Comp & DV> inquiry character
Abfragedienst m <Telekom> answering service
Abfrageklinke f <Telekom> jack
Abfragemethode f <Comp & DV> lookup
Abfragemodus m <Elektronik> interrogation mode
abfragen v 1. <Comp & DV> interrogate, query, poll; 2. <Telekom> scan (Adresse); sense
Abfragerate f <Regelung> revisit rate
Abfrageschalter m <Comp & DV> sampler
Abfragesprache f <Comp & DV> query language, QL, structured query language, SQL
Abfragestation f <Comp & DV> inquiry station
Abfragesteuerung f <Comp & DV> inquiry control
Abfragesystem n 1. <Künstl Int> query system; 2. <Telekom> polling system
Abfrageverarbeitung f <Comp & DV> inquiry processing, query processing
Abfühlen n <Comp & DV> sensing
Abfuhr f <Abfall> collection

Abfuhr f der Nachwärme <Kerntech> afterheat release
abführen v <Bau> carry off (Wasser)
Abführmechanismus m <Textil> take-away mechanism
Abführung f 1. <Fertig> clearance (Späne); 2. <Heiz & Kälte> eduction
Abfuhrwagen m <Abfall> skip, tipper truck
Abfüll... <Verpack> filling
Abfüllanlage f <Verpack> filling machine
Abfüllautomat m <Verpack> bottling machine
Abfüllbehälter m <Lebensmittel> bottling tank
abfüllen v <Fertig> fill, fill in; bottle (in Flaschen)
Abfüllen n <Lebensmittel> racking (von Getränken)
Abfüllen n auf Flaschen <Lebensmittel> filling bottles
Abfüllgerät n <Maschinen> filler, filling machine
Abfülllinie f <Verpack> filling line
Abfüllmaschine f <Fertig> filler, filling machine
Abfüllpackung f <Verpack> dosing packing
Abfüll- und Dosierautomat m <Verpack> filling and dosing machine
Abfüll- und Kappenaufsetzmaschine f <Verpack> filling and capping machine
Abfüll- und Siegelmaschine f <Verpack> filling and sealing machine
Abfüll- und Versiegelungseinheit f für Beutel <Verpack> sachet form fill seal unit
Abfüllung f in Kartons <Verpack> bag-in-a-box packaging
Abfüllung f in Säcke von Hand <Verpack> hand bagging
Abfüllvorrichtung f <Lebensmittel> dispenser
Abfüllwaage f 1. <Gerät> bag-filling scale, dispensing scale; 2. <Verpack> checkweighing machine
Abgabeseite f <Maschinen> discharge side
Abgang m 1. <Bau> (AE) raveling; 2. <Kfztech> output; 3. <Telekom, Trans> originating, outgoing, output
Abgangskanal m <Telekom> outgoing channel
Abgangsverkehr m 1. <Telekom> originating traffic, outgoing traffic; 2. <Trans> originating traffic
Abgangswelle f <Kfztech> output shaft
Abgas n 1. <Fertig> waste gas; 2. <Heiz & Kälte> flue gas; 3. <Kerntech> off-gas, waste gas; 4. <Kfztech> emission, exhaust gas; 5. <Lufttrans> emission; exhaust (Triebwerk, Motor); 6. <Mechan> exhaust; 7. <Phys> flue gas; 8. <Thermod> exhaust gas, flue gas; 9. <Umweltschmutz> exhaust gas, waste gas
Abgasanlage f <Kfztech> exhaust system
abgasarmes Gemisch n <Kfztech> lean mixture
Abgasausströmöffnung f <Mechan> exhaust gate
abgasbeheizt adj <Mechan> exhaust-operated
Abgasdüse f <Lufttrans> exhaust nozzle
Abgasdüsenverschlussstücke npl <Lufttrans> exhaust nozzle breeches
Abgasentschwefelung f <Abfall> (AE) waste gas desulfurization, (BE) waste gas desulphurization
Abgasfilterung f <Thermod> exhaust gas cleaning
Abgasgehäuse n <Lufttrans> exhaust case
abgasgetrieben adj <Mechan> exhaust-operated
Abgasgewicht n <Mechan> exhaust weight
Abgaskatalysator m <Kfztech> (BE) catalytic converter, (AE) catalytic muffler, (BE) catalytic silencer
Abgaskondensator m <Kerntech> off-gas condenser
Abgaskonus m <Lufttrans> exhaust cone
Abgaskrümmer m <Mechan> exhaust manifold
Abgasleitung f 1. <Fertig> exhaust duct; 2. <Kfztech, Lufttrans> exhaust pipe; 3. <Mechan> exhaust conduit, exhaust pipe; 4. <Thermod, Wassertrans> exhaust pipe
Abgasmessgerät n <Gerät> waste gas meter
Abgasmessstrecke f <Heiz & Kälte> flue gas test section
Abgasprüfgerät n <Mechan> (BE) exhaust gas analyser, (AE) exhaust gas analyzer

Abgasreiniger m <Umweltschmutz> exhaust gas cleaner
Abgasreinigung f 1. <Abfall> waste gas cleaning; 2. <Thermod> exhaust gas cleaning
Abgasreinigungsanlage f 1. <Sicherheit> flue gas cleaning installation; 2. <Umweltschmutz> exhaust gas cleaner
Abgasreinigungsfilter n <Abfall, Sicherheit> exhaust purification filter
Abgasrohr n 1. <Kerntech> flue tube; 2. <Kfztech> exhaust pipe, tail pipe *(Motor)*; 3. <Lufttrans, Mechan, Thermod, Wassertrans> exhaust pipe
Abgasrohrkrümmer m <Kfztech, Lufttrans> exhaust manifold
Abgasrückführung f <Kfztech> exhaust gas recirculation
Abgasrückführung f **mit Lufteinblasung** <Kfztech> exhaust gas recirculation with air injection
Abgassammler m <Kfztech, Lufttrans, Thermod> exhaust manifold
Abgasschalldämpfer m <Kfztech> *(AE)* exhaust muffler, *(BE)* exhaust silencer, exhaust nozzle
Abgasschubrahmen m <Lufttrans> exhaust nozzle
Abgassensor m <Kfztech> lambda probe
Abgasstutzen m <Mechan> exhaust stack
Abgastemperatur f 1. <Kfztech> exhaust gas temperature; 2. <Lufttrans> exhaust gas temperature, jet pipe temperature
Abgastemperaturanzeige f <Kfztech, Lufttrans> exhaust gas temperature indicator
Abgasturbine f 1. <Lufttrans> exhaust gas turbine; 2. <Mechan> exhaust-driven turbine
Abgasturbolader m 1. <Kfztech> exhaust turbocharger, turbosupercharger; 2. <Mechan> exhaust-turbine supercharger
Abgasventil n <Kfztech> exhaust valve
Abgasverbrennung f <Kfztech> exhaust gas combustion
Abgaswärme f <Thermod> waste gas heat
Abgaswärmerückgewinnung f <Thermod> waste gas heat recovery
Abgaswiederverwertung f <Kfztech> exhaust recycling
abgebaut adj <Bau> struck
abgeben v <Thermod> emit, give off
Abgeben n <Chemie> liberation *(Verbindungen)*
abgeblättert adj <Fertig> scaled
abgeblendeter Scheinwerfer m <Kfztech> *(AE)* dipped headlight
abgebrochen darstellen v <Konstzeich> represent broken
abgebrochene Prüfung f <Qual> curtailed inspection
abgebrochener Anflug m <Lufttrans> discontinued approach
abgebunden adj <Fertig> hardened *(Kleber)*
abgedacht adj <Fertig> pointed *(Zahnrad)*
abgedichtet adj <Fertig> gasketed
abgefahrener Reifen m <Kfztech> *(AE)* bald tire, *(BE)* bald tyre
abgefangene Landung f <Lufttrans> flared landing
abgefaste Kante f <Bau, Maschinen> bevel edge, *(AE)* beveled edge, *(BE)* bevelled edge, chamfered edge
abgefaste Scheibe f <Maschinen> *(AE)* beveled washer, *(BE)* bevelled washer
abgeführte Wärme f <Heiz & Kälte> heat removed
abgegebene Leistung f <Maschinen> power output
abgegebene Strahlungsmenge f <Strahlphys> radiated output
abgegebene Wärmedichte f <Kerntech> heat output density
abgeglichene Leitung f <Elektriz> balanced line, balancing coil
abgehängte Decke f <Bau> suspended ceiling

abgehend adj <Telekom> outgoing
abgehende Fernleitung f <Telekom> outgoing trunk circuit
abgehende Leitung f <Telekom> outgoing circuit, outgoing line
abgehende Rufe gesperrt <Telekom> outgoing calls barred
abgehende Verbindung f <Telekom> outgoing call
abgehender Anruf m <Telekom> outgoing call
abgehender Kanal m <Telekom> go channel *(Richtfunk)*
abgehender Ruf m <Telekom> call request, connection request *(CQ)*
abgehender Verkehr m <Telekom> outgoing traffic, outward traffic
abgehendes Bündel n <Telekom> outgoing group
abgehoben adj <Telekom> off hook *(Hörer)*
abgehörte Telefonleitung f <Telekom> tapped telephone line
abgeklungene Radioaktivität f <Strahlphys> cooled-down radioactivity
abgekühlte Radioaktivität f <Strahlphys> cooled-down radioactivity
abgelagertes Holz n <Wassertrans> seasoned timber, seasoned wood *(Schiffbaumaterial)*
abgelaufene Zeit f <Comp & DV> elapsed time
abgelegenes Verteilungssystem n <Nichtfoss Energ> remote network
abgelegter Satz m <Druck> dead matter, dead type
abgeleitete Anströmgeschwindigkeit f <Lufttrans> derived gust velocity
abgeleitete Einheit f <Maschinen, Phys> derived unit
abgeleitete Schaltung f <Elektriz> derived circuit
abgeleiteter Kanal m <Telekom> tributary channel
abgeleiteter Strom m <Elektriz> derived current
abgelenkte Bohrung f <Erdöl> deviated well *(Bohrtechnik)*
abgelenkter Strahl m <Fernseh> deflected beam
abgelenktes Bohren n <Erdöl> deviated drilling *(Bohrtechnik)*
abgelesener Messwert m <Gerät> reading
abgelesener Wert m <Gerät> reading
abgemeldete Endeinrichtung f <Telekom> de-affiliated terminal *(Telefon)*
abgenommen adj <Telekom> off hook *(Hörer)*
abgenutzt adj 1. <Mechan> worn; 2. <Textil> worn out
abgenutztes Werkzeug n <Sicherheit> worn tool *(Unfallursache)*
abgeplattet adj <Geom> oblate
abgeplatteter Kern m <Kerntech> oblate nucleus
abgeplattetes Ellipsoid n <Geom, Phys> oblate ellipsoid
abgeplattetes Sphäroid n <Geom> oblate spheroid
abgeplatzte Ecke f <Ker & Glas> chipped corner
abgeplatzte Kante f <Ker & Glas> chipped edge
abgereicherter Kernbrennstoff m <Kerntech> depleted nuclear fuel
abgerundete Anfahrdüse f <Hydraul> rounded approach orifice
abgerundete Kante f <Ker & Glas> rounded edge
abgeschaltet adj 1. <Elektrotech> off, switched off; 2. <Telekom> disabled *(Gerät)*
abgeschaltete Leitung f <Telekom> dead line
abgeschalteter Thyristor m <Elektronik> off thyristor
abgeschalteter Transistor m <Elektronik> off transistor
abgeschaltetes Handy n <Mobilkom> deactivated mobile handheld phone, deactivated mobile phone, deactivated radio phone
abgeschert adj <Fertig> sheared
abgeschirmt adj <Bau, Telekom> shielded
abgeschirmte Ader f <Elektriz> screened core

abgeschirmte

abgeschirmte Antenne f <Fernseh> screened aerial
abgeschirmte Doppelleitung f <Telekom> screened pair
abgeschirmte symmetrische Doppelader f <Telekom> shielded balanced pair
abgeschirmte symmetrische Leitung f <Elektrotech> shielded symmetric pair of wires
abgeschirmte Übertragung f <Elektrotech> shielded transmission
abgeschirmter Ausgang m <Elektrotech> shielded output, screened output
abgeschirmter Draht m <Elektrotech> shielded wire
abgeschirmter Eingang m <Elektrotech> shielded input, screened input
abgeschirmter Transformator m <Elektrotech> shielded transformer
abgeschirmtes Gehäuse n <Elektrotech> shielded enclosure
abgeschirmtes Kabel n <Aufnahme, Comp & DV, Elektriz, Elektrotech, Fernseh, Phys> screened cable, shielded cable
abgeschlagenes Garn n <Ker & Glas> sloughed yarn
abgeschliffen adj <Fertig> attrite
abgeschlossen adj 1. <Phys> self-contained; 2. <Umweltschmutz> sealed
abgeschlossen adj/in sich <Heiz & Kälte> self-contained
abgeschlossene Deponie f <Abfall> complete fill
abgeschlossene Leitung f/durch Impedanz <Phys> line terminated by an impedance
abgeschlossene Schale f <Kerntech> closed shell (Atom)
abgeschlossener Behälter m <Labor> closed vessel
abgeschlossener Innenraum m <Phys> enclosure
abgeschlossenes Intervall n <Math> closed interval
abgeschlossenes System n <Phys> isolated system
abgeschnitten adj <Fertig> sheared
abgeschnittene Pyramide f <Geom> truncated pyramid
abgeschnittener Kegel m <Geom> truncated cone
abgeschrägt adj <Bau> splayed (Zimmerhandwerk)
abgeschrägte Gehrungsfuge f <Bau> splayed mitre joint
abgeschrägte Kante f <Maschinen> bevel edge, (AE) beveled edge, (BE) bevelled edge
abgeschrägter Meißel m <Maschinen> (AE) beveled chisel, (BE) bevelled chisel
abgeschrägter Stein m <Ker & Glas> skew block
abgeschreckte Scherben fpl <Ker & Glas> quenched cullet
abgesetzte Bohrung f <Fertig> shouldered hole
abgesetzte Konzentratoreinheit f <Telekom> RCU, remote concentration unit
abgesetzte Vermittlung f <Telekom> stand-alone exchange
abgesetzter Konzentrator m <Telekom> remote concentrator, remote line concentrator
abgesetzter Meißel m <Fertig> offset cutting tool
abgesetztes Abwasser n <Wasserversorg> settled sewage
abgesetztes Endgerät n <Telekom> remote terminal, RT
abgesetztes Luftschraubenblatt n <Lufttrans> offset blade (Hubschrauber)
abgesetztes Schlaggelenk n <Lufttrans> offset flapping hinge (Hubschrauber)
abgesichert adj <Elektrotech, Sicherheit> fuse-protected
abgesichertes Gerät n <Sicherheit> fail-safe device
abgesoffene Turbine f <Hydraul> drowned turbine
abgespeckter PC-Client m <Comp & DV> thin PC client
abgesteifter Schacht m <Bau> timbered shaft

abgestimmte Leitung f <Elektronik> resonant line
abgestimmte Pistenlänge f <Lufttrans> balanced field length
abgestimmter Schwingkreis m <Wellphys> tuned circuit
abgestimmtes Filter n <Elektronik> tuned filter
abgestimmtes Reed-Relais n <Elektrotech> matched reed relay
abgestrahlte Leistung f <Telekom> radiated power
abgestuft adj <Bau> stepped; screened (Körnung)
abgestufte Toleranz f <Qual> stepped tolerance
abgestufter Grenzwert m <Qual> stepped limiting value
abgestufter Höchstwert m <Qual> stepped upper limiting value
abgestufter Mindestwert m <Qual> stepped lower limiting value
abgestufter Steigflug m <Lufttrans> stepped climb
abgestumpft adj <Mechan> dull
abgestumpfte Pyramide f <Geom> truncated pyramid
abgetastetes Signal n <Telekom> sampled signal
abgetragen adj <Textil> worn out
abgetreppt adj <Bau> benched, stepped
abgewalmtes Mansardendach n <Bau> double pitch roof
abgewickelter Verkehr m <Telekom> handled traffic
abgewinkelter Schraubenzieher m <Maschinen> offset screwdriver
abgießen v <Chemie> decant
Abgießen n <Chemtech, Lebensmittel> decantation
Abgleich m 1. <Fernseh> alignment (der Videoköpfe); 2. <Gerät> trimming (Messbereich); 3. <Mechan> equalization; 4. <Phys> balancing (Brückenabgleich); 5. <Telekom> tuning (Frequenzempfänger)
Abgleicharbeiten fpl <Qual> adjusting operations
Abgleichband n <Fernseh> alignment tape
Abgleichbereich m <Metall> levelling line
Abgleichbesteck n <Gerät> alignment tool set, trimming kit
Abgleicheinrichtung f <Gerät, Qual> calibration equipment
Abgleichelement n <Gerät> adjusting element, trimming element
abgleichen v 1. <Bau> trim; level (Mauern); 2. <Fertig> parfolicalize; 3. <Funktech> align; 4. <Mechan> trim; 5. <Telekom> tune (Frequenz)
Abgleichen n 1. <Funktech> alignment; adjustment (Empfänger); 2. <Telekom> balancing
Abgleicher m <Bau> planisher
Abgleichfehler m <Telekom> alignment fault
Abgleichfrequenz f <Elektronik> tie-down point (bei Überlagerungsempfänger)
Abgleichmechanismus m <Gerät> balancing mechanism
Abgleichpunkt m <Gerät> balance point (Brücke)
Abgleichverstärker m <Elektronik> (AE) leveling amplifier, (BE) levelling amplifier
Abgleichwiderstand m 1. <Elektriz> balancing resistor; 2. <Elektrotech> adjustable resistor; 3. <Gerät> trimming resistor; balancing resistor (Bauelement)
Abgleisen n <Eisenbahn> derailing
Abgleitung f <Elektrotech> slip
abgraten v 1. <Bau> chip; 2. <Fertig> clip, snag
Abgraten n in Trommelmaschine <Fertig> barrel deburring
Abgratfehler m <Fertig> mistrimming • **mit Abgratfehler** <Fertig> mistrimmed
Abgratpresse f 1. <Fertig> stripping press; 2. <Maschinen> trimming machine
Abgratwerkzeug n <Fertig> trimming tool

abgreifen v <Maschinen> (AE) caliper, (BE) calliper
Abgreifen n <Maschinen> (AE) calipering, (BE) callipering
Abgreifpunkt m <Elektrotech> tapping point
abgrenzen v <Elektrotech> insulate
Abgriff m <Phys, Telekom> tapping
Abguss m 1. <Fertig> pouring; 2. <Metall> casting
abhalten v <Comp & DV> hold
Abhämmern n <Fertig> peening
Abhang m <Bau> downhill slope, slope
abhängig adj <Patent> dependent
abhängige Gleichungen fpl <Math> dependent equations
abhängige Variable f <Math> dependent variable
abhängiger Datensatz m <Comp & DV> member set
abhängiger Patentanspruch m <Patent> dependent claim
abhängiges Patent n <Patent> dependent patent
Abhängigkeit f <Fertig, Sicherheit> interlock, interlocking
Abhängigkeitsschaltung f <Fertig> interlock
Abhaspelmaschine f <Kunststoff> reeling machine
Abhebebewegung f <Fertig> relief motion (Hobelmeißel)
Abhebegeschwindigkeit f <Lufttrans> liftoff speed
Abhebegewicht n <Raumfahrt> liftoff weight
abheben v 1. <Fertig> skim; cut off (Späne); clear (Werkzeug); 2. <Lufttrans> lift off; 3. <Telekom> lift (Hörer) • **Hörer abheben** <Telekom> remove the handset (Telefon)
Abheben n 1. <Fertig> skimming; relieving (Hobelmeißel); 2. <Raumfahrt> lift-off
Abhebeöse f <Fertig> lifting lug (Gießen)
Abhilfe f 1. <Qual> remedy; 2. <Telekom> corrective measure
Abhilfemaßnahme f <Qual> corrective action, corrective measure, remedy
Abhitze f <Heiz & Kälte, Maschinen, Thermod> waste heat
Abhitzekessel m <Heiz & Kälte, Thermod> waste heat boiler
Abhitzerückgewinnung f <Thermod> waste heat recovery
Abhitzeverwerter m <Metall> regenerator
Abholen n <Abfall> collection
Abhör... <Aufnahme> control
Abhördienst m <Telekom> interception service, radio monitoring
abhören v <Telekom> tap (Telefon)
Abhören n <Telekom> interception, tapping (von Telefongesprächen)
Abhörkabine f <Aufnahme> control cubicle
Abhörlautsprecher m <Aufnahme> control loudspeaker
abhörsicheres Telefon n <Telekom> interception-proof telephone, tap-proof telephone
Abhörsicherheit f <Telekom> safety from interception, security against interception
Abietan n <Chemie> abietate
Abietin n <Chemie> coniferin
Abietinsäure f <Papier> abietic acid
abisolierbar adj <Fertig> strippable
abisolieren v <Kfztech> strip
Abkant... <Fertig, Maschinen> edging, folding
abkanten v 1. <Bau> bend at right angles; 2. <Fertig> fold (Blech)
Abkanten n <Fertig, Maschinen> edging, folding
Abkantmaschine f <Fertig, Maschinen> edging machine, folding machine
Abkantpresse f <Fertig> brake, press brake
Abkantung f <Bau> chamfer
Abkantwinkel m <Fertig, Maschinen> angle of bend (Blechabkantmaschine)
abkapseln v <Anstrich> encapsule

Abkipp... 1. <Bau> dumping; 2. <Lufttrans> stall, stalling
abkippen v <Bau> shoot, tip
Abkippen n <Lufttrans> dive (des Flugzeugs nach vorne)
Abkippförderkorb m <Bau> dump skip
Abkipptrudelverhalten n <Lufttrans> stall spin characteristics
abklären v 1. <Chemie> clarify; 2. <Chemtech> clarify, elutriate
Abklärgefäß n <Chemtech> decantation vessel, decanter, decanting glass
Abklärung f <Chemie, Patent, Qual> clarification
abklemmen v 1. <Elektrotech, Kontroll> disconnect; 2. <Telekom> disconnect (Draht)
Abkling... <Kerntech> decay
Abklingbecken n <Kerntech> neutralization pond
Abklingcharakteristik f <Elektrotech> decay characteristic
abklingen v <Strahlphys> decay
Abklingen n 1. <Elektrotech> decay; 2. <Fernseh> damping, decay; 3. <Kerntech> decay, neutralization; cooling (Reaktor); 4. <Strahlphys> decay
Abklingen n **des angeregten Zustandes** <Kerntech, Strahlphys> excited-state deactivation
abklingende Schwingung f <Telekom> ringing (Oszillation)
abklingende Welle f <Phys> decaying wave
abklingendes Feld n <Optik, Telekom> evanescent field
Abklingfaktor m <Elektrotech> decay factor
Abklingkonstante f <Phys, Strahlphys, Teilphys> decay constant
Abklingrate f <Akustik> decay rate
Abklingteich m <Kerntech> discharge pond (für verbrauchte Brennelemente)
Abklingzeit f 1. <Elektronik, Elektrotech> decay time; 2. <Kerntech> cooling-down period (von radioaktivem Material); 3. <Optik> settling time; 4. <Phys> decay time (Impuls)
Abklingzeitkonstante f <Gerät> damping time constant
abklopfen v <Fertig> rap
Abklopfen n <Bau> picking
Abklopfer m <Fertig> rapper (Gießen)
Abklopfhammer m <Maschinen> boilermaker's hammer
abknallen v <Fertig> pop (Flamme)
Abknallen n <Heiz & Kälte> backfiring (Brenner)
Abknickung f <Konstzeich> offset, zigzag
Abkochmittel n <Chemie> scouring agent (Rohseide)
Abkochung f <Lebensmittel> decoction
abkohlen v <Kohlen> break coal
A/B-Konverter m <Telekom> a/b converter (ISDN-Anschluss für Analogtelefon)
abkoppeln v <Kontroll> disconnect
Abkopplung f <Kerntech> disconnect rod (für Abschaltstab)
Abkreiden n <Kunststoff> chalking
abkühlen v 1. <Heiz & Kälte> chill; 2. <Ker & Glas> cool down; 3. <Mechan> chill, quench; 4. <Textil> cool; 5. <Thermod> chill
Abkühlen n <Kunststoff> chilling
Abkühlphase f <Ker & Glas> cooling-down period
Abkühlung f 1. <Heiz & Kälte> chilling; 2. <Textil, Verpack> cooling
Abkühlungsgeschwindigkeit f 1. <Heiz & Kälte> cooling rate; 2. <Thermod> rate of cooling
Abkühlungskurve f <Metall, Qual> cooling curve
Abkühlzeit f <Kerntech> cooling-down period
abkuppeln v 1. <Maschinen> throw out; 2. <Mechan> disengage
abkürzen v <Qual> curtail
Abkürzung f <Trans> shortcut

Abladestation

Abladestation f <Bau> dumping station
Ablage f 1. <Comp & DV> filing; 2. <Verpack> deposit
ablagern v <Abfall> tip (Müll)
Ablagern n <Verpack> deposition (von Karton)
Ablagerung f 1. <Bau> alluviation, deposit, sediment; 2. <Chemie> deposit; 3. <Chemtech> sediment; desiccation (Holz); 4. <Fertig> deposit; 5. <Umweltschmutz> deposition (Bergbau); 6. <Wasserversorg> accretion, deposit, sediment
Ablagerungsbecken n <Wasserversorg> settling basin
Ablagerungsgeschwindigkeit f <Umweltschmutz> deposition velocity (radioaktive Partikel)
Ablagerungsplatz m <Abfall, Bau> storage area
Ablagerungswert m <Umweltschmutz> deposition value
ablandig adj <Wassertrans> offshore
ablandiger Wind m <Wassertrans> offshore wind
Ablass m 1. <Bau> bibcock (eines Ausgussbeckens); 2. <Kohlen> blow-off; 3. <Maschinen> drain; 4. <Wasserversorg> outlet
ablassbarer unnutzbarer Treibstoff m <Lufttrans> drainable unusable fuel
ablassen v 1. <Fertig> bleed (Flüssigkeit); 2. <Maschinen> bleed, drain (Flüssigkeit); deflate (Luft aus Reifen); 3. <Raumfahrt> jettison (Raumschiff); 4. <Sicherheit> lower; 5. <Umweltschmutz> discharge (Wasser)
Ablassen n <Kerntech, Mechan> bleeding
Ablassen n **der Luft** <Kfztech> deflation (Reifen)
Ablassgraben m <Wasserversorg> outlet channel
Ablasshahn m 1. <Bau, Fertig> drain cock; 2. <Kfztech> drain cock (Kühler); 3. <Maschinen> drain cock, draw-off tap, pet cock
Ablasskühlung f <Raumfahrt> dump cooling (Raumschiff)
Ablassöffnung f 1. <Maschinen, Metall> discharge aperture, discharge opening; 2. <Sicherheit> relief (Druckentlastung)
Ablassrohr n <Wasserversorg> outlet pipe
Ablassstollen m <Wasserversorg> tailrace tunnel
Ablassventil n 1. <Eisenbahn> overflow valve; 2. <Hydraul> discharge valve, poppet valve; 3. <Maschinen> drain valve; 4. <Raumfahrt> jettison valve (Raumschiff); 5. <Wassertrans> drain valve
Ablassventilsteuerung f <Hydraul> poppet valve gear (Dampfmaschine)
Ablation f <Comp & DV, Kerntech> ablation
Ablationsimpuls m <Kerntech> ablating momentum
Ablationsion n <Kerntech> ablated ion
Ablationskühlung f <Raumfahrt> ablative cooling (Raumschiff)
Ablationsschild m <Raumfahrt> ablation shield (Raumschiff)
ablativ adj <Raumfahrt> ablative (Raumschiff)
Ablauf m 1. <Comp & DV> flow, run; 2. <Fertig> operating sequence, work cycle; 3. <Kerntech> outlet; 4. <Kontroll> routine; 5. <Meerschmutz, Nichtfoss Energ> runoff; 6. <Phys> course (Experiment); 7. <Wasserversorg> outflow
Ablauf m **des Zeitgebers** <Telekom> expiry of timer
Ablaufbacke f <Kfztech> trailing shoe (Bremse)
Ablaufbahnhof m <Eisenbahn> classification yard with hump
Ablaufbogen m <Bau> arch of discharge
Ablaufdiagramm n <Comp & DV> flow diagram, flowchart
ablaufen v 1. <Comp & DV> pass; 2. <Fertig, Wasserversorg> run off
ablaufen lassen v <Comp & DV> run (Programm)
Ablaufen n <Kunststoff> curtaining
ablaufende Bandspule f <Fernseh> supply reel
ablaufende Filmrolle f <Fernseh> supply roll
ablaufendes Wasser n <Wassertrans> falling tide (Gezeiten)
Ablauffolge f <Kontroll> sequence
Ablaufgerinne n <Wasserversorg> tail race (eines Wasserrads)
Ablaufgeschwindigkeit f <Qual> rate of progress
Ablaufhaspel f <Fertig> pay-off reel
Ablaufkanal m <Nichtfoss Energ, Wasserversorg> flume
Ablaufplansymbol n <Comp & DV> flowchart symbol
Ablaufprogrammierung f <Comp & DV> run-off programming
Ablauframpe f <Bau> gravity incline (Eisenbahn)
Ablaufrangierbetrieb m <Eisenbahn> hump shunting
Ablaufrichtung f <Comp & DV> flow direction
Ablaufrohr n 1. <Fertig> outlet pipe (Kunststoffinstallationen); 2. <Wasserversorg> delivery pipe
Ablaufschritt m <Comp & DV> step
Ablaufsteuerung f 1. <Comp & DV> sequencer, sequencing; 2. <Telekom> sequence control
ablaufsynchroner Betrieb m <Comp & DV> real-time operation
Ablaufüberwachungssystem n <Kontroll> supervising system
Ablaufverfolgung f <Comp & DV> trace
Ablaufverfolgungsprogramm n <Comp & DV> trace program
Ablauge f <Abfall> waste lye
Ablauger m <Bau> paint stripper
Abläutesignal n <Eisenbahn> train-announcing signal
ablegen v 1. <Erdöl> lay down (Leitung vom Verlegeschiff); 2. <Wassertrans> cast off (festmachen)
• **Datensicherungsbänder ablegen** <Comp & DV> file data storage tapes
Ablegen n **auf Gips** <Ker & Glas> laying on plaster
Ablegen n **auf Stoff** <Ker & Glas> laying on cloth
Ablegen n **von Rohglas** <Ker & Glas> laying (zum Schleifer)
Ablegeplatz m <Ker & Glas> laying yard
Ablegesatz m <Druck> dead matter, dead type
Ablehngrenze f <Qual> limiting quality
ableichtern v <Meerschmutz> lighten
ableiten v 1. <Bau> carry off (Wärme); 2. <Fertig> leak
Ableitstrom m <Elektrotech> leakage current
Ableitung f 1. <Elektrotech> dissipation, leakage; 2. <Funktech> leak; 3. <Geom> differentiation (trigonometrischer Funktionen); 4. <Math> derivative, differential coefficient, differential derivative; derivative (einer Funktion); 5. <Telekom> leakage; 6. <Wasserversorg> offtake; diverting (eines Flusses)
Ableitung f **trigonometrischer Funktionen** <Geom> differentiation of trigonometrical functions
Ableitungskanal m <Wasserversorg> discharge canal
Ableitungsverlust m <Elektrotech> dissipative loss
Ableitvorrichtung f <Elektrotech> sink
Ableitwiderstand m <Elektrotech> bleeder, bleeder resistor
Ablenk... 1. <Bau> baffle; 2. <Gerät> deflecting, deflection
Ablenkamplitude f <Fernseh, Gerät> amplitude of deflection
Ablenkblech n 1. <Bau> baffle plate; 2. <Chemtech> deflector; 3. <Heiz & Kälte> deflector plate; 4. <Ker & Glas> baffle (Ofen); 5. <Kfztech> baffle (Auspufftopf); 6. <Lufttrans> deflector; 7. <Maschinen> baffle, deflector plate
Ablenkbohren n <Erdöl> sidetrack drilling
Ablenkelektrode f <Elektrotech, Gerät, Kerntech> deflecting electrode, deflection electrode
Ablenkempfindlichkeit f <Elektronik> deflection sensitivity

ablenken v 1. <Fernseh> deflect; 2. <Gerät> deflect *(Elektronenstrahl)*; 3. <Maschinen> deflect
Ablenker m <Fertig> deflector
Ablenkfaktor m <Elektronik> deflection factor
Ablenkjoch n 1. <Elektronik> deflection yoke *(Katodenstrahlröhre)*; 2. <Fernseh> deflection yoke, scanning yoke
Ablenkkeil m <Erdöl> whipstock
Ablenkmagnet m 1. <Fernseh> deflection magnet; 2. <Teilphys> deflecting magnet
Ablenknormen fpl <Fernseh> scanning standards
Ablenkplatte f 1. <Chemtech> baffle; 2. <Elektronik> deflection plate *(Ionenfalle)*; 3. <Fernseh> deflection plate; 4. <Fertig> baffle board; 5. <Gerät> deflecting electrode; 6. <Heiz & Kälte> deflector plate; 7. <Ker & Glas> baffle plate; 8. <Maschinen> baffle; 9. <Raumfahrt> deflection plate *(Raumschiff)*; 10. <Strahlphys> deflection plate
Ablenkrohr n <Strahlphys> deflection tube
Ablenksieb n <Papier> baffle wire
Ablenkspule f 1. <Elektrotech> deflection coil; 2. <Fernseh> scanning coil; 3. <Kerntech, Phys> deflecting coil
Ablenkstrahl m <Fernseh> scanning beam
Ablenkstromgenerator m <Fernseh> scan current generator
Ablenksystem n <Fernseh> deflection system
Ablenkung f 1. <Erdöl> sidetracking *(Tiefbohrtechnik)*; 2. <Gerät> scan *(Oszilloskop)*; 3. <Maschinen, Mechan, Optik> deflection; 4. <Phys> deviation *(Lichtstrahlen)*; 5. <Raumfahrt> deflection; 6. <Strahlphys> deflection *(elektrisches oder magnetisches Feld)*; 7. <Telekom> deviation *(eines Strahls)*; 8. <Wassertrans> deviation *(Kompass)*
Ablenkung f im Magnetfeld <Kerntech> magnetic deflection
Ablenkventil n <Hydraul> deflecting valve
Ablenkverstärker m 1. <Elektronik> deflection amplifier; 2. <Gerät> sweep deflection amplifier *(Oszilloskop)*
Ablenkvorrichtung f <Maschinen, Telekom> deflector
Ablenkweite f <Fernseh> trace interval
Ablenkwinkel m 1. <Fernseh> deflection angle; 2. <Phys> angle of deviation
Ablesebereich m <Fertig> reading range
ablesen v <Gerät> read *(Messergebnis)*
Ablesen n <Bau> reading
Ablesesystem n <Maschinen> read-out system
Ablesewert m <Gerät> reading
Ablesung f 1. <Gerät> reading; 2. <Telekom> readout
Ablieferungsprüfung f <Qual> user's inspection
Ablieferungszeichnung f <Qual> as-delivered condition
abliegende Zunge f <Eisenbahn> open point, open switch
ablöhnen v <Wassertrans> pay off *(Besatzung)*
Ablösbarkeit f <Papier> release
ablöschen v <Chemie> slake *(Kalk)*
Ablöseeinrichtung f <Sicherheit> blow-off system
ablösen v <Math> commute
Ablöseschaltung f <Telekom> combining circuit
Ablösung f <Maschinen> parting
Ablösung f der Verdichterströmung <Lufttrans> compressor stall *(Turbinenmotoren)*
Ablösungsmannschaft f <Lufttrans> relief crew
Abluft f 1. <Abfall> cleaned gas, scrubbed gas; 2. <Heiz & Kälte> vitiated air; 3. <Papier> exhaust
Abluftleistung f <Heiz & Kälte> extracted-air flow rate
Ablufttreiniger m <Textil> exhaust cleaning installation
Abluftstrom m <Heiz & Kälte> exhaust airstream
Abluftsystem n <Sicherheit> exhaust vent installation
Abluftventilator m <Papier> exhaust fan
Abluftvolumenstrom m <Heiz & Kälte> extracted-air flow rate

Abmaß n 1. <Maschinen> deviation, permissible allowance; 2. <Qual> deviation; 3. <Textil> tolerance
abmelden v <Telekom> de-affiliate
abmelden v/sich 1. <Comp & DV> log off, log out; sign off *(vom System)*; 2. <Mobilkom> sign off
Abmelden n <Comp & DV> logoff, logout
Abmesseinheit f <Verpack> volumetric filling unit
abmessen v <Kontroll, Textil, Wasserversorg> *(AE)* gage, *(BE)* gauge
Abmesspumpe f <Verpack> dosing pump
Abmessung f 1. <Comp & DV> physical dimension; 2. <Druck> dimension; 3. <Elektronik> measurement
Abmessung f von Hand <Verpack> hand dosing
Abmessungskompatibilität f <Telekom> dimensional compatibility
abmontiert adj <Bau> struck *(Gerüst)*
abmustern v <Wassertrans> discharge *(Schiffsbesatzung)*
Abmusterung f <Wassertrans> paying-off *(Besatzung)*
Abnäher m <Textil> dart
Abnahme f 1. <Maschinen> inspection; 2. <Math> decrease; 3. <Mechan, Qual> acceptance
Abnahmebericht m <Mechan> acceptance report
Abnahmebescheinigung f <Qual> acceptance certificate
Abnahmekriterium n <Qual> acceptance criterion
Abnahmelehre f <Maschinen> *(AE)* inspection gage, *(BE)* inspection gauge
Abnahmemodell n <Raumfahrt> qualification model
Abnahmeprobebrand m <Raumfahrt> acceptance firing test
Abnahmeprobeflug m <Lufttrans> acceptance flight
Abnahmeprotokoll n <Qual> acceptance certificate, acceptance report
Abnahmeprüfprotokoll n <Patent, Qual> certificate of acceptance
Abnahmeprüfung f 1. <Comp & DV> acceptance test; 2. <Fertig> receiving inspection; 3. <Kohlen> acceptance test; 4. <Qual> acceptance checking, acceptance inspection, acceptance test; 5. <Telekom> acceptance test
Abnahmeprüfungen fpl <Wassertrans> acceptance trials
Abnahmeprüfzeugnis n <Qual> acceptance test certificate, inspection report
Abnahmepunkt m <Qual> witness point
Abnahmetest m 1. <Maschinen> acceptance inspection; 2. <Mechan> acceptance testing; 3. <Telekom> acceptance test
Abnahmeverfahren n <Qual> acceptance procedure
Abnahmeverweigerung f <Elektronik> rejection *(Computertechnik)*
Abnahmevorschriften fpl <Qual, Telekom> acceptance test specification
Abnahmewalze f <Papier> pick-up roll
Abnahmezeichnung f <Heiz & Kälte, Qual> acceptance drawing
abnehmbar adj <Foto, Maschinen, Optik> detachable, removable
abnehmbare Andruckplatte f <Foto> detachable pressure plate
abnehmbare Backen fpl <Maschinen> detachable jaws
abnehmbare Kupplungslasche f <Eisenbahn> removable coupling link
abnehmbare Rückwand f <Foto> removable back
abnehmbare Tastatur f <Comp & DV> detachable keyboard
abnehmbarer Griff m <Maschinen> detachable handle
abnehmen v 1. <Comp & DV> decay; 2. <Math> decrease; 3. <Qual> accept; 4. <Telekom> unhook
Abnehmen n 1. <Ker & Glas> take-down; 2. <Papier> pick-up

Abnehmer *m* <Ker & Glas> doff
Abnehmerzähler *m* <Gerät> consumer meter *(für Energieverbrauch)*
abnormale Beendigung *f* <Comp & DV> abnormal termination
abnormale Struktur *f* <Metall> abnormal structure
abnutzen *v* 1. <Anstrich> erode; fret *(durch Korrosion oder Reiben)*; 2. <Textil> wear out
Abnutzung *f* 1. <Anstrich> erosion; 2. <Elektriz> wearout; 3. <Kerntech> wastage; 4. <Maschinen> wear; 5. <Mechan> abrasion, fretting, wear; 6. <Meerschmutz, Textil> wear
Abnutzungsausgleich *m* <Fertig> wear compensation
Abnutzungsfaktor *m* <Mechan> abrasion factor
Abnutzungsfläche *f* <Maschinen, Textil> wearing surface
Abnutzungsgrenze *f* <Fertig> wear limit
Abonnementfernsehen *n* <Fernseh> subscription television, subscription TV
Abonnement-Service *m* <Comp & DV> subscription-based service
Abpackung *f* **in Steigen** <Verpack> tray packing
abpausen *v* <Maschinen> trace
abplatzen *v* 1. <Bau> spall; 2. <Ker & Glas> chip
Abplatzen *n* 1. <Anstrich, Bau> spalling; 2. <Ker & Glas> chipping
Abprall *m* <Fertig> ricochet
abprallen *v* <Fertig> ricochet
abpumpen *v* <Maschinen, Wassertrans> pump off, pump out
abputzen *v* <Fertig> scour
Abquetsch... 1. <Kunststoff> flash; 2. <Textil> quetsch
abquetschen *v* <Foto> squeegee
Abquetschfläche/mit <Fertig> landed
Abquetschgrat *m* <Kunststoff> flash
Abquetschvorrichtung *f* <Textil> *(AE)* quetsch unit
Abquetschwalze *f* 1. <Papier> squeeze roll; 2. <Textil> *(AE)* quetsch roller
Abquetschwerkzeug *n* <Kunststoff> *(AE)* flash mold, *(BE)* flash mould
Abraham'scher Impuls *m* <Kerntech> Abraham momentum
Abrasion *f* <Anstrich, Kohlen> abrasion
Abrasionswiderstand *m* <Anstrich> abrasion resistance
abrasiver Verschleiß *m* <Maschinen> abrasive wear
Abraum... *adj* <Kohlen> overburden
Abraumbau *m* <Kohlen> open-pit mining
Abraumbohrer *m* <Kohlen> overburden drill
Abraumdruck *m* <Kohlen> overburden pressure
abräumen *v* <Bau> clear
Abräumen *n* <Kohlen> stripping
Abraumkippe *f* <Bau> spoil area
Abraumschicht *f* <Kohlen> overburden
Abrechnungsdatei *f* <Comp & DV> accounting file
abreiben *v* 1. <Bau> rub *(Putz)*; 2. <Fertig> waste; grind *(Werkzeug)*; 3. <Maschinen, Mechan, Papier> abrade, abrase
abreibend *adj* <Papier> abradant
abreichern *v* <Kerntech> deplete
Abreicherung *f* <Kerntech> depletion
abreißen *v* 1. <Bau> break down; demolish *(Gebäude)*; 2. <Elektriz> interrupt *(Lichtbogen)*; 3. <Maschinen> tear down
Abreißen *n* <Fertig> interruption *(Lichtbogen)*
Abreißstab *m* **für Schreiberstreifen** <Gerät> chart paper tear-off bar
Abreißstartstrom *m* <Elektriz> breakaway starting current
Abreißverschluss *m* <Verpack> snap-off closure, tear-off closure

Abreißzündung *f* <Kfztech> make-and-break ignition
Abrichtdiamant *m* <Fertig> diamond dresser
abrichten *v* <Maschinen> dress, true, true up
Abrichten *n* 1. <Ker & Glas> planing; 2. <Maschinen> dressing, truing
Abrichten *n* **durch Profilrolle** <Fertig> crush dressing
Abrichter *m* <Fertig> truer; dresser *(für Schleifscheibe)*
Abrichtgerät *n* <Maschinen> dressing device
abrichthobeln *v* <Maschinen> surface
Abrichthobeln *n* <Maschinen> surface planing, surfacing
Abrichtrolle *f* <Fertig> block truer
Abrichtscheibe *f* <Maschinen> truing wheel
Abrichtvorrichtung *f* <Maschinen> dressing equipment
Abrichtvorrichtung *f* **für Schleifscheiben** <Maschinen> grinding wheel dressing equipment
Abrieb *m* 1. <Anstrich> abrasion; 2. <Bau> attrition; 3. <Druck> abrasion; 4. <Fertig> abrasion *(spanende Bearbeitung)*; 5. <Ker & Glas> attrition; 6. <Kerntech> galling; 7. <Kohlen> fines; 8. <Kunststoff> abrasion; 9. <Maschinen> abrasion, attrition; 10. <Mechan, Papier> abrasion
Abriebbeständigkeit *f* <Papier> abrasion resistance
Abriebeigenschaften *fpl* <Papier> abrasiveness
abriebfest *adj* 1. <Druck> abrasion-resistant; 2. <Mechan> abrasion-proof
Abriebfestigkeit *f* <Anstrich, Bau, Kunststoff, Maschinen, Mechan> abrasion resistance
Abriebkorrosion *f* <Chemie> fretting corrosion
Abriebmarkierung *f* <Ker & Glas> scuff mark
Abriebprüfmaschine *f* 1. <Kunststoff> abrasion tester; 2. <Papier> abrasion tester, adrader
Abriebprüfung *f* <Maschinen> abrasion test
Abriebsprotektor *m* <Lufttrans> certification weight, chafing strip
Abriebtest *m* <Maschinen> attrition test
Abriebverschleiß *m* <Bau> abrasive wear
Abriegelung *f* <Sicherheit, Wasserversorg> blocking
Abriss *m* **der Verdichterströmung** <Lufttrans> compressor stall *(Turbinenmotor)*
Abrisshäufigkeit *f* <Papier> breakage rate
Abrollabrichten *n* <Fertig> crushing *(Schleifscheibe)*
abrollen *v* 1. <Aufnahme> roll off; 2. <Comp & DV> scroll
Abrollen *n* <Papier> reeling off
Abrollhaspel *f* <Fertig> uncoiler
Abruf *m* 1. <Comp & DV> fetch; 2. <Ker & Glas> call down; 3. <Qual> requisition; 4. <Telekom> polling, request
Abruf *m* **auf Aufforderung** <Comp & DV> demand fetching
Abrufanweisung *f* <Comp & DV> fetch instruction
Abrufbefehl *m* <Comp & DV> fetch instruction
Abrufbusdienst *m* <Trans> demand-scheduled bus service
abrufen *v* 1. <Comp & DV> fetch, request, retrieve; 2. <Telekom> request, select
Abrufen *n* **von Nachrichten** <Comp & DV> message retrieval
Abrufen *n* **von Systemmeldungen** <Comp & DV> message retrieval
Abrufen *n* **von Text** <Comp & DV> text retrieval
Abrufphase *f* <Comp & DV> fetch phase
Abrufsignal *n* <Comp & DV> fetch signal
Abruftaste *f* <Comp & DV> attention key
Abruftechnik *f* <Comp & DV> polling mode
Abrufunterbrechung *f* <Comp & DV> attention interrupt
Abrufzyklus *m* <Comp & DV> fetch cycle
Abrundeisen *n* <Bau> arrissing tool
abrunden *v* 1. <Comp & DV> round down, round off; 2. <Math> round
Abrunden *n* 1. <Fertig> radii-forming *(Spanung)*; 2. <Math> rounding

Abrundung f <Math> truncation
abrüsten v <Bau> dismantle
Abrutschen n <Ker & Glas> slough
ABS 1. <Kfztech> *(Antiblockiersystem)* ABS, antiblocking system, antilock braking system; 2. <Kunststoff> *(Acrylnitril-Butadien-Styrol)* ABS, acrylonitrile butadiene styrene *(Copolymer)*
Absacken n <Bau> subsidence
Absacklinie f <Verpack> sack-filling line
Absackung f <Bau> slump
Absackwaage f <Gerät> sacking balance
Absatz m 1. <Bau> bench, berm; 2. <Comp & DV> paragraph; 3. <Druck> break; 4. <Mechan> offset
Absatzdrehen n <Fertig> shoulder turning
absatzfähig adj <Textil> marketable
Absatzmaß n <Konstzeich> stepped dimension
Absatzwechsel m <Druck> new paragraph
absaufen v <Kfztech> flood
Absaug... 1. <Chemtech> filtering; 2. <Heiz & Kälte> extract; 3. <Sicherheit> suction
Absauganlage f <Lebensmittel, Sicherheit> exhauster, exhaust system
absaugbarer unnutzbarer Treibstoff m <Lufttrans> drainable unusable fuel
Absaugbehälter m <Lebensmittel> exhaustion box
absaugen v <Heiz & Kälte> extract
Absauger m 1. <Labor> aspirator; 2. <Mechan> exhaust pump
Absaugeschweißtisch m <Sicherheit> welding table
Absaugflasche f <Chemtech> filtering flask
Absauggebläse n <Heiz & Kälte> extract fan
Absauglüftung f <Fertig> extract ventilation
Absaugmittel n <Mechan> absorbent
Absaugpumpe f <Hydraul> aspiration pump, aspiring pump
Absaug- und Filtervorrichtung f <Sicherheit> suction and filter installation *(für Staub und Späne)*
Absaugung f 1. <Fertig> extraction *(Kunststoffinstallationen)*; 2. <Sicherheit> welding; 3. <Textil> suction
Absaugventil n <Labor> exhaust valve
Absaugvorrichtung f <Fertig> eductor
Abschäler m <Bau> scarifier
Abschalt... 1. <Maschinen> shut-off; 2. <Raumfahrt> shutdown
Abschaltanweisung f <Raumfahrt> shutdown procedure *(Raumschiff)*
Abschaltdruck m <Maschinen> shut-off pressure
abschalten v 1. <Comp & DV> disconnect; 2. <Elektrotech> switch off, de-energize, disable, disconnect, isolate; turn off *(Stromversorgung abstellen)*; 3. <Kontroll> switch off; 4. <Maschinen> stop; 5. <Raumfahrt> shut down, shut off; 6. <Telekom> disconnect *(Gerät)*
Abschalten n <Raumfahrt> shutdown procedure *(Raumschiff)*
Abschaltfühler m <Raumfahrt> shutdown sensor *(Raumschiff)*
Abschaltimpuls m <Elektronik> turn-off pulse
Abschaltkreis m <Regelung> de-energizing circuit
Abschaltrelais n <Elektriz> cutoff relay
Abschaltschütz n <Elektriz> cutoff relay
Abschaltspannung f <Elektrotech> interrupting voltage
Abschaltstrom m <Elektriz> breaking capacity, breaking current, cutoff current
Abschaltstromkreis m <Elektrotech> shutdown circuit
Abschaltung f 1. <Elektrotech> cutoff, de-energization, disconnection; turn-off *(Stromversorgung)*; 2. <Fernseh> cutoff; 3. <Hydraul> cutoff *(Dampfzufuhr)*; 4. <Kerntech> shutdown *(Reaktor)*; 5. <Raumfahrt> shutdown *(Raumschiff)*; 6. <Telekom> disconnection *(Teilnehmer)*

Abschaltverzögerung f 1. <Kerntech> scram delay; 2. <Kontroll> turn-off delay
Abschaltvorgang m <Elektrotech> power down operation, switch off operation
Abschaltvorrichtung f 1. <Elektrotech> power-down feature *(Halbleiterspeichertechnik)*; 2. <Hydraul> cutoff device *(Dampfzufuhr)*
Abschaltzeit f 1. <Elektriz, Elektrotech> opening time, turn-off time; 2. <Kerntech> disable time, switched-off time; *(AE)* turnaround time, *(BE)* turnround time *(bei Reinigung)*; *(AE)* turnaround time, *(BE)* turnround time *(einer Wiederaufbereitungsanlage)*; 3. <Kontroll> turn-off time; 4. <Mechan> downtime
Abschattung f 1. <Fernseh> shadowing; 2. <Kerntech> corner cutting *(Bildschirm)*; 3. <Telekom> shadowing
abschätzen v <Bau> rate
Abschätzung f <Maschinen, Qual> assessment
Abschätzung f **der Folge mit der größten Wahrscheinlichkeit** <Telekom> maximum likelihood sequence estimation
Abschäum... <Ker & Glas> skim, skimming
Abschäumbalken m <Ker & Glas> skim bar
abschäumen v <Ker & Glas> skim
Abschäumer m <Ker & Glas> skimmer
Abschäumloch n <Ker & Glas> skimming hole
Abschäumvorbau m <Ker & Glas> skim pocket
Abscheide... <Chemtech> separation
abscheiden v 1. <Abfall> segregate; 2. <Chemtech> separate, settle
Abscheiden n <Chemtech> separation *(Komponenten)*
Abscheider m 1. <Bau> interceptor; 2. <Chemtech> separator, settling tank, trap; 3. <Erdöl> catchpot *(Rohrleitungen)*; 4. <Fertig> strainer *(Kunststoffinstallationen)*; 5. <Heiz & Kälte> trap; 6. <Maschinen, Meerschmutz> separator
Abscheiderzyklon m <Lebensmittel> cyclone *(Fliehkraftscheider)*
Abscheidevorrichtung f <Chemtech> separator
Abscheidewirkung f <Chemtech> separation effect
Abscheidung f 1. <Abfall> segregation; 2. <Chemie> deposit
Abscherbolzen m <Fertig, Maschinen> shear pin
abscheren v 1. <Bau> shear; 2. <Fertig> shear off *(Kunststoffinstallationen)*; 3. <Wassertrans> sheer off *(Navigation)*
Abscheren n <Ker & Glas> shearing-off
Abscherfestigkeit f 1. <Bau, Maschinen, Metall> shearing tenacity; 2. <Mechan> shear strength
Abscherstift m <Maschinen> shear pin
Abscherung f <Bau, Maschinen, Metall> shearing
Abscherversuch m <Maschinen, Qual> shear test *(Nieten)*
abscheuern v <Wassertrans> hog
abschießen v <Raumfahrt> launch
Abschilferung f <Mechan> scale
Abschirm... 1. <Elektriz> shielding; 2. <Lufttrans> deflector
Abschirmblech n <Lufttrans> deflector
Abschirmeffekt m <Elektriz> shielding effect
abschirmen v 1. <Bau> shield; 2. <Chemtech> baffle; 3. <Funktech, Raumfahrt> shield
Abschirmen n <Fernseh> screening
Abschirmfaktor m <Kerntech> screen factor
Abschirmgitter n <Sicherheit> screen grid
Abschirmkabel n <Elektrotech> shielding cable
Abschirmkonstante f <Kerntech> screening constant
Abschirmleiter m <Elektriz> shielding conductor
Abschirmplatte f <Kerntech> screen plate
Abschirmrelais n <Elektrotech> shielding relay, screening relay

Abschirmung

Abschirmung f 1. <Aufnahme, Bau> screening, shielding; 2. <Comp & DV> screening, shielding *(Schutz)*; 3. <Elektriz, Elektrotech> screening, shielding *(eines Netzwerkes)*; 4. <Fernseh, Funktech, Kerntech, Phys, Raumfahrt> screening, shielding; 5. <Sicherheit> safeguard, safeguarding device; screen guard, screening, shielding; 6. <Strahlphys> screening, shielding; 7. <Telekom> screen *(Kabel)*
Abschirmungseffekt m 1. <Lufttrans> shielding effect; 2. <Telekom> screen effect
Abschirmungszahl f <Kerntech> screening number
Abschirmvorrichtung f <Sicherheit> screen guard
abschlacken v <Fertig> flush
Abschlacken n <Fertig> deslagging
Abschlagen n <Bau> spalling
abschlämmen v 1. <Chemie> clarify; 2. <Chemtech> elutriate; 3. <Wasserversorg> blow down
Abschlämmen n 1. <Chemie> clarification; 2. <Chemtech> elutriation; 3. <Wasserversorg> blowdown
Abschlämmer m <Erdöl> desilter *(Bohrtechnik)*
abschleifen v 1. <Bau> sand *(mit Sandpapier)*; 2. <Ker & Glas> cut off *(der Glaskanten)*; 3. <Maschinen, Mechan, Papier> abrade, abrase
Abschleifen n 1. <Fertig> abrading; 2. <Maschinen> grinding-off
Abschleifen n **von Porzellanemail** <Ker & Glas> stoning
abschleifend adj <Mechan> abrasive
Abschlepp... <Trans> towing
abschleppen v <Trans> tow
Abschleppen n <Trans> debogging *(eines Fahrzeuges aus Schlamm)*
Abschlepphaken m <Kfztech> tow hook *(für Anhänger)*
Abschleppöse f <Kfztech> towing bracket
Abschleppstange f <Kfztech> bullbar
Abschleppwagen m <Kfztech> *(AE)* salvage car, *(BE)* salvage lorry, towing vehicle, *(AE)* wrecker
abschleudern v <Chemtech> centrifuge
abschließbar adj <Sicherheit> lockable
abschließbarer Zugang m <Sicherheit> lockable access door
abschließen v <Telekom> terminate *(Kabel)*
Abschließen n <Elektrotech> terminating
Abschliff m <Fertig> abraded material
Abschluss m 1. <Comp & DV> closedown *(des aktiven Betriebszustandes)*; 2. <Kontroll> termination *(eines Prozesses)*
Abschlussanweisung f <Comp & DV> close statement
Abschlussblende f <Foto> front diaphragm
Abschlussblock m <Kontroll> terminal block
Abschlusscheck m <Lufttrans> postflying check
Abschlussdeckel m <Fertig> end plate
Abschlusselement n <Elektrotech> terminating element
Abschlusselement n **für Lichtleitfasern** <Optik> *(AE)* optical fiber pigtail, *(BE)* optical fibre pigtail
Abschlussflansch m <Maschinen> blank flange, blind flange
Abschlussimpedanz f <Elektrotech> terminating impedance
Abschlusskappe f <Elektrotech> end cap
Abschlussmast m <Elektriz> terminal tower
Abschlussmauer f <Bau> head wall
Abschlussplatte f <Foto> stop plate
Abschlussprozedur f <Kontroll> termination procedure
Abschlussseite f <Comp & DV> trailer page
Abschlussstecker m <Comp & DV> terminator
Abschlussstein m <Ker & Glas> seal block, tuckstone
Abschlussstopfen m <Fertig> end plug *(Kunststoffinstallationen)*
Abschlussstück n **für Lichtleitfasern** <Optik> *(AE)* optical fiber pigtail, *(BE)* optical fibre pigtail
Abschlusswiderstand m 1. <Comp & DV> terminator; 2. <Elektrotech> terminating resistor
Abschlusszeichen n <Comp & DV> terminator
Abschmelzbug m <Raumfahrt> ablating cone *(Raumschiff)*
Abschmelzen n <Phys> ablating
abschmelzende Schweißelektrode f <Bau> consumable welding
Abschmelzleistung f 1. <Fertig> deposition rate *(Schweißen)*; 2. <Umweltschmutz> deposition rate
Abschmelzung f <Elektrotech> smelting
Abschmier... <Maschinen> grease, lubrication
Abschmierfett n <Maschinen> lubricating grease
Abschmiergrube f <Maschinen> grease pit, greasing pit
Abschnappkupplung f <Maschinen> impulse coupling
Abschneide... 1. <Comp & DV> truncation; 2. <Telekom> cutback
Abschneidefehler m <Comp & DV> truncation error
Abschneidemethode f <Telekom> cutback technique
abschneiden v 1. <Comp & DV> truncate; 2. <Fertig> draw *(Eisen)*; 3. <Geom> subtend, truncate; 4. <Ker & Glas> *(BE)* break off, cap; 5. <Maschinen> shear
Abschneiden n 1. <Bau> cutting-off; 2. <Comp & DV> clipping, scissoring, truncation; 3. <Fertig> cropping; 4. <Geom> truncating; 5. <Ker & Glas> cutting-off *(Kanten)*; 6. <Künstl Int> pruning
Abschnellen n <Textil> picking
Abschnitt m 1. <Bau, Comp & DV, Druck, Eisenbahn> section; 2. <Geom> intercept, segment; 3. <Kerntech> leg *(Rohr)*; 4. <Kohlen> cuttings; 5. <Maschinen, Telekom> section; 6. <Textil> cutting
Abschnitt m **auf Geraden** <Geom> line segment
abschnittsweise Beladung f <Kerntech> zoned fuel loading *(Reaktor)*
abschnittsweise Leitweglenkung f <Telekom> link-by-link traffic routing
Abschöpf... <Meerschmutz> skimming
Abschöpfbarke f <Meerschmutz> skimming barge
Abschöpfeinrichtung f <Meerschmutz> skimmer
abschöpfen v <Fertig, Umweltschmutz> skim off
Abschöpfölsperre f <Meerschmutz> skimming barrier
abschrägen v 1. <Bau> bevel, slope, splay; 2. <Fertig, Maschinen> chamfer
Abschrägen n 1. <Bau> *(AE)* beveling, *(BE)* bevelling; 2. <Ker & Glas> *(AE)* beveling, *(BE)* bevelling, siding
Abschrägung f 1. <Bau> bevel, cant, chamfer; 2. <Elektrotech> ramp; 3. <Fertig> bevel, splay; 4. <Ker & Glas, Maschinen> bevel
Abschraubbohrrohr n <Erdöl> unscrewing pipe *(Bohrtechnik)*
abschrauben v 1. <Maschinen> unscrew; 2. <Mechan> loosen
Abschrauben n <Maschinen> unscrewing
Abschraubrohr n <Erdöl> unscrewing pipe *(Bohrtechnik)*
Abschreck... 1. <Kerntech> quench; 2. <Metall> quench, quenching
Abschreckalterung f <Kerntech, Metall> *(BE)* quench ageing, *(AE)* quench aging
Abschreckbad n <Metall> quenching bath
abschrecken v 1. <Bau> quench; 2. <Heiz & Kälte> chill; 3. <Mechan> chill, quench; 4. <Metall> quench; 5. <Thermod> chill; quench *(Stahl)*
Abschrecken n 1. <Ker & Glas> chilling; 2. <Metall> quench
Abschreckflüssigkeit f <Mechan> quenchant, quenching liquor
Abschreckhärten n <Kerntech, Metall> quench hardening

Abschreckmittel n <Fertig, Mechan> quenchant
Abschreckprüfung f <Werkprüf> thermal shock test (Glas)
Abschreckrissempfindlichkeit f <Thermod> heat treatment crack sensitivity
Abschreckversuch m <Maschinen, Metall, Qual> quench hardening test
Abschreibung f <Fertig> depreciation
Abschrot m 1. <Fertig> anvil cutter, hardie; 2. <Maschinen> anvil chisel, anvil cutter, hardie
abschroten v <Fertig> chop
abschwächen v 1. <Elektriz> damp; 2. <Ergon> attenuate; 3. <Foto> reduce; 4. <Maschinen> deaden
Abschwächer m <Foto> reducer
Abschwächung f 1. <Comp & DV> damping attenuation; 2. <Phys> attenuation; 3. <Raumfahrt> de-emphasis (Weltraumfunk); 4. <Wellphys> attenuation
Abschwächungsmittel n <Chemie> diluent
abschwefeln v <Chemie> (AE) desulfurize, (BE) desulphurize
Abschwefelung f <Chemie> (AE) desulfurization, (BE) desulphurization
ABS-Copolymer n <Elektriz> ABS copolymer
abseihen v <Chemtech> filter
Abseilgerät n <Sicherheit> rope-grab fall protection device
Absender m <Telekom> sender
absenkbare Rumpfnase f <Lufttrans> droop nose (Concorde)
absenken v 1. <Bau> sink; 2. <Bau> sink (Bohrloch); 3. <Kohlen> depress; 4. <Sicherheit> lower
Absenken n <Fertig> driving
Absenkung f <Erdöl> subsidence; drawdown (Bohrtechnik)
Absetz... 1. <Abfall> sedimentation, settling; 2. <Chemtech> precipitation, sedimentation, settling; 3. <Wasserversorg> sedimentation, settling
Absetzbecken n <Abfall, Bau, Chemtech, Wasserversorg> sedimentation basin, settling basin
Absetzbehälter m <Chemtech> precipitation tank, sedimentation tank
Absetzbottich m <Chemtech> settling tub, settling vat
Absetzdauer f <Chemtech> settling time
absetzen v 1. <Chemtech> settle; 2. <Fertig> shoulder; offset (Schneidmeißel); 3. <Kohlen> retreat; 4. <Wassertrans> plot (Peilung)
Absetzen n 1. <Kunststoff> sedimentation; 2. <Phys> settling
Absetzer m <Kohlen> boom stacker
Absetzgefäß n 1. <Abfall> settling vessel; 2. <Chemtech> thickener (Elektrochemie)
Absetzgeschwindigkeit f 1. <Abfall> settling velocity; 2. <Chemtech> settling speed
Absetzgrube f <Wasserversorg> cess pit, cess pool
Absetzkammer f <Abfall> settling chamber
Absetzklärung f <Abfall> decantation, settling
Absetzkonus m <Chemtech> settling cone
Absetzprobe f <Bau, Kohlen> sedimentation test
Absetztank m 1. <Chemtech> settling reservoir; 2. <Erdöl> settling tank (Bohrtechnik)
Absetzzisterne f <Chemtech> settling cistern
absieben v <Meerschmutz> sift; screen (Öl)
Absieben n <Chemtech> sieving
absinken v 1. <Chemtech> settle; 2. <Wasserversorg> draw down
Absinken n 1. <Erdöl> subduction (Geologie); 2. <Thermod> fall; 3. <Wasserversorg> drawdown
Absinkzone f <Erdöl> subduction zone (Geologie)
ABS-Kunststoff m <Verpack> ABS plastic

absolut adj <Comp & DV> absolute
absolut stetig adv <Math> absolute continuous
Absolut... <Gerät, Math, Phys> absolute
Absolutbetrag m <Math> modulus (einer komplexen Zahl)
Absolutbewegung f <Phys> absolute motion
Absolutdruck m <Heiz & Kälte> absolute pressure
Absolutdruck-Manometer n <Gerät> (AE) absolute pressure gage, (BE) absolute pressure gauge
absolute Adresse f <Comp & DV> absolute address, direct address, specific address
absolute Dielektrizitätskonstante f 1. <Elektriz> absolute permittivity; 2. <Elektrotech> absolute permittivity, dielectric constant
absolute Feuchte f <Heiz & Kälte, Phys> absolute humidity
absolute Geometrie f <Geom> absolute geometry
absolute Geschwindigkeitsänderung f <Elektriz> absolute speed variation
absolute Härte f <Bau> absolute hardness
absolute Häufigkeit f <Math> absolute frequency
absolute Instruktion f <Comp & DV> absolute instruction
absolute Kapazität f <Trans> absolute capacity
absolute Konstanz f <Telekom> absolute stability
absolute Permeabilität f <Elektriz, Elektrotech> absolute permeability
absolute Spannungsänderung f <Elektrotech> absolute voltage changing
absolute Temperatur f 1. <Heiz & Kälte> absolute temperature; 2. <Hydraul> absolute temperature (θ); 3. <Labor> absolute temperature (T); 4. <Lebensmittel, Phys> absolute temperature; 5. <Raumfahrt, Thermod> Kelvin temperature; 6. <Thermod> absolute temperature, thermodynamic temperature (θ)
absolute Temperaturskale f <Phys> perfect gas scale of temperature
absoluter Betrag m <Math> absolute value (komplexe Zahl)
absoluter Brechungsindex m <Strahlphys> absolute refractive index
absoluter Code m <Comp & DV> absolute code, specific code
absoluter Druck m <Heiz & Kälte> absolute pressure
absoluter Druckwandler m <Bau> absolute pressure transducer
absoluter Fehler m <Comp & DV> absolute error
absoluter Helligkeitsschwellwert m <Fernseh> absolute threshold of luminance
absoluter Nullpunkt m <Heiz & Kälte, Phys, Thermod> absolute zero
absoluter wasserfreier Alkohol m <Lebensmittel> dehydrated alcohol
absoluter Wert m <Comp & DV> absolute value
absolutes Blocksystem n <Eisenbahn> absolute block system
absolutes Maßsystem n <Werkprüf> absolute system
absolutes Moment n <Math> absolute moment (Verteilungsmerkmal)
absolutes Vakuum n <Heiz & Kälte> absolute vacuum
Absolutierung f <Chemtech> dehydration
Absolutkursdarstellung f <Funkort> true motion display
Absolutmaßsystem n <Gerät> absolute measure system (bei numerischer Steuerung)
Absolutmesssystem n <Elektronik> absolute measuring system
Absolutwert m 1. <Heiz & Kälte> absolute value; 2. <Math> modulus (komplexe Zahlen); absolute value (reelle Zahl)
absondern v 1. <Chemie> abstract; exude (Harz); 2. <Thermod> emit, give off

absondern

absondern v/sich <Chemie> exude
Absondern n <Chemtech> separation *(Komponenten)*
Absorber m <Heiz & Kälte> absorber
Absorberelement n <Heiz & Kälte, Kerntech> absorber element
Absorberelement n **mit Gelenkverbindung** <Kerntech> articulated absorber
Absorberglied n <Kerntech> absorber member
Absorberplatte f <Kerntech> absorber plate
Absorberschalldämpfer m <Sicherheit> *(AE)* absorption muffler, *(BE)* absorption silencer
absorbierbar adj <Papier> absorbable
absorbieren v <Kohlen, Maschinen, Papier, Sicherheit> absorb
absorbierendes Förderband n <Umweltschmutz> absorbent belt skimmer *(zur Ölaufnahme)*
absorbierte Dosis f *(D)* <Strahlphys> absorbed dose *(D)*
absorbierte Dosisrate f <Strahlphys> absorbed dose rate
absorbierte Energie f <Metall, Strahlphys> absorbed energy
Absorption f 1. <Elektrotech> absorption *(von Leistung, Energie, Kraft)*; 2. <Erdöl> gas absorption *(trockene Gasreinigung)*; 3. <Funktech, Heiz & Kälte, Kohlen, Kunststoff, Lebensmittel, Optik, Papier, Strahlphys, Thermod, Wasserversorg> absorption
Absorption f **ionisierender Strahlung** <Strahlphys> absorption of ionizing radiation
Absorption f **von Röntgenstrahlen** <Strahlphys> X-ray absorption
Absorptionsabschwächer m <Telekom> absorptive attenuator
Absorptionsanlage f <Erdöl> absorption plant *(Raffinerie)*
Absorptionsband n <Phys, Strahlphys> absorption band
Absorptionsbehälter m <Chemtech> absorption cell
Absorptionsdämpfungsglied n <Elektronik> absorptive attenuator
Absorptionsdosis f *(D)* <Kerntech> absorbed dose *(D)*
Absorptions-Dynamometer n <Gerät> absorption dynamometer
Absorptionseinrichtung f <Sicherheit> absorption plant
absorptionsfähig adj <Chemie> absorbent
Absorptionsfähigkeit f <Verpack> absorbency
Absorptionsfaktor m (α) <Optik> absorption factor (α)
Absorptionsfalle f <Chemtech> absorber trap
Absorptionsfiltern n <Aufnahme> absorption filtering
Absorptionsfrequenzmessgerät n <Gerät> absorption frequency meter
Absorptionsgefäß n 1. <Chemtech> absorption vessel; 2. <Papier> absorber
Absorptionsgeometrie f <Strahlphys> geometry of absorption
Absorptionsgrad m 1. <Optik> absorption capacity; 2. <Papier> absorption factor; 3. <Phys, Qual> coefficient of absorption; 4. <Telekom> absorption factor
Absorptions-Hygrometer n <Gerät> absorption hygrometer
Absorptionskälteanlage f <Heiz & Kälte> absorption refrigeration system
Absorptionskältekreislauf m <Heiz & Kälte> absorption refrigerating cycle
Absorptionskältemaschine f 1. <Heiz & Kälte> absorption refrigeration machine; 2. <Thermod> absorption type refrigerator
Absorptionskante f <Kerntech, Strahlphys> absorption edge
Absorptionskoeffizient m (α) <Akustik, Funktech, Phys, Strahlphys> absorption coefficient (α)

Absorptionskolonne f 1. <Chemtech> absorber column; 2. <Erdöl> absorption column *(Raffinerie)*
Absorptionskühlschrank m <Heiz & Kälte> absorption refrigerator
Absorptionskühlung f <Nichtfoss Energ> absorption cooling
Absorptionsküvette f <Chemtech> absorption cell *(analytische Chemie)*
Absorptionslinie f <Phys> absorption line
Absorptionsmaximum n <Elektrotech> absorption peak
Absorptionsmessgerät n <Papier> absorptionmeter
Absorptionsmesstechnik f <Gerät> absorptiometry
Absorptionsmessung f <Kerntech> absorptiometry
Absorptionsmittel n 1. <Chemtech> absorber; 2. <Kohlen, Papier, Umweltschmutz> absorbent
Absorptionsmodulator m <Elektronik> absorptive modulator
Absorptionsplatte f <Nichtfoss Energ> absorption plate
Absorptionspumpe f <Phys> absorption pump
Absorptionsquerschnitt m <Kerntech, Strahlphys> absorption cross-section
Absorptionsrate f <Wasserversorg> rate of absorption
Absorptionsrohr n 1. <Chemtech> absorber tube; 2. <Labor> absorption tube
Absorptionsröhrchen n <Labor> absorption tube
Absorptionssäule f 1. <Chemtech> absorber column; 2. <Kohlen> absorption tower; 3. <Papier> absorption column
Absorptionsschaltung f <Elektriz> absorption circuit
Absorptionsschwund m <Funktech> absorption fading
Absorptions-Spektralanalyse f <Strahlphys> absorption spectroanalysis
Absorptions-Spektrometer n <Labor> absorption spectrometer
Absorptions-Spektrometrie f <Telekom> absorption spectrometry
Absorptions-Spektrophotometer n <Strahlphys> absorption spectrophotometer
Absorptions-Spektroskopie f <Phys, Strahlphys> absorption spectroscopy
Absorptionsspektrum n <Optik, Papier, Phys, Raumfahrt, Strahlphys> absorption spectrum
Absorptionsstrom m <Elektriz> absorption current
Absorptionsturm m 1. <Chemtech> absorption tower; 2. <Erdöl> absorption tower *(Raffinerie)*; 3. <Kohlen, Lebensmittel> absorption tower
Absorptionsverlust m <Elektrotech, Fernseh, Funktech> absorption loss
Absorptionsvermögen n 1. <Heiz & Kälte> absorptivity; 2. <Kohlen> absorptance; 3. <Optik> absorptance; 4. <Papier> absorbency; 5. <Phys> absorptance; 6. <Wasserversorg> absorption capacity
Absorptionswärme f <Nichtfoss Energ, Thermod> heat of absorption
Absorptionswert m <Mechan> absorbing capacity
Absorptionszahl f 1. <Fernseh> absorptivity; 2. <Maschinen> absorption coefficient
Abspann... <Bau, Elektrotech, Erdöl> guy
Abspannanker m <Erdöl> guy anchor
Abspanndraht m <Bau> stay wire
abspannen v 1. <Bau> anchor *(Mast)*; 2. <Wassertrans> rig
Abspannen n <Bau> anchoring *(Mast)*
Abspannisolator m <Elektrotech> shackle insulator; guy insulator *(bei Abspanndraht)*
Abspannmast m <Elektriz> dead-end tower, span pole
Abspannring m <Erdöl> guy ring
Abspannseil n 1. <Bau> guy rope; 2. <Erdöl> guy anchor; 3. <Fertig> guy; 4. <Funktech> guy wire

Abspanntransformator m <Elektrotech> step-down transformer
Abspannung f 1. <Bau> guy, guying; 2. <Fertig, Funktech, Mechan> guy
Abspannungs-Unterwerk n <Elektriz> step-down station
Absperr... <Bau, Labor, Maschinen> cutoff, shut-off
Absperrarmatur f <Maschinen> shut-off valve
absperren v 1. <Bau> close, shut, stop; 2. <Fertig> shut off *(Kunststoffinstallationen)*; 3. <Hydraul> cut off *(Dampfzufuhr)*; 4. <Wassertrans> block off; 5. <Wasserversorg> shut off *(des Wassers)*
Absperren n <Bau> stopping
Absperrhahn m 1. <Bau> cutoff cock, *(AE)* faucet, stopcock; 2. <Labor, Maschinen> stopcock
Absperrklappe f 1. <Bau> flap; 2. <Fertig> butterfly valve *(Kunststoffinstallationen)*; 3. <Heiz & Kälte> butterfly damper; 4. <Maschinen, Wasserversorg> shut-off valve
Absperrorgan n 1. <Fertig> cock; valve *(Kunststoffinstallationen)*; 2. <Heiz & Kälte> obturator; 3. <Hydraul> valve *(von Rohrleitungen)*; 4. <Maschinen> obturator
Absperrplatte f <Hydraul> cutoff plate *(Dampfzufuhr)*
Absperrschieber m <Maschinen> shut-off slide
Absperrung f 1. <Bau> barrier *(Straße)*; 2. <Hydraul> cutoff, cutting-off *(Dampfzufuhr)*; 3. <Sicherheit> barricade, crush barrier
Absperrventil n 1. <Bau> check valve, stop valve; 2. <Fertig> check valve, stop valve *(Kunststoffinstallationen)*; 3. <Maschinen, Mechan, Nichtfoss Energ, Papier> check valve, stop valve; 4. <Raumfahrt> check valve, stop valve *(Raumschiff)*; 5. <Sicherheit> shutdown valve; 6. <Wassertrans> check valve, stop valve *(Motor)*; 7. <Wasserversorg> check valve, stop valve
Absperrvorrichtung f 1. <Heiz & Kälte> barrier *(gegen Brandübertragung)*; 2. <Hydraul> cutoff device *(Dampfzufuhr)*; 3. <Ker & Glas> shut-off
Abspiel... <Aufnahme> replay
abspielen v <Aufnahme> replay
Abspielen n <Aufnahme> playback
Abspielgerät n <Aufnahme> play-only recorder
Abspielgeräusch n <Aufnahme> needle noise, surface noise
Abspielkopf m <Aufnahme> replay head
Abspielnadel f <Aufnahme> needle
Abspielzeit f <Aufnahme> playing time *(einer CD, Kassette oder Schallplatte)*
Abspitzen n <Bau> picking *(von Stein)*
absplitten v <Bau> gravel
Absplittern n <Bau> spalling
Abspreng... <Ker & Glas> crack-off, wetting-off
Absprengen n <Ker & Glas> burning-off, cracking-off
Absprenghaken m <Ker & Glas> *(AE)* crack-off iron, *(BE)* wetting-off iron
Absprengkappe f <Ker & Glas> moil
Absprengring m <Ker & Glas> cracking ring
Abspreng- und Kantenschmelzmaschine f <Ker & Glas> *(BE)* burning-off and edge-melting machine, *(AE)* remelting machine
Absprengverschluss m <Ker & Glas> pry-off finish
Absprengwerkzeug n <Ker & Glas> cracking tool
absprießen v <Bau> brace *(Rahmen)*
abspringen v 1. <Fertig> rebound; 2. <Raumfahrt> bail out
abspulen v 1. <Fertig> run off; 2. <Foto> unspool
Abspulen n <Maschinen> unwinding
Abstand m 1. <Comp & DV> gap; 2. <Elektriz> pitch; 3. <Elektrotech> clearance; gap, spacing *(bei Relais)*; 4. <Gerät> range; 5. <Hydraul> space *(Kolben, Zylinderwand)*; 6. <Kerntech> clearance; 7. <Lufttrans> spacing clearance *(Propeller, Flügel)*; 8. <Maschinen, Mechan> clearance; 9. <Raumfahrt> separation *(Raumschiff)*; 10. <Telekom> spacing *(Kanal, Frequenz)*; 11. <Trans> spacing; 12. <Wassertrans> clearance, spacing
Abstand m **zwischen Brennstoff und Hülse** <Kerntech> clad-fuel clearance
Abstand m **zwischen Energiebändern** <Kerntech> energy band gap
Abstand m **zwischen Funkzonen** <Mobilkom> distance ratio *(Frequenzgleichlage)*
Abstandscheibe f <Maschinen> shim
Abstandsensor m <Gerät> proximity sensor
abstandsgleich adj <Maschinen> equally-spaced
Abstandshalter m 1. <Bau> distance piece, spacer block; 2. <Comp & DV> filler plate
Abstandskontrolle f <Trans> headway control
Abstandsleiste f <Maschinen> bumper rod
Abstandsmaske f <Elektronik> proximity mask
Abstandsmesser m <Metrol> *(AE)* gap gage, *(BE)* gap gauge
Abstandsring m 1. <Fertig> spacer; 2. <Wassertrans> calibration ring *(Radar)*
Abstandssensor m <Fertig> distance sensor
Abstandsstange f <Maschinen> distance bar
Abstandsstück n 1. <Ker & Glas> spacer; 2. <Maschinen> distance piece
Abstandsverlust m <Aufnahme> spacing loss
Abstandswarnanzeiger m <Lufttrans> proximity warning indicator
Abstandswarnvorrichtung f <Trans> headway warning device
Abstapeln n <Verpack> destacking *(von Paletten)*
Abstech... <Maschinen> cutoff, cutting-off
Abstechdrehmaschine f <Maschinen> cutoff machine, cutting-off lathe
abstechen v <Fertig> truncate
Abstechen n 1. <Fertig> parting *(Spanung)*; 2. <Lebensmittel> racking; 3. <Maschinen> parting-off
Abstechmaschine f <Maschinen> cutting-off machine
Abstechmeißel m <Maschinen> cutting-off tool
Abstechmeißelhalter m <Maschinen> cutting-off tool holder
Abstechschlitten m <Maschinen> cutting-off slide
Abstechstahl m <Maschinen> parting tool
Abstech- und Formdrehmaschine f <Maschinen> cutting-off and forming lathe
Abstechwerkzeug n <Maschinen> cutting-off tool, parting tool
abstecken v <Bau> peg out, set out *(Vermessung)*
Abstecken n <Bau> setting out, staking *(Vermessung)*
Absteckkette f <Bau> surveyor's chain
Absteckpfahl m <Bau> peg, surveyor's staff; picket, stake *(Vermessung)*
Absteh... <Ker & Glas> conditioning, soaking
Abstehen n <Ker & Glas> conditioning
Abstehofen m <Ker & Glas> soaking pit *(Guss und optisches Glas)*
Abstehzone f <Ker & Glas> conditioning zone
absteifen v 1. <Bau> brace, prop, shore, shore up; 2. <Fertig> truss; 3. <Kohlen> shore, shore up; 4. <Textil> stiffen
Absteifen n <Bau, Kohlen> shoring
Absteifung f 1. <Bau> sheeting; 2. <Fertig> stiffening
absteigender Ast m <Comp & DV> descendant
Abstell... 1. <Eisenbahn> storage; 2. <Lufttrans> parking
abstellen v 1. <Bau> stock; 2. <Lufttrans> shut down *(Motor und Triebwerk)*; 3. <Textil> stop
Abstellen n <Eisenbahn> stabling
Absteller m <Bau> deadbolt *(Schloss)*
Abstellfläche f <Lufttrans> parking area

Abstellgleis

Abstellgleis n <Eisenbahn> storage siding
Abstell-Lamelle f <Textil> drop pin
absteppen v <Textil> quilt, stitch down
Absterben n <Kfztech> stall *(Motor)*
Abstich m 1. <Fertig, Ker & Glas> tapping; 2. <Lebensmittel> racking
Abstichgraben m <Fertig> sow
Abstichloch n 1. <Fertig> taphole; 2. <Metall> notch
Abstichpfanne f <Fertig> tap ladle, tapping ladle
Abstieg m <Raumfahrt> descent
Abstiegstriebwerk n <Raumfahrt> descent engine *(Raumschiff)*
Abstimm... <Elektronik> tuning
Abstimmanzeige f 1. <Elektronik> tuning indicator; 2. <Funktech> zero beat indicator
Abstimmauge n <Elektronik> magic eye
abstimmbarer Breitband-Oszillator m <Elektronik> wideband tunable oscillator
abstimmbarer Oszillator m <Elektronik> tunable oscillator
abstimmbarer Schwingungskreis m <Elektronik> tuned circuit
abstimmbares Klystron n <Raumfahrt> tunable klystron
abstimmbares Magnetron n <Elektronik> tunable magnetron
abstimmbares Voltmeter n <Elektriz> selective voltmeter
Abstimmbereich m <Strahlphys> tuning range
Abstimmeigenschaften fpl <Strahlphys> tuning characteristics
abstimmen v 1. <Elektronik> tune; 2. <Funktech> tune *(Frequenz)*; 3. <Papier> adjust *(Farbton)*; 4. <Telekom> tune *(Empfänger)*; 5. <Wellphys> tune *(auf eine Frequenz)*
Abstimmen n <Elektronik> tuning
Abstimmkondensator m <Elektrotech> tuning capacitor
Abstimmkreis m <Elektrotech, Telekom> tuning circuit
Abstimmschaltung f <Elektronik> tuning circuit
Abstimmschraube f <Elektrotech, Phys> tuning screw
Abstimmschraubendreher m <Funktech> alignment tool
Abstimmung f 1. <Elektrotech> tuning; 2. <Funktech, Telekom> tuning *(Empfänger)*
Abstimmung f des Durchlassbereichs <Funktech> passband tuning
abstoppen v <Wassertrans> check, stopper *(Tauwerk)*
Abstoß m <Elektriz, Phys> repulsion
abstoßende Kraft f <Phys> force of repulsion, repulsive force
Abstoßkraft f <Elektriz> repulsive force
Abstoßrangierbetrieb m <Eisenbahn> fly shunting
Abstoßung f <Phys> repulsion
Abstoßungskraft f <Elektrotech> repulsive force
Abstrahlen n <Fertig> release
abstrahlendes Kabel n <Telekom> radiating cable
Abstrahlkeulenbreite f <Funktech> beamwidth *(Richtantenne)*
Abstrahlung f 1. <Elektronik> radiation; 2. <Optik> radiant emittance
abstrakt adj <Comp & DV> abstract
abstraktes Symbol n <Comp & DV> abstract symbol
Abstraktion f <Comp & DV> abstraction
abstreichen v 1. <Bau> strickle; 2. <Fertig> strickle; level *(Gießen)*
Abstreicher m <Bau> striker
Abstreichlineal n <Fertig> strike
Abstreichplatte f <Bau> strickle board
Abstreichvorrichtung f <Lebensmittel> scraper
abstreifen v 1. <Bau> wipe; 2. <Comp & DV> truncate; 3. <Metall> strip

Abstreifer m 1. <Bau> wiper; 2. <Ker & Glas> squeegee *(zum Emaillieren)*; 3. <Kohlen> stripper; 4. <Maschinen> scraper, stripper plate; 5. <Mechan, Papier> scraper
Abstreifmesser n <Textil> knife
Abstreifring m <Kfztech> oil control ring *(Kolben)*
Abstreusplitt m <Bau> blotter material
Abstrich m <Fertig> scum
Abström... <Lufttrans> trailing
Abströmkante f <Lufttrans> trailing edge *(Flügel)*
Abströmkante f des Leitwerks <Lufttrans> fin leading edge
Abströmkegel m <Heiz & Kälte> diffuser cone
abstufen v <Bau> grade, graduate
Abstufen n <Comp & DV> staging
Abstufung f <Druck> graduation
abstumpfen v <Druck, Maschinen> blunt
Absturz m <Comp & DV> crash
abstürzen v <Comp & DV> crash
Absturzsicherung f <Sicherheit> fall arrrester, fall-arresting device, fall harness
Abstützbohle f <Bau> raking shore
Abstützbrett n <Bau, Kohlen> forepole *(Tunnelbau)*
abstützen v <Bau> prop, strut
Abstützstrebe f <Lufttrans> brace
absuchen v <Funkort> scan *(Radar)*
Absuchen n <Telekom> scanning *(Radar)*
Abszisse f <Comp & DV, Elektriz, Fertig, Math> abscissa
abtakeln v <Wassertrans> unrig *(Segeln)*
Abtast... <Elektriz, Gerät> scanning
Abtastbereich m 1. <Comp & DV> scan area; 2. <Gerät> scanning range
Abtasteinrichtung f <Fernseh> scanning device
Abtastelektrode f <Elektrotech> sensing electrode
Abtastelektronenmikroskop n <Elektriz, Phys> scanning electron microscope
Abtastelement n 1. <Gerät> sampling element; 2. <Kerntech> sensing element
abtasten v 1. <Comp & DV> read, sample, sense, scan; 2. <Elektronik> sample; 3. <Fernseh, Fertig, Lufttrans> scan *(Radar)*; 4. <Maschinen> *(AE)* caliper; *(BE)* calliper; 5. <Telekom> sample; scan *(Signal, Bild)*; sense; 6. <Wassertrans, Wellphys> scan *(Radar)*
Abtasten n 1. <Comp & DV> sampling, scanning, sensing; 2. <Fernseh> scanning; 3. <Gerät> scanning; data sampling *(von Daten)*; measuring data sampling, measuring data scanning *(von Messdaten)*; 4. <Kerntech> sensing; 5. <Maschinen> *(AE)* calipering, *(BE)* callipering; 6. <Phys, Strahlphys> scanning
Abtaster m 1. <Akustik> pick-up; 2. <Comp & DV> sampler, scanner; 3. <Elektriz, Elektronik> scanner; 4. <Fernseh> analyser *(Bild)*; sampler; 5. <Gerät, Kontroll, Strahlphys> scanner; 6. <Telekom> sampler, scanner
Abtastfehler m 1. <Aufnahme> tracking error; 2. <Comp & DV, Fernseh> scanning error
Abtastfilterung f <Elektronik> sampled data filtering
Abtastfläche f 1. <Fernseh> scanning area; 2. <Raumfahrt> scan platform *(Raumschiff)*
Abtastfrequenz f 1. <Elektronik> digitizing rate, sample rate, sampling frequency, sampling rate, scanning frequency; 2. <Telekom> sampling frequency, sampling rate
Abtastgerät n 1. <Comp & DV> scanner, scanning device; 2. <Wellphys> scanner *(medizinisch)*
Abtastgeschwindigkeit f 1. <Comp & DV> sampling rate, scanning rate, scanning speed; 2. <Fernseh> scanning speed; 3. <Lufttrans, Wassertrans> scanning rate, scanning speed *(Radar)*
Abtastglied n <Gerät> sampling element, scanner
Abtastimpuls m <Gerät> sample pulse

Abtastintervall n 1. <Comp & DV, Telekom> sampling interval; 2. <Gerät> scan interval
Abtastkopf m 1. <Fernseh> scanning head; 2. <Gerät, Optik> pick-up head
Abtastlaser m <Optik> read laser, scanning laser
Abtastlaserstrahl m <Optik> scanning laser beam
Abtastlichtstrahl m <Aufnahme> scanning light beam *(für optische Aufzeichnung auf Film)*
Abtastlücke f <Fernseh> scanning gap
Abtastoszilloskop n <Gerät> sampling oscilloscope
Abtastpunkt m 1. <Comp & DV> scanning spot; 2. <Fernseh> scanning dot
Abtastpunktsteuerung f <Fernseh> scanning spot control
Abtastrate f 1. <Comp & DV> scan rate; 2. <Telekom> sampling interval, sampling rate
Abtastregelung f <Regelung> sampling control
Abtastregler m <Regelung> sampling controller
Abtastschalter m <Elektrotech> scanning switch
Abtastschaltung f <Telekom> sampler
Abtastschlitz m <Aufnahme> scanning slit *(für optische Tonspur auf Film)*
Abtastsignal n <Elektronik> sampled signal
Abtast-Spektralanalysator m <Elektronik> (BE) sampling spectrum analyser, (AE) sampling spectrum analyzer
Abtastspektrometer n <Strahlphys> scanning spectrometer
Abtaststrahl m <Elektronik, Fernseh> scanning beam
Abtastsystem n <Gerät> sampled data system
Abtasttheorem n <Comp & DV, Gerät, Telekom> sampling theorem
Abtasttrommel f <Fernseh> scanning drum
Abtastumfang m <Elektronik> sampled data size
Abtastumsetzer m <Elektronik> scan converter
Abtast- und Haltekreis m <Elektrotech> sample-and-hold circuit
Abtast- und Halteschaltung f <Telekom> sample-and-hold circuit
Abtastung f 1. <Comp & DV> sampling, scanning; 2. <Elektronik, Fernseh, Fertig, Gerät, Lufttrans, Telekom, Wassertrans, Wellphys> sampling, scan, scanning
Abtastung f **im Gegentakt** <Optik> push-pull scanning
Abtastung f **mit variabler Geschwindigkeit** <Fernseh> variable-speed scanning
Abtastverfahren n <Fernseh> scanning process
Abtastverhalten n <Regelung> sampling action
Abtastvertikalverstärker m <Elektronik> sampling vertical amplifier
Abtastverzerrung f <Aufnahme> tracking distortion
Abtastwert m 1. <Elektronik> sample, sampled value; 2. <Telekom> sample
Abtastzeile f <Comp & DV, Elektronik, Fernseh> scanning line
Abtastzyklus m 1. <Fernseh> scanning cycle; 2. <Gerät> sampling cycle
abtauen v 1. <Heiz & Kälte> thaw; 2. <Lebensmittel> defrost
Abtauen n <Lebensmittel> defrosting
Abteil n <Eisenbahn> compartment
Abteilung f 1. <Bau> bay *(Werkstatt)*; 2. <Mechan> bay
abteilungsübergreifendes Qualitätssicherungssystem n *(TQMS)* <Qual> Total Quality Management System *(TQMS)*
Abteilventil n <Fertig> block valve *(Kunststoffinstallationen)*
abteufen v 1. <Bau> bore; 2. <Kohlen> bore, sink
Abteufgerüst n <Kohlen> sinking trestle
Abteufpumpe f <Kohlen, Wasserversorg> borehole pump

abtönen v <Bau> tint
Abtönen n <Kunststoff> shading, tinting
Abtönung f <Kunststoff> shading, tinting
Abtrag m 1. <Bau> cut *(Erdreich)*; 2. <Bau, Eisenbahn, Kohlen> excavated material
Abtragemethode f <Optik> ablative method
abtragen v 1. <Anstrich> erode; 2. <Bau> skim, wreck; clear out, cut *(Erdreich)*; transfer *(Lasten)*; 3. <Fertig> erode *(Zerspanung)*
Abtragen n 1. <Maschinen> removal; 2. <Phys> ablating
abtragende Oberfläche f <Anstrich> abrasive surface
abtragendes Mittel n <Anstrich> abrasive surface
abtragendes Verfahren n <Fertig> metal-removing method
Abtragung f 1. <Anstrich, Fertig> erosion *(Zerspanung)*; 2. <Meerschmutz> wear
Abtragungsrate f 1. <Fertig> surface removal rate, workpiece removal rate; 2. <Raumfahrt> erosion rate *(Raumschiff)*
Abtransport m <Bau> removal
Abtransport m **von Korrosionsprodukten** <Kerntech> carry-off of corrosion products
abtreiben <Wassertrans> be making leeway *(nach Lee; Schiffsbewegung)*
abtrennbarer Nasenkegel m <Lufttrans> detachable nose cone
abtrennbares Düsenaggregat n <Lufttrans> detachable pod *(Hubschrauber)*
abtrennen v 1. <Bau> partition; 2. <Chemie> isolate; 3. <Kontroll> disconnect; 4. <Strömphys> bail out
abtrennen v/sich <Chemtech> separate out
Abtrennen n <Fertig> separating *(Stoffteilchen)*
Abtrennen n **der Blaskappe** <Ker & Glas> bursting-off
Abtrennen n **des Nietkopfes** <Fertig> rivet washing
Abtrennende n <Fertig> crop end
Abtrennung f 1. <Abfall> separation; 2. <Bau> partitioning *(durch Trennwände)*; 3. <Chemie> dissociation; 4. <Phys, Raumfahrt> separation *(Raumschiff)*
abtreppen v <Bau> bench
Abtretender m <Patent> assignor
Abtretungsempfänger m <Patent> assignee
Abtrieb m <Maschinen> output, power takeoff
Abtriebsdrehmoment n <Fertig> output torque
Abtriebsdrehzahl f <Maschinen> output speed
Abtriebsglied n <Maschinen> output link
Abtriebskupplung f <Maschinen> output drive clutch
Abtriebsrad n <Maschinen> following gear
Abtriebsseite f <Maschinen> output end, power takeoff side
Abtriebsteilchen n <Fertig> abraded particle
Abtriebswelle f 1. <Kfztech> output shaft, third motion shaft; 2. <Maschinen> output shaft
Abtriftmesser m <Lufttrans> driftmeter
Abtropf... 1. <Labor> draining, dripping; 2. <Maschinen> drip
Abtropfblech n <Maschinen> drain pan, drip plate
Abtropfbrett n <Fertig> drain board
Abtropfflüssigkeit f <Lebensmittel> drip
Abtropfgestell n <Labor, Lebensmittel> draining rack
Abtropfröhrchen n <Labor> dripping tube
Abtropfschale f <Maschinen> drip pan, drip tray
Abtrudeln n <Raumfahrt> spin-down *(Raumschiff)*
abvieren v <Bau> timber *(Holz)*
Abwälzfräsen n <Maschinen> gear hobbing, hob cutting, hobbing
Abwälzfräser m <Maschinen> gear hob, generating cutter, hob
Abwälzfräsmaschine f <Maschinen> hob, hobber, hobbing machine

A/B-Wandler

A/B-Wandler m <Telekom> a/b converter *(ISDN-Anschluss für Analogtelefon)*
Abwärme f 1. <Abfall> thermal discharge, waste heat; 2. <Heiz & Kälte, Kerntech> waste heat; 3. <Thermod> offheat, waste heat; 4. <Umweltschmutz> waste heat
abwärts gehende Destillation f <Thermod> distillation by descent
abwärts gehender Kolbenhub m <Kfztech> downstroke *(Motor)*
Abwärts... <Mechan> downstroke
Abwärtsbewegung f <Mechan> downstroke
Abwärtsfrequenz f <Funktech> downlink frequency *(Satellitenfunk)*
Abwärtshub m 1. <Fertig> downstroke; 2. <Hydraul> instroke; 3. <Kfztech> downstroke *(Motor)*; 4. <Maschinen> downstroke
abwärtskompatibel adj <Comp & DV> downward compatible
Abwärtskompatibilität f <Comp & DV> downward compatibility
Abwärtsmischer m <Telekom> down-converter *(Frequenzwechselschalter)*
Abwärtsmodulation f <Elektronik> downward modulation
Abwärtsstrategie f <Künstl Int> top-down strategy
Abwärtsstrecke f 1. <Funktech> downlink *(Satellitenfunk)*; 2. <Telekom> downlink feeder link
Abwärtstransformator m <Elektriz, Phys> step-down transformer
Abwärtsübertrager <Telekom> step-down transformer
Abwärtsumsetzer m <Telekom> down-converter
Abwärtsunterwerk n <Elektriz> step-down station
Abwärtsverbindung f <Mobilkom> downlink
Abwärtswandler m <Phys> step-down transformer
abwaschbar adj <Papier> washable
Abwaschbecken n <Bau> sink
Abwaschfestigkeit f <Konstzeich> resistance to washing
Abwasser n 1. <Abfall> drain water, effluent, wastewater; 2. <Bau> drain water, sewage; 3. <Erdöl> effluent; 4. <Wasserversorg> sewage, sewage wastewater
Abwasser n **aus Sanitäranlagen** <Wasserversorg> sanitary wastewater
Abwasserablauf m <Wasserversorg> sewage effluent
Abwasseranalyse f 1. <Abfall> wastewater analysis; 2. <Wasserversorg> sewage analysis
Abwasseranfall m <Abfall> sewage flow, volume of sewage
Abwasseraufbereitung f 1. <Abfall> sewage treatment; 2. <Umweltschmutz> wastewater treatment
Abwasserbecken m <Abfall> stabilization pond
Abwasserbehandlung f 1. <Abfall> sewage treatment; 2. <Wasserversorg> wastewater purification
Abwasserbehandlung f **mittels aerober Reinigung** <Abfall> aerobic sewage treatment
Abwasserbehandlungsanlage f <Abfall, Bau> clarification plant, sewage treatment plant
Abwasserbehandlungsverfahren n <Abfall> sewage treatment process
Abwasserbeseitigung f 1. <Abfall> sewage disposal, sewage water disposal; 2. <Bau> sewage disposal; 3. <Umweltschmutz> wastewater disposal
Abwassereinlauf m <Wasserversorg> wastewater outfall
Abwassereinleitung f 1. <Abfall> effluent discharge, sewage discharge; 2. <Meerschmutz> marine sewage disposal *(ins Meer)*; 3. <Umweltschmutz> wastewater discharge; 4. <Wasserversorg> effluent discharge, sewage effluent; underground wastewater disposal *(in den Untergrund)*
Abwassereinleitung f **ins Meer** <Abfall> marine sewage disposal

Abwassereinleitungsstelle f <Abfall, Bau> sewage outfall
Abwasserentsorgung f <Abfall> sewage disposal
Abwasserfaulraum m <Abfall> hydrolizing tank, privy tank, septic tank
Abwasserfischteich m <Abfall> wastewater fishpond
Abwasserkanal m <Bau, Wasserversorg> sewer
Abwasserkanalisation f <Bau, Wasserversorg> sewerage
Abwasserkanalreinigung f <Wasserversorg> sewer cleaning
Abwasserkläranlage f 1. <Abfall> clarification plant, sewage treatment plant; 2. <Bau> clarification plant, sewage treatment plant, sewage disposal plant; 3. <Wasserversorg> sewage disposal plant
Abwasserklärung f 1. <Abfall> sewage purification, sewage treatment; 2. <Wasserversorg> sewage clarification
Abwasserkontrolle f <Umweltschmutz> wastewater control
Abwasserleitung f 1. <Bau> sewer; 2. <Wasserversorg> canalization
Abwassermenge f <Abfall> sewage flow, volume of sewage
Abwasserpilz m <Abfall> sewage fungus
Abwasserreinigung f 1. <Abfall> sewage purification, sewage treatment; 2. <Wasserversorg> sewage clarification, wastewater purification
Abwasserreinigungsanlage f <Abfall, Bau> sewage treatment works, wastewater purification plant
Abwasserrohr n <Bau> drainpipe, waste pipe
Abwasserrückführung f <Kerntech> wastewater recycling operation
Abwassersammeltank m <Abfall> wastewater collection tank
Abwassersammler m <Abfall> interceptor sewer
Abwassersanierung f <Abfall> wastewater renovation
Abwasserschlamm m 1. <Abfall> effluent sludge; sewage sludge *(von Haushalten)*; 2. <Chemie> digested sludge
Abwasserstripper m <Abfall> wastewater stripper
Abwasserverrieselung f <Wasserversorg> broad irrigation
Abwasserversenkung f <Wasserversorg> underground wastewater disposal
Abwasserzusammensetzung f <Umweltschmutz> sewage composition
abwechselnd adj <Maschinen> alternating
abwedeln v <Foto> dodge
Abwehrring m <Elektriz> guard ring
Abwehrsystem n <Sicherheit> suppression system
abweichen v <Elektriz> fluctuate
Abweichung f 1. <Comp & DV> variance; jitter *(eines Übertragungssignals)*; 2. <Elektriz> fluctuation; 3. <Fernseh, Funktech> deviation; 4. <Maschinen> deviation, variation; 5. <Math> deviation; 6. <Nichtfoss Energ, Phys> declination; 7. <Qual> deviation, discrepancy, fluctuation; 8. <Raumfahrt> drift *(Raumschiff)*; deviation *(Weltraumfunk)*; 9. <Telekom> deviation, drift; 10. <Wassertrans> declination *(astronomische Navigation)*
Abweichungsanalyse f <Comp & DV> analysis of variance
Abweichungsanzeige f 1. <Gerät> deviation indication, error indication; 2. <Qual> error indication
Abweichungsbericht m <Qual> nonconformance report
abweichungsfreie Stirnräder npl <Fertig> zero-deviation cylindrical gears
Abweichungskreis m <Optik> circle of declination
Abweichungswinkel m <Lufttrans> declination angle *(Navigation)*

abweisen v <Telekom> reject
Abweiseplatte f <Chemtech> deflector plate
Abweiser m <Raumfahrt> jet deflector *(Raumschiff)*
Abweisstein m <Bau> baffle brick
abwerfbar adj <Lufttrans> jettisonable
abwerfen v <Raumfahrt> jettison *(Raumschiff)*
Abwerfen n <Raumfahrt> drop *(Raumschiff)*
Abwesenheitsfunktion f <Comp & DV> vacation switch
Abwickel... <Fertig, Maschinen> decoiling, unwinding
Abwickeleinrichtung f <Fertig> decoiler
abwickeln v 1. <Foto> unwind; 2. <Geom> develop; 3. <Telekom> carry
Abwickeln n 1. <Fernseh> batting down; 2. <Maschinen> unwinding
Abwickelspule f <Gerät> supply reel *(Bandregistriergerät)*
Abwickelvorrichtung f <Textil> unwinder
Abwicklung f 1. <Bau, Geom> development; 2. <Telekom> handling *(Verkehr)*
Abwind m <Lufttrans> (AE) downdraft, (BE) downdraught, downwash, downwind
abwischen v <Bau> wipe
Abwracken n <Wassertrans> ship breaking
Abwurf m <Raumfahrt> drop *(Raumschiff)*
Abwurfbremsschirm m <Lufttrans> message chute *(Hubschrauber)*
Abwurfeinrichtung f <Maschinen> tripper
Abwurfgebiet n <Raumfahrt> drop zone *(Raumschiff)*
Abwurfhöhe f <Lufttrans> free drop height
Abwurfkapsel f <Raumfahrt> ejectable capsule *(Raumschiff)*
Abwurfschacht m <Bau> chute
Abwurfspitze f <Raumfahrt> ejectable nose cone *(Raumschiff)*
abwürgen v <Kfztech> stall *(Motor)*
Abwürgen n 1. <Kfztech> stall *(des Motors)*; 2. <Lufttrans> stall *(des Kompressors, des Turbomotors)*
abzählbar adj <Math> countable
Abzählmaschine f <Fertig> counter/dispenser
Abzapf... <Elektriz> tapping
Abzapfbreite f <Elektriz> tapping range
abzapfen v <Fertig, Maschinen> bleed *(Flüssigkeit)*
Abzapfintervall n <Elektriz> tapping step
Abzapfpunkt m <Elektriz> tapping
Abzapfpunkt m **für reduzierte Leistung** <Elektriz> reduced power tapping
Abzapfstrom m **aus einer Wicklung** <Elektriz> tapping current of winding
abzapfungslos adj <Elektriz> untapped *(Transformator)*
Abzapfwechsel m **bei Last** <Elektriz> load-tap-changer
Abzapfwechsler m **unter Last** <Elektriz> on-load tapchanger
Abzieh... 1. <Bau> screed; 2. <Ker & Glas> (BE) transfer
Abziehapparate mpl <Textil> doffing devices
abziehbare Kupplung f <Elektriz> pull-off coupling
abziehbarer Schutzbelag m <Verpack> peelable protective coating
Abziehbild n <Ker & Glas> (AE) decal, (BE) transfer
Abziehbohle f <Bau> screed board
abziehen v 1. <Bau> rub, smooth; finish *(Beton)*; level *(Mauern)*; strike off *(Oberfläche)*; 2. <Bau> draw; 3. <Comp & DV> disconnect *(Stecker, Kabel)*; 4. <Fertig> hone; 5. <Foto> print; 6. <Lebensmittel> skin *(Haut, Fell)*; 7. <Maschinen> hone, withdraw; 8. <Math> subtract; 9. <Mechan> hone
Abziehen n 1. <Bau, Fertig> stripping *(Kokille)*; 2. <Foto> printing; 3. <Lebensmittel> racking
Abziehen n **von Schlacke** <Metall> skimming off the dross

Abzieher m <Kfztech> gear puller *(Werkzeug)*
Abziehfestigkeit f <Werkprüf> peel strength
Abziehfilmverpackung f <Verpack> peel-off wrapping
Abziehhülse f <Maschinen> withdrawal sleeve
Abziehstein m 1. <Fertig> oilstone; 2. <Mechan> honing stone
Abziehsystem n <Verpack> peelable system
Abziehvorrichtung f <Maschinen> extractor, withdrawal tool
Abzisse f 1. <Math> horizontal axis, x-axis; 2. <Phys> x-coordinate
abzuführende Verlustleistung f <Heiz & Kälte> amount of heat to be dissipated
abzuführende Wärmemenge f <Heiz & Kälte> amount of heat to be dissipated
Abzug m 1. <Comp & DV> print; 2. <Foto> print *(eines Negativs)*; 3. <Heiz & Kälte> eduction; 4. <Labor> fume hood; 5. <Mechan> vent; 6. <Papier> exhaust; 7. <Qual> penalty *(bei Folgestichprobenprüfung)* • **Abzüge machen** 1. <Druck> pull proofs; 2. <Foto> print • **mit Abzug** <Maschinen> flued *(Erhitzer)*
Abzugsgas n 1. <Mechan> exhaust gas; 2. <Phys> flue gas
Abzugsgewicht n <Textil> take-down weight
Abzugsgraben m <Wasserversorg> drainage ditch
Abzugshaube f 1. <Bau, Sicherheit> exhaust hood, hood; 2. <Umweltschmutz> exhaust hood
Abzugskanal m <Wasserversorg> delivery channel
Abzugsleine f <Wassertrans> tripping line
abzugslos adj <Heiz & Kälte, Maschinen> flueless
Abzugsöse f <Fertig> withdrawal eye *(Kunststoffinstallationen)*
Abzugsrohr n 1. <Bau> discharge pipe, drain pipe, vent pipe; 2. <Fertig> eduction pipe; offtake; 3. <Heiz & Kälte, Ker & Glas, Kerntech, Kunststoff, Labor, Maschinen, Mechan> vent pipe
Abzugsschrank m <Labor> fume cupboard
Abzugvorrichtung f <Fertig> eductor
Abzweig m 1. <Elektriz> branch; 2. <Elektrotech> branch, stub, tapping; 3. <Mechan, Phys, Telekom> branch
Abzweigdose f 1. <Bau> joint box; 2. <Elektriz> branch box; 3. <Elektrotech> conduit box, distribution box, junction box; 4. <Heiz & Kälte> junction box
Abzweigelement n <Optik> branching device
abzweigen v 1. <Bau> branch, branch off; 2. <Telekom> branch off, tap; drop *(Kanal)*
Abzweigen n <Telekom> tapping *(Kanal)*
Abzweigfilter n <Elektronik, Telekom> ladder filter
Abzweigkabel n <Telekom> branch cable
Abzweigkasten m 1. <Bau, Elektriz> branch box; 2. <Elektrotech> junction box
Abzweigklemme f <Elektriz, Telekom> branch terminal
Abzweigkreis m <Elektriz> branched circuit
Abzweigleitung f 1. <Comp & DV> branch *(Schaltung)*; 2. <Elektriz> branch line; 3. <Fertig> shunt line
Abzweigmuffe f 1. <Fertig> Y-joint, branch tee, paralleljoint sleeve; 2. <Telekom> branch T
Abzweigpunkt m <Elektronik> branch point *(in einer Schaltung)*
Abzweigregler m <Elektrotech> tapped control
Abzweigrohr n 1. <Bau> branch pipe; 2. <Fertig> branch; 3. <Maschinen, Mechan> branch pipe; 4. <Wasserversorg> branch pipe *(zur Feuerlöschung)*
Abzweigspule f <Elektrotech> tapped coil
Abzweigstromkreis m <Elektrotech> branched circuit
Abzweigstück n <Bau> Y-branch
Abzweigung f 1. <Bau> junction, turnout, branch *(Sanitärbereich)*; 2. <Elektriz, Elektrotech> branch, spur; 3. <Elektronik> branch *(einer Schaltung)*; tapping; 4.

AC

<Mechan, Phys> branch; 5. <Telekom> leg *(Kabel)*; tap *(ISDN)*; 6. <Wasserversorg> branching
AC 1. <Elektrotech> *(Wechselstrom)* AC *(alternating current)*; 2. <Fertig> *(Adaptivsteuerung)* AC *(adaptive control)*
AC-Adapter m <Elektrotech> AC adapter
AC-Animeter n <Elektrotech> AC animeter
AC-Ausgang m <Elektrotech> AC output
AC-Beschichten n <Chemtech> AC bed coating
AC-Betrieb m <Elektrotech> AC mode, AC operation
AC-Brücke f <Elektrotech> AC bridge
ACC *(automatische Chrominanzregelung)* <Fernseh> ACC *(automatic chrominance control)*
Account m **mit minimalen Berechtigungen** <Comp & DV> minimally privileged account
AC-Dickfilm-Elektrolumineszenzanzeige f <Elektronik> AC thick-film electroluminescent display
AC-Eingang m <Elektrotech> AC input
AC-Entladung f <Elektrotech> AC discharge
AC-Erregung f <Elektrotech> AC excitation
AC-Erzeugung f <Elektrotech> AC current generation, AC generation
Acetal n <Kunststoff, Lebensmittel> acetal
Acetaldehyd m 1. <Chemie> ethanal; 2. <Lebensmittel> acetaldehyde
Acetaldol n 1. <Chemie> acetaldol; 2. <Lebensmittel> aldol
Acetanhydrid n <Lebensmittel> acetic anhydride
Acetat n 1. <Chemie, Lebensmittel> acetate; 2. <Textil> acetate, cellulose acetate
Acetatfaser f <Textil> acetate, cellulose acetate
Acetatfaserstoff m <Textil> acetate, cellulose acetate
Acetatfolie f <Verpack> acetate film
Acetatkleber m <Verpack> acetate adhesive, acetate glue
Acetatlaminat n <Verpack> acetate laminate
Acetatseide f <Kunststoff> rayon
Acetatverbundmaterial n <Verpack> acetate laminate
Acetessig... <Chemie> diacetic
Acetin n <Lebensmittel> acetin
Acetoglycerid n <Lebensmittel> acetoglyceride
Acetolyse f <Lebensmittel> acetolysis
Aceton n 1. <Chemie> acetone, propanone; 2. <Druck, Kunststoff> acetone
Acetonextraktion f <Kunststoff> acetone extraction
Acetonharz n <Fertig, Kunststoff, Verpack> acetone resin
Acetonitril n <Chemie> acetonitrile, ethanenitrile
Acetophenon n <Chemie> acetophenone
Acetoxygruppe f <Lebensmittel> acetoxy group
Acetylcellulose f <Kunststoff> acetate, cellulose acetate
AC-Feld n <Elektriz> AC field
AC-gekoppelt adj <Kontroll> AC-coupled
AC-Generator m <Elektriz, Phys> AC generator
AC-GS *(Wechselstrom-Gleichstrom)* <Elektrotech> AC-DC
AC-GS-Umsetzer m <Elektrotech> AC-DC converter
AC-GS-Umsetzung f <Elektrotech> AC-DC conversion
AC-GS-Wandler m <Elektrotech> AC-DC converter
AC-GS-Wandlung f <Elektrotech> AC-DC conversion
Achat m <Ker & Glas> agate
Achatsteingut n <Ker & Glas> agate ware
Achromat m <Druck, Foto, Optik> achromatic lens
achromatisch adj <Ergon, Optik, Phys> achromatic
achromatische Linse f <Druck, Foto> achromatic lens
achromatische Ringe mpl <Phys> achromatic fringes
achromatische Streifen mpl <Phys> achromatic fringes
achromatisches Dublett n <Phys> achromatic doublet
achromatisches Objektiv n <Optik> achromatic lens
achromatisieren v <Optik> achromatize
Achromatisierung f <Optik> achromatization

Achromatismus m <Optik> achromatism
Achs... <Bau, Kfztech> axle
Achsabstand m <Fertig, Kfztech, Maschinen> *(AE)* center distance, *(BE)* centre distance
Achsaggregat n <Eisenbahn> *(BE)* bogie
Achsantrieb m <Kfztech> final drive, wheel and axle drive
Achsantriebverhältnis n <Kfztech> final drive ratio
Achsbaum m <Kfztech> axletree
Achsbolzen m <Kfztech> axle pin
Achsbuchsenführung f <Kfztech> axle box guide, axle guide
Achsbürste f <Eisenbahn> axle brush
Achsdichtung f <Fertig> axial seal *(Kunststoffinstallationen)*
Achsdruck m <Bau, Eisenbahn> axle load
Achsdrucklager n <Maschinen> axial thrust bearing
Achse f 1. <Bau, Eisenbahn> axle; 2. <Geom> axis; 3. <Kfztech, Maschinen> axle; 4. <Mechan> trunnion; axis *(einer Riemenscheibe oder eines Kräftepaars)*; 5. <Papier, Phys> axis • **Achse verschieben** <Bau> *(AE)* move the center line, *(BE)* move the centre line • **von Achse zu Achse** <Maschinen> *(AE)* from center to center, *(BE)* from centre to centre
Achse f **des Zugsattelzapfens** <Kfztech> fifth-wheel kingpin axis *(bei Sattelschlepper)*
Achselhöhe f <Druck> shoulder height
Achsen... <Geom> axial
Achsenabstand m <Maschinen> distance between axles
Achsenbeschriftung f <Geom> axis labelling
Achsendrehmaschine f <Maschinen> axle lathe
achsenparalleler Strahl m <Optik> paraxial ray
Achsenschenkel m <Fertig> spindle
Achsenskala f <Geom> axis scale
Achsensymmetrie f <Geom> axial symmetry
Achsensystem n <Geom> system of coordinates
Achsenversatz m <Fertig> axial misalignment, misalignment *(Faseroptik)*
Achsenwinkel m 1. <Fertig> shaft angle *(Kegelräder)*; 2. <Maschinen> axis angle
achsfern adj <Fertig> abaxial
Achsflansch m <Kfztech> axle flange *(Räder)*
Achsführung f <Eisenbahn> cylindrical axle guide
Achsgabel f 1. <Eisenbahn> pedestal; 2. <Kfztech> axle guard; 3. <Trans> pedestal
Achsgehäuse n <Kfztech> axle casing
Achsgehäusedeckel m <Kfztech> axle box cover, axle box lid
Achsgehäuse-Enwässerungsbehälter m <Kfztech> axle box cellar, axle box sponge-box
Achsgelenkverbindung f <Kfztech> axle articulation
achsgerade einstellen v <Fertig> align
Achshalterstreg m <Kfztech> axle guide stay
Achskappe f <Kfztech> axle cap
Achslagefehler m <Fertig> error of alignment caused by deflection of the shafts *(Zahnrad)*
Achslager n 1. <Eisenbahn> *(BE)* axle box, *(AE)* journal box; 2. <Kerntech> journal bearing; 3. <Kfztech> axle box bearing; 4. <Trans> *(BE)* axle box, *(AE)* journal box *(Wagen, Fahrgestell)*
Achslagergehäuse n <Maschinen> journal box
Achslast f 1. <Bau, Eisenbahn> axle load; 2. <Maschinen> axle weight
Achslaufbuchse f <Kfztech> axle bush, axle bushing
Achslenker m <Kfztech> axle guide
Achsmitte f <Kfztech> *(AE)* axle center, *(BE)* axle centre
Achsmotor m <Eisenbahn> direct drive motor
Achsmutter f <Kfztech> axle nut
Achsparallelität f <Kfztech> tracking

Achsrichtung f <Fertig> end • **in Achsrichtung** <Fertig> endwise
Achsschälmaschine f <Fertig> axle peeling lathe
Achsscheibe f <Bau> axle pulley *(Beschläge)*
Achsschenkel m <Kfztech> control arm *(Federung, Aufhängung)*; steering knuckle *(Lenkung)*; stub axle *(Rad)*
Achsschenkelbolzen m <Maschinen, Mechan> *(AE)* kingbolt, *(BE)* kingpin
Achsschenkelbolzenspreizung f <Kfztech> *(AE)* kingbolt inclination, *(BE)* kingpin inclination, steering axis inclination
Achsschenkelfederbein n <Kfztech> McPherson strut
Achssitz m <Maschinen> axle seat
Achsstand m <Eisenbahn, Kfztech> wheelbase
Achssturzwinkel m <Lufttrans> angle of wing setting, toe-in angle *(Fahrwerk)*
Achsübersetzungsverhältnis n <Kfztech> axle ratio
Achsversatz m <Fertig> axial misalignment, misalignment *(Faseroptik)*
Achswelle f <Kfztech> axle shaft
Achswellenkegelrad n <Kfztech> differential side gear
Achszapfen m <Kfztech> journal; spindle *(Rad)*
achtbindig adj <Chemie> octavalent
Acht-Bit <Comp & DV, Elektronik> eight-bits
Acht-Bit-Byte n <Comp & DV> eight-bit byte
Acht-Bit-Genauigkeit f <Elektronik> eight-bit accuracy
Acht-Bit-Umsetzer m <Elektronik> eight-bit converter
Acht-Bit-Umsetzung f <Elektronik> eight-bit conversion
Achteck n <Geom, Metall> octagon
achteckig adj <Geom> octagonal
achteckige Mutter f <Maschinen> octagonal nut
achteckige Reibahle f <Maschinen> octagonal reamer
Achtelkreisfehler m <Funkort> octantal error *(Funkpeilung)*
Achtelmeile f <Metrol> furlong
Achter... 1. <Elektronik> eight-...; 2. <Math> octa...; 3. <Wassertrans> aft, stern
Achteralphabet n <Elektronik> eight-level code
achteraus adv <Wassertrans> astern; abaft *(hinter dem Schiff)* • **genau achteraus** <Wassertrans> dead astern • **achteraus laufen** <Wassertrans> go astern
Achterdeck n <Wassertrans> afterdeck, quarter deck
Achtergruppe f <Chemie> octet
achterlastig adj <Wassertrans> trimmed by the stern
achterlastig adv <Wassertrans> down by the stern *(Schiff)*
achterlich adj <Wassertrans> abaft, abaft the beam
achterlicher Wind m <Wassertrans> following wind
achtern adv 1. <Raumfahrt> aft; 2. <Wassertrans> astern; abaft *(hinter mittschiffs)* • **nach achtern** <Wassertrans> aft
Achterschiff n <Wassertrans> aft section
Achterspring f <Wassertrans> backspring aft *(Mooring)*
Achterstag n <Wassertrans> aft stay *(Tauwerk)*
Achtersteven m <Wassertrans> stern frame, sternpost *(Schiffbau)*
achtflächig adj <Chemie, Geom> octahedral
Achtflächner m <Geom> octahedron
achtförmig verbogenes Rad n <Trans> buckled wheel
Achtkant m <Maschinen> octagon
Achtkanteisen n <Metall> octagon iron
achtkantig adj <Maschinen> octagonal
Achtkantstahl m <Metall> octagon bar
Achtknoten m <Wassertrans> figure-of-eight knot
Achtpol m <Elektrotech> octupole
Achtpolröhre f <Elektronik> octode
Achtspur-Recorder m <Aufnahme> eight-track recorder
Achtungssignal n <Aufnahme> audio cue *(Tonstudio)*

achtwertige Phasenumtastung f <Telekom> eight-phase phase shift keying
Achtzylinder-V-Motor m <Kfztech> V-eight engine
ACIA *(Asynchron-Übertragungs-Schnittstellenanpasser, asynchronischer Übertragungs-Schnittstellenanpasser)* <Kontroll> ACIA *(asynchronous communications interface adaptor)*
ACIA-Schaltkreis m <Kontroll> ACIA switching circuit
Acidimeter n 1. <Chemie> acidimeter, acidometer; 2. <Chemtech, Lebensmittel> acidimeter
Acidimetrie f <Chemie, Chemtech, Lebensmittel, Papier> acidimetry
acidimetrisch adj <Chemie, Papier> acidimetric
Acidität f <Chemie> acidity
Acidolyse f <Lebensmittel> acidolysis
Acidometer n 1. <Chemie> acidimeter, acidometer; 2. <Chemtech> acidimeter
aci-Form f <Chemie> aci-form
AC-Josephson-Effekt m <Elektronik> AC Josephson effect
AC-Kompensator m <Elektriz> AC potentiometer
AC-Komponente f <Elektriz> AC component
AC-Kondensator m <Elektrotech> AC capacitor
AC-Koppler m <Telekom> AC coupler
AC-Kraft f <Elektrotech> AC electromotive force
AC-Kreis m <Elektriz> AC circuit, AC network
AC-Last f <Elektrotech> AC load
AC-Leistung f <Elektrotech> AC power
AC-Leitung f <Elektrotech> AC line
AC-Lichtbogen m <Elektrotech> AC arc
AC-Lichtbogenschweißen n <Elektrotech> AC arc welding
AC-Marker m <Wassertrans> AC marker *(Radar)*
AC-Maschine f <Elektrotech> AC machine
AC-Messbrücke f <Elektriz> AC bridge
AC-Messinstrument n <Elektriz> AC meter
Acmetrapezgewinde n 1. <Fertig> acme standard screw thread; 2. <Maschinen> trapezoidal thread
AC-Motor m <Elektriz, Elektrotech, Fertig, Phys> AC motor
ACN *(automatische Himmelsnavigation)* <Raumfahrt> ACN *(automatic celestial navigation)*
ACNA *(Analogrechner für Netzabgleich)* <Comp & DV> ACNA *(analog computer for net adjustment)*
AC-Netz n 1. <Elektriz> AC network; 2. <Elektrotech> AC network, AC power line
AC-Netzausfall m <Elektrotech> AC power failure
AC-Netzleitung f <Elektrotech> AC power line
ACO *(Anpassungssteuerung mit Optimierung)* <Labor> ACO *(adaptive control optimization)*
Aconit... <Chemie> aconitic
Aconitase f <Chemie> aconitase
Aconitat n <Chemie> aconitate
Aconitin n <Lebensmittel> aconitine
AC-Quelle f <Elektrotech> AC source, current source
ACR *(Anflugradar)* <Raumfahrt> ACR *(approach control radar)*
AC-Relais n <Elektriz, Elektrotech> AC armature relay, AC relay
Acryl... <Anstrich> acrylic
Acrylat n <Kunststoff> acrylate
Acrylfarbe f <Bau> acrylic paint
Acrylfaserstoff m <Textil> acrylic
Acrylgewebe n <Textil> acrylic
Acrylglas n <Verpack> acrylic plastic
Acrylharz n <Fertig, Kunststoff, Mechan, Verpack> acrylic resin
Acrylkautschuk m <Verpack> acrylic rubber
Acrylkunststoff m <Fertig> acrylic plastic
Acryllack m <Kunststoff> acrylic paint

Acrylnitril-Butatien-Styrol

Acrylnitril-Butatien-Styrol n 1. <Elektriz> acrylonitrile butadiene styrene; 2. <Kunststoff> acrylonitrile butadiene styrene *(Copolymer)*
Acrylnitrilgummi m <Mechan> acrylonitrile rubber
Acrylnitrilkautschuk m <Kunststoff> acrylonitrile rubber
Acrylschlichte f <Textil> acrylic size
Acrylstoff m <Textil> acrylic
AC-Schaltkreis m <Elektrotech> AC circuit
AC-Schaltung f <Elektrotech> AC circuit
AC-Schweißlichtbogen m <Fertig> AC welding arc
AC-Servomotor m <Elektrotech> AC servomotor
AC-Spannung f <Elektriz> AC voltage
ACSR *(Einseitenband mit kompandierter Amplitude)* <Funktech> ACSS *(amplitude-compandered single sideband)*
AC-Stellmotor m <Elektrotech> AC servomotor
Actin n <Lebensmittel> actin
Actinidenelement n 1. <Phys> actinide; 2. <Strahlphys> actinide element, actinoid
Actinidenreihe f <Strahlphys> actinide series, actinium series
Actomyosin n <Lebensmittel> actomyosin
ACU *(automatische Anrufeinheit, automatisches Rufgerät)* <Raumfahrt, Telekom> ACU *(automatic calling unit)*
AC-Übertragungsleitung f <Elektriz> AC transmission line
AC-Versorgung f 1. <Elektriz> AC supply; 2. <Elektrotech> AC current source, AC power source
AC-Versorgungssystem n <Raumfahrt> AC power system *(Raumschiff)*
AC-Verstärker m <Elektrotech> AC amplifier
AC-Voltmeter n <Elektriz, Elektrotech> AC voltmeter
AC-Vorspannung f <Aufnahme> AC bias
AC-Widerstand m <Elektrotech> AC resistance
acyclisch adj <Chemie> acyclic
acyclisches Diacylamin n <Chemie> imide
acyclisches Säureamid n <Chemie> imide
Acyl n <Chemie> acyl
acylieren v <Chemie> acylate
A/D *(Analog-Digital-...)* <Elektronik, Fernseh, Fertig> A/D *(analog-digital)*
Adaptationsniveau n <Ergon> adaptation level
Adapter m 1. <Comp & DV, Elektriz, Elektrotech> adaptor; 2. <Fertig> adaptor *(Kunststoffinstallationen)*; 3. <Funktech, Labor, Maschinen, Mechan, Phys, Telekom, Textil> adaptor
Adapterplatte f <Maschinen> adaptor plate
Adaption f <Ergon> adaptation
adaptiv adj 1. <Comp & DV> adaptive; 2. <Künstl Int> adaptive *(Programm, System)*; 3. <Mechan, Phys> adaptive
adaptive Abstimmung f <Elektronik> adaptive tuning
adaptive Abtastung f <Gerät> adaptive sampling
adaptive Antenne f <Telekom> adaptive antenna
adaptive Codierung f <Telekom> adaptive coding
adaptive Differenz-Pulscodemodulation f *(ADPCM)* <Telekom> adaptive differential pulse code modulation *(ADPCM)*
adaptive Entzerrung f <Elektrotech> AC distortion device, AC filter, adaptive equalization
adaptive Kanalzuordnung f <Comp & DV, Telekom> adaptive channel allocation
adaptive Prädiktionscodierung f <Telekom> adaptive predictive coding
adaptive Signalverarbeitung f <Elektronik> adaptive signal processing
adaptiver Entzerrer m <Elektronik> adaptive equalizer *(Übertragungstechnik)*
adaptiver Prozess m <Comp & DV> adaptive process

adaptives Filter n <Elektronik, Telekom> adaptive filter
adaptives Filtern n <Comp & DV> adaptive filtering
adaptives Kippen n <Elektronik> adaptive sweep *(Katodenstrahlröhren)*
adaptives Regelungssystem n <Elektrotech, Mechan, Phys> adaptive control system
Adaptivsteuerung f 1. <Elektrotech> AC, adaptive control; 2. <Fertig> adaptive control *(Spannung)*
Adcock-Peiler m <Funkort> Adcock direction finder
Addendum n <Druck> addendum
Adder m <Math> adder
Addier... <Comp & DV, Elektriz, Math> adding
Addiereinrichtung f <Comp & DV> adder
Addierer m <Math> adder; adding machine *(Rechenmaschine)*
Addierer m **mit Parallelübertrag** <Comp & DV> look-ahead
Addiermaschine f <Comp & DV, Math> adding machine
Addierschaltkreis m <Math> adder
Addierschaltung f 1. <Elektriz> adding network; 2. <Elektronik> adder *(Schaltkreistechnik)*
Addierzähler m <Elektronik> adding counter
Addition f <Math> addition
Additionspolymer n <Kunststoff> addition polymer
Additionspolymerisat n <Kunststoff> addition polymer
Additionspolymerisation f <Kunststoff> addition polymerization
Additionsstelle f <Regelung> summing point
Additionsvorgang m <Elektronik> additive process
Additionszähler m <Gerät> accumulating counter
Additiv n 1. <Abfall> additive; 2. <Chemie> dope *(Mineralöl)*; 3. <Druck, Kunststoff, Maschinen> additive
additiv adj <Math> additive *(Term)*
additive Mischung f <Elektronik> additive mixing *(Hochfrequenztechnik)*
additives Rauschen n <Telekom> additive noise
Additivsynthese f <Foto> additive synthesis
Adduktion f <Chemie, Ergon> adduction
Adenin n 1. <Chemie> adenine, aminopurine; 2. <Lebensmittel> adenine
Adenosin n <Chemie> adenosine
Adenosintriphosphat n *(ATP)* <Lebensmittel> adenosine triphosphate *(ATP)*
Ader f 1. <Elektriz> conductor, core; 2. <Elektrotech> conductor, wire, core; core *(eines Drahtseils oder elektrischen Kabels)*; 3. <Kohlen> seam; 4. <Phys> conductor; 5. <Telekom> core *(Kabel)*
Ader f **zum Stöpselhals** <Telekom> R-wire, ring wire
Ader f **zur Stöpselspitze** <Telekom> tip wire
Adernabschirmung f <Elektriz> core screen
ADF *(Radiokompass, Funkkompass, automatischer Funkpeiler)* <Funkort, Lufttrans, Telekom> ADF *(automatic direction finder)*
Adhäsiometer n <Kunststoff> adherometer
Adhäsion f <Kunststoff> adhesion
Adhäsionseisenbahn f <Eisenbahn> *(AE)* adhesion railroad, *(BE)* adhesion railway
Adhäsionsfestigkeit f <Kunststoff> adhesive strength
Adhäsionskraft f <Verpack> adherence
Adhäsionsmesser m <Kunststoff> adherometer
Adhäsionssystem n <Trans> adhesion system
Adhäsionsverbesserer m <Kunststoff> adhesion promoter
Adhäsionszug m <Eisenbahn> total adherence train
ADI *(duldbare tägliche Aufnahmemenge)* <Lebensmittel> ADI *(acceptable daily intake)*
adiabatisch adj <Mechan, Phys, Strömphys, Thermod> adiabatic
adiabatisch adv <Strömphys, Thermod> adiabatically

adiabatische Änderung f <Strömphys, Thermod> adiabatic change
adiabatische Ausdehnung f <Strömphys, Thermod> adiabatic expansion
adiabatische Entmagnetisierung f <Strömphys, Thermod> adiabatic demagnetization
adiabatische Invariante f <Strömphys, Thermod> adiabatic invariant
adiabatische Kompressibilität f <Phys> isentropic compressibility
adiabatische Kompression f <Strömphys, Thermod> adiabatic compression
adiabatische Schallwellen fpl <Strömphys, Thermod> adiabatic sound waves
adiabatische Stoßwelle f <Strömphys, Thermod> adiabatic shock wave
adiabatische Transformation f <Strömphys, Thermod> adiabatic transformation
adiabatische Wand f <Strömphys, Thermod> adiabatic wall
adiabatische Zustandsänderung f <Strömphys, Thermod> adiabatic change
adiabatischer Beiwert m <Strömphys, Thermod> adiabatic coefficient, adiabatic curve
adiabatischer Druckabfall m <Strömphys, Thermod> adiabatic pressure drop
adiabatischer Prozess m <Strömphys, Thermod> adiabatic process
adiabatischer Temperaturgradient m <Strömphys, Thermod> adiabatic temperature gradient
adiabatischer vertikaler Gradient m <Thermod> adiabatic lapse rate
adiabatischer Wirkungsgrad m <Strömphys, Thermod> adiabatic efficiency
adiabatisches System n <Strömphys, Thermod> adiabatic system
adiabatisches Temperaturgefälle n <Thermod> adiabatic lapse rate
adiabatisches Verhalten n <Strömphys, Thermod> adiabatism
Adipinsäureester m <Kunststoff> adipic ester
Adiuretin n <Chemie> vasopressin
Adjazenz f <Geom> adjacency
Administrator m **des technischen Dienstes** <Comp & DV> technical services supervisor
Admittanz f <Akustik, Elektrotech, Phys> admittance
ADPCM (adaptive Differenz-Pulscodemodulation) <Telekom> ADPCM (adaptive differential pulse code modulation)
Adrenocorticotropin n <Chemie> corticotrophin
Adress... <Comp & DV> address, sequence
Adressabbildung f <Comp & DV> address mapping
Adressat m <Trans> consignee
Adressbus m <Comp & DV> address bus
Adresse f <Comp & DV, Funktech> address
Adressen-Auflösungsprotokoll n <Telekom> address resolution protocol
Adressendatei f <Telekom> address file
Adressenende n <Comp & DV> end of address
Adressenformat n <Comp & DV> address format
Adressengenerierung f <Comp & DV> address generation
Adressenmatrize f <Verpack> address stencil
Adressenregister n <Comp & DV> address register
Adressenschild n <Verpack> address label
Adressenvielfachleitung f <Comp & DV> address highway
Adressfolgeregister n <Comp & DV> sequence control register, sequence counter, sequence register
adressierbar adj <Comp & DV> addressable
adressierbare Speicherstelle f <Comp & DV> addressable location
adressieren v <Comp & DV> address
Adressiersystem n <Comp & DV> addressing system
Adressierung f <Comp & DV, Telekom> addressing
Adressierung f **über Basisadresse** <Comp & DV> base displacement address
Adressierungsart f <Comp & DV> addressing mode
Adressliste f <Comp & DV> mailing list
Adressmodifikation f <Comp & DV> address modification
Adressposition f <Comp & DV> address position
Adressraum m <Comp & DV> address space
Adsorbens n <Chemie> adsorbent
adsorbierbar adj <Chemie> adsorbable
adsorbieren v <Kohlen> adsorb
adsorbiertes Wasser n <Bau> adsorbed water
Adsorption f <Kohlen, Kunststoff, Lebensmittel> adsorption
Adsorptionsfalle f <Kerntech> adsorption trap
Adsorptionsisotherme f <Kerntech> adsorption isotherm
Adsorptionskohle f <Chemie> active carbon
Adsorptionsmittel n <Lebensmittel> adsorbent
Adsorptionswärme f <Kerntech> adsorption heat
Adsorptionswirkung f <Abfall> adsorption efficiency
ADU (Analog-Digital-Umsetzer) <Comp & DV, Fernseh, Fertig> ADC (analog-digital converter)
A/D-Umsetzer m <Comp & DV, Elektriz, Elektronik, Fertig, Gerät, Phys, Telekom> A/D converter
A/D-Umsetzung f <Aufnahme, Comp & DV, Elektriz, Elektronik, Fertig> A/D conversion
ADV (automatische Datenverarbeitung) <Comp & DV> ADP (automatic data processing)
Advektion f <Phys, Strömphys> advection
A/D-Wandler m <Comp & DV, Elektriz, Elektronik, Fertig, Labor, Phys, Telekom> A/D converter
A/D-Wandlung f <Aufnahme, Comp & DV, Elektriz, Elektronik, Fertig> A/D conversion
AE (astronomische Einheit) <Labor> AU (astronomical unit)
AEC (Amerikanischer Atomenergieverband) <Kerntech> AEC (Atomic Energy Commission)
aerob adj <Ergon, Lebensmittel> aerobic
aerob stabilisierter Schlamm m <Abfall> aerobically digested sludge
aerobe Bakterien npl <Abfall> aerobic bacteria
aerobe Gärung f <Abfall, Lebensmittel> aerobic fermentation
aerobe Schlammfaulung f <Wasserversorg> aerobic sludge digestion
aerobe Schlammstabilisierung f <Abfall> aerobic sludge stabilization
aerobe Zersetzung f <Abfall> aerobic decomposition
aerober Abbau m <Abfall> aerobic degradation
aerober Metabolismus m <Ergon> aerobic metabolism
aerobes Behandlungsverfahren n <Abfall> aerobic treatment process
Aerobier m <Lebensmittel> aerobe
Aerobus m <Lufttrans> aerobus
Aerodynamik f <Maschinen, Phys, Trans> aerodynamics
Aerodynamikkoeffizient m <Raumfahrt> aerodynamic coefficient
aerodynamisch adj <Maschinen, Phys, Trans> aerodynamic
aerodynamische Form f <Kfztech> aerodynamic shape (Karosserie)
aerodynamische Geräusche npl <Lufttrans> aerodynamic noise, airframe noise

aerodynamische Kraft

aerodynamische Kraft f <Nichtfoss Energ> aerodynamic power
aerodynamische Last f <Lufttrans> aerodynamic load
aerodynamische Stabilisierungsflosse f <Raumfahrt> aerodynamic stabilizing fin
aerodynamische Verwindung f <Lufttrans> aerodynamic twist
aerodynamische Verzögerung f <Lufttrans> aerodynamic lag
aerodynamische Waage f 1. <Lufttrans> aerodynamic balance, wind tunnel balance; 2. <Wassertrans> wind tunnel balance
aerodynamischer Auftrieb m <Lufttrans> aerodynamic lift
aerodynamischer Druck m <Lufttrans> aerodynamic pressure
aerodynamischer Faktor m <Lufttrans> aerodynamic factor
aerodynamischer Mittelpunkt m <Lufttrans> (AE) aerodynamic center, (BE) aerodynamic centre
aerodynamischer Mittelpunkt m **des Blattes** <Lufttrans> (AE) blade aerodynamic center, (BE) blade aerodynamic centre
aerodynamischer Übergang m <Lufttrans> fairing (Flugzeug); fillet (Flugwerk)
aerodynamischer Windkanal m <Lufttrans, Wassertrans> wind tunnel
aerodynamischer Wirkungsgrad m <Lufttrans> aerodynamic efficiency
aerodynamisches Bremsen n <Raumfahrt> aerodynamic braking (Raumschiff)
aerodynamisches Luftkissenfahrzeug n <Trans> aerodynamic-type air cushion vehicle
aerodynamisches Schweben n <Trans> aerodynamic levitation
aerodynamisches Verhalten n <Kfztech> aerodynamic stance
Aeroelastizität f <Kerntech, Nichtfoss Energ> aeroelasticity
Aerograph m <Druck, Ker & Glas> aerograph
Aerographie f <Druck, Ker & Glas> aerography
aeromagnetischer Zug m <Eisenbahn> aeromagnetic train
Aerometer n 1. <Metrol> aerometer; 2. <Papier> aerometer (Luftdichtemesser); 3. <Phys> aerometer
Aerometrie f <Papier, Phys> aerometry
aerometrisch adj <Papier> aerometric
aeronautische Werkstoffnorm f (AMS) <Raumfahrt> aeronautical material standard (AMS)
aeronautischer Informationsdienst m (AIS) <Raumfahrt> aeronautical information service (AIS)
Aerosol n 1. <Phys, Sicherheit> aerosol; 2. <Umweltschmutz> aerosol (schwebstoffhaltige Luft); 3. <Verpack> aerosol
Aerosolbehälter m <Maschinen> aerosol container
Aerosoldose f <Verpack> aerosol container
Aerosolpackung f <Abfall> aerosol dispenser
Aerosolsprühdose f <Maschinen> aerosol spray container
Aerosoltreibgas n <Erdöl> aerosol propellant (Petrochemie)
Aerosolventil n <Verpack> aerosol valve
Aerosolverschluss m <Verpack> aerosol cap
Aerostatik f <Maschinen> aerostatics
aerostatisches Luftkissenfahrzeug n <Trans> aerostatic-type air cushion vehicle
AES (Auger'sche Elektronenspektroskopie) <Phys, Strahlphys> AES (Auger electron spectroscopy)
Aesculin n <Chemie> aesculin

affine Geometrie f <Geom> affine geometry
Affinität f (A) <Chemie> affinity (A)
Affintransformation f <Metall> affine transformation
AFGC (automatische Frequenz- und Verstärkungsregelung) <Elektronik, Fernseh, Funktech> AFGC (automatic frequency and gain control)
AFI (automatische Fahrzeugidentifikation) <Trans> AVI (automatic vehicle identification)
Aflatoxin n <Lebensmittel> aflatoxin (Toxikologie)
AFO (automatische Fahrzeugortung) <Trans> AVL (automatic vehicle location)
A4-Format n <Papier> A4 size (internationale Papiergröße)
AFR (automatische Frequenzregelung) <Elektronik, Fernseh, Funktech> AFC (automatic frequency control)
AFS (fester Flugfunkdienst) <Lufttrans, Telekom> AFS (aeronautical-fixed service)
AFT (automatische Scharfabstimmung) <Funktech> AFT (automatic fine tuning)
AFTN (festes Flugfunknetz) <Lufttrans, Telekom> AFTN (aeronautical-fixed telecommunication network)
Ag (Silber) <Chemie> Ag (silver)
AG (amerikanisches Maß) <Maschinen> AG (American gage)
Agar n <Lebensmittel> agar (aus Polysacchariden bestehender Rotalgenextrakt)
Agar-Agar n <Lebensmittel> agar-agar (aus Polysacchariden bestehender Rotalgenextrakt)
Agatlinie f <Druck> agate line
AGCA (automatische Anflugsteuerung vom Boden) <Raumfahrt> AGCA (automatic ground-controlled approach)
AGCL (automatische Landesteuerung vom Boden) <Raumfahrt> AGCL (automatic ground-controlled landing)
AGE (Allylglycidether) <Kunststoff> AGE (allyl glycidyl ether)
Agene n <Lebensmittel> agene
Agens n <Chemie> agent
Agglomerat n <Fertig> agglomerate
Agglomeratbildung f <Chemtech> agglomeration
Agglomeration f <Chemtech, Kunststoff> agglomeration
Agglomerationsverfahren n <Metall> agglomerating process
agglomerieren v <Chemtech> agglomerate
Agglutination f <Chemtech, Lebensmittel> agglutination
agglutinierend adj <Chemie> agglutinant
Agglutinin n <Lebensmittel> agglutinin (spezifischer Antikörper)
Aggradation f <Wasserversorg> aggradation
Aggregat n <Fertig> bank
Aggregat n 1. <Bau> aggregate, set; 2. <Comp & DV> unit; 3. <Fertig> aggregate, package, unit; 4. <Kfztech> unit; 5. <Kohlen> aggregate; 6. <Maschinen> assembly, unit; generator set, genset (Antrieb); 7. <Metall> aggregate; 8. <Wassertrans> generator set, genset (Antrieb)
Aggregatkratzer m <Trans> aggregate scraper
Aggregatschrappförderer m <Trans> aggregate scraper
Aggregatzustand m <Maschinen> state of aggregation
aggressiv adj <Fertig> corrosive (Kunststoffinstallationen)
aggressive Mittel npl <Heiz & Kälte> corrosive media
aggressives Wasser n <Wasserversorg> aggressive water
AGR (fortgeschrittener Gas-Graphit-Reaktor) <Kerntech> AGR (advanced gas-cooled reactor)
Ahle f <Druck> bodkin
Ahminges fpl <Wassertrans> (AE) draft marks, (BE) draught marks (Schiffkonstruktion)

ähnlich sein v <Math> assimilate
ähnliche Dreiecke npl <Geom> similar triangles
ähnliche Figuren fpl <Geom> similar figures
ähnliche Ladungsmengen fpl <Elektriz> like charges
ähnliche Pole mpl <Elektriz> like poles
Ähnlichkeitsverhältnis n <Geom> similarity relation
AI (Schallimpedanz, Schallwellenwiderstand, akustische Impedanz, akustischer Scheinwiderstand) <Akustik> ZA (acoustic impedance)
AIA (Amerikanischer Luft- und Raumfahrtverband) <Raumfahrt> AIA (Aerospace Industries Association)
Airbag m <Kfztech> air bag
Airbag-Haltesystem n <Kfztech> air bag restraint system
Airbus m <Lufttrans> airbus
Air-Terminal m <Lufttrans> airways terminal
Air-Terminal m **für Frachtflugzeuge** <Lufttrans> freight terminal, cargo terminal
Airy-Scheibchen n <Phys> (BE) Airy disc, (AE) Airy disk
AIS (aeronautischer Informationsdienst) <Raumfahrt> AIS (aeronautical information service)
AK (akustische Kapazität) <Akustik> AC (acoustic capacitance)
Akklimatisierung f <Ergon> acclimatization
Akkomodation f <Ergon> accommodation
Akkord m <Akustik> chord
Akkordarbeit f <Bau> job work
Akku m (Akkumulator) 1. <Comp & DV, Elektriz, Elektrotech> accumulator, rechargeable battery, storage battery; 2. <Funktech> accumulator, rechargeable battery, storage battery (wiederaufladbare Batterie); 3. <Heiz & Kälte, Hydraul, Kfztech, Papier, Phys> accumulator, rechargeable battery, storage battery
Akkubatterie f <Comp & DV, Elektriz, Elektrotech, Funktech, Heiz & Kälte, Hydraul, Kfztech, Papier, Phys> accumulator, rechargeable battery, storage battery
Akkuladeschaltung f <Elektriz> charging circuit
Akkumulator m (Akku) <Comp & DV, Elektriz, Elektrotech> accumulator, rechargeable battery, storage battery; 2. <Funktech> accumulator, rechargeable battery, storage battery (wiederaufladbare Batterie); 3. <Heiz & Kälte, Hydraul, Kfztech, Papier, Phys> accumulator, rechargeable battery, storage battery
Akkumulatorbatterie f <Elektrotech, Kfztech, Telekom> accumulator battery, rechargeable battery, storage battery
Akkumulatorelektrode f <Elektriz> battery electrode
Akkumulatorentladung f <Elektriz> accumulator discharge
Akkumulatorfahrzeug n <Trans> accumulator vehicle
Akkumulatorkasten m <Papier> accumulator box
Akkumulatorladung f <Elektriz> accumulator charge
Akkumulatorleistungsanzeige f <Papier> accumulator capacity indicator
Akkumulatorplatte f <Elektriz> accumulator plate, battery plate
Akkumulatorregister n <Comp & DV> accumulator register
Akkumulatorsäure f 1. <Chemie, Elektriz> electrolyte; 2. <Elektrotech> battery acid
Akkumulatortriebwagen m <Trans> accumulator railcar
Akkumulatorzelle f <Elektriz, Elektrotech> accumulator cell, storage cell
akkumulierte Dosis f/**von Strahlenarbeitern** <Strahlphys> dose accumulated by workers
akkumulierte Energiedosis f <Kerntech> accumulated dose, cumulative dose
akkumulierter Fehler m <Gerät> accumulated error
Akkuplatte f <Elektriz> accumulator plate

A-Kohle f (Aktivkohle) <Chemie, Kohlen, Kunststoff, Lebensmittel, Papier, Wasserversorg> activated carbon, activated charcoal, active carbon
Akquisition f <Künstl Int> acquisition (des Wissens)
Akronym n <Comp & DV> acronym
Akteneinsicht f <Patent> inspection of files
Aktentaschencomputer m <Comp & DV> laptop computer, portable
aktinisch adj <Foto> actinic
aktinische Strahlen mpl <Foto> actinic rays
Aktinität f <Druck> actinic effect
Aktiniumemanation f <Strahlphys> actinium emanation
Aktinometrie f 1. <Phys> actinometry; 2. <Strahlphys> actinometry (hauptsächlich Licht)
Aktion f <Comp & DV, Ergon, Kfztech> action
Aktions... 1. <Comp & DV> action, drop-down; 2. <Trans> working
Aktionseintrag m <Comp & DV> action entry
Aktionsfenster n <Comp & DV> drop-down menu, pull-down menu
Aktionspotenzial n <Ergon> action potential
Aktionsradius m 1. <Kfztech> cruising range (Elektrofahrzeuge); 2. <Trans> useful working range
Aktionsturbine f 1. <Hydraul> action turbine; 2. <Nichtfoss Energ> impulse turbine
aktiv adj <Comp & DV, Elektrotech, Fertig> active
Aktiv... <Comp & DV, Elektrotech, Fertig> active
Aktivabfallverdampfer m <Kerntech> radioactive waste evaporator
Aktivation f <Künstl Int> activation (von Neuronen)
Aktivator m 1. <Elektrotech> activator; 2. <Kohlen> activating agent, activator; 3. <Kunststoff> activator
Aktivdatei f <Mobilkom> active subscribers file, GSM
aktive Aufhängung f <Eisenbahn> active suspension
aktive Datei f <Comp & DV> active file
aktive Emanation f <Kerntech> active emanation
aktive Fahrzeugsicherheit f <Trans> active motor vehicle safety
aktive Feldzeit f <Fernseh> active field period
aktive Flanke f 1. <Fertig> active profile (Getriebelehre); 2. <Maschinen> active face (eines Zahns)
aktive Führung f <Raumfahrt> active guidance (Raumschiff)
aktive Gruppenantenne f <Funktech> active array
aktive Instandhaltungsdauer f <Qual> active maintenance time
aktive integrierte Mikrowellenschaltung f <Elektronik> active microwave integrated circuit
aktive Kamera f <Fernseh> hot camera
aktive Länge f <Kerntech> active length (eines Brennelementes)
aktive Leitung f <Elektronik, Fernseh> active line
aktive Schicht f <Elektronik> active layer
aktive Sicherheitsfunktion f <Kfztech> active safety feature
aktive Sonnenenergie f <Nichtfoss Energ> active solar energy
aktive Spiegelmaschine f <Kerntech> active mirror
aktive Steuerung f <Raumfahrt> active control (Raumschiff)
aktive Übertragungsleitung f <Telekom> active line
aktive vorbeugende Wartungsdauer f <Qual> active preventative maintenance time
aktive Zielsuchlenkung f <Lufttrans> homing active guidance
aktiver Bandpass m <Elektronik> active band-pass filter
aktiver Bereich m <Elektronik> active region (Halbleitersubstrat)
aktiver Dipol m <Elektrotech> active dipole

aktiver

aktiver Equalizer m <Aufnahme> active equalizer *(Entzerrer)*
aktiver Erddruck m <Kohlen> active earth pressure
aktiver Füllstoff m <Kunststoff> active filler *(verstärkend wirkend)*
aktiver Infrarotdetektor m <Trans> active infrared detector
aktiver Kreislauftest m <Kerntech> active test loop
aktiver Prozessor m <Telekom> active processor
aktiver Richtstollen m <Kohlen> active mine heading
aktiver Schaltkreis m <Phys> active circuit
aktiver Spiegel m <Kerntech> active mirror
aktiver Strahler m <Funktech> active antenna, primary source *(Antenne)*
aktiver Transducer m <Elektriz, Elektrotech> active transducer
aktiver Vierpol m <Elektrotech> active quadripole
aktiver Wandler m <Elektriz, Elektrotech> active current transformer, active transducer, active voltage transformer
aktives Anfahren n **eines Reaktors** <Kerntech> energetic start-up *(Atomkraftwerk)*
aktives Bandpassfilter n <Elektronik> active band-pass filter, active notch filter
aktives Bauelement n <Elektriz, Elektrotech> active component
aktives Element n <Elektrotech> active element
aktives Filter n <Elektronik, Telekom> active filter
aktives Filter n **dritter Ordnung** <Elektronik> third order active filter
aktives Filtern n <Elektronik> active filtering
aktives Glied n <Regelung> active element, final controlling element
aktives I-Element n <Elektronik> active integrator *(Automatisierungstechnik)*
aktives Integrierglied n <Elektronik> active integrator
aktives Lasermedium n <Optik, Telekom> active laser medium
aktives Leuchtfeuer-Kollisionswarnsystem n <Lufttrans> active beacon collision avoidance system
aktives Lösemittel n <Kunststoff> active solvent, true solvent
aktives Lösungsmittel n <Verpack> active solvent
aktives Metall n <Metall> active metal
aktives Netzwerk n <Elektrotech> active network
aktives Sonnenenergiesystem n <Nichtfoss Energ> active solar energy system
aktives Sonnensystem n <Nichtfoss Energ> active solar system
aktives System n <Akustik> active system
aktives Wasser n <Wasserversorg> active water
aktivieren v 1. <Comp & DV> enable, execute *(Rechner)*; 2. <Kohlen, Kontroll, Papier> activate; 3. <Textil> boost
aktiviert adj <Elektrotech> activated
aktivierte Holzkohle f <Lebensmittel> activated charcoal
aktivierte Kohle f <Lebensmittel> activated carbon, active carbon
aktivierte Tonerde f <Lebensmittel> activated alumina
aktivierter Komplex m <Metall> activated complex
aktivierter Zustand m <Metall> activated state
aktiviertes Aluminiumoxid n <Lebensmittel> activated alumina
aktiviertes Molekül n <Strahlphys> activated molecule
Aktivierung f 1. <Kerntech, Kohlen> activation; 2. <Künstl Int> activation *(von Neuronen)*; 3. <Metall, Papier, Strahlphys, Telekom> activation
Aktivierung f **durch Gammastrahlen** <Strahlphys, Teilphys, Wellphys> gamma photon activation
Aktivierungsanalyse f 1. <Phys> activation analysis; 2. <Strahlphys> activation analysis, radioactivation analysis
Aktivierungsanalyse f **mit Hilfe geladener Teilchen** <Kerntech, Strahlphys, Teilphys> charged-particle activation analysis
Aktivierungsbereich m <Metall> activation area
Aktivierungsenergie f 1. <Kerntech> activation; 2. <Metall, Strahlphys> activation energy
Aktivierungsentropie f <Metall> activation entropy
Aktivierungslog n <Erdöl> activation log *(Bohrlochmessung)*
Aktivierungsmittel n <Papier> activator
Aktivierungsparameter n <Metall> activation parameter
Aktivierungswärme f <Nichtfoss Energ, Thermod> heat of activation
Aktivität f 1. <Comp & DV> activity; 2. <Kerntech> activity *(A)*; 3. <Phys> activity *(A)*
Aktivität f **eines Atomkerns** <Strahlphys> nuclear activity
Aktivitätsbeiwert m <Raumfahrt> activity factor *(Weltraumfunk)*
Aktivitätsgrenzwerte mpl <Kerntech> activity threshold
Aktivitätskoeffizient m <Phys> activity coefficient
Aktivitätsüberspannung f <Raumfahrt> activity overvoltage *(Raumschiff)*
Aktivitätsverzeichnis n <Kerntech> activity inventory
Aktivkohle f *(A-Kohle)* <Chemie, Kohlen, Kunststoff, Lebensmittel, Papier, Wasserversorg> activated carbon, activated charcoal, active carbon
Aktivkohle-Absorption f <Umweltschmutz> active carbon absorption
Aktivkohlebehandlung f <Abfall> activated carbon treatment
Aktivkohlebett n <Kerntech> activated charcoal bed
Aktivkohlefilter n <Kerntech> activated carbon filter
Aktivruß m <Kunststoff> activated carbon black
Aktor m <Künstl Int> actor
aktualisieren v 1. <Comp & DV> refresh, update; 2. <Maschinen> bring up to date, update
aktualisiert anzeigen v <Comp & DV> refresh
Aktualisierung f 1. <Comp & DV> refresh, update; 2. <Telekom> updating
Aktualisierungsdatei f <Comp & DV> update file
Aktualisierungslauf m <Comp & DV> update run
Aktualisierungsmodus m <Comp & DV> update mode
Aktualisierungsrate f <Comp & DV> refresh rate
Aktualisierungsspeicher m <Comp & DV> refresh memory
Aktualisierungszyklus m <Comp & DV> refresh cycle
Aktuator m <Elektriz> actuator *(Stellglied)*
aktuelle Datei f <Comp & DV> update
aktueller Adressschlüssel m <Comp & DV> actual key *(Programmiersprache)*
Akuphonie f <Akustik> acouphony
Akustik f 1. <Aufnahme> acoustics *(Lautlehre)*; 2. <Ergon, Funktech, Phys> acoustics
Akustikentwurf m <Aufnahme> acoustical design
Akustikkoppler m <Comp & DV, Elektronik> acoustic coupler *(Peripheriegerät)*
Akustiklog n <Erdöl> acoustic log *(Bohrlochmessung)*; sonic log *(Messtechnik)*
Akustikprozessor m <Künstl Int> acoustic processor
Akustikspektrum n <Aufnahme> acoustic spectrum
Akustikverstärker m <Aufnahme> acoustic amplifier
akustisch adj <Aufnahme, Erdöl, Phys> acoustic, sonic
akustische Abschirmung f <Aufnahme> acoustic screen, acoustic shielding
akustische Absorption f 1. <Akustik> total acoustic absorption *(a)*; 2. <Aufnahme> acoustic absorption *(Schallschlucken)*
akustische Admittanz f *(YA)* <Akustik, Elektrotech> acoustic admittance *(YA)*

akustische Ausbreitungskonstante f <Akustik> acoustic propagation constant
akustische Beugung f <Akustik> acoustic diffraction
akustische Blendung f <Akustik> aural dazzling
akustische Dämpfung f <Elektronik> acoustic damping
akustische Diffraktion f <Akustik> acoustic diffraction
akustische Dispersion f <Akustik> acoustic dispersion
akustische Emission f <Kerntech> acoustic emission
akustische Energie f <Elektrotech> acoustic energy
akustische Federung f <Aufnahme> acoustic compliance
akustische Impedanz f 1. <Akustik> (AI) acoustic impedance (ZA); 2. <Aufnahme, Elektrotech, Phys> acoustic impedance (ZA)
akustische Kapazität f (AK) <Akustik> acoustic capacitance (AC)
akustische Kapselung f <Sicherheit> acoustic enclosure
akustische Kernimpedanz f <Akustik> transfer acoustic impedance
akustische Kopplung f <Elektronik> acoustic coupling
akustische Masse f 1. <Akustik> inertance; 2. <Akustik> acoustic mass (AM); 3. <Phys> acoustic inertance
akustische Mobilität f <Akustik> acoustic mobility
akustische Oberflächenwelle f (AOW) <Elektronik, Telekom> surface acoustic wave (SAW)
akustische Perspektive f <Aufnahme> acoustic perspective
akustische Reaktanz f <Akustik, Elektrotech> acoustic reactance
akustische Resistanz f <Akustik> acoustic resistance
akustische Rückkopplung f <Aufnahme> acoustic feedback
akustische Steifheit f <Akustik, Phys> acoustic stiffness
akustische Streuung f 1. <Akustik> acoustic dispersion; 2. <Aufnahme> acoustic scattering
akustische Suszeptanz f <Akustik> acoustic susceptance
akustische Trägheitsmasse f <Akustik> inertance
akustische Übertragungslinie f <Elektrotech> acoustic transmission line
akustische Verzögerungsleitung f <Comp & DV> acoustic delay line
akustische Verzögerungsstrecke f <Elektronik> acoustic delay line
akustische Warnanlage f <Gerät, Sicherheit> audio alarm system
akustischer Absorptionskoeffizient m <Phys> acoustic absorption coefficient
akustischer Absorptionsverlust m <Aufnahme> acoustic absorption loss
akustischer Akzeptanzpegel m (AAP) <Akustik> acoustic comfort index (ACI)
akustischer Alarm m <Telekom> audible alarm
akustischer Anrufmelder m <Telekom> tone pager
akustischer Blindleitwert m (AB) <Akustik> acoustic susceptance (BA)
akustischer Blindwiderstand m <Akustik, Elektrotech, Phys> acoustic reactance
akustischer Frequenzbereich m <Strahlphys> audio range
akustischer Leitwert m (AL) <Akustik> acoustic conductance (GA)
akustischer Oszillator m <Strahlphys> audio oscillator
akustischer Resonator m <Elektronik> acoustic resonator
akustischer Scheinwiderstand m 1. <Akustik> (AI) acoustic impedance (ZA); 2. <Aufnahme, Elektrotech, Phys> acoustic impedance (ZA)

akustischer Speicher m 1. <Comp & DV> acoustic memory, acoustic store; 2. <Elektronik> acoustic delay line
akustischer Tonabnehmer m <Aufnahme> acoustic pick-up
akustischer Träger m <Elektronik> acoustic carrier
akustischer Widerstand m <Akustik> acoustic resistance
akustischer Wirkungsgrad m <Aufnahme> acoustic efficiency
akustischer Zweig m <Phys> acoustic branch (Festkörperphysik)
akustisches Filter n 1. <Akustik, Aufnahme> acoustic filter; 2. <Elektronik> acoustic filter, acoustic-wave filter; 3. <Maschinen> acoustic filter
akustisches Interferometer n <Akustik> acoustic interferometer
akustisches Radiometer n <Akustik> acoustic radiometer
akustisches Rufzeichen n <Telekom> audible signal
akustisches Signal n 1. <Eisenbahn> sound signal; 2. <Elektronik, Maschinen> acoustic signal; 3. <Telekom> sound signal (Telefon)
akustisches System n <Akustik> acoustic system
akustisches Zentrum n <Akustik> (AE) effective acoustic center, (BE) effective acoustic centre
akustooptische Modulation f <Elektronik> acousto-optic modulation
akustooptischer Effekt m <Optik> acousto-optic effect
akustooptischer Modulator m <Elektronik, Optik> acousto-optic modulator
akustooptischer Prozessor m <Elektronik> acousto-optic processor
Akut m <Druck> acute accent
akute Wirkung f <Umweltschmutz> acute effect
Akzent m <Druck> accent
Akzentbuchstabe m <Druck> accented letter
akzeptabler Qualitätspegel m (AQL) <Qual> acceptable quality level (AQL)
akzeptabler Zuverlässigkeitspegel m (ARL) <Qual> acceptable reliability level (ARL)
Akzeptanz f <Raumfahrt, Telekom> acceptance
Akzeptanzkriterium n <Raumfahrt> acceptance criterion (Raumschiff)
Akzeptanzwinkel m 1. <Fertig> acceptance angle, angle of acceptance (Faseroptik); 2. <Telekom> acceptance angle
Akzeptor m <Comp & DV, Elektronik> acceptor (Halbleiter)
Akzeptoratom n <Elektronik, Phys> acceptor atom
Akzeptorniveau n <Elektronik> acceptor level (Halbleiter)
Akzeptorverunreinigung f <Elektronik> acceptor impurity
Akzidenz f <Druck> job, jobbing
Akzidenzsatz m <Druck> job composition
Akzidenzschriften fpl <Druck> jobbing types
Al (Aluminium) <Chemie> Al (aluminium)
AL 1. <Akustik> (akustischer Leitwert) GA (acoustic conductance); 2. <Telekom> (Anschlussleitung) subscriber's line
Alabasterglas n <Ker & Glas> alabaster glass
Alanin n <Chemie> alanine
Alantin n <Chemie> alantin, dahlin, helenin, inulin, sinistrin
Alantstärke f <Chemie> alant starch, alantin, inulin
Alarm m <Kerntech, Maschinen, Sicherheit> alarm
• **Alarm auslösen** 1. <Sicherheit> give the alarm; 2. <Telekom> trigger an alarm
Alarmanlage f <Sicherheit> alert system, safety alarm
Alarmblinker m <Sicherheit> alarm flashing light

Alarmdrucker *m* <Telekom> alarm print-out facility
Alarmeinstellung *f* <Telekom> alarm setting
Alarmglas *n* <Ker & Glas> signal glass
Alarmglocke *f* 1. <Elektriz, Papier> alarm bell; 2. <Sicherheit> alarm bell, warning bell
Alarmierung *f* <Telekom> alerting
Alarmleitung *f* <Telekom> alarm circuit
Alarmmeldelampe *f* <Telekom> alarm indication lamp
Alarmmelder *m* <Gerät> alarm annunciator
Alarmmeldesignal *n* <Elektronik> blue signal *(Nachrichtenübertragungstechnik)*
Alarmpatrone *f* <Eisenbahn> torpedo
Alarmrelais *n* <Elektriz> alarm relay
Alarmruf *m* <Telekom> alarm call
Alarmschalter *m* <Gerät> alarm switch
Alarmschaltungskarte *f* <Telekom> alarm card
Alarmschütz *n* <Elektriz> alarm relay
Alarmschwimmer *m* <Papier> alarm float
Alarmsicherung *f* <Elektriz> alarm fuse
Alarmsignal *n* 1. <Eisenbahn> danger signal; 2. <Kerntech> alarm; 3. <Maschinen, Wassertrans> alarm signal
Alarmsirene *f* <Sicherheit> auditory signal
Alarmsystem *n* <Sicherheit> alarm system, alert system
Alarmtafel *f* <Lufttrans, Sicherheit> general warning panel
Alaun *m* <Chemie, Lebensmittel, Papier> alum
alaunen *v* <Chemie> aluminate
Alaunerde *f* <Chemie> alumina
alaunhaltig *adj* <Chemie> aluminiferous
Alban *n* <Chemie> alban
Albedo *f* 1. <Chemie> pith; 2. <Raumfahrt> albedo
Albit *m* <Ker & Glas> albite
Albumin *n* <Lebensmittel> albumin
albuminartig *adj* <Chemie> albuminoid, albuminous
Albuminat *n* <Chemie, Lebensmittel> albuminate
Albuminpapier *n* <Druck, Papier> albumenized paper
Albuminverfahren *n* <Foto> albumen process
Albumose *f* <Lebensmittel> albumose
ALC 1. <Funktech> *(automatische Pegelregelung)* ALC *(automatic level control)*; 2. <Kfztech> *(automatischer Niveauausgleich)* ALC *(automatic level control)*
Älchen *n* <Lebensmittel> eelworm *(Phytopathologie)*
Aldehyd *m* <Chemie, Kunststoff, Lebensmittel> aldehyde
aldehydhaltig *adj* <Chemie> aldehydic
aldehydisch *adj* <Chemie> aldehydic
Aldehydsäure *f* <Lebensmittel> aldehyde acid
Aldohexose *f* <Chemie, Lebensmittel> aldohexose
Aldol *n* 1. <Chemie> acetaldol; 2. <Lebensmittel> aldol
Aldose *f* <Lebensmittel> aldose
Aldosteron *n* <Lebensmittel> aldosterone
aleatorisch *adj* <Math> aleatoric, aleatory
aleatorische Reihe *f* <Math> aleatory series
Alfapapier *n* <Druck> esparto paper
Algebra *f* <Comp & DV, Math> algebra
algebraisch *adj* <Math> algebraic
algebraische Geometrie *f* <Geom> algebraic geometry
algebraische Gleichung *f* <Math> literal equation
algebraische Zahl *f* <Math> algebraic number
algebraischer Ausdruck *m* <Math> algebraic expression
algebraisches Symbol *n* <Math> algebraic symbol
algebraisches Zeichen *n* <Math> algebraic symbol
Alginat *n* <Kunststoff> alginate
Alginsäure *f* <Lebensmittel> alginic acid
Algorithmik *f* <Comp & DV> algorithmics
algorithmisch *adj* <Comp & DV> algorithmic
algorithmische Sprache *f* <Comp & DV> algorithmic language
Algorithmus *m* <Comp & DV, Ergon, Math> algorithm

Alhidade *f* 1. <Eisenbahn, Kfztech, Lufttrans, Trans> sight rule; 2. <Wassertrans> alidade, index bar, sight rule
Alias *m* <Comp & DV> alias
Alias-Effekt *m* <Elektronik> aliasing
Aliasing-Frequenz *f* <Elektronik> aliased frequency
aliphatisch *adj* 1. <Chemie> acyclic, aliphatic; 2. <Erdöl> aliphatic *(Petrochemie)*
aliphatischer Kohlenwasserstoff *m* <Kunststoff> aliphatic hydrocarbon
aliphatisches Lösungsmittel *n* <Erdöl> aliphatic solvent *(Raffinerie)*
aliphatisches Polyamin *n* <Kunststoff> aliphatic polyamine
aliphatisches Solvent *n* <Erdöl> aliphatic solvent
Alitieren *n* <Fertig> aluminizing, calorizing; alitizing *(Stahl)*
alitieren *v* <Fertig> aluminize; alitize *(Stahl)*
Alizarinviolett *n* <Chemie> gallein
Alkali *n* 1. <Chemie> alkali; 2. <Erdöl> alkali *(Petrochemie)*; 3. <Kohlen, Papier, Textil> alkali
Alkali-Akkumulator *m* <Papier> alkaline accumulator
Alkalibatterie *f* <Foto> alkaline cell
alkalibeständig *adj* <Papier> alkali-proof
alkalibeständiges Papier *n* <Verpack> alkali-proof paper
Alkalibeständigkeit *f* <Kunststoff> alkali resistance
Alkalicellulose *f* <Papier> alkali cellulose
Alkaligehalt *m* <Chemie> alkalinity
alkalihaltig *adj* <Chemie> alkaline
Alkalimesser *m* <Lebensmittel> alkalimeter
Alkalimessung *f* <Chemie> alkalimetry
Alkalimetall *n* <Fertig, Metall> alkali metal
Alkalimeter *n* <Papier> alkalimeter *(Laugenmesser)*
Alkalimetrie *f* <Chemie> alkalimetry
Alkalinität *f* <Chemie> alkalescence, alkalinity
alkalisch *adj* <Anstrich, Chemie, Funktech> alkaline
alkalische Batterie *f* <Elektrotech> alkaline battery
alkalische Zelle *f* <Elektrotech> alkaline cell, alkaline storage cell
alkalischer Akkumulator *m* <Elektriz, Elektrotech> alkaline accumulator, alkaline storage battery
alkalischer Sammler *m* <Elektrotech> alkaline storage battery
Alkalisierung *f* <Chemie> alkalization
Alkalität *f* <Lebensmittel, Umweltschutz> alkalinity *(Laugengrad)*
Alkalität vor der Ansäuerung *f* <Umweltschutz> preacidification alkalinity
Alkalizelle *f* <Foto> alkaline cell
Alkaloid *n* <Chemie> alkaloid
Alkan *n* 1. <Chemie> alkane, methane series; 2. <Erdöl> alkane *(Petrochemie)*
Alkaptonurie *f* <Chemie> alkaptonuria
Alken *n* 1. <Chemie> alkene, olefin, olefine; 2. <Erdöl> alkene *(Petrochemie)*
Alken... <Chemie> olefinic
Alkengehalt *m* <Chemie> olefinic content
Alkine *npl* <Chemie> alkynes *(Petrochemie)*
Alkohol *m* <Chemie, Lebensmittel> alcohol
Alkoholat *n* <Chemie, Lebensmittel> alcoholate
Alkoholbrenner *m* <Chemie, Lebensmittel> distiller
Alkoholbrennstoffe *mpl* <Kohlen> alcohol fuels
alkoholisch *adj* <Chemie, Lebensmittel> alcoholic
alkoholische Gärung *f* <Chemie, Lebensmittel> alcoholic fermentation
Alkoholthermometer *n* <Heiz & Kälte> alcohol thermometer
Alkyd *n* <Chemie> alkyd, alkyd resin
Alkydharz *n* 1. <Chemie> alkyd, alkyd resin; 2. <Fertig, Kunststoff> alkyd resin
Alkyl *n* <Chemie, Erdöl> alkyl *(Petrochemie)*

Alkylaromaten npl <Erdöl> alkyl aromatics *(Petrochemie)*
Alkylenimin n <Chemie> imine
Allanit m <Kerntech> orthite
Allantoin n <Chemie> allantoin
Alleinflugzeit f <Lufttrans> solo time
alleiniger Erfinder m <Patent> sole inventor
allelotrop adj <Chemie> allelotropic
Allen n <Chemie> allene, propadiene
Allen'sche Schleifenmethode f <Elektriz> Allen's loop test *(Kabelwiderstandsmessung)*
Allen'sche Stromschlingenprüfung f <Elektriz> Allen's loop test *(Kabelwiderstandsmessung)*
Alleskleber m <Verpack> all-purpose adhesive
Alles-oder-Nichts-Reaktion f <Ergon> all-or-none response, all-or-nothing response
allgemeine Eingrenzung f <Telekom> general localization
allgemeine Kreisgleichung f <Geom, Math> general equation of the circle
allgemeine Relativitätstheorie f <Phys> general theory of relativity
allgemeine Wartung f <Kerntech> general maintenance
allgemeines Fernsprechwählnetz n <Telekom> general switched telephone network
allgemeines Paketfunksystem n <Funktech, Telekom> general packet radio system
Allgeschwindigkeitsquerruder n <Trans> all-speed aileron
Allglasfaser f 1. <Elektrotech> *(AE)* all-glass optical fiber; 2. <Optik> *(AE)* all-glass fiber, *(BE)* all-glass fibre
Alligator... 1. <Elektriz> crocodile; 2. <Maschinen> alligator
allmählicher Ausfall m <Telekom> gradual failure
Allmenge f <Math> universal set
Allomorphie f <Chemie> allotropism
allotriomorph adj <Fertig> allotriomorphic *(Kristall)*
Allotropie f 1. <Chemie> allotropism, allotropy; 2. <Fertig> allotropy
allotropisch adj <Chemie, Fertig> allotropic
Allpass m <Elektronik> all-pass filter
Allpassfilter n <Elektronik> all-pass filter
Allplastfaser f <Optik> *(AE)* all-plastic fiber, *(BE)* all-plastic fibre
Allradantrieb m <Kfztech> all-wheel drive, AWD, four-wheel drive
Allradbremse f <Kfztech> all-wheel brake, four-wheel brake
Allrichtungsmikrofon n 1. <Akustik> omnidirectional microphone; 2. <Aufnahme> astatic microphone
Allroundterminal n <Wassertrans> multipurpose terminal
allseitig bearbeitet adj <Mechan> machined all over
Allstrommotor m <Elektriz> *(AE)* all-current motor, *(BE)* all-mains motor
All-Tantal-Kondensator m <Elektrotech> all-tantalum capacitor
Allterrain-Reifen m <Kfztech> *(AE)* town-and-country tire, *(BE)* town-and-country tyre
Alluvialschicht f <Wasserversorg> alluvial bed
Alluvion f <Bau> alluvion
Alluvium n 1. <Bau> alluvium; 2. <Wasserversorg> alluvial deposit
Allwellenantenne f <Phys> multiband antenna
Allwellenempfänger m <Funktech, Telekom> all-wave receiver
Allwetter... <Lufttrans> all-weather
Allwetterflüge mpl <Lufttrans> all-weather operations
Allwetterhubschrauber m <Lufttrans> all-weather helicopter
Allyl n <Chemie> allyl
Allylalkohol m <Chemie> allyl alcohol
Allylen n <Chemie> allylene, methylacetylene, propyne
Allylglycidether m *(AGE)* <Kunststoff> allyl glycidyl ether *(AGE)*
Allylmethylendioxybenzen n <Chemie> allylmethylenedioxybenzene, safrole
Allzweck-... <Elektrotech> GP, general-purpose, all-purpose
Allzweck-Laminat n <Elektronik> general-purpose laminate
Aloebitter m <Chemie> aloetic gum, aloin
Aloebitterstoff m <Chemie> aloetic gum, aloin
ALOHA-Verfahren n <Telekom> ALOHA system
ALOHA-Zugriffssystem n **mit festen Zeitschlitzen** <Telekom> slotted ALOHA system
Aloin n <Chemie> aloin, barbaloin
Alpha n (α) <Geom> alpha (α)
Alphaablauf m <Fernseh> alpha wrap
Alphabet n <Comp & DV> alphabet
alphabetisch adj <Comp & DV> alphabetic
alphabetischer Code m <Comp & DV> alphabetic code
alphabetischer Schlüssel m <Comp & DV> alphabetic code
Alpha-Cellulose f <Papier> alpha cellulose
Alphaeisen n <Metall> alpha iron
Alphageometrie f <Fernseh> alphageometry *(Videotex)*
alphageometrische Anzeige f <Telekom> alphageometric display
alphageometrisches Bildschirmgerät n <Telekom> alphageometric display
Alpha-Ionisierungsgasanalyse f <Kerntech> alpha ionization gas analysis
Alphamosaik-Verfahren n <Comp & DV> alphamosaic mode
alphanumerisch adj <Comp & DV> alphanumeric
alphanumerische Anzeige f <Telekom> alphanumeric display
alphanumerische Sortierung f <Comp & DV> alphanumeric sort
alphanumerischer Code m <Comp & DV> alphanumeric code
alphanumerisches Bildschirmgerät n <Telekom> alphanumeric display
alphanumerisches Zeichen n <Comp & DV> alphanumeric character
Alpha-Profil n <Optik> alpha profile
Alpha-Spektrometrie f <Phys, Strahlphys, Teilphys> alpha ray spectrometry
Alpha-Strahlen mpl <Elektriz, Optik, Phys, Strahlphys, Teilphys> alpha rays
Alpha-Strahler m <Kerntech, Phys, Strahlphys, Teilphys> alpha emitter
Alpha-Teilchen n <Elektriz, Phys, Strahlphys, Teilphys> alpha particle
Alpha-Zerfall m <Kerntech, Phys, Strahlphys, Teilphys> alpha decay, radioactive transmutation
Alpha-Zerfallsenergie f <Phys, Strahlphys, Teilphys> alpha disintegration energy
ALR *(automatische Lautstärkeregelung)* <Funktech> AVC *(automatic volume control)*
Alt... <Abfall, Maschinen> waste
Altbackenwerden n <Lebensmittel> staling *(Brot)*
Altblei n <Abfall> scrap lead
altern v 1. <Anstrich> cure; 2. <Bau> weather; mature *(Farbe, Bitumen, Beton)*; 3. <Fertig> age-harden *(Leichtmetalle)*; 4. <Papier> age
Altern n 1. <Anstrich> curing; 2. <Bau> maturing *(Farbe, Bitumen, Beton)*; 3. <Fertig> age hardening *(Leichtmetalle)*

Alternativ...

Alternativ... <Maschinen, Telekom> alternative
alternative Betriebsart f <Elektronik> alternate mode
alternative Brennstoffe mpl <Kohlen> alternative fuels
alternative Energiequellen fpl <Kohlen, Nichtfoss Energ> alternative energy sources
alternative Leitweglenkung f <Telekom> alternative routing
alternative Prüfmethode f <Telekom> alternative test method
alternative Technologie f <Nichtfoss Energ> soft technology
alternative Verkehrslenkung f <Telekom> alternative routing
alternativer Netzbetreiber m <Telekom> competitive access provider
Alternativmaterial n <Maschinen> alternative material
Alternativname m <Comp & DV> alias
Alternator m 1. <Elektriz> alternating-current generator, alternator; 2. <Elektrotech, Phys> alternator
Alternatorfeldspannung f <Elektriz> alternator field voltage
Alternieren n <Maschinen> alternation
alternierend adj 1. <Elektriz> alternating (Strom); 2. <Maschinen> alternating
alternierende Bewegung f <Maschinen> alternating motion, alternation of a movement
alternierende Komponente f <Elektriz> alternating component
alternierende Reihe f <Math> alternating series
alternierender Fluss m <Elektriz> alternating flux
Altersbestimmung f **mit Hilfe von Urantochternukliden** <Kerntech> dating by uranium daughters
Altersbestimmung f **mittels Radiokohlenstoff** <Phys> radiocarbon dating
Alterung f 1. <Fertig> age hardening (Leichtmetalle); 2. <Kunststoff, Lebensmittel, Papier, Telekom, Verpack> (BE) ageing, (AE) aging
Alterungsausfall m <Kontroll> wearout failure
alterungsbeständig adj <Verpack> (BE) ageing-resistant, (AE) aging-resistant
alterungsfähig adj <Fertig> age-hardenable (Leichtmetalle)
Alterungsfähigkeit f <Fertig> age hardenability (Leichtmetalle)
Alterungshärtung f <Elektriz> intentional accelerated curing, intentional component curing, intentional normal curing
Alterungstest m <Verpack> (BE) ageing test, (AE) aging test
Alterungsuntersuchung f <Raumfahrt> (BE) ageing studies, (AE) aging studies
Altfahrzeug n <Abfall> end-of-life vehicle
Altgerät n <Elektriz> second-hand appliance
Altglas n <Abfall> waste glass
Altglascontainer m <Abfall> bottle bank, waste glass container
Altglasscherben fpl <Ker & Glas> ecology cullet
Altglasverwertung f <Abfall> glass recycling
Altimeter n <Trans> altimeter
Altlast f <Abfall> abandoned site, old landfill site, problem site
Altlastaufarbeitung f <Umweltschmutz> processing of an old site
Altmaterial n <Fertig, Textil> junk • **aus Altmaterial gewinnen** <Umweltschmutz> recover
Altmedikamente npl <Abfall> pharmaceutical waste
Altmetall n <Abfall> scrap metal
Altocumulus m <Lufttrans> altocumulus (Meteorologie)

Altöl n 1. <Abfall> residual oil; 2. <Umweltschmutz> used oil, waste oil
Altölaufbereitung f <Abfall> waste oil preparation
Altölaufbereitungsbetrieb m <Abfall> oil regeneration plant
Altölrückgewinnung f <Abfall> waste oil recovery
Altölschmierung f <Maschinen> waste oil lubrication
Altöltank m <Abfall> slop tank
Altölwiederverwertung f <Abfall> waste oil recovery
Altostratus m <Lufttrans> altostratus (Meteorologie)
Altpapier n <Abfall, Papier> waste paper
Altpapieraufbereitung f <Abfall> waste paper preparation
Altpapierkompressor m <Abfall> waste paper compressing press
Altpapierrecycling n <Abfall> waste paper recycling
Altpapiersammlung f <Abfall> collection of waste paper, paper collection, waste paper collection
Altreifen m <Abfall> (AE) scrap tire, (BE) scrap tyre
Altsand m <Fertig> floor sand, used foundry sand (Gießen)
Altweiberknoten m <Wassertrans> granny knot
Alufolie f (Aluminiumfolie) <Lebensmittel, Metall> (BE) aluminium foil, (AE) aluminum foil
Alu-Knetlegierung f <Metall> alclad
Alumetieren n <Fertig> (BE) aluminium coating by spraying, (AE) aluminum coating by spraying
Alumetierung f <Fertig> alumetizing
Aluminat n <Chemie> aluminate
aluminieren v 1. <Fertig> (BE) aluminium-coat, (AE) aluminum-coat, (BE) aluminium-plate, (AE) aluminum-plate, aluminize; 2. <Metall> aluminize; 3. <Papier> aluminate
Aluminieren n <Fertig> aluminizing
Aluminierung f <Papier> alumination
aluminisieren v <Künstl Int, Metall> aluminize
Aluminisieren n <Metall> aluminization
Aluminium n (Al) <Chemie> (BE) aluminium, (AE) aluminum (Al)
Aluminiumammoniumsulfat n <Chemie> ammonia alum
aluminiumangereichert adj <Anstrich> (BE) aluminium-filled, (AE) aluminum-filled
aluminiumangereicherte Chromat-Phosphat-Beschichtung f <Anstrich> (BE) aluminium-filled chromate/phosphate coat, (AE) aluminum-filled chromate/phosphate coat
Aluminiumanode f <Elektrotech> (BE) aluminium anode, (AE) aluminum anode
aluminiumbehandelt adj <Fertig> (BE) aluminium-killed, (AE) aluminum-killed
Aluminiumblech n <Fertig> (BE) aluminium sheet, (AE) aluminum sheet
Aluminiumbronze f <Mechan, Metall> (BE) aluminium bronze, (AE) aluminum bronze
Aluminiumdose m <Verpack> metal can (Getränke)
Aluminiumelektrolytkondensator m <Elektrotech> (BE) aluminium electrolytic capacitor, (AE) aluminum electrolytic capacitor
Aluminiumfolie f 1. <Lebensmittel> (BE) aluminium foil, (AE) aluminum foil (Alufolie); 2. <Metall> (BE) aluminium foil, (AE) aluminum foil
Aluminiumgatter n <Elektronik> (BE) aluminium gate, (AE) aluminum gate
aluminiumhaltig adj <Chemie> aluminiferous
aluminiumhaltiger Uranbrennstoff m <Kerntech> uranium aluminide fuel
Aluminiumhülse f <Kerntech> (BE) aluminium can, (AE) aluminum can (Brennelement)
Aluminiumhydroxid n <Kunststoff> (BE) aluminium hydroxide, (AE) aluminum hydroxide

Aminopurin

Aluminium-Keramik f <Anstrich> *(BE)* aluminium ceramic, *(AE)* aluminum ceramic
Aluminiumkondensator m **mit Festelektrolyt** <Elektrotech> *(BE)* solid aluminium capacitor, *(AE)* solid aluminum capacitor
Aluminiumlegierung f <Maschinen> *(BE)* aluminium alloy, *(AE)* aluminum alloy
Aluminiumleiter m <Elektriz> *(BE)* aluminium conductor, *(AE)* aluminum conductor
Aluminiumlot n <Metall> *(BE)* aluminium solder
Aluminiummessing n <Metall> *(BE)* aluminium brass, *(AE)* aluminum brass
Aluminiumoxid n 1. <Chemie, Fertig> alumina, *(BE)* aluminium oxide, *(AE)* aluminum oxide; 2. <Papier> alumina
Aluminiumoxidgehalt m <Kohlen> alumina content
Aluminiumoxidschneide f <Fertig> *(BE)* aluminium oxide tool tip, *(AE)* aluminum oxide tool tip
Aluminiumpellet n <Kerntech> *(BE)* aluminium pellet, *(AE)* aluminum pellet
Aluminiumschrott m <Abfall> *(BE)* aluminium scrap, *(AE)* aluminum scrap
Aluminiumsilikatfaser f <Heiz & Kälte> *(BE)* aluminium silicate fibre, *(AE)* aluminum silicate fiber
Aluminiumstahl m <Metall> *(BE)* aluminium steel, *(AE)* aluminum steel
Aluminiumsulfat n <Lebensmittel, Papier> alum
aluminiumüberzogenes Teflon® n <Raumfahrt> aluminized Teflon®
aluminiumverspiegelter Bildschirm m <Elektronik> aluminized screen
Alumino... <Chemie> aluminic, alumino
aluminothermisch adj <Fertig> aluminothermic
aluminothermisches Schweißen n 1. <Bau, Eisenbahn> thermit welding; 2. <Fertig> *(AT-Schweißen)* aluminothermic welding *(AT welding)*
AM 1. <Akustik> *(akustische Masse)* AM *(acoustic mass)*; 2. <Aufnahme, Comp & DV, Elektriz, Elektronik, Fernseh, Funktech, Phys, Telekom, Wellphys> *(Amplitudenmodulation)* AM *(amplitude modulation)*
Amalgam n <Fertig, Kohlen, Metall> amalgam
Amalgamation f <Fertig> amalgamation
Amalgamationsplatte f <Kohlen, Metall> amalgamation plate
Amalgamator m <Maschinen> amalgamator
Amalgambildung f <Chemie> amalgamation, mercurification
amalgamieren v <Fertig, Kohlen, Metall> amalgamate
Amalgamierplatte f <Kohlen, Metall> amalgamation plate
Amalgamierstisch m <Kohlen> amalgamating table
Amalgamierung f <Kohlen, Metall> amalgamation
Amarin n <Chemie> amarine
Amateur m <Fernseh, Funktech, Telekom> amateur
Amateurfernsehen n *(ATV)* <Fernseh> amateur television *(ATV)*
Amateurfunkdienst m <Funktech, Telekom> amateur radio service
Amateurfunker m <Telekom> radio amateur
Amboss m 1. <Fertig> stake *(Blech)*; 2. <Maschinen, Mechan> anvil
Ambossbett n <Maschinen> anvil bed
Ambossblock m <Maschinen> anvil block, block of an anvil
Ambosseinsatz m <Fertig> holdfast
Ambosshahn m <Fertig> anvil pallet face
Ambosshorn n <Maschinen> beak iron
Ambossschlacke f <Metall> anvil dross
Ambossvierkantloch n <Fertig> anvil hardie hole, hardie hole
Ambulanzwagen m <Trans> ambulance

AME *(Atommasseneinheit)* <Kerntech> AWU *(atomic weight unit)*
Ameisensäure f 1. <Fertig> formic acid *(Kunststoffinstallationen)*; 2. <Lebensmittel> formic acid *(Konservierungsstoff)*
Ameisensäure... <Chemie> formic
Ameisensäurealdehyd n <Chemie> formaldehyde
Ameisensäureamid n <Chemie> formamide, methanamide
Amerikanische Datenübertragungs-Codenorm f *(ASCII)* <Comp & DV, Druck, Telekom> American Standard Code for Information Interchange *(ASCII)*
Amerikanische Einheit f **für Drahtdurchmesser** *(AWG)* <Labor> *(AE)* American wire gage, *(BE)* American wire gauge *(AWG)*
amerikanische Holzvolumeneinheit f <Metrol> *(AE)* cord
amerikanische Projektion f <Fertig> third angle system
Amerikanischer Fernsehnormungsausschuss m *(NTSC)* <Fernseh> National Television Standards Committee *(NTSC)*
amerikanisches Maß n *(AG)* <Maschinen> *(AE)* American gage, *(BE)* American gauge *(Drahtdurchmesser, Gewinde)*
Amerikanisches NPT-Rohrgewinde n <Maschinen> American NPT, American National Pipe Taper *(Faden)*
AM-Gang m <Elektronik> AM response
AMI *(bipolare Schrittinversion)* <Telekom> AMI *(alternate mark inversion)*
Amici-Prisma n <Phys> Amici prism, direct-vision prism, roof prism
Amid n <Kunststoff> amide
amidischer Härter m <Kunststoff> amide hardener
Amidogen n <Chemie> amidogen
Amidogruppe f <Chemie> amido group
Amidoschwefel m <Chemie> *(AE)* amido-sulfuric, *(BE)* amido-sulphuric, *(AE)* sulfamic, sulphamic *(Säure)*
Amin n <Kunststoff> amine
Aminierung f <Chemie> amination
aminisch gehärtetes Epoxidharz n <Kunststoff> amine cured epoxy
aminischer Härter m <Kunststoff> amine curing agent
Amino... <Chemie> amino
Aminoanisol n <Chemie> anisidine
Aminoazo... <Chemie> aminoazo
Aminobenzensulfonamido-Pyridin n <Chemie> *(AE)* aminobenzenesulfamidopyridine, *(AE)* sulfapyridine, *(BE)* sulphapyridine *(Pharmazie)*
Aminobenzol n <Chemie> aminobenzene, aniline, phenylamine
Aminocarbonsäure f <Chemie> amino acid
Aminodimethylbenzol n <Chemie> aminodimethylbenzene, dimethylaniline
Aminoessigsäure f <Lebensmittel> glycine
Aminoethan n <Chemie> aminoethane, ethylamine
Aminoethanol n <Chemie> ethanolamine
Aminoharz n <Kunststoff> amino resin
Aminomethan n <Chemie> aminomethane, methylamine
Aminomethanamidin n <Chemie> guanidine
Aminonaphthalin n <Chemie> naphthylamine
Aminonitrobenzol n <Chemie> nitroaniline
Aminophenetol n <Chemie> ethoxyaniline, phenetidine
Aminophenolethylether m <Chemie> ethoxyaniline, phenetidine
Aminophenylmethylether m <Chemie> anisidine
Aminophenylsulfonamid n <Chemie> *(AE)* sulfanilamide, *(BE)* sulphanilamide *(Pharmazie)*
Aminoplast m <Fertig> amino-plast, amino-plastic
Aminopurin n <Chemie> adenine, aminopurine

Aminosäure

Aminosäure f <Chemie, Lebensmittel> amino acid
Aminotoluol n <Chemie> aminotoluene, toluidine
Aminoxylen n <Chemie> xylidine
Aminoxylol n <Chemie> dimethylaniline
Ammin n <Chemie> ammine
Amminverbindung f <Chemie> ammine
Ammonal n <Chemie> ammonal
Ammonalaun m <Papier> ammonial alum
Ammoniak n 1. <Chemie, Elektronik> ammonia *(flüssig)*; 2. <Fertig, Papier, Raumfahrt, Umweltschutz> ammonia
ammoniakalisch adj <Chemie> ammoniacal
ammoniakhaltig adj <Chemie> ammoniacal
Ammoniak-Maser f <Elektronik> ammonia maser
Ammoniakwasser n <Papier> ammonia liquor
Ammonit n <Chemie> ammonia dynamite, ammonite
Ammonium n <Chemie> ammonium
Ammoniumchlorid n <Chemie, Lebensmittel> ammonium chloride
Ammoniumgruppe f <Chemie> ammonium, ammonium radical, ammonium residue
Ammoniumhexachlorostannat n <Chemie> ammonium hexachlorostannate
Ammoniumhydrat n <Papier> ammonia hydrate
Ammoniumhydroxid n <Erdöl> ammonium hydroxide *(Petrochemie)*
Ammoniumperchlorat n <Raumfahrt> ammonium perchlorate *(Raumschiff)*
amorph adj <Anstrich, Fertig, Kohlen> amorphous
amorphe Schicht f <Elektronik> amorphous layer
amorpher Halbleiter m <Elektronik> amorphous semiconductor
amorphes Gefüge n <Kunststoff> amorphous structure
amorphes Silizium n <Elektronik> amorphous silicon
amorphes Trägermaterial n <Elektronik> amorphous substrate
Ampere n *(A)* <Elektriz, Elektrotech, Fertig, Funktech, Metrol, Phys> ampere *(A)*
Ampere-Laplace-Satz m <Elektriz> Ampere-Laplace theorem
Amperemeter n 1. <Elektriz, Elektrotech> ammeter, amperemeter; 2. <Fernseh, Funktech> ammeter; 3. <Gerät> ammeter, amperemeter; 4. <Kfztech, Labor, Phys> ammeter
Ampere'sche Molekularströme mpl <Phys> amperian currents
Ampere'sche Schwimmerregel f <Elektriz> Ampere's rule, amplitude modulation
Ampere'sches Gesetz n 1. <Elektriz> Ampere's law, Ampere's theorem; 2. <Phys> Ampere's law
Amperesekunde f <Elektriz> ampere-second *(Einheit)*
Amperestunde f 1. <Elektriz> ampere-hour *(Einheit der Ladung)*; 2. <Metrol, Phys> ampere-hour
Amperestundenzähler m <Gerät> ampere-hour meter
Amperewindung f <Elektriz, Phys> ampere-turn
Amperewindungszahl f <Elektriz> ampere-turn
Amperezahl f <Elektriz, Elektrotech> amperage
Amphibienfahrzeug n <Wassertrans> amphibian vehicle
amphibisch adj <Wassertrans> amphibian
Amphibolasbest m <Fertig> amphibole
Amplidyne f <Elektrotech> amplidyne
Amplitron n <Elektronik, Phys> amplitron
Amplitude f *(A)* 1. <Akustik, Aufnahme, Comp & DV, Elektriz, Elektronik, Funktech, Phys> amplitude, A; 2. <Wassertrans> amplitude, A *(eines Himmelskörpers)*; 3. <Wellphys> amplitude, A
amplitudenabhängige Folgesteuerung f <Regelung> signal amplitude sequencing control
Amplituden-Amplitudengang m <Elektronik> amplitude-amplitude response
Amplituden-Amplitudengangkurve f <Elektronik> amplitude-amplitude response curve
Amplituden-Amplituden-Verzerrung f <Elektronik> amplitude-amplitude distortion
Amplitudenbegrenzer m <Elektronik, Funktech> amplitude limiter
Amplitudenbegrenzerschaltung f <Fernseh> amplitude limiter circuit
Amplitudendemodulation f <Elektronik, Funktech> amplitude demodulation
Amplitudeneichung f <Elektronik> amplitude calibration
Amplitudenentzerrer m <Telekom> amplitude equalizer
Amplitudenfilter n <Elektronik> amplitude filter
Amplitudenfrequenzgang m <Elektronik, Funktech> amplitude-frequency response
Amplitudenfrequenzgangkurve f <Elektronik, Funktech> amplitude-frequency response curve
Amplitudenfrequenzverzerrung f <Elektronik, Telekom> amplitude-frequency distortion
Amplitudengang m <Elektronik, Telekom> amplitude response
Amplitudeninformation f <Elektronik, Telekom> amplitude information
Amplitudenjustierung f <Elektronik> amplitude adjustment
Amplitudenkorrektur f <Fernseh> amplitude corrector
Amplitudenmodulation f *(AM)* <Aufnahme, Comp & DV, Elektriz, Elektronik, Fernseh, Funktech, Phys, Telekom, Wellphys> amplitude modulation *(AM)*
Amplitudenmodulationsgang m <Elektronik, Funktech> amplitude modulation response
Amplitudenmodulationsrauschen n <Aufnahme, Elektronik, Funktech> amplitude modulation noise
Amplitudenmodulationsträger m <Elektronik> amplitude modulation carrier
Amplitudenmodulator m <Elektronik, Funktech, Telekom> amplitude modulator
amplitudenmoduliert adj <Elektronik, Funktech, Telekom> amplitude-modulated
amplitudenmodulierter Träger m <Elektronik, Telekom> amplitude-modulated carrier
amplitudenmoduliertes Fernsehen n *(AM-TV)* <Fernseh> amplitude modulation television *(AM-TV)*
amplitudenmoduliertes Signal n <Elektronik, Telekom> amplitude modulation signal
Amplitudenraster m <Elektronik> amplitude grid
Amplitudenresonanz f <Phys, Wellphys> amplitude resonance
Amplitudenschrift f <Akustik> variable area recording
Amplitudenschwelle f <Elektronik> amplitude threshold
Amplitudensieb n 1. <Aufnahme> clipper *(Video)*; 2. <Fernseh> sync separator
Amplitudenspektrum n <Funktech, Phys, Wellphys> amplitude spectrum
Amplitudenspitze f <Elektronik> peak amplitude
Amplitudensteuerung f <Phys, Wellphys> amplitude control
Amplitudentastung f <Telekom> amplitude keying
Amplitudenteilung f <Phys> amplitude division
Amplitudenumtastung f *(ASK)* <Elektronik, Telekom> amplitude-shift keying *(ASK)*
Amplitudenverzerrung f 1. <Elektronik> amplitude distortion, amplitude filter; 2. <Phys, Telekom, Wellphys> amplitude distortion
Amplitudenwahrscheinlichkeitsverteilung f <Telekom> amplitude probability distribution
AM-PM Übertragungskoeffizient m <Raumfahrt> AM-PM transfer coefficient

AM-PM Umwandlungskoeffizient *m* <Raumfahrt> AM-PM conversion coefficient
Ampulle *f* 1. <Ker & Glas> *(BE)* ampoule, *(AE)* ampule; 2. <Labor> *(BE)* ampoule, *(AE)* ampule, *(BE)* phial; 3. <Verpack> *(BE)* ampoule, *(AE)* ampule
AM-Rauschen *n* <Aufnahme, Elektronik, Funktech> AM noise
AMS *(aeronautische Werkstoffnorm)* <Raumfahrt> AMS *(aeronautical material standard)*
AM-Signal *n* <Elektronik, Funktech, Telekom> AM signal
Amt *n* <Telekom> exchange • **zwischen Ämtern verlaufend** <Telekom> interoffice
Amt *n* **mit Leitungsvermittlung** <Telekom> circuit-switched exchange
Amt *n* **zweiter Ordnung** <Telekom> *(AE)* secondary center, *(BE)* secondary centre
amtlich gemessene Tonne *f* <Wassertrans> measured ton
amtliche Güteprüfung *f* <Qual> government inspection
amtliche Qualitätssicherung *f* <Qual> government quality assurance
AM-Träger *m* <Elektronik, Funktech> AM carrier
Amtsanruf *m* <Telekom> exchange line call, external call
amtsberechtigt *adj* <Telekom> unrestricted *(Telefonie)*
Amtsberechtigung *f* <Telekom> exchange access *(Telefonie)*
Amtsklappenschrank *m* <Telekom> exchange switchboard
Amtsleitung *f* <Telekom> *(BE)* central exchange trunk, *(AE)* central office trunk, exchange line, trunk
Amtsverbindungsleitung *f* <Telekom> trunk
Amtszeichen *n* <Telekom> dialling tone, exchange tone
amu *(atomare Masseneinheit)* <Kerntech> amu *(atomic mass unit)*
Amygdalose *f* <Chemie> amygdalose, gentiobiose, isomaltose
Amyl *n* <Papier> amyl
Amylacetat *n* <Kunststoff, Papier> amyl acetate
Amylalkohol *m* 1. <Chemie> amyl alcohol, pentanol, pentyl alcohol; 2. <Papier> amyl alcohol
Amylase *f* <Chemie> amylase, diastase
Amylen *n* <Chemie> amylene, pentene
Amylin *n* <Chemie> amylin
Amylopektin *n* <Textil> amylopectin
anachromatisches Objektiv *n* <Foto> anachromatic lens
anaerob *adj* <Ergon, Lebensmittel> anaerobic
anaerob härtender Klebstoff *m* <Kunststoff> anaerobic adhesive
anaerobe Faulung *f* <Abfall> anaerobic digestion
anaerobe Gärung *f* <Abfall> anaerobic digestion, anaerobic fermentation
anaerober Teich *m* <Abfall> anaerobic lagoon
Anaerobier *m* <Lebensmittel> anaerobe
anakustische Zone *f* <Aufnahme> anacoustic zone
anallaktisch *adj* <Optik> anallactic
Anallaktismus *m* <Optik> anallactism
analog *adj* 1. <Comp & DV> analog; 2. <Elektronik, Funktech, Telekom> *(AE)* analog, *(BE)* analogue
analog arbeitendes Messgerät *n* <Gerät> analog instrument
Analog... <Comp & DV, Elektronik, Gerät> analog
Analog-Amperemeter *n* <Elektrotech> analog ammeter
Analoganzeige *f* <Gerät> analog read-out
Analogaufzeichnung *f* <Telekom> analog recording
Analogchip *m* <Elektronik> analog chip
Analogdaten *npl* <Elektronik> analog data
Analog-Digital-... *(A/D)* <Elektronik, Fernseh, Fertig, Telekom> analog-digital *(A/D)*

Analog-Digital-Parallelumsetzer *m* <Elektronik> flash analog-digital converter
Analog-Digital-Umsetzer *m (ADU)* <Comp & DV, Elektronik, Fernseh, Fertig, Telekom> analog-digital converter *(ADC)*
Analog-Digital-Umsetzerausrüstung *f* <Fertig> analog-digital conversion equipment
Analog-Digital-Wandler *m* <Comp & DV, Elektronik, Fernseh, Fertig, Telekom> analog-digital converter
Analog-Digital-Wandlerausrüstung *f* <Fertig> analog-digital conversion equipment
Analog-Digital-Wandlung *f* <Aufnahme, Comp & DV, Elektriz, Elektronik, Fertig, Telekom> analog-digital conversion
analoge Anzeige *f* <Gerät> analog read-out
analoge Bipolarschaltung *f* <Elektronik> analog bipolar integrated circuit
analoge integrierte Schaltung *f* <Elektronik> analog integrated circuit
analoge Mietleitung *f* <Telekom> analog private wire
analoge Teilnehmeranschlussleitung *f* <Telekom> analog subscriber line, subscriber loop
analoge Verzögerungsleitung *f* <Fernseh> analog delay line
analoger Leitungsverstärker *m* <Comp & DV, Telekom> analog line driver
analoger Port *m* <Telekom> analog port
analoger Vergleicher *m* <Elektronik> analog comparator
analoger Vermittlungsprozessor *m* <Telekom> analog call processor
analoger Wert *m* <Comp & DV> analog quantity
analoges Fernsprechnetz *n* <Telekom> analog telephone network
analoges Filter *n* <Elektronik> analog filter
analoges Filtern *n* <Elektronik> analog filtering
analoges Gatter *n* <Elektronik> analog gate
analoges Schieberegister *n* <Comp & DV> analog shift register
analoges Schnurlostelefon *n* <Mobilkom> analog cordless telephone, analog CT
analoges Trägerfrequenzsystem *n* <Telekom> analog carrier system
analoges Vermittlungssystem *n* <Telekom> analog switching system
Analoggerät *n* <Telekom> analog device
Analoginstrument *n* <Gerät> analog instrument
Analogkanal *m* <Telekom> analog channel
Analogkarte *f* <Elektronik> analog board
Analogkreis *m* <Elektrotech> analog circuit
Analogleitung *f* <Telekom> analog circuit
Analogmenge *f* <Comp & DV> analog quantity
Analogmessgerät *n* 1. <Elektrotech> analog meter; 2. <Gerät> analog instrument, analog measuring instrument
Analogmessinstrument *n* <Elektriz, Metrol> analog measuring instrument
Analogmesssystem *n* <Comp & DV> analog measuring system
Analogmodulation *f* <Elektronik, Phys> analog modulation
Analogrechner *m* <Comp & DV> analog computer
Analogrechner *m* **für Netzabgleich** *(ACNA)* <Comp & DV> analog computer for net adjustment *(ACNA)*
Analogrechnung *f* <Comp & DV> analog calculation
Analogschaltung *f* <Elektronik, Telekom> analog circuit
Analogschnittstelle *f* <Telekom> analog interface
Analogschreiber *m* <Gerät> analog data recorder
Analogsignal *n* 1. <Comp & DV> analog signal; 2. <Elektronik, Phys, Telekom> analog signal

Analogsignalgenerator

Analogsignalgenerator m <Elektronik> analog signal generator
Analogsignalverarbeitung f <Elektronik> analog signal processing
Analogspannungsmessgerät n <Elektrotech> analog voltmeter
Analogstellglied n <Elektrotech> analog actuator
Analogstromkreis m <Elektrotech> analog circuit
Analogstrommesser m <Elektrotech> analog ammeter
Analogsystem n <Telekom> analog system
Analogübertragung f <Telekom> analog transmission
Analogverbindung f <Telekom> analog circuit
Analogvoltmeter n <Elektrotech> analog voltmeter
Analogwert m <Comp & DV> analog quantity
Analogwertschreiber m <Gerät> analog data recorder
Analysator m <Comp & DV, Elektriz, Elektronik, Metall, Phys, Telekom> *(BE)* analyser, *(AE)* analyzer
Analysator m **für Einschwingungsvorgänge** <Elektronik> *(BE)* transient network analyser, *(AE)* transient network analyzer *(Netzwerke)*
Analysatorschaltung f <Elektronik> *(BE)* analyser circuit, *(AE)* analyzer circuit
Analyse f 1. <Anstrich> analysis; 2. <Künstl Int> analysis *(von Bildern)*; 3. <Maschinen, Textil, Wasserversorg> analysis
Analyse f **von Einschwingvorgängen** <Elektronik> transient analysis
Analyseausrüstung f <Labor> analytical kit
Analyseautomat m <Gerät> *(BE)* autoanalyser, *(AE)* autoanalyzer
Analysefehler m <Kohlen, Qual> analysis error
Analysegerät n 1. <Gerät> analytical instrument; 2. <Telekom> *(BE)* analyser, *(AE)* analyzer
Analysemesseinrichtung f <Gerät> analysis measuring equipment
Analysemessgerät n <Gerät> *(BE)* analyser, *(AE)* analyzer, analytical instrument
Analysemessung f <Gerät> analytical measurement
Analyseprobe f <Kohlen, Qual> analysis sample
Analysewaage f 1. <Gerät> analytical balance; 2. <Labor> chemical balance; 3. <Phys> analytical balance
Analysis f <Math> calculus
analytische Geometrie f <Geom, Math> analytical geometry
analytische Mechanik f <Mechan, Phys> analytical mechanics
anamorphotisches Verfahren n <Telekom> anamorphosis
Anastigmat m <Foto, Phys> anastigmat
anastigmatische Einschlaglupe f <Fertig> anastigmatic folding magnifier
anastigmatisches Objektiv n <Foto> anastigmatic lens
Anastigmatlinse f <Foto> anastigmat lens, anastigmatic lens
anatomisch gestaltet adj <Fertig> contoured *(Kunststoffinstallationen)*
anatomische Achsen fpl <Ergon> anatomical axes
Anattofarbstoff m <Lebensmittel> anatto, bixin
An-/Ausschalter m <Elektrotech> on-off switch
anbändseln v <Wassertrans> seize
Anbau m <Bau> extension *(Gebäude)*
anbauen v <Wassertrans> fit *(Schiffteile)*
Anbaurakete f <Raumfahrt> kick rocket *(Raumschiff)*
Anbauteil n <Maschinen> attaching part
anbieten v 1. <Bau> tender; 2. <Comp & DV> serve
Anbieten n <Telekom> offer
Anbieter m <Comp & DV> player; provider, service provider
anbinden v <Aufnahme> lace up

Anblaskühlung f <Heiz & Kälte> air blast cooling
Anblaswinkel m <Phys> angle of attack
anblatten v <Bau, Fertig> halve
Anblatten n <Fertig> halving
anbohren v <Bau> tap
Anbohren n <Maschinen> spot drilling
Anbohrer m <Fertig, Maschinen> spotting drill
Anbohrschelle f <Fertig> tapping saddle, tapping tee *(Kunststoffinstallationen)*
Anbohrung f 1. <Fertig> dimple; 2. <Maschinen> dimpled hole
Anbordnahme f <Wassertrans> shipping; boarding *(Ladung)*
Anböschung f <Abfall, Bau> ramp landfill, slope landfill, slope method
anbringen v 1. <Bau> fix; 2. <Comp & DV> mount; 3. <Maschinen> attach
ANC *(Luftfahrt-Navigationsausschuss)* <Lufttrans> ICAO *(International Civil Aviation Organization)*
Anchorit n <Metall> anchorite
andämmen v <Bau> bank up
andauernder Fehler m <Comp & DV> hard error
ändern v 1. <Comp & DV> modify; 2. <Maschinen, Patent, Qual> amend
Anderson'sche Messbrücke f <Elektriz> Anderson Bridge
Änderung f 1. <Akustik> alteration; 2. <Maschinen, Patent, Qual> amendment, change
Änderung f **der Betrachtungsweise** <Comp & DV> viewing transformation
Änderung f **der Induktanz** <Elektriz> self-inductance variation
Änderungsanweisung f <Comp & DV> update statement
Änderungsdatei f <Comp & DV> amendment file, change file, movement file, transaction file
Änderungsnachweis m <Qual> record of changes
Änderungssatz m <Comp & DV> amendment record, change record
Änderungsvermerk m <Konstzeich> modification note
Änderungsverzeichnis n <Qual> table of revision
Änderungszustand m <Konstzeich> modification status
Andock... <Raumfahrt> docking
andocken v <Raumfahrt> dock *(Raumschiff)*
Andocken n <Raumfahrt> docking *(Raumschiff)*
Andocköffnung f <Raumfahrt> docking port
Andockzwischenstück n <Raumfahrt> docking piece
Andrehkurbel f <Maschinen> starting handle
Andruck m <Aufnahme, Foto> pressure
Andrückdeckel m <Verpack> snap cap
Andrucke mpl <Druck> press proofs
Andrücketikette f <Verpack> self-adhesive label
Andruckfenster n <Foto> pressure gate
Andruckplatte f <Foto> pressure plate
Andruckrolle f 1. <Aufnahme> pressure roller; 2. <Fernseh> tape roller
Andruckwalze f <Ker & Glas> backup roll
aneinander fügen v <Maschinen> join
aneinander gekoppelt adj <Fertig> coupled *(Maschinen)*
Anelastizität f <Phys> anelasticity
Anemometer n 1. <Gerät, Labor> anemometer; 2. <Lufttrans> wind velocity indicator; 3. <Nichtfoss Energ, Papier, Wassertrans> anemometer
anemometrisch adj <Papier> anemometric
Anerkennung f <Patent, Qual> approval
Aneroidbarometer n 1. <Gerät> *(AE)* bellows gage, *(BE)* bellows gauge; 2. <Labor, Phys> aneroid barometer
Anethol n <Chemie> anethole
Aneurin n <Chemie> aneurin, thiamin
Anfahrbrenner m <Ker & Glas> start-up burner

anfahren v 1. <Comp & DV> start; 2. <Kontroll, Maschinen> start, start up
Anfahren n 1. <Ker & Glas> starting; 2. <Maschinen> start-up, starting
Anfahren n **mit reduzierter Spannung** <Elektriz> partial voltage starting
Anfahrtransformator m <Elektrotech> starting transformer
Anfahrvorgang m <Maschinen> start-up procedure
Anfall m <Fertig> yield
Anfall m **von Abfällen** <Abfall> waste formation, waste production, waste stream
anfällig adj <Anstrich> prone
Anfang m **der Nachricht** *(SOM)* <Comp & DV> start of message *(SOM)*
anfangen v 1. <Comp & DV> start; 2. <Comp & DV> log in, log on, sign on, start
anfängliche Rissausbreitung f <Kerntech> initial crack growth
anfängliche Rückspannung f <Elektriz> initial inverse voltage
anfängliche Überschussreaktivität f <Kerntech> built-in reactivity
Anfangs... <Comp & DV, Lufttrans> initial
Anfangsadresse f <Comp & DV> initial address
Anfangsanflug m <Lufttrans> initial approach
Anfangsanweisung f <Comp & DV> header statement
Anfangsbeladung f **mit spaltbarem Material** <Kerntech> initial fissile charge
Anfangsbeurteilung f <Qual> initial evaluation
Anfangsbit n <Comp & DV, Telekom> start bit
Anfangsbruttogewicht n <Lufttrans> initial gross weight
Anfangsdruck m <Maschinen> initial pressure
Anfangsempfindlichkeit f <Gerät> minimum scale sensitivity
Anfangsfehler m <Comp & DV> initial error
Anfangsflugbahn f <Lufttrans> initial approach path
Anfangsflugposition f <Lufttrans> initial approach fix
Anfangsgeschwindigkeit f <Raumfahrt> initial velocity
Anfangskapazität f <Elektriz> initial capacity
Anfangskennsatz m <Comp & DV> header label
Anfangskrängungswinkel m <Wassertrans> angle of loll *(Schiffkonstruktion)*
Anfangskritikalität f <Kerntech> initial criticality
Anfangsladung f <Lufttrans> initial forming charge
Anfangsmagnetisierung f <Phys> initial magnetization curve
Anfangsmodus m <Raumfahrt> fundamental mode *(Raumschiff)*
Anfangspunkt m <Geom> starting point *(einer Linie)*
Anfangsschwinden n <Bau> initial drying shrinkage
Anfangssetzung f <Kohlen> initial settlement
Anfangsspannung f <Maschinen> initial tension
Anfangsstabilität f <Wassertrans> initial stability *(Schiffbau)*
Anfangsstadium n <Bau, Metall> initial stage
Anfangsstatus m <Comp & DV> initial state
Anfangsstrom m <Elektriz> initial current, starting current
Anfangstemperatur f <Heiz & Kälte> initial temperature
Anfangstest m **ohne Last** <Kerntech> start-up zero power test
Anfangsvorzündung f <Lufttrans> initial advance
Anfangswertobjekt n <Comp & DV> initial value object
anfärben v <Textil> dye
Anfärben n <Textil> dyeing
Anfärbung f <Textil> dyeing
anfasen v <Maschinen> chamfer
Anfaswerkzeug n <Bau, Maschinen> chamfering tool
Anfertigung f **von Probedrucken** <Druck> proof printing

anfeuchten v 1. <Bau> damp, dampen, moisten, temper, wet; temper *(Sand)*; 2. <Fertig> temper *(Sand)*; 3. <Lebensmittel, Thermod> moisten
Anfeuchten n <Papier> dampening
Anfeuchten n **des Gemenges** <Ker & Glas> batch wetting
Anfeuchten n **von Ton** <Ker & Glas> clay wetting
Anfeuchter m 1. <Fertig> damper; 2. <Papier> moistener
anflanschen v <Maschinen> flange
Anflanschen n <Maschinen> flanging, flanging-on
anfliegen v <Kohlen> settle
Anflug m <Lufttrans> approach
Anflug m **auf Startbahnende** <Lufttrans> approach end of runway
Anflug m **ohne Verfahrenskurve** <Lufttrans> straight-in approach
Anflugbahn f <Lufttrans> approach path
Anflugbefeuerung f <Lufttrans> approach lighting system
Anflugende n **der Landebahn** <Lufttrans> approach end of runway
Anflugfeuer n <Lufttrans> homing beacon
Anflugfolge f <Lufttrans> approach sequence
Anflugfreigabe f <Lufttrans> approach clearance
Anflugführung f <Raumfahrt> approach guidance
Anflugführungssender m <Lufttrans> radio homing beacon
Anflugfunkfeuer n <Lufttrans> homing beacon, radio marker
Anfluggeschwindigkeit f <Lufttrans, Raumfahrt> approach speed
Anfluggleitwinkel m <Lufttrans> approach path
Anflugkontrolldienst m <Lufttrans> approach control service
Anflugkontrolle f <Lufttrans> approach control
Anflugkontrollstelle f <Lufttrans> approach control office
Anflugkontrollstufe f <Lufttrans> approach control rating
Anfluglärmmesspunkt m <Lufttrans> approach noise measurement point
Anflugleitstrahl m <Lufttrans> approach light beacon
Anflugnavigationskarte f <Lufttrans> approach chart
Anflugphase f <Lufttrans> approach phase
Anflugposition f <Lufttrans> approach fix
Anflugpräzisionsradarstufe f <Lufttrans> approach surveillance radar rating
Anflugradar m *(ACR)* <Funkort, Raumfahrt> approach control radar *(ACR)*
Anflugradaranlage f <Funkort> radar approach control equipment
Anflugradareinstufung f <Lufttrans> approach radar rating
Anflugradarstufe f <Lufttrans> approach radar rating, precision radar rating
Anflugschneise f <Lufttrans> landing lane
Anflugstraße f <Raumfahrt> descent path *(Raumschiff)*
Anflugüberwachungsradareinstufung f <Lufttrans> approach surveillance radar rating
Anflugverfahren n <Raumfahrt> approach procedure
Anflugzeit f <Lufttrans> approach time
anfordern v 1. <Comp & DV> query, request; 2. <Telekom> request
Anforderung f 1. <Comp & DV> demand, enquiry, query, request; 2. <Patent> requirement; 3. <Qual> requirement, requisition; 4. <Telekom> request
Anforderung f **nach Datentransfer** <Telekom> data transfer request
Anforderungskanal m <Raumfahrt> request channel *(Weltraumfunk)*
Anforderungsstapel m <Comp & DV> request stack
Anfrage f 1. <Comp & DV> inquiry, request; 2. <Patent> inquiry; 3. <Telekom> enquiry, inquiry, request

anfrageintensiv

anfrageintensiv *adj* <Comp & DV> query-intensive
anfressen *v* <Fertig> pit
Angabenliste *f* <Comp & DV> option list
angeben *v* <Comp & DV> identify
angeblasene Klappe *f* <Lufttrans> blown flap *(Grenzschichtsteuerung)*
Angebot *n* **und Verwaltung** *f* **von HTML** <Comp & DV> HTML delivery and management
angebotene Belegung *f* <Telekom> offered call
angefachte aperiodische Seitenbewegung *f* <Lufttrans> lateral divergence
angeflanscht *adj* <Mechan> flanged
angefressen *adj* <Fertig> pitted
angegebene Flugbahn *f* <Lufttrans> indicated flight path
angegossen *adj* <Mechan> integrally cast
angegossener Steckverbinder *m* <Elektrotech> one-piece connector
angehren *v* <Fertig> *(AE)* miter, *(BE)* mitre
angeklammerte Halterung *f* <Verpack> clip-on carrier
angekoppelte Stromkreise *mpl* <Elektrotech> coupled circuits
angekräuseltes Garn *n* <Textil> abraded yarn
Angel *f* 1. <Fertig> buckle *(Gattersäge)*; 2. <Maschinen> fang, tang; 3. <Mechan> hinge
angelegte Druckgradienten *mpl* <Strömphys> imposed pressure gradients *(Strömungen)*
angelegte EMK *f* <Phys> applied emf
angelegte Spannung *f* <Elektriz> impressed voltage
angelenkt *adj* <Mechan> articulated
angelenkte Hinterkantenklappe *f* <Lufttrans> trailing edge flap
Angelwurzel *f* <Fertig> heel; shoulder *(Feile)*
angemacht *adj* <Lebensmittel> dressed *(Salat)*
angenäherter Knotenpunkt *m* <Akustik> partial node
angenommener Schiffsort *m* <Wassertrans> estimated position *(Navigation)*
angeordnete Spule *f/nahebei* <Elektriz> adjacent coil
angepasst *adj* <Phys> matched
angepasste Impedanz *f* <Elektrotech> matched impedance
angepasste Last *f* <Elektrotech, Phys> matched load
angepasste Ummantelung *f* <Optik> matched cladding
angepasste Widerstände *mpl* <Elektrotech> matched resistors
angepasster Mantel *m* <Telekom> matched cladding
angepasster Scheinwiderstand *m* <Elektrotech> matched impedance
angepasster technischer Service *m* <Comp & DV> personalized technical service
angepasster Wellenleiter *m* <Elektrotech, Funktech> matched waveguide
angepasstes Filter *n* <Elektronik> matched filter
angepasstes Filtern *n* <Elektronik> matched filtering
angepresste Dichtung *f/durch Flüssigkeitsdruck* <Fertig> automatic packing
angeregt *adj* <Kerntech> excited *(Atomkern, Molekül)*
angeregte Strahlung *f* <Telekom> stimulated emission
angeregter Zustand *m* <Kerntech> excited state
angeregtes Atom *n* <Kerntech> excited atom
angereichert *adj* <Lebensmittel> fortified
angereicherter Brennstoff *m* <Kerntech> enriched fuel
angereicherter Kernbrennstoff *m* <Kerntech> enriched nuclear fuel
angereicherter Reaktor *m* <Kerntech> enriched reactor
angereichertes Material *n* <Strahlphys> enriched material
angereichertes Uran *n* <Strahlphys> enriched uranium
angeriebenes Garn *n* <Textil> abraded yarn
angesäuert *adj* <Chemie> acidulous, sourish
angesäuerte Bodenfläche *f* <Umweltschmutz> acidic area
angeschaltetes Handy *n* <Mobilkom> activated mobile hand-held telephone, activated mobile phone, activated radio cellphone
angeschlämmt *adj* <Anstrich> slurried
angeschliffener Grat *m* <Fertig> shoulder
angeschuhte Stange *f* <Bau> shoed bar
angesetztes Holz *n* <Bau> pieced wood
angestoßenes Atom *n* <Kerntech> knocked-on atom
angestoßenes Teilchen *n* <Kerntech> knock-on
angetriebene Scheibe *f* <Maschinen> follower
angetriebene Spindel *f* <Fertig> idler impeller *(Schraubenpumpe)*
angetriebene Welle *f* <Maschinen> driven shaft
angetriebenes Element *n* <Maschinen> driven element
angetriebenes Rad *n* <Maschinen> driven wheel
angetriebenes Werkzeug *n* <Mechan> power tool
angetriebenes Zahnrad *n* <Maschinen> follower
angewandte Mathematik *f* <Math> applied mathematics
angewandte Thermodynamik *f* <Maschinen> applied thermodynamics
angezapfter Fluss *m* <Wasserversorg> beheaded river
angezeigt *adj* <Maschinen> indicated
angezeigte Eigengeschwindigkeit *f* <Lufttrans> indicated airspeed
angezeigte Fluggeschwindigkeit *f* <Lufttrans> IAS, indicated airspeed
angezeigter Anstellwinkel *m* <Lufttrans> indicated pitch angle *(Hubschrauber)*
angezeigter Messwert *m* <Gerät> indicated value, reading
angezeigter Steigungswinkel *m* <Lufttrans> indicated pitch angle *(Hubschrauber)*
angezeigter Wert *m* 1. <Gerät> reading; 2. <Kerntech> indicated value
Angießkanal *m* <Kunststoff> feed
angleichen *v* <Bau> match
Angleichung *f* <Regelung> adaptation *(Regelgrößen)*
angreifen *v* <Bau> catch
angreifend *adj* <Fertig> active *(Kraft)*
angrenzen *v* <Bau> abut
Angrenzen *n* <Geom> adjacency; adjacency *(zweier Seiten)*
angrenzend *adj* 1. <Comp & DV> contiguous; 2. <Geom> adjacent
angrenzende Dateien *fpl* <Comp & DV> contiguous files
angrenzende Winkel *mpl* <Geom> contiguous angles
Angriff *m* <Kohlen> attack
Angriffspunkt *m* <Maschinen> contact point
Angström *n* *(Å)* <Metrol> angstrom *(Å)*
Angström-Einheit *f* <Lebensmittel> angstrom unit
Anguss *m* 1. <Kunststoff> gate, sprue; 2. <Maschinen> sprue
Angussbuchse *f* <Maschinen> feed bush, sprue bush
Angussdruckstift *m* <Maschinen> sprue ejector pin
Angussfarbe *f* <Ker & Glas> *(AE)* colored clay, *(BE)* coloured clay
Angusskanal *m* <Fertig> runner *(Kunststoffe)*
Angusskegel *m* <Kunststoff> feed
angusslose Form *f* <Maschinen> *(AE)* runnerless mold, *(BE)* runnerless mould
Angussöffnung *f* <Kunststoff> gate, sprue opening
Angusszieher *m* <Maschinen> sprue puller
Anhall *m* <Akustik> attack
anhalten *v* 1. <Comp & DV> hold, stop; 2. <Textil> stop
Anhalten *n* 1. <Nichtfoss Energ> stall; 2. <Textil> stop
Anhaltestift *m* <Eisenbahn> catch pin
Anhaltleistung *f* *(S)* <Kerntech> stopping power *(S)*

Anhang *m* <Druck> appendix
Anhängelast *f* <Trans> trailing load
anhängen *v* 1. <Druck> run on; 2. <Maschinen> attach
Anhänger *m* 1. <Bau, Eisenbahn> *(BE)* bogie, *(AE)* trailer; 2. <Kfztech> *(BE)* bogie, *(BE)* lorry, *(AE)* trailer, *(AE)* truck, trail car; 3. <Textil> tag; 4. <Trans> trailer wagon; 5. <Verpack> hangtag, label
Anhängerbremse *f* <Kfztech> trailer brake
Anhängsel *n* <Verpack> swing ticket
Anhäufung *f* <Metall> clustering
Anhebeeinrichtung *f* <Fertig> elevator *(Räumen)*
anheben *v* 1. <Bau> hoist; 2. <Fertig> jack; 3. <Mechan> lift
Anheben *n* <Sicherheit> lifting
Anhebeschlitten *m* <Fertig> broach-handling slide; elevating slide *(Räumen)*
Anhebung *f* <Bau> elevation
anheizen *v* <Thermod> heat up
Anheizen *n* <Eisenbahn> preheating
Anheizklappe *f* <Heiz & Kälte> start-up flap
anholen *v* <Wassertrans> haul *(Tauwerk)*
Anholtau *n* <Wassertrans> messenger *(Tauwerk)*
Anhub *m* <Kerntech> lift
Anhubstange *f* <Mechan> push rod
Anhydrid *n* <Chemie, Papier> anhydride
anhydridischer Härter *m* <Kunststoff> anhydride hardener
anhydrisch *adj* <Chemie> anhydrous
Anhydrit *n* 1. <Erdöl> anhydrite *(Mineral)*; 2. <Lebensmittel> *(AE)* calcium sulfate, *(BE)* calcium sulphate
Anilin *n* 1. <Chemie> aminobenzene, aniline, phenylamine; 2. <Papier> aniline
Anilinfarbe *f* <Druck> aniline ink
Anilinfarbstoff *m* 1. <Chemie> *(AE)* aniline color, *(BE)* aniline colour, coal-tar dye; 2. <Foto> aniline dye
Anilinformaldehydharz *n* <Fertig> aniline formaldehyde resin, aniline resin
Anilingummidruck *m* <Druck> aniline rubber-plate printing
Anilinharz *n* <Fertig> aniline formaldehyde resin, aniline resin
Anilinpunkt *m* <Kunststoff> aniline point
Anion *n* <Elektriz, Elektrotech, Erdöl, Kohlen, Lebensmittel, Phys, Strahlphys> anion
Anionenaustauscher *m* <Kohlen> anion exchanger
anionisch *adj* <Kohlen> anionic
Anionotropie *f* <Chemie> anionotropy
Anisaldehyd *n* <Chemie> anisaldehyde, methoxybenzaldehyde
anisentropisch *adj* <Thermod> anisentropic
Anisidin *n* <Chemie> anisidine
anisochron *adj* <Comp & DV> anisochronous
anisochrone Übertragung *f* <Comp & DV, Telekom> anisochronous transmission
anisoelastisch *adj* <Raumfahrt> anisoelastic *(Gyroskop)*
anisoelastische Verschiebung *f* <Raumfahrt> anisoelastic drift *(Gyroskop)*
Anisoelastizitätsfaktor *m* <Raumfahrt> anisoelasticity factor *(Gyroskop)*
Anisol *n* 1. <Chemie> anisole, methoxybenzene; 2. <Lebensmittel> anisole
anisotherm *adj* <Chemie> athermal
anisotrop *adj* <Kunststoff, Mechan, Optik, Telekom> anisotropic
Anisotropie *f* 1. <Fertig> directionality; 2. <Kunststoff> anisotropy
Anisotropie *f* **der Turbulenz** <Strömphys> anisotropy of turbulence
Anker *m* 1. <Comp & DV> member set; 2. <Elektriz, Elektrotech> armature *(Dynamo, Wechselstromgenerator)*; 3. <Fertig> stay, tie rod; belt *(Gießen)*; keeper *(Magnet)*; 4. <Kohlen> roof bolt *(Tunnelabdeckung)*; 5. <Mechan> anchor, guy; 6. <Phys> armature; 7. <Wassertrans> anchor • **Anker lichten** <Wassertrans> weigh anchor • **Anker schlippen** <Wassertrans> trip anchor • **Anker über Grund schleifen** <Wassertrans> club • **Anker vom Grund lösen** <Wassertrans> trip anchor • **den Anker fallen lassen** <Wassertrans> cast anchor, drop anchor *(Festmachen)* • **vor Anker gehen** <Wassertrans> anchor • **vor Anker legen** <Wassertrans> anchor; bring up to anchor *(des Schiffes)* • **vor Anker liegen** <Wassertrans> ride at anchor; lie at anchor *(Festmachen)* • **vor Anker treiben** <Wassertrans> drag the anchor *(Festmachen)* • **vor dem Anker aufgedreht** <Wassertrans> wind-rode *(Festmachen des Schiffes)*
Anker *m* **mit geschlossenen Nuten** <Elektriz> closed slot armature
Ankerarm *m* <Wassertrans> anchor arm
Ankerball *m* <Wassertrans> anchor ball *(Signal)*
Ankerblechpaket *n* <Elektriz> armature core
Ankerblindwiderstand *f* <Elektriz> armature reactance
Ankerboje *f* 1. <Meerschmutz> mooring buoy; 2. <Wassertrans> anchor buoy
Ankerbolzen *m* 1. <Bau> lag screw, rockbolt; 2. <Maschinen> tie bolt; 3. <Mechan> anchor bolt, tie rod
Ankerbuchse *f* <Elektriz> armature spider
Ankerdeck *n* <Wassertrans> anchor deck *(Schiff)*
Ankereisen *n* 1. <Bau> tie bar *(Klammer)*; T-cramp *(Werksteinarbeiten)*; 2. <Elektrotech> armature iron
Ankerfeld *n* <Elektriz> armature field
Ankerflügel *m* <Wassertrans> anchor fluke
Ankerflügelspitze *f* <Wassertrans> anchor bill
Ankerflunke *f* <Wassertrans> anchor fluke; fluke *(Festmachen)*
Ankerfußschuh *m* <Labor> horseshoe foot *(Mikroskop)*
Ankergegenwirkung *f* <Elektriz, Elektrotech> armature reaction
Ankergehäuse *n* <Elektrotech> armature casing
Ankergeschirr *n* <Wassertrans> ground tackle *(Festmachen)*
ankergesteuerter Motor *m* <Elektriz> armature-controlled motor
Ankergrund *m* <Wassertrans> anchoring ground
Ankerhals *m* <Wassertrans> anchor crown
Ankerhand *f* <Wassertrans> fluke *(Festmachen)*
Ankerhub *m* <Fertig> escapement
Ankerinduktion *f* <Elektriz> armature induction
Ankerkern *m* <Elektriz, Elektrotech> armature core
Ankerkette *f* <Wassertrans> anchor chain, anchor cable, cable, chain cable; cable chain *(Festmachen)*
Ankerkettenbefestigung *f* <Wassertrans> anchor cable attachment
Ankerkettenschlipper *m* <Wassertrans> senhouse slip *(Decksausrüstung)*
Ankerklüse *f* <Wassertrans> hawse pipe *(Schiffbau)*
Ankerkreis *m* <Elektriz> armature circuit
Ankerkrone *f* <Wassertrans> anchor crown
Ankerlaterne *f* <Wassertrans> anchor light, riding light *(Signal)*
Ankerleiter *m* <Elektriz> armature conductor
Ankerlicht *n* <Wassertrans> anchor light, riding light *(Signal)*
Ankermanöver *n* <Wassertrans> anchoring
ankern *v* 1. <Wassertrans> anchor; 2. <Wassertrans> drop anchor
Ankern *n* <Wassertrans> anchoring
Ankerpeilung *f* <Wassertrans> anchor bearing
Ankerplatte *f* 1. <Bau> anchoring plate; bearing plate *(Spannbeton)*; 2. <Eisenbahn, Kohlen> anchor plate, anchoring plate; 3. <Fertig> foundation plate

Ankerplatz

Ankerplatz m <Wassertrans> anchorage
Ankerprüfgerät n <Elektriz> armature tester
Ankerreaktanz f <Elektriz> armature reactance
Ankerregelung f <Elektriz> armature control
Ankerrelais n <Elektrotech> armature relay
Ankerring m <Wassertrans> anchor ring
Ankerrückwirkung f <Elektriz, Elektrotech> armature reaction
Ankerrückwirkungsausgleich m <Elektrotech> armature reaction compensation
Ankersäule f <Ker & Glas> buckstay
Ankerschall m <Akustik> reference sound
Ankerschallbeschleunigung f <Akustik> reference sound acceleration
Ankerschild m <Elektriz> armature end plate
Ankerschildanschlüsse mpl <Elektriz> armature end connections
Ankerschildklemmen fpl <Elektriz> armature end connections
Ankerschraube f 1. <Bau> anchor bolt, lag screw; 2. <Maschinen> anchor bolt
Ankerschraubenrohr n <Fertig> anchor bolt tube
Ankerspule f 1. <Elektriz> armature winding; 2. <Elektrotech> armature coil
Ankerstab m 1. <Bau> anchor bar, stay rod; 2. <Eisenbahn, Kohlen> anchor rod; 3. <Elektriz, Elektrotech> armature bar
Ankerstange f 1. <Elektriz> armature bar; 2. <Lufttrans> dog bone, tie bar *(Hubschrauber)*
Ankerstern m <Elektriz, Elektrotech> armature spider
Ankerstich m <Wassertrans> cable clinch *(Festmachen)*; clinch *(Knoten)* • **mit Ankerstich befestigen** <Wassertrans> clinch *(Tauwerk)*
Ankerstock m <Wassertrans> anchor stock
Ankerstrom m <Elektriz, Elektrotech> armature current
Ankertasche f <Wassertrans> anchor pocket
Ankertrosse f <Wassertrans> anchor cable
Ankervorrichtung f <Bau> anchorage
Ankerwelle f <Elektriz> armature shaft
Ankerwicklung f 1. <Elektriz> armature coil, armature winding; 2. <Elektrotech> armature winding
Ankerwiderstand m <Elektriz, Elektrotech> armature resistance
Ankerwinde f 1. <Bau> windlass; 2. <Mechan> capstan; 3. <Wassertrans> windlass *(Deckausrüstung)*
Ankerwulst m <Wassertrans> anchor boss
Ankerzahn m <Elektriz> armature tooth
Ankerzweig m <Elektrotech> armature branch
anklagen v <Patent> prosecute
anklammern v 1. <Maschinen> cramp; 2. <Verpack> clip to
Anklang m <Akustik> concord
ankleben v <Fertig> freeze
Anklopf... <Telekom> call waiting
Anklopfanzeige f <Telekom> call waiting indication
Anklopfzeichen n <Telekom> call waiting signal
ankochen v <Lebensmittel> parboil
ankommend adj <Telekom> incoming
ankommende Fernleitung f <Telekom> incoming trunk circuit
ankommende Leitung f <Telekom> incoming circuit, incoming line
ankommende Nachricht f <Telekom> incoming message *(Fax)*
ankommende Verbindung f <Telekom> terminating connection *(ISDN)*
ankommender Anruf m <Telekom> incoming call
• **ankommende Anrufe gesperrt** <Telekom> ICB, incoming-calls-barred

ankommender Kanal m <Telekom> incoming channel, return channel
ankommender Ruf m <Telekom> incoming call *(Signal)*
ankommender Verkehr m 1. <Telekom> incoming traffic; 2. <Trans> *(AE)* incoming traffic, *(BE)* inward traffic
ankommendes Bündel n <Telekom> incoming group
ankommendes Signal n <Elektronik> incoming signal
Ankoppelschaltung f <Gerät> coupling network
Ankoppelung f <Elektrotech> coupling
Ankoppelungsführung f <Lufttrans> docking guidance system
Ankoppelungsverlust m <Optik> coupling loss
Ankoppelungswirkungsgrad m <Optik> coupling efficiency
ankörnen v <Fertig> mark
Ankörnen n <Fertig> marking, punching
Ankörner m 1. <Maschinen> *(AE)* center punch, *(BE)* centre punch, prick punch, punch; 2. <Mechan> *(AE)* center punch, *(BE)* centre punch
Ankörnmaschine f <Maschinen> *(AE)* centering machine, *(BE)* centring machine
Ankörnung f <Fertig> mark
Ankreis m <Fertig, Geom> escribed circle
Ankündigungssignal n <Eisenbahn> distant caution signal, distant signal
Ankunftsflughafen m <Lufttrans> airport of arrival
ankuppeln v 1. <Eisenbahn> couple, hook; 2. <Kfztech> hitch, hook; 3. <Maschinen> attach
Ankuppeln n <Kfztech> hitching
ankurbeln v <Fertig, Lufttrans> crank
Anlage f 1. <Bau> equipment, plant, set; 2. <Comp & DV> *(BE)* station; 3. <Druck> lay; 4. <Fertig> abutting piece, block *(Werkstückaufspannung)*; plant; 5. <Maschinen> installation, unit; 6. <Mechan> layout; 7. <Telekom> installation; system *(Gerät)*
Anlage f **mit hohem Lärmpegel** <Sicherheit> noisy equipment
Anlage f **zur Flugerprobung** <Lufttrans> flying test bench
Anlage f **zur Schwefelrückgewinnung** <Abfall> *(AE)* sulfur recovery plant, *(BE)* sulphur recovery plant
Anlage f **zur Trennung von Uranisotopen** <Kerntech> uranium isotope separation plant
Anlagenausfall m <Sicherheit> breakdown of equipment, equipment fault, equipment outage; plant downtime; plant failure
Anlagenbau m <Fertig> plant manufacturing *(Kunststoffinstallationen)*
Anlagenberechtigung f <Telekom> installation barring level
Anlagengestaltung f <Ergon, Sicherheit> equipment design
Anlagenlärm m <Sicherheit> equipment noise
Anlagenleistung f <Kerntech> unit output
Anlagenprojektierung f <Ergon, Sicherheit> equipment design
Anlagensicherheit f <Sicherheit> equipment safety; plant safety
Anlagensteuerung f <Fertig, Maschinen> plant control
Anlagentechnik f <Comp & DV> systems engineering
Anlagenüberwachung f <Sicherheit> equipment monitoring; plant surveillance
Anlagenzuverlässigkeit f <Sicherheit> system reliability
anlagern v <Fertig> age
Anlagerung f <Fertig> *(BE)* ageing, *(AE)* aging
Anlandung f <Wasserversorg> alluvial deposit
Anlass... 1. <Elektriz> starting; 2. <Lufttrans> primer; 3. <Maschinen> starting; 4. <Metall> tempering
Anlassbad n <Metall> tempering bath

Anlassdrehmoment n <Maschinen> starting torque
Anlasseinrichtung f <Elektriz> starting device
Anlasseinspritzpumpe f <Lufttrans> primer pump
anlassen v 1. <Fertig> temper; draw *(Stahl)*; 2. <Lufttrans> crank; 3. <Maschinen> start, start up; 4. <Metall> temper
Anlassen n 1. <Comp & DV> start; 2. <Fertig> drawing, temper *(Wärmebehandlung von Stahl)*; 3. <Lufttrans> cranking *(Kolbenmotor)*
Anlasser m 1. <Elektriz> starter; 2. <Elektrotech> motor starter, starter; 3. <Kfztech, Kontroll> starter; 4. <Maschinen> starter motor, starting motor
Anlasser m **mit Verzögerung** <Elektriz> time delay starter
Anlasserantriebseinheit f <Kfztech> starter drive assembly
Anlasserbatterie f <Elektriz, Elektrotech> starter battery
Anlasserdüse f <Kfztech> starter jet
Anlasserelektrode f <Elektrotech> starter electrode
Anlasserfeldspule f <Kfztech> starter field coil
Anlasserfeldwicklung f <Kfztech> starter field winding
Anlasserkabel n <Kfztech> starter cable
Anlasserknopf m <Kfztech> starter button
Anlasserkohlebürste f <Kfztech> starter brush
Anlasserkollektor m <Kfztech> starter commutator
Anlasserkurbel f 1. <Kfztech> starting crank; 2. <Lufttrans> crank switch
Anlassermotor m <Elektrotech, Kfztech> starter motor
Anlasserpolschuh m <Kfztech> starter pole shoe
Anlasserrheostat m <Elektriz> starting rheostat
Anlasserritzel n 1. <Kfztech> drive pinion; starter motor pinion *(Motor)*; 2. <Maschinen> starting gear
Anlasserritzelwelle f <Kfztech> drive pinion shaft
Anlasserschalter m <Elektriz> starter
Anlasserschleifring m <Kfztech> starter collector ring, starter slip ring
Anlasserstarterzug m <Kfztech> starter control
Anlasserzahnkranz m <Kfztech> starter ring gear *(Motor)*
Anlassfarbe f <Metall> *(AE)* tempering color, *(BE)* tempering colour
Anlasshebel m <Maschinen> starting handle, starting lever
Anlasskondensator m <Elektriz> starting capacitor
Anlasskraftstoffdüse f <Kfztech> starting jet
Anlassluft f <Wassertrans> starting air *(Schiffsmotor)*
Anlassmotor m 1. <Elektriz> starting motor; 2. <Kfztech> starter
Anlassofen m <Metall> tempering furnace
Anlassprobe f <Metall> temper test
Anlassschalter m <Elektrotech> starter, starter switch
Anlassschaltung f <Telekom> start-up circuit
Anlassspartransformator m <Elektriz> autotransformer starter
Anlasssprödigkeit f <Metall> temper brittleness
Anlassspule f <Kfztech> booster coil
Anlassumschalter m <Elektriz> starting changeover switch
Anlasswiderstand m <Elektriz> starting rheostat
Anlasszahnkranz m <Kfztech> flywheel starter ring gear *(Motor)*
Anlauf m <Maschinen> starting
Anlaufdrehmoment n 1. <Fertig> starting torque, starting torque *(Kunststoffinstallationen)*; 2. <Heiz & Kälte> starting torque
anlaufen v 1. <Fertig> tarnish *(Metallflächen)*; 2. <Kontroll> start up
anlaufen lassen v <Maschinen> start, start up
Anlaufen n 1. <Kunststoff> blushing; 2. <Maschinen> start-up

Anlaufflanke f <Fertig> lifting flank *(Kurve)*
Anlaufhafen m <Wassertrans> port of call
Anlaufkupplung f <Maschinen> centrifugal clutch
Anlaufmoment n <Heiz & Kälte> starting torque
Anlaufprozedur f <Comp & DV> initialization
Anlaufreibung f <Maschinen> starting friction
Anlaufscheibe f <Maschinen> thrust washer
Anlaufstrom m <Heiz & Kälte> starting current
Anlaufwert m <Regelung> reaction value, transition value *(einer Strecke)*
Anlaufzeit f 1. <Comp & DV> rise time, start time; 2. <Maschinen> response time; 3. <Papier> start-up
Anlege... <Wassertrans> docking, landing
Anlegebrücke f 1. <Bau> jetty; 2. <Wassertrans> landing pier
Anlegekante f <Druck> lay edge
Anlegemanöver n <Wassertrans> *(AE)* docking maneuver, *(BE)* docking manoeuvre *(Hafen)*
anlegen v 1. <Comp & DV> create; 2. <Elektrotech, Fertig> feed *(Spannung)*; 3. <Wassertrans> berth, land
Anlegen n **an die Wand** <Strömphys> wall attachment
Anlegeplatte f <Druck> feed board
Anlegeplatz m <Wassertrans> landing place, landing
Anlegesteg f <Wassertrans> jetty; pier *(Hafen)*
Anlegestelle f <Wassertrans> landing place
Anlegethermometer n <Gerät> contact thermometer, surface temperature sensor
Anlegewinkel m <Metrol> square
Anlegewinkelmesser m <Metrol> bevel protractor
Anleimen n <Verpack> sizing
Anleimmaschine f 1. <Fertig> glueing machine; 2. <Verpack> gluing machine
Anlieferungszustand m <Qual> as-received condition
anliegend adj <Geom> adjacent
anliegende Seiten fpl <Geom> adjacent sides
anliegende Wirbel mpl <Strömphys> attached eddies
Anliegestrich m <Lufttrans> lubber's line *(Kompass)*
anlösen v <Fertig> bite *(Kunststoffe)*
anmachen v <Bau> temper; mix *(Beton)*; temper *(Mörtel)*
Anmachen n <Bau> mixing
Anmachwasser n <Bau> mixing water
anmelden v 1. <Comp & DV> log in, log on, sign on; 2. <Mechan> apply for *(zum Patent)*
anmelden v/sich 1. <Comp & DV> log in, log on *(in einem System)*; 2. <Mobilkom> sign on
Anmelden n <Comp & DV> login, logon
Anmelder m <Patent> applicant
Anmeldetag m <Patent> date of filing, date of registration
Anmeldung f 1. <Comp & DV> login; 2. <Patent> application • **Anmeldung einreichen** <Patent> file an application • **Anmeldung ist anhängig** <Patent> application is pending
Anmerkung f <Comp & DV> annotation
annähern v <Kontroll> approximate
Annäherung f <Wassertrans> approach
Annäherung f **auf geringsten Abstand** <Wassertrans> closest approach *(Navigation)*
Annäherungsbahn f <Raumfahrt> rendezvous trajectory
Annäherungsbeleuchtung f <Eisenbahn> approach lighting *(Signal)*
Annäherungsgeschwindigkeit f 1. <Trans> approach speed; 2. <Wassertrans> closing speed *(Navigation)*
Annäherungslog n <Erdöl> proximity log
Annäherungsradar m <Funkort, Raumfahrt> rendezvous radar
Annäherungsschalter m <Gerät> approximating pick-up
Annäherungssensor m 1. <Gerät> proximity sensor; 2. <Sicherheit> proximity detector
Annäherungsverfahren n <Raumfahrt> rendezvous procedure

Annäherungsverschluss

Annäherungsverschluss *m* <Lufttrans> approach locking
Annahme *f* 1. <Math> assumption; 2. <Mechan, Qual> acceptance; 3. <Patent> presumption
Annahme *f* **von Langadressen** <Telekom> long-address acceptance
Annahmekennlinie *f* <Qual> operating characteristic curve
Annahmeprüfprotokoll *n* <Qual> inspection certificate
Annahmeprüfung *f* 1. <Gerät> acceptance inspection; 2. <Qual> acceptance test
Annahmestichprobenplan *m* <Qual> acceptance sampling plan
Annahmestichprobenprüfung *f* <Qual> acceptance sampling inspection
annahmetauglich *adj* <Qual> acceptable
Annahmetauglichkeit *f* <Qual> acceptability
Annahme- und Rückweisungskriterien *npl* <Qual> acceptance and rejection criteria
Annahmeverfahren *n* <Qual> acceptance procedure
Annahmezahl *f* <Qual> acceptance number
annässen *v* <Bau> wet
Annattofarbstoff *m* <Lebensmittel> annatto
annehmbare Qualitätsgrenzlage *f (AQL)* <Qual> acceptable quality level *(AQL)*
annehmbare Qualitätslage *f (AQL)* <Qual> acceptable quality level *(AQL)*
annehmbare Spannungs- und Frequenzgrenzen *fpl* <Nichtfoss Energ> acceptable voltage and frequency boundaries
Annehmbarkeitsnachweis *m* <Patent, Qual> evidence of acceptability
annehmen *v* 1. <Math> assume; 2. <Qual> accept
Annihilation *f* <Teilphys> annihilation
Anode *f* 1. <Elektrotech> drain; plate *(Elektroplattierung, Galvanisierung)*; 2. <Elektrotech, Fernseh, Funktech, Phys, Wassertrans> (BE) anode, (AE) plate *(Elektrik)*
Anodenbasisschaltung *f* <Elektrotech> cathode follower
Anodenbasisverstärker *m* <Elektrotech> cathode follower amplifier
Anodengalvanisierung *f* <Metall> anodic coating
Anodenhemmstoff *m* <Metall> anodic inhibitor
Anodenkennlinie *f* <Elektrotech> anode characteristic
Anodenkorrosion *f* <Metall> anode corrosion
Anodenkreis *m* <Elektrotech> anode circuit, plate circuit
Anodenmodulation *f* <Elektronik> anode modulation
Anodensättigung *f* <Elektrotech> anode saturation
Anodenspannung *f* <Elektrotech> anode voltage
Anodenstrahlen *mpl* <Elektrotech, Strahlphys> anode rays
Anodenstrom *m* <Elektrotech> anode current
anodisch *adj* <Elektriz> anodic
anodisches Polieren *n* <Metall> electropolishing
anodisieren *v* <Chemie> anodize
Anodisieren *n* <Metall> anodizing
anomale Dispersion *f* <Phys> anomalous dispersion
anomaler Zeeman-Effekt *m* <Phys> anomalous Zeeman effect
Anomalie *f* <Thermod> anomaly *(des Wassers)*
anordnen *v* 1. <Bau> locate; 2. <Comp & DV> order, set-up; 3. <Elektrotech> locate, set-up; 4. <Mechan> set up
Anordnung *f* 1. <Comp & DV> array, layout, setup; 2. <Elektriz> layout; 3. <Elektrotech> arrangement, grouping, set-up; 4. <Maschinen> arrangement, layout; 5. <Mechan> layout; 6. <Telekom> array • **mit fester Anordnung der Spindeln** <Fertig> *(AE)* fixed center, *(BE)* fixed centre
Anordnungszeichnung *f* <Konstzeich> arrangement drawing

anorganisch *adj* <Anstrich> anorganic, inorganic
anorganischer Flüssigkeitslaser *m* <Elektronik> inorganic liquid laser
Anorthit *m* <Ker & Glas> anorthite
anpassbar *adj* <Comp & DV> customizable
anpassen *v* 1. <Fernseh, Funktech> match; 2. <Mechan> adjust; 3. <Raumfahrt> adapt
anpassen *v/sich* <Raumfahrt> adapt *(den Weltraumbedingungen)*
Anpassen *n* **einer Kurve** <Math> curve fitting *(an gegebene Datenwerte)*
Anpassglied *n* 1. <Elektronik> transforming section; 2. <Elektrotech> adapter, matching element, matching pad
Anpassung *f* 1. <Ergon> adaptation, adjustment; 2. <Funktech> adjustment; 3. <Heiz & Kälte> acclimatization; 4. <Metrol> adjustment
Anpassungsdämpfung *f* <Elektronik> matching attenuation
anpassungsfähig *adj* 1. <Künstl Int> adaptive *(Programm, System)*; 2. <Mechan> adjustable
anpassungsfähiges System *n* <Comp & DV> adaptive system
Anpassungsfähigkeit *f* <Ergon> adaptability, flexibility
Anpassungsglied *n* <Funktech, Telekom> adapter
Anpassungsimpedanz *f* <Funktech> matching impedance
Anpassungsnetzwerk *n* <Telekom> impedance matching network
Anpassungsschicht *f* <Telekom> adaptation layer *(ISDN)*
Anpassungssteuerung *f* **mit Optimierung** *(ACO)* <Labor> adaptive control optimization *(ACO)*
Anpassungstransformator *m* <Elektriz, Elektrotech> matching transformer
Anpassungsübertrager *m* <Telekom> matching transformer
Anpassungsunterbrechung *f* <Fernseh> match cut
Anpassungsverstärker *m* <Elektronik> matching amplifier
anpeilen *v* <Wassertrans> locate *(Schiff, Seezeichen)*
ANPN-System *n* <Trans> Army Navy Performance Number System *(Brennstoff-Oktanzahl)*
Anprall *m* <Chemie> impingement
Anpressdruck *m* <Maschinen> contact pressure
Anpressdruck *m* **des Stromabnehmers** <Eisenbahn> pantograph pressure
anquellen *v* <Textil> swell
Anquellen *n* <Textil> swelling
Anrauen *n* <Bau> deadening
Anregelzeit *f* <Comp & DV> rise time
anregen *v* 1. <Mechan> actuate; 2. <Phys> actuate, energize
Anregung *f* 1. <Akustik> incitation; 2. <Elektriz, Kerntech> excitation; 3. <Strahlphys> excitation; energization *(Atomkern, Atom)*; 4. <Telekom> excitation
Anregungsenergie *f* <Kerntech, Strahlphys> excitation energy
Anregungsfunktion *f* <Kerntech, Strahlphys> excitation function
Anregungspegel *m* <Akustik> critical band level
Anregungsquelle *f* <Kerntech, Strahlphys> excitation source
Anregungszustand *m* <Metall> excited state
anreichern *v* 1. <Chemtech> concentrate; 2. <Kohlen, Lebensmittel> enrich
Anreichern *n* <Kohlen> enrichment
Anreicherung *f* 1. <Elektronik, Kohlen> enrichment; 2. <Lebensmittel> enrichment, fortification; 3. <Phys> enrichment *(von Uran)*; 4. <Telekom> enhancement *(Halbleiter)*

Anreicherung f **mit Leuchtstoff** <Elektronik> phosphorus doping
Anreicherung f **mit Sauerstoff** <Chemie> aeration
Anreicherungsabfall m <Kerntech> enrichment tails
Anreicherungsanlage f **mit Ultrazentrifuge** <Kerntech> ultracentrifuge enrichment plant
Anreicherungsbecken n <Abfall> infiltration basin
Anreicherungs-Isolierschicht-Feldeffekttransistor m <Elektronik> enhancement-mode FET
Anreicherungstyp m <Elektronik> enhancement mode
Anreiß... 1. <Bau> plotting; 2. <Fertig> marking
anreißen v 1. <Bau> plot, score, scribe; 2. <Fertig> mark, snap, whitewash
Anreißen n 1. <Elektronik> scribing *(mit Lasern)*; 2. <Fertig> layout, marking; 3. <Kunststoff> tear initiation
Anreißer m 1. <Elektronik> scriber *(mit Lasern)*; 2. <Maschinen> scriber
Anreißkasten m <Fertig> box angle plate
Anreißkörner npl <Fertig> prick punch
Anreißplatte f 1. <Maschinen> marking-out table, surface plate; 2. <Mechan> surface plate
Anreißschritt m <Elektronik> scribing step *(mit Lasern)*
Anreiz m 1. <Ergon> incentive; 2. <Telekom> event *(eines Kontrollverfahrens)*
Anreizen v <Comp & DV> start
Anreizverarbeitung f <Telekom> event processing
Anriss m 1. <Fertig> initial cracking; incipient crack *(Dauerbruchzone)*; 2. <Kerntech, Lufttrans> incipient crack; 3. <Mechan> cracking, flaw
Anrollstart m <Kfztech> rolling start
Anruf m <Telekom> call, ringing, telephone call • **einen Anruf entgegennehmen** answer a call
Anruf m **an alle** <Telekom> general call
Anruf m **in Warteschleife** <Telekom> call queue
Anrufablehnung f <Telekom> disregard incoming call
Anrufbeantworter m <Telekom> answering machine
Anrufdurchschaltung f **zu besetztem Teilnehmer** <Telekom> completion of calls to busy subscriber
anrufen v 1. <Comp & DV> poll; 2. <Telekom> make a call, ring
anrufender Teilnehmer m <Telekom> calling customer, calling party, calling subscriber
Anrufer m <Telekom> caller
Anruffilter n <Telekom> call filter
Anrufglocke f <Elektrotech> call bell
Anrufmelder m <Telekom> pager
Anrufsignal n <Telekom> calling signal
Anrufsperreinrichtung f <Telekom> call barring equipment
Anrufsucher m <Elektronik, Telekom> line finder
Anruftaste f <Elektrotech> call button *(elektrische Klingel)*
Anrufumleiter m <Telekom> call diverter
Anrufumleitung f <Telekom> call diversion, diversion service
Anrufverarbeitung f <Telekom> call processing
Anrufversuch m <Telekom> call attempt
Anrufversuche mpl **zur Hauptverkehrsstunde** <Telekom> BHCA, busy hour call attempts
Anrufverteiler m <Telekom> call distributor
Anrufwarteschleife f <Telekom> call hold
Anrufweiterschaltung f <Telekom> call transfer, terminal call forwarding
Anrufweiterschaltung f **bei Teilnehmer besetzt** <Telekom> call forwarding busy, CBF *(ISDN)*
Anrufzeichen n <Eisenbahn> calling-on signal
Anrufzustand m <Telekom> alerting
Anrundung f <Fertig> initial curvature
Ansage f 1. <Lufttrans> PA, public address; 2. <Telekom> announcement

Ansageanlage f <Lufttrans> PA system, public address system
Ansagemaschine f <Telekom> announcement machine, recorded announcement machine
Ansammlung f <Elektrotech> collection *(Elektronen, Strom)*
Ansatz m 1. <Bau> deposit, nose; 2. <Fertig> lug; 3. <Kfztech> neck; 4. <Kunststoff> mix; 5. <Maschinen> lug, nose, shoulder; 6. <Mechan> batch, neck
Ansatzbohrung f <Fertig> hole with shoulder, stepped hole
Ansatzdrehen n <Fertig> shoulder turning
Ansatzflansch m <Maschinen> neck flange
Ansatzrahmen m <Kfztech> stub frame *(Karosserie)*
Ansatzrohr/mit <Maschinen> tubulated
Ansatzsäge f <Bau> tenon saw
Ansatzstück n <Fertig> lateral
ansäuerbar adj <Chemie> acidifiable
ansäuern v 1. <Anstrich, Chemie> acidify; 2. <Lebensmittel> acidify, acidulate; 3. <Papier, Textil, Umweltschmutz> acidify
ansäuernd adj <Umweltschmutz> acidifying
Ansäuerung f <Umweltschmutz> acidification
Ansaug... 1. <Hydraul> suction; 2. <Kohlen> inlet
Ansaugbehälter m <Hydraul> suction tank
Ansaugdruck m 1. <Kohlen> inlet pressure; 2. <Maschinen> inlet pressure, intake pressure
ansaugen v 1. <Fertig> prime; take *(Vorschub)*; 2. <Heiz & Kälte> aspirate
ansaugen lassen v <Wasserversorg> prime *(Pumpe)*
Ansaugen n 1. <Elektrotech> absorption *(von Elektronen, Gasen)*; 2. <Heiz & Kälte> suction; 3. <Kfztech> inlet *(Luft-Kraftstoff-Gemisch)*; 4. <Mechan> intake, suction
Ansaugenlassen n <Maschinen, Wasserversorg> priming *(Pumpe)*
Ansauggrube f <Erdöl> suction pit
Ansaughöhe f <Heiz & Kälte> suction head
Ansaughub m <Kfztech> induction stroke
Ansaugkanal m <Kfztech> inlet port
Ansaugkrümmer m 1. <Kfztech> induction manifold, inlet manifold *(Motor)*; 2. <Lufttrans> intake manifold *(Motor, Triebwerk)*
Ansaugleistung f <Heiz & Kälte> suction capacity
Ansaugluftkammer f <Lufttrans> plenum chamber
Ansaugmenge f <Maschinen> intake capacity
Ansaugöffnung f <Lufttrans> intake *(Motor, Triebwerk)*
Ansaugpumpe f <Hydraul> suction pump
Ansaugring m **am Flugzeugrumpf** <Lufttrans> nacelle intake ring
Ansaugrohr n 1. <Hydraul> induction pipe; 2. <Kfztech> suction pipe; induction manifold *(Motor)*; 3. <Lufttrans> intake manifold *(Triebwerk)*; 4. <Wassertrans> suction pipe; 5. <Wasserversorg> priming pipe *(Pumpe)*
Ansaugschlitz m <Kfztech> port *(Motor)*
Ansaugseite f <Maschinen> intake side
Ansaugspinne f <Kfztech> induction manifold *(Motor)*
Ansaugstutzen m <Maschinen> inlet manifold
Ansaugtank m <Hydraul> suction tank
Ansaugung f 1. <Kfztech> induction; 2. <Maschinen> indraught, intake, suction; 3. <Papier> aspiration, suction
Ansaugventil n 1. <Fertig> intake valve; 2. <Heiz & Kälte, Hydraul> suction valve; 3. <Maschinen> priming valve
anschäkeln v <Wassertrans> shackle on
Anschaltekoppler m <Telekom> access matrix
anschalten v 1. <Elektrotech> turn on *(Stromversorgung)*; 2. <Telekom> enable, join
Anschalten n <Elektrotech> turn-on *(Stromversorgung)*; turn-on *(Übergang in den An-Zustand)*
Anschellen n <Mechan> clamping

Anschlag

Anschlag m 1. <Akustik> attack; 2. <Bau> back stop, stop; rabbet *(Fenster, Tür)*; 3. <Fertig> cupping operation; end stop *(Kunststoffinstallationen)*; 4. <Kfztech> stop; 5. <Maschinen> end stop, stop; 6. <Mechan> dog, lug, snubber, stop
Anschlagbolzen m 1. <Fertig> T-slot bolt; 2. <Maschinen> trip dog
Anschlagbund m <Maschinen> stop collar
anschlagdrehen v <Fertig> trip
Anschlagdrehen n <Fertig> tripping
Anschlagdrucker m <Telekom> impact printer
anschlagen v 1. <Comp & DV> hit *(Taste)*; 2. <Wassertrans> bend; 3. <Wasserversorg> meet
anschlagfreier Drucker m <Comp & DV> nonimpact printer *(Laserdrucker)*
Anschlagmittel pl 1. <Bau> lifting tackle; 2. <Wassertrans> cargo-handling gear
Anschlagplatte f <Fertig> stop plate *(Kunststoffinstallationen)*
Anschlagschleifen n <Fertig> shoulder grinding
Anschlagschraube f <Foto> stop screw
Anschlagstift m 1. <Fertig> pilot; 2. <Foto> stop pin
Anschlagwinkel m <Maschinen> try square
anschleifen v <Ker & Glas> start a cut
Anschließbarkeit f <Telekom> connectivity
anschließen v 1. <Bau> connect; 2. <Comp & DV> attach, connect; 3. <Elektrotech> connect, connect up, plug in; connect *(Komponenten)*; 4. <Telekom> connect, connect up, join, plug in
Anschluss m 1. <Bau> connection, joint; 2. <Comp & DV> attachment, connector, input port; 3. <Druck> port; 4. <Eisenbahn> junction; 5. <Elektriz> connection; 6. <Elektrotech> contact; connection *(von Stromleitern)*; 7. <Fertig> connection; coupling *(Kunststoffinstallationen)*; 8. <Telekom> port, connection; access *(Ausrüstung)*; line *(Telefon)* • **kein Anschluss unter dieser Nummer** <Telekom> number-unobtainable tone, NUT *(NU-Ton)*
Anschluss m **der Saugleitung** <Fertig> inlet *(Pumpe)*
Anschluss m **integrierter Schaltungen** <Elektrotech> integrated-circuit connection
Anschluss m **zum Grundtarif** <Telekom> BRA, basic rate access
Anschlussbereich m exchange area, service area *(Telefon)*
Anschlussbewehrungsstab m <Bau> starter bar
Anschlussbuchse f 1. <Comp & DV> connector, port; 2. <Elektriz> connector socket *(Anhängerwagen)*; 3. <Elektrotech> connector socket
Anschlusschip m <Elektronik> companion chip
Anschlussdose f 1. <Elektrotech> connection box, jack; 2. <Heiz & Kälte> junction box
Anschlussdraht m 1. <Elektronik> lead; 2. <Elektrotech> pigtail
Anschlusseinheit f <Comp & DV> interface unit
Anschlusserweiterung f <Comp & DV> terminal extension
Anschlussfaser f <Telekom> *(AE)* optical fiber pigtail, *(BE)* optical fibre pigtail; pigtail *(Lichtwellenleiter)*
Anschlussfeld n <Elektrotech> connecting terminal, connection terminal
Anschlussflansch m 1. <Maschinen> connecting flange; 2. <Mechan, Phys> coupling flange
Anschlussfleck m <Elektronik> bonding pad *(metallisiert)*
Anschlussgas n <Erdöl> connection gas *(Förderung)*
Anschlussgebühr f <Mobilkom> activation charge *(Mobiltelefon)*
Anschlussgerät n <Comp & DV> peripheral equipment, peripheral, peripheral unit
Anschlussgleis n <Bau> siding

Anschlusshahn m <Bau> union cock
Anschlusskabel n <Elektrotech> connecting cable
Anschlusskanal m <Telekom> access channel
Anschlusskasten m <Elektriz> *(BE)* joint box, junction box, terminal box
Anschlusskennung f **der gerufenen Station** <Telekom> called line identification
Anschlusskennung f **der rufenden Station** <Telekom> calling line identification *(CLI)*
Anschlussklemme f 1. <Elektriz, Elektrotech> connecting terminal, connection terminal, terminal; 2. <Kfztech> terminal; 3. <Telekom> block terminal
anschlusskompatibel adj <Telekom> pin compatible
Anschlusskontaktstelle f <Kontroll> terminal pad
Anschlussleiste f 1. <Elektrotech> terminal block; 2. <Telekom> connection strip
Anschlussleitung f 1. <Elektrotech> connecting lead; 2. <Telekom> subscriber's line, branch line
Anschlusslinie f <Eisenbahn> branch line, feeder line
Anschlusslitze f <Kerntech> pigtail
Anschlussmaß n 1. <Fertig> mounting dimension; 2. <Maschinen> connecting dimension, fitting dimension, mating dimension
Anschlussmodul n <Comp & DV> input/output switching module
Anschlussmöglichkeit f <Telekom> connectivity
Anschlussnetz n <Telekom> access network, connection network
Anschlussplatte f <Elektrotech> socket board
Anschlusspunkt m <Elektrotech> junction point
Anschlussspannung f <Gerät> connection voltage
Anschlussspeicher m <Telekom> subscriber's store
Anschlussstecker m 1. <Comp & DV> connector; 2. <Elektriz> coupler connector
Anschlussstelle f 1. <Bau> *(AE)* interchange, *(BE)* junction *(Autobahn)*; 2. <Comp & DV> port; 3. <Eisenbahn> rail junction point
Anschlussstift m <Kontroll> terminal pin
Anschlussstrecke f <Telekom> access link *(ISDN)*
Anschlussstück n 1. <Bau> nipple, union; 2. <Elektrotech> connector, coupling; 3. <Mechan> gooseneck
Anschlussstutzen m <Fertig> spigot *(Kunststoffinstallationen)*
Anschlussteil n <Elektrotech, Funktech> connector *(Stecker)*
Anschlussverriegelung f <Fernseh> slavelock
Anschlussweiche f <Eisenbahn> junction points
Anschlusswert m <Fertig> input power
Anschlusswiderstand m <Elektrotech> ferrule resistor
Anschlusswinkel m <Bau> angle bracket
Anschlusszug m <Eisenbahn> connecting train, feeder train
Anschnitt m 1. <Fertig> chamfer edge, ingate; gate *(Kunststoffe)*; 2. <Kunststoff> gate • **mit kegeligem Anschnitt** <Fertig> tapered-ended
Anschnittstechnik f <Fertig> gating
Anschnittwinkel m <Fertig> angle of chamfer *(Spiralbohrer)*
Anschrägen n <Fertig> chamfer *(Kunststoffinstallationen)*
anschrauben v <Maschinen> screw
Anschrauben n <Maschinen> bolting, screwing
anschreiben v <Fertig> escribe
anschütten v <Bau> bank up, slope
Anschüttung f <Kohlen> backfill
Anschwänzen n <Chemie> sparging *(Brauwesen)*
Anschweiß... <Maschinen> weld-on
Anschweißmutter f <Maschinen> weld nut, welding nut
anschwellen v 1. <Bau, Kohlen> belly out; 2. <Bau> bulge; 3. <Phys> swell

Anschwemmung f <Bau> alluviation
Anseilschutz m <Sicherheit> fall-arresting device
ansenken v 1. <Maschinen> spot-face; 2. <Mechan> countersink
ansetzen v <Comp & DV, Fernseh, Telekom> schedule
Ansetzen n **von Bädern** <Foto> making-up of baths
Anspannung f <Fertig> stretch
Ansprache f <Akustik> designation
Ansprech... <Aufnahme, Elektriz, Elektrotech, Gerät, Telekom> threshold
Ansprecheigenschaft f <Telekom> response characteristic
Ansprechempfindlichkeit f 1. <Aufnahme, Elektrotech> threshold sensitivity; 2. <Optik, Telekom> responsiveness
Ansprechgeschwindigkeit f <Telekom> speed of response
Ansprechgrenze f 1. <Elektronik> response threshold; 2. <Gerät> threshold limit
Ansprechrelais n <Elektrotech> operate relay, response relay
Ansprechschwelle f <Gerät> threshold limit
Ansprechsignal n <Gerät> threshold signal
Ansprechspannung f <Elektrotech> response voltage, operate voltage
Ansprechstrom m <Elektrotech> response current, operate current
Ansprechtemperatur f <Heiz & Kälte> critical temperature
Ansprechverhalten n 1. <Aufnahme, Elektronik> response; 2. <Telekom> response characteristic
Ansprechvermögen n <Elektrotech> response sensitivity
Ansprechverzögerung f <Elektrotech> response lag, operate lag
Ansprechwert m <Gerät> threshold limit
Ansprechzeit f 1. <Aufnahme> attack time (eines Begrenzers); 2. <Elektronik> response time; 3. <Elektrotech> operate time, response time; 4. <Gerät> answering time (Messgerät); 5. <Metrol, Telekom> response time
ansprengen v <Meerschmutz> wring
Anspruch m 1. <Patent> claim, entitlement; 2. <Qual> entitlement • **in Anspruch nehmen** <Patent> claim
Ansprüche mpl **der gleichen Kategorie** <Patent> claims in the same category
Ansprüche mpl **verschiedener Kategorien** <Patent> claims in different categories
Anspruchsniveau n 1. <Ergon> level of aspiration; 2. <Qual> grade
anstauchen v <Fertig> squeeze; head (Köpfe)
Anstauchen n 1. <Fertig> heading (Köpfe); 2. <Maschinen> heading
Anstauchwerkzeug n <Maschinen> heading tool
anstechen v <Lebensmittel> tap (Fass)
Anstechteil n <Fertig> looping piece (Gießen)
anstehende Ader f 1. <Bau> outburst; 2. <Kohlen> apex, outburst
ansteigen v <Bau> rise
Ansteigen n 1. <Erdöl> rising; 2. <Strömphys> surge
ansteigend adj <Bau> uphill
Anstellen n <Chemtech> setting (Gärung)
Anstellwinkel m 1. <Lufttrans> attack angle, blade angle; 2. <Maschinen> setting angle
Anstellwinkel m **des Luftschraubenblattes** <Lufttrans> blade attack angle
Anstellwinkel m **für Nullauftrieb** <Lufttrans> zero-lift angle
Anstellwinkelanschlag m <Lufttrans> pitch stop (Hubschrauber)
Anstellwinkelanzeiger m <Lufttrans> angle of attack indicator

ansteuern v 1. <Comp & DV> select; 2. <Funktech> drive; 3. <Telekom> select (Adresse); 4. <Wassertrans> approach, steer for (Navigation)
Ansteuern n <Comp & DV> gating
Ansteuerung f 1. <Comp & DV> selection; 2. <Elektronik> control (eines Geräts); 3. <Funktech> driving (Transistor); 4. <Telekom> excitation (eines Transmitters); 5. <Wassertrans> approach
Ansteuerung f **eines Leitstrahls** <Lufttrans> beam interception
Ansteuerungssignal n <Elektronik> gating signal
Ansteuerungstonne f <Wassertrans> sea buoy
Anstich machen v/den <Ker & Glas> tap (am Glasschmelzofen)
Anstieg m 1. <Bau> elevation (Wasser); 2. <Erdöl> rising (Ölspiegel in Bohrung); 3. <Phys> slope
Anstiegsrate f <Elektrotech> rate of rise
Anstiegszeit f 1. <Aufnahme> rise time; attack time (eines Verstärkers); 2. <Comp & DV, Phys> rise time; 3. <Regelung> ramp response
Anstiegszeit f **der Impulsvorderflanke** <Fernseh, Telekom> leading-edge pulse time
Anstoß m 1. <Anstrich> impact; 2. <Mechan> impulse; 3. <Raumfahrt> nudging
anstoßen v <Bau> abut
anstoßend adj <Geom> contiguous
anstoßende Winkel mpl <Geom> adjacent angles
Anstoßmechanismus m <Raumfahrt> kick-off mechanism
Anstrich m 1. <Anstrich> paint; 2. <Papier> coat
Anstrichmittelwanne f <Labor> sump
Anstrichstoff m **zum Spritzen** <Kunststoff> spraying paint
Anstrichsystem n <Anstrich> paint system
Anström... <Lufttrans> leading
Anströmgeschwindigkeit f **an der Rotorspitze** <Lufttrans> rotor tip velocity (Hubschrauber)
Anströmkante f <Lufttrans> leading edge
Anströmkantenrippe f <Lufttrans> leading-edge rib
Anströmwinkel m <Lufttrans> angle of incidence
Anteil m 1. <Elektriz> component; 2. <Kunststoff> content; 3. <Mechan> component; 4. <Metrol> rate (verhältnismäßig)
Anteil m **erfolgreich abgewickelter Anrufe** <Telekom> call success rate
Anteil m **fehlerhafter Einheiten** <Qual> fraction defective, fraction nonconforming (in einer Stichprobe)
Anteil m **von Ausgangsatomen** <Kerntech> parent fraction
Antenne f 1. <Elektrotech, Fernseh, Funktech> aerial, antenna; 2. <Kfztech> aerial, antenna (Zubehör); 3. <Phys, Raumfahrt, Telekom> aerial, antenna
Antenne f **für Schwerewellen** <Strahlphys> gravitational wave aerial
Antenne f/im **Brennpunkt erregte** <Funktech> focal point feed antenna
Antenne f **mit periodischer Strahlschwenkung** <Funktech> sweep antenna
Antenne f **mit Reflektor** <Raumfahrt> reflector antenna (Weltraumfunk)
Antenne f **mit Richtwirkung** <Funktech> directive array
Antenne f **mit schwenkbarer Charakteristik** <Funktech, Telekom> steerable antenna
Antennenabstimmspule f (ATI) <Funktech> aerial-tuning inductance, antenna-tuning inductance (ATI)
Antennenanpassgerät n <Funktech> aerial-tuning unit
Antennenanpassung f <Funktech> antenna matching device (Gerät); aerial-tuning unit (ATU); aerial matching
Antennenanschluss m <Fernseh, Funktech> aerial terminal

Antennenaufzug

Antennenaufzug m <Lufttrans> halyard
Antennenaufzugseil n <Lufttrans> halyard
Antennenbündelung f <Funktech> antenna directivity
Antennendiversity n <Funktech> antenna diversity
Antenneneinführung f <Funktech> lead-in
Antennenfeld n <Funktech> antenna array
Antennengewinn m 1. <Aufnahme, Elektronik> power gain; 2. <Fernseh, Funktech, Phys> aerial gain; 3. <Raumfahrt> antenna gain
Antennengewinnfunktion f <Elektronik> gain function
Antennengruppe f <Funktech> antenna array
Antennenhauptkeule f <Funktech> antenna main lobe
Antennenkabel n <Elektrotech> aerial cable, antenna cable
Antennenkuppel f 1. <Lufttrans> blister *(Hubschrauber)*; 2. <Raumfahrt, Telekom> radome
Antennenleitung f <Elektriz> aerial line
Antennenmast m 1. <Fernseh> aerial mast *(Funk)*; 2. <Funktech> aerial mast, antenna pole; 3. <Telekom, Wassertrans> aerial mast *(Funk)*
Antennennachführsystem n <Funktech> antenna tracking system
Antennenrelaissystem n ohne Verstärkung <Funktech, Telekom> nonboosted antenna repeater system
Antennenrichtwirkung f <Fernseh, Funktech> aerial directivity, antenna directivity
Antennenschüssel f <Raumfahrt> dish antenna
Antennenspeiseleitung f <Funktech> antenna feeder, antenna feeder line
Antennenspiegel m <Funktech> antenna reflector
Antennenstrahlungswiderstand m <Funktech> aerial radiation resistance
Antennensystem n <Funktech, Raumfahrt> antenna system
Antennentrennverstärker m <Funktech> antenna multicoupler *(AtWe)*
Antennenverlängerungsspule f <Funktech> aerial loading coil
Antennenverstärker m <Elektronik, Funktech> antenna booster, booster
Antennenverstärkung f <Phys> aerial gain
Antennenweiche f 1. <Fernseh> combiner; 2. <Funktech> antenna multicoupler *(AtWe)*; combiner
Antennenwirkfläche f 1. <Elektronik> capture area *(Antennentechnik)*; 2. <Funktech> absorption cross-section, effective aperture
Antennenwirkungsgrad m 1. <Fernseh> aerial efficiency; 2. <Funktech> aerial efficiency, antenna efficiency
Antennenwirkwiderstand m <Funktech, Phys> aerial resistance
Antennenzuleitung f 1. <Fernseh> aerial lead; 2. <Funktech> aerial lead, antenna feeder, antenna feedline
Anthracen n <Chemie> anthracene
Anthracenfarbstoff m <Chemie> anthracene dye
Anthracenöl n <Chemie> anthracene oil
Anthragallol n <Chemie> anthragallol
Anthrazit m <Kohlen> anthracite, hard coal
Anthrazitkohle f <Kohlen> anthracite coal
anthropogen bedingte Übersäuerung f <Umweltschmutz> anthropogenic acidification
anthropogen verursachte Bodenerschütterung f <Umweltschmutz> man-made earth tremor
anthropogen verursachtes Erdbeben n <Umweltschmutz> man-made earthquake
Anthropometrie f <Ergon> anthropometry
anthropomorpher Roboter m <Künstl Int> anthropomorphic robot
anthropotechnisch adj <Ergon> anthropotechnical
Antialbumose f <Lebensmittel> anti-albumose

Antialiasing f <Elektronik, Fernseh, Telekom> anti-aliasing
Antialiasing-Filter n <Elektronik, Fernseh, Telekom> anti-aliasing filter
Antiausschwimmmittel n <Kunststoff> antiflooding agent
Antibackmittel n <Lebensmittel> anticaking agent
Antibaryon n <Teilphys> antibaryon
Antibindung f <Strahlphys> antibonding
Antibindungsbahn f <Strahlphys> antibonding atomic orbital
Antibindungselektronen npl <Strahlphys> antibonding electrons
Antiblockiersystem n *(ABS)* 1. <Kfztech> antiblocking system, antilock system, anti-lock braking system, ABS; antiskid braking system, ASBS *(Bremsung)*; 2. <Lufttrans> antiskid unit
Antiblockmittel n <Kunststoff> antiblocking agent
Antichlor n <Chemie, Papier> antichlor
Antidot n <Chemie> antidote
Antienzym n <Lebensmittel> anti-enzyme
antiferromagnetisch adj <Phys> antiferromagnetic
Antiferromagnetismus m <Elektriz, Phys> antiferromagnetism
Antifouling-Anstrichfarbe n <Kunststoff> antifouling paint
Antifriktionsmittel n <Papier> antifriction
Antigen n <Chemie> antigen
Antihautmittel n <Kunststoff> antiskinning agent
Antikatode f <Elektriz, Phys> anticathode
Antikglas n <Ker & Glas> antique glass
Antiklinale f <Erdöl> anticline *(Geologie)*
antiklinale Falle f <Erdöl> anticlinal trap *(Geologie)*
Antiklopfmittel n 1. <Erdöl> antiknock *(Raffinerie)*; 2. <Kfztech> antiknock additive; antiknock agent *(Kraftstoff)*; 3. <Umweltschmutz> antiknock additive
Antikoinzidenz f <Gerät, Strahlphys> anticoincidence
Antikoinzidenzschaltung f <Phys> anticoincidence circuit
Antikoinzidenzzähler m <Gerät> anticoincidence counter
Antikollisionslicht n <Lufttrans> anticollision light
Antikondensationsbeutel m <Verpack> desiccant bag
Antilichthofbelag m <Foto> antihalation backing
antimagnetisch adj <Chemie> antimagnetic, nonmagnetic
Antimaterie f <Phys> antimatter
Antimon n *(Sb)* <Chemie> antimonic, antimonous, antimony, stibic, stibium
antimonartig adj <Chemie> antimonial
Antimonat n <Chemie> antimoniate, stibate, antimonite
Antimonglanz m <Chemie> antimonite, antimony glance, stibnite
antimonhaltig adj <Chemie> antimonial
Antimonid n <Chemie> antimonide, stabide
antimonig adj <Chemie> stibious
Antimonit m <Chemie> antimonite, antimony glance, stibnite
Antimontetroxid n <Chemie> antimony tetroxide
Antimykotikum n <Chemie> fungicide *(pilztötendes Mittel)*
Antineutrino n <Phys> antineutrino
Antineutron n <Phys> antineutron
Antioxidans n <Kunststoff, Lebensmittel> antioxidant
Antioxidationsmittel n <Kunststoff, Lebensmittel> antioxidant
Antioxidationswirkstoff m <Kunststoff> antioxidant
antiparallel adj <Geom> antiparallel
Antiparallel... <Elektrotech, Geom> antiparallel

antiparallele Anordnung f <Elektrotech> antiparallel arrangement; back-to-back arrangement *(Kondensatoren)*
Antiparallelgelenkviereck n <Fertig> anti-ager-parallel four-bar *(Verbindungen)*
Antipassat m <Wassertrans> antitrades *(Windart)*
Antipodenpunkte mpl <Geom> antipodal points
antippen v <Comp & DV> identify *(Auswahl auf Tablett)*
Antiproton n <Phys, Teilphys> antiproton
Antiprotonenring m **mit geringer Energie** *(LEAR)* <Teilphys> Low-Energy Antiproton Ring *(LEAR)*
Antiquark n <Phys, Teilphys> antiquark
Antiquaschrift f <Druck> Roman type
Antireaktivität f <Kerntech> antireactivity
Antireflexbelag m <Ker & Glas, Telekom> antireflection coating
Antireflexionsüberzug m <Optik> antireflection coating
Antiresonanz f <Akustik, Elektronik> antiresonance
Antiresonanzfrequenz f *(fA)* <Akustik, Elektronik> antiresonant frequency *(fA)*
Antirutschsohle f <Sicherheit> slip-resistant sole *(Sicherheitsschuhwerk)*
Antisatellit-Laser m 1. <Elektronik> antisatellite laser; 2. <Papier> anti froth
Antischaummittel n 1. <Chemie> defoaming agent; 2. <Erdöl> antifoam agent; 3. <Kohlen, Kunststoff, Maschinen> antifoaming agent; 4. <Papier> antifoam, antifroth
Antischleiermittel n <Foto> antifogging agent
Antischrumpfbehandlung f <Textil> antishrink treatment
Antischwappdämpfer mpl <Raumfahrt> antislosh baffles *(Einbauten)*
Antischwerkraft f <Raumfahrt> antigravity
Antiskidsystem n <Lufttrans> antiskid unit
Antispritzmittel n <Lebensmittel> antispattering agent
Antistatik f <Fertig, Maschinen, Textil> antistatic
Antistatikausrüstung f <Maschinen> antistatic protection
Antistatikgerät n <Textil> static eliminator
Antistatikmittel n <Fertig, Textil> antistatic agent
Antistatikschuhe mpl <Sicherheit> antistatic footwear
Antistatikspray n <Comp & DV> antistatic spray
Antistatikum n <Kunststoff, Textil> antistatic agent
antistatisch adj <Elektriz, Foto, Maschinen> antistatic
antistatische Matte f <Comp & DV> antistatic mat
antistatische Schutzkleidung f <Sicherheit> antistatic protective clothing
antistatischer Belag m <Foto> antistatic backing
antistatisches Material n <Sicherheit> antistatic material
antisymmetrische Wellenfunktion f <Phys> antisymmetric wave function
antisymmetrischer Tensor m <Strömphys> antisymmetric tensor
Antiteilchen n <Phys, Teilphys> antiparticle
Antivalenz f 1. <Comp & DV> symmetric difference; 2. <Elektronik> anticoincidence *(Exklusiv-ODER-Verknüpfung)*
Antivalenzfunktion f <Comp & DV> nonequivalence function
Antivalenzglied n <Comp & DV> nonequivalence gate
Antivalenztor n <Comp & DV> nonequivalence gate
Antivalenzverknüpfung f <Comp & DV> nonequivalence operation
Antivereisungsschieber m <Lufttrans> engine anti-icing gate valve
Antivibrationsgriff m <Ergon, Sicherheit> antivibration handle
antizyklonale Generation f <Metrol> anticyclonic generation
Antizyklone f <Wassertrans> anticyclone
Antonit m <Kerntech> antonite

Antrag m <Patent> request
antreiben v 1. <Maschinen> drive; 2. <Mechan> actuate, drive; 3. <Papier> drive; 4. <Phys> actuate; 5. <Raumfahrt> drive *(Raumschiff)*; 6. <Textil> drive
Antrieb m 1. <Ergon> incentive; 2. <Fernseh> transport mechanism; 3. <Fertig> actuation; 4. <Kfztech> propulsion; drive *(Triebstrang)*; 5. <Kontroll> drive *(Plattenspielwerk)*; 6. <Maschinen> drive, impulsion, propulsion; 7. <Mechan> actuator, drive; 8. <Papier> drive; 9. <Raumfahrt> propulsion *(Raumschiff)*; 10. <Textil> drive; 11. <Wassertrans> propulsion *(Schiffantrieb)* • **mit elektrischem Antrieb** <Foto> electrically-driven • **ohne Antrieb** <Fertig> undriven
Antrieb m **durch feste Antriebsräder** <Kfztech, Wassertrans> propulsion by stationary drive wheels
Antrieb m **durch Luftdruck** <Kfztech, Wassertrans> propulsion by air pressure
Antrieb m **durch Spiralantrieb mit wechselnder Steigung** <Kfztech, Wassertrans> propulsion by spiral drive with varying pitch
Antrieb m **mit konstanter Drehzahl** <Maschinen> constant-speed drive
Antrieb m **mit Ritzel und Zahnstange** <Kerntech> rack-and-pinion drive gear
Antrieb m **ohne Nachbrenner** <Lufttrans> dry power *(Triebwerk, Motor)*
Antrieb m **ohne Rutschkupplung** <Fertig> positive drive *(Gewindebohrer)*
Antriebsachse f 1. <Fertig> actuator shaft *(Kunststoffinstallationen)*; 2. <Kfztech> driving axle, live axle *(Triebstrang)*; 3. <Maschinen> driving axle, live axle
Antriebsaggregat n 1. <Fertig> driving package; 2. <Maschinen> drive unit, mover, power unit, prime mover
Antriebsbatterie f <Kfztech> drive battery
Antriebsdrehzahl f <Kfztech> engine speed *(Motor)*
Antriebseinheit f <Raumfahrt> propulsion unit *(Raumschiff)*
Antriebselement n <Maschinen> driving element
Antriebsgehäuse n <Fertig> actuator housing *(Kunststoffinstallationen)*
Antriebsglied n 1. <Fertig> driving link, follower, input member *(Getriebelehre)*; 2. <Maschinen> driving member
Antriebskegelrad n <Kfztech> drive pinion, pinion gear
Antriebskette f 1. <Kfztech> drive chain *(Motor, Triebstrang)*; 2. <Maschinen> driving chain, transmission chain
Antriebskette f **der Nockenwelle** <Kfztech> camshaft drive chain
Antriebskettenrad n 1. <Fertig> driving sprocket; 2. <Kfztech> sprocket *(Motorradgetriebe)*
Antriebskraft f 1. <Kfztech> drive power; 2. <Maschinen> drive power, propulsive force; 3. <Wassertrans> motive force *(Schiffkonstruktion)*
Antriebskurbel f <Fertig> driving crank
Antriebsleistung f 1. <Maschinen> driving power; 2. <Wassertrans> propulsion power
Antriebsmagnet m <Kfztech, Wassertrans> propulsion magnet
Antriebsmaschine f 1. <Kfztech> motor; 2. <Lufttrans, Maschinen> prime mover
Antriebsmechanismus m <Kontroll> drive mechanism
Antriebsmotor m 1. <Aufnahme, Elektriz> drive motor; 2. <Fertig> mover; 3. <Foto> drive motor; 4. <Kfztech> propulsion engine, propulsion motor; 5. <Kontroll> driving motor; 6. <Wassertrans> propulsion engine, propulsion motor *(Schiffantrieb)*
Antriebspotenzial n <Elektrotech> driving potential
Antriebspropeller m <Lufttrans> driving propeller
Antriebsrad n 1. <Fertig> impeller; 2. <Kfztech> driving wheel; drive wheel *(Rad, Kraftübertragung)*; 3. <Maschi-

nen> driver, driving gear, driving wheel, leader; 4. <Papier> leader

Antriebsriemen *m* 1. <Fertig> driving belt; 2. <Maschinen> belt, driving belt, transmission belt

Antriebsriemenscheibe *f* <Kfztech> drive pulley *(Drehstromlichtmaschine)*

Antriebsritzel *n* 1. <Kfztech> drive pinion, pinion gear; 2. <Maschinen> driving pinion

Antriebsritzelwelle *f* <Kfztech> drive pinion shaft

Antriebsrolle *f* 1. <Comp & DV> capstan; 2. <Fernseh> drive sprocket; 3. <Maschinen> driving pulley

Antriebsschale *f* <Kfztech> input shell

Antriebsscheibe *f* 1. <Fertig> driver pulley; 2. <Maschinen> *(BE)* driving disc, *(AE)* driving disk

Antriebsseite *f* 1. <Elektrotech> drive end; 2. <Maschinen> driving end; 3. <Papier> drive side

Antriebsspindel *f* <Maschinen> drive spindle

Antriebsspule *f* <Elektrotech> drive coil

Antriebsstange *f* <Mechan> push rod

Antriebsstrang *m* <Maschinen> train *(eines Antriebssystems)*

Antriebssystem *n* 1. <Maschinen> drive system, transmission system; 2. <Raumfahrt> propulsion system *(Raumschiff)*; 3. <Textil> drive system

Antriebstrommel *f* <Kfztech> driving drum

Antriebswelle *f* 1. <Elektrotech> capstan, shaft; 2. <Fertig> drive, driver; drive shaft *(Kunststoffinstallationen)*; 3. <Kfztech> axle shaft; input shaft *(Kupplung, Getriebe)*; drive shaft, half shaft *(Triebstrang)*; 4. <Maschinen> drive shaft, driving shaft, engine shaft, transmission shaft; 5. <Wassertrans> driving shaft *(Motor)*

Antriebszahnrad *n* **der Antriebswelle** <Kfztech> clutch gear

Antrocknungszeit *f* <Kunststoff> tack free time

Antwort *f* 1. <Comp & DV> response; 2. <Elektronik> reply; 3. <Ergon, Telekom> response

antwortabhängige Nachricht *f* <Comp & DV> response message

Antwortfunkfeuer *n* <Lufttrans> responder beacon

Antwortmodus *m* 1. <Comp & DV> response mode; 2. <Telekom> auto answer *(Modem)*

Antwortsender *m* 1. <Lufttrans> responder *(Kommunikationswesen)*; 2. <Phys> transponder

Antwortsignal *n* <Telekom> answer signal

Antwortwimpel *m* <Wassertrans> answering pennant *(Signal)*

Antwortzeit *f* <Comp & DV, Ergon, Telekom> response time

An- und Abfuhr *f* <Trans> conveying

Anvisieren *n* <Bau> sighting

anvisierte Alhidade *f* <Bau> sighted alidade

anvisierte Höhe *f* <Bau> sighted level

Anvulkanisation *f* <Kunststoff> scorch

Anvulkanisationsverhinderer *m* <Kunststoff> anti-scorching agent

anvulkanisieren *v* <Thermod> scorch

Anwärm... 1. <Bau> heating; 2. <Thermod> warm-up

Anwärmbrenner *m* <Bau> heating blowpipe

anwärmen *v* <Thermod> heat up, warm up

Anwärmloch *n* <Ker & Glas> glory hole

Anwärmzeit *f* <Thermod> warm-up time

anweisen *v* <Comp & DV> order

Anweisung *f* 1. <Comp & DV> directive, instruction, order; statement *(in Programm)*; 2. <Kontroll> instruction

Anweisungskennsatz *m* <Comp & DV> statement label

Anweisungsnummer *f* <Comp & DV> statement number

anwendbar *adv* applicable; practicable; usable

anwenden *v* <Mechan> apply

Anwender *m* <Comp & DV, Maschinen, Qual, Telekom> user

Anwenderauswahl-eingebuchtes Handy *n* <Mobilkom> user selected log-in

Anwender-Betriebsumgebung *f* <Comp & DV> user-operating environment

anwenderfreundliches System *n* <Sicherheit> user-friendly system

Anwenderschnittstelle *f* <Telekom> customer interface

Anwendung *f* 1. <Anstrich> application; 2. <Comp & DV> application *(Software)*; data application; 3. <Mechan> application; 4. <Maschinen> operation, use • **eine Anwendung ausführen** <Comp & DV> run an application

anwendungsbezogene Sprache *f* <Comp & DV> application-oriented language

Anwendungsdatei *f* <Comp & DV> application file

Anwendungseinheit *f* <Fertig> unit *(Kunststoffinstallationen)*

Anwendungsentwicklung *f* <Comp & DV> application development

Anwendungshandbuch *n* <Comp & DV> application manual

Anwendungsinstanz *f* <Telekom> application entity

Anwendungsmodell *n* <Comp & DV> usage model

anwendungsorientierte Programmiersprache *f* <Comp & DV> application oriented language

Anwendungsprogramm *n* <Comp & DV> application, application program, application software, business software, user program

Anwendungsprogrammpaket *n* <Comp & DV> application package

Anwendungsschicht *f* 1. <Comp & DV> application layer *(ISO-Referenzmodell)*; 2. <Telekom> application layer

Anwendungsschnittstelle *f (API)* <Telekom> application program interface, application programming interface, API

Anwendungsteil *n* **intelligentes Netz** <Telekom> intelligent network application part

Anwendungs- und Datenlogik *f* <Comp & DV> business and data logic

Anwerfen *n* **des Außenputzes** <Bau> rendering

Anwurf *m* <Bau> roughcast

Anzahl *f* 1. <Gerät> count; 2. <Metrol> rate; 3. <Qual> number, quantity

Anzahl *f* **der Arbeitsgänge** <Bau, Fertig> number of passes

Anzahl *f* **der Proben** <Werkprüf> number of specimens

anzapfen *v* 1. <Bau> blend; 2. <Fertig> tap *(Punktschweißen)*; 3. <Funktech> bug; 4. <Phys> tap

Anzapftransformator *m* <Elektrotech> tapped transformer

Anzapfung *f* 1. <Elektriz, Funktech> tap; 2. <Elektrotech, Phys, Telekom> tapping

Anzapfungsanzeige *f* <Elektriz> tap position indicator

Anzapfungswähler *m* <Elektriz> tap selector, tap switch

Anzapfungswechsel *m* <Elektriz> tap change operation

Anzapfungswechsler *m* <Elektriz> tap changer

Anzapfwiderstand *m* <Elektrotech> tapped resistor

Anzeichen *n* <Kohlen, Metall> prospect *(Erz)*

anzeichnen *v* <Bau, Fertig> scribe

Anzeige *f* 1. <Comp & DV> display, readout; 2. <Druck> display; 3. <Druck> advertisement; 4. <Elektriz> display, readout; 5. <Gerät> reading; 6. <Hydraul> indicator; 7. <Kontroll> display; 8. <Maschinen> telltale; 9. <Raumfahrt> display *(Bildschirm)*; 10. <Telekom> display, readout

Anzeige *f*/**achtzig Spalten breite** <Comp & DV> eighty-column screen

Anzeige *f* **der gerufenen Nummer** <Telekom> called number display

Anzeige *f* **der Grenzwertüberschreitung** <Gerät> out-of-limits indication

Anzeige *f* **der Nummer des rufenden Teilnehmers** <Telekom> CLIP, calling line identification presentation
Anzeige *f* **der rufenden Leitung** <Telekom> CLID, calling line identification display
Anzeige *f* **der Rufnummer des erreichten Teilnehmers** <Telekom> connected line identification presentation *(COLP)*
Anzeige *f* **des Blatteinstellwinkels** <Lufttrans> *(AE)* blade angle check gage, *(BE)* blade angle check gauge *(Hubschrauber)*
Anzeigebereich *m* <Gerät> indicating range, indication range
Anzeigeeinheit *f* <Gerät> display device
Anzeigeeinrichtung *f* 1. <Gerät> read-out device; 2. <Maschinen> telltale
Anzeigeformat *n* <Comp & DV, Kerntech> display format
Anzeigegenauigkeit *f* <Gerät> accuracy of indication
Anzeigegerät *n* 1. <Comp & DV> display device; 2. <Gerät> display device, read-out meter; 3. <Kontroll> display unit, monitor
Anzeigegerät *n* **mit Kreisskale** <Gerät> round scale indicator
Anzeigehintergrund *m* <Comp & DV> background display
Anzeigelampe *f* <Elektriz> indicator lamp
Anzeigeleuchte *f* <Kfztech> pilot light *(Zubehör)*
Anzeigemessgerät *n* <Hydraul> indicator
Anzeigemittel *n* <Fertig> detecting agent
Anzeigemodus *m* <Comp & DV> display mode
anzeigen *v* 1. <Comp & DV, Druck, Elektronik, Kontroll> display; 2. <Metrol> read; 3. <Raumfahrt> display *(Bildschirm)*
Anzeigenabbild *n* <Comp & DV> screen image
Anzeigenabteilung *f* <Druck> advertising department
Anzeigenadel *f* <Foto> indicator needle
anzeigendes Messgerät *n* <Gerät> read-out meter
Anzeigenfahne *f* <Druck> ad galley
Anzeigenlayout *n* <Druck> advertisement layout
Anzeigenschrift *f* <Druck> ad face
Anzeigenseite *f* <Druck> advertisement page
Anzeigensetzer *m* <Druck> advertisement setter
Anzeigensetzerei *f* <Druck> advertisement composing room
anzeigepflichtig *adj* <Qual> notifiable
Anzeiger *m* 1. <Comp & DV> flag, indicator; 2. <Kfztech> *(AE)* gage, *(BE)* gauge; 3. <Maschinen> indicator
Anzeiger *m* **für spannungstragende Leitung** <Elektriz, Sicherheit> live line indicator
Anzeigeregister *n* <Comp & DV> display register
Anzeigeröhre *f* <Elektronik> indicator tube
Anzeigesäule *f* <Labor> cup *(eines Barometers)*
Anzeigeskale *f* <Gerät> indicating scale
Anzeigestelle *f* <Kerntech> indicator bay *(eines Massenspektrometers)*
Anzeigetafel *f* <Mechan> index table
Anzeigethermometer *n* <Heiz & Kälte> indicating thermometer
Anzeigewert *m* <Gerät> indicated value, reading
Anzeigezeit *f* <Gerät> display time
Anziehdrehmoment *n* <Maschinen> tightening torque
anziehen *v* 1. <Maschinen> fasten, tighten; 2. <Phys> attract
Anziehen *n* <Eisenbahn, Maschinen> tightening
anziehend *adj* <Phys> attractive
Anziehung *f* <Phys> attraction
Anziehungseffekt *m* <Trans> attractive effect
Anziehungskraft *f* 1. <Elektriz> attractive force; 2. <Phys> attractive force, force of attraction, gravitation; 3. <Raumfahrt> pull

Anzug *m* 1. <Elektrotech> pick-up; 2. <Maschinen> tightening
Anzugmutter *f* <Maschinen> tightening nut
Anzugschraube *f* <Fertig> draw-in bolt *(Frässpindel)*
Anzugsmoment *n* <Kfztech> pick-up
Anzugsspannung *f* <Elektrotech> pick-up voltage
Anzugstoff *m* <Textil> suiting
anzünden *v* <Thermod> kindle
Anzünder *m* <Heiz & Kälte> lighter
AOCS *(Fluglage- und Umlaufbahnkontrollsystem)* <Raumfahrt> AOCS *(attitude and orbit control system)*
AOQ *(durchschnittliche Fertigproduktqualität)* <Qual> AOQ *(average outgoing quality)*
AOQL *(durchschnittlicher Fertigproduktqualitätsgrenzwert)* <Qual> AOQL *(average outgoing quality limit)*
AOS *(automatisches Signal zur Mikrofonübergabe)* <Telekom> AOS *(automatic over signal)*
AOW *(akustische Oberflächenwelle)* <Elektronik, Telekom> SAW *(surface acoustic wave)*
AOW-Bauelement *n* <Elektronik, Telekom> SAW device
AOW-Expansionsfilter *n* <Elektronik> SAW expansion filter
AOW-Filterung *f* <Elektronik> SAW filtering
AOW-Kompressionsfilter *n* <Elektronik> SAW compression filter
AOW-Laufzeitleitung *f* <Elektronik> SAW delay line
Apatit *m* <Chemie> apatite, phosphate of lime
APD *(Avalanchephotodiode)* <Elektronik, Optik> APD *(avalanche photodiode)*
aperiodisch *adj* 1. <Fertig> dead *(Schwingung)*; 2. <Phys> aperiodic
aperiodisch gedämpft *adj* <Aufnahme> dead beat
aperiodisch gedämpftes Galvanometer *n* <Elektriz> dead beat galvanometer
aperiodisch gedämpftes Instrument *n* <Metrol> aperiodic instrument
aperiodische Leitung *f* 1. <Elektronik> nonresonant line; 2. <Funktech> flat line
aperiodische Schaltung *f* <Elektronik> aperiodic circuit *(frequenzunabhängig)*
aperiodischer Stromkreis *m* <Comp & DV> aperiodic circuit
aperiodisches Filter *n* <Funktech> aperiodic filter
aperiodisches Galvanometer *n* <Elektriz> aperiodic galvanometer
Apertur *f* <Comp & DV, Funktech, Mechan, Telekom> aperture *(Antenne)*
Aperturantenne *f* <Funktech, Telekom> aperture antenna
Aperturblende *f* 1. <Foto> aperture diaphragm; 2. <Phys> aperture stop
Aperturgitter *n* <Elektronik> aperture grill
Aperturverzerrung *f* <Telekom> aperture distortion
Apfelsäure *f* <Lebensmittel> malic acid
Apfelsinenschale *f* <Ker & Glas> orange peel
Apfelsinenschaleneffekt *m* <Kunststoff> orange peel
Aphel *n* <Phys> aphelion
Aphongetriebe *n* <Kfztech> helical gear
API *(Amerikanisches Erdölinstitut)* <Erdöl> API *(American Petroleum Institute)*
API-Dichte *f* <Erdöl> API gravity *(Öl)*
Apionol *n* <Chemie> apionol, phenetrol
Aplanasie *f* <Optik> aplanatism
Aplanat *m* 1. <Foto> aplanatic lens; 2. <Optik> aplanat
Apochromasie *f* <Optik> apochromatism
Apochromat *m* <Fertig, Foto> apochromatic lens
apochromatisch *adj* <Optik> apochromatic
apochromatische Korrektur *f* <Foto> apochromatic correction
Apogäum *n* <Phys, Raumfahrt> apogee

Apogäumsmanöver

Apogäumsmanöver n <Raumfahrt> (AE) apogee maneuver, (BE) apogee manoeuvre
Apogäumstriebwerk n <Raumfahrt> apogee motor (Raumschiff)
Apostilb n (asb) <Optik> apostilb (asb)
Apparat m <Labor, Maschinen> apparatus, device, instrument • **am Apparat bleiben** <Telekom> hold the line
Apparate mpl <Fertig, Maschinen> equipment
Apparatebau m <Fertig> apparatus construction; instruments engineering
Apparatesatz m <Elektrotech, Maschinen> set (Gruppe von Geräten)
Apparateschnur f <Telekom> instrument cord
Appleton-Schicht f <Phys> Appleton layer, F layer
Appret n <Textil> finish
Appreteur m <Textil> finisher
appretieren v 1. <Fertig> finish; 2. <Textil> size
Appretieren n <Textil> sizing
Appretur f 1. <Fertig> finish; 2. <Textil> finish (Baumwolle)
Appreturflotte f <Textil> finishing bath
Approximation f <Math> approximation
Approximationsfehler m <Gerät> approximation error
APR 1. <Elektronik> (automatische Phasenregelung) APC (automatic phase control); 2. <Fernseh> (automatische Phasensteuerung) APC (automatic phase control)
aprotisches Lösemittel n <Chemie> aprotic solvent
APT (programmierte Werkzeuge) <Comp & DV> APT (automatically programmed tools)
apyrisch adj <Chemie> apyrous
AQL (akzeptabler Qualitätsspegel, annehmbare Qualitätsgrenzlage, annehmbare Qualitätslage) <Qual> AQL (acceptable quality level)
Aquädukt m <Bau, Nichtfoss Energ, Wasserversorg> aqueduct
Aquakultur f <Wasserversorg> aquiculture
Aquaplaning n <Kfztech> aquaplaning
aquatisches System n <Wasserversorg> aquatic system
Äquator m <Wassertrans> equator (Geographie)
äquatoriale Brennlinie f <Phys> sagittal focal line
äquatoriale Fokuslinie f <Phys> sagittal focal line
äquatoriale synchrone Umlaufbahn f <Raumfahrt> equatorial synchronous orbit
äquatoriale Umlaufbahn f <Raumfahrt> equatorial orbit
Äquatortaufe f <Wassertrans> crossing the line ceremony
Äquatorüberflug m <Raumfahrt> equatorial crossing
Äquatorüberquerung f <Wassertrans> crossing the line
Aquiclude f <Wasserversorg> aquiclude
äquidistant adj <Bau, Geom> equidistant
Äquidistante f <Geom> equidistant line
äquidistante Linie f <Geom> equidistant line
Äquidistanzlinie f <Erdöl> median line (Seekartierung)
Aquifer m <Nichtfoss Energ, Wasserversorg> aquifer
äquimolar adj <Chemie> equimolecular
Äquinoktialgezeit f <Nichtfoss Energ> equinoctial tide
äquipotenzial adj <Elektriz, Elektrotech> equipotential
Äquipotenzial n <Phys> equipotential
Äquipotenzialfläche f <Elektriz, Elektrotech, Funktech, Phys, Raumfahrt> equipotential surface
Äquipotenzialkatode f <Elektrotech> unipotential cathode
Äquipotenziallinie f <Raumfahrt> equipotential line
äquivalent adj <Math> equivalent
Äquivalent n <Math> equivalent
Äquivalent n je Million (EPM) <Umweltschmutz> equivalent per million (EPM)
Äquivalentdosis f <Phys> dose equivalent
äquivalente Absorptionsfläche f <Akustik> equivalent absorption area
äquivalente Dichte f <Erdöl> equivalent density (Bohrtechnik)
äquivalente Fluggeschwindigkeit f (EAS) <Lufttrans> equivalent airspeed (EAS)
äquivalente isotrope Strahlungsleistung f (EIRP) <Funktech, Raumfahrt> effective isotropically-radiated power, EIRP (Weltraumfunk)
äquivalente Rauschleistung f <Telekom> noise equivalent power (NEP)
äquivalente Rauschtemperatur f <Raumfahrt> equivalent noise temperature (Weltraumfunk)
äquivalente Rauschzahl f <Raumfahrt> equivalent noise temperature (Weltraumfunk)
äquivalente Steigbögeschwindigkeit f <Lufttrans> equivalent vertical gust speed
äquivalente Strahlungsleistung f <Telekom> equivalent radiated power
äquivalente Stufenindex-Brechzahldifferenz f <Telekom> ESI refractive index difference
äquivalente Tiefe f <Erdöl> equivalent depth (Bohrtechnik)
äquivalente Zirkulationsdichte f <Erdöl> ECD, equivalent circulating density (Bohrtechnik)
äquivalenter Dauerschallpegel m <Sicherheit> equivalent continuous sound level
äquivalenter Stufenindex m <Telekom> ESI, equivalent step index
äquivalenter Zufallsverkehrswert m <Telekom> equivalent random traffic intensity
äquivalentes Bohrschlammgewicht n <Erdöl> EMW, equivalent mud weight (Bohrtechnik)
äquivalentes Stufenprofil n <Telekom> ESI profile, equivalent step index profile
Äquivalentleitwert m <Thermod> equivalent conductance
Äquivalentschaltung f <Elektriz> equivalent circuit
Äquivalentvolumen n <Kerntech, Phys, Strahlphys> atomic volume
Äquivalentwiderstand m <Elektriz> equivalent resistance
Äquivalenz f <Comp & DV> equivalence
Äquivalenzfunktion f <Comp & DV> equivalence function, equivalence operation
Äquivalenzglied n 1. <Comp & DV> equivalence gate; 2. <Elektronik> exclusive NOR gate
Äquivalenzklasse f <Math> equivalence class
Äquivalenznormalruß m <Umweltschmutz> equivalent standard smoke
Äquivalenzprinzip n <Phys> principle of equivalence
Äquivalenzverknüpfung f 1. <Comp & DV> equivalence operation; 2. <Elektronik> exclusive NOR circuit
Ar (Argon) <Chemie> Ar (argon)
Ar n (a) <Metrol> are (a)
AR (Ausgangsregister) <Mobilkom> HLR (home location register)
Arabinose f <Chemie> arabinose
arabische Zahl f <Math> cipher (Ziffer)
arabische Zahlen fpl <Math> arabic numerals
arabische Ziffern fpl <Math> arabic numerals
Arabit m <Chemie> arabitol
Arabitol f <Chemie> arabitol
Arachidonsäure f <Lebensmittel> arachidonic acid
Arachin... <Chemie> arachic
Arachinalkohol m <Chemie> arachic alcohol, eicosyl alcohol
A-Rahmen m <Bau> A-frame (Dach)
Aramid n <Kunststoff> aramid
Aräometer n 1. <Erdöl> hydrometer; 2. <Gerät> areometer; 3. <Labor> hydrometer; 4. <Lebensmittel> densimeter, hydrometer; 5. <Phys> areometer

Aräometrie f <Phys> araeometry
Arbeit f 1. <Elektriz> energy; 2. <Ergon> job, labour, task, work; 3. <Maschinen> operation, work
Arbeit f mit Gefahrstoffen <Sicherheit> handling of dangerous materials
arbeiten v 1. <Ergon> work; 2. <Maschinen> function, operate, run, work
arbeitender Analog-Digital-Umsetzer m/nach Zählmethode <Gerät> analog-digital counter-type converter
arbeitendes Büro n/im Rechnerverbund <Comp & DV> integrated office system
arbeitendes Teil n <Maschinen> working part
Arbeitsablauf m 1. <Comp & DV> workflow; 2. <Fertig> machining cycle, operating cycle
Arbeitsabschnitt m <Comp & DV> session
Arbeitsakt m <Fertig> machining cycle
Arbeitsanforderung f <Ergon> job demand
Arbeitsaufgabe f <Ergon> work task
Arbeitsauftrag m <Comp & DV> job
Arbeitsband n <Comp & DV> scratch tape
Arbeitsbegleitpapier n <Qual> process sheet
Arbeitsbegleitpapiere npl <Qual> accompanying papers
Arbeitsbelastung f <Comp & DV, Ergon> workload
Arbeitsbereich m 1. <Comp & DV> work area, working area, workspace; 2. <Kerntech> operation area
Arbeitsbereicherung f <Ergon> job enrichment
Arbeitsbeschreibung f <Ergon> job description, job specification
Arbeitsbewertung f <Ergon> job evaluation
Arbeitsbrücke f <Bau> staging
Arbeitsbühne f 1. <Bau> platform, stage, working platform; 2. <Erdöl> drill floor *(Bohrtechnik)*; 3. <Ker & Glas> working platform
Arbeitsdatei f <Comp & DV> scratch file, work file
Arbeitsdatensatz m <Comp & DV> work record
Arbeitsdatenträger m <Comp & DV> work volume
Arbeitsdruck m 1. <Heiz & Kälte> operating pressure; 2. <Maschinen> operating pressure, working pressure
Arbeitsebene f <Fertig> working plane
Arbeitseingriff m <Fertig> working engagement
Arbeitselement n <Comp & DV> work item
Arbeitsende n <Ker & Glas> working end *(des Wannenofens)*
Arbeitsenergieumsatz m <Ergon> work energy expenditure
Arbeitserweiterung f <Ergon> job enlargement
Arbeitsfläche f 1. <Ergon> work surface; 2. <Fertig> machined surface, working area, working surface; 3. <Maschinen> face
Arbeitsfolgeregler m <Elektrotech> sequencer
Arbeitsfrequenz f <Maschinen> operating frequency
Arbeitsfuge f <Bau> construction joint *(Beton)*
Arbeitsfunktion f <Elektrotech> work function
Arbeitsgang m 1. <Comp & DV> operation, pass, transaction; 2. <Fertig> working traverse; 3. <Maschinen> cutting stroke
Arbeitsgeschwindigkeit f 1. <Metall> operating rate; 2. <Maschinen> working speed
Arbeitsgestalter m <Ergon> job designer
Arbeitsgestaltung f <Ergon> job design, work design
Arbeitsgruppe f <Comp & DV> workgroup
Arbeitshandschuhe mpl <Sicherheit> working gloves
Arbeitshub m 1. <Fertig> working stroke; 2. <Kfztech> power stroke; 3. <Maschinen> working stroke
Arbeitsinformationsmittel n <Ergon> job aid
Arbeitsinhalt m <Ergon> job content, work content
Arbeitskanal m <Telekom> working channel
Arbeitskarte f 1. <Ergon> instruction card; 2. <Qual> job card; work card *(Laufkarte)*

Arbeitskennlinie f 1. <Elektriz> dynamic characteristic; 2. <Fertig> operating characteristic; performance characteristic
Arbeitskleidung f <Sicherheit, Textil> working clothes
Arbeitskontakt m 1. <Elektriz> normally open contact; 2. <Elektrotech> make contact, normally open contact
Arbeitsleben n <Elektrotech> service life
Arbeitslehre f 1. <Maschinen> *(AE)* manufacturing gage, *(BE)* manufacturing gauge; 2. <Phys> work standard
Arbeitsleistung f 1. <Comp & DV> performance; 2. <Fertig> output
Arbeitsmaschine f <Maschinen> machine
Arbeitsmechanismus m <Maschinen> working mechanism
Arbeitsmittel npl <Ergon> work equipment
Arbeitsmodus m <Comp & DV> work mode
Arbeitsnorm f <Fertig> output quota
Arbeitsnormal n <Qual> working standard
Arbeitsplan m <Phys> flowchart
Arbeitsplatte f <Comp & DV> work disk
Arbeitsplatz m 1. <Comp & DV> workplace; workstation *(vom Netzwerk abhängiger Rechner)*; 2. <Ergon> job, workplace; 3. <Maschinen> work station; 4. <Telekom> operating position
Arbeitsplatzanalyse f <Ergon> job analysis
Arbeitsplatzbeschreibung f <Ergon> job description
Arbeitsplatzgestaltung f <Ergon> design of workplace, job design, job layout, layout of workplaces, workplace design, workplace layout
Arbeitsplatzlärm m <Sicherheit> workplace noise
Arbeitsplatzrechner m <Comp & DV> work station
Arbeitsplatzüberprüfung f <Sicherheit> inspection of the workplace, survey of workplaces
Arbeitsprüfung f <Qual> operating duty test
Arbeitspuffer m <Comp & DV> scratch pad
Arbeitspunkt m <Elektrotech> operating point
Arbeitspunkt-Drift f <Regelung> point drift
Arbeitsraum m 1. <Ergon> workspace; 2. <Fertig> driving free length, driving side; 3. <Mechan> workshop
Arbeitsschiff n <Erdöl> work barge *(Offshore-Arbeiten)*
Arbeitsschuhwerk n <Sicherheit> industrial footwear, work shoes
Arbeitsschutzbrille f <Sicherheit> protective goggles
Arbeitsschutzfachmann m <Sicherheit> safety engineer, safety expert
arbeitsschutzgerechte Konstruktion f <Ergon, Sicherheit> design for safeguarding
Arbeitsschutzingenieur m <Sicherheit> safety engineer
Arbeitsschutzkleidung f <Sicherheit> industrial protective clothes, safety clothing, safety protective apparel, workers' protective clothing
Arbeitsschutzschuhe mpl <Sicherheit> safety footwear
arbeitsschutztechnische Gestaltung f <Sicherheit> safety design
Arbeitssicherheit f <Sicherheit> occupational safety, work safety
Arbeitssicherheitseinrichtung f <Sicherheit> safety device
Arbeitssicherheitsgerät n <Sicherheit> safety device
Arbeitssicherheitstechnik f <Sicherheit> safety engineering
Arbeitssitzung f <Comp & DV> work session
Arbeitsspannung f 1. <Elektriz> working voltage; 2. <Elektrotech> closed-circuit voltage, operating voltage
Arbeitsspeicher m <Comp & DV> working memory, working storage
Arbeitsspeicherbank f <Comp & DV> memory bank
Arbeitsspeicherbereich m <Comp & DV> partition
Arbeitsspiel n 1. <Kfztech> working cycle; 2. <Maschinen> cycle, working cycle; 3. <Mechan> duty cycle

Arbeitsstation

Arbeitsstation f <Verpack> work station
Arbeitsstätte f <Fertig> work site; workplace; workshop; workstation
Arbeitsstelle f <Fertig> work station *(Maschine)*
Arbeitsstiefel m <Sicherheit> work boot
Arbeitsstrom m <Elektrotech> operating current
Arbeitsstromrelais n <Elektriz> working current relay
Arbeitsstück n <Mechan> workpiece
Arbeitsstunde f <Bau> manhour
Arbeitssystem n <Ergon> work system
arbeitstägliche Abdeckung f <Abfall> daily cover *(einer Deponie)*
Arbeitstakt m 1. <Fertig> work cycle; 2. <Hydraul> stroke; 3. <Kfztech> power stroke
Arbeitstemperatur f <Heiz & Kälte> operating temperature
Arbeitstisch m 1. <Fertig> workholding table; live pass *(Walzen)*; 2. <Maschinen> workbench
Arbeitsturbine f <Lufttrans> free turbine
Arbeitsumgebung f <Ergon> work environment
Arbeitsunfall m <Sicherheit> accident at work, occupational accident
Arbeitsvermögen n 1. <Maschinen> working capacity; 2. <Nichtfoss Energ> available power
Arbeitsverwaltung f <Verpack> work handling *(verschiedener Aufgaben)*
Arbeitsvorgang m <Maschinen> operation
Arbeitsvorschub m <Fertig> working feed
Arbeitswechsel m <Ergon> job rotation
Arbeitsweise f 1. <Comp & DV> mode; 2. <Elektrotech> operation; 3. <Ergon> mode of operation, operation, procedure
Arbeitswissenschaft f <Ergon, Sicherheit> ergonomics, human engineering, work science
Arbeitszyklus m 1. <Maschinen> duty cycle, working cycle; 2. <Trans> operating cycle
Arbeitszylinder m <Kfztech> power cylinder
arbiträre Achse f <Geom> arbitrary axis
Arcatomschweißen n <Fertig> hydrogen arc-welding
archäologische Metallurgie f <Metall> archeological metallurgy
Archimedes'sche Schraube f <Hydraul> Archimedean screw *(Fördertechnik)*
Archimedes'scher Körper m <Geom> Archimedean solid
Archimedes'sches Prinzip n <Phys> Archimedes' principle
Archipel m <Wassertrans> archipelago *(Geographie)*
Architektur f <Comp & DV> architecture *(Hardware)*
Architektur offener Systeme <Telekom> OSA, open systems architecture
Archiv n <Comp & DV> archive
Archivbild n <Foto> file picture
Archivdatei f <Comp & DV> archived file
archivieren v <Comp & DV> archive
Archivieren n <Comp & DV> archiving
Archivierungssystem n <Comp & DV> archivist
Archivton m <Aufnahme> stock sound
Arecain n <Chemie> arecaine
Argentan n <Metall> argentan
Argentit n <Chemie> argentite
Arginase f <Chemie> arginase
Arginin n <Chemie> arginine
Argon n *(Ar)* <Chemie> argon *(Ar)*
Argongaslaser m <Strahlphys> argon gas laser
Argon-Ionen-Laser m <Druck> argon-ion laser
Argonlaser m <Elektronik> argon laser
Argon-Sauerstoff-Entkohlen n *(AOD)* <Metall> argon-oxygen decarburization *(AOD)*

Argonschutzgas n <Kerntech> argon gas blanket
ARGOS *(automatische Satellitenerfassung von geomagnetischen Daten)* <Wassertrans> ARGOS *(Automatic Remote Geomagnetic Observatory System)*
Argument n <Comp & DV, Math> argument *(unabhängige Variable einer Funktion oder Winkel-Koordinate einer komplexen Zahl)*
Argyrit m <Chemie> argentite
Arithmetik f <Comp & DV, Math> arithmetic
arithmetisch adj <Comp & DV, Math> arithmetic
arithmetische Folge f <Fertig> arithmetic progression
arithmetische Operationen fpl <Math> arithmetic operations
arithmetische Reihe f <Math> arithmetic progression, arithmetic series
arithmetische Stellenverschiebung f <Comp & DV> arithmetic shift
arithmetischer Mittelrauwert m <Maschinen> CLA height, arithmetic average height, *(AE)* center line average height, *(BE)* centre line average height
arithmetischer Mittelwert m <Math, Qual> arithmetic mean
arithmetischer Operator m <Comp & DV> arithmetic operator *(Programmiersprache)*
arithmetisches Mittel n <Comp & DV, Math, Qual> arithmetic mean
Arkade f <Bau> arcade
Arkansas-Abziehstein m <Bau> Arkansas oilstone
Arkansas-Polierstein m <Bau> Arkansas oilstone
Arkose f <Ker & Glas> arkose
ARL *(akzeptabler Zuverlässigkeitspegel)* <Qual> ARL *(acceptable reliability level)*
Arm m 1. <Elektrotech> bracket *(von elektrischer Leuchte)*; 2. <Fertig> arm; bracket *(Umlaufgetriebe)*; 3. <Maschinen> bracket; 4. <Papier> arm
Armatur f 1. <Erdöl> valve *(Absperrung, Regelorgan)*; 2. <Fertig> valve *(Kunststoffinstallationen)*
Armaturen fpl <Maschinen> fittings
Armaturenbrett n <Kfztech> dash, dashboard; instrument panel *(Zubehör)*
Armauflage f 1. <Bau> elbow rail; 2. <Ergon> arm pad, arm support, pad
Armausladung f <Fertig> throat *(Punktschweißmaschine)*
armiert adj <Elektriz, Telekom> *(AE)* armored, *(BE)* armoured
armierter Kunststoff m <Verpack> reinforced plastic
armierter Schlauch m <Bau> *(AE)* armored hose, *(BE)* armoured hose
armiertes Kabel n <Elektriz, Elektrotech, Maschinen, Telekom> *(AE)* armored cable, *(BE)* armoured cable
Armierung f 1. <Bau> reinforcement; 2. <Elektriz> *(AE)* armor, *(BE)* armour, reinforcement, sheath; 3. <Elektrotech> *(AE)* armor, *(BE)* armour *(Kabel)*
Armierungsschelle f <Elektriz> *(AE)* armor clamp, *(BE)* armour clamp
Armkreuzring m <Papier> backing wire
Armlehne f 1. <Ker & Glas> chair arm; 2. <Kfztech> armrest *(Fahrzeuginnenausstattung)*
Armleuchter m <Elektriz> chandelier
Armozement m <Bau> ferrocement
Aroma n <Lebensmittel> *(AE)* flavor, *(BE)* flavour
Aromaten npl <Chemie> aromatic compounds, aromatics
aromatische Verbindung f <Lebensmittel> aromatic compound
aromatischer Kohlenwasserstoff m <Chemie, Kunststoff> aromatic hydrocarbon
ARPA *(automatische Radaraufnahmehilfe)* <Wassertrans> ARPA *(automatic radar plotting aid)*

ARQ *(automatische Wiederholanforderung)* <Funktech, Telekom> ARQ *(automatic repeat request)*
Arrest *m* <Wassertrans> arrest
Arretier... 1. <Fertig> lock; 2. <Foto> locating
Arretierfeder *f* <Fertig> check
Arretierhebel *m* <Fertig> catch, lock lever
Arretierschraube *f* <Bau> stop screw
Arretierstift *m* 1. <Fertig> location pin; 2. <Foto> locating pin
Arretierung *f* 1. <Comp & DV> latch; 2. <Fertig> detent; 3. <Maschinen> detent, lock, stop; 4. <Sicherheit> safety locking device
ARRL *(Amerikanischer Amateurdachverband)* <Funktech> ARRL *(American Radio Relay League)*
Arrowrootstärke *f* <Lebensmittel> arrowroot
Arsen *n (As)* <Chemie> arsenic *(As)*
Arsenat *n* <Chemie> arsenate, arsenite
Arsenid *n* <Chemie> arsenide
Arsenigsäureanhydrid *n* <Chemie> arsenic oxide, arsenic trioxide
Arsen-Implantation *f* <Elektronik> arsenic implantation
Arsenosulfid *n* <Chemie> *(AE)* sulfarsenide, *(BE)* sulpharsenide
Arsen-Oxid *n* <Chemie> arsenic oxide, arsenic trioxide
Arsentrioxid *n* <Chemie> arsenic oxide, arsenic trioxide
Arsin *n* <Chemie> arsane
Arsphenamin *n* <Chemie> arsphenamine, salvarsan
ARSR *(Flugüberwachungsradar)* <Funkort, Raumfahrt> ARSR *(air route surveillance radar)*
Art *f* <Comp & DV> type
Art. Nr. *f (Artikelnummer)* <Verpack> part number
Artefakt *n* <Ergon> artifact
artesisch *adj* 1. <Erdöl> artesian *(Brunnen)*; 2. <Wasserversorg> artesian
artesische Quelle *f* <Wasserversorg> artesian spring
artesischer Brunnen *m* <Wasserversorg> artesian well
artesisches Wasser *n* <Kohlen> artesian water
Artikel *m* <Maschinen> item
Artikelnummer *f* 1. <Maschinen> item number; 2. <Verpack> part number *(Art. Nr.)*
ARU 1. <Comp & DV> *(Sprachausgabe-Einheit)* ARU *(audio response unit)*; 2. <Funktech> *(automatische Rauschunterdrückung)* ANL *(automatic noise limiter)*
Aryl... <Chemie> aryl
Arylamin *n* <Chemie> aminoarene, arylamine
As *(Arsen)* <Chemie> As *(arsenic)*
asb *(Apostilb)* <Optik> asb *(apostilb)*
Asbest *m* 1. <Fertig> asbestos; asbestos *(Kunststoffinstallationen)*; 2. <Ker & Glas, Kunststoff, Mechan, Papier, Sicherheit, Textil> asbestos
Asbestbelastungsgrenzwert *m* <Sicherheit> asbestos exposure limit
Asbestdichtung *f* <Fertig> asbestos gasket
Asbesteinlage *f* <Bau> asbestos-plaited packing
Asbestfaser *f* 1. <Bau> asbestos thread; 2. <Kerntech> asbestos wool
asbestfreie Schutzkleidung *f* <Sicherheit> asbestos-free protective clothing
Asbestgarn *n* <Textil> asbestos yarn
Asbestin *n* <Papier> asbestine
Asbestpappe *f* <Papier> asbestos board
Asbestplatte *f* <Fertig, Papier> asbestos sheet
Asbestzement *m* <Bau> asbestos cement
Asbestzementplatten *fpl* <Bau> asbestos cement sheeting
Asche *f* 1. <Kohlen, Kunststoff, Papier> ash; 2. <Thermod> ashes; 3. <Umweltschmutz> ash
aschearmes Filterpapier *n* <Labor> ashless filter paper
Aschebrei *n* <Maschinen> slurry

aschefrei *adj* 1. <Papier> ashless; 2. <Umweltschmutz> ash-free
aschefreies Filterpapier *n* <Lebensmittel> ashless filter paper
Aschegehalt *m* <Kohlen, Lebensmittel, Papier> ash content
Aschenkasten *m* <Heiz & Kälte> ash box
Ascherückstand *m* <Maschinen> ash residue
Asche- und Verbrennungsrückstand *m* <Umweltschmutz> ash and combustion residue
ASCII *(Amerikanische Datenübertragungs-Codenorm)* <Comp & DV, Druck, Telekom> ASCII *(American Standard Code for Information Interchange)*
Ascorbin... <Chemie> ascorbic
ASD *(Rutschsicherung)* <Kfztech> ASD *(antiskid device)*
ase *(Flugnormwirkungsgrad)* <Raumfahrt> ase *(air standard efficiency)*
ASE 1. <Künstl Int> *(automatische Spracherkennung)* ASR *(automatic speech recognition)*; 2. <Trans> *(Selbststabilisierungsgerät)* ASE *(automatic stabilizing equipment)*
aseptisch *adj* <Lebensmittel, Verpack> aseptic
aseptische Abfüllung *f* <Lebensmittel> aseptic filling
ASG *(Flugnormengruppe)* <Raumfahrt> ASG *(aeronautical standards group)*
ASI 1. <Lufttrans> *(Eigengeschwindigkeitsanzeiger, Geschwindigkeitsmesser)* ASI *(airspeed indicator)*; 2. <Raumfahrt> *(Geschwindigkeitsanzeiger)* ASI *(airspeed indicator)*
ASK *(Amplitudenumtastung)* <Elektronik, Telekom> ASK *(amplitude-shift keying)*
Askarel *n* <Elektrotech> Askarel *(Isolieröl)*
Äskulin *n* <Chemie> esculin
ASL *(atomare Sicherheitslinie)* <Kerntech> ASL *(atomic safety line)*
ASLT *(fortschrittliche Festkörperlogik)* <Elektronik> ASLT *(advanced solid logic technology)*
ASME *(Amerikanische Gesellschaft der Maschinenbau-Ingenieure)* <Qual> ASME *(American Society of Mechanical Engineers)*
ASME-Code *m* <Mechan> ASME code
Asparagin... <Chemie> aspartic
Asparaginsäure *f* <Lebensmittel> aspartic acid
Aspartam *n* <Lebensmittel> aspartame *(Süßstoff)*
Aspektverhältnis *n* <Phys> aspect ratio
Asphalt *m* <Bau> asphalt, bitumen
Asphaltbeton *m* <Bau> asphalt concrete
Asphaltbinderschicht *f* <Bau> binder course
Asphaltdecke *f* <Bau> asphalt surfacing
Asphaltemulsion *f* <Maschinen> bitumen emulsion
asphaltieren *v* <Bau> asphalt, bituminize
Asphaltieren *n* <Bau> asphalting, bituminization
asphaltiert *adj* <Bau> bituminized
Asphaltkaltgemisch *n* <Bau> cold mix
Asphaltkocher *m* <Bau> asphalt boiler
Asphalttragschicht *f* <Bau> bituminous base course
Asphaltwerk *n* <Bau> asphalt plant *(Tiefbau)*
asphärische Korrekturplatte *f* <Fernseh> aspheric corrector plate
Aspiration *f* <Hydraul, Papier> aspiration
Aspirations-Porosimeter *n* <Papier> aspiration porosimeter
Aspirationsporositätsprüfer *m* <Papier> aspiration porosity tester
Aspirations-Psychrometer *n* 1. <Gerät> Aßmann psychrometer; 2. <Gerät, Papier> aspiration psychrometer *(Feuchtigkeitsmesser)*
Aspirationspumpe *f* <Hydraul> aspiration pump

Aspirator

Aspirator m <Labor> aspirator
ASR 1. <Comp & DV> *(automatischer Sender-Empfänger)* ASR *(automatic send-receive)*; 2. <Comp & DV> *(automatisches Senden und Empfangen)* ASR *(automatic send and receive)*; 2. <Lufttrans> *(Flughafen-Überwachungsradar)* ASR *(airport surveillance radar)*
Assembler m <Comp & DV> assembler *(Programmiersprache)*
Assemblerinstruktion f <Comp & DV> assembler directive, assembler instruction
Assemblerprogramm n <Comp & DV> assembler
Assemblersprache f <Comp & DV> assembly language
Assemblierer m <Comp & DV> assembler
Assistent m <Comp & DV> wizard
Aßmann'sches Psychrometer n 1. <Gerät> Aßmann psychrometer *(Aspirations-Psychrometer)*; 2. <Gerät> aspiration psychrometer
Assoziationsliste f <Künstl Int> association list
assoziativ adj <Künstl Int> associative *(Suche)*
Assoziativ... <Comp & DV, Math> associative
Assoziativadressierung f <Comp & DV> associative addressing
Assoziativdatei f <Comp & DV> content-addressable file
assoziative Zentraleinheit f <Telekom> associative processor
assoziatives Netz n <Künstl Int> associative network, semantic net, semantic network
Assoziativgesetz n <Math> associative law
Assoziativspeicher m *(CAM)* 1. <Comp & DV> associative memory, *(AE)* associative storage, *(BE)* associative store, content-addressable memory, *(AE)* content-addressable storage, *(BE)* content-addressable store; 2. <Künstl Int> associative memory, content-addressable memory *(AM)*
assoziierte Zeichengabe f <Telekom> *(AE)* associated signaling, *(BE)* associated signalling
Ast m 1. <Bau> knot *(Holz)*; 2. <Geom> branch
astabile Kippschaltung f <Elektronik> astable multivibrator
astabile Schaltung f <Elektronik> astable circuit
astatisch adj <Elektriz, Phys> astatic
astatischer Kreisel m <Raumfahrt> free gyroscope
astatisches Amperemeter n <Elektriz> astatic ammeter
astatisches Galvanometer n <Elektrotech, Phys> astatic galvanometer
astatisches Spannungsmessgerät n <Elektriz> astatic voltmeter
Asteroid m <Raumfahrt> asteroid
astfrei adj <Bau> knotless
Asthma-Papier n <Papier> asthma paper
astigmatisch adj <Fertig, Optik, Phys> astigmatic
Astigmatismus m <Fernseh, Optik, Phys> astigmatism
ASTM *(Amerikanische Gesellschaft für Werkstoffprüfung)* <Maschinen, Qual> ASTM *(American Society for Testing Materials)*
Astro... <Raumfahrt> astro...
Astrodynamik f <Raumfahrt> astrodynamics
Astrofixierung f <Raumfahrt> astro fix
Astroführung f <Raumfahrt> stellar guidance
Astrokompass m <Raumfahrt> astrocompass
Astrolenkung f <Raumfahrt> celestial guidance
Astromechanik f <Raumfahrt> celestial mechanics
Astrometrie f <Raumfahrt> astrometry
Astronavigation f <Wassertrans> astronavigation, astronomical navigation
astronomische Einheit f *(AE)* <Metrol> astronomical unit *(AU)*
astronomische Kamera f <Foto> astronomical camera
astronomische Navigation f 1. <Lufttrans> celestial navigation; 2. <Wassertrans> astronavigation, astronomical navigation, celestial navigation
astronomisches Besteck n <Wassertrans> astronomical position
astronomisches Fernrohr n <Phys> astronomical telescope
Astrophysik f <Raumfahrt> astrophysics
Asymmetrie f <Math> asymmetry
Asymmetriefehler m <Phys> coma
asymmetrisch adj 1. <Comp & DV> asymmetric; 2. <Geom, Math> asymmetrical, asymmetric
asymmetrische Ablenkung f <Fernseh> asymmetric deflection
asymmetrische digitale Teilnehmer-Anschlussleitung f <Telekom> asymmetric digital subscriber line
asymmetrische digitale Teilnehmerleitung f <Telekom> asymmetric digital subscriber line *(ADSL)*
asymmetrische Doppelleiterspeisung f <Funktech> asymmetric twin feed
asymmetrische Schaltung f <Elektriz> asymmetric circuit
asymmetrischer Fehlerbereich m <Elektrotech> asymmetric error range
asymmetrisches Gewinde n <Maschinen> asymmetric thread
asymmetrisches Trapezgewinde n <Maschinen> asymmetric trapezoidal screwthread
Asymptote f <Math> asymptote
asymptotisch adj <Geom> asymptotic
asymptotische Näherung f <Telekom> asymptomatic approximation
asymptotisches Verhalten n <Gerät> *(AE)* asymptotic behavior, *(BE)* asymptotic behaviour
asynchron adj <Comp & DV, Eisenbahn, Elektriz, Fernseh, Funktech, Phys> asynchronous *(Satelliten, Umlaufbahnen)*
Asynchron... <Comp & DV, Elektriz, Telekom> asynchronous
Asynchronbetrieb m <Telekom> asynchronous mode, asynchronous operation
asynchrone Schaltung f <Comp & DV> asynchronous circuit
asynchrone serielle Schnittstelle f <Comp & DV> asynchronous serial interface, start-stop interface
asynchrone Übertragung f 1. <Comp & DV> asynchronous communication, asynchronous transmission, asynchronous mode; 2. <Telekom> asynchronous transmission
asynchrone Verbindung f <Elektriz> asynchronous link *(Leitungsnetze)*
asynchroner Betrieb m <Elektriz> asynchronous running; asynchronous operation *(Energienetz)*
asynchroner Betriebsmodus m <Comp & DV> asynchronous capability
asynchroner Lauf m <Elektrotech> asynchronous operation
asynchroner Linearmotor m <Elektrotech> asynchronous linear motor
asynchroner Transfermodus m *(ATM)* <Telekom> asynchronous transfer mode, ATM
asynchroner Übertragungsmodus m *(ATM)* <Telekom> asynchronous transfer mode *(ATM)*
asynchrones Netz n <Telekom> asynchronous network
asynchrones Zeitmultiplexverfahren n <Telekom> ATDM, asynchronous time-division multiplexing
Asynchrongenerator m <Elektriz, Elektrotech> asynchronous alternator, asynchronous generator

asynchronischer Übertragungsschnittstellenanpasser m *(ACIA)* <Kontroll> asynchronous communications interface adaptor *(ACIA)*
Asynchronlinearmotor m <Trans> asynchronous linear induction motor
Asynchronmaschine f <Elektrotech> asynchronous machine
Asynchronmodem n <Elektronik> asynchronous modem
Asynchronmotor m 1. <Elektriz> asynchronous motor, induction motor; 2. <Elektrotech> asynchronous motor; 3. <Fertig> induction motor; 4. <Trans> asynchronous motor
Asynchronmotor m **mit Kompensationswicklung** <Elektriz> compensated induction motor
Asynchron-Übertragungsschnittstellenanpasser m *(ACIA)* <Kontroll> asynchronous communications interface adaptor *(ACIA)*
Aszension f <Fertig> elevating *(Kapillar)*
aT *(Thermodiffusionskonstante)* <Phys> aT *(thermal diffusion constant)*
AT *(fortschrittliche Technologie)* <Comp & DV> AT *(advanced technology)*
ataktisches Polymer n <Kunststoff> atactic polymer
ATC 1. <Eisenbahn> *(automatische Zugsteuerung)* ATC, ATO *(automatic train operation)*; 2. <Funktech> *(kapazitive Antennenanpassung)* ATC *(aerial-tuning capacitor)*; 3. <Metall> *(automatischer Werkzeugwechsler)* ATC *(automatic tool changer)*
ATE *(automatische Prüfeinrichtung)* <Comp & DV> ATE *(automatic test equipment)*
Atebrin n <Chemie> atebrin, quinacrine
Atelier n <Foto> studio
Atelierarbeit f <Foto> studio work
Atelierkamera f <Foto> studio camera
Atembeutel m <Sicherheit> respirator storage bag *(eines Atemschutzgerätes)*
Atemgerät n <Sicherheit> breathing apparatus
Atemgerät n/**von der Außenluft unabhängiges** <Sicherheit> powered respirator
Atemgrenzwert m <Umweltschutz> breathing capacity
Atemschutz m <Sicherheit> respiratory protection
Atemschutzausrüstung f <Sicherheit> respiratory protective equipment
Atemschutzbehälter m <Sicherheit> respirator canister
Atemschutzfilter n <Sicherheit> respirator filter, respiratory filter
Atemschutzgerät n <Sicherheit> respiratory protection apparatus, protective respirator, respiratory protective device, respiratory protective equipment
Atemschutzgerät n **mit Luftzuführung** <Sicherheit> air-supplied respirator
Atemschutzgerät n/**von der Umgebungsluft unabhängiges** <Sicherheit> self-contained air supply respirator
Atemschutzgesichtsmaske f **mit Vollschichtscheibe** <Sicherheit> full-facepiece respirator
Atemschutzleichtmaske f <Sicherheit> respirator facepiece
Atemschutzmaske f <Sicherheit> full-mask respirator, respirator facepiece
Atemschutzsystem n <Umweltschutz> breathing protection system
Atemventil n <Maschinen> breather
ATF *(Automatikgetriebeöl)* <Kfztech> ATF *(automatic transmission fluid)*
ätherisch adj <Chemie> volatile
ätherisches Öl n <Lebensmittel> essential oil
athermisch adj <Chemie> athermal
Athodyd n <Lufttrans> athodyd *(Staustrahltriebwerk)*
ATI *(Antennenabstimmspule)* <Funktech> ATI *(aerial-tuning inductance)*

ATM 1. <Optik> *(Azimutal-Transversal-Mode)* ATM, azimuthal transversal mode *(optische Fasern)*; 2. <Telekom> *(asynchroner Transfermodus)* ATM, asynchronous transfer mode
Atmolyse f <Chemie> atmolysis
Atmosphäre f <Phys> atmosphere
atmosphärisch adj <Phys> atmospheric
atmosphärisch bedingte Spektrallinie f <Phys> atmospheric line
atmosphärische Absorption f <Strahlphys> atmospheric absorption
atmosphärische Entladung f <Elektrotech> lightning discharge
atmosphärische Erscheinung f <Umweltschmutz> atmospheric phenomenon
atmosphärische Korrosion f <Metall> atmospheric corrosion
atmosphärische Leitschicht f <Funktech> atmospheric duct *(Wellenausbreitung)*
atmosphärische Luftbelastung f <Umweltschmutz> atmospheric loading
atmosphärische Masse f <Nichtfoss Energ> air mass
atmosphärische Säurekapazität f <Umweltschmutz> atmospheric acidity
atmosphärische Störung f 1. <Aufnahme, Elektronik> atmospheric noise; 2. <Elektrotech> atmospherics; 3. <Funktech> atmospheric noise; 4. <Wassertrans> atmospheric disturbance
atmosphärische Überspannung f <Elektriz> atmospheric overvoltage, overvoltage of atmospheric origin
atmosphärische Verdunklung f <Umweltschmutz> atmospheric obscurity
atmosphärische Welle f <Phys> sky wave
atmosphärischer Auswaschvorgang m <Umweltschmutz> atmospheric scrubbing
atmosphärischer Druck m <Wassertrans> atmospheric pressure
atmosphärischer Dukt m <Funktech> atmospheric duct *(Wellenausbreitung)*
atmosphärischer Niederschlag m <Umweltschmutz> atmospheric fallout, atmospheric precipitation, precipitation
atmosphärischer Schwefel m <Umweltschmutz> *(AE)* atmospheric sulfur, *(BE)* atmospheric sulphur
atmosphärisches Rauschen n 1. <Aufnahme> atmospheric noise; 2. <Elektrotech> atmospheric noise, atmospherics; 3. <Funktech> atmospheric noise
ATMOS-Spur f <Aufnahme> atmos track
atmungsaktiv adj <Kunststoff> breathable
Atmungsaktivität f <Kunststoff> breathability
Atmungsapparat m <Kohlen> breathing apparatus
Atmungsferment n <Chemie> oxygenase *(Biochemie)*
Atoll n <Wassertrans> atoll *(Geographie)*
Atom n <Comp & DV, Phys, Teilphys> atom
Atomabfallbeseitigung f **durch Kernumwandlung** <Kerntech> waste disposal by nuclear transmutation
Atomabsorptions-Spektrometer n <Labor> atomic absorption spectrometer
Atomabsorptions-Spektroskopie f <Kerntech, Phys, Strahlphys> atomic absorption spectroscopy
Atomanlage f <Elektrotech, Kerntech> nuclear power plant
Atomantrieb m <Wassertrans> marine nuclear plant • **mit Atomantrieb** <Kerntech, Wassertrans> nuclear-powered *(Schiff, U-Boot)*
atomar adj <Phys, Teilphys> atomic
atomare Absorptionsanalyse f <Kerntech, Phys, Strahlphys> atomic absorption analysis
atomare Elektronenumlaufbahn f <Kerntech, Phys, Strahlphys> atomic orbital

atomare

atomare Fluoreszenzanalyse f <Kerntech, Phys, Strahlphys> atomic fluorescence analysis *(Spektrometrie)*
atomare Masseneinheit f *(amu)* <Kerntech> atomic mass unit *(amu)*
atomare Polarisation f <Kerntech, Phys, Strahlphys> atomic polarization
atomare Sicherheitslinie f *(ASL)* <Kerntech> atomic safety line *(ASL)*
atomare Strahlung f <Kerntech, Phys, Strahlphys> atomic radiation
atomare Streuung f <Kerntech, Phys, Strahlphys> atomic scattering
atomare Struktur f <Kerntech> atomistic structure
atomare Wärmekapazität f <Kerntech, Phys, Strahlphys> atomic heat capacity
atomarer Querschnitt m <Kerntech, Phys, Strahlphys> atomic cross-section
atomarer Urandampf m <Kerntech> *(AE)* atomic uranium vapor, *(BE)* atomic uranium vapour
atomarer Wasserstoffmaser m <Elektronik> atomic hydrogen maser
atomarer Zwischenraum m <Kerntech, Phys, Strahlphys> atomic interspace
atomares Absorptions-Spektrophotometer n <Kerntech, Phys, Strahlphys> atomic absorption spectrophotometer
atomares Energieniveau n <Kerntech, Phys, Strahlphys> atomic energy level
Atomforschung f <Kerntech> atomic research
Atomfrequenznormal n <Telekom> atomic frequency standard
Atom-Gas-Laser m <Elektronik> atomic gas laser
atomgetrieben adj <Kerntech, Wassertrans> nuclear-powered *(Schiff, U-Boot)*
Atomgewicht n 1. <Kerntech, Maschinen> atomic weight; 2. <Phys, Strahlphys> atomic mass, atomic weight
Atomgewichtseinheit f <Kerntech> atomic weight unit
Atomhülle f <Kerntech, Phys, Strahlphys> atomic electron shell
Atomisator m <Chemtech> atomizer
Atomistik f <Phys> atomicity
Atomizität f <Comp & DV> atomicity
Atomkern m 1. <Kerntech, Phys, Strahlphys> atomic core, atomic nucleus; 2. <Teilphys> nucleus
Atomkraftwerk n <Elektrotech> nuclear power station
Atommasse f *(Ma)* <Kerntech> atomic mass *(Ma)*
Atommasseneinheit f *(AME)* <Kerntech> atomic weight unit *(AWU)*
Atommassenkonstante f *(mu)* <Kerntech> unified atomic mass constant *(mu)*
Atommeiler m <Kerntech> pile
Atommüll m <Abfall, Kerntech, Strahlphys> nuclear waste, radioactive waste
Atommüll m **mit langer Halbwertzeit** <Abfall, Kerntech> long-life radioactive waste
Atomphysik f <Phys> atomic physics
Atomradius m <Kerntech, Phys, Strahlphys> atomic radius
Atomradiustheorie f <Kerntech> theory of effective radius
Atomrakete f <Kerntech, Raumfahrt> atomic rocket
Atomspektroskopie f <Kerntech, Phys, Strahlphys> atomic spectroscopy
Atomspektrum n <Kerntech, Phys, Strahlphys> atomic spectrum
Atomstrahl m <Kerntech, Phys, Strahlphys> atomic beam
Atomstruktur f 1. <Kerntech> atomic structure; 2. <Phys> atomic structure, structure of the atom; 3. <Strahlphys> atomic structure
Atomtest m <Kerntech, Werkprüf> nuclear test

Atom-U-Boot n <Wassertrans> nuclear-powered submarine, nuclear submarine *(Marine)*
Atomuhr f <Kerntech, Phys, Strahlphys, Telekom> atomic clock
Atomumordnen n <Metall> atomic shuffling
Atomverschiebung f <Metall> atomic displacement
Atomversuch m <Kerntech, Werkprüf> nuclear test
Atomvolumen n <Kerntech, Phys, Strahlphys> atomic volume
Atomwärme f 1. <Kerntech, Phys, Strahlphys> atomic heat capacity; 2. <Thermod> atomic heat
Atomzahl f *(Z)* <Kerntech> atomic number *(Z)*
ATP *(Adenosintriphosphat)* <Lebensmittel> ATP *(adenosine triphosphate)*
Atropa... <Chemie> atropic
AT-Schweißen n *(aluminothermisches Schweißen)* <Fertig> AT welding *(aluminothermic welding)*
Attapulgit n <Erdöl> attapulgite *(Adsorptionsmittel)*
Atterberg'sche Konsistenzgrenzen fpl <Bau> Atterberg limits *(Tiefbau)*
Attrappe f <Wassertrans> mock-up *(Schiffkonstruktion)*
Attribut n 1. <Comp & DV> attribute; 2. <Künstl Int> attribute *(eines Objekts)*; 3. <Patent, Qual> attribute; 4. <Telekom> attribute *(Dokument)*
Attributprüfung f 1. <Gerät> acceptance inspection; 2. <Qual> inspection by attributes
ATU *(Antennenanpassung)* <Funktech> ATU *(aerial-tuning unit)*
ATV *(Amateurfernsehen)* <Fernseh> ATV *(amateur television)*
Atwater-Faktoren mpl <Lebensmittel> Atwater factors
Atwater-Tabelle f <Lebensmittel> Atwater table
Atwood'sche Fallmaschine f <Phys> Atwood's machine
Ätz... 1. <Chemie> caustic; 2. <Fertig> etching
Ätzbad n <Fertig> etching bath
Ätzdruck m <Textil> discharge printing
ätzen v 1. <Bau> bite, etch; 2. <Chemie> attack, etch; 3. <Metall> etch *(Platinen)*
Ätzen n 1. <Bau> pickling; 2. <Chemie> attack, etching; 3. <Druck> etching
ätzend adj 1. <Chemie> caustic; 2. <Fertig> etching
Ätzfigur f <Fertig> etch figure
Ätzflüssigkeit f <Metall> etching solution
Ätzgrube f <Fertig> etching pit
Ätzkali n <Chemie> potassium hydroxide
Ätzkalk m <Chemie, Meerschmutz> quicklime
Ätzkraft f <Chemie> causticity, corrosiveness
Ätzmaschine f <Druck> etching machine
Ätzmittel n <Fertig> etchant
Ätznatron n <Chemie> caustic soda, sodium hydrate, sodium hydroxide
ätzpolieren v <Metall> attack-polish
Ätzung f 1. <Druck, Elektronik> etching; 2. <Fertig> bite
Ätzwasser n <Chemie> caustic water, nitric acid
Audio... <Akustik, Aufnahme> audio
Audio-CD f <Optik> *(BE)* audio compact disc, *(AE)* audio compact disk
Audio-CD-Spieler m <Optik> *(BE)* audio compact disc player, *(AE)* audio compact disk player
Audio-Eingang m <Aufnahme> audio input
Audiofrequenzbereich m <Strahlphys> audio range
Audiogramm n <Akustik, Ergon> audiogram
Audiogramm-Maskierung f <Akustik> audiogram masking
Audiokonferenz f <Telekom> telephone conference
Audiometer n <Akustik, Ergon, Gerät> audiometer
Audiometrie f <Akustik, Aufnahme, Ergon> audiometry
audiometrische Kabine f <Aufnahme> audiometric booth

Audiosignal *n* <Elektronik> audio signal
Audio-Videotext *m* <Telekom> audio videotex
Auditbeweis *m* <Qual> audit evidence
Auditfeststellung *f* <Qual> audit finding
Aufarbeitung *f* 1. <Kerntech> reprocessing; 2. <Maschinen> regeneration
Aufarbeitungsvorrichtung *f* <Umweltschmutz> recovery device
Aufarbeitungszentrum *n* **für wiederverwertbare feste Abfallmaterialien** <Umweltschmutz> *(AE)* processing center for recyclable solid waste materials, *(BE)* processing centre for recyclable solid waste materials
Aufbau *m* 1. <Bau> assembly, erection; 2. <Comp & DV> architecture, layout, setup; 3. <Erdöl> rigging up *(Bohrtechnik)*; 4. <Fertig> assembly *(Kunststoffinstallationen)*; 5. <Kontroll> structure; 6. <Mechan> assembly; 7. <Papier> erection; 8. <Wassertrans> superstructure
Aufbau *m* **einer Nachricht** <Comp & DV> message structure
aufbauen *v* 1. <Bau> build up, set up; 2. <Comp & DV> set-up; 3. <Kontroll> configure; 4. <Maschinen, Mechan> erect; 5. <Telekom> set up
Aufbauenergie *f* <Metall> formation energy
Aufbaufaktor *m* <Kerntech> build-up factor
Aufbaufeld *n* <Konstzeich> add-on block
Aufbäumung *f* <Textil> beaming
Aufbauorganisation *f* <Qual> organizational structure
Aufbauschneide *f* <Maschinen> built-up edge
aufbereiten *v* 1. <Abfall> treat; 2. <Bau> prepare; 3. <Chemtech> concentrate *(Erz)*; 4. <Fertig> dress *(Erz)*; 5. <Kohlen> condition, table
Aufbereiten *n* <Kohlen> milling
Aufbereitung *f* 1. <Chemtech> breaking *(Hadern)*; 2. <Erdöl> treatment *(Erdgas)*; 3. <Fertig> dressing *(Erz)*; 4. <Ker & Glas> beneficiation; 5. <Kohlen> conditioning, processing; 6. <Telekom> synthesis *(Frequenz)*; 7. <Wasserversorg> make-up
Aufbereitung *f* **des Eingangssignals** <Elektronik> input signal conditioning
Aufbereitungsanlage *f* 1. <Abfall> preparation plant, processing plant, reprocessing plant; 2. <Kohlen> cleaning plant; preparation plant; ore dressing plant
Aufbereitungsbehälter *m* <Kohlen> conditioning tank
Aufbereitungsprodukt *n* <Lebensmittel> concentrate
Aufbereitungsverfahren *n* <Abfall> conditioning process, preparation process, treatment process
aufbewahren *v* <Bau> keep, store
Aufbewahrung *f* 1. <Eisenbahn, Kerntech> storage; 2. <Qual> filing, retention
Aufbewahrungsfrist *f* 1. <Patent> filing period; 2. <Qual> filing period, retention period
Aufbewahrungskanister *m* <Kerntech> storage canister
aufbiegen *v* <Bau> bend up *(Bewehrungsstahl)*
Aufbiegung *f* <Bau> bend
Aufblasanschluss *m* <Meerschmutz> inflation cuff *(einer Ölsperre)*
aufblasbar *adj* <Meerschmutz> inflatable
aufblasbare Dichtung *f* <Kerntech> inflatable seal
aufblasbare Pontons *mpl* <Lufttrans, Wassertrans> inflatable pontoons
aufblasbare Seenotrutsche *f* <Lufttrans> inflatable slide
aufblasbares Boot *n* <Wassertrans> inflatable dinghy
aufblasen *v* <Phys> inflate
Aufblasen *n* <Meerschmutz> inflation
Aufblättern *n* <Phys> exfoliation
Aufblitzen *n* 1. <Elektrotech, Foto> flashing; 2. <Metall> fulguration
aufbocken *v* <Kfztech> jack *(Verwendung von Werkzeug, Karosserie)*

Aufbocken *n* <Maschinen> jacking
aufbohren *v* 1. <Kfztech> rebore *(Motor, Zylinder)*; 2. <Maschinen> counterbore, rebore
Aufbohren *n* <Maschinen> boring, counterboring, reboring
Aufbohren *n* **tiefer Bohrungen** <Fertig> deep-hole boring
Aufbohrer *m* **für vorgegossene Löcher** <Fertig> core drill *(Spanung)*
Aufbohrmeißel *m* <Erdöl> borer bit *(Bohrtechnik)*
Aufbohrung *f* <Maschinen> counterbore
Aufbrauch *m* <Metall> exhaustion
aufbrechen *v* 1. <Bau> break; break open *(eine Tür)*; 2. <Sicherheit> break open
Aufbrechen *n* <Mechan> breaking up
aufbringen *v* 1. <Anstrich> apply; 2. <Wassertrans> fit *(Schiffinstandhaltung)*
Aufbringen *n* 1. <Elektronik> deposition *(Leiterplatte)*; 2. <Wassertrans> seizure *(Schiff)*
Aufbringen *n* **einer neuen Schotterschicht** <Bau> *(AE)* remetaling, *(BE)* remetalling *(Straßenbau)*
Aufbringen *n* **von Schotter** <Bau, Trans> macadamization
Aufbringung *f* <Wassertrans> seizure *(Schiff)*
Aufbuchtung *f* <Wassertrans> hog
aufdämmen *v* <Bau> bank up
aufdampfen *v* <Fertig> *(AE)* vapor-deposit, *(BE)* vapour-deposit
Aufdampfen *n* 1. <Fertig> *(AE)* vapor depositing, *(BE)* vapour depositing; 2. <Thermod> *(AE)* vapor deposition, *(BE)* vapour deposition
Aufdampfverfahren *n* <Chemtech> *(AE)* vapor deposition technique, *(BE)* vapour deposition technique
Aufdornen *n* 1. <Fertig> piercing; 2. <Maschinen> drifting
aufdornen *v* 1. <Fertig> drift, pierce; 2. <Mechan> ream
Aufdorner *m* <Mechan> reamer
Aufdornversuch *m* <Fertig> expanding test
aufdrehen *v* <Wasserversorg> turn on; turn on *(einen Hahn)*
Aufdrehen *n* <Maschinen> untwisting
Aufdruck *m* <Textil> impression
aufeinander abgestimmte Dioden *fpl* <Elektronik> matched diodes
aufeinander abgestimmte Röhren *fpl* <Elektronik> matched tubes
aufeinander abgestimmte Transistoren *mpl* <Elektronik> matched transistors
aufeinander folgender Fehlerblock *m* <Telekom> CEB, consecutive error block
Aufenthaltsdatei *f* <Mobilkom> visitor location register
Aufenthaltsmodul *n* <Raumfahrt> habitation module *(bemannte Raumstation)*
Aufenthaltsrufnummer *f* <Mobilkom> mobile station roaming number *(zellulares Netz)*
Aufenthaltswahrscheinlichkeit *f* <Strahlphys> probability of presence *(eines Teilchens)*
Auffächern *n* <Kerntech> unfolding *(eines Spektrums)*
Auffahren *n* **der Weiche** <Eisenbahn> forcing the points
Auffahrt *f* <Bau> access road
Auffahrunfall *m* <Sicherheit, Trans> rear end collision
Auffang *m* 1. <Anstrich> strip; 2. <Heiz & Kälte> drip; 3. <Meerschmutz> collection, interception
Auffangbecken *n* 1. <Bau, Eisenbahn> catchpit; 2. <Fertig, Kfztech> sump *(Öl)*
Auffangbehälter *m* <Erdöl> receiver
Auffangeinrichtung *f* <Meerschmutz> collection device *(für Öl)*
auffangen *v* 1. <Meerschmutz> collect *(Öl)*; 2. <Telekom> pick up *(Signal)*

Auffangen

Auffangen *n* eines Leitstrahls <Lufttrans> beam interception
Auffangschale *f* <Heiz & Kälte> drip tray
Auffangtank *m* <Anstrich> strip tank
Auffangtank *m* **für Reaktorkühlmittel** <Kerntech> reactor coolant drain tank
Auffangvorrichtung *f* <Fertig> collecting device, collector
Auffangwanne *f* <Anstrich> strip tank
aufflammen lassen *v* <Thermod> flare up
aufflanschen *v* <Erdöl> flange up
auffordern *v* 1. <Comp & DV> prompt, request; 2. <Telekom> request
Aufforderung *f* 1. <Comp & DV> demand, enquiry, query; 2. <Qual> requisition; 3. <Telekom> request
Aufformziehen *n* <Textil> boarding
auffrischen *v* 1. <Elektronik> refresh *(Daten)*; 2. <Fertig> recuperate; 3. <Wassertrans> rise *(Wind)*
Auffrischung *f* <Akustik> recruitment
aufführen *v* <Comp & DV> list
auffüllen *v* 1. <Bau> fill up *(Graben)*; 2. <Comp & DV> pad
Auffüllen *n* 1. <Comp & DV> padding; 2. <Ker & Glas> making-up
Auffüllladung *f* <Elektrotech> topping charge
Auffüllung *f* 1. <Bau> made ground *(Erdreich)*; pugging *(zur Schalldämmung)*; 2. <Umweltschmutz> backfill
Aufgabe *f* 1. <Comp & DV> task; 2. <Patent> problem *(technische Aufgabe in Erfindungsbeschreibung)*
Aufgabegut *n* <Fertig> charge
Aufgabenanalyse *f* <Ergon> task analysis
Aufgabenbereich *m* <Regelung> range of desired variable
Aufgabenbeschreibung *f* <Ergon> task description
Aufgabenerfüllung *f* <Ergon> accomplishment of task
Aufgabengröße *f* <Regelung> indirectly controlled variable
Aufgabenhierarchie *f* <Ergon> task hierarchy
aufgabenkritisch *adj* <Comp & DV> business-critical
aufgabenkritische Umgebung *f* <Comp & DV> business-critical environment
aufgabenorientiertes Wartesystem *n* <Telekom> task-oriented queuing
Aufgabensteuerung *f* <Comp & DV> task management
Aufgabetrichter *m* <Bau> feed hopper
Aufgabevorrichtung *f* <Maschinen> feeder, feeding device
Aufgangszeit *f* <Lufttrans> rise time *(Schallknall)*
aufgearbeitet *adj* <Elektronik> regenerated
aufgebrachte Epitaxialschicht *f*/**mit Dampfphase** <Elektronik> *(AE)* vapor phase grown epitaxial layer, *(BE)* vapour phase grown epitaxial layer
aufgebrachte Schicht *f* <Elektronik> deposited layer
aufgedampfte Schicht *f* 1. <Chemtech> *(AE)* vapor-deposited layer, *(BE)* vapour-deposited layer; 2. <Elektronik> evaporated layer
aufgedampfte Schicht *f*/**im Vakuum** <Elektronik> vacuum-deposited film
aufgedrückte Spannung *f* <Metall> applied stress
aufgegebene Bohrung *f* <Erdöl> lost hole *(Bohrtechnik)*
aufgehängte Steuerung *f* <Sicherheit> pendant switch control
aufgeheizt *adj* <Ker & Glas> fired-on
aufgeklebte Dichtung *f* <Verpack> bonded seal, glued seal
aufgeladener Motor *m* <Eisenbahn> supercharged engine
aufgelaufen *adj* <Wassertrans> aground *(Schiff)*
aufgelaufener Fehler *m* <Gerät> accumulated error
aufgelegt *adj* <Wassertrans> laid-up *(Schiff)*

aufgelegte elektromotorische Kraft *f* <Elektriz> impressed electromotive force *(EMK)*
aufgelegte Spannung *f* <Elektriz> impressed voltage
aufgelegte Tonnage *f* <Wassertrans> idle shipping *(Seehandel)*
aufgelöst *adj* <Chemtech> dissolved
aufgelöste Scherspannung *f* <Metall> resolved shear stress
aufgelöster Stoff *m* <Lebensmittel> solute
aufgerahmter Latex *m* <Kunststoff> creamed latex
aufgesattelte Treppe *f* <Bau> cutstring staircase, open string stairs
aufgeschrumpft *adj* <Thermod> heat-shrunk
aufgeschüttet *adj* <Bau> heaped; made-up *(Erdreich)*
aufgesetzter spannungsgeregelter Oszillator *m* <Elektronik> set-on voltage-controlled oscillator
aufgespannt *adj* <Fertig> set-up *(Werkzeug)*
aufgeständerte Fahrbahn *f* <Eisenbahn> elevated track beam
aufgestecktes Filter *n* <Foto> mounted filter
aufgeteilt *adj* <Comp & DV> partitioned
aufgetriebener Flaschenhals *m* <Ker & Glas> flared neck
aufgetriebenes Ende *n* <Ker & Glas> flared end
aufgewickeltes Garn *n* <Textil> beamed yarn
aufgezeichnete Dosis *f* <Strahlphys> dose recorded
aufgezeichnete Wellenlänge *f* <Akustik> recorded wavelength
Aufglasur *f* <Ker & Glas> overglazing
Aufgliederung *f* <Bau> breakdown
aufglühen *v* <Textil> blaze
Aufgrundlaufen *n* <Wassertrans> *(BE)* earthing, *(AE)* grounding *(eines Schiffes)*
Aufguss *m* <Lebensmittel> infusion
aufhalden *v* <Kohlen> stockpile
Aufhängehaken *m* <Maschinen> suspension hook
Aufhänger *m* <Maschinen> hanger
Aufhängung *f* 1. <Elektrotech> suspension *(in Messwerken)*; 2. <Kfztech, Maschinen> suspension; 3. <Phys> suspension *(Montage)*
Aufhängung *f* **eines Trimm-Abschaltstabes** <Kerntech> shim safety rod suspension
Aufhängungen *fpl* <Maschinen> hanger fixtures
Aufhängungsteile *npl* <Maschinen> hanger fixtures
aufhäufen *v* <Bau> bank
Aufhäufen *n* <Bau> piling up
aufheben *v* <Telekom> set off *(Alarm)*
Aufheben *n* <Trans> lifting
Aufhebung *f* <Patent> revocation
Aufhebungszeichen *n* <Akustik> natural
Aufheiz... <Kunststoff, Thermod> heating
aufheizen *v* <Ker & Glas, Textil, Thermod> heat, heat up
Aufheizen *n* <Ker & Glas> firing-on, heating-up
Aufheizkurve *f* 1. <Kunststoff> heating curve; 2. <Thermod> heating-up curve
Aufheiztiefe *f* <Thermod> heating depth
Aufheizzeit *f* <Thermod> heating-up time
Aufhell... <Foto> fill-in
Aufhellblitz *m* <Foto> fill-in flash
aufhellen *v* 1. <Bau> tint; 2. <Textil> raise *(Färben)*
Aufhelllicht *n* <Foto> fill-in light
Aufhellschirm *m* <Foto> reflecting screen
Aufhellungsschirm *m* <Foto> reflecting screen
aufheulen lassen *v* <Kfztech> rev up *(Motor)*
Aufhöhung *f* <Bau> raising *(einer Mauer)*
Aufkimmung *f* <Wassertrans> deadrise *(Schiffskonstruktion)*; rise of floor *(Schiffkonstruktion)*
Aufklappmenü *n* <Comp & DV> drop-down menu, pop-up menu

Aufklärungssatellit m <Raumfahrt> observation satellite
Aufkleber m 1. <Comp & DV> external label; 2. <Papier> paster
Aufklebezettel m <Qual> sticker
aufklotzen v <Textil> pad *(Färben)*
Aufklotzung f <Wassertrans> chock *(Schiffbau)*
Aufkochen n <Chemtech> boiling
Aufkochneigung f <Ker & Glas> tendency to reboil
aufkohlen v 1. <Fertig> carburize, cementite; 2. <Metall> cement, cementite
Aufkohlen n 1. <Fertig> acierage *(Stahl)*; 2. <Maschinen> carburizing; 3. <Metall> cementation
Aufkohlung f <Fertig, Funktech> carburization
Aufkohlungsmittel n <Fertig> carburizer
aufkommen v <Wassertrans> rise *(Sturm)*
aufkratzen v <Bau> scratch
aufladbare Zelle f <Phys> rechargable cell
Auflade... <Elektrotech> charging
Aufladegebläse n 1. <Fertig> booster; 2. <Kfztech> supercharger
aufladen v 1. <Bau> saddle *(Last)*; 2. <Elektrotech> boost, charge *(Batterie)*
Auflader m <Mechan> supercharger
Aufladestelle f <Trans> charging point
Aufladevorrichtung f <Trans> charger
Aufladezeit f <Foto> recharge time
Aufladung f 1. <Elektrotech> charge, charging; 2. <Kfztech> supercharging; 3. <Lufttrans> supercharge
Aufladung f **unter Last** <Kerntech> (AE) on-load refueling, (BE) on-load refuelling
Auflage f 1. <Bau> seat; 2. <Fertig> heel; 3. <Maschinen> rest
Auflagefläche f 1. <Bau> bearing area, seat; 2. <Maschinen> bearing surface, seat, seating; 3. <Trans> bearing surface
Auflageholz n <Bau> pole plate
Auflagekonsole f <Mechan> bearing pad
Auflagennummer f <Verpack> batch number
Auflageplatte f <Maschinen> base plate
Auflageplattieren n <Mechan> overlay cladding
Auflagepuffer m <Mechan> bearing pad
Auflagepunkt m <Maschinen> point of support
Auflager m 1. <Bau> bearing, saddle, support; 2. <Comp & DV, Maschinen> support; 3. <Trans> bearing
Auflagerfläche f <Bau> bedding surface
auflagern v <Bau> rest
Auflagerplatte f 1. <Bau> bearing plate, bed plate; 2. <Fertig> bed plate *(Spanung)*
Auflagerung f <Bau> support
Auflageschiene f <Trans> bearing rail
Auflageschiene f **für das Werkstück** <Maschinen> work plate
auflandiger Wind m <Wassertrans> onshore wind
Auflast f <Kohlen> surcharge load
Auflaufbremsbacke f <Kfztech> primary shoe
auflaufen v <Wassertrans> beach *(absichtlich, auf den Strand)*
auflaufendes Wasser n <Wassertrans> rising tide *(Gezeiten)*
auflegen v 1. <Telekom> hang up *(Hörer)*; 2. <Wassertrans> lay up *(Schiff)*
Auflicht n 1. <Fertig> incident light; 2. <Foto> reflected light
Auflichtmikroskop n <Metrol> microscope for reflected light
aufliegen v 1. <Bau> rest; 2. <Bau> seat
auflisten v <Comp & DV> list
Auflistung f <Comp & DV> listing
auflockern v <Fertig> open; aerate *(Formsand)*
Auflockern n <Fertig> opening *(Formsand)*
auflodern v <Thermod> blaze up
Auflösbarkeit f <Chemie> dissolubility
Auflösebehälter m <Chemtech> dissolver
auflösen v <Meerschmutz> foul
Auflösen n <Chemtech> dissolution *(Vorgang)*
Auflösung f 1. <Akustik> resolution; 2. <Chemie> decomposition, dissolution; 3. <Comp & DV, Elektriz> definition, resolution; 4. <Fernseh> resolution *(Grafik)*; 5. <Math> solution; 6. <Patent> cancellation *(eines Vertrages)*; 7. <Phys> suspension
Auflösung f **der Verzahnung** <Elektronik> de-interleaving
Auflösungsmittel n <Chemie> dissolvent
Auflösungsvermögen n 1. <Comp & DV> resolution; 2. <Elektronik, Foto> resolving power; 3. <Funktech> discrimination; 4. <Metall, Phys> resolving power; 5. <Telekom> resolution
Auflösungsvermögen n **des Ohres** <Akustik> aural resolving power
Aufmachung f <Wasserversorg> make-up
Aufmaß n <Heiz & Kälte> allowance
Aufmaße npl <Wassertrans> offsets *(Schiffbau)*
aufmauern v <Bau> brick up, cope, mason
Aufmauern n <Bau> coping
Aufnahme f 1. <Akustik> detection, record, recording; 2. <Aufnahme> recording; constant-velocity recording *(mit konstanter Geschwindigkeit)*; 3. <Bau> survey *(Gebäude, Bausubstanz)*; 4. <Elektrotech> absorption *(von Leistung, Energie, Kraft)*; 5. <Fertig> seat; 6. <Foto> shot; 7. <Hydraul> IP, input; 8. <Lebensmittel> absorption; 9. <Maschinen> accommodation; 10. <Telekom> recording *(Akustik)*; slot *(Karte)*
Aufnahme f **aus der Froschperspektive** <Foto> low-angle shot
Aufnahme f **des Imprägniermittels** <Textil> dip pick-up
Aufnahme f **für Werkstücke** <Fertig> workholder
Aufnahme f **und Orientierung** <Fertig> mounting and location
Aufnahme f **von Streuspannungen** <Gerät> stray signal pick-up
Aufnahme f **von unten** <Foto> low-angle shot
Aufnahme f **zum sofortigen Abspielen** <Akustik, Aufnahme> instantaneous recording
Aufnahme-/Abspielgerät n <Aufnahme> recorder-player
Aufnahmeabstellknopf m <Aufnahme> record defeat tab
Aufnahmebandspule f <Fernseh> take-up spool
Aufnahmebehälter m <Kerntech> receiving assembly
Aufnahmeblase f <Ker & Glas> gathering bubble
Aufnahmebühne f <Aufnahme> recording stage
Aufnahmebunker m <Abfall> receiving bunker
Aufnahmecharakteristik f <Aufnahme> directional characteristic *(des Mikrofons)*
Aufnahmeeigenschaft f <Fernseh> recording characteristic
Aufnahmeeisen n <Ker & Glas> gathering iron
Aufnahmeende n <Ker & Glas> gathering end
Aufnahmeentfernung f <Foto> shooting distance
Aufnahmeentzerrung f <Aufnahme> pre-equalization
Aufnahmefähigkeit f 1. <Papier> absorptiveness; 2. <Verpack> holding capacity; 3. <Wasserversorg> absorption capacity
Aufnahmeflansch m <Fertig> holding flange, mounting flange
Aufnahmegerät n <Aufnahme, Comp & DV, Telekom> recorder
Aufnahmegeräusch n <Aufnahme> recording noise
Aufnahmekabine f <Aufnahme> recording booth
Aufnahmekanal m <Aufnahme> recording channel

Aufnahmekette

Aufnahmekette f <Fernseh> recording chain
Aufnahmeknopf m <Fernseh> record button
Aufnahmekopf m <Fernseh> recording head
Aufnahmeloch n <Ker & Glas> gathering hole
Aufnahmemagnetkopf m <Akustik> recording magnetic head
Aufnahmenzähler m <Foto> exposure counter
Aufnahmeobjektiv n <Foto> taking lens
Aufnahmepegel m <Aufnahme> recording level
Aufnahmepunkt m <Konstzeich> location point
Aufnahmeraum m <Aufnahme> recording room
Aufnahmerohr n <Maschinen> pilot hole
Aufnahmeröhre f <Fernseh> pick-up tube
Aufnahmeschuh m <Ker & Glas> gathering shoe
Aufnahmesender m <Fernseh> pick-up transmitter
Aufnahmesession f <Aufnahme> recording session
Aufnahme-/Speicherröhre f <Elektronik> recording storage tube
Aufnahmespitze f <Fertig> (AE) work center, (BE) work centre
Aufnahmespur f <Akustik> recording track
Aufnahmestrom m <Fernseh> record current
Aufnahmestudio n <Aufnahme> recording studio
Aufnahmetaste f <Aufnahme> record button
Aufnahmetemperatur f <Ker & Glas> gathering temperature
Aufnahmetreiber m <Fernseh> record driver
Aufnahmevorgang m <Aufnahme> recording process
Aufnahmevorrichtung f <Fertig> workholding device, workholding fixture
Aufnahme/Wiedergabe-Magnetkopf m <Akustik> recording/reproducing magnetic head
aufnehmen v 1. <Aufnahme> record; 2. <Bau> house; 3. <Bau> conduct a survey (Vermessung); 4. <Chemie> imbibe (Flüssigkeiten); 5. <Fernseh> record; 6. <Ker & Glas> gather; 7. <Sicherheit> absorb; 8. <Telekom> pick up (Ton); 9. <Umweltschutz> ingest (Biologie)
Aufnehmen n 1. <Fernseh> recording; 2. <Ker & Glas> gathering
aufnehmendes Außenteil n <Fertig> female part
Aufnehmer m 1. <Elektronik> sensor (für Messwerte); 2. <Ergon, Kerntech, Maschinen, Trans> pick-up
Aufnehmersignal n <Elektronik> sensor signal
Aufprall m 1. <Kohlen> impact; 2. <Maschinen> impact, impingement; 3. <Mechan> impact; 4. <Metall> impingement
Aufprallbruchstelle f <Kerntech> impact fracture
aufprallendes Teilchen n <Kerntech> impinging particle
Aufprallfestigkeit f <Kerntech> impact strength
Aufprallgeräusch n <Sicherheit> impact noise
aufprallsichere Motorhaube f <Kfztech> (BE) crush-proof safety bonnet, (AE) crush-proof safety hood
Aufpralltest m <Kerntech, Qual> impact check
Aufprallversuch m <Kfztech> collision test, impact test
aufpressen v <Maschinen> force on
Aufpumpen n <Meerschmutz> inflation
Aufputzmontage f <Comp & DV> surface mounting
Aufquellen n <Kunststoff> swelling
aufquellender Boden m <Kohlen> swelling soil
Aufrahmen n <Kunststoff> creaming
Aufrahmmittel n <Kunststoff> creaming agent
Aufrahmung f <Kunststoff> creaming
aufrauen v 1. <Bau> key, scarify, score (streichen); 2. <Fertig> rag; sharpen (Schleifscheibe); boss (Walzen); 3. <Maschinen> roughen; 4. <Textil> nap; raise (Tuch)
Aufrauen n 1. <Bau> keying, scarification (Streichen); 2. <Fertig> ragging; 3. <Textil> napping; raising (Tuch)
Aufrauungszone/mit <Fertig> scuffed (Getriebelehre, Zahnflanke)

aufrecht gehaltener Diamant m <Ker & Glas> diamond held upright
aufrecht transportieren v <Verpack> keep upright
Aufrechterhaltung f <Bau> retention
Aufrechterhaltung f **eines Patents** <Patent> maintenance of a patent
Aufreibdorn m <Bau> reaming iron
Aufreiben n <Maschinen> reaming, reaming-out
Aufreiß... <Verpack> pull-off, tear-off
Aufreißdeckel m <Verpack> tear tab lid
aufreißen v <Bau> break open, rip, scarify; break open (Straße)
Aufreißen n <Bau> ripping; scarification (Straßen)
Aufreißen n **des Druckteiles** <Sicherheit> vessel bursting
Aufreißer m 1. <Bau> scarifier; ripper (Straßenbau); 2. <Trans> scarifier
Aufreißlasche f <Verpack> pull-off closure
Aufreißpackung f <Verpack> tear-off pack
Aufreißstreifen m <Verpack> tear strip
aufrichten v <Verpack> erect
aufrichtender Hebelarm m <Wassertrans> righting lever arm (Schiffbau)
aufrichtendes Moment n <Wassertrans> righting moment (Schiffbau)
Aufrichter m <Ker & Glas> uprighter
Aufrichtungsmarkierung f <Lufttrans> (AE) leveling mark, (BE) levelling mark
Aufriss m 1. <Bau> elevation (eines Gebäudes); 2. <Fertig> shear draft (Zeichnung); 3. <Maschinen> front elevation; 4. <Mechan, Wassertrans> elevation (Schiffkonstruktion)
Aufroll... 1. <Foto> winder; 2. <Papier> wind-up
aufrollen v 1. <Comp & DV> scroll; 2. <Maschinen> roll, roll up
Aufrollen n <Papier> reeling
Aufrollmaschine f <Papier> winder
Aufrollvorrichtung f 1. <Foto> rewinder; 2. <Papier> wind-up stand
Aufruf m <Comp & DV> call, polling
Aufrufantwortmodus m <Comp & DV> normal response mode
Aufrufbefehl m <Comp & DV> call instruction
Aufrufbetrieb m 1. <Comp & DV> polling selection; 2. <Telekom> polling mode, selecting mode
aufrufen v <Comp & DV> poll
Aufrufliste f <Comp & DV> polling list
aufrühren v <Chemtech> agitate
aufrunden v 1. <Comp & DV> round off, round up; 2. <Math> round
aufrüsten v <Comp & DV> upgrade
Aufrüstposition f <Lufttrans> rigging position (Luftfahrzeug)
Aufrüstung f <Comp & DV> upgrade (Hardware)
aufsatteln v <Bau> saddle (Stufen)
Aufsatz m <Mechan> cowl
Aufsatzkranz m <Bau> (AE) curb, (BE) kerb
Aufsatzventil n <Maschinen> yoke valve
Aufsatzzusammenbau m <Nichtfoss Energ> top assembly
aufsaugen v <Heiz & Kälte> aspirate
Aufsaugen n 1. <Chemie> occlusion; 2. <Elektrotech> absorption (von Elektronen, Gasen); 3. <Lebensmittel> absorption
aufschalten v <Mechan, Phys> actuate
Aufschalten n <Telekom> trunk offer
Aufschaltzeit f <Telekom> switching time (Vermittlung)
Aufschaltwert m <Phys> modulation factor
Aufschäumblasen fpl <Ker & Glas> reboil bubbles

aufschäumen v 1. <Chemtech> foam; 2. <Fertig> froth
Aufschäumen n 1. <Ker & Glas> reboil; 2. <Kunststoff> frothing
Aufschäumung f <Fertig> expansion *(Kunststoffe)*
aufschichten v <Bau> pile, stack
aufschießen v <Wassertrans> coil up *(Tauwerk)*
Aufschlag m 1. <Anstrich> impact; 2. <Bau> strike; surcharge *(Preis)*; 3. <Fertig> deposit; 4. <Kohlen> strike; 5. <Maschinen> impact
Aufschlagen n <Wassertrans> slamming *(Schiffbewegung)*
Aufschlagkrater m <Raumfahrt> impact crater
Aufschlagleuchten n <Fernseh> impact fluorescence
aufschlämmen v <Anstrich> suspend
Aufschlämmung f <Fertig, Kohlen, Lebensmittel> slurry
aufschließen v 1. <Chemie> macerate *(Pflanzenteile)*; 2. <Lebensmittel> macerate *(Nahrungsmittel)*
Aufschließen n 1. <Abfall> digestion, fouling; 2. <Chemie> maceration *(Pflanzenteile)*
aufschlitzen v <Textil> split
Aufschluss m 1. <Abfall> digestion, fouling; 2. <Chemie> dissociation *(Mineralerze)*; pulping *(chemisch)*; 3. <Papier> digestion
Aufschlussapparat m <Labor> digestion apparatus
Aufschlussbohren n <Erdöl> exploration drilling, exploratory drilling
Aufschlussbohrung f <Erdöl> exploration well, wildcat drilling
Aufschmelzverfahren n <Fertig> fusing method
aufschrauben v 1. <Kerntech> unbolt, unscrew; 2. <Maschinen> unscrew
Aufschrauben n <Maschinen> unscrewing
Aufschraubfräser m <Fertig> screw-on cutter
aufschrumpfbar adj <Thermod> heat-shrinkable
Aufschrumpfen n 1. <Fertig> shrinking-on; 2. <Maschinen> shrinkage, shrinking-on; 3. <Thermod> heat shrink fitting, heat shrinking
aufschütten v <Bau> bank up, raise
Aufschüttung f 1. <Kohlen> fill; 2. <Metall> debris
aufschweißen v <Fertig> deposit *(Schweißen)*
Aufschweißlegierung f <Fertig> hard-facing alloy, surfacing alloy
Aufschweißung f <Fertig> deposition
aufschwemmen v <Anstrich> suspend
aufschwimmen lassen v <Wassertrans> float off *(gestrandetes Schiff)*
Aufschwimmen n 1. <Kfztech> aquaplaning; 2. <Kunststoff> flooding *(von Farbstoffen)*
Aufsetz... <Lufttrans, Raumfahrt> touchdown
Aufsetzbauelement n *(SMC)* <Telekom> surface-mounted component *(SMC)*; surface mounting device *(SMD)*
aufsetzen v <Raumfahrt> touch down *(Raumschiff)*
Aufsetzen n <Lufttrans> touchdown
Aufsetzen n **von Flicken** <Fertig> patching *(Kessel)*
Aufsetzen n **von Verschlusskappen** <Verpack> capping
Aufsetzgeschwindigkeit f <Lufttrans> touchdown speed
Aufsetzpunkt m <Raumfahrt> touchdown point *(Raumschiff)*
Aufsetztechnologie f <Telekom> surface mounting technology
Aufsetzzone f <Lufttrans> touchdown zone
Aufsetzzonenbefeuerung f <Lufttrans> runway touchdown zone
Aufsicht f 1. <Papier> lookdown; 2. <Sicherheit, Telekom> supervision *(Telefon)*
Aufsichtsbehörde f <Patent, Qual> regulatory authority
Aufsichtssucher m **mit Mattscheibe** <Foto> reflex viewfinder

aufslippen v <Wassertrans> haul up
aufspalten v <Bau> split
Aufspalten n <Mechan> cracking
Aufspaltung f 1. <Anstrich> analysis; 2. <Bau> splitting; 3. <Chemie> breakdown
Aufspaltung f **eines Multipletts** <Phys> splitting of multiplet
Aufspann... 1. <Fertig> setting; 2. <Maschinen> chucking, clamping
aufspannen v 1. <Fertig> set *(Werkzeug)*; 2. <Maschinen> chuck
Aufspannen n <Textil> tentering
Aufspannen n **der Werkzeuge** <Fertig> tooling
Aufspannfläche f <Maschinen> clamping surface
Aufspannplatte f 1. <Fertig> bolster; 2. <Maschinen> adaptor plate, backing plate
Aufspanntisch m <Fertig> worktable
Aufspannung f <Fertig> setting *(Werkzeug)*
Aufspannvorrichtung f <Maschinen> chuck, chucking device, fixture
Aufspannwinkel m <Maschinen, Metrol> angle plate
aufsplittern v <Bau> split
Aufspringbildwand f <Foto> self-erecting screen
Aufspritzung f <Phys> *(AE)* metalization, *(BE)* metallization
aufspröden v <Anstrich, Fertig> embrittle
Aufsprödung f <Anstrich> embrittlement
aufspulen v <Phys> wind
Aufspulen n <Phys> winding
Aufstampfen n <Fertig> tamping *(Formen)*
aufstapeln v <Verpack> stack up
Aufstäuben n <Fertig> deposition by sputtering
Aufsteck... 1. <Foto> push; 2. <Gerät> clip; 3. <Maschinen> shell
Aufsteckfassung f <Foto> push-on mount, slip-on sleeve
Aufsteckfräser m 1. <Fertig> arbor cutter, arbor-type mill, hole-type cutter; 2. <Maschinen> shell mill
Aufsteckgatter n <Textil> creel
Aufsteckreibahle f <Maschinen> shell reamer
Aufsteckschlüssel m <Mechan> *(BE)* socket spanner, socket wrench
Aufstecksenker m <Maschinen> arbor-mounted counterbore, shell drill
Aufsteckskale f <Gerät> clip-on scale
Aufsteckwerkzeug n <Fertig> shell tool
aufsteigende Destillation f <Thermod> distillation by ascent
aufsteigende Sortierung f <Comp & DV> ascending sort
aufstellen v 1. <Fertig> install, rig; 2. <Maschinen> erect; 3. <Telekom> set up; 4. <Wassertrans> install *(Technik)*
Aufstell- und Abstellgleis n <Eisenbahn> siding
Aufstellung f 1. <Bau> assembly; erection; setting; 2. <Comp & DV> installation; 3. <Fernseh> setup; 4. <Fertig> assembly; erection; setting; 5. <Kontroll> positioning; 6. <Patent> drawing-up; 7. <Telekom> installation
Aufstellungszeichnung f <Heiz & Kälte> installation drawing
Aufsticken n <Chemie> nitridation
Aufstieg m 1. <Erdöl> rising; 2. <Lufttrans> lift
Aufstiegsmodus m <Raumfahrt> ascending mode
Aufstiegsstufe f <Raumfahrt> ascent stage *(Raumschiff)*
Aufstoß m <Chemie> impingement
aufstoßen v <Druck> jog
aufstreichbare Schlämmbeschichtung f <Anstrich> paint-on slurry coating
aufstreichen v <Anstrich> apply
Aufstreichmaschine f <Fertig> spreader
Aufstreuen n <Fertig> dusting
Aufstrich m <Anstrich> application

Aufströmung

Aufströmung f <Lufttrans> upwash *(Lufttüchtigkeit)*
Aufstromvergaser m <Kfztech> *(AE)* updraft carburetor, *(BE)* updraught carburettor
Aufstützring m <Fertig> seal support ring *(Kunststoffinstallationen)*
Aufsuchung f <Erdöl> exploration
Aufsummierzähler m <Gerät> totalizing counter
auftakeln v <Wassertrans> rig
auftanken v 1. <Kfztech> fill up, refuel; 2. <Lufttrans, Wassertrans> refuel
Auftanken n <Maschinen, Meerschmutz> *(AE)* refueling, *(BE)* refuelling
Auftanken n **in der Luft** <Lufttrans> *(AE)* refueling in flight, *(BE)* refuelling in flight
Auftankfahrzeug n <Lufttrans> *(AE)* refueling tanker, *(BE)* refuelling tanker
auftasten v <Kontroll> strobe
auftauchen v <Wassertrans> surface *(U-Boot)*
auftauen v 1. <Heiz & Kälte> defrost; 2. <Lebensmittel> thaw
Auftauen n <Lebensmittel> defrosting
Auftauversuch m <Heiz & Kälte, Maschinen> defrosting test
aufteilen v 1. <Bau> split into; 2. <Comp & DV> decouple; partition
Aufteilung f 1. <Bau> division; 2. <Chemie, Telekom> partitioning
Aufteilungskabel n *(AtK)* <Telekom> distribution cable
Auftragefläche f <Fertig> base *(Meißelschaft)*
auftragen v 1. <Anstrich> apply; 2. <Bau> deposit, plot, spread; distribute *(Farbe)*; 3. <Kunststoff> coat
Auftragen n <Kunststoff> coating
Auftragmenge f <Bau> application rate
Auftragnehmer m <Bau> contractor
Auftragschweißung f <Maschinen> build-up welding
auftragsgebunden adj <Konstzeich> order-tied *(Zeichnungen, Bestellliste)*
Auftragslinie f <Fertig> abscissa
Auftragsmaschine f <Kunststoff, Verpack> coating machine
Auftragsmetall n <Fertig> deposited metal
Auftragsschreiben n <Bau> notice of award
auftragsschweißen v <Fertig> pad
Auftragsschweißen n <Fertig> hard-facing by welding
Auftragsüberwachung f <Comp & DV> order tracking
Auftragsvergabesystem n <Comp & DV> order placement system
Auftragswalze f 1. <Fertig> spreading roll; 2. <Papier> applicator roll, spread roll
Auftragswerkstatt f <Fertig> job shop
Auftragung f <Anstrich> application
Auftragwalze f <Metall> spreader roll
Auftragwalze f **für Klebstoff** <Verpack> adhesive applicator
Auftreff... 1. <Lufttrans> ram; 2. <Metall> impact
auftreffen v <Mechan> impinge on
auftreffende Welle f <Phys, Wellphys> incident wave
auftreffender Strahl m <Phys, Wellphys> incident beam, incident ray
auftreffendes Teilchen n <Kerntech> incident particle
Auftreffenergie f <Metall> impact energy
Auftreffgeschwindigkeit f <Metall> impact velocity
Auftreffwucht f <Lufttrans> ram effect *(Aerodynamik)*
Auftreiben n 1. <Ker & Glas> flaring; 2. <Lebensmittel> swell
Auftreiber m <Ker & Glas> reamer
Auftrennen n **des Zylinders in Längsrichtung** <Ker & Glas> splitting of the cylinder
auftretender Fehler m/**durch Näherung** <Gerät> approximative error

Auftrieb m 1. <Fertig> lifting pressure *(Gießen)*; 2. <Hydraul> buoyancy; 3. <Lufttrans> buoyancy, lifting, lift; 4. <Nichtfoss Energ> lift *(Aerodynamik)*; 5. <Phys> upthrust; buoyancy *(Flüssigkeit)*; 6. <Strömphys> buoyancy, uplift; 7. <Wassertrans> buoyancy *(Schiff, U-Boot)*
Auftrieb m **am Luftschraubenblatt** <Lufttrans> blade lift *(Hubschrauber)*
Auftriebsbeiwert m 1. <Hydraul> lift coefficient, CL; 2. <Lufttrans> lift coefficient; 3. <Nichtfoss Energ, Phys, Qual> lift coefficient, CL; 4. <Wassertrans> lift coefficient, CL *(Schiffbau)*
auftriebserhöhende Vorrichtungen fpl <Lufttrans> high-lift devices
Auftriebskomponente f <Lufttrans> lift component
Auftriebskraft f <Phys, Strömphys> buoyancy force
Auftriebskurve f <Wassertrans> buoyancy curve *(Schiffskonstruktion)*
Auftriebsmittelpunkt m <Lufttrans> *(AE)* lift center, *(BE)* lift centre
Auftriebsparameter m <Phys, Strömphys> buoyancy parameter
Auftriebsreserve f <Wassertrans> reserve buoyancy
Auftriebssteigung f <Lufttrans> lift curve slope
Auftriebsverteilung f <Lufttrans> lift distribution
Auftriebszahl f *(CL)* <Hydraul, Nichtfoss Energ, Phys, Qual, Wassertrans> lift coefficient, CL *(Schiffbau)*
Auftrittsbreite f <Bau> foothold
Auftrudeln n <Raumfahrt> spin-up *(Raumschiff)*
auftuchen v <Wassertrans> furl *(Segeln)*
Auf- und Abbewegung f <Maschinen> reciprocating motion, up-and-down motion
auf- und abgehend adj <Maschinen> reciprocating
Auf- und Abwickelmaschine f <Verpack> unwinding machine
Auf- und Abwickelvorrichtungen fpl <Verpack> wind/unwind equipment
Aufwachsen n <Elektronik> epitaxial growth *(von Kristallen)*
Aufwallen n <Chemtech> boiling
Aufwallung f <Phys> bubbling
Aufwältigung f <Erdöl> workover *(Bohrtechnik)*
Aufwärmloch n <Ker & Glas> warming-in hole
Aufwärts... 1. <Comp & DV> upward; 2. <Elektriz, Elektrotech> step-up; 3. <Kfztech> upward; 4. <Maschinen> upstroke
Aufwärtsbewegung f 1. <Maschinen> rise, upward movement; 2. <Wassertrans> scend, up-movement
Aufwärtsblock m <Telekom> uplink block
Aufwärtsbohren n <Ker & Glas> upward drilling
Aufwärtsgang m <Kfztech> upstroke
Aufwärtshub m 1. <Hydraul> outstroke *(Kolben)*; 2. <Kfztech> upstroke; upstroke *(Motor, Kolben)*; 3. <Maschinen> upstroke
aufwärtskompatibel adj <Comp & DV> upward compatible
Aufwärtskompatibilität f <Comp & DV> upward compatibility
Aufwärtsstrategie f <Künstl Int> bottom-up strategy
Aufwärtsströmung f <Kerntech> upflow, upward flow
Aufwärtstakt m <Maschinen> upstroke
Aufwärtstransformator m <Elektriz, Elektrotech, Phys, Telekom> step-up transformer
Aufwärtsübertrager m <Telekom> step-up transformer
Aufwärtsverdampfung f <Heiz & Kälte> vertical evaporation
Aufwärtswandler m <Phys> step-up transformer
Aufwärtszähler m <Elektronik> up counter
Aufweichen n <Abfall> maceration
Aufweite... 1. <Maschinen> sizing; 2. <Metall> expanding

aufweiten v <Bau> bulge, expand
Aufweiten n <Fertig> flaring
Aufweitewalzwerk n <Metall> expanding mill
Aufweitewerkzeug n <Maschinen> sizing tool
Aufwerfen n <Foto> buckling *(Emulsion)*
aufwerten v <Bau> grade up *(Gebäude)*
Aufwickelkassette f <Foto> take-up cassette, take-up spool
Aufwickelmaschine f 1. <Foto> film winder; 2. <Verpack> rewind machine, winding machine, winding-on machine
aufwickeln v 1. <Maschinen> roll, roll up, wind, wind up; 2. <Papier> wind
Aufwickelspule f 1. <Aufnahme> take-up spool; 2. <Comp & DV> take-up reel
Aufwickelvorrichtung f <Fertig> roll-up device
Aufwind m <Lufttrans> upwash *(Lufttüchtigkeit)*
Aufwinde... <Textil> take-up
aufwinden v <Maschinen> jack up
Aufwindesystem n <Textil> take-up system
Aufwindevorrichtung f <Textil> take-up motion
Aufwölbung f <Bau> camber
Aufzählung f <Comp & DV> enumeration
Aufzählungstyp m <Comp & DV> enumeration type
aufzeichnen v 1. <Aufnahme> record; 2. <Bau> plot; 3. <Comp & DV> record, register; 4. <Druck> record; 5. <Fernseh> video; 6. <Foto> register; 7. <Kontroll, Phys> record
Aufzeichnen n <Comp & DV> logging, recording
Aufzeichnen n **mit doppelter Schreibdichte** <Comp & DV> double-density recording
Aufzeichnung f 1. <Akustik> recording; 2. <Phys> record; 3. <Telekom> recording
Aufzeichnung f **auf Videomagnetband** <Fernseh> videotaping
Aufzeichnung f **der Flugerprobungsdaten** <Lufttrans> flight test recorder
Aufzeichnung f **mittels Elektronenstrahl** <Elektronik> electron beam recording *(direkt auf Mikrofilm)*
Aufzeichnungscharakteristik <Akustik> recording characteristic
Aufzeichnungsdichte f <Comp & DV> recording density, storage density
aufzeichnungsfähige CD f <Optik> *(BE)* recordable optical disc, *(AE)* recordable optical disk
Aufzeichnungsfrequenzkurve f <Aufnahme> recording characteristic
Aufzeichnungsgerät n <Comp & DV> recording instrument
Aufzeichnungsmedium n <Comp & DV> recording medium
Aufzeichnungsmodus m <Comp & DV> recording mode
Aufzeichnungsoberfläche f <Comp & DV> recording surface *(eines Datenträgers)*
Aufzeichnungsspur f <Comp & DV> recording track
Aufzeichnungsstufe f <Comp & DV> recording level
Aufzeichnungsverlust m <Aufnahme> recording loss
aufziehen v 1. <Textil> handle; 2. <Wassertrans> hoist *(Segel, Flagge)*
Aufziehen n 1. <Chemtech> absorption *(Farbstoffe)*; 2. <Textil> attachment *(Farbstoffe)*
Aufziehkarton m <Foto> mount *(zum Aufziehen eines Fotos)*
Aufzug m 1. <Bau> escalator, *(BE)* lift; 2. <Maschinen> *(AE)* elevator, *(BE)* lift; 3. <Mechan> elevation; 4. <Trans> *(AE)* elevator, *(BE)* lift
Aufzuggewinde n <Lufttrans> *(AE)* elevator hoist, *(BE)* lift hoist
Aufzugkolben m <Lufttrans> hoisting block

Aufzugmaschine f <Lufttrans> engine hoist
Aufzugring m <Lufttrans> hoisting ring
Auf-Zu-Klappe f <Heiz & Kälte> on/off butterfly valve
Aufzupfen n <Textil> picking *(Wolle)*
Auge n <Maschinen, Wassertrans> eye
Augenabstand m <Ergon> interocular distance
augenblickliches elektrisches Dipolmoment n <Strahlphys> instantaneous electric dipole moment
Augenblicksbelastung f <Werkprüf> instantaneous load
Augenblicksfrequenz f <Elektronik> instantaneous frequency
Augenblicksfrequenzmessung f <Elektronik> instantaneous frequency measurement
Augenblickskapazität f <Trans> momentary capacity
Augenblickswert m <Gerät> instantaneous value
Augenbolzen m <Mechan> eyebolt
Augendiagramm n <Telekom> eye diagram, eye-shape pattern
Augendusche f <Sicherheit> emergency eyewash unit, emergency eyewashing unit
Augenfilter n <Sicherheit> eye filter *(gegen Laserstrahlen)*
Augenhöhe f <Bau> eye level
Augenkreis m <Optik> eye ring
Augenlinse f <Foto> eyepiece lens
Augenmuschel f <Foto> eyecup
Augenreizstoff m <Chemie> *(BE)* lachrymator, *(AE)* lacrimator
Augenscheinprüfung f <Qual> visual examination
Augenschraube f 1. <Fertig> eyebolt; 2. <Kfztech> eyebolt *(Kupplung)*; 3. <Maschinen> eyebolt
Augenschutz m 1. <Raumfahrt> eyepiece; 2. <Sicherheit> eye protector; personal eye protector *(für Schweißarbeiten)*
Augenspülbrunnen m <Sicherheit> rinsing station
Augenspüleinrichtung f <Sicherheit> emergency eyewash unit, eye-rinsing station, eyewash safety station, eye-washing safety station, rinsing station
Augenspülmittelflasche f <Sicherheit> eye-rinse bottle
Augentropfflasche f <Ker & Glas> eye-drop bottle
Auger-Ausbeute f <Phys, Strahlphys> Auger yield
Auger-Effekt m <Phys, Strahlphys> Auger effect
Auger-Elektron n <Phys, Strahlphys> Auger electron
Auger'sche Elektronenspektroskopie f *(AES)* <Phys, Strahlphys> Auger electron spectroscopy *(AES)*
Augplatte f <Wassertrans> eye plate *(Deckbeschläge)*
Augspleiß m <Wassertrans> eye splice *(Tauwerk)*
Augspliss m <Maschinen> eye splice
Augstropp m <Wassertrans> snotter
Aurantiin n <Chemie> naringin
Aurichlorid n <Chemie> auric chloride, gold chloride, gold trichloride
aus adv <Maschinen> off
ausbaggern v <Wassertrans> dredge
Ausbaggern n 1. <Bau> digging; 2. <Wassertrans> dredging
ausbalancieren v <Mechan> trim
Ausbau m <Bau> completion, development
Ausbauchung f 1. <Bau> belly, bulge; 2. <Kohlen> belly; 3. <Mechan> expansion
ausbauen v <Fertig> dismantle *(Kunststoffinstallationen)*
Ausbauen n <Bau> removing
ausbäumen v <Wassertrans> boom out *(Segeln)*
Ausbessern n 1. <Druck> touching-up; 2. <Textil> mending
ausbeulen v 1. <Bau, Fertig, Maschinen> buckle, bulge, cripple, flatten, planish; 2. <Metall> planish
Ausbeulen n 1. <Fertig, Maschinen> buckling, crippling, planishing; 2. <Metall> planishing

Ausbeulhammer

Ausbeulhammer *m* <Maschinen> planisher, planishing hammer
Ausbeulung *f* 1. <Kerntech> buckling *(in einem Brennelement)*; 2. <Maschinen> bulge; 3. <Metall> buckling
Ausbeute *f* 1. <Chemie> yield *(Reaktion)*; 2. <Comp & DV, Elektronik, Erdöl, Fertig, Kerntech, Papier, Textil> yield
Ausbeute *f* **an Thermoionen** <Kerntech> thermal neutron yield
Ausbeutefaktor *m* <Erdöl> recovery factor *(Förderung)*
Ausbeutung *f* **der geothermalen Energie** <Nichtfoss Energ, Umweltschmutz> geothermal energy exploitation
Ausbildung *f* 1. <Phys> development, formation; 2. <Qual> training
Ausbildung *f* **durch Lernmaschine** <Comp & DV> machine learning
Ausbildung *f* **einer Grenzschicht** <Strömphys> boundary layer formation
Ausbildungsreaktor *m* <Kerntech> training reactor
Ausbildungsschiff *n* <Wassertrans> training ship *(Marine)*
Ausbinden *n* <Aufnahme> lacing *(eines Kabelstrangs)*
Ausbissbelastung *f* <Kohlen> crop load
Ausblaseleitung *f* <Mechan> exhaust line
ausblasen *v* <Elektriz, Heiz & Kälte> blow down, blow out
Ausblasen *n* <Elektriz, Heiz & Kälte> blowing down, blowing out
Ausblaseventil *n* <Hydraul> blow valve
ausbleichen *v* <Foto> bleach out
ausblenden *v* 1. <Aufnahme> fade down, fade out; 2. <Telekom> blank
Ausblenden *n* 1. <Aufnahme> fading down, fading out; 2. <Comp & DV> gating, reverse clipping; masking *(von Programmen)*; 3. <Elektronik> gating
Ausblenden in der Bildspur <Elektronik> trace blanking
ausblühen *v* <Bau, Kohlen, Metall> bloom, effloresce
Ausblühen *n* <Kunststoff> blooming
Ausblühung *f* 1. <Bau> blooming, efflorescence; 2. <Ker & Glas> scum; bloom *(durch Sulfatbildung während Kühlung)*; 3. <Kohlen, Metall> blooming, efflorescence
ausbluten *v* 1. <Kunststoff> bleed; 2. <Textil> bleed off
ausbohren *v* 1. <Fertig> enlarge; 2. <Kfztech> rebore *(Motor, Zylinder)*; 3. <Maschinen> counterbore, rebore
Ausbohren *n* <Maschinen> boring, counterboring, reboring
Ausbohrmeißel *m* <Maschinen> boring tool
Ausbohr- und Stirnmaschine *f* <Fertig> boring and facing mill
ausbrechen *v* 1. <Erdöl> gush *(Öl aus Bohrung)*; 2. <Fertig> chip *(Schneide bei Spannung)*; 3. <Kfztech> break away; 4. <Sicherheit> break out
Ausbrechen *n* 1. <Kfztech> breakaway *(Lastwagen)*; 2. <Thermod> outbreak
Ausbrechen *n* **der Ladung** <Lufttrans> cargo swing *(Hubschrauber)*
Ausbrechen *n* **der Schneide** <Fertig> edge chipping
ausbreiten *v/sich* 1. <Meerschmutz> spread; 2. <Wellphys> propagate
Ausbreiten *n* <Fertig> hammering
ausbreitender Riss *m/sich* <Kohlen> progressive failure
Ausbreiteprobe *f* <Phys> flattening test
Ausbreiter *m* <Textil> stretcher
Ausbreitmaß *n* <Bau> slump *(Beton)*
Ausbreitmaßprüfung *f* <Bau> slump test
Ausbreitung *f* 1. <Bau> diffusion; 2. <Comp & DV, Funktech, Künstl Int> propagation; 3. <Raumfahrt> spread; 4. <Telekom> propagation; 5. <Umweltschmutz> diffusion
Ausbreitung *f* **der Emission** <Phys> evolution
Ausbreitung *f* **über Meteorschweife** <Funktech> meteor trail propagation
Ausbreitung *f* **über sporadische E-Schicht-Reflektionen** <Funktech> sporadic E
Ausbreitung *f* **von Funkwellen** <Strahlphys> radiowave propagation
Ausbreitung *f* **von Schallwellen** <Elektrotech> acoustic-wave propagation
Ausbreitungsanomalien *fpl* **in der Ionosphäre** <Funktech> ionospheric propagation anomalies
ausbreitungsbedingte Verzögerung *f* <Raumfahrt> propagation delay *(Weltraumfunk)*
Ausbreitungsdämpfung *f* <Funktech> propagation loss
Ausbreitungsfunktion *f* <Phys> propagator
Ausbreitungsgeschwindigkeit *f* 1. <Bau> rate of spread; 2. <Telekom> propagation velocity; 3. <Wellphys> speed of propagation
Ausbreitungskoeffizient *m* <Optik> propagation coefficient
Ausbreitungskonstante *f* <Optik, Phys> propagation constant
Ausbreitungslaufzeit *f* <Telekom> propagation delay
Ausbreitungsmode *f* <Optik> propagation mode *(elektromagnetische Welle)*
Ausbreitungsmodus *m* <Funktech> propagation mode
Ausbreitungsschlauch *m* <Raumfahrt> heat pipe *(Führung)*
Ausbreitungsschlauch *m* **in der Atmosphäre** <Funktech> atmospheric duct
Ausbreitungsverluste *mpl* <Phys> propagation losses
Ausbreitungsweg *m* 1. <Funktech> path; 2. <Sicherheit> propagation path *(Lärm)*
ausbringen *v* <Wasserversorg> put out *(Fender)*
Ausbringen *n* 1. <Druck> quad; 2. <Fertig> output
Ausbruch *m* 1. <Bau> outpouring; 2. <Erdöl> blowout; breakout *(Bohrtechnik, Fördertechnik)*; gush *(plötzlicher Ölausbruch aus Bohrung)*; 3. <Fertig> chip *(Schneide)*; 4. <Kfztech> deflection *(bei Reifen)*; 5. <Konstzeich> auxiliary section; 6. <Thermod> outbreak
Ausbruchsicherung *f* <Erdöl> blowout preventer
ausbuchen *v* <Mobilkom> log out
ausbuchsen *v* 1. <Fertig> bush *(Schleifscheibe)*; 2. <Maschinen> bush
ausbüchsen *v* <Fertig> rebush
Ausbüchsen *n* <Maschinen> bushing
Ausbuchten *n* <Kerntech> ballooning
Ausbuchtung *f* **des Kötzers** <Textil> bulge
ausdehnen *v* 1. <Bau> expand; 2. <Fertig> flare
ausdehnende Erde *f/sich* **durch Wärme** <Kohlen> dilatant soil
Ausdehnung *f* 1. <Geom> dilation; 2. <Heiz & Kälte, Kunststoff, Maschinen, Mechan, Metall> expansion; 3. <Phys> extension; 4. <Raumfahrt> expansion *(Raumschiff)*; 5. <Thermod> expansion
Ausdehnungsbogen *m* <Heiz & Kälte> expansion bend
Ausdehnungsgefäß *n* 1. <Heiz & Kälte> expansion tank; 2. <Maschinen> expansion vessel
Ausdehnungshub *m* <Thermod> expansion stroke
Ausdehnungsknie *n* <Mechan> expansion bend
Ausdehnungskoeffizient *m* 1. <Kunststoff> coefficient of expansion; 2. <Mechan, Qual> expansion coefficient
Ausdehnungskupplung *f* <Maschinen> expansion coupling, slip joint
Ausdehnungsnocke *f* <Mechan> expansion cam
Ausdehnungsrohrverbindungen *fpl* <Maschinen> bellow expansion joints
Ausdehnungsstoß *m* <Bau> expansion joint
Ausdehnungsvermögen *f* **der Gase** <Thermod> expansibility of gases
Ausdehnungswärme *f* <Bau, Nichtfoss Energ, Thermod> heat of expansion

Ausdehnungszahl f <Bau> coefficient of expansion
Ausdrehen n <Maschinen> boring
Ausdrehmeißel m <Maschinen> boring tool
Ausdrehschneidstahl m <Maschinen> boring cutter
Ausdruck m 1. <Comp & DV, Telekom> hard copy, printout; 2. <Math> expression, term
ausdrücken v <Papier> squeeze
Ausdünstung f <Sicherheit> fume
Ausdünstungsventil n <Lufttrans> exhalation valve
Ausecken n <Fertig> notching
auseinander brechen v <Bau> break up
auseinander gezogene Darstellung f <Mechan> exploded view
auseinander laufend adj <Geom> divergent
auseinander liegende Maße npl <Konstzeich> dimensions at different locations
Auseinanderspreizen n der Ziehschleifsteine <Fertig> expansion of honing stones
Ausfächerung f <Elektrotech> fan-out
Ausfachung f <Bau> nogging
Ausfädeln n <Trans> leaving a traffic stream
ausfahrbar adj <Fertig> retractable
ausfahrbare Antenne f <Phys> periscope aerial, periscope antenna
ausfahrbare Hilfsstütze f <Lufttrans> outrigger
ausfahrbares Langsieb n <Papier> roll-out Fourdrinier
ausfahrender Verkehr m <Trans> outbound traffic, outward traffic
Ausfall m 1. <Bau, Comp & DV, Elektriz> breakdown, failure; 2. <Elektrotech> breakdown, failure; drop-out (Stromnetz); 3. <Fernseh, Funktech, Kerntech, Maschinen, Mechan> breakdown, failure; 4. <Raumfahrt> blackout; 5. <Sicherheit> breakdown, failure (einer Maschine); 6. <Telekom> breakdown, failure (Netz); outage; 7. <Wassertrans> flare (Schiffkonstruktion)
Ausfall m des Vorstevens <Wassertrans> stem rake (Schiffkonstruktion)
Ausfall m mit Datenverlust <Maschinen> gang tuning capacitor
Ausfall m mit Zerstörung <Elektronik> destructive breakdown
Ausfallart f <Qual> failure mode
Ausfallart-, Ausfallauswirkungs- und Ausfallbedeutungsanalyse f <Qual> failure mode, effects and criticality analysis
Ausfallart- und Wirkungsanalyse f <Qual> failure mode and effects analysis
Ausfallbild n <Fernseh> early-finish video
Ausfalldatenkarte f <Lufttrans> failure data card
Ausfalldauer f 1. <Kerntech> outage time; 2. <Qual> outage duration; 3. <Telekom> outage time, unavailability time
Ausfalldichte f <Qual> failure density
Ausfalldichteverteilung f <Qual> failure density distribution
ausfallen v 1. <Chemtech> precipitate; 2. <Telekom> go down
ausfällen v <Abfall> precipitate
Ausfallerkennung f des Datenträgers <Telekom> data carrier failure detector
Ausfallhäufigkeit f <Qual> failure frequency
Ausfallhäufigkeitsdichte f <Qual> failure density
Ausfallkriterium n <Qual> failure criterion
Ausfallmodus m <Qual> failure mode
Ausfallmuster n <Qual> outfall sample
Ausfallquote f 1. <Maschinen> failure rate; 2. <Qual> failure quota; 3. <Raumfahrt> failure rate (Raumschiff)
Ausfallrate f <Comp & DV, Elektrotech, Qual, Telekom> failure rate

Ausfallratengewichtung f <Qual> failure rate weighting
Ausfallratenniveau n <Qual> failure rate level
Ausfallrisiko n <Qual> failure risk
ausfallsanft adj <Kontroll> fail-soft
ausfallsicher adj <Comp & DV, Elektrotech, Kontroll, Mechan, Qual, Raumfahrt, Sicherheit, Telekom> fail-safe
ausfallsicherer Betrieb m <Comp & DV> fail-safe operation
ausfallsicheres System n <Comp & DV> fail-safe system
ausfalltolerierend adj <Kontroll, Sicherheit> fail-soft
Ausfallton m <Fernseh> early-finish audio
Ausfällung f <Abfall> coagulation, precipitation
Ausfallursache f <Qual> failure cause
Ausfallverhalten n <Qual> failure mode
Ausfallwahrscheinlichkeit f <Qual> failure probability
Ausfallwahrscheinlichkeitsdichte f <Qual> failure probability density
Ausfallwahrscheinlichkeitsverteilung f <Qual> failure probability distribution
Ausfallwinkel m <Phys, Wellphys> angle of reflection
Ausfallzeit f 1. <Comp & DV> downtime, fault time; 2. <Erdöl> downtime (einer Anlage); 3. <Fernseh> downtime; 4. <Kerntech> oscillating electron, unavailability time; 5. <Lufttrans> downtime; 6. <Telekom> out-of-service time; downtime (Versagen)
ausfaltbare Antenne f <Telekom> unfurlable aerial, unfurlable antenna
Ausfiltern n <Comp & DV> gating
Ausflanschen n <Fertig> notching (Werkstofftrennen)
ausflecken v <Foto> spot
Ausflecken n <Foto> spotting
Ausfließen n <Wasserversorg> efflux
ausflocken v 1. <Chemie> clot, flocculate; 2. <Chemtech> coagulate, flocculate
Ausflockung f 1. <Abfall> flocculation; 2. <Chemie> flocculation; scavenging (Mineralaufbereitung); 3. <Chemtech, Erdöl, Kohlen, Kunststoff, Lebensmittel> flocculation • **zur Ausflockung bringen** <Chemtech> coagulate
Ausflockungsmittel n <Chemtech> coagulator
Ausflockungspunkt m <Chemtech> flocculation point
ausfluchten v <Bau> line out (Linie)
Ausfluchtung f <Bau> alignment
Ausflugsdampfer m <Wassertrans> excursion steamer; pleasure boat (Boottyp)
Ausfluss m 1. <Kerntech> issue; 2. <Meerschmutz> runoff (Vorgang); 3. <Phys> effusion, outward flux; 4. <Wasserversorg> outflow; Wehr discharge (Fluss)
Ausflusskoeffizient m 1. <Hydraul> coefficient of efflux; 2. <Hydraul> discharge coefficient (C); 3. <Qual> coefficient of efflux
Ausflussmenge f <Kerntech> issue
Ausflussseite f <Kerntech> outlet side; delivery side (einer Pumpe); outlet edge (einer Turbine)
Ausflussstrahl m <Fertig> issuing jet
Ausflussventil n <Hydraul> discharge valve
Ausflusswehr n für Aktivabfall <Kerntech> effluent weir
Ausfräsen n <Bau> routing
ausfugen v <Bau> joint
Ausfugung f <Bau> pointing (Aktion)
Ausfuhr f <Wassertrans> export
ausführbare Anweisung f <Comp & DV> executable statement
ausführbarer Befehl m <Comp & DV> executable instruction
ausführen v 1. <Comp & DV> execute, run (Programm); 2. <Patent, Qual> carry out; 3. <Telekom> perform • **eine Anwendung ausführen** <Comp & DV> run an application

Ausfuhrgenehmigung

Ausfuhrgenehmigung f <Wassertrans> *(BE)* export licence, *(AE)* export license *(Dokumente)*
Ausfuhrhafen m <Wassertrans> shipping port *(Hafen)*
ausführliche Darstellung f <Konstzeich, Patent> detailed representation
Ausführung f 1. <Bau> design, workmanship; 2. <Comp & DV> execution, running; 3. <Druck> finish; 4. <Fertig> grade; make; model *(Kunststoffinstallationen)*; 5. <Kontroll> execution; 6. <Maschinen> version; 7. <Patent> carrying out; 8. <Qual> carrying out, workmanship
Ausführung f **für Fahrzeugeinbau** <Mobilkom> car mounting version
Ausführung f **links** <Konstzeich> left-hand version
Ausführung f **rechts** <Konstzeich> right-hand version
Ausführungsanweisung f <Comp & DV> execute statement
Ausführungsausfallart- und -wirkungsanalyse f *(DFMEA)* <Qual> design failure mode and effects analysis *(DFMEA)*
Ausführungseinheit f <Comp & DV> run unit
Ausführungsform f <Patent> embodiment *(einer Erfindung)*
Ausführungsmodus m <Comp & DV> execute mode
Ausführungsphase f <Comp & DV> execute phase, execution phase
Ausführungsprogramm n <Comp & DV> executable
Ausführungssignal n <Comp & DV> execute signal
Ausführungszeichnung f <Heiz & Kälte, Qual> as-built drawing
Ausführungszeit f <Comp & DV> execution time
Ausfüllungsgrad m <Lufttrans> solidity *(Propeller)*
Ausgabe f 1. <Comp & DV> output, output data; 2. <Druck, Funktech, Kontroll> output
Ausgabeanschluss m <Comp & DV> output port
Ausgabeanschlusspunkt m <Comp & DV> output port
Ausgabeanzeige f <Comp & DV> output display
Ausgabebereich m <Comp & DV> output area
Ausgabeblock m <Comp & DV> output block
Ausgabedatei f <Comp & DV, Druck> output file
Ausgabedaten npl <Comp & DV> output data
Ausgabedrucker m <Comp & DV> terminal printer
Ausgabeeinheit f <Druck> output device
Ausgabeelement n <Comp & DV> output element
ausgabegebunden adj <Comp & DV> output-limited
ausgabegebundener Prozess m <Comp & DV> output-limited process
Ausgabegerät n <Comp & DV, Druck> output device
Ausgabegeschwindigkeit f <Comp & DV> output rate
Ausgabegröße f <Elektrotech> output quantity
Ausgabekanal m <Comp & DV> output channel
Ausgabekapazität f <Comp & DV> output capacity
Ausgabekennsatz m <Comp & DV> output label
Ausgabeklasse f <Comp & DV> output class
Ausgabekonfiguration f <Comp & DV> output configuration
Ausgabeleistung f <Elektriz> output
Ausgabeleitung f <Fernseh> outgoing line
Ausgabemedium n <Comp & DV> output medium
Ausgabeöffnung f <Comp & DV> gate
Ausgabepuffer m <Comp & DV> output buffer
Ausgaberücksetzung f <Raumfahrt> output backoff *(Weltraumfunk)*
Ausgabesatz m <Comp & DV> output record
Ausgabesteuerzeichen n <Comp & DV> output control character
Ausgabewarteschlange f <Comp & DV> output queue
Ausgang m 1. <Comp & DV> exit, output, outlet; 2. <Fernseh, Funktech> output; 3. <Kerntech> issue; 4. <Kontroll> output; 5. <Telekom> outlet

Ausgangsadmittanz f <Elektrotech> output admittance
Ausgangsanschluss m <Telekom> output port
Ausgangsbelastbarkeit f <Elektrotech> fan-out
Ausgangscode m <Comp & DV> source code
Ausgangsdämpfung f <Elektronik> output attenuation
Ausgangsdaten npl <Comp & DV> raw data
Ausgangsdose f <Elektriz> outlet box
Ausgangsdraht m <Elektriz> leading-out wire
Ausgangselektrode f <Elektrotech> output electrode
Ausgangsfeld n <Comp & DV> parent field
Ausgangsformat n <Konstzeich> starting size
Ausgangsgleichstrom m <Elektrotech> DC output, direct current output
Ausgangsgröße f <Elektrotech> output quantity • **zurück zur Ausgangsgröße** <Raumfahrt> anisoelastic *(Gyroskop)*
Ausgangsimpedanz f <Elektriz, Elektrotech, Fernseh, Telekom> output impedance
Ausgangskammer f <Elektronik> output cavity
Ausgangskanal m <Telekom> outgoing channel
Ausgangskapazität f <Elektrotech> output capacitance
Ausgangskasten m <Elektriz> outlet box
Ausgangsklemme f 1. <Elektriz> output terminal *(Kontakt)*; 2. <Elektrotech> output terminal
Ausgangskondensator m <Elektrotech> output capacitor
Ausgangskreis m <Elektrotech, Telekom> output circuit
Ausgangsladung f <Elektrotech> output charge
Ausgangsleistung f 1. <Elektriz> output power; 2. <Elektrotech> output power, output, power output; 3. <Telekom> output power
Ausgangsleitung f 1. <Aufnahme> line out; 2. <Elektriz> outgoing circuit; 3. <Fernseh> line out
Ausgangsleitwert m <Elektrotech> output admittance
Ausgangsmaterial n <Kohlen> raw material
Ausgangsmonitor m <Fernseh> output monitor
Ausgangsnuklid n <Kerntech> parent nuclide
Ausgangspegel m <Aufnahme, Fernseh, Telekom> output level
Ausgangsphase f <Metall> parent phase
Ausgangsregister n *(AR)* <Mobilkom> home location register *(HLR)*
Ausgangsregler m <Fernseh> output control
Ausgangsschaltung f <Elektrotech> output circuit
Ausgangsseite f <Fertig> outlet side *(Kunststoffinstallationen)*
Ausgangssignal n <Elektrotech, Fernseh, Telekom> output signal
Ausgangsspalte f <Druck> last column
Ausgangsspannung f <Elektriz, Elektrotech, Fernseh, Telekom> output voltage
Ausgangssprache f <Comp & DV> source language
Ausgangsstecker m <Elektriz> plug-type outlet
Ausgangssteckverbinder m <Telekom> outconnector
Ausgangsstelle f <Comp & DV> outconnector
Ausgangsstellung bringen v/in <Comp & DV> restore
Ausgangsstoff m **einer Reaktion** <Chemie> reactant
Ausgangsstrahl m <Kerntech> ejected beam
Ausgangsstrom m <Elektriz> output current
Ausgangsstufe f **ohne Transformator** <Aufnahme> transformerless output stage
Ausgangsteilchen n <Kerntech> initiating particle
Ausgangsteiler m <Elektronik> output attenuator
Ausgangstemperatur f <Heiz & Kälte> initial temperature
Ausgangstext m <Comp & DV> corpus
Ausgangstor n <Telekom> output port
Ausgangstransformator m <Aufnahme, Elektriz, Elektrotech, Phys> output transformer

Ausgangsübertrager m <Phys> output transformer
Ausgangsverstärker m <Elektronik, Funktech, Telekom> output amplifier
Ausgangsverzeichnis n <Comp & DV> root directory
Ausgangsverzweigung f <Elektrotech> fan-out, output branch
Ausgangswandler m <Elektrotech> output transducer
Ausgangswerte mpl <Qual> benchmarks
Ausgangswicklung f <Elektriz> output winding
Ausgangszeile f <Druck> break line
Ausgasung f <Kohlen> degasifying *(Bergbau)*
ausgebildetes Personal n 1. <Sicherheit> trained staff; 2. <Telekom> occupational forces
ausgeblendetes Signal n <Elektronik> gated signal
ausgebrochene Bohrung f <Erdöl> wild well *(Förderung von Erdöl und Erdgas)*
ausgebundener Kabelfächer m <Elektrotech> laced cablefan
ausgedehnt adj <Raumfahrt> expanded *(Raumschiff)*
ausgefächertes Kabel n <Elektrotech> fanned cable
ausgefahrene Straße f <Bau> heavy road
ausgefallen adj <Telekom> out of order
ausgeflossenes Öl n <Meerschmutz> spill
ausgeformtes Kabelende n <Elektrotech, Telekom> cable form
ausgefräst adj <Mechan> milled
ausgeführte A/D-Wandlung f /auf dem Chip <Elektronik> on-chip analog-to-digital conversion
ausgeführte D/A-Wandlung f /auf dem Chip <Elektronik> on-chip digital-to-analog conversion
ausgefüllt adj <Bau> filled
ausgeglichene Trittstufenfläche f <Bau> balance step
ausgeglichenes Querruder n <Lufttrans> balanced aileron
ausgeheilt adj <Thermod> annealed *(Halbleiter)*
ausgehen v <Elektrotech> go out *(Licht)*
ausgekehlt adj <Fertig> recessed
ausgekleideter Kanal m <Bau> lined canal
ausgekolkt adj <Fertig> grooved
ausgelaufen adj <Fertig> attrite
ausgelaufener Block m <Metall> bled ingot
ausgelaufenes Pleuellager n <Kfztech> run bearing
ausgelaugt adj <Kohlen> barren *(Erz)*
ausgelaugtes Gangerz n <Kohlen> barren gangue
ausgelegte Konzeption f /als Digitalschaltung <Elektronik> digital-circuit design
ausgenutzter Entwickler m <Foto> exhausted developer
ausgeprägte Seitenkeulen fpl <Elektronik> high sidelobes *(Antenne)*
ausgeprägter Pol m <Elektriz, Elektrotech> salient pole
ausgerückt adj <Maschinen> out of gear
ausgeschaltete Stellung f <Maschinen> off position
ausgeschlossen adj <Druck> justified
ausgeschlossener Text m <Druck> justified text
ausgesetzt adj 1. <Anstrich> exposed; 2. <Comp & DV> suspended
ausgesetzte Adressierung f <Comp & DV> deferred addressing
ausgespitzt adj <Fertig> pointed *(Bohrer)*
ausgestellt adj <Textil> flared
ausgestellter Abschnitt m <Raumfahrt> flared section *(Raumschiff)*
ausgewählter optischer Rohling m <Ker & Glas> selected chunk
ausgewähltes Amt n <Patent> elected office
ausgewalzt adj <Fertig, Metall> sheeted
Ausgewogenheit f <Aufnahme> balance
ausgewuchtet adj <Fertig> balanced
ausgezogen adj <Anstrich> exposed

ausgießen v <Fertig, Maschinen> bush
Ausgießverschluss m <Verpack> pour spout closure, pour spout seal
Ausgleich m 1. <Elektriz> balance; 2. <Elektronik, Elektrotech> compensation, equalization; 3. <Maschinen> make-up; 4. <Mechan> equalization; 5. <Phys> compensation
Ausgleich m **des Parallaxenfehlers** <Gerät> compensation of parallax *(Ablesefehler)*
Ausgleichbecken n 1. <Nichtfoss Energ> surge tank; 2. <Wasserversorg> equalizing tank
Ausgleichbehälter m 1. <Bau> make-up tank; 2. <Heiz & Kälte> equalizer tank; 3. <Kfztech> expansion tank; header tank *(Motor)*; 4. <Mechan> expansion tank; 5. <Wasserversorg> compensator reservoir
Ausgleichbereich m <Kerntech> range of compensation
Ausgleicheinrichtung f <Phys> compensator
ausgleichen v 1. <Bau> average out, planish; trim *(Straße)*; 2. <Kohlen> compensate; 3. <Papier> balance; 4. <Phys> compensate; 5. <Qual> average out
Ausgleichen n 1. <Bau> planishing; 2. <Papier> balancing
Ausgleichen n **mit Scheiben** <Maschinen> shimming
Ausgleicher m <Heiz & Kälte> equalizer
Ausgleichfläche f **am Ruder** <Lufttrans> balance tab
Ausgleichgetriebe n 1. <Fertig, Kfztech> differential; 2. <Maschinen> balance gear, differential gear, equalizing gear
Ausgleichgewicht n 1. <Bau> counterweight; 2. <Maschinen> balance weight
Ausgleichhebel m <Maschinen> balance lever
Ausgleichhorn n <Lufttrans> horn balance
Ausgleichimpulse mpl <Fernseh> equalizing pulses
Ausgleichkegelrad n <Kfztech> differential pinion, differential spider pinion; pinion gear *(Teil des Ausgleichsgetriebes)*; differential bevel gear *(Triebstrang)*
Ausgleichkolben m 1. <Fertig> dummy piston; 2. <Hydraul> balancing piston; 3. <Maschinen> balance piston
Ausgleichkraft f <Lufttrans> equalizing
Ausgleichkupplung f <Maschinen> compensating coupling, flexible coupling, resilient coupling
Ausgleichkurve f <Elektronik> equalization curve
Ausgleichluftdüse f <Kfztech> air correction jet
Ausgleichmagnetstreifen m <Fernseh> balancing magnetic stripe
Ausgleichpunkt m <Lufttrans> balance point
Ausgleichrad n <Maschinen> compensating gear
Ausgleichriet n <Textil> spacing reed
Ausgleichritzel n <Kfztech> differential pinion, pinion gear
Ausgleichruder n <Lufttrans> balance tab, balanced control surface
Ausgleichschacht m <Nichtfoss Energ> surge shaft
Ausgleichschaltung f 1. <Fernseh> shaping network; 2. <Gerät> compensating circuit
Ausgleichscheibe f <Maschinen> shim
Ausgleichscheibenventil n <Nichtfoss Energ> *(BE)* balanced disc valve, *(AE)* balanced disk valve
Ausgleichschiene f <Kfztech> equalizer bar
Ausgleichschienen-Drehgestell n <Eisenbahn> equalizer bar bogie
Ausgleichschwingung f <Phys> transient oscillation
Ausgleichsdüse f <Kfztech> compensating jet *(Vergaser)*
Ausgleichslage f <Bau> regulating course
Ausgleichsmodulation f <Telekom> dither
Ausgleichstellerventil n <Nichtfoss Energ> *(BE)* balanced disc valve, *(AE)* balanced disk valve
Ausgleichstern m <Kfztech> differential spider
Ausgleichstromversorgungsleitung f <Elektriz> equalizing feeder

Ausgleichswicklung

Ausgleichswicklung f <Elektriz> compensating winding
Ausgleichsystem n <Elektrotech> balancer
Ausgleichträger m <Kfztech> differential casing
Ausgleichunterlegscheibe f <Lufttrans> balance washer
Ausgleichventil n <Fertig> pressure-maintaining valve
Ausgleichverhalten n <Gerät> (AE) transient behavior, (BE) transient behaviour
Ausgleichwert m <Regelung> compensation value
Ausgleichzeit f 1. <Gerät> settling time; 2. <Regelung> compensation time
ausglühen v 1. <Mechan> anneal; 2. <Metall> anneal, temper; 3. <Phys, Thermod> anneal (Stahl)
Ausglühen n 1. <Elektronik, Heiz & Kälte, Metall> annealing; 2. <Thermod> annealing; anneal (Stahl)
Ausguck m <Wassertrans> lookout (bei Navigation für Person) • **Ausguck halten** <Wassertrans> keep a lookout
Ausguss m 1. <Bau> lip, sewer; 2. <Fertig> lining; 3. <Ker & Glas> lip; 4. <Labor> spout; 5. <Maschinen> antifriction lining, nozzle; 6. <Umweltschmutz> sink; 7. <Verpack> pour spout; 8. <Wasserversorg> spout (Pumpe)
Ausgussmetall n <Maschinen> bush metal, bushing metal
Ausgussröhre f <Wasserversorg> delivery pipe
Ausgussstutzen m <Maschinen> pouring sleeve
Aushaken n <Maschinen> unhooking
aushämmern v 1. <Fertig> pane; 2. <Maschinen> batter
aushärtbar adj <Mechan, Thermod> thermosetting
aushärten v 1. <Anstrich> cure; 2. <Bau> cure, mature (Mörtel, Beton); 3. <Kunststoff> cure; 4. <Metall> temper
Aushärten n 1. <Anstrich> curing; 2. <Bau> cure; 3. <Fertig> curing; 4. <Kunststoff> hardening; 5. <Metall> age hardening
Aushärtezeit f <Bau> curing period
Aushärtung f 1. <Fertig> (BE) quench ageing, (AE) quench aging; 2. <Metall> dispersion hardening, precipitation hardening; 3. <Wassertrans> curing (Schiffbau)
aushärtungsfähig adj <Fertig> age-hardenable (Stahl)
Aushärtungsfähigkeit f <Fertig> age hardenability (Stahl)
Aushärtungstemperatur f <Anstrich> cure temperature
Aushärtungszeit f <Bau> cure rate
ausheben v 1. <Bau> cut; sink (Baugrube); 2. <Fertig> draw; lift (Modell)
Ausheben n 1. <Bau> digging; 2. <Fertig> withdrawing; drawing (Gießen)
Aushebestift m <Mechan> ejector pin
ausheilen v <Mechan, Phys> anneal
Ausheilen n <Thermod> annealing
Ausheizofen m <Thermod> firebox
aushöhlen v <Bau> hollow out, hollow
Aushöhlen n <Fertig> hollowing
Aushöhlung f <Bau> hollow
Ausholer m <Wassertrans> outhaul (Segeln)
Aushub m 1. <Bau> excavation; 2. <Fertig> relief motion, retraction (Tieflochbohrer)
Aushubboden m <Bau> spoil
auskehlen v 1. <Bau> channel; hollow (Holz); 2. <Fertig> hollow, hollow out, (AE) mold, (BE) mould; 3. <Maschinen> recess
Auskehlung f 1. <Bau> hollowing; 2. <Fertig> (BE) channelling; 3. <Maschinen> recess; 4. <Mechan> flute
auskippen v <Trans> dump
Auskippen n <Bau> dumping
auskitten v <Bau> stop with putty
auskleiden v <Bau> line
Auskleidung f 1. <Bau> coating, lining; 2. <Metall, Papier> lining; 3. <Wasserversorg> lining (einer Pumpe)

Auskleidung f des Fahrzeughimmels <Kfztech> head lining (Innenausstattung)
Ausklink... <Maschinen> trip
ausklinken v 1. <Bau> notch; cope (Träger); 2. <Fertig> interrupt (Gewindebohrer); cope (Träger); 3. <Maschinen> release, throw out; 4. <Mechan> disengage; 5. <Raumfahrt> drop (Raumschiff)
Ausklinken n 1. <Fertig> coping (eines Trägers); 2. <Kerntech> uncoupling (eines Brennelementes); unlatching (eines Elementes bei Schnellabschaltung); 3. <Raumfahrt> decoupling (Weltraumfunk)
Ausklinken n der Ladung <Lufttrans> load release (Flugwesen)
Ausklinkhaken m <Bau> releasing hook
Ausklinkmechanismus m <Maschinen> trip, trip gear
Ausklinkstelle f <Lufttrans> knock-out station
ausknicken v <Fertig> (AE) buckle, (BE) collapse
Ausknicken n <Fertig> crippling
auskohlen v <Fertig> crater (Spanung)
Auskohlung f <Fertig> crater (Spanung)
auskolken v <Fertig> pit; erode (Spanfläche)
Auskolken n 1. <Fertig> cupping; 2. <Maschinen> pitting
Auskolkung f 1. <Fertig> crater wear; erosion (Spanfläche); pit (Spanung); 2. <Maschinen> crater
Auskopier... <Druck, Foto> printing-out
Auskopieremulsion f <Foto> printing-out emulsion
Auskopierpapier n <Druck> printing-out paper
Auskoppel... <Elektronik> catcher
auskoppeln v <Telekom> tap
Auskoppelraum m <Elektronik> catcher cavity (Klystron)
Auskoppelspalt m <Elektronik> catcher space
Auskopplung f <Eisenbahn> uncoupling
auskragen v <Bau> project
auskragend adj <Bau> overhanging
Auskragung f <Bau> cantilever, overhang, projection
Auskreiden n <Kunststoff> chalking
Auskreidung f <Kunststoff> chalking
Auskreuzen n <Telekom> transposition (Telefonleitung)
auskristallisieren v <Chemtech> crystallize out
Auskristallisieren n <Ker & Glas> (AE) sulfuring, (BE) sulphuring
Auskunftsdienst m <Telekom> inquiry facility
Auskunftsdienste mpl <Telekom> directory enquiries
auskuppeln v 1. <Eisenbahn> uncouple; 2. <Kfztech> put out of gear, unclutch; 3. <Kfztech> declutch; 4. <Maschinen> declutch, unclutch
Auskuppeln n 1. <Kfztech> clutch throwout, declutching, disengagement; 2. <Maschinen> unclutching
auskurbelbares Mittelrohr n <Foto> (AE) geared center column, (BE) geared centre column (Stativ)
ausladen v 1. <Comp & DV> roll out; 2. <Wassertrans> unship (Ladung)
Ausladen n <Wassertrans> landing
ausladend adj 1. <Bau> overhanging; 2. <Mechan> flared
Ausladung f 1. <Bau> radius (eines Kranes); 2. <Fertig> overhang, throat distance; 3. <Maschinen> overhang • **ohne Ausladung** <Fertig> throatless
auslagern v <Comp & DV> roll out, swap
Auslagern n <Comp & DV> swap-out
Auslagerungsfunktion f <Comp & DV> swapping
Auslandsfernwahl f <Telekom> IDD, IDDD, (AE) international direct dialing, (BE) international direct dialling, (AE) international direct distance dialing, (BE) international direct distance dialling
Auslands-Kopfvermittlungsstelle f <Telekom> IGN, international gateway node
Auslandsvermittlungsstelle f <Telekom> international gateway exchange
Auslängung f <Kunststoff> sag

Auslass m 1. <Heiz & Kälte> discharge; 2. <Hydraul> gating; 3. <Kerntech, Kfztech, Maschinen, Wasserversorg> outlet
Auslassdampf m <Hydraul> exhaust steam
Auslassdeckung f <Hydraul> exhaust cover, exhaust lap *(Steuerschieber)*
Auslassdurchflussregelung f <Wasserversorg> outlet flow control
auslassen v <Hydraul> gate
Auslasshub m <Kfztech> exhaust stroke
Auslasskanal m 1. <Hydraul> exhaust port; 2. <Kfztech> exhaust passage; exhaust port *(Viertaktmotor)*
Auslasskante f <Hydraul> exhaust edge *(Steuerschieber)*
Auslassnocken m <Mechan> exhaust cam
Auslassöffnung f 1. <Hydraul> exhaust port; 2. <Ker & Glas> outlet port; 3. <Kfztech> exhaust port *(Motor, Auspuff)*; 4. <Maschinen> outlet
Auslassschlitz m <Hydraul, Kfztech> exhaust port *(Zweitakt- und Kreiskolbenmotor)*
Auslasstemperatur f <Heiz & Kälte, Phys, Thermod> discharge temperature
Auslassüberdeckung f <Hydraul> exhaust cover, exhaust lap *(Steuerschieber)*
Auslassung f 1. <Comp & DV> skip; 2. <Druck> ellipsis, out
Auslassungszeichen n 1. <Comp & DV> ignore character; 2. <Druck> caret
Auslassventil n 1. <Heiz & Kälte> discharge valve; 2. <Hydraul> delivery valve, exhaust valve, outlet valve; 3. <Kfztech> outlet valve; 4. <Maschinen> discharge valve; 5. <Mechan, Papier> exhaust valve
Auslastung f <Heiz & Kälte> capacity utilization • **mit voller Auslastung arbeiten** <Maschinen> work to full capacity
Auslauf m 1. <Bau> mouth; 2. <Fertig> drainage *(Kunststoffinstallationen)*; 3. <Ker & Glas> spout *(beim Walzverfahren)*; 4. <Maschinen> run-out, runout
Auslaufabdeckung f <Ker & Glas> *(AE)* spout cover
Auslaufanschluss m <Papier> outlet
Auslaufbecher m <Kunststoff> cup, flow cup
auslaufen v 1. <Bau> leak; 2. <Kohlen> bleed off; 3. <Kunststoff> bleed; 4. <Textil> bleed off *(Farbe)*; 5. <Wassertrans> get under way, put to sea, sail away; set out *(Schiff)*; 6. <Wasserversorg> run out
Auslaufen n 1. <Meerschmutz> leakage; 2. <Wassertrans> sailing
auslaufend adj <Wassertrans> outward bound *(Schiff)*
Auslaufrille f 1. <Akustik> lead-out groove; 2. <Aufnahme> concentric groove, lead-out groove; 3. <Maschinen> undercut
Auslaufrückstand m <Ker & Glas> spout
Auslaufschacht m <Ker & Glas> running-out pit
Auslaufstein m <Ker & Glas> tap out block
Auslaufventil n 1. <Fertig> outlet valve *(Kunststoffinstallationen)*; 2. <Maschinen> plug cock, plug valve
Auslaufzähler m <Gerät> outflow meter
auslaugen v 1. <Abfall> lixiviate; 2. <Chemie> extract, leach; 3. <Fertig> lixiviate; 4. <Meerschmutz> leach
Auslaugen n 1. <Abfall> leaching; 2. <Chemie> extraction, leaching; 3. <Kohlen, Lebensmittel, Umweltschmutz> leaching
Auslaugkoeffizient m <Kerntech> leaching coefficient
Auslaugmittel n <Kerntech> leachant, leaching agent
Auslaugtest m <Kerntech> leachability test, leaching test
Auslaugung f 1. <Abfall> elutriation, lixiviation; 2. <Fertig> lixiviation; 3. <Ker & Glas> dealkalization
Auslaugungsgraben m <Wasserversorg> leaching trench
Auslaugverfahren n <Abfall> leaching property

Auslegearm m <Eisenbahn> post bracket
auslegen v 1. <Bau> inlay *(mit Parkett)*; 2. <Mechan> design
Auslegen n **von Seezeichen** <Wassertrans> seamarking
Ausleger m 1. <Bau> cantilever, flange; boom *(Kran)*; 2. <Eisenbahn> cantilever; 3. <Erdöl> boom *(Kran)*; 4. <Fertig> boom, jib, radial arm; outrigger *(Kran)*; arm *(Radialbohrer)*; 5. <Kerntech> jib *(Kran)*; 6. <Lufttrans> jib *(Hubschrauber)*; 7. <Maschinen> boom, cantilever, jib, outrigger, radial arm; 8. <Papier> beam, cantilever; 9. <Phys> cantilever; 10. <Raumfahrt> boom *(Raumschiff)*
Auslegerarm m <Bau, Fertig> jib
Auslegerbalken m <Bau, Phys> cantilever beam
Auslegerbrücke f <Bau> cantilever bridge
Auslegerklemmung f <Fertig> arm clamping; arm-clamping mechanism *(Radialbohrer)*
Auslegerkran m 1. <Bau, Kerntech> jib crane; 2. <Lufttrans> jib *(eines Hubschraubers)*; 3. <Maschinen> jib crane; 4. <Mechan> derrick
Auslegung f <Maschinen> design
Auslegung f **als integrierte Schaltung** <Elektronik> integrated-circuit design
Auslegung f **einer Analogschaltung** <Elektronik> *(AE)* analog circuit design, *(BE)* analogue circuit design
Auslegung f **einer integrierten Schaltung** <Elektronik> integrated-circuit layout
Auslegungsabbrand m <Kerntech> design burnup
Auslegungsbeben n <Kerntech> safe shutdown earthquake *(größtes verzeichnetes Erdbeben)*
Auslegungsbericht m <Qual> design report
Auslegungsbestimmungen fpl <Qual> design specifications
Auslegungsdruck m <Heiz & Kälte, Qual> design pressure
Auslegungsgeschwindigkeit f <Trans> design speed
Auslegungsgrenzen fpl <Qual> design limits
Auslegungskriterium n <Kerntech> design criterion
Auslegungsleistungsabgabe f <Kerntech> designed power required output
Auslegungsstörfall m <Kerntech> design basis accident, design basis event
Auslegungsstrahlenpegel m <Kerntech> design irradiation level
Auslenkung f <Kfztech> deflection *(bei einem Reifen)*
auslesen v 1. <Comp & DV> read out *(Informationen)*; 2. <Kohlen> cull
Auslesen n <Comp & DV> readout
Ausleuchtung f 1. <Elektrotech> illumination *(Ziel)*; 2. <Funktech> illumination *(Antenne)*; 3. <Raumfahrt> luminance *(Weltraumfunk)*
Ausleuchtung f **einer Hemisphäre** <Telekom> hemispherical coverage
Ausleuchtungswirkungsgrad m <Raumfahrt> illumination efficiency *(Weltraumfunk)*
Ausleuchtzone f <Telekom> footprint
Ausleuchtzone f **auf der Erde** <Telekom> Earth coverage area *(Satellitenkommunikation)*
Auslieferungsband m <Comp & DV> product shipment tape
ausloggen v <Mobilkom> log out
auslöschen v <Thermod> quench *(Feuer)*
Auslöschung f <Phys> extinction
Auslöse... <Aufnahme, Comp & DV, Telekom> release
Auslöseanforderung f <Comp & DV, Telekom> clear request
Auslöseanschlag m <Fertig> trip dog
Auslösedauer f <Aufnahme> release time
Auslöseeinrichtung f <Sicherheit> trip bar, trip device
Auslösehandgriff m **mit Verschlussauslöser** <Foto> pistol grip with shutter release

Auslösehebel

Auslösehebel m <Bau> trip lever
Auslöseimpuls m 1. <Comp & DV> trigger; 2. <Telekom> triggering lead pulse
Auslöseknopf m <Foto> shutter release button
Auslösekontakte mpl <Kfztech> trigger contacts
auslösen v 1. <Comp & DV, Elektriz> trigger; 2. <Mechan> release; 3. <Phys> trigger, trip
Auslösen n 1. <Comp & DV> triggering; 2. <Elektriz> tripping; 3. <Telekom> clear down; release *(Telefon)*
auslösendes Teilchen n <Kerntech> initiating particle
Auslöseprozedur f <Telekom> clearing procedure
Auslöser m 1. <Elektriz> shutter release, trigger; 2. <Elektrotech> release; 3. <Fertig> detent, tripper; 4. <Foto> release, trigger *(Kameraverschluss)*; 5. <Maschinen> release, trigger; 6. <Sicherheit> trip device
Auslöserrelais n 1. <Elektriz> tripping relay; 2. <Foto> trigger relay
Auslöserstromkreis m <Foto> triggering circuit
Auslöseschalter m 1. <Elektronik> trigger switch; 2. <Sicherheit> release, trip switch
Auslöseschaltung f <Elektriz> tripping circuit
Auslöseschaltvorrichtung f <Sicherheit> trip switch device, trip switch
Auslösesicherung f <Sicherheit> safety trip control
Auslösesignal n <Kerntech> actuating signal *(in automatischer Steuerung)*
Auslösespule f <Elektriz> tripping coil
Auslösestrom m <Elektrotech> release current *(bei Schalter)*
Auslöseverzögerung f <Elektrotech> release lag
Auslösevorrichtung f 1. <Elektriz> trip gear; 2. <Maschinen> tripper, tripping device, tripping mechanism
Auslösezeichen n <Telekom> release signal
Auslösezeit f <Elektrotech> release time
Auslösung f 1. <Elektrotech> firing, realizing; 2. <Maschinen> release *(durch Nocken)*; 3. <Telekom> cleardown
Auslösungsanforderungspaket n <Telekom> clear-request packet
auslösungsfreies Abschalten n <Elektriz> trip-free release
ausloten v <Bau> lead
Ausmahlungsgrad m <Lebensmittel> extraction rate *(bei Getreide)*
Ausmauerung f <Bau> nogging
Ausmeißeln n <Fertig> gouging
Ausmessgerät n <Bau> measuring apparatus
Ausmündung f <Hydraul> opening *(in Trennwand, Platte)*
Ausnadeln n <Textil> unpinning
Ausnahme f <Comp & DV> exception
Ausnahmebedingung f <Comp & DV> exception
Ausnahmebehandlungsroutine f <Comp & DV> exception handler
Ausnahmeverwaltung f <Comp & DV> exception management
ausnehmen v <Maschinen> recess
ausnutzen v <Patent> take advantage of
Ausnutzung f <Mechan, Qual> efficiency
Ausnutzungsfaktor m <Maschinen> utilization factor
Ausnutzungskurve f <Nichtfoss Energ> utilization curve
auspacken v <Comp & DV> unpack
auspflanzen v <Bau> bed out
auspolstern v <Textil> pad
Ausprägung f <Künstl Int> instance
auspressen v <Bau> grout *(Spannbeton)*
ausprobieren v <Bau> try
Ausprüfen n <Kerntech> checkout
Auspuff m 1. <Kfztech> exhaust, exhaust system; 2. <Lufttrans> exhaust *(Triebwerk, Motor)*; 3. <Mechan> exhaust, *(AE)* muffler, *(BE)* silencer

Auspuffanlage f <Kfztech> exhaust system
Auspuffdampf m <Thermod> dead steam
Auspuffgas n <Kfztech, Mechan> exhaust gas
Auspuffgegendruck m <Lufttrans, Mechan> exhaust backpressure
Auspuffgehäuse n <Kfztech> *(AE)* muffler shell, *(BE)* silencer shell
Auspuffhub m <Kfztech, Mechan> exhaust stroke
Auspuffkrümmer m <Kfztech> exhaust manifold
Auspuffrohr n <Kfztech, Lufttrans, Mechan, Thermod, Trans, Wassertrans> exhaust pipe
Auspuffsammelleitung f <Mechan> exhaust manifold
Auspuffsammler m <Mechan> exhaust collector, exhaust manifold
Auspuffstutzen m <Mechan> exhaust stack
Auspuffsystem n <Mechan> exhaust arrangement
Auspufftopf m 1. <Kfztech> *(AE)* exhaust muffler, *(BE)* exhaust silencer; *(AE)* muffler, *(BE)* silencer *(Auspuffanlage)*; 2. <Mechan> *(AE)* exhaust muffler, *(BE)* exhaust silencer
Auspuffummantelung f <Kfztech> *(AE)* muffler jacket, *(BE)* silencer jacket
Auspuffventil n <Mechan> exhaust valve
Auspuffverkleidung f <Lufttrans> exhaust case
auspumpen v 1. <Elektronik> evacuate *(Röhren)*; 2. <Wassertrans> evacuate *(Schiff)*; 3. <Wasserversorg> pump out
Auspumpen n <Wasserversorg> pumping-out
ausrechnen v <Math> calculate
Ausregelzeit f <Gerät> transient time
Ausreiben n **fluchtender Bohrungen** <Fertig> line reaming
Ausreiben n **von Grundbohrungen** <Fertig> blind-hole reaming
ausreichend adj <Kontroll> satisfactory
Ausreißer m 1. <Gerät> outlier *(Messwert im unnormalem Streubereich)*; 2. <Qual> maverick, outlier
ausreiten v <Wassertrans> hike *(Segeln)*
Ausreitgurt m <Wassertrans> hiking strap *(Segeln)*
ausrichten v 1. <Bau> align, take out of wind; orient *(Instrument)*; 2. <Foto> sight *(Kamera)*; align *(Projektor)*; 3. <Funktech> point *(Antenne)*; 4. <Ker & Glas> true up; 5. <Maschinen> align
Ausrichten n 1. <Bau> boning; 2. <Fertig> lining-up; 3. <Funktech> alignment *(Antenne)*; 4. <Kerntech> deconvolution; 5. <Raumfahrt> pointing *(Weltraumfunk)*
Ausrichtepassstift m <Fertig> location dowel
Ausrichtgenauigkeit f <Funktech> pointing accuracy *(Antenne)*
Ausrichtung f 1. <Bau, Druck> alignment; 2. <Maschinen> lining-in, location; 3. <Mechan> alignment, *(AE)* leveling, *(BE)* levelling; 4. <Metall> ordering *(Moleküle)*
Ausrichtung f am Rand <Comp & DV> justification
Ausrichtungsfehler m 1. <Raumfahrt> pointing error *(Weltraumfunk)*; 2. <Telekom> alignment fault *(Antenne)*
Ausrichtungsgenauigkeit f <Raumfahrt> pointing accuracy *(Weltraumfunk)*
Ausrichtungsverlust m <Raumfahrt> pointing loss *(Weltraumfunk)*
Ausrück... 1. <Kfztech> clutch; 2. <Maschinen> clutch, disengaging, release
ausrücken v 1. <Fertig> demesh; unclutch *(Kupplung)*; 2. <Maschinen> throw out of action, throw out of gear; 3. <Mechan> disengage
Ausrücken n <Trans> disconnecting
Ausrücken n der Kupplung <Kfztech> declutching
Ausrücker m <Maschinen> release, stop motion
Ausrückfeder f <Maschinen> clutch spring

Ausrückgabel f 1. <Kfztech> clutch release fork *(Kupplung)*; 2. <Maschinen> clutch fork, fork
Ausrückhebel m <Maschinen> disengaging lever, release lever
Ausrücklager n <Kfztech> clutch release bearing *(Kupplung)*
Ausrundung f <Fertig> internal radius
ausrüsten v 1. <Bau> fit, outfit; 2. <Maschinen> fit out; 3. <Telekom> equip; 4. <Textil> finish; 5. <Wassertrans> equip, fit out *(Schiff)*
Ausrüsten n 1. <Bau> fitting-out; 2. <Wassertrans> fitting out *(Schiff)*
Ausrüstung f 1. <Bau> plant; 2. <Fertig> outfit; 3. <Foto> apparatus; 4. <Kerntech> kit; 5. <Maschinen> equipment, gear, outfit; 6. <Papier> finishing; 7. <Textil> finish
Ausrüstungsbecken n <Wassertrans> fitting-out berth
Aussägen n <Ker & Glas> sawing out
Aussagenlogik f <Künstl Int, Math> propositional logic
Ausschachtbarkeit f <Bau, Kohlen> excavatability
ausschachten v <Bau, Kohlen> excavate
Ausschachten n <Bau> digging
Ausschachtung f <Bau, Kohlen> excavation
ausschalen v 1. <Bau> strike, strip; strike *(Beton)*; 2. <Bau> strip formwork
Ausschalen n <Bau> stripping
Ausschalt... 1. <Elektrotech> cutoff, turn-off; 2. <Kontroll> turn-off
ausschalten v 1. <Comp & DV> disable, disconnect; 2. <Elektrotech> disconnect; turn off *(Lampe)*; 3. <Fernseh, Funktech, Kontroll> switch off; 4. <Maschinen> stop; 5. <Telekom> switch off
Ausschalten n <Elektrotech> turn-off *(Lampe)*
ausschaltende Windgeschwindigkeit f <Nichtfoss Energ> cutout wind speed
Ausschalter m 1. <Elektrotech> switch, cutout, interrupter; 2. <Phys> single throw switch
Ausschaltrille f <Akustik, Aufnahme> lead-out groove
Ausschaltung f <Elektrotech> cutoff, disconnection
ausschaltverzögertes Relais n <Elektrotech> off-delay relay
Ausschaltverzögerung f <Kontroll> turn-off delay
Ausschaltverzug m <Elektrotech> opening time
Ausschaltzeit f 1. <Elektrotech> off period *(Schaltung)*; 2. <Kontroll> turn-off time
Ausschaltzustand m <Elektrotech> off-state
Ausschalung f <Bau> release
Ausschankgerät n <Verpack> dispenser
Ausschärfung f <Fertig> scarfing; bevel *(Blech)*
Ausschäumen n <Chemtech> foaming *(Hohlräume)*
Ausscheidung f 1. <Chemie> exudation; 2. <Chemtech> sediment; 3. <Metall> precipitation
Ausscheidungsglühen n <Metall, Thermod> precipitation anneal
ausscheidungshärten v <Metall, Thermod> precipitation-harden
Ausscheidungshärten n 1. <Fertig> *(BE)* ageing, *(AE)* aging; age hardening *(Stahl)*; 2. <Metall> age hardening
Ausscheidungshärtung f 1. <Elektriz> intentional accelerated curing, intentional component curing, intentional normal curing; 2. <Fertig, Thermod> precipitation hardening
Ausscheidungsmittel n <Chemtech> separating agent
ausscheren v <Textil> nap the pile
Ausscheren n 1. <Fertig> crippling; 2. <Trans> leaving a line of traffic
ausschieben v <Fertig> exhaust *(Motor)*
ausschießen v <Druck> impose
Ausschießen n <Druck> imposition
ausschiffen v <Wassertrans> disembark, land, put ashore *(Passagiere)*

Ausschiffung f <Wassertrans> landing; disembarkation *(Passagiere)*
Ausschlachten n <Kfztech> cannibalizing *(Autoverwertung)*
ausschlacken v <Fertig> slag
Ausschlag... <Gerät> deflection
Ausschlagbecken n <Kohlen> slurry pond
ausschlagen v <Gerät> deflect *(Zeiger)*
ausschlagen lassen v <Maschinen> deflect *(Zeigernadel)*
Ausschlagen n <Fertig> stuffing
Ausschläger m <Textil> finisher scutcher
Ausschlagmethode f <Gerät> deflection method
Ausschlagwinkel m <Gerät> deflection angle *(Zeiger)*
ausschlämmen v <Wasserversorg> cleanse
Ausschleifen n <Kfztech> reboring
Ausschleudermaschine f <Chemtech> centrifuge
ausschleudern v <Chemtech> centrifuge
ausschließen v <Druck> justify
Ausschließen n <Druck> justification
ausschließlich zugeordnet adj <Comp & DV> dedicated
ausschließliche Lizenz f <Patent> *(BE)* exclusive licence, *(AE)* exclusive license
ausschließliches Recht n <Patent> exclusive right
Ausschlusstaste f <Druck> justification key
Ausschlusstrommel f <Druck> justifying scale
Ausschmelzmodell n <Fertig> investment pattern *(Gießen)*
Ausschmelzverfahren n <Maschinen> investment casting • **im Ausschmelzverfahren genauigkeitsgegossen** <Fertig> investment-cast
ausschmieden v <Fertig> draw out
Ausschmieden n <Fertig> drawing-out
ausschneiden v 1. <Fertig> rout; 2. <Maschinen> cut out
Ausschneiden n <Fertig> routing
Ausschnitt m 1. <Comp & DV> window; 2. <Elektrotech> cutout; 3. <Fertig> blanking; 4. <Ker & Glas> cutout; 5. <Maschinen> blank, blanking, cutout; 6. <Mechan> cutout; 7. <Phys> aperture
Ausschnittsvergrößerung f <Foto> cutout photograph, section enlargement
ausschöpfen v <Wassertrans> bail
Ausschuss m 1. <Bau> rejects; spoilage; 2. <Druck> broke; 3. <Fertig> refuse, rejects; 4. <Kohlen> scrap; 5. <Mechan> discard, scrap; 6. <Papier> broke; 7. <Qual> scrap, waste
Ausschuss... <Fertig, Maschinen> no-go
Ausschussgrenze f <Qual> limiting quality, limiting quality level, lot tolerance percentage of defectives, rejectable quality level
Ausschusslehre f <Maschinen> *(AE)* no-go gage, *(BE)* no-go gauge
Ausschusspapier n <Papier> refuse
Ausschussporzellan n <Ker & Glas> outshot of porcelain
Ausschussquote f <Qual> rejects rate
Ausschussseite f <Maschinen> no-go end
Ausschussteil n <Qual> reject, rejected item
ausschütten v <Bau> tip
Ausschütten n <Bau> dumping
Ausschwefeln n <Kunststoff> *(AE)* sulfur blooming, *(BE)* sulphur blooming
Ausschweifung f <Fertig> flaring
Ausschwimmen n <Kunststoff> floating *(von Farbstoffen)*
Ausschwimmen n **des Pigments** <Kunststoff> pigment floating
Ausschwing... <Aufnahme, Elektrotech> decay
Ausschwingen n <Aufnahme, Funktech> decay
Ausschwingkurve f <Aufnahme> decay curve

Ausschwingzeit

Ausschwingzeit f <Elektrotech> decay time *(Messanzeige)*
ausschwitzen v 1. <Chemie> exude; 2. <Lebensmittel> sweat *(Gießerei)*; sweat *(Ofen)*; 3. <Metall> sweat, sweating
Ausschwitzen n <Chemie, Kunststoff> exudation
Aussehen n 1. <Textil> look; 2. <Werkprüf> appearance, visual appearance
ausseigern v <Fertig> liquate
Außen... 1. <Elektriz> exterior, external; 2. <Elektronik> external; 3. <Verpack> exterior, outside
Außenabmessungen fpl <Maschinen> outside dimensions
Außenabnahme f beim Zulieferanten <Qual> subcontractor source inspection
Außenantrieb m <Lufttrans> power takeoff
Außenaufnahme f <Fernseh> field pick-up
Außenbord... <Maschinen, Wassertrans> outboard
Außenborder m <Wassertrans> outboard motorboat
Außenborderschlauchboot n <Wassertrans> outboard inflatable
Außenbordmotor m <Wassertrans> outboard motor
außenbords adv <Wassertrans> outboard
Außenbordschnellboot n <Wassertrans> out-board speedboat
aussenden v <Comp & DV> emit
Außendienstunterstützung f <Comp & DV> field support
Außendrehen n <Fertig> external turning, outside turning
Außendurchmesser m 1. <Fertig> OD, outside diameter *(Kunststoffinstallationen)*; 2. <Maschinen, Mechan> OD, outside diameter
Außeneinbau m <Fertig> external mounting
Außenfläche f 1. <Geom> outside face; 2. <Verpack> exterior surface
Außenflächenräumen n <Fertig> surface broaching
Außenflächenräummaschine f <Fertig> surface broaching machine
außengekühlt adj <Kerntech> externally cooled
Außengewinde n <Maschinen> external screw thread, male thread, outside screw thread
Außengewindeschneiden n <Fertig, Maschinen> external threading
Außengewindeschneider m <Maschinen> outside-threading tool
Außengewindeschneidmaschine f <Fertig> bolt cutter
Außengewindeschraube f <Maschinen> male screw
Außenhafen m <Wassertrans> *(AE)* outer harbor, *(BE)* outer harbour
Außenhaut f 1. <Lufttrans> skin *(Luftfahrzeug)*; 2. <Wassertrans> outer hull, outer skin, shell, skin *(Schiffbau)*
Außenhautbeplattung f <Wassertrans> shell plating *(Schiffbau)*
Außenhautplan m <Wassertrans> drawing of shell expansion *(Schiffkonstruktion)*
Außenhülle f <Fertig> coating *(Faseroptik)*
Außenkabel n <Fertig> outdoor cable *(Faseroptik)*
Außenkegel m <Fertig> external taper
aussenken v <Fertig> boss *(Naben)*
Aussenken n <Fertig> spotting
Außenkühlung f <Heiz & Kälte> surface cooling
Außenlastträger m <Lufttrans> pylon
Außenläufermotor m <Elektriz> external rotor motor
Außenleiter m <Telekom> outer conductor
Außenlinie f <Fertig> contour
Außenluft f <Heiz & Kälte> outdoor air, outside air
Außenlufttemperaturanzeige f <Lufttrans> outside air temperature indicator *(Flugwesen)*
Außenlufttemperaturfühler m <Lufttrans> outside air temperature probe *(Flugwesen)*

außenluftunabhängiges Atemschutzgerät n <Sicherheit> powered respirator
außenluftunabhängiges Filtergerät n <Sicherheit> powered air-purifying respirator
Außenmantel m <Fertig> coating *(optical fibres)*
Außenmaße npl <Verpack> outside dimensions
Außenmikrometer n <Maschinen> external micrometer
Außenpackmaschine f <Verpack> exterior packaging machine
Außenplanetenmission f <Raumfahrt> outer planet mission
Außenpolgenerator m <Elektriz> exterior pole generator, external pole generator
Außenrad n <Kfztech> annulus
Außenraster n <Elektronik> external grid
Außenräumen n <Fertig> external broaching, surface broaching
Außenräummaschine f 1. <Fertig> surface broaching machine; 2. <Maschinen> external broaching machine
Außenräumwerkzeug n 1. <Fertig> surface broach; 2. <Maschinen> external broach
Außenreportage f <Fernseh> OB, outside broadcast
Außenring m 1. <Fertig> cup *(Kegelrollenlager)*; outer race *(Kugellager)*; 2. <Maschinen> outer race
Außenrundschleifen n <Maschinen> cylindrical grinding, external cylindrical grinding
Außen-Rundschleifmaschine f <Fertig> external cylindrical grinding machine
Außenrüttler m <Bau> external vibrator
außenschleifen v <Fertig> surface grind
Außenschleifen n <Fertig> surface grinding
Außenschürze f <Wassertrans> peripheral skirt
Außenseite f 1. <Bau> face; 2. <Verpack> exterior surface
außenseitige elektrische Einrichtung f <Elektriz> outdoor electrical installation
Außenspiegel m <Kfztech> side mirror *(Zubehör)*
außenstehender Pfeil m <Konstzeich> outward-positioned arrowhead
Außenströmung f <Strömphys> free stream
Außentaster m <Maschinen> outside calipers
Außenteil n <Maschinen> outer member
Außentreppe f <Bau> fliers
Außenübertragung f <Fernseh> OB, outside broadcast
Außenübertragungsgruppe f <Fernseh> OB unit
Außenübertragungskabel n <Fernseh> outside plant cable
Außen- und Innentaster m <Maschinen> *(AE)* outside-and-inside calipers, *(BE)* outside-and-inside callipers
Außenverpackung f <Verpack> outer case
außenverzahnte Zahnscheibe f <Fertig> external tooth lock washer
außenverzahntes Rad n <Maschinen> external gear
Außenverzahnung f <Maschinen> external toothing, outside gearing
Außenvoreinströmung f <Hydraul> outside lead
Außenwand f <Bau> enclosing wall
Außenwandverkleidung f <Bau> siding
Außenwiderstand m <Elektrotech> external resistor
Außenwinkel m <Geom> exterior angle, outward angle
Außerband... <Elektronik, Telekom> out-of-band
Außerbandfilterung f <Elektronik> out-of-band filtering
Außerbandsignalisierung f <Telekom> *(AE)* out-of-band signaling, *(BE)* out-of-band signalling
Außerbandzeichengabe f <Telekom> *(AE)* out-band signaling, *(BE)* out-band signalling
äußere Beanspruchung f <Werkprüf> external strain
äußere Elektronenschalen fpl <Strahlphys> outer orbital complex

äußere Form f <Werkprüf> geometry
äußere Induktanz f <Elektriz> external inductance
äußere Isolierung f <Elektriz> external insulation, outer insulation
äußere Klappe f <Verpack> outer flap
äußere Kraft f <Metall> external force
äußere Mauerschale f <Bau, Kohlen> mantle
äußere Reibung f <Maschinen> external friction
äußere Schaltung f <Elektriz> external circuit
äußere Spaltzone f <Kerntech> *(AE)* outer-fueled zone, *(BE)* outer-fuelled zone
äußere Störung f <Elektriz> external disturbance, external interference *(von außen verursachte Störung)*
äußere Überdeckung f <Hydraul> outside lap; outside lap *(Steuerschieber)*
äußere Umhüllung f <Elektriz> oversheath
äußere Umkleidung <Elektriz> oversheath *(Armierung)*
äußere Voreinströmung f <Hydraul> outside lead
äußere Wechselwinkel mpl <Geom> alternate exterior angles
äußerer Befund m <Qual> visual inspection result
äußerer Druck f <Druck> outer form
äußerer Photoeffekt m 1. <Elektronik> photoelectric emission; 2. <Optik> photoemissive effect
äußerer photoelektrischer Effekt m 1. <Optik> external photoelectric effect; 2. <Telekom> external photoelectric effect, photoemissive effect
äußerer Planet m <Raumfahrt> outer planet
äußerer Radkasten m <Kfztech> external wheel case
äußerer Spalt m <Hydraul> outside clearance
äußerer Weichmacher m <Kunststoff> external plasticizer
äußerer Winkel m <Geom> outward angle
äußeres Einfahrtsignal n <Eisenbahn> outer home signal
äußeres Einflugzeichen n <Lufttrans> outer marker *(Startbahn)*
äußeres Elektron n <Metall> valence electron
äußeres Gestell n <Kfztech> perimeter frame
äußeres Nukleon n <Kerntech> peripheral nucleon
äußeres Spiel n <Hydraul> clearance
äußeres Vektorprodukt n <Math, Phys> vector product
äußeres Zwischenstück n <Raumfahrt> external interface *(Raumschiff)*
außerirdische Störung f <Funktech> extraterrestrial noise
außerirdisches Leben n <Raumfahrt> extra-terrestrial life
äußerlich adj <Umweltschmutz> external
außermittig adj <Maschinen> *(AE)* off-center, *(BE)* off-centre
Außermittigdrehen n <Fertig, Maschinen> eccentric turning
außermittige Bohrung f <Fertig> eccentric bore
außermittiges Spannen n <Fertig> eccentric chucking
Außermittigkeit f <Papier> *(AE)* off-center, *(BE)* off-centre
außerordentlicher Strahl m <Phys> extraordinary ray
außerphasig adj <Elektronik> out-of-phase
außerplanmäßiges Ausfahren n <Kerntech> unscheduled withdrawal *(eines Steuerstabes)*
äußerst starkes Licht n <Wellphys> intense light
aussetzbare Antenne f <Raumfahrt> deployable aerial, deployable antenna *(Weltraumfunk)*
aussetzen v 1. <Anstrich> expose; 2. <Phys> break down
• **Witterungseinflüssen aussetzen** <Bau> weather
Aussetzen n 1. <Fernseh> dropout; 2. <Raumfahrt> drop *(Raumschiff)*; 3. <Wassertrans> launching *(Boot)*
Aussetzen n **gegen saure Halogenide** <Anstrich> acid halide exposure
aussetzende Belastung f <Elektriz, Fertig> intermittent load
aussetzender Betrieb m <Elektriz> intermittent duty
aussetzender Betrieb m **mit veränderlicher Belastung** <Elektriz> variable intermittent duty
Aussickern n <Meerschmutz> seepage
aussieben v <Meerschmutz> screen *(Öl)*
aussondern v <Qual> segregate
Aussonderung f <Qual> segregation *(von fehlerhaften Einheiten)*
ausspachteln v <Bau> grout, smooth
ausspannen v <Maschinen> unclamp
aussparen v 1. <Bau> block out, box out, recess; 2. <Fertig> recess; 3. <Maschinen> relieve
Aussparen n <Bau, Maschinen> recessing
Aussparung f 1. <Bau> notch, pocket, recess; 2. <Fertig> pocket, recess; 3. <Maschinen> clearance, recess, relief; 4. <Mechan> clearance, relief
ausspeichern v <Comp & DV> roll out
Ausspeicherung f <Comp & DV> roll out
Aussperrung f <Comp & DV> lockout
ausspitzen v <Fertig> thin *(Spiralbohrer)*
Ausspitzen n <Fertig> pointing *(Bohrer)*; thinning *(Spiralbohrer)*
ausspülen v 1. <Nichtfoss Energ> scour; 2. <Wasserversorg> flush *(mit Wasser)*
Ausspülen n 1. <Mechan> flushing; 2. <Wasserversorg> cleaning out
Ausspülung f 1. <Chemtech> washing out; 2. <Papier> baring
ausstanzen v 1. <Fertig> blank; 2. <Mechan> punch out
ausstatten v 1. <Bau> furnish; 2. <Maschinen> fit out; 3. <Telekom> equip
Ausstattung f 1. <Bau> equipment, set; 2. <Maschinen> equipment
Ausstattung f **mit Geräten** <Gerät> instrumentation
Ausstattungsgrad m <Kfztech> trim level
aussteifen v 1. <Bau> buttress; 2. <Maschinen> brace, buttress; 3. <Mechan> buttress
Aussteifung f <Bau> stiffening
ausstellen v <Textil> flare
Aus-Stellung f <Maschinen> off position
ausstemmen v <Bau> mortice, mortise
Aussteuerung f 1. <Comp & DV, Elektronik> modulation *(Radio)*; 2. <Elektrotech> excitation
Aussteuerungsanzeige f 1. <Aufnahme> level indicator; 2. <Fernseh> VI meter
Aussteuerungsbereich m **des Vertikalverstärkers** <Elektronik> vertical-amplifier dynamic range
Aussteuerungsgrenze f 1. <Akustik> maximum recording level; 2. <Aufnahme, Elektrotech> overload level
Aussteuerungsmesser m <Aufnahme, Fernseh, Funktech> *(AE)* peak program meter, *(BE)* peak programme meter
Aussteuerungsmessgerät n <Gerät> volume unit meter
ausstöpseln v 1. <Bau> unstop; 2. <Funktech, Telekom> unplug *(Stecker)*
Ausstoß m 1. <Erdöl> expulsion *(Kapillarwasser in Schieferschichten)*; 2. <Kerntech> discharge; 3. <Kfztech, Lufttrans> emission; 4. <Maschinen> output, production; 5. <Mechan> ejection
Ausstoß m **von Auspuffgasen** <Kfztech> exhaust gas emission
Ausstoßdüse f <Kerntech> discharge nozzle
ausstoßen v 1. <Fertig> knock out; 2. <Strömphys> jet; 3. <Thermod> give off
Ausstoßer m <Fertig> lifter
Ausstoßrate f <Erdöl> expulsion rate *(Kapillarwasser in Schieferschichten)*

Ausstoßung

Ausstoßung f <Chemie> expulsion *(Gase, Flüssigkeiten)*
Ausstoßvorrichtung f <Kunststoff, Maschinen> ejector
ausstrahlen v 1. <Comp & DV> emit; 2. <Thermod> radiate
Ausstrahlung f 1. <Chemie> emanation; 2. <Funktech> emission
Ausstrippen n **mit Dampf** <Abfall> steam stripping *(Sickerwasserbehandlung)*
Ausstrippen n **mit Luft** <Abfall> air stripping
Ausström... 1. <Heiz & Kälte> discharge; 2. <Phys> stream
Ausströmdruck m <Heiz & Kälte> discharge pressure
ausströmen v <Thermod> emit
ausströmen lassen v <Heiz & Kälte> discharge
Ausströmen n 1. <Heiz & Kälte> discharges; 2. <Wasserversorg, Elektrotech> leakage, efflux
Ausströmgeschwindigkeit f <Phys> free stream velocity
Ausströmraum m <Wasserversorg> volute chamber *(einer Kreisel- oder Zentrifugalpumpe)*
Ausströmung f 1. <Mechan> exhaust; 2. <Sicherheit> effluent, exhaust
Ausströmungsaushöhlung f <Hydraul> exhaust cavity *(Abdampfkavitation)*
Ausströmvorgang m <Mechan> exhaust process
Austast... <Elektronik, Fernseh> blanking
austasten v <Telekom> blank
Austasten n 1. <Comp & DV, Elektronik, Fernseh> gating; 2. <Gerät> blanking
Austastgenerator m <Elektronik> blanking generator
Austastimpuls m <Fernseh> blanking pulse
Austastkreis m <Fernseh> blanking circuit
Austastlücke f <Fernseh> blanking interval
Austastpegel m <Fernseh> blanking level
Austastsignal n <Elektronik, Fernseh> blanking signal
Austastspannung f <Fernseh> blanking voltage
Austast- und Synchronisiersignal n <Fernseh> blanking and sync signal
Austast- und Synchronisiersignalmischer m <Fernseh> blanking and sync signal mixer
Austastung f <Fernseh> blanking
Austastverstärker m <Elektronik> blanking amplifier *(Fernsehtechnik)*
Austausch m <Comp & DV, Maschinen> exchange
austauschbar adj <Foto, Maschinen> interchangeable
austauschbare Logik f <Elektronik> compatible logic
austauschbares optisches Medium n <Optik> alterable optical medium
austauschbares Teil n <Maschinen> interchangeable part
Austauschbarkeit f <Maschinen> interchangeability
austauschen v <Comp & DV, Telekom> swap
Austauschenergie f <Metall> exchange energy
Austauschsprungbefehl m <Comp & DV> exchange jump
Austauschsteuerungsnummer f <Comp & DV> interchange control number
Austauschsteuerungsreferenz f <Comp & DV> interchange control reference
Austauschwerkstoff m <Maschinen> alternative material
Austenit m <Fertig, Metall> austenite
austenitisch adj <Fertig, Mechan> austenitic
austenitischer Stahl m <Mechan> austenitic steel
austenitisieren v <Fertig> austenitize
Austenitisierung f <Fertig> austenizing
Austenitstahl m <Metall> austenic steel
Austenitstahlrohr n <Maschinen> austenitic stainless steel tube
Austernbagger m <Wassertrans> oyster dredge, oyster dredger *(Schifftyp)*

Austrag m <Kohlen> discharge
Austreiben n <Druck> quad
Austreiber m 1. <Fertig> taper key; 2. <Maschinen> *(AE)* center key, *(BE)* centre key, drift bolt, pin punch
Austreiberlappen/mit <Fertig> tanged
Austreibung f <Chemie> expulsion *(Gase, Flüssigkeiten)*
austreten v <Heiz & Kälte> discharge
austreten lassen v <Heiz & Kälte> discharge
austretender Fluss m <Phys> outward flux
austretendes Teilchen n <Kerntech> outcoming particle
austretendes Tragflächenboot n <Wassertrans> emerging foil craft
Austrieb m 1. <Erdöl> expulsion; 2. <Kunststoff> flash
Austriebrate f <Erdöl> expulsion rate
Austritt m 1. <Elektrotech> outlet *(elektrische Schwingungen)*; 2. <Erdöl> spillage *(Öl)*; 3. <Heiz & Kälte> discharge; 4. <Maschinen> outlet
Austrittsarbeit f 1. <Elektrotech> work function *(bei elektrischen Röhren)*; 2. <Phys> work function *(des Elektrons)*
Austrittsdivergenz f <Telekom> output divergence
Austrittsfläche f <Mechan> exhaust area
Austrittsgeschwindigkeit f 1. <Fertig, Hydraul> exit velocity; 2. <Umweltschutz> efflux velocity
Austrittshöhe f <Kerntech> discharge head *(einer Pumpe)*
Austrittskonus m <Lufttrans> exhaust cone
Austrittskonus m **einer Düse** <Phys> exit cone of nozzle
Austrittsluke f <Optik> exit port
Austrittsöffnung f <Maschinen> outlet
Austrittspfosten m <Bau> newel
Austrittspupille f 1. <Optik> eye ring; 2. <Phys> exit pupil
Austrittsschaufelrad n <Wassertrans> exducer
Austrittsseite f <Textil> downstream
Austrittstemperatur f <Umweltschutz> outlet temperature
Austrittsventil n <Hydraul> escape valve
Austrittsverlust m <Mechan> exhaust loss
Austrittswinkel m 1. <Hydraul> exit angle; 2. <Optik, Telekom> output angle
austrocknen v <Bau, Thermod> dry out
Austrocknen n <Thermod> drying-out
Austrocknung f <Chemie> desiccation
austrommeln v <Meerschmutz> reel out *(Sperre)*
ausüben v 1. <Fertig> impart *(Druck)*; 2. <Phys> exert *(Kraft)*
Auswahl f 1. <Comp & DV> menu selection, selection; 2. <Math> choice, sampling, selection; 3. <Qual> applicability
Auswahl f **der Prüfschärfe** <Qual> procedure for normal, tightened and reduced inspection
auswählbarer Name m <Comp & DV> generic name
Auswahlcode m <Comp & DV> option code
Auswahleinheit f <Qual> sampling unit
auswählen v <Comp & DV> select
Auswahlmenü n <Comp & DV> menu
Auswahlmöglichkeit f <Comp & DV> available choice, option
Auswahlregel f <Kerntech, Phys, Strahlphys> selection rule
Auswahlsortierung f <Comp & DV> selective sort
Auswahlverfahren n <Qual> sampling
auswalzen v 1. <Bau> sheet out *(Tiefbau)*; 2. <Fertig> bloom *(Luppen)*; get down *(Walzen)*; 3. <Metall> sheet out
Auswalzen n <Fertig> blooming-down *(im Blockwalzwerk)*
Auswanderung f <Telekom> drift *(auf null)*
Auswärtsdrehung f <Ergon> eversion
auswaschen v 1. <Bau, Kohlen> erode, leach; 2. <Textil> launder

Auswaschen n <Chemtech> washing out
Auswaschung f 1. <Bau> erosion; 2. <Lebensmittel> elutriation; 3. <Umweltschmutz> washout
auswattieren v <Textil> pad
auswechselbar adj 1. <Foto> interchangeable; 2. <Maschinen> interchangeable, removable
auswechselbare Ambossbahn f <Fertig> anvil pallet
auswechselbare Mattscheibe f <Foto> interchangeable focusing screen
auswechselbare Platte f <Comp & DV> removable disk
auswechselbarer Aufsichtsucher m <Foto> interchangeable waist-level finder
auswechselbarer Einsatz m <Maschinen> removable insert
auswechselbares Kation n <Umweltschmutz> exchangeable cation
Auswechselbarkeit f <Maschinen> interchangeability
Auswechselbeutel m <Verpack> liner bag
auswechseln v <Bau> trim (Balken)
Ausweich... <Eisenbahn, Lufttrans, Raumfahrt, Wassertrans> alternate
ausweichen v <Wassertrans> give way (Navigation)
Ausweichflughafen m <Lufttrans> alternate airport
Ausweichgleis n <Eisenbahn> passing track, siding, turnout
Ausweichlandung f <Raumfahrt> alternate landing
Ausweichmanöver n <Wassertrans> emergency turn
Ausweichrangiergleis n <Eisenbahn> classification siding
Ausweichstelle f 1. <Bau> turnout; 2. <Eisenbahn> passing point, shunt, turnout
Ausweichsystem n <Comp & DV> backup system
Ausweis m <Comp & DV> badge
Ausweisanhänger m <Sicherheit> sling identification tag
Ausweiskontrolle f <Trans> passport check
Ausweisleser m <Comp & DV> badge reader
Ausweitdorn m <Fertig> drift
auswerfen v <Fertig> spew
Auswerfer m <Kunststoff, Mechan> ejector
Auswerferbolzen m <Mechan> ejector pin
Auswerferbuchse f <Maschinen> ejector sleeve
Auswerferplatte f <Maschinen> ejector plate
Auswerferstift m <Maschinen> ejection pin, ejector pin, knock-out pin
Auswerfvorrichtung f <Maschinen> ejector
auswerten v 1. <Comp & DV> interpret; 2. <Math> evaluate, exploit
Auswertung f 1. <Comp & DV> interpretation; 2. <Math> evaluation, exploitation
Auswertung f **von Prüfergebnissen** <Qual> evaluation of test results
Auswertungs-Session f <Aufnahme> scoring session
auswiegen v <Fertig> poise
Auswirkung f <Mechan> effect
auswölben v <Bau> vault
Auswucht... <Lufttrans, Maschinen> balancing
auswuchten v 1. <Fertig> balance, true; 2. <Heiz & Kälte> balance
Auswuchten n 1. <Fertig> truing; 2. <Maschinen> balancing, counterbalancing
Auswuchtgewicht n <Lufttrans> balance weight
Auswuchtmaschine f <Maschinen> balancing machine
Auswuchtung f <Fertig> balance
Auswurf m 1. <Kerntech> washback; 2. <Metall> spittings; 3. <Umweltschmutz> effluent, efflux
Auswurfkraft f <Raumfahrt> ejection force (Raumschiff)
Auszeichnung f <Verpack> (AE) labeling, (BE) labelling
Auszeichnungsschrift f <Druck> display face, display type

ausziehbar adj <Maschinen> telescopic, telescoping
ausziehbare Leitung f <Phys> line stretcher
ausziehbarer Bohrmeißel m <Erdöl> collapsible bit (Bohrtechnik)
ausziehen v <Anstrich> expose
Auszieher m <Bau> extractor
Ausziehleiter f <Bau> extension
Auszug m <Foto> extension (Balg)
Auszugsfilter n <Foto> (AE) color separation filter, (BE) colour separation filter
Auszugsring m <Foto> extension ring
Authentifizierung f <Comp & DV> authentication
Authentifizierungsprozedur f <Mobilkom> authentication procedure
Authentifizierungsverfahren n <Comp & DV> authentication policy
Auto n (Automobil) <Kfztech> auto, automobile, car, passenger car (Fahrzeugart)
Autobahn f <Trans> (AE) expressway, (BE) motorway, (AE) superhighway
Autobahnparkplatz m <Trans> lay-by
Autobahnzubringerkontrolle f <Trans> (BE) slip road control
Autobahnzubringerverkehrszählung f <Trans> (BE) slip road census, (BE) slip road count, (BE) slip road metering
Autobandspanner m <Fernseh> autotension
Autocue n <Fernseh> autocue
Autodyn n <Elektronik> autodyne
Autoedieren n <Fernseh> autoediting
Autoedit n <Fernseh> automatic editing
Autoempfänger m <Funktech> car radio receiver
Autoequalizer m <Fernseh> autoequalization
Autoersatzteillager n <Kfztech> auto parts store
Autofähre f <Wassertrans> ferry
Autofrettage f <Mechan> cold drawing
Autofriedhof m <Kfztech> car dump
autogen adj <Kohlen, Maschinen> autogenous
Autogen... <Kohlen, Maschinen> autogenous
Autogenbrenner m <Bau> oxyacetylene blowpipe, oxyacetylene blowtorch
autogene Schweißung f <Maschinen> autogenous welding
autogenes Brennschneiden n <Thermod> flame cutting
autogenes Mahlen n <Kohlen> autogenous milling
Autogengas n <Bau, Fertig> oxyacetylene gas
Autogenhärten n <Fertig> flame hardening
Autogenmühle f <Kohlen> autogenous mill
Autogenschweißbrenner m <Mechan> acetylene-oxygen torch
Autogenschweißen n 1. <Bau> oxyacetylene welding; 2. <Maschinen> autogenous welding, gas welding; 3. <Mechan> oxyacetylene welding; 4. <Thermod> gas welding
Autogiro n <Lufttrans> autogyro
Autoheizung f <Kfztech> car heater
Autoionisation f <Phys, Strahlphys> autoionization
autokatalytisch adj <Kerntech> autocatalytic
autokatalytische Wirkung f <Metall> autocatalytic effect
Autoklav m 1. <Chemie, Ker & Glas, Kohlen, Lebensmittel> autoclave; 2. <Thermod> digester
Autoklavieren n <Chemie> retorting
Autokollimationsfernrohr n <Metrol> autocollimator
Autokompensator m <Gerät> automatic potentiometer
Autokorrelation f <Elektronik, Telekom> autocorrelation
Autokorrelationsanalyse f <Gerät> autocorrelation analysis
Autokran m <Bau> mobile crane
Autolock n <Fernseh> automatic lock

Autolyse

Autolyse f <Lebensmittel> autolysis *(Biochemie)*
Automat m 1. <Elektriz> automaton; 2. <Maschinen> automatic; 3. <Verpack> dispensing machine *(für Getränke, Snacks)*
Automatenmessing n <Maschinen> free-cutting brass
Automatenpackung f <Verpack> package for vending machine
Automatenstahl m 1. <Fertig> free machine steel; 2. <Maschinen> free-cutting steel, machining steel; 3. <Metall> free-cutting steel
Automatentheorie f <Comp & DV> automatic theory, automaton theory
Automatik f <Verpack> automatic control
Automatikgetriebe n <Kfztech> automatic transmission
Automatikgetriebeöl n *(ATF)* <Kfztech> automatic transmission fluid *(ATF)*
Automatikmeißel m <Bau> self-coring chisel *(Holzbau)*
Automatikschalter m <Regelung> automatic control switch
Automatikschalthebel m <Kfztech> selector lever *(Getriebe)*
Automatiksystem n <Raumfahrt> automatic system
Automatiktür f <Bau> self-closing door
Automation f <Bau, Comp & DV, Ergon, Maschinen, Raumfahrt> automation
Automationsunterstützung f <Comp & DV> automation support
automatisch adj 1. <Comp & DV, Funktech> automatic; 2. <Maschinen> self-acting; 3. <Mechan> automatic; 4. <Telekom> unattended
automatisch arbeitende Armatur f <Erdöl> automatic valve *(Rohrleitungen)*
automatisch eingebuchtes Handy n <Mobilkom> automatically logged-in mobile, preprogrammed log-in mobile
automatisch getaktete Farbkorrekturschaltung f <Fernseh> *(AE)* color automatic time-base corrector, *(BE)* colour automatic time-base corrector
automatisch herabfallende Sauerstoffmaske f <Lufttrans> quick-downing oxygen mask
automatische Abgriffverbindung f <Elektrotech> automated tap bonding
automatische Abschaltung f <Sicherheit> automatic shutdown
automatische Absperrarmatur f <Erdöl> automatic shutoff valve *(Rohrleitungen)*
automatische Abstandskontrolle f <Kfztech> automatic control of headway
automatische Abstimmung f <Telekom> automatic tuning
automatische Amplitudenregelung f <Elektronik, Funktech, Telekom> automatic gain control
automatische Amtsholung f <Telekom> automatic exchange line seizure
automatische Anflugkontrolle f <Lufttrans> automatic approach control *(Navigation)*
automatische Anflugsteuerung f **vom Boden** *(AGCA)* <Raumfahrt> automatic ground-controlled approach *(AGCA)*
automatische Anlagen f <Lebensmittel> CIP
automatische Anlagenreinigung f <Lebensmittel> cleaning in place
automatische Anrufumlegung f <Telekom> automatic call transfer
automatische Antwort f <Comp & DV> auto-reply
automatische Auslösung f <Kerntech> automatic release
automatische Ausschaltung f <Aufnahme> automatic stop

automatische Außerbetriebsetzung f <Sicherheit> automatic shutdown
automatische Autobahn f <Trans> *(AE)* automatic highway, *(BE)* automatic motorway
automatische Bahnführung f <Wassertrans> track control *(Navigation)*
automatische Bahnregelung f <Wassertrans> track control *(Navigation)*
automatische Brandmeldeanlage f <Bau, Eisenbahn> automatic fire alarm
automatische Bremse f <Kfztech> automatic brake
automatische Bürette f <Labor> automatic burette
automatische Chrominanzsteuerung f <Fernseh> automatic chrominance control
automatische Datenkonvertierung f <Comp & DV> automatic data conversion
automatische Datenumwandlung f <Comp & DV> automatic data conversion
automatische Datenverarbeitung f *(ADV)* <Comp & DV> automatic data processing *(ADP)*
automatische Diagnose f <Künstl Int> automatic diagnosis
automatische Dosieranlage f <Bau> automatic weighbatcher
automatische Drehbank f <Mechan> automatic lathe
automatische Drehzahlregelung f <Kfztech> automatic speed control
automatische Dynamikdrängung f <Aufnahme> automatic volume compression
automatische Eichungskontrolle f <Metall> automatic gauge control
automatische Erfassung f **der Ortsveränderungen** <Mobilkom> automatic roaming
automatische Ersatzschaltung f <Telekom> automatic change-over *(Standby-Betätigung)*
automatische Expansionsvorrichtung f <Maschinen> automatic expansion gear
automatische Fahr- und Bremssteuerung f <Trans> automatic running and braking control, automatic speed control
automatische Fahrzeugidentifikation f *(AFI)* <Funkort, Trans> automatic vehicle identification *(AVI)*
automatische Fahrzeugortung f *(AFO)* <Trans> automatic vehicle location *(AVL)*
automatische Fehlerberichtigung f <Lufttrans> automatic error correction
automatische Fehlererkennung f <Comp & DV, Telekom> automatic error detection
automatische Fehlerkorrektur f <Telekom> automatic error correction
automatische Fernwahl f <Telekom> *(AE)* automatic trunk dialing, *(BE)* automatic trunk dialling
automatische Freigabe f <Fertig> automatic release
automatische Frequenzregelung f *(AFR)* <Elektronik, Fernseh, Funktech> automatic frequency control *(AFC)*
automatische Frequenzumtastung f <Elektronik, Funktech, Telekom> automatic frequency shift keying *(Fernschreiben)*
automatische Frequenz- und Verstärkungsregelung f *(AFGC)* <Elektronik, Fernseh, Funktech> automatic frequency and gain control *(AFGC)*
automatische Geschwindigkeitssteuerung f <Trans> automatic speed control
automatische Gesprächsumlegung f <Telekom> automatic call transfer
automatische Gittervorspannung f <Elektrotech> self-bias
automatische Haltlichtanlage f <Eisenbahn> automatic flashing light signals

automatische Handschrifterkennung f <Künstl Int> automatic handwriting recognition
automatische Helligkeitsregelung f (ABC) <Fernseh> automatic brightness control (ABC)
automatische Helligkeitssteuerung f <Fernseh> automatic brightness control
automatische Himmelsnavigation f (ACN) <Raumfahrt> automatic celestial navigation (ACN)
automatische KFZ-Identifizierung f (automatische Kraftfahrzeug-Identifizierung) <Kfztech> ACI (automatic car identification)
automatische Kopierdrehmaschine f <Maschinen> automatic copying lathe
automatische Kopiermaschine f <Foto> automatic printer
automatische Kraftfahrzeug-Identifizierung f (automatische KFZ-Identifizierung) <Kfztech> (AE) automatic car identification, (BE) automatic wagon identification (automatic car identification)
automatische Kraftübertragung f <Mechan> automatic transmission
automatische Landesteuerung f vom Boden (AGCL) <Raumfahrt> automatic ground-controlled landing (AGCL)
automatische Lasttraverse f <Trans> spreader
automatische Lastübergabe f <Elektriz> automatic load transfer
automatische Lautstärkeregelung f (ALR) <Funktech> automatic volume control (AVC)
automatische Lichtwellenleiterausrichtung f <Telekom> automatic fiber alignment
automatische Messbereichsumschaltung f <Gerät> automatic ranging
automatische Müllsortierung f <Abfall> mechanical separation
automatische Nachformdrehmaschine f <Maschinen> automatic copying lathe
automatische Nachführung f <Raumfahrt> autotracking
automatische Nachrichten-Speichervermittlungsstelle f <Telekom> (AE) automatic message switching center, (BE) automatic message switching centre
automatische Nebenstellenanlage f <Telekom> private branch exchange
automatische Nullstellung f <Kontroll> automatic reset (Rückstellung)
automatische Ortswählvermittlung f <Telekom> regional automatic circuit exchange
automatische Ortung f bewegter Objekte <Funktech, Telekom> automatic location of moving objects, moving target indication (MTI)
automatische Ortung f und Registrierung f <Telekom> automatic location-registration (von Schiffen)
automatische Papierzuführung f <Comp & DV> feeder (Drucker)
automatische Pegelregelung f (ALC) <Funktech> automatic level control (ALC)
automatische Phasenregelung f (APR) <Elektronik> automatic phase control (APC)
automatische Phasensteuerung f (APR) <Fernseh> automatic phase control (APC)
automatische Probeentnahmevorrichtung f <Labor> automatic sampling device
automatische Prüfeinrichtung f (ATE) <Comp & DV> automatic test equipment (ATE)
automatische Prüfung f <Comp & DV> automatic check
automatische Querbewegung f <Maschinen> automatic transverse movement (des Werkzeugs)
automatische Radaraufnahmehilfe f (ARPA) <Wassertrans> automatic radar plotting aid (ARPA)
automatische Rauschunterdrückung f (ARU) <Funktech> automatic noise limiter (ANL)
automatische Regelung f <Elektriz> automatic control
automatische Reparatur f <Textil> automatic repair
automatische Rückspulvorrichtung f <Foto> automatic rewinder
automatische Rückstellung f <Elektriz> automatic reset
automatische Satellitenerfassung f von geomagnetischen Daten (ARGOS) <Wassertrans> Automatic Remote Geomagnetic Observatory System (ARGOS)
automatische Scharfabstimmung f 1. <Elektronik> automatic frequency control; 2. <Fernseh> automatic frequency control (Empfänger); 3. <Funktech> automatic frequency control; 4. <Funktech> automatic fine tuning (AFT)
automatische Scharfeinstellung f <Foto> automatic focusing
automatische Schlauchbeutelfüllanlage f <Verpack> automatic flexible-bag-filling machine
automatische Schneidvorrichtung f <Ker & Glas> automatic cutter
automatische Schraubendrehbank f <Maschinen, Mechan> automatic screw machine
automatische Schrifterkennung f <Künstl Int> automatic writing recognition
automatische Schubsteuerung f <Lufttrans> autothrottle
automatische Schweißung f <Maschinen> automatic welding
automatische Seitennummerierung f <Comp & DV> automatic page numbering
automatische Sendersuche f und Speicherung f <Fernseh> automatic tuner search and storage (ATS)
automatische Silbentrennung f <Druck> automatic hyphenation
automatische Spannungsregelung f <Elektriz> automatic voltage control
automatische Spracherkennung f (ASE) <Künstl Int> automatic speech recognition (ASR)
automatische Sprachverarbeitung f <Künstl Int> automatic speech processing
automatische Standorterfassung f <Funktech, Telekom> self-location
automatische Starrkupplung f <Trans> rigid automatic coupling
automatische Steuerung f <Bau, Elektriz, Erdöl, Maschinen, Verpack> automatic control
automatische Stirnradfräsmaschine f <Maschinen> automatic spur-gear-cutting machine
automatische Telefonzentrale f (ATZ) <Telekom> automatic telephone exchange
automatische Thermoformung f <Verpack> automatic thermoforming (von aufgerolltem Material)
automatische Titration f <Labor> automatic titration
automatische Übersetzung f <Künstl Int> automatic translation
automatische Umschaltung f 1. <Elektriz> automatic change-over switching; 2. <Telekom> automatic change-over
automatische Vakuumbremse f <Maschinen> automatic vacuum brake
automatische Verkehrsumschaltung f <Telekom> automatic change-over
automatische Vermittlungsstelle f <Telekom> automatic switchboard
automatische Verstärkungsregelung f (AVR) <Elektronik, Funktech, Telekom> automatic gain control (AGC)
automatische Verstellung f <Lufttrans> autofeathering (Propeller)

automatische

automatische Videokassettenaufzeichnung f <Aufnahme, Fernseh> automatic video cassette recording, automatic video recording
automatische Vorrichtung f <Telekom> automatic device
automatische Waggonidentifikation f <Trans> ACI, (AE) automatic car identification, (BE) automatic wagon identification
automatische Wahlwiederholung f <Telekom> automatic call repetition, automatic redialling; last number redial (LR)
automatische Weiterleitung f <Comp & DV> autoforwarding
automatische Werkzeugmaschine f <Maschinen> automatic machine tool
automatische Wetterstation f <Wassertrans> automatic weather station
automatische Wiederholanforderung f (ARQ) <Funktech, Telekom> automatic repeat request (ARQ)
automatische Zeichenerkennung f <Comp & DV, Metrol> automatic character recognition
automatische Zielaufschaltung f <Lufttrans> lock-on (Flugwesen)
automatische Zielverfolgung f <Funkort> automatic tracking (Radar)
automatische Zuführung f 1. <Erdöl> automatic feed (Bohrtechnik); 2. <Maschinen> automatic feed, power feed; 3. <Verpack> automatic feeding system
automatische Zugdeckung f <Eisenbahn> (BE) ATP, automatic train protection
automatische Zugsteuerung f (ATC) <Eisenbahn> automatic train control, automatic train operation (ATO)
automatische Zugsteuerung f per Minicomputer 1. <Eisenbahn> automatic train operation by mini computer; 2. <Trans> ATOMIC
automatische Zugüberwachung f <Eisenbahn> automatic train monitoring
automatische Zündung f <Lufttrans> autoignition
automatische Zurückstellung f <Elektriz> automatic reset
automatische Zusammentragung f <Verpack> automatic collation
automatische Zustellung f <Maschinen> automatic down-feed, automatic feed
automatischer Abruf m <Telekom> automatic recall
automatischer Abspanntransformator m <Elektrotech> step-down autotransformer
automatischer Anrufverteiler m <Telekom> ACD, automatic call distributor
automatischer Aufwärtstransformator m <Elektrotech> step-up autotransformer
automatischer Ausschalter m <Elektriz> automatic switch, cutout switch
automatischer Azimutanzeiger m <Lufttrans> OBI, omnibearing indicator
automatischer Betrieb m 1. <Bau> automatic operation; 2. <Comp & DV> hands-off operation, unattended operation
automatischer Diawechsler m <Foto> automatic slide changer
automatischer Dienst m <Telekom> automatic operation
automatischer Fernbetrieb m <Telekom> automatic trunk working
automatischer Feueralarm m <Sicherheit> automatic fire alarm
automatischer Feuermelder m <Sicherheit> automatic fire detection system
automatischer Filmdruck m <Textil> automatic screen printing

automatischer Fliehkraftregler m <Nichtfoss Energ> automatic governor
automatischer Funkkompass m (ADF) <Funkort, Lufttrans> automatic direction finder (ADF)
automatischer Kompensator m <Gerät> automatic potentiometer
automatischer Kreditkartendienst m <Telekom> automatic credit card service
automatischer Niveauausgleich m (ALC) <Kfztech> automatic level control (ALC)
automatischer Peilempfänger m <Funkort, Lufttrans> automatic direction finder
automatischer Personenschnellverkehr m <Trans> automated personal rapid transit
automatischer Prober m <Kohlen> automatic sampler
automatischer Prüfcode m <Comp & DV> self-checking code
automatischer Rauschbegrenzer m (ANL) <Funktech> automatic noise limiter (ANL)
automatischer Regler m <Maschinen> automatic regulator
automatischer Rückruf m **bei Besetzt** <Telekom> automatic call-back on busy; completion of calls to busy subscribers, CCBS (Dienstmerkmal im Euro-ISDN)
automatischer Rufnummerngeber m (ACU) <Telekom> automatic calling unit (ACU)
automatischer Schalter m <Elektrotech> automatic switch, self-acting switch
automatischer Schnellverkehr m <Trans> rapid automatic transport
automatischer Schussspulenwechsel m <Textil> automatic pirn change
automatischer Sender-Empfänger m (ASR) <Comp & DV> automatic send-receive (ASR)
automatischer Spitzenbegrenzer m <Fernseh> automatic peak limiter
automatischer Steuerschalter m <Elektriz> automatic control switch
automatischer Stromregler m <Elektriz> automatic current controller
automatischer Transformator m **zur Spannungserhöhung** <Elektrotech> step-up autotransformer
automatischer Trennschalter m <Elektriz> automatic cut-off switch
automatischer Vergrößerer m <Foto> automatic enlarger
automatischer Verkehrsbereichwechsel m <Mobilkom> automatic roaming
automatischer Vorschub m <Maschinen> power feed
automatischer Wassersprühnebel-Feuerlöscher m <Sicherheit> automatic fire-fighting system
automatischer Weißabgleich m <Fernseh> automatic white balance
automatischer Werkzeugwechsler m (ATC) <Metall> automatic tool changer (ATC)
automatischer Zentrifugalregler m <Nichtfoss Energ> automatic governor
automatisches Abfangen n <Telekom> automatic intercept system
automatisches Ausrücken n <Maschinen> automatic throwing out of action
automatisches Formen n <Ker & Glas> automatic forming
automatisches Getriebe n <Maschinen> automatic transmission
automatisches Identifikationssystem n <Wassertrans> automatic identification system, AIS (Navigation)
automatisches Kurssteuerungssystem n <Lufttrans> automatic flight-control system

automatisches Laden n <Comp & DV> autoload
automatisches Ladesystem n <Foto> automatic loading system
automatisches Leitsystem n <Elektrotech> automatic guidance system
automatisches Lösen n <Maschinen> automatic release
automatisches Palettenstapel- und Strecksystem n <Verpack> robotic palletizing and stretch system
automatisches Pressen n <Ker & Glas> automatic pressing
automatisches Programmentwicklungssystem n <Comp & DV> automatic programming tool
automatisches Rohrventil n <Bau> self-closing cock
automatisches Rufgerät n (ACU) <Raumfahrt, Telekom> automatic calling unit (ACU)
automatisches Schalten n <Trans> automatic switching
automatisches Senden n **und Empfangen** n <Telekom> automatic send and receive, ASR (Modemverfahren)
automatisches Senden-Empfangen n <Comp & DV> automatic send-receive
automatisches Speisen n <Ker & Glas> automatic feeding
automatisches Sperren n <Comp & DV> automated locking
automatisches Steuerelement n <Kerntech> automatic control assembly
automatisches Transportsystem n <Trans> automatic transportation system
automatisches Trennen n <Maschinen> automatic release
automatisches Verfolgungsradar n <Funkort> automatic tracking radar
automatisches Wählen n <Telekom> automatic calling, automatic dialling, one-button dialling, one-touch dialling
automatisches Zielfluggerät n <Funkort, Lufttrans> automatic direction finder
automatisches Zurückklappen n **der Luftschraubenblätter** <Lufttrans> automatic blade-folding system (Hubschrauber)
automatisieren v 1. <Comp & DV> automate; 2. <Fertig> automate, automatize
automatisierte Maschinenwerkzeugprogrammierung f (AUTOPROMT) <Fertig> automated programming of machine tools (AUTOPROMT)
automatisierte Werkzeugpositionierung f (AUTOSPOT) <Fertig> automated system for positioning tools (AUTOSPOT)
automatisierte Wetterstation f <Nichtfoss Energ> automated weather station
automatisierter Fahrkartenverkauf m (AFC) <Eisenbahn> automatic fare collection, AFC
automatisierter Kugelhahn m <Fertig> actuated ball valve (Kunststoffinstallationen)
automatisiertes Entscheiden n <Künstl Int> automated decision-making
Automatisierung f <Comp & DV, Ergon, Fertig, Maschinen> automation
Automatverschluss m <Foto> self-cocking shutter
Autometamorphismus m <Nichtfoss Energ> autometamorphism
Automobil n (Auto) <Kfztech> automobile, car
autonom adj 1. <Comp & DV> stand-alone; 2. <Kontroll> autonomous; 3. <Raumfahrt> self-contained (Raumschiff)
autonome Arbeitsgruppe f <Ergon> autonomous work group
autonome Aufzeichnung f <Nichtfoss Energ> spontaneous log
autonome Navigationshilfe f <Lufttrans> self-contained navigational aid
autonomes System n <Raumfahrt> stand-alone system (Weltraumfunk)
auto-orthogonale Faltungscodierung f <Telekom> self-orthogonal convolutional coding
Autopilot m 1. <Lufttrans> automatic pilot, gyropilot, autopilot; 2. <Raumfahrt> autopilot; 3. <Wassertrans> autopilot, gyropilot
Autopilot-Drehknopf m <Lufttrans> autopilot turn knob
Autopilot-Steigungsfeineinstellung f <Lufttrans> autopilot pitch sensitivity system
AUTOPROMT (automatisierte Maschinenwerkzeugprogrammierung) <Fertig> AUTOPROMT (automated programming of machine tools)
Autoradio n <Funktech> car radio receiver
Autoradiographie f <Phys, Strahlphys> autoradiography
Autoradiolyse f <Phys, Strahlphys> autoradiolysis
Autoreisezug m <Eisenbahn> car sleeper train, motorail
Autorenkorrektur f <Druck> AA, author's alterations
Autorotationsflug m <Lufttrans> autorotation flight; autorotative flight (Hubschrauber)
Autorotationsübergangszeit f <Lufttrans> autorotation transition time
Autoservobetrieb m <Fernseh> autoservo mode
AUTOSPOT (automatisierte Werkzeugpositionierung) <Fertig> AUTOSPOT (automated system for positioning tools)
Autospurlageeinstellung f <Fernseh> autotracking
Autostraße f <Trans> (AE) expressway, (BE) motorway, (AE) superhighway
Autotelefon n 1. <Kfztech> car phone, in-car telephone; 2. <Mobilkom> car phone, car telephone, in-car telephone; 3. <Trans> car telephone
Autothermikkolben m <Kfztech> autothermic piston
Autotransformator m <Elektrotech> autotransformer
autotrophe Ernährung f <Lebensmittel> autotrophy
autotropher Mikroorganism m <Lebensmittel> autotroph
Autotrophie f <Lebensmittel> autotrophy (Biochemie)
Autotypie f <Druck> halftone process
Autotypieraster n <Druck> halftone screen
Autoverschrottungsanlage f <Abfall> car fragmentation plant
Autowerkstatt f <Trans> garage
Autowrack n <Abfall> scrap motorcar
Autoxidation f <Lebensmittel> autoxidation
Autozubehör n <Kfztech> car accessories
Auxochrom n <Chemie> auxochrome
auxochrome Gruppe f <Chemie> auxochrome
AV (Säurewert) <Chemie> AV (acid value)
Avalanchediode f <Phys> avalanche diode
Avalanchedurchbruch m <Elektriz> avalanche breakdown
Avalanchedurchbruchspannung f <Elektriz> avalanche voltage
Avalanchephotodiode f (APD) <Elektronik, Optik> avalanche photodiode (APD)
Avenin n <Chemie> avenin
Aventurin n <Ker & Glas> aventurine
A-Vermittlungsstelle f <Telekom> originating exchange
A-Verstärker m <Phys> class A amplifier
Avionik f <Lufttrans> avionics
Avionikkonsole f <Raumfahrt> avionics console
Avogadro'sche Konstante f <Phys, Thermod> Avogadro's constant
Avogadro'sche Zahl f (NA) <Phys, Thermod> Avogadro's number (NA)
Avogadro'sches Gesetz n <Phys, Thermod> Avogadro's hypothesis
Avoirdupois-Gewicht n <Metrol> avoirdupois weight

AVR *(automatische Verstärkungsregelung)* <Elektronik, Funktech, Telekom> AGC *(automatic gain control)*
AWACS *(Überwachungs- und Leitsystem im Flugzeug)* <Lufttrans> AWACS *(airborne warning and control system)*
AWG *(Amerikanische Einheit für Drahtdurchmesser)* <Metrol> AWG *(American wire gage)*
AWS *(Automatisches Warnsystem)* <Eisenbahn> AWS, Automatic Warning System *(induktive Zugbeeinflussung)*
axial *adj* <Geom, Papier> axial
axial zusammenschiebbare Lenksäule *f* <Trans> axially collapsing steering column
Axial... <Elektriz, Fertig, Ker & Glas, Maschinen, Optik, Telekom> axial
Axialablagerung *f* <Ker & Glas> axial deposition
Axialabscheideverfahren *n* **aus Dampfphase** *(VAD)* 1. <Optik> *(AE)* vapor phase axial deposition technique, *(BE)* vapour phase axial deposition technique *(VAD)*; 2. <Telekom> *(AE)* vapor phase axial deposition technique, *(BE)* vapour-phase axial deposition technique *(VAD)*
Axialabschirmung *f* <Kerntech> axial shield
Axialanker *m* <Elektriz> axial armature
Axialbeanspruchung *f* 1. <Fertig> thrust load; 2. <Maschinen> axial load, axial strain
Axialdauer *f* <Raumfahrt> axial period
Axialdruck *m* <Fertig> axial thrust
Axialdrucklager *n* 1. <Maschinen> axial thrust bearing, end-thrust bearing, thrust bearing; 2. <Mechan> thrust bearing
axiale Abtastung *f* <Optik> axial scanning
axiale Ausbreitungskonstante *f* <Telekom> axial propagation coefficient
axiale Platteninterferometrie *f* <Optik> axial slab interferometry
axiale Temperaturverteilung *f* <Kerntech> axial temperature distribution
Axialebene *f* <Fertig> axial plane
Axialempfindlichkeit *f* <Akustik> axial sensitivity
axialer Ausbreitungskoeffizient *m* <Optik> axial propagation coefficient
axialer Strahl *m* <Optik, Telekom> axial ray
axialer Tischvorschub *m* <Fertig> table feed
axialer Turboverdichter *m* <Lufttrans> axial compressor
axialer Vorschub *m* <Maschinen> axial feed
axiales Austauschen *n* <Kerntech> axial shuffling *(Brennelementen)*
axiales Flächenträgheitsmoment *n* <Fertig> geometrical moment
axiales Plasma-Abscheideverfahren *n* <Telekom> axial plasma deposition
Axialgebläse *n* <Maschinen> axial fan, propeller fan
Axialgeschwindigkeit *f* <Nichtfoss Energ> axial velocity
Axialgeschwindigkeitsmelder *m* <Raumfahrt> axial velocity sensor
Axialhonen *n* <Fertig> axial honing
Axialinterferenzmikroskopie *f* <Optik> axial interference microscopy
Axialität *f* <Fertig> alignment
Axialkolbenpumpe *f* <Mechan> axial piston pump
Axialkompressor *m* <Mechan> axial compressor
Axialkompressortriebwerk *n* <Lufttrans> gas turbine engine
Axialkugellager *n* <Maschinen> thrust ball-bearing
Axiallager *n* <Maschinen, Nichtfoss Energ> thrust bearing
Axiallast *f* <Metall> axial load
Axiallüfter *m* 1. <Heiz & Kälte> axial flow fan; 2. <Sicherheit> axial blower
Axialpendelrollenlager *n* <Maschinen> self-aligning roller thrust bearing

Axialpumpe *f* <Maschinen> axial flow pump, axial pump
Axialrad *n* <Maschinen> axial flow wheel
Axial-Radial-Verdichter *m* <Maschinen> axial centrifugal compressor
Axialrillenkugellager *n* <Maschinen> deep-groove ball thrust bearing
Axialrollenlager *n* <Maschinen> thrust roller bearing
Axialschlag *m* <Maschinen> axial eccentricity
Axialschub *m* <Fertig, Maschinen> axial thrust
Axialspiel *n* 1. <Fertig> amount of axial freedom, end play; 2. <Mechan> axial clearance, end play
Axialstellglied *n* <Raumfahrt> axial actuator *(Raumschiff)*
Axialstrahl *m* <Optik> axial ray
Axialstrom *m* <Lufttrans> axial flow
Axialströmung *f* 1. <Kfztech> axial flow *(Motor)*; 2. <Maschinen> axial flow
Axialströmungshubgebläse *n* <Lufttrans> axial flow lift fan
Axialteilung *f* <Maschinen> axial pitch
Axialturbine *f* <Hydraul> axial flow turbine
Axialventilator *m* 1. <Elektriz> axial fan; 2. <Sicherheit> axial ventilator
Axialvergrößerung *f* <Mechan> axial magnification
Axialverhältnis *n* <Raumfahrt> axial ratio
Axialversatz *m* 1. <Fernseh> axial displacement; 2. <Fertig> axial misalignment *(Faseroptik)*
Axialvorschub *m* <Maschinen> axial feed
Axialzylinderrollenlager *n* <Maschinen> axial cylindrical roller bearing, thrust cylindrical roller bearing
Axiom *n* <Geom, Math> axiom
axiomatisch *adj* <Geom> axiomatic
Axonometrie *f* <Geom> axonometry
axonometrische Projektion *f* <Konstzeich> axonometric projection
Axt *f* <Bau> *(AE)* ax, *(BE)* axe
Azelain... <Chemie> azelaic
Azeotrop *n* <Chemie> azeotrope
azeotrop *adj* <Chemie> azeotropic
azeotrope Destillation *f* <Lebensmittel> azeotropic distillation
azeotropes Gemisch *n* 1. <Chemie> azeotrope; 2. <Lebensmittel> azeotropic mixture
Azetozon *n* <Lebensmittel> acetyl benzoyl peroxide
Azetyl *n* <Papier> acetyl
Azetylen *n* 1. <Bau> acetylene; 2. <Chemie> acetylene, ethine, ethyne; 3. <Fertig> acetylene, acetylene gas; 4. <Mechan> acetylene
Azetylbrenner *m* <Bau> acetylene blowpipe
Azetylenbrennschneiden *n* <Fertig> acetylene cutting
Azetylenbrennschneider *m* <Fertig> acetylene cutter
Azetylendruck *m* <Fertig> acetylene pressure
Azetylendruckminderer *m* <Fertig> acetylene pressure regulator, acetylene regulator
Azetylenentwickler *m* <Bau, Chemtech, Fertig, Maschinen> acetylene generator
Azetylenerzeuger *m* <Chemtech> acetylene generator
Azetylenerzeugung *f* <Fertig> acetylene generation
Azetylenerzeugungsanlage *f* <Fertig> acetylene gas generating plant, acetylene generator station, acetylene producing plant
Azetylenflamme *f* <Fertig> acetylene flame
Azetylenflasche *f* <Fertig, Maschinen, Mechan, Sicherheit> acetylene cylinder
Azetylengas *n* <Chemie> acetylene gas
azetylenisch *adj* <Chemie> acetylenic
Azetylenleitung *f* <Fertig> acetylene line
Azetylensauerstoffbrenner *m* <Bau> oxyacetylene blowpipe, oxyacetylene blowtorch
Azetylensauerstoffschweißen *n* <Fertig> oxyacetylene welding

Azetylenschlauch m <Fertig> acetylene hose
Azetylenschweißbrenner m <Bau, Fertig> acetylene blowpipe
Azetylenschweißen n <Fertig> acetylene welding, oxyacetylene welding
Azetylenstahlflasche f <Fertig> acetylene cylinder
Azetylenüberschuss m <Fertig> acetylene excess
Azetylenventil n <Fertig> acetylene valve
Azetylgruppe f <Lebensmittel> acetyl group
Azetylid n <Chemie> acetylide
azetylieren v <Papier> acetylate
Azetylierung f <Lebensmittel, Papier> acetylation
Azetylsalicyl... <Chemie> acetylsalicylic
Azetylzahl f <Kunststoff> acetyl value
Azetylzellulosefilm m <Verpack> cellulose acetate film
Azid n <Chemie> azide, hydrazoate
azid adj <Chemie> acidic
Azidität f <Druck> acidity
Azimino... <Chemie> azimino
Aziminoverbindung f <Chemie> azimino compound
Azimut m <Bau, Funkort, Funktech, Phys, Raumfahrt, Wassertrans> azimuth (Navigation)
Azimutabweichung f <Aufnahme> azimuth deviation
Azimutalbefestigung f <Raumfahrt> Az-El mount
azimutale Führung f <Lufttrans> back azimuth guidance (Navigation)
azimutale Quantenzahl f <Phys> azimuthal quantum number
azimutaler Kurvenanflug m <Lufttrans> curved azimuth approach path (Mikrowellenlandesystem)
azimutales Steuerwerk n <Lufttrans> azimuthal control
Azimutalschub m <Raumfahrt> azimuth thrust
Azimutal-Transversal-Mode f (ATM) <Optik> azimuthal transversal mode (ATM)
Azimutalverlust m <Fernseh> azimuth loss
Azimutalverzeichnung f <Fernseh> azimuth distortion
Azimuteinstellung f <Fernseh> azimuth adjustment
Azimut-Elevationsmontierung f <Funkort, Funktech> azimuth-elevation mount (Antenne)
Azimutkreisel m <Raumfahrt> azimuth gyro
Azobenzol n <Chemie> azobenzene
Azofarbstoff m <Textil> azoic dye
Azophenylen n <Chemie> azophenylene, phenazine
Azotometer n <Chemie> azotometer, nitrometer
Azoverbindung f <Chemie> azo compound, azo derivative
Azulen n <Chemie> azulene
Azulminsäure f <Chemie> azulmin
Azurit n <Chemie> azurite, chessylite

B

b 1. <Kohlen, Labor> (Bar) b, bar (Luftdruckeinheit); 2. <Raumfahrt> (Raumbreite) b (galactic latitude)
B 1. <Akustik, Elektrotech, Phys> (Bel) B, bel; 2. <Aufnahme> (magnetischer Scheinwiderstand) B (magnetic induction); 3. <Chemie> (Bor) B (boron); 4. <Elektriz, Elektrotech, Phys, Telekom> (Magnetinduktion) B (magnetic induction); 5. <Kerntech> (Bindungsenergie) B (binding energy); 6. <Strahlphys, Teilphys> (Bindungsenergie, Kernbindungsenergie) B (binding energy); 7. <Thermod> (Volumenelastizitätsmodul) B (modulus of volume elasticity)
Ba (Barium) <Chemie> Ba (barium)
Babbitmetall n <Fertig> babbitt
Babinet'scher Kompensator m <Phys> Babinet compensator
Babinet'sches Prinzip n <Phys> Babinet's principle
Babinet'sches Theorem n <Phys> Babinet's principle
Back f <Wassertrans> forecastle (Schiffbau)
Backbone m <Comp & DV> backbone
backbord adv 1. <Lufttrans, Raumfahrt> port (Raumschiff); 2. <Wassertrans> port (Schiff)
Backbord n <Lufttrans, Wassertrans> port
Backbord achteraus adv <Wassertrans> on the port quarter
Backbord voraus adv <Wassertrans> on the port bow
Backdeck n <Wassertrans> raised deck (Schiff)
Backe f 1. <Kerntech, Kohlen> jaw; 2. <Maschinen> cheek, jaw; 3. <Mechan> jaw; 4. <Wassertrans> cheek (Deckausrüstung)
Backeigenschaft f <Lebensmittel> baking quality
backen v <Kohlen, Lebensmittel> bake
Backen n 1. <Fertig> caking; 2. <Kfztech, Maschinen> baking (Bohrtechnik); 3. <Metall> baking
Backenbohrmeißel m <Erdöl> Mother Hubbard bit (Bohrtechnik)
Backenbrecher m 1. <Kohlen> jaw crusher; 2. <Maschinen> jaw breaker, jaw crusher
Backenbremse f <Kfztech, Maschinen> shoe brake
backend adj 1. <Chemtech> agglutinative; 2. <Kohlen> caking
backende Kohle f <Kohlen> caking coal
Back-End-Kernsystem n <Comp & DV> back-end core system
Back-End-System n <Comp & DV> back-end system
Backenfänger m <Bau> casing spears
Backenfutter n <Maschinen> dog chuck, jaw chuck
Backenklemme f <Trans> clip with jaws
Backenklemmvorrichtung f <Labor> clamp with jaws
Backenwerkzeug n <Kunststoff> (AE) split mold, (BE) split mould
backholen v <Wassertrans> back (Segel)
backlegen v <Wassertrans> lay aback (Segel)
Backstagswind m <Wassertrans> quarter wind
backstehen v <Wassertrans> aback (Segeln)
Backtracking n <Künstl Int> backtracking
Bad n 1. <Metall> bath; 2. <Textil> bath, dip
Badeanzugstoff m <Textil> swimsuit fabric
Badewannenkurve f <Math> bath-tube curve (Statistik, z. B. der Fehlerhäufigkeit)
Badthermometer n <Foto> tray thermometer
Baffle n <Chemtech> baffle (Vakuumtechnik)
Bagassenwalze f <Lebensmittel> bagasse roller
BA-Gewinde n <Maschinen> British Association screw thread
Bagger m 1. <Bau, Kfztech> excavator; 2. <Kohlen> digger
Baggerarbeiten fpl <Wassertrans> dredging operations
Baggereimer m 1. <Trans> digging bucket; 2. <Wassertrans> dredge bucket
Baggereimerzähne mpl <Trans> digging bucket teeth
Baggerkette f 1. <Bau> bucket chain; 2. <Maschinen> excavator chain
Baggerkorb m <Bau> grab
Baggern n <Kohlen> digging
Baggerpumpe f <Wassertrans> dredge pump
Baggersand m <Bau> dredging sand
Baggerschaufel f <Bau> bucket
Baggersumpf m <Kohlen> sedimentation pond

Bahn

Bahn f 1. <Phys> orbit, path, trajectory; 2. <Raumfahrt> path, trajectory; 3. <Telekom> path *(Elektronen)*; 4. <Textil> sheeting, width; 5. <Wassertrans> path
Bahn f **eines Teilchens** <Phys> path of a particle
Bahnanlage f <Eisenbahn> *(AE)* railroad system, *(BE)* railway system
Bahnantrieb m <Elektrotech> traction drive
Bahnaufzeichnung f <Strahlphys> trajectography
Bahnbenutzer m <Eisenbahn> *(AE)* railroad user, *(BE)* railway user
Bahnbetrieb m <Eisenbahn> service
Bahnbildung f <Papier> web formation
Bahnbreite f <Druck> web width
Bahndrehimpuls m <Phys, Strahlphys> orbital angular momentum
Bahndrehimpulsquantenzahl f <Phys> orbital angular momentum quantum number
Bahneigenschaften fpl <Elektronik> bulk properties *(Mikroelektronik)*
Bahnelektronen npl <Strahlphys> orbital electrons
Bahngeschwindigkeit f <Phys> orbital velocity *(Umlaufbahn)*
Bahnhalbleiter m <Elektronik> bulk semiconductor
Bahnhof m <Eisenbahn> *(AE)* railroad depot, *(BE)* railway depot, *(AE)* railroad station, *(BE)* railway station, station
Bahnhofsanlage f <Eisenbahn> station area
Bahnhofswagen m <Eisenbahn> wagon
Bahnimpuls m <Strahlphys> orbital momentum
Bahnkorrektur f <Raumfahrt> path correction
Bahnmotor m <Eisenbahn> traction motor
Bahnnetz n <Eisenbahn> *(AE)* railroad system, *(BE)* railway system
Bahnoberbau m <Eisenbahn> permanent way
Bahnpostwagen m <Eisenbahn> *(AE)* mail car, *(AE)* mail van, *(BE)* mailcoach, *(BE)* post wagon
Bahnquantenzahl f <Phys, Strahlphys> orbital quantum number
Bahnrakete f <Raumfahrt> orbital rocket
Bahnrendezvous n <Raumfahrt> earth orbit rendezvous *(im Erdorbit)*
Bahnriss m **in der Nasspartie** <Papier> wet break
Bahnsatellit m <Raumfahrt> orbiting satellite
Bahnsatellit m **für Amateurfunkzwecke** *(OSCAR)* <Funktech, Raumfahrt> Orbiting Satellite Carrying Amateur Radio, OSCAR *(Weltraumfunk)*
Bahnscheibe f <Elektronik> bulk wafer
Bahnschranke f <Eisenbahn> *(AE)* railroad gate, *(BE)* railway gate
Bahnschreibung f <Raumfahrt> trajectography
Bahnstation f <Raumfahrt> orbiting station
Bahnsteig m <Eisenbahn> platform
Bahnsteigkarte f <Eisenbahn> platform ticket
Bahnstromnetz n <Eisenbahn> traction network
Bahnstromsystem n <Eisenbahn> traction network
Bahnstromversorgung f <Eisenbahn> traction power supply
Bahntransport m <Eisenbahn> rail transport, *(AE)* railroad transport, *(BE)* railway transport
Bahnüberführung f <Eisenbahn> bridge, *(AE)* road over railroad, *(BE)* road over railway
Bahnübergangsfahrzeug n <Raumfahrt> orbital transfer vehicle
Bahnverfolgung f 1. <Funktech> tracking *(Satelliten)*; 2. <Phys> ray tracing
Bahnverkehr m <Eisenbahn> service
Bahnvermessung f <Raumfahrt> tracking *(Raumschiff)*
Bahnwärter m <Eisenbahn> lineman
Bai f <Nichtfoss Energ> bay *(Geographie)*
Bailey-Behelfsbrücke f <Bau> Bailey bridge

Bainit m <Fertig, Metall> bainite
Bainitferrit m <Metall> bainitic ferrite
Bajonett n <Elektrotech> bayonet
Bajonettfassung f 1. <Bau> bayonet fitting; 2. <Elektriz> bayonet socket; bayonet holder, bayonet socket *(für Leuchtstofflampe)*; 3. <Foto> bayonet mount; 4. <Maschinen> bayonet socket
Bajonettkupplung f <Elektrotech> bayonet coupling
Bajonett-Lampenfassung f <Elektriz> bayonet lamp holder
Bajonettsockel m 1. <Elektriz> bayonet cap *(Leuchtstofflampe)*; 2. <Elektrotech, Foto> bayonet base
Bajonettsteckverbinder m **mit Überwurfmutter** *(BNC-Stecker)* <Elektroniks> bayonet nut connector *(BNC)*
Bajonettverbindung f <Elektriz> bayonet joint
Bajonettverschluss m 1. <Fertig> bayonet fastening *(Kunststoffinstallationen)*; 2. <Foto> *(BE)* bayonet socket, *(AE)* quarter-turn fastener; 3. <Ker & Glas> bayonet cap finish; 4. <Kerntech> bayonet closure; 5. <Mechan> bayonet locking; 6. <Verpack> bayonet catch
Bake f 1. <Funkort, Funktech> beacon; 2. <Raumfahrt> beacon; beacon generator *(Raumschiff)*; 3. <Wassertrans> beacon *(Seezeichen)*
Bakelit n <Fertig> bakelite
Bakensignal n <Raumfahrt> beacon *(Weltraumfunk)*
Bakterien fpl <Chemie, Lebensmittel, Umweltschutz> bacteria
Bakterientoxin n <Chemie> bacterial toxin, bacteriotoxin
Bakterienvermehrungsgefäß n <Lebensmittel> bacteria propagation tank
Bakterienzahl f <Wasserversorg> bacterial count *(im Wasser)*
Bakterienzuchtbehälter m <Lebensmittel> bacteria propagation tank
bakteriologische Reinigung f <Abfall> bacteriological treatment, bacteriological purification
bakteriologischer Trockenschrank m <Labor> bacteriological oven
Bakteriolyse f <Lebensmittel> bacteriolysis
Bakteriophage m <Lebensmittel> bacteriophage
Bakteriostatikum n <Lebensmittel> bacteriostat
bakteriostatisches Mittel n <Lebensmittel> bacteriostat
Bakterizid n <Chemie, Lebensmittel> bactericide *(Bakteriengift)*
Balance f <Aufnahme> balance
Balancefeder f <Maschinen> equalizer spring
Balanceregelung f <Aufnahme> balance control
Balanceruder n <Wassertrans> balanced rudder
Balanciermaschine f <Maschinen> beam engine
Balata f <Fertig, Kunststoff> balata
Balatariemen m <Maschinen> balata belt
Balg m <Maschinen> bellows
Balgen m <Foto, Maschinen> bellows
Balgenansatz m <Foto> bellows attachment
Balgenauszug m <Foto> bellows extension
Balgendichtung f <Foto> bellows covering
Balgenkamera f <Foto> bellows camera
Balgenklappkamera f <Foto> bellows-type folding camera
Balgenrahmen m <Foto> bellows frame
Balgenverschluss m <Foto> bellows shutter
Balgfeder f <Kfztech> cushion spring *(Kupplung)*
Balgfederdurchflussmessgerät n <Gerät> bellows flowmeter
Balgfedermanometer n <Gerät> *(AE)* bellows gage, *(BE)* bellows gauge
Balgmanometer n <Fertig> bellows gauge
Balken m 1. <Bau> *(BE)* baulk, summer; 2. <Fernseh> bar, flagpole; 3. <Mechan> girder

Balkenanzeige f <Gerät> bar graph display
Balkenauflage f <Metrol> beam support *(Waage)*
Balkenauflagerplatte f <Bau> wall plate
Balkenaussparung f <Bau> wall box
Balkenbiegung f <Bau> beam bending
Balkenbrücke f <Bau> girder bridge
Balkenbucht f <Wassertrans> round of beam; camber *(Schiffbau)*
Balkencode m <Kontroll> bar code
Balkendiagramm n <Comp & DV, Math> bar chart, bar diagram, bar graph
Balkengenerator m <Fernseh> bar generator
Balkenknie n <Wassertrans> beam knee *(Schiffbau)*
Balkenlehre f <Metrol> *(AE)* beam caliper, *(AE)* beam caliper gage, *(BE)* beam calliper, *(BE)* beam calliper gauge
Balkenleiter m <Elektronik> beam lead *(Mikroelektronik)*
Balkenleitergerät n <Elektronik> beam lead device
Balkenleitertechnik f <Elektronik> beam lead technique
Balkenmuster n <Fernseh> bar pattern
Balkenstein m <Bau> corbel
Balkenträger m <Bau> beam
Balkenwaage f 1. <Gerät> beam-type scale; 2. <Metrol> beam and scales, beam balance, beam scales
Balkenwerk n <Bau> framework
Ball m <Labor> ball *(Kamera)*
Ballast m 1. <Elektrotech> ballast; 2. <Mechan> load; 3. <Trans> ballast *(Schiff)* • **in Ballast fahren** <Wassertrans> sail in ballast
Ballastkiel m <Wassertrans> ballast keel *(Schiffbau)*
Ballastkreis m <Elektrotech> ballasting circuit
Ballaststoffe mpl <Lebensmittel> *(AE)* crude fiber, *(BE)* crude fibre, *(AE)* dietary fiber, *(BE)* dietary fibre, roughage
ballaststoffreich adj <Lebensmittel> *(AE)* high-fiber, *(BE)* high-fibre
Ballasttank m <Wassertrans> ballast tank *(Schiff)*
Ballastwiderstand m 1. <Elektriz> ballast resistor, barretter; 2. <Elektrotech> ballast resistor
Ballastzustand m <Wassertrans> ballast condition
ballen v 1. <Kohlen> ball; 2. <Verpack> bale
Ballen m 1. <Fertig> body, pack *(Walze)*; 2. <Papier> bale; 3. <Textil> bale, packet; 4. <Verpack> bale • **in Ballen verpacken** <Verpack> bale
Ballengriff m <Maschinen> ball knob
Ballenlader m <Trans> bale loader
Ballenlänge f <Fertig> barrel length, body length; surface length *(Walze)*
Ballenpresse f <Papier, Verpack> baling press
Ballen-Pulper m <Papier> bale pulper
Ballenumreifung f <Verpack> bale hoop
ballig adj <Fertig> convex
ballig bearbeiten v <Fertig> crown
ballige Fläche f <Fertig> spherical surface
ballige Riemenscheibe f <Fertig> crown-face pulley
Balligkeit f 1. <Fertig> crowning *(Zahnrad)*; 2. <Maschinen> camber • **ohne Balligkeit** <Fertig> noncrowned *(Getriebelehre)*
balligtragendes Zahnrad n <Fertig> gear with localized tooth bearing
ballistisch adj <Raumfahrt> ballistic
ballistische Auslesevorrichtung f <Abfall> ballistic sorter
ballistische Bahn f <Raumfahrt> ballistic path
ballistische Flugbahn f <Raumfahrt> ballistic trajectory
ballistische Kurve f <Maschinen> ballistic curve
ballistische Rakete f <Raumfahrt> ballistic missile
ballistische Sichtung f <Abfall> ballistic separation
ballistische Sortierung f <Abfall> ballistic sorting *(von Müll)*

ballistisches Galvanometer n <Elektriz, Gerät, Phys> ballistic galvanometer
Ballon m 1. <Labor> bulb *(Gebläseballon)*; 2. <Lebensmittel> carboy
Ballonkorb m <Mechan> car
Ballonreifen m <Kfztech> *(AE)* balloon tire, *(BE)* balloon tyre
Ballung f 1. <Elektronik> bunching *(Klystron)*; 2. <Lufttrans> bunching
Ballungsraum m <Elektronik> bunching space *(Klystron)*
Balmer'sche Gleichung f <Phys> Balmer's formula
Balmer'sche Serie f <Phys, Teilphys> Balmer series
Balun m <Funktech> balanced-to-unbalanced transformer, balun *(zum Anschluss symmetrischer Antennen an Koaxialkabel)*
Baluster m <Bau> baluster
Bambus m <Ker & Glas> bamboo
Bambuseffekt m <Kerntech> bamboo effect
Bananenbahn f <Kerntech> banana trajectory
Bananenbuchse f <Elektrotech> banana jack
Bananenstecker m <Elektrotech, Foto> banana plug
Bananenumlaufbahn f <Kerntech> banana orbit
Banbury-Innenmischer m <Kunststoff> Banbury mixer
Band n 1. <Akustik> band; 2. <Bau> band; hinge *(Baubeschlag)*; 3. <Comp & DV> band, ribbon, tape; 4. <Elektriz> tape *(zum Isolieren)*; 5. <Funktech> band *(Frequenz)*; 6. <Ker & Glas> ribbon; 7. <Kontroll> line; 8. <Kunststoff> tape; 9. <Maschinen> band, hoop; 10. <Metrol> ribbon, tape line; 11. <Papier> strap; 12. <Phys> band *(Frequenz)* • **auf Band aufnehmen** <Aufnahme> tape • **auf Band aufzeichnen** <Fernseh> tape • **Band begrenzen** <Elektronik> band-limit
Band n **in Kehrlage** <Telekom> reversed band, reversed CF-band
Band n **in Regellage** 1. <Funktech> erect band *(Frequenzband)*; 2. <Telekom> erect band, erect CF-band
Bandabdichtung f <Verpack> tape sealer
Bandabschöpfgerät n <Meerschmutz> rope skimmer
Bandabspielgerät n <Aufnahme> tape player
Bandanfang m *(BOT)* <Comp & DV> beginning of tape *(BOT)*
Bandanfangsmarke f <Comp & DV> BOT marker, tape mark, beginning-of-tape marker
Bandantrieb m 1. <Aufnahme> capstan drive, capstan servo, tape drive; 2. <Comp & DV> capstan, tape transport; 3. <Fernseh> tape drive
Bandantriebsachse f <Aufnahme, Elektrotech> capstan *(Tonbandgerät)*
Bandantriebswelle f <Akustik, Fernseh> capstan
Bandaufnahme f <Aufnahme> tape recording
Bandaufnahmegerät n <Aufnahme> tape recorder
Bandausnutzung f <Telekom> band efficiency
Bandauszug m <Comp & DV> tape dump
Bandbasis f <Aufnahme> tape base
Bandbefeuchtungsvorrichtung f <Verpack> tape-moistening device
bandbegrenztes Signal n <Elektronik> band-limited signal
Bandbelag m <Aufnahme> tape backing
Bandbeschichtungsmaterial n <Aufnahme> tape-coating material
Bandbibliothek f <Comp & DV> tape library
Bandbiegung f <Fernseh> tape curvature
Bandbreite f *(BW)* 1. <Akustik> band; 2. <Aufnahme, Comp & DV, Elektronik, Fernseh, Funktech> bandwidth, BW; 3. <Optik> bandwidth, BW *(Wellenlänge)*; 4. <Telekom> bandwidth
Bandbreite f **des Vertikalverstärkers** <Elektronik> vertical-amplifier bandwidth

Bandbreite

Bandbreite *f* **nach Carson** <Raumfahrt> Carson's rule bandwidth *(Weltraumfunk)*
Bandbreite-Länge-Produkt *n* <Telekom> length-bandwidth product *(Faseroptik)*
bandbreitenbegrenzter Betrieb *m* <Telekom> bandwidth-limited operation
bandbreitenbegrenzter Vorgang *m* <Optik> bandwidth-limited operation
Bandbreitenkompression *f* <Fernseh, Telekom> bandwidth compression
Bandbreitenökonomie *f* <Telekom> bandwidth economy
Bandbreitenreduzierung *f* <Fernseh> bandwidth compression
Bandbreite-Reichweite-Produkt *n* <Telekom> length-bandwidth product *(Faseroptik)*
Bandbremse *f* 1. <Maschinen> band brake, strap brake; 2. <Mechan> band brake
Bändchen *n* <Akustik, Aufnahme, Optik> ribbon
Bändchenkabel *n* <Optik> ribbon cable
Bändchenlautsprecher *m* <Akustik, Aufnahme> ribbon loudspeaker
Bändchenmikrofon *n* <Akustik, Aufnahme> ribbon microphone
Banddatei *f* <Comp & DV> tape file
Banddatenauszug *m* <Comp & DV> tape dump
Banddehnung *f* <Elektronik> bandwidth expansion
Banddiagramm *n* <Qual> band chart
Banddistribution *f* <Comp & DV> tape distribution
Banddrucker *m* <Comp & DV> band printer, belt printer
Banddurchlauf-Geschwindigkeit *f* <Akustik> tape speed
Bande *f* 1. <Phys> band *(Spektrum)*; 2. <Textil> stripe
Bandeinführung *f* 1. <Aufnahme> tape threading; 2. <Fernseh> tape input guide
Bandeinheit *f* <Comp & DV> tape unit
Bandeisen *n* 1. <Fertig> hoop; 2. <Maschinen, Verpack> band iron
Bandende *n* 1. <Comp & DV> end of reel; 2. <Comp & DV> end of tape
Bandendemarke *f* <Comp & DV> EOT marker, end-of-tape marker
Bandenspektrum *n* <Phys, Strahlphys> band spectrum
Bänder *npl* <Wellphys> bands *(aufgrund von Interferenz)*
Bändermodell *n* <Phys, Strahlphys> band theory, band theory of solids
Banderole *f* <Verpack> band label, banderole
Banderolendruck- und Klebemaschine *f* <Verpack> *(AE)* print and apply labeling machine, *(BE)* print and apply labelling machine
Banderolieren *n* <Verpack> banding
Banderoliermaschine *f* <Verpack> *(AE)* labeling machine, *(BE)* labelling machine, strapping machine
Bandetikett *n* <Comp & DV> tape label
Bandfahrsteig *m* <Trans> *(BE)* belt-type moving pavement, *(AE)* belt-type moving sidewalk
Bandfahrung *f* <Bau> man-riding
Bandfeder *f* <Maschinen> volute spring
Band-Film-Übertragung *f* <Fernseh> tape-to-film transfer
Bandfilter *n* **mit flachem Kurvenverlauf** <Elektronik, Funktech> flat band-pass filter
Bandfilterkurve *f* <Telekom> band-pass filter response, band-pass response
Bandfiltersperrbereich *m* <Telekom> band-pass filter stopband
Bandfiltersperrbereichsdämpfung *f* <Telekom> band-pass filter stopband attenuation
Bandfluss-Frequenzgang *m* <Akustik> recording characteristic

Bandförderer *m* 1. <Fertig> belt conveyor, ribbon conveyor; 2. <Maschinen> band conveyor; 3. <Papier> belt conveyor
Bandförderer *m* **mit Zugseil** <Papier> cable conveyor
Bandformat *n* <Comp & DV> tape format
Bandführung *f* 1. <Aufnahme, Comp & DV> tape guide; 2. <Fernseh> ribbon guide, tape alignment guide, tape guide, tape output guide
Bandführungssystem *n* <Aufnahme> vacuum guide system
Bandgerät *n* 1. <Akustik> tape recorder; 2. <Comp & DV> tape unit
Bandgeschwindigkeit *f* <Aufnahme, Fernseh> tape speed
Bandgeschwindigkeitsregelung *f* <Fernseh> tape speed control
bandgesteuerter Zeilenguss *m* <Druck> tape-controlled linecasting
bandgewickelter Kern *m* <Elektrotech> tape-wound core
Bandhaspel *f* <Aufnahme> tape reel
Bandheber *m* <Aufnahme, Fernseh> tape lifter
Bandholzschleifmaschine *f* <Fertig> band sander
Bandiermaschine *f* <Verpack> paper-banding machine
Bandkabel *n* 1. <Elektrotech> flat cable; 2. <Telekom> ribbon cable
Bandkassette *f* <Aufnahme> tape cassette
Bandkennsatz *m* <Comp & DV> tape label
Bandkennzeichnung *f* <Comp & DV> tape label
Bandkühlofen *m* <Ker & Glas> conveyor belt lehr
Bandkupplung *f* <Maschinen> band clutch, band coupling
Bandlängenanzeige *f* <Fernseh> tape length indicator
Bandlängsschlupf *m* <Aufnahme, Fernseh> cinching
Bandlauf *m* <Aufnahme, Fernseh> tape run
Bandlaufwerk *n* <Comp & DV> tape deck, tape drive
Bandleiter *m* <Elektrotech, Phys> strip line
Bandleitung *f* <Elektrotech, Phys> strip line
Bandlochung *f* <Foto> perforation *(Film)*
Bandlücke *f* <Phys> gap
Bandmarke *f* <Comp & DV> tape mark
Bandmaschine *f* <Ker & Glas> ribbon machine
Bandmaß *n* 1. <Mechan> tape measure; 2. <Metrol> tape, tape measure
Bandmechanismus *m* **mit Servosteuerung** <Fernseh> servo-controlled tape mechanism
Bandmittenfrequenz *f* <Elektronik, Funktech, Telekom> midband frequency
Bandmittenverstärkung *f* <Elektronik, Funktech> midband gain
Bandmodell *n* <Phys, Strahlphys> band theory of solids
bandolierte Bauteile *npl* <Elektrotech> bandoliered components
bandolierte Komponenten *fpl* <Elektrotech> bandoliered components
Bandoxidschicht *f* <Aufnahme, Fernseh> tape oxide layer
Bandpass *m* <Funktech> band-pass
Bandpassfilter *n* *(BPF)* <Aufnahme, Elektronik, Fernseh, Funktech, Phys, Telekom> band-pass filter *(BPF)*
Bandpassfilter *n* **dritter Ordnung** <Elektronik> third order band-pass filter
Bandpassfilter *n* **zweiter Ordnung** <Elektronik> second order band-pass filter
Bandpassformung *f* <Elektronik> band-pass filter shaping
Bandpassverstärker *m* <Aufnahme, Elektronik> band-pass amplifier
Bandring *m* <Verpack> coil
Bandrolle *f* <Aufnahme> tape roller

Bandrückseite f <Aufnahme> tape backing
Bandsäge f <Maschinen> band saw
Bandsägemaschine f <Maschinen> band-sawing machine
Bandsalat m <Aufnahme> tape spill
Bandschaben n <Aufnahme> tape scrape
Bandschalldruckpegel m <Aufnahme> band pressure level
Bandscharnier n <Bau> strap hinge
Bandscheibe f <Maschinen> band wheel
Bandscheider m <Kohlen> belt separator
Bandschlaufe f <Fernseh> tape loop
Bandschlaufencassette f <Fernseh> tape loop cassette
Bandschleife f <Aufnahme> tape loop (Endlosband)
Bandschleifen n 1. <Fertig> abrasive band grinding, abrasive belt grinding; 2. <Maschinen> belt grinding
Bandschleifmaschine f 1. <Fertig> abrasive band grinding machine; 2. <Maschinen> abrasive belt grinder, belt grinder
Bandschlupf m <Fernseh> tape slippage
Bandschräglauf m <Aufnahme> tape skew
Bandschreiber m 1. <Labor> chart recorder; 2. <Lufttrans> continuous chart recorder
Bandseil n <Maschinen> flat rope
Bändsel aufsetzen v <Wassertrans> seize
Bandsortieren n <Comp & DV> tape sorting
Bandspannarm m <Aufnahme> tape tension arm
Bandspannung f 1. <Aufnahme, Fernseh> tape tension; 2. <Fertig> belt tension (Schleifband)
Bandspannungsregelung f <Aufnahme, Fernseh> tape tension control
Bandspannungsregler m <Aufnahme, Fernseh> tape tension control
Bandsperre f 1. <Elektronik> active band-stop filter; 2. <Elektrotech> blocking network; 3. <Funktech, Telekom> band-rejection filter
Bandsperrfilter n 1. <Aufnahme> band-stop filter; 2. <Elektronik> active band-stop filter, band-stop filter; 3. <Funktech, Telekom> band-rejection filter, band-stop filter
Bandspleißer m <Aufnahme> tape splicer
Bandspreizung f 1. <Funktech, Phys, Strahlphys> band-spread; 2. <Telekom> band spreading
Bandspreizverfahren n <Wassertrans> spread spectrum technique (Elektronik)
Bandsprosse f <Elektronik> frame (Platz, den Zeichen auf Magnetband einnimmt)
Bandspule f 1. <Aufnahme> reel spindle; 2. <Comp & DV> tape reel; 3. <Fernseh> spool
Bandstahl m 1. <Fertig> strip; 2. <Metall> hoops, skelp
Bandstahlbeschichtung f <Kunststoff> coil coating
Bandstand m <Aufnahme, Fernseh> tape count
Bandstreifigkeit f <Textil> barriness in the weft
Bandströmung f <Heiz & Kälte> laminar flow
Bandstruktur f <Phys> band structure (Festkörperphysik, Emissionsspektrum)
Bandtransport m <Aufnahme, Comp & DV> tape transport
Bandtransportgeometrie f <Aufnahme> tape transport geometry
Bandtransportrolle f <Aufnahme, Elektrotech> capstan
Bandtrieb m <Maschinen> belt drive
Bandtrockner m <Chemtech> belt drier, belt dryer
Bandumwickelung f <Optik> tape wrap
Bandverdehnung f <Aufnahme> tape curvature
Bandverspleißung f <Aufnahme> tape splice
Bandvorschub m 1. <Aufnahme> tape advance; 2. <Comp & DV> tape skip
Bandwalze f <Fertig> belt drum (Bandschleifen)
Bandware f <Textil> narrow fabric

Bandweite f (BW) 1. <Aufnahme> bandwidth, BW (eines aufgezeichneten Signals); 2. <Comp & DV, Elektronik, Fernseh, Funktech, Optik, Telekom> bandwidth, BW
Bandwickeln n <Fertig> taping
Bandwölbung f <Aufnahme> tape cupping
Bandzuführer m <Chemtech> belt feed
Bandzugempfindlichkeit f <Akustik> tension sensitivity
Bandzugregler m <Fernseh> tension servo
Banjo n <Eisenbahn> banjo
Banjogehäuse n <Kfztech> banjo-type housing (Hinterachse)
Bank f 1. <Bau> bank (Sand, Fels); 2. <Elektrotech> bank (Gruppe von Kondensatoren); bank (Tasten, Kontakte); 3. <Labor> bench; 4. <Maschinen> table; 5. <Wassertrans> bank (Geographie)
Bankett n <Bau> banquette, bench, berm; flank (einer Straße)
Banknotenpapier n <Papier> (BE) banknote paper, (AE) onionskin paper
Bankplatte f <Ker & Glas> seat
Bankpostpapier n <Druck> bank paper
Bankschraube f <Bau> bench screw
Bankschraubstock m 1. <Bau> (BE) bench vice, (AE) bench vise; 2. <Maschinen> (BE) table vice, (AE) table vise (höhenverstellbarer Tisch)
Bankverkehr m **über das Internet** <Telekom> on-line banking
Banyan-Koppelnetz n <Telekom> Banyan net (Durchschaltung von ATM-Zellen)
Bar n (b) <Kohlen, Metrol> bar, b (Luftdruckeinheit)
Bär m <Fertig> hammer head; salamander (Gießen)
Baracke f <Bau> barrack
Barbaloin n <Chemie> aloin, barbaloin
Barbitur... <Chemie> barbituric
Barbiturat n <Chemie> barbiturate
Barcode m <Verpack> bar code
Barcodelesegerät n <Verpack> bar code reader
Barcodeleser m <Comp & DV> bar code reader
Barcode-Scanner- und Decoder-Logik f <Verpack> bar code scanner and decoder logic
Bardeen-Cooper-Schrieffer-Theorie f (BCS-Theorie) <Phys> Bardeen-Cooper-Schrieffer-Theory (BCS theory)
Bareboat-Charter m <Wassertrans> bareboat charter
barettförmige Feile f <Maschinen> barrette file
bargeldloses Telefon n <Telekom> cashless telephone
BARITT-Diode f <Phys> BARITT diode, barrier injection transit-time diode
Barium n (Ba) <Chemie> barium (Ba)
Bariummonosulfid n <Druck> (AE) barium monosulfide, (BE) barium monosulphide
Bariumoxid n <Chemie> bariumoxide, baryta
Barkhausen-Effekt m <Phys> Barkhausen effect
Barlow-Rad n <Phys> Barlow's wheel
Barn n <Phys, Teilphys> barn (Einheit des Wirkungsquerschnittes)
Barnett-Effekt m <Phys> Barnett effect
Barodiffusion f <Kerntech> barodiffusion
Barograph m 1. <Gerät, Labor> barograph (Luftdruck); 2. <Phys> barograph, recording barometer; 3. <Wassertrans> barograph (Wetterkunde)
Barometer n <Labor, Phys, Wassertrans> barometer (Wetterkunde)
Barometerstand m 1. <Gerät> barometer reading; 2. <Maschinen> barometric reading; 3. <Wassertrans> barometer reading
barometrisch adj <Phys> barometric
barometrische Steuerung f <Lufttrans> barometric controller

barometrischer 84

barometrischer Druckhöhenregler m <Lufttrans> barometric altitude controller
barometrischer Höhenmesser m <Lufttrans> pressure altimeter
Baroskop n <Phys> baroscope
Barotrauma n <Akustik> barotrauma
barotrope Flüssigkeit f <Strömphys> barotropic fluid
Barovakuum-Meter n <Gerät> compound pressure-and-vacuum gauge
Barre f <Wassertrans> bar *(Geographie)*
Barrel npl **pro Tag** *(B/d)* <Erdöl> barrels per calendar day, BCD *(Fördermenge Öl)*
Barren m 1. <Ker & Glas> bullion; 2. <Kohlen> ingot; 3. <Maschinen> bullion; 4. <Mechan> ingot
Barreneisen n <Metall> bar iron
Barrenzinn n <Metall> bar tin
Barretter m 1. <Elektriz> ballast resistor, barretter; 2. <Elektrotech> barretter *(Stabilisatorröhre)*; 3. <Phys> barretter
Barriere f <Sicherheit> fixed barrier guard
Barrierewirkung f <Kunststoff> hold-out
Bartlett-Kraft f <Kerntech> Bartlett force
Barwagen m <Eisenbahn> *(BE)* bar coach
Baryon n <Phys, Teilphys> baryon
Baryonenzahl f <Phys, Teilphys> baryon number
Baryt m <Ker & Glas, Kunststoff> baryte
Baryterde f <Chemie> baryta
Baryzentrum n <Raumfahrt> *(AE)* barycenter, *(BE)* barycentre *(Raumschiff)*
Basalt m <Bau, Nichtfoss Energ> basalt
Base f <Chemie> base
Basengehalt m <Chemie> alkalinity *(Boden)*
Basenkation n <Umweltschutz> base cation
Basis f 1. <Aufnahme> base *(Tonband)*; 2. <Chemie> base, basis; 3. <Comp & DV> base, radix; 4. <Elektronik, Elektrotech> base; 5. <Funktech> base *(Transistor)*; 6. <Geom> base *(einer geometrischen Figur)*; 7. <Maschinen, Telekom> base
Basisabdichtung f <Abfall> bottom sealing; base sealing *(einer Deponie)*
Basisadresse f <Comp & DV> base address
Basisadressenregister n <Comp & DV> base address register
Basisanschluss m 1. <Elektrotech> base contact *(Transistor)*; 2. <Telekom> base access *(ISDN)*
Basisband n 1. <Comp & DV, Telekom, Elektrotech> baseband *(BB)*; 2. <Fernseh> base band
Basisbandcharakteristik f <Telekom> baseband response function
Basisbandmodem n <Telekom> baseband modem
Basisbandsignal n <Telekom> baseband signal
Basisbandübertragungsfunktion f <Telekom> baseband transfer function
Basisbereich m <Elektronik> base region
Basisbreite f <Elektronik> base thickness one, base zone thickness
basisch adj <Chemie> alkaline, basic
basisch Bleicarbonat n <Chemie> basic lead carbonate, ceruse, cerussa
basisch Carbonat n <Chemie> basic carbonate, subcarbonate
basisch Chlorid n <Chemie> basic chloride, subchloride
basisch Erz n <Metall> base ore
basisch Nitrat n <Chemie> basic nitrate, subnitrate
basisch Salz n <Chemie> basic salt, subsalt
basisch stellen v <Chemie> basify
basischer Farbstoff m <Textil> basic dye
basischer Martinofen m <Ker & Glas> basic openhearth furnace

basischer Stahl m <Metall> basic steel
basisches Acetat n <Chemie> basic acetate
Basisdiffusion f <Elektronik> base diffusion *(Transistortechnik)*
Basis-Dotierung f <Elektronik> base doping *(Transistortechnik)*
Basiseinheit f 1. <Comp & DV> primitive; 2. <Telekom> base unit *(BU)*
Basiselektrode f <Elektrotech> base electrode
Basiserweiterung f <Elektronik> base widening
Basisformat n <Comp & DV> native format
Basisgröße f <Phys> elementary magnitude
Basiskanal m <Telekom> basic access channel *(ISDN)*
Basiskomplement n <Comp & DV> radix complement
Basiskreis m <Geom> base circle
Basislinie f <Wassertrans> base line *(Navigation)*
Basismodulation f <Elektronik> base modulation
Basismodus m <Comp & DV> native mode
Basispunkt m <Geom> base point
Basisraum m <Elektronik> base region
Basisregister n <Comp & DV> base register *(CPU)*
Basisschaltung f <Elektrotech> *(BE)* earthed-base connection, *(AE)* grounded-base connection; common-base connection *(bei Transistoren)*
Basisschicht f <Fernseh> base film
Basis-Sende/Empfangs-Station f <Mobilkom> base transceiver station *(BTS)*
Basisstation f <Mobilkom> base station
Basisstationssteuerung f <Telekom> base station controller
Basisstörstelle f <Elektronik> base impurities
Basisvektor m <Phys> basis vector
Basisvolumen n <Trans> base volume
Basisweite f <Elektronik> base thickness two, base width
Basiswiderstand m <Elektrotech> base resistance
Basiswissen n <Künstl Int> basic knowledge
Basov-Diagramm n <Kerntech> Basov diagram
Bass m <Akustik> bass
Bassanhebung f 1. <Aufnahme> bass compensation; 2. <Elektronik> bass boost *(Radiotechnik)*
Bassfilter n <Aufnahme> bass-cut filter
BAS-Signal n <Fernseh> composite signal
Bassreflexgehäuse n <Aufnahme> bass-reflex enclosure
Bassregelung f <Aufnahme> bass control
Bassstimme f <Akustik> bass
Basswiedergabe f <Aufnahme> bass response
Bastard... <Maschinen, Mechan> bastard
Bastardfeile f 1. <Maschinen> bastard file, bastard-cut file; 2. <Mechan> bastard file
Bastardhieb m <Maschinen> bastard cut
Batch n <Kunststoff> batch
Batch-Betrieb m <Comp & DV> batch mode, batch processing, batch window
Batchdestillation f <Kerntech> batch distillation
Batch-Prozess m <Comp & DV> batch process
Bathymetrie f <Nichtfoss Energ> bathymetry
Batterie f 1. <Elektriz> battery; 2. <Elektrotech> battery *(Monozelle)*; battery *(Speicherzelle)*; 3. <Fertig> group; 4. <Foto, Funktech, Kfztech, Maschinen, Phys, Telekom, Wassertrans> battery
Batterie f **von Messkondensatoren** <Elektrotech> capacitance box *(im geschlossenen Gehäuse)*
Batterieanschluss m <Elektriz, Elektrotech> battery terminal
Batterieanschlussklemme f <Kfztech> battery terminal
Batterieantrieb m <Fertig> accumulator driver
Batterieaufladefähigkeit f <Nichtfoss Energ> battery charging capability
Batteriebetrieb m <Elektrotech> battery operation

batteriebetrieben *adj* <Elektrotech> battery-powered, self-powered
batteriebetriebener Diabetrachter *m* <Foto> battery viewer
batteriebetriebener Lastkraftwagen *m* <Trans> battery truck
batteriebetriebener Zaun *m* <Elektrotech> battery-operated electric fence controller
batteriebetriebenes Elektrofahrzeug *n* <Trans> battery-powered electric vehicle
Batterieblitzlicht *n* <Foto> battery-powered flash unit
batterieelektrische Zugförderung *f* <Elektrotech> battery-electric traction
batterieelektrischer Bus *m* <Elektrotech> battery-electric road vehicle
batterieelektrisches Fahrzeug *n* <Kfztech> battery-electric road vehicle
Batterieelektrode *f* <Elektriz> battery electrode
Batterieelement *n* <Elektriz> battery cell
Batteriefach *n* <Foto> battery chamber
Batteriefahrzeug *n* <Elektrotech> battery vehicle
Batteriegehäuse *n* <Kfztech> battery box
batteriegepufferter Transferbus *m* <Raumfahrt> battery transfer bus *(Raumschiff)*
batteriegespeist *adj* <Elektrotech> battery-powered
Batteriegestell *n* 1. <Elektriz> battery frame, battery framework; 2. <Kfztech> battery cradle
batteriegestützte Reservestromversorgung *f* <Kontroll> stand-by battery power supply
Batterieglas *n* <Ker & Glas> battery jar
Batteriehauptschalter *m* <Kfztech> battery master switch
Batteriekammer *f* <Foto> battery chamber
Batteriekasten *m* <Elektriz, Kfztech> battery box
Batterieklemme *f* 1. <Elektriz> battery terminal; 2. <Elektrotech> battery clip *(federnd)*; 3. <Foto> battery grip
Batteriekontakt *m* <Foto> battery terminal
Batterieladegerät *n* <Elektrotech> battery charger
batterieladende Windturbine *f* <Nichtfoss Energ> battery charging wind turbine
Batterieladestelle *f* <Kfztech> battery loading point
Batterieladung *f* <Elektrotech> battery charge
Batterielokomotive *f* <Eisenbahn> battery locomotive
Batterienotstromversorgung *f* <Kontroll> stand-by battery power supply
Batteriepack *m* <Foto> powerpack unit
Batteriepackung *f* <Funktech> battery pack
Batterieplatte *f* <Elektrotech, Kfztech> battery plate
Batteriepol *m* <Elektrotech, Kfztech> battery terminal
Batteriesammelschiene *f* <Raumfahrt> battery transfer bus *(Raumschiff)*
Batteriesäure *f* <Chemie> battery acid, electrolyte
Batterieschutzschalter *m* <Foto> battery switch
Batterieteil *n* <Foto> battery pack *(Blitzlicht)*
Batterieumschaltrelais *n* <Kfztech> battery change-over relay
Batteriewechselstelle *f* <Kfztech> battery exchange point
Batteriezelle *f* 1. <Elektrotech> battery cell; 2. <Kfztech> accumulator cell; 3. <Raumfahrt> battery cell *(Raumschiff)*
Batteriezündung *f* <Kfztech> battery ignition
Batteriezündunterbrecher *m* <Kfztech> breaker
Batteriezustand *m* <Telekom> battery condition
Bau *m* <Bau> construction
Bauablaufgeschwindigkeit *f* <Bau> rate of progress
Bauablaufplan *m* <Bau> construction schedule, progress chart
Bauabschnitt *m* <Bau> stage

Bauarbeiten *fpl* <Bau> construction work
Bauarbeiten *fpl* **der öffentlichen Hand** <Bau> public works
Bauart *f* 1. <Bau> style; 2. <Fertig> build; construction; design; make; model; type; version • **in offener Bauart** <Maschinen> open-type
Bauaufsicht *f* <Wassertrans> survey *(Schiffbau)*
Bauaufzug *m* <Bau> hoist; mechanical platform
Bauausführung *f* <Bau> building construction, construction
Baubedingungen *fpl* <Bau> specifications
Baubeschläge *mpl* <Bau> builder's hardware
Baubeschränkung *f* <Bau> building restriction
Baubeschreibung *f* <Bau> specification
Baubreite *f* <Bau> overall width
Bauch *m* 1. <Akustik> antinode; 2. <Fertig> loop *(Welle, Schwingung)*; 3. <Wassertrans> belly *(eines Segels)*; 4. <Wellphys> antinode
bauchen *v/sich* <Maschinen> bulge
Bauchgurt *m* <Sicherheit> waist belt *(Fallschutz)*
Bauchigkeit *f* <Wassertrans> belly *(eines Bootes)*
Bauchlandung *f* <Lufttrans> belly landing
Baud *n* <Comp & DV, Telekom> baud, Bd *(Einheit der Schrittgeschwindigkeit)*
Baudienst *m* <Trans> Way and Structures Department
Baudrate *f* <Comp & DV, Telekom> baud rate, line digit rate
Baueinheit *f* <Elektriz> assembly
Baueisen *n* <Bau, Metall> structural iron
Bauelement *n* <Telekom> component *(elektrisches Bauelement)*
Bauelement *n* **einer integrierten Schaltung** <Elektronik> integrated-circuit element
Bauelementebene *f* <Elektronik> component level
Bauelementeseite *f* <Telekom> component side
bauen *v* <Bau> construct
Bauentwurf *m* <Bau> construction plan
Bauflucht *f* <Bau> alignment
Bauform *f* <Maschinen> design, version
Baufortschrittsbericht *m* <Bau> progress report
Baugelände *n* 1. <Abfall> zoning site; 2. <Bau> building area, building site, site
Baugenehmigung *f* <Bau> building permit, planning permission
Bauglas *n* <Ker & Glas> structural glass
Baugrenzlinie *f* <Bau> building line
Baugröße *f* <Bau> overall dimensions
Baugrube *f* 1. <Bau> pit, trench; 2. <Kohlen> excavation
Baugrund *m* <Bau> site, subsoil
Baugruppe *f* 1. <Bau, Elektriz> assembly; 2. <Funktech> assembly, module; 3. <Kfztech> unit; 4. <Mechan, Qual> assembly; 5. <Telekom> module, package
Baugruppe *f* **zur Peilung und Kraftübertragung** <Raumfahrt> BAPTA, bearing and power transfer assembly
Baugruppenträger *m* <Fernseh, Funktech, Telekom> card cage
Baugutachten *n* <Bau> survey
Bauhof *m* <Bau> timber yard
Bauhöhe *f* <Bau> overall height
Bauholz *n* <Bau> builder's timber, timber
Bauingenieur *m* <Bau> civil engineer
Baukasten *m* <Comp & DV> modular
Baukastenbauweise *f* <Fertig> modular construction, unit construction
Baukastenprinzip *n* <Fertig> *(AE)* building-block construction, modular principle; assembly of unit parts *(Kunststoffinstallationen)*
Baukastensystem *n* 1. <Fertig> unit assembly system; 2. <Maschinen> modular system

Baukonstruktion

Baukonstruktion f <Bau> building construction
Baulänge f <Fertig> length *(Kunststoffinstallationen)*
Bauleitplan m <Bau> land use plan
Baum m 1. <Fertig, Raumfahrt, Wassertrans> boom; 2. <Künstl Int> tree; 3. <Textil> beam
Baumangel m <Bau> constructional defect
Baumaß n <Fertig> dimension *(Kunststoffinstallationen)*
Bäumen n <Textil> beaming
Baumé-Skale f <Lebensmittel> Baumé scale
Baumfärbeapparat m <Textil> beam-dyeing machine
Baumfärben n <Textil> beam dyeing
Baumfärbung f <Textil> beam dyeing
Baumgei f <Wassertrans> boom guy
baumkantig adj <Bau> rough-hewn
Baumkurre f <Wassertrans> beam trawl *(Fischerei)*
Bäummaschine f <Textil> beamer
Baumnetz n <Telekom> tree network
Baumnetzwerk n <Comp & DV> tree network
Baumniederhalter m <Wassertrans> boom vang, kicking strap *(Segeln)*
Baumstruktur f 1. <Comp & DV> tree, tree structure; 2. <Telekom> tree structure
Baumstumpf m <Bau> stub, stump
Baumtopologie f <Comp & DV> tree topology
Baumusterprüfung f <Qual> prototype test
Baumwollatlas m <Textil> satin
Baumwolle f <Textil> cotton • **doppelt mit Baumwolle umhüllt** <Elektriz> double-cotton-covered, DCC • **einfach mit Baumwolle isoliert** <Elektriz> single-cotton-covered *(SCC)* • **einschichtig mit Baumwolle bedeckt** <Elektriz> single cotton covered
Baumwollentkörnungsmaschine f <Textil> gin
Baumwollgarn n <Textil> cotton
Baumwollgeflecht n <Elektrotech> cotton braid
baumwollisoliert adj <Elektrotech> cotton-covered *(Stromdraht)*
baumwollisolierter Draht m <Elektriz> double-covered cotton wire
Baumwollisolierung f <Elektriz> cotton insulation
Baumwollnummerierung f <Textil> cotton count
Baumwollöffner m <Textil> opener
Baumwollsaatöl n <Lebensmittel> cottonseed oil
Baumwollsatin m <Textil> satin
Baumwollspinnerei f <Textil> cotton spinning
Bauordnung f <Bau> building code
Baupappe f <Papier> felt
Bauplan m 1. <Bau> construction plan; 2. <Kerntech> assembly plan
Bauplatz m <Bau> construction site, job site
Bauprogramm n <Bau> *(AE)* construction program, *(BE)* construction programme
Baureihe f 1. <Maschinen> series; 2. <Telekom> range *(Herstellung)*
Bausachverständiger m <Bau> building expert
Bausatz m <Funktech, Kerntech> assembly kit, kit
Bausch m <Textil> bulk *(Faser)*
Bauschreiner m <Bau> joiner
Bauschreinerei f <Bau> joinery
Bauschutt m 1. <Abfall> construction waste, demolition waste; 2. <Bau> demolition waste, waste
Bauschuttcontainer m <Abfall, Bau> waste skip
Bauschvermögen n <Textil> bulk
Bauspantenriss m <Wassertrans> frame plan *(Schiffkonstruktion)*
Baustahl m 1. <Bau, Metall> structural steel; 2. <Fertig> mild steel
Baustatik f <Bau> structural analysis, structural engineering

Baustein m 1. <Comp & DV> module, unit; 2. <Elektronik, Fertig, Hydraul> module
Bausteinsystem n <Comp & DV> modularity
Bausteintechnik f <Fertig> packaging technique
Baustelle f <Bau> construction site, field, job site, site, work site • **auf der Baustelle** <Bau> on site
Baustellenabfall m <Abfall> construction waste
Baustellenbesprechung f <Bau> site meeting
Baustelleneinrichtung f <Bau> job site installations, site installations
Baustelleningenieur m <Mechan> field engineer
Baustellenroden n <Bau> grubbing
Baustellenschweißen n <Kerntech> field weld
Baustellenüberwachung f <Bau> site monitoring
Baustoffe mpl <Bau> materials
Bautafel f <Bau> lath *(Putzträger)*
Bautagebuch n <Bau> builder's diary
Bautechniker m <Bau> builder
bautechnische Richtlinien fpl <Bau, Qual> code of practice
Bauteil n 1. <Bau> component, member; 2. <Elektriz, Hydraul> element; 3. <Kfztech> part; 4. <Maschinen> component, member; 5. <Patent, Telekom> component
Bauteilauswahl f <Qual, Raumfahrt> component selection *(Raumschiff)*
Bauteilbeschaffung f <Raumfahrt> component procurement *(Raumschiff)*
Bauteileprüfung f <Qual, Werkprüf> component testing
Bautischler m <Bau> joiner
Bautischlerei f <Bau> join-ery
Bauvorhaben n <Bau> construction project
Bauvorrichtung f 1. <Mechan> jig; 2. <Wassertrans> erection jig *(Schiffbau)*
Bauvorschriften fpl <Bau> building regulations
Bauweise f <Bau> construction, design
Bauwerft f <Wassertrans> construction yard *(Schiffbau)*
Bauwerk n <Bau> construction, structure
Bauxit m <Fertig, Ker & Glas> bauxite
Bauzeichnung f <Bau, Mechan> layout drawing
Bauzeit f <Bau> construction time
Bayes'sches Verfahren n <Künstl Int> Bayesian decision method, Bayesian inference theory
Bazin'sche Formel f <Hydraul> Bazin's formula
Bazooka-Balun m <Funktech> bazooka balun *(Antennen)*
BB *(Basisband)* <Elektrotech> BB *(baseband)*
BBD 1. <Elektrotech> *(Eimerkettenspeicher)* BBD *(bucket brigade device)*; 2. <Telekom> *(Eimerkettenschaltung)* BBD *(bucket brigade device)*
B-Bild n <Fernseh> B-frame, bidirectionally predictively coded frame
BBL *(Blindlandung mit Bakenunterstützung)* <Raumfahrt> BBL *(beacons and blind landing)*
BBV *(Breitbandverstärker)* <Elektronik> wideband amplifier
BCC *(Blockprüfzeichen)* <Telekom> BCC *(block check character)*
BCD 1. <Comp & DV> *(binärcodierte Dezimalzahl, binärcodierte Drehzahl)* BCD *(binary-coded decimal)*; 2. <Funktech> *(binärcodierte Dezimalzahl)* BCD *(binary-coded decimal)*
BCI *(Rundfunkstörung)* <Telekom> BCI *(broadcast interference)*
BCS *(Bardeen-Cooper-Schrieffer)* <Comp & DV> BCS *(Bardeen-Cooper-Schrieffer)*
BCS-Theorie f <Phys> BCS theory *(Supraleitung)*
B/d *(Barrel pro Tag)* <Erdöl> BCD *(barrels per calendar day)*
BDP *(Verbundpapier)* <Papier> BDP *(bonded double paper)*
Be *(Beryllium)* <Chemie> Be *(beryllium)*

BE 1. <Elektriz> *(elektrischer Blindleitwert)* BE *(electric susceptance)*; 2. <Telekom> *(Basiseinheit)* BU *(base unit)*
beabsichtigte Flugbahn f <Lufttrans> intended flight path
beabsichtigte Störung f <Telekom> jamming
beanspruchen v 1. <Bau> load, strain; 2. <Maschinen> stress; 3. <Patent> claim; 4. <Qual> load; 5. <Telekom> engage
beanspruchtes Bauteil n <Sicherheit> load-bearing structure
Beanspruchung f 1. <Bau> stress; 2. <Ergon, Ker & Glas> strain; 3. <Kohlen> stress; 4. <Mechan> strain, stress; 5. <Metall> stress; 6. <Phys, Qual> strain; 7. <Wassertrans> stress *(Werkstoff)*; 8. <Werkprüf> stress
Beanspruchungsbeginn m <Qual> beginning of stress
Beanspruchungskombination f <Qual> load combination
Beanspruchungsschwankung f <Metall> fluctuating stress
beanstandet adj <Qual> nonconforming
Beanstandung f <Patent, Qual> objection
beantragen v 1. <Mechan> apply for; 2. <Qual> submit
Bearbeitbarkeit f <Kunststoff, Maschinen> machinability
bearbeiten v 1. <Bau> handle; 2. <Comp & DV> edit; 3. <Maschinen> work; 4. <Mechan> machine; 5. <Qual> handle
bearbeitet adj <Maschinen, Mechan> machined
bearbeitete Oberfläche f <Mechan> machined surface
Bearbeitung f 1. <Druck> adaptation; 2. <Maschinen, Mechan> machining
Bearbeitung f **des geblasenen Glaspostens** <Ker & Glas> working on blown post
Bearbeitungsfehler m <Maschinen> machining defect
Bearbeitungsmaschinen fpl **im Serienbetrieb** <Verpack> in-line finishing equipment
Bearbeitungsspuren fpl <Fertig> lay
Bearbeitungstoleranz f <Maschinen> machining allowance
Bearbeitungsvorschub m <Fertig> drilling feed *(beim Bohren)*
Bearbeitungszeit f <Comp & DV> productive time, target phase, run time
Bearbeitungszugabe f <Fertig> allowance for machining, tooling allowance
Bearbeitungszyklus m <Maschinen> machining cycle, working cycle
Beatmungsgerät n <Sicherheit> breathing apparatus *(Sicherheitsausrüstung)*; respirator
Beaufort-Skale f <Wassertrans> Beaufort scale
Beaufschlagung f <Fertig> discharge • **mit äußerer Beaufschlagung** <Fertig> inward flow
beauftragen v <Patent, Qual> appoint
beauftragen v **mit** <Patent, Qual> entrust with
Beaumé-Skale f <Phys> Beaume scale *(Aräometrie)*
bebaute Fläche f <Bau> architectural area, building area, built-up area
Bebauungsgelände n <Abfall> zoning site
bebildern v <Druck> illustrate
bebunkern v <Wassertrans> bunker
Becher m <Labor, Lebensmittel> beaker *(Laborgerät)*
Becher m **mit Schnabel** <Labor> beaker with spout
Becherbruch m <Metall> cup and cone fracture
Becherglas n <Ker & Glas> beaker
Becherhalter m <Labor> beaker holder
Becherleiter f <Bau> bucket ladder
Becherrelais n <Elektriz> sealed relay
Becherschließzeit f <Kunststoff> cup-closing time
Becherschöpfrad n <Hydraul> tympanus
Becherwerk n <Bau> bucket elevator

Becken n 1. <Bau> pan; 2. <Erdöl> basin *(Geologie)*; 3. <Kohlen, Labor> basin; 4. <Papier> pond; 5. <Wasserversorg> basin, lagoon
Beckenplatte f <Ker & Glas> pool tablet
Beckenstein m <Ker & Glas> pool block
Beckentiefe f <Papier> pond depth
Beckmann'sches Flüssigkeitsthermometer n <Phys> Beckmann thermometer
Beckschicht f <Wasserversorg> capping
Becquerel n *(Bq)* <Metrol, Phys> becquerel *(Bq)*
Becquereleffekt m <Elektriz> Becquerel effect
Becquerelzelle f <Elektriz> Becquerel cell
bedachen v <Bau> roof
Bedarf m <Bau, Kohlen, Qual> requirement
bedarfsgesteuert adj <Comp & DV> demand-responsive
bedarfsgesteuerter Vielfachzugriff m *(DAMA)* <Telekom> demand-assigned multiple access *(DAMA)*
bedarfsgesteuertes System n <Trans> demand-responsive system
bedarfsmäßiges Einspeichern n <Comp & DV> demand staging
Bedarfswartung f 1. <Comp & DV> remedial maintenance; 2. <Telekom> corrective maintenance
bedecken v 1. <Bau> cope, put under cover, top; 2. <Fertig> cope
Bedecken n <Fertig> coping
bedeckt adj <Elektriz> covered
Bedeckung f <Funktech, Telekom> coverage
Bedielung f <Bau> deck
Bedienelement n <Ergon, Maschinen> actuator
bedienen v 1. <Bau> attend to; 2. <Mechan> handle
Bediener m 1. <Comp & DV, Druck, Maschinen> operator; 2. <Telekom> server
Bedienerantwort f <Comp & DV> operator response
Bedienerbefehl m <Comp & DV> operator command
Bedienerführung f <Telekom> prompting
Bedienerkonsole f 1. <Comp & DV> control panel, operator console; 2. <Eisenbahn> control panel
Bedienermeldung f <Comp & DV> operator message
Bedienernachricht f <Comp & DV> operator message
Bedienerschulung f <Comp & DV> operator training
Bedienerstation f <Comp & DV> operator terminal
Bedienfeld n <Comp & DV> front panel, panel
Bedienhebel m <Maschinen> control lever
Bedienpodest n <Lufttrans> control pedestal
Bedienpult n <Gerät, Lufttrans> console
Bedientafel f <Comp & DV> subpanel
bediente Maschine f/**von Hand** <Verpack> hand-operated machine
Bedienteil n <Ergon> control element
Bedienung f 1. <Elektrotech> operation; 2. <Fertig> operating; 3. <Maschinen> control, operation, operation; 4. <Papier> operating; 5. <Raumfahrt> handling *(des Raumschiffs)*; 6. <Telekom> control, operation
Bedienung f **am Gerät** <Mechan> local control
Bedienungs... adj <Fertig> operating
Bedienungsanleitung f 1. <Fertig> operating instructions; 2. <Maschinen> instruction book, operating instructions
Bedienungsfehler m <Kohlen> operating error
Bedienungsfreundlichkeit f <Mechan> ease of operation
Bedienungsgang m <Kerntech> gangway
Bedienungsgleis n <Eisenbahn> line serving a siding
Bedienungshandbuch n <Maschinen> instruction book
Bedienungshandgriff m <Lufttrans> box-type stiffener
Bedienungshebel m <Maschinen> operating lever
Bedienungsknopf m <Elektrotech> knob
Bedienungsperson f <Ergon> human operator

Bedienungspersonal

Bedienungspersonal n <Kerntech> service staff
Bedienungspult n 1. <Comp & DV> system control panel; 2. <Eisenbahn> control desk; 3. <Gerät> console; 4. <Mechan> control panel; 5. <Telekom> operating console, operator's console
Bedienungsreep n <Wassertrans> lanyard
Bedienungsschalter m <Elektriz> operating switch
Bedienungsspannung f <Fertig> operating voltage
Bedienungsstand m <Fertig> platform
Bedienungssystem n <Telekom> operator system
Bedienungstafel f 1. <Elektriz> control board; 2. <Wassertrans> control panel (Elektrik)
Bedienungszeit f <Fertig> machine-handling time
bedingt adj <Comp & DV, Kontroll> conditional
bedingt lösbare Verbindung f <Fertig> semipermanent connection (optical fibres)
bedingte Anweisung f <Comp & DV> conditional instruction
bedingte Erwartung f <Math> conditional expectation
bedingte Löslichkeit f <Metall> restricted solubility
bedingte Verzweigung f <Comp & DV> conditional branch
bedingter Befehl m <Comp & DV> conditional instruction
bedingter Programmstopp m <Comp & DV> checkpoint
bedingter Schutz m <Patent> conditional protection
bedingter Sprungbefehl m <Comp & DV> conditional jump
bedingtes Gegensprechen n <Telekom> alternating duplex communication, semiduplex
Bedingung f 1. <Comp & DV, Math> condition, constraint; 2. <Patent, Qual> requirement
Bedingung f **für gleich bleibenden Zustand** <Elektrotech> steady state condition
Bedingungsausdruck m <Comp & DV> conditional expression
Bedingungsbühne f <Labor> bench
Bedruckbarkeit f <Papier> printability
bedrucken v <Textil> impress, print
Bedrucken n <Textil> printing
bedruckte Faltschachtel f <Verpack> printed folding carton
bedruckte Seite f <Druck> printed sheet
bedruckter Stoff m <Textil> printed fabric
bedrucktes Etikett n <Verpack> printed label
beenden v 1. <Comp & DV> exit; 2. <Comp & DV> sign off
Beerben n <Künstl Int> inheritance, subtyping
befahrene Straße f/wenig <Bau> low-traffic road
Befehl m 1. <Comp & DV> command, instruction, order; 2. <Kontroll> command
Befehl m **zur Speicherung nach der LIFO-Methode** <Comp & DV> push instruction, push operation
Befehlsablauf m <Comp & DV> instruction cycle
Befehlsabruf m <Comp & DV> instruction fetching
Befehlsadressregister n <Comp & DV> IAR, instruction address register
Befehlsausführung f <Comp & DV> instruction execution
Befehlscode m <Comp & DV> instruction code, order code, program counter
Befehlscodeprozessor m <Comp & DV> OCP, order code processor
Befehlsdecodierer m <Comp & DV> instruction decoder
Befehlseingabeaufforderung f <Comp & DV> command input request
Befehlseingabezeile f <Comp & DV> command input line
Befehlseinheit f <Comp & DV> control station
Befehlsfolge f <Comp & DV> instruction sequence, instruction stream, sequence of instructions
Befehlsfolgeregister n <Comp & DV> sequence control register, sequence counter, sequence register
Befehlsformat n <Comp & DV> instruction format
befehlsgesteuerte Schnittstelle f <Comp & DV> command-driven interface
Befehlskette f <Comp & DV> pipeline
Befehlslänge f <Comp & DV> instruction length
Befehlsparameter m <Comp & DV> switch
Befehlsprozessor m <Comp & DV> instruction processor
Befehlsregister n <Comp & DV> instruction register
Befehlsschlüssel m <Comp & DV> program counter
Befehlssequenz f <Comp & DV> instruction sequence
Befehlsspeicherbereich m <Comp & DV> instruction area
Befehlssprache f (CL) <Comp & DV> command language (CL)
Befehlssystem n <Comp & DV> command system
Befehls- und Steuersystem n <Comp & DV> command-and-control system
Befehlsvorrat m <Comp & DV> instruction repertoire, instruction set
Befehlswort n <Comp & DV> instruction word
Befehlszähler m <Comp & DV> program counter
Befehlszeile f <Comp & DV> command line
Befehlszwischenspeicher m <Comp & DV> instruction cache
Befehlszyklus m <Comp & DV> instruction cycle
befeilen v <Maschinen> file
befestigen v 1. <Bau> fasten, mount, pave, secure, tack, tail; batten (mit Latten); 2. <Mechan> fasten; 3. <Meerschmutz> moor; 4. <Papier> attach; 5. <Verpack> clip to; 6. <Wassertrans> secure
Befestigen n 1. <Bau> sealing; 2. <Maschinen, Mechan> fastening
Befestigung f 1. <Bau> fastening; revetment (Ufer); (BE) pavement, (AE) sidewalk (von Wegen oder Straßen); 2. <Kfztech> panel mounting (für Karosserieblech); 3. <Maschinen> attachment, fixing, fastening; 4. <Mechan> clamping, fastening; 5. <Papier> attachment; 6. <Raumfahrt> fastener (Raumschiff)
Befestigung f **ohne Bolzen** <Nichtfoss Energ> no-bolt fixing
Befestigungsband n <Verpack> clip band
Befestigungsbügel m 1. <Foto> mounting bracket (Blitzlicht); 2. <Maschinen> mounting bracket
Befestigungsfeder f <Maschinen> mounting spring
Befestigungsfläche f 1. <Kerntech> seating; 2. <Maschinen> seat
Befestigungsfuß m <Foto> mounting foot
Befestigungsgeschirr n <Meerschmutz> mooring bracket
Befestigungsglied n <Maschinen> attachment link
Befestigungsmittel n <Maschinen> fastener, fastening device
Befestigungsschraube f 1. <Bau> fixing screw; 2. <Fertig> holding-down bolt; 3. <Maschinen> attaching screw, fixing bolt
Befestigungsschrauben fpl <Maschinen> threaded fasteners
Befestigungsstück n <Maschinen> attachment fitting
Befestigungssystem n <Raumfahrt> strapdown system (Raumschiff)
Befestigungswinkel m 1. <Bau> angle bracket; 2. <Maschinen> angle bracket, mounting bracket
befeuchten v 1. <Bau> moisten, water; 2. <Chemtech> water; 3. <Heiz & Kälte> humidify; 4. <Lebensmittel> moisten; 5. <Papier> humidify; 6. <Phys> wet; 7. <Thermod> humidify, moisten
Befeuchter m 1. <Heiz & Kälte, Lufttrans, Papier, Thermod> humidifier; 2. <Verpack> moistening device
Befeuchtung f <Heiz & Kälte, Thermod> humidification

Befeuchtungsvorrichtung f <Verpack> moistening equipment
Befeuerung f <Wassertrans> light
befinden v **in/sich** <Comp & DV> reside
Beflechtung f <Fertig> braiding
Befolgungsgrad m <Trans> obedience level
Beförderer m <Trans> conveyor
befördern v 1. <Bau, Kohlen> haul; 2. <Textil, Trans> carry; 3. <Wassertrans> ship
Beförderung f <Trans> carriage, conveyance, transport
Beförderung f **von Handelsgütern** <Trans> merchant haulage
Befrachter m <Wassertrans> charterer *(Seehandel)*
Befrachtung f <Wassertrans> charterage, chartering *(Seehandel)*
Befrachtungsagent m <Trans> freight agent
Befrachtungsmakler m <Wassertrans> chartering broker *(Seehandel)*
Befragung f <Ergon, Künstl Int> interview
Befragungswerkzeug n <Künstl Int> consultation tool
Befüllung f <Abfall> waste injection *(einer Kaverne)*
Befund m <Qual> findings
Befundbericht m <Qual> report of findings
Befundprüfung f <Qual> as-found test
BEG *(Bodeneffektgerät)* <Kfztech> GEM *(ground effect machine)*
begast adj <Verpack> gas-flushed
Begasungsextruder m <Kunststoff> gas injection extruder
begehbare Gefrieranlage f <Heiz & Kälte> walk-in freezer
begehbarer Gitterrost m <Heiz & Kälte> step-on grille
begichten v <Fertig> fill *(Hochofen)*
Begichtung f <Fertig, Metall> charging, filling
Beginn m <Comp & DV> start
Beginn m **des Sinkflugs** <Lufttrans> top of descent
beginnende kritische Wärmestromdichte f <Kerntech> departure from nuclear boiling
beginnender Ermüdungsbruch m <Lufttrans> incipient fatigue failure
Beginner's All-purpose Symbolic Instruction Code m *(BASIC)* <Druck> beginner's all-purpose symbolic instruction code *(BASIC)*
Beginnzeichen n <Telekom> answer signal
Beglaubigung f <Qual> certificate
Beglaubigungsschein m <Patent, Qual> certificate of approval
Beglaubigungszeichen n <Qual> certification mark
Begleit... <Konstzeich, Raumfahrt> track
Begleitkarte f <Qual> job card
Begleitlinie f <Konstzeich> tracer line
Begleitmineral n <Bau> accessory mineral
Begleitradar m <Raumfahrt> tracking radar
Begleitstraße f <Bau> frontage road
Begleittextdatei f <Comp & DV> narrative file
Begleitung f <Lufttrans> tracking
begradigen v <Bau> rectify, straighten; level *(Straßenbau)*
begrenzen v 1. <Bau> border; 2. <Elektronik> limit; 3. <Elektrotech> clip; 4. <Math> confine
Begrenzer m 1. <Aufnahme, Elektronik, Elektrotech> limiter; 2. <Funktech> clipper, limiter *(Sprechfunk)*; 3. <Kontroll, Maschinen> limiter; 4. <Raumfahrt> limiter *(Weltraumfunk)*; 5. <Telekom> limiter
Begrenzer m **für Integralanteil** <Regelung> integral action limiter
Begrenzerdiode f 1. <Elektronik> clipper diode, limiter diode; 2. <Fernseh> limiter diode
Begrenzerschaltung f 1. <Elektronik> clipper, clipper circuit, limiter; 2. <Elektrotech> limiter

begrenzte Sichtweite f <Trans> limit of visibility
begrenzte Warteschlange f <Telekom> limited waiting queue
begrenzter Betrieb m/**durch Quantenrauschen** <Optik, Telekom> quantum-noise-limited operation
begrenzter Fluss m **durch ein Schaltelement** <Phys> flux cut by a circuit element
begrenzter Fluss m **durch einen Leiter** <Phys> flux cut by a conductor
Begrenzung f 1. <Aufnahme, Elektronik> limiting; 2. <Erdöl> confinement; 3. <Geom> boundary; 4. <Qual> limiting
Begrenzung f **des Schadstoffausstoßes** <Maschinen> emission control
Begrenzungslinie f <Bau> boundary line
Begrenzungsregelung f <Qual, Regelung> limiting control
Begrenzungsregelung f **nach oben** <Regelung> high-limiting control
Begrenzungsregelung f **nach unten** <Regelung> low-limiting control
Begrenzungsschalter m <Elektriz, Elektrotech, Kontroll, Mechan> limit switch
Begrenzungsschaltkreis m <Fernseh, Telekom> limiter
Begrenzungssymbol n <Comp & DV> delimiter, separator symbol
Begrenzungsverstärker m 1. <Aufnahme> limiter amplifier; 2. <Elektronik> limiting amplifier
Begrenzungswert m <Elektronik> clipping level
Begrenzungszeichen n <Comp & DV> delimiter, separator
Begriffsabhängigkeit f <Künstl Int> conceptual dependency, CD
Begründungsverwaltung f <Künstl Int> consistency maintenance, truth maintenance
Begrünen n <Bau> grassing
begutachten v <Patent, Qual> appraise, evaluate, survey
Begutachtung f <Qual> appraisal, expert valuation, survey
Behaglichkeit f <Ergon> comfort
Behälter m 1. <Bau, Erdöl> tank; 2. <Fertig> receiver; tank *(Kunststoffinstallationen)*; 3. <Funktech> receptacle; 4. <Heiz & Kälte> tank; 5. <Hydraul> drum; 6. <Ker & Glas> chest; 7. <Kfztech> reservoir *(Öl, Kraftstoff)*; 8. <Kohlen> container; 9. <Kontroll> enclosure; 10. <Labor> basin, bucket, vessel; 11. <Mechan> case, tank; 12. <Meerschmutz> scoop; 13. <Textil> vessel; 14. <Umweltschmutz> tank
Behälter m **für Messeinheit** <Verpack> unit dose container
Behälter m **mit Plane** <Trans> tarpaulin-covered container
Behälteranschluss m <Fertig> tank adaptor *(Kunststoffinstallationen)*
Behälterbau m <Fertig> pressure vessel construction
Behälterglas n <Ker & Glas> container glass
Behälterkammer f <Wasserversorg> chamber *(einer Kanalschleuse)*
Behältermessanlage f <Gerät> *(AE)* tank gaging system, *(BE)* tank gauging system
Behältermesssystem n <Gerät> *(AE)* tank gaging system, *(BE)* tank gauging system
Behälterwagen m <Eisenbahn> *(AE)* tank car, *(BE)* tank wagon
Behälterwand f <Fertig> tank wall *(Kunststoffinstallationen)*
Behälterzerknall m <Sicherheit> vessel bursting
behandeln v 1. <Fertig> treat, process; 2. <Lebensmittel> process

Behandlung

Behandlung f **der Anmeldung** <Patent> processing of an application
Behandlung f **fehlerhafter Einheiten** <Qual> control of non-conforming items
Behandlung f **von Klärschlamm** <Abfall> treatment of sewage sludge
Beharrung f <Heiz & Kälte, Maschinen> inertia
Beharrungsbremse f <Eisenbahn> continuous brake
Beharrungsgeschwindigkeit f <Eisenbahn> balancing speed
Beharrungsgesetz n <Maschinen> law of inertia
Beharrungsmoment n <Maschinen> rotational inertia
Beharrungsregler m <Maschinen> inertia governor
Beharrungsvermögen n <Heiz & Kälte, Maschinen, Mechan, Phys, Umweltschmutz> inertia
Beharrungszustand m <Elektrotech> steady state, steady state condition
behauen v 1. <Bau> axe, mill; adze *(Holzbau)*; square *(Stein)*; 2. <Fertig> cut *(Feile)*
Behauen n <Bau> dressing *(von Stein)*
behauene Größe f <Bau> dressed size
behauener Naturstein m <Bau> dressed stone
behaupten v <Math> assert, contend
Behauptung f <Math> assertion, contention
behebbarer Fehler m <Comp & DV> recoverable error
beheben v 1. <Qual> remedy *(Fehler)*; correct *(Mängel)*; 2. <Telekom> recover *(Fehler)*
Behebung f <Qual> remedy
beheizte Windschutzscheibe f <Lufttrans> *(BE)* heated windscreen pane, *(AE)* heated windshield pane
beheizter Container m <Trans> heated container
Beheizung f <Fertig> heating
Behelfs... <Fertig> makeshift, temporary
Behelfsbau m <Bau> temporary structure
Behelfsbrücke f <Bau> temporary bridge
Behelfsruder n <Wassertrans> jury rudder
Behen... <Chemie> behenic
beherrschter Prozess m <Qual> controlled process, process in control
behindertengerechte Gestaltung f <Ergon> design for disabled persons
Behörde f <Patent> authority
Behörde f **für internationale Vorprüfungen** <Patent> international preliminary examining authority
beibehalten v <Mechan> maintain *(Werkzeuge, Geräte)*
Beibehaltung f 1. <Bau, Chemie> retention; 2. <Optik> conservation
Beibehaltung f **der Helligkeit** <Optik> conservation of brightness
Beibehaltung f **der Strahlungsintensität** <Optik> conservation of radiance
Beiboot n <Wassertrans> ship's boat
beidäugig adj <Optik, Phys> binocular
beiderseitige Belüftung f <Heiz & Kälte> double-ended ventilation
Beidhändigkeit f <Ergon> ambidexterity
beidohrig adj <Akustik> binaural
beidrehen v <Wassertrans> heave to *(Schiff)*
beidseitig bedrucken v <Druck> perfect
beidseitig beschreibbarer Datenträger m <Comp & DV> double-sided disk
beidseitiger Gehrungswinkel m <Ker & Glas> *(AE)* miter bevel both sides, *(BE)* mitre bevel both sides
Beifahrer m <Kfztech> front-seat passenger
Beifahrerseite f <Kfztech> near side
beigedreht adj <Wassertrans> ahull *(Schiff)*
beigedreht legen v <Wassertrans> lay ahull *(Schiff)*
Beijing Electron Positron Collider m *(BEPC)* <Teilphys> Beijing Electron Positron Collider *(BEPC)*

Beil n <Bau> *(AE)* ax, *(BE)* axe
Beilage f 1. <Fertig> shim, spacer; 2. <Druck> supplement
Beilageblatt n <Verpack> leaflet insert *(Druckschrift als Anlage)*
Beilagefolie f <Mechan> shim
Beilagscheibe f <Maschinen> shim
Beilegering m <Fertig> packing *(Nutenfräser)*
beiliegen v <Wassertrans> be hove to, lie, try *(Schiffsbewegung)*
Beimengung f 1. <Chemie> impurity; 2. <Ker & Glas> admix; 3. <Kunststoff> admixture
Beimischung f 1. <Bau> admixture; 2. <Chemie> impurity; 3. <Kunststoff, Textil> admixture
Beinahezusammenstoß m <Lufttrans, Raumfahrt, Sicherheit> near miss
Beinfreiheit f <Ergon, Kfztech> legroom
Beinfreiraum m <Ergon> legroom
Beinschutz m <Sicherheit> leg protector
Beinschützer m <Sicherheit> leg protector
beißen v <Bau> catch
beißender Geruch m <Umweltschmutz> *(AE)* acrid odor, *(BE)* acrid odour
Beißzange f 1. <Elektriz> pliers; 2. <Maschinen> pincers, pliers, tongs
Beitel m <Mechan> chisel
Beiwagen m <Kfztech> side car
Beiwagendrehgestell n <Eisenbahn> trailer bogie
Beizbrüchigkeit f <Fertig> acid brittleness
Beize f 1. <Chemie> caustic; 2. <Fertig> etchant, mordant, pickle, stain
beizen v 1. <Bau> etch; 2. <Fertig> etch, mordant
Beizen n <Fertig> pickling
beizend adj <Chemie, Qual> caustic
Beizfärben n <Foto> mordant dyeing
Beizkraft f <Chemie> causticity
Beizmittel n <Bau> stripper
Bekämpfung f **der Umweltverschmutzung** <Umweltschmutz> environmental pollution control
bekiesen v <Bau> grit
Bekleidung f 1. <Hydraul> clothing; 2. <Textil> apparel
bekohlen v <Kohlen> coal
Bekohlungsvorrichtung f <Kohlen> coal-handling plant
Bel n *(B)* <Akustik, Elektrotech, Phys> bel *(B)*
Belademaschine f <Kerntech> fuel-charging machine
beladen v 1. <Mechan> load; 2. <Thermod> charge *(Brennofen, Hochofen)*; 3. <Trans> load, stuff; 4. <Wassertrans> load *(Schiff)*
Beladen n <Trans> loading, stuffing
Beladerohr n <Kerntech> feeder pipe
Beladeseite f <Kerntech> charge face
Beladezone f <Kerntech> charge area
Beladung f 1. <Elektrotech, Maschinen> loading; 2. <Raumfahrt> loading *(Raumschiff)*; 3. <Trans> loading
Beladung f **bei abgeschalteter Last** <Kerntech> off-load charging
Beladung f **mit Brennmaterial** <Kerntech> fuel charge
Beladung f **unter Last** <Kerntech> on-load charging, *(AE)* on-load fueling, *(BE)* on-load fuelling
Belag m 1. <Bau> cover, overlay, overlaying, paving; 2. <Kfztech> pad; lining *(Bremse, Kupplung)*; 3. <Lebensmittel> bloom *(auf Früchten)*; 4. <Maschinen> facing; 5. <Mechan> lining
Belagdicke f <Maschinen> thickness of lining
Belagverschleiß m <Kfztech> lining wear
Belassung f <Qual> accept as is
Belastbarkeit f 1. <Maschinen, Qual> load rating, loading capacity; 2. <Sicherheit> permissive load
Belastbarkeit f **im Gebrauch** <Metall> working stress

belasten v 1. <Bau, Qual> burden, load; weight; 2. <Phys> stress; 3. <Sicherheit> burden, pollute; weight
belastendes Mittel n <Sicherheit> polluting agent
belastetes Kabel n <Elektrotech> loaded cable
Belastung f 1. <Aufnahme> loading *(eines Lautsprechers)*; 2. <Bau> load, stress; 3. <Comp & DV> load; 4. <Elektrotech> loading; 5. <Ergon> stress; 6. <Fertig> load, loading; 7. <Funktech> load; 8. <Hydraul> load *(Ventil)*; 9. <Kerntech> exposure; 10. <Kohlen> stress; 11. <Maschinen, Mechan, Metall, Papier, Qual> load; 12. <Raumfahrt> battery drain, loading *(Raumschiff)*; 13. <Telekom> load
Belastung f an Drittperson <Telekom> third party charging
Belastung f außerhalb der Spitzenzeit <Elektrotech> off-peak load
Belastung f durch Widerstand <Elektrotech> resistive load
Belastungsanalyse f <Ergon, Raumfahrt> stress analysis *(Raumschiff)*
Belastungsbild n **einer Verbindung** <Trans> desire line
Belastungscharakteristik f <Lufttrans> load characteristic
Belastungsdiagramm n 1. <Elektrotech> load curve; 2. <Mechan> load diagram
Belastungsfähigkeit f 1. <Mechan> carrying capacity; 2. <Wasserversorg> rated capacity, rated load
Belastungsfaktor m 1. <Mechan> load factor; 2. <Qual, Raumfahrt> loading factor *(Weltraumfunk)*
belastungsfreies Anfahren n <Elektriz> no-load start
Belastungsfunktion f <Metall> loading function
Belastungsgeschwindigkeit f <Metall> rate of loading
Belastungsgrenze f <Maschinen> load limit
Belastungskapazität f 1. <Elektriz> peaking capacity; 2. <Elektrotech> input capacitance *(bei Voltmetern)*
Belastungskennlinie f <Elektrotech> load line
Belastungskoeffizient m <Elektrotech> load factor
Belastungskraft f <Fertig> active force
Belastungskurve f <Elektrotech> load curve
Belastungslinie f <Maschinen> load line
Belastungsprüfung f <Bau, Qual> loading test
Belastungsregelung f <Elektrotech> load regulation
Belastungsregler m <Papier> load governor
Belastungsschaubild n <Mechan> load diagram
Belastungsschutz m <Sicherheit> load protector
Belastungsspannung f <Elektrotech> on-load voltage
Belastungsspitze f <Elektrotech> peak load
Belastungsspule f <Phys> loading coil
Belastungstal n <Elektrotech> off-peak load
Belastungstestanlage f <Eisenbahn> static load test bed
Belastungsveränderungsmuster n <Raumfahrt> load fluctuation pattern *(Raumschiff)*
Belastungswert m <Anstrich> strain rate
Belastungswiderstand m 1. <Elektrotech, Phys> load resistance; 2. <Telekom> load impedance
belatten v <Bau> lath *(Wand)*
Belebtschlamm m 1. <Abfall, Umweltschmutz> activated sludge; 2. <Wasserversorg> mixed liquor
Belebtschlammanlage f <Abfall> activated sludge plant
Belebtschlammbecken n <Abfall> activated sludge tank, aeration tank
Belebtschlammverfahren n <Abfall> activated sludge process
Belebungsbecken n <Wasserversorg> aeration basin
Beleg m <Comp & DV> document
Belegbearbeitung f <Comp & DV> document processing
belegen v 1. <Bau> occupy *(Gebäude)*; tile *(mit Fliesen)*; 2. <Fertig> charge, coat *(Schleifmittel)*; 3. <Telekom> seize *(Leitung, Telefon)*; 4. <Wassertrans> belay *(Tauwerk)*

Belegklampe f <Wassertrans> belaying cleat, cleat *(Deckausrüstung)*
Beleglesen n <Comp & DV> document reading
Belegleser m <Comp & DV> document reader, mark reader
Belegmuster n <Qual> known-good device
Belegnagel m <Wassertrans> belaying pin *(Deckausrüstung)*
Belegsortierer m <Comp & DV> document sorter
belegt adj <Telekom> engaged *(Telefonie)*
Belegung f 1. <Comp & DV> load; 2. <Fertig> charging *(Schleifscheibe)*; 3. <Künstl Int> binding *(einer Variablen)*; 4. <Telekom> distribution, occupancy, seizure; usage *(Kanal)*
Belegung f des Spektrums <Telekom> spectral occupancy
Belegungsdauer f <Telekom> holding time
Belegungsdetektor m <Trans> occupancy detector
Belegungsrate f <Telekom, Trans> occupancy rate
Belegungsschleife f <Trans> presence loop
Belegungsversuch m <Telekom> bid
Belegungsversuche mpl **pro Stunde** <Telekom> call attempts per hour *(CA/h)*
Belegungswahrscheinlichkeit f **eines Leitungsbündels** <Telekom> seizing probability of a trunk line
Belegungszählung f <Telekom> peg count *(BZ)*
Belegungszeit f <Telekom> holding time
Belegungszeitaufzeichnung f <Telekom> call duration recording
Belegwiederherstellung f <Comp & DV> document recovery
beleuchtete Einstellscheibe f <Foto> illuminated dial
Beleuchtung f 1. <Kfztech> light; 2. <Metall> illumination; 3. <Phys> illuminance, irradiance
Beleuchtungsanlage f <Elektrotech> lighting system
Beleuchtungsausrüstung f <Foto> lighting equipment
Beleuchtungsdämpferabdeckung f <Lufttrans> dimmer cap
Beleuchtungsglas n <Ker & Glas> lighting glass
Beleuchtungskabel n <Elektrotech> light cable
Beleuchtungskörper m 1. <Elektrotech> light fitting; 2. <Heiz & Kälte> light fitting, light fixture
Beleuchtungsmesser m <Elektriz> luxmeter
Beleuchtungsnetzschaltung f <Elektriz> lighting circuit
Beleuchtungsöffnung f <Mechan> lightening hole
Beleuchtungsstärke f 1. <Elektriz> illumination, intensity of illumination; 2. <Elektrotech> illumination *(Lichtstromstärke)*; 3. <Ergon> illuminance, illumination; 4. <Phys> brightness; 5. <Telekom> irradiance *(Quotient)*
Beleuchtungsstärkemesser m <Metrol> footcandle meter
Beleuchtungsstärkemessgerät n <Gerät> luxmeter
Beleuchtungsstativ n <Foto> lighting stand
Beleuchtungstärke f <Strahlphys> brightness
Beleuchtungs- und Bildregieraum m <Fernseh> lighting and vision control room
belichten v <Foto> expose *(Emulsion)*
Belichtung f 1. <Druck, Foto> exposure; 2. <Phys> exposure, light exposure
Belichtungsanlage f <Druck> exposure unit
Belichtungsapparat m <Labor> illuminating apparatus
Belichtungsautomat m <Gerät> automatic exposure timer
Belichtungsautomatik f <Foto> automatic exposure, automatic timer, exposure timer
Belichtungsfaktor m <Foto> exposure factor
Belichtungsmesser m 1. <Foto> light meter; exposure meter *(mit Nachführsystem)*; 2. <Phys> exposure meter, light meter

Belichtungsmesserskale

Belichtungsmesserskale f <Foto> light meter scale
Belichtungsmessersonde f <Foto> light meter probe
Belichtungsmesszelle f <Foto> light meter cell
Belichtungsschaltuhr f <Foto> darkroom timer
Belichtungsspielraum m <Foto> exposure latitude
Belichtungsstärke f <Phys> exposure rate
Belichtungstabelle f <Foto> exposure-calculating chart; exposure scale *(am Apparat)*
Belichtungszeit f <Foto> exposure time
Belichtungszeitautomat m <Gerät> automatic exposure timer
beliebig adv <Math> arbitrarily
beliebig verstellbar adj <Maschinen> adjustable at will
Belleville-Feder f <Maschinen> Belleville spring
Belsazzar m <Ker & Glas> belshazzar
Beltramiströmungen fpl <Strömphys> Beltrami flows
belüften v 1. <Bau> aerate, ventilate, vent; 2. <Heiz & Kälte> ventilate; 3. <Maschinen> vent; 4. <Sicherheit> air *(Werkstatt)*; 5. <Thermod> ventilate
Belüfter m 1. <Abfall> aerator; 2. <Chemtech> diffuser *(Druckbelüftung)*
belüftet adj <Heiz & Kälte, Thermod> ventilated
belüftete Bremsen fpl <Kfztech> ventilated brakes
belüftete Wasserlamelle f <Wasserversorg> ventilated nappe
belüfteter Bohrschlamm m <Erdöl> aerated mud *(Bohrtechnik)*
belüfteter Propeller m <Lufttrans, Wassertrans> ventilated propeller
Belüftung f 1. <Abfall> bio-aeration; 2. <Bau> aeration, airing; 3. <Heiz & Kälte, Kfztech, Maschinen, Thermod> ventilation; 4. <Verpack> aeration; 5. <Wassertrans> ventilation
Belüftungsart f <Heiz & Kälte> method of ventilation
Belüftungsbecken n 1. <Abfall> activated sludge tank, aeration tank; 2. <Wasserversorg> aeration basin
Belüftungsbehälter m <Wasserversorg> aeration basin
Belüftungseinrichtung f <Lufttrans> aerator
Belüftungshaube f <Lufttrans> air scoop
Belüftungshaubenabdeckung f **für Seitenflosse** <Lufttrans> blanking cover for fin air scoop
Belüftungsklappe f <Maschinen> *(AE)* louver, *(BE)* louvre
Belüftungsloch n <Maschinen> vent hole
Belüftungsschacht m <Sicherheit> ventilation shaft
Belüftungsstörung f <Heiz & Kälte> ventilation breakdown
Belüftungsventil n <Heiz & Kälte, Maschinen> ventilation valve
Belüftungsweg m <Heiz & Kälte> ventilating passage
bemannen v <Wassertrans> man *(Besatzung)*
bemannte Arbeitsstation f <Raumfahrt> manned workshop
bemannte Raumfahrt f <Raumfahrt> manned space flight, manned space travel
bemannte Weltraumforschung f <Raumfahrt> manned space research
bemannter Betrieb m <Telekom> attended operation
bemannter Flug m <Raumfahrt> manned flight
bemannter Hubschrauber m <Lufttrans> manned helicopter
bemannter Orbitalflug m <Raumfahrt> manned orbital space flight *(auf Umlaufbahnen)*
bemanntes Amt n <Telekom> manned exchange
bemanntes Fahrzeug n <Raumfahrt> *(AE)* manned maneuvering unit, *(BE)* manned manoeuvring unit
bemanntes Orbitallabor n <Raumfahrt> MOL, manned orbiting laboratory
bemanntes Raumfahrzeug n <Raumfahrt> manned spacecraft

bemaßen v <Comp & DV> dimension *(Konstruktionszeichnung mit Bemaßungen versehen)*
bemaßte Darstellung f <Konstzeich> dimension representation, dimensioned representation
Bemaßung f <Maschinen> dimensioning
Bemerkung f <Comp & DV> comment
bemessen v <Bau> batch, design, rate, size
Bemessung f <Bau, Maschinen> design, rating
Bemessungsdaten npl **für Wasserkühlung** <Heiz & Kälte> water-cooled rating
Bemessungstemperatur f <Heiz & Kälte> design temperature
Bemessungsverkehrsstärke f <Trans> design volume
Bemusterung f <Fertig> *(AE)* sample molding, *(BE)* sample moulding *(Kunststoffinstallationen)*
benachbarte Schicht f <Kohlen, Metall> *(AE)* neighboring layer, *(BE)* neighbouring layer, *(AE)* neighboring stratum, *(BE)* neighbouring stratum
Benadelung f <Metall> pinning
Bendix-® <Kfztech> Bendix®, Bendix-type
Bendixanlasser m <Kfztech> Bendix-type starter; Bendix starter *(Motor)*
Bendixantrieb m <Kfztech> Bendix-type starter
benennen v <Comp & DV> declare
Benennung f <Patent> designation *(eines Vertragsstaates)*
Benetzbarkeit f <Kunststoff> wettability
benetzen v <Phys, Thermod> wet
Benetzen n <Maschinen> wetting
benetzte Oberfläche f <Wassertrans> wetted surface *(Schiffkonstruktion)*
Benetzungsmittel n 1. <Anstrich> surfactant; 2. <Ker & Glas, Kunststoff, Maschinen, Meerschmutz> wetting agent
Benioff-Zone f <Nichtfoss Energ> Benioff zone
benötigte Startabbruchstrecke f <Lufttrans> accelerate-stop distance required
benötigte Startstrecke f <Lufttrans> takeoff distance required
benötigte Zeit f **von einem Gate zum anderen** <Lufttrans> ramp-to-ramp time
Bentonit m 1. <Erdöl> bentonite *(Bohrtechnik)*; 2. <Fertig, Kohlen> bentonite
Benummerung f <Druck> numbering
benutzen v <Comp & DV> use
Benutzen n <Comp & DV> use
Benutzer m 1. <Comp & DV> user; 2. <Kontroll> operator; 3. <Telekom> user
Benutzerabfrage f <Comp & DV> user query
Benutzeranmeldung f <Comp & DV> user log-on
Benutzerattribut n <Comp & DV> user attribute
Benutzerautorisierung f <Comp & DV> user authorization
Benutzerberechtigung f <Comp & DV> user authorization
Benutzerbereich m <Comp & DV> user area
Benutzerdaten npl <Comp & DV> user data
benutzerdefiniert adj <Comp & DV> user-defined
benutzerdefinierte Systemumgebung f <Comp & DV> user environment, user-operating environment
benutzerfreundlich adj <Comp & DV> user-friendly
Benutzerfreundlichkeit f <Telekom> usability *(Software)*
Benutzerführung f 1. <Comp & DV> prompt, user guidance; 2. <Telekom> user guide
Benutzergruppe f <Comp & DV> user group
Benutzerhandbuch n <Comp & DV> user manual
Benutzerkennung f <Telekom> user ID
Benutzerkennzeichen n <Comp & DV> user ID
Benutzerklasse f 1. <Comp & DV> access category; 2. <Telekom> class of service, user class of service

Benutzermenü n <Comp & DV> user menu
Benutzername m <Comp & DV> user name
Benutzeroberfläche f <Comp & DV> desk top, user interface
Benutzerprogramm n <Comp & DV> user program
Benutzerschnittstelle f <Comp & DV, Telekom> user interface
Benutzerunterstützung f <Comp & DV> support staff
benutzerverwaltetes Account n <Comp & DV> user managed account
Benutzerverwaltung f <Comp & DV> account management, user administration
Benutzerzugang m <Telekom> user access
Benutzerzugriffsmodus m <Comp & DV> user access mode
Benutzung f von Schutzgas <Kerntech> inert gas blanketing
Benutzungsbeweis m <Patent> evidence of use
Benutzungsdauer f <Qual> operating time
Benutzungsgrad m <Patent, Trans> degree of utilization
Benzal n <Chemie> benzal
Benzalacetophenon n <Chemie> chalcone
Benzaldehyd m <Chemie> benzaldehyde
Benzaldoxim n <Chemie> benzaldehyde oxime, benzaldoxime
Benzamid n <Chemie> benzamide
Benzanilid n <Chemie> benzanilide
Benzen n <Chemie> benzene
Benzidin n <Chemie> benzidine
Benzil n <Chemie> benzil
Benzin n 1. <Chemie> benzin, naphtha *(für technische Zwecke oder als Reformingstock)*; 2. <Erdöl> *(AE)* gas, *(AE)* gasoline, *(BE)* petrol *(Destillationsprodukt)*; 3. <Kfztech> fuel, *(AE)* gas, *(AE)* gasoline, *(BE)* petrol; 4. <Thermod> *(AE)* gas, *(AE)* gasoline, *(BE)* petrol; 5. <Trans> motor gasoline
Benzin n mit geringem Bleigehalt <Kfztech> *(AE)* low-lead gasoline, *(BE)* low-lead petrol
Benzinbeständigkeit f <Kunststoff> *(AE)* gas resistance, *(AE)* gasoline resistance, *(BE)* petrol resistance
Benzindampfrückgewinnungsanlage f <Umweltschmutz> *(AE)* gas vapor recovery plant, *(AE)* gasoline vapor recovery plant, *(BE)* petrol vapour recovery plant
Benzinfilter n <Kfztech> *(AE)* gas filter, *(AE)* gasoline filter, *(BE)* petrol filter *(Kraftstoff)*
Benzingemisch n <Kfztech> *(AE)* gas mixture, *(AE)* gasoline mixture, *(BE)* petrol mixture *(Zweitaktmotor)*
Benzingewinnung f <Erdöl> *(AE)* recovery of gasoline, *(BE)* recovery of petrol *(aus Raffinerierohgas, Erdgas)*
Benzingsicherung f <Maschinen> circlip
Benzingsicherungsring m <Maschinen> circlip
Benzinkanister m <Trans> jerry can
Benzinleitung f <Kfztech> fuel line *(Kraftstoff)*
Benzinmotor m 1. <Kfztech> *(AE)* gas motor, *(AE)* gasoline motor, *(BE)* petrol motor; *(AE)* gas engine, *(AE)* gasoline engine, *(BE)* petrol engine *(Motor)*; 2. <Thermod, Wassertrans> *(AE)* gas engine, *(AE)* gasoline engine, *(BE)* petrol engine *(Verbrennungsmotor)*
Benzinpumpe f <Kfztech> fuel pump, *(AE)* gas pump, *(AE)* gasoline pump, *(BE)* petrol pump *(Kraftstoff)*
Benzinschlauch m <Kfztech, Kunststoff> *(AE)* gas hose, *(AE)* gasoline hose, *(BE)* petrol hose
Benzintank m <Kfztech> fuel tank; *(AE)* gas tank, *(AE)* gasoline tank, *(BE)* petrol tank *(Kraftstoff)*
Benzinuhr f <Kfztech> *(AE)* fuel gage, *(BE)* fuel gauge, fuel indicator
benzin- und ölbeständiger Schlauch m <Kunststoff> *(AE)* gas-and-oil-resisting hose, *(AE)* gasoline-and-oil-resisting hose, *(BE)* petrol-and-oil-resisting hose

Benzinverbrauch m <Kfztech> *(AE)* gas consumption, *(AE)* gasoline consumption, *(BE)* petrol consumption *(Motor)*
Benzoat n <Chemie> benzoate
Benzocarbonitril n <Chemie> benzocarbonitrile, benzonitrile
Benzochinolin n <Chemie> phenanthridine
Benzochinon n <Chemie> benzoquinone, quinone
Benzocyclopentadien n <Chemie> indene
benzoehaltig adj <Chemie> benzoic
Benzoesäure f <Lebensmittel> benzoic acid
Benzoesäuresalz n <Chemie> benzoate
benzoid adj <Chemie> benzenoid
Benzoin n <Chemie> benzoin
Benzol n 1. <Chemie> benzene; 2. <Erdöl> benzene *(Petrochemie)*; 3. <Fertig> benzene
Benzoldiol n <Chemie> hydroquinone, quinol
Benzolhexachlorid n <Lebensmittel> benzene hexachloride
Benzonaphthol n <Chemie> benzonaphthol
Benzonitril n <Chemie> benzonitrile
Benzophenanthren n <Chemie> chrysene
Benzophenon n <Chemie> benzophenone
Benzopyrazin n <Chemie> quinoxaline
Benzopyren n <Chemie> benzopyrene
Benzopyridin n <Chemie> benzopyridine, quinoline
Benzothiophen n <Chemie> benzothiophene, thionaphthene
Benzoylbenzol n <Chemie> benzophenone
Benzoylperoxid n <Kunststoff, Lebensmittel> benzoyl peroxide
Benzoylsalicin n <Chemie> benzoylsalicin, populin
Benzoylsulfonimid n <Chemie> saccharin
Benzyl n <Chemie> benzyl
Benzylalkohol m <Chemie> benzyl alcohol, phenylcarbinol
Benzylcinnamat n <Lebensmittel> benzyl cinnamate
Benzyliden n <Chemie> benzylidene
Benzylzellulose f <Fertig> benzyl cellulose
Beobachterfehler m <Gerät> personal error
beobachtete Schwelle f <Kerntech> observed threshold *(einer Kernreaktion)*
beobachteter Standort m <Wassertrans> observed position *(Navigation)*
Beobachtung f <Qual> observation, surveillance
Beobachtung f durch Radarsonde <Lufttrans> radarsonde observation
Beobachtungsbrunnen m 1. <Abfall> monitoring well, observation well; 2. <Wasserversorg> observation well
Beobachtungsgröße f <Elektronik> measurand
Beobachtungskammer f <Erdöl> observation chamber *(Tauchtechnik)*
Beobachtungspunkt m <Ergon> point of observation
Beobachtungsraster m <Kerntech> observation grid
Beobachtungssatellit m <Raumfahrt> observation satellite
Beobachtungswert m <Gerät, Qual> observed value
BEPC (Beijing Electron Positron Collider) <Teilphys> BEPC *(Beijing Electron Positron Collider)*
Beplankung f <Bau> veneering
beplatten v <Wassertrans> plate *(Schiffbau)*
Beplattung f <Wassertrans> plating *(Schiffbau)*
Beprobung f <Wasserversorg> sampling
BER (Bitfehlerquote, Bitfehlerrate) <Comp & DV, Telekom> BER *(bit error rate)*
Berandung f <Strömphys> boundary wall
Beratungssystem n <Künstl Int> advisory system
berechenbare Funktion f <Comp & DV, Math> calculable function

berechenbarer

berechenbarer Kondensator *m* <Phys> calculable capacitor
berechnen *v* 1. <Bau> design; 2. <Math> calculate, compute, evaluate; 3. <Mechan> design
Berechnung *f* 1. <Bau> design; 2. <Comp & DV, Lufttrans> computation *(Schwerpunkt)*; 3. <Math> account, calculation, computation, evaluation
Berechnungsdruck *m* <Heiz & Kälte> design pressure
Berechnungstemperatur *f* <Heiz & Kälte> design temperature
berechtigt *adj* <Patent, Qual> entitled
Berechtigung *f* 1. <Comp & DV> authorization, clearance, privilege; 2. <Mobilkom> authentication
Berechtigung *f* **zu IFR-Flügen** <Lufttrans> instrument rating
Berechtigungscode *m* <Mobilkom> authorization code
Berechtigungsebene *f* <Comp & DV> clearance level
Berechtigungsmarke *f* <Telekom> token
Berechtigungszentrum <Mobilkom> *(AE)* authentication center *(GSM)*
Bereich *m* 1. <Akustik> band; 2. <Anstrich, Bau> area; 3. <Comp & DV> area, range, array; 4. <Elektronik> domain, range; 5. <Funktech, Gerät> range; 6. <Künstl Int> domain *(des Wissens)*; 7. <Papier, Wassertrans> range; 8. <Phys> domain; 9. <Telekom> domain *(Internet)*
Bereich *m* **eines Fernamts** <Telekom> trunk switching exchange area
Bereich *m* **mit angehobenem Nullpunkt** <Regelung> elevated-zero range
Bereich *m* **niedriger Ausbeute** <Elektronik> low-yield region
Bereichsänderung *f* <Gerät> change in range
Bereichsaufspaltung *f* <Regelung> signal amplitude sequencing control, split ranging
Bereichsschalter *m* 1. <Elektriz> scale switch; 2. <Elektrotech> selector switch
Bereichsentstörungsfilter *n* <Comp & DV> band-rejection filter
Bereichskennung *f* <Telekom> area identification signal
Bereichskennzahl *f* <Telekom> area code
Bereichsmodell *n* <Telekom> compartmental model
Bereichsschalter *m* <Funktech> band switch
Bereichssperrfilter *n* <Comp & DV> band-stop filter
Bereichsstruktur *f* <Metall> domain structure
Bereichsumschalter *m* <Funktech> band switch
Bereichsumschaltung *f* 1. <Fernseh, Funktech> waveband switching; 2. <Gerät> automatic ranging
Bereichsunterdrückung *f* <Gerät> suppression of range
Bereichsunterschreitung *f* <Comp & DV> underflow
Bereichsverhältnis *n* <Regelung> rangeability
Bereichswahl *f* <Gerät> automatic ranging
Bereichswähler *m* <Gerät> range selector
bereifen *v* <Mechan> hoop
bereinigen *v* <Comp & DV> debug
bereit *adj* <Comp & DV> ready • **bereit für den Einsatz im Netzwerk** <Comp & DV> network ready
Bereitschaft *f* 1. <Comp & DV, Telekom> stand-by; 2. <Raumfahrt> availability
Bereitschaftsaggregat *n* <Elektrotech> stand-by set
Bereitschaftsbetrieb *m* 1. <Kontroll> stand-by, stand-by mode; 2. <Mobilkom> stand-by mode; 3. <Telekom> stand-by working
Bereitschaftsgerät *n* <Elektrotech> stand-by unit
Bereitschaftsstellung *f* <Elektrotech> on position
Bereitschaftssystem *n* <Sicherheit> back-up system
Bereitschaftszeit *f* <Comp & DV> stand-by time
Bereitstellung *f* <Telekom> provision
Bereitzustand *m* <Comp & DV> ready state
Berg *m* <Bau, Kohlen> hill

Bergarbeiter *m* <Kohlen> miner
Bergbau *m* <Kohlen> mining
Bergbauproduktion *f* <Kohlen> mine yield
Bergeklein *n* <Kohlen> refuse
bergen *v* <Wassertrans> salvage
Berglandwirtschaft *f* <Lebensmittel> hill farming
Bergmittel *n* <Kohlen> stone band
Bergreinigerzelle *f* <Kohlen> scavenger cell
Bergrutsch *m* <Trans> landslide, landslip
Bergteich *m* <Kohlen> tailing pond
Bergung *f* 1. <Lufttrans> recovery, rescue *(Flugmanöver)*; 2. <Raumfahrt> recovery *(Raumschiff)*; 3. <Sicherheit, Wassertrans> recovery, salvage; rescue *(Notfall)*
Bergungsgerät *n* <Sicherheit> rescue device
Bergungskran *m* <Kfztech> salvage crane
Bergungspaket *n* <Raumfahrt> recovery package *(Raumschiff)*
Bergungsschiff *n* <Wassertrans> salvage vessel, wrecker
Bergungsschlepper *m* <Trans, Wassertrans> salvage tug
Bergwerk *n* <Kohlen> mine
Bericht *m* <Comp & DV> report
Berichterstattung *f* **über Satellit** <Telekom> satellite news gathering
berichtigen *v* <Qual> rectify
berichtigte angezeigte Eigengeschwindigkeit *f* <Lufttrans> calibrated airspeed *(Flugwesen)*
berichtigte angezeigte Flugmindestgeschwindigkeit *f* <Lufttrans> minimum calibrated speed in flight time *(bei normaler Luftströmung)*
Berichtigung *f* 1. <Comp & DV> upgrade; 2. <Math> adjustment, correction; 3. <Patent, Qual> correction
Berichtigung *f* **der Anzeige der rufenden Nummer** <Telekom> CLIR, calling line identification rectification
Berichtigungsfaktor *m* <Lufttrans> correction factor *(Auftriebswiderstand)*
Berichtserstellung *f* <Comp & DV> report generation
Berieselung *f* 1. <Chemtech> scrubbing; 2. <Fertig> flooding; 3. <Wasserversorg> border irrigation, irrigation
Berieselungskanal *m* <Wasserversorg> catch-feeder
Berieselungskanone *f* <Elektronik> flooding gun
Berieselungskühler *m* <Heiz & Kälte> spray cooler
Berieselungsturm *m* <Chemtech> scrubber, washing column
Berkelium *n* *(Bk)* <Chemie> berkelium *(Bk)*
Berme *f* 1. <Abfall> segregation berm; 2. <Bau> berm; 3. <Eisenbahn> bank; 4. <Kohlen> berm; 5. <Wasserversorg> offset
Berner Vierkantschlüssel *m* <Eisenbahn> Berne key
Bernoulli'sche Gleichung *f* 1. <Phys> Bernoulli's equation *(Geschwindigkeits- und Druckänderungen entlang Stromlinie)*; 2. <Strömphys> Bernoulli's equation
Bernoulli'sches Theorem *n* <Phys, Strömphys> Bernoulli's theorem
Berst... <Kunststoff, Papier, Textil, Verpack> bursting
Berstdruck *m* 1. <Fertig> bursting pressure *(Kunststoffinstallationen)*; 2. <Ker & Glas, Papier, Verpack> bursting pressure
bersten *v* 1. <Bau, Papier> burst; 2. <Thermod> detonate
Bersten *n* <Papier> burst
Berstfestigkeit *f* <Kunststoff, Papier, Textil, Verpack> bursting strength
Berstfestigkeitsprüfer *m* <Papier> bursting-strength tester
Berstindex *m* <Papier> burst ratio
Berstscheibe *f* <Maschinen, Mechan, Sicherheit> *(BE)* bursting disc, *(AE)* bursting disk, *(BE)* safety disc, *(AE)* safety disk

Berstversuch m <Sicherheit> bursting trial
Berücksichtigung f <Math> consideration
berufen v <Patent, Qual> appoint
berufsbedingte Lärmbelastung f <Sicherheit> occupational noise exposure
Berufsverkehr m <Trans> business traffic, home-to-work traffic
Berufung f <Patent> appeal
beruhigen v 1. <Fertig> quiet (Stahl); kill (Walzen); 2. <Metall> kill
beruhigt adj <Fertig> dead (Stahl); killed (Walzen)
beruhigt vergossen adj <Fertig> solid (Stahl)
beruhigter Stahl m <Metall> killed steel
Beruhigung f <Fertig> quieting (Stahl)
Beruhigungsbecken n <Bau> stilling basin
Beruhigungskammer f <Abfall> settling chamber
Beruhigungszeit f 1. <Gerät> damping period, transient time, transition time; 2. <Metrol> response time (Messinstrument)
berührender Fühler m <Gerät> contact sensor
berührender Sensor m <Gerät> contact sensor
Berührgerade f <Geom, Math> tangent
Berührung f <Geom> tangency
Berührungsbogen m <Fertig> arc of conduct (Schleifscheibe)
Berührungsebene f <Geom> tangent plane
Berührungseingabe f <Kontroll> touch input
berührungsempfindlicher Bildschirm m <Comp & DV> touch screen, touch-sensitive screen
berührungsempfindlicher Klebstoff m **mit hoher Klebfestigkeit** <Verpack> high-tack pressure-sensitive adhesive
Berührungsfläche f <Maschinen> contact area, surface of contact
berührungsfrei adj <Elektriz> contactless, noncontact
berührungsfreie Dichtung f <Maschinen> noncontacting seal
berührungsfreier Abnehmer m <Elektriz> contactless pick-up
berührungsfreies Trägerfahrzeug n <Kfztech> contactless support vehicle
Berührungsfühler m <Gerät> contact sensor
Berührungslinie f 1. <Geom> tangent; 2. <Maschinen> line of contact; 3. <Math> tangent
berührungslos wirkende Schutzvorrichtung f <Sicherheit> contactless safety device, non-contact guard, non-contact safety device
berührungslos wirkender Detektor m <Sicherheit> proximity detector
berührungsloses Oberflächenthermometer n <Gerät> contactless surface thermometer
Berührungs-Maus-Kissen n <Comp & DV> sensor mouse pad, touch pad
Berührungsmessgerät n <Gerät> contact-measuring instrument
Berührungsmesskopf m <Gerät> contact head
Berührungspunkt m 1. <Geom> point of tangency, tangent point; 2. <Phys> point of contact
Berührungsschalter m 1. <Elektriz> touch contact switch; 2. <Elektrotech, Kontroll> touch switch
Berührungsschutz m <Sicherheit> accidental contact protection
Berührungssensor m 1. <Gerät> tactile sensor, touch sensor; 2. <Gerät> contact sensor; 3. <Sicherheit> presence sensing device, pressure-sensitive device
berührungssichere Buchse f <Elektriz> shockproof socket
berührungssichere Steckverbindung f <Elektriz> shockproof socket

Berührungsspannung f <Elektriz> shock hazard voltage
Berührungstaste f <Gerät> touching key
Berührungsthermometer n <Gerät> contact thermometer, surface temperature sensor
Beryll m <Ker & Glas> beryl
Beryllerde f <Ker & Glas> beryllia
Beryllium n (Be) <Chemie> beryllium (Be)
berylliummoderierter Reaktor m <Kerntech> beryllium-moderated reactor
Besan m <Wassertrans> mizzen (Segeln)
Besanden n <Ker & Glas> sanding
besänftigen v <Anstrich> temper (Legierungen)
Besanmast m <Wassertrans> mizzen mast (Segeln)
Besatz m 1. <Lebensmittel> dockage; 2. <Textil> trimming
Besatztuch n <Textil> facing
Besatzungskabine f <Lufttrans> crew compartment, flight deck
Besatzungsliste f <Wassertrans> crew list
Besatzungsraum m <Lufttrans> cockpit
Besäum... <Bau, Fertig> edging, squaring
besäumen v <Bau> trim (Holz)
Besäumen n 1. <Bau> squaring (Holz); 2. <Fertig> edging (Holz)
Besäumkreissäge f <Fertig> edger
Besäummaschine f <Maschinen> trimmer
besäumt adj <Bau> square-edged (Holz)
beschädigt adj 1. <Kerntech> defective (Brennstab); 2. <Sicherheit> damaged, defective; 3. <Qual, Telekom> damaged
beschädigter Wagen m <Eisenbahn> (AE) damaged car, (BE) damaged wagon
beschädigtes Brennelement n <Kerntech> damaged fuel assembly
beschädigtes Garn n <Ker & Glas> damaged yarn
Beschädigung f <Fertig, Qual> damage, scoring
Beschaffungsspezifikation f <Raumfahrt> procurement specifications (Raumschiff)
Beschaufelung f 1. <Maschinen> blades; 2. <Mechan> blading
Bescheid m <Telekom> interception
Bescheidansage f <Telekom> intercept announcer
Bescheiddienst m <Telekom> changed-number interception
bescheinigter Werkstoff-Prüfbericht m <Qual> certified material test report, CMTR
Bescheinigung f <Patent> certificate
beschichten v 1. <Bau> coat, overcoat; surface (Material); 2. <Kunststoff, Lebensmittel> coat
Beschichten n 1. <Anstrich> plating; 2. <Kunststoff> coating; 3. <Phys> cladding; 4. <Verpack> coating (mit Kunststoff)
beschichtet adj <Fertig> coated (Kunststoffinstallationen)
beschichtete Packung f <Verpack> laminated pack
beschichteter Lichtwellenleiter m <Optik, Telekom> coated optical fiber, jacketed optical fiber
beschichtetes Band n <Anstrich> film
beschichtetes Gewebe n <Kunststoff> coated fabric
beschichtetes Klebeband n <Verpack> self-adhesive laminated tape
beschichtetes Kunstdruckpapier n <Verpack> coated synthetic paper
Beschichtung f 1. <Bau> coating; 2. <Optik> sheath; 3. <Phys, Textil> coating; 4. <Verpack> (AE) enameling, (BE) enamelling
Beschichtungsloch n <Aufnahme> drop-out (Magnetband)
Beschichtungsmaschine f <Kunststoff> coating machine
Beschichtungssystem n <Anstrich> coating system

Beschichtungsverfahren

Beschichtungsverfahren n 1. <Anstrich> coating process; 2. <Fertig> coating technique
beschicken v <Trans> fill
Beschicken n 1. <Ker & Glas> ladling *(mit einem Löffel)*; 2. <Textil> feeding
Beschicktür f <Fertig> charging door
Beschickung f 1. <Fertig> round *(Hochofen)*; 2. <Kerntech> supply *(während des Betriebs)*; 3. <Maschinen> loading; 4. <Textil> batch
Beschickung f **bei abgeschalteter Last** <Kerntech> off-load charging
Beschickung f **unter Last** <Kerntech> on-load charging, *(AE)* on-load fueling, *(BE)* on-load fuelling, *(AE)* on-load refueling, *(BE)* on-load refuelling
Beschickungsarmatur f <Erdöl> feed tank
Beschickungseinrichtung f 1. <Abfall> charging facility, loading mechanism; 2. <Fertig> loading device
Beschickungskammer f <Abfall> charging chamber *(Atommülllagerung)*
Beschickungskopf m <Kerntech> feeder head
Beschickungsmaschine f <Verpack> case loader *(für Kästen)*
Beschickungsrohr n <Kerntech> feeder pipe
Beschickungsseite f <Kerntech> reactor charging face *(eines Reaktors)*
Beschickungstisch m <Verpack> feeding table
Beschickungstrichter m 1. <Fertig> hopper; 2. <Mechan> feed hopper
Beschickungstür f <Fertig> charging door
Beschickungsventil n <Erdöl> feed valve
Beschickungsverfahren n <Kerntech> feeding process
Beschickungsverriegelung f <Kerntech> feeder lock *(bei kugelförmigen Brennelementen)*
Beschickungszone f <Kerntech> charge area
Beschlag m 1. <Ker & Glas> bloom *(Fleck)*; 2. <Mechan> ferrule, plate
beschlagen v <Fertig> clout
beschleunigen v 1. <Eisenbahn, Kfztech> accelerate; 2. <Papier, Phys> accelerate, speed up
beschleunigend adj <Papier> accelerant, accelerative
beschleunigende Kraft f <Phys> accelerating force
Beschleuniger m 1. <Bau, Chemtech> accelerator; 2. <Elektrotech> accelerator, activator; 3. <Kfztech> accelerator pedal; 4. <Kunststoff, Lebensmittel, Mechan, Papier, Teilphys> accelerator; 5. <Wassertrans> accelerator *(Bootbau)*; catalyst *(im Kunststoffbootbau)*
Beschleunigerdüse f <Kfztech> accelerator jet *(Vergaser)*
Beschleuniger-Hohlraumresonator m <Teilphys> accelerator cavity
Beschleunigerpumpe f <Kfztech> accelerating pump; accelerator pump *(Vergaser)*
beschleunigt adj <Phys> accelerated
beschleunigte Bewegung f <Mechan> accelerated motion
beschleunigte Filtration f <Chemtech> accelerated filtration
beschleunigte Kommutation f <Elektriz> accelerated commutation
beschleunigte Kompostierung f <Abfall> accelerated composting, mechanical composting, rapid fermentation
beschleunigte Lebensdauerprüfung f <Gerät> accelerated life test
beschleunigte Leitwegumlenkung f <Telekom> forced rerouting
beschleunigte Prüfung f 1. <Qual> accelerated test; 2. <Werkprüf> accelerated testing
beschleunigter Abschaltstab m <Kerntech> accelerated scram rod
beschleunigter Leichtwasserreaktor m <Kerntech> accelerator-driven light-water reactor
beschleunigter Salzschmelzenbrüter m <Kerntech> accelerator molten-salt breeder
beschleunigter Test m <Maschinen> accelerated test
beschleunigtes Kriechen n <Metall> accelerated creep
Beschleunigung f 1. <Eisenbahn, Kfztech> acceleration; 2. <Kontroll> speed-up; 3. <Papier> acceleration
Beschleunigungsabbruch m <Kfztech> acceleration stop
Beschleunigungsabbruchstrecke f <Kfztech> accelerate-stop distance
Beschleunigungsanode f 1. <Elektrotech> accelerating anode, second anode; 2. <Strahlphys> accelerating anode
Beschleunigungsanzeiger m <Lufttrans> acceleration detector
Beschleunigungsaufnehmer m 1. <Ergon> acceleration pick-up; 2. <Heiz & Kälte> acceleration sensor
Beschleunigungsbelastung f <Raumfahrt> G-force *(Raumschiff)*
Beschleunigungsdüse f <Kfztech> acceleration jet
Beschleunigungselektrode f 1. <Kerntech, Strahlphys> accelerating electrode; 2. <Telekom> accelerator
Beschleunigungskammer f <Kerntech> accelerating chamber
Beschleunigungskraft f <Ergon> acceleration force
Beschleunigungsleistung f <Kfztech> accelerating power
Beschleunigungsmesser m <Elektrotech, Ergon, Lufttrans, Mechan> accelerometer
Beschleunigungsmessgerät n <Gerät, Papier, Phys> accelerometer
Beschleunigungspegel m <Sicherheit> acceleration level *(Lärm)*
Beschleunigungsrakete f <Raumfahrt> acceleration rocket *(zum Treibstoffsammeln)*
Beschleunigungsrelais n <Elektriz> acceler-ation relay
Beschleunigungsröhre f <Chemtech> accelerating tube
Beschleunigungsschreiber m <Gerät> accelerograph
Beschleunigungssensor m <Gerät> acceleration pick-up
Beschleunigungsspannung f <Elektriz, Elektrotech> accelerating voltage
Beschleunigungsspur f <Trans> acceleration lane
Beschleunigungssteuereinheit f <Lufttrans> acceleration control unit
Beschleunigungsvermögen n <Kfztech> pick-up
Beschleunigungsvorrichtung f <Trans> acceleration device
Beschleunigungsweg m <Kfztech> acceleration distance
Beschleunigungszeit f <Comp & DV, Papier> acceleration time
Beschneidemaschine f <Maschinen> trimmer, trimming machine
beschneiden v 1. <Druck> cut, trim; 2. <Elektrotech> clip; 3. <Papier> trim
Beschneiden n <Fernseh> cropping
Beschnitt m <Druck> trimming edge
Beschnittbreite f <Druck> trim width
beschnittene Lichtpause f <Konstzeich> trimmed blueprint sheet
beschnittene Zeichnung f <Konstzeich> trimmed drawing sheet
beschnittenes Format n 1. <Druck> trim size, trimmed size; 2. <Verpack> trimmed size *(Papier)*
Beschnittmarken fpl <Druck> trim marks
Beschnittrand m 1. <Druck> trimmed edges; 2. <Konstzeich> trimming edge

Beschnittzeichen n <Druck> crop mark, cutting mark
beschottern v <Bau> gravel; metal (Tiefbau)
Beschottern n <Bau, Trans> macadamization
beschränken v <Math> confine
Beschränker m <Kontroll> limiter
beschränkt adj <Math> confined, limited, restricted
beschränkter Informationsübermittlungsdienst m <Telekom> restricted information transfer service
Beschränktheit f <Math> dullness
Beschränkung f 1. <Künstl Int> constraint; 2. <Math> limitation; 3. <Phys> constringence
beschreibbar mehrfach lesbar adj/nur einmal <Comp & DV, Optik> write-once read many times
beschreibbare CD f 1. <Comp & DV> CD-R, CD recordable, compact disk recordable; 2. <Optik> (BE) compact disc-interactive, (AE) compact disk-interactive, writable optical disk
beschreibbare Scheibe f <Optik> writable disk
beschreibend adj <Patent, Qual> descriptive
beschreibendes Modell n <Ergon> descriptive model
Beschreiber m <Comp & DV> descriptor
Beschreibung f <Patent, Qual> description
Beschreibzeit f <Elektronik> writing time (Photolack)
beschrieben adj <Elektronik> written-state
beschriebene Oberfläche f <Comp & DV> recorded surface (eines Datenträgers)
Beschriften n <Comp & DV> (AE) labeling, (BE) labelling
Beschriftung f 1. <Comp & DV> annotation; 2. <Druck> lettering; 3. <Ker & Glas> lettering (auf Gefäßboden); 4. <Konstzeich> inscription, lettering
Beschriftungsanlage f <Verpack> marking equipment
Beschriftungsfeld n <Akustik> label area
Beschriftungsmaschine f <Verpack> marking machine
Beschwerde f <Patent> appeal
Beschwerdeschrift f <Patent> notice of appeal
beschweren v <Bau> burden
beseitigen v <Bau, Qual> cure (Mängel)
Beseitigung f 1. <Bau> removal; 2. <Kerntech> decommissioning (einer Reaktoranlage); 3. <Meerschmutz> disposal; 4. <Umweltschmutz> removal (von organischen Bestandteilen); removal (von Schwebstoffen durch Sedimentablagerung)
Beseitigungssystem n <Sicherheit> suppression system
besetzen v <Textil> face
besetzt adj <Telekom> busy, engaged (Telefonie)
Besetzt... <Comp & DV, Telekom> busy
besetzte Leitung f <Kontroll> (AE) busy line, (BE) engaged line
besetzte Rufnummer f <Telekom> (AE) busy number, (BE) engaged tone
besetzte Signalleitung f <Kontroll> (AE) busy line, (BE) engaged line
besetztes Band n <Phys> full band (Festkörpermaterial)
Besetztstatus m <Telekom> busy status
Besetztton m <Telekom> (AE) busy number, (BE) engaged tone
Besetztzeichen n <Comp & DV> (AE) busy signal, (BE) engaged signal
Besetztzustand m <Telekom> busy state
Besetzung f <Phys, Strahlphys> population
Besetzungsdichte f angeregter Atome <Strahlphys> population density of excited atoms
Besetzungsinversion f <Phys> population inversion
Besetzungsumkehr f <Strahlphys> population inversion
Besichtigung f 1. <Bau> surveying; 2. <Mechan, Qual> inspection; 3. <Raumfahrt> walkaround inspection (Raumschiff)
Bespannung f <Papier> clothing
bespanten v <Wassertrans> frame (Schiffbau)

Bespantung f <Wassertrans> timber (Schiffbau)
bespieltes Band n <Fernseh> prerecorded tape
besprengen v 1. <Chemtech> water; 2. <Wasserversorg> sprinkle
besprühen v <Meerschmutz, Papier> spray
bespultes Kabel n <Telekom> loaded cable
Bespulung f <Telekom> coil loading
Bessemerbirne f 1. <Fertig> acid Bessemer converter, acid converter; 2. <Maschinen> Bessemer converter
Bessemerroheisen n <Fertig> acid Bessemer pig, acid pig
Bessemerstahl m 1. <Fertig> acid Bessemer steel, acid converter steel; 2. <Maschinen> Bessemer steel (BS); 3. <Metall> basic Bessemer steel
Bessemerverfahren n <Fertig> acid Bessemer process, acid converter process
bessern v/sich <Wassertrans> abate (Wetter)
Bestand m 1. <Anstrich> residue; 2. <Qual> survivals
beständig adj <Kunststoff> lasting, resistant
Beständigkeit f 1. <Anstrich> durability; 2. <Kunststoff> durability, resistance; 3. <Maschinen> durability; 4. <Textil> resistance
Beständigkeit f gegen Sonnenlicht <Kunststoff> sunlight resistance
Bestandteil m 1. <Anstrich> component; 2. <Kunststoff, Lebensmittel> ingredient; 3. <Mechan> component
bestätigen v 1. <Comp & DV, Kerntech> acknowledge (Notabschaltung); 2. <Raumfahrt> validate
Bestätigung f 1. <Comp & DV> ACK, acknowledgement, authentication, verification; 2. <Kontroll, Maschinen> verification; 3. <Patent> certification; 4. <Qual> certification, verification; 5. <Telekom> ACK, acknowledgement
Bestätigungssignal n <Telekom> acknowledgement signal, confirmation signal
bestäuben v 1. <Lebensmittel> dust; 2. <Papier> powder
Bestäuben n <Ker & Glas> dry spray
Besteck n 1. <Bau> set of instruments; 2. <Maschinen> set (von Instrumenten) • **Besteck absetzen** <Wassertrans> plot the position (Navigation)
Besteckabsetzen n <Wassertrans> plotting (Navigation)
Bestellerrisiko n <Qual> consumer's risk
Bestellung angefertigt adj/auf <Verpack> custom-made
Bestensuche f <Künstl Int> best-first search
Best-First-Suche f <Künstl Int> best-first search
bestimmbare Ursache f <Qual> assignable cause
bestimmen v <Qual> determine, rate; specify (näher bestimmen)
bestimmter Flammpunkt m/im geschlossenen Raum <Kunststoff> closed-cup flash point
bestimmtes Integral n <Math> definite integral
Bestimmung f 1. <Sicherheit> determination; 2. <Thermod> determination (des Wärmewertes)
Bestimmung f über Höhenstaffelung in bestimmten Quadranten <Lufttrans> quadrantal height rule
Bestimmungsamt n <Patent> designated office
Bestimmungskurve f <Geom> defining curve
Bestimmungspunkte mpl <Geom> control points
Bestmarke f <Comp & DV> benchmark
bestrahlt adj <Phys> irradiated
Bestrahlung f 1. <Kunststoff> exposure to radiation; 2. <Strahlphys, Teilphys> irradiation; 3. <Umweltschmutz> irradiation, radiation
Bestrahlung f mit Kobalt 60 <Kerntech> cobalt 60 gamma irradiation
Bestrahlungsdichte f <Nichtfoss Energ> irradiance
Bestrahlungsdosis f <Strahlphys> exposure dose (gemessen in Röntgen)
Bestrahlungsfolie f <Strahlphys> beam foil
Bestrahlungsgeometrie f <Kerntech> geometry of irradiation

Bestrahlungshärten

Bestrahlungshärten n <Metall> irradiation hardening
Bestrahlungskammer f 1. <Kerntech> radiation chamber; 2. <Strahlphys> irradiation chamber
Bestrahlungsreaktor m <Kerntech> radiation reactor
Bestrahlungsrisiken npl <Strahlphys> exposure risks
Bestrahlungsschaden m <Kerntech> radiation damage
Bestrahlungsstärke f 1. <Optik> irradiance *(H)*; 2. <Phys> exposure rate, irradiance, irradiation, radiation intensity
Bestrahlungsversuch m **innerhalb des Reaktors** <Kerntech> in-pile test
bestreichen v <Anstrich> paint
bestücken v 1. <Fertig> tip *(Spanung)*; 2. <Maschinen> hard-face
Bestücken n <Kerntech> hard-facing
bestückt adj <Maschinen> hard-faced
Bestückung f <Fertig> tip, tooling
Bestückungsdichte f <Elektronik> packing density *(bei ICs)*
Bestwertsteuerung f 1. <Elektrotech> AC, adaptive control; 2. <Fertig> adaptive control
Bestzeit... <Comp & DV> minimum-access, minimum-delay
Bestzeitcode m <Comp & DV> minimum-access code, minimum-delay code
Bestzeitprogramm n <Comp & DV> minimum-access routine
Bestzeitprogrammierung f <Comp & DV> minimum-access programming
Besucherdatei f 1. <Mobilkom> visitor location register; 2. <Telekom> visitor data base, VDB *(Mobilfunk, für das Besucherregister im GSM)*
besultes Kabel n <Telekom> coil-loaded cable
Beta n *(β)* <Geom> beta *(β)*
Betaabsorptionsanalyse f <Phys, Strahlphys, Teilphys> beta particle absorption analysis
Betaamylase f <Lebensmittel> beta-amylase
Betadichtemesser m <Kerntech> *(AE)* beta density gage, *(BE)* beta density gauge
betakeln v <Wassertrans> rig
Betamessing n <Fertig> beta brass
Betanken n <Kfztech, Lufttrans> *(AE)* fueling, *(BE)* fuelling
Betankung f <Lufttrans> fuel transfer, *(AE)* refueling, *(BE)* refuelling
Betankungsausleger m <Lufttrans> *(AE)* refueling boom, *(BE)* refuelling boom
Betarückstreuanalyse f <Phys, Strahlphys, Teilphys> beta particle backscattering analysis
Betarückstreumesser m <Kerntech> *(AE)* beta backscatter gage, *(BE)* beta backscatter gauge
Betaspektrum n <Strahlphys, Teilphys> beta ray spectrum
Betastabilitätsinsel f <Kerntech> beta stability island
Betastrahlen mpl 1. <Elektriz> beta rays; 2. <Phys, Strahlphys, Teilphys> beta radiation, beta rays
Betastrahlenspektrum n <Phys, Strahlphys, Teilphys> beta ray spectrum
Betastrahler m <Phys, Strahlphys, Teilphys> beta emitter
Betastrahlung f <Phys, Strahlphys, Teilphys> beta emission, beta radiation
Betateilchen n 1. <Elektriz> beta particle; 2. <Phys> beta particle *(Elektron)*; 3. <Strahlphys, Teilphys> beta particle
Betatest m <Comp & DV> beta test • **einen Betatest durchführen** <Comp & DV> beta test
Betateststandort m <Comp & DV> beta site
betätigen v <Maschinen> actuate, operate, work
Betätigung f <Maschinen> actuation, control, operation
Betätigungsbügel m <Fertig> actuator *(Kunststoffinstallationen)*

Betätigungseinrichtung f <Maschinen> actuator attachment
Betätigungselement n <Comp & DV, Elektrotech, Maschinen> actuator
Betätigungsglied n <Elektrotech> actuator
Betätigungsknopf m <Gerät> control button
Betätigungsplatte f <Lufttrans> actuating plate
Betätigungsspannung f <Elektriz> actuating voltage
Betätigungsspindel f <Fertig> actuating screw
Betätigungsstange f <Lufttrans> actuating rod
Betätigungsstrom m <Elektriz> operating current
Betätigungstaste f <Gerät> operating key
Betatron n <Phys, Strahlphys> betatron
Betatronbewegung f <Phys> betatron motion
Betazerfall m 1. <Kerntech> beta decay, beta disintegration; 2. <Phys, Strahlphys, Teilphys> beta decay
Betazerfallsenergie f <Phys, Strahlphys, Teilphys> beta disintegration energy
Bethe-Goldstone-Gleichung f <Kerntech> Bethe-Goldstone equation
Betitelung f <Fernseh> titling
Beton m <Bau, Ker & Glas> concrete • **Beton einbringen** <Bau> pour concrete
Beton m **mit Steineinlagen** <Bau> cyclopean concrete
Betonarbeiten fpl <Bau> concrete work
Betonaufbruchhammer m <Bau> concrete breaker
Betonauskleidung f <Bau> concrete lining
Betonbau m <Bau> concrete structure
Betonblockstein m <Bau> concrete block
Betondachziegel m <Bau> concrete roofing tile
Betondecke f <Bau> *(BE)* concrete pavement, *(AE)* concrete sidewalk *(Straße)*
Betondosieranlage f <Bau> concrete-batching plant
Betoneisen n <Bau> reinforcing bar
Betonfertigteile npl <Bau> precast concrete
Betonfertigteilwerk n <Bau> precasting plant
Betonhaltbarkeit f <Bau> concrete durability
betonieren v 1. <Bau> concrete; 2. <Bau> pour concrete
Betonieren n <Bau> concreting
betoniert adj/**vor Ort** <Bau> cast-in-situ
Betonkern m <Bau> core
Betonmauerwerk n <Bau> concrete masonry
Betonmischanlage f <Bau> batch plant, concrete mixing plant
Betonmischer m <Bau> batch mixer, concrete mixer
betonnen v <Wassertrans> buoy *(Seezeichen)*
Betonnest n <Bau> honeycomb
Betonnung f <Wassertrans> buoyage *(Navigation)*
Betonpfahl m <Kohlen> concrete pile
Betonplattform f <Erdöl> concrete platform *(Offshore-Technik)*
Betonring m <Bau> concrete ring
Betonrippenkonstruktion f <Heiz & Kälte> waffle slab
Betonrohr n <Bau> concrete pipe
Betonrüttler m <Bau> concrete vibrating machine, concrete vibrator
Betonsäge f <Bau> concrete saw
Betonschrapper m <Trans> concrete scraper
Betonschutt m <Trans> concrete scrap
Betonschütttrichter m <Bau> tremie
Betonschwelle f <Bau, Eisenbahn> *(BE)* concrete sleeper, *(AE)* concrete tie *(Eisenbahn)*
Betonschwellenverlegegerät n <Bau, Eisenbahn> *(BE)* concrete sleeper layer, *(AE)* concrete tie layer
Betonschwellenvorspannung f <Bau, Eisenbahn> *(BE)* concrete sleeper prestressing, *(AE)* concrete tie prestressing
Betonsohle f <Heiz & Kälte> concrete floor
Betonstein m <Ker & Glas> concrete block

betonummantelt *adj* <Bau> haunched
Betonverflüssiger *m* <Bau> plastifying admixture
Betrachtung *f* 1. <Foto> viewing; 2. <Math> consideration
Betrachtungseinheit *f* 1. <Qual> item; 2. <Telekom> entity
Betrachtungsobjektiv *n* <Foto> viewing lens
Betrachtungsvergrößerer *m* <Foto> viewing magnifier
Betrag *m* 1. <Comp & DV> sum; 2. <Math> amount
Betrag *m* **eines Vektors** 1. <Math> norm of a vector; 2. <Phys> magnitude of vector, modulus of vector
betrauen *v* **mit** <Patent, Qual> entrust with
betreiben *v* 1. <Fertig> run *(eine Anlage)*; 2. <Kerntech> operate; 3. <Maschinen> operate, work
Betreiber *m* 1. <Maschinen, Telekom> operator; 2. <Telekom> operator, carrier
Betreiber-Netzkennzahl *f* <Telekom> carrier prefix
Betrieb *m* 1. <Comp & DV> operation; running *(von Software)*; 2. <Elektrotech> operation *(einer Maschine)*; 3. <Fertig> duty, operation, plant, run, service, working; 4. <Maschinen> duty, operation, run, service, working; 5. <Mechan> mill, shop • **außer Betrieb** 1. <Elektriz> out of operation; 2. <Elektrotech> off; 3. <Kontroll> inoperative; 4. <Maschinen> idle, out of action; 5. <Telekom> disabled *(Gerät)* • **außer Betrieb befindlich** <Lufttrans> rest period • **außer Betrieb nehmen** <Bau, Eisenbahn> take out of service • **außer Betrieb setzen** 1. <Bau> put out of service; 2. <Lufttrans> shut down *(Motor, Triebwerk)* • **in Betrieb** <Elektriz> in operation, in-service • **in Betrieb befindlich** <Elektrotech, Maschinen> active, operating, operative • **in Betrieb nehmen** <Maschinen> bring into service, put into operation, set into operation • **in Betrieb sein** 1. <Comp & DV> run; 2. <Maschinen> function, operate, run • **in Betrieb setzen** 1. <Kontroll> activate; 2. <Kerntech, Maschinen> set into operation • **nicht in Betrieb** <Telekom> idle
Betrieb *m* **auf einer Frequenz** <Elektronik, Funktech, Telekom> single frequency operation
Betrieb *m* **auf zwei Ebenen** <Elektrotech> bilevel operation
Betrieb *m* **bei konstantem Druck** <Raumfahrt> constant-pressure operation *(Raumschiff)*
Betrieb *m* **mit Ersatzschaltung** <Telekom> stand-by working *(Übertragung)*
Betrieb *m* **mit geregeltem Schlupf** <Elektrotech> operation with controlled slip
Betrieb *m* **mit gleich bleibender Zeilenzahl** <Fernseh> constant line number operation
Betrieb *m* **mit konstantem Luftspaltfluss** <Elektrotech> constant air-gap flux operation, operation with constant air-gap flux
Betrieb *m* **mit konstantem Strom** <Elektrotech> operation with constant current
Betrieb *m* **mit konstanter Leistung** <Elektrotech> operation with constant power
Betrieb *m* **mit konstanter Spannung** <Elektrotech> operation with constant voltage
betriebliches Fernsehen *n* <Fernseh> business television
Betriebmessbrücke *f* <Gerät> field bridge
Betriebs... <Maschinen> operating
Betriebsanleitung *f* <Fertig, Maschinen> operating instructions, operating manual
Betriebsart *f* 1. <Comp & DV> mode, mode of operation; 2. <Elektronik> mode; 3. <Elektrotech, Telekom> duty type, mode, mode of operation, operating mode, type of duty; 4. <Regelung> step setting mode
Betriebsartbefehlsformat *n* <Comp & DV> mode instruction code
Betriebsartenextraktor *m* <Raumfahrt> mode extractor *(Weltraumfunk)*
Betriebsartenschalter *m* <Elektrotech> selector switch

Betriebsartenwahlschalter *m* <Lufttrans> mode selector switch
Betriebsartregister *n* <Comp & DV> mode register
Betriebsartwechsel *m* <Comp & DV> mode change
Betriebsaufnahme *f* <Kontroll> initial operation phase
Betriebsausfall *m* <Elektrotech, Sicherheit> breakdown, failure, shutdown
Betriebsbeauftragter *m* **für Abfall** *(BfA)* <Abfall> Waste Management Officer
betriebsbedingte Verzögerung *f* <Trans> operational delay
betriebsbedingter Defekt *m* <Kerntech> operations-related defect
betriebsbedingter Fehler *m* <Umweltschmutz> operational error
Betriebsbedingungen *fpl* 1. <Elektriz, Elektrotech, Metrol> operating conditions; 2. <Telekom> environmental conditions
betriebsbereit *adj* 1. <Comp & DV> online, ready for operation, ready; 2. <Maschinen> in working order
Betriebsbereitschaft *f* 1. <Fernseh> stand-by; 2. <Maschinen, Qual> operability, operational readiness, readiness for operation; 3. <Telekom> data set ready *(Modem)*
Betriebsbereitschaftssignal *n* <Comp & DV> enabling signal
Betriebsbremse *f* <Kfztech, Maschinen> service brake
Betriebsdaten *npl* <Elektrotech, Phys> rating
Betriebsdauer *f* <Kerntech> operating lifetime
Betriebsdrehzahl *f* <Maschinen> operating speed
Betriebsdruck *m* 1. <Fertig> working pressure *(Kunststoffinstallationen)*; 2. <Maschinen> operating pressure, working pressure; 3. <Raumfahrt> operating pressure *(Raumschiff)*
Betriebs-Durchlassbereich *m* <Elektronik> composite passband
betriebseigener Standard *m* <Kerntech> in-house standard
Betriebserfahrung *f* <Qual> operational experience
Betriebserlaubnis *f* 1. <Kfztech> certification *(Rechtsvorschriften)*; 2. <Lufttrans> operating permit
Betriebserwartung *f* <Elektrotech> service life
Betriebsflughöhe *f* <Raumfahrt> operating altitude
Betriebsfrequenz *f* <Elektriz> power frequency
Betriebsführung *f* <Kerntech> operative management
Betriebsgeschwindigkeit *f* <Lufttrans> operating speed
Betriebsgewicht *n* <Raumfahrt> operating weight
Betriebsgipfelhöhe *f* <Lufttrans> operating ceiling, service ceiling
Betriebshandbuch *n* 1. <Kontroll> manual; 2. <Lufttrans> operational manual; 3. <Maschinen> operations manual
Betriebshandrad *n* <Maschinen> operating hand-wheel
Betriebsinstrument *n* <Gerät> plant instrument
betriebsinterne Software *f* <Comp & DV> in-house software
Betriebskontrolle *f* <Elektrotech> process control
Betriebslast *f* 1. <Bau> rolling load; 2. <Maschinen> working load
Betriebslasten-Simulation *f* <Werkprüf> testing under service loading conditions
Betriebslebensdauer *f* <Elektrotech> service life
Betriebsmessgerät *n* <Gerät> field instrument, plant instrument
Betriebsmessinstrument *n* <Gerät> field instrument
Betriebsmittel *npl* 1. <Comp & DV> resource; 2. <Ergon> work equipment; 3. <Fertig> equiment; working materials; 4. <Telekom> ressources, system resources
Betriebsmittelzuteilung *f* <Comp & DV> resource allocation

Betriebsmittelzuweisung

Betriebsmittelzuweisung f <Comp & DV> resource allocation
Betriebsmodus m <Comp & DV, Telekom> mode of operation
Betriebspausenzeit f <Qual> nonrequired time
Betriebsprüfung f 1. <Qual> functional test, in-service test; 2. <Werkprüf> functional test
Betriebsraum m <Telekom> operations room
Betriebsrechner m <Comp & DV> scheduling computer
Betriebssaal m <Telekom> operations room *(Telefon)*
betriebssicher adj 1. <Elektrotech, Kfztech> fail-safe; 2. <Sicherheit> fail-safe, safe to operate
betriebssichere Steuerung f <Sicherheit> reliable control
Betriebssicherheit f 1. <Elektrotech> reliability; 2. <Fertig, Sicherheit> operational safety
Betriebsspannung f 1. <Elektriz> operating voltage *(eines Systems)*; 2. <Elektrotech> operating voltage; 3. <Fertig> voltage *(Kunststoffinstallationen)*; 4. <Kontroll> operating voltage
Betriebssprache f *(CL)* <Comp & DV> command language *(CL)*
Betriebssteuerung f <Elektrotech> process control
Betriebsstörung f 1. <Elektrotech> interruption; 2. <Fernseh, Funktech, Telekom> breakdown *(System)*
Betriebsstrom m <Elektriz, Elektrotech> operating current
Betriebsstunde f <Fertig> service hour
Betriebsstundenanzeiger m <Wassertrans> operating hours indicator *(Motor)*
Betriebsstundenzähler m 1. <Gerät> operating hour meter, time meter; 2. <Mechan> elapsed time counter; 3. <Sicherheit> working hours counter
Betriebssystem n 1. <Comp & DV> OS, operating system, operation system; 2. <Telekom> OS, operating system
Betriebssystemkern m <Comp & DV> kernel, operating system kernel
Betriebssystemplatte f <Comp & DV> master disk
Betriebstemperatur f <Heiz & Kälte, Kfztech, Kontroll, Lufttrans, Raumfahrt> operating temperature • **auf Betriebstemperatur aufheizen** <Ker & Glas> fire over *(Schmelzofen in Leerzustand)*
Betriebstest m <Metrol> operational test
Betriebsüberwachung f *(ISM)* <Telekom> in-service monitoring *(ISM)*
Betriebsumgebung f <Comp & DV> operating environment
Betriebs- und Datenserver m <Telekom> administration and data server
Betriebs- und Wartungszentrum n <Telekom> OMC, *(AE)* operations and maintenance center, *(BE)* operations and maintenance centre
Betriebsunterbrechung f <Telekom> outage
Betriebsverhalten n 1. <Comp & DV> performance; 2. <Maschinen> operating characteristic
Betriebsverwaltung f <Kerntech> operative management
Betriebswasser n <Wasserversorg> industrial water, process water
Betriebsweise f 1. <Comp & DV> mode, mode of operation; 2. <Maschinen> operation; 3. <Telekom> mode of operation
Betriebszeit f <Comp & DV> operating time, uptime
Betriebszentrale f <Raumfahrt> *(AE)* operation center, *(BE)* operation centre
Betriebszentrum n *(BZ)* <Telekom> *(AE)* operations center, *(BE)* operations centre
Betriebszustand m 1. <Elektrotech> operating conditions; 2. <Fertig> working order
Betriebszustandsanzeiger m <Comp & DV> mode indicator
Betriebszuverlässigkeit f <Elektrotech, Sicherheit> reliability
Betriebszyklus m <Kontroll> duty cycle
betrügerische Nutzung f <Telekom> fraudulent use
Bett n <Maschinen> bed *(einer Drehmaschine)*
Bett n **mit Einsatzbrücke** <Maschinen> gap bed
Bettdecke f <Textil> bedspread, blanket
Bettführung f <Maschinen> ways
Bettführungsbahn f <Maschinen> ways
Bettschlitten m <Maschinen> bed carriage, bed slide
Betttuchstoff m <Textil> sheeting
Bettung f 1. <Bau> ballast; bed *(Mörtel)*; 2. <Trans> ballast
bettungslose Schiene f <Eisenbahn> ballastless track
Bettungsmörtel m <Bau> bedding mortar
Bettungsmörtelschicht f <Bau> bedding course
Bettungsreinigung f <Eisenbahn> ballast screening
Bettungsrückstand m <Eisenbahn> ballast residue
Beuchen n <Textil> kier boiling
Beuchfass n <Textil> kier
beugen v <Fertig> diffract
Beugung f 1. <Elektrotech> diffraction; 2. <Ergon> flexion; 3. <Fertig, Foto, Funktech, Optik, Telekom, Wellphys> diffraction
Beugung f **von Atomstrahlen** <Kerntech, Phys, Strahlphys> atomic beam diffraction
Beugungsbild n <Metall> diffraction pattern
Beugungsgitter n <Optik, Phys, Telekom, Wellphys> diffraction grating
Beugungsspektrum n <Optik, Strahlphys> diffraction spectrum
Beugungswinkel m <Fertig> angle of deflection
Beule f 1. <Maschinen> bulge; 2. <Mechan> dent
beulen v <Mechan, Metall> buckle
beulen v/sich <Maschinen> bulge
Beulen n <Maschinen> buckling
Beullast f <Mechan> buckling load
Beurteilung f <Ergon> assessment, estimation, evaluation, rating
Beurteilungsfehler m <Qual> error
Beutel m <Verpack> bin liner, pouch, sachet
Beutel m **mit Innenausfütterung** <Verpack> lined bag
Beutel m **mit Reißverschluss** <Verpack> zip lock bag
Beutel m **mit Seitenfalten** <Verpack> square bag with gussets
Beutel mpl **von der Rolle** <Verpack> reel feed bags
Beutelabfüllanlage f <Verpack> bag-filling machine
Beutelchen n **für Maßeinheit** <Verpack> unit dose sachet
Beutelfilter n <Kohlen> bag filter
Beutelhalter m <Verpack> bag holder
Beutelhersteller m <Verpack> bagmaker
Beutelherstellungsmaschine f <Verpack> pouch-making machine
Beutelöffnungsmaschine f <Verpack> bag opener
Beutelpackmaschine f <Verpack> bag-loading machine
Beutelpapier n <Verpack> bag paper
Beutelpositioniersystem n <Verpack> bag-placing system
Beutelrollenhülse f <Verpack> bag reel
Beutelverpackung f <Verpack> bag packaging
Beutelverschließmaschine f <Verpack> bag-sealing equipment, sack-closing machine
Bevölkerungsdichte f <Bau> population density
Bevölkerungsgesamtdosis f <Umweltschmutz> population dose
Bevölkerungsteildosis f <Umweltschmutz> subpopulation collective dose

bevorrechtigte Operation f <Comp & DV> privileged operation
bevorrechtigter Befehl m <Comp & DV> privileged instruction
bevorrechtigter Bereich m <Comp & DV> privileged account
bevorzugte Annahmegrenzen fpl <Qual> preferred acceptable quality levels
bewachsen v <Wassertrans> foul *(Schiff)*
bewachsener Schiffsboden m <Wassertrans> foul bottom
bewässerbar adj <Bau> irrigable
bewässerbares Gebiet n <Bau> irrigable area
bewässern v <Chemtech> water
Bewässerung f <Wasserversorg> irrigation
Bewässerungsgraben m <Wasserversorg> catch-feeder
Bewässerungskanal m <Wasserversorg> irrigation canal
beweglich adj <Maschinen, Mobilkom, Telekom> mobile
bewegliche Arbeitsbühne f <Bau> moving platform
bewegliche Backe f <Maschinen> chop, movable jaw
bewegliche Brücke f <Wassertrans> movable bridge *(Schleusen, Binnenwasserstraßen)*
bewegliche Druckgießformhälfte f <Fertig> ejector die
bewegliche Einrichtung f <Telekom> mobile installation
bewegliche Funkstelle f <Mobilkom> mobile station
bewegliche Komponente f <Umweltschmutz> mobile component
bewegliche Kraftübertragung f <Mechan> flexible drive
bewegliche Ladung f <Elektrotech> moving charge
bewegliche Last f <Bau> moving load
bewegliche Rolle f <Maschinen> idle pulley
beweglicher Anschlag m <Maschinen> movable stop
beweglicher Backen m <Fertig> movable die *(Stauchen)*
beweglicher Flugfunksatellitendienst m <Raumfahrt> aeronautical mobile satellite service *(Weltraumfunk)*
beweglicher Kern m <Elektriz> movable core
beweglicher Kontakt m <Elektrotech> moving contact
beweglicher Landfunkdienst m <Mobilkom> land mobile radio service
beweglicher Rost m 1. <Fertig> movable grate, travelling grate; 2. <Heiz & Kälte> moving grate; 3. <Maschinen> movable grate, travelling grate
beweglicher Satellitenfunkdienst m <Raumfahrt> mobile satellite service *(Weltraumfunk)*
beweglicher Satelliten-Seefunkdienst m <Telekom, Wassertrans> maritime mobile satellite service *(Satellitenfunk)*
beweglicher Seefunkdienst m (bS) 1. <Mobilkom> marine mobile service; 2. <Telekom, Wassertrans> maritime mobile service
beweglicher Seefunk-Fernsprechdienst m **über Satellit** <Funktech, Mobilkom, Wassertrans> maritime mobile satellite telephone service
beweglicher Sitz m <Hydraul> dancing seat
beweglicher Spindelstock m <Maschinen> sliding headstock, sliding poppet
beweglicher Tisch m <Ker & Glas> moving table
bewegliches Brechwerk n <Kohlen> mobile crusher
bewegliches Gelenk n <Mechan> flexible drive
bewegliches Herzstück n <Eisenbahn> switch diamond
bewegliches Rotorblatt n <Lufttrans> movable rotor blade *(Hubschrauber)*
bewegliches Schaltstück n <Elektrotech> moving contact
bewegliches Teil n <Maschinen> moving part
bewegliches Wehr n <Wasserversorg> bar weir
Beweglichkeit f <Phys> mobility
Beweglichkeitsgrad m <Maschinen> degree of mobility
Bewegtbild n <Fernseh> full-motion picture
Bewegtbild-Videokonferenz f <Telekom> full-motion videoconferencing
Bewegung f 1. <Maschinen> motion, movement; 2. <Mechan> motion • **die Bewegung umkehren** <Maschinen> reverse the motion • **in Bewegung halten** <Foto> agitate *(Bad)* • **in Bewegung versetzen** <Maschinen> impart motion to, set in motion
Bewegungsablaufdetektor m <Trans> motion detector
Bewegungsamplitude f <Gerät> amplitude of movement *(Messgerät)*
Bewegungsbereich m 1. <Ergon> random observation method; 2. <Kerntech> limit of travel
Bewegungsdatei f <Comp & DV> movement file, transaction file
Bewegungsebene/mit versetzter <Fertig> drunken
Bewegungsenergie f 1. <Maschinen> energy of motion; 2. <Mechan> kinetic energy; 3. <Nichtfoss Energ> energy of motion; 4. <Phys> kinetic energy
Bewegungsfuge f <Bau> movement joint
Bewegungsgeschwindigkeit f <Kerntech> rate of travel *(eines Steuerstabs)*
Bewegungsgleichung f <Maschinen> motion equation
Bewegungsgröße f 1. <Maschinen> kinetic quantity; 2. <Phys> momentum
Bewegungsimpedanz f <Akustik> motional impedance
Bewegungslehre f 1. <Mechan> kinematics, kinetics; 2. <Metall, Phys> kinetics
Bewegungsmittelpunkt m <Mechan, Phys> *(AE)* center of motion, *(BE)* centre of motion
Bewegungsrichtung f <Maschinen> direction of motion
Bewegungssatz m <Comp & DV> transaction record
Bewegungsschaufel f <Lufttrans> impeller blade *(Turbomotor)*
Bewegungssteuerung f <Kontroll> motion control
Bewegungsstudie f <Ergon> motion study
Bewegungsumkehr f <Maschinen> reversing the motion
Bewegungsunschärfe f <Foto> motion blur
bewehren v <Sicherheit> cage in
Bewehren n <Bau> reinforcing
bewehrt adj 1. <Elektrotech> *(AE)* armored clad, *(BE)* armoured clad; 2. <Telekom> *(AE)* armored, *(BE)* armoured
bewehrte Länge f <Bau> embedment length
bewehrter Beton m <Bau> reinforced concrete
bewehrtes Kabel n 1. <Elektrotech> *(AE)* armored cable, *(BE)* armoured cable, sheathed cable; 2. <Maschinen, Telekom> *(AE)* armored cable, *(BE)* armoured cable
Bewehrung f 1. <Bau> armoring, reinforcement; 2. <Fertig> armoring; 3. <Elektrotech> *(AE)* armor, *(BE)* armour *(Kabel)*
Bewehrungsarbeiten fpl <Bau> steel fixing
Bewehrungsdraht m <Optik> *(AE)* armor wire, *(BE)* armour wire *(Lichtleitkabel)*
Bewehrungsmatte f <Bau> reinforcement mat, reinforcement mesh, wire mesh reinforcement
Bewehrungsnetz n <Bau> mat reinforcement
Bewehrungsstahl m <Bau> reinforcing bar
Beweis m <Künstl Int, Math> proof
beweisen v 1. <Math> prove; 2. <Qual> verify
Beweisfindungsstrategie f <Künstl Int> proof strategy
Beweistheorie f <Künstl Int> proof theory
Beweisverfahren n <Künstl Int> theorem proving
bewerten v 1. <Ergon> assess, evaluate, rate; 2. <Lebensmittel> grade; 3. <Qual> evaluate, grade, rate; 4. <Patent, Sicherheit> evaluate; 5. <Textil> assess
bewerteter Schalldruckpegel m <Akustik> weighted sound level
bewerteter Störabstand m <Telekom> weighted signal-to-noise ratio

Bewertung

Bewertung f 1. <Bau, Künstl Int> evaluation; 2. <Ergon, Patent, Qual, Sicherheit> assessment, estimation, evaluation
Bewertungsfaktor m 1. <Qual> weighting factor; 2. <Sicherheit> quality factor *(Strahlenschutz)*
Bewertungsgröße f <Ergon, Qual> quantity of assessment
Bewertungsprogramm n <Comp & DV> benchmark program
bewettern v <Kohlen> aerate
Bewetterung f <Kohlen> aeration
Bewicklung f <Phys> winding
Bewicklungsbreite f <Textil> dressed width of warp
Bewilligung f 1. <Erdöl> *(BE)* production licence; 2. <Trans> approval
Bewitterung f <Kunststoff, Textil> weathering
Bewitterungsbeanspruchung f <Kunststoff> exposure to weather
Bewitterungshaut f <Kunststoff> alligatoring, crazing
bewuchsverhindernde Farbe f <Wassertrans> antifouling paint *(Schiffinstandhaltung)*
bezeichnen v 1. <Comp & DV> identify, label; 2. <Telekom> identify
Bezeichner m <Comp & DV, Telekom> identifier
bezeichnete Frequenz f <Telekom> designated frequency
Bezeichnung f 1. <Akustik> designation; 2. <Comp & DV> label; 3. <Patent> designation *(des Gegenstandes der Erfindung)*
Bezeichnung f der Erfindung <Patent> title of the invention
beziehen v **auf/sich** <Patent> reduce to, refer to *(Patentanspruch)*
Beziehung f <Math> relation
Beziehungsgraph m <Künstl Int> relation graph
Beziehungssymbol n <Künstl Int> relation symbol, relational operator
Bézier-Fläche f <Geom> Bézier surface
Bezug m 1. <Druck> blanket; 2. <Metrol> reference
Bezugsadresse f <Comp & DV> base address, reference address
Bezugsatmosphäre f <Raumfahrt> reference atmosphere
Bezugsband n 1. <Akustik> reference tape; 2. <Aufnahme> standard play tape
Bezugsbedingungen fpl <Metrol> reference conditions
Bezugsbemaßung f <Konstzeich> reference dimensioning
Bezugsebene f <Bau> datum plane
Bezugselektrode f <Labor> reference electrode *(Elektrochemie)*
Bezugsempfindlichkeit f <Telekom> reference sensibility
Bezugsenergie f <Akustik> reference energy
Bezugsfläche f <Fertig> reference surface
Bezugsfrequenz f <Telekom> reference frequency
Bezugsgerät n <Qual> known-good device
Bezugsgeräusch n <Telekom> reference noise
Bezugsgeräuschquelle f <Aufnahme> reference noise source
Bezugsgesamtheit f <Qual> standard population
Bezugsgröße f <Comp & DV> datum
Bezugshaken m <Konstzeich> tick
Bezugshöhe f <Bau> datum, datum level
Bezugshorizont m <Mechan> datum line
Bezugskante f 1. <Fernseh> reference edge; 2. <Fertig> datum edge
Bezugsklotz m <Fertig> datum block
Bezugskopplung f <Elektrotech> reference coupling

Bezugskreis m <Maschinen> reference circle
Bezugslandeanfluggeschwindigkeit f <Lufttrans> reference landing approach speed *(ein Triebwerk außer Kraft)*
Bezugslautstärke f <Aufnahme> reference volume
Bezugslinie f 1. <Bau> datum line; 2. <Geom> reference line; 3. <Konstzeich> witness line; 4. <Maschinen> datum line, leader line, reference line
Bezugsmarkensensor m <Kfztech> reference sensor
Bezugsmaß n 1. <Konstzeich> datum dimension; 2. <Maschinen> reference dimension
Bezugsmaßsystem n <Gerät> absolute measure system *(bei numerischer Steuerung)*
Bezugsmenge f <Qual> basic size
Bezugsmessmethode f <Telekom> reference test method
Bezugsnormal n 1. <Metrol> reference standards; 2. <Qual> reference standard
Bezugspegel m 1. <Akustik> reference volume; 2. <Telekom> reference level
Bezugsphase f <Fernseh> reference phase
Bezugspotenzial n <Regelung> signal common
Bezugsprofil n <Maschinen> reference profile
Bezugspunkt m 1. <Akustik> reference point; 2. <Bau> datum point; reference mark *(Vermessung)*; 3. <Comp & DV> datum, reference mark, reference; 4. <Geom, Maschinen, Telekom> reference point
Bezugspunkt m für die Messung des Anfluglärms <Lufttrans> approach reference noise measurement point
Bezugspunktdaten npl <Qual> benchmark data
Bezugsrauschwert m <Elektronik> reference noise
Bezugsreaktor m <Kerntech> reference reactor
Bezugsreibungsbedingungen fpl <Lufttrans> reference friction condition *(Start- und Landebahn)*
Bezugsschall m <Akustik> reference sound
Bezugsschalldruck m <Akustik, Umweltschmutz> reference sound pressure
Bezugsschalleistung f <Akustik> reference sound power
Bezugsschallgeschwindigkeit f <Akustik> reference sound velocity
Bezugsschallintensität f <Akustik> reference sound intensity
Bezugsschallpegel m <Fernseh> reference audio level
Bezugsschwarz n <Fernseh> reference black
Bezugssignal n <Elektronik, Telekom> reference signal
Bezugssignaleingang m <Elektronik> reference signal input
Bezugssignalphase f <Elektronik> reference signal phase
Bezugsspannung f <Elektrotech> reference voltage
Bezugsstation f <Raumfahrt> reference station *(Weltraumfunk)*
Bezugssteigung f <Lufttrans> standard pitch *(Propeller)*
Bezugsstück n 1. <Fertig> master; 2. <Maschinen> reference piece
Bezugssystem n 1. <Ergon> benchmark system; 2. <Maschinen> reference system; 3. <Phys> frame of reference, standard
Bezugstabelle f <Comp & DV> reference table
Bezugstaktgeber m <Telekom> reference clock
Bezugstemperatur f 1. <Heiz & Kälte> reference temperature, standard temperature; 2. <Metrol> reference temperature
Bezugston m <Akustik> reference tone
Bezugstonpegel m <Aufnahme> reference audio level
Bezugsweiß n <Fernseh> reference white, white reference
Bezugswerkstück n <Maschinen> master piece
Bezugszahnstange f <Maschinen> basic rack

Bezugszeichen n <Patent> reference sign
Bezugszeit f <Comp & DV> reference time
Bezugszylinder m <Fertig> reference cylinder
BfA m (Betriebsbeauftragter für Abfall) <Abfall> Waste Management Officer (Berufsbezeichnung)
BFO (Schwebungsfrequenzoszillator) <Elektronik, Funktech, Phys> BFO (beat frequency oscillator)
Bg (geometrisches Buckling) <Kerntech> Bg (geometric buckling)
BHA (Butylhydroxyanisol) <Lebensmittel> BHA (butylated hydroxyanisole)
B/H-Schleife f <Elektriz> B/H loop
BHT (Butylhydroxytoluol) <Lebensmittel> BHT (butylated hydroxytoluene)
Bi (Bismut) <Chemie> Bi (bismuth)
biaxial adj <Geom, Metall> biaxial
biaxial orientierte Folie f <Kunststoff> biaxially oriented film
biaxiale Belastung f <Metall> biaxial loading
biaxiale Orientierung f <Kunststoff> biaxial orientation
Bibeldruckpapier n <Druck, Papier> Bible paper, thin paper
Biberschwanz m <Bau> plain tile
Bibliothek f <Comp & DV> library
Bibliotheksautomation f <Comp & DV> library automation
Bibliotheksfunktion f <Comp & DV> library function
Bibliotheksmusik f <Aufnahme> library music
Bibliotheksprogramm n <Comp & DV> library program
Bibliotheksroutine f <Comp & DV> library routine
Bibliotheksverwaltungsprogramm n <Comp & DV> librarian program
Bi-CMOS-Transistor m <Comp & DV> BiCMOS transistor
Bicyclo-Decan n <Chemie> biclodecane, decahydronaphthalene
bidirektional adj <Comp & DV, Telekom> bidirectional
bidirektional prädiktiv codiertes Bild n <Fernseh> bidirectionally predictively coded frame
bidirektionale Suche f <Künstl Int> bidirectional search
bidirektionale Triggerdiode f (Diac) <Elektronik, Funktech> diode alternating-current switch, diac (Zweiweg-Schaltdiode)
bidirektionale Übertragung f <Telekom> bidirectional transmission
bidirektionale Verbindung f <Telekom> bidirectional connection
bidirektionaler Datenfluss m <Comp & DV> bidirectional flow
bidirektionaler Koppler m <Elektrotech> bidirectional coupler
bidirektionaler Schalter m <Elektrotech> bidirectional switch
bidirektionaler Transducer m <Elektrotech> bidirectional transducer
bidirektionaler Übertragungsweg m <Comp & DV> bidirectional bus
bidirektionaler Verkehr m <Telekom> bidirectional traffic
bidirektionaler Wandler m <Elektrotech> bidirectional transducer
bidirektionales Mikrofon n <Akustik> bidirectional microphone
bidirektionales Netzwerk n <Elektrotech> bidirectional network
Biege... <Kunststoff, Metall, Wassertrans> bending, flexural
Biegeapparat m <Bau> power bender
Biegebeanspruchung f 1. <Kunststoff, Metall, Wassertrans> bending stress; 2. <Qual, Werkprüf> flexural stress

Biegebelastung f <Mechan> bending stress
Biegedorn m <Fertig> internal mandrel
Biegeeigenschaft f <Maschinen> bending property
Biege-Elastizitätsmodul m (Biege-E-Modul) <Kunststoff> flexural modulus of elasticity
Biege-E-Modul m (Biege-Elastizitätsmodul) <Kunststoff> flexural modulus of elasticity
Biegefeder f <Maschinen> flexion spring, spring subjected to bending
Biegefestigkeit f 1. <Kerntech, Kunststoff> flexural strength; 2. <Maschinen> resistance to bending, ultimate bending strength; 3. <Mechan> fatigue strength; 4. <Papier> bending strength; 5. <Phys> flexural rigidity; 6. <Qual> resistance to bending; 7. <Telekom> flexibility strength; 8. <Werkprüf> flexural strength
Biegeform f 1. <Ker & Glas> (AE) bending mold, (BE) bending mould; 2. <Maschinen> former
Biegekante f <Fertig> forming edge
Biegemaschine f 1. <Maschinen> bending machine, bending press, machine for bending; 2. <Mechan> bending machine
Biegemodul n <Werkprüf> flexural modulus
Biegemoment n 1. <Bau> flexural moment; 2. <Maschinen, Mechan, Wassertrans> bending moment (Schiffkonstruktion)
biegen v 1. <Bau> bend, camber, inflect; 2. <Maschinen> fold; 3. <Metall, Papier> bend
Biegen n 1. <Bau> bending; 2. <Ker & Glas> bend; bending (Flachglas); bending (von Glasrohren); twisting (von Rohren); 3. <Metall> bending
Biegepresse f <Fertig> bending press
Biegeprisma n <Fertig> vee die
Biegeprobe f <Metall> bend test piece
Biegeprüfgerät n <Kunststoff> bending tester
Biegeprüfmaschine f <Gerät> bending tester
Biegeprüfstab m <Metall> bend test piece
Biegeradius m 1. <Maschinen> bend radius; 2. <Mechan> bending radius
Biegerichten n <Maschinen> straightening by bending
Biegeriss m <Fertig, Metall> flexural crack
Biegeschablone f <Fertig> (AE) form
Biegeschenkel m <Fertig> flexible section (Kunststoffinstallationen)
Biegeschwellspannung f <Fertig> fluctuating bending, fluctuating bending stress
Biegeschwinger m <Elektronik> flexure-mode resonator
Biegeschwingung f <Maschinen, Qual> flexure
Biegespannung f <Kunststoff, Maschinen, Wassertrans> bending stress
Biegesteiferanschluss m <Bau> rigid joint
Biegesteifigkeit f 1. <Maschinen> bending strength; 2. <Mechan, Qual> flexural strength
Biegesteifigkeitsprüfer m <Papier> bending stiffness tester
Biegestempel m <Maschinen> bending die
Biegeverlust m <Fertig> bend loss, bending loss (Faseroptik)
Biegeverschluss m <Ker & Glas> bent finish
Biegeversuch m 1. <Bau> flexural test; 2. <Fertig> beam test; 3. <Maschinen> bend test, bending test; 4. <Metall, Qual> bend test; 5. <Werkprüf> bending test
Biegewalze f <Maschinen> bending roll, bending rollers
Biegewalzmaschine f <Maschinen> bending rollers
Biegewerkzeug n <Maschinen> bending tool
Biegewiderstand m <Bau> flexural resistance
Biegewinkel m <Fertig, Maschinen> angle of bend
Biegezange f <Fertig> bending pliers, claw
Biegezugfestigkeit f <Anstrich> tensile bond strength
biegsam adj <Bau, Elektrotech, Maschinen> flexible

biegsame

biegsame Antriebswelle *f* <Maschinen> flexible drive shaft
biegsame Leitung *f* <Elektrotech> flexible wire
biegsame Metallrohrleitung *f* <Maschinen> flexible metal conduit
biegsame Verbindung *f* <Maschinen> flexible joint
biegsame Welle *f* <Maschinen> flexible shaft
biegsamer Hohlleiter *m* <Elektrotech, Funktech> flexible waveguide
biegsamer Leiter *m* <Elektrotech> flexible conductor
biegsamer Schlauch *m* <Maschinen> hose pipe
biegsames Kabel *n* <Elektrotech> flexible cable
biegsames Rohr *n* <Bau> flexible hose
Biegsamkeit *f* <Kunststoff, Textil, Wassertrans> flexibility *(Material)*
Biegung *f* 1. <Eisenbahn> flexion; 2. <Elektrotech> bend *(Glasfaseroptik)*; bend *(Wellenleiter)*; 3. <Kerntech> deflection *(eines Brennelementes)*; 4. <Maschinen, Mechan> bend; 5. <Metall> deflection; 6. <Wassertrans> bend
Biegungsverlust *m* <Elektrotech> bending loss
Bienenkorbkühler *m* <Kfztech> honeycomb radiator
Bi-Ergol-Technologie *f* <Raumfahrt> bi-ergol technology
Biertreber *mpl* <Lebensmittel> brewer's grain
Bierwürze *f* <Lebensmittel> wort *(Brauerei)*
Biesennaht *f* <Textil> cording seam
bifilar *adj* <Elektriz, Funktech, Phys> bifilar
bifilar gewickelter Transformator *m* <Elektrotech> double-wound transformer
Bifilardrahtwicklung *f* <Elektriz> bifilar winding
Bifilarwicklung *f* <Elektriz> bifilar winding
Bifokalglas *n* <Ker & Glas> bifocal lens
Bifokuslinse *f* <Optik> bifocal lens *(mit runder Einfügung)*
Bi-Fuel-Fahrzeug *n* <Kfztech, Kohlen, Nichtfoss Energ> bi-fuel vehicle
bifunktionelle Verbindung *f* <Chemie> bifunctional compound
BIGFET *(Feldeffekttransistor mit bipolarisoliertem Gatter)* <Elektronik> BIGFET *(bipolar-insulated gate field-effect transistor)*
Biguanid *n* <Chemie> biguanide
bijektiv *adj* <Geom, Math> bijective *(injektiv und surjektiv)*
bikonkav *adj* <Optik> biconcave
Bikonkavlinse *f* <Optik, Phys> biconcave lens
bikonvex *adj* <Optik> biconvex, convexo-convex
Bikonvexlinse *f* <Optik, Phys> biconvex lens
bikubisch *adj* <Geom> bicubic
bilaterale Symmetrie *f* <Geom> bilateral symmetry
bilateraler Transducer *m* <Elektrotech> bilateral transducer
bilateraler Wandler *m* <Elektrotech> bilateral transducer
Bild *n* 1. <Comp & DV> picture; 2. <Druck> face, figure, image; 3. <Ergon> image; 4. <Fernseh> frame, image, picture; 5. <Fertig> pattern; 6. <Foto> image, picture; 7. <Funktech, Ker & Glas, Phys> image
Bild *n* **aus ASCII-Zeichen** <Telekom> emoticon, graphic character icon
Bildablenkung *f* <Fernseh> vertical sweep
Bildabtaster *m* 1. <Comp & DV> scanner; 2. <Fernseh> *(BE)* image analyser, *(AE)* image analyzer, image scanner; 3. <Telekom> *(BE)* image analyser, *(AE)* image analyzer, image scanner
Bildabtastgerät *n* <Kontroll> scanner
Bildabtastung *f* 1. <Fernseh> image analysis; 2. <Telekom> image analysis, scanning
Bildanalyse *f* <Künstl Int> image analysis
Bildanpassung *f* <Fernseh> picture match
Bildauflösung *f* <Comp & DV, Telekom> image resolution, picture definition, resolution

Bildaufnahmemodul *m* <Kontroll> vision input module
Bildaufnahmeröhre *f* 1. <Elektronik> pick-up tube, vidicon tube; 2. <Fernseh> camera tube
Bildaufnehmer *m* 1. <Fernseh> imager; 2. <Telekom> image sensor
Bild-Austast-Synchronsignal *n* <Fernseh> composite signal, composite video signal
Bildcodierung *f* <Fernseh, Telekom> coding of images, picture coding
Bilddarstellung *f* <Math> pictogram
Bilddatei *f* <Comp & DV, Elektronik> image file
Bilddatenkompression *f* <Fernseh> image data compression
Bilddigitalisierer *m* <Elektronik> image digitizer
Bilddigitalisierung *f* <Elektronik, Fernseh, Telekom> image digitization
Bildebene *f* <Akustik> focal plane, image plane
Bildelement *n* <Comp & DV, Fernseh> pixel, picture element
Bilden *n* **einer Fahrgemeinschaft** <Trans> car pooling
Bilderfassung *f* <Telekom> image grab *(Digitalverarbeitung)*
Bilderglas *n* <Ker & Glas> picture glass
Bilderkennen *n* <Künstl Int> image recognition
Bilderzeugung *f* <Fernseh> image formation
Bildfernsprechen *n* <Telekom> video telephony
Bildfernsprecher *m* <Telekom> videophone
Bildfläche *f* <Fernseh> scanning field
Bildflackern *n* <Fernseh> image flicker, picture flutter
Bildflattern *n* <Fernseh> fluttering video level
Bildfolge *f* <Fernseh> image sequence
Bildformat *n* <Phys> aspect ratio
Bildfreiraum *m* <Druck> placeholder
Bildfrequenz *f* 1. <Elektronik> image frequency *(Faksimile)*; 2. <Fernseh> frame frequency, image frequency, vertical frequency
Bildgröße *f* <Foto> picture size
Bild-im-Bild *n* <Fernseh> picture in picture
Bildinversion *f* <Fernseh> inversion of image
Bildkanal *m* <Fernseh> video channel, video-frequency channel
Bildkommunikation *f* <Telekom> image communication, videophone, visual communication
Bildkompression *f* 1. <Comp & DV> image compression, image data compression; 2. <Fernseh> picture compression; 3. <Telekom> image compression, image data compression
Bildkonservierung *f* <Fernseh> image retention
Bildkontur *f* <Fernseh> image edge
Bildlaufregler *m* <Fernseh> hold control
Bildlegende *f* <Druck> caption
Bildleuchtdichte *f* <Elektronik, Strahlphys> luminance
bildlich anzeigen *v* <Telekom> display
Bildmaske *f* 1. <Fernseh> framing mask; 2. <Foto> mask
Bildmaßstab *m* <Foto> scale of image
Bildmischer *m* <Fernseh> vision mixer
Bildmonitor *m* <Fernseh> picture monitor
Bildmuster *n* <Fernseh> pattern
Bildmustererkennung *f* <Elektronik> pattern recognition
Bildmustererzeugung *f* <Fernseh> pattern generation
Bildmustergenerator *m* <Fernseh> pattern generator
Bildnachlauf *m* <Fernseh> image lag
Bildpegelanzeige *f* <Fernseh> video level indicator
Bildphasenwinkel-Koeffizient *m* <Elektronik> image phase-change coefficient
Bildplatte *f* 1. <Aufnahme, Comp & DV> optical disk; 2. <Fernseh> *(BE)* videodisc, *(AE)* videodisk
Bild-Preemphase *f* <Fernseh> video pre-emphasis
Bildprojektion *f* <Elektronik> image projection *(Lithographie)*

Bildpunkt m <Comp & DV, Fernseh> pixel, picture element
Bildraster m <Elektronik> raster
Bildrasterabtastung f <Elektronik> raster scanning
Bildrasterscannen n <Elektronik> raster scanning
Bildrasterwandler m <Fernseh> scan converter
Bildreaktor m <Kerntech> image reactor
Bildregieraum m <Fernseh> video control room, vision control room
Bildretusche f <Druck> image retouching
Bildröhre f 1. <Elektronik> metal-cone tube, picture tube; 2. <Fernseh> picture tube, kinescope
bildsam adj <Metall, Qual> ductile
Bildscanner m <Fernseh> image scanner
Bildschärfe f 1. <Foto> definition; 2. <Optik> sharpness
Bildschicht <Telekom> picture layer
Bildschirm m 1. <Comp & DV> display screen, display, monitor, screen; 2. <Elektronik> display screen, screen; 3. <Fernseh> monitor, scope, screen; 4. <Funktech> screen; 5. <Telekom> display screen
Bildschirm m **mit starker Grundhelligkeit** <Elektronik> high-brightness screen
Bildschirmadapter m <Comp & DV> display adaptor, monitor adaptor
Bildschirmanschluss m <Comp & DV> display adaptor
Bildschirmanzeige f 1. <Comp & DV> panel; 2. <Telekom> display, visual display unit
Bildschirmanzeigeformat n <Comp & DV> screen format
Bildschirmauflösung f <Comp & DV> display resolution
Bildschirmausdruck m <Comp & DV> screen dump
Bildschirmausgabe f <Comp & DV> soft copy
Bildschirmausgabegerät n <Gerät> output display terminal
Bildschirmauszug m <Comp & DV> screen dump
Bildschirmblättern n <Comp & DV> scrolling
Bildschirmcontroller m <Comp & DV> display controller
Bildschirmdarstellung f <Comp & DV> soft copy
Bildschirmeditor m <Comp & DV> screen editor
Bildschirmeinstellung f <Comp & DV> display setting
Bildschirm-Endgerät n <Telekom> screen terminal
Bildschirmgerät n 1. <Comp & DV> VDU, visual display unit; 2. <Kontroll, Phys> VDU; 3. <Telekom> VDU, video display unit, video terminal
Bildschirmgestaltung f <Ergon> display design
Bildschirmgröße f <Comp & DV> display size
Bildschirmhelligkeit f <Wassertrans> display brilliance (Radar)
Bildschirminhalt m <Comp & DV> screen content
• **Bildschirminhalt löschen** <Comp & DV> clear screen
Bildschirmkonsole f <Comp & DV> display console
Bildschirmmenü n <Comp & DV> display menu
Bildschirmprozessor m <Comp & DV> display processor
Bildschirmregler m <Elektronik> CRT controller
Bildschirmschoner m <Comp & DV> screen saver
Bildschirmspeicher m <Comp & DV> screen memory
Bildschirmsteuereinheit f <Comp & DV> display controller
Bildschirmtext m (Btx) 1. <Fernseh> teletex, Videotex®; 2. <Telekom> Teletext®, interactive videotex
Bildschirmtreiber m <Comp & DV> display driver
Bildschirmtyp m <Comp & DV> monitor type
Bildschirmübersichtsseite f <Fernseh> survey
Bildschwankungen fpl <Fernseh> jitter
Bildseitenverhältniseinstellung f <Fernseh> aspect-ratio adjustment
bildseitiger Brennpunkt m <Foto> rear focus
Bildsequenz f <Telekom> image sequence

Bildsignal n 1. <Elektronik> image signal, picture signal; 2. <Fernseh, Phys> video signal; 3. <Telekom> picture signal
Bildsignal n **mit Austastung** <Fernseh> video signal with blanking
Bildsignalcodierung f <Fernseh> composite signal coding
Bildsignalimpuls m <Fernseh> video signal pulse
Bildspeicher m 1. <Elektronik> image storage; 2. <Fernseh> frame store
Bildspeicherröhre f <Elektronik> image storage tube, storage tube
Bildspeicherung f 1. <Elektrotech> image storage; 2. <Fernseh> picture storage
Bildspur f <Elektronik> trace
Bildspur-Integration f <Elektronik> trace integration
Bildspur-Intensivierung f <Elektronik> trace intensification
Bildsteuerung f <Fernseh> framing control
Bildstörung f 1. <Elektronik> long-line effect; 2. <Fernseh> image interference, picture defect
Bildstrich m <Fernseh> frame line
Bildstricheinstellung f <Fernseh> framing
Bildsynchronimpuls m <Fernseh> frame sync pulse
Bildsynchronisierung f <Fernseh> frame synchronization control
Bildsynthese f 1. <Elektronik> imaging; 2. <Künstl Int> image synthesis
Bildsynthese-Anordnung f <Elektronik> imaging array
Bildsynthese-Chip m <Elektronik> imaging chip
Bildtelefon n <Telekom> picture phone, videophone
Bildtelefonvermittlungssystem n <Telekom> videophone switching system
Bildtext m <Druck> caption
Bildtiefe f 1. <Fernseh> depth of field; 2. <Telekom> pixel word length (Bildcodierung)
Bildträger m 1. <Elektronik> picture carrier; 2. <Fernseh> image carrier, video carrier, vision carrier
Bildtransfer m <Telekom> image transfer
Bildübermittlung f <Telekom> video messaging (ISDN)
Bildüberschrift f <Verpack> caption
Bildübertragung f 1. <Comp & DV> picture transmission; 2. <Telekom> image transmission, phototransmission, picture transmission
Bildübertragung f **über das Internet** <Telekom> phototransmission via Internet
Bildumlauf m <Comp & DV> wraparound
Bildumsetzung f <Elektronik> image conversion
Bildung f <Akustik, Mechan, Phys, Thermod> formation
Bildung f **der Lautheit** <Akustik> formation of loudness
Bildung f **von Oberflächenerhebungen** <Papier> blistering (Fehler im Papier)
Bildungsenthalpie f <Mechan, Phys, Thermod> enthalpy of formation
Bildungswärme f <Thermod> heat of formation
Bildunterschrift f <Druck, Verpack> caption
Bildverarbeitung f 1. <Comp & DV> picture processing; 2. <Elektronik, Künstl Int, Telekom> image processing
Bildverbesserung f <Fernseh> image enhancement
Bildverdichtung f <Elektronik> image compression
Bildverriegelung f <Fernseh> pixlock
Bildverschiebung f <Fernseh> roll-over
Bildversetzung f <Metall> image dislocation
Bildverständnis n (BV) <Künstl Int> image comprehension, image understanding
Bildverstärker m 1. <Elektronik> image intensifier; image intensifier (Röntgenstrahlen); 2. <Fernseh> image intensifier, video amplifier
Bildverstärkerröhre f <Elektronik> image intensifier tube

Bildverstärkung

Bildverstärkung f <Elektronik> image enhancement
Bildverstehen n (BV) <Künstl Int> image comprehension, image understanding
Bildverteilung f <Fernseh> split image
Bildvorlage f <Konstzeich> master illustration
Bildwandler m 1. <Comp & DV> imager; 2. <Elektronik, Fernseh> image converter
Bildwandlerröhre f <Elektronik, Fernseh> image converter tube
Bildwandlung f <Telekom> image conversion
Bildwechselfrequenz f <Fernseh> field frequency, frame rate, vertical frequency
Bildwechselfrequenzabweichung f <Fernseh> frame slip
Bildwechselimpuls m <Fernseh> frame pulse
Bildwechselschalter m <Fernseh> field shift switch
Bildwechselsynchronisierung f <Fernseh> field sync alignment
Bildwiederholfrequenz f <Comp & DV> refresh rate
Bildwiederholmodus m <Elektronik> refresh mode
Bildwiederholsignal n <Elektronik> refresh signal
Bildwiederholung f <Fernseh> refresh
Bildzähler m <Foto> frame counter
Bildzeichen n 1. <Konstzeich> pictorial symbol; 2. <Patent> figurative mark
Bildzeile f 1. <Comp & DV> display line; 2. <Fernseh> scanning line
Bildzerfall m <Fernseh> picture breakup
Bildzerlegung f <Telekom> scanning
Bild-zu-Bild-Codierung f <Telekom> interframe coding
Bildzwischenfrequenzfilter n <Elektronik> picture carrier filter
Bilge f <Wassertrans> bilge (Schiffbau)
Bilgegebläse n <Wassertrans> bilge blower
Bilgenpumpe f <Meerschmutz> gulley sucker
Bilgewasser n <Wassertrans> bilge water
Bilirubin n <Chemie> bilirubin
Billet'sche Halblinse f <Phys> Billet's split lens
Billigflagge f <Wassertrans> flag of convenience (Seehandel)
billigstes Angebot n <Bau> lowest bid
Billion f <Math> (AE) trillion, (BE) billion (eine Million Millionen)
Bimetall n <Fertig> bimetal
Bimetall... <Metall> bimetallic
Bimetalldraht m <Elektrotech> bimetallic wire
Bimetallinstrument n <Gerät> bimetallic instrument
bimetallisch adj <Metall> bimetallic
Bimetallkolben m <Kfztech> bimetal piston
Bimetallkontakt m <Elektrotech> bimetallic contact
Bimetallschalter m <Elektriz> bimetallic switch
Bimetallstreifen m 1. <Elektriz> bimetallic strip, bimetallic switch; 2. <Metall, Phys, Thermod> bimetallic strip
Bimetallthermometer n 1. <Gerät> bimetallic thermometer; 2. <Thermod> bimetallic strip thermometer
bimodale Verteilung f <Qual> bimodal distribution
bimodale Wahrscheinlichkeitsverteilung f <Qual> bimodal probability distribution
bimodaler Bus m <Kfztech> dual-mode bus
bimolekular adj <Chemie> bimolecular
Bimsstein m <Fertig, Ker & Glas> pumice
Binaphthalin n <Chemie> binaphthyl
binär adj <Chemie, Comp & DV, Elektriz, Math, Metall> binary
Binäraddierer m <Elektronik> binary adder
Binärarithmetik f <Comp & DV, Elektronik> binary arithmetic
Binärcode m <Comp & DV> binary code
binärcodierte Dezimalzahl f (BCD) <Comp & DV, Telekom> binary-coded decimal (BCD)

binärcodierte Drehzahl f (BCD) <Comp & DV> binary-coded decimal (BCD)
binärcodiertes Signal n <Elektronik> binary-coded signal
Binärcodierung f <Telekom> binary coding
Binärdividierer m <Elektronik> binary divider
binäre Addition f <Elektronik> binary addition
binäre Division f <Elektronik> binary division
binäre Legierung f <Elektriz> binary
binäre Multiplikation f <Elektronik> binary multiplication
binäre Phasenumtastung f (BPSK) <Telekom> binary phase shift keying (BPSK)
binäre Schaltkette f <Regelung> binary switching chain
binäre Schaltung f 1. <Comp & DV> binary circuit; 2. <Elektronik> divider
binäre Subtraktion f <Elektronik> binary subtraction
binäre Suche f <Comp & DV> binary search
binäre Verzögerungsleitung f <Fernseh> binary delay line
binäre Wasserstoffverbindung f <Chemie> hydride
binärer Abschaltkreis m <Regelung> binary de-energizing circuit
binärer Baum m <Comp & DV> binary tree, heap, tree
binärer Speicherauszug m <Comp & DV> binary dump
binärer Suchbaum m <Künstl Int> binary search tree
binäres Ladeprogramm n <Comp & DV> binary loader
binäres Signal n <Comp & DV, Elektronik> binary signal
binäres Sortieren n <Comp & DV> binary sort
binäres Suchen n <Comp & DV> dichotomizing search
binäres Verknüpfungsglied n <Regelung> binary combinational element
Binärexponent m <Comp & DV> binary exponent
Binärfolge f <Telekom> binary sequence
Binärlogik f <Comp & DV> binary logic
Binärmuster n <Comp & DV> bit pattern
Binäroperation f <Comp & DV> binary operation
Binärpunkt m <Comp & DV> binary point
Binärschreibweise f <Comp & DV> binary notation, binary representation
Binärsequenz f <Comp & DV> binary sequence
Binärsignal n <Comp & DV, Elektronik> binary signal
Binärspalte f <Comp & DV> binary column
Binärspeicherelement n <Telekom> two-state register
binärsynchrone Übertragung f (BSC-Übertragung) <Comp & DV> binary synchronous communication (BSC)
Binärsystem n <Comp & DV> binary system
Binäruntersetzer m <Elektronik> binary scaler
Binärzahl f 1. <Comp & DV> binary number; 2. <Math> binary, binary number
Binärzähler m 1. <Comp & DV, Elektronik> binary counter; 2. <Gerät> binary counter, dual counter
Binärzeichen n 1. <Comp & DV> binary character, bit; 2. <Math> binary digit
Binärzeichenfolge f <Telekom> binary sequence
Binärziffer f <Comp & DV> binary digit
binaural adj <Ergon> binaural
Binde... 1. <Bau> cementing; 2. <Comp & DV> link, linking; 3. <Maschinen> link; 4. <Metall> binding
Bindedraht m <Elektrotech> tie wire
Bindeeisen n <Ker & Glas> stowing tool
bindefest adj <Anstrich> well-bonded
Bindeglied n <Maschinen> link
Bindelader m <Comp & DV> link loader, linking loader
Bindemaschine f <Verpack> bundle-tying machine
Bindemittel n 1. <Abfall> binder, binding agent; 2. <Aufnahme> binder; 3. <Bau> binder, cement, cementing material; 4. <Chemie> agglutinant; 5. <Fertig> agglomerant, agglutinant; 6. <Ker & Glas, Kunststoff> binder; 7. <Lebensmittel, Meerschmutz> binding agent; 8. <Metall> binding agent, cement; 9. <Papier> fixing agent

Bindemittelemulsion f <Chemtech> emulsion binder
binden v 1. <Bau> bind *(Zement)*; 2. <Druck> bind *(Bücher)*
Binden n 1. <Druck> bookbinding; 2. <Kohlen> binding; 3. <Künstl Int> binding *(einer Variablen)*
Binder m 1. <Anstrich> binder; 2. <Bau> binder, girder; 3. <Comp & DV> link editor, linker; 4. <Fertig> matrix; 5. <Meerschmutz> binding agent
Binderbalken m <Bau> binding beam, main girder
Binderei f <Druck> bindery
Binderfarbe f <Bau> water-based paint
Binderlage f <Bau> base course, header course
binderlose Dachkonstruktion f <Bau> untrussed roof
Binderprogramm n <Comp & DV> linkage editor
Binderschicht f <Bau> binder
Bindersparren m <Bau> common rafter, principal rafter
Binderstein m <Bau> header; binding stone *(Mauerwerk)*; through stone *(in Wandstärke)*
Binderziegel m <Bau> bonder, bondstone
Bindesäule f <Bau> clustered column
Bindeschicht f 1. <Anstrich> tie coat; 2. <Bau> tack coat *(Straßenbau)*
Bindestrich m <Druck> hyphen
Bindeton m <Bau> pipe clay
Bindewort n <Comp & DV> connective
Bindfaden m <Textil> cord, twine
bindig adj <Bau> cohesive
bindiger Boden m <Kohlen> cohesive soil
bindiges Erdmaterial n <Bau, Kohlen> binder soil
Bindigkeit f 1. <Bau> cohesion; 2. <Kohlen> cohesion, cohesion strength
Bindung f 1. <Erdöl> bond *(Petrochemie)*; 2. <Ker & Glas> bond; 3. <Kunststoff> linkage; 4. <Künstl Int> binding *(einer Variablen)*; 5. <Maschinen> bond
Bindungsenergie f (B) <Kerntech, Strahlphys, Teilphys> binding energy (B)
Bindungsenergiekurve f <Kerntech> binding energy curve
Bindungsmuster n <Textil> weave
Bindungswärme f <Nichtfoss Energ, Thermod> heat of absorption
Bindungswertigkeit f <Kerntech, Metall> covalence
Binnen... <Wassertrans, Wasserversorg> home, inland, river
Binnengewässer npl <Wasserversorg> inland waters
Binnenhafen m <Wassertrans> inner port
Binnenquelle f <Umweltschmutz> internal source
Binnenschiff n <Wassertrans> canal boat *(Schifftyp)*; inland vessel, inland waterway craft
Binnenschifffahrt f <Wassertrans> inland navigation, inland water transport
Binnenschifffahrtsstraße f <Wassertrans> inland waterway
Binnenstrecke f <Wassertrans> inland haulage
Binnenvorsteven m <Wassertrans> apron *(Schiffbau)*
Binnenwasserstraße f <Wassertrans> inland waterway
binokular adj <Ergon, Optik, Phys> binocular
binokulare Konkurrenz f <Ergon> binocular rivalry
binokulare Rivalität f <Ergon> binocular rivalry
Binokular-Mikroskop n <Labor> binocular microscope
Binom n <Math> binomial
Binomial... <Phys, Qual> binomial
Binomialkoeffizient m <Math> binomial coefficient
Binomialverteilung f <Phys> binomial distribution
Binominal... <Qual> binomial
binominale Grundgesamtheit f <Qual> binominal population
Binominalverteilung f <Math, Qual> binominal distribution

Binominalwahrscheinlichkeit f <Qual> binominal probability
binomischer Lehrsatz m <Math> binomial theorem
Bioabbaubarkeit f <Verpack> biodegradation
Bioabfall m <Umweltschmutz> biological waste
biochemischer Indikator m <Umweltschmutz> biochemical tracer
Biofilter n 1. <Chemtech> biological filter; 2. <Umweltschmutz> biofilter
Biogas n 1. <Abfall> biogas, digester gas, fermentation gas, manure gas; 2. <Umweltschmutz> fermentation gas
Biogasreinigungsanlage f <Umweltschmutz> biogas digester
biogen adj <Erdöl> biogenic
biogenetisch adj <Erdöl> biogenic *(Kohlenwasserstoff-Formation)*
Bioglas n <Ker & Glas> bioglass
Bioindikator m <Umweltschmutz> bioindicator, biological indicator
Biokraftstoff m <Kfztech> biodiesel, biodiesel fuel, biofuel
biologisch abbaubar adj <Anstrich, Umweltschmutz> biodegradable
biologisch abbaubare Substanz f <Umweltschmutz> biodegradable substance
biologisch abbaubarer Abfall m <Abfall> biodegradable waste
biologisch abbaubarer Schadstoff m <Umweltschmutz> biodegradable pollutant
biologisch nicht abbaubarer Abfall m <Abfall> non-biodegradable waste
biologische Abbaubarkeit f <Umweltschmutz> biodegradability
biologische Abschirmung f <Kerntech> biological shield
biologische Abwasserreinigung f <Wasserversorg> biological water treatment
biologische Affinität f <Kerntech> biological affinity *(von Radioisotopen)*
biologische Behandlung f <Umweltschmutz> biological treatment
biologische Gefahr f <Sicherheit> biological hazard
biologische Kläranlage f <Umweltschmutz> biological clarification plant
biologische Kriegsführung f <Umweltschmutz> biological warfare
biologische Nachreinigung f <Abfall> secondary sewage treatment
biologische Umwandlung f <Abfall> biological energy conversion
biologische Zersetzung f <Meerschmutz> biodegradation, biological degradation
biologischer Abbau m 1. <Meerschmutz, Umweltschmutz> biodegradation, biological degradation; 2. <Verpack> biodegradation
biologischer Effekt m **ionisierender Strahlung** <Strahlphys> biological effect of ionizing radiation
biologischer Körper m <Chemtech> biological filter
biologischer Rasen m 1. <Chemtech> biological filter; 2. <Wasserversorg> bacteria bed
biologischer Sauerstoffbedarf m (BSB) <Abfall, Lebensmittel, Umweltschmutz> biological oxygen demand (BOD)
biologischer Schild m <Kerntech> biological shield
biologisches Agens n <Umweltschmutz> biological agent
biologisches Filter n <Abfall> biological filter
biologisches Gleichgewicht n <Umweltschmutz> biological equilibrium

biologisch-schützendes

biologisch-schützendes Kühlsystem n <Kerntech> biological protection cooling system
Biomasse f <Nichtfoss Energ> biomass
Biomechanik f <Ergon> biomechanics
Biometrie f <Ergon> biometry
Biomüllkompost m <Umweltschutz> biowaste compost
Biomüllkompostierung f <Umweltschutz> biological waste composting
Biophysik f <Phys> biophysics
Biopolartechnik f <Elektronik> bipolar technology
biorientierte Folie f <Kunststoff> biaxially oriented film
Biose f <Chemie> biose
Biosolarzelle f <Nichtfoss Energ> biosolar cell
biosozial adj <Ergon> biosocial
Biosphäre f <Nichtfoss Energ, Umweltschutz> biosphere
Biostabilisator m <Abfall> biostabilizer
Biotechnologie f <Lebensmittel, Chemie> biotechnology
Biotin n <Lebensmittel> biotin
Biotit m <Ker & Glas> biotite
Biot-Savart'sches Gesetz n <Phys> Biot-Savart law, Laplace's law
Bioumwandlung f <Kohlen, Nichtfoss Energ> bioconversion
Biozönose f <Umweltschutz> biocoenosis
Biphenyl n <Chemie> biphenyl, phenylbenzene
bipolar adj <Comp & DV, Funktech, Phys> bipolar
Bipolar... <Comp & DV, Elektronik, Funktech, Phys> bipolar
Bipolarcode m <Telekom> bipolar code
bipolare Diode f <Elektronik> bipolar diode
bipolare integrierte Schaltung f <Comp & DV> bipolar integrated circuit
bipolare integrierte Siliziumschaltung f <Elektronik> silicon bipolar integrated circuit
bipolare Logik f <Elektronik> bipolar logic
bipolare Schrittinversion f (AMI) <Telekom> alternate mark inversion (AMI)
bipolare Stromversorgung f <Elektrotech> bipolar power supply
bipolarer Feldeffekttransistor m **mit isoliertem Gatter** (BIGFET) <Elektronik> bipolar insulated-gate field-effect transistor (BIGFET)
bipolarer Leistungstransistor m <Elektronik> bipolar power transistor
bipolares Signal n <Telekom> bipolar signal
Bipolar-Mischtechnik f <Elektronik> merged bipolar technology
Bipolar-Technologie f <Elektronik> bipolar technology
Bipolartransistor m 1. <Elektronik> bipolar transistor (mit P- und mit N-Halbleitern aufgebaut); 2. <Phys> bipolar transistor, junction transistor; 3. <Telekom> bipolar transistor
Biprisma n <Fertig, Optik> biprism
biquinär adj <Comp & DV> biquinary
Biquinärcode m <Comp & DV> biquinary code
biquinäre Zahl f <Comp & DV> biquinary number
Birne f 1. <Fertig> Bessemer converter; 2. <Metall> converter
Birnenblitz m <Foto> flash bulb
birnenförmiges Gefäß n <Labor> pear-shaped vessel
Bisazimethylen n <Chemie> ketazine
Bisazofarbstoff m <Chemie> bis-azo dye
B-ISDN-Dienst m <Telekom> broadband ISDN service
B-ISDN-Protokollreferenzmodell n <Telekom> B-ISDN protocol reference model
bisheriger Stand m **der Technik** <Patent> background art
Biskuit... <Ker & Glas> biscuit

Biskuiteintauchvorrichtung f <Ker & Glas> biscuit dipper
Biskuitporzellan n <Ker & Glas> biscuit-baked porcelain
Biskuitware f <Ker & Glas> biscuit ware (Keramik)
Bismut n (Bi) <Chemie> bismuth (Bi)
Bisphenol A n <Kunststoff> bisphenol A
Bisswinkel m <Kohlen> angle of nip
bistabil adj <Comp & DV, Elektronik, Funktech> bistable
bistabile Kippschaltung f <Elektronik> flip-flop (mit zwei stabilen Ausgangszuständen)
bistabile Kippschattung f <Phys> flip-flop
bistabile Schaltung f <Comp & DV> flip-flop
bistabiler Multivibrator m <Elektronik> bistable multivibrator, scaling circuit
bistabiles Kippglied n <Elektronik> bistable circuit
bistabiles Relais n <Elektrotech> bistable relay
Bistabilität f <Telekom> bistability
Bisulfit n <Chemie> (AE) bisulfite, (BE) bisulphite
Bit n <Comp & DV, Telekom> binary digit, bit
Bit n **mit höchster Wertigkeit** <Comp & DV> high-order bit
Bitabbildung f <Comp & DV> bitmap
Bitbearbeitung f <Comp & DV> bit handling
Bitbündelübertragung f <Telekom> bulk transmission
Bitbündelvermittlung f <Telekom> burst switching (ITG)
Bitebene f <Comp & DV> bit plane
Bitelement n <Elektronik> bit slice
Bitfehlerquote f (BER) <Comp & DV, Telekom> binary error rate, bit error rate (BER)
Bitfehlerrate f (BER) <Comp & DV, Telekom> binary error rate, bit error rate (BER)
Bitfluss m <Telekom> bit stream
Bitfolgefrequenz f <Telekom> bit rate
Bit-für-Bit-Codierung f <Telekom> bit-by-bit encoding
Bitgruppe f 1. <Comp & DV> byte; 2. <Telekom> bit group, burst, byte
Bitkette f <Comp & DV, Elektronik> bit string
Bitmap f <Comp & DV> bitmap, paper feed, paper throw (Dateiformat)
Bitmapping n <Elektronik> bit mapping (Speicherung als digitales Muster)
Bitmuster n <Comp & DV, Kontroll, Telekom> bit pattern
bitorientierte Datenübertragungssteuerung f <Comp & DV, Telekom> high-level data link control
bitorientierte Übertragungssteuerungsverfahren n <Comp & DV, Telekom> high-level data link control
bitparallel adj <Telekom> parallel by bit
bitparallele Datenübertragung f <Comp & DV, Telekom> bit-parallel transfer
Bitrate f <Telekom> bit rate
Bitratenanpassungseinheit f <Telekom> rate adaption unit
Bitratenhalbierung f <Mobilkom> halfrate (GSM: 6,5 kb/s; codierte Sprache)
Bitratenumschalter m <Telekom> bit switch
Bitreihe f <Comp & DV> bit stream
Bits npl **pro Sekunde** (bps) <Comp & DV, Telekom> bits per second, bps (Maß für die Datenübertragungsgeschwindigkeit)
Bits npl **pro Zoll** (bpi) <Comp & DV> bits per inch, bpi (Maß für die Aufzeichnungsdichte)
bit-serielle Übertragung f <Comp & DV, Telekom> bit-serial transfer
Bit-Slice m <Elektronik> bit slice (Prozessorelement)
Bit-Slice-Prozessor m <Elektronik> bit slice processor (Prozessor zur Erzielung bestimmter Wortlängen)
Bitstehlen n <Telekom> bit stealing
Bitstrom m <Telekom> bit stream
Bitter-Magnet m <Phys> Bitter magnet

Bitterstoff m <Chemie> bittern
Bitter-Streifen m <Phys> Bitter pattern
Bitterwert m <Lebensmittel> bittering value
Bittiefe f <Telekom> resolution in bits *(video coding)*
Bitübertragungsschicht f <Comp & DV, Telekom> physical layer *(OSI)*
Bitumen n 1. <Bau> bitumen; 2. <Erdöl> bitumen *(Destillationsprodukt)*; asphalt *(Raffinerie)*; 3. <Kunststoff> bitumen
Bitumenanstrich m <Fertig> bituminous paint
Bitumenanstrichfarbe f <Kunststoff> bituminous paint
Bitumenbeton m <Bau> asphalt concrete
Bitumenemulsion f <Bau> bitumen emulsion
Bitumenfolie f <Bau> bituminous membrane
bitumenhaltig adj <Bau> bituminous
Bitumenlack m <Fertig> bitumen varnish
Bitumenmembran f <Bau> bituminous membrane
Bitumenpapier n <Papier> tarred brown paper
Bitumenpappe f <Bau> bitumen board
bituminieren v <Bau> bituminize
Bituminieren n <Bau> bituminization
bituminiertes Papier n <Verpack> bitumen-coated paper
bituminös adj <Bau> bituminous
bituminöse Kohle f <Kohlen> bituminous coal
Bitverarbeitung f <Comp & DV> bit manipulation
Bitvollgruppe f <Telekom> envelope *(Datenvermittlung)*
bivalent adj <Chemie> bivalent, divalent
bivalentes Heizystem n <Heiz & Kälte> fuel/electric heating system
Bivalenz f <Chemie> bivalence, divalence
Bivinyl n <Chemie> bivinyl, divinyl
Bixin n <Lebensmittel> anatto, annatto, bixin
Bk *(Berkelium)* <Chemie> Bk *(berkelium)*
B-Kanal m <Telekom> B-channel
Blachenstoff m <Textil> canvas
Blackbox f <Comp & DV, Lufttrans> black box
Black-Matrix-Bildröhre f <Fernseh> black matrix tube
Blähmittel n <Chemtech> foaming agent
blanchieren v <Lebensmittel> blanch
blank adj 1. <Bau> bare, sound; 2. <Elektrotech> bare; 3. <Fertig> bright-finished; bare *(Draht)*; 4. <Maschinen> bright, plain
blank reiben v <Bau> scour
Blank... <Fertig> bright; bare *(Draht)*
Blankätzbad n <Ker & Glas> clear etching bath
Blankätzen n <Ker & Glas> clear etching
Blankbrennen n <Fertig> bright dip finishing
Blankdraht m 1. <Fertig> naked wire; 2. <Metall> bright wire; 3. <Telekom> bare wire
blanke Kette f <Metall> bright chain
blanke Schraube f <Maschinen> bright bolt
blanker Aluminiumdraht m <Elektrotech> BAW, *(BE)* bare aluminium wire, *(AE)* bare aluminum wire
blanker Draht m <Telekom> bare wire
blanker Leitungsdraht m <Elektrotech> bare wire
blanker Stahl m <Metall> bright steel
blanker Stahldraht m <Metall> bright steel wire
Blanket n <Kerntech> breeding blanket
blankgeglüht adj <Fertig> bright-annealed
blankgeglühter Draht m <Maschinen> bright-annealed wire
blankglühen v <Fertig> bright-anneal
Blankglühen n <Metall> bright annealing
Blankschleifen n <Ker & Glas> smooth grinding; smoothing *(von Spiegelglas)*
Blankziehen n <Fertig> brightdrawing
Blas... <Ker & Glas, Verpack> blow
Blas-, Abfüll- und Verschließsystem n <Verpack> blow fill seal system

Blas-Blas-Verfahren n <Ker & Glas> blow-and-blow process
Bläschen n 1. <Ker & Glas> seed; 2. <Optik> bubble
Bläschenbildungspotenzial n <Ker & Glas> seeding potential
bläschenfrei adj <Ker & Glas> seed-free
Blase f 1. <Chemie> void; 2. <Chemtech> bubble; 3. <Fertig> flaw *(Guss)*; 4. <Kohlen> bubble; 5. <Kunststoff> blister, bubble; 6. <Metall> blister, blow hole; 7. <Papier, Phys> bubble; 8. <Wassertrans, Werkprüf> blister
Blase f mit eingeschlossenem Fremdkörper <Ker & Glas> cat's eye
Blasebalg m 1. <Fertig> bellows, smith's bellows; 2. <Mechan> bellows
Blasebalgpumpe f <Labor> bellows pump
blasen v <Elektriz, Papier, Raumfahrt> blow *(Raumschiff)*
Blasen n <Fertig> air blasting • **Blasen bilden** <Papier> bubble • **Blasen werfen** <Werkprüf> blister
Blasen n im Konverter <Fertig> Bessemer blow
Blasenauftrieb m <Ker & Glas> train of bubbles
Blasenbildung f <Papier> bubbling
Blasendestillation f <Kerntech> batch distillation
Blasendestillationsanlage f <Erdöl> batch still *(Raffinerie)*
Blasenfolie f <Verpack> bubble film
Blasenglocke f <Lebensmittel> bubble cap *(einer Glockenbodenkolonne)*
Blasenkammer f <Kerntech, Phys, Strahlphys, Teilphys> bubble chamber
Blasenloch n <Fertig> pit hole
Blasenmodell n <Metall> bubble model
Blasenprobe f <Papier> bubble test
blasenreiches Glas n <Ker & Glas> very seedy glass
Blasenschleier m <Ker & Glas> feather
Blasensieden n <Heiz & Kälte> nucleate boiling
Blasenspeicher m <Comp & DV> bubble memory, magnetic bubble memory
Blasenstahl m <Metall> blister steel
Blasenverpackung f <Verpack> air bubble wrap, bubble pack
Blasenzähler m <Chemtech> *(AE)* bubble gage, *(BE)* bubble gauge
Blasflasche f <Verpack> blown bottle
Blasfolie f <Kunststoff> blown film
Blasform f 1. <Fertig> blow die *(Kunststoffe)*; 2. <Maschinen> *(AE)* blow mold, *(BE)* blow mould
Blasformen n <Kunststoff> *(AE)* blow molding, *(BE)* blow moulding
Blasformverfahren n <Verpack> *(AE)* blow-molding process, *(BE)* blow-moulding process
Blaskopf m 1. <Ker & Glas> blow head; 2. <Kunststoff> parison die
Blasloch n <Fertig> heating gate *(Thermitschweißen)*
Blasmagnet m <Elektrotech> magnetic blowout
Blasöl n <Maschinen> blown oil
Blasring m <Ker & Glas> blowing ring
Blasrotor m <Lufttrans> jet-flapped rotor *(eines Hubschraubers)*
Blasspule f <Elektrotech> blowout coil
Blastisch m <Ker & Glas> blow table
Blasvorgang m <Metall> shot blasting
Blaswalze f <Papier> blow roll
Blaswirkung f <Fertig> arc blow; magnetic arc blow *(Lichtbogenschweißen)*
Blatt n 1. <Bau> blade; 2. <Comp & DV> leaf; 3. <Druck> leaf, sheet; 4. <Fertig> band; blade *(einer Säge)*; 5. <Künstl Int> end node, terminal node *(eines Baumes)*; 6. <Maschinen> blade, leaf; 7. <Textil> reed
Blatt n der Luftschraube <Lufttrans> blade *(Hubschrauber)*

Blatt

Blatt *n* **einer Metallbandsäge** <Maschinen> metal-cutting bandsaw blade
Blatt *n* **Papier** <Papier> slip
Blattanstellwinkel *m* <Lufttrans> blade pitch angle *(Hubschrauber)*
Blattanstellwinkel-Übertragungsgerät *n* <Lufttrans> blade pitch transmitter *(Hubschrauber)*
Blattbelastung *f* <Lufttrans> blade loading *(Hubschrauber)*
Blattbildungseinheit *f* <Papier> former *(Maschine)*
Blattbohrmeißel *m* <Erdöl> drag bit *(Bohrtechnik)*; spudding bit *(Flachbohrtechnik)*
Blattbreite *f* <Textil> width of reed
Blättchen *n* <Kunststoff> platelet
Blatter *f* <Ker & Glas> blister
Blätteranker *m* <Elektriz> laminated armature
Blätterbürste *f* <Elektriz> laminated brush
Blättergips *m* <Chemie> selenite
Blätterkern *m* <Elektriz> laminated core
Blätterknauf *m* <Bau> finial
Blätterkohle *f* <Kohlen> paper coal, papyraceous lignite, slate-foliated lignite
Blätterlack *m* <Fertig> shellac
Blätterleiste *f* <Comp & DV> scroll bar
Blättermagnet *m* 1. <Elektriz> laminated magnet; 2. <Phys> compound magnet
Blättermodus *m* <Comp & DV> scroll mode
blättern *v* <Comp & DV> browse, scroll
Blättern *n* <Comp & DV> browsing, scroll, scrolling
Blattfeder *f* 1. <Kfztech> leaf spring *(Motor)*; 2. <Maschinen> laminated spring, leaf spring, plate spring
Blattfederstarrachse *f* <Kfztech> Hotchkiss drive *(Triebstrang)*
Blattfilm *m* <Foto> sheet film
Blattindex *m* <Umweltschmutz> leaf-area index
Blattkante *f* <Konstzeich> sheet border, sheet edge
Blattkräuselkrankheit *f* <Lebensmittel> leaf-curl *(Pflanzenkrankheitslehre)*
Blattoberfläche *f* <Umweltschmutz> foliar surface
Blattroller *m* <Lebensmittel> leaf-roller *(Pflanzenkrankheitslehre)*
Blattspitzengeräusch *n* <Lufttrans> blade slap *(Hubschrauber)*
Blattspitzen-Geschwindigkeitsverhältnis *n* <Nichtfoss Energ> tip speed ratio
Blattspitzenverlustfaktor *m* <Lufttrans> blade tip loss factor *(Hubschrauber)*
Blattspurprüfung *f* <Lufttrans> blade tracking *(Hubschrauber)*
Blattsteigung *f* **beim Propeller** <Lufttrans> pitch *(Lage eines Flugzeugs oder Schiffs)*
Blattsteigungs-Synchronismusanzeiger *m* <Lufttrans> pitch throttle synchronizer
Blattsteigungsverriegelung *f* <Lufttrans> pitch locking system *(Hubschrauber)*
Blattung *f* <Bau> splice joint; halved joint *(Holzbau)*
Blattwerkstoffe *mpl* <Nichtfoss Energ> blade materials *(Propeller)*
Blattwespe *f* <Lebensmittel> sawfly *(Schädlingsbekämpfung)*
Blattwinkelanzeigersynchro *f* <Lufttrans> pitch detector synchro
Blattwinkelradius *m* <Lufttrans> pitch radius
Blattwinkelverstellung *f* <Lufttrans> pitch control *(Hubschrauber)*
Blattwinkelverstellungsarm *m* <Lufttrans> pitch control arm *(Hubschrauber)*
Blattwinkelverstellungsbelastung *f* <Lufttrans> pitch control load *(Hubschrauber)*
Blattwinkelverstellungshebel *m* <Lufttrans> pitch control lever *(Hubschrauber)*
Blattwinkelverstellungs-Stangenwinkel *m* <Lufttrans> pitch control rod angle
Blattziffer *f* <Druck> folio
Blaubruchgebiet *n* <Fertig> blue-brittle range
blaubrüchig *adj* <Fertig> blue-brittle
Blaubrüchigkeit *f* <Metall> blue-brittleness
Blaudruck *m* <Textil> blueprint
blaues Kieselgel *n* <Verpack> blue silica gel
blaues Vitriol *n* <Chemie> *(AE)* copper sulfate, *(BE)* copper sulphate
blaugeglüht *adj* <Fertig, Metall> open-annealed
Blaugitter *n* <Fernseh> blue screen-grid
Blauglasquarz *n* <Chemie> blue hyaline-quartz
blauglühen *v* <Fertig> open-anneal
Blauglühen *n* <Metall> blue annealing, open annealing
Blauglühung *f* <Fertig> open annealing
blaugrüner Laser *m* <Elektronik> blue-green laser
Blaugrünfiltereinstellung *f* <Foto> minus red filter adjustment
Blaukanone *f* <Fernseh> blue gun
Blaumischer *m* <Fernseh> blue adder
Blaupause *f* <Bau, Druck, Maschinen> blueprint
Blau-Schwarz-Pegel *m* <Fernseh> blue-black level
blauspröde *adj* <Fertig> blue-short
Blausprödigkeit *f* 1. <Fertig> blue-shortness; 2. <Metall> blue-brittleness
Blaustein *m* <Chemie> blue vitriol, copper vitriol, bluestone
Blaustrahl *m* <Fernseh> blue beam
Blaustrahlmagnet *m* <Fernseh> blue-beam magnet
Blauton *m* <Ker & Glas> blue clay
Blauvitriol *n* <Chemie> blue vitriol, *(AE)* copper sulfate, *(BE)* copper sulphate
Blauwärme *f* <Fertig> blue heat
Blauwertspitze *f* <Fernseh> blue peak level
Blech *n* 1. <Fertig> metal sheet, sheet metal; 2. <Kerntech> sheet; 3. <Kfztech> panel *(Karosserieteil)*; 4. <Mechan> sheet; 5. <Metall> plate, sheet, sheet metal; 6. <Wassertrans> sheet *(Metall)*
Blechbiegemaschine *f* 1. <Fertig> plate-bending rolls, sheet metal-bending machine, sheet-bending machine; 2. <Maschinen> plate-bending press; 3. <Metall> sheet-bending machine, sheet metal-bending machine
Blechbiegen *n* 1. <Fertig> sheet bending; 2. <Maschinen> plate bending; 3. <Metall> sheet-bending
Blechbiegewalzen *fpl* <Maschinen> plate-bending rollers, plate-bending rolls
Blechbohrer *m* <Fertig> hole cutter, sheet drill
Blechbördelmaschine *f* <Fertig> sheet-bordering machine
Blechen *n* <Elektrotech> lamination *(von Eisenkernen)*
Blechfalz- und Biegemaschine *f* <Maschinen> plate folding and bending machine
Blechhalter *m* <Fertig> blank holder
Blechkantenhobelmaschine *f* <Fertig> plate planer
Blechkantennachformen *n* <Fertig> plate-edge profiling
Blechkettenläufer *m* <Elektrotech> chain rim motor, segmental rim motor
Blechlehre *f* 1. <Fertig> *(AE)* plate gage, *(BE)* plate gauge, *(AE)* sheet gage, *(BE)* sheet gauge; 2. <Maschinen> *(AE)* plate gage, *(BE)* plate gauge, *(AE)* sheet gage, *(BE)* sheet gauge, *(AE)* sheet iron gage, *(BE)* sheet iron gauge; 3. <Metrol> *(AE)* plate gage, *(BE)* plate gauge
Blechmaterial *n* <Fertig, Metall> sheet stock
Blechpaket *n* 1. <Elektrotech> core stack, laminated core, stack of sheets *(Transformator)*; 2. <Fertig> packet, stacked sheets

Blechpaketbohren n <Fertig> packet drilling
Blechrichtmaschine f <Eisenbahn, Fertig> plate-straightening machine
Blechrohr n <Bau> sheet iron pipe
Blechronde f <Elektrotech> circular lamination, circular punching
Blechrundbiegemaschine f <Fertig, Metall> sheet metal-bending roll
Blechschablone f <Konstzeich> sheet metal stencil
Blechschere f 1. <Fertig> tinman's shear; 2. <Maschinen> metal shears
Blechschornstein m <Bau> steel chimney
Blechschraube f <Maschinen> self-tapping screw, sheet metal screw, tapping screw
Blechschraubengewinde n <Maschinen> tapping screw thread
Blechschrott m <Fertig> sheet scrap
Blechstegträger m <Bau> plate web girder
Blechtafelschere f <Fertig> plate shear
Blechträger m <Fertig> plate girder
Blechtrommel f <Verpack> metal drum
Blechwalze f <Maschinen> plate roll
Blechwalzen n <Fertig> plate rolling
Blechwalzwerk n <Maschinen> plate mill
Blechzange f <Fertig> dog
Blechzuschnitt m <Fertig> blank
Blei n (Pb) <Chemie> lead
Bleiakkumulator m <Elektriz, Phys> lead accumulator
Bleiansatzstück n <Labor> extension lead
bleiarmes Benzin n <Kfztech> (AE) low-lead gasoline, (BE) low-lead petrol
Bleibatterie f <Elektriz> lead accumulator
bleibende Dehnung f <Kunststoff> tension set
bleibende Regeldifferenz f <Regelung> steady state deviation from the desired value
bleibende Verformung f 1. <Kunststoff> permanent set, set; 2. <Maschinen> permanent set; 3. <Metall> permanent deformation
bleibender Fehler m <Comp & DV> solid error
Bleiblech n <Bau, Metall> sheet lead
Bleich... 1. <Foto> bleach; 2. <Ker & Glas, Papier> bleaching
Bleichbad n <Foto> bleach bath
Bleichbütte f <Papier> bleaching chest
bleichen v <Lebensmittel, Papier, Textil> bleach
Bleichen n <Lebensmittel> bleaching
Bleicherde f <Ker & Glas> bleaching clay
Bleichholländer m <Papier> bleaching engine
Bleichkalk m <Textil> bleaching powder
Bleichlösung f <Papier> bleaching liquor
Bleichmittel n <Lebensmittel> bleaching agent
Bleichpulver n 1. <Lebensmittel> bleaching powder, chlorinated lime; 2. <Papier, Textil> bleaching powder
Bleichturm m <Papier> bleaching tower
Bleichverfahren n <Papier> bleaching
Bleidichtung f 1. <Bau> lead joint; 2. <Fertig> lead packing
Bleifarbe f <Bau> lead paint
Bleifeile f <Bau> shave hook
Bleifilter n <Kfztech> lead filter
bleifrei adj <Erdöl> lead-free (Vergaserkraftstoffe)
bleifreies Benzin n 1. <Kfztech> (AE) lead-free gasoline, (BE) lead-free petrol, (AE) unleaded gasoline, (BE) unleaded petrol; 2. <Umweltschmutz> (AE) lead-free gasoline, (BE) lead-free petrol
Bleiglaszähler m <Strahlphys> lead glass counter
Bleiglätte f <Kunststoff> lead oxide, litharge
bleihaltig adj <Chemie> plumbic, plumbous
Bleikammerkristalle mpl <Chemtech> chamber crystals

Bleikammerverfahren n <Chemtech> chamber process
Bleikristallglas n <Ker & Glas> lead crystal glass; full lead crystal glass (mit 30 % Bleigehalt)
Bleilot n <Elektriz, Maschinen> soft solder
Bleimantel m 1. <Elektriz> lead sheath; 2. <Elektrotech> lead sheath (bei Kabel)
Bleimantelkabel n <Elektrotech> lead-covered cable, lead-sheathed cable
Bleimennige f <Ker & Glas, Kunststoff> red lead
Bleinaphthenat n <Kunststoff> lead naphthenate
Blei-Orthoplumbat n <Chemie> lead tetroxide
Bleioxid n <Chemie> lead dioxide, lead oxide, litharge
Bleisatz m <Druck> hot-metal typesetting
Bleischwamm m <Kfztech> sponge lead
Bleisicherung f an der Fahrwerksfelge <Lufttrans> landing-gear wheel rim fusible plug
Bleisilikat n <Ker & Glas> lead silicate
Bleistiftkeule f <Telekom> pencil beam
Bleistiftstrahl m <Telekom> pencil beam
Bleisulfat n <Chemie> (AE) lead sulfate, (BE) lead sulphate
Bleisulfid n <Chemie> (AE) lead sulfide, (BE) lead sulphide
Bleitetraethyl n 1. <Chemie> lead tetraethyl, tetraethyl lead, tetraethylplumbane; 2. <Kfztech> tetraethyl lead
Bleitype f <Druck> lead printing letter
bleiumhülltes Kabel n <Elektriz> lead-covered cable, lead-sheathed cable
Blei- und Zinnlegierungen fpl <Maschinen> lead and tin alloys
bleiverglastes Fenster n <Ker & Glas> leaded light
Bleiversiegelung f <Verpack> lead seal
Bleizinnlegierung f <Metall> terne metal
Bleizucker m <Lebensmittel> sugar of lead
Bleizusatzstoffe mpl <Umweltschmutz> lead additives
Blende f 1. <Aufnahme> diaphragm; 2. <Fernseh> aperture; 3. <Fertig> shutter, slit; 4. <Foto> F stop, aperture, diaphragm; 5. <Funktech> diaphragm; 6. <Gerät> measuring orifice (Pneumatik, Optik); 7. <Mechan> diaphragm; 8. <Phys> aperture • **Blende schließen** <Foto> stop down
Blendeneinstellknopf m <Foto> aperture setting knob
Blendeneinstellring m <Foto> aperture setting ring
Blendeneinstellung f <Foto> aperture stop, stop
Blendenöffnung f 1. <Comp & DV> aperture; 2. <Foto> focal aperture
Blendenrevolver m <Optik> revolving diaphragm
Blendenring m <Foto> aperture ring
Blendenrotor m <Kfztech> trigger wheel (mit gleicher Anzahl von Blenden wie Zylinder am Motor)
Blendenskale f <Foto> aperture scale
Blendensteuerung f <Fernseh> iris control button
Blendenvorwähler m <Foto> aperture priority camera
Blendenwert m <Foto> f-number, lens stop
Blendlicht n <Sicherheit> glare
Blendmauer f <Bau> screen wall
Blendschutz m <Lufttrans> glare shield (gegen heiße Abgase)
Blendschutzanstrich m <Kunststoff> shading paint
Blendschutzglas n <Ker & Glas> antidazzle glass
Blendschutzschicht f <Telekom> antiglare coating
Blendschutzvisier n <Kfztech> antidazzle visor (Motorradhelm)
Blendung f <Optik> glare
blendungsfrei adj 1. <Sicherheit> glare-free; 2. <Trans> antidazzling
Blendziegel m <Bau> facing brick
blicken v <Metall> brighten
Blickfeldblende f <Phys> field stop

blind

blind adj <Fertig> false (Walzen, Kaliber)
Blind... 1. <Bau> blank, blind; 2. <Elektriz> reactive; 3. <Lufttrans> blind; 4. <Maschinen> blank
Blindachse f <Eisenbahn> blind axle
Blindband m <Druck> mock-up
Blindbelegung f <Telekom> false seizure (Telefon)
Blindboden m <Fertig> false bottom
Blindbohrung f <Bau, Kohlen> blind hole
Blindelement n <Elektrotech> reactive element
Blindenergie f <Elektriz> reactive energy
blinder Gewindebolzen m <Maschinen> blind stud bolt
Blindfarben fpl <Druck> dropout colours
Blindfenster n <Bau> blank window, blind window
Blindflansch m <Maschinen> blank flange, blind flange
Blindflug m 1. <Lufttrans> blind flight; 2. <Raumfahrt> blind navigation
Blindfluglandesystem n durch Eigenpeilung (ILS) <Lufttrans, Raumfahrt> instrument landing system (ILS)
Blindfluglandung f <Lufttrans> blind landing
blindgebohrte Welze f <Papier> blind-drill roll
Blindkaliber n <Fertig> dummy pass (Walzen)
Blindkomponente f 1. <Elektriz> wattless component; 2. <Elektronik> quadrature component; 3. <Elektrotech> idle component, reactive component
Blindlandung f mit Bakenunterstützung (BBL) <Raumfahrt> beacons and blind landing (BBL)
Blindlast f 1. <Elektrotech, Phys, Telekom> reactive load; 2. <Trans> dummy load
Blindleistung f 1. <Elektriz, Elektrotech, Phys> reactive power; 2. <Telekom> idle power
Blindleistungseinheit f (var) <Labor> volt-amperes reactive (var)
Blindleistungskomponente f <Elektriz> quadrature power
Blindleistungsverbesserung f <Elektrotech> power factor correction
Blindleistungsverbrauch m <Elektriz> reactive energy
Blindleitung f <Elektrotech> adjustable short, stub
Blindleitwert m <Elektrotech, Phys> susceptance
Blindniet m <Maschinen> blind rivet
Blindprägung f <Druck> blind blocking, blind embossing
Blindschacht m 1. <Bau> blind shaft; 2. <Bau, Kohlen> blind pit
Blindschaltkreis m <Elektrotech> reactive circuit
Blindspannung f <Elektriz, Phys> reactive voltage
Blindspannungskomponente f <Elektriz> quadrature voltage
Blindstich m <Fertig> blind pass; dead pass, dummy pass (Walzen)
Blindstrom m 1. <Elektriz> idle current, reactive current, wattless current; 2. <Elektrotech> reactive current, wattless current
Blindstromkomponente f <Elektriz> quadrature component
Blindstrommaschine f <Elektriz> compensator
Blindtür f <Bau> blank door, blind door
Blindverbrauchszähler m <Gerät> varhour meter
Blindwiderstand m 1. <Elektriz> reactance (X); 2. <Elektrotech, Funktech> reactance; 3. <Phys> inductive reactance, reactance
Blindwiderstandsrelais n <Elektrotech> reactance relay
Blindwiderstandsschaltung f <Elektriz> reactance circuit
Blink... <Elektriz, Verpack, Wassertrans> flashing
blinken v 1. <Comp & DV> blink; 2. <Metall> brighten
Blinken n <Comp & DV> blinking
Blinker m <Kfztech> direction indicator; flasher, indicator (Zubehör)
Blinkfeuer n <Wassertrans> flashing light (Seezeichen)

Blinklampe f <Elektriz> flashlight (Morselampe)
Blinkleuchte f <Kfztech> flashing light
Blinklicht n <Sicherheit> flashing light
Blinkzeichen n <Wassertrans> flashing signal (Signal)
Blip m <Wassertrans> radar blip (Radar)
Blisterkarte f <Verpack> blister card
Blisterpack n <Verpack> shrink pack
Blisterpackautomat m <Verpack> blister packaging machine
Blisterpackung f <Verpack> blister pack
Blisterrand- und Klarsichtfolienverpackungsmaschine f <Verpack> blister edge and foil machine
Blitz m 1. <Elektrotech> lightning; 2. <Foto> flash
Blitzableiter m 1. <Elektriz> lightning arrester; 2. <Elektrotech> arrester, lightning rod, lightning conductor, lightning arrester, surge arrester; 3. <Sicherheit> lightning conductor, lightning rod
Blitzableiterstab m <Elektrotech> lightning rod
Blitzableiterstange f <Elektriz> lightning rod
Blitzdauer f <Foto> flash duration
Blitzeinschlag m <Elektriz> lightning stroke
Blitzeinschlagstelle f <Elektriz> lightning strike position
Blitzen n 1. <Elektriz> lightning; 2. <Foto> flashing
Blitzentladung f <Elektriz, Elektrotech> lightning discharge
Blitzfeuer n <Raumfahrt> blinking light
Blitzfolgezeit f <Foto> recycle time, recycling time
blitzgeschützt adj <Elektrotech, Sicherheit> lightning-proof, lightning-resistant
blitzgeschützte Starkstromleitung f <Elektriz> lightning-resistant power line
blitzgeschützter Transformator m <Elektriz> lightning-proof transformer
Blitzkontakt m <Foto> hot-shoe flash contact
Blitzleiste f <Foto> flash bar
Blitzlicht n <Elektrotech, Foto, Mechan> flashlight
Blitzlichtaufnahme f <Foto> flash picture
Blitzlichtkontakt m <Foto> flash contact
Blitzlicht-Steckverbindung f <Foto> flash socket
Blitzröhre f <Elektronik> flash tube
Blitzrohrzange f <Maschinen> grip pipe-wrench
Blitzschalter m <Foto> flash switch
Blitzschlag m 1. <Elektriz> lightning strike; 2. <Elektrotech> lightning discharge; 3. <Phys> stroke
Blitzschuh m <Foto> flash shoe
Blitzschutz m 1. <Elektriz> lightning protection, lightning arrester; 2. <Elektrotech> arrester, lightning arrester, lightning protection; 3. <Funktech> lightning protection; 4. <Raumfahrt> lightning arrester; 5. <Sicherheit> lightning arrester, lightning protection, lightning protector
Blitzschutz m mittels Luftstrecke <Elektriz> air gap protector
Blitzschutzanlage f <Sicherheit> lightning protective system
Blitzschutzschalter m <Sicherheit> lightning switch
Blitzschutzsicherung f <Telekom> surge arrester
Blitzstoß m <Elektriz> lightning surge
Blitzstrecke f <Elektriz> lightning path
Blitzstrom m <Elektrotech> lightning current
Blitzwürfel m <Foto> flash cube
Block m 1. <Comp & DV> block, pad; 2. <Erdöl> block (Aufsuchen; Gewinnung); (BE) licence block, (AE) license block (Recht); 3. <Fertig> ingot, pad; 4. <Kohlen> ingot; 5. <Maschinen> bar, block, ingot; 6. <Mechan> ingot; 7. <Wassertrans> block
Block m mit fester Länge <Comp & DV> fixed-length block
Blockade f <Wassertrans> blockade

Blockanweisung m <Comp & DV> block statement
Block-Copolymer n <Kunststoff> block copolymer
Blockdiagramm n <Comp & DV, Elektrotech, Maschinen> block diagram
Blockdruck m <Textil> block printing
Blockeisen n <Metall> ingot iron
Blocken n <Eisenbahn, Kunststoff, Telekom> blocking *(OSI)*
Blockende n *(EOB)* <Comp & DV, Telekom> end of block *(EOB)*
Blockfehlerrate f <Comp & DV, Telekom> block error rate
Blockform f <Fertig> *(AE)* ingot mold, *(BE)* ingot mould
Blockgerüst n <Fertig> blooming stand
Blockgeschwindigkeit f <Lufttrans, Trans> block speed
Blockgröße f <Comp & DV, Telekom> block size
Blockheizkraftwerk n <Bau> block heating power plant
Blockheizung f <Kfztech> block heater
blockierbar adj <Sicherheit> lockable
blockieren v 1. <Elektrotech> interlock; 2. <Hydraul> bar *(Dampfmaschine)*; jam *(Ventil)*; 3. <Kfztech> jam *(Bremse)*; 4. <Maschinen> jam; 5. <Maschinen> lock; 6. <Wassertrans> block off, blockade *(Schifffahrt, Hafen)*
Blockieren n 1. <Lufttrans> stall *(Kompressor, Turbomotor)*; 2. <Maschinen> blocking, jam, jamming
Blockierregler m <Kfztech> antiblocking system, anti-skid braking system
blockiert adj <Maschinen> blocked
blockierte Tragschere f <Ker & Glas> stuck shank
Blockierung f 1. <Eisenbahn> blocking; 2. <Elektrotech> interlock; blocking *(der Leitfähigkeit)*; 3. <Lufttrans> interlock; 4. <Metall> locking; 5. <Telekom> blocking
Blockierungseffekte mpl <Telekom> blockage effects
blockierungsfreie Übertragung f <Telekom> non-blocking transmission
blockierungsfreier Koppler m <Telekom> nonblocking switch *(Vermitteln)*
blockierungsfreies Netz n <Telekom> nonblocking network
Blockierungswahrscheinlichkeit f <Telekom> blocking probability
Blockierverhinderer m <Lufttrans> antiskid unit
Blockierzustand m <Elektrotech> off-state *(von Thyristoren)*
Blocking n <Kunststoff> blocking
Blockingeffekt m <Phys> blocking effect
Blockkaliber n <Fertig> blooming pass, cogging pass
Blockkette f <Maschinen> block chain
Blockkokille f <Fertig> *(AE)* ingot mold, *(BE)* ingot mould
Blockkondensator m <Elektrotech> blocking capacitor
Blocklänge f <Comp & DV, Telekom> block length, block size
Blocklehm m <Kohlen> till
Blocklöscher m <Fernseh> bulk eraser
Blockmeißelhalter m <Fertig> multiple-tool block
Blockmetall n <Metall> ingot metal
Blocknummer f <Erdöl> block number
Blockparität f <Comp & DV> horizontal parity
Blockprüfung f <Comp & DV, Telekom> LRC, longitudinal redundancy check
Blockprüfzeichen n *(BCC)* <Telekom> block check character *(BCC)*
Blockprüfzeichenfolge f <Telekom> FCS, frame-checking sequence
Blockquantisierung f <Telekom> block quantization
Blockreaktanz f <Elektrotech> block reactance
Blockrolle f <Wassertrans> pulley
Blocksatz m <Comp & DV> justification • **im Blocksatz setzen** <Druck> block
Blockschaltbild n 1. <Comp & DV, Elektrotech> block diagram; 2. <Kontroll> schematic unit diagram; 3. <Maschinen> block diagram, kinematic diagram, mimic diagram; 4. <Telekom> block diagram
Blockscheibe f 1. <Maschinen> sheave, sheave block; 2. <Wassertrans> pulley
Blockschema n <Telekom> block diagram
Blockschere f 1. <Fertig> bloom shears; 2. <Maschinen> billet shears, bloom shears
Blocksortierung f <Comp & DV> block sort
Blockstahl m <Metall> ingot steel
Blockstein m <Bau> block
Blockstraße f <Fertig> cogging train
Blockstruktur f <Comp & DV> block structure
Blockstufe f <Bau> flyer
Blockübertragung f 1. <Comp & DV> block transfer; 2. <Telekom> block transmission
Blockungsfaktor m <Comp & DV> blocking factor
Blockverband m <Bau> old English bond *(Mauerwerk)*
Blockverdichtung f <Comp & DV> block compaction
Blockverschluss m <Eisenbahn> block signal interlocking
Blockwagen m <Maschinen> block carriage
Blockwalze f <Fertig> blooming roll, cogging roll
Blockwalzgerüst n <Fertig> cogging stand
Blockwalzwerk n <Fertig> blooming mill, cogging mill
Blockzugriff m <Comp & DV> block retrieval
Blockzwischenraum m <Comp & DV> IBG, interblock gap
Bloop-Lampe f <Aufnahme> bloop lamp
Blow-Down Druckbeaufschlagung f <Raumfahrt> blow-down pressurization
Blowout-Preventer m *(BOP)* <Erdöl> blowout preventer, BOP *(Bohrtechnik, Fördertechnik)*
Blutdruckmessgerät n <Gerät> blood pressure meter
Blutgift n <Chemie> *(BE)* haemotoxin, *(AE)* hemotoxin
Blutschwarz n <Lebensmittel> blood black
Blutserumalbumin n <Lebensmittel> blood albumin
B-Modulation f <Elektronik> B-modulation
BMOSFET *(Halbleiter-Feldeffekttransistor mit Rückgatter)* <Elektronik> BMOSFET *(back-gate metal-oxide semiconductor field-effect transistor)*
BNC-Stecker m *(Bajonettsteckverbinder mit Überwurfmutter)* <Elektronik> BNC *(bayonet nut connector)*
Bö f <Wassertrans> gust, squall
boardseitige Diagnostik f *(OBD)* <Kfztech> on board diagnostics *(OBD)*
Bobine f <Textil> bobbin
Bock m <Fertig> trestle; stand *(Auswuchten)*
Bockbrücke f <Bau> trestle bridge
Bockkran m <Eisenbahn> gantry crane
Bocklager n <Fertig> pedestal bearing
Bockschere f <Bau> bench shears
Bocksprungtest m <Comp & DV, Qual> leapfrog test
Bockstütze f <Bau> trestle shore
Bockwalzen n <Fertig> roll cogging
Boden m 1. <Bau, Kohlen> *(BE)* earth, *(AE)* ground, soil; 2. <Textil> blotch *(Textildruck)* • **vom Boden gesteuert** <Raumfahrt> ground-controlled
Bodenabfertigung f <Lufttrans> ground handling services
Bodenabstand m <Kfztech> road clearance
Bodenabzugskanal m <Wasserversorg> bottom culvert
Bodenadresse f <Telekom> *(BE)* earth address, *(AE)* ground address
Bodenantenne f <Funktech> ground antenna, terrestrial antenna
Bodenart f <Kohlen> type of soil
Bodenaushub m <Abfall> excavation
Bodenbefeuerung f <Lufttrans> ground lighting
Bodenbelag m 1. <Bau> decking; 2. <Ker & Glas> bottom paving; 3. <Sicherheit> flooring

Bodenbelastung

Bodenbelastung f <Umweltschutz> impact of soil
Bodenbeplattung f <Wassertrans> bottom plating
Bodenbetrieb m 1. <Lufttrans> ground operation *(Flughafen)*; 2. <Raumfahrt> ground operation *(Raumschiff)*
Bodenbewegung f 1. <Bau> earthwork; 2. <Lufttrans> *(AE)* ground maneuver, *(BE)* ground manoeuvre
Bodenbewegung f **nach oben** <Kerntech> upward heave of ground
Bodenblech n <Lufttrans> floor panel
Bodenbohranlage f <Erdöl> ground rig *(Bohrtechnik)*
Boden-Bord Funkverkehr m <Funktech, Lufttrans> ground-to-air communication
Bodencodierer m <Verpack> bottom coder
Bodendichte f <Kohlen> bulk density
Bodendiffusion f <Elektronik> top-bottom diffusion
Bodendruck m 1. <Bau, Kohlen> bearing load, earth pressure; 2. <Eisenbahn> ground pressure
Bodeneffekt m <Lufttrans> ground effect
Bodeneffektfahrzeug n <Kfztech> surface effect vehicle
Bodeneffektgerät n *(BEG)* <Kfztech> ground effect machine *(GEM)*
Bodeneinfluss m <Lufttrans> ground effect
Bodeneinrichtungen fpl 1. <Lufttrans> ground installations; 2. <Raumfahrt> ground facilities
Bodenempfangsstelle f 1. <Funktech> *(BE)* earth receiving station, *(AE)* ground receiving station; 2. <Raumfahrt> *(BE)* earth receiving station, *(AE)* ground receiving station *(Weltraumfunk)*
Bodenenergieversorgung f <Raumfahrt> ground power system
Bodenentleererkübel m <Bau> drop-bottom bucket
Bodenfalz m <Verpack> bottom fold
Bodenfalz- und Nähmaschine f <Verpack> bottom folding and seaming machine
Bodenfilter n <Abfall> soil filter
Bodenfließen n <Bau> mudflow
Bodenfräse f <Trans> pulvimixer
Bodenfreiheit f 1. <Fertig> underclearance; 2. <Kfztech> ground clearance; road clearance *(Karosserie)*
bodenfremde Substanz f <Umweltschutz> allochthonous matter
Bodenfüllung f <Verpack> bottom filling
Bodenfunkfeuer n <Funktech, Lufttrans> ground beacon, ground radio beacon
Bodenfunkstation f <Funktech, Lufttrans> ground radio station
Bodenfunkstelle f <Funktech> ground station
Bodengang m **neben dem Kiel** <Wassertrans> keel strake
bodengebundener Hochgeschwindigkeitstransport m <Trans> high-speed ground transportation
bodengeleiteter Anflug m <Lufttrans> ground-controlled approach
bodengestützte Seefunknavigation f <Funktech, Wassertrans> maritime terrestrial radio navigation
Bodenglas n <Ker & Glas> bottom glass
Bodenhaftung f <Kfztech> ground adhesion, road adherence, road adhesion
Bodenheizkabel n <Fertig> floor-warming cable
Bodenkippe f <Bau> tip
Bodenklappe f 1. <Heiz & Kälte> floor damper; 2. <Kfztech> drop bottom; 3. <Verpack> bottom flap
Bodenkolonne f <Erdöl> plate column *(Destillationstechnik)*
Bodenkunde f 1. <Bau, Kohlen> pedology, soil science; 2. <Wasserversorg> soil science
Bodenluftkonzentration f <Umweltschutz> soil atmosphere concentration

Bodenluke f 1. <Bau> trap door; 2. <Lufttrans> floor hatch
Bodenmechanik f <Bau, Kohlen> soil mechanics
Bodenniveau n <Kerntech> ground level
Bodenplatte f 1. <Bau> bed plate, bottom, tile; 2. <Elektronik> base *(des Quarzhalters)*; 3. <Fertig> loam plate; 4. <Mechan> floor plate; 5. <Raumfahrt> sole plate *(Raumschiff)*
Bodenpressung f <Bau, Kohlen> bearing load
Bodenprobe f <Kohlen> core sample
Bodenprüfung f <Lufttrans> ground test
Bodenpunkt m <Mechan> benchmark
Bodenresonanz f <Lufttrans> ground resonance
Bodenriss m <Ker & Glas> bottom tear
Bodenrückstand m <Chemie> residue *(Öltank)*
Bodensatz m 1. <Bau> deposit; 2. <Chemtech> precipitate; 3. <Lebensmittel> sludge; 4. <Verpack> deposit
Bodensau f <Fertig> salamander *(Gießen)*
Bodensäulen-Abscheider m <Umweltschutz> plate column scrubber
Bodenschicht f <Wasserversorg> bottom deposit
Bodensenkung f 1. <Bau> subsidence; 2. <Kohlen> settlement
Bodensetzung f <Abfall> settlement, settling
Bodensicht f <Lufttrans> ground visibility
Bodenstabilisierung f <Bau> cementation
Bodenstation f 1. <Fernseh> earth station; 2. <Funktech> ground station; 3. <Raumfahrt> earth station, land station *(Weltraumfunk)*; 4. <Wassertrans> local user terminal *(LUT)*
Bodenstaubsauger m <Elektrotech> floor vacuum cleaner
Bodenstein m 1. <Fertig> hearth bottom; 2. <Ker & Glas> bottom block
Bodenstoß m <Bau> thrust
Bodenströmung f <Wasserversorg> bottom current, bottom flow
Bodenstromversorgung f <Lufttrans> ground power supply
Bodentasse f <Verpack> base cup
Boden- und Seitenbeladungsanlage f <Verpack> automatic side and bottom loading machine
Bodenuntersuchung f 1. <Bau> subsoil exploration, soil exploration; 2. <Kohlen> soil exploration
Bodenventil n <Maschinen> bottom valve, foot valve
Bodenverdichtungsmesser m <Kohlen> *(AE)* settlement gage, *(BE)* settlement gauge
Bodenvermörtelung f <Bau> soil stabilization
bodenverschmutzender Stoff m <Umweltschutz> soil pollutant
Bodenversiegelung f <Abfall> surface sealing
bodenverunreinigender Stoff m <Umweltschutz> land pollutant
Bodenverunreinigung f 1. <Abfall> ground surface contamination; 2. <Umweltschutz> land pollution, soil pollution
Bodenwanne f <Kfztech> floor pan *(Karosserie)*
Bodenwasser n <Wasserversorg> soil water
Bodenwelle f <Funktech, Telekom, Wassertrans> ground wave *(Funk)*
Bodenwellenausbreitung f <Funktech> ground wave propagation
Bodenwrange f <Wassertrans> floor, floor plate *(Schiffbau)*
Böen fpl <Lufttrans> gust
Böenabminderungsfaktor m <Lufttrans> gust alleviation factor
Böenbelastungsgrenze f <Lufttrans> gust load limit
Böenbildungszeit f <Lufttrans> gust formation time
Böenlastvielfaches n <Lufttrans> gust load factor

Böennenngeschwindigkeit f <Lufttrans> nominal gust velocity
Böentiefe f <Lufttrans> gust gradient distance
Böenverriegelung f <Lufttrans> gust lock
Böen-V-n-Diagramm n <Lufttrans> gust V-n diagram, gust envelope
Bogen m 1. <Bau> arch, bow; 2. <Druck> sheet, signature; 3. <Künstl Int> arc *(Graph)*; edge, link *(zwischen Knoten in einem Graph)*; 4. <Maschinen> bend, bow; 5. <Raumfahrt> flute *(Raumschiff)*
Bogenanfang m <Bau> springing
Bogenanlage/mit <Druck> sheet-fed
Bogenbildung f <Fertig> arcing
Bogenblende f <Bau> blind arch *(Architektur)*
Bogenbrücke f <Bau> arched beam bridge
Bogenbrücke f **mit Zugband** <Bau> bowstring bridge
Bogendickenmesser m <Bau> *(AE)* bow calipers, *(BE)* bow callipers
Bogenentladung f <Elektriz> arc discharge
Bogenentladungsröhre f <Elektrotech> arc discharge tube
Bogenfeder f <Maschinen> bow spring
Bogenfederzirkel m <Bau> bow spring compasses
bogenförmiger Anschnitt m <Ker & Glas> scalloped bevel
Bogengang m <Bau> arcade
Bogengewölbe n <Bau> arched vault
Bogenherzstück n <Eisenbahn> curved common crossing
Bogenkalender m <Papier> sheet calender
Bogenlampe f <Elektriz> arc lamp
Bogenlampenkohle f <Maschinen> arc lamp carbon
Bogenlänge f <Geom> arc length
Bogenleibung f <Bau> intrados
Bogenlicht n <Elektriz> arc light
Bogenlinie f <Konstzeich> curved line
Bogenmaß n 1. <Geom> circular measure; 2. <Maschinen> radian measure
Bogenmauer f <Wasserversorg> arch dam
Bogenminute f <Phys> arc minute
Bogenpresse f <Druck> sheet-fed machine
Bogenrückschlag m <Elektrotech> arc back
Bogensäge f 1. <Bau> backsaw; 2. <Maschinen> bow saw
Bogenschenkel m <Bau> haunch
Bogenschneider m <Verpack> sheet-cutting machine
Bogenschweißen n <Mechan> arc welding
Bogensekunde f 1. <Metrol> second; 2. <Phys> arc second; second of arc *(Winkelmaß)*
Bogensignatur f <Druck> signature number
Bogenspektrum n <Phys> arc spectrum
Bogenspitzzirkel m <Metrol> dividers with quadrant
Bogenstaumauer f <Bau, Wasserversorg> arch dam
Bogenstein m <Bau> arched tile
Bogenträger m **mit Zugband** <Bau> bowstring girder
Bogenverband m <Bau> arch bond
Bogenverzahnung f <Fertig> hyphoid teeth
Bogenzahnkupplung f <Maschinen> curved-tooth gear coupling
Bogenziegel m <Bau> arch brick, *(AE)* gage brick, *(BE)* gauge brick, voussoir
bogenziehende Kontakte mpl <Elektriz> arcing contacts
Bogenzirkel m 1. <Bau> bow compass; 2. <Maschinen> wing compasses
Bogenzuführmaschine f <Verpack> sheet machine
Bogenzuführungsapparat m <Druck> sheet feeder
Bogheadkohle f <Kohlen> boghead coal
Bogie m 1. <Eisenbahn> *(BE)* bogie, *(AE)* trailer; 2. <Kfztech> *(BE)* bogie *(Anhänger)*

Bohle f <Bau> batten, board, deal, plank • **Bohlen hochkant verlegen** <Bau> set boards edgewise
Bohlenbelag m <Bau> planking
Bohlengang m <Bau> duck board, strake
Bohlwand f <Mechan> bulkhead
Böhmit n <Ker & Glas> boehmite
Bohr... 1. <Erdöl> drilling; 2. <Fertig> boring, drilling; 3. <Mechan> drill
Bohranlage f <Erdöl> drilling rig, rig
Bohranlagenetage f <Erdöl> rig floor *(Bohrtechnik)*
Bohransatzpunkt m <Bau, Kohlen> boring site
Bohrarbeiter m <Erdöl> floorman, toolpusher; driller *(Bohrtechnik)*; derrick man *(Tiefbohrtechnik)*
Bohrarm m <Kohlen> cutter arm
Bohrausschuss m <Fertig> drilling rejects
Bohrbank f 1. <Maschinen> boring bench; 2. <Mechan> drill press
bohrbarer Stopfen m <Erdöl> drillable plug
Bohrbarkeit f <Erdöl> drillability
Bohrbedingungen fpl <Erdöl> drilling conditions
Bohrbetrieb m <Erdöl> drilling operations *(Tiefbohrtechnik)*
Bohrbrunnen m <Wasserversorg> bored well
Bohrbuchse f <Maschinen> drill bushing
Bohrbüchse f <Maschinen> jig bush
Bohreinrichtung f <Maschinen> drilling attachment
Bohreinsatz m <Maschinen> drill
bohren v 1. <Bau> bore, drill; 2. <Erdöl> bore; 3. <Kohlen> bore, prospect; 4. <Maschinen, Mechan> bore, drill; 5. <Metall> prospect; 6. <Papier> drill
Bohren n 1. <Bau> boring, drilling; 2. <Erdöl, Kohlen> boring; 3. <Maschinen, Mechan> boring, drilling; 4. <Papier> drilling
Bohren m **mit geraden Nuten** <Maschinen> straight-fluted drill
Bohren m **mit Tiefenanschlag** <Maschinen> stop drill
Bohren m **mit Vierkantschaft** <Maschinen> square shank drill
Bohren n **ohne Bohrlochverrohrung** <Erdöl> open-hole drilling *(Bohrtechnik)*
Bohren n **von Rohknüppeln** <Fertig> billet drilling
Bohren n **von Sprenglöchern** <Kohlen> drilling of blast holes
Bohrer m 1. <Bau> bit, borer; 2. <Erdöl> driller; 3. <Kohlen> borer; 4. <Maschinen> drill • **Bohrer betreiben** <Maschinen> run a drill *(mit Druckluft)*
Bohrer m **für Bohrwinden** <Maschinen> brace bit
Bohrer m **mit geraden Nuten** <Maschinen> straight-fluted drill
Bohrer m **mit Tiefenanschlag** <Maschinen> stop drill
Bohrer m **mit Vierkantschaft** <Maschinen> square shank drill
Bohrereinsatz m <Bau> bit
Bohrerfutter n <Maschinen> drill socket
Bohrergrößen fpl <Maschinen> *(AE)* gage numbers, *(BE)* gauge numbers
Bohrerhalter m <Maschinen> drill holder
Bohrerlehre f <Maschinen> *(AE)* drill gage, *(BE)* drill gauge
Bohrerschleifgerät n <Maschinen> drill sharpener
Bohrerschleifmaschine f <Maschinen> drill grinder
Bohrfortschritt m <Erdöl> rate of penetration; drilling rate *(Bohrtechnik)*
Bohrfutter n 1. <Fertig> drill chuck; 2. <Maschinen> drill chuck, drill head; 3. <Mechan> drill bushing, jig, pad
Bohrgarnitur f <Erdöl> drill string *(Bohrtechnik)*
Bohrgeschwindigkeit f 1. <Bau> drilling rate; 2. <Erdöl> rotation rate *(Bohrtechnik)*; rate of penetration *(Tiefbohrtechnik)*

Bohrgestängeeinbau

Bohrgestängeeinbau m <Erdöl> snubbing (Bohrtechnik)
Bohrgestell n <Nichtfoss Energ> drilling rig
Bohrgut n <Erdöl> cuttings, drill cuttings
Bohrhammer m <Bau> rock drill
Bohrhubinsel f <Erdöl> jack-up rig (Offshore-Technik)
Bohrhülse f <Maschinen> boring sleeve
Bohringenieur m <Erdöl> drilling engineer
Bohrinsel f <Wassertrans> drilling platform, oil rig
Bohrinselversorgungsschiff n <Wassertrans> offshore drilling rig supply vessel
Bohrkäfer m <Lebensmittel> borer (Pflanzenschädlinge)
Bohrkasten m <Fertig> box drill jig
Bohrkern m 1. <Bau, Erdöl> core; 2. <Fertig> core, pin, plug (Hohlbohren)
Bohrkernanalyse f <Erdöl> core analysis
Bohrkernrohr n <Erdöl> core barrel
Bohrkette f <Kohlen> cutter chain
Bohrklein n <Erdöl> cuttings; drill cuttings (Bohrtechnik)
Bohrkleinausfall m <Erdöl> cuttings dropping out
Bohrkleingas n <Erdöl> cuttings gas (Tiefbohrtechnik)
Bohrknarre f <Maschinen> ratchet, ratchet brace, ratchet drill
Bohrkopf m <Maschinen> boring head, cutter head
Bohrkopf m **mit verstellbarem Bohrbild** <Fertig> adjustable centre head
Bohrkratzer m 1. <Kohlen> scraper; 2. <Wasserversorg> cleaner-up
Bohrkrone f <Kohlen> drill bit
Bohrkronenbolzen mpl <Erdöl> drill bit studs
Bohrkronenzähne mpl <Erdöl> drill bit studs (Tiefbohrtechnik)
Bohrkurbel f <Maschinen> bit brace, bit stock, breast drill brace, crank brace
Bohrleier f <Fertig> hand brace
Bohrleistung f <Fertig> drilling capacity
Bohrloch n 1. <Bau> borehole, hole; 2. <Erdöl> well; 3. <Kohlen> borehole; 4. <Maschinen, Wasserversorg> bore, borehole • **aus dem Bohrloch fahren** <Erdöl> coming out of hole (Bohrwerkzeug)
Bohrlochkopf m 1. <Bau> casing head; 2. <Erdöl> wellhead
Bohrlochmessung f <Erdöl> downhole measurements (Bohrtechnik); well logging (Messtechnik)
Bohrlochproduktivitätstest m <Erdöl> drill stem test
Bohrlochpumpe f <Bau, Kohlen, Wasserversorg> borehole pump
Bohrlochsohle f <Erdöl> bottom hole (Bohrtechnik)
Bohrlochsohlenausrüstung f (BSA) <Erdöl> bottom hole assembly (BHA)
Bohrlochsohlenbedingungen fpl <Erdöl> bottom hole conditions
Bohrlochsohlen-Zementstopfen m <Erdöl> bottom cementing plug
Bohrlochtemperatur-Logging n <Erdöl> temperature well logging
Bohrlochtemperatur-Messverfahren n <Erdöl> temperature well logging
Bohrlochvermessung f <Erdöl> logging
Bohrlochwandung f <Wasserversorg> casing
Bohrmannschaft f <Erdöl> drilling crew
Bohrmaschine f 1. <Kohlen> drill; 2. <Maschinen> boring machine, drill, drill press, drilling machine; 3. <Mechan> boring machine, drill, drilling machine
Bohrmaschine f **mit Hebelvorschub** <Maschinen> lever feed drilling machine
Bohrmaschinenfutter n <Mechan> chuck
Bohrmeißel m 1. <Bau> bore bit, boring bit, jumper; 2. <Erdöl> bit (Bohrtechnik); drill bit (Tiefbohrtechnik); 3. <Fertig> drill bit, single point boring tool; 4. <Kohlen> cutting bit; 5. <Maschinen> boring tool

Bohrmeißelabnutzung f <Erdöl> bit wear
Bohrmeißelauflast f (WOB) <Erdöl> weight on bit (WOB)
Bohrmeißelbrecher m <Erdöl> bit breaker
Bohrmeißelverschleiß m <Erdöl> bit wear
Bohrmeißelwechsel m <Erdöl> BC, bit change
Bohrmesser n <Fertig> bit
Bohrplattform f <Erdöl> drilling platform (Bohrtechnik)
Bohrprogramm n <Erdöl> (AE) drilling program, (BE) drilling programme (Tiefbohrtechnik)
Bohrrechteck n <Fertig> area of drilling
Bohrrohr n 1. <Erdöl> drill pipe; 2. <Wasserversorg> casing pipe
Bohrsäule f <Maschinen> drilling pillar
Bohrschablone f 1. <Fertig> jig; 2. <Maschinen> drill template, drilling jig, drilling template
Bohrschacht m <Kohlen> pit
Bohrschachthebeschwinge f <Erdöl> pit lever (Getriebe)
Bohrschappe f <Kohlen> spoon sampler
Bohr'scher Radius m (a0) <Phys> Bohr radius (a0)
Bohr'sches Magneton n <Phys> Bohr magneton
Bohrschiff n 1. <Erdöl> drill barge (Bohrtechnik); 2. <Wassertrans> drill ship
Bohrschlamm m **auf Ölbasis** <Erdöl> oil-base mud (Tiefbohrtechnik)
Bohrschlamm m **auf Wasserbasis** <Erdöl> water-based mud
Bohrschlammpumpe f <Erdöl> slush pump
Bohrschlitten m <Fertig> headstock; drilling head (Radialbohrer)
Bohr-Sommerfeld-Modell n <Phys> Bohr-Sommerfeld model
Bohrspan m <Fertig> drill chip
Bohrspindel f 1. <Fertig> drill spindle; 2. <Maschinen> boring spindle, drill spindle, drilling spindle; 3. <Mechan> boring spindle
Bohrspindelkopf m <Fertig> drilling head
Bohrspitzenhalter m <Maschinen> bit holder
Bohrstahl m <Maschinen> drill steel
Bohrstange f 1. <Bau> bore rod, boring rod; 2. <Maschinen> boring bar
Bohrstangenführungslager n <Fertig> steady bearing
Bohrstelle f <Bau, Kohlen> boring site
Bohrstift m <Bau> drill pin
Bohrstrang m <Erdöl> drilling line (Tiefbohrtechnik)
Bohrtiefe f <Maschinen> drill depth, drilling depth
Bohrtiefenbegrenzung f <Fertig> depth control
Bohrtisch m 1. <Erdöl> drilling table (Tiefbohrtechnik); 2. <Kerntech> jig table; 3. <Maschinen> drilling table
Bohrtour f <Erdöl> trip (Bohrtechnik)
Bohrturm m 1. <Bau> derrick; 2. <Erdöl> derrick (Bohrtechnik); 3. <Mechan> derrick; 4. <Nichtfoss Energ> drilling rig • **den Bohrturm auf Schlitten versetzen** <Erdöl> skid the rig (Bohrtechnik)
Bohrturm m **mit Gewinde** <Kerntech> threaded hole
Bohrturmkeller m <Erdöl> derrick cellar
Bohrturmkopf m <Erdöl> crown, derrick crown
Bohrturmkran m <Erdöl> derrick crane
Bohrturmsohle f <Erdöl> derrick floor (Bohrtechnik)
Bohr- und Drehwerk n <Maschinen> boring and turning mill
Bohr- und Fräsmaschine f <Maschinen> boring and milling machine
Bohr- und Plandrehmaschine f <Fertig> boring and facing lathe
Bohr- und Schrämmaschine f <Kohlen> holing and shearing machine
Bohrung f 1. <Erdöl> well; 2. <Fertig> hole (Kunststoffinstallationen); 3. <Hydraul> opening (in Trennwand,

Platte); 4. <Ker & Glas> *(BE)* bore, *(AE)* corkage; 5. <Kfztech> bore *(Motor, Zylinder)*; 6. <Kohlen> bore, drilling; 7. <Maschinen> bore, borehole, passage, port; 8. <Nichtfoss Energ, Raumfahrt> bore • **Bohrung vorgießen** <Fertig> core
Bohrungsdurchmesser *m* 1. <Erdöl, Kohlen> well bore *(Bohrtechnik)*; 2. <Maschinen> bore, size of bore
Bohrungslehre *f* <Maschinen> *(AE)* caliper gage, *(BE)* calliper gauge, *(AE)* internal cylindrical gage, *(BE)* internal cylindrical gauge
Bohrungsmessgerät *n* <Metrol> *(AE)* bore gage, *(BE)* bore gauge
Bohrvorrichtung *f* <Maschinen> drilling jig
Bohrvorschacht *m* <Erdöl> derrick cellar
Bohrvorschub *m* <Fertig> drilling feed
Bohrwelle *f* <Maschinen> boring bar
Bohrwerk *n* <Maschinen, Mechan> boring mill
Bohrwerkständer *m* <Maschinen> boring-mill column
Bohrwerkzeug *n* 1. <Bau> drilling tool; 2. <Maschinen> drill
Bohrwinde *f* 1. <Bau> brace; 2. <Fertig> bit brace
Boiler *m* 1. <Heiz & Kälte> hot-water tank; 2. <Maschinen> hot-water heater, water heater; 3. <Mechan, Thermod> boiler
Boje *f* <Wassertrans> buoy *(Seezeichen)* • **an die Boje gehen** <Wassertrans> pick up moorings
Bojenreep *n* <Wassertrans> buoy rope *(Tauwerk)*
Bolometer *n* 1. <Elektriz, Elektrotech, Fertig> bolometer; 2. <Gerät> bolometric instrument; 3. <Heiz & Kälte, Phys, Raumfahrt, Thermod> bolometer
bolometrisch *adj* <Fertig> bolometric
bolometrisches Messinstrument *n* <Gerät> bolometric instrument
Boltzmann'sche Funktion *f* <Phys, Thermod> Boltzmann function
Boltzmann'sche Gleichung *f* <Phys, Thermod> Boltzmann equation of particle conservation, Boltzmann equation
Boltzmann'sche Konstante *f (k)* 1. <Phys> Boltzmann constant *(k)*; 2. <Thermod> Boltzmann constant *(Boltzmann'sche Zahl, k)*
Boltzmann'sche Zahl *f (k)* 1. <Phys> Boltzmann constant *(k)*; 2. <Thermod> Boltzmann constant *(Boltzmann'sche Konstante, k)*
Bolzen *m* 1. <Bau> bolt, fang bolt, pintle, pin, screw, stay, stud; 2. <Fertig> bolt; pin *(Kette)*; 3. <Kfztech, Maschinen> bolt; 4. <Mechan> hob
Bolzen *m* **mit Splint** <Fertig> cotter bolt
Bolzen *m* **mit versenktem Kopf** <Bau> countersunk head-bolt
Bolzenanker *m* <Bau> rag bolt
Bolzenaufschweißen *n* <Fertig> stud welding
Bolzenkopf *m* <Maschinen> bolt head
Bolzenkupplung *f* <Maschinen> bolt coupling, pin coupling
Bolzenloch *n* <Maschinen> bolt hole
Bolzenlochbohrmaschine *f* <Fertig> stud-inserting machine
Bolzenschere *f* <Maschinen> bolt clipper
Bolzenschneider *m* 1. <Kfztech> bolt cutter; 2. <Maschinen> bolt cropper, bolt cutter; 3. <Mechan> bolt cutter
Bolzenschraube *f* <Bau, Maschinen> bolt
Bolzenschweißen *n* <Bau> stud welding
Bolzenschweißpistole *f* <Bau> stud welding gun
Bolzenstange *f* <Bau> stud bolt
Bolzentreiber *m* <Maschinen> pin drift
Bolzenverbindung *f* 1. <Fertig> bolted union; 2. <Maschinen> bolt fastening; 3. <Raumfahrt> bolted connection

Bombage *f* <Lebensmittel> swell
Bombenkalorimeter *n* <Gerät, Phys> bomb calorimeter
Bombieren *n* 1. <Fertig> crowning *(Walzen)*; 2. <Lebensmittel> blowing *(Konservendosen)*
bombierter Siebsauger *m* <Papier> cambered suction box
Bombierung *f* <Papier> camber
Bombierungsstange *f* <Papier> camber bar
Boole'sche Abfrage *f* <Comp & DV> Boolean query
Boole'sche Algebra *f* 1. <Comp & DV> Boolean algebra, Boolean logic; 2. <Math> Boolean algebra
Boole'sche Komplementierung *f* <Comp & DV> NOT operation
Boole'sche Variable *f* <Comp & DV> Boolean variable
Boole'sche Verknüpfung *f* <Regelung> Boolean term
Boole'scher Datentyp *m* <Comp & DV> logical-type
Boole'scher Operator *m* <Comp & DV> Boolean operator
Boole'scher Primärausdruck *m* <Comp & DV> Boolean primary
Boole'scher Ring *m* <Comp & DV> Boolean ring
Boole'scher Sekundärausdruck *m* <Comp & DV> Boolean secondary
Boole'scher Wert *m* <Comp & DV> Boolean value, logical value
Booster *m* <Maschinen> booster
Boosterpumpe *f* <Maschinen> booster pump
Booster-Verstärker *m* <Funktech> booster amplifier
Boot *n* <Wassertrans> boat • **Boote aussetzen** <Wassertrans> lower the boats *(Notfall)*
Booten *n* <Comp & DV> boot-up
Bootsanhänger *m* <Wassertrans> boat trailer
Bootsdeck *n* <Wassertrans> boat deck
Bootshaken *m* <Wassertrans> boathook
Bootshaus *n* <Trans> boat house
Bootshissstropp *m* <Wassertrans> boat sling
Bootsklampe *f* <Wassertrans> boat chock *(Deckausrüstung)*
Bootskörper *m* <Wassertrans> hull
Bootskurs *m* <Wassertrans> boat's heading
Bootsladung *f* <Wassertrans> boat load
Bootslagerung *f* <Wassertrans> cradle
Bootslift *m* <Wassertrans> boat elevator, boat lift
Bootsmann *m* <Wassertrans> boatswain
Bootsmannsstuhl *m* <Wassertrans> boatswain's chair
Bootsrolle *f* <Wassertrans> boat stations bill *(Notfall)*
Bootstaljenläufer *m* <Wassertrans> boat fall *(Deckausrüstung)*
Bootstank *m* <Wassertrans> boat tank
Bootstransport *m* <Wassertrans> boat carriage
Bootstrap *m* 1. <Fernseh> bootstrap; 2. <Math> bootstrap *(statistisches Schätzverfahren)*
Bootstrapping *n* <Math> bootstrapping *(statistisches Schätzverfahren)*
Bootsverdeck *n* <Wassertrans> canopy
BOP *(Blowout-Preventer)* <Erdöl> BOP (blowout preventer)
Bor *n (B)* <Chemie> boron *(B)*
Bor... <Chemie> boracic, boric
Boral *n* <Kerntech> boral
Boran *n* <Chemie> borane
borartig *adj* <Chemie> boric
Borat *n* <Chemie> borate
Borax *m* 1. <Chemie> borax, disodium tetraborate decahydrate; 2. <Ker & Glas> borax
Boraxperle *f* <Bau> borax bead
Boraxperlenversuch *m* <Chemie> borax bead test
Boraxperlprüfung *f* <Metall> borax bead test
Bord *m* <Lufttrans, Wassertrans> board *(Schiff)* • **an Bord befindlich** 1. <Lufttrans> inboard; 2. <Wassertrans> a-

Bord

board, on board • **an Bord bringen** <Wassertrans> put on board • **an Bord gehen** <Wassertrans> board • **an Bord nehmen** <Wassertrans> embark *(Passagiere, Ladung)*; ship *(Passagiere)* • **über Bord** <Wassertrans> overboard • **über Bord fallen** <Wassertrans> fall overboard • **über Bord spülen** <Wassertrans> wash overboard • **über Bord werfen** <Wassertrans> jettison • **von Bord gehen** <Wassertrans> disembark *(Passagiere)*
Bordabstandswarnanzeiger *m* <Lufttrans> airborne proximity warning indicator
Borda-Mundstück *n* <Hydraul> Borda mouthpiece
Borda-Mündung *f* <Hydraul> Borda mouthpiece
Bord-Boden-Funkverbindung *f* <Funktech, Telekom> air-to-ground radio communication
Bord-Bord-Alarmierung *f* <Telekom> ship-to-ship alerting
Bord-Bord-Funkverbindung *f* <Funktech, Telekom> air-to-air radio communication
Bord-Bord-Verbindung *f* <Telekom> air-to-air communication
Bord-Bord-Verkehr *m* <Telekom> air-to-air communication
Bordcomputer *m* <Raumfahrt> on-board computer *(Raumschiff)*
Borde *f* <Textil> trimming
Bördel… 1. <Bau> flanged; 2. <Fertig> flanging; 3. <Maschinen> crimping; 4. <Textil> flanging
Bördelblech *n* <Bau> flanged plate
Bördeleisen *n* <Fertig> bordering tool
Bördelmaschine *f* 1. <Fertig> flanging machine; 2. <Maschinen> seaming machine; 3. <Verpack> flanging machine
bördeln *v* 1. <Fertig> border, clinch; 2. <Maschinen> seam; 3. <Mechan> crimp; 4. <Verpack> falten
Bördeln *n* 1. <Fertig> flanging; 2. <Maschinen> crimping
Bördelpresse *f* <Maschinen> crimping machine, flanging press
Bördelrand *m* <Mechan> bead
Bördelversuch *m* <Fertig> flanging test
Bördelwerkzeug *n* <Maschinen> crimping tool
Bördelzange *f* <Maschinen> crimping pliers
Bordempfangsschein *m* <Wassertrans> mate's receipt *(Handelsmarine)*
Bordenergieversorgung *f* <Elektrotech> on-board energy supply *(Magnetschwebetechnik)*
Bordfernsehen *n* <Fernseh> airborne television
Bordkollisionswarnsystem *n* <Lufttrans> airborne collision avoidance system
Bordkommunikationsstation *f* <Telekom> on-board communication station
Bordküche *f* <Lufttrans> galley
Bord-Land-Alarmierung *f* <Telekom> ship-to-shore alerting
Bord-Land-Funkverbindung *f* <Funktech, Wassertrans> ship-to-shore radio communication *(Funk)*
Bordnetz *n* 1. <Kfztech> vehicle electrical distribution system; 2. <Wassertrans> shipboard electrical system
Bordnetzumrichter *m* <Elektrotech> on-board converter *(Magnetschwebetechnik)*
Bordradar *n* <Elektronik, Funkort, Lufttrans> airborne radar *(Cockpit)*
Bordradargerät *n* <Elektronik, Funkort, Lufttrans> airborne radar equipment
Bordrand *m* 1. <Fertig> flange *(Riemenscheibe)*; 2. <Maschinen> flange
Bordrechner *m* <Kfztech> car computer
Bordschaltung *f* <Raumfahrt> on-board switching *(Weltraumfunk)*
Bordsprechanlage *f* 1. <Aufnahme> intercom; 2. <Telekom> interphone

Bordstein *m* 1. <Bau> *(AE)* curb, *(BE)* kerb, *(AE)* curbstone, *(BE)* kerb, *(BE)* kerbstone; 2. <Ker & Glas> fluxline block
Bordsysteme *npl* <Raumfahrt> on-board systems *(Raumschiff)*
Bordterminal *m* **für Satellitenfunk** *(SES)* <Wassertrans> ship earth station *(SES)*
Borduhrrücksetzung *f* <Raumfahrt> clock recovery *(Weltraumfunk)*
Bordüre *f* <Druck> border
Bordverarbeitung *f* <Raumfahrt> on-board processing *(Weltraumfunk)*
Bore *f* <Wassertrans> bore *(Fluss)*
Borhydrid *n* <Chemie> borane
Borid *n* <Chemie> boride
Borneocampher *m* <Chemie> borneol
Borneol *n* <Chemie> borneol, camphol
Borneolacetat *n* <Chemie> bornyl acetate
Bornyl… <Chemie> bornyl
Bornylacetat *n* <Chemie> bornyl acetate
Bornylalkohol *m* <Chemie> borneol, bornyl alcohol
Borosilikat *n* <Chemie> borosilicate
Borosilikatglas *n* <Ker & Glas, Labor> borosilicate glass
Borsäure *f* 1. <Chemie> boric acid, orthoboric acid; 2. <Ker & Glas> boric acid
Borsäuremischpumpe *f* <Kerntech> boric acid blender
Borstahl *m* <Metall> boron steel
Borstahlabsorber *m* <Kerntech> boronated steel absorber
Bort *m* <Fertig> bort
Borte *f* 1. <Ker & Glas> braid; 2. <Textil> braid, braiding
Borte *f* **der Glasscheibe** <Ker & Glas> edge of the sheet
Bortenführungen *fpl* <Ker & Glas> edge guides
Bortenhalter *m* <Ker & Glas> edge holder
Bortrioxid *n* <Ker & Glas> boric oxide
Böschung *f* 1. <Bau> bank, slope; 2. <Kohlen> slope
Böschungsabdeckung *f* <Bau> facing
Böschungsabsatz *m* <Wasserversorg> offset
Böschungserdhobel *m* <Bau> angledozer
Böschungsfuß *m* <Bau, Kohlen> slope toe
Böschungshobel *m* <Bau> backsloper
Böschungskrone *f* <Bau, Kohlen> slope top
Böschungsmauer *f* <Bau> retaining wall
Böschungsneigung *f* <Bau> batter
Böschungssicherung *f* <Bau> slope protection
Böschungsstandfestigkeit *f* <Bau, Kohlen> slope stability
Böschungsversagen *n* <Bau, Kohlen> slope failure
Bose-Einstein'sche Kondensation *f* <Phys> Bose-Einstein condensation
Bose-Einstein'sche Statistik *f* <Phys> Bose-Einstein statistics
Bose-Einstein'sche Verteilung *f* <Phys> Bose-Einstein distribution
Boson *n* <Phys, Teilphys> boson
Bosse *f* <Maschinen> boss
Bossieren *n* <Ker & Glas> palleting
Bossierholz *n* <Ker & Glas> pallet
BOT *(Bandanfang)* <Comp & DV> BOT *(beginning of tape)*
Bote *m* <Trans> carrier
Bottich *m* 1. <Fertig> vat; 2. <Labor> trough; 3. <Lebensmittel> vat; 4. <Metall> pot; 5. <Textil> vat; 6. <Verpack> tub *(für Papierbeleimung)*
Bottomonium *n* <Teilphys> bottomonium
Bottom-Up-Strategie *f* <Künstl Int, Teilphys> bottom-up strategy
Botulinus *m* <Lebensmittel> botulinus
Botulismus *m* <Lebensmittel> botulism
Bouilleurkessel *m* <Hydraul> elephant boiler

Bourdon'sche Röhre f <Erdöl> (AE) Bourdon gage, (BE) Bourdon gauge
Bourdon'sches Manometer n 1. <Gerät> (AE) Bourdon tube gage, (BE) Bourdon tube gauge, (AE) boundary tube gage, (BE) boundary tube gauge; 2. <Phys> (AE) Bourdon gage, (BE) Bourdon gauge
Bourdonsches-Rohr n <Phys> (AE) Bourdon gage, (BE) Bourdon gauge
Bowdenzug m 1. <Kfztech> Bowden cable (Kupplung, Bremse); 2. <Mechan> Bowden cable
Box f <Foto, Verpack> box
Boxcar-Oszilloskop n <Gerät> boxcar oscilloscope (Sampling-Oszilloskop)
Boxermotor m <Kfztech> flat engine
Boxkamera f <Foto> box camera
Boxpalette f <Verpack> box pallet with sidewalls
Boyle-Mariotte'sches Gesetz n 1. <Phys> Boyle's law; 2. <Thermod> Boyle's law, Mariotte's law
Boyle'sche Temperatur f <Phys> Boyle temperature
B-Papierformat n <Druck> B-size
BPF (Bandpassfilter) <Aufnahme, Elektronik, Fernseh, Funktech, Phys, Telekom> BPF (band-pass filter)
bpi (Bits pro Zoll) <Comp & DV> bpi (bits per inch)
bps (Bits pro Sekunde) <Comp & DV> bps (bits per second)
BPS (Bremspferdestärke) <Eisenbahn, Fertig, Kfztech, Mechan> BHP (brake horsepower)
BPSK (Zweiphasenumtastung, binäre Phasenumtastung) <Telekom> BPSK (binary phase shift keying)
Bq (Becquerel) <Metrol, Phys> Bq (becquerel)
Br (Brom) <Chemie> Br (bromine)
Brackett'sche Serie f <Phys> Brackett series (Atomphysik)
Brackwasser n <Bau, Kohlen, Wasserversorg> brackish water
Brackwassermoor n <Wasserversorg> salt swamp
Bradford-Brecher m <Kohlen> Bradford breaker
Bragg-Beugungsgitterlinse f <Telekom> Bragg grating lens (optische Kommunikationstechnik)
Bragg'sche Regel f <Phys> Bragg rule
Bragg'sche Zelle f <Elektrotech> Bragg cell
Bragg'scher Winkel m <Phys> Bragg angle
Brailtau n <Wassertrans> bow line
Brain-Drain m <Trans> brain drain
Bramme f 1. <Fertig> plate slab; slab (Halbzeug); 2. <Metall> slab
Brammenherstellung f <Fertig> slabbing
Brammenschere f <Fertig> slab shears
Brammenwalzwerk n <Fertig> slabbing mill
Brand m 1. <Ker & Glas> baking, burning; 2. <Lebensmittel> rust; mildew (am Getreide); 3. <Sicherheit> fire
Brand m **an der Oberfläche** <Sicherheit> surface fire
Brandausbreitungsverhütung f <Sicherheit> fire spread prevention
Brandbekämpfung f <Bau, Lufttrans, Sicherheit, Wassertrans> firefighting (Notfall)
Brandbekämpfungseinrichtung f <Sicherheit> firefighting equipment
Brandende n <Raumfahrt> flameout (Raumschiff)
Brandfleck m <Fertig> burning (Schleifen)
Brandgefahr f <Lufttrans, Sicherheit, Wassertrans> fire hazard
Brandlast f <Thermod> fire load
Brandmarke f <Ker & Glas> burn mark
Brandmauer f 1. <Bau> fireproof wall, party wall; 2. <Sicherheit> fireproof wall
Brandmeldeanlage f <Sicherheit> fire alarm system
Brandmeldesystem n <Sicherheit> fire alarm system
Brandriss m <Fertig> crazing (Krokille)

Brandsatz m <Sicherheit> incendiary agent, incendiary device
Brandschott n 1. <Lufttrans> fire wall; 2. <Wassertrans> fire bulkhead (Schiffkonstruktion)
Brandschottring m <Lufttrans> fireseal
Brandschutz m 1. <Sicherheit> fire engineering (als Wissenschaft); fire protection; 2. <Thermod> fire prevention, fireproofing
Brandschutzanlage f <Sicherheit> fire equipment, fire protection equipment
Brandschutzausrüstung f <Sicherheit> fire protection equipment
Brandschutzbeschichtung f <Kunststoff> fireproof coating
Brandschutzkabel n <Lufttrans, Sicherheit> fire wire
Brandschutzkleidung f <Sicherheit> fire-protective garment
Brandschutzkonstruktion f <Sicherheit> fire-resistant construction
Brandschutzleitung f <Lufttrans, Sicherheit> fire wire
Brandschutztechnik f <Sicherheit> fire engineering, fire protection engineering (als Wissenschaft); fire equipment, fire protection equipment
brandschutztechnische Anlage f <Sicherheit> fire-fighting installation
brandschutztechnische Mittel npl <Sicherheit> fire protection equipment
Brandschutztür f <Sicherheit> fire door, fireproof door
Brandschutz- und Rettungsgerät n <Sicherheit> firefighting and rescue equipment
Brandschutzvorschrift f <Sicherheit> fire regulation
Brandstiftung f <Thermod> arson
Brandung f <Wassertrans> breakers (Meer)
Brandverhalten n 1. <Heiz & Kälte> (AE) fire behavior, (BE) fire behaviour; 2. <Sicherheit> (AE) burning behavior, (BE) burning behaviour (von Textilprodukten)
Brandverhütung f <Sicherheit> preventive fire protection
Brandwache f <Thermod> fire station
Branntkalk m <Chemie, Ker & Glas, Meerschmutz> quicklime
Brasse f <Wassertrans> brace (Tauwerk)
Brau... <Lebensmittel> brewing, malting
brauchbares Bildfeld n <Foto> effective image field
Brauchwasser n <Wasserversorg> process water
Brauchwassererwärmer m <Heiz & Kälte> service water calorifier
Brauen n <Lebensmittel> brewing
Braugerste f <Lebensmittel> malting barley
Brauindustrie f <Lebensmittel> brewing industry
Braun... <Chemie, Kohlen> brown
Braunbeizen n <Metall> bluing of iron
Brauneisen n <Chemie> (BE) brown haematite, (AE) brown hematite, brown iron ore
Braunfleckigkeit f <Lebensmittel> scald mark (an Obst)
Braunkohle f <Kohlen, Thermod> brown coal, lignite
Braun'sche Röhre f <Comp & DV, Druck, Elektriz, Elektronik, Fernseh, Funktech> cathode-ray tube
Braunschliffpappe f <Papier> brown mechanical pulp board
Bräunung f <Lebensmittel> bloom (Brot)
Brausesieb n <Kohlen> spraying screen
Brauwasser n <Lebensmittel> brewing liquor
Breadth-First-Suchverfahren n <Künstl Int> breadth-first search
Breakpointbetrieb m <Kontroll> breakpoint operation
Breakpointschalter m <Kontroll> breakpoint switch
Brech... <Chemie, Kohlen> crushing
Brechanlage f <Chemtech, Kohlen> crushing plant
Brechbacke f 1. <Chemtech> crusher jaw; 2. <Fertig> breaker jaw; 3. <Labor> jaw crusher (Präparierung)

Brechberge

Brechberge pl <Kohlen> broken rocks
Brecheisen n 1. <Bau> crowbar, pinch bar; 2. <Elektrotech> crowbar; 3. <Fertig> crowbar, pry; 4. <Maschinen> jimmy bar
brechen v 1. <Bau> break; 2. <Chemtech> separate out *(Emulsion)*; 3. <Fertig> refract *(Schall, Licht)*; 4. <Papier> break; 5. <Phys> refract *(Lichtstrahl)*; 6. <Wassertrans> break
Brechen n 1. <Chemtech> milling *(Erz)*; 2. <Kohlen> crushing
Brechen n **in einem Durchgang** <Kohlen> open circuit crushing
brechend adj <Raumfahrt> refractory *(Raumschiff)*
brechender Winkel m <Phys> angle of a prism *(Prisma)*
Brecher m 1. <Chemtech> crushing machine; 2. <Fertig> crusher; 3. <Kohlen> crusher, crushing mill, grinding mill; breaker *(Bergbau)*; 4. <Maschinen> breaker, crusher; 5. <Wassertrans> breaker *(Seezustand)*
Brecherplatte f <Kohlen> jaw plate
Brechgut n <Bau> crushed material
Brechkraft f 1. <Phys> converging power, power *(Linse)*; 2. <Wellphys> converging power *(Linse)*
Brechprisma n <Optik> refracting prism
Brechpunkt m <Bau> breakpoint, break
Brechstange f 1. <Bau> pinch bar, wrecking bar; claw bar *(mit Finne)*; 2. <Elektrotech> crowbar; 3. <Fertig> wrecking bar; 4. <Maschinen> handspike, jim crow; 5. <Mechan> crowbar
Brechung f 1. <Fertig> refraction *(Schall, Licht)*; 2. <Optik> refraction, refringency; refringence *(Strahlen)*; 3. <Phys> refraction; 4. <Raumfahrt> refractivity *(Weltraumfunk)*; 5. <Wellphys> refraction *(Lichtwelle)*
Brechungsachse f <Optik> axis of refraction
Brechungsbrennkurve f <Optik> caustic by refraction
Brechungsebene f <Optik> plane of refraction
Brechungsgesetze npl <Phys> laws of refraction
Brechungsindex m 1. <Foto> refractive index; 2. <Optik> index of refraction; refractive index *(eines Mediums)*; 3. <Phys> refractive index; 4. <Raumfahrt> refractive index *(Weltraumfunk)*; 5. <Wellphys> refractive index
Brechungsindex m **der Luft** <Mechan, Phys> air refractive index
Brechungsindexkontrast m <Optik> refractive index contrast
Brechungsindexprofil n <Optik> refractive index profile
Brechungskoeffizient m 1. <Foto> refractive index; 2. <Optik> relative refractive index; 3. <Telekom> refractive index
Brechungsmesser m <Labor> refractometer
Brechungsverlust m <Aufnahme> refraction loss
Brechungsvermögen n <Funktech> refractivity
Brechungswinkel m (r) <Optik, Phys> angle of refraction (r)
Brechungszahl f <Optik> relative refractive index
Brechvermögen n <Phys> refractivity
Brechwalze f <Kohlen> cracker, crusher roll, crushing roll
Brechwalzwerk n 1. <Kohlen> breaker, crushing mill; 2. <Maschinen> crushing roll
Brechwerk n <Kohlen> crusher, stamp mill
Brechwerkzeuge npl <Ker & Glas> pinching tools
Brechwirkungsgrad m <Kohlen> crushing efficiency
Brechwurz f <Lebensmittel> ipecac, ipecacuanha *(Pharmakologie)*
Brechzahl f 1. <Fertig> index of refraction; 2. <Optik> relative refractive index; 3. <Telekom> refractive index
Brechzahleinbruch m <Telekom> index dip
Brechzahlprofil n <Telekom> index profile
Brechzahlunterschied m <Telekom> refractive index contrast

Brei m <Abfall, Fertig, Kohlen, Lebensmittel> pulp, slurry
breit hergestellter Teppich m <Textil> broadloom carpet
Breit... 1. <Bau> broad; 2. <Fertig> broad, wide
Breitband n 1. <Comp & DV> broadband, wideband; 2. <Fernseh> broadband; 3. <Fertig> wide strip; 4. <Telekom> broadband, wideband
Breitbandantenne f 1. <Fernseh> broadband aerial; 2. <Funktech> broadband aerial, wideband antenna; 3. <Raumfahrt> wideband antenna
Breitband-Bandpassfilter n <Elektronik> wideband band-pass filter
Breitband-Codemultiplexzugriff m <Telekom> wideband code division multiple access
Breitbandempfänger m <Telekom> wideband receiver
Breitbandfernsehen n *(FSTV)* <Fernseh> fast-scan television *(FSTV)*
Breitbandfilter n <Elektronik> wideband filter
Breitbandfiltern n <Elektronik> wideband filtering
Breitband-Hochpassfilter n <Elektronik> wideband high-pass filter
breitbandig adj <Telekom> broadband, wideband
breitbandiges lokales Glasfasernetz n <Telekom> fiber optic broadband LAN, optical fiber high-speed LAN
breitbandiges Rauschen n <Telekom> broadband noise, wideband noise
Breitbandimpuls m <Fernseh> broad pulse
Breitband-ISDN n <Telekom> broadband ISDN, wideband ISDN
Breitband-ISDN-Dienst m <Telekom> broadband ISDN service
Breitbandkommunikation f <Telekom> broadband communication
Breitbandkoppelfeld n <Telekom> broadband switching network
Breitbandkoppelnetz n <Telekom> wideband switching network
Breitbandkoppelpunkt m <Telekom> broadband crosspoint
Breitbandkoppler m <Telekom> broadband switch
Breitband-Leistungsverstärker m <Elektronik> wideband power amplifier
Breitbandmehrfachzugriff m *(SSMA)* <Raumfahrt> spread spectrum multiple access, SSMA *(Weltraumfunk)*
Breitbandmessung f <Elektronik> wideband measurement
Breitbandmodem n <Elektronik, Telekom> wideband modem
Breitband-Modulation f <Elektronik> wideband modulation
Breitband-Paketvermittlung f <Telekom> fast packet switching
Breitbandrauschen n 1. <Aufnahme> broadband noise; 2. <Elektronik> wideband noise
Breitbandröhre f <Elektronik> wideband tube
Breitband-Schaltkreis m <Telekom> wideband circuit *(Komponente)*
Breitbandsignal n <Elektronik, Telekom> wideband signal
Breitbandstahl m <Fertig> wide strip
Breitbandstörung f <Elektronik> wideband interference
Breitbandstrahlen mpl <Strahlphys> wideband beams
Breitband-Tiefpassfilter n <Elektronik> wideband low-pass filter
Breitbandübertragung f <Telekom> wideband transmission
Breitbandverbindung f <Telekom> wideband circuit
Breitbandverstärker m 1. <Elektronik> *(BBV)* broadband amplifier; wideband amplifier; 2. <Funktech> broadband amplifier

Breitbandwalzwerk n <Fertig> wide-strip mill
Breitband-Zugangsnetz n <Telekom> broadband access network
Breitbeil n <Bau> *(AE)* adz, *(BE)* adze *(Holzverarbeitung)*
Breitbildformat n <Fernseh> letterbox format *(16:9)*
Breitbrenner m <Labor> flat-flame burner
Breite f 1. <Comp & DV, Druck, Geom> width; 2. <Papier> breadth, width
Breite f **der reduzierten Spanfläche** <Fertig> width of reduced face
Breite f **des Bandes** <Aufnahme> tape width
Breite f **über alles** <Maschinen> overall width
Breitenausdehnung f <Ker & Glas> spread
Breitende n <Bau> poll *(eines Hammers)*
Breitengrad m <Wassertrans> latitude *(Navigation)*
Breitenjitter m <Elektronik> width jitter
Breitenmetazentrum n <Wassertrans> *(AE)* transverse metacenter, *(BE)* transverse metacentre *(Schiffkonstruktion)*
breitenmoduliertes Impulssignal n <Regelung> width-modulated pulse signal
Breitensuche f <Künstl Int> breadth-first search
breitenvariable Tonspur f <Aufnahme> variable-width sound track
breiter Drehmeißel m <Fertig> wide-face square-nose tool
Breitflansch-Scharnierband n <Bau> H hinge
Breitflanschträger m <Bau> H-beam, H-girder, broad-flange girder
Breitfußfahrbalken m <Trans> inverted-T-shaped track girder *(Hängebahn)*
Breithacke f <Bau> mattock
Breithalter m <Textil> stretcher, temple
breitkantiges Wehr n <Wasserversorg> broad-crested weir
Breitschlichten n <Fertig> wide finishing
Breitschlichtmaschine f <Fertig> broad-nose machine
Breitschlichtmeißel m <Fertig> broad-finishing tool, wide-finishing tool
Breitschlitzdüse f <Kunststoff> sheet die
Breitschrift f <Druck> expanded type, extended type
Breitspur f <Eisenbahn> *(AE)* broadgage, *(BE)* broadgauge
Breitstrahler m <Funktech> broadside array
Breitstuhlteppich m <Textil> broadloom carpet
Breit-Wigner-Resonanz f <Kerntech> Breit-Wigner resonance
Brems... <Fertig, Kfztech, Mechan> brake
Bremsaggregat n <Mechan> brake load
Bremsankerplatte f <Kfztech> brake anchor plate, brake carrier plate, brake shield
Bremsanlage f <Kfztech> brake system
Bremsausgleichgestänge n <Kfztech> brake compensator
Bremsbacke f 1. <Fertig> brake shoe; 2. <Kfztech> brake shoe; 3. <Maschinen> brake jaw, brake shoe; 4. <Mechan> brake shoe
Bremsbacken-Bremsweg m <Trans> braking distance less brake lag distance
Bremsbackenrückzugfeder f <Kfztech> brake release spring
Bremsband n 1. <Kfztech> brake band *(Kupplung)*; 2. <Mechan> brake band
Bremsbaugruppe f <Kfztech> brake assembly
Bremsbelag m 1. <Ker & Glas> brake lining; 2. <Kfztech> brake friction pad, brake lining; brake lining *(Bremsanlage)*; brake pad *(der Scheibenbremse)*; 3. <Maschinen, Mechan> brake lining • **Bremsbeläge erneuern** <Kfztech> reline the brakes

Bremsbelagverschleißanzeige f <Kfztech> brake lining wear indicator
Bremsbelastung f <Eisenbahn, Kfztech, Mechan> brake load
Bremsbereich m <Kfztech> brake area
Bremsberg m <Kohlen> running jig
Bremsbetätigung f <Kfztech> brake application
Bremsbetätigungshebel m <Eisenbahn> brake lever
Bremsdichte f <Kerntech> slowing-down density
Bremsdrehkraft f <Kfztech> brake torque
Bremsdrehmoment n <Kfztech> brake torque
Bremsdruck m <Eisenbahn, Kfztech> brake pressure
Bremsdruckregler m <Maschinen> brake pressure regulator
Bremsdüse f <Raumfahrt> resistojet *(Raumschiff)*
Bremsdynamo m <Elektriz> brake dynamo, dynamometer, dynamometric dynamo
Bremsdynamometer n <Gerät, Maschinen> absorption dynamometer
Bremse f <Kfztech, Maschinen, Mechan> brake • **die Bremse anziehen** <Kfztech> put on the brakes • **die Bremse fest anziehen** <Kfztech> put the brakes on hard • **die Bremse ganz durchtreten** <Kfztech> put the brakes on full
Bremse f **mit Bremskraftverstärker** <Maschinen> power brake
Bremselektrode f <Elektrotech> reflecting electrode
bremsen v 1. <Eisenbahn> brake; 2. <Kfztech> apply the brake, brake, put on the brake; 3. <Mechan> decelerate
Bremsen n <Eisenbahn, Maschinen, Trans> braking
Bremsentlüftungsgerät n <Kfztech> brake bleeder unit
Bremsfading n <Kfztech> brake fade *(Bremsklötze, Bremsbeläge)*
Bremsfallschirm m 1. <Lufttrans> brake parachute, deceleration parachute, drag chute, drag parachute; 2. <Raumfahrt> drag chute *(Raumschiff)*
Bremsfallschirmhülle f <Lufttrans> drag chute cover
Bremsfläche f <Kerntech> slowing-down area
Bremsflüssigkeit f 1. <Kfztech> brake fluid *(Bremsanlage)*; 2. <Maschinen> brake fluid
Bremsflüssigkeitsbehälter m 1. <Kfztech> brake-fluid reservoir; 2. <Maschinen> brake-fluid tank
Bremsfutter n <Mechan> brake lining
Bremsgestänge n 1. <Eisenbahn> brake rigging; 2. <Kfztech> brake linkage *(Bremsanlage)*; 3. <Maschinen> brake linkage
Bremsgewicht n <Mechan> counterbalance
Bremsgitter n 1. <Elektronik> suppressor grid *(Elektronenröhren)*; 2. <Elektrotech> suppression grid
Bremshebel m <Maschinen> brake lever
Bremskeil m <Kfztech> brake wedge
Bremskette f <Maschinen> braking chain
Bremskissen n <Mechan> brake pad
Bremsklotz m 1. <Eisenbahn> brake block; 2. <Kfztech> brake pad *(Bremsanlage)*; pad *(Scheibenbremse)*; 3. <Maschinen> brake block; 4. <Mechan> chock
Bremskolben m <Kfztech> brake piston
Bremskombination f <Eisenbahn> brake blending
Bremskontrollleuchte f <Kfztech> brake-warning light *(Bremsanlage)*
Bremskraft f 1. <Eisenbahn> braking power; 2. <Kfztech> brake effort; 3. <Maschinen> brake power; 4. <Trans> braking power
Bremskraftregler m <Maschinen> brake-power control facility
Bremskraftverstärker m <Kfztech> brake servo *(Bremsanlage)*
Bremskraftverteiler m <Kfztech> brake-power distributor
Bremskreisaufteilung f <Kfztech> L-split system

Bremskupplung

Bremskupplung f <Eisenbahn, Mechan> brake coupling
Bremslänge f <Kerntech> slowing-down length
Bremslauf m <Lufttrans> run-up *(Motor und Triebwerk)*
Bremsleistung f 1. <Eisenbahn> braking power; 2. <Fertig, Kfztech, Mechan> brake horsepower; 3. <Trans> braking power
Bremsleitung f <Kfztech> brake line
Bremsleitungsverbindung f <Eisenbahn> brake-pipe connection
Bremsleuchte f <Kfztech> stop lamp
Bremslicht n <Kfztech> stop lamp
Bremslösefeder f <Kfztech> brake release spring
Bremslüfter m <Maschinen> brake motor
Bremsluftschraube f <Trans> braking airscrew
Bremsmaschine f <Elektriz> brake motor
Bremsmasse f <Eisenbahn> brake weight
Bremsmoment n <Maschinen> brake torque, braking moment
Bremsmotor m <Elektriz, Maschinen> brake motor
Bremsnocken m 1. <Kfztech> brake bar, brake cam; 2. <Mechan> brake cam
Bremsnutzung f <Kerntech> resonance escape probability
Bremspedal n <Kfztech> brake pedal
Bremspferdestärke f *(BPS)* <Fertig, Kfztech, Mechan> brake horsepower *(BHP)*
Bremspotenzial n <Phys> stopping potential
Bremsprobe f <Kfztech, Qual> brake testing
Bremspropeller m <Wassertrans> reversible pitch propeller
Bremsprozent n <Eisenbahn> percentage of brake power
Bremsprüfstand m <Mechan> brake-test stand
Bremsrad n <Mechan> brake wheel
Bremsrakete f <Raumfahrt> retro rocket
Bremsreaktion f <Trans> brake reaction
Bremsregler m <Trans> braking governor
Bremsring m 1. <Kfztech> brake plate *(Automatikgetriebe)*; 2. <Maschinen> brake ring
Bremssattel m <Kfztech> *(AE)* brake caliper, *(BE)* brake calliper, *(AE)* caliper, *(BE)* calliper
Bremssattel m **der Scheibenbremse** <Kfztech> *(AE)* caliper, *(BE)* calliper
Bremsscheibe f <Kfztech, Maschinen, Mechan> *(BE)* brake disc, *(AE)* brake disk
Bremsscheibenzentriervorrichtung f <Kfztech> *(BE)* brake disc alignment jig, *(AE)* brake disk alignment jig
Bremsschild n <Kfztech> brake carrier plate, brake shield
Bremsschirm m <Lufttrans> brake parachute
Bremsschlauch m <Kfztech> brake hose
Bremsschlauch m **der Druckluftbremse** <Maschinen> air brake hose
Bremsschub m <Lufttrans> reverse thrust
Bremsschuh m 1. <Kfztech> shoe; 2. <Mechan> brake shoe
Bremsschwund m <Kfztech> braking fading; brake fade *(Bremsklötze, Bremsbeläge)*
Bremsseil n <Kfztech, Maschinen> brake cable
Bremssichtweite f <Trans> stopping sight distance
Bremsspiel n <Kfztech> brake clearance
Bremsspindelstütze f <Eisenbahn> brakescrew support
Bremsspurkranz m <Eisenbahn> brake flange
Bremsstange f <Kfztech> brake rod
Bremsstellung f <Lufttrans> brake pitch; braking pitch *(Hubschrauber)*
Bremssteuerung f <Mechan> brake control
Bremsstrahlung f <Phys, Strahlphys, Teilphys> bremsstrahlung
Bremsstrahlungsquelle f <Kerntech> bremsstrahlung source

Bremsstrecke f <Eisenbahn, Kfztech> braking distance
Bremssubstanz f <Kerntech> moderator
Bremssystem n 1. <Eisenbahn> brake system, braking system; 2. <Kfztech> braking system; 3. <Maschinen> brake system, braking system; 4. <Mechan> brake system
Bremstest m <Maschinen> brake test
Bremsträger m <Kfztech> brake anchor plate, brake shield
Bremsträgerplatte f <Kfztech> brake carrier plate
Bremstrommel f 1. <Kfztech> brake drum *(Bremsanlage)*; 2. <Mechan> brake drum
Bremsung f 1. <Eisenbahn> braking; 2. <Ergon> deceleration; 3. <Kfztech> braking; 4. <Mechan> braking, deceleration; 5. <Trans> braking
Bremsung f **auf Halt** <Eisenbahn> braking to a stop
Bremsventil n <Eisenbahn, Kfztech> brake valve
Bremsvermögen n 1. <Kerntech> slowing-down power; 2. <Phys> stopping power
Bremsversagen n <Eisenbahn, Kfztech> brake failure
Bremsversuch m <Maschinen> brake test
Bremsverzögerung f <Trans> braking deceleration
Bremsvorrichtung f <Trans> deceleration device
Bremswagen m <Eisenbahn> brake van, brake wagon
Bremsweg m 1. <Eisenbahn, Kfztech> braking distance; 2. <Trans> stopping distance
Bremsweg m **ohne Bremsverzögerungsabstand** <Trans> braking distance less brake lag distance
Bremswelle f <Kfztech, Maschinen> brake shaft
Bremswiderstand m <Eisenbahn, Trans> braking resistance
Bremswirkungsverlust m <Kfztech> braking fading
Bremszeit f <Qual, Trans> braking time
Bremszugstange f <Kfztech> brake connecting rod
Bremszustand m <Maschinen> brake state
Bremszylinder m 1. <Fertig> dashpot; 2. <Kfztech, Maschinen> brake cylinder
Brenn... 1. <Chemtech> distilling; 2. <Elektronik> focal; 3. <Phys> caustic, focal
Brennapparat m <Chemtech> distilling apparatus
brennbar adj 1. <Chemie, Fertig> combustible; 2. <Kunststoff> flammable; 3. <Thermod> combustible
brennbarer Abfall m <Umweltschmutz> combustible waste
brennbarer Stoff m <Sicherheit> combustible material
brennbares Material n <Umweltschmutz> combustible material
Brennbarkeit f 1. <Kunststoff> flammability; 2. <Sicherheit, Thermod, Verpack> combustibility
Brennbarkeitstest m <Sicherheit> fire test *(für Möbel)*
Brennebene f 1. <Elektronik> focal plane; 2. <Phys> caustic, focal plane
Brenneisen n <Verpack> branding iron
Brennelement n **in Rasterbauweise** <Kerntech> grid-spaced fuel assembly
Brennelementhülse f <Kerntech> fuel cladding
Brennelementüberprüfung f **mit Gammastrahlen** <Kerntech> gamma heating
brennen v 1. <Ker & Glas> bake *(von Ton)*; 2. <Textil> bake *(Textildruck)*; 3. <Thermod> bake *(Keramik)*
Brennen n <Kohlen> calcination, roasting
brennend adj <Heiz & Kälte, Ker & Glas, Metall, Thermod> burning
brennende Kohle f <Kohlen> burning coal
Brenner m 1. <Bau> blowpipe, burner, torch; 2. <Fertig> blow-pipe, torch; 3. <Maschinen, Mechan> blowpipe, burner, torch; 4. <Telekom, Thermod> burner *(Bunsenbrenner, von Gasherd)*
Brennerdüse f 1. <Fertig> blowpipe nozzle; 2. <Heiz & Kälte> burner head, burner mouth; 3. <Maschinen> burner nozzle

Brennerei f <Lebensmittel> distillery
Brennereinsatz m <Fertig> torch head
Brennerflamme f <Fertig> torch flame
Brennerkopf m <Heiz & Kälte> burner head
Brennerlöten n <Bau> torch brazing
Brennermaul n <Ker & Glas> port mouth
Brennermaulwange f <Ker & Glas> port side wall
Brennermundstück n 1. <Bau> tip; 2. <Fertig> burner head
Brenneröffnung f <Ker & Glas> eye (des Hafenofens)
Brennerschlauch m <Fertig> torch hose
Brennerspitze f <Bau> tip
Brennfläche f <Optik> caustic
Brennfleck m 1. <Elektronik> focal spot; 2. <Fertig> arcing end
Brenngasflasche f <Fertig> fuel-gas cylinder
brenngeschnitten adj <Fertig> gas-cut
Brenngut n <Abfall> incinerator charge
Brennhärtemaschine f <Fertig> flame-hardening machine
brennhärten v <Thermod> flame-harden
brennhobeln v <Fertig> hog (Stahlblöcke)
Brennintervall n <Ker & Glas> firing range
Brennkammer f 1. <Heiz & Kälte> fire box; 2. <Raumfahrt> combustion chamber, combustor (Raumschiff); 3. <Thermod> firebox
Brennkammerauskleidung f <Heiz & Kälte> burner liner
Brennkapsel f <Ker & Glas> sagger
Brennkapselstapel m <Ker & Glas> bung of saggers
Brennkraftmaschine f 1. <Kfztech> combustion engine, explosion engine, explosion motor; 2. <Lufttrans> combustion engine; 3. <Maschinen> internal combustion engine; 4. <Trans, Wassertrans> combustion engine
Brennkurve f <Optik> caustic
Brennlinie f <Optik> caustic
Brennofen m <Bau, Fertig, Lebensmittel, Metall, Thermod> kiln
Brennpunkt m 1. <Foto> focal point; 2. <Geom> focus; 3. <Heiz & Kälte> fire point; 4. <Phys> focus; 5. <Sicherheit> fire point
brennputzen v <Fertig> torch-deseam
Brennputzen n <Fertig> flame chipping, flame descaling, flame deseaming, flame scarfing
Brennraum m <Kfztech, Lufttrans, Thermod, Wassertrans> combustion chamber
Brennschluss m <Raumfahrt> flameout; burnout (Raumschiff)
brennschneiden v <Fertig> oxygen-cut, torch-cut
Brennschneiden n 1. <Fertig> gas cutting, oxycutting, oxygen cutting; 2. <Maschinen, Mechan> acetylene cutting, flame cutting; 3. <Metall> flame cutting; 4. <Thermod> gas cutting
Brennstab m <Kerntech> fuel rod
Brennstempel m <Verpack> branding iron
Brennstoff m <Heiz & Kälte, Kfztech, Maschinen, Thermod> fuel
Brennstoffanreicherung f <Kerntech> feed enrichment
Brennstoffaufbereitung f <Kerntech> make-up fuel
Brennstoffaufbereitung f durch Wäsche <Kerntech> scrubbing
Brennstoffauffrischung f <Kerntech> fuel regeneration
Brennstoffdichte f <Kerntech> fuel density
Brennstoffdosierung f <Gerät> fuel metering
Brennstoffelement n 1. <Maschinen> fuel element; 2. <Thermod> fuel cell
Brennstoffersparnis f <Thermod> fuel economy
Brennstoff-Fördertisch m <Kerntech> fuel transfer table
Brennstoffhülse f <Kerntech> fuel cladding
Brennstoffinventar n <Kerntech> activity inventory, fuel inventory
Brennstoffkanal m <Kerntech> fuel channel
Brennstoffkreislauf m **im Reaktorkern** <Kerntech> in-core fuel cycle
Brennstoffkühlung f <Thermod> fuel cooling
Brennstofflebensdauer f <Kerntech> fuel life
Brennstoffmatrix f <Kerntech> matrix fuel
Brennstoff-Pellett n **mit Uranoxid** <Kerntech> uranium oxide pellet
Brennstoffraster m <Kerntech> grid-spaced fuel assembly
Brennstoffschäden mpl <Kerntech> fuel detriment (Wachsen, Beulenbildung, Risse)
brennstoffsparend adj <Thermod> fuel-efficient
Brennstoffzelle f <Elektriz> fuel cell
Brennstoffzellenantrieb m <Wassertrans> fuel cell drive
Brennstoffzinsen mpl <Kerntech> fuel rates
Brennstumpfschweißen n <Fertig> flash welding
Brennstütze f <Ker & Glas> (AE) buck, (BE) dot, (BE) point bar
Brenntemperatur f <Ker & Glas> firing temperature
Brennverhalten n <Kunststoff> flammability
Brennweite f <Foto, Phys> focal length
Brennweitenbereich m <Foto> focusing range
Brennwert m 1. <Heiz & Kälte> gross calorific value, useful heat; 2. <Lebensmittel, Phys, Thermod> calorific value
Brenzcatechin-Monomethylether m <Chemie> guaiacol
Brenztraubensäure f <Chemie> pyruvic acid
Brett n <Bau> board
Bretterverkleidung f 1. <Bau> boarding; 2. <Kohlen> planking
Brettschaltungsmodell n <Elektrotech> breadboard model (Versuchsschaltung)
Brewster'scher Effekt m <Funktech> Brewster effect (Polarisationsdrehung)
Brewster'scher Einfallswinkel m <Phys> Brewster incidence
Brewster'scher Winkel m <Optik, Phys> Brewster angle
Bride f <Maschinen> buckle
Brief m 1. <Comp & DV> letter; mail message (E-Mail); 2. <Druck> letter
Briefkopf m <Druck> letterhead
Briefqualität f <Comp & DV> letter-quality
Briggs-Gewinde n <Maschinen> Brigg's pipe thread
Brikett n <Kohlen> briquette
Brikettfett n <Maschinen> block grease
brikettiertes Eisen n (HBI) <Metall> hot briquetted iron, HBI
brikettiertes Karbid n <Fertig> cake of carbide
Brillantschliff m <Ker & Glas> brilliant cutting
Brillantsucher m <Foto> reflecting viewfinder
Brillanz f 1. <Elektronik> brilliance, brilliancy (Fernsehtechnik); 2. <Metall> brilliance, brilliancy
Brille f 1. <Fertig> gland (Stopfbuchse); 2. <Maschinen> gland
Brillenglas n **mit 3½″ Wölbungsradius** <Ker & Glas> coquille
Brillenglas n **mit 7″ Wölbungsradius** <Ker & Glas> micoquille
Brillouin'sche Zone f <Phys> Brillouin zone (Festkörperphysik)
Brinell... <Fertig, Maschinen> Brinell
Brinelleffekt m <Maschinen> Brinell effect
Brinellhärte f 1. <Fertig> Brinell hardness, Brinell hardness number; 2. <Maschinen> BHN, Brinell hardness number; 3. <Mechan> Brinell hardness
Brinellhärteprüfung f <Fertig> ball indentation test
Brinellhärtetest m <Maschinen> Brinell hardness testing machine

Brinellieren 124

Brinellieren n <Maschinen> brinelling
Brinellkugel f <Fertig> ball penetrator
Brinellkugelprüfung f <Maschinen> Brinell ball test
Brinellprüfung f <Maschinen> Brinell test
Britanniametall n <Metall> Britannia, Britannia metal
Britische Norm f <Qual> British Standard, BS
Britische Normenspezifikation f (BSS) <Maschinen> British Standard Specification (BSS)
Britische Normvorschrift f (BSS) <Maschinen> British Standard Specification (BSS)
Britische Wärmeeinheit f (BTU, BThU) <Labor, Maschinen> British Thermal unit (BThU, BTU)
Britischer Bushel m <Lebensmittel> bushel
Britisches BA-Gewinde n <Maschinen> BA screw thread
Britisches SF-Gewinde n <Maschinen> British Standard fine thread
Brix-Skala f <Lebensmittel> Brix scale (für Zucker)
bröckelig adj 1. <Chemtech> pulverulent; 2. <Kunststoff> friable
Brodeln n <Phys> bubbling
Brokat m <Textil> brocade
Brokatgewebe n <Textil> brocade
Brom n (Br) <Chemie> bromine (Br)
Brom... <Chemie> bromic
Bromaceton n <Chemie> bromoacetone
Bromal n <Chemie> bromal
Bromat n <Chemie> trioxobromate
Brombenzol n <Chemie> bromobenzene
Bromelin n <Lebensmittel> bromelin
Bromgelatine-Verfahren n <Foto> gelatino-bromide process
Bromid n <Chemie> bromide
Bromoform n <Chemie> bromoform
Bromölabzug m <Foto> bromoil print
Bromphenol n <Chemie> bromophenol
Bromsalz n <Chemie> bromide
Bromsilberdruck m <Druck> bromide, bromide print
Bromsilberkollodiumplatte f <Foto> silver-bromide collodion plate
Bromsilberpapier n <Druck, Foto> bromide paper
Bronze f <Mechan> gunmetal
Bronzebüchse f <Maschinen> gunmetal bush
Bronzeführungsbuchse f <Maschinen> bronze guide bush
Bronzelager n <Maschinen> gunmetal bearing
Bronzeschweißen n <Bau, Metall> bronze welding
Bronzierung f <Metall> bronzing
Brookfield-Viskosität f <Kunststoff> Brookfield viscosity
broschiert adj <Druck> stitched
Brot n <Lebensmittel> bread
Brotgärung f <Lebensmittel> panary fermentation
Brotschrift f <Druck> body type
Brown'sche Molekularbewegung f <Phys, Strahlphys, Thermod> Brownian molecular movement, Brownian motion, Brownian movement
Brown-und-Sharpe-Kegel m <Fertig> BS taper, Brown and Sharpe taper
Browser m <Comp & DV, Telekom> browser (Internet)
Bruch m 1. <Druck> fraction; 2. <Erdöl> fault; 3. <Fertig> crushing; 4. <Kohlen> break, crack; 5. <Maschinen> fracture; 6. <Math> fraction; 7. <Mechan> crack, fracture; 8. <Metall> crack, failure, fracture, rupture; 9. <Papier> break; 10. <Textil> breakage, break, burst • **einen Bruch kürzen** <Math> reduce a fraction
Bruchbeanspruchung f <Maschinen, Mechan, Qual> breaking stress
Bruchbelastung f <Papier, Verpack> breaking load
Bruchbild n <Mechan> breaking pattern

Bruchbildung f 1. <Erdöl> fracturing (Geologie); 2. <Metall, Qual> fracturing
Bruchdehnung f 1. <Bau> flexural strength; 2. <Fertig> elongation; 3. <Kunststoff> elongation at break, ultimate elongation; 4. <Mechan> flexural strength; 5. <Papier> stretch at breaking point; 6. <Qual> elongation, elongation at break, flexural strength; 7. <Werkprüf> elongation at break
Bruchdruck m <Erdöl> fracture pressure (Geologie)
Bruchelement n <Heiz & Kälte> rupture member
Bruchfestigkeit f 1. <Bau> ultimate strength; 2. <Fertig> breaking strain (Kunststoffinstallationen); 3. <Kunststoff> breaking strength; 4. <Maschinen> ultimate breaking strength; 5. <Mechan> breaking strength; 6. <Metall> rupture strength; 7. <Textil> breaking strength; 8. <Werkprüf> fracture strength
Bruchglas n <Ker & Glas> cullet
Bruchgradient m <Erdöl> fracture gradient (Geologie)
Bruchgrenze f 1. <Maschinen> breaking point, ultimate breaking strength; 2. <Wassertrans> breaking strain
Bruchhefe f <Lebensmittel> flocculating yeast (Brauerei)
brüchig adj 1. <Anstrich> brittle; 2. <Chemtech> pulverulent; 3. <Kunststoff> brittle, friable; 4. <Mechan> brittle
brüchige Kreide f <Bau> fractured chalk
Brüchigkeit f <Kunststoff, Qual> brittleness
Bruchkanten fpl <Mechan> break edges
Bruchkegel m <Ker & Glas> fracture cone
Bruchkreis m <Bau> slip circle
Bruchkriterium n <Metall, Qual> fracture criterion
Bruchkriterium n **nach Griffith** <Kerntech> Griffith's fracture criterion
Bruchlast f 1. <Bau> breaking load; 2. <Fertig> collapse load; 3. <Kohlen> breaking load, failure load; 4. <Kunststoff> breaking load; 5. <Maschinen> breaking load, rupture load; 6. <Mechan> breaking load; 7. <Qual> collapse load; 8. <Verpack, Wassertrans> breaking load
Bruchlinie f <Konstzeich> break line
Bruchlochwicklung f <Elektrotech> fractional slot winding
Bruchmechanik f <Mechan> fracture mechanics
Bruchmechanikversuch m <Werkprüf> fracture mechanics test
Bruchmuster n <Ker & Glas> fracture pattern
Bruchplatte f <Mechan> (BE) bursting disc, (AE) bursting disk
Bruchpunkt m <Phys> yield point
Bruchrechnen n <Math> fractional arithmetic
Bruchreis m <Lebensmittel> broken rice
bruchsicher adj <Verpack> shockproof
Bruchsicherheit f <Werkprüf> fracture strength
Bruchspannung f <Bau, Maschinen, Mechan, Qual> breaking stress
Bruchspiegel m <Ker & Glas> fracture mirror
Bruchstein m <Bau> broken stone, rubblestone
Bruchsteinmauerwerk n <Bau> rubble masonry
Bruchstelle f <Nichtfoss Energ> fracture
Bruchstrich m <Math> bar, fraction stroke
Bruchstück n <Sicherheit> fragment (von abgenutztem Werkzeug)
Bruchteil m 1. <Comp & DV> fractional part; 2. <Math> fraction
bruchteilige Frequenzabweichung f <Elektronik> fractional frequency deviation
Bruchtest m <Mechan, Qual> breaking test
Bruchursprung m <Werkprüf> fracture origin
Bruchverhalten n <Metall, Qual> (AE) fracture behavior, (BE) fracture behaviour
Bruchversuch m 1. <Bau> breaking test; 2. <Maschinen> fracture test

Bruchwertdesign *n* <Raumfahrt> design to buckling strength *(konstruiert bis zur Beulgrenze)*
Bruchzähigkeit *f* <Kunststoff> fracture toughness
Bruchzähigkeitsfaktor *m* <Werkprüf> fracture toughness factor
Bruchzahl *f* <Math> fractional, fractional number, fractional numeral
Bruchzeit *f* <Metall> time to rupture
Bruchzone *f* <Kohlen> mat *(über dem Abbau)*
Bruchzugkraft *f* <Bau> ultimate strength
Brucin *n* <Lebensmittel> brucine
Brücke *f* 1. <Bau> bridge *(eines Laufkranes)*; 2. <Comp & DV, Eisenbahn> bridge; 3. <Elektriz> jumper; 4. <Elektrotech> bridge; 5. <Funktech> bridge *(Schaltung)*; 6. <Gerät, Ker & Glas> bridge; 7. <Maschinen> bridge, runner, *(AE)* traveler, *(BE)* traveller; 8. <Telekom> bridge *(zwischen Netzen)*; 9. <Wassertrans> navigation bridge *(Handelsmarine)*; bridge *(Schiff)* • **Brücke über einen Fluss schlagen** <Bau> throw a bridge over river
Brücke *f* **mit Gleitkontakt** <Elektrotech> slide bridge
Brücke *f* **mit oben liegender Fahrbahn** <Bau> deck bridge
Brückenabgleich *m* <Elektrotech> bridge balancing
Brückenarm *m* <Elektriz> bridge arm *(Brückenzweig)*
Brückenaufbau *m* <Wassertrans> bridge superstructure *(Schiffbau)*
Brückenaufnahme *f* <Bau, Eisenbahn> bridge survey
Brückenausgleich *m* <Gerät> bridge balance
Brückenblech *n* <Bau, Eisenbahn> bridge plate
Brückenbogen *m* <Bau> arch
Brückendeck *n* <Wassertrans> bridge deck *(Schiffbau)*
Brückendiagonalspannung *f* <Gerät> unbalance voltage *(Messbrücke)*
Brückendrehkran *m* <Bau> jib crane
Brückendurchfahrtshöhe *f* <Wassertrans> air draught *(Schiffkonstruktion)*
Brückenfachwerkträger *m* <Bau, Eisenbahn> bridge truss
Brückenfahrbahn *f* <Bau> bridge deck
Brückenfilter *n* <Elektronik, Telekom> lattice filter
Brückengeländer *n* <Bau, Eisenbahn> bridge railing
Brückengleichgewicht *n* <Gerät> bridge balance
Brückengleichrichter *m* <Elektriz, Elektrotech> bridge rectifier
Brückenhaus *n* <Wassertrans> bridge house *(Schiffbau)*
Brückenkahn *m* <Wassertrans> pontoon
Brückenkran *m* 1. <Bau> bridge crane, overhead crane; 2. <Maschinen> overhead crane
brückenlose Kontakte *mpl* <Elektrotech> non-bridging contacts
Brückenmessung *f* <Gerät> bridge measurement
Brückenmischer *m* 1. <Elektronik> push-pull mixer; 2. <Elektrotech> balanced mixer
Brückenpfeiler *m* <Bau> pylon
Brücken-Router *m* <Telekom> bridge router, B-router, brouter
Brückenschaltung *f* 1. <Elektriz> bridge circuit, bridge connection; 2. <Elektrotech> bridge circuit
Brückenschaltung *f* **aus Ohm'schen Elementen** <Gerät> resistive bridge
Brückenstecker *m* <Kontroll> strapping plug
Brückenüberbau *m* <Bau> bridge deck
Brückenverhältnisarm *m* <Elektriz> ratio arm
Brückenverstärker *m* <Elektronik> bridge amplifier
Brückenwaage *f* <Bau> weighbridge
Brückenwärter *m* <Wassertrans> bridge keeper *(Fluss)*
Brückenweiche *f* <Funktech> bridge diplexer *(Antenne)*
Brückenwiderstand *m* <Elektriz> bridge resistor *(Bauelement)*; bridge resistance *(physikalische Größe)*

Brückenzweig *m* <Gerät> ratio arm *(Brückenschaltung)*
Brüdenverdichtung *f* <Chemtech> *(AE)* vapor compression, *(BE)* vapour compression
brühen *v* <Lebensmittel, Thermod> scald
brühend heiß *adj* <Thermod> scalding
Brumm *m* <Akustik> hum, ripple
Brummabstand *m* <Elektronik> signal-to-hum ratio
Brummeinkopplung *f* <Aufnahme> hum pickup
brummen *v* <Funktech> hum
Brummen *n* <Akustik, Elektrotech, Funktech> hum
Brummfaktor *m* <Elektronik> ripple factor
Brummfilter *n* <Strahlphys> ripple filter
Brummfrequenz *f* <Strahlphys> ripple frequency
Brummkompensationsspule *f* <Elektrotech> humbucking coil
Brummspannung *f* 1. <Elektronik> ripple voltage; 2. <Elektrotech> hum voltage
Brummstreifen *mpl* <Fernseh> hum bars
brünieren *v* <Fertig> black-finish, brown; blue *(Stahl)*
Brünieren *n* 1. <Fertig> black finishing, blackening *(Stahl)*; 2. <Metall> bronzing *(Stahl)*
Brunnen *m* <Bau, Kohlen> spring, well
Brunnen *m* **vom Beobachtungsnetz** <Wasserversorg> observation well
Brunnenbau *m* <Kohlen, Nichtfoss Energ> well sinking
Brunnenbohren *n* <Kohlen, Nichtfoss Energ> well boring, well drilling
Brunnenkopf *m* <Nichtfoss Energ> well head
Brunnenkopfdruck *m* <Nichtfoss Energ> wellhead pressure
Brunnenkopftemperatur *f* <Nichtfoss Energ> wellhead temperature
Brunnenkopfventil *n* <Nichtfoss Energ> wellhead valve
Brunnenring *m* <Kohlen, Nichtfoss Energ> well casing
Brust *f* 1. <Fertig> breast; 2. <Maschinen> breast, chest
Brustbohrer *m* <Maschinen> breast drill, chest drill
Brustbohrmaschine *f* <Maschinen> breast drill, chest drill
Brusthölzer *npl* <Bau> walings
Brustleier *f* 1. <Fertig> breast drill; 2. <Maschinen> breast drill, chest drill
Brustplatte *f* <Maschinen> breast plate
Brüstung *f* <Bau> balustrade, parapet, railing
Brüstungsgeländer *n* <Hydraul> breasting parapet
Brüstungsmauer *f* <Bau> breast wall *(brusthoch)*
Brustwalze *f* <Papier> breast roll
Brustzapfenaufwölbung *f* <Bau> tusk
Brustzapfenverbindung *f* <Bau> tusk tenon joint
Brut *f* <Kerntech> breeding
Brutabschnitt *m* <Kerntech> breeding section *(eines Brennstabes)*
Brüter *m* <Kerntech> breeding reactor
Brüterreaktor *m* <Kerntech, Phys> breeder reactor
Brutgewinn *m* 1. <Elektronik> conversion gain *(Atomphysik)*; 2. <Kerntech> breeding gain
Brutkreislauf *m* <Kerntech> breeding cycle
Brutmantel *m* <Kerntech> breeding blanket
Brutreaktor *m* <Kerntech> breeding reactor
Brutschrank *m* <Labor> incubator *(Mikrobiologie)*
Brutto... <Metrol> gross
Bruttogewicht *n* <Metrol, Verpack> gross weight
Bruttoleistung *f* <Kerntech> gross installed capacity
Bruttoraumgehalt *m* <Wassertrans> gross registered tonnage, gross tonnage
Bruttoraumzahl *f* *(BRZ)* <Wassertrans> gross tonnage *(Raummaß, ersetzt die Angabe in BRT)*
Bruttoregistertonne *f* 1. <Metrol> displacement ton; 2. <Wassertrans> gross ton
Bruttotonnage *f* <Erdöl> gross tonnage *(Schifffahrt)*

Bruttovolumen *n* <Verpack> gross volume
Bruttowärmeverlust *m* <Heiz & Kälte> gross heat loss
Brutverfahren *n* <Kerntech> breeding process
Brutvorgang *m* <Kerntech> breeding process
Brutvorgangswirkungsgrad *m* <Kerntech> breeding process efficiency
Brutzyklus *m* <Kerntech> breeding cycle
BRZ *(Bruttoraumzahl)* <Wassertrans> GT, gross tonnage *(Raummaß, ersetzt die Angabe in BRT)*
BS 1. <Fertig> *(Windfrischstahl)* BS *(Bessemer steel)*; 2. <Maschinen> *(Bessemerstahl)* BS *(Bessemer steel)*
BSA *(Bohrlochsohlenausrüstung)* <Erdöl> BHA *(bottom hole assembly)*
BSB *(biologischer Sauerstoffbedarf)* <Abfall, Lebensmittel, Umweltschmutz> BOD *(biological oxygen demand)*
BSC-Übertragung *f* *(binärsynchrone Übertragung)* <Comp & DV, Telekom> BSC *(binary synchronous communication)*
BSF-Gewinde *n* *(Gewinde nach britischem Standard)* <Maschinen> BSF *(British standard fine screw thread)*
BSP-Gewinde *n* *(Gewinde nach britischem Standard)* <Maschinen> BSP *(British standard pipe thread)*
BSS *(Britische Normenspezifikation, Britische Normvorschrift)* <Maschinen> BSS *(British Standard Specification)*
BThU *(Britische Wärmeeinheit)* <Maschinen, Metrol> BThU, British Thermal unit *(Energie)*
BTU *(Britische Wärmeeinheit)* 1. <Labor> BThU, British Thermal unit; 2. <Maschinen> *(AE)* BTU, British Thermal unit
Btx *(Bildschirmtext)* 1. <Fernseh> Videotex®; 2. <Telekom> Teletext®
Btx-Anbieter *m* <Fernseh> videotex information provider
Bubble Jet Drucker *m* <Druck> bubble-jet printer
Bubblesort *n* <Comp & DV> bubble sort *(einfaches Sortierverfahren)*
Bubbling *n* <Ker & Glas> bubbling
Buch *n* <Druck> book
Buch *n* **mit festem Einband** <Druck> casebound book, cased book
Buchbinden *n* <Druck> bookbinding
Buchbinder *m* <Druck> bookbinder
Buchbinderei *f* <Druck> bindery, bookbinding
Buchbindernadel *f* <Druck> bookbinder's needle
Buchbinderpappe *f* <Druck> millboard
Buchbinderstempel *m* <Druck> bookbinder's brass
Buchdecke *f* <Druck> book case, case
Buchdruck *m* <Druck> letterpress, letterpress printing
Buchdruckmaschine *f* <Druck> letterpress-printing machine
Buchdruckpresse *f* <Druck> letterpress-printing machine
Büchner'sche Nutsche *f* <Labor> Büchner funnel *(Filtrieren)*
Büchner'scher Kolben *m* <Labor> Büchner flask *(Filtrieren)*
Büchner'scher Trichter *m* <Labor> Büchner funnel *(Filtrieren)*
Buchrücken *m* <Druck> shelfback
Buchsatz *m* <Druck> book composition
Buchse *f* 1. <Aufnahme, Comp & DV> jack; 2. <Elektriz> bearing, female connector; 3. <Elektrotech> jack; 4. <Fernseh> socket *(Stecker)*; 5. <Funktech> jack, receptacle, socket; 6. <Maschinen> bush, bushing, liner, sleeve, bushing; 7. <Mechan> bushing, sleeve; 8. <Telekom> jack, socket • **mit Buchse versehen** <Elektrotech> jacked
Buchse *f* **aus Kupferlegierung** <Maschinen> copper alloy bush
Buchse *f* **mit berührungsgeschützten Kontakten** <Elektriz> socket with shrouded contacts

Büchse *f* 1. <Fertig> liner; 2. <Maschinen> bushing, liner; 3. <Verpack> *(AE)* can, *(BE)* tin • **in Büchsen** <Lebensmittel> *(AE)* canned, *(BE)* tinned
Buchsenfeld *n* <Aufnahme, Telekom> *(BE)* jack panel, *(AE)* patch panel
Buchsenkette *f* <Maschinen> bush chain
Buchsenkontakt *m* <Elektrotech> socket contact
Buchsenleiste *f* <Elektrotech> socket board
Buchsenstecker *m* <Elektriz> socket plug
Buchsensteckverbinder *m* <Elektrotech> female connector
Buchsenteil *m* <Elektrotech> female contact
Buchsenverbindung *f* <Elektriz> socket coupler
Buchsenzieher *m* <Maschinen> bush extractor
Buchstabe *m* <Comp & DV, Druck> letter
Buchstabe *m* **mit Oberlänge** <Druck> ascending letter
Buchstabe *m* **ohne Ober- und Unterlängen** <Druck> short letter
Buchstabenalgebra *f* <Math> literal algebra
Buchstabenbezeichnung *f* <Maschinen> *(AE)* letter gage, *(BE)* letter gauge *(für Stahldraht)*
Buchstabencode *m* <Comp & DV> mnemonic code *(Kurzbefehl)*
Buchstabenfehler *m* <Druck> literal error
Buchstabengenerator *m* <Fernseh> character generator
Buchstabengleichung *f* <Math> literal equation
Buchstabenkennung *f* <Konstzeich> letter code
Buchstabenumschaltung *f* <Comp & DV, Telekom> letter shift
Bucht *f* <Wassertrans> bay *(Geographie)*; bight *(Tauwerk)*
Buchzeichen *n* <Druck> bookmark
Buckel *m* <Maschinen> boss
buckelgeschweißt *adj* <Fertig> projection-welded
buckelschweißen *v* <Fertig> projection-weld
Buckelschweißen *n* <Maschinen> projection welding
Buckelschweißung *f* <Bau, Fertig> projection welding
Buckeye-Kupplung *f* <Eisenbahn> *(AE)* buckeye coupler
Buckram *m* <Druck> buckram
Bufotoxin *n* <Chemie> bufotoxine
Bug *m* 1. <Bau> strut; 2. <Raumfahrt> nose cone *(Raumschiff)*; 3. <Wassertrans> bow, prow • **mit Bug voran sinken** <Wassertrans> go down by the bows *(Schiff)* • **mit vollem Bug** <Wassertrans> bluff-bowed • **vor dem Bug** <Wassertrans> across the bow
Buganker *m* <Wassertrans> bow anchor, bower anchor
Bugaufklotzung *f* <Wassertrans> bow chock
Bugdruckwelle *f* <Raumfahrt> bow shock
Bügel *m* 1. <Bau> bow, fastening, shackle, stirrup, yoke; 2. <Fertig> stirrup; pin *(Kunststoffinstallationen)*; C-frame *(Messschraube)*; 3. <Maschinen> bow, shackle, strap, yoke; 4. <Mechan> clevis, yoke; 5. <Meerschmutz> shackle
Bügelbolzen *m* <Bau> U-bolt
bügelfrei *adj* <Textil> drip-dry
Bügelmessschraube *f* <Fertig> outside micrometer
Bügelsäge *f* <Fertig, Maschinen, Mechan> hacksaw
Bügelschraube *f* 1. <Bau> U-bolt; 2. <Bau> strap bolt
Bugfahrwerk *n* <Lufttrans> nose gear
Bugfahrwerksbein *n* <Lufttrans> nose gear leg
Bugfahrwerksklappe *f* <Lufttrans> nose gear door
Bugfahrwerkslenkung *f* <Lufttrans> nose gear steering
Bugfahrwerkssattel *m* <Lufttrans> nose gear saddle
Bugfahrwerks-Steuerungsverriegelung *f* <Lufttrans> nose gear steer lock
Bugfender *m* <Wassertrans> noseband; bow fender *(Deckausrüstung)*
Bugholz *n* <Bau> angle tie; angle brace *(Holzbau)*
Bugkettenstopper *m* <Wassertrans> bow stopper *(Deckausrüstung)*

Bugklappe f <Wassertrans> bow door
Bugpforte f <Wassertrans> bow door
Bugradfahrwerkszahnrad n <Lufttrans> nose gear wheel
Bugradlenkstange f <Lufttrans> nose wheel steering bar
Bugradsteuerrad n <Lufttrans> nose wheel steering control wheel
Bugradsteuerung f <Lufttrans> nose wheel steering
Bugsee f <Wassertrans> bow wave
bugsieren v <Wassertrans> tow *(Schiff)*
Bugsieren n <Wassertrans> towage *(Schiff)*
Bugspriet n <Wassertrans> bowsprit *(Schiffbau)*
Bugstrahlruder n <Wassertrans> bow thruster *(Schiffantrieb)*
Bugverdichtungsstoß m <Raumfahrt> bow shock
Bugwelle f <Lufttrans, Wassertrans> bow wave
Buhne f 1. <Wassertrans> *(AE)* groin, *(BE)* groyne; 2. <Wasserversorg> breakwater, *(AE)* groin, *(BE)* groyne
Buline f <Wassertrans> bowline *(Tauwerk)*
Bullauge n 1. <Raumfahrt> viewing port *(Raumschiff)*; 2. <Wassertrans> porthole
Bulldozer m <Bau, Trans> bulldozer
Bullentalje f <Wassertrans> boom vang, kicking strap *(Segeln)*
Bumerang m <Strömphys> boomerang
Bund m 1. <Druck> gutter; 2. <Fertig> rap; barrel *(Walze)*; 3. <Maschinen> collar, flange, set collar; 4. <Mechan> collar, flange; 5. <Textil> waist band
Bundbolzen m <Maschinen> flanged bolt
Bundbuchse f <Fertig> flange adaptor *(Kunststoffinstallationen)*
Bündel n 1. <Fernseh> burst *(Fibern, Fehlern)*; 2. <Ker & Glas, Papier> bundle; 3. <Phys> bunch; 4. <Telekom> bundle, fibre bundle, group, group of lines; 5. <Strahlphys> beam, cone; 6. <Verpack> bundle
Bündel n **von Absorberelementen** <Kerntech> absorber element bundle
Bündelanordnung f <Kerntech> banked configuration *(von Brennelementen)*
Bündelbreite f <Optik> beamwidth *(Strahlen)*
Bündelelement n <Kerntech> block-shaped fuel element
Bündelentkopplung f <Phys> bunch decoupling
Bündeler m <Phys> collimator
Bündelfunk m <Funktech, Mobilkom> trunking
Bündelfunknetz n <Mobilkom> trunked mobile radio network
Bündelkabel n <Elektrotech> *(BE)* bunched cable, *(AE)* bundled cable
Bündelknoten m <Fernseh> crossover
Bündelmaschine f 1. <Fertig> bundler, bundling machine; 2. <Verpack> bundle-tying machine
bündeln v 1. <Comp & DV> multiplex; 2. <Papier> bundle
Bündelung f 1. <Eisenbahn> grouping; 2. <Elektronik> collimation; 3. <Lufttrans> bunching; 4. <Metall> clustering
Bündelungselektrode f <Fernseh> beam-forming plate, focusing electrode
Bündelungsfaktor m <Nichtfoss Energ> directionality factor
Bündelungsgrad m <Akustik> sound power concentration
Bündelungsmagnet m <Fernseh> focusing magnet
Bündelungsspule f <Fernseh> focusing coil
Bundes-Immissionsschutzgesetz n <Abfall> *(AE)* Federal Clean Air Act *(in Deutschland)*
Bundesmarine f <Wassertrans> Federal Navy *(Marine)*

Bundesstraße f <Trans> *(BE)* trunk road
Bundfalz m <Druck> back fold
bündig adj <Bau, Druck, Maschinen> flush
bündig machen v 1. <Fertig> flush; 2. <Maschinen> make flush
bündige Überlappverbindung f <Fertig> flush joint
bündiger Stoß m <Fertig> flushing
Bundlager n <Maschinen> flange bearing
Bundmutter f <Maschinen> collar nut, flanged nut
Bundschraube f <Maschinen> collar screw
Bundsteg m <Druck> gutter
Bungeegurt m <Kfztech> bungee cord
Bunker m 1. <Bau> bunker, hopper; 2. <Erdöl> bunker *(Produktlagerung)*; 3. <Kohlen, Lufttrans, Wassertrans> bunker
Bunkerfüllstandsmesser m <Gerät> bin level meter
Bunkerkohle f <Kohlen> bunker coal
Bunkeröle npl <Lufttrans, Wassertrans> bunker oil
Bunkertank m <Umweltschmutz> bunker tank
Bunsenbrenner m <Labor> Bunsen burner
Bunsenelement n <Elektriz> Bunsen cell
bunt glasierte Kachel f <Ker & Glas> encaustic tile
Bunt... 1. <Ker & Glas> stained; 2. <Kunststoff> *(AE)* colored, *(BE)* coloured
Buntbartschloss n <Bau> warded lock
Buntglas n <Ker & Glas> tinted glass
Buntglasfenster n <Ker & Glas> stained glass window
Buntpapier n <Ker & Glas> paper stain
Buntpigment n <Kunststoff> *(AE)* colored pigment, *(BE)* coloured pigment
Buntsandstein m <Erdöl> bunter *(Formation des Trias)*
Bürette f <Labor> burette
Bürettenständer m <Labor> burette stand
Bürgersteig m <Bau> *(BE)* pavement, *(AE)* sidewalk
Burgunderflasche f <Ker & Glas> burgundy bottle
Büro n <Comp & DV, Telekom> office
Büroabfall m <Abfall> office waste
Büroautomatisierung f <Comp & DV> OA, office automation
Bürodruckmaschine f <Druck> office printing machine
Bürofernschreiben n <Comp & DV> teletext
Bürogebäude n <Bau> office building
Bürokommunikation f <Telekom> office communication
Bürokommunikationsprotokoll n *(TOP)* <Telekom> technical and office protocol *(TOP)*
Bürokopiergerät n <Comp & DV> office copier
Burrus-Diode f <Optik, Telekom> Burrus diode
Burst m <Elektriz> burst *(Übertragung, Fehler)*
Bürste f 1. <Bau> wiper; 2. <Elektriz, Mechan> brush, contact brush; 3. <Papier> brush
Bürsten n <Elektrotech, Ker & Glas, Kfztech, Textil> brushing
Bürstenabheber m <Elektriz> brush lifting device
Bürstenbrücke f <Elektrotech> brush rocker
Bürstenentladung f <Elektriz> brush discharge
Bürstenfeuer n 1. <Elektrotech> brown-out, brush discharge, sparking; 2. <Phys> brush discharge
Bürstenglättmaschine f <Papier> brush polishing machine
Bürstenhalter m <Elektriz, Elektrotech> brush holder
Bürstenhalterarm m <Elektrotech> brush rod
Bürstenjoch n <Elektriz> brush yoke
Bürstenkontaktwiderstand m <Elektriz> brush contact resistance
bürstenlose Maschine f <Elektrotech> brushless machine
bürstenloser Generator m <Elektriz> brushless generator motor

bürstenloser

bürstenloser Gleichstrommotor *m* <Elektrotech> permanent brushless d.c. motor, permanent magnet synchronous motor, PMBLDC motor
bürstenloser Induktionsmotor *m* **mit gewickeltem Läufer** <Elektrotech> brushless wound-rotor induction motor
bürstenloser Motor *m* <Elektriz> brushless generator motor
Bürstennacheilwinkel *m* <Elektriz> angle of brush lag
Bürstensatinage *f* <Papier> brush glazing
Bürstensprühen *n* <Elektriz> brush sparking
Bürstenstellung *f* <Elektrotech> brush position
Bürstenstreichverfahren *n* <Papier> brush coating
Bürstenstriche *mpl* <Ker & Glas> brush lines
Bürstenträger *m* <Elektriz> brush holder
Bürstenverlustwiderstand *m* <Elektriz> brush contact resistance
Bürstenvoreilung *f* <Elektriz> angle of lead of brushes
Bürstenvoreilwinkel *m* <Elektriz> angle of lead of brushes
Bürstenwähler *m* <Elektriz> brush selector
Bürstenwalze *f* <Papier> brush roller
Bürstenwaschmaschine *f* <Textil> brush washer
Bürstenwinkel *m* <Elektriz> brush angle
Burstoszillator *m* <Fernseh> burst-locked oscillator
Burst-Switching *n* <Telekom> burst switching *(ITG)*
Burstverstärker *m* <Fernseh> burst amplifier
Bus *m* 1. <Comp & DV> *(BE)* bus, highway, *(AE)* trunk; 2. <Elektrotech> bus; 3. <Kfztech> bus, coach; 4. <Telekom> bus
Bus *m* **auf Eisenbahngleisen** <Eisenbahn, Trans> *(AE)* bus on railroad tracks, *(BE)* bus on railway tracks
Bus *m* **auf Eisenbahnwaggon** <Eisenbahn, Trans> *(AE)* bus on railroad wagon, *(BE)* bus on railway wagon
Busabschlussstecker *m* <Comp & DV> bus terminator
Büschel *n* 1. <Metall> pencil; 2. <Optik> bundle; 3. <Textil> cluster
Büschelentladung *f* 1. <Elektriz> brush discharge; 2. <Elektrotech> brown-out, brush discharge; 3. <Phys> brush discharge
Büschungsmauer *f* <Bau> toe wall
Bushel *n* <Metrol> bushel
Buskollision *f* <Comp & DV> bus collision
Buskonfiguration *f* <Telekom> bus configuration
Busleitung *f* <Telekom> highway, bus
Buslinie *f* <Trans> bus line
Bus-Master *m* <Comp & DV> bus master
Bus-Maus-Adapter *m* <Comp & DV> bus-mouse adaptor
Busnetz *n* <Comp & DV> bus network
Busnetzwerk *n* <Comp & DV> bus network
Busplatine *f* <Comp & DV> bus board
Busschnittstelle *f* <Comp & DV> bus interface
Bussolenlinse *f* <Ker & Glas> compass lens
Busspur *f* <Trans> bus lane
Busspur *f* **mit Leitsystem** <Trans> bus lane equipped with guiding device
Bustaxidienst *m* <Trans> on-call bus system
Bustopologie *f* <Comp & DV> bus topology
Buszuteiler *m* <Telekom> bus arbitrator
Butadien *n* <Chemie, Erdöl> butadiene *(Petrochemie)*
Butadien-Acrylnitril-Kautschuk *m* <Kunststoff> butadiene acrylonitrile rubber
Butadienkautschuk *m* 1. <Fertig> bivinyl rubber; 2. <Kunststoff> butadiene rubber
Butadien-Styrol-Copolymerisat *n* <Erdöl> butadiene-styrene copolymer *(Petrochemie)*

Butan *n* <Erdöl> butane *(Petrochemie)*
Butandion *n* <Chemie> butanedione, diacetyl
Butangastanker *m* <Wassertrans> butane carrier; butane gas tanker *(Schifftyp)*
Butan-Tanker *m* <Erdöl> butane tanker
Butan-Tankschiff *n* <Erdöl> butane tanker *(Schifffahrt)*
Butan-Tankwagen *m* <Erdöl> butane carrier *(Eisenbahn, Straße)*
Butenluv *m* <Wassertrans> bumpkin *(Schiffbau)*
Bütte *f* <Papier> chest
Büttenofen *m* <Ker & Glas> Butten furnace
Büttenrand *m* <Papier> deckle edge • **mit Büttenrand versehen** <Verpack> deckle
Büttenrandschneider *m* <Foto> jagged edge trimmer
Butter... <Chemie> butyric
Butterfass *n* <Lebensmittel> firkin
Butterherstellung *f* <Lebensmittel> churning
buttern *v* <Lebensmittel> churn
Buttern *n* <Lebensmittel> churning
Buttersäure *f* 1. <Chemie> butyric; 2. <Lebensmittel> butyric acid
Butterworth-Filter *n* <Elektronik> Butterworth filter
Butyl... <Kunststoff, Lebensmittel> butyl, butylated
Butylacetat *n* <Kunststoff> butyl acetate
Butylether *n* <Lebensmittel> butyl ether
Butylhydroxyanisol *n* (BHA) <Lebensmittel> butylated hydroxyanisole, BHA *(Antioxidans)*
Butylhydroxytoluol *n* (BHT) <Lebensmittel> butylated hydroxytoluene *(BHT)*
Butylkautschuk *m* 1. <Fertig> butyl rubber *(Kunststoffinstallationen)*; 2. <Kunststoff> butyl rubber
Butylkresol *n* <Lebensmittel> BHT, butylated hydroxytoluene
Butylphthalat *n* <Kunststoff> butyl phthalate
Butyrat *n* <Chemie> butyrate
Butyrin *n* <Chemie> butyrin
Butzen *m* 1. <Fertig> sludge *(Lochen)*; 2. <Kunststoff> flash
BV *(Bildverstehen, Bildverständnis)* <Künstl Int> IU *(image understanding)*
BW *(Bandbreite, Bandweite)* <Aufnahme, Comp & DV, Elektronik, Fernseh, Funktech, Optik, Telekom> BW *(bandwidth)*
B-Y-Achse *f* <Fernseh> B-Y axis
B-Y-Signal *n* <Fernseh> B-Y signal
Bypass *m* 1. <Erdöl> bypass *(Rohrleitungsbau)*; 2. <Hydraul> bypass
Bypass-Bohrung *f* <Kfztech> bypass bore
Bypass-Filter *n* <Maschinen> bypass filter
Bypass-Luftstrom *m* **im Triebwerk** <Lufttrans> engine bypass air
Bypass-Schalter *m* <Elektriz> bypass switch
Bypass-Triebwerk *n* 1. <Lufttrans> bypass engine, ducted fan; 2. <Thermod> bypass engine
Bypass-Turbinensystem *n* <Kerntech> turbine bypass system
Bypass-Verhältnis *n* <Lufttrans> bypass ratio *(Turbolüfter, Turbojet)*
Byte *n* <Comp & DV, Telekom> byte
byteparallele Übertragung *f* <Telekom> byte-parallel transmission
byteserielle Übertragung *f* <Telekom> byte-serial transmission
Byte-Umschalter *m* <Telekom> byte switch
byteweise *adj* <Telekom> byte-by-byte
byzantinischer Bogen *m* <Bau> stilted arch
BZ *(Betriebszentrum)* <Telekom> *(AE)* operations center, *(BE)* operations centre
B-Zeit *f* <Kunststoff> B-stage time

C

c 1. <Elektronik, Elektriz, Kohlen, Kunststoff, Telekom, Umweltschmutz> *(Konzentration)* c *(concentration)*; 2. <Hydraul> *(Wellenausbreitungsgeschwindigkeit)* c *(wave celerity)*; 3. <Metrol> *(Zenti...)* c *(centi)*; 4. <Metrol> *(Lichtgeschwindigkeit)* c *(velocity of light)*; 5. <Optik> *(Lichtgeschwindigkeit)* c *(speed of light in empty space)*; 6. <Phys> *(spezifische Wärme)* c *(specific heat capacity)*; 7. <Phys> *(Schallgeschwindigkeit)* c *(speed of sound)*; 8. <Thermod> *(spezifische Wärmekapazität)* c *(specific heat capacity)*
C 1. <Bau, Elektriz, Elektrotech, Erdöl, Heiz & Kälte> *(Kapazität)* C *(capacity)*; 2. <Chemie> *(Kohlenstoff)* C *(carbon)*; 3. <Elektriz, Elektrotech, Metrol, Phys> *(Coulomb)* C *(coulomb)*; 4. <Hydraul> *(Cauchy'sche Zahl)* C *(Cauchy coefficient)*; 5. <Hydraul> *(Chezy-Koeffizient)* C *(Chezy coefficient)*; 6. <Hydraul> *(Ausflusskoeffizient, Durchflusskoeffizient)* C *(discharge coefficient)*; 7. <Labor> *(Celsius)* C *(centigrade)*; 8. <Nichtfoss Energ> *(Schüttkoeffizient)* C *(discharge coefficient)*; 9. <Funktech, Phys, Telekom> *(Kapazität)* C *(capacitance)*
Ca *(Calcium)* <Chemie, Lebensmittel> Ca *(calcium)*
CA 1. <Heiz & Kälte> *(kontrollierte Atmosphäre)* CA *(controlled atmosphere)*; 2. <Kunststoff, Textil> *(Celluloseacetat)* CA *(cellulose acetate)*
Cabalglas n <Ker & Glas> cabal glass
Cabrio n <Kfztech> cabriolet, convertible
Cache m <Comp & DV> cache
Cache-Speicher m <Comp & DV> cache memory
C-Achse f <Geom> C axis *(definiert Drehbewegung um die Z-Achse)*
CAD *(computergestützte Konstruktion, computergestützter Entwurf)* <Comp & DV, Elektriz, Kontroll, Mechan, Telekom, Trans> CAD *(computer-aided design)*
Cadaverin n <Chemie> cadaverin, pentamethylenediamine
CADCAM *(computergestützte Konstruktion und Fertigung)* <Comp & DV> CADCAM *(computer-aided design and manufacturing)*
Caisson m <Wassertrans> caisson
Caissonkrankheit f <Wassertrans> caisson disease
CAL 1. <Comp & DV> *(computergestützter Unterricht)* CAI *(computer-aided instruction)*; 2. <Comp & DV> *(computergestütztes Lernen)* CAL *(computer-aided learning)*
Calabarin n <Chemie> calabarine, eserine, physostigmine *(ein Alkaloid)*
CA-Lagerung f <Lebensmittel> controlled-atmosphere storage
Calciferol n <Chemie> calciferol
calcifizieren v <Chemie> calcify
calcinieren v <Chemie> calcine
Calcium n *(Ca)* <Chemie, Lebensmittel> calcium *(Ca)*
• **Calcium entziehen** <Chemie> decalcify *(Medizin)*
Calciumacetylid n <Chemie> calcium acetylide, calcium carbide
Calciumcarbid n <Chemie> calcium acetylide, calcium carbide
Calciumcarbonat n <Lebensmittel> calcium carbonate
Calciumchlorid n <Chemie, Lebensmittel> calcium chloride
Calciumcyanamid n <Chemie> calcium carbide, calcium cyanamide
Calciumhydroxid n <Chemie, Lebensmittel> calcium hydroxide
Calciumnaphthenat n <Kunststoff> calcium naphthenate
Calciumpantothenat n <Lebensmittel> calcium pantothenate
Calciumphosphat n <Lebensmittel> calcium phosphate
Calciumsulfat n <Lebensmittel> *(AE)* calcium sulfate, *(BE)* calcium sulphate
Californium n *(Cf)* <Chemie> californium *(Cf)*
CAM 1. <Comp & DV, Elektriz> *(computergestützte Fertigung, computergestützte Produktion)* CAM *(computer-aided manufacturing)*; 2. <Comp & DV, Künstl Int> *(Assoziativspeicher, inhaltsadressierbarer Speicher)* CAM *(content-addressable memory)*
Camcorder m <Fernseh> camcorder
Camera-Lucida f <Foto> camera lucida
Campbell-Stokes-Aufzeichner m <Nichtfoss Energ> Campbell-Stokes recorder
Camper m <Kfztech> *(AE)* camper, *(BE)* caravan
Camphen n <Chemie> camphene
Camphorat n <Chemie> camphorate
Candela f *(cd)* <Elektrotech, Metrol, Optik, Phys> candela *(cd)*
Cantharidin n <Chemie> cantharidine
Cantilever-Langsieb n <Papier> cantilever foudrinier
CAP *(computergestütztes Publizieren)* <Druck> CAP *(computer-aided publishing)*
CAPI CAPI, common application programming interface *(Software-Schnittstelle zwischen ISDN und PC)*
Caproin n <Chemie> caproin
Caprolactam n <Chemie> caprolactam
Caproyl... <Chemie> octanoyl
Capryl... <Chemie> hexyl
Capryliden n <Chemie> caprylidene, octyne
Capsaicin n <Chemie> capsaicin
Capsicin n <Chemie> capsicin
Capture-Effekt m <Telekom> capture effect
CAR *(Zivilflugvorschriften)* <Raumfahrt> CAR *(Civil Air Regulations)*
Caravan m <Kfztech> *(AE)* camper, *(BE)* caravan, *(AE)* trailer
Carbamat n <Chemie> carbamate
Carbamid n <Chemie> carbamide, urea
Carbamid... <Chemie> carbamic
Carbamidsäurehydrazid n <Chemie> semicarbazide
Carbamoyl n <Chemie> carbamoyl
Carbanil n <Chemie> carbanil, phenyl isocyanate
Carbanilid n <Chemie> carbanilide
Carbanion n <Chemie> carbanion
Carbazid n <Chemie> carbazide
Carbazol n <Chemie> carbazole, dibenzopyrrole
Carben n <Chemie> carbene
Carbeniat-Anion n <Chemie> carbanion
Carbid n <Chemie> carbide
Carbidkohle f <Chemie> carbide carbon
carbocyclisch adj <Chemie> carbocyclic, homocyclic, isocyclic
Carbohydrase f <Lebensmittel> carbohydrase
Carbol... <Chemie> carbolic
Carbolyase f <Chemie> decarboxylase
Carbonatation f <Chemie> carbonation *(Entkalkung des Zuckerrübensaftes)*
carbonisieren v <Lebensmittel> aerate *(Getränke mit Kohlensäure sättigen)*
Carbonisierung f <Chemie> charring *(Brennverhalten von Textilien)*
Carbonohydrazid n <Chemie> carbazide, carbonohydrazide
Carbonylchlorid n <Chemie> carbonyl dichloride, phosgene

Carbonylsulfid

Carbonylsulfid n <Umweltschmutz> *(AE)* carbonyl sulfide, *(BE)* carbonyl sulphide
Carborundum® n <Chemie> Carborundum®, silicon carbide
Carbostyril n <Chemie> carbostyril
Carboxyl n <Chemie> carboxyl
Carboxylation f <Chemie> carbonation *(Entkalkung des Zuckerrübensaftes)*
carboxyliertes Polymer n <Kunststoff> carboxylated polymer
Carboxymethylcellulose f *(CMC)* <Kunststoff, Lebensmittel> carboxymethylcellulose, CMC *(Verdickungsmittel, Emulgator)*
Carbroabzug m <Foto> *(AE)* carbro color print, *(BE)* carbro colour print
Carbrodrucken n <Foto> carbro printing
Cardew-Spannungsmessgerät n <Elektriz> Cardew voltmeter
Carmin n <Chemie> carmine
Carminfarbe f <Lebensmittel> carmine
Carnitin n <Chemie> carnitine, novain
Carnot'sche Maschine f <Phys, Thermod> Carnot engine
Carnot'scher Kreisprozess m <Phys, Thermod> Carnot cycle
Carnot'sches Theorem n <Phys> Carnot's theorem *(Wärmekraftmaschine)*
Caron n <Chemie> carone
Caroten n <Chemie> carotene
Carotin n <Lebensmittel> carotene
Carrageen n <Lebensmittel> carrageen
Carrier m <Textil> carrier *(Färberei)*
Carrier-Auswahlnummer f <Telekom> carrier prefix
Carrier-Vorwahl f <Telekom> carrier prefix
CASE *(computergestützte Softwareentwicklung)* <Comp & DV> CASE *(computer-aided software engineering)*
Casein n <Kunststoff> casein
Cashewnussschalenöl n <Kunststoff> cashew-nut shell oil
Cashey-Kasten m <Ker & Glas> cashey box
Casing n <Erdöl> casing *(Bohrtechnik)*
Cäsium n *(Cs)* <Chemie> *(BE)* caesium, *(AE)* cesium *(Cs)*
Cassegrain'sche Antenne f 1. <Funktech, Phys> Cassegrain aerial, Cassegrain antenna; 2. <Raumfahrt> *(BE)* Cassegrain aerial, *(AE)* Cassegrain antenna *(Weltraumfunk)*
Cassegrain'sches Teleskop n <Phys> Cassegrain telescope
Cassettenradio n <Kfztech> radio-cassette *(Zubehör)*
Castorin n <Chemie> castorin
CAT 1. <Comp & DV> *(computerunterstützte Übersetzung)* CAT *(computer-assisted translation)*; 2. <Elektronik> *(Senderöhre mit gekühlter Anode)* CAT *(cooled-anode transmitting valve)*; 3. <Lufttrans> *(Kaltluftturbulenzen)* CAT *(cold air turbulence)*
Cat-Cracker m <Erdöl> cat cracker *(Raffinerie)*
Catechin n <Chemie> catechinic acid
Catecholamin n <Chemie> catecholamine
Catechugerb... <Chemie> catechutannic
Caterpillar-Planierraupe f <Trans> caterpillar bulldozer
Catforming n <Chemie> catforming
CATV *(Fernsehen über Gemeinschaftsantenne)* <Fernseh> CATV *(community antenna television system)*
CATVI 1. <Fernseh> *(Kabelfernsehstörung)* CATVI *(cable television interference)*; 2. <Fernseh> *(störende Beeinflussung des Kabelfernsehdienstes)* CATVI *(cable television interference)*

Cauchy-Green'scher Verzerrungstensor m <Strömphys> Cauchy-Green strain tensor
Cauchy'sche Zahl f *(C)* <Hydraul> Cauchy coefficient *(C)*
Cauer-Filter n <Phys> Cauer filter
CAV *(computerunterstütztes Sehen)* <Künstl Int> CAV *(computer-aided vision)*
Cavendish-Experiment n <Phys> Cavendish experiment
CA-Verpackung f *(Verpackung in geregelter Atmosphäre)* <Verpack> CAP *(controlled-atmosphere packaging)*
CB-Funk m <Funktech, Telekom> CB, CB radio, CB radio communication, citizen's band radio
CBR-Wert m <Bau> California Bearing Ratio
CCD *(Ladungsgekoppeltes Bauelement, CCD-Element)* <Elektronik, Fernseh, Phys, Telekom> CCD *(charge-coupled device)*
CCD-Bildaufnehmer-Zeile f <Fernseh> CCD line imager
CCD-Bildwandler m <Fernseh> CCD image sensor, CCD imager
CCD-Filter n <Elektronik> CCD filter
CCD-Signalverarbeitung f <Elektronik> CCD signal processing
CCD-Zeile f <Fernseh> CCD line imager
CCITT *(Internationaler Beraterausschuss für den Fernschreib- und Telefondienst)* <Telekom> CCITT *(International Telegraph and Telephone Consultative Committee)*
C-Compiler m <Comp & DV> C-compiler
CCTV *(angewandtes Fernsehen, industrielles Fernsehen)* <Fernseh> CCTV *(closed-circuit television)*
cd *(Candela)* <Elektrotech, Metrol, Optik, Phys> cd *(candela)*
Cd *(Kadmium)* <Chemie> Cd *(cadmium)*
CD 1. <Comp & DV, Elektronik, Telekom> *(Trägerdetektion, Trägererkennung)* CD *(carrier detection)*; 2. <Comp & DV, Telekom> *(Kollisionserkennung)* CD *(collision detection)*; 3. <Comp & DV, Optik> *(Compact-Disk)* CD *(compact disk)*
CD-Abspielgerät n <Comp & DV, Optik> CD player
CD-Archivierungssystem n <Comp & DV, Optik> optical disk filing system
CD-Bibliothek f <Comp & DV, Optik> optical disk library
CD-Festwertspeicher m <Comp & DV, Optik> optical disk read-only memory
CD-i *(beschreibbare CD)* <Optik> CD-I *(compact disk-interactive)*
CD-Kassette f <Comp & DV, Optik> optical disk cassette
CD-Laufwerk n 1. <Comp & DV> optical disk drive; 2. <Optik> optical disk drive, optical drive
CD-Leser m <Comp & DV, Optik> optical disk reader
CD-Lese/Schreib-Laufwerk n <Comp & DV> CD read/write drive, CD R/W-drive
CDM *(kompandierte Deltamodulation)* <Telekom> CDM *(companded delta modulation)*
CD-Platte f <Optik> *(AE)* disk platter
CD-ROM 1. <Comp & DV> *(Compact-Disk ohne Schreibmöglichkeit)* CD-ROM *(compact disk read-only memory)*; 2. <Optik> *(Compact-Disk-Speicher ohne Schreibmöglichkeit)* CD-ROM *(compact disk read-only memory)*
CD-ROM-Diskettenlaufwerk n <Comp & DV> CD-ROM disk drive
CD-ROM-Festplattenlaufwerk n <Comp & DV> CD-ROM hard disk drive
CD-ROM-Spieler m <Optik> CD-ROM player
CD-Spieler m 1. <Comp & DV> optical disk player; 2. <Optik> CD player, optical disk player
CD-Spieler m **für Ton** <Optik> audio CD player
CD-Wechsler m <Comp & DV, Optik> optical disk exchanger

Ce *(Cerium)* <Chemie> Ce *(cerium)*
CE *(elektrische Kapazität)* <Akustik> CE *(electric capacitance)*
CEBAF *(Gleichstromelektronenbeschleuniger)* <Teilphys> CEBAF *(continuous electron beam facility)*
Cedren *n* <Chemie> cedrene
Cedrol *n* <Chemie> cedrol
Ceiling-Temperatur *f* <Kunststoff> ceiling temperature
Celluloid *n* 1. <Chemie> celluloid; 2. <Kunststoff> xylonite
Cellulose *f* <Kunststoff, Lebensmittel> cellulose
Celluloseacetat *n* *(CA)* <Kunststoff, Textil> acetate, cellulose acetate *(CA)*
Celluloseacetobutyrat *n* <Kunststoff> cellulose acetobutyrate
Celluloseanstrichfarbe *f* <Kunststoff> cellulose paint
Celluloseglykolsäure *f* <Lebensmittel> carboxymethylcellulose
Cellulosenitrat *n* <Chemie, Kunststoff> cellulose nitrate
Cellulosepropionat *n* <Kunststoff> CP, cellulose propionate
Cellulosetriacetat *n* *(CTA)* <Kunststoff> cellulose triacetate *(CTA)*
Celsius *n* *(C)* <Metrol> centigrade *(C)*
Celsius-Grad *m* <Phys, Thermod> centigrade
Celsius-Temperaturskale *f* <Phys> centigrade
CEN *(Comité Européen de Normalisation)* <Qual> CEN *(Europäischer Normungsausschuss)*
Centrex-Vermittlung® *f* <Telekom> Centrex system®
Centronics-Schnittstelle® *f* <Druck> Centronics interface®
Cephalosporin *n* <Chemie> cephalosporin
Cer... <Chemie> ceric, cerium
Ceran *n* <Chemie> cerane, hexacosane
Cerenkov'sche Strahlung *f* <Strahlphys, Teilphys> Cerenkov radiation
Cerenkov'scher Detektor *m* <Strahlphys, Teilphys> Cerenkov detector
Cerenkov'scher Effekt *m* <Strahlphys, Teilphys> Cerenkov effect
Cerenkov'scher Zähler *m* <Strahlphys, Teilphys> Cerenkov counter
Cerfluorit *m* <Chemie> yttrocerite
Cerium *n* *(Ce)* <Chemie> cerium *(Ce)*
Cermet *n* <Kerntech> cermet *(Metallkeramik-Werkstoff für Brennelemente)*
CERN *(Europäisches Kernforschungszentrum)* <Teilphys> CERN *(European Organization for Nuclear Research)*
Cerussa *n* <Chemie> ceruse, cerussa
Ceten *n* <Chemie> hexadecanol
Cetylalkohol *m* <Chemie> ethal, hexadecanol
Cevadin *n* <Chemie> cevadine
Cevin *n* <Chemie> cevine
Cf *(Californium)* <Chemie> Cf *(californium)*
C-Feder *f* <Maschinen> C spring
C-Format-Videorecorder *m* <Fernseh> C format videotape recorder
C-förmiges Gestell *n* <Fertig> C-frame *(Presse)*
CGA *(Farbgrafikadapter)* <Comp & DV> CGA *(colour graphics adaptor)*
CGS-System *n* *(Zentimeter-Gramm-Sekunde-System)* <Metrol> CGS system *(centimetre-gramme-second system)*
Chalcon *n* <Chemie> chalcone
Chalkanthit *m* <Chemie> blue vitriol, bluestone, chalcanthite
Chalkogenidglas *n* <Ker & Glas> chalcogenide glass
Chambray *m* <Textil> chambray
Chaos *n* <Strömphys> chaos *(Stabilität)*
chaotische Bewegung *f* <Strömphys> chaotic motion

Chaptalisierung *f* <Lebensmittel> chaptalization
Charakteristik *f* <Elektriz, Maschinen, Math> characteristic *(Merkmal)*
charakteristisch *adj* characteristic
charakteristische Empfindlichkeit *f* <Akustik> characteristic sensitivity
charakteristische Frequenz *f* <Elektronik, Fernseh, Funktech> characteristic frequency
charakteristische Impedanz *f* <Elektriz> surge impedance
charakteristischer Leitungswiderstand *m* <Elektrotech, Phys> characteristic impedance
charakteristisches Röntgenspektrum *n* <Strahlphys> characteristic X-ray spectrum
Charge *f* 1. <Anstrich> batch; 2. <Elektrotech> charge; 3. <Fertig> melt; 4. <Lebensmittel> batch; 5. <Papier> batch *(Verarbeitung)*; 6. <Telekom, Textil> batch
Chargenbeschickung *f* <Kerntech> batch fuel loading
Chargenbetrieb *m* <Chemtech> batch processing
Chargenentladung *f* <Kerntech> batch extraction
Chargenmischer *m* <Chemtech, Ker & Glas, Lebensmittel> batch mixer
Chargenmischung *f* <Bau> batch mix
Chargenofen *m* <Chemtech> batch furnace
Chargenstreuung *f* <Qual> batch variation
Chargenumfang *m* <Qual> batch size
Chargenverarbeitung *f* <Papier> batch processing
Charles-Gesetz *n* <Phys> Charles's law
Charm *m* <Phys, Teilphys> charm
Charmeuse *f* <Textil> charmeuse, locknit
Charmonium *n* <Phys> charmonium
Charpy'sche Spitzkerbprobe *f* <Kerntech> Charpy V-notch test
Charpy'scher Kerbschlagbiegeversuch *m* <Metall> Charpy impact test
Charpy'scher Kerbschlagversuch *m* 1. <Qual> Charpy test, Charpy impact test; 2. <Phys> Charpy test; 3. <Werkprüf> Charpy impact test
Charpy'scher Pendelschlagversuch *m* <Anstrich> Charpy impact test
Charpy'scher Rundkerbversuch *m* <Mechan> Charpy V-notch test, Charpy impact test
Charpy'scher V-Kerbtest *m* <Mechan> Charpy V-notch test
Charpy'sches Schlagzähigkeitsprüfgerät *n* <Kunststoff> Charpy impact tester
Charter *f* <Wassertrans> charter *(Schiff)*
Charter *f* **eines bloßen Schiffes** <Wassertrans> bareboat charter
Charterbuchung *f* **im Voraus** <Lufttrans> advance booking charter *(ABC)*
Chassis *n* <Kfztech> chassis *(Karosserie)*
Chassislängsträger *m* <Kfztech> chassis member *(Karosserie)*
chatten *v* <Telekom> chat *(Internet)*
Chaulmoograöl *n* <Chemie> chaulmoogra oil
Chavibetol *n* <Chemie> chavibetol
Chavicol *n* <Chemie> chavicol
Checkliste *f* <Patent, Qual> check list
Chefapparat *m* <Telekom> boss extension *(Telefon)*
Chef-Sekretär-Anlage *f* <Telekom> manager/secretary station
Chef-Sekretärin-Anlage *f* <Telekom> executive-secretary system
Chelat *n* <Chemie> chelate • **Chelate bilden** <Chemie> chelate
Chelatbildner *m* <Chemie> chelating agent
Chelatbildung *f* <Chemie, Kerntech> chelation
chelatisieren *v* <Chemie> chelate

CHEMFIX-Verfahren

CHEMFIX-Verfahren n <Abfall> CHEMFIX process
chemiebeständig adj <Verpack> chemically resistant
Chemiefaser f <Chemtech, Fertig> (AE) chemical fiber, (BE) chemical fibre
Chemiereaktor m <Kerntech> chemonuclear fuel reactor
Chemiezellstoff m <Papier> dissolving pulp
Chemigraph m <Druck> process engraver
Chemikalien fpl <Papier> chemicals
Chemikalienbeständigkeit f <Sicherheit> resistance to chemicals
Chemikalienschutz m <Sicherheit> chemical protection
Chemikalienschutzkleidung f <Sicherheit> chemical protection clothing
Chemikalientanker m <Wassertrans> chemical carrier
Chemilumineszenz f <Phys, Strahlphys> chemiluminescence
chemisch adj <Textil> chemical
chemisch gebundenes Wasser n <Wasserversorg> combined water
chemisch neutrales Öl n <Erdöl> chemically neutral oil
chemisch reinigen v <Textil> dry-clean
chemisch resistent adj <Verpack> chemically resistant
chemisch träge adj <Anstrich> chemically inert
chemisch widerstandsfähiges Glas n <Ker & Glas> chemically-resistant glass
chemisch wirksam adj <Phys> actinic
chemisch-atomare Masseneinheit f <Kerntech> chemical atomic mass unit
chemische Abfälle mpl <Abfall> chemical waste
chemische Abscheidung f aus der Dampfphase <Telekom> (AE) chemical vapor deposition technique, (BE) chemical vapour deposition technique, (AE) vapor phase chemical deposition, (BE) vapour phase chemical deposition
chemische Abwasserreinigung f <Wasserversorg> chemical water treatment
chemische Analyse f <Kohlen> chemical analysis
chemische Aufbereitung f <Kohlen> chemical treatment
chemische Behandlung f <Kohlen> chemical treatment
chemische Beschichtung f <Kerntech> chemical coating
chemische Beständigkeit f 1. <Ker & Glas> chemical durability; 2. <Kohlen> chemical stability; 3. <Kunststoff> chemical resistance
chemische Bindung f <Erdöl> chemical bond (Petrochemie)
chemische Dosimetrie f <Strahlphys> chemical dosimetry (Radioaktivität)
chemische Entwicklung f <Foto> chemical development
chemische Fällung f <Abfall> chemical precipitation
chemische Gefahr f <Sicherheit> chemical hazard
chemische Kapselung f <Sicherheit> encapsulating chemical protection
chemische Kohlereinigung f <Kohlen, Umweltschmutz> chemical coal cleaning
chemische Oberflächenbearbeitung f <Mechan> chemical machining
chemische Pulpe f <Abfall> chemical pulp
chemische Reaktion f <Anstrich> reaction
chemische Rückstände mpl <Kerntech> chemical drains
chemische Sicherheitstechnik f <Sicherheit> chemical safety engineering
chemische Trimmung f <Kerntech> chemical shimming
chemische Verstärkung f <Foto> chemical intensification
chemische Waage f <Labor> chemical balance
chemische Wechselwirkung f zwischen Pellet und Hülse <Kerntech> pellet-clad chemical interaction

chemische Wiederaufbereitungsanlage f <Kerntech> chemical reprocessing plant
chemischer Kampfstoff m <Umweltschmutz> toxic agent (Militär)
chemischer Laser m <Elektronik> chemical laser
chemischer Prozess m in der Atmosphäre <Umweltschmutz> atmospheric chemical process
chemischer Sauerstoffbedarf m (CSB) <Umweltschmutz> chemical oxygen demand (COD)
chemischer Schadstoff m <Sicherheit> hazardous chemical
chemischer Wirkstoff m <Umweltschmutz> chemical agent
chemisches Bedampfungsverfahren n <Elektronik> (AE) chemical vapor deposition technique, (BE) chemical vapour deposition technique
chemisches Enthülsen n <Kerntech> chemical decanning, chemical decladding (von Brennmaterial)
chemisches Gleichgewicht n <Phys> chemical balance
chemisches Härten n <Metall> chemical hardening
chemisches Nachweismittel n <Chemie> reagent (anorganische Chemie)
chemisches Polieren n <Metall> chemical polishing
chemisches Potenzial n <Phys> chemical potential
chemisches Raketentriebwerk n <Maschinen> chemical rocket engine
chemisches Treibmittel n <Lebensmittel> chemical leavening
Chenillegarn n <Textil> chenille yarn
Chessylith m <Chemie> chessylite
Chezy-Koeffizient m (C) <Hydraul> Chezy coefficient (C)
chiffrieren v <Telekom> cipher, encipher, encode
Chiffrierschlüssel m <Telekom> key
Chiffrierung f <Telekom> ciphering, encryption
Chill-Roll-Coextrusion f <Kunststoff> chill roll coextrusion
Chill-Roll-Extrusion f <Kunststoff> chill roll extrusion
China... <Chemie> quinic
China-Blau n <Ker & Glas> China blue
Chinacrin n <Chemie> atebrin, quinacrine
Chinaldin n <Chemie> methylquinoline, quinaldine
Chinalizarin n <Chemie> quinalizarin
Chinamin n <Chemie> quinamine
Chinarinde f <Chemie> cinchona bark
Chinazin n <Chemie> quinoxaline
Chinhydron n <Chemie> quinhydrone
Chinicin n <Chemie> chinicine, quinicine
Chinidin n <Chemie> conchinine, quinidine
Chinin n <Chemie> quinine
Chininum n <Chemie> quinine (Pharmazie)
Chinit m <Chemie> cyclohexanediol, quinitol
chinoid adj <Chemie> quinoid
Chinolin n <Chemie> quinoline
Chinolyl... <Chemie> quinolyl
Chinon n <Chemie> quinone
Chinonphenolimin n <Chemie> indophenol
Chinovabitter n <Chemie> chinovin, quinova bitter, quinovin
Chinovin n <Chemie> chinovin, quinova bitter, quinovin
Chinoxalin n <Chemie> quinoxaline
Chip m <Comp & DV, Elektronik, Telekom> chip, microchip, wafer • **außerhalb des Chips gelegen** <Elektronik> off chip • **auf Chip generiert** <Elektronik> generated on chip • **nicht auf dem Chip befindlich** <Elektronik> off chip
Chipauslegung f <Elektronik> chip layout
Chipentwurf m <Elektronik> chip design
Chip-Fläche f <Elektronik> chip area

Chipkarte f 1. <Comp & DV, Telekom> chip card, smart card; 2. <Telekom> smart card
Chipkartenleser m <Telekom> chip-card reader, smart card reader
Chipkomplexität f <Elektronik> chip complexity
Chipkondensator m <Elektrotech> on-chip capacitor
Chiprate f <Telekom> chip rate
Chipsatz m <Elektronik> chip set
Chipträger m <Elektronik> chip carrier
Chirpen n <Telekom> chirping
Chirping n <Telekom> chirping
Chitosamin n <Chemie> chitosamine, glucosamine
Chlor n (Cl) <Chemie> chlorine (Cl)
Chloracetat n <Chemie> chloroacetate
Chloral... <Chemie> chloral
Chloralformamid n <Chemie> chloral formamide
Chloralose f <Chemie> chloralose
Chloranil n <Chemie> chloranil
Chlorat n <Chemie> chlorate
chloren v <Chemie> chlorinate
Chloressig... <Chemie> chloracetic
Chlorfasern fpl <Textil> (AE) chlorofibers, (BE) chlorofibres
Chlorhydrat n <Chemie> chlorine hydrate
Chlorhydrin n <Chemie> chlorohydrin
Chlorid n <Chemie> chloride
Chloridglas n <Ker & Glas> chloride glass
Chloridpapier n <Foto> chloride paper
chlorieren v <Chemie> chlorinate
chloriertes Polyethylen n (PE-C) <Kunststoff> chlorinated polyethylene (CPE)
chloriertes Polyvinylchlorid n (PVC-C) <Kunststoff> chlorinated polyvinyl chloride (CPVC)
Chlorierung f <Abfall, Chemie> chlorination
chlorig adj <Chemie> chlorous
Chlorit n 1. <Chemie> chlorite; 2. <Erdöl> chlorite (Mineral); 3. <Kohlen> chlorite
Chlorkalk m 1. <Lebensmittel> bleaching powder, chlorinated lime; 2. <Textil> bleaching powder
Chlorkautschuk m <Kunststoff> chlorinated rubber
Chloroplatinat n <Chemie> tetrachloroplatinate
Chloropren n <Chemie> chloroprene
Chloroprenkautschuk m (CPK) <Kunststoff> chloroprene rubber (CR)
Chlorphenol n <Chemie> chlorophenol
Chlorpikrin n <Chemie> aquinite, nitrochloroform
chlorsauer adj <Chemie> chloric
Chlorsäure f <Lebensmittel> chloric acid
Chlorschwefelisocyanat n (CSI) <Umweltschmutz> chlorosulfonyl isocyanate (CSI)
Chlorung f <Chemie> chlorination
Choke m <Kfztech, Mechan> choke
Chol... <Chemie> cholic
Cholesten n <Chemie> cholestene, cholesterol
Cholesterin n 1. <Chemie> cholestene, cholesterol; 2. <Lebensmittel> cholesterol
Cholin n <Lebensmittel> choline
Cholinesterase f <Chemie> cholesterase, choline esterase
Cholsäureester m <Chemie> cholate
Chondrin n <Chemie> chondrin
chopperstabilisierter Verstärker m <Elektronik> chopper stabilized amplifier
Chopper-Verstärker m <Elektronik> chopper amplifier
Chrom n (Cr) <Chemie> chromium (Cr)
Chromalaun m <Chemie> (AE) ammonium chromic sulfate, (BE) ammonium chromic sulphate
Chromat n <Chemie> chromate
chromatieren v <Metall> chromate
chromatisch adj <Optik> chromatic

chromatische Aberration f <Optik, Strahlphys> chromatic aberration
chromatische Aberration f **längs der optischen Achse** <Phys> longitudinal chromatic aberration
chromatische Abweichung f <Foto> chromatic aberration
chromatische Dispersion f <Optik> chromatic dispersion
chromatische Tonleiter f <Akustik> chromatic scale
chromatische Verzeichnung f <Optik> chromatic distortion
chromatische Verzerrung f <Telekom> chromatic distortion
chromatische Zerlegung f <Optik> chromatic dispersion (Licht)
chromatischer Abbildungsfehler m <Fernseh> chromatic aberration
chromatischer Fehler m <Optik> chromatic distortion
chromatischer Halbton m <Akustik> chromatic semitone
Chromatograph m <Gerät> chromatograph
Chromatographie f <Chemie> chromatography
Chromatographiepapiere npl <Labor> chromatography papers
Chromatographiesäule f <Labor> chromatography column
Chromattank m <Labor> chromatography tank
Chromdioxidband n <Aufnahme, Fernseh> chrome dioxide tape
Chromel n <Metall> chromel
Chromerz n <Ker & Glas> chrome ore
chromgrün adj <Ker & Glas> chrome green
Chrominanz f <Elektronik> chrominance (Fernsehtechnik)
Chrominanzbandbreite f <Fernseh> chrominance bandwidth
Chrominanzdemodulator m <Fernseh> chrominance demodulator
Chrominanzhilfsträgersignal n <Fernseh> chrominance subcarrier signal
Chrominanzsignal n <Elektronik, Fernseh, Raumfahrt> chrominance signal (Weltraumfunk)
Chrominanzträger m <Fernseh> chrominance carrier
Chrominanzträgerleistung f <Fernseh> chrominance carrier output
Chrominanzunterträger m <Fernseh> chrominance subcarrier
Chrominanzverstärker m <Elektronik, Fernseh> chrominance amplifier
Chromit m <Ker & Glas> chromite
Chromium n (Cr) <Chemie> chromium (Cr)
Chromnickelstahl m <Metall> nickel chrome steel, nickel chromium steel
Chromodynamik f <Teilphys> chromodynamics
Chromogen n <Chemie> chromogen
Chromosphäre f <Funktech, Raumfahrt> chromosphere
Chromotrop... <Chemie> chromotropic
Chromoxid n <Ker & Glas> chromic oxide
Chromsäuresalz n <Chemie> chromate
Chromstahl m <Metall> chrome steel, chromium steel
Chromvanadiumstahl m <Metall> chrome vanadium steel
Chromverstärker m <Foto> chrome intensifier
Chronometer n <Gerät, Phys, Wassertrans> chronometer (Navigation)
Chronometergang m <Wassertrans> chronometer rate (Navigation)
Chrysen n <Chemie> chrysene
Chrysoidin n <Chemie> chrysoidine, diaminoazobenzene
Chrysophan... <Chemie> chrysophanic
Ci (Curie) <Phys, Strahlphys> Ci (curie)

CIM

CIM 1. <Comp & DV> *(CompuServe® Information Manager)* CIM *(CompuServe® Information manager)*; 2. <Comp & DV> *(computerintegrierte Fertigung)* CIM *(computer-integrated manufacture)*
cim *(Kubikzoll pro Minute)* <Labor> cim *(cubic inches per minute)*
Ciminit n <Erdöl> ciminite *(Mineral)*
Cinchonidin n <Chemie> chinidine, cinchonidine
Cinchonin n <Chemie> cinchonine
Cinch-Stecker m <Fernseh> cinch, cinch connector, cinch plug
Cineol n <Chemie> cineole
Cineol... <Chemie> cineolic
Cinnamyl... <Chemie> cinnamyl
Cinnolin n <Chemie> cinnoline
Circular-Pitch m <Maschinen> CP, circular pitch
CISC *(Prozessor mit komplettem Befehlssatz, konventioneller Rechner)* <Comp & DV> CISC *(complex instruction set computer)*
cis-ständig adj <Chemie> cis
Cis-Trans... <Chemie> cis-trans
Citral n <Chemie> citral
Citrat n <Chemie> citrate
Citronell n <Lebensmittel> citronella
Citronellal n <Chemie> citronellal, dimethyloctenal
Citronellaldehyd m <Chemie> citronellal, dimethyloctenal
Citronellaöl n <Chemie> citronella oil, citronyl
Citronellöl n <Lebensmittel> citronella oil *(Bestandteil von Lebensmittelaromen)*
Citrullin n <Chemie> aminoureidovaleric acid, citrulline
cl *(Geschwindigkeit von Längswellen)* <Akustik> cl *(velocity of longitudinal waves)*
Cl *(Chlor)* <Chemie> Cl *(chlorine)*
CL 1. <Comp & DV> *(Befehlssprache, Betriebssprache)* CL *(command language)*; 2. <Hydraul, Nichtfoss Energ, Phys, Wassertrans> *(Auftriebsbeiwert, Auftriebszahl)* CL *(lift coefficient)*; 3. <Lufttrans> *(Auftriebszahl)* CL *(lift coefficient)*
Claisen-Destillierkolben m <Chemtech> Claisen flask
Claisen-Kolben m <Chemtech> Claisen flask
Clausius-Clapeyron'sche Gleichung f <Phys> Clapeyron's equation
Clausius-Mosotti'sche Gleichung f <Phys> Clausius-Mosotti formula *(Lichtbrechung)*
Clausius-Rankine-Prozess-Motor m <Kfztech> Rankine cycle engine
Clausius'sche Formulierung f <Phys> Clausius statement *(des zweiten Hauptsatzes der Thermodynamik)*
Clearscan-Verfahren n <Wassertrans> clearscan *(Radar)*
cleaven v <Fertig> cleave *(Faseroptik)*
Cleaver m <Fertig> cleaver *(Faseroptik)*
Cleveit m <Kerntech> cleveite *(Pechblendeart)*
Client-seitiges Zertifikat n <Comp & DV> client-side certificate
Client/Server-Kommunikationssystem n <Comp & DV> client/server messaging system
Client/Server-System n <Comp & DV> client-server system
Clip-On-Kältesatz m <Heiz & Kälte> clip-on refrigerating machine
Clip-On-Messinstrument n <Elektriz> clip-on instrument
Clipper m <Elektronik, Lufttrans, Telekom> clipper
Clipperverstärker m <Elektronik> clipper amplifier
Clon m <Comp & DV> clone
Closed-Loop-Verkehrsregelung f <Lufttrans> closed-loop traffic control system
Closed-Shop-Betrieb m <Comp & DV> hands-off operation

Clupein n <Chemie> clupein
Cluster m <Comp & DV, Kerntech, Kontroll> cluster
Clustermodell n <Kerntech> cluster model *(Atomkern)*
Cm *(Curium)* <Chemie> Cm *(curium)*
CM *(mechanische Auslenkung)* <Akustik> CM *(mechanical compliance)*
CMC *(Carboxymethylcellulose)* <Kunststoff, Lebensmittel> CMC *(carboxymethylcellulose)*
CM-Cellulose f <Lebensmittel> carboxymethylcellulose
CMD-Spieler m <Optik> CMD player
CMOS *(Komplementär-Metalloxid-Halbleiter)* <Elektronik> CMOS *(complementary metal oxide semiconductor)*
CMOS-Halbleiterelement n <Comp & DV> CMOS semiconductor
CMOS-Koppelpunkt m <Telekom> CMOS crosspoint
CMOS-Logik f <Elektronik> CMOS logic
CMOS-Transistoren mpl <Elektronik> CMOS transistors
C-Multiplex m *(Codemultiplex)* <Telekom> CDM *(code-division multiplexing)*
CNC *(computernumerische Steuerung)* <Maschinen> CNC *(computerized numeric control)*
Co *(Cobalt, Kobalt)* <Chemie> Co *(cobalt)*
Coanda'scher Effekt m <Strömphys> Coanda effect *(Strömungsverhalten)*
coax *(koaxial)* <Funktech> coax *(coaxial)*
Cobalamin n <Chemie> cobalamin
Cobalt n *(Co)* <Chemie> cobalt *(Co)*
Cobalt-60 n <Chemie> cobalt-60, radiocobalt
Cobaltammin n <Chemie> cobaltammine
Cobaltchlorid n <Ker & Glas> cobalt chloride
Cobaltiak n <Chemie> cobaltammine
Cobaltnaphthenat n <Kunststoff> cobalt naphthenate
COBOL *(problemorientierte Programmiersprache für Geschäftsbetrieb)* <Comp & DV> COBOL *(common business oriented language)*
Cochenille m <Lebensmittel> cochineal
Cockpit n 1. <Kfztech> cockpit; 2. <Lufttrans> cockpit, flight deck; 3. <Wassertrans> cockpit *(Schiff)*
Cockpittonbandgerät n <Lufttrans> cockpit voice recorder
Code m 1. <Comp & DV> code; 2. <Elektronik> code *(Nachrichtentechnik)*; code *(Regelwerk zur Darstellung von Informationen)*; 3. <Telekom> code *(Übergang)*
Code m **der Zielvermittlungsstelle** <Telekom> destination point code
Codeänderung f <Comp & DV> code change
Codeausgaben fpl <Qual> code editions
Codebereich m <Comp & DV> code area
Codec m *(Codierer-Decodierer)* <Comp & DV, Elektronik, Telekom> codec *(coder-decoder)*
Code-Element n <Comp & DV, Telekom> code element
Codeerweiterungssteuerzeichen n <Telekom> code extension character
codegeteilter Mehrfachzugriff m <Telekom> code-division multiple access *(CDMA)*
Codein n <Chemie> codeine, methylmorphine
Codeklasse f <Patent, Qual> code class
Codekonverter m <Elektrotech> code converter
Codemultiplex m *(C-Multiplex)* <Telekom> code-division multiplexing *(CDM)*
Codeschlüssel m <Elektrotech> key
codespezifisches Problem n <Comp & DV> code-related problem
Codestempel m <Qual> code symbol stamp
Codesteuerzeichen n <Comp & DV> code extension character
Codetaste f <Comp & DV> code key
Codethylin n <Chemie> ethylmorphine
codetransparent adj <Telekom> code-transparent

Codeumsetzer m 1. <Comp & DV> encoder; 2. <Elektrotech, Telekom> code converter
Codeumsetzung f 1. <Comp & DV> code conversion; 2. <Telekom> transcoding, code conversion
codeunabhängiges Steuerungsverfahren n <Comp & DV, Telekom> high-level data link control
Codewandler m 1. <Elektrotech> code converter; 2. <Telekom> transcoder, code converter
Codier... <Elektronik, Fernseh, Kontroll, Telekom> coding, encoding
Codierblatt n <Comp & DV> coding sheet
Codier/Decodier-Baustein m <Telekom> encoding/decoding device *(codec)*
codieren v 1. <Elektronik> encode; 2. <Fernseh> code, encode; 3. <Telekom> encode
Codieren n 1. <Comp & DV> coding; 2. <Elektronik, Telekom> coding, encoding
Codierer m 1. <Elektronik, Fernseh> coder, encoder; 2. <Kontroll> encoder; 3. <Telekom> coder, encoder; 4. <Trans> coding device
Codierer-Decodierer m *(Codec)* <Comp & DV, Elektronik, Telekom> coder-decoder *(codec)*
Codierfehler m <Comp & DV, Elektronik> coding error
Codierhöhenmesser m <Lufttrans> encoding altimeter
Codiersystem n <Trans> code-decode system
Codiersystem n **mit Einzelkopf** <Verpack> single-head coding system
Codiersystem n **mit vier Köpfen** <Verpack> four-head coding system
codiert adj <Telekom> coded
Codiertabelle f <Comp & DV> coding table
codierte Impulse mpl <Fernseh> encoded pulses
codierte Kennzeichenübersicht f <Trans> precoded tag survey
codierte Stereophonie f <Aufnahme> coded stereo
codierte Übertragung f <Telekom> ciphered transmission, coded transmission, encrypted transmission
codierter Exponent m <Comp & DV> biased exponent
codiertes Signal n <Elektronik, Telekom> encoded signal
codiertes Videosignal n <Fernseh> coded TV-signal
Codiertheorie f <Comp & DV> coding theory
Codierung f 1. <Elektronik> coding, encoding; 2. <Funktech> encoding; 3. <Telekom> coding, encoding; 4. <Verpack> encoding
Codierungsfeld n <Comp & DV> code field
Codierungspotenziometer n <Raumfahrt> encoding potentiometer *(Weltraumfunk)*
Codierverfahren n <Telekom> coding scheme
Coenzym n <Lebensmittel> coenzyme
Coeruleum n <Chemie> ceruleum
Coil n <Verpack> coil
Coil-Coating n <Kunststoff> coil coating
CO_2-Laser n *(Kohlendioxidlaser)* <Elektronik> CO_2 laser *(carbon dioxide laser)*
Colchicin n <Chemie> colchicine
Colcothar m <Chemie> colcothar
Coliforme npl <Lebensmittel> coliform bacteria
Collargol n <Chemie> collargol
Collider m <Teilphys> collider
Collodinlösung f <Chemie> colloxylin
Collodium n <Chemie> collodion, nitrated cellulose
Colloxylin n <Chemie> colloxylin
Colophen n <Chemie> colophene
Colophonium n <Chemie> colophony, rosin
colorieren v <Druck> *(AE)* color, *(BE)* colour
Colpitts-Oszillator m <Elektronik> Colpitts oscillator
Columbit m <Chemie> columbite
Comen... <Chemie> comenic

Compact-Disk f *(CD)* 1. <Comp & DV> compact disk *(CD)*; 2. <Optik> *(BE)* compact disc, *(AE)* compact disk *(CD)*
Compact-Disk m **ohne Schreibmöglichkeit** *(CD-ROM)* <Comp & DV> compact disk read-only memory *(CD-ROM)*
Compact-Disk-Festwertspeicher m <Comp & DV> compact disk read-only memory
Compact-Disk-Speicher m **ohne Schreibmöglichkeit** *(CD-ROM)* <Optik> *(BE)* compact disc read-only memory, *(AE)* compact disk read-only memory *(CD-ROM)*
Compound n <Kunststoff> compound
Compoundkern m <Kerntech, Phys, Strahlphys> compound nucleus
Compound-Zustand m <Kerntech> compound state
Compton'sche Streuung f <Phys, Strahlphys, Teilphys> Compton scattering
Compton'sche Wellenlänge f <Kerntech, Phys, Strahlphys, Teilphys> Compton wavelength
Compton'scher Effekt m <Phys, Strahlphys, Teilphys> Compton effect
Compton'sches Kontinuum n <Phys, Strahlphys, Teilphys> Compton continuum
Compton'sches Spektrometer n <Phys, Strahlphys, Teilphys> Compton spectrometer
Compurverschluss m <Foto> Compur shutter
CompuServe® Information Manager m *(CIM)* <Comp & DV> CompuServe® Information Manager *(CIM)*
Computer m 1. <Comp & DV> computer, machine; 2. <Elektriz, Wassertrans> computer
Computer m **mit reduziertem Befehlsvorrat** <Comp & DV> RISC, reduced instruction set computer
Computer m **mit seriellem Anschluss** <Comp & DV> serial computer
Computer m **mit variabler Wortlänge** <Comp & DV> variable-word-length computer
computerabhängig adj <Comp & DV> machine-dependent
Computeranimation f <Comp & DV, Fernseh> computer animation
Computerbediener m <Comp & DV> computer operator
Computerfachkenntnis f <Comp & DV> computer literacy
Computergeometrie f <Geom> computational geometry, computer geometry
computergesteuert adj <Telekom> computer-controlled
computergesteuertes Echtzeitfahrzeugsystem n <Kfztech> real-time locating system
computergesteuertes Fahrzeugsystem n *(AGVS)* <Kfztech> automated guided vehicle system *(AGVS)*
computergestützt adj 1. <Comp & DV> computer-aided, computer-assisted; 2. <Kontroll> computer-aided
computergestützte Abwicklung f <Raumfahrt> computerized management *(Weltraumfunk)*
computergestützte Fertigung f *(CAM)* <Comp & DV, Elektriz> computer-aided manufacturing *(CAM)*
computergestützte Konstruktion f *(CAD)* <Comp & DV, Elektriz, Kontroll, Mechan, Telekom, Trans> computer-aided design *(CAD)*
computergestützte Konstruktion f **und Fertigung** f *(CADCAM)* <Comp & DV> computer-aided design and manufacturing *(CADCAM)*
computergestützte Produktion f *(CAM)* <Comp & DV, Elektriz> computer-aided manufacturing *(CAM)*
computergestützte Softwareentwicklung f *(CASE)* <Comp & DV> computer-aided software engineering *(CASE)*
computergestützter Entwurf m *(CAD)* <Comp & DV, Elektriz, Kontroll, Mechan, Telekom, Trans> computer-aided design *(CAD)*

computergestützter

computergestützter Unterricht m (CAL) <Comp & DV> computer-aided instruction, computer-assisted instruction (CAI)
computergestütztes Lernen n (CAL) <Comp & DV> computer-aided learning, computer-assisted learning (CAL)
computergestütztes Publizieren n (CAP) <Druck> computer-aided publishing (CAP)
computergestütztes Training n <Telekom> computer-based training
Computergrafik f <Comp & DV, Fernseh> computer graphics
computerintegrierte Fertigung f (CIM) <Comp & DV> computer-integrated manufacture (CIM)
Computerintelligenz f <Künstl Int> computer intelligence, machine intelligence
Computerkenntnisse fpl <Comp & DV> computer literacy
Computerkommunikationsnetz n <Telekom> computer communications network
Computerkunst f <Comp & DV> computer art
Computerlauf m <Comp & DV> machine run
computerlesbar adj <Comp & DV> machine-readable
computerlesbare Daten npl <Comp & DV> machine-readable data
Computerlogik f <Comp & DV> computer logic
Computer-Maus-Kissen n <Comp & DV> mouse pad
Computernetz n <Comp & DV, Telekom> computer network
Computernetzarchitektur f <Comp & DV> computer network architecture
computernumerische Steuerung f (CNC) <Maschinen> computerized numeric control (CNC)
Computerpapier n <Druck> printout paper
Computerraum m <Aufnahme> machine room
Computersatz m <Comp & DV> computer setting
Computerschnittstelle f <Telekom> computer interface
Computerschnittstelle f **für elektronischen Bankverkehr** <Telekom> home banking computer interface
Computersicherheit f <Comp & DV> computer security
Computersystem n <Comp & DV, Elektriz> computer system
Computersystem n **mit Haupt- und Nebenrechner** <Comp & DV> master-slave system
Computertechnik f <Comp & DV> computer technology
Computer-to-Plate-System n <Druck> direct-to-plate system, computer-to-plate system, CTP system
computerunabhängig adj <Comp & DV> machine-independent
computerunterstützte Übersetzung f (CAT) <Comp & DV> computer-assisted translation (CAT)
computerunterstützte Überwachung f <Qual> computer-aided inspection
computerunterstütztes Problemlösen n <Künstl Int> computer-aided problem solving
computerunterstütztes Sehen n (CAV) <Künstl Int> computer-aided vision (CAV)
Computervision f <Künstl Int> artificial vision, computational vision, computer vision, CV
Computerwesen n <Comp & DV> computing
ComSat m 1. <Raumfahrt> comsat, communication satellite (Nachrichtensatellit); 2. <Wassertrans> comsat, communication satellite (Fernmeldesatellit)
Conche f <Lebensmittel> conche (Schokoladenfabrikation)
conchieren v <Lebensmittel> conche
Conchinin n <Chemie> conchinine, quinidine
Cone f <Textil> cone
Confinement n <Phys, Teilphys> confinement
Coniferin n <Chemie> coniferin
Constraint m <Künstl Int> constraint

Container m 1. <Kohlen> container; 2. <Meerschmutz> scoop; 3. <Trans, Verpack> container (Entsorgung)
Container m **für den kombinierten Verkehr** <Trans> intermodal container
Container m **mit festen Rädern** <Trans> container with fixed wheels
Container m **mit Seitenwandvorhang** <Eisenbahn> curtain-sided container
Container m **mit zu öffnender Oberseite** <Trans> container with opening top
Container Freight Station f (CFS) <Wassertrans> container freight station, CFS
Containeraufstellfläche f <Trans> (AE) marshaling area, (BE) marshalling area
Containerentladung f <Trans> container destuffing, container stripping, container unpacking
Containerfracht f <Wassertrans> (AE) containerized cargo, (BE) containerised cargo
containerisieren v <Trans> containerize
Containerisierung f <Trans, Verpack> containerization
Containerkai m <Wassertrans> container wharf
Containerkapsel f <Trans> container capsule
Containerladefähigkeit f <Bau, Wassertrans> container capacity
Containerleichter m <Wassertrans> container lighter
Containerleichter-Mutterschiff-System n <Trans> CLASS, containerized lighter aboard ship system
Container-LKW m <Trans> (BE) container carrier lorry, (AE) container carrier truck
Containerpackstation f <Wassertrans> container freight station, CFS (Transport und Logistik)
Containerschiff n (CTS) <Wassertrans> container ship (CTS)
Containerspülanlage f <Verpack> container rinsing equipment
Containerstandplatz m <Trans> container berth
Containerstation f 1. <Telekom> containerized station; 2. <Trans> container terminal
Containertragwagen m <Eisenbahn> (BE) container car, (AE) container truck
Containertransportschiff n <Wassertrans> container transport ship
Containerumschlagkran m <Trans> transtainer crane
Containerumschlagplatz m <Trans> container station
Containerwagen m <Trans> (BE) container car, (AE) container truck
Containerzug m <Eisenbahn> freightliner train
Controller m <Comp & DV> controller
Convolver m <Elektronik> convolver (militärische Nachrichtentechnik)
Convolvulin n <Chemie> convolvulin, rhodeorhetin
Cooper-Paare npl <Phys> Cooper pairs (Supraleitung)
Copolymer n <Chemie, Erdöl, Kunststoff, Textil> copolymer
Copolymerisat n <Chemie, Erdöl, Kunststoff, Textil> copolymer
Copolymerisation f <Kunststoff> copolymerization
Coprozessor m <Comp & DV> coprocessor
Copyright-Vermerk m <Druck> copyright notice
COR (Druckausgabeverkleinerung) <Comp & DV> COR (character output reduction)
Coracit m <Kerntech> coracite
Cordierit m <Ker & Glas> cordierite
Coriandrol n <Chemie> linalool
Coriolis-Beschleunigung f <Mechan, Raumfahrt> Coriolis acceleration
Coriolis-Kraft f <Phys, Raumfahrt, Strömphys> Coriolis force
Coronen n <Chemie> coronene
Corticoid n <Chemie> corticoid, corticosteroid

Corticosteroid n <Chemie> corticoid, corticosteroid
Corticosteron n <Chemie> corticosterone
Corticotropin n <Chemie> corticotrophin
Cos-Austrittsgesetz n <Optik> cosine emission law
COSMOS (Komplementär-Symmetrischer Metalloxid-Halbleiter) <Elektronik> COSMOS (complementary-symmetrical metal oxide semiconductor)
cot (Kotangens) <Geom> cot (cotangent)
Cotton-Mouton'scher Effekt m <Phys> Cotton-Mouton effect
Cotton'sche Waage f <Phys> Cotton balance (Magnetismus)
Couette'sche Strömung f <Strömphys> Couette flow
Coulomb n (C) <Elektriz, Elektrotech, Metrol, Phys> coulomb, C (SI-Einheit der Elektrizitätsmenge oder elektrischen Ladung)
Coulomb'sche Abstoßung f <Phys> Coulomb repulsion
Coulomb'sche Barriere f <Strahlphys> Coulomb barrier
Coulomb'sche Energie f <Strahlphys> Coulomb energy
Coulomb'sche Fließbedingung f <Phys> Coulomb's theorem
Coulomb'sche Torsionswaage f <Phys> Coulomb's torsion balance
Coulomb'sche Waage f <Phys> (AE) Coulomb gage, (BE) Coulomb gauge
Coulomb'sches Gesetz n <Elektriz, Phys> Coulomb's law
Coulometer n <Chemie, Elektrotech, Phys> coulometer, voltameter
Coumaronharz n <Kunststoff> coumarone resin
Countdown m <Raumfahrt> countdown (Rückwärtszählung zum Startzeitpunkt)
Cowper'scher Winderhitzer m <Ker & Glas> Cowper stove
Cp (Wärmekapazität bei konstantem Druck) <Labor> Cp (heat capacity at constant pressure)
CPFSK (phasenkontinuierliche Frequenzumtastung) <Elektronik, Funktech, Telekom> CPFSK (continuous phase frequency shift keying)
CPK (Chloroprenkautschuk) <Kunststoff> CR (chloroprene rubber)
CPM (Methode des kritischen Weges) <Comp & DV> CPM (critical path method)
CPT-Theorem n <Phys> CPT theorem, charge conjugation parity operation time reversal theorem (Elementarteilchenphysik)
CPU (Zentraleinheit, zentrale Rechnereinheit) <Comp & DV, Telekom> CPU (central processing unit)
CQR-Anker n (Danford-Anker) <Wassertrans> CQR anchor (coastal quick release anchor)
Cr (Chrom, Chromium) <Chemie> Cr (chromium)
CR (Rotationsauslenkung) <Akustik> CR (rotational compliance)
Cracken n <Erdöl> cracking (Raffinerie)
Craquelé eglas n <Ker & Glas> crackled glass
Crashsensor m <Kfztech> collision sensor
Craze-Bildung f <Kunststoff> crazing
Crazing-Effekt m <Kunststoff> crazing
CRC (zyklische Blockprüfung, zyklische Blocksicherung, CRC-Prüfung) <Comp & DV, Elektronik, Labor, Telekom> CRC (cyclic redundancy check)
CRCA (kaltgewalzt und ausgeglüht) <Metall> CRCA (cold-rolled and annealed)
Crêpe m <Textil> crepe
Crêpe-Kautschuk m <Kunststoff> crepe rubber
Cresolharz n <Kunststoff> cresol resin
Crestfaktor m <Elektrotech> crest factor
crimpbar adj <Fertig> crimpable (Faseroptik)
Crimpen n <Fertig> crimping (Faseroptik)
Crocein... <Chemie> crocein

Crocin n <Chemie> crocin
Crookesglas n <Ker & Glas> Crookes glass
Crookes'sche Röhre f <Elektronik> Crookes tube
Crookes'scher Dunkelraum m <Phys> Crookes dark space
Crossbar-Selektor m <Elektrotech> cross coupling, crossbar selector
Crossbar-System n <Telekom> crossbar system
Crossbar-Wähler m <Telekom> crossbar selector
Cross-Connector m <Telekom> cross-connect
Cross-Track-Error m <Wassertrans> cross-track error (Satellitennavigation)
Croton... <Chemie> crotonic
Crotonaldehyd m <Chemie> crotonaldehyde, methylacrolein
Crotyl n <Chemie> butenyl, crotyl
Crown n <Papier> crown
Crude-Oil Washing n (COW) <Wassertrans> crude-oil washing, COW (Tankreinigungsverfahren)
Crusher m <Kohlen> crusher
Cryotron n Kryotron
Cs (Cäsium) <Chemie> Cs (caesium)
CS (Durchschaltevermittlung) <Comp & DV, Telekom> CS (circuit switching)
CSB (chemischer Sauerstoffbedarf) <Umweltschmutz> COD (chemical oxygen demand)
CSI (Chlorschwefelisocyanat) <Umweltschmutz> CSI (chlorosulfonyl isocyanate)
CSM (Kommando- und Servicemodul) <Raumfahrt> CSM, command and service module (Raumschiff)
CSMA (Mehrfachzugriff mit Trägerkennung) <Comp & DV, Telekom> CSMA (carrier sense multiple access)
CSMA/CD (CSMA/CD-Verfahren) <Comp & DV, Telekom> CSMA/CD (carrier sense multiple access with collision detection)
CSMA/CD-Verfahren n (CSMA/CD) <Comp & DV> carrier sense multiple access with collision detection (CSMA/CD)
CSMA-Zugriffsverfahren n (CSMA) <Comp & DV, Telekom> carrier sense multiple access (CSMA)
CSN (Durchschalte-Vermittlungsnetz) <Comp & DV, Telekom> CSN (circuit-switched network)
CSPDN (leitungsvermitteltes öffentliches Datennetz) <Telekom> CSPDN (circuit-switched public data network)
C-Stahl m <Anstrich> carbon steel
ct (Geschwindigkeit von Transversalwellen) <Labor> ct (velocity of transversal waves)
CTA (Cellulosetriacetat) <Kunststoff> CTA (cellulose triacetate)
CTCSS (Hilfsträgergeräuschsperre) <Funktech> CTCSS (continuous tone-coded squelch system)
CTD 1. <Elektrotech> (Ladungsverschiebeschaltung) CTD, charge transfer device (Halbleiter); 2. <Phys> (ladungsgekoppeltes Bauelement) CTD, charge transfer device; 3. <Raumfahrt> (Ladungsübertragungsgerät) CTD, charge transfer device; 4. <Telekom> (Ladungstransferelement) CTD, charge transfer device
CTOL-Flugzeug n (konventionell startendes und landendes Flugzeug) <Lufttrans> CTOL aircraft (conventional takeoff and landing aircraft)
CTP-Belichtungssystem n <Druck> direct-to-plate system, computer-to-plate system, CTP system
CTP-System n <Druck> direct-to-plate system, computer-to-plate system, CTP system
CTS (Containerschiff) <Wassertrans> CTS (container ship)
CT-Standard m <Telekom> cordless telephone standard (CEPT Standard für Cordless-Telefone)
Cu (Kupfer) <Chemie, Metall> Cu (copper)
Cubebin n <Chemie> cubebin

CUG

CUG *(geschlossene Benutzergruppe, geschlossener Benutzerkreis)* <Comp & DV, Telekom> CUG *(closed user group)*
Cumalin *n* <Chemie> coumalin
Cumalin... <Chemie> coumalic
Cumar... <Chemie> coumaric
Cumaran *n* <Chemie> coumaran
Cumarin *n* <Chemie> benzopyrone, coumarin, cumarin
Cumarinsäureanhydrid *n* <Chemie> benzopyrone, coumarin, cumarin
Cumin... <Chemie> cumic
Cuminaldehyd *m* <Chemie> cumic aldehyde
Cumol *n* <Chemie> cumene, cumol, isopropylbenzene
Cumyl... <Chemie> cumyl
Cuprat *n* <Chemie> cuprate
Cuprit *m* <Chemie> cuprite, red copper ore
Cupromangan *n* <Chemie> cupromanganese, manganese copper
Curcuma *f* <Chemie> turmeric
Curcumin *n* <Chemie> curcumin
Curie *n (Ci)* 1. <Phys> curie, Ci; 2. <Strahlphys> curie, Ci *(alte Einheit der Radioaktivität)*
Curie'sche Konstante *f* <Phys, Strahlphys> Curie constant
Curie'sche Temperatur *f* <Phys, Strahlphys> Curie point, Curie temperature
Curie'scher Punkt *m* 1. <Elektriz> Curie point; 2. <Phys, Strahlphys> Curie point, Curie temperature
Curie'sches Gesetz *n* <Phys, Strahlphys> Curie's law
Curie-Weiss-Gesetz *n* <Phys, Strahlphys> Curie-Weiss law
Curium *n (Cm)* <Chemie> curium *(Cm)*
Curium-Reihe *f* <Strahlphys> curium series
Curryklemme *f* <Wassertrans> cam cleat *(Beschläge)*
Cursor *m* <Comp & DV, Druck> cursor, indicator
Cursorausgangsstellung *f* <Comp & DV> cursor home
Cursorsteuerungsfeld *n* <Comp & DV> touchpad
Cursortaste *f* <Comp & DV> cursor key
Curtain-Coater *m* <Kunststoff> curtain coater
CVD *(Gasphasenabscheidung)* <Elektronik, Telekom> CVD *(chemical vapour deposition)*
C-Verstärker *m* <Elektronik> amplifier class C
CVS *(Teilstromentnahme nach Verdünnung)* <Umweltschmutz> CVS *(constant volume sampling)*
CW *(Dauerstrich, ungedämpfte Welle)* <Aufnahme, Elektronik, Elektrotech, Funktech, Telekom> CW *(continuous wave)*
CW-Betrieb *m* <Teilphys> CW mode
CW-Radar *n* <Wassertrans> CW radar
CW-Wert *m* <Kfztech> drag coefficient
Cyan... <Chemie> cyanic
Cyanamid *n* <Chemie> cyanamide
Cyanat *n* <Chemie> cyanate
Cyaneinstellung *f* <Foto> cyan filter adjustment
Cyanhärtung *f* <Metall> cyanide hardening
Cyanid *n* <Chemie> cyanide
Cyanidlaugerei *f* <Chemie> cyanide lixiviation process *(Gold- und Silbergewinnung)*
Cyanisierung *f* <Chemie> cyanidation
Cyanoaurat *n* <Chemie> dicyanoaurate
Cyanoferrat *n* <Chemie> hexacyanoferrate
Cyantoluol *n* <Chemie> cyanotoluene, tolunitrile
Cyanwasserstoff *m* <Chemie> hydrocyanic
cyclisch *adj* <Erdöl> cyclic
cyclische Kette *f* <Chemie> closed chain
cycloaliphatisches Amin *n* <Kunststoff> cycloaliphatic amine
Cycloalkan *n* 1. <Chemie> cyclane, cycloalkane, naphthene; 2. <Erdöl> cycloalkane *(Petrochemie)*
Cyclobutan *n* <Chemie> cyclobutane, tetramethylene
Cycloheptadecenon *n* <Chemie> civetone, cycloheptadecenone
Cycloheptanon *n* <Chemie> cycloheptanone, suberone
Cyclohexadien *n* <Chemie> cyclohexadiene, dihydrobenzene
Cyclohexan *n* 1. <Chemie> cyclohexane, hexahydrobenzene; 2. <Erdöl> cyclohexane *(Petrochemie)*
Cyclohexancarbonsäure *f* <Chemie> hexahydrobenzoic acid
Cyclohexandiol *n* <Chemie> cyclohexanediol, quinitol
Cyclokautschuk *m* <Kunststoff> cyclized rubber
Cyclonit *n* <Chemie> cyclonite, hexogen
Cycloolefin *n* <Erdöl> cycloolefin *(Petrochemie)*
Cycloparaffin *n* <Erdöl> cycloparaffin *(Petrochemie)*
Cyclopentan *n* <Chemie> cyclopentane, pentamethylene
Cyclopropan *n* <Chemie> cyclopropane, trimethylene

D

d 1. <Chemie, Phys, Teilphys> *(Deuteron)* d *(deuteron)*; 2. <Labor> *(Dezi)* d *(deci...)*
D 1. <Akustik> *(Schwärzung)* D *(optical density)*; 2. <Chemie> *(Deuterium)* D *(deuterium)*; 3. <Elektriz> *(Verschiebung)* D *(displacement)*; 4. <Elektronik, Funktech, Phys> *(Diffusionskoeffizient)* D *(diffusion coefficient)*; 5. <Fertig, Geom, Maschinen> *(Durchmesser)* D *(diameter)*; 6. <Fertig, Phys> *(Versetzung)* D *(displacement)*; 7. <Kerntech> *(Absorptionsdosis)* D *(absorbed dose)*; 8. <Optik> *(optische Dichte)* D *(optical density)*; 9. <Strahlphys> *(absorbierte Dosis)* D *(absorbed dose)*; 10. <Thermod> *(vierter Virialkoeffizient)* D *(fourth virial coefficient)*
da *(Deka...)* <Labor> da *(deca...)*
D/A *(Digital-Analog-...)* <Aufnahme, Comp & DV, Elektronik, Labor, Telekom> D/A *(digital-analog)*
DA *(direkter Zugriff)* <Comp & DV> DA *(direct access)*
Dach *n* <Bau, Eisenbahn, Kfztech, Kohlen> roof
Dachantenne *f* <Mobilkom> rooftop antenna
Dachbinder *m* <Bau> roof frame, roof truss, truss
Dachboden *m* <Bau> garret
Dachdeckung *f* <Bau> roofing
Dachfenster *n* 1. <Bau> roof light, skylight; 2. <Ker & Glas> roof light
Dachfirst *m* <Bau> crest
Dachgepäckträger *m* <Kfztech> roof rack
Dachhammer *m* <Bau> roofer's hammer, *(BE)* slate axe, slate knife
Dachkehle *f* <Bau> valley, valley gutter
Dachknick *m* <Bau> break, *(AE)* curb, *(BE)* kerb; breakpoint *(eines Mansardendaches)*
Dachlandeplatz *m* **für Hubschrauber** <Lufttrans> rooftop heliport
Dachlatte *f* <Bau> batten
Dachleiter *m* <Sicherheit> roof conductor
Dachmanschette *f* <Fertig> gland, seal *(Kunststoffinstallationen)*
Dachneigung *f* <Bau> roof pitch
Dachpappe *f* <Bau> roofing felt
Dachpappenrandstreifen *m* <Bau> selvage, selvedge
Dachpfanne *f* <Bau> pantile, roofing tile
Dachprisma *n* <Phys> Amici prism, roof prism
Dachrahmen *m* <Bau> purlin

Dachrinne f <Bau> eaves gutter, eaves trough, gutter
Dachrinne f **hinter einer Brüstungsmauer** <Bau> parapet gutter
Dachrinnenhalter m <Bau> gutter bracket
Dachschalung f <Bau> roofing
Dachschieferlatte f <Bau> (AE) slate ax, (BE) slate axe
Dachschirmplatte f <Kerntech> roof shielding plate
Dachschräge f 1. <Bau> roof pitch; 2. <Elektronik> pulse tilt; 3. <Telekom> tilt
Dachspriegel m <Eisenbahn> carline
dachstabilisiertes Schnellverkehrssystem n <Trans> top-stabilized rapid transit system
Dachstein m <Bau> roofing tile, saddle stone
Dachstuhl m <Bau> principal, roof truss, truss
Dachstuhl-Auflageplatte f <Bau> roof plate
Dachterassenwohnung f <Bau> penthouse
Dachziegel m <Bau> roofing tile, tile
Dachziegelende n <Bau> tail
Dachzierleiste f <Kfztech> (AE) drip molding, (BE) drip moulding (Karosserie)
Dacron® n <Wassertrans> (AE) Dacron®, (BE) Terylene® (Segeln)
Daguerreotypie f <Foto> daguerreotype
Dahlin n <Chemie> dahlin, helenin, inulin, sinistrin
Dalbe f <Wassertrans> dolphin (Festmachen)
Dalton'sches Partialdruckgesetz n <Phys, Thermod> Dalton's law
DAMA (bedarfsgesteuerter Vielfachzugriff) <Telekom> DAMA (demand-assigned multiple access)
Damköhlerzahlen fpl <Strömphys> Damköhler numbers
Damm m <Wasserversorg> (AE) dike, (BE) dyke, weir
Dämm... <Aufnahme, Heiz & Kälte, Verpack> insulating
Dammausspülung f <Eisenbahn> embankment erosion, embankment washout
Dammbalkenwehr n <Wasserversorg> stoplog weir
dämmen v 1. <Bau> insulate; 2. <Wasserversorg> dam
Dammkrone f <Bau, Wasserversorg> crest
Dämmmatte f <Bau> blanket, insulating mat
Dämmplatte f 1. <Aufnahme> acoustic tile; 2. <Heiz & Kälte, Verpack> insulating board
Dämmschicht f <Elektrotech> insulating layer (bei Akustik)
Dammschüttung f <Eisenbahn> embanking
Dammstein m <Ker & Glas> skimmer block
Dämmstein m <Heiz & Kälte> insulating brick
Dämmstoff m 1. <Bau> insulating material, insulator; 2. <Chemie> insulator; 3. <Elektrotech> insulating material (bei Akustik); 4. <Fertig> lag (Wärmeisolierung); 5. <Heiz & Kälte> insulant, insulator
Dammstraße f <Bau> causeway
Dämmung f 1. <Bau> insulation; 2. <Erdöl> lagging (von Rohren, Behältern); 3. <Kunststoff, Telekom, Wassertrans> insulation
Dämmungswert m <Akustik> SRI, Sound Reduction Index, TL, transmission loss
Dämmzahl f <Akustik> SRI, Sound Reduction Index, TL, transmission loss
Dampf m 1. <Chemie, Chemtech> steam, (AE) vapor, (BE) vapour; 2. <Elektronik, Erdöl> (AE) vapor, (BE) vapour; 3. <Hydraul> steam; 4. <Ker & Glas> (AE) vapor, (BE) vapour; 5. <Kerntech> steam, (AE) vapor, (BE) vapour; 6. <Kfztech> (AE) vapor, (BE) vapour; 7. <Maschinen> steam, (AE) vapor, (BE) vapour; 8. <Metall> (AE) vapor, (BE) vapour; 9. <Nichtfoss Energ, Papier> steam; 10. <Phys> steam, (AE) vapor, (BE) vapour; 11. <Textil> steam; 12. <Thermod> (AE) vapor, (BE) vapour • **mit Dampf behandeln** <Textil> steam • **mit Dampf entfetten** <Anstrich> (AE) vapor-degrease, (BE) vapour-degrease • **unter Dampf** <Maschinen> under steam

Dampfabblaseventil n <Hydraul> steam relief valve
Dampfabgabesystem n <Kerntech> steam dumping system
Dampfabscheider m <Hydraul, Maschinen> steam separator
Dampfabscheidung f <Elektronik> (AE) vapor deposition, (BE) vapour deposition
Dampfabschrecken n <Metall> (AE) vapor quenching, (BE) vapour quenching
Dampfabsperrventil n <Hydraul> steam stop valve
Dampfaufmachen n <Hydraul> steam raising
Dampfaufsaugzeit f <Bau> soaking period (Autoklav)
Dampfausdehnungszeit f <Hydraul> duration of steam, expansion
Dampfausgleich m <Hydraul> steam balance
Dampfauslass m <Hydraul, Maschinen> steam outlet
Dampfauslassöffnung f <Hydraul> steam port
Dampfaustritt m <Hydraul, Maschinen> steam outlet
Dampfaustrittsöffnung f <Hydraul> eduction port (Dampfzylinder)
Dampfbad n <Chemtech> (AE) vapor bath, (BE) vapour bath
Dampfbehälter m <Hydraul> steam case
Dampfbehandlung f <Papier, Textil> steaming
Dampfblase f <Chemtech> (AE) vapor bubble, (BE) vapour bubble
Dampfblasen n <Ker & Glas> steam blowing
Dampfblasenbildung f <Kfztech> (AE) vapor lock, (BE) vapour lock (Störung des Flüssigkeitszuflusses)
Dampfboiler m <Hydraul> steam boiler
Dampfbremse f <Hydraul> steam brake
Dampfbüchse f <Hydraul> steam chest
Dampfdestillation f <Chemtech> steam distillation
dampfdicht adj <Heiz & Kälte, Hydraul, Maschinen> steamtight
Dampfdichte f <Chemie, Phys, Thermod> (AE) vapor density, (BE) vapour density
dampfdichte Kleidung f <Sicherheit> vapour-proof garment
Dampfdichteschreiber m <Gerät> (AE) vapor density recorder, (BE) vapour density recorder
Dampfdichtung f <Hydraul> steam packing
Dampfdom m <Hydraul> dome, steam dome
Dampfdomnieter m <Hydraul> dome riveter
Dampfdruck m 1. <Erdöl> (AE) vapor pressure, (BE) vapour pressure; 2. <Maschinen> steam pressure; 3. <Thermod> (AE) vapor pressure, (BE) vapour pressure; 4. <Thermod> steam pressure, (AE) vapor pressure, (BE) vapour pressure
Dampfdruckkochtopf m <Elektrotech> rapid cooking pot, vapour pressure cooking pot
Dampfdruckkurve f <Thermod> (AE) vapor pressure diagram, (BE) vapour pressure diagram
Dampfdruckmanometer n <Phys> (AE) steam gage, (BE) steam gauge
Dampfdrucksterilisator m <Labor> autoclave
Dampfdruckthermometer n <Heiz & Kälte, Metrol> (AE) vapor pressure thermometer, (BE) vapour pressure thermometer
Dampfdrucktopf m <Phys> autoclave, vapour pressure pot, Papin's pot
Dampfdurchflussmesser m <Papier> steam flowmeter
Dampfdurchlässigkeit f <Heiz & Kälte, Thermod> (AE) vapor permeability, (BE) vapour permeability
Dampfdüse f 1. <Hydraul> steam jet; 2. <Maschinen> steam nozzle
Dampfeinlass m <Hydraul, Maschinen> steam inlet
Dampfeinlassöffnung f <Hydraul> steam port
Dampfeinlassventil n <Maschinen> steam throttle

Dampfeintritt

Dampfeintritt m <Hydraul, Maschinen> steam inlet
Dampfeintrittskanal m <Hydraul> steam admission port
Dampfeintrittsventil n <Hydraul> eduction valve
Dampfemissionen fpl <Umweltschutz> steam-laden emissions
dämpfen v 1. <Aufnahme> attenuate; 2. <Bau> damp, muffle (Schall); 3. <Elektriz, Elektronik> damp; 4. <Ergon> attenuate; 5. <Hydraul> cushion; 6. <Lebensmittel> steam (Speisen); 7. <Maschinen> absorb, cushion, deaden; 8. <Sicherheit> absorb; 9. <Telekom> damp (Signal); 10. <Textil> steam
dämpfende Zwischenschicht f <Fertig> dolly
dämpfendes Element n <Elektronik> attenuating element
Dampfentladungslampe f <Elektrotech, Thermod> (AE) vapor discharge lamp, (BE) vapour discharge lamp
Dampfentlastungsventil n <Hydraul> steam relief valve
Dampfentnahme f <Maschinen> steam extraction
Dampfentspannungszeit f <Hydraul> duration of steam expansion
Dampfer m <Wassertrans> steamer; steamboat (Schifftyp)
Dämpfer m 1. <Akustik> damper; 2. <Eisenbahn> (BE) bumper, damper, (AE) fender; 3. <Elektrotech> damper; 4. <Heiz & Kälte> attenuator; 5. <Hydraul> cushion; 6. <Mechan> dashpot
Dämpferspule f <Elektrotech> damping coil
Dämpferwicklung f <Elektrotech> amortisseur winding, damper winding, damping winding
Dampferzeuger m 1. <Heiz & Kälte> steam generator; 2. <Hydraul> boiler; 3. <Kerntech> (AE) vapor generator, (BE) vapour generator; 4. <Mechan, Thermod> boiler
Dampferzeugung f 1. <Hydraul> steam raising; 2. <Wassertrans> steam generation, steam raising (Motor)
Dampfextraktion f <Lebensmittel> steam extraction
Dampffahne f <Abfall> (AE) vapor plume, (BE) vapour plume
dampffixieren v <Textil> steam-set
dampfförmig adj <Chemie> (AE) vaporous, (BE) vapourous
dampfförmige Phase f <Thermod> (AE) vapor phase, (BE) vapour phase
Dampfgenerator m <Heiz & Kälte, Kerntech> steam generator
Dampfhahn m <Hydraul> steam cock
dampfhaltig adj <Thermod> humid
Dampfhammer m <Maschinen> steam hammer
Dampfheizschlange f <Heiz & Kälte> steam coil
Dampfheizung f <Heiz & Kälte> steam heating
Dampfheizungsanlage f <Heiz & Kälte> steam heating
Dämpfkalander m <Textil> steam calender
Dampfkammer f <Hydraul> steam chamber
Dampfkanal m <Hydraul> steam port
Dampfkante f <Hydraul> steam edge
Dampfkasten m <Hydraul> steambox, steam case, steam chest
Dampfkessel m <Heiz & Kälte, Hydraul, Labor, Maschinen, Thermod> steam boiler
Dampfkesselanlage f <Maschinen> steam boiler plant
Dampfkesselmanometer n <Gerät> (AE) boiler gage, (BE) boiler gauge
Dampfkesselspeisung f <Heiz & Kälte> boiler feed
Dampfkochtopf m <Thermod> digester
Dampfkolben m <Hydraul> steam piston
Dampfkondensierung f <Kerntech> valve off
Dampfkraft f <Hydraul> steam power
Dampfkraftgenerator m <Elektrotech> steam electric generating set
Dampfkraftwerk n <Elektrotech> steam electric power plant, steam electric power station
Dampfkreis m <Hydraul> steam loop

Dampflampe f <Elektrotech, Thermod> (AE) vapor discharge lamp, (BE) vapour discharge lamp
Dampfmantel m <Heiz & Kälte, Hydraul> steam jacket
Dampfmaschine f <Hydraul, Kfztech, Maschinen, Phys, Thermod> steam engine
Dampfmaschinenanzeigegerät n <Hydraul> steam engine indicator
Dampfmaschinenanzeiger m <Hydraul> steam engine indicator
Dampfmaschinenindikator m <Hydraul> steam engine indicator (Druckverlaufaufzeichnungsgerät in Kolbenmaschinenzylindern)
Dampfmessgerät n <Hydraul> (AE) steam gage, (BE) steam gauge
Dampföffnungen fpl <Maschinen> steam ports
Dampfomnibus m <Kfztech> steambus
Dampfpackung f <Hydraul> steam packing
Dampfphase f <Chemtech, Elektronik> (AE) vapor phase, (BE) vapour phase
Dampfphasenepitaxie f <Chemtech, Elektronik> (AE) vapor phase epitaxy, (BE) vapour phase epitaxy
Dampfphasennitrierung f <Chemtech> (AE) vapor phase nitration, (BE) vapour phase nitration
Dampfphasenreaktion f <Elektronik> (AE) vapor phase reaction, (BE) vapour-phase reaction
Dampfpipeline f <Hydraul> steam pipeline
Dampfpunkt m <Maschinen> steam point
Dampfraum m 1. <Erdöl> ullage (in Tank); 2. <Hydraul> steam space
Dampfregler m <Hydraul> steam governor
Dampfregulator m <Hydraul> steam governor
Dampfrohr n <Hydraul, Maschinen> steam pipe
Dampfrohrleitung f <Hydraul> steam pipeline
Dampfrückleitung f <Kfztech> (AE) vapor return line, (BE) vapour return line
Dampfsammler m <Hydraul> steam accumulator
Dampfschiff n <Wassertrans> steamer, steamship
Dampfschlange f <Heiz & Kälte> steam coil
Dampfschlauch m <Hydraul> steam hose
Dampfseparator m <Hydraul> steam separator
Dampfspannung f 1. <Heiz & Kälte> steam pressure; 2. <Maschinen> steam tension
Dampfspeicher m <Hydraul, Papier> steam accumulator
Dampfsperre f 1. <Bau> vapour membrane, (AE) water vapor barrier, (BE) water vapour barrier; 2. <Kfztech> (AE) vapor lock, (BE) vapour lock
Dampfstabilität f <Hydraul> steam balance
Dampfstrahl m <Maschinen, Umweltschmutz> steam jet
Dampfstrahlbrenner m <Heiz & Kälte> steam jet burner
Dampfstrahlen n <Meerschmutz> steam jet cleaning
Dampfstrahlpumpe f 1. <Hydraul> injector, steam ejector; 2. <Maschinen> steam jet pump
Dampfstrahlverdichter m <Hydraul> ejector condenser
Dampfstutzen m <Hydraul> steam nozzle
Dampftechnik f <Maschinen> steam engineering
Dampftopf m <Hydraul> steam trap
Dampftrockner m <Hydraul> steam drier, steam dryer
Dampfturbine f <Hydraul, Maschinen, Thermod, Wassertrans> steam turbine (Motor)
Dampfüberdeckung f <Hydraul> steam lap
Dampfüberhitzer m <Heiz & Kälte> steam superheater
Dampfüberlappung f <Hydraul> steam lap
Dämpfung f 1. <Akustik> muffling, muting; 2. <Aufnahme> attenuation; 3. <Bau> absorption (Stößen); 4. <Comp & DV> damping, damping attenuation; 5. <Elektriz> damping attenuation; 6. <Elektronik> attenuation; 7. <Elektrotech> damping; absorption (bei Stoß); 8. <Ergon> damping; 9. <Fernseh> attenuation; 10. <Funktech> absorption, attenuation; 11. <Gerät> damping; 12. <Maschinen>

absorption; 13. <Metall> damping; 14. <Optik> attenuation; coupling loss *(Lichtwellenleiter)*; 15. <Phys> attenuation, damping; loss *(Licht)*; extenuation *(Schwingkreis)*; 16. <Raumfahrt> attenuation *(Weltraumfunk)*; 17. <Telekom> attenuation, loss; roll-off *(Filter)*; 18. <Wellphys> attenuation • **mit niedriger Dämpfung** low-loss
Dämpfung *f* **an einer Kante** <Funktech> loss around a corner
Dämpfung *f* **durch Ausrichtfehler** <Funktech> pointing loss *(Antenne)*
Dämpfung *f* **durch Ausrichtungsfehler** <Telekom> angular misalignment loss, misalignment loss
Dämpfung *f* **durch Längsversatz** <Telekom> gap loss
Dämpfung *f* **durch Makrobiegungen** <Telekom> macrobend loss
Dämpfung *f* **durch Mikrokrümmungen** <Telekom> microbend loss
Dämpfung *f* **durch seitlichen Versatz** <Telekom> lateral offset loss, transverse offset loss
dämpfungsarm *adj* <Funktech, Telekom> low-loss
Dämpfungsband *n* <Elektronik> attenuation band *(Filter)*
dämpfungsbegrenzter Betrieb *m* <Telekom> attenuation-limited operation
dämpfungsbegrenzter Vorgang *m* <Optik> attenuation-limited operation
Dämpfungsbelag *m* <Telekom> attenuation coefficient *(Kabel)*
Dämpfungsdiode *f* <Elektronik> attenuator diode *(Mikrowellentechnik)*
Dämpfungseinrichtung *f* <Elektrotech> damper
Dämpfungselement *n* <Gerät> attenuating element
Dämpfungsfähigkeit *f* <Fertig> absorbability *(Stoß)*
Dämpfungsfaktor *m* 1. <Aufnahme> attenuation factor; 2. <Elektriz> damping factor; 3. <Elektrotech> decay factor; 4. <Phys> damping factor
Dämpfungsfilter *n* <Aufnahme> attenuating filter
Dämpfungsglied *n* 1. <Akustik> attenuator, pad; 2. <Elektronik> attenuator, pad; absorptive attenuator *(Automatisierungstechnik)*; 3. <Elektrotech> pad; 4. <Ergon> attenuator; 5. <Funktech> attenuator *(Schaltkreis, Baugruppe)*; 6. <Gerät> attenuating element, attenuator pad; 7. <Telekom> attenuator
Dämpfungsglied *n* **aus Ohm'schen Widerständen** <Telekom> resistive attenuator
Dämpfungsgrad *m* <Gerät> attenuation ratio
Dämpfungskoeffizient *m* 1. <Akustik> attenuation coefficient; 2. <Kerntech> linear attenuation coefficient; 3. <Optik> attenuation coefficient; 4. <Phys> damping coefficient; attenuation coefficient *(Licht)*; 5. <Telekom> linear attenuation coefficient
Dämpfungskondensator *m* <Elektrotech> damping capacitor
Dämpfungskonstante *f* 1. <Aufnahme> decay factor; 2. <Elektronik, Phys> attenuation constant, damping coefficient; attenuation coefficient *(Licht)*; 3. <Telekom> attenuation coefficient, attenuation constant
Dämpfungskontur *f* <Elektronik> attenuation contour
Dämpfungsmaß *n* <Gerät> attenuation ratio
Dämpfungsmesser *m* <Elektrotech> decremeter
Dämpfungsmessgerät *n* <Gerät> decibel meter
Dämpfungsmoment *n* <Lufttrans> damping moment
Dämpfungsnetzwerk *n* <Elektriz> damping resistor
Dämpfungsregler *m* <Lufttrans> damping gyro regulator
Dämpfungssystem *n* <Raumfahrt> surge baffle system *(Raumschiff)*
Dämpfungsventil *n* <Maschinen> dashpot valve
Dämpfungsverhältnis *n* 1. <Ergon> damping ratio; 2. <Phys> decrement

Dämpfungsverlauf *m* <Aufnahme, Elektronik> response curve
Dämpfungsverzerrung *f* <Aufnahme, Phys> attenuation distortion
Dämpfungsvorrichtung *f* <Maschinen> damping device
Dämpfungswicklung *f* <Elektriz> damping coil
Dämpfungswiderstand *m* 1. <Elektriz> damping resistor, damping resistance; 2. <Phys> damping resistance
Dämpfungswinkel *m* <Phys> loss angle
Dämpfungszeitkonstante *f* <Gerät> damping time constant
Dämpfungsziffer *f* <Akustik> attenuation coefficient
Dampfventil *n* <Hydraul, Maschinen> steam valve
Dampfverbindung *f* <Papier> steam joint
Dampfverbrauchszähler *m* <Gerät> steam consumption meter
Dampfverteiler *m* <Papier> steam header
Dampfwagen *m* <Kfztech> steam car
Dampfweg *m* <Hydraul> steam way
Dampfzufuhrrohr *n* <Hydraul> steam supply pipe
Dampfzustand *m* <Thermod> *(AE)* vapor phase, *(BE)* vapour phase
Dampfzylinder *m* <Hydraul> steam cylinder
Danait *n* <Hydraul> danaide
Dandyroller *m* <Verpack> dandy roll
Danford-Anker *m* *(CQR-Anker)* <Wassertrans> Danford anchor, coastal quick release anchor *(CQR anchor)*
Danksagung *f* <Druck> ACK, acknowledgement
Daphnetin *n* <Chemie> daphnetin
Daphnin *n* <Chemie> daphnin
dargestellte Wellenform *f* <Elektronik> displayed waveform *(auf dem Oszilloskop)*
Darlingtonleistungstransistor *m* <Elektronik> power Darlington transistor
Darm *m* <Lebensmittel> gut
darren *v* 1. <Fertig> dry, kiln-dry; 2. <Lebensmittel> cure, kiln-dry *(Brauerei)*
Darrmalz *n* <Lebensmittel> kiln malt *(Brauereiwesen)*; cured malt, kiln-dried malt *(Destillation, Gärung)*
Darrofen *m* <Lebensmittel> drying kiln, drying oven, kiln
Darrschrank *m* <Lebensmittel> drying cupboard
d'Arsonval-Galvanometer *n* <Elektrotech> d'Arsonval galvanometer
darstellen *v* 1. <Bau> plot; 2. <Elektronik> display; 3. <Telekom> display; display *(auf dem Bildschirm)*
darstellende Geometrie *f* <Geom> descriptive geometry
Darstellung *f* 1. <Bau> plot; 2. <Comp & DV> representation; 3. <Konstzeich> display, presentation; 4. <Math> account; 5. <Telekom> display
Darstellung *f* **der Erfindung** <Patent> disclosure of the invention *(in der Beschreibung)*
Darstellung *f* **im Frequenzbereich** <Elektrotech> frequency domain representation
Darstellung *f* **im Zeitbereich** <Elektrotech> time domain representation
Darstellungsfeld *n* <Konstzeich> display area, presentation area
Darstellungsfläche *f* <Comp & DV> display space
Darstellungsgrafik *f* <Comp & DV> presentation graphics
Darstellungsschicht *f* <Comp & DV> presentation layer
Darstellzeit *f* <Gerät> display time
Dasermantel *m* <Optik> *(AE)* fiber coating, *(BE)* fibre coating
DAT *(Digital-Audio-Tape)* <Aufnahme> DAT *(digital audio tape)*
Datagramm *n* <Comp & DV, Telekom> datagram
Data-Mining-Technologie *f* <Comp & DV> data mining technology

Datarom®

Datarom® *n* <Optik> Datarom®
Data-Warehouse *n* <Comp & DV> data warehouse
Data-Warehousing-Technologie *f* <Comp & DV> data warehousing technology
DAT-Cassette *f* <Aufnahme> DAT cassette
Datei *f* 1. <Comp & DV> data file, data set document, file; 2. <Telekom> file
Datei *f* **für den Direktzugriff** <Comp & DV> direct file
Datei *f* **für wahlfreien Zugriff** <Comp & DV> random file
Dateiabfrage *f* <Comp & DV> file interrogation
Dateiaktualisierung *f* <Comp & DV> file updating
Dateiaufbau *m* <Comp & DV> file layout
Dateibediener *m* <Telekom> file server
Dateibeschreibung *f* <Comp & DV> file description
Dateidefinition *f* <Comp & DV> data set definition
Dateiende *n* (EOF) <Comp & DV, Telekom> end of document, end of file
Dateiendekennsatz *m* <Comp & DV> trailer
Dateieröffnungsroutine *f* <Comp & DV> open routine
Dateierstellung *f* <Comp & DV> file creation, file preparation
Dateigeneration *f* <Comp & DV> generation data set
Dateigenerierung *f* <Comp & DV> generation data set
Dateigröße *f* <Comp & DV> file size
Dateigruppe *f* <Comp & DV> file set
Dateiindex *m* <Comp & DV> file index
Dateikennsatz *m* <Comp & DV> file label
Dateikennung *f* <Comp & DV> file identifier, file name
Dateilöschung *f* <Comp & DV> file deletion, file purge
Dateiname *m* <Comp & DV> file identification, file name
Dateinamenerweiterung *f* <Comp & DV> file extension, file name extension
Dateiorganisation *f* <Comp & DV> file organization
Dateipflege *f* <Comp & DV> file maintenance
Dateiprüfung *f* <Comp & DV> file validation
Dateisäuberung *f* <Comp & DV> file cleanup
Dateischutz *m* <Comp & DV> file protection, file security
Dateiserver *m* <Comp & DV> file server
Dateispeicherung *f* <Comp & DV> file storage
Dateistruktur *f* <Comp & DV> file organization, file structure
Dateitransfer *m* <Comp & DV, Telekom> file transfer
Dateitrenner *m* <Comp & DV> file separator
Dateiübertragung *f* <Comp & DV, Telekom> file transfer
Dateiumfang *m* <Comp & DV> file extent
Dateiumwandlung *f* <Comp & DV> file conversion
Dateiverarbeitung *f* <Comp & DV> file processing
Dateiverwaltung *f* <Comp & DV> file handling, file management
Dateiverwaltungsroutine *f* <Comp & DV> file-handling routine
Dateiverzeichnis *n* <Comp & DV> file directory
Dateivorbereitung *f* <Comp & DV> file preparation
Dateiwiederherstellung *f* <Comp & DV> file restore
Dateizugriff *m* <Comp & DV> file access
Dateizuordnung *f* <Comp & DV> file allocation
Daten *npl* <Comp & DV, Elektronik, Elektrotech, Funktech, Labor, Patent, Qual, Telekom> data • **Daten aus Speicher aufrufen** <Comp & DV> download • **Daten eingeben** <Comp & DV> enter data • **Daten erfassen** <Comp & DV> capture data
Daten *npl* **in maschinenlesbarer Form** <Comp & DV> script
Datenabfrage *f* <Comp & DV> data query, data retrieval
Datenabfrageschlüssel *m* <Telekom> polling key
Datenabnahmezustand *m* <Comp & DV> accept data state
Datenabruf *m* <Comp & DV> data retrieval
Datenabstraktion *f* <Comp & DV> data abstraction
Datenadressenkettung *f* <Comp & DV> data chaining

Datenaggregat *n* <Comp & DV> data aggregate
Datenanschlussgerät *n* <Telekom> data interface unit
Datenanzeigestation *f* <Raumfahrt> display terminal (Weltraumfunk)
Datenauflösung *f* <Comp & DV> data resolution
Datenaufnehmer *m* <Elektriz> logger, recorder
Datenaufzeichnung *f* <Comp & DV, Kerntech, Telekom> data logging, data recording
Datenaufzeichnungsgerät *n* <Comp & DV> data recorder
Datenaufzeichnungsmedium *n* <Gerät> data recording medium
Datenausgabe *f* **auf Mikrofilm** <Comp & DV> computer output on microfilm
Datenaustausch *m* 1. <Comp & DV, Elektronik, Funktech> data communication, handshake; 2. <Telekom> data communication, data exchange, information exchange
Datenaustauschvermittlung *f* <Comp & DV> DSE, data switching exchange
Datenautobahn *f* 1. <Comp & DV> communication highway, information highway, information super-highway; 2. <Telekom> information highway, information super-highway
Datenbank *f* <Comp & DV> computer data base, data bank, database, repository
Datenbankabfrage *f* <Comp & DV> database query
Datenbankabfragesprache *f* <Comp & DV> database query language
Datenbankadministrator *m* (DBA) <Comp & DV> database administrator, DBA, senior database technician
Datenbankanschluss *m* <Comp & DV> database connectivity
Datenbankaufgliederung *f* <Comp & DV> database mapping
Datenbankbetreiber *m* <Comp & DV> host
Datenbanknetz *n* <Comp & DV> database network
Datenbankservice *m* <Comp & DV> database services
Datenbanksprache *f* <Comp & DV> QL, query language, database language
Datenbankverwalter *m* (DBA) <Comp & DV> database administrator (DBA)
Datenbankverwaltung *n* <Comp & DV> DBM, database management
Datenbankverwaltungssystem *n* <Comp & DV> database management system (DBMS)
Datenbasis *f* <Telekom> database
Datenbasisverwaltungssystem *n* <Telekom> database management system
Datenbehandlungssprache *f* <Comp & DV> data manipulation language
Datenbereich *m* <Elektronik> data domain
Datenbereinigung *f* <Comp & DV> data cleansing
Datenbericht *m* <Patent, Qual> data report
Datenbeschreibung *f* <Comp & DV> data description
Datenbeschreibungssprache *f* <Comp & DV> data description language
Datenbetrieb *m* <Comp & DV> data mode
Datenbewegung *f* <Comp & DV> transaction
Datenblatt *n* <Comp & DV, Trans> data sheet
Datenbündel *n* <Comp & DV> data burst
Datenbus *m* 1. <Comp & DV> data bus; 2. <Kontroll> bus, data highway; 3. <Raumfahrt> bus, data bus (Raumschiff); 4. <Telekom> data highway
Datencodierung *f* <Comp & DV> data encoding
Datendarstellung *f* <Comp & DV> data representation
Datendefinition *f* <Comp & DV> data definition
Datendurchlauf *m* <Comp & DV> (BE) throughput, (AE) thruput

Datensatz

Datendurchsatz *m* <Comp & DV> *(BE)* throughput, *(AE)* thruput
Dateneingabe *f* <Comp & DV> data entry, data input
Dateneingabe *f* **über Tastatur** <Comp & DV> keyboarding
Dateneingabestelle *f* <Comp & DV> data entry
Dateneingabeterminal *n* <Telekom> data entry terminal
Dateneinheit *f* <Comp & DV> entity
Dateneintrag *m* <Elektronik> data entry
Datenelement *n* 1. <Comp & DV> data element, data item; 2. <Elektronik> data element
Datenempfänger *m* <Elektrotech> data sink
Datenende *n (EOD)* <Comp & DV, Telekom> end of data *(EOD)*
Datenendeinrichtung *f (DEE)* <Comp & DV, Telekom> data terminal, data terminal equipment *(DTE)*
Datenendgerät *n* <Telekom> data terminal
Datenerfasser *mpl* <Comp & DV> data input people
Datenerfassung *f* 1. <Comp & DV> data acquisition, data capture, data collection, data entry, data gathering, data logging; 2. <Elektronik, Gerät> data acquisition; 3. <Kerntech> data logging; 4. <Telekom> data collection
Datenerfassungsstation *f* <Comp & DV> data collection platform
Datenerfassungssystem *n* <Comp & DV> data acquisition system, data logger
Datenerstellung *f* <Comp & DV> data origination
Datenextraktion *f* <Elektronik> data extraction
Daten-Fax-Modem *n* <Telekom> data fax modem
Datenfehler *m* <Comp & DV> data error
Datenfeld *n* <Comp & DV> data field, data item, item
Datenfeld *n* **mit konstanter Länge** <Comp & DV> constant length field
Datenfernsprecher *m* <Comp & DV> data phone
Datenfernübertragung *f* 1. <Comp & DV, Elektronik> data communication; 2. <Funktech> data communication, digital communication; 3. <Telekom> data communication
Datenfernverarbeitung *f* 1. <Comp & DV> remote data processing, teleprocessing; 2. <Telekom> teleinformatics
Datenfernverarbeitungsnetz *n* <Telekom> teleprocessing network
Datenfernverarbeitungverbindung *f* <Comp & DV> teleprocessing connection
Datenfestnetz *n* <Telekom> dedicated circuit data network
Datenfluss *m* 1. <Comp & DV> data flow, data stream; 2. <Elektrotech> data stream
Datenfluss *m* **in Rückwärtsrichtung** <Comp & DV> reverse direction flow
Datenflussdiagramm *n* <Comp & DV> data flow diagram
Datenflussplan *m* <Comp & DV> data flow chart
Datenformat *n* <Comp & DV> data format
Datenfunk *m* <Telekom> radio data transmission
datengesteuertes System *n* <Künstl Int> data-driven system
datengetriebenes System *n* <Künstl Int> data-driven system
Datengruppierung *f* <Comp & DV> data aggregate
Datenhierarchie *f* <Comp & DV> data hierarchy
Dateninfrastruktur *f* <Comp & DV> record infrastructure
Datenintegrität *f* <Comp & DV> data integrity
Datenkanal *m* 1. <Comp & DV> data channel; 2. <Telekom> information channel, data channel
Datenkanalmultiplexer *m* <Comp & DV, Telekom> data channel multiplexer
Datenkasse *f (POS-Terminal)* <Comp & DV, Verpack> point of sale terminal *(POS terminal)*
Datenkassette *f* <Comp & DV> data cartridge

Datenkommunikation *f* <Comp & DV, Elektronik, Funktech, Telekom> data communication
Datenkommunikationsnetz *n* <Comp & DV> data communication network
Datenkommunikationsstation *f* <Comp & DV> data communication terminal
Datenkompression *f* <Comp & DV, Telekom> data compaction, data compression, data reduction
Datenkonverter *m* <Elektronik> data converter *(Hardware oder Software zur Datenumsetzung)*
Datenkonvertierung *f* 1. <Elektronik> data conversion *(des Datenformats)*; 2. <Gerät> data conversion
Datenkonzentrator *m* 1. <Comp & DV> data concentrator; 2. <Mobilkom> input/output controller *(DKZ)*
Datenkorrektur *f* <Comp & DV> data cleaning
Daten-Landeskennzahl *f* <Telekom> data country code *(öffentliches Datennetz)*
Datenleitung *f* 1. <Lufttrans> data link; 2. <Telekom> data circuit
Datenleitungssteuerung *f* <Comp & DV> data link control
Datenleitungs-Steuerungsprotokoll *n* <Comp & DV> data link control protocol
Datenlink *m* <Telekom> data link
Datenmigration *f* <Comp & DV> data migration
Datenmodell *n* <Comp & DV> data model
Datenmodem *n* <Telekom> data modem
Datenmodulation *f* <Telekom> data modulation
Datenmodus *m* <Comp & DV> data mode
Datenmonitor *m* <Fernseh> viewdata terminal
Datenmultiplexer *m* <Comp & DV, Elektronik, Telekom, Trans> data multiplexer
Datenmultiplexing *n* <Elektronik> data multiplexing
Datenmultiplexor *m* <Comp & DV, Elektronik, Telekom, Trans> data multiplexer
Datenname *m* <Comp & DV> data name
Datennetz *n* <Comp & DV, Telekom> data network, data communication network
Datennetzkennung *f* <Telekom> data network identification code *(DNIC)*
Datennetzkennzahl *f* <Telekom> data network identification code *(DNIC)*
Datenpaket *n* 1. <Elektrotech> data packet; 2. <Telekom> data packet, packet
Datenpaket-Endstelle *f* <Telekom> packet mode terminal
Datenpaketübertragung *f* <Comp & DV> packet transmission
Datenpaketvermittlung *f* <Comp & DV, Elektrotech, Telekom> data packet switching
Datenpfad *m* <Comp & DV> data path
Datenpflege *f* <Comp & DV> data management
Datenphase *f* <Telekom> data phase
Datenprotokollfunktion *f* <Comp & DV> data logger
Datenprotokollierung *f* <Comp & DV> data logging
Datenprüfung *f* <Comp & DV, Gerät> data check, data checking, data validation, data verification
Datenquelle *f* <Comp & DV, Elektrotech, Telekom> data source
Datenrate *f* <Telekom> bit rate, data rate
Datenrecorder *m* <Elektriz, Labor> data recorder
Datenregister *n* <Comp & DV> data register
Datenregistriergerät *n* <Gerät> data recorder
Datenreihe *f* <Comp & DV> stream
Datenrückwand *f* <Foto> data back
Datensammlung *f* <Comp & DV> data gathering
Datensatz *m* <Comp & DV> data record, record *(Datenelement)*; data set *(geordnete Datenmenge)*; frame *(kurzer Datensatz)*

Datensatzaktualisierung

Datensatzaktualisierung f <Comp & DV> record updating
Datensatzbereich m <Comp & DV> record area
Datensatzdefinition f <Comp & DV> data set definition
Datensatzerstellung f <Comp & DV> record creation
Datensatzformat n <Comp & DV> record format, record layout
Datensatzklasse f <Comp & DV> record class
Datensatzkopf m <Comp & DV> record head
Datensatzlänge f <Comp & DV> record length
Datenschnittstelle f <Comp & DV> data interface
Datenschub m <Raumfahrt> data burst *(Weltraumfunk)*
Datenschutz m <Comp & DV> *(BE)* data privacy, *(AE)* data security, data protection, privacy
Datensender m <Elektrotech> data source
Datensenke f <Elektrotech, Telekom> data sink *(Empfangsstelle)*
Datensicherheit f 1. <Comp & DV> *(BE)* data privacy, *(AE)* data security, data reliability, security; 2. <Telekom> information security
Datensicherung f 1. <Comp & DV> backup, data protection; 2. <Elektrotech> backup; 3. <Telekom> data protection
Datensicherungsband n <Comp & DV> data storage tape • **Datensicherungsbänder ablegen** <Comp & DV> file data storage tapes • **Datensicherungsbänder wiederauffinden** <Comp & DV> retrieve data storage tapes
Datensicherungsschicht f <Comp & DV, Telekom> data link layer, link layer
Datensicherungsverfahren n <Comp & DV> backup procedure
Datensichtgerät n 1. <Comp & DV> VDU, display, visual display unit; 2. <Gerät> data display unit; 3. <Kontroll, Phys> VDU; 4. <Telekom> data display terminal, display terminal, VDU, video display unit, video terminal
Datenspeicher m 1. <Comp & DV> data memory, data storage, memory, store; 2. <Elektrotech> data storage; 3. <Telekom> information storage
Datenspeicherkarte f <Comp & DV> memory card
Datenspeichern n <Kerntech, Telekom> data recording
Datenspeicherung f <Kerntech, Telekom> data storage
Datensperre f <Comp & DV> lock
Datenspur f <Comp & DV> data track
Datenstation f 1. <Comp & DV> data station, data terminal, terminal, workstation; 2. <Telekom> terminal station
Datenstation f **für Datenpaketbetrieb** <Comp & DV> packet mode terminal
Datenstation f **ohne Plattenspeicher** <Comp & DV> diskless workstation
Datenstationsadapter m <Comp & DV> terminal adaptor
Datenstationsanschluss m <Comp & DV> terminal port
Datenstationsentwicklung f <Comp & DV> workstation development
Datenstationsjob m <Comp & DV> terminal job
Datenstationskomponente f <Comp & DV> terminal component
Datenstationsmodus m <Comp & DV> terminal mode
Datenstationsprotokoll n <Comp & DV> terminal log
Datenstationsserver m <Comp & DV> terminal server
Datenstationssteuerung f <Comp & DV> terminal control
Datenstau m <Telekom> data congestion
Datenstelle f <Regelung> data position
Datenstrecke f <Telekom> data link
Datenstrom m 1. <Comp & DV> data stream, stream; 2. <Elektrotech, Telekom> data stream
Datenstrombandlaufwerk n <Comp & DV> streaming tape drive
Datenstromeinheit f <Comp & DV> streamer
Datenstruktur f <Comp & DV> data structure, record

Datenteil n <Comp & DV> data division
Datentelefon n <Elektronik> data phone *(mit Modem, Netzwerk)*
Datenträger m 1. <Aufnahme> recording medium; 2. <Comp & DV> data carrier, data medium, *(AE)* disk, recording medium, storage medium, volume; 3. <Telekom> data carrier, storage medium
Datenträgerdetektor m *(DCD)* <Telekom> data carrier detector *(DCD)*
Datenträgerende n <Comp & DV> end of medium
Datenträgerkennsatz m <Comp & DV> header label, volume label
Datenträgeroberfläche f <Comp & DV> data surface
Datentransfergeschwindigkeit f <Telekom> data transfer rate, information transfer rate
Datentransfersystem n <Telekom> data transfer system
Datentransportnetz n <Telekom> data transport network
Datentyp m <Comp & DV> data type
Datenübermittlung f <Comp & DV, Elektronik, Funktech, Telekom> data communication
Datenübermittlungsabschnitt m <Comp & DV, Elektronik, Telekom> data communication link
Datenübermittlungskanal m *(DCC)* <Telekom> data communication channel *(DCC)*
Datenübertragung f 1. <Comp & DV> communication, data communication, data transfer, data transmission; 2. <Elektronik> communication, data communication; 3. <Funktech> data communication; 4. <Telekom> data communication, data transfer, data transmission
Datenübertragung f **oberhalb des Sprachbandes** <Telekom> data above voice
Datenübertragung f **über Satellit** <Telekom> satellite data transmission *(DASAT)*
Datenübertragungsabschnitt m <Comp & DV, Elektronik, Telekom> data communication link
Datenübertragungsblock m <Comp & DV> frame *(Datenfernverarbeitung)*
Datenübertragungsebene f <Comp & DV> data link level
Datenübertragungseinrichtung f *(DÜE)* 1. <Comp & DV> data communication terminating equipment *(DCE)*; 2. <Telekom> data circuit terminating equipment, data communication terminating equipment *(DCE)*
Datenübertragungsgeschwindigkeit f 1. <Comp & DV> baud rate, bit rate, data rate, data transfer rate; 2. <Optik> data transfer rate; 3. <Telekom> data rate, data transfer rate
Datenübertragungskanal m 1. <Comp & DV> data transmission channel; 2. <Kontroll> data highway; 3. <Telekom> data communication channel *(DCC)*; data highway, data transmission channel
Datenübertragungsleitung f <Comp & DV> scheduled circuit
Datenübertragungsphase f <Telekom> data phase
Datenübertragungssignal n <Elektronik> communications signal
Datenübertragungssteuereinheit f <Comp & DV, Elektronik> multiplexer
Datenübertragungssteuerung f 1. <Comp & DV> data link control, line control; 2. <Telekom> data link control
Datenübertragungssystem n <Telekom> data transmission system
Datenübertragungssystem n **für GSM-Netze** <Mobilkom, Telekom> high-speed circuit switched data, high-speed mobile data *(HSMD)*
Datenübertragungssystemsteuerung f <Lufttrans> data link
Datenübertragungsumschaltung f <Comp & DV> data link escape

Datenübertragungsverbindung f <Raumfahrt> data link *(Weltraumfunk)*
Datenübertragungsvorrechner m <Telekom> front-end processor
Datenübertragungsweg m <Comp & DV> data bus, data highway
Datenumlagerung f <Comp & DV> data migration
Datenumsetzung f <Elektronik> data conversion
Datenunabhängigkeit f <Comp & DV> data independence
Datenunterbrechung f <Comp & DV> data break
Datenursprung m <Comp & DV> data origin
Datenverarbeitung f *(DV)* 1. <Comp & DV> data processing, DP; 2. <Elektronik> data processing, DP *(Verarbeitung analoger oder digitaler Daten)*; 3. <Kontroll, Telekom> data processing, DP
Datenverarbeitung f **an Bord** <Raumfahrt> on-board processing *(Weltraumfunk)*
Datenverarbeitungskosten fpl <Comp & DV> computing cost
Datenverarbeitungssystem n <Comp & DV, Kerntech, Telekom> dataprocessing system
Datenverarbeitungsumgebung f <Comp & DV> computing environment
Datenverbindung f <Telekom> data circuit
Datenverdichter m <Telekom> compressor
Datenverdichtung f <Comp & DV, Telekom> data compaction, data compression, data reduction
Datenvereinbarung f <Comp & DV> data declaration
Datenverkehr m <Comp & DV> traffic
Datenverlust m **durch Überlauf** <Druck> overrun
Datenverlustausfall m <Elektrotech> overrun
Datenvermittlung f <Telekom> data switching
Datenvermittlungsstelle f <Telekom> DSE, data switch, data switching exchange
Datenvermittlungssystem n <Telekom> data switching system
Datenverschlüsselung f <Comp & DV, Telekom> data encryption
Datenverstärker m <Elektronik> data amplifier
Datenverwaltung f <Comp & DV> data control, data management
Datenverzeichnis n <Comp & DV> data dictionary, data directory
Datenvielfachleitung f <Comp & DV> data highway
Datenvorbereitung f <Comp & DV> data preparation, preprocessing
Datenwählnetz n <Telekom> switched data network *(Datex-L-Netz)*
Datenwandler m <Gerät> data converter
Datenwandlung f <Elektronik> data conversion
Datenwarteschlange f <Comp & DV> data queue
Datenweg m <Elektronik, Telekom> data path
Datenwiederauffindesystem n <Strahlphys> data recovery system
Datenwort n <Comp & DV> data word
Datenwörterbuch n <Comp & DV> data dictionary
Datenzentrale f <Comp & DV> *(AE)* data center, *(BE)* data centre
Datex-L n <Telekom> circuit-switched data network *(CSDN)*
Datexnetz n <Telekom> datex network
Datierung f <Phys> dating
Datiscagelb n <Chemie> datiscin, datisosid
Datiscin n <Chemie> datiscin
Datum n <Math> date, datum
Datumscodierung f <Verpack> date code
Datumsgrenze f 1. <Raumfahrt> datum line; 2. <Wassertrans> date line
Datumsstempel m <Verpack> date code

DAU *(Digital-Analog-Umsetzer)* <Comp & DV, Elektronik, Telekom> DAC *(digital-analog converter)*
Daube f 1. <Fertig> stave; 2. <Mechan> hoop
Dauer f **der Überfahrt** <Wassertrans> crossing time
Dauerablauf m <Fertig> continuous cycling
Dauerbeanspruchung f <Elektriz> continuous duty, uninterrupted duty
Dauerbelastung f 1. <Kohlen, Maschinen> permanent load; 2. <Metall> repeated loading
Dauerbetrieb m 1. <Comp & DV> steady state; 2. <Elektriz> uninterrupted duty; 3. <Elektrotech> steady state; 4. <Raumfahrt> continuous spectrum
Dauerbetriebsbedingung f <Elektrotech> steady state condition
Dauerbiegefestigkeit f <Kunststoff> flexing endurance
Dauerbruch m <Fertig> fatigue durability, fatigue fracture, progressive fracture
Dauerbruchbeginn m <Lufttrans> incipient crack
Dauerbruchrastlinie f <Fertig> fatigue crescent
Dauerbruchzone f <Fertig> fatigue nucleus
Dauerdehngrenze f <Metall> creep strength
Dauerdrehzahl f <Maschinen> continuous speed
Dauerdruckfeuerlöscher m <Sicherheit> stored pressure fire extinguisher
Dauerdrucklöscher m <Sicherheit> stored pressure fire extinguisher
Dauereingriff m <Fertig> permanent mesh *(Getriebelehre)*
Dauerelektrode f <Elektriz> nonconsumable electrode, permanent electrode
Dauererhitzung f <Lebensmittel> batch-type pasteurization *(Molkereiwesen)*
Dauerfestigkeit f 1. <Fertig> fatigue strength; fatigue limit *(im Schwellbereich)*; 2. <Kunststoff> durability; 3. <Lufttrans, Mechan, Metall> fatigue strength; 4. <Qual> fatigue limit
Dauerfestigkeitsdiagramm n <Phys> Smith chart
Dauerfestigkeitsprüfmaschine f <Fertig> alternate strength testing machine
Dauerfettschmierung f <Fertig> permanent lubrication *(Kunststoffinstallationen)*
Dauerfeuerbeständigkeit f <Fertig> refractoriness under load
Dauerfluss m <Phys> steady flow
Dauerform f <Fertig> *(AE)* permanent mold, *(BE)* permanent mould
Dauerfrost m <Erdöl, Kohlen> permafrost
Dauergießform f <Fertig> *(AE)* permanent mold, *(BE)* permanent mould
Dauergleichgewicht n <Phys> secular equilibrium
dauerhaft adj <Bau> permanent
dauerhafte Wasserkraftquelle f <Nichtfoss Energ> continuous hydro-source
Dauerhaftigkeit f <Anstrich, Kunststoff, Metrol, Textil> durability
Dauerhaltbarkeit f <Fertig> fatigue, fatigue durability
Dauerhub m <Fertig> continuous stroking
Dauerinspektion f <Lufttrans> fatigue inspection
Dauerkurzschlussprüfung f <Elektrotech> sustained short-circuit test
Dauerkurzschlussstrom m <Elektriz> steady-state short circuit current, sustained short-circuit current
Dauerkurzschlussversuch m <Elektrotech> sustained short-circuit test
Dauerlast f <Elektriz> constant load, continuous load
Dauerleistung f 1. <Aufnahme> continuous power output; 2. <Elektriz> constant duty, continuous output; 3. <Kontroll> continuous duty

Dauerleistungsgrenze

Dauerleistungsgrenze f <Ergon> permanent performance limit
Dauermagnet m <Elektriz, Elektrotech, Maschinen, Phys, Telekom, Trans> permanent magnet
Dauermagnetgenerator m <Elektrotech> permanent magnet generator
dauermagnetische Substanz f <Phys> hard magnetic material
Dauermagnetlöschung f <Aufnahme> permanent magnet erasing
Dauermagnetrelais n <Elektriz> permanent magnet relay
Dauermagnetsynchronmotor m <Elektrotech> permanent magnet synchronous motor
Dauermessung f <Gerät> long-term measurement
dauernd adj 1. <Kontroll> continuous; 2. <Phys> steady-state
dauernde Magnetzentrierung f <Fernseh> (AE) permanent magnet centering, (BE) permanent magnet centring
Dauernennlast f <Elektriz> nominal continuous load, rated continuous load
Dauernennleistung f <Elektriz> continuous output, continuous power, continuous rating
dauerplissiert adj <Textil> permanently pleated
Dauerprüfung f <Lufttrans, Werkprüf> fatigue test
Dauerruf m <Elektrotech> continuous ringing (Telefonklingel)
Dauerschwingbeanspruchung f 1. <Fertig> cycling, fatigue, fatigue loading; 2. <Qual> fatigue loading
Dauerschwingfestigkeit f 1. <Fertig> endurance limit, endurance, fatigue limit; 2. <Maschinen> fatigue limit; 3. <Qual> endurance limit, fatigue limit; 4. <Werkprüf> fatigue strength
dauerschwingungsbeansprucht adj <Fertig> fatigue-loaded
Dauersignal n <Elektronik> continuous signal
Dauerspannung f 1. <Elektrotech> constant voltage; 2. <Ker & Glas> permanent stress
Dauerspeicher m 1. <Comp & DV> permanent memory, permanent storage; 2. <Elektrotech> permanent memory
Dauerstandfestigkeit f <Fertig> limiting creep stress
Dauerstandkriechgrenze f <Kohlen, Metall> creep strength
Dauerstrahl m <Elektronik> continuous beam
Dauerstreckgrenze f <Metall> repeated yield point
Dauerstrich m (CW) <Aufnahme, Elektronik, Elektrotech, Funktech, Telekom> continuous wave (CW)
Dauerstrichbetrieb m <Teilphys> continuous-wave mode
Dauerstrichgaslaser m <Elektronik> CW gas laser, continuous-wave laser
Dauerstrichlaser m 1. <Elektronik> CW laser; 2. <Wellphys> CW laser, continuous-wave laser
Dauerstrichlaserstrahl m <Elektronik> CW laser beam, continuous-wave beam
Dauerstrichradar n <Funkort, Wassertrans> continuous wave radar
Dauerstrichradardetektor m <Funkort, Trans> CW radar detector, continuous-wave radar detector
Dauerstrichsignal n <Funktech> continuous-wave signal
Dauerstrichultraschalldetektor m <Trans> CW ultrasonic detector, continuous-wave ultrasonic detector
Dauerstrom m 1. <Elektrotech> constant current, contactor, continuous current; 2. <Phys> steady current
Dauertauchen n <Erdöl> saturation diving (Tauchtechnik)
Dauertaucher m <Erdöl> saturation diver (Tauchtechnik)
Dauerüberwachungsgerät n <Sicherheit, Umweltschmutz> continuous monitoring device
Dauerumschaltung f <Comp & DV> SO, shift out
Dauerverbindung f <Telekom> full time circuit (Vermitteln)
Dauerversuch m 1. <Maschinen> endurance test, fatigue test; 2. <Metrol, Werkprüf> endurance test
Dauerversuchmaschine f <Maschinen> fatigue-testing machine
Dauerversuchsprobe f <Fertig> fatigue specimen
Dauerwärmebeständigkeit f 1. <Kunststoff> heat stability; 2. <Thermod> thermal stability
Dauerwerbung f <Fernseh> back-to-back commercials
Dauerzählstelle f <Trans> continuous counting station
Dauerzugfestigkeit f <Maschinen> endurance tensile strength
Dauerzustand m 1. <Elektrotech> steady state condition; 2. <Phys> steady state, steady state condition
Daumen m <Fertig, Maschinen> cog
Daumeneinschnitt m <Druck> side index
Daumenglashals m <Ker & Glas> danny neck
Daumenindex m <Druck> thumb index
Daumenradschalter m <Funktech> thumbwheel switch
Daumenregister n <Druck> thumb index
Daumensteuerung f <Fertig> cam gear
D/A-Umsetzer m <Comp & DV, Elektronik, Telekom> D/A converter
D/A-Umsetzung f <Aufnahme, Comp & DV, Elektronik, Telekom> D/A conversion
Davit m <Mechan, Wassertrans> davit
D/A-Wandler m <Comp & DV, Elektronik, Telekom> D/A converter
D/A-Wandlung f <Aufnahme, Comp & DV, Elektronik, Telekom> D/A conversion
Daycruiser m <Wassertrans> day cruiser (Schiffstyp)
dB (Dezibel) <Akustik, Aufnahme, Elektronik, Funktech, Phys, Strahlphys, Umweltschmutz> dB (decibel)
DBA (Datenbankadministrator, Datenbankverwalter) <Comp & DV> DBA (database administrator)
DBCS-Endezeichen n <Comp & DV> SI character, shift-in character
DBCS-Startzeichen n <Comp & DV> SO character, shift-out character
D-Beiwert m <Regelung> derivative action coefficient
dBi (Dezibel über Isotropstrahler) <Strahlphys> dBi (decibels over isotropic)
DC-AC-Spannungswandler m <Elektrotech> DC-AC converter
DC-AC-Spannungswandlung f <Elektrotech> DC-AC conversion
DCC (Datenübermittlungskanal) <Telekom> DCC (data communication channel)
DCD (Datenträgerdetektor) <Telekom> DCD (data carrier detector)
DC-Josephson-Effekt m <Elektrotech> DC Josephson effect
DCTL (direkt gekoppelte Transistorlogik) <Elektronik> DCTL (direct-coupled transistor logic)
DC-Transducer m <Elektrotech> DC transducer
DD 1. <Hydraul> (Strömungswiderstand) DD (coefficient of drag); 2. <Nichtfoss Energ> (Luftwiderstandsbeiwert, Widerstandsbeiwert) DD (coefficient of drag)
DDE (direkte Dateneingabe, direkter Dateneintrag) <Comp & DV> DDE (direct data entry)
DDP (Doppeldiodenpentode) <Elektronik> DDP (double diode pentode)
DDT (Dichlordiphenyltrichlorproäthan) <Chemie> DDT (dichlordiphenyltrichlorproethane)
deaktivieren v 1. <Maschinen> deactivate; 2. <Raumfahrt> passivate (Raumschiff); 3. <Telekom> take off-line
Deaktivierung f <Chemie> deactivation, inactivation
Deaktivierung f des angeregten Zustandes <Kerntech, Strahlphys> excited-state deactivation

Dean-und-Stark-Apparat m <Labor> Dean and Stark apparatus
Debricin n <Chemie> ficin *(Milchsaft in Ficusarten)*
de-Broglie'sche Welle f <Elektrotech, Phys, Strahlphys> de Broglie wave
Debugger m <Comp & DV> debugger, debugging program, software debugger
Debye'sche Frequenz f <Phys> Debye frequency
Debye'sche Temperatur f (θD) <Phys, Thermod> Debye temperature (θD)
Debye'sches Modell n <Phys> Debye model
Decahydronaphthalin n <Chemie> decahydronaphthalene, decalin, naphthane
Decalin n <Chemie> decahydronaphthalene, decalin, naphthane
Decan n <Chemie> decane
Decan... <Chemie> capric
Decanol n <Chemie> decanol
Decarbonisierung f <Chemie> decarbonization
Decarboxylase f <Chemie> decarboxylase
dechiffrieren v <Elektronik, Telekom> decipher *(Nachrichtentechnik)*
Dechiffrieren n <Raumfahrt> descrambling *(Weltraumfunk)*
Dechiffriervorrichtung f <Trans> decoding device
Dechsel f 1. <Bau> *(AE)* adz *(Holzbau)*; 2. <Eisenbahn> adz, adze
Deck n <Wassertrans> deck *(Schiffbau)*
Deckanstrich m <Bau, Kunststoff> finishing coat, top coat
• **mit Deckanstrich versehen** <Bau> finish
Deckasphaltschicht f <Bau> asphalt surfacing
Deckbalken m <Wassertrans> beam *(Schiffbau)*
Deckbalkenknieblech n <Wassertrans> beam bracket *(Schiffbau)*
Deckbogen m <Druck> top blanket
Decke f 1. <Bau> roof; paving, veneer *(Straße)*; 2. <Fertig> roof *(Schmelzherd)*; 3. <Kohlen> roof; 4. <Textil> blanket; 5. <Wassertrans> deckhead • **Decke einziehen** <Bau> ceil
Deckel m 1. <Druck> tympan; 2. <Elektrotech> cap; 3. <Fertig> roof *(Lichtbogenofen)*; 4. <Hydraul> cap *(Dampfbüchse)*; cover *(Dampfzylinder)*; 5. <Kfztech> cap; 6. <Labor> lid; 7. <Maschinen> flap, head, lid; 8. <Verpack> lid *(Verschlusskappe)*
Deckellager n <Maschinen> cap bearing
Deckelpackstoff m <Verpack> cap-sealing compound
Deckelring m <Verpack> lever ring
Deckelsiegelmasse f <Verpack> lid sealing compound
Deckelstanze f <Verpack> capping press
Deckelverschluss m <Fertig> lid cover
Deckelversiegelung f <Verpack> cap sealing
decken v <Bau> tile *(Dach)*
Deckenabzweigdose f <Elektrotech> ceiling tapping box
Deckenbalken m <Bau> ceiling joist
Deckenbemessung f <Bau> *(BE)* pavement design, *(AE)* sidewalk design *(Straßenbau)*
Deckengeräte npl <Heiz & Kälte> ceiling-hung equipment
Deckenkanalsystem n <Heiz & Kälte> ceiling-mounted ducting
Deckenkran m 1. <Mechan> overhead crane; 2. <Verpack> *(AE)* overhead traveling crane, *(BE)* overhead travelling crane
Deckenkühlschlange f <Heiz & Kälte> ceiling coil
Deckenleuchte f 1. <Elektriz> ceiling fitting, ceiling luminaire *(Lampe)*; 2. <Wassertrans> deckhead light
Deckenoberlicht n 1. <Bau> laylight; 2. <Lufttrans> overhead light
Deckenpapier n <Papier> liner

Deckenrose f <Elektrotech> ceiling rose *(elektrische Leuchte)*
Deckenschallplatte f <Aufnahme> ceiling baffle
Deckenschalttafel f <Lufttrans> overhead panel
Deckenspannung f <Elektrotech> nominal exciter ceiling voltage *(Erregermaschine)*; ceiling level, ceiling voltage *(im Regelkreis)*
Deckensteckdose f <Elektrotech> ceiling outlet box
Deckenstrahler m <Elektrotech> ceiling spotlight *(Licht)*
Deckenträger m <Bau> joist
Deckentransportband n <Verpack> overhead conveyor
Deckenunterzug m <Bau> floor joist
Deckenventilator m <Maschinen> ceiling fan
Deckenvorgelegewelle f <Maschinen> ceiling countershaft
Deckenzwischenraum m <Bau> plenum
Deckfähigkeit f <Chemie> opacity
Deckfarbe f <Kunststoff> covering paint
Deckflachstahl m <Metall> boom plate
Deckgebirge n 1. <Erdöl> overburden *(Geologie)*; 2. <Nichtfoss Energ> cap rock
Deckgestein n <Nichtfoss Energ> cap rock
Deckglas n 1. <Ker & Glas> cover glass; 2. <Labor> cover glass, cover slip
Deckhubschrauber m <Lufttrans> carrier-borne helicopter
Deckkraft f <Kunststoff> covering power, hiding power
Decklage f 1. <Bau> decking; 2. <Fertig> coincidence *(Getriebelehre)*
Deckleiste f 1. <Bau> batten, fillet; 2. <Maschinen> cover strip
Deckmasse f <Ker & Glas> cover coat
Deckmessglas n <Ker & Glas> *(AE)* cover glass gage, *(BE)* cover glass gauge
Deckmetall n <Fertig> deposited metal
Deckpeilung f <Wassertrans> alignment bearing
Deckplatte f 1. <Bau> flange tile; 2. <Fertig> coping; 3. <Metrol> surface plate; 4. <Papier> crown
Deckplatte f der Wannensteine <Ker & Glas> top course of tank blocks
Deckplatte f mit Rand <Ker & Glas> lipped cover tile
Deckring m <Ker & Glas> plunger ring
Decksaufbau m <Wassertrans> superstructure; deck superstructure *(Schiffbau)*
Decksbalken m <Wassertrans> deck beam, transverse beam *(Schiffbau)*
Decksbelag m <Wassertrans> deck covering, decking *(Schiffbau)* • **mit Decksbelag** <Wassertrans> sheathed deck
Decksbeplattung f <Wassertrans> deck plating *(Schiffbau)*
Decksbeschläge mpl <Wassertrans> deck fittings *(Ausrüstung)*
Decksbucht f <Wassertrans> camber *(Schiffbau)*
Deckschicht f 1. <Abfall> covering layer *(einer Deponie)*; 2. <Anstrich> top coat; 3. <Bau> covering; wearing course *(Straße)*; 4. <Erdöl> cap rock *(Geologie)*; 5. <Kunststoff> finishing coat; 6. <Verpack> paper liner
Decksfracht f <Wassertrans> deck cargo *(Ladung)*
Deckshaus n <Wassertrans> deckhouse
Deckskran m <Wassertrans> deck crane *(Ladung)*
Decksladung f <Wassertrans> deck cargo
Deckslängsbalken m <Wassertrans> deck longitudinal *(Schiffbau)*
Decksmannschaft f <Wassertrans> deck crew
Decksmaschinen pl <Wassertrans> deck machinery
Decksplan m <Wassertrans> deck plan *(Schiffbau)*
Decksplatte f <Wassertrans> deck plate *(Ausrüstung)*
Decksplitt m <Bau> blotter material

Decksquerverband

Decksquerverband *m* <Wassertrans> deck transverse structure *(Schiffbau)*
Decksstütze *f* <Wassertrans> deck pillar *(Schiffbau)*
Deckstrāger *m* <Wassertrans> deck girder *(Schiffbau)*
Deckstrak *m* <Wassertrans> deck line, sheer line *(Schiffbau)*
Deckstütze *f* <Wassertrans> stanchion
Decksunterzug *m* <Wassertrans> deck girder *(Schiffbau)*
deckungsgleich *adj* <Geom> congruent
deckungsgleiche Dreiecke *npl* <Geom> congruent triangles
Deckungsgleichheit *f* <Geom> congruence
Deckvermögen *n* <Kunststoff> covering power, hiding power
Deckziegel *m* <Ker & Glas> cover tile
Decoder *m* <Comp & DV, Elektronik, Fernseh, Kontroll, Telekom> decoder
decodieren *v* <Elektronik, Telekom> decode
Decodieren *n* <Elektronik> decoding
decodieren-codieren *v* <Fernseh, Telekom> decode-encode
Decodierer *m* <Comp & DV, Elektronik, Fernseh, Kontroll, Telekom> decoder
Decodiermatrix *f* <Fernseh, Telekom> decoding matrix
Decodierung *f* <Fernseh, Funktech, Raumfahrt, Telekom> decoding
Decodierung *f* **mit harter Entscheidung** <Telekom> hard decision decoding
Decodierung *f* **mit weicher Entscheidung** <Telekom> soft decision decoding
Decodiervorrichtung *f* <Trans> decoding device
Deconfinement-Impuls *m* <Kerntech> deconfining momentum
Decyl... <Chemie> decyl
Decylalkohol *m* <Chemie> decanol
dediziert *adj* <Comp & DV> dedicated
dedizierte Umgebung *f* <Comp & DV> dedicated environment
dedizierter Computer *m* <Comp & DV> dedicated computer
dedizierter Kanal *m* <Comp & DV> dedicated channel
dedizierter Modus *m* <Comp & DV> dedicated mode
Deduktion *f* 1. <Comp & DV> inference; 2. <Math> deduction
deduktive Argumentationsführung *f* <Math> deductive reasoning
DEE *(Datenendeinrichtung)* <Comp & DV, Telekom> DTE *(data terminal equipment)*
Defekt *m* <Elektrotech, Kerntech, Metall> defect, flaw
Defektbogenausstoß *m* <Verpack> faulty sheet ejection
defekte Lettern *fpl* <Druck> broken types
Defektelektron *n* <Fertig> hole
Defektstruktur *f* <Metall> defect structure
definieren *v* 1. <Comp & DV> set, set up; 2. <Math> define
definierte Frequenz *f* <Elektronik> discrete frequency
Definition *f* 1. <Comp & DV> setup; 2. <Elektronik> definition *(Anzahl der pro Fläche unterscheidbaren Bildpunkte)*; definition *(Datenverarbeitung)*; 3. <Math> definition
Definitionstestbild *n* <Fernseh> definition test pattern
deflagrieren *v* <Chemie> deflagrate
Deflektor *m* 1. <Chemtech> deflector; 2. <Lufttrans> deflector, load hook up; 3. <Nichtfoss Energ> deflector; 4. <Raumfahrt> jet deflector *(Raumschiff)*; 5. <Telekom> deflector
Deflektorplatte *f* <Lufttrans> deflector plate
Defokussierung *f* <Elektronik> defocusing
Defoliationsmittel *n* <Chemie> defoliant
Defoliator *m* <Chemie> defoliant

Deformation *f* <Kerntech, Kunststoff, Maschinen, Mechan, Phys, Strahlphys> deformation
Deformationsgradient *m* <Phys> deformation gradient
Deformationsmesser *m* <Mechan> extensometer
dehnbar *adj* 1. <Kunststoff> extensible; 2. <Maschinen> tensile; 3. <Mechan> elastic; 4. <Metall> ductile; 5. <Papier> stretchable; 6. <Qual> ductile; 7. <Textil> stretchy
Dehnbarkeit *f* 1. <Kunststoff> extensibility; 2. <Maschinen> tensility; 3. <Mechan> elasticity; 4. <Metall> ductility; 5. <Papier> elasticity; 6. <Textil> stretch
dehnen *v* 1. <Bau> strain, stretch; 2. <Papier> stretch
Dehner *m* <Telekom> expander
Dehnfähigkeit *f* <Werkprüf> high elongation
Dehnfuge *f* 1. <Bau> contraction joint, expansion joint; 2. <Heiz & Kälte> expansion joint
Dehngrenze *f* 1. <Fertig> proof stress; 2. <Maschinen> limit of elasticity
Dehnpassung *f* <Maschinen> expansion fit
Dehnschlupf *m* <Fertig> creep *(Riemen)*
Dehnung *f* 1. <Elektronik> expansion; 2. <Kunststoff> elongation, strain, stretch; 3. <Labor> strain; 4. <Maschinen> expansion, extension; 5. <Mechan> expansion; 6. <Metall> elongation, strain; 7. <Papier> elongation, stretch; 8. <Phys> elongation, strain; 9. <Telekom> extension; 10. <Wassertrans> strain; 11. <Werkprüf> expansion
Dehnungsangleicher *m* <Erdöl> compensator *(Formstück für Rohrleitungsbau)*
Dehnungsausgleich *m* <Maschinen> expansion compensation
Dehnungsausgleicher *m* <Hydraul> expansion joint
Dehnungsausgleichskupplung *f* <Hydraul> expansion coupling
Dehnungsband *n* <Heiz & Kälte> expansion loop
Dehnungsbeanspruchung *f* <Metall> tensile stress
Dehnungsbogen *m* <Heiz & Kälte> expansion bend, expansion loop
Dehnungsfilter *n* <Elektronik> expansion filter
Dehnungsfuge *f* 1. <Bau> running joint *(Straßenbau)*; 2. <Maschinen, Mechan> expansion joint
Dehnungsgrenze *f* <Mechan> elastic limit
Dehnungskoeffizient *m* 1. <Bau> strain modulus; 2. <Maschinen> expansion coefficient; 3. <Qual> strain modulus
Dehnungsmessbrücke *f* <Elektrotech> *(AE)* strain gage bridge, *(BE)* strain gauge bridge
Dehnungsmesser *m* 1. <Bau, Elektrotech> *(AE)* strain gage, *(BE)* strain gauge; 2. <Fertig> extensometer; 3. <Gerät> dilatometer; 4. <Kunststoff, Maschinen> *(AE)* strain gage, *(BE)* strain gauge; 5. <Mechan> extensometer; 6. <Metrol> extensometer *(Materialprüfung)*; 7. <Papier> extensometer; 8. <Phys> dilatometer
Dehnungsmessfühler *m* <Elektrotech, Gerät> *(AE)* strain gage, *(BE)* strain gauge
Dehnungsmessstreifen *m* 1. <Bau> horizontal clip gauge; 2. <Elektrotech, Kunststoff, Mechan, Metrol> *(AE)* strain gage, *(BE)* strain gauge
Dehnungsrest *m* <Fertig> set
Dehnungsriss *m* <Maschinen, Mechan, Werkprüf> expansion crack
Dehnungsspannungsmessgerät *n* <Phys> *(AE)* strain gage, *(BE)* strain gauge
Dehnungsstreifenbrücke *f* <Gerät> *(AE)* strain gage bridge, *(BE)* strain gauge bridge
Dehnungstensor *m* <Phys> strain tensor
Dehnungstest *m* <Phys> tensile test
Dehnungsverbindung *f* <Mechan> expansion joint
Dehnvermögen *n* <Fertig, Qual> ductility

Dehnwechselprüfanlage f <Werkprüf> dynamic strain testing system
Dehydratation f <Chemie, Chemtech> dehydration
dehydratisieren v 1. <Chemie> desiccate; dehydrate (einer Verbindung); 2. <Heiz & Kälte, Lebensmittel> dehydrate
Dehydratisierung f <Chemie, Chemtech, Heiz & Kälte> dehydration
dehydrieren v <Chemie> dehydrogenate, dehydrogenize
dehydriert adj <Chemie> dehydrated
Dehydro-Gefrieren n <Heiz & Kälte> dehydrofreezing
Dehydrogenase f <Lebensmittel> dehydrogenase (dehydrierendes Enzym)
Deich m <Bau, Nichtfoss Energ, Wasserversorg> (AE) dike, (BE) dyke
Deichsel f <Kfztech> drawbar (Anhänger)
Deichselbolzen m <Kfztech> drawbar bolt (Anhänger)
Deinking n <Papier> de-inking, ink removal
Deinkinganlage f <Abfall> de-inking unit
deinkter Faserstoff m <Papier> de-inked paper stock
Deiodothyroxin n <Chemie> thyronine
Deka... (da) <Metrol> deca... (da)
Dekadendämpfer m <Elektronik> decade attenuator
Dekadengehäuse n <Lufttrans> decade box
Dekadenkasten m <Lufttrans> decade box
dekadisch einstellbare Selbstinduktivität f <Elektrotech> decade inductance box
dekadisch einstellbarer Kondensator m <Elektrotech> decade capacitor
dekadisch einstellbarer Widerstand m <Elektrotech> decade resistor
dekadische Einstellung f <Elektrotech> decade box
dekadische Extinktion f <Phys, Strahlphys> absorbance
dekadischer Oszillator m <Elektronik> decade oscillator
Dekaeder n <Geom> decahedron
Dekagon n <Geom> decagon
dekagonal adj <Geom> decagonal
Dekagramm n <Metrol> decagram
Dekaliter m <Metrol> (AE) decaliter, (BE) decalitre
Dekameter m <Metrol> (AE) decameter, (BE) decametre
Dekameterwellen fpl <Funktech> decametric waves
Dekanter m <Chemtech, Umweltschutz> decanter
dekantieren v <Chemie, Chemtech, Umweltschutz> decant
Dekantieren n 1. <Chemtech> decantation (Vorgang); 2. <Erdöl, Lebensmittel> decantation
Dekantiergefäß n <Chemtech> decantation glass, decantation vessel
Dekantierglas n <Chemtech> decanting glass; precipitation vessel (Labor)
Dekapieren n <Fertig> pickling
dekatieren v <Textil> decatize, steam
Dekatiermaschine f <Textil> decatizing machine
Dekatur f <Textil> decatizing
deklarativ adj <Comp & DV> declarative
deklaratives Wissen n <Künstl Int> declarative knowledge
deklarieren v <Comp & DV> declare
Deklination f 1. <Lufttrans, Nichtfoss Energ, Phys> declination; 2. <Wassertrans> declination (astronomische Navigation)
Deklinationsabweichung f <Phys> angle of magnetic declination
Deklinationskreis m <Phys> declination circle
Deklinationsmesser m <Phys> declinometer
Deklinationswinkel m <Lufttrans> declination angle (Navigation)
Dekommutation f <Elektronik> decommutation (Prüfung und Entschlüsselung)

Dekommutator m <Elektronik> decommutator
Dekompressionskammer f 1. <Erdöl> decompression chamber (Tieftauchtechnik); 2. <Wassertrans> decompression chamber
Dekompressionsunfall m <Kerntech> depressurization accident
Dekontamination f <Chemie, Chemtech> decontamination
Dekontaminationsbrause f <Sicherheit> decontamination shower
Dekontaminationsfaktor m <Strahlphys> decontamination factor
dekontaminieren v <Kerntech> decontaminate
Dekontaminierung f <Chemie, Kerntech, Strahlphys, Umweltschmutz> decontamination
Dekontaminierungssystem n <Kerntech> decontamination system
Dekoration f **bei Herstellung** <Ker & Glas> decoration during production
Dekorationsflachglas n <Ker & Glas> spandrel glass
Dekorationslack m <Kunststoff> decorative varnish
Dekorationsstoff m <Textil> furnishing fabric
Dekorrelation f <Telekom> decorrelation
Dekorschliff m <Ker & Glas> decorative cutting
Dekrement n <Phys> decrement
Dekrepitationsprüfung f <Ker & Glas> decrepitation test
delaminieren v <Kunststoff> delaminate
Delaminierung f <Kunststoff> delamination
Deleaturzeichen n <Druck> deletion mark
D-Elektrode f <Phys> dee (beim Zyklotron)
Delorenzit m <Kerntech> delorenzite
Delphi-Detektor m <Teilphys> Delphi detector
Delphinin n <Chemie> delphinin
Delta n <Elektrotech, Erdöl> delta
Delta n **x** <Math> delta x
Delta-Anpassung f <Funktech> delta match (Antennenanpassung)
Delta-Dreieck-Schaltung f <Elektriz> mesh connection
Delta-Eisen n <Metall> delta iron
Deltaflügel m <Lufttrans> delta wing
deltaförmig adj <Erdöl> deltaic
Deltametall n <Metall> delta metal
Deltamodulation f (DM) 1. <Elektronik> delta modulation, DM; 2. <Raumfahrt> delta modulation, DM (Weltraumfunk); 3. <Telekom> delta modulation, DM
Deltamodulation f **mit variablen Flanken** <Raumfahrt> variable-slope delta modulation (Weltraumfunk)
Deltaschaltung f <Elektriz> delta connection
Deltaschleife f <Funktech> delta loop (Antenne)
Delta-Sternschaltung f <Elektriz> delta star connection
Delta-Sternumformung f <Elektriz> delta-to-star conversion
Deltastrahlen mpl <Strahlphys> delta rays
Deltastrahlung f <Strahlphys> delta rays
Demethylation f <Chemie> demethylation
demethylieren v <Chemie> demethylate
Demineralisierung f <Chemtech> demineralizing
Demodulation f 1. <Akustik> detection; 2. <Comp & DV, Elektronik, Fernseh, Funktech, Phys> demodulation; 3. <Raumfahrt> demodulation, detection (Weltraumfunk); 4. <Telekom> demodulation
Demodulator m 1. <Akustik, Comp & DV, Elektronik, Fernseh, Funktech, Phys> demodulator; 2. <Raumfahrt> demodulator (Weltraumfunk); 3. <Telekom> demodulator
Demodulator m **mit Bestimmung momentaner Frequenz** <Raumfahrt> instantaneous frequency estimation demodulator
Demodulator m **mit einspeisungsstabilisiertem Oszillator** <Raumfahrt> injection-locked oscillator demodulator (Weltraumfunk)

Demodulator

Demodulator *m* **mit Momentanfrequenz-Abschätzung** <Funktech> instantaneous frequency estimation demodulator
Demodulator *m* **mit Schwellwerterhöhung** <Raumfahrt> threshold extension demodulator *(Weltraumfunk)*
demodulieren *v* <Elektronik, Fernseh, Funktech, Telekom> demodulate
demoduliertes Signal *n* <Elektronik> demodulated signal
Demontage *f* <Maschinen> disassembly, dismantling
demontierbar *adj* <Bau> collapsible
demontieren *v* 1. <Bau> dismantle; 2. <Kerntech> strip; 3. <Maschinen> disassemble, dismantle, dismount, take down
Demulgator *m* <Umweltschmutz> demulsifier, emulsion breaker
Demulgierprodukt *n* <Umweltschmutz> demulsifying product
Demulgierung *f* <Chemtech> breaking
demultiplexen *v* <Elektronik> demultiplex
Demultiplexen *n* <Elektronik> demultiplexing
Demultiplexer *m* 1. <Comp & DV, Elektronik> demultiplexer; 2. <Telekom> demultiplexer, demux
demultiplexieren *v* <Telekom> demultiplex
Demultiplexieren *n* <Telekom> demultiplexing
Demultiplexing *n* <Comp & DV> demultiplexing
Demux *m* <Telekom> demux
denaturierter Alkohol *m* 1. <Chemie> denatured alcohol, methylated spirit; 2. <Lebensmittel> denatured alcohol
Dendrit *m* 1. <Fertig> fir tree crystal, pine crystal; dendrite *(Gefüge)*; 2. <Metall> dendrite *(Gefüge)*
dendritisch *adj* 1. <Fertig> dendritic *(Gefüge)*; 2. <Metall> dendritic
Denier *n* <Textil> denier
Denitrierung *f* **des Abfalls** <Abfall> waste denitrification
Denitrifikation *f* <Umweltschmutz> denitrification
Densimeter *n* <Labor> hydrometer
Densitometer *n* <Foto, Gerät, Labor, Optik, Phys> densitometer
Densitometrie *f* <Foto, Labor, Optik, Phys> densitometry
Dentalkeramik *f* <Ker & Glas> dental ceramic
Denudation *f* <Metall> denudation
denudierte Zone *f* <Metall> denuded zone
Depaketierer *m* <Telekom> depacketizer
Dephlegmator *m* <Chemie> dephlegmator *(Destillation)*
 • **mit Dephlegmator behandeln** <Chemie> dephlegmate
Dephlegmiersäule *f* <Chemie> dephlegmator
Deplacement *n* <Wassertrans> displacement *(Schiffskonstruktion)*
Depolarisation *f* <Chemie, Elektrotech> depolarization
Depolarisationsfeld *n* <Aufnahme> depolarizing field
Depolarisator *m* <Chemie> depolarizer
depolarisieren *v* <Chemie, Elektrotech, Raumfahrt> depolarize
depolarisierendes Feld *n* <Phys> depolarizing field
Depolarisierer *m* <Chemie> depolarizer
Depolarisierung *f* <Raumfahrt> depolarization *(Weltraumfunk)*
Depolarisierungsmittel *n* <Elektriz> depolarizing agent
Depolymerisation *f* <Chemie, Kunststoff> depolymerization
Depolymerisierung *f* <Chemie> depolymerization
Deponie *f* 1. <Abfall> dumping site, landfill, waste dump, waste site, waste tip, storage site, waste; 2. <Kohlen> landfill; 3. <Umweltschmutz> dumping ground, landfill, repository, waste disposal site
Deponiebetrieb *m* <Abfall> waste site operation
Deponieeinrichtung *f* <Abfall, Sicherheit> waste disposal facility

Deponieentgasung *f* <Abfall, Sicherheit> landfill degasification
Deponiegelände *n* <Abfall> tipping site
Deponiegut *n* <Abfall> fill mass, waste mass
Deponieoberfläche *f* <Abfall> operating face, working face
Deponieplatz *m* <Abfall> waste site
Deponieschluss *m* <Abfall> waste site closure
Deponiesickerwasser *n* <Bau> leachate
Deponiesickerwasserbehandlung *f* <Abfall> leachate treatment
Deponiestandort *m* <Abfall> landfill site, waste disposal site
Deponietyp *m* <Abfall> landfill design, landfill type
Depression *f* <Wassertrans> depression *(Tiefdruckgebiet)*
Depth-First-Suchverfahren *n* <Künstl Int> depth-first search
Derating *n* <Elektrotech> derating
Derrick *m* <Bau> derrick
Derrickkran *m* 1. <Bau> derrick crane, derrick; 2. <Lufttrans> derrick *(Hubschrauber)*
Derrickmast *m* <Bau> derrick post
Derrin *n* <Chemie> derrin, rotenone
Derriswurzelextrakt *m* <Chemie> derrin, rotenone
Desaminase *f* <Chemie> desaminase
desaxierte Pleuelstange *f* <Kfztech> offset connecting rod
Descrambler *m* <Telekom> descrambler
Desensibilisator *m* <Foto> desensitizer
Desensibilisierung *f* <Foto, Telekom> desensitization
Desensibilisierungsbad *n* <Foto> desensitizing bath
Design *n* <Druck, Ergon, Maschinen, Qual> design
Designbestätigung *f* <Qual> design validation
Design-Ergonomie *f* <Ergon> design ergonomics
Designverifizierung *f* <Qual> design verification
Desinfektionsmittel *n* <Sicherheit> disinfectant
desinfizieren *v* <Sicherheit> disinfect
Desintegratormühle *f* <Lebensmittel> disintegrator
deskriptives Modell *n* <Ergon> descriptive model
Deskriptor *m* <Comp & DV> descriptor
Desktop-Publishing *n* (DTP) <Comp & DV, Druck> desktop publishing *(DTP)*
Desmotropie *f* <Chemie> desmotropy
Desodoriermittel *n* <Kunststoff> deodorant
desorbieren *v* <Chemie> desorb
Desorption *f* <Chemie> desorption
Desoxidation *f* <Chemie> deoxidization
Desoxidationsmittel *n* 1. <Chemie> deoxidizer, reducing agent; 2. <Fertig> scavenger
Desoxidieren *n* <Chemie> deoxidization
Destillat *n* 1. <Erdöl> distillate *(Raffinerie, Destillationsprodukt)*; 2. <Thermod> distillate
Destillation *f* 1. <Erdöl> distillation *(Raffinerietechnik)*; 2. <Thermod> distillation
Destillation *f* **im Vakuum** <Chemtech> vacuum distillation
Destillation *f* **mittels Sonnenenergie** <Nichtfoss Energ> solar distillation
Destillationsanlage *f* <Chemtech> still
Destillationsapparat *m* 1. <Chemtech> distiller, distilling apparatus, still; 2. <Labor> still
Destillationsbereich *m* <Thermod> distillation range
Destillationsbetrieb *m* <Chemtech> distillery
Destillationsgas *n* <Chemtech, Thermod> distillation gas
Destillationsgerät *n* <Chemtech> distiller, distilling apparatus, still
Destillationskolben *m* <Chemtech, Labor> distilling flask
Destillationskolonne *f* 1. <Chemtech> distillation column, distilling column; 2. <Erdöl> distillation tower, frac-

tionating tower *(Raffinerie)*; 3. <Thermod> distillation tower
Destillationsrohr *n* <Chemtech> distilling tube
Destillationsrückstand *m* <Lebensmittel> distillery residue
Destillationsturm *m* <Chemtech> distilling tower
Destillieranlage *f* 1. <Chemtech> distillery; 2. <Labor> still; 3. <Lebensmittel> distillery
Destillierapparat *m* <Labor> distillation apparatus
destillieren *v* <Thermod> distil
Destillieren *n* <Thermod> distillation
Destillierkolben *m* 1. <Chemtech> distilling flask; 2. <Thermod> distillation flask
Destillierkolonne *f* <Chemtech> distilling tower
Destillierofen *m* <Metall> distillation furnace, distillation retort
destilliertes Wasser *n* <Kfztech> distilled water *(Batterie)*
destruktive Interferenz *f* <Phys, Wellphys> destructive interference
Desulfonierung *f* <Chemie> *(AE)* desulfonation, *(BE)* desulphonation
Desulfurierung *f* <Chemie> *(AE)* desulfurization, *(BE)* desulphurization
DESY *(Deutsches Elektronensynchroton)* <Teilphys> DESY
Detail *n* <Druck, Fernseh> detail
detaillierte Gebührenberechnung *f* <Telekom> detailed billing
Detailmontage *f* <Druck> separate make-up
Detailwiedergabe *f* <Fernseh> detail rendition
Detektion *f* <Telekom> detection
Detektionsempfindlichkeit *f* <Telekom> detectivity *(Empfänger)*
Detektionsschwelle *f* <Telekom> detection threshold
Detektor *m* <Elektriz, Elektronik, Elektrotech, Funktech, Kerntech, Optik, Phys, Telekom> detector
Detektor *m* **für ungeladene Teilchen** <Kerntech> neutral particle detector
Detektordiode *f* <Elektronik> detector diode
Detektordiode *f* **für Schottky-Barriere** <Elektronik> Schottky barrier detector diode
Detektorröhre *f* **für Kurzzeitprobenentnahme** <Sicherheit> detector tube for short-term sampling
Detektorschaltung *f* <Elektronik> detector circuit
Detektorschleife *f* <Trans> detection loop
Detektorsignal *n* <Elektronik> detector signal
Detergens *n* 1. <Erdöl> detergent; 2. <Meerschmutz> surface active agent, surfactant
Detergensöl *n* <Kfztech> detergent oil *(Schmierung)*
Detergentzusatz *m* <Maschinen> detergent additive
Determinante *f* <Comp & DV, Math> determinant
deterministisch *adj* <Künstl Int> deterministic
deterministisches Zugriffsverfahren *n* <Telekom> deterministic access mode
Detonation *f* <Thermod> detonation
Detonationsknall *m* <Raumfahrt> boom *(Raumschiff)*
Detonator *m* <Eisenbahn> detonator
detonieren *v* <Thermod> detonate
Deuterium *n* (D) <Chemie> deuterium (D)
Deuteriumoxid *n* (D_2O) <Chemie> deuterium oxide (D_2O)
Deuteron *n* (d) <Chemie, Phys, Teilphys> deuteron (d)
Deutlichkeit *f* <Telekom> clarity
Deutsche Gesellschaft *f* **zur Rettung Schiffbrüchiger** *(DGzRS)* <Wassertrans> German Lifeboat Institution, GLI
Deutsche Norm *f* <Maschinen, Qual> German Standard
Deutsches Elektronensynchroton *n* *(DESY)* <Teilphys> DESY
Deutsches *n* **Institut für Normung** *(DIN)* <Maschinen> German Standards Institution

Devastierung *f* <Umweltschmutz> land degradation, land disturbance
Deviation *f* <Lufttrans, Wassertrans> deviation *(Kompass)*
Deviationsanzeige *f* <Lufttrans> deviation indicator
Deviationsgeber *m* <Lufttrans> deviation detector
Deviationssignal *n* <Lufttrans> deviation signal
Dewar-Gefäß *n* 1. <Chemtech, Labor> Dewar flask *(Isolierung)*; 2. <Thermod> Dewar flask, Dewar vessel
Dextran *n* <Chemie> dextran
Dextrin *n* 1. <Chemie> amylin, dextrin; 2. <Kunststoff, Lebensmittel> dextrin *(Stärkeabbauprodukt)*
Dextrose *f* <Lebensmittel> dextrose, grape sugar
dezentral *adj* 1. <Comp & DV> distributed; 2. <Telekom> decentralized
dezentrale Datenbank *f* <Comp & DV> distributed database
dezentrale Prozesse *mpl* <Comp & DV> distributed processes
dezentraler Aufbau *m* <Bau, Comp & DV> distributed architecture
dezentraler Vektorenrechner *m* <Comp & DV> distributed array processor
dezentrales Informationssystem *n* <Comp & DV> distributed information system
dezentrales Netz *n* <Comp & DV> distributed network
dezentrales System *n* <Telekom> decentralized system
dezentralisiert *adj* <Comp & DV> decentralized
Dezi... (d) <Metrol> deci... (d)
Dezibel *n* (dB) 1. <Akustik, Aufnahme> decibel, dB; 2. <Elektronik> decibel, dB *(logarithmisches Dämpfungsmaß)*; 3. <Funktech> decibel, dB; 4. <Phys> decibel, dB *(Einheit der Dämpfung)*; 5. <Strahlphys, Umweltschmutz> decibel, dB
Dezibel *npl* **über Isotropstrahler** *(dBi)* <Strahlphys> decibels over isotropic *(dBi)*
Dezibelskale *f* <Aufnahme> decibel scale
Dezigramm *n* <Metrol> decigram
Deziliter *m* (dl) <Metrol> *(AE)* deciliter, *(BE)* decilitre (dl)
dezimal *adj* <Comp & DV> decimal
Dezimal... <Comp & DV, Elektronik, Labor, Math> decimal
Dezimal-Binär-Umsetzer *m* <Elektronik, Gerät> decimal-to-binary converter
Dezimal-Binär-Umsetzung *f* <Elektronik, Labor> decimal-to-binary conversion
Dezimalbruch *m* <Math> decimal fraction
Dezimalbruchentwicklung *f* <Math> expansion of decimal fraction
Dezimal-Dual-Umsetzer *m* <Elektronik, Gerät> decimal-to-binary converter
Dezimalkomma *n* <Comp & DV, Math> decimal point
Dezimalschreibweise *f* <Comp & DV, Math> decimal notation
Dezimalskale *f* <Gerät> decimal scale
Dezimalstelle *f* 1. <Comp & DV> decimal point; 2. <Math> decimal place
Dezimalsystem *n* <Math> decimal system
Dezimalteiler *m* <Elektronik> decade scaler
Dezimalwaage *f* <Gerät> decimal balance
Dezimalzahl *f* <Math> decimal, decimal number, decimal numeral
Dezimeter *m* <Metrol> *(AE)* decimeter, *(BE)* decimetre
Dezimeterwelle *f* 1. <Elektronik, Fernseh, Funktech, Telekom> ultrahigh frequency; 2. <Wellphys> decimetric wave, ultrahigh frequency wave
Dezineper *n* <Akustik> decineper
Dezitonne *f* <Metrol> *(AE)* quintal
Dezi-Verstärker *m* <Elektronik> microwave amplifier
Dezi-Verstärkerröhre *f* <Elektronik> microwave amplifier tube

Dezi-Verstärkung

Dezi-Verstärkung f <Elektronik> microwave amplification
D-Flipflop n <Elektrotech> delay flip-flop, D-type flip-flop
DFT *(diskrete Fourier-Transformation)* <Elektronik> DFT *(discrete Fourier transform)*
DFV *(Datenfernverarbeitung)* <Comp & DV> TP *(teleprocessing)*
DFV-Verbindung f <Comp & DV> TP connection, communication link
D-Glied n <Regelung> derivative element
D-Glucosamin n <Chemie> D-glucosamine
DGPS *(Differenzial-GPS)* <Wassertrans> DGPS, differential global positioning system *(Satellitennavigation)*
DGzRS f *(Deutsche Gesellschaft zur Rettung Schiffbrüchiger)* <Wassertrans> GLI *(German Lifeboat Institution)*
Di *(Richtwirkungsindex)* <Akustik> Di *(directivity index)*
Dia n <Foto> slide, *(AE)* transparency
Diac *(bidirektionale Triggerdiode)* 1. <Elektronik> diac, diode alternating-current switch *(Wechselstromdiodenschalter)*; 2. <Funktech> diac, diode alternating-current switch
Diacetyl n <Chemie> diacetyl
Diacetylen n <Chemie> diacetylene
Diagenese f 1. <Erdöl> diagenesis *(Geologie)*; 2. <Nichtfoss Energ> diagenesis
diagenetisch adj 1. <Erdöl> diagenetic *(Geologie)*; 2. <Nichtfoss Energ> diagenetic
Diagnose f <Comp & DV> diagnosis
Diagnosecode m <Comp & DV> flag code
Diagnoseexpertensystem n <Künstl Int> diagnostic expert system
Diagnosehilfe f <Telekom> diagnostic aid
Diagnoseprogramm n <Comp & DV> diagnostic program
Diagnosetest m <Comp & DV> diagnostic test
Diagnostik f <Comp & DV> diagnostics
diagonal adj <Geom, Kerntech, Math> diagonal
diagonal adv <Geom> diagonally
Diagonal... <Kerntech, Kfztech, Konstzeich> diagonal
Diagonale f 1. <Math> diagonal; 2. <Wassertrans> diagonal line
diagonale Spannung f <Bau> diagonal tension
diagonale Stegknickung f <Bau> diagonal web buckling
diagonale Strebe f <Kerntech> diagonal member rod
Diagonalkreuz n <Konstzeich> diagonal cross
Diagonalreifen m 1. <Kfztech> *(AE)* bias ply tire, *(BE)* bias ply tyre, *(AE)* fabric-laminated thread tire, *(BE)* fabric-laminated thread tyre; 2. <Kunststoff> *(AE)* cross ply tire, *(BE)* cross ply tyre, *(AE)* diagonal ply tire, *(BE)* diagonal ply tyre
Diagonalschneidemaschine f <Mechan> angle cutter
Diagonalstab m <Bau> brace
Diagonalstrebe f 1. <Hydraul> diagonal stay; 2. <Kerntech> diagonal; 3. <Kfztech> antiroll bar
Diagonalstütze f <Hydraul> diagonal stay *(Kessel)*
Diagonalverband m <Bau> diagonal bracing
Diagonalverstrebung f <Wasserversorg> diagonal brace *(eines Schleusentors)*
Diagramm n <Comp & DV, Maschinen, Telekom> diagram • **Diagramm aufnehmen** <Maschinen> take a chart
Diagrammabreißstab m <Gerät> chart paper tear-off bar
Diagrammantriebsmotor m <Gerät> chart motor
Diagrammstreifen m <Gerät> chart, continuous diagram, recording chart, strip chart
Diagrammtransport m <Gerät> chart transport
Diagrammtrommel f <Gerät> chart drum
Diakasten m <Foto> slide box
Diakaustik f <Optik> diacaustic
diakaustisch adj <Optik> diacaustic
Diakopieraufsatz m <Foto> slide copying attachment
Diakopieren n <Foto> slide copying, slide duplication

Diakopiergerät n <Foto> slide copying device
diakritisch adj <Comp & DV, Druck, Konstzeich> diacritical
diakritisches Zeichen n 1. <Druck> diacritical mark; 2. <Konstzeich> diacritical sign
Dialin n <Chemie> dialin, dihydronaphthalene
Diallylphthalat-... <Kunststoff> diallylphthalate
Diallylphthalat-Formmasse f <Kunststoff> *(AE)* diallylphthalate molding compound, *(BE)* diallylphthalate moulding compound
Diallylphthalat-Pressmasse f <Kunststoff> *(AE)* diallylphthalate molding compound, *(BE)* diallylphthalate moulding compound *(PDAP)*
Dialog m <Comp & DV> *(AE)* dialog, *(BE)* dialogue
Dialogbetrieb m <Comp & DV> conversational mode, interactive mode • **im Dialogbetrieb arbeitend** <Comp & DV> interactive
Dialogbetriebmodus m <Comp & DV> interactive mode
Dialogentzerrer m <Aufnahme> *(AE)* dialog equalizer, *(BE)* dialogue equalizer
dialogfähige Datenstation f <Comp & DV> interactive terminal
dialogfähige Grafikverarbeitung f <Comp & DV> interactive graphics
Dialogfenster n <Comp & DV> pop-down, pop-up window
Dialogfenstermenü n <Comp & DV> pop-up menu
Dialoggerät n <Telekom> interactive terminal
Dialogkommunikation f <Comp & DV, Telekom> conversational communication
Dialognetz n <Telekom> interactive network
dialogorientiert adj <Künstl Int> *(AE)* dialog-oriented, *(BE)* dialogue-oriented
Dialogsprache f <Comp & DV> conversational language
Dialogspur f <Fernseh> *(AE)* dialog track, *(BE)* dialogue track
Dialogsystem n <Telekom> interactive system
Dialursäure f <Chemie> dialuric acid, hydroxybarbituric acid, tartronylurea
Dialysat n <Chemie> dialyzate
Dialyse f <Chemie> dialysis • **durch Dialyse trennen** <Chemie> dialyze
dialysieren v <Chemie> dialyze
dialytisch adj <Chemie> dialytic, dialytical
diamagnetisch adj <Elektriz, Phys, Strahlphys> diamagnetic
diamagnetische Abschirmung f **des Atomkerns** <Strahlphys> diamagnetic shielding of the nucleus
diamagnetische Anisotropie f <Strahlphys> diamagnetic anisotropy
diamagnetischer Stoff m <Elektriz> diamagnetic material
Diamagnetismus m 1. <Chemie, Elektriz, Elektrotech, Phys> diamagnetism; 2. <Strahlphys> diamagnetics, diamagnetism
Diamant m 1. <Maschinen> diamond; 2. <Mechan> *(AE)* diamond • **durch Diamant gerissen** <Fertig> diamond-scribed *(Messoptik)*
diamantbesetzt adj <Mechan> *(AE)* jeweled, *(BE)* jewelled
Diamantbohrmeißel m <Erdöl> diamond bit *(Bohrtechnik)*
Diamantgitter n <Metall> diamond lattice
Diamantglanz m <Fertig> *(AE)* adamantine luster, *(BE)* adamantine lustre
diamanthart adj <Fertig> adamantine
Diamanthohlbohrmeißel m <Erdöl> diamond core drill *(Bohrtechnik)*
Diamantkegel m <Fertig> brale

Diamantkronen bohren v/mit <Erdöl> diamond drilling *(Bohrtechnik)*
Diamantmarkierungsstift m <Labor> diamond-tipped pen
Diamantmeißel m <Maschinen> diamond nose chisel, diamond point chisel
Diamantmörser m <Labor> percussion mortar *(Schleifen)*
Diamantnadel f <Aufnahme> diamond stylus
Diamantpaste f <Maschinen, Metall> diamond paste
Diamantpunktierung f <Ker & Glas> diamond point engraving
Diamantsäge f <Maschinen> diamond saw
Diamantschleifscheibe f <Maschinen> diamond-grinding wheel
Diamantschneidrad n <Ker & Glas> diamond-slitting wheel
Diamantspitze f <Bau, Maschinen> diamond point
Diamantstahl m <Mechan> carbon steel
Diamantwerkzeug n <Mechan> diamond tool
diametral adj <Geom> diametric, diametrical
diametral entgegengesetzt adj <Geom> diametrically opposed
Diametral-Pitch m *(DP)* <Maschinen> diametral pitch *(DP)*
diametrische Projektion f <Konstzeich> diametric projection
Diamid n <Chemie> diamide, diazane, hydrazine
Diamin n <Chemie> diamine, diazane, hydrazine
Diaminodiphenylmethan n <Kunststoff> diaminodiphenylmethane
Diamontriffelung f <Wasserversorg> diamond riffle
Diaper m <Textil> diaper
Diaphon n <Wassertrans> diaphone *(Navigation)*
Diaphragmapumpe f <Wasserversorg> diaphragm pump
Diapir m <Erdöl> diapir *(Geologie)*
Diapositiv n <Foto> slide, *(AE)* transparency
Diapositivabtaster m <Fernseh> slide scanner
Diapositivaufnahme f <Fernseh> slide pick-up
Diaprojektor m <Foto> slide projector
Diarähmchen n <Foto> slide holder
Diarahmung f <Foto> slide mounting
Diarsenat n <Chemie> diarsenate, pyroarsenate
Diastase f <Lebensmittel> diastase
Diastereomer n <Chemie> diastereomer, epimer
Diät f <Lebensmittel> diet
diathermisch adj <Thermod> diathermanous, diathermic
diatonische Tonleiter f <Akustik> diatonic scale, gamut
diatonischer Halbton m <Akustik> diatonic semitone
diatonischer Vierklang m <Akustik> diatonic tetrachord
Diätsalz n <Lebensmittel> salt substitute
Diätzucker m <Lebensmittel> dietary sugar
Diavorlage f <Konstzeich> original for slides
Diawechsler m <Foto> slide changer
Diazan n <Chemie> diazane, hydrazine
Diazo... <Chemie> diazo *(Diazoverbindung)*
Diazobenzol n <Chemie> diazobenzene
Diazoessig... <Chemie> diazoacetic
Diazoimid n <Chemie> diazoimide, hydrogen azide
Diazol n <Chemie> diazole
Diazonium... <Chemie> diazonium
Diazoschicht f <Druck> diazo coating
diazotieren v <Chemie> diazotize
Dibenzanthracen n <Chemie> dibenzanthracene, naphthophenanthrene
Dibenzoparadiazin n <Chemie> dibenzopyrazine, phenazine
Dibenzopyran n <Chemie> xanthene
Dibenzopyrazin n <Chemie> dibenzopyrazine, phenazine
Dibenzopyron n <Chemie> xanthone

Dibenzopyrrol n <Chemie> carbazole
Dibenzoyl... <Chemie> dibenzoyl
Dibenzylamin n <Chemie> dibenzylamine
Dibrom... <Chemie> dibromo
Dibrombenzol n <Chemie> dibromobenzene
Dibromhydrin n <Chemie> dibromohydrin
Dibutylphthalat n <Kunststoff> dibutylphthalate
Dichlor... <Chemie> dichloro
Dichloraceton n <Chemie> dichloroacetone
Dichlorbenzol n <Chemie> dichlorobenzene
Dichlordiphenyltrichlorproäthan n *(DDT)* <Chemie> dichlordiphenyltrichlorproethane *(DDT)*
Dichloressig... <Chemie> dichloroacetic
Dichlorid n <Chemie> bichloride, dichloride
Dichroismus m 1. <Chemie> dicroism; 2. <Phys> dichroism
dichroitisch adj <Foto, Phys, Telekom> dichroic
dichroitischer Schleier m <Foto> dichroic fog
dichroitischer Spiegel m <Optik> dichroic mirror
Dichromat n <Chemie> bichromate, dichromate
dicht adj <Erdöl> impervious *(Geologie)* • **dicht am Wind** <Wassertrans> close-hauled *(Segeln)*
dicht angeholt adj <Wassertrans> close-hauled *(Segeln)*
dicht gelagerter Kies m <Bau> tight gravel
dicht gepackt adj <Anstrich> closely-packed
dicht gepacktes Gitter n <Metall> close-packed lattice
Dicht... 1. <Fertig> sealing; 2. <Maschinen> packing, sealing
Dichte f 1. <Akustik> optical density; 2. <Bau, Chemie, Comp & DV, Erdöl, Kohlen, Kunststoff, Mechan, Phys> density; 3. <Math> density, density function; 4. <Strahlphys, Textil> density, specific gravity
Dichte f der freien Elektronen <Strahlphys> free-electron density
Dichte f der kinetischen Energie <Akustik> kinetic energy density
Dichteänderung f <Strömphys> density variation
Dichtefühler m <Gerät> density probe
Dichtefunktion f <Math> density, density function, probability density function, pdf *(Wahrscheinlichkeitstheorie)*
Dichtelog n <Erdöl> density log *(Bohrlochmessung)*
Dichtemesser m 1. <Labor, Metrol> densimeter; 2. <Phys> densimeter, viscosimeter
Dichtemessfühler m <Gerät> density probe
Dichtemessgerät n 1. <Metrol> densimeter; 2. <Phys> densimeter, gravimeter
Dichtemessung f 1. <Gerät> densimetry, density measurement; 2. <Labor> density measurement; 3. <Lebensmittel, Metrol> densimetry; 4. <Phys> densimetry, density measurement
Dichtemodifikator m <Lebensmittel> density modifier
Dichtemodulation f <Elektronik, Funktech> density modulation *(Mikrowellen)*
dichten v <Bau> stop
Dichten n <Bau> stopping
Dichtepegel m <Akustik> density level
dichtes Bohrloch n <Erdöl> tight hole
dichtes Erz n <Kohlen> compact ore
dichtes Wellenlängenmultiplex n <Telekom> dense wavelength division multiplex *(fiber optics)*
Dichteumfang m <Druck> density range, dynamic range, tonal range
Dichteverhältnis n 1. <Erdöl> specific gravity; 2. <Kohlen> relative density
Dichtewaage f <Gerät> density balance
Dichtfilz m <Maschinen> packing felt
Dichtfläche f 1. <Fertig> sealing surface *(Kunststoffinstallationen)*; 2. <Ker & Glas, Maschinen> sealing surface
Dichtgas n <Kerntech> seal gas

Dichtheit 154

Dichtheit f 1. <Fertig> leakproof closure *(Kunststoffinstallationen)*; 2. <Kerntech> leak tightness
Dichtheitsprüfung f <Kerntech> leak test
dichtholen v <Wassertrans> set taut *(Tauwerk)*
Dichtigkeit f 1. <Kohlen> density; 2. <Maschinen> tightness; 3. <Nichtfoss Energ> solidity; 4. <Sicherheit> tightness
Dichtigkeitsprüfung f <Kerntech> leak test
Dichtigkeitstest m *(DT)* <Erdöl> leak-off test, LOT *(Bohrtechnik, Flüssigkeiten, Gase)*
Dichtkante f <Ker & Glas> sealing edge
Dichtlippe f <Maschinen> sealing lip
Dichtmanschette f <Maschinen> cup
Dichtmasse f <Chemie, Kunststoff> sealant
Dichtmaterial n <Maschinen> packing material
Dichtmittel n <Anstrich, Ker & Glas, Kfztech> sealant *(Karosserie)*
Dichtnut f <Maschinen> sealing groove
Dichtprüfung f <Bau, Qual> density test
Dichtring m 1. <Kfztech> sealing ring; 2. <Maschinen> packing ring, sealing ring; 3. <Mechan> gasket; 4. <Papier> sealing ring
Dichtscheibe f <Maschinen> gasket
Dichtschweißen n <Fertig> caulking
Dichtsitz m <Sicherheit> respirator fit *(eines Atemschutzgerätes)*
Dichtstoff m <Heiz & Kälte> sealant
Dichtstreifen m <Verpack> band sealer
Dichtung f 1. <Bau> packing *(zwischen beweglichen Teilen)*; 2. <Elektriz> seal; 3. <Erdöl> sealing, seal; 4. <Fertig> gasket; gasket, seal *(Kunststoffinstallationen)*; 5. <Ker & Glas> sealing, seal; 6. <Kerntech> packing seal; 7. <Kfztech> gasket; 8. <Labor> seal; 9. <Maschinen> gasket, jointing, seal; 10. <Mechan> gasket, packing; 11. <Verpack, Wassertrans> seal *(Schiffbau)*
Dichtung f **durch luftleeren Raum** <Verpack> chamber-type vacuum sealing
Dichtungsbahn f <Abfall> liner sheet
Dichtungseinheit f <Kerntech> seal unit
Dichtungsgraben m <Bau> cutoff ditch
Dichtungsgummiring m <Bau> washer
Dichtungsmanschette f 1. <Maschinen> cup; 2. <Mechan> gasket
Dichtungsmasse f <Bau> sealing compound
Dichtungsmaterial n <Kerntech> sealing material
Dichtungsmittel n <Bau> curing compound *(zur Nachbehandlung)*
Dichtungsmontage f <Kerntech> seal assembly
Dichtungsring m 1. <Fertig> toroidal ring, washer; 2. <Kerntech> joint ring; 3. <Maschinen> packing ring
Dichtungsscheibe f 1. <Fertig> packing washer; 2. <Maschinen> round washer; 3. <Mechan> gasket
Dichtungsschleier m <Wasserversorg> cutoff wall *(eines Wehres)*
Dichtungsschweißnaht f <Kerntech> sealing weld
Dichtungsstutzen m <Wassertrans> gland
Dichtungswand f <Abfall> slurry wall
Dichtungszange f <Maschinen> sealing pliers
Dichtwerkstoff m <Fertig> sealing material *(Kunststoffinstallationen)*
dick adj <Ker & Glas> heavy *(Schulter, Boden, Ecke eines Behälters)*
dick umhüllt adj/**sehr** <Fertig> shielded *(Elektrode)*
Dick... <Fertig> thick
Dickdruckpapier n <Papier> bulking paper
dicke Blase f <Ker & Glas> heavy seed
dicke Platte f <Ker & Glas> heavy panel
Dicke f 1. <Anstrich> thickness; 2. <Druck> set, set width, width; 3. <Kunststoff> thickness; 4. <Maschinen> *(AE)* gage, *(BE)* gauge; 5. <Mechan> thickness; 6. <Papier> thickness; *(AE)* caliper, *(BE)* calliper *(des Papiers)*
Dicke f **des Luftschraubenblattes** <Lufttrans> blade depth *(Hubschrauber)*
Dicke f **des Ziehbalkens** <Ker & Glas> depth of the drawbar
Dickenabnahme f <Fertig> reduction *(Walzen)*
Dickenlehre f <Gerät> *(AE)* thickness gage, *(BE)* thickness gauge
Dickenmesseinrichtung f <Gerät> *(AE)* thickness gage, *(BE)* thickness gauge
Dickenmesser m <Kunststoff, Papier> *(AE)* thickness gage, *(BE)* thickness gauge
Dickenmessgerät n **mit Gammastrahlen** <Gerät> gamma thickness meter
Dickenmesslehre f <Metrol> *(AE)* caliper, *(BE)* calliper
Dickensensor m <Gerät> *(AE)* thickness gage, *(BE)* thickness gauge
Dickenverhältnis n <Lufttrans> thickness ratio
Dickenverlust m <Aufnahme, Fernseh> thickness loss
Dickfilmschaltung f <Phys> printed circuit
dickflüssig adj <Fertig> viscid
Dickflüssigkeit f 1. <Fertig> viscidity; 2. <Lebensmittel> ropiness
Dickoxid n <Elektronik> thick oxide
Dickoxid-Metallgate-MOS-Schaltung f <Elektronik> thick oxide metal-gate MOS circuit
Dickschicht f <Elektronik, Raumfahrt> thick film
Dickschichtbauelement n <Elektronik> thick film device
Dickschichthybridschaltung f <Elektronik> thick film hybrid circuit
Dickschichtkondensator m <Elektronik> thick film capacitor
Dickschichtleiter m <Elektronik> thick film conductor
Dickschichtmaterial n <Elektronik> thick film material
Dickschichttechnik f <Elektronik, Raumfahrt> thick film technology *(Raumschiff)*
Dickschichtwiderstand m <Elektronik> thick film resistor
Dickspiegelglas n <Ker & Glas> thick polished plate glass
Dicktafelglas n <Ker & Glas> *(AE)* crystal sheet glass, *(BE)* thick sheet glass
Dickten... 1. <Druck> width; 2. <Maschinen> thickness; 3. <Metrol> feeler
Dicktenhobelmaschine f <Maschinen> planing and thicknessing machine, thicknessing machine
Dicktenhobeln n <Maschinen> thicknessing
Dicktenlehre f <Metrol> *(AE)* feeler gage, *(BE)* feeler gauge
Dicktenschablone f <Metrol> *(AE)* feeler gage, *(BE)* feeler gauge
Dicktentabelle f <Druck> width table
Dicyan n <Chemie> cyanogen
Didotsystem n <Druck> Didot system
Didymium n <Chemie> didymium
Diebstahlalarmanlage f <Sicherheit> theft alarm installation
diebstahlsicher adj <Verpack> pilfer-proof
Diebstahlsicherung f <Comp & DV> lock
Diebstahlsicherung f **mit Trickschaltung** <Kfztech> antitheft ignition lock
Diebstahlverhütung f <Sicherheit> theft prevention device
Dieldrin n <Chemie, Lebensmittel> dieldrin *(Insektizid)*
Diele f <Bau> plank, vestibule
Dielektrikum n <Elektriz, Elektrotech, Funktech, Phys, Telekom> dielectric • **mit Luft als Dielektrikum ausgestattet** <Elektriz> air-dielectric
Dielektrikum n **mit Verlust** <Elektriz> lossy dielectric

dielektrisch *adj* <Chemie, Elektriz, Elektrotech, Funktech, Maschinen, Phys, Raumfahrt, Telekom> dielectric
dielektrische Absorption *f* <Elektrotech> dielectric absorption
dielektrische Antenne *f* 1. <Raumfahrt> dielectric antenna *(Weltraumfunk)*; 2. <Telekom> dielectric antenna
dielektrische Eigenschaften *fpl* <Elektriz> dielectric properties
dielektrische Erwärmung *f* <Elektriz, Elektrotech, Heiz & Kälte, Kunststoff> dielectric heating
dielektrische Hysterese *f* <Elektriz, Elektrotech> dielectric hysteresis
dielektrische Isolierung *f* <Elektrotech> dielectric isolation
dielektrische Ladung *f* <Telekom> dielectric charge
dielektrische Leitfähigkeit *f* <Raumfahrt> permittivity
dielektrische Polarisation *f* <Elektrotech> dielectric polarization
dielektrische Prüfung *f* <Elektrotech> dielectric test
dielektrische Stärke *f* <Elektriz, Elektrotech> dielectric strength
dielektrische Suszeptibilität *f* <Elektriz> dielectric susceptibility
dielektrischer Durchbruch *m* <Elektriz> dielectric breakdown
dielektrischer Resonator *m* <Raumfahrt, Telekom> dielectric resonator
dielektrischer Stoff *m* <Elektrotech> dielectric material
dielektrischer Verlust *m* <Elektriz, Elektrotech, Phys> dielectric loss
dielektrischer Verlustwinkel *m* <Elektriz> dielectric loss angle
dielektrischer Wellenleiter *m* <Telekom> dielectric waveguide
dielektrisches Anschwellen *n* <Kerntech> dielectric swelling
dielektrisches Medium *n* <Elektrotech> dielectric medium
dielektrisches Resonanzfilter *n* <Raumfahrt> dielectric resonator filter *(Weltraumfunk)*
Dielektrizität *f* <Elektrotech> dielectricity
Dielektrizitätskonstante *f* 1. <Elektriz> permittivity, relative permittivity; 2. <Elektrotech> dielectric constant, permittivity; 3. <Funktech> permittivity; 4. <Kunststoff> dielectric constant, permittivity; 5. <Phys, Telekom> permittivity
Dielektrizitätskonstante *f* **der Luft** <Elektrotech> permittivity of air
Dielektrizitätskonstante *f* **des Vakuums** <Elektrotech, Phys> permittivity of free space
Dielektrizitätszahl *f* <Elektrotech, Phys> relative permittivity
Dielektronen-Spektrometer *n* **mit hoher Akzeptanz** *(HADES)* <Teilphys> high acceptance di-electron spectrometer *(HADES)*
dielen *v* <Bau> floor
Dielung *f* <Bau> flooring, planking
Dien *n* <Chemie> diene
Dienst *m* <Telekom> service • **außer Dienst gestellt** <Wassertrans> decommissioned *(Schiff)* • **in Dienst** <Elektrotech> in operation, in-service • **in Dienst befindlich** <Wassertrans> in commission *(Schiff)* • **in Dienst stellen** <Wassertrans> commission *(Schiff)*
Dienst *m* **der unteren Ebene** <Telekom> lower level service
Dienst *m* **zum Grundtarif** <Telekom> basic rate service
Dienstanbieter *m* 1. <Comp & DV> services provider; 2. <Telekom> server *(im Internet)*; service provider
Dienstanforderung *f* <Telekom> request for service

Dienstbit *n* <Telekom> service bit
diensteintegrierende Selbstwählnebenstelle *f* <Telekom> integrated services PABX
diensteintegrierende Vermittlungsstelle *f* <Telekom> integrated services exchange
diensteintegrierendes digitales Breitbandnetz *n* <Telekom> wideband integrated services digital network
diensteintegrierendes digitales Netz *n* <Telekom> integrated services digital network *(ISDN)*
diensteintegrierendes Gebäude-Verkabelungssystem *n* <Telekom> integrated communications cabling system
Dienstekonvergenz *f* <Telekom> services convergence
Dienstelement *n* <Telekom> service primitive
Diensterbringer *m* <Telekom> service provider
Dienste-Steuerungsknoten *m* <Telekom> service control point *(Intelligentes Netzwerk)*
Dienstewechsel *m* <Telekom> swap
Dienstgipfelhöhe *f* <Lufttrans> service ceiling
Dienstgüte *f* <Telekom> grade of service, quality of service
diensthabender Operator *m* <Kerntech> on shift operator
dienstintegrierendes Digitalnetz *n* <Telekom> integrated services digital network
Dienstleistungsmarke *f* <Patent> service mark
Dienstleitung *f* <Telekom> order wire, service line, traffic circuit; engineer's order wire *(Telefon)*
Dienstmerkmal *n* <Telekom> facility *(ISDN)*
Dienstpersonal *n* <Kerntech> service staff
Dienstprogramm *n* <Comp & DV> service program, utility program
Dienstprogramm *n* **für Shell-Prozeduren** <Comp & DV> shell script utility
Dienstschicht *f* <Telekom> service layer
dienstspezifisches verbindungsorientiertes Protokoll *n* <Telekom> service-specific connection-oriented protocol
Dienststammelement *n* <Telekom> primitive
Dienstunterbrechung *f* <Telekom> service disruption
Dienstvermittlungspunkt *m* <Telekom> service switching point *(SSP)*
Dienstzeit *f* <Telekom> service time
Dienstzugangskennung *f* <Telekom> service access point identifier *(SAPI)*
Diesel *m* <Erdöl> diesel fuel • **auf Diesel umstellen** <Kfztech> convert to diesel
Dieselaggregat *n* 1. <Elektrotech> diesel-driven generating set; 2. <Wassertrans> diesel generator *(Antrieb)*
dieselbetriebenes Notstromaggregat *n* <Kerntech> diesel generator standby power plant
dieselelektrische Lokomotive *f* <Eisenbahn> diesel electric locomotive
dieselelektrische Rangierkleinlokomotive *f* <Eisenbahn> diesel-electric shunting motor tractor
dieselelektrischer Motor *m* <Kfztech> diesel electric engine
dieselelektrischer Triebwagen *m* <Eisenbahn> diesel electric railcar
dieselelektrisches Kraftwerk *n* <Elektrotech> diesel electric power station
dieselelektrisches Triebfahrzeug *n* <Eisenbahn, Elektrotech> diesel-electric motor vehicle
Dieselelektroantrieb *m* <Kfztech> diesel electric drive
Diesel-Gas... *adj* <Kfztech, Nichtfoss Energ> dual-fuel
Diesel-Gasfahrzeug *n* <Kfztech, Nichtfoss Energ> dual-fuel vehicle
Dieselgenerator *m* <Wassertrans> diesel generator *(Antrieb)*

Dieselhammer *m* <Kohlen> diesel hammer
Dieselhorst-Martin-Kabel *n* <Elektrotech> multiple-twin quad
dieselhydraulische Lokomotive *f* <Eisenbahn> diesel hydraulic locomotive
dieselhydraulischer Motor *m* <Maschinen> diesel hydraulic engine
Dieselkraftstoff *m* 1. <Erdöl> diesel fuel; 2. <Kfztech> diesel fuel, fuel
Dieselkraftwerk *n* <Elektrotech> diesel electric power station
Diesellokomotive *f* <Eisenbahn> diesel locomotive
Dieselmannschaftswagen *m* <Eisenbahn> diesel generator unit crew car
Dieselmotor *m* 1. <Kfztech, Maschinen> compression-ignition engine, diesel engine; 2. <Wassertrans> diesel engine • **mit Dieselmotor** <Maschinen> diesel-powered
Dieselmotor *m* **in Kreuzkopfbauart** <Wassertrans> crosshead engine
Dieselmotor *m* **mit indirekter Einspritzung** <Kfztech> indirect injection diesel engine
Dieselöl *n* 1. <Eisenbahn> diesel oil; 2. <Kfztech> diesel fuel, diesel oil
Dieselstromerzeuger *m* <Nichtfoss Energ> diesel generator set, diesel genset
Dieseltriebwagenzug *m* *(DMU)* <Eisenbahn> diesel motorcoach, diesel multiple unit, DMU
Diesel- und Windbatteriekraft *f* <Nichtfoss Energ> diesel and wind-battery power
diesig *adj* <Wassertrans> hazy *(Wetter)*
Diethen *n* <Chemie> bivinyl
Diethylen *n* <Chemie> bivinyl, diethene
Diethylendiamin *n* <Chemie> diethylenediamine, piperazine
Differenz *f* <Math> difference *(Ergebnis einer Subtraktion)*
Differenzdetektor *m* <Telekom> differential detector
Differenzdruck *m* 1. <Erdöl> differential pressure *(Messtechnik)*; 2. <Heiz & Kälte> differential pressure, pressure differential; 3. <Labor, Mechan, Nichtfoss Energ> differential pressure
Differenzdruck-Durchflussmesser *m* <Gerät> differential pressure flowmeter
Differenzdruckhöhe *f* <Nichtfoss Energ> differential head
Differenzdruckmesser *m* <Mechan> *(AE)* differential pressure gage, *(BE)* differential pressure gauge
Differenzdruck-Messumformer *m* <Gerät> differential pressure transducer, differential head pressure transducer
Differenzdruck-Messwandler *m* <Gerät> differential pressure transducer
Differenzdruck-Messzelle *f* <Gerät> differential pressure cell
Differenzdruckwandler *m* <Gerät> differential pressure transducer
Differenzeingang *m* <Elektronik> differential input *(am Verstärker)*
Differenzfrequenz *f* <Funktech> difference frequency
Differenzial *n* 1. <Aufnahme, Elektriz, Elektronik, Elektrotech, Fernseh, Kfztech, Kunststoff, Lufttrans, Maschinen> differential; 2. <Math> differential *(infinitesimale Variablenänderung)*; 3. <Mechan, Thermod> differential
Differenzial *n* **mit regelbarem Schlupf** <Mechan> controlled slip differential
Differenzial-Amperemeter *n* <Elektriz> differential ammeter
Differenzialbetrieb *m* <Elektronik> differential mode
Differenzialbremsung *f* <Lufttrans> differential braking

differenzialerregter Doppelschlussgenerator *m* <Elektriz> differentially excited compound generator
differenziales Magnetometer *n* <Elektriz> differential magnetometer
Differenzialflaschenzug *m* 1. <Kerntech> differential chain block; 2. <Mechan> differential pulley, differential chain block
Differenzial-Galvanometer *n* <Elektrotech> differential galvanometer
Differenzialgeber *m* <Metrol, Regelung> differential pick-up, differential transducer
Differenzialgehäuse *n* <Kfztech> differential case
Differenzialgetriebe *n* 1. <Kfztech> differential; 2. <Maschinen> differential gear
Differenzialgewinn *m* <Fernseh> differential gain
Differenzialgleichung *f* <Math> differential equation
Differenzial-GPS *f (DGPS)* <Wassertrans> differential global positioning system *(DGPS)*
Differenzial-Kalorimeter *n* <Thermod> differential scanning calorimeter
Differenzial-Kalorimetrie *f* <Thermod> differential scanning calorimetry
Differenzialkondensator *m* <Elektrotech> differential capacitor
Differenzialmikrofon *n* <Aufnahme> differential microphone
Differenzialquotient *m* <Math> differential coefficient, differential quotient
Differenzialrechnung *f* <Math> differential calculus
Differenzialrelais *n* <Elektriz, Elektrotech> differential relay
Differenzial-Scanningkalorimeter *n* <Thermod> differential scanning calorimeter
Differenzial-Scanningkalorimetrie *f* <Kunststoff, Thermod> differential scanning calorimetry
Differenzialschraube *f* <Maschinen> compound screw, differential screw
Differenzialschutzrelais *n* <Elektriz> differential protection relay
Differenzialspannung *f* <Elektrotech> differential voltage
Differenzialsperre *f* <Kfztech> differential lock *(Triebstrang)*
Differenzialspule *f* <Elektriz> differential coil
Differenzialsteuerung *f (PID-Steuerung)* <Elektriz> derivative control
Differenzialteilen *n* <Maschinen> differential indexing
Differenzialthermoanalyse *f (DTA)* <Kunststoff, Thermod, Umweltschmutz> differential thermal analysis *(DTA)*
Differenzialtransducer *m* <Elektrotech> differential transducer
Differenzialtransformator *m* <Elektriz, Regelung> differential transformer
Differenzialverhältnis *n* <Kfztech> differential ratio
Differenzialverstärker *m* <Elektronik, Mechan> differential amplifier
Differenzialverzögerung *f* <Elektronik> differential delay
Differenzialwandler *m* <Elektrotech> differential transducer
Differenzialwelle *f* <Kfztech> differential shaft
Differenzialwicklung *f* <Elektrotech> differential winding
Differenzialwirkung *f* <Lufttrans> differential effect
Differenzialzeit *f* <Elektronik> differential time
Differenzialzwischenrad *n* <Kfztech> differential pinion
differenzielle Modendämpfung *f* <Optik, Telekom> differential mode attenuation
differenzielle Modenverzögerung *f* <Optik> differential mode delay
differenzielle Phase *f* <Fernseh, Funktech> differential phase

differenzielle Quantenausbeute f <Telekom> differential quantum efficiency
differenzieller Quantenwirkungsgrad m <Optik, Telekom> differential quantum efficiency
Differenzierbeiwert m <Regelung> derivative action coefficient
differenzieren v <Math> differentiate
Differenzieren n <Math> differentiation
differenzierende Regelung f <Regelung> derivative control
differenzierendes Glied n <Regelung> derivative element
differenzierendes Verhalten n <Regelung> derivative action
Differenzierer m <Regelung> derivative unit
Differenzierschaltung f <Elektronik> differentiating circuit
differenziertes Signal n <Elektronik> differentiated signal
Differenzierzeit f <Regelung> derivative action time
Differenzkanal m <Aufnahme> difference channel
Differenzkomparator m <Elektronik> differential comparator
Differenzmodulation f <Elektronik> differential modulation
Differenznote f <Aufnahme> difference note
Differenzphasenumtastung f (DPSK) <Telekom> differential phase shift keying (DPSK)
Differenz-Pulscodemodulation f (DPCM) <Elektronik, Telekom> differential pulse code modulation (DPCM)
Differenzsignal n <Elektronik> difference signal, differential signal, differential mode signal
Differenzsignalquelle f <Elektronik> differential signal source
Differenzspannungsmessgerät n <Elektriz> differential voltmeter
Differenztemperatur f <Erdöl> differential temperature (Messtechnik)
Differenzthermoanalyse f <Umweltschmutz> differential thermal analysis
Differenzthermoelement n <Thermod> differential thermocouple
Differenzton m <Akustik> difference tone
Differenzverstärker m 1. <Elektronik> differential amplifier, instrumentation amplifier; 2. <Funktech, Telekom> differential amplifier
Differenzverstärkung f <Elektronik> differential gain
Diffraktion f <Elektrotech, Telekom> diffraction
Diffraktionstechnik f <Erdöl, Phys> diffractometry (Lichtwellen)
diffundierte Schicht f <Elektronik> diffused layer
diffundierter Emitter-Kollektor-Transistor m <Elektronik> diffused emitter-collector transistor
diffundierter Übergang m <Elektronik> diffused junction (Halbleiter)
diffuse Reflexion f <Strahlphys> diffuse reflection
diffuse Strahlung f <Nichtfoss Energ, Raumfahrt> diffuse radiation
diffuser Nebel m <Raumfahrt> diffuse nebula
diffuser Schallpegel m <Akustik> diffuse sound level
Diffusion f 1. <Bau> diffusion; 2. <Chemie> diffusion (von Flüssigkeiten); 3. <Chemtech, Elektriz, Funktech, Kerntech, Kohlen, Maschinen, Phys, Umweltschmutz> diffusion
Diffusion f **im Magnetfeld** <Kerntech> diffusion across the magnetic field
Diffusionsapparat m <Chemtech> diffusion apparatus, diffusion cell; diffuser (Lebensmittel)
Diffusionsdotierung f <Elektronik> diffusion doping
Diffusionsfehlerstelle f <Elektronik> diffusion defect

Diffusionsfläche f <Heiz & Kälte> diffusion area
Diffusionsgalvanisierung f <Anstrich> diffused plating
Diffusionsglühen n <Chemtech> diffusion annealing
Diffusionskern m <Kerntech> diffusion kernel
Diffusionskoeffizient m (D) <Elektronik, Funktech, Phys> diffusion coefficient (D)
Diffusionslänge f 1. <Elektronik> diffusion length; 2. <Kerntech> diffusion length
diffusionslegierter Transistor m <Elektronik> diffused alloy transistor (Legierung kombiniert mit Diffusion)
diffusionslose Rückwirkung f <Metall> diffusionless reaction
Diffusionsmethode f <Kerntech> diffuse scattering method
Diffusionsofen m <Elektronik> diffusion oven
Diffusionsphotodiode f <Elektronik> diffused photodiode
Diffusionspumpe f <Maschinen, Phys> diffusion pump
Diffusionsspannung f <Phys> diffusion voltage (Halbleiterphysik)
Diffusionsstrom m <Elektriz> diffusion current
Diffusionswiderstand m <Lufttrans> spray drag (Luftfahrzeug)
Diffusionszelle f <Chemtech> diffusion cell
Diffusor m 1. <Nichtfoss Energ> diffuser; 2. <Raumfahrt> diffuser (für gasförmigen Sauerstoff); 3. <Telekom> diffuser; 4. <Wasserversorg> volute chamber
Diffusor m **für Auflichtmessung** <Foto> diffuser for incident measurement
Diffusverstärker m <Fernseh> matting amplifier
digital adj <Aufnahme, Comp & DV, Elektriz, Elektronik, Fernseh, Funktech, Labor, Phys, Raumfahrt, Telekom> digital • **digital darstellen** <Elektronik> digitize (Signale)
digital codierte Videoplatte f <Optik> (BE) digitally-encoded videodisc, (AE) digitally-encoded videodisk
digital übertragene Berichterstattung f **via Satellit** <Telekom> digital satellite news gathering
digital umgesetztes Signal n <Elektronik> digitized signal
Digital... <Aufnahme, Comp & DV, Elektronik, Fernseh, Labor, Telekom> digital
Digitalabstimmung f <Elektronik> digital tuning
Digitalabtastvoltmeter n <Metrol> sample-and-hold digital voltmeter
Digital-Analog-... (D/A) <Aufnahme, Comp & DV, Elektronik, Fernseh, Labor, Telekom> digital-analog (D/A)
Digital-Analog-Umsetzer m (DAU) 1. <Comp & DV> digital-analog converter (DAC); 2. <Elektronik, Telekom> digital-analog converter (DAC)
Digital-Analog-Umsetzung f 1. <Aufnahme> digital-analog conversion; 2. <Comp & DV> digital-analog conversion; 3. <Elektronik, Gerät, Telekom> digital-analog conversion
Digital-Analog-Wandler m 1. <Comp & DV> digital-analog converter; 2. <Elektronik, Telekom> digital-analog converter
Digital-Analog-Wandlung f 1. <Aufnahme> digital-analog conversion; 2. <Comp & DV> digital-analog conversion; 3. <Elektronik, Gerät, Telekom> digital-analog conversion
Digitalanzeige f 1. <Elektronik> digital display; 2. <Gerät> digital display, digital readout; 3. <Telekom> digital readout
Digital-Audio-Tape n (DAT) <Aufnahme> digital audio tape (DAT)
Digitalaufzeichnung f <Aufnahme, Fernseh, Telekom> digital recording
Digitalausgabe f <Comp & DV> digital output
Digitalausgang m <Elektronik> digital output
Digitalbereich m <Elektronik> digital domain

Digitalchip

Digitalchip m <Elektronik> digital chip *(Speicherbaustein)*
Digitalcode m <Elektronik> digital code
Digitalcodierung f <Elektronik> digital coding
Digitaldaten npl <Elektronik> digital data
Digital-Digital-Umsetzung f <Gerät> digital-digital conversion
Digitaldruck m <Druck> digital printing
Digitaldruckmaschine f <Druck> digital press, digital printing machine
digitale Anzeige f <Elektriz, Gerät> digital readout
digitale Audiocassette f <Aufnahme> digital audio tape cassette
digitale Auffüllung f <Telekom> digital filling
digitale Aufzeichnung f <Aufnahme, Fernseh, Telekom> digital recording
digitale Bildverarbeitung f <Elektronik, Fernseh, Telekom> digital image processing
digitale CD f <Optik> *(BE)* digital optical disc, *(AE)* digital optical disk
digitale Codierung f <Telekom> digital coding
digitale Darstellung f <Comp & DV, Elektronik> digital representation • **in digitale Darstellung umsetzen** <Gerät> digitalize
digitale Datenfernverarbeitung f <Comp & DV> digital communications
digitale Druckmaschine f <Druck> digital press, digital printing machine
digitale Fernmeldeleitung f <Telekom> digital telecommunication circuit *(ISDN)*
digitale Fernvermittlung f <Telekom> digital trunk exchange
digitale Filterung f <Telekom> digital filtering
digitale Funkverbindung f <Funktech, Telekom> digital radio link
digitale Hauptvermittlungsstelle f <Telekom> DMNSC, *(AE)* digital main network switching center, *(BE)* digital main network switching centre
digitale Hierarchie f <Telekom> digital hierarchy
digitale Impulsfolge f <Telekom> digital pulse stream
digitale Integration f <Elektronik> digital integration
digitale integrierte Schaltung f <Elektronik> digital-integrated circuit
digitale internationale Mietleitung f <Telekom> international digital leased circuit, dIML
digitale Kamera f <Foto> digital camera
digitale Kommunikation f <Telekom> digital communications
digitale Koppelmatrix f <Telekom> digital switching matrix
digitale Leitung f <Telekom> digital circuit
digitale Lichtwellenleiterübertragung f <Telekom> digital optical fiber communication, digital optical fiber transmission
digitale Logik f <Elektronik> digital logic *(Schaltalgebra)*
digitale Millimeterwellen-Richtverbindung f <Telekom> millimetrical digital link
digitale Modulation f <Elektronik, Phys, Telekom> digital modulation
digitale Momentanfrequenzmessung f <Elektronik> digital instantaneous frequency measurement
digitale Multiplikation f <Elektronik> digital multiplication
digitale Nachrichtentechnik f <Telekom> digital telecommunication technology
digitale Nachrichtenübertragung f <Telekom> digital transmission of information
digitale Nebenstellenanlage f <Telekom> digital branch exchange, digital PABX, digital private automatic branch exchange
digitale optische Platte f <Comp & DV> digital optical disk

digitale Ortsvermittlung f <Telekom> digital local exchange
digitale Ortsvermittlungsstelle f <Telekom> digital local exchange, digital local office *(DIVO)*
digitale Phasenmodulation f <Comp & DV, Telekom> digital phase modulation
digitale Phasenverschiebung f <Elektronik> digital phase shifting
digitale Pseudorauschfolge f <Telekom> digital pseudo noise sequence
digitale Rahmenstruktur f <Telekom> digital frame structure
digitale Regelung f <Regelung> digital control
digitale Regenerierung f <Elektronik> digital regeneration
digitale Richtfunkstrecke f <Funktech, Telekom> digital radio link
digitale Rückkopplung f <Telekom> digital feedback
digitale Satelliteneinrichtungssteuerung f <Fernseh> digital satellite equipment control *(DiSEqC)*
digitale Satellitenverbindung f <Fernseh> digital satellite link
digitale Schaltung f <Comp & DV> digital circuit
digitale Schnittstelle f <Telekom> digital interface
digitale Schnittstelle f **für Musikinstrumente** <Telekom> musical instrument digital interface, MIDI
digitale Schrift f <Druck> digital font
digitale Sichtanzeige f <Comp & DV> digital readout
digitale Signalverarbeitung f <Comp & DV, Elektronik> digital signal processing
digitale Sprache f <Elektronik, Telekom> digital speech
digitale Sprachinterpolation f *(DSI)* <Raumfahrt, Telekom> digital speech interpolation *(DSI)*
digitale Sprachsynthese f <Elektronik> digital speech synthesis
digitale Steuerung f <Fernseh, Telekom> digital control
digitale Störung f <Telekom> digital interference
digitale Teilnehmeranschlusseinheit f <Telekom> digital subscriber access unit, digital subscriber loop
digitale Teilnehmer-Anschlussleitung f <Telekom> digital subscriber line *(DSL)*
digitale Teilnehmer-Anschlussleitung f **mit hoher Datenrate** <Telekom> high data rate digital subscriber line
digitale Teilnehmer-Anschlussleitung f **mit sehr hoher Datenrate** <Telekom> very high data rate digital subscriber line
digitale Teilnehmer-Einzel-Anschlussleitung f <Telekom> single digital subscriber line
digitale Teilnehmerleitung f **mit sehr hoher Bitrate** <Telekom> very-high-bit-rate digital subscriber line *(VHDSL)*
digitale Tonaufzeichnung f <Aufnahme> digital audio recording
digitale Transitsteuerung f <Telekom> digital transit command
digitale Übertragung f <Telekom> digital transmission
digitale Verarbeitung f <Telekom> digital processing
digitale Verbindung f <Telekom> digital connection
digitale Vermittlungsanlage f <Elektrotech> digital switching equipment
digitale Vermittlungsstelle f <Telekom> digital exchange, digital switch, *(AE)* digital switching center, *(BE)* digital switching centre
digitale Vermittlungsstelle f **für den Fernverkehr** <Telekom> digital trunk exchange
digitale Vermittlungsstelle f **für den Ortsverkehr** <Telekom> digital local exchange

digitale Video-CD f <Aufnahme, Fernseh> digital versatile disc, digital video disc *(DVD)*
digitale Videoeffekte mpl <Fernseh> DVE, digital video effects
digitale Videoplatte f <Optik> *(BE)* digital videodisc, *(AE)* digital videodisk
Digitalein n <Chemie> digitalein
Digitaleingabe f <Elektronik> digital input
Digitalelement n <Telekom> digit
digitaler Addierer m <Elektronik> digital adder
digitaler Anrufbeantworter m <Telekom> digital answerer
digitaler Computer m <Comp & DV> digital computer
digitaler Fernseher m <Fernseh> digital TV receiver
digitaler Fernsehfunk m <Fernseh> digital video broadcasting *(DVB)*
digitaler Fernseh-Rundfunk m <Fernseh> digital television broadcasting
digitaler Flugdatenschreiber m <Lufttrans> digital flight data recorder
digitaler Frequenzzähler m <Gerät> counter
digitaler Halbbilderzeuger m <Fernseh> digital framer
digitaler Hörfunk m <Funktech> digital audio broadcasting *(DAB)*; digital radio broadcasting
digitaler Kennungsrahmen m <Telekom> digital identification frame
digitaler Konzentrator m <Telekom> digital concentrator
digitaler Messschritt m <Regelung> digital measuring step
digitaler Messwert m <Gerät> digital reading
digitaler Multiplizierer m <Elektronik> digital multiplier
digitaler Phasenregelkreis m <Telekom> digital phase-locked loop *(DPLL)*
digitaler Plotter m <Comp & DV> digital plotter
digitaler Rahmenaufbau m <Telekom> digital frame structure
digitaler Regenerator m <Elektronik> digital regenerator
digitaler Satellitenempfänger m <Fernseh> digital satellite receiver
digitaler Satellitenhörfunk m <Fernseh, Funktech> satellite digital audio broadcasting
digitaler Satelliten-Hörrundfunk m <Funktech> digital satellite radio
digitaler Satellitenrundfunk m <Fernseh, Funktech> digital satellite broadcasting, digital satellite radio *(DSR)*
digitaler Satz m <Druck> digital typesetting
digitaler Schnurlostelefonstandard m <Mobilkom> DECT standard of ETSI, digital European cordless telephone standard
digitaler Selektivruf m <Telekom> DSC, digital selective calling
digitaler Signalprozessor m <Comp & DV, Elektronik> digital signal processor *(DSP)*
digitaler Stromkreis m <Comp & DV> digital circuit
digitaler terrestrischer Hörfunk m <Fernseh, Funktech> terrestrial digital audio broadcasting *(DAB-T)*
digitaler Trägerbaustein m <Telekom> digital carrier module
digitaler Übertragungsweg m <Telekom> digital circuit
digitaler Videocamerarecorder m <Aufnahme, Fernseh> digital camcorder
digitaler Videoplattenspieler m <Fernseh> digital video disc player, DVD-player
digitaler Videorecorder m <Fernseh> digital videotape recorder *(DVTR)*
digitaler Workflow m <Druck> digital workflow
digitaler Zähler m <Gerät> digital counter
digitales angepasstes Filter n <Elektronik> digital-matched filter

digitales Aufzeichnungsgerät n <Kerntech> digital recorder
digitales Ausgabesignal n <Elektronik> digital output signal
digitales Befehlszeichen n <Telekom> digital command signal *(DCS)*
digitales Computersystem n <Kerntech> digital process computer system
digitales Dämpfungsglied n <Elektronik> digital attenuator
digitales diensteintegrierendes Netz n <Telekom> integrated services digital network *(ISDN)*
digitales Drehzahlmessgerät n <Gerät> counter
digitales Einbaugerät n <Gerät, Metrol> digital panel meter
digitales Eingangssignal n <Elektronik, Regelung> digital input signal
digitales europäisches Schnurlostelefon n <Mobilkom> digital European cordless telephone *(DECT)*
digitales Fernsehen n <Fernseh> digital video broadcasting
digitales Fernsprechnetz n <Telekom> digital telecommunication network
digitales Filtern n <Comp & DV, Elektronik> digital filtering
digitales Gerät n <Elektronik> digital device
digitales Instrument n <Elektriz> digital instrument
digitales Integrierglied n <Elektronik> digital integrator
digitales Kabelfernsehen n <Fernseh> cable digital video broadcasting
digitales Koppelelement n <Telekom> digital switching element
digitales Koppelvielfach n <Telekom> digital switching matrix
digitales Lichtwellenleitersystem n <Telekom> digital fiber-optic system
digitales Messgerät n <Gerät> digital measuring instrument
digitales Messinstrument n 1. <Gerät> digital measuring instrument; 2. <Metrol> digital readout measuring instrument
digitales Mobilfunknetz n <Mobilkom> GSM-net, mobile digital radio communications
digitales Mobilfunksystem n **der USA** <Mobilkom> digital advanced mobile phone system, Digital AMPS
digitales Multimeter n <Elektrotech> digital multimeter
digitales Multiplexverfahren n <Elektronik, Telekom> digital multiplexing
digitales paneuropäisches Mobilfunksystem n <Mobilkom> global system for mobile communications *(GSM)*
digitales Punkt-zu-Mehrpunkt-Funksystem n <Funktech, Telekom> digital multipoint system
digitales Richtfunkgerät n <Funktech> digital radio relay equipment
digitales Richtfunksystem n <Funktech, Telekom> digital radio relay system
digitales Rundfunksystem n <Fernseh, Funktech> digital broadcast system
digitales Satellitenfernsehen n <Fernseh> satellite digital video broadcasting *(DVB-S)*
digitales Satelliten-Radio n <Funktech> digital satellite radio, DSR
digitales schnurloses Telefon n <Mobilkom> DECT-standard telephone, digital cordless telephone
digitales Signal n 1. <Comp & DV, Phys> digital signal; 2. <Elektronik, Telekom> digital signal, discrete signal
digitales Stellglied n <Elektrotech> digital actuator
digitales Steuergerät n <Kfztech> digital control box
digitales Tonband n <Comp & DV> digital audio tape

digitales

digitales Vermitteln *n* <Telekom> digital switching
digitales Vermittlungsnetz *n* <Telekom> digital switching network
digitales Vermittlungssystem *n* <Telekom> digital switch, digital switching system
digitales Voltmeter *n* <Elektrotech> digital voltmeter
digitales Wohnungsnetz *n* <Telekom> inhome digital network
Digitalfehler *m* <Telekom> digital error
Digitalfernsehen *n* <Fernseh> digital television
Digitalfernsehen *n* **mit Standardauflösung** <Fernseh> standard definition television
Digitalfilter *m* <Elektronik, Telekom> digital filter
Digitalfotografie *f* <Foto> digital photography
Digitalfüllzeichen *n* <Telekom> digital pad
Digitalgeometrie *f* <Geom> digital geometry
Digitalhierarchie *f* <Telekom> digital hierarchy
Digitalin *n* <Chemie, Lebensmittel> digitalin
digitalisieren *v* 1. <Aufnahme, Comp & DV, Elektriz, Fernseh> digitize; 2. <Maschinen> digitize *(Modell)*; 3. <Phys> digitize; 4. <Telekom> digitalize, digitize
Digitalisierer *m* <Comp & DV, Elektronik> digitizer
Digitalisiergerät *n* <Comp & DV, Funktech, Telekom> digitizer
digitalisiert *adj* 1. <Druck> digitized; 2. <Raumfahrt> digitalized *(Weltraumfunk)*
Digitalisiertablett *n* <Comp & DV, Elektronik> digitizing tablet
digitalisierte Daten *npl* <Elektronik, Telekom> digitized data
digitalisierte Sprache *f* <Telekom> digitized speech
digitalisiertes Bild *n* <Elektronik> digital image, digitized image
Digitalisierung *f* 1. <Comp & DV> digitization; 2. <Elektronik> digitizing *(Umsetzung von Grafiken in digitale Daten)*; 3. <Kerntech, Phys> digitization; 4. <Telekom> digitalization, digitization
Digitalisierung *f* **von Signalen** <Elektronik> signal digitization
Digitalisierungstablett *n* <Comp & DV, Elektronik> digitizing tablet
Digitalkamera *f* <Fernseh, Foto> digital camera
Digitalkanal *m* <Telekom> data channel, digital channel, digital transmission channel, PC channel
Digitalkassette *f* <Comp & DV> digital cassette
Digitalkompaktkassette *f* <Aufnahme, Comp & DV> digital audio tape *(DAT)*; digital compact cassette *(DCC)*
Digitalkonverter *m* <Gerät> digital converter
Digitalmagnetbandkassette *f* <Aufnahme, Comp & DV> DAT cassette, digital audio tape cassette
Digitalmessgerät *n* <Gerät, Metrol> digital measuring equipment, digital measuring instrument
Digitalmikrometer *n* <Metrol> digital readout micrometer
Digitalmultimeter *n* <Metrol> digital multimeter *(DMM)*
Digital-Multiplexeinrichtung *f* <Telekom> digital multiplex equipment
Digitalrechner *m* <Comp & DV> digital computer
Digitalschaltung *f* <Elektronik> digital circuit
Digitalsignal *n* 1. <Comp & DV> digital signal; 2. <Elektronik> digital signal, discrete signal; 3. <Phys> digital signal, time and value-discrete signal; 4. <Telekom> digital signal, discrete signal, time and value-discrete signal
Digitalsignalanalysator *m* <Elektronik> *(BE)* digital signal analyser, *(AE)* digital signal analyzer
Digitalsignalanalyse *f* <Elektronik> digital signal analysis
Digitalsignal-Grundleitung *f* <Telekom> digital line link, DSGL
Digitalsignalverbindung *f (DSV)* <Telekom> digital connection; digital path *(DSV)*

Digitalsignalverteiler *m* <Telekom> digital distribution frame
Digitalstrommessgerät *n* <Elektriz> digital ammeter
Digitalsubtrahierer *m* <Elektronik> digital subtractor
Digitaltechnik *f* <Comp & DV, Elektronik> digital technique
Digitalübertragung *f* <Telekom> digital transmission
Digitaluhr *f* <Comp & DV> digital clock
Digitalumsetzer *m* <Elektronik, Gerät> digital converter
Digitalverarbeitung *f* <Elektronik> digital processing
Digitalvermittlung *f* <Telekom> digital switching
Digitalvideorecorder *m* <Fernseh> DVTR, digital videotape recorder
Digitalvoltmeter *n* <Elektriz> digital voltmeter
Digitalzähler *m* <Gerät> digital counter
Diglycidether *m* <Chemie> diglycidyl ether
Diglycidylether *m* <Chemie> diglycidyl ether
Diglykololeat *n* <Lebensmittel> diglycol oleate
Digraph *m* <Künstl Int> digraph, directed graph, oriented graph
Dihydro... <Chemie> dihydro...
Dihydrobenzol *n* <Chemie> cyclohexadiene, dihydrobenzene
Dihydrodioxonaphthalin *n* <Chemie> dihydrodiketonaphthalene, naphthoquinone
Dihydroergotamin *n* <Chemie> dihydroergotamine
Dihydronaphthalin *n* <Chemie> dialin, dihydronaphthalene
Dihydrostreptomycin *n* <Chemie> dihydrostreptomycin
Dihydrotachysterin *n* <Chemie> dihydrotachysterol
Dihydrothiazol *n* <Chemie> dihydrothiazole, thiazoline
Dihydroxyaceton *n (DHA)* <Chemie> dihydroxyacetone
Dihydroxy-α-Carotin *n* <Chemie> dihydroxy-α-carotene, lutein
Dihydroxypropanon *n (DHA)* <Chemie> dihydroxyacetone
Diiodmethan *n* <Chemie> diiodomethane, methylene iodide
Diisopropylidenaceton *n* <Chemie> diisopropylidene acetone, phorone
Diktaphon *n* <Aufnahme> dictaphone, voice recorder
Diktiergerät *n* <Aufnahme> dictaphone, dictation machine
Dilatanz *f* <Kunststoff> dilatancy
DIL-Gehäuse *n (Dual-in-Line-Gehäuse)* <Elektronik, Elektrotech, Funktech, Telekom> DIP *(dual-in-line package)*
Dimension *f* <Comp & DV, Druck, Maschinen, Math> dimension
Dimension *f* **einer Größe** <Phys> dimension of a quantity
Dimensionalität *f* <Ergon> dimensional characteristic
dimensionieren *v* <Bau> size
Dimensionierung *f* 1. <Comp & DV, Ergon> dimensioning; 2. <Maschinen> dimensioning, sizing
Dimensionierungsstabilität *f* <Verpack> dimensional stability
Dimensionsgleichung *f* <Phys> dimensional equation
dimensionslose Darstellung *f* <Konstzeich> dimensionless representation
Dimensionsstabilität *f* <Kunststoff> dimensional stability
Dimer *n* <Chemie> dimer
dimer *adj* <Chemie> dimeric
Dimeres *n* <Chemie> dimer
Dimethoxyphthalid *n* <Chemie> dimethoxyphthalide, meconin, opianyl
Dimethyl... <Chemie> dimethyl...
Dimethylamin *n* <Chemie> dimethylamine
Dimethylanilin *n* <Chemie> dimethylaniline, xylidine
Dimethylarsan *n* <Chemie> dimethylarsane
Dimethylarsin *n* <Chemie> dimethylarsane

Dimethylbenzol n <Chemie> dimethylbenzene, xylol
Dimethylbutanon n <Chemie> dimethylbutanone
Dimethylessig... <Chemie> dimethylacetic
Dimethylhydroxybenzol n <Chemie> hydroxydimethylbenzene, xylenol
Dimethylketon n <Chemie> acetone, propanone
Dimethylmorphin n <Chemie> dimethylmorphine, paramorphine, thebaine
Dimethyloctadienol n <Chemie> dimethyloctadienol, nerol
Dimethyloctenal n <Chemie> citronellal, dimethyloctenal
Dimethylpyridin n <Chemie> lutidine
Dimethylxanthin n <Chemie> dimethylxanthine, theobromine
dimetrisch adj <Geom> dimetric
dimetrische Projektion f <Geom, Konstzeich> dimetric projection
Dimmer m <Elektriz, Elektrotech, Lufttrans> dimmer (Beleuchtung)
Dimmerabdeckung f <Lufttrans> dimmer cap
Dimmerschalter m <Elektriz, Elektrotech> dimmer switch
dimolekular adj <Chemie> bimolecular
DIN (Deutsches Institut für Normung) <Maschinen> DIN, German Standards Institution
Dinasstein m <Ker & Glas> dinas brick
Dinatriumtetraborat-Dekahydrat n <Chemie> borax, disodium tetraborate decahydrate
DIN-Format n <Druck> DIN size
Dingi n <Wassertrans> dinghy (Boot)
DIN-Größe f <Druck> DIN size
Dinitro... <Chemie> dinitro...
Dinitrobenzol n <Chemie> dinitrobenzene
Dinitrogenoxid n <Chemie> dinitrogen oxide
Dinitronaphthalin n <Chemie> dinitronaphthalene
Dinitrophenol n <Chemie> dinitrophenol
Dinitrotoluol n <Chemie> binitrotoluene, dinitrotoluene
Dioctylphthalat n (DOP) <Kunststoff> dioctylphthalate (DOP)
Diode f 1. <Comp & DV, Elektriz> diode; 2. <Elektronik> diode (zweipoliges Halbleiterbauelement mit nichtlinearer Strom-Spannungskennlinie); 3. <Funktech, Kerntech> diode; 4. <Kfztech> diode (Elektrikzündung); 5. <Phys, Telekom> diode
Diode f mit einfachem pn-Übergang <Elektronik> p-n homojunction diode
Diode f mit hoher Trägerbeweglichkeit <Elektronik> hot carrier diode
Diode f mit niedriger Verlustleistung <Elektronik> low-power diode
Diode f mit pn-Übergang <Elektronik> p-n junction diode
Diodenbegrenzer m <Elektronik> diode limiter
Dioden-Entstörbaugruppe f <Elektronik> diode suppressor
Diodenentstörung f <Elektronik> diode suppression
Diodenfolge f <Elektronik> diode string
Diodenfrequenzvervielfacher m <Elektronik> diode frequency multiplier
Diodengatter n <Elektronik> diode gate
Diodengleichrichter m <Elektronik> diode rectifier
Diodenkennlinie f <Elektronik> diode characteristic
Diodenkoppelpunkt m <Telekom> diode crosspoint
Diodenlaser m <Elektronik> diode laser
Diodenlogik f <Elektronik> diode logic
Diodenmatrix f <Telekom> diode array
Diodenmischer m <Elektronik> diode mixer
Diodenmodulation f <Elektronik> diode modulation
Diodenmodulator m <Elektronik> diode modulator
Diodenphasenschieber m <Elektronik> diode phase shifter

Diodenprüfgerät n <Elektronik> diode tester
Diodenschalter m <Fernseh> diode switch
Diodenspannung f <Elektronik> diode voltage
Dioden-Transistor-Logik f (DTL) <Elektronik> diode transistor logic (DTL)
Diodenumsetzer m <Elektronik> diode modulator
Diodenverstärker m <Elektronik> diode amplifier
Diode-Triode f <Elektronik> diode triode
Diolefin n 1. <Chemie> dialkene, diene, diolefin; 2. <Erdöl> diolefin (Petrochemie)
D-Ionosphärenschicht f <Funktech, Strahlphys> D-layer
Diopter n 1. <Bau> vane (Vermessung); 2. <Phys> diopter (optisches Instrument); 3. <Raumfahrt> diopter (Raumschiff)
Dioptrie f (dpt) <Optik> (AE) diopter, (BE) dioptre (dpt)
Dioptrik f <Optik> dioptrics
Dioxid n <Chemie> dioxide
Dioxobor... <Chemie> metaboric
Dioxoborat n <Chemie> dioxoborate, metaborate
Dipalmitin n <Chemie> dipalmitin
Diphenyl... <Chemie> diphenyl
Diphenylenimid n <Chemie> carbazole
Diphenylether m <Chemie> diphenyl ether, phenoxybenzene
Diphenylglyoxal n <Chemie> benzil, bibenzoyl
Diphenylharnstoff m <Chemie> carbanilide
Diphenylimid n <Chemie> carbazole
Diphenylketon n <Chemie> benzophenone, diphenyl ketone
Diphenylmethandiisocyanat n (MDI) <Kunststoff> diphenylmethane diisocyanate (MDI)
Diphenylsulfoharnstoff m <Chemie> thiambutosine (Pharmazie)
Diphosphat n <Chemie> pyrophosphate
diphosphorig adj <Chemie> pyrophosphorous
Dipikrylamin n <Chemie> dipicrylamin, hexanitrodiphenylamine
Diplexer m <Telekom> diplexer
Dipol m 1. <Elektriz, Elektrotech> dipole; 2. <Funktech> dipole (Antenne); doublet; 3. <Metall, Telekom> dipole
Dipolantenne f 1. <Fernseh> dipole aerial; 2. <Funktech> dipole, doublet; 3. <Telekom> dipole; 4. <Wassertrans> (BE) dipole aerial (Funk)
Dipol-Dipol-Wechselwirkung f <Strahlphys> dipole-dipole interaction
Dipolmoment n <Elektrotech, Phys> dipole moment
Dipolreihe f <Elektronik> linear array
dippen v <Wassertrans> dip (Flagge)
DIP-Relais n <Elektrotech> DIP relay
DIP-Schalter m <Elektriz, Elektrotech> DIP switch, dual-in-line package switch
Dirac-Konstante f <Phys> Dirac constant (h-quer)
Dirac-Maß n <Math> Dirac measure (Einpunktmaß)
direkt angetriebene Pumpe f <Wasserversorg> direct-acting pump, direct-action pump
direkt angetriebener Propeller m <Lufttrans> direct drive propeller
direkt anzeigendes Instrument n <Gerät> direct-reading instrument
direkt anzeigendes Messgerät n <Gerät> direct-reading instrument
direkt anzeigendes Messinstrument n <Elektriz, Gerät> direct-reading instrument
direkt empfangbarer Fernsehsatellit m <Fernseh> direct broadcast satellite
direkt empfangbarer Satellit m <Fernseh> direct broadcasting satellite
direkt geerdet adj <Elektrotech> directly-earthed

direkt 162

direkt geheizte Katode f <Elektrotech> directly-heated cathode
direkt gekoppelt adj 1. <Comp & DV> close-coupled; 2. <Elektriz> direct-coupled
direkt gekoppelte Transistorlogik f (DCTL) <Elektronik> direct-coupled transistor logic (DCTL)
direkt gekuppelt adj 1. <Comp & DV> close-coupled, direct-coupled; 2. <Elektriz> direct-coupled
direkt netzbetriebener Motor m <Elektriz> across the line motor
direkt reduziertes Eisen n (DRI) <Metall> direct reduced iron (DRI)
direkt übertragene Musik f <Aufnahme> live music (nicht im Studio)
direkt übertragener Ton m <Aufnahme> live sound
direkt wirkende oben liegende Nockenwelle f <Kfztech> direct-acting overhead camshaft
direkt wirkender Propeller m <Lufttrans> direct drive propeller
direkt wirkender Temperaturregler m <Heiz & Kälte> thermostatic valve
Direktablesung f <Maschinen> direct reading
Direktanlassen n <Elektriz> direct starting (Elektromotor)
Direktantrieb m 1. <Elektrotech, Kfztech> direct drive, gearless drive (Getriebe); 2. <Maschinen> direct drive, gearless drive
Direktausgabe f <Elektronik> direct output
Direktbefehl m <Comp & DV> immediate instruction
Direktbelichtung f 1. <Druck> direct imaging; 2. <Elektronik> direct writing
Direktbetrieb m <Elektrotech> direct operation (Magnetschwebetechnik)
Direktdampf m <Hydraul> live steam
Direktdampfinjektor m <Hydraul> live steam injector
Direktdaten npl <Comp & DV> immediate data
Direktdestillat n <Erdöl> straight run product
Direkt-Duplikatfilm m <Foto> direct duplicating film
direkte Adresse f 1. <Comp & DV> immediate address; 2. <Elektronik> direct address (Angabe im Befehlsadressteil der Speicheradresse)
direkte Adressierung f <Comp & DV> direct addressing
direkte Brennstoffzelle f <Kfztech> direct cell
direkte Datei f <Comp & DV> direct file
direkte Dateneingabe f (DDE) <Comp & DV> direct data entry (DDE)
direkte Destillatfraktion f <Erdöl> straight run product
direkte Digitalsteuerung f <Comp & DV> direct digital control
direkte Einwahl f <Telekom> (BE) DDI, (AE) direct dialing-in, (BE) direct dialling-in
direkte Elektronenstrahl-Belichtung f <Elektronik> direct electron beam writing
direkte Energieumwandlung f <Elektrotech> direct energy conversion
direkte Kaltwasserstoffbrennstoffzelle f <Kfztech> direct cold hydrogen cell
direkte Komponente f <Elektriz> direct component
direkte Kupplung f <Nichtfoss Energ> direct coupling
direkte Rundfunkübertragung f über Satellit <Telekom> direct broadcasting by satellite
direkte Strahlung f <Nichtfoss Energ> direct radiation
direkte Überstromauslösung f <Elektriz> direct overcurrent release
direkte Umlaufbahn f <Raumfahrt> direct orbit
direkte Verarbeitung f <Comp & DV> random processing
direkte Verdrahtung f <Elektriz> point-to-point wiring
direkte Wahl f <Telekom> (AE) direct dialing, (BE) direct dialling

direkte Wasserstoff-Sauerstoff-Brennstoffzelle f <Trans> direct hydrogen-oxygen cell
direkte zyklische Verstellung f <Lufttrans> primary cyclic variation (Hubschrauber)
Direkteingabe f <Elektronik> direct input
Direkteinspritzung f <Kfztech, Maschinen> direct injection
direkter Abbruch m <Comp & DV> immediate cancel
direkter AC-Umformer m <Elektriz> direct AC converter
direkter Dateneintrag m (DDE) <Comp & DV> direct data entry (DDE)
direkter Durchbruch m <Kerntech> direct breakthrough
direkter Gang m <Kfztech> direct drive (Getriebe)
direkter Kernphotoeffekt m <Kerntech> direct photonuclear effect
direkter Lichtbogenofen m <Fertig> direct arc furnace
direkter Mustervergleich m <Comp & DV> template matching, templet matching
direkter Piezoeffekt m <Elektrotech> direct piezoelectric effect
direkter Schallpegel m <Aufnahme> direct sound level
direkter Solargewinn m <Nichtfoss Energ> direct solar gain
direkter Speicherzugriff m <Comp & DV> direct memory access
direkter Vielfachzugriff m <Telekom> random multiple access
direkter Zugriff m 1. <Comp & DV> random access; 2. <Comp & DV> direct access (DA); 3. <Telekom> direct access
direktes Auffangen n <Umweltschmutz> direct interception
direktes Licht n <Foto> direct light
direktes Methylalkohol-Luftsauerstoffelement n <Trans> direct methanol air cell
direktes Spinnen n von Glasseidensträngen <Ker & Glas> direct roving
Direktfarbstoff m <Textil> direct dyestuff
Direktflug m <Lufttrans> direct flight
Direktkopplung f 1. <Comp & DV> close coupling; 2. <Elektriz> direct coupling
Direktleitung f <Telekom> direct line
Direktmodulation f <Elektronik> direct modulation
Direktmodus m <Comp & DV> immediate mode
Direktmustererzeugung f <Elektronik> direct pattern generation
Direktor m <Raumfahrt> director (Antennenteil)
Direktregler m <Gerät> primary controller
Direktrix f <Geom> directrix
Direktschreiben n <Elektronik> direct writing (IC-Herstellung durch Direktbelichtung)
Direktsichtbildröhre f <Elektronik> direct view storage tube
Direkt-Speicherzugriff m (DMA) <Comp & DV> direct memory access (DMA)
Direktsteuerung f <Maschinen> DC, direct control
Direktstrom m <Ker & Glas> direct current (der Glasströmung im Ofen)
Direktumwandlung f <Nichtfoss Energ> direct conversion
Direktwahl f <Telekom> one-touch dialling
Direktwahl f aus dem PC <Telekom> direct dialling from the PC
Direktzugriff m <Comp & DV, Elektrotech> direct access, random access
Direktzugriffsdatei f <Comp & DV> random access file
Direktzugriffsspeicher m 1. <Comp & DV> random access storage; 2. <Comp & DV> direct access storage, memory random access (DMA); 3. <Comp & DV, Elekt-

ronik> random access memory *(RAM)*; 4. <Elektrotech> direct access memory

Disassemblierer *m* <Comp & DV> disassembler

Disazofarbstoff *m* <Chemie> bis-azo dye

Dischwefel... <Chemie> *(AE)* pyrosulfuric, *(BE)* pyrosulphuric

Dischwefelsäure *f* <Chemie> *(AE)* hydrosulfurous acid, *(BE)* hydrosulphurous acid

Disilan *n* <Chemie> disilane

Disilicoethan *n* <Chemie> disilane

Disilikat *n* <Chemie> disilicate

disjunkte Mengen *fpl* <Math> disjoint sets

Disjunktion *f* <Comp & DV> disjunction

Disk *f (Diskette)* <Comp & DV> disk

Diskette *f* 1. <Comp & DV> floppy disk, disk, diskette; 2. <Druck> diskette, floppy disk; 3. <Telekom> floppy disk

Diskettenlaufwerk *n* <Comp & DV> disk drive, diskette drive, floppy disk drive

Diskettenlesegerät *n* <Telekom> floppy disk reader

diskontinuierliche Arbeitsweise *f* <Chemtech> batch processing

diskontinuierliche Belastung *f* <Elektriz> intermittent load

diskontinuierlicher Betrieb *m* <Elektriz> intermittent duty

diskontinuierlicher Fehler *m* <Elektriz> intermittent fault

diskontinuierlicher Kocher *m* <Papier> batch digester

diskontinuierlicher Mischer *m* <Chemtech> batch mixer

diskontinuierlicher Pulper *m* <Papier> batch pulper

diskontinuierlicher Speiseeisbereiter *m* <Lebensmittel> batch freezer

Diskordanz *f* <Erdöl> unconformity *(Geologie)*

Diskordanzfalle *f* <Erdöl> unconformity trap *(Geologie)*

diskret *adj* <Comp & DV, Elektronik, Qual, Telekom> discrete

diskrete Fourier-Transformation *f (DFT)* <Elektronik> discrete Fourier transform *(DFT)*

diskrete Fourier-Transform-Technik *f* <Elektronik> discrete Fourier transform

diskrete Zufallsgröße *f* <Math> discrete random variable *(mit diskretem Bereich)*

diskreter Bipolar-Transistor *m* <Elektronik> discrete bipolar transistor

diskreter Kanal *m* <Telekom> discrete channel

diskreter N-Kanal-Feldeffekttransistor *m* <Elektronik> n-channel discrete FET

diskreter Verstärker *m* <Elektronik> discrete amplifier

diskretes Bauelement *n* <Telekom> discrete component

diskretes Filter *m* <Elektronik> discrete filter

diskretes Halbleiterbauelement *n* <Elektronik> discrete semiconductor device

diskretes Merkmal *n* <Qual> discrete characteristic

Diskretisierung *f* <Math> discretization, truncation

Diskriminator *m* 1. <Elektronik> discriminator *(Entscheider)*; 2. <Fernseh> discriminator; 3. <Funktech> discriminator *(Schaltung, Schaltkreis)*; 4. <Telekom> discriminator

Diskriminierungsvermögen *n* <Ergon> sensory discrimination

Diskussionsgruppe *f* <Comp & DV> discussion group

disparat *adj* <Comp & DV> disparate

Dispatcher *m* <Comp & DV, Kfztech> dispatcher *(Schmieranlage, Öl)*

Dispergator *m* 1. <Kunststoff> deflocculating agent; 2. <Meerschmutz> dispersant, dispersing agent

Dispergens *n* 1. <Chemie> deflocculant; 2. <Chemtech> dispersive medium; 3. <Erdöl> dispersant; 4. <Meerschmutz, Umweltschmutz> dispersant, dispersing agent

dispergieren *v* <Chemie> deflocculate

Dispergiermittel *n* 1. <Chemie> deflocculant; 2. <Chemtech> dispersive medium; 3. <Kunststoff> deflocculating agent, dispersant, dispersing agent; 4. <Lebensmittel> dispersing agent

dispergierter Brennstoff *m* <Kerntech> dispersion fuel

Dispergierung *f* <Chemie> deflocculation

Dispergierungsmittel *n* <Kohlen> dispersing agent

Dispersant *n* 1. <Fertig> dispersant; 2. <Meerschmutz, Umweltschmutz> dispersant, dispersing agent

Dispersantbehälter *m* <Meerschmutz> bucket *(unter einem Helikopter befestigter Behälter zum Ausbringen von Dispersionsmitteln)*

disperse Phase *f* <Umweltschmutz> disperse phase

Dispersion *f* 1. <Anstrich> dispersion; 2. <Chemtech> dispersion, dispersivity; 3. <Comp & DV, Elektriz, Kohlen, Kunststoff, Phys, Telekom, Wellphys> dispersion

Dispersion *f* **von Farben** <Phys, Wellphys> *(AE)* dispersion of colors, *(BE)* dispersion of colours *(aufgrund von Brechung)*

dispersionsbegrenzt *adj* <Telekom> dispersion-limited *(Faseroptik)*

Dispersionseinrichtung *f* <Meerschmutz> dispersing equipment *(zur Bekämpfung von Ölverschmutzungen)*

Dispersionsfarbe *f* <Bau> water-based paint

Dispersionsfarbstoff *m* <Textil> disperse dyestuff

dispersionsgekühlter Reaktor *m* <Kerntech> dispersion-cooled reactor

Dispersionsgleichung *f* <Phys, Wellphys> dispersion equation

Dispersionshärten *n* <Metall> precipitation hardening

Dispersionshärtung *f* <Metall> dispersion hardening

Dispersionskleber *m* <Fertig> dispersion adhesive

Dispersionskneter *m* <Chemtech> dispersion kneader

Dispersionskompensator *m* <Telekom> dispersion compensation module *(Faseroptik)*

Dispersionsmedium *n* <Phys> dispersion medium

Dispersionsmittel *n* 1. <Kerntech> dispersion agent; 2. <Kunststoff, Meerschmutz> dispersant, dispersing agent

Dispersionsrelation *f* <Wellphys> dispersion relation

Dispersionsteilchen *n* <Chemtech> particulate *(Schmutzteilchen)*

dispersionsverschoben *adj* <Telekom> dispersion-shifted *(Faseroptik)*

Dispersoid *n* <Chemtech> dispersoid

Display *n* 1. <Comp & DV> display, display unit, screen; 2. <Druck, Verpack> display

Displayschachtel *f* <Verpack> presentation box

Dispositionsgleis *n* <Eisenbahn> relief track

dissipatives Medium *n* <Phys> dissipative medium

Dissonanz *f* <Akustik> discordance, dissonance

Dissousgasflasche *f* <Fertig> acetylene cylinder

Dissoziation *f* <Kohlen> dissociation

Dissoziationswärme *f* <Thermod> heat of dissociation

dissoziierbar *adj* <Chemie> dissociable

dissoziieren *v* <Chemie> dissociate

Dissoziierung *f* <Chemie> dissociation

Dissymmetrie *f* <Chemie, Geom> dissymmetry

dissymmetrisch *adj* <Geom> dissymmetric, dissymmetrical

distal *adj* <Ergon> distal

Distanz *f* 1. <Fertig> distance; 2. <Maschinen> spacer; 3. <Telekom> distance

Distanzadresse *f* <Comp & DV> displacement address

Distanzblech *n* <Erdöl> stabilizer

Distanzblock *m* <Maschinen> spacer block

Distanzbuchse *f* <Fertig> distance bush, distance piece *(Kunststoffinstallationen)*

Distanzscheibe *f* <Maschinen> shim

Distickoxid *n* <Chemie> nitrous oxide

Distickstoffmonoxid

Distickstoffmonoxid n <Chemie> dinitrogen monoxide
Distickstoffoxid n <Umweltschmutz> nitrous oxide
Distributivgesetz n <Math> distributive law
Disulfat n <Chemie> (AE) disulfate, (BE) disulphate, (AE) pyrosulfate, (BE) pyrosulphate
Disulfid n <Chemie> (AE) disulfide, (BE) disulphide
Disulfit n <Chemie> (AE) pyrosulfite, (BE) pyrosulphite
Disulfuryl... <Chemie> (AE) pyrosulfuryl, (BE) pyrosulphuryl
Dithion... <Chemie> dithionic
Dithionat n <Chemie> dithionate
dithionige Säure f <Chemie> dithionous acid, (AE) tetraoxodisulfuric acid, (BE) tetraoxodisulphuric acid
Dithionit n <Chemie> dithionite, (AE) hyposulfite, (BE) hyposulphite
ditonisches Komma n <Akustik> Didyme comma
diurnal adj <Ergon> diurnal
divergent adj <Geom, Optik> divergent
divergente Düse f <Phys> divergent nozzle (im Windkanal)
divergente Reihe f <Math> divergent series
Divergenz f <Phys> divergence (eines Vektorfeldes)
Divergenzwinkel m <Fertig> divergence angle, exit angle
divergierende Reihe f <Math> divergent series
Diversity n <Funktech, Telekom> diversity
Diversityaustausch m <Kohlen> diversity exchange
Diversityfaktor m <Kohlen> diversity factor
DIVF (digitale Vermittlungsstelle für den Fernverkehr) <Telekom> digital trunk exchange
Dividend m <Comp & DV, Math> dividend
dividieren v 1. <Geom> divide; 2. <Math> divide
Divinyl n <Chemie> bivinyl, divinyl
Division f <Comp & DV, Math> division (Grundrechenart)
Divisor m <Comp & DV, Math> divisor
DIVO (digitale Vermittlungsstelle für den Ortsverkehr) <Telekom> digital local exchange
D-Kanal m <Telekom> D-channel
dl (Deziliter) <Labor> dl (deciliter)
DM (Deltamodulation) <Elektronik, Raumfahrt, Telekom> DM (delta modulation)
DMA (Direkt-Speicherzugriff, Direktzugriffsspeicher) <Comp & DV> DMA (direct memory access)
DMC (kittartige Formmasse) <Kunststoff> DMC (dough-moulding compound)
DM-Vierer m <Elektrotech> multiple-twin quad
D-Netz n <Telekom> D-net
D-Netz, E-Netz n <Telekom> digital cellular mobile radio system, digital cellular radio system, DCS (Mobilfunknetz nach GSM-Standard)
Dock n <Wassertrans> dock
Dockadaptor m <Raumfahrt> docking adaptor
Docke f <Textil> batch (Warendocke)
Dockgeld n <Wassertrans> dockage
Docking n <Raumfahrt> docking
Docking-Fenster n <Raumfahrt> docking port
Docking-Stutzen m <Raumfahrt> docking probe
Docking-Tunnel m <Raumfahrt> docking tunnel
Dockmöglichkeiten fpl <Wassertrans> dockage
Dockstück n <Raumfahrt> docking piece
Dodecan n <Chemie> bihexyl, dihexyl, dodecane
Dodecanoyl... <Chemie> lauryl
Dodecawolframatophosphat n <Chemie> phosphatododecatungstate, phosphotungstate
Dodecyl... <Chemie> dodecyl
Dodekaeder n <Geom> dodecaeder, dodecahedron (Zwölfflächner)
Dodekagon n <Geom> dodecagon (Zwölfeck)
DOHC-Motor m (Querstromkopfmotor) <Kfztech> DOHC engine (direct-acting overhead camshaft engine)

doktornegativ adj <Chemie> sweet (Erdöl)
Dokument n <Comp & DV> document, text file
Dokument n **mit langer Zykluszeit** <Comp & DV> long life-cycle document
Dokumentation f <Comp & DV, Patent, Qual> documentation
Dokumentation f **der Zusatzeinrichtungen** <Comp & DV> feature documentation
Dokumentenabruf m <Comp & DV> document retrieval
Dokumenten-Archivierungssystem n **mit CD** <Comp & DV, Optik> optical data disk document filing system
Dokumentenaustauschprotokoll n <Telekom> document interchange protocol
Dokumentenbaum m <Comp & DV> document tree
Dokumentenfilm m <Foto> document film
Dokumentenglas n <Ker & Glas> document glass
Dokumentlayout n <Comp & DV> document layout
Dollbord n <Wassertrans> gunnel, gunwale (Schiffbau)
Dolle f <Wassertrans> (BE) rowlock; (AE) oarlock (Bootszubehör)
Dolomit m 1. <Erdöl> dolomite (Geologie); 2. <Ker & Glas> dolomite
Dolomitstein m <Bau> dolomite brick
Domäne f 1. <Künstl Int> domain (des Wissens); 2. <Telekom> domain (Internet)
Domänenname m <Comp & DV> domain name
Domänenname m **für kundenspezifischen Postbereich** <Comp & DV> custom mail domain name
Domänenname m **für Postbereich** <Comp & DV> mail domain name
Domänenstruktur f <Phys> domain structure
Domänenwissen n <Künstl Int> domain knowledge
Dominante f <Akustik> dominant
dominantes Anion n <Umweltschmutz> dominant anion
dominantes Kation n <Umweltschmutz> dominant cation
Donator m 1. <Comp & DV> donor; 2. <Elektronik> donator, donor (elektronenabgebendes Atom oder Störstelle)
Donatoratom n <Elektronik, Phys> donor atom
Donatorniveau n <Elektronik> donor level
Donatorverunreinigung f <Elektronik> donor impurity
DOP (Dioctylphthalat) <Kunststoff> DOP (dioctylphthalate)
Dopamin n <Chemie> dopamine
Dopans n <Optik> dopant
Doppel... 1. <Akustik, Druck> double; 2. <Elektrotech> twin; 3. <Kfztech> tandem; 4. <Lufttrans, Verpack> twin
Doppelabsacksystem n <Verpack> twin bagging system
Doppelabzweigdose f <Elektrotech> biforcating box
Doppelachse f 1. <Kfztech> tandem axle (LKW); 2. <Maschinen> double axle
Doppelader f <Telekom> pair (Telefon)
doppeladrig adj <Funktech> bifilar
doppeladriges Kabel n <Elektriz> double-core cable
Doppelankerrelais n <Elektriz> double-armature relay
Doppelanzeige f <Fertig> dual indicator
doppelatomig adj <Chemie> biatomic, diatomic
Doppel-b n <Akustik> double flat
Doppelbackenbremse f <Eisenbahn> (AE) clasp brake
Doppelbandpolieren n <Ker & Glas> twin polishing
Doppelbandpoliermaschine f <Ker & Glas> twin polisher
Doppelbandschleifmaschine f <Ker & Glas> twin grinder
doppelbasisch adj <Chemie> bibasic, dibasic
Doppelbasisdiode f <Elektronik> double base diode
Doppelbelegung f <Telekom> double seizure
Doppelbelichtung f <Foto> double exposure
Doppelbeschichtung f <Telekom> double layer coating

Doppelboden *m* 1. <Bau> double floor, raised floor; 2. <Heiz & Kälte> raised floor; 3. <Wassertrans> double bottom *(Schiffbau)*
Doppelbohrung *f* <Maschinen> twin bore
doppelbrechend *adj* <Optik> birefringent
doppelbrechendes Medium *n* <Optik> birefringent medium
Doppelbrechung *f* <Optik, Strahlphys> birefringence
Doppelbruch *m* <Math> complex fraction
Doppelbuchstabe *m* <Druck> double letter
Doppelbus *m* **mit verteilten Warteschlangen** <Telekom> distributed queue dual bus *(DQDB)*
Doppelchlorid *n* <Chemie> bichloride, dichloride
Doppeldeckpalette *f* 1. <Trans> reversible pallet; 2. <Verpack> double-decked pallet
Doppeldeltaflügel *m* <Lufttrans> double delta wing
Doppeldiodenpentode *f (DDP)* <Elektronik> double diode pentode *(DDP)*
Doppeldrahtschaltung *f* <Elektriz> two-wire circuit
Doppeldruckregler *m* <Heiz & Kälte> dual pressure controller
Doppelelementrelais *n* <Elektriz> two-element relay
Doppelfadenaufhängung *f* <Elektriz, Phys> bifilar suspension
doppelfädig *adj* <Elektriz, Phys> bifilar
doppelfädiges Elektrometer *n* <Elektriz> bifilar electrometer
Doppelfarbigkeit *f* <Chemie> dicroism
Doppelfensterfaser *f* <Optik> *(AE)* double-window fiber, *(BE)* double-window fibre
Doppelflugzeug *n* <Lufttrans> composite aircraft
Doppelflüssigtreibstoffantrieb *m* <Raumfahrt> liquid bipropellant propulsion *(Raumschiff)*
Doppelformatkamera *f* <Foto> dual-format camera
Doppelfußboden *m* <Bau> framed floor
doppelgängiges Gewinde *n* <Maschinen> two-start thread
Doppelgarn *n* <Textil> two-ply yarn
Doppelgehäuse *n* <Maschinen> double casing
Doppelgelenk *n* 1. <Kfztech> constant-velocity universal joint *(Antriebswelle, Kardanwelle)*; 2. <Maschinen> double joint
doppelgewickelter Generator *m* <Elektriz> double-wound generator
Doppelgewindeschraube *f* <Bau> double-threaded screw
Doppelhaken *m* <Maschinen> clip hooks
Doppelhebel *m* <Fertig> double lever handle *(Kunststoffinstallationen)*
Doppelherzstück *n* <Eisenbahn> diamond crossing, double diamond crossing
Doppelhiebfeile *f* <Fertig> double-cut file
doppelhiebige Feile *f* <Maschinen> double-cut file
doppelhörig *adj* <Akustik> binaural
Doppelhub *m* 1. <Fertig> cycle, reciprocation; 2. <Maschinen> double stroke
Doppelhülle *f* <Wassertrans> double hull *(Schiffbau, U-Boote)*
Doppelimpuls *m* <Elektriz> double impulse, double pulse
Doppelisolator *m* <Elektrotech> double insulator
Doppelisolierung *f* <Elektrotech> double insulation
Doppel-I-Trägeraufhängung *f* <Kfztech> twin I-beam suspension
Doppeljersey *m* <Textil> double jersey
Doppelkabel *n* <Elektriz> duplex cable, foto double cable
Doppelkabelauslöser *m* <Foto> double cable release
Doppelkäfigmotor *m* <Elektriz> double-squirrel cage motor

Doppelkäfigwicklung *f* <Elektrotech> double-squirrel cage winding
Doppelkammlinearmotor *m* <Elektrotech> double-sided linear motor
Doppelkamm-Magnetron *n* <Elektronik, Phys> interdigital magnetron
Doppelkapselmikrofon *n* <Akustik> differential microphone
Doppelkassette *f* <Foto> twin magazine
Doppelkegelhälfte *f* <Geom> nappe
Doppelkeilriemen *m* <Maschinen> double-V belt
Doppelkettenstichnaht *f* <Textil> two-thread chainstitch seam
Doppelkettfäden *mpl* <Textil> twin ends
Doppelklicken *n* <Comp & DV> double click *(mit Maus)*
Doppelklinke *f* <Maschinen> double ratchet
Doppelkolben *m* <Kfztech, Maschinen> double piston, twin piston
Doppelkolbenkompressor *m* <Hydraul> duplex compressor
Doppelkolbenmotor *m* 1. <Kfztech> twin-piston engine; 2. <Maschinen> double-piston engine
Doppelkolbenpumpe *f* <Hydraul> duplex pump
Doppelkondensator-Motor *m* <Elektriz> dual capacitor motor
Doppelkontaktrelais *n* <Elektriz> two-element relay
Doppelkonusantenne *f* <Telekom> biconical antenna
Doppelkonuslautsprecher *m* <Aufnahme> dual-cone loudspeaker
Doppelkopfniete *f* <Bau> bullhead rivet
Doppelkopfschiene *f* <Eisenbahn> bull-headed rail, double-headed rail
Doppelkörpervergaser *m* <Kfztech> *(AE)* twin-barreled carburetor, *(BE)* twin-barrelled carburettor
Doppelkreisdiagramm *n* <Funktech> figure-of-eight diagram
Doppelkreuz *n* 1. <Akustik> double sharp; 2. <Optik> double crucible
Doppelkreuzgelenk *n* <Maschinen> double universal joint
Doppelkreuzmethode *f* <Optik> double crucible method
Doppelkreuzungsweiche *f* <Eisenbahn> double slip
Doppelkreuzverfahren *n* <Optik> double crucible technique
Doppelkristall *m* <Metall> twin
Doppelkurbel *f* <Maschinen> double crank, duplex crank
Doppelkurbelpresse *f* <Maschinen> double-crank press
Doppellaschennietung *f* <Maschinen> double-strap butt joint
Doppellattenkiste *f* <Verpack> double-battened case
Doppelleinen *n* <Druck> buckram
Doppelleitung *f* <Telekom> pair *(Telefon)*
Doppelleitwerk *n* <Lufttrans> twin-tail unit
Doppellinie *f* 1. <Druck> double rule; 2. <Optik> doublet *(Spektralanalyse)*
Doppellinse *f* <Ker & Glas> twins
Doppelmast *m* <Wassertrans> derrick mast
doppelmäulig *adj* <Fertig> double-end
Doppelmesserschalter *m* <Elektriz> double-throw knife switch
Doppelmesserschneider *m* <Papier> dual knife cutter
Doppelmesserumschalter *m* <Elektriz> double-pole double-throw knife switch
Doppelmikrofon *n* <Aufnahme> double-button microphone
Doppelmodulation *f* 1. <Elektronik> compound modulation, double modulation; 2. <Funktech> double modulation
Doppelmotor *m* <Kontroll> tandem motor

Doppelmuffe

Doppelmuffe f <Labor> bosshead *(Feststellvorrichtung)*
doppeln v <Wassertrans> double
Doppelnetzspannung f <Elektrotech> dual supply voltage
Doppelobjektiv n <Foto> doubled lens, doublet lens
Doppel-Öffner-Kontakt m <Elektriz> break-break contact
Doppelöffnerschalter m <Elektriz> double break switch
Doppelpackung f <Verpack> twin pack
Doppelpeilung f <Wassertrans> running fix *(Navigation)*
doppelphasig adj <Elektriz> biphase
Doppelplattformpalette f <Verpack> double-platform pallet
doppelpolige Elektrode f <Elektriz> bipolar electrode
doppelpolige Leitung f <Elektriz> bipolar line
Doppelpoller m <Wassertrans> bitts
Doppelpolwicklung f <Elektriz> bipolar winding
Doppelprisma n <Optik> biprism
Doppelprofilglas n <Ker & Glas> double-bended glass
Doppelquerlenkeraufhängung f <Kfztech> double-wishbone suspension
Doppelquerlenkervorderachse f <Kfztech> double-wishbone
Doppelräder fpl <Lufttrans> dual wheel
Doppelrampe f <Lufttrans> dual platform
Doppelreflektorantenne f <Telekom> double reflector antenna
Doppelregistervergaser m <Kfztech> *(AE)* four-barrel carburetor, *(BE)* four-barrel carburettor, *(AE)* quad carburetor, *(BE)* quad carburettor
Doppelringschlüssel m <Maschinen> *(BE)* double open-ended spanner, double open-ended wrench, *(BE)* double-ended box spanner, double-ended box wrench, *(BE)* double-ended open-jaw spanner, double-ended open-jaw wrench, *(BE)* double-ended ring spanner, double-ended ring wrench, *(BE)* double-ended spanner, double-ended wrench
Doppelrumpfboot n <Wassertrans> catamaran *(Boottyp)*
Doppelschaufelabscheider m <Umweltschmutz> double bucket collector
doppelschichtige Wellfaserplatte f <Verpack> *(AE)* double-wall corrugated fiberboard, *(BE)* double-wall corrugated fibreboard
Doppelschichtwicklung f <Elektriz> double layer winding
Doppelschlichtfeile f <Fertig> dead smooth cut file
Doppelschlussmaschine f <Elektrotech> compound-characteristic machine
Doppelschlussmotor m <Elektrotech> compound motor
Doppelschlusswicklung f <Elektrotech> compound winding
Doppelschneckenextruder m <Abfall, Kunststoff> twin-screw extruder
Doppelschrauben fpl <Wassertrans> twin propellers *(Schiffantrieb)*
Doppelschraubendampfer m <Wassertrans> twin-screw steamer
Doppelschwinge f <Maschinen> double rocker
Doppelseitenband n *(DSB)* <Elektronik, Funktech> double sideband *(DSB)*
doppelseitig adj 1. <Bau> double-faced; 2. <Druck> back-to-back; 3. <Elektronik> double-sided; 4. <Optik> two-sided; 5. <Telekom> double-sided; 6. <Verpack> double-faced, double-sided
doppelseitig kaschierte Leiterplatte f <Telekom> double-sided printed circuit
doppelseitig klebendes Band n <Verpack> double-sided tape
doppelseitig platiert adj <Fertig> duo-clad
doppelseitige Scheibe f <Optik> *(BE)* two-sided disc, *(AE)* two-sided disk
doppelseitige Wellpappe f <Verpack> double-faced corrugated board
doppelseitiger Amplitudenbegrenzer m <Elektronik> amplitude gate
doppelseitiger Verteiler m <Telekom> double-sided distribution frame
doppelseitiges Bedrucken n <Druck> back-to-back printing
doppelseitiges Krepppapier n <Papier, Verpack> double-faced crepe paper
doppelseitiges Trägermaterial n <Elektronik> double-sided substrate *(für Leiterplatten)*
doppelseitiges Wachspapier n <Papier, Verpack> double-faced wax paper
Doppelsilicat n <Chemie> bisilicate
Doppelsitzventil n <Maschinen> double-seat valve
Doppelspalt-Löschkopf m <Aufnahme> double-gap erase head
Doppelspielband n <Aufnahme> double-play tape
Doppelspinsatellit m <Raumfahrt> dual spin satellite *(Raumschiff)*
Doppelspinstabilisierung f <Raumfahrt> dual spin stabilization *(Raumschiff)*
Doppelspitzhacke f <Kohlen> coal pick
Doppelspur f 1. <Akustik> dual track; 2. <Aufnahme> twin track
Doppelspur-Aufnahmegerät n <Aufnahme> twin-track recorder
Doppelstatoranordnung f <Elektrotech> double stator device *(Linearmotor)*
Doppelstatordrehkondensator m <Elektrotech> split stator variable capacitor
Doppelstatorlinearmotor m <Elektrotech> double-sided linear motor
Doppelstecker m <Fertig> biplug
Doppelsteghohlleiter m <Elektrotech> double ridge waveguide
Doppelsteuerschulung f <Lufttrans> dual instruction
Doppelsteuerstange f <Lufttrans> dual rod
Doppelstichprobenentnahme f <Qual> double sampling
Doppelstichprobenprüfplan m <Qual> double-sampling plan
Doppelstift-Steckverbindung f <Elektriz> two-contact connector
Doppelstrahl-Katodenstrahlröhre f <Elektronik> dual-beam cathode-ray tube
Doppelstrangpolymer n <Kunststoff> double-strand polymer, ladder polymer
Doppelstrich m <Druck> double rule
Doppelstrombetrieb m <Telekom> double current operation
Doppelstromrichter m <Elektrotech> double converter
Doppelstromtor-Strahlsteuerungsröhre f <Fernseh> gated beam tube
Doppelstromversorgung f <Elektrotech> dual power supply
doppelt gekröpfte Kurbelwelle f <Maschinen> two-throw crankshaft
doppelt gerichtete Leitung f <Telekom> both-way circuit, both-way line
doppelt gerichtetes Bündel n <Telekom> both-way group
doppelt gespeister Motor m <Elektriz> double-fed motor
doppelt kautschukbedeckt adj <Elektriz> double-pure-rubber-covered
doppelt wirkend adj 1. <Fertig> double-acting *(Kunststoffinstallationen)*; 2. <Maschinen> double-acting

doppelt wirkende Presse f <Maschinen> double-acting press
doppelt wirkende Pumpe f <Maschinen> double-acting pump
doppelt wirkender Kompressor m <Heiz & Kälte> double-acting compressor
doppelt wirkender Servomotor m <Nichtfoss Energ> double-acting servomotor
doppelt wirkender Stoßdämpfer m <Kfztech> double-acting shock absorber
doppelt wirkender Verdichter m <Heiz & Kälte> double-acting compressor
doppelt wirkender Zylinder m <Maschinen> double-acting cylinder
Doppel-T-Anker m <Elektriz> shuttle armature
doppelte bituminöse Oberflächenbehandlung f <Bau> double bituminous surface treatment
doppelte D-Schaltung f <Elektriz> double delta connection
doppelte Flachspule f <Elektriz> (BE) double disc winding, (AE) double disk winding
doppelte Gleisverbindung f <Eisenbahn> double crossover
doppelte Kreuzungsweiche f <Bau> double crossover
doppelte kreuzweise Bewehrung f <Bau> four-way reinforcement
doppelte Linie f <Druck> double rule
doppelte oben liegende Nockenwelle f <Kfztech> double overhead camshaft (Motor)
doppelte Pufferung f <Comp & DV> double buffering
doppelte Schreibdichte f <Comp & DV> DD, double density
doppelte Speicherkapazität f <Comp & DV> double density
doppelte Symmetrie f <Geom> twofold symmetry
Doppel-T-Eisen n <Bau> H-iron
doppelter Anschnitt m <Ker & Glas> double bevel
doppelter Erdfehler m <Elektriz> (BE) double-earth fault, (AE) double-ground fault
doppelter Erdschluss m <Elektriz> (BE) double-earth fault, (AE) double-ground fault
doppelter Kameraauszug m <Foto> double camera extension
doppelter Netzanschluss m <Elektriz> duplicate supply
doppelter oben liegender Nockenwellenmotor m <Kfztech> dual overhead cam engine
doppelter Scheitelwert m <Akustik, Phys> peak-to-peak value
doppelter Sicherheitsschutz m <Comp & DV> double protection security
doppeltes Tandemradfahrgestell n <Lufttrans> dual tandem wheel undercarriage
Doppeltiegelverfahren n <Telekom> double crucible technique
Doppeltischmaschine f <Ker & Glas> two-table machine
Doppel-T-Netz n <Elektrotech> twin-T network
Doppeltreibstoffdruckmesser m <Lufttrans> (AE) dual-fuel pressure gage, (BE) dual-fuel pressure gauge
Doppeltrommelkessel m <Heiz & Kälte> bi-drum boiler
Doppel-T-Träger m 1. <Bau> H-beam, H-girder, I-beam; 2. <Fertig> I-beam; 3. <Metall> I beam
Doppelüberlagerung f <Elektronik, Funktech> double conversion
Doppelüberlagerungsempfänger m <Funktech> double conversion receiver
Doppelumschalter m <Phys> double-throw switch
Doppelunterbrecher m <Kfztech> dual point breaker
Doppelvergaser m 1. <Kfztech> (AE) dual carburetor, (BE) dual carburettor, (AE) twin carburetor, (BE) twin carburettor, (AE) twin-choke carburetor, (BE) twin-choke carburettor; 2. <Mechan> (AE) dual carburetor, (BE) dual carburettor
Doppelverglasung f <Bau, Heiz & Kälte> double glazing
Doppelverglasungseinheit f <Ker & Glas> double glazing unit
Doppelverschlüsselung f <Telekom> concatenated encryption, double encryption
Doppelvoltmeter n <Elektriz, Metrol> double-range voltmeter, double voltmeter
doppelwandig adj 1. <Heiz & Kälte> double-skin (Rolle); 2. <Wassertrans> double-skin (Schiffkonstruktion)
Doppelwattmeter n <Elektriz, Metrol> double wattmeter
Doppelweggleichrichten n 1. <Elektriz> bridge rectification; 2. <Elektrotech> full-wave rectification
Doppelweggleichrichter m 1. <Elektriz> bridge rectifier; 2. <Elektrotech> full-wave rectifier
Doppelwegspeisung f <Elektriz> two-way feed
Doppelwendel f 1. <Elektriz> coiled coil; 2. <Maschinen> double helix
Doppelwendelfaden m <Elektriz> coiled coil filament
Doppelwendelglühfaden m <Elektriz> coiled coil filament
Doppelwendellampe f <Elektriz> coiled coil lamp
Doppelwicklungsanker m <Elektriz> double-wound armature
Doppelwicklungstransformator m <Elektriz> double-wound transformer
Doppelzeilensprungabtastung f <Fernseh> twin-interlaced scanning
Doppelzellenkondensator m <Elektriz> two-cell capacitor
Doppelzirkel m <Bau> (AE) double calipers, (BE) double callipers
Döpper m 1. <Bau> rivet set, rivet snap, riveting set; 2. <Fertig> holding-up snap, snap; 3. <Maschinen> (AE) header, snap die
Doppler m <Comp & DV> duplicator, duplicating punch
Doppler-Bandbreite f <Elektronik> Doppler bandwidth
Doppler-Breite f <Strahlphys> Doppler width (im optischen Frequenzbereich)
Doppler-Effekt m <Akustik, Elektronik, Funktech, Wellphys> Doppler effect
Doppler-Filter n <Elektronik> Doppler filter
Doppler-Filterung f <Elektronik> Doppler filtering
Doppler-Frequenz f <Elektronik, Funktech> Doppler frequency
Doppler-Modulation f <Elektronik> Doppler modulation
Doppler-Navigation f 1. <Funkort, Lufttrans> Doppler navigation; 2. <Raumfahrt> Doppler navigation (Raumschiff)
Doppler-Navigationssystem n <Elektronik, Funktech, Raumfahrt> Doppler navigation system (Raumschiff)
Doppler-Peiler m 1. <Funkort> Doppler direction finder; 2. <Funktech> commutated-antenna direction finder (CADF); Doppler direction finder
Doppler-Radar n 1. <Funkort> continuous wave radar, cw-radar, Doppler radar (Verkehrsradar); 2. <Wassertrans> Doppler radar
Doppler'sche Frequenz f <Elektronik> Doppler frequency
Doppler-Trägheitsnavigation f <Raumfahrt> Doppler inertial navigation (Raumschiff)
Doppler-Verbreiterung f <Strahlphys> Doppler broadening
Doppler-Verschiebung f <Aufnahme, Funktech, Raumfahrt> Doppler shift
Dorn m 1. <Bau> bolt, pin drift, spur; broach (eines Schlosses); 2. <Fertig> core bar, pritchel, tongue; 3. <Ker & Glas> spike; 4. <Kunststoff> mandrel, mandril; 5. <Ma-

Dorndurchmesser

schinen> *(AE)* arbor, *(BE)* arbour, gudgeon, mandrel, mandril
Dorndurchmesser *m* <Fertig> arbor diameter
Dornpresse *f* <Maschinen> arbor press, mandrel press, mandril press
Dornschaft *m* <Fertig> arbor shank
Dornschraubzwinge *f* <Fertig> arbor clamp
Dornstange *f* <Fertig> mandrel supporting rod, mandril supporting rod *(Bending)*; bar *(Stopfenzug)*
Dorntraglager *n* <Fertig> arbor bracket
DOS®-Betriebssystem *n* <Comp & DV> DOS®, disk operating system
Dose *f* <Verpack> *(AE)* can, *(BE)* tin • **in Dosen konserviert** <Lebensmittel> *(AE)* canned, *(BE)* tinned
Dosenabfüllautomat *m* <Verpack> *(AE)* can filling machine, *(BE)* tin filling machine
Dosenabfüllinie *f* <Verpack> *(AE)* can filling line, *(BE)* tin filling line
Dosenbarometer *n* <Labor, Lufttrans> aneroid barometer
Dosendichtungsmasse *f* <Verpack> *(AE)* can sealing compound, *(BE)* tin sealing compound
Dosenhüllenentfernung *f* <Verpack> *(AE)* can delabeling, *(BE)* tin delabelling
Dosenkontakt *m* <Elektrotech> female contact
Dosenneuetikettierung *f* <Verpack> *(AE)* can relabeling, *(BE)* tin relabelling
Dosenring *m* <Verpack> pull ring
Dosenverpackungsmaschine *f* <Verpack> *(AE)* can packing machine, *(BE)* tin packing machine
Dosenverschließmaschine *f* <Verpack> *(AE)* can closing machine, *(BE)* tin closing machine
Dosier... 1. <Gerät> metering, proportioning; 2. <Verpack> dosing
Dosieranlage *f* <Bau> batch plant
Dosierautomat *m* <Gerät> automatic proportioner
Dosierbandwaage *f* <Gerät> metering conveyor balance
Dosiereinrichtung *f* <Gerät> proportioning device
dosieren *v* 1. <Bau> batch, meter; 2. <Metrol> measure out
Dosierer *m* <Verpack> dosing feeder
Dosiergerät *n* <Gerät> proportioning device
Dosierkugelhahn *m* <Fertig> metering ball valve *(Kunststoffinstallationen)*
Dosiermaschine *f* <Verpack> dosing machine
Dosierpumpe *f* 1. <Gerät> metering pump; 2. <Labor> metering pump *(Weiterleitung von Flüssigkeiten)*; 3. <Maschinen> metering pump; 4. <Verpack> dosing pump
Dosiertank *m* <Kerntech> batching tank
Dosierung *f* 1. <Fertig> proportioning *(Flüssigkeiten)*; 2. <Kohlen> dosage; 3. <Maschinen> metering
Dosierungsöffnung *f* <Maschinen> metering hole
Dosierventil *n* 1. <Kfztech> metering valve, proportioning valve; 2. <Maschinen> metering valve
Dosiervorrichtung *f* 1. <Fertig> batcher; 2. <Verpack> dosing apparatus, metering equipment
Dosierwaage *f* <Gerät> gravimetric meter, metering balance
Dosierzähler *m* <Gerät> batching counter
Dosimeter *n* 1. <Ergon> dosimeter; 2. <Gerät> radiation dosimeter; 3. <Strahlphys> dosimeter
Dosimeterglas *n* <Ker & Glas> dosimeter glass
Dosimetrie *f* <Phys, Strahlphys> dosimetry
Dosimetrie f bei hohem Strahlungspegel <Strahlphys> high-level dosimetry
Dosis *f* 1. <Kerntech> dose rate; 2. <Strahlphys> dose
Dosisgrenzwert *m* **für berufliche Bestrahlung** <Umweltschutz> occupational dose limit
Dosisleistung *f* <Strahlphys> R, dose rate
Dosisleistungseffekt *m* <Strahlphys> dose rate effect

Dosismesser *m* <Gerät> radiation dosimeter
Dosismessgerät *n* <Gerät, Phys> dosemeter
Dosisrate *f* <Strahlphys> R, dose rate
Dosiswirkung *f* <Umweltschutz> dose response
Dosis-Wirkungsbeziehung *f* <Umweltschutz> dose response relationship
Dotier... 1. <Elektronik> implant; 2. <Phys> dope
Dotierdosis *f* <Elektronik> implant dose
Dotierelement *n* <Phys> dopant
dotieren *v* 1. <Elektronik> implant; 2. <Optik, Phys> dope
Dotierstoff *m* <Phys> dopant
dotiert *adj* <Comp & DV, Elektronik, Optik, Phys> doped
dotierte Silicafaser *f* <Optik> *(AE)* doped silica fiber, *(BE)* doped silica fibre
dotierter Halbleiter *m* <Elektronik, Phys> doped semiconductor
Dotierung *f* 1. <Elektronik> implantation; doping *(Verunreinigung zur Veränderung von Halbleitereigenschaften)*; 2. <Phys> doping *(von Halbleitern mit Fremdelementen)*
Dotierungsausgleich *m* <Elektronik> doping compensation
Dotierungskonzentration *f* <Fertig> dopant concentration *(Faseroptik)*
Dotierungsmaterial *n* <Elektronik> dopant, doping agent
Dotierungsniveau *n* <Elektronik> doping level
Dotierungsprofil *n* <Elektronik> doping profile, impurity concentration profile
Dotierungsstoff *m* <Comp & DV> dopant
Doublet *n* <Foto> doubled lens, doublet
Dove'sches Umkehrprisma *n* <Optik> Dove prism
Dozer m mit neigbarem Schild <Trans> tilting dozer
DP *(Diametral-Pitch)* <Maschinen> DP *(diametral pitch)*
DPCM *(Differenz-Pulscodemodulation)* <Elektronik, Telekom> DPCM *(differential pulse code modulation)*
DPDT *(zweipoliger Umschalter, zweipoliger Wechselschalter)* <Elektrotech> DPDT *(double-pole double-throw)*
DPSK *(Phasendifferenzmodulation, Phasendifferenzumtastung)* <Elektronik, Telekom> DPSK *(differential phase shift keying)*
DPST *(zweipoliger Ein/Aus-Schalter)* <Elektrotech> DPST *(double-pole single-throw)*
dpt *(Dioptrie)* <Optik> dpt *(dioptre)*
Drachenviereck *n* <Geom> kite
Dragganker *m* <Wassertrans> grapnel *(Festmachen)*
Draggen *m* <Wassertrans> grapnel *(Festmachen)*
dragiert *adj* <Lebensmittel> panned *(Süßwaren)*
Draht *m* 1. <Elektrotech> wire; 2. <Fertig> *(AE)* rod *(Schweißen)*
Draht m mit Flussmittelkern <Bau> flux-cored wire *(Schweißen)*
Drahtantenne *f* <Telekom> wire aerial
Drahtauslöser *m* <Foto> cable release
Drahtauslösernippel *m* <Foto> cable release socket
Drahtbonden *n* <Elektrotech> wire bonding
Drahtbündel *n* <Elektrotech> wire bundle
Drahtbürste *f* 1. <Bau> scratch brush; 2. <Fertig> wire brush
Drahtdicke *f* <Maschinen> *(AE)* wire gage, *(BE)* wire gauge
Drahtelektrode *f* <Elektriz> wire electrode
Drahtende *n* <Elektrotech> wire end
Drahtfunk *m* <Telekom> cablecast
Drahtgeflecht *n* <Bau> wire netting
Drahtgeflechtbrille *f* <Sicherheit> wire mesh goggles
Drahtgeflechtschutzgitter *n* <Sicherheit> wire mesh screen
Drahtgewebe *n* 1. <Fertig> wire cloth; 2. <Kunststoff> wire mesh
drahtgewickelte Spule *f* <Elektriz> wire-wound coil

drahtgewickelter Anker m <Elektriz> wire-wound armature
drahtgewickelter Widerstand m <Elektrotech> wire-wound resistor
Drahtgitter n <Ker & Glas> wire mesh
Drahtgittercontainer m <Trans> lattice-sided container, skeleton container
Drahtgitterverstärkung f <Ker & Glas> wire mesh reinforcement
Drahtglas n 1. <Bau> (AE) armored glass, (BE) armoured glass; 2. <Ker & Glas> (AE) armored glass, (BE) armoured glass, wire glass, wired glass
Drahtgussglas n <Ker & Glas> wired cast glass
Drahtheftklammer f 1. <Bau> wire staple; 2. <Verpack> staple
Drahtheftmaschine f <Verpack> stapling machine
Drahtkammer f <Teilphys> wire chamber
Drahtkern m <Elektrotech> wire core
Drahtkontaktieren n <Elektrotech> wire bonding
Drahtkugellager n <Maschinen> wire race ball bearing
Drahtlehre f <Maschinen> SWG, (AE) standard wire gage, (BE) standard wire gauge, (AE) wire gage, (BE) wire gauge
Drahtlitze f <Maschinen> strand wire
Drahtlitzenleiter m <Elektriz> tinsel conductor
drahtlos ausgesandt adj <Phys> radio
drahtlose Anwendungssprache f <Mobilkom> wireless application protocol, WAP (Internet-Zugang über Handy)
drahtlose Datenpaketübertragung f <Mobilkom> cellular digital packet data
drahtlose Kommunikationstechnik f <Comp & DV> wireless access technology
drahtlose Kopfgarnitur f <Aufnahme> wireless headset (Kopfhörer und Mikrofon)
drahtlose Steuerungsautomatik f <Lufttrans, Trans, Wassertrans> radio telecontrol
drahtloser Breitbandanschluss m <Telekom> broadband wireless access
drahtloser Breitband-Codemultiplexzugriff m <Funktech, Telekom> wireless wideband code multiplex access, wireless wideband CDMA
drahtloser Infrarot-Kopfhörer m <Aufnahme> wireless infrared headphones
drahtloser Internet-Zugang m <Funktech, Telekom> wireless Internet access
drahtloser Teilnehmeranschluss m <Telekom> radio in the loop (RITL); wireless local loop
drahtloser Web-Zugang m <Telekom> wireless web access
drahtloses intelligentes Netz n <Funktech, Telekom> wireless intelligent network
drahtloses lokales Netz n <Funktech, Mobilkom, Telekom> radio LAN, wireless LAN, wireless local area network
drahtloses Schwerhörigengerät n <Funktech, Telekom> wireless hearing aid receiver
drahtloses Telefon n <Mobilkom> cell phone, (AE) cellular phone, cordless telephone, mobile, (BE) mobile phone
Drahtmodelldarstellung f <Comp & DV> wire frame representation
Drahtnachlaufschweißen n <Fertig> backward welding
Drahtnagel m 1. <Bau> brad; 2. <Maschinen> wire nail
Drahtnetz n <Labor> wire gauze
Drahtpotenziometer n <Elektriz> wire-wound potentiometer
Drahtquerschnitt m <Metrol> (AE) wire gage, (BE) wire gauge
Drahtrolle f <Bau> wire reel
Drahtrollenlager n <Maschinen> wire race roller bearing
Drahtseil n <Fertig, Maschinen, Sicherheit> wire rope

Drahtseilbahn f <Trans> (AE) cable railroad, (BE) cable railway, cable road, ropeway
Drahtseilschlaufe f <Sicherheit> wire rope sling
Drahtspeichenrad n <Kfztech> wire wheel
Drahtspeichergerät n <Aufnahme> wire recorder
Drahtstärke f <Maschinen> (AE) wire gage, (BE) wire gauge
Drahtsteigung f <Elektriz> pitch
Drahtstöpselverschluss m <Ker & Glas> (AE) wired stopper finish
Drahtstraße f <Maschinen> wire mill
Drahttarget n <Strahlphys> wire mesh target
Drahttauwerk n <Wassertrans> wire rope
Drahtübertragungsweg m <Elektrotech> metallic circuit
Drahtumschnürungsapparat m <Verpack> wire strapping equipment
Drahtverschnürung f **für Beutel** <Verpack> wire bag tie
drahtverstärkter Schlauch m <Kunststoff, Maschinen> wire-reinforced hose
drahtverstärktes Brandschutzglas n <Ker & Glas> Georgian-wired glass
Drahtwalzwerk n <Kohlen> rod mill
Drahtwellenleitung f <Telekom> Goubau line
Drahtwiderstand m <Elektriz, Phys> wire-wound resistor
Drahtzange f <Maschinen> tongs
Drahtzange f **mit flachem Maul** <Mechan> flat-nosed pliers
Drahtzaun m <Bau> wire fence
Drahtziehbank f <Fertig> wire-drawing bench
Drahtzieheisen n <Maschinen> wire-drawing die
Drahtziehen n <Maschinen> wire drawing
Drahtzuführung f <Ker & Glas> wire guide
Drahtzug m <Fertig> wire drawing
Drahtzusammenstapelmaschine f <Verpack> wire stacking machine
Drain m <Elektronik, Elektrotech, Phys> drain
Drainage f 1. <Erdöl> drainage (Geologie); 2. <Wasserversorg> drainage
Drainanschluss m <Elektrotech> drain terminal
Drainelektrode f <Phys> drain (beim Feldeffekttransistor)
Drainschalter m <Elektrotech> drain contact
Drainschaltung f <Elektrotech> drain connection
Drainstrom m <Elektrotech> drain current
Drainverstärker m <Elektronik> drain amplifier
Drainvorspannung f <Elektrotech> drain bias
Drall m 1. <Elektriz> lay; 2. <Fertig> helix, moment of impulse, spiral; 3. <Maschinen> twist; 4. <Raumfahrt> spin; 5. <Textil> twist
drallfreies Tau n <Wassertrans> nonkinking rope
Dralllänge f 1. <Elektriz> length of lay (Kabelherstellung); 2. <Elektrotech> lay
Drallnut f <Fertig> helical broaching
Drallnuträumen n <Fertig> helical broaching
Drallrichtung f 1. <Elektriz> direction of lay; 2. <Fertig> hand of helix, hand of spiral; 3. <Textil> direction of twist
Drallverhältnis n <Elektriz> lay ratio
Drallwinkel m 1. <Fertig> angle of the tooth helix; 2. <Mechan> angle of twist (Drahtseiltechnik)
DRAM (dynamischer RAM) <Comp & DV> DRAM (dynamic random access memory)
drapieren v <Textil> drape
Drapierung f <Textil> draping
Draufsicht f <Geom, Konstzeich> top view
Drechselbank f <Maschinen> lathe
drechseln v <Maschinen> turn
Drechseln n <Maschinen> turning
Drechsler m <Maschinen> turner, wood turner
Drechslerbank f <Bau> wood-turning lathe
Drechslerwerkzeug n <Bau> wood-turning tools

D-Regeleinrichtung *f* <Regelung> derivative control system
dreggen *v* <Wassertrans> drag
Dreggnetz *n* <Wassertrans> dredge net *(Fischerei)*
D-Region *f* <Funktech> D-region *(Wellenausbreitung)*
Dreh... 1. <Maschinen> rotary; 2. <Mechan> torsional
Drehachse *f* 1. <Fertig> fulcrum, hinge; 2. <Maschinen> axis of rotation, swivel axis; 3. <Mechan> axis of revolution, hinge; 4. <Phys> axis of revolution
Drehanker *m* 1. <Elektriz> rotary armature, rotating armature; 2. <Elektrotech> pivoted armature, revolving armature
Dreharbeit *f* <Fertig, Maschinen> lathe work
Dreharm *m* 1. <Kfztech> torque arm; 2. <Mechan> swivel; 3. <Meerschmutz> rotatable arm
Drehautomat *m* <Maschinen> automatic lathe
Drehbank *f* <Bau, Fertig, Maschinen, Mechan> lathe
Drehbankarbeit *f* <Maschinen> lathe work
Drehbankbett *n* <Maschinen, Mechan> lathe bed
Drehbankfutter *n* <Maschinen, Mechan> lathe chuck
Drehbankschlitten *m* <Maschinen, Mechan> lathe slide
Drehbankspindel *f* <Maschinen> lathe spindle
Drehbankspindelstock *m* <Maschinen> lathe headstock
Drehbankspitze *f* <Maschinen> *(AE)* lathe center, *(BE)* lathe centre
Drehbankwerkzeughalter *m* <Maschinen> lathe toolpost
drehbar *adj* 1. <Fertig> hinged; 2. <Funktech> revolving; 3. <Labor> revolving, rotary; 4. <Mechan> pivoted; 5. <Raumfahrt, Telekom> rotatable
drehbare Antenne *f* <Telekom> rotatable antenna
drehbare Düse *f* <Raumfahrt> rotatable nozzle *(Raumschiff)*
drehbare Radarreflektorantenne *f* <Funkort> revolving radar reflector
drehbare Richtantenne *f* <Funktech> rotatable beam antenna, rotatable directional antenna
drehbare Stellung *f* <Labor> revolving stage *(Mikroskop)*
drehbarer Pfeifenkopf *m* <Labor> revolving nose piece *(Mikroskop)*
drehbarer Tisch *m* <Metrol> rotary table
drehbarer Zeichenkopf *m* <Metrol> protractor
Drehbeanspruchung *f* <Metall> twisting strain
Drehbewegung *f* 1. <Fertig> rotation *(Kunststoffinstallationen)*; 2. <Maschinen, Phys> rotation
Drehbohren *n* <Bau, Kohlen> rotary drilling
Drehbohrer *m* 1. <Bau> rotary drill, twist gimlet; 2. <Fertig> rotary tool
Drehbohrmaschine *f* <Kohlen> rotary machine
Drehbolzen *m* <Fertig> pivot
Drehbrett *n* <Fertig> loam board
Drehbrücke *f* 1. <Bau> pivot bridge, swing bridge, turn bridge, turning bridge; 2. <Maschinen> turn bridge; 3. <Wassertrans> swing bridge
Drehdiamanten *mpl* <Maschinen> turning diamonds
Drehdurchmesser *m* <Maschinen> swing
Dreheiseninstrument *n* <Elektrotech, Gerät, Metrol> electromagnetic instrument, ferrodynamic instrument, moving-iron instrument, soft iron instrument
Dreheiseninstrument *n* **mit Magnet** <Gerät, Metrol> permanent magnet moving-iron instrument
Dreheisenmessgerät *n* <Elektrotech> moving-iron instrument
Dreheisenmesswerk *n* <Gerät, Metrol> iron-vane movement instrument, moving-iron instrument
Dreheisenoszillograph *m* <Metrol> soft-iron oscillograph
Dreheisenvoltmeter *n* <Elektriz, Metrol> moving-iron voltmeter
drehelastisch *adj* <Maschinen> torsionally elastic
Drehelastizität *f* <Maschinen> torsional elasticity

drehen *v* 1. <Comp & DV> rotate; 2. <Ker & Glas> throw; 3. <Maschinen> rotate, turn; 4. <Wassertrans> turn *(Schiff)*
drehen *v*/sich <Wassertrans> veer *(Wind)*
Drehen *n* 1. <Bau> *(BE)* slewing, *(AE)* sluing; turning *(eines Krans)*; 2. <Maschinen> turning
Drehen *n* **von der Stange** <Maschinen> bar turning
drehend *adj* <Phys> rotatory *(Molekül)*
drehend *adj*/im Uhrzeigersinn <Lebensmittel> clockwise-rotating
Dreher *m* 1. <Ker & Glas> thrower; 2. <Maschinen> turner; 3. <Mechan> lathe operator
Drehfeld *n* 1. <Elektriz> rotating field; 2. <Elektrotech, Labor> rotary field; 3. <Telekom> rotating field
Drehfeldinstrument *n* <Gerät> rotating field instrument
Drehfeldumformer *m* <Elektriz> induction frequency converter, rotating field converter, rotary field converter
Drehfenster *n* <Bau> pivot-hung window
Drehfensterflügel *m* <Bau> pivot-hung sash
Drehfeuer *n* <Lufttrans> rotating beacon
Drehfilter *n* <Kohlen, Wasserversorg> rotary filter
Drehflügelfenster *n* <Bau> side-hung window
Drehflügelflugzeug *n* 1. <Lufttrans> gyroplane, rotary wing aircraft, rotating wing aircraft, rotor aircraft; 2. <Trans> gyroplane
Drehflügelzähler *m* <Gerät> rotating blade meter
Drehformblasverfahren *n* <Ker & Glas> paste mould blowing
Drehform-Press-Blasverfahren *n* <Ker & Glas> paste mould press-and-blow process
Drehfunkfeuer *n (DFF)* <Funkort, Lufttrans> omnirange indicator
drehgelagerter Anker *m* <Elektrotech> pivoted armature
Drehgelenk *n* 1. <Lufttrans> blade pitch change hinge *(Hubschrauber)*; 2. <Maschinen> hinge, hinge joint • **über Drehgelenk verbunden** <Fertig> pin-connected
drehgelenkig *adj* <Fertig> pivoting
drehgelenkig angeordnet *adj* <Fertig> pivoted
drehgelenkig anordnen *v* <Fertig> pivot
drehgelenkige Anordnung *f* <Fertig> pivoting
Drehgelenkverbindung *f* <Maschinen> swivel joint
Drehgeschwindigkeit *f* 1. <Lufttrans> rotative speed; 2. <Maschinen> rotating speed
Drehgestell *n* 1. <Fertig> dolly; 2. <Telekom> rotator
Drehgestellgüterwagen *m* <Kfztech> bogie wagon *(mit Schwenkdach)*
Drehgestellmotor *m* <Elektrotech> bogie-mounted motor
Drehgestellpendel *n* <Eisenbahn> swing link
Drehgestellrahmen *m* <Eisenbahn> bogie frame
Drehgestellwagen *m* <Maschinen> bogie car
Drehgriff *m* <Kfztech> twist grip *(Motorrad)*
Drehhaken *m* <Wassertrans> swivel hook *(Takelage, Beschläge)*
Drehherz *n* <Fertig, Maschinen> driver, driving dog, lathe carrier, lathe dog
Drehimpuls *m* <Phys, Teilphys> angular momentum
Drehimpulsquantenzahl *f* <Teilphys> spin
Drehkippfenster *n* <Bau> *(AE)* center-hung window, *(BE)* centre-hung window *(horizontal oder vertikal)*
Drehknopf *m* 1. <Elektrotech> knob, rotary knob; 2. <Maschinen> knob
Drehkolben *m* <Maschinen> rotary piston
Drehkolbengasmessgerät *n* <Gerät> lobed-impeller gas meter
Drehkolbenmotor *m* 1. <Kfztech> rotating piston engine; 2. <Maschinen> rotary engine
Drehkolbenpumpe *f* <Kontroll, Maschinen, Wasserversorg> rotary pump
Drehkolbenverdichter *m* <Maschinen> rotary compressor

Drehkondensator m 1. <Elektriz> rotary capacitor, variable capacitor; 2. <Elektrotech> variable capacitor
Drehkondensatorbereich m <Elektrotech> variable capacitor sector
Drehkopf m <Mechan> turret
Drehkörper m <Kontroll> rotor
Drehkraft f <Phys> rotatory power
Drehkran m 1. <Bau> *(BE)* all-round swing crane, rotary crane, *(BE)* slewing crane, *(AE)* sluing crane; 2. <Kerntech> *(BE)* slewing crane, *(AE)* sluing crane
Drehkranz m <Kontroll> turntable
Drehkreis m <Wassertrans> turning circle *(Schiff)*
Drehkreuz n 1. <Fertig> capstan, turnstile; 2. <Lufttrans> spider unit *(Hubschrauber)*; 3. <Maschinen> capstan wheel, spider wheel
Drehkristallmethode f <Strahlphys> rotating crystal method *(Röntgenbeugung)*
Drehkuppel f <Fertig> turret
Drehlager n 1. <Eisenbahn> pivot bearing; 2. <Fertig> rotary bearing *(Kunststoffinstallationen)*; 3. <Kerntech> pivot bearing
Drehling m <Fertig> tool holder bit
Drehmagnetcassette f <Aufnahme> moving-magnet cartridge
Drehmagnetgalvanometer n <Elektrotech, Gerät, Metrol> moving-magnet galvanometer
Drehmagnetmedium n <Elektrotech> moving-magnet medium
Drehmagnetmesswerk n <Gerät> moving-magnet movement
Drehmantel m <Fertig> outer sleeve
Drehmaschine f 1. <Fertig> duplex lathe, gear lathe; 2. <Maschinen> turning lathe
Drehmaschine f mit Brücke <Maschinen> break lathe, gap lathe
Drehmaschine f mit Sechskantrevolverkopf <Maschinen> hexagon turret lathe
Drehmeißel m 1. <Fertig> lathe tool; 2. <Maschinen> lathe tool, turning tool
Drehmelder m für Drehmomente <Gerät> synchro torque transmitter
Drehmoment n 1. <Elektriz, Elektrotech, Erdöl> torque; 2. <Fertig> radial force; 3. <Fertig> torque *(Kunststoffinstallationen)*; 4. <Kfztech> torque *(Motor)*; 5. <Maschinen> turning moment; 6. <Maschinen> torque; 7. <Mechan> angular momentum; 8. <Mechan> torque; 9. <Phys> rotational moment • **für hohes Drehmoment geeignet** <Mechan> high-torque
Drehmoment n des Motors <Kfztech, Lufttrans> engine torque
Drehmoment n des Triebwerks <Kfztech, Lufttrans> engine torque
Drehmoment n eines blockierten Läufers <Elektriz> locked rotor torque
Drehmoment n eines blockierten Rotors <Elektriz> locked rotor torque
Drehmomentanzeiger m <Gerät> torque indicator
Drehmomentausgleichsluftschraube f <Lufttrans> antitorque rotor
Drehmomentausgleichspropeller m <Lufttrans> antitorque propeller
Drehmomentausgleichsvorrichtung f <Lufttrans> antitorque device
Drehmomentendiagramm n <Elektrotech, Metrol> speed-torque curve
Drehmomentenstückegabel f <Lufttrans> torque link *(Fahrwerk)*
Drehmomentmesser m 1. <Fertig> dynamometer; 2. <Maschinen> torquemeter, torsiometer

Drehmomentminderer m <Fertig> automatic torque limiting device
Drehmomentmotor m <Elektrotech> torque motor
Drehmomentschlüssel m 1. <Kerntech> torque spanner; 2. <Kfztech> torque wrench *(Werkzeug)*; 3. <Maschinen> torque wrench
Drehmomentskoeffizient m <Nichtfoss Energ, Qual> coefficient of torque
Drehmomentwandler m 1. <Kfztech> converter *(Automatikgetriebe)*; torque converter *(Getriebe)*; 2. <Maschinen> torque converter
Drehmomentwandlergehäuse n <Kfztech> torque converter housing
Drehofen m <Kohlen> rotary kiln
Drehpfanne f <Kfztech> *(BE)* bogie pivot, *(AE)* truck pivot *(Anhänger)*
Drehplatte f <Kontroll> turntable
Drehpol m <Maschinen> *(AE)* center of revolution, *(BE)* centre of revolution
Drehpositionsbestimmung f <Comp & DV> rotation position sensing, rotational position sensing
Drehpotenziometer n <Elektriz, Elektrotech> rotary potentiometer
Drehprisma n <Optik> rotating prism
Drehpunkt m 1. <Fertig> fulcrum; 2. <Geom> pivot point; 3. <Maschinen> pivot, pivot point; 4. <Mechan> *(AE)* center of motion, *(BE)* centre of motion, pivot; 5. <Phys> *(AE)* center of motion, *(BE)* centre of motion, fulcrum
Drehrahmen m <Foto> revolving back *(Kamera)*
Drehrahmenpeiler m <Funkort> rotating frame antenna direction finder
Drehregelventil n <Bau> plug cock, plug tap
Drehregler m <Elektriz> induction regulator
Drehrichtung f 1. <Fertig> hand of rotation; 2. <Maschinen> sense of rotation
Drehrichtungsanzeiger m <Gerät> rotation indicator
drehrichtungsunabhängiger Lüfter m <Heiz & Kälte> bidirectional fan
Drehring m <Mechan> swivel
Drehrohrofen m 1. <Abfall> rotary furnace, rotary kiln; 2. <Ker & Glas> rotary kiln
Drehsäule f <Wasserversorg> quoin post *(einer Schleuse)*
Drehschalter m 1. <Elektriz> rotary-type switch; 2. <Elektrotech> rotary switch; 3. <Kontroll> rotary wafer switch
Drehscheibe f 1. <Eisenbahn> turntable; 2. <Fertig> swivel; 3. <Ker & Glas> throwing wheel; 4. <Maschinen> *(BE)* rotating disc, *(AE)* rotating disk
Drehscheibenfluglehrapparat m <Lufttrans> link trainer
Drehscheibenschalter m <Elektrotech> rotary wafer switch
Drehscheibenventil n <Kfztech> *(BE)* rotary disc valve, *(AE)* rotary disk valve *(Zweitaktmotor)*
Drehschieber m 1. <Kfztech> *(BE)* rotary disc valve, *(AE)* rotary disk valve *(Zweitaktmotor)*; 2. <Maschinen, Nichtfoss Energ> rotary valve
Drehschlauch m <Erdöl> rotary hose *(Bohrtechnik)*
Drehschraubstock m <Maschinen> *(BE)* swivel vice, *(AE)* swivel vise
Drehschüssel f <Ker & Glas> rotating bowl
Drehschwingung f <Maschinen> rotary oscillation
Drehseil n <Erdöl> spinning line *(Bohrtechnik)*
Drehspäne mpl <Maschinen> turnings
Drehspiegel m <Phys> rotating mirror
Drehspindel m <Maschinen> lathe spindle
Drehspulamperemeter n <Elektriz, Metrol> moving-coil ammeter
Drehspule f <Elektrotech> moving coil *(Galvanometer)*
Drehspulgalvanometer n <Elektriz, Elektrotech, Phys> moving-coil galvanometer

Drehspulinstrument *n* <Elektriz, Elektrotech, Gerät, Metrol> moving-coil instrument, permanent magnet moving-coil instrument
Drehspulmessgerät *n* <Elektriz, Elektrotech, Gerät, Metrol> moving-coil instrument, moving-coil meter
Drehspulmesswerk *n* <Elektriz, Elektrotech, Metrol> d'Arsonval movement, moving-coil mechanism, permanent-magnet moving-coil mechanism
Drehspulrelais *n* <Elektriz> moving-coil relay
Drehspulspannungsmessgerät *n* <Elektriz, Metrol> moving-coil voltmeter
Drehspulspiegelgalvanometer *n* <Metrol> moving-coil mirror galvanometer
Drehspulstrommessgerät *n* <Elektriz, Metrol> moving-coil ammeter
Drehspultonabnehmer *m* <Aufnahme> moving-coil pickup
Drehspulvibrationsgalvanometer *n* <Elektriz, Metrol> moving-coil vibration galvanometer
Drehspulvoltmeter *n* <Elektriz, Metrol> moving-coil voltmeter
Drehspulzeigergalvanometer *n* <Elektriz, Metrol> moving-coil pointer galvanometer
Drehstab *m* <Kfztech> torsion bar; stabilizer bar *(Federung, Aufhängung)*
Drehstabstabilisator *m* <Kfztech> torque stabilizer, anti-roll bar *(Federung)*
Drehstahl *m* <Maschinen> lathe tool
Drehstift *m* <Bau> hinge pin, turn pin
Drehstrom *m* <Elektriz, Elektrotech> three-phase current
Drehstrom *m* **entgegen dem Uhrzeigersinn** <Maschinen> rotation anticlockwise
Drehstrom *m* **im Uhrzeigersinn** <Maschinen> rotation clockwise
Drehstromalternator *m* <Elektriz> three-phase alternator
Drehstromanker *m* <Elektriz> three-phase current armature
Drehstrombrückenschaltung *f* <Elektronik> three-phase bridge circuit, three-phase bridge connection
Drehstromeinspeisung *f* <Elektriz> three-phase supply
Drehstromerregermaschine *f* <Elektrotech> phase advancer
Drehstromgenerator *m* 1. <Elektriz> alternator, three-phase generator; 2. <Elektrotech, Phys> alternator
Drehstromgleichrichterbrücke *f* <Elektriz> three-phase rectifier
Drehstrominduktionsmotor *m* <Kfztech> three-phase induction motor
Drehstromkommutatorkaskade *f* <Elektrotech> cascaded induction and commutator machine
Drehstromkommutatormaschine *f* <Elektrotech> three-phase commutator machine
Drehstromlichtmaschine *f* <Kfztech, Nichtfoss Energ> alternator
Drehstrommotor *m* <Elektriz> three-phase motor
Drehstromnetz *n* <Elektriz> three-phase supply network
Drehstromschaltung *f* <Elektriz> polyphase circuit
Drehstromschrittmotor *m* <Elektrotech> three-phase stepper motor
Drehstromschutzleiterdrossel *f* <Elektriz> three-phase neutral reactor
Drehstromsynchronmotor *m* <Elektrotech> three-phase synchronous motor
Drehstromtransformator *m* <Elektrotech> three-phase transformer
Drehstromtriebfahrzeug *n* <Elektrotech> three-phase a.c. motor vehicle
Drehstromversorgung *f* <Elektrotech> three-phase supply
Drehstromzugförderung *f* <Elektrotech> three-phase a.c. traction
Drehsupport *m* <Maschinen> swivel head, swivel slide rest
Drehteil *n* 1. <Fertig> lathe work; harp *(Waagerechtstoßmaschine)*; hob swivel head *(Wälzfräsen)*; 2. <Maschinen> turned part
Drehteller *m* <Akustik> turntable
Drehtisch *m* 1. <Erdöl> rotary table *(Bohrtechnik)*; 2. <Kontroll> turntable; 3. <Maschinen> revolving table, rotary table, rotating table; 4. <Mechan> turntable
Drehtischzuführung *f* <Verpack> turntable feed
Drehtor *n* <Bau> swing gate
Drehtransformator *m* <Elektrotech> rotary transformer
Drehtrennschalter *m* <Elektrotech> centre-break disconnector
Drehtrommel *f* <Abfall> revolving drum
Drehtrommelzuführung *f* <Maschinen> rotary drum feeder
Drehtür *f* <Bau> swing door, swinging door
Drehumformer *m* <Elektriz> inverted rotary converter, rotary converter
Drehung *f* 1. <Elektrotech> turn *(Rotation)*; 2. <Maschinen> revolution, rotation, turn; 3. <Textil> twist, twisting
• **in Drehung versetzen** <Maschinen> impart a rotary motion *(Welle)*
Drehungen *fpl* **pro Meter** <Textil> *(AE)* turns per meter, *(BE)* turns per metre
Drehungsbeiwert *m* <Textil> twist factor
drehungsfrei *adj* <Phys> irrotational, nonkinking
Drehungsgeschwindigkeit *f* <Lufttrans> rotation speed
Drehvektor *m* <Elektrotech> spiral vector
Drehventil *n* <Hydraul> rotary valve
Drehverbindung *f* <Elektrotech> rotary joint
Drehversuch *m* <Metall> torsion test
Drehvorrichtung *f* 1. <Elektrotech> barring gear, turning gear; 2. <Kerntech, Wassertrans> turning gear *(Motor)*
Drehwähler *m* <Elektrotech> uniselector *(Fernmeldewesen)*
Drehwählersystem *n* <Telekom> rotary system
Drehwählervermittlungsstelle *f* <Telekom> rotary exchange
Drehwechselfestigkeit *f* <Fertig> alternate torsional strength
Drehwerkzeug *n* <Maschinen> lathe tool
Drehwinkel *m* 1. <Fertig> angle of rotation *(Kunststoffinstallationen)*; 2. <Gerät> angle of rotation; 3. <Maschinen> angle of twist
Drehwinkelbegrenzung *f* <Fertig> limitation of rotation angle *(Kunststoffinstallationen)*
Drehwinkelrate *f* <Lufttrans> angular roll rate
Drehzahl *f* 1. <Elektriz> number of revolutions, revolutions per minute *(rpm)*; 2. <Gerät> rotational speed, speed of rotation; 3. <Kfztech> revs *(Motor)*; 4. <Maschinen> rotational speed, speed; 5. <Phys> speed, speed of rotation
Drehzahl *f* **der Schleifscheibe** <Fertig> grinding wheel RPM
Drehzahl *f* **des Rotors** <Lufttrans> rotor speed *(Hubschrauber)*
Drehzahlanzeige *f* <Maschinen> revolution indication
Drehzahlanzeigegerät *n* <Gerät> speed indicator
Drehzahlanzeiger *m* <Maschinen> revolution indicator
Drehzahlbegrenzer *m* 1. <Kfztech> speed limiter; 2. <Maschinen> overspeed protection
Drehzahlbereich *m* 1. <Kfztech> rev range; 2. <Maschinen> range of speeds, speed range
Drehzahldrehmomentkennlinie *f* <Heiz & Kälte> speed torque characteristic
Drehzähler *m* <Fertig> counter

Drehzahlmesser m 1. <Kfztech> revolution counter *(Motor, Zubehör)*; 2. <Maschinen> revolution counter, speed counter
Drehzahlminderer m <Maschinen> speed reducer
Drehzahlregelung f <Elektriz, Textil> speed control
Drehzahlregler m 1. <Elektrotech, Kfztech> actuator; governor *(Motordrehzahlbegrenzer)*; 2. <Maschinen> speed controller, speed governor; 3. <Mechan> governor; 4. <Nichtfoss Energ> speed control device; 5. <Optik> controller *(für CD-Spieler)*; 6. <Trans> actuator *(KFZ)*
Drehzahlreihe f <Fertig> group of speeds
Drehzahlsichtanzeigesystem n <Trans> displayed speed system
Drehzahlüberschreitungsprüfung f <Elektriz> overspeed test
Drehzahlverhältnis n <Maschinen> speed ratio
Drehzahlwächter m <Fertig> overspeed control, speed control
Drehzahn m <Fertig> tool bit
Drehzapfen m 1. <Fertig> gudgeon; 2. <Kerntech> king journal, trunnion; 3. <Kfztech> *(BE)* bogie pin, *(AE)* truck pin, pivot pin *(Anhänger)*; trunnion *(Universalgelenk)*; 4. <Kohlen> trunnion; 5. <Maschinen> fulcrum pin, gudgeon pin, hinge pin, piston pin, pivot pin, swivel pin; 6. <Mechan> journal, pivot
Drehzapfenlager n <Eisenbahn> *(AE)* center plate, *(BE)* centre plate
Drehzentrum n <Geom> centre of rotation
Dreiachs... <Raumfahrt> three-axis
Dreiachsenanzeige f <Raumfahrt> three-axis indicator *(Raumschiff)*
Dreiachsenkreisel m <Raumfahrt> three-axis gyro unit *(Raumschiff)*
Dreiachsen-Pinchversuch m <Kerntech> triaxial pinch experiment
Dreiachsenstabilisierung f <Raumfahrt> three-axis stabilization *(Raumschiff)*
dreiachsige Spannung f <Metall> triaxial stress
Dreiachs-Winkeldrehsensor m <Lufttrans> angular three-axis rate sensor
Dreiadern-Kabel n <Elektriz> triple core cable
Dreiadressbefehl m <Comp & DV> three-address instruction
dreiadriges Kabel n <Elektrotech> three-conductor cable
Dreiaxialprüfung f 1. <Bau> triaxial test; 2. <Maschinen> three-jaw chuck
Dreibackensetzstock m <Maschinen> three-jaw steady, three-jaw steadyrest
dreibasige Säure f <Chemie> triacid
Dreibein n <Bau> gin
Dreibeinbohrbühne f <Erdöl> shear leg *(Bohrtechnik)*
Dreibeinfahrwerk n <Lufttrans> tricycle landing gear *(Fahrwerk)*
Dreibeinkran m <Bau> shear leg
Dreidecker m <Lufttrans> triplane
dreidimensional *adj (3-D)* 1. <Maschinen> three-dimensional; 2. <Phys> three-dimensional *(3-D)*
dreidimensional fräsen v <Fertig> pocket *(Gesenke)*
dreidimensionale Geometrie f <Geom> geometry of three dimensions
dreidimensionale Grafik f <Comp & DV> three-dimensional graphics
dreidimensionale Integration f <Elektronik> three-dimensional integration
dreidimensionale Koordinaten fpl <Geom> x-y-z coordinates
dreidimensionaler IC m <Elektronik> three-dimensional integrated circuit

dreidimensionales Bild n <Telekom> three-dimensional image
dreidimensionales Nachformfräsen n <Fertig> kellering
Dreidimensionalität f <Geom> three dimensionality
Dreidrahtgenerator m <Elektriz> three-wire generator
Dreidrahtsystem n <Elektriz> three-wire system
Dreieck n 1. <Elektrotech> delta; 2. <Geom> triangle
Dreieckbogen m <Bau> triangular arch
dreieckig *adj* <Geom> triangular
dreieckiger Nocken m <Maschinen> triangular cam
Dreieckigkeit f <Ker & Glas> triangularity
Dreieckschaltung f 1. <Elektriz> delta connection; 2. <Elektrotech> delta connection; mesh connection *(Speicherröhrenanschlüsse)*
Dreiecksmatrix f <Comp & DV> triangular matrix
Dreiecksschaltung f <Phys> delta connection
Dreieck-Sternschaltung f <Elektriz> delta star connection
Dreieck-Sternumformung f <Elektriz> delta-to-star conversion
Dreieckstest m <Lebensmittel> triangle test, triangle testing
Dreiecksungleichung f <Math> triangle inequality
Dreiecksvermessung f <Bau> triangulation
Dreieckszahn m <Maschinen> fleam tooth
Dreielektrodenröhre f 1. <Chemie> three-electrode valve, triode *(Elektrizität)*; 2. <Elektronik> three-electrode tube
Dreierkonferenz f <Telekom> add-on third party, flexible add-on, three-way conversation *(ISDN)*
Dreierkonferenzschaltung f <Telekom> three-party call
Dreierpackung f <Verpack> triple pack
Dreierverbindung f <Telekom> three-party call, three-way call
dreifach *adj* <Chemie> ternary, triple
dreifach beschichtetes Teilchen n <Kerntech> triplex-coated particle
dreifach gekröpfte Kurbel f <Maschinen> three-throw crank
dreifach gekröpfte Kurbelwelle f <Maschinen> three-throw crankshaft
dreifach gewickelter Transformator m <Elektriz> triple-wound transformer
Dreifach... <Chemie> triple
dreifacher Alphaprozess m <Kerntech> triple alpha process
Dreifachform f <Ker & Glas> triple cavity mould
Dreifachkabel n <Elektrotech> three-conductor cable
Dreifachnorm f <Fernseh> triple standard
Dreifarben... 1. <Elektronik> three-beam; 2. <Foto> *(AE)* three-color, *(BE)* three-colour
Dreifarben-Bildröhre f <Elektronik> *(AE)* three-beam color picture tube, *(BE)* three-beam colour picture tube
Dreifarbenfotografie f <Foto> *(AE)* three-color photography, *(BE)* three-colour photography
Dreifarbenplatte f <Foto> *(AE)* three-color plate, *(BE)* three-colour plate
Dreifingerregel f <Elektriz> left-hand rule
dreiflächig *adj* <Geom> trihedral
Dreifuß m <Labor> tripod
dreigängig *adj* <Fertig, Maschinen> three-start
dreigängiges Gewinde n <Maschinen> three-start thread
Dreigelenkbogen m <Bau> three-hinged arch
dreigestaltig *adj* <Chemie> trimorphic, trimorphous
Dreigestaltigkeit f <Chemie> trimorphism
Dreigitterröhre f <Elektronik> three-grid tube
dreigliedrig *adj* <Math> trinomial
Dreihalskolben m <Labor> three-necked flask
Dreikanalzweitaktmotor m <Kfztech> three-port two-stroke engine

Dreikant...

Dreikant... 1. <Bau> triangular; 2. <Maschinen> three-square
Dreikantfeile f <Maschinen> three-square file, tri square file, triangular file
Dreikantleiste f <Bau> triangular fillet
Dreikantprisma n <Phys> roof prism
Dreiklauenfutter n <Maschinen> three-pronged chuck
Dreikomponentenlegierung f <Metall> three component alloy
Dreikreis-Kernkraftwerk n <Kerntech> three-circuit nuclear power plant
Dreilagenholz n <Bau> three-ply wood
Dreileiterkabel n <Elektrotech> three-conductor cable
Dreileiternetz n <Elektrotech> three-wire mains
Dreilochbrenner m <Heiz & Kälte> treble jet burner
Dreimastmarssegelschoner m <Trans> (BE) barque schooner
Dreimessermaschine f <Druck> three-sided cutting machine
Dreiniveaulaser m <Elektronik> three-level laser
Dreiniveaumaser m <Elektronik> three-level maser
Dreinullstellen-Antialiasing-Filter n <Elektronik> three-zeros antialiasing filter (zur Verhinderung von Faltungsfrequenzen)
Dreiphasen... <Elektriz, Elektrotech> three-phase
Dreiphasen-Erdungstransformator m <Elektriz> (BE) three-phase earthing transformer, (AE) three-phase grounding transformer
Dreiphasen-Induktionsmotor m <Elektrotech> three-phase induction motor
Dreiphasenkommutatormaschine f <Elektrotech> three-phase commutator machine
Dreiphasenläufer m <Elektrotech> three-phase rotor
Dreiphasenlinearmotor m <Elektrotech> three-phase linear motor
Dreiphasenmaschine f <Elektriz, Elektrotech> three-phase machine
Dreiphasenmotor m <Elektrotech> three-phase motor
Dreiphasennetz n <Elektrotech> three-wire mains
Dreiphasenrotor m <Elektrotech> three-phase rotor
Dreiphasenrotorwicklung f <Elektrotech> three-phase rotor winding
Dreiphasenschaltung f <Telekom> three-phase circuit
Dreiphasenstator m <Elektrotech> three-phase stator
Dreiphasenstatorwicklung f <Elektrotech> three-phase stator winding
Dreiphasenstrom m <Elektrotech> three-phase current
Dreiphasenstromversorgung f <Elektrotech> three-phase supply
Dreiphasenstromwendermaschine f <Elektrotech> three-phase commutator machine
Dreiphasensynchronmotor m <Elektrotech> three-phase synchronous motor
Dreiphasensystem n <Elektrotech> three-phase system, three-wire system
Dreiphasen-Wanderfeld-Wicklung f <Elektrotech> three-phase traveling field winding (Magnetschwebetechnik)
Dreiphasenwechselstrommotor m <Kfztech> three-phase alternomotor, three-phase motor
dreiphasig adj <Chemie, Elektriz, Elektrotech, Phys> three-phase
dreipoliger Schalter m <Elektrotech> three-pole switch
dreipoliges Filter n <Elektrotech> three-pole filter
dreiprotonige Säure f <Chemie> triacid
Dreipunkt... 1. <Lufttrans, Maschinen> three-point; 2. <Regelung> three-step
Dreipunktbiegen n <Metall> three-point bending

Dreipunktbiegeprobe f <Kerntech> three-point bending specimen
Dreipunktgurt m <Kfztech> safety belt (Sicherheitszubehör)
Dreipunktlagenwinkel m <Lufttrans> ground angle
Dreipunktlager n <Maschinen> three-point support
Dreipunktlandung f <Lufttrans> three-point landing
Dreipunktregelung f <Regelung> three-step control
Dreipunktsicherheitsgurt m <Kfztech> three-point seat belt
Dreipunktsignal n 1. <Elektronik> three-level signal; 2. <Regelung> three-step signal
dreirädrig adj <Maschinen> three-wheeled
Dreirad-Zweikreisbremsanlage f <Kfztech> L-split system
dreiringig adj <Chemie> tricyclic
Dreiröhrenkamera f <Fernseh> three-tube camera
Dreirollenbohrmeißel m <Erdöl> three-cone bit; tricone bit (Bohrtechnik)
Dreisatz m <Math> rule of three
dreischäftig adj <Wassertrans> three-stranded (Tauwerk)
Dreischichtenfilm m <Foto> tripack film
dreischichtige Wellpappe f <Verpack> triple wall corrugated board
Dreischritt-Relais n <Elektriz> three-step relay
Dreiseitenbeschneidemaschine f <Druck> three-sided cutting machine
dreiseitig adj <Geom> three-sided
dreiseitige Matrix f <Comp & DV> triangular matrix
Dreispur-Stereo n <Aufnahme> three-track stereo
Dreistellungsschalter m <Elektriz> three-position switch, three-way switch
Dreistift-Steckbüchse f <Elektriz> three-pin socket
Dreistrahl-... 1. <Elektronik> three-gun; 2. <Kerntech> triple beam
Dreistrahl-Farbfernsehröhre f <Elektronik> (AE) three-gun color picture tube, (BE) three-gun colour picture tube
Dreistrahl-Koinzidenzspektrometer n <Kerntech> triple beam coincidence spectrometer
dreistufiger Verstärker m <Elektronik> three-stage amplifier
Dreitastenmaus f <Comp & DV> three-button mouse
dreiteiliger Ölabstreifring m <Kfztech> three-piece oil control ring
Drei-Teilnehmer-Gespräch n <Telekom> three-way call
Dreiteilung f <Geom> trisection
Dreiviertelwind m <Wassertrans> quarter wind
Dreiwege... 1. <Aufnahme, Bau> three-way; 2. <Elektriz, Kunststoff, Maschinen> trifurcate; 3. <Mechan> three-way
Dreiwegehahn m <Bau, Maschinen> three-way cock
Dreiwegekugelhahn m <Fertig> three-way ball valve (Kunststoffinstallationen)
Dreiwegesystem n <Aufnahme> three-way system
Dreiwegeventil n <Hydraul, Maschinen> three-way valve
Dreiwegexpander m <Elektriz> trifurcator
Dreiwegexpansionskasten m <Elektriz> trifurcating box
Dreiwegexpansionsspleiß m <Elektriz> trifurcating joint
Dreiwegezapfluftventil n <Lufttrans> air cross bleed valve
dreiwertig adj <Chemie> tribasic, trivalent
dreiwertige Logik f <Elektronik> tristate logic
dreiwertiger Alkohol m <Chemie> trihydric acid, triol
Dreiwertigkeit f <Chemie> tervalence, trivalence, trivalency
Dreiwicklungstransformator m <Elektrotech> three-winding transformer
Dreizustand... <Elektronik> three-state
Dreizustandsausgang m <Elektronik> three-state output
Dreizustandsgatter n <Elektronik> three-state gate
Dreizustandslogik f <Elektronik> three-state logic

Dreizylindermotor m <Maschinen> three-cylinder engine
Drempel m 1. <Bau> (AE) miter sill, (BE) mitre sill; 2. <Wasserversorg> clap sill
dressieren v <Fertig> kill (Blech)
dressiert adj <Fertig> killed, nonkinking, pinch-passed (Blech)
Drexon-Karte f <Optik> Drexon card
Drift f 1. <Akustik, Elektriz> drift; 2. <Elektronik> droop; 3. <Raumfahrt> drift (Raumschiff); 4. <Telekom> drift; 5. <Wassertrans> drift current, drift
Driftanzeige f <Lufttrans> drift indicator
driftarmer Oszillator m <Elektronik> low-drift oscillator
Driftausgleich m <Gerät> drift compensation
driften v <Metrol> drift
Driften n <Bau> drifting
Driftfehler m <Lufttrans> drift error (Höhenmesser)
Driftgeschwindigkeit f <Metall> drift velocity
Driftkammer f <Teilphys> drift chamber
Driftkompensation f <Gerät> drift compensation
Driftstabilisierung f <Fernseh> driftlock
Driftströmung f <Wassertrans> drift current
Drillbohrer m <Maschinen> Archimedean drill
Drillen n <Ker & Glas> lacing
Drillings... <Metall> triple
Drillingspresspumpe f <Wasserversorg> three-throw pump
Drillingsverbindung f <Metall> triple junction
Drillschraubendreher m <Maschinen> spiral ratchet screwdriver
Drillstem-Test m <Nichtfoss Energ> drill stem test
dringendes Gespräch n <Telekom> urgent call
Dringlichkeit f <Comp & DV, Qual> precedence
dritte Harmonische f <Elektronik> third harmonic
dritte Oberwelle f <Elektronik> third harmonic
dritte Partei f <Patent> third party
dritte Potenz f <Fertig, Math> cube • **in dritte Potenz erheben** <Math> cube
dritte Reinigungsstufe f <Abfall> tertiary sewage treatment
dritte Schiene f <Eisenbahn> live rail, third rail
dritte Wurzel f <Math> cube root
Drittelgeviert n <Druck> three-to-em space
Drittelspatium n <Druck> thick space
dritten Grades <Math> cubic
dritter Brand m <Ker & Glas> third firing
dritter Hauptsatz m **der Thermodynamik** <Phys> third law of thermodynamics
Drop-Frame-Anzeige f <Fernseh> drop-frame indicator
Drop-Frame-Betrieb m <Fernseh> drop-frame mode
Drossel f 1. <Elektriz, Elektrotech> choke, choking coil, inductor, reactor; 2. <Erdöl> jet (Bohrtechnik); choke (um Rohrströmung zu verhindern); 3. <Fertig> damper, restrictor; 4. <Funktech> choke; 5. <Lufttrans> throttle (Luftfahrzeug); 6. <Telekom> choke • **ohne Drossel** chokeless
Drosselanschlagschraube f <Kfztech> throttle stop screw (Vergaser)
Drosselblende f 1. <Erdöl, Heiz & Kälte> orifice plate; 2. <Maschinen> orifice meter
Drosselklappe f 1. <Bau> damper; 2. <Fertig> choke; throttle valve (Kunststoffinstallationen); 3. <Heiz & Kälte> butterfly damper, damper flap; 4. <Kfztech> throttle plate, throttle valve; throttle (Vergaser); 5. <Maschinen> butterfly valve, throttle valve
Drosselklappendämpfer m <Kfztech> throttle dashpot (Vergaser)
Drosselklappenhebel m <Kfztech> throttle control lever
Drosselklappenschalter m <Kfztech> throttle valve switch
Drosselklappenventil n 1. <Hydraul> butterfly throttle-valve; 2. <Lufttrans> throttle (Luftfahrzeug)
Drosselkolben m <Elektrotech> choke plunger
Drosselladedruckventil n <Kfztech> throttle boost pressure valve
Drosselluftklappe f **des Vergasers** <Mechan> choke
drosseln v 1. <Fertig> throttle (Kunststoffinstallationen); 2. <Kfztech> choke; 3. <Lufttrans> throttle back; 4. <Maschinen> throttle
Drosseln n 1. <Eisenbahn, Maschinen> throttling; 2. <Lufttrans> choking
Drosselplatte f <Kfztech> restrictor plate
Drosselschaltung f <Elektriz> choke circuit
Drosselschieber m 1. <Heiz & Kälte> damper slide; 2. <Kfztech> throttle slide (Vergaser)
Drosselspeisung f <Elektriz> choke feed
Drosselspule f 1. <Elektriz> choke coil; 2. <Elektrotech> choke, choke coil, inductor, reactance coil, reactor; 3. <Telekom> choke
Drosselstoß m **der Schienen** <Eisenbahn> impedance bond, reactance bond
Drosselung f 1. <Akustik> partial masking; 2. <Eisenbahn> throttling; 3. <Lufttrans> choking; 4. <Maschinen> throttling
Drosselventil n 1. <Fertig> throttle valve; 2. <Heiz & Kälte> restrictor valve, throttle valve; 3. <Hydraul> butterfly valve, throttle valve; 4. <Kfztech> throttle valve; 5. <Kohlen> choker valve; 6. <Maschinen> butterfly valve
Drosselventilhebel m <Hydraul> throttle lever
Drosselventilstange f <Hydraul> throttle rod, throttle stem
Drosselwirkung f <Raumfahrt> throttle control (Raumschiff)
Drosselzapfendüse f <Kfztech> throttle pintle nozzle
Druck m 1. <Bau> thrust; 2. <Comp & DV> print, printout; 3. <Elektrotech> pressure; 4. <Hydraul> head, pressure delivery, pressure head; pressure (Dampf); head (zum Gegenpumpen); 5. <Mechan> pressure, push; 6. <Meerschmutz> thrust; 7. <Papier> pressure; 8. <Phys> sound pressure; 9. <Raumfahrt> pressure (Raumschiff); 10. <Textil, Thermod> pressure • **auf normalen Druck bringen** <Kerntech> depressurize • **mit Druck beaufschlagen** <Heiz & Kälte> pressurize • **unter Druck gesetzt** <Mechan> pressurized • **unter Druck halten** <Heiz & Kälte> pressurize • **unter Druck setzen** <Heiz & Kälte> pressurize • **unter Druck verformbar** <Fertig> malleable
Druck m **auf volle Bogen** <Comp & DV, Druck> even working
Druck... <Fertig> compressed
Druckabbau m <Raumfahrt> depressurization (Raumschiff)
Druckabdichtung f <Hydraul> pressure seal
Druckabfall m 1. <Fertig> friction pressure drop, pressure drop; head loss (Hydraulik); 2. <Heiz & Kälte, Hydraul, Nichtfoss Energ> head loss, pressure drop; 3. <Papier> pressure drop
druckabhängig adj <Fertig> pressure-dependent
druckabhängiges Atemschutzgerät n <Sicherheit> pressure-demand respirator
Druckabhängigkeit f <Fertig> pressure dependence
Druckablassventil n <Kfztech> relief valve (Schmierung)
Druckamplitude f <Akustik> pressure amplitude
Druckänderung f <Wellphys> pressure variation
Druckänderungs-Geschwindigkeitsregulierung f <Lufttrans> pressure rate-of-change regulating
Druckänderungs-Geschwindigkeitsschalter m <Lufttrans> pressure rate-of-change switch
Druckangleichsunfall m <Kerntech> blowdown accident
Druckanschluss m <Elektrotech> pressurized connection

Druckanstieg

Druckanstieg *m* 1. <Heiz & Kälte> pressure rise; 2. <Maschinen> pressure build-up
Druckanzeiger *m* 1. <Gerät> pressure indicator; 2. <Heiz & Kälte> *(AE)* indicating pressure gage, *(BE)* indicating pressure gauge
Druckanzug *m* 1. <Raumfahrt> G-suit, pressure suit *(Raumschiff)*; 2. <Sicherheit> positive-pressure suit
Druckaufbau *m* <Lufttrans> pressure build-up
Druckaufbauventil *n* <Lufttrans> pressurizing valve
Druckauflage *f* <Comp & DV> print run
Druckaufnehmer *m* <Metrol> barometric sensor, pressure gauge, pressure pick-up, pressure sensor, pressure transducer
Druckauftrag *m* <Druck> print job
Druckausbreitung *f* <Phys> pressure broadening
Druckausgabe *f* 1. <Comp & DV> hard copy; 2. <Druck> printer output; 3. <Telekom> printout
Druckausgabepuffer *m* <Druck> print buffer
Druckausgabeverkleinerung *f (COR)* <Comp & DV> character output reduction *(COR)*
Druckausgleich *m* 1. <Fertig> pressure balance, pressure compensation; 2. <Heiz & Kälte> expansion
Druckausgleichbehälter *m* <Nichtfoss Energ> surge tank
Druckausgleichdose *f* <Mechan> expansion bellows
Druckausgleichkolben *m* <Hydraul> balancing piston
Druckausgleichventil *n* <Hydraul> balanced valve
Druckausrichtung *f* <Comp & DV> print position
Druckausrüstung *f* <Erdöl> pressure equipment *(Tieftauchtechnik)*
Druckautomat *m* <Druck> flat-bed cylinder press
Druckbalg *m* <Gerät> pressure bellows
Drückbank *f* 1. <Fertig> lathe *(Metalldrücken)*; 2. <Maschinen> spinning lathe
Druckbeanspruchung *f* 1. <Maschinen, Mechan, Metall> compression stress; 2. <Qual> compression load, compressive strength; 3. <Verpack> compression load; 4. <Werkprüf> compressive strength
Druckbefehl *m* <Comp & DV> print command
Druckbegrenzungsventil *n* <Maschinen> pressure relief valve
Druckbehälter *m* <Fertig, Kerntech, Maschinen> pressure vessel
druckbelastet *adj* <Sicherheit> pressurized
druckbelüfteter Motor *m* <Elektriz> forced-ventilation motor
Druckbelüftung *f* <Lufttrans> pressurization
druckbeständig *adj* <Qual, Werkprüf> compression-proof
Druckbogen *m* <Druck> sheet, signature
Druckbolzen *m* <Kfztech> nozzle holder spindle
Druckbuchstabe *m* <Druck> printing letter
Druckbügelregler *m* <Gerät> chopper bar controller
Druckcharakteristik *f* <Wassertrans> pressure characteristic *(Wetterkunde)*
Druckdatei *f* <Druck> print file *(Digitaldruck)*
Druckdecke *f* <Textil> blanket
druckdicht *adj* <Fertig> pressure-sealed, pressure-tight; leakproof under pressure *(Kunststoffinstallationen)*
druckdichter Boden *m* <Lufttrans> pressurized floor
druckdichter Druckluftverteiler *m* <Lufttrans> pressurized manifold
druckdichter Wagen *m* <Eisenbahn> pressure-sealed wagon
Druckdichtung *f* <Eisenbahn> pressure sealing
Druckdifferenz *f* 1. <Hydraul> pressure difference; 2. <Kfztech> pressure differential
Druckdifferenzmelder *m* <Regelung> pneumatic limit operator
Druckdifferenzregelventil *n* <Kfztech> pressure differential warning valve
Druckdomäne *f* <Comp & DV> printing domain

Druckdose *f* <Mechan> *(AE)* crusher gage, *(BE)* crusher gauge
Druckecho *n* <Aufnahme> printing echo
drucken *v* <Comp & DV, Druck, Papier, Textil> print
Drucken *n* 1. <Comp & DV> printing; 2. <Druck> printing, printing process; 3. <Ker & Glas, Papier, Textil> printing
Drucken von Hand <Druck> hand printing
drücken *v* 1. <Comp & DV> push *(Taste)*; 2. <Fertig> drag *(Schneidenrücken)*; 3. <Kunststoff> pressure-form; 4. <Maschinen> spin; 5. <Mechan> push
Drücken *n* 1. <Comp & DV> push; 2. <Fertig> dragging *(Spanung)*; 3. <Kunststoff> pressure forming; 4. <Maschinen> spinning
druckentlastetes Ventil *n* <Hydraul> balanced valve
Druckentlastungseinrichtung *f* <Sicherheit> pressure-relief device
Druckentnahmebohrung *f* <Fertig> pressure tap
Druckentspannungsventil *n* <Hydraul> pressure relief valve
Drucker *m* 1. <Comp & DV> printer; 2. <Druck> printer *(Maschine)*; printer *(Person)*; 3. <Telekom> printer
Drucker mit fliegendem Abdruck <Comp & DV> hit-on-the-fly printer
Drucker mit Typenradwalze <Comp & DV> *(BE)* barrel printer, *(AE)* drum printer
Drucker und Setzer *m* <Druck> printer and typesetter
Drücker *m* <Fertig> spinner
Druckerauftrag *m* <Comp & DV> print job *(in Druckerwarteschlange)*
Druckerausgabe *f* <Druck> printer output
Drückerauslöser *m* <Foto> trigger release *(Kameraverschluss)*
Druckerbefehle *mpl* <Druck> printer commands
Druckerdgas *n (CNG)* <Kohlen> compressed natural gas, CNG
Druckerei *f* <Druck> print shop, printing shop, printing trade, printing works
Druckereiarbeiter *m* <Druck> printing trade worker
Druckereibedarf *m* <Druck> printer's supply
Druckereibetrieb *m* <Druck> printing shop
Druckereigeräte *npl* <Druck> printing equipment
Druckereizubehör *n* <Druck> printing accessories
Druckerfarbe *f* <Druck> printer's ink, printing ink
Druckergewerbe *n* <Druck> printing trade
Druckerhöhungspumpe *f* 1. <Erdöl> booster pump; 2. <Raumfahrt> boost pump
Druckerkunst *f* <Druck> art of printing
Druckerlaubnis erteilen *v* <Druck> pass for press
Druckerlehrling *m* <Druck> printer's devil
Druckerpapier *n* <Druck> printing paper
Druckerpool *m* <Comp & DV> printer pool
Druckerpresse *f* <Druck> press
Druckerpuffer *m* <Druck> print buffer
Druckerschnittstelle *f* <Comp & DV> hard copy interface
Druckerschwärze *f* <Druck> printer's ink, printing ink
Druckerserver *m* <Comp & DV> print server
Druckersprache *f* <Druck> printer language
Druckertreiber *m* <Comp & DV> printer driver
Druckerzeichen *n* <Druck> printer's mark
Druckerzeuger *m* <Fertig> pressure generator
Druckerzeugung *f* <Trans> pressurization
Druckfarbe *f* 1. <Kunststoff> printing ink; 2. <Papier> ink, printing ink
druckfärben *v* <Textil> dye under pressure
Druckfeder *f* <Maschinen> compression spring
Druckfehler *m* <Druck> misprint, typo
druckfest *adj* <Qual, Werkprüf> compression-proof
druckfeste Kapselung *f* <Sicherheit> flame-proof enclosure

Druckfestigkeit f 1. <Bau, Kunststoff> compressive strength; 2. <Maschinen, Metall, Qual, Verpack> compression strength
Druckfestigkeitsversuch m <Bau> crushing test
Drückfett n <Fertig> spinning lubricate
Druckfilter n <Abfall, Chemtech, Kohlen, Wasserversorg> pressure filter
Druckfirnis m <Druck> printing varnish
Druckfläche f <Druck> printing area • **über die Druckfläche hinausgehen** <Druck> bleed
Druckflasche f **für Helium** <Raumfahrt> compressed-helium bottle *(Raumschiff)*
Druckflüssigkeit f <Fertig> pressure liquid
Druckform f 1. <Comp & DV> form; 2. <Druck> *(BE)* forme, *(AE)* printing form, *(BE)* printing forme, *(AE)* type form, *(BE)* type forme
Druckformat n 1. <Comp & DV> print format; 2. <Druck> print format, printing format
Druckformträger m <Druck> carriage
Druckfreistrahlgebläse n **zum Gussputzen** <Fertig> air blast cleaning unit
Druckfüllung f <Verpack> press filling *(für Sprühdosen)*
Drückfutter n <Maschinen> spinning mandrel
Druckgang m <Druck> print cycle, printing cycle
Druckgas n 1. <Kohlen> compressed gas; 2. <Raumfahrt> pressurizing gas *(Raumschiff)*
Druckgasflasche f <Fertig> pressure cylinder *(Schweißen)*
druckgasisoliertes Kabel n <Elektriz> external gas pressure cable, internal gas pressure cable
Druckgaskabel n <Telekom> gas-cushion cable
Druckgas-Leistungsschalter m <Elektrotech> gas-blast circuit-breaker
Druckgastank m <Raumfahrt> pressurizing gas tank *(Raumschiff)*
Druckgefälle n 1. <Erdöl, Fertig> pressure gradient; 2. <Heiz & Kälte> pressure drop
Druckgefäß n 1. <Bau, Heiz & Kälte> pressure vessel; 2. <Labor> autoclave; 3. <Mechan> pressure vessel; 4. <Phys> reactor pressure vessel
druckgegossen *adj* <Fertig> die-cast *(Metall)*
druckgekühlter Motor m <Elektriz> forced-ventilation motor
Druckgeschwindigkeit f 1. <Comp & DV> print speed; 2. <Druck> print rate, printing rate, print speed, printing speed
druckgießen v <Maschinen> die-cast
Druckgießform f <Fertig> die-casting die
Druckglied n <Bau> strut
Druckgradient m <Erdöl, Strömphys> pressure gradient
Druckgradient m **null** <Strömphys> zero pressure gradient *(Untersuchung von Grenzschichten)*
Druckgradientendiagramm n <Erdöl> pressure vs depth plot *(Geologie: Darstellung von Druck in Abhängigkeit zur Tiefe)*
Druckgradientenmikrofon n <Akustik> velocity microphone
Druckgrenze f <Hydraul> head limit
Druckguss m 1. <Fertig> die casting *(Metall)*; 2. <Maschinen> die casting
Druckgussautomat m <Fertig> automatic die-casting machine
Druckgussform f <Maschinen> die-casting die, pressure die-casting die
Druckgusslegierung f <Fertig> die-casting alloy
Druckgussmaschine f <Maschinen> die-casting machine
Druckhammer m <Druck> printing hammer
Druckhebewinde f <Fertig> hydraulic jack

Druckhöhe f 1. <Heiz & Kälte> head, pump head; 2. <Hydraul> pressure head, static head; 3. <Kerntech> elevation head *(einer Pumpe)*; 4. <Kohlen> pressure head; 5. <Lufttrans> pressure altitude, pressure height; 6. <Maschinen> elevation head; 7. <Nichtfoss Energ> effective head
Druckhöhenmessgerät n <Gerät> head meter
Druckimpuls m <Akustik> pressure impulse
Druckindustrie f <Druck> printing industry
Druckjob m <Druck> print job
Druckkabel n <Elektriz> pressure cable
Druckkabine f 1. <Raumfahrt> cabin *(Raumschiff)*; 2. <Sicherheit> pressurized cabin
Druckkammer f <Fertig> casting chamber, pressure chamber
Druckkammerlautsprecher m <Akustik> pneumatic loudspeaker
Druckkappe f <Kfztech> pressure cap
Druckkessel m <Mechan> pressure vessel
Druckkissen n <Verpack> pressure pad *(zur Unterlage)*
Druckkleidung f <Sicherheit> pressurized clothing
Druckknopf m 1. <Comp & DV, Elektriz> push button; 2. <Elektrotech> button; 3. <Kontroll> button, push button; 4. <Lufttrans> push button *(für das Wiederanstellen des Triebwerks im Flug)*; 5. <Maschinen, Mechan> push button; 6. <Telekom> button, press button, push button
Druckknopfanlasser m <Elektriz> push-button starter
Druckknopfschalter m <Trans> pedestrian push button
Druckknopfsteuerung f <Mechan> push-button control
Druckknoten m <Fertig> pressure node
Druckkochtopf m <Maschinen> pressure cooker
Druckkoeffizient m <Maschinen, Nichtfoss Energ> pressure coefficient
Druckkolben m <Bau> ram
Druckkopf m 1. <Comp & DV> print head; 2. <Druck, Papier> printing head
Druckkraft f 1. <Fertig> tonnage; 2. <Maschinen, Qual> compressive force
Druckkraftfestigkeit f <Mechan> compression strength
Druckkufe f <Aufnahme> pressure pad
Druckkühlung f <Maschinen> compression refrigeration
Druckkurve f <Hydraul> pressure curve
Drucklager n 1. <Maschinen> thrust bearing; 2. <Wassertrans> thrust bearing, thrust block *(Motor)*
Drucklagerwelle f <Wassertrans> thrust shaft *(Antriebsanlage)*
Drucklegung f <Druck> going to press
Druckleiste f <Fertig> thrust strip
Druckleistung f <Druck> printing rate *(einer Druckmaschine)*
Druckleitung f 1. <Hydraul> head pipe; 2. <Kfztech> delivery pipe; 3. <Nichtfoss Energ> penstock; 4. <Wasserversorg> delivery pipe, penstock
Druckleitungsanschluss m <Fertig> outlet
Drucklinie f <Kohlen> compression curve
drucklos geöffnet *adj* <Fertig> fail-safe to open *(Kunststoffinstallationen)*
drucklos geschlossen *adj* <Fertig> fail-safe to close *(Kunststoffinstallationen)*
drucklose Leitung f <Kerntech> unpressurized line
druckloser Abschnitt m **des Flugzeugrumpfes** <Lufttrans> fuselage non-pressurized section
druckloser Refiner m <Papier> atmospheric refiner
Druckluft f <Anstrich, Maschinen> compressed air
Druckluftakkumulator m <Fertig> hydropneumatic accumulator
Druckluftanhebung f <Fertig> airlift
Druckluftatemgerät n <Sicherheit> breathing apparatus *(für industrielle Einsätze)*

druckluftbetätigte

druckluftbetätigte Spanneinrichtung f <Fertig> air clamp
druckluftbetätigtes Ventil n <Maschinen> pneumatically operated valve
Druckluftbohren n <Erdöl> air flooding *(Bohrtechnik)*
Druckluftbohrer m 1. <Bau> pneumatic drill; 2. <Maschinen> air drill, compressed air drill; 3. <Mechan, Phys> air drill
Druckluftbohrmaschine f <Fertig> pneumatic drill
Druckluftbremse f 1. <Kfztech> air brake, air pressure brake, compressed air brake, pneumatic brake; 2. <Maschinen> air brake, compressed air brake, pneumatic brake; 3. <Mechan, Phys> air brake
Druckluftbremsschlauch m <Kunststoff> air brake hose
Druckluftbremssystem n <Kfztech, Maschinen> compressed air braking system
Druckluft-Differenzialkolbenakkumulator m <Fertig> differential accumulator
Drucklufteinbringung f <Erdöl> air repressuring *(Öllagerstätte)*
Drucklüfter m 1. <Heiz & Kälte> *(AE)* forced-draft fan, *(BE)* forced-draught fan; 2. <Maschinen> pressure fan
Druckluftflasche f 1. <Fertig, Heiz & Kälte> air bottle; 2. <Mechan, Phys> air cylinder
Druckluftförderer m <Fertig> pneumatic conveyor
Druckluftförderung f 1. <Fertig> pneumatic conveying *(von Schüttgut)*; 2. <Kohlen> pneumatic handling
Druckluftfutter n 1. <Fertig> air-actuated chuck; 2. <Maschinen> air chuck, air-operated chuck
Druckluftgebläse n <Fertig> air lance
druckluftgekühlter Transformator m <Elektriz> air blast transformer
Druckluftgeräte npl <Maschinen> compressed air equipment
druckluftgesteuerte Vakuumbremse f <Eisenbahn> air-pressure-controlled vacuum brake
Druckluftgussputzen n <Fertig> air blasting
Drucklufthammer m <Maschinen> pneumatic hammer
Drucklufttheber m 1. <Chemtech> airlift pump; 2. <Maschinen> air hoist, pneumatic hoist
Druckluft-Kolbenakkumulator m <Fertig> air-loaded accumulator with piston
Druckluftkraftwerk n <Elektrotech> compressed-air power station
Druckluftkreisel m <Lufttrans> aerogyro *(Navigation)*
Druckluftleistungsschalter m <Elektrotech> air blast circuit breaker
Druckluftleitung f <Kfztech, Maschinen> compressed air line
Druckluftmeißel m <Fertig> pneumatic chipping hammer
Druckluftmotor m <Maschinen> air engine, air motor, compressed air engine
Druckluft-Niethammer m <Maschinen> pneumatic riveter
Druckluft-Pressformmaschine f <Fertig> air-operated squeezer
Druckluftprobe f <Fertig> pneumatic test
Druckluft-Putzstrahlen n <Fertig> air blast cleaning, compressed air blast-cleaning
Druckluftschalter m <Elektriz, Elektrotech> air-blast breaker, air-blast circuit breaker, air-blast switch, air breaker, air-pressure circuit breaker, air-pressure switch, pneumatic switch
Druckluftschlagbohrer m <Bau> pneumatic hammer drill
Druckluftschlauch m <Bau, Fertig, Maschinen, Mechan, Phys> compressed air hose
Druckluftschleuse f <Wasserversorg> lock
Druckluftschrauber m <Fertig> pneumatic screw driver

Druckluftspanndorn m <Fertig> pneumatic expanding mandrel
Druckluftspannung f <Fertig> air chucking
Druckluftspeicher m <Fertig> air accumulator, air hydraulic accumulator
Druckluftstampfer m <Fertig> pneumatic rammer *(Gießen)*
Druckluftstrom m <Fertig> air blast
Druckluftstutzen n <Maschinen> compressed air socket
Druckluftsystem n 1. <Lufttrans> air pressure system; 2. <Maschinen> compressed air system
Drucklufttechnik f <Kfztech> pneumatics
Drucklufttest m <Fertig> pneumatic test
Drucklufttturbine f <Maschinen> air turbine
Drucklüftungssystem n <Heiz & Kälte> plenum system
Druckluftventil n <Maschinen> pneumatic valve
Druckluftwerkzeug n <Maschinen> air tool, pneumatic tool
Druckluftzange f <Fertig> pneumatic collet
Druckluftziehkissen n <Fertig> pneumatic die cushion
Druckluftzylinder m 1. <Lufttrans> pneumatic cylinder *(Hubschrauber)*; 2. <Maschinen> compressed air cylinder, pneumatic cylinder; 3. <Mechan, Phys> air cylinder
Druckmaschine f 1. <Druck> press printing machine, printing press; 2. <Textil> printing machine
Drückmaschine f <Maschinen> spinning lathe
Druckmedien npl <Druck> print media
Druckmessdose f 1. <Gerät> load cell; 2. <Kohlen> pressure cell
Druckmesseinrichtung f <Gerät> pressure-measuring equipment
Druckmesser m 1. <Bau> *(AE)* pressure gage, *(BE)* pressure gauge; 2. <Fertig> manometer, *(AE)* pressure gage, *(BE)* pressure gauge; 3. <Heiz & Kälte, Hydraul> *(AE)* pressure gage, *(BE)* pressure gauge; 4. <Kohlen> compressometer; 5. <Kontroll> *(AE)* pressure gage, *(BE)* pressure gauge; 6. <Labor> manometer; 7. <Papier> *(AE)* pressure gage, *(BE)* pressure gauge; 8. <Phys> manometer
Druckmessgerät n 1. <Bau> pressure meter; 2. <Erdöl> manometer, *(AE)* pressure gage, *(BE)* pressure gauge; 3. <Gerät> head meter; 4. <Hydraul, Kohlen, Labor, Phys> *(AE)* pressure gage, *(BE)* pressure gauge
Druckmessung f <Raumfahrt> pressure measurement *(Raumschiff)*
Druckmikrofon n <Akustik> pressure microphone
Druckminderer m <Heiz & Kälte, Lufttrans, Maschinen> pressure reducer
Druckminderung f <Maschinen> depressurization
Druckminderungsventil n 1. <Heiz & Kälte> pressure-reducing valve; 2. <Maschinen> reducing valve
Druckminderventil n 1. <Bau, Heiz & Kälte> pressure-reducing valve; 2. <Fertig> pressure relief valve, pressure-reducing valve; 3. <Maschinen> depressurization valve, pressure-reducing valve, reducing valve
Druckmittelpunkt m 1. <Lufttrans> *(AE)* center of pressure, *(BE)* centre of pressure *(Aerodynamik)*; 2. <Phys> *(AE)* center of pressure, *(BE)* centre of pressure
Druckmittelpunkt m des Blattes der Luftschraube <Lufttrans> *(AE)* blade center of pressure, *(BE)* blade centre of pressure *(Hubschrauber)*
Druckmodul m <Kunststoff> compressive modulus
druckölbetätigt adj <Fertig> oil-actuated
Druckölbetätigung f <Fertig> oil actuation
Druckölbrenner m <Heiz & Kälte> pressure jet oil burner
Druckölschmierung f <Maschinen> forced-oil cooling
Druckpapier n 1. <Foto> printing paper; 2. <Verpack> printings
Druckplatte f 1. <Druck> plate, printing plate; 2. <Fertig> pressure plate *(Kunststoffinstallationen)*; 3. <Kfztech>

pressure plate; driven plate *(Kupplung)*; 4. <Maschinen> thrust plate
Druckplatte f **auf Polyesterbasis** <Druck> polyester-based plate
Druckplatte f **für wasserlosen Offsetdruck** <Druck> waterless plate
Druckplattenantriebsriemen m <Kfztech> pressure plate drive strap
Druckplattenausrückgabel f <Kfztech> pressure plate release lever *(Kupplung)*
Druckplattenausrückhebel m <Kfztech> pressure plate release lever
Druckplattenfeder f <Kfztech> pressure plate spring
Druckplattenherstellung f <Druck> platemaking
Druckposition f <Comp & DV> print position
Druckpresse f <Druck> press, press printing machine, printing press
Druckpresse f **für einzelne Bogen** <Druck> sheet-fed machine
Druckprobe f 1. <Maschinen> pressure test; 2. <Mechan> compression test
Druckpropeller m <Lufttrans> pusher propeller
Druckprüfmaschine f <Verpack> compression test machine
Druckprüfung f <Kohlen, Qual, Verpack, Werkprüf> compression test
Druckpuffer m <Druck> print buffer
Druckpumpe f 1. <Hydraul, Kohlen> pressure pump; 2. <Maschinen> double-acting pump; 3. <Wasserversorg> force pump, pressure pump, ram pump
Druckpunkt m <Comp & DV> key force
Druckraumabschluss m <Fertig> pressure seal
Druckreduzierer m <Raumfahrt> pressure reducer *(Raumschiff)*
Druckreduzierung f <Phys> pressure reduction
Druckreduzierventil n <Heiz & Kälte, Hydraul, Sicherheit> pressure-reducing valve, safety valve
Druckreflexionsfaktor m <Akustik> pressure reflection coefficient
Druckregelung f <Heiz & Kälte> pressure control
Druckregelventil n <Fertig> pressure control valve, pressure-regulating valve
Druckregler m 1. <Bau, Elektrotech> pressure regulator; 2. <Heiz & Kälte> pressure controller; 3. <Hydraul> pressure regulator; 4. <Lufttrans> barostat; 5. <Papier> pressure controller
druckreife Vorlage f <Druck> CRC, camera ready copy
Druckring m 1. <Fertig> reaction ring; 2. <Kfztech, Maschinen> thrust collar
Druckriss m 1. <Bau> compression crack; 2. <Ker & Glas> pressure check
Druckrohr n 1. <Fertig> gun; 2. <Kfztech> pressure tube; 3. <Maschinen> pressure pipe
Druckröhrenreaktor m <Kerntech> pressure tube reactor
Druckrohrleitung f 1. <Nichtfoss Energ> penstock; 2. <Wasserversorg> penstock, pentrough
Druckrohrtemperatur f <Phys, Thermod> discharge temperature
Druckrolle f <Maschinen> pressure roller
Drucksache f <Druck> printed matter
Drucksammler m <Papier> pressure accumulator
Druckschacht m <Wasserversorg> pressure well
Druckschalter m <Elektrotech, Fertig, Hydraul> pressure switch
Druckschaltungsverbinder m <Elektrotech> printed circuit connector
Druckscheibe f <Fertig> thrust washer; washer *(Kunststoffinstallationen)*

Druckschlauchgerät n <Sicherheit> airline respirator, pressure-demand respirator
Druckschmieden n <Maschinen> press forging
Druckschmierung f 1. <Fertig> force-feed lubrication, pressure-feed lubrication; 2. <Heiz & Kälte> forced lubrication; 3. <Kfztech> pressure lubrication; 4. <Maschinen> pressure-feed lubrication
Druckschott n <Lufttrans> pressure bulkhead
Druckschraube f 1. <Fertig> clamping nut; 2. <Maschinen> pressure screw
Druckschreiber m <Gerät> pressure recorder, recording manometer
Druckschwankung f <Fertig> pressure fluctuation
Druckschwankungs-Ausgleichsakkumulator m <Fertig> alleviator
Druckschweißen n <Thermod> pressure welding
Druckschweißung f 1. <Fertig> upset welding; 2. <Thermod> pressure welding
Druckseite f 1. <Druck> page, printed sheet; 2. <Fertig> discharge side; 3. <Hydraul> pressure side
Drucksensordetektor m <Eisenbahn> pressure-sensitive detector
Druckserver m <Comp & DV> print server
Drucksintern n <Chemtech> sintering under pressure
Druckspalte f <Druck> column, print column
Druckspannung f **parallel zum Korn** <Bau> compressive stress parallel to grain
Druckspannung f **senkrecht zum Korn** <Bau> compressive stress perpendicular to grain
Druckspant m <Raumfahrt> pressure bulkhead *(Raumschiff)*
Druckspeicher m 1. <Fertig> pressure accumulator; 2. <Heiz & Kälte> pressurized hot water tank; 3. <Kunststoff, Maschinen> hydraulic accumulator
Druckstab m 1. <Bau> strut; 2. <Fertig> compression member; 3. <Mechan> strut
Druckstelle f <Comp & DV> print position
Druckstempel m <Fertig> thrust die
Drucksterilisator m <Lebensmittel> autoclave
Drucksteuerung f <Heiz & Kälte> pressure control
Drucksteuerzeichen n <Comp & DV> print control character
Druckstift m <Maschinen> pressure pin
Druckstock m <Druck> block, printing block
Druckstockherstellung f <Druck> blockmaking
Druckstockhöhenprüfer m <Druck> *(AE)* type height gage, *(BE)* type height gauge
Druckstoß m <Erdöl> surge; water hammer *(Rohrleitungen)*
Druckstoßdrossel f <Lufttrans> antisurge baffle
Druckstoßventil n <Lufttrans> antisurge valve
Druckstück n <Fertig> compressor *(Kunststoffinstallationen)*
Druckstufe f <Fertig, Hydraul> pressure stage
Drucksturz m <Thermod> explosive decompression
Drucksystem n <Lufttrans> pressure system
Drucktaste f 1. <Comp & DV> push button; 2. <Elektrotech> key; 3. <Fertig> punchbutton key; 4. <Maschinen> push button; 5. <Telekom> press button, push button
drucktastengeschaltet adj <Fertig> punchbutton-operated
Drucktastenschalttafel f <Fertig> punchbutton panel
Drucktastensteuerung f <Fertig> punchbutton control
Drucktastentafel f <Fertig> punchbutton panel
Drucktaster m <Elektrotech> push-button switch
Drucktechnik f <Druck> printing technology
Drückteil m <Fertig> spun part
Druck-Temperatur-Schreiber m <Phys> barothermograph

Drucktendenz

Drucktendenz f <Wassertrans> pressure tendency *(Wetterkunde)*
Drucktest m <Hydraul> pressure test
Drucktisch m <Druck> printing table
Drucktransmitter m <Hydraul> pressure transmitter
Drucktuch n <Textil> blanket
Druckturbine f <Hydraul> pressure turbine
Drucktype f <Druck> movable type, printing type
Druckübertragung f <Fertig> pressure transmission
Druckübertragungsgerät n <Hydraul> pressure transmitter
Druckumformmaschine f <Verpack> pressure-forming machine
Druckumlaufschmierung f 1. <Fertig> pressure circulation lubrication, pressure lubrication; 2. <Heiz & Kälte> forced lubrication; 3. <Kfztech> forced-feed lubrication; 4. <Maschinen> force-feed lubrication
Druckunterlegscheibe f <Kfztech> thrust washer
Druckunterschied m <Heiz & Kälte> pressure differential
Druckventil n 1. <Heiz & Kälte> discharge valve; 2. <Maschinen> delivery valve, head valve
Druckverbindung f <Elektrotech> pressurized connection
Druckverbreiterung f <Strahlphys> pressure broadening *(von Spektrallinien)*
Druckverformung f <Verpack> compression damage
Druckverformungsrest m <Kunststoff> compression set, residual set
Druckverhältnis n <Fertig> pressure ratio
Druckverhältnis n **im Triebwerk** <Lufttrans> engine pressure ratio
Druckverlust m 1. <Erdöl, Heiz & Kälte> pressure drop; 2. <Hydraul> loss of pressure; 3. <Kfztech> deflation *(Reifen)*; 4. <Maschinen> pressure loss; 5. <Nichtfoss Energ> pressure drop
Druckvermerk m <Konstzeich> notation
Druckverringerung f <Maschinen> depressurization
Druckverstärker m <Hydraul> intensifier
Druckverstärkerpumpe f <Mechan> booster pump
Druckversuch m <Fertig, Kohlen, Maschinen, Metall, Verpack> compression test
Druckversuch m **an Rohren** <Fertig, Qual> collapse test
Druckverteilung f <Fertig> pressure distribution
Druck-Volumen-Diagramm n *(P/V-Diagramm)* <Thermod> pressure volume diagram
Druckvorbereitung f <Druck> prepress
Druckvorgang m <Druck> printing process
Druckvorlage f <Konstzeich> photomaster
Druckvorlagenhersteller m <Druck> process engraver
Druckvorrichtung f <Druck> printing apparatus
Druckvorstufe f <Druck> prepress
Druckvorstufenbetrieb m <Druck> prepress plant, prepress shop
Druckwaage f <Fertig> pressure-maintaining valve
Druckwächter m 1. <Fertig> pressure control device; 2. <Heiz & Kälte> pressure switch
Druckwalze f 1. <Comp & DV> print drum; 2. <Druck> printing roller, roller
Druckwalzenmasse f <Druck> printing roller composition
Druckwandler m <Gerät> pressure transducer
Druckware f <Textil> printed fabric
Druckwasser n <Wasserversorg> pressurized water, water under pressure
Druckwasserkolben m <Bau> hydraulic piston
Druckwasserkühlung f <Heiz & Kälte> pressurized water cooling
Druckwasserreaktor m <Kerntech, Phys> pressurized water reactor
Druckwasserspeicher m <Fertig, Papier> hydraulic accumulator
Druckwassersystem n <Hydraul> hydraulic system
Druckwelle f 1. <Akustik, Aufnahme> compressional wave; 2. <Hydraul> pressure surge; 3. <Telekom> pressure wave; 4. <Wassertrans> thrust shaft *(Antriebsanlage)*
Druckwellenfront f <Sicherheit> shock front *(Explosion)*
Drückwerkzeug n <Maschinen> spinning tool
Druckwiderstand m <Lufttrans> pressure drag
Druckzentrum n <Maschinen> *(AE)* center of pressure, *(BE)* centre of pressure
Druckzone f <Comp & DV> print position
Druckzuführung f <Kfztech> pressure feed
Druckzug m <Heiz & Kälte> *(AE)* forced draft, *(BE)* forced draught
Druckzylinder m 1. <Druck> cylinder, printing cylinder; 2. <Kunststoff> impression cylinder; 3. <Lufttrans> master cylinder *(Bremse)*; 4. <Papier> printing cylinder
Drüse f <Lebensmittel> gland
drusenreich adj <Erdöl> vuggy, vugular
drusig adj <Erdöl> vuggy, vugular *(Geologie)*
DSB *(Zweiseitenband)* <Elektronik, Funktech> DSB *(double sideband)*
D-Schicht f 1. <Funktech> D-region; D-layer *(Wellenausbreitung)*; 2. <Phys> D-layer
DSI *(digitale Sprachinterpolation)* <Raumfahrt, Telekom> DSI *(digital speech interpolation)*
D-Steuerung f <Elektriz> derivative control
DSV *(Digitalsignalverbindung)* <Telekom> digital connection
DT *(Dichtigkeitstest)* <Erdöl> LOT *(leak-off test)*
DTA *(Differenzialthermoanalyse)* <Kunststoff, Thermod, Umweltschutz> DTA *(differential thermal analysis)*
DTL *(Dioden-Transistor-Logik)* <Elektronik> DTL *(diode transistor logic)*
DTP *(Desktop-Publishing)* <Comp & DV, Druck> DTP *(desktop publishing)*
Dual... 1. <Comp & DV> dual; 2. <Elektronik, Telekom> binary
Dual-Band-Handy n <Mobilkom> db-mobile, dual-band hand-held telephone, dual-band radio telephone
Dualdividierer m <Elektronik> binary divider
duale Schaltung f <Phys> dual network
duales Kartenbild n <Telekom> binary image
Dual-in-Line-Gehäuse n *(DIL-Gehäuse)* <Elektronik, Elektrotech, Funktech> dual-in-line package *(DIP)*
Dualinput m <Fernseh> dual input
Dualmultiplizierer m <Elektronik, Telekom> binary multiplier
Dual-Port-Speicher m <Comp & DV> dual port memory
Dualprozessor-Lastteilungssystem n <Telekom> dual-processor load-sharing system
Dualprozessorsystem n <Telekom> dual-processor system
Dualsubtrahierer m <Elektronik> binary subtractor
Dualzähler m <Gerät> binary counter, dual counter
Dualzahlmultiplizierer m <Elektrotech> binary multiplier
Dübel m 1. <Bau> dowel, dowel pin, joggle, plug, tenon, trenail; key *(Holzbau)*; 2. <Fertig> dowel pin, dowel, slip; 3. <Mechan> key, peg
Dübelbohrer m 1. <Bau> bradawl; 2. <Maschinen> pin drill
Dübelloch n <Bau> dowel hole
dübeln v 1. <Bau> joggle, plug; 2. <Fertig> dowel pin
Dübelstein m <Bau> anchorage block
Dublett n 1. <Nichtfoss Energ> doublet *(elektronisch)*; 2. <Phys> doublet *(Spektroskopie)*
Dublette f <Druck> doublet
Dublettstruktur f <Strahlphys> doublet structure *(von Spektrallinien)*

Ducht f <Wassertrans> thwart
Dückdalbe f <Wassertrans> mooring pile *(Festmachen)*
Dückdalben m <Wassertrans> mooring post *(Festmachen)*
DüE *(Datenübertragungseinrichtung)* 1. <Comp & DV> DCE *(data communication terminating equipment)*; 2. <Telekom> DCE *(data circuit terminating equipment)*
Düker m 1. <Bau> culvert; 2. <Telekom> underwater pipe *(Kabelführung)*
Dukt m <Funktech> duct; wave duct *(Wellenausbreitung)*
Duktausbreitung f <Funktech> duct propagation
duktil adj <Metall, Qual> ductile
Dulcin n <Chemie> dulcin, dulcine
Dulcit n <Chemie> ducite, ducitol, melampyrit
Dulcitol n <Chemie> ducite, ducitol, melampyrit
duldbare tägliche Aufnahmemenge f *(ADI)* <Lebensmittel> acceptable daily intake *(ADI)*
Dunkeladaptation f <Ergon> dark adaptation
Dunkelfeldbeleuchtung f 1. <Fertig> dark field illumination; 2. <Phys> dark field illumination, dark ground illumination
Dunkelkammer f <Foto> darkroom
Dunkelkammerbeleuchtung f <Foto> safelight
Dunkelkorrektur f <Fernseh> shading corrector
Dunkelleitung f <Elektrotech> dark conduction
Dunkellinienspektrum n <Raumfahrt> dark line spectrum
Dunkelschriftschirm m <Elektronik> dark trace screen
Dunkelsignal n <Fernseh> shading signal
dunkelste Rotglühhitze f <Metall> dark red heat
dunkelsteuern v <Fernseh> blank
Dunkelsteuerung f 1. <Elektronik> blanking *(End- und Peripheriegeräte)*; 2. <Fernseh> shading
Dunkelsteuerungssignal n <Elektronik> blanking signal
Dunkelstrom m <Elektrotech, Optik, Phys, Telekom> dark current
Dunkeltastung f <Fernseh> blanking
Dunkelwiderstand m <Elektrotech, Kerntech> dark resistance
dunkle Interferenzringe mpl <Phys> dark fringe
dunkle Rothitze f <Metall> blood red heat
dunkler Interferenzstreifen m <Phys> dark fringe
Dünn... 1. <Druck, Elektronik> thin; 2. <Fertig> *(AE)* light gage, *(BE)* light gauge; 3. <Papier> lightweight; 4. <Telekom> thin
Dunnage f <Wassertrans> dunnage
Dünnblech n <Fertig> *(AE)* light gage sheet metal, *(BE)* light gauge sheet metal
Dünndruckpapier n <Druck, Papier> airmail paper, *(BE)* banknote paper, Bible paper, India paper, *(AE)* onionskin paper, thin paper
Dünne f <Chemie> tenuity
dünne Linse f <Phys> thin lens
dünne Magnetschicht f <Elektronik> magnetic thin film
dünne Schicht f <Chemie> film, lamina
dünne transparente Goldschicht f <Ker & Glas> starved gold
dünne Verblendung f <Bau> veneer
dünnes Blech n <Metall> thin plate
dünnes Präparat n <Kerntech> thin source *(radioaktive Quelle)*
dünnes Spatium n <Druck> thin space
dünnflüssiger Schlamm m <Kerntech> liquid slurry
Dünnpapier n <Papier> lightweight paper
Dünnschaftschraube f <Maschinen> reduced shaft bolt
Dünnschicht f <Elektronik, Telekom> thin film, thin layer
Dünnschichtbauelement n <Elektronik> thin film device
Dünnschicht-Chromatographie f <Elektronik> thin layer chromatography

Dünnschicht-Elektrolumineszenz f *(TFEL)* <Elektronik> thin film electroluminescence *(TFEL)*
Dünnschichthybridschaltung f <Elektronik> thin film hybrid circuit
Dünnschichtkondensator m 1. <Elektronik> thin film capacitor; 2. <Telekom> thin layer capacitor
Dünnschichtleiter m <Elektronik> thin film conductor
Dünnschichtmaterial n <Elektronik> thin film material
Dünnschichtspeicher m <Comp & DV> thin film memory
Dünnschichttechnik f <Elektronik> thin film technology
Dünnschichttransistor m <Elektronik> thin film transistor
Dünnschicht-Wärmeaustauscher m <Heiz & Kälte> scraped-surface heat exchanger
Dünnschichtwellenleiter m <Telekom> thin film waveguide
Dünnschichtwiderstand m <Elektronik> thin film resistor
Dünntafelglas n <Ker & Glas> thin sheet glass
dünnwandig adj <Maschinen> thin-walled
dünnwandige Lagerschale f <Maschinen> thin-walled half-bearing
dünnwandiger Zylinder m <Maschinen> thin-walled cylinder
Dünnware f <Ker & Glas> thin ware
Dunst m 1. <Elektrotech> damp; 2. <Foto> atmospheric haze *(Atmosphäre)*; 3. <Kohlen> damp; 4. <Umweltschmutz> haze; 5. <Wassertrans> mist *(leichter Nebel)*
dünsten v <Lebensmittel> steam
Dunstkoeffizient m <Qual, Umweltschmutz> coefficient of haze
Dunstventil n <Lufttrans> exhalation valve
Dünung f <Wassertrans> swell *(Seezustand)*
Duodezbogen m <Druck> duodecimo
Duodezimalsystem n <Math> duodecimal system *(Zahlensystem zur Basis zwölf)*
Duotriode f <Elektronik> double triode
Duowalzwerk n <Fertig> two-high mill
Duplex n <Comp & DV, Elektrotech, Funktech, Telekom> duplex, full duplex
Duplexbetrieb m 1. <Comp & DV> duplex operation, duplexing; 2. <Funktech> full duplex *(Sprechfunk)*; 3. <Telekom> duplex operation
Duplexbrenner m <Lufttrans> duplex burner
Duplexkabel n <Elektriz> duplex cable, foto double cable
Duplexkarton m <Verpack> duplex board
Duplexkette f <Kfztech> double roller chain
Duplexkompressor m <Hydraul> duplex compressor
Duplexpappe f <Verpack> duplex board
Duplexpumpe f <Hydraul> duplex pump
Duplexschmelzverfahren n <Fertig> duplex process, duplexing
Duplex-Sprechweg m <Telekom> two-frequency channel
Duplexübertragung f <Comp & DV> duplex
Duplexverfahren n <Fertig> duplex process, duplexing
Duplexverkehr m <Funktech> duplex
Duplikat n <Foto> duplicate
Duplikatfilm m <Foto> duplicating film
duplizieren v <Anstrich> forge
Dur-Akkord m <Akustik> major chord
Duralumin n <Metall> duralumin
Duraluminium n <Mechan> duralumin
durchbiegen v <Fertig> form
Durchbiegen n 1. <Bau> bending; 2. <Fertig> forming
Durchbiegung f 1. <Bau> bending, camber, deflection, flexion; 2. <Eisenbahn> sagging; 3. <Kerntech> sag; 4. <Maschinen> deflection; 5. <Papier> bending; 6. <Wassertrans> sagging *(Schiffbau)*
Durchbiegung f unter Last <Maschinen> deflection under load

Durchblättern 182

Durchblättern n <Comp & DV> browsing *(einer Datenbank, eines Textes)*
durchbohren v <Bau> hole, pierce
durchbrechen v 1. <Bau> break through, hole, pierce; 2. <Fertig> hole
Durchbrechen/mit <Fertig> holed
durchbrennen v 1. <Elektriz> blow *(Sicherung)*; 2. <Funktech> fuse
Durchbrennen n <Elektriz> blowing, blowing out *(einer Sicherung)*
durchbrochene weiße Linie f <Bau> broken white line *(Straßenmarkierung)*
durchbrochenes Mauerwerk n <Bau> trellis work
Durchbruch m 1. <Bau> breakpoint, breakthrough; opening *(für Fenster, Tür)*; 2. <Elektriz> breakdown; 3. <Hydraul> opening *(in Trennwand, Platte)*; 4. <Kohlen, Textil> breakthrough
Durchbruchfeldstärke f <Elektrotech> disruptive strength
Durchbruchspannung f <Elektriz, Phys, Telekom> breakdown voltage *(Halbleiter)*
Durchbruchstelle f <Bau, Kohlen> breakthrough point
Durchbruchszeichnung f <Konstzeich> penetration drawing
Durchdrehen n 1. <Kfztech> wheel spin; 2. <Lufttrans> belting-in
Durchdrehen n **des Rades** <Kfztech> spinning of the wheel
durchdringbar adj <Chemie> permeable
Durchdringbarkeit f <Nichtfoss Energ> permeability
durchdringen v 1. <Anstrich> penetrate; 2. <Bau> intersect, penetrate
durchdringend adj 1. <Anstrich> penetrant; 2. <Fertig> hard *(Strahl)*
Durchdringung f 1. <Elektronik, Erdöl> penetration; 2. <Geom, Kerntech> interpenetration; 3. <Mechan> penetration
Durchdringungsfestigkeit f <Sicherheit> resistance to penetration *(z. B. Schutzhelm)*
Durchdringungsgrad m <Erdöl> penetration rate *(Tiefbohrtechnik)*
Durchdringungspotenzial n <Kerntech> penetration potential
Durchdringungswahrscheinlichkeit f <Kerntech> penetration factor
Durchdrückpackung f <Verpack> push-through pill pack
Durchdrückpackungsfolie f <Verpack> push-through packaging sheet
Durchdrückverpackung f <Verpack> blister pack
Durchfahrt f 1. <Eisenbahn> transit; 2. <Wassertrans> pass *(Navigation)*
Durchfahrtstraße f <Bau, Trans> *(BE)* thoroughfare, *(AE)* thruway
durchfärben v <Textil> penetrate
Durchfeuchten n <Bau> soaking
Durchfluss m 1. <Bau> flow; 2. <Fertig> *(BE)* throughput, *(AE)* thruput *(Kunststoffinstallationen)*; 3. <Hydraul, Ker & Glas> flow; 4. <Kfztech> passage; 5. <Kohlen, Wasserversorg> flow
Durchflussanzeiger m <Maschinen> flow indicator
Durchflussbegrenzer m <Fertig> flow restrictor
Durchflussdichtemessgerät n <Gerät> *(BE)* continuous flow density analyser, *(AE)* continuous flow density analyzer
Durchflussgeschwindigkeit f 1. <Fertig> speed of flow *(Kunststoffinstallationen)*; 2. <Heiz & Kälte, Kohlen, Phys> rate of flow
Durchflusskalorimeter n <Thermod> continuous flow calorimeter
Durchflusskennlinie f <Maschinen> flow characteristic

Durchflusskoeffizient m 1. <Hydraul> discharge coefficient; 2. <Nichtfoss Energ, Qual> flow coefficient
Durchflusskorrekturrechner m <Gerät> correcting flow calculator
Durchflussmenge f 1. <Erdöl> flow rate; 2. <Heiz & Kälte, Kunststoff> flow rate, rate of flow
Durchflussmengenmesser m <Metrol> rate-of-flow meter
Durchflussmengenmessung f <Gerät> flow measurement
Durchflussmengenregler m <Maschinen> flow rate controller
Durchflussmessblende f <Gerät> orifice plate
Durchflussmesser m 1. <Bau, Kohlen, Labor, Maschinen, Papier, Phys, Thermod> flowmeter; 2. <Wasserversorg> flowmeter; instant flowmeter *(für Wasser)*
Durchflussmesser m **mit Permanentmagnet** <Kerntech> permanent-magnet flowmeter
Durchflussmesser m **mit Tauchglockenwirkdruckgeber** <Gerät> bell flowmeter
Durchflussmessgerät n 1. <Elektrotech> flowmeter; 2. <Gerät> aneroid flowmeter; pipe flowmeter *(für Rohre)*
Durchflussmessgerät n **mit Balgfeder als Wirkdruckgeber** <Gerät> bellows flowmeter
Durchflussmesszelle f <Gerät> continuous flow cell, flow-through cell, flow-through-type cell
Durchflussquerschnitt m <Raumfahrt> bore *(Raumschiff)*
Durchflussrate f 1. <Heiz & Kälte> flow rate, rate of flow; 2. <Wasserversorg> discharge, rate of flow
Durchflussrechner m **für den Volumenstrom** <Gerät> volumetric flow calculator
Durchflussrefraktometer n <Gerät> continuous flow refractometer
Durchflussregelung f <Heiz & Kälte, Qual> flow control
Durchflussregelventil n <Maschinen> flow control valve
Durchflussrichtung f <Fertig> flow direction *(Kunststoffinstallationen)*
Durchflussrohr n <Maschinen> flow pipe
Durchflussschalter m <Elektriz> flow switch
Durchflussvolumen n <Phys> volume rate
Durchflusszähler m 1. <Gerät> flow-counting device, volumetric flow meter; 2. <Heiz & Kälte> liquid flow counter; 3. <Thermod> flowmeter
Durchflusszählung f <Gerät> flow counting
Durchflusszelle f <Gerät> flow-through cell, flow-through-type cell
Durchflutung f <Elektriz> magnetomotive force
Durchführbarkeit f <Bau, Comp & DV, Patent, Qual> feasibility
Durchführbarkeitsbericht m <Comp & DV, Patent> feasibility report
Durchführbarkeitsstudie f <Bau, Patent> feasibility study
durchführen v 1. <Comp & DV> implement; 2. <Patent, Qual> carry out; 3. <Telekom> pass through *(Kabel)*
• **einen Betatest durchführen** <Comp & DV> beta test
Durchführung f 1. <Comp & DV> execution, implementation; 2. <Elektrotech> feed-through, feed-through lead, leading through, lead-through; outdoor bushing *(Freiluftausführung)*; bushing, grommet *(für Kabel)*; 3. <Fertig> duct *(Kunststoffinstallationen)*; 4. <Kontroll> execution; 5. <Mechan> duct, feed-through; 6. <Patent> carrying out; 7. <Phys> feedthrough; 8. <Qual> carrying out; 9. <Raumfahrt> feed-through; 10. <Telekom> inlet; bushing *(Kabel)*
Durchführungseingang m <Elektrotech> feed-through input
Durchführungshülse f <Elektrotech> grommet

Durchführungsisolator m <Elektrotech> feed-through insulator
Durchführungskondensator m <Elektrotech, Phys> feed-through capacitor
Durchgang m 1. <Bau> passage; 2. <Comp & DV> walk-through; 3. <Fertig> throat; sag *(Riemen)*; 4. <Kerntech> transit *(eines geladenen Teilchens)*; 5. <Maschinen> undersize • **in einem Durchgang hergestellt** <Fertig> one-holed
Durchgangsamt n <Telekom> tandem exchange
Durchgangsbahnhof m <Trans> through station
Durchgangsbohrung f 1. <Fertig> through hole; 2. <Maschinen> clearance hole, through hole
Durchgangsdämpfung f 1. <Akustik> transmission loss; 2. <Telekom> transmission loss *(Vierpol)*
Durchgangsdämpfung f **ins Gebäude** <Telekom> building penetration loss
Durchgangsdrehzahl f <Nichtfoss Energ> runaway speed
Durchgangsdurchmesser m <Fertig> diametral capacity *(Spanung)*
Durchgangseilgüterzug m <Eisenbahn> *(AE)* through freight train, *(BE)* through goods train
Durchgangsfernamt n <Telekom> *(AE)* transit switching center, *(BE)* transit switching centre
Durchgangshafen m <Wassertrans> port of transit
Durchgangsknotenamt n <Telekom> *(BE)* junction tandem exchange
Durchgangskontrolle f <Gerät> conduction test
Durchgangsleitung f <Telekom> through line
Durchgangsprüfer m 1. <Elektriz> continuity tester; 2. <Gerät> continuity tester; circuit continuity tester *(für elektrische Leitungen)*
Durchgangsprüfung f <Gerät> conduction test
Durchgangsrichtung f <Elektriz> forward-conducting direction
Durchgangsschalter m <Elektriz> continuity switch
Durchgangsschraube f <Maschinen> through bolt
Durchgangsunterbrechung f <Telekom> continuity fault
Durchgangsventil n 1. <Heiz & Kälte> two-way valve; 2. <Maschinen> full-way valve, straight-way valve; 3. <Nichtfoss Energ> straight flow valve
Durchgangsverkehr m 1. <Telekom> transit traffic; 2. <Trans> bypassing traffic, external-external traffic, through traffic, transient currents
Durchgangsvermittlungsstelle f <Telekom> tandem exchange, transit exchange
Durchgangszeit f 1. <Elektrotech> transit time; 2. <Kerntech> transit time *(von geladenen Teilchen)*
durchgebrannte Sicherung f 1. <Elektriz> open fuse; 2. <Elektrotech> blown fuse
durchgebrannter Kontakt m <Kfztech> burnt contact *(Ventil)*
Durchgehen n <Maschinen> racing *(eines Motors)*
durchgehend geschweißte Schienen fpl <Eisenbahn> continuous-welded rail
durchgehende bewehrte Fundamentplatte f <Bau, Fertig> raft foundation
durchgehende digitale Verbindungsmöglichkeit f <Telekom> end-to-end digital connectivity
durchgehende Neigung f <Eisenbahn> continuous gradient
durchgehende weiße Linie f <Trans> continuous white line *(Straßenmarkierung)*
durchgehender Zug m <Eisenbahn> nonstop train
durchgehendes Gleis n <Eisenbahn> main track
durchgehendes Loch n <Maschinen> through hole
durchgehendes Mauerwerk n <Bau> blank wall, blind wall

durchgehendes Protokoll n <Comp & DV> end-to-end protocol
durchgehendes Zeitband n <Trans> through band
durchgelassene Welle f <Phys> transmitted wave
durchgelassener Strahl m <Phys> transmitted beam
durchgestrichene Null f <Druck> slashed zero
durchgezogene Maßlinie f <Konstzeich> continuous dimension line
durchgießen v <Fertig> strain
Durchgreifspannung f <Elektronik> punch-through
Durchgriff m <Elektronik> punch-through
Durchhang m 1. <Bau> dip, sag *(von Leitungen)*; 2. <Wassertrans> slack *(Tauwerk)*
Durchhang m **einer Schwärzungskurve** <Foto> toe region of characteristic curve
durchhängen v <Bau> sag
durchkohlen v <Kohlen, Thermod> carbonize
durchkontaktiertes Loch n 1. <Elektriz> *(AE)* metalized hole, *(BE)* metallized hole; 2. <Elektronik> plated-through hole
Durchkontaktierung f <Kontroll, Telekom> through connection
Durchlass m 1. <Bau> culvert; 2. <Fertig> passage *(Werkstück)*; 3. <Ker & Glas> throat; 4. <Kfztech> passage; 5. <Mechan> feedthrough, port
Durchlassabdeckstein m <Ker & Glas> throat cover
Durchlassband n <Comp & DV, Funktech, Telekom> passband
Durchlassbereich m 1. <Aufnahme, Comp & DV> passband; 2. <Elektronik> filter pass band, passband; 3. <Elektrotech> conducting zone; 4. <Funktech, Phys> passband
Durchlasscharakteristik f <Telekom> transmission characteristic *(Filter)*
Durchlassdämpfung f <Elektronik, Telekom> passband attenuation
Durchlassdauerspannung f <Elektriz> continuous on-state voltage
Durchlassfunktion f <Kerntech> transmission function
Durchlassgrad m <Telekom> transmittance
durchlässig adj 1. <Akustik, Anstrich> porous; 2. <Bau> pervious; 3. <Chemie, Kohlen, Kunststoff> permeable; 4. <Metall> porous; 5. <Nichtfoss Energ> pervious, porous; 6. <Telekom> transparent; 7. <Textil, Werkprüf> permeable
durchlässige Scheibe f <Optik> *(BE)* transmissive disc, *(AE)* transmissive disk
durchlässiger Boden m <Bau> pervious soil
Durchlässigkeit f 1. <Akustik> porosity, transmission; 2. <Anstrich> porosity; 3. <Chemie, Erdöl, Kohlen, Kunststoff> permeability; 4. <Metall> porosity; 5. <Nichtfoss Energ> perviousness, porosity; 6. <Optik> transmission; 7. <Phys> transmission power, transmittance; 8. <Telekom> transmission, transparency; 9. <Textil> permeability; 10. <Wellphys> transmittance; 11. <Werkprüf> permeability
Durchlässigkeitsbeiwert m <Nichtfoss Energ> hydraulic conductivity
Durchlässigkeitsbereich m <Maschinen> passband
Durchlässigkeitsfaktor m 1. <Akustik> diffuse density; 2. <Phys> transmission coefficient *(Schall)*
Durchlässigkeitsgrad m <Optik> transmittance
Durchlässigkeitskoeffizient m <Abfall, Qual> coefficient of permeability
Durchlässigkeitsspektrum n <Kerntech> transmission spectrum
Durchlässigkeitswert m <Abfall, Qual> coefficient of permeability
Durchlassintervall n <Fernseh> forward-stroke interval

Durchlassöffnung

Durchlassöffnung f <Kerntech> transfer port *(an Handschuhkasten)*
Durchlassrichtung f <Elektronik> conducting direction
Durchlassseitenstein m <Ker & Glas> throat cheek
Durchlassspannung f 1. <Elektriz> continuous on-state voltage; 2. <Phys> forward voltage
Durchlassstrom m 1. <Elektronik> conducting-state current, continuous on-state current, on-state current; 2. <Elektrotech> cut-off current, forward current, let-through current, on-state current
Durchlassstromkennlinie f <Elektrotech> cut-off current characteristic, let-through current characteristic
Durchlasstransistor m <Elektronik> pass transistor
Durchlassverhalten n <Elektronik, Telekom> passband response
Durchlassvorspannung f <Elektrotech> forward bias
Durchlasszeit f <Elektrotech> conduction angle *(bei Thyristor)*
Durchlasszustand m 1. <Elektronik> conducting state *(Halbleiter)*; 2. <Elektrotech> on-state *(Thyristor)*
Durchlauf m <Comp & DV> pass, run, run-through
Durchlaufanzeiger m <Maschinen> flow indicator
Durchlaufbalkenträger m <Bau> continuous beam
Durchlaufbelichter m <Verpack> continuous printer
Durchlaufbetrieb m **mit Aussetzbelastung** <Elektrotech> continuously running duty with intermittent loading, continuously operation duty-type
Durchlaufdampfgenerator m <Kerntech> OTSG, once-through steam generator
Durchlaufdosenreiniger m <Lebensmittel> *(AE)* straight-through can washer, *(BE)* straight-through tin washer
durchlaufen v <Telekom> pass through
Durchlaufen n <Chemie> percolation
durchlaufender Träger m <Elektronik> continuous beam *(Hochbau)*
Durchlauf-Entwicklungsmaschine f <Foto> continuous processing machine
Durchläufer m <Qual> passed component
Durchlauferhitzer m <Heiz & Kälte> flow-type heater, hot-water heater, instantaneous water heater
Durchlaufglühen n <Fertig> continuous annealing
Durchlaufkochanlage f <Lebensmittel> continuous cooker
Durchlaufofen m <Heiz & Kälte> continuous kiln
Durchlaufplatte f <Bau> continuous slab
Durchlaufschleifen n <Fertig> through-feed grinding
Durchlaufschmierung f <Maschinen> once-through lubrication
Durchlaufträger m <Fertig> continuous beam
Durchlaufvernetzung f <Kunststoff> continuous vulcanization
Durchlaufzeit f <Comp & DV, Kontroll, Telekom> *(AE)* turnaround time, *(BE)* turnround time
Durchleuchtung f <Elektrotech> fluoroscopy *(Röntgen)*
Durchlicht n <Foto, Labor, Strahlphys> transmitted light
Durchlichtmikroskop n <Metrol> microscope for transmitted light
Durchlüfter m <Lufttrans> aerator
Durchmesser m *(D)* <Fertig, Geom, Maschinen> diameter *(D)*
Durchmesser m **der Blattsteigungsachse** <Lufttrans> *(AE)* pitch center diameter, *(BE)* pitch centre diameter
Durchmessermaß n <Konstzeich> diametral dimension
Durchmesserreduktion f <Fertig> breaking down *(Walzen)*
Durchmessersymbol n <Geom> diameter symbol
Durchmesserteilung f <Maschinen> diametral pitch
Durchmessertoleranz f <Maschinen> tolerance on the diameter
Durchmesserverjüngung f <Fertig> longitudinal clearance *(Bohrer)*
Durchmesserwicklung f <Elektrotech> full-pitch diametrical winding
Durchmesserzeichen n <Konstzeich> diameter symbol
durchperlen v <Phys> bubble through
durchplattiertes Loch n <Elektronik> plated-through hole *(in Leiterplatten)*
durchpoltern v <Fertig> dish
Durchprüfung f <Comp & DV> program checkout *(eines Programms auf Fehler)*
Durchreiben n <Kerntech> chafing
durchreißen v <Fertig> *(AE)* louver, *(BE)* louvre *(Blechbearbeitung)*
Durchrutschen n <Kfztech> clutch slip
Durchrutschen n **des Bandes** <Aufnahme> tape slippage
durchsacken v <Wassertrans> sag *(Schiff)*
Durchsacken n 1. <Fertig> sag; 2. <Lufttrans> stall *(Strömungsabriss eines Flugzeugs)*
Durchsackwarngerät n <Lufttrans> stall warning device
Durchsageanlage f <Lufttrans> PA system, public address system
Durchsatz m 1. <Comp & DV> *(BE)* throughput, *(AE)* thruput; 2. <Erdöl> flow rate, *(BE)* throughput, *(AE)* thruput; 3. <Heiz & Kälte> rate of flow; 4. <Kontroll, Maschinen> *(BE)* throughput, *(AE)* thruput; 5. <Mechan> flow rate; 6. <Telekom> *(BE)* throughput, *(AE)* thruput
Durchsatzgeschwindigkeit f <Kontroll> throughput rate
Durchsatzklasse f <Telekom> throughput class
Durchsatzrate f 1. <Kontroll> throughput rate; 2. <Telekom> *(BE)* throughput, *(AE)* thruput
Durchsatzzeit f <Kontroll> throughput time
Durchschalte... <Telekom> switching
Durchschalteeinheit f <Telekom> switching unit
Durchschaltesystem n <Telekom> circuit switching system
Durchschaltevermittlung f 1. <Comp & DV> circuit switching *(CS)*; 2. <Telekom> circuit switch; 3. <Telekom> circuit switching *(CS)*
Durchschalte-Vermittlungseinrichtung f <Telekom> circuit switching unit
Durchschalte-Vermittlungsnetz n *(CSN)* <Comp & DV, Telekom> circuit-switched network *(CSN)*
Durchschaltnetzwerk n <Elektrotech> switching network *(Fernmeldewesen)*
Durchschaltung f <Telekom> through connection
Durchscheinen n <Strahlphys> translucence
durchscheinend adj <Textil> sheer
durchscheinende Materialien npl <Strahlphys> translucent substances
durchscheinendes Medium n <Phys> translucent medium
Durchscheuern n <Wassertrans> chafing *(Segeln, Tauwerk)*
durchschießen v 1. <Druck> lead out; 2. <Textil> interline *(Druck)*
Durchschießen n <Comp & DV> interleaving
Durchschlag m 1. <Elektriz, Elektronik, Elektrotech> breakdown; puncture *(im Kondensatorwickel)*; breakdown, disruptive breakdown, disruptive discharge, insulation breakdown *(von Gasröhren)*; breakdown *(bei p-n-Anschlussisolator)*; 2. <Kunststoff> breakdown; 3. <Maschinen> drift, drift punch, piercer, puncture; 4. <Papier> blind carbon copy, carbon, carbon copy, copy, duplicate; 5. <Telekom> blind carbon copy, carbon copy *(E-Mail)*

durchschlagen v 1. <Bau> hole; 2. <Kunststoff> bleed; 3. <Mechan> overshoot
durchschlagend adj <Elektrotech> disruptive
Durchschläger m <Maschinen> drift, solid punch
Durchschlagfestigkeit f <Phys> dielectric strength
Durchschlagprüfung f <Elektriz> puncture test
Durchschlagpunkt m <Papier> puncture point
durchschlagsicher adj <Sicherheit> puncture-resistant
Durchschlagspannung f 1. <Elektriz> breakdown voltage, disruptive voltage; 2. <Elektrotech> disruptive voltage; breakdown voltage (Gas); breakdown voltage (bei p-n-Anschlussisolator); breakdown voltage (Ölleitung); 3. <Maschinen, Telekom> breakdown voltage
Durchschlagspannungsprüfung f <Elektriz, Metrol> disruptive discharge test
Durchschlagsprüfer m <Papier> puncture tester
Durchschlagstoßspannung f <Elektriz> breakdown impulse voltage, impulse breakdown voltage
Durchschlagversuch m <Elektriz, Metrol> breakdown test, insulation breakdown test
Durchschlämmung f <Chemie> percolation (Boden)
durchschleifen v <Fernseh> pipe
Durchschleusen n 1. <Nichtfoss Energ> sluicing; 2. <Wasserversorg> sluicing; lockage (eines Schiffs)
Durchschlupf m <Qual> average outgoing quality
Durchschnitt m <Math> intersection (Mengenlehre); average (Statistik)
durchschnittlich adj 1. <Comp & DV> mean; 2. <Kerntech, Maschinen, Nichtfoss Energ, Phys, Qual, Trans> average
durchschnittliche Anhaltezeit f <Trans> average stopped time
durchschnittliche Anzahl f **der geprüften Einheiten je Los** <Qual> average total inspection
durchschnittliche Energie f (W) <Kerntech> average energy, W (pro Ionenpaar)
durchschnittliche Fahrzeuglänge f <Trans> average vehicle length
durchschnittliche Fertigproduktqualität f (AOQ) <Qual> average outgoing quality (AOQ)
durchschnittliche kinetische Molekularenergie f (ε) <Phys> average molecular kinetic energy (ε)
durchschnittliche mittlere Temperatur f <Heiz & Kälte> average mean temperature
durchschnittliche Momentgeschwindigkeit f <Trans> average spot speed
durchschnittliche Reisegeschwindigkeit f <Eisenbahn, Trans> average overall travel speed
durchschnittliche Reisezeit f <Eisenbahn, Trans> average journey time
durchschnittliche Stichprobengröße f <Qual> average sample number
durchschnittliche Tagesleistung f <Kohlen, Nichtfoss Energ> average daily output
durchschnittliche Windgeschwindigkeit f <Nichtfoss Energ> average wind speed
durchschnittlicher Fertigproduktqualitätsgrenzwert m (AOQL) <Qual> average outgoing quality limit (AOQL)
durchschnittlicher Gesamtprüfumfang m <Qual> average total inspection
durchschnittlicher Gleitwegfehler m <Lufttrans> mean glide path error
durchschnittlicher Stichprobenumfang m <Qual> average sample number
durchschnittlicher Wasserspiegel m <Wasserversorg> intermediate water level
durchschnittliches Tagesverkehrsaufkommen n <Trans> ADT, average daily traffic

durchschnittliches Tagesverkehrsaufkommen n **pro Jahr** (AADT) <Trans> annual average daily traffic (AADT)
Durchschnittsdichte f <Trans> average density
Durchschnittsgeschwindigkeit f 1. <Eisenbahn, Kfztech, Lufttrans> mean speed; 2. <Trans> average running speed; 3. <Wassertrans> mean speed
Durchschnittsleistung f <Kohlen, Phys> average output
Durchschnittsprobe f <Qual> average sample
Durchschnittswert m <Comp & DV, Math> mean, mean value
Durchschnittszeitintervall n <Trans> average time interval
durchschossen adj <Comp & DV> interleaved
Durchschreibmöglichkeit f <Elektronik> write-through capability
Durchschuss m 1. <Comp & DV, Druck> leading; 2. <Elektronik> lead (Druckwesen); 3. <Textil> pick • **ohne Durchschuss** <Druck> solid
durchsetzende Stelle f <Patent, Qual> enforcement authority
Durchsetzung f <Patent> enforcement
Durchsicht f <Papier> lookthrough
durchsichtig adj 1. <Anstrich> glassy; 2. <Comp & DV> transparent; 3. <Ker & Glas> seethrough; 4. <Phys, Strahlphys> transparent
durchsichtige Materialien npl <Strahlphys> translucent substances, transparent substances
durchsichtiger Fleck m <Ker & Glas> clear spot
durchsichtiger Vorhangstoff m <Textil> casement cloth
durchsichtiges Material n <Phys> transparent medium
durchsichtiges Medium n <Phys> transparent medium
Durchsichtigkeit f <Wellphys> transmittance
Durchsichtprisma n <Phys> Amici prism
durchsickern v 1. <Bau> leak, pass through; 2. <Bau, Kohlen> seep
Durchsickern n <Bau, Kohlen, Meerschmutz> seepage
durchsieben v 1. <Bau> screen (Erde); 2. <Kohlen> screen; 3. <Meerschmutz> sift
Durchsieben n 1. <Chemtech> sieving; 2. <Meerschmutz> screening
durchspülen v <Bau, Maschinen, Wasserversorg> flush
Durchspülung f <Mechan> scavenging
Durchsteckschraube f 1. <Bau> bolt and nut; 2. <Fertig> bolt
Durchsteckstromwandler m <Elektrotech> bar-type current transformer
durchstellen v <Telekom> put through
Durchstoß m 1. <Abfall> puncture; 2. <Geom> penetration
Durchstoßfestigkeit f <Abfall> puncturability, puncture resistance
Durchstoßpunkt m <Geom> penetration point
Durchstoßungspunkt m <Geom> penetration point
Durchstoßwicklung f <Elektriz> push-through winding
Durchstrahlung f <Elektronik, Kerntech, Labor, Phys> transmission
Durchstrahlungselektronenmikroskop n <Elektronik, Kerntech, Labor, Phys> transmission electron microscope
Durchstrahlungs-Katodenstrahlröhre f <Elektronik> penetration CRT
Durchstrahlverfahren n <Kerntech> transmission technique
durchströmen v <Bau> pass
Durchströmungskanal m <Phys> port
durchsuchen v <Comp & DV> browse, scan
durchtränken v <Lebensmittel> soak, steep
durchtreiben v <Fertig> drift

Durchtreiber

Durchtreiber m 1. <Bau> pin punch; 2. <Maschinen> drift punch
durchtrittfest adj <Sicherheit> puncture-resistant *(Schuhwerk)*
durchtrittsicher adj <Sicherheit> puncture-resistant *(Schuhwerk)*
durchverbinden v 1. <Funktech> interconnect; 2. <Telekom> put through
Durchwahl f <Telekom> *(AE)* direct inward dialing, *(BE)* direct inward dialling *(Nebenstelle)*
Durchwahl zur Endstelle f <Telekom> direct dialling in, DDI *(Dienstmerkmal im Euro-ISDN)*
Durchwahlrufnummer f <Telekom> direct dial-in number
Durchwärmdauer f <Thermod> heat penetration time
durchwärmen v <Ker & Glas, Metall> soak
durchweicht adj <Papier> soggy
Durchziehvorrichtung f <Elektriz> pull box
Durchziehwicklung f <Elektrotech> pull-through winding
Durchzugspule f <Elektriz> pull-through winding
Dur-Dreiklang m <Akustik> major common chord
Duren n <Chemie> durene
Durometer n <Kunststoff> durometer
Duroplast m 1. <Fertig> thermosetting plastic *(Kunststoffinstallationen)*; 2. <Kunststoff> thermoset, thermosetting plastic
duroplastische Kunststoffmasse f <Kunststoff> thermosetting compound
Dur-Tonart f <Akustik> major key
Dur-Tonleiter f <Akustik> major scale
Düse f 1. <Bau> nozzle; 2. <Druck> air vent; 3. <Fertig> nozzle, orifice, tue iron, tuyere; aperture *(Matrize)*; tip *(Schneidbrenner)*; 4. <Hydraul> orifice *(Unterwasser)*; 5. <Kfztech> jet, nozzle; 6. <Kohlen, Kunststoff, Labor, Lufttrans> nozzle; 7. <Maschinen> mouth, nozzle, opening, orifice, port, porthole; 8. <Mechan, Nichtfoss Energ, Papier> nozzle; 9. <Phys, Raumfahrt> jet, nozzle *(Raumschiff)* • **Düse am Rohrende anbringen** <Maschinen> fit nozzle on end of pipe
Düse f an der Spitze des Luftschraubenblattes <Lufttrans> blade tip nozzle *(Hubschrauber)*
Düse f des Strahltriebwerks <Lufttrans> jet nozzle
Düse f mit Strahlumlenkung <Lufttrans> thrust vectoring nozzle
Düsen fpl <Maschinen> dies
Düsenansatz m <Maschinen> nozzle adaptor
Düsenantrieb m <Lufttrans, Raumfahrt> jet propulsion *(Raumschiff)*
Düsendorn m <Fertig> core *(Extruder)*
Düsendurchmesser m <Nichtfoss Energ> jet diameter
Düsenfächer m <Lufttrans> ducted fan
Düsenfläche f <Raumfahrt> nozzle area *(Raumschiff)*
Düsenflugzeug n <Lufttrans> jet, jet plane
Düsengeschwindigkeit f <Raumfahrt> jet velocity *(Raumschiff)*
Düsengeschwindigkeitskoeffizient m <Nichtfoss Energ> nozzle velocity coefficient
Düsenhals m <Maschinen> nozzle throat
Düsenhalsquerschnitt m <Raumfahrt> nozzle throat, throat of nozzle *(Raumschiff)*
Düsenhalter m <Kfztech> nozzle holder
Düsenhaubenverkleidung f <Lufttrans> nozzle cowl *(Flugwesen)*
Düsenlastflugzeug n <Trans> heavy jet
Düsenlippen fpl <Ker & Glas> slot lips
Düsenmund m <Phys> nozzle exit
Düsenpropeller m 1. <Eisenbahn> ducted propeller; 2. <Trans> carinated propeller, shrouded propeller
Düsenschwefelbrenner m <Papier> *(AE)* jet sulfur burner, *(BE)* jet sulphur burner

Düsenstein m <Ker & Glas> burner block
Düsenstock m <Fertig> penstock
Düsenstrahl m <Lufttrans> jet wash; jet *(eines Vergasers)*
Düsenstrahlschutzwand f <Lufttrans> blast fence *(Fangeinrichtung für Flugzeuge)*
Düsenstrahltriebwerk n <Lufttrans> jet engine
Düsentemperaturanzeige f <Lufttrans> nozzle temperature indicator *(Flugwesen)*
Düsentragflächenboot n <Wassertrans> jetfoil *(Schifftyp)*
Düsentriebwerk n 1. <Lufttrans> jet engine; 2. <Raumfahrt> jet engine *(Raumschiff)*; 3. <Thermod> jet engine
Düsenverkehrsflugzeug n <Lufttrans> *(BE)* jet aeroplane, *(AE)* jet airplane
Düsenwirkungsgrad m <Raumfahrt> nozzle efficiency *(Raumschiff)*
Dutch-Roll f <Lufttrans> Dutch roll *(Flugzeug mit gepfeiltem Flügel)*
Dutch-Tropfen m <Ker & Glas> Dutch drop
Dutzend n <Math> dozen
DV *(Datenverarbeitung)* <Comp & DV, Elektronik, Kontroll, Telekom> DP *(data processing)*
DVD *(digital versatile disk)* DVD
DVD-Abspielgerät n <Comp & DV, Telekom> digital video disc player, DVD-player
D-Ventil n <Hydraul> D-valve
dwars adv <Wassertrans> abeam
Dwars... <Wassertrans> beam, breast
Dwarsbalken m <Wassertrans> crossbeam, crosspiece *(Schiffbau)*
Dwarsfeste f <Wassertrans> breast line
dwarsschiffs adv <Wassertrans> athwartships
Dwarssee f <Wassertrans> beam sea
Dy *(Dysprosium)* <Chemie> Dy *(dysprosium)*
Dyn n <Metrol> dyn, dyne
Dynamik f <Aufnahme, Fernseh, Funktech, Maschinen, Mechan, Phys> dynamics
Dynamikbereich m 1. <Aufnahme> volume range; 2. <Fernseh, Funktech> dynamic range
Dynamikdehner m <Aufnahme> expander
Dynamikdehnung f <Funktech> expansion
Dynamikkompression f <Elektronik> volume compression
Dynamikpresser m <Aufnahme> volume compressor
Dynamikpresser m **und -dehner** m <Aufnahme> compander
Dynamikpressung f <Aufnahme> compression
Dynamikregelung f <Telekom> companding
Dynamikumfang m <Akustik> volume range
dynamisch adj <Akustik, Aufnahme, Bau, Comp & DV, Elektrotech, Fernseh, Funktech, Kohlen, Kunststoff, Maschinen, Raumfahrt, Telekom> dynamic
dynamische Adressumsetzung f <Comp & DV> dynamic address translation
dynamische Ähnlichkeit f <Strömphys> dynamic similarity *(von zwei Strömungen)*
dynamische Arbeit f <Ergon> dynamic effort
dynamische Auftriebskraft f <Phys> lifting force
dynamische Auswuchtung f <Fertig> dynamic balancing
dynamische Bedingungen fpl <Elektrotech> dynamic conditions
dynamische Belastung f 1. <Bau> dynamic loading; 2. <Lufttrans> impact load; 3. <Metall> dynamic loading; 4. <Werkprüf> dynamic load
dynamische Dichtung f <Maschinen> dynamic seal
dynamische Druckhöhe f <Heiz & Kälte> velocity head
dynamische Eigenschaften fpl <Kunststoff> dynamic properties
dynamische Entkopplung f <Raumfahrt> dynamic decoupling *(Raumschiff*

dynamische Erholung f <Metall> dynamical recovery
dynamische Ersatzplatten fpl/zwei <Comp & DV> double dynamic spare disks
dynamische Erwärmung f <Thermod> dynamic heating
dynamische Fokussierung f <Fernseh> dynamic focusing
dynamische Komponente f <Lufttrans> dynamic component
dynamische Konvergenz f <Fernseh> dynamic convergence
dynamische Last f <Bau> dynamic loading
dynamische Leistungsaufnahme f <Elektrotech> dynamic power consumption
dynamische Lotung f <Kohlen> dynamic sounding
dynamische Muskelarbeit f <Ergon> dynamic effort
dynamische Positionierung f <Erdöl> dynamic positioning (Nautik)
dynamische Programmierung f <Comp & DV> dynamic programming
dynamische Prüfung f <Werkprüf> dynamic test
dynamische Rauschunterdrückung f <Aufnahme> dynamic noise suppressor
dynamische Reibung f <Phys> dynamic friction
dynamische Stabilität f <Lufttrans> dynamic stability
dynamische Überspannung f <Elektrotech> dynamic overvoltage
dynamische Verzerrung f <Akustik> dynamic distortion
dynamische Viskosität f 1. <Fertig> absolute viscosity; 2. <Kunststoff, Maschinen, Nichtfoss Energ, Phys, Strömphys> dynamic viscosity
dynamische Wechselwirkung f <Metall> dynamic interaction
dynamische Zuordnung f <Comp & DV> dynamic allocation
dynamische Zuweisung f <Comp & DV> dynamic allocation
dynamischer Ausgleich m <Lufttrans> dynamic balancing
dynamischer Belegungsdetektor m <Lufttrans> dynamic presence detector
dynamischer Bereich m <Strahlphys> dynamic range
dynamischer Bewegungsdetektor m <Lufttrans> dynamic movement detector
dynamischer Dauerspeicher m <Comp & DV> permanent dynamic memory
dynamischer Druck m 1. <Fertig> dynamic head; 2. <Lufttrans> dynamic pressure
dynamischer Eluationstest m <Abfall> dynamic leaching test
dynamischer Haltestrom m <Elektrotech> latching current
dynamischer Kegelpenetrometer m (DCP) <Bau> dynamic cone penetrometer (DCP)
dynamischer Lautsprecher m <Akustik, Aufnahme> dynamic loudspeaker
dynamischer Parameter m <Comp & DV> dynamic parameter
dynamischer Programmaustausch m <Comp & DV> swapping
dynamischer RAM m (DRAM) <Comp & DV> dynamic random access memory (DRAM)
dynamischer Speicher m 1. <Comp & DV> dynamic memory; 2. <Kerntech> delay line storage
dynamischer Speicherauszug m <Comp & DV> dynamic dump, snapshot dump
dynamischer Speicherbereich m <Comp & DV> dynamic storage area
dynamischer Spurwinkel m <Kfztech> dynamic toe angle (Lenkung)

dynamischer Widerstand m 1. <Elektriz> dynamic resistance (Resonanzzustand eines Schwingkreises); 2. <Funktech> dynamic resistance
dynamisches Abschöpfgerät n <Meerschmutz> dynamic skimmer
dynamisches Auswuchten n <Fertig, Lufttrans> dynamic balancing
dynamisches Gleichgewicht n <Maschinen> dynamic balance
dynamisches Loten n <Kohlen> dynamic sounding
dynamisches Modell n <Raumfahrt> dynamic model (Raumschiff)
dynamisches Trimmen n <Elektrotech> dynamic trimming
Dynamo m <Elektriz, Elektrotech, Kfztech, Wassertrans> dynamo (Elektrik)
Dynamo m **mit englischer Rahmenmontierung** <Elektriz> cradle dynamo
Dynamo m **mit konstanter Spannung** <Elektriz> constant-voltage dynamo
Dynamoeffekt m <Raumfahrt> dynamo effect
dynamoelektrisch adj <Elektrotech> dynamo-electric
Dynamograph m <Mechan> dynamograph
Dynamometamorphose f <Nichtfoss Energ> dynamic metamorphism
Dynamometer n 1. <Elektriz, Ergon> dynamometer; 2. <Labor> dynamometer (Energie); 3. <Maschinen, Mechan> dynamometer
Dynamometer-Leistungsmesser m <Elektriz> dynamometer wattmeter
Dynamotor m <Elektrotech> dynamotor
Dynode f <Phys> dynode
Dysprosium n (Dy) <Chemie> dysprosium (Dy)
D-Zug m <Eisenbahn> express train, fast train

E

e (Elektron) <Elektriz, Elektrotech, Funktech, Phys, Teilphys> e (electron)
E 1. <Druck, Erdöl, Heiz & Kälte, Hydraul, Maschinen, Nichtfoss Energ, Phys, Thermod> (Evaporation, Verdampfung) E (evaporation); 2. <Elektriz, Elektrotech, Phys> (elektrische Feldstärke) E (electric field strength); 3. <Elektriz, Elektrotech, Kerntech, Mechan, Metrol, Phys, Thermod> (Energie) E (energy); 4. <Elektrotech> (elektrischer Feldvektor) E (electric field vector); 5. <Erdöl, Hydraul, Kohlen, Kunststoff, Metall, Phys> (Young'scher Modul) E (Young's modulus); 6. <Optik> (Energie) power
E/A 1. <Comp & DV> (Eingabe/Ausgabe) I/O (input/output); 2. <Elektriz> (Eingabe/Ausgabe, Eingang/Ausgang) I/O (input/output)
E/A-Prozessor m <Comp & DV> input/output processor
Early-Effekt m <Elektronik> Early effect (Transistoren)
EAS (äquivalente Fluggeschwindigkeit) <Lufttrans> EAS (equivalent airspeed)
Easy-Gleitbereich m <Metall> easy-glide region
E/A-System n <Comp & DV> input/output system
Ebbe f 1. <Nichtfoss Energ> ebb, falling tide, low tide; 2. <Wassertrans> ebb, ebb tide, low tide
Ebbekrafterzeugung f <Nichtfoss Energ> ebb generation
ebben v <Nichtfoss Energ, Wassertrans> ebb (Gezeiten)
Ebbeströmung f <Nichtfoss Energ> ebb tide
Ebbetor n <Wasserversorg> aft gate, tail gate

Ebbstrom *m* <Nichtfoss Energ, Wassertrans> ebb stream
EBCDIC-Code *m* (erweiterter BCD-Code für Datenaustausch) <Comp & DV, Elektronik, Telekom> EBCDIC (extended binary-coded decimal interchange code)
EBCS-System *n* (europäisches Leichterträgersystem) <Wassertrans> EBCS (European barge carrier system)
eben *adj* 1. <Bau> flush, level; 2. <Fertig> coplanar (Kräfte); 3. <Geom> planar • **eben mit der Oberfläche eingebaut** <Fertig> flush-mounted
eben polarisierte Welle *f* <Wellphys> plane-polarized wave
Ebene *f* 1. <Comp & DV, Elektronik> level (Fläche); 2. <Geom> plane; 3. <Mechan> flat • **nicht in einer Ebene liegend** <Fertig> noncoplanar
ebene Figuren *fpl* <Geom> plane figures
ebene Fläche *f* <Maschinen> plain surface
ebene Geometrie *f* <Geom> plane geometry
ebene parallele Wellen *fpl* <Wellphys> plane parallel waves (von weit entfernter Quelle)
ebene Welle *f* <Akustik, Elektrotech, Funktech, Optik, Phys, Telekom> plane wave
Ebenengleichung *f* <Geom> plane equation (eines Polygons)
ebener Kiel *m* <Wassertrans> even keel (Schiffkonstruktion) • **auf ebenem Kiel** <Wassertrans> even keel
ebener Spiegel *m* <Phys> plane mirror
ebener Winkel *m* <Geom> plane angle
ebenerdig *adj* <Bau> at grade, even with the ground
ebenerdige Führungsschiene *f* <Bau, Eisenbahn> guideway at grade
ebenes Dreieck *n* <Geom> plane triangle
ebenes Polygon *n* <Geom> planar polygon, plane polygon
ebenes Schallfeld *n* <Akustik> free sound field
ebenflächig *adj* <Fertig> planar
Ebenheit *f* <Papier> flatness
Ebenheit *f* **der Straßendecke** <Bau> (BE) pavement surface evenness, (AE) sidewalk surface evenness
Ebenheitstoleranz *f* <Maschinen> flatness tolerance
ebnen *v* 1. <Bau> level (Straßenbau); 2. <Maschinen> flatten; 3. <Mechan> plane
E-Bogen *m* <Elektrotech> E-plane bend (Hohlleiter); E bend, E plane (Wellenleiter)
Ebonit *n* <Kunststoff> ebonite, vulcanite
e-Business-Anwendung *f* <Comp & DV> e-business application
EC (Ethylcellulose) <Kunststoff> EC (ethyl cellulose)
Ecdyson *n* <Chemie> ecdysone
Ecgonin *n* <Chemie> ecgonine
Echelette-Gitter *n* <Phys> echelette grating
Echinochrom *n* <Chemie> echinochrome
Echinus *m* <Bau> ovolo
Echo *n* 1. <Akustik, Aufnahme, Comp & DV> echo; 2. <Elektronik> echo (mit deutlichem Verzug wahrgenommene Schallreflexion); 3. <Funktech, Phys> echo; 4. <Wellphys> echo (reflektierter Schall)
Echoanzeige *f* 1. <Funkort> radar pip; 2. <Wassertrans> radar blip, radar pip
Echoeffekt *m* <Mobilkom> echo effect
Echoimpuls *m* 1. <Funkort> blip, reflected pulse (Radar); 2. <Telekom> echo impulse, echo pulse, reflected pulse
Echoimpulsgerät *n* <Fertig> reflectorscope
Echokanal *m* <Telekom> E-channel
Echokompensation *f* <Elektronik, Telekom> echo cancellation
Echokompensator *m* <Elektronik, Telekom> (AE) echo canceler, (BE) echo canceller (zur Einspeisung des kompensierenden Gegensignals)

Echokompensator-Chip *m* <Elektronik> (AE) echo-canceling chip, (BE) echo-cancelling chip
Echokontrolle *f* <Comp & DV> read back check
Echolot *n* 1. <Strahlphys> sonar, supersonic radar; 2. <Wassertrans> echo sounder, echo sounding, sonar; echo depth finder (Navigation); 3. <Wellphys> echo sounder
Echolotung *f* <Funktech> sounding (Navigation, Ionosphäre)
echometrische Messung *f* <Erdöl> acoustic well logging (Bohrlochmessung)
Echoortung *f* <Akustik, Funktech> echo ranging, echolocation
Echoplex *n* <Comp & DV> echoplex
Echoprüfung *f* <Comp & DV> echo check
Echoschreiber *m* <Phys> echograph
Echosignal *n* <Elektronik, Telekom> echo signal
Echosperre *f* <Comp & DV, Elektronik> echo suppressor (Telefon)
Echostrom *m* <Elektrotech> echo return current
Echounterdrückung *f* <Elektronik, Telekom> echo suppression (Telefon)
Echounterdrückungsschaltung *f* <Raumfahrt> echo suppressor (Weltraumfunk)
Echoverzerrung *f* <Elektronik, Telekom> echo distortion
Echozeichen *n* <Funkort> pip (Radar)
echt *adj* 1. <Comp & DV> real; 2. <Textil> actual
echte Adresse *f* <Comp & DV> real address
echte Farbe *f* <Textil> (AE) fast color, (BE) fast colour
echte Halbwertsbreite *f* <Strahlphys> true half-width
echte Klassengrenzen *fpl* <Qual> true class limits
echte Untermenge *f* <Comp & DV> proper subset
echter Bruch *m* <Math> proper fraction
echter Mehltau *m* <Lebensmittel> powdery mildew (Pflanzenkrankheitslehre)
echtes Labkraut *n* <Lebensmittel> lady's bedstraw
echtes Lösemittel *n* <Kunststoff> true solvent
Echtheit *f* <Textil> fastness
Echtheitsprüfungscode *m* <Telekom> authentication code (AC)
Echtzeit *f* <Comp & DV, Elektronik, Kontroll, Telekom> real time • **in Echtzeit** <Telekom> real-time
Echtzeitanalysator *m* <Akustik> (BE) real-time analyser, (AE) real-time analyzer
Echtzeitanalyse *f* <Elektronik> real-time analysis
Echtzeitausgabe *f* <Comp & DV> real-time output
Echtzeitbetrieb *m* <Comp & DV> real-time operation
Echtzeitbetriebssystem *n* <Comp & DV> real-time operating system
Echtzeitdaten-Registriergerät *n* <Nichtfoss Energ> real-time data logger
Echtzeiteingabe *f* <Comp & DV> real-time input
Echtzeithochsprache *f* <Telekom> CCITT high-level language, high-level programming language, high-level real-time language
Echtzeitprogrammiersprache *f* <Telekom> real-time programming language
Echtzeit-Signalverarbeitung *f* <Elektronik> real-time signal processing
Echtzeitsimulation *f* <Elektronik> real-time simulation
Echtzeitsimulator *m* <Elektronik> real-time simulator
Echtzeit-Spektralanalysator *m* <Elektronik> (BE) real-time spectral analyser, (AE) real-time spectral analyzer
Echtzeit-Spektralanalyse *f* <Elektronik> real-time spectral analysis
Echtzeitsprache *f* <Comp & DV> real-time language
Echtzeitsteuerung *f* <Trans> real-time control
Echtzeitsystem *n* <Comp & DV> real-time system
Echtzeitübertragung *f* <Telekom> streaming (Daten)

Echtzeituhr f <Comp & DV> real-time clock
Echtzeitumsetzer m <Telekom> real-time conversion facility
Echtzeit-Umsetzersatellit m <Fernseh> real-time repeater satellite
Echtzeitverarbeitung f <Comp & DV> real-time processing
Eckband n <Bau> corner band
Eckblech n <Fertig> gusset plate
Ecke f 1. <Bau> corner; 2. <Geom> corner, vertex; 3. <Maschinen> nose; 4. <Phys> edge
Eckenheftmaschine f <Verpack> corner stapling machine
Eckenheftung f <Verpack> corner stapling
Eckenmaß n <Maschinen> width across corners
Eckenversteifung f <Verpack> corner reinforcement
Eckfrequenz f <Elektronik, Funktech, Telekom> crossover frequency
Eckhahn m <Maschinen> right-angle stop cock
eckige Klammer f 1. <Comp & DV> bracket; 2. <Math> square bracket
Ecklautsprecher m <Aufnahme> corner loudspeaker
Ecknaht f <Fertig> corner weld *(Schweißen)*
Eckpfeiler m <Bau> corner pillar, jamb stone
Eckpfosten m <Bau> corner post
Eckrohrzange f <Maschinen> multiple pliers
Eckschiene f <Fertig> angle bar
Eckstab m <Kerntech> corner rod *(in Brennelementkonfiguration)*
Eckstab m **eines Brennelementbündels** <Kerntech> bundle corner rod
Eckstab m **in Bündelelement** <Kerntech> fuel assembly corner rod
Eckstein m 1. <Bau> corner block, pillar stone, quoin; 2. <Ker & Glas> corner block
Ecksteuerelement n <Kerntech> edge control assembly, edge control element
Eckventil n 1. <Eisenbahn> angle cock; 2. <Fertig> right angle valve *(Kunststoffinstallationen)*; 3. <Maschinen> angle valve
ECL *(emittergekoppelte Logik)* <Comp & DV, Elektronik> ECL *(emitter-coupled logic)*
ECL-Gatteranordnung f <Elektronik> ECL gate array
e-Commerce-Transaktion f <Comp & DV> e-commerce transaction
Edelgas n <Kerntech> inert gas
Edelgasröhre f <Elektronik> rare gas tube
Edelgasschutzmantel m <Fertig> inert gas shield • **mit Edelgasschutzmantel** <Fertig> inert gas-shielded
Edelgasschutzmantel... <Fertig> inert gas-shielded
Edelgasschweißen n <Fertig> inert gas welding
Edelkohle f <Kohlen> pure coal
Edelmetall n <Metall, Umweltschmutz> noble metal
Edelmetallthermoelement n <Gerät> noble metal thermocouple
Edelpassung f <Maschinen> force fit
Edelrost m <Fertig> patina • **mit Edelrost überziehen** <Fertig> patinate
Edelstahl m 1. <Fertig> high-alloyed steel *(Kunststoffinstallationen)*; 2. <Kfztech> stainless steel; 3. <Metall> high-alloyed steel, refined steel, special steel
Edelstahlbecher m <Labor> stainless steel beaker
Edestin n <Chemie> edestin
Edieren n **von Hand** <Fernseh> manual editing
Edison-Batterie f <Elektriz> Edison cell
Edison-Sockel m <Elektrotech> screw cap *(Glühbirnen)*
Edison-Zelle f <Elektriz> Edison cell
Editbetrieb m <Fernseh> edit mode
editgesteuerte Synchronisierung f <Fernseh> edit sync

editieren v <Comp & DV> edit
Editimpuls m <Fernseh> edit pulse
Editor m <Comp & DV, Fernseh> editor
edles Metall n <Metall, Umweltschmutz> noble metal
edles Thermoelement n <Gerät> noble metal thermocouple
EDTV *(hochauflösendes Fernsehen)* <Fernseh> EDTV *(extended definition television)*
EDV *(elektronische Datenverarbeitung)* <Comp & DV, Elektriz, Elektronik, Kontroll> EDP *(electronic data processing)*
EEB *(elektroerosive Bearbeitung)* <Maschinen> EDM *(electro-discharge machining)*
E-Ebene f <Funktech> E-plane
EEPROM *(elektrisch löschbarer programmierbarer Lesespeicher)* <Elektronik> EEPROM *(electrically-erasable programmable read-only memory)*
EEROM *(elektronisch löschbarer Festwertspeicher, elektronisch löschbarer Lesespeicher)* <Comp & DV> EEROM *(electronically erasable read-only memory)*
Effekt m <Elektriz> effect
Effektbogen m <Elektrotech> flame arc
Effektbus m <Fernseh> effects bus
Effekte mpl <Fernseh, Qual> effects
Effektgarn n <Textil> fancy yarn
Effektgenerator m <Fernseh> effects generator
effektiv adj <Akustik> effective
effektive Adresse f <Comp & DV> effective address
effektive Baustellenfläche f <Bau> net site area
effektive Datenübertragungsgeschwindigkeit f <Comp & DV, Telekom> effective data transfer rate
effektive Druckhöhe f <Hydraul> effective head
effektive Fallhöhe f <Kohlen> effective drop height
effektive Haftreibung f <Elektrotech> true adhesion
effektive Korngröße f <Kohlen> effective grain size
effektive Leistung f <Mechan> actual power
effektive Masse f <Akustik> effective mass
effektive Modenamplitude f <Optik> effective mode volume
effektive Neutronenlebensdauer f *(l)* <Kerntech> effective neutron lifetime *(l)*
effektive Neutronen-Multiplikationskonstante f *(keff)* <Kerntech> effective neutron multiplication constant
effektive Pferdestärke-Stunde f *(effektive PS-Stunde)* <Mechan> actual horsepower hour
effektive PS-Stunde f *(effektive Pferdestärke-Stunde)* <Mechan> actual horsepower hour
effektive Schlitzbreite f <Fernseh> effective slit width
effektive Spaltbreite f <Akustik, Aufnahme> effective gap length
effektive Teilchendichte f <Kerntech> effective particle density
effektive Verdunstung f <Wasserversorg> effective evaporation
Effektiv-EMK f <Elektriz> effective electromotive force
effektiver Dampfdruck m <Hydraul, Kerntech> effective steam pressure
effektiver Schalldruck m <Akustik, Umweltschmutz> effective sound pressure
effektiver Schutz m <Sicherheit> positive protection
effektiver Temperaturbereich m <Qual, Thermod> effective temperature range
effektiver Widerstand m <Phys> effective resistance
Effektivleistung f **in Pferdestärke** *(Effektivleistung in PS)* <Mechan> actual horsepower
Effektivleistung f **in PS** *(Effektivleistung in Pferdestärke)* <Mechan> actual horsepower
Effektivspannung f <Elektriz> root mean square voltage
Effektivstrom m <Elektrotech> rms current

Effektivwert *m* <Aufnahme, Elektronik, Optik, Phys> root mean square value
Effektivwert *m* **des Stromes** <Elektrotech> rms current
Effektivwertanzeiger *m* <Metrol> root-mean-square value detector
Effektlautsprecher *m* <Aufnahme> effects loudspeaker
Effektmikrofon *n* <Aufnahme> effects microphone
Effektor *m* <Fertig> effector *(eines Roboters)*
effektorisches Handeln *n* <Ergon> effector process
Effektspeicher *m* <Fernseh> effects bank
effloreszierend *adj* <Chemie> efflorescent
Effusion *f* <Kerntech> effusion
Effusionsofen *m* <Elektronik> effusion oven *(Lava-Ausflussofen)*
EFS *(essenzielle Fettsäure)* <Lebensmittel> EFA *(essential fatty acid)*
EFuRD *(Europäischer Funkrufdienst)* <Telekom> European radio-paging system
Egalisierer *m* <Textil> *(AE)* leveling agent, *(BE)* levelling agent
Egalisiermittel *n* <Textil> *(AE)* leveling agent, *(BE)* levelling agent
Egalisierwalze *f* <Papier> evener roll
Egoutteur *m* <Papier, Verpack> dandy roll
Egoutteurwalze *f* <Papier> forming roll
EHF *(Millimeterwellen)* <Funktech> EHF *(extremely high frequency)*
Ehrenfest-Gleichung *f* <Phys> Ehrenfest's equation
E.h.t. *(Hochspannung, Höchstspannung)* <Fernseh> EHT *(extremely high tension)*
Eichamt *n* <Metrol> calibration service
Eichanordnung *f* <Gerät, Qual> calibration set-up
Eichbergmotor *m* <Elektrotech> repulsion motor with fixed single set of brushes
Eichbericht *m* <Metrol> calibrating report
Eichdiagramm *n* <Metrol> calibrating plot
Eicheinrichtung *f* <Gerät, Qual> calibration equipment
Eichelzucker *m* <Chemie> quercite, quercitol
eichen *v* 1. <Bau> *(AE)* gage, *(BE)* gauge; 2. <Elektriz, Labor, Phys, Strahlphys> calibrate
Eichen *n* <Erdöl> *(AE)* gaging, *(BE)* gauging
Eichengerb... <Chemie> quercitannic
Eichfehlergrenze *f* <Gerät, Qual> calibration limit
Eichflug *m* <Lufttrans> calibration flight
Eichfrequenz *f* 1. <Akustik> SF, standard frequency; 2. <Funktech> calibration frequency
Eichgenauigkeit *f* <Gerät> accuracy of calibration
Eichgerät *n* 1. <Funktech> calibrator; 2. <Gerät> calibration instrument; 3. <Qual> calibration instrument, calibrator
Eichgewicht *n* <Labor> calibration weight
Eichimpuls *m* <Elektronik> calibration pip *(zur Entfernungseichung eines Radargeräts)*
Eichinstrument *n* <Gerät, Mechan> calibration instrument
Eichinvarianz *f* <Mechan> *(AE)* gage invariance, *(BE)* gauge invariance
Eichmaß *n* 1. <Fertig, Kfztech> *(AE)* gage, *(BE)* gauge; 2. <Metrol, Phys, Telekom> standard
Eichmikrofon *n* <Akustik> standard microphone
Eichnormal *n* <Metrol> calibration standard, standard measure
Eichsignal *n* <Elektronik> calibration signal *(Datenkommunikation)*
Eichstelle *f* <Metrol> calibrating facility
Eichtank *m* <Erdöl> *(AE)* gaging tank, *(BE)* gauging tank
Eichton *m* <Akustik> reference tone
Eichtrafo *m* <Kerntech> calibrating transformer
Eichtransformator *m* <Kerntech> calibrating transformer

Eichung *f* 1. <Aufnahme, Comp & DV, Elektriz, Elektronik> calibration; 2. <Erdöl> *(AE)* gaging, *(BE)* gauging; 3. <Labor, Metrol, Phys, Qual> calibration
Eichversuch *m* <Maschinen, Qual> calibration test
Eichwert *m* <Gerät, Qual> calibration value
eiförmig *adj* <Metall> oval
eigen *adj* <Anstrich> inherent
Eigen... <Anstrich, Chemie, Gerät, Strahlphys> self-...
Eigenabgleich *m* <Gerät> self-balance
Eigenabsorption *f* <Chemie> self-absorption
Eigenabsorption *f* **von Strahlung** <Strahlphys> self-absorption of radiation *(durch angeregte Atome)*
Eigenantrieb/mit 1. <Kfztech> automotive; 2. <Wassertrans> self-propelled
Eigenbedarfsgenerator *m* <Elektrotech> ancillary generator, auxiliary generator, house generator; station service generator *(Kraftwerk)*
Eigenbedarfsschalttafel *f* <Elektrotech> auxiliary suppliers board
Eigenbedarfstransformator *m* <Elektrotech> auxiliary transformer, house-service transformer; station service transformer *(Kraftwerk)*
eigenbetrieben *adj* <Elektrotech> self-powered
Eigendämpfung *f* <Lufttrans> internal damping
Eigenempfindlichkeit *f* <Akustik> characteristic sensitivity
Eigenerwärmung *f* <Elektrotech> self-heating
Eigenerweichungstemperatur *f* <Ker & Glas> self-sagging temperature
Eigenfeedback *n* <Kerntech> inherent feedback
Eigenfluss *m* <Phys> self-flux
Eigenfrequenz *f* 1. <Akustik, Elektronik, Fernseh> characteristic frequency, natural frequency; 2. <Funktech> characteristic frequency, natural frequency; 3. <Phys> natural frequency; 4. <Raumfahrt> eigenfrequency *(Raumschiff)*; 5. <Strahlphys> eigenfrequency
Eigenfrequenzen *fpl* <Wellphys> natural harmonics
Eigenfrequenzschwingung *f* <Akustik, Elektronik, Phys> natural frequency oscillation
Eigenfunktion *f* <Phys, Strahlphys> eigenfunction
Eigengeschwindigkeit *f* <Lufttrans> airspeed
Eigengeschwindigkeitsanzeiger *m* *(ASI)* <Lufttrans> airspeed indicator *(ASI)*
Eigengewicht *n* <Fertig> dead weight *(Kunststoffinstallationen)*
Eigenhalbleiter *m* 1. <Comp & DV, Elektronik> i-type semiconductor, intrinsic semiconductor; 2. <Phys> intrinsic semiconductor
Eigenidentifizierung *f* <Telekom> self-identification
Eigenionisierung *f* <Phys, Strahlphys> autoionization
Eigenjustierung *f* <Gerät> automatic adjustment
Eigenkapazität *f* 1. <Elektriz, Elektrotech> distributed capacitance, self-capacitance; 2. <Phys> self-capacitance
Eigenkapital *n* <Erdöl> equity capital *(Finanzen)*
Eigenkonvektion *f* <Heiz & Kälte> natural convection
Eigenkopie *f* <Telekom> local copy
Eigenkorrelation *f* <Elektronik> autocorrelation
Eigenkühlung *f* <Heiz & Kälte> self-cooling
Eigenlast *f* <Bau> dead load, permanent weight
Eigenleiterschichtdiode *f* <Elektronik> intrinsic-barrier diode
Eigenleitfähigkeit *f* <Elektrotech> intrinsic conductivity
Eigenleitungsdichte *f* <Elektronik> intrinsic density
Eigenlüftung *f* <Gerät> self-ventilation
Eigenmasse *f* <Bau> permanent weight
Eigenmodulation *f* <Telekom> automodulation
Eigenmodus *m* <Comp & DV> native mode
Eigennachführung *f* <Telekom> autotracking
Eigenpeilung *f* <Funktech> self-bearing

Eigenperiode f <Elektronik> natural period
Eigenpotenzialog n <Erdöl> self-potential log; spontaneous potential log *(Messtechnik)*
Eigenrauschen n 1. <Akustik> inherent noise pressure *(Mikrofon)*; 2. <Aufnahme> inherent noise; 3. <Elektronik> ground noise; 4. <Telekom> inherent noise
Eigenregelung f <Elektronik> inherent regulation *(bei gleich bleibender Drehzahl)*
Eigenreibung f <Ker & Glas> internal friction
eigenrelative Adressierung f <Comp & DV> self-relative addressing
Eigenrückkopplung f <Elektronik> inherent feedback
Eigenschaft f <Künstl Int> feature
Eigenschaft f **der Quarks** <Phys, Teilphys> charm
Eigenschaften fpl 1. <Ergon, Geom> properties; 2. <Telekom> characteristics
Eigenschaften fpl **von Winkeln** <Geom> properties of angles
Eigenschatten m <Raumfahrt> eigenshadow *(Raumschiff)*
Eigenscherben fpl <Ker & Glas> factory cullet
Eigenschwingung f 1. <Funktech> natural oscillation; 2. <Maschinen> self-oscillation
Eigenschwingungen fpl <Fertig> self-exited vibrations, self-induced vibrations
Eigenschwingungszustand m <Akustik> natural mode of vibration
Eigensetzung f <Bau> inherent settlement
eigensicher adj <Elektrotech, Sicherheit> intrinsically safe
eigensichere Anlage f <Sicherheit> intrinsic safety system, intrinsically safe equipment
Eigenspannung f <Maschinen, Metall> internal stress
eigenständig adj <Comp & DV> stand-alone
eigenständige Substanz f <Umweltschmutz> autochthonous matter
eigenständige Zeichnung f <Konstzeich> independent drawing
Eigensynchronisierung f <Telekom> autosynchronization
Eigentemperatur f <Elektronik> intrinsic temperature
Eigentemperaturbereich m <Elektronik> intrinsic temperature range
Eigentumstelefon n <Telekom> own-a-phone
Eigenvektor m <Comp & DV, Elektronik, Phys> eigenvector
Eigenverbrauch m <Elektrotech> power consumption
Eigenverlust m **an der Verbindungsstelle** <Optik> intrinsic joint loss
Eigenverständigungsanlage f <Telekom> interphone
Eigenverstärkung f <Elektronik> internal gain
Eigenverzerrung f <Telekom> inherent distortion
Eigenvolumen n <Phys> specific volume
Eigenwärme f <Heiz & Kälte> specific heat
Eigenwert m <Comp & DV, Elektronik, Phys, Strahlphys> eigenvalue
Eigenzeit f 1. <Phys> proper time; 2. <Regelung> inherent delay
Eigenzündung f <Kfztech> compression ignition *(Zündanlage bei Dieselmotor)*
Eightball-Mikrofon n <Aufnahme> eightball mike
Eigner m <Comp & DV> owner
Eignung f **zur Korrektur von Büschelfehlern** <Elektrotech> burst error correcting capability
Eignungsprüfung f 1. <Kerntech> qualification test; 2. <Qual> ability testing, aptitude test, performance test
Eignungstest m <Ergon> ability test, aptitude test, normal arm's reach
Eiisolator m <Elektrotech> egg insulator, egg-shaped insulator, strain insulator

Eikosylalkohol m <Chemie> eicosyl alcohol
Eilauftrag m <Druck> rush order
Eilgang m <Maschinen> fast traverse
Eilgangwelle f <Maschinen> quick-motion shaft
Eilgutbahnhof m <Eisenbahn> parcels depot
Eilgüterzug m <Eisenbahn> express parcels train
Eilrücklauf m <Maschinen> quick return
Eilvorschub m <Maschinen> quick feed
Eilwartung f <Comp & DV> emergency maintenance
Eimer m <Bau, Labor, Mechan, Wassertrans> bucket
Eimerbagger m <Bau, Wassertrans> bucket dredge, bucket dredger
Eimerkette f <Wassertrans> bucket chain *(Baggern)*
Eimerkettenaufzug m <Bau> bucket elevator
Eimerkettenbagger m 1. <Bau> bucket excavator; 2. <Wassertrans> ladder dredge, ladder dredger
Eimerkettenschaltung f (BBD) <Telekom> bucket brigade device (BBD)
Eimerleiter f <Wassertrans> ladder *(Baggern)*
Einadressbefehl m <Comp & DV> one-address instruction, single address instruction
Einadresscode m <Comp & DV> single address code
Einadressrechner m <Comp & DV> one-address computer
einadriger Hohlleiter m <Elektrotech> uniconductor waveguide
einadriges Kabel n 1. <Elektriz> single conductor cable, single-core cable; 2. <Elektrotech> single conductor cable
einander durchdringende Ebenen fpl <Geom> intersecting planes
Einanker-Frequenzumformer m <Elektriz> rotary frequency converter
Einankerumformer m <Elektrotech> rotary converter, single-armature converter, synchronous converter
Einanodengleichrichter m <Elektrotech> single anode rectifier
Einanodenröhre f <Elektronik> single anode tube
einarbeiten v <Kunststoff> incorporate
Einarbeiten n <Fertig> sinking
Einarbeitung f 1. <Qual> indoctrination; 2. <Textil> contraction
Einarmhebel m <Fertig> single-armed lever handle *(Kunststoffinstallationen)*
einarmige Spindelpresse f <Maschinen> swan neck fly press, swan neck screw press
Einarmzapfverbindung f <Bau> housed joint *(Holzbau)*
einatomig adj <Chemie> monoatomic
einatomiges Gas n <Phys> monatomic gas
einäugige Spiegelreflexkamera f (SLR) <Foto> single lens reflex camera (SLR)
ein/aus adv <Maschinen> on/off
Ein-/Aus-... <Comp & DV, Elektriz, Hydraul> on/off
Ein-/Ausgabe f **parallel zu Rechenprogramm** *(Spool-Programm)* <Comp & DV> simultaneous peripheral operations on-line (SPOOL)
Ein-/Ausgabewerk n <Telekom> interface processor
Ein-/Ausgangsschnittstelle f **für serielle Anschlüsse** <Telekom> asynchronous communications interface adaptor (ACIA)
Ein-/Aushebelschalter m <Elektriz> lever on-off switch
Ein-/Auslagerungsrate f <Comp & DV> swap rate
Ein-/Auslassöffnung f <Hydraul> port opening; gate *(Dampfturbine)*
Ein-/Auslassöffnungsfläche f <Hydraul> port face
Ein-/Auslassschieber m <Hydraul> gate valve
Ein-/Auslassschiebersteuerung f <Hydraul> gate gear
Ein-/Auslassschiebestange f <Hydraul> gate stem
Ein-/Auslasssteuerung f <Hydraul> gate gear
Ein-Ausschalt-Eigenzeit f <Elektrotech> close-open time

Ein-/Ausschalter

Ein-/Ausschalter *m* <Elektriz, Elektrotech, Foto, Maschinen, Phys> on-off switch
Ein-Ausschalt-Zeit *f* <Elektrotech> make-break time
Ein-/Ausspeicherung *f* <Comp & DV> roll in/roll out
Ein-/Aussteuerung *f* <Elektriz> on-off control
Einbadentwickler *m* <Foto> single bath developer
Einband *m* <Druck> cover
einbasisch *adj* <Chemie> monobasic
Einbau *m* 1. <Elektronik> mounting; 2. <Maschinen> installation, mounting
Einbauanleitung *f* <Maschinen> fitting instructions
Einbaubrenner *m* <Heiz & Kälte> built-in burner
einbauen *v* 1. <Bau, Elektronik> build in, embed, fit in, house, mount; 2. <Wassertrans> install
Einbauhöhe *f* <Fertig> overall height *(Kunststoffinstallationen)*
Einbauinstrument *n* <Gerät, Metrol> flush instrument, flush-mounting instrument, panel meter, panel-type meter
Einbaulage *f* 1. <Fertig> assembling position; positioning *(Kunststoffinstallationen)*; 2. <Heiz & Kälte> mounting position
Einbaulautsprecher *m* <Aufnahme> cabinet loudspeaker
Einbaumaß *n* 1. <Fertig> tool fitting dimension; 2. <Maschinen> fitting dimension, mounting dimension
Einbaumaße *npl* <Fertig> overall dimensions *(Kunststoffinstallationen)*
Einbaumikrofon *n* <Aufnahme> built-in microphone
Einbaumodem *n* <Comp & DV> built-in modem, integrated modem
Einbaumotor *m* <Elektrotech> built-in motor
Einbaureibahle *f* <Fertig> block-type reamer
Einbausatz *m* <Maschinen> installation kit
Einbauspuler *m* <Textil> built-in bobbin winder
Einbaustück *n* <Fertig> chuck *(Walzen)*
Einbautank *m* <Maschinen> built-in tank
Einbautechnik *f* <Abfall> refuse deposition technique
Einbauten *mpl* <Bau> fixed equipment
Einbauvorrichtung *f* <Kerntech> fitting stand
Einbechern *n* <Kunststoff> potting
Einbereichinstrument *n* <Gerät> single range instrument
einbetonieren *v* <Bau> set in concrete
einbetten *v* <Bau> bed in, embed
Einbetten *n* <Kunststoff> embedding, encapsulation
Einbettmasse *f* <Kunststoff> encapsulant
Einbettung *f* 1. <Elektriz> bedding; 2. <Geom> immersion; 3. <Maschinen> embedding
Einbettungsvermögen *n* <Fertig> embeddability
einbeulen *v* <Mechan> dent
Einbeulung *f* 1. <Fertig> dome; 2. <Maschinen> buckle
einbinden *v* <Bau> fix, fix in, tail in
Einbindung *f* 1. <Abfall> grain encapsulation *(Sondermüll)*; 2. <Bau> embedment *(einer Stütze)*
einblatten *v* <Fertig> adze
einblenden *v* 1. <Aufnahme> fade in; 2. <Fernseh> superimpose
Einblenden *n* <Fernseh> inlay
Einblicklinse *f* <Optik> ocular
Einblicksöffnung *f* <Foto> eyepiece
Einbootungsdeck *n* <Wassertrans> embarkation deck *(Zugang zu Rettungsbooten)*
einbrechen *v* <Anstrich> disrupt
einbrennbares Abziehbild *n* <Ker & Glas> ceramic transfer
Einbrenne *f* <Lebensmittel> roux
einbrennen *v* 1. <Bau> bake; 2. <Fernseh> burn in; 3. <Fertig> bake
Einbrennen *n* 1. <Druck, Elektronik> burning-in *(Dickschicht-Leiterplatten)*; 2. <Kunststoff> baking
Einbrennetikett *n* <Verpack> hot-transfer label
Einbrennlack *m* 1. <Bau> baking varnish; 2. <Kunststoff> stoving enamel, stoving finish, stoving varnish; 3. <Verpack> *(AE)* enameling, *(BE)* enamelling
Einbrennmuffel *f* <Ker & Glas> decorating lehr
Einbrennofen *m* 1. <Ker & Glas> decorating kiln; 2. <Textil> baking stove
Einbringen *n* **der ersten Rohrtour** <Erdöl> spudding in *(Bohrtechnik)*
Einbruch *m* 1. <Erdöl> caving *(Bohrtechnik)*; 2. <Mechan> breaking in; 3. <Wasserversorg> irruption
Einbruchsalarmanlage *f* <Sicherheit> intruder alarm equipment
einbruchssicher *adj* <Sicherheit> burglar-proof
einbuchen *v* <Mobilkom> log in, sign on
Einbuchen *n* **des Teilnehmer-Aufenthaltsortes** <Mobilkom> mobile location registration *(zellulares Netz)*
Einbuchung *f* 1. <Mobilkom> log in, registration; 2. <Telekom> log in
eindampfen *v* <Chemie> evaporate *(bis zur Trockne)*
• **zur Trockne eindampfen** <Chemtech> evaporate to dryness
Eindampfkessel *m* <Chemtech> vaporizer
Eindampfung *f* <Chemie> inspissation
eindecken *v* <Bau> roof
Eindeck-Flachpalette *f* <Trans> single-faced pallet
Eindeckpalette *f* <Trans> single-decked pallet
Eindeckschiff *n* <Wassertrans> single-decked ship
eindeutig *adv* <Math> unique
eindeutige Worterkennung *f* <Raumfahrt> unique word detection *(Weltraumfunk)*
eindeutiges Wort *n* <Raumfahrt> UW, unique word *(Weltraumfunk)*
Eindickapparat *m* <Chemtech> thickener *(Gerät)*
eindicken *v* 1. <Chemtech> condense; 2. <Fertig> body, body up, inspissate; 3. <Lebensmittel> boil down
Eindicker *m* <Kunststoff, Lebensmittel> thickener
Eindickkegel *m* <Kohlen> thickening cone
Eindickung *f* 1. <Chemie> inspissation; 2. <Fertig> inspissation; bodying *(Öl)*
Eindickzylinder *m* <Papier> concentrator
eindimensional *adj* <Phys> one-dimensional
eindocken *v* <Wassertrans> dock, drydock
Eindockung *f* <Wassertrans> docking *(Schiff)*
Eindosen *n* <Lebensmittel> canning
Eindrahtspeisung *f* <Elektriz> single supply
Eindrahtsystem *n* <Elektriz> single wire system
eindrehen *v* <Fertig> neck *(kreisförmige Nut)*
Eindrehung *f* <Maschinen> groove, recess
eindringen *v* <Anstrich, Bau> penetrate
Eindringen *n* **von Wasser in ein Brennelement** <Kerntech> water logging
Eindringkörper *m* 1. <Fertig> penetrator; 2. <Maschinen> indenter; 3. <Metrol> penetrator
Eindringkraft *f* <Strahlphys> penetrating power
Eindringprüfung *f* <Bau> static penetration test
Eindringtiefe *f* <Bau, Comp & DV, Elektronik, Kerntech, Maschinen, Werkprüf> penetration depth
Eindringtiefenmesser *m* <Bau> penetrometer
Eindringung *f* <Mechan> penetration
Eindringungsversuch *m* <Kohlen> penetration test
Eindringverfahren *n* <Kerntech> penetration method *(bei Werkstoffprüfung)*
Eindruck *m* <Maschinen, Metall> indentation
Eindrückdeckel *m* <Verpack> lever lid
Eindruckhärte *f* 1. <Kunststoff> indentation hardness; 2. <Maschinen> indentation hardness, penetration hardness
Eindruckkalotte *f* <Fertig> ball impression
einebnen *v* <Fertig> flush

Einebnen n 1. <Bau> (AE) leveling, (BE) levelling; 2. <Mechan> equalization, (AE) leveling, (BE) levelling
eineindeutig adv <Math> injective
Einerkomplement n <Comp & DV> one's complement
einfach adj 1. <Comp & DV> single; 2. <Kontroll> straightforward; 3. <Textil> plain
einfach gerichtet adj <Telekom> one-way
einfach gerichteter Verstärker m <Elektronik, Telekom> one-way repeater
einfach übersetztes Getriebe n <Elektronik> single reducting gearing
einfach wirkend adj <Heiz & Kälte> single-acting
einfach wirkende Maschine f <Maschinen> single action engine, single-acting engine
einfach wirkende Pumpe f <Hydraul> single-acting pump
einfach wirkender Servomotor m <Nichtfoss Energ> single-acting servomotor
einfach wirkender Verdichter m <Heiz & Kälte> single-acting compressor
Einfach... <Comp & DV> single
Einfachabdeckstruktur f <Elektronik> single level masking structure
Einfachbettdrehmaschine f <Maschinen> plain-bed lathe
Einfachdrehbank f <Maschinen> plain lathe
Einfachdrehmaschine f <Maschinen> plain lathe
einfache Baumwolleisolierung f <Elektriz> single cotton cover (SCC)
einfache bituminöse Oberflächenbehandlung f <Bau> single bituminous surface treatment
einfache Einspeisung f <Elektriz> single feeder
einfache Erde f <Elektriz> (BE) single earth, (AE) single ground
einfache Gegentaktmischstufe f <Elektronik> single-balanced mixer
einfache Genauigkeit f <Comp & DV> single precision
einfache Gleiskreuzung f <Eisenbahn> common crossing
einfache harmonische Bewegung f <Phys, Wellphys> simple harmonic motion
einfache Hybridschaltung f <Elektronik> simple hybrid circuit
einfache Scherspannung f <Metall> simple shear stress
einfache Speiseleitung f <Elektriz> single feeder
einfacher Bootskörper m <Wassertrans> single hull
einfacher Flammofen m <Metall> air furnace
einfacher Kameraauszug m <Foto> single camera extension
einfacher Öffnungskontakt m <Elektrotech> single break contact
einfacher Polykristallinprozess m <Elektronik> single level polysilicon process
einfacher Strebebogen m <Bau> flying buttress
einfacher Taljereepsknoten m <Wassertrans> single Matthew Walker (Knoten)
einfacher Zeilenabstand m <Comp & DV> monospacing
einfacher Zwirn m <Textil> twine (Spinnen)
einfaches Einstecken n <Nichtfoss Energ> plug-in simplicity
einfaches Kreuz n <Textil> end-and-end lease, one-and-one lease
einfaches Objektiv n <Foto> single lens
einfaches Pendel n <Phys> simple pendulum
einfaches schritthaltendes Paket n (SIP-Paket) <Comp & DV, Elektronik> single in-line package (SIP)
einfaches schritthaltendes Speichermodul n (SIMM) <Comp & DV> single in-line memory module (SIMM)
Einfachexpansionsmaschine f <Maschinen> simple expansion engine, single expansion engine

Einfachfahrschein m <Trans> (AE) one-way ticket, (BE) single ticket
Einfachfilter n <Elektronik> single section filter
Einfachform f <Maschinen> (AE) single impression mold, (BE) single impression mould
Einfachfräsmaschine f <Maschinen> plain-milling machine
Einfachimpuls m <Elektronik> single pulse
Einfachimpulssignal n <Elektronik> single pulse signal
Einfachkammlinearmotor m <Elektrotech> single-sided linear motor
Einfachraketentreibstoff m <Chemie, Raumfahrt> monopropellant
Einfachschlüssel m <Maschinen> single-ended spanner
Einfachschraubenschlüssel m <Maschinen> single-ended spanner
Einfachstapelfasergarn n <Ker & Glas> (AE) single staple-fiber yarn, (BE) single staple-fibre yarn
Einfachstatorlinearmotor m <Elektrotech> single-sided linear motor
Einfachstichprobe f <Qual> single sample
Einfachstichprobenentnahme f <Qual> single sampling
Einfachstichprobenprüfplan m <Qual> single sampling plan
Einfachstichprobenprüfung f <Qual> single sampling inspection
Einfachtafelglas n <Ker & Glas> single thickness sheet glass
Einfachteilen n <Maschinen> single indexing
Einfachüberlappung f <Maschinen> single overlap
Einfachverstärker m <Elektronik> monolithic amplifier
Einfädeln n von Papier <Papier> threading of paper
Einfädelung f 1. <Kfztech> (AE) weaving maneuver, (BE) weaving manoeuvre; 2. <Trans> joining a traffic system
Einfädelungsaufkommen n <Trans> merge volume (Verkehrsströme)
Einfadenaufhängung f <Elektriz> unifilar suspension
einfahren v <Kfztech> run in (Motor)
Einfahren n 1. <Erdöl> going-in hole (Bohrtechnik); stabbing (Zapfen in Muffengewinde, Gestänge); 2. <Maschinen> running-in
einfahrender Verkehr m <Trans> entering traffic
Einfahrt f <Wassertrans> mouth (Geographie, Hafen); inlet (Hafen)
Einfahrvorsignal n <Eisenbahn> entry warning signal
Einfall m 1. <Kerntech> fall-back; 2. <Optik> incidence (Strahl); 3. <Phys> incidence
Einfallen n 1. <Erdöl> dip; dipping (Geologie); 2. <Nichtfoss Energ> dip (Geologie)
einfallend adj 1. <Optik> incident (Strahl); 2. <Telekom> incoming (Welle)
einfallende Betondecke f <Bau> drop panel
einfallende Welle f <Phys, Wellphys> incident wave
einfallender Strahl m 1. <Funktech> incident ray; 2. <Phys, Wellphys> incident beam, incident ray
einfallendes Licht n <Foto, Phys, Wellphys> incident light
Einfallsebene f <Phys> plane of incidence
Einfallstelle f <Fertig> shrink mark (Gießen)
Einfallstellentropfen m <Ker & Glas> (BE) crater drip, (AE) top tin
Einfallswinkel m 1. <Elektrotech, Fertig, Funktech, Lufttrans, Optik, Phys> angle of incidence; 2. <Telekom> angle of arrival; 3. <Wellphys> angle of incidence
Einfallwinkelsonde f <Lufttrans> incidence probe
einfalzen v 1. <Bau> join; 2. <Fertig> rabbet
Einfang m <Kerntech> capture
einfangen v <Raumfahrt> capture (Raumschiff)
Einfangen n 1. <Fertig> aligning; alignment (Messgerät); 2. <Raumfahrt> capture (eines Satelliten)

Einfangstrahlung

Einfangstrahlung f <Kerntech> capture radiation
einfärben v <Bau> (AE) color, (BE) colour
Einfarbenpunktschreiber m <Gerät> (AE) single color point recorder, (BE) single colour point recorder
einfarbig adj 1. <Comp & DV, Druck, Foto> monochrome; 2. <Textil> plain
Einfärbung f <Papier> inking
Einfaserkabel n <Elektrotech, Telekom> (AE) single fiber cable, (BE) single fibre cable
Einfaserleitung f <Elektrotech> (AE) single fiber line, (BE) single fibre line
Einfassborte f <Textil> braid
einfassen v 1. <Bau> border; 2. <Textil> face
Einfassung f 1. <Bau> skirting; 2. <Kfztech> rim (Scheinwerfer); 3. <Meerschmutz> skirt
Einfeldträger m <Bau> simple beam
Einfettmittel n <Lebensmittel> greasing agent (Trennmittel für Backbleche)
Einfingerwahl f <Telekom> (AE) single digit dialing, (BE) single digit dialling
Einfliegen n <Lufttrans> flight test
einfluchten v <Fertig> flush
Einfluchten n 1. <Bau> running; 2. <Ker & Glas> marking out
Einflugsteuerkurs m <Lufttrans> inbound heading
Einflussfunktion f <Kerntech> importance function
Einflussgebiet n <Wasserversorg> area of influence
Einflusslinie f <Bau> influence line (Brückenbau)
einförmig adj <Chemie> monotonic, uniform
Einfressung f <Kerntech> corrosion
Einfriedung f <Bau> boundary fence, enclosure
einfrieren v 1. <Heiz & Kälte> freeze; 2. <Lebensmittel, Papier> freeze
einfrieren lassen v/sich <Heiz & Kälte> freeze
Einfrieren n 1. <Heiz & Kälte> freezing; 2. <Verpack> deep-freezing
Einfriertemperatur f <Kunststoff> glass transition temperature
Einfügedämpfungsmessmethode f <Telekom> reference test method (Lichtleitfaser)
einfügen v 1. <Bau> join; 2. <Comp & DV> paste, patch; 3. <Fertig> rabbet; 4. <Math> adapt, insert
Einfügen n <Fernseh> edit-in, insert edit, insert editing
Einfügung f <Comp & DV> insert
Einfügungsdämpfung f 1. <Phys> insertion loss; 2. <Telekom> extrinsic joint loss, insertion loss
Einfügungsverlust m <Optik> insertion loss
Einfügungsverstärkung f <Phys> insertion gain
einführen v 1. <Elektrotech> plug in; 2. <Telekom> insert (Steckkarte)
Einführkabel n <Elektriz> lead-in cable
Einfuhrlizenz f <Wassertrans> (BE) import licence, (AE) import license (Dokument)
Einführungsdraht m <Elektrotech> lead-in wire
Einführungsrille f <Aufnahme> lead-in groove
Einfülladapter m <Kfztech> filler adaptor
Einfülldeckel m <Mechan> filler cap
Einfüllöffnung f <Raumfahrt> filling hole (Raumschiff)
Einfüllstutzen m <Kfztech> filler neck
Einfüllstutzen m **für Reaktorkühlmittel** <Kerntech> reactor coolant inlet nozzle
Einfülltrichter m <Textil> hopper
Einfüllventil n 1. <Kfztech> filler valve, filling valve; 2. <Raumfahrt> filling valve (Raumschiff)
Einfüllverlust m <Raumfahrt> spillover loss
Einfüllverschluss m <Kfztech> radiator cap
Eingabe f 1. <Comp & DV> entry, IP, input; 2. <Elektriz, Fernseh, Funktech, Hydraul, Kontroll> IP, input; 3. <Patent> entry; 4. <Phys> IP, input

Eingabe f **am Steuerpult** <Kontroll> console input
Eingabe f **über Tastatur** <Comp & DV> keyboard entry
Eingabe f **von Hand** <Comp & DV> manual input
Eingabeanschluss m <Comp & DV> input port
Eingabeanweisung f <Comp & DV> input statement
Eingabearbeitswarteschlange f <Comp & DV> input work queue
Eingabeaufforderung f <Comp & DV> prompt
Eingabe/Ausgabe f (E/A) <Comp & DV, Elektriz> input/output (I/O)
Eingabe/Ausgabeanforderung f <Comp & DV> input/output request
Eingabe/Ausgabeanschluss m <Comp & DV> input/output port
Eingabe/Ausgabebefehl m <Comp & DV> input/output instruction
eingabe-/ausgabebegrenzt adj <Comp & DV> input/output limited
Eingabe/Ausgabebus m <Comp & DV> input/output bus
Eingabe/Ausgabedatei f <Comp & DV> input/output file
Eingabe/Ausgabeeinheit f <Comp & DV> input/output device
Eingabe/Ausgabegerät n <Comp & DV> input/output device
Eingabe/Ausgabeinstruktion m <Comp & DV> input/output instruction
Eingabe/Ausgabeinterrupt m <Comp & DV> input/output interrupt
Eingabe/Ausgabekanal m <Comp & DV> input/output channel
Eingabe/Ausgabeprozessor m <Comp & DV> IOP, input/output processor
Eingabe/Ausgabepuffer m <Comp & DV> input/output buffer
Eingabe/Ausgaberegister n <Comp & DV> input/output register
Eingabe/Ausgabesteuerung f <Comp & DV> input/output control
Eingabe/Ausgabesystem n <Comp & DV> IOS, input/output system
Eingabe/Ausgabeunterbrechung f <Comp & DV> input/output interrupt
Eingabebedingung f <Comp & DV, Patent> entry condition
Eingabebefehl m <Comp & DV, Patent> entry instruction
eingabebegrenzt adj <Comp & DV> input-limited
Eingabebereich m <Comp & DV> input area
Eingabeblock m <Comp & DV> input block
Eingabedatei f <Comp & DV> input file
Eingabedaten npl <Comp & DV> input data
Eingabedatensatz m <Comp & DV> input record
Eingabeeinheit f <Comp & DV> input device, input unit
Eingabefehler m <Telekom> keying error
Eingabefolge f <Comp & DV> input sequence
Eingabegerät n 1. <Comp & DV> input device, input unit; 2. <Elektrotech> input device
Eingabegerät n **für Alphanumerik** <Mobilkom> input device for alphanumerics, EGA (Funkpersonenruf)
Eingabeglied n <Regelung> input element
Eingabekanal m <Kontroll> input port
Eingabeleitung f <Comp & DV> input lead
Eingabemodus m <Comp & DV> input mode
Eingabeprogramm n <Comp & DV> input routine, reader
Eingabepuffer m <Comp & DV> input buffer
Eingabepufferverstärker m <Elektronik> input buffer amplifier
Eingaberegister n <Comp & DV> input register
Eingaberoutine f <Comp & DV> input routine

Eingaberückstellung f <Raumfahrt> input back-off (Wanderwellenröhre)
Eingabesatz m <Comp & DV> input record
Eingabespeicherbereich m <Comp & DV> input storage
Eingabesteuerung f <Elektronik> input control
Eingabetablett n <Comp & DV> data tablet
Eingabetaste f <Comp & DV> return, return key
Eingabevorgang m <Comp & DV> transaction
Eingabewarteschlange f <Comp & DV> entry queue, input queue, input work queue
Eingabezeile f <Comp & DV> entry line
Eingang m 1. <Elektriz, Elektronik, Fernseh> IP, input; 2. <Fertig> mouth (Kunststoffinstallationen); 3. <Funktech, Kontroll> IP, input; 4. <Telekom> inlet
Eingang/Ausgang m (E/A) <Elektriz> input/output (I/O)
eingängig adj <Maschinen> single-start, single-thread
eingängiges Gewinde n <Maschinen> single thread
Eingangsabgriff m <Elektrotech> input tapping
Eingangsablauf m <Telekom> incoming procedure
Eingangsadmittanz f <Elektrotech> input admittance
Eingangs-Ausgangs-Schnittstelle f <Telekom> input/output interface
Eingangsbestand m <Verpack> accepted stock
Eingangsdämpfungsglied n <Elektronik> input attenuator
Eingangsdruck m <Kohlen> inlet pressure
Eingangselektrode f <Elektrotech> input electrode
Eingangselement n <Regelung> receiver element
Eingangsfächerung f <Elektrotech> fan-in
Eingangsfehler m <Comp & DV> inherited error
Eingangsfilter n 1. <Elektronik, Funktech> input filter; 2. <Raumfahrt> input filter (Weltraumfunk)
Eingangsfilter n **mit Drossel** <Funktech> choke-input filter
Eingangsfilterung f <Elektronik> input filtering
Eingangsflughafen m <Lufttrans> airport of entry
Eingangsfolgeliste f <Comp & DV> push-up list
Eingangsfolgestapel f <Comp & DV> push-up stack
Eingangsgatter n <Elektronik> input gate
Eingangsgleichstrom m <Elektrotech> DC input
Eingangsglied n <Gerät> receiving element (einer Messeinrichtung)
Eingangshohlraum m <Elektronik> input cavity
Eingangsimpedanz f 1. <Akustik> loaded impedance (bei Nennbelastung); 2. <Elektrotech, Fernseh, Telekom> input impedance
Eingangsimpuls m <Elektronik> input pulse
Eingangskanal m <Kontroll> input port
Eingangskapazität f <Elektrotech> input capacitance
Eingangskegelritzelwelle f <Lufttrans> input bevel pinion shaft (Hubschrauber)
Eingangsklemme f <Elektriz, Elektrotech> input terminal
Eingangskontrolle f 1. <Abfall> weigh office (einer Deponie); 2. <Qual> incoming inspection
Eingangskreis m <Elektrotech> input circuit
Eingangsleistung f 1. <Elektrotech, Maschinen> input power; 2. <Fernseh> input, input power; 3. <Funktech> input; 4. <Telekom> input, input power
Eingangsleitung f <Aufnahme, Fernseh> line in
Eingangsmesswandler m <Elektrotech> input transductor
Eingangsmobilvermittlungsstelle f <Mobilkom> (AE) gateway mobile switching center, (BE) gateway mobile switching centre
Eingangs-MSC f <Mobilkom> gateway MSC
Eingangspegel m <Fernseh, Telekom> input level
Eingangspegel m **für Tonfrequenzsignal** <Aufnahme> audio signal input level
Eingangsport m <Telekom> input port

Eingangsprüfung f <Qual> incoming inspection, on-receipt inspection, receiving inspection
Eingangspunkt m <Comp & DV, Patent> entry point
Eingangsregister n <Telekom> incoming register
Eingangsresonator m <Elektronik> input resonator (Klystron); buncher resonator (verursacht Geschwindigkeitsmodulation)
Eingangsschaltung f <Elektriz> input circuit
Eingangsscheinleitwert m <Elektrotech> input admittance
Eingangsscheinwiderstand m <Elektrotech> input impedance
Eingangssignal n <Elektronik, Fernseh, Telekom> input signal, incoming signal
Eingangssignalleistung f <Elektronik> input signal power
Eingangssignalquantelung f <Elektronik> input signal quantization
Eingangs-Signal-Rausch-Abstand m <Funktech> input signal-to-noise ratio
Eingangssignal-Rausch-Verhältnis n <Elektronik> input signal-to-noise ratio
Eingangsspannung f <Elektriz, Elektrotech> input voltage
Eingangsstelle f <Patent> receiving section
Eingangssteuerung f <Elektronik> input control
Eingangsstrom m <Elektriz> input current
Eingangsstufe f 1. <Elektronik> input stage; 2. <Funktech> premixer
Eingangsstufenverstärkung f <Funktech> input stage gain
Eingangstransformator m <Elektrotech, Phys> input transformer
Eingangsübertrager m <Funktech> input transformer
Eingangsverhalten n <Elektronik> input response
Eingangsverstärker m <Elektronik> input amplifier
Eingangswelle f 1. <Kfztech> input shaft (Kupplung, Getriebe); 2. <Maschinen> input shaft
Eingangswiderstand m 1. <Elektrotech> input resistance; 2. <Phys> input impedance
Eingangszuleitung f <Elektrotech> input lead
eingebaut adj 1. <Bau> encastré, built-in; 2. <Elektrotech, Maschinen, Mechan> built-in
eingebaute Alterung f <Abfall> built-in obsolescence
eingebaute Barriere f <Kerntech> engineered barrier (zur Sicherheit in Atomkraftwerken)
eingebaute Batterie f <Elektrotech> internal battery
eingebaute Freispracheinrichtung f <Mobilkom> built-in handsfree unit
eingebaute Prüfeinrichtung f <Gerät> built-in test equipment
eingebaute Schutzvorrichtung f <Sicherheit> built-in guard
eingebaute Standardfunktion f <Comp & DV> built-in function
eingebaute Stromversorgung f <Elektrotech> built-in power supply
eingebauter Belichtungsmesser m <Foto> built-in exposure meter
eingebauter Motor m <Elektrotech> built-in motor
eingebauter Oszillator m <Phys> local oscillator
eingebautes Ladegerät n <Trans> built-in charger
eingeben v 1. <Comp & DV> type; enter, key in (Daten, Befehl); 2. <Elektronik> input
Eingeben n <Elektronik> inputting
eingebettet adj <Comp & DV, Maschinen> embedded
eingebetteter Befehl m <Comp & DV> embedded command
eingebetteter Code m <Comp & DV> embedded code

eingebetteter

eingebetteter Dehnmessstreifen m <Bau> embedded strain gauge
eingebettetes System n <Comp & DV> embedded computer, embedded system
eingebrannter Zeitcode m <Fernseh> burnt-in time code
eingebranntes Ventil n <Kfztech> burnt valve *(Motor)*
eingebuchtes Handy n <Mobilkom> logged-in mobile
eingebundenes Post-Script n <Comp & DV> encapsulated PostScript
eingebundenes Programm n <Comp & DV> encapsulated program
eingedämmt adj <Thermod> banked-up
eingedampfter Latex n <Kunststoff> evaporated latex
eingedickter Schlamm m <Kohlen> thickened sludge
eingedost adj <Chemie> *(AE)* canned, *(BE)* tinned
eingedrückter Mantel m <Optik> depressed cladding
eingeebnet adj <Bau> *(AE)* leveled, *(BE)* levelled
eingeengter Sattelpunkt m <Metall> constricted node
eingefahrener Notabschaltstab m <Kerntech> inserted scram rod
eingefahrener Schnellabschaltstab m <Kerntech> scrammed rod
eingefroren adj <Wassertrans> icebound *(Schiff)*
eingegangener Bestand m <Verpack> accepted stock
eingehen v <Textil> contract, shrink
Eingehen n <Textil> shrinkage
eingehende Mail f <Comp & DV> incoming mail
eingehende Meldung f <Comp & DV> incoming message
eingehende Nachricht f <Comp & DV> incoming message
eingekapselte Strahlungsquelle f <Kerntech> encapsulated source
eingekehlt adj <Fertig> necked-down *(Schmieden)*
eingekerbte Düse f <Lufttrans> notched nozzle
eingeklebt adj <Druck> tipped-in
eingelagert adj <Maschinen> embedded
eingelagerter Abfall m <Abfall> emplaced waste
eingelassen adj <Elektriz> flush-mounting
eingelassene Schleife f <Trans> embedded loop
eingelassener Heißring m <Wassertrans> flush lifting ring *(Deckbeschläge)*
eingelassener Schalter m <Elektriz> flush switch
eingelegte Drähte mpl <Elektriz> flush wiring
eingeleitet adj <Phys> induced
eingenähtes Etikett n <Textil> sewn-in label
eingeprägte Kraft f <Fertig> active force *(auf einen Körper wirkend)*
eingerastet adj <Fertig> engaged
eingereichter Flugplan m <Lufttrans> air-filed flight plan
eingerissen adj <Fertig> ragged
eingeschalt adj <Bau> timbered
eingeschaltet adj <Elektrotech> active, closed, on
eingeschalteter Stromkreis m <Elektrotech> closed circuit
eingeschalteter Zustand m <Elektrotech> on-state
eingeschliffen adj <Maschinen> ground-in
eingeschliffener Glasstopfen m <Labor> ground stopper
eingeschlossene Luft f <Kunststoff, Verpack> entrapped air
eingeschlossener Winkel m <Geom> included angle
eingeschlossenes Teilchen n <Kerntech> trapped particle
eingeschnittener Synchronisationsimpuls m <Fernseh> serrated pulse
eingeschnittenes Walmdach n <Bau> hip and valley roof
eingeschränkter Belegungsdetektor m <Trans> limited presence detector

eingeschränktes Progressivsystem n <Trans> limited progressive system
eingeschriebene Kugel f <Fertig> insphere
eingeschriebener Kreis m <Geom> inscribed circle
eingeschriebener Winkel m <Geom> inscribed angle
eingeschriebenes Quadrat n <Geom> inscribed square
eingeschwungener Zustand m <Comp & DV, Telekom> steady state *(Oszillation)*
eingesenkter Faden m <Ker & Glas> depressed thread
eingesetzt adj <Maschinen> inserted
eingesetzter Stahl m <Metall> case-hardened steel
eingesetzter Zahn m <Fertig> cog
eingespannt adj <Bau> encastré
eingespannter Träger m <Bau> encastré beam
eingespülte Füllung f <Bau> hydraulic fill
eingestellt adj <Papier> adjusted
eingestochen adj <Fertig> recessed
eingetaucht adj <Phys> immersed
eingetragene Marke f <Patent> registered trade mark
eingetragener Benutzer m <Patent> registered user
eingetragenes Warenzeichen n <Patent> registered trade mark
Eingeweide npl <Lebensmittel> gut
Eingewöhnung f <Heiz & Kälte> acclimatization
eingezapfter Mauerstein m <Bau> tusk
Eingießen n <Kunststoff> potting
Eingitterröhre f <Elektronik> single grid tube
Einglasen n <Ker & Glas> glazing *(Einbauen von Fenstern)*
eingleisige Gleisabzweigung f <Eisenbahn> single line turnout
eingleisige Strecke f <Eisenbahn> single track line
eingreifen v <Maschinen> engage, mesh
Eingreifen n 1. <Maschinen> meshing; 2. <Mechan> mating
Eingreifswinkel m <Fertig> angle of obliquity of action *(Getriebelehre)*
eingrenzen v 1. <Comp & DV> localize; 2. <Meerschmutz> corral *(Ölverschmutzung)*
Eingrenzung f 1. <Comp & DV> isolation *(einer Fehlerquelle)*; 2. <Telekom> locating; sectionalization *(Fehler)*
Eingriff m 1. <Comp & DV> intervention; 2. <Fertig> mesh; contact *(Getriebelehre)*; 3. <Maschinen> contact, engagement, gearing, intermeshing, meshing • **außer Eingriff bringen** <Fertig> disengage *(Zahnräder)* • **gegen Eingriff sichern** <Sicherheit, Verpack> tamperproof • **im Eingriff** 1. <Fertig> engaged; 2. <Maschinen> engaged, meshed • **in Eingriff bringen** <Maschinen> mesh • **in Eingriff kommen** <Maschinen> come into gear • **nicht im Eingriff** <Maschinen> out of gear • **voll im Eingriff** <Fertig> fully meshed • **wieder in Eingriff bringen** <Maschinen> re-engage
Eingriffsbogen m <Lufttrans> arc of contact *(des Gurtes, Riemens)*
Eingriffsbogen m **vor dem Wälzpunkt** <Fertig> arc of action *(Getriebelehre)*
Eingriffsfeld n <Fertig> zone of action
Eingriffsgrenze f <Qual> action limit
Eingriffslinie f <Maschinen> line of action, line of engagement
Eingriffssteuerung f <Lufttrans> override control
Eingriffsstörung f <Fertig> tooth interference
Eingriffsteilung f <Maschinen> contact pitch
Eingriffstiefe f <Maschinen> depth, working depth of teeth
Eingriffswinkel m 1. <Fertig> angle of approach *(Getriebelehre)*; 2. <Maschinen> angle of obliquity, angle of pressure, pressure angle
Einguss m <Maschinen> sprue

Eingussabschneider m <Fertig> sprue cutter
Eingussbolzen m <Maschinen> sprue pin
Einhakdeckel m <Verpack> hooked lid
einhaken v <Fertig> bind (Bohren)
Einhaken n <Fertig> binding (Spanung)
Einhaltung f <Fertig> maintenance (Toleranz)
einhängbarer Sack m <Verpack> insertable sack
Einhängefeld n <Bau> suspended span (Brücke)
Einhängemaschine f <Druck> casing-in machine
einhängen v 1. <Bau> hinge; 2. <Druck> case in; 3. <Telekom> hang up (Hörer)
einhausen v <Sicherheit> cage in, enclose
Einhausung f <Sicherheit> acoustic enclosure, acoustical enclosure, enclosure
Einheftkante f <Druck> binding edge
einheimische Energiequelle f <Kohlen, Nichtfoss Energ> indigenous energy resource
Einheit f 1. <Comp & DV> device, unit; 2. <Elektriz, Kfztech, Maschinen, Optik, Phys, Telekom> component, set, unit; 3. <Künstl Int> artificial neuron, cell, neuron, unit (neurales Netzwerk); 4. <Metrol> unit (Maßeinheit)
Einheit f der absorbierten Strahlungsdosis <Strahlphys> unit of absorbed dose
Einheit f der Bestrahlung <Strahlphys> unit of exposure
Einheit f der Entropie <Thermod> unit of entropy
Einheit f des Dosisäquivalents <Strahlphys> unit of dose equivalent (als Maß für biologische Wirkung)
Einheit f im Raumwinkelmaß <Geom> steradian
Einheit f mit einem oder mehreren Hauptfehlern <Qual> major defective
Einheit f mit einem oder mehreren Nebenfehlern <Qual> minor defective
Einheit f/zu prüfende <Qual> unit under test
Einheitencode m <Comp & DV> device code
Einheitenfolge f <Comp & DV> unit string
Einheitensteuerung f <Comp & DV> device control
Einheitensteuerzeichen n <Comp & DV> device control character
Einheitensystem n <Elektriz, Maschinen> system of units
einheitliche Rufnummer f <Telekom> universal number
einheitlicher Aufbau m <Comp & DV> unified architecture
einheitliches Feld n <Phys> uniform field
einheitliches Kabel n <Optik> unit-type cable
Einheitlichkeit f der Erfindung <Patent> unity of invention
Einheitsbohrung f <Fertig, Maschinen> basic bore
Einheitselement n <Math> identity element
Einheitsfläche/pro <Phys> per unit area
Einheitsimpuls m <Elektriz> Delta-function, unit impulse, unit pulse
Einheitslänge/pro <Phys> per unit length
Einheitsmasse/pro <Phys> per unit mass
Einheitsmatrix f <Math> identity matrix
Einheitsmessumformer m für Mischungsabweichungen <Regelung> composition deviation transmitter
Einheitssatz m <Bau, Patent> flat-rate fee
Einheitsschritt m 1. <Comp & DV> unit element; 2. <Telekom> unit interval
Einheitsschub m <Raumfahrt> unit thrust (Raumschiff)
Einheitssignal n <Regelung> standard signal
Einheitssprungantwort f <Elektriz> unit step response, transient response
Einheitsvektor m 1. <Math> unit vector; 2. <Phys> standard vector, unit vector
Einheitswelle f <Maschinen> basic shaft
Einhiebfeile f <Maschinen> single-cut file
einhiebig adj <Fertig> float-cut (Feile)

einhiebige Feile f 1. <Fertig> float; 2. <Maschinen> float, float-cut file, single-cut file
einhieven v <Wassertrans> heave in (Tauwerk)
einholen v <Wassertrans> hoist (Schiff); haul in (Tauwerk)
Einholen n auf der Umlaufbahn <Raumfahrt> orbital catchup
Einholtau n <Wassertrans> dead man (Deckausrüstung)
Einhüllende f <Math> envelope (Hüllkurve für eine Kurvenschar)
Einhüllung f <Phys> enclosure
Einimpfen n <Kerntech> seeding
Einkanalprotokoll n <Comp & DV> single channel protocol
Ein-Kanal-pro-Träger m (SCPC) 1. <Raumfahrt> single channel per carrier, SCPC (Weltraumfunk); 2. <Telekom> single channel per carrier, SCPC
Ein-Kanal-pro-Träger-System n (SCPC) 1. <Raumfahrt> single channel per carrier, SCPC (Weltraumfunk); 2. <Telekom> single channel per carrier, SCPC
Einkanalträger m <Raumfahrt> single channel carrier (Weltraumfunk)
Einkanalverstärker m <Elektronik> single channel amplifier
Einkapselung f 1. <Abfall> sealing (Deponie); 2. <Kerntech> encapsulation
Einkapselung f des Verfahrens <Sicherheit> process enclosure
Einkaufsbedingungen fpl <Raumfahrt> procurement specifications (Raumschiff)
einkerben v 1. <Bau> score; 2. <Fertig> incise, nick, serrate
Einkerben n <Fertig> brinelling
Einkerbung f 1. <Bau> indentation; 2. <Fertig> incision, indentation; 3. <Maschinen> dent, indentation
einklammern v <Druck> bracket
Einklang m <Akustik> concord, consonance
einkleben v <Druck> tip in
einklinken v <Erdöl> latch on (Bohrtechnik)
Einklinken n <Elektrotech> latching
einkochen v 1. <Chemtech> thicken by boiling; 2. <Lebensmittel> boil down; bottle (Früchte oder Gemüse)
Einkoppelstrecke f <Elektronik> buncher space (Klystron); input gap (Wellenleiter)
Einkoppelungsfaser f <Optik> (AE) launching fiber, (BE) launching fibre
Einkopplung f <Telekom> launching
Einkopplungsbedingung f bei stationärer Modenverteilung <Telekom> steady state launching condition
Einkörperschiff n <Wassertrans> single hull ship
Einkörperverdampfer m <Lebensmittel> single effect evaporator
Einkreisreaktor m <Kerntech> one-cycle reactor
Einkristall m <Elektronik> single crystal (Monokristall)
Einkristallfaden m <Kunststoff> whisker
Einkristallhalbleiter m <Elektronik> single crystal semiconductor
Einkristallzüchtung f <Elektronik> single crystal growth
einkuppeln v <Kfztech> engage
Einkuppeln n <Maschinen> meshing
einladen v <Comp & DV> roll in
Einlage f 1. <Bau> core; 2. <Fertig> insert (Kunststoffinstallationen); 3. <Kerntech, Kohlen> liner; 4. <Kunststoff> insert; 5. <Verpack> inlay, insert, liner
einlagern v <Comp & DV> roll in
Einlagern n <Comp & DV> swap-in
Einlagerung f <Werkprüf> inclusion
Einlagestoff m <Textil> interlining
einlagig adj 1. <Bau> single-layer; 2. <Fertig> single-ply (Riemen)

Einlass

Einlass m 1. <Hydraul> gating; 2. <Ker & Glas> inlet port; 3. <Kfztech> induction, inlet; 4. <Maschinen> admission, inlet, intake; 5. <Mechan, Telekom> inlet
einlassen v <Hydraul> gate
Einlassgeschwindigkeit f <Hydraul> inlet velocity
Einlasshub m <Kfztech> induction stroke
Einlasskanal m 1. <Hydraul> induction port *(Dampfzylinder)*; 2. <Kfztech> inlet port
Einlasskrümmer m <Kfztech> induction manifold *(Motor)*
Einlasslufttrichter m <Lufttrans> inlet throat
Einlassöffnung f 1. <Hydraul> induction port *(Dampfzylinder)*; 2. <Kfztech> inlet port; 3. <Maschinen> inlet, intake
Einlassring m **am Flugzeugrumpf** <Lufttrans> nacelle intake ring
Einlassrohr n <Hydraul> induction pipe
Einlassseite f <Maschinen> inlet side, intake side
Einlasssonde f <Kerntech> injection well
Einlassspinne f <Kfztech> induction manifold, inlet manifold *(Motor)*
Einlassstutzen m <Fertig> inlet connection
Einlasssystem n <Maschinen> intake system
Einlasstür f <Bau> wicket
Einlassüberdeckung f <Hydraul> outside lap *(Steuerschieber)*
Einlassventil n 1. <Hydraul> induction valve, inlet valve; intake valve *(Dampf)*; 2. <Kfztech> induction valve, suction valve; inlet valve *(Motor)*; 3. <Maschinen> admission valve
Einlassverbindungskabel n <Kerntech> inlet jumper
Einlauf m 1. <Bau> hopper head *(Regenrohr)*; 2. <Fertig> gate *(Gießen)*; 3. <Mechan> intake; 4. <Wasserversorg> inlet, intake
Einlaufanschluss m <Papier> inlet
einlaufen v <Maschinen> run in
Einlaufen n 1. <Maschinen> running-in; 2. <Textil> shrinkage
einlaufend adj <Wassertrans> inward-bound
einlaufender Verkehr m <Trans> inbound traffic, *(AE)* incoming traffic, *(BE)* inward traffic
Einlaufrille f <Akustik> lead-in groove
Einlaufschacht m <Umweltschmutz> sink
Einlaufzeit f <Thermod> warm-up time
einlegbarer Schalter m <Elektriz> recessed switch
Einlegeende n <Ker & Glas> charging end
Einlegekeil m <Maschinen> sunk key
einlegen v 1. <Aufnahme> thread *(eines Bandes in Gerät)*; 2. <Bau> insert; inlay *(Holz)*; 3. <Fertig> inject *(Spanung)*; insert *(Werkstück)*
Einlegen n <Ker & Glas> charging
Einlegen n **von nur einem Glasposten** <Ker & Glas> single gob feeding
Einleger m <Fertig> feeder *(Gießen)*
Einlegering m <Fertig> spacing ring *(Kunststoffinstallationen)*
Einlegestreifen m <Textil> paper collar
Einlegeteil n <Fertig> valve end *(Kunststoffinstallationen)*
Einlegevorbau m <Ker & Glas> dog house, filling end
einleiten v 1. <Comp & DV> initialize; 2. <Phys> induce
Einleiten n <Meerschmutz, Umweltschmutz, Wasserversorg> discharge
Einleiterkabel n <Elektrotech> single conductor cable
Einleitung f <Abfall, Meerschmutz, Umweltschmutz, Wasserversorg> discharge *(von Abwässern)*
Einleitungsanlage f <Abfall> discharge system
Einleitungsbremse f <Eisenbahn> single pipe brake
Einleitungskanal m <Wasserversorg> effluent channel
einlesen v <Gerät> read *(Programm)*
Einlocheinspritzdüse f <Kfztech> single jet injection nozzle

einloggen v <Mobilkom> log in, log on
einloten v <Bau> plumb
Einmachglas n 1. <Ker & Glas> *(AE)* canning jar, *(BE)* preserving jar; 2. <Verpack> glass jar
Einmaischen n <Lebensmittel> mashing
einmal beschreibbare Platte f <Comp & DV> write-once disk, write-once read many times disk
einmal beschreibbare Scheibe f/nur <Optik> *(BE)* write-once disc, *(AE)* write-once disk
einmal beschreibbare Speicher mpl/nur <Optik> write-once optical storage
einmal gestrichenes Papier n <Papier> single-coated paper
Einmalartikel m **aus gepresstem Zellstoff** <Verpack> *(AE)* molded pulp article, *(BE)* moulded pulp article *(in Krankenhaus)*
Einmalbeschreibung-Mehrfachlesen n (WORM) <Optik> write once read many times *(WORM)*
einmalig beschreibbare CD f <Comp & DV> WO-CD, write-once compact disk
einmalige Beladung f <Kerntech> once-through charge
einmalige Kosten fpl <Raumfahrt> non-recurrent cost
einmaliger Impuls m <Elektronik> nonrecurrent pulse, nonrecursive pulse
Einmalkarbon-Farbband n <Konstzeich> once-only ribbon
Ein-Mann-Brücke f <Wassertrans> one-man bridge
einmanteln v <Bau> box
einmauern v <Bau> brick in
einmäulig adj <Fertig> single-end *(Lehre)*
Einmessen n <Metrol> calibration
Einmesszeichen n <Telekom> calibration signal
Einmoden... <Elektronik, Elektrotech, Fernseh, Optik, Phys, Telekom> single mode
Einmodenfaser f 1. <Optik> *(AE)* single mode fiber, *(BE)* single mode fibre; 2. <Telekom> *(AE)* monomode fiber, *(BE)* monomode fibre, *(AE)* single mode fiber, *(BE)* single mode fibre
Einmodenglasfaser f <Elektrotech> *(AE)* single mode optical fiber, *(BE)* single mode optical fibre
Einmodenkabel n <Elektrotech> single mode fiber cable
Einmodenlaser m <Elektronik, Funktech, Telekom> single frequency laser
Einmoden-Lichtleiter m <Phys> *(AE)* monomode fiber, *(BE)* monomode fibre
Einmoden-Lichtleitfaser f <Optik, Telekom> *(AE)* single-mode optical fiber, *(BE)* single-mode optical fibre
Einmoden-Lichtwellenleiter m <Telekom> *(AE)* monomode fiber, *(BE)* monomode fibre, *(AE)* single-mode fiber, *(BE)* single-mode fibre
Einmoden-Lichtwellenleiterkabel n <Telekom> single mode fiber cable
Einmoden-LWL-Faser f <Telekom> *(AE)* single mode optical fiber, *(BE)* single mode optical fibre
einmolekular adj <Chemie> monomolecular, unimolecular
Einmotorenantrieb m <Elektrotech> monomotor drive, single-motor drive
einohrig adj <Akustik, Aufnahme> monaural
einordnen v <Lebensmittel, Qual> grade
einparametrisches digitales Signal n <Regelung> single parameter digital signal
einpassen v 1. <Bau> fit in, fit, seat; 2. <Maschinen> fit in, fit into; 3. <Mechan> fit
Einpassen n 1. <Comp & DV> sizing; 2. <Maschinen, Mechan> fitting
Einpegeln n <Telekom> level adjustment
Einpegelungston m <Aufnahme> line-up tone
einpeilen v <Bau> locate
Einpendeln n <Elektrotech> hunting *(Regler)*

Einpflügen *n* **eines Kabels** <Telekom> plough burial of a cable
Einphasen *n* <Elektronik, Kontroll> single-phase
Einphasenbrücke *f* <Elektronik> single-phase bridge
Einphasenbrückenschaltung *f* <Elektronik> single-phase bridge connection
Einphasenmaschine *f* <Elektrotech> single-phase machine
Einphasenmotor *m* **mit Anlaufkondensator** <Elektrotech> capacitor-start motor
Einphasenmotor *m* **mit Hilfswicklung** <Elektrotech> split-phase motor
Einphasenmotor *m* **mit Hilfswicklung und Drosselspule** <Elektrotech> reactor start split-phase motor
Einphasenmotor *m* **mit Hilfswicklung und Widerstand** <Elektrotech> resistance start split-phase motor
Einphasenmotor *m* **mit Kondensatorhilfsphase** <Elektrotech> capacitor-start motor
Einphasenreaktion *f* <Metall> monophase reaction
Einphasensteuergerät *n* <Trans> one-phase controller
Einphasenstrom *m* 1. <Elektriz> single phase electric current; 2. <Elektrotech> single phase current
Einphasenwechselstrom-Motor *m* <Elektrotech> single-phase motor
Einphasenwechselstrom-Triebfahrzeug *n* <Elektrotech> single-phase a.c. motor vehicle
einphasig *adj* 1. <Elektriz> monophase, single-phase, uniphase; 2. <Elektrotech> one phase, single-phase, uniphase; 3. <Kontroll> single-phase
einphasige Leitung *f* <Elektriz> monopolar line
einphasige Maschine *f* <Elektrotech> single phase machine
einphasige Stromversorgung *f* <Elektriz, Elektrotech> single phase supply
einphasige Wicklung *f* <Elektrotech> single phase winding
einphasiger Brückengleichrichter *m* <Elektriz> single phase bridge rectifier
einphasiger Gleichrichter *m* <Elektriz> single wave rectifier
einphasiger Induktionsmotor *m* <Elektrotech> single phase induction motor
einphasiger Motor *m* <Elektriz, Elektrotech> single phase motor
einphasiger Strom *m* <Elektrotech> single phase current
einphasiger Transformator *m* <Elektrotech> single phase transformer
einplanen *v* <Comp & DV, Fernseh, Telekom> schedule
Einplatinenrechner *m* <Comp & DV> single board computer
Einplattenruder *n* <Wassertrans> single plate rudder
einpolar *adj* <Elektrotech> unipolar
einpolig *adj* <Comp & DV, Elektriz, Elektronik> monopolar, unipolar
einpoliger Ein-/Ausschalter *m* *(SPST-Schalter)* 1. <Elektriz> single pole single-throw switch, single toggle switch *(SPST switch)*; 2. <Elektrotech, Kontroll> single pole single-throw switch *(SPST switch)*; single toggle switch
einpoliger Schalter *m* <Elektriz, Elektrotech> single pole switch
einpoliger Umschalter *m* *(SPDT-Schalter)* <Elektrotech> single pole double-throw switch *(SPDT switch)*
einpoliger Wechselschalter *m* *(SPDT-Schalter)* <Elektriz> single pole double-throw switch, SPDT switch *(mit mittlerer Ruhelage)*
einpoliges Einschaltrelais *n* <Elektrotech> SPST relay
einpoliges Umschaltrelais *n* <Elektrotech> SPDT relay
Einprägen *n* <Druck> stamping

einprägsame Darstellung *f* <Konstzeich> expressive representation
Einpressbohrung *f* <Erdöl> injection well *(Tiefbohrtechnik)*
einpressen *v* <Bau> grout *(Zementmörtel)*
Einpressen *n* 1. <Bau> grouting *(Spannbeton)*; 2. <Kohlen> grouting
Einpressmutter *f* <Maschinen> press nut
Einpresspumpe *f* <Bau> injection pump
einprofilig *adj* <Fertig> single-edge *(Schleifscheibe)*
Einprogrammsystem *n* <Comp & DV> monoprogramming system
einprotonig *adj* <Chemie> monobasic *(Säure)*
Einprozessor-Server *m* <Comp & DV> single-processor server
Einpunktmaß *n* <Math> Dirac measure, one-point measure *(Dirac-Maß)*
Einpunktverbindung *f* <Elektriz> single point bonding
Einrahmung *f* <Bau> casing *(eines Fensters, einer Tür)*
einrammen *v* <Bau> drive in *(Nägel)*
Einrammen *n* 1. <Bau> ramming *(Pfähle)*; 2. <Fertig> driving; 3. <Kohlen> ramming
einrasten *v* 1. <Fertig> pawl; insert *(Stift)*; 2. <Kfztech> catch; 3. <Maschinen> engage *(Klinke)*; 4. <Mechan> latch, lock, lock in place, snap; 5. <Telekom> lock
Einrasten *n* 1. <Fertig> insertion; 2. <Telekom> acquisition *(Synchronisation)*
einrastender Elektromagnet *m* <Elektriz> latching electromagnet
Einrastknopf *m* <Bau> lock knob
Einrastpunkt *m* <Fertig> detent point
Einrastung *f* <Maschinen> stop
Einreichung *f* <Patent> filing
einreihig *adj* <Maschinen> single-row
einreihige Nietüberlappung *f* <Bau> single-riveted lap joint
einreihige Nietverbindung *f* <Bau> single-riveted joint
einreißen *v* <Anstrich, Bau, Metall> crack
Einreißen *n* 1. <Ker & Glas> rip-in; 2. <Kunststoff> tear
Einreißfestigkeit *f* <Kunststoff> tear resistance, tear strength
einrichten *v* 1. <Bau> set; 2. <Comp & DV> initialize; set-up *(ein System)*; 3. <Maschinen> set, set-up; 4. <Telekom> establish *(Netzwerk)*; set up *(Verbindung)*
Einrichten *n* 1. <Lufttrans> setting *(eines Instrumentes)*; 2. <Maschinen> setting, setup
Einrichtprozedur *f* <Comp & DV> initial set-up procedure
Einrichtung *f* 1. <Bau> facility; 2. <Fertig> facility *(Kunststoffinstallationen)*; 3. <Maschinen> installation, device; 4. <Telekom> device, equipment, facility; unit *(Ausrüstung)*
Einrichtung *f* **für Probebetrieb** <Maschinen> try-out facility
Einrichtung *f* **zur Hochfrequenzschweißung** <Verpack> high-frequency welding equipment
Einrichtung *f* **zur Verfahrensüberwachung** <Elektrotech> process controller
Einriegelschloss *n* <Bau> deadlock
einringen *v* <Bau> place *(Beton)*
einrollen *v* 1. <Fertig> crush; 2. <Textil> curl
einrotorig *adj* <Lufttrans> single rotor *(Hubschrauber)*
einrücken *v* <Maschinen> mesh, throw into action, throw into gears
Einrücken *n* <Maschinen> engagement, engaging, gearing
Einrüsten *n* <Bau> scaffolding
Eins *f* <Math> one
Einsabstoff *m* <Erdöl> feedstock *(Raffinerie, Vergasung)*
Einsackapparat *m* <Verpack> bagging machine

einsacken

einsacken v <Lebensmittel, Verpack> bag
Einsacken n <Papier> bagging
einsalzen v <Lebensmittel> cure
Einsammeln n <Abfall> collection
Einsatz m 1. <Akustik> attack; 2. <Fertig> application *(Kunststoffinstallationen)*; case *(Stahl)*; 3. <Kfztech, Kunststoff> insert; 4. <Maschinen> insert, socket; 5. <Mechan> insert, lining; 6. <Textil> batch; 7. <Verpack> insert
Einsatz m **in Fertigungsreihe** <Fertig> in-line operation
Einsatzbereich m <Gerät> range
Einsatzbereich m **eines Geräts** <Gerät> instrument range
einsatzbereiter Wartebetrieb m <Raumfahrt> hot standby *(Raumschiff)*
Einsatzbohrer m <Erdöl> insert bit *(Bohrtechnik)*
Einsatzbrücke f <Maschinen> gap bridge, gap piece
Einsatzentfernung f <Lufttrans> operating range *(Flugwesen)*
Einsatzfilter n <Heiz & Kälte> cartridge-type filter
einsatzgehärteter Stahl m <Mechan> case-hardened steel
Einsatzgeschwindigkeit f <Lufttrans> operating speed
einsatzhärten v <Fertig> carburize; case-harden, cement *(Stahl)*
Einsatzhärten n 1. <Fertig> carbon case hardening, pack hardening; 2. <Maschinen> case hardening
Einsatzhärteofen m <Maschinen> carburizing furnace
Einsatzhärtung f 1. <Fertig> carburization; 2. <Metall> case hardening
Einsatzlaufbüchse f <Mechan> liner
Einsatzlebensdauer f <Raumfahrt> service life *(Raumschiff)*
Einsatzmeißel m <Fertig> bit insert
Einsatzmittel n <Comp & DV> resource
Einsatzofen m <Chemtech> batch furnace
Einsatzpause f <Lufttrans> rest period
Einsatzprüfung f <Raumfahrt> field trial
Einsatzreichweite f <Lufttrans> operating range *(Flugwesen)*
Einsatzspirale f <Foto> tank reel
Einsatztiefe f <Maschinen> case depth
Einsatzversuch m <Raumfahrt> field trial
Einsatzzentrale f <Sicherheit> *(AE)* emergency center, *(BE)* emergency centre
einsaugen v 1. <Fertig> inspire; 2. <Heiz & Kälte> aspirate
Einsaugen n <Heiz & Kälte, Mechan> suction
Einsaugung f <Fertig> inspiration
einsäurig adj <Chemie> monoacidic
Einschalt... <Comp & DV, Elektrotech, Fernseh, Fertig> turn-on
Einschaltantwort f <Elektrotech> transient response
Einschaltdauer f <Elektrotech, Fertig> duty cycle
einschalten v 1. <Comp & DV> connect; enable *(Rechner)*; 2. <Elektriz> switch on; 3. <Elektrotech> make, power up; turn on *(Lampe)*; 4. <Fernseh, Funktech, Kontroll> switch on; 5. <Phys> connect, enable, energize, switch on; 6. <Telekom> switch on
Einschalten n 1. <Elektrotech> turn-on *(Lampe)*; closure *(eines elektrischen Stromkreises)*; 2. <Strahlphys> onset *(eines Magnetfeldes)*
einschaltende Windgeschwindigkeit f <Nichtfoss Energ> cut-in wind speed
Einschalter m <Elektrotech> circuit closer
Einschaltimpuls m 1. <Elektronik> turn-on pulse; 2. <Elektrotech> make pulse
Einschaltlaststrom m <Elektrotech> on-load current
Einschaltprellen n <Elektriz> pull-in bouncing
Einschaltstellung f <Elektrotech> on position
Einschaltstoß m <Elektrotech, Phys> transient
Einschaltstoßstrom m <Elektrotech> inrush current
Einschaltstoßstrom-Begrenzer m <Elektrotech> inrush current limiter
Einschaltstrom m 1. <Elektriz> inrush current; 2. <Elektrotech> make current; 3. <Heiz & Kälte> starting current
Einschalttor n <Regelung> normally open gate
Einschaltung f <Elektrotech> power up
einschaltverzögertes Relais n <Elektrotech> on-delay relay
Einschaltverzögerung f <Kontroll> turn-on delay
Einschaltwiderstand m <Elektrotech> on resistance
Einschaltzeit f 1. <Elektronik, Elektrotech> closing time, make time, on-time, turn-on time; on-period *(Schaltung)*; gate-controlled rise time *(Thyristor)*; 2. <Kontroll> turn-on time
Einschaltzustand m <Elektrotech> on-state
einschätzen v 1. <Math> appreciate; 2. <Qual> assess, estimate; rate; value
Einschätzung f 1. <Math> appreciation; 2. <Qual> assessment, estimate; rating; valuation
Einscheiben... <Maschinen> single pulley
Einscheibenantrieb m <Maschinen> single pulley drive
Einscheibenfensterglas n <Ker & Glas> single thickness window glass
Einscheibentrockenkupplung f 1. <Kfztech> single dry plate clutch; 2. <Maschinen> *(BE)* dry single-disc clutch, *(AE)* dry single-disk clutch
einscheren v <Wassertrans> reeve *(Tauwerk)*
einschichtig adj <Bau> single-layer • **einschichtig mit Baumwolle bedeckt** <Elektriz> single cotton covered
Einschicht-Keramikkondensator m <Elektrotech> single layer ceramic capacitor
Einschichtwicklung f <Elektrotech> one-position winding, single-layer winding
Einschiebeinheit f <Elektriz> plug-in unit
einschieben v 1. <Comp & DV> insert *(Absatz, Zeichen)*; 2. <Telekom> insert *(Steckkarte)*
Einschieben n <Ker & Glas> setting in
Einschieben n **der Scherben** <Ker & Glas> pushing down the cullet
Einschiebtreppe f <Bau> folding staircase
Einschienen... <Eisenbahn> monorail
Einschienenbahn f <Bau, Eisenbahn> monorail *(System)*
Einschienenbahn f **mit asymmetrischer Aufhängung** <Eisenbahn> monorail with asymmetric suspension
Einschienenbahn f **mit Pendelfahrzeugaufhängung** <Eisenbahn> monorail with pendulum vehicle suspension
Einschienenbahn f **mit pneumatischer Aufhängung** <Eisenbahn> monorail with pneumatic suspension
Einschienengreiferlaufkatze f <Eisenbahn> monorail grab trolley
Einschienenhängebahn f <Eisenbahn> monorail conveyor, monorail with hanging cars, suspended monorail
Einschienenhochbahn f <Eisenbahn> elevated monorail
Einschienensattelbahn f <Eisenbahn> supported monorail
Einschiffung f <Wassertrans> boarding *(Passagiere)*
einschlagen v 1. <Bau> drive in *(Nagel)*; 2. <Fertig> pocket *(Ventilsitz)*; 3. <Maschinen> drive, drive in • **im Brandfall Glas einschlagen** <Sicherheit> in case of fire, break the glass *(Feuermelder)*
Einschlagfaden m <Textil> pick
Einschlagmaschine f <Verpack> envelope machine, wrapping machine
Einschlagszentrum n <Mechan, Phys> *(AE)* center of impact, *(BE)* centre of impact
Einschleifen n 1. <Ker & Glas> truing; grinding *(eines Stopfens)*; 2. <Maschinen> grinding-in

Einschleifpaste f <Fertig> grinding paste
einschließen v 1. <Anstrich> encapsule; 2. <Bau> enclose, include; 3. <Chemie> occlude; 4. <Meerschmutz> entrap *(einen Ölteppich)*
Einschließen n 1. <Kerntech> containment *(von radioaktiven Stoffen)*; encapsulation *(von Aktivabfall)*; 2. <Umweltschmutz> entrapment
Einschließen n **in Beton** <Kerntech> embedding in concrete
Einschließung f 1. <Bau> housing; 2. <Kerntech> containment *(von radioaktiven Stoffen)*
Einschluss m 1. <Abfall> grain encapsulation *(von Sondermüll)*; 2. <Chemie> occlusion; 3. <Kerntech> enclosure *(von radioaktivem Material)*; inclusion *(von Verunreinigungen)*; 4. <Metall> inclusion; 5. <Phys> confinement, containment; 6. <Raumfahrt> envelope; 7. <Sicherheit> enclosure, entrapment; 8. <Teilphys> confinement; 9. <Umweltschmutz> entrapment; 10. <Werkprüf> inclusion
Einschluss m **in Beton** <Kerntech> embedding in concrete
Einschluss m **von Medien** <Anstrich> media entrapment
Einschlusstiefe f <Elektronik> cavity depth, void depth
einschmelzen v 1. <Metall> smelt; 2. <Textil> melt
Einschmelzen n <Textil> oiling
Einschmelztiefe f <Fertig> depth of fusion *(Schweißen)*
Einschnappverschluss m <Verpack> snap-on closure
einschneiden v <Bau> notch
einschneidig adj 1. <Fertig> single-edge *(Werkzeug)*; 2. <Maschinen> single-point
einschneidiges Werkzeug n <Maschinen> single point cutting tool
Einschnitt m 1. <Bau> indentation; 2. <Eisenbahn> cutting; 3. <Wasserversorg> notch
einschnittig adj <Maschinen> single-shear
Einschnüreffekt m <Elektronik, Phys> pinch effect
Einschnürung f 1. <Fertig> formation of neck; choking *(Kunststoffe)*; 2. <Maschinen> contraction, necking; 3. <Metall, Phys> necking
Einschnurvermittlungsschrank m <Telekom> single cord switchboard
einschränkend adj <Elektronik> limiting
Einschränkung f 1. <Künstl Int> constraint; 2. <Qual> curtailment; 3. <Telekom> restriction *(Dienste)*
einschrauben v <Bau> drive
Einschraubenschiff n <Wassertrans> single screw ship
Einschraubteil m <Fertig> union bush *(Kunststoffinstallationen)*
Einschraubtiefe f <Fertig> screw penetration
einschreiben v <Geom> inscribe
Einschritt... <Comp & DV> single step
Einschrittbetrieb m <Comp & DV> single step operation
Einschrittoperation f <Comp & DV> single step operation
Einschrumpfen n <Verpack> shrink wrapping
Einschub m 1. <Elektrotech> plug-in, plug-in unit, removable part, slide-in module; 2. <Telekom> plug-in unit; 3. <Verpack> compartmented tray, insert
Einschubmodul n <Elektronik> card module
Einschuss m <Textil> pick; weft *(Weben)*
Einschussfaden m <Textil> weft
einschwalben v <Bau> dovetail
Einschweißfolie f <Druck> blister pack
einschwenkbarer Filter n <Foto> swing-in filter
Einschwing... <Elektriz, Fernseh> transient
Einschwingbedingungen fpl <Elektrotech> transient conditions
Einschwingen n 1. <Phys> transient oscillation; 2. <Telekom> pull-in *(PLL)*
Einschwingspannung f <Elektriz> transient voltage, prospective transient recovery voltage *(Energieerzeugung)*

Einschwingverhalten n 1. <Elektrotech, Gerät> transient response, *(AE)* transient behavior, *(BE)* transient behaviour; 2. <Telekom> transient response
Einschwingverzögerung f <Telekom> delay *(Stufenfunktion)*
Einschwingvorgang m <Akustik, Kontroll, Phys> transient
Einschwingzeit f 1. <Gerät> damping period, settling time, transient time; 2. <Telekom, Werkprüf> response time
Einschwingzustand m <Elektriz> transient state
Einschwungprellen n <Elektriz> pull-in bouncing
Einseitenband n *(SSB)* <Elektronik, Fernseh, Funktech, Telekom, Wassertrans> single sideband *(SSB)*
Einseitenband n **mit kompandierter Amplitude** *(ACSR)* <Funktech> amplitude-compandered single sideband *(ACSS)*
Einseitenband-Amplitudenmodulation f <Elektronik> single-sideband amplitude modulation *(SSB-AM)*
Einseitenband-Demodulation f <Telekom> single-sideband demodulation
Einseitenband-Empfang m <Funktech> single-sideband reception, SSB reception
Einseitenbandfilter n <Elektronik> single sideband filter
Einseitenband-Frequenzmodulation f <Funktech> single-sideband frequency modulation
Einseitenbandmodulation f <Elektronik> single sideband modulation
Einseitenbandmodulator m <Elektronik> single sideband modulator
Einseitenbandsender m <Fernseh> single-sideband transmitter, SSB transmitter
Einseitenbandsendung f <Fernseh> single sideband transmission
Einseitenbandspektrum n <Telekom> single-sideband spectrum, spectral density of a single-sideband signal
Einseitenbandübertragung f <Phys> single sideband transmission
Einseitenbandunterdrückung f <Telekom> single-sideband suppression
einseitig beschreibbare Diskette f <Comp & DV> single-sided disk, single-sided diskette
einseitig geglättete Pappe f <Druck, Papier> MG board, machine-glazed board
einseitig geglättetes Papier n <Druck, Papier> MG paper, machine-glazed paper
einseitig gelagerte Kurbel f <Maschinen> outside crank
einseitig gerichtet adj <Telekom> unidirectional
einseitig gerichtete Verbindung f <Telekom> unidirectional connection
einseitig gerichteter Strom m <Elektriz> unidirectional current
einseitig gerichteter Transducer m <Elektrotech> unidirectional transducer
einseitig gestrichene Pappe f <Verpack> one-sided coated board
einseitige Diskette f <Optik> *(BE)* single-sided disc, *(AE)* single-sided disk
einseitige Leiterplatte f <Elektronik> single-sided printed circuit
einseitige Palette f <Kfztech> single platform pallet
einseitiger Feldmagnet m <Elektrotech> single-sided field system
einseitiger Strom m <Elektriz> unidirectional current
einseitiges Getriebe n <Elektrotech> unilateral gearing
einseitiges Verteilergestell n <Telekom> single-sided distribution frame
Eins-Element n <Math> identity element
Einsenken n <Maschinen> recessing

Einsenkpresse

Einsenkpresse f <Maschinen> hobbing press
Einsenkung f 1. <Eisenbahn> subsidence; 2. <Maschinen> counterbore
Einsenkung f **des Index** <Optik> index dip
einsetzbares Fach n <Verpack> compartmentable insert
einsetzen v 1. <Bau> insert, set; 2. <Comp & DV> insert, paste; set *(Variable)*; 3. <Math> insert *(Wert in Formel, Lösung in Gleichung)*; 4. <Metall> case-harden; 5. <Wassertrans> set in *(von Gezeiten)*
Einsetzen n 1. <Bau> sealing; 2. <Math> insertion; 3. <Strahlphys> onset *(eines Magnetfeldes)*
Einsetzkamera f <Fernseh> insert camera
einsickern v <Bau> penetrate
Einsickern n <Kohlen> infiltration
Einsickerung f <Wasserversorg> infiltration
Einsinken n <Umweltschmutz> subsidence
Einsinkpunkt m <Ker & Glas> sinking point
einsinnigbewehrte Platte f <Bau> one-way slab
einsinnige Bewehrung f <Bau> one-way reinforcement
einsitziges Ventil n <Hydraul> single-seated valve
einspannen v 1. <Bau> clamp; fix *(Balken, Träger)*; 2. <Comp & DV> load *(Papier)*; 3. <Fertig> chuck *(Spannung)*; 4. <Maschinen> chuck, clamp, mount; 5. <Mechan> hitch
Einspannen n 1. <Bau> stretching; 2. <Maschinen> mounting; 3. <Mechan> chucking
Einspannfutter n <Mechan> chuck
Einspannkopf m <Fertig> gripping head *(Zerreißstab)*
Einspannkraft f <Bau> fixed-end force
Einspannmoment n <Bau> fixed-end moment
Einspannstelle f <Maschinen> bearing point
Einspannung f 1. <Fertig> constraint *(Balken)*; 2. <Maschinen> fixing; 3. <Mechan> clamping
Einspannvorrichtung f 1. <Bau> shackle; 2. <Maschinen> chucking device, clamping device, clamping fixture, gripping device; 3. <Mechan> jig
einspeichern v <Comp & DV> roll in
Einspeicherung f <Comp & DV> roll in
Einspeicherungsimpuls m <Comp & DV> write pulse
Einspeicherungszeit f <Comp & DV> write time
Einspeisarmatur f <Erdöl> feed tank
einspeisen v 1. <Bau> supply; 2. <Fernseh, Mechan> feed
Einspeisepegel m <Elektronik> injection level
Einspeiseschlauchverbindung f <Erdöl> feed hose union
Einspeiseschleuse f <Papier> airlock feeder
Einspeiseventil n <Erdöl> feed valve
Einspeisung f 1. <Elektronik> injection *(eines Signals in Schaltung)*; 2. <Fernseh> feed, incoming feed; 3. <Hydraul, Kfztech, Mechan> feed
einspeisungsstabilisierter Oszillator m <Raumfahrt> injection-locked oscillator *(Weltraumfunk)*
einspielen v <Fernseh> play in
Einspielzeit f <Werkprüf> response time
Einspindelautomat m <Maschinen> single spindle automatic
Einspindelbohrmaschine f <Maschinen> single spindle boring machine
Einspindeldrehmaschine f <Maschinen> single spindle lathe
Eins-plus-Eins-Adressbefehl m <Comp & DV> one-plus-one address instruction
Eins-plus-Eins-Trägersystem n <Telekom> one-plus-one carrier system
einspringend adj <Geom> reentrant
Einspritzaggregat n <Maschinen> fuel injector
Einspritzbohrung f <Kerntech> injection borehole
Einspritzdruck m <Maschinen> injection pressure

Einspritzdüse f 1. <Chemtech, Kfztech> injection nozzle *(Kraftstoff)*; 2. <Mechan> injector
Einspritzdüse f **mit Halter** <Kfztech> injector *(Kraftstoff)*
Einspritzdüsenhalter m <Kfztech> injection nozzle holder
einspritzen v <Raumfahrt> inject
Einspritzen n **von Anlasskraftstoff** <Lufttrans> priming
Einspritzhahn m <Hydraul> injection cock
Einspritzkniehebel m <Lufttrans> mixer bellcrank *(Hubschrauber)*
Einspritzkompressor m <Hydraul> injection compressor
Einspritzkondensator m <Hydraul> injection condenser, jet condenser
Einspritzleitung f <Kfztech> delivery pipe
Einspritzmenge f <Raumfahrt> priming charge *(Raumschiff)*
Einspritzpumpe f <Kfztech, Maschinen, Wassertrans> injection pump *(Motor)*
Einspritzrohr n <Hydraul> injection pipe
Einspritz- und Zündungsrechner m <Kfztech> digital control box
Einspritzung f 1. <Kfztech> injection *(Kraftstoff)*; 2. <Raumfahrt> primer *(Raumschiff)*
Einspritzventil n <Labor> injection valve *(Gaschromatographie)*
Einspritzzapfendüse f <Kfztech> pintle injection nozzle
Einspruch m <Patent> appeal, opposition
Einspruchsgründe mpl <Patent> grounds for opposition
Einspruchsschrift f <Patent> notice of opposition
Einspruchsverfahren n <Patent> opposition proceedings
einspuliges Stromstoßrelais n <Elektrotech> single coil latching relay
einspuliges Stützrelais n <Elektrotech> single coil latching relay
Einspülschiff n <Erdöl> bury barge *(Schiff zum Einspülen von Seeleitungen)*
Einspuraufnahme f <Aufnahme> single track recording
Einspuraufzeichnung f <Akustik> one-track recording
einspurige Fahrbahn f <Kfztech> one-groove track
Einstahlschneidwerkzeug n <Maschinen> single point cutting tool
Einstampfgerät n <Verpack> repulping equipment
Einständerblechkanten-Hobelmaschine f <Fertig> open-side plate planing machine
Einständerhobelmaschine f <Maschinen> open-side planing machine
Einstandspreis m <Kunststoff> cost price *(Lagerwert)*
Einstaubewässerung f <Wasserversorg> subsurface irrigation
Einstechdrehen n <Maschinen> recessing
Einstechdrehmaschine f <Maschinen> grooving machine
einstechen v <Maschinen> recess
Einstechgewindeschleifen n <Maschinen> plunge-cut thread grinding
Einstechmeißel m <Maschinen> recessing tool
Einstechschleifen n <Maschinen> plunge grinding, plunge-cut grinding
Einstechwerkzeug n <Maschinen> plunging tool
Einsteck... <Bau, Elektrotech> plug-in
einsteckbarer SIM-Chip m <Mobilkom> plug-in-SIM *(für Handys)*
Einsteckbaugruppe f <Elektrotech> plug-in unit
Einsteckeinheit f <Elektriz> plug-in unit
einstecken v 1. <Elektriz, Elektrotech> plug, plug in; 2. <Fertig> plug *(Niet)*; 3. <Funktech> plug; 4. <Telekom> plug in, insert *(Stecker)*
einstecken v **und betreiben** v <Telekom> plug-and-play
Einsteckende n <Bau> spigot *(einer Muffenrohrverbindung)*

Einsteckfeder f <Bau> loose tongue *(Holzbau)*
Einsteckgruppe f <Elektrotech> plug-in unit
Einsteck-Hörgerät n <Sicherheit> insert-type hearing device
Einsteckmeißel m <Fertig> boring-bar cutter, inserted tool, tool holder bit
Einsteckmodul m <Elektrotech> plug-in module
Einsteckrelais n <Elektrotech> plug-in relay
Einsteckschloss n <Bau> mortise dead lock, mortise lock
Einsteckspule f <Elektrotech> plug-in coil
Einsteckteil n <Elektrotech> plug-in component
Einsteigeplattform f <Eisenbahn> boarding platform
Einsteigöffnung f <Maschinen> access hole
Einstein-de-Haas'scher Effekt m <Phys, Thermod> Einstein-de Haas effect
Einsteinium n *(Es)* <Chemie, Strahlphys> einsteinium *(Es)*
Einstein'sche Koeffizienten mpl <Phys, Strahlphys> Einstein coefficients
Einstein'sche Temperatur f 1. <Phys> Einstein temperature; 2. <Thermod> Einstein temperature *(θ K)*
Einstell... <Maschinen> adjusting
Einstellanschlag m <Mechan> adjustable stop
einstellbar adj 1. <Elektriz> variable; 2. <Kontroll, Maschinen> adjustable
einstellbare Ansicht f <Comp & DV> customized view
einstellbare Kurzschlussbrücke f <Phys> adjustable short-circuit bridge
einstellbare Schutzvorrichtung f <Sicherheit> adjustable guard
einstellbarer Generator m <Raumfahrt> pointable generator *(Raumschiff)*
einstellbarer Schraubenschlüssel m <Mechan> *(BE)* adjustable spanner, *(AE)* adjustable wrench
einstellbarer Spannungserzeuger m <Elektriz> variable voltage generator
einstellbarer Spannungsteiler m <Elektriz> adjustable voltage divider
einstellbarer Thermoschalter m <Heiz & Kälte> adjustable thermostatic switch
einstellbarer Wellenleiter m <Funktech> adjustable waveguide
einstellbarer Widerstand m <Elektriz> adjustable resistor
einstellbares Kontaktthermometer n <Gerät> adjustable contact thermometer
einstellbares Relais n <Telekom> variable relay
einstellbares Schneidmesser n <Bau> *(AE)* cutting gage, *(BE)* cutting gauge *(für Furniere)*
Einstellbrett n <Foto> enlarger baseboard *(des Vergrößerungsapparates)*
Einstelldruck m <Heiz & Kälte> setting pressure
Einstellelement n <Gerät> setting device
einstellen v 1. <Bau> adjust, set; 2. <Comp & DV, Kontroll> set; 3. <Metrol> adjust *(Mikroskop)*; 4. <Papier> adjust, set up • **auf unendlich einstellen** <Foto> focus for infinity
Einstellen n 1. <Bau> setting; 2. <Lufttrans> setting *(von Instrumenten)*; 3. <Regelung> positioning
Einstellen n **des Luftschraubenblattes** <Lufttrans> blade setting *(Hubschrauber)*
Einstellfilter f <Foto> red swing filter
Einstellgenauigkeit f <Gerät> accuracy of adjustment
Einstellglied n <Gerät> adjusting element
Einstellhülse f <Kfztech> adjusting sleeve
einstellig adj <Comp & DV> unary
einstellige Operation f <Comp & DV> unary operation
Einstellknopf m 1. <Elektriz> control knob; 2. <Mechan> adjusting knob

Einstelllehre f <Maschinen> *(AE)* setting gage, *(BE)* setting gauge
Einstellmaß n <Konstzeich> setting dimension
Einstellmutter f <Maschinen> regulating nut, set nut
Einstellscheibe f <Fertig> adjustment dial *(Kunststoffinstallationen)*
Einstellschraube f 1. <Elektrotech> tuning screw; 2. <Maschinen> adjusting screw, regulating screw, set screw, temper screw; 3. <Mechan> adjusting screw; 4. <Phys> tuning screw; 5. <Trans> adjustable pitch propeller
Einstellskala f **am Messgerät** <Gerät> meter dial
Einstellspannung f <Elektriz> adjusting voltage *(Abgleichspannung)*
Einstellstift m 1. <Maschinen> set pin; 2. <Mechan> alignment pin
Einstellung f 1. <Bau> sight *(Vermessung)*; 2. <Comp & DV> computer setting; 3. <Elektronik> adjustment; 4. <Ergon> attitude, set; 5. <Funktech> adjustment; 6. <Kunststoff> formulation; 7. <Maschinen> adjustment, regulation, setting, setup; 8. <Metrol> adjustment; 9. <Papier> adjusting, adjustment; 10. <Trans> synchronization *(auf konstant Grün)*
Einstellung f **auf null** <Maschinen> zeroizing
Einstellungsgesundheitsuntersuchung f <Sicherheit> pre-employment health screening
Einstellvorrichtung f <Papier> adjuster
Einstellwert m 1. <Gerät> setting value; 2. <Kerntech> index value; 3. <Maschinen> setting
Einstellwiderstand m <Elektrotech> adjustable resistor
Einstellwinkel m 1. <Fertig> cutting tool angle; 2. <Maschinen> entering angle
Einstellwinkel m **der Luftschraube** <Lufttrans> blade angle *(Hubschrauber)*
Einstellwinkel m **des Blattes** <Lufttrans> blade angle *(Hubschrauber)*
Einstellwinkel m **des Luftschraubenblattes** <Lufttrans> blade angle, blade-setting angle *(Hubschrauber)*
Einstellwinkelschwankung f <Lufttrans> incidence oscillation
Einstellzeit f 1. <Comp & DV> response time; 2. <Gerät> settling time; balance time *(Brücke)*
Einstellzeit f **einer Frequenzdekade** <Telekom> synthesizer setting time
einstemmen v <Fertig> mortice, mortise
Einstemmen n 1. <Bau> *(BE)* morticing, *(AE)* mortising; 2. <Fertig, Mechan> caulking
Einstich m 1. <Fertig> neck *(Spiralbohrer)*; 2. <Maschinen> plunge cut, recess
Einstichboden m <Ker & Glas> push-up, pushed punt
Einstieg m <Erdöl> manway
Einstiegloch n <Maschinen> access hole
Einstiegsluke f <Lufttrans> hatch
Einstiegsöffnung f <Bau, Kerntech> manhole
Einstiegsstelle f <Fernseh> in point
Einstrahl-Katodenstrahlröhre f <Elektronik> single beam cathode ray tube
Einstrahloszilloskop n <Gerät> single beam oscilloscope
Einstrahlröhre f <Elektronik> single beam tube
Einstrahlspektrometer n <Strahlphys> single beam spectrophotometer
Einstrahlung f <Raumfahrt> irradiation *(Weltraumfunk)*
Einstreichen n **mit Flussmittel** <Bau> fluxing
einstrippen v <Druck> strip in
Einströmen n <Maschinen> inlet
Einströmkanal m <Fertig> sprue hole
Einströmöffnung f <Maschinen> inlet
Einströmrate f <Lufttrans> inflow ratio

Einströmung

Einströmung f <Lufttrans> inflow
Einströmungsöffnung f <Kerntech> inlet end *(einer Turbine)*
Einströmwinkel m <Lufttrans> inflow angle
Einstufen... <Fertig> one-blow *(Stauchautomat)*
Einstufenkompressor m <Maschinen> single stage compressor
Einstufenrückführung f <Kerntech> rabbit
Einstufenverdampfer m <Lebensmittel> single effect evaporator
Einstufenverstärker m <Elektronik, Kerntech, Lebensmittel, Maschinen> single stage amplifier
einstufig adj <Comp & DV> single-level
einstufige Turbine f <Hydraul> single stage turbine
einstufiger Geradeausempfänger m **mit Rückkopplung** <Funktech> feedback audion, one-tube radio and audio-amplification and detection, regenerative receiver
einstufiger Kompressor m <Hydraul> single stage compressor
einstufiger Verdichter m <Heiz & Kälte, Maschinen> single stage compressor
einstufiger Verstärker m <Elektriz> single stage amplifier
Einstufung f <Ergon> rating, score
Einsturz m 1. <Bau, Kohlen> collapse; 2. <Erdöl> caving
einstürzen v <Bau> fail, fall in
einstweilig adj <Bau, Patent, Qual> provisional
einstweilige Verfügung f <Patent> interim injunction
einstweiliger Schutz m <Patent> provisional protection
Einsumpfen n <Bau> soaking *(Kalk)*
Einsumpfzeit f <Bau> soaking period
Eins-Verstärker m <Elektronik> unity gain amplifier *(Verstärker mit Verstärkungsfaktor Eins)*
Eins-Verstärkung f <Elektronik> unity gain
Eins-zu-eins Abbildung f <Math> injective mapping
Eins-Zustand m <Elektronik> one state
Eintagstide f <Wassertrans> diurnal tide
Eintaktröhre f <Elektronik> single-ended tube
Eintaktverstärker m <Elektronik> single-ended amplifier
eintasten v 1. <Druck> key in, keyboard; 2. <Telekom> key in *(Nummer)*
Eintasten n <Telekom> keying
eintauchen v 1. <Metall> dip; 2. <Papier> immerse; 3. <Raumfahrt> dive *(Raumschiff)*; 4. <Textil> steep; 5. <Verpack> dip
Eintauchen n 1. <Bau> plunging; 2. <Chemie> immersion; 3. <Ker & Glas> dip *(des Fangeisens bei der Herstellung von Tafelglas)*; 4. <Maschinen> dipping; 5. <Papier> immersion; 6. <Raumfahrt> dive
Eintauchmesszelle f <Gerät> immersion cell
eintauchmetallisieren v <Fertig> whiten
Eintauchobjektiv n <Phys> immersion objective *(Mikroskop)*
Eintauch-Thermostat m <Heiz & Kälte> immersion-type thermostat
Eintauchtiefe f <Wassertrans> depth of immersion
einteilen v <Mechan> index
einteilig adj <Fertig> solid
einteilige Bifokallinsen fpl <Ker & Glas> solid bifocals
einteilige Schneidkluppe f <Maschinen> one-part screw plate
einteiliger Schieber m <Heiz & Kälte> single leaf damper
einteiliges Lager n <Maschinen> solid bearing
Einteilung f <Comp & DV> partitioning, grouping *(in Gruppen)*
Einteilung f der Funkfrequenzen <Funktech> classification of radio frequencies
Einthoven'sches Galvanometer n <Elektriz> Einthoven galvanometer

Eintonnenvertäuung f <Erdöl> single buoy mooring
Eintrag m 1. <Comp & DV> entry; item *(in Liste)*; 2. <Patent> entry
eintragen v 1. <Comp & DV> enter, register; 2. <Telekom> log
Einträgersystem n <Trans> monobeam system
Eintraggabel f <Ker & Glas> carrying-in fork
Eintragung f <Lufttrans, Patent, Trans, Wassertrans> registration
eintreiben v <Bau> pile *(von Pfählen)*
Eintreten n **in Umlaufbahn** <Raumfahrt> entry into orbit *(Raumschiff)*
Eintritt m 1. <Lufttrans> intake *(Motor, Triebwerk)*; 2. <Maschinen> inlet, intake
Eintrittskanal m <Fertig> inlet port
Eintrittsleitschaufel f <Lufttrans> intake guide vane
Eintrittsleitschaufel-Staudruck m <Lufttrans> intake guide vane ram
Eintrittsöffnung f 1. <Bau> throat; 2. <Maschinen> inlet, intake
Eintrittsorbit m <Raumfahrt> injection orbit
Eintrittspupille f <Phys> entrance pupil
Eintrittsschalldämpfer m <Mechan> *(AE)* inlet muffler, *(BE)* inlet silencer
Eintrittstemperatur f <Umweltschmutz> inlet temperature
Eintrittswahrscheinlichkeit f <Comp & DV> probability
Eintrittswinkel m 1. <Elektronik> acceptance angle; 2. <Phys> angle of contact
eintrommeln v <Meerschmutz> reel in *(eine Sperre)*
Ein- und Ausstiegsluke f <Raumfahrt> manhole
Einwahl f **ins Netz** <Telekom> *(AE)* direct outward dialing, *(BE)* direct outward dialling
einwählen v <Mobilkom> log on
Einwählvorgang m <Telekom> access via switched lines
Einwalzen n <Bau> rolling
einwandern v <Fertig> diffuse
Einwanderung f <Fertig> diffusion
einwandfrei adj <Bau, Qual> sound
Einwärtsfluss m <Phys> inward flux
einwässern v <Fertig> macerate
einwecken v <Lebensmittel> bottle *(Früchte oder Gemüse)*
Einweg... 1. <Abfall, Heiz & Kälte, Ker & Glas, Kfztech, Lebensmittel, Phys, Sicherheit, Trans, Verpack> half-wave; 2. <Elektriz> half wave, one-way; 3. <Telekom> one-way
Einwegbehälter m <Verpack> one-way container
Einwegfilter n <Heiz & Kälte> disposable filter
Einwegflasche f 1. <Abfall> disposable bottle, nonreturnable bottle, one-way bottle; 2. <Ker & Glas> single trip bottle; 3. <Verpack> disposable bottle, nonreturnable bottle, one-way bottle
Einweggleichrichter m <Elektriz, Phys> half-wave rectifier
Einwegkupplung f <Kfztech> free engine clutch
Einwegmaske f <Sicherheit> disposable respirator
Einwegölfilter n <Kfztech> throw-away oil filter
Einwegpackung f <Verpack> one-way pack
Einwegpalette f 1. <Trans> expendable pallet, nonreusable pallet, one-way pallet; 2. <Verpack> expandable pallet, nonreturnable pallet, one-way pallet
Einwegpolythenverpackung f <Verpack> expanded polythene packaging
Einwegprodukt n <Abfall> throw-away product
Einwegschaltung f <Elektronik> half-wave circuit, one-way circuit, single-wave circuit, single-way circuit, single-way connection
Einwegschutzkleidung f <Sicherheit> disposable protective clothing

Einwegspiegel m <Ker & Glas> seethrough mirror
Einwegverpackung f 1. <Abfall> disposable container, nonreturnable container, one-way pack, throw-away pack; 2. <Lebensmittel, Verpack> nonreturnable packaging
einweichen v 1. <Chemie> macerate; 2. <Lebensmittel> macerate, soak, steep; 3. <Papier> batch, soak; 4. <Textil> soak through, soak, steep
Einweichen n <Papier, Textil> soaking
einwellig adj <Maschinen> single-shaft
einwellige integrierte Optikschaltung f <Elektronik> single mode optical integrated circuit
einwelliges Licht n <Metall> monochromatic light
einwertig adj <Chemie> monobasic, monovalent, univalent
einwertige Säure f <Chemie> primary acid
Einwertigkeit f <Chemie> monovalence, univalence, monovalency, univalency
Einwickelmaschine f <Fertig> wrapper, wrapping machine
Einwicklungstransformator m <Elektriz> one-coil transformer
einwirken v auf <Phys> act upon
Einwohnergleichwert m <Abfall, Wasserversorg> inhabitant equivalent, population equivalent
Einzahnschlagfräser m <Fertig> single point cutter
einzapfen v <Fertig> mortice, mortise
einzäunen v <Bau> fence in
Einzeilenfenster n <Comp & DV> strip window
Einzel... <Comp & DV, Fernseh, Fertig, Kontroll, Telekom, Textil, Verpack> single
Einzeladresscode m <Comp & DV> single address code, single address instruction
Einzeladressnachricht f <Comp & DV> single address message
Einzelalarm m <Telekom> minor alarm
Einzelantrieb m 1. <Fertig> individual drive, self-contained drive; 2. <Maschinen> individual drive
Einzelbauelement n <Elektrotech> discrete component
Einzelbenutzersystem n <Comp & DV> single user system
Einzelbenutzerzugriff m <Comp & DV> single user access
Einzelberechnung f <Telekom> detailed billing
Einzelbild n 1. <Comp & DV> frame; 2. <Fernseh> frame by frame, picture still
Einzelbitfehler m <Comp & DV> single bit error
Einzelblattzuführung f <Comp & DV> sheet feeding
Einzelbuchstabensetz- und -gießmaschine f <Druck> single-type composing and casting machine
Einzeldiffusionsvorgang m <Elektronik> single diffusion process
Einzeldruck m <Konstzeich> one-off print
Einzelempfang m <Funktech> individual reception (Rundfunksatellit)
Einzelentwickler m <Comp & DV> single developer
Einzelfaden-Schlichten n <Textil> single end sizing
Einzelfahrzeug n ohne Fahrer <Kfztech> driverless single car
Einzelfaserkabel n <Telekom> (AE) single-fiber cable, (BE) single-fibre cable
Einzelfaserleitung f <Telekom> (AE) single-fiber line, (BE) single-fibre line
Einzelfederung f <Maschinen> individual suspension
Einzelfertigung f <Fertig> jobbing
Einzelflotation f <Kohlen> single flotation
Einzelfundament n <Bau> footing, foundation block, single footing
Einzelgerät n <Kontroll> stand-alone device
Einzelgussteil n <Fertig> jobbing casting

Einzelhandelpackung f <Verpack> retail package
Einzelhub m <Maschinen> single stroke
Einzelkläranlage f <Wasserversorg> separate sewerage system
Einzelkondensator m <Elektrotech> discrete capacitor
Einzelkontrolle f <Trans> individual control
Einzelkristalldiodenmischer m <Elektronik> single-ended crystal mixer
Einzelleitung f <Elektriz> independent feeder
Einzellinearinduktionsmotor m <Trans> SLIM, single linear inductor motor
Einzellinse f <Kerntech> Einzel lens
Einzelmesswert m <Gerät> individual measuring value, measurement value
einzeln adj <Comp & DV> single
einzelner Escape-Peak m <Strahlphys> single escape peak
einzelnes Anführungszeichen n <Druck> turned comma
Einzelnetzgerät n <Elektrotech> single supply device, single supply unit
Einzeloperation f <Comp & DV> single operation
Einzelpackung f <Verpack> unit pack
Einzelplatzsystem n <Comp & DV> single user system
Einzelplatzzugriff m <Comp & DV> single user access
Einzelpreis m <Maschinen> break-up price
Einzelprobe f <Qual> increment
Einzelprotokoll-Router m <Telekom> single protocol router
Einzelradaufhängung f <Kfztech> independent suspension
Einzelresonanz f <Kerntech> single level resonance
Einzelschicht-Wellfaserplatte f <Verpack> (AE) single wall-corrugated fiberboard, (BE) single wall-corrugated fibreboard
Einzelschrittbetrieb m <Kontroll> step-by-step operation
Einzelschrittsignal n <Elektronik> one-shot signal
Einzelspeicher m <Telekom> individual store
Einzelstaubabscheider m <Sicherheit> individual dust removal apparatus
Einzelsteuerung f <Trans> individual control
Einzelsteuerung f des Durchflusses in Kühlkanälen <Kerntech> individual channel flow control
Einzelstrahlbrechung f <Optik> monorefringence
Einzelstromelement n <Elektrotech> discrete power component
Einzelstromversorgung f <Elektrotech> single power supply
Einzelstromversorgungsspannung f <Elektrotech> single supply voltage
Einzelteil n 1. <Elektronik> component; 2. <Fertig> component part (Kunststoffinstallationen); 3. <Kfztech> component part, part; 4. <Konstzeich> component; 5. <Maschinen> component; separate parts (der Drehmaschine)
Einzelteilzeichnung f <Konstzeich, Patent> component drawing
Einzeltonaudiogramm n <Akustik> pure tone audiogram
Einzeltonnenfestmachen n <Erdöl> single buoy mooring (Schifffahrt)
Einzeltreibstoffschubtriebwerk n <Raumfahrt> monopropellant thruster (Raumschiff)
Einzeltype f <Druck> movable type
Einzelventilator m <Kfztech> single fan
Einzelverbindungsnachweis m <Telekom> itemized billing
Einzelverfahren n <Patent, Qual> detailed procedure
Einzelwahlverfahren n <Telekom> call-by-call (Privatanbieter)

Einzelwiderstand

Einzelwiderstand m <Elektrotech> discrete resistor
Einzelworterkennung f <Künstl Int> isolated words recognition
einziehbar adj <Mechan> retractable
einziehbare Antenne f <Mobilkom> retractable antenna
einziehbare Räder npl <Kfztech> retractable wheels
einziehbares Filter n <Foto> retractable filter
einziehen v <Wassertrans> reeve *(Tauende in Block)*
Einziehen n 1. <Papier> draw; 2. <Telekom> pulling-in *(Kabel)*
Einzifferaddierglied n <Elektronik> one-digit adder
Einziffersubtrahierglied n <Elektronik> one-digit subtractor
Einzonenreaktor m <Kerntech> one-zone reactor
einzufügender Text m <Druck> inside forme
Einzug m <Textil> draft *(Weben)*
Einzugsbereich m 1. <Funktech> capture area; 2. <Raumfahrt> range
Einzugsbereich m **eines Knotenamts** <Telekom> *(AE)* group-switching center catchment area, *(BE)* group-switching centre catchment area
Einzugsgebiet n 1. <Bau, Kohlen, Nichtfoss Energ> catchment area; 2. <Wasserversorg> catchment area, drainage area, drainage basin
Einzugsgebiet n **mit Binnenentwässerung** <Wasserversorg> blind drainage area
Einzugsrolle f <Fertig> feed roll *(Draht)*
Einzweckmaschine f <Maschinen> single purpose machine
Einzylindermotor m 1. <Kfztech> one-cylinder engine; 2. <Maschinen> one-cylinder engine, single cylinder engine
Einzylindertrockner m <Papier> yankee dryer
EIRP *(äquivalente isotrope Strahlungsleistung)* <Funktech, Raumfahrt> EIRP *(effective isotropically-radiated power)*
eisbehindert adj <Wassertrans> icebound *(Schiff, Hafen)*
Eisblumenglas n <Ker & Glas> ice-patterned glass
Eisbrecher m 1. <Lufttrans> ice guard; 2. <Wassertrans> ice guard, icebreaker
Eisbrecherfrachtschiff n <Trans> icebreaking cargo ship
Eisbrechertanker m <Trans> icebreaking oil tanker
Eisen n 1. <Chemie> iron *(Fe)*; 2. <Metall> iron
Eisen... <Chemie> ferric
Eisenbahn f <Eisenbahn> *(AE)* railroad, *(BE)* railway
Eisenbahnanschnitt m <Eisenbahn> rail cutting
Eisenbahnbetrieb m <Eisenbahn> *(AE)* railroad operation, *(BE)* railway operation
Eisenbahnbrücke f <Eisenbahn> *(AE)* railroad bridge, *(BE)* railway bridge
Eisenbahndamm m <Eisenbahn> embankment
Eisenbahneinschnitt m <Eisenbahn> *(AE)* railroad cutting, *(BE)* railway cutting
Eisenbahnfährdock n <Trans> train ferry dock
Eisenbahnfähre f <Wassertrans> ferry
Eisenbahnfahrplan m <Eisenbahn> *(AE)* railroad schedule, *(BE)* railway schedule, *(AE)* railroad timetable, *(BE)* railway timetable
Eisenbahnfahrzeuge npl <Eisenbahn> *(AE)* railroad vehicles, *(BE)* railway vehicles
Eisenbahnfrachtterminal m <Eisenbahn> *(AE)* railroad freight terminal, *(BE)* railway freight terminal
Eisenbahnknotenpunkt m <Eisenbahn> *(AE)* railroad center, *(BE)* railway centre, *(AE)* railroad junction, *(BE)* railway junction
Eisenbahnmaterial n <Eisenbahn> *(AE)* railroad material, *(BE)* railway material, *(AE)* railroad stock, *(BE)* railway stock
Eisenbahnschiene f <Eisenbahn> *(AE)* railroad track, *(BE)* railway track

Eisenbahnschwelle f <Eisenbahn> *(AE)* cross tie, *(BE)* sleeper, tie
Eisenbahnsicherheit f <Sicherheit> rail safety
Eisenbahnstrecke f <Eisenbahn> *(AE)* railroad line, *(BE)* railway line
Eisenbahntransport m <Eisenbahn> *(AE)* railroad transport, *(BE)* railway transport
Eisenbahnüberführung f <Eisenbahn> *(AE)* railroad overbridge, *(BE)* railway overbridge *(für Straßen)*
Eisenbahnunterführung f <Eisenbahn> *(AE)* railroad underbridge, *(BE)* railway underbridge *(für Straßen)*
Eisenbahnverkehr m <Eisenbahn> *(AE)* railroad traffic, *(BE)* railway traffic
Eisenbahnverkehrsordnung f <Eisenbahn> *(BE)* railroad regulations, *(AE)* railway regulations
Eisenbahnwagen m <Eisenbahn> *(AE)* freight car, *(BE)* wagon
Eisenbahnwaggon m <Eisenbahn> *(AE)* railroad car, *(BE)* railway carriage
Eisenband n <Mechan> ferrule
Eisenbandcutter m <Verpack> iron band cutter
Eisenbeschläge mpl <Bau> ironwork
Eisenblech n <Fertig, Metall> sheet iron
Eisenbrücke f <Bau> iron bridge
Eisenfeilspäne mpl <Phys> iron filings
eisengeschirmtes Messwerk n <Gerät> iron-screened movement
Eisengießerei f <Fertig> iron foundry
Eisenglimmer m <Kunststoff> micaceous iron oxide
Eisenglut f <Metall> red-hot iron
Eisengraupe f <Metall> granular iron
Eisenguss m <Metall> iron casting
Eisenhüttenkunde f <Metall> metallurgy
Eisenkarbid n <Metall> cementite, iron carbide
Eisenkern m <Elektrotech> iron core • **ohne Eisenkern** <Elektrotech> air core *(Spule)*
Eisenkernfüllfaktor m <Elektriz> core space factor
Eisenkernspannungsmessgerät n <Elektriz> iron core voltmeter
Eisenkernstrommessgerät n <Elektriz> iron core ammeter
Eisenkerntransformator m <Elektrotech> iron core transformer
Eisenlegierung f <Anstrich> ferrous alloy
eisenlos adj <Elektrotech> air core *(Spule)*
Eisenmetalle npl <Metall> ferrous metals
Eisennadelinstrument n <Gerät> permanent magnet moving-iron instrument
Eisen-Nickel-Akkumulator m 1. <Elektriz> Edison cell; 2. <Raumfahrt> Ni-Fe battery *(Raumschiff)*
Eisennippel m <Mechan> ferrule
Eisenoxid n 1. <Aufnahme> ferric oxide; 2. <Ker & Glas> ferrous oxide; 3. <Chemie> iron oxide; 4. <Metall> black iron oxide
Eisenrohr n <Bau> iron pipe
Eisensalzlicht-Pausverfahren n <Konstzeich> ferrosalt method of reproduction
Eisensättigung f <Elektrotech> magnetic saturation
Eisenschlacke f <Metall> iron slag
Eisenschlechten fpl <Ker & Glas> iron slips
Eisenschrott m <Abfall> ferrous scrap, junk iron, scrap iron
Eisenschrottpresse f <Abfall> junk press, scrap-baling press
Eisenstab m <Metall> iron bar
Eisensulfat-Heptahydrat n <Chemie> *(AE)* iron sulfate, *(BE)* iron sulphate
Eisenträger m <Bau> iron girder
Eisenverlust m <Elektriz, Phys> iron loss

Eisenverluste *mpl* <Elektriz, Phys> core losses
Eisenverlustmessung *f* <Elektriz, Metrol> core-loss measurement, core-loss test, iron-loss measurement, iron-loss test
Eisenvitriol *n* <Chemie> copperas, iron vitriol
Eisenwaren *fpl* <Maschinen> hardware
Eisenwasserstoffröhre *f* <Elektrotech> barretter
Eisenwasserstoffwiderstand *m* <Elektrotech> barretter
eiserner Vorhang *m* <Thermod> fire curtain *(zum Feuerschutz im Theater)*
Eisessig *m* <Chemie, Lebensmittel> glacial acetic acid
Eisfach *n* <Maschinen> ice-making compartment
eisig *adj* <Thermod> icy
eiskalt *adj* <Thermod> ice-cold
Eiskondensator *m* <Heiz & Kälte, Kerntech> ice condenser
Eislast *f* <Elektrotech> ice loading *(Freileitung)*
Eislinse *f* <Kohlen> ice lens
Eismaschine *f* <Thermod> ice-making machine
Eismessfühler *m* <Lufttrans, Wassertrans> ice probe
Eispack *n* <Lufttrans, Wassertrans> ice pack
Eispunkt *m* <Heiz & Kälte, Phys> ice point
Eisscholle *f* <Wassertrans> ice floe
Eisstücke *npl* <Heiz & Kälte> broken ice
Eistarget *n* <Kerntech> ice target
Eiswarnzeichen *n* <Lufttrans, Wassertrans> ice-warning sign
Eiswürfel *m* <Thermod> ice cube
Eiweiß *n* <Lebensmittel> albumen
Eiweißfasern *fpl* <Textil> *(AE)* protein fibers, *(BE)* protein fibres
eiweißreich *adj* <Lebensmittel> high-protein
Eiweißschlichte *f* <Textil> protein size
Ejektor *m* 1. <Labor> syphon; 2. <Mechan> ejector
ejektorartiges Schneidabfallausstoßsystem *n* <Verpack> ejector-type trim exhaust system
Ejektorpumpe *f* <Maschinen> ejector pump
Ekliptik *f* <Raumfahrt> ecliptic *(scheinbare Sonnenbahn)*
E-Krümmer *m* <Elektrotech> E bend, E plane *(Wellenleiter)*
EL *(Elektrolumineszenz-Anzeige)* <Elektronik> EL *(electroluminescent display)*
Elaidin *n* <Chemie> elaidin
Elaidin... <Chemie> elaidic
Elain *n* <Chemie> olein
Elast *n* <Chemie> elastomer
elastisch *adj* 1. <Fertig> resilient; mechanical *(Hysterese)*; 2. <Kunststoff, Mechan> elastic; 3. <Textil> flexible
elastisch zurückfedern *v* <Fertig> resile
elastische Ausdehnung *f* <Qual, Verpack> elastic elongation
elastische Dehnung *f* <Kunststoff> stretch
elastische Durchbiegung *f* <Qual, Verpack> elastic deformation
elastische Eigenschaften *fpl* <Strömphys> elastic properties *(Flüssigkeiten)*
elastische Konstante *f* <Strömphys> elastic constant
elastische Kupplung *f* <Heiz & Kälte> flexible coupling
elastische Masse *f* <Chemie> elastomer
elastische Nachwirkung *f* <Kunststoff> retarded elasticity
elastische Nachwirkungen *fpl* <Strömphys> elastic aftereffects *(Turbulenz)*
elastische Schiene *f* <Eisenbahn> resilient rail
elastische Spannung *f* <Phys> stress
elastische Straßendecke *f* <Bau> flexible pavement
elastische Streuung *f* <Phys, Strahlphys, Teilphys> elastic scattering
elastische Wellen *fpl* <Phys> elastic waves

elastischer Aufprall *m* <Kerntech> elastic impact
elastischer Bereich *m* <Qual, Mechan> elastic range
elastischer Modus *m* <Raumfahrt> elastic mode
elastischer Schlupf *m* <Fertig> stretch
elastischer Stoß *m* <Kerntech, Phys, Strahlphys> elastic collision
elastisches Gelenk *n* <Maschinen> flexible joint
elastisches Rad *n* <Trans> elastic wheel
elastisches Seil *n* <Wassertrans> sandow
elastisches Vorspannen *n* <Fertig> prespringing
Elastizität *f* 1. <Kunststoff> elasticity; 2. <Maschinen> elasticity, resilience, resiliency; 3. <Mechan, Metall> elasticity; 4. <Papier> resiliency; 5. <Phys, Qual> elasticity; 6. <Textil> flexibility, resilience; 7. <Verpack, Wassertrans> elasticity *(Holz, Metall)*
Elastizitätsbereich *m* <Mechan, Qual> elastic range
Elastizitätsgebiet *n* <Mechan, Qual> elastic range
Elastizitätsgrenze *f* 1. <Bau> elastic limit; 2. <Kerntech> yield point; 3. <Kunststoff, Mechan, Phys, Qual, Verpack> elastic limit
Elastizitätskoeffizient *m* <Kunststoff, Metall, Qual> coefficient of elasticity
Elastizitätskonstante *f* <Maschinen, Metall, Qual> elastic constant
Elastizitätsmodul *m* 1. <Bau> modulus of elasticity, Young's modulus (E); 2. <Fertig> Young's modulus *(Faseroptik)*; 3. <Hydraul> bulk modulus of elasticity (K); 4. <Kohlen, Kunststoff, Lufttrans, Maschinen, Metall> modulus of elasticity, Young's modulus (E); 5. <Phys> bulk modulus, modulus of elasticity; 6. <Werkprüf> modulus of elasticity
Elastizitätsmodus *m* <Raumfahrt> elastic mode
Elastomer *n* 1. <Chemie> elastomer; 2. <Erdöl> elastomer *(Kunststoff)*; 3. <Fertig> elastomer, elastometric material *(Kunststoffinstallationen)*; 4. <Kunststoff> elastomer
Elastomeres *n* <Chemie> elastomer
Elastomerverschnitt *m* <Kunststoff> elastomer blend
elastoplastisch *adj* <Kunststoff> elastoplastic
Electronic-Publishing *n* <Comp & DV, Druck, Elektronik> electronic publishing *(Erstellen von Druckerzeugnissen auf Großrechnern)*
ELED *(Kantenemitter-Lumineszenzdiode, kantenstrahlende Lumineszenzdiode)* <Telekom> ELED *(edge-emitting light-emitting diode)*
Elefantenhaut *f* <Kunststoff> alligatoring, crazing
Elektret *n* <Elektriz, Phys> electret
Elektretfolienmikrofon *n* <Akustik, Aufnahme> electret-foil microphone
elektrifizieren *v* <Elektrotech> electrify
Elektrifizierung *f* <Elektriz, Phys> electrification
elektrisch *adj* <Elektriz, Elektronik> electric, electrical
elektrisch abgestimmter Oszillator *m* <Elektronik> electrically-tuned oscillator
elektrisch angetrieben *adj* <Foto, Mechan> electrically-driven
elektrisch betrieben *adj* <Fertig> electric-powered
elektrisch gehaltener Koppelpunkt *m* <Telekom> electrically-held crosspoint
elektrisch leitende Versiegelung *f* <Kerntech> electrical conductor seal
elektrisch löschbarer programmierbarer Lesespeicher *m* *(EEPROM)* <Elektronik> electrically-erasable programmable read-only memory *(EEPROM)*
elektrisch neutral *adj* <Chemie> uncharged *(Punkt)*
elektrisch vorgespanntes Relais *n* <Elektriz> biased relay
elektrische Abstimmung *f* <Elektronik> electric tuning
elektrische Anlage *f* <Elektriz, Elektrotech, Wassertrans> electrical installation, electrical plant

elektrische

elektrische Ausgabe f <Elektronik> electrical output
elektrische Beleuchtung f <Elektriz> electric lighting
elektrische Beschaltung f <Elektrotech> electrical wiring
elektrische Bewegungsimpedanz f <Akustik> electrical motional impedance
elektrische Bogenentladung f <Phys> electric arc
elektrische Bohrmaschine f 1. <Bau> power drill; 2. <Maschinen> electric drill
elektrische Bremsung f <Eisenbahn> electric braking
elektrische Durchgangsprüfung f <Elektriz> continuity test
elektrische Eigenschaft f <Elektronik> electrical characteristic
elektrische Eindringtiefe f <Kerntech> electric penetration
elektrische Eingangsimpedanz f <Akustik> free electrical impedance *(bei unbelastetem Ausgang)*
elektrische Einrichtung f <Elektriz> electrical installation
elektrische Energie f 1. <Elektriz, Elektrotech> electric power, electrical power, electric energy, electrical energy; 2. <Phys> electric energy
elektrische Entstaubung f <Sicherheit> electrical dust removal installation
elektrische Erhitzung f <Kerntech> joule heating
elektrische Feldkonstante f <Phys> permittivity of free space
elektrische Feldstärke f *(E)* <Elektriz, Elektrotech, Phys> electric field strength *(E)*
elektrische Feuerung f <Elektriz> electric lighting
elektrische Größe f <Elektriz> electric variable
elektrische Heizung f <Elektriz, Heiz & Kälte> electric heating
elektrische Impedanz f <Aufnahme> electrical impedance
elektrische Installation f <Wassertrans> electrical installation
elektrische Installationsarbeit f <Elektriz> electrical installation work
elektrische Isolierplatte f <Elektriz> electrical-insulating board
elektrische Isolierung f <Elektrotech> electrical insulation
elektrische Kapazität f *(CE)* <Akustik> electric capacitance *(CE)*
elektrische Kennlinie f <Elektrotech> electrical characteristic
elektrische Komponente f <Elektrotech> electrical component
elektrische Konstante f <Phys> electric constant
elektrische Kontinuität f <Elektrotech> electrical continuity
elektrische Kraftstoffpumpe f <Kfztech> electric fuel pump *(Kraftstoffzufuhr)*
elektrische Kupplung f <Elektrotech> electric coupler
elektrische Ladung f <Elektriz, Phys, Teilphys> electric charge
elektrische Laufzeitkette f <Elektronik> electric delay line
elektrische Leistung f 1. <Elektriz> electric power, electrical power; 2. <Kerntech, Telekom> electrical power
elektrische Leitfähigkeit f <Elektrotech> electrical conduction, electrical conductivity
elektrische Lokomotive f <Elektriz> electric locomotive
elektrische Maschine f <Elektrotech> electrical machine
elektrische Messung f <Elektrotech> electrical measuring
elektrische Mittelführerstandlok f **mit langen und flachen Aufbauten** <Eisenbahn> crocodile
elektrische Pferdestärke f <Elektriz> electric horsepower

elektrische Pferdestärkestunde f <Elektriz> electric horsepower hour
elektrische Polarisation f <Elektriz, Phys> electric polarization
elektrische Prüfung f <Elektrotech, Werkprüf> electrical test
elektrische Säge f <Elektriz> electric saw
elektrische Schaltdifferenz f <Aufnahme> electric hysteresis
elektrische Schaltuhr f <Gerät> automatic electric timer
elektrische Schaltung f <Elektriz, Elektronik, Elektrotech> electric circuit, electric wiring, electrical wiring
elektrische Schutzeinrichtung f <Sicherheit> electrical safeguard
elektrische Schwingung f <Elektronik> electric oscillation
elektrische Schwingungen fpl <Wellphys> electrical oscillations
elektrische Sicherheitseinrichtung f <Sicherheit> electrical safeguard
elektrische Spannung f 1. <Elektriz> electric potential; 2. <Phys> voltage
elektrische Strahlungskochplatte f <Elektrotech> electric radiant type of hot-plate
elektrische Stromdichte f <Elektriz> electric current density
elektrische Suszeptanz f <Elektriz> electric susceptance
elektrische Suszeptibilität f <Elektrotech, Phys> electric susceptibility
elektrische Technologie f <Elektriz> electrotechnology
elektrische Übertragungsleitung f <Elektrotech> electrical transmission line
elektrische Übertragungsschaltung f <Elektrotech> electric transducer
elektrische Variable f <Elektriz> electric variable
elektrische Verbindung f <Elektriz, Elektrotech> electric linkage, electrical connection
elektrische Verdrahtung f <Elektrotech> electric wiring, electrical wiring
elektrische Verlustleistung f <Elektriz> electric losses
elektrische Verriegelung f <Sicherheit> power interlocking
elektrische Verschiebung f <Elektrotech, Phys> electric flux, electric displacement
elektrische Vierpolübergänge mpl <Strahlphys> electric quadrupole transitions
elektrische Welle f <Elektrotech, Phys, Telekom> electric wave
elektrische Zündung f <Elektriz> electric lighting
elektrischer Anlasser m <Elektriz> electric starter
elektrischer Anschluss m <Elektrotech> electrical connection, electrical connector
elektrischer Ausgang m <Elektrotech> electrical output
elektrischer Bewegungsscheinwiderstand m <Akustik> electrical motional impedance
elektrischer Blindleitwert m *(BE)* <Elektriz> electric susceptance *(BE)*
elektrischer Bohrer m <Elektriz> electric drill
elektrischer Brennofen m <Elektriz> electric furnace
elektrischer Bus m <Trans> battery bus
elektrischer CO$_2$-Entladungslaser m <Elektronik> electric-discharge CO$_2$ laser
elektrischer Dipol m <Phys> electric dipole
elektrischer Eingang m <Elektrotech> electrical input
elektrischer Entladungslaser m <Elektronik> electric discharge laser
elektrischer Fehler m <Elektrotech> electrical breakdown
elektrischer Feldvektor m *(E)* <Elektrotech> electric field vector *(E)*

elektrischer Fluss m <Elektriz, Phys> electric flux
elektrischer Funke m <Elektriz> electric spark
elektrischer Hauteffekt m <Akustik> electrodermal effect
elektrischer Heizapparat m <Elektriz> electric heater
elektrischer Impuls m <Elektrotech> electric pulse
elektrischer Induktionsofen m <Elektrotech> electric induction furnace
elektrischer Kleinmotor m <Elektrotech> fractional horsepower motor
elektrischer Kontakt m <Elektriz, Elektrotech> electrical contact
elektrischer Konvektionsofen m <Thermod> electric convector
elektrischer Leiter m 1. <Elektrotech> electrical conductor; 2. <Phys> electric conductor
elektrischer Leitungsbruchalarm m <Elektriz> electric wire-break alarm
elektrischer Lichtbogen m <Elektriz> electric arc
elektrischer Lichtbogenschmelzofen m <Elektriz> EAF, electric-arc furnace
elektrischer Nullpunkt m <Elektriz> electrical zero
elektrischer Nullsteller m <Elektriz> electrical zero adjuster
elektrischer Ofen m <Elektriz, Thermod> electric oven
elektrischer Pol m <Elektrotech> electric pole
elektrischer Resonator m <Elektronik> electrical resonator
elektrischer Schienenwagen m <Elektriz> electric railcar
elektrischer Schock m <Elektriz> electric shock
elektrischer Schwingungserzeuger m <Elektronik> electrical oscillator
elektrischer Signalwandler m <Elektrotech> electric transducer
elektrischer Stecker m <Elektrotech, Labor> electric plug
elektrischer Steckverbinder m <Elektrotech> electrical connector
elektrischer Strom m <Elektriz, Elektrotech, Phys> electric current
elektrischer Transducer m <Elektrotech> electric transducer
elektrischer Triebwagenzug m *(EMU)* <Eisenbahn> electric multiple unit, EMU
elektrischer Trockner m <Elektriz> electric dryer
elektrischer Verdrahtungsplan m <Wassertrans> electrical-wiring diagram
elektrischer Vierpol m <Elektrotech> electric quadrupole
elektrischer Wandler m <Elektrotech> electric transducer
elektrischer Widerstand m <Phys> electric resistance *(physikalische Größe)*
elektrischer Wirkungsgrad m <Elektrotech> electrical efficiency
elektrischer Zünder m <Elektriz> electric lighter
elektrisches Bauteil n <Elektrotech> electrical component
elektrisches Bild f <Elektrotech> electric image
elektrisches Dipolmoment n <Phys> electric dipole moment
elektrisches Feld n <Aufnahme, Elektriz, Elektrotech, Fernseh, Phys> electric field
elektrisches Filter n <Elektronik> electric filter, electrical filter
elektrisches Handwerkzeug n <Sicherheit> portable power tool
elektrisches Haushaltgerät n <Elektriz> domestic appliance, electrical household appliance
elektrisches Heizelement n <Elektriz> electric heater
elektrisches Heizgerät n <Maschinen> electric heater
elektrisches Heizkissen n <Thermod> electric heating pad
elektrisches Kraftwerk n <Elektriz, Elektrotech> electric power station
elektrisches Leuchten n <Raumfahrt> electroglow
elektrisches Lieferfahrzeug n <Kfztech> electrovan
elektrisches Messgerät n <Sicherheit> electric-measuring apparatus
elektrisches Mixgerät n <Lebensmittel> electric mixer
elektrisches Netzwerk n <Phys> electric network
elektrisches Potenzial n <Elektriz, Elektrotech, Phys> electric potential, electrical potential
elektrisches Quadrupolmoment n <Phys> quadrupole electrical moment
elektrisches Relais n <Elektriz, Elektrotech> electric relay, electrical relay
elektrisches Schneiden n <Thermod> electric-arc cutting
elektrisches Schutzgerät n <Sicherheit> electrical protection equipment
elektrisches Signal n 1. <Elektronik> electrical signal; 2. <Elektrotech> electric signal
elektrisches Stellglied n <Raumfahrt> electric actuator *(Raumschiff)*
elektrisches Triebfahrzeug n <Elektriz> electric locomotive
elektrisches Verriegelungssystem n <Elektriz> electric-interlocking system
elektrisches Versorgungsunternehmen n <Elektrotech> electric utility
elektrisches Wärmeäquivalent n <Elektriz> Joule's equivalent
elektrisches Wechselfeld n <Elektrotech> alternating electric field
elektrisches Zählgerät n <Gerät> counting instrument
elektrisieren v <Elektrotech> electrify
Elektrisiermaschine f <Elektrotech> static electrical machine
Elektrizität f <Elektriz, Elektrotech, Funktech, Phys> electricity
Elektrizitätsausgangsleistung f <Kerntech> electrical output *(eines Reaktors)*
Elektrizitätserzeugung f <Elektriz> generation of electricity
Elektrizitätskontrolltafel f <Kerntech> electrical control board
Elektrizitätskraftwerk n *(E-Werk)* <Elektronik> electric power station
Elektrizitätsmessgerät n <Elektriz> electricity meter
Elektrizitätsnetz n 1. <Elektriz> grid *(großes oder flächendeckendes Versorgungsnetz)*; 2. <Elektrotech> grid *(Verteilung der E-Energie)*
Elektrizitätsschalttafel f <Kerntech> electrical control board
Elektrizitätsschaltwarte f <Kerntech> electrical control room
Elektrizitätssystem n **mit ungeerdetem Mittelleiter** <Elektriz> isolated neutral system, isolated system
Elektrizitätstransformierung f <Elektriz> transformation of electricity
Elektrizitätsübertragung f <Elektriz> transmission of electricity
Elektrizitätsunterwerk n <Elektriz> electric power substation
Elektrizitätsversorgung f <Elektriz> electricity supply
Elektrizitätsversorgungsunternehmen n *(EVU)* <Elektriz> electricity supply company
Elektrizitätswerk n <Bau, Elektrotech> electricity generation station, generating plant

Elektrizitätswirtschaft

Elektrizitätswirtschaft f <Elektriz> electricity sector economics
Elektrizitätszähler m <Elektriz> electricity meter
Elektro... <Akustik, Elektronik, Elektrotech, Maschinen, Mechan, Phys> electrical
Elektroakustik f <Elektrotech> electroacoustics
elektroakustisch adj <Aufnahme> electroacoustic
elektroakustische Kette f <Akustik> electroacoustic chain
elektroakustischer Frequenzgang m <Aufnahme> electroacoustical frequency response
elektroakustischer Reziprozitätskoeffizient m <Akustik> electroacoustical reciprocity coefficient
elektroakustischer Wandler m <Akustik, Elektrotech> electroacoustic transducer
Elektroantrieb m 1. <Foto> electric drive; 2. <Kerntech> electrical drive
Elektroblech n <Metall> silicone steel sheet
Elektrobohrer m <Maschinen, Mechan> electric drill
Elektrobus m <Kfztech> electric bus, electrobus
Elektrochemie f <Elektriz> electrochemistry
elektrochemisch überziehen v <Anstrich> galvanize
elektrochemische Bearbeitung f <Fertig> electrochemical machining
elektrochemische Energie f <Elektrotech> electrochemical energy
elektrochemischer Sekundärgenerator m <Kfztech> secondary electrochemical generator
Elektrocochleographie f <Akustik> electrocochleography
Elektrocureverfahren n <Kerntech> electron beam curing
Elektrodampfkessel m <Maschinen> electric steam boiler
Elektrode f <Elektriz, Elektrotech, Fernseh, Labor, Mechan, Metall, Phys> electrode plate
Elektrodenabstand m 1. <Elektrotech> electrode gap, spark gap; 2. <Kfztech> spark plug gap
Elektrodenanordnung f <Telekom> electrode configuration
Elektrodenhalter m <Bau, Elektrotech> electrode holder
Elektrodenkapazität f <Elektrotech> inter-electrode capacitance
Elektrodenkennlinie f <Elektrotech> electrode characteristic
Elektrodenkessel m <Heiz & Kälte> electrode boiler
Elektrodenkohle f <Elektrotech> electrode carbon
Elektrodenleitwert m <Elektriz> electrode admittance
Elektrodenmantel m <Fertig> electrode coating (Schweißen)
Elektrodenpotenzial n <Elektriz, Mechan, Phys> electrode potential
Elektrodenschweißen n <Bau> electrode welding
Elektrodenspannung f <Elektriz> electrode potential
Elektrodenspitze f <Elektriz> electrode tip
Elektrodenstrahloszilloskop n <Elektronik> cathode-ray oscilloscope (zur Bildschirmdarstellung von schnell schwankenden, periodischen Spannungswerten)
Elektrodenvorspannung f <Elektrotech> electrode bias, electrode bias voltage
Elektrodialyse f <Chemie> electrodialysis
Elektrodiesel-Lokomotive f <Eisenbahn> electro-diesel locomotive
Elektrodynamik f <Elektriz, Phys, Teilphys> electrodynamics
elektrodynamische Gleisbremse f <Eisenbahn> eddy current rail brake
elektrodynamischer Lautsprecher m <Akustik> electrodynamic loudspeaker

elektrodynamischer Schwingungsaufnehmer m <Gerät> electrodynamic vibration pick-up
elektrodynamisches Instrument n <Elektriz> electrodynamic instrument
elektrodynamisches Messinstrument n <Elektriz> electrodynamic instrument
elektrodynamisches Messwerk n <Gerät> electrodynamic movement
elektrodynamisches Mikrofon n <Akustik, Aufnahme> electrodynamic microphone
elektrodynamisches Relais n <Elektriz> electrodynamic relay
elektrodynamisches Schwebesystem n <Eisenbahn> electrodynamic levitation system
Elektrodynamometer n 1. <Elektriz> electrodynamometer (Messung mechanischer Leistung); 2. <Phys> electrodynamometer
elektrodynamometrisch adj <Elektriz> electrodynamometric
Elektroeingang m <Elektronik> electrical input
Elektroenergieübertragung f <Elektrotech> electric power transmission
elektroerosive Bearbeitung f (EEB) 1. <Fertig> electroerosion machining; 2. <Maschinen> electro-discharge machining (EDM)
Elektrofahrzeug n (EV) 1. <Elektrotech> battery vehicle; 2. <Kfztech, Nichtfoss Energ> electric vehicle, EV
Elektrofahrzeug n **für den Stadtverkehr** <Trans> urban electric vehicle
Elektrofilter n <Elektronik, Heiz & Kälte> electrostatic filter
Elektroformen n <Maschinen> electroforming
elektrofotografischer Drucker m <Comp & DV> electrophotographic printer
Elektrogerät n <Lebensmittel> electrical appliance
elektrographischer Drucker m <Comp & DV> electrographic printer
Elektrohebezeug n <Mechan> electric hoist
Elektroheizung f <Heiz & Kälte> electric heating
elektrohydraulisches Bohren n <Kohlen> electrodrilling
Elektroingenieur m <Elektrotech> electrical engineer
Elektroinstallation f <Elektrotech> electrical installation
Elektrokinetik f <Elektrotech, Phys> electrokinetics
elektrokinetische Energie f <Elektriz> electrokinetic energy
Elektrokochplatte f <Thermod> electric hot plate
Elektro-Kuppelvorrichtung f <Eisenbahn> jumper head
Elektroleitung f <Thermod> electric power line
Elektrolieferfahrzeug n <Kfztech> electric pickup, (BE) float
Elektrolieferwagen m <Kfztech> electric delivery truck
Elektrolog n <Erdöl> electric log (Bohrlochmesstechnik)
Elektrolumineszenz f 1. <Elektronik> electroluminescence (durch elektrische Felder induziert); 2. <Kerntech> electrofluorescence, electroluminescence; 3. <Optik, Phys, Telekom> electroluminescence
Elektrolumineszenz-Anzeige f (EL) <Elektronik> electroluminescent display (EL)
Elektrolyse f <Druck, Elektriz, Elektrotech, Phys> electrolysis
Elektrolyse f **von Wasser** <Kerntech> water electrolysis
Elektrolyseapparat m <Elektrotech> electrolytic unit
Elektrolysebad n 1. <Elektriz, Elektrotech> electrolytic bath; 2. <Phys> electrolytic cell
Elektrolyseur m <Chemie> electrolyzer
Elektrolysezelle f 1. <Elektrotech> electrolytic cell; 2. <Fertig> pot; 3. <Phys> electrolytic unit, electrolytic cell
elektrolysieren v <Chemie, Phys> electrolyze
Elektrolysierung f <Elektriz> electrolyzation
Elektrolyt m <Chemie, Elektriz, Elektrotech, Kunststoff, Phys, Telekom> electrolyte

Elektrolyt... <Chemie> electrolytic
Elektrolytätzen n <Kerntech> electrolytic etching
Elektrolytglas n <Ker & Glas> copper light
Elektrolytgleichrichter m <Elektrotech> electrolytic rectifier
elektrolytisch adj <Chemie, Elektriz, Metall, Phys> electrolytic
elektrolytisch behandeln v <Wassertrans> anodize (Metall)
elektrolytisch plattieren v <Elektrotech> electroplate
elektrolytisch zerlegen v <Chemie> electrolyze
elektrolytische Abscheidung f <Elektrotech, Fertig> electrodeposition
elektrolytische Korrosion f <Metall> electrolytic corrosion
elektrolytische Leitfähigkeit f <Elektriz> electrolytic conductivity
elektrolytische Reinigung f <Metall> electrolytic cleaning
elektrolytische Zelle f <Chemie> electrolyzer
elektrolytischer Gleichrichter m <Elektrotech> electrolytic rectifier
elektrolytisches Härten n <Fertig> electrolytic hardening
elektrolytisches Ventil n <Elektrotech> electrolytic rectifier
Elektrolytkondensator m 1. <Elektrotech> electrolytic capacitor (Elkom); 2. <Funktech, Phys, Telekom> electrolytic capacitor, electrochemical capacitor
Elektrolytstand m <Kfztech> acid level
Elektromagnet m <Chemie, Elektriz, Elektrotech, Fernseh, Phys> electromagnet
elektromagnetisch adj 1. <Elektriz> electromagnetic; 2. <Fertig> solenoid; 3. <Funktech, Phys> electromagnetic
elektromagnetisch betrieben adj <Elektrotech> electromagnetically-operated
elektromagnetische Ablenkung f <Elektrotech> electromagnetic deflection
elektromagnetische Abschirmung f <Phys, Teilphys, Wellphys> electromagnetic screen
elektromagnetische Abstimmung f <Elektronik> electromagnetic tuning
elektromagnetische Aussendung f <Phys, Teilphys, Wellphys> electromagnetic emission
elektromagnetische Beeinflussbarkeit f <Phys, Teilphys, Wellphys> electromagnetic sensitivity
elektromagnetische Beeinflussung f zwischen Systemen <Phys> intersystem interference
elektromagnetische Dämpfung f <Elektriz> electromagnetic damping (Schwingungen)
elektromagnetische Einheit f <Elektriz> electromagnetic unit
elektromagnetische Energie f <Elektrotech, Phys, Wellphys> electromagnetic energy
elektromagnetische Fokussierung f <Elektrotech> electromagnetic focusing
elektromagnetische Induktion f <Elektrotech, Phys, Teilphys, Wellphys> electromagnetic induction
elektromagnetische Induktionsbremse f <Elektrotech> electromagnetic induction brake
elektromagnetische Interferenz f <Elektrotech, Funktech> electromagnetic interference (EMI)
elektromagnetische Isolierung f <Elektrotech> electromagnetic isolation
elektromagnetische Kompatibilität f <Elektrotech> electromagnetic compatibility
elektromagnetische Kopplung f <Elektriz, Elektrotech> electromagnetic coupling
elektromagnetische Kraft f <Elektriz, Elektrotech, Phys, Teilphys, Wellphys> electromagnetic force

elektromagnetische Kupplung f 1. <Eisenbahn, Kfztech> electromagnetic clutch, electromagnetic coupling; 2. <Mechan> electromagnetic clutch
elektromagnetische Linse f <Elektrotech, Fernseh> electromagnetic lens
elektromagnetische Pulsfestigkeit f <Elektriz, Phys, Teilphys, Wellphys> electromagnetic pulse hardening
elektromagnetische Pumpe f <Kerntech> electromagnetic pump
elektromagnetische Schirmung f <Elektriz> electromagnetic shielding
elektromagnetische Störempfindlichkeit f <Elektriz, Elektrotech> electromagnetic susceptibility
elektromagnetische Störmatrix f <Elektriz, Elektrotech> electromagnetic interference matrix
elektromagnetische Störung f <Comp & DV, Elektrotech, Funktech, Raumfahrt> EMI, electromagnetic interference
elektromagnetische Strahlen mpl <Elektriz, Elektrotech, Optik, Phys, Teilphys, Telekom, Wellphys> electromagnetic radiation
elektromagnetische Strahlung f <Elektriz, Elektrotech, Optik, Phys, Teilphys, Telekom, Wellphys> electromagnetic radiation
elektromagnetische Strahlungswirkung f <Elektriz, Phys, Teilphys, Wellphys> electromagnetic radiation effect
elektromagnetische Verletzbarkeit f <Elektriz> electromagnetic vulnerability
elektromagnetische Verträglichkeit f (EMV) <Elektriz, Funktech, Raumfahrt> electromagnetic compatibility (EMC)
elektromagnetische Wechselwirkung f <Phys, Teilphys, Wellphys> electromagnetic interaction
elektromagnetische Welle f <Elektriz, Elektrotech, Phys, Teilphys, Telekom, Wellphys> electromagnetic wave
elektromagnetische Wellengleichungen fpl <Phys, Teilphys, Wellphys> electromagnetic-wave equations
elektromagnetische Zündung f <Eisenbahn, Kfztech> electromagnetic ignition
elektromagnetischer Auslöser m <Foto> electromagnetic shutter release
elektromagnetischer Energieimpuls m <Elektrotech> electromagnetic energy pulse (EMP)
elektromagnetischer Impuls m <Elektrotech, Funktech, Telekom> electromagnetic pulse
elektromagnetischer Lautsprecher m <Akustik, Aufnahme> electromagnetic loudspeaker
elektromagnetischer Resonator m <Elektronik> electromagnetic resonator
elektromagnetischer Schrittmotor m <Elektrotech> variable reluctance stepper motor
elektromagnetischer Umwelteinfluss m <Elektriz, Phys> electromagnetic environmental effect
elektromagnetisches Drehmoment n <Elektriz> electromagnetic moment
elektromagnetisches Feld n <Elektriz, Elektrotech, Phys, Telekom, Wellphys> electromagnetic field
elektromagnetisches Futter n <Maschinen> electromagnetic chuck
elektromagnetisches Kalorimeter n <Strahlphys, Teilphys> electromagnetic calorimeter
elektromagnetisches Mikrofon n <Akustik, Aufnahme> electromagnetic microphone
elektromagnetisches Moment n <Elektriz, Phys, Teilphys, Wellphys> electromagnetic moment
elektromagnetisches Querführungssystem n <Eisenbahn, Kfztech> electromagnetic lateral guidance system
elektromagnetisches Relais n <Elektriz, Elektrotech> electromagnetic relay

elektromagnetisches

elektromagnetisches Schweben n <Eisenbahn, Kfztech> electromagnetic levitation
elektromagnetisches Spektrum n <Elektronik, Elektrotech, Phys, Teilphys, Wellphys> electro-magnetic spectrum
elektromagnetisches Umfeld n <Raumfahrt> electromagnetic environment
elektromagnetisches Ventil n <Hydraul> solenoid valve
Elektromagnetismus m <Elektriz, Phys> electromagnetism
Elektromagnetlautsprecher m <Akustik> electromagnet loudspeaker
Elektromaschine f <Elektrotech> electric machine
Elektromechanik f <Elektriz> electromechanics
elektromechanische Tonaufzeichnung f <Aufnahme> electromechanical recording
elektromechanische Vermittlung f <Telekom> electromechanical switching
elektromechanische Vermittlungseinrichtung f <Telekom> electromechanical switching unit
elektromechanische Vermittlungsstelle f <Telekom> electromechanical exchange
elektromechanischer Kopplungsfaktor m <Akustik> electromechanical coupling factor
elektromechanischer Transducer m <Elektrotech> electromechanical transducer
elektromechanischer Wandler m <Akustik, Elektrotech> electromechanical transducer
elektromechanisches Filter n <Elektronik> electromechanical filter
elektromechanisches Gerät n <Elektrotech> electromechanical device
elektromechanisches Relais n <Elektrotech> electromechanical relay
elektromechanisches Vermittlungssystem n <Telekom> electromechanical switching system
elektromerer Effekt m <Chemie> mesomeric effect
Elektrometallurgie f <Elektriz> electrometallurgy
Elektrometer n 1. <Elektriz, Elektrotech> electrometer; 2. <Labor> electrometer (elektrisches Laden); 3. <Phys> electrometer
Elektrometerröhre f <Elektronik, Elektrotech> electrometer tube
Elektrometerverstärker m <Elektronik> electrometer amplifier
Elektrometrie f <Elektrotech> electrometry
elektrometrische Titration f <Chemie> electrometric titration
Elektromotor m 1. <Elektrotech> electric motor, electromotor; 2. <Kfztech> electric motor
elektromotorische Gegenkraft f <Eisenbahn> back electromotive force
elektromotorische Kraft f (EMK) <Bau, Eisenbahn, Elektriz, Elektrotech, Fernseh, Funktech, Phys> electromotive force (EMF)
Elektron n (e) 1. <Elektriz, Elektrotech, Funktech> electron, e; 2. <Phys> electron, e (Elementarteilchen); 3. <Teilphys> electron, e
Elektronenabbildung f <Elektronik> electron imaging
elektronenabgebendes Atom n <Phys> donor atom
Elektronenabspaltung f **durch Photoeffekt** <Kerntech> photodetachment
Elektronenabtaststrahl m <Fernseh> electron scanning beam
Elektronenabtastung f <Kerntech> electron scanning
Elektronenanlagerung f <Kerntech> electron attachment
elektronenanziehend adj <Chemie> electrophilic
Elektronenapparatur f <Elektrotech> electronic device
Elektronenbahn f <Kerntech> electron path

Elektronenbeschleuniger m <Teilphys> electron accelerator
Elektronenbeschuss m <Elektronik, Teilphys> electron bombardment
Elektronenbeschusstriebwerk n <Raumfahrt> electron bombardment thruster (Raumschiff)
Elektronenbeugung f <Strahlphys> electron diffraction, electron spectroscopic diffraction
Elektronenbild n <Elektronik> electron image
Elektronenbildwandler m <Elektronik> electron image tube (Bildwandlerröhre)
Elektronendichte f 1. <Phys> electron density, electron population; 2. <Raumfahrt> electron population (Raumschiff)
Elektroneneinfang m 1. <Phys> electron capture (Kernphysik); 2. <Strahlphys> electron capture
Elektroneneinfangdetektor m <Umweltschmutz> electron capture detector
Elektroneneinstrahlung f <Raumfahrt> electron irradiation
Elektronenemission f <Elektronik, Teilphys> electron emission
Elektronenfahrplan m <Elektronik> Applegate diagram (Klystron)
Elektronenfelderzeugung f <Fernseh> electronic field production
Elektronenfluss m <Fernseh> electron stream
Elektronenflutlithographie f <Elektronik> electron flood lithography
Elektronengas n <Phys> electron gas
Elektronengerät n <Elektrotech> electron device
Elektronenhülle f <Kerntech> electron shell
Elektronenkanone f <Elektronik, Fernseh, Phys, Strahlphys> electron gun
Elektronenkaskade f <Kerntech> electron cascade
Elektronenkonfiguration f <Phys, Strahlphys, Teilphys> electronic configuration
Elektronenkontinuum n <Kerntech> electron continuum
elektronenkoppelnder Oszillator m <Elektrotech> electron-coupling oscillator
Elektronenkopplung f <Elektronik, Elektrotech> electron coupling
Elektronenkühlung f <Teilphys> electron cooling
Elektronenlaufzeit f <Elektrotech> transit time (von Katode zu Anode)
elektronenleitend adj <Elektronik> n-type
elektronenleitende Epitaxialschicht f <Elektronik> n-type epitaxial layer
elektronenleitende Störstelle f <Elektronik> n-type impurity
elektronenleitendes Silizium n <Elektronik> n-type silicon
elektronenleitendes Trägermaterial n <Elektronik> n-type substrate
Elektronenleitfähigkeit f <Strahlphys> electron conductivity
Elektronenleitvermögen n <Strahlphys> electron conductivity
Elektronenlinearbeschleuniger m <Teilphys> electron linear accelerator
Elektronenlinse f <Elektronik, Fernseh, Phys> electron lens
Elektronenlochrekombination f <Kerntech> electron hole recombination
Elektronenmasse f (me) <Chemie, Kerntech, Teilphys> electron mass (me)
Elektronenmikrofon n <Aufnahme> electronic microphone
Elektronenmikroskop n <Elektronik, Labor, Metall, Phys, Strahlphys> electron microscope

Elektronenmikroskopbild n <Strahlphys> electron micrograph image
Elektronenmikroskopie f <Elektronik> electron microscopy
Elektronenniederschlag m <Bau> electrodeposition
Elektronenoptik f <Phys> electron optics
Elektronenpaar n <Teilphys> electron pair
Elektronenpolarisation f <Phys, Strahlphys, Teilphys> electronic polarization
Elektronenpolarisierung f <Phys, Strahlphys, Teilphys> electronic polarization
Elektronenpopulation f <Raumfahrt> electron population (Raumschiff)
Elektronen-Positronen-Kollideranlage f (LEP) <Teilphys> large electron-positron collider (LEP)
Elektronenquelle f 1. <Elektronik> electron source; electron gun (Physik); 2. <Strahlphys> electron source
Elektronenradiographie f <Kerntech> electron radiography
Elektronenradius m (re) <Kerntech> electron radius (re)
Elektronenröhre f 1. <Elektronik> electron tube, electronic tube, electronic valve, thermionic tube, thermionic valve; 2. <Raumfahrt> electron tube (Weltraumfunk)
Elektronenröhrenansatz m <Elektronik> electron tube neck
Elektronenröhrengitter n <Elektronik> electron tube grid
Elektronenröhrenhalter m <Elektronik> electron tube holder
Elektronenröhrenheizung f <Elektronik> electron tube heater
Elektronenröhrenkolben m <Elektronik> electron tube envelope
Elektronenröhrenoszillator m <Elektronik> electron tube oscillator
Elektronenröhrensockel m <Elektronik> electron tube base
Elektronenschale f <Phys, Strahlphys, Teilphys> electron shell, electronic subshell
Elektronenschauer m <Kerntech> electron shower
Elektronensenke f <Kerntech> electron sink
Elektronensonde f <Kerntech> electron probe
Elektronenspeicher m <Elektrotech> electronic memory
Elektronenspeicherring m <Teilphys> electron storage ring
Elektronenspektroskopie f 1. <Phys> electron spectroscopy; 2. <Strahlphys> electron energy loss spectroscopy, electron spectroscopic imaging
elektronenspendend adj <Chemie> nucleophilic (Valenz)
Elektronenspiegel m <Elektronik> electron mirror
Elektronenspinresonanz f (ESR) <Phys, Strahlphys, Teilphys> electron spin resonance (ESR)
Elektronenspinresonanz-Magnetometer n <Kerntech> electron spin resonance magnetometer
Elektronenstoßionentriebwerk n <Raumfahrt> electron impact ion engine (Raumschiff)
Elektronenstrahl m (ES) <Comp & DV, Elektriz, Elektronik, Fernseh, Kerntech, Strahlphys, Telekom, Wellphys> electron beam, electronic beam (EB)
Elektronenstrahl m **für Direktbelichtung** <Elektronik> direct-write electron beam
Elektronenstrahlabdecklack m <Elektronik> electron beam resist
Elektronenstrahlabtastung f <Elektronik> electron beam scanning
Elektronenstrahlaufzeichnung f <Gerät> electron beam recording, electronic beam recording
Elektronenstrahlausrichtmethode f <Elektronik> electron beam alignment method

Elektronenstrahlbearbeitung f <Elektronik> electron beam machining
Elektronenstrahlbildung f <Elektronik> electronic beam forming
Elektronenstrahlbündelung f <Elektronik> electron beam focusing
Elektronenstrahldirektbelichtung f <Elektronik> electron beam direct writing
Elektronenstrahlhärtungsanlage f <Fertig> electron-beam hardening equipment (Oberflächentechnik)
Elektronenstrahlkolonne f <Elektronik> electron beam column
Elektronenstrahllaser m <Elektronik> electron beam laser
Elektronenstrahllithographie f <Elektronik> electron beam lithography
Elektronenstrahl-Lithographiebearbeitungsmaschine f <Elektronik> electron beam lithography machine
Elektronenstrahlmaske f <Elektronik> electron beam mask
Elektronenstrahlnachbeschleunigung f <Elektronik> electron beam acceleration
Elektronenstrahl-Projektionslithographie f <Elektronik> projection electron-beam lithography
Elektronenstrahl-Projektionsschreiber m <Elektronik> electron beam projection printer
Elektronenstrahlresist m <Elektronik> electron beam resist
Elektronenstrahlröhre f <Elektronik> electron beam tube
Elektronenstrahlschmelzen n <Kerntech> electron beam melting
Elektronenstrahlschneiden n <Elektronik> electron beam cutting
Elektronenstrahlschweißen n <Bau, Elektriz, Kerntech> electron beam welding
Elektronenstrahlspannung f <Fernseh> electron beam voltage
Elektronenstrahlsystem n <Elektronik> (AE) color gun, (BE) colour gun
Elektronenstrahlung f <Metall> electron beam
Elektronenstrahlverarbeitung f <Elektronik> electron beam processing
Elektronenstrahlvergütung f <Elektronik> electron beam annealing
Elektronenstreuung f <Kerntech> electron scattering
Elektronenstrom m <Elektrotech> electron current
Elektronenstruktur f <Kerntech> electronic structure
Elektronensynchrotron n <Teilphys> electron synchrotron
Elektronentheorie f **der Metalle** <Strahlphys> electron theory of metals
Elektronentrajektorie f <Kerntech> electron trajectory
Elektronentransportdiode f <Elektronik> transferred-electron diode
Elektronenüberlauf m <Telekom> spillover
Elektronenverbindung f <Metall> electron compound
Elektronenverdampfung f <Phys> thermionic emission
Elektronenvervielfacher m <Elektronik, Strahlphys> electron multiplier
Elektronenvervielfacherröhre f <Elektronik> electron multiplier tube
Elektronenvolt n (eV) <Elektriz, Elektrotech, Phys, Strahlphys, Teilphys> electronvolt (eV)
Elektronenwanderung f <Kerntech> electron drift
Elektronenweg m <Fernseh> electron path
Elektronenwellenmagnetron n <Elektronik> electron wave magnetron
Elektronenwellenröhre f <Elektronik> electron wave tube

Elektronenwolke

Elektronenwolke f <Fernseh, Kerntech, Strahlphys> electron cloud
Elektronenzusammenstoß m <Telekom> electron collision
Elektronenzyklotron n <Teilphys> microtron
Elektronik f <Comp & DV, Eisenbahn, Elektronik, Kfztech, Lufttrans, Wassertrans> electronic engineering *(als Fachgebiet)*; electronic equipment *(als Technik)*; electronics *(Steuereinrichtungen)*
Elektronikbremse f <Kfztech> electronic-braking control
Elektronikmessgerät n <Metrol> *(AE)* electronic gage, *(BE)* electronic gauge
Elektronikmodul n <Elektronik> electronic module
Elektronikschablone f <Elektronik> electronic stencil
elektronisch adj <Elektriz, Elektronik> electronic
elektronisch abgestimmter Oszillator m <Elektronik> electronically-controlled oscillator
elektronisch abgestimmtes Filter n <Elektronik> electronically-tuned filter
elektronisch gesteuert adj <Elektronik> electronically-controlled
elektronisch gesteuerte Einspritzung f <Kfztech> electronic injection
elektronisch gesteuertes Ventil n <Bau> electronically-controlled valve
elektronisch gesteuertes Wählsystem n <Telekom> electronically controlled switching system *(ESS)*
elektronisch löschbarer Festwertspeicher m *(EEROM)* <Comp & DV> electronically erasable read-only memory *(EEROM)*
elektronisch löschbarer Lesespeicher m *(EEROM)* <Comp & DV> electronically erasable read-only memory *(EEROM)*
elektronisch löschbarer ROM m <Comp & DV> electronically erasable ROM
elektronische Abstimmung f 1. <Elektronik> electronic tuning *(Klystron)*; 2. <Telekom> electronic tuning
elektronische Abtastung f <Elektronik, Funktech, Telekom> frequency scanning
elektronische Antenne f <Funktech> active antenna
elektronische Aufklärung f <Elektronik> electronic intelligence
elektronische automatische Nebenstellenvermittlung f <Telekom> electronic private automatic branch exchange *(EPABX)*
elektronische Benzineinspritzung f <Kfztech> electronic metering of fuel injection
elektronische Berichterstattung f 1. <Comp & DV> electronic news reporting; 2. <Fernseh> electronic news gathering
elektronische Bilderzeugung f <Elektronik> electronic imaging
elektronische Bremsschlupfregelung f <Kfztech> antiblocking system, antiskid braking system
elektronische Bremssteuerung f <Kfztech> electronic-braking control
elektronische Datenverarbeitung f *(EDV)* <Comp & DV, Elektriz, Elektronik, Kontroll> electronic data processing *(EDP)*
elektronische Frequenzsteuerung f <Elektronik> electronic frequency control
elektronische Gegenmaßnahmen fpl <Raumfahrt> electronic countermeasures *(Weltraumfunk)*
elektronische Geldanweisung f <Telekom> EFT, electronic funds transfer
elektronische Geldüberweisung f **vom Kassenterminal** <Telekom> electronic fund(s) transfer at point of sale, EFTPOS
elektronische Geschwindigkeitskontrolle f <Kfztech> electronic speed control
elektronische integrierte Schaltung f <Elektronik> electronic integrated circuit
elektronische Karte f <Wassertrans> electronic chart, electronic map *(Navigation)*
elektronische Kassenanzeige f *(POS-Anzeige)* <Verpack> point of sale display *(POS display)*
elektronische Kontenführung f <Telekom> homebanking, on-line banking
elektronische Mailbox f <Comp & DV> electronic mailbox
elektronische Morsetaste f <Funktech> El bug
elektronische Nachrichtenspeichervermittlung f <Telekom> electronic message switch
elektronische Nachrichtenübermittlung f <Comp & DV, Elektronik, Telekom> electronic messaging
elektronische Post f *(E-Mail)* <Comp & DV, Elektronik, Telekom> electronic mail *(e-mail)*
elektronische Prüfmuster npl <Elektronik> electronic test patterns
elektronische Regelung f <Kfztech> electronic control
elektronische Schaltuhr f <Foto> electronic timer
elektronische Schaltung f 1. <Phys> electronic network; 2. <Telekom> electronic circuit
elektronische Schutzvorrichtung f <Sicherheit> electronic guard, electronic protective device
elektronische Signalverarbeitung f <Elektronik> electronic signal processing
elektronische Sperre f <Comp & DV> electronic lock
elektronische Sprachsynthese f <Elektronik> electronic speech synthesis
elektronische Steuerung f <Wassertrans> electronic control
elektronische Stromversorgung f <Elektrotech> electronic power supply
elektronische Technik f <Elektronik> electronic engineering
elektronische Überwachung f <Telekom> electronic surveillance
elektronische Uhr f <Telekom> electronic clock
elektronische Unterschrift f <Telekom> electronic signature
elektronische Verkehrshilfen fpl <Trans> electronic traffic aids
elektronische Vermittlung f <Telekom> electronic switching
elektronische Vermittlung f **mit Reed-Relais** <Telekom> reed relay electronic exchange
elektronische Vermittlungsstelle f <Telekom> electronic exchange
elektronische Verstimmempfindlichkeit f <Elektronik> electronic-tuning sensitivity
elektronische Verstimmung f <Elektronik> electronic tuning *(Oszillator)*
elektronische Verteilungsfunktion f <Kerntech> electronic partition function
elektronische Waage f 1. <Labor> electronic balance; 2. <Verpack> electronic-weighing scales
elektronische Wärmekapazität f <Phys, Strahlphys, Teilphys> electronic heat capacity
elektronische Wärmeleitfähigkeit f <Kerntech> electronic heat conductivity
elektronische Zündung f <Kfztech> electronic ignition
elektronischer Abstimmbereich m <Elektronik> electronic-tuning range *(Klystron)*
elektronischer Bankverkehr m <Telekom> homebanking, on-line banking, telebanking
elektronischer Bankverkehr m **per Handy** <Telekom> mobile banking, mobile homebanking *(bei Mobilfunk mit Internetzugang)*

elektronischer Bleistift m <Elektronik> electronic pencil
elektronischer Briefkasten m 1. <Comp & DV> electronic mailbox; 2. <Telekom> electronic mailbox, mailbox, user agent
elektronischer Datenaustausch m 1. <Comp & DV> electronic data exchange; 2. <Telekom> electronic data exchange, electronic data interchange
elektronischer Datenaustausch m **für Verwaltung, Handel und Transport** <Telekom> electronic data interchange for administration, commerce and transport, EDIFACT *(Standard)*
elektronischer Drehzahlregler m 1. <Elektronik> electronic speed controller, solid-state speed controller; 2. <Kfztech> electronic speed controller
elektronischer Fahrtrichtungsschalter m <Kfztech> electronic direction reverser
elektronischer Geschäftsverkehr m <Telekom> E-commerce
elektronischer Halbleiter m <Kerntech> electronic semiconductor
elektronischer Handel m <Comp & DV> e-commerce
elektronischer Impulsgenerator m <Elektrotech> electronic impulse generator, electronic pulse generator
elektronischer Kommutator m <Kfztech> electronic commutation
elektronischer Koppelpunkt m <Telekom> electronic crosspoint
elektronischer Lichtsatz m <Druck> electronic photocomposition
elektronischer Modum m <Kerntech> electronic instrument module
elektronischer Postdienst m <Comp & DV> electronic mail service
elektronischer Rauschgenerator m *(RG)* <Elektronik, Gerät> electronic noise generator *(ENG)*
elektronischer Rechner m <Comp & DV> electronic calculator
elektronischer Regler m <Kfztech> transistorized regulator
elektronischer Schaltkreis m <Elektronik> electronic circuit
elektronischer Speicher m <Elektrotech> electronic memory
elektronischer Taktgeber m <Elektronik> electronic clock
elektronischer Theodolit m **mit Digitalanzeige** <Bau> electronic digital theodolite *(Vermessung)*
elektronischer Vergaser m <Kfztech> *(AE)* electronic carburetor, *(BE)* electronic carburettor
elektronischer Verstimmbereich m <Elektronik> electronic-tuning range *(Oszillator)*
elektronischer Zähler m <Elektronik> electronic counter *(mit elektronischen Schaltungen aufgebaut)*
elektronischer Zahlungsverkehr m <Comp & DV> EFT, electronic funds transfer
elektronischer Zeitschalter m <Elektronik> electronic timer
elektronisches Ablagesystem n <Comp & DV> electronic filing
elektronisches Antiblockiersystem n <Kfztech> electronic antiskid system, electronic antilocking device
elektronisches Arbeitsblatt n <Comp & DV> spreadsheet
elektronisches Bauelement n 1. <Elektronik> electronic component, electronic device; 2. <Kerntech> electronic instrument module; 3. <Mechan> chip; 4. <Telekom> electronic component
elektronisches Bauteil n <Comp & DV> electronic component

elektronisches Büro n <Comp & DV> electronic office
elektronisches Datenvermittlungssystem n <Telekom> electronic data switching system *(EDS)*
elektronisches digitale Wählsystem n <Telekom> digital electronic switching system *(EWSD)*
elektronisches Diskussionsforum n <Comp & DV> bulletin board, bb *(E-mail)*
elektronisches Editieren n <Comp & DV> electronic editing
elektronisches Fahrzeug n <Kfztech> electronic car
elektronisches Gerät n <Elektronik> electronic equipment
elektronisches Getriebe n <Kfztech> electronic transmission
elektronisches Gravieren n <Druck> electronic engraving
elektronisches Kassenterminal n <Comp & DV, Telekom> EPS, electronic point-of-sale
elektronisches Messgerät n 1. <Elektronik, Gerät> electronic instrument; 2. <Metrol> *(AE)* electronic gage, *(BE)* electronic gauge
elektronisches Mitteilungssystem n <Telekom> electronic message system
elektronisches Postfach n <Telekom> electronic mailbox, e-mail account
elektronisches Publizieren n <Comp & DV, Druck, Elektronik, Telekom> electronic publishing
elektronisches Relais n <Elektriz> electronic relay *(elektronisch gesteuerter Schalter)*
elektronisches Scannen n <Druck> electronic scanning
elektronisches Schlüsselsystem n <Telekom> electronic key system
elektronisches Steuergerät n <Kfztech> electronic control unit
elektronisches Steuerungssystem n <Elektronik> electronic control system
elektronisches Telefonbuch n <Telekom> electronic telephone directory
elektronisches Testbild n <Fernseh> electronic test pattern
elektronisches Umbruchterminal m <Druck> electronic makeup terminal
elektronisches Vermittlungssystem n <Telekom> electronic switching system
elektronisches Zählen n <Elektronik> electronic counting
Elektronlochpaar n <Phys> electron hole pair
Elektronneutrino n <Phys> electron neutrino
Elektronstrahlpumpen n <Elektronik> electron beam pumping
Elektronutzfahrzeug n <Kfztech> commercial electric vehicle
Elektroofen m 1. <Elektriz> electric oven; 2. <Fertig, Kohlen> electric furnace; 3. <Thermod> electric oven
elektrooptisch adj 1. <Elektronik> electro-optical; 2. <Telekom> electro-optic
elektrooptische Schutzvorrichtung f <Sicherheit> electro-optical safety device
elektrooptische Signalverarbeitung f <Elektronik> electro-optical signal processing
elektrooptischer Effekt m <Optik> electro-optic effect
elektrooptischer Modulator m <Elektronik> electro-optical modulator
elektrooptischer Schalter m <Telekom> electro-optic switch
elektrooptischer Tonabnehmer m <Aufnahme> light beam pickup
Elektroosmose f <Chemie> electro-osmosis
elektroosmotisch adj <Chemie> electro-osmotic

elektrophil

elektrophil adj <Chemie> electrophilic
Elektrophor m <Elektrotech> electrophorus
Elektrophorese f <Elektriz, Labor> electrophoresis
Elektrophoresekammer f <Labor> electrophoresis cell
Elektrophoresezelle f <Labor> electrophoresis cell
elektrophoretisch adj <Fertig> electrophoretic
elektrophoretische Wanderung f <Kerntech> electrophoretic migration
elektrophoretisches Lackieren n <Fertig> electrophoretic coating
elektroplattieren v <Elektrotech> electroplate
Elektroplattieren n <Elektrotech, Fertig> electroplating
Elektroplattierung f <Elektrotech> electroplating, plating
elektropneumatische Bremse f <Eisenbahn, Kfztech> electropneumatic brake
elektropositive Elemente npl <Phys, Strahlphys, Teilphys> electropositive elements
Elektroproduktion f <Phys, Strahlphys, Teilphys> electroproduction
Elektrorauschen n <Elektronik> electric noise, electrical noise
Elektrorund m <Fertig> manufactured corundum
Elektrosägen n <Kerntech> electrical sawing
Elektroschlackeschweißen n <Bau, Kerntech> electroslag welding
Elektroschlepper m <Kfztech> electric truck
elektroschwache Theorie f <Teilphys> electroweak theory
elektroschwache Wechselwirkung f <Teilphys> electroweak interaction
Elektroschweißen n 1. <Elektriz> electric welding; 2. <Mechan> arc welding; 3. <Thermod> electric-arc welding
Elektroschweißer m <Mechan> arc welder
Elektroschweißmaschine f <Mechan> arc-welding machine
elektrosensitiver Drucker m <Comp & DV> electrosensitive printer
elektrosensitives Papier n <Comp & DV> electrosensitive paper
Elektrosicherheitssystem n <Sicherheit> electrosensitive safety system
Elektroskop n 1. <Elektriz> electroscope (Instrument zur Messung elektrostatischer Ladungen); 2. <Phys, Strahlphys, Teilphys> electroscope
Elektrosprayen n <Kerntech> electrospraying
Elektrostatik f <Elektriz, Elektrotech, Phys> electrostatics
Elektrostatikladungsmessgerät n <Elektriz> electrostatic meter
elektrostatisch adj <Elektriz, Telekom> electrostatic
elektrostatische Abschirmung f <Elektriz, Phys> electrostatic screen, static screen
elektrostatische Anziehung f <Elektrotech> electrostatic attraction
elektrostatische Anziehungskraft f <Elektriz> electrostatic attraction force
elektrostatische Elektronenbündelung f <Elektrotech> electrostatic focusing
elektrostatische Flussdichte f <Elektriz> electrostatic flux density
elektrostatische Induktion f <Elektriz, Phys> electrostatic induction
elektrostatische Ionenoszillation f <Kerntech> electrostatic ion oscillation
elektrostatische Kraft f <Elektrotech> electrostatic force
elektrostatische Ladung f <Elektriz> electrostatic charge
elektrostatische Linse f 1. <Elektronik> electrostatic CRT (Katodenstrahlröhren); 2. <Elektrotech> electrostatic lens, focusing electrode; 3. <Phys> electrostatic lens

elektrostatischer Drucker m <Comp & DV> electrographic printer, electrostatic printer
elektrostatischer Fluss m <Elektriz> electrostatic flux
elektrostatischer Generator m <Elektrotech> electrostatic generator
elektrostatischer Kollektor m <Kerntech> electrostatic collector
elektrostatischer Lautsprecher m <Akustik, Aufnahme> electrostatic loudspeaker
elektrostatischer Plotter m <Comp & DV> electrostatic plotter
elektrostatischer Schirm m <Elektronik> electrostatic screen
elektrostatischer Staubabscheider m (ESA) <Umweltschmutz> electrostatic precipitator (ESP)
elektrostatisches Feld n <Elektriz, Phys> electrostatic field
elektrostatisches Filter n 1. <Elektronik> electrostatic filter (Kohlekraftwerk); 2. <Heiz & Kälte> electrostatic filter
elektrostatisches Instrument n <Elektriz, Elektrotech, Labor, Phys> electrometer
elektrostatisches Luftfilter n <Sicherheit> electrostatic air filter
elektrostatisches Mikrofon n <Akustik, Aufnahme> condenser microphone, electrostatic microphone
elektrostatisches Pulverbeschichten n <Kunststoff> electrostatic powder coating
elektrostatisches Relais n <Elektriz> electrostatic relay
Elektrostauchen n <Fertig> metal gathering
Elektrostraßenfahrzeug n <Kfztech> electric road vehicle
Elektrostriktion f <Elektriz, Phys> electrostriction
Elektrosynthese f <Chemie> electrosynthesis
elektrothermal adj <Elektriz> electrothermal
elektrothermisch adj <Elektriz> electrothermal, electrothermic
Elektrowärmegerät n <Heiz & Kälte> electric-heating appliance
Elektrowelle f (E-Welle) <Elektrotech> electric wave (E wave)
Element n 1. <Bau> member; 2. <Comp & DV> element, item; 3. <Elektriz, Elektrotech> element, cell; 4. <Fertig> detail (Kraftmaschine); 5. <Funktech> element, member; 6. <Hydraul> element; 7. <Künstl Int> artificial neuron, cell, neuron, unit (neurales Netzwerk); slot (eines Schemas); 8. <Maschinen> element, link; 9. <Math, Optik, Patent, Qual> element, member; 10. <Telekom> chip, element
Element n der Actinidenreihe <Strahlphys> actinide element
elementar adj <Math> elementary
elementare Rechenzeit f <Comp & DV> basic machine time
elementarer Anreicherungsfaktor m <Kerntech> elementary enrichment factor
elementarer Trenneffekt m <Kerntech> elementary separation effect
elementares Trennvermögen n <Kerntech> elementary separative power
Elementarfaden m <Kunststoff> monofilament
Elementarladung f <Phys, Teilphys> elementary charge
Elementarlautsprecher m <Akustik> elementary loudspeaker
Elementarteilchen n <Phys, Teilphys> charm, charmed quark, elementary particle
Elementarteilchenphysik f <Phys> particle physics
Elementarwelle f <Fertig> wavelet
Elementarzelle f <Kerntech> unit cell
Elementenfamilie f <Strahlphys> family of elements

Elementenreihe f <Strahlphys> family of elements
elementfremde Mengen fpl <Math> disjoint sets
Elementtriade f <Chemie> triad
Elementzeichen n <Math> element symbol
Eleostearin... <Chemie> elaeostearic
Elevation f <Wassertrans> elevation (Navigation)
Elevationswinkel m <Raumfahrt> elevation angle
Elko m (Elektrolytkondensator) <Elektrotech> electrolytic capacitor
Ellag... <Chemie> ellagic
Ellagengerbstoff m <Chemie> ellagitannin
Ellbogengelenk n <Kerntech> elbow (eines Manipulators)
Ellbogenschutz m <Sicherheit> elbow pad
Ellipse f <Geom> ellipse
Ellipsenbogen m 1. <Bau> elliptical arch; 2. <Geom> elliptical arc
Ellipsenzirkel m 1. <Fertig> trammel point; 2. <Ker & Glas> trammel; 3. <Maschinen> (AE) egg calipers, (BE) egg callipers, trammel
Ellipsoid n <Geom, Phys> ellipsoid
ellipsoid adj <Geom> ellipsoidal
Ellipsometer n <Phys> ellipsometer
Elliptikfeder f <Maschinen> elliptic spring
elliptisch adj <Geom, Math> elliptic, elliptical (Integral)
elliptisch polarisierte Welle f <Akustik, Funktech, Phys> elliptical-polarized wave
elliptische Bahn f <Phys> elliptical orbit
elliptische Frequenzgangkurve f <Elektronik> elliptic frequency response curve
elliptische Geometrie f <Geom> elliptical geometry (nichteuklidische Geometrie zweiter Art)
elliptische Polarisation f 1. <Funktech, Phys> elliptical polarization; 2. <Raumfahrt> elliptical polarization (Weltraumfunk)
elliptischer Raum m <Geom> elliptical space
elliptischer Spiegel m <Phys> elliptical mirror
elliptisches Gewölbe n <Bau> (AE) three-centered arch, (BE) three-centred arch
elliptisches Zahnrad n <Maschinen> elliptical gear
Elliptizität f <Raumfahrt> ellipticity (Weltraumfunk)
Ellipton n <Chemie> elliptone
Ellsworthit m <Kerntech> ellsworthite
Elmsfeuer n <Wassertrans> St Elmo's fire, corposant (Wetterkunde)
Elongation f <Fertig> displacement (Schwingung)
Eloxalaluminium n <Heiz & Kälte> (BE) anodized aluminium, (AE) anodized aluminum
Eloxalqualität f <Metall> anodizing quality
Eloxalverfahren n 1. <Fertig> (AE) aluminite process; 2. <Metall> anodizing
eloxieren v <Chemie, Fertig, Wassertrans> anodize (Metall)
Eloxierung f <Fertig> anodization
ELSBM (ungeschützte Einzeltonnenvertäuung) <Erdöl> ELSBM (exposed location single buoy mooring)
Eluat n <Chemie> eluate
Eluationsversuch m <Abfall> leaching test
Eluent m <Chemie> eluant, elution agent
eluieren v <Chemie> elute
Eluieren n <Chemie, Chemtech> elution
Elution f <Chemie> elution
Elutionsmittel n <Chemie> eluant, elution agent
Elutionsversuch m <Abfall> leachability test
E-Mail f (elektronische Post) <Comp & DV, Elektronik, Telekom> e-mail (electronic mail)
E-Mail-Adresse f <Telekom> domain address, e-mail address
E-Mail-Funktion f <Comp & DV> E-mail facility
Emaille f <Bau, Ker & Glas> enamel

Emailledraht m <Elektriz, Elektrotech> (AE) enameled wire, (BE) enamelled wire
Emaillefarbe f <Ker & Glas> (AE) enamel color, (BE) enamel colour
Emaillelack m <Bau> baking varnish
Emaillemischung f <Fertig> slip
Emaillierung f <Verpack> (AE) enameling, (BE) enamelling
E-Mail-Mitteilung f <Comp & DV> mail message
Emanation f <Kerntech, Strahlphys> emanation
Embelin n <Chemie> embelin
EMC-Kollaboration f <Teilphys> European collaboration for muon physics
Emetin n <Chemie> emetin, emetine
EMI-Filterung f (Filterung gegen elektromagnetische Beeinflussung) <Elektronik> EMI filtering
Emission f 1. <Comp & DV, Kfztech, Lufttrans, Phys> emission; 2. <Strahlphys> emission (spontane, induzierte Emission); 3. <Teilphys, Umweltschmutz> emission
Emission f **in die Luft** <Phys, Strahlphys> emission-into-the-air
Emissionsbande f <Phys, Strahlphys> emission band
Emissionsbegrenzer m <Kerntech> gag
Emissionsdaten npl <Umweltschmutz> emission data
Emissionselektrode f <Elektrotech> emitter
emissionsfreies Fahrzeug n (ZEV) <Nichtfoss Energ> zero emission vehicle, ZEV
Emissionslinie f <Phys, Strahlphys> emission line
Emissionsmikroskop n <Metall> emission microscope
Emissionsort m <Umweltschmutz> emission point
Emissionsphotoschicht f <Elektronik> photoemissive layer
Emissionsquelle f <Umweltschmutz> emission source, pollution emitter
Emissionsspektralanalyse f <Phys, Strahlphys> emission spectral analysis
Emissionsspektrum n <Phys, Strahlphys, Telekom> emission spectrum
Emissionsstandard m <Umweltschmutz> emission standard, level of emission
Emissionsstärke f <Optik, Phys, Strahlphys> emissivity
Emissionsverhinderung f <Phys> blocking effect
Emissionsvermögen n <Heiz & Kälte, Phys, Strahlphys, Telekom> emissivity
Emissionsverzeichnis n <Umweltschmutz> emission inventory
Emittanz f <Nichtfoss Energ> emittance
Emitter m 1. <Comp & DV, Elektrotech> emitter; 2. <Phys> emitter (elektronenliefernde Transistorelektrode)
Emitter m **des Transistors** <Elektronik> transistor emitter
Emitterbasisanschluss m <Elektrotech> (BE) earthed-emitter connection, (AE) grounded-emitter connection
Emitterbasisdurchschlag m <Elektronik> emitter-base breakdown
Emitterbasissperrschicht f <Elektronik> emitter-base junction
Emitterelektrode f <Elektrotech> emitter electrode
Emitterfolger m <Elektrotech, Phys> emitter follower
Emittergebiet n <Elektrotech> emitter region
emittergekoppelte Logik f (ECL) <Comp & DV, Elektronik> emitter-coupled logic (ECL)
Emitterkontakt m <Elektrotech> emitter contact
Emitterschaltung f <Elektrotech> common-emitter connection
Emitterverstärker m <Elektronik> common-emitter amplifier
Emitterzone f <Elektrotech> emitter region
emittierende Diode f <Elektronik> emissive diode

emittierte Strahlung f <Strahlphys> emitted radiation
EMK (elektromotorische Kraft) <Bau, Eisenbahn, Elektriz, Elektrotech, Fernseh, Funktech, Phys> EMF (electromotive force)
Emodin n <Chemie> emodin
Emodin... <Chemie> emodic
Emodol n <Chemie> emodin
E-Modul m 1. <Bau> modulus of elasticity; 2. <Fertig> Young's modulus (Faseroptik); 3. <Kohlen, Kunststoff, Lufttrans, Maschinen, Metall, Phys, Werkprüf> modulus of elasticity
E-Modus m 1. <Elektrotech> E mode, TM mode; 2. <Optik> E Mode, TM mode; 3. <Telekom> E mode, TM mode
Emoticon n <Telekom> emoticon, graphic character icon
Empfang m 1. <Comp & DV> receipt, reception; 2. <Funktech> reception • **Empfang bestätigen** <Telekom> acknowledge
Empfang m mit faseroptischem Endgerät <Optik> (AE) receive fiberoptic terminal device, (BE) receive fibreoptic terminal device
empfangen v <Comp & DV> receive (Daten)
Empfangen n von Kurznachrichten <Mobilkom, Telekom> short message service-mobile terminated (SMS-MT)
empfangenes Signal n <Elektronik> received signal
Empfänger m 1. <Comp & DV> addressee; 2. <Elektronik, Fernseh, Funktech> receiver; 3. <Raumfahrt> receiver (Weltraumfunk); 4. <Telekom> receiver; 5. <Trans> consignee; 6. <Umweltschmutz> receptor; 7. <Wassertrans> receiver (Satellitenfunk)
Empfänger m für Lichtwellenleiterübertragung <Telekom> (AE) fiberoptic receiver, (BE) fibreoptic receiver
Empfänger m mit Stummabstimmung <Funktech> muting receiver
Empfängerbandpass m <Funktech> receiver bandpass
Empfängerbereich m <Umweltschmutz> receptor region
Empfängerdiode f <Elektronik> receiver diode
Empfängerfeinabstimmung f (RIT) <Funktech> receiver incremental tuning (RIT)
Empfängerkarte f <Elektronik> receiver board
Empfängermesssender m <Funktech> standard signal generator
Empfängermesssender m mit Frequenzdekade <Funktech> synthesized signal generator
Empfängerröhre f <Elektronik> receiving tube
Empfängersperröhre f <Elektronik> TR tube
Empfängerverstärkung f <Elektronik> receiver gain
empfänglich adj <Anstrich> prone
Empfangsanlage f <Comp & DV> receive-only equipment
Empfangsantenne f 1. <Fernseh> receiving aerial, receiving antenna; 2. <Funktech> receiving aerial, receiving antenna, receive antenna; 3. <Phys> receiving aerial, receiving antenna
Empfangsbereich m 1. <Fernseh> receiving range; 2. <Funktech> range, receiving range
empfangsbereite Schnittstelle f <Comp & DV> listen port
empfangsbereites Handy n <Mobilkom> mobile ready for receive, mobile RR
Empfangsbescheinigung f <Patent> receipt (für Unterlagen)
Empfangsbestätigungszeichen n <Comp & DV> acknowledgement character
Empfangs-Bodenstation f <Fernseh, Telekom> receiving earth station
Empfangsfilter n <Elektronik> receive filter
Empfangsfrequenz f <Elektronik, Fernseh, Funktech, Telekom> reception frequency
Empfangsgerät n <Telekom> receive machine

Empfangsgleichrichtung f <Phys> demodulation (Funkwellen)
Empfangskanal m <Telekom> return channel (Richtfunk)
Empfangskarte f <Elektronik> receiver board
Empfangsloch n <Fernseh, Mobilkom> shadow
Empfangslücke f <Gerät> dead spot
Empfangsoszillator m <Elektronik, Funktech> local oscillator
Empfangsoszillator m mit Frequenzaufbereitung <Elektronik> synthesized local oscillator
Empfangsoszillatorfrequenz f <Elektronik, Funktech, Telekom> local oscillator frequency
Empfangsoszillatorröhre f <Elektronik> local oscillator tube
Empfangsoszillatorsignal n <Elektronik> local oscillator signal
Empfangs-Parabolantenne f <Fernseh> receiving dish antenna
Empfangspegel m <Elektronik> reception level
Empfangsquarz m <Elektronik> receive crystal
Empfangsquittung f <Comp & DV> receipt
Empfangssignal n <Telekom> incoming signal (Empfänger)
Empfangssucher m <Elektronik> ranger finder
Empfehlung f recommendation (z. B. in Standards)
Empfehlungen fpl **für die Nährstoffzufuhr** <Lebensmittel> RDA, recommended dietary allowances
empfindlich adj **für alle Farben** 1. <Chemie> panchromatic (Fotografie); 2. <Druck, Foto> panchromatic
empfindlich machen v <Foto> sensitize
empfindliche Emulsion f <Foto> orthochromatic emulsion
empfindliches Papier n <Foto> sensitive paper
Empfindlichkeit f 1. <Akustik> response responsivity, responsiveness; 2. <Comp & DV, Elektriz, Elektrotech, Funktech, Gerät, Kohlen, Kontroll, Optik> sensitivity; 3. <Phys> sensitiveness, sensitivity, susceptibility; 4. <Raumfahrt> sensitivity (Weltraumfunk); 5. <Telekom> sensitivity
Empfindlichkeit f **am Skalenanfangswert** <Gerät> minimum scale sensitivity
Empfindlichkeit f **nach DIN** <Foto> DIN speed
Empfindlichkeitsindex m <Foto> exposure index
Empfindlichkeitsschwelle f <Gerät> threshold of sensitivity
Empfindung f <Akustik> sensation
Empfindungsfunktion f <Akustik> stimulus-sensation relation
Empfindungsgrößen fpl <Akustik> values of sensation
Empfindungsschwelle f <Ergon> threshold of feeling
Empfindungsstufen fpl <Akustik> sensation steps
empfundener Lärmpegel m <Sicherheit> perceived noise level
empirisch adv <Math> empirical (auf Beobachtungen basierend)
empirische Methode f <Bau, Qual> trial-and-error
empirische Verteilungsfunktion f <Math> empirical distribution function, sample distribution function
EMS (Expansionsspeicher-Spezifikation) <Comp & DV> EMS (expanded memory specification)
EMS-Speicherverwalter m <Comp & DV> expanded memory manager
Emulation f <Comp & DV, Elektronik> emulation (Simulation eines anderen Computersystems mit Hardware- oder Softwaremitteln)
Emulator m 1. <Comp & DV> emulator; 2. <Elektronik> emulator (Hardware-Zusatz); emulator (Programm)
Emulgator m <Chemtech, Kerntech, Kunststoff, Lebensmittel, Papier, Verpack> emulsifier, emulsifying agent
Emulgatorflüssigkeit f <Chemtech> emulsifying liquid

Emulgier... <Chemtech> emulsifying
emulgierbar adj <Chemtech> emulsifiable
Emulgierbarkeit f <Chemtech> emulsifiability
emulgieren v <Chemtech, Foto, Kunststoff, Papier, Umweltschutz> emulsify
Emulgieren n <Chemtech> emulsification
Emulgiermaschine f <Chemtech, Kerntech> emulsifying machine
Emulgiermittel n 1. <Chemtech, Kerntech> emulsifier; 2. <Kunststoff> emulsifier, emulsifying agent; 3. <Lebensmittel, Papier> emulsifier; 4. <Verpack> emulsifying agent
Emulgierung f <Chemie> emulsification
Emulgierungs... <Chemtech> emulsifying
emulieren v <Comp & DV, Elektronik> emulate *(Verhalten eines anderen Programms simulieren)*
Emulsin n <Chemtech> emulsin
Emulsion f 1. <Bau> emulsion; 2. <Chemtech> emulsion *(Binder)*; 3. <Erdöl, Foto, Kerntech, Kunststoff, Lebensmittel, Meerschmutz, Phys> emulsion
emulsionbeschichtet adj <Verpack> emulsion-coated
Emulsionieren n <Chemie> emulsification
Emulsionsanlage f <Kerntech> emulsifying machine
Emulsionsbeständigkeit f <Chemtech> emulsion persistence
Emulsionschargennummer f <Foto> emulsion batch number
Emulsionsfarbe f <Bau, Kunststoff> emulsion paint
Emulsionsflüssigkeit f <Chemtech, Sicherheit> emulsifying liquid
Emulsionspolymerisation f <Kunststoff> emulsion polymerization
Emulsionsspalter m 1. <Lebensmittel> de-emulsifying agent; 2. <Meerschmutz> demulsifier, emulsion breaker; 3. <Umweltschmutz> emulsion breaker
Emulsionstest m <Chemtech> emulsion test
EMV *(elektromagnetische Verträglichkeit)* <Elektriz, Funktech, Raumfahrt> EMC, electromagnetic compatibility *(Raumfahrt)*
EMV-gerecht adj <Elektriz> EMC-compatible
EN *(Europäische Norm)* <Elektriz> European Standard
Enantiomer n <Chemie> enantiomer, optical isomer
enantiomorph adj <Chemie> enantiomorphic, enantiomorphous
Enantiomorphie f <Chemie> enantiomorphism
enantiotrop adj <Chemie> enantiotropic
End... <Abfall, Fertig, Gerät, Kerntech, Lufttrans, Telekom> final, ultimate
Endabbrand m <Kerntech> ultimate burn up; final fuel burnup *(von Brennelementen)*
Endabdeckung f <Abfall> final covering, final cover *(einer Deponie)*
Endablesung f <Gerät> final reading
Endabnahme f <Qual> final acceptance
Endabschaltung f <Fertig> end switch *(Kunststoffinstallationen)*
Endabschnitt m <Kerntech> end section *(eines Brennelementes)*
Endamt n <Telekom> end exchange, terminal exchange, terminating exchange
Endanflug m <Lufttrans> final approach
Endanflugbahn f <Lufttrans> final approach path
Endanflugpunkt m <Lufttrans> final approach fix, final approach point
Endanode f <Fernseh> final anode
Endanschlag m <Fertig> end stop
Endanwender m <Comp & DV> end user
Endanwendung f <Textil> end use
Endauflager n <Bau> abutment *(Architektur)*
Endauslösevorrichtung f <Kerntech> final trip assembly

Endaustastung f <Fernseh> final blanking
Endbahnhof m <Eisenbahn> terminal station
Endbearbeitung f <Fertig, Ker & Glas> finishing
Endbedingung f <Comp & DV> end condition
Endbegrenzungsleuchte f <Kfztech> end outline marker lamp
Endbenutzer m <Telekom> end user
Endbogenstück n <Lufttrans> edge box member
Endcode m <Comp & DV> tail
Enddeckel m <Papier> end deckle
Enddruck m <Maschinen> final pressure
Ende n <Comp & DV, Papier> end • **am Ende herausgezogen** <Ker & Glas> drawn out at end • **Ende werfen** <Wassertrans> throw a line *(Tauwerk)*
Ende n/**das äußerste** <Wassertrans> bitter end *(Tauwerk)*
Ende n **der Blankschmelze** <Ker & Glas> seed-free time
Ende n **des Übertragungsblocks** <Comp & DV, Telekom> end-of-transmission block
Ende n **ohne Kuppe** <Maschinen> as-rolled end
Endeinheit f <Comp & DV> terminal device
Endeinrichtung f <Comp & DV, Telekom> terminal, terminal device; terminal equipment *(TE)*; terminal installation
Endemarke f <Fertig> blockmark *(Datenverarbeitung)*
Endflansch m <Maschinen> end flange
Endformat n <Papier> trimmed size
Endgehalt m <Kerntech> tail assay *(von Natururan)*
Endgerät n 1. <Comp & DV> terminal, terminal device; terminal equipment *(TE)*; terminal installation; 2. <Telekom> terminal, terminal device; terminal equipment *(TE)*; terminal installation, user equipment
Endgerät n **für Fernbetriebsführung** <Telekom> remote operating terminal
Endgeräteanpassung f <Telekom> terminal adaptor
Endgeräte-Anschlusssteuerung f <Telekom> medium access control
Endgerätesubadressierung f <Telekom> terminal subaddressing
Endgeschwindigkeit f <Raumfahrt> all-burnt velocity *(Raumschiff)*
Endglied n <Maschinen> end link
endgültige Tonmischung f <Aufnahme> final mix
endgültiger Befehl m <Comp & DV> effective instruction
endgültiges Verwerfen n <Qual> final rejection
Endinstallation f <Kerntech> ultimate installation
Endkeil m <Kohlen> end cleat
Endknoten m 1. <Künstl Int> end node, terminal node *(eines Baumes)*; 2. <Telekom> terminating junction
Endkontrolle f <Qual> final inspection
Endlage f 1. <Fertig> end position *(Kunststoffinstallationen)*; 2. <Gerät> end-point position *(Zeiger)*; 3. <Kerntech> limit of travel *(eines Regel- oder Moderatorstabes)*
Endlageneinstellung f <Kerntech> final position setting
Endlager n 1. <Abfall> disposal zone, vitrification process; 2. <Maschinen> end bearing
Endlagerstätte f <Umweltschmutz> repository *(für radioaktiven Abfall)*
Endlagerung f 1. <Abfall> final dumping *(von Atommüll oder radioaktivem Abfall)*; final storage, ultimate storage *(von Müll)*; 2. <Kerntech> ultimate waste disposal *(von Atommüll)*
Endlagerung f **von Abfällen** <Abfall> permanent waste storage
Endlagerungsstätte f <Abfall> disposal zone, vitrification process
endlich adj <Math> finite
endliche Impulsdauer f <Elektriz> finite impulse duration, finite pulse duration

endliche Reihe f <Math> finite series
endlos adj <Papier, Textil> endless
Endlos... <Papier, Textil> endless
Endlosantrieb m <Aufnahme> closed-loop drive
endloser Breitkeilriemen m <Maschinen> endless wide V-belt
Endlosfaden m <Textil> filament
Endlosfaser f 1. <Kunststoff> filament; 2. <Metall> (AE) continuous fiber, (BE) continuous fibre
Endlosfasermatte f <Kunststoff> continuous strand mat
Endloskabel n <Kfztech> endless cable
Endloskette f <Kfztech, Papier> endless chain
Endlosmagnetbandkassette f <Akustik> endless magnetic loop cartridge
Endlosmatte f <Kunststoff> continuous strand mat
Endlospapier n 1. <Comp & DV> (AE) continuous forms, (BE) continuous stationery, (AE) continuous-feed paper, fanfold, fanfold stationery; 2. <Druck> (AE) continuous forms, (BE) continuous stationery, (AE) continuous-feed paper
Endlospapiereinzug m <Comp & DV> continuous feed
Endlosriemen m <Kfztech> endless belt
Endlosschleife f <Comp & DV> infinite loop
Endlosspinnfaden m <Ker & Glas> continuous filament
Endlosverfahren n <Ker & Glas> continuous-drawing process
Endmarke f <Comp & DV> terminator
Endmaß n 1. <Maschinen> (AE) end gage, (BE) end gauge; 2. <Metrol> end measure, (AE) gage block, (BE) gauge block, length bar
Endmaßvergleichsmesser m <Metrol> (AE) gage block comparator, (BE) gauge block comparator
Endmast m <Elektriz> terminal tower
Endmontage f <Kerntech> final assembly, ultimate installation
Endoenzym n <Chemie> endo-enzyme
endogene Variable f <Math> endogen variable, response variable (Regressionsanalyse)
Endoskop n <Phys> endoscope
Endoskopie f <Kerntech> endoscopy
Endosperm n <Lebensmittel> endosperm (Nährgewebe im Samen)
endotherm adj 1. <Ker & Glas, Raumfahrt> endothermic (Raumschiff); 2. <Thermod> endothermal, endothermic
endotherme Reaktion f <Nichtfoss Energ> endothermic reaction
endothermer Prozess m <Thermod> endothermic process
endothermisch adj 1. <Ker & Glas, Raumfahrt> endothermic (Raumschiff); 2. <Thermod> endothermal, endothermic
endothermische Reaktion f <Nichtfoss Energ> endothermic reaction
endothermischer Prozess m <Thermod> endothermic process
Endpflock m <Kohlen> end cleat
Endplatte f <Maschinen> end plate
Endprodukt n <Kohlen> commercial coal (Kohle)
Endprüfung f <Qual> final inspection
Endpunkt m 1. <Comp & DV> exit point; 2. <Fernseh> out point; 3. <Geom> ending point (einer Linie)
Endpunkt m einer Achse <Geom> axis end point
Endpunktkoordinaten fpl <Geom> end-point coordinates
Endrahmenstück n <Verpack> end frame member
Endreinigungsvorgang m <Kerntech> tail end process
Endrille f <Akustik> finishing groove
Endrippe f <Lufttrans> end rib
Endrohr n <Kfztech> tail pipe
Endrohrverlängerung f <Kfztech> tail pipe extension

Endschalter m 1. <Elektriz, Elektrotech> limit switch, proximity switch; 2. <Fertig> depth-control limit switch; 3. <Kontroll> end position switch, limit switch; 4. <Mechan> limit switch; 5. <Sicherheit> overtravel switch (einer Maschine)
Endserienverpackung f <Verpack> end-of-line packaging
Endspalte f <Comp & DV> end column
Endspeisung f <Funktech> end feeding (Antenne)
Endspiel n <Maschinen, Mechan> end play
endspiraliger Bohrer m <Fertig> high-helix drill
Endstation f <Eisenbahn> terminal
Endstelle f 1. <Comp & DV> (BE) station; 2. <Telekom> terminal station
Endstellung f <Gerät> end-point position, ultimate position (Zeiger)
Endstück n 1. <Fertig> butt; 2. <Kerntech> end fitting
Endstufe f 1. <Elektrotech, Funktech> power amplifier (PA); 2. <Telekom> terminating stage
Endsymbol n <Comp & DV> terminal symbol
Endsystemteil n <Telekom> user agent
Endtetrode f mit Elektronenbündelung <Elektronik> beam power tube
Endumsetzer m <Elektronik> final modulator (Trägerfrequenz)
Endvakuum n <Thermod> ultimate vacuum
Endverbraucher m <Comp & DV> end user
Endverkehr m <Telekom> terminating traffic
Endvermittlungsstelle f (EVS) <Telekom> terminating office, terminal exchange
Endverschluss m <Telekom> termination (Kabel)
Endverstärker m 1. <Elektronik, Elektrotech> final amplifier, output amplifier, power amplifier; 2. <Funktech, Telekom> output amplifier
Endverstärkung f <Elektronik> final amplification (Radio)
Endverzweiger m <Telekom> block terminal; terminal box (Telefon)
Endwert m <Qual> target
Endzurückweisung f <Qual> final rejection
Endzusammenbau m <Kerntech> final assembly
energetische Verwertung f <Abfall> energy recovery
energetischer Wirkungsgrad m <Nichtfoss Energ, Thermod> energy efficiency
Energie f (E) 1. <Elektriz, Elektrotech, Kerntech, Labor, Mechan, Nichtfoss Energ> energy, E; power, P; 2. <Optik> power; 3. <Phys, Thermod> energy, E • **Energie abschalten** <Elektrotech> de-energize
Energie f aus Abfall <Abfall> residue derived energy
Energieabbau m <Kerntech> energy degradation
energieabsorbierend adj <Nichtfoss Energ> energy-absorbing
Energieabsorption f <Nichtfoss Energ, Telekom> energy absorption
Energieanschluss m <Elektrotech> power supply
energiearmer Laser m <Elektronik> low-energy laser
energiearmer Strahl m <Elektronik> low-energy beam
Energieaustauschreaktion f <Kerntech> energy exchange reaction
Energiebedarf m <Nichtfoss Energ, Thermod> energy demand
Energiebereich m <Nichtfoss Energ, Strahlphys> energy range
Energiebilanz f <Wasserversorg> energy balance
Energie/Brennstoff-Diversität f <Kohlen> energy/fuel diversity
Energiedichte f <Telekom> power density
Energiedichte f einer Strahlung <Strahlphys> energy density of radiation
Energiedosis f <Phys, Strahlphys> absorbed dose of ionizing radiation

Energieeinheit f <Thermod> unit of energy
Energieerhaltung f 1. <Nichtfoss Energ, Phys> conservation of energy; 2. <Thermod> energy conservation
Energieersparnis f <Nichtfoss Energ, Thermod> energy saving
Energieerzeugung f <Elektrotech, Kohlen, Nichtfoss Energ> power generation
Energieerzeugungsanlage f <Elektrotech> power plant
Energiefluss m <Nichtfoss Energ, Phys> energy fluence
Energieflussbild n <Nichtfoss Energ, Thermod> energy flow chart
Energieflussdiagramm n <Nichtfoss Energ, Thermod> energy flow chart
Energieflussdichte f 1. <Kerntech> energy flux density (I); 2. <Optik> power flux density
Energieflussrate f <Nichtfoss Energ, Phys> energy fluence rate
Energiegehalt m <Nichtfoss Energ, Thermod> energy content
Energiegewinnung f <Nichtfoss Energ> energy extraction
Energiegleichgewicht n 1. <Nichtfoss Energ, Strömphys> energy balance (bei turbulenter Bewegung); 2. <Thermod> energy balance
Energiehaushalt m <Nichtfoss Energ, Umweltschutz> energy budget
Energieinhalt m <Nichtfoss Energ, Thermod> energy content
Energie-intensiv adj <Nichtfoss Energ, Thermod> energy-intensive (Verfahren, Industrie)
Energiekaskade f <Strömphys> energy cascade
Energiekonverter m <Elektrotech, Nichtfoss Energ> energy converter
Energiekrise f <Nichtfoss Energ, Thermod> energy crisis
Energieleitung f <Elektrotech> transmission line
Energieleitungsnetz n <Elektriz> transmission line network
energielos adj <Elektriz> wattless
Energielücke f <Nichtfoss Energ, Phys> energy gap (Halbleiter)
Energiemessgerät n <Nichtfoss Energ> energy meter
Energiemusterfaktor m <Nichtfoss Energ> energy pattern factor
Energieniveau n 1. <Nichtfoss Energ, Phys> energy level; 2. <Sicherheit> power level
Energiepegel m <Sicherheit> power level
Energieprodukt n <Elektriz> energy product; BH product (des Magnetmaterials)
Energiequelle f 1. <Elektriz, Elektrotech> power source, power supply, source; 2. <Nichtfoss Energ> energy source, power source
Energiereflexionskoeffizient m <Optik> power reflection coefficient
energiereicher Strahl m <Elektronik> high-energy beam
energiereiches Elektron n <Elektronik> high-energy electron
energiereiches Ion n <Elektronik> high-energy ion
energiereiches Teilchen n <Elektronik> high-energy particle
Energieressourcen fpl <Abfall, Nichtfoss Energ> energy resources
Energierückgewinnung f 1. <Abfall> energy recovery; 2. <Nichtfoss Energ, Thermod> energy recovery, energy regeneration
Energierückgewinnungsfaktor m <Nichtfoss Energ> energy recovery factor
Energierückspeisung f <Elektrotech> energy regeneration (Elektrotraktion)
energiesparend adj <Nichtfoss Energ, Thermod> energy-saving

energiesparende Technologie f <Abfall, Nichtfoss Energ> energy-saving technology
Energiespeicher m <Nichtfoss Energ, Raumfahrt> energy storage device (Raumschiff)
Energiespeicherung f <Elektrotech, Nichtfoss Energ, Thermod> energy storage
Energiespeicherung f **in der Schwachlastzeit** <Kerntech> off-peak energy storage
Energiespektrum n <Raumfahrt> energy spectrum
Energiestreuung f 1. <Nichtfoss Energ> energy dispersal; 2. <Raumfahrt> energy dispersal (Weltraumfunk)
Energietal n <Kerntech> energy valley
Energietechnik f <Nichtfoss Energ, Umweltschutz> energy technology
Energieträger m <Elektriz, Nichtfoss Energ> energy carrier
Energietransportkoeffizient m <Nichtfoss Energ, Phys, Qual> energy transfer coefficient
Energieübertragung f 1. <Elektrotech> energy transmission, power transmission; 2. <Nichtfoss Energ, Thermod> energy transfer, energy transmission
Energieübertragung f **durch mechanische Schwingung** <Nichtfoss Energ, Wellphys> energy transfer by vibration (Körperschall)
Energieumformung f <Elektriz, Nichtfoss Energ> energy transformation
Energieumsatz m 1. <Ergon> energy expenditure, metabolic rate; 2. <Nichtfoss Energ> energy expenditure
Energieumwandler m <Elektrotech, Nichtfoss Energ, Thermod> energy converter
Energieumwandlung f <Elektrotech, Nichtfoss Energ, Thermod> energy conversion
Energieumwandlungskoeffizient m <Kerntech> mass energy transfer coefficient
energieunabhängiger Speicher m <Comp & DV> non-volatile memory, permanent memory, permanent storage
Energieverbrauch m 1. <Elektriz> energy consumption, power consumption; 2. <Nichtfoss Energ, Phys, Thermod> energy consumption
Energieverengung f <Metall> constriction energy
Energieverlust m 1. <Elektriz, Elektrotech> energy loss, power loss; 2. <Nichtfoss Energ> energy loss; 3. <Phys> degradation of energy; 4. <Thermod> energy loss
Energieverlust m **von Elektronen** <Strahlphys> electron energy loss
Energieversorgung f 1. <Eisenbahn, Elektrotech> power supply; 2. <Nichtfoss Energ> energy supply, power supply; 3. <Telekom> power feeding; 4. <Thermod> energy supply
Energieversorgungsnetz n <Elektronik, Elektrotech> electric power supply system, electric distribution system
Energieversorgungsstation f <Elektriz, Elektrotech> energy-supply station, pre-conditioning supply, pre-heating supply
Energieversorgungsunternehmen n <Elektriz> power utility, utility
Energieversorgungswagen m <Eisenbahn> power source car
Energiewandler m <Elektrotech, Nichtfoss Energ, Telekom, Thermod> energy converter
Energiewiedergewinnung f <Nichtfoss Energ, Thermod> energy recuperation
Energiewirkungsgrad-Verhältnis n (EER) <Kohlen, Nichtfoss Energ> energy efficiency ratio, EER
Energiezähler m <Nichtfoss Energ> energy meter
Energiezustand m **eines Atoms** <Kerntech, Phys, Strahlphys> atomic state
E-Netz n <Mobilkom> digital cellular system, E-net
ENF (extrem tiefe Frequenz) <Funktech> ELF (extremely low frequency)

eng gruppierte Wohngebäude *npl* <Bau> cluster housing
eng toleriert *adj* <Maschinen> close-tolerance
enge Bohrung *f* <Erdöl> slim hole *(Bohrtechnik)*
enge Kopplung *f* <Elektrotech, Phys> tight coupling
enge Passung *f* <Maschinen> close fit
enge Toleranz *f* <Maschinen> close tolerance
Engel *mpl* <Elektronik> clutter suppression *(Radartechnik)*
Engelecho *n* <Funkort> angel echo *(Radar)*
enger ausschließen *v* <Druck> close up, keep in
Enghalsflasche *f* <Labor> narrow-necked bottle
Enghalspackung *f* <Ker & Glas> narrow neck container
Engländer *m* <Maschinen> coach wrench, *(BE)* shifting spanner
Englergrad *m* <Fertig, Maschinen> Engler degree
englische Schreibschrift *f* <Druck> script type
Engpass *m* <Mechan, Telekom, Trans> bottleneck
Engschrift *f* <Konstzeich> close-spaced characters, close-spaced lettering
Engspaltschweißen *n* <Mechan> narrow-gap welding
enharmonische Noten *fpl* <Akustik> enharmonic notes
Enol *n* <Chemie> enol
Enol... <Chemie> enolic
Enolase *f* <Chemie> enolase
enolisch *adj* <Chemie> enolic
Enolisierung *f* <Chemie> enolization
entartet *adj* <Elektronik, Metall, Phys> degenerate
entarteter Halbleiter *m* <Elektronik> degenerate semiconductor
entartetes Elektronengas *n* <Strahlphys> degenerate electron gas
Entartung *f* 1. <Elektronik> degeneracy, degeneration; 2. <Kerntech> degradation *(von Teilchen, Energieniveaus)*; 3. <Phys> degeneracy; 4. <Strahlphys> degeneration *(von Energieniveaus)*
Entbastungsmittel *n* <Chemie> scouring agent *(Textil, Seide)*
Entbasung *f* <Kohlen> desorption
entbehrlich *adj* <Raumfahrt> expendable *(Raumschiff)*
Entbindung *f* <Patent> release
Entbituminieren *n* <Kohlen> debituminization
Entblätterungsmittel *n* <Chemie> defoliant
entblocken *v* <Comp & DV> deblock
Entblocken *n* <Comp & DV> deblocking
Entblößung *f* <Metall> denudation
Entbrummspule *f* <Aufnahme, Elektrotech, Fernseh> humbucking coil
Entbündeln *n* <Elektronik> debunching
Entdröhnmittel *n* <Sicherheit> vibration deadener
enteisen *v* <Heiz & Kälte, Raumfahrt> de-ice *(Raumschiff)*
Enteisenung *f* 1. <Abfall> deferrization; 2. <Ker & Glas> de-ironing
Enteiser *m* 1. <Funktech> de-icer *(Antenne)*; 2. <Kfztech, Lufttrans, Raumfahrt> de-icer *(Raumschiff)*
Enteiserhaube *f* <Lufttrans> de-icer boot, de-icer trunk
Enteiserleitung *f* <Lufttrans> de-icing duct
Enteiserluftauslass *m* <Lufttrans> de-icing air outlet
Enteiserpumpe *f* <Lufttrans> de-icing pump
Enteisung *f* 1. <Heiz & Kälte> de-icing; 2. <Raumfahrt> anti-icing, de-icing
Enteisung *f* **des Motors** <Lufttrans> engine de-icing *(des Triebwerks)*
Enteisungsflüssigkeit *f* <Anstrich> de-icing fluid
Enteisungsluft *f* <Lufttrans> de-icing air
Enteisungsstiefel *m* <Raumfahrt> de-icer boot
Enteisungssystem *n* <Funktech, Raumfahrt> anti-icing system *(für Antennen)*
Entemulgator *m* <Lebensmittel> demulsifier
Entenflugzeug *n* <Lufttrans> canard wing aircraft, tail first configuration aircraft

Enterhaken *m* <Wassertrans> grapple
Entermannschaft *f* <Wassertrans> boarding party *(Seeräuberei)*
entern *v* <Wassertrans> board
entfaltbare Antenne *f* <Raumfahrt> unfurlable antenna *(Weltraumfunk)*
Entfaltung *f* <Telekom> deconvolution
entfärben *v* <Textil> bleach
Entfärber *m* <Ker & Glas> *(AE)* decolorizer, *(BE)* decolourizer
Entfärbungspulver *n* <Textil> bleaching powder
entfeinter Beton *m* <Bau> no-fines concrete
entfernbar *adj* <Optik> removable
entfernbares Teil *n* <Elektriz> removable part
Entfernen *n* <Bau> removing, removal
Entfernen *n* **von Randbeschnitt** <Papier> trim removal
entfernt *adj* <Comp & DV> remote
entfernte Quelle *f* <Umweltschmutz> distant source
Entfernung *f* 1. <Eisenbahn> removal *(der Stangenkupplung)*; 2. <Elektronik> range; 3. <Funktech> distance; 4. <Gerät> range; 5. <Telekom> distance, range; 6. <Wassertrans> range *(Navigation, Radar, Funk)*
Entfernung *f* **bewegungsunfähiger Lutfahrzeuge** <Lufttrans> disabled aircraft removal *(Flughafen)*
Entfernungseinstellring *m* <Foto> focusing ring
Entfernungsmesseinrichtung *f* <Funkort, Gerät, Lufttrans> distance-measuring equipment *(DME)*
Entfernungsmessen *n* **mit einer Kette** <Bau> chaining
Entfernungsmesser *m* 1. <Bau> odometer; 2. <Foto, Metrol> range finder; 3. <Wassertrans> RF, range finder; distance finder *(Navigation)*
Entfernungsmessgerät *n* <Gerät> RF, range finder
Entfernungsmessung *f* 1. <Elektronik> range finding; 2. <Raumfahrt> ranging
Entfernungsradar *m* <Funkort> radar range finder
Entfernungsring *m* 1. <Foto> focusing ring; 2. <Wassertrans> calibration ring *(Radar)*
Entfernungsskale *f* <Foto> distance scale
Entfernungstaste *f* *(Taste Entf)* <Comp & DV> delete key *(DEL key)*
entfetten *v* 1. <Bau, Elektriz, Maschinen, Mechan> degrease; 2. <Textil> scour *(Wolle)*
Entfetten *n* 1. <Maschinen, Mechan> degreasing; 2. <Textil> scouring
Entfetter *m* <Anstrich> degreaser
Entfettung *f* <Bau, Ker & Glas> degreasing
Entfettungsbehälter *m* <Lebensmittel> degreasing tank
Entfettungseinrichtung *f* <Elektriz> degreaser
Entfettungsmittel *n* 1. <Mechan> degreasing agent; 2. <Verpack> degreasing compound
entfeuchten *v* <Heiz & Kälte> dehumidify
Entfeuchter *m* 1. <Chemie, Chemtech> desiccator; 2. <Heiz & Kälte> dehydrator; 3. <Kerntech> dehumidifier
Entfeuchtung *f* <Heiz & Kälte> dehumidification
Entfeuchtungsgerät *n* <Heiz & Kälte> dehumidifier
Entfeuchtungsmittel *n* <Lebensmittel> desiccant
entflammbar *adj* <Kunststoff, Sicherheit> flammable
entflammbare Flüssigkeit *f* <Sicherheit> flammable liquid
entflammbarer Dampf *m* <Sicherheit> *(AE)* flammable vapor, *(BE)* flammable vapour
entflammbarer Werkstoff *m* <Sicherheit> flammable material
Entflammbarkeit *f* 1. <Chemie> ignitability; 2. <Sicherheit> flammability
entflammen *v* <Thermod> fire up
Entflammungspunkt *m* 1. <Heiz & Kälte> flash point; 2. <Thermod> kindling point
Entflechtung *f* <Elektriz> layout *(Leiterplatte)*

Entflocken n <Chemie> deflocculation
Entformen n 1. <Ker & Glas> take-out *(Herausnehmen des Gegenstands aus Form)*; 2. <Kunststoff> *(AE)* demolding, *(BE)* demoulding
Entformen n **mit Einstechen** <Ker & Glas> take-out with push-up
Entformer m <Ker & Glas> take-out *(Vorrichtung zum Entformen)*
Entformungsmittel n <Kunststoff> *(AE)* mold release agent, *(BE)* mould release agent, release agent
Entformungsvorrichtung f <Kunststoff> extractor
Entfrittung f <Metall> decohesion
entfrosten v <Heiz & Kälte> de-ice, defrost
Entfrosten n <Heiz & Kälte, Maschinen> defrosting
Entfroster m <Kfztech> demister; defroster *(Zubehör)*
Entfrostung f <Heiz & Kälte> de-icing
entführen v <Lufttrans> hijack, skyjack *(Flugzeug)*
entgasen v 1. <Abfall> degasify; 2. <Chemtech> degas; 3. <Elektronik> degas *(Elektronenröhre)*; 4. <Thermod> free from gas
Entgasen n 1. <Chemtech> degassing; 2. <Elektronik> degassing *(Elektronenröhre)*; 3. <Kunststoff> degassing
Entgaser m 1. <Erdöl> deaerator; 2. <Heiz & Kälte> degasser
entgast adj <Thermod> free from gas
entgaster Brennstab m <Kerntech> vented fuel rod
entgastes Brennelement n <Kerntech> vented fuel assembly
entgastes Öl n <Erdöl> dead oil
Entgasung f <Abfall> degassing; gas drainage *(Deponie)*; degasification *(einer Deponie)*
Entgasungsindex m <Ker & Glas> outgassing index
entgegengerichteter Gradient m <Strömphys> adverse gradient
entgegengesetzt adj <Math> opposite
entgegennehmen v <Telekom> answer *(Anruf)* • **einen Anruf entgegennehmen** answer a call
entgegenwirkendes Feld n <Elektriz> opposing field
Entgiftung f 1. <Kerntech> decontamination, depoisoning; 2. <Strahlphys> decontamination *(Kernspaltprodukte)*
entglasen v <Ker & Glas> devitrify
Entglasung f <Ker & Glas> devitrification
Entglasungssteinchen n <Ker & Glas> devitrification stone
Entgleisanzeiger m <Eisenbahn> dragging detector
entgleisen v <Eisenbahn> derail
Entgleisung f <Eisenbahn> derailment
Entgleisungsweichen fpl <Eisenbahn> catch points, *(BE)* derailing points, *(AE)* derailing switch
entgraten v 1. <Fertig> deburr *(Kunststoffinstallationen)*; deflash *(Kunststoffe)*; 2. <Kunststoff> deflash; 3. <Maschinen, Mechan> deburr
Entgraten n <Ker & Glas, Maschinen> deburring
Entgrat- und Abfasmaschine f <Fertig> deburring and chamfering machine
Entgratwerkzeug n <Maschinen> burr remover
Enthacken n <Raumfahrt> descrambling *(Weltraumfunk)*
Enthacker m <Raumfahrt> descrambler
Enthalpie f (H) <Heiz & Kälte, Kohlen, Mechan, Nichtfoss Energ, Phys, Raumfahrt, Thermod> enthalpy (H)
enthalten v <Bau> include
Enthärten n <Metall> softening
Enthärter m <Textil> softener *(Wasser)*
enthärtetes Wasser n <Wasserversorg> soft water
Enthärtungsanlage f <Fertig> water-softening plant *(Kunststoffinstallationen)*
Enthärtungsmittel n <Textil> softening agent *(Wasser)*
entharzen v <Fertig> deresinify
Enthitzer m <Heiz & Kälte> desuperheater

enthülsen v <Lebensmittel> husk
Enthülsen n <Kerntech> decanning, decladding *(von Brennelementen)*
enthülstes Brennelement n <Kerntech> uncanned fuel element
entionisiertes Wasser n <Elektriz> de-ionized water
Entionisierungsgitter n <Elektronik> de-ionizing grid
Entionisierungsmittel n <Labor> de-ionizer *(Wasser)*
Entität f <Künstl Int> entity
entkalken v <Chemie> decalcify
entkarbonisieren v <Chemie, Kohlen> decarbonate
Entkarbonisieren n <Chemie> decarbonization
entkeimt adj <Lebensmittel> sterilized
entkernen v <Lebensmittel> *(AE)* pit, *(BE)* stone *(Früchte)*
Entkerner m <Erdöl> corer *(Bohrtechnik)*
entkoffeiniert adj <Lebensmittel> decaffeinated
entkohlen v 1. <Fertig> decarburize; 2. <Thermod> decarbonize
entkohlt adj <Fertig> soft *(Randzone)*
entkohlte Schicht f <Fertig> bark
Entkohlung f 1. <Fertig> decarburization; 2. <Thermod> decarbonization
entkomprimieren v <Comp & DV> unpack *(data)*
entkoppelt adj <Comp & DV> decoupled
entkoppelte Mehrgrößenregelung f <Regelung> non-interacting control
Entkoppelungskreis m <Lufttrans> antiresonant circuit
Entkopplung f 1. <Elektrotech, Funktech> decoupling; 2. <Raumfahrt> decoupling *(Weltraumfunk)*; 3. <Telekom> decoupling
Entkopplungsfilter n <Elektronik> decoupling filter
Entkopplungskondensator m <Elektriz, Elektrotech> decoupling capacitor, decoupling condenser
Entkorkmaschine f <Verpack> uncorking machine
entkrusten v <Mechan> descale
Entkupplungsstange f <Eisenbahn> shunter's pole
Entlade... <Elektrotech, Lufttrans, Raumfahrt> discharge
Entladebunker m <Abfall> unloading hopper
Entladehaken m <Lufttrans> cargo release hook
Entladekondensator m <Elektrotech> discharge capacitor
Entladekreis m <Elektrotech> discharge circuit
entladen v 1. <Bau> dump; 2. <Comp & DV> unload; 3. <Maschinen, Trans> discharge, unload, unstuff
Entladen n 1. <Comp & DV> unloading; 2. <Trans> unstuffing
Entladeregler m <Raumfahrt> discharge regulator *(Raumschiff)*
Entladeschaltung f <Elektrotech> discharge circuit
Entladestab m <Kerntech> unloading rod
Entladestation f <Lufttrans> discharging station
Entladestrom m <Elektrotech> discharge current
Entladung f 1. <Elektriz, Elektrotech> discharge; 2. <Metall> unloading; 3. <Phys> unloading; discharge *(elektrisch)*; 4. <Telekom> discharge; 5. <Trans> unloading
Entladungskanal m <Kerntech> transfer canal
Entladungskreis m <Elektrotech> discharge circuit
Entladungslampe f <Elektrotech> discharge lamp
Entladungsmikrofon n <Aufnahme> discharge microphone
Entladungsröhre f 1. <Elektronik> discharge tube *(Glimmlampe)*; 2. <Phys> discharge tube
Entladungsspitzenspannung f <Elektriz> peak arc voltage
Entladungsstrom m <Elektriz> discharge current
Entladungswiderstand m <Elektriz> discharge resistor *(Bauelement)*; discharge resistance *(physikalische Größe)*
entlasten v 1. <Bau, Qual> ease; 2. <Telekom> relieve

Entlasten *n* <Bau> removing
entlastendes Wasser *n* <Umweltschmutz> deballasting water
entlastet *adj* <Fertig> pressure-balanced *(Dichtung)*
entlasteter Schieber *m* <Maschinen> balanced slide valve
entlastetes Querruder *n* <Lufttrans> balanced aileron
Entlastung *f* 1. <Maschinen> relief; 2. <Phys> unloading
Entlastungsanlage *f* <Bau> spillway
Entlastungsbogen *m* <Bau> relieving arch, safety arch; discharging arch *(Mauerwerk)*
Entlastungseinrichtung *f* <Sicherheit> relief device
Entlastungsgerinne *n* <Wasserversorg> discharge flume
Entlastungskanal *m* <Hydraul> spillway canal *(Hydraulik)*
Entlastungsöffnung *f* <Mechan> lightening hole
Entlastungsschleuse *f* <Wasserversorg> discharge sluice
Entlastungsstrecke *f* <Eisenbahn> bypass line
Entlastungsventil *n* 1. <Heiz & Kälte> relief valve; 2. <Maschinen> unloading valve; 3. <Mechan, Nichtfoss Energ> relief valve
Entlastungswehr *n* <Wasserversorg> spillway, waste weir
Entlastungszug *m* <Eisenbahn> relief train
Entlaufen *n* <Eisenbahn> runaway *(von Wagen)*
entlaugen *v* <Fertig> wash
Entlaugen *n* <Fertig> washing
entleeren *v* 1. <Bau> drain; 2. <Elektronik> evacuate *(Röhren)*; 3. <Kohlen> drain; 4. <Wassertrans> evacuate *(Schiff)*; 5. <Wasserversorg> blow off
Entleeren *n* <Fertig> drainage *(Kunststoffinstallationen)*
Entleerung *f* 1. <Erdöl> drain; 2. <Wasserversorg> depletion
Entleerungshahn *m* <Erdöl> drain tap
Entleerventil *n* <Nichtfoss Energ> purging valve
entlieschen *v* <Lebensmittel> husk *(Mais)*
entlöten *v* <Maschinen> unsolder
Entlöten *n* <Maschinen> unsoldering
entlüften *v* 1. <Bau> vent; 2. <Fertig> bleed *(Luft)*; 3. <Heiz & Kälte> vent; 4. <Kfztech> bleed *(Bremse)*; 5. <Maschinen> bleed, vent *(Luft)*
Entlüften *n* 1. <Erdöl> bleeding; 2. <Heiz & Kälte, Kerntech> venting
Entlüfter *m* 1. <Bau> extractor fan; 2. <Erdöl> air exhaust; 3. <Heiz & Kälte> air bleeder, vent; 4. <Kfztech> breather *(Kurbelgehäuse des Motors)*; 5. <Mechan> exhaust pump
Entlüftung *f* 1. <Bau> airing; 2. <Erdöl> air vent; 3. <Fertig> air drain, air escape; 4. <Heiz & Kälte> ventilation; 5. <Ker & Glas> vent; 6. <Kerntech> air drain, vent; 7. <Kfztech> bleeding, ventilation; 8. <Kunststoff, Labor> vent; 9. <Maschinen> bleeding, vent, venting, ventilation; 10. <Mechan> vent; 11. <Thermod, Wassertrans> ventilation
Entlüftungsanlage *f* <Lebensmittel> exhauster
Entlüftungsarmatur *f* <Erdöl> bleed valve
Entlüftungsbehälter *m* <Lebensmittel> exhaustion box
Entlüftungsbohrung *f* <Fertig> air vent *(Ziehwerkzeug)*
Entlüftungshahn *m* <Fertig> air drain petcock
Entlüftungshaubenventil *n* <Lufttrans> air vent valve
Entlüftungsklappe *f* 1. <Bau> vent cap; 2. <Kerntech> air vent
Entlüftungsleitung *f* <Mechan> exhaust line
Entlüftungsöffnung *f* 1. <Bau> vent; 2. <Fertig> whistler *(Formen)*; 3. <Heiz & Kälte> vent, vent port; 4. <Kerntech, Kunststoff, Labor> vent; 5. <Maschinen> vent, vent hole
Entlüftungsrohr *n* 1. <Bau, Heiz & Kälte, Ker & Glas, Kerntech> vent pipe; 2. <Kfztech> breather; 3. <Kunststoff, Labor, Maschinen, Mechan> vent pipe; 4. <Raumfahrt> standpipe

Entlüftungsschraube *f* <Kfztech, Maschinen> bleeder screw
Entlüftungsstopfen *m* <Maschinen> bleed plug
Entlüftungsventil *n* 1. <Fertig> air drain valve, air relief valve, breather; 2. <Kerntech> air drain valve; 3. <Kfztech> bleed valve; 4. <Lebensmittel> air bleed valve; 5. <Lufttrans> air bleed valve; vent valve *(für luftführende Leitungen)*; 6. <Maschinen> air vent valve, bleed valve
entmagnetisieren *v* 1. <Aufnahme, Comp & DV> degauss; 2. <Fernseh> degauss, demagnetize; 3. <Kohlen> demagnetize; 4. <Maschinen> degauss, demagnetize; 5. <Phys, Telekom> demagnetize
Entmagnetisieren *n* <Comp & DV, Fernseh, Maschinen> degaussing
Entmagnetisierer *m* <Aufnahme> degausser
Entmagnetisiergerät *n* 1. <Fernseh> degausser; 2. <Maschinen> demagnetizer
Entmagnetisierspule *f* <Fernseh> degaussing coil
Entmagnetisierung *f* 1. <Aufnahme> demagnetization; 2. <Comp & DV> degaussing; 3. <Elektriz> demagnetization; 4. <Fernseh> degaussing, demagnetization; 5. <Maschinen> degaussing; 6. <Mechan, Phys> demagnetization
Entmagnetisierungsfeld *n* <Elektriz, Phys> demagnetizing field
Entmagnetisierungsgerät *n* <Comp & DV> degausser
Entmagnetisierungsverlust *m* <Aufnahme> demagnetization loss
entmasten *v* <Wassertrans> dismast *(Schiff)*
entmineralisieren *v* <Chemtech> demineralize
Entmineralisierung *f* <Chemtech> demineralization
Entmineralisierungsanlage *f* <Kerntech> demineralizing plant
Entmischen *n* 1. <Bau> settlement *(Frischbeton)*; 2. <Kerntech> segregation *(von Legierungen)*; 3. <Kunststoff> segregation
Entmischungseffekt *m* <Chemtech> separation effect *(Phasen oder Gemische)*
Entmischungsvorgang *m* <Chemtech> separation process
Entnahme *f* 1. <Fertig> withdrawal; 2. <Maschinen> discharge; 3. <Wasserversorg> intake
Entnahme *f* **von Wasser** <Wasserversorg> tapping
Entnahmegerät *n* <Abfall> extractor
Entnahmematerial *n* <Bau> borrow
Entnahmestelle *f* <Bau> borrow pit
Entnahmeversuch *m* <Wasserversorg> pumping test
Entnebeler *m* <Lufttrans> demister
Entnebelungsventilator *m* <Lufttrans> defogging fan
entnehmen *v* 1. <Elektriz> abstract; 2. <Papier> pick
Entnickelung *f* <Fertig> denickelfication
entölen *v* <Bau, Elektriz, Maschinen, Mechan> degrease
Entölen *n* <Abfall> oil removal, oil separation
Entölung *f* <Bau> degreasing
entpacken *v* <Comp & DV> unpack *(Daten)*
entpaketieren *v* <Comp & DV> unpack
Entpalettisiermaschine *f* <Verpack> depalletizer
Entphosphoren *n* <Chemie> dephosphorization
Entphosphorungsverfahren *n* <Metall> dephosphorizing process
Entpolymerisation *f* <Verpack> depolymerization
Entrahmungszentrifuge *f* <Lebensmittel> cream separator
entrastern *v* <Druck> descreen
entregen *v* <Phys> de-energize
Entriegeln *n* <Comp & DV> unlocking
Entriegelung *f* <Kerntech> unlatching
Entriegelung *f* **der Ladung** <Lufttrans> load release *(Flugwesen)*

Entriegelungs... <Comp & DV, Kontroll> nonlocking
Entriegelungstaste f <Kontroll> unlock key
Entrinden n <Papier> barking
Entrindungstrommel f <Papier> barking drum
Entropie f <Comp & DV, Fertig, Mechan, Nichtfoss Energ, Phys, Telekom, Thermod> entropy
Entropiecode m <Telekom> variable-length code
Entropiefluss m <Thermod> entropic flux
Entrosten n <Maschinen> derusting
entsalzen v 1. <Chemtech> desalinate, desalinize, desalt; 2. <Wasserversorg> desalinate, desalinize
Entsalzen n <Chemtech> desalination, desalinization, desalting
Entsalzung f 1. <Chemtech> desalination, desalinization; demineralizing *(Hydrochemie)*; 2. <Fertig> desalinization; desalination *(Kunststoffinstallationen)*; 3. <Maschinen> desalting; 4. <Wasserversorg> desalination, desalinization
Entsalzungsanlage f 1. <Bau, Chemtech> desalination plant, desalinization plant; 2. <Wasserversorg> desalination plant, salt water plant
Entsalzungsreaktor m <Kerntech> desalination reactor
Entsalzungsverfahren n <Abfall> mineralization technique *(Abfälle werden so behandelt, dass sie wie Kies oder Sand abgelagert werden)*
Entsander m <Erdöl> desander *(Bohrtechnik)*
entsättigen v <Fernseh> desaturate
entsättigte Farben fpl <Fernseh> *(AE)* desaturated colors, *(BE)* desaturated colours
Entsäuerung f <Chemie> neutralization *(Ölraffination)*
entschalen v <Bau> dismantle
Entschärfung f von Hand <Raumfahrt> manual disarming *(Raumschiff)*
entschäumen v <Fertig> scum, skim
Entschäumen n <Fertig> scumming, skimming
Entschäumer m 1. <Lebensmittel> defoaming agent; 2. <Papier> defoamer
Entscheidung f <Comp & DV> decision
Entscheidung f **stimmhaft/stimmlos** <Telekom> voiced/unvoiced decision *(Sprachcodierung)*
Entscheidungsbaum m <Comp & DV, Künstl Int> decision tree, DT
Entscheidungsgehalt m <Comp & DV> decision content
Entscheidungsgeschwindigkeit f <Lufttrans> decision speed
Entscheidungsgraph m <Künstl Int> decision graph
Entscheidungshilfesystem n <Künstl Int> decision-support system, DSS
Entscheidungshöhe f <Lufttrans> decision height
Entscheidungssymbol n <Comp & DV> decision box
Entscheidungstabelle f <Comp & DV, Künstl Int> decision table
entscheidungsunterstützendes System n <Künstl Int> decision-support system, DSS
Entscheidungszeitpunkt m <Telekom> decision instant *(eines Digitalsignals)*
Entschieferung f <Erdöl> deslating
entschlammen v 1. <Kohlen> deslurry; 2. <Wasserversorg> scour
Entschlammer m <Erdöl> desilter
Entschlämmsieb n <Kohlen> depulping screen, desliming screen
Entschlämmung f <Kohlen> deslurrying
Entschleimung f <Lebensmittel> degumming
entschlichten v <Textil> desize
Entschlichten n <Textil> desizing
Entschlichtung f <Ker & Glas, Textil> desizing
entschlüsseln v 1. <Comp & DV> decode; 2. <Elektronik> decipher, decrypt; 3. <Telekom> decipher

Entschlüsselung f 1. <Comp & DV> decoding, decryption; 2. <Elektronik> deciphering; 3. <Raumfahrt> decoding *(Weltraumfunk)*; 4. <Telekom> deciphering
Entschlüsselungsvorlage f <Raumfahrt> filter mask *(Weltraumfunk)*
Entschrottung f <Abfall> scrap metal separation
Entschwärzen n **von Schlamm** <Abfall> de-inking
Entschwefelung f <Kohlen, Umweltschmutz> *(AE)* desulfurization, *(BE)* desulphurization
Entschweißen n <Textil> scouring *(Wolle)*
entscrambeln v <Telekom> unscramble
Entscrambler m <Telekom> unscrambler
entseuchen v 1. <Kerntech> decontaminate; 2. <Sicherheit> decontaminate, disinfest
Entseuchung f <Kerntech, Sicherheit> decontamination
Entseuchungsgrad m <Kerntech> degree of decontamination
entsilbern v <Metall> desilver
Entsilberung f <Metall> desilverization
Entsorgung f <Abfall, Wasserversorg> disposal
Entsorgung f **von Kernreaktoren** <Kerntech> nuclear reactor poison removal
Entsorgungsanlage f **an Land** <Meerschmutz> shore reception facility
Entsorgungseinrichtung f <Abfall> disposal facility
Entsorgungslogistik f <Umweltschmutz> logistics of disposal
Entsorgungstank m <Raumfahrt> disposal tank *(Raumschiff)*
Entsorgungsweg m <Abfall> disposal route
entspannen v 1. <Hydraul> expand *(Dampf)*; 2. <Textil> relax; 3. <Thermod> relieve stress
entspannte Faser f <Textil> *(AE)* relaxed fiber, *(BE)* relaxed fibre
entspannte Luft f <Heiz & Kälte> expanded air
Entspannung f 1. <Hydraul> expansion *(Dampf)*; 2. <Ker & Glas> stress relaxation; 3. <Metall> relaxation; 4. <Werkprüf> stress relief *(mechanisch)*
Entspannungsbehälter m <Sicherheit> blow-down tank
Entspannungsglühen n <Thermod> stress-relieving anneal
Entspannungsmittel n <Anstrich> surfactant *(Wasser)*
Entspannungsunterkühlung f <Kerntech> flash subcooling, flash undercooling
Entspannungsventil n <Heiz & Kälte, Hydraul> expansion valve
Entspannungsverdampfung f 1. <Kerntech> flash evaporation; 2. <Thermod, Wasserversorg> flash distillation
Entspannungszeit f <Metall> relaxation time
Entspannungszentrum n <Metall> *(AE)* relaxation center, *(BE)* relaxation centre
entspelzen v <Lebensmittel> husk *(Reis)*
Entsperren n <Comp & DV> unlocking
entspiegelt adj <Ker & Glas> bloomed
entspiegeltes Glas n <Ker & Glas> coated glass, nonreflecting glass
Entspiegelung f <Foto> blooming
entsprechen v <Math> agree
entsprechende Lufttüchtigkeitsanforderung f <Lufttrans> appropriate airworthiness requirement
entstandene Helligkeit f <Strahlphys> developed luminosity
entstauben v <Kohlen> dedust
Entstauber m 1. <Abfall> dust separator; 2. <Chemtech, Kohlen> dust collector
Entstaubungsanlage f 1. <Kunststoff> dust collector; 2. <Sicherheit, Umweltschmutz> dust collection equipment, dust exhaust system

Entstaubungsgerät

Entstaubungsgerät n 1. <Bau> dedusting unit; 2. <Sicherheit> dust exhaust appliance
Entstaubungssystem n <Sicherheit, Verpack> dust removal system
Entstaubungsvorrichtung f <Sicherheit> dust exhaust appliance
entstearinisieren v <Chemie> winterize
Entstehung f <Papier> formation
Entstipper m <Papier> deflaker
Entstördrossel f <Elektriz> suppressor choke
entstörend adj <Elektriz> anti-interference
Entstörer m <Elektriz, Elektronik, Elektrotech, Telekom> suppressor
Entstörfilter n <Fernseh, Telekom> interference filter
Entstörkondensator m <Elektriz, Elektrotech> decoupling capacitor, decoupling condenser, suppressor capacitor
Entstörung f 1. <Elektronik> interference rejection; 2. <Funktech> interference elimination; 3. <Telekom> fault clearance
Entstörvorrichtung f <Elektriz, Elektronik, Elektrotech, Telekom> suppressor
Enttonung f <Erdöl> deslating *(Bohrtechnik)*
enttrichtern v <Fertig> sprue *(Gießen)*
Enttrichtern n <Fertig> spruing
Enttrübung f <Wassertrans> anticlutter control *(Radar)*
Entwässerer m <Kerntech> dephlegmator
entwässern v 1. <Bau> drain; 2. <Fertig, Heiz & Kälte> dehydrate; 3. <Hydraul> dewater *(Pumpe)*; 4. <Kohlen, Papier> drain; 5. <Textil> dewater; 6. <Wasserversorg> drain
Entwässern n <Textil> dewatering
entwässernd adj/zum Ozean <Wasserversorg> exorheic
entwässert adj 1. <Chemie> anhydrous, dehydrated; 2. <Erdöl> anhydrous
entwässerter Abfall m <Abfall, Kerntech> dewatered waste
entwässerter Schlamm m <Abfall> dewatered sludge
Entwässerung f 1. <Bau> dewatering; 2. <Chemie> dehydration; 3. <Erdöl> water knock-out *(Gasförderung)*; 4. <Kohlen> drain; 5. <Maschinen, Papier> drainage; 6. <Wasserversorg> drainage, draining
Entwässerungsbauwerk n <Bau> drainage structure
Entwässerungsgefällestufe f <Lufttrans> drainage terrace
Entwässerungsgerinne n <Bau, Kohlen> drainage channel
Entwässerungsgraben m 1. <Bau> drain, drainage channel; 2. <Kohlen> drainage channel; 3. <Wasserversorg> drainage ditch
Entwässerungsgraben m mit Böschung <Wasserversorg> berm ditch
Entwässerungshahn m <Bau> drip cock
Entwässerungskanal m 1. <Nichtfoss Energ> *(AE)* dike, *(BE)* dyke; 2. <Wasserversorg> drainage channel, sewer
Entwässerungsloch n <Bau> weephole
Entwässerungsmittel n <Chemtech, Heiz & Kälte> dehydrator
Entwässerungspresse f <Papier> dewatering press
Entwässerungspumpe f <Wasserversorg> drainage pump, draining engine, draining pump
Entwässerungsrohr n <Bau> drain
Entwässerungsschicht f 1. <Abfall> drainage layer *(einer Deponie)*; 2. <Bau> pervious blanket
Entwässerungssieb n <Kohlen> draining screen
Entwässerungsstollen m <Wasserversorg> water adit
Entwässerungsventil n <Eisenbahn, Maschinen, Wassertrans> drain valve
Entwässerungsvorrichtung f <Wassertrans> drain

Entwässerungswalze f <Papier> dewatering roll
Entweichen n <Elektrotech> leakage
Entweichgeschwindigkeit f <Phys> escape velocity
entwerfen v 1. <Mechan> design; 2. <Textil> style
Entwerfen n 1. <Mechan> designing; 2. <Textil> styling
entwickeln v <Anstrich, Bau, Maschinen> develop
entwickeltes Bild n <Foto> developed picture
Entwickler m 1. <Anstrich> developer; 2. <Fertig> generator *(Schweißen)*; 3. <Foto> developer
Entwicklerflüssigkeit f <Foto> developer
Entwicklerrahmen m <Foto> tray
Entwicklerschale f <Foto> tray
Entwicklerzange f <Foto> print tongs
Entwicklung f 1. <Bau> development; 2. <Foto> processing; 3. <Maschinen> development; 4. <Phys> evolution
Entwicklungsbad n <Foto> developing bath
Entwicklungsbohrung f <Erdöl> development well, extension well
Entwicklungsinstrument n <Comp & DV> development toolkit
Entwicklungsklammer f <Foto> developing clip
Entwicklungsmaschine f <Foto> processing machine
Entwicklungsmuster n <Telekom> prototype
Entwicklungsrahmen m <Foto> developing frame
Entwicklungsspirale f <Foto> developing spiral
Entwicklungstank m <Foto> developing tank
Entwicklungstankthermometer n <Foto> developing tank thermometer
Entwicklungstrommel f <Foto> processing drum
Entwicklungswerkzeug n <Künstl Int> development tool
Entwicklungszange f <Foto> developing tongs
Entwurf m 1. <Comp & DV> design, layout; 2. <Heiz & Kälte> design
Entwurf f und Entwicklung f <Qual> design and development
Entwurfbüro n <Bau> design office
entwürfeln v <Telekom> descramble, unscramble
Entwürfeln n <Telekom> descrambling
Entwürfler m <Telekom> descrambler, unscrambler
Entwurfsautomatisierung f <Comp & DV> design automation
Entwurfshandbuch n <Telekom> designer handbook
Entwurfsprogramm n <Comp & DV> design aid
Entwurfsprüfung f <Qual> design review
Entwurfszeichnung f 1. <Heiz & Kälte> draft drawing; 2. <Konstzeich> preliminary drawing
Entzerrer m 1. <Aufnahme> equalizer; 2. <Comp & DV> equalizer, repeater; 3. <Elektronik> equalizer; 4. <Elektrotech> balancer, compensator; 5. <Telekom> equalizer
Entzerrer m mit Vorabfühlung <Aufnahme> presence equalizer
Entzerrerschaltung f <Telekom> equalizer circuit
Entzerrung f 1. <Akustik> de-emphasis, equalization; 2. <Aufnahme, Comp & DV, Elektronik> equalization; 3. <Fertig> correction *(Messgerät)*; 4. <Telekom> equalization
Entzerrung f durch quantisierte Rückkopplung <Telekom> quantized feedback egalization
Entzerrungskreis m <Lufttrans> antiresonant circuit
Entzerrungsschaltung f <Elektrotech> balancing network
entzinken v <Chemie> dezincify
Entzinkung f <Fertig> dezincification
Entzinnen n <Fertig, Metall> detinning
Entzinnung f <Abfall> detinning
entzündbar adj <Chemie, Fertig, Thermod> combustible
entzünden v <Umweltschutz> ignite
Entzünden n <Kunststoff> ignition
Entzunderer m <Metall> descaler

entzundern v 1. <Fertig> descale, scour; 2. <Mechan> descale
Entzundern n <Fertig> descaling, scouring
Entzündlichkeit f 1. <Chemie> ignitability; 2. <Thermod> combustibility
Entzündungsexperiment n <Kerntech> ignition experiment
Entzündungspunkt m <Thermod> flash point
Entzündungstemperatur f <Thermod> inflammation point, inflammation temperature
Enveloppe f <Mechan> envelope curve
Enzianblau n <Strömphys> gentian violet *(Strömungsvisualisierung)*
Enzym n <Lebensmittel> enzyme
Enzyminhibitor m <Lebensmittel> anti-enzyme
EOB *(Blockende)* <Comp & DV, Telekom> EOB *(end of block)*
EOD *(Datenende)* <Comp & DV, Telekom> EOD *(end of data)*
EOF *(Dateiende)* <Comp & DV, Telekom> EOF *(end of file)*
E-Ofen m <Fertig> electric furnace
EOM *(Nachrichtenende)* <Comp & DV, Telekom> EOM *(end of message)*
Eosin n <Chemie> eosin, tetrabromofluoresceine
EOT *(Bandende)* <Comp & DV> EOT *(end of tape)*
Eötvös-Gravitationswaage f <Phys> Eotvos balance
EP *(Höchstdruck)* <Maschinen> EP *(extreme pressure)*
Ephedrin n <Chemie> ephedrine
Ephemeriden fpl <Raumfahrt> ephemerides
Epichlorhydrin n <Chemie> epichlorhydrin
Epidiaskop n <Foto> epidiascope
Epikoprostanol n <Chemie> epicoprostanol
Epimer n <Chemie> epimer
Epimerisierung f <Chemie> epimerization
EPIRB *(Seenot-Funkbake mit Positionsmeldung)* <Funktech, Telekom, Wassertrans> emergency position-indicating radio beacon *(Funk)*
epitaktisch adj <Elektronik> epitactic
epitaktische Ablagerung f <Elektronik> epitaxial diffusion-junction transistor
epitaktisches Abscheiden n **aus Dampfphase** <Chemtech> *(AE)* vapor phase epitaxy, *(BE)* vapour phase epitaxy
Epitaxial... <Elektronik, Metall, Telekom> epitaxial
epitaxiale Siliziumschicht f <Elektronik> epitaxial silicon film
epitaxialer Siliziumplanartransistor m <Elektronik> silicon epitaxial planar transistor
epitaxiales Aufwachsen n <Elektronik> epitaxial layer deposition, epitaxial overgrowth
Epitaxialplanartransistor m <Elektronik> epitaxial diffusion-junction transistor
Epitaxialreaktor m <Elektronik> epitaxy reactor
Epitaxialschicht f <Elektronik, Telekom> epitaxial layer
Epitaxialtransistor m <Elektronik> epitaxial transistor
Epitaxialversetzung f <Metall> epitaxial dislocation
Epitaxialwafer m <Elektronik> epitaxial wafer
Epitaxie f 1. <Elektronik> epitaxy *(Aufwachstechnik)*; 2. <Metall, Strahlphys> epitaxy
epitaxisch adj <Elektronik> epitactic
Epitrochoide f <Geom> epitrochoid
Epizentrum n <Phys> *(AE)* epicenter, *(BE)* epicentre
epizyklisch adj <Fertig> epicycloidal
epizykloid adj <Geom> epicycloidal
epizykloidal adj <Mechan> epicycloidal
Epizykloide f <Fertig, Geom> epicycloid
epizykloidisch adj <Maschinen> epicycloidal
EPM *(Äquivalent je Million)* <Umweltschutz> EPM *(equivalent per million)*

EPNS *(versilberte Gegenstände)* <Metall> EPNS *(electroplated nickel silver)*
Epoxidharz n 1. <Bau, Chemie, Elektriz> epoxy resin; 2. <Fertig> epoxy resin *(Kunststoffinstallationen)*; 3. <Kunststoff, Verpack> epoxy resin
epoxidieren v <Chemie> peroxidize
epoxidiertes Öl n <Kunststoff> epoxidized oil
Epoxidmatrix f <Raumfahrt> epoxy matrix *(Raumschiff)*
Epoxy... <Bau, Chemie, Elektriz, Fertig, Kunststoff, Telekom, Verpack> epoxy
EPROM *(löschbarer programmierbarer Lesespeicher)* <Comp & DV> EPROM *(erasable programmable read-only memory)*
EP-Schmierstoff m <Maschinen> EP lubricant
Equalizer m <Elektronik, Fernseh> equalizer
Equalizerverstärker m <Fernseh> equalizing amplifier
Equilenin n <Chemie> equilenin
Er *(Erbium)* <Chemie> Er *(erbium)*
Erbeulen n <Mechan> buckling
Erbium n *(Er)* <Chemie> erbium *(Er)*
erbiumdotierter Faserverstärker m <Telekom> erbium-doped fiber amplifier *(Faseroptik)*
erblindet adj <Ker & Glas> struck
Erbsenkette f <Maschinen> beaded chain
Erbskohle f <Kohlen> pea coal, smalls
Erdanschluss m <Elektriz, Elektrotech> *(BE)* earth, *(AE)* ground, *(BE)* earth connection, *(AE)* ground connection, *(BE)* earth terminal, *(AE)* ground terminal
Erdantenne f <Funktech> ground antenna, terrestrial antenna
Erdanziehungskraft f <Phys> gravitation
Erdanziehungspotenzial n <Phys> gravitational potential
Erdarbeit f <Bau> digging
Erdaufschüttung f <Bau> earth fill
Erdausleuchtungsgebiet n <Lufttrans, Raumfahrt> earth coverage area
Erdbau m <Bau> earthwork
Erdbauarbeiten fpl <Bau> earthwork
Erdbaumaschinen fpl <Bau> earthworking machinery
Erdbeben n <Bau> earthquake
Erdbebenbelastung f <Bau, Kohlen> seismic load
Erdbebenkunde f <Phys> seismology
Erdbebenmesser m <Bau, Kohlen> seismograph
Erdbebenregistriergerät m <Phys> seismograph
erdbebensicher adj <Sicherheit> earthquake-resistant
erdbebensichere Bauweise f <Bau, Sicherheit> seismic design
erdbebensichere Bemessung f <Bau> seismic design
Erdbebensicherheitsuntersuchung f <Werkprüf> earthquake safety study
Erdbebensimulator m <Werkprüf> earthquake simulator
Erdbebenwarte f <Sicherheit> seismologic station
Erdbebenwelle f 1. <Akustik> Rayleigh wave; 2. <Wellphys> seismic wave
Erdbehälter m <Wasserversorg> earth reservoir
Erdbeobachtungssatellit m <Raumfahrt> earth observation satellite
Erdbeschleunigung f 1. <Mechan, Phys> acceleration; 2. <Raumfahrt> gravitational acceleration *(g)*
Erdbewegungsmaschinen fpl <Maschinen> earthmoving machinery
Erdbohrer m <Bau> auger
Erddamm m <Bau> bank
Erddammpfahl m <Kohlen> embankment pile
Erddammpfahltreiben n <Kohlen> embankment piling
Erddruck m 1. <Bau> earth pressure, soil pressure; 2. <Erdöl> geopressure *(Geologie)*; 3. <Kohlen> earth pressure, soil pressure

Erddruckbeiwert

Erddruckbeiwert m <Bau, Kohlen> earth pressure coefficient
Erddruckmessdose f <Bau> soil pressure gauge
Erde f <Funktech, Telekom> (BE) earth, (AE) ground • **an Erde gelegt** <Elektrotech> (BE) connected to earth, (AE) connected to ground, (BE) earthed, (AE) grounded • **an Erde legen** 1. <Elektriz> (BE) earth, (AE) ground; 2. <Telekom> (AE) ground • **mit Erde verbunden** <Elektrotech> (BE) connected to earth, (AE) connected to ground, (BE) earthed, (AE) grounded
Erdefunkstelle f 1. <Funktech> (BE) earth receiving station, (AE) ground receiving station; 2. <Raumfahrt> (BE) earth receiving station, (AE) ground receiving station (Weltraumfunk)
Erdeinsturz m <Bau> fall of earth
Erdelektrode f <Elektriz> (BE) earth rod, (AE) ground rod
erden v 1. <Elektriz, Elektrotech> (BE) earth, (AE) ground; 2. <Telekom> (AE) ground
Erderkundungssatellit m <Funktech, Telekom> earth exploration satellite
Erde-Satellit-Verbindung f <Lufttrans, Raumfahrt> earth-to-satellite link, uplink
Erdfehler m <Elektriz> (BE) earth fault, (AE) ground fault (an elektrischer Leitung)
Erdfehlerschutz m <Elektriz> (BE) earth fault protection, (AE) ground fault protection
Erdfehlerschutzeinrichtung f <Elektriz> (BE) earth fault protection, (AE) ground fault protection
Erdfehlerstrom m <Sicherheit> earth fault current
erdfeuchter Beton m <Bau> dry-packed concrete
Erdfixpunktumlaufbahn f <Raumfahrt> earth-parking orbit
Erdfluchtgeschwindigkeit f <Raumfahrt> earth escape velocity (zweite kosmische Geschwindigkeit)
Erdfluchtstufe f <Raumfahrt> earth escape stage (Raketenstufe mit zweiter kosmischer Geschwindigkeit)
erdfrei adj <Elektrotech> floating
Erdfunkenstrecke f <Elektriz> earth terminal arrester
Erdfunkstelle f 1. <Funktech> ground station; 2. <Raumfahrt> CES, coast earth station; earth station, land station (Weltraumfunk)
Erdfunkstelle f auf einem Schiff <Raumfahrt> shipborne earth station (Weltraumfunk)
Erdgas n <Erdöl, Heiz & Kälte, Thermod, Umweltschmutz> natural gas
Erdgasaustauschgas n (SNG) <Erdöl> synthetic natural gas (SNG)
Erdgasfahrzeug n (NGV) <Kfztech> natural gas vehicle, NVG
Erdgasfeld n <Thermod> gas field
Erdgaskondensat n <Erdöl> NGL, natural gas liquid
Erdgastanker m 1. <Trans> liquid natural gas carrier; 2. <Wassertrans> methane carrier
Erdgleiche f <Kerntech> ground level
Erdhobel m <Bau> scraper
Erdhügel m <Bau, Kohlen> knoll
Erdinduktionskompass m <Lufttrans> gyrosyn compass
Erdinduktionskompassanzeige f <Lufttrans> gyrosyn compass indicator
Erdkabel n 1. <Elektriz, Elektrotech> (BE) earth lead, (AE) ground lead, underground cable; 2. <Telekom> buried cable
Erdkegel m <Bau> dumpling (Straßenbau)
Erdkern m <Nichtfoss Energ> earth's core
Erdklemme f <Elektriz> (BE) earth clamp, (AE) ground clamp, (BE) earth clip, (AE) ground clip, (BE) earth terminal, (AE) ground terminal
Erdklumpen m <Bau> clod
Erdkommandostation f <Raumfahrt> command earth station

Erdkriechstrom m <Elektriz> (BE) earth leakage current, (AE) ground leakage current
Erdkrümmung f <Raumfahrt> earth curvature
Erdkruste f <Nichtfoss Energ> earth's crust
Erdlader m <Bau> scraper
Erdlast f <Fertig> backfill material (Kunststoffinstallationen)
Erdleiter m 1. <Elektriz> (BE) earth conductor, (AE) ground conductor; 2. <Elektrotech> (BE) earth line, (AE) ground line, (BE) earth wire, (AE) ground wire; 3. <Funktech, Telekom> (BE) earth wire, (AE) ground wire
Erdleitung f <Elektrotech> (BE) earth line, (AE) ground line, underground line
Erdmagnetfeld n <Nichtfoss Energ, Phys, Raumfahrt> earth's magnetic field
erdmagnetisches Feld n <Elektriz> earth's magnetic field
Erdmagnetismus m <Phys> geomagnetism, terrestrial magnetism
Erdnähe f <Phys> perigee
erdnahe Umlaufbahn f <Raumfahrt> low-earth orbit (LEO); near-earth orbit
Erdnetz n <Telekom> (BE) earth network, (AE) ground network
Erdoberfläche f <Raumfahrt> terrestrial surface
Erdöl n <Erdöl> crude, crude oil
Erdölanalyse f <Erdöl> crude assay, crude oil analysis
Erdölbegleitgas n <Erdöl> (AE) associated gas, (BE) associated petrol (Ölförderung)
Erdölbetrieb m <Erdöl> petroleum workings
Erdölbohrlocheinrichtung f <Erdöl> oil well appliance
Erdölbohrturm m <Erdöl> oil well derrick
Erdölbohrung f <Erdöl> oil well (Ölförderung)
Erdölerzeugnis n <Umweltschmutz> petroleum product
Erdölexploration f <Erdöl> oil exploration (Erdölsuche)
Erdölfalle f <Erdöl> oil trap (Erdölgeologie)
Erdölförderpumpe f <Erdöl> oil well pump (Ölförderung)
Erdölförderung f <Erdöl> drawing petroleum
erdölführend adj <Erdöl> oil-bearing
Erdölgas n <Erdöl> (AE) associated gas, (BE) associated petrol
Erdölgeologie f <Erdöl> petroleum geology
erdölhaltig adj <Erdöl> oil-bearing (Erdölgeologie)
Erdölingenieur m <Erdöl> petroleum engineer (Person)
Erdöllagerstätte f <Erdöl> oilfield, pool of petroleum; oil reservoir, petroleum reservoir (Erdölexploration)
Erdölraffinerie f <Erdöl> refinery; oil refinery, petroleum refinery (Ölindustrie)
Erdölverarbeitung f <Erdöl> oil refining (Ölindustrie, Raffinerie)
Erdorbit/im <Raumfahrt> earth-orbiting
Erdparkorbit m <Raumfahrt> earth-parking orbit
Erdpech n <Bau> mineral pitch
Erdplanum n <Bau> grade
Erdrückleitung f 1. <Elektrotech> ground return; 2. <Telekom> earth return
Erdruhedruck m <Bau, Kohlen> earth pressure at rest
Erdrutsch m <Trans> landslide, landslip
Erdsatellit m <Raumfahrt> earth satellite (Erforschung von Bodenschätzen)
Erdsatellit m zur Ferndatenaufnahme <Raumfahrt> earth remote-sensing satellite
Erdschalter m <Elektriz> earth switch
Erdschein m <Raumfahrt> earthshine
Erdschelle f <Elektriz> (BE) earth clamp, (AE) ground clamp
Erdschiene f <Elektriz> (BE) earth bus, (AE) ground bus, (BE) earthing bus, (AE) grounding bus
Erdschluss m 1. <Elektriz> (BE) earth fault, (AE) ground fault, (BE) earth leakage, (AE) ground leakage; (BE) earth

fault *(Fehler an elektrischer Leitung)*; 2. <Telekom> *(BE)* earth fault, *(AE)* ground fault
Erdschlussanzeiger *m* 1. <Elektriz> *(BE)* earth leakage detector, *(AE)* ground leakage detector; 2. <Elektrotech> *(BE)* earth indicator, *(AE)* ground indicator
Erdschlusslöschspule *f* <Elektrotech> arc suppression coil
Erdschlussmessgerät *n* <Elektriz> *(BE)* earth leakage meter, *(AE)* ground leakage meter
Erdschlussmeter *m* <Metrol> earth-leakage meter
Erdschlussortung *f* <Metrol> earth-fault location
Erdschlussprüfer *m* <Elektrotech> *(BE)* earth detector, *(AE)* ground detector, *(BE)* earth leakage indicator, *(AE)* ground leakage indicator
Erdschlussprüfung *f* <Elektriz, Metrol> earth-fault test, earth-leakage testing
Erdschlussreaktanz *f* <Elektrotech> neutral compensator
Erdschlussschutz *m* <Elektriz> *(BE)* earth-fault protection, earth-fault relaying, *(AE)* ground-fault protection, leakage protective system
Erdschlussstrom *m* <Elektriz> *(BE)* earth leakage current, *(AE)* ground leakage current
Erdschlussstromunterbrecher *m* <Elektriz> *(BE)* earth leakage circuit breaker, *(AE)* ground leakage circuit breaker
Erdschüttdamm *m* <Wasserversorg> earth dam
Erdsegment *n* 1. <Raumfahrt> earth segment *(Weltraumfunk)*; 2. <Telekom> earth segment
Erdsieb *n* <Kohlen> screen
Erdspannung *f* <Elektriz> earth potential
Erdspiralbohrer *m* <Kohlen> earth drill
Erdstab *m* <Elektriz> *(BE)* earth rod, *(AE)* ground rod
Erdstampfer *m* <Bau> earth rammer
Erdstaudamm *m* <Wasserversorg> earth dam
Erdstoffstruktur *f* <Bau, Kohlen> soil structure
Erdstoffzementgemisch *n* <Bau> soil-cement
Erdstrom *m* <Elektriz> *(BE)* earth current, *(AE)* ground current
Erdstromrelay *n* <Elektriz> earth fault relay
Erdstromschutzdrossel *f* <Elektriz> earthing reactor
erdsymmetrisch *adj* <Elektrotech> balanced to earth
erdsymmetrische Spannung *f* <Elektriz> balanced to earth voltage
erdsynchrone Umlaufbahn *f* <Raumfahrt> earth synchronous orbit
erdsynchroner Satellit *m* <Raumfahrt> earth synchronous satellite
Erdtaste *f (ET)* <Telekom> grounding key *(Telefon)*
Erdteer *m* <Bau> mineral tar
Erdumlaufbahn *f* <Raumfahrt> earth orbit
Erdumlaufbahntreffen *n* <Raumfahrt> earth orbit rendezvous
Erdumlaufecho *n* <Funkort> round-the-world echo
Erdumsegelung *f* <Wassertrans> circumnavigation
Erdung *f* 1. <Bau, Eisenbahn, Elektriz, Elektrotech, Funktech, Kfztech, Kohlen> *(BE)* earth, *(BE)* earthing, *(AE)* ground, *(AE)* grounding; 2. <Sicherheit> earthing installation; 3. <Wassertrans> *(BE)* earth, *(BE)* earthing, *(AE)* ground, *(AE)* grounding
Erdung *f* **des Flugzeugrumpfes** <Lufttrans> fuselage ground connection
Erdung *f* **des Flugzeugs** <Lufttrans> grounding of aircraft
Erdungsanschluss *m* <Eisenbahn, Kfztech, Wassertrans> *(BE)* earth connector, *(AE)* ground connector
Erdungsdraht *m* <Fernseh> *(BE)* earth wire, *(AE)* ground wire
Erdungseinstellung *f* <Elektriz> *(BE)* earthing position, *(AE)* grounding position *(Schalterstellung zur Erdung)*
Erdungselektrode *f* <Elektriz> *(BE)* earth electrode, *(AE)* ground electrode, *(BE)* earth rod, *(AE)* ground rod, *(BE)* earthing rod, *(AE)* grounding rod
Erdungskabel *n* 1. <Eisenbahn> *(BE)* earth cable, *(AE)* ground cable; 2. <Elektrotech> *(BE)* earth cable, *(BE)* earth lead, *(AE)* ground cable, *(AE)* ground lead; 3. <Kfztech, Wassertrans> *(BE)* earth cable, *(AE)* ground cable
Erdungsklemme *f* <Elektriz> earth clamp, earth terminal, earthing clamp, *(BE)* earthing clip, earthing terminal, *(AE)* grounding clip
Erdungskontakt *m* <Elektrotech> earth contact, earthing contact, earth-return brush *(Radsatzkontakt)*
Erdungsleiter *m* <Elektrotech, Funktech, Telekom> *(BE)* earth wire, *(AE)* ground wire
Erdungsleitung *f* <Elektrotech> *(BE)* earth line, ground cable, *(AE)* ground line
Erdungsmanschette *f* <Elektrotech> ground sleeve
Erdungsmessgerät *n* <Elektriz, Metrol> earth tester
Erdungsplatte *f* <Phys> *(BE)* earth plate, *(AE)* ground plate
Erdungsschalter *m* <Elektriz> earth switch, *(BE)* earthing switch, *(AE)* grounding switch
Erdungsschelle *f* <Elektriz> *(BE)* earthing clip, *(AE)* grounding clip
Erdungsschiene *f* <Elektriz> *(BE)* earth bar, *(AE)* ground bar, *(BE)* earth bus, *(AE)* ground bus
Erdungsschütz *n* <Elektriz> earthing contactor
Erdungsseil *n* <Elektriz> *(BE)* earth wire, *(AE)* ground wire
Erdungsstab *m* <Elektriz> *(BE)* earth electrode, *(AE)* ground electrode, *(BE)* earthing rod, *(AE)* grounding rod
Erdungsstange *f* <Lufttrans> *(BE)* earthing bar, *(AE)* grounding bar
Erdverkabelung *f* <Bau> underground cabling
erdverlegt *adj* <Bau> buried, underground
erdverlegte Schleife *f* <Trans> buried loop
erdverlegtes Kabel *n* <Telekom> buried cable
Erdverlegung *f* 1. <Bau> underground laying *(Leitungen)*; 2. <Erdöl> burial *(Rohrleitung)*
Erdvermessungskunde *f* <Geom> geodesy
Erdwärme *f* <Erdöl, Nichtfoss Energ> geothermics
Erdwiderstandsmessgerät *n* <Elektriz> *(BE)* earth resistance meter, *(AE)* ground resistance meter
Ereignis *n* 1. <Comp & DV> event; 2. <Phys> event *(in der Relativitätstheorie)*; 3. <Math> event *(Teilmenge des Stichprobenraumes)*; 4. <Teilphys> event
Ereignisbehandlung *f* <Comp & DV> event handling
ereignisbezogene Potenziale *npl* <Ergon> event-related potentials
Ereignisbit *n* <Comp & DV> event bit
ereignisgesteuert *adj* <Künstl Int> event-driven
Ereignisschreiber *m* 1. <Elektriz> event recorder; 2. <Gerät> event recorder, time recorder
Ereignisverfolgung *f* <Comp & DV> event trapping
Ereigniszähler *m* 1. <Elektriz> operation counter; 2. <Gerät> event counter
Ereigniszeit *f* <Phys> time of event
Erfahrungswissen *n* <Künstl Int> experiential knowledge
Erfassungsbereich *m* <Gerät> coverage
Erfassungssystem *n* <Kfztech> detection system
erfinden *v* <Patent> invent
Erfinder *m* <Patent> inventor
erfinderisch *adj* <Patent> inventive
erfinderische Leistung *f* <Patent> inventive merit
Erfindernennung *f* <Patent> designation of the inventor
Erfindung *f* <Patent> invention
erfolglose Verbindung *f* <Telekom> unsuccessful call, unsuccessful connection

erfolgloser Anruf *m* <Telekom> ineffective call
erfolgloser Verbindungsversuch *m* <Telekom> unsuccessful call attempt
erfolgreich abgewickelter Anruf *m* <Telekom> successful call
erfolgreiche Verbindung *f* <Telekom> successful call, successful connection
erfolgreicher Verbindungsversuch *m* <Telekom> successful call attempt
Erfolgsrate *f* <Comp & DV> yield
erforderliche Frequenz *f* <Nichtfoss Energ> required frequency
erforderliche Schlagzahl *f* <Bau> blow count
erforderliche Zulaufhöhe *f* <Kerntech> net positive suction head *(einer Pumpe)*
erforschen *v* <Math> explore
Erfüllung *f* **der Zielvorstellung** <Künstl Int> goal satisfaction
Erg *n* <Metrol> erg
ergänzend *adj* <Patent> supplementary
Ergänzung *f* <Comp & DV> attribute *(Programmiersprache)*
Ergänzungskegel *m* <Fertig> complementary cone
Ergänzungswinkel *m* <Math> conjugate angle
Ergänzungszeichnung *f* <Konstzeich> supplementary drawing
Ergebnis *n* 1. <Kerntech> issue; 2. <Maschinen> output; 3. <Math> result *(einer Berechnung)*; outcome *(eines Zufallsexperimentes)*; 4. <Qual> result
Ergebnis *n* **eines Zufallsexperiments** <Math> outcome *(Statistik)*
Ergebnisabweichung *f* <Qual> error of result
ergebnisloser Versuch *m* <Phys> inconclusive test
Ergol *n* <Raumfahrt, Thermod> ergol
Ergometer *n* <Metrol> ergometer
Ergonomie *f* <Comp & DV, Ergon, Maschinen, Raumfahrt, Sicherheit> ergonomics
ergonomisch *adj* 1. <Ergon> ergonomic; 2. <Maschinen> ergonomical; 3. <Verpack> ergonomic
ergonomisch gestaltet *adj* <Ergon> ergonomical, ergonomically designed
ergonomisch gestaltete Schutzvorrichtung *f* <Ergon, Sicherheit> ergonomic safety device
ergonomisch gestalteter Arbeitsplatz *m* <Ergon> ergonomical workplace, ergonomically designed workplace
Ergot-Alkaloid *n* <Chemie> ergot alkaloid
Ergotinin *n* <Chemie> ergotinine
Erguss *m* <Phys> effusion
Ergussgestein *n* <Nichtfoss Energ> extrusive rocks, igneous rocks
erhabene Fuge *f* <Ker & Glas> prominent joint
erhalten *v* 1. <Maschinen> maintain *(Werkzeug, Geräte)*; 2. <Mechan> maintain *(Werkzeug, Geräte)*; 3. <Sicherheit> receive *(Geldbetrag)*; 4. <Telekom> retain *(Daten)*
• **Netzwerkbandbreite erhalten** <Comp & DV> conserve network bandwidth
Erhalten *n* <Raumfahrt> acquisition
Erhaltung *f* 1. <Elektrotech> holding *(bei Thyristoren)*; 2. <Erdöl, Phys, Strömphys, Thermod> conservation
Erhaltung *f* **der rotationsfreien Bewegung** <Strömphys> permanence of irrotational motion
Erhaltung *f* **der Strahlungsdichte** <Telekom> conservation of radiance
Erhaltungsstützen *fpl* <Thermod> conservation laws
erhärten *v* <Bau> harden; freeze *(Beton)*
Erhärten *n* <Bau> hardening
Erhärtung *f* <Bau> seasoning *(Beton)*
erheben *v* <Bau> rise
Erhebung *f* <Fertig> prominence

Erhebungswinkel *m* 1. <Mechan> quadrant angle; 2. <Nichtfoss Energ> angle of incidence; 3. <Telekom> elevation angle
Erhebungswinkel *m* **der Sonne** <Nichtfoss Energ> solar altitude angle
erhitzen *v* <Textil, Thermod> heat
Erhitzen *n* <Textil, Thermod> heating
Erhitzer *m* <Mechan> heater
Erhitzung *f* <Chemie> calefaction
Erhitzung *f* **durch Luftreibung** <Raumfahrt> air friction heating *(Raumschiff)*
Erhitzungsgeschwindigkeit *f* <Thermod> rate of heating
Erhitzungskurve *f* <Thermod> heating curve
erhöhen *v* 1. <Bau> raise; 2. <Fernseh> bump up *(Zuschauerrate)*
Erhöhen *n* <Bau> heightening; keying-up *(Angebot)*
erhöhte Sicherheit *f* <Sicherheit> increased safety
erhöhte Straße *f* <Trans> (BE) flyover, (AE) skyway
erhöhte Temperatur *f* <Thermod> excess temperature
erhöhter Durchlass *m* <Ker & Glas> lifted throat
erhöhter Fußweg *m* <Wassertrans> causeway
erhöhter Wasserspiegel *m* <Bau, Kohlen, Nichtfoss Energ> raised water table
Erhöhung *f* <Telekom> enhancement
Erhöhungswinkel *m* <Nichtfoss Energ> angle of incidence
Erhöhungszeichen *n* <Akustik> sharp
Erholspannung *f* <Elektriz> recovery voltage
Erholung *f* 1. <Elektrotech> recovery *(von dynamischen Lasten)*; 2. <Metall, Textil> recovery
Erholungsgrad *m* <Metall> recovery rate
Erichsen'scher Tiefziehversuch *m* <Fertig> Erichsen-type ductility test
Erinnerungsalarmdienst *m* <Telekom> reminder alarm service
Erinnerungsanruf *m* <Telekom> reminder call
erkanntes Signal *n* <Elektronik> detected signal
erkennen *v* 1. <Elektronik> detect; 2. <Kontroll> identify
Erkennen *n* 1. <Ergon> cognition, recognition; 2. <Raumfahrt> detection
Erkennen *n* **einzelner Wörter** <Künstl Int> isolated words recognition
Erkennen *n* **fließend gesprochener Sprache** <Künstl Int> connected-speech recognition, continuous speech recognition
Erkennen *n* **fließender Sprache** <Künstl Int, Telekom> connected speech recognition, continuous speech recognition
Erkennen *n* **gebundener Rede** <Künstl Int> connected-speech recognition, continuous speech recognition
Erkennen *n* **kontinuierlicher Sprechsprache** <Künstl Int> connected-speech recognition, continuous speech recognition
Erkenntnis *n* <Künstl Int> cognition
Erkennung *f* 1. <Akustik> identification, recognition; 2. <Comp & DV> recognition; 3. <Elektronik> detection; 4. <Kontroll> identification; 5. <Künstl Int> recognition
Erkennung *f* **kontinuierlicher Sprechsprache** <Künstl Int> isolated words recognition
Erkennungsfunkbake *f* <Funktech> identification beacon
Erkennungssystem *n* <Künstl Int> recognition system
Erkennungsteil *n* <Comp & DV> identification division
Erkennungsvorrichtung *f* **für fehlende Kappen** <Verpack> missing cap detector
Erkennweite *f* <Eisenbahn> sighting distance
Erkerfenster *n* <Bau> bay window
erklären *v/***für gültig** <Comp & DV> validate
erklären *v/***für nichtig** <Patent> revoke

erklärende Anweisung f <Comp & DV> narrative statement
Erklärung f <Comp & DV> narrative
Erklärungskomponente f <Künstl Int> explanation subsystem, explanation component *(eines Expertensystems)*
Erklärungsteil n <Künstl Int> explanation subsystem, explanation component *(eines Expertensystems)*
Erl *(Erlang)* <Telekom> Erl *(Erlang)*
Erlang n *(Erl)* <Telekom> Erlang, Erl *(Einheit des Verkehrswertes)*
erlaubter Elektrondipolübergang m <Strahlphys> allowed electron dipole transition
erlaubter Übergang m <Strahlphys> allowed transition
erlaubtes Energieband n <Strahlphys> allowed energy band
erleichtern v <Bau, Qual> ease
Erleichterungsloch n <Lufttrans> lightening hole
Erlenmeyer-Kolben m <Labor> Erlenmeyer flask, conical flask
erlöschen v <Elektrotech> go out *(Licht)*
Ermächtigungsbescheinigung f <Patent, Qual> certificate of authorization
Ermangelungsschließen n <Künstl Int> default reasoning
ermäßigte Gesprächsgebühr f <Telekom> cheap call rate
ermäßigter Tarif m <Telekom> reduced rate
ermittelbar adj <Math> ascertainable
ermitteln v 1. <Math> ascertain; 2. <Qual> identify; investigate
Ermitteln n böswilliger Anrufe <Telekom> malicious call tracing
Ermittlung f 1. <Comp & DV> identification; 2. <Math> ascertainment
Ermittlung f der Wortzahl <Comp & DV> word count
Ermittlungsergebnis n <Qual> result of determination
ermüden v <Anstrich, Metall> fatigue
Ermüdung f <Anstrich, Bau, Kunststoff, Maschinen, Mechan, Metall, Werkprüf> fatigue
Ermüdung f bei niedriger Lastspielzahl <Metall> low-cycle fatigue
Ermüdungsanriss m <Metall> fatigue crack
Ermüdungsausfall m <Kontroll> wearout failure
Ermüdungsbruch m <Metall> fatigue failure
Ermüdungseigenschaften fpl <Mechan, Qual> fatigue properties
Ermüdungsfestigkeit f 1. <Qual> fatigue resistance; 2. <Werkprüf> fatigue resistance, fatigue strength
Ermüdungshärten n <Metall> fatigue hardening
Ermüdungsprüfung f <Werkprüf> fatigue test
Ermüdungsriss m <Werkprüf> fatigue crack
Ermüdungsverhalten n <Maschinen, Metall> *(AE)* fatigue behavior, *(BE)* fatigue behaviour
Ermüdungsverschleiß m <Maschinen> fatigue wear
Ermüdungsversuch m 1. <Metall> fatigue test; 2. <Werkprüf> endurance test
Ermüdungsvorriss m <Kerntech> fatigue precrack
Ermüdungsweichmachen n <Metall> fatigue softening
Ernährung f <Lebensmittel> nutrition
Ernährungsergänzungsstoff m <Lebensmittel> nutritional supplement
Ernährungsstörung f <Lebensmittel> nutritional disorder
Ernährungsumstellung f <Lebensmittel> dietary change
erneuerbar adj <Nichtfoss Energ> renewable
erneuerbare Energiequelle f <Elektriz> renewable energy source
Erneuerung f der Qualifikationen <Qual> requalification
erneut anzeigen v <Comp & DV> refresh

erneut speichern v <Comp & DV> resave
erneute Qualifizierung f <Qual> requalification
erneute Verbindung f <Comp & DV> session restart
erneuter Schreibvorgang m <Comp & DV> rewrite
Erniedrigungszeichen n <Akustik> flat *(Musik)*
erodieren v <Anstrich, Bau, Elektriz, Kohlen> erode
Erodiermaschine f <Fertig> eroding machine
Erosion f <Bau, Kohlen, Nichtfoss Energ> erosion
Erosionsabbrand m <Raumfahrt> erosive burning
Erosionsrate f <Raumfahrt> erosion rate *(Raumschiff)*
erproben v <Qual> test
Erprobung f <Gerät, Kerntech> trial
Erprobungsflug m <Lufttrans> proving flight, test flight
errechnete Adresse f <Comp & DV> generated address
errechnete Flügelfläche f <Lufttrans> design wing area
errechnete Fluggeschwindigkeit f <Lufttrans> design airspeed
errechnete Geschwindigkeit f bei größter Böenintensität <Lufttrans> design speed for maximum gust intensity *(Lufttüchtigkeit)*
errechnete Landegeschwindigkeit f <Lufttrans> design-landing speed
errechnete Landeklappengeschwindigkeit f <Lufttrans> design flap speed *(Lufttüchtigkeit)*
errechnete Last f <Lufttrans> design load
errechnete Radbelastung f <Lufttrans> design wheel load
errechnete Reisefluggeschwindigkeit f <Lufttrans> design-cruising speed
errechnete Startmasse f <Lufttrans> design takeoff mass
errechnete Sturzfluggeschwindigkeit f <Lufttrans> design-diving speed
errechnete Tragflügelfläche f <Lufttrans> design wing area
errechneter Raddruck m <Lufttrans> design wheel load
errechnetes Fluggewicht n <Lufttrans> design flight weight
errechnetes Gewicht n <Lufttrans> design weight
errechnetes Landegewicht n <Lufttrans> design-landing weight
errechnetes Rollgewicht n <Lufttrans> design taxi weight
erregen v <Elektrotech> energize, excite
Erreger m <Elektriz, Elektrotech, Nichtfoss Energ> exciter
Erregeranode f <Elektrotech> excitation anode, keep-alive electrode
Erregerdynamo m <Elektriz> exciting dynamo
Erregerfeld n <Elektrotech> exciting field
Erregerfrequenz f <Werkprüf> excitation frequency
Erregerkreis m <Elektriz, Elektrotech> field circuit, energizing circuit
Erregermaschine f <Elektrotech> exciter
Erregerspule f <Elektriz> field coil
Erregerstrom m <Elektriz, Elektrotech> excitation current, field current, energizing current, induction current
Erregerstufe f <Funktech> exciter *(Sender)*
Erregerwicklung f <Elektriz> excitation winding, field coil
Erregerwiderstand m <Elektrotech> field rheostat
erregt adj <Elektrotech> energized
Erregung f 1. <Akustik, Elektriz> excitation; 2. <Elektrotech> energization, excitation; 3. <Ergon> arousal; 4. <Telekom> excitation
Erregung f durch Strahlungskopplung <Funktech> parasitic excitation
Erregungsgeschwindigkeit f <Elektriz> excitation response
Erregungspegel m <Akustik> excitation level
erreichen v <Telekom> reach *(Teilnehmer)*

Erreichen

Erreichen n <Raumfahrt> acquisition
Erreichen n **der Fluglage** <Raumfahrt> acquisition of attitude *(Raumschiff)*
Erreichen n **der Umlaufbahn** <Raumfahrt> acquisition of orbit *(Raumschiff)*
Erreichen n **des Normalmodus** <Raumfahrt> acquisition of normal mode *(Raumschiff)*
erreichter Messwert m <Gerät> achieved-measuring value
erreichter Wert m <Gerät> achieved-measuring value
errichten v 1. <Bau> construct, put up; 2. <Telekom> establish *(Netzwerk)*
Errichtung f <Bau> erection
Ersatz m 1. <Lufttrans> reserves; 2. <Telekom> stand-by
Ersatzboot n <Erdöl> stand-by boat *(Schifffahrt)*
Ersatzglied n <Maschinen> repair link
Ersatzkreis m <Elektrotech> *(AE)* analog circuit, *(BE)* analogue circuit
Ersatzlampe f <Foto> spare bulb
Ersatzlast f <Funktech> dummy load
Ersatzleitweg m <Telekom> alternative route
Ersatzmittel n <Lebensmittel> surrogate
Ersatzplatte f <Comp & DV> spare disk
Ersatzrad n <Kfztech> spare wheel
Ersatzrädersatz m <Maschinen> set of change wheels
Ersatzschaltgerät n *(EsG)* <Comp & DV> stand-by equipment
Ersatzschaltung f 1. <Elektriz, Elektronik> equivalent circuit; 2. <Elektrotech> analog circuit, equivalent circuit; 3. <Funktech, Phys, Telekom> equivalent circuit
Ersatzstromkreis m <Elektrotech> equivalent circuit
Ersatzteil n 1. <Eisenbahn> spare part; 2. <Fertig> spare part *(Kunststoffinstallationen)*; 3. <Kfztech, Lufttrans, Maschinen, Wassertrans> spare part
Ersatzteile npl <Maschinen> spares
Ersatzweg m <Telekom> alternative route
Ersatzwerkzeug n <Maschinen> spare tool
Ersatzwiderstand m <Elektriz> equivalent resistance
Ersatzzeichen n <Comp & DV> substitute character
erschließen v <Bau> develop *(von Bauland)*
Erschließung f 1. <Bau> development *(Bauland)*; 2. <Erdöl> development *(Lagerstätte)*
Erschließungsphase f <Erdöl> development phase *(Lagerstätte)*
Erschließungsvorhaben n <Erdöl> development project *(Lagerstätte)*
erschmolzenes Glas n/**aus Scherben** 1. <Ker & Glas> glass melted from cullet; 2. <Ker & Glas> body
erschmolzenes Glas n/**nur aus Gemenge** <Ker & Glas> glass melted from batch only
erschmolzenes Opalglas n/**im Hafenofen** <Ker & Glas> pot opal
erschmolzenes Rubinglas n/**im Hafenofen** <Ker & Glas> pot ruby
erschöpfbare Energiequelle f <Kohlen, Nichtfoss Energ> depletable energy source
erschöpfend adj <Math> exhaustive, sufficient
erschöpfende Schätzfunktion f <Math> sufficient estimator *(enthält alle Informationen der Stichprobe über einen Parameter)*
erschöpfte Batteriekapazität f <Comp & DV> low battery charge
erschöpfter Kernbrennstoff m <Kerntech> depleted nuclear fuel
Erschöpfung f <Wasserversorg> depletion
Erschütterung f 1. <Phys> vibration; 2. <Raumfahrt> judder, shock *(Raumschiff)*
Erschütterungsdämpfer m <Lufttrans> vibration damper
Erschütterungsfestigkeit f <Kunststoff> shock resistance

Erschütterungsprobe f <Verpack> jarring test
erschwerte Prüfung f <Qual> tightened inspection
ersetzen v <Math> substitute
ersoffene Turbine f <Hydraul> drowned turbine
Erspinnen n <Textil> spinning
Erst... <Raumfahrt> initial
erstarren v 1. <Abfall, Bau> set; 2. <Fertig> freeze *(Metall)*; 3. <Fertig> congeal
Erstarrung f 1. <Abfall, Bau> setting; 2. <Chemtech> coagulation, gelation
Erstarrungsbad n <Chemtech> coagulating bath
Erstarrungsflüssigkeit f <Chemtech> coagulation liquid
Erstarrungsgeschwindigkeit f 1. <Bau> rate of curing; 2. <Ker & Glas> setting rate
Erstarrungspunkt m 1. <Strömphys> congealing point; 2. <Thermod> solidification point
Erstarrungszeit f <Bau, Kunststoff> setting time
Erstattung f <Patent> reimbursement
Erstausbau m <Telekom> initially-installed capacity *(Amt)*
Erstausrüstung f <Maschinen> original equipment
Erstbelegung f <Comp & DV> initialization *(Programm)*
Erstbündel n <Telekom> first-choice group
erste Harmonische f <Elektronik> first harmonic
erste Ionisationsstufe f <Phys> first ionization potential
erste Lage f **der Y-Schweißnaht** <Mechan> root pass
erste Projektionsebene f <Geom> ground plane
erste Schutzschicht f <Fertig> primary coating
erste Zwischenfrequenz f <Elektronik, Funktech> first intermediate frequency
Erste-Hilfe-Kasten m <Sicherheit> first-aid box
Erste-Hilfe-Raum m <Sicherheit> first-aid treatment room
Erste-Hilfe-Schrank m <Sicherheit> first-aid cupboard
erstellen v <Comp & DV> create, load
Ersteller m <Konstzeich> *(AE)* draftsman, *(BE)* draughtsman
Erstellung f <Comp & DV> creation, preparation
Erstellung f **von Modellen** <Comp & DV, Elektronik, Geom> *(AE)* modeling, *(BE)* modelling
Erstellungsnummer f <Comp & DV> generation number
erster Detektor m <Elektronik> first detector, first mixer *(in Überlagerungsempfängern)*
erster Hauptsatz m **der Wärmelehre** <Phys> first law of thermodynamics
erster Mischer m <Elektronik, Funktech> first mixer
erster Nassfilz m <Papier> first dryer
erster Oxidationsbrand m <Ker & Glas> first oxidizing firing
erster Summand m <Comp & DV> augend
erster Überlagerungsoszillator m <Elektronik, Funktech> first local oscillator
erster Vergaserlufttrichter m <Kfztech> primary barrel
erster ZF-Verstärker m *(erster Zwischenfrequenzverstärker)* <Elektronik, Funktech> first IF amplifier *(first intermediate frequency amplifier)*
erster Zwischenfrequenzverstärker m *(erster ZF-Verstärker)* <Elektronik, Funktech> first intermediate frequency amplifier *(first IF amplifier)*
erstes Anfahren n <Kerntech> beginning of life
erstes kritisches Experiment n <Kerntech> first critical experiment
erstes Kritischwerden n <Kerntech> first criticality, first divergence
erstes Transmissionsfenster n <Optik> *(AE)* first fiber window, *(BE)* first fibre window *(Lichtleitfaser)*
erstes Walzenpaar n <Ker & Glas> first pair of rollers
ersticken v <Thermod> smother *(Feuer)*
Erstickung f <Sicherheit> suffocation *(Dämpfe)*
Erstmahlen n <Kohlen> primary grinding
Erstneutron n <Kerntech> first-flight neutron

Erstprobe f <Kohlen> primary sample
Erstprüfung f <Qual> original inspection
erstrecken v **von/sich** <Telekom> range from
Erstsetzung f <Kohlen> primary settlement
Erstziehen n <Maschinen> first drawing
erteilen v <Wassertrans> admit
Erteilung f <Patent> grant
Ertrag m <Chemie, Elektronik, Kohlen, Nichtfoss Energ, Textil> yield
Eruca... <Chemie> erucic
Eruption f <Erdöl> breakout; blowout (Bohrtechnik)
erwärmen v <Textil, Thermod> heat
Erwärmen n <Textil, Thermod> heating
Erwärmung f <Thermod> temperature rise
Erwärmungsprüfung f <Elektriz, Metrol> heating test, temperature rise test
erwartete Betriebsbedingungen fpl <Lufttrans> anticipated-operating conditions
erwarteter Wert m <Math> expected value
erwartetes Potenzial n <Phys> advanced potential
Erwartung f <Math> expectation (Mittelung)
Erwartungsbohrung f <Erdöl> appraisal well (Lagerstättenerkundung)
erwartungstreu adv <Math> unbiased (unverzerrte Schätzfunktion)
Erwartungswert m <Qual> expected value
erweichen v <Fertig> plasticize
Erweichung f 1. <Fertig> plasticization; 2. <Ker & Glas> sagging (Keramik)
Erweichungsintervall n <Ker & Glas> softening range
Erweichungsofen m <Ker & Glas> softening furnace
Erweichungspunkt m 1. <Heiz & Kälte> softening point; 2. <Ker & Glas> deformation point; 3. <Kunststoff, Textil> softening point
Erweichungstemperatur f 1. <Kunststoff> softening point; 2. <Werkprüf> heat distortion temperature (Kunststoffe)
erweiterbar adj 1. <Bau> extensible; 2. <Comp & DV> upgradable; 3. <Fertig> extensible
erweiterbare Adressierung f <Comp & DV> extensible addressing
erweiterbare Sprache f <Comp & DV> extensible language
erweiterbares Etikettiersystem n <Verpack> (AE) modular labeling system, (BE) modular labelling system
Erweiterbarkeit f <Comp & DV> extensibility
erweitern v 1. <Bau> expand; 2. <Comp & DV> extend, scale; upgrade; 3. <Math> extend (von Brüchen); 4. <Mechan> ream
erweitert adj <Mechan> flared
erweiterte Adressierung f <Comp & DV> extended addressing
erweiterte Bohrung f <Erdöl> development well
erweiterte IDE-Schnittstelle f <Telekom> AT attachment packet interface, ATAPI (PC-Bus-Schnittstelle); extended IDE, EIDE (Schnittstelle für PC-Plattenspieler)
erweiterte indizierte Zugriffsmöglichkeit f **für sequenzielle Dateien** (QUISAM) <Comp & DV> Queued Unique Index Sequential Access Method
erweiterte Stummelwelle f/**fasenringartig nach außen** <Lufttrans> bevel ring-flared stub shaft
erweiterte Transaktionssysteme npl <Comp & DV> extended transaction systems
erweiterte Zugriffsmethode f <Comp & DV> queued access method
erweiterte Zugriffsmöglichkeit f **für sequenzielle Dateien** (QSAM) <Comp & DV> queued sequential access method (QSAM)

erweiterter BCD-Code m **für Datenaustausch** (EBCDIC-Code) <Comp & DV, Elektronik, Telekom> extended binary-coded decimal interchange code (EBCDIC)
erweiterter Dienst m <Telekom> enhanced service
Erweiterung f 1. <Bau> development; 2. <Comp & DV> upgrade (Hardware); 3. <Druck> expansion; 4. <Textil> flare
Erweiterungsanweisung f <Comp & DV> option instruction
Erweiterungsbohren n <Erdöl> appraisal drilling (Lagerstättenerkundung)
Erweiterungsbohrer m <Erdöl> reaming bit, shell (Bohrtechnik)
Erweiterungsbohrmeißel m <Erdöl> eccentric bit; expansion bit (Bohrtechnik)
Erweiterungsbohrung f <Erdöl> extension well, outpost well (Bohrtechnik); outpost well (Lagerstättenabbau)
Erweiterungsfähigkeit f <Telekom> open-endedness (System)
Erweiterungsfeld n <Konstzeich> extension block
Erweiterungskarte f 1. <Comp & DV> expansion card, feature expansion card; 2. <Elektronik> extension card
Erweiterungsmöglichkeit f <Telekom> scalability
Erweiterungsnetz n <Telekom> expansion network
Erweiterungsplatine f <Comp & DV> expansion board, expansion card
Erweiterungsposition f <Comp & DV> expansion slot
Erweiterungsring m <Erdöl> expansion ring (Bohrtechnik)
Erweiterungsspeicher m (XMS) <Comp & DV> extended memory specification (XMS)
Erweiterungsspeicherröhre f <Elektronik> expansion storage tube
Erweiterungssteckkarte f <Comp & DV> plug-in board
Erweiterungssteckplatz m <Comp & DV> expansion slot
Erwerb m <Künstl Int> acquisition (des Wissens)
Erythrin n <Chemie> erythrine, tetraiodofluorescein
Erythrit m <Chemie> erythritol
Erythrittetranitrat n <Chemie> tetranitrol
Erythrose f <Chemie> erythrose
Erythrosin n <Chemie> erythrosine
Erythrulose f <Chemie> erythrulose
Erz n <Fertig, Kohlen> ore
Erzanalyse f <Kerntech> ore assaying, ore testing
Erzanreicherungsanlage f <Kerntech> ore enrichment plant
Erzaufbereitung f <Kohlen> mineral processing
Erz-Bulk-Öl-Frachter m <Wassertrans> combination bulk carrier
erzen v <Fertig> ore down
Erzen n <Fertig> oreing down
erzeugen v 1. <Bau> prepare; 2. <Geom> generate; 3. <Heiz & Kälte> develop (Druck)
Erzeugende f <Geom> generatrix
Erzeuger m <Funktech> generator
Erzeugnis n 1. <Comp & DV, Fertig> product; 2. <Papier> make
erzeugte Elektrizität f <Kerntech> generated electricity
erzeugter elektromagnetischer Impuls m/**vom System** <Elektrotech, Elektronik> system generated electromagnetic pulse
Erzeugungskopie f <Fernseh> generation copy
Erzfrachter m <Wassertrans> ore carrier (Schifftyp)
erzielte Qualität f <Qual> quality achievement
Erz-Kohle-Öl-Frachtschiff n <Trans> ore-coal-oil carrier
Erzlager n <Kerntech> ore deposit
Erzniere f <Kohlen> nodule
Erz-Öl-Tanker m <Wassertrans> ore-oil carrier (Schiff)
Erz-Schlamm-Öl n (OSO) <Wassertrans> ore-slurry-oil (OSO)

Erz-Schlamm-Öl-Tanker *m* <Wassertrans> ore-slurry-oil tanker
Erz-Schüttgut-Öl *n (OBO)* <Wassertrans> ore-bulk oil *(OBO)*
Erz-Schüttgut-Öl-Frachter *m* <Wassertrans> ore-bulk-oil carrier *(Schiff)*
Erzvorkommen *n* <Kerntech> ore deposit
erzwungene Emission *f* <Elektronik, Phys> stimulated emission
erzwungene Kommutierung *f* <Elektrotech> forced commutation
erzwungene Konvektion *f* <Phys> forced convection
erzwungene Schwingung *f* 1. <Akustik> forced vibration; 2. <Elektronik> forced oscillation; 3. <Phys> forced oscillation *(meist elektrisch)*; forced vibration *(meist mechanisch)*
Es *(Einsteinium)* <Chemie, Strahlphys> Es *(einsteinium)*
ES 1. <Elektronik, Kerntech> *(Elektronenstrahl)* EB *(electronic beam)*; 2. <Künstl Int> *(Expertensystem)* ES *(expert system)*
ESA 1. <Raumfahrt> *(Europäische Raumfahrtbehörde)* ESA *(European Space Agency)*; 2. <Umweltschmutz> *(elektrostatischer Staubabscheider)* ESP *(electrostatic precipitator)*
Esaki-Diode *f* 1. <Elektronik> Esaki diode; 2. <Phys> Esaki diode, tunnel diode
ESCA *(Photoelektronenspektroskopie)* <Phys> ESCA *(electron spectroscopy for chemical analysis)*
Escape-Peak *m* <Strahlphys> escape peak *(bei Gamma-Strahlung)*
Escape-Sequenz *f* <Comp & DV> escape sequence
Escape-Zeichen *n* <Comp & DV> escape character
Escape-Zeichenfolge *f* <Comp & DV> escape sequence
E-Schicht *f* <Funktech, Phys> E-layer, Heaviside layer
E-Schmelzung *f* <Elektrotech> electric smelting
Eserin *n* <Chemie> eserine, physostigmine
ESPRIT <Elektriz> ESPRIT, European Semiconductor Production Research Initiative
ESR *(Elektronenspinresonanz)* <Phys, Strahlphys, Teilphys> ESR *(electron spin resonance)*
essenzielle Fettsäure *f (EFS)* <Lebensmittel> essential fatty acid *(EFA)*
essigartig *adj* <Chemie> acetic, acetous
Essigerzeuger *m* <Lebensmittel> vinegar generator
Essigester *m* <Lebensmittel> ethyl acetate
essigsauer *adj* <Chemie> acetous
Essigsäure *f* 1. <Fertig> acetic acid *(Kunststoffinstallationen)*; 2. <Kunststoff, Lebensmittel> acetic acid
Essigsäureanhydrid *n* <Lebensmittel> acetic anhydride
Essigsäurebacterium *n* <Lebensmittel> acetobacter
Essigsäurebakterien *fpl* <Lebensmittel> acetic bacteria
Essigsäurebildung *f* <Chemie> acetification
Essigsäureester *m* <Chemie> acetate
Essigsäureethylester *m* <Lebensmittel> acetic ether
Essigsäuregärung *f* <Lebensmittel> acetic fermentation
Essigsäuremethylester *m* <Chemie> methyl acetate
Ester *m* <Chemie> ester
Ester *m* **der Arsensäure** <Chemie> arsenate
Estergummi *m* <Kunststoff> ester gum
Esterharz *n* <Kunststoff> ester gum
Estrich *f* <Bau> floor paving
Estrichstärke *f* <Bau> screed height
Estron *n* <Chemie> oestrone, theelin
ETA *(voraussichtliche Ankunftszeit)* <Lufttrans, Wassertrans> ETA *(estimated time of arrival)*
Eta-Faktor *m* <Phys> eta-factor *(Neutronenausbeute bei Absorption)*
Etage *f* <Bau> floor
Etagenbogen *m* <Erdöl> dogleg *(Rohrleitung)*

Etagenhöhe *f* <Kunststoff> daylight
Etagenofen *m* <Abfall> multiple-hearth incinerator
Etagenpresse *f* <Kunststoff> daylight press, multiple-daylight press
Etagenschalter *m* <Elektriz> floor switch, landing switch
Etagentrockenpartie *f* <Papier> stacked dryer section
Etagenwagen *m* <Lufttrans> service trolley *(Werkstatt)*
Etagenwohnung *f* <Bau> apartment
Etalon *n* <Phys> etalon, primary standard, standard
Eta-Mason *n* <Phys> eta mason
ETD *(voraussichtliche Abflugzeit)* <Lufttrans, Wassertrans> ETD *(estimated time of departure)*
Ethal *n* <Chemie> ethal
Ethan *n* 1. <Chemie> ethane, hexadecanol; 2. <Erdöl> ethane *(Petrochemie)*
Ethanal *n* 1. <Chemie> ethanal; 2. <Lebensmittel> acetaldehyde
Ethandiamid *n* <Chemie> oxamide
Ethandinitril *n* <Chemie> cyanogen
Ethandioyl... <Chemie> oxalyl
Ethandisäuremonoureid *n* <Chemie> oxaluric acid
Ethannitril *n* <Chemie> acetonitrile, ethanenitrile
Ethanol *n* 1. <Chemie, Erdöl> ethanol *(Petrochemie)*; 2. <Lebensmittel> ethanol, ethyl alcohol
Ethanolat *n* <Chemie> ethanolate
Ethanolyse *f* <Chemie> ethanolysis
Ethansäure *f* <Lebensmittel> acetic acid
Ethanthiol *n* <Chemie> ethanethiol
Ethen *n* <Chemie> ethene, ethylene
Ethenyl... <Chemie> vinyl
Ethenyliden... <Chemie> vinylidene
Ether *m* <Chemie> ether
etherartig *adj* <Chemie> ethereal
etherisch *adj* <Chemie> ethereal
Ethin *n* <Chemie> acetylene, ethine, ethyne
Ethion... <Chemie> ethionic
Ethlen *n* <Erdöl> ethene
Ethoxyacetanilid *n* <Chemie> ethoxyacetanilide, phenacetin
Ethoxyanilin *n* <Chemie> ethoxyaniline, phenetidine
Ethoxybenzol *n* <Chemie> ethoxybenzene, phenetole
Ethyl *n* <Erdöl> ethyl *(Petrochemie)*
Ethylacetat *n* <Kunststoff, Lebensmittel> ethyl acetate
Ethylaldehyd *m* <Lebensmittel> acetaldehyde
Ethylalkohol *m* 1. <Erdöl> ethyl alcohol *(Petrochemie)*; 2. <Foto> ethyl alcohol; 3. <Lebensmittel> ethanol, ethyl alcohol
Ethylamin *n* <Chemie> aminoethane, ethylamine
Ethylanilin *n* <Chemie> ethylaniline
Ethylat *n* <Chemie> ethylate
Ethylcellulose *f (EC)* <Kunststoff> ethyl cellulose *(EC)*
Ethyldimethylmethan *n* <Chemie> isopentane
Ethylen *n* 1. <Chemie, Erdöl> ethene, ethylene *(Petrochemie)*; 2. <Kunststoff, Lebensmittel> ethylene
Ethylen... <Chemie> ethylenic
Ethylenkohlenwasserstoff *m* <Chemie> alkene, olefin, olefine
Ethylen-Propylen-Kautschuk *m* <Kunststoff> ethylene propylene rubber
Ethylenvinylacetat *n (EVA)* <Kunststoff> ethylene vinyl acetate *(EVA)*
Ethylhydrosulfid *n* <Chemie> ethanethiol
Ethyliden *n* <Chemie> ethylidene
Ethylidenradikal *n* <Chemie> ethylidene
ethylieren *v* <Chemie> ethylate
Ethylierung *f* <Chemie> ethylation
Ethylmercaptan *n* <Chemie> ethanethiol
Ethylmorphin *n* <Chemie> ethylmorphine

Ethylphenylether m <Chemie> ethoxybenzene, phenetole
Ethylschwefel... <Chemie> *(AE)* ethylsulfuric, *(BE)* ethylsulphuric; vinic *(Säure)*
Ethylthioethanol n <Erdöl> ethylthioethanol *(Petrochemie)*
Ethylurethan n <Chemie> ethyl urethane
Ethylvanillin n <Lebensmittel> ethyl-vanillin *(Aromastoff)*
Etikett n 1. <Comp & DV> external label, label; 2. <Textil> label, tag; 3. <Verpack> label
Etikettausgabe f <Verpack> label dispenser
Etikettenaufdruckmaschine f <Verpack> label-overprinting machine
Etikettenauszeichnungsmaschine n <Verpack> label-coding machine
Etikettenfolie f <Verpack> label film
etikettieren v <Textil> label, tag
Etikettieren n <Textil> *(AE)* labeling, *(BE)* labelling, tagging
Etikettieren n **mit Druckluft** <Verpack> *(AE)* air-blast labeling, *(BE)* air-blast labelling
Etikettieren n **mit Luftdruck** <Verpack> *(AE)* air-jet labeling, *(BE)* air-jet labelling
Etikettierer m <Verpack> *(AE)* labeler, *(BE)* labeller
Etikettiermaschine f <Verpack> *(AE)* labeling machine, *(BE)* labelling machine
Etikettiermaschine f **für Packungsoberseiten** <Verpack> *(AE)* front of pack labeler, *(BE)* front of pack labeller
Etikettierung f <Verpack> *(AE)* labeling, *(BE)* labelling
Etikettrückseite f <Verpack> back
Eu *(Europium)* <Chemie> Eu *(europium)*
Eucalyptol n <Chemie> eucalyptol
Eudiometer n <Chemie> eudiometer
Eudiometrie f <Chemie> eudiometry
Eugenol n <Chemie> eugenol
euklidische Geometrie f <Geom> Euclidean geometry
euklidischer Raum m <Geom, Math, Phys> Euclidean space *(Punktraum mit euklidischer Norm)*
euklidischer Vektorraum m <Math> Euclidian vector space *(reeller Vektorraum mit Skalarprodukt)*
Eule fangen v/eine <Wassertrans> broach to *(Segeln)*
Euler'sche Bewegungsgleichungen fpl <Strömphys> Eulerian equations
Euler'sche Kurve f <Math> Euler circles
Euler'sche Winkel mpl <Phys> Euler angles
Euler'scher Graph m <Math> Eulerian graph *(Graphentheorie)*
Europäische Atomgemeinschaft f *(EURATOM)* <Teilphys> European Organization for Nuclear Research *(EURATOM)*
Europäische Norm f *(EN)* <Elektriz> European Standard
europäische Patentanmeldung f <Patent> European patent application
Europäische Raumfahrtbehörde f *(ESA)* <Raumfahrt> European Space Agency *(ESA)*
Europäischer Funkrufdienst m *(EFuRD)* <Telekom> European radio-paging system
Europäischer Normungsausschuss m *(CEN)* <Qual> European Committee for Standardization *(CEN)*
Europäisches Funkrufsystem n **der zweiten Generation** <Funktech, Telekom> enhanced radio message system *(ERMES)*
Europäisches Institut n **für Telekom-Standards** <Telekom> European Telecommunications Standards Institute *(ETSI)*
Europäisches Kernforschungszentrum n *(CERN)* <Teilphys> European Organization for Nuclear Research *(CERN)*

europäisches Leichterträgersystem n *(EBCS-System)* <Wassertrans> European barge carrier system *(EBCS)*
europäisches Nachrichtensatellitensystem n <Telekom> European communication satellite system *(ECS)*
europäisches Patent n <Patent> European patent
europaweite GSM-Standards mpl <Telekom> pan-European GSM standards
europaweites Bündelfunksystem n <Funktech> Trans-European Trunked Radio *(digital, abhörsicher)*
Europiep m <Telekom> European radio-paging system
Europium n *(Eu)* <Chemie> europium *(Eu)*
Euroschlitz m <Verpack> euroslot
Euro-Testumgebung f <Comp & DV> euro test environment
eustatisch adj <Nichtfoss Energ> eustatic
Eutektikum n <Chemie, Ker & Glas> eutectic
eutektisch adj <Chemie> eutectic
eutektische Legierung f <Metall> eutectic alloy
eutektische Reaktion f <Metall> eutetic reaction
eutektische Transformation f <Metall> eutetic transformation
eutektischer Stahl m <Metall> eutectoid steel
Eutektoid n <Metall> eutectoid
Eutrophierung f <Umweltschutz> eutrophication
eV *(Elektronenvolt)* <Elektriz, Elektrotech, Phys, Strahlphys, Teilphys> eV *(electronvolt)*
EVA *(Ethylenvinylacetat)* <Kunststoff> EVA *(ethylene vinyl acetate)*
evakuieren v 1. <Elektronik> evacuate *(Röhren)*; 2. <Wassertrans> evacuate *(Schiff)*
evakuiert adv <Phys, Thermod> under vacuum
Evaluierung f <Bau, Künstl Int, Patent, Qual> evaluation
Evaporation f *(E)* <Druck, Erdöl, Heiz & Kälte, Hydraul, Maschinen, Nichtfoss Energ, Phys, Thermod> evaporation *(E)*
Evaporationskühlung f <Heiz & Kälte, Thermod> evaporative cooling
Evaporator m <Chemtech> evaporator
Evaporimeter n <Chemtech> evaporimeter
Evaporit n <Erdöl> evaporite *(Geologie)*
E-Verhüttung f <Elektrotech> electric smelting
Evolute f <Geom> evolute
Evolvente f 1. <Fertig> involute curve; 2. <Geom> involute
Evolventenkerbverzahnung f <Maschinen> involute serrations, involute spline
Evolventenrad n <Maschinen> involute gear
Evolventenverzahnung f <Maschinen> involute gearing
evolventisch adj <Fertig> involute
evozieren v <Ergon> evoke
evozierte Reaktion f <Ergon> evoked response
EVS *(Endvermittlungsstelle)* <Telekom> terminating exchange
EVSt *(Endvermittlungsstelle)* <Telekom> terminal exchange
EVU *(Elektrizitätsversorgungsunternehmen)* <Elektriz> electricity supply company
E-Welle f *(Elektrowelle)* 1. <Elektrotech> E wave, TM wave; 2. <Phys, Telekom> E wave, TM wave *(electric wave)*
E-Werk n <Elektrotech> station *(Kraftwerk)*; electric power station
exakt adj <Mechan, Phys> accurate
exakte Logik f <Künstl Int> exact logic
Exemplar n <Druck> copy
Exhaustor m 1. <Chemtech> dust catcher; 2. <Maschinen> exhauster; 3. <Mechan> exhaust pump
Exiton n <Phys> exciton
exklusive ODER-Verknüpfung f <Comp & DV> non-equivalence operation

exklusives-ODER-Glied n <Elektronik> exclusive OR circuit, exclusive OR gate
Ex-Motor m (explosionsgeschützter Motor) <Elektrotech, Thermod> explosion proof motor
exogen adj <Math, Sicherheit> exogenous
exogene Variable f <Math> exogenous variable (Regressor, unabhängige Variable)
exotherm adj <Thermod> exothermal, exothermic
exothermer Prozess m <Thermod> exothermic process
exothermisch adj <Raumfahrt> exothermic (Raumschiff)
exotischer Chip m <Elektronik> exotic chip
exotisches Signal n <Elektronik> exotic signal
Expander m <Telekom> expander
expandieren v <Hydraul> expand (Dampf)
expandiert adj <Raumfahrt> expanded (Raumschiff)
expandierter Graph m <Künstl Int> expanded graph
Expansion f 1. <Hydraul> expansion (Dampf); 2. <Maschinen> expansion; 3. <Raumfahrt> expansion (Raumschiff); 4. <Thermod> expansion
Expansionsbeanspruchung f <Hydraul> expansion stress
Expansionsbohrmeißel m <Erdöl> expansion bit
Expansionsdüse f <Lufttrans> expansion nozzle (Strahltriebwerk)
Expansionsfalle f <Hydraul> expansion trap
Expansionshub m <Maschinen, Mechan, Thermod> expansion stroke
Expansionskammer f <Maschinen> expansion chamber
Expansionskasten m <Hydraul> expansion box
Expansionskoeffizient m <Maschinen> expansion coefficient
Expansionskupplung f <Maschinen> slip joint
Expansionskurve f <Fertig> expansion curve, expansion line
Expansionslinie f <Fertig> expansion curve, expansion line
Expansionsmaschine f <Maschinen> expansion engine
Expansionsperiode f <Hydraul> expansion period
Expansionsplatte f <Hydraul> expansion plate
Expansionspunkt m <Hydraul> expansion point
Expansionsraum m <Ker & Glas> expansion space
Expansionsring m <Erdöl> expansion ring
Expansionsrohr n <Labor> expansion tube
Expansionsschalter m <Elektriz> air breaker
Expansionsschaltkulisse f <Hydraul> expansion notch (Quadrant)
Expansionsschaltnocke f <Hydraul> expansion notch (Quadrant)
Expansionsschieber m <Hydraul> expansion slide
Expansionsspannung f <Hydraul> expansion stress
Expansionsspeicher m <Comp & DV> expanded memory
Expansionsspeicher-Spezifikation m (EMS) <Comp & DV> expanded memory specification (EMS)
Expansionssteuernocke f <Hydraul> expansion cam (Dampfmaschine)
Expansionsstufe f <Telekom> expansion stage
Expansionsturbine f <Heiz & Kälte> expansion turbine
Expansionsventil n <Heiz & Kälte, Hydraul, Maschinen> expansion valve
Expansionswelle f <Lufttrans> expansion wave (Überschallknall)
Expansionszeitraum m <Hydraul> expansion period
Experiment n 1. <Kohlen> trial; 2. <Phys> experiment
Experiment n innerhalb des Reaktors <Kerntech> in-reactor experiment
Experimentalphysiker m <Teilphys> experimental physicist
Experimentalsicherheitsauto n <Kfztech> ESV, experimental safety vehicle
Experimente npl **mit hochenergetischem Teilchenstrahl** <Strahlphys> fast beam experiments
experimentell adj <Phys> experimental
experimentelles Fernsehen n <Fernseh> experimental television
Experimentenpaket n <Raumfahrt> experiment package
Experimentiermodul n <Raumfahrt> experiment module
Expertensystem n 1. <Comp & DV> (XPS) expert system (XPS); 2. <Künstl Int> (ES, XPS) expert system (XPS)
Expertensystem-Shell f <Künstl Int> expert system shell
explodieren v <Thermod> detonate
Exploration f <Erdöl> exploration (Lagerstätten)
Explorationsbohranlage f <Erdöl> exploration rig (Lagerstätten)
Explorationsbohren n <Erdöl> exploration drilling, exploratory drilling (Lagerstätten)
Explorationsphase f <Erdöl> exploration phase (Lagerstätten)
Explosion f 1. <Erdöl> explosion (Petrochemie, Physik); 2. <Sicherheit> explosion; 3. <Thermod> detonation
explosionsartiges Feuer n <Thermod> flash fire
Explosionsbekämpfung f <Sicherheit> explosion control
Explosionsbürette f <Chemie> eudiometer
explosionsfähiges Gemisch n <Sicherheit> explosive mixture
Explosionsformgebung f <Mechan> explosive forming
Explosionsformung f <Thermod> explosive forming
explosionsgefährdeter Bereich m 1. <Elektriz> explosive gas atmosphere; 2. <Sicherheit> explosive range
explosionsgefährlicher Abfall m <Abfall> explosive waste
Explosionsgemisch n <Kfztech> explosive mixture
explosionsgeschützt adj 1. <Elektriz, Lufttrans, Mechan, Sicherheit> explosion-proof; 2. <Thermod> flameproof; 3. <Verpack> explosion-proof
explosionsgeschützte elektrische Anlage f <Sicherheit> explosion-proof electric equipment
explosionsgeschützte Elektrogeräte npl <Sicherheit> flameproof electrical equipment
explosionsgeschützter Motor m (Ex-Motor) <Elektrotech, Thermod> flameproof motor
explosionsgeschütztes elektrisches Betriebsmittel n <Sicherheit> explosion-proof electric equipment
Explosionsmesser m <Labor> explosimeter (brennbare Gase)
Explosionsmotor m <Mechan> internal combustion machine
Explosionsschutz m <Sicherheit> explosion control, explosion protection
Explosionsschutzeinrichtung f <Sicherheit> explosion safeguard
explosionssicher adj <Elektriz, Lufttrans, Mechan, Sicherheit, Verpack> explosion-proof
explosionssichere Ausführung f <Sicherheit> explosion-proof construction
explosionssichere Konstruktion f <Sicherheit> explosion-proof construction
explosionssichere Verglasung f <Ker & Glas> explosion-proof glazing
Explosionssicherheit f <Sicherheit> resistance to explosions
Explosionsunterdrückung f <Sicherheit> explosion suppression
Explosionszeichnung f <Maschinen> exploded view
explosiv adj 1. <Kfztech> explosive; 2. <Thermod> detonatable, explosive
explosive Luft f <Sicherheit> explosive atmosphere
explosiver Abfall m <Abfall> explosive waste
Explosivstoffe mpl <Thermod> explosives

Explosivumformung f <Fertig, Metall> explosive forming
Exponent m 1. <Comp & DV> exponent; 2. <Math> exponent, index, superscript
Exponential... <Akustik, Elektriz, Elektronik, Math> exponential
Exponentialfunktion f <Math> exponential function
Exponentialkurve f <Elektriz> exponential curve
Exponentialröhre f <Elektronik> exponential tube
Exponentialtrichter m <Akustik> exponential horn
Exponentialverstärker m <Elektronik> exponential amplifier
Exponentialverteilung f <Comp & DV> exponential distribution
exponentiell adj <Math> exponential
exponentielle Glättung f <Math> exponential smoothing
exponentieller Zerfall m <Elektronik> exponential decay (Kerntechnik)
exponentielles Brechzahlprofil n <Telekom> power law index profile
Exportgüterverpackung f <Verpack> export packaging
Exportlizenz f <Wassertrans> (BE) export licence, (AE) export license (Dokumente)
Expositionszeit f <Ergon> exposure time
Exschutzeinrichtung f <Sicherheit> explosion safeguard
Exsikkator m 1. <Chemie, Chemtech> desiccator; 2. <Labor> desiccator (Trocknen); 3. <Lebensmittel> desiccator
Extender m <Kunststoff> extender
Extensionsgröße f <Phys> extensive quantity
extern adj <Comp & DV, Maschinen, Sicherheit> external
extern erregter Motor m <Elektriz> separately-excited motor
externe Aufrufschnittstelle f (ECI) <Comp & DV> external call interface, ECI
externe Blockierung f <Telekom> external blocking
externe Datei f <Comp & DV> external data file
externe Eingabe f <Comp & DV> external input
externe Einspritzung f <Kfztech> external injection
externe Modulation f <Elektronik> external modulation
externe Quelle f <Elektronik> external source
externe Sortierung f <Comp & DV> external sort
externe Steuereinheit f (PCU) <Comp & DV> peripheral control unit (PCU)
externe Unterbrechung f <Comp & DV, Elektronik> external interrupt
externer Radioaktivitätspegel m <Strahlphys> level of external radioactivity
externer Schnittstellenadapter m <Comp & DV> peripheral interface adaptor
externer Speicher m 1. <Comp & DV> external memory, external storage, external store, external memory; 2. <Elektrotech> external memory
externer Taktgeber m <Comp & DV> external clock
externer Ursprung m <Elektronik> external source
externer Verstärker m <Elektronik> remote amplifier
externer Widerstand m <Elektriz> external resistance
externes Drehmoment n <Maschinen> external torque
externes Gerät n <Comp & DV> external device
externes Interface n <Raumfahrt> external interface (Raumschiff)
externes Magnetfeld n <Elektriz> external magnetic field
externes Signal n <Elektronik> external signal
externes Vorspannen n <Bau> external prestressing (Beton)
Externumschaltung f <Mobilkom> intercell hand-off, interhandoff (Zellenwechsel)
Extinktionsmessgerät n <Gerät> haze meter
extradünnes Tafelglas n <Ker & Glas> extra-thin sheet glass
extragalaktisch adj <Raumfahrt> extra-galactic

extrahartes Papier n <Foto> extra-hard paper
extrahierbarer Schwefel m <Kunststoff> (AE) extractable sulfur, (BE) extractable sulphur
extrahieren v 1. <Chemie> abstract, extract, leach; 2. <Papier> extract
Extrahochspannungskabel n <Elektriz> extra-high voltage cable
Extrakt n <Papier> extract
Extrakteur m <Chemtech> extractor
Extraktion f <Chemtech, Labor> extraction
Extraktionsapparat m <Chemtech> extractor
Extraktionshaube f <Labor> extraction hood
Extraktionshülse f 1. <Labor> extraction thimble (Soxhlet-Apparat); 2. <Lebensmittel> extraction thimble
Extraktionsmittel n 1. <Chemie> menstruum (für Drogenauszüge); 2. <Kerntech> eluant, eluting agent; 3. <Lebensmittel> extraction solvent
Extraktionssäule f <Papier> extractor
Extraktionsventilator m <Labor> extraction fan
extraktive Destillation f <Lebensmittel> extractive distillation
Extrapolation f <Math> extrapolation
extrapolieren v <Math> extrapolate
extraterrestrische Forschung f <Phys> extraterrestrial investigation, extraterrestrial research
extraweiches Papier n <Foto> extra-soft paper
extrem hohe Frequenz f (EHF) <Funktech> extremely high frequency (EHF); microwave frequency
extrem tiefe Frequenz f (ENF) <Funktech> extremely low frequency (ELF)
Extrempunkt m <Geom> turning point
Extremum n <Math> extreme
Extremwert m <Math, Qual> extreme value
Extremwertauswahleinheit f <Regelung> high-low signal selector
Extrinsic-Halbleiter m <Phys> extrinsic semiconductor
Extrudat n <Fertig> extrudate (Faseroptik)
Extruder m <Kunststoff> extruder
Extruderdüse f <Kunststoff> extrusion die
Extruderfolie f <Kunststoff> extruded film
Extrudierbarkeit f <Kunststoff> extrudability
extrudieren v <Kunststoff> extrude
Extrudieren n <Kunststoff> extrusion
extrudiert adj <Fertig> (AE) extrusion molded, (BE) extrusion moulded
extrudierte Folie f <Kunststoff> extruded film
Extrusion f 1. <Fertig> (AE) extrusion molding, (BE) extrusion moulding; 2. <Kunststoff, Papier> extrusion
Extrusionsblasen n <Verpack> (AE) extrusion blow molding, (BE) extrusion blow moulding
Extrusionsmaschine f <Kunststoff> extrusion machine
Extrusionswerkzeug n <Kunststoff> extrusion die
Exzenter m <Maschinen, Textil> eccentric
Exzenterbohrmeißel m <Erdöl> eccentric bit (Bohrtechnik)
Exzenterbolzen m <Maschinen> eccentric bolt
Exzenterbuchse f <Maschinen> eccentric bush
Exzenterbügel m <Maschinen> eccentric strap
Exzenterdrehen n <Fertig> eccentric turning
Exzenternocken m <Maschinen> eccentric cam
Exzenterpresse f <Fertig, Maschinen> eccentric press
Exzenterscheibe f <Maschinen> (BE) eccentric disc, (AE) eccentric disk, eccentric sheave
Exzenterstange f <Mechan> eccentric rod
Exzenterstift m <Fertig> calm pin
Exzenterwelle f <Fertig, Maschinen> eccentric shaft
Exzenterzapfen m <Maschinen> eccentric pin
exzentrisch adj <Mechan> eccentric
exzentrische Anomalie f <Raumfahrt> eccentric anomaly

exzentrische Belastung f 1. <Bau> eccentric loading; 2. <Maschinen> eccentric load
exzentrisches Futter n <Maschinen> eccentric chuck
Exzentrizität f 1. <Akustik> eccentricity; 2. <Maschinen> eccentricity, throw; 3. <Mechan> eccentricity; 4. <Optik> core-cladding concentricity error *(Lichtleiter, von Kern zu Mantel)*; 5. <Raumfahrt> eccentricity
Exzentrizitätsmaß n <Maschinen> throw
Exzitonenübergang m <Elektronik> exciton transition
Exzitron n <Elektrotech> excitron

F

F 1. <Akustik, Aufnahme, Comp & DV, Elektronik, Funktech, Phys> *(Frequenz)* f *(frequency)*; 2. <Chemie> *(Fluor)* F *(fluorine)*; 3. <Elektriz, Elektrotech, Metrol, Phys> *(Farad)* F *(farad)*; 4. <Elektronik, Funktech> *(Rauschzahl)* F *(noise figure)*; 5. <Hydraul, Phys> *(Frouden'sche Zahl)* F *(Froude number)*; 6. <Kerntech> *(hyperfeine Quantenzahl)* F *(hyperfine quantum number)*; 7. <Metall, Phys> *(Kraft)* F *(force)*; 8. <Metall, Phys> *(freie Energie)* F *(free energy)*; 9. <Metrol> *(Fahrenheit)* F *(Fahrenheit)*; 10. <Metrol> *(Femto…)* f *(femto…)*
fA *(Antiresonanzfrequenz)* <Akustik, Elektronik> fA *(antiresonant frequency)*
Fabrik f <Mechan> factory, mill, plant, works
Fabrikanlage f <Maschinen> industrial plant, industrial unit
Fabrikanschlussgleis n <Eisenbahn> factory siding
Fabrikationsnummer f <Maschinen> serial number
fabrikfertig adj <Heiz & Kälte> factory-assembled
fabrikfertige Anlage f <Heiz & Kälte> factory-assembled system
fabrikfertige Verpackung f <Verpack> prefabricated package
Fabrikhalle f <Mechan> fabricating shop
Fabrikprüfung f <Qual> shop test
Fabrikschiff n <Wassertrans> factory ship
Fabry-Pérot'sches Interferometer n <Phys, Raumfahrt> Fabry-Pérot interferometer
Faceplatte f <Wassertrans> face plate *(Schiffbau)*
Facette f <Ker & Glas> arrissed edge
facettiert adj <Ker & Glas, Metall> facetted
facettierter Ring m <Kerntech> facetted ring
facettiertes Bläschen n <Kerntech> facetted bubble
Fach n 1. <Maschinen> compartment; 2. <Mechan, Telekom> bay
Fachbereichswissen n <Künstl Int> domain knowledge
Fächer m <Heiz & Kälte> fan, ventilator
fächerförmiges Lichtbündel n <Elektronik> fan beam *(Beleuchtungstechnik)*
Fächerfunkfeuer n <Funkort, Wassertrans> fan marker beacon *(Seezeichen)*
Fächerkasten m <Verpack> compartment case
Fächerkeule f <Elektronik> fan beam *(Antennentechnik)*
Fächermaschine f <Papier> harper machine
Fächermethode f <Abfall> cell method
Fächerplatte f <Wassertrans> gusset plate *(Schiffbau)*
Fächerscheibe f <Maschinen> serrated lock washer
Fächerwand f <Verpack> partition wall
Fachgrundnorm f <Elektrotech, Fertig, Maschinen> generic standard

Fachwerk n <Bau> framework, lattice, truss, trussing
Fachwerkbinder m <Bau> open-web girder
Fachwerkbinderdach n <Bau> trussed roof
Fachwerkbrücke f <Bau> frame bridge, truss bridge
Fachwerkhaus n <Bau> frame house, half-timbered house
Fachwerkholzträger m <Bau> trussed wooden beam
Fachwerkträger m <Bau> lattice girder, trussed beam, trussed girder
Fachwerkwand f <Bau> stud wall
Fackel f 1. <Erdöl> flare *(Raffinerie, Bohrfeld)*; 2. <Mechan> torch
fackeln v <Textil> blaze
Fackelrohr n <Erdöl> flare stack *(Raffinerie, Bohrfeld)*
fade adj <Lebensmittel> *(AE)* unflavored, *(BE)* unflavoured
Faden m 1. <Bau> filament; 2. <Ker & Glas> string, thread; 3. <Metrol> fathom; 4. <Papier> thread; 5. <Textil> thread, yarn; 6. <Wassertrans> fathom *(Maßeinheit)* • **mit Faden heften** <Druck> thread
Fadenabschneider m <Textil> thread trimmer
Fadenanzahl f <Textil> ply *(Garn)*
Fadenauflage f <Ker & Glas> applied thread
Fadenbruch m <Textil> broken end
Fadenelektrometer m <Elektrotech, Metrol> filament electrometer, thread electrometer
Fadenende n <Textil> yarn end
fadenförmiger Knoten m <Ker & Glas> stringy knot
Fadenführer m <Textil> feeder, yarn carrier, yarn guide; carrier *(Spinnen)*
Fadenkreuz n 1. <Comp & DV> crosshair; 2. <Elektronik, Elektrotech> axis intersection, reticle; 3. <Fertig> hair cross; 4. <Optik> spider lines
Fadenkreuzlinie f <Optik> webspider line
Fadenkreuzokular n <Optik> eyepiece with cross-wires
Fadenlieferer m <Textil> guide
Fadenmikrometer n <Metrol> filar micrometer
Fadennetz n <Fertig> reticle
Fadenscheinigkeit f <Textil> scratching
Fadenschluss m <Textil> cover
Fadenspanner m <Textil> yarn-tensioning device
Fadenverdickung f <Textil> slub
Fadenwächter m am Gatter <Textil> stop motion on creel
Fadenzähler m 1. <Druck> line tester; 2. <Papier> thread counter
Fadenziehen n <Lebensmittel> ropiness *(Brot)*
Fadenzufuhrregelung f <Textil> yarn feed control
Fading n <Akustik, Funktech, Telekom> fade, fading
Fadingregelung f <Funktech> automatic volume control
Fähigkeit f 1. <Ergon> ability; 2. <Phys> power
Fähigkeit f zu Was-wäre-wenn-Folgerungen <Künstl Int> what-if capability *(eines Expertensystems)*
Fähigkeit f zur Korrektur von Fehlerbursts <Telekom> burst error-correcting capability
Fähigkeiten fpl <Comp & DV> capabilities *(eines Programms, Computers, Benutzers)*
Fahne f 1. <Textil> ply *(Wolle&)*; 2. <Wassertrans> flag *(Flagge)*
Fahnenabzug m <Druck> galley, galley proof, slip proof
Fahr… <Trans> driving
Fahrabschaltventil n <Hydraul> riding cutoff valve
Fahrbahn f 1. <Bau> runway *(eines Krans)*; 2. <Trans> roadway
Fahrbahn f für allgemeinen Verkehr <Trans> nonreserved space
Fahrbahndurchbiegung f <Elektrotech> track deflection *(Magnetschwebetechnik)*
Fahrbahndynamik f <Elektrotech> guideway deflection *(Magnetschwebetechnik)*

Fahrbahnlinienlast f <Elektrotech> linear distributed load on guideway *(Magnetschwebetechnik)*
Fahrbahnmarkierung f <Bau> road painting
Fahrbahnunstetigkeit f <Eisenbahn> unevenness of trackway
Fahrbalken m <Trans> track girder
fahrbare Hubbühne f <Bau> portable hoisting platform
fahrbarer Portalkran m <Wassertrans> *(AE)* traveling gantry crane, *(BE)* travelling gantry crane *(Hafen)*
fahrbarer Schraubstock m <Bau> portable vice
fahrbarer Verdichter m <Abfall> compactor vehicle, compression vehicle, packer body
fahrbarer Wagenheber m <Trans> mobile jack
Fährbetrieb m <Wassertrans> ferry service, ferrying
Fahrdraht m <Elektrotech> catenary wire, collector wire, overhead contact wire, train line, trolley line, trolley wire
Fahrdrahtaufhängung f <Elektrotech> catenary suspension
Fähre f 1. <Raumfahrt> shuttle *(Raumschiff)*; 2. <Wassertrans> ferry
fahren v 1. <Elektrotech> drive, run *(ein Kraftwerk)*; 2. <Eisenbahn> drive; 3. <Fertig> run *(eine Anlage)*; 4. <Kfztech> drive, ride; 5. <Trans> carry; drive *(Fahrzeug)*; 6. <Wassertrans> sail; run *(Schiff)*
Fahren n <Eisenbahn> running
Fahrenheit n *(F)* <Metrol> Fahrenheit *(F)*
Fahrenheit-Temperaturskale f <Phys> Fahrenheit scale
Fahrer m <Trans> driver
Fahrerbereich m <Kfztech> cockpit
Fahrerhaus f <Eisenbahn, Kfztech> cab *(Karosserie)*
Fahrerlaubnis f <Kfztech> *(BE)* driver's licence, *(AE)* driver's license *(Rechtsvorschriften)*
fahrerloser Zug m <Eisenbahn> unmanned train
Fahrerseite f <Kfztech> offside
Fahrfläche f <Lufttrans> tread *(Fahrwerk)*
Fahrgast m <Trans> passenger
Fahrgastflug m <Trans> passenger flight
Fahrgastraum m <Eisenbahn> passenger compartment
Fahrgastsitz m <Trans> passenger seat
Fahrgemeinschaft f <Trans> car pool
Fahrgeschwindigkeit f <Trans> progression speed, running speed
Fahrgestell n 1. <Fertig> bogie; 2. <Kfztech> undercarriage; *(BE)* bogie *(Anhänger)*; chassis *(Karosserie)*; 3. <Maschinen> bogie; 4. <Mechan> carriage; 5. <Trans> carriage; *(BE)* bogie, *(AE)* bogie truck, *(AE)* trailer *(Kran)*
Fahrgestell n **mit Kabine** <Kfztech> chassis-cab *(LKW)*
Fahrgestell-A n **mit Motor** <Kfztech> carriage A containing the motor
Fahrgestellachswelle f <Lufttrans> landing-gear shaft
Fahrgestellauskreuzungseinrichtung f <Lufttrans> landing-gear bracing installation
Fahrgestelleinfahrverriegelungskasten m <Lufttrans> landing-gear up-lock box
Fahrgestellhaubenhalterung f <Lufttrans> *(BE)* landing-gear boot retainer, *(AE)* landing-gear trunk retainer
Fahrgestellkompensierungsstange f <Lufttrans> landing-gear compensation rod
Fahrgestellquerbock m <Lufttrans> landing-gear diagonal truss
Fahrgestellquerstrebe f <Lufttrans> landing-gear diagonal truss
Fahrgestellspur f <Lufttrans> landing-gear track
Fahrgestellstoßdämpfer m <Lufttrans> landing-gear bumper
Fahrgestellstütze f <Lufttrans> landing leg support
Fahrgestellverschlusshalterung f <Lufttrans> *(BE)* landing-gear boot retainer, *(AE)* landing-gear trunk retainer

Fahrkarte f <Eisenbahn> *(AE)* railroad ticket, *(BE)* railway ticket
Fahrkartenautomat m <Eisenbahn> add-value machine *(zur Aufwertung existierender Karten)*
Fahrkarteneinzugsautomat m <Eisenbahn> *(AE)* faregate
Fahrkorb m <Trans> car *(eines Aufzugs)*
Fahrkran m 1. <Bau> portable crane; 2. <Mechan> *(AE)* traveling crane, *(BE)* travelling crane
Fährlandungsbrücke f <Wassertrans> ferry-landing stage
Fahrleitung f 1. <Elektrotech> catenary, collector wire, contact line, overhead contact wire, overhead wire, train line, trolley line, trolley wire; 2. <Mechan, Phys> catenary
Fahrleitungsmast m <Eisenbahn> catenary support
Fahrmanipulator m <Kerntech> *(AE)* traveling manipulator, *(BE)* travelling manipulator
Fahrmotor m <Elektrotech> traction motor
Fahrmotorengruppenschaltung f <Elektrotech> motor combination
Fahrmotortrennschalter m <Elektrotech> disconnecting switch reserver, traction motor isolating switch
Fahrpedal n <Kfztech> *(BE)* accelerator, *(AE)* gas pedal
Fahrplan m <Eisenbahn> *(AE)* schedule, *(BE)* timetable
Fahrplankonstruktion f <Eisenbahn> *(AE)* schedule compilation, *(BE)* timetable compilation
Fahrplantrasse f <Eisenbahn> train path
Fahrrad n <Maschinen> bicycle
Fahrraddynamo m <Elektrotech> bicycle dynamo
Fahrradergometer n <Ergon> bicycle ergometer
Fahrradpumpe f <Maschinen> bicycle pump
Fahrradscheinwerferlampe f <Elektrotech> bicycle headlight lamp
Fahrradventil n <Maschinen> bicycle valve
Fahrradwerkzeuge npl <Maschinen> bicycle tools
Fahrrinne f <Wassertrans> ship canal; channel *(Navigation)*
Fahrschiene f <Eisenbahn> runner
Fahrschwingung f <Sicherheit> ride vibration
Fährseil n <Wassertrans> ferry cable
Fahrsicherheit f <Sicherheit> ride safety
Fahrspur f <Trans> lane
Fahrspurrichtungssignal n <Trans> lane direction control signal
Fahrspurwechsel m <Trans> lane switching
Fahrsteig m <Trans> *(BE)* moving pavement, *(AE)* moving sidewalk
Fahrstrahl m <Phys> position vector
Fahrstraße f <Trans> road
Fahrstraßeneinstellung f **durch Rechner** <Eisenbahn> computer route setting
Fahrstraßenfestlegung f <Eisenbahn> holding of a route
Fahrstraßenmatrix f <Eisenbahn> routing diagram
Fahrstraßensperre f <Eisenbahn> route locking
Fahrstrecke f <Lufttrans, Trans> distance covered *(Auto, Bahn)*
Fahrstrom m <Elektrotech> traction current *(Elektrotraktion)*
Fahrstromkreis m <Elektrotech> traction circuit *(Elektrotraktion)*
Fahrstromrückleitung f <Elektrotech> return circuit *(Elektrotraktion)*
Fahrstuhl m <Bau, Elektriz, Mechan, Trans> *(AE)* elevator, *(BE)* lift
Fahrt f 1. <Trans> running; 2. <Wassertrans> way of ship • **Fahrt achteraus machen** <Wassertrans> be making sternway *(Schiffsbewegung)* • **Fahrt voraus** <Wassertrans> headway • **Fahrt voraus machen** <Wassertrans> be making headway *(Schiffsbewegung)* • **freie Fahrt ge-**

Fahrt

ben <Eisenbahn, Wassertrans> clear the line • **halbe Fahrt voraus** <Wassertrans> half ahead • **halbe Fahrt zurück** <Wassertrans> half astern
Fahrt *f* **durch das Wasser** <Wassertrans> speed through the water
Fahrt *f* **in Vielfachsteuerung** *f* <Elektrotech> multiple unit operation *(Zugsteuerung)*
Fahrt *f* **über Grund** <Wassertrans> speed made good over the ground
Fahrtbericht *m* <Trans> trip report analysis
Fahrtdauer *f* <Trans> journey time
Fahrtenregler *m* <Kfztech> drive control unit
Fahrtenschreiber *m* 1. tachograph, trip recorder *(Elektrotraktion)*; 2. <Kfztech> tachograph, trip recorder
Fahrtgeschwindigkeit *f* <Wassertrans> rate of sailing
Fahrtmessanlage *f* <Wassertrans> speedometer
Fahrtmesser *m* <Lufttrans, Wassertrans> log
Fahrtregler *m* <Kfztech> cruise control
Fahrtrichtungsanzeiger *m* 1. <Kfztech> direction indicator, indicator; 2. <Trans> trafficator
Fahrtsignal *n* <Eisenbahn, Wassertrans> clear signal
Fahrtstörungslaterne *f* <Wassertrans> not-under-command light *(Signal)*
Fahrttrimm *f* <Wassertrans> squat
fahrtüchtig *adj* <Wassertrans> navigable
Fahrtüchtigkeit *f* <Wassertrans> navigability
Fahrtwender *m* <Eisenbahn> braking switchgroup, reversing switchgroup
Fahrtwind *m* <Wassertrans> headwind
Fahrtzweck *m* <Trans> trip purpose
Fahrvorrichtung *f* <Kerntech> traversing mechanism *(an Dickenmessgerät)*
Fahrwasser *n* <Wassertrans> channel, fairway *(Navigation)* • **in Fahrwasser einlaufen** <Wassertrans> enter a channel *(Navigation)*
Fahrwassermarkierungen *fpl* <Wassertrans> channel markings; fairway markings *(Seezeichen)*
Fahrwasserzeichen *n* <Wassertrans> fairway mark
Fahrwerk *n* 1. <Lufttrans> landing gear, undercarriage; 2. <Maschinen> bogie assembly, running gear
Fahrwerkabwerftest *m* <Lufttrans> landing-gear drop test
Fahrwerkanzeige *f* <Lufttrans> landing-gear position indicator
Fahrwerkausfahren *n* <Lufttrans> landing-gear extension
Fahrwerkbein *n* <Lufttrans> landing-gear leg
Fahrwerkeinfahrverriegelung *f* <Lufttrans> landing-gear up-lock, landing-gear retraction lock
Fahrwerkentriegelung *f* <Lufttrans> landing-gear unlocking
Fahrwerkgabelstange *f* <Lufttrans> landing-gear fork rod
Fahrwerkgelenkträger *m* <Lufttrans> landing-gear hinge beam
Fahrwerkgelenkträgeranschluss *m* <Lufttrans> landing-gear hinge beam fitting
Fahrwerkgrube *f* <Lufttrans> landing-gear well
Fahrwerkhauptbremszylinder *m* <Lufttrans> landing-gear master brake cylinder
Fahrwerkhauptfederbein *n* <Lufttrans> landing-gear main shock strut
Fahrwerkklappenentriegelung *f* <Lufttrans> landing-gear door unlatching
Fahrwerkklappenverschluss *m* <Lufttrans> landing-gear door latch
Fahrwerkklappenverschlusskasten *m* <Lufttrans> landing-gear door latching box
Fahrwerkschieberventil *n* <Lufttrans> landing-gear sliding valve
Fahrwerksicherheitssteuerung *f* <Lufttrans> landing-gear safety override

Fahrwerksicherheitsverschluss *m* <Lufttrans> landing-gear safety lock
Fahrwerksperre *f* <Lufttrans> landing-gear down latch
Fahrwerksteuergerät *n* <Lufttrans> landing-gear control unit
Fahrwerksteuerung *f* <Lufttrans> landing-gear control unit
Fahrwerkverriegelung *f* <Lufttrans> landing-gear down latch
Fahrwerkverriegelungsbolzen *m* <Lufttrans> landing-gear lock pin
Fahrwiderstand *m* <Kfztech> road resistance *(Reifen)*
Fahrzeit *f* <Trans> running time
Fahrzeug *n* 1. <Kfztech> car; 2. <Mechan> vehicle; 3. <Meerschmutz> lightening vessel *(in das geleichtert wird)*; 4. <Wassertrans> craft
Fahrzeug *n* **innerhalb der Erdgravitation** <Raumfahrt> earth capture vehicle *(Raumschiff)*
Fahrzeug *n* **mit Allradantrieb** <Kfztech> four-wheel drive vehicle
Fahrzeug *n* **mit inhärent geringer Emission** *(ILEV)* <Nichtfoss Energ> inherently low emission vehicle, ILEV
Fahrzeug *n* **mit Unterdruckaufhängung** <Kfztech> suction-suspended vehicle
Fahrzeug *n* **mit Vergasermotor** <Kfztech, Umweltschmutz> *(AE)* gasoline engine vehicle, *(BE)* petrol engine vehicle
Fahrzeug *n* **mit verschiedenen Brennstoffmöglichkeiten** <Kohlen, Nichtfoss Energ> flexible fuel vehicle, FFV
Fahrzeugabstand *m* <Trans> vehicular gap
Fahrzeugantenne *f* <Kfztech> car antenna, vehicle antenna
Fahrzeugantriebsdifferenzial *n* <Kfztech> traction differential
Fahrzeugausstattung *f* <Kfztech> car accessories
Fahrzeugbatterie *f* 1. <Elektrotech> traction battery *(Elektrotraktion)*; 2. <Kfztech> vehicle battery
Fahrzeugbeleuchtung *f* <Kfztech> autocar lighting, automobile lighting, vehicle lighting
Fahrzeugbremsweg *m* <Kfztech> braking distance
Fahrzeugbrief *m* <Kfztech> engine logbook *(KFZ)*
Fahrzeugerfassung *f* <Trans> vehicle intercept survey
Fahrzeugfähre *f* <Wassertrans> vehicle ferry
Fahrzeugflotte *f* <Kfztech> fleet of vehicles
Fahrzeugfluss *m* **zur Hauptverkehrszeit** <Trans> vehicular flow at the peak hour
Fahrzeugfolgeabstand *m* <Trans> headway *(Verkehr)*
Fahrzeugfolgezeit *f* <Trans> gap, time headway, vehicle extension period
Fahrzeugfolgezeitanalyse *f* <Trans> headways distribution analysis
Fahrzeugfolgezeitdetektor *m* <Trans> gap detector
Fahrzeughalogenlampe *f* <Kfztech> tungsten-halogen auto lamp
Fahrzeug-Identifizierungsnummer *f (FIN)* <Kfztech> vehicle identification number, VIN
Fahrzeuginnenraum *m* <Kfztech> interior of car
Fahrzeuginventar *n* <Kfztech> vehicle inventory
Fahrzeuglärm *m* <Sicherheit> vehicle noise
Fahrzeugmarkierung *f* <Trans> vehicle tagging
fahrzeugmontierter kurzer Primärlinearmotor *m* <Kfztech> vehicle-mounted short primary linear motor
Fahrzeugortsbestimmung *f* <Funktech, Kfztech, Telekom> vehicle location
Fahrzeugortung *f* <Funktech, Kfztech> vehicle localization, vehicle tracking
Fahrzeugortungs-Subsystem *n* <Telekom> vehicle location subsystem

Fahrzeugpark m <Eisenbahn, Kfztech> rolling stock
Fahrzeugschein m <Kfztech> engine logbook
Fahrzeugschlange f <Trans> (AE) line, (BE) queue
Fahrzeugschlangendetektor m <Trans> (AE) line detector, (BE) queue detector
Fahrzeugschwebesystem n <Eisenbahn> SVS, suspended vehicle system
Fahrzeugschwingung f <Sicherheit> transport vibration
Fahrzeugstation f <Mobilkom> vehicular station
Fahrzeugtanker m <Wassertrans> vehicle tanker
Fahrzeugturbine f <Kfztech> vehicle turbine
Fahrzeugverkehr m <Lufttrans> (BE) aerodrome vehicle operations, (AE) airdrome vehicle operations
Fahrzeugzufahrt f <Trans> vehicle ramp
Fail-safe-Technik f <Sicherheit> fail-safe technology
Faksimile n (Fax) <Comp & DV, Funktech, Telekom> facsimile (fax)
Faksimile-Mitteilung f <Comp & DV, Telekom> facsimile message
Faktenwissen n <Künstl Int> factual knowledge
Faktor m <Comp & DV, Math> factor • **in Faktoren zerlegen** <Math> decompose to factors
Faktor m **der schnellen Spaltung** <Phys> fast fission factor (Kernphysik)
Faktoren... <Math> factorial
Faktorenanalyse f <Ergon> factor analysis
faktoriell adj <Math> factorial
faktorielle Gestaltung f <Ergon> factorial design
Faktorisierung f <Math> factorization (Produktdarstellung)
Fakultät f <Math> factorial
fakultativer Aerobier m <Lebensmittel> facultative aerobe
Fall m 1. <Comp & DV> instance; 2. <Metrol> case; 3. <Textil> draping properties; 4. <Wassertrans> rake; halyard (Tauwerk)
Fallbär m <Mechan> ram
fallbasiertes Inferieren n <Künstl Int> case-based reasoning, CBR, exemplar-based reasoning
fallbasiertes Schließen n <Künstl Int> case-based reasoning, CBR, exemplar-based reasoning
Fallbasierung f <Künstl Int> case-based reasoning, CBR, exemplar-based reasoning
Fallbetankung f <Lufttrans> (AE) gravity refueling, (BE) gravity refuelling
Fallbetankungshahn m <Lufttrans> gravity filler plug
Fallbirne f <Bau> breaker ball
Fallbremse f <Sicherheit> fall-arresting device, rope-grab fall protection device
Fallbremseinrichtung f <Sicherheit> fall-arresting equipment, fall stopper, rope-grab fall protection device, shock absorber
Fallbügelpunktschreiber m <Gerät> chopper bar dot recorder
Fallbügelregler m <Gerät> chopper bar controller
Fallbügelschreiber m <Gerät> hoop drop recorder
Falldurchbiegungsmesser m <Bau> falling weight deflectometer, FWD
Falle f 1. <Erdöl> trap (Geologie); 2. <Funktech> trap
fallen lassen v <Patent> abandon (Anmeldung)
Fallen n 1. <Erdöl> dip (Geologie); 2. <Thermod> fall
fallende Tide f <Wassertrans> falling tide (Gezeiten)
fallende Widerstandscharakteristik f <Elektrotech> negative resistance characteristic
fallender Guss m <Fertig> top pouring
fallendes Gießen n <Fertig> direct casting
Fallfüllung f <Raumfahrt> gravity filling (Raumschiff)
Fallgesetz n <Maschinen> law of gravitation
Fallgewicht n <Maschinen> drop weight

Fallgewichtsprüfung f <Maschinen> hammer test
Fallhammer m 1. <Bau> monkey; 2. <Fertig> drop stamp, tup; 3. <Kohlen, Maschinen> drop hammer
Fallhärteprüfung f <Metall> ball test
Fallhöhe f 1. <Bau> head; 2. <Hydraul> head, head of water; 3. <Kohlen> drop height; 4. <Verpack> drop height, height of fall
Fallkasten m <Wasserversorg> drip pump, drop box
Fallklappentafel f <Elektrotech> annunciator
Fallkugel f <Maschinen> drop ball
Fallleitung f <Kerntech> penstock
Fallmauer f <Wasserversorg> lift wall (Kanalschleuse)
Fällmittel n <Abfall> coagulant
Fallout m <Kerntech, Umweltschmutz> fallout
Fallprobe f <Maschinen> drop test, falling-weight test
Fallreepstreppe f <Wassertrans> accommodation ladder
Fallrichtung f <Erdöl> down dip (Geologie)
Fallrinne f <Mechan> chute
Fallrohr n 1. <Bau> downspout, leader, soil pipe; rainwater pipe (Dachrinne); band (Kaminschacht); 2. <Erdöl> downcomer (Destillationstechnik); 3. <Kerntech> downcomer; 4. <Papier> drop leg; 5. <Wasserversorg> soil pipe
Fallrohrauslauf m <Bau> shoe
Fallrohreindicker m <Papier> gravity thickener
Fallschacht m <Wasserversorg> pressure well
Fallschirm m <Lufttrans> parachute • **mit Fallschirm abwerfen** <Lufttrans> airdrop
Fallschirmaufziehleine f <Lufttrans> parachute release handle
Fallschirmbremsung f <Raumfahrt> parabrake (Raumschiff)
Fallschirmlicht n <Wassertrans> parachute flare (Signal)
Fallschloss n <Bau> spring lock
Fallschmierung f <Maschinen> gravity lubrication
Fallschnecke f <Fertig> drop worm
Fallschutz m <Sicherheit> fall protection, overhead protection, protection against falls
Fallschutzeinrichtung f <Sicherheit> fall arrest equipment
Fallschutzmittel n <Sicherheit> fall arrester, fall-arresting device, fall protection equipment
Fallseil n <Sicherheit> fall-arresting lanyard
Fallsicherungsanlage f <Sicherheit> fall arrest equipment
Fallstrom m <Maschinen> (AE) downdraft, (BE) downdraught
Fallstromkühler m <Kfztech> upright radiator
Fallstromvergaser m <Kfztech> (AE) downdraft carburetor, (BE) downdraught carburettor
Fallstudie f <Ergon, Künstl Int> case study
Falltür f <Bau> trap door
Fallüberlaufdamm m <Lufttrans> gravity spillway dam
Fällungsanalyse f <Chemtech> precipitation analysis
Fällungsmittel n <Chemtech> precipitating agent
Fallverschluss m <Foto> guillotine shutter
Fallversuch m 1. <Maschinen> drop test, falling-weight test; 2. <Verpack> drop test
Fallwind m <Lufttrans> down gust, downwash, downwind
falsch adj <Comp & DV> false
falsch beschneiden v <Fertig> mistrim
falsch verbunden adj <Telekom> wrongly-connected
falsche akustische Wahrnehmung f <Akustik> paracusis
falsche Nummer f <Telekom> wrong number
falscher Rücken m <Druck> false back
falsches Körpergehalt m <Ker & Glas> false body
falsches Wiederauffinden n <Comp & DV> false retrieval

Falschfahrt

Falschfahrt f <Eisenbahn> running on wrong line
Falschluft f <Kfztech> air leak
Falschsignal n <Elektronik> false signal
Falt... <Funktech, Konstzeich, Raumfahrt, Verpack> folding
Faltantenne f <Raumfahrt> collapsible antenna *(Weltraumfunk)*
Faltart f <Konstzeich> folding mode
faltbar adj <Maschinen> collapsible
faltbares und wiederzuverwertendes Verpackungssystem n <Verpack> collapsible and reusable packaging system
Faltbett n <Sicherheit> folding bed *(Erste-Hilfe)*
Faltboot n <Wassertrans> collapsible boat
Faltdipol m <Funktech> folded dipole
Falte f 1. <Anstrich> lap; 2. <Fertig> pucker; flopper *(Blech)*; 3. <Ker & Glas> crimp, lap; 4. <Papier> wrinkle; 5. <Textil> crease, pleat
falten v 1. <Fertig> crease; 2. <Papier> fold; 3. <Textil> pleat
Falten n 1. <Papier> folding; 2. <Textil> pleating
Faltenbalg m <Maschinen, Mechan> bellows
Faltenbildung f <Fertig> puckering
faltenfrei adj <Anstrich> unlapped
Faltenhalter m <Fertig> blank holder
Faltenhalterkraft f <Fertig> blank-holder force
Faltenrohr n <Maschinen> quill tube
Faltflügelflugzeug n <Lufttrans> folding-wing aircraft
Faltgut n <Konstzeich> folded material
Faltkante f <Konstzeich> folding edge
Faltkarton m <Verpack> collapsible case, folding carton
Faltprospekt m <Druck> folder
Faltschachtel f 1. <Papier> folding box; 2. <Verpack> folding cardboard box
Falttür f <Bau> accordion door, flexible door, folding door
Falt- und Aufrichtmaschine f <Verpack> large case erector
Faltung f 1. <Elektronik> folding; 2. <Telekom> convolution
Faltungscode m <Telekom> convolution code, convolutional code
Faltungscodierung f mit halber Geschwindigkeit <Telekom> rate one-half convolutional coding
Faltungsfrequenz f <Comp & DV> aliasing
Faltungsoperation f <Elektronik> convolution *(Nachrichtenverarbeitung)*
Faltungsprodukt n 1. <Elektronik> convolution integral, convolution product *(Nachrichtenverarbeitung)*; 2. <Math> convolution product
Faltverschluss m <Verpack> tuck-in closure
Faltversuch m 1. <Maschinen> folding test; 2. <Metall> cold bend test, doubling-over test; bend test *(Platten, Stäben)*; 3. <Verpack> folding test
Faltwerkdach n <Bau> folded-plate roof
Falz m 1. <Bau> joggle, lap, mortise, *(BE)* plough, *(AE)* plow, rabbet, seam; 2. <Fertig> lock seam *(Blech)*; rabbet; 3. <Papier> fold
Falzbarkeit f <Fertig> foldability
Falzbodenkarton m <Verpack> folded-bottom box
falzen v 1. <Bau> bead, bend, rabbet; 2. <Fertig> bend, welt; seam *(Blech)*; 3. <Mechan> crimp; 4. <Verpack> crease
Falzen n 1. <Bau> bending; 2. <Fertig> bending, folding, seaming
Falzfestigkeit f <Verpack> folding strength
Falzfuge f <Bau> rebated joint
Falzhobel m <Bau> fillister, rabbet plane, rebate plane
Falzmarken fpl <Druck> folding marks
Falzmaschine f 1. <Druck> folder unit; 2. <Maschinen> folding machine, seaming machine

Falzmeißel m <Fertig> groover
Falztrichter m <Papier> former
Falz- und Anleimmaschine f <Verpack> crease and glueing machine
Falz- und Verschließmaschine f <Verpack> folding and seaming machine
Falzung f <Bau> feather edge
Falzwalzen fpl <Druck> folding rollers
Falzzudrücken n <Fertig> grooving
Familienpackung f <Verpack> economy size pack, family packet
Fan m <Lufttrans> fan *(Mantelstromtriebwerk)*
Fang... <Elektronik> capture
Fangarbeit f <Erdöl> fishing job
Fangbereich m 1. <Elektronik> lock-in range; capture range *(AFC)*; 2. <Funktech> lock-in range, locking range
Fangdamm m 1. <Bau> batardeau, coffer, cofferdam; 2. <Wasserversorg> cofferdam; coffer *(einer Schleuse)*
Fangeisen n <Ker & Glas> bait
Fangen n <Telekom> call interception, call trace; pull-in *(PLL)*
Fanggerät n <Erdöl> fishing tool
Fanggitter n <Elektrotech> suppression grid
Fangglocke f <Erdöl> overshot
Fanghilfeverstärker m <Elektronik> lock-in amplifier
Fangleine f <Wassertrans> painter *(Festmachen)*
Fangleitung f <Sicherheit> roof conductor
Fangort m <Elektrotech> trapping site
Fangstelle f 1. <Comp & DV> trap; 2. <Elektrotech> trapping site
Fangstoff m <Elektrotech> getter
Fangverhalten n <Telekom> acquisition behaviour
Fangvorrichtung f <Sicherheit> safety gear
Fangwerkzeug n <Erdöl> fishing tool
Fan-Triebwerk n <Lufttrans> fan jet engine
Farad n 1. <Elektriz, Elektrotech> farad *(F)*; 2. <Funktech> farad *(Einheit der Kapazität)*; 3. <Metrol, Phys> farad *(F)*
Faraday'sche Drehung f <Raumfahrt> Faraday rotation
Faraday'sche Gesetze npl <Elektriz, Phys> Faraday's laws
Faraday'sche Konstante f <Phys> Faraday constant, faraday
Faraday'sche Scheibe f <Elektrotech> *(BE)* Faraday disc, *(AE)* Faraday disk
Faraday'sche Schlitzscheibe f <Phys> *(BE)* Faraday disc, *(AE)* Faraday disk
Faraday'scher Dunkelraum m <Elektronik, Phys> Faraday dark space
Faraday'scher Effekt m <Elektrotech, Phys> Faraday effect
Faraday'scher Käfig m 1. <Elektriz> Faraday cage; 2. <Elektrotech> Faraday cage, Faraday screen; 3. <Funktech> Faraday cage; 4. <Kerntech> Faraday cage, Faraday shield; 5. <Phys> Faraday cage
Faraday'scher Zylinder m <Phys> Faraday cylinder
Faraday'sches Gefäß n <Elektriz> Faraday ice pail
Faraday'sches Induktionsgesetz n <Elektriz, Phys> Faraday's law
Farb... <Anstrich, Fernseh> *(AE)* color, *(BE)* colour
Farbabbrennlampe f <Bau> paint-burning lamp
Farbabgleich m <Druck> *(AE)* color matching, *(BE)* colour matching
Farbabgleichung f <Ergon> *(AE)* color matching, *(BE)* colour matching
Farbabmusterung f <Druck> *(AE)* color matching, *(BE)* colour matching
Farbabstimmung f <Comp & DV> *(AE)* color balance, *(BE)* colour balance
Farbabweichung f <Fernseh> chromaticity aberration

Farbabzug m <Foto> (AE) color print, (BE) colour print
Farbadapter m <Comp & DV> (AE) color adaptor, (BE) colour adaptor
Farbaffinität f <Textil> dyeing affinity
Farbanalysator m <Foto> (AE) color analyzer, (BE) colour analyser
Farbanteile mpl <Fernseh> chrominance components
Farbanzeige f <Comp & DV> (AE) color display, (BE) colour display
Farbanzeige f für Strahlungsdosis <Strahlphys> (AE) dose color indicators, (BE) dose colour indicators
Farbart-, Bild-, Austast-, Synchronsignal n <Fernseh> (AE) composite color signal, (BE) composite colour signal
Farbauflösungsvermögen n <Phys> chromatic resolving power
Farbaufnahme f <Foto> (AE) color picture, (BE) colour picture
Farbaufrollen n <Bau> roller painting
Farbaufsatz m <Foto> (AE) color head, (BE) colour head
Farbauszugsfilter n <Foto> (AE) color separation filter, (BE) colour separation filter
Farbauszugsnegativ n <Foto> (AE) color separation negative, (BE) colour separation negative
Farbauszugsüberlagerung f <Fernseh> (AE) chromakey, (AE) color separation overlay, (BE) colour separation overlay
Farbauszugverfahren n <Druck> (AE) color separation, (BE) colour separation
Farbbalken mpl <Fernseh> (AE) color bars, (BE) colour bars
Farbbalkengenerator m <Fernseh> (AE) color bar generator, (BE) colour bar generator
Farbband n <Comp & DV> inked ribbon, ribbon
Farbbandführung f <Aufnahme, Gerät> ribbon guide (Registriergerät)
Farbbeizen n <Ker & Glas> staining
Farbbeständigkeit f <Kunststoff> (AE) color fastness, (BE) colour fastness
Farbbezugssignal n <Fernseh> (AE) color reference signal, (BE) colour reference signal
Farbbildröhre f <Fernseh> colour picture tube
Farbbildschirm m <Comp & DV> (AE) color display, (AE) color monitor, (BE) colour display, (BE) colour monitor
Farbburst m <Fernseh> (AE) color burst, (BE) colour burst
Farbcode m für Feuerlöscher <Sicherheit> (AE) fire extinguisher color code, (BE) fire extinguisher colour code
Farbdecoder m <Fernseh> (AE) color decoder, (BE) colour decoder
Farbdecodierer m <Comp & DV> (AE) color decoder, (BE) colour decoder
Farbdichtemesser m <Papier> (BE) colour densitometer
Farbdifferenz f <Fernseh> (AE) color difference, (BE) colour difference
Farbdifferenzsignal n <Fernseh> (AE) color difference signal, (BE) colour difference signal
Farbdisplay n <Telekom> colour display
Farbdreieck n <Phys> (AE) color triangle, (BE) colour triangle
Farbdruck m <Druck> (AE) color printing, (BE) colour printing, (AE) colorwork, (BE) colourwork
Farbdrucker m <Comp & DV> colour printer
Farbe f 1. <Anstrich> paint; 2. <Druck> (AE) color, (BE) colour; 3. <Fernseh> chroma; 4. <Lebensmittel, Papier, Teilphys, Telekom> (AE) color, (BE) colour; 5. <Textil> (AE) color, (BE) colour, dye • **die Farbe abstimmen** <Textil> match the shade • **empfindlich für alle Farben** 1. <Chemie> panchromatic (Fotografie); 2. <Druck, Foto> panchromatic

Färbeapparat m <Fertig> dyeing equipment
Färbebad n <Textil> bath
Färbebaum m <Textil> beam
Färbebeschleuniger m <Textil> carrier
farbecht adj <Verpack> (AE) colorfast, (BE) colourfast
Farbechtheit f <Kunststoff> (AE) color fastness, (BE) colour fastness
Färbeflotte f <Fertig> dyeing solution (Kunststoffinstallationen)
Färbehülse f <Textil> cone tube
Farbeindringprüfung f <Mechan> dye penetrant test
färben v <Foto, Textil> dye • **in einem Stück färben** <Textil> dye in the piece
Farbenanalyse f <Fernseh> (AE) color analysis, (BE) colour analysis
Farbendruck m <Druck> (AE) color printing, (BE) colour printing, (AE) colorwork, (BE) colourwork
Farbendruckmaschine f <Druck> (AE) color printing machine, (BE) colour printing machine
Farbenform f <Druck> (AE) colorform, (BE) colour form
Farbenlichtdruck m <Druck> (AE) color collotype, (BE) colour collotype
Farbenmesser m <Phys> colorimeter
Farbenmessung f <Phys, Strahlphys> colorimetry
Farbenphase f <Fernseh> (AE) color phase, (BE) colour phase
Farbenschwund m <Textil> fading
Farbensehen n <Ergon> (AE) color vision, (BE) colour vision
Farbenspektrum n <Foto> chromatic spectrum
Farbentrennung f <Fernseh> (AE) color separation, (BE) colour separation
Farbentwickler m <Foto> (AE) color developer, (BE) colour developer
Farbentwicklung f <Foto> (AE) color development, (BE) colour development
Farbenzerlegung f <Fernseh> (AE) color break-up, (BE) colour break-up
Färberei f <Textil> dyeing
Färbewickel m <Textil> mock cake
Farbfehler m 1. <Elektronik> (AE) color artefact, (BE) colour artefact (Nachrichtentechnik); 2. <Fernseh> chromatic aberration, (AE) color error, (BE) colour error; 3. <Optik> chromatic distortion; chromatic aberration (Linsenfehler); 4. <Strahlphys> chromatic aberration
Farbfeld n <Fernseh> (AE) color field, (BE) colour field
Farbfeldkorrektur f <Fernseh> (AE) color field corrector, (BE) colour field corrector
Farbfernkopieren n <Telekom> colour telefax
Farbfernsehempfänger m <Fernseh> (AE) color television receiver, (BE) colour television receiver
Farbfernsehen n <Fernseh> (AE) color television, (BE) colour television
Farbfernseher m <Fernseh> (BE) colour TV set
Farbfernsehgerät n <Fernseh> (AE) color television receiver, (BE) colour television receiver, (BE) colour TV set
Farbfernsehnorm f <Fernseh> (AE) color television standard, (BE) colour television standard
Farbflimmern n <Fernseh> chromatic flicker
Farbfotografie f <Foto> (AE) color photography, (BE) colour photography
Farbfotoverfahren n <Foto> (AE) color printing process, (BE) colour printing process
Farbgarn n <Textil> dyed yarn
Farbgebung f <Comp & DV> painting
Farbgenauigkeit f <Druck> colour accuracy
Farbglas n <Ker & Glas> (AE) colored glass, (BE) coloured glass
Farbgleichgewicht n 1. <Fernseh> chromatic balance; 2. <Foto> (AE) color balance, (BE) colour balance

Farbgrafik

Farbgrafik f <Comp & DV> (AE) color graphics, (BE) colour graphics
Farbgrafikadapter m (CGA) <Comp & DV> (AE) color graphics adaptor, (BE) colour graphics adaptor (CGA)
Farbhilfsträgerbezug m <Fernseh> chrominance subcarrier reference
Farbhintergrundgenerator m <Fernseh> (AE) color background generator, (BE) colour background generator
farbig adj <Telekom> (AE) colored, (BE) coloured
farbig verzieren v <Ker & Glas> (AE) put down in color work, (BE) put down in colour work
farbige Ränder mpl <Druck> (AE) colored edges, (BE) coloured edges
farbiges Paketband n <Verpack> (AE) colored strapping, (BE) coloured strapping
farbiges Rauschen n <Phys> pink noise
farbiges Umführungsband n <Verpack> (AE) colored strapping, (BE) coloured strapping
Farbkomponente f <Fernseh> chromatic component
Farbkontrollstreifen m <Druck> colour control strip
Farbkoordinaten fpl 1. <Phys> chromaticity coordinates; chromatic coordinates (Farbdreieck); 2. <Strahlphys> (AE) color coordinates, (BE) colour coordinates
Farbkörper m <Kunststoff, Textil> pigment
Farbkorrektur f <Fernseh> (AE) color correction, (BE) colour correction
Farbkorrekturfilter n <Elektronik> (AE) color correction filter, (BE) colour correction filter
farbkorrigiertes Objektiv n <Foto> (AE) color-corrected lens, (BE) colour-corrected lens
farbkundlich adj <Anstrich> chromate
farbliches Blinklicht n <Trans> coloured flashing light (Verkehrsampel)
farblos adj <Lebensmittel> (AE) colorless, (BE) colourless
farblose Masse f <Ker & Glas> (AE) colorless flux, (BE) colourless flux
farbloses Glas n <Ker & Glas> (AE) colorless glass, (BE) colourless glass
farbloses Produkt n <Meerschmutz> white product
Farbmaßzahl f <Fernseh> chromaticity
Farbmessung f <Wasserversorg> colorimetry
Farbmetallographie f <Metall> (AE) color metallography, (BE) colour metallography
Farbmischung f 1. <Comp & DV> dithering; 2. <Ker & Glas> (AE) color mix, (BE) colour mix
Farbmodulator m <Fernseh> (AE) color modulator, (BE) colour modulator
Farbmonitor m <Comp & DV> (AE) color monitor, (BE) colour monitor
Farbmühle f <Ker & Glas> paint mill
Farbnachlauf m <Fernseh> chroma delay
Farboszillator m <Elektronik, Fernseh> chrominance subcarrier oscillator
Farbpalette f <Comp & DV> (AE) color palette, (BE) colour palette
Farbphase f <Fernseh> chrominance phase
Farbphasendiagramm n <Fernseh> (AE) color phase diagram, (BE) colour phase diagram
Farbpigment n <Anstrich> pigment
Farbpyrometer n 1. <Strahlphys> (AE) color pyrometer, (BE) colour pyrometer; 2. <Thermod> colorimetric pyrometer
Farbrad n <Optik> (AE) color wheel, (BE) colour wheel
Farbrahmen m <Fernseh> (AE) color framing, (BE) colour framing
Farbraster n <Druck, Foto> (AE) color screen, (BE) colour screen
Farbrauschen n <Fernseh> (AE) color noise, (BE) colour noise, (AE) cross-color noise, (BE) cross-colour noise

244

Farbsättigung f <Fernseh> (AE) color saturation, (BE) colour saturation
Farbschaltung f <Aufnahme> (AE) color sampling, (BE) colour sampling
Farbschaltungsabfolge f <Fernseh> (AE) color sampling sequence, (BE) colour sampling sequence
Farbschaltungsgeschwindigkeit f <Fernseh> (AE) color sampling rate, (BE) colour sampling rate
Farbschattierungstreifenbildung f <Fernseh> banding on hue
Farbschlieren fpl <Ker & Glas> (AE) color streaks, (BE) colour streaks
Farbschlüssel m <Fernseh> (AE) chromakey, (AE) color separation overlay, (BE) colour separation overlay
Farbschwankung f <Fernseh> chroma flutter
Farbschwellwert m <Fernseh> (AE) color threshold, (BE) colour threshold
Farbsignal n <Fernseh> (AE) color signal, (BE) colour signal
Farbsplitten n <Fernseh> chromatic splitting
Farbsplitter m <Anstrich> paint chip
Farbspritzen n <Anstrich> spray painting
Farbspritzgerät n <Anstrich> paint-spraying apparatus
Farbspritzroboter m <Sicherheit> paint-spraying robot
Farbsprühen n <Anstrich> spray painting
Farbstärke f <Kunststoff> (AE) color strength, (BE) colour strength
Farbsteuergitter n <Fernseh> (AE) color grid, (BE) colour grid
Farbsteuerung f <Fernseh> chroma control, chroma pilot
Farbstich m <Foto> (AE) color cast, (BE) colour cast
Farbstoff m 1. <Druck> (AE) coloring matter, (BE) colouring matter; 2. <Ker & Glas> (AE) coloring agent, (BE) colouring agent, pigment; 3. <Papier> stain; 4. <Textil> dye, dyestuff
Farbstoff m **zur Flächenrissprüfung** <Fertig> dye penetrant
Farbstoffaufnahme f <Textil> dye uptake
Farbstoffherstellung f <Ker & Glas> (AE) color striking, (BE) colour striking
Farbstofflaser m <Phys> dye laser
Farbsynchrongatter n <Fernseh> burst gate
Farbsynchronisierung f <Fernseh> (AE) color lock, (BE) colour lock
Farbsynchronisierungssignal n <Fernseh> (AE) color sync signal, (BE) colour sync signal
Farbsynchronphase f <Fernseh> burst phase
Farbsynchronsignal n <Fernseh> burst, color burst
Farbsynchrontrennung f <Fernseh> burst separator
Farbsynthesizer m <Fernseh> (AE) color synthesiser, (BE) colour synthesizer
Farbtafel f <Fernseh> chromaticity diagram, (AE) color chart, (BE) colour chart
Farb-Telefaxen n <Telekom> (AE) telefax, (BE) color colour telefax
Farbtemperaturmessgerät n <Foto> (AE) color temperature meter, (BE) colour temperature meter
Farbtestbild n <Fernseh> (AE) color pattern, (BE) colour pattern
Farbtestbild n **mit Balken** <Fernseh> (AE) color bar test pattern, (BE) colour bar test pattern
Farbtiefe f <Textil> depth of shade
Farbton m 1. <Druck> hue, tint; 2. <Fernseh> chromaticity; 3. <Foto> hue, tint; 4. <Ker & Glas> tint; 5. <Kunststoff> (AE) color tone, (BE) colour tone; 6. <Textil> shade
Farbtonregelung f <Fernseh> hue control
Farbtönung f <Kunststoff> (AE) color tone, (BE) colour tone

Farbträger m <Fernseh> chrominance carrier, *(AE)* color subcarrier, *(BE)* colour subcarrier
Farbträger-Demodulation f <Elektronik, Fernseh> chrominance subcarrier demodulation
Farbträger-Demodulator m <Elektronik, Fernseh> chrominance subcarrier demodulator
Farbträger-Modulation f <Elektronik, Fernseh> chrominance subcarrier modulation
Farbträger-Modulator m <Elektronik, Fernseh> chrominance subcarrier modulator
Farbträger-Unterdrückung f <Fernseh> *(AE)* color kill, *(BE)* colour kill
Farbtrennverfahren n <Druck> *(AE)* color separation, *(BE)* colour separation
Farbtüchtigkeit f <Ergon> *(AE)* acuity of color perception, *(BE)* acuity of colour perception
Farbumkehrentwicklung f <Foto> *(AE)* color reversal process, *(BE)* colour reversal process
Farbumkehrfilm m <Foto> *(AE)* color reversal film, *(BE)* colour reversal film, *(AE)* reversal-type color film, *(BE)* reversal-type colour film
Farbumrandung f <Fernseh> *(AE)* color fringing, *(BE)* colour fringing
Färbung f <Lebensmittel> *(AE)* coloring, *(BE)* colouring
Farbunterscheidung f <Qual> *(AE)* color discrimination, *(BE)* colour discrimination
Farbunterscheidungsvermögen n <Ergon> *(AE)* color discrimination, *(BE)* colour discrimination
Farbveränderung f <Ker & Glas> *(AE)* color change, *(BE)* colour change
Farbverbrauch m <Papier> ink coverage
Farbverfahrenchemikalien fpl <Foto> *(AE)* color processing chemicals, *(BE)* colour processing chemicals
Farbverschiebung f <Comp & DV> *(AE)* color shift, *(BE)* colour shift
Farbwert m 1. <Elektronik> chrominance; 2. <Phys> tristimulus value
Farbzerlegung f <Optik> chromatic dispersion *(Licht)*
Farbzuordnungstabelle f <Comp & DV> *(AE)* color map, *(BE)* colour map
Farinograph m <Lebensmittel> farinograph
Farmer'scher Abschwächer m <Foto> Farmer's reducer
Fase f 1. <Bau> chamfer; scarf *(Holz)*; 2. <Fertig> bevel, heel; 3. <Maschinen> land • **mit Fase** <Fertig> landed
fasen v <Fertig> chamfer
fasenartiger Anschliff m <Fertig> ridge
fasenartiger Anschliff m **an der Schneide** <Fertig> primary land
Fasenfreiwinkel m <Fertig> secondary clearance angle
Fasenhöhe f <Fertig> depth of body clearance *(Spiralbohrer)*
Fasenring m <Maschinen> bevel ring
Fasenwinkel m <Maschinen> angle of bevel
Faser f <Optik, Papier, Telekom, Textil> *(AE)* fiber, *(BE)* fibre • **Fasern trennen** <Fertig> cleave *(optische Fasern)* • **gegen die Faser arbeiten** <Bau> work against the grain
Faser f **mit einheitlichem Brechungsindex** <Optik> *(AE)* uniform-index fiber, *(BE)* uniform-index fibre
Faser f **mit parabolischem Indexprofil** <Optik, Telekom> *(AE)* parabolic-index fiber, *(BE)* parabolic-index fibre
Faser f **mit zweitem Transmissionsfenster** <Optik> *(AE)* second window fiber, *(BE)* second window fibre
Faser f **ohne verschobenen Dispersionsnullpunkt** <Telekom> non zero dispersion shifted fiber, NZ-DSF
Faserabschluss m <Optik> *(AE)* fiber buffer, *(BE)* fibre buffer

Faserachse f <Optik> *(AE)* fiber axis, *(BE)* fibre axis
Faserband n <Textil> sliver
Faserbart m <Textil> tuft *(Spinnen)*
Faserbeschichtung f <Telekom> *(AE)* fiber coating, *(BE)* fibre coating
faserbewehrt adj <Bau> *(AE)* fiber-reinforced, *(BE)* fibre-reinforced
faserbewehrter Beton m <Bau> *(AE)* fiber-reinforced concrete, *(BE)* fibre-reinforced concrete
Faserbündel n <Ker & Glas, Optik, Telekom> *(AE)* fiber bundle, *(BE)* fibre bundle *(Lichtwellenleiter)*
Faserdämpfung f <Telekom> *(AE)* fiber loss, *(BE)* fibre loss
Faserflor m <Textil> pile
Fasergehalt m <Textil> *(AE)* fiber content, *(BE)* fibre content
Faserholz n <Abfall, Papier> pulpwood
Faserhülle f <Telekom> *(AE)* fiber buffer, *(BE)* fibre buffer, *(AE)* fiber jacket, *(BE)* fibre jacket
Faserhülse f <Telekom> ferrule *(Lichtleitfaser)*
faserig adj <Papier> fibrous
faserige Mikrostruktur f <Metall> fibrous microstructure
Faserisolierung f <Heiz & Kälte> fibrous insulation
Faserkalk m <Papier> agalite *(Asbest)*
Faserkern m <Telekom> *(AE)* fiber core, *(BE)* fibre core
Faserkerngröße f <Elektronik> *(AE)* fiber core size, *(BE)* fibre core size
Faserlänge f <Textil> staple length
Fasermantel m <Optik> *(AE)* fiber cladding, *(AE)* fiber coating, *(BE)* fibre coating, *(AE)* fiber jacket, *(BE)* fibre cladding, *(BE)* fibre jacket
Faseroptik f <Comp & DV, Elektrotech, Ker & Glas, Phys, Telekom> *(AE)* fiber optics, *(BE)* fibre optics
Faseroptik f **zur Lichtübertragung** <Optik> *(AE)* transit fiberoptic, *(BE)* transit fibreoptic
Faseroptik-Ausrüstung f <Labor> *(AE)* fiber optics equipment, *(BE)* fibre optics equipment
faseroptische Technologie f <Elektrotech> *(AE)* fiberoptic technology, *(BE)* fibreoptic technology
faseroptische Übertragung f <Optik> *(AE)* fiberoptic transmission, *(BE)* fibreoptic transmission
faseroptischer Empfänger m 1. <Elektrotech> *(AE)* fiberoptic receiver, *(BE)* fibreoptic receiver; 2. <Telekom> *(AE)* receive fiberoptic terminal device, *(BE)* receive fibreoptic terminal device
faseroptischer Kreisel m <Wassertrans> *(BE)* fibre optic gyro *(Navigation)*
faseroptischer Messwandler m <Elektrotech> *(AE)* fiberoptic transducer, *(BE)* fibreoptic transducer
faseroptischer Sender m <Elektrotech> *(AE)* fiberoptic transmitter, *(BE)* fibreoptic transmitter
faseroptischer Steckverbinder m <Elektrotech> *(AE)* fiberoptic connector, *(BE)* fibreoptic connector
faseroptischer Transducer m <Elektrotech> *(AE)* fiberoptic transducer, *(BE)* fibreoptic transducer
faseroptisches Endgerät n <Optik> *(AE)* fiberoptic terminal device, *(BE)* fibreoptic terminal device
faseroptisches Kabel n <Optik> *(AE)* fiberoptic cable, *(BE)* fibreoptic cable
faseroptisches Kabelnetz n <Elektrotech> *(AE)* fiberoptic network, *(BE)* fibreoptic network
faseroptisches Terminal n <Optik> *(AE)* fiberoptic terminal device, *(BE)* fibreoptic terminal device
faseroptisches Übertragungssystem n <Elektrotech> *(AE)* fiberoptic transmission system, *(BE)* fibreoptic transmission system
Faserquetschung f <Fertig> *(AE)* ruptured fiber structure, *(BE)* ruptured fibre structure
Faserrichtung/quer adv zur <Fertig> *(AE)* across the fiber grain, *(BE)* across the fibre grain

Faserschichtglas

Faserschichtglas n <Ker & Glas> ply glass *(Flachglas)*
Faserschlaufe f <Sicherheit> *(AE)* fiber-type sling, *(BE)* fibre-type sling
Faserschnittmatte f <Kunststoff> chopped-strand mat
Faserspeiser m <Ker & Glas> *(AE)* fiber feeder, *(BE)* fibre feeder
Faserspleiß m <Telekom> *(AE)* optical fiber splice, *(BE)* optical fibre splice
Faserstippe f <Papier> flake
Faserstoff m 1. <Papier> pulp; 2. <Textil> fibrous material
Faserstoffdichtung f <Mechan> *(AE)* fiber gasket, *(BE)* fibre gasket
Faserstoffriemen m <Maschinen> *(AE)* fiber belt, *(BE)* fibre belt
Faserstoffschicht f <Papier> furnish layer
Faserstoffzusammensetzung f <Papier, Verpack> *(AE)* fiber composition, *(BE)* fibre composition
Faserstreuung f <Optik> *(AE)* fiber scattering, *(BE)* fibre scattering
Faserstruktur f <Metall> *(AE)* fiber texture, *(BE)* fibre texture
Fasertaper m <Telekom> *(AE)* tapered fiber, *(BE)* tapered fibre
Fasertauwerk n <Wassertrans> *(AE)* fiber rope, *(BE)* fibre rope
Fasertorf m <Kohlen> fibrous peat
Fasertrenngerät n <Fertig> cleaver *(Faseroptik)*
Faserüberlänge f <Optik> *(AE)* fiber excess length, *(BE)* fibre excess length
Faserumhüllung f <Optik> fibre jacket, secondary coating
Faserummantelung f <Optik> *(AE)* fiber buffet, *(BE)* fibre buffet
Faserverbinder m <Optik> joint
Faserverstärkung f <Bau> *(AE)* fiber reinforcement, *(BE)* fibre reinforcement
Faserwendel f <Optik> *(AE)* fiber helix, *(BE)* fibre helix
Faserziehen n 1. <Fertig> drawing into fibers *(Faseroptik)*; 2. <Telekom> *(AE)* fiber drawing, *(BE)* fibre drawing
Fass n 1. <Bau, Kohlen> barrel; 2. <Kunststoff> drum; 3. <Lebensmittel> barrel, vat; 4. <Mechan, Phys, Trans> barrel • **frisch vom Fass gezapft** <Lebensmittel> drawn from the wood
Fass n **für Atommüll** <Kerntech> waste drum
Fassade f <Bau> façade, face, front face, frontage *(Gebäude)*
Fassadenfarbe f <Kunststoff> house paint
Fassadenverkleidung f <Bau> curtain-walling
fassen v 1. <Bau> catch; 2. <Fertig> hold *(Flüssigkeit)*; bite *(Werkzeug)*; 3. <Wassertrans> grip
Fassonschmieden n <Fertig> swaging
Fassreifen m <Fertig, Verpack> hoop
Fasstonne f <Wassertrans> barrel buoy *(Navigation)*
Fassung f 1. <Elektrotech> holder *(Glühlampe)*; fuse base *(Sicherung)*; snubber resistor, socket *(elektrische Lampen)*; 2. <Fertig> receptacle; 3. <Foto> mounting *(Kamera, Licht)*; 4. <Funktech> socket; 5. <Maschinen> mount, setting; 6. <Telekom> socket
Fassung f **der Frontlinse** <Foto> mount of front element
Fassungsvermögen n 1. <Bau, Elektriz> capacitance, capacity *(Behälter)*; 2. <Fertig> holding capacity *(von Zuführvorrichtungen)*; 3. <Heiz & Kälte> freezer capacity *(Tiefkühlgerät)*; 4. <Maschinen> capacity; 5. <Wassertrans> volume *(Maßeinheit)*
Fasswicklung f <Elektrotech> barrel winding
Fastsenkrechtstarter m *(STOL-Flugzeug)* <Lufttrans> short takeoff and landing aircraft *(STOL aircraft)*
Fastzusammenstoß m <Lufttrans, Sicherheit, Trans> near collision, near miss

Faszie f <Bau> fascia
Faulbaumbitter n <Chemie> frangulin
Faulbecken n <Wasserversorg> septic tank
Faulbehälter m <Abfall> digester, digestion tank, digestion sump
Fäule f <Wassertrans> rot *(Holz)*
faulfähiger Schlamm m <Abfall> putrescible sludge
Faulfähigkeit f <Abfall> putrescibility
Faulgas n 1. <Abfall> biogas, digester gas, fermentation gas; 2. <Thermod> digester gas
Faulgrube f <Wasserversorg> septic tank
Fäulnisalkaloid n <Chemie> ptomaine
fäulnisfähiger Stoff m <Abfall> putrescible matter
fäulnissicher adj <Papier> rotproof
Faulraum m 1. <Abfall> digestion tank, digestion sump; 2. <Thermod> digestion tank
Faulschlamm m <Abfall, Bau, Chemie> digested sludge
Faultank m <Abfall> anaerobic sludge digestor
Faulteich m <Abfall> anaerobic lagoon
Faulturm m <Abfall> anaerobic sludge digester
Faulung f <Abfall> fouling
Faustachse f <Kfztech> stub axle *(Rad)*
Fäustel m 1. <Fertig> hammer; 2. <Maschinen> club hammer
Fax n 1. <Comp & DV> fax, facsimile *(Faksimile, Faxgerät, Fernkopieren, Fernkopierer)*; 2. <Funktech> fax, facsimile *(Faksimile, Fernkopieren, Telefax)*; 3. <Telekom> fax, facsimile *(Faksimile, Faxgerät, Fernkopieren, Fernkopierer, Telefax)*
Fax-Abruf m <Telekom> fax on demand
faxen v <Telekom> fax, send by fax
Faxgerät n *(Fax)* <Comp & DV, Telekom> facsimile equipment, facsimile machine, fax equipment
Faxkombi n <Telekom> fax-phone
Fax-Modem n <Comp & DV, Telekom> fax modem
Faxvorlage f <Comp & DV> fax template
Faxweiche f <Telekom> fax switch *(Telefon/Fax)*
FB *(Flughafenbake)* <Lufttrans, Raumfahrt> *(BE)* aerodrome beacon, *(AE)* airdrome beacon
FBAS-Signal n <Fernseh> *(AE)* composite color signal, *(BE)* composite colour signal
FCKW *(Fluorchlorokohlenwasserstoff)* <Chemie, Umweltschmutz> CFC *(chlorofluorocarbon)*
FCNE *(Flugüberwachungs- und Navigationsausrüstung)* <Lufttrans> FCNE *(flight control and navigational equipment)*
FdW *(Fahrt durchs Wasser)* <Wassertrans> speed through the water
Fe *(Eisen)* <Chemie> Fe *(iron)*
F&E *(Forschung und Entwicklung)* <Chemtech, Comp & DV, Kfztech> R&D *(research & development)*
Feder f 1. <Bau> feather, spring, tongue; 2. <Kfztech> spring; 3. <Maschinen> key, spring; 4. <Mechan, Phys> spring • **Feder kalibrieren** <Maschinen> scale
Feder f **mit geschlossener Wicklung** <Maschinen> close-coil spring
Feder f **mit gleich bleibender Federkraft** <Fertig> constant-force spring
Feder f **und Nut** f 1. <Bau> tongue and groove; 2. <Fertig> key and slot; 3. <Maschinen> key and feather
Federantrieb m <Elektrotech> spring action, spring drive
Federauflage f <Kfztech> spring seat
Federbandkupplung f <Maschinen> coil clutch, spring band clutch
Federbein n <Kfztech> strut *(Fahrwerk)*
federbelastetes Ventil n <Hydraul> spring-loaded valve
Federbewegung f <Maschinen> oscillation of a spring
Federbogen m <Fertig> swing arm *(Kunststoffinstallationen)*

Federbolzen m <Bau> spring bolt, spring hanger pin
Federbuchse f <Elektrotech> spring jack
Federbügel m <Maschinen> spring band, spring buckle, spring shackle, strap
Federcharakteristik f <Maschinen> spring characteristic
Federdruckkörper m <Fertig> (AE) automatic center punch, (BE) automatic centre punch
Federdynamometer n <Metrol> spring balance
Federfalle f <Bau> latch bolt
Federfassung f <Elektrotech> snap-in socket
Federgalvanometer n <Metrol> spring galvanometer
federgespannter Spannhebel m <Foto> spring-tensioned pressure lever
Federhaken m <Maschinen> spring hook
Federhaltebügel m <Kfztech> spring retainer
Federkeil m 1. <Bau> feather, feather tongue (Holz); 2. <Maschinen> feather, feather key
Federkennlinie f <Maschinen> spring characteristic
Federklammer f <Kfztech> rebound clip
Federklemme f **am Ende des Hefteisens** <Ker & Glas> gadget
Federkommutator m <Elektriz> spring commutator
Federkonstante f <Maschinen> spring constant
Federkorbpresse f <Ker & Glas> spring cage press
Federkraft f 1. <Maschinen> spring force; 2. <Mechan> spring
Federkraftregler m <Maschinen> spring governor
Federleiste f **für die gedruckte Schaltung** <Elektrotech> printed circuit connector
Federlösemechanismus m <Raumfahrt> spring release device (Raumschiff)
Federmanometer n <Phys> spring manometer
federn v <Maschinen> be resilient
federnd adj <Kunststoff> resilient
federnde Dichtung f <Maschinen> resilient seal
federnde Windung f <Fertig> active coil
federnder Anschlag m <Maschinen> spring stop
Federnut f <Fertig> spring groove
Federopazität f <Ker & Glas> plume opacity
Federpaket n 1. <Fertig> spring unit (Kunststoffinstallationen); 2. <Maschinen> spring assembly
Federpuffer m <Eisenbahn> spring buffer
Federrate f <Maschinen> spring rate
Federring m 1. <Fertig> lock washer; 2. <Maschinen> lock washer, split washer, spring clip; 3. <Mechan> lock washer
Federringdichtung f <Maschinen> spring lock washer
Federrollenlager n <Maschinen> flexible roller bearing
Federrückstellschalter m <Elektriz> spring return switch
Federrückstellung f <Fertig> spring return mechanism (Kunststoffinstallationen)
Federsatz m <Maschinen> nest of springs
Federschalter m <Elektriz, Elektrotech> spring switch, snap-action switch
Federscheibe f <Maschinen> split lock washer, spring washer
Federschienen fpl <Eisenbahn> spring points
Federschloss n <Bau> spring lock
Federsitz m <Kfztech> spring seat
Federspannmotor m <Mechan> clockwork
Federstahl m <Metall> spring steel
Federteilzirkel mpl <Maschinen> spring dividers
Federteller m <Maschinen> spring plate
Federung f 1. <Fertig> cushioning; 2. <Maschinen> spring suspension
Federungs- und Stoßdämpfereinstellung f <Kfztech> spring and damper setting
Federventil n <Hydraul> spring valve

Federverbindung f <Bau> (BE) ploughed-and-feathered joint, (AE) plowed-and-feathered joint, slip tongue joint, tongue-and-groove joint
Federvorsteckstift m <Maschinen> spring cotter
Federwaage f 1. <Maschinen> spring balance; 2. <Metrol> spiral balance, spring balance; 3. <Phys> spring balance
Fehlalarm m <Telekom> false alarm
Fehlalarmwahrscheinlichkeit f <Telekom> false alarm probability
Fehlanflugsverfahren n <Lufttrans> missed approach procedure
fehlangepasst adj <Fernseh, Funktech, Phys> mismatched
Fehlanruf m <Telekom> false call
Fehlanrufhäufigkeit f <Telekom> false calling rate
Fehlauslösung f <Elektriz> false trip
Fehlaustrag m <Kohlen> outsize
Fehlbohrung f <Erdöl> dry hole (Tiefbohrtechnik)
fehleingestellt adj <Lufttrans> out-of-pitch (Hubschrauber)
fehleingestelltes Blatt n <Lufttrans> out-of-pitch blade (Hubschrauber)
Fehler m 1. <Comp & DV> bug, defect, error, fault; 2. <Elektriz, Elektronik, Elektrotech> defect, error, failure, fault; 3. <Ker & Glas, Kerntech> defect (Material); 4. <Lebensmittel> blemish; 5. <Maschinen> fault; 6. <Math> error, shortcoming; 7. <Mechan> fault, flaw; 8. <Metall> defect, flaw, 9. <Phys> error; 10. <Qual> defect, error, nonconformity, nonconformance; 11. <Telekom> fault; 12. <Textil> flaw
Fehler m **bei der Ausführung** <Comp & DV> run-time error
Fehler m **durch Synchronisationsverlust** <Telekom> out-of-synchronization error, out-of-sync error
Fehler m **im Flankendurchmesser** <Metrol> pitch diameter error
Fehler m **im Teilkreisdurchmesser** <Metrol> pitch diameter error
Fehlerabschätzung f <Math, Qual> error estimation
Fehleranalyse f 1. <Comp & DV, Math, Qual> error analysis; 2. <Strahlphys> failure analysis
fehleranfällig adj <Sicherheit> error-prone
Fehleranteil m <Qual> fraction defective
Fehleranzahl f **pro Einheit** <Qual> defects per unit
Fehleranzeige f <Telekom> fault display
Fehleranzeiger m <Elektriz> fault detector
Fehlerbedingung f 1. <Comp & DV> error condition, fault; 2. <Qual> error condition
fehlerbehaftete Sekunde f <Telekom> errored second
Fehlerbehandlungsprogramm n <Comp & DV> failure routine
Fehlerbehandlungsroutine f <Qual> failure routine
Fehlerbehebung f 1. <Comp & DV> error handling, error management, error recovery, error trapping; 2. <Qual> error handling, error management, error recovery; 3. <Telekom> fault maintenance
Fehlerbehebungsverfahren n <Comp & DV> error recovery procedure
Fehlerbericht m 1. <Comp & DV> error report; 2. <Qual> defect note, error report
Fehlerberichterstattung f <Qual> nonconformance reporting
Fehlerberichtigung f <Comp & DV> error correction, fix
Fehlerbeseitigung f <Comp & DV, Telekom> debugging
Fehlerbestimmung f <Comp & DV> problem determination
Fehlercode m <Comp & DV> error code
Fehlerdiagnose f <Comp & DV, Elektriz, Qual, Telekom> error diagnosis, fault diagnosis

Fehlerdiagnostik

Fehlerdiagnostik *f* <Comp & DV> error diagnostics
Fehlerdichte *f* <Elektronik> defect density
Fehlerempfindlichkeit *f* <Qual, Telekom> error susceptibility
Fehlerentdeckung *f* <Comp & DV> fault detection
fehlererkennende Codierung *f* <Telekom> error detection coding
Fehlererkennung *f* <Comp & DV, Elektronik, Qual, Telekom> error detection, fault detection, fault identification, problem diagnosis
Fehlererkennungscode *m* 1. <Comp & DV, Elektronik> error-detecting code; 2. <Telekom> error detection code
Fehlererkennungseinrichtung *f* <Qual> error detector
Fehlererkennungsprogramm *n* <Comp & DV> fault location program
Fehlerfeststellung *f* <Fernseh> error detection
Fehlerfortpflanzung *f* <Comp & DV, Qual> error propagation
fehlerfrei *adj* <Textil> faultless
fehlerfreie Einheit *f* <Qual> conforming item
fehlerfreundliches System *n* <Sicherheit> error-tolerated system
Fehlergrenze *f* <Metrol> limit of error
fehlerhaft *adj* 1. <Comp & DV> corrupt *(Daten)*; 2. <Elektriz> faulty, out of operation; 3. <Mechan> faulty; 4. <Qual> nonconforming; defective *(Werkstoff)*; 5. <Textil> faulty
fehlerhafte Datei *f* <Comp & DV> corrupt file
fehlerhafte Eingabe *f* <Comp & DV> garbage
fehlerhafte Einheit *f* <Qual> nonconforming item
fehlerhafte Funktion *f* **des Lesekopfes** <Aufnahme, Comp & DV> head crash
fehlerhafte Funktion *f* **des Schreib-/Leseschreibkopfes** <Aufnahme, Comp & DV> head crash
fehlerhafte Funktion *f* **des Schreibkopfes** <Aufnahme, Comp & DV> head crash
fehlerhafter Block *m* <Telekom> erroneous block
fehlerhafter Zeitabschnitt *m* <Telekom> erroneous period
fehlerhaftes Datenbit *n* <Comp & DV> bad data bit
Fehlerhäufigkeit *f* 1. <Comp & DV> error rate; 2. <Math> frequency of errors; 3. <Telekom> error rate
Fehlerhäufung *f* 1. <Comp & DV> burst, error burst; 2. <Qual, Telekom> error density
Fehlerklassifizierung *f* <Qual> classification of nonconformance, classification of nonconformities
Fehlerkontrolle *f* <Qual, Telekom> error check
Fehlerkontrollzeichen *n* <Telekom> error check character, error check signal
Fehlerkorrektur *f* 1. <Comp & DV, Elektronik, Qual> error correction; 2. <Telekom> error correction, error recovery
Fehlerkorrekturcode *m* 1. <Comp & DV> error-correcting code, error correction code, self-checking code; 2. <Elektronik, Telekom> error-correcting code; 3. <Raumfahrt> error-correcting code *(Weltraumfunk)*
Fehlerkorrekturschlüssel *m* <Raumfahrt> error-correcting code *(Weltraumfunk)*
fehlerkorrigierende Codierung *f* <Telekom> error correction coding
fehlerkorrigierender Code *m* <Telekom> error correction code
Fehlerliste *f* <Comp & DV, Qual> error list
Fehlerlokalisierung *f* <Elektriz, Telekom> fault location
fehlerloses Programm *n* <Comp & DV> star program
Fehlermeldung *f* 1. <Comp & DV> error message; 2. <Qual> defect note; 3. <Telekom> error message
Fehlermuster *n* <Qual, Telekom> error pattern
Fehlernachricht *f* <Comp & DV> error message
Fehlerortsmessung *f* <Metrol> determination of fault location, fault location
Fehlerortung *f* 1. <Elektriz, Telekom> fault location; 2. <Kontroll> troubleshooting
Fehlerortungsfahrzeug *n* <Elektrotech> fault location van *(Kabelfehlerortung)*
Fehlerortungsgerät *n* <Gerät> fault location instrument
Fehlerprogramm *n* <Comp & DV> error program
Fehlerprotokollierung *f* 1. <Comp & DV> error logging, failure logging; 2. <Qual> error logging
Fehlerprüfcode *m* <Comp & DV> error-checking code
Fehlerprüfung *f* <Comp & DV, Qual> error checking
Fehlerquote *f* <Comp & DV, Qual, Telekom> error rate
Fehlerquotenmessung *f* <Qual, Telekom> error rate measurement
Fehlerrate *f* 1. <Comp & DV, Elektronik> error rate, failure rate; 2. <Qual> error rate, failure rate, outage rate; 3. <Telekom> error rate
Fehlerroutine *f* <Comp & DV, Qual> error routine
Fehlerrückführung *f* <Künstl Int> backpropagation, BP, backward error propagation, error back propagation *(neurales Netzwerk)*
Fehlerschutz *m* <Qual, Telekom> error protection
Fehlerschutzcode *m* <Qual, Telekom> error protection code
Fehlersicherung *f* <Qual, Telekom> error protection
Fehlersicherungsgerät *n* <Qual, Sicherheit, Telekom> ECD, error control device
Fehlersignal *n* <Elektronik> error signal
Fehlerspanne *f* <Comp & DV, Qual> margin of error
Fehlerspannungs-Stromunterbrecher *m* <Elektriz> fault voltage circuit breaker
Fehlerstrom *m* <Elektrotech> leakage current
Fehlersuche *f* 1. <Elektrotech> fault finding; 2. <Kontroll> troubleshooting; 3. <Telekom> debugging
Fehlersuchprogramm *n* <Comp & DV> debugger
Fehlersuchtabelle *f* <Maschinen> fault-finding table
fehlertolerant *adj* <Comp & DV, Elektrotech> fault-tolerant
fehlertolerantes System *n* <Comp & DV, Telekom> fault-tolerant system
Fehlertoleranz *f* <Comp & DV, Elektrotech, Telekom> fault tolerance
fehlertolerierte Technik *f* <Sicherheit> fault-tolerated engineering
Fehlerüberwachung *f* <Comp & DV, Qual> error control
Fehlerumgehung *n* <Comp & DV> workaround
Fehlerwahrscheinlichkeit *f* <Künstl Int, Qual, Telekom> error probability
Fehlerwiderstand *m* <Elektrotech> fault resistance
Fehlfarbe *f* <Comp & DV> false colour
Fehlfunktion *f* <Elektriz, Metrol, Raumfahrt> malfunction *(Raumschiff)*
fehlgeleiteter Anruf *m* <Telekom> misdirected call
fehlgeordnet *adj* 1. <Fertig> disordered *(Kristall)*; 2. <Qual> disordered
fehlgeschlagener Job *m* <Comp & DV> failed job
Fehlguss *m* <Ker & Glas> waste
Fehlinhalt *m* <Ker & Glas> off-content
Fehlkontakt *m* <Elektrotech> bad contact
Fehlmenge *f* <Lebensmittel> ullage *(Flüssigkeit)*
Fehlmessung *f* <Gerät> faulty measurement
Fehlprodukt *n* <Qual> nonconforming product
Fehlregistrierung *f* <Fernseh> misregistration
Fehlschalten *n* <Elektriz> false switching
fehlschlagen *v* <Bau> fail
Fehlschließung *f* <Elektrotech> false closure
Fehlsignal *n* <Elektronik> false signal
Fehlstelle *f* 1. <Aufnahme> drop-out *(Magnetbandaufzeichnung)*; 2. <Chemie> defect,

imperfection, lattice defect, void; vacancy *(Kristall)*; 3. <Comp & DV> blemish *(Speicher)*; 4. <Elektronik> electron hole, hole; 5. <Elektrotech> cavity, void; dry spot *(durch Kriechstrom)*; 6. <Fertig> let-go *(Lösen von Plastschichten)*; 7. <Metall> void, blow hole *(Gießharz)*
Fehlstellenleitfähigkeit f <Elektrotech> p-type conductivity
Fehlstellenstreuung f <Kerntech> defect scattering
fehlsynchronisiert adj <Fernseh> out of sync
Fehlweisung f <Lufttrans, Wassertrans> compass error *(Navigation)*
Fehlzündung f 1. <Elektronik> mode jump *(Magnetron)*; 2. <Kfztech> backfire, misfire, misfiring
Feile f 1. <Kfztech> file *(Werkzeug)*; 2. <Maschinen, Mechan> file
feilen v <Maschinen> file
Feilen n <Maschinen> filing
Feilenbürste f <Maschinen> file card
Feilenhärte f <Maschinen> file hardness
Feilenhärteprüfung f <Maschinen> file test
Feilenhieb m <Maschinen> cut, file cut
Feilkolben m 1. <Bau> *(BE)* pin vice, *(AE)* pin vise; 2. <Maschinen> *(BE)* filing vice, *(AE)* filing vise, *(BE)* tail vice, *(AE)* tail vise, *(BE)* hand vice, *(AE)* hand vise
Feilmaschine f <Maschinen> filing machine
Feilscheibe f <Maschinen> circular-cut file
Feilspan m <Fertig> filing
fein mahlen v <Chemtech> pulverize
Fein... <Elektronik, Gerät, Telekom> fine
Feinabstimmung f <Elektronik, Gerät, Telekom> fine tuning
Feinanalyse f <Ergon> fine analysis
Feinausrichtung f <Funktech> precision pointing *(Antenne)*
Feinbearbeitung f 1. <Fertig> finishing; 2. <Maschinen> fine machining, precision machining
Feinblech n <Metall> sheet, thin sheet
Feinblechwalzen n <Fertig> roller sheet, sheet rolling
Feinblechwalzwerk n <Fertig> sheet mill
Feinbohren n 1. <Fertig> *(AE)* borizing, fine boring; 2. <Maschinen> fine boring, precision boring
Feinbrechen n <Kohlen> fine crushing
Feinbrecher m <Chemtech> fine-crushing mill
Feindosierventil n <Gerät> fine-metering valve
Feindrehmaschine f <Maschinen> precision lathe
Feindruckmessgerät n <Gerät> micromanometer
Feine f <Metall> fineness
feine Garnnummer f <Textil> fine count
feine Skalenmarke f <Gerät> hairline
Feineinstellschraube f <Optik> fine adjustment screw
Feineinstellskale f <Gerät> vernier scale
Feineinstellung f 1. <Elektronik, Fernseh> fine adjustment; 2. <Fertig> fine adjustment, precision setting, vernier adjustment; 3. <Funktech> fine adjustment; 4. <Gerät> fine setting, vernier adjustment; 5. <Kerntech> fine adjustment; 6. <Optik> fine adjustment *(mittels Mikrometerschraube)*
Feinen n <Fertig> refining
feiner Furnace-Ruß m *(FF-Ruß)* <Kunststoff> fine furnace carbon black *(FF carbon black)*
feiner Skalenstrich m <Gerät> hairline
feiner Strich m <Druck> fine line
Feinerde f <Kohlen> fine soil
feines Garn n <Textil> fine-count yarn
Feinfilter n 1. <Maschinen> fine-mesh filter; 2. <Wasserversorg> microfilter
Feinfiltration f <Chemie> clarification
Feinfolie f <Kunststoff> film
Feinfräsen n <Maschinen> fine milling, precision milling

Feingehalt m <Metall> fineness
Feingemisch n <Lufttrans> lean mixture
F&E-Ingenieur m *(Forschungs- und Entwicklungsingenieur)* <Maschinen> research-development engineer
Feingewindeschraube f <Bau> fine-pitch screw
Feingleiten n <Metall> fine slip
Feingold n <Metall> fine gold, gold of standard fineness
Feingut npl <Maschinen> fine sizes
Feinheit f <Bau, Papier> fineness
Feinheitanalyse f <Chemtech> particle size analysis
Feinheitsfestigkeit f <Textil> breaking length
Feinheitsgrad m <Maschinen> fineness ratio
Feinheitsmodul n <Bau> fineness modulus
Feinhöhenmesser m <Lufttrans> sensitive altimeter
Feinhonen n <Fertig> precision honing
Feinjustierung f <Kerntech> fine adjustment
Feinkeramikmaschine f <Ker & Glas> machine for fine ceramics
Feinkies m <Bau> fine gravel
Feinkohle f <Kohlen> culm
Feinkorn n 1. <Bau> close grain; 2. <Metall> close grain, fine grain
Feinkornbild n <Foto> fine-grain image
Feinkornentwickler m <Foto> fine-grain developer
feinkörnig adj 1. <Fertig> sappy; even *(Bruchfläche)*; 2. <Bau> close-grained; 3. <Metall> close-grained, fine-grained
feinkörniger Kies m <Bau> fine gravel
feinkörniger Sand m <Bau> fine sand
feinkörniger Stahl m <Kerntech> fine-grained steel
Feinkornstahl m <Kerntech> fine-grained steel
feinkristallin adj <Chemie> microcrystalline
Feinlinie f <Elektronik> fine line
Feinlinienleiterplatte f <Elektronik> fine-line printed circuit
Feinlunker m <Optik> pinhole
Feinmahlen n <Kohlen> fine grinding
Feinmahlung f <Kohlen> pulverization
Feinmessmanometer n <Gerät> *(AE)* precision gage, *(BE)* precision gauge
Feinmühle f <Bau> pulverizer
Feinraster n <Druck> fine screen
Feinrechen m <Abfall> fine screen
feinregeln v <Heiz & Kälte> control finely
Feinsand m <Kohlen> fine sand
Feinschicht f <Kunststoff, Wassertrans> gel coat *(Schiffbau)*
Feinschlämme f <Bau> laitance
Feinschliff m <Anstrich> microsection
Feinschnitt m <Anstrich> microsection
Feinschraubengewinde n **nach britischem Standard** <Maschinen> British standard fine screw thread
Feinsieb n <Maschinen> fine screen
Feinsilber n <Metall> fine silver
Feinstbohren n <Maschinen> precision boring
Feinstdreharbeit f <Fertig> superfinish turning
Feinstdrehmaschine f <Fertig> superfinisher
Feinsteinstellung f <Fertig> metal adjustment • **mit Feinsteinstellung** <Fertig> micrometer-adjustable
Feinstellskale f <Gerät> micrometric scale
Feinsteuerstab m <Kerntech> fine control member
Feinsteuerung f <Kerntech> fine control *(Reaktor)*
Feinstoff m <Bau, Papier> fines
Feinstruktur f 1. <Phys> fine structure *(Atomphysik)*; 2. <Strahlphys> fine structure
Feinstrukturaufspaltung f <Kerntech> fine-structure splitting
Feinstrukturkonstante f <Phys> fine-structure constant
feinstufig adj <Fertig> sensitive

Feinstvermahlung f <Kerntech> comminution
feinstzerteilt adj <Kunststoff> micronized
Feintaster m <Fertig> (AE) precision dial gage, (BE) precision dial gauge
Feinungsschlacke f <Fertig> finishing slag
Feinverteilen n <Chemtech> dispersion
Feinwaage f <Metrol> special accuracy weighing machine
Feinwerktechnik f <Maschinen> precision engineering
Feinzuschlagstoff m <Bau> fine aggregate
Feld n 1. <Akustik> field; 2. <Bau> bay; span (Trägern); 3. <Comp & DV, Elektriz, Elektrotech> field (Leiter); 4. <Erdöl> block, field; 5. <Funktech> array (Antennen); 6. <Konstzeich> block, panel; field; 7. <Telekom> array, section
Feld n **für Maßstabsangaben** <Konstzeich> panel provided for scale particulars
Feld n **mit abklingender Stärke** <Optik, Telekom> evanescent field
Feldbahnlokomotive f <Eisenbahn> pug
Feldbegrenzung f <Kfztech> field frame
Feldbuch n <Bau> field book (Vermessung)
Feldbussystem n <Elektronik> field area network
Felddesorptions-Massenspektrometer n <Kerntech> field desorption mass spectrometer
Feldeffekt m <Elektrotech, Telekom> field effect
Feldeffekttransistor m (FET) <Comp & DV, Elektronik, Optik, Phys, Raumfahrt> field effect transistor (FET)
Feldeffekttransistor m **mit Verarmungsschicht** <Elektronik> depletion mode FET
Feldeffektverstärker m <Elektronik> field effect amplifier
Feldeinteilung f <Konstzeich> block subdivision
Feldemission f 1. <Elektronik> field emission (Elektronenröhre); 2. <Elektrotech> cold emission; 3. <Phys> field emission
Feldemissionsmikroskop n <Phys> field emission microscope
Feldendezeichen n <Comp & DV> field delimiter
Felderprobung f <Telekom> field trial
Felderregung f <Elektriz> field excitation
Feldflackern n <Kerntech> field flutter
feldfreier Emissionsstrom m <Kerntech> field-free emission current
Feldgehäuse n <Kfztech> field frame
Feldgenerator m <Fernseh> safe area generator
Feldgradient m <Elektrotech> electric field gradient
Feldgruppe f <Comp & DV> array
Feldgruppenelement n <Comp & DV> array element
Feldionenemissionsmikroskop n <Phys> field ion microscope
Feldkabel n <Elektrotech> field cable, field wire
Feldkennung f <Telekom> label (Daten)
Feldkrümmung f <Phys> curvature of the field
Feldleitung f <Erdöl> flow line (Leitung von Bohrung zu Sammelstation)
Feldleitungstemperatur f <Erdöl> flow line temperature
Feldlinie f <Elektrotech, Phys> field line
Feldlinse f <Foto> front element (Objektiv); field lens (Okular)
Feldmagnet m <Aufnahme, Elektrotech> field magnet; alternator magnet (im Generator); field system (Elektrotraktion)
Feldmesskette f <Bau> land measuring chain
Feldmessung f <Bau> land measuring
Feldmikrofon n <Aufnahme> field microphone
Feldoxid n <Elektronik> field oxide
Feldpol m <Elektrotech> field pole
feldprogrammierbar adj <Telekom> field-programmable
Feldregler m 1. <Elektriz, Elektrotech> field regulator, field rheostat; 2. <Kerntech> field regulator

Feldreglerrheostat n <Elektriz> field rheostat
Feldrichtung f <Elektrotech> field direction
Feldschaltung f <Elektriz> field circuit
Feldschwächgrad m <Elektrotech> weakening ratio
Feldschwächung f <Elektrotech, Regelung> field weakening
Feldschwächung f **durch Anzapfung** <Elektrotech> field weakening by tapping
Feldschwächung f **durch Nebenschluss** <Elektrotech> field shunting
Feldschwächungsschalter m <Elektrotech> field-weakening switchgroup
Feldspannung f <Elektriz> field voltage
Feldspat m <Ker & Glas> feldspar
Feldspule f <Aufnahme, Elektriz, Elektrotech> field coil
Feldstärke f 1. <Elektriz, Elektrotech> field intensity, field strength, magnetizing force; 2. <Funktech> field intensity
Feldstärkelinie f <Elektrotech> line of force
Feldstärkemesser m <Funktech> field strength meter
Feldstärkemessgerät n <Elektriz> field strength meter
Feldstrom m <Elektriz, Elektrotech> field current
Feldübertragungsfaktor m <Akustik> free-field tension sensitivity
Feldunterbrecher m <Kerntech> field discharge switch, field-breaking switch
Feldunterdrücker m <Elektriz> field suppressor
Felduntersuchung f <Kohlen> field investigation
Feldvektor m <Elektrotech> field vector
Feldvermesser m <Bau> Ordnance Surveyor
Feldversuch m 1. <Maschinen> field test; 2. <Telekom> field trial
Feldwellenwiderstand m <Elektriz, Elektrotech> field impedance, free-space impedance
Feldwicklung f <Elektriz, Elektrotech> field coil, field winding
Felge f 1. <Fertig> felloe; 2. <Kfztech> rim (Rad)
Felgenflansch m <Kfztech> rim flange (Rad)
Felgenschulter f <Kfztech> flange (Autofelge)
Fels m <Bau, Kohlen> rock
Felsband n <Kohlen> rock ledge
Felsbohrmeißel m <Erdöl> rock bit (Bohrtechnik)
Felsdruck m <Bau, Kohlen> rock pressure
Felsen m <Wassertrans> rock (Geographie)
Felsenmehl n <Bau> rock flour
Felsgründung f <Bau, Kohlen> rock foundation
Felshöhle f <Kohlen> rock cut
Felsmechanik f <Bau, Kohlen> rock mechanics
FEM (Finite-Elemente-Methode) <Maschinen> FEM (finite elements method)
Femto... (F) <Metrol> femto... (f)
Femtometer n <Labor> femtometer
Fenchen n <Chemie> fenchene
Fenchon n <Chemie> fenchone
Fenchyl... <Chemie> fenchyl
Fender m 1. <Meerschmutz> fender; 2. <Wassertrans> (BE) bumper; fender (Deckausrüstung) • **Fender ausbringen** <Wassertrans> put out (Festmachen)
Fenske-Ringe mpl <Labor> Fenske helices (Destillation)
Fenster n <Comp & DV, Kfztech, Maschinen, Wassertrans> window (Schiffbau)
Fenster n **der Atmosphäre** <Phys, Strahlphys> atmospheric window
Fensterblende f **gegen Seeschlag** <Wassertrans> deadlight (Schiffbau)
Fensterblendrahmen m <Bau> window frame
Fensterbrüstung f <Bau> spandrel
Fensterdichtung f <Kfztech> window seal
Fensterfilter n <Elektronik> window filter
Fensterflügel m <Bau, Ker & Glas> casement

Fensterflügelrahmen *m* <Bau> sash
Fenstergitter *n* <Bau> window bars
Fensterglas *n* <Bau> window glass
Fensterglasflügel *m* <Bau> glazed sash
Fenstergummi *n* <Kfztech> window seal; pane rabbet *(Windschutzscheibe)*
Fensterheber *m* <Kfztech> window regulator
Fensterkitt *m* <Bau> bedding putty
Fensterladen *m* <Bau> folding shutter, shutter
Fensteröffnung *f* <Bau> window opening
Fensterrahmen *m* <Kfztech> window frame
Fensterriegel *m* <Bau> sash rail, window catch
Fensterschließer *m* <Bau> window fastener
Fenstersprosse *f* <Bau> sash bar, window bar
Fensterstab *m* <Bau> sash bar
Fenstersturz *m* <Bau> lintel *(Fenster)*
Fenstertechnik *f* <Comp & DV> window clipping, windowing *(Bildschirmunterteilung)*
Fenstertransformation *f* <Comp & DV> window transformation
Fenstertür *f* <Bau> French casement, glazed door
Fermate *f* <Akustik> rest
Fermat'sche Primzahl *f* <Math> Fermat prime number
Fermat'sche Vermutung *f* <Math> Fermat's last theorem
Fermat'sches Prinzip *n* <Phys> Fermat's principle
Fermentation *f* <Chemie> fermentation, zymosis
Fermi-Dirac'sche Statistik *f* <Phys> Fermi-Dirac statistics
Fermi-Dirac'sche Verteilung *f* <Phys> Fermi-Dirac distribution *(Quantenstatistik)*
Fermion *n* <Phys, Teilphys> fermion
Fermi'sche Energie *f* <Phys> Fermi energy
Fermi'sche Grenze *f* <Phys> Fermi limit
Fermi'sche Kugel *f* <Phys> Fermi sphere
Fermi'sche Oberfläche *f* <Phys> Fermi surface
Fermi'scher Wellenvektor *m* <Phys> Fermi wave vector
Fermi'sches Niveau *n* <Phys> Fermi level
Fermium *n* *(Fm)* <Chemie> fermium *(Fm)*
Fern... <Comp & DV, Kontroll> remote
Fernabfrage *f* <Telekom> remote access, remote inquiry *(von Anrufbeantwortern)*
Fernabtastung *f* <Comp & DV> remote sensing
Fernamt *n* <Telekom> *(AE)* toll exchange, *(AE)* toll switch, *(BE)* trunk exchange
Fernamtssystem *n* <Telekom> four-wire exchange system, trunk exchange system
Fernanschlusskabel *n (FAsk)* <Telekom> long-distance cable
Fernanzeige *f* <Gerät> remote indication
Fernanzeigegerät *n* <Gerät> remote indicating instrument
Fernaufnahmepunkt *m* <Fernseh> remote pickup point
fernbedient *adj* <Kfztech, Maschinen> remote, remote-operated, remotely controlled
fernbediente Autoverriegelung *f* <Kfztech> remote keyless entry
fernbediente Weiche *f* <Eisenbahn> automatic switch
Fernbedienung *f* 1. <Foto, Funktech> remote control; 2. <Sicherheit> remote-handling device; 3. <Telekom> remote control
Fernbedienungsgerät *n* <Aufnahme> remote control device
Fernbedienungsinstrument *n* <Kerntech> remote handling tool
Fernbedienungsvorrichtung *f* <Kerntech> remote handling tool
Fernbeförderung *f* <Trans> long-haul carriage
fernbetätigtes Zeichen *n* <Trans> remote control sign
Fernbetrieb *m* <Telekom> remote operation

Ferndrucken *n* <Comp & DV> remote printing
ferne Datenstation *f* <Comp & DV> remote terminal
ferne Infrarotstrahlung *f* <Strahlphys> far infrared
ferne Stapeldatenstation *f* <Comp & DV> remote batch terminal
ferne UV-Strahlung *f* <Strahlphys> far ultraviolet
ferne zweite Kopie *f* <Comp & DV> remote secondary copy
Ferneinkauf *m* <Telekom> teleshopping
ferner Drucker *m* <Comp & DV> remote printer
ferner Host *m* <Comp & DV> remote host
ferner Test *m* <Comp & DV> remote test
Fernerfassung *f* <Funkort> remote detection
Fernerkundung *f* <Meerschmutz> remote sensing
fernes Infrarot *n* <Strahlphys> far infrared
fernes Laden *n* <Comp & DV> remote loading
fernes Ultraviolett *n* <Phys, Strahlphys> far ultraviolet
Fernfahrer *m* <Kfztech> *(BE)* long-haul lorry driver, *(AE)* long-haul truck driver
Fernfehlerdiagnose *f* <Telekom> remote fault diagnosis *(vom Wartungszentrum)*
Fernfehlerortung *f* <Telekom> remote fault localization
Fernfeld *n* <Telekom> distant field
Fernfeldanalyse *f* <Funktech, Telekom> far-field analysis
Fernfeldbereich *m* <Optik, Telekom> far-field region
Fernfeldbeugungsdiagramm *n* <Funktech> far-field diffraction pattern
Fernfeldbrechungsmuster *n* <Optik> far-field diffraction pattern
Fernfeldmuster *n* <Optik> far-field pattern
Fernfeldstrahlungsdiagramm *n* 1. <Funktech> far-field pattern, far-field radiation pattern; 2. <Telekom> far-field pattern
Fernfeldstrahlungsmuster *n* <Optik> far-field radiation pattern
Fernflug *m* <Lufttrans> long-distance flight
Ferngas *n* <Heiz & Kälte> grid gas
Ferngasleitung *f* <Thermod> gas pipeline
Ferngasnetz *n* <Thermod> gas grid
Ferngastransport *m* <Trans> long-distance gas transport
Ferngeber *m* <Gerät> retransmitting slide wire
Ferngespräch *n* <Telekom> *(AE)* toll call, *(BE)* trunk call
Ferngespräch *n* **zu Ortsgebühr** <Telekom> local-charge-rate trunk call
ferngesteuert *adj* <Fernseh, Maschinen> remote-controlled
ferngesteuerte Kamera *f* <Fernseh> remote-controlled camera
ferngesteuerte Sendung *f* <Fernseh> remote broadcast
ferngesteuerte Vermittlung *f* <Telekom> remote switching
ferngesteuerte Vermittlungseinheit *f* <Telekom> remote switching unit *(RSU)*
ferngesteuerter Oszillator *m* <Elektronik> remote controlled oscillator
ferngesteuertes Entsicherungs- und Sicherungsgerät *n* <Raumfahrt> remote arming and safety unit *(Raumschiff)*
ferngesteuertes Vermittlungssystem *n* <Telekom> remote-switching system
Fernglas *n* 1. <Ker & Glas> field glass magnifier; 2. <Phys, Wassertrans> binoculars
Fernheizung *f* <Bau, Heiz & Kälte, Kohlen> district heating
Fernheizwerk *n* <Kohlen, Thermod> district-heating station
Fernhörer *m* <Telekom> receiver *(Telefon)*
Ferninstandhaltung *f* <Telekom> remote maintenance
Fernkabel *n* 1. <Elektrotech> long-distance cable; 2. <Telekom> long-distance cable, trunk cable

Fernkontrolle 252

Fernkontrolle f <Elektriz> remote control
fernkopieren v <Telekom> facsimile, fax
Fernkopieren n (Fax) 1. <Comp & DV> facsimile, fax; remote copying; 2. <Funktech, Telekom> facsimile, fax
Fernkopierer m (Fax) <Comp & DV, Telekom> facsimile equipment, facsimile machine, fax equipment
Fernladen n <Comp & DV, Telekom> remote loading
Fernlastzug m <Kfztech> long-distance road train
Fernleitung f <Elektrotech, Telekom> long-distance line
Fernleitungsbetrieb m <Elektrotech> trunking
Fernleitungshauptverteiler m (FHV) <Telekom> trunk distribution frame (TDF)
Fernleitungsnetz n <Telekom> trunk network
Fernmanagement n <Telekom> remote management
Fernmanipulation f <Sicherheit> remote-handling device
Fernmeldeamt n <Telekom> telecommunications office
Fernmeldeanlage f <Eisenbahn> remote signalling
Fernmeldegeheimnis n <Telekom> secrecy of telecommunications
Fernmeldegesellschaft f <Telekom> telecommunication operator
Fernmeldeingenieur m <Telekom> engineer for telecommunications, telecommunications engineer
Fernmeldekabel n <Telekom> telecommunication cable
Fernmeldeleitung f <Telekom> telecommunication circuit, telecommunications line
Fernmeldelinie f <Telekom> trunk
Fernmeldenetz n 1. <Comp & DV> communication network; 2. <Telekom> communication network, telecommunication network
Fernmeldesatellit m (ComSat) <Raumfahrt, Telekom> communication satellite (comsat)
Fernmeldesatellitensystem n NASA <Telekom> communication satellite system NASCM
Fernmeldetechnik f <Telekom> telecommunications
Fernmeldeübertragungsweg m <Telekom> telecommunication circuit
Fernmeldeverbindung f <Telekom> telecommunications line
Fernmeldevermittlungsanlage f <Telekom> telecommunications exchange
Fernmeldewesen n <Telekom> telecommunication, telecommunications
Fernmessgerät n <Maschinen> telemeter
Fernmesssystem n <Gerät> telemetering system
Fernmesstechnik f <Comp & DV> telemetry
Fernmessung f 1. <Elektriz> telemetry; 2. <Funktech> remote measuring, telemetering, telemetry; 3. <Gerät> remote sensing; 4. <Kerntech> remote metering, telemetry; 5. <Meerschmutz> remote sensing; 6. <Metrol, Telekom> remote measuring, telemetering, telemetry
Fernnebensprechen n <Telekom> far-end crosstalk
Fernnetz n 1. <Nichtfoss Energ> remote network; 2. <Telekom> trunk network, trunking network
Fernnetzgerät n <Elektrotech> remote power supply
Fernordnungsstoß n <Kerntech> distant collision
Fernproofs fpl <Druck> remote proofs pl (über Netz übertragen)
Fernreisebus m <Kfztech> long-distance bus
Fernrohr n 1. <Phys> refracting telescope, refractor; 2. <Raumfahrt, Wassertrans> telescope
Fernrohrlinse f <Foto> telescopic lens
Fernsatz m <Druck> teletypesetting
Fernschnellzug m <Eisenbahn> express train
Fernschreibamt n <Telekom> telegraph office
Fernschreiben n <Telekom> telex message
Fernschreibendgerät n <Telekom> teletype terminal, TTY terminal; teletype machine, TTY machine (Fernschreiber)

Fernschreib-Entzerrer m <Elektronik> regenerative repeater (Telegrafie)
Fernschreiber m (FS) <Comp & DV, Telekom> Teletype®, (BE) teleprinter, (AE) teletypewriter
Fernschreiberschnittstelle f <Telekom> teleprinter interface
Fernschreibmaschine f <Comp & DV, Telekom> (BE) teleprinter, (AE) teletypewriter
Fernschreibtelegrafie f <Telekom> teletype-telegraphy
Fernseh... <Fernseh> television
Fernsehaufnahmewagen m <Fernseh> television camera truck
Fernsehaufzeichnung f <Fernseh> telerecording
Fernsehbildröhre f <Fernseh> picture tube, television picture tube, television tube
Fernsehbildschärfe f <Fernseh> television image sharpness
Fernsehbildübertragung f <Fernseh> television (TV); transmission of moving pictures, videophone transmission
Fernsehempfänger m <Fernseh> television receiver
Fernsehempfangsstörung f (TVI) <Fernseh> television interference (TVI)
Fernsehen n (TV) <Fernseh> television (TV) • **im Fernsehen senden** <Fernseh> telecast
Fernsehen n **auf Abruf** <Fernseh> video-on-demand (Kabelfernsehen)
Fernsehen n **gegen Bezahlung** <Fernseh> pay per view, pay-TV
Fernsehen n **mit bedingtem Programmabruf** <Fernseh> near video-on demand
Fernsehen n **mit erhöhter Auflösung** <Fernseh> enhanced definition TV; high definition TV (HDTV)
Fernsehen n **mit erhöhter Bildqualität** <Fernseh> enhanced quality [definition] television
Fernsehen n **mit höherer Bildqualität** <Fernseh> high-quality TV (HQ TV)
Fernsehen n **mit langsamer Abtastung** (SSTV) <Fernseh> slow scan television (SSTV)
Fernsehen n **mit Restseitenbandmodulation** <Fernseh> television with vestigial-sideband modulation
Fernsehen n **über Gemeinschaftsantenne** (CATV) <Fernseh> community antenna television system (CATV)
Fernsehfunk m <Fernseh> television broadcasting
Fernsehgerät n <Fernseh> television set
Fernsehkabel n <Fernseh> television cable
Fernsehkabelnetz n <Fernseh> television cable network
Fernsehkamera f <Fernseh> television camera
Fernsehkamera-Röhre f <Fernseh> television camera tube
Fernsehkanal m <Elektrotech> tv-channel, television channel
Fernsehmonitor m <Fernseh> television monitor
Fernsehnorm f <Fernseh> television standard, television system • **für mehrere Fernsehnormen ausgelegt** <Fernseh> multistandard
Fernsehrelaisstation f <Fernseh> television relay
Fernsehröhre f <Fernseh, Ker & Glas> television tube
Fernsehrundfunk m <Fernseh> television broadcasting
Fernsehrundfunksignal n <Fernseh> broadcast television signal
Fernsehsender m <Fernseh> television transmitter
Fernsehsendung ausstrahlen v/als <Fernseh> televise
Fernsehstörung f (TVI) <Fernseh> television interference (TVI)
Fernsehübertragungsrechte npl <Fernseh> television rights
Fernseh-Umsetzer m <Fernseh> relay station, translator station

Fernseh- und Infrarrotbeobachtungssatellit *m* <Fernseh, Telekom> television and infrared observation satellite
Fernspeisesystem *n* <Telekom> power feeding system
Fernspeisung *f* <Telekom> remote power supply *(Telefon)*
Fernsprech... <Telekom> telephone
Fernsprechamt *n* <Telekom> telephone exchange
Fernsprechansagedienst *m* <Telekom> recorded public information service
Fernsprechapparat *m* <Telekom> telephone instrument
Fernsprechapparat *m* **mit Nummernschalterwahl** <Telekom> rotary dial telephone set
Fernsprechapparat *m* **mit Wählscheibe** <Telekom> rotary dial set, rotary dial telephone set
Fernsprechauftragsdienst *m* <Telekom> absent subscriber service
Fernsprechauskunft *f* <Telekom> directory enquiries
Fernsprechband *n* <Telekom> telephony band, voice band
Fernsprechbasisband *n* <Telekom> telephony baseband
Fernsprechbuch *n* <Telekom> telephone book, telephone directory
Fernsprechen *n* <Telekom> telephony
Fernsprechendgerät *n* <Telekom> telephone terminal
Fernsprecher *m* <Telekom> telephone
Fernsprechgebühr *f* <Telekom> call charge
Fernsprechkabel *n* <Elektrotech> telephone line
Fernsprechkonferenz *f* <Telekom> audio conference
Fernsprechleitung *f* <Elektrotech> telephone line
Fernsprechnebenstelle *f* <Telekom> telephone extension
Fernsprechnetz *n* <Telekom> telephone network, voice network
Fernsprechschrank *m* <Telekom> telephone switchboard
Fernsprechschrank *m* **mit Stöpselschnüren** <Telekom> plug and cord switchboard
Fernsprechübertrager *m* <Elektrotech> repeating coil
Fernsprechverbindung *f* <Phys> telecommunication
Fernsprechvermittlung *f* **mit ZB-Betrieb** <Telekom> central battery switchboard
Fernsprechvermittlungsamt *n* <Telekom> telephone switching center
Fernsprechvermittlungsstelle *f* <Telekom> telephone exchange
Fernsprechvermittlungstechnik *f* <Telekom> telephone switching
Fernsprechverzeichnis *n* <Telekom> telephone directory
Fernsprechwählnetz *n* <Telekom> switched telephone network, switched network
Fernsprechwählvermittlung *f* <Telekom> automatic telephone switch
Fernsprechwesen *n* <Telekom> telephony
Fernsprechzelle *f* <Telekom> telephone box
Fernsteuerung *f* <Elektriz, Kontroll, Mechan, Telekom> remote control, supervisory control, telecontrol
Fernsteuerung *f* **durch Fernseh-Kamera** <Fernseh, Trans> remote control by television camera
Fernstoß *m* <Kerntech> distant collision
Fernstrecke *f* <Eisenbahn> trunk line
Fernstromversorgung *f* <Elektrotech> remote power supply
Ferntest *m* <Comp & DV> remote test
Fernthermometer *n* 1. <Heiz & Kälte> remote thermometer; 2. <Thermod> *(AE)* remote temperature gage, *(BE)* remote temperature gauge

Fernüberwachung *f* <Fernseh, Telekom> remote monitoring
Fernverarbeitung *f* <Comp & DV> teleprocessing *(von Daten)*
Fernverbindungskabel *n* <Elektrotech, Telekom> trunk cable
Fernverkehrsdienst *m* **zu Ortsgebühren** <Telekom> extended-area service
Fernverkehrssystem *n* <Telekom> wide-area system
Fernvermittlung *f* <Telekom> *(AE)* toll exchange, *(BE)* trunk exchange, trunk switching
Fernvermittlungsanlage *f* <Telekom> *(AE)* toll switch
Fernvermittlungsstelle *f* *(FVSt)* <Telekom> *(AE)* toll exchange, *(BE)* trunk exchange, *(AE)* trunk switching center, *(BE)* trunk switching centre *(toll exchange)*
Fernverwaltung *f* 1. <Comp & DV> remote administration; 2. <Telekom> remote management
Fernwahl *f* <Telekom> STD, *(AE)* subscriber trunk-dialing, *(BE)* subscriber trunk-dialling
Fernwahlzugangskennzahl *f* <Telekom> *(AE)* direct distance-dialing access code, *(BE)* subscriber trunk-dialling access code
Fernwärme *f* <Heiz & Kälte, Thermod> district heating
Fernwärmekraftwerk *n* <Kohlen, Thermod> district-heating station
Fernwasserversorgung *f* <Wasserversorg> distant water supply
Fernwirkungsfeldstärke *f* <Raumfahrt> far-field intensity
Fernzähler *m* <Gerät> telecounter
Fernzeichnen *n* <Telekom> telescript, telewriting
Fernzugriff *m* <Comp & DV> remote access
Fernzugriff-Server *m* <Telekom> remote access server
Ferrat *n* <Chemie> ferrate
Ferredoxin *n* <Chemie> ferredoxin
Ferricyan *n* <Chemie> ferricyanogen
ferrimagnetisch *adj* <Phys> ferrimagnetic
Ferrimagnetismus *m* <Phys> ferrimagnetism
Ferrit *n* <Chemie, Elektriz, Elektrotech, Phys> ferrite
Ferrit *n* **mit rechteckiger Magnetisierungsschleife** <Elektrotech> square loop ferrite
Ferritantenne *f* <Funktech> ferrite antenna
Ferritbegrenzer *m* <Elektrotech> ferrite limiter
Ferritin *n* <Chemie> ferritin
ferritisch *adj* <Metall, Phys> ferritic
ferritischer rostfreier Stahl *m* <Phys> ferritic stainless steel
Ferritisolator *m* <Elektrotech> ferrite isolator
Ferritkern *m* 1. <Comp & DV> ferrite core, magnetic core; 2. <Elektriz, Elektrotech, Telekom> ferrite core
Ferritkopf *m* <Aufnahme, Fernseh> ferrite head
Ferritmodulator *m* <Elektronik> microwave amplitude modulator
Ferritphasenregler *m* <Elektrotech> ferrite phase shifter
Ferritphasenschieber *m* <Elektrotech> ferrite phase shifter
Ferritspulenkern *m* <Funktech> ferrite slug
Ferritstab *m* <Elektrotech, Funktech, Phys> ferrite rod
Ferritstabantenne *f* <Funktech, Wassertrans> ferrite rod antenna
Ferritzirkulator *m* <Funktech> ferrite circulator *(Mikrowellen)*
Ferrochrom *n* <Metall> ferrochrome
Ferrocyan *n* <Chemie> ferrocyanogen
ferrodynamischer Leistungsmesser *m* <Elektriz> ferrodynamic wattmeter
ferrodynamisches Wattmeter *n* <Elektriz> ferrodynamic wattmeter
Ferroelektrikum *n* <Elektrotech> ferroelectricity
ferroelektrisch *adj* <Phys> ferroelectric

ferroelektrischer

ferroelektrischer Kristall *m* <Elektrotech> ferroelectric crystal
Ferroelektrizität *f* <Phys> ferroelectricity
Ferrolegierung *f* <Metall> ferroalloy
Ferromagnetikum *n* <Elektrotech> ferromagnetic material
ferromagnetisch *adj* <Aufnahme, Phys> ferromagnetic
ferromagnetischer Verstärker *m* <Elektronik> ferromagnetic amplifier
ferromagnetisches Material *n* <Elektriz> ferromagnetic material
Ferromagnetismus *m* <Aufnahme, Elektriz, Phys> ferromagnetism
Ferroresonanz *f* <Elektrotech> ferroresonance
Ferroresonanzkreis *m* <Elektrotech> ferroresonance circuit
Ferrozement *m* <Bau> ferrocement
Ferse *f* <Ker & Glas> heel
fertig bearbeitetes Teil *n* <Maschinen> finished part
fertig montiert *adj* <Heiz & Kälte> factory-assembled
fertig stellen *v* <Bau> finish
Fertigbeton *m* <Bau> ready-mixed concrete
Fertigblasen *n* <Ker & Glas> final blow, settle blow
Fertigbohren *n* <Maschinen> finish boring
Fertigbohren mit Diamanten <Fertig> diamond boring
Fertigbohrkopf *m* <Fertig> boring head
Fertigdrehen *n* <Maschinen> final turning
fertigen *v* <Mechan> fabricate
Fertiger *m* <Bau> finisher *(Tiefbau)*
Fertiggesenk *n* <Maschinen> finishing die
fertiggesenkschmieden *v* <Fertig> finish-stamp
Fertigkaliber *n* <Fertig> finishing groove
Fertigkeit *f* <Ergon> skill
Fertigkeitsanalyse *f* <Ergon> skills analysis
Fertigläppen *n* <Fertig> finish lap
Fertigmachen *n* <Druck> finishing
Fertigmaß *n* <Maschinen> actual size
Fertigmehl *n* <Lebensmittel> *(BE)* self-raising flour
Fertigpackung *f* <Abfall> prepackaging
Fertigproduktelager *n* <Verpack> finished goods store
fertigputzen *v* <Fertig> rumble
Fertigreibahle *f* <Maschinen> finishing reamer
fertigschmieden *v* <Fertig> finish-stamp
Fertigschneideeisen *n* <Fertig> bottoming die
Fertigschneider *m* 1. <Fertig> final tap, finishing tap *(Gewinde)*; 2. <Maschinen> third tap
Fertigschnitt *m* <Maschinen> finishing cut
Fertigstellung *f* <Bau, Qual> completion
Fertigstellungsdatum *n* <Bau, Qual> completion date
Fertigteil *n* <Bau> precast unit
Fertigteilbau *m* <Bau> system building construction
Fertigung *f* <Fertig> fabrication, manufacturing, production
Fertigungsautomation *f* <Elektriz> manufacturing automation
Fertigungsfehler *m* <Fertig> error of gear cutting *(Getriebelehre)*
fertigungsgerechte Konstruktion *f (DFM)* <Qual> design for manufacturing, DFM
fertigungsgerechte Maßeinteilung *f* <Konstzeich> production-oriented dimensioning
Fertigungsmittelzeichnung *f* <Konstzeich> production facility drawing
Fertigungsplan *m* <Comp & DV, Fertig> production schedule
Fertigungsprüfer *m* <Qual> in-process inspector
Fertigungsprüfplan *m* <Qual> in-process inspection plan
Fertigungsprüfung *f* <Qual> in-process inspection, manufacturing inspection, process inspection

254

Fertigungsregel *f* <Comp & DV> production rule
Fertigungsroboter *m* <Mechan> assembly robot
Fertigungsstätte *f* <Fertig> *(AE)* production center, *(BE)* production centre
Fertigungssteuerung *f* 1. <Comp & DV> production control; 2. <Kontroll, Maschinen> manufacturing control
Fertigungsstraße *f* <Maschinen, Verpack, Wassertrans> production line *(Schiffbau)*
Fertigungsstückliste *f* <Konstzeich> production parts list
Fertigungstechnik *f* 1. <Elektronik> manufacturing technique; 2. <Fertig, Maschinen> production engineering
fertigungstechnischer Arbeitsvorgang *m* <Fertig> machining operation
Fertigungsüberwachung *f* <Qual> process control, production surveillance
Fertigungsüberwachungsunterlagen *fpl* <Qual> process-controlling documents
fertigungs- und montagegerechte Konstruktion *f* <Qual> design for manufacturability and assembly
Fertigungsverfahren *n* <Fertig> operating procedure
Fertigungszeichnung *f* 1. <Heiz & Kälte, Konstzeich> production drawing; 2. <Maschinen> working drawing
Fertigungszentrum *n* <Fertig> *(AE)* production center, *(BE)* production centre
Fertigungszyklus *m* <Kunststoff> *(AE)* molding cycle, *(BE)* moulding cycle
Fertigwalzen *n* <Fertig> finish roll-forming
Fertigware *f* <Textil> finished goods
Fertigzeichnung *f* <Maschinen> final drawing
Fertigziehen *n* <Maschinen> final drawing, finish drawing
Ferula... <Chemie> ferulic
Fessel *f* <Raumfahrt> tether *(Raumschiff)*
Fesselflug *m* <Raumfahrt> captive flight
fest *adj* 1. <Anstrich, Bau> solid *(Untergrund)*; 2. <Comp & DV> static; 3. <Maschinen> fixed, permanent; 4. <Mechan, Telekom> fixed; 5. <Textil> firm; 6. <Verpack> moisture-proof
fest abgestimmter Hohlraumresonator *m* <Elektronik, Funktech> fixed-tuned cavity resonator
fest angeschlossen *adj* <Elektrotech> hard-wired
fest eingebaute Wohnungsgegenstände *mpl* <Bau> fixtures and fittings
fest gekoppelt *adj* <Elektriz> close-coupled
fest gewordenes Fett *n* <Lebensmittel> solidified fat
fest verkabeltes Netz *n* <Telekom> fixed network
fest verlegt *adj* <Bau> permanent
fest zugeordnete Frequenz *f* <Telekom> dedicated frequency
Festabstimmung *f* <Elektronik, Funktech> fixed tuning
Festaderkabel *n* <Telekom> tight-jacketed cable
Festamplitude *f* <Elektronik> fixed amplitude
Festanschlag *m* <Fertig> positive stop
Festanschluss *m* <Telekom> tie line
festbacken *v* <Chemtech> cake
festbinden *v* <Wassertrans> lash
festbrennen *v* <Textil> bake *(Verschmutzung)*
Festdaten *npl* <Comp & DV> fixed data
Festdielektrikum *n* <Elektrotech> solid dielectric
feste Ablaufsteuerung *f* <Comp & DV, Telekom> fixed sequencer
feste Absperrung *f* <Sicherheit> fixed barrier guard
feste drahtlose Anschlussleitung *f* <Telekom> fixed wireless loop *(Telefon)*
feste Entfernungsbefeuerung *f* <Lufttrans> fixed distance lights *(Flughafen)*
feste Feuerlöschanlage *f* <Sicherheit> stationary fire-fighting installation
feste Flugfunkstelle *f* <Lufttrans, Telekom> aeronautical-fixed station

Festkörpergeometrie

feste Flugmeldeverbindung f <Lufttrans, Telekom> aeronautical-fixed circuit
feste Kapselung f <Sicherheit> fixed enclosure
feste Kopplung f <Elektriz, Elektrotech> close coupling, tight coupling
feste Kupplung f <Maschinen> permanent coupling, solid coupling
feste Länge f <Comp & DV> fixed length
feste Phase f <Thermod> solid phase
feste Riemenscheibe f <Maschinen> fixed pulley
feste Rolle f <Maschinen> fixed pulley
feste Schürze f <Wassertrans> rigid skirt
feste Schutzvorrichtung f <Sicherheit> built-in guard
feste Sperre f <Sicherheit> fixed guard
feste Spitze f 1. <Fertig> (AE) dead center, (BE) dead centre; 2. <Maschinen> (AE) dead center, (BE) dead centre, (AE) fixed center, (BE) fixed centre
feste Versetzung f <Metall> immobile dislocation
feste virtuelle Verbindung f <Comp & DV, Telekom> PVC, permanent virtual circuit
feste Wortlänge f <Comp & DV, Telekom> fixed word length
fester Abfall m <Abfall> solid waste
fester Abfallstoff m <Abfall> solid waste
fester Aggregatzustand m <Elektronik, Teilphys> solid state
fester Anschlag m <Maschinen> fixed stop, hard stop
fester Block m <Wassertrans> standing block (Beschläge)
fester Brennstoff m <Heiz & Kälte, Maschinen> solid fuel
fester Einband m <Druck> hardback
fester Fehler m <Lufttrans> fixed error (Funkhöhenmesser)
fester Flugfunkdienst m (AFS) <Lufttrans, Telekom> (AE) aeronautical-fixed network, (BE) aeronautical-fixed service, (AE) aeronautical-fixed system (Bodenfunkstationen)
fester Griff m <Textil> firm handle
fester Kernbrennstoff m <Kerntech> solid nuclear fuel
fester Kohlenstoff m <Chemie> fixed carbon
fester Rost m <Heiz & Kälte> fixed grate
fester Schenkel m <Fertig> blade, stock (Winkelmesser)
fester Setzstock m <Maschinen> fixed steadyrest
fester Siedlungsabfall m <Abfall> MSW, municipal solid waste
fester virtueller Schaltkreis m <Comp & DV> PVC, permanent virtual circuit
fester Zahnkranz m <Lufttrans> fixed ring gear
fester Zustand m <Teilphys> solid state
festes Aufspulen n <Foto> tight spooling
festes Bullauge n <Wassertrans> deadlight (Schiffbau)
festes Dämpfungsglied n 1. <Elektronik> fixed attenuator; 2. <Telekom> pad
festes Dielektrikum n <Elektrotech> solid dielectric
festes Feld n <Comp & DV> fixed field
festes Flugfunknetz n (AFTN) <Lufttrans, Telekom> aeronautical-fixed telecommunication network (AFTN)
festes Gehäuse n <Sicherheit> fixed enclosure
festes Gehrungsdreieck n <Bau> (AE) miter square, (BE) mitre square
festes Material n <Phys> solid
festes Rad n <Maschinen> fixed wheel
festfahren v <Nichtfoss Energ> stall
Festfeuer n <Wassertrans> fixed light (Seezeichen)
Festformat n <Comp & DV> fixed format
Festfrequenzempfänger m <Funktech> fixed-frequency receiver
Festfrequenzgenerator m <Elektronik> fixed-frequency synthesizer

Festfrequenzmagnetron n <Elektronik> fixed-frequency magnetron
Festfrequenzoszillator m 1. <Elektronik> fixed-frequency oscillator; 2. <Raumfahrt> local oscillator
Festfrequenzsender m <Funktech> fixed-frequency transmitter
festfressen v <Fertig> bind (Werkzeug)
festfressen v/sich <Kerntech> seize
Festfressen n 1. <Fertig> galling; 2. <Kfztech> jamming (Kolbenringe); seizing (Lager, Kolben)
festgefressen adj <Kfztech> jammed
festgelegter Fahrweg m <Wassertrans> lane
festgelegtes Verfahren n <Qual> routine
Festgenerator m <Raumfahrt> fixed generator
festgeschaltete Leitung f <Comp & DV, Telekom> dedicated line
festgeschaltete Verbindung f <Telekom> fixed connection, permanent circuit
festgeschalteter Anschluss m <Telekom> dedicated port
festgeschalteter Kanal m <Telekom> dedicated channel
festgeschalteter Zeichengabekanal m <Telekom> (AE) dedicated signaling channel, (BE) dedicated signalling channel
festgezurrt und aufgehängt adj <Raumfahrt> strap-down-mounted (Raumschiff)
festhaken v 1. <Maschinen> clasp, hook; 2. <Mechan> hitch
festhaken v/sich 1. <Fertig> clog (Säge); 2. <Wassertrans> grip (Festmachen)
Festhalten n der Darstellung <Elektronik> display retention
Festhöhe f <Fertig> fixed height
Festigkeit f 1. <Bau> strength; 2. <Kerntech> toughness; 3. <Kohlen> strength; 4. <Lebensmittel> consistency; 5. <Maschinen> strength; 6. <Mechan> stress; 7. <Metall> strength; 8. <Nichtfoss Energ> solidity; 9. <Papier, Phys> strength; 10. <Telekom> stability; 11. <Textil> strength, tenacity
Festigkeitsberechnung f <Maschinen> stability calculation
Festigkeitseigenschaften fpl <Werkprüf> physical properties
Festigkeitsgrenze f <Mechan> breaking point
Festigkeitsprüfung f <Raumfahrt> stress analysis (Raumschiff)
festkleben v <Bau> stick
festklemmen v 1. <Bau> clamp; 2. <Fertig> choke; 3. <Maschinen> lock; 4. <Mechan, Wassertrans> jam
festklemmen v/sich <Mechan> jam
Festklemmen n <Fertig> sticking (Bohrer)
Festklopfen n <Bau> ramming
Festkomma n 1. <Comp & DV> fixed point; 2. <Math> fixed decimal point; 3. <Telekom> fixed point
Festkommabetrieb m <Comp & DV> fixed-point operation
Festkommarechnung f <Comp & DV> fixed-point arithmetic
Festkommaschreibweise f <Comp & DV> fixed-point notation
festkommen v <Wassertrans> run aground (Schiff)
Festkondensator m <Elektrotech, Phys> fixed capacitor
Festkontakt m <Elektriz, Elektrotech> fixed contact
Festkörper m <Comp & DV, Phys, Strahlphys, Teilphys> solid
Festkörperbauelement n <Elektriz, Phys, Telekom> solid state device
Festkörpereffekt m <Kerntech> solid state effect
Festkörpergeometrie f <Geom> CSG, constructive solid geometry

Festkörperphysik

Festkörperphysik f <Phys, Teilphys> solid state physics
Festkörperrelais n <Elektrotech> solid state relay
Festkörperspannungsableiter m <Elektrotech> solid state surge arrester
Festkörperspeicher m <Comp & DV, Elektrotech> solid state memory device
Festkörperverstärker m <Telekom> solid state amplifier
Festkörperzähler m <Strahlphys> solid state detector
Festlandsockel m <Erdöl, Wassertrans> continental shelf *(Geologie)*
Festlast f <Elektrotech> fixed load
festlegen v <Comp & DV> declare, set
Festlegung f <Fertig> convention
Festlegung f **der Prioritätsfolge** <Comp & DV> priority sequencing
Festluftkissen n <Wassertrans> static air cushion
festmachen v 1. <Meerschmutz> moor *(Schiff)*; 2. <Wassertrans> make fast, moor, secure; furl *(Segeln)*
Festmachen n <Wassertrans> mooring
Festmachleine f <Wassertrans> mooring line *(Tauwerk)*
Festnetztelefon n <Telekom> fixed telephone
Festoxidbrennstoffzelle f <Elektriz> solid oxide fuel cell
Festplatte f 1. <Comp & DV> Winchester disk, fixed disk, *(AE)* disk; 2. <Comp & DV, Telekom> hard disk, HD
Festplattencontroller m <Comp & DV> disk controller
Festplattenlaufwerk n <Comp & DV> disk drive; hard disk drive *(HDD)*
Festplattentreiber-Schnittstelle f <Comp & DV> integrated device equipment *(IDE)*
Festpol m <Elektriz> fixed pole
Festpropeller m <Wassertrans> fixed-pitch propeller
Festpunkt m 1. <Bau> datum, fixed point; benchmark *(Vermessung)*; 2. <Eisenbahn> midpoint anchor *(Fahrleitung)*; 3. <Kerntech> fixed point
Festpunkteinstellung f <Regelung> terminal-based conformity
Festpunktnetz n <Bau> observation grid *(Vermessung)*
Festrad n <Maschinen> fixed wheel
Festsattel m <Kfztech> *(AE)* fixed caliper, *(BE)* fixed calliper
Festsattelscheibenbremse f <Kfztech> *(AE)* fixed caliper disk brake, *(BE)* fixed calliper disc brake
Festscheibe f <Fertig> fast pulley
Festschelle f <Fertig> fixed-point bracket *(Kunststoffinstallationen)*
Festschrauben n <Maschinen> screwing
festschrauben v 1. <Bau> bolt; 2. <Maschinen> screw
festsetzen v <Math> assign
festsetzen v/sich <Chemtech> settle on
Festsetzen n <Fertig> seizing *(Kunststoffinstallationen)*
Festsitz m (FS) <Maschinen> driving fit, force fit
festspannen v <Maschinen> clamp, mount
Festspannen n 1. <Bau> stretching; 2. <Maschinen> mounting
Festspeicherinstruktion f <Comp & DV> read-only instruction
feststampfen v <Bau, Eisenbahn> tamp
feststehend adj <Mechan> fixed
feststehende Achse f 1. <Geom> fixed axis; 2. <Maschinen> dead axle
feststehende Richtungsschaufel f <Lufttrans> fixed stator vane
feststehende Spule f <Elektriz> fixed coil
feststehender Flügel m <Lufttrans> fixed wing
feststehender Querbalken m <Fertig> fixed rail
feststehender Tisch m <Ker & Glas> fixed table *(Spiegelherstellung)*
feststehendes Teil n <Phys> stator
feststellbar adj <Math> ascertainable

Feststellbremse f <Kfztech, Mechan> parking brake
Feststelleinrichtung f <Fertig> retainer
feststellen v 1. <Bau> lock; 2. <Math> ascertain; 3. <Qual> determine, identify
Feststellen n **des Bildschirms** <Comp & DV> screen locking *(Gegensatz zu Blättern)*
Feststellen n **des Luftschraubenblattes** <Lufttrans> blade setting *(Hubschrauber)*
Feststeller m 1. <Fertig> locking pin *(Kunststoffinstallationen)*; 2. <Raumfahrt> fastener *(Raumschiff)*
Feststellring m <Maschinen> lock ring
Feststellung f 1. <Math> ascertainment; 2. <Qual> determination
Feststellvorrichtung f <Maschinen> blocking device
Feststoff m <Anstrich, Maschinen> solid
Feststoffabfall m <Umweltschmutz> solid waste
Feststoffaustauscher m <Kohlen> solid exchanger
Feststoffextraktion f <Chemie> leaching
Feststoffgehalt m <Abfall> solids content
Feststoffpumpe f <Maschinen> solids pump
Feststoffrakete f 1. <Maschinen> solid propellant rocket engine; 2. <Raumfahrt> dry-fuelled rocket
Feststoffschmierung f <Maschinen> solid lubrication
Feststoffschubtriebwerk n <Raumfahrt> solid fuel booster
Feststoffsystem n <Raumfahrt> solid propellant system
Feststoffteilchen n <Chemtech, Umweltschmutz> solid particle
Festtantalkondensator m <Elektrotech> solid tantalum capacitor
Festtreibstoff m <Raumfahrt> solid propellant *(Raumschiff)*
festverdrahtet adj <Comp & DV, Elektrotech> hard-wired
festverdrahtete Logik f <Comp & DV> hard-wired logic
festverdrahtete Verbindung f <Comp & DV> hard-wired connection
festverdrahtetes Programm n <Comp & DV> hard-wired program
festverdrahtetes programmierbares Vermittlungssystem n <Telekom> hard-wired programmable switching system
Festvermittlung f <Telekom> cross-connect
Festwertregelung f 1. <Elektrotech> regulation; 2. <Regelung> fixed command control
Festwertspeicher m *(ROM)* <Comp & DV, Elektriz, Elektrotech, Funktech> read-only memory *(ROM)*; non-erasable storage
Festwiderstand m <Elektriz, Elektrotech, Phys> fixed resistor
festzeitgesteuertes Signal n <Trans> pretimed signal
Festzeitrelais n <Elektriz> definitive-time relay
festziehen v <Maschinen> fasten, tighten
Festziehen n <Maschinen> tightening
festzuzurrende Gerätschaft f <Raumfahrt> strapdown equipment *(Raumschiff)*
festzuzurrende Trägheitsplattform f <Raumfahrt> strapdown inertial platform *(Raumschiff)*
FET *(Feldeffekttransistor)* <Comp & DV, Elektronik, Optik, Phys, Raumfahrt> FET *(field effect transistor)*
FET-Eingang m <Elektronik> FET front end, FET input
fett adj 1. <Chemie> fatty; 2. <Druck> bold
Fett n 1. <Druck, Kfztech> grease; 2. <Lebensmittel> fat; 3. <Maschinen, Mechan> grease; 4. <Papier> fat, grease
Fettabscheider m 1. <Abfall> grease separator, grease trap; 2. <Meerschmutz> skimming tank
fettabweisendes Papier n <Verpack> grease-resistant paper
fettaffin adj <Chemie> lipophile, lipophilic
fettähnlich adj <Chemie> lipoid

fettarm adj <Lebensmittel> low-fat
fettartig adj <Chemie> lipoid
Fettbeständigkeit f <Kunststoff> grease resistance
Fettbüchse f 1. <Fertig> grease cup, greaser; 2. <Maschinen> grease cup
fettdicht adj <Papier, Verpack> greaseproof
fettdichtes Papier n 1. <Lebensmittel> greaseproof paper; 2. <Papier> grease-resistant paper
Fettdichtigkeitsprobe f <Papier> greaseproof proof
Fettdruck m 1. <Comp & DV> bold face, bold print; 2. <Druck> bold face
Fettdrucken n <Druck> emboldening
Fettdurchlässigkeit f <Verpack> permeability to grease
fette Schrift f <Druck> heavy type
fetter Beton m <Bau> fat concrete, rich concrete
fetter Druck m <Druck> bold face
fetter Ton m 1. <Erdöl> fat clay *(Mineral)*; 2. <Ker & Glas> rich clay
fettes Gemisch n <Kfztech> rich mixture
fettes Öl n <Chemie> fat oil, fatty oil, fixed oil
Fettfänger m <Meerschmutz> skimming tank
fettfrei adj <Lebensmittel> nonfat
fettgeschmiert adj <Fertig> grease-lubricated
Fettharz n <Chemie> oleoresin
fettiges Alkyd n <Kunststoff> long oil alkyd
Fettkalk m <Papier> fat lime
Fettkante f <Kunststoff> fat edge
Fettkappe f <Kfztech> grease cap
Fettkennzahl f <Lebensmittel> lipid value
Fettkohle f <Kohlen> fat coal, rich coal
fettlöslich adj <Chemie> liposoluble
Fetton m <Chemie> smectite
Fettpackung f <Maschinen> grease packing
Fettpistole f <Kfztech> grease gun *(Werkzeug)*
Fettpresse f <Bau, Maschinen> grease gun
Fettreif m <Lebensmittel> bloom *(Schokolade)*
Fettsäure f <Chemie, Lebensmittel, Papier> fatty acid
Fettsäureglycerid n <Lebensmittel> fatty acid glyceride
fettundurchlässiges Papier n <Lebensmittel> greaseproof paper
FET-Verstärker m <Elektronik> FET amplifier
FET-Verstärker m **in Gateschaltung** <Funktech> common gate FET amplifier
feucht adj 1. <Bau> damp; 2. <Papier> damp, moist; 3. <Phys> hygroscopic; 4. <Textil> wet; 5. <Thermod> humid, moist
Feuchte f <Bau, Phys, Textil, Verpack, Werkprüf> moisture
Feuchteanzeiger m <Phys> hygroscope
Feuchtegehalt m <Bau, Phys, Textil, Verpack, Werkprüf> moisture content
feuchtehaltig adj <Phys> hygroscopic
Feuchtemesser m 1. <Heiz & Kälte> moisture meter; 2. <Phys> moisture meter, psychrometer
Feuchtemessfühler m <Gerät> moisture head
Feuchtemessgerät n <Gerät> moisture content meter
Feuchtemessung f <Phys> psychrometry
Feuchter m <Papier> dampener
feuchtgeglättetes Papier n <Druck> water-finished paper
Feuchtglättwerk n <Papier> breaker stack
Feuchtigkeit f 1. <Anstrich, Elektrotech> humidity, damp; 2. <Fertig> humidity *(Kunststoffinstallationen)*; 3. <Heiz & Kälte> humidity; 4. <Kohlen> damp; 5. <Papier> moisture; 6. <Phys, Thermod> humidity • **Feuchtigkeit entziehen** <Heiz & Kälte> dehumidify
feuchtigkeitsabweisend adj <Verpack> moisture-repellent
Feuchtigkeitsanzeige f <Verpack> humidity indicator

Feuchtigkeitsaufnahme f 1. <Papier> moisture regain; 2. <Verpack> humidity absorber; 3. <Werkprüf> moisture regain
Feuchtigkeitsaufnahme f **im Normalklima** <Werkprüf> moisture regain in the standard atmosphere
feuchtigkeitsbeständig adj <Verpack> moisture-proof
Feuchtigkeitsbestimmung f <Verpack> moisture determination
Feuchtigkeitsdetektor m <Elektronik, Metrol> moisture sensor
feuchtigkeitsfest adj <Bau, Verpack> damp-proof
Feuchtigkeitsgehalt m <Heiz & Kälte, Kohlen, Lebensmittel, Papier, Textil, Wasserversorg, Werkprüf> moisture content
Feuchtigkeitsmesser m <Heiz & Kälte, Phys> moisture meter
Feuchtigkeitsmessung f <Wassertrans> humidity measurement
Feuchtigkeitsprobe f <Verpack> moisture test
Feuchtigkeitsregler m <Heiz & Kälte> humidistat
Feuchtigkeitssperre f <Bau> damp-proof course, dpc
Feuchtigkeitssperrmembrane f <Bau> damp-proof membrane, dpm
Feuchtigkeit- und Temperaturfühler m <Trans> moisture and temperature detector
Feuchtkugelthermometer n <Thermod> wet bulb thermometer
Feuchtraum m <Heiz & Kälte> damp location
Feuchtraumisolator m <Elektrotech> mushroom insulator
Feuchtstaubabsaugung f <Sicherheit> wet dust removal installation
Feuchtverfahren n <Konstzeich> semidry-method
Feuchtwalze f <Verpack> damping roll
Feuer n 1. <Sicherheit, Thermod> fire; 2. <Wassertrans> loom *(unter der Kimm)*
Feueralarm m <Sicherheit, Thermod> fire alarm
Feueranzünder m <Heiz & Kälte> fire lighter
feuerbeständig adj 1. <Heiz & Kälte, Sicherheit> fire-resistant; 2. <Thermod> fire-resisting
feuerbeständige Beschichtung f <Thermod> fire-resisting coating
feuerbeständige Farbe f <Thermod> fire-resisting paint
feuerbeständige Trennwand f <Thermod> fire-resisting bulkhead
feuerbeständige Tür f <Sicherheit> fire-resistant door
feuerbeständige Wand f <Sicherheit> fire-resistant partition
Feuerbeständigkeit f 1. <Fertig> refractoriness; 2. <Kunststoff> fire resistance
Feuerbrücke f <Thermod> fire stop
Feuerbüchse f <Eisenbahn> firebox
Feuerfalle f <Sicherheit> firetrap, flame trap
feuerfest adj 1. <Bau, Heiz & Kälte> refractory; 2. <Mechan, Sicherheit> fireproof; 3. <Thermod> fire-resisting, fireproofed, fireproof, refractory; 4. <Verpack> fireproof
feuerfest machen v <Sicherheit, Thermod> fireproof
Feuerfestbetonerzeugnis n <Ker & Glas> castable
feuerfeste Auskleidung f <Bau, Heiz & Kälte> refractory lining
feuerfeste Ausmauerung f <Bau, Heiz & Kälte> refractory lining
feuerfeste Beschichtung f <Thermod> fire-resisting coating
feuerfeste Farbe f <Thermod> fire-resisting paint
feuerfeste Töpferwaren fpl <Ker & Glas> fireproof pottery
feuerfeste Trennwand f <Lufttrans> fire wall

feuerfester

feuerfester Handschuh m <Labor> heat-resistant glove
feuerfester Sandstein m <Ker & Glas> fire resisting sandstone
feuerfester Stoff m <Bau, Fertig> refractory
feuerfester Ton m 1. <Ker & Glas> chamotte; 2. <Thermod> fireclay
feuerfester Ziegel m <Bau> fire brick
Feuerfesterzeugnis n <Ker & Glas> refractory
feuerfestes Geschirr n <Ker & Glas> oven-to-table ware
feuerfestes Glas n <Thermod> oven proof glass
feuerfestes Material n <Thermod> refractory
feuerfestes Metall n <Metall> refractory metal
feuerfestes Papier n <Papier> flameproof paper
feuerfestes Schott n <Raumfahrt> fireproof bulkhead (Raumschiff)
Feuerfestglas n <Ker & Glas> flameproof glass
Feuerfestigkeit f 1. <Optik> refractoriness (Materialwissenschaft); 2. <Sicherheit> resistance to fire
Feuerfestmaterial n <Bau, Kohlen> refractory material
Feuerflecken mpl <Ker & Glas> fire marks
Feuergefahr f <Thermod> fire hazard
Feuergefährlichkeit f <Sicherheit, Thermod> flammability
feuerhemmend adj 1. <Heiz & Kälte, Sicherheit> fire-retardant; 2. <Thermod> fire-retarding
feuerhemmende Beschichtung f <Kerntech> fire-retardant coat
Feuerholz n <Thermod> firewood
Feuerklappe f <Heiz & Kälte> fire damper
Feuerleiter f 1. <Bau> fire escape; 2. <Sicherheit> fire ladder; 3. <Thermod> fire escape
Feuerleitradar n <Funkort> fire control radar
Feuerleitung f <Wassertrans> fire control (Marine)
Feuerlöschanlage f <Sicherheit> fire-extinguishing installation, fire-fighting equipment
Feuerlöschboot n <Wassertrans> fire boat
Feuerlöschbrause f <Sicherheit> safety shower (Feuer)
Feuerlöschdecke f <Sicherheit> fire blanket
Feuerlöscheinrichtung f 1. <Lufttrans> firefighting equipment; 2. <Sicherheit> fixed fire extinguisher; 3. <Wassertrans> firefighting equipment
Feuerlöscher m <Kfztech, Lufttrans, Sicherheit, Thermod, Wassertrans> fire extinguisher
Feuerlöscherladung f <Sicherheit> fire extinguisher filling
Feuerlöscherschlagbolzen m <Lufttrans> extinguisher striker
Feuerlöscherzuschläger m <Lufttrans> extinguisher striker
Feuerlöschfahrzeug n <Thermod> fire engine
Feuerlöschgeräte pl <Sicherheit> fire-extinguishing equipment
Feuerlöschhydrant m <Thermod, Wassertrans> fire hydrant
Feuerlöschkanone f <Wassertrans> water gun (Brandbekämpfung)
Feuerlöschmittel n <Sicherheit> extinguishant, fire-extinguishing agent
Feuerlöschmonitor m <Meerschmutz> fire monitor
Feuerlöschpumpe f <Thermod> fire pump, fire extinguishing pump
Feuerlöschsteigleitung f <Thermod> fire riser, fire-rising main
Feuermeldeanlage f 1. <Lufttrans, Wassertrans> fire detection system (Notfall); 2. <Sicherheit> fire alarm system, fire detection system
Feuermelder m <Elektriz, Sicherheit> fire detector
Feuermelderdraht m <Lufttrans, Wassertrans> fire-detecting wire
Feuermelderkabelbaum m <Lufttrans, Wassertrans> fire detection harness

258

Feuermeldersystem n <Sicherheit> fire detection and alarm system
Feuermeldeschleife f <Raumfahrt> fire detection loop (Raumschiff)
feuern v 1. <Künstl Int> fire (Regel); 2. <Thermod> fuel
Feuerplatte f <Ker & Glas> deadplate
Feuerpolierer m <Ker & Glas> fire finisher
feuerpoliert adj <Ker & Glas> fire-polished
feuerpolierte Kante f <Ker & Glas> fire-polished edge
Feuerpolitur f <Ker & Glas> fire finish, fire polishing
Feuerraum m 1. <Heiz & Kälte> combustion chamber; 2. <Hydraul> combustion chamber, fire chamber (Dampfkessel)
Feuerrisiko n **durch elektrische Ursache** <Sicherheit> electric fire risk
Feuerrohr n <Heiz & Kälte> smoke tube
Feuerschaden m <Sicherheit> fire damage
Feuerschein m <Thermod> firelight
Feuerschiff n 1. <Thermod> fireship; 2. <Wassertrans> lightship, lightvessel
Feuerschneise f <Thermod> firebreak
Feuerschott n <Wassertrans> fire bulkhead (Schiffkonstruktion)
Feuerschutz m <Sicherheit, Thermod> fire guard, protection against fire
Feuerschutzgitter n <Sicherheit> fire screen
Feuerschutzisolierglas n <Sicherheit> insulating glass for fire protection
Feuerschutzschott n <Raumfahrt> fire bulkhead (Raumschiff)
Feuerschutztür f <Thermod> fire door
Feuerschutzwand f <Raumfahrt> fire wall (Raumschiff)
Feuerschweißen n <Fertig> forge welding
Feuersektor m <Wassertrans> sector of a light (Seezeichen)
feuersicher adj <Elektriz, Mechan> fireproof
feuersichere Telefonanlage f <Bau> fireproof telephone system
Feuersicherheit f <Sicherheit> fire safety (in Gebäuden)
Feuersirene f <Sicherheit> fire siren
Feuerspritze f <Thermod> fire hose
Feuerton m <Ker & Glas> pipe clay
Feuertür f 1. <Eisenbahn> firebox door; 2. <Thermod> fire door
Feuerübung f <Sicherheit> fire drill
Feuer- und Hitzeschutzkleidung f <Sicherheit> protective clothing against heat and fire
Feuerung f 1. <Elektrotech> firing; 2. <Fertig> furnace; 3. <Maschinen> firing system; 4. <Mechan> heating
Feuerungsanlage f <Maschinen> firing plant
Feuerungsdecke f <Fertig> crown (Ofen)
Feuerungsstoß m <Raumfahrt> pyrotechnical shock (Raumschiff)
Feuerungssystem n <Maschinen> firing system
Feuerungstechnik f <Raumfahrt> pyrotechnics
Feuerungsventil n <Raumfahrt> pyrotechnic valve (Raumschiff)
Feuerungswandung f <Maschinen> furnace wall
Feuerverhinderung f <Thermod> fire prevention
Feuerverhütung f <Sicherheit> fire prevention
feuerverzinken v <Anstrich> galvanize
feuerverzinkt adj <Heiz & Kälte> hot-dip galvanized
Feuerverzinkung f <Fertig> pot galvanizing
Feuervorhang m <Thermod> fire curtain
Feuerwache f <Thermod> fire station
Feuerwehr f <Sicherheit, Thermod> fire brigade
Feuerwehrauto n <Thermod> fire engine
Feuerwehraxt f <Sicherheit> firefighting axe
Feuerwehrfahrzeug n <Sicherheit> firefighting vehicle

Feuerwehrgerät n <Sicherheit> firefighting equipment
Feuerwehrhelm m <Sicherheit> fireman's helmet
Feuerwehrschlauch m <Sicherheit, Thermod> fire hose
Feuerwehrwagen m <Sicherheit> fire engine
Feuerwerk n <Thermod> firework display
Feuerwerkskörper m <Thermod> firework
Feuerwiderstandsdauer f <Heiz & Kälte> fire-resistance time
feuerwiderstandsfähig adj <Sicherheit> fire-resistant
feuerwiderstandsfähige bauliche Ausführung f <Sicherheit> fire-resistant construction
feuerwiderstandsfähige Wand f <Sicherheit> fire-resistant partition
Feuerwiderstandsfähigkeit f <Kunststoff> fire resistance
Feuerwiderstandsklasse f <Heiz & Kälte> fire rating
Feuerzelle f <Kerntech> fire cell
Feuerzone f <Kerntech> fire area *(durch Feuersperren abgetrennt)*
Feynman'sches Diagramm n <Phys> Feynman diagram
FFS *(flexibles Fertigungssystem)* <Künstl Int> FMS *(flexible manufacturing system)*
FFT *(schnelle Fourier-Transformation)* <Elektronik> FFT *(fast Fourier transform)*
FHV *(Fernleitungshauptverteiler)* <Telekom> TDF *(trunk distribution frame)*
Fiberdichtung f <Mechan> *(AE)* fiber gasket, *(BE)* fibre gasket
Fiberglas n 1. <Bau> *(AE)* glass fiber, *(BE)* glass fibre; 2. <Wassertrans> *(AE)* fiberglass, *(BE)* fibreglass *(Schiffbau)*
Fibonacci-Verfahren n <Comp & DV> Fibonacci search
Fibonacci-Zahlen fpl <Math> Fibonacci numbers
Fibrillierung f <Papier> fibrillating
Fibroin n <Chemie> fibroin
Ficim n <Chemie> ficin
Fick'sches Gesetz n <Phys> Fick's law
Fiedelbohrer m <Maschinen> bow drill
fieren v 1. <Wassertrans> pay out *(Kette)*; lower *(Segeln)*; slacken *(Tauwerk)*; 2. <Wassertrans> veer *(Tauwerk)*
FIFA *(Spaltstoffabbrand, Spaltstoffverbrauch)* <Kerntech> FIFA *(fissions per initial fissile atom)*
FIFA-Wert m <Kerntech> FIFA value
FIFO-Prinzip n *(zuerst Abgelegtes wird als Erstes bearbeitet)* <Comp & DV> FIFO *(first-in-first-out)*
Figur f <Druck, Patent> figure *(Zeichen)*
figurative Konstante f <Comp & DV> figurative constant
figuratives Element n <Patent> figurative element *(Zeichen)*
fiktive Bezugsverbindung f <Telekom> HRC, hypothetical reference connection
fiktive Bindungsenergie f <Kerntech> fictitious binding energy
Filament n <Kunststoff, Textil> filament
Filamentdenier n <Textil> filament denier
Filicin... <Chemie> filicic
Film m 1. <Foto> film; 2. <Hydraul> film *(Wasser)*; 3. <Kunststoff, Mechan> film
Filmabtaster m <Fernseh> film scanner
Filmabtastung f <Fernseh> telecine scan
Filmandruckplatte f <Foto> pressure pad *(Kamera)*
Filmanzeiger m <Foto> film type indicator
Filmaufnahme f <Fernseh> film pick-up
filmbildend adj <Kunststoff> film-forming
Film-Coating n <Verpack> film coating
Filmdicke f <Kunststoff> film thickness
Filmdosimeter n <Kerntech, Strahlphys> film badge, film dosimeter
Filmdosimetrie f <Kerntech> film dosimetry

Filmdruck m <Druck> serigraphy
Filmführung f <Foto> film advance leader
Filmgelenk n <Kunststoff> integral hinge, living hinge
Filmgestell n <Foto> film rack
Filmhaltevorrichtung f <Foto> film holder
Filmkartonierung f <Verpack> film cartoning
Filmkassette f <Foto> cassette
Filmklammer f <Foto> film clip
Filmladegerät n <Foto> bulk film loader
Filmmontage f <Druck> film mounting
Filmpatrone f <Foto> cartridge *(Kleinbildkamera)*
Filmplakette f <Kerntech> film badge
Filmprojektor m <Elektriz> movie projector
Filmrissbildung f <Kunststoff> mud cracking
Filmrolle f <Foto> spool of films
Filmrückschicht f <Foto> film backing
Filmrückspulgabel f <Foto> film rewind handle
Filmscharnier n <Kunststoff> integral hinge, living hinge
Filmschlüsselschalter m <Elektrotech> membrane keyswitch
Filmschrumpfung f <Foto> film shrinkage
Filmsender m <Fernseh> film transmitter
Filmsieden n <Heiz & Kälte, Kerntech> film boiling
Filmspeicher m <Comp & DV> thin film memory
Filmspule f <Foto> film spool *(Kleinbildfilm)*
Filmtastatur f <Elektrotech> membrane keyboard
Filmtransport m <Foto> film transport
Filmtransporthebel m <Foto> film transport lever
Filmtransportkurbel f <Foto> film advance crank, film transport crank
Filmtransportzahntrommel f <Foto> film transport sprocket
Filmtrockengerät n <Foto> film dryer
Filmvorspann m <Foto> film leader *(Projektor)*
Filter n 1. <Chemtech, Comp & DV, Elektriz, Elektronik> filter; 2. <Erdöl> filter, strainer; 3. <Fertig> strainer; 4. <Foto, Funktech> filter; 5. <Kfztech> filter *(Vergaser, Öl)*; 6. <Kohlen, Labor, Mechan, Papier, Phys, Telekom> filter
Filter n/**auf dem Chip befindliches** <Elektronik> on-chip filter
Filter n **dritter Ordnung** <Elektronik> third-order filter
Filter n **erster Ordnung** <Elektronik> first-order filter
Filter n **für elliptisch polarisiertes Licht** <Elektronik> elliptic filter
Filter n **für relativ konstante Bandbreite** <Elektronik> constant-percentage bandwidth filter
Filter n **gegen elektromagnetische Beeinflussung** <Elektronik> electromagnetic-interference filter
Filter n **geradzahliger Ordnung** <Elektronik> even-order filter
Filter n **höherer Ordnung** <Elektronik> high-order filter
Filter n **mit drei Nullstellen** <Elektronik> three-zeros filter
Filter n **mit geringer Durchlassbreite** <Elektronik> narrow-band filter
Filter n **mit induktivem Eingang** <Elektrotech> choke-input filter
Filter n **mit unbegrenztem Impulsansprechverhalten** <Elektronik> infinite impulse response filter
Filter n **niedriger Ordnung** <Elektronik> low-order filter
Filter n **ungerader Ordnung** <Elektronik> odd-order filter
Filter n **zweiter Ordnung** <Elektronik> second-order filter
Filter-Abtast-Detektor m <Telekom> filter-and-sample detector
Filteramplitudenfrequenzgang m <Elektronik> filter amplitude response
Filteranlage f 1. <Sicherheit> filter plant; 2. <Umweltschmutz> filtering unit; 3. <Wasserversorg> filter plant
Filteraufnahme f <Fernseh> filter holder

Filterbanksystem

Filterbanksystem n <Telekom> filter-bank system
Filterbelag m <Chemtech> filter cake
Filterbett n <Chemtech, Kerntech, Wasserversorg> filter bed
Filterbeutel m <Chemtech> filter bag
Filterbohrung f <Chemtech> filtering well
Filterbrunnen m <Chemtech, Wasserversorg> filtering well
Filterdämpfung f <Elektronik> filter attenuation
Filterdeckel m <Kfztech> filter plug
Filterdrossel f 1. <Elektriz, Elektrotech> filter choke; 2. <Kfztech> filter choke, filter choke unit
Filterdurchlauf m <Kerntech> filter run
Filtereigenfunktion f <Elektronik> filter characteristic function
Filtereinsatz m 1. <Fertig> strainer; 2. <Heiz & Kälte, Kfztech, Mechan> filter cartridge
Filterelement n <Kfztech> filter cartridge, filter element
Filterfaktor m <Foto> filter factor
Filterflanke f <Elektronik> filter slope
Filterflasche f <Chemtech> filter flask, filtering flask
Filterfrequenz f <Elektronik> filter frequency
Filterfrequenzgang m <Elektronik> filter frequency response
Filtergehäuse n <Kfztech> filter housing; filter bowl *(Vergaser)*
Filtergerät n <Sicherheit> air filtering respirator
Filtergestaltung f <Elektronik> filter shaping
Filtergewebe n <Chemtech> filter cloth
Filtergrenzfrequenz f <Elektronik> filter cut-off frequency
Filtergruppe f <Elektronik> bank of filters
Filterhalter m <Labor> filter support
Filterhalterung f <Fernseh> filter holder
Filterhaus n <Kerntech> filter house
Filterhilfsmittel n <Chemtech> filter aid
Filterkammer f <Wasserversorg> filter gallery
Filterkartusche f <Chemtech> filter cartridge
Filterkerze f 1. <Kerntech> filtering candle; 2. <Mechan> filter cartridge
Filterkies m <Chemtech, Wasserversorg> filter gravel
Filterkohle f <Kohlen> filtering charcoal
Filterkondensator m <Elektrotech> filter capacitor
Filterkuchen m <Kohlen, Papier> filter cake
Filtermanschette f <Labor> filter funnel
Filtermasse f <Chemtech> filter pulp, filter stuff
Filtermembrane f <Labor> filter membrane
filtern v 1. <Chemtech> filter *(Gasfeststoff)*; 2. <Comp & DV, Elektronik, Foto> filter
Filtern n <Chemie, Chemtech, Comp & DV, Telekom> filtering
Filternullpunkt m <Elektronik> filter zero
Filternutsche f nach Büchner <Labor> Büchner funnel *(Filtrieren)*
Filterordnung f <Elektronik> filter order
Filterpapier n <Labor> filter paper
Filterpassbereich m <Elektronik> filter pass band
Filterpatrone f 1. <Kfztech> filter cartridge; 2. <Maschinen> cartridge filter
Filterphasenverhalten n <Elektronik> filter phase response
Filterplattenhahn m <Fertig> filter plate valve *(Kunststoffinstallationen)*
Filterpol m <Elektronik> filter pole
Filterpresse f 1. <Abfall, Chemtech> filter press; 2. <Ker & Glas> clay press, filter press; 3. <Kohlen, Labor, Lebensmittel, Papier> filter press
Filterpressenstoff m <Chemtech> filter-press cloth
Filterpumpe f <Kfztech, Labor> filter pump
Filterquarz m <Elektronik> filter crystal

Filterrahmen m <Chemtech> filter frame
Filterreihe f <Elektronik> filter bank
Filterrest m <Anstrich> residue
Filterrückstand m <Chemtech> filter cake
Filtersatz m <Foto> filter set
Filterschicht f 1. <Chemtech> filter bed, filtering layer; 2. <Wasserversorg> filter bed
Filtersieb n <Kunststoff> filter screen
Filterstoff m <Ker & Glas, Wasserversorg> filter cloth
Filtersynthese f <Elektronik> filter synthesis
Filtertrennschärfe f <Elektronik> filter discrimination
Filtertrichter m 1. <Chemtech> filtering cone; 2. <Labor> filter funnel
Filtertrog m <Kohlen> filter feed trough
Filtertrommel f <Chemtech> filter drum
Filtertuch n 1. <Textil> bolting fabric; 2. <Wasserversorg> filter cloth
Filterung f <Elektronik, Telekom> filtering
Filterung f gegen elektromagnetische Beeinflussung *(EMI-Filterung)* <Elektronik> electromagnetic-interference filtering
Filterverdickungsmittel n <Kohlen> filter thickener
Filterverhalten n <Elektronik> filter response
Filterverschmutzung f <Heiz & Kälte> filter fouling
Filterverstärker m <Elektronik> filter amplifier
Filtrat n <Chemtech, Kohlen> filtrate
Filtration f <Abfall, Erdöl, Papier> filtration
Filtrationsanlage f <Kohlen> filtration plant
Filtrationsprüfer m <Papier> filtration tester
Filtrationsverhältnis n <Anstrich> strain rate
Filtrierapparat m <Chemtech> filter
Filtrierbarkeit f <Kohlen> filterability
filtrieren v <Chemtech> filter
Filtrierkolben m <Labor> filtration flask
Filtriermasse f <Papier> filter mass
Filtrierpapier n <Chemtech> filter paper
Filtriersieb n <Chemtech> filtering screen
Filtrierstativ n <Labor> funnel stand
Filtriertuch n <Chemtech> filter cloth
Filtrierung f <Abfall, Erdöl> filtration
Filz m 1. <Druck> blanket; 2. <Fertig, Papier, Textil> felt
Filzdichtung f <Maschinen> felt packing
Filzfilter n <Maschinen> felt filter
Filzkonditionierer m <Papier> felt conditioner
Filzleitwalze f <Papier> felt-carrying roll
Filzmarke f <Papier> felt mark
Filzpolierer m <Ker & Glas> felt polisher
Filzreinigungsvorrichtung f <Papier> felt whipper
Filzscheibe f <Maschinen> felt washer
Filzseite f <Druck> felt side
Filzspannvorrichtung f <Papier> felt stretcher
Filztrockner m <Papier> felt dryer
FIN *(Fahrzeug-Identifizierungsnummer)* <Kfztech> VIN *(Vehicle Identification Number)*
Fingerhut m <Ker & Glas> thimble
Fingerhutrohr n <Kerntech> thimble
Fingerling m <Wassertrans> pintle
Fingerschutz m <Sicherheit> finger stall
Finger-Server m <Comp & DV> finger server
Finish-Presse f <Textil> press finishing machine
finite Impulsantwort f *(FIR)* <Telekom> finite impulse response *(FIR)*
Finite-Elemente-Methode f *(FEM)* <Maschinen> finite elements method *(FEM)*
finites Element n <Maschinen, Telekom> finite element
Finne f <Fertig, Maschinen> pane, peen
Finsternisdauer f <Raumfahrt> eclipse period
FIR *(finite Impulsantwort)* <Telekom> FIR *(finite impulse response)*

Firewall-Funktion f <Comp & DV> firewall facility
FIR-Filter n <Elektronik, Telekom> finite impulse response filter, FIR filter, nonrecursive filter
Firmenname m <Patent> trade name
Firmennetz n <Telekom> company network, corporate network (CN); enterprise network
Firmenportal n <Telekom> enterprise portal (Internet)
Firmware f <Comp & DV> firmware
Firnis m 1. <Fertig> boiled oil; 2. <Mechan> lacquer
First m <Bau> ridge (Dach) • **mit First versehen** <Bau> ridge
Firstabdeckung f <Bau> ridge capping
Firstbalken m <Bau> ridge beam
Firstbrett n <Bau> ridge piece
Firstlinie f <Bau> ridge line
Firstpfosten m <Bau> crown post; king post (Dachstuhl)
Firststein m <Bau> crown tile, ridge tile
Firststück n <Bau> ridge piece
Firstziegel m 1. <Bau> crest tile, head, ridge tile; 2. <Ker & Glas> ridge tile
FIR-System n <Elektronik> finite impulse response
Fischauge n <Kunststoff> fish eye
Fischaugenobjektiv n <Foto> fish-eye lens
Fische mpl <Fernseh> spikes (Bildstörung)
Fischerboot n <Wassertrans> fishing boat, fishing smack
Fischerei f <Wassertrans> fishery
Fischereifahrzeug n <Wassertrans> fishing boat, fishing vessel
Fischereihafen m <Wassertrans> fishery harbour, fishing harbour, fishing port
Fischereischutzboot n <Wassertrans> fishery protection vessel
Fischernetz n <Wassertrans> fishing net
Fischfanggebiet n <Wassertrans> fishing area, fishing grounds
Fischgerinne n <Nichtfoss Energ> fish pass
Fischgraben n <Nichtfoss Energ> fish pass
Fischgrätenantenne f <Funktech> fishbone antenna
Fischgrätenbildung f <Fernseh> herringboning
Fischgrätenmuster n <Bau> herringbone pattern (Parkettbodenbelag)
Fischgrätenstoff m <Textil> herringbone
Fischgrätenverkrümmung f <Ker & Glas> herringbone distortion
Fischgrätkühlrippen fpl <Kerntech> herringbone fins
Fischgrund m <Wassertrans> fishing ground
Fischleim m 1. <Druck> fish glue; 2. <Lebensmittel> isinglass
Fischplanke f <Wassertrans> king plank (Schiffbau)
Fischschwanzbohrer m <Erdöl> fishtail bit (Bohrtechnik)
Fischschwanzbohrmeißel m <Erdöl> fishtail bit
Fischzucht f <Lebensmittel> fish breeding
Fisetin n <Chemie> fisetin
Fisher-Information f <Math> Fisher information (Parameterabschätzung)
Fissium n <Kerntech> fissium
Fitting n <Maschinen> fitting, screwed fitting
Fix m <Wassertrans> position fix, fix (Navigation)
Fixationspunkt m <Ergon> point of fixation
fixer Kohlenstoff m <Chemie> fixed carbon
Fixfocus-Kamera f <Foto> fixed-focus camera
Fixfocus-Objektiv n <Foto> fixed-focus lens
Fixierbolzen m <Maschinen> locating stud
Fixiereinrichtung f <Fertig> locating device
Fixieren n 1. <Foto> fixing; 2. <Textil> boarding
Fixierfaden m <Foto> fixing thread
Fixiermittel n 1. <Foto> fixing agent; 2. <Textil> fastener
Fixierschraube f <Maschinen> locating screw
Fixierstift m <Maschinen> alignment pin, locating pin

fixiertes Walzenwehr n <Nichtfoss Energ> fixed roller sluice gate
Fixierung f 1. <Fertig> location (eines Werkstückes); 2. <Konstzeich> fixing; 3. <Textil> curing
Fixpunkt m <Mechan> benchmark
Fixpunkte mpl <Thermod> fixed reference points (Temperaturskale)
Fizeau'sche Ringe mpl <Phys> Fizeau fringes
Fjord m <Wassertrans> fjord
FKO (Fließkommaoperation) <Comp & DV> FLOP (floating-point operation)
FKP (Fließkommaprozessor) <Comp & DV> FPP (floating-point processor)
flach adj 1. <Bau> level; 2. <Elektronik, Funktech> flat; 3. <Ker & Glas, Kohlen> shallow; 4. <Telekom> flat; 5. <Wassertrans> shallow (Wasser)
flach geneigt adj <Bau> low-gradient
flach liegende Hülsen fpl <Verpack> lay flat tubing
Flachantenne f 1. <Funktech> planar array antenna; 2. <Telekom> flat antenna
Flachbatterie f <Telekom> flat battery
Flachbauelement n (SMD) <Elektriz> surface mounting device (SMD)
Flachbettdruck m <Druck> flat-bed printing
Flachbettgerät n <Comp & DV> flat bed
Flachbettplotter m <Comp & DV> flat bed plotter
Flachbettscanner m <Comp & DV> flat bed scanner
Flachbildschirm m <Comp & DV, Elektronik, Fernseh> flat display screen, flat screen, flat panel display, liquid crystal display, LCD
Flachboden m <Bau> flat top
Flachbodenätzgrübchen n <Metall> flat-bottomed etch pit
flachbödig adj <Wassertrans> flat-bottomed (Schiff)
Flachbogen m <Bau> flat arch, scheme arch, segmental arch
Flachbohrer m <Maschinen> flat drill
Flachdach n <Bau> cut roof, decking
Flachdeck n <Wassertrans> flat top
Flachdichtung f 1. <Fertig> flat gasket (Kunststoffinstallationen); 2. <Kerntech> flat gasket, flat-packing gasket; 3. <Kfztech> gasket; 4. <Maschinen> flat packing, gasket; 5. <Wassertrans> gasket
Flachdraht m <Elektrotech> flat wire
Flachdruck m <Druck> flat printing, planography
flache Beleuchtung f <Foto> flat lighting
flache Gruppenantenne f <Funktech> planar array antenna
flache integrierte Schaltung f <Elektronik> planar integrated circuit
flache Kante f <Ker & Glas> flat edge
flache Kante f und Schräge f <Ker & Glas> flat edge and bevel
flache Kehre f <Lufttrans> flat turn
flache Kümpelvertiefung f <Kerntech> dishing shallow depression (Brennstoffpellets)
flache Kurve f <Geom> flat curve
flache ringförmige Kammer f <Kerntech> pancake-shaped annular chamber
flache Wiedergabe f <Aufnahme, Elektronik> flat response
Fläche f 1. <Anstrich> area; 2. <Geom> area, face, surface; 3. <Papier, Phys> area
Fläche f eines Kreises <Geom> area of a circle
Fläche f/mit Kühlrippen versehene <Heiz & Kälte> finned surface
Fläche f zweiter Ordnung <Geom> quadric surface
Flacheisen n 1. <Maschinen> flat bar; 2. <Metall> flat
Flächen n <Fertig> slabbing • **Flächen bearbeiten** <Fertig> slab • **Flächen fräsen** <Fertig> face

Flächenabtaster

Flächenabtaster *m* <Gerät> area scanner
Flächenbearbeitung *f* <Ker & Glas> surface working
Flächenbelegung *f* <Funktech> illumination *(Antenne)*
Flächenberechnung *f* <Fertig> mensuration
Flächendeckung *f* <Telekom> coverage
Flächendiode *f* <Elektronik> junction diode
Flächendruckkraft *f* <Eisenbahn> surface pressure
Flächeneinheit *f* <Metrol> unit of area
Flächenemitter-Lumineszenzdiode *f* <Telekom> surface-emitting light-emitting diode
Flächenfräsen *n* <Maschinen> surface milling
Flächenfräsmaschine *f* <Maschinen> surface-milling machine
Flächengewicht *n* <Papier> basis weight, substance
Flächengitter *n* <Metall> plane lattice
Flächengründung *f* <Kohlen> pad foundation
Flächenheizkörper *m* 1. <Heiz & Kälte> panel heater; 2. <Labor> heating mantle
Flächenheizung *f* <Heiz & Kälte> radiant panel heating, radiant-heating system
Flächenhelle *f* <Elektrotech> luminance
Flächenhelligkeit *f* <Optik> brightness
Flächenhelligkeitsmesser *m* <Gerät> brightness meter
Flächeninhalt *m* <Bau, Geom> area
Flächeninhaltsbestimmung *f* <Nichtfoss Energ> quadrature
Flächenkontaktdiode *f* <Elektronik> junction diode
Flächenkrümmung *f* <Geom> curvature of surfaces
Flächenladungsdichte *f* <Elektrotech> surface density
Flächenlast *f* <Maschinen> surface load
Flächenmaße *npl* <Metrol> square measures
Flächenmassenmessung *f* <Gerät> area mass measurement
Flächennutzungsplan *m* <Wasserversorg> zoning plan
Flächenpressung *f* 1. <Fertig> pressure intensity; 2. <Maschinen> surface pressure
Flächenprofil *n* <Maschinen> surface profile
Flächenregel *f* <Lufttrans> area rule
Flächenschleifmaschine *f* <Fertig> surface-grinding machine
Flächenschwerpunkt *m* <Phys> centroid
Flächenträgheitsmoment *n* <Maschinen> geometrical moment of inertia
Flächentransistor *m* <Elektronik> junction transistor
Flächenverlust *m* <Ker & Glas> loss of sheet
Flächenversetzung *f* <Kerntech> face gap
Flächenwiderstand *m* <Elektrotech> sheet resistance
Flächenwirkungsgrad *m* <Funktech> aperture efficiency *(Antenne)*
flacher Auswerferstift *m* <Maschinen> flat ejector pin
flacher Bildschirm *m* 1. <Comp & DV> flat screen; 2. <Elektronik> flat panel display; 3. <Telekom> flat screen
flacher Bipolartransistor *m* <Elektronik> planar bipolar transistor
flacher Dreiphasenlinearmotor *m* <Elektrotech> flat three-phase linear motor
flacher Linearmotor *m* <Elektrotech> flat linear motor
flacher Plattenkollektor *m* <Nichtfoss Energ> flat plate collector
flacher Schlauchbeutel *m* <Verpack> lay flat film bag
flacher Sinkflug *m* <Lufttrans> shallow descent
flacher Stab *m* <Metall> flat bar
flacher Treibriemen *m* <Maschinen> flat transmission belt
flaches Display *n* <Fernseh, Telekom> flat display, flat panel, flat screen
flaches Gehäuse *n* <Elektronik, Telekom> flatpack
flaches Optikwerkzeug *n* <Ker & Glas> flat optical tool
flaches Überlaufwehr *n* <Hydraul> flat-crested weir

Flachfacette *f* <Ker & Glas> flat facet
Flachfeder *f* <Maschinen> flat spring
Flachfeile *f* <Maschinen> flat file
Flachformzylinderpresse *f* <Druck> flat-bed cylinder press
Flachführung *f* <Fertig> square guide
flachgängig *adj* <Fertig> square *(Gewinde)*
Flachgehäuse *n* <Elektronik, Telekom> flatpack
flachgehendes Schiff *n* <Wassertrans> shallow-draught vessel
Flachgewinde *n* <Maschinen> flat thread, square thread
Flachgewindemeißel *m* <Fertig> square thread tool
Flachgewölbe *n* <Ker & Glas> jack arch
Flachglas *n* 1. <Bau> plate glass; 2. <Ker & Glas> flat glass
Flachgurt *m* <Maschinen> flat belt
Flachheizkörper *m* <Heiz & Kälte> flat radiator
Flachinstrument *n* <Gerät> flat-face instrument
Flachkabel *n* <Elektriz, Elektrotech, Telekom> flat cable, ribbon cable
Flachkämmmaschine *f* <Textil> rectilinear-combing machine
Flachkarton *m* <Verpack> flat pack
Flachkeil *m* 1. <Fertig> flat key, parallel key; 2. <Maschinen> flat key, key on flat, rectangular key
Flachkopf *m* <Maschinen> pan head
Flachkopfniet *m* <Maschinen> pan head rivet
Flachkopfschraube *f* <Maschinen> pan head screw
Flachlitzenseil *n* <Fertig> flattened strand rope
Flachmaterial *n* <Ker & Glas, Maschinen> flats
Flachmeißel *m* 1. <Fertig> square-nosed tool; 2. <Maschinen> flat chisel
Flachmetall *n* <Metall> flat
Flachpackung *f* <Bau, Verpack> flat pack
Flachpalette *f* <Verpack> flat pallet
Flachpresse *f* <Bau> flat jack *(Stahlbeton)*
Flachprofilgerät *n* <Gerät> flat edgewise pattern instrument
Flachrelais *n* <Elektrotech> flat relay
Flachriemen *m* <Maschinen> flat belt
Flachriementrieb *m* <Maschinen> flat-belt drive
Flachringdynamo *m* <Elektriz> flat ring dynamo
Flachrücken *m* <Druck> flat back
Flachrundkopf *m* 1. <Fertig> truss head; 2. <Maschinen> mushroom head
Flachrundkopfniet *m* <Bau> cup head rivet
Flachrundkopfschraube *f* <Maschinen> mushroom-head bolt
Flachrundschraube *f* <Maschinen> saucer head screw
Flachsauger *m* <Papier> flat box
Flachschieber *m* 1. <Hydraul> plain slide valve; 2. <Maschinen> flat slide valve
Flachschiene *f* <Eisenbahn> strap rail
Flachschleifen *n* <Maschinen> surface grinding
Flachschleifmaschine *f* 1. <Ker & Glas> flat-grinding machine; 2. <Maschinen> surface grinder, surface-grinding machine
Flachseil *n* <Maschinen> flat rope
Flachsenken *n* <Maschinen> end facing, spot facing
Flachspan *m* <Fertig> flake
flachspritzen *v* <Fertig> quench *(Kunststoffe)*
Flachspritzen *n* <Fertig> quenching *(Kunststoffe)*
Flachspule *f* <Elektriz, Elektrotech> loop coil, pancake coil
Flachspulinstrument *n* <Gerät> flat-coil instrument
Flachstab *m* <Maschinen> flat rod
Flachstahl *m* <Metall> flat bar, flats
Flachstampfer *m* <Fertig> flat rammer *(Formen)*
Flachstößel *m* <Fertig> flat follower *(Getriebelehre)*

Flachstrickerei f <Textil> flat knitting
Flachstrickmaschine f <Textil> flat-knitting machine
Flachstromvergaser m <Kfztech> (AE) horizontal carburetor, (BE) horizontal carburettor; (AE) sidedraft carburetor, (BE) sidedraft carburettor (veraltetes Prinzip)
Flachtrudeln n <Lufttrans> flat spin
Flachwagen m <Eisenbahn, Kfztech> flat wagon
flachwalzen v <Fertig> slab
Flachwasser n <Wassertrans> shallows
Flachwicklung f <Elektriz> (BE) disc winding, (AE) disk winding
Flachzange f <Fertig> flat-nosed pliers
Flachziegel m 1. <Bau> plain tile; 2. <Ker & Glas> flat tile
Flackerfrequenz f <Fernseh> flicker frequency
flackern v 1. <Elektronik> flicker; 2. <Fertig> flare
Flackern n 1. <Comp & DV, Fernseh> flicker; 2. <Telekom> flickering
Flagge f 1. <Telekom> flag; 2. <Wassertrans> (AE) colors, (BE) colours, flag
Flaggenkasten m <Wassertrans> flag locker, signal locker
Flaggenknopf m <Wassertrans> truck
Flaggenmast m <Wassertrans> signal mast
Flaggensignal n <Wassertrans> flag signal (Kommunikation)
Flaggenspind m <Wassertrans> flag locker, signal locker
Flaggenstock m <Wassertrans> flagstaff
Flaggenzeichen n <Wassertrans> flag signal (Kommunikation)
Flagge-Q f <Wassertrans> Q flag
Flaggleine f <Lufttrans> halyard
Flaggschiff n <Wassertrans> flagship (Marine)
Flamm... <Textil> blazing
flammbar adj <Thermod> flammable
Flammbarkeit f 1. <Thermod> flammability; 2. <Verpack> combustibility, flammability
flammbeständig adj <Sicherheit> flame-resistant
Flamme f <Sicherheit, Thermod> fire, flame
flämmen v <Fertig> scarf
Flämmen n <Fertig> scarfing
Flammenabsorptionsphotometrie f <Gerät> absorption flame photometry
Flammenausbreitungsgeschwindigkeit f <Thermod> rate of spread of flame
Flammenaussetzer m <Thermod> flame failure
flammenbeständig adj <Elektriz> flameproof
flammenbeständige Beleuchtungseinrichtung f <Elektriz> flameproof lighting installation
flammenbeständiger Schalter m <Elektriz> flameproof switch
flammend adj 1. <Textil> blazing; 2. <Thermod> blazing, flaming
Flammendämpfung f <Ker & Glas> flame attenuation
flammendurchschlagsichere elektrische Einrichtung f <Sicherheit> flameproof electrical equipment
Flammenemissionsspektroskopie f <Phys> flame emission spectroscopy
flammenfest adj <Verpack> flameproof
flammengehärtet adj <Fertig> flame-hardened
Flammenhärtemaschine f <Fertig> flame-hardening machine
Flammenhärtung f <Fertig> flame hardening
Flammenhemmstoff m <Lufttrans, Sicherheit> flame retardant
Flammenhemmung f <Sicherheit, Werkprüf> flame retardancy (Kunststoffe)
Flammenhydrolyse f <Optik> flame hydrolysis
Flammenlöscher m <Raumfahrt> flame arrester
Flammenlötung f <Fertig> torch brazing

Flammenmelder m <Heiz & Kälte> flame detector
Flammenphotometer n <Labor> flame photometer
flammenphotometrischer Detektor m <Umweltschmutz> flame photometric detector
Flammenplatieren n <Fertig> flame plating
Flammenrückschlag m 1. <Bau> flashback; backfire (Schweißen); 2. <Kfztech> backfire; 3. <Thermod> flashback
Flammenrückschlagsicherung f <Kfztech> flame trap (Motor)
flammensicher adj <Fertig, Lufttrans, Verpack> flameproof
flammensicherer Schalter m <Elektriz> flameproof switch
flammensicheres Gehäuse n <Sicherheit> flameproof enclosure
Flammenspektroskopie f <Thermod> flame spectroscopy
Flammenspektrum n <Phys, Thermod> flame spectrum
Flammensperre f 1. <Erdöl> flame arrester (Sicherheitstechnik); 2. <Lufttrans, Thermod> flame trap
Flammenstabilisator m <Lufttrans> flame holder
Flammenstrahlbohren n <Kohlen> jet drilling
Flammenwächter m <Heiz & Kälte> flame detector
Flammenwerfer m <Thermod> flame thrower
flammfest adj <Werkprüf> nonflammable
Flammfugenhobeln n <Bau> flame gouging
Flammhärten n <Metall> flame hardening
flammhemmend adj <Kunststoff> flame-retardant
flammhemmend eingestellt adj <Kunststoff> fire-retardant
flammhemmendes Zusatzmittel n <Kunststoff> flame-retardant
Flammkohle f <Kohlen> flaming coal
Flammofen m 1. <Fertig> air furnace, reverberatory furnace; 2. <Heiz & Kälte> air furnace; 3. <Metall> reverberatory furnace
flammofenfrischen v <Fertig> puddle
Flammofenfrischen n <Fertig> puddling
Flammpunkt m 1. <Erdöl> flash point (Destillation); 2. <Heiz & Kälte, Kfztech, Kunststoff, Lebensmittel, Maschinen, Thermod> flash point; 3. <Wassertrans> flash point (Material)
Flammpunktprüfer m <Maschinen> flash tester
Flammpunktprüfgerät n <Labor> flash point apparatus
Flammrohr n 1. <Fertig, Heiz & Kälte> flue; 2. <Kerntech> flame tube (Gasturbine)
Flammrohrkessel m 1. <Heiz & Kälte> flue boiler; 2. <Wassertrans> Scotch boiler (Dampferzeugung)
Flammrückschlagsicherung f <Lufttrans, Sicherheit> flame trap
Flammschutz m <Sicherheit> flame protection
Flammschutzmittel n <Kunststoff, Sicherheit> flame-retardant
Flammschweißen n <Kunststoff> flame welding
flammsicher adj <Sicherheit> flameproof
flammsicherer Rührer m <Labor> flameproof stirrer
Flammsperre f <Bau, Sicherheit> flame arrester
Flammspritzen n <Fertig, Kerntech> flame spraying
Flammstrahlen n 1. <Fertig> flame blasting; 2. <Ker & Glas> scarfing
Flammstrahlen n zum Entzundern <Fertig> flame cleaning
Flammstrahlreinigen n <Bau> flame cleaning
flammwidrig adj 1. <Heiz & Kälte> flame-retardant; 2. <Kunststoff> fire-retardant
Flanke f 1. <Bau> flank; 2. <Fertig> side (Getriebelehre); flank (Gewinde); 3. <Maschinen> flank, side; 4. <Mechan> flank

Flankenabfall

Flankenabfall *m* <Telekom> roll-off *(Merkmal)*
Flankendurchmesser *m* <Maschinen> effective diameter, minor diameter, pitch diameter
Flankenformfehler *m* <Maschinen> flank form error
flankengesteuerter Multivibrator *m* <Elektronik> edge-triggered flip-flop
Flankenlinie *f* <Maschinen> flank line
Flankenrate *f* <Elektronik> edge rate
Flankenspiel *n* 1. <Fertig> backlash *(Getriebelehre)*; 2. <Maschinen> flank clearance; 3. <Mechan> backlash
Flankensteilheit *f* <Elektronik> edge steepness
Flankenverriegelung *f* <Elektronik> edge latching
Flankenwinkel *m* 1. <Fertig> thread angle; profile angle *(Gewinde)*; 2. <Maschinen> thread angle
Flansch *m* 1. <Akustik> flange; 2. <Bau> boom; 3. <Elektriz, Elektrotech> coupling, flange; 4. <Erdöl> flange *(Rohrleitungen)*; 5. <Fertig> flange; 6. <Kfztech> flange; web *(Kurbelwelle)*; 7. <Labor, Maschinen, Mechan, Meerschmutz, Raumfahrt> flange; 8. <Telekom> connector, coupling *(Wellenleiter)*
Flanscharmaturen *fpl* <Maschinen> flanged fittings
Flanschbefestigung *f* <Maschinen, Raumfahrt> flange mounting
flanschen *v* 1. <Erdöl> flange up; 2. <Maschinen> flange
Flanschen *n* <Fertig, Maschinen> flanging
Flanschengussrohr *n* <Bau> flanged cast-iron pipe
Flanschfläche *f* <Maschinen> flange facing
Flanschhälfte *f* <Maschinen> half-flange
Flanschkupplung *f* <Maschinen> flange coupling, half-coupling
flanschlose Bremssohle *f* <Eisenbahn> flangeless brake shoe
Flanschmotor *m* <Elektriz, Mechan> flange motor
Flanschpressenbearbeitung *f* <Ker & Glas> flange press finish
Flanschrohr *n* 1. <Bau> flange pipe; 2. <Maschinen> flanged pipe
Flanschstahl *m* <Metall> flange steel
Flanschverbindung *f* 1. <Maschinen> flange connection, flange coupling, flange joint, flange union, flanged connection, flanged coupling, flanged joint, flanged union; 2. <Telekom> flanged joint
Flanschwelle *f* <Maschinen> flange shaft, flanged shaft
Fläschchen *n* <Labor> *(BE)* phial, *(AE)* vial
Flasche *f* 1. <Fertig, Ker & Glas> flask; 2. <Maschinen> pulley • **auf Flaschen abfüllen** <Lebensmittel> bottle
Flasche *f* **mit formgepresstem Hals** <Labor> *(AE)* bottle with molded neck, *(BE)* bottle with moulded neck
Flasche *f* **mit Schraubverschluss** <Verpack> screw cap bottle
Flasche *f* **mit unebenem Boden** <Ker & Glas> rocker
Flasche *f* **ohne Pfand** <Verpack> nonreturnable bottle
Flaschenabfüllmaschine *f* <Lebensmittel> bottling machine
Flaschenabpackungsanlage *f* <Verpack> bottle-packing machine
Flaschenausrichter *m* <Verpack> bottle unscrambler
Flaschenbrennofen *m* <Ker & Glas> bottle kiln *(Keramik)*
Flaschenetikettaufdruck *m* <Verpack> graphics coordinated with bottle labels
Flaschenfüllapparat *m* <Lebensmittel> bottle filler
Flaschengas *n* <Thermod> bottled gas
Flaschenglas *n* <Ker & Glas> bottle glass
Flaschenhals *m* 1. <Ker & Glas> neck of a bottle; 2. <Lebensmittel, Mechan> bottleneck
Flaschenhülle *f* <Verpack> bottle jacket
Flaschenhülse *f* <Verpack> bottle sleeve
Flaschenindustrie *f* <Ker & Glas> bottle industry
Flaschenkappenaufsetzer *m* <Verpack> bottle-capping machine
Flaschenkapsel *f* <Verpack> bottle capsule
Flaschenkasten *m* <Verpack> bottle crate, crate
Flaschenkorkmaschine *f* <Verpack> bottle-corking machine
Flaschenlecksensor *m* <Verpack> bottle leak detector
Flaschenpfand *n* <Verpack> bottle deposit, deposit
Flaschenspülmaschine *f* <Lebensmittel> bottle washer
Flaschenstopfen *m* <Verpack> bottle stopper
Flaschenträger *m* <Verpack> bottle carrier
Flaschenummantellungsmaschine *f* <Ker & Glas> bottle-casing machine
Flaschenverschließmaschine *f* <Verpack> bottle-closing machine, bottle-sealing machine
Flaschenverschluss *m* 1. <Ker & Glas> closure for bottles; 2. <Verpack> bottle closure
Flaschenverschlusskappe *f* <Verpack> bottle cap
Flaschenwagen *m* <Fertig> trolley *(Schweißen)*
Flaschenwaschmaschine *f* <Verpack> bottle-rinsing machine, bottle-washing machine
Flaschenwinde *f* <Maschinen> bottle jack
Flaschenzähler *m* <Gerät> bottle counter
Flaschenzählgerät *n* <Gerät> bottle counter
Flaschenzug *m* 1. <Bau> tackle; 2. <Fertig> block and tackle, pulley, tackle; 3. <Maschinen> block and pulley, block and tackle, lifting block, lifting tackle, pulley block; 4. <Mechan> chain block, tackle; 5. <Phys> pulley; 6. <Sicherheit> hoist; 7. <Wassertrans> pulley block *(Deckausrüstung)*
Flaschenzuggehäuse *n* <Maschinen> pulley shell
Flaschenzughaken *m* <Maschinen> pulley block hook
Flashdestillation *f* <Thermod> flash distillation
Flatpack *m* <Telekom> flat pack
Flatpack-Gehäuse *n* <Telekom> flat pack
Flatter… <Lufttrans> flutter
Flatterdämpfer *m* <Lufttrans> shimmy damper *(Luftfahrzeug)*
Flatterecho *n* <Akustik> flutter echo
Flattereffekt *m* <Fernseh> flutter effect
Flatterfaktor *m* <Aufnahme> flutter factor
flattern *v* <Maschinen> chatter, knock
Flattern *n* 1. <Aufnahme> flutter *(schnelle Tonhöhenschwankungen)*; 2. <Comp & DV> jitter; 3. <Fernseh, Funktech> flutter; 4. <Lufttrans> buffeting; flutter *(Aerodynamik)*; 5. <Maschinen> knocking
Flattersatz *m* **rechts** <Druck> ragged right setting
Flatterschwingung *f* <Lufttrans> flutter *(Aerodynamik)*
Flattersitz *m* <Hydraul> fluttering seat
flaue Kühle *f* <Wassertrans> light airs
Flaute *f* <Wassertrans> calm *(Wind)*
Flavan *n* <Chemie> flavan
Flavanon *n* <Chemie> flavanone
Flavin *n* <Chemie> flavin
Flavon *n* 1. <Chemie> flavone, phenylchromone; 2. <Lebensmittel> flavone *(Pflanzenfarbstoff)*
Flavonoid *n* <Lebensmittel> flavonoid
Flavonol *n* <Chemie> flavonol
Flavoprotein *n* <Lebensmittel> flavoprotein
Flavopurpurin *n* <Chemie> flavopurpurin
Flechtdraht *m* <Elektrotech> braided wire
Flechte *f* <Textil> braid
flechten *v* <Bau> bend *(Bewehrungsstahl)*
Flechten *n* <Textil> braiding
Flechtenrot *n* <Chemie> orcein *(Farbstoff)*
Flechttechnik *f* <Textil> braiding technique
Flechtwerk *n* <Bau> trellis, trellis work
Fleck *m* 1. <Druck> speckle; 2. <Ker & Glas> freak; 3. <Textil> blotch, speckle
Fleck *m* **unter Oberfläche** <Ker & Glas> pip under finish

Fleckempfindlichkeitsklasse f <Ker & Glas> staining class (zur Charakterisierung von optischem Glas)
Fleckenbildung f <Ker & Glas> specking; staining (atmosphärischer Angriff)
fleckenfrei adj <Papier> stainless
Fleetwagen m <Kfztech> program car
Fleisch n 1. <Druck> beard; 2. <Lebensmittel> meat
Fleischbeschaugesetz n <Abfall> Meat Inspection Act
Fleischmilch f <Chemie> sarcolactic (Säure)
Fleischseite f <Fertig> flesh side (Riemen)
Fleischwolf m <Lebensmittel> meat grinder, mincer, mincing machine
Fleißspannung f <Phys> yield stress
FLEI-Verkehr m (Flugzeug-Eisenbahn-Verkehr) <Trans> rail-air-rail service
Fleming'sche Dreifingerregel f <Elektriz> Fleming's rules
Fleming'sche rechtshändige Dreifingerregel f <Elektriz> right-hand rule
flexibel adj <Elektriz> flexible
Flexibilität f <Kunststoff> flexibility
flexible Feldgruppe f <Comp & DV> flexible array
flexible Flachbaugruppe f <Elektronik> flexible printed circuit
flexible gedruckte Schaltung f <Elektronik> flexible printed circuit
flexible Isolierplatte f <Fertig> blanket insulator
flexible Kupplung f <Maschinen> flexible coupling
flexible Leiterplatte f <Elektronik> flexible printed circuit
flexible Stahlschlauchleitung f <Maschinen> flexible steel piping
flexible Verbindung f <Eisenbahn> flexible connection
flexibler Datenträger m <Comp & DV> flexible disk
flexibler Einband m <Druck> soft cover
flexibler Hohlleiter m <Elektrotech, Funktech> flexible waveguide (Weltraumfunk)
flexibler Leiter m <Elektriz> flexible conductor
flexibler Reflektor m <Raumfahrt> flexible reflector (Raumschiff)
flexibler Schlauch m <Meerschmutz> flexible hose
flexibler Schwimmtank m <Meerschmutz> floating flexible tank (zur Aufnahme von Öl)
flexibler Umschlag m <Druck> limp binding
flexibler Widerstand m <Elektrotech> flexible resistor
flexibles Fertigungssystem n (FFS) <Künstl Int> flexible manufacturing system (FMS)
flexibles Kabel n <Elektriz> flexible cable
flexibles Produktionssteuerungssystem n <Künstl Int> flexible manufacturing system
Flexion f <Ergon> flexion
Flexodruck m 1. <Druck> aniline print, flexographic printing, flexography; 2. <Kunststoff> flexographic printing
Flexofaltschachtel-Klebmaschine f <Verpack> flexofolder gluer
flicken v <Textil> mend
Flicken n <Textil> mending
Flicken n **des Ofenfutters** <Ker & Glas> patching
Flickstein m <Ker & Glas> patch block
fliegen v <Lufttrans> pilot
fliegend aufgespannter Dorn m <Fertig> (AE) stub arbor, (BE) stub arbour
fliegende Brücke f <Bau, Wassertrans> flying bridge
fliegende Fertigungsprüfung f <Qual> patrol inspection
fliegende Lagerung f <Fertig> cantilever support (Walze)
fliegende Verbindung f <Telekom> jumper
fliegender Akzent m <Druck> floating accent, piece accent
Fliehgewicht n <Maschinen> centrifugal weight, flyweight
Fliehkraft f <Chemtech, Maschinen, Phys, Raumfahrt, Strömphys, Umweltschmutz> centrifugal force

Fliehkraftabsauggebläse n <Kerntech> centrifugal extractor
Fliehkraftabscheider m 1. <Chemtech> centrifugal separator; 2. <Umweltschmutz> cyclone-recovery skimmer
Fliehkraftabschneider m <Erdöl> cyclone
Fliehkraftbeschleunigung f <Lufttrans> centrifugal acceleration
Fliehkraftgebläse n <Kfztech> centrifugal supercharger
Fliehkraftkupplung f <Kfztech, Maschinen> centrifugal clutch
Fliehkraftlüfter m <Heiz & Kälte> centrifugal fan
Fliehkraftregelung f <Lufttrans> governor control link
Fliehkraftregler m 1. <Kfztech> governor (Motordrehzahlbegrenzer); 2. <Nichtfoss Energ> governor
Fliehkraftreiniger m <Chemtech> centrifugal cleaner
Fliehkraftschalter m 1. <Gerät> tachometric relay; 2. <Maschinen> centrifugal switch
Fliehkraftschmierung f <Maschinen> centrifugal lubrication
Fliehkraft- und Vakuumregler m <Lufttrans> centrifugal and vacuum governor
Fliehkraftversteller m <Kfztech> centrifugal advance mechanism (Zündung)
Fliehkraftzündversteller m <Kfztech> centrifugal advance mechanism
Fliehpendeltachometer n <Gerät> flyweight tachometer
Fliese f <Bau> slab, tile
fliesen v <Bau> tile
Fliesenboden m <Ker & Glas> tiling
Fliesenfußboden m <Bau> tile floor, tile flooring
Fließ... <Comp & DV, Elektrotech, Erdöl> flow
Fließband n 1. <Kontroll> line; 2. <Maschinen> conveyor line; 3. <Mechan> assembly line
Fließbandarchitektur f <Kontroll> pipelined architecture
Fließbandstraße f <Maschinen> assembly line
Fließbandverarbeitung f <Telekom> pipelining
Fließbett n <Chemtech, Umweltschmutz> fluidized bed
Fließbetttrockner n <Chemtech> fluidized-bed dryer
Fließbettverfahren n <Chemtech> fluid-catalyst process
Fließbild n <Comp & DV> flowchart
Fließdiagramm n <Erdöl, Qual> flow sheet
fließen v <Comp & DV, Phys, Strömphys, Textil, Wassertrans> Gezeiten flow (Fluss)
Fließen n 1. <Bau> creep; 2. <Elektrotech, Kunststoff, Maschinen, Textil> flow
fließend adj <Phys> floating
fließender Erdstoff m <Kohlen> running soil
fließender Verkehr m <Bau, Trans> moving traffic
fließendes Gewässer n <Wasserversorg> running water
fließendes Wasser n <Wasserversorg> running water
fließfähiges Produkt n <Verpack> free-flow product
Fließfertigung f <Verpack> continuous production line
Fließfigur f <Fertig> worm
fließgepresst adj <Maschinen> extruded
Fließgeschwindigkeit f 1. <Heiz & Kälte> flow velocity; 2. <Kunststoff> flow rate
Fließgrenze f 1. <Anstrich> yield strength; 2. <Bau> liquid limit; 3. <Kerntech> yield point; 4. <Kohlen> liquid limit; 5. <Mechan> yield point, yield strength; 6. <Phys> yield point; 7. <Werkprüf> yield strength
Fließgrenzgerät n <Kohlen> liquid limit device
Fließkommamodus m <Comp & DV> noisy mode
Fließkommaoperation f (FKO) <Comp & DV> floating-point operation (FLOP)
Fließkommaprozessor m (FKP) <Comp & DV> floating-point processor (FPP)
Fließlehre f <Kohlen, Metall> rheology
Fließlinien-Schrumpfverpackungen fpl <Verpack> shrink flow line wrappers

Fließlöten

Fließlöten n <Elektrotech> reflow soldering
Fließpresse f 1. <Maschinen> extruder; 2. <Metall> extrusion press
fließpressen v <Maschinen, Metall> extrude
Fließpressen n <Maschinen, Metall> extrusion
Fließpressteil n <Maschinen> extruded part
Fließpunkt m 1. <Erdöl> flow point; 2. <Heiz & Kälte> pour point; 3. <Ker & Glas> flow point; 4. <Kfztech> pour point (Öl); 5. <Strömphys> flow point
Fließreibung f <Bau> hydraulic friction
Fließsand m <Kohlen> quicksand
Fließsatz m <Druck> body matter, solid matter
Fließschaumverpackung f <Verpack> flow foam wrap
Fließschema n <Erdöl, Kohlen, Qual> flow sheet
Fließschweißen n <Fertig> flow welding (Kunststoffe)
Fließspan m <Maschinen> continuous chip
Fließton m <Kohlen> quick clay
Fließvariable f <Kohlen, Metall> rheological variable
Fließverhalten n <Kunststoff> flow characteristics
Fließvermögen n 1. <Kohlen> rheological properties; 2. <Maschinen, Mechan> fluidity; 3. <Metall> plasticity, rheological properties; 4. <Phys, Strömphys> fluidity
Fließvermögen bei Kälte <Mechan> cold flow
Fließwasser n <Wasserversorg> running water
Fließwiderstand m <Phys> flow stress
Fließzeit f <Kunststoff> scorch time (Kautschukmischung)
Fließzonenbildung f <Kunststoff> crazing
flimmerfrei adj 1. <Comp & DV> flicker-free; 2. <Elektronik> jitter-free (Leuchtdioden)
flimmern v <Elektronik> flicker
Flimmern n 1. <Comp & DV> flicker; 2. <Raumfahrt> scintillation (Weltraumfunk); 3. <Telekom> scintillation; flickering (Video)
Flimmerphotometer n <Phys> flicker photometer
Flimmerrauschen n <Raumfahrt> scintillation noise (Weltraumfunk)
flinke Sicherung f <Elektrotech> fast-acting fuse
Flip-Flop m <Comp & DV> flip-flop, toggle
Flip-Flop-Schaltung f <Comp & DV, Phys> flip-flop
Flip-Flop-Vorgang m <Kerntech> flop-over process
Floatglas n <Ker & Glas> float glass
Flocke f 1. <Fertig> shatter crack, thermal burst; 2. <Kunststoff> flock
flockenförmig adj <Chemtech> flocculent
flockengefärbt adj <Textil> stock-dyed
Flockenriss m <Fertig> fish eye (Stahl)
Flockigkeit f <Chemtech> flocculence
Flockpunkt m <Chemtech> flocculation point
Flockspritzen n <Verpack> flock spraying
Flocktest m <Chemtech> flocculation test
Flockung f 1. <Abfall, Chemie, Chemtech> flocculation; 2. <Erdöl> flocculation (Aufbereitungstechnik); 3. <Kohlen, Kunststoff, Lebensmittel> flocculation
Flockungschemikalie f <Chemtech> flocculent
Flockungseinrichtung f <Chemtech> flocculator
Flockungshilfsmittel n <Kunststoff> flocculant, flocculating agent
Flockungsmittel n 1. <Abfall> coagulant, flocculant; 2. <Chemtech> flocculent; 3. <Kohlen> flocculant; 4. <Kunststoff> flocculant, flocculating agent
Flockungsreaktor m <Chemtech> flocculator
Floppy f 1. <Comp & DV> FD, floppy disk; 2. <Druck> FD, diskette, floppy disk
Flor m <Textil> pile
Flordichte f <Textil> density of pile
Florgewicht n <Textil> pile weight
Florhöhe f <Textil> pile height
Florteiler m <Textil> tape condenser
Floß n <Wassertrans> raft

Floßbrücke f <Wassertrans> floating bridge
Flosse f <Wassertrans> fin (Schiffbau)
Flossenkiel m <Wassertrans> fin keel (Schiffbau)
Flossenstabilisator m <Wassertrans> (AE) fin stabilizer, (BE) fin stabiliser
Flotation f <Kohlen> flotation
Flotationsanlage f <Chemtech> flotation plant
Flotationsflüssigkeit f <Chemtech> flotation liquid
Flotationskammer f <Meerschmutz> flotation chamber (Ölsperre)
Flotationsschaum m <Chemtech> flotation froth
Flotationsverfahren n <Kohlen> flotation process
Flotationszelle f mit Luftrührung <Kohlen> pneumatic flotation cell
flott adj <Wassertrans> afloat (Schiff)
Flotte f 1. <Textil> bath; 2. <Wassertrans> fleet, navy
Flottenbasis f <Wassertrans> naval base
Flowmeter m <Wasserversorg> flowmeter
Flöz n <Kohlen> seam
Flözarbeit f <Kohlen> seam work
Flucht f <Fertig> alignment • **außer Flucht** <Bau> out-of-line
fluchten v <Bau> align
Fluchten n 1. <Fertig> aligning; 2. <Maschinen> alignment, lining-in
fluchtend adj 1. <Fertig> lining-up; 2. <Maschinen> in-line
Fluchtendmachen n <Maschinen> lining-up
Fluchtgerät n <Sicherheit> emergency equipment, escape apparatus, escape respirator, personnel escape set
Fluchtgeschwindigkeitsstartmotor m <Raumfahrt> launch escape motor
flüchtig adj 1. <Chemie, Comp & DV> volatile; 2. <Druck> fugitive; 3. <Elektrotech> transient; 4. <Kunststoff> volatile; 5. <Phys> transient; 6. <Textil> volatile
flüchtige Emissionen fpl <Umweltschmutz> fugitive emissions
flüchtige Masse f <Kohlen> volatile body
flüchtige Spannung f <Elektriz> transient voltage
flüchtiger Netzvorgang m <Elektriz> line transient
flüchtiger Speicher m <Comp & DV, Elektrotech> volatile memory
Flüchtigkeit f <Chemie, Kohlen, Kunststoff, Textil> volatility
Flüchtigkeit f von Gasen <Thermod> fugacity of gases
Fluchtkapsel f <Erdöl> escape capsule (Offshore)
Fluchtlinie f <Mechan> alignment
Fluchtpunkt m <Geom, Konstzeich> vanishing point
Fluchtpunktperspektive f <Geom> true perspective, vanishing point perspective
Fluchtpunktprojektion f <Konstzeich> vanishing point projection
Fluchtstab m <Bau> range pole, range rod (Vermessung)
Fluchtstange f <Bau> rod, target (Vermessung)
Fluchtungsfehler m 1. <Fertig> malalignment, misalignment; 2. <Maschinen> alignment error, misalignment
Fluchtungsfernrohr n <Fertig> optical alignment-testing telescope
Fluchtweg m <Sicherheit> emergency escape route, fire rescue path
Fluenz f <Phys> fluence
Flug m <Lufttrans> flight (mit Reisegeschwindigkeit)
Flug m mit doppelter Besatzung <Lufttrans> double-crew operation
Flugasche f 1. <Kohlen> flue dust; 2. <Papier, Umweltschmutz> fly ash
Flugbahn f 1. <Lufttrans> flight path, path; 2. <Maschinen, Raumfahrt> trajectory
Flugbahnabfangen n <Lufttrans> flight path levelling
Flugbahnkurskreisel m <Lufttrans> directional gyro

Flugbegleitpersonal n <Lufttrans> flight crew
Flugbenzin n 1. <Erdöl> (AE) aviation gasoline, (BE) aviation petrol (Destillationsprodukt); 2. <Lufttrans> (BE) aviation fuel
Flugbereich m <Lufttrans> flight envelope
Flugbetrieb m <Raumfahrt> in-flight operation (Raumschiff)
Flugbetriebsmeldung f <Lufttrans> dismantling, flight regularity message
Flugbetriebsplanung f <Lufttrans> in-flight operational planning
Flugbewegung f <Lufttrans> aircraft movement
Flugbewegungsleitung f <Lufttrans> flight controls
Flugblatt n <Papier> fly sheet
Flugboot n <Trans> flying boat
Flugbordpeiler m <Funktech, Lufttrans> airborne DF equipment, airborne direction finding equipment
Flugbordwetterradar n <Funkort, Lufttrans> airborne weather radar
Flugcomputer m <Lufttrans> flight computer
Flugdaten npl <Lufttrans> flight data
Flugdatenschreiber m <Lufttrans> flight data recorder
Flugdeck n <Lufttrans> flight deck
Flügel m 1. <Fertig> lobe, vane; blade (Rührwerk); 2. <Heiz & Kälte> blade, vane; 3. <Maschinen> blade, impeller, vane, wing; 4. <Nichtfoss Energ> vane; 5. <Phys> wing
Flügel m in Mitteldeckeranordnung <Lufttrans> midwing
Flügelanode f <Elektrotech> vane-type anode
Flügelausrundung f <Lufttrans> wing fillet
Flügelbohrmeißel m <Erdöl> blade bit (Bohrtechnik)
Flügelbremse f <Maschinen> fan brake
Flügelfenster n <Bau> French window, casement window
Flügelgeschwindigkeit f <Nichtfoss Energ> vane velocity
Flügelhahn m <Nichtfoss Energ> butterfly cock, butterfly valve
Flügelhinterkante f <Lufttrans> trailing edge
Flügelholm m <Fertig> spar
Flügelklappe f <Lufttrans> wing flap
Flügelmauer f <Bau> wing wall; head wall (Brücken)
Flügelmutter f 1. <Bau> butterfly nut, fly nut, thumb nut, wing nut; 2. <Kfztech> butterfly nut, finger nut, fly nut, thumb nut, wing nut; 3. <Maschinen> butterfly nut, fly nut, thumb nut, wing nut
Flügeloberseite f <Phys> upper surface of wing
Flügelpumpe f <Wasserversorg> semirotary pump, vane pump
Flügelrad n 1. <Heiz & Kälte> vane wheel; 2. <Kfztech> impeller (Pumpe); 3. <Lufttrans> impeller (Turbomotor); 4. <Maschinen, Mechan> impeller
Flügelradanemometer n 1. <Gerät, Metrol> rotating cup anemometer, rotating vane anemometer, vane anemometer; 2. <Heiz & Kälte> windmill-type anemometer
Flügelradströmungsmesser m <Metrol> vane current meter
Flügelradwattmessgerät n <Elektriz, Nichtfoss Energ> vane-type wattmeter
Flügelradzähler m <Gerät, Nichtfoss Energ> vane meter
Flügelrelais n <Elektriz> vane-type relay
Flügelschiene f <Eisenbahn> wing rail
Flügelschraube f <Bau, Fertig, Maschinen> butterfly screw, thumb bolt, thumb screw, wing bolt, wing screw
Flügelspalt m <Lufttrans> wing slot
Flügelspitze f <Lufttrans> wing tip
Flügelspitzenwirbel m <Lufttrans> wing tip vortex
Flügelstreckung f <Lufttrans> wing aspect ratio (Tragfläche)
Flügelstützschwimmer m <Lufttrans> wing tank
Flügeltiefe f 1. <Lufttrans> chord; 2. <Nichtfoss Energ> chord length (Aerodynamik)
Flügeltür f <Bau> folding door
Flügelübergang m <Lufttrans> wing fillet
Flügelvorderklappenkante f <Lufttrans> slat of the leading edge
Flügelwand f <Bau> return wall
Flügelzellenpumpe f <Maschinen> vane pump
Flugerlaubnis entziehen v <Lufttrans> (BE) earth, (AE) ground
Flugerprobungszentrum n <Lufttrans> (AE) flight test center, (BE) flight test centre
Flugfeld n <Lufttrans> airfield
Flugfeldüberwachungssystem n <Funkort, Lufttrans> aerodrome surface surveillance, airport surveillance radar (ASR)
Flugfreigabe f <Lufttrans> flight clearance
Flugfreigabeüberprüfung f <Raumfahrt> FRR, flight readiness review
Flugfunkdienst m <Telekom> aircraft radio service
Flugfunke m <Sicherheit> flying spark
Fluggast m <Trans> passenger
Fluggastbrücke f <Lufttrans> aerobridge, air bridge, boarding bridge, jetway, passenger bridge
Fluggastraum m <Lufttrans> passenger cabin
Fluggeschwindigkeit f <Lufttrans> airspeed
Fluggesellschaft f <Lufttrans> airline
Flughafen m <Lufttrans> (BE) aerodrome, (AE) airdrome, airport
Flughafenabfertigungsgebäude n <Trans> air terminal
Flughafenbake f 1. <Lufttrans> (BE) aerodrome beacon, (AE) airdrome beacon; 2. <Raumfahrt> (BE) aerodrome beacon, (AE) airdrome beacon, airport beacon
Flughafengebühr f <Lufttrans> airport fee
Flughafenleuchtfeuer n <Funktech, Lufttrans> airport beacon
Flughafenradar n <Funkort, Lufttrans> airport radar, terminal radar
Flughafenüberwachungsradar n (ASR) <Funkort, Lufttrans> aerodrome surface surveillance, airport surveillance radar (ASR)
Flughalle f <Lufttrans> air terminal
Flughandbuch n <Lufttrans> flight manual
Flughindernisbefeuerung f <Lufttrans> aeronautical warning lights
Flughöhe f 1. <Lufttrans> flight altitude, flight level; 2. <Raumfahrt> altitude
Fluginformationsdienst m <Lufttrans> flight information service
Fluginformationszentrale f <Lufttrans> (AE) flight information center, (BE) flight information centre
Flugkabine f <Lufttrans> flight compartment
Flugkabinenbeleuchtung f <Lufttrans> flight compartment lights
Flugkabinenzugangstreppe f <Lufttrans> flight compartment access stairway
Flugkommandoanlage f <Lufttrans> flight director
Fluglage f <Raumfahrt> attitude, pitch attitude
Fluglagebezugsystem n <Raumfahrt> attitude reference unit
Fluglagekontrollsystem n <Raumfahrt> attitude control unit
Fluglage- und Umlaufbahnkontrollsystem n (AOCS) <Raumfahrt> attitude and orbit control system (AOCS)
Flugleistungsstörung f <Lufttrans> flight technical error
Flugleiter m <Lufttrans> flight controller
Fluglogbuch n <Lufttrans> flight log
Fluglotse m <Lufttrans> air traffic controller

Flugmasse

Flugmasse *f* <Lufttrans> gross weight
Flugnachbesprechung *f* <Lufttrans> postflying check
Flugnavigationsfunkdienst *m* <Funkort, Lufttrans> aeronautical radio navigation service
Flugnormengruppe *f (ASG)* <Raumfahrt> aeronautical standards group *(ASG)*
Flugnormwirkungsgrad *m (ase)* <Raumfahrt> air standard efficiency *(ase)*
Flugparameter *mpl* <Lufttrans> flight data
Flugplan *m* <Lufttrans> aerial timetable
Flugplandaten *npl* <Lufttrans> flight plan data
Flugplatz *m* <Lufttrans> *(BE)* aerodrome, *(AE)* airdrome
Flugradar *n* <Funkort> airborne radar
Flugrechnersystem *n* <Raumfahrt> flight data system
Flugrettung *f* <Lufttrans> air emergency
Flugsand *m* <Bau> flying sand
Flugschneise *f* <Lufttrans> lane
Flugschrauber *m* <Lufttrans> gyrodyne
Flugschreiber *m* <Lufttrans> black box, flight recorder
Flugschubvektorierung *f* <Raumfahrt> in-flight thrust vectoring *(Raumschiff)*
Flugsequenz *f* <Raumfahrt> in-flight sequence *(Raumschiff)*
Flugsicherung *f (FS)* <Lufttrans> air traffic control *(ATC)*
Flugsicherungszentrale *f* <Lufttrans> *(AE)* air traffic control center, *(BE)* air traffic control centre
Flugsicht *f* <Lufttrans> flight visibility
Flugsimulator *m* <Comp & DV> flight simulator
Flugspektrum *n* <Lufttrans> flight spectrum
Flugstaub *m* <Metall> flue dust
Flugsteig *m* <Lufttrans> gate, ramp
Flugsteuerungen *fpl* <Lufttrans> flight controls
Flugstrecke *f* <Lufttrans, Trans> distance covered *(Flugzeug)*
Flugstreckeneinrichtungen *fpl* <Lufttrans> air route facilities
Flugstützpunkt *m* <Trans> air base
flugtechnisch bedingter Ausfall *m* <Lufttrans> flight technical error
Flugtestanlage *f* <Lufttrans> flying test bench
Flugturbinenkerosin *n* <Erdöl> ATK, aviation turbine kerosene *(Destillationsprodukt)*
Flugüberwachungsinstrumente *npl* <Lufttrans> flight instruments
Flugüberwachungsradar *n (ARSR)* <Raumfahrt> air route surveillance radar *(ARSR)*
Flugüberwachungs- und Navigationsausrüstung *f (FCNE)* <Lufttrans> flight control and navigational equipment *(FCNE)*
Flugverkehr *m* <Lufttrans> air traffic
Flugverkehr-Überwachungsradar *n* <Funkort, Lufttrans> air-traffic control radar
Flugverlaufsplan *m* <Lufttrans> flight progress board
Flugvorbereitung *f* <Lufttrans> preflight planning
Flugvorbereitungsinformationen *fpl* <Lufttrans> preflight information
Flugweg *m* <Lufttrans> flight path
Flugwegzeichnen *n* <Lufttrans> radar plotting
Flugwerk *n* <Lufttrans> airframe *(ohne Motor)*
Flugwerksleitebene *f* <Lufttrans> airframe reference plane
Flugwetterrechner *m* <Lufttrans> air data computer
Flugwetterwarte *f* <Lufttrans> aeronautical meteorological station
Flugzeitdatenanalyse *f* <Strahlphys> time-of-flight data analysis
Flugzeitmethode *f* <Strahlphys> time-of-flight method
Flugzeug *n* <Lufttrans> *(BE)* aeroplane, *(AE)* airplane, aircraft, plane

Flugzeug *n* **für Raketenstarts** <Raumfahrt> launching aircraft
Flugzeug *n* **mit absenkbarer Rumpfspitze** <Lufttrans> droop-nose aircraft
Flugzeug *n* **mit nur einem Mittelgang** <Lufttrans> single aisle aircraft
Flugzeug *n* **mit variabler Geometrie** <Lufttrans> variable-geometry aircraft
Flugzeug *n* **mit vier Strahltriebwerken** <Lufttrans> four-engine jet aircraft
Flugzeugabgas *n* <Umweltschmutz> aircraft waste gas
Flugzeugabstellplatz *m* <Lufttrans> aircraft-parking position *(Flughafen)*
Flugzeugachse *f* <Lufttrans> aircraft axis
Flugzeugaufzug *m* <Lufttrans> aircraft lift
Flugzeugausrüstung *f* <Lufttrans> aircraft equipment
Flugzeugbeleuchtung *f* <Lufttrans> aircraft light
Flugzeug-Eisenbahn-Verkehr *m (FLEI-Verkehr)* <Trans> rail-air-rail service
Flugzeugentführung *f* <Lufttrans> hijack, skyjack
Flugzeuggleichgewicht *n* <Lufttrans> aircraft balance
Flugzeughalle *f* <Lufttrans> hangar
Flugzeugkategorie *f* <Lufttrans> aircraft category
Flugzeugkennung *f* <Lufttrans> aircraft identification
Flugzeuglängsfeld *n* <Lufttrans> aircraft longitudinal field *(Magnetfeld)*
Flugzeugnutzleistung *f* <Lufttrans> aircraft effectivity
Flugzeugquerfeld *n* <Lufttrans> aircraft lateral field *(Magnetfeld)*
Flugzeugradar *n* <Funkort> airborne radar
Flugzeugrumpf *m* <Lufttrans> fuselage, nacelle
Flugzeugrumpfbasis *f* <Lufttrans> fuselage datum line
Flugzeugrumpfkasten *m* <Lufttrans> fuselage box
Flugzeugrumpfmittelkasten *m* <Lufttrans> *(AE)* fuselage center box, *(BE)* fuselage centre box
Flugzeugschlepper *m* 1. <Lufttrans> aircraft tug; 2. <Trans> aircraft tractor
Flugzeugschleppstart *m* <Lufttrans> *(BE)* aeroplane tow launch, *(AE)* airplane tow launch
Flugzeugschwerpunkt *m* <Lufttrans> aircraft balance
Flugzeugtelefon *n* <Mobilkom> in-flight phone
Flugzeugträger *m* 1. <Lufttrans> carrier; 2. <Wassertrans> aircraft carrier *(Marine)*
Flugzeugtreppe *f* <Lufttrans> air stairs
Flugzeugtriebwerksemissionen *fpl* <Lufttrans> aircraft engine emissions
Flugzeugüberholungsdaten *npl* <Lufttrans> aircraft overhaul rating
Flugzeugvereisungsanzeiger *m* <Lufttrans> aircraft icing indicator
Flugzeugwartungsdaten *npl* <Lufttrans> aircraft maintenance rating
Flugzustand *m* <Lufttrans> flight status
fluid *adj* <Maschinen> fluidal, fluidic
Fluid *n* <Chemie> fluid
Fluidantrieb *m* <Mechan> fluid drive
Fluidelement *n* <Maschinen> fluidic device
Fluidgetriebe *n* <Maschinen> fluidic transmission
Fluidik *f* <Phys> fluidics
Fluidität *f* <Maschinen, Mechan, Phys, Strömphys> fluidity
Fluidkompass *m* <Wassertrans> liquid compass
fluktuierendes Rauschen *n* <Akustik> fluctuating noise
Fluor *n (F)* <Chemie> fluorine *(F)*
Fluoranthen *n* <Chemie> fluoranthene
Fluorchlorokohlenwasserstoff *m (FCKW)* <Umweltschmutz, Verpack> chlorofluorocarbon *(CFC)*
Fluoren *n* <Chemie> fluorene
Fluorenon *n* <Chemie> fluorenone
Fluorescein *n* <Chemie> fluorescein

Fluoreszenz f <Chemie, Phys> fluorescence
Fluoreszenz... <Chemie> fluorescent
Fluoreszenzanalyse f <Phys, Strahlphys, Wellphys> fluorescence analysis
Fluoreszenzanregungsspektrum n <Phys, Strahlphys, Wellphys> fluorescence excitation spectrum
Fluoreszenzlampe f <Strahlphys> fluorescent lamp
Fluoreszenzschirm m <Strahlphys> fluorescent screen
fluoreszierend adj <Chemie, Druck> fluorescent
Fluorid n <Chemie> fluoride
Fluoridierung f <Chemie> fluoridation
Fluorid-Opalglas n <Ker & Glas> fluoride opal glass
Fluoridzusatz m <Chemie> fluoridation (Trinkwasser)
Fluorkältemittel n <Heiz & Kälte> fluorocarbon refrigerant
Fluorkiesel... <Chemie> fluosilicic
Fluorkohlenstoff m <Kunststoff> fluorocarbon resin
Fluoroaluminat n <Chemie> fluoaluminate
Fluorobor... <Chemie> fluoboric
Fluoroborat n <Chemie> borofluoride
fluoroborsaueres Salz n <Chemie> fluoborate
Fluoroform n <Chemie> fluoroform
Fluorophosphat n <Chemie> fluophosphate
Fluorosilicat n <Chemie> hexafluorosilite
Fluorozirconat n <Chemie> zirconifluoride
Fluorpolymer n <Kunststoff> fluorocarbon resin
Fluorsiliciumverbindung f <Chemie> hexafluorosilite
Fluorsulfon... <Chemie> (AE) fluosulfonic, (BE) fluosulphonic
fluorwasserstoffsauer adj <Chemie> hydrofluoric
Flur m <Bau> vestibule
Flurplatte f <Wassertrans> floor plate (Schiffbau)
Fluss m 1. <Comp & DV> flow; 2. <Elektriz, Elektrotech> current, flow, flux; 3. <Kontroll, Maschinen> flow; 4. <Mechan> flux; 5. <Metall, Nichtfoss Energ> flow; 6. <Phys> flow, flux; 7. <Wassertrans> river
Flussablagerung f <Wasserversorg> aggradational deposit
flussabwärts adv <Bau> downstream
Flussbagger m <Wasserversorg> river dredge
Flussbau m <Wasserversorg> river works
Flussbett n <Wasserversorg> river bed
Flussboot n <Wassertrans> river boat
Flüsschen n <Wassertrans> (AE) creek
Flussdampfer m <Wassertrans> water bus
Flussdelta n <Erdöl> delta
Flussdiagramm n <Comp & DV, Phys> flowchart
Flussdiagrammbeschriftung f <Comp & DV> flowchart text
Flussdiagrammsymbol n <Comp & DV> flowchart symbol
Flussdichte f <Elektriz, Phys> flux density
Flussdichtemessgerät n <Gerät> flux meter
Flussdichtewölbung f <Kerntech> Bg, geometric buckling
Flusseinzugsgebiet n <Wassertrans, Wasserversorg> river basin
Flusseisen n <Bau, Metall> structural iron
Flussfähre f <Wassertrans> river ferry
Flussgebiet n <Wasserversorg> river basin
Flussgeschwindigkeit f des Reaktorkernkühlmittels <Kerntech> core coolant flow rate
Flusshafen m <Wassertrans> river port
flüssig adj 1. <Strömphys> fluid; 2. <Thermod> liquid
Flüssig... <Umweltschmutz> fluid
Flüssigantriebsystem n <Raumfahrt> liquid propellant system (Raumschiff)
Flüssigbett n <Umweltschmutz> fluidized bed
Flüssigchromatographie f <Umweltschmutz> liquid chromatography

flüssige Brennstoffe mpl <Heiz & Kälte> liquid fuels
flüssige Katode f <Elektrotech> pool cathode
flüssige Luft f <Thermod> liquid air
flüssiger Abfall m <Abfall> liquid waste
flüssiger Ammoniak m <Thermod> liquid ammonia
flüssiger Anlasswiderstand m <Elektriz> liquid starter resistance
flüssiger Kautschuk m <Kunststoff> liquid rubber
flüssiger Stickstoff m <Thermod> liquid air
flüssiger Wasserstoff m <Thermod> liquid hydrogen
Flüssigerdgas n (LNG) <Erdöl, Thermod> liquefied natural gas (LNG)
Flüssigerdgasbus m <Kfztech> liquid natural gas bus
Flüssigerdgastanker m <Wassertrans> liquefied natural gas tanker
Flüssigerdgastransporter m <Thermod> liquid natural gas carrier
flüssiges Begleitprodukt n <Erdöl> associated liquids (Erdgasförderung)
flüssiges Chlor n <Thermod> liquid chlorine
flüssiges Helium n <Thermod> liquid helium
flüssiges Kühlmittel n <Maschinen> liquid coolant
flüssiges Monergol n <Thermod> liquid monopropellant
flüssiges Paraffin n <Thermod> liquid paraffin
flüssiges Triergol n <Thermod> liquid tripropellant
Flüssiggas n (LPG) <Erdöl, Heiz & Kälte, Kfztech, Thermod, Wassertrans> liquefied petroleum gas (LPG)
Flüssiggasbus m <Kfztech> liquefied petroleum gas bus
Flüssiggasmotor m <Kfztech> liquefied petroleum gas engine
Flüssiggastanker m <Wassertrans> liquefied petroleum gas tanker
Flüssiggastransporter m <Thermod> liquefied petroleum gas carrier
flüssiggekühlter Motor m <Kfztech> liquid-cooled engine
Flüssiggut n <Wassertrans> liquid bulk cargo
Flüssigkeit f 1. <Heiz & Kälte> liquid; 2. <Phys, Strömphys> fluid; 3. <Thermod> liquid • **Flüssigkeit rückführen** <Fertig> vent (Hydraulik)
Flüssigkeiten fpl **in rotierenden Systemen** <Strömphys> fluids in rotating systems
Flüssigkeitsanlasser m <Elektrotech> liquid starter
Flüssigkeitsbremse f <Maschinen> hydraulic brake
Flüssigkeitsdichtemessgerät n <Gerät> areometer
Flüssigkeitsdruck m 1. <Heiz & Kälte> hydrostatic pressure; 2. <Strömphys> fluid pressure
Flüssigkeitsdruckmessdose f <Gerät> hydraulic capsule
Flüssigkeitseinlass m <Maschinen> fluid inlet
flüssigkeitsfest adj <Verpack> liquidproof
flüssigkeitsfester Karton m <Verpack> liquidproof carton
flüssigkeitsgekühlt adj <Thermod> liquid-cooled
Flüssigkeitsgetriebe n 1. <Maschinen> hydraulic transmission; 2. <Mechan> fluid drive
Flüssigkeitsguss m <Raumfahrt> liquid slosh (Raumschiff)
Flüssigkeitshonen n <Maschinen> liquid honing
Flüssigkeitsindex m <Kohlen> liquidity index
Flüssigkeitskatode f <Elektrotech> pool cathode
Flüssigkeitskühler m <Heiz & Kälte> liquid chiller
Flüssigkeitskühlung f <Thermod> liquid cooling
Flüssigkeitskupplung f 1. <Hydraul> hydraulic clutch; 2. <Kfztech> fluid coupling (Kraftübertragung); 3. <Maschinen> fluid coupling
Flüssigkeitslaser m <Elektronik> liquid laser
Flüssigkeitslasermedium n <Elektronik> liquid laser medium

Flüssigkeitsleitung

Flüssigkeitsleitung f <Maschinen> fluid pipeline
Flüssigkeitslöscher m <Sicherheit> nonflammable liquid extinguisher
Flüssigkeitsmaß n <Metrol> liquid measure
Flüssigkeitsmassengut n (OBO) <Wassertrans> ore-bulk oil (OBO)
Flüssigkeitsmassengutfrachter m <Wassertrans> ore-bulk oil carrier
Flüssigkeitsregler m <Elektrotech> liquid controller
Flüssigkeitsreibung f 1. <Heiz & Kälte> fluid friction; 2. <Lebensmittel, Maschinen, Phys> viscous friction
Flüssigkeitssäule f <Maschinen> liquid column
Flüssigkeitsschwund m <Raumfahrt> ullage (Raumschiff)
Flüssigkeitsspiegel betätigt adj/durch <Fertig> float-operated
Flüssigkeitsstand m <Verpack> liquid level
Flüssigkeitsstandanzeiger m 1. <Erdöl> (AE) gage glass, (BE) gauge glass (Messtechnik); 2. <Gerät> liquid level indicator; 3. <Maschinen> level indicator; 4. <Verpack> liquid level indicator
Flüssigkeitsstandglas n <Heiz & Kälte> (AE) gage glass, (BE) gauge glass
Flüssigkeitsstandregler m <Verpack> liquid level control
Flüssigkeitsstrahl m <Fertig> jet of liquid
Flüssigkeitsstrahlbohren n <Fertig> abrasive jet drilling
Flüssigkeitsstrom m <Kerntech> liquid flow
Flüssigkeitsströmung f <Strömphys> fluid flow, liquid flow
Flüssigkeitsteilchen n <Strömphys> fluid particle
Flüssigkeitsthermometer n <Thermod> liquid expansion thermometer, liquid thermometer
Flüssigkeitstropfenmodell n <Phys> liquid drop model (Kernphysik)
Flüssigkeitsverlust m <Chemie> ullage
Flüssigkeitsverschluss m <Maschinen> liquid seal
Flüssigkeitswaage f <Lebensmittel> hydrometer
Flüssigkeitswiderstand m <Elektriz> liquid rheostat
Flüssigkeitszerstäuber m <Wasserversorg> pulverizer
Flüssigkristall m <Comp & DV, Elektriz, Elektronik, Kerntech, Telekom> liquid crystal (LC)
Flüssigkristallanzeige f (LCD) <Comp & DV, Elektriz, Elektronik, Fernseh, Gerät, Telekom, Thermod> liquid crystal display (LCD)
Flüssigmetall n <Metall> liquid metal
Flüssigmetallionenquelle f <Strahlphys> liquid metal ion source
Flüssigmetallkühlung/mit <Thermod> liquid metal cooled
Flüssigmetallreaktor m (FMR) <Kerntech> flowable solids reactor (FSR)
Flüssigmetall-Wärmeaustauscher m <Thermod> liquid metal heat exchanger
Flüssigmist m <Abfall> slurry
Flüssigphase f <Thermod> liquid phase
Flüssigphasenepitaxie f <Elektronik> liquid phase epitaxy
Flüssigrakete f <Thermod> liquid fuel rocket, liquid propellant rocket
Flüssigsauerstoff m (LOX) <Raumfahrt, Thermod> liquid oxygen (lox)
Flüssigschlamm m <Abfall> liquid sludge, slurry
Flüssigschmiere f <Raumfahrt> liquid slosh (Raumschiff)
Flüssigstickstoff m <Raumfahrt> liquid nitrogen
Flüssigtreibstoff m 1. <Raumfahrt> liquid propellant; 2. <Thermod> liquid fuel, liquid propellant
Flüssigwasserstoff m <Raumfahrt> liquid hydrogen
Flusskabel n <Telekom> subfluvial cable
Flusskanalisierung f <Wasserversorg> river training

Flusslauf m **von Oberwasser** <Wasserversorg> headwater reach
Flusslinie f <Elektrotech> flux line, line of flux
Flussmessgerät n 1. <Gerät> flux meter; 2. <Phys> fluxmeter
Flussmeter n <Gerät> flux meter
Flussmittel n 1. <Elektrotech> flux, soldering flux; 2. <Funktech> flux (Löthilfe); 3. <Ker & Glas> flux; 4. <Kohlen> flux powder; 5. <Maschinen> flux additive; 6. <Metall> flux • **Flussmittel zusetzen** <Fertig> flux • **in Flussmittel tauchen** <Metall> flux
Flussmittel n **zum Lichtbogenschweißen** <Fertig> arc flux
flussmittelgefülltes Lot n <Elektriz> cored solder
Flussmündung f 1. <Nichtfoss Energ> estuary; 2. <Wasserversorg> river mouth
Flussprojektierung f <Nichtfoss Energ> run-of-river scheme
Flussquantum n <Phys> flux quantum
Flussrate f <Phys> flow rate
Flussregulierung f <Wasserversorg> river training
Flussrichtung f <Comp & DV> flow direction
Flussröhre f <Phys> tube
flusssaueres Salz n <Chemie> fluoride
Flussschiff n <Wassertrans> river boat
Flussschifffahrt f <Wassertrans> river navigation, river traffic
Flussschlepper m <Wassertrans> river tug
Flussspat m <Ker & Glas> fluorspar
Flussstahl m <Metall> mild steel
Flusssteuerung f <Comp & DV, Telekom> flow control
Flussstrom m <Elektrotech> forward current
Flussufer n <Wassertrans> river bank • **am Flussufer lebend** <Wasserversorg> riparian
Flussufermauer f <Wasserversorg> river wall
Flussventil n <Lufttrans> flux valve
Flusswechseldichte f <Comp & DV> packing density
Flusswölbung f <Kerntech> buckling, geometric buckling
Flusszeit f <Elektrotech> on period
Flut f 1. <Nichtfoss Energ> high tide, rising tide; 2. <Wassertrans> flood tide, flood, high tide, high water, tide
Flutbecken n <Nichtfoss Energ> tidal basin
Flutbeleuchtung f <Elektriz> floodlighting
fluten v 1. <Kfztech> flood (Motor); 2. <Wassertrans> flood (Schleusen)
Fluthafen m <Wassertrans> (AE) tidal harbor, (BE) tidal harbour, tidal port
Flutkraftwerk n <Nichtfoss Energ> tidal power station
Flutlicht n <Elektrotech> floodlight
Flutöffnung f <Wasserversorg> flood arch
Flutstrom m <Wassertrans> flood stream (Navigation, Gezeiten)
Flutströmung f <Nichtfoss Energ> flowing tide
Flutstunde f <Nichtfoss Energ> lagging of the tide
Fluttor n <Wassertrans> floodgate (Schleuse)
Flutung f <Ker & Glas> flooding
Flutventil n 1. <Wassertrans> seacock (Schiffbau); 2. <Wasserversorg> flooding cock
Flutverlust m <Nichtfoss Energ> flood loss
Flutwechsel m <Wassertrans> turn of the tide
Flutwelle f <Wassertrans> tidal wave
fluviales Schwemmland n <Wasserversorg> fluvial alluvium
Fluxmeter n <Elektrotech> fluxmeter
Fly-By-Punkt m <Raumfahrt> fly-by point (Raumschiff)
Flyer m <Textil> flyer spinning frame, roving frame
Flyergarn n <Textil> rove
Fm (Fermium) <Chemie> Fm (fermium)
FM (Frequenzmodulation) <Comp & DV, Elektronik, Funktech, Phys, Telekom> FM (frequency modulation)

FM-Aufzeichnung f <Aufnahme> FM recording
FM-Modem n <Elektronik, Funktech> FM modem
FMR (Flüssigmetallreaktor) <Kerntech> FSR (flowable solids reactor)
FM-Rundfunk m <Funktech> FM broadcasting
FM-Signal n <Elektronik, Funktech> FM signal
FM-Stereophonie f <Aufnahme> FM stereo
FM-Träger m <Elektronik, Funktech> FM carrier
Fock f <Wassertrans> forestaysail
Fockfall n <Wassertrans> jib halyard (Segeln, Taue)
Fockmast m <Wassertrans> foremast (Segeln)
Fockstag n <Wassertrans> forestay
Fogging n <Comp & DV> fogging
Fokalebene f <Foto> focal plane
Fokalfläche f <Phys> caustic surface (Optik)
Fokallinie f <Optik> caustic curve
Fokus m 1. <Foto> focal point; 2. <Geom> focus
Fokuseinstellung f <Foto> focus setting
Fokuslampe f <Elektriz> focusing lamp
Fokus-Servo m <Optik> focus servo
Fokussieraufsatz m <Foto> focusing stage
fokussieren v 1. <Elektronik, Foto> focus; 2. <Gerät, Labor> focus (Elektronenstrahl); 3. <Phys> focus
Fokussierhilfe f <Foto> focusing aid
Fokussierknopf m <Foto> focusing knob
Fokussiermagnet m <Elektrotech> focusing magnet
Fokussierpunkt m <Fernseh> crossover
Fokussierspule f <Elektrotech> focusing coil
fokussiert adj 1. <Foto> in focus; 2. <Phys> focused
fokussierter Ionenstrahl m <Phys> focused ion beam
fokussierter Strahl m <Phys> focused beam
Fokussierung f <Elektronik> focusing (Katodenstrahlröhre, Teilchenstrahl)
Fokussierungsanlage f für Ionenstrahlen <Strahlphys> ion beam-focusing column
Fokussierungsanode f <Elektrotech> focusing anode
Fokussierungselektrode f <Elektrotech> focusing electrode
Fokussierungsmagnet m <Raumfahrt> focusing magnet (Weltraumfunk)
Foley-Spur f <Aufnahme> Foley track
Folge f 1. <Comp & DV> series, suite; 2. <Kontroll> routine, sequence; 3. <Math> sequence (Zahlen); 4. <Telekom> sequence
Folgeausfall m <Qual> dependent failure, subsequent failure
Folgebereich m 1. <Comp & DV> overflow area; 2. <Elektronik> follow range
Folgeblatt n <Konstzeich> continuation sheet
Folgedrosselung f <Akustik> temporal partial masking
Folgefeld n <Phys> wake field
folgegedrosselte Lautheit f <Akustik> temporally partial-masked loudness
Folgegerät n <Kontroll> slave (führungsabhängig)
folgegesteuert adj <Comp & DV> sequence-controlled
folgegesteuerter Ablauf m <Comp & DV> sequencing
Folgenummer f <Comp & DV> sequence number
Folgenutzung f <Abfall> after use
Folgeprüfung f 1. <Comp & DV> sequence control; 2. <Qual> sequential test
Folgeregelung f 1. <Elektrotech> servomechanism; 2. <Kontroll> servo control; 3. <Regelung> sequence control
Folgeregelungssystem n 1. <Elektrotech> servomechanism; 2. <Kontroll> servo control system, servo system
Folgerelais n <Elektrotech> sequence relay
folgern v <Math> deduce
Folgerung f <Math> implication
Folgesatz m <Math> corollary
Folgeschäden mpl <Kerntech> consequential damage

Folgesteuerung f <Comp & DV, Elektrotech, Regelung> sequencing control, sequential control, sequential phase control
Folgesteuerungseinheit f <Telekom> sequencer
Folgesteuerungsmechanismus m <Comp & DV> servomechanism
Folgestichprobenentnahme f <Qual> sequential sampling
Folgestichprobenplan m <Qual> sequential-sampling plan
Folgewerkzeug n <Maschinen> follow-on tool, progressive tool
Folie f 1. <Chemie> film (hauchdünn); 2. <Fertig> leaf; 3. <Kohlen> foil; 4. <Kunststoff> film, sheet; 5. <Metall, Papier> foil
Folienabpressmaschine f <Verpack> foil-backing machine
Folienblasen n <Kunststoff> film blowing
Folieneinschweißung f <Lebensmittel> shrink-wrap (in Folie)
Folienextrudieranlage f <Verpack> film extrusion equipment
Folienextrusion f <Kunststoff> film extrusion
Foliengießen n <Kunststoff> film casting
Folienkondensator m <Elektriz> capacitor film
Folienpapier n <Papier> foil paper
Folienversiegelung f <Verpack> foil sealing
Folin... <Chemie> folinic
Folio n <Druck> folio
Folioformat n <Druck> folio
Folsäure f <Lebensmittel> folic acid
Fond m <Textil> blotch (Nassverfahren)
Foolscap n <Druck> foolscap
Förde f <Wassertrans> firth
Förder... <Erdöl, Fertig> production
Förderanlage f 1. <Bau> conveyor system; 2. <Kerntech> conveyor; 3. <Kohlen> conveyor system; 4. <Maschinen> transporter; 5. <Trans> conveying plant
Förderapparat m <Trans> conveyor
Förderband n <Abfall, Maschinen, Mechan, Trans, Verpack> band conveyor, belt conveyor, conveying belt
Förderband zur Ölaufnahme 1. <Meerschmutz> belt skimmer; 2. <Umweltschmutz> belt skimmer
Förderbandfilter n <Kerntech> TBF, (AE) traveling belt filter, (BE) travelling belt filter
Förderbandskimmer m <Umweltschmutz> conveyor belt skimmer
Förderbandtrockner m <Chemtech> belt drier, belt dryer
Förderbandwaage f <Gerät> belt balance
Förderbandwerkstoff m <Textil> conveyor belting
Förderbohrung f <Erdöl> production well
Förderdruck m 1. <Heiz & Kälte> delivery pressure; 2. <Hydraul> head; 3. <Wasserversorg> delivery pressure
Förderer m <Bau, Kohlen, Trans, Verpack> conveyor
Förderer m **mit veränderlicher Geschwindigkeit** <Trans> variable-speed conveyor belt
Fördererz n <Kohlen> crude ore
Fördergerät n <Bau, Kohlen, Trans> conveyor
Fördergrus m <Kohlen> dross
Förderhöhe f 1. <Bau> head (Pumpe); 2. <Erdöl> head; 3. <Fertig> discharge head; 4. <Heiz & Kälte> delivery head, pump head; 5. <Hydraul> head, static head; 6. <Kerntech> delivery head (Pumpe); 7. <Maschinen> discharge head, lift; 8. <Wasserversorg> delivery head, delivery lift, discharge head
Förderhöhengrenze f <Hydraul> head limit
Förderhorizont m <Erdöl> pay zone, production horizon (Fördertechnik)
Förderkasten m <Fertig> tote box

Förderkette

Förderkette f <Maschinen> conveyor chain
Förderkohle f <Kohlen> pit coal, rough coal, run of mine coal, unscreened coal
Förderkohlesieb n <Kohlen> raw coal screen
Förderkolonne f <Erdöl> production string
Förderkopf m <Erdöl> wellhead
Förderkorb m <Trans> skip
Förderleistung f 1. <Bau> capacity; 2. <Heiz & Kälte> delivery rate, discharge capacity; 3. <Maschinen> carrying capacity, delivery rate, discharge; 4. <Verpack> carrying capacity (Transportband); 5. <Wasserversorg> delivery (Pumpe)
Fördermenge f 1. <Fertig> discharge; 2. <Maschinen> delivery, discharge, output; 3. <Wasserversorg> capacity • **mit gleich bleibender Fördermenge** <Fertig> constant-flow
Fördermittel n <Fertig> conveyance, haulage means
fördern v 1. <Bau, Kohlen> haul; 2. <Trans> convey, transport; hoist (senkrecht)
Förderphase f <Erdöl> production phase
Förderplattform f <Erdöl> production platform
Förderpumpe f <Fertig, Heiz & Kälte, Maschinen> feed pump
Förderrichtung f <Heiz & Kälte> discharge direction
Förderrohr n <Bau> screw elevator
Förderrutsche f <Maschinen> chute, slide
Förderschnecke f 1. <Hydraul> Archimedean screw; 2. <Maschinen, Verpack> screw, screw conveyor, worm, worm conveyor
Förderseil n 1. <Eisenbahn> traction cable; 2. <Maschinen> hoisting rope
Förderseite f 1. <Fertig> delivery side, discharge side; 2. <Hydraul> pressure side; 3. <Maschinen> discharge side
Förderstrecke f <Kohlen> adit
Förderstrom m 1. <Fertig> delivery, oil flow; 2. <Heiz & Kälte> flow of discharge
Fördersystem n <Wassertrans> conveyor system (Ladungsumschlag)
Fördertechnik f <Fertig> handling engineering, hoisting and conveying
Fördertrommel f <Fertig> hoisting drum
Forderung f <Bau, Kohlen> requirement
Förderung f 1. <Maschinen> delivery, discharge (Pumpe); 2. <Trans> conveyance, transport; 3. <Wasserversorg> offtake
Fördervolumen n <Maschinen> delivery, displacement
Fördervorrichtung f <Papier, Trans> conveyor
Förderwagen m 1. <Kohlen, Labor> trolley; 2. <Trans> (AE) cart, (BE) truck
Form f 1. <Bau> form; 2. <Druck> (BE) forme, (AE) printing form; 3. <Fertig> chase, die; 4. <Kunststoff, Maschinen> (AE) mold, (BE) mould; 5. <Metall> (AE) mold, (BE) mould, shape; 6. <Patent> form (Zusammenfassung); 7. <Textil> shape • **in verdünnter Form** <Anstrich> slurried
Form f **für thermoplastischen Guss** <Maschinen> (AE) thermoplastic mold, (BE) thermoplastic mould
Form f **für Wachsausschmelzverfahren** <Maschinen> (AE) lost wax mold, (BE) lost wax mould
Formabweichung f <Metrol> form errors
Formaldehyd m 1. <Chemie> formaldehyde, methanal; 2. <Kunststoff, Textil> formaldehyde
Formaldehydsulfoxylat n <Lebensmittel> (AE) formaldehyde sulfoxylate, (BE) formaldehyde sulphoxylate
formale Logik f <Comp & DV> formal logic
formale Sprache f <Comp & DV> formal language
formales Parameter n <Comp & DV> formal parameter
Formalin n <Chemie> formalin
Formamid n <Chemie> formamide, methanamide

Formänderung f 1. <Akustik> warp; 2. <Kunststoff> strain; 3. <Maschinen> deformation, yield; 4. <Metall> shape change
Formänderungsarbeit f <Maschinen> deformation work
Formänderungsfestigkeit f <Maschinen> yield strength
Formänderungswiderstand m <Fertig> consistency
Formant m <Akustik> formant
Formant-Vocoder m <Telekom> formant vocoder
Format n <Comp & DV, Druck, Fernseh> format
Formatbreite f <Druck> measure
formatieren v 1. <Comp & DV> format (Diskette, Text); 2. <Druck> format
Formatierung f <Telekom> formatting
Formatierungsanweisung f <Comp & DV> tag
Formatierungsprogramm n <Comp & DV> formatter
Formatierungssequenz f <Comp & DV> tag
Formation f <Erdöl, Kohlen, Lufttrans> formation
Formationsbewertung f <Erdöl> formation evaluation (Geologie)
Formationsdruck m <Erdöl> formation pressure (Geologie)
Formationsdruckgradient m <Erdöl> formation pressure gradient (Geologie)
Formationsflug m <Lufttrans> formation flight
Formationskunde f <Kohlen> stratigraphy
Formationstest m <Erdöl> formation test (Geologie)
Formationstester m <Papier> formation tester
Formationswasser n <Erdöl> formation water (Geologie)
Formatleiste f <Papier> deckle board
Formazyl n <Chemie> formazyl
Formbarkeit f <Metall> plasticity
Formbeständigkeit f <Kerntech> dimensional stability (Brennelement)
Formbetätigungseinrichtung f <Fertig> die-actuating mechanism
Formblasen n <Ker & Glas> (AE) mold-blowing, (BE) mould-blowing
Formdeckel m <Papier> deckle strap
Formdrehen n <Maschinen> contouring
Formdrehmaschine f <Maschinen> forming lathe
Formeinsatz m <Kunststoff> (AE) mold insert, (BE) mould insert
Formel f 1. <Comp & DV> rule; 2. <Math> formula
formen v 1. <Anstrich> forge; 2. <Bau, Ker & Glas> (AE) mold, (BE) mould (Ton)
Formen n 1. <Ker & Glas> forming (mit Tafeln); 2. <Maschinen, Mechan> forming
Formen n **mit Tonplatten** <Ker & Glas> (AE) molding with clay sheets, (BE) moulding with clay sheets
Formenbauer m <Ker & Glas> (AE) mold maker, (BE) mould maker
Formenbrecher m <Kohlen> breaker
Formentleerer m <Ker & Glas> (AE) mold emptier, (BE) mould emptier
Formentrocknen n <Fertig> (AE) mold drying, (BE) mould drying
Former m <Fertig> (AE) molder, (BE) moulder
Formerei f <Fertig, Metall> (AE) molding shop, (BE) moulding shop
Formfaktor m <Teilphys> form factor
Formfräsen n <Maschinen> form milling, profile milling
Formfräser m 1. <Fertig> formed circular cutter, multiple-tooth gear cutter; 2. <Maschinen> form cutter, form-milling cutter, formed-milling cutter
Formfräsmaschine f 1. <Fertig> form-milling machine; 2. <Ker & Glas> shape-cutting machine
Formgebung f 1. <Maschinen> design; 2. <Metall> forming
formgeschäumtes Polystyrol n **und PET** n <Verpack> (AE) form-molded polystyrene and PET, (BE) form-moulded polystyrene and PET

formgewickelte Spule f <Elektriz> form-wound coil
Formgleichheit f <Chemie> isomorphism
Formgravieren n <Maschinen> (AE) mold engraving, (BE) mould engraving
Formgrube f 1. <Fertig> (AE) molding hole, (BE) moulding hole; 2. <Ker & Glas> pit
Formgrubentür f <Ker & Glas> pit door
Formhalter m <Ker & Glas> (AE) mold holder, (BE) mould holder
Formholz n <Fertig> wood die; former (Streckziehen)
Formiat n <Chemie> formate
Formkachel f <Bau> trimmer
Formkasten m 1. <Fertig> (AE) molding box, (BE) moulding box, (AE) molding flask, (BE) moulding flask; 2. <Metall> (AE) molding box, (BE) moulding box
Formkonduktor m <Elektriz> sector-shaped conductor, shaped conductor
Formkurven fpl <Wassertrans> hydrostatic curves (Schiffkonstruktion)
Formlehre f <Metrol> (AE) receiving gage, (BE) receiving gauge
Formleiter m <Elektriz> sector-shaped conductor
Formling m <Ker & Glas> briquette
formlos adj <Anstrich> amorphous
Formmarkierung f <Ker & Glas> (AE) mold mark, (BE) mould mark
Formmaschine f <Fertig> (AE) molding machine, (BE) moulding machine (Gießen)
Formmaske f <Fertig> (AE) molding shell, (BE) moulding shell
Formmaskenverfahren n <Fertig> C-process; (AE) shell-molding process, (BE) shell-moulding process (Gießen)
Formmassepulver n <Kunststoff> (AE) molding powder, (BE) moulding powder
Formmeißel m <Maschinen> forming cutter, forming tool
Formmodell n <Fertig> (AE) mold pattern, (BE) mould pattern
Formpappe f <Papier, Verpack> (AE) molded board, (BE) moulded board
Formpresse f <Kunststoff> (AE) compression-molding machine, (BE) compression-moulding machine
Formpressen n 1. <Fertig, Ker & Glas> (AE) compression molding, (BE) compression moulding; 2. <Kunststoff> (AE) compression molding, (BE) compression moulding, (AE) molding, (BE) moulding; 3. <Verpack> (AE) compression molding, (BE) compression moulding
Formrippe f <Lufttrans> false rib
Formsand m <Fertig> (AE) molding sand, (BE) moulding sand
Formsatz m <Druck> runaround
Formscheibe f 1. <Ker & Glas> former; 2. <Maschinen> form shim
Formschleifen n <Maschinen> form grinding, profile grinding, profiling
Formschlichte f 1. <Anstrich> (AE) mold coating, (BE) mould coating; 2. <Fertig> dressing; 3. <Ker & Glas> (AE) mold coating, (BE) mould coating
Formschließen n <Druck> lock-up
formschlüssig adj <Fertig> positive
formschlüssige Kupplung f <Maschinen> positive clutch
formschlüssige Ventildrehvorrichtung f <Kfztech> positive-type valve rotator
formschlüssiges Kuppeln n <Kfztech> jaw clutching
Formschräge f <Ker & Glas> shaped bevel
Formschuh m <Papier> forming shoe
Formsignal n <Eisenbahn> semaphore signal
Formstahl m <Maschinen> forming tool
Formstahlhalter m <Maschinen> forming tool holder

Formstanzen n <Kunststoff> pressure forming
Formsteifigkeit f <Kerntech> inherent stability
Formstempel m <Papier> (AE) mold stamp, (BE) mould stamp
Formstich m <Fertig> former (Walzen)
Formstück n <Maschinen> plain fitting
Formteil n 1. <Elektrotech> (AE) molding, (BE) moulding; 2. <Fertig> compact
Formteilgrat m <Kunststoff> flash
Formteilherstellung f <Kunststoff> (AE) molding, (BE) moulding
Formtoleranz f <Maschinen> form tolerance
Formtrennfuge f <Fertig> cutoff (Kunststoffe)
Formtrennmittel n <Kunststoff> (AE) mold release agent, (BE) mould release agent, release agent
Formtrocknen n <Fertig> (AE) mold drying, (BE) mould drying
Formular n 1. <Comp & DV> form; 2. <Druck> printed form; 3. <Patent> form
Formularmaske f <Comp & DV> form overlay
Formularmodus m <Comp & DV> form mode
Formularname m <Comp & DV> form name
Formularressource f <Comp & DV> form resource
Formulartransport m <Druck> form feed
Formularvorschub m <Comp & DV> form feed
Formularzufuhr f <Druck> form feed
Formulierung f <Kunststoff> formulation
Formung f <Textil> boarding
Formungs-, Abfüll- und Siegelanlage f <Verpack> form, fill and seal machine
Formungs-, Abfüll- und Siegelmaschine f <Verpack> form, fill and seal machine (waagerecht und senkrecht arbeitend)
Formungstemperatur f <Verpack> forming temperature (Wärmeformung)
Formungsverstärker m <Elektronik> shaping amplifier
Formwalze f <Ker & Glas> former roller
Formwerkzeug n 1. <Fertig> forming tool; 2. <Kunststoff> (AE) mold, (BE) mould
Formwiderstand m 1. <Fertig> form drag; body drag (Strömung); 2. <Lufttrans> form drag; 3. <Wassertrans> hull resistance (Schiffkonstruktion)
Formyl n <Chemie> formyl
Formylgruppe f <Chemie> formyl
Formzyklus m <Kunststoff> (AE) molding cycle, (BE) moulding cycle
Forschung f und Entwicklung f (F&E) <Chemtech, Comp & DV, Kfztech> research & development, R&E
Forschungsreaktor m <Kerntech> laboratory reactor, research reactor
Forschungsschiff n <Wassertrans> research ship, research vessel
Forschungs- und Entwicklungsingenieur m (F&E-Ingenieur) <Maschinen> research-development engineer
Forschungszentrum n <Telekom> (AE) research center, (BE) research centre
Forstwirtschaftspolitik f <Umweltschmutz> forestry policy
fortgeschritten adj <Maschinen, Telekom> sophisticated (Gerät)
Fortgeschrittene Telekommunikationstechnologien fpl **und Dienste** mpl <Telekom> Advanced Communications Technologies and Services, ACTS (Programm der Europäischen Union)
fortgeschrittener Gas-Graphit-Reaktor m (AGR) <Kerntech> advanced gas-cooled reactor (AGR)
fortgeschrittenes intelligentes Netz n <Telekom> advanced intelligent network, AIN
Fortin'sches Barometer n <Phys> Fortin barometer

fortlaufend

fortlaufend *adj* <Kontroll> successive
fortlaufend nummerieren *v* <Patent> number consecutively
fortlaufende Abtastung *f* <Fernseh> sequential scanning
fortlaufende Nummer *f* <Patent> serial number
fortlaufende Qualifikationsprüfung *f* <Kerntech> ongoing qualification test
fortlaufende Zeilensprungabtastung *f* <Fernseh> progressive interlace
fortlaufender Text *m* <Druck> running text
fortlaufender Zeilensprung *m* <Fernseh> sequential interlace
Fortluft *f* <Heiz & Kälte> exhaust air, outgoing air
Fortluftstrom *m* <Heiz & Kälte> exhaust airstream
Fortpflanzung *f* <Comp & DV, Künstl Int> propagation
Fortpflanzung f von Fehlern <Künstl Int> backward error propagation, error back propagation
Fortpflanzung f von Schallwellen <Elektrotech> acoustic-wave propagation
Fortpflanzungsgleichung *f* <Phys> propagation equation
Fortpflanzungskonstante *f* 1. <Akustik> propagation coefficient; 2. <Telekom> propagation coefficient, propagation constant
Fortschaltrelais *n* <Elektriz> multiposition relay, stepping relay, stepping switch
Fortschaltung *f* <Elektronik> increment
Fortschaltungsadressierung *f* <Comp & DV> implied addressing
fortschreiben *v* <Comp & DV> update
Fortschreibung *f* <Comp & DV> file maintenance, file updating
fortschreitende Welle *f* 1. <Akustik> *(AE)* traveling wave, *(BE)* travelling wave; 2. <Telekom> outward-propagating wave, *(AE)* traveling wave, *(BE)* travelling wave; 3. <Wellphys> *(AE)* traveling wave, *(BE)* travelling wave
fortschreitende Wellen *fpl* <Wellphys> progressive waves
fortschrittliche Festkörperlogik *f (ASLT)* <Elektronik> advanced solid logic technology *(ASLT)*
fortschrittliche Luftunterstützung *f* <Lufttrans> advanced airborne fire support system *(Militär)*
fortschrittliche Technologie *f (AT)* <Comp & DV> advanced technology *(AT)*
Fortschrittsdiagramm *n* <Qual> progress chart
Fortschrittsstadium *n* <Patent, Qual> stage
Fortschrittsüberwachung *f* <Qual> progress control
Fortsetzungsblatt *n* <Druck> continuation sheet
fossile Strahlung *f* <Raumfahrt> fossil radiation
fossiler Brennstoff *m* <Thermod, Umweltschmutz> combustible fossil fuel, fossil fuel
Fotikon *n* <Elektronik> photicon
Foto *n* <Foto> picture
Fotoabzug *m* <Druck> photographic print, photoprint
Fotoapparat *m* <Foto> camera
Fotobox *f* <Foto> box camera
Fotoecken *fpl* <Foto> corner mounts
Fotoeffekt *m* 1. <Elektronik> photoelectric effect; 2. <Raumfahrt> photovoltaic effect *(Raumschiff)*; 3. <Strahlphys> photoelectric effect, photoemissive effect
Fotogerät *n* <Foto> photographic apparatus
Fotograf *m* <Foto, Phys> photographer
Fotografie *f* <Foto> photograph
fotografieren *v* <Foto> photograph
fotografische Belichtung *f* <Akustik> photographic exposure
fotografische Platte *f* <Druck> photographic plate
fotografischer Abzug *m* <Druck> photographic print, photoprint

Fotolabor *n* <Foto> photographic laboratory
Fotopapier *n* 1. <Druck> bromide paper, photopaper; 2. <Foto> photographic paper
Fotoprobeabzüge *mpl* <Druck> photographic proofs
Fotothek *f* <Foto> picture library
Foucault-Pendel *n* <Phys> Foucault pendulum
Foucault-Strom *m* <Elektriz> eddy current
Foulard *m* <Textil> pad mangle; pad *(Färben)*
Foulardfärbung *f* <Textil> pad dyeing
foulardieren *v* <Textil> pad
Foulardieren *n* <Textil> padding
Fourier-Analyse *f* <Elektronik, Ergon, Math, Phys, Telekom> Fourier analysis, harmonic analysis
Fourier-Entwicklung *f* <Math> Fourier expansion, Fourier series *(Darstellung einer periodischen Funktion)*
Fourier-Integral *n* <Phys> Fourier integral
Fourier-Reihe *f* <Phys> Fourier series
Fourier-Transformation *f* 1. <Elektronik> Fourier transform, Fourier transformation; 2. <Math> Fourier transformation; 3. <Phys> Fourier transform, Fourier transformation
Fourier-Transformationsspektroskopie *f* <Phys> Fourier transform spectroscopy
Fourier-Transformierte *f* <Elektronik> Fourier transform
FPS *(schnelle Paketvermittlung)* <Telekom> FPS *(fast packet server)*
fR *(Resonanzfrequenz)* <Akustik, Elektronik, Telekom, Wellphys> fR *(resonant frequency)*
Fr *(Francium)* <Chemie> Fr *(francium)*
Frac-Behandlung *f* <Erdöl> fracturing *(Förderung)*
Fracht *f* 1. <Trans> carriage; 2. <Wassertrans> freight, load; cargo *(Ladung)*
Frachtabfertigung *f* <Lufttrans> cargo handling
Frachtabfertigungsterminal *n* <Lufttrans> cargo terminal
Frachtbehälter *m* <Meerschmutz> scoop
Frachter *m* <Wassertrans> freighter
Frachtflugzeug *n* <Lufttrans> cargo plane, freighter
Frachtgut *n* <Eisenbahn> ordinary freight
Frachtgutwagenladung *f* <Trans> *(AE)* carload, *(BE)* wagonload
Frachthubschrauber *m* <Lufttrans> cargo helicopter
Frachtkabine *f* <Lufttrans> cargo compartment *(Flugwesen)*
Frachtkabineneinrichtung *f* <Lufttrans> cargo compartment equipment *(Flugwesen)*
Frachtkabinentür *f* <Lufttrans> cargo compartment door *(Flugwesen)*
Frachtkahn *m* <Wassertrans> cargo barge
Frachtkosten *pl* <Eisenbahn, Verpack> freight rate
Frachtraum *m* 1. <Lufttrans> hold *(Flugzeug)*; 2. <Raumfahrt> cargo hold *(Raumschiff)*
Frachtsatzanzeiger *m* <Eisenbahn, Wassertrans> scale rates
Frachtschiff *n* 1. <Trans> cargo ship; 2. <Wassertrans> cargo ship, dry-cargo ship, freight barge, freighter
Frachtschiff n mit Fahrgastbeförderung <Trans> passenger cargo ship
Frachtschlinge *f* <Lufttrans> cargo sling *(Hubschrauber)*
Frachttarif *m* <Eisenbahn, Verpack> freight rate
Frachtterminal *n* <Lufttrans> cargo terminal
Frachttonne *f* <Metrol> freight ton
Frachtumschlagplatz *m* <Lufttrans, Wassertrans> cargo-handling berth
Fracht- und Passagierschiff *n* <Wassertrans> cargo and passenger ship
Frage/Antwort-Gerät *n* <Telekom> interrogator-transponder
Fragebogen *m* <Ergon> questionnaire

Fragmentierung f <Comp & DV> fragmentation
Fraktal n <Comp & DV> fractal
fraktale Geometrie f <Geom> fractal geometry
Fraktaltheorie f <Geom> fractal theory
Fraktal-Transformation f <Fernseh> fractal transform *(Bildcodierung)*
Fraktil n <Qual> fractile
Fraktion f <Erdöl> fraction *(Destillation)*
Fraktionierapparat m <Chemtech> fractionating apparatus
Fraktionieren n 1. <Chemie> fractionation; 2. <Chemtech> fractional distillation
fraktionierende Destillation f 1. <Chemie> fractionation; 2. <Chemtech> fractional distillation
Fraktionierkolben m <Chemtech> distillation flask
Fraktionierkolonne f <Chemtech> fractionating column
Fraktioniersäule f <Labor> fractionation column *(Destillation)*
fraktionierte Destillation f <Erdöl> fractional distillation
fraktionierte Trennung f <Chemie> fractionation
Frame m <Künstl Int> frame
Francis-Turbine f <Nichtfoss Energ> Francis turbine
Francium n (Fr) <Chemie> francium *(Fr)*
Franck-Condon'sches Prinzip n <Phys> Franck-Condon principle
Franck-Hertz'scher Versuch m <Phys> Franck-Hertz experiment
Frangulin n <Chemie> frangulin
Franki-Pfahl m <Bau> Franki pile
Französische Norm f <Maschinen> French standard
Französische Prägung f <Ker & Glas> French embossing
Franzstandort m <Lufttrans> pinpoint
Fräsbohrer m <Bau> router
Fräsdorn m <Maschinen> cutter arbor, milling cutting arbor
Fräsdornmutter f <Fertig> arbor nut
Fräsdornstützlager n <Fertig> arbor support
Fräsdorntraglager n 1. <Fertig> arbor yoke, arbor-supporting bracket; 2. <Maschinen> arbor support
fräsen v <Fertig, Maschinen> mill
Fräsen n <Fertig, Maschinen> milling
Fräsen n **schräger Flächen** <Fertig> angle milling
Fräser m *(Fertig, Maschinen)* cutter, mill, milling cutter, rotary multipoint cutter
Fräser m **für Scheibenfedern** <Maschinen> Woodruff key cutter
Fräser m **mit eingesetzten Zähnen** <Maschinen> inserted tooth-milling cutter
Fräser m **mit geraden Zähnen** <Maschinen> milling cutter with straight teeth
Fräser m **mit grober Zahnteilung** <Fertig> coarse-pitch cutter
Fräser m **mit hinterdrehten Zähnen** <Maschinen> relieved-milling cutter
Fräser m **mit sägeförmigen Zähnen** <Maschinen> sawtooth cutter
Fräser m **mit Spiralzähnen** <Maschinen> milling cutter with spiral teeth
Fräserabhebung f <Fertig> cutter lift, cutter relief
Fräserdorn m <Maschinen> milling machine arbor
Fräsermesser n <Ker & Glas> cutter blade
Fräsersatz m <Fertig> set of cutters
Fräserschaft f <Fertig> cutter shank
Fräserschleifmaschine f <Maschinen> cutter grinder
Fräserstandzeit f <Fertig> cutter life
Fräserwiege f <Fertig> cutter cradle
Fräskopf m <Maschinen> cutter head, milling head
Fräslader m <Kohlen> continuous miner

Fräsmaschine f 1. <Fertig> milling machine, offset-milling machine; 2. <Maschinen, Mechan> milling machine
Fräsmaschine f **mit Handvorschub** <Fertig> router
Fräsmaschinendorn m <Fertig> milling machine arbor
Fräsmaschinenständer m <Fertig> milling machine column
Fräsmaschinentisch m <Maschinen> milling machine table, milling table
Frässchnitt m <Fertig> milling cut
Frässpan m <Fertig> milling
Frässpindel f 1. <Fertig> cutter spindle, milling spindle; 2. <Maschinen> cutter mandrel, cutter spindle, milling spindle
Frässpindelstock m <Fertig> cutter head
Frästisch m <Maschinen> milling machine table, milling table
Fräsvorrichtung f <Maschinen> milling attachment, milling jig
Fräsvorschub m <Fertig> milling feed
Fräswerkzeug n 1. <Fertig> milling tool; 2. <Maschinen> cutter, milling cutter
Fraunhofer'sche Beugungserscheinung f <Phys> Fraunhofer diffraction
Fraunhofer'sche Linien fpl <Phys> Fraunhofer lines
Fraunhofer'scher Bereich m <Raumfahrt> Fraunhofer region *(Weltraumfunk)*
Fraunhofer'sches Beugungsmuster n <Optik> Fraunhofer diffraction pattern
Freeware f <Comp & DV> freeware package *(Programm)*
frei adj <Telekom> idle; unoccupied *(Leitung)*
frei aufliegend adj <Bau> simply-supported
frei belegbare Funktionstaste f <Comp & DV> soft key
frei belegbare Tastatur f <Comp & DV> soft keyboard
frei geblasen adj <Ker & Glas> free-blown
frei stehend adj 1. <Bau> self-supporting; 2. <Verpack> freestanding
frei stehende Zwischenwand f <Bau> self-supporting partition
frei stehender Kran m <Bau> independent crane
frei stehendes Gerät n <Kontroll> stand-alone device
frei verstellbar adj <Maschinen> adjustable at will
frei werden v <Telekom> to be released
frei werdende Wärme f <Heiz & Kälte> released heat
frei werdendes Neutron n <Kerntech> nascent neutron
Freiarmmaschine f <Textil> arm-bed machine
Freibiegeversuch m <Fertig> free bend test *(Schweißen)*
Freibord m 1. <Erdöl> ullage; 2. <Wassertrans> freeboard *(Schiffbau)*
Freibordhöhe f <Wassertrans> depth for freeboard *(Schiffkonstruktion)*
Freibordmarke f <Wassertrans> plimsoll line
Freibordzuschlag m <Wassertrans> freeboard allowances *(Schiffkonstruktion)*
freie Baustelle f <Bau> delivered site
freie elektrische Bewegungsimpedanz f <Akustik> free electrical motional impedance
freie Elektronendichte f <Strahlphys> free-electron density
freie Energie f *(F)* <Metall, Phys> free energy *(F)*
freie Energie f **nach Helmholtz** <Phys> Helmholtz free energy, Helmholtz function
freie Flüssigkeitsoberfläche f <Phys> free surface of liquid
freie Konvektionsströmung f <Strömphys> free convection flow
freie Ladung f <Phys> free charge
freie Liste f <Comp & DV> free list
freie Mode f <Optik> unbound mode

freie

freie Oberfläche f <Wassertrans> free surface (Schiffkonstruktion)
freie Schwingung f 1. <Akustik> free vibration; 2. <Elektriz, Elektronik, Fertig> free oscillation; 3. <Phys> free oscillation; free vibration (mechanisch)
freie Strahlenquelle f <Kerntech> free source
freie Variable f <Künstl Int> free variable
freie Wärme f <Thermod> free heat
freier Betriebskanal m <Mobilkom> idle-working channel
freier Kanal m <Telekom> clear channel
freier Platz m <Comp & DV> empty slot
freier Raum m für den Kuppler <Eisenbahn> Berne rectangle
freier Rotor m <Lufttrans> free rotor (Hubschrauber)
freier Schwefel m <Kunststoff> (AE) free sulfur, (BE) free sulphur
freier Träger m <Bau> free beam
freier Zugriff m <Raumfahrt> random access (Weltraumfunk)
freies Elektron n <Elektriz, Phys, Teilphys> free electron
freies Format n <Comp & DV> free format
freies Grundwasser n <Wasserversorg> free groundwater
freies Radikal n <Lebensmittel> free radical
freies Schweben n <Phys> levitation
freies Walzenwehr n <Nichtfoss Energ> free roller sluice gate
freifahrende Turbine f <Lufttrans> free turbine
Freifall m <Verpack> free fall
Freifallbohrung f <Bau> free-fall boring, free-fall drilling
Freifall-Rettungsboot n <Wassertrans> freefall lifeboat
Freifallstanze f <Bau> free-falling stamp
Freifeld n <Comp & DV> free field
Freifeld-Bedingungen fpl <Aufnahme> free-field conditions
Freifeld-Übertragungsfaktor m <Aufnahme> free-field response
Freifläche f 1. <Bau> open area; 2. <Fertig> tool flank; flank, relief face (Spanung); 3. <Lufttrans> clearway (Flughafen); 4. <Maschinen> flank
Freifläche f am Umfang <Fertig> body clearance, diametral clearance (Spiralbohrer)
Freifläche f an Fase <Fertig> land clearance
Freiflächenfase f <Fertig> tool flank chamfer
Freiflächen-Orthogonalebene f <Fertig> tool flank orthogonal plane
freiformgeschmiedet adj <Fertig, Metall> hammer-forged, hand-forged
Freiformkurve f <Geom> free-form curve
Freiformschmieden n 1. <Fertig> hammer forging, hammering; 2. <Maschinen> open die forging; 3. <Metall> hammer forging
freiformschmieden v <Fertig> hammer-forge, hand-hammer
Freifräsung f am Umfang <Fertig> land clearance (Spiralbohrer)
Freigabe f 1. <Bau, Comp & DV> release; 2. <Elektronik> enabling; 3. <Lufttrans, Wassertrans> clearance (Genehmigung); 4. <Patent, Telekom> release
Freigabe f durch die Flugsicherung <Lufttrans> air traffic control clearance
Freigabe f zur Ausführung von Arbeiten <Qual> permit to work
Freigabeauflage f <Erdöl> relinquishment requirement (Recht)
Freigabedauer f <Lufttrans, Wassertrans> clearance period
Freigabeimpuls m <Comp & DV, Elektronik> enable pulse

Freigabeknopf m <Fernseh> release button
Freigabesignal n 1. <Comp & DV, Sicherheit, Telekom> enabling signal; 2. <Elektronik> enable signal
Freigabestand m <Comp & DV> release level
Freigabevorrichtung f <Sicherheit> enabling device
Freigabezeit f <Trans> green period
freigeben v 1. <Elektronik, Kontroll> enable; 2. <Mechan> release; 3. <Sicherheit> approve; 4. <Mobilkom> enable
freigegebenes Gatter n <Elektronik> enabled gate
freigelegt adj <Anstrich> exposed
freigelegter Schussfaden m <Textil> loose pick
freigeschalteter Kanal m <Telekom> blanked channel
freigesetzte Energie f <Nichtfoss Energ, Phys> released energy
freigesetzte kinetische Energie f geladener Teilchen in Materie (K) <Phys> kerma (K); kinetic energy released mass (Kerma)
freigesetzte Wärme f <Heiz & Kälte, Nichtfoss Energ> released heat
freigestellte Prüfung f <Qual> optional test
Freihafen m <Wassertrans> free port
freihalten v <Telekom> keep clear
freihalten v von/sich <Wassertrans> steer clear of (Navigation)
freihändig erstellte Zeichnung f <Konstzeich> freehand drawing
Freihandlinie f <Konstzeich> continuous irregular line, freehand line
Freihandschleifen n <Maschinen> freehand grinding
freihandschmieden v <Fertig> hammer
Freiheit f von Einbrandkerben <Fertig> absence of undercutting (Schweißen)
Freiheitsgrad m <Ergon, Maschinen, Phys, Qual, Thermod> degree of freedom
Freiheitsgradzahl f <Phys> variance
Freikolbengasturbine f <Maschinen> free-piston gas turbine
Freikolbenmaschine f <Maschinen> free-piston engine
freikreuzen v/sich <Wassertrans> claw off (absegeln vom Legerwall)
Freilager n <Wassertrans> bonded warehouse (Zoll)
Freilauf m 1. <Maschinen> freewheel; 2. <Mechan> free-wheel mechanism
Freilaufdiode f <Elektronik> freewheeling diode
freilaufender Rotor m <Lufttrans> free rotor (Hubschrauber)
freilaufendes Signal n <Elektronik> free-running signal
Freilaufkupplung f 1. <Kfztech> free engine clutch, overrunning clutch, roller clutch; 2. <Maschinen> freewheel clutch; 3. <Mechan> roller clutch
Freilauf- und Kuppelungseinheit f <Lufttrans> freewheel and clutch unit
Freilaufvorrichtung f <Maschinen> freewheel mechanism
freilegen v <Anstrich> expose
Freileitung f <Eisenbahn, Elektriz, Elektrotech, Telekom> overhead line, aerial line, open-wire line
Freileitungsableiter m <Elektrotech> intermediate line-type arrester, line-type arrester
Freileitungskabel n 1. <Elektrotech> overhead cable, overhead-line cable; 2. <Telekom> aerial cable, overhead cable
Freileitungsmodell n <Elektrotech> transmission-line model
Freileitungsnetz n <Telekom> overhead network
Freileitungssystem n <Elektrotech> overhead system
Freileitungsübertragung f <Elektrotech> open-wire transmission line
Freiluft f <Bau> open air

Freiluftausführung f <Elektrotech> outdoor design
Freiluftgerät n <Elektrotech> outdoor apparatus, outdoor device
Freiluftkabel n <Elektriz> outdoor cable, overhead cable
Freiluftkegelresonanz f <Aufnahme> free-air cone resonance
Freiluftleitung f <Elektriz> overhead line
Freiluftprüffeld n <Elektrotech, Metrol> open area test site
Freiluftquarzoszillator m <Elektronik> free-air crystal oscillator
Freiluftschaltgerät n <Elektriz> outdoor switchgear
Freilufttransformator m <Elektrotech> outdoor transformer
freimachen v <Bau> unstop *(Rohr)*
Freimachung f der Baustelle <Bau> clearing and grubbing
Freipfeiler m <Bau> pillar
freiprogrammierbares logisches Feld n <Comp & DV> field programmable device, field programmable logic array
Freiraumausbreitung f 1. <Funktech> free propagation, free space propagation; 2. <Telekom> free propagation
Freiraumdämpfung <Funktech> free space loss
Freiraumgrunddämpfung f <Telekom> free space basic loss
freischalten v <Mobilkom> enable
Freischalten n <Telekom> clear down
freischwebend adj <Mechan> cantilevered
freischwimmend adj <Wassertrans> hull borne
freischwingender Oszillator m <Elektronik, Funktech> free-running oscillator
Freischwingfrequenz f <Elektronik> free-running frequency
Freisetzung f <Abfall> discharge, release
Freispeicher m <Comp & DV> heap
Freisprecheinrichtung f <Mobilkom> handsfree device *(Autotelefon)*
freisteuern v von <Wassertrans> steer clear of *(Navigation)*
Freistich m <Maschinen> undercut
Freistrahl m <Fertig> free jet
Freistrahlturbine f <Mechan> impulse turbine
Freistrahlwindkanal m <Phys> open-jet wind tunnel
Freitextabfrage f <Comp & DV> free text query
freitragend adj <Bau> overhanging
freitragende Treppe f <Bau> fliers
freitragende Wand f <Bau> cantilevered wall
freitragende Windung f <Raumfahrt> air core winding *(Raumschiff)*
freitragendes Pultdach n <Bau> single pitch roof
Freiträger m <Fertig> cantilever beam
Freiträgertreppe f <Bau> hanging stairs
Freiwahl f <Telekom> hunting
Freiwange f <Bau> face string, outside string
Freiwasser n <Wasserversorg> surplus water
Freiwinkel m 1. <Fertig> back-off clearance, clearance angle, relief angle; 2. <Maschinen> clearance angle, orthogonal clearance, relief angle
Freiziehen n <Trans> debogging *(Schlamm)*
Freizustand m 1. <Comp & DV> idle condition; 2. <Telekom> idle state, idle condition
Fremdantrieb m <Lufttrans> belting-in run
Fremdatom n <Elektronik> impurity
fremdbewegtes Kühlmittel n <Heiz & Kälte> forced-circulated coolant
Fremddatei f <Mobilkom> visitor location register
Fremdemissionen fpl <Umweltschmutz> foreign emissions

fremderregter Dynamo m <Elektrotech> separate-excited dynamo
fremderregter Generator m <Elektrotech> separate-excited generator
fremderregter Magnet m <Chemie> electromagnet
Fremdgeräusch n 1. <Aufnahme> extraneous noise; 2. <Telekom> external noise
Fremdgeschmack m <Lebensmittel> off-flavour
Fremdkapazität f <Elektrotech> parasitic capacitance
Fremdkomponente f <Elektrotech> parasitic component
Fremdkörper m 1. <Fertig> fish; 2. <Lebensmittel> foreign body
Fremdkraftbremse f <Kfztech> power brake
Fremdkühlung f <Heiz & Kälte> separate cooling
Fremdkühlung f mit Luft <Heiz & Kälte> air blast cooling
Fremdnetz n <Mobilkom> visited network
Fremdorganisation f <Qual> outside agency
Fremdquelle f 1. <Fertig> outside source; 2. <Umweltschmutz> foreign source
Fremdscherben fpl <Ker & Glas> foreign cullet
Fremdspannung f <Elektrotech> external voltage
Fremdspannungsquelle f <Elektrotech> external voltage source
Fremdstelle f <Qual> outside agency
Fremdstoff m 1. <Anstrich> contaminant; 2. <Lebensmittel> foreign matter
Fremdstoffniederschlag m <Umweltschmutz> contamination fallout
Fremdstrom m <Elektrotech> parasitic current
Fremdsynchronisierung f <Elektronik> external synchronization
Fremdsynchronisierungseinrichtung f <Fernseh> slaving unit
Fremdteil n <Elektrotech> parasitic component
Fremdteilzeichnung f <Konstzeich> foreign part drawing
Fremdvermittlungsstelle f <Mobilkom> visited mobile switching center
Fremdzufuhr f <Umweltschmutz> external input
Frequenz f (F) <Akustik, Aufnahme, Comp & DV, Elektronik, Funktech, Phys> frequency (f)
Frequenz f des übernächsten Kanals <Elektronik, Fernseh, Funktech, Telekom> second channel frequency
Frequenzabfall m <Elektronik, Funktech, Telekom> frequency fall-off
Frequenzabgleich m <Elektronik, Fernseh, Funktech, Telekom> frequency adjustment, frequency alignment
frequenzabhängig adj <Funktech, Telekom> frequency-dependent, frequency-sensitive
Frequenzabstand m 1. <Elektronik> frequency separation, mode separation; 2. <Funktech, Telekom> frequency separation
Frequenzabstimmung f <Elektronik, Funktech, Telekom> frequency tuning
Frequenzabtastung f <Elektronik> frequency scanning
Frequenzabweichung f 1. <Comp & DV> frequency deviation; 2. <Elektronik, Funktech, Telekom> frequency departure
Frequenzanalysator m <Elektronik, Funktech> (BE) wave analyser, (AE) wave analyzer
Frequenzanalyse f <Telekom> harmonic analysis
Frequenzänderung f <Elektronik, Funktech> frequency change
Frequenzauflösungsvermögen n <Elektronik, Funktech, Telekom> frequency resolving power
Frequenzaufnahme f <Elektronik, Funktech, Telekom> frequency record
Frequenzauswanderung f 1. <Elektronik> frequency pulling *(bei Belastung)*; 2. <Funktech, Telekom> frequency pulling, frequency drift

Frequenzband

Frequenzband n 1. <Elektronik, Fernseh> frequency band; 2. <Funktech> frequency band, waveband; 3. <Telekom> frequency band
Frequenzbandentstörungsfilter n <Comp & DV> band-rejection filter
Frequenzbändertausch m <Funktech, Telekom> frequency frogging
Frequenzbandsperrfilter n <Comp & DV> band-stop filter
Frequenzbandumsetzung f <Elektronik, Funktech, Telekom> frequency conversion
Frequenzbereich m 1. <Akustik> frequency band, frequency domain, frequency range; 2. <Aufnahme> frequency band, frequency domain, frequency range (von Mikrofonen); 3. <Comp & DV, Elektronik, Funktech, Telekom> frequency band, frequency domain, frequency range; 4. <Wellphys> waveband
Frequenzbeweglichkeit f <Elektronik, Funktech> frequency agility
Frequenzcharakteristik f <Telekom> frequency characteristic
Frequenzdekade f 1. <Elektronik, Funktech> frequency synthesizer; 2. <Telekom> synthesizer, frequency synthesizer
Frequenzdemodulation f <Elektronik, Funktech, Telekom> frequency demodulation
Frequenzdemodulator m <Elektronik, Funktech, Telekom> frequency demodulator, frequency detector
Frequenzdiversity n <Elektronik, Funktech, Telekom> frequency diversity
Frequenzdrift f <Elektronik, Funktech, Telekom> frequency drift
Frequenzeichung f <Elektronik, Fernseh, Funktech> frequency calibration
Frequenzeinstellung f <Elektronik, Fernseh, Funktech, Telekom> frequency adjustment, frequency setting
Frequenzfenster n **in Erdatmosphäre** <Phys, Strahlphys> atmospheric window
Frequenzgang m 1. <Akustik> response; 2. <Comp & DV, Elektronik, Funktech, Telekom, Wellphys> frequency characteristic, frequency response, harmonic characteristic, harmonic response, response
frequenzgesteuerte Antenne f <Elektronik, Funktech, Telekom> frequency scanner
frequenzgeteiltes Vermittlungssystem n <Elektronik, Telekom> frequency division switching system
Frequenzgrenze f <Elektronik, Funktech, Telekom> frequency cut-off
Frequenzgruppe f <Akustik> critical band
Frequenzgruppenbreite f <Akustik> crit-ical band width
Frequenzgruppenintensität f <Akustik> critical band intensity
Frequenzgruppenpegel m <Akustik> critical band level
Frequenzgruppentausch m <Akustik> frequency frogging
Frequenzhalbierschaltung f <Elektronik> binary counter (Schaltkreistechnik)
Frequenzhub m 1. <Elektronik> frequency deviation, frequency sweep; 2. <Raumfahrt> deviation (Weltraumfunk); 3. <Telekom> frequency deviation, frequency shift (Frequenzmodulation)
Frequenzkanal m <Elektrotech> channel
Frequenzkompensation f <Aufnahme, Elektronik> frequency compensation
Frequenzkomponente f <Aufnahme, Elektronik> frequency component
Frequenzkonstanthaltung f <Elektronik, Funktech, Telekom> frequency stabilization
Frequenzkontrolle f <Elektronik, Funktech, Telekom> frequency monitoring

Frequenzkontrollgerät n <Phys> wavemeter
Frequenzlücke f <Elektronik, Funktech, Telekom> frequency gap
Frequenzmanagement n <Mobilkom> frequency management
Frequenzmessbrücke f <Gerät> frequency-measuring bridge
Frequenzmesser m 1. <Elektronik, Funktech> frequency meter; 2. <Phys> wavemeter; 3. <Telekom> frequency meter; 4. <Wassertrans> wavemeter (Funk)
Frequenzmodulation f (FM) <Comp & DV, Elektronik, Funktech, Phys, Telekom> frequency modulation (FM)
Frequenzmodulationsrauschen n <Elektronik, Funktech, Telekom> frequency modulation noise
Frequenzmodulator m <Elektronik, Funktech, Telekom> frequency modulator
frequenzmoduliert adj <Elektronik> frequency-modulated
Frequenzmultiplexverfahren n <Elektronik, Funktech, Telekom> frequency division multiplexing (FDM)
Frequenznachführung f <Elektronik, Funktech, Telekom> frequency tracking
Frequenznormal n <Elektronik, Funktech, Telekom> frequency standard
Frequenznutzung f <Funktech> usage of frequencies
Frequenzökonomie f <Telekom> spectrum efficiency
Frequenzquelle f <Elektronik, Funktech, Telekom> frequency source
Frequenzraster n <Funktech, Telekom> frequency raster
Frequenzregelung f <Elektronik, Funktech, Telekom> frequency control, frequency regulation
Frequenzrelais n <Elektronik, Funktech> frequency relay
Frequenzschwankungen fpl <Akustik> wow
frequenzselektive Flüssigkristallanzeige f <Elektronik> dichroic LCD
frequenzselektiver Schwund m <Elektronik, Funktech, Telekom> frequency-selective fading
frequenzselektiver Verstärker m <Elektronik, Funktech, Telekom> frequency-selective amplifier
frequenzselektives Fading n <Elektronik, Funktech, Telekom> frequency-selective fading
Frequenzskala f <Elektronik, Funktech, Telekom> frequency scale
Frequenzspektrum n <Aufnahme, Comp & DV, Elektronik, Funktech, Telekom> frequency spectrum
Frequenzsprungoszillator m <Elektronik, Funktech, Telekom> frequency-hopping oscillator
Frequenzsprungverfahren n <Elektronik, Funktech, Telekom> frequency hopping
Frequenzstabilisierung f <Elektronik, Funktech, Telekom> frequency stabilization
Frequenzstufe f <Akustik> step difference limen
Frequenzsynthese f <Elektronik, Funktech, Telekom> frequency synthesis
Frequenzsynthese generator m <Elektronik, Funktech, Telekom> frequency synthesizer
Frequenzteilband n <Elektronik, Funktech, Telekom> frequency subband
Frequenzteiler m <Elektriz, Funktech, Telekom> frequency divider
Frequenztoleranz f <Funktech, Telekom> frequency tolerance
Frequenzüberdeckung f <Elektronik, Funktech, Telekom> frequency coverage
Frequenzüberlappung f <Elektronik, Funktech, Telekom> frequency overlap, frequency overlapping
Frequenzüberwachung f <Elektronik, Funktech, Telekom> frequency monitoring
Frequenzumformer m <Comp & DV> frequency changer

Frequenzumformer-Unterwerk n <Elektriz> substation for frequency conversion
Frequenzumkehrung f <Elektronik, Funktech, Telekom> frequency inversion
Frequenzumrichter m <Elektronik> frequency converter, variable-frequency inverter
Frequenzumsetzer m 1. <Elektronik, Funktech> frequency converter, frequency changer, frequency translator; 2. <Telekom> frequency converter, frequency translator
Frequenzumsetzung f <Elektronik, Funktech, Telekom> frequency conversion, frequency translation, frequency transposition
Frequenzumtastung f (FSK) <Comp & DV, Elektronik, Funktech, Telekom> frequency shift keying (FSK)
Frequenzunsicherheit f <Elektronik, Funktech, Telekom> frequency uncertainty
Frequenzunterdrückung f <Elektronik, Funktech, Telekom> frequency rejection
Frequenzverdoppler m <Elektronik, Funktech, Telekom> frequency doubler
Frequenzvergabe f <Funktech, Telekom> frequency allocation
Frequenzverlauf m <Elektronik, Funktech, Telekom, Wellphys> frequency response
Frequenzversatz m 1. <Comp & DV> frequency shift; 2. <Elektronik, Funktech, Telekom> frequency departure, frequency offset
Frequenzverschachtelung f <Elektronik, Funktech, Telekom> frequency interlace
Frequenzverschiebung f 1. <Comp & DV> frequency drift; 2. <Elektronik, Funktech, Telekom> frequency displacement, frequency shift
Frequenzverschiebung f **durch Dopplereffekt** <Funktech> Doppler shift
Frequenzverteilung f <Comp & DV, Elektronik> frequency distribution
Frequenzvervielfacher m 1. <Elektronik> frequency multiplier, harmonic generator; 2. <Funktech, Telekom> frequency multiplier
Frequenzvervielfachung f 1. <Elektronik> frequency multiplication, harmonic generation; 2. <Funktech, Telekom> frequency multiplication
Frequenzvervielfachungsklystron n <Elektronik> frequency multiplier klystron
Frequenzvervielfachungs-Reaktanzdiode f <Elektronik> harmonic generator varactor
Frequenzverzerrung f <Elektronik, Funktech, Telekom> frequency distortion
Frequenzvielfach-Zugriffsverfahren n <Elektronik, Funktech, Telekom> FDMA, frequency division multiple access
Frequenzwahlschalter m <Elektronik, Funktech, Telekom> frequency selector
Frequenzwanderung f **eines Oszillators** <Elektronik> oscillator drift
Frequenzwandler m 1. <Elektronik, Funktech> frequency converter, frequency changer; 2. <Telekom> frequency converter
Frequenzwandlung f <Elektronik, Funktech, Telekom> frequency conversion
Frequenzwechsel m <Elektronik> frequency change
Frequenzwechsler m <Elektronik> frequency changer
Frequenzweiche f 1. <Aufnahme> crossover network; 2. <Elektronik, Funktech> frequency-separating filter; 3. <Telekom> diplexer, frequency-separating filter
Frequenzwiederbenutzung f <Elektronik, Funktech, Telekom> frequency reuse
Frequenzwiederverwendung f <Elektronik, Funktech, Telekom> frequency reuse
Frequenzwobbelung f 1. <Elektronik> frequency sweep, wobbling; 2. <Funktech, Telekom> frequency sweep
Frequenzzähler m 1. <Elektronik, Funktech> frequency counter; 2. <Gerät> counter; 3. <Telekom> frequency counter
Frequenzziehen n <Elektronik, Funktech, Telekom> frequency pulling
Frequenzzuteilung f 1. <Funktech> frequency allocation (Frequenzplan); 2. <Telekom> frequency allocation
Frequenzzuweisung f <Elektronik, Fernseh> frequency allocation
Fresnel'sche Beugung f <Phys> Fresnel diffraction
Fresnel'sche Beugungszone f <Funktech, Telekom> Fresnel zone
Fresnel'sche Gleichungen fpl <Phys> Fresnel's formulae
Fresnel'sche Linse f <Foto, Phys> Fresnel lens
Fresnel'sche Reflexion f <Optik, Telekom> Fresnel reflection
Fresnel'sche Reflexionsmethode f <Telekom> Fresnel reflection method
Fresnel'sche Zone f 1. <Funktech> Fresnel zone; 2. <Telekom> Fresnel region, Fresnel zone
Fresnel'scher Spiegel m <Phys> Fresnel mirrors
Fresnel'sches Beugungsdiagramm n <Telekom> Fresnel diffraction pattern
Fresnel'sches Beugungsmuster n <Optik> Fresnel diffraction pattern
Fresnel'sches Biprisma n <Phys> Fresnel biprism
Fresnel-Zone f 1. <Funktech> Fresnel zone; 2. <Telekom> Fresnel region, Fresnel zone
fressen v 1. <Anstrich> fret (Korrosion, Reiben); 2. <Fertig> scuff; 3. <Maschinen> seize
Fressen n <Maschinen> seizing, seizure
Frettbohrer m <Bau> auger gimlet
Freund-Feind-Erkennung f <Funktech> friend-foe-identification
friemeln v <Maschinen> cross-roll
Friemelwalzwerk n <Fertig> reeler
frieren v <Bau> freeze
frierend adj <Thermod> freezing
Fries m <Bau> (AE) molding, (BE) moulding
Friktion f <Maschinen, Papier> friction
friktioniert adj <Papier> friction-glazed
Friktionsfeder f <Maschinen> friction spring
Friktionskalander m <Papier> friction-glazing calender
Friktionsrolle f <Papier> friction reel
Frisch... <Lebensmittel> fresh
Frischbeton m <Bau> green concrete
Frischdampf m <Hydraul, Kerntech, Maschinen, Thermod> live steam
Frischdampfinjektor m <Hydraul> live steam injector
Frischdampfturbine f <Maschinen> live-steam turbine
frischen v 1. <Fertig> decarburize; 2. <Metall> refine
Frischen n <Fertig> oxidation of impurities (Stahl)
frischer Brennstoff m <Kerntech> fresh fuel
frischer Uran-Brennstoff m <Kerntech> fresh uranium
Frischhaltefolie f <Verpack> cling film
Frischhaltemittel n <Lebensmittel> antistaling agent (Brot)
Frischluft f <Heiz & Kälte> fresh air
Frischluftdruckschlauchgerät n <Sicherheit> continuous line respirator
Frischlüfter m <Heiz & Kälte> (AE) forced-draft fan, (BE) forced-draught fan
Frischluftkühlung f <Heiz & Kälte> fresh-air cooling
Frischölschmierung f <Maschinen> total loss lubrication
Frischschlamm m <Abfall> fresh sludge
Frischwasser n <Abfall, Wassertrans, Wasserversorg> freshwater

Frischwassererzeuger

Frischwassererzeuger m <Wassertrans> freshwater condenser
Frischwasserfreibord m <Wassertrans> freshwater freeboard *(Schiffkonstruktion)*
Frist f <Patent> time limit
Fritte f <Ker & Glas> frit
Fritten n <Ker & Glas> fritting *(der Charge)*
Frittenglasur f <Ker & Glas> fritted glaze
Frittezone f <Ker & Glas> fritting zone
Frittung f <Chemie> vitrification *(Geologie)*
Front f <Comp & DV, Konstzeich> front
Frontabschnitt m <Konstzeich> frontal section
Frontalbereich m <Maschinen> frontal area
Frontalzusammenstoß m 1. <Kfztech> head-on collision; 2. <Raumfahrt> head-on collision *(Raumschiff)*; 3. <Sicherheit> head-on collision
Frontantrieb m <Mechan> front wheel drive
Front-End-Anwendung f <Comp & DV> front-end application
Frontgabelstapler m <Fertig> front-end forklift truck
Frontispiz n <Druck> frontispiece
Frontlader-Tieflöffelkombination f <Bau> loader backhoe *(Straßenbau)*
Frontlenker m <Kfztech> cab over engine *(Karosserie)*
Frontlinse f <Foto> front element *(Objektiv)*
Frontmotor m <Kfztech> front engine, front-mounted engine
Frontplatte f 1. <Fernseh> face plate; 2. <Foto> lens panel; 3. <Gerät> front panel *(Gerät)*; 4. <Maschinen> breast plate, front panel, panel
Frontschaufellader m <Meerschmutz> front-end loader
Frontscheibe f <Kfztech, Raumfahrt> *(BE)* windscreen, *(AE)* windshield *(Raumschiff)*
Frontscheibenheizung f <Kfztech> windshiel heater
Frontschott n <Trans> break-bulkhead
Frontseite f <Maschinen> face
Frontseite/Reversseite f <Comp & DV> face-side/revers-side
Frontspalt m <Akustik> front gap
Fronttafel f <Gerät> front panel *(Gerät)*
Frontverkleidung f <Kfztech> front fairing *(Fahrzeug)*
Frontwalze f <Papier> face roll
Frontwand f <Bau> face wall
Froschbeinwicklung f <Elektrotech> frog-leg winding
Frost m <Kohlen> frost
Frostanfälligkeit f <Kohlen> frost susceptibility
Frostaufbruch m <Kohlen> frost heave
frostbeständiger Boden m <Kohlen> frost-resistant soil
Frosteindringtiefe f <Kohlen> frost penetration depth
frostfreie Tiefe f <Kohlen> frost-free level
Frostgrenze f <Kohlen> frost limit
Frostpunkt m <Heiz & Kälte> ice point
Frostschadstelle f <Kohlen> frost boil
Frostschutz m <Bau, Kfztech> antifreeze *(Kühlsystem)*
Frostschutz... <Heiz & Kälte> anti-freeze
Frostschutzmittel n 1. <Heiz & Kälte> antifreeze agent; 2. <Kfztech> antifreeze; 3. <Verpack> frost-preventive agent
Frostschutzpapier n <Papier> antifreeze paper
Frostschutzschicht f <Bau> subbase *(Tiefbau)*
Frostschutzwächter m <Heiz & Kälte> antifreeze detector
froststabilisierter Latex m <Kunststoff> freeze-thaw resistant latex
Frost-Tau-Beständigkeit f <Heiz & Kälte> freeze-thaw resistance
Frouden'sche Zahl f *(F)* <Hydraul, Phys> Froude number, F *(Strömungslehre)*
Fruchtfleisch n <Lebensmittel> pulp
Fruchtzucker m <Lebensmittel> fructose, *(BE)* laevulose, *(AE)* levulose
Fructosan n <Chemie> fructosan, inulin
Fructose f <Chemie, Lebensmittel> fructose
Frühausfall m 1. <Elektrotech> early failure; 2. <Kontroll> wear-in failure
frühe Veröffentlichung f <Patent> early publication
frühere Anmeldung f <Patent> earlier application
frühere Priorität f <Patent> earlier priority
frühhochfester Beton m <Bau> high-early-strength concrete
frühhochfester Zement m <Bau> high-early-strength cement, rapid-hardening cement
Frühjahrs-Tagundnachtgleiche f <Raumfahrt> vernal equinox
Frühlingsmaximum n <Umweltschutz> spring maximum of fallout
Frühlingssäureschock m <Umweltschutz> spring acid shock
Frühstart m <Fernseh> early start
Frühstartbild n <Fernseh> early-start video
Frühstartton m <Fernseh> early-start audio
Frühwarnradar n 1. <Funkort> early-warning radar; 2. <Funktech, Wassertrans> distant early-warning radar, early-warning radar
Frühzündung f 1. <Kfztech> advance, advanced ignition, premature ignition; 2. <Lufttrans> ignition advance, spark advance
FS 1. <Comp & DV, Telekom> *(Fernschreiber)* TTY *(teletypewriter)*; 2. <Lufttrans> *(Flugsicherung)* ATC *(air traffic control)*; 3. <Maschinen> *(Festsitz)* force fit
F-Schicht f 1. <Funktech> F layer; 2. <Phys> Appleton layer, F layer
FSK *(Frequenzumtastung)* <Comp & DV, Elektronik, Funktech, Telekom> FSK *(frequency shift keying)*
FSK-Modem n <Elektronik, Telekom> FSK modem
FSTV *(Breitbandfernsehen)* <Fernseh> FSTV *(fast-scan television)*
Fuchs m <Ker & Glas> flue
Fuchson n <Chemie> fuchsone
Fuchsschwanz m 1. <Bau> handsaw, tenon saw; 2. <Maschinen> handsaw
Fucose f <Chemie> fucose
Fucosterin n <Chemie> fucosterol
Fucoxanthin n <Chemie> fucoxanthin
FüG *(Fahrt über Grund)* <Wassertrans> speed made good over the ground
Fuge f 1. <Bau> joint, meeting, mortise, seam; 2. <Fertig> interstice; 3. <Kerntech> joint *(Schweißen)*; 4. <Maschinen, Mechan> gap • **Fuge auf Fuge** <Bau> straight joint *(Mauerwerk)*
Fugeisen n <Bau> jointer
Fugendichtung f <Bau> waterstop
Fugenflanke f <Fertig> joint face
Fugenhobel m <Bau> gouging plane
fugenhobeln v <Fertig> torch-gouge
Fugenhobeln n <Fertig> gas gouging, gouging, groove cutting
Fugenmasse f <Bau> joint sealer
Fugenmörtelbrett n <Bau> hawk
Fugennaht f <Fertig> groove weld
Fugenschneidegerät n <Bau> joint cutter
Fugenvergussmasse f <Bau> joint-sealing compound, sealing compound
Fugenvorbereitung f <Fertig> edge preparation, plate-edge preparation *(Schweißen)*
Fügung f <Fertig> mating
fühlbare Kühlwirkung f <Heiz & Kälte> sensible-cooling effect

fühlbare Wärme f <Heiz & Kälte> sensible heat
fühlbare Wärmelast f <Heiz & Kälte> sensible heat load
Fühler m 1. <Elektriz> pick-up, probe; 2. <Fertig> stylus *(zum Abtasten)*; 3. <Gerät, Kohlen> probe; 4. <Maschinen> feeler, probe, tracer pin; 5. <Optik> detector *(Messglied)*; 6. <Raumfahrt> sensor *(Weltraumfunk)*; 7. <Wassertrans> sensor *(Messinstrumente)*
Fühler m **und Geber** m <Telekom> scanner distributor
Fühlerkabel n <Nichtfoss Energ> sensor cable
Fühlerlehre f 1. <Kerntech> *(AE)* feeler gage, *(BE)* feeler gauge; 2. <Maschinen> *(AE)* feeler gage, *(BE)* feeler gauge, *(AE)* thickness gage, *(BE)* thickness gauge
Fühlglied n <Raumfahrt> sensor *(Weltraumfunk)*
Fühllehre f 1. <Kfztech> *(AE)* feeler gage, *(BE)* feeler gauge *(Werkzeug)*; 2. <Metrol> *(AE)* feeler gage, *(BE)* feeler gauge
Fühlschiene f <Eisenbahn> locking bar
Fühlsensor m <Gerät> tactile sensor
Fühluhr f <Metrol> *(AE)* dial indicating gage, *(BE)* dial indicating gauge
führen v 1. <Hydraul> gate; 2. <Lufttrans> pilot, steer; 3. <Trans> steer; 4. <Wassertrans> steer; con, handle, navigate *(Schiff)*
Führerschein m <Kfztech> *(BE)* driver's licence, *(AE)* driver's license *(Rechtsvorschriften)*
Führersicherheitsvorrichtung f <Eisenbahn> driver's safety device, DSD
Führerstand m 1. <Eisenbahn> cab, driver's cab; 2. <Kfztech> cab *(Karosserie)*
Führerstandssignal n <Eisenbahn> *(AE)* cab signaling, *(BE)* cab signalling
Fuhrlohn m <Trans> cartage
Fuhrpark m <Bau> rolling stock
Führschraube f <Mechan> lead screw
Führung f 1. <Bau> forcing *(Schloss)*; 2. <Elektriz> conduit *(Kabelrohr, Schutzrohr)*; 3. <Hydraul> guide *(Turbine)*; 4. <Kontroll> guide; 5. <Maschinen> guide, guidance, guiding, slide bar; 6. <Papier> guide
Führung f **für Luftschraubenblatt** <Lufttrans> blade sleeve *(Hubschrauber)*
Führungen fpl 1. <Ker & Glas> guides; 2. <Maschinen> ways
Führungsantenne f <Raumfahrt> guidance antenna *(Raumschiff)*
Führungsbahn f 1. <Bau, Eisenbahn> guideway; 2. <Fertig> bearing; 3. <Maschinen> guideway, ways
Führungsbereich m <Regelung> range of command
Führungsblock m <Maschinen> guide block
Führungsbohrung f <Fertig> guide hole
Führungsbuchse f 1. <Fernseh> female guide; 2. <Fertig> adaptor bushing; 3. <Kerntech> guide bushing; 4. <Kfztech> pilot bushing; 5. <Maschinen> guide bush
Führungsempfänger m <Raumfahrt> guidance receiver *(Raumschiff)*
Führungsfase f <Fertig> land *(Spiralbohrer)*
Führungsfehler m <Fernseh> guide errors
Führungsgestell n <Maschinen> die set
Führungsgröße f <Regelung> reference input variable, reference variable
Führungshülse f <Fertig> ferrule *(Faseroptik)*
Führungslager n 1. <Kfztech> pilot bearing; 2. <Maschinen> guide bearing
Führungsleiste f 1. <Fertig> plate *(Spitzenlosschleife)*; 2. <Maschinen> guide rail, work rest blade
Führungslenker m <Kfztech> radius arm *(Aufhängung)*
Führungslicht n <Foto> key light
Führungslineal n 1. <Fertig> fence; 2. <Maschinen> gib
Führungsloch n <Comp & DV> feed hole, sprocket hole
Führungsmagnet m <Trans> guidance magnet

Führungsmeißel m <Erdöl> pilot bit *(Bohrtechnik)*
Führungsnase f <Mechan> locating key
Führungsnavigationssystem n <Raumfahrt> guidance navigation system *(Raumschiff)*
Führungsplatte f <Maschinen> guide plate
Führungspolster n <Trans> guidance cushion
Führungsprismen npl <Fertig> vees
Führungspunkte mpl <Druck> dot leaders
Führungsräder npl <Trans> guide wheels
Führungsring m <Hydraul> guide ring
Führungsrohr n <Maschinen> guide tube
Führungsrolle f 1. <Fertig> guide roller *(Bandsäge)*; 2. <Maschinen> jockey, jockey pulley, jockey roller, jockey wheel; 3. <Mechan> jockey pulley; 4. <Textil> guide roller; 5. <Wassertrans> fairlead *(Decksausrüstung)*
Führungssäule f <Maschinen> guide pillar
Führungsschaft m <Fertig> rear pilot; back pilot *(Reibahle)*
Führungsscheibe f <Maschinen> guide, guide pulley
Führungsschiene f 1. <Eisenbahn> guide rail; 2. <Fertig> beam *(Messschieber)*; 3. <Maschinen> guide bar, work rest blade
Führungsschiff n <Wassertrans> command ship *(Marine)*
Führungsschlitten m <Maschinen> pilot carriage
Führungsschnitt m <Fertig> subpress die
Führungsstange f 1. <Bau> guide pole *(Ramme)*; 2. <Maschinen> guide bar, slide bar; 3. <Textil> guide bar
Führungsstein m <Maschinen> slide block
Führungsstift m 1. <Maschinen> box pin, guide pin; 2. <Mechan> alignment pin, guide pin
Führungsverhalten n <Regelung> command action
Führungswalze f <Textil> guide roller
Führungswelle f <Mechan> idler
Führungszapfen m 1. <Elektrotech> aligning plug *(Röhre)*; 2. <Fertig> *(AE)* teat; pilot *(Reibahle)*; 3. <Maschinen> guide pin, pilot, pilot pin
Führungszentrum n <Telekom> *(AE)* management center, *(BE)* management centre
Füll... <Elektrotech, Kohlen, Verpack> filling
Füllapparat m <Verpack> feeder
Füllart f **der Deponie** <Abfall> tip-filling method
Fülldamm m <Bau> fill dam
Fülldichte f <Kohlen> bulk density
Fülldruck m <Sicherheit> inflation pressure
Fülldüse f <Raumfahrt> nozzle *(Raumschiff)*
Fülleinrichtung f <Verpack> feeding device, filling device
Füllelement n <Elektrotech> wet cell
Füllelemente npl <Heiz & Kälte> infill panels
füllen v <Ker & Glas, Trans> fill
Füllen n **des Ofens** <Ker & Glas> filling the furnace *(vor Beginn des Schmelzvorgangs)*
Fuller-Bonnot-Luftstromkegelmühle f <Kohlen> Fuller-Bonnot mill
Fullererde f <Bau> Fuller's earth
Fullerkreide f <Bau> Fuller's chalk
Füllfaktor m <Telekom> filling coefficient
Füllgrad m **der Schallplatte** <Aufnahme> groove spacing *(Rillenabstand)*
Füllhahn m <Fertig> feed cock
Füllhöhe f <Maschinen> filling level
Füllholz n <Ker & Glas> packing piece
fülliger Griff m <Textil> full handle
Fülligkeit f <Textil> bulk *(Garn)*
Füllkit m <Fertig> beaumontage
Füllkokssäule f <Kohlen> coke bed
Füllkörperabscheider m <Umweltschmutz> packed-bed scrubber

Füllkörperkolonne

Füllkörperkolonne f 1. <Chemtech> packed column, packed tower *(Destillation, Absorption)*; 2. <Erdöl> packed column *(Raffinerie)*
Füllkörpersäule f <Chemtech> packed column, packed tower
Füllloch n <Verpack> filling hole
Füllmaschine f <Fertig, Maschinen> filler, filling machine
Füllmaschine f **bis zur Dichtheitsgrenze** <Verpack> density-filling machine
Füllmaterial n 1. <Kerntech> filling, filling material; 2. <Textil> filling
Füllöffnung f <Mechan, Verpack> filling hole
Füllpunkt m <Ker & Glas> filling point
Füllraum m <Fertig> pot *(Kunststoffe)*
Füllraum-Abquetschwerkzeug n <Kunststoff> *(AE)* semipositive mold, *(BE)* semipositive mould
Füllraumwerkzeug n <Kunststoff> *(AE)* positive mold, *(BE)* positive mould
Füllrohr n <Wassertrans> filling pipe
Füllschlauchverbindung f <Erdöl> feed hose union
Füllsender m <Fernseh> booster station, fill-in transmitter
Füllspant m <Wassertrans> filling frame
Füllstand m 1. <Bau> level; 2. <Erdöl> liquid level *(Behälter, Tanks)*; 3. <Gerät> level; 4. <Verpack> fill level, level of filling
Füllstandsanzeige f <Kerntech> level meter
Füllstandsanzeigegerät n <Gerät> indicating level meter
Füllstandsanzeiger m 1. <Bau> level indicator; 2. <Gerät> filling level indicator, level indicator, liquid level indicator
Füllstandsmessgerät n <Gerät> *(AE)* level gage, *(BE)* level gauge
Füllstandsmessung f <Gerät> level measurement
Füllstandswächter m <Gerät> level indicator
Füllstandsüberwachung f <Verpack> liquid level control
Füllstoff m <Elektrotech, Ker & Glas, Kunststoff> filler
Füllstück n <Elektriz> filler
Füllstutzen m <Verpack> filling nozzle
Fülltrichter m <Mechan> hopper
Füllung f 1. <Elektrotech> charge; 2. <Ker & Glas> fill; 3. <Lebensmittel> stuffing *(Fleisch)*; 4. <Maschinen> admission; 5. <Mechan> batch
Füllvorrichtung f <Fertig> filling device
Füllzeichen n 1. <Comp & DV> filler, filler character, gap character, pad; 2. <Druck> filler; 3. <Telekom> filler, idle character
Füllziffer f <Comp & DV> gap digit
Füllzylinder m <Fertig> hopper
Fulminat n <Chemie> fulminate
Fulven n <Chemie> fulvene
Fumar... <Chemie> fumaric
Fundament n 1. <Bau> base, base plate, foundation; 2. <Druck> bed, forme bed; 3. <Kohlen> foundation; 4. <Papier> basement
fundamentale Kraft f <Teilphys> fundamental force *(Wechselwirkung)*
Fundamentalmode f <Optik> fundamental mode
Fundamentanker m <Bau> anchor bolt
Fundamentblock m <Bau> footing block, foundation block
Fundamentgewölbe n <Bau> inverted arch
Fundamentkonstruktion f <Bau> substructure
Fundamentplatte f 1. <Bau> bottom plate, foundation plate, raft; 2. <Kohlen> raft; 3. <Maschinen> bed plate
Fundamentrost m <Bau> grillage foundation
Fundamentschraube f <Bau> foundation bolt, lag screw
Fundamentschwelle f <Kohlen> foot wall
Fundamentzeichnung f <Konstzeich> foundation drawing

fünfatomig adj <Chemie> pentatomic
Fünfeck n <Geom> pentagon
fünfeckig adj <Geom> pentagonal
Fünfermarkierung f <Qual> tally
fünfflächig adj <Geom> pentahedral
Fünfgittermischröhre f <Elektrotech> pentagrid converter
Fünfgitterröhre f <Elektronik> pentode
Fünfkant m <Maschinen> pentagon
fünfkantiges Räumwerkzeug n <Maschinen> five-sided broach
Fünfkantmutter f <Maschinen> pentagon nut
fünfschichtige Sperrfolie f <Verpack> five-layer barrier film
fünfstufige Ganztonleiter f <Akustik> pentatonic scale
fünfwertig adj <Chemie> pentavalent, quinquevalent
fünfwertiges Element n <Chemie> pentad
Fünfwertigkeit f <Chemie> pentavalence, quinquevalence
Fungistatikum n <Lebensmittel> fungistat
Fungizid n <Chemie, Umweltschmutz> fungicide
Funk m <Funktech> radio
Funkamateur m 1. <Funktech> ham; 2. <Telekom> radio amateur
Funkanflughilfen fpl <Lufttrans> radio approach aids
Funkantenne f 1. <Funktech> *(BE)* radio aerial, *(AE)* radio antenna; 2. <Kfztech> *(BE)* radio aerial, *(AE)* radio antenna *(Zubehör)*; 3. <Trans> *(BE)* radio aerial, *(AE)* radio antenna
Funkarmbanduhr f <Funktech> radio-controlled wrist watch
Funkbake f <Funkort, Lufttrans, Wassertrans> radio beacon *(Seezeichen)*
Funkbereich m <Funktech, Lufttrans, Trans> radio range
Funkbereichskennzahl f <Telekom> area code
Funkbetrieb m <Raumfahrt> operation *(Weltraumfunk)*
Funkbrücke f <Lufttrans, Trans, Wassertrans> radio link
Funkdienst m 1. <Raumfahrt> service *(Weltraumfunk)*; 2. <Wassertrans> radio duties *(Wachdienst)*
Funke m <Elektriz, Elektrotech, Ker & Glas, Phys> spark
Funkeinrichtung f <Funktech, Lufttrans, Trans, Wassertrans> radio facility
Funkeleffekt m <Elektronik> Schottky effect
Funkelfeuer n <Wassertrans> quick-flashing light *(Seezeichen)*
Funkeln n <Textil> scintillation
Funkelrauschen n 1. <Elektronik> flicker noise *(Halbleiterrauschen)*; 2. <Funktech> flicker noise
Funkempfänger m 1. <Funktech, Phys> radio receiver; 2. <Telekom> receiver *(Radio)*
Funken m <Elektriz> spark
Funken n <Funktech> radio
Funkenbildung f <Elektrotech, Fertig> sparking
Funkenentladung f <Phys> spark discharge
funkenerodieren v <Elektriz> spark-erode
Funkenerosion f <Maschinen> spark erosion
Funkenerosionsbearbeitung f <Maschinen> electro-spark machining, spark machining
funkenfrei adj <Elektrotech> nonarcing
funkenfreie Kommutierung f <Elektrotech> sparkless commutation
Funkeninduktor m <Elektrotech> spark coil
Funkenkammer f <Phys, Teilphys> spark chamber
Funkenkondensator m <Elektrotech> spark capacitor
Funkenlänge f <Elektrotech> sparking distance
Funkenlöscher m <Elektriz, Elektrotech> spark arrester, spark blow out, spark extinguisher, spark quencher, spark suppressor
Funkenlöschspule f <Elektrotech> blowout coil

Funkenlöschung f <Elektrotech> spark quenching, spark suppression
Funkenprüfung f <Werkprüf> spark test, spark testing
Funkenregen m <Bau> shower of sparks
funkensicher adj <Sicherheit> non-sparking, spark-proof (Werkzeug)
funkensichere Ausrüstung f <Sicherheit> spark-proof equipment
funkensicheres Werkzeug n <Sicherheit> non-sparking tool, spark-proof tool
Funkensieb n <Eisenbahn> chimney netting
Funkenspektrum n <Strahlphys> spark spectrum
Funkenstrecke f 1. <Elektriz, Elektrotech> arcing air gap, discharge gap, spark absorber, spark gap, sparking distance; 2. <Kfztech> spark plug gap; 3. <Phys> spark gap
Funkenstreckendurchschlag m <Elektriz, Elektrotech> spark gap breakdown
Funkenstreckenmodulation f <Elektriz, Elektrotech> spark gap modulation
Funkenstreckenüberschlag m <Elektriz, Elektrotech> spark gap flash-over
Funkentstörung f <Aufnahme, Elektrotech> noise suppression
Funkenüberschlag m 1. <Elektrotech> arc-over; 2. <Thermod> flashover
Funkenunterdrückung f <Elektrotech> spark suppression
Funkenzähler m <Strahlphys> spark counter
Funkenzündung f <Kfztech> spark ignition
Funker m <Wassertrans> radio operator
Funkfehlweisung f <Lufttrans> quadrantal error
Funkfelddämpfung f <Telekom> transmission loss (Richtfunk)
Funkfeldern/mit zwei <Funktech, Telekom> two-hop
Funkfeldlänge f <Funktech, Telekom> hop length
Funkfernschreiben n (RTTY) <Funktech> radioteletype (RTTY)
Funkfernsprechen n 1. <Funktech> wireless telephony; 2. <Telekom> radiotelephony, wireless telephony
Funkfernsprechteilnehmer m <Mobilkom, Telekom> mobile radio subscriber, wireless subscriber
funkfernsteuern v <Funktech, Trans> radioguide
Funkfernsteuerung f <Funktech, Lufttrans, Trans, Wassertrans> radio control, radio remote control, radio steering
Funkfeststation f <Mobilkom> base station
Funkfeuer n 1. <Funkort> radio beacon; 2. <Funktech> beacon, radio range; 3. <Lufttrans> beacon, radio beacon, radio range; 4. <Strahlphys> radio beacon; 5. <Trans> radio range; 6. <Wassertrans> radio range; radio beacon (Seezeichen)
Funkfrequenz f (HF) <Aufnahme, Elektronik, Fernseh, Funktech, Telekom> radio frequency (RF)
Funkfrequenzbereich m <Funktech, Lufttrans, Trans> radio range
Funkfrequenzspektrum n <Elektronik, Funktech> radio spectrum
Funkgerät n <Funktech, Lufttrans, Trans, Wassertrans> radio equipment
Funkgeräusch n <Elektronik> radio noise
funkgesteuert adj <Funktech, Trans> radio-controlled
Funkhorizont m <Funktech> quasi-optical horizon, radio horizon
Funkkanal m <Lufttrans> radio channel
Funkkompass m 1. <Funkort, Lufttrans> radio compass; 2. <Phys, Strahlphys> radiogoniometer; 3. <Wassertrans> radio compass
Funkkontakt m <Wassertrans> radio contact

Funkkonzentrator m <Mobilkom> base station
Funk-LAN n <Telekom> radio LAN, radio local area network, wireless LAN, wireless local area network
Funkleitstrahl m 1. <Funkort> glide path beam, localizer beam, radio beacon; 2. <Lufttrans> glide path beam, radio beacon; localizer beam (Flugwesen); 3. <Strahlphys> radio beam; 4. <Wassertrans> radio beacon
Funkleitung f <Lufttrans> radio guidance
Funkloch n <Funktech> radio fade-out
Funkmast m <Funktech, Telekom> radio transmission mast
Funkmessstation f <Funkort> reconnaissance radar station
Funkmessverfahren n <Funkort> radio detecting and ranging
Funknavigation f <Funkort, Lufttrans, Trans, Wassertrans> radio navigation
Funknavigationssystem n <Funktech, Wassertrans> long-range navigation system, long-range radio navigation system
Funknetz n <Telekom> radio network
Funknetz n **mit Pulsamplitudenmodulation** <Telekom> pulse amplitude modulation network
Funknotrufsystem n <Telekom> distress radio call system
Funkortung f 1. <Funkort> radar control, radiolocation, radio position fixing; 2. <Lufttrans> radar control; 3. <Strahlphys> radiolocation; 4. <Trans> radar control
Funkpeilantenne f <Funkort, Trans, Wassertrans> radio direction finding antenna, RDF antenna
Funkpeiler m <Funkort, Phys, Trans, Wassertrans> radio direction finder
Funkpeiler m **mit Sichtanzeige** <Funkort> visual display direction finder
Funkpeilgerät n <Funkort, Lufttrans, Phys, Trans, Wassertrans> automatic direction finder (ADF); direction finder, radio compass, radio direction finder, radiogoniometer
Funkpeilgerätantenne f <Funkort, Lufttrans, Trans, Wassertrans> radio direction finder antenna
Funkpeilkompass m <Funkort, Lufttrans, Wassertrans> radio compass
Funkpeilrahmen m <Funkort> radio direction finder frame
Funkpeilstelle f <Funkort, Lufttrans, Wassertrans> radio direction finding station
Funkpeiltechnik f <Phys, Strahlphys> radiogoniometry
Funkpeilung f 1. <Funkort> direction finding, radio bearing, radio direction finding; 2. <Funktech, Lufttrans, Trans, Wassertrans> bearing, radio bearing; radio direction finding (RDF); radio position fixing
Funkraum m <Wassertrans> radio room
Funkreichweite f <Funktech, Lufttrans, Trans> radio range
Funkruf m <Telekom> paging, radiopaging
Funkrufempfänger m **mit alphanumerischer Anzeige** <Telekom> alphanumeric pager
Funkruffunktion f <Telekom> paging facility
Funkrufsystem n **mit Tonfolgecodierung** <Telekom> sequential tone coded radio-paging system
Funkseitenpeilung f 1. <Funkort> relative bearing; 2. <Lufttrans> curved azimuth approach path, relative bearing
Funksender m <Funktech, Phys> radio transmitter
Funksignal n 1. <Elektronik, Funktech> radio signal; 2. <Raumfahrt> (AE) signaling, (BE) signalling (Weltraumfunk)
Funksonde f <Telekom> radiosonde
Funkspektrum n <Elektronik, Funktech> radio spectrum
Funksprechanlage f <Lufttrans, Trans, Wassertrans> radio link

Funksprechgerät

Funksprechgerät *n* <Funktech, Mobilkom, Wassertrans> radiotelephone
Funkstandort *m* <Lufttrans, Trans, Wassertrans> radio fix *(Navigation)*
Funkstation *f* <Raumfahrt> station *(Weltraumfunk)*
Funksteuerung *f* <Lufttrans> radio guidance
Funkstrecke *f* <Funktech> radio path, radio link
Funkstreckendämpfung *f* <Funktech> path attenuation
Funkstreckenprofil *n* <Funktech> path profile
Funkstreifenwagen *m* <Trans> radio patrol car
Funktagebuch *n* <Funktech> log
Funktaxi *n* <Kfztech> radio taxicab
Funktechnik *f* <Funktech, Lufttrans, Trans, Wassertrans> radio engineering
Funkteilnehmer *m* <Telekom> radio subscriber • **vom Funkteilnehmer abgehend** <Mobilkom> mobile-originated
Funkteilsystem *n* <Telekom> radio subsystem
Funktelefon *n* <Funktech, Trans> radiophone, radio telephone
Funktelefon *n* **mit dekadischer Frequenzeinstellung** <Funktech> synthesized radio telephone
Funktelefondienst *m* *(FuTelD)* <Mobilkom> mobile telephone service
funktelefonisch anrufen *v* <Funktech, Trans> radiophone
Funktelegrafie *f* <Funktech> radio telegraphy
Funktion *f* 1. <Comp & DV> function; feature *(Programm)*; 2. <Elektronik> function; 3. <Maschinen> action; 4. <Math, Patent, Qual> function • **mit vielen Funktionen** <Comp & DV> feature-rich
Funktion *f* **auf zwei Ebenen** <Elektrotech> bilevel operation
Funktion *f* **dritten Grades** <Math> cubic function
Funktional *n* <Math> functional *(Funktionalanalysis)*
funktional aufgeteiltes System *n* <Telekom> functionally divided system
Funktionalanalysis *f* <Math> functional analysis
funktionale Sprache *f* <Comp & DV> functional language
Funktionalität *f* <Comp & DV> functionality, capability
funktionelle Nachprüfung *f* <Qual> functional verification
funktioneller Entwurf *m* <Bau, Comp & DV> functional design
funktionelles Design *n* <Bau, Comp & DV> functional design
funktionieren *v* <Maschinen> operate, work
Funktionsablauf *m* <Fertig> operational sequence
funktionsbedingtes Maß *n* <Konstzeich> functionally significant dimension
funktionsbeeinflussende Wartung *f* <Telekom> function-affecting maintenance
Funktionsbereich *m* <Comp & DV> functional area
funktionsbezogene Bezugsebene *f* <Konstzeich> functionally important datum plane
funktionsbezogene Maßeintragung *f* <Konstzeich> function-related dimensioning
Funktionsblock *m* <Comp & DV> functional block
Funktionscode *m* <Comp & DV> function code, function digit
Funktionsdiagramm *n* <Comp & DV> functional diagram
Funktionseinheit *f* 1. <Comp & DV> functional unit; 2. <Telekom> element, module
Funktionsempfänger *m* <Elektrotech> synchro transformer
funktionsfähige Windgeschwindigkeit *f* 1. <Funktech> survival wind speed *(Richtantenne)*; 2. <Nichtfoss Energ> survival wind speed
Funktionsgeber *m* <Elektronik> function generator

Funktionsgraph *m* <Künstl Int> function graph
Funktionskontrolle *f* <Telekom> functional test
Funktionsplan *m* 1. <Konstzeich> function plan; 2. <Maschinen> functional diagram
Funktionsprüfung *f* 1. <Comp & DV, Qual> functional test; 2. <Telekom> performance test
Funktionsschema *n* 1. <Konstzeich> function diagram; 2. <Maschinen> functional diagram
funktionssicher *adj* <Fertig, Sicherheit> reliable
funktionssicheres Bauteil *n* <Sicherheit> reliable component
Funktionstabelle *f* <Comp & DV> truth table
Funktionstaste *f* <Comp & DV> dead key, function key
Funktionsteilung *f* <Telekom> function sharing
Funktionstüchtigkeit *f* <Kerntech> serviceability *(eines Brennstabs)*
Funktionsverlauf *m* <Regelung> course of the function
Funktionswähler *m* <Elektriz> function selector
Funktionsweise *f* <Maschinen> mode of operation
Funktionszeichen *n* <Comp & DV> functional character
Funktotalschwund *m* <Funktech> radio fade-out
Funkübertragung *f* <Funktech, Telekom> transmission by radio
Funküberwachungsdienst *m* <Funktech, Telekom> interception service, radio monitoring
Funkuhr *f* <Funktech> radio clock, radio-controlled clock
Funkverbindung *f* 1. <Funktech, Lufttrans> radio link; 2. <Telekom> radiocommunication; 3. <Trans> radio link; 4. <Wassertrans> radio contact, radio link
Funkverbindung *f* **Bord-Boden** <Telekom> air-ground communication
Funkverbindung *f* **Erde-Weltraum** <Funktech> Earth-space radio link
Funkverbindung *f* **über Meteorschwarmreflektion** <Raumfahrt> meteor burst communication, meteor scatter communication *(Weltraumfunk)*
Funkverkehr *m* 1. <Funktech> radio communication; 2. <Telekom> radio communication; 3. <Trans, Wassertrans> radio communication
Funkverkehrsleitung *f* <Lufttrans, Trans, Wassertrans> radio link
Funkvermittlungsstelle *f* *(FuVst)* <Mobilkom> *(AE)* mobile services switching center, *(BE)* mobile services switching centre
Funkversorgung *f* <Telekom> coverage *(Radio)*
Funkwelle *f* 1. <Elektronik> radiowave; 2. <Funktech, Phys, Telekom> radio wave
funkwellenundurchlässig *adj* <Strahlphys> radio-opaque
Funkzelle *f* <Mobilkom> cell *(zellulares Netz)*
Funkzellenwechsel *m* <Mobilkom> change-over
Funkzielanflug *m* <Funktech, Lufttrans> radio homing
Funkzone *f* <Mobilkom> cell *(zellulares Netz)*
Funkzonengrenze *f* <Mobilkom> cell boundary
Funkzonengruppe *f* <Mobilkom, Telekom> cluster
Funkzonenwechsel *m* <Mobilkom> cell change
Funkzugangskanal *m* <Telekom> radio access bearer
Funkzugangsnetz *n* <Telekom> radio access network
Furaldehyd *m* <Kunststoff> furfuraldehyde
Furan *n* <Chemie> furan
Furche *f* 1. <Maschinen> ridge; 2. <Mechan, Optik> groove *(Beugungsgitter)*
furchen *v* <Bau, Fertig, Maschinen> ridge
Furfural *n* <Kunststoff> furfural
Furfuran *n* <Chemie> furan
Furfuryl... <Chemie> furfuryl
Furil *n* <Chemie> furil
Furil... <Chemie> furilic
Furling-Geschwindigkeit *f* <Nichtfoss Energ> Furling speed

Furnier n <Bau> veneer
Furnieren n <Bau> veneering
Furnierholz n <Bau> veneer
Furnierplatte f <Fertig> ply
Furnierpresse f <Maschinen> veneering press
Furniersäge f <Fertig> veneer saw
Fusarc-Verfahren n <Fertig> fusarc process
Fuselöl n <Lebensmittel> fusel oil
Fusion f <Ergon, Phys> fusion
Fusionenergie f <Kohlen> fusion energy
Fusionsprozesse mpl 1. <Kerntech> nuclear fusions; 2. <Teilphys> nuclear fusion
Fusionsreaktion f <Kerntech> thermonuclear reaction
Fusionsreaktor m **mit Linearbeschleuniger** <Kerntech> LADR, linear accelerator-driven reactor
Fusions-Spleißmaschine f <Fertig> fusion splicer (Faseroptik)
Fusionswelle f <Kerntech> thermonuclear combustion wave
Fuß m 1. <Bau> foot (Maßeinheit); bottom (eines Hügels); toe; 2. <Druck> base; 3. <Fertig> patten (Säule, Schiene); 4. <Ker & Glas> root; 5. <Maschinen> base, leg, root; 6. <Metrol, Wassertrans> foot (Maßeinheit)
Fuß m **pro Sekunde** <Metrol> foot per second
Fuß m **pro Sekunde-Quadrat** <Metrol> foot per second squared
Fußausrundung f <Maschinen> fillet
Fußbalken m <Bau> sole plate
Fußbedienung f <Maschinen> foot-operated control
Fußbetätigung f <Maschinen> foot-operated control
fußbetriebener Rollenschneider m <Ker & Glas> foot-operated score
Fußboden m <Bau> floor
Fußbodenbelag m <Bau> flooring
Fußbodenfliese f <Bau> flooring tile
Fußbodenheizung f 1. <Bau> screed heating; 2. <Heiz & Kälte> floor heating, underfloor heating; 3. <Thermod> underfloor heating
Fußbodennagel m <Bau> flooring nail
Fußbodenplatte f 1. <Bau> flooring tile; 2. <Ker & Glas> floor tile
Fußbodenquerträger m <Eisenbahn> floor beam
Fußbodenspeicherheizung f <Heiz & Kälte> thermal storage floor heating
Fußbremse f <Kfztech> brake pedal, foot brake
Fußbrett n <Bau> toeboard
fusselfreie Kleidung f <Mechan> lint-free cloth
Fußfläche f <Maschinen> root surface
Fußgänger m <Trans> pedestrian • **durch Fußgänger gelenkt** <Trans> pedestrian-controlled
fußgängerbetätigtes Signal n <Trans> pedestrian-actuated signal
Fußgängerbrücke f <Trans> pedestrian bridge
Fußgängerphase f <Trans> pedestrian phase
Fußgängerschutzinsel f <Bau> refuge
Fußgängerschutzstreifen m <Bau> refuge
Fußgängertunnel m <Trans> pedestrian underpass
Fußgängerüberführung f <Eisenbahn> footbridge
Fußgängerverkehr m <Trans> pedestrian traffic
Fußgangschaltung f <Kfztech> foot change
Fußhebel m 1. <Ker & Glas> foot boards; 2. <Maschinen> treadle
Fußhebeldämpfungsgestänge n <Lufttrans> pedal damper assembly (Flugzeug)
Fußhöhe f 1. <Fertig> dedendum (Getriebelehre); 2. <Maschinen> dedendum
Fußholz n <Bau> bottom rail, ground plate, groundsill, sole piece; pole plate (Dachkonstruktion)
Fußkreis m <Maschinen> root line

Fußkreisdurchmesser m <Maschinen> dedendum circle, root diameter
Fußkreislinie f <Maschinen> dedendum line
Fußlager n <Maschinen> footstep bearing
Fußleiste f <Bau> base, base board, (AE) baseboard, (AE) mopboard, (BE) skirting board, skirting
Fußpedal n <Kfztech> footpedal
Fußpfette f <Bau> wall plate
Fußplatte f 1. <Bau> sole plate; 2. <Wassertrans> heel plate (Schiffbau)
Fußpumpe f <Maschinen> foot pump
Fußpunkt m 1. <Funktech> base (Antenne); 2. <Metrol> foot; 3. <Telekom> base (Antenne)
Fußraste f <Kfztech> foot rest (Motorrad)
Fußraum m <Kfztech> footwell
Fußrohr n <Wasserversorg> tail pipe
Fußschalterkontakt m <Elektriz> floor contact switch
Fußschalthebel m <Kfztech> foot change lever
Fußschicht f <Ker & Glas> eaves course
Fußschraube f <Maschinen> foot screw
Fußträger m <Ker & Glas> foot carrier
Fußventil n <Maschinen> foot valve
Fußweg m <Bau> footway
Futter n 1. <Bau> lining; 2. <Kerntech> liner; 3. <Maschinen> chuck; 4. <Mechan> lining
Futter n **mit einzeln verstellbaren Backen** <Maschinen> independent chuck
Futter n **mit vier Einzelverstellbacken** <Maschinen> four-jaw independent chuck
Futterautomat m <Maschinen> automatic chucking lathe, chucker, chucking automatic lathe
Futterholz n <Bau> furring, furring piece
Futterplatte f <Maschinen> back plate, chuck plate
Futterrohr n <Erdöl> casing • **Futterrohre absetzen** <Erdöl> set casing
Futterscheibe f <Fertig> back plate (Spannfutter)
Futterschutz m <Maschinen> chuck guard
Futterspannung f <Fertig> chucking (Metallspanung)
Futterstoff m <Textil> lining fabric
Fuzzy-Logik f <Comp & DV, Künstl Int> fuzzy logic
FVSt (Fernvermittlungsstelle) <Telekom> (AE) toll exchange, (BE) trunk exchange, (AE) trunk switching center, (BE) trunk switching centre
f-Zahl f <Phys> f-number

G

g 1. <Chemie, Labor, Phys> (Gramm) g (gram); 2. <Kerntech, Phys> (gyromagnetisches Verhältnis) g (gyromagnetic ratio); 3. <Phys> (statistisches Gewicht) g (statistical weight); 4. <Raumfahrt> (Erdbeschleunigung) g (gravitational acceleration)
G 1. <Aufnahme, Elektriz> (Gauß) G (gauss); 2. <Elektronik, Ergon, Fernseh, Funktech, Raumfahrt, Telekom> (Gewinn) G (gain); 3. <Maschinen, Phys> (Schermodul) G (shear modulus); 4. <Metrol> (Giga…) G (giga); 5. <Phys, Thermod> (Gibbs'sche Funktion) G (Gibbs function)
Ga (Gallium) <Chemie> Ga (gallium)
GaAs (Galliumarsenid) <Elektronik, Funktech, Optik, Phys> GaAs (gallium arsenide)
GaAs-Laser m <Phys> GaAs laser

Gabardine

Gabardine *m* <Textil> gaberdine
Gabel *f* 1. <Hydraul> gab; 2. <Ker & Glas> fork; 3. <Kfztech> fork *(Motorradgetriebe)*; yoke *(Universalgelenk)*; 4. <Maschinen> fork; 5. <Mechan> bracket
Gabel *f* mit Bremsnickausgleich <Kfztech> antidive fork *(Motorrad)*
Gabelbein *n* <Kfztech> fork leg *(Motorrad)*
Gabelgelenk *n* <Maschinen> knuckle joint
gabelgesteuerte Ventilbewegung *f* <Hydraul> hook gear valve motion
Gabelhebel *m* <Maschinen> fork lever, forked lever
Gabelkabelschuh *m* <Elektrotech> fork-type cable lug
Gabelkopf *m* 1. <Maschinen> fork head, yoke; 2. <Mechan> clevis, yoke
Gabellehre *f* <Metrol> *(AE)* caliper gage, *(BE)* calliper gauge
gabeln *v* <Fertig> bisect
Gabelpfanne *f* <Fertig> shank ladle *(Gießen)*
Gabelpleuel *n* <Maschinen> fork-end connecting rod
Gabelpleuelstange *f* <Maschinen> connecting rod with fork end
Gabelpunkt *m* <Telekom> two-to-four wire transition point
Gabelrohr *n* 1. <Bau> forked pipe; 2. <Mechan> Y-tube
Gabelrückholfeder *f* <Kfztech> fork return spring *(Kupplung)*
Gabelschaltung *f* <Telekom> hybrid circuit
Gabelschlüssel *m* 1. <Fertig> *(BE)* spanner, wrench *(Kunststoffinstallationen)*; 2. <Maschinen> engineer's wrench, *(BE)* face spanner, face wrench, fork wrench, *(BE)* open spanner, open wrench, *(BE)* open-end spanner, open-end wrench
Gabelschraubenschlüssel *m* <Mechan> *(BE)* open-end spanner, open-end wrench
Gabelschweißung *f* <Fertig> cleft weld
Gabelspanneisen *n* <Fertig> U-clamp
Gabelstapler *m* <Bau, Trans, Verpack> forklift, forklift truck
Gabelstapler-an-Gabelstapler-Umschlag *m* <Trans> *(AE)* truck-to-truck handling, *(AE)* truck-to-truck operation
Gabelstaplerbetrieb umstellen *v/auf* <Trans> palletize
Gabelstößel *m* <Kfztech> fork push rod *(Kupplung)*
Gabelstück *n* <Bau> Y branch
Gabelübertrager *m* <Telekom> hybrid transformer *(z. B. Telefon)*
Gabelumschalter *m* <Telekom> cradle switch; switch hook *(Telefon)*
Gabelunsymmetrieverlust *m* <Telekom> balance return loss, line balancing error loss
Gabelverbindung *f* <Elektriz> forked connection
Gabelzinke *f* <Maschinen> fork arm
Gablonzer-Ware *f* <Ker & Glas> Gablonz glassware
Gadolinium *n* (Gd) <Chemie> gadolinum *(Gd)*
Gaffel *f* <Wassertrans> gaff *(Segeln)*
Gaffelklaue *f* <Wassertrans> crutch *(Segeln)*
Galactan *n* <Chemie> galactan, gelose
Galacton... <Chemie> galactonic
Galactosamin *n* <Chemie> aminodeoxy-D-galactose, galactosamine
Galactosan *n* <Chemie> galactan, gelose
galaktische Wolke *f* <Raumfahrt> galactic cloud
Galaktose *f* <Lebensmittel> galactose *(Milchsaccharid)*
Galaxienhaufen *m* <Raumfahrt> galaxy cluster
Galilei'sche Transformation *f* <Phys, Strömphys> Galilean transformation
Galilei'sches Bezugssystem *n* <Phys> Galilean frame
Galilei'sches Fernrohr *n* <Phys> Galilean telescope
Galilei'sches Teleskop *n* <Phys> Galilean telescope
Gallat *n* <Chemie> gallate
Galle *f* <Ker & Glas> gall
Gallein *n* <Chemie> gallein
Gallert *f* <Verpack> gelatine

gallertartiger Bohrschlamm *m* <Erdöl> colloidal mud
Gallertbildung *f* <Chemtech> gelation, jellification
Gallium *n* (Ga) <Chemie> gallium *(Ga)*
Galliumarsenid *n* (GaAs) <Elektronik, Funktech, Optik, Phys> gallium arsenide *(GaAs)*
Galliumarsenid-Chip *m* <Elektronik> gallium arsenide chip
Galliumarsenid-Diode *f* <Elektronik> gallium arsenide diode
Galliumarsenid-Laser *m* <Strahlphys> gallium arsenide laser
Galliumarsenid-Logik *f* <Elektronik> gallium arsenide logic
Galliumarsenid-MOS-Transistor *m* <Elektronik> gallium arsenide MOS transistor
Galliumarsenid-Solarzelle *f* <Elektronik> gallium arsenide solar cell
Galliumarsenid-Trägermaterial *n* <Elektronik> gallium arsenide substrate
Gall-Kette *f* <Maschinen> plate link chain
Gallone *f* <Phys> gallon *(Volumeneinheit)*
Gallonenflasche *f* <Ker & Glas> gallon jug
Gallotannin *n* <Chemie> gallotannin
Gallusgerbsäure *f* <Chemie> gallotannic acid
Gallussäure *f* <Chemie> gallic acid
Galvanisation *f* 1. <Chemie> voltaization; 2. <Elektrotech> electroplating
galvanisch *adj* <Chemie, Elektriz> galvanic *(Elektrochemie; z. B. Strom)*; voltaic; electroplated *(Überzug)*
galvanisch aktiv *adj* <Anstrich> galvanically sacrificial, sacrificial
galvanisch gekoppelter Verstärker *m* <Elektronik> DC-coupled amplifier, direct-coupled amplifier
galvanisch getrennt *adj* <Elektriz> indirect connected, indirect-coupled
galvanisch isolierte Stromversorgung *f* <Elektriz> floating supply
galvanisch leitende Verbindung *f* <Nichtfoss Energ> ohmic contact
galvanisch verbunden *adj* direct-connected, direct-coupled
galvanisch versilbern *v* <Elektrotech> electrosilver
galvanisch verzinken *v* <Fertig> electrogalvanize
galvanische Anode *f* <Erdöl> sacrificial anode
galvanische Isolierung *f* <Elektrotech> galvanic isolation
galvanische Kette *f* <Chemie> galvanic cell
galvanische Kopplung *f* 1. <Elektrotech> direct coupling; 2. <Elektrotech> direct coupling, galvanic couple
galvanische Versilberung *f* <Elektrotech> electrosilvering
galvanische Zelle *f* <Elektrotech> galvanic cell
galvanischer Strom *m* <Elektrotech> galvanic current
galvanischer Überzug *m* <Fertig> electroplate • mit galvanischem Überzug versehen <Metall> plate
galvanischer Vorgang *m* <Anstrich> galvanic action
galvanisches Element *n* <Chemie, Elektriz, Elektrotech> galvanic cell, voltaic cell
galvanisches Verbleien *n* <Fertig> lead plating, terne plating
galvanisches Vernickeln *n* <Fertig> nickel plating
galvanisches Verzinken *n* <Fertig> electrogalvanizing
galvanisieren *v* <Elektrotech> electroplate
Galvanisieren *n* 1. <Anstrich> plating; 2. <Elektrotech> electroplating, plating; 3. <Fertig> electroplating
galvanisiert *adj* 1. <Anstrich, Elektriz> electroplated; 2. <Fertig> plated; 3. <Mechan> galvanized • im galvanisierten Zustand <Fertig> as-deposited
Galvanisierung *f* 1. <Bau> electrodeposition; 2. <Elektriz> electroplating
Galvano *n* <Druck> electro, electrotype
Galvanometer *n* <Elektriz, Elektrotech, Labor, Phys> galvanometer

Galvanometermesswerk n <Gerät> galvanometer movement
Galvanometer-Nebenschlusswiderstand m <Elektriz> galvanometer shunt
Galvanometer-Shunt m <Elektrotech> galvanometer shunt
Galvanoplastik f 1. <Elektrotech> electroforming; 2. <Fertig> electroform
galvanoplastisch adj <Fertig> galvanoplastic
Galvanotechnik f <Fertig> electroplating industry (Kunststoffinstallationen)
Gamaschen fpl <Sicherheit> gaiters
Gambir m <Chemie> gambier
Gamma n <Fernseh> gamma
Gamma-Abschirmung f <Kerntech> gamma shield
Gamma-Anpassung f <Funktech> gamma match (Antenne)
Gamma-Aufheizung f <Kerntech> gamma heating
Gamma-Eigenschaft f <Elektronik> gamma characteristic (Bildaufnahmeröhre)
Gamma-Eisen n <Metall> gamma iron
Gamma-Erzbreikonzentrationsmessgerät n <Kerntech> gamma ore pulp content meter
Gamma-Escape-Peak n <Strahlphys, Teilphys, Wellphys> gamma ray escape peak
Gamma-Fehler m <Fernseh> gamma error
Gamma-Gamma Log n <Erdöl> gamma-gamma log (Bohrlochmessung)
Gamma-Korrektur f <Fernseh> gamma corrector
Gamma-Kurve f <Elektronik> gamma characteristic (Bildaufnahmeröhre)
Gamma-Log n <Erdöl> gamma-gamma log, gamma ray log (Bohrlochmessung)
gammametrische Qualitätsbestimmung f des Erzes <Kerntech> gammametric ore assaying
Gammaquant n <Strahlphys, Teilphys, Wellphys> gamma particle, gamma quantum
Gammaradiographie f <Strahlphys, Teilphys, Wellphys> gamma radiation
Gamma-Rückstreumethode f <Kerntech> gamma backscatter method
Gamma-Spektrometer n <Strahlphys, Teilphys, Wellphys> gamma ray spectrometer
Gammaspektrum n <Strahlphys, Teilphys, Wellphys> gamma ray spectrum
Gammastrahl m 1. <Elektronik> gamma ray; 2. <Strahlphys, Teilphys, Wellphys> gamma beam, gamma ray
Gammastrahlen mpl <Strahlphys, Teilphys, Wellphys> gamma radiation
Gammastrahlenabsorptionsanalyse f <Strahlphys, Teilphys, Wellphys> gamma ray absorption analysis
Gammastrahlenaktivierungsanalyse f <Strahlphys> photoactivation analysis
Gammastrahlenaufheizung f <Strahlphys, Teilphys, Wellphys> gamma ray heating
Gammastrahlenaufzeichnung f <Nichtfoss Energ> gamma ray log
Gammastrahlenbohrlochmessung f <Erdöl> gamma ray well logging (Messtechnik)
Gammastrahlendetektor m <Gerät> gamma radiation detector
Gammastrahlendickenmessgerät n <Gerät> gamma thickness meter
Gammastrahlenfilm m <Strahlphys, Teilphys, Wellphys> gamma film (im Strahlenschutz)
Gammastrahlenkonstante f <Strahlphys, Teilphys, Wellphys> gamma constant
Gammastrahlenspektrometer n <Strahlphys, Teilphys, Wellphys> gamma ray spectrometer

Gammastrahlung f 1. <Elektriz, Elektronik> gamma radiation, gamma ray; 2. <Strahlphys> gamma emission, gamma radiation; 3. <Teilphys, Wellphys> gamma emission, gamma radiation, gamma ray
Gammastrahlungsüberwachung f <Kerntech> gamma ray survey
Gammastrahlverfahren n <Strahlphys, Teilphys, Wellphys> gamma radiation
Gammaübergang m <Strahlphys, Teilphys, Wellphys> gamma ray transformation
Gammaumwandlung f <Chemie> vitrification (Gummiherstellung)
Gammexan n <Chemie> gammexane
Gang m 1. <Comp & DV> running; 2. <Fertig> channel (Extruderschnecke); start (Schnecke); flight (Schraubenlinie); 3. <Maschinen> action, operation, run, running, work, working; 4. <Mechan> gear; 5. <Telekom> response • **Gang einlegen** <Kfztech> put into gear • **in Gang setzen** <Ergon, Kerntech, Maschinen> set into operation
Gangerz n <Kohlen> gangue mineral
Ganghöhe f 1. <Elektriz, Elektrotech> pitch; 2. <Fertig> screw pitch; 3. <Maschinen> lead, pitch; 4. <Mechan> pitch
Ganglinie f <Wasserversorg> hydrograph
Gangschalthebel m <Kfztech> (BE) gear lever, (AE) gear shift (Getriebe)
Gangschaltung f 1. <Kfztech> gear change, (BE) gear lever, (AE) gear shift (Getriebe); 2. <Maschinen> speed-changing device, speed-changing mechanism; 3. <Mechan> (BE) gear lever, (AE) gear shift
Gangwechsel m <Kfztech> gear change (Getriebe)
Gangwechselgeschwindigkeit f <Kfztech> gear change rate
Gangwerk n <Maschinen> motion
Gangzahl f <Maschinen> number of starts
Gänseaugenstoff m <Textil> diaper (Jacquardgewebe)
ganze Zahl f <Comp & DV, Math> integer
ganzer Ton m <Akustik> whole tone
Ganzglasfaser f <Telekom> (AE) all-glass fiber, (BE) all-glass fibre
Ganzjahresöl n <Kfztech> multigrade oil
ganzjährige Klimatisierung f <Heiz & Kälte> all-year air conditioning
Ganzkörper... <Kerntech> whole-body
Ganzkörperbestrahlung f <Kerntech> whole-body irradiation
Ganzkörperschutzmittel n <Sicherheit> whole-body suit
Ganzkörperschwingung f <Sicherheit> whole-body vibration
Ganzkörperspektrometer n <Kerntech> whole-body counter
Ganzkörperzähler m <Kerntech> HBC, human-body counter, whole-body counter
Ganzlochwicklung f <Elektrotech> integer-slot winding, integral-slot winding
Ganzpolteilwicklung f <Elektriz> full-pitch winding
Ganzschutzanzug m <Sicherheit> encapsulating suit, whole-body suit
Ganzseitenmontagesystem n <Druck> page composition system
Ganzstoffkasten m <Papier> stuff chest
Ganzzahl f <Comp & DV> integer
ganzzahliger Datentyp m <Comp & DV> integer type
ganzzahliger Spin m <Phys> integral spin
Ganzzeichen n <Druck> solid letter
Garage f <Trans> garage
Garantie f <Qual> guarantee
Garantiegewicht n <Lufttrans> guaranteed weight
Garantiekappe f <Verpack> guarantee cap

garantiert farbecht *adj* <Textil> guaranteed not to fade
garantierte Entnahme *f* <Nichtfoss Energ> guaranteed draw-off
garantierte Flugbahn *f* <Lufttrans> guaranteed flight path
garantierter Schub *m* <Lufttrans> guaranteed thrust
Garantieverschluss *m* <Verpack> guarantee closure
Gärbehälter *m* <Lebensmittel> fermenter
Gardine *f* <Textil> curtain
Gardinenbildung *f* <Kunststoff> sagging
gären *v* <Chemie> ferment *(Flüssigkeit)*
Garn *n* <Ker & Glas, Kunststoff, Papier, Textil> yarn • **im Garn färben** <Textil> yarn-dye
Garnapplikator *m* <Ker & Glas> yarn applicator
Garnelentrawler *m* <Wassertrans> shrimp trawler
Garnfärben *n* <Textil> yarn dyeing
Garnier *n* <Wassertrans> dunnage
Garnitur *f* 1. <Bau> set; 2. <Druck> series; 3. <Fertig> fittings *(Kunststoffinstallationen)*; 4. <Textil> set
Garnkörper *m* <Textil> package
Garnnummer *f* <Textil> size
Garnspule *f* <Ker & Glas, Textil> pirn
Garnsträhne *f* <Textil> hank
Garnstrang *m* <Textil> hank
Garnträger *m* <Textil> bobbin
Gartenblankglas *n* <Ker & Glas> horticultural glass
Gartenhäcksler *m* <Gerät> garden shredder
Gartenklarglas *n* <Ker & Glas> horticultural cast glass
Gärung *f* 1. <Chemie> fermentation *(Vorgang)*; 2. <Lebensmittel> fermentation
Gärungsgase *npl* <Abfall> fermentation gases
Gas *n* 1. <Heiz & Kälte, Phys> gas; 2. <Thermod> *(AE)* gas, *(AE)* gasoline • **Gas geben** 1. <Kfztech> accelerate; 2. <Lufttrans> advance throttle • **Gas wegnehmen** 1. <Kfztech> decelerate; 2. <Lufttrans> throttle back
Gasabsaugbohrung *f* <Kohlen> methane-draining boring
Gasabscheider *m* <Kerntech> gas stripper *(für Kühlwasser)*
Gasabsorption *f* <Erdöl> absorption of gases
Gasabzug *m* <Kerntech> gas vent
Gasanalyse *f* <Erdöl> gas analysis *(Messtechnik)*
Gasanalysegerät *n* <Erdöl> *(BE)* gas analyser, *(AE)* gas analyzer *(Messtechnik)*
Gasanalysengerät *n* **mit Wärmeleitfähigkeitszelle** <Gerät> *(BE)* conduction-of-heat gas analyser, *(AE)* conduction-of-heat gas analyzer
Gasanreicherung *f* <Thermod> gas enrichment
Gasanzeichen *n* <Erdöl> gas show *(Exploration)*
Gasanzünder *m* <Heiz & Kälte> gas lighter
gasarm *adj* <Kfztech> lean
gasarmes Gemisch *n* <Kfztech> poor mixture
Gasaufkohlen *n* <Thermod> gas carburizing
Gasaufkohlung *f* <Fertig> gas carburization
Gasausgang *m* <Kfztech> gas outlet
Gasausscheidung *f* <Metall> gas precipitate
Gasaustritt *m* <Erdöl> trip gas *(Bohrtechnik)*
Gasaustrittsöffnung *f* <Kfztech> gas outlet
Gasaustrittstemperatur *f* <Kerntech> gas outlet temperature
Gasbadeofen *m* <Heiz & Kälte> gas geyser
Gasbehälter *m* <Thermod> gas holder
gasbeheizt *adj* <Heiz & Kälte, Maschinen> gas-fired
gasbeheizter Ofen *m* <Heiz & Kälte, Thermod> gas-fired furnace
gasbenzinelektrisches Fahrzeug *n* <Kfztech> *(AE)* gas electric vehicle
gasbeständig *adj* <Thermod> gasproof
gasbetriebener Bus *m* <Kfztech> *(AE)* gas-fueled bus, *(BE)* gas-fuelled bus

gasbetriebener Motor *m* <Maschinen> gas engine
gasbetriebenes Auto *n* <Kfztech> *(AE)* gas-fueled car, *(BE)* gas-fuelled car
Gasbildungsvermögen *n* <Lebensmittel> gassing power *(beim Teig)*
Gasblase *f* 1. <Fertig> blister, gas cavity; 2. <Ker & Glas> air bell, bubble
Gasblasendurchflussmesser *m* <Labor> bubble flow meter
Gasblasenströmungsmesser *m* <Chemtech> *(AE)* bubble gage, *(BE)* bubble gauge
Gasboiler *m* <Thermod> gas boiler
Gasbrenner *m* 1. <Heiz & Kälte> gas burner; 2. <Thermod> gas burner, gas ring
Gasbrenner *m* **ohne Gebläse** <Thermod> atmospheric burner
gasbrenngeschnitten *adj* <Fertig> flame cut
Gasbrennschnitt *m* <Fertig> flame-cut
Gasbürette *f* <Labor> gas burette
Gaschromatograph *m* <Labor> gas chromatograph
Gaschromatographie *f* <Lebensmittel, Thermod> gas chromatography
Gasdetektor *m* <Thermod> gas leak detector
gasdicht *adj* 1. <Fertig> gastight; gastight *(Kunststoffinstallationen)*; 2. <Maschinen> gasproof; 3. <Mechan> gastight; 4. <Thermod> gasproof, gastight
gasdichte Brille *f* <Sicherheit> gas-tight goggles
gasdichte Schutzbrille *f* <Sicherheit> respirator spectacles
Gasdichteschreiber *m* <Gerät> gas density recorder
Gasdichtewaage *f* <Gerät> buoyancy gas balance
Gasdrossel *f* 1. <Kfztech> *(BE)* accelerator, *(AE)* gas pedal *(Vergaser)*; 2. <Mechan> gooseneck
Gasdrosselung *f* <Kerntech> gas baffle
Gasdruck *m* <Phys, Thermod> gas pressure
Gasdruckminderungsventil *n* <Thermod> gas pressure-reducing valve
Gasdruckregler *m* <Heiz & Kälte, Maschinen> gas pressure regulator
Gasdurchlässigkeit *f* <Kunststoff, Thermod> gas permeability
Gasdurchsatz *m* <Raumfahrt> gas flow *(Raumschiff)*
Gasdynamik *f* <Maschinen> gas dynamics
Gaseinpressung *f* <Erdöl> gas injection
Gaseinschluss *m* <Kerntech> gas cavity
gasen *v* <Textil> gas
Gasen *n* <Textil> singeing
Gasentladung *f* <Elektronik> gas discharge
Gasentladungslampe *f* <Elektriz> gas discharge lamp
Gasentladungsrelais *n* <Elektronik, Elektrotech> gas-filled relay
Gasentladungsröhre *f* <Elektronik> gas discharge tube, gas tube
Gasentladungsspalt *m* <Kerntech> gas discharge gap
Gasentladungsstrecke *f* <Kerntech> gas discharge gap
Gasentladungsventil *n* <Elektrotech> gas-filled rectifier
Gasentölungstrieb *m* <Erdöl> solution gas drive *(Fördertechnik)*
Gasentschwefelung *f* <Umweltschmutz> *(AE)* gas desulfurization, *(BE)* gas desulphurization
Gasentwickler *m* **nach Kipp** <Labor> Kipp's apparatus *(Generator)*
Gaserzeuger *m* <Thermod> gas generator
Gasexplosion *f* <Kohlen, Sicherheit> gas explosion
Gasfackel *f* <Thermod> gas flare *(zum Abfackeln)*
Gasfeder *f* <Maschinen> gas spring
Gasfeld *n* <Erdöl, Thermod> gas field
Gas-Fest-Chromatographie *f* <Umweltschmutz> gas-solid chromatography

Gasfeuerung f <Heiz & Kälte> gas-heating system
Gasfeuerungsautomat m <Heiz & Kälte> automatic gas-firing unit
Gasflamme getrocknet adj/mit <Fertig> skin-dried (Form)
Gasflasche f 1. <Kerntech> gas bottle; 2. <Maschinen> gas cylinder; 3. <Sicherheit> transportable gas container; 4. <Thermod> gas bottle, gas cylinder
Gas-Flüssigkeit-Chromatographie f <Thermod, Umweltschutz> gas-liquid chromatography
Gasfokussierung f <Elektronik> gas focusing (Strahlkonzentrierung durch Gasfüllung)
gasförmig adj <Phys, Thermod> gaseous
gasförmige Phase f <Thermod> gaseous phase
gasförmiger Abfall m <Umweltschutz> gaseous waste
gasförmiger Brennstoff m <Erdöl, Heiz & Kälte> gaseous fuel
gasförmiger Luftschadstoff m <Umweltschutz> gaseous air pollutant
gasförmiger luftverunreinigender Stoff m <Umweltschutz> gaseous air pollutant
gasförmiger Schadstoff m <Umweltschutz> gaseous pollutant
gasförmiger Zustand m <Thermod> gaseous phase
gasförmiges aktives Medium n <Elektronik> gaseous active medium
gasförmiges Medium n <Umweltschutz> gaseous medium
gasförmiges Verbrennungsprodukt n <Umweltschutz> gaseous combustion product
gasfrei adj <Thermod> free from gas
gasgefeuert adj <Heiz & Kälte> gas-fired
gasgefüllt adj <Thermod> gas-filled
gasgefüllte Diode f <Elektronik> gas diode
gasgefüllte Gleichrichterröhre f <Elektriz, Elektronik> gas-filled detector tube, gas-filled rectifier
gasgefüllte Photozelle f <Elektronik> gas phototube
gasgefüllte Röhre f <Elektronik> gas-filled tube
gasgefüllte Schaltröhre f <Elektronik> gas-filled switching tube
Gasgehalt m <Thermod> gas content
Gasgehalt m im Bohrschlamm <Erdöl> gas-cut mud (Tiefbohrtechnik)
gasgekohlt adj <Fertig> gas-carburized
gasgekühlt adj <Thermod> gas-cooled
gasgekühlter Brutreaktor m <Kerntech> GCBR, gas-cooled breeder reactor
gasgekühlter Reaktor m <Kerntech> gas-cooled nuclear power plant
Gasgenerator m <Trans> gas generator
gasgeschmiertes Lager n <Kerntech> gas-lubricated bearing
Gasgestänge n <Kfztech> throttle control rod; accelerator linkage, throttle linkage (Vergaser)
Gasgewinde n <Maschinen> gas threads
Gasgewinnung f aus Mülldeponien <Umweltschutz> landfill gas extraction
Gasgleichrichterröhre f <Elektrotech> gas-filled rectifier
Gasgleichung f <Thermod> gas equation
Gasgrenzschicht f <Elektrotech> gas boundary layer
gashaltig adj <Thermod> gassy
gashaltiges Öl n <Thermod> live oil
Gashebel m <Raumfahrt> throttle control (Raumschiff)
Gasheizung f 1. <Heiz & Kälte> gas-heating system; 2. <Thermod> gas fire (Gerät); gas heating (Heizen mit Gas) • **mit Gasheizung** <Maschinen> gas-fired
Gashinweis m <Erdöl> gas show
Gashydrat n <Chemie> gas hydrate (Erdgasaufbereitung)
gasiertes Garn n <Textil> gassed yarn

Gasinjektion f <Erdöl> gas injection (Ölgewinnung)
gasisoliertes Kabel n <Elektriz> gas-insulated line
Gasisolierung f <Elektrotech> gas insulation
Gaskältemaschine f <Heiz & Kälte> gas-refrigerating machine
Gaskanal m <Fertig> vent hole
Gaskappe f <Erdöl> gas cap (Lagerstättenkunde)
Gaskappentrieb m <Erdöl> gas cap drive (Ölförderung)
Gaskissen n <Kerntech> gas cushion
Gaskohle f <Kohlen> gas coal
Gaskoks m <Kohlen> gas-coke
Gaskonstante f (R) <Phys, Thermod> gas constant (R)
Gaskreislauf m 1. <Heiz & Kälte> gas circuit; 2. <Kerntech> gas circulation loop
Gaskühler m <Heiz & Kälte> gas cooler
Gaskühlschrank m <Thermod> gas refrigerator
Gaslagerstätte f <Erdöl> gas field
Gaslagerung f <Lebensmittel> gas storage
Gaslaser m 1. <Elektronik> gas dynamic laser, gas laser; 2. <Strahlphys> gas laser
Gasleitung f 1. <Bau, Erdöl> gas pipeline (Erdgastransport); 2. <Thermod> gas pipe
Gasleitungsleck n <Thermod> gas leak
Gasleitungsnetz n <Erdöl> gas grid
Gaslift m <Erdöl> gaslift (Erdölgewinnung)
Gaslöten n <Bau> torch brazing
Gasluftgemisch n <Phys, Thermod> gas-air mixture
Gas-Maser m <Elektronik> gas maser
Gasmessung f <Thermod> gasometry
Gasmotor m 1. <Maschinen> gas engine; 2. <Thermod> (AE) gas engine, (AE) gasoline engine, (BE) petrol engine
Gasmultiplikationsfaktor m <Elektronik> gas multiplication factor
Gasnitrieren n <Thermod> gas nitriding
Gasofen m <Thermod> gas fire
Gasöl n 1. <Erdöl> gas oil (Raffinerie, Destillationsprodukt); 2. <Thermod> gas oil
Gas-Öl-Verhältnis n (GÖV) <Erdöl> gas-to-oil ratio, GOR (Erdölgewinnung)
Gasometrie f <Thermod> gasometry
Gaspatrone f <Erdöl> gas bottle
Gaspedal n <Kfztech> (BE) accelerator, accelerator pedal, (AE) gas pedal (Vergaser)
Gasphasenabscheidung f (CVD) 1. <Elektronik> (AE) chemical vapor deposition, (BE) chemical vapour deposition, CVD (nach chemischem Verfahren); 2. <Telekom> (AE) chemical vapor deposition, (BE) chemical vapour deposition, CVD
Gasphasennitrierung f <Chemtech> (AE) vapor phase nitration, (BE) vapour phase nitration
Gasphasenpfropfen m <Kerntech> gas phase grafting
Gasphotozelle f <Elektronik> gas phototube
Gaspore f <Optik> pinhole
Gasprüfer m <Chemie> eudiometer
Gasrauschen n <Elektronik> gas noise
gasreich adj <Thermod> gassy
Gasreiniger m <Sicherheit> gas-cleaning equipment
Gasreinigung f <Sicherheit> gas cleaning
Gasreinigungsanlage f <Kerntech> gas purifiers
Gasrelais n <Elektronik, Elektrotech> gas-filled relay
Gasrohr n <Bau, Kfztech, Thermod> gas pipe
Gasröhre f <Elektronik> gas tube
Gasrohr-Innengewinde n <Fertig> BSP parallel female thread (Kunststoffinstallationen)
Gasrohrzange f <Bau> gas pliers
Gasrückgewinnung f <Umweltschutz> gas recovery
Gasrückstand m <Elektronik, Umweltschutz> residual gas
Gasruß m <Kunststoff> channel black

Gassammler

Gassammler *m* <Fertig> gas holder *(Schweißen)*
Gasschieber *m* <Maschinen> gas valve
Gasschleuse *f* <Heiz & Kälte, Kerntech> gas lock
Gasschweißbrenner *m* <Bau> oxyacetylene blowpipe, oxyacetylene blowtorch
Gasschweißen *n* 1. <Maschinen> gas welding; 2. <Thermod> oxyacetylene welding
Gasschweißung *f* <Maschinen> gas welding
Gassen *fpl* <Druck> rivers of white
gassenbesetzt *adj* <Telekom> all trunks busy, ATB
Gassenbesetztzustand *m* <Telekom> congestion *(Telefonverkehr)*
Gasspürgerät *n* <Labor> gas detector, leak detector
Gasspürröhrchen *n* <Umweltschutz> gas detection tube
Gasstrahl *m* <Maschinen> gas jet
Gasstrom *m* <Thermod> gas flow
Gasströmung *f* <Thermod> gas flow
Gastanker *m* <Wassertrans> gas tanker
Gasteer *m* <Kohlen> gas tar
Gastetrode *f* <Elektronik> gas tetrode *(Elektronenröhre)*
Gastgeber *m* <Comp & DV> host
Gasthermometer *n* 1. <Gerät> gas-filled thermometer; 2. <Heiz & Kälte, Phys, Thermod> gas thermometer
Gastriode *f* <Elektronik> gas triode
Gasturbine *f* <Fertig, Maschinen, Mechan, Thermod, Wassertrans> gas turbine *(Motor)*
Gasturbine *f* mit geschlossenem Kreislauf <Trans> closed cycle gas turbine
Gasturbine *f* mit offenem Kreislauf <Trans> open-cycle gas turbine
Gasturbinenbus *m* <Kfztech> gas turbine bus
Gasturbinen-Elektrizitätswerk *n* <Elektriz> gas turbine power station
Gasturbinen-Kraftwerk *n* <Elektriz> gas turbine power station
Gasturbinen-Reisebus *m* <Kfztech> gas turbine motor coach
Gasturbinen-Triebwagen *m* <Eisenbahn> gas turbine railcar
Gasturbinen-Triebwerk *n* <Lufttrans> gas turbine engine
Gasturbinen-Zug *m* <Eisenbahn> gas turbine train
Gasverflüssigung *f* <Maschinen, Thermod> liquefaction of gases
gasverpackt *adj* <Lebensmittel> gas-packed
Gaswaschanlage *f* 1. <Kerntech> gas washer; 2. <Thermod> gas-scrubbing plant
Gaswäsche *f* <Thermod> gas scrubbing
Gaswaschen *n* <Chemtech> scrubbing
Gaswaschflasche *f* <Labor> wash bottle
Gaswaschflasche *f* nach Dreschel <Labor> Dreschel gas-washing bottle
Gaswaschturm *m* <Chemtech> scrubber
Gaswegnahme *f* <Lufttrans> engine shut-off stop
Gaszähler *m* <Labor, Thermod> gas meter
Gaszug *m* <Kfztech> *(AE)* carburetor control cable, *(BE)* carburettor control cable
Gaszylinder *m* <Kerntech> gas cylinder
Gate *n* 1. <Comp & DV, Elektrotech> gate *(Thyristorelektrode)*; gate *(Transistorelektrode)*; 2. <Phys> gate *(Elektrode beim Feldeffekttransistor)*
Gate-Array *n* 1. <Elektronik> gate array *(anwendungsspezifisch verdrahtbare Standard-Anordnung von Gattern)*; 2. <Telekom> gate array
Gate-Dielektrikum *n* <Elektrotech> gate dielectric
Gate-Drain-Sperrkapazität *f* <Elektrotech> gate-to-drain capacitance
Gate-Fehlerstrom *m* <Elektrotech> gate leakage current
Gate-Katoden-Widerstand *m* <Elektrotech> gate-to-cathode resistor

Gate-Kontakt *m* <Elektrotech> gate contact
Gate-Schaltung *f* <Elektrotech> common-gate connection
Gate-Senke-Sperrkapazität *f* <Elektrotech> gate-to-drain capacitance
Gate-Source-Spannung *f* <Elektrotech> gate-to-source voltage
Gate-Source-Sperrkapazität *f* <Elektrotech> gate-to-source capacitance
Gate-Spannung *f* <Elektrotech> gate voltage
Gate-Substrat-Sperrkapazität *f* <Elektrotech> gate-to-substrate capacitance
Gateverstärker *m* <Elektronik> common-gate amplifier
Gateway *m* 1. <Comp & DV> gateway; 2. <Telekom> gateway; network gateway *(zwischen ungleichen Netzen)*
Gateway-Host *m* <Comp & DV> gateway host
Gattchen *n* <Wassertrans> grommet *(Segeln)*
Gatter *n* 1. <Comp & DV, Elektronik> gate; 2. <Fertig> gate *(Säge)*; 3. <Phys> gate
Gatteranordnung *f* 1. <Elektronik> gate array, logic array; 2. <Phys> gate array
Gatteransteuerungssignal *n* <Elektronik> gate-drive signal
Gatterausgang *m* <Elektrotech> gate output
Gatterdichte *f* <Elektronik> gate density
Gattereingang *m* <Elektronik> gate input
Gatterfeld *n* <Telekom> gate array
Gatterführung *f* <Maschinen> saw guide
Gatterkurzschluss *m* <Elektronik> gate short
Gattermatrix *f* <Comp & DV> gate array
Gatterschaltung *f* <Elektronik> logic gate
Gatterverbindung *f* <Elektronik> gate interconnect
Gatterverzögerung *f* <Elektronik> gate delay
gattieren *v* <Metall> burden
Gattierung *f* <Fertig> composition
Gattungsname *m* <Comp & DV> generic name
GAU *(größter anzunehmender Unfall)* <Kerntech> MCA *(maximum credible accident)*
Gaufrage *f* <Papier> embossing
gaufrieren *v* <Papier> emboss
Gaufrierkalander *m* <Papier> embossing calender
Gaufrierwalze *f* <Papier> embossing roll
Gauge *n* <Textil> *(AE)* gage, *(BE)* gauge
Gault-Ton *m* <Bau> Gault Clay *(Geologie)*
Gauß *n* (G) 1. <Aufnahme> gauss, G *(Einheit der magnetischen Induktion)*; 2. <Elektriz> gauss, G
Gauß-Filter *m* <Telekom> Gaussian filter *(bei GMFSK)*
Gauß-Filter-Minimalphasenumtastung *f* <Telekom> GMSK, Gaussian filtered minimum shift keying
Gauß-Impuls *m* <Telekom> bell-shaped pulse, Gaussian pulse
Gaußmeter *m* <Metrol> gaussmeter
Gauß'sche Fehlerverteilung *f* <Aufnahme, Math> Gaussian distribution
Gauß'sche Frequenzumtastung *f* <Telekom> Gaussian frequency shift keying *(GFSK)*
Gauß'sche Glockenkurve *f* <Math> bell-shaped curve, Gaussian bell-shaped curve, Gaussian curve
Gauß'sche Krümmung *f* <Geom> Gaussian curvature
Gauß'sche Kurve *f* <Geom> Gaussian curve
Gauß'sche Quadratur *f* <Comp & DV, Math> Gaussian quadrature
Gauß'sche Verteilung *f* <Comp & DV, Elektriz, Phys> Gaussian distribution
Gauß'sche Zahlenebene *f* <Math> Gaussian plane, Gauss number plane *(Veranschaulichung der komplexen Zahlen)*
Gauß'scher Impuls *m* <Telekom> Gaussian pulse
Gauß'scher Lehrsatz *m* <Elektriz> Gauss's theorem

Gauß'scher Satz *m* 1. <Math> Gauss's theorem *(Vektoranalysis)*; 2. <Phys> Gauss's law *(Elektrostatik)*; Gauss's theorem *(Vektoranalysis)*
Gauß'scher Strahl *m* <Telekom> Gaussian beam
Gauß'sches Rauschen *n* <Elektronik, Telekom> Gaussian noise
gautschen *v* <Papier> couch
Gautschpresse *f* <Papier> couch press
Gautschwalze *f* <Papier> couch roll
Gautschwalzenbezug *m* <Papier> couch roll jacket
Gautschwalzenfilzbürste *f* <Papier> jacket brush
Gay-Lussac'sches Gesetz *n* <Maschinen, Phys> Charles's law, Gay-Lussac's law
Gazepapier *n* <Verpack> reinforced paper
GB *(Gigabyte)* *n* <Comp & DV, Optik, Telekom> GB *(gigabyte)*
GCA-System *n* <Lufttrans> GCA system, ground-controlled approach system *(Landesystem)*
Gd *(Gadolinium)* <Chemie> Gd *(gadolinum)*
gealtert *adj* <Thermod> aged
Gebäckfehler *m* <Lebensmittel> baking fault
Gebälk *n* <Bau> beams
Gebäude *n* <Eisenbahn, Kontroll> structure
Gebäudeanschlussleitung *f* <Bau> branch line
Gebäudebrand *m* <Sicherheit> structural fire
Gebäudeflügel *m* <Bau> wing
Gebäudehülle *f* <Bau> building envelope
Gebäudeinstallation *f* <Bau> building services
Gebäudeleitung *f* <Bau> building line
Gebäuderäumung *f* <Sicherheit> evacuation of buildings
Gebäudeschutz *m* <Sicherheit> safeguarding of buildings
Gebäudeverkabelung *f* <Telekom> building cabling
Geber *m* <Telekom> sender *(Telegrafie)*
gebeugte Welle *f* <Phys> diffracted wave
Gebiet *n* 1. <Akustik> field; 2. <Bau> area, zone; 3. <Kohlen> zone; 4. <Künstl Int> domain *(des Wissens)*; 5. <Metall> zone
Gebiet *n* **mit schlechtem Empfang** <Fernseh> poor reception area
Gebietsauslass *m* <Wasserversorg> outfall
Gebietsrückhalt *m* <Wasserversorg> retention
Gebietsspeicherung *f* <Wasserversorg> retention
Gebietsüberdeckung *f* <Telekom> area coverage
Gebilde *n* <Bau> form
Gebirgsbahn *f* <Eisenbahn> *(AE)* mountain railroad, *(BE)* mountain railway
Gebirgsdruck *m* <Erdöl> overburden pressure *(Geologie)*
Gebirgsmasse *f* <Kohlen> mountain mass
Gebirgsschlag *m* <Bau> rockburst
Gebläse *n* 1. <Bau> fan, ventilating fan; 2. <Elektriz> fan; 3. <Fertig> fanner; 4. <Kfztech> fan; blower *(Kühlanlage)*; 5. <Lebensmittel> blower; 6. <Lufttrans> defogging fan, fan; 7. <Maschinen> blast engine, blower; 8. <Mechan, Papier> blower
Gebläsebrenner *m* <Bau> *(BE)* blowlamp
Gebläsebürste *f* <Foto> blower brush
Gebläseflügel *m* <Heiz & Kälte, Mechan, Thermod> fan blade
Gebläseflügelrad *n* <Maschinen> fan wheel
gebläsekühlen *v* <Heiz & Kälte, Mechan, Thermod> fan-cool
Gebläsekühlung *f* <Heiz & Kälte, Mechan, Thermod> fan cooling
Gebläsekupolofen *m* <Fertig> blast cupola
Gebläselaufrad *n* <Maschinen> blower wheel
Gebläsemaschine *f* <Maschinen> blast engine
geblasene Glasröhre *f* <Labor> blown-glass tube
geblasenes Glasrohr *n* <Labor> blown-glass tube

geblasenes Tafelglas *n* <Ker & Glas> *(AE)* blown sheet, *(BE)* cylinder glass
Gebläseofen *m* <Heiz & Kälte, Thermod> blast furnace
Gebläserad *n* <Kerntech, Kohlen> impeller
geblaut *adj* <Metall> blued
gebläutes Blech *n* <Metall> blued sheet
gebleichter Papierbrei *m* <Verpack> bleached pulp
gebleichter Zellstoff *m* <Verpack> bleached pulp
gebleichtes Mehl *n* <Lebensmittel> bleached flour
gebogen *adj* 1. <Bau> arched; 2. <Ker & Glas, Metall> bent
gebogene Linie *f* <Geom> curved line
gebogener Abschnitt *m* <Metall> bent section
gebogener Meißel *m* <Maschinen> cranked tool
gebogener Schraubenschlüssel *m* <Maschinen> *(BE)* S-shaped spanner, S-shaped wrench, *(BE)* curved spanner, curved wrench
gebördelte Kappe *f* <Verpack> flanged cap
gebördelter Boden *m* <Ker & Glas> flanged bottom
gebördelter Papierbecher *m* <Verpack> crimp paper cup
gebördelter Rand *m* <Maschinen> flanged edge
gebrannte Form *f* <Ker & Glas> *(AE)* burnt mold, *(BE)* burnt mould
gebrannter Kalk *m* <Bau> quicklime
gebrannter Ton *m* <Ker & Glas> burnt clay
gebrannter Ziegel *m* <Bau> burnt brick
gebranntes Eisen *n* <Metall> burnt iron
gebranntes Steingut *n* <Ker & Glas> burnt earthenware
Gebrauch *m* <Comp & DV> use
Gebrauchsanweisungen *fpl* <Verpack> directions for use, instructions for use
Gebrauchsdauer *f* 1. <Fertig> operational life *(Kunststoffinstallationen)*; 2. <Kunststoff> pot life; 3. <Maschinen> working life
gebrauchsfertig *adj* <Foto> ready-made *(Lösung)*
gebrauchsgetestet *adj* <Mechan> field-tested
Gebrauchsgrafik *f* <Druck> commercial art
Gebrauchsmuster *n* <Patent> utility model
Gebrauchsnormal *n* <Qual> service standard, working standard
Gebrauchsschrift *f* <Druck> body type
Gebrauchstauglichkeit *f* <Patent, Qual> fitness for use
Gebrauchs- und Installationsanweisungen *fpl* <Verpack> handling and installation instructions
Gebrauchszertifikat *n* <Patent> utility certificate
gebrochen *adj* 1. <Bau> broken; 2. <Fertig> stepped *(Härtung)*; 3. <Kohlen> broken; 4. <Math> broken, fractional *(Zahl)*
gebrochene Linie *f* <Geom> broken line
gebrochene Zahl *f* <Math> fractional
gebrochener Exponent *m* <Math> fractional exponent
gebrochener Strahl *m* 1. <Optik> *(AE)* refracted rayoptical fiber, *(BE)* refracted rayoptical fibre *(Lichtleiter)*; 2. <Phys, Telekom> refracted ray
gebrochener Zuschlagstoff *m* <Bau> broken aggregate
gebrochenes Dach *n* <Bau> *(BE)* gambrel roof, *(AE)* mansard roof
gebrochenes Licht *n* <Optik> refracted light
gebuchter Zusteigeverkehr *m* <Trans> pick-up traffic
Gebühr *f* 1. <Patent> fee; 2. <Telekom> call charge, charge
Gebührenabrechnungssystem *n* <Telekom> call-accounting system
Gebührenangabe *f* <Telekom> charging information
Gebührenansage *f* <Telekom> advice of charge
Gebührenanzeige *f* <Telekom> advice of charge, AOC *(Dienstmerkmal im EURO-ISDN)*

Gebührenanzeiger 292

Gebührenanzeiger m **beim Teilnehmer** <Telekom> subscriber's private meter
Gebührenerfassung f <Telekom> call metering, metering
Gebührenfernsehen n <Fernseh> pay per view, pay-TV; pay-per-channel TV, pay-per-packet TV *(Entgelt programmkanalweise)*
Gebührenfernsehnetz n <Fernseh, Telekom> paytelevision network
gebührenfrei adj <Telekom> free of charge
gebührenfreie Nummer f <Telekom> freephone number
gebührenfreie Rufnummer f <Telekom> *(BE)* freefone number, *(AE)* toll-free number
gebührenfreie Schnellstraße f <Trans> *(AE)* freeway
gebührenfreie Verbindung f <Telekom> free call
gebührenfreier Anruf m <Telekom> free call, *(BE)* freephone call, no-charge call, *(AE)* toll-free call, green number service
gebührengünstige Zeit f <Telekom> low-charge period
Gebühreninformation f **im laufenden Gespräch** <Telekom> call-in-progress cost information
gebührenpflichtige Ansprüche mpl <Patent> claims incurring fees
gebührenpflichtige Autobahn f <Trans> toll road, *(AE)* turn-pike
gebührenpflichtige Straße f <Trans> *(AE)* turnpike
gebührenpflichtiges Fernsehen n <Fernseh> pay television
gebührenpflichtiges Kabel n <Fernseh> pay cable
Gebührenstelle f <Telekom> *(AE)* billing center, *(BE)* billing centre
Gebührenübernahme f <Comp & DV> reverse charge
Gebührenverrechnung f <Telekom> billing
Gebührenzähler m <Telekom> call-charging equipment, charge meter, subscriber's meter *(Telefon)*
Gebührenzone f <Telekom> charging area
gebündelter Strahl m <Elektronik> collimated beam
gebunden adj <Ker & Glas> bonded
gebundene Schleifmittel fpl <Maschinen> bonded abrasive products
gebundene Variable f <Künstl Int> bound variable
gebundene Wärme f <Bau, Erdöl, Heiz & Kälte, Thermod> latent heat
gebundener Schwefel m <Kunststoff> *(AE)* combined sulfur, *(BE)* combined sulphur
gebundenes Buch n <Druck> bound book
gebundenes Elektron n <Teilphys> bound electron
gebundenes Glasfaservlies n <Ker & Glas> bonded mat
gebundenes Wasser n <Lebensmittel> bound water
gebürstet adj <Elektrotech, Ker & Glas, Kfztech, Mechan> brushed
gedämmt adj <Bau> insulated
gedämpft adj 1. <Phys> damped; 2. <Thermod> banked-up
gedämpft schwingendes Instrument n <Gerät> damped periodic instrument
gedämpfte Schwingung f <Aufnahme, Elektronik, Funktech> damped oscillation
gedämpfte Schwingungen fpl <Wellphys> damped vibrations
gedämpfte Sinusgröße f <Elektrotech> damped sinusoidal quantity
gedämpfter Schallmessraum m <Raumfahrt> reverberation chamber
gedämpftes System n <Gerät> attenuated system
Gedankenflussanalyse f <Ergon> thought-stream analysis
gedeckt adj <Wassertrans> decked *(Schiffbau)*
gedeckter Güterwagen m <Eisenbahn> *(AE)* boxcar, *(AE)* freight car, *(BE)* wagon

gedeckter Hafen m <Ker & Glas> closed port
gedeckter Wagen m <Eisenbahn> *(AE)* covered car, *(BE)* covered wagon, *(AE)* freight van, *(BE)* goods van
gedehnt adj <Druck> elongated
gedehnte x-Ablenkung f <Elektronik> expanded sweep *(Radar)*
gedornt adj <Fertig> indented
gedrehte Scheibe f <Maschinen> turned washer
gedrehter Rand m <Ker & Glas> turned rim
gedrehtes Teil n <Maschinen> turned part
gedrosselte Lautheit f <Akustik> partially-masked loudness
gedruckt adj <Patent> printed
gedrückt adj <Fertig> spun *(Metall, Blech)*
gedruckte Mikrowellenschaltung f <Elektronik, Funktech> microwave-printed circuit
gedruckte Schaltkarte f <Elektronik> printed board
gedruckte Schaltung f 1. <Comp & DV, Elektronik> printed circuit board *(PCB)*; 2. <Phys, Telekom> printed circuit
gedruckte Verdrahtung f <Elektronik> printed wiring
gedrückter Spitzbogen m <Bau> *(AE)* four-centered arch, *(BE)* four-centred arch
gedrucktes Buch n <Druck> printed book
geeichte Unterlegscheibe f <Maschinen> calibrated spacer
geeichter Eimer m <Bau> stamped bucket
geeichtes Wasserrückhaltebecken n <Umweltschmutz> calibrated watershed *(Landwirtschaft)*
geeignet adv <Math> appropriate
geeigneter Speicher m/**vorwiegend zum Lesen** <Comp & DV> read-mostly memory
geerdet adj <Elektriz, Elektrotech> *(BE)* earthed, *(AE)* grounded
geerdeter Schalter m <Elektriz> *(BE)* earthed switch, *(AE)* grounded switch
Gefahr f <Sicherheit> danger, hazard
Gefahr f **durch Hochfrequenzwellen** <Sicherheit> radiowave hazard
Gefahr f **durch Laserstrahlung** <Sicherheit> laser radiation hazard
Gefahr f **durch Ultraschall** <Sicherheit> ultrasonic hazard
gefährden v <Sicherheit> be a danger to, endanger, threaten
gefahrdrohend adj <Sicherheit> dangerous, hazardous
gefahrdrohende Menge f <Sicherheit, Umweltschmutz> dangerous concentration, hazardous concentration
Gefährdung f <Sicherheit> danger, hazard, risk
Gefährdungsfaktor m <Sicherheit> hazardous agent
gefährdungsfrei adj <Sicherheit> hazardous-free, nonhazardous, riskless, safe
gefährdungsfreie Konstruktion f <Sicherheit> safety design
gefährdungsfreie Technologie f <Sicherheit> fail-safe technology, safe technology
Gefahrenabwehr f <Sicherheit> hazard control
Gefahrenbake f <Lufttrans> hazard beacon
Gefahrenbereich m <Sicherheit> danger area, danger zone, high-risk area
Gefahrenbeseitigung f <Sicherheit> hazard control
Gefahrenfaktor m <Sicherheit> hazardous agent
Gefahrenfeuer n <Lufttrans> hazard beacon
Gefahrenkennzeichnung f <Sicherheit> identification of hazards
Gefahrenpunkt m <Sicherheit> danger point
Gefahrenschalter m <Sicherheit> emergency switching device
Gefahrensignal n <Sicherheit> danger signal

Gefahrenstelle f <Sicherheit> danger point
Gefahrenverhütung f <Sicherheit> hazard prevention
Gefahrgut n <Sicherheit, Umweltschutz> dangerous goods, hazardous goods, hazardous material
Gefahrgutbeauftragter m <Umweltschutz> authorized person for hazardous goods
Gefahrgutumschlag m <Sicherheit, Umweltschutz> transshipment of hazardous goods
gefährlich adj <Sicherheit, Umweltschutz> dangerous, hazardous
gefährliche Ansammlung f <Sicherheit, Umweltschmutz> dangerous built-up (von Schadstoffen)
gefährliche Frachtgüter npl <Lufttrans> dangerous goods
gefährliche Konzentration f <Umweltschutz> dangerous concentration, hazardous concentration
gefährliche Ladungen fpl <Sicherheit, Trans> dangerous loads
gefährliche Luftschadstoffe mpl <Kohlen, Umweltschmutz> hazardous air pollutants pl, HAPs
gefährliche Maschine f <Sicherheit> dangerous machine
gefährlicher Abfall m <Abfall> hazardous waste
gefährlicher Arbeitsstoff m <Sicherheit> hazardous substance in the workplace
gefährlicher Baustoff m <Sicherheit> hazardous material
gefährlicher chemischer Stoff m <Sicherheit> hazardous chemical
gefährlicher Luftschadstoff m <Umweltschutz> hazardous air pollutant
gefährlicher Stoff m <Sicherheit> dangerous material, harmful substance, hazardous material; harmful agent, hazardous agent
gefährlicher Werkstoff m <Sicherheit> hazardous material
gefährliches Material n <Sicherheit> hazardous material
gefahrlose Konzentration f 1. <Kerntech> safe concentration (von Kernbrennstoff); 2. <Sicherheit> safe concentration
gefahrlose Technik f <Sicherheit> fail-safe technology, safe technology
Gefahrmeldeeinrichtung f <Sicherheit> alarm unit
Gefahrstoff m <Sicherheit> dangerous material, dangerous substance, hazardous substance
Gefahrstoffeindämmung f <Sicherheit, Umweltschmutz> containment of dangerous substances
Gefahrstoffkataster n <Umweltschutz> register of hazardous substances
Gefahrstofflager n <Sicherheit> storage of dangerous materials
Gefälle n 1. <Bau> downward gradient, downward slope, falling gradient, gradient, incline, pitch, slant, slope; grade (Gelände); 2. <Geom> grade, slope; 3. <Hydraul> head, head of water; slope (S)
Gefälle n einer Geraden <Geom> gradient of a straight line
Gefällewinkelmesser m <Metrol> clinometer
gefaltet adj <Textil> pleated
gefalzte Verbindung f <Bau> rabbeted joint
gefärbt adj <Telekom> (AE) colored, (BE) coloured
gefärbtes Verbundglas n <Ker & Glas> tinted-laminated glass
Gefäß n <Labor> vessel
Gefäßdurchdringen n <Kerntech> vessel penetration
Gefäßförderanlage f <Kohlen> skip extraction
Gefäßreinigungsanlage f <Bau> tank-cleaning plant
Gefechtskopf m <Raumfahrt> warhead (Raumschiff)
gefedert adj <Maschinen> spring-loaded, springborne
gefederte Andruckplatte f <Foto> spring-mounted pressure plate

gefederte Verbindung f <Bau> feather joint (Holzbau)
gefederter Dielenfußboden m <Bau> tongued flooring
gefedertes Gewicht n <Kfztech> sprung weight (Karosserie)
gefedertes Rad n <Kfztech> sprung gear (Getriebe)
gefesselt adj <Mechan> captive
gefilterte Quadraturphasenumtastung f <Telekom> filtered QPSK
gefilterter Schlamm m <Abfall> filtration sludge
geflanscht adj <Elektriz> flanged
geflanschtes Rohr n <Maschinen> flanged pipe
Geflecht n 1. <Bau> netting; 2. <Elektrotech> braid (Isolierung von elektrischer Leitung); braid (Kupferdraht-Abschirmung)
geflutet adj <Wassertrans> flooded (Schleuse)
geforderte Genauigkeit f <Patent, Qual> required accuracy
geförderte Luftmenge f <Heiz & Kälte> rate of air delivered
geforderte Verfügbarkeitszeit f <Qual> required time
geformt adj 1. <Maschinen> shaped; 2. <Papier> (AE) molded, (BE) moulded
geformter Impuls m <Elektronik> shaped pulse
geformter Reflektor m <Raumfahrt> shaped reflector (Weltraumfunk)
gefräst adj <Mechan> milled
Gefrier… <Heiz & Kälte, Lebensmittel, Thermod> freezing, frozen
Gefrieranlage f <Heiz & Kälte> freezing plant, refrigerating plant
Gefrierbrand m 1. <Lebensmittel> freezerburn; 2. <Thermod> humidity loss by sublimation
Gefriercontainer m <Heiz & Kälte> refrigerated container
Gefriereffekt m <Thermod> mechanothermal effect (von Helium)
gefrieren v 1. <Bau> freeze; 2. <Heiz & Kälte> freeze (von Wasser)
Gefrieren n <Heiz & Kälte, Lebensmittel, Metall, Papier> freezing
Gefrierfach n 1. <Heiz & Kälte> freezer compartment (eines Kühlschranks); 2. <Maschinen> household freezer compartment
gefrierfest adj <Heiz & Kälte, Lufttrans> antifreezing
Gefriergerät n <Heiz & Kälte, Maschinen> freezer
Gefriergerät n für Lebensmittel <Maschinen> food freezer
gefriergetrocknet adj <Chemtech, Heiz & Kälte, Lebensmittel, Thermod> freeze-dried
gefriergetrocknetes Produkt n <Verpack> freeze-dried product
Gefrierkette f <Lebensmittel> freezer chain (Weg vom Hersteller zum Endverbraucher)
Gefrierkonzentration f 1. <Chemtech> freeze concentration; 2. <Heiz & Kälte> freeze-concentration
Gefrierkost f <Verpack> deep-frozen food
Gefrierkurve f <Metall> freezing curve
Gefriermaschine f <Heiz & Kälte> freezer
Gefriermischung f <Heiz & Kälte> freezing mixture
Gefriermittel n <Heiz & Kälte> freezing medium
Gefriermöbel n <Heiz & Kälte> display case
Gefrierpunkt m <Heiz & Kälte, Thermod, Verpack, Wassertrans> freezing point
Gefrierpunktserniedrigung f <Thermod> depression of freezing point
Gefrierraum m <Heiz & Kälte> chill room, freezing room
Gefrierschachtverfahren n <Thermod> low-temperature sinking
Gefrierschrank m <Lebensmittel> deep freeze (Verpackung)

Gefrierschutzmittel

Gefrierschutzmittel n <Heiz & Kälte> antifreeze agent
Gefriertheke f <Heiz & Kälte> display case
gefriertrocknen v <Lebensmittel, Thermod> freeze-dry
Gefriertrocknen n <Chemtech, Thermod, Verpack> freeze-drying
Gefriertrockner m <Chemtech, Heiz & Kälte> freeze-drier
Gefriertrocknung f 1. <Chemtech, Heiz & Kälte> freeze-drying, lyophilization; 2. <Lebensmittel> dehydrofreezing, freeze-drying; 3. <Verpack> freeze-drying
Gefriertrocknungskolben m <Ker & Glas> lyophilization flask
Gefriertruhe f 1. <Heiz & Kälte> chest freezer; 2. <Lebensmittel> freezer chest; 3. <Thermod> freezer
Gefriertunnel m <Heiz & Kälte> freezing tunnel, lyophilization tunnel
Gefrierverpackung f <Verpack> deep freeze packaging
gefrorener Boden m <Kohlen> frozen ground
Gefüge n <Fertig> spacing *(Schleifscheibe)* • **von lunker- und fehlerfreiem Gefüge** <Fertig, Metall> metallurgically sound
Gefügeausbildung f <Kohlen> crystal structure
Gefügebeständigkeit f <Bau, Werkprüf> structural stability *(Kunststoffe)*
Gefügeumwandlung f <Fertig> structural change *(Stahl)*
Gefühlssimulator m <Raumfahrt> feel simulator
geführte optische Welle f <Telekom> optical guided wave
geführte Welle f <Telekom> guided wave
geführter Modus m <Telekom> bound mode, guided mode
gefüllt adj <Bau> filled
gefülltes PTFE n <Fertig> PTFE and graphite *(Kunststoffinstallationen)*
gefunkt adj <Phys> radio
gefurcht adj <Fertig> grooved
gegebene Größe f <Comp & DV> datum
Gegenbegrenzer m <Elektronik> inverse limiter
Gegenbewegung f <Maschinen> alternation of a motion, counter-motion
Gegenbogen m <Bau> inflected arch, reversed arch
Gegendruck m 1. <Hydraul> back pressure *(Zylinder)*; 2. <Maschinen, Papier> back pressure
Gegendruckturbine f 1. <Maschinen> back-pressure turbine; 2. <Mechan> impulse turbine
Gegendruckventil n <Hydraul> back-pressure valve
Gegendruckzylinder m <Kunststoff> impression cylinder
gegeneinander geschaltet adj <Elektriz> back-to-back connected
gegeneinander geschaltete Anordnung f <Elektrotech> back-to-back arrangement *(Dioden)*
gegenelektromotorische Kraft f *(Gegen-EMK)* <Eisenbahn, Elektriz, Elektrotech, Lufttrans> back electromotive force *(bemf)*; counter electromotive force *(cemf)*
Gegen-EMK f *(gegenelektromotorische Kraft)* <Eisenbahn, Elektriz, Elektrotech, Lufttrans> bemf *(back electromotive force)*; cemf *(counter electromotive force)*
Gegenfeder f <Fertig> return spring
Gegenflansch m 1. <Kerntech> counter flange, mating flange; 2. <Maschinen> companion flange, counterflange
Gegenflutung f <Wassertrans> counterflooding
Gegengewicht n 1. <Bau> counterweight; 2. <Fertig> dolly; 3. <Funktech> counterpoise *(Viertelwellenantenne)*; 4. <Lufttrans> balance weight, counterbalance; 5. <Maschinen, Mechan> balance weight, counterbalance, counterweight
Gegengewicht n **des Luftschraubenblattes** <Lufttrans> blade balance weight
Gegengewichtskasten m <Fertig> balance box
Gegengewölbe n <Bau> inverted arch

Gegengift n <Chemie> antidote
Gegengrünphase f <Trans> opposing green
gegenhalten v 1. <Fertig> dolly *(Nieten)*; 2. <Maschinen> hold up
Gegenhalten n <Maschinen> holding-up
Gegenhalter m 1. <Bau> rivet dolly; 2. <Fertig> dolly *(Nieten)*; hold-on; overarm, overhanging arm *(Fräsmaschine)*; 3. <Maschinen> holder-up, holding-up hammer, overarm, overhanging arm; 4. <Sicherheit> load backrest *(Gabelstapler)*
Gegenhalterschere f <Fertig> arbor brace; brace *(Fräsmaschine)*
Gegeninduktion f 1. <Elektriz> mutual inductance, mutual induction; 2. <Phys> mutual induction
Gegeninduktionskoeffizient m <Elektriz, Elektrotech, Qual> coefficient of mutual inductance
Gegeninduktivität f <Elektrotech, Phys> mutual inductance; mutual inductor *(Spule)*
Gegeninduktivitätskopplung f <Elektrotech> mutual inductance coupling
Gegeninduktoren mpl <Elektriz> mutual inductors
gegeninduzierter Anker m <Elektriz> inverse-induced armature
Gegenkolbenmotor m <Maschinen> opposed piston engine
Gegenkopplung f 1. <Aufnahme, Elektronik, Elektrotech> negative feedback, inverse feedback; 2. <Funktech> inverse feedback; 3. <Telekom, Wellphys> negative feedback
Gegenkopplungsfilter n <Elektronik> inverse feedback filter
Gegenkraft f <Maschinen> counteracting force
Gegenkurs m <Wassertrans> reciprocal course, reciprocal track *(Navigation)*
Gegenlauf m 1. <Fernseh> reverse action; 2. <Fertig> reverse rotation; 3. <Maschinen> counter-rotation *(eines Propellers)*
gegenlaufende Welle f <Elektrotech> *(AE)* reverse traveling-wave, *(BE)* reverse travelling-wave
Gegenlauffräsen n 1. <Fertig> conventional milling, outcut milling, up-milling, upcut milling; 2. <Maschinen> conventional milling, standard milling, upcut milling
Gegenlauffrässchnitt m <Fertig> conventional cut
gegenläufige Bewegung f <Maschinen> alternation of a motion
gegenläufige Propeller mpl <Lufttrans> counter-rotating propellers
gegenläufige Treppe f <Bau> dogleg stairs
gegenläufige Walzenstreichmaschine f <Papier> reverse roll coater
gegenläufiger Axiallüfter m <Lufttrans> counter-revolving axial fan
Gegenlaufschleifen n <Fertig> up-grinding
Gegenlaufwirbeln n <Fertig> opposed whirling
Gegenlicht beleuchten v/im <Foto> backlight
Gegenlicht/im <Foto> backlit
Gegenlichtaufnahme f <Foto> backlighted photo
Gegenlichtblende f <Foto> hood, lens hood
Gegenmodulation f <Elektronik> inverse modulation
Gegenmutter f 1. <Maschinen> jam nut, locknut; 2. <Mechan> locknut
Gegenparallelschaltung f <Elektrotech> anti-parallel connection
Gegenpassat m <Wassertrans> antitrades *(Windart)*
gegenphasig adj <Elektriz, Elektronik> in-phase opposition, opposite phase
gegenphasig adv <Fernseh> antiphase
gegenphasige Antenne f <Raumfahrt> endfire antenna *(Weltraumfunk)*

Gegenprobe f 1. <Kohlen> check sample; 2. <Maschinen> countertest; 3. <Qual> check sample
Gegenprofil n <Maschinen> mating profile
Gegenrad n <Maschinen> mating gear
Gegenschaltung f <Elektriz> bucking circuit
Gegenschiene f <Eisenbahn> safety rail
Gegenschlaghammer m <Fertig> impacter
Gegensee f <Wassertrans> head sea
gegenseitig anziehen v <Phys> attract each other
gegenseitig versetzen v <Comp & DV> interleave
gegenseitige Abhängigkeit f <Kontroll> interplay
gegenseitige Blockierung f <Kontroll> deadlock
gegenseitige Induktivität f <Elektrotech> mutual inductance
gegenseitige Peilung f <Lufttrans> reciprocal bearing (Navigation)
gegenseitige Störung f der Kanäle <Elektrotech> interchannel interference
gegenseitige Störung f von Diensten <Telekom> interservice interference
gegenseitige Synchronisierung f <Telekom> mutual synchronization
gegenseitiges Authentifizieren n <Comp & DV> mutual authentication
gegenseitiges Sperren n <Comp & DV> deadlock, deadly embrace
Gegenspannung f <Elektrotech> reverse voltage
Gegenspannungsschutz m <Elektrotech> reverse voltage protection
Gegensprechanlage f 1. <Aufnahme> talkback; 2. <Fernseh, Funktech> talkback circuit
Gegensprechen n 1. <Aufnahme> talkback; 2. <Funktech, Telekom> duplex
Gegensprechfunkverkehr m <Telekom> duplex radiotelephony
Gegensprechmikrofon n <Aufnahme> talkback microphone
Gegenstand m <Patent> subject matter
Gegenständer m <Maschinen> back rest
Gegenstelle f <Telekom> distant end, remote terminal
Gegenstempel m <Maschinen> counterpunch
Gegenstößel m <Mechan> plunger
Gegenstrom m 1. <Elektrotech> countercurrent; 2. <Kohlen, Maschinen> counterflow
Gegenstrombremsung f <Elektrotech> breaking by plugging, countercurrent breaking, plug breaking, plugging (Asynchronmotor); regenerative braking (Maschine)
Gegenstromdestillation f <Chemie> rectification
Gegenstromdiffusionsanlage f <Kerntech> countercurrent diffusion plant
Gegenstromfilter n <Abfall> reverse flow filter
Gegenstromklassierer m <Kohlen> countercurrent classifier
Gegenstromkondensation f <Heiz & Kälte> counterflow condensation
Gegenstromkühlturm m <Heiz & Kälte> counterflow-cooling tower
Gegenstromkühlung f <Heiz & Kälte> counterflow cooling
Gegenstromlinie f <Hydraul> counter streamline
Gegenstromwärmeaustauscher m <Heiz & Kälte, Lebensmittel> counterflow heat exchanger
Gegentakt m <Mechan> push-pull
Gegentaktbetrieb m <Funktech> push-pull operation
Gegentaktmischer m 1. <Elektronik> balanced mixer; 2. <Funktech> double-balanced mixer
Gegentaktmodulator m <Elektronik> balanced modulator

Gegentaktschalter m <Elektriz> push-pull switch
Gegentaktschaltung f <Aufnahme> push-pull circuit
Gegentaktumsetzer m <Elektronik> push-pull modulator
Gegentaktverstärker m 1. <Aufnahme> balanced amplifier; 2. <Elektronik> balanced amplifier, paraphrase amplifier, push-pull amplifier
gegenüberliegend adj <Geom> opposite
gegenüberliegende Seiten fpl <Geom> opposite sides (eines Quadrats)
gegenüberliegende Zylinder mpl <Kfztech> opposed cylinders
Gegenüberwicklung f <Elektriz> diametrical winding
Gegenvakuumventil n <Nichtfoss Energ> antivacuum valve
Gegenverkehr m <Trans> opposing traffic
Gegenwelle f <Maschinen> second motion shaft
Gegenwind m <Lufttrans, Wassertrans> headwind
Gegenwinkel mpl <Geom> corresponding angles
gegenwirkend adj <Maschinen> reactive
gegenwirkende Drehkraft f <Lufttrans> antagonistic torque
Gegenwirkleitwert m <Elektrotech> transconductance
gegisstes Besteck n <Wassertrans> dead-reckoning position, estimated position (Navigation)
geglättete Kante f <Ker & Glas> smoothed edge
geglätteter Verkehr m <Telekom, Trans> smooth traffic
gegliedert adj <Bau> jointed
geglüht adj <Metall, Thermod> annealed (Stahl)
gegorener Malztrank m <Lebensmittel> malt extract
gegossene Type f <Druck> metal type
gegossene Zeile f <Druck> slug
gegossenes Loch n <Maschinen> cast hole
Gehalt m 1. <Anstrich, Kohlen, Kunststoff> content; 2. <Maschinen> tenor; 3. <Metall, Papier> content; 4. <Textil> analysis, content
gehaltene Verbindung f <Telekom> call held
gehaltener Diamant m/**mit festem Griff** <Ker & Glas> diamond held with firm grip
gehämmert adj <Ker & Glas> hammered (Glas)
Gehänge n 1. <Fertig> ball (Gießen, Gießkran); 2. <Kfztech> shackle (Blattfeder); 3. <Maschinen> hanger
gehärtet adj 1. <Chemie> hydrogenated (Fett, Öl); 2. <Lebensmittel> hydrogenated; 3. <Metall> hardened (Stahl); 4. <Thermod> heat-treated
Gehäuse n 1. <Bau, Elektriz, Elektrotech> box, cabinet, case, casing, cubicle, enclosure, housing, shell; frame (eines elektronischen Geräts); 2. <Fertig> body, housing (Kunststoffinstallationen); 3. <Heiz & Kälte> enclosure; 4. <Hydraul> casing (Dampfzylinder, Turbine, Zentrifugalpumpe); 5. <Kfztech> casing (Motor, Getriebe); 6. <Kohlen> shell (Maschine); 7. <Kontroll> enclosure; 8. <Lufttrans> case (Flugwesen); 9. <Maschinen> box, cage, case, casing, housing, shell; 10. <Mechan> case, casing; 11. <Nichtfoss Energ> housing; 12. <Raumfahrt> pod (Raumschiff); 13. <Telekom> cabinet, package; 14. <Wassertrans> casing
Gehäuseabstrahlung f <Telekom> cabinet radiation
Gehäusedeckel m 1. <Foto> body cap; 2. <Maschinen> casing cover
Gehäusedichtung f <Fertig> body seal (Kunststoffinstallationen)
geheftet adj <Druck> stitched
geheftete Schachtel f <Verpack> stitched box
Geheimnummer f <Telekom> unlisted number (Telefon)
Geheimtür f <Bau> jib door
gehemmt adj <Fertig> held back (Getriebelehre)
Gehlinie f <Bau> walking line (Treppe)
gehobelte Schalung f <Bau> wrought shuttering
gehobene Last f <Lufttrans> lifted load

Gehör

Gehör n 1. <Akustik> audio, auditory; 2. <Ergon> audition
Gehörempfindlichkeit f <Akustik> aural sensitivity
Gehörempfindlichkeitsmessgerät n <Akustik> audiometer
Gehörempfindlichkeitsmessung f <Akustik> audiometry
Gehörempfindlichkeitsskale f <Akustik> aural sensation scale
Gehörempfindung f <Akustik> auditory sensation
Gehörermüdung f <Akustik> hearing fatigue
Gehörgang m <Akustik> auditory canal
Gehörmesser m <Aufnahme> audiometer
Gehörmessgerät n <Wellphys> sonometer
Gehörprüfkabine f <Aufnahme> audiometric booth
Gehörprüfraum m <Akustik> audiometric test room
Gehörschärfemessung f <Akustik> audiometry
Gehörschutz m <Akustik, Ergon, Sicherheit> ear protection, ear protector, hearing conservation
Gehörschutzmittel n <Sicherheit> ear protector, hearing protection device, hearing protector
Gehörschutzstopfen m <Sicherheit> insert-type protector, noise protective plug
Gehörschutzstöpsel m <Sicherheit> insert-type protector, noise protective plug
Gehörschwankungen fpl <Akustik> aural flutter (in der Tonhöhe)
gehren v <Fertig> (AE) miter, (BE) mitre
Gehrstoß m <Bau> (AE) miter joint, (BE) mitre joint
Gehrung f 1. <Bau> bevel, (AE) miter, (BE) mitre; 2. <Druck, Fertig, Ker & Glas> (AE) miter, (BE) mitre
Gehrungsfuge f <Bau> (AE) miter joint, (BE) mitre joint
Gehrungskrümmung f <Ker & Glas> (AE) miter return, (BE) mitre return
Gehrungsschere f <Fertig> (AE) beveling shear, (BE) bevelling shear
Gehrungsschleifmaschine f <Ker & Glas> (AE) miter-grinding machine, (BE) mitre-grinding machine
Gehrungsschnitt m <Fertig> (AE) miter cut, (BE) mitre cut
Gehrungsschnittlehre f <Bau> (AE) miter board, (AE) miter box, (BE) mitre board, (BE) mitre box
Gehrungsstanzmaschine f <Bau> (AE) miter-cutting machine, mitre-cutting machine
Gehrungswinkel m 1. <Bau> (AE) miter square, (BE) mitre square; 2. <Fertig> bevel; 3. <Ker & Glas> (AE) miter bevel, (BE) mitre bevel; 4. <Metrol> bevel
Gehweg m <Bau> banquette, (BE) pavement, (AE) sidewalk
Gehwegplatte f <Bau> flag, paving stone
Gehwegplattenbelag m <Bau> (BE) flagstone pavement, (AE) flagstone sidewalk
Gei f <Wassertrans> guy (Tauwerk)
Geigenharz n <Chemie> rosin
Geiger-Müller-Zählrohr n <Phys, Strahlphys, Teilphys> Geiger counter, Geiger-Müller counter, Geiger-Muller tube, Geiger tube
Geigerzähler m <Fertig, Phys, Strahlphys, Teilphys, Umweltschmutz> Geiger counter, contamination meter
Geiger-Zählrohr n <Phys, Strahlphys, Teilphys> Geiger tube
Geiser m <Nichtfoss Energ> geyser
Geißler-Röhre f <Elektronik> Geissler tube
Geisterbild n 1. <Fernseh> ghost image, multipath signals; 2. <Foto> ghost
Geisterecho n <Funkort, Wassertrans> ghost echo (Radar)
geistiges Eigentum n <Patent> intellectual property
geistiges Eigentumsrecht n <Patent> intellectual property rights, IPR

Geitau n <Wassertrans> guy (Tauwerk)
gekämmt adj <Textil> combed
gekapselt adj <Maschinen, Sicherheit> enclosed
gekapselte Anlage f <Sicherheit> enclosed equipment
gekapselte Induktanz f <Elektriz> sealed reactor
gekapselte Kontakte mpl <Elektrotech> sealed contacts
gekapselte Sicherung f <Elektriz, Elektrotech> enclosed fuse, sealed fuse
gekapselter Motor m 1. <Elektriz> closed motor, enclosed motor, sealed motor; 2. <Elektrotech, Maschinen> enclosed motor; 3. <Mechan> canned motor
gekapselter Transformator m <Elektriz> sealed transformer
gekehlt adj <Mechan> keyed
gekennzeichneter Name m <Comp & DV> qualified name
gekerbt adj 1. <Fertig> notched, scalloped, serrated; 2. <Maschinen> notched
gekerbte Mutter f <Maschinen> notched nut
gekettete Baumstruktur f <Comp & DV> threaded tree
gekettete Datei f 1. <Comp & DV> chained file, threaded file; 2. <Telekom> threaded file
gekettete Liste f <Comp & DV> chained list
gekettete Programmiersprache f <Comp & DV> threaded language
gekettetes Programm n <Comp & DV> thread
geklebt adj 1. <Druck> pasted; 2. <Elektriz> laminated
geklebte Bindung f <Druck> perfect binding
geklebtes Glasleinen n <Ker & Glas> bonded glass cloth
geknickte Freifläche f <Fertig> offset tool flank
geknickte Spanfläche f <Fertig> offset tool face
geknickter Linienzug m <Konstzeich> zigzag line
gekniffene Fadenauflage f <Ker & Glas> pinched thread
gekochte Stärke f <Lebensmittel> boiled starch
gekohlter Stahl m <Maschinen> carbon steel
gekoppelt adj 1. <Elektriz, Elektrotech> coupled; 2. <Maschinen> connected
gekoppelte Gleichungen fpl <Math> simultaneous equations
gekoppelte Kontakte f <Elektrotech> mated contacts
gekoppelte Moden fpl <Optik, Telekom> coupled modes
gekoppelte Motoren mpl <Fertig> coupled engines
gekoppelte Oszillatoren mpl <Phys> coupled oscillators
gekoppelte Stromkreise mpl <Elektriz> coupled circuits
gekoppelte Systeme npl <Phys> coupled systems
gekoppelte Verschluss- und Blendeneinstellung f <Foto> coupled speed and F-stop setting
gekoppelter Entfernungsmesser m <Foto> coupled rangefinder
gekoppeltes Potenziometerpaar n <Elektriz> dual-ganged potentiometer
gekräuselt adj <Textil> crimped
gekreppt adj <Papier> creped
gekreuzt adj <Maschinen> crossed
gekreuzte Nicol-Prismen npl <Phys> crossed Nicols
gekröpft adj <Maschinen> cranked, swan-necked
gekröpfte Achse f 1. <Kfztech> dropped axle (Triebstrang); 2. <Maschinen> dropped axle
gekröpfte Spule f <Elektrotech> bent coil
gekröpfter Gewindemeißel m <Fertig> offset single-point threading tool
gekröpfter Meißel m <Maschinen> goosenecked tool
gekröpfter Schraubenschlüssel m <Maschinen> (BE) bent spanner, gooseneck wrench, offset wrench
gekröpftes Drehherz n <Maschinen> bent-tail lathe dog
gekröpftes Werkzeug n <Maschinen> bent tool
gekrümmt adj 1. <Geom> curved; 2. <Ker & Glas> bent
gekrümmte Fläche f <Geom> curved surface
gekrümmte Linie f <Geom> curved line
gekrümmte Wand f <Bau> bulged wall

gekrümmter Flaschenhals m <Ker & Glas> bent neck
gekrümmter Lichtwellenleiter m <Optik, Telekom> bent optical fiber, curved optical fiber
gekrümmtes Schaufelblatt n <Hydraul> curved vane (Turbine, Kreiselpumpe)
gekühlt adj 1. <Heiz & Kälte> refrigerated; 2. <Lebensmittel> chilled; 3. <Thermod> chilled, cooled
gekühlte Abstellfläche f <Heiz & Kälte> refrigerated shelf area
gekümpelt adj <Fertig> dished
gekümpelter Boden m <Kerntech> dished bottom (eines Reaktordruckgefäßes)
gekuppelt adj 1. <Fertig> engaged, interlocked; 2. <Maschinen> coupled
gekuppelter verstellbarer Nutenfräser m <Fertig> interlocking mill
gekürzt zeichnen v <Konstzeich> draw in shortened form
gekürzte Fassung f <Druck> abridged edition
gekürzter Bruch m <Math> basic fraction
Gel n <Erdöl, Kunststoff> gel
geladene Kapazität f <Eisenbahn, Kohlen> loaded capacity
geladenes Teilchen n <Elektronik, Elektrotech, Kerntech, Strahlphys, Teilphys> charged particle
Gelände n <Bau> field
Gelände n für Schleudertraining <Kfztech> skidpad (Straßenfahrt)
Geländefahrzeug n <Kfztech> all-terrain vehicle
geländegängig adj <Kfztech> off-the-road, off-the-highway
geländegängiger LKW m <Kfztech> (BE) cross-country lorry, (AE) cross-country truck
geländegängiges Fahrzeug n <Kfztech> off-highway vehicle
Geländehöhe f 1. <Bau> level; 2. <Eisenbahn> ground level
Geländemotorrad n <Kfztech> cross-country motorcycle, scrambling motor cycle
Geländeneigung f <Bau> fall of ground
Geländer n 1. <Bau> balustrade, guard rail, handrail, railing; 2. <Mechan> handrail; 3. <Sicherheit> guard rail
Geländerausfachung f <Bau> paling
Geländerpfosten m <Bau> baluster
Geländespiegel m <Phys> periscope
Gelatine f <Druck, Verpack> gelatine
Gelatinekapsel f <Verpack> gelatine capsule
Gelatinieren n <Kunststoff> gelation
Gelatinierung f <Kunststoff> gelation
Gelatinierungsmittel n <Kunststoff> gelling agent
Gelb n <Foto> yellow
Gelbbrennsäure f <Chemie> aqua fortis
gelbes Katechu n <Chemie> gambier
Gelbfiltereinstellung f <Foto> minus blue filter adjustment, yellow filter adjustment
Gelbguss m 1. <Fertig> high brass; 2. <Metall> brass casting, high brass
Gelbkörperhormon n <Chemie> progesterone
Gelbsignaldauer f <Trans> hour of yellow signal indication
Gelbstich m <Textil> yellowness
gelbstichig adj <Textil> yellowish
Gelbstrohpapier n <Papier> yellow straw paper
Gelbstrohstoff m <Papier> yellow straw pulp
Gelchromatographie f <Labor> gel permeation chromatography
Gelcoat m <Kunststoff, Wassertrans> gel coat (Schiffbau)
Geldausgabeautomat m <Comp & DV> ATM, automatic teller machine
Geldkarte f <Telekom> payment card

Geldkarte f für Kartentelefon <Telekom> calling card, telephone card
Geldkartenfernsprecher m <Telekom> card-operated payphone, card payphone, card phone, chip-card telephone
geleiten v <Wassertrans> convoy (Schiff)
geleitete Welle f <Optik> guided wave
Gelenk n 1. <Bau> joint; 2. <Comp & DV> link; 3. <Ergon> joint; 4. <Kfztech> knuckle; 5. <Lufttrans> cardan; 6. <Maschinen> joint; 7. <Mechan> knuckle; 8. <Trans> articulated • **mit Gelenk verbunden** <Mechan> articulated
Gelenkarm m <Fertig> articulated arm
Gelenkbolzen m 1. <Bau> joint bolt; 2. <Maschinen> hinged bolt, pintle
Gelenkbus m <Trans> articulated bus
Gelenkfahrsteig m <Trans> (BE) articulated-type moving pavement, (AE) articulated-type moving sidewalk
Gelenkfahrzeug n <Kfztech> articulated vehicle, twister
Gelenkflügel m der Luftschraube <Lufttrans> articulated blade (Hubschrauber)
Gelenkhebebühne f <Fertig> articulated elevating platform
gelenkig adj <Bau> jointed
gelenkig verbunden adj <Mechan> linked
Gelenkkette f <Mechan> sprocket chain
Gelenkketteneinstellung f <Kfztech> sprocket and chain timing
Gelenkkreuz n <Kfztech> spider (Universalgelenk)
Gelenkkupplung f 1. <Maschinen> universal joint coupling; 2. <Trans> articulated coupling
Gelenklager n 1. <Maschinen> swing support; 2. <Mechan> ball-and-socket joint
Gelenkoberleitungsbus m <Kfztech> articulated trolleybus
Gelenkpunkt m <Phys> fulcrum
Gelenkrotor m <Lufttrans> articulated rotor (Hubschrauber)
Gelenkschienenfahrzeug n <Eisenbahn> articulated railcar
Gelenkstellung f <Ergon> joint posture
Gelenkstraßenbahn f <Kfztech> articulated tramway
gelenkter unbemannter Hubschrauber m <Lufttrans> drone helicopter
gelenktes Rad n <Kfztech> steered wheel (Lenkung)
Gelenktriebwagen m <Kfztech> (AE) articulated streetcar, (BE) articulated tramcar
Gelenkverbindung f 1. <Bau> hinge joint; 2. <Ergon> articulation; 3. <Maschinen> articulation, link joint
Gelenkwagen m <Kfztech> articulated car, (BE) articulated lorry, (AE) articulated truck
Gelenkwelle f 1. <Fertig> universal shaft; 2. <Kfztech> propeller shaft; 3. <Mechan> cardan shaft
Geleucht n <Sicherheit> miner's safety lamp
Gelierdauer f <Verpack> gel time
gelieren v <Chemtech> coagulate
Gelieren n <Chemtech> jellification
Geliermittel n <Kunststoff, Lebensmittel, Meerschmutz> gelling agent
Gelierung f <Chemtech> jellification
Gelierzeit f <Verpack> gel time
gelinde Hitze f <Thermod> gentle heat
gelinde kochen v <Thermod> boil slowly
gelinde Wärme f <Thermod> gentle heat
gelitzter Draht m <Elektriz, Elektrotech> stranded cable, stranded conductor
gelitzter Leiter m <Elektriz> stranded conductor
gelocht adj <Comp & DV> perforated
gelochte Dämmplatte f <Aufnahme> perforated absorbent tile

gelochte

gelochte Platte f <Metall> perforated plate
geloggte Entfernung f <Wassertrans> distance logged (Navigation)
gelöschter Kalk m <Lebensmittel> slaked lime
gelöst adj <Chemtech> dissolved
gelöste organische Substanz f <Umweltschmutz> dissolved organic matter
gelöster Sauerstoff m <Chemie> dissolved oxygen
gelotet adj <Fertig> cast
gelötete Blechverbindung f <Bau> plumb joint
Gel-Permeations-Chromatographie f (GPC) <Kunststoff, Labor> gel permeation chromatography (GPC)
Gelsemin n <Chemie> gelsemine
gelten v <Math> apply
Geltendmachung f <Patent> enforcement
Geltungsbereich m <Maschinen> scope
Geltungsdauer f <Bau> validity period
gelüftet adj <Heiz & Kälte, Thermod> ventilated
Gel-Zeit f <Kunststoff> gel time
Gel-Zelle f <Elektrotech> gel cell
gemahlen adj <Papier> beaten
gemallte Breite f <Wassertrans> (AE) molded breadth, (BE) moulded breadth (Schiffkonstruktion)
gemallte Seitenhöhe f <Wassertrans> (AE) molded depth, (BE) moulded depth (Schiffbau)
gemeiner Bruch m <Math> common fraction, vulgar fraction
gemeinsam benutzbarer virtueller Bereich m (SVA) <Comp & DV> shared virtual area (SVA)
gemeinsam benutzen v <Comp & DV> share
gemeinsam benutzte Datei f <Comp & DV> shared file
gemeinsam benutzter Speicher m <Comp & DV> shared memory
gemeinsam benutztes Speichersystem n <Comp & DV> shared memory system
gemeinsam genutzte Anwendung f <Comp & DV> collaborative application
gemeinsam nutzen v <Comp & DV> share
gemeinsame Abfrage f <Comp & DV> partitioning sensing
gemeinsame Account-Datenbank f <Comp & DV> common account database
gemeinsame Anmelder mpl <Patent> joint applicants
gemeinsame Benennung f <Patent> joint designation
gemeinsame Busleitung f <Telekom> common highway
gemeinsame Eigenschaften fpl <Strahlphys> common properties (elektromagnetischer Wellen)
gemeinsame Fußzeile f <Comp & DV> common footer
gemeinsame Katode f <Elektrotech> common cathode
gemeinsame Luftschnittstelle f <Telekom> common air interface
gemeinsame Nutzung f <Comp & DV> sharing
gemeinsame Nutzung f der Betriebsmittel <Comp & DV, Telekom> resource sharing
gemeinsame Rückleitung f <Eisenbahn> common return
gemeinsame Schnittstelle f <Fernseh> common interface, CI (Chipkartenschnittstelle)
gemeinsame Zahnhöhe f <Fertig> depth of engagement (Getriebelehre)
gemeinsame Zweigleitung f <Elektriz> common branch (elektrisches Versorgungsnetz)
gemeinsamer Anodenanschluss m <Elektrotech> common anode connection
gemeinsamer Bereich m <Comp & DV> common area
gemeinsamer Dateizugriff m <Comp & DV> file sharing
gemeinsamer Highway m <Telekom> common highway
gemeinsamer Nenner m <Math> common denominator
gemeinsamer Signalisierungskanal m <Telekom> centralized control signalling, common signalling channel (CSC); outslot signalling

gemeinsamer Speicher m <Telekom> common store
gemeinsamer Teiler m <Math> common factor, common ratio
gemeinsamer Träger m <Fernseh, Telekom> common carrier
gemeinsamer Zugriff m <Comp & DV> shared access
gemeinsamer Zweig m <Elektrotech> common branch (bei Strompfad)
gemeinsames Phänomen n <Phys> cooperative phenomenon
gemeinschaftlich stranggepresste Folie f <Verpack> coextruded film
gemeinschaftliches Fahrzeug n <Trans> public automobile
Gemeinschaftsantenne f 1. <Fernseh> community antenna; 2. <Telekom> collective aerial, collective antenna, common aerial, community aerial
Gemeinschaftseinrichtung f <Telekom> common equipment
Gemeinschaftskläranlage f <Wasserversorg> community sewage works, public water supply
Gemeinschaftsleitung f <Telekom> party line, shared service line
Gemeinschaftssystem n <Comp & DV> multiuser system
Gemenge n <Ker & Glas> batch (Glasherstellung)
Gemengecharge f ohne Scherben <Ker & Glas> batch charge without cullet
gemengefrei adj <Ker & Glas> batch-free
Gemengehaus n <Ker & Glas> batch house
Gemengehaut f <Ker & Glas> batch crust
Gemengemischung f <Ker & Glas> batch mixing
Gemengesatz m <Ker & Glas> batch pile
Gemengeschmelzgrenze f <Ker & Glas> (AE) batch-melting line, (BE) silica scum line
Gemengespeiser m <Ker & Glas> batch charger
Gemengestaub m <Ker & Glas> batch dust
Gemengestein m <Ker & Glas> batch stone
Gemengeturm m <Ker & Glas> batch tower
gemessene Schwelle f <Kerntech> observed threshold (einer Kernreaktion)
gemessene Spannung f <Elektriz> measured voltage
gemessener Stromwert m <Elektriz> measured current
gemessener Überdruck m <Heiz & Kälte> (AE) gage pressure, (BE) gauge pressure (Überdruck über atmosphärischem Druck)
gemessener Wert m <Gerät> measured value
Gemisch n 1. <Fertig> mix; 2. <Kunststoff> blend; 3. <Lebensmittel, Maschinen> mixture
Gemischregler m <Kfztech> mixture control unit
Gemischregulierung f <Lufttrans> mixture control (Motor und Triebwerk)
Gemischschmierung f <Maschinen> oil-in-gasoline lubrication
gemischt algebraischer Ausdruck m <Math> mixed algebraic expression
gemischt betriebene Leitung f <Telekom> both-way circuit
gemischtadriges Kabel n <Elektrotech> composite cable
Gemischtbasisschreibweise f <Comp & DV> mixed-base notation, mixed-radix notation
Gemischtbremsventil n <Eisenbahn> brake-blending valve
Gemischtdruck m <Druck> intermixed characters
gemischte Ablagerung f <Abfall> codeposition, codisposal
gemischte Datenaufzeichnung f <Comp & DV> MDR, miscellaneous data recording

gemischte Energieversorgung f <Trans> mixed power supply
gemischte Feldschwächung f <Elektriz> combined field-weakening
gemischte Strahlung f <Strahlphys> mixed radiation
gemischte Versetzung f <Metall> mixed dislocation
gemischter Prozess m <Elektronik> mixed process
gemischter Satz m <Druck> mixed styles
gemischtes Produkt n <Math> triple scalar product *(von Vektoren)*
gemischtpaariges Kabel n <Elektrotech> composite cable
genähert adj <Math> approximated
genau adj <Math, Mechan, Metrol, Phys> accurate • **genau achteraus** <Wassertrans> dead astern • **genau voraus** <Wassertrans> dead ahead
Genaubohren n <Maschinen> precision drilling
genaue Druckmarkierung f <Verpack> accurate print registration
genaue Karte f <Metrol> accurate map
genaue Mitte f <Mechan> *(AE)* dead center, *(BE)* dead centre
genaue Registrierung f <Verpack> accurate print registration
genaue Überprüfung f <Bau> meticulous inspection
genaue Wiedergabe f <Foto> accurate reproduction
Genauigkeit f 1. <Comp & DV, Maschinen, Mechan, Metrol, Qual, Phys> accuracy, precision; 2. <Wassertrans> accuracy *(des Schiffsstandortes)*
Genauigkeit f des Sollwertes <Kerntech> set point accuracy
Genauigkeitsgießform f <Fertig, Maschinen> *(AE)* investment mold, *(BE)* investment mould
Genauigkeitsklasse f <Gerät> class *(Messgerät)*
Genauigkeitsmaß n <Math> accuracy measure *(Schätzung)*
Genauigkeitsprüfung f <Maschinen> accuracy test
Genauigkeitsspiralbohrer m <Fertig> cylinder bit
genehmigen v <Qual, Sicherheit> approve
genehmigt adj <Patent, Qual, Sicherheit> approved
Genehmigung f 1. <Elektriz> approval; 2. <Funktech> *(BE)* licence, *(AE)* license *(Funkverwaltung)*; 3. <Patent, Qual, Trans> approval
geneigt adv **um einen Winkel von** <Geom> inclined at an angle of
geneigte Ebene f <Phys> inclined plane
geneigte Fläche f 1. <Geom> inclined plane; 2. <Maschinen> cant
geneigte Linie f <Konstzeich> sloping line
geneigter Dipol m <Funktech> drooping dipole
General... <Elektrotech, Wassertrans> master
Generalplan m <Wassertrans> general arrangement plan *(Schiffbau)*
Generalruf m <Telekom> global call
Generalschalter m <Elektrotech> master switch
Generalschlüssel m <Bau> master key
Generalüberholung f <Trans> overhaulage
Generalunternehmer m <Bau> general contractor
Generalverkehrsplan m <Trans> traffic master plan
Generator m 1. <Bau, Comp & DV, Elektriz, Elektrotech> alternator, electric generator, generator, lighting dynamo; 2. <Funktech, Hydraul> generator; 3. <Kerntech> generation time; 4. <Maschinen> producer; 5. <Phys> generator, voltage generator; 6. <Telekom> generator; 7. <Wassertrans> generator *(Elektrik)*
Generator m **für die Gittervorspannung** <Elektrotech> bias generator
Generator m **mit ausgeglichener Verbunderregung** <Elektrotech> *(BE)* level-compounded generator, *(BE)* flat-compounded generator
Generator m **mit ausgeprägten Polkanten** <Elektriz> salient pole generator
Generator m **mit Doppelwicklung** <Elektriz> double-wound generator
Generator m **mit erweitertem Luftspalt** <Elektriz> salient pole generator
Generator m **mit Über-Verbunderregung** <Elektrotech> over-compounded generator
Generator m **mit Unter-Verbunderregung** <Elektrotech> under-compounded generator
Generatoranlage f <Elektrotech> generating plant
Generatorbürste f <Kfztech> generator brush *(KFZ-Elektrik)*
Generatorgeschwindigkeit f <Nichtfoss Energ> generator speed
Generatorgruppe f <Elektrotech> generating set
generatorisches Bremsen n <Eisenbahn> dynamic braking
Generatorkohle f <Kohlen> generator coal, producer coal
Generatorleistung f <Elektrotech> generator output power
Generatorsatz m <Elektrotech> generating set
Generator-Signalgabe f <Elektrotech> *(AE)* generator signaling, *(BE)* generator signalling
Generatrix f <Geom> generatrix
generelles Problemlösen n <Künstl Int> GPS, general problem solving
generelles Zugangsprofil n <Telekom> generic access profile, GAP *(Cordless-Telefonie-Standard)*
generieren v <Geom> generate
generierter Fehler m <Comp & DV> generated error
Generierung f <Comp & DV> generation
Generierungsprogramm n <Comp & DV> generating program, generator
generisch adj <Comp & DV> generic
generische Kaskade f <Kerntech> generic cascade
generischer Name m <Comp & DV> generic name
generisches Zugangsprofil n <Telekom> generic access profile, GAP *(Cordless-Telefonie-Standard)*
genetisch wichtige Dosis f <Umweltschutz> genetically significant dose
genietete Überlappung f <Maschinen> riveted lap joint
Genistein n <Chemie> genistein, prunetol
genormte Bemessungswerte mpl <Heiz & Kälte> standard ratings
genormte Schnittstelle f <Telekom> standard interface, standardized interface
Gentex n *(Telegrammwähldienst)* <Telekom> gentex *(general telegraph exchange)*
Gentianin n <Chemie> gentianin
Gentiobiose f <Chemie> gentiobiose
Gentiopikrin n <Chemie> gentiopicrin
Gentiopikrosid n <Chemie> gentiopicrin
Gentisin n <Chemie> gentisin
Gentisin... <Chemie> gentisic
Genua f <Wassertrans> genoa *(Segel)*
Genusssäure f <Lebensmittel> edible acid
genutet adj 1. <Bau> slotted; 2. <Fertig> grooved, slotted
genuteter Anker m <Elektrotech> slotted armature
genutetes Kabel n <Optik> grooved cable
Geodäsie f <Geom> geodesy
Geodäte f <Geom> geodesic
geodätisch adj <Wassertrans> geodesic, geodetic
geodätische Linie f <Geom> geodesic
Geodimeter n <Bau> geodimeter
geoelektrisches Log n <Erdöl> electric log
geöffnet adj <Phys> open *(Stromkreis)*
geöffnete Schaltung f <Elektrotech> broken circuit
geöffneter Kontakt m <Elektriz> open contact

geographische Breite f <Wassertrans> latitude *(Navigation)*
geographische Flugbahn f <Lufttrans> required flightpath
geographische Länge f <Wassertrans> longitude *(Navigation)*
geographische Sichtweite f <Wassertrans> geographical range *(Navigation)*
Geologie f <Erdöl, Kohlen> geology
geologische Vermessung f <Erdöl, Kohlen> geological survey
geölte Lager npl <Maschinen> oiled bearings
geomagnetisch adj <Raumfahrt> geomagnetic
geomagnetische Abrissenergie f <Raumfahrt> geomagnetic cut-off energy
geomagnetische Albedo f <Raumfahrt> geomagnetic albedo *(Raumschiff)*
Geomagnetismus m <Phys, Raumfahrt> geomagnetism
Geometer n <Geom> geometer, geometrician
Geometrie f <Elektrotech, Fernseh, Funktech, Kerntech> geometry
Geometrie-Eichung f <Fernseh> geometric calibration
Geometriefaktor m <Kerntech> geometry factor
Geometriefehler m <Fernseh> geometric error
Geometrieverhältnis n <Elektrotech> aspect ratio *(MOSFET)*
geometrisch adj <Geom> geometric, geometrical
geometrische Auflösungslänge f <Kerntech> geometrical resolution length
geometrische Datenverarbeitung f <Geom> computational geometry, computer geometry
geometrische Eigenschaften fpl <Geom> geometric properties
geometrische Figur f <Geom> geometric figure
geometrische Fläche f <Geom> geometric surface
geometrische Form f <Geom> geometric shape
geometrische Optik f 1. <Optik> geometric optics; 2. <Phys> geometrical optics; 3. <Telekom> geometric optics
geometrische Reihe f 1. <Geom> geometric progression; 2. <Math> geometric series
geometrische Steigung f <Lufttrans> geometrical pitch *(Propeller)*
geometrische Strahlauflösung f <Kerntech> geometric beam resolution
geometrische Trennung f <Kerntech> separation by geometry
geometrische Werte mpl <Bau> geometrical data
geometrischer Körper m <Geom> geometric solid
geometrischer Mittelwert m <Math, Qual> geometric mean
geometrischer Ort m <Geom> geometric locus, locus
geometrischer Werkzeugspanwinkel m <Fertig> tool geometrical rake
geometrisches Buckling n (Bg) <Kerntech> geometric buckling (Bg)
geometrisches Mittel n <Comp & DV, Geom, Math, Qual> geometric mean
Geomorphologie f <Bau> geomorphology
Geophon n <Erdöl, Kohlen> geophone
Geophysik f <Erdöl, Kohlen, Phys> geophysics
geophysikalisch adj <Bau, Kohlen> geophysical
geophysikalische Aufnahme f <Bau> geophysical survey
geophysikalische Untersuchung f <Erdöl> geophysical survey *(Geologie)*
geophysikalisches Ingenieurwesen n <Bau> geophysical engineering
geophysikalisches Log n <Erdöl> geophysical log *(Messtechnik)*
geophysikalisches Schürfen n <Bau> geophysical prospecting
geordnete Ablagerung f <Abfall> proper disposal, sanitary landfilling, sound disposal
geordnete Beseitigung f <Abfall> safe disposal
geordnete Datenmenge f <Comp & DV> data set document
geordnete Deponie f <Abfall> controlled dumping, controlled tipping, sanitary landfill
geordnete Festlösung f <Metall> ordered solid solution
geordnete Legierung f <Metall> ordered alloy
geordnete Liste f <Comp & DV> ordered list
geordnete Menge f <Math, Metall> ordered set *(Menge mit Ordnungsrelation)*
geordnete Suche f <Künstl Int> ordered search
geordneter Baum m <Comp & DV> ordered tree
geordnetes Ablagern n <Abfall> landfilling
geordnetes Paar n <Math> ordered pair
geostationär adj <Funktech> geostationary
geostationäre Bahn f <Phys> geostationary orbit
geostationäre Erdumlaufbahn f <Raumfahrt> geostationary earth orbit *(GEO)*
geostationäre Satellitenrundfunk-Raumstation f <Fernseh, Raumfahrt> broadcasting geostationary satellite space station
geostationäre Satellitenumlaufbahn f <Raumfahrt> geostationary satellite orbit *(GSO)*
geostationärer Satellit m <Fernseh, Phys, Raumfahrt, Telekom, Wassertrans> geostationary satellite *(Navigation, Satellitenfunk)*
geostationärer Wettersatellit m <Telekom> geostationary meteorological satellite *(GMS)*
geostatischer Druck m <Erdöl> geostatic pressure *(Geologie)*
geosynchron adj <Funktech, Raumfahrt> geosynchronous
Geosynklinale f <Erdöl> geosyncline *(Geologie)*
Geotechnik f <Bau> geotechnics
geothermale Energie f <Umweltschmutz> geothermal energy
Geothermik f <Erdöl, Nichtfoss Energ> geothermics *(Erdformationswärme)*
geothermisch adj <Nichtfoss Energ> geothermal
geothermische Anlage f <Nichtfoss Energ> geothermal plant
geothermische Bohrausrüstung f <Nichtfoss Energ> geothermal drilling equipment
geothermische Energie f 1. <Nichtfoss Energ> geothermal energy, geothermal power; 2. <Phys> geothermal energy
geothermische Quellen fpl <Nichtfoss Energ> geothermal resources
geothermische Tiefenstufe f <Nichtfoss Energ> geothermal gradient
geothermischer Dampf m <Nichtfoss Energ> geothermal steam
geothermischer Gradient m <Erdöl> geothermal gradient *(Geophysik)*
geothermischer Kreislauf m <Nichtfoss Energ> geothermal circuit
geothermisches Feld n <Nichtfoss Energ> geothermal field
geothermisches Kraftwerk n <Elektriz, Elektrotech> geothermal power station
geothermisches Log n <Erdöl, Nichtfoss Energ> geothermal log *(Messtechnik)*
gepaartes Kabel n <Elektrotech> paired cable
Gepäck n <Eisenbahn> baggage, luggage
Gepäckabteil n <Eisenbahn> baggage compartment, *(AE)* baggage room, *(BE)* luggage compartment

Gepäckannahmestelle f <Trans> parcels office
Gepäckausgabeband n <Lufttrans> baggage claim belt
Gepäckfördereinrichtung f <Trans> bag conveyor
Gepäckkarren m <Trans> barrow
Gepäcklader m <Lufttrans> baggage loader
Gepäcknetz n <Eisenbahn> rack
Gepäckraum m <Eisenbahn> vestibule
Gepäckschlepper m <Eisenbahn> luggage trolley
gepackte Dezimalzahl f <Comp & DV> packed decimal
Gepäckterminal n <Trans> baggage terminal
gepacktes Format n <Comp & DV> packed format
Gepäckwagen m <Eisenbahn> (AE) baggage car, (BE) luggage van
gepanzert adj <Elektrotech> (AE) armored clad, (BE) armoured clad
gepanzerter Transformator m <Elektrotech> shielded transformer
gepanzertes Kabel n <Elektrotech> shielded cable
gepanzertes Transportfahrzeug n <Trans> AT vehicle, (AE) armored transport vehicle, (BE) armoured transport vehicle
geparkt adj <Telekom> parked
geparkte Leitung f <Telekom> parked line
gepastete Form f <Ker & Glas> (AE) paste mold, (BE) paste mould
gepfeilter Flügel m <Lufttrans> back-swept wing
gepfeilter Flügel m/nach vorn <Lufttrans> forward-swept wing
geplant beschleunigte Bauelemente-Alterung f <Elektriz> (BE) intentional accelerated component ageing, (AE) intentional accelerated component aging, (BE) intentional component ageing, (AE) intentional component aging, (BE) intentional normal component ageing, (AE) intentional normal component aging
geplante Nichtverfügbarkeit f <Qual> scheduled outage
geplatztes Brennelement n <Kerntech> burst can, burst slug
gepolstert adj <Mechan> padded
gepolter Elektrolytkondensator m <Elektrotech> polarized electrolytic capacitor
gepolter Kondensator m <Elektrotech> polarized capacitor
gepolter Stecker m <Elektrotech> polarized plug
gepolter Verbinder m <Elektrotech> polarized connector
gepoltes Relais n <Elektriz, Elektrotech> polarized relay
gepoltes Relais n ohne Vorspannung <Elektriz> unbiased-polarized relay
geprägt adj <Maschinen> embossed
geprüft adj <Qual> tested
geprüfte Sicherheit f <Sicherheit> proved safety
geprüfter Sicherheitsbereich m <Sicherheit> approved safety area
gepulste Elektronenkanone f <Strahlphys> pulsed electron gun
gepulste Elektronenquelle f <Strahlphys> pulsed electron gun
gepulste Strömung f <Strömphys> pulsating flow
gepulster Laser m <Elektronik> pulsed laser
gepulster Laser m/mit kurzen Impulsen <Wellphys> short-pulsed laser
gequantelt adj <Phys> quantized
Geraddrehen n <Fertig> straight turning
gerade adj 1. <Bau> square-headed (Türöffnung); 2. <Comp & DV> even; 3. <Druck> regular; 4. <Math> even (Zahl, Funktion)
Gerade f <Geom> line, straight line
gerade Destillation f <Thermod> distillation by ascent
gerade Frequenz f <Elektrotech> straight line frequency
gerade Funktion f <Math> even function (Graph ist Ordinate-symmetrisch)

gerade Kante f <Ker & Glas> straight edge
gerade Linie f <Phys> straight line
gerade Packung f <Ker & Glas> straight packing
gerade Parität f <Comp & DV> even parity (Prüfung auf gerade Bitzahl)
gerade Schleifscheibe f <Maschinen> straight wheel, straight-grinding wheel
gerade Seite f <Druck> verso
gerade Stirn f <Fertig> square end (Schaftfräser)
gerade Zahl f <Math> even number, even numerate
Geradeausanflug m <Lufttrans> straight-in approach
Geradeausempfang m <Funktech> straight reception
Geradeaustest m <Telekom> quick test
Geradeausverkehr m <Trans> straight-through traffic
Geradeausverstärkung f <Funktech> straight amplification
geradeeinfaches Herzstück n <Eisenbahn> straight common crossing
geradegedrehte Riemenscheibe f <Maschinen> straight-faced pulley
Geradengleichung f <Geom> line equation
gerader Durchlass m <Ker & Glas> straight throat
gerader Frequenzgang m <Aufnahme, Funktech> flat frequency response
gerader Kreiskegel m <Geom> right circular cone
gerader Kreiszylinder m <Geom> right circular cylinder
gerader Rücken m <Druck> square back
gerader Spleiß m <Elektriz> straight joint
gerader Strang m <Eisenbahn> straight track
gerader Treppenlauf m <Bau> flyers
gerader Vorsteven m <Wassertrans> straight stem (Bootbau)
gerades Band n <Bau> T-hinge, cross-garnet hinge
geradestehend adj <Druck> regular
geradestoßen v <Druck> jog
Geradführung f <Maschinen> slide bar, straight guide
Geradheit f <Fertig> straightness
Geradheitstoleranz f <Maschinen> straightness tolerance
geradlinig adj <Mechan> rectilineal
geradlinige Abtastung f <Fernseh> rectilinear scanning
geradlinige Antenne f <Telekom> rectilinear antenna
geradlinige Ausbreitung f <Wellphys> rectilinear propagation (von Lichtwellen)
geradlinige Bewegung f <Mechan, Phys> rectilineal motion
Geradlinigkeitsprüfinstrument n <Metrol> straightness-measuring instrument
Geradsichtprisma n <Phys> Amici prism, direct-vision prism
Geradstirnrad n <Maschinen> spur gear, spur wheel
geradverzahnt adj <Maschinen> straight
geradverzahntes Rad n <Maschinen> straight tooth wheel
Geradverzahnung f 1. <Fertig> straight gear cutting; 2. <Maschinen> spur teeth
geradzahlige harmonische Schwingungen fpl <Phys> even harmonic vibrations
geradzahlige Oberwelle f <Elektronik, Funktech> even harmonic
geradzahlige Parität f <Phys> even parity
geradzahlig-geradzahliger Kern m <Phys> even-even nucleus
geradzahlig-ungeradzahliger Kern m <Phys> even-odd nucleus
geradzahnig adj <Maschinen> straight-tooth
Geradzahnrad n <Maschinen> spur wheel
gerafft adj <Textil> gathered
gerahmtes Fenster n <Comp & DV> tiled window

gerammter Bereich m <Kerntech> rammed area *(als Lager)*
gerändelt adj <Mechan> knurled
Geraniol n <Chemie> geraniol
Geraniumaldehyd n <Chemie> citral, geranialaldehyde
Geranyl... <Chemie> geranyl
Geranylalkohol m <Chemie> geraniol
Gerät n 1. <Bau> equipment, utensil; 2. <Comp & DV> device, instrument; 3. <Kontroll> device; 4. <Lebensmittel> appliance; 5. <Maschinen> appliance, device, tackle; 6. <Telekom> device, equipment, unit; 7. <Textil> appliance
Gerät n **mit seriellem Zugriff** <Comp & DV> serial access device
Gerät n **mit wahlfreiem Zugriff** <Comp & DV> random access device
Gerät n **zur Signaleingabe** <Regelung> device for signal input
Geräte npl <Gerät, Maschinen> equipment
Geräte npl **der Unterhaltungselektronik** <Elektrotech> consumer electronic equipment
Geräteanordnung f <Comp & DV, Telekom> hardware configuration
Geräteausfall m 1. <Gerät> instrument malfunction; 2. <Sicherheit> equipment outage
Geräteausrüstung f <Fertig> instrumentation
Geräteausstattung f <Fertig> instrumentation
Geräteerkennung f <Elektronik, Gerät> device identification *(im Automatisierungssystem)*
Gerätefehler m <Gerät> instrumental error
Gerätegehäuse n <Gerät> instrument cabinet
Gerätegruppe f <Comp & DV> cluster
Gerätekennung f <Comp & DV> device code, device flag
Gerätekennungsdatei f <Mobilkom> equipment identity register
Gerätekonfiguration f <Comp & DV, Telekom> hardware configuration
Gerätekonstruktion f <Ergon, Sicherheit> equipment design
Gerätelärm m <Sicherheit> equipment noise
Gerätelebenserwartung f <Raumfahrt> commercial life
Gerätemotor m <Kontroll> system motor
Geräteprüfung f 1. <Comp & DV> hardware check; 2. <Sicherheit> equipment check
Gerätesatz m <Elektrotech, Gerät> set
Geräteschicht f <Telekom> equipment layer
Geräteschnittstelle f <Comp & DV> terminal interface
Geräteschuppen m <Bau> tool shed
Gerätesicherheit f <Sicherheit> equipment safety, mechanical equipment safety, mechanical safeguarding
Gerätestecker m <Elektrotech> connector, power line connector
Gerätesteckvorrichtung f <Elektrotech> coupler
Gerätesteuerprogramm n <Comp & DV> peripheral software driver
Gerätesteuerung f <Comp & DV, Telekom> DC, device controller
Gerätetechnik f <Telekom> hardware
gerätetechnisch adj <Gerät, Regelung> device-related
gerätetechnische Ausrüstung f <Gerät> instrumentation
Geräteträger m <Gerät> instrument rack
Gerätetreiber m <Comp & DV> device driver
Gerätetyp m <Elektrotech, Gerät> type of device
Geräteverbundgruppe f <Kontroll> cluster
Gerätezuverlässigkeit f <Sicherheit> equipment reliability
Geräusch n <Akustik, Elektronik, Phys, Telekom> noise
Geräuschabstand m <Telekom> signal-to-noise ratio
geräuscharm adj <Sicherheit> low-noise

geräuscharm startendes und landendes Flugzeug n <Lufttrans> QTOL aircraft, quiet takeoff and landing aircraft
geräuscharme Hupe f <Kfztech> low-tone horn
geräuscharmer Modus m <Druck> quiet mode
geräuscharmes Aufnahmesystem n <Aufnahme> noiseless-recording system
geräuscharmes Kurzstart- und Landeflugzeug n <Lufttrans> QSTOL aircraft, quiet short takeoff and landing aircraft
Geräuschbewertungsfaktor m <Telekom> psophometric weighting factor
geräuschdämpfend adj <Akustik> antinoise
Geräuschdämpfer m <Heiz & Kälte> *(AE)* muffler, *(BE)* silencer
Geräuscheffekte mpl <Fernseh> sound effects
Geräuschemission f <Sicherheit> noise emission
Geräuschentwicklung f <Sicherheit> noise emission
Geräuschfilter n <Aufnahme> scratch filter
Geräusch-Instrumentarium n <Aufnahme> sound equipment
Geräuschkulisse f <Aufnahme> sound effects
geräuschlos adj <Sicherheit> noiseless, silent
geräuschlose Steuerkette f <Kfztech> noiseless timing chain
geräuschloser Lauf m <Sicherheit> noiseless running
geräuschloser Motor m <Sicherheit> noiseless motor
Geräuschmessgerät n <Gerät> interference level meter, sound level meter
Geräuschmodulation f <Aufnahme> noise modulation
Geräuschpegel m <Sicherheit, Telekom, Umweltschmutz> noise level
Geräuschpegelanzeiger m <Umweltschmutz> weighted noise level indicator
Geräuschsperre f <Telekom> squelch
Geräuschspur f <Aufnahme> buzz track *(auf Band)*
Geräuschverhalten n <Sicherheit> noise characteristics
geräuschvolle Anlage f <Sicherheit> noisy equipment
geraut adj <Fertig> raised
gerbstoffartig adj <Chemie> tannic
gereckte Polypropylenfolie f <Verpack> orientated polypropylene film
gereckte Salzblasen fpl <Ker & Glas> *(AE)* gray blibes, *(BE)* grey blibes
gerecktes Polypropylen-Etikett n <Verpack> orientated polypropylene label
geregelt adj <Kontroll> regulated, stabilized
geregelte Ablagerung f <Umweltschmutz> regulated deposition
geregelte Temperatur f <Heiz & Kälte> controlled temperature
geregelter Druck m <Maschinen> controlled pressure
geregeltes Bussystem n <Raumfahrt> regulated bus system *(Raumschiff)*
geregeltes Netzgerät n <Elektrotech> regulated power supply
geregeltes Verbrennungssystem n <Lufttrans> controlled combustion system
gereinigtes Wasser n <Umweltschmutz> depolluted water
gerichtete Bohrung f <Erdöl> directional well *(Bohrtechnik)*
gerichteter Graph m <Künstl Int> digraph, directed graph, oriented graph
gerichteter Lautsprecher m <Akustik> directional loudspeaker
gerichteter Strahl m <Elektronik> shaped beam
gerichtetes Glied n <Regelung> directional element
gerieft adj <Fertig> serrated, striated

geriffelt *adj* 1. <Fertig> corrugated; 2. <Mechan> fluted, knurled
geringe Steigung *f* <Lufttrans> fine pitch *(Luftschraube)*
geringfügiges Gleiten *n* <Metall> banal slip
geringpermeabler Grundwasserleiter *m* <Wasserversorg> aquiclude, aquitard
geringwertige Kohle *f* <Kohlen> inferior coal
Gerinne *n* 1. <Bau> flume, gutter, launder; 2. <Nichtfoss Energ> flume, sluice box; 3. <Wasserversorg> raceway, race, sluice box
gerinnen *v* 1. <Chemtech> coagulate; 2. <Fertig> congeal
gerinnen lassen *v* <Chemtech> coagulate
Gerinnen *n* <Chemie> clotting
Gerinnung *f* <Lebensmittel> coagulating
Gerinnungsmittel *n* <Chemtech> coagulator
Gerippe *n* 1. <Bau> stud *(Fachwerk)*; 2. <Lufttrans> hull
gerippt *adj* 1. <Heiz & Kälte> finned; 2. <Mechan, Textil> ribbed
gerippte Fläche *f* <Heiz & Kälte> finned surface
gerippte Platte *f* <Bau> ribbed slab
geripptes Glas *n* <Ker & Glas> reeded glass
gerissen *adj* <Bau, Mechan, Metall> cracked
gerissene Blase *f* <Ker & Glas> broken seed
gerissene Oberfläche *f* <Ker & Glas> checked finish, crizzled finish
gerissene Schwelle *f* <Eisenbahn> *(BE)* split sleeper, *(AE)* split tie
gerissener Faden *m* <Textil> broken end
Germanatglas *n* <Ker & Glas> germanide glass
Germanium-Avalanche-Photodiode *f* <Elektronik> germanium avalanche photodiode
Germaniumdiode *f* <Elektriz> germanium diode
Germaniumgleichrichter *m* <Elektriz, Elektrotech> germanium rectifier
Geröll *m* <Bau> pebbles
Geröllblock *m* <Bau> boulder *(Geologie)*
geronnen *adj* <Chemtech> coagulated
geronnene Milch *f* <Lebensmittel> curd
Geruch *m* <Abfall, Chemtech> *(AE)* odor, *(BE)* odour
geruchlos *adj* <Verpack> nonodorous
Geruchsbeeinflussung *adj*/**frei von** <Verpack> *(AE)* odor proof, *(BE)* odour proof
Geruchsbekämpfung *f* <Abfall, Chemtech, Umweltschmutz> *(AE)* odor control, *(BE)* odour control
Geruchsbelästigung *f* <Abfall> *(AE)* odor nuisance, *(BE)* odour nuisance
Geruchsemission *f* <Umweltschmutz> *(AE)* odor emissions, *(BE)* odour emissions
geruchsintensiver Stoff *m* <Sicherheit, Umweltschmutz> *(AE)* odor-bearing substance, *(BE)* odour-bearing substance
Geruchsstoff *m* <Sicherheit> odorous agent
geruchsverbessernder Stoff *m* <Umweltschmutz> deodorant
Geruchsverschluss *m* 1. <Bau> drain trap, stench trap, stink trap, trap; 2. <Chemtech> trap
gerufener Teilnehmer *m* <Telekom> called customer, called party, called subscriber
gerufenes Telefon *n* <Telekom> called telephone
Gerümpel *n* <Abfall> litter
gerundete Zahlen *fpl* <Math> rounded numbers
Gerüst *n* 1. <Bau> gantry, scaffold, scaffolding; 2. <Maschinen> skeleton; 3. <Mechan> frame • **Gerüst abbauen** <Bau> take down scaffolding
Gerüstbau *m* <Bau> staging
Gerüstbock *m* <Bau> trestle
Gerüstboden *m* <Bau> stage
Gerüstbohle *f* <Bau> scaffold board
Gerüsteiweiß *n* <Lebensmittel> scleroprotein

Gerüsteiweißstoff *m* <Lebensmittel> albuminoid
Gerüstfangnetz *n* <Sicherheit> scaffolding protective net
Gerüstpfosten *m* <Bau> standard
Gerüststange *f* <Bau> putlog, scaffold pole
gesamt *adj* <Bau, Maschinen, Qual> overall, total
Gesamt... 1. <Bau, Maschinen> total; 2. <Trans> overall
Gesamtabfallaufkommen *n* <Umweltschmutz> total appearance of waste
Gesamtablagerung *f* <Umweltschmutz> total deposition
Gesamtabmessungen *fpl* <Maschinen> overall dimensions
Gesamtabnahme *f* <Bau> general acceptance *(eines Projektes)*
Gesamtachsstand *m* <Eisenbahn> total wheelbase
Gesamtauftrieb *m* <Lufttrans> total lift
Gesamtausfall *m* 1. <Sicherheit> catastrophic failure; 2. <Telekom> total failure
Gesamtaußenmaß *n* <Qual> overall dimensions
Gesamtbaustellenbereich *m* <Bau> gross site area
Gesamtbearbeitungszeit *f* <Comp & DV> processing time
Gesamtbeladung *f* <Kerntech> total charge
Gesamtbetrag *m* <Comp & DV> sum
Gesamtbildschirmeditor *m* <Comp & DV> full screen editor
Gesamtbreite *f* <Maschinen> overall width
Gesamtdämpfungsverlauf *m* <Akustik> overall response curve
Gesamtdrehimpuls *m* <Phys> total angular momentum
Gesamtdrehimpulsquantenzahl *f* <Phys> total angular momentum quantum number
Gesamtdruck *m* <Maschinen, Qual> overall pressure
gesamte mittlere freie Weglänge *f* <Kerntech> total mean free path
Gesamtenergie *f* <Maschinen> total energy
Gesamtenergiedichte *f* <Akustik> total energy per unit volume
gesamter Stahlverbrauch *m* <Metall> steel intensity
gesamtes atomares Bremsvermögen *n* <Phys> total-atomic stopping power
gesamtes lineares Bremsvermögen *n* <Phys> total-linear stopping power
Gesamtfahrzeit *f* <Eisenbahn, Trans> overall travel time
Gesamtfehler *m* <Gerät> total error
Gesamtfläche *f* 1. <Bau> total area; 2. <Nichtfoss Energ> gross area *(eines Kollektors)*; 3. <Papier> overall face
Gesamtfluggewicht *n* <Lufttrans> AUW, all-up weight
Gesamtflugmasse *f* <Lufttrans> AUW, all-up weight
Gesamtfluss *m* <Kerntech> gross flow
Gesamtgewicht *n* **des Fahrzeugs** <Kfztech> gross vehicle weight
Gesamtgröße *f* <Metrol> total size
Gesamtheizfläche *f* <Heiz & Kälte> aggregate heating surface
Gesamthöhe *f* 1. <Kerntech> total height; 2. <Maschinen, Qual> overall height
Gesamtkonfiguration *f* <Telekom> total configuration
Gesamtlänge *f* 1. <Bau> overall length; 2. <Maschinen> length overall, overall length; 3. <Metrol, Qual, Textil> overall length; 4. <Wassertrans> length overall *(Schiff)*
Gesamtlautheit *f* <Akustik> total loudness
Gesamtmassenbremsvermögen *n* <Phys> total mass stopping power
Gesamtölvolumen *n* <Erdöl> oil in place *(Erdölgeologie)*
Gesamtreisegeschwindigkeit *f* <Eisenbahn, Trans> overall travel speed
Gesamtsäurezahl *f* *(GSZ)* <Chemie> total acid number *(TAN)*
Gesamtschallpegel *m* <Sicherheit> overall noise level

Gesamtschattierung

Gesamtschattierung f <Textil> overall shade
Gesamtschub m <Lufttrans> gross thrust
Gesamtschutz m <Sicherheit> all-round protection, overall protection, total protection
Gesamtschutzanzug m <Sicherheit> whole-body suit
Gesamtschwefelgehalt m <Chemie> (AE) total sulfur, (BE) total sulphur
Gesamtstrahlungs-Pyrometer n <Gerät, Thermod> total radiation pyrometer
Gesamtsumme f 1. <Math> amount; 2. <Qual> sum total, total
Gesamtteilung f <Maschinen> total pitch
Gesamttiter m <Chemie, Textil> total denier
Gesamtüberdeckungsgrad m <Fertig> total contact ratio (Getriebelehre)
Gesamtverfügbarkeit f <Telekom> overall availability
Gesamtverluste mpl <Elektriz> total losses
Gesamtvolumenstrom m <Heiz & Kälte> total volumetric flow
Gesamtwiderstand m 1. <Elektriz> total resistance; 2. <Lufttrans> total drag
Gesamtwirkungsgrad m <Heiz & Kälte, Kohlen, Nichtfoss Energ, Qual> overall efficiency
Gesamtzeitintervall n <Trans> overall time interval
Gesamtzonenbeanspruchung f <Kohlen> area load
gesättigt adj <Heiz & Kälte> saturated
gesättigt alicyclisch adj <Chemie> naphthenic
gesättigte Erde f <Kohlen> saturated soil
gesättigte Logik f <Elektronik> saturated logic
gesättigte Luft f <Bau> saturated air
gesättigter Dampf m 1. <Phys> (AE) saturated vapor, (BE) saturated vapour; 2. <Thermod> saturated steam
gesättigter Kern m <Elektrotech> saturated core
gesättigter Kohlenwasserstoff m <Erdöl> alkane; saturated hydrocarbon (Geologie)
gesättigter Ringkerntransformator m <Elektrotech> saturated toroidal transformer
gesättigter Transformator m <Elektrotech> saturated transformer
geschachtelte Intervalle npl <Math> nest of intervals
Geschäftsdrucksachen fpl <Druck> business stationery
Geschäftsflugzeug n <Lufttrans> business aircraft, executive aircraft
Geschäftshubschrauber m <Lufttrans> executive helicopter
Geschäftskommunikationssystem n <Telekom> business communication system
Geschäftsnetz n <Telekom> business system
Geschäftsobjektbibliothek f <Comp & DV> business object's library
Geschäftsprozessablauf m <Comp & DV> business process flow
Geschäftssystemmanagement n <Comp & DV> business system management
geschälter Reis m <Lebensmittel> hulled rice
geschaltet adj <Telekom> switched
geschaltete Röhre f/als Diode <Elektronik> diode-connected tube
geschaltete Strecken fpl/**in Reihe** <Telekom> series of links
geschalteter Gleichstrommotor m/**in Reihe** <Elektrotech> series dc motor
geschalteter Strom m <Elektriz> switched current
geschalteter Transistor m/**als Diode** <Elektronik> diode-connected transistor
geschätzte verstrichene Zeit f <Lufttrans> estimated elapsed time
geschäumtes Polystyrol n (Schaum-PS) <Verpack> expanded polystyrene (ep)

geschichtet adj 1. <Kohlen> stratified; 2. <Strömphys> laminar
geschichtete Probe f <Sicherheit> stratified sampling
geschichtete Probenahme f <Kohlen> stratified sampling
geschichtete Stichprobe f <Qual> stratified sample
geschichtete Strömung f <Strömphys> stratified flow
geschichtete Unterlegscheibe f <Maschinen> peel shim
geschichtete Zufallsstichprobe f <Qual> stratified random sample
geschichteter Kunststoff m <Elektriz> laminated plastic
Geschiebe n <Kohlen> boulder
Geschiebeboden m <Kohlen> boulder soil
Geschiebefracht f <Wasserversorg> sediment discharge
Geschiebelehm m <Kohlen> boulder clay, glacial clay, till
Geschiebemergel m <Kohlen> glacial clay
Geschirr n 1. <Fertig> tackle; 2. <Raumfahrt> safety harness
Geschirreinschiebevorrichtung f <Ker & Glas> ware pusher
Geschirrkeramik- und Haushaltsglasindustrie f <Ker & Glas> tableware and domestic glass industry
geschlepptes Wasserfahrzeug n <Wassertrans> tow
geschlichtete Kette f <Textil> sized warp
geschliffene Kante f <Ker & Glas> polished edge
geschliffenes Drahtglas n <Ker & Glas> polished-wired glass
geschliffenes Glas n <Ker & Glas> cut glass
geschlitzt adj <Fertig> notched
geschlitzter Wellenleiter m <Elektrotech> slotted waveguide
geschlossen adj 1. <Abfall> sealed; 2. <Fertig> compact
geschlossen adj/**in sich** <Bau, Phys> self-contained
geschlossene Benutzergruppe f (CUG) <Comp & DV, Telekom> closed user group (CUG)
geschlossene Heizungsanlage f <Heiz & Kälte> closed-type heating system
geschlossene Kompostierung f <Abfall> mechanical composting, rapid fermentation
geschlossene Kurve f <Geom, Math> closed curve
geschlossene Maßkette f <Konstzeich> closed chain dimensioning, closed chain dimension
geschlossene Membran f <Akustik> closed diaphragm
geschlossene Schleife f <Kontroll> closed loop
geschlossene Stellung f <Elektriz> closed position
geschlossene Zelle f <Kunststoff> closed cell
geschlossene Zugeinheit f <Eisenbahn> unsplittable train
geschlossener Benutzerkreis m (CUG) <Comp & DV, Telekom> closed user group (CUG)
geschlossener Betrieb m <Comp & DV> closed shop operation
geschlossener Brennstoffkreislauf m <Kerntech> closed fuel cycle
geschlossener Container m <Trans> closed container, covered container
geschlossener Getriebekasten m <Maschinen> enclosed gears
geschlossener Gleichrichter m <Elektrotech> sealed rectifier
geschlossener Hohlkastenträger m <Trans> closed box girder
geschlossener Kreislauf m 1. <Lufttrans> closed loop; 2. <Maschinen> closed circuit, closed loop
geschlossener Kühlkreislauf m <Kerntech> closed cycle cooling system
geschlossener Regelkreis m <Elektriz, Elektronik, Regelung> closed loop

geschlossener Scheibendrehschalter m <Elektrotech> sealed wafer rotary switch
geschlossener Spind m <Wassertrans> closed locker
geschlossener Stromkreis m <Elektriz, Elektrotech> closed circuit
geschlossener Überdruck-Windkanal m <Nichtfoss Energ> pressure tunnel
geschlossener Wirkungskreis m <Fertig> feedback loop
geschlossener Zug m <Eisenbahn> block train
geschlossenes Druckgaskabel n <Elektriz> self-contained pressure cable
geschlossenes Farbfernsehsignal n <Fernseh> (AE) composite color signal, (BE) composite colour signal
geschlossenes Gefäß n <Labor> closed vessel
geschlossenes Gesenk n <Maschinen> closed die
geschlossenes Kaliber n <Fertig> close pass (Walzen)
geschlossenes Kühlsystem n <Kfztech> closed and sealed cooling system
geschlossenes Messgerät n <Gerät> self-contained instrument
geschlossenes Polygon n <Geom> closed polygon
geschlossenes Regelsystem n <Regelung> closed loop
geschlossenes Servolenksystem n/in sich <Kfztech> self-contained power steering system
geschlossenes System n <Thermod> closed system
geschlossenzelliger Schaumkunststoff m <Kunststoff> closed-cell cellular plastic
geschlossenzelliger Schaumstoff m 1. <Heiz & Kälte> closed-cell foamed plastic; 2. <Verpack> closed-cell foam
Geschmack m <Lebensmittel, Teilphys> (AE) flavor, (BE) flavour
Geschmacksstoff m <Lebensmittel> (AE) flavoring, (BE) flavouring
Geschmacksverstärker m <Lebensmittel> (AE) flavor enhancer, (BE) flavour enhancer, (AE) flavor potentiator, (BE) flavour potentiator
geschmiedet adj <Maschinen, Mechan> forged, wrought
geschmiedeter Stahl m <Metall> forge steel
geschmiertes Band n <Aufnahme> lubricated tape
geschmolzen adj 1. <Metall> molten; 2. <Thermod> fused, melted, molten
geschmolzener Kern m <Raumfahrt> molten core
geschmolzener Quarz m <Optik> fused quartz
geschmolzenes Metall n <Metall, Sicherheit> molten metal
geschmolzenes Silica n <Optik> fused silica
geschmolzenes Silikatglas n <Telekom> fused silica
geschnitten adj <Kohlen> cutoff
geschnittene Viskose-Filamentfasern fpl <Textil> cut staple
geschobener Zug m <Eisenbahn> backup train
Geschoss n <Bau> floor
Geschossdecke f <Bau> floor
Geschossquerbalken m <Bau> summer, summer tree
geschraubte Verbindung f <Kerntech> screwed joint
geschrumpfter Abschnitt m <Hydraul> contracted section
Geschützbronze f <Fertig> gunmetal
Geschützmetall n <Fertig> gunmetal
geschützt adj <Bau, Sicherheit> guarded, screened
geschützte Anlage f im Freien <Heiz & Kälte> sheltered installation
geschützte Zahnräder npl <Maschinen> guarded gears
geschützter Speicher m <Comp & DV> protected storage
geschützter Speicherplatz m <Comp & DV> protected location
geschütztes Feld n <Comp & DV> protected field
geschütztes Getriebe n <Sicherheit> guarded gears
Geschützturm m <Wassertrans> turret (Marinefahrzeug)
geschwärzt adj <Thermod> blackened
geschwärzt darstellen v <Konstzeich> show in black
geschweifte Klammern fpl <Math> braces (Notierung in der Mengenlehre)
geschweißt adj <Maschinen, Thermod> welded
geschweißte Hülse f <Kerntech> weld-deposited cladding (eines Brennelementes)
Geschwindigkeit f 1. <Akustik> velocity; 2. <Comp & DV> rate; 3. <Elektronik, Elektrotech, Maschinen> speed, velocity; 4. <Mechan> velocity; 5. <Metrol> rate; 6. <Papier> speed; 7. <Phys> speed, velocity; 8. <Textil, Trans> speed • **Geschwindigkeit verringern** 1. <Eisenbahn, Maschinen> reduce speed; 2. <Trans> slow down
Geschwindigkeit f der Außenströmung <Strömphys> free-stream velocity (außerhalb der Grenzschicht)
Geschwindigkeit f über Grund 1. <Lufttrans> ground speed; 2. <Trans> group speed; 3. <Wassertrans> ground speed
Geschwindigkeit f von Längswellen (cl) <Akustik> velocity of longitudinal waves (cl)
Geschwindigkeit f von Transversalwellen (ct) <Labor> velocity of transversal waves (ct)
geschwindigkeitsabhängige Frequenz f <Akustik> speed variation frequency
Geschwindigkeitsabnahme f <Kfztech> deceleration (bei einem Motor)
Geschwindigkeitsanzeige f <Maschinen> speed indicator
Geschwindigkeitsanzeiger m 1. <Maschinen> speedometer; 2. <Papier> speed indicator; 3. <Raumfahrt> airspeed indicator (ASI)
Geschwindigkeitsbegrenzer m <Kfztech> speed limiter
Geschwindigkeitsdetektor m <Trans> speed detector
Geschwindigkeitsdiagramm n <Maschinen, Nichtfoss Energ> velocity diagram
Geschwindigkeitsdruckhöhe f <Hydraul> velocity head
Geschwindigkeitseinheit f <Raumfahrt> velocity increment
Geschwindigkeitsempfänger m <Aufnahme> velocity microphone
Geschwindigkeitsfeld n <Lufttrans> wing velocity field
Geschwindigkeitsgefälle n <Kunststoff> shear rate
Geschwindigkeitskoeffizient m <Nichtfoss Energ> velocity coefficient
Geschwindigkeitskonstante f <Metall> rate constant
Geschwindigkeitsmesser m 1. <Kfztech> tachometer; speedometer (Zubehör); 2. <Lufttrans> airspeed indicator (ASI); 3. <Phys> tachometer
Geschwindigkeitsmesserantrieb m <Kfztech> speedometer drive gear
Geschwindigkeitsmessgerät n 1. <Gerät> rate-measuring instrument, speed indicator; 2. <Phys> velocimeter
Geschwindigkeitsmessung f <Phys> velocimetry
Geschwindigkeitsmodulation f 1. <Elektronik, Phys> speed modulation, velocity modulation; 2. <Telekom> velocity modulation
Geschwindigkeitspotenzial n <Akustik> velocity potential
Geschwindigkeitsprofil n <Nichtfoss Energ, Phys> velocity profile
Geschwindigkeitsregelung f <Eisenbahn, Elektriz, Kontroll, Trans> speed control
Geschwindigkeitsregler m <Fernseh, Nichtfoss Energ> velocity control servo
Geschwindigkeitsschreiber m <Eisenbahn, Kfztech> speed recorder

Geschwindigkeitsschwankung

Geschwindigkeitsschwankung f <Fernseh> wow
Geschwindigkeitsschwankungen fpl <Strömphys> velocity fluctuations
Geschwindigkeitssensor m <Nichtfoss Energ> speed sensor
Geschwindigkeitssteuerung f 1. <Kontroll> speed control; 2. <Phys> velocity modulation
Geschwindigkeitsstufenturbine f <Hydraul> velocity stage turbine
Geschwindigkeitstiefenkurve f 1. <Erdöl> velocity depth curve (Seismik); 2. <Nichtfoss Energ> velocity depth curve
Geschwindigkeitsüberwachung f <Eisenbahn> speed supervision
Geschwindigkeitsverlust m <Kerntech, Nichtfoss Energ> velocity loss
Geschwindigkeitswähler m **nach Flugzeitmessung** <Kerntech> time-of-flight velocity selector
Geschwindigkeitswählschalter m <Eisenbahn> speed selector
Geschwindigkeitszunahme f <Raumfahrt> velocity increment
Geschwindigkeit-Verkehrsaufkommen-Kurve f <Trans> speed volume curve
geschwungener Deckel m <Ker & Glas> swung baffle
Gesenk n 1. <Fertig> swage; die (Schmieden); 2. <Maschinen> die, forging die, swage; 3. <Mechan> die, swage • **im Gesenk geschmiedet** <Maschinen> drop-forged
Gesenkformen n <Mechan> die-stamping
Gesenkfräsen n <Fertig, Maschinen> die-sinking
Gesenkfräser m <Fertig> cherry
Gesenkfräsmaschine f <Maschinen> die-sinking machine
gesenkgeschmiedeter Stahl m <Metall> drop-forged steel
gesenkgeschmiedet adj <Fertig, Maschinen> drop-forged
Gesenkhalter m <Fertig> holding shoe
Gesenkoberteil n <Maschinen> top swage
Gesenkplatte f <Maschinen> swage block
Gesenkpresse f <Fertig> stamping press
gesenkpressschmieden v <Fertig> iron (Gießen)
Gesenkschmiedeanteil n <Fertig> stamping plant
Gesenkschmiedehammer m <Fertig> drop stamp, stamp
Gesenkschmieden n 1. <Fertig> die-forging, drop-forging, impact die forging; die-stamping (Schmieden); 2. <Maschinen, Mechan> drop-forging
Gesenkschmiederohling m <Fertig> dummy
Gesenkschmiedeteil n <Fertig> drop-forging, stamping
Gesetz n <Sicherheit> act, law
Gesetz n **der korrespondierenden Zustände** <Phys> law of corresponding states
Gesetz n **von der Erhaltung der Teilchenzahl** <Kerntech> particle number conservation law
Gesetz n **von der Paritätserhaltung** <Strahlphys> parity conservation law
gesetzliche Längeneinheiten fpl <Metrol> legal units of length
gesetzt adj/**ohne Zeildurchschuss** <Druck> set solid
Gesetztes n <Druck> matter
gesichert adj <Mechan> fused
gesicherte Übertragung f <Telekom> secure transmission
Gesichtsfeld n <Ergon> visual field
Gesichtsmaske f <Sicherheit> mask (Schutzkleidung)
Gesichtsschlagschutz m <Sicherheit> filtering facepiece (mit Schutzglas)
Gesichtsschutz m <Sicherheit> face shield, face visor, visor

Gesichtsschutzschale f <Sicherheit> face protector
Gesiebe n <Chemtech> sieve cloth
gesiebt adj <Bau> screened
gesiebter Poliersand m <Anstrich> mesh abrasive grit
Gesims n <Bau> cornice, (AE) molding, (BE) moulding
Gesimsband n <Bau> string
gesintert adj <Chemtech, Maschinen, Mechan> sintered
gespaltener Kern m <Elektrotech> gapped core
Gespannguss m 1. <Fertig> group casting, group teeming; 2. <Metall> group casting
Gespannplatte f <Fertig> bottom plate, bottom-pouring plate (Getriebelehre)
gespannter Grundwasserleiter m <Wasserversorg> confined aquifer
gespanntes Grundwasser n <Kohlen, Wasserversorg> confined ground water
gespeicherte Ansage f <Telekom> prerecorded message
gespeicherte Energie f <Metall, Nichtfoss Energ, Phys, Strahlphys> stored energy
gespeichertes Programm n <Comp & DV> stored program
gespeichertes Rufnummernverzeichnis n <Mobilkom> call number memory, number finder, stored call number directory (bei Handys)
Gesperre n <Maschinen> locking mechanism
gesperrt adj 1. <Kontroll> disabled; 2. <Telekom> barred (Teilnehmerstation); barred (aus- und eingehende Telefongespräche)
gesperrte Flächen fpl <Lufttrans> unserviceable areas (Flughafen)
gesperrte Kapazität f <Aufnahme> clamped capacitance
gesperrte Leitung f/**für abgehende Anrufe** <Telekom> outgoing calls barred line
gesperrte Leitung f/**für ankommende Anrufe** <Telekom> incoming calls barred line
gesperrter Eingabebereich m <Comp & DV> dead zone
gespleißtes Kabel n <Telekom> spliced cable
gespleißtes Tau n <Wassertrans> spliced rope
Gespräch n <Telekom> call • **ein Gespräch anmelden** <Telekom> book a call
Gespräch n **in Wartestellung** <Telekom> camp-on call
Gespräch n **mit ermäßigter Gebühr** <Telekom> cheap call
Gespräch n **mit Voranmeldung** <Telekom> booked call
Gespräch n **zu Ortsgebühr** <Telekom> local-charge-rate call
Gesprächsabwicklung f <Telekom> call handling (Telefon)
Gesprächsaufzeichnung f **im Cockpit** <Lufttrans> voice recorder
Gesprächsdaten pl <Telekom> call data
Gesprächsdatenaufzeichnung f <Telekom> call data recording
Gesprächsdatenerfassung f <Telekom> call data acquisition, call logging
Gesprächsdatenregistrierung f <Telekom> call data recording
Gesprächsdatenspeicher m <Telekom> call data store
Gesprächsdauer f <Telekom> call duration, duration of call (Telefon)
Gesprächsendezeichen n <Telekom> end-of-communication signal
Gesprächsgebühr f <Telekom> call charge (Telefon)
Gesprächsgebühreneinheit f <Telekom> call charge unit
Gesprächsgebührenzähler m <Telekom> call charge meter, call-charging equipment, charge meter (Telefon)
Gesprächsgeheimhaltung f <Telekom> voice privacy

Gesprächsmodus m <Comp & DV> conversational mode
Gesprächsverbindung f <Telekom> call *(Telefon)*
Gesprächsweiterleitung f <Mobilkom> automatic hand-off
Gesprächsweiterverbindung f <Telekom> call transfer
Gesprächszähler m <Telekom> message register, subscriber's meter
Gesprächszeit f <Mobilkom> talk time; airtime
gesprengtes Gestein n <Kohlen> blasted stone
gesprenkelt adj <Verpack> mottled
gesprungen adj <Bau, Mechan, Metall> cracked
gespundet adj <Bau> tongued-and-grooved
Gestaltänderungsenergie f <Fertig> distortion energy
gestalten v <Bau> develop
Gestaltfestigkeit f <Maschinen> fatigue strength
Gestaltung f 1. <Druck, Ergon, Maschinen> design; 2. <Telekom> configuration
Gestaltungsergonomie f <Ergon> design ergonomics
Gestänge n 1. <Kfztech> linkage; 2. <Maschinen> bar linkage, rod, rod linkage; 3. <Mechan> linkage, rod system
Gestängeanheber m <Erdöl> elevator
gestanzter Stahl m <Metall> stamped steel
gestanztes Blech n <Elektrotech, Metall> stamping
gestanztes Trafoblech n <Elektrotech> core lamination
Gestein n <Bau, Kohlen> rock
Gesteinsabbau m <Bau, Kohlen> stoneworking
Gesteinsanker m <Bau> rockbolt
Gesteinsbohrer m 1. <Bau> rock borer, rock drill, tapped valve drill; 2. <Kohlen> rock borer, rock drill
Gesteinsbohrmaschine f <Kohlen> rock drill
Gesteinsdruck m <Bau, Kohlen> rock pressure
Gesteinsmehl n <Kohlen> stone dust
Gesteinsprallmühle f <Kohlen> impact breaker
Gesteinsschotter m <Bau, Hydraul> rock rubble
Gesteinsstaub m <Kohlen> stone dust
Gesteinsstrecke f <Kohlen> stone drift
Gesteinsverwitterung f <Kohlen> rock decay
Gestell n 1. <Bau> trestle; 2. <Ker & Glas> horse; 3. <Maschinen> column, cradle, frame, support; 4. <Mechan> cradle, rack; 5. <Papier> rack; 6. <Telekom> bay, rack; 7. <Textil, Verpack> frame
gestellbefestigt adj <Verpack> rack-mount *(auf Gestell)*
Gestellrahmen m <Telekom> rack
Gestellsäge f <Maschinen> frame saw
Gestellverdrahtung f <Telekom> cable harness
gesteuerte Leitwegumlenkung f <Telekom> controlled routing
gesteuerte Verzögerungsschaltung f <Fernseh> controlled delay lock
gesteuerter Oszillator m <Elektronik> controlled oscillator
gesteuerter Thermonuklearreaktor m (TNR) <Kerntech> controlled thermonuclear reactor (CTR)
gesteuertes Fräsen n <Sicherheit> jig routing
gestockte Antennengruppen fpl <Funktech> baying aerials
gestopft adj <Fertig> blind *(Windform)*
gestört adj <Maschinen, Telekom> out of order
gestörte Bodenprobe f <Kohlen> (AE) remolded sample, (BE) remoulded sample
gestörte Frequenz f <Strahlphys> perturbed frequency
gestörte Leitung f <Telekom> faulty line
gestörte Übertragung f <Telekom> disturbed transmission
gestörte Verbindung f <Telekom> disturbed connection, faulty connection
gestörte Verdichterförderung f <Lufttrans> surge *(Turbomotor)*
gestörte Welle f <Phys> distorted wave

gestörter Betrieb m <Maschinen> faulty operation
gestörter Boden m <Bau, Kohlen> remoulded soil
gestörter Kompass m <Wassertrans> disturbed compass *(Navigation)*
gestörtes Gespräch n <Telekom> faulty call
gestoßene Verbindung f <Bau> abutting joint *(Holzbau)*
gestrandet adj <Wassertrans> stranded; shipwrecked *(Schiff)*
gestrandetes Schiff n <Wassertrans> shipwreck
gestreckt adj 1. <Bau> square-headed; 2. <Geom> prolate; 3. <Mechan> elongated
gestreckte kleine Blase f <Ker & Glas> blibe
gestreckte Länge f <Konstzeich> developed view
gestreckte Spule f <Phys> solenoid
gestreckter Winkel m <Geom> flat angle
gestrecktes Rotationsellipsoid n 1. <Geom> prolate spheroid; 2. <Phys> prolate ellipsoid
gestreut laden v <Comp & DV> scatter-load
gestreut lesen v <Comp & DV> scatter-read
gestreute Speicherungsform f <Comp & DV> random organization
gestreutes Laden n <Comp & DV> scatter load, scattered load
gestreutes Lesen n <Comp & DV> scatter read, scattered read
gestreutes Licht n <Optik> scattered light
gestrichene Pappe f <Verpack> enamel board
gestrichenes Papier n <Papier, Verpack> coated paper
Gestrick n <Textil> knitted fabric
Gestübbe n <Kohlen, Metall> brasque
gestuft adj <Maschinen> stepped
gestülpt adj <Fertig> inside-out redrawn *(Ziehen)*
gestutzte Tragfläche f <Lufttrans> clipped wing
gesundheitsschädlich adj <Sicherheit> harmful to health
gesundheitsschädlicher Arbeitsstoff m <Sicherheit> harmful substance in the workplace
Gesundheitstechnik f <Sicherheit> health engineering, sanitary engineering
gesundheitstechnische Anlage f <Sicherheit> health equipment
gesüßt adj <Lebensmittel> sweetened
getaktete Arbeit f <Ergon> paced work
getaktete Schaltung f <Elektronik> clocked circuit
getaktetes System n <Kontroll> clocked system
geteertes Tau n <Wassertrans> tarred rope *(Tauwerk)*
geteilte Anflugbahn f <Lufttrans> segmented approach path
geteilte Felge f <Maschinen> split rim
geteilte Muffe f <Maschinen> split collar, split sleeve
geteilte Reibscheibe f <Eisenbahn> (BE) split friction disc, (AE) split friction disk
geteilte Riemenscheibe f <Maschinen> split pulley
geteilte Skala f <Maschinen> divided dial
geteilte Stromschiene f <Elektriz> sectionalized busbar
geteilter Balken m <Metrol> divided beam *(Waage)*
geteilter Bildschirm m <Comp & DV> split screen
geteilter Trennschalter m <Elektrotech> divided support disconnecting
geteiltes Gehäuse n <Kfztech> split housing *(Hinterachsgruppe)*
geteiltes Lager n <Maschinen> split bearing
getempert adj <Mechan> tempered
getestet adj <Qual> tested
getöntes Normalglas n <Ker & Glas> neutral-tinted glass
Getränkeverpackung f <Abfall> beverage container
getränktes Papier n <Konstzeich> impregnated paper
Getreidebrand m <Lebensmittel> smut
Getreidekäfer m <Lebensmittel> weevil

Getreidemotte

Getreidemotte f <Lebensmittel> grain moth *(Pflanzenkrankheitslehre)*
getrennt erstellte Zeichnung f <Konstzeich> separately-elaborated drawing
getrennte Abfalllagerung f <Abfall> waste segregation
getrennte Aufbewahrung f <Qual> quarantining
getrennte Empfängerfeinabstimmung f <Funktech> receiver incremental tuning
getrennte Klärung f **von Abwässern** <Wasserversorg> separate sewerage system
getrennte Müllabfuhr f <Abfall> selective collection, separate collection
getrennte Müllsammlung f <Abfall> selective collection, separate collection
getrennter Perlit m <Metall> divorced pearlite
getrennter Stromkreis m <Elektrotech> open circuit
getreue Wiedergabe f <Aufnahme> faithful reproduction
Getriebe n 1. <Fertig> gear; 2. <Kfztech> transmission; 3. <Maschinen> driving gear, gear, gearbox, gearing, mechanism, transmission gear; 4. <Mechan, Papier, Wassertrans> gear *(Motor)*
Getriebe n **für die Schaltung des Vorschubs** <Maschinen> change-feed box
Getriebe n **für Hilfseinrichtungen** <Lufttrans> accessory gearbox
Getriebeaggregat n <Maschinen> gear assembly
Getriebeantriebswelle f <Kfztech> gearbox drive shaft, primary shaft
Getriebeautomatik f <Maschinen> automatic transmission
Getriebebremse f <Maschinen> transmission brake
Getriebedeckel m <Maschinen> gear cover
Getriebeeingangswelle f <Maschinen> gearbox input shaft
Getriebegehäuse n <Kfztech> gear case, gearbox, gearbox housing
Getriebehauptwelle f <Kfztech> mainshaft, third motion shaft
Getriebekasten m 1. <Fertig> gearbox; 2. <Maschinen> gear case, gear casing, gearbox; 3. <Mechan> gearbox
Getriebekopf m <Maschinen> gear head
Getriebelehre f <Fertig> kinematics
getriebelos adj <Mechan> gearless
Getriebemotor m <Mechan> back geared motor, geared motor
getriebene Scheibe f <Maschinen> *(BE)* driven disc, *(AE)* driven disk
Getrieberad n <Maschinen> gear, gearwheel
Getrieberäder npl <Maschinen> gearing
Getrieberadsatz m <Kfztech> set of gears
Getriebeturbine f <Mechan> geared turbine
Getriebeuntersetzung f <Kfztech> transmission reduction
Getriebewelle f <Maschinen> gear shaft
Getriebezahnrad n <Maschinen> gear
getrocknet adj <Lebensmittel, Thermod> desiccated, dried
Getter n <Elektriz, Elektrotech> getter
Getterstoff m <Elektrotech> getter
getunter Motor m <Kfztech> hotted-up engine
Geübtheit f <Ergon> proficiency
Geviert n <Druck> em, mutton
gewachsener Boden m <Umweltschutz> unspoilt land
gewachsener Übergang m <Elektronik> grown junction *(Halbleiter)*
gewählte virtuelle Verbindung f <Telekom> switched virtual circuit
gewährleistete Flugbahn f <Lufttrans> guaranteed flight path

gewährleistetes Gewicht n <Lufttrans> guaranteed weight
Gewährleistungsmarke f <Patent> certification mark
gewalzt adj <Elektriz, Metall> rolled • **mit gewalztem Gewinde** <Fertig> roll-threaded
Gewände n <Bau> jamb
gewartet adj <Maschinen> serviced
gewaschene Kohle f <Kohlen> cleaned coal
gewaschene und ausgepresste Pappartikel mpl <Verpack> washed and squashed consumer waste cartons *(aus dem Handel zur Entsorgung)*
Gewässer npl <Wassertrans> waters
Gewässerbelastung f <Umweltschutz> impact of waters
Gewässergüte f <Wasserversorg> quality of water
Gewässerkunde f 1. <Nichtfoss Energ, Wassertrans> hydrography; 2. <Wasserversorg> hydrology
gewässerkundlich adj <Wassertrans> hydrographic
Gewässerreinigungsschiff n <Umweltschutz> depolluting ship
Gewässerschutz m <Umweltschutz> river and lake protection, water protection
Gewässerversauerung f <Umweltschutz> water acidification
Gewässerverunreinigung f <Wasserversorg> water pollution
Gewässerwissenschaft f <Wasserversorg> hydroscience
Gewebe n 1. <Papier> fabric; 2. <Textil> cloth, fabric, web, woven fabric; fabric *(in Strangform)*
Gewebe n **mit Lichtpausschicht** <Konstzeich> cloth with diazo coating
Gewebebandpresse f <Papier> fabric press
Gewebebindung f <Textil> weave of a fabric
Gewebefilter n <Abfall> fabric filter
Gewebekonstruktion f <Textil> fabric construction
Gewebepackung f <Fertig> fabric packing
Geweberiemen m <Maschinen> fabric belt
Gewebestrang m <Textil> rope
gewellt adj 1. <Fertig> corrugated; 2. <Ker & Glas> wavy; 3. <Maschinen> corrugated; 4. <Papier> corrugated, fluted
gewellte Ausdehnungsverbindung f <Kerntech> corrugated expansion joint
gewellte Federscheibe f <Maschinen> corrugated spring washer, crinkle washer
gewellte Kante f <Ker & Glas> curled edge
gewellte Scheibe f <Maschinen> corrugated washer
gewellter Rohrausgleicher m <Kerntech> corrugated tube compensator
Gewerbeabfall m <Abfall> industrial waste
Gewerbehygiene f <Ergon> industrial hygiene
Gewerbekühlschrank m <Heiz & Kälte> commercial refrigerator
gewerblich anwendbar sein v <Patent> to be susceptible of industrial application
gewerbliche Anwendbarkeit f <Patent> industrial application
gewerbliche Anwendung f <Patent> exploitation in industry
gewerbliche Funkdienste mpl <Funktech, Raumfahrt, Telekom> business services
gewerblicher Abfall m <Abfall> commercial waste, industrial waste, process waste, trade waste
gewerblicher Müll m <Abfall> industrial waste
gewerblicher Verkehr m <Trans> commercial traffic
gewerbliches Abwasser n <Abfall> industrial effluent
gewerbliches Eigentum n <Patent> industrial property
Gewerk n <Bau> trade

gewetzte Freifläche f <Fertig> primary clearance *(Reibahle)*
Gewicht n 1. <Druck, Maschinen> weight; 2. <Math> weight *(Gewichtungsfaktor)*; 3. <Metrol, Phys> weight
Gewicht n **des Seils** <Lufttrans> cable weight
Gewicht n **pro Quadratmeter** <Druck> *(AE)* grams per square meter, *(BE)* grams per square metre
Gewichte npl **und Maße** npl <Metrol> weights and measures
gewichten v <Ergon> weight
gewichteter Durchschnitt m <Qual> weighted average
gewichteter Graph m <Künstl Int> weighted graph
gewichteter Mittelwert m <Qual> weighted average
gewichtetes Mittel n <Math, Phys> weighted mean
Gewichtsanteil m <Kerntech> weight fraction
Gewichtsausgleich m **des Luftschraubenblattes** <Lufttrans> blade balance weight *(Hubschrauber)*
gewichtsbelastetes Manometer n <Gerät> *(AE)* dead-weight gage, *(BE)* dead-weight gauge
Gewichtseinheit f <Metrol> weight unit
Gewichtskraft f <Metrol, Phys> weight
Gewichtslosigkeit f <Phys> weightlessness
gewichtsmolar adj <Chemie> molal
Gewichtsoptimierung f <Raumfahrt> weight optimization *(Raumschiff)*
Gewichtsplattform f <Erdöl> gravity platform *(Offshore-Technik)*
Gewichtsschwerpunkt m <Wassertrans> *(AE)* center of gravity, *(BE)* centre of gravity *(Schiffkonstruktion)*
Gewichtsskale f <Metrol> weighing scale
Gewichtsversuch m <Kohlen> weight penetration test
Gewichtung f 1. <Comp & DV> weighting; 2. <Labor, Math> weighting
Gewichtungsfaktor m 1. <Math> weight; 2. <Raumfahrt> weighting factor *(Weltraumfunk)*
gewickelter Dynamo m/**als Nebenschluss** <Elektrotech> shunt-wound dynamo
gewickelter Motor m/**als Nebenschluss** <Elektrotech> shunt-wound motor
gewickelter Stator m <Elektrotech> wound stator
Gewinde n 1. <Fertig> thread; 2. <Maschinen> thread; thread *(einer Mutter)*; 3. <Mechan> coil • **Gewinde fertigen** <Fertig, Maschinen> thread • **Gewinde herstellen** <Fertig, Maschinen> thread • **mit Gewinde** <Fertig, Maschinen> threaded • **mit gewalztem Gewinde** <Fertig> roll-threaded • **mit Gewinde versehen** <Maschinen> screw • **mit kegeligem Gewinde** <Maschinen> taper-threaded
Gewinde n **nach britischem Standard** <Maschinen> British standard fine screw thread, BSF *(BSF-Gewinde)*; British standard pipe thread, BSP *(BSP-Gewinde)*
Gewindeachse f <Fertig> thread axis
Gewindeanschluss m <Fertig> threaded joint *(Kunststoffinstallationen)*
Gewindebearbeitungsmaschine f <Maschinen> screwing machine
Gewindebemaßung f <Konstzeich> thread dimensioning
Gewindebohreinheit f <Maschinen> tapping unit
Gewindebohreinrichtung f <Maschinen> tapping attachment
Gewindebohren n <Maschinen> tapping
Gewindebohrer m 1. <Bau> screw tap, *(BE)* tap; 2. <Fertig> tap; acme thread tap *(für Acme-Trapezgewinde)*; 3. <Maschinen> screw tap, tap
Gewindebohrerhalter m <Maschinen> tap holder
Gewindebohrmaschine f <Maschinen> tapping machine
Gewindebohr- und Schneidmaschine f <Maschinen> screwing and tapping machine
Gewindebohrung f <Fertig> taphole; tapped hole *(Innengewinde)*
Gewindebolzen m 1. <Kfztech> threaded bolt; 2. <Maschinen> stud, stud bolt, threaded fastener
Gewindebolzenabschneider m <Fertig> bolt clipper
Gewindebuchse f <Fertig> screwed bush; threaded bush *(Kunststoffinstallationen)*
Gewindedarstellung f <Konstzeich> representation of a thread
Gewindedrehen n <Maschinen> screw cutting, thread turning
Gewindedreher m <Bau> pipe threader
Gewindedrehmaschine f <Maschinen> threading lathe
Gewindedrücken n <Maschinen> thread bulging
Gewindedurchmesser m <Maschinen> thread diameter
Gewindeeinsatz m <Maschinen> thread insert
Gewindeflanke f <Maschinen> thread flank
Gewindeform f <Maschinen> thread form
gewindeformende Schraube f <Maschinen> thread-forming screw
Gewindefräsen n <Fertig, Maschinen> thread milling
Gewindefräser m <Fertig, Maschinen> thread-milling cutter
Gewindefräsmaschine f <Fertig, Maschinen> thread-milling machine
Gewindefreistich m <Maschinen> thread undercut
Gewindefurchen n <Mechan> thread ridging
Gewindeganganzeiger m <Maschinen> thread dial indicator
Gewindeherstellung f <Fertig> threading
Gewindeinstallation f <Maschinen> screwing
Gewindelänge f <Maschinen> thread length
Gewindelehrdorn m 1. <Maschinen> *(AE)* plug gage for threads, *(BE)* plug gauge for threads; 2. <Metrol> *(AE)* plug thread gage, *(BE)* plug thread gauge
Gewindelehre f 1. <Maschinen> *(AE)* screw pitch gage, *(BE)* screw pitch gauge, *(AE)* screw thread gage, *(BE)* screw thread gauge, *(AE)* thread gage, *(BE)* thread gauge; 2. <Metrol> *(AE)* center gage, *(BE)* centre gauge, *(AE)* screw thread gage, *(BE)* screw thread gauge
Gewindeloch n 1. <Fertig> taphole; threaded hole *(Kunststoffinstallationen)*; 2. <Maschinen> taphole, tapped hole
Gewindemeißel m <Fertig, Maschinen> threading tool
Gewindemessschraube f <Fertig> thread micrometer
Gewindemesszylinder m <Metrol> screw-thread measuring cylinder
Gewindemuffe f 1. <Fertig> threaded socket end *(Kunststoffinstallationen)*; 2. <Maschinen> threaded sleeve
Gewindemutter f <Maschinen> threaded nut
Gewindenachbohrer m <Bau, Maschinen> plug tap
Gewindenachschneiden n <Fertig, Maschinen> rethreading
Gewindenippel m <Maschinen> threaded nipple
Gewindeprofil n <Fertig, Maschinen> thread profile
Gewinderille f <Fertig> thread groove
Gewindering m <Maschinen> ring nut
Gewinderohr n <Maschinen> threaded pipe
Gewinderollen n <Maschinen> thread rolling
Gewindeschablone f <Maschinen> *(AE)* screw pitch gage, *(BE)* screw pitch gauge
Gewindeschälen n 1. <Fertig> thread peeling; 2. <Maschinen> thread whirling
Gewindeschleifen n <Fertig, Maschinen, Mechan> thread grinding
Gewindeschleifmaschine f <Fertig> thread-grinding wheel
Gewindeschneidbacke f <Fertig> chaser, threading die

Gewindeschneidbohrer *m* <Fertig> tap
Gewindeschneideinrichtung *f* 1. <Fertig> thread-cutting attachment; 2. <Maschinen> screw-cutting attachment, thread-cutting attachment
Gewindeschneideisen *n* <Maschinen> screwing die
Gewindeschneiden *n* 1. <Fertig> thread cutting, threading; 2. <Maschinen> screw cutting, thread cutting
gewindeschneidende Schraube *f* <Maschinen> thread-cutting screw
gewindeschneidender Bohrer *m* <Maschinen> tapping drill
Gewindeschneider *m* <Maschinen> threader
Gewindeschneidkluppe *f* <Maschinen> screw plate, tap plate
Gewindeschneidkluppe *f* **mit Ratsche** <Maschinen> ratchet-screwing stock
Gewindeschneidkopf *m* 1. <Fertig> die head; 2. <Maschinen> die head, die stock, screw stock, screwing chuck, screwing head
Gewindeschneidkopf *m* **für Außengewinde** <Fertig> bolt die head
Gewindeschneidmaschine *f* <Maschinen> machine tapper, threading machine
Gewindeschneidschraube *f* <Maschinen> self-cutting screw
Gewindespitzenabrundung *f* <Fertig> crest truncation
Gewindesteigung *f* <Kfztech> thread pitch
Gewindesteigungswinkel *m* <Maschinen> thread lead angle
Gewindestift *m* 1. <Bau> grub screw; 2. <Fertig> grub screw *(Kunststoffinstallationen)*; 3. <Maschinen> grub screw, headless pin; 4. <Mechan> headless screw
Gewindestopfen *m* <Bau> screw plug
Gewindestrehlen *n* 1. <Fertig> thread chasing; 2. <Maschinen> screw chasing, thread chasing
gewindestrehlen *v* <Fertig> chase
Gewindestrehler *m* <Fertig, Maschinen> screw chaser, thread chaser
Gewindeteilung *f* <Maschinen> screw pitch, thread pitch
Gewindetiefe *f* <Bau, Fertig, Maschinen> thread depth
Gewindetiefenmesser *m* <Metrol> *(AE)* height gage, *(BE)* height gauge
Gewindewalzautomat *m* **mit Backen** <Fertig> automatic flat-die thread-rolling machine
Gewindewalzen *n* <Maschinen> thread rolling
Gewindewälzen *n* <Fertig> thread hobbing
Gewindewälzfräser *m* <Fertig> thread-milling hob
Gewindewirbeln *n* <Maschinen> thread whirling
Gewindezahn *m* <Fertig> thread ridge
Gewindezapfen *m* <Fertig> threaded journal • **mit Gewindezapfen** 1. <Fertig> threaded *(Spindel)*; 2. <Maschinen> threaded
Gewinn *m* 1. <Elektronik, Ergon, Fernseh> gain, G; 2. <Funktech> gain, G *(Antenne)*; 3. <Kohlen, Nichtfoss Energ> yield; 4. <Raumfahrt> gain, G *(Weltraumfunk)*; 5. <Telekom> gain, G *(Antenne)*
gewinnbare Reserven *fpl* <Erdöl> recoverable reserves
gewinnen *v* 1. <Chemie> obtain; 2. <Kohlen> exploit, get, mine
gewinnender Strom *m* <Wasserversorg> gaining stream
gewinngeführt *adj* <Elektronik> gain-controlled *(Laser)*
Gewinnung *f* 1. <Erdöl> extraction *(Bohrtechnik)*; 2. <Fertig, Kohlen> extraction
Gewinnveränderung *f* <Elektronik> gain change
Gewirk *n* <Textil> knitted fabric
Gewissheit *f* <Math> assurance
Gewissheitsfaktor *m* <Künstl Int> certainty factor, confidence factor
gewogenes Mittel *n* <Math> weighted mean

gewöhnliche Sechskantmutter *f* <Maschinen> ordinary hexagonal nut
gewöhnlicher Portlandzement *m* <Bau> ordinary Portland cement
Gewöhnung *f* <Ergon> habituation
Gewölbe *n* 1. <Bau> vault; 2. <Ker & Glas> arch *(des Glashafens)*
Gewölbefläche *f* <Bau> intrados
Gewölberücken *m* <Bau> extrados
Gewölberückenfläche *f* <Bau> back
Gewölbesperrmauer *f* <Bau> arch dam
Gewölbestein *m* <Bau> arch stone
Gewölbestromwender *m* <Elektrotech> arch-bound commutator
Gewölbewicklung *f* <Elektrotech> barrel winding
gewölbt *adj* 1. <Bau> arched, bonneted; 2. <Maschinen> dished
gewölbte Federscheibe *f* <Maschinen> curved spring washer
gewölbte Oberfläche *f* <Ker & Glas> warped finish
gewölbter Boden *m* <Mechan> dished head
gewölbter Luftansaugekanal *m* <Lufttrans> variable geometry inlet
gewölbter Verschluss *m* <Ker & Glas> bulged finish
gewölbtes Deck *n* <Wassertrans> cambered deck
gewölbtes Tafelglas *n* <Ker & Glas> *(AE)* bow and warp, *(BE)* warped sheet
gewunden *adj* <Bau> meandering
gewürfelt *adj* <Math> aleatoric
Geysir *m* <Nichtfoss Energ> geyser
gezackt *adj* 1. <Fertig> scalloped; 2. <Mechan, Sicherheit> jagged
gezackte Kante *f* <Maschinen> pinked edge
gezackte Schaufel *f* <Bau> pronged shovel
gezackte Schneide *f* <Sicherheit> jagged edge *(einer Klinge)*
gezackter Riss *m* <Ker & Glas> hackle
gezählter Impuls *m* <Gerät> count
gezahnt *adj* 1. <Maschinen> geared, toothed; 2. <Mechan> cogged
gezahnter Keilriemen *m* <Maschinen> cogged V belt
gezahnter Treibriemen *m* <Maschinen> toothed drive belt
Gezeit *f* <Nichtfoss Energ, Wassertrans> tide
Gezeiten *fpl* <Wassertrans> tides
Gezeitenaufzeichnung *f* <Nichtfoss Energ> marigram
Gezeitenbereich *m* <Nichtfoss Energ> tidal range
Gezeitenbewegung *f* <Nichtfoss Energ> tidal movement
Gezeitenenergie *f* 1. <Nichtfoss Energ> tidal power, tidal energy; 2. <Phys> tidal energy
Gezeitenhub *m* <Wassertrans> range of tide
Gezeitenkraft *f* <Nichtfoss Energ> tidal power
Gezeitenkraftwerk *n* 1. <Elektrotech> tidal power plant; 2. <Nichtfoss Energ> tidal power plant, tidal power station
Gezeitenmesser *m* <Nichtfoss Energ> *(AE)* tide gage, *(BE)* tide gauge
Gezeitenmühle *f* <Nichtfoss Energ> tide mill
Gezeitenprisma *n* <Nichtfoss Energ> tidal prism
Gezeitensignale *npl* <Wassertrans> tidal signals, tide signals
Gezeitenstrom *m* 1. <Nichtfoss Energ> tidal current; 2. <Wassertrans, Wasserversorg> tidal current, tidal flow, tidal stream
Gezeitenstromatlas *m* <Wassertrans> tidal stream atlas
Gezeitenstromschnelle *f* <Wassertrans> tide race
Gezeitentabelle *f* <Wassertrans> tide chart; tidal chart *(Navigation)*
Gezeitentafel *f* <Wassertrans> tide chart; tidal chart *(Navigation)*

Gezeitentafeln *fpl* <Wassertrans> tide tables
Gezeitenwechsel *m* <Wassertrans> turn of the tide
Gezeitenzone *f* <Wassertrans> littoral
gezielte Sendung *f* <Fernseh> narrowcasting
gezogener Draht *m* <Maschinen> drawn wire
gezogener Stiel *m* <Ker & Glas> drawn stem
gezogener Zonenübergang *m* <Elektronik> drawn junction *(Halbleiter)*
gezogener Zucker *m* <Lebensmittel> pulled sugar *(Süßwaren)*
gezündete Röhre *f* <Elektronik> fired tube
g-Faktor *m* <Phys> g-factor
GFK 1. <Kunststoff> *(Glasfaserkunststoff, glasfaserverstärkter Kunststoff)* GRP, glass fibre-reinforced plastic; 2. <Verpack> *(glasfaserverstärkter Kunststoff)* GRP, glass fibre-reinforced plastic; 3. <Wassertrans> *(glasfaserverstärkter Kunststoff)* GRP, glass fibre-reinforced plastic *(Schiffbau)*
ggT *(größter gemeinsamer Teiler)* <Math> HCF *(highest common factor)*
Gibbs'sche freie Energie *f* <Phys> Gibbs free energy
Gibbs'sche Funktion *f (G)* <Phys, Thermod> Gibbs function *(G)*
Gibbs'sche Phasenregel *f* <Kerntech, Thermod> Gibbs phase rule
Gicht *f* <Fertig> batch, top
Gichtbühne *f* <Fertig> charging platform
Gichtgas *n* 1. <Fertig> blast furnace gas; 2. <Kerntech> stack gas
Gichtglocke *f* <Fertig> bell
Gichtverschluss *m* <Fertig> bell and hopper
Giebel *m* <Bau> gable
Giebelbogen *m* <Bau> triangular arch
Giebeldach *n* <Bau> gable roof
Giebelfußstein *m* <Bau> foot skew, skew
Giebelmauer *f* <Ker & Glas> gable wall
Giebelwand *f* <Bau> flank wall
Gientalje *f* <Wassertrans> winding tackle *(Deckausrüstung)*
Gier... <Lufttrans, Maschinen, Nichtfoss Energ, Raumfahrt> yaw
Gierachse *f* <Lufttrans> yaw axis
Gierbewegung *f* <Lufttrans> yaw
gieren *v* 1. <Nichtfoss Energ> yaw; 2. <Raumfahrt> yaw *(Raumschiff)*; 3. <Wassertrans> yaw *(Navigation)*
Gieren *n* 1. <Lufttrans, Nichtfoss Energ, Raumfahrt> yaw *(Raumschiff)*; 2. <Wassertrans> yawing *(Schiffsbewegung)*
Gierjustierung *f* <Nichtfoss Energ> yaw adjustment
Gierkontrolle *f* <Nichtfoss Energ> yaw control
Giermoment *n* <Lufttrans, Maschinen, Nichtfoss Energ> yawing moment
Gierschwingung *f* <Lufttrans> snaking *(Aerodynamik)*
Gierung *f* 1. <Phys> yaw; 2. <Raumfahrt> yaw rate *(Quantität)*
Gierungsachse *f* <Raumfahrt> yaw axis *(Raumschiff)*
Gierungswinkel *m* <Raumfahrt> yaw angle *(Raumschiff)*
Gierwinkelgeschwindigkeit *f* <Lufttrans> angular yaw rate
Gieß... <Druck, Ker & Glas, Maschinen> casting
Gießakt *m* <Fertig> shot *(Gießen)*
Gießansatz *m* <Ker & Glas> casting scar
Gießaufsatz *m* <Fertig> feeder *(Gießen)*
Gießbäche *mpl* <Druck> rivers of white
gießen *v* 1. <Bau> pour *(Beton)*; 2. <Druck> cast; 3. <Lebensmittel> pour; 4. <Maschinen, Papier> cast
Gießen *n* 1. <Fertig> casting, founding; 2. <Ker & Glas> casting *(von Feuerfesterzeugnissen)*; casting *(von Spiegelglas)*; 3. <Kunststoff, Maschinen, Mechan, Metall> casting; 4. <Sicherheit> pouring

Gießen *n* **in Mehrfachform** <Fertig> gating
Gießer *m* 1. <Druck> composition caster; 2. <Fertig> founder; 3. <Ker & Glas> ladler
Gießerei *f* <Fertig, Kohlen, Metall> foundry
Gießereiformschwärze *f* <Fertig> blackening *(Gießen)*
Gießereikran *m* <Ker & Glas> casting crane
Gießereisand *m* <Bau, Metall> foundry sand
Gießfolie *f* <Kunststoff> cast film
Gießform *f* 1. <Druck> *(AE)* casting mold, *(BE)* casting mould, *(AE)* mold, *(BE)* mould; 2. <Maschinen> *(AE)* casting mold, *(BE)* casting mould, *(AE)* mold for casting, *(BE)* mould for casting; 3. <Mechan> *(AE)* ingot mold, *(BE)* ingot mould; 4. <Metall> *(AE)* mold, *(BE)* mould
Gießharz *n* <Kunststoff> casting resin
Gießharzdurchführung *f* <Elektrotech> cast-resin bushing
gießharzisoliert *adj* <Elektrotech> cast-resin insulate
Gießharzstromwandler *m* <Elektrotech> cast resin block-type current transformer
Gießharzstützer *m* <Elektrotech> cast-resin post insulator
Gießharztransformator *m* <Elektrotech> cast-resin transformer
Gießkasten *m* <Papier> casting box
Gießlippe *f* <Ker & Glas> casting lip
Gießloch *n* <Ker & Glas> orifice
Gießlochring *m* <Ker & Glas> orifice ring
Gießlöffel *m* 1. <Bau> ladle *(Maurerwerkzeug)*; 2. <Fertig> handladle, ladle; 3. <Maschinen> pouring spoon
Gießmaschine *f* 1. <Druck> caster, casting machine, type caster; 2. <Ker & Glas> casting unit; 3. <Papier> casting machine
Gießpfanne *f* <Fertig, Metall> foundry ladle
Gießpfannenkran *m* <Fertig> ladle crane
Gießrinne *f* 1. <Fertig> spout; 2. <Ker & Glas> trough *(bei der Herstellung von Walzglas)*
Gießrinnenlippe *f* <Ker & Glas> trough lip
Gießschlicker *m* <Ker & Glas> casting slip
Gießspirale *f* <Fertig> *(AE)* fluidity mold, *(BE)* fluidity mould
Gießteil *n* <Kunststoff> casting
Gießtisch *m* <Ker & Glas> casting table
Gießtrichter *m* 1. <Fertig> cast gate, ingate, pouring gate, skim gate; 2. <Metall> sprue
Gießtrichteransatz *m* <Fertig> sprue
Gießtrichtermodell *n* <Fertig> gate pin, gate stick
Gießtümpel *m* <Fertig> runner basin
Gießverfahren *n* <Kunststoff> casting
Gießwalzen *fpl* <Ker & Glas> casting rollers
Gießwanne *f* <Fertig> tundish
giftig *adj* <Chemie, Sicherheit, Umweltschmutz> toxic
Giftigkeit *f* <Chemie, Sicherheit, Umweltschmutz> toxicity
Giftmüll *m* <Abfall> poisonous waste, toxic waste
Giftmüllentsorgungsanlage *f* <Umweltschmutz> toxic waste disposal plant
Giftschrank *m* <Labor> poisons cupboard
Giftstoff *m* <Chemie, Umweltschmutz> toxic agent, toxicant
Giftstoffentfernung *f* **aus Kernreaktoren** <Kerntech> nuclear reactor poison removal
Giga... *(G)* <Metrol> giga... *(G)*
Gigabyte *n* *(GB)* <Comp & DV, Optik, Telekom> gigabyte, GB *(Informationseinheit)*
Gigahertz *n* <Funktech> gigahertz
Gigahöchstintegration *f* <Elektronik> gigascale integration
Gigaplatte *f* <Comp & DV, Optik> gigadisk
GIGO *(Müll rein, Müll raus)* <Comp & DV> GIGO *(garbage in, garbage out)*

GII *(globale Informations-Infrastruktur)* <Comp & DV> GII *(Global Information Infrastructure)*
Gilbert *n* <Elektrotech> gilbert *(CGS-Einheit)*
Gill *n* <Metrol> gill
Gillung *f* <Wassertrans> counter *(Schiffbau)*
Gillungsheck *n* <Wassertrans> counter stern *(Schiffbau)*
Gimpe *f* <Textil> gimp
Ginn'sche Gleichung *f* <Kerntech> Ginn equation
Giorgi'sches Einheitssystem *n* <Metrol> Giorgi system of units
Gipfel *pl* <Geom> apices
Gipfelstation *f* <Trans> top station
Gips *m* <Erdöl> gypsum *(Mineral)*
Gips… <Chemie> selenitic *(Kalk)*
Gipsbauplatte *f* <Bau> plasterboard
Gipsgestein *n* <Bau> plaster rock, plaster stone
Gipskartonplatte *f* <Bau> plasterboard
Gipsmörtel *m* <Bau> plaster
Gipsputzunterlage *f* <Bau> furring *(aus Streckmetall)*
Gipsschwefelzement *m* <Bau> supersulphated cement
Girlande *f* <Papier> festoon
Gischt *f* <Wassertrans> foam, spray *(Meer)*
Gispe *f* <Ker & Glas> seed *(Blase)*
Gitter *n* 1. <Bau> lattice, screen, trellis; 2. <Elektriz, Elektronik, Elektrotech> grid, lattice; grid *(Elektronenröhre)*; 3. <Fernseh> grating; 4. <Fertig> grid, lattice; 5. <Funktech> grid; 6. <Kerntech> grid *(Elektrode)*; 7. <Kohlen> grid; 8. <Math, Metall> lattice; 9. <Phys> grid; 10. <Raumfahrt> lattice *(Raumschiff)*
Gitterableitwiderstand *m* <Phys> grid-leak resistor
Gitterabstand *m* <Metall> lattice spacing
Gitterbalken *m* <Bau> lattice truss
Gitterboxpalette *f* 1. <Trans> crate pallet; 2. <Verpack> box pallet with sidewalls
Gitterbrücke *f* <Bau> lattice bridge
Gittercharakteristik *f* <Elektronik> grid characteristic
Gitterfehler *m* 1. <Elektronik> lattice defect *(Halbleiter)*; 2. <Metall> lattice defect
gitterförmig *adj* <Fertig> latticed
Gitterfurche *f* <Optik> groove *(Beugungsgitter)*
gittergesteuerte Röhre *f* <Elektronik> grid-controlled tube
gittergesteuerter Lichtbogen-Quecksilbergleichrichter *m* <Elektrotech> grid-controlled mercury arc rectifier
Gitterkammer *f* <Ker & Glas> checker chamber
Gitterkatodenkapazität *f* <Elektrotech> grid-cathode capacitance
Gitterkisteneinlage *f* <Papier> crate liners
Gitterkondensator *m* <Elektrotech> grid capacitor
Gitterkonstante *f* <Metall> lattice constant
Gitterlücke *f* <Metall> vacancy *(Kristalle)*
Gitterluftschraubenblätter *npl* <Lufttrans> cascade blades, cascade vanes
Gittermast *m* 1. <Bau> pylon, tower; 2. <Funktech> lattice mast
Gittermauer *f* <Bau> screen wall
Gittermauerwerk *n* <Ker & Glas> checkers
Gittermodulation *f* <Elektronik, Fernseh> grid modulation
Gitternetz *n* <Raumfahrt> mesh *(Sieb)*
Gitterpfosten *m* <Bau> trellis post
Gitterplatte *f* <Elektriz> battery electrode, battery plate
Gitterpunkt *m* <Metall> lattice point
Gitterrippe *f* <Lufttrans> lattice rib
Gitterrost *n* 1. <Bau> grate; 2. <Heiz & Kälte> grating, grille
Gitterschaufeln *fpl* <Lufttrans> cascade vanes *(Turbine)*
Gittersonde *f* <Kerntech> grid probe
Gitterspektrograph *m* <Kerntech> diffraction spectrograph, grating spectrograph
Gitterspiegel *m* <Raumfahrt> reticulated mirror *(Raumschiff)*
Gitterstab *m* <Bau> screen bar
Gitterstein *m* <Fertig, Ker & Glas> checker brick
Gittersteuerleistung *f* <Elektrotech> grid-driving power
Gitterstrich *m* <Optik> groove *(Beugungsgitter)*
Gitterstrom *m* <Elektrotech> grid current
Gitterträger *m* 1. <Bau> lattice beam, lattice girder; 2. <Mechan, Raumfahrt> lattice girder *(Raumschiff)*
Gitterträgerbogen *m* <Bau> lattice girder arch
Gitterträgerplatte *f* <Kerntech> grid support plate
Gitterturbulenz *f* <Strömphys> grid turbulence
Gitterumformer *m* <Elektrotech> grating converter
Gitterverformung *f* <Metall> lattice deformation
Gittervorspannung *f* 1. <Elektrotech> bias, gate bias, grid bias; gate bias *(Thyristoren)*; 2. <Funktech> grid bias
Gittervorspannung *f* **durch Katodenwiderstand** <Elektrotech> self-bias
Gittervorwiderstand *m* <Phys> bias resistor *(Elektronenröhre)*
Gitterwandler *m* <Elektrotech> grating converter
Gitterwerk *n* <Bau> latticework
Gitterwerkmast *m* <Elektriz> lattice tower
Gitterwiderstand *m* <Funktech> grid-leak resistor
Gitterzerkleinerer *m* <Abfall> grid crusher
Gitterziegel *m* <Bau> perforated brick
Gitterzuordnung *f* <Metall> lattice correspondence
Glanz *m* 1. <Druck, Kunststoff> gloss; 2. <Metall> brightness, brilliance, burnish, sheen; 3. <Papier> gloss; 4. <Textil> glaze, *(AE)* luster, *(BE)* lustre, radiance, sheen; 5. <Verpack> gloss
Glanzappretur *f* <Textil> glazing
glänzen *v* <Textil> glaze
glänzend *adj* 1. <Kunststoff> glossy; 2. <Textil> bright, glossy, lustrous; 3. <Verpack> glossy
Glanzfarbe *f* <Bau> gloss paint
Glanzgarn *n* <Textil> glazed yarn
Glanzgold *n* <Ker & Glas> bright gold, burnishing gold
Glanzkarton *m* <Verpack> glazed board
Glanzkohle *f* <Kohlen> glance coal
Glanzlack *m* **für Anstrich** <Verpack> gloss
glanzlos *adj* 1. <Bau> dull, flat *(Farbe)*; 2. <Textil> *(AE)* lusterless, *(BE)* lustreless
Glanzmaschine *f* <Verpack> glazing machine
Glanzmesser *m* <Kunststoff> gloss meter
Glanzpapier *n* 1. <Druck> glossy paper; 2. <Papier> glazed paper; 3. <Verpack> glazed paper, glossy paper
Glanzpappe *f* <Verpack> glazed board
Glanzschleifen *n* <Papier> rubbing
Glanzsilber *n* <Ker & Glas> bright silver, burnishing silver
Glanzwinkelgitter *n* <Phys> blazed grating
Glas *n* 1. <Anstrich, Chemie> glass; 2. <Verpack> glass container, glass jar • **im Brandfall Glas einschlagen** <Sicherheit> in case of fire, break the glass *(Feuermelder)* • **in Gläser einmachen** <Lebensmittel> bottle *(Früchte, Gemüse)*
Glas *n* **mit hohem Brechungsindex** <Ker & Glas> flint glass
Glasanalyse *f* <Ker & Glas> glass analysis
Glasandruckplatte *f* <Foto> glass pressure plate
glasartig *adj* 1. <Anstrich> glassy; 2. <Chemie, Fertig> vitreous; 3. <Ker & Glas> glassy
Glasausbeute *f* <Ker & Glas> glass yield
Glasauskleidung *f* <Elektrotech> glass cladding
Glasballon *m* <Ker & Glas, Labor> carboy
Glasbaustein *m* <Ker & Glas> glass block
Glasbehälter *m* 1. <Abfall> bottle bank; 2. <Verpack> glass container
Glasbetonplatte *f* <Ker & Glas> glass concrete panel

Glasbildner m <Ker & Glas> glass former
Glasblasen n <Ker & Glas> blowing, glassblowing
Glasbläser m <Ker & Glas> blower, glassblower
Glasblock m <Kerntech> glass block
Glasbruch m <Abfall> cullet
Glascontainer m <Abfall> bottle bank
Glasdach n <Ker & Glas> glass roof
Glasdachziegel m <Ker & Glas> glass roof tile
Glaselektrode f <Labor> glass electrode
Glasemail n <Ker & Glas> vitreous enamel
Glasemailschild n <Ker & Glas> vitreous enamel label
Glasendmaß n <Metrol> optical flat
Glas-Epoxid-Laminat n <Elektronik> glass-epoxy laminate
Glas-Epoxid-Leiterplatte f <Elektronik> glass-epoxy printed circuit board
Glaser m <Bau> glazier
Glaserdiamant m 1. <Bau> diamond pencil, glass cutter, glazier's diamond; 2. <Ker & Glas> diamond pencil
Glaserkitt m 1. <Bau> putty, sash putty; 2. <Ker & Glas> putty
Glaserzange f <Ker & Glas> glazier's pliers
Glasfarbe f <Ker & Glas> *(AE)* glass color, *(BE)* glass colour
Glasfaser f 1. <Bau> *(AE)* glass fiber, *(BE)* glass fibre; 2. <Comp & DV> *(AE)* optical fiber, *(BE)* optical fibre; 3. <Elektrotech> *(AE)* optical fiber, *(BE)* optical fibre; 4. <Kunststoff> *(AE)* glass fiber, *(BE)* glass fibre; 5. <Raumfahrt> *(AE)* optical fiber, *(BE)* optical fibre *(zur Signalübertragung)*; 6. <Telekom> *(AE)* glass fiber, *(BE)* glass fibre; 7. <Wassertrans> *(AE)* glass fiber, *(BE)* glass fibre *(Schiffbaumaterial)*
Glasfaser f **bis ans Haus** <Telekom> fibre to the curb, fibre to the home
Glasfaser f **in der Anschlussleitung** <Telekom> fibre in the loop
Glasfaseranschluss m <Elektrotech> *(AE)* fiberoptic connection, *(BE)* fibreoptic connection
glasfaserbewehrter Beton m <Bau> glass-fiber reinforced concrete *(GRC)*
Glasfaserkabel n <Comp & DV, Elektrotech, Telekom> *(AE)* fiberoptic cable, *(BE)* fibreoptic cable, optical cable, *(AE)* optical fiber cable, *(BE)* optical fibre cable
Glasfaserkabel n **für mehrere Einsatzarten** <Elektrotech> *(AE)* multimode optical fiber, *(BE)* multimode optical fibre
Glasfaserkabel n **mit großem Durchmesser** <Elektrotech> *(AE)* large-core glass fiber, *(BE)* large-core glass fibre
Glasfaserkabel-Steckverbinder m <Elektrotech> *(AE)* fiberoptic connector, *(BE)* fibreoptic connector
Glasfaserkabelübertragung f <Telekom> *(AE)* optical fiber cable transmission, *(BE)* optical fibre cable transmission
Glasfaserkommunikationssystem n <Telekom> *(AE)* optical fiber communication system, *(BE)* optical fibre communication system
Glasfaserkreisel m <Raumfahrt> *(AE)* fiberoptic gyrometer, *(BE)* fibreoptic gyrometer *(Raumschiff)*
Glasfaserkunststoff m *(GFK)* <Kunststoff> *(AE)* glass fiber-reinforced plastic, *(BE)* glass fibre-reinforced plastic *(GRP)*
Glasfaserlaminat n <Verpack> *(AE)* glass fiber laminate, *(BE)* glass fibre laminate
Glasfasernachrichtentechnik f <Telekom> *(AE)* optical fiber communications, *(BE)* optical fibre communications
Glasfaserprofil n <Telekom> *(AE)* optical fiber profile, *(BE)* optical fibre profile
Glasfaserschichtkunststoff m <Verpack> *(AE)* glass fiber laminate, *(BE)* glass fibre laminate

Glasfaserschweißstelle f <Telekom> *(AE)* optical fiber connection, *(AE)* optical fiber junction, *(BE)* optical fibre connection, *(BE)* optical fibre junction
Glasfaser-Seekabel n <Telekom> submarine optical fiber cable, undersea fiber-optic cable
Glasfaserspleiß m <Telekom> *(AE)* optical fiber splice, *(BE)* optical fibre splice
Glasfaserstoff m <Textil> *(AE)* fiberglass, *(BE)* fibreglass
Glasfaserstrang m <Ker & Glas> strand
Glasfasertechnik f 1. <Comp & DV, Elektrotech, Telekom> *(AE)* fiber optics, *(BE)* fibre optics; 2. <Fertig> *(BE)* optical fibre technology
Glasfasertechnologie f <Elektrotech> *(AE)* fiberoptic technology, *(BE)* fibreoptic technology
Glasfasertemperatursensor m <Metrol> *(AE)* fiber-optical thermometer, *(BE)* fibre-optic thermometer
Glasfaser-Tiefseekabel n <Telekom> deep submarine optical-fiber cable
Glasfaserübertragung f <Telekom> *(AE)* transmission by optical fiber, *(BE)* transmission by optical fibre
Glasfaserübertragungssystem n <Telekom> *(AE)* fiber-optic transmission system, *(BE)* fibreoptic transmission system
Glasfaserverbindung f <Fertig> *(BE)* optical fibre link
glasfaserverstärkt adj <Verpack> *(AE)* glass fiber-reinforced, *(BE)* glass fibre-reinforced
glasfaserverstärkter Kunststoff m *(GFK)* <Kunststoff, Verpack, Wassertrans> *(AE)* glass fiber-reinforced plastic, *(BE)* glass fibre-reinforced plastic *(GRP)*
Glasfaserverstärkung f <Ker & Glas> *(AE)* glass fiber reinforcement, *(BE)* glass fibre reinforcement
Glasfaservlies n <Ker & Glas> chopped-strand mat, *(AE)* glass fiber mat, *(BE)* glass fibre mat
Glasfilament n <Ker & Glas> glass filament
Glasfilm m <Ker & Glas> glass film
Glasfläschchen n <Ker & Glas> *(BE)* phial, *(AE)* vial
Glasformzange f <Ker & Glas> pucella
Glasfritte f <Ker & Glas> glass frit
Glasgefäß n <Verpack> glass jar
Glasgestrick n <Ker & Glas> knitted glass fabric
Glasgewebe n <Verpack> glass fabric
Glasglocke f 1. <Ker & Glas> bell jar *(Uhr)*; 2. <Labor> bell jar
Glashafenton m <Ker & Glas> pot clay
Glashafenträger m **mit Schwanenhals** <Ker & Glas> goosenecked pot carriage
Glashalbleiter m <Elektronik> amorphous semiconductor
Glashalter m <Elektrotech> glass holder
Glashärte f <Metall> glass hardness
Glashartgewebe n <Kerntech> glass-reinforced laminate
Glasheizkörper m <Ker & Glas> glass heating panel
glasieren v 1. <Fertig, Foto> glaze; 2. <Ker & Glas> glaze *(Substrat)*; 3. <Lebensmittel> glaze
Glasiermaschine f <Ker & Glas> glazing machine
glasiert adj <Ker & Glas> glazed *(Töpferwaren)*
glasierte Ofenkachel f <Ker & Glas> Dutch tile
glasierte Töpferwaren fpl <Ker & Glas> glazed pottery
glasiertes Steingut n <Ker & Glas> glazed earthenware
glasig adj <Anstrich> glassy
glasiger Feldspat m <Ker & Glas> glassy feldspar
Glasisolator m <Elektriz, Ker & Glas> glass insulator
Glaskamee f <Ker & Glas> glass cameo
Glaskappe f <Ker & Glas> blank seam
Glaskeramik f <Ker & Glas, Kerntech> glass ceramic
Glaskolben m 1. <Ker & Glas> bulb; 2. <Lebensmittel> flask *(Laborgerät)*
Glaskondensator m <Elektrotech> glass capacitor
Glaskrug m <Ker & Glas> glass jug
Glaskugel f <Ker & Glas> bead

Glaslaser

Glaslaser m <Elektronik> glass laser
Glasleiste f <Bau> window bar
Glaslot n <Ker & Glas> solder glass
Glaslüftungsstein m <Ker & Glas> glass ventilating brick
Glasmacherlehrling m <Ker & Glas> taker-in
Glasmacherstuhl m <Ker & Glas> chair
Glasmacherwerkzeug n <Ker & Glas> glassmaker's tool
Glasmalerei f <Ker & Glas> painting on glass
Glasmarmor m <Ker & Glas> glass marble (zur Herstellung von Glasfasern)
Glasmikrosphäre f <Ker & Glas> glass microsphere
Glasoberfläche f <Ker & Glas> overlay, overlaying
Glasofenreise f <Ker & Glas> campaign
Glaspassivierung f <Elektronik> glass passivation; glassivation (Halbleiter)
Glaspegelregler m <Ker & Glas> glass level controller
Glasperle f 1. <Anstrich> glass bead; 2. <Ker & Glas> bead, glass bead
Glaspflasterplatte f <Ker & Glas> glass-paving slab
Glasplatte f <Labor> glass plate
Glasposten m <Ker & Glas> gob
Glaspostenende n <Ker & Glas> gob tail
Glaspostentemperatur f <Ker & Glas> gob temperature
Glaspostenverarbeitung f <Ker & Glas> gobbing
Glaspostenverteiler m <Ker & Glas> gob distributor
Glaspressen n <Ker & Glas> pressing
Glaspunkt m <Kunststoff> glass transition temperature
Glasrecycling n <Abfall> glass recycling
Glasrohr n <Elektronik, Ker & Glas> glass tube
Glasröhre f <Elektronik> glass tube
Glasroving n <Textil> roving
Glasrührstab m <Labor> glass-stirring rod
Glassatz m <Ker & Glas> batch formula
Glasschale f <Ker & Glas> glass dish
Glasscheibe f <Bau> pane
Glasscherben fpl <Abfall, Ker & Glas> cullet
Glasschleifer m <Ker & Glas> glass cutter (Person)
Glasschmuckperle f <Ker & Glas> prunt
Glasschneidediamant m <Ker & Glas> diamond for glass cutting
Glasschneider m 1. <Bau> glass cutter; 2. <Ker & Glas> vitrea cutter; glass cutter (Werkzeug)
Glasschneiderdiamant m <Ker & Glas> cutting diamond
Glasschneidertisch m <Ker & Glas> cutter's table
Glasschneidertischlineal n <Ker & Glas> cutter's table ruler
Glasschneidezange f <Ker & Glas> cutter's pliers
Glasseidenmatte f <Wassertrans> chopped-strand mat (Schiffbau)
Glasseidenvlies n <Ker & Glas> continuous strand mat
Glasseidenzwirn m <Ker & Glas> spun roving
Glassinterfiltertiegel m <Labor> sintered glass filter crucible
Glassinterfiltertrichter m <Labor> sintered glass filter funnel
Glasspinnfadengarn n <Ker & Glas> glass continuous filament yarn
Glaspulver n <Ker & Glas> powdered glass
Glasstab m <Labor> glass rod
Glasstahlbeton m <Ker & Glas> glass-reinforced concrete
Glasstapelfasergarn n <Ker & Glas> (AE) glass staple-fiber yarn, (BE) glass staple-fibre yarn
Glasstaub m <Ker & Glas> glass dust
Glas-Substrat n <Elektronik> glass substrate
Glastasche f <Ker & Glas> glass pocket
Glastuch n <Wassertrans> glass cloth (Schiffbaumaterial)
Glasübergangstemperatur f <Kunststoff> glass transition temperature
Glasumwandlungstemperatur f <Kunststoff> glass transition temperature
Glasur f 1. <Bau> enamel; 2. <Ker & Glas> glaze, glossing; 3. <Lebensmittel> frosting, icing; 4. <Textil> glazing
Glasurerz n <Ker & Glas> potter's ore
Glasurofen m <Ker & Glas> glaze kiln
Glasurqualität f <Ker & Glas> glazing quality
Glasurschleifer m <Ker & Glas> glaze grinder
Glasurstein m <Bau> glazed brick
Glasurziegel m <Bau> glazed brick
Glaswalzendämpfer m <Ker & Glas> glass roll dampener
Glaswaren fpl <Ker & Glas> glassware
Glaswaschapparat m <Labor> glass washer
Glaswatte f <Verpack> glass wadding
Glaswolle f 1. <Ker & Glas> spun glass; 2. <Kunststoff> glass wool; 3. <Wassertrans> (AE) fiberglass, (BE) fibreglass (Schiffbau)
Glaswollefilter n <Ker & Glas> glass wool filter
Glaszange f <Ker & Glas> chipping tool
Glasziegel m <Ker & Glas> glass brick
Glaszustand m <Ker & Glas> vitreous state
glatt adj 1. <Anstrich> smooth; 2. <Fertig> slick; 3. <Ker & Glas> plain (ohne Dekor); 4. <Math> smooth; 5. <Papier> even, smooth
glatt hobeln v <Bau> surface; shoot (Holzkanten)
glatt streichen v 1. <Bau> strike (Fugen); 2. <Fertig> sleek (Gießen)
Glatt... <Ker & Glas, Papier, Wassertrans> flush, straight
Glattbeplattung f <Wassertrans> flush plating (Schiffbau)
Glattbrand m <Ker & Glas> sharp fire
Glattdeck n <Wassertrans> flush deck (Schiffbau)
Glätte f 1. <Papier> smoothness; 2. <Trans> slipperiness
glätten v 1. <Bau> even, flush, skim, smooth, trowel; level (Mauern); 2. <Fertig> slick (Gießform); 3. <Maschinen> flat; 4. <Mechan> polish; 5. <Metall> burnish; 6. <Textil> glaze, iron out, surface
Glätten n 1. <Bau> planing; 2. <Comp & DV> smoothing (von Rundungen); 3. <Ker & Glas> smoothing (von Hohlglas); 4. <Maschinen, Metall> burnishing
Glätteprüfer m <Papier> smoothness tester
glatter Satz m <Druck> straight text matter
glattes Gewebe n <Textil> plain fabric
Glattgarn n <Textil> flat yarn
Glattmantelwalze f <Bau> smooth roller
Glattnegativ n <Foto> straight negative
Glattpresse f <Papier> plain press
Glättpresse f <Papier> smoothing press
Glattpressenwalze f <Papier> plain press roll
Glattsandstrahlen n <Ker & Glas> plain sandblast
Glattschachtpackung f <Ker & Glas> smooth plain packing
Glattstoßmaschine f <Druck> jogging machine
Glattstreichen n <Fertig> sleeking (Gießen)
Glättung f 1. <Elektronik, Gerät> smoothing; 2. <Mechan> burnishing; 3. <Papier> glaze, polish, smooth finish
Glättungsdrossel f <Elektriz, Elektrotech> smoothing choke, smoothing inductor, smoothing reactor, ripple-filter choke
Glättungskondensator m <Elektriz, Elektrotech> smoothing capacitor, smoothing filter capacitor
Glättungskreis m <Elektriz, Elektrotech> smoothing circuit
Glättungsschaltung f <Elektriz, Elektronik> smoothing filter
Glättungswiderstand m <Elektriz, Elektrotech> smoothing resistor
Glättwalze f <Papier> smoothing roll
Glattwalzenstuhl m <Fertig> smooth roller mill
Glättwerk n <Papier> thickness calender

Glättwerkzeug n <Maschinen> burnishing tool
Glättzahn m 1. <Fertig> button (Räumwerkzeug); 2. <Maschinen> burnishing tooth
Glättzylinder m <Papier> MG cylinder, machine-glazing cylinder, glazing cylinder
Glaubensmaß n <Künstl Int> MB, measure of belief
Glaukonitmergel m <Bau> glauconite marl
Gleason-Verzahnung f <Maschinen> Gleason gear teeth
gleich adj <Math> equal
gleich bleibende Peilung f <Wassertrans> steady bearing (Navigation)
gleich bleibender Differenzialdruck m <Lufttrans> constant differential pressure
gleich bleibender Schaltschritt m <Comp & DV> monospacing
gleich schwere Kerne mpl <Kerntech> isobar
gleich wahrscheinlich adv <Math> equally likely
gleich weit entfernt adj <Bau, Geom> equidistant
gleichachsiger gegenläufiger Propeller m <Lufttrans> coaxial propeller
gleichachsiges Korn n <Metall> equiaxed grain
gleicharmige Brücke f <Elektriz> equal arm bridge
gleichberechtigter Spontanbetrieb m <Telekom> balanced mode
Gleichdruck m <Heiz & Kälte> balanced pressure
Gleichdruckturbine f 1. <Hydraul> action turbine; 2. <Nichtfoss Energ> impulse turbine
Gleichdruckvergaser m <Kfztech> (AE) suction carburetor, (BE) suction carburettor
gleiche Mengen fpl <Math> equal sets (Mengenlehre)
gleichen v <Geom> equal
gleichförmig adj <Metall> homogeneous
gleichförmig angeregte Gassäule f <Strahlphys> uniformly-excited column of gas
gleichförmige Bewegung f <Phys> uniform motion
gleichförmiger Schall m <Sicherheit> steady state noise
gleichgerichteter Ausgang m <Elektrotech> rectified output
gleichgerichteter Strom m <Elektrotech> rectified current, unidirectional current
gleichgerichteter Wechselstrom m <Elektriz> rectified alternating current
Gleichgewicht n 1. <Fertig> poise; 2. <Lufttrans> aerodynamic balance; 3. <Mechan> equilibrium; 4. <Metrol, Papier> balance; 5. <Phys> balance, equilibrium • **aus dem Gleichgewicht** <Metrol> out of balance • **aus dem Gleichgewicht bringen** <Sicherheit> overbalance • **nicht im Gleichgewicht** <Phys> not in equilibrium
Gleichgewicht n **des Luftschraubenblattes** <Lufttrans> blade balance (Hubschrauber)
Gleichgewichtsanflug m <Lufttrans> steady approach
Gleichgewichtsbedingung f 1. <Mechan> equation of equilibrium; 2. <Optik> steady state condition
Gleichgewichtsdichte f <Erdöl> equilibrium density (Bohrtechnik)
Gleichgewichtsfehler m <Mechan> unbalance
Gleichgewichtskonstante f (K) <Thermod> equilibrium constant (K)
Gleichgewichtskraft f <Lufttrans> equalizing
Gleichgewichtskurve f <Metall> equilibrium curve
Gleichgewichtslänge f <Telekom> equilibrium-mode distribution length, equilibrium length
Gleichgewichtsmodenverteilung f <Telekom> equilibrium mode distribution
Gleichgewichtspunkt m 1. <Bau> (AE) center of gravity, (BE) centre of gravity; 2. <Gerät> balance point (Brücke); 3. <Nichtfoss Energ> (AE) center of gravity, (BE) centre of gravity

Gleichgewichtsstrahlungsdiagramm n <Telekom> equilibrium radiation pattern
Gleichgewichtsthermodynamik f <Thermod> thermostatics
Gleichgewichtsventil n <Hydraul> equilibrium valve
Gleichgewichtszustand m 1. <Phys> steady state; 2. <Thermod> state of equilibrium
Gleichgewichtszustand m **des Gezeitenwechsels** <Nichtfoss Energ> equilibrium tide
Gleichheit f <Comp & DV, Math> equality
Gleichheitsprüfer m <Phys> comparator
Gleichkanal... <Telekom> cochannel
Gleichkanalfunkstörung f <Funktech, Telekom> co-channel radio interference, common channel radio interference (CCI)
Gleichkanalschutzabstand m <Funktech, Telekom> co-channel protection ratio
Gleichkanalstörer m <Funktech, Telekom> cochannel interferer
Gleichkanalstörung f <Aufnahme, Fernseh, Funktech, Telekom> cochannel interference
Gleichkanalwiederholabstand m <Funktech, Telekom> cochannel reuse distance
Gleichklang m <Akustik> consonance, unison
gleichkörnige Erde f <Kohlen> even-grained soil
gleichkörniger Kieszuschlagstoff m <Bau> single size gravel aggregate
gleichlastig adj <Wassertrans> on-even-keel
Gleichlauf m 1. <Elektronik> synchronism; 2. <Fernseh> tracking, sync; 3. <Kfztech> synchronization; 4. <Kontroll, Maschinen> synchronism; 5. <Telekom> synchronism, synchronization, sync • **Gleichlauf wiederherstellen** <Telekom> resynchronize
gleichlaufend adj 1. <Elektrotech> in step; 2. <Telekom> synchronous
gleichlaufend existent adj <Kontroll> concurrent
Gleichlauffehler m 1. <Elektronik> clocking error; 2. <Telekom> synchronization error
Gleichlauffräsen n 1. <Fertig> climb-feed milling, climb milling, cutting down, down-cut milling; 2. <Maschinen> down-cut milling
Gleichlaufgelenk n <Kfztech> constant-velocity universal joint (Antriebswelle, Kardanwelle)
Gleichlaufgenerator m <Fernseh> tracking generator
Gleichlaufgetriebe n <Maschinen> synchromesh gear
Gleichlaufimpuls m <Elektronik> clocking pulse
Gleichlaufkreise mpl <Elektrotech, Fernseh, Funktech> ganged circuits
Gleichlaufregler m <Fernseh> tracking control
Gleichlaufschleifen n <Fertig> down grinding
Gleichlaufschwankung f <Akustik, Telekom> flutter (Aufzeichnung)
Gleichlaufschwankungen fpl <Fernseh> wow and flutter (Tonbandgerät)
Gleichlaufsteuerung f <Kontroll> synchronization
Gleichlichtphotometer n <Gerät> continuous-light photometer
Gleichmaß n <Akustik> time
gleichmäßig adj 1. <Fertig> even (Druck); 2. <Math> monotone, uniformly
gleichmäßig anregendes Rauschen n <Akustik> uniform-exciting noise
gleichmäßig verdeckendes Rauschen n <Akustik> uniform-masking noise
gleichmäßig verteilt adj <Anstrich> smooth
gleichmäßige Beschleunigung f <Papier> uniform acceleration
gleichmäßige Korrosion f <Metall> uniform corrosion

gleichmäßige

gleichmäßige Lagenwicklung f <Elektriz> uniform-layer winding
gleichmäßiger Anflug m <Lufttrans> steady approach
gleichmäßiger Lauf m <Maschinen> smooth running
gleichmäßiges Rauschen n <Akustik> steady noise
Gleichmäßigkeitskoeffizient m <Kohlen> uniformity coefficient
Gleichmäßigkeitsprüfung f <Metall, Qual> homogeneity test
gleichmöglich adv <Math> equally likely
gleichmolar adj <Chemie> equimolecular
gleichnamige Pole mpl <Phys> like poles
gleichphasig adj <Elektriz, Elektronik, Telekom> in-phase
gleichphasige Fläche f <Elektrotech> equiphase surface
gleichphasige Komponente f <Elektronik> in-phase component
gleichphasiger Strom m <Elektriz> in-phase current
gleichphasiges Signal n <Elektronik> in-phase signal
Gleichpolgenerator m <Elektrotech> homopolar generator
Gleichpotenzial-Linse f <Kerntech> unipotential lens
gleichrichten v <Elektriz, Elektrotech, Funktech> rectify
Gleichrichter m 1. <Aufnahme> rectifier *(elektrisches Ventil)*; 2. <Elektriz, Elektronik, Elektrotech> detector, rectifier; 3. <Funktech> demodulator, rectifier; detector, signal detector *(im Empfänger)*; 4. <Kfztech> detector, rectifier *(KFZ-Elektrik)*; 5. <Phys> detector, rectifier; 6. <Telekom> demodulator, rectifier; detector, signal detector *(im Empfänger)*
Gleichrichter m **als Brücke geschaltet** <Elektrotech> bridge rectifier
Gleichrichter m **zur Batterieladung** <Elektriz> charging rectifier
Gleichrichteranlage f <Elektrotech> rectifier equipment, rectifier plant
Gleichrichteranode f <Elektrotech> rectifier anode
Gleichrichteranschluss m <Elektrotech> rectifying junction
Gleichrichterbrücke f <Elektrotech> bridge rectifier, rectifier bridge
Gleichrichterdiode f <Elektronik> rectifier diode
Gleichrichterelement n <Elektrotech> rectifier element
Gleichrichterfilter n <Aufnahme> rectifier filter
Gleichrichtergerät n <Telekom> rectifier unit
Gleichrichterinstrument n <Gerät> rectifier instrument
Gleichrichterkreis m <Elektrotech> rectifying circuit
Gleichrichterlokomotive f <Eisenbahn> rectifier locomotive
Gleichrichtermessgerät n <Gerät> rectifier instrument
Gleichrichterrauschen n <Elektronik> rectifier detector noise
Gleichrichterröhre f <Elektrotech> rectifier tube
Gleichrichterschaltung f <Elektrotech> rectifying circuit
Gleichrichterspannung f <Elektronik, Elektrotech> undulating voltage
Gleichrichterstation f <Elektrotech> rectifier substation
Gleichrichtertrafo m <Elektrotech> rectifier transformer
Gleichrichtertransformator m <Elektrotech> rectifier transformer
Gleichrichterunterwerk n <Elektrotech> rectifier substation
Gleichrichterwandler m <Elektrotech> rectifier transformer
Gleichrichterzelle f <Elektrotech> rectifier cell
Gleichrichtgrad m <Elektrotech> rectification factor
Gleichrichtung f 1. <Akustik> detection; 2. <Elektriz, Elektrotech> rectification; 3. <Raumfahrt> detection *(Weltraumfunk)*
gleichschenklig adj <Geom> isosceles

gleichschenkliges Dreieck n <Geom> isosceles triangle
gleichschenkliges Winkeleisen n <Bau> equal-sided angles
Gleichschlagseil n <Maschinen> long-lay rope
gleichschwebende Temperatur f <Akustik> equal temperament
gleichseitig adj <Geom> equilateral
gleichseitiges Giebeldach n <Bau> span roof
Gleichspannung f <Elektrotech> DC voltage
Gleichspannungsmesser m <Elektrotech> DC voltmeter
Gleichspannungsquelle f <Elektrotech> DC voltage source
Gleichspannungsteiler m <Elektriz> DC potentiometer
Gleichspannungswandler m <Elektrotech> DC-DC converter, direct-current converter
Gleichspannungswandlung f <Elektrotech> DC-DC conversion
Gleichspulenwicklung f <Elektrotech> diamond winding
Gleichstrom m *(GS)* <Aufnahme, Comp & DV, Eisenbahn, Elektriz, Elektrotech, Fernseh, Funktech, Fertig, Phys, Telekom> direct current *(DC)*
Gleichstromamperemeter n <Elektrotech> DC ammeter
Gleichstromanteil m **im Stoßkurzschlussstrom** <Elektrotech> aperiodic component
Gleichstrom-Ausgleichsmaschinensatz m <Elektrotech> direct current balancer
Gleichstrombremsung f <Elektrotech> d.c. braking, d.c. injection braking *(Drehstrommotor)*
Gleichstrombrücke f <Elektrotech> DC bridge
Gleichstromdirektumformer m <Elektriz> direct d-c converter
Gleichstromdynamo m <Elektriz> direct-current generator
Gleichstromelektronenbeschleuniger m *(CEBAF)* <Teilphys> continuous electron beam facility *(CEBAF)*
Gleichstromerzeuger m <Elektrotech> DC generator
Gleichstromerzeugung f <Elektrotech> DC generation
gleichstromfrei adj d.c. free, direct-current-free
gleichstromgekoppelt adj <Elektriz> direct-coupled
Gleichstromgenerator m 1. <Elektriz, Elektrotech> DC generator, direct-current generator, dynamo; 2. <Phys> DC generator
Gleichstromisolation f <Elektrotech> DC isolation
Gleichstrom-Josephson-Effekt m <Phys> DC Josephson effect
Gleichstromkompensator m <Elektriz> DC potentiometer, direct-current potentiometer
Gleichstromkomponente f <Elektriz, Telekom> DC component, direct-current component
Gleichstromkonverter m <Elektrotech> direct-current converter
Gleichstromkoppler m <Telekom> DCC, direct-current coupler
Gleichstromkopplung f <Elektriz> direct coupling
Gleichstromkreis m 1. <Eisenbahn> track circuit; 2. <Elektrotech> DC circuit
Gleichstromlichtbogenschweißen n <Fertig> DC welding
Gleichstromlöschkopf m <Aufnahme> DC erase head
Gleichstrommaschine f <Elektrotech> DC machine
Gleichstrommesser m <Elektriz> DC meter
Gleichstrommittenabgleich m <Fernseh> *(AE)* DC centering, *(BE)* DC centring
Gleichstrommotor m <Elektriz, Elektrotech, Fertig, Phys> DC motor, direct-current motor
Gleichstrommotor m **in Bürstenbauweise** <Elektrotech> brush-type DC motor
Gleichstromnetz n <Elektriz, Elektrotech> DC network, direct-current network

Gleichstromnetzwerk n <Elektrotech> DC network, direct-current network
Gleichstrompegel m <Fernseh> DC level
Gleichstrompotenziometer n <Elektriz> DC potentiometer, direct-current potentiometer
Gleichstromprüfzeile f <Fernseh> dc insertion
Gleichstromquelle f <Elektrotech> DC supply, direct-current supply
Gleichstromrauschen n <Aufnahme> DC noise
Gleichstromregelung f <Elektriz> DC regulation, direct-voltage regulation
Gleichstromrelais n <Elektriz, Elektrotech> DC relay, direct-current relay
Gleichstromschaltvorrichtung f <Elektrotech> DC switching
Gleichstromschweißen n <Fertig> DC welding
gleichstromseitiges Filter n <Elektrotech> d.c. filter
Gleichstromservometer n <Elektrotech> DC servometer
Gleichstromsignalisierung f <Elektronik> *(AE)* DC signaling, *(BE)* DC signalling
Gleichstromsteller m <Elektronik> d.c. chopper, d.c. motor controller
Gleichstromtrafo m <Elektrotech> DC transformer
Gleichstromtransformator m <Elektrotech> DC transformer
Gleichstromtriebfahrzeug n <Elektrotech> d.c. motor vehicle
Gleichstromturbine f <Hydraul> parallel flow turbine
Gleichstromumformer m <Elektriz> DC converter
Gleichstromumrichter m <Comp & DV> inverter
Gleichstromverdichter m <Heiz & Kälte> uniflow compressor
Gleichstromversorgung f <Elektriz, Elektrotech> DC supply, direct-current supply
Gleichstromverstärker m <Elektronik> DC amplifier
Gleichstromverstärkung f <Elektronik> DC amplification, DC current gain
Gleichstromverzerrung f <Kerntech> direct-current distortion
Gleichstromvoltmeter n <Elektrotech> DC voltmeter
Gleichstromvorspannung f <Fernseh> DC biasing
Gleichstromvorwärmer m <Heiz & Kälte> uniflow preheater
Gleichstromwandler m 1. <Elektriz, Elektrotech> direct-current transformer, DC transducer; 2. <Raumfahrt> DC-AC converter *(Raumschiff)*
Gleichstromwärmeaustauscher m <Heiz & Kälte, Kerntech> parallel-flow heat exchanger
Gleichstrom-Wechselstrom-Konverter m <Elektrotech> DC-AC-inverter
Gleichstromwiderstand m <Elektrotech> DC resistance, ohmic resistance
Gleichstrom-Zugförderungsmotor m <Trans> direct-current traction motor
Gleichtakt m <Elektronik> common mode *(Differenzialverstärker)*
Gleichtaktfeuer n <Wassertrans> isophase light
Gleichtaktsignal n <Elektronik, Telekom> common-mode signal
Gleichtaktspannung f <Elektrotech> common-mode voltage
Gleichtaktunterdrückung f <Elektronik, Telekom> common-mode rejection
Gleichtaktunterdrückungsverhältnis n <Elektronik, Telekom> common-mode rejection ratio
Gleichtaktverstärkung f <Elektronik> common-mode gain
gleichtemperiert adj <Mechan, Phys, Thermod> isothermal

Gleichung f <Chemie, Math, Strömphys> equation
Gleichung f **dritten Grades** <Fertig, Math> cubic equation
Gleichung f **ersten Grades** f <Math> equation of first degree
Gleichungen fpl **für reibungsbehaftete Strömungen** <Phys> viscous flow equations
Gleichungsseiten fpl <Math> equation members
Gleichverteilung f **der Energie** <Phys> equipartition of energy
Gleichwellensignal n <Elektrotech> continuous-wave signal, CW signal
Gleichwertigkeit f <Comp & DV> equivalence
gleichwinklig adj <Geom> equiangular, isogonal
gleichwinkliges Dreieck n <Geom> equiangular triangle
gleichzeitig adj <Comp & DV, Kontroll> simultaneous
gleichzeitig ablaufend adj <Comp & DV, Qual> concurrent
gleichzeitig ablaufendes Kopieren n <Comp & DV> concurrent copy
gleichzeitige Ausführung f <Comp & DV, Qual> concurrent execution
gleichzeitige Ausführung f **mehrerer Jobs** <Comp & DV> multitasking
gleichzeitige Führung f <Raumfahrt> iterative guidance
gleichzeitige Verarbeitung f <Comp & DV> concurrent processing
gleichzeitige Verarbeitung f **von zwei Kübeln** <Ker & Glas> double gobbing
gleichzeitiger Ablauf m **mehrerer Programme** <Comp & DV> multitasking
gleichzeitiger Benutzer m <Comp & DV> concurrent user
Gleichzeitigkeit f <Phys> simultaneity
Gleis n <Eisenbahn> line, rail track, track • **Gleis frei** <Eisenbahn> line-clear
Gleis n **außer Betrieb** <Eisenbahn> track out of service
Gleis n **in Betrieb** <Eisenbahn> track in service
Gleis n **mit dritter Schiene** <Eisenbahn> *(AE)* mixed-gage track, *(BE)* mixed-gauge track
Gleisabschnitt m <Eisenbahn> track section
Gleisanschluss m <Bau> siding
Gleisarbeiten n <Eisenbahn> track laying
Gleisarbeiter m <Eisenbahn> track layer
Gleisbild n <Eisenbahn> track diagram
Gleisbremse f <Eisenbahn> rail brake
Gleisdrossel f <Eisenbahn> impedance bond
Gleise npl <Eisenbahn> metals
Gleisinstabilität f <Eisenbahn> instability of track
Gleisjochlegemaschine f <Eisenbahn> track-panel laying machine
Gleiskette f <Mechan> track
Gleiskettenschrappwagen m <Trans> caterpillar-hauling scraper
Gleislage f <Eisenbahn> track bed
Gleislagerung f <Eisenbahn> track bed
Gleismontage f <Eisenbahn> track assembly
Gleisrelais n <Eisenbahn> track relay
Gleisschotter m <Eisenbahn> track ballast
Gleissenkung f <Eisenbahn> depression of track
Gleisstrang m **aus endlos zusammengeschweißten Schienen** <Eisenbahn> long-welded rail
Gleistafel f <Eisenbahn> track diagram
Gleisüberführung f <Eisenbahn> *(BE)* flyover, *(AE)* skyway
Gleisverbindung f <Eisenbahn> crossover
Gleisverformung f <Eisenbahn> distortion of the track
Gleisverlegung f <Bau> track laying
Gleisverwerfung f <Eisenbahn> distortion of the track, warping of track

Gleisvorarbeiter

Gleisvorarbeiter m <Eisenbahn> track-laying foreman *(Person)*
Gleiszwischenraum m <Eisenbahn> space between rails
Gleit... <Mechan> antifriction
Gleitaxiallager n <Maschinen> plain thrust bearing
Gleitbacke f <Fertig> flat die
Gleitbahn f 1. <Lufttrans> glide path; 2. <Maschinen> slideway, track
Gleitbahnkurs-Funkfeuer n <Funkort, Lufttrans> glide path localizer
Gleitband n <Metall> glide band
Gleitbereich m <Comp & DV> floating area
Gleitbewegung f <Elektrotech> slip
Gleitboot n <Wassertrans> gliding boat, hydroplane
Gleitbruch m <Kerntech> gliding fracture, sliding fracture
Gleitbügel m <Eisenbahn> pantograph
Gleitebene f 1. <Fertig> sliding plane; 2. <Kerntech> glide plane
gleiten v <Maschinen, Mechan> slide
Gleiten n 1. <Elektrotech> slip; 2. <Maschinen> slide, sliding
gleitende Last f <Elektrotech> sliding load
gleitender Löschkopf m <Fernseh> flying erase head
Gleiter m <Trans> skidder
Gleitfeder f <Maschinen> sliding key
Gleitfenster-Protokoll n <Telekom> sliding-window protocol *(Datenübertragung)*
Gleitfestigkeit f <Maschinen, Qual> resistance to sliding
Gleitfläche f 1. <Fertig> rubbing surface; 2. <Kohlen> slip surface; 3. <Metall> gliding plane
Gleitflächendichtung f <Maschinen> floating seal
Gleitflosse f <Lufttrans> chine *(Wasserflugzeug)*
Gleitflug m <Lufttrans> gliding flight
Gleitflugentfernung f <Lufttrans> gliding distance
Gleitflugwinkel m <Lufttrans> gliding angle
Gleitfunke m <Elektrotech> creepage discharge spark, surface discharge spark
Gleitfunkenanordnung f <Elektrotech> creepage discharge arrangement
Gleitfunkenoberfläche f <Elektrotech> creepage discharge surface
Gleitgelenk n <Maschinen> slip joint
Gleitgeometrie f <Metall> geometry of glide
Gleitgeschwindigkeit f <Eisenbahn> sliding speed *(Rad)*
Gleithemmung f <Akustik> antiskating
Gleitkomma n 1. <Comp & DV> floating point; 2. <Math> floating decimal point
Gleitkomma-Arithmetik f <Comp & DV> floating-point arithmetic
Gleitkommabetrieb m <Comp & DV> FLOP, floating-point operation
Gleitkommaoperation f <Comp & DV> FLOP, floating-point operation
Gleitkommaschreibweise f <Comp & DV> floating-point notation
Gleitkommazahl f <Comp & DV> floating-point number
Gleitkontakt m <Elektrotech> gliding contact, rubbing contact, sliding contact
Gleitkreis m <Bau> slip circle *(Bodenmechanik)*
Gleitkufe f <Fertig> skid
Gleitlager n 1. <Kfztech> plain bearing; 2. <Maschinen> bearing, friction-type bearing, plain bearing
Gleitmarkierung f <Metall> slip marking
Gleitmittel n <Ker & Glas, Kunststoff> lubricant
Gleitmodul n 1. <Kohlen> rigidity modulus; 2. <Maschinen> modulus of transverse elasticity
Gleitpfad m <Lufttrans> glide path, glide slope
Gleitpfadantenne f <Funkort, Lufttrans> glide aerial
Gleitpotenziometer n <Elektrotech> slide potentiometer

Gleitreibung f <Ergon, Maschinen, Phys> sliding friction
Gleitreibungszahl f <Mechan, Phys, Qual> coefficient of sliding friction
Gleitring m <Elektrotech> slip ring
Gleitringdichtung f <Fertig> face seal; mechanical seal *(Kunststoffinstallationen)*
Gleitschaftkolben m <Kfztech> full slipper piston
Gleitschalung f <Bau> moving form, sliding formwork, sliding shuttering
Gleitschicht f <Maschinen> antifriction layer
Gleitschiene f 1. <Eisenbahn> slide rail; 2. <Fertig> skid; 3. <Kohlen> slide; 4. <Maschinen> crosshead guide, slide bar, slide rail
Gleitschienenträger m <Maschinen> slide bar carrier
Gleitschirm m <Raumfahrt> parafoil *(Raumschiff)*
Gleitschuh m 1. <Fertig> shoe, slipper; 2. <Eisenbahn> slipper; 3. <Maschinen> crosshead shoe, sliding block
Gleitschuhkolben m <Kfztech> full slipper piston
Gleitschutz m <Elektrotech> contact slipper, wheel slide protection device *(Elektrotraktion)*
Gleitschutz-Deckfarbe f <Wassertrans> nonslip deck paint *(Schiff- und Bootsbau)*
Gleitschutzeinrichtung f <Elektrotech> wheel slide protection device *(Elektrotraktion)*
Gleitschutzkette f <Maschinen> non-skid chain
gleitsicher adj <Sicherheit> antislip, non-slip
gleitsicherer Fußbodenbelag m <Sicherheit> antislip floor coating
gleitsicheres Schuhwerk n <Sicherheit> antislip footwear
Gleitsicherheit f <Sicherheit> antislip protection, slip resistance
Gleitsitz m 1. <Maschinen> sliding fit; 2. <Mechan> close-sliding fit, sliding fit
Gleitstange f <Fertig> tiller
Gleitstein m <Eisenbahn> link-block guide
Gleitstrahl m <Funkort, Lufttrans> glide path beam
Gleitstück n <Maschinen> slide, slide block, slipper
Gleitstufenhöhe f <Metall> slip step height
Gleit- und Widerstandsbeiwert m *(LD-Beiwert)* <Lufttrans> lift and drag ratio, LD ratio *(Nutzeffekt des Luftfahrzeugs)*
Gleitung f 1. <Bau> shear; 2. <Elektrotech> slip; 3. <Fertig> slipping
Gleitverhältnis n <Lufttrans> glide ratio
Gleitwagen m <Kfztech> skid car
Gleitwegfunkfeuer n <Lufttrans> glide path beacon
Gleitwegleitstrahler m <Funkort, Lufttrans> glide path beam
Gleitzahl f 1. <Lufttrans> lift-to-drag ratio; 2. <Maschinen> skid number; 3. <Wassertrans> lift-drag ratio *(Bootbau)*
Gleitzylinder m <Metall> slip cylinder
Glied n 1. <Comp & DV> member; term *(Reihe)*; 2. <Elektriz> link *(eines Systems)*; 3. <Maschinen> link, member; 4. <Telekom> section *(Filter)*; 5. <Wassertrans> link
Glied n **mit Zweipunktverhalten** <Regelung> element with two-step action
Gliederbandförderer m <Trans> apron conveyor
Gliederkessel m <Heiz & Kälte> sectional boiler
Gliederkette f <Maschinen> link chain
Gliederriemen m 1. <Kunststoff> link belting; 2. <Maschinen> link belt
Gliederschiff n <Wassertrans> articulated ship
Gliederung f <Bau> division
Gliederzug m <Eisenbahn> articulated train
Glimmeinsatzprüfung f <Elektrotech> partial discharge inception test
glimmen v <Thermod> *(AE)* smolder, *(BE)* smoulder *(Glut, schwelendes Feuer)*

Glimmentladung f 1. <Elektriz> glow discharge, corona discharge; 2. <Elektronik, Elektrotech, Phys, Strahlphys> glow discharge
Glimmentladungslampe f <Elektriz> glow discharge lamp
Glimmentladungsröhre f <Elektronik> glow discharge tube
Glimmentladungsventil n <Elektrotech> glow discharge rectifier
Glimmer m <Elektriz, Fertig, Ker & Glas, Kunststoff> mica
Glimmerkondensator m <Elektriz, Elektrotech> mica dielectric capacitor, mica capacitor
Glimmkatode f <Elektrotech> glow discharge cathode
Glimmlampe f <Elektrotech> glow lamp
Glimmlampe f mit Neonfüllung <Elektrotech> neon lamp
Glimmlichtgleichrichter m <Elektrotech> glow discharge rectifier
Glimmschalter m <Elektrotech> glow switch
global adj <Comp & DV> global
globale Informations-Infrastruktur f (GII) <Comp & DV> Global Information Infrastructure (GII)
globale Variable f <Comp & DV> global variable
globaler Positionsbestimmungssatellit m <Kfztech> global positioning satellite
globales Ersetzen n <Comp & DV> global change (Funktion bei Textprogrammen)
globales Hyperbelnavigationsverfahren n <Funktech> global hyperbolic navigation system
globales Mobilfunksystem n <Mobilkom> global system for mobile communication (GSM)
globales Netz n <Telekom> global area network (GAN)
globales Positionsbestimmungssystem n (GPS) <Wassertrans> global position finding system, global-positioning system, GPS (Satellitennavigation)
globales Satelliten-Navigationssystem n <Funktech> Global Navigation Satellite System (GLONASS)
globales Seenot- und Sicherheitssystem n <Telekom> global marine distress and safety system (GMDSS)
globales Suchen n und Ersetzen <Comp & DV> global search and replace
globales Umweltüberwachungssystem n <Umweltschmutz> Global Environment Monitoring System
Globalstrahl m <Raumfahrt> global beam (Weltraumfunk)
Globoid n <Maschinen> globoid
Globoidgetriebe n <Maschinen> enveloping worm drive, globoid gear, globoidal gear, globoidal worm gear
Globoidschnecke f <Maschinen> double-enveloping worm, enveloping tooth wheel
Glocke f 1. <Bau> cap (eines Dachfirstes); 2. <Chemtech> bubble; 3. <Elektriz> bell; 4. <Erdöl> bubble cap (Destillationstechnik); bubble cap (Raffinerie, Fraktioniertechnik)
Glockenbatterie f <Elektriz> bell battery
Glockenboden m 1. <Chemtech, Erdöl> bubble tray (Fraktioniertechnik); 2. <Kerntech> bubble cap
Glockenbodenkolonne f 1. <Chemtech, Erdöl> bubble-cap tray column; 2. <Erdöl> bubble-cap tower (Fraktioniertechnik)
glockenförmige Kurve f <Geom> bell-shaped curve
glockenförmiger Aufnahmestutzen m <Raumfahrt> bell mouth
Glockengut n <Metall> bell metal
Glockenhaube f <Kerntech> bubble hood
Glockenisolator m <Elektrotech> bell-shaped insulator, petticoat insulator
Glockenkappe f <Erdöl> bell cap (Raffinerie)
Glockenkegel m <Ker & Glas> bell cone
Glockenkurve f <Geom, Math> bell-shaped curve (normal distribution curve)
Glockenmanometer n <Gerät> bell-type manometer
Glockenmetall n <Metall> bell metal

Glockenrock m <Textil> flared skirt
Glockenspinnmaschine f <Textil> cap-spinning frame
Glockentonne f <Wassertrans> bell buoy (Seezeichen)
Glockentransformator m <Elektriz> bell transformer
Glockentrichter m <Labor> thistle funnel
Glockenventil n 1. <Hydraul> bell valve; 2. <Maschinen> bell-shaped valve, bellows valve
Glockenzentriervorrichtung f <Maschinen> (AE) bell-centering punch, (BE) bell-centring punch
Glosskalander m <Papier> gloss calender
Glove-Box f <Kerntech> (AE) glove box, (BE) glove compartment
Glucagon n <Chemie> glucagon
Glucamin n <Chemie> glucamine
Glucar... <Chemie> glucaric
Glucin n <Chemie> glucin
Glucon... <Chemie> gluconic
Glucopyranose f <Chemie> glucopyranose
Glucosamin n <Chemie> glucosamine
Glucosan n <Chemie> glucosan
Glucose f <Lebensmittel> grape sugar
Glucosid n <Chemie> glucoside
Glüh... <Fertig, Thermod> annealing
Glühanlage f <Metall> annealing plant
Glühbad n <Fertig, Thermod> annealing bath
Glühbehandlung f <Fertig> heat exchange
Glühbirne f 1. <Elektriz> bulb, incandescent lamp; 2. <Elektrotech> bulb (Lampe)
Glühdraht m <Elektriz> filament
Glühdrahthalterung f <Elektriz> anchor (Lampenkolben)
glühelektrische Elektronenemission f <Phys> thermionic emission
Glühemission f <Elektrotech> thermionic emission
Glühemissionskonverter m <Elektrotech> thermionic converter
Glühemissionswandlung f <Elektrotech> thermionic conversion
glühen v 1. <Thermod> bake (Stahl); 2. <Thermod> glow
glühend adj 1. <Metall> red-hot; 2. <Thermod> glowing
glühende Kohle f <Kohlen> live coal
Glühentladung f <Elektriz, Elektronik, Phys, Strahlphys> glow discharge
Glühfaden m <Elektrotech> filament
Glühfadenpyrometer n <Phys> disappearing filament pyrometer
glühfrischen v <Fertig> malleablize
Glühfrischen n <Fertig> malleablizing
Glühhaube f <Fertig> annealing bell
Glühhitze f <Thermod> glowing heat
Glühkatode f <Elektrotech> hot cathode, thermionic cathode
Glühkatodenentladung f <Elektrotech> glow discharge
Glühkatodenröhre f <Elektronik> hot-cathode tube, thermionic tube
Glühkatodentriode f <Elektronik> thermionic triode
Glühkerze f 1. <Kfztech> glow plug; 2. <Thermod> heat plug; glow plug (im Motor)
Glühkopfmotor m <Kfztech> hot-bulb engine
Glühlampe f <Elektrotech> incandescent lamp, lamp
Glühofen m <Fertig, Kerntech, Metall> annealing furnace
Glühpunkt m (GP) <Metall> annealing point (AP)
Glühschiffchen n <Labor> combustion boat
Glühschmelzer m <Bau> heating melter (Tiefbau)
Glühspan m <Mechan> scale
Glühstrumpf m <Thermod> gas mantle
Glühverlust m <Kohlen, Umweltschmutz> ignition loss
Glühzyklus m <Thermod> heat cycle
Gluon n 1. <Phys> gluon (Elementarteilchen); 2. <Teilphys> gluon

Glut

Glut *f* <Thermod> burning heat
Glutamin *n* <Chemie> glutamine
Glutamin... <Chemie> glutamic
Glutaminat *n* <Chemie> glutamate
Glutaraldehyd *n* <Chemie> glutaraldehyde
Glutardialdehyd *n* <Chemie> glutaraldehyde
Glutathion *n* <Chemie> glutathione
Glutbeständigkeit *f (Metall, Metall)* resistance to glow heat
glutenfrei *adj* <Lebensmittel> gluten-free
Gluthitze *f* <Thermod> glowing heat
Glyceraldehyd *m* <Chemie> glyceraldehyde
Glycerid *n* <Chemie, Lebensmittel> glyceride
Glycerin *n* 1. <Chemie> glyceric; 2. <Chemie> glycerine, glycerol
Glycerinacetat *n* <Lebensmittel> acetin
Glycerinaldehyd *m* <Chemie> glyceraldehyde
Glycerindibutyrat *n* <Chemie> dibutyrin
Glycerinmonoacetat *n* 1. <Chemie> monoacetin; 2. <Lebensmittel> glycerol monoacetate
Glycerintriacetat *n* <Chemie> triacetin
Glycerintributyrat *n* <Chemie> tributyrin
Glycerintrimyristat *n* <Chemie> myristin
Glycerintrinitrat *n* <Chemie> nitroglycerin
Glycerintrioleat *n* <Chemie> triolein
Glycerintripalmitin *n* <Chemie> tripalmitin
Glycerintripalmitinsäureester *m* <Chemie> tripalmitin
Glycerol *n* <Chemie> glycerine, glycerol
Glycerolmonoacetat *n* <Chemie> monoacetin
Glycerolmonostearat *n* <Chemie> monostearin
Glyceroltrinitrat *n* <Chemie> nitroglycerin
Glyceroltrioleat *n* <Chemie> olein
Glycerophosphat *n* <Chemie> glycerophosphate
Glycerophosphor... <Chemie> glycerophosphoric
Glyceryl... <Chemie> glyceryl
Glycerylmonostearat *n* <Chemie> monostearin
Glyceryltetradecanoat *n* <Chemie> myristin
Glyceryltripalmitat *n* <Chemie> palmitin
Glyceryltristearat *n* <Chemie> glyceryl tristeate, tristearin
Glycid... <Chemie> glycidic
Glycogen *n* <Chemie> glycogen
Glycol *n* <Chemie> glycol
Glycol... <Chemie> glycolic
Glycolylharnstoff *m* <Chemie> hydantoin
Glycosidase *f* <Chemie> carbohydrase
Glycyl... <Chemie> glycyl
Glycyrrhizinsäure *f* <Chemie> glycyrrhizine
Glykogen *n* <Lebensmittel> animal starch, glycogen
Glykokoll *n* <Lebensmittel> glycine
Glykol... <Chemie> glycollic
Glykolid *n* <Chemie> glycolide
Glykolipid *n* <Chemie> glycolipid
Glykolyse *f* <Lebensmittel> glycolysis
Glykoprotein *n* <Chemie> glucoprotein
Glykosid *n* <Lebensmittel> glycoside
Glykuron... <Chemie> glycuronic
Glyoxal *n* <Chemie> glyoxal
Glyoxalin *n* <Chemie> glyoxaline
Glyoxim *n* <Chemie> glyoxime
Glyoxyl... <Chemie> glyoxylic
Glyoxyldiureid *n* <Chemie> allantoin, glyoxydiureide
Glyptalharz *n* <Kunststoff> glyptal resin
g/m *(Gramm pro Quadratmeter)* <Druck> gsm *(grams per square metre)*
GMDSS *(System zur Rettung von Menschenleben bei Seenotfällen)* <Wassertrans> GMDSS *(global marine distress and safety system)*
Gneis *m* <Bau, Erdöl> gneiss *(Mineralogie)*
gnomonische Projektion *f* <Funktech, Wassertrans> gnomonic projection *(Navigation)*

Golay-Zelle *f* <Phys> Golay cell
Gold *n* <Metall> gold
Goldamalgam *n* <Metall> gold amalgam
Goldbeschichtung *f* <Raumfahrt> gold plating *(Raumschiff)*
Goldblatt-Elektroskop *n* <Elektriz, Elektrotech, Phys> gold leaf electroscope
Goldchlorid *n* <Chemie> gold chloride, gold trichloride
Goldcyanid *n* <Chemie> gold cyanide
golddotierte Diode *f* <Elektronik> gold-doped diode
Golddotierung *f* <Elektronik> gold doping
Goldener Schnitt *m* 1. <Druck> golden rectangle, golden section; 2. <Geom> golden section
Goldepoxid *n* <Elektronik> gold epoxy
Goldfolie *f* <Druck> gold foil
Goldgehalt *m* <Metall> gold content
goldplattieren *v* <Metall> plate
Goldpulver *n* <Ker & Glas> powdered gold
Goldschnitt *m* <Druck> gilt edges
Goldsondenmethode *f* <Kerntech> gold probe method *(der Eichung)*
Goldtonbad *n* <Foto> gold toning
Goldwaschen *n* <Chemie> panning
Golf *m* <Wassertrans> gulf
Gondel *f* <Eisenbahn> gondola
Gongtonne *f* <Wassertrans> gong buoy *(Seezeichen)*
Goniometer *n* 1. <Funkort, Geom> goniometer; 2. <Metrol> angulometer
Goniometerantenne *f* <Funktech> cross-coil antenna, crossed-loop antenna
Goniometerpeiler *m* <Funktech> cross-coil direction finder, crossed-loop direction finder
Gooch-Tiegel *m* <Labor> Gooch crucible *(Filtrieren)*
Göpel *m* <Bau> gin *(Bohrarbeiten)*
Görtlerzahl *f* <Strömphys> Görtler number *(Grenzschichttheorie)*
Gösch *f* <Wassertrans> jack flag *(Flagge)*
Gotisch *n* <Druck> gothic
Goubau-Leitung *f* <Telekom> Goubau line
GÖV *(Gas-Öl-Verhältnis)* <Erdöl> GOR, gas-to-oil ratio *(Lagerstättentechnik)*
GP *(Glühpunkt)* <Metall> AP *(annealing point)*
GPC *(Gel-Permeations-Chromatographie)* <Kunststoff, Labor> GPC *(gel permeation chromatography)*
GPF-Ruß *m* <Kunststoff> GPF carbon black, general-purpose furnace carbon black
GPG *(Grundprimärgruppe)* <Telekom> basic group
GPS 1. <Funkort> *(weltweites Ortungssystem)* GPS, global-positioning system; 2. <Wassertrans> *(globales Positionsbestimmungssystem)* GPS, global-positioning system *(Satellitennavigation)*
GPS-Empfänger *m* <Wassertrans> GPS receiver *(Satellitennavigation)*
graben *v* 1. <Bau> dig, excavate; 2. <Kohlen> excavate
Graben *m* 1. <Bau> trench; ditch *(Entwässerung)*; 2. <Erdöl> trench *(Leitungsbau)*; 3. <Kohlen> ditch, trench; 4. <Nichtfoss Energ> *(AE)* dike, *(BE)* dyke; 5. <Wasserversorg> ditch line, ditch race
Graben im Silizium <Elektronik> silicon trench
Grabenarbeiten *fpl* <Bau> trenchwork
Grabenätzverfahren *n* <Elektronik> trench etch technique
Grabenaushub *m* <Bau> trenching
Grabenbagger *m* <Bau> trench excavator
Grabenbewässerung *f* <Wasserversorg> ditch irrigation
Grabenbildung *f* <Elektronik> trench formation
Grabenbreite *f* <Elektronik> trench width
Grabenherstellung *f* <Bau> trenching
Grabenisolationstechnik *f* <Elektronik> trenchh isolation technology

Grabenkondensator m <Elektronik> trench capacitor
Grabenmethode f <Abfall> trench method
Graben-MOS-Transistor m <Elektronik> trench metal-oxide semiconductor transistor *(TMOS)*
Grabentransistorzelle f <Elektronik> trench transistor cell *(TTC)*
Grabenverbau m <Bau> trench sheeting
Grabenverfüllgerät n <Trans> back filler
Grabenzelle f <Elektronik> trench cell
Grabzähne mpl <Bau> cutting teeth
Grad m 1. <Fertig> order; 2. <Geom> degree; 3. <Metrol> rate; 4. <Phys, Thermod> degree
Grad m **Celsius** <Metrol> degree Celsius • **in Grad Celsius geteilt** <Mechan> centigrade
Grad m **Kelvin** m <Thermod> degree Kelvin
Gradbogen m 1. <Maschinen> protractor; 2. <Mechan> quadrant scale
Gradeinteilung f 1. <Comp & DV> calibration; 2. <Konstzeich> angular spacing; 3. <Labor> graduation
Gradient m 1. <Elektrotech> graded index, gradient; 2. <Geom> gradient *(einer Kurve)*; 3. <Künstl Int, Math> gradient; 4. <Phys> gradient *(Vektorrechnung)*; 5. <Strömphys> gradient *(Temperatur- und Druckgradient)*
Gradient m **des elektrischen Feldes** <Elektrotech> electric field gradient
Gradient m **einer Geraden** <Geom> gradient of a straight line
Gradientenfaser f <Telekom> *(AE)* gradient index fiber, *(BE)* gradient index fibre
Gradientenindex m <Optik, Telekom> graded index
Gradientenindexfaser f <Optik, Telekom> *(AE)* graded index fiber, *(BE)* graded index fibre *(Lichtleitfaser mit kontinuierlich veränderdetem Brechungsindex)*
Gradientenindexprofil n <Optik, Telekom> graded index profile
Gradientenkern m <Elektrotech, Optik, Telekom> graded index core
Gradientenlichtleitfaser f <Elektrotech> *(AE)* graded index multimode optical fiber, *(BE)* graded index multimode optical fibre *(mit mehreren Arbeitsweisen)*
Gradientenlichtwellenleiter m <Telekom> *(AE)* gradient index fiber, *(BE)* gradient index fibre
Gradientenlichtwellenleiter m **mit Potenzprofil** <Optik, Telekom> *(AE)* power law index fiber, *(BE)* power law index fibre
Gradientenzug m <Bau> line of levels
Gradientmikrofon n <Akustik> gradient microphone
Gradteiler m <Mechan> vernier
Grafik f 1. <Comp & DV> graphics; 2. <Druck> art work, graphic arts
Grafikadapter m <Comp & DV> graphic display adaptor
Grafikarbeitsplatz m <Comp & DV> graphics workstation
Grafikbildschirmadapter m <Comp & DV> graphic display adaptor
Grafikblock m <Comp & DV> Rand tablet, graphics pad, graphics tablet
Grafikdrucker m <Comp & DV> graphics printer
Grafikkarte f <Comp & DV> graphic display adaptor
Grafikkurvenschreiber m <Comp & DV> graphics plotter
Grafikmodus m <Comp & DV> graphics mode, plotting mode
Grafikplotter m <Comp & DV> graphics plotter
Grafikprozessor m <Comp & DV> graphics processor
Grafiksoftwarepaket n <Comp & DV> graphic software package
Grafiktableau n <Fernseh> graphic tablet
Grafiktablett n <Comp & DV> graphics tablet
Grafikvorlage f <Druck> art work

Grafikzeichen n <Comp & DV> graphic character, graphics character
grafisch adj <Druck> graphic, graphical
grafisch darstellen v 1. <Comp & DV> plot; 2. <Math> figure, graph
grafische Benutzeroberfläche f <Comp & DV> graphical interface
grafische Benutzerschnittstelle f *(GUI)* <Comp & DV> graphical user interface *(GUI)*
grafische Darstellung f 1. <Comp & DV> graphical representation; 2. <Lufttrans> plot *(Navigation)*
grafische Industrie f <Druck> graphic arts industry
grafische Lösung f <Bau> graphical solution
grafische Massenberechnung f <Nichtfoss Energ> mass diagram
grafische Methoden fpl <Math> graphical methods
grafische Symbole npl <Sicherheit> graphical symbols *(für Brandschutzpläne)*
grafischer Arbeitsplatz m <Comp & DV> graphics workstation
grafisches Editieren n <Comp & DV> graphical editing
grafisches Entwurfspaket n <Comp & DV> graphic design package
grafisches Gerät n <Gerät> display device
grafisches Gewerbe n <Druck> printing industry
Graham'sches Diffusionsgesetz n <Phys> Graham's law of diffusion
Gramm n *(g)* <Chemie, Metrol, Phys> gram *(g)*
Gramm n **pro Quadratmeter** *(g/m)* <Druck> *(AE)* grams per square meter, *(BE)* grams per square metre *(gsm)*
grammatische Markierung f <Comp & DV> grammatical tagger
Gramme-Ring m <Elektriz> Gramme winding
Grammion n <Metrol> gram-ion
Grammkalorie f <Metrol> gram calorie, *(AE)* gram centimeter heat-unit, *(BE)* gram centimetre heat-unit
Grammophon n <Aufnahme> *(BE)* gramophone, *(AE)* phonograph
Gran n <Metrol> grain
Granalie f <Metall> shot
Grand-Tourisme-Prototyp m *(GTP)* <Kfztech> grand touring prototype *(GTP)*
Granit m <Bau> granite
Granulat n <Chemtech> granulate
Granulation f <Chemtech> granulation
Granulationsrauschen n <Telekom> speckle noise
Granulator m 1. <Chemtech> granulating machine, granulator *(Zucker)*; 2. <Verpack> granulating machine
Granulierapparat m <Chemtech> granulator
granulieren v 1. <Chemtech> granulate; 2. <Fertig> grain; 3. <Verpack> granulate
Granulieren n <Kunststoff> pelletizing
Granulierextruder m <Kunststoff> pelletizer *(Kautschuk)*
Granuliermühle f <Chemtech> granulating crusher
Granulit n <Erdöl> granulite *(Mineralogie)*
Graph m 1. <Comp & DV, Künstl Int> graph; 2. <Math> graph *(einer Funktion)* • **einen Graph zeichnen** <Math> plot a graph
Graphensuche f <Künstl Int> graph search
Graphensuchverfahren n <Künstl Int> graph search
Graphentheorie f <Künstl Int, Math> graph theory
Graphit m <Chemie, Fertig> plumbago
Graphitabschirmung f <Kerntech> graphite shielding
graphitbeschichtetes Brennelement n <Kerntech> graphite-clad fuel element
Graphitbeschichtung f <Kerntech> graphite coating
Graphitblock m <Bau> graphite block
Graphitbürste f <Elektriz> graphite brush

Graphitformschwärzung

Graphitformschwärzung f <Fertig> graphite blacking (Gießen)
Graphitführungsrohr n <Kerntech> graphite guide tube
Graphitisation f <Metall> graphitization
Graphitkugel f <Kerntech> graphite pebble
graphitmoderiert adj <Kerntech> graphite-moderated
graphitmoderierter gasgekühlter Reaktor m <Kerntech> graphite-moderated gas-cooled reactor
Graphitpackung f <Maschinen> graphite packing
Graphitschmierung f <Mechan> graphite grease
Graphitschwund m <Kerntech> graphite shrinkage
Graphitstaub m <Chemie> plumbago (Gießerei)
Graphitstruktur f <Kerntech> graphite structure
Graphittiegel m <Fertig> graphite crucible, plumbago crucible
graphitumhülltes Brennelement n <Kerntech> graphite-clad fuel element
Gras n <Kerntech> grass (Störung auf Oszilloskop)
Grat m 1. <Bau> groin (Architektur); 2. <Fertig> rag; 3. <Kohlen> rib; 4. <Maschinen> burr, wire edge; 5. <Mechan> burr; 6. <Phys> fin line (Werkstück) • **Grat abscheren** <Fertig> trim • **mit Grat** <Fertig, Maschinen, Mechan> burred
Gratbildung f <Fertig> formation of burrs; finning (Walzen)
Gräting f <Wassertrans> grating (Schiffbau)
Gratsparren m <Bau> angle rafter, hip rafter
Gratsparrendach n <Bau> hip and ridge roof
Grätzgleichrichter m <Elektriz> bridge rectifier
Gratziegel m <Bau> hip tile
Grau n <Druck, Fernseh, Foto> (AE) gray, (BE) grey
grauer Körper m <Fernseh> (AE) gray body, (BE) grey body
grauer Star m **bei Glasmachern** <Ker & Glas> glassworker's cataract
Grauglasscheibe f <Fernseh> black screen
Grauguss m 1. <Fertig> cast iron (Kunststoffinstallationen); 2. <Mechan> (AE) gray-cast iron, (BE) grey-cast iron
Graugussimpfungszusatz m <Fertig> inoculant
Graugussrohr n <Maschinen> (AE) gray iron pipe, (BE) grey iron pipe
Graugusszusatz m <Fertig> inoculant
Graukeil m <Fernseh> staircase
Graukeilsignal n <Fernseh> staircase signal
Graupappe f <Druck, Verpack> millboard
Grauschatten m <Fernseh> black shading
Grauskala f <Fernseh> (AE) gray scale, (BE) grey scale
Grauskalenwert m <Fernseh> (AE) gray scale value, (BE) grey scale value
Graustufe f <Comp & DV> (AE) gray scale, (BE) grey scale
Graustufung f <Foto> (AE) gray scale, (BE) grey scale
Grautönung f <Konstzeich> (AE) gray shading, (BE) grey shading, (AE) gray toning, (BE) grey toning
Grauwertstauchung f <Fernseh> black crush
Graveur m <Ker & Glas> engraver
gravieren v <Druck> engrave
Gravieren n <Druck, Elektronik> engraving
Gravierfräsen n <Maschinen> engraving
Gravimetrie f <Phys> gravimetry
gravimetrisch adj <Kohlen> gravimetric
gravimetrische Messung f <Erdöl> gravimetric analysis (Prospektion)
Gravitation f <Umweltschmutz> gravity
Gravitationskonstante f <Phys, Raumfahrt> gravitation constant
Gravitationskraft f <Phys> force of inertia, gravitation
Gravitationswellen fpl 1. <Phys, Strahlphys> gravity waves; 2. <Wellphys> gravity waves (theoretische Ausbreitung der Gravitationskraft)

Graviton n <Phys, Teilphys> graviton
Gravur f <Druck, Ker & Glas> engraving • **Gravuren einsenken** <Fertig> type
Gravurmaschine f <Ker & Glas> engraving lathe
Gray n (gy) 1. <Phys> (AE) gray, gy (Einheit der Energiedosis); 2. <Teilphys> (AE) gray, gy
gregorianisches Teleskop n <Phys> Gregorian telescope (Spiegelteleskop)
Greifanker m <Wassertrans> grappling hook (Festmachen)
Greifbagger m 1. <Bau> grab crane; 2. <Wassertrans> grab dredge, grab dredger
Greifbereich m <Fertig> gripping range (von Robotern)
greifen v 1. <Bau> bite (Schrauben); 2. <Ergon> grab; 3. <Fertig, Maschinen> bite (Feile)
Greifen n <Maschinen> biting
Greifen n **der Walzen** <Maschinen> biting of the rolls
Greifer m 1. <Bau> grab; 2. <Druck> gripper; 3. <Ergon> gripper, pantograph; 4. <Kerntech> gripper; finger action tool (eines Manipulators); 5. <Maschinen> clutch, gripper; 6. <Metall, Wassertrans> grab (Baggern)
Greifer mpl **und Garnträger** mpl <Textil> gripper and yarn carriers
Greiferauflage f <Druck> gripper pad
Greiferkorb m <Bau> grab bucket
Greiferkübel m <Maschinen> grab bucket
Greiferscheibe f <Trans> clip pulley
Greiferseilscheibe f <Trans> clip pulley
Greifhaken m <Kerntech> grip hook
Greifkraft f <Ergon> grip strength
Greifraum m <Ergon> reaching space
Greifvermögen n <Fertig> bite (Walzen)
Greifvorrichtung f <Maschinen> gripping device
Greifvorrichtung f **eines Roboters** <Maschinen> robot gripping device
Greifwerkzeug n 1. <Ergon> gripper; 2. <Kerntech> gripper tool
Greifwinkel m <Fertig> angle of contact, angle of nip (Walzen)
Greifzange f <Kerntech> grapple
Greifzirkel m <Metrol> (AE) caliper, (BE) calliper
Grellweiß n <Metall> dazzling white
Grenz… <Math, Qual> limiting, marginal
Grenzabmessungen fpl <Maschinen> limit dimensions
Grenzabweichung f <Qual> limiting deviation
Grenzanteil m <Qual> limiting proportion
Grenzbrennstoffkassette f <Kerntech> limiting fuel assembly
Grenzdauer f <Akustik> critical duration
Grenze f 1. <Bau> boundary; 2. <Maschinen> limit; 3. <Math> boundary, limit; 4. <Metall> boundary
Grenzempfindlichkeit f 1. <Gerät> threshold of sensitivity; 2. <Phys> tangential signal sensitivity
Grenzfläche f 1. <Fertig> interface; 2. <Heiz & Kälte, Lufttrans> boundary layer; 3. <Metall> interface boundary; interface (zwischen Medien); 4. <Optik> core-cladding interface (Lichtleiter, zwischen Kern und Mantel); 5. <Phys> surface; 6. <Telekom> interface
grenzflächenaktiv adj <Chemie> surface-active
grenzflächenaktiver Stoff m <Umweltschmutz> surfactant
grenzflächenaktives Mittel n <Chemie> surfactant
Grenzflächenchemie f <Chemie> boundary surface chemistry
Grenzflächenenergie f <Metall> interface energy
Grenzflächenspannung f 1. <Kohlen> surface tension; 2. <Meerschmutz> interfacial tension
Grenzfrequenz f 1. <Elektronik> cutoff frequency, threshold frequency; 2. <Funktech, Telekom> cutoff frequency, critical frequency

Grenzgatter n <Elektronik> threshold gate
Grenzkonzentration f 1. <Kunststoff> critical concentration; 2. <Umweltschmutz> limiting concentration
Grenzkorn n <Kohlen> near-mesh material
Grenzlast f 1. <Elektriz> limit load; 2. <Lufttrans> limit load (eines Luftschraubenblattes); 3. <Maschinen> limit load
Grenzlastfaktor m <Lufttrans> limit load factor
Grenzlehrdorn m <Maschinen> (AE) internal limit gage, (BE) internal limit gauge
Grenzlehre f 1. <Maschinen> (AE) go and no-go limit gage, (BE) go and no-go limit gauge, (AE) limit gage, (BE) limit gauge; 2. <Metrol> (AE) limit gage, (BE) limit gauge
Grenzleistung f bei Schnellabschaltung <Kerntech> emergency shutdown power
Grenzlinie f <Trans> cordon line
Grenzlinienübersicht f <Trans> cordon line survey
Grenzmarkierung f <Bau> boundary mark
Grenzmaß n 1. <Maschinen> boundary dimensions, limit size, size limit; 2. <Mechan> limit size
Grenzmauer f <Bau> boundary wall
Grenzpfosten m <Bau> boundary post
Grenzpunkt m <Hydraul> cutoff point
Grenzqualität f <Qual> limiting quality
Grenzrachenlehre f <Maschinen> (AE) external caliper gage, (BE) external caliper gauge
Grenzreibung f <Maschinen> boundary friction
Grenzrelais n <Elektriz> limit value relay
Grenzschalter m 1. <Elektriz, Elektrotech> check switch, limit switch; 2. <Kontroll, Mechan> limit switch
Grenzschaufelturbine f <Hydraul> limit turbine
Grenzschicht f 1. <Fertig> boundary layer (Strömung); 2. <Heiz & Kälte, Maschinen, Mechan, Nichtfoss Energ> boundary layer; 3. <Optik> barrier layer; 4. <Phys> interface; boundary layer (in Wandnähe eines Flüssigkeitscontainers oder Kanals); 5. <Strömphys> boundary layer
Grenzschichtablösung f <Strömphys> boundary layer separation
Grenzschichtbeeinflussung f <Strömphys, Trans> boundary layer control
Grenzschichtdicke f 1. <Hydraul> boundary layer momentum thickness; 2. <Strömphys> boundary layer thickness
Grenzschichteinfluss m <Strömphys> boundary layer effect
Grenzschichtgleichung f <Strömphys> boundary layer equation
Grenzschichtkapazität f <Elektrotech> junction capacitance
Grenzschichtschmierung f <Lufttrans> boundary lubrication
Grenzschmierung f <Maschinen> boundary lubrication
Grenzspannung f <Elektrotech> critical voltage
Grenzstein m <Bau> boundary stone, landmark
Grenzsteinmarkierung f <Bau> monument (Vermessung)
Grenzstrom m <Elektriz> limiting current
Grenztaster m <Elektriz> microswitch
grenzübergreifende Systeme npl <Mobilkom, Telekom> cross-border systems
grenzüberschreitende Abfallverbringung f <Abfall> transboundary movement of waste
grenzüberschreitender Datenfluss m <Telekom> transborder data flow
Grenzverteilung f <Math> limiting distribution
Grenzviskosität f <Strömphys> limiting viscosity number
Grenzwellenlänge f 1. <Elektronik> threshold wavelength; 2. <Optik> cutoff wavelength (einer Einzelmode); cutoff wavelength (einer Schwingungsmode); 3. <Phys> cutoff wavelength (Hohlleiter); 4. <Telekom> cutoff wavelength
Grenzwert m 1. <Akustik> threshold value; 2. <Ergon> limen, threshold; 3. <Math> limes, limiting value; 4. <Qual> limiting value, tolerance limit; 5. <Sicherheit> exposure limit (für Exposition gegenüber gefährlichen Stoffen); 6. <Telekom> limiting value
Grenzwertanzeiger m <Gerät> limit indicator
Grenzwertdesign n <Raumfahrt> design to breaking strength (konstruiert bis zur Bruchlast)
Grenzwerteinstellung f <Kerntech> limit setting
Grenzwertmelder m <Gerät> limit indicator
Grenzwertprüfung f <Comp & DV> MC, marginal check, marginal test
Grenzwertsensor m <Gerät> threshold detector
Grenzwertüberwachung f 1. <Comp & DV> marginal check; 2. <Sicherheit> control of limit values, surveillance of limit values
Grenzwertvergleich m <Gerät> limit comparison
Grenzwinkel m <Fertig, Optik, Phys, Telekom> critical angle
Grenzwinkel m **der Totalreflexion** <Optik> angle of total reflection
Grenzzustand m <Bau> limit state
Grenzzustandsdesign n <Bau> limit state design
Grieß m 1. <Kerntech> grass (Störung auf Oszilloskop); 2. <Lebensmittel> middlings
Grießkleie f <Lebensmittel> middlings bran
Grießkohle f <Kohlen> smalls
Grießkohle f ohne Feinkohle <Kohlen> small coal without fines
Griff m 1. <Ergon> grip; 2. <Fertig> hilt; 3. <Foto> handgrip; 4. <Ker & Glas> grip; 5. <Kfztech> handle; 6. <Kohlen> grip; 7. <Maschinen> grip, handle; 8. <Mechan> handle, knob; 9. <Papier> handle; 10. <Raumfahrt> handle (Raumschiff); 11. <Textil> handle (Stoff); 12. <Verpack> handle
griffig adj <Fertig> open (Schleifscheibe)
griffig machen v <Fertig> sharpen
Griffith'sche Risse mpl <Ker & Glas> Griffith flaw
Griffkreuz n <Maschinen> star handle
Griffrippe f <Fertig> reinforcing rib (Kunststoffinstallationen)
Griffschalter m <Elektrotech> lever switch
Grignard-... <Chemie> organomagnesium
Grignard-Verbindung f <Chemie> organomagnesium compound
grob adj <Maschinen> coarse, unfinished
grob entwerfen v <Bau> sketch
Grob... <Textil> coarse-count
Grobabgleich m <Gerät> coarse balance (Brücke)
Grobabtastung f <Fernseh> coarse scanning
Grobbeton m <Bau> no-fines concrete
Grobblech n 1. <Metall> heavy plates, plate; 2. <Wassertrans> plate (Schiffbau)
Grobbrechen n <Fertig> coarse crushing
Grobbrecher m <Kohlen> primary crusher
grobdendritisches Gefüge n <Fertig> ingotism
grobe Teilung f <Maschinen> coarse-pitch
grobe Übereinstimmung f <Comp & DV> fuzzy match
grobe Unterlegscheibe f <Maschinen> coarse washer
Grobeinstellung f <Elektriz, Fertig> coarse adjustment
Grobeinstellung f mittels Gestell und Zahnrad <Optik> coarse adjustment by rack and pinion (Mikroskop)
groberes Gewebe n <Textil> coarser woven fabric
grobes Filter n <Wasserversorg> coarse filter
grobes Sieb n <Kohlen> riddle
Grobfeile f <Maschinen> coarse file, rough file, rough-cut file

Grob-Fein-Einstellung

Grob-Fein-Einstellung *f* <Gerät> coarse-fine adjustment
Grobfilter *n* <Maschinen> coarse filter
Grobflyer *m* <Textil> slubbing frame
grobgängige Schraube *f* <Bau> coarse-pitch screw
Grobgarn *n* <Textil> coarse yarn
Grobgewinde *n* <Maschinen> coarse thread
Grobhieb *m* 1. <Fertig> coarse cut *(Feile)*; 2. <Maschinen> rough cut
Grobholz *n* <Bau> rough wood
Grobkeramik *f* <Ker & Glas> ordinary ceramic
Grobkies *m* <Bau> rubble
grobkiesiger Sand *m* <Bau> coarse gravelly sand
Grobklärbecken *n* <Wasserversorg> roughing tank
Grobkohle *f* <Kohlen> lump coal
grobkonzentrieren *v* <Fertig> rag
Grobkorn *n* <Bau, Fertig, Kohlen, Metall> coarse grain
Grobkorngefüge *n* <Fertig> pebbles
Grobkornglühen *n* <Fertig> coarse-grain annealing
grobkörnig *adj* 1. <Bau> rough-grained, coarse-grained; 2. <Kohlen, Metall> coarse-grained
grobkörnige Erde *f* <Bau, Kohlen> coarse soil
grobkörniges Bild *n* <Foto> coarse-grain image
grobporige Schleifscheibe *f* <Fertig> open-structure wheel
Grobraster *n* <Druck> coarse screen
Grobrechen *n* <Abfall> coarse screen
Grobsand *m* <Anstrich, Bau, Mechan> grit
Grobschleifen *n* <Maschinen> rough grinding
Grobschleifscheibe *f* <Maschinen> rough-grinding wheel
grobschmieden *v* <Fertig> rough-forge
Grobschmieden *n* <Fertig> preliminary drawing, rough forging
Grobsieb *n* <Kohlen> scalping screen
Grobsieben *n* <Kohlen> scalping
Grobstichnaht *f* <Textil> rope-stitch seam
Grobstoff *m* <Chemie> screenings *(Papier)*
Grobstoffentferner *m* <Papier> junk remover
Grobstofffänger *m* <Wasserversorg> roughing tank
Grobstraße *f* <Fertig> blooming train
Grobstruktur *f* <Chemie> macrostructure
Grobvakuum *n* <Phys> coarse vacuum
Grobverstellung *f* <Maschinen> coarse feed
Grobvorschub *m* <Fertig> coarse feed
Grobvorschubreihe *f* <Fertig> coarse-feed series
Grobwalze *f* <Maschinen> breaking-down roll
Grobwalzen *n* 1. <Fertig> roughing; 2. <Metall> blooming
Grobwalzwerk *n* <Maschinen> blooming mill
Grobzerkleiner *m* <Kohlen> coarse crusher
Grobzerkleinerung *f* 1. <Kohlen> coarse crushing, coarse grinding; 2. <Papier> breaking
Grobzerkleinerungsmaschine *f* <Chemtech, Kohlen> coarse crushing mill
Grobzug *m* <Fertig> roughing block *(Draht)*
Grobzuschlagstoff *m* <Bau> ballasting material, coarse aggregate
Großaufnahme *f* <Foto> close-up
Großbaum *m* <Wassertrans> main boom *(Segeln)*
Großbehälterumschlag *m* <Trans> containerization
Großbildschirm *m* <Comp & DV> LSD, large-screen display
Großbohrloch *n* <Kohlen> well drill hole
Großboot *n* <Wassertrans> longboat *(Beiboot)*
Großbrasse *f* <Wassertrans> main brace *(Tauwerk)*
Großbuchstabe *m* <Druck> upper case • **alles in Großbuchstaben** <Druck> all caps • **in Großbuchstaben** <Comp & DV> in upper case
Großcontainer *m* <Verpack> large-size container
große Bahnhalbachse *f* <Raumfahrt> semimajor axis *(eines Orbits)*

große Menge *f* <Bau> bulk
große Reynoldszahl *f* <Strömphys> high Reynolds number
große Sekunde *f* <Akustik> major second
große Septime *f* <Akustik> major seventh
große Sext *f* <Akustik> major sixth
große Spantiefe *f* <Maschinen> heavy cut
große Steigung *f* <Lufttrans> high pitch *(Hubschrauber)*; coarse pitch *(am Propeller)*
große Sturzwelle *f* <Wassertrans> roller *(Seezustand)*
große Terz *f* <Akustik> major third
große Tonne *f* <Metrol> long ton
große Vereinigung *f* <Teilphys> grand unification
Größe *f* 1. <Druck> weight; 2. <Elektronik> quantity; 3. <Fertig> quantity, value; 4. <Metrol> *(AE)* gage, *(BE)* gauge; 5. <Textil> size
Größe *f* **des Spanraums** <Fertig> amount of chip space
Größendarstellung *f* <Comp & DV> signed magnitude representation
Größenklasse *f* <Phys> magnitude *(Stern)*
Größenordnung *f* 1. <Comp & DV, Phys> order of magnitude; 2. <Strahlphys> order *(des Hintergrundrauschens)*
Größenordnung *f* **einer physikalischen Größe** <Phys> magnitude of quantity
Größenschwelle *f* <Ergon> size threshold
Größenwandler *m* <Comp & DV> quantizer
Größenzusammenhang *m* <Regelung> relationship between variables, relationship between quantities
größer als *adj* <Math> greater than
großer Ganzton *m* <Akustik> major whole tone
großer Kondensator *m* <Elektrotech> large-value capacitor
großer Streuwinkel *m* <Kerntech> large-angle scattering
Größergleichzeichen *n* <Math> chevron
Größerzeichen *n* <Math> chevron
großes Pleuelauge *n* <Kfztech> connecting rod big end
Großfall *n* <Wassertrans> main halyard *(Segeln, Taue)*
Großflächenbasisstation *f* <Mobilkom> umbrella site
Großflächenwärmetauscher *m* <Maschinen> extended-surface heat exchanger
großflächige Bestrahlungsnorm *f* <Kerntech> large-area radiation standard
Großflughafen *m* <Lufttrans> air terminal
Großformatdruck *m* <Druck> large-format print, large-format printing, LFP
großformatiger Druck *m* <Druck> large-format printing *(LFP)*
großformatiges Druckerzeugnis *n* <Druck> large-format print *(LFP)*
Großformatklappkamera *f* <Foto> large-format folding camera
Großintegration *f (LSI)* <Comp & DV, Elektronik, Phys, Telekom> large-scale integration *(LSI)*
Groß-/Kleinschreibung *f* <Comp & DV> case height
Großkreis *m* 1. <Funktech, Phys> great circle; 2. <Raumfahrt> meridian
großkreisabhängig *adj* <Raumfahrt> meridional
Großkreisfunkstrecke *f* <Funktech> great circle radio path
Großkreiskarte *f* <Wassertrans> great circle chart *(Navigation)*
Großkreiskurs *m* <Wassertrans> great circle route; great circle course *(Navigation)*
Großkreisroute *f* <Raumfahrt> great circle path
Großlochbohrung *f* <Kohlen> large-hole boring
Großmast *m* <Wassertrans> mainmast *(Segeln)*
Großmotorrad *n* <Kfztech> large-capacity motorcycle
Großoberflächenfehler *m* <Fertig> rat
Großplatten *fpl* <Metall> large plates

Großraum... <Lufttrans> wide-bodied
Großraumcontainer m (HC) <Trans> High Cube (HC)
Großraumdüsenflugzeug n <Lufttrans> jumbo jet
Großraumfrachter m <Wassertrans> bulk carrier
Großraumjet m <Lufttrans> wide-bodied aircraft
Großraum-LKW m <Kfztech> large-capacity truck
Großraumwagen m <Eisenbahn> saloon coach (für Personen)
Großrechner m <Comp & DV> large-scale computer, mainframe computer, supercomputer
Großring... <Chemie> macrocyclic
Großsack m <Verpack> multiply sack
Großschot f <Wassertrans> mainsheet (Segeln)
Großschotwagen m <Wassertrans> (AE) traveler, (BE) traveller (Segeln)
Großschreibung f <Druck> capitalization
Großsegel n <Wassertrans> mainsail (Segeln)
Großsegler m <Wassertrans> tall ship (Segeln)
Großserienfertigung f <Fertig> large-batch production, quantity production
Großsignal n <Elektronik> large signal
Großsignal-Bandbreite f <Elektronik> large-signal bandwidth
Großsignal-Bedingungen fpl <Elektronik> large-signal conditions
Großsignal-Betrieb m <Elektronik> large-signal operation
Großspeicher m <Comp & DV> mass storage
Großspeicher-Chip m <Elektronik> random logic chip
Großspeicherschaltung f <Elektronik> random logic circuit
Großstag n <Wassertrans> mainstay (Takelage)
größte Abmessungen fpl <Wassertrans> extreme dimensions (Schiffbau)
größte Breite f <Wassertrans> extreme breadth (Schiffkonstruktion)
größte Schiffsbreite f <Wassertrans> maximum beam (Schiff- und Bootsbau)
größte zulässige Nutzleistung f <Elektrotech> overload level
größter anzunehmender Unfall m (GAU) <Kerntech> maximum credible accident (MCA)
größter Durchschlupf m <Qual> average outgoing quality limit
größter gemeinsamer Teiler m (ggT) <Math> highest common factor (HCF)
größter Leistungseingang m <Elektrotech> maximum power input
größter Tiefgang m <Wassertrans> extreme draught (Schiffkonstruktion)
größter Wassergehalt m <Kohlen> liquid limit
größter zulässiger Fehler m <Metrol> maximum permissible error
Größtmaß n <Maschinen, Qual> maximum size
größtmögliche Schlepptiefe f <Wasserversorg> full-dredging depth
größtmögliche Signalempfindlichkeit f <Elektronik> minimum detectable signal
Größtspiel n <Maschinen, Qual> maximum clearance
Größtübermaß n <Maschinen> maximum allowance
Großvaterdatei f <Comp & DV> grandfather file
Großwant f <Wassertrans> main shroud (Tauwerk)
Großzelle f <Mobilkom> macrocell (zellulares Netz)
Groteskschrift f <Druck, Konstzeich> sans serif
Groupware f <Comp & DV> groupware
Grübchen n <Metall> dimple
Grübchenbildung f 1. <Ker & Glas> pitting; 2. <Raumfahrt> potting (Raumschiff)
Grube f 1. <Abfall> trench landfill; 2. <Erdöl, Kohlen, Wasserversorg> pit

Grubenarbeiter m <Kohlen> collier, miner
Grubenbahn f <Eisenbahn, Kohlen> (AE) mine railroad, (BE) mine railway
Grubenbau m <Kohlen> mine opening, mine working, underground excavation, underground working
Grubenbetrieb m <Kohlen> underground mining
Grubengas n <Kohlen, Thermod> firedamp
Grubengasprüfgerät n <Kohlen> firedamp-proof machine
Grubenholz n <Kohlen> timber
Grubenlampe f 1. <Kohlen> Davy lamp; 2. <Sicherheit> miner's safety lamp
Grubenlokomotive f <Eisenbahn> hauling engine
Grubenwagen m <Eisenbahn, Kohlen> tram
Grubenwand f <Abfall> embankment (einer Deponie)
Grubenzimmerung f <Kohlen> timbering
Grund m 1. <Maschinen> root; 2. <Wassertrans> bottom, ground • **auf Grund** <Wassertrans> aground • **auf Grund laufen** <Wassertrans> go aground, run ashore; ground, run aground (Schiff) • **auf Grund setzen** <Wassertrans> ground, run aground (Schiff) • **den Grund berühren** <Wassertrans> touch bottom (Schiff)
Grundablass m <Wasserversorg> bottom outlet
Grundarmierung f <Bau> main reinforcement
Grundbacke f <Fertig> actual chuck jaw
Grundbaustein m <Elektriz> basic module
Grundbefehl m <Comp & DV> basic instruction
Grundbezug m **der Session** <Telekom> basic session reference
Grundbindung f <Textil> plain weave
Grundbitrate f <Telekom> basic bit rate
Grundbogen m <Bau> reversed arch
Grundbohrung f <Fertig> blind bore, blind hole
Grundbruch m <Bau> base failure
Grundbüchse f 1. <Fertig> liner bushing (Bohren); 2. <Maschinen> bottom brass
Grundcodierung f <Comp & DV> basic coding
Grundebene f 1. <Fertig> basal plane; 2. <Geom> ground plane
Grundeinheit f 1. <Elektriz, Maschinen> elementary unit, fundamental unit (einer Struktur); 2. <Metrol> base unit (eines Einheitensystems)
gründen v <Bau> lay (Gebäudefundament)
Grunderregung f <Elektriz> basic excitation
Grundfalte f <Konstzeich> basic fold
Grundfarbe f 1. <Comp & DV, Druck> (AE) primary color, (BE) primary colour; 2. <Ker & Glas> (AE) ground color, (BE) ground colour; 3. <Optik, Phys> (AE) primary color, (BE) primary colour
Grundfarben fpl 1. <Druck> (AE) fundamental colors, (BE) fundamental colours; 2. <Strahlphys> (AE) prime colors, (BE) prime colours
Grundfarbstoff m <Papier> basic dye
Grundfehler m <Metrol> intrinsic error (Messinstrument)
Grundfernsprechgebühr f <Telekom> basic call charge
Grundfläche f 1. <Bau> area, floor space; 2. <Metall> basal plane
Grundflächengleiten n <Metall> basal slip
Grundformen fpl <Geom> basic shapes
Grundfrequenz f 1. <Elektronik, Fernseh> BF, basic frequency, fundamental frequency; 2. <Funktech, Telekom> BF, base frequency, basic frequency, fundamental frequency
Grundgebühr f <Telekom> basic call charge
Grundgefüge n <Fertig> matrix
Grundgerät n <Gerät> basic instrument
Grundgeräusch n 1. <Akustik> ground noise; 2. <Bau> random noise; 3. <Elektronik, Telekom> basic noise (Telefontechnik); 4. <Werkprüf> background noise

Grundgeräuschpegel

Grundgeräuschpegel *m* <Sicherheit> background noise level
Grundgesamtheit *f* <Qual> population
Grundgeschwindigkeit *f* <Kontroll> basic speed
Grundgestein *n* <Wasserversorg> bedrock
Grundgewebe *n* <Textil> ground cloth, ground fabric; backing fabric *(Frottierware)*; backing *(Teppich)*
Grundgruppe *f* 1. <Elektrotech> basic group; 2. <Telekom> basic group, primary group
Grundhieb *m* <Fertig> undercut *(Feile)*
grundieren *v* 1. <Bau> prime; precoat *(Farbe)*; 2. <Papier> stain
Grundierfarbe *f* <Ker & Glas> *(AE)* flat color, *(BE)* flat colour
Grundiermasse *f* <Fertig> size
Grundiermittel *n* <Kunststoff> primer
grundierte Rohkarosserie *f* <Kfztech> body in white
Grundierung *f* 1. <Anstrich> base coat; 2. <Bau> prime coat; undercoat *(Farbe)*; 3. <Fertig> priming; 4. <Kunststoff, Raumfahrt> primer *(Raumschiff)*
Grundkapazität *f* <Comp & DV> base capacity
Grundkegel *m* <Maschinen> base cone
Grundkomponente *f* <Phys> fundamental component *(Schwingung)*
Grundkonzentration *f* <Umweltschutz> background concentration, instantaneous concentration
Grundkörper *m* <Fertig> main casting, stock
Grundkreis *m* <Maschinen> base circle, root circle
Grundladung *f* <Raumfahrt> priming charge *(Raumschiff)*
Grundlagenforschung *f* <Comp & DV> basic research
Grundlagenwissen *n* <Comp & DV> knowledge base
Grundlänge *f* **der Start- und Landebahn** <Lufttrans> runway basic length
Grundlast *f* 1. <Kohlen> base load; 2. <Nichtfoss Energ> base load, off-peak generation; 3. <Trans> basic load
Grundlastanlage *f* <Kohlen, Nichtfoss Energ> base-load plant
Grundlastbetrieb *m* <Elektrotech> base-load duty *(eines Kraftwerks)*
Grundlastbündel *n* <Telekom> first-choice group *(Schaltung)*
Grundlasteinheit *f* <Kohlen, Nichtfoss Energ> base-load unit
Grundlastgeneratorsatz *m* <Elektrotech> base-load set
Grundlastkapazität *f* <Kohlen, Nichtfoss Energ> base-load capacity
Grundlastkessel *m* <Kerntech> base-load boiler
Grundlastkraftwerk *n* <Elektrotech> base-load power station
Grundlastmaschine *f* <Elektrotech> base-load machine *(z. B. Generator)*
grundlegende Leistungsfähigkeit *f* <Trans> basic capacity
grundlegende Verschmutzung *f* <Umweltschutz> background pollution
Grundleistung *f* <Kerntech> base power
gründliche Überholung *f* <Kerntech, Sicherheit> major overhaul
Grundlinie *f* 1. <Bau> base, baseline *(Vermessung)*; 2. <Druck> baseline; 3. <Geom> base *(einer geometrischen Figur)*
Grundlinienkontrolle *f* <Mechan> baseline inspection
Grundloch *n* <Maschinen> blind hole, bottom hole
Grundlochgewindebohrer *m* <Maschinen> bottoming tap, plug tap
Grundlochreibahle *f* <Maschinen> bottoming reamer
Grundmasse *f* <Bau> matrix
Grundmaterial *n* <Anstrich> substrate
Grundmauer *f* <Bau> base wall

Grundmetall *n* 1. <Maschinen> base metal, parent metal; 2. <Metall> base metal, basis metal
Grundmode *f* <Optik> fundamental mode
Grundmodell *n* <Kfztech> base model
Grundmodus *m* <Phys> dominant mode *(Hohlleiter)*
Grundnahrungsmittel *n* <Lebensmittel> staple food
Grundnorm *f* <Fertig, Maschinen> basic specification, basic standard
Grundpegel *m* <Telekom> floor, noise floor *(z. B. eines verrauschten Signals)*
Grundpetrochemikalien *fpl* <Erdöl> basic petrochemicals
Grundplatine *f* <Comp & DV> motherboard
Grundplatte *f* 1. <Bau> bed plate *(Hoch- und Tiefbau)*; 2. <Eisenbahn> bed plate; 3. <Erdöl> bed plate *(Offshore)*; 4. <Elektronik, Elektrotech> case back; 5. <Fertig> die bed, die shoe; shoe *(Schnitt)*; 6. <Kfztech> backing plate *(Bremsanlage)*; 7. <Maschinen> base plate, bed plate; 8. <Mechan> base plate; 9. <Wassertrans> bed plate *(Maschine)*
Grundprimärgruppe *f* *(GPG)* <Telekom> basic group, basic primary group
Grundprüfung *f* <Elektrotech> basic test
Grundrahmen *m* <Kerntech> base frame
Grundrauschen *n* 1. <Funktech, Gerät> basic noise; 2. <Telekom> background noise
Grundriss *m* 1. <Bau> sketch; 2. <Maschinen> plan view
Grundschaltbild *n* <Funktech, Kerntech, Telekom> basic circuit diagram
Grundschicht *f* 1. <Anstrich> base coat; 2. <Fertig> primer
Grundschleier *m* <Foto> base fog
Grundschleppnetz *n* 1. <Meerschmutz> trawl net; 2. <Wassertrans> bottom trawl *(Fischerei)*
Grundschriftfeld *n* <Konstzeich> basic title block
Grundschwelle *f* 1. <Bau> ground plate, groundsill, sill, sill plate; 2. <Wassertrans> sill *(Hafen)*; 3. <Wasserversorg> sill
Grundschwingung *f* 1. <Elektronik, Elektrotech> first harmonic, fundamental frequency, funnel; dominant mode *(Wellenleiter)*; 2. <Funktech> fundamental frequency; 3. <Phys> fundamental mode; 4. <Telekom> fundamental frequency
Grundschwingungsleistung *f* <Elektriz> fundamental power
Grundschwingungsleistungsfaktor *m* <Elektriz> displacement factor, fundamental power factor, power factor of the fundamental
Grundschwingungsmode *f* 1. <Akustik> fundamental vibration mode; 2. <Optik> fundamental mode
Grundsee *f* <Wassertrans> ground swell *(Seezustand)*
grundsohlig *adj* <Erdöl> bottom-hole *(Bohrtechnik)*
Grundstellung *f* <Comp & DV> initial state, reset
Grundstruktur *f* <Comp & DV> framework
Grundstück *n* <Bau> plot
Grundstücksbegrenzungsmauer *f* <Bau> party wall
Grundstücksleitung *f* <Bau> supply pipe
Grundstücksmakler *m* <Bau> estate agent
Grundtakt *m* <Elektronik> basic clock rate, basic pulse rate
Grundtaktgenerator *m* <Comp & DV> master clock
Grundtertiärgruppe *f* <Telekom> basic master group, basic tertiary group
Grundton *m* 1. <Akustik> fundamental tone, keynote; 2. <Wellphys> fundamental tone
Grundtyp *m* 1. <Elektrotech> dominant mode *(Wellenleiter)*; fundamental mode *(einer Welle)*; 2. <Phys> dominant mode
grundüberholen *v* <Wassertrans> refit

Grundüberholung f 1. <Kerntech> major overhaul; 2. <Wassertrans> refit
Gründung f <Bau> footing *(Gebäude)*
Gründungspfahl m <Bau> pier
Gründungssohle f <Bau> foundation
Gründungswanne f <Bau> tank *(Grundbau)*
Grundverbindung f <Comp & DV> basic linkage
Grundwasser n 1. <Kohlen> ground water; 2. <Umweltschmutz> underground water; 3. <Wasserversorg> ground water, subsoil water, subterranean water
Grundwasserangebot n <Wasserversorg> ground water resources
Grundwasserbelastung f <Umweltschmutz> impact on ground water
Grundwassereinzugsgebiet n <Wasserversorg> ground water basin
Grundwassererneuerung f <Wasserversorg> natural ground water recharge
grundwasserführende Schicht f <Wasserversorg> water-bearing stratum
Grundwasserkunde f <Kohlen> geohydrology
Grundwasserleiter m <Wasserversorg> aquifer
Grundwasserschließung f <Wasserversorg> capture of ground water
Grundwasserschutz m <Abfall> ground water protection
Grundwasserspiegel m 1. <Bau> ground water level, water table; 2. <Erdöl> water table; 3. <Kohlen> ground water table, water table; 4. <Wasserversorg> ground water level, phreatic water level
Grundwasserstand m <Wasserversorg> ground water level, ground water table
Grundwasserstauer m <Wasserversorg> aquifuge
Grundwasserstrom m <Umweltschmutz> underground water flow
Grundwassertiefe f <Wasserversorg> ground water depth
Grundwasserversorgung f <Kohlen> ground water supply
Grundwehr n <Hydraul, Wasserversorg> drowned weir, submerged weir
Grundwelle f 1. <Akustik> fundamental wave; 2. <Elektrotech> fundamental mode; 3. <Phys> carrier; 4. <Wassertrans> ground wave *(Seezustand)*; 5. <Wellphys> carrier
Grundwelle f **des Flusses** <Elektriz> fundamental of flux
Grundwelle f **des Stromes** <Elektriz> fundamental of current
Grundwerk n <Papier> bed plate
Grundwerkkasten m <Papier> bed plate box
Grundwerkstoff m 1. <Bau> base material; 2. <Fertig> backing metal, base metal *(Schweißen)*
Grundwerkwalze f <Papier> bed roll
Grundzahl f <Math> radix
Grundzahlen fpl <Math> radices
Grundzeit f <Comp & DV> productive time
Grundzustand m 1. <Comp & DV> stable state; 2. <Phys> ground state; 3. <Strahlphys> basic state; 4. <Teilphys> ground state
Grundzyklus m <Comp & DV> basic machine time
grüne Welle f <Trans> progressive signal system; *(BE)* phased traffic lights, *(AE)* synchronized lights *(Ampelschaltung)*; linked lights *(KFZ-Verkehr)*
grünes Elektronenstrahlsystem n <Elektronik, Fernseh> green gun
Grüne-Welle-Verkehrssignalsteuerung f <Trans> linked traffic signal control
Grünfläche f <Umweltschmutz> green area, green space, park area
Grüngitter n <Fernseh> green screen-grid
Grünlaser m <Elektronik> green beam laser

Grünling m 1. <Bau> sun-dried brick; 2. <Kerntech> green compact
Grünmischer m <Fernseh> green adder
Grünpellet n <Kerntech> green pellet
Grünphase f <Trans> green phase
Grün-Schwarz-Pegel m <Fernseh> green-black level
Grünsignaldauer f <Trans> hour of green signal indication
Grünspan m <Chemie, Fertig> verdigris
Grünspitzenwert m <Fernseh> green peak level
Grünstein m <Bau> green stone
Grünstrahl m <Fernseh> green beam
Grünstrahllaser m <Elektronik> green beam laser
Grünstrahlsystem n <Elektronik, Fernseh> green gun
Grünzeit f <Trans> green period, green time
Grünzeitverschiebung f <Trans> offset
Gruppe f 1. <Akustik> field; 2. <Chemie> residue; 3. <Comp & DV> cluster, group, set; 4. <Elektrotech> group; bank *(Tasten, Kontakte)*; bank *(von Kondensatoren)*; 5. <Funktech> array *(Antennen)*; 6. <Maschinen> battery, cluster; 7. <Math> group; 8. <Telekom> array, group • **in Gruppen zusammenfassen** <Comp & DV> cluster
Gruppenanrufkennung f <Telekom> group call identity
Gruppenantenne f <Funktech, Telekom> array antenna
Gruppenauswahl f <Qual> stratified sampling
Gruppenbrechungsindex m <Optik> group index
Gruppenbrechzahl f <Telekom> group index
gruppencodiertes Aufzeichnen n <Comp & DV> GCR, group code recording
Gruppenfahrkarte f <Eisenbahn> group ticket
Gruppengeschwindigkeit f 1. <Akustik> group velocity; 2. <Optik> group velocity *(Wellenausbreitung)*; 3. <Phys> group velocity; 4. <Telekom> envelope velocity, group velocity
Gruppenindex m <Elektriz> group index *(Licht)*
Gruppenlaufzeit f 1. <Comp & DV, Elektrotech> group delay; 2. <Telekom> group delay, group transmission delay
Gruppenlaufzeitdifferenz f <Telekom> differential group delay
Gruppenlaufzeitdifferenz f **durch Modendispersion** <Telekom> differential mode delay, multimode group delay
Gruppenlaufzeitverzerrung f <Telekom> envelope delay distortion
Gruppenlaufzeitverzögerung f 1. <Telekom> group delay distortion; 2. <Raumfahrt> group delay distortion *(Weltraumfunk)*
Gruppenmarke f <Comp & DV> group mark, group marker
Gruppenschalter m <Elektrotech> gang switch
Gruppenschaltung f **der Fahrmotoren** <Elektrotech> motor combination
Gruppensteuereinheit f <Comp & DV> cluster controller
Gruppensteuerung f <Regelung> group control
Gruppentheorie f <Math> group theory
Gruppenvermittlungsstelle f *(GrVST)* <Telekom> *(AE)* group-switching center, *(BE)* group-switching centre *(GSC)*
Gruppenverschlüsselung f <Comp & DV> GCR, group code recording
Gruppenverzögerung f <Elektrotech, Telekom> group delay
Gruppenzeichnung f <Konstzeich> group drawing
gruppiert adj <Comp & DV> ganged
Gruppierung f 1. <Elektriz, Elektrotech> layout, grouping; 2. <Trans> consolidation
Gruppierungsschalter m <Elektrotech> grouping switch
Grus m <Kohlen> slack coal

Gruskohle

Gruskohle f <Kohlen> slack coal
GrVST *(Gruppenvermittlungsstelle)* <Telekom> GSC *(group-switching centre)*
GS *(Gleichstrom)* <Aufnahme, Comp & DV, Eisenbahn, Elektriz, Elektrotech, Fernseh, Fertig, Funktech, Phys, Telekom> DC *(direct current)*
GS-Löschkopf m <Aufnahme> DC erase head
GSM-Netz n <Telekom> mobile digital radio communications
GS-Motor m **mit Bürsten** <Elektrotech> brush-type DC motor
GS-Vormagnetisierung f <Aufnahme> DC bias
GS-WS-Wandler m <Phys> DC-AC converter
GS-WS-Wandlung f <Phys> DC-AC conversion
GSZ *(Gesamtsäurezahl)* <Chemie> TAN *(total acid number)*
G/T-Rauschzahl f <Funktech, Raumfahrt> gain-to-noise temperature ratio, G/T *(Satellitenkommunikation)*
Guajacol n <Chemie> guaiacol
Guajacon... <Chemie> guaiaconic
Guajakharz n <Lebensmittel> guaiac resin, gum guaiacum
Guajakholz n <Mechan> lignum vitae
Guajaret... <Chemie> guaiaretic
Guanidin n <Chemie> guanidine
Guanidinabspaltung f <Chemie> depurination
Guanin n <Chemie> guanine
Guano m <Chemie> guano
Guanosin n <Chemie> guanosine, vernine
Guanyl... <Chemie> guanyl
Guanylguanidin n <Chemie> biguanide
Guckloch n 1. <Maschinen> spyhole; 2. <Raumfahrt> peep hole *(Raumschiff)*
GUI-Darstellungsschicht f <Comp & DV> GUI presentation layer
GUI-Schnittstelle f <Comp & DV> GUI interface
Gülle f <Abfall> slurry
Gully m <Umweltschmutz> sink
Gulon... <Chemie> gulonic
Gulose f <Chemie> gulose
gültig adj <Comp & DV, Qual> effective, operative, valid
 • **für gültig erklären** <Comp & DV> validate
Gültigkeit f <Comp & DV, Telekom> validity
Gültigkeitsbereich m <Comp & DV, Qual> scope
Gültigkeitsbereich m **eines Namens** <Telekom> title domain *(OSI)*
Gültigkeitsprüfung f <Comp & DV, Telekom> validation, validity check
Gumbo n <Erdöl> gumbo *(Geologie)*
Gummi n 1. <Elektriz> rubber; 2. <Kunststoff> india rubber, rubber • **einfach mit Gummi isoliert** <Elektriz> single-rubber-covered
Gummiabfall m <Abfall> rubber waste, waste rubber
Gummiabfallverwertung f <Kunststoff> rubber scrap recycling, scrap recycling
Gummiarabikum n <Lebensmittel> gum arabic *(Klebstoff)*
Gummiboot n <Wassertrans> rubber boat, rubber dinghy
Gummibuchse f <Fertig> rubber bush *(Kunststoffinstallationen)*
Gummidichtring m <Elektrotech> grommet
Gummidichtung f 1. <Elektrotech> grommet; 2. <Fertig, Maschinen> rubber gasket
Gummidichtungsring m <Maschinen> grommet
Gummidruckschlauch m <Wasserversorg> rubber delivery hose
gummielastisch adj <Fertig> rubber-like
gummieren v <Druck> gum
Gummiermaschine f <Verpack> glue-gumming machine, gumming machine

gummiert adj <Kunststoff> rubberized
gummierter Rand m <Verpack> gummed edge
gummiertes Etikett n <Verpack> gummed label
gummiertes Gewebe n <Kunststoff> coated fabric
gummiertes Papier n <Verpack> gummed paper
Gummifeder f <Maschinen> rubber spring
Gummiförderschlauch m <Wasserversorg> rubber delivery hose
Gummiform f <Maschinen> rubber mould
Gummigurt m <Maschinen> rubber belt
Gummigurtbandförderer m <Fertig> rubber-belt conveyor
Gummihandschuh m 1. <Kunststoff> india rubber glove; 2. <Sicherheit> rubber glove *(für elektrische Einsätze)*
gummiisoliertes Kabel n <Elektrotech> rubber-insulated cable
Gummikabel n <Elektrotech> rubber cable, rubber-insulated cable
Gummikissen n <Kfztech> rubber pad
Gummiklotz m <Kfztech> rubber pad
Gummikolben m <Labor> rubber bulb *(Pipette)*
Gummilager n 1. <Kfztech> rubber mounting *(Motor)*; 2. <Maschinen> rubber mounting
Gummimembran f <Maschinen> rubber diaphragm
Gummimotoraufhängung f <Kfztech> rubber engine mounting
Gummipuffer m 1. <Foto> rubber tip; 2. <Maschinen> rubber buffer
Gummiquetschwalze f <Foto> squeegee
Gummiradwalze f <Bau> *(AE)* rubber-tired roller, *(BE)* rubber-tyred roller; *(AE)* pneumatic-tired roller, *(BE)* pneumatic-tyred roller *(Straßenbau)*
Gummiriemen m <Maschinen> rubber belting
Gummisaugschlauch m <Wasserversorg> rubber suction-hose
Gummischlauch m 1. <Kunststoff> india rubber hose, rubber hose; 2. <Labor> rubber tubing; 3. <Maschinen> rubber hose
Gummistoff m <Elektriz> rubberized material
Gummistopfen m <Labor> rubber stopper
Gummituch n <Druck> rubber blanket
Gummiwalze f 1. <Fertig> squeegee; 2. <Ker & Glas> squeegee *(für optisches Glas)*
Gummiwendelantenne f <Funktech> rubber ducky antenna
Gummizylinder m <Druck> blanket cylinder
Gundiode f <Elektronik> gun diode
Gunn-Diode f 1. <Funktech> Gunn-effect diode; 2. <Phys> electron transfer diode; Gunn diode *(Halbeiterphysik)*
Gunn-Effekt m <Elektronik> Gunn effect
günstigste Geschwindigkeit f <Trans> optimum speed
Gunverstärker m <Elektronik> gun amplifier
Gurt m 1. <Bau> boom, fascia; 2. <Kohlen> wale, waling; 3. <Lufttrans> brace; 4. <Maschinen> band, strap; 5. <Nichtfoss Energ> chord
Gurtband n <Textil> webbing
Gurtbandförderer m <Maschinen> band conveyor, belt conveyor
Gurtblech n <Bau> boom plate
Gurte mpl <Raumfahrt> safety harness
Gürtel m <Kfztech> belt *(Gürtelreifen)*; ply *(Reifen)*
Gürtelreifen m 1. <Kfztech> *(AE)* radial tire, *(BE)* radial tyre *(Reifen)*; 2. <Kunststoff> *(AE)* radial ply tire, *(BE)* radial ply tyre
Gurtholz n <Bau> waling
Gurtplatte f <Bau> flange plate *(eines Trägers)*
Gurtsims m <Bau> fascia
Gurttrommel f <Fertig> belt fastener

Gurtwerkstoff m <Kunststoff> belting
Gurtzuführer m <Chemtech> belt feed
Guss m <Metall, Papier> casting
Guss... <Metall> cast-iron
Gussabzweigdose f <Elektrotech> cast iron conjunction box
Gussblase f <Fertig> blister
Gussbronze f <Mechan, Metall> cast bronze
Gusseisen n <Fertig, Maschinen, Mechan, Metall, Papier> cast iron
Gusseisengelenk n <Bau> cast-iron joint
Gusseisenrohr n <Bau> cast-iron pipe
gusseisern adj <Metall> cast-iron
gusseiserner Krümmer m <Bau> cast-iron elbow
Gussfehler m <Fertig> shift
Gussform f 1. <Maschinen> (AE) mold for casting, (BE) mould for casting; 2. <Wassertrans> (AE) mold, (BE) mould
gussgekapselt adj <Elektrotech> cast-metal clad
Gussglas n <Ker & Glas> cast glass
Gusshaut f <Fertig> scale
Gussmessing n <Metall> cast brass
Gussmodell n <Mechan> casting pattern
Gussnaht f <Fertig, Maschinen> burr
Gussnarbe f <Mechan> flaw
Gussputzen n <Fertig> tumbling
Gussputztrommel f <Fertig> rumble, shaker barrel, tumbler
Gussrohrleitung f <Maschinen> cast-iron pipeline
Gussschablone f <Papier> casting template
Gussspiegelglas n <Ker & Glas> cast plate glass
Gussstahl m 1. <Mechan> cast steel; 2. <Metall> casting steel
Gussstück n <Metall> casting
Gussteil n <Fertig> casting
Gussverteiler m <Elektrotech> cast-iron box-type distribution board (Kabeltechnik)
Gusszinnbronze f <Metall> cannon metal, gunmetal
Gut n 1. <Patent, Qual> commodity; 2. <Wassertrans> rigging (Tauwerk)
gut erreichbar adj <Maschinen> conveniently-placed (Hebel)
gut freihalten v von <Wassertrans> give a wide berth (Schiffsführung)
gut funktionierend adj <Comp & DV> well-behaving (Programm)
gut leitende Diode f <Elektronik> high-conductance diode
gut moderierter Reaktorkern m <Kerntech> well-moderated core
gut sichtbare Kleidung f <Sicherheit> highly visible clothing
gut verbunden adj <Anstrich> well-bonded
Gutachten n <Qual> expert opinion, expert's report, expertise, survey
gute Herstellungstechnik f <Qual> good manufacturing practice, GMP
Güte f 1. <Akustik> quality; 2. <Fertig, Metall> grade; 3. <Qual> quality
Güte f **der Oberfläche** <Anstrich> surface finish
Güte f **des Zuschlagstoffs** <Bau> quality of aggregate
Güteanforderung f <Textil> spec, specification
Gütebestätigungsstufe f <Qual> assessment level
Gütebestätigungssystem n <Qual> system of quality assessment
Gütebewertung f <Qual> quality appraisal
Gütefaktor m 1. <Elektronik, Telekom> figure of merit; 2. <Funktech, Mechan, Phys, Qual, Umweltschmutz> (Q-Faktor) quality factor (Q factor)

Gütefaktor m **einer Spule** <Elektriz> coil Q-factor
Gütefaktormesser m (Q-Meter) <Phys> Q meter
gütegeschalteter Laser m <Kerntech> Q-switched laser
Gütegrad m <Mechan> rating
Güteklasse f 1. <Gerät> class (Messgerät); 2. <Mechan, Qual> quality class
Gütekontrolle f <Fertig> inspection and quality control
Gütekriterium n <Chemie, Telekom> performance index
Gütemerkmal n <Qual> quality criterion
Güteminderung f <Qual> deterioration
Gutenberg'sche Diskontinuität f <Nichtfoss Energ> Gutenberg discontinuity
Güteprüfung f **durch den öffentlichen Auftraggeber** <Qual> government quality assurance
Güter npl <Eisenbahn> goods
Güterannahme f <Eisenbahn> (AE) freight inwards, (BE) goods inwards
Güterbahnhof m 1. <Eisenbahn> (AE) freight station, (BE) goods station, (AE) freight yard, (BE) goods yard; 2. <Trans> (AE) freight depot, (BE) goods depot
Güterboden m <Eisenbahn> (AE) freight shed, (BE) goods shed
Güterfernverkehr m <Trans> long-distance goods traffic, road haulage, road transport
Güterlastkraftwagen m (Güter-LKW) <Kfztech> (AE) freight truck, (BE) goods lorry (freight truck)
Güter-LKW m (Güterlastkraftwagen) <Kfztech> (AE) freight truck, (BE) goods lorry (goods lorry)
Güterlokomotive f <Eisenbahn> freight locomotive
Güterrutsche f <Eisenbahn> (AE) freight chute, (BE) goods chute
Güterschuppen m 1. <Eisenbahn> (AE) freight shed, (BE) goods shed; 2. <Trans> (AE) freight depot, (BE) goods depot
Güterträger m <Eisenbahn> (AE) freight porter, (BE) goods porter
Güterwagen m <Eisenbahn> (AE) railroad freight car, (BE) railway freight car, van
Güterwagendrehgestell n <Eisenbahn> (AE) freight truck, (BE) goods lorry
Güterwaren fpl <Verpack> goods
Güterzug m <Eisenbahn> (AE) freight train, (BE) goods train
Güteschaltbetrieb m <Elektronik> Q switching (Laser-Sperrung)
Gütezahl f <Mechan> quality index
Gütezeichen n 1. <Mechan> quality sign; 2. <Patent> certification mark; 3. <Qual> mark of conformity; 4. <Verpack> quality mark
Guthaben n <Kohlen> asset
Guthabenkarte f <Mobilkom, Telekom> prepaid card
Gutlehrdorn m <Fertig> go-screw plug
Gutlehre f 1. <Fertig> (AE) go gage, (BE) go gauge; 2. <Maschinen> (AE) go gage, (BE) go gauge, (AE) pass gage, (BE) pass gauge
Gutquittung f <Telekom> positive acknowledgement
Gutrachenlehre f <Fertig> (AE) go snap-gage, (BE) go snap-gauge
Gut-/Schlecht-Entscheidung f <Qual> go/no-go decision, pass/fail decision
Gutseite f 1. <Fertig> go end (Lehre); 2. <Maschinen> go end
Gutseitelehrung f <Fertig> (AE) go end gaging, (BE) go end gauging
Gutstoff m <Papier> accepted stock
Guttapercha f <Chemie, Kunststoff> guttapercha
Gut- und Ausschusslehre f <Maschinen> (AE) go and no-go gage, (BE) go and no-go gauge
Gx (Systemkonstante) <Akustik> Gx (system-rating constant)

gy *(Gray)* 1. <Phys> gy, gray *(Einheit der Energiedosis)*; 2. <Teilphys> gy, gray *(Einheit der Strahlendosis)*
Gynocard... <Chemie> gynocardic
Gyrator *m* <Phys> gyrator
Gyrobus *m* <Kfztech> gyrobus
Gyrograph *m* <Raumfahrt> gyrograph
gyromagnetischer Effekt *m* <Phys> gyromagnetic effect
gyromagnetisches Verhältnis *n (g)* <Kerntech, Phys> gyromagnetic ratio *(g)*
Gyroskop *n* <Phys> gyroscope
Gyrostat *m* <Chemie> gyrostat
gyrostatisch *adj* <Chemie> gyrostatic
Gyrotron *n* <Telekom> gyrotron

H

h 1. <Comp & DV, Funktech, Geom> *(Höhe)* h *(height)*; 2. <Metrol> *(Hekto...)* h *(hecto...)*; 3. <Metrol> *(Stunde)* h *(hour)*; 4. <Phys, Strahlphys, Teilphys> *(Planck'sche Konstante, Planck'sches Wirkungsquantum)* h *(Planck's constant)*
h quer *adj* <Phys> h-bar *(Dirac-Konstante)*
H 1. <Chemie> *(Wasserstoff)* H *(hydrogen)*; 2. <Elektriz, Elektrotech, Funktech, Metrol, Phys> *(Henry)* H *(henry)*; 3. <Elektriz, Elektrotech> *(magnetische Feldstärke)* H *(magnetic field strength)*; 4. <Heiz & Kälte, Kohlen, Mechan, Nichtfoss Energ, Phys, Raumfahrt, Thermod> *(Enthalpie)* H *(enthalpy)*; 5. <Hydraul> *(Hamilton'sche Funktion)* H *(Hamiltonian function)*; 6. <Optik> *(Bestrahlungsstärke)* H *(irradiance)*; 7. <Phys> *(Magnetfeldstärke)* H *(magnetic field strength)*
ha *(Hektar)* <Metrol> ha *(hectare)*
Haar *n* <Textil> hair
Haarfeuchtigkeitsmesser *m* <Heiz & Kälte> hair hygrometer
Haargarnteppich *m* <Textil> haircord carpet
Haarhygrometer *n* <Heiz & Kälte, Labor, Phys> hair hygrometer *(Feuchtigkeitsmesser)*
Haarkristall *m* <Elektronik, Kunststoff, Metall> whisker
Haarlineal *n* <Maschinen> knife edge straight edge, straight edge
Haarlinie *f* <Druck> hairline
Haarnadelfeder *f* <Maschinen> hairpin spring
Haarnadeltrockner *m* <Ker & Glas> hairpin cooler
Haarriss *m* 1. <Fertig> shatter crack; 2. <Ker & Glas> fire crack *(in Glasuren)*; 3. <Kerntech> capilliary crack, hairline crack; 4. <Mechan> fine crack, hair crack, hairline crack, microflaw, tiny crack; 5. <Textil> craze
Haarrissbildung *f* 1. <Fertig> crazing *(Gießen)*; 2. <Ker & Glas> crazing *(auf Ziegeln, Glas)*
Haar-Roving *n* <Ker & Glas> hairy roving
Haarseite *f* <Fertig> grain side *(Riemen)*
Haarspatium *n* <Druck> hairline space
Haarstrich *m* <Gerät> hairline *(auf Skale)*
Habitus *m* <Metall> habit *(Kristalle)*
Habitusfläche *f* <Metall> habit plane
Hacker *m* <Textil> comb *(Spinnen)*
Hackerkamm *m* <Textil> doffer comb
Hackmaschine *f* <Papier> chipper
Hackmesser *n* <Papier> chipper knife
Hackschnitzel *m* <Papier> chip
Häcksler *m* <Papier> chopper

Hadamard-Ableitung *f* <Math> Hadamard derivative *(Differenzialgleichung)*
Hadamard-Differenzial *n* <Math> Hadamard-differential
Ha-Dec-Befestigung *f* <Raumfahrt> Ha-Dec mount
Hadernpapier *n* <Druck, Papier> rag paper
Hadernsortierer *m* <Papier> rag sorter
HADES *(Dielektronen-Spektrometer mit hoher Akzeptanz)* <Teilphys> HADES *(high acceptance di-electron spectrometer)*
Hadron *n* <Phys, Teilphys> hadron
Hadron-Elektron-Ring-Anlage *f (HERA)* <Teilphys> hadron-electron ring collider *(HERA)*
hadronisch *adj* <Teilphys> hadronic
hadronisches Kalorimeter *n* <Strahlphys, Teilphys> hadronic calorimeter
Hadronkollideranlage *f (LHC)* <Phys> large hadron collider *(LHC)*
Hafen *m* 1. <Bau, Kohlen, Ker & Glas> glas-melting pot, pot; 2. <Wassertrans> *(AE)* harbor, *(BE)* harbour, port
Hafenabstich *m* <Ker & Glas> pot mouth
Hafenanlagen *fpl* <Wassertrans> port installations
Hafenausrüstung *f* <Wassertrans> port equipment
Hafenbahnhof *m* <Eisenbahn> maritime terminal
Hafenbecken *n* <Wassertrans> basin
Hafendamm *m* 1. <Bau> breakwater; 2. <Wassertrans> mole
Hafeneinrichtungen *fpl* <Wassertrans> port facilities
Hafeneinsatz *m* <Ker & Glas> pot setting
Hafenfähre *f* <Wassertrans> *(AE)* harbor ferry, *(BE)* harbour ferry
Hafenfahrzeug *n* <Wassertrans> *(AE)* harbor craft, *(BE)* harbour craft
Hafengelände *n* <Wassertrans> *(AE)* harbor area, *(BE)* harbour area, port area
Hafengewölbe *n* <Ker & Glas> pot arch
Hafenkran *m* <Wassertrans> quay crane
Hafenkühlung *f* <Ker & Glas> pot cooling
Hafenlogistik *f* <Wassertrans> port logistics
Hafenofen *m* <Ker & Glas> pot furnace
Hafenschlepper *m* <Wassertrans> *(AE)* harbor tug, *(BE)* harbour tug
Hafenschlitten *m* <Ker & Glas> pot carriage
Hafenspeicher *m* <Wassertrans> dock warehouse
Hafenspeiserkopf *m* <Ker & Glas> pot spout
Hafensperre *f* <Wassertrans> boom
Hafnium *n (Hf)* <Chemie> hafnium *(Hf)*
HAF-Ruß *m* *(hochabriebfester Furnace-Ruß)* <Kunststoff> HAF carbon black *(high abrasion furnace carbon black)*
Haftband *n* <Verpack> pressure-sensitive tape
Haftbedingung *f* <Strömphys> no-slip condition
Haftbestückung *f* <Elektronik> surface mounting
haften *v* 1. <Bau> cleat, stick; 2. <Kunststoff, Papier, Verpack> adhere
Haften *n* 1. <Elektrotech> sticking; 2. <Fertig> holding; 3. <Ker & Glas> sticking
haftend *adj* <Papier> adherent
Haftenergie *f* <Metall> bond energy
Haftfähigkeit *f* <Papier> adhesiveness
haftfest *adj* <Anstrich> well-bonded
Haftfestigkeit *f* 1. <Anstrich> bond strength; 2. <Kerntech> bonding strength; 3. <Kunststoff> adhesive strength, bond strength; 4. <Verpack> adherence
Haftfestigkeitsprüfung *f* <Kunststoff> adhesion test
Haftfestigkeitsversuch *m* <Maschinen> adhesion strength test
Haftgewicht *n* <Eisenbahn> adhesive weight
Haftgrund *m* <Kunststoff> primer, wash primer
Haftklebepapier *n* <Verpack> pressure-adhesive paper, pressure-sensitive paper

Haftmarkierung f <Ker & Glas> sticking mark
Haftmittel n <Kunststoff> bonding agent
Haftprüfung f <Verpack> bonding test
Haftpulver n <Fertig> blotter powder *(Fluoreszenzverfahren)*
Haftreibung f 1. <Ergon> static friction; 2. <Maschinen> static friction, stiction
haftreibungsfreier Oszillator m <Elektronik> antistiction oscillator
Haftscheibe f <Verpack> *(BE)* adhesive disc, *(AE)* adhesive disk
Haftschicht f <Anstrich> bonding layer
Haftschweißen n <Mechan> tack welding
Haftstoff m <Raumfahrt> bonding agent *(Raumschiff)*
Haftung f 1. <Anstrich> bond; 2. <Bau> adhesion, bond, bonding; 3. <Fertig> stick; 4. <Kohlen, Kunststoff> adhesion; 5. <Meerschmutz> liability
Haftungsgrenze f <Kfztech> limit of adhesion
Haftverbindung f <Fertig> joint
Haftvermittler m <Kunststoff> coupling agent, primer
Haftvermögen n 1. <Bau> adhesion; 2. <Fertig> grip *(Riemen)*; 3. <Papier> adherence, tackiness; 4. <Werkprüf> peel strength
Haftverstärker m <Kunststoff> adhesion promoter
Haftwasser n 1. <Erdöl> interstitial water *(Geologie)*; 2. <Kohlen> pellicular water
Haftwert m <Kfztech> adhesion coefficient
Hagen-Poiseuille'sches Gesetz n <Phys> Hagen-Poiseuille law
Hager-Schleuderscheibe f <Ker & Glas> *(BE)* Hager disc, *(AE)* Hager disk
Hahn m 1. <Bau> cock; 2. <Fertig> plug valve, valve *(Kunststoffinstallationen)*; 3. <Kerntech> valve; 4. <Labor> *(AE)* faucet, *(BE)* tap *(Bedienung)*; 5. <Maschinen> cock, *(AE)* faucet, spigot, *(BE)* tap; 6. <Mechan> cock, *(AE)* faucet, *(BE)* tap, valve; 7. <Wassertrans> cock
Hahnepot f <Wassertrans> bridle *(Tauwerk)*
Hahnhülse f <Labor> barrel *(Spritzgerät)*
Hahnkegel m <Bau> plug
Hahnküken n 1. <Fertig> taper plug; 2. <Hydraul> plug
Hahnventil n 1. <Labor> cock; 2. <Maschinen> cock valve; 3. <Mechan> cock
Haidinger'sche Ringe mpl <Phys> Haidinger fringes
Haken m 1. <Bau> hook; 2. <Fertig> crochet, holdfast; 3. <Hydraul> gab; 4. <Ker & Glas> hook, pick, snap; 5. <Maschinen> hook; 6. <Mechan> hitch, hook; 7. <Textil> hook • **an Haken befestigen** <Maschinen> hook
Haken m **zum Lösen der Fracht** <Lufttrans> cargo release hook *(Hubschrauber)*
Hakenbügel m <Bau> U-bolt
Hakeneisen n <Ker & Glas> trying iron
Hakenkeil m <Fertig> dog key
Hakenlast f <Erdöl> hook load *(Bohrtechnik)*
Hakenmarkierung f <Ker & Glas> hook mark
Hakennagel m <Bau> spike
Hakenschalter m <Elektrotech> gravity switch
Hakenschlag m <Wassertrans> blackwall hitch
Hakenschlüssel m <Maschinen> hook spanner
Hakenschraube f 1. <Bau> screw hook; 2. <Maschinen> hook bolt
Hakenstift m <Bau> sprig bolt
Hakenverschluss m <Verpack> hooked lock
Hakenzahn m <Maschinen> peg tooth
Halbacetal n <Chemie> hemiacetal
Halbachse f <Geom> semiaxis
Halbaddierer m 1. <Comp & DV> half-adder; 2. <Elektronik> half-adder *(vernachlässigte Überträge)*
halbaktive Zielsuchlenkung f <Lufttrans> homing semiactive guidance

halbaktives Fahrwerk n <Lufttrans> semiactive landing gear
halbamtsberechtigt adj <Telekom> semirestricted *(Nebenstelle)*
halbamtsberechtigte Nebenstelle f <Telekom> partially restricted extension
Halbanthrazit m <Kohlen> semianthracite
Halbautomatik f <Kfztech> semiautomatic transmission *(Triebstrang, Getriebe)*
halbautomatisch adj <Ker & Glas, Maschinen, Telekom, Verpack> semiautomatic
halbautomatische Etikettiermaschine f <Verpack> *(AE)* semiautomatic labeling machine, *(BE)* semiautomatic labelling machine
halbautomatische Paketiermaschine f <Verpack> semiautomatic strapping machine
halbautomatische Signalgebung f <Eisenbahn> semiautomatic signalling
halbautomatischer Fernbetrieb m <Telekom> semiautomatic trunk working
halbautomatischer Vermittlungsschrank m <Telekom> automanual switchboard
halbautomatisches Pressen n <Ker & Glas> semiautomatic pressing
halbautomatisches System n <Telekom> semiautomatic system
Halbband m <Druck> quarter binding
Halbbaum m <Textil> half beam *(für Kettenwirkerei)*
halbberuhigt adj <Fertig> balanced *(Stahl)*
halbbeweglich adj <Maschinen> semiportable
Halbbild n <Fernseh> field, frame
Halbbildabschluss m <Fernseh> interfield cut
Halbbinder m <Bau> half-truss
Halbbogen m <Bau> haunch
Halbbyte n <Comp & DV> nibble, nybble
Halbcellulose f <Chemie> hemicellulose
Halbdach n <Bau> pent roof
Halbdieselmotor m <Kfztech, Maschinen> semidiesel engine
Halbdunkelaufnahme f <Foto> twilight shot
halbduplex adj *(HD)* <Comp & DV> half-duplex *(HDX)*
Halbduplex m <Comp & DV, Telekom> half-duplex, semiduplex, two-way alternate
Halbduplexbetrieb m <Comp & DV, Funktech> half-duplex operation
Halbduplex-Funkverkehr m <Telekom> half-duplex radiocommunication
Halbduplexmodus m <Comp & DV> half-duplex mode
Halbduplex-Operation f <Comp & DV> half-duplex operation
halbdurchlässige Farbe f <Ker & Glas> *(AE)* semitransparent color, *(BE)* semitransparent colour
halbdurchlässige Membran f <Phys> semipermeable membrane
halbdurchlässige Photokatode f <Elektrotech> semitransparent photocathode
halbdurchlässige Platte f <Phys> semireflecting plate
halbdurchlässige Scheibe f <Phys> semireflecting plate
halbe Hubhöhe f <Maschinen> half-travel
Halbellipse f <Geom> semi-ellipse
halbellipsenförmiger Boden m <Kerntech> hemiellipsoidal bottom
halbellipsenförmiger Deckel m <Kerntech> hemiellipsoidal head
Halbelliptikfeder f <Maschinen> half-elliptic spring, semielliptic spring
halber Balken m <Wassertrans> half beam *(Schiffbau)*
halber Hub m <Maschinen> midtravel
halber Schritt m <Comp & DV> half-space

halbfester

halbfester brennbarer Abfall m <Umweltschmutz> semisolid combustible waste
Halbfett n <Druck> medium face
halbfette Kohle f <Kohlen> semibituminous coal
halbfette Schrift f <Druck> medium face
Halbfeuchttrennung f <Abfall> semiwet sorting
halbfliegende Achse f <Kfztech> semifloating axle
halbformatige Leiterplatte f <Elektronik> half-sized board
Halbfranzband n <Druck> half-bound
halbgebleichter Zellstoff m <Papier> semibleached pulp
halbgedrehte Treppe f <Bau> half-turn stairs
halbgekapselter Motor m <Elektriz> semienclosed motor
halbgerichtetes Mikrofon n <Aufnahme> semidirectional microphone
halbgetauchtes Tragflächenboot n <Wassertrans> surface piercing craft
Halbgeviert n <Druck> en
Halbgeviert-Zwischenraum m <Druck> en space
Halbhohlniet m <Maschinen> semitubular rivet
halbieren v 1. <Bau> halve; 2. <Geom> bisect
halbierend adj <Geom> bisecting
Halbierung f <Geom> bisection
Halbierungslinie f <Geom> bisector
Halbierungspunkt m <Geom> midpoint
Halbierungssuchverfahren n <Comp & DV> dichotomizing search
Halbimpuls m <Elektronik> half-pulse *(Daten)*
Halbinsel f <Wassertrans> peninsula
halbisolierendes Trägermaterial n <Elektronik> semi-insulating substrate
halbkalibrierte Walze f <Fertig> half-roll
Halbkanten fpl <Geom> half edges
Halbkartonaufrichter m <Verpack> tray erector
Halbkettenfahrzeug n <Trans> half-track vehicle
halbklassische Näherung f <Kerntech> semiclassical approximation
halbkompiliert adj <Comp & DV> semicompiled
halbkontinuierliches Gießen n <Ker & Glas> semicontinuous casting
Halbkreis m <Bau, Geom> semicircle
Halbkreisbogen m <Bau> round arch, semicircular arch
halbkreisförmig adj <Geom, Kerntech> semicircular
halbkreisförmiges Betaspektrometer n <Kerntech> semicircular beta spectrograph
Halbkugel f <Geom> half sphere, hemisphere
Halbkugelendspant m <Raumfahrt> hemispherical end rib *(Raumschiff)*
halbkugelförmiger Brennraum m <Kfztech> hemispherical combustion chamber
halbleitend adj <Anstrich> semiconductive
Halbleiter m <Comp & DV, Elektriz, Elektronik, Funktech, Phys> semiconductor, solid
Halbleiter m **unter Elektronenbeschuss** <Elektronik> electron-bombarded semiconductor
Halbleiter m **vom Typ n–** <Elektronik> n–-type semiconductor
Halbleiter m **vom Typ n+** <Elektronik> n+-type semiconductor
Halbleiterband-Elementekonstruktion f <Elektronik> VMOS, vertical metal oxide semiconductor
Halbleiterbauelement n 1. <Comp & DV, Elektronik> semiconductor device, solid state device; 2. <Telekom> solid state device
Halbleiterbauteil n <Comp & DV, Elektronik> solid state device, semiconductor component
halbleiterbestückt adj <Fernseh, Funktech, Telekom> solid state

Halbleiterchip m <Elektronik> semiconductor chip, solid-state device
Halbleiterdatenspeicher m **mit wahlfreiem Zugriff** <Comp & DV> data RAM, random access data memory
Halbleiterdatenspeicher m/**nur lesbarer** <Comp & DV> data ROM, read only data memory
Halbleiterdehnungsmessstreifen m <Gerät> *(AE)* semiconductor strain gage, *(BE)* semiconductor strain gauge
Halbleiterdiode f <Comp & DV, Elektronik> semiconductor diode
Halbleiterdotierung f <Elektronik> semiconductor doping
Halbleitereinkristall m <Elektronik> semiconductor single crystal
Halbleiter-Feldeffekttransistor m **mit Rückgatter** *(BMOSFET)* <Elektronik> back-gate metal-oxide semiconductor field-effect transistor *(BMOSFET)*
Halbleitergleichrichter m <Elektriz, Elektrotech> semiconductor rectifier, solid-state rectifier
Halbleiterherstellung f <Elektronik> semiconductor fabrication
Halbleiterkamera f <Elektronik> solid-state camera
Halbleiterkomponente f <Elektronik> semiconductor component, solid-state component
Halbleiterkoppelpunkt m <Telekom> semiconductor crosspoint
Halbleiterkristall m <Elektronik> semiconductor crystal, solid-state crystal
Halbleiterkristallscheibe f <Elektrotech> wafer
Halbleiterlaser m 1. <Elektronik> injection laser, semiconducting laser, semiconductor laser, solid-state laser; 2. <Optik, Strahlphys, Telekom> semiconductor laser
Halbleitermaser m <Elektronik> solid-state maser
Halbleitermaterial n <Elektronik> semiconductor material, solid-state material
Halbleitermessrelais n <Elektriz, Elektrotech> solid-state measuring relay
Halbleitermikrofon n <Aufnahme> semiconductor microphone
Halbleitermotoranlasser m <Elektrotech> solid-state motor starter
Halbleiterpapier n <Foto> semiconductor layer paper
Halbleiterphotodetektor m <Elektronik> semiconductor photodetector
Halbleiterrelais n <Elektriz, Elektrotech> semiconductor relay
Halbleiterschaltelement n <Elektriz> semiconductor switching device
Halbleiterschalter m <Elektrotech> semiconductor switch, solid-state switch
Halbleiterscheibe f <Comp & DV, Elektronik> semiconductor wafer
Halbleiterschicht f <Elektronik> semiconductor layer
Halbleiterspeicher m <Comp & DV, Elektrotech> semiconductor memory
Halbleiterstromrichter m <Elektrotech> semiconductor converter
Halbleitertechnik f <Elektronik> semiconductor technology, solid-state technology • **vollständig in Halbleitertechnik** <Elektronik> all-solid state *(Ausführung)*
Halbleiterthermoelement n <Gerät> semiconductor thermocouple
Halbleiterträgermaterial n <Elektronik> semiconductor substrate
Halbleiterverstärker m <Telekom> semiconductor amplifier
Halbleiterwerkstoff m <Elektronik> semiconductor material

Halbleiterwiderstand m <Elektrotech> semiconductor resistor
Halbleiterzähler m <Strahlphys> semiconductor counter
Halbmaske f <Sicherheit> half-mask, respirator facepiece
halbmatt adj <Textil> semimatt
Halbmesserlehre f <Fertig> (AE) radius gage, (BE) radius gauge
halbmondförmig adj <Geom> crescent-shaped
Halbparabelbrücke f <Bau> hogbacked bridge
Halbpodest n <Bau> quarter-landing
Halbportalkran m <Bau> semigantry crane
Halbradiallüfter m <Heiz & Kälte> mixed-flow fan
Halbring m <Maschinen> half-ring
Halbrücke f <Elektrotech> half-bridge
Halbrücken-Anordnung f <Elektrotech> half-bridge arrangement
Halbrund n <Fertig, Maschinen> half round
halbrunde Kante f <Ker & Glas> half-round edge
Halbrundfeile f <Maschinen> half-round file
Halbrundkopf m 1. <Elektriz> button; 2. <Maschinen> cup head, round head
Halbrundkopfschraube f <Maschinen> (AE) button-headed screw, cup head bolt, half-round screw
Halbrundniet m 1. <Fertig> button-headed rivet; 2. <Maschinen> round-head rivet
Halbrundschraube f 1. <Fertig> securing screw (Kunststoffinstallationen); 2. <Maschinen> (AE) button-headed screw, cup head bolt, half-round screw
Halbrundzange f <Maschinen> half-round pliers
Halbschale f <Maschinen> half-liner
Halbschatten m <Optik, Phys> penumbra • **Halbschatten erzeugend** <Optik> penumbrous • **im Halbschatten** <Optik> penumbral
Halbscheibe f <Maschinen> half-washer
Halbschnitt m <Fertig, Konstzeich> half-section
Halbschnittzeichnung f <Konstzeich> half-section drawing
Halbschwingachse f <Kfztech> de Dion axle
halbseitige Spur f <Akustik> unilateral track
halbspröder Bruch m <Metall> semibrittle fracture
Halbstahl m <Metall> half-steel
halbstarre automatische Kupplung f <Kfztech> semirigid automatic coupling
halbstocks setzen v <Wassertrans> half-mast (Flagge)
Halbstrahl m <Geom> half-line
Halbsubtrahierer m <Comp & DV, Elektronik> half subtractor
Halbsubtrahierglied n <Comp & DV> half subtractor
Halbsubtrahiersignal n <Comp & DV> half subtractor
Halbtagstide f <Wassertrans> semidiurnal tide (Gezeiten)
Halbtaucher m <Erdöl> semisubmersible rig (Offshore-Technik)
Halbtide f <Wassertrans> half-tide (Gezeiten)
Halbton m 1. <Akustik> semitone; 2. <Druck> continuous tone, halftone; 3. <Phys> semitone
Halbtonbild n <Fernseh> halftone image
halbweißes Glas n <Ker & Glas> half-white glass
Halbwelle f 1. <Elektriz> half-wave; 2. <Kfztech> half shaft (Triebstrang); 3. <Phys, Strahlphys> half-wave
Halbwellendipol m <Phys> half-wave dipole
Halbwellendipolantenne f <Funktech, Strahlphys> half-wave dipole aerial, half-wave dipole antenna
Halbwellengleichrichter m <Elektrotech> half-wave rectifier
Halbwellengleichrichtung f <Elektrotech> half-wave rectification
Halbwellenleitung f <Phys> half-wave line
Halbwertdicke f <Phys> half-thickness, half-value thickness

Halbwertsbreite f 1. <Elektronik> half-power width; 2. <Optik> FWHM, full width at half maximum, full duration at half maximum; 3. <Phys, Strahlphys> half-width; 4. <Telekom> full width at half height, FWHH (Impuls)
Halbwertspunkt m <Telekom> three-dB point
Halbwertszeit f 1. <Kerntech> half-life, operating lifetime; 2. <Phys, Strahlphys, Teilphys> half-life, half-time, half-value period, radioactive half-life; 3. <Telekom> (HWZ) full duration half maximum (FDHM)
Halbwertszeitbereich m <Strahlphys> range of half-life
Halbwort n <Comp & DV> half-word
halbzählig adj <Fertig> half-integer (Spin)
halbzähliger Spin m <Phys> half-integral spin
Halbzellstoff m <Papier> semichemical pulp
Halbzeug n <Elektriz, Kunststoff> semifinished product
Halbzeugholländer m <Papier> breaker
Halbzyklus m <Elektrotech> half-cycle
Halde f 1. <Abfall> waste site (Bergbau); 2. <Bau> spoil heap, stockpile; 3. <Kohlen> dump, tip • **auf Halde lagern** <Bau> stockpile
Haldenabfall m <Kohlen> tails
Haldenabfallbeseitigung f <Kohlen> tail disposal
Haldengelände n <Abfall> dumping site (Bergbau)
Haldenkohle f <Kohlen> stock coal
Half-Plate-Kamera f <Foto> half-plate camera
Halle f <Bau> shed
Hall-Effekt m 1. <Elektriz> Hall effect (durch quer liegenden Strom und Magnetfeld erzeugte Spannung in Halbleitern); 2. <Phys, Raumfahrt, Strahlphys> Hall effect
Hallenschiff n <Bau> span
Hallfeld n <Sicherheit> reverberant field (Lärm)
Hall-Generator m <Funktech> Hall generator
halliges Studio n <Aufnahme> live studio
Halligkeit f <Akustik> liveness
Hallraum m 1. <Akustik> echo chamber, reverberation room; 2. <Aufnahme> reverberant room, reverberation chamber, reverb; 3. <Sicherheit> live room, reverberant room, reverberation chamber, reverberation room (Lärm)
Hallraumprüfung f <Metrol> reverberant-field test
Hall'sche Beweglichkeit f <Phys> Hall mobility
Hall'sche Mobilität f (μ H) <Funktech> Hall mobility (μ H)
Hall'sche Sonde f <Phys, Strahlphys> Hall probe
Hall'sche Spannung f <Phys> Hall voltage
Hall'scher Geber m <Kfztech> Hall generator
Hall'scher Generator m <Kfztech> Hall generator
Hall'scher IC m <Kfztech> Hall IC
Hall'scher Koeffizient m (RH) <Phys> Hall coefficient (RH)
Hall'scher Widerstand m <Phys> Hall resistance
Hall'sches Feld n <Phys> Hall field
Hall'sches Ionentriebwerk n <Raumfahrt> Hall-ion thruster (Raumschiff)
Hall'sches Magnetometer n <Phys> Hall magnetometer
Hall-Sonde f <Funktech> Hall probe
Hallwachs-Effekt m <Optik> photoemissive effect
Halo m <Raumfahrt> halo
Haloantenne f <Raumfahrt> halo
Halogen n <Chemie> halogen
halogenartig adj <Chemie> haloid
Halogenation f <Chemie> halogenation
Halogenid n <Chemie> halide, halogenide
Halogenisierung f <Chemie> halogenation
Halogenkohlenwasserstoff n <Raumfahrt> halon (Raumschiff)
Halogenkohlenwasserstoff-Kältemittel n <Heiz & Kälte> halocarbon refrigerant
Halogenlampe f <Elektriz> halogen lamp
Halogensilberemulsion f <Foto> silver halide emulsion
Halogensilberpapier n <Foto> silver halide paper

Halographie

Halographie f <Chemie> halography
Halokinese f <Erdöl> halokinesis *(Salztektonik)*
Halon n <Raumfahrt> halon *(Raumschiff)*
Halonfeuerlöscher m <Sicherheit> halon fire extinguisher
Haloumlaufbahn f <Raumfahrt> halo orbit
Hals m 1. <Elektronik> neck *(einer Katodenstrahlröhre)*; 2. <Ker & Glas, Kfztech, Labor> neck; 3. <Maschinen> collar, neck, throat; 4. <Wassertrans> tack *(Segeln)*
Halsausweitung f <Maschinen> flared neck
halsen v 1. <Wassertrans> wear *(Segeln)*; 2. <Wassertrans> jibe *(Segeln)*
Halslager n <Maschinen> collar bearing, neck bearing
Halsring m <Ker & Glas> neck ring; spout *(der Owens-Maschine)*
Halsringhalter m <Ker & Glas> neck ring holder
Halsstück n <Bau> throat
Halszapfen m <Mechan> (BE) gudgeon pin
Halt m 1. <Comp & DV> halt; 2. <Kfztech, Maschinen, Mechan> stop
haltbar adj 1. <Kunststoff> lasting; 2. <Textil> fast • **haltbar bis zum** <Verpack> best before • **haltbar machen** <Lebensmittel> preserve; cure *(Räuchern, Salzen, Pökeln)*
Haltbarkeit f 1. <Bau> service life; 2. <Kunststoff> durability, shelf life; 3. <Lebensmittel> keeping quality; 4. <Maschinen, Textil> durability; 5. <Verpack> shelf impact
Haltbarkeitsdatum n <Verpack> sell-by date
Haltbarkeitsdauer f <Lebensmittel> shelf life
Haltbarkeitsprüfung f <Verpack> shelf life test
Haltbarkeitstest m <Verpack> durability test
Halteanode f <Elektrotech> holding anode, keep-alive electrode
Haltearm m <Lufttrans> hold *(Luftfahrzeug)*
Haltebedingung f <Comp & DV> halt condition, stop condition
Haltebefehl m <Comp & DV> halt instruction, stop instruction
Haltebremse f <Eisenbahn> holding brake
Haltebucht f <Trans> bus bay
Haltecode m <Comp & DV> stop code
Halteeinrichtung f <Sicherheit> stop device
Haltefahrstreifen m für Busse <Trans> bus stopping lane
Haltegestell n für Reagenzgläser <Labor> rack for test tubes
Halteimpuls m <Comp & DV> hold
Halteklemme f für Mikroskopstellung <Labor> clip of microscope stage
Haltekraft f <Bau> cohesion
Haltelasche f <Bau> cleat
halten v <Telekom> put on hold
Halten n <Elektrotech> holding *(Quecksilberdampfröhren)*
Halten n einer Verbindung <Telekom> call hold
Haltenetz n <Lufttrans> mooring harness
Halteplatte f <Maschinen> retaining plate
Halteplattform f <Kerntech> holding pedestal
Haltepodest n <Kerntech> holding pedestal
Haltepunkt m 1. <Kontroll> breakpoint; 2. <Mechan> stop; 3. <Phys> recalescence; 4. <Qual> hold point
Haltepunkt m bei Abkühlung <Fertig> Ar point, aspiration point temperature
Haltepunkt m bei Erwärmung <Fertig> decalescence point
Halter m 1. <Bau> clamp; 2. <Fertig> (AE) arbor, (BE) arbour *(Reibahle)*; cap *(Schneidbacke)*; 3. <Kfztech> handle; 4. <Maschinen> holder; 5. <Mechan> clamp, fastener, handle
Haltering m 1. <Lufttrans> mooring ring; 2. <Maschinen> holding ring, retaining ring; 3. <Mechan> retaining ring

Halterung f 1. <Fertig> fastening; bracket, holder *(Kunststoffinstallationen)*; 2. <Ker & Glas> mechanical boy; 3. <Kfztech> panel mounting *(für Karosserieblech)*; 4. <Kohlen> attachment; 5. <Lufttrans> fairlead; 6. <Maschinen> attachment, holder, holding fixture, support
Halterungsklammer f <Raumfahrt> launcher release gear *(für Rakete beim Start)*
Halteschelle f am Luftschraubenblatt <Lufttrans> blade retention strap *(Hubschrauber)*
Halteseil n <Maschinen> holding rope
Haltespeicher m <Telekom> control memory, hold latch
Haltestelle f 1. <Textil> stopping mark; 2. <Trans> halt, stop *(Station)*
Haltestift m <Fertig> retaining pin
Haltestock m <Bau> bench stop *(Zimmermannsarbeiten)*
Haltestrahl m 1. <Comp & DV> holding beam; 2. <Elektronik> holding beam *(Computer)*; 3. <Strahlphys> holding beam
Haltestrom m <Elektriz, Elektrotech> holding current
Halteturm m mit Fluchtmöglichkeit <Raumfahrt> emergency escape tower *(Raumschiff)*
Haltewicklung f <Elektriz> holding winding *(eines Relais)*
Haltezeit f 1. <Comp & DV> hold time; 2. <Funktech> lock-up time; 3. <Trans> halt
Haltezustand/im <Telekom> on-hold
Haltgliedsteuerung f <Regelung> holding element control
Haltscheibe f <Eisenbahn> stop signal *(für Rangierfahrten)*
Haltsignal n <Eisenbahn> stop board, stop signal
Haltung f <Ergon> attitude, posture
Hämatein n <Chemie> (BE) haematein, (AE) hematein
Hämatin n <Chemie> (BE) haematin, (AE) hematin
Hämatin... <Chemie> (BE) haematic, (AE) hematic
Hämatit m <Fertig> (BE) haematite, (AE) hematite
Hämatoporphyrin n <Chemie> (BE) haematoporphyrin, (AE) hematoporphyrin
Hämatoxylin n <Chemie> (BE) haematoxylin, (AE) hematoxylin
Hamilton-... <Teilphys> Hamiltonian
Hamilton-Jacobi'sche Gleichung f <Phys> Hamilton-Jacobi equation
Hamilton'sche Funktion f (H) <Hydraul> Hamiltonian function (H)
Hamilton'sche Gleichungen fpl <Phys> Hamilton's equations
Hamilton'scher Operator m <Phys> Hamiltonian operator
Hammer m 1. <Bau, Ker & Glas, Kohlen, Maschinen> hammer; 2. <Mechan> hammer, mallet
Hammer m mit gerader Finne <Bau> straight-pane hammer, straight-peen hammer
Hammer m mit Kugelfinne <Maschinen> ball-pane hammer, ball-peen hammer
Hammerbacken m <Fertig> rotary swaging die
Hammerbär m <Fertig> hammer tup
hämmerbar adj <Mechan, Metall> malleable
Hämmerbarkeit f <Metall> ductility
Hammerbrecher m <Kohlen, Maschinen> hammer crusher
Hammerfinne f <Fertig> hammer peen
hammergeschmiedetes Teil n <Fertig> hammered forging
Hammerhärten n <Metall> hammer hardening
Hammerkopf m <Maschinen> hammer head
Hammerkopfschraube f <Maschinen> T-head bolt, hammer-head bolt, hammer-head screw
Hammermühle f 1. <Abfall> hammer mill, hammer-mill crusher; 2. <Lebensmittel> hammer mill

hämmern v 1. <Bau> beat, sledge; 2. <Fertig> chase, forge, peen, rotary-swage; 3. <Maschinen> hammer; 4. <Metall> beat
Hämmern n 1. <Bau> hammering; 2. <Fertig> forging, swaging; 3. <Kerntech> hammering; 4. <Maschinen> hammering, peening
hammernieten v <Fertig> hammer-rivet
Hammernieten n <Fertig> hammer riveting
Hammeröhr n <Fertig> hammer eye
Hammerschlag m 1. <Fertig> forge scale, hammer scale; 2. <Maschinen> hammer blow; 3. <Mechan> scale
Hammerschlaganstrich m <Kunststoff> hammer finish
Hammerschlaglack m <Kunststoff> hammer finish paint
Hammerschraube f <Maschinen> T-head bolt, hammer-head bolt, hammer-head screw
Hammerschweißen n <Bau> forge welding
Hammerunterbrecher m <Elektrotech> trembler
Hammerzerkleinerer m <Abfall> hammer mill, hammer-mill crusher
Hamming-Abstand m <Telekom> Hamming distance
Hämoglobin n <Chemie> (BE) haemoglobin, (AE) hemoglobin
Hämolyse f <Chemie> (BE) haemolysis, (AE) hemolysis
Hämolysin n <Chemie> (BE) haemolysin, (AE) hemolysin
Hämopyrrol n <Chemie> (BE) haemopyrrole, (AE) hemopyrrole
Hämosiderin n <Chemie> (BE) haemosiderin, (AE) hemosiderin
Hämotoxin n <Chemie> (BE) haemotoxin, (AE) hemotoxin
Hand mischen v/von <Bau> spade (Beton)
Hand stampfen v/von <Fertig> hand-ram
Hand/von <Mechan> manual
Hand zusammengestellt adj/von <Verpack> hand-assembled
Hand... <Druck, Fertig, Kerntech, Telekom> manual
Handabschaltung f <Kerntech> manual shutdown
Handapparat m <Telekom> hand-held receiver, handset
Handapparateschnur f <Telekom> handset cord
Handarbeit f 1. <Druck> handwork; 2. <Ker & Glas> off-hand working
Handarmatur f <Fertig> hand-operated valve (Kunststoffinstallationen)
Hand-Arm-System n <Ergon> hand-arm system
Handauflage f <Maschinen> handrest, turning rest
Handauflegeverfahren n <Kunststoff> hand lay-up
Handauslösesignal n <Regelung> manually-operated releasing signal
Handauslösung f <Elektriz> manual reset
Handautomatikschalter m <Kerntech> hand-automatic switch
handbedient adj <Mechan> hand-operated, manually-operated
handbedienter Drehzahlmesser m <Verpack> hand tachometer
handbedienter Zug m <Kerntech> hand-operated pull
handbedientes Fernsprechsystem n <Telekom> manual system
Handbedienung f 1. <Comp & DV> manual operation; 2. <Mechan> manual control; 3. <Telekom> manual working
handbetätigt adj <Kontroll, Maschinen> hand-operated, manually-operated
handbetätigter Schalter m <Elektriz> hand-operated switch
handbetätigtes Werkzeug n <Mechan> handtool
Handbetätigung f 1. <Comp & DV> manual operation; 2. <Fertig> manual override (Kunststoffinstallationen)
Handbetrieb m <Comp & DV> manual operation
handbetrieben adj <Kontroll, Mechan, Raumfahrt> hand-operated, manually-operated

Handbetriebsanzeiger m <Foto> manual control indicator
Handblasebalg m <Ker & Glas> hand bellow
Handblechschere f <Fertig> handshears
Handbohrer m 1. <Bau> gimlet; 2. <Maschinen> hand brace, handdrill
Handbohrmaschine f 1. <Maschinen> crank brace; 2. <Mechan> handdrill
Handbremse f 1. <Kfztech> handbrake, parking brake; 2. <Maschinen> handbrake; 3. <Mechan> parking brake
Handbuch n 1. <Comp & DV> documentation, manual; 2. <Funktech, Kontroll> manual; 3. <Maschinen> handbook; 4. <Patent, Qual> documentation
Handdrahtzug m <Kerntech> hand wire pull
Handdruck m <Textil> hand-block printing
Handdruckmaschine f <Druck> hand-printing machine
Handdurchschläger m <Bau> nail set, set
Handelsflotte f <Wassertrans> commercial fleet, merchant fleet
Handelshafen m <Wassertrans> trading port, commercial port
Handelsschiff n <Wassertrans> trade ship, commercial ship
Handelsschifftonnage f <Wassertrans> merchant tonnage
Handelstoluol n <Chemie> toluol
handelsüblicher Verstärker m <Elektronik> commercial amplifier
Hand-Endgerät n <Telekom> hand-held terminal
Handetikettiermaschine f <Verpack> hand labeller
Handfackel f <Wassertrans> handflare (Signal)
Handfeuerlöscher m <Sicherheit> portable fire extinguisher
Handform f <Kunststoff> (AE) portable mold, (BE) portable mould
Handfunkfernsprecher m <Mobilkom> hand-held radio telephone, mobile (Handy)
Handfunksprechgerät n 1. <Funktech> hand-carried transceiver; walkie-talkie (Bündelfunk); 2. <Mobilkom> walkie-talkie (Bündelfunk); 3. <Telekom> hand-carried transceiver
Handgaszugsteuerung f <Kfztech> hand throttle control
handgeführt adj <Fertig> hand-held (Läppwerkzeug)
handgeführtes Elektrowerkzeug n <Sicherheit> portable power tool
handgehaltenes Nivellier n <Bau> hand level
Handgelenkblutdruckanzeiger m <Elektronik, Gerät> wrist blood pressure monitor
Handgelenkschutz m <Sicherheit> wrist protector
handgemischter Beton m <Bau> hand-mixed concrete
Handgepäck n <Eisenbahn> hand luggage
handgeregelt adj <Elektrotech> manually-controlled
Handgeschicklichkeit f <Ergon> manual dexterity
handgeschöpftes Papier n <Druck> handmade paper
Handgetriebe n <Fertig> manual drive unit, reduction gear (Kunststoffinstallationen)
Handgewindebohrer m <Fertig> handtap
Handgießpfanne f <Fertig> handladle
Handgriff m 1. <Bau> handle; 2. <Maschinen> grip, handgrip
handhaben v 1. <Bau> handle; 2. <Ergon> manipulate; 3. <Mechan, Qual, Wassertrans> handle (Schiff)
Handhabung f 1. <Ergon> manipulation; 2. <Telekom, Textil, Verpack> handling
Handhabung f von Gütern <Verpack> handling of goods
Handhabung f von Mehrwegflaschen <Verpack> handling of returnables
Handhabungsgerät n <Ergon> manipulator
Handhebel m <Maschinen> hand lever

Handhebelbohrer

Handhebelbohrer *m* <Maschinen> sensitive drill
Handhebelpresse *f* <Ker & Glas> side lever press
Handhebel-Reihenbohrmaschine *f* <Maschinen> sensitive gang drill
Handhebelschere *f* 1. <Fertig> handshears; 2. <Maschinen> crocodile shears
Handhebelvorschub *m* 1. <Fertig> sensitive feed; 2. <Maschinen> hand lever feed • **mit Handhebelvorschub** <Fertig> sensitive
Handhobel *m* <Fertig> manual plane
Handhubwagen *m* <Verpack> manual lift truck
Händigkeit *f* <Ergon> handedness
Handkarren *m* <Bau> handbarrow
Handkette *f* <Bau> handchain
Handklassierung *f* <Kohlen> hand screening
Handkontrolle *f* <Elektriz> manual control
Handkreuz *n* <Fertig, Maschinen> pilot wheel
Handkurbel *f* 1. <Fertig> ball handle crank; 2. <Maschinen> manual crank; 3. <Mechan> crank
Handlaminat *n* <Kunststoff> hand lay-up laminate
Handlaminieren *n* <Kunststoff> hand lay-up
Handlampe *f* <Elektriz, Elektrotech> portable lamp, inspection lamp
Handlauf *m* 1. <Bau, Kerntech> handrail; 2. <Wassertrans> handrail; top rail *(Schiffsausrüstung)*
Handlaufkrümmling *m* <Bau> wreath *(Treppe)*
Handleuchte *f* <Bau> portable light
handlich *adj* <Mechan> handy
Handlocher *m* <Comp & DV> keypunch
Handluftschieber *m* <Heiz & Kälte> main valve, manual damper
Handlung *f* <Ergon, Qual> action
Handlungsanalyse *f* <Ergon> activity analysis
Handlungsfolge *f* <Ergon> *(AE)* serial behavior, *(BE)* serial behaviour
Handlungsträger *m* <Künstl Int> actor
Handmaschinenvibration *f* <Sicherheit> vibration emitted by portable hand-held machines
Handmatrizen *fpl* <Druck> sorts
Handmikrofon *n* <Aufnahme> hand microphone
Handnachbildung *f* <Ergon> artificial hand
Handpeilkompass *m* <Wassertrans> hand-bearing compass
Handpfanne *f* <Fertig> handladle *(Gießen)*
Handpresse *f* <Druck> handpress
Handpressenrähmchen *n* <Druck> frisket
Handpumpe *f* <Fertig, Sicherheit> handpump *(Feuerlöschgerät)*
Handrad *n* 1. <Fertig> handwheel *(Kunststoffinstallationen)*; 2. <Maschinen, Mechan> handwheel
Handregelung *f* 1. <Maschinen> manual setting; 2. <Regelung> manual control
Handreglerschalter *m* <Regelung> manual control switch
Handreibahle *f* 1. <Kfztech> handreamer, reamer *(Werkzeug)*; 2. <Maschinen> handreamer, reamer
Handrückstellung *f* <Elektriz> manual reset
Handsäge *f* <Bau, Maschinen> handsaw
Handsatz *m* <Druck> hand composition, hand setting, manual typesetting
Handsauger *m* <Ker & Glas> handsucker
Handschachten *n* <Bau> shovel work
Handschalten *n* <Elektrotech> manual switching
Handschalter *m* <Regelung> manual control switch
Handschere *f* <Fertig> handshears
Handschleifen *n* <Maschinen> hand grinding
Handschliff *m* <Fertig> offhand grinding
Handschmierung *f* <Maschinen> hand lubrication, manual lubrication
Handschmiervorrichtung *f* <Maschinen> manual-lubricating device
Handschneidbrenner *m* <Fertig> hand flame-cutting torch
Handschraube *f* <Maschinen> handscrew
Handschraubspindel *f* <Maschinen> handscrew
Handschraubstock *m* <Maschinen> *(BE)* hand vice, *(AE)* hand vise
Handschrifterkennung *f* <Künstl Int> handwriting recognition
Handschuh *m* <Fertig> gauntlet *(Schweißen)*
Handschuhe *mpl* **mit hoher Schutzwirkung** <Sicherheit> heavy-duty gloves
Handschuheingriff *m* <Kerntech> glove port
Handschuhfach *n* <Kfztech> *(AE)* glove box, *(BE)* glove compartment *(Innenausstattung)*
Handschuhkasten *m* <Kerntech> *(AE)* glove box, *(BE)* glove compartment
Handschuhöffnung *f* <Kerntech> glove port
Handschutz *m* <Sicherheit> handshield
Handschutzschild *m* <Bau, Sicherheit> handshield, hand-shield type protector, hand-type protector, welding handshield *(Schweißen)*
Handshake *m* <Comp & DV, Kontroll, Telekom> handshake
Handsiebdruck *m* <Textil> hand screen printing
Handsignal *n* <Trans> handsignal
Handsortierung *f* <Abfall> hand sorting, manual sorting *(von Müll)*
Handspindelbremse *f* <Eisenbahn> screw brake with crank handle
Handstellteil *n* <Ergon> hand control
Handsteuerung *f* 1. <Bau, Elektriz> manual control; 2. <Kerntech> manual handling; 3. <Lufttrans> manual control
Handstichprobe *f* <Kohlen> hand sampling
Handtalje *f* <Wassertrans> handy billy *(Deckausrüstung)*; jigger *(Tauwerk)*
Handtuchstoff *m* <Textil> towelling
handübertragene Schwingung *f* <Sicherheit> hand-transmitted vibration
Handvermittlung *f* <Telekom> manual board
Handvermittlungsschrank *m* <Telekom> cord switchbord
Handvermittlungsstelle *f* <Telekom> manual exchange
Handverpacken *n* <Verpack> hand packing
Handvorschub ausfräsen *v/mit* <Fertig> rout *(Blech)*
Handwagen *m* <Trans> trolley
Handwasserspritze *f* <Sicherheit> stirrup pump for water
Handwerkzeug *n* <Bau> set of tools
Handy *n* <Mobilkom> cell phone, *(AE)* cellular phone, hand-held radio telephone, mobile, mobile phone
Handy *n* **für das D- und E-Netz** <Mobilkom> db-mobile, dual-band handheld telephone, dual-band radio telephone
Handy *n*, **gleichzeitig Schnurlostelefon zum Festnetz** <Mobilkom> dual mode mobile
Handy *n*/**in ein Netz eingebuchtes** <Mobilkom> network logged-in mobile
Handy *n* **mit Eintastenbedienung** <Mobilkom> one-touch mobile
Handy *n* **mit Hörfreisprechgarnitur und FM-Radio** <Mobilkom> mobile with headset and FM receiver
Handy *n* **mit integriertem Organizer** <Mobilkom> mobile organizer
Handy *n* **mit MP3-Abspielmöglichkeit** <Mobilkom> mobile with MP3 player
Handy *n* **mit Taschenrechner mit alphanumerischer Tastatur und Display** <Mobilkom> mobile organizer

Handy-Alarm-Kugelschreiber m <Mobilkom> mobile pen
Handy-Anruf m <Mobilkom> mobile radio call
Handy-Autohalterung f <Mobilkom> mobile phone car-holding device, mobile phone car-mounting
Handybanking n **über das Internet** <Mobilkom> mobile banking, mobile homebanking *(bei Mobilfunk mit Internetzugang)*
Handy-Benutzer m <Mobilkom> mobile phone user, mobile radio subsriber *(Mobilfunkteilnehmer)*
Handy-Gürtelclip m <Mobilkom> mobile belt-clip
Handy-Mailbox f <Mobilkom> mobile radio telephone mailbox, SMS storage
Handy-Nummer f <Mobilkom> mobile subsriber number
Handy-Telefonguthaben-Karte f <Mobilkom> mobile telephone credit card, Xtra Cash
Handy-Wörterbuch n <Mobilkom> mobile phone dictionary
Handzeichen n <Sicherheit> hand signal
Handzentrifuge f <Labor> hand centrifuge *(Trennen)*
Handzufuhr f <Verpack> hand feed
Handzuführung f <Maschinen> hand feed • **mit Handzuführung** <Druck> hand-fed
Handzuführungslocher m <Comp & DV> hand-feed punch
Handzustellung f <Maschinen> hand feed
Hanf m <Fertig, Wassertrans> hemp *(Tauwerk)*
Hanfdichtung f <Fertig> hemp packing
Hanfseil n <Maschinen, Verpack> hemp rope
Hang m <Kohlen> slope
Hangar m <Lufttrans> hangar
Hängebahn f <Eisenbahn> *(AE)* suspended railroad, *(BE)* suspended railway
Hängebaugerüst n <Bau> boat scaffold
Hängebrücke f <Bau> cable-stayed bridge, suspension bridge
Hängebügel m <Maschinen> stirrup hanger
Hängebühne f <Bau> cradle, hanging scaffold
Hängedecke f <Ker & Glas> flying arch
Hängeeisen n <Bau> hanger
Hängegerüst n <Bau> flying scaffold, hanging scaffold, hanging stage, suspended scaffold, *(AE)* traveling cradle, *(BE)* travelling cradle
Hängehaken m <Maschinen> suspension hook
Hangeinschnitt m <Bau> sidehill cut
Hängeisolator m <Elektrotech> suspension insulator
Hängekran m 1. <Bau> suspension crane; 2. <Maschinen> overhead crane, suspension crane; 3. <Mechan> overhead crane
Hängelager n 1. <Bau> hanger; 2. <Maschinen> hanging bearing
Hängelampe f <Elektriz, Elektrotech> hanging lamp, luminaire
Hängelaufkran m <Maschinen> *(AE)* overhead traveling crane, *(BE)* overhead travelling crane
Hängematte f <Wassertrans> hammock
hängender Drucktastenschalter m/**am Schwenkarm** <Fertig> punchbutton pendant
hängender Einzug m <Comp & DV> reverse indention
hängender Torpfosten m <Bau> hanging post, hinge post
hängendes Ventil n 1. <Hydraul> drop valve *(Dampfmaschine)*; 2. <Kfztech> OHV, overhead valve
Hanger m <Wassertrans> topping lift *(Tauwerk)*
Hängeruder n <Wassertrans> underhung rudder
Hängesäule f 1. <Maschinen> joggle piece, joggle post; 2. <Wasserversorg> heelpost *(einer Schleuse)*
Hängetrockner m <Papier> festoon dryer
Hängeventil n <Hydraul> drop valve *(Dampfmaschine)*

Hängeventilsteuerung f <Hydraul> drop valvegear
Hängevorlagespeicher m <Hydraul> inverted pattern accumulator
Hangkanal m 1. <Nichtfoss Energ> headrace canal; 2. <Wasserversorg> headrace
Hank n <Textil> hank
Hantel f <Metall> dumbbell
haptisch adj <Ergon> haptic *(Tastsinn)*
Hardcopy f <Comp & DV> hard copy
Hardenit m <Metall> hardenite
Hardware f <Comp & DV, Telekom> hardware *(Gerätetechnik einer Datenverarbeitungsanlage)*
Hardware-Aufrüstung f <Comp & DV> hardware upgrade
Hardware-Ausrüstung f <Comp & DV> hardware resources
Hardware-Betriebsmittel npl <Comp & DV> hardware resources
Hardwarefehler m <Comp & DV> hard failure
Hardware-Fehler m <Comp & DV> machine error
Hardware-Kompatibilität f <Comp & DV> hardware compatibility
Hardware-Konfiguration f <Comp & DV, Telekom> hardware configuration
hardwarenah adj <Comp & DV> machine-intimate
Hardware-Service m <Comp & DV> hardware maintenance
Hardware-Sicherheit f <Comp & DV> hardware security
Hardware-Steuerung f <Comp & DV> hardware control
Hardware-Unterbrechung f <Comp & DV> hardware interrupt
Hardy-Scheibe f <Maschinen> *(BE)* Hardy disc, *(AE)* Hardy disk
Harmalin n <Chemie> harmaline
Harmin n <Chemie> harmine, yageine
Harmonie f <Akustik> harmony
Harmonik f <Akustik> harmonics
Harmonikatrennwand f <Bau> slip partition
harmonisch adj <Akustik> harmonic
harmonische Analyse f 1. <Math> harmonic analysis *(Bestimmung einer Fourier-Reihe)*; 2. <Mechan, Phys> harmonic analysis
harmonische Molltonleiter f <Akustik> harmonic minor scale
harmonische Ordnung f <Elektronik> harmonic order
harmonische Reihe f <Akustik> harmonic series
harmonische Schwingung f 1. <Elektronik> harmonic oscillation; 2. <Phys> harmonic vibration; harmonic oscillation *(Oberschwingungen)*
harmonische Verzerrung f <Aufnahme, Elektronik> harmonic distortion
harmonische Wellen fpl <Wellphys> harmonic waves
Harmonische f <Akustik, Funktech, Mechan> harmonic
Harmonische f **höherer Ordnung** f <Elektronik> high-order harmonic
Harmonische f **niedriger Ordnung** f <Elektronik> low-order harmonic
harmonischer Analysator m <Phys> *(BE)* harmonic analyser, *(AE)* harmonic analyzer
harmonischer Mischer m <Elektronik> harmonic mixer
harmonischer Mittelwert m <Math> harmonic mean
harmonischer Oszillator m <Phys> harmonic oscillator
harmonisches Mittel n <Math> harmonic mean
Harn... <Chemie> uric *(Säure)*
Harnstoff m <Chemie> carbamide, urea
Harnstoffadditionsverbindung f <Lebensmittel> urea adduct
Harnstoffaddukt n <Lebensmittel> urea adduct
harnstoffausscheidend adj <Chemie> ureotelic

Harnstoff

Harnstoff-Formaldehydharz n (UFH) <Elektriz, Fertig, Kunststoff> urea formaldehyde resin (UFR)
Harnstoffharz n <Elektriz, Fertig, Kunststoff> urea resin
Harpune f <Wassertrans> harpoon (Walfang)
hart aufgelötetes Plättchen n <Fertig> brazed-on tip
hart eingelötet adj <Fertig> sandwich-brazed
Hart... <Verpack> rigid
härtbar adj 1. <Maschinen> hardenable; 2. <Thermod> heat-treatable
Härtbarkeit f <Anstrich, Maschinen, Metall> hardenability
Hartblei n <Fertig> antimonial lead
harte Begrenzung f <Elektronik, Funktech> hard limiting
harte Bodenschicht f <Bau> hardpan
harte Landung f 1. <Lufttrans> hard landing, rough landing; 2. <Raumfahrt> hard landing
harte Röntgenstrahlen mpl <Phys, Strahlphys> hard X-rays
harte Straßendecke f <Bau> rigid pavement
Härte f <Kunststoff, Maschinen, Mechan, Metall, Papier> hardness
Härtebad n 1. <Foto> hardener; 2. <Metall> hardening bath, quenching bath
Härtegeschwindigkeit f <Kunststoff> cure rate
Härtegrad m 1. <Ker & Glas> temper; 2. <Maschinen> grade, hardness; 3. <Metall> temper
Harteisen n <Metall> hard iron
Härtekomponente f <Kunststoff> hardener
Härtemesser m <Kunststoff, Labor> durometer
härten v 1. <Anstrich> temper (Legierungen); 2. <Bau> condition; 3. <Chemie> hydrogenate (Fett, Öl); 4. <Kunststoff> cure; 5. <Lebensmittel> hydrogenate (Fett, Öl); 6. <Maschinen, Metall> harden
Härten n 1. <Kohlen, Kunststoff, Maschinen, Metall> hardening; 2. <Textil> baking
härtendes Säurebad n <Foto> acid-hardening bath
Härtenorm f <Fertig> hardness standard
Härteofen m 1. <Fertig> quench furnace; 2. <Metall> hardening furnace
Härteöl n <Chemie> tallow oil
Härteprüfer m 1. <Labor> durometer, hardness tester; 2. <Maschinen, Papier> hardness tester
Härteprüfgerät n <Kunststoff> durometer
Härter m 1. <Ker & Glas> hardening (auf der Glasur); 2. <Kunststoff> hardener
harter Begrenzer m <Elektronik, Funktech> hard limiter
harter Griff m <Textil> hard handle
harter Kunststoff m <Kunststoff> rigid plastic
Härterei f <Metall> hardening shop
Härteriss m <Thermod> heat treatment crack
hartes Bromsilberpapier n <Foto> hard bromide paper
Härteskala f <Mechan> hardness scale
Härtetest m <Mechan, Phys> hardness test
Härtetester m <Metrol> hardness tester
Härtezeit f <Kunststoff> curing time, setting time
Hartfaserplatte f <Bau> hardboard
Hartfett n <Lebensmittel> hydrogenated fat
hartformatiert adj <Comp & DV> hard-sectored
hartformatierte Platte f <Comp & DV> hard-sectored disk
Hartformatierung f <Comp & DV> hard-sectoring
Hartgaslastschalter m <Elektrotech> hard-gas evolving switch
Hartgasschalter m <Elektrotech> auto-blast interrupter switch
hartgelötet adj 1. <Fertig> brazed, hard-soldered; 2. <Thermod> brazed
Hartgesteinbohrmeißel m <Erdöl> hard formation bit (Bohrtechnik)
hartgezogen adj <Thermod> cold-drawn

Hartglas n 1. <Bau> toughened glass; 2. <Ker & Glas> fused silica, hard glass, tempered glass; 3. <Metall> silica glass
Hartgummi n 1. <Elektriz> vulcanite; 2. <Kunststoff> ebonite, vulcanite
Hartgusseisen n <Metall> hard cast iron
Hartgussstrahlmittel n <Metall> chill cast shot
Hartholz n <Bau> hardwood
Hartimpuls m <Elektronik> hard pulse
Hartkautschuk m <Kunststoff> ebonite, vulcanite
Hartlandung f 1. <Lufttrans> hard landing, rough landing; 2. <Raumfahrt> hard landing
Hartlegierung f <Fertig> hard alloy
Hartlegierungsauflage f <Fertig> alloy facing • **mit Hartlegierungsauflage** <Fertig> alloy-faced
Hartley-Oszillator m <Elektronik> Hartley oscillator
Härtling m <Fertig> hard head
Hartlot n <Maschinen> brazing solder
Hartlötbrenner m <Bau> brazing blowpipe
hartlöten v 1. <Fertig> braze, hard-solder; 2. <Maschinen, Mechan, Thermod> braze
Hartlöten n 1. <Elektriz, Fertig> hard-soldering; 2. <Thermod> brazing
Hartlötflussmittel n <Bau> brazing flux
Hartlötung f 1. <Fertig> hard solder; brazing (Lötmittel); braze (Verfahren); 2. <Maschinen> brazing; 3. <Metall> hard solder, hard-brazing solder (Lötmittel); 4. <Sicherheit> brazing (Verfahren)
hartmagnetischer Werkstoff m <Mechan, Phys> hard magnetic material
Hartmessinglot n <Metall> hard brass solder
Hartmetall n 1. <Fertig> carbide, cemented carbide; 2. <Maschinen> tungsten carbide; 3. <Metall> cemented carbide
Hartmetallauflage f <Mechan> hard-facing
hartmetallbestückt adj <Maschinen> carbide-tipped
hartmetallbestückter Drehstahl m <Maschinen> turning tool with carbide tip
hartmetallbestückter Meißel m <Maschinen> carbide-tipped tool
Hartmetallgesenk n <Fertig> carbide die
Hartmetallmeißel m <Fertig, Maschinen> carbide tool
Hartmetallscheibe f <Fertig> carbide tip (Bohrer)
Hartmetallschneidplättchen n <Maschinen> carbide tip
Hartmetallspitze f <Maschinen> tungsten carbide tip
Hartmetall-Wendeschneidplatte f <Maschinen> indexable hard metal insert
Hartmetallwerkzeug n <Maschinen> tungsten carbide tool
Hartpappe f <Verpack> hardboard, millboard
Hartperlit m <Metall> troostite
Hartporzellan n <Ker & Glas> hard porcelain
Hart-PVC n <Verpack> rigid PVC
Hartroheisen n <Metall> hard pig iron
Hartrudersignal n <Elektronik> hard-over signal (Luftfahrt)
Hartschmiedeeisen n <Metall> hard iron
Hartschweißen n <Bau> hard-facing
Hartsektorieren n <Comp & DV> hard-sectoring
hartsektoriert adj <Comp & DV> hard-sectored
hartsektorierte Platte f <Comp & DV> hard-sectored disk
Hartstrahlung f <Strahlphys> hard radiation
Hartumpolung f <Telekom> line reversal
Härtung f 1. <Kunststoff> curing; 2. <Lebensmittel> hydrogenation (Fett, Öl); 3. <Metall> hardening
Härtung f durch Diffusion <Kerntech> diffusion hardening (des Neutronenspektrums)
Hartverchromung f 1. <Fertig> industrial chromium plating; 2. <Maschinen> hard chromium plating

Hartweizen m <Lebensmittel> durum wheat
Hartzinn n <Fertig> pewter
Harz n 1. <Fertig> gum; 2. <Kunststoff, Textil> resin
Harz n **für Formmaskenverfahren** <Kunststoff> (AE) shell-molding resin, (BE) shell-moulding resin
Harz n **im A-Zustand** <Kunststoff> A-stage resin, resol
Harz n **im B-Zustand** <Kunststoff> B-stage resin
Harz n **im C-Zustand** <Kunststoff> C-stage resin, resite
Harz n **im Resolzustand** <Kunststoff> A-stage resin, resol
harzartig adj <Fertig> gummy
harzfrei adj <Fertig> nonresinous
harzgebundenes Sperrholz n <Kunststoff> resin-bonded plywood
Harzmittel n <Fertig> binder
Harzrückstand m <Fertig> gum (Öl)
Harzträger m <Fertig> filler
Hash m <Comp & DV> hash
Hash-Code m <Comp & DV> hash code
Hash-Funktion f <Comp & DV> hash function
Hashing n <Comp & DV> hashing (Abbildung mit Hilfe der Streuspeichertechnik)
Hash-Tabelle f <Comp & DV> hash table
Hash-Verfahren n <Comp & DV> hashing (Abbildung mit Hilfe der Streuspeichertechnik)
Hash-Zeichen n <Comp & DV> hashmark
Haspe f <Bau> knuckle (Fenster, Tür); hasp, staple (Schloss)
Haspel f 1. <Aufnahme> reel; 2. <Bau> winder; 3. <Fertig> reel, roll, spool; 4. <Ker & Glas> winder; 5. <Papier> reel; 6. <Textil> winch
haspeln v <Textil> wind
Haspeln n <Bau> winding
Haspeltrommel f <Textil> swift
Haspen m <Fertig> catch
Hatchettin m 1. <Kerntech> hatchettite; 2. <Phys> adipocerite (Mineralogie)
Haube f 1. <Bau> coping; 2. <Comp & DV> hood; 3. <Elektrotech> cap; 4. <Fertig> bell (Wärmebehandlung); 5. <Labor> head (eines Destillationsapparats); 6. <Maschinen> (BE) bonnet, (AE) hood, shroud; 7. <Mechan> cap; 8. <Papier> hood; 9. <Sicherheit> cover
Haubenschloss n <Kfztech> (BE) bonnet catch, (AE) hood catch
Haubenverkleidung f <Mechan> cowl
Haueisen n <Bau> mattock
hauen v <Kohlen> hew
Hauen n <Maschinen> cutting
Haufen m <Bau> heap
Haufenwolke f <Lufttrans, Wassertrans> cumulus
Häufigkeit f 1. <Math> frequency (Statistik); 2. <Phys> abundance
Häufigkeit f **der Elemente** <Strahlphys> nuclear abundance
Häufigkeitsdichte f 1. <Elektronik> frequency density; 2. <Math> frequency density (Statistik)
Häufigkeitskoeffizient m 1. <Math> frequency factor (Statistik); 2. <Phys, Qual> coefficient of abundance
Häufigkeitsverteilung f 1. <Math> frequency distribution (Statistik); 2. <Telekom> frequency distribution
haufwerksporiger Beton m <Bau> no-fines concrete
Haupt n <Maschinen> head
Haupt... <Comp & DV, Funktech, Telekom> master
Hauptabrechnungsstelle f <Telekom> major account holder
Hauptachse f 1. <Geom> major axis; 2. <Maschinen> main axle; 3. <Optik> principal axis (eines gewölbten Spiegels oder einer Linse); 4. <Phys> principal axis (eines Festkörperkristalls)

Hauptachsentransformation f <Math> major axis transformation (Kegelschnitte)
Hauptalarm m <Telekom> major alarm
Hauptamt n <Telekom> central exchange, host exchange, main exchange
Hauptanode f <Elektrotech> main anode
Hauptanschlussbereich m **einer Fernvermittlungsstelle** <Telekom> (AE) main trunk-switching center area, (BE) main trunk-switching centre area
Hauptanschlussbereich m **eines Fernamtes** <Telekom> main trunk exchange area
Hauptanschlussklemme f <Kfztech> main terminal
Hauptanschlussleitung f <Telekom> main line
Hauptansicht f <Konstzeich> principal view
Hauptantrieb m <Maschinen> main drive, master drive
Hauptantriebswelle f 1. <Kfztech> mainshaft (Getriebe); 2. <Lufttrans> main drive shaft (Hubschrauber)
Hauptantriebszahnrad n <Kfztech> main drive gear
Hauptanwendung f <Comp & DV> core application, prime application
Hauptanzapfstelle f <Elektriz> principal tapping
Hauptauftragnehmer m <Bau> general contractor
Hauptausfall m <Lufttrans> basic failure (einer technischen Anlage)
Hauptbalken m <Bau> main beam
Hauptband n <Comp & DV> master tape
Hauptbatterie f <Lufttrans> main battery (eines Luftfahrzeugs)
Hauptbehälterdruck m <Eisenbahn> main air reservoir pressure (Bremse)
Hauptbeschichtung f <Fertig> secondary coating (Faseroptik)
Hauptbewehrung f <Bau> main reinforcement
Hauptblatt n <Kfztech> main leaf, top leaf (der Blattfeder)
Hauptbremsleitung f <Eisenbahn> main brake pipe
Hauptbremsschlauch m <Eisenbahn> main brake hose
Hauptbremszylinder m <Kfztech> brake master cylinder, master cylinder
Hauptbrenner m <Heiz & Kälte> main burner
Hauptdampf m <Kerntech> mainsteam
Hauptdampfleitung f <Hydraul> mainsteam pipe
Hauptdatei f <Comp & DV> master file
Hauptdatenstation f <Comp & DV> master terminal
Hauptdeck n <Wassertrans> main deck
Hauptdomäne f <Telekom> top level domain (Internet)
Hauptdüse f <Kfztech> main jet (Vergaser)
Hauptebene f 1. <Foto> nodal plane, principal plane; 2. <Optik> principal plane
Haupteinflugzeichen n <Lufttrans> middle marker (ILS)
Haupteintrag m <Comp & DV> primary entry
Hauptentladungsstrecke f <Elektrotech> main gap
Hauptentwässerungsleitung f <Bau> main sewer
Haupterzeugnis n <Lebensmittel> staple
Hauptfahrwerk n <Lufttrans> main-landing gear (Hubschrauber)
Hauptfahrwerkachsenträger m <Lufttrans> main gear axle beam
Hauptfahrwerkschiebeklappe f <Lufttrans> main gear-sliding door
Hauptfahrwerksklappe f <Lufttrans> main landing gear door
Hauptfahrwerksstützstrebe f <Lufttrans> main landing gear brace strut
Hauptfaktor m <Textil> chief factor
Hauptfeder f <Maschinen> mainspring
Hauptfehler m <Qual> major defect, major failure
Hauptförderstrecke f <Kohlen> gangway
Hauptfreifläche f <Fertig> major flank
Hauptgasleitung f <Thermod> gas main

Hauptgebälk

Hauptgebälk n <Bau> principal
Hauptgleis n <Eisenbahn> main track
Haupthahn m <Maschinen> main tap
Hauptintensität f <Optik> principal maxima
Hauptkabel n <Elektrotech> main cable
Hauptkarte f <Comp & DV> master card
Hauptknotenpunkt m <Eisenbahn> *(AE)* major railroad junction, *(BE)* major railway junction
Hauptkonsole f <Comp & DV> master console
Hauptkontakte mpl <Elektriz> main contacts
Hauptkrümmung f <Ker & Glas> principal curvature
Hauptlager n <Bau, Kfztech, Maschinen, Wassertrans> main bearing
Hauptlagerbuchse f <Kfztech> main-bearing bushing
Hauptlagerhülse f <Kfztech> main-bearing bushing
Hauptlandevorrichtung f <Lufttrans> main-landing gear *(Hubschrauber)*
Hauptlast f <Bau> main load
Hauptleitung f 1. <Bau> main line; main *(Wasser, Strom)*; 2. <Comp & DV> ethyne; 3. <Elektrotech> bus, main; 4. <Heiz & Kälte> main; 5. <Wasserversorg> head pipe
Hauptluftbehälter m <Eisenbahn> main air reservoir *(Bremse)*
Hauptmaske f <Elektronik> master mask
Hauptmaßstab m <Konstzeich> principal scale
Hauptmast m <Wassertrans> mainmast *(Segeln)*
Hauptmaximum n <Phys> principal maxima
Hauptmenge f <Math, Phys> bulk
Hauptmischregler m <Fernseh> master control fader
Hauptmodus m <Raumfahrt> fundamental mode *(Raumschiff)*
Hauptmonitor m <Fernseh> master monitor
Hauptmotor m <Lufttrans> master engine
Hauptnenner m <Math> common denominator
Hauptnetzleitung f <Elektriz> trunk main
Hauptnormal n <Qual> master standard, primary standard
Hauptplatine f 1. <Comp & DV> motherboard, wraparound; 2. <Elektronik> motherboard
Hauptplatte f <Comp & DV> master disk
Hauptpleuel n <Maschinen> main rod
Hauptpleuelstange f <Maschinen> main rod
Hauptpol m <Elektriz, Elektrotech> main pole, field pole
Hauptpresse f <Papier> main press
Hauptprogramm n <Comp & DV> main program, main routine, master program, MP; background program *(bei interruptfähigem System)*
Hauptprozessor m <Comp & DV, Telekom> main processor, master processor, regional processor
Hauptpunkte mpl 1. <Optik> cardinal points; 2. <Phys> principal points
Hauptquantenzahl f 1. <Lufttrans> main quantum number; 2. <Phys> principal quantum number
Hauptquelle f <Umweltschmutz> major source
Hauptrechner m <Comp & DV> host computer, master computer
Hauptregelventil n <Kfztech> main regulator valve
Hauptregieraum m <Fernseh> central control room
Hauptregler m <Fernseh> master control
Hauptrichtantenne f <Raumfahrt> main beam *(Weltraumfunk)*
Hauptrippe f <Lufttrans> main rib
Hauptriss m <Metall> main crack
Hauptrohrleitung f <Umweltschmutz> pipeline
Hauptrotor m <Lufttrans> main rotor *(Hubschrauber)*
Hauptrotorblatt n <Lufttrans> main rotor blade
Hauptrotorkopf m <Lufttrans> main rotor head
Hauptrotornabe f <Lufttrans> main rotor hub
Hauptrotorwelle f <Lufttrans> main rotor shaft

Hauptroutine f <Comp & DV> main routine, master routine
Hauptrücksetzsignal n <Kontroll> master reset signal
Hauptsammelkanal m 1. <Abfall> interceptor sewer; 2. <Bau> main sewer
Hauptsammelschiene f <Elektrotech, Telekom> main bar, main busbar
Hauptsammler m 1. <Abfall> interceptor sewer, main collector, main sewer; 2. <Bau> main drain *(Abwasser)*
Hauptsatz m <Comp & DV> master record
Hauptsatz m **der Thermodynamik** <Thermod> law of thermodynamics
Hauptschalter m 1. <Elektriz, Elektrotech> main switch, master switch; 2. <Fernseh> master switch
Hauptschaltgetriebe n <Lufttrans> main gearbox *(Hubschrauber)*
Hauptschaltgetriebeaufnahme f <Lufttrans> main gearbox support
Hauptschaltgetriebegehäusearm m <Lufttrans> main gearbox support
Hauptschaltgetriebehalterung f <Lufttrans> main gearbox support
Hauptschaltkontakte mpl <Elektriz> main-switching contacts
Hauptschenkel m <Kerntech> main leg *(eines Transformatorkernes)*
Hauptschiene f <Elektrotech> main bar
Hauptschifffahrtsroute f <Wassertrans> main-trading route *(Seehandel)*
Hauptschlüssel m <Bau> master key
Hauptschlussmaschine f <Elektriz> series-excited machine
Hauptschlussmotor m <Elektriz> series motor
Hauptschneide f 1. <Fertig> active-cutting edge, lip, major-cutting edge, working cutting edge; 2. <Maschinen> major-cutting edge
Hauptschnittdruck m <Fertig> main tool thrust
Hauptschnittfläche f <Fertig> work surface
Hauptschnittkraft f <Fertig> main-cutting force
Hauptsignal n <Eisenbahn> home signal
Hauptsignalsteuergerät n <Eisenbahn> master controller *(Verkehrsregelung)*
Hauptskalenteilung f <Gerät> major graduation
Hauptsolargenerator m <Raumfahrt> main solar generator *(Raumschiff)*
Hauptspant n <Wassertrans> midship frame, midship section *(Schiffbau)*
Hauptsparren m <Bau> principal rafter
Hauptspeicher m <Comp & DV> central memory, main memory, main store, primary memory, primary storage, primary store
Hauptspeicherabbild n <Comp & DV> memory map
Hauptspeicherauszug m <Comp & DV> memory dump
Hauptspeichererweiterung f <Comp & DV> memory expansion
Hauptspeicherkapazität f <Comp & DV> memory capacity
Hauptspeicherstruktur f <Comp & DV> memory map
Hauptspeicherzugriff m <Comp & DV> memory access
Hauptspindel f <Maschinen> main spindle
Hauptspindelgetriebe n <Lufttrans> headgear
Hauptstart- und Landebahn f <Lufttrans> main runway, primary runway
Hauptstation f <Comp & DV> master station
Hauptstellenimpuls m <Elektronik> master pulse
Hauptsteuerpult n <Fernseh> master control panel
Hauptstrahl m 1. <Funktech> main beam *(Antenne)*; 2. <Optik, Phys> principal ray; 3. <Raumfahrt> main beam *(Weltraumfunk)*

Hauptstrecke f 1. <Eisenbahn> (AE) arterial railroad, (BE) arterial railway, main line, (AE) main-line railroad, (BE) main-line railway; 2. <Elektrotech> main gap
Hauptstrom m <Elektriz, Elektrotech> main current
hauptstromgeregeltes Netzgerät n <Elektrotech> series-regulated power supply
Hauptstromkreis m <Elektriz> main circuit
Hauptstromölfilter n <Kfztech> full-flow oil filter (Schmierung)
Hauptstromregelung f <Elektrotech> series current control
Hauptstromregler m <Elektrotech> series controller
Haupttakt m <Telekom> master clock
Haupttaktgeber m <Comp & DV> master clock
Hauptteil m **der IP-Mitteilung** <Telekom> body of the IP-Message
Hauptteilstrich m <Gerät> major graduation
Hauptterminal n <Comp & DV> central terminal
Hauptträger m <Bau> main girder
Hauptträgheitsachse f <Maschinen> main axis of inertia
Haupttrennventil n <Kerntech> main-isolating valve
Haupttriebwerk n <Lufttrans> master engine
Haupttyp m <Elektrotech> dominant mode (Wellenleiter)
Hauptuhr f <Comp & DV> master clock
Haupt- und Nebenkanäle mpl <Heiz & Kälte> main and branch ductwork
Hauptverbindung f 1. <Comp & DV> busbar; 2. <Eisenbahn> main line
Hauptverbindungsstraße f <Trans> main road
Hauptverkehr m <Eisenbahn> peak hour traffic
Hauptverkehrsader f <Bau, Trans> (BE) thoroughfare, (AE) thruway
Hauptverkehrsstraße f <Trans> (AE) arterial highway, (BE) arterial motorway, arterial road, main road
Hauptverkehrsstunde f (HVStd) <Telekom> busy hour
Hauptverkehrszeit f 1. <Eisenbahn> peak hours; 2. <Telekom> busy period; 3. <Trans> peak hour
Hauptvermittlungsstelle f <Telekom> regional switch
Hauptvermittlungsstelle f **im Fernwahlnetz** (HVSt) <Telekom> central exchange, central office (CO); main exchange
Hauptverstärker m (PA) <Phys> power amplifier (PA)
Hauptverstärkerröhre f <Elektronik> power amplifier tube
Hauptverstärkertransistor m <Elektronik> power amplifier transistor
Hauptverstärkungsregler m <Aufnahme, Elektronik> master gain control
Hauptverteiler m (HVt) <Telekom> main distribution frame (MDF)
Hauptverteiler m **für Verstärkerämter** <Telekom> main repeater distribution frame
Hauptverteilung f <Telekom> (AE) distribution center, (BE) distribution centre (Starkstromtechnik)
Hauptverteilungsnetz n <Nichtfoss Energ> main grid network
Hauptvorschubbewegung f <Maschinen> main feed motion
Hauptwärmeaustauscher m <Lufttrans> primary heat exchanger
Hauptwarteschlange f <Telekom> master queue
Hauptwasserleitung f <Wasserversorg> delivery main, (BE) mains, (AE) supply network, water main
Hauptwasserrohr n <Wasserversorg> water main
Hauptwelle f <Kfztech, Lufttrans, Maschinen, Mechan> mainshaft
Hauptwellenlager n <Wassertrans> mainshaft bearing (Schiffbau)
Hauptwert m <Math> main value
Hauptzeichnung f <Konstzeich> general arrangement drawing
Hauptzeit f **beim Schleifen** <Fertig> actual-grinding time
Hauptzylinder m <Kfztech, Lufttrans> master cylinder (Bremse)
Haus n <Bau> house
Hausanschlusskabel n <Fernseh> drop cable (Kabelfernsehen)
Hausanschlusskasten m <Bau> branch box (Elektroversorgung)
Hausbau m <Bau> house building
Hausbrandkohle f <Kohlen> domestic coal
hauseigenes Netz n <Telekom> in-house network
Hausenblase f <Lebensmittel> isinglass (Klärmittel)
Haushaltabfall m <Abfall, Wasserversorg> domestic refuse, domestic waste
Haushaltabwasser n <Wasserversorg> domestic sewage, domestic waste water
Haushaltabwässer npl <Abfall> household wastewater
Haushaltboiler m <Heiz & Kälte, Maschinen> domestic boiler
Haushaltbrennstoff m <Kohlen> household fuel
Haushaltelektronik f <Elektrotech> domestic electronic equipment
Haushaltgasgerät n <Heiz & Kälte> domestic gas appliance
Haushaltkeramik f <Ker & Glas> crockery ware
Haushaltkühlschrank m <Maschinen> domestic refrigerator, household refrigerator
Haushaltnetzinstallation f <Elektrotech> domestic electric installation
Haushaltporzellan n <Ker & Glas> household porcelain
Haushaltroboter m <Künstl Int> domestic robot
Haushalttextilien fpl <Textil> household textiles
Haushaltung f <Raumfahrt> housekeeping
Haushaltverbraucher m <Elektriz> domestic consumer
Haushaltwasser n <Wasserversorg> domestic water
Haushaltwasserversorgung f <Wasserversorg> domestic water supply
Hausinstallation f <Elektriz> indoor installation
Hausinstallationskabel n <Elektriz> indoor cable
Hausinstallationsschalter m <Elektriz> house-wiring switch
Hausisolierung f <Elektriz> indoor insulation
Hausleiterkabel n <Elektriz> indoor cable
Hausmüll m 1. <Abfall> consumer waste, domestic waste; 2. <Umweltschmutz> consumer waste, domestic waste, municipal waste; 3. <Wasserversorg> municipal waste
Hausmülldeponie f <Abfall> municipal waste landfill
Hausmüllzusammensetzung f <Abfall> waste composition
Hausnebenstelle f **mit Wählbetrieb** <Telekom> private automatic exchange
Haussprechanlage f <Aufnahme> intercom
Hausvermittlungssystem n <Telekom> house exchange system
Hausvorschriften fpl <Druck> house style
Hauswirtschaftslehre f <Lebensmittel> home economics
Haut f <Kohlen, Kunststoff, Lebensmittel> skin
Hautblase f <Ker & Glas> skin blister
Hauteffekt m <Elektriz, Elektrotech, Phys, Werkprüf> skin effect
Hauttiefe f <Phys> skin depth
Häutungshormon n <Chemie> ecdyson
Hautverhütungsmittel n <Kunststoff> antiskinning agent
Hautwiderstand m <Kohlen> skin resistance
Hautwiderstandsänderung f <Ergon> galvanic skin response
Havarie f 1. <Kerntech> average; 2. <Fernseh, Funktech, Telekom> breakdown (System); 3. <Wassertrans> average; damage by sea (Schiff)

havariebedingter

havariebedingter Ausfluss m <Umweltschmutz> accidental discharge *(Öl, Chemikalien)*
Havarieschutz m <Kerntech> RPS, reactor protection system
H-Bahn f *(hochgeständerte Einschienenhängebahn)* <Eisenbahn> cabin taxi, overhead monorail
H-Band n <Bau> H hinge
H-Bogen m <Elektrotech> H-plane bend
HD 1. <Comp & DV> *(Festplatte)* HD *(hard disk)*; 2. <Comp & DV> *(halbduplex)* HDX *(half-duplex)*; 3. <Maschinen> *(Hochleistung)* HD *(heavy duty)*
HDLC-Prozedur f <Comp & DV, Telekom> HDLC, high-level data link control
HDLC-Verfahren n <Comp & DV> HDLC, high-level data link control
HD-Öl n *(Heavy-Duty-Öl, Hochleistungsöl)* <Maschinen> HD oil *(heavy-duty oil)*
HDTV *(hochauflösendes Fernsehen, hochzeiliges Fernsehverfahren)* <Fernseh> HDTV *(high-definition television)*
He *(Helium)* <Chemie> He *(helium)*
Head-Up-Display n <Raumfahrt> head-up display *(Frontscheibensichtanzeige, Blickfelddarstellung)*
Hebb-Regel f <Künstl Int> Hebbian rule
Hebb'sche Lernregel f <Künstl Int> Hebbian rule
Hebdrehwählersystem n <Telekom> Strowger system
Hebearm m <Lufttrans> hoist arm *(Hubschrauber)*
Hebeauge n <Lufttrans> hoisting eye
Hebeausleger m <Lufttrans> hoist boom
Hebebaum m <Mechan> handspike
Hebebock m <Maschinen> jack
Hebebremsklotz m <Lufttrans> hoisting block
Hebebrücke f <Bau> lift bridge
Hebebühne f 1. <Eisenbahn, Lufttrans> lifting platform; 2. <Maschinen> elevating platform; 3. <Mechan> hydraulic jack; 4. <Wassertrans> lifting platform
Hebedaumen m 1. <Fertig> wiper; 2. <Maschinen> lifter
Hebegerät n 1. <Maschinen> lifting apparatus, lifting device, lifting equipment; 2. <Sicherheit> lifting appliance
Hebehaken m <Maschinen> lifting hook
Hebekette f <Sicherheit> lifting chain
Hebekran m <Maschinen> hoisting crane
Hebel m <Fertig, Maschinen> lever
Hebelarm m 1. <Bau> lever arm; 2. <Maschinen> lever arm, moment arm
Hebelausschalter m <Elektrotech> single throw switch
Hebelbremse f <Maschinen> lever brake
Hebeldrehpunkt m <Fertig> *(AE)* leveling fulcrum, *(BE)* levelling fulcrum
Hebelgestänge n <Maschinen> leverage
Hebelgriff m <Bau> lever handle
Hebelhammer m <Fertig> tilt hammer
Hebelkraft f <Ergon, Fertig, Maschinen> leverage
hebeln v <Maschinen> lever
Hebelpresse f <Maschinen> lever press
Hebelpunkt m <Phys> fulcrum
Hebelrelais n <Elektrotech> single throw relay
Hebelschalter m 1. <Elektriz> toggle switch; 2. <Elektrotech> lever switch
Hebelschere f <Maschinen> lever shears
Hebelstrecker m <Wassertrans> forestay release lever
Hebelsystem n <Maschinen> lever system
Hebelübersetzung f <Maschinen> leverage
Hebelventil n <Hydraul> lever valve
Hebelvorschub m <Maschinen> lever feed
Hebelwechslerschalter m <Elektriz> double-throw switch
Hebelwerk n <Maschinen> compound lever
Hebelwirkung f <Ergon> leverage

Hebemagnet m <Elektriz> lifting magnet
Hebemaschine f <Maschinen> hoisting machine
heben v 1. <Maschinen> hoist, jack up, lift; 2. <Wassertrans> heave in
heben v **und senken** v/sich <Wassertrans> heave
Heben n 1. <Bau> raising; 2. <Maschinen> hoisting, lifting; 3. <Sicherheit> lifting
hebendes Kissen n <Trans> height-on cushion
H-Ebene f <Elektrotech, Funktech> H-plane
Hebeponton m <Wassertrans> camel
Hebepumpe f <Lufttrans> hoist pump
Heber m 1. <Bau> lifter; 2. <Labor> siphon, syphon; 3. <Maschinen> jack; 4. <Phys> siphon, syphon
hebern v 1. <Labor> siphon, syphon; 2. <Nichtfoss Energ> syphon; 3. <Phys> siphon, syphon
Hebeschaft m <Maschinen> lifting shaft
Hebeseilschlaufe f <Lufttrans> hoisting sling
Hebeseiltrenner m <Lufttrans> hoist cable cutter *(Hubschrauber)*
Hebesystem n <Fertig> hoisting system
Hebetisch m <Mechan> elevating table
Hebevorrichtung f 1. <Eisenbahn> lifting gear, lifting tackle; 2. <Labor> jack; 3. <Lufttrans> lifting gear, lifting tackle; 4. <Maschinen> lifting apparatus; 5. <Wassertrans> lifting gear, lifting tackle
Hebewerk n 1. <Bau> lift; 2. <Erdöl> draw works *(Bohrtechnik)*; 3. <Maschinen> elevator
Hebewinde f <Bau> windlass
Hebezeug n 1. <Bau> *(AE)* elevator, hoist, *(BE)* lift, lifting table, gin, tackle; 2. <Elektriz, Fertig> *(AE)* elevator, *(BE)* lift, gin, lifting table, tackle; 3. <Lufttrans> gin, lift, lifting table, tackle; 4. <Maschinen> *(AE)* elevator, *(BE)* lift, gin, lifting table, tackle, purchase; 5. <Mechan> gin, *(BE)* lift, tackle, lifting table; 6. <Wassertrans> *(AE)* elevator, *(BE)* lift, gin, lifting table, tackle
hecheln v <Textil> hackle
Hecheln n <Textil> hackling
Heck n 1. <Kfztech> rear end, tail *(Karosserie)*; 2. <Raumfahrt> afterbody *(Raumschiff)*; 3. <Wassertrans> stern *(Schiffbau)*
Heckaufreißer m <Trans> rear-mounted ripper
Heckfenster n <Kfztech> rear window *(Karosserie)*
Heckkanzel f <Wassertrans> stern pulpit *(Decksausrüstung)*
Heckklappe f 1. <Kfztech, Mechan, Trans> tailgate; 2. <Wassertrans> stern door *(Ro-Ro-Schiff)*
Heckkorb m <Wassertrans> pushpit *(Deckausrüstung)*
Hecklaterne f <Wassertrans> stern light *(Navigation)*
Heckleine f <Wassertrans> stern line *(Festmachen)*
Heckleuchte f <Kfztech> rear lamp, tail light, tail lamp *(Beleuchtung)*
Heckluftschraube f <Lufttrans> tail propeller
Heckmotor m <Kfztech> rear engine, rear-mounted engine
Heckreling f <Wassertrans> taffrail *(Schiffbau)*
Heckrotor m <Lufttrans> tail rotor *(Hubschrauber)*
Heckscheibe f <Kfztech> back window, rear window *(Karosserie)*
Heckschraube f <Lufttrans> tail rotor *(Hubschrauber)*
Heckspiegel m <Wassertrans> transom *(Schiffbau)*
Hecksporn m <Lufttrans> tail skid
Heckstoßwelle f <Lufttrans> tail shock wave
Heckstrahlpropeller m <Wassertrans> stern thruster *(Schiffantrieb)*
Hecktür f 1. <Kfztech> hatchback, tailgate; 2. <Trans> tailgate
Hecküberhang m <Wassertrans> aft rake *(Schiffkonstruktion)*; fantail *(Schiffbau)*
Hede f <Textil> tow

hedonischer Maßstab *m* <Lebensmittel> hedonic scale
Heft *n* <Fertig> handle, hilt
Heftahle *f* <Maschinen> elevating machinery, hoist, hoisting gear, jack, lift
Heftapparat *m* 1. <Fertig> stitcher; 2. <Verpack> stapling equipment
Heftbolzen *m* <Mechan> locating pin
Heftdraht *m* <Verpack> stapling wire, stitching wire
Hefteisen *n* <Ker & Glas> punty, sticking up iron • **mit Hefteisen aufnehmen** <Ker & Glas> put on the punty
heften *v* <Bau> tack
Heften *n* 1. <Bau> stitching; 2. <Druck> sewing
Heftfaden *m* <Druck> binding thread
Heftfalte *f* <Konstzeich> filling fold
heftiges Sieden *n* <Kerntech> violent boiling
Heftklammer *f* <Verpack> staple
Heftmaschine *f* <Druck> sewing machine
Heftniet *m* <Fertig> dummy rivet
Heftrand *m* <Konstzeich> binding margin; filling margin *(einer Zeichnung)*
heilen *v* <Anstrich> cure
Heimarbeit *f* **am Computer** <Comp & DV> telecommuting
Heimarbeiter *m* <Comp & DV> telecommuter
Heimatdatei *f* <Mobilkom> home location register
Heimat-Funkvermittlungsstelle *f* <Mobilkom> home exchange
Heimathafen *m* <Wassertrans> home port, port of documentation, port of registration, port of registry
Heimcomputer *m* <Comp & DV> home computer
Heimempfang *m* <Funktech> individual reception *(Rundfunksatellit)*
heimkehrend *adj* <Wassertrans> inward-bound *(Schiff)*
Heimroboter *m* <Künstl Int> domestic robot
Heimtextilien *npl* <Textil> home textiles
Heimvideo-Aufzeichnungssystem *n* (VHS-C) <Fernseh> video home system-compact *(VHS-C)*
Heimvideosystem *n* (VHS) <Fernseh> video home system *(VHS)*
Heimvideosystem *n* **für Videokamera** <Fernseh> video home system for camcorder *(VHS-C)*
Heimvideosystem *n* **mit höherer Bildqualität** <Fernseh> super video home system *(S-VHS)*
Heisenberg'sche Unbestimmtheitsrelation *f* <Teilphys> Heisenberg uncertainty principle
heiß *adj* <Heiz & Kälte, Kerntech, Phys, Thermod> hot
Heiß... <Heiz & Kälte, Kerntech, Maschinen, Phys, Thermod, Verpack> hot
Heißabfüllung *f* <Verpack> hot-filling
Heißaufladung *f* <Kerntech> *(AE)* hot-refueling, *(BE)* hot-refuelling
Heißauge *n* <Wassertrans> lifting eye *(Schiffbau, Deckbeschläge)*
Heißbeschickung *f* <Kerntech> *(AE)* hot-refueling, *(BE)* hot-refuelling
Heißdampf *m* <Heiz & Kälte, Maschinen, Phys, Thermod> superheated steam
Heißdampfkühler *m* <Heiz & Kälte> desuperheater
heiße Chemie *f* <Kerntech> hot chemistry
heiße Sohle *f* <Nichtfoss Energ> liquid brine
heiße Zelle *f* <Kerntech> hot cell
Heißendvergütung *f* <Ker & Glas> hot end coating
heißes brikettiertes Eisen *n* (HBI) <Metall> hot briquetted iron, HBI
heißes Eisenbrikettieren *n* <Metall> hot iron briquetting
heißes Labor *n* <Kerntech> hot laboratory
heißes Triebwerksteil *n* <Lufttrans> hot section *(eines Triebwerkes)*
heißes Wasser *n* <Thermod> hot strength, hot water

heißfixierbar *adj* <Thermod> heat-setting
Heißfolien-Kartoncodierer *m* <Verpack> hot foil carton coder
Heißform *f* <Ker & Glas> *(AE)* hot mold, *(BE)* hot mould
Heißgas-Entlastungsventil *n* <Heiz & Kälte> hot-gas by-pass valve
heißgelaufenes Lager *n* <Fertig> overheating bearing
Heißgemisch *n* <Bau> hot mix *(Bitumen)*
heißgesiegelt *adj* <Thermod> heat-sealed
Heißglasdraht *m* <Ker & Glas> hot glass wire
Heißglasdrahtschneiden *n* <Ker & Glas> hot glass wire cutting
Heißkalandrieren *n* <Papier> hot calendering
Heißkanalfaktor *m* <Kerntech> hot channel factor
Heißkanalverfahren *n* <Fertig> hot runner moulding *(Kunststoffe)*
Heißkanalwerkzeug *n* <Kunststoff> *(AE)* hot runner mold, *(BE)* hot runner mould
Heißkaschieren *n* <Verpack> heat lamination
Heißklebeband *n* <Verpack> heat-fix tape
Heißklebeetikett *n* <Verpack> heat seal label, heat-activated label
Heißklebefolie *f* <Verpack> heat-sealing tape
Heißkleben *n* <Verpack> thermal sealing
Heißlabor *n* <Kerntech> hot laboratory
heißlaufen *v* 1. <Kfztech> overheat; 2. <Maschinen> run hot
Heißlaufen *n* <Kfztech> overheating
Heißläufer *m* <Eisenbahn> hot box
Heißläufersuchgerät *n* <Eisenbahn> hot-box detector
Heißleim *m* <Verpack> hot-setting glue
Heißleiter *m* <Phys, Telekom> thermistor
Heißlötstelle *f* <Elektriz> hot junction
Heißluft *f* <Kfztech, Labor, Maschinen, Sicherheit> hot-air
Heißluftgebläse *n* <Labor> hot-air blower
Heißluftheizung *f* <Sicherheit> hot-air radiation heating system
Heißluftmaschine *f* <Maschinen> caloric engine
Heißluftmotor *m* <Kfztech, Maschinen> hot-air engine
Heißluftofen *m* <Kunststoff> air oven
Heißluftschlichtmaschine *f* <Textil> hot-air sizing machine
Heißluftstrom *m* <Textil> hot-air stream
Heißlufttrockner *m* <Textil> hot-air dryer
Heißluftturbine *f* **mit geschlossenem Kreislauf** <Trans> closed cycle hot-air turbine
Heißluft- und Flammenausstoß *m* **des Ofens** <Ker & Glas> sting-out
Heißluftventil *n* <Lufttrans> hot-air valve
Heißluftventilator *m* <Heiz & Kälte> hot-air fan
Heißprägen *n* <Verpack> hot-stamping *(von Folien)*
Heißpresse *f* <Maschinen> hot press
Heißpressen *n* <Verpack> hot-pressing
Heißräucherung *f* <Lebensmittel> hot-smoking
Heißring *m* <Wassertrans> lifting ring *(Deckbeschläge)*
Heißschmelzbeschichter *m* <Papier> hot-melt coating
Heißschmelzkleber *m* <Kunststoff> hot-melt adhesive
Heißschrumpfsitz *m* <Maschinen> hot-shrink fit
Heißsiegelanlage *f* <Verpack> heat-sealing equipment
Heißsiegelbeschichtung *f* <Verpack> heat seal coating
heißsiegelfähig *adj* 1. <Fertig> hot-sealing; 2. <Papier> heat-sealing; 3. <Thermod> heat-sealable
heißsiegeln *v* 1. <Fertig> hot-seal; heat-seal *(Kunststoffe)*; 2. <Thermod> heat-seal
Heißsiegeln *n* 1. <Kunststoff, Thermod> heat-sealing; 2. <Verpack> hot-blade sealing
Heißsiegelpapier *n* <Verpack> heat-sealable paper
Heißsiegel- und Verschweißmaschine *f* <Verpack> heat-sealing and welding machine *(für Schrumpfpackungen)*

Heißsiegelverpackungen

Heißsiegelverpackungen *fpl* <Thermod> heat-sealed wrappings
Heißspülung *f* <Anstrich> hot rinse
Heißstrahltriebwerk *n* <Lufttrans> thermal jet engine
Heißverkleben *n* <Thermod> heat-sealing
heißverklebt *adj* <Thermod> heat-sealed
Heißverschweißgerät *n* <Labor> heat seal apparatus *(Polyäthylenbeutel)*
heißversiegelt *adj* <Fertig> hot-sealed
heißverstreckte Faser *f* <Textil> *(AE)* heat-stretched fiber, *(BE)* heat-stretched fibre
Heißvulkanisation *f* <Thermod> hot creep, hot-curing
Heißwäsche *f* <Anstrich> hot wash
Heißwasser *n* <Thermod> hot water
Heißwasserboiler *m* <Thermod> hot-water boiler
Heißwassererzeuger *m* <Heiz & Kälte> high-temperature water heating appliance
Heißwassergerät *n* <Thermod> hot-water boiler
Heißwasserheizungsanlage *f* <Heiz & Kälte> high-temperature water heating system
Heißwasserspeicher *m* <Heiz & Kälte> thermal storage water heater
Heißwassertrichter *m* <Labor> funnel heater
Heißwasservulkanisation *f* <Kunststoff> hot-water vulcanization
Heißwasserwaschen *n* <Meerschmutz> hot-water washing
Heiz... <Heiz & Kälte, Kfztech> heating
Heizanlage *f* <Kfztech> heating system *(Zubehör)*
Heizbalg *m* <Kunststoff> bladder
Heizband *n* <Labor> heating tape
Heizdraht *m* 1. <Elektrotech> filament; 2. <Heiz & Kälte> fire bar; 3. <Mechan> filament; 4. <Metall> resistance wire
Heizeinsatz *m* <Heiz & Kälte> heating element
Heizelement *n* 1. <Elektriz, Elektrotech> heating element, heater; 2. <Foto> element heater; 3. <Heiz & Kälte> fire bar, heater, heating element
Heizen *n* <Thermod> heating
Heizer *m* <Heiz & Kälte> heater
Heizergebläse *n* <Heiz & Kälte> heater fan
Heizfaden *m* <Elektrotech, Funktech> heater filament *(Elektronenröhre)*
Heizfadenspannung *f* <Elektrotech> heater voltage
Heizfadentemperatur *f* <Elektriz> filament temperature
Heizfadenwiderstand *m* <Elektriz> filament resistance, filament resistor
Heizfähigkeit *f* <Heiz & Kälte, Thermod> heating capacity
Heizfläche *f* 1. <Heiz & Kälte> flat radiator, heating surface; 2. <Kerntech> heating surface; 3. <Thermod> effective heating surface, heating surface
Heizflächenrohr *n* <Kerntech> heating surface tube
Heizgebläse *n* 1. <Heiz & Kälte> fan-assisted air heater; 2. <Kfztech> heater *(Zubehör)*
Heizgerät *n* 1. <Heiz & Kälte> heater, heating appliance; 2. <Maschinen> heating device; 3. <Thermod> heater
Heizgürtel *m* <Heiz & Kälte> heating belt
Heizkabel *n* <Elektriz> heating cable
Heizkammer *f* <Thermod> heating chamber
Heizkanal *m* <Verpack> heating channel
Heizkatode *f* <Elektrotech> heating cathode
Heizkessel *m* <Heiz & Kälte> heating and hot water boiler
Heizkörper *m* 1. <Elektrotech> heater; 2. <Heiz & Kälte> heater, radiator; 3. <Thermod> radiator
Heizkraft *f* 1. <Mechan> heating; 2. <Thermod> heating power
Heizkraftwerk *n* <Thermod> CHPS, combined heat and power station
Heizkreis *m* <Heiz & Kälte> heating circuit

Heizleistung *f* 1. <Heiz & Kälte> heat output, heating capacity; 2. <Phys, Thermod> calorific output, calorific power
Heizleiter *m* <Elektrotech> heater
Heizlüfter *m* 1. <Heiz & Kälte> fan heater; 2. <Lufttrans> heater blower; 3. <Maschinen, Mechan> fan heater; 4. <Sicherheit> fan-assisted air heater; 5. <Thermod> fan heater
Heizmantel *m* 1. <Labor> heating mantle; 2. <Thermod> heating jacket
Heizofen *m* 1. <Mechan> heater; 2. <Thermod> heating furnace; 3. <Verpack> heating tunnel
Heizöl *n* 1. <Bau> fuel oil; 2. <Erdöl> fuel oil *(Destillationsprodukt)*; 3. <Heiz & Kälte> fuel oil; 4. <Thermod> heating oil, oil fuel
Heizplatte *f* 1. <Kunststoff> heated plate; 2. <Labor> hotplate
Heizraum *m* <Heiz & Kälte> boiler room, combustion chamber
Heizrohr *n* 1. <Kerntech> fire tube; 2. <Wassertrans> heat pipe
Heizrohrkessel *m* <Heiz & Kälte> multitubular boiler
Heizrohrschlange *f* <Bau> coil
Heizschlange *f* <Heiz & Kälte> calorifier, coil, heating coil
Heizspule *f* <Thermod> heating coil
Heizstab *m* 1. <Heiz & Kälte> heating element, immersion heater; 2. <Kerntech> heater rod
Heizstift *m* <Kerntech> heating pin
Heizstrahler *m* <Heiz & Kälte, Maschinen, Strahlphys, Thermod> radiant heater
Heizstrom *m* 1. <Elektriz> filament current *(Glühkatodenröhre)*; 2. <Elektrotech, Thermod> filament current, heating current
Heiztechniker *m* <Maschinen> heating technician
Heiztransformator *m* <Elektrotech> heating transformer
Heizung *f* 1. <Fertig, Heiz & Kälte> heating; 2. <Kfztech> heater *(Zubehör)*; 3. <Thermod> heating
Heizungs-, Lüftungs- und Klimatechnik *f* <Heiz & Kälte> heating, ventilation and air conditioning
Heizungsanlage *f* <Thermod> heating installation, heating plant
Heizungsbau *m* <Heiz & Kälte> heating installation
Heizungskanal *m* <Heiz & Kälte> heating duct
Heizungsnetzteil *n* <Elektrotech> heater power supply
Heizungssystem *n* <Thermod> heating system
Heizungstechnik *f* <Raumfahrt> heat transfer engineer
Heizvorrichtung *f* <Elektrotech> heater
Heizwert *m* 1. <Abfall, Fertig, Heiz & Kälte, Phys> calorific value; 2. <Thermod> calorific value, thermal power
Heizwertmesser *m* <Maschinen> calorimeter
Heizwicklung *f* <Thermod> heating coil
Heizzug *m* <Thermod> flue
Hektar *m (ha)* <Metrol> hectare *(ha)*
Hektarzähler *m* <Metrol> acremeter
Hekto... *(h)* <Metrol> hecto... *(h)*
Hektogramm *n* <Metrol> hectogram
Hektographie *f* <Druck> hectography
Hektoliter *m (hl)* <Labor> *(AE)* hectoliter, *(BE)* hectolitre *(hl)*
Hektowatt *n* <Elektrotech> hectowatt
Helianthin *n* <Chemie> helianthin, helianthine
Helicin *n* <Chemie> helicin
Helikoid *n* <Geom> helicoid
helikoid *adj* <Geom> helicoid
Heliostat *m* <Nichtfoss Energ> heliostat
heliothermisch *adj* <Nichtfoss Energ> heliothermal
Heliotropin *n* <Chemie> heliotropin, piperonal
heliotropisch *adj* <Nichtfoss Energ> heliotropic
Helipot *n* <Elektriz> helical potentiometer *(Mehrdrehungspotentiometer)*

Helium n (He) <Chemie> helium (He)
Helium-Entwässerungsanlage f <Kerntech> helium dehydrator unit
Helium-Gasflasche f <Raumfahrt> compressed-helium bottle (Raumschiff)
Helium-Lecksortung f <Kerntech> helium leak detection
Helium-Lecktest m <Kerntech> helium leak test
Helium-Lösungskältemaschine f <Phys> helium dilution refrigerator
Helium-Neonlaser m <Strahlphys> helium neon laser
Heliumschutz-Gasschweißen n <Fertig> heliarc welding
Helix f <Geom> helix
Helixantenne f <Raumfahrt> helix antenna (Weltraumfunk)
hell adj 1. <Optik> light; 2. <Textil> bright
helle Flamme f <Thermod> blaze
helle Hinterlegung f <Comp & DV> highlighting (auf dem Bildschirm)
helle Kante f <Phys> bright edge
Hellegatt n <Wassertrans> storeroom (Ladung)
heller Lichtschein m <Thermod> blaze
helleres Zentrum n <Foto> hot spot (des Lichtkegels einer Atelierlampe)
helles Weizenmehl n <Lebensmittel> patent flour
Hellfeld n <Metall> bright field
Hellfeldbeleuchtung f <Phys> bright field illumination
Helligkeit f 1. <Elektronik> brightness, intensity, luminance (Fernsehtechnik); 2. <Ergon, Fernseh, Optik> brightness; 3. <Phys> luminosity; brightness (nicht allgemein definierbare Größe); 4. <Strahlphys> brightness
Helligkeitskurve f <Fernseh> brightness curve
Helligkeitsmessgerät n <Gerät> brightness meter
Helligkeitsmodulation f <Elektronik> intensity modulation
Helligkeitspegel m <Fernseh> bright level
Helligkeitsregler m 1. <Elektrotech> dimmer switch; 2. <Fernseh> brightness control
Helligkeitsschwankung f <Fernseh> fluttering of brightness level
Helligkeitssteuerung f <Elektronik> brightness modulation
Helligkeitsumfang m <Foto> brightness range
Helligkeitsverhältnis n <Fernseh, Phys, Strahlphys> brightness ratio
Helligkeitswert m <Fernseh> brightness value
Helling f <Wassertrans> building berth, building slip; slip, slipway (Schiffbau)
Hellmarke f <Elektronik> intensify pip (Katodenstrahlröhre)
Hellrotglühhitze f <Metall> bright red heat
Hellsteuerung f <Elektronik> unblanking (Katodenstrahlröhre)
Helltastimpulse mpl <Fernseh> unblanking pulses
Helltastschaltung f <Fernseh> unblanking circuit
Helltastung f <Fernseh> unblanking
Helmholtz'sche Spulen fpl <Phys> Helmholtz coils
Helmholtz'scher Resonator m <Akustik, Phys> Helmholtz resonator
Helmholtz'sches Galvanometer n <Elektriz> Helmholtz galvanometer
Helvetica f <Druck> Helvetica (Schrift)
Hemdenstoff m <Textil> shirting
Hemiacetal n <Chemie> hemiacetal
Hemicellulose f <Chemie> hemicellulose
Hemipin... <Chemie> hemipinic
Hemisphäre f <Geom> hemisphere
hemmen v 1. <Maschinen> inhibit; 2. <Mechan> jam
Hemmrad n <Maschinen> escape wheel

Hemmschiene f <Bau> skid track
Hemmschuh m 1. <Maschinen> scotch; 2. <Mechan> shoe, skid
Hemmstoff m <Anstrich, Lebensmittel> inhibitor
Hemmung f 1. <Fertig> escapement; 2. <Maschinen> escapement mechanism
Hemmwerk n <Fertig> escapement
Hennegatt n <Wassertrans> rudder port
Henry n (H) <Elektriz, Elektrotech, Funktech, Metrol, Phys> henry, H (Einheit der Induktivität)
Hentriacontanon n <Chemie> hentriacontanone, palmitone
Heparin n <Chemie> heparin
Heptaeder n <Geom> heptahedron
Heptagon n <Geom> heptagon
heptagonal adj <Geom> heptagonal
Heptan n <Chemie, Erdöl> heptane (Kohlenwasserstoff)
Heptan-1-al n <Chemie> oenanthal
heptavalent adj <Chemie> heptavalent
Hepten n <Chemie> heptene, heptylene
Hept-1-in n <Chemie> heptyne
Heptode f <Elektronik> heptode (Elektronenröhre)
Heptose f <Chemie> heptose
Heptyl... <Chemie> heptyl, heptylic
Heptylen n <Chemie> heptene, heptylene
HERA (Hadron-Elektron-Ring-Anlage) <Teilphys> HERA (hadron-electron ring collider)
herablassen v <Sicherheit> lower
herabsetzen v <Elektrotech> drop, decrease (Spannung)
herabsetzende Äußerungen fpl <Patent> disparaging statements
heranführen v <Telekom> link up (eine Leitung)
Herausheber m <Bau> extractor
herauslösen v 1. <Chemie> elute (adsorbierte Stoffe aus festen Adsorptionsmitteln); 2. <Chemtech> dissolve away, dissolve out
herausnehmen v <Telekom> drop (Kanal)
herausschmelzen v <Fertig> eliquate
Herausschmelzen n <Fertig> eliquation
Herausschneiden n <Fernseh> edit-out
herausspringen v <Fertig> jump (Kette)
herausspritzen v <Fertig> squirt
herauszeichnen v <Konstzeich> draw separately
Herauszeichnen n **von Einzelheiten** <Konstzeich> separate drawing of details
herausziehen v <Bau> draw; pull out (Nagel)
Herausziehen n 1. <Bau> drawing; extraction (eines Nagels); 2. <Fertig> retraction (Bohrer)
Herd m <Fertig> hearth
Herdformerei f <Fertig> (AE) open-sand molding, (BE) open-sand moulding
Herdfrischverfahren n <Metall> Siemens-Martin process
Herdglas n <Ker & Glas> slag glass
Herdofen m <Kohlen> open hearth furnace
Herdraum m <Bau> hearth
Herdraum m **des Hafenofens** <Ker & Glas> pot room
Herdwanderung f <Fertig> bottom bank
hereinkommender Fluss m <Phys> inward flux
hergestellt adj <Lebensmittel> prepared
hergestellt adj/im Schleudergussverfahren <Fertig> spun
hergestellte Verbindung f <Telekom> connected call (CC)
herkömmliches Telefonnetz n <Telekom> traditional telephone network
herkömmliches Telefonsystem n <Telekom> plain old telephone system (POTS)
Herkunft f <Comp & DV, Elektrotech> source
Herkunftsangabe f <Patent> indication of source

herleiten

herleiten v <Math> deduce
hermetisch abgeschlossen adj 1. <Elektrotech> hermetically-sealed; 2. <Heiz & Kälte, Mechan> airtight, hermetically-sealed; 3. <Phys> airtight
hermetisch abgeschlossene Baugruppe f <Elektrotech> hermetically-sealed unit
hermetisch abgeschlossener Verdichtersatz m <Heiz & Kälte> hermetically-sealed compressor unit
hermetisch abgeschlossenes Gerät n <Elektrotech> hermetically-sealed unit
hermetische Versiegelung f 1. <Kerntech> hermetic sealing; 2. <Verpack> hermetic seal
hermetischer Luftabschluss m <Kerntech> hermetic sealing
hermetischer Verdichter m <Maschinen> hermetic refrigerant compressor
hermetischer Verschluss m 1. <Telekom> hermetic sealing; 2. <Verpack> hermetic closure
herstellen v 1. <Bau, Mechan> fabricate; 2. <Telekom> establish, set up *(Verbindung)*
Herstellen n **der Mater** <Optik> mastering
Herstellen n **einer Satelliten-Querverbindung** 1. <Telekom> intersatellite link acquisition; 2. <Raumfahrt> intersatellite link acquisition *(Weltraumfunk)*
Herstellen n **U-förmiger Biegeteile** <Fertig> channel bending
herstellerspezifische Lösung f <Comp & DV> proprietary solution
Herstellgrenzqualität f <Qual> manufacturing quality limit
Herstellung f 1. <Fertig> fabrication, manufacturing, production; 2. <Ker & Glas> manufacture; 3. <Lebensmittel> preparation
Herstellung f **auf einem Glasposten** <Ker & Glas> making on a post
Herstellung f **balligtragender Flächen** <Fertig> crowning *(Spanung)*
Herstellung f **des Gleichgewichtes** <Kerntech> balancing
Herstellung f **integrierter Schaltung** <Elektronik> integrated-circuit fabrication
Herstellung f **mit Glasmacherpfeife** <Ker & Glas> making on blow-pipe
herstellungsbedingter Brennelementdefekt m <Kerntech> fabrication-related fuel defect
Herstellungsdatum n <Verpack> date of manufacture
Herstellungswert m <Qual> objective value
Hertz n *(Hz)* <Elektriz, Elektrotech, Fernseh, Funktech, Metrol, Phys> hertz *(Hz)*
Hertz-... <Elektriz> Hertzian
Hertz-Oszillator m <Elektriz> Hertzian oscillator
Hertz'sche Pressung f <Fertig> Hertz-calculated stresses, hertz equation
Hertz'scher Bruch m <Ker & Glas> Hertzian fracture
Hertz'scher Dipol m 1. <Funktech> Hertzian dipole; 2. <Raumfahrt> Hertzian dipole *(Weltraumfunk)*
Hertz'scher Strahl m <Funktech> Hertzian beam
herunterdrücken v <Verpack> press down
herunterfahren v 1. <Comp & DV> shut down; 2. <Fertig> lower, move down
Herunterfahren n **des Servers** <Comp & DV> shutting down the server
Heruntergehen n <Lufttrans> letdown
Herunterschalten n <Telekom> fall-back *(Bitrate des Modems)*
herunterwalzen v <Fertig> cog *(Walzen)*
Herunterziehen n **von Glasfasern** <Ker & Glas> picking down
hervorholen v <Kontroll> retrieve

Hervorströmen n <Bau> outpouring
Herz n <Maschinen> dog
Herzblatt-Polierschaufel f <Bau> heart trowel
herzförmiger Nocken m <Maschinen> heart-shaped cam
Herzkurve f 1. <Aufnahme> cardioid diagram; 2. <Maschinen> cardioid
Hesperidin n <Lebensmittel> hesperidin
Hesse'sche Normalform f <Comp & DV> normal form, Hesse's standard form
Hessian m <Textil> hessian
heteroatomig adj <Chemie> heteroatomic
Heteroauxin n <Chemie> heteroauxin
heterocyclisch adj <Chemie> heterocyclic
Heterodynempfang m <Telekom> heterodyne reception
Heterodyn-Lichtwellenleiter-Verbindung f <Telekom> heterodyne optical-fiber communications
heterogen adj <Metall> heterogeneous
heterogene Strahlung f <Kerntech> heteroradiation
heterogener Reaktor m <Wassertrans> heterogeneous reactor
Heterojunktion f <Optik> heterojunction
heteropolar adj <Elektriz, Elektrotech> heteropolar
Heterosid n <Chemie> heteroside
Heteroübergang m 1. <Elektronik> heterojunction *(Halbleiter)*; 2. <Optik> heterojunction
Heteroübergangs-Feldeffekttransistor m <Elektronik> heterojunction FET
Heteroxanthin n <Chemie> heteroxanthine
Heuler m <Telekom> howler
Heultonne f <Wassertrans> whistle buoy *(Seezeichen)*
Heuristik f <Künstl Int> heuristics
heuristisch adj <Comp & DV, Künstl Int> heuristic
heuristisches Inferieren n <Künstl Int> heuristic reasoning
heuristisches Wissen n <Künstl Int> heuristic knowledge
Heusler'sche Legierung f <Fertig> magnet alloy
HEX 1. <Comp & DV, Geom> *(hexadezimal)* hex *(hexadecimal)*; 2. <Geom> *(Hexagon)* hex *(hexagon)*
Hexacontan n <Chemie> hexacontane
Hexacosan n <Chemie> cerane, hexacosane
Hexacyanoferrat n <Chemie> prussiate
Hexadecan n <Chemie> hexadecane
Hexadecanoat n <Chemie> hexadecanoate, palmitate
Hexadecanol n <Chemie> ethal, hexadecanol
Hexadecyl n <Chemie> cetyl, hexadecyl
Hexadecylalkohol m <Chemie> hexadecanol
Hexadecylen n <Chemie> cetene, hexadecylene
hexadezimal adj *(HEX)* 1. <Comp & DV> hexadecimal, sexadecimal; 2. <Geom> hexadecimal *(hex)*
Hexadezimal-Darstellung f <Comp & DV> hexadecimal notation
hexadezimale Zahlendarstellung f <Comp & DV> hexadecimal notation
Hexaeder n <Geom> hexahedron
hexaedrisch adj <Geom, Kerntech> hexahedral
Hexafluorsilicat n <Chemie> hexafluorosilicate
Hexagon n *(HEX)* <Geom> hexagon *(hex)*
hexagonal adj <Geom, Kerntech> hexagonal
Hexahydropyrazin n <Chemie> diethylenediamine, hexahydropyrazine, piperazine
Hexahydropyridin n <Chemie> hexahydropyridine, piperidine
Hexamethylendiisocyanat n <Kunststoff> hexamethylene diisocyanate
Hexamethylentetramin n <Chemie> methenamine, urotropine
Hexamin n <Chemie> methenamine, urotropine
Hexamincobalt n <Chemie> luteocobaltic
Hexan n <Chemie, Erdöl> hexane *(Kohlenwasserstoff)*

Hexan... <Chemie> caproic
Hexanitrodiphenylamin n <Chemie> dipicrylamin, hexanitrodiphenylamine
Hexanol n <Chemie> hexyl alcohol
hexavalent adj <Chemie> hexavalent
Hexen n <Chemie> hexene, hexylene
Hexenspiegel m <Ker & Glas> witch mirror
Hexin n <Chemie> hexyne
Hexode f <Elektronik> hexode
Hexogen n <Chemie> cyclonite, hexogen
Hexosan n <Chemie> hemicellulose
Hexose f <Chemie> hexose
Hexyl... <Chemie> capryl, hexylic
Hexylalkohol m <Chemie> hexyl alcohol
Hexylen n <Chemie> hexene, hexylene
Hf (Hafnium) <Chemie> Hf (hafnium)
HF 1. <Aufnahme, Elektronik, Fernseh, Funktech, Telekom> (Funkfrequenz, Hochfrequenz) HF (high frequency); RF (radio frequency); 2. <Elektriz> (Hochfrequenz) HF (high frequency); 3. <Wassertrans> (Hochfrequenz) HF (high frequency); RF (radio frequency)
HF-Abschirmung f <Aufnahme, Fernseh> RF shielding
HF-Abschnitt m <Elektronik> RF section
HF-Abschnittsgenerator m <Elektronik> RF section generator
HF-Filter n <Aufnahme, Elektronik, Funktech> high-frequency filter
HF-Generator m <Elektrotech> RF alternator, RF generator
HF-Leistungsgenerator m <Funktech> high-frequency power generator
HF-Linearität f <Funktech> RF linearity
HF-Löschkopf m <Aufnahme> HF erase head
HF-Messsender m <Funktech, Gerät> high-frequency signal generator
HF-Mikrofon n <Aufnahme> RF microphone
HF-Nachsynchronisierung f <Fernseh> RF dub
HF-Oszillator m <Elektronik> RF oscillator
HF-Puls m <Funktech> RF pulse
H-Frequenz f <Fernseh> horizontal frequency
HF-Schweißgerät n <Verpack> high-frequency welding equipment
HF-Signal n <Elektronik, Funktech> HF signal, high-frequency signal
HF-Signalgenerator m <Elektronik, Gerät> HF signal generator
HF-Sonde f <Raumfahrt> RF sensor (Weltraumfunk)
HF-Spannung f <Funktech> high-frequency voltage
HF-Spektrum n <Elektronik, Funktech> HF spectrum, high-frequency spectrum
HF-Spule f <Elektronik> RF coil
HF-Störung f <Elektronik, Fernseh> radio interference, RF interference
HF-Strecke f <Elektronik> RF section
HF-Streckengenerator m <Elektronik> RF section generator
HF-Strom m <Elektrotech> RF current
HF-Stromquelle f <Elektrotech> RF current source
HF-Stufe f 1. <Elektronik> RF stage; 2. <Funktech> front end (Empfänger)
HF-Träger m <Elektronik> RF carrier
HF-Transformator m <Elektrotech> RF transformer
HF-Transistor m <Elektronik> RF transistor
HF-Übertrager m <Elektrotech, Funktech> high-frequency transformer
HF-Verstärker m <Elektronik, Telekom> RF amplifier
HF-Verstärkung f <Elektronik> RF amplification
HF-Verteiler m <Telekom> high-frequency distribution frame
HF-Vorstufe f <Elektronik> RF stage (des Empfängers)
Hg (Quecksilber) <Chemie> Hg (mercury)
H-Glied n <Elektrotech> H-network
Hieb m <Maschinen> cut
Hiebteilung f <Fertig> coarseness (Feile)
Hierarchie f <Comp & DV> hierarchy
Hierarchie f einer Regelung <Regelung> control hierarchy
Hierarchie f einer Steuerung <Regelung> control hierarchy
hierarchischer Code m <Telekom> embedded code
hierarchischer objektorientierter Entwurf m <Comp & DV> hierarchical object-oriented design
hierarchisches Modell n <Comp & DV> hierarchical model
hierarchisches Programmieren n <Künstl Int> hierarchical programming
hierarchisches System n <Telekom> hierarchical system
hieven v <Wassertrans> haul up
Hifi-Klang m <Akustik, Aufnahme, Fernseh> hi-fi sound
Higgs-Teilchen n <Teilphys> Higgs particle
Highway m <Telekom> highway
Hilfe f <Comp & DV> help
Hilfeanzeige f <Comp & DV> help display
Hilfebereich m <Comp & DV> help area
Hilfebildschirm m <Comp & DV> help screen
Hilfedatei f <Comp & DV> help file
Hilfefunktion f <Comp & DV> help function
Hilfemeldung f <Comp & DV> help message
Hilfemenü n <Comp & DV> help menu
Hilfenachricht f <Comp & DV> help message
Hilfeprogramm n <Comp & DV> help program
Hilfe-System n <Comp & DV> help system
Hilfs... <Maschinen> auxiliary
Hilfsanlage f <Telekom> emergency installation
Hilfsansteuerungsfunkfeuer n <Lufttrans, Wassertrans> compass locator
Hilfsantrieb m <Lufttrans> accessory drive
Hilfsausrüstung f <Fertig> ancillary equipment
Hilfsbit n <Comp & DV> service bit
Hilfsbremsanlage f <Kfztech> emergency brake system
Hilfsdynamo m <Elektriz> booster dynamo
Hilfseinrichtung f <Mechan> ancillary equipment
Hilfselektrode f <Ker & Glas> auxiliary electrode
Hilfsfahrstraße f <Trans> auxiliary route
Hilfsfahrzeug n <Wassertrans> auxiliary vessel
Hilfsflügel m <Lufttrans> flap (Flugzeug)
Hilfsflügelrollenlager n <Lufttrans> flap roller carriage
Hilfsflügelspurrippe f <Lufttrans> flap track rib
Hilfsgerät n <Gerät> autoranging, auxiliary device
Hilfsgitter n <Elektronik> injection grid (zur zusätzlichen Steuerung des Elektronenstroms)
Hilfshammer m <Fertig> anvil top tool
Hilfskanal m <Aufnahme> return channel
Hilfskessel m <Heiz & Kälte> auxiliary boiler
Hilfskompass m <Lufttrans> stand-by compass
Hilfskontakt m <Elektriz, Elektrotech> auxiliary contact
Hilfskoordinatensystem n <Geom> auxiliary coordinate system
Hilfskoppelgruppe f <Telekom> auxiliary switching unit
Hilfskoppelstelle f <Telekom> auxiliary switching point
Hilfskraftbremse f <Kfztech> servo brake
Hilfskraftlenkung f <Kfztech> power steering, power-assisted steering
Hilfskraftmodulatorventil n <Kfztech> servo modulator valve
Hilfskran m <Eisenbahn> wrecking crane
Hilfskreisbogen m <Konstzeich> auxiliary arc

Hilfslinien *fpl* <Druck> feint rules
Hilfsmaschine *f* <Wassertrans> auxiliary engine *(Schiffantrieb)*
Hilfsmaschinen *fpl* <Wassertrans> auxiliary machinery
Hilfsmaß *n* <Konstzeich> auxiliary dimension
Hilfsmaßstab *m* <Gerät> auxiliary scale
Hilfsmaßzahl *f* <Math> ancillary statistic
Hilfsmittel *n* 1. <Fertig> resource; 2. <Kohlen> adjuvant; 3. <Textil> appliance
Hilfsmotor *m* 1. <Elektriz, Elektrotech> auxiliary motor, servomotor; 2. <Maschinen, Wassertrans> auxiliary engine *(Schiffantrieb)*
Hilfsmühle *f* <Nichtfoss Energ> booster mill
Hilfsoperation *f* <Comp & DV> auxiliary operation
Hilfsoszillator *m* <Elektronik> keep-alive oscillator, local oscillator
Hilfsplattform *f* <Bau> relieving platform
Hilfsprogramm *n* 1. <Comp & DV> tool, utility; 2. <Telekom> utility program
Hilfsprozessor *m* <Comp & DV> auxiliary processor
Hilfspumpe *f* <Mechan> booster pump
Hilfsrelais *n* <Comp & DV, Elektriz, Elektrotech> all-or-nothing relay, secondary relay, slave relay
Hilfsrippe *f* <Lufttrans> false frame
Hilfsrotor *m* <Lufttrans> auxiliary rotor *(Hubschrauber)*
Hilfsschalter *m* 1. <Elektriz> auxiliary switch; 2. <Fertig> auxiliary control *(Kunststoffinstallationen)*; 3. <Telekom> auxiliary switch
Hilfsschiff *n* <Wassertrans> support vessel *(Schifftyp)*
Hilfsschnitt *m* <Ker & Glas> auxiliary cut
Hilfsseil *n* <Trans> emergency cable
Hilfsservosteuerung *f* <Maschinen> auxiliary servo control
Hilfsskale *f* <Gerät> auxiliary scale
Hilfsspeicher *m* <Comp & DV> secondary memory, secondary storage
Hilfsständer *m* <Fertig> auxiliary housing *(Einständerhobelmaschine)*
Hilfssteuerungsventil *n* <Maschinen> pilot valve
Hilfsstromversorgung *f* <Elektriz> stand-by supply
Hilfsstütze *f* <Bau> flying shore
Hilfsträger *m* <Funktech, Fernseh> subcarrier
Hilfsträgerfrequenz *f* <Elektronik, Fernseh, Funktech, Telekom> subcarrier frequency
Hilfsträgergenerator *m* <Fernseh, Funktech> subcarrier generator
Hilfsträgergeräuschsperre *f (CTCSS)* <Funktech> continuous tone-coded squelch system *(CTCSS)*
Hilfsträgermodulation *f* <Fernseh, Funktech, Telekom> subcarrier modulation
Hilfsträger-Offset *n* <Fernseh, Funktech> subcarrier offset
Hilfstransformator *m* <Elektriz> auxiliary transformer
Hilfstriebwerk *n* <Lufttrans> auxiliary power unit
Hilfsventil *n* <Maschinen> auxiliary valve
Hilfszentrale *f* <Telekom> satellite exchange *(Telefon)*
Hilfszufallsgröße *f* <Math> ancillary statistic
Hilfszug *m* <Eisenbahn> breakdown train, *(AE)* breakdown car, *(BE)* breakdown wagon
Hilfszugriffsspeicher *m* <Comp & DV> intermediate access memory
Himmelskörper *m* <Raumfahrt> heavenly body
Himmelstemperatur *f* <Nichtfoss Energ> sky temperature
hin und her *adv* <Maschinen> to and fro
hinausfahren *v* **über** <Eisenbahn> overshoot *(ein Haltesignal)*
hinausschießen *v* **über** <Lufttrans, Metall> overshoot
Hinausschießen *n* 1. <Lufttrans> overshooting *(Flugwesen)*; 2. <Metall> overshooting

hindern *v* <Raumfahrt> retard *(Raumschiff)*
Hindernis *n* 1. <Bau, Sicherheit> barricade, obstruction; 2. <Qual> obstacle
Hindernis *n* **in geschichteter Flüssigkeit** <Strömphys> obstacle in stratified fluid *(Turbulenzuntersuchung)*
Hindernis *n* **in rotierendem Fluid** <Strömphys> obstacle in rotating fluid *(Turbulenzuntersuchung)*
Hindernisbegrenzungsfläche *f* <Lufttrans> obstacle limitation surface
Hindernisbeseitigung *f* <Sicherheit> clearing
Hindernisgewinn *m* <Elektronik, Funktech> obstacle gain
hindurchfließen *v* <Telekom> pass through *(Strom)*
hindurchgehen *v* <Telekom> pass through
Hineinfegen *n* <Ker & Glas> insweep
hineinpassen *v* <Maschinen> fit into
Hineinschneiden *n* <Fernseh> in-edit
hinlaufende Welle *f* <Elektrotech> forward wave
hinreichend *adv* <Math> sufficient
Hinterachsantriebswelle *f* <Kfztech> rear axle drive shaft *(Triebstrang)*
Hinterachsbrücke *f* <Kfztech> differential casing
Hinterachse *f* <Kfztech> back axle, rear axle *(Triebstrang)*
Hinterachse *f* **mit doppelter Untersetzung** <Kfztech> double-reduction rear axle
Hinterachsgehäuse *n* <Kfztech> rear axle housing *(Triebstrang)*
Hinterachsgehäusekörper *m* <Kfztech> rear axle housing assembly
Hinterachskörper *m* <Kfztech> rear axle assembly
Hinterachstrichter *m* <Kfztech> rear axle flared tube
Hinterachswelle *f* <Kfztech> rear axle shaft
Hinterachswellenrad *n* <Kfztech> differential side gear
hinterarbeiten *v* 1. <Fertig> machine-relieve; clear *(Spanung)*; 2. <Maschinen> back off, relieve *(Maschine)*
Hinterarbeiten *n* 1. <Fertig> backing-off, relief, relieving; 2. <Maschinen> backing-off, relief
Hinterbohren *n* <Maschinen> backing-off boring
hinterdrehen *v* 1. <Fertig> relieve by turning; 2. <Maschinen> back off, relieve by turning
Hinterdrehen *n* <Maschinen> backing-off, relief, relieving
Hinterdrehmaschine *f* 1. <Fertig> relieving lathe; 2. <Maschinen> backing-off lathe, relieving lathe
hinterdrehte Zähne *mpl* <Maschinen> backed-off teeth, relieved teeth
hinterdrehter Fräser *m* <Maschinen> relieved-milling cutter
hintere Brennebene *f* <Foto> rear focal plane
hintere Ebene *f* <Geom> back plane *(Ebene parallel zur Draufsicht)*
hintere Einzelradaufhängung *f* <Kfztech> independent rear suspension
hintere Flanke *f* <Elektronik> tail *(des Impulses)*
hintere Führung *f* <Fertig> rear pilot *(Reibahle)*
hintere Kantenplatte *f* <Ker & Glas> rear lip tile
hintere Scheibe *f* <Maschinen> tail pulley
hintere Schwarzschulter *f* <Fernseh> back porch
hintere Spannung *f* <Aufnahme> back tension *(Tonband)*
hintereinander geschaltet *adj* 1. <Elektriz, Elektrotech> series-connected; 2. <Gerät> series-connected; 3. <Kontroll> serial
hintereinander geschalteter Transformator *m* <Elektriz> series-connected transformer
hintereinander geschalteter Widerstand *m* <Elektriz> series-connected resistance
Hintereinanderschaltung *f* <Elektriz, Elektrotech> series arrangement, series connection, tandem connection
hinterer Kolben *m* <Kfztech> secondary piston
hinterer Kontakt *m* <Elektriz> back contact
hinterer Spalt *m* <Akustik, Aufnahme> back gap

hinterer Verkleidungskonus m <Raumfahrt> aft skirt *(Raumschiff)*
hinterer Verschlussstein m <Ker & Glas> back tweel
hinterer Zellenring m <Raumfahrt> aft frame section *(Raumschiff)*
hinteres Drehmoment n <Kfztech> rear end torque
hinteres Ende n <Elektronik> tail
hinteres Lot n <Wassertrans> aft perpendicular *(Schiffkonstruktion)*
Hinterflanke f <Elektronik, Fernseh, Phys> trailing edge
Hinterfräsen n <Maschinen> relief milling
hinterfüllen v <Bau, Kohlen> backfill
Hinterfüllung f <Bau, Kohlen> backfill
Hintergrund m <Comp & DV, Druck, Foto> background
Hintergrundabsorption f <Phys, Strahlphys, Teilphys> background absorption
Hintergrundbeleuchtung f <Elektrotech> backlighting
Hintergrundgeräusch n <Akustik, Aufnahme, Werkprüf> background noise
Hintergrundmusik f <Aufnahme> background music
Hintergrundprogramm n <Comp & DV> background program
Hintergrundrauschen n 1. <Akustik> background noise; 2. <Elektronik> background noise, grass; 3. <Raumfahrt> background noise
Hintergrundsrauschtemperatur f <Raumfahrt> sky noise temperature *(Weltraumfunk)*
Hintergrundstrahlung f <Phys, Strahlphys, Teilphys> background radiation
Hintergrundverarbeitung f <Comp & DV> background processing
Hinterkante f **des Luftschraubenblattes** <Lufttrans> blade trailing edge *(Hubschrauber)*
Hinterkantenwirbel mpl <Strömphys> trailing vortices
Hinterkipper m <Bau> end dump truck
Hinterkipperanhänger m <Trans> rear tipping trailer
Hinterkippung f <Bau> end dump *(Kfz)*
Hinterlagedichtung f <Fertig> backing seal *(Kunststoffinstallationen)*
Hinterlegung f <Patent> deposit
hintermauern v <Bau> back up
Hintermauerungsmaterial n <Bau> backing
Hinterrad n <Kfztech> rear wheel
Hinterradachse f <Mechan> rear axle
Hinterradantrieb m <Kfztech> rear wheel drive
Hinterradantrieb m **beim Heckmotor** <Kfztech> rear engine rear wheel drive
Hinterradaufhängung f <Kfztech> rear suspension
hinterschleifen v <Fertig> relief-grind
Hinterschleifen n <Maschinen> backing-off, relieving
Hinterschleifwinkel m <Fertig> relief angle
Hinterschliff m 1. <Fertig> back-off clearance, relief; 2. <Maschinen> relief
Hinterschliff m **der Fase** <Fertig> primary clearance *(Spanung)*
Hinterschlifffläche f <Fertig> flank *(Bohrer)*
hinterschneiden v <Maschinen> undercut
Hinterschneidung f <Kunststoff, Maschinen> undercut
Hinterschraube f <Lufttrans> rear propeller
Hinterseite f <Geom> back face *(eines Objektes)*
Hintersetzwinkel m <Fertig> radial relief *(Reibahle)*
Hintersteven m <Wassertrans> stern frame, sternpost *(Schiffbau)*
Hinterwand f <Maschinen> back-end plate
hinübernehmen v <Druck> overrun
Hin- und Herbewegung f <Maschinen> alternating motion, reciprocating motion
hin- und hergehend adj <Maschinen> reciprocating
hin- und herlaufend adj <Mechan> reciprocating

hin- und herschalten v <Comp & DV> toggle
Hin- und Rückleitung f <Telekom> go-and-return line
Hinweis m **auf Spezifikationsänderungen** *(SCN)* <Trans> specification change notice *(SCN)*
Hinweisansagegerät n <Telekom> interception equipment
hinweisend adj <Phys> point
Hinweislinie f <Konstzeich> leader line
Hinweissymbol n <Comp & DV> sentinel
Hinweistafel f <Bau> signboard
Hinweiston m <Telekom> alerting tone, number-unobtainable tone
Hinweiszeichen n <Comp & DV> reference mark
hinzufügen v <Comp & DV> add
hissen v <Wassertrans> hoist *(Segeln, Flagge)*
Histogramm n <Comp & DV, Ergon, Math, Phys, Qual, Telekom> histogram
Histon n <Chemie> histone
Hittorf'scher Dunkelraum m 1. <Elektronik> Hittorf dark space; 2. <Phys> Crookes dark space
Hitzdraht m <Gerät> hot wire
Hitzdrahtdurchflussmessgerät n <Gerät> hot-wire flowmeter
Hitzdrahtinstrument n 1. <Gerät> thermal expansion instrument; 2. <Phys> hot-wire anemometer
Hitzdrahtleistungsmesser m <Elektriz> hot-wire wattmeter
Hitzdrahtmesswerk n <Gerät> expansion movement
Hitzdrahtmikrofon n <Akustik> hot-wire microphone
Hitzdrahtrelais n <Elektriz> hot-wire relay
Hitzdrahtstromgerät n <Elektriz> hot-wire ammeter, thermal ammeter
Hitze f 1. <Heiz & Kälte, Kohlen, Phys, Textil> heat; 2. <Thermod> burning heat, heat
hitzeabweisend adj <Sicherheit> heat-resistant, heatproof
Hitzeausgleich m <Raumfahrt> heat transfer engineer
Hitzebarriere f <Anstrich> thermal barrier
Hitzebehandlung f 1. <Anstrich, Metall> heat treatment; 2. <Textil> baking
hitzebeständig adj 1. <Erdöl> thermostable; 2. <Heiz & Kälte, Phys> heat-resistant; 3. <Raumfahrt> heat-resistant *(Raumschiff)*; 4. <Verpack> heat-resistant
hitzebeständiger Stahl m <Metall> heat-resisting steel
hitzebeständiges Glas n <Ker & Glas> heat-resisting glass
Hitzebeständigkeit f 1. <Qual> resistance to heat; 2. <Thermod> resistance to heat, temperature resistance
Hitzefarbe f <Metall> heat tint
Hitzefärben n <Metall> heat-tinting
Hitzefluss m <Raumfahrt> heat flux
Hitzekachel f <Raumfahrt> tile
Hitzemantel m <Raumfahrt> heat shroud
Hitzemauer f <Anstrich, Lufttrans> thermal barrier
Hitzeriss m <Mechan> hot tear
Hitzeschild m 1. <Lufttrans> heat shield, heating shield; 2. <Raumfahrt> heat shield, thermal protection shield *(Raumschiff)*; 3. <Sicherheit> heat shield, thermal protection shield; 4. <Thermod> heat shield
Hitzeschutz m <Sicherheit> overheat protection
Hitzeschutzanzug m <Sicherheit> heat protective suit, proximity suit
Hitzeschutzhandschuh m <Sicherheit> heat-resistant glove
Hitzeschutzkleidung f <Sicherheit> heat-protective clothing, heatproof clothing
Hitzeschutzmaterial n <Sicherheit> heat-protective material
Hitzeschutzstiefel mpl <Sicherheit> heat protective boots

Hitzeschutzwand

Hitzeschutzwand f <Bau, Lufttrans> heat-insulating wall
Hitzestrahlung f <Raumfahrt> radiative heat transfer *(Raumschiff)*
Hitzestrahlungsmesser m <Thermod> pyrometer
hitzesuchend adj <Raumfahrt> heat-seeking *(Raumschiff)*
hitzetrocknen v <Thermod> dry by heat
Hitze- und UV-Bestrahlungstest m <Ker & Glas> bake and UV-irradiation test
Hitzeverschweißen n <Verpack> heat welding
Hitzeversiegelmaschine f <Verpack> heat-sealing machine
Hitzeversiegelung f <Verpack> heat induction seal, heat sealing
Hitzeversiegler m <Verpack> heat-sealing device
Hitzewelle f <Thermod> heat wave
Hitzewiderstand m <Lufttrans> heating resistor
H-Krümmer m <Elektrotech> H-plane bend
hl *(Hektoliter)* <Labor> hl *(hectoliter)*
HMF-Ruß m <Kunststoff> HMF carbon black
H-Modus m <Elektrotech, Telekom> H-mode, TE mode, transverse electric mode
HNF *(höchste nutzbare Frequenz)* <Funktech> MUF *(maximum usable frequency)*
Ho *(Holmium)* <Chemie> Ho *(holmium)*
Hobel m <Bau, Fertig> plane *(Werkzeug)*
Hobelbank f <Bau> carpenter's bench, joiner's bench
Hobeleisen n <Bau, Maschinen> plane iron
Hobelkasten m <Bau> plane stock
Hobelmaschine f <Bau, Fertig, Maschinen, Mechan> planer, planing machine
Hobelmeißel m 1. <Bau> paring chisel; 2. <Maschinen> planer tool
Hobelmesser n <Bau> plane iron
hobeln v <Fertig, Maschinen, Mechan> plane
Hobeln n <Bau, Maschinen> planing
Hobelspan m 1. <Bau> shaving; 2. <Fertig> planing chip
Hobelspäne mpl <Bau> wood shavings
Hobeltisch m <Maschinen> planer table
Hobelwerkzeug n <Fertig> planing tools
hoch adj 1. <Maschinen, Papier> high; 2. <Phys> high *(Temperatur)*; 3. <Umweltschutz> high *(Konzentration)*
• **zu hoch ansteuern** <Raumfahrt> overdrive *(Weltraumfunk)*
hoch dosiert adj <Elektronik> high-dose
hoch entwickelt adj <Telekom> advanced, sophisticated *(Gerät)*
hoch entwickeltes Berichten n <Comp & DV> advanced reporting
hoch entwickeltes Protokollieren n <Comp & DV> advanced logging
hoch liegender Wellenleiter m <Funktech> elevated duct *(Wellenausbreitung)*
hoch radioaktiv adj <Kerntech> highly-radioactive
hoch stapeln v <Verpack> stack up
hochabriebfester Furnace-Ruß m *(HAF-Ruß)* <Kunststoff> high abrasion furnace carbon black *(HAF carbon black)*
Hochachse f <Lufttrans> normal axis
Hochachsewindturbine f <Nichtfoss Energ> vertical-axis wind turbine
hochaktiver Abfall m <Kerntech> HAW, highly-active waste
hochangereichertes Uran n <Kerntech> highly-enriched uranium
hochauflösend adj <Comp & DV> high-resolution
hochauflösende Abtastung f <Strahlphys> high-resolution scan
hochauflösende Untersuchung f <Strahlphys> high-resolution study *(Linienprofile)*

hochauflösender Scan m <Strahlphys> high-resolution scan
hochauflösendes Fernsehen n 1. <Fernseh> extended definition television *(EDTV)*; 2. <Fernseh> high-definition television *(HDTV)*
Hochausbeute-Faserstoff m <Papier> high-yield pulp
Hochbahn f 1. <Eisenbahn> *(AE)* elevated railroad, *(BE)* elevated railway, *(AE)* overhead railroad, *(BE)* overhead railway; 2. <Trans> overhead track
Hochbau m <Bau> building construction
Hochbauschgarn n <Textil> high-bulk spun yarn
Hochbehälter m <Bau> tower tank
hochbelastbar adj 1. <Elektriz, Maschinen> high-load; 2. <Sicherheit> heavy-duty
hochbelastbarer Schutzhelm m <Sicherheit> heavy-duty helmet
Hochdach-Kastenwagen m <Kfztech> raised roof van
Hochdach-Transporter m <Kfztech> raised roof van
Hochdruck m 1. <Druck> letterpress, letterpress printing; 2. <Phys> high performance, high pressure
Hochdruckadditiv n <Erdöl> extreme-pressure additive *(Bohrtechnik)*
Hochdruckanlage f <Sicherheit> high-pressure system
Hochdruckaufnehmer m <Heiz & Kälte> high-pressure pickup
Hochdruckdampfhärten n <Bau> high-pressure steam-curing
Hochdruckdirekteinspritzen n <Kfztech> high pressure direct injection *(HTBI)*
Hochdruckflüssigchromatographie f *(HPLC)* <Labor, Lebensmittel> high-pressure liquid chromatography *(HPLC)*
Hochdruckgebiet n <Wassertrans> high-pressure area *(Wetterkunde)*
Hochdruckheizung f <Heiz & Kälte> high-pressure heating system
Hochdruckkeil m <Wassertrans> ridge *(barometrischer Druck)*
Hochdruckkessel m <Heiz & Kälte> high-pressure boiler
Hochdruckkolbenverdichter m <Maschinen> high-pressure piston compressor
Hochdruckkompressor m <Maschinen> high-pressure compressor
Hochdruckkraftstoffpumpe f <Kfztech> high-pressure fuel pump
Hochdruckmanometer n <Gerät> *(BE)* high-range gauge, *(AE)* high-range gage
Hochdruckmesser m <Heiz & Kälte> *(AE)* high-pressure gage, *(BE)* high-pressure gauge
Hochdruckquecksilberdampflampe f <Elektriz> high-pressure mercury lamp
Hochdruckregler m <Heiz & Kälte> high-pressure controller
Hochdruckreifen m <Kfztech> *(AE)* high-pressure tire, *(BE)* high-pressure tyre
Hochdruckreinigung f <Meerschmutz> high-pressure water blasting
Hochdruckrotationsmaschine f <Druck> letterpress rotary
Hochdruckschmierstoff m <Maschinen> extreme-pressure lubricant, high-pressure lubricant
Hochdruckschwimmerventil n <Heiz & Kälte> high-pressure float valve
Hochdrucksicherheitsventil n <Sicherheit> high-pressure safety valve
Hochdruckspülen n <Meerschmutz> high-pressure flushing
Hochdrucktank m <Raumfahrt> high-pressure tank *(Raumschiff)*

Hochdruckumgebung f <Sicherheit> high-pressure atmosphere
Hochdruckvakuumpumpe f <Maschinen> high-pressure vacuum pump
Hochdruckventil n <Maschinen> high-pressure valve
Hochdruckzylinder m <Maschinen> high-pressure cylinder
Hochebene f <Erdöl> plateau
hochenergetische Strahlung f <Strahlphys> high-level radiation
Hochenergieband n <Fernseh> high-energy tape
Hochenergiefusion f <Kerntech> high-energy fusion
Hochenergiemetallumformung f <Maschinen> high-energy metal forming
Hochenergiephysik f <Phys> high-energy physics
Hochenergieproton n <Raumfahrt> high-energy proton
Hochenergiestrahlung f <Strahlphys> high-energy radiation
Hochfahrbahn f <Eisenbahn> overhead trackway
hochfahren v <Aufnahme> bring up
Hochfahrzeit f <Aufnahme> run-up time
hochfeste Faser f <Textil> (AE) high-tenacity fiber, (BE) high-tenacity fibre
hochfester Bohrschlamm m <Erdöl> high-solid mud (Bohrtechnik)
hochfester Stahl m <Metall> high-tensile steel
Hochflussreaktor m <Kerntech> high flux reactor
Hochformat n 1. <Comp & DV> portrait format, portrait representation; 2. <Druck> portrait format
hochfrequent adj <Elektriz, Elektronik, Funktech> high-frequency, radio-frequency
hochfrequente Erwärmung f <Elektrotech> high-frequency heating
Hochfrequenz f (HF) <Aufnahme, Elektriz, Elektronik, Fernseh, Funktech, Telekom, Wassertrans> high frequency (HF); radio frequency (RF)
Hochfrequenzeisenkern m <Elektrotech> powdered iron core
Hochfrequenzerwärmung f <Kunststoff> dielectric heating, radio-frequency heating
Hochfrequenzfilter n <Aufnahme, Elektronik, Funktech> high-frequency filter
Hochfrequenzgenerator m <Elektriz, Elektrotech, Funktech> high-frequency generator, radio-frequency alternator
Hochfrequenzheizung f <Elektrotech> high-frequency heating
Hochfrequenzinduktionserwärmung f <Elektriz> dielectric heating, eddy current heating, high-frequency heating
Hochfrequenzinduktionslöten n <Bau> high-frequency induction brazing
Hochfrequenzkabel n <Elektriz> high-frequency cable
Hochfrequenzleistungsgenerator m <Funktech> high-frequency power generator
Hochfrequenzlitze f <Elektrotech> stranded conductor
Hochfrequenz-Messsender m <Funktech, Gerät> high-frequency signal generator
Hochfrequenzofen m <Elektrotech> high-frequency furnace
Hochfrequenz-Rundumreichweite f <Elektrotech> doppler very high-frequency omnidirectional radio range
Hochfrequenzschweißung f <Verpack> high-frequency welding (von Folien)
Hochfrequenzsignal n <Elektronik, Funktech> high-frequency signal, radio signal
Hochfrequenzspektrum n <Elektronik, Funktech> high-frequency spectrum
Hochfrequenzstörung f 1. <Aufnahme> radio-frequency interference; 2. <Elektronik> radio interference, radio noise, radio-frequency interference; 3. <Funktech, Telekom> radio-frequency interference
Hochfrequenzstrom m <Elektriz, Elektrotech> radio-frequency current
Hochfrequenzstufe f <Funktech> high-frequency stage
Hochfrequenzträger m <Elektronik> radio-frequency carrier
Hochfrequenztransformator m <Elektrotech, Funktech> high-frequency transformer
Hochfrequenztransistor m <Elektronik> high-frequency transistor
Hochfrequenzverstärkung f 1. <Elektronik> high-frequency amplification; 2. <Elektronik, Funktech> high-frequency amplification, microwave amplification
Hochgebirgsbahn f <Eisenbahn> (AE) mountain railroad, (BE) mountain railway
hochgeleimtes Papier n <Druck> hard-sized paper
hochgemahlener Zellstoff m <Papier> wet-beaten pulp
hochgenaue Abwiegung f <Verpack> ultrahigh accuracy weighing
Hochgeschwindigkeit f 1. <Maschinen> high speed; 2. <Wasserversorg> high velocity
Hochgeschwindigkeitsabtastung f <Fernseh> high-velocity scanning
Hochgeschwindigkeits-CMOS-Logik f <Elektronik> high-speed CMOS logic (HCMOS, HSCMOS)
Hochgeschwindigkeits-Datenkommunikation f <Telekom> high-speed data communication
Hochgeschwindigkeits-Ethernet n <Telekom> fast Ethernet, high-speed Ethernet
Hochgeschwindigkeitsfahrzeug n <Kfztech> HHSV, high hypothetical speed vehicle
Hochgeschwindigkeitsgasturbinenreisebus m <Kfztech> high-speed gas turbine motor coach
Hochgeschwindigkeitsgasturbinentriebwagen m <Eisenbahn> high-speed gas turbine railcar
Hochgeschwindigkeitskopie f <Fernseh> high-speed duplication
Hochgeschwindigkeits-Lagerungssystem n <Kfztech> high-speed sortation system
Hochgeschwindigkeits-LAN n <Telekom> high-speed LAN, high-speed local area network (HSLAN)
Hochgeschwindigkeitsmodem n <Telekom> high data rate modem, high-speed modem
Hochgeschwindigkeitsschienenfahrzeug n <Eisenbahn> super high-speed rail vehicle
Hochgeschwindigkeitsschiff n <Wassertrans> high-speed craft (HSC)
Hochgeschwindigkeitsschleifmaschine f <Maschinen> high-speed grinding machine
Hochgeschwindigkeitsschütteln n <Lufttrans> high-speed buffeting
Hochgeschwindigkeitsstrecke f <Kfztech> super-speedway
Hochgeschwindigkeitsteilchen n <Kerntech> high-speed particle
Hochgeschwindigkeitsübertragung f <Telekom> high-speed transmission
Hochgeschwindigkeitsverkehr m <Trans> super high-speed traffic
Hochgeschwindigkeitszug m <Eisenbahn> APT, (BE) advanced passenger train, high-speed train
hochgeständerte Einschienenhängebahn f (H-Bahn) <Eisenbahn> overhead monorail
hochgestelltes Zeichen n <Comp & DV, Druck> superscript
hochgiftig adj <Sicherheit> extremely toxic, high-toxic, highly toxic
hochgiftiger Stoff m <Sicherheit> high-toxic substance

Hochglanz

Hochglanz *m* <Kunststoff> high gloss • **auf Hochglanz bringen** <Anstrich> polish
Hochglanzfolie *f* <Verpack> high-gloss foil
Hochglanzfoto *n* <Foto> glossy print
Hochglanzmaschine *f* <Foto> glazing machine
Hochglanzpapier *n* 1. <Foto> glossy paper; 2. <Verpack> high-gloss paper
Hochglanzplatte *f* <Foto> glazing sheet
hochglanzpolieren *v* <Anstrich> burnish
Hochglanzpolitur *f* <Maschinen> mirror finish
Hochglanztrockenpresse *f* <Foto> dryer-glazer
Hochgleis *n* <Eisenbahn> elevated track
Hochglühen *n* <Metall> full annealing
hochgradige Hitze *f* <Kerntech> high-grade heat
hochhebeln *v* <Maschinen> lever up
hochheben *v* <Maschinen> hoist
Hochheben *n* <Bau> raising
hochintegrierte Schaltung *f* <Elektronik> high-density integrated circuit
hochintegrierter logischer Schaltkreis *m* <Elektronik> high-density logic
hochintegrierter Schaltkreis *m (LSI-Kreis)* <Elektronik, Phys, Telekom> large-scale integrated circuit *(LSI circuit)*
Hochintensitätslichtbogen *m* <Elektriz> high-intensity electric arc
Hochkanal *m* <Funktech> elevated duct *(Wellenausbreitung)*
hochkanaliges Übertragungssystem *n/sehr* <Telekom> very-large-capacity transmission system
hochkant biegen *v* <Bau> bend on edge
hochkanten *v* <Bau> raise on edge
hochkantiges Wachstum *n* <Metall> edgewise growth
Hochkomma *n* <Druck> turned comma
Hochkurzzeiterhitzung *f* <Lebensmittel> HTST, high-temperature short time pasteurization
hochladen *v* <Comp & DV> upload
Hochladen *n* <Comp & DV> uploading
Hochlastwiderstand *m* <Elektrotech> power resistor
hochlaufen *v* <Lufttrans> run-up *(Motor und Triebwerk)*
Hochlaufen *n* <Lufttrans> engine run-up *(Leistung des Motors)*; run-up *(Motor und Triebwerk)*
Hochleistung *f* 1. <Fertig, Maschinen> heavy duty; 2. <Maschinen> heavy duty *(HD)*; 3. <Mechan> heavy duty; 4. <Phys> high performance
Hochleistungsband *n* <Aufnahme> high-output tape
Hochleistungsbatterie *f* <Trans> high-performance battery
Hochleistungsbipolartransistor *m* <Elektronik> high-power bipolar transistor
Hochleistungsbrenner *m* <Bau> high-pressure blowpipe
Hochleistungs-Caching *n* <Comp & DV> high performance caching
Hochleistungsdrehmaschine *f* <Maschinen> heavy-duty lathe
Hochleistungsdüse *f* <Kfztech> high-speed auxiliary jet
Hochleistungsfilterung *f* <Abfall> high-rate filtration
Hochleistungs-Flüssigkeitschromatographie *f* <Umweltschmutz> high-performance liquid chromatography
Hochleistungsgleichrichter *m* <Elektrotech> high-power rectifier
Hochleistungskontakt *m* <Elektrotech> heavy-duty contact
Hochleistungslaser *m* 1. <Elektronik> high-energy laser; 2. <Kerntech> high-power laser
Hochleistungslift *m* <Bau> heavy duty lift
Hochleistungslüfter *m* <Heiz & Kälte> high-performance fan
Hochleistungsmanipulator *m* **im Reaktorkern** <Kerntech> in-core power manipulator

Hochleistungsöl *n (HD-Öl)* <Maschinen> heavy-duty oil *(HD oil)*
Hochleistungsplattenlaufwerk *n* <Comp & DV> high performance disk drive
Hochleistungsröhre *f* <Elektronik> high-power tube
Hochleistungsthyristor *m* <Elektrotech> high-power SCR
Hochleistungstransformator *m* <Elektriz> high-power transformer
Hochleistungsverbraucher *m* <Elektrotech> high-power load
Hochleistungsverstärker *m* <Raumfahrt> high-power amplifier *(Weltraumfunk)*
Hochlicht *n* <Druck> highlight
Hochlinie *f* <Eisenbahn> elevated line
Hochmodul-Furnace-Ruß *m* <Kunststoff> high-modulus furnace carbon black
hochmolekular *adj* <Chemie> macromolecular
Hochofen *m* <Fertig, Ker & Glas, Kohlen, Maschinen, Thermod> blast furnace
Hochofenausmauerung *f* <Fertig> blast furnace lining, *(AE)* shirt
Hochofengas *n* <Fertig> blast furnace gas
Hochofenmetall *n* <Metall> hot metal
Hochofenzement *m* <Bau> blast furnace cement
hochohmig *adj* 1. <Funktech> high-impedance; 2. <Telekom> high-resistance
hochohmiger Zustand *m* <Elektrotech> high-impedance state
Hochohmwiderstand *m* <Telekom> high resistance
Hochpass *m* <Telekom> high-pass
Hochpassfilter *n* <Aufnahme, Comp & DV, Elektriz, Elektronik, Fernseh, Phys, Telekom> high-pass filter
Hochpassfilter *n* **zweiter Ordnung** <Elektronik> second order high-pass filter
Hochpassfilterung *f* <Elektronik> high-pass filtering
Hochprägung *f* <Druck> embossing, relief printing
hochratiges Faksimile *n* <Telekom> high-speed facsimile
hochreines Pigment *n* <Kunststoff> high-purity pigment
hochrot *adj* <Metall> bright red
Hochschulterlager *n* <Maschinen> rigid deep-groove ball bearing
Hochsee *f* <Wassertrans> high seas
Hochsee… <Wassertrans> deep-sea, seagoing
Hochseebagger *m* <Meerschmutz> marine dredge
Hochseebergungsschlepper *m* <Trans> seagoing salvage tug
Hochseefischerei *f* <Wassertrans> deep-sea fishing *(Fischerei)*
Hochseekreuzer *m* <Wassertrans> ocean-going cruiser
Hochseenavigation *f* <Wassertrans> deep-sea navigation
Hochseeschiff *n* <Wassertrans> ocean-going ship, sea-going vessel
Hochseeschifffahrt *f* <Wassertrans> deep-sea navigation
Hochseeschlepper *m* <Wassertrans> seagoing tug
hochseetaugliches Schiff *n* <Wassertrans> ocean-going ship, ocean-going vessel
Hochseevermessungsschiff *n* <Wassertrans> ocean survey vessel
Hochsetzsteller *m* <Elektrotech> boost chopper, step-up chopper *(Gleichstromsteller)*
Hochsicherheitsverglasung *f* <Kfztech> high-safety glazing
Hochsiedendes *n* <Thermod> distillation tail
Hochspannung *f* 1. <Elektriz> high tension *(HT)*; high voltage *(h.v., HV)*; 2. <Fernseh> EHT, extra-high tension

Hochspannungsanlage f <Elektrotech> high-voltage facility, high-voltage plant
Hochspannungsanschlussklemme f <Kfztech> high-tension terminal
Hochspannungsbauelement n <Elektronik> high-voltage component, high-voltage device *(Hsp)*
Hochspannungsblitzschutz m <Sicherheit> lightning arrester for high voltage
Hochspannungseinrichtung f <Elektriz> high-voltage equipment
Hochspannungserzeuger m <Elektriz> high-voltage generator
Hochspannungsfreileitung f <Elektriz> high-voltage overhead line, high-voltage overhead transmission line
Hochspannungsgleichrichter m <Elektriz> high-voltage rectifier
Hochspannungsgleichstromübertragung f <Elektrotech> DC high-tension power transmission, HVDC transmission, high-voltage direct current transmission
Hochspannungsimpulsgenerator m <Elektriz> high-voltage impulse generator, high-voltage pulse generator
Hochspannungsisolierung f <Elektriz> high-voltage insulation
Hochspannungskabel n <Elektriz> high-voltage cable
Hochspannungsleitung f 1. <Bau> transmission line; 2. <Elektriz> high-voltage transmission line
Hochspannungsmotor m <Elektriz> high-voltage motor
Hochspannungsnetz n <Elektrotech> high-tension power supply, high-voltage power supply
Hochspannungspol m <Kfztech> high-tension terminal
Hochspannungsporzellanisolator m <Elektriz> high-voltage porcelain insulator
Hochspannungsprüfgerät n <Elektriz> high-voltage tester
Hochspannungsprüfkreis m <Elektriz> high-voltage testing circuit
Hochspannungsprüfschaltung f <Elektriz> high-voltage testing circuitry
Hochspannungsprüftechnik f <Elektriz> high-voltage testing technique
Hochspannungsprüfung f <Elektriz> high-voltage test, high-voltage testing
Hochspannungsprüfverfahren n <Elektriz> high-voltage testing procedure
Hochspannungsschaltanlage f <Elektriz> high-voltage switchgear
Hochspannungsseekabel n <Elektriz> high-voltage undersea cable
Hochspannungssicherung f <Elektriz> high-voltage fuse
Hochspannungssteckdose f <Kerntech> high-potential socket
Hochspannungssteuerung f <Elektriz> high-voltage control *(bei Wechselstrom-Triebfahrzeugen)*
Hochspannungsstoßgenerator m <Elektriz> high-voltage impulse generator
Hochspannungsstromunterbrecher m <Elektriz> high-voltage circuit breaker
Hochspannungsstromversorgung f <Elektriz> high-voltage power supply
Hochspannungsstromversorgungsnetz n <Elektriz> high-voltage grid
Hochspannungstransformator m <Elektriz> high-voltage transformer
Hochspannungstransmissionselektronenmikroskopie f <Elektronik> high-voltage transmission electron microscopy *(HVTEM)*
Hochspannungsübertragungsnetz n <Elektriz> high-voltage transmission network

Hochspannungsversorgung f <Fernseh> EHT supply
Hochspannungswicklung f <Elektriz> high-voltage winding
höchst eben adj <Anstrich> ultrasmooth
höchst glatt adj <Anstrich> ultrasmooth
Höchst... <Elektronik, Kfztech, Lufttrans, Maschinen> maximum
hochstabiler Bohrschlamm m <Erdöl> high-solid mud
hochstabiler Oszillator m <Elektronik> highly stable oscillator
Höchstanteil m <Qual> upper limiting proportion
Höchstauftrieb m <Lufttrans> maximum lift
Höchstbelastung f 1. <Elektrotech> peak load; 2. <Lufttrans> maximum load
Höchstbelastungsgrenze f <Sicherheit> maximum exposure limit
Höchstbetriebshöhe f <Lufttrans> maximum operating altitude
Höchstdrehzahl f 1. <Kfztech> peak revs *(Motor)*; 2. <Maschinen> maximum speed
Höchstdrehzahl f des Motors <Kfztech> peak engine speed
Höchstdruck m *(EP)* <Maschinen> extreme pressure *(EP)*
höchste Amplitude f <Elektronik> peak amplitude
höchste brauchbare Übertragungsfrequenz f <Elektrotech> maximum usable frequency
höchste Dauerleistung f <Lufttrans> maximum continuous power
höchste für längere Zeit entnehmbare Leistung f <Lufttrans> METO power, maximum except takeoff power
höchste nutzbare Frequenz f *(HNF)* <Funktech> maximum usable frequency *(MUF)*
höchste Rotordrehzahl f <Lufttrans, Nichtfoss Energ> maximum rotor speed
höchste Schlagbelastung f <Lufttrans> flapping stress peak *(Hubschrauber)*
hochstegig adj <Fertig> high-webbed
Hochstellen n <Kerntech> setup *(Regeln)*
Hochstellung f <Comp & DV> superscript
höchster Außenluftüberdruck m <Lufttrans> free-air peak overpressure *(Überschallknall)*
höchster Punkt m vor Sinkflug <Lufttrans> top of descent *(künstlicher Flugsimulator)*
Höchstflugdauer f <Lufttrans> endurance, maximum flying time
Höchstfrequenz f 1. <Elektronik> superhigh frequency; 2. <Funktech> microwave
Höchstfrequenzgenerator m 1. <Elektronik> microwave generator, microwave synthesizer, super-high frequency generator; 2. <Funktech> microwave generator, microwave synthesizer
Höchstfrequenzoszillator m <Elektronik, Funktech> microwave oscillator
Höchstfrequenzschaltung f <Funktech> microwave circuit
Höchstfrequenzsignal n <Elektronik, Telekom> microwave signal
Höchstfrequenzsignalgenerator m <Elektronik> microwave signal generator
Höchstfrequenzsignalquelle f <Elektronik> microwave signal source
Höchstgeschwindigkeit f 1. <Eisenbahn, Kfztech, Lufttrans, Maschinen> maximum speed; 2. <Trans> speed limit *(Geschwindigkeitsbegrenzung)*; 3. <Wassertrans> maximum speed • **mit Höchstgeschwindigkeit** <Trans> at full speed
Höchstgeschwindigkeit f **in Normalfluglage mit Nennleistung** <Lufttrans> maximum speed in level flight with rated power

Höchstgewicht *n* <Verpack> maximum weight
Höchstintegration *f (VLSI)* <Comp & DV, Elektronik, Telekom> very large-scale integration *(VLSI)*
Höchstintegrationsschaltkreis *m* <Phys> very large-scale integrated circuit
Höchstlast *f* 1. <Elektrotech, Papier> peak load; 2. <Wassertrans> maximum load
Höchstlautstärke-Geschwindigkeit *f* <Aufnahme> peak volume velocity
Höchstleistung *f* <Elektriz, Elektrotech> maximum output, maximum power, peak power
Höchstleistung *f* **bei Nennwindgeschwindigkeit** <Nichtfoss Energ> maximum power at rated wind speed
Höchstleistungsmischungsverhältnis *n* <Kfztech> maximum output mixture ratio
Höchstleistungsrechner *m* <Comp & DV> number cruncher, supercomputer
Höchstnennstrom *m* <Elektrotech> maximum current rating
Höchstnutzlast *f* <Lufttrans> maximum payload
Höchstquantil *n* <Qual> upper limiting quantile
Hochstraße *f* 1. <Bau> elevated highway, elevated motorway; 2. <Trans> *(BE)* flyover, *(AE)* skyway
Hochstromdiode *f* <Elektronik> high-current diode
Hochstromtransistor *m* <Elektronik> high-current transistor
höchstschmelzendes Metall *n* <Anstrich> refractory metal
Höchstschweißstrom *m* <Bau> maximum welding current
Höchstspannung *f* 1. <Elektriz, Elektrotech> extra-high tension; extra-high voltage *(e.h.v., EHV)*; maximum voltage, peak voltage; ultrahigh voltage *(uhv, UHV)*; 2. <Fernseh> extremely high tension
Höchstspannungsrelais *n* <Elektriz> maximum voltage relay
Höchststromstärke *f* <Elektrotech> peak current
Höchsttemperatur *f* **in der Brennelementhülse** <Kerntech> PCT, peak cladding temperature
Höchstverbrauch *m* <Elektriz> maximum demand
Höchstverbrauchszähler *m* <Elektrotech> demand meter
Höchstvergrößerung *f* <Metall> ultimate magnification
Höchstwert *m* 1. <Elektriz, Elektronik> maximum value, peak value; 2. <Kerntech> maximum; 3. <Papier> peak; 4. <Qual> upper limiting value
Höchstwertanzeiger *m* <Elektronik> peak indicator
höchstwertige Binärstelle *f* <Comp & DV> MSB, most significant bit
höchstwertige Ziffer *f* <Comp & DV> most significant digit
höchstwertiges Bit *n* <Comp & DV> MSB, most significant bit
höchstwertiges Zeichen *n* <Comp & DV> most significant character
höchstzulässige Betriebsgeschwindigkeit *f* <Lufttrans> maximum permissible operating speed
höchstzulässige Dosis *f* <Strahlphys> maximum permissible dose *(ionisierende Strahlung)*
höchstzulässige Fahrwerkbetriebsgeschwindigkeit *f* <Lufttrans> maximum landing-gear operating speed
höchstzulässige Geschwindigkeit *f* <Kfztech> maximum design speed
höchstzulässige Geschwindigkeit *f* **bei ausgefahrenem Fahrwerk** <Lufttrans> maximum landing-gear extended speed
höchstzulässige Geschwindigkeit *f* **bei ausgefahrenen Klappen** <Lufttrans> maximum flap extended speed

höchstzulässige Konzentration *f (HZK)* <Umweltschmutz> maximum allowable concentration *(MAC)*; threshold limit value *(TLV)*
höchstzulässige Konzentration *f* **in der Umwelt** <Umweltschmutz> threshold limit value in the free environment
höchstzulässige Machzahl *f* <Lufttrans> maximum permissible Mach number
höchstzulässige Riemenspannung *f* <Maschinen> maximum allowable belt tension
höchstzulässige Überdrehzahl *f* **des Motors** <Lufttrans> maximum engine overspeed
höchstzulässige Überspannung *f* <Elektriz> assigned maximum overvoltage
höchstzulässiges Gesamtgewicht *n* <Eisenbahn, Kfztech, Lufttrans, Wassertrans> maximum total weight
Hochtank *m* <Trans, Wassertrans> deep tank
Hochtemperatur *f* <Metall, Kerntech, Phys> high temperature
hochtemperaturbeständige Isolierung *f* <Elektriz> high-temperature insulation
Hochtemperatur-Festigkeitsprüfung *f* <Werkprüf> high-temperature strength test
Hochtemperaturfett *n* <Mechan> high-temperature grease
Hochtemperaturkriechen *n* <Metall> high-temperature creep
Hochtemperaturreaktor *m (HTR)* <Kerntech> high-temperature reactor *(HTR)*
Hochtemperaturreaktor *m* **mit Bündelelementen** <Kerntech> *(AE)* block-type element-fueled high temperature reactor, *(BE)* block-type element-fuelled high temperature reactor
Hochtemperatur-Salzschmelztreibstoffzelle *f* <Kfztech> high-temperature molten salts fuel battery
Hochtemperatursupraleitfähigkeit *f* <Phys> high-temperature superconductivity
Hochtemperaturtreibstoffzelle *f* <Kfztech> high-temperature fuel cell
Hochtemperaturtrocknung *f* <Fertig> drying at high temperature
Hochtemperaturwindkanal *m* <Lufttrans> hot shot wind tunnel
Hoch-Tief-Verhalten *n* <Regelung> high-low action
Hochtonausgleich *m* <Aufnahme> treble compensation
hochtonig *adj* <Aufnahme> treble
Hochtonlautsprecher *m* <Aufnahme> high-frequency loudspeaker, high-frequency speaker, treble loudspeaker, tweeter, tweeter loudspeaker
hochtoxische Substanz *f* <Sicherheit> high-toxic substance
Hochvakuum *n* <Elektronik, Mechan, Phys> high vacuum
Hochvakuumbildröhre *f* <Fernseh> high-vacuum cathode ray tube
Hochvakuum-Katodenstrahlröhre *f* <Gerät> high-vacuum cathode ray tube
Hochvakuumofen *m* <Maschinen> high-vacuum furnace
Hochvakuumphotozelle *f* <Elektronik> vacuum phototube
Hochvakuumröhre *f* <Elektronik> high-vacuum tube
Hochvakuum-Ventilbauelement *n* <Elektronik> high-vacuum valve device
hochverdichteter Motor *m* <Maschinen> supercompression engine
hochverfügbar *adj* <Comp & DV> fault-tolerant
Hochwasser *n* <Wassertrans> high tide, high water
Hochwasserentlastungsanlage *f* <Wasserversorg> flood spillway, spillway
Hochwassermarke *f* <Wassertrans> high-water mark

Hochwasserschutz m 1. <Nichtfoss Energ> flood control; 2. <Wasserversorg> flood abatement, flood control, flood prevention
Hochwasserspiegel m <Bau> high-water level
Hochwasserüberlauf m <Wasserversorg> high-water overflow
Hochwasserüberschwemmungsgebiet n <Wasserversorg> flood plain
Hochwasserüberwachung f <Wasserversorg> flood control
hochwertig adj 1. <Kohlen> high-grade; 2. <Mechan> high-tensile; 3. <Qual> high-grade
hochwertiger Stahl m <Mechan> high-tensile steel
hochwertiges Benzin n <Kfztech> (AE) high-test gasoline, (BE) high-test petrol
hochwertiges Erz n <Kohlen> high-grade ore
hochwinden v 1. <Bau> hoist; 2. <Fertig> hoist, jack; 3. <Wassertrans> winch up
Hochzahl f <Math> exponent
hochzeiliges Fernsehverfahren n (HDTV) <Fernseh> high-definition television (HDTV)
hochziehen v <Mechan> hoist
Hochziehen n <Lufttrans> pitch-up
höffiges Gebiet n <Erdöl> zone of petroleum accumulation
Hoffman-Elektrometer n <Elektriz> Hoffman electrometer
hohe Auflösung f <Comp & DV> high resolution
hohe Dämpfung f <Telekom> high loss (Transmission)
hohe Dichtigkeit f/sehr (VHD) <Optik> very high density (VHD)
hohe Drehzahl f <Maschinen> high speed
hohe Geschwindigkeit f <Maschinen> high speed • **für hohe Geschwindigkeit** <Mechan> high-speed • **mit hoher Geschwindigkeit** at high velocity, high-speed • **mit hoher Geschwindigkeit laufen lassen** <Fertig> overspeed
hohe Losgröße f <Fertig> high-quantity lot
hohe See f <Wassertrans> open sea; heavy swell (Seezustand)
hohe Wassersäule f <Nichtfoss Energ> high head
hohe Welle f <Wassertrans> billow (Seezustand)
Höhe f 1. <Bau> elevation, grade; 2. <Comp & DV, Funktech> height; 3. <Geom> altitude; height; 4. <Hydraul> head (zum Gegenpumpen); 5. <Phys> altitude; 6. <Wassertrans> elevation (Navigation); altitude (astronomische Navigation) • **auf gleiche Höhe bringen** <Druck> level • **auf gleicher Höhe** <Wassertrans> abreast • **auf gleicher Höhe mit** <Bau> level with
Höhe f der Dünung <Wassertrans> height of the swell
Höhe f der Gezeiten <Wassertrans> height of the tide
Höhe f über alles <Maschinen> overall height
Höhe f über dem durchschnittlichen Geländeniveau <Funktech> height above average terrain
Höhe f über dem Meeresspiegel 1. <Bau> height above sea level; 2. <Nichtfoss Energ> elevation above sea level; 3. <Wassertrans> elevation above sea level (Navigation)
Hoheitsgewässer npl <Wassertrans> territorial waters
Höhenanhebung f <Aufnahme> treble boost
Höhenaufnahme f <Bau> (AE) leveling, (BE) levelling
Höhendämpfung f <Aufnahme> treble roll-off
Höheneinstellung f <Kerntech> height position (eines Kontrollstabs)
Höhenfehler m <Lufttrans> height-keeping error
Höhenflosse f <Lufttrans> horizontal stabilizer, tailplane
Höhenförderer m <Fertig> elevator
Höhenfries m <Bau> stile (einer Tür)
höhengleicher Bahnübergang m <Bau, Eisenbahn> (AE) grade crossing, (BE) level crossing

Höhenkammeraufstieg m <Lufttrans> chamber ascent
Höhenkorrektureinrichtung f <Kfztech> ride height corrector
Höhenleitwerk n <Lufttrans> horizontal stabilizer, tailplane
Höhenlinie f <Bau> contour, contour line
Höhenmarke f <Mechan> benchmark
Höhenmaßstab m <Metrol> (AE) height gage, (BE) height gauge
Höhenmesser m <Phys, Trans> altimeter
Höhenmessereinstellung f <Lufttrans> altimeter setting
Höhenmesserkontrollorte mpl <Lufttrans> preflight altimeter check locations
Höhenmessgerät n <Gerät> altimeter
Höhenplan m <Bau> longitudinal section; contour map (Landvermessung)
Höhenquerruder n <Lufttrans> elevon
Höhenregler m <Aufnahme> treble control
Höhenreißer m 1. <Maschinen> scribing block; 2. <Metrol> (AE) height gage, (BE) height gauge, (AE) vernier height gage, (BE) vernier height gauge, (AE) surface gage, (BE) surface gauge, surface geometry meter
Höhenrichtwerk n <Lufttrans> elevator
Höhenruder n <Lufttrans> elevator
Höhenruderausschlag m <Lufttrans> elevator deflection
Höhenruderservosteuerung f <Lufttrans> elevator follow-up
Höhenrudersteuerung f <Lufttrans> elevator control
Höhenrudertrimmklappe f <Lufttrans> elevator trim
Höhenschnitt m <Konstzeich> vertical section
Höhenschnittpunkt m <Geom> (AE) orthocenter, (BE) orthocentre
Höhenschreiber m <Phys> barograph
Höhensteuer n <Lufttrans> elevator
Höhensteuerung f <Lufttrans> altitude controller
Höhenstrahlung f 1. <Phys> cosmic rays; 2. <Strahlphys> cosmic ray background
Höhen- und Breitenverhältnis n <Druck> aspect ratio
höhenverstellbar adj <Fertig> elevating (Radialbohrer)
höhenverstellbarer Tisch m <Maschinen> height-adjustable table
Höhenwinkel m 1. <Bau> elevation angle; 2. <Funkort> angle of elevation (Radar); 3. <Nichtfoss Energ> angle of incidence; 4. <Raumfahrt, Telekom> elevation angle
hoher Anstellwinkel m <Lufttrans> high pitch (Hubschrauber)
hoher Anzeigewert m <Gerät> high reading
hoher Bahnsteig m <Eisenbahn> elevated platform
hoher Bildpegel m <Fernseh> high picture level
hoher Integrationsgrad m (LSI) 1. <Comp & DV, Elektronik> large-scale integration, LSI; 2. <Phys> large-scale integration, LSI (mehr als etwa 1000 Bauelemente); 3. <Telekom> large-scale integration, LSI
hoher Schwefelgehalt n <Erdöl> (AE) high sulfur content, (BE) high sulphur content (Erdöl, Erdgas, Raffinerieprodukte)
hoher Widerstand m <Phys, Telekom> high resistance
höhere Ableitungen fpl <Math> higher derivatives
höhere Berechtigung f <Comp & DV> elevated priviledge • **mit höheren Berechtigungen laufen** <Comp & DV> run with elevated priviledges
höhere Dienste mpl <Telekom> higher-level services
höhere Energieleistung f <Abfall> UCV, upper calorific value
höhere Logik f <Elektronik> high-level logic
höhere Mathematik f <Math> higher mathematics
höhere Programmiersprache f (HPS) 1. <Comp & DV> high-order language, high-level language (HLL); 2. <Künstl Int> advanced language, high-level language

höhere

(HLL); 3. <Telekom> high-level language *(HLL)*; high-order language
höhere zyklische Steigung f <Lufttrans> high-order cyclic pitch *(Hubschrauber)*
höherer Grad m <Geom> higher degree
höheres Elementenpaar n <Fertig> higher pair *(Getriebelehre)*
höherwertig adj <Comp & DV> high-order
hohes Einheitsdrehzahlrad n <Nichtfoss Energ> high specific speed wheel
hohl adj <Bau> hollow
Hohladerkabel n <Telekom> loose cable structure
Hohladerstruktur f <Telekom> loose tube structure
Hohlanode f <Elektriz> hollow anode
Hohlbohrer m <Maschinen> hollow drill
Hohleisen n <Bau> gouge
hohler Flaschenboden m <Ker & Glas> punt
hohles Gegenstück n <Fertig> female part
Hohlfahrbalken m <Trans> hollow-type track girder
Hohlfeder f <Maschinen> hollow spring
hohlflächig adj <Fertig> dished
Hohlfräser m <Maschinen> running-down cutter
hohlgeschliffen adj <Mechan> dished
Hohlglasblock m <Ker & Glas> hollow glass block
Hohlhals m <Ker & Glas> hollow neck
Hohlheit f <Bau> hollowness
Hohlkammerplatte f <Kunststoff> cellular sheet
Hohlkatoden-Ionenquelle f <Kerntech> hollow cathode ion source
Hohlkehle f 1. <Bau> fillet, gorge, quirk; 2. <Eisenbahn> *(AE)* tire groove, *(BE)* tyre groove; 3. <Fertig> fillet
hohlkehlen v <Fertig> rebate
Hohlkeil m <Maschinen> saddle key
Hohlkörper m <Bau> hollow
Hohlkugel f <Fertig> hollow sphere
Hohlleiter m 1. <Elektriz, Elektrotech> hollow conductor, wave duct, waveguide; 2. <Funktech> wave duct, waveguide; 3. <Telekom> waveguide
Hohlleiter m **mit nur einem Leiter** <Elektrotech> uniconductor waveguide
Hohlleiterabschnitt m <Telekom> waveguide section
Hohlleiterantenne f <Telekom> waveguide antenna
Hohlleiterbauelement n <Elektrotech> waveguide component
Hohlleiterbereich m <Elektrotech> waveguide section
Hohlleiterbereich m **mit Schlitz** <Elektrotech> waveguide slotted section
Hohlleiterfestlast f <Elektrotech> waveguide fixed load
Hohlleiterflansch m <Telekom> waveguide flange
Hohlleitergleitlast f <Elektrotech> waveguide sliding load
Hohlleiterisolator m <Elektrotech> waveguide isolator
Hohlleiterkolben m <Elektrotech> waveguide plunger
Hohlleiterkomponente f <Elektrotech> waveguide component
Hohlleiterkopplung f <Elektrotech> waveguide coupling
Hohlleiterlast f <Elektrotech> waveguide load
Hohlleiterphasenregler m <Elektrotech> waveguide phase shifter
Hohlleitertransformator m <Elektrotech> waveguide transformer
Hohlleiterübergang m <Elektrotech> waveguide transition
Hohlmaß n <Fertig, Metrol> liquid measure
Hohlmeißel m <Bau> gouge *(Holzbearbeitung)*
Hohlniet m <Maschinen> hollow rivet
Hohlpfanne f <Ker & Glas> gutter tile
Hohlprägen n <Fertig, Maschinen> embossing
Hohlprofil n <Heiz & Kälte> hollow section
Hohlrad n <Kfztech> annulus

Hohlraum m 1. <Bau> cavity, cell; core *(eines Ziegels)*; 2. <Elektronik> cavity; 3. <Fertig> cavity, hollow; 4. <Funktech> cavity; 5. <Kerntech> cavity *(eines Druckgefäßes)*; 6. <Kohlen> cavity, cell, void; 7. <Mechan> cavity; 8. <Metall> cavity, void; 9. <Phys, Strömphys> cavity
Hohlraumbildung f 1. <Bau, Kohlen, Mechan, Metall, Nichtfoss Energ, Phys> cavitation; 2. <Raumfahrt> cavity *(Weltraumfunk)*; 3. <Strömphys> cavitation
hohlraumfreier Stahl m <Metall> interstitial free steel
Hohlraumgehalt m <Bau> porosity
Hohlraumgitter n <Elektronik> resonator grid
Hohlraummagnetron n <Elektronik> cavity magnetron
Hohlraumresonanz f <Elektronik> cavity resonance
Hohlraumresonanzeffekt m <Aufnahme> cavity resonance effect
Hohlraumresonator m 1. <Elektronik> cavity oscillator, cavity resonator; 2. <Phys, Teilphys, Telekom> cavity resonator
Hohlraumvolumen n <Textil> void volume
hohlschleifen v <Fertig> dish
Hohlschlüssel m <Mechan> *(BE)* box spanner, box wrench
Hohlschraube f <Maschinen> banjo bolt, hollow bolt
Hohlschraubenverbindung f 1. <Fertig> banjo fitting; 2. <Maschinen> banjo union
Hohlspiegel m <Phys> concave mirror
Hohlsteindecke f <Bau> hollow pot flooring, pot floor
Hohlstelle f <Fertig> rag
Hohlstift m <Maschinen> hollow pin
Hohltarget n <Kerntech> hollow target
Hohlträger m <Bau> box girder
Hohlträgerbrücke f <Bau> box girder bridge
Höhlung f 1. <Bau> hole; 2. <Maschinen> cavity
Hohlwelle f 1. <Eisenbahn> quill shaft; 2. <Fertig, Maschinen> hollow shaft
Hohlzeug n <Ker & Glas> hollow ware
Hohlzeugpressmaschine f <Ker & Glas> hollow-ware presser
Holden'scher Effekt m <Kerntech> Holden effect
Holländermesser n <Papier> beater bar, rag knife
Holländertropfen m <Ker & Glas> Dutch drop
Holländerwalze f <Papier> beater roll
Hollerith-Code m <Comp & DV> Hollerith code
Hollerith-Karte f <Comp & DV> Hollerith card
Holm m 1. <Lufttrans> spar *(Luftfahrzeug)*; 2. <Mechan> boom; 3. <Raumfahrt> spar *(Raumschiff)*
Holmbeschlag m **des Luftschraubenblattes** <Lufttrans> blade attachment fitting *(Hubschrauber)*
Holmenkasten m **des Leitwerks** <Lufttrans> fin spar box
Holmenwand f **des Flugzeugrumpfkastens** <Lufttrans> fuselage box beam wall
Holmium n *(Ho)* <Chemie> holmium *(Ho)*
holoedrisch adj <Fertig> holohedral
Hologramm n 1. <Comp & DV> holographic image; 2. <Phys, Wellphys> hologram
Holographie f 1. <Comp & DV, Phys> holography; 2. <Raumfahrt> holography *(Raumschiff)*; 3. <Strahlphys, Wellphys> holography
holographischer Scanner m <Comp & DV> holographic scanner
holographischer Speicher m 1. <Comp & DV> holographic memory, holographic storage; 2. <Telekom> holographic memory
Holz n <Ker & Glas, Papier> wood • **mit Holz verkleiden** <Bau> timber
Holzabfall m <Abfall> wood waste
holzähnliche Kohle f <Kohlen> xyloid coal
Holzarbeiten f <Bau> woodwork

Holzbalkenzug-Ankerverbindung f <Bau> haunched mortise and tenon joint
Holzbau m <Bau> building in wood, woodwork
Holzbock m <Bau> timber jack
Holzbolzen m <Bau> needle
Holzbrücke f <Bau> timber bridge
Holzbuchstaben mpl <Druck> woodtypes
Holzdrehbank f <Bau> wood-turning lathe
hölzerne Typen fpl <Druck> woodtypes
Holzfachwerk n <Bau> timber framing
Holzfachwerkträger m <Bau> timber truss
Holzfaserbruch m 1. <Fertig> woody fracture; 2. <Kerntech> fibrous fracture
Holzfloß n <Bau> timber raft
holzfrei adj <Verpack> woodfree
holzfreies Papier n <Druck, Papier> woodfree paper
Holzgurtgesims n <Bau> stringer
holzhaltig adj <Papier> woody
holzhaltiges Papier n <Druck, Papier> wood-containing paper
Holzhammer m 1. <Bau> beetle (Plasterarbeiten); 2. <Fertig> wood mallet; 3. <Ker & Glas, Maschinen, Mechan> mallet
Holzklammer f <Bau> timber dog
Holzkohle f <Fertig, Kohlen> charcoal, wood charcoal
Holzkohlefilter n <Labor> charcoal filter
Holzkohlenstaub m <Kohlen> pulverized charcoal
Holzmehl n <Kunststoff> wood flour
Holznagel m <Bau> trenail
Holzpfahl m 1. <Bau> pale, spile; 2. <Kohlen> wooden pile
Holzpflock m 1. <Bau> spile; 2. <Fertig> runner stick
Holzplatte f <Ker & Glas> nog plate
Holzplatte f **für Anwärmgefäß der Pfeife** <Ker & Glas> shoe nog plate
Holzplatz m <Bau> timber yard
Holzrahmen m <Bau> timber frame
Holzriegel m <Bau> nogging piece
Holzrost m <Bau> pontoon
Holzrutsche f <Bau> log chute
Holzsäge f <Bau> wood saw
Holzschlaghammer m <Bau> bossing mallet
Holzschleifen n <Fertig> glasspapering, sanding
Holzschleifer m <Papier> grinder
Holzschleifmaschine f <Fertig> sander, sandpapering machine
Holzschliff m 1. <Papier> groundwood pulp, mechanical wood pulp; 2. <Verpack> mechanical wood pulp
Holzschliffkarton m <Verpack> mechanical pulp board
Holzschliffpappe f <Papier> mechanical pulp board
Holzschraube f <Bau> woodscrew
Holzschutzmittel n <Bau> wood preservative
Holzschwelle f <Bau> abutment, wooden sleeper
Holzspaltkeil m <Bau> timber splitting wedge
Holzspanplatte f <Verpack> chipboard
Holzsparren m <Bau> timber rafter
Holzspiritus m <Thermod> wood alcohol
Holzstoff m <Verpack> (AE) molded pulp article, (BE) moulded pulp article
Holzstopfen m <Wassertrans> wooden plug (Schiffbau)
Holztäfelung f <Bau> wainscot
Holzunterbau m <Bau> crib
Holzverkleidung f 1. <Bau> timbering; 2. <Hydraul> wood lagging
Holzverlattung f <Bau> battening, lathing
Holzverschalung f <Hydraul> wood lagging
Holzverstärkung f <Wassertrans> wood reinforcement (Schiffbau)
Holzwolle f <Verpack> wood wool

Holzzellstoff m 1. <Druck> chemical wood pulp; 2. <Papier> wood pulp
Holzziegel m <Bau> wood brick
Homebanking-Computerschnittstelle f <Telekom> home banking computer interface
Homepage f <Comp & DV> home page
Home-Verzeichnis n <Comp & DV> home directory
Homobrenzcatechin n <Chemie> homopyrocatechol
homocyclisch adj <Chemie> carbocyclic, homocyclic, isocyclic
Homodynoszillator m <Elektronik> homodyne oscillator
homogen adj <Math, Qual> homogeneous
homogene Anregung f <Strahlphys> homogeneous stimulus
homogene isotrope Turbulenz f <Strömphys> homogeneous isotropic turbulence
homogene Strahlung f <Phys> homogeneous radiation
homogener Mantel m 1. <Optik> homogeneous cladding (Lichtleiter); 2. <Telekom> homogeneous cladding
homogener Reaktor m <Kerntech, Trans> homogeneous reactor
homogener Stimulus m <Strahlphys> homogeneous stimulus
homogenes Medium n <Phys> homogeneous medium
homogenes System n <Thermod> heating zone, homogeneous system
Homogenisator m <Labor> homogenizer (Präparation)
Homogenisieren n <Kunststoff> homogenization, homogenizing
Homogenisierung f <Metall> homogenization, homogenizing
Homojunktion f <Optik> homojunction
homologe Reihe f <Erdöl> homologous series (Petrochemie)
homologe Temperatur f <Metall> homologous temperature
homopolar adj <Elektriz> homopolar, unipolar
Homopolymer n <Kunststoff> homopolymer
Homopolymerisat n <Kunststoff> homopolymer
Homopolymerisation f <Kunststoff> homopolymerization
Homoterephtal... <Chemie> homoterephthalic
Homoübergang m 1. <Elektronik> homojunction (einfacher Übergang); 2. <Optik> homojunction
homozyklisch adj <Chemie> carbocyclic, homocyclic, isocyclic
Honahle f <Fertig, Maschinen> hone, honing tool
honen v <Fertig, Maschinen, Mechan> hone
Honen n <Fertig, Maschinen> honing
Honmaschine f <Fertig, Maschinen> honing machine
Honstein m <Maschinen> honestone, honing stone
Hooke'sches Gesetz n <Bau, Phys> Hooke's law
Hopcalit n <Chemie> hopcalite
hopfenähnlich adj <Chemie> lupuline
Hopperbagger m <Wassertrans> hopper dredge, hopper dredger
Hopperschute f <Wassertrans> hopper barge
Hör... 1. <Akustik> aural; 2. <Comp & DV> audio
hörbar adj <Ergon> audible
hörbarer Frequenzbereich m <Akustik> audible frequency range
hörbarer Schall m <Sicherheit> audible sound
hörbares Signal n <Aufnahme> audible signal
Hörbarkeitsgrenze f <Akustik> hearing threshold level
Hörbarkeitsmesser m <Aufnahme> audibility meter
Hörbarkeitsschwelle f <Umweltschmutz> threshold of audibility
Hörbereich m <Aufnahme, Ergon> audible range
Horchgerät n <Akustik> sound locator
Horchortung f <Wellphys> sound ranging

Hordein

Hordein n <Chemie> hordein
Hörer m <Telekom> handset, receiver, hand-held receiver
• **Hörer abgehoben** <Telekom> receiver off hook • **Hörer abheben** <Telekom> remove the handset *(Telefon)*
• **Hörer aufgelegt** <Telekom> on-hook condition, receiver on hook
Hörer m **mit Verstärker** <Telekom> amplified handset
Hörerempfindlichkeit f <Telekom> receive loudness rating *(Telefon)*
Hörerschnur f <Telekom> handset cord
Hörfeld f <Akustik> auditory sensation area
Hörfläche f <Akustik> auditory sensation area
hörfrequent adj <Aufnahme> audio-frequency
Hörfrequenz f 1. <Akustik, Elektronik> acoustic frequency, audio frequency; 2. <Aufnahme> acoustic frequency
Hörfrequenzmesser m <Strahlphys> audiometer
Hörfunk m <Funktech> sound broadcasting
Hörgerät n 1. <Akustik> hearing aid; 2. <Ergon> artificial ear
Hörhilfe f 1. <Akustik> hearing aid; 2. <Ergon> artificial ear
Horizont m <Telekom> optical horizon
horizontal adj <Geom> horizontal
horizontal eingebauter Motor m <Kfztech> horizontal engine
horizontal und topladende Kartoniereinrichtung f <Verpack> horizontal and top loader cartoner
Horizontal... <Elektronik, Fernseh, Verpack> horizontal
Horizontalablenkplatte f 1. <Elektronik> horizontal deflection plate; 2. <Fernseh> X-plate
Horizontalablenkplatten fpl <Phys> horizontal-deflecting plates
Horizontalablenkspule f <Elektrotech> horizontal deflection coil
Horizontalablenkung f <Fernseh> horizontal sweep
Horizontalablenkungssteuerung f <Fernseh> horizontal deflection control
Horizontalablenkverstärker m <Gerät> sweep deflection amplifier *(Oszilloskop)*
Horizontalabtastfrequenz f <Elektronik, Fernseh, Funktech> horizontal-scanning frequency
Horizontalabtastung f <Fernseh> horizontal scanning
Horizontalachse f <Bau> horizontal axis
Horizontalauflösung f <Fernseh> horizontal resolution
Horizontalaufzug m <Trans> horizontal elevator
Horizontalaustastintervall n <Fernseh> horizontal-blanking interval
Horizontalaustastung f <Fernseh> horizontal blanking
Horizontalbalken m <Fernseh> horizontal bar
Horizontalbrunnen m <Bau> shallow well
Horizontaldynamikkonvergenz f <Fernseh> horizontal dynamic convergence
Horizontale f 1. <Geom> horizontal; 2. <Mechan> datum line
horizontale Ablenkung f <Elektronik> horizontal deflection
horizontale Achse f *(X-Achse)* <Math> horizontal axis *(x-axis)*
horizontale Bildzentrierung f <Fernseh> *(AE)* horizontal-centering control, *(BE)* horizontal-centring control
horizontale Ebene f <Geom> horizontal plane
horizontale Einwicklung f <Verpack> horizontal wrapping
horizontale Kartonfüllmaschine f <Verpack> horizontal case loader
horizontale Komponente f <Phys> horizontal component
horizontale Luftströmung f <Phys, Strömphys> advection

horizontale Polarisation f <Elektrotech, Phys, Telekom> horizontal polarization
horizontale Schichten fpl <Strömphys> horizontal layers
horizontale und vertikale Linien fpl <Lufttrans> horizontal and vertical bars *(flugrichtungsanzeigend)*
Horizontaleinfangen n <Fernseh> horizontal lock
Horizontalentfernung f <Funkort> ground range *(Radar)*
horizontaler Bilddurchlauf m <Comp & DV> scrolling
horizontaler Reiseflug m <Lufttrans> level cruise
horizontaler Windgradient m <Lufttrans> horizontal wind shear
horizontales Richtdiagramm n <Funktech> horizontal directivity diagram, horizontal directivity pattern
Horizontalflug m <Lufttrans> level flight
Horizontalformat n <Druck> landscape format
Horizontalfrequenz f <Fernseh> horizontal frequency
Horizontalkartoniermaschine f <Verpack> horizontal-cartoning machine
Horizontalkraft f <Bau> horizontal thrust
Horizontalmühle f <Sicherheit> horizontal-milling machine
Horizontalregler n <Fernseh> horizontal hold control
Horizontalschnitt m <Konstzeich> horizontal section
Horizontalsynchronisierimpuls m <Fernseh> breezeway
Horizontalsynchronisierung f <Fernseh> horizontal synchronization
Horizontaltabulator m <Comp & DV> HT, horizontal tabulator
Horizontal- und Vertikaleinschlagmaschine f <Verpack> horizontal and vertical wrapping machine
Horizontalvergaser m <Kfztech> *(AE)* horizontal carburetor, *(BE)* horizontal carburettor
Horizontalverstärker m <Elektronik> horizontal amplifier
horizontieren v <Lufttrans> level out
Horizontierungseinheit f <Lufttrans> *(AE)* leveling unit, *(BE)* levelling unit
Horizontmelder m <Raumfahrt> horizon sensor *(Raumschiff)*
Hörkapsel f <Telekom> receiver inset
Hörmelder m <Gerät> acoustic alarm device, audible alarm device
Hörminimummethode f <Funkort> aural null method *(Funkpeilung)*
Horn n 1. <Fertig> beak *(Schmieden)*; 2. <Maschinen> beak
Hornantenne f <Funktech, Raumfahrt> horn, horn antenna
Hornausgleich m <Lufttrans> balance horn
Hornklausel f <Künstl Int> Horn clause
Hornschiene f <Eisenbahn> wing rail
Hornstein n <Erdöl> chert *(Geologie)*
Hornstrahler m <Lufttrans> horn *(Funkwesen)*
Hörpegel m <Ergon> hearing level
Hörpeilung f <Funkort> auditory direction finding
Hörprothese f <Ergon> auditory prosthesis
Hörrundfunk m <Funktech> sound broadcasting
Hörrundfunk m über Antenne <Funktech> audio broadcasting by antenna, terrestrial audio broadcasting
Hörrundfunkempfänger m <Funktech> sound broadcast receiver
Hörrundfunksender m <Funktech> sound broadcast transmitter
Hörsaal m <Aufnahme> auditorium
Hörschall m <Sicherheit> audible sound
Hörschärfe f 1. <Akustik> auditory acuity; 2. <Ergon> hearing acuity
Hörschwelle f 1. <Akustik> aural threshold, hearing threshold, threshold of audibility; 2. <Ergon> hearing

threshold, hearing threshold level; 3. <Phys> hearing threshold
Hörschwellenabwanderung f <Ergon> hearing threshold shift (Lärm)
Hörschwellendifferenz f <Akustik> hearing threshold difference
Hörschwellenmessgerät n <Akustik, Gerät> audiometer
Hörschwellenpegel m <Akustik> hearing threshold level
Hör-Sprech-Schalter m <Aufnahme, Funktech> talk-listen switch
Hörtest m <Akustik> hearing test
Hörverlustfaktor m <Akustik> hearing loss factor
Hörvermögen n <Ergon> audition, hearing
Hörverständigung f im Fahrzeug <Kfztech> in-vehicle aural communication
Hosenboje f <Wassertrans> breeches buoy
Hosenrohr n 1. <Fertig> Y-pipe; 2. <Maschinen, Mechan> breeches pipe
Host m <Comp & DV> host
Hostcomputer m <Comp & DV> host computer
Host-Integration f <Comp & DV> host integration
Hostsystem n <Comp & DV> host system
Hot-Carrier-Diode f <Elektronik> hot carrier diode
Hot-Spot m <Kerntech> hot spot
Hovercraft n <Wassertrans> (BE) hovercraft (Schifftyp)
h-Parameter m (Hybrid-Parameter) <Elektronik> hybrid parameter
HPC-Ruß m (schwer verarbeitbarer Kanalruß) <Kunststoff> HPC carbon black (hard-processing channel carbon black)
HPLC (Hochdruckflüssigchromatographie) <Labor, Lebensmittel> HPLC (high-pressure liquid chromatography)
H-Profil m <Fertig> H-section
HPS (höhere Programmiersprache) <Comp & DV, Telekom> HLL (high-level language)
H-Rahmen-Drehgestell n <Eisenbahn> H frame bogie
HTML-Formular n <Comp & DV> HTML form
HTR (Hochtemperaturreaktor) <Kerntech> HTR (high-temperature reactor)
HTTP-fähig adj <Comp & DV> HTTP-compliant
HTTP-Web-Servermaschine f <Comp & DV> HTTP Web server engine
Hub m 1. <Bau> lift; 2. <Erdöl> heave (Seegang); 3. <Fertig> stroke (Kunststoffinstallationen); blow (Presse, Schere); 4. <Funktech> swing (Frequenz); 5. <Hydraul> (AE) elevator, (BE) lift; stroke (Steuerkolben); 6. <Kerntech> lift; 7. <Kfztech> stroke; 8. <Lufttrans> lift; 9. <Maschinen> daylight, displacement, stroke, throw, travel; 10. <Phys> stroke; 11. <Telekom> hub, shift; 12. <Textil> traverse
Hubanzeigestift m <Fertig> position indicator (Kunststoffinstallationen)
Hubbegrenzer m 1. <Fertig> stop; 2. <Sicherheit> overrun stop (für Werkzeuge); travel limiting device (für Krane)
Hubbegrenzung f 1. <Fertig> lift limiter, stroke limiter (Kunststoffinstallationen); 2. <Sicherheit> travel limiting device (für Krane)
Hubbereich m <Kerntech> range of movement (eines Steuerstabes)
Hubbrücke f 1. <Bau> lifting bridge; 2. <Trans> lift bridge; 3. <Wassertrans> lifting bridge
Hubdruck m <Trans> lifting pressure
Hubel m <Ker & Glas> blank
Hubende n 1. <Fertig> bottom stroke; 2. <Maschinen> end of stroke, end of travel
Hubfenster n <Bau> sash window
Hubgebläse n <Lufttrans> lift fan
Hubgerät n <Nichtfoss Energ> lift-type device

Hubgeschwindigkeit f 1. <Kerntech> rate of travel (eines Steuerstabes); 2. <Maschinen> hoisting speed
Hubglied n <Fertig> follower (Kurve)
Hubgriff m <Bau> lift
Hubhöhe f 1. <Kerntech> stroke (eines Elementes); 2. <Maschinen> lift
Hubkarren m <Verpack> lifting vehicle
Hubkette f <Bau> lifting chain
Hubklinke f <Bau> lift latch
Hubkolbenmotor m <Kfztech, Lufttrans, Maschinen, Wassertrans> piston engine, reciprocating engine
Hubkolbenverbrennungsmaschine f <Kfztech> reciprocating internal combustion engine
Hubkolbenverdichter m <Maschinen> reciprocating compressor, reciprocating piston compressor
Hubkolbenzähler m 1. <Gerät> reciprocating piston-type meter; 2. <Regelung> ball prover flow measuring device
Hubkraft f <Maschinen, Trans> lifting power
Hublader m <Bau> loading shovel (Tiefbau)
Hublänge f <Maschinen> length of stroke
Hublänge f des Kolbens <Maschinen> piston stroke
Hubmagnet m <Maschinen> lifting magnet
Hubmessung f <Telekom> deviation measurement
Hubmitte f <Fertig> midstroke
Hubmotor m <Maschinen> hoisting motor, lift motor
Hubplattform f 1. <Erdöl> jack-up rig; 2. <Maschinen> lift platform; 3. <Mechan> elevating table
Hubrad n <Maschinen> elevating wheel, lifting wheel
Hubraum m 1. <Kfztech> displacement; capacity (Motor); 2. <Maschinen> displacement; 3. <Mechan, Phys> cubic capacity
Hubrohr n <Bau> lift pipe
Hubscheibe f <Fertig> main driving gear (Getriebelehre)
Hubschrauber m <Lufttrans> helicopter • mit Hubschrauber befördern <Lufttrans> helicopter
Hubschrauber m mit Doppelfunktion <Lufttrans> dual-role helicopter
Hubschrauberavionikausrüstung f <Lufttrans> helicopter avionics package
Hubschrauberflughafen m <Trans> heliport
Hubschrauberlandedeck n <Lufttrans> helicopter landing deck
Hubschrauberlandefläche f <Lufttrans> helicopter landing surface, heliport deck
Hubschrauberlandeplattform f 1. <Lufttrans> helicopter landing platform; 2. <Wassertrans> helicopter pad
Hubschrauberlandeplatz m 1. <Erdöl> heliport; helipad (auf Plattform); 2. <Lufttrans> helipad, helistop; spot (auf Schiffen); 3. <Trans> heliport
Hubschrauberstation f <Lufttrans> helicopter station
Hubschrauberverhalten n <Lufttrans> (AE) helicopter behavior, (BE) helicopter behaviour
Hubschrauberzubringerdienst m <Lufttrans> helicopter shuttle service
Hubseil n 1. <Eisenbahn> fall rope; 2. <Maschinen> hoisting rope
Hubspiegel m <Foto> swing-up mirror
Hubspindel f 1. <Fertig> raising screw; 2. <Maschinen> elevating screw
Hubstange f <Hydraul> lifting rod
Hubstapler m 1. <Fertig> stacker truck; 2. <Mechan> lifting truck
Hubtor n <Bau> lift gate
Hubtragschrauber m <Trans> heligyro
Hubventil n 1. <Hydraul> lifting valve; 2. <Maschinen> lift valve
Hubverhältnis n <Elektronik> deviation ratio (Modulation)
Hubverlagerung f <Fertig> ram positioning

Hubvermögen

Hubvermögen *n* **mit Haken** <Kerntech> lifting capacity with hook
Hubvolumen *n* <Fertig> piston displacement
Hubwagen *m* 1. <Mechan> lifting truck; 2. <Verpack> lift truck
Hubweg *m* <Maschinen> stroke
Hubwelle *f* <Lufttrans> lift shaft
Hubwerk *n* 1. <Erdöl> elevator; 2. <Fertig> hoisting gear
Hubwinde *f* <Lufttrans> hoist
Hubwindeanschlussstück *n* <Lufttrans> hoist fitting *(Hubschrauber)*
Hubwindehebel *m* <Lufttrans> hoist lever
Hubwindenfahrgestell *n* <Lufttrans> hoisting carriage
Hubzähler *m* <Gerät> stroke counter
Hubzählgerät *n* <Gerät> stroke counter
Hubzapfen *m* <Fertig> crankpin
Hubzeit *f* <Elektriz> *(AE)* traveling time, *(BE)* travelling time
Huckepackbahn *f* <Eisenbahn> piggyback rail
Huckepackreaktor *m* <Kerntech> package reactor, transportable reactor
Huckepackverkehr *m* 1. <Eisenbahn> piggyback traffic, piggyback transport, rail transport of road trailers; 2. <Lufttrans> piggyback traffic, piggyback transport; 3. <Trans> trailers on flat cars, TOFC
Huf *m* <Elektriz, Phys> horse shoe
Hufeisengewölbe *n* <Bau> horseshoe arch
Hufeisenmagnet *m* <Elektriz, Phys> horseshoe magnet
Hufeisenofen *m* <Ker & Glas> horseshoe-fired furnace
Hufeisenprofile *npl* <Maschinen> horseshoe sections
Hufeisenrettungsboje *f* <Wassertrans> horseshoe lifebuoy
Hufeisenwirbel *m* <Strömphys> horseshoe vortex
Hula-Hoop-Antenne *f* <Funktech> hula hoop aerial, hula hoop antenna
Hülle *f* 1. <Comp & DV> hood; 2. <Elektrotech> envelope; 3. <Fertig, Labor> jacket *(Glasartikel)*; 4. <Lebensmittel> skin; 5. <Mechan> cover; 6. <Raumfahrt> case, jacket, shell *(Raumschiff)*
Hüllengewebe *n* <Fertig> casing *(Keilriemen)*
Hüllkurve *f* 1. <Fernseh, Geom> envelope; 2. <Mechan> envelope curve; 3. <Raumfahrt> envelope *(Weltraumfunk)*; 4. <Telekom> envelope
Hüllkurvendemodulation *f* <Funktech> envelope detection
Hüllkurvendetektor *m* <Funktech> envelope detector, peak rectifier
Hüllkurvengleichrichter *m* <Funktech> envelope detector, peak rectifier
Hüllkurvenverzögerung *f* <Raumfahrt> envelope delay *(Weltraumfunk)*
Hüllmaterial *n* <Kerntech> cladding material
Hüllschnitt *m* <Maschinen> profiling cut
Hülltemperaturgrenze *f* <Kerntech> cladding temperature limit
Hülse *f* 1. <Bau> tube; barrel *(eines Schlosses)*; 2. <Fertig> quill, sleeve, thimble; shell *(Kunststoffinstallationen)*; 3. <Kerntech> cladding; 4. <Maschinen> bush, bushing, collar, collet, runner, sleeve; 5. <Mechan> collar, sheath; 6. <Papier> core *(der Papierrolle)*; 7. <Telekom> ferrule; 8. <Textil> bobbin; 9. <Verpack> sleeve
Hülse *f* **mit Außengewinde** <Maschinen> threaded bush
Hülsenfrucht *f* <Lebensmittel> legume, pulse
Hülsenkette *f* <Maschinen> bushed roller chain
Hülsenlehre *f* <Ker & Glas, Metrol> *(AE)* ring gage, *(BE)* ring gauge
hülsenloses Brennelement *n* <Kerntech> canal-ray discharge
Hülsenrohr *n* <Bau> socket pipe

Hülsenschiebermotor *m* <Kfztech> sleeve valve engine
Hülsenüberwachung *f* <Kerntech> cladding monitoring
Humulen *n* <Chemie> humulene
Humus *m* <Kohlen> humus
Humusbehälter *m* <Wasserversorg> humus tank
Hundekoje *f* <Wassertrans> quarter berth
Hundert-Jahr-Sturm *m* <Erdöl> hundred year storm *(Offshore)*
Hundert-Jahr-Welle *f* <Erdöl> hundred year wave *(Offshore)*
Hundertprozentprüfung *f* <Qual> one-hundred-percent inspection
Hundertstel unterteilt *adj/in* <Mechan> centigrade
Hund'sche Regeln *fpl* <Phys> Hund's rules
Hutmutter *f* 1. <Fertig> acorn nut; handwheel nut *(Kunststoffinstallationen)*; 2. <Lufttrans> acorn nut; 3. <Maschinen> cap nut, dome nut; 4. <Mechan> cap nut
Hutschraube *f* <Maschinen> cap bolt
Hütte *f* <Metall> smeltery
Hütteneis *n* <Ker & Glas> granulated glass
Hüttenkunde *f* <Fertig> metallurgy, process metallurgy
hüttenmännisch *adj* <Fertig, Metall> metallurgical
Hüttenmetall *n* <Metall> virgin metal
Hüttenofen *m* <Maschinen, Metall> metallurgical furnace
Hüttenwerk *n* <Metall> smeltery
Hüttenzement *m* <Metall> slag cement
Hüttenzink *n* <Metall> spelter
Huygens'sches Okular *n* <Phys> Huygens' eyepiece
Huygens'sches Prinzip *n* <Phys> Huygens' principle
HV-Naht *f* <Fertig> single bevel groove weld *(Schweißen)*
HVSt *(Hauptvermittlungsstelle)* <Telekom> main exchange
HVStd *(Hauptverkehrsstunde)* <Telekom> busy hour
HVt *(Hauptverteiler)* <Telekom> MDF *(main distribution frame)*
H-Welle *f* <Elektrotech, Telekom> H-wave, TE wave, transverse electric wave
HWZ *(Halbwertszeit)* 1. <Kerntech, Phys, Strahlphys, Teilphys> T½ *(half-life)*; 2. <Telekom> FDHM *(full duration half maximum)*
Hyacinthenaldehyd *m* <Chemie> hyacinthin, phenylacetaldehyde
Hyacinthin *n* <Chemie> hyacinthin, phenylacetaldehyde
Hybrid... <Kfztech, Telekom, Thermod> hybrid
Hybridantrieb *m* 1. <Kfztech> hybrid propulsion; 2. <Wassertrans> hybrid drive
Hybridbus *m* <Kfztech> hybrid bus
Hybride *f* <Elektronik> hybrid junction *(Mikrowellen)*
hybride Anwendung *f* <Comp & DV> hybrid application
hybrider Treibstoff *m* <Thermod> lithergol
hybrider Vermittlungsprozessor *m* <Telekom> hybrid call processor
hybrides Glasfaser-Funkverbindungsnetz *n* <Telekom> hybrid fiber wireless network
Hybridfahrzeug *n* <Kfztech> hybrid vehicle
Hybridkabel *n* <Telekom> hybrid cable *(optische Fasern/Kupferdrähte)*
Hybridkonfiguration *f* <Nichtfoss Energ> hybrid configuration
Hybridlager *n* <Maschinen> hybrid bearing
Hybrid-Mikroschaltung *f* <Elektronik> hybrid microcircuit
Hybridmode *f* <Optik> hybrid mode *(Wellenausbreitung)*
Hybridmodus *m* <Telekom> hybrid mode
Hybridmotor *m* <Kfztech> hybrid engine
Hybridnetz *n* **Glasfaser-Koaxkabel** <Telekom> hybrid fiber coax network
Hybrid-Parameter *m* *(h-Parameter)* <Elektronik> hybrid parameter
Hybridplattform *f* <Erdöl> hybrid platform *(Offshore-Betrieb)*

Hybridrechnen n <Comp & DV> digitally implemented analogue processing
Hybridrechner m <Comp & DV> hybrid computer
Hybridschaltkreis m 1. <Phys> hybrid circuit; 2. <Telekom> hybrid integrated circuit
Hybridschaltung f 1. <Comp & DV> hybrid circuit; 2. <Elektronik> hybrid circuit, hybrid integrated circuit
Hybridschnittstelle f <Comp & DV> hybrid interface
Hybridschrittmotor m <Comp & DV> hybrid stepping motor
Hybridstation f <Telekom> balanced station
Hybridsystem n <Kfztech, Telekom> hybrid system
Hybridtragflächenboot n <Wassertrans> hybrid foil craft
Hybridwerkzeug n <Künstl Int> hybrid tool *(Expertensysteme)*
Hydantoin n <Chemie> hydantoin
Hydantoinsäure f <Chemie> hydantoic acid
Hydracryl... <Chemie> hydracrylic
Hydrant m 1. <Sicherheit> fire hydrant; 2. <Wasserversorg> hydrant, *(AE)* plug, water hydrant, *(AE)* water plug
Hydrast... <Chemie> hydrastic
Hydrastin n <Chemie> hydrastine
Hydrat n <Chemie> hydrate • **Hydrat bilden** <Chemie> hydrate
Hydratation f 1. <Bau> hydration *(Zement)*; 2. <Chemie> hydration
Hydratationswärme f 1. <Bau> heat of hydration *(Zement)*; 2. <Thermod> heat of hydration
Hydratbildung f <Chemie> hydration
hydratisch adj <Chemie> hydrous
hydratisieren v <Chemie> hydrate
hydratisiert adj <Chemie> hydrous
hydratisiertes Proton n <Chemie> oxonium
Hydratropa... <Chemie> hydratropic
Hydraulik f <Erdöl, Hydraul, Maschinen> hydraulics
Hydraulikaggregat n <Fertig> power pack
Hydraulikanlage f <Wassertrans> hydraulic system
Hydraulikantrieb m <Maschinen> hydraulic drive
Hydraulikarmaturen fpl <Hydraul> hydraulic fittings
Hydraulikbehälter m <Hydraul> hydraulic reservoir
Hydraulikdetektor m <Kfztech> hydraulic detector
Hydraulikflüssigkeit f <Anstrich, Hydraul, Maschinen> hydraulic fluid
Hydraulikflüssigkeitsbehälter m <Maschinen> hydraulic fluid reservoir
Hydraulikgenerator m <Hydraul> hydraulic generator
Hydraulikgetriebe n <Maschinen> hydraulic transmission
Hydraulikheber m <Maschinen> hydraulic jack
Hydraulikkolben m 1. <Fertig> hydraulic jack; 2. <Maschinen> hydraulic ram
Hydraulikkraft f <Hydraul> hydraulic power
Hydraulikkupplung f <Hydraul> hydraulic clutch
Hydrauliköl n <Hydraul> hydraulic fluid
Hydraulikpresse f 1. <Bau, Hydraul> hydraulic jack; 2. <Labor> hydraulic press
Hydraulikpumpe f <Maschinen, Meerschmutz> hydraulic pump
Hydraulikschaltbild n <Kontroll> hydraulic diagram
Hydraulikschlauch m <Kunststoff> hydraulic hose
Hydraulikspeicher m <Kunststoff> hydraulic accumulator
Hydrauliksystem n <Hydraul, Maschinen> hydraulic system
Hydraulikverriegelung f <Hydraul> hydraulic locking
Hydraulikverschluss m <Hydraul> hydraulic locking
Hydraulikwinde n <Hydraul> hydraulic jack
Hydraulikzylinder m <Maschinen> hydraulic cylinder
hydraulisch adj <Mechan, Wassertrans> hydraulic
hydraulisch angetriebener Zylinder m <Hydraul> hydraulic actuating cylinder
hydraulisch betätigter Zylinder m <Hydraul> hydraulic actuating cylinder
hydraulisch betätigtes Ventil n <Heiz & Kälte> hydraulically-operated valve
hydraulisch betriebenes Gerät n <Maschinen> hydraulically-operated device
hydraulisch unterstützte Bremse f <Kfztech> hydraulically-assisted brake
hydraulisch unterstützte Kupplung f <Kfztech> hydraulically-assisted clutch
hydraulisch vorgesteuert adj <Fertig> pilot-operated
hydraulische Bremse f <Kfztech, Maschinen> hydraulic brake
hydraulische Druckquelle f <Hydraul> hydraulic pressure source
hydraulische Druckversorgung f <Hydraul> hydraulic pressure supply
hydraulische Förderhöhe f <Erdöl> hydraulic head
hydraulische Geräte npl <Maschinen> hydraulic equipment
hydraulische Kopierfräsmaschine f <Maschinen> hydraulic copy mill
hydraulische Kupplung f 1. <Kfztech> hydraulic clutch *(Kraftübertragung)*; 2. <Maschinen> hydraulic coupling
hydraulische Maschinen fpl <Maschinen> hydraulic machinery
hydraulische Presse f <Kunststoff> hydraulic press
hydraulische Pumpe f <Maschinen> hydraulic pump
hydraulische Rissbildung f <Nichtfoss Energ> hydraulic fracturing
hydraulische Setzmaschine f <Kohlen> plunger-type jig
hydraulische Strangpresse f <Maschinen> hydraulic extruder
hydraulische Wasserkraft f <Hydraul> hydraulic power
hydraulischer Antrieb m <Maschinen> hydraulic drive
hydraulischer Aushub m <Bau> hydraulic excavation
hydraulischer Bagger m <Bau> hydraulic excavator
hydraulischer Bremskraftverstärker m <Kfztech> hydraulic brake servo
hydraulischer Heber m <Maschinen, Mechan> hydraulic jack
hydraulischer Kalk m <Bau> hydraulic lime
hydraulischer Kompensator m <Erdöl> hydraulic compensator *(Plattform)*
hydraulischer Regler m <Kfztech> hydraulic control system
hydraulischer Sohlenauftrieb m <Kohlen> hydraulic bottom heave
hydraulischer Stoßdämpfer m <Kfztech> dashpot *(Vergaser)*
hydraulischer Strahlantrieb m <Kfztech> hydraulic jet propulsion
hydraulischer Ventilfederheber m <Kfztech> hydraulic tappet, hydraulic valve lifter
hydraulischer Ventilstößel m <Kfztech> hydraulic tappet, hydraulic valve lifter
hydraulischer Verlust m <Kerntech> hydraulic loss
hydraulischer Widder m <Nichtfoss Energ> hydraulic ram
hydraulischer Wirkungsgrad m <Nichtfoss Energ> hydraulic efficiency
hydraulisches Bremssystem n <Kfztech, Maschinen> hydraulic brake system
hydraulisches Gefälle n <Bau> hydraulic gradient
hydraulisches Gestänge n <Kfztech> hydraulic linkage
hydraulisches Getriebe n <Maschinen> hydraulic transmission
hydraulisches Zirkulationssystem n <Erdöl> hydraulic circulation system *(Bohrtechnik)*

Hydrazid

Hydrazid n <Chemie> hydrazide
Hydrazin n 1. <Chemie> diamide, diazane, hydrazine; 2. <Raumfahrt> hydrazine
Hydrazinantrieb m <Raumfahrt> hydrazine propulsion *(Raumschiff)*
Hydrazinantriebssystem n <Raumfahrt> hydrazine propulsion system *(Raumschiff)*
Hydrazobenzol n <Chemie> phenylhydrazine
Hydrid n <Chemie, Metall> hydride
Hydrierapparat m <Lebensmittel> hydrogenator
hydrieren v <Chemie, Lebensmittel> hydrogenate
Hydrieren n <Chemie> hydrogenation
hydrierende Spaltung f <Erdöl> hydroracking *(Raffinerie)*
hydriert adj 1. <Chemie> hydrous; 2. <Lebensmittel> hydrogenated
hydrierte Schicht f <Ker & Glas> hydrated layer
hydriertes Fett n <Lebensmittel> hydrogenated fat
Hydrierung f 1. <Chemie> hydrogenation; 2. <Erdöl> hydrogenation *(Raffinerie)*; 3. <Lebensmittel> hydrogenation
Hydrinden n <Chemie> hydrindene, indan
Hydro… <Chemie> hydro…
hydroaromatisch adj <Chemie> hydroaromatic
Hydrobatterie-Dieselsystem n <Nichtfoss Energ> hydro-battery-diesel system
Hydrobilirubin n <Chemie> hydrobilirubin, urobilin
Hydrobrom… <Chemie> hydrobromic
Hydrobromid n <Chemie> hydrobromide
Hydrocellulose f <Chemie> hydrocellulose
Hydrochinon n <Chemie> hydroquinone
Hydrochlorid n <Chemie> hydrochloride
Hydrocortison n <Chemie> hydrocortisone
Hydrocotarnin n <Chemie> hydrocotarnine
Hydrodynamik f <Erdöl, Maschinen, Phys, Strömphys, Wassertrans> hydrodynamics *(Schiffkonstruktion)*
hydrodynamisch adj 1. <Chemie> hydrodynamic; 2. <Erdöl> hydrodynamic *(Hydraulik)*; 3. <Phys> hydrodynamic
hydrodynamische Faserverwirbelung f <Kunststoff> hydraulic entanglement process, hydroentanglement process
hydrodynamische Kupplung f 1. <Kfztech> fluid coupling *(Kraftübertragung)*; 2. <Maschinen> hydrodynamic clutch
hydrodynamische Peilung f <Raumfahrt> hydrodynamic bearing *(Raumschiff)*
hydrodynamische Schmierung f <Maschinen> hydrodynamic lubrication
hydrodynamischer Auftrieb m <Wassertrans> hydrodynamic lift
hydrodynamischer Dämpfungsfaktor m <Nichtfoss Energ> hydrodynamic damping factor
hydrodynamischer Strömungswiderstand m <Wassertrans> hydrodynamic drag
hydrodynamisches Modell n <Nichtfoss Energ> hydrodynamic model
hydroelastische Radaufhängung f <Kfztech> hydroelastic suspension
hydroelektrische Umformung f <Maschinen> hydroelectric forming
hydroelektrischer Generator m <Elektrotech> hydroelectric generator
hydroelektrisches Umformen n <Maschinen> electrohydraulic forming
Hydroelektrizität f <Elektriz> hydroelectricity
Hydrofluorid n <Chemie> hydrofluoride
Hydrogen… <Chemie> hydrogen, monohydric
Hydrogenazid n <Chemie> diazoimide

Hydrogencarbonat n <Chemie> hydrocarbonate
Hydrogenperoxid n <Chemie> hydrogen peroxide
Hydrogensalz n <Lebensmittel> acid salt
Hydrogensulfat n <Chemie> *(AE)* hydrosulfate, *(BE)* hydrosulphate
Hydrogensulfid n <Chemie> *(AE)* hydrogen sulfide, *(BE)* hydrogen sulphide, *(AE)* hydrosulfide, *(BE)* hydrosulphide, *(AE)* sulfhydrate, *(BE)* sulphydrate
Hydrogensulfit n <Chemie> *(AE)* hydrogen sulfite, *(BE)* hydrogen sulphite, *(AE)* hydrosulfite, *(BE)* hydrosulphite
Hydrogeologie f <Wasserversorg> hydrogeology
Hydrographie f <Nichtfoss Energ, Wassertrans> hydrography
hydrographisch adj <Wassertrans> hydrographic
Hydrolastikfederung f <Kfztech> Hydrolastic suspension
Hydrologie f <Wasserversorg> hydrology
hydrologische Bilanz f <Wasserversorg> hydrological balance
hydrologische Untersuchung f <Bau, Kohlen> hydrological study
hydrologischer Zyklus m <Wasserversorg> hydrologic cycle
Hydrolyse f <Kunststoff, Lebensmittel> hydrolysis
Hydromechanik f <Strömphys> hydromechanics
hydromechanische Diesellokomotive f <Eisenbahn> diesel-hydromechanical locomotive
hydromechanische Kupplung f <Maschinen> hydromechanical clutch
hydromechanischer Fliehkraftregler m <Nichtfoss Energ> hydromechanical governor
hydromechanischer Zentrifugalregler m <Nichtfoss Energ> hydromechanical governor
Hydrometallurgie f <Kohlen, Metall> hydrometallurgy
Hydrometer n 1. <Elektriz> hydrometer; 2. <Erdöl> hydrometer *(Messtechnik)*; 3. <Kohlen> hydrometer; 4. <Qual> hydrometer *(Batterie)*
Hydrometrie f <Chemie, Wasserversorg> hydrometry
Hydromotor m <Maschinen> hydraulic motor
Hydroniumion n <Chemie> hydroxonium ion
hydrophil adj <Kohlen> hydrophilic
hydrophob adj <Chemie, Kohlen> hydrophobic
Hydrophobiermittel n <Textil> water repellent
Hydrophobierung f <Textil> hydrophobizing
Hydrophon n 1. <Erdöl> hydrophone *(Offshore-Seismik)*; 2. <Telekom> hydrophone
hydropneumatische Aufhängung f <Kfztech> hydropneumatic suspension
hydropneumatische Bremse f <Kfztech> hydropneumatic brake
hydropneumatischer Speicher m <Maschinen> hydropneumatic accumulator
hydropneumatischer Stoßdämpfer m <Kfztech> oleopneumatic shock absorber
Hydropulsor m <Nichtfoss Energ> hydraulic ram
Hydropumpe f <Meerschmutz> hydraulic pump
Hydrosilicat n <Chemie> hydrosilicate
Hydrosol n <Chemie> hydrosol
Hydrosphäre f <Umweltschmutz> hydrosphere
Hydrostatik f <Bau, Maschinen, Mechan, Phys, Strömphys, Wasserversorg> hydrostatics
hydrostatisch adj <Maschinen, Phys> hydrostatic
hydrostatische Höhe f <Erdöl> hydrostatic head
hydrostatische Lagerung f <Mechan> hydrostatic bearing
hydrostatische Schmierung f <Maschinen> hydrostatic lubrication
hydrostatische Spannung f <Metall> hydrostatic stress
hydrostatische Übertragung f <Mechan> hydrostatic transmission

hydrostatischer Druck m 1. <Erdöl> hydrostatic pressure *(Geologie)*; 2. <Heiz & Kälte, Kohlen> hydrostatic pressure; 3. <Nichtfoss Energ> hydraulic thrust; 4. <Strömphys> hydrostatic pressure
hydrostatisches Gleichgewicht n 1. <Strömphys> hydrostatic balance; 2. <Thermod> hydrostatic equilibrium
hydrostatisches Lager n <Maschinen> hydrostatic bearing
Hydrosulfid n <Chemie> *(AE)* hydrogen sulfide, *(BE)* hydrogen sulphide, *(AE)* sulfhydrate, *(BE)* sulphydrate, *(AE)* hydrosulfide, *(BE)* hydrosulphide
Hydrosystem n <Maschinen> fluid-power system
hydrothermale Prozesse mpl <Nichtfoss Energ> hydrothermal processes
Hydroventil n <Maschinen> hydraulically-operated valve
Hydroxozinkat n <Chemie> zincate
Hydroxybarbitursäure f <Chemie> dialuric acid, hydroxybarbituric acid, tartronylurea
Hydroxycarbonsäure f <Chemie> hydroxycarboxylic acid
Hydroxycholin n <Chemie> muscarine
Hydroxydinitrobenzol n <Chemie> dinitrophenol
Hydroxyethylzellulose f <Erdöl> hydroxyethylcellulose *(Petrochemie)*
Hydroxyindol n <Chemie> hydroxyindole, indolol, indoxyl
Hydroxyketon n <Chemie> ketol
hydroxyliert adj <Lebensmittel> hydroxylated
Hydroxynaphthalin n <Chemie> naphthol
Hydroxynaphthochinon n <Chemie> hydroxynaphthoquinone, juglone
Hydroxyphenanthren n <Chemie> phenanthrol
Hydroxytryptamin n <Chemie> serotonin
Hydrozimt... <Chemie> hydrocinnamic
Hydrozyklon m 1. <Abfall, Chemtech> hydrocyclone; 2. <Erdöl> hydrocyclone *(Bohrtechnik)*; 3. <Kohlen> hydrocyclone
Hydrozylinder m <Maschinen> fluid-power cylinder
Hygrometer n 1. <Erdöl> hygrometer *(Messtechnik)*; 2. <Gerät, Heiz & Kälte, Maschinen, Phys, Thermod> hygrometer
Hygrometrie f <Wasserversorg> hygrometry
hygroskopisch adj <Bau, Mechan, Wasserversorg> hygroscopic
Hyochol... <Chemie> hyocholic
Hyoscin n <Chemie> hyoscine
hyperabrupter Übergang m <Elektronik> hyperabrupt junction
hyperballistisch adj <Raumfahrt> hyperballistic
Hyperbel f <Geom> hyperbola
Hyperbelfunkortung f 1. <Funkort> hyperbolic position finding; 2. <Funktech> hyperbolic radio position finding
Hyperbelnavigation f <Funkort, Wassertrans> hyperbolic navigation, hyperbolic radio navigation
Hyperbelorbit m <Raumfahrt> hyperbolic orbit
Hyperbelräder npl <Maschinen> hyperbolical wheels
Hyperbelverfahren n **zur Positionsbestimmung** <Wassertrans> hyperbolic position-fixing system *(Navigation)*
hyperbolisch adj <Geom> hyperbolic
hyperbolische Geometrie f <Geom> hyperbolic geometry
hyperbolische Spirale f <Geom> hyperbolic spiral
hyperbolischer Raum m <Geom> hyperbolic space
Hyperboloid n <Geom> hyperboloid
Hyperboloidgetriebe n <Maschinen> hyperbolical gear
Hyperebene f <Geom> hyperplane
hypereutektisch adj <Metall> hypereutectic
hypereutektoidischer Stahl adj <Metall> hypereutectoid steel

hyperfeine Quantenzahl f *(F)* <Kerntech> hyperfine quantum number *(F)*
Hyperfeinstruktur f <Phys, Strahlphys> hyperfine structure
Hyperfläche f <Geom> hypersurface
Hypergol... <Raumfahrt> hypergol *(Raumschiff)*
hypergolisch adj <Raumfahrt> hypergolic *(Raumschiff)*
hypergolische Eigenschaft f <Raumfahrt> hypergolic property
Hyperkardioid-Mikrofon n <Akustik> hypercardioid microphone
Hyperladung f <Phys> hypercharge
Hyperlink m <Telekom> hyperlink
Hyperniere f <Funktech> hypercardioid *(Richtcharakteristik)*
Hyperon n <Phys, Teilphys> hyperon
Hyperoxid n <Chemie> hyperoxide
Hyperschallströmung f <Strömphys> hypersonic flow
Hypertext m <Comp & DV> hypertext
Hypertext-Übertragungsprotokoll n <Telekom> hypertext transfer protocol, HTTP *(Internet)*
hyperthermische Felder npl <Nichtfoss Energ> hyperthermal fields
Hypochlorid n <Chemie> oxychloride
hypochlorige Säure f <Chemie> hypochlorous acid
Hypochlorit n <Chemie> hypochlorite
Hypochlorsäure f <Chemie> hypochlorous acid
Hypodiphosphor... <Chemie> hypophosphoric
hypoeutektisch adj <Metall> hypoeutectic
Hypoidgetriebe n 1. <Kfztech> hypoid gearing; 2. <Maschinen> hypoid bevel gears
Hypoidkegelgetriebe n <Kfztech> hypoid bevel gearing
Hypoidkegelrad n <Maschinen, Mechan> hypoid bevel gear
Hypoidkegelschraubgetriebe n <Maschinen> hypoid bevel gears
Hypoidrad n <Maschinen> hypoid gear
Hypoidwälzfräsautomat m <Fertig> automatic hypoid generator
Hyponitrit n <Chemie> hyponitrite
Hypophosphat n <Chemie> hypophosphate
hypophosphorisch adj <Chemie> hypophosphorous
Hypotenuse f <Geom> hypotenuse
Hypothermie f <Wassertrans> hypothermia
Hypothese f <Künstl Int> hypothesis
Hypothese f **des elastischen Grenzzustandes** <Fertig> Mohr's strength theory
Hypothesenbildung f <Künstl Int> hypothesis generation
Hypothesengenerierung f <Künstl Int> hypothesis generation
hypothesengesteuert adj <Künstl Int> hypothesis-driven
hypothesengetrieben adj <Künstl Int> hypothesis-driven
hypothetische Referenzverbindung f <Telekom> hypothetical reference connection *(HRC)*
hypotonisch adj <Chemie> hypotonic
Hypotrochoide f <Geom> hypotrochoid
Hypoxanthin n <Chemie> hypoxanthine, sarkine
Hypozykloide f <Geom> hypocycloid
Hystazarin n <Chemie> hystazarin
Hysterese f <Aufnahme, Elektriz, Elektrotech, Fernseh, Kunststoff, Maschinen, Metall, Phys> hysteresis
Hysteresefehler m <Lufttrans> hysteresis error *(Höhenmesser)*
Hysteresekurve f <Funktech> hysteresis
Hystereseschleife f 1. <Elektriz, Elektrotech> B/H loop, hysteresis loop; 2. <Kunststoff, Maschinen, Metall, Phys> hysteresis loop
Hystereseverlust m <Elektrotech, Kunststoff> hysteresis loss

Hystereseverlustzahl

Hystereseverlustzahl f <Elektriz> hysteresis coefficient
Hysteresis f <Maschinen> hysteresis
Hysteresiskupplung f <Elektrotech> hysteresis coupling
Hysteresismotor m <Elektrotech> hysteresis motor
Hz *(Hertz)* <Elektriz, Elektrotech, Fernseh, Funktech, Metrol, Phys> Hz *(hertz)*
HZK *(höchstzulässige Konzentration)* <Umweltschmutz> MAC *(maximum allowable concentration)*; TLV *(threshold limit value)*
HZK f **am Arbeitsplatz** <Umweltschmutz> MAC in the workplace, TLV in the workplace
HZK f **in Umwelt** <Umweltschmutz> MAC in the free environment, TLV in the free environment

I

I 1. <Akustik, Elektriz> *(Intensität, Stärke)* I *(intensity)*; 2. <Chemie> *(Iod, Jod)* I *(iodine)*; 3. <Elektriz, Phys, Telekom> *(elektrischer Strom)* I *(electric current)*; 4. <Elektriz, Optik> *(Intensität)* I *(intensity)*; 5. <Kerntech> *(Energieflussdichte)* I *(energy flux density)*; 6. <Kerntech> *(nukleare Spinquantenzahl)* I *(nuclear spin quantum number)*
I-Achse f <Fernseh> I axis
I-Anker m <Elektriz> shuttle armature
I-Beiwert m <Regelung> integral action coefficient
ICAS *(Kommerzieller und Amateurfunkdienst)* <Funktech> ICAS *(Intermittent Commercial and Amateur Services)*
IC-Maske f <Elektronik> IS mask
Icosan n <Chemie> icosane
ICRP *(Internationale Strahlenschutzkommission)* <Strahlphys> ICRP *(International Commission on Radiological Protection)*
IC-Verfahren n *(Spritzprägeverfahren)* <Kunststoff> injection compression process
ID *(Identifikation, Kennung)* <Comp & DV> identification
Ideal... <Elektrotech, Trans> ideal
ideale Flüssigkeit f <Phys> perfect fluid
ideale Phasenfokussierung f <Elektronik> ideal bunching
idealer Leiter n <Elektronik, Elektrotech> ideal conductor, perfect conductor
ideales Filter n <Elektronik> ideal filter
ideales Gas n <Phys, Thermod> ideal gas, perfect gas
Idealgeschwindigkeit f <Nichtfoss Energ> ideal velocity
Idealgleichrichter m <Elektrotech> ideal rectifier
Idealtransformator m <Elektrotech> ideal transformer
I-Demodulator m <Fernseh> I demodulator
Identifikation f 1. <Comp & DV> identity; 2. <Comp & DV, Maschinen> identification
Identifikationscode m <Comp & DV> authentication code, identification code
Identifikationskennzeichen n <Comp & DV> tag, identification character
identifizieren v <Comp & DV, Telekom> identify
Identifizierer m <Comp & DV, Telekom> identifier
identifizierte Quellen fpl <Nichtfoss Energ> identified resources
Identifizierung f 1. <Comp & DV> identity; 2. <Telekom> identification
Identifizierung f **des rufenden Anschlusses** <Telekom> CLI, calling line identification

Identifizierung f **unerwünschter Anrufe** <Telekom> malicious call identification, MCID *(Dienstmerkmal im Euro-ISDN)*
Identität f <Math, Patent, Qual> identity
Identität f **des lokalen Jobs** <Comp & DV> local job identity
Identkarte f <Qual> identity card
Identnummer f <Patent, Qual> identity number
Ideogramm n <Comp & DV> ideogram
idiomorpher Kristall m <Metall> idiomorphic crystal
Idit n <Chemie> idite, iditol
Iditol n <Chemie> idite, iditol
Idler-Frequenz f <Elektronik, Funktech, Telekom> idler frequency
Idon... <Chemie> idonic
Idose f <Chemie> idose
Idozucker... <Chemie> idosaccharic
IFR *(Instrumentenflugregeln)* <Lufttrans> IFR *(instrument flight rules)*
IFRB *(Internationaler Ausschuss für Frequenzregistrierung)* <Raumfahrt, Telekom> IFRB *(International Frequency Registration Board)*
IFU *(Internationale Fernmeldeunion)* <Telekom, Wassertrans> ITU *(International Telecommunication Union)*
IGFET *(Isolierschicht-Feldeffekttransistor)* <Elektronik> IGFET *(insulated gate field-effect transistor)*
I-Glied n *(Integralelement)* <Regelung> integral element
Iglu-Container m <Lufttrans> igloo container
Ignitron n <Elektriz, Elektrotech> ignitron
Ignitronlokomotive f <Eisenbahn> ignitron locomotive
Ignitronröhre f <Elektriz, Elektrotech> ignitron
I-Halbleiter m <Comp & DV> intrinsic semiconductor
IIR *(unbegrenztes Ansprechen auf Impuls)* <Elektronik> IIR *(infinite impulse response)*
IIR-Digitalfilter n <Telekom> IIR digital filter
IIR-Filter n <Telekom> IIR filter
Ikonoskop n <Elektronik> iconoscope
Ikosaeder n <Geom> icosahedron
ikosaedrisch adj <Geom> icosahedral
Illit m <Erdöl, Kohlen> illite
IL-Logik f *(integrierte Injektionslogik)* <Elektronik> IL-Logik *(integrated injection logic)*
Illumination f <Elektriz> illumination
Illustration f <Druck> illustration
Illustrationsfarbe f <Druck> halftone ink
illustrieren v <Druck> illustrate
ILS *(Blindfluglandesystem durch Eigenpeilung, Instrumentenlandesystem)* <Funkort, Lufttrans, Raumfahrt> ILS *(instrument landing system)*
imaginäre Einheit f <Math> imaginary unit
imaginäre Zahl f <Math> imaginary number
Imaginärkomponente f <Math> imaginary component, imaginary part *(einer komplexen Größe)*
Imaging-System n <Comp & DV> imaging system
Imband-Zeichengabe f <Telekom> *(AE)* in-band signaling, *(BE)* in-band signalling
Imhoffbrunnen m <Abfall> Imhoff tank
Imid n <Chemie> imide
Imidazo-Pyrimidin n <Chemie> purine
Imido... <Chemie> imido
Imidogruppe f <Chemie> imido group
Imidoxanthin n <Chemie> aminohypoxanthine, guanine, imidoxanthine
Iminoharnstoff m <Chemie> guanidine
imitieren v <Math> mimic *(Bootstrapverfahren)*
Immediatanalyse f <Chemie> proximate analysis
Immersion f <Chemie> immersion
Immersionslinse f 1. <Fernseh> immersion electron lens; 2. <Labor> immersion lens *(Mikroskop)*

Immersionsobjektiv n <Metall> immersion objective
Immersionsöl-Linse f <Labor> oil immersion lens *(Mikroskop)*
Immissionsgrenzwert m **der Luft** <Umweltschmutz> ambient air emission standard, ambient air quality standard
Impaktion f <Umweltschmutz> impaction
Impatt-Diode f *(Lawinenlaufzeitdiode)* <Elektronik, Phys> impatt diode *(impact ionization avalanche transit-time diode)*
Impatt-Oszillator m <Elektronik> impatt oscillator
Impedanz f 1. <Aufnahme> impedance; 2. <Elektriz, Elektrotech> apparent resistance *(Größe)*; impedance; impedor *(Bauelement)*; 3. <Funktech, Phys, Telekom> impedance
Impedanzanpassung f <Elektrotech, Phys, Telekom> impedance matching
Impedanzanpassungsnetz n <Elektrotech> impedance matching network
Impedanzausgleicher m <Elektriz> impedance corrector
Impedanzbrücke f <Gerät> impedance measuring bridge
Impedanzfehlanpassung f <Elektrotech, Telekom> impedance mismatch
Impedanzkupplung f <Elektrotech> impedance coupling
Impedanzkurve f <Elektriz, Elektrotech, Elektronik> impedance characteristic
Impedanzkurve f **eines blockierten Läufers** <Elektriz> locked rotor impedance characteristic
Impedanzmessbrücke f <Elektrotech, Gerät> impedance bridge, impedance measuring bridge
Impedanzrelais n <Elektriz> impedance relay
Impedanzspannung f <Elektrotech> impedance voltage
Impedanzspannungsabfall m <Elektriz> impedance voltage drop
Impedanzspule f <Elektrotech> impedance coil
Impedanzverhältnis n <Aufnahme, Telekom> impedance ratio
Impedanzwandler m <Phys> impedance transformer
Impedanzwandlung f <Elektrotech, Telekom> impedance conversion
Impfen n <Kerntech> inoculation
Impfkompostrückführung f <Abfall> recycling of inoculated compost
Impfkristall m <Fertig, Thermod> seed crystal
Impfung f <Kerntech> inoculation
Implantation f **radioaktiver Ionen** <Teilphys> radioactive ion implantation
Implantationstiefe f <Elektronik> implant depth
Implantationszeit f <Elektronik> implant time
implantierte Basis f <Elektronik> implanted base
implantierte Diode f <Elektronik> implanted diode
implantierte Zenerdiode f <Elektronik> implanted Zener diode *(IZD)*
implantierter Transistor m <Elektronik> implanted transistor
implementieren v <Comp & DV, Maschinen> implement
Implementierung f <Comp & DV> implementation
Implikation f <Math> implication *(Folgerung)*
implizieren v <Math> implicate
implizierte Adressierung f <Comp & DV> implied addressing
implizit adj <Math> implicit
implizite Adressierung f <Comp & DV> inherent addressing
implodieren v <Elektronik> implode
Implosion f <Elektronik> implosion
Implosion f **der schwarzen Strahlung** <Kerntech> implosion of black body radiation
imprägnieren v 1. <Bau> proof, temper; 2. <Textil> dip, impregnate

Imprägnieren n 1. <Bau> soaking; 2. <Kunststoff> impregnation
Imprägnierlack m <Elektriz> impregnating varnish *(Tränklack)*
Imprägniermaschine f <Verpack> impregnating machine
Imprägniermasse f <Verpack> impregnating agent
Imprägniermittel n 1. <Chemie> impregnant, saturant; 2. <Verpack> impregnating agent
imprägniert adj <Papier> impregnated
imprägnierte Katode f <Elektrotech> impregnated cathode
imprägnierte Spule f <Elektrotech> impregnated coil
imprägniertes Gewebe n <Verpack> impregnated fabric
imprägniertes Kabel n <Elektriz> impregnated cable
imprägniertes Papier n <Druck, Verpack> impregnated paper
imprägniertes papierisoliertes Kabel n <Elektriz> impregnated paper insulated cable
Imprägnierung f <Bau, Papier, Textil> impregnation
Imprägnierwachs n <Verpack> impregnating wax
imprimaturbereit adj <Druck> passed for press
Improved-Ruß m <Kunststoff> improved carbon black
Impuls m 1. <Aufnahme> pulse; burst *(Tonimpuls, Rauschimpuls)*; 2. <Comp & DV> pulse, signal; 3. <Elektriz, Elektronik, Elektrotech> pulse; 4. <Fernseh> spike; 5. <Funktech, Kontroll> pulse; 6. <Maschinen> impetus, impulse, impulsion, momentum; 7. <Mechan> impulse, momentum; 8. <Optik> pulse; 9. <Papier> impulse; 10. <Phys> momentum; impulse *(Zeitintegral der Kraft)*; 11. <Telekom> impulse, pulse; 12. <Wassertrans> pulse *(Radar)*
Impuls m **eines Schaltkennzeichens** <Elektronik> signal pulse
Impuls m **mit großer Anstiegsgeschwindigkeit** <Elektronik> fast rise time impulse, fast rise time pulse
Impuls m **mit hoher Amplitude** <Elektronik> high-amplitude pulse
Impuls m **mit kurzer Anstiegszeit** <Elektronik> fast-rise pulse
Impulsabfall m <Elektronik> impulse decay, impulse fall, pulse decay, pulse fall, pulse tilt
Impulsabfallzeit f <Aufnahme, Telekom> pulse decay time
Impulsabfrage f <Telekom> impulse scanning, pulse scanning
Impulsabstand m <Elektronik> impulse interval, impulse separation, impulse spacing, pulse interval, pulse separation, pulse spacing, pulse-digit spacing
Impulsamplitude f <Aufnahme, Elektronik> pulse amplitude
Impulsanstiegszeit f <Fernseh, Telekom> leading-edge pulse time
Impulsantwort f <Telekom> impulse response, pulse response
impulsartig rauschen v <Comp & DV> burst
Impulsbeschleuniger m <Strahlphys> impulse accelerator
Impulsbetrieb m <Elektronik> pulsed operation
Impulsbildung f <Elektronik> bunching *(in Laufzeitröhren)*
Impulsbreite f <Comp & DV, Elektronik, Fernseh, Phys, Telekom> pulse duration, pulse width
Impulsbreitemodulation f <Elektronik> impulse-width modulation, pulse-width modulation
Impulsbreitenmodulator m <Elektronik> impulse-width modulator, pulse-width modulator
Impulsbreitenregelung f <Elektronik> impulse-width control, pulse-width control
Impulsdach n <Elektronik> impulse top, pulse top
Impulsdachschräge f <Elektronik> impulse droop, impulse tilt, pulse droop, pulse tilt

Impulsdaten

Impulsdaten pl <Elektronik> impulse data, pulse data
Impulsdauer f <Comp & DV, Elektronik, Fernseh, Kontroll, Phys, Telekom> impulse duration, impulse length, impulse width, on-time of pulse, pulse duration, pulse length, pulse width
Impulsdauermodulation f <Elektronik> impulse-duration modulation, pulse-duration modulation (PDM)
Impulsdehner m <Elektronik> impulse stretcher, pulse stretcher
Impulsdemodulator m <Elektronik> impulse detector, pulse detector
Impulsdichte f <Elektronik> impulse rate, pulse rate
Impulsdiskriminator m <Funktech> pulse count discriminator
Impulsdispersion f <Optik, Telekom> pulse dispersion
Impuls-Doppler-Radar n <Funkort> impulse Doppler radar, pulse Doppler radar
Impulse mpl **je Sekunde** <Telekom> pulses per second
Impulsecho n <Funktech> impulse echo, pulse echo
Impulsechospannung f <Funktech> impulse echo voltage, pulse echo voltage
Impulsechoverfahren n <Funktech> impulse echo voltage, pulse echo voltage
Impulseingang m <Elektronik> impulse input, pulse input
Impulselement n <Elektronik> impulse element, pulse element
Impulsempfänger m <Elektronik> impulse receiver, pulse receiver
Impulsenergie f <Elektronik> impulse energy, pulse energy
Impulsentladung f <Elektronik> impulse discharge, pulse discharge
Impulsentladungslampe f <Elektrotech> impulse discharge lamp, pulse discharge lamp
Impulsentladungszeit f <Elektronik> impulse spark-over time, pulse spark-over time
Impulsentzerrer m <Elektronik> impulse corrector, impulse regenerator, pulse corrector, pulse regenerator
Impulsentzerrung f <Elektronik, Fernseh> impulse correction, impulse regeneration, pulse correction, pulse regeneration, pulse restoration
Impulserhaltung f <Phys> conservation of momentum
Impulserzeuger m <Elektronik, Fertig> impulse generator, pulse generator, pulser
Impulserzeugung f <Elektronik> impulse generation, pulse generation
Impulsfernmessgerät n <Metrol> impulse-type telemeter, pulse-type telemeter
Impulsfolge f <Comp & DV, Elektronik> pulse sequence, pulse train
Impulsfolgefrequenz f 1. <Comp & DV, Elektronik, Phys, Telekom> impulse recurrence frequency, impulse repetition frequency, impulse repetition rate, pulse recurrence frequency; pulse repetition frequency (PRF); pulse repetition rate; 2. <Elektriz> discharge repetition rate, impulse repetition rate, pulse repetition rate (bei Teilentladungen)
Impulsfolgefrequenzsignal n <Telekom> impulse repetition frequency signal, pulse repetition frequency signal
Impulsfolgeperiode f <Elektronik> pulse repetition period
Impulsform f <Elektronik> pulse shape
Impulsformung f <Comp & DV> pulse shaping
Impulsfrequenz f <Elektronik> impulse frequency
Impulsfunktion f <Elektronik> impulse function
Impulsgang m <Optik> impulse response
Impulsgeber m <Elektronik, Kerntech> impulse generator, impulse initiator, pulse initiator, pulse generator, pulser, pulsing device, surge generator
Impulsgeberrad n <Kfztech> trigger wheel

Impulsgebertaste f <Telekom> impulse sending key, pulse sending key
Impulsgenerator m <Comp & DV, Elektronik, Elektrotech, Kerntech, Telekom> impulse generator; pulse generator (PG); pulse synthesizer, pulser
Impulsgerät n <Elektronik, Elektrotech> impulse device, pulse device, pulser unit
Impulsgeräusch n <Elektronik> impulse noise, pulse noise
impulsgesteuert adj <Elektronik> impulse-controlled, pulsed, pulse-controlled, impulse-operated, pulse-operated
Impulsgruppenfrequenz f <Elektronik> impulse frame repetition rate, pulse frame repetition rate
impulshaltig adj <Akustik> impulsive
impulshaltige Schwingung f <Sicherheit> impact vibration
impulshaltiges Rauschen n <Akustik> impulsive noise
Impulshammer m <Elektronik> impact hammer (Modalanalyse)
Impulshärten n <Elektrotech> impulse hardening, pulse hardening (z. B. durch Laser)
Impulshitzesiegler m <Verpack> impulse heat sealer
Impulshöhe f 1. <Comp & DV> pulse height; 2. <Elektronik> pulse amplitude
Impulskennlinien fpl <Elektronik> pulse characteristics
Impulskette f <Comp & DV> pulse train
Impulskompression f <Telekom> pulse compression
Impulskondensator m <Elektriz> pulse capacitor
Impulslänge f <Telekom, Wassertrans> pulse length (Radar)
Impulslaser m <Elektronik, Strahlphys> pulsed laser
Impulsmagnetron n <Elektronik> pulsed magnetron
Impulsmaser m <Elektronik> pulsed maser
Impulsmodulierung f <Elektriz> pulse modulation
Impulsmoment n <Lufttrans> angular momentum
Impulsoszillator m <Elektronik> pulsed oscillator
Impulspause f <Elektronik> pulse spacing
Impulspegel m <Akustik> burst level
Impulspeilung f <Funktech> impulse direction finding, pulse direction finding
Impulsperiodendauer f <Elektronik> pulse duration
Impulsprofil n <Telekom> pulse profile
Impulsradar n <Funkort> pulse radar
Impulsradarsender m <Funkort> pulse-radar transmitter
Impulsrand m <Elektronik> impulse margin, pulse margin
Impulsrauschen n <Raumfahrt> pulse noise (Weltraumfunk)
Impulsregenerierung f <Fernseh, Telekom> pulse regeneration
Impulsrelais n <Elektriz, Elektrotech> impulse relay, pulse relay
Impulsrücken m <Elektriz, Elektrotech> tail of an impulse, tail of a pulse
Impulsrückflanke f <Telekom> pulse trailing edge
Impulsrückstrahlverfahren n <Funktech> impulse reflection method, pulse reflection method
Impulsschalter m <Elektrotech> impulse switch, pulse switch, semiflush switch, semirecessed switch, semisunk switch
Impulsschaltung f <Elektronik> pulse circuit
Impulsschaltung f **mit Lawinentransistor** <Elektronik> avalanche transistor impulse circuit, avalanche transistor pulse circuit
Impulsschwankung f <Elektriz, Elektronik, Elektrotech> impulse oscillation, pulse oscillation
Impulsschweißen n <Fertig> impulse welding, pulse welding
Impulssignal n <Elektronik> pulse signal

Impulsspannung f <Elektriz> impulse voltage
Impulsspitze f <Elektriz> pulse spike
Impulsspitzenbegrenzer m <Telekom> pulse clipper
Impulsstopfen n <Telekom> justification
Impulsstörung f <Akustik, Comp & DV, Telekom> impulse noise
Impulsstörungsbegrenzer m <Funktech> noise pulse limiter
Impulsstrom m <Elektriz> impulse current
Impulssynchronisierung f <Fernseh> pulse sync
Impulstachometer n <Metrol> impulse tachometer, pulse tachometer
Impulstaktfrequenz f <Elektronik, Telekom> impulse repetition frequency, impulse repetition rate, pulse repetition frequency, pulse repetition rate
Impulstaktschwankung f <Telekom> jitter
Impulstastverhältnis n 1. <Aufnahme> burst duty factor (bei Tonimpulsen); 2. <Comp & DV> mark-to-space ratio; 3. <Elektronik> pulse duty factor
Impulstechnik f <Werkprüf> impulse circuit technique, pulse technique, pulsing technique
Impulsträger m <Telekom> pulse carrier
Impulstrennstufe f <Fernseh> sync separator
Impulstrennung f <Fernseh> pulse separator
Impulstriggern n <Fernseh> pulse triggering
Impulsübertrager m <Telekom> PT, pulse transformer
Impulsübertragungsgang m <Optik> impulse response
Impulsultraschalldetektor m <Trans> pulsed ultrasonic detector
Impulsverbreiterung f 1. <Optik> pulse broadening, pulse spreading; 2. <Telekom> pulse broadening, pulse spreading, pulse widening
Impulsverkehrsradar n <Trans> pulsed radar detector
Impulsverschlüssler m <Strahlphys> pulse coder
Impulsverstärker m <Aufnahme, Elektronik> pulse amplifier
Impulsverstärkung f <Elektronik> pulse amplification
Impulsvorderflanke f <Telekom> pulse leading edge
Impulswahl f <Telekom> impulse action, pulse action
Impulswähler m <Telekom> impulse selector, pulse selector
Impulswahlfernsprecher m <Telekom> dial-impulse telephone, dial-pulse telephone
Impulswahlverfahren n <Telekom> loop disconnect pulsing, pulse dialling, pulse selection
Impulswandler m <Telekom> impulse converter, impulse transformer, pulse converter, pulse transformer
Impulswiederholer m <Elektronik> pulse regenerator
Impulszähler m <Elektriz, Elektronik> impulse counter, pulse counter
Impulszähltechnik f <Telekom> pulse-counting technique
Impulszeichengabe f <Telekom> (AE) impulse signaling, (BE) impulse signalling
Impulszentrale f <Telekom> central pulse distributor
Impulszittern n <Telekom> time jitter (PCM)
In (Indium) <Chemie> In (indium)
inaktinisches Glas n <Ker & Glas> inactinic glass
inaktiv adj <Chemie, Comp & DV> inactive
inaktiver Reaktor m <Kerntech> cold reactor
inaktivieren v <Comp & DV> disable
Inaktivierung f <Comp & DV> disarmed state
inakzeptable Qualität f <Qual, Telekom> unacceptable quality
Inanspruchnahme f <Patent> claiming
Inbetriebnahme f 1. <Comp & DV> installation; 2. <Kontroll> initial operation phase; 3. <Mechan> commissioning
 • **für die Inbetriebnahme vorbereiten** <Hydraul> prime (Dampfkessel)

Inbetriebnahmeprüfung f <Telekom> commissioning test
Inbetriebnahmeventil n <Hydraul> equilibrium valve
Inbetriebnahmevorbereitung f <Hydraul> priming
Inbusschlüssel m <Maschinen> (BE) Allen key, (AE) Allen wrench
Inbusschraube f <Maschinen> Allen screw, socket head screw
Inchgewinde n <Maschinen> inch thread
Inchscrap-Verfahren n <Abfall> cryogrinding, freeze-grinding
In-Circuit-Test m <Elektriz> in-circuit test (z. B. von Schaltungen)
Incore-... <Kerntech> in-core
Incore-Brennstoffkreislauf m <Kerntech> in-core fuel cycle
Incore-Lebensdauer f des Brennstoffes <Kerntech> in-core fuel life
Indan n <Chemie> hydrindene, indan
Indanon n <Chemie> indanone
Indanthron n <Chemie> indanthrene
Indazin n <Chemie> indazine
Indazol n <Chemie> indazole
Inden n <Chemie> indene
Index m 1. <Comp & DV, Maschinen, Math> exponent, index, subscript, superscript; 2. <Telekom> index
Indexfehler m <Comp & DV> index error
indexieren v <Comp & DV> index
Indexieren n <Comp & DV, Maschinen> indexing
indexierte Datei f <Comp & DV> indexed file
indexierte Variable f <Comp & DV> subscripted variable
indexierter Befehl m <Comp & DV> indexed instruction
Indexierung f <Telekom> indexing
Indexliste f <Comp & DV> subscript (FORTRAN)
Indexloch n <Comp & DV> index hole
Indexprofil n <Optik, Telekom> index profile
Indexregister n <Comp & DV> index register
Indexsequenz f <Comp & DV> IS, indexed sequence
indexsequenzielle Datei f <Comp & DV> indexed sequential file
indexsequenzielle Zugriffsmethode f <Telekom> indexed sequential access method
indexsequenzieller Zugriff m <Comp & DV> indexed sequential access
indexsequenzierte Datei f <Comp & DV> indexed sequential file
indexsequenzierter Zugriff m <Comp & DV> indexed sequential access
Indican n <Chemie> indicane
Indienststellung f <Mechan, Wassertrans> commissioning (Schiff)
indifferentes Lösemittel n <Chemie> aprotic solvent
Indikator m 1. <Hydraul> indicator; 2. <Ker & Glas> tracer; 3. <Maschinen, Wasserversorg> indicator
Indikatoratom n <Kerntech> (AE) labeled atom, (BE) labelled atom
Indikatordiagramm n <Maschinen> indicator diagram
Indikatorfunktion f <Math> indicator function (einer Menge)
Indikatorpapier n <Foto, Verpack> indicator paper
indirekt beeinflusste Regelstrecke f <Regelung> indirectly controlled system
indirekt beeinflusste Steuerstrecke f <Regelung> indirectly controlled system
indirekt geheizte Katode f <Elektrotech> heater-type cathode, indirectly heated cathode
indirekt gesteuertes System n <Telekom> indirect-control system
indirekte Adressierung f <Comp & DV> deferred addressing, indirect addressing

indirekte

indirekte Beleuchtung f <Elektrotech> indirect illumination
indirekte Photoleitfähigkeit f <Elektronik> indirect photoconductivity
indirekte Regelung f <Elektrotech> indirect control
indirekte Überstromabschaltung f <Elektriz> indirect over-current release
indirekter Gleichstromsteller m <Elektrotech> d.c. chopper converter, indirect d.c. converter
Indium n (In) <Chemie> indium (In)
Individual-Section-Flaschenblasmaschine f (IS-Maschine) <Ker & Glas> individual section machine (IS machine)
individuelle Abnehmerleitung f <Fernseh> individual trunk
indizieren v <Comp & DV> index
indizierte Adresse f <Comp & DV> indexed address
indizierte Adressierung f <Comp & DV> indexed addressing
indizierte Datei f <Comp & DV> indexed file
indizierte Instruktion f <Comp & DV> indexed instruction
Indol n <Chemie> indole
Indolol n <Chemie> hydroxyindole, indolol, indoxyl
Indolylessig... <Chemie> indolylacetic
Indonaphthen n <Chemie> indene
Indophenin n <Chemie> indophenine, induline
Indophenol n <Chemie> indophenol
Indoxyl n <Chemie> hydroxyindole, indolol, indoxyl
Indoxyl... <Chemie> indoxylic
Indoxylschwefel... <Chemie> (AE) indoxylsulfuric, (BE) indoxylsulphuric
Induktanz f <Elektriz, Elektrotech> inductance
Induktanzbrücke f <Elektriz> inductance bridge (Messbrücke)
Induktanzmessgerät n <Elektriz> inductance meter
Induktion f <Elektriz, Elektrotech, Maschinen, Telekom> induction, magnetic field density
Induktionsfeld n <Elektrotech, Fernseh, Telekom> induction field
Induktionsfluss m <Elektrotech> induction flux, magnetic flux
induktionsfrei adj <Elektriz> noninductive
induktionsfreie Last f <Elektriz> noninductive load
induktionsfreier Stromkreis m <Elektriz> noninductive circuit
induktionsfreier Widerstand m <Elektriz> noninductive resistor
Induktionsfrequenzumformer m <Elektriz> induction frequency converter
Induktionsgenerator m <Elektriz, Elektrotech> induction generator
Induktionshärten n <Fertig> induction hardening
induktionshartlöten v <Fertig> induction-braze
Induktionshartlöten n <Fertig> induction brazing
Induktionshärtung f <Elektrotech> induction hardening
Induktionsheizgerät n <Elektrotech> induction heater
Induktionsheizung f <Elektrotech, Mechan, Thermod> induction heating (Verfahren)
Induktionsinstrument n <Elektrotech> induction instrument
Induktionskoeffizient m <Phys, Qual> coefficient of induction
Induktionskopplung f <Elektriz> induction coupling
induktionslöten v <Fertig> induction-braze
Induktionslöten n <Fertig> induction brazing
Induktionsmotor m <Elektriz, Elektrotech, Fertig, Kfztech, Phys> induction motor
Induktionsmotor m **mit Repulsionsanlauf** <Elektrotech> repulsion start induction motor
Induktionsmotor m **mit Schleifringläufer** <Elektrotech> wound-rotor induction motor

Induktionsmotor/Repulsionsmotor m <Elektrotech> repulsion induction motor
Induktionsofen m <Elektrotech, Phys> induction furnace
Induktionspumpe f <Elektrotech> induction pump
Induktionsregler m <Elektrotech> induction voltage regulator
Induktionsrelais n <Elektriz, Elektrotech> induction relay
Induktionsrinnenofen m <Heiz & Kälte> channel induction furnace
Induktionsschleifendetektor m <Kfztech> induction loop detector
Induktionsschweißen n <Elektriz, Fertig> induction welding
Induktionssiegler m <Verpack> induction sealer
Induktionsspannung f 1. <Elektrotech> induced electromotive force, induction voltage; 2. <Telekom> induced voltage
Induktionsspannungsregler m <Elektrotech> induction voltage regulator
Induktionsspule f <Elektriz, Elektrotech, Kfztech> inductance coil, induction coil, inductor
Induktionsspule f **mit Luftspaltkern** <Elektriz> air gap induction coil
Induktionsstrom m <Telekom> induced current
Induktionssystem n 1. <Kfztech> induction system; 2. <Phys> inducing system
Induktionswiderstand m <Phys> inductive reactance
Induktionszeit f <Metall> induction period
induktive Abstimmung f <Elektrotech> inductive tuning
induktive Belastung f <Telekom> inductive load
induktive Erwärmung f <Kunststoff> induction heating
induktive Kopplung f <Elektriz, Elektrotech> flux linkage, inductive coupling, magnetic coupling
induktive Last f <Elektriz, Elektrotech> inductive load
induktive Reaktanz f <Elektriz, Elektrotech> inductive reactance
induktive Rückkopplung f <Elektrotech> inductive feedback
induktive Schaltung f <Elektriz> inductive circuit
induktive Zugbeeinflussung f <Eisenbahn> automatic warning system, AWS
induktiver Blindwiderstand m (XL) <Elektriz> inductive reactance (XL)
induktiver Drahtwicklungswiderstand m <Elektrotech> inductive wirewound resistor
induktiver Durchflussmesser m <Elektrotech> electromagnetic flowmeter
induktiver Kondensator m <Elektrotech> inductive capacitor
induktiver Näherungsschalter m <Elektrotech> inductive proximity switch
induktiver Spannungsteiler m <Elektriz> inductive potential divider
induktiver Widerstand m <Elektrotech> inductive reactance
induktives AC-Potenziometer n <Elektriz> inductive potential divider
induktives Durchflussmengenmessgerät n <Gerät> inductive flow-meter
Induktivität f (L) 1. <Aufnahme> inductance, L (Spule); 2. <Elektriz, Elektrotech, Funktech, Labor, Phys, Telekom> inductance, L
induktivitätbehafteter Widerstand m <Elektrotech> inductive resistor
Induktivitätsmessbrücke f <Elektrotech> inductance bridge
Induktivitätsmessgerät n <Gerät> inductance measuring instrument
Induktometer n <Elektriz, Elektrotech> inductometer

Induktor *m* <Elektrotech> inductor
Induktorfrequenzumformer *m* <Elektrotech> inductor frequency converter
Induktorgenerator *m* <Elektrotech> inductor generator
Induktormaschine *f* <Elektrotech> inductor machine
Induktorsynchronmotor *m* <Elektrotech> inductor-type synchronous motor
Indulin *n* <Chemie> induline
Industrieabfall *m* <Abfall> industrial waste
Industrieabwasser *n* 1. <Abfall> industrial waste water; 2. <Wasserversorg> industrial effluent
Industriealkohol *m* <Lebensmittel> industrial alcohol *(Lösungsmittel)*
Industriearbeitskleidung *f* <Sicherheit> industrial clothing
Industrieautomation *f* <Maschinen> industrial automation
Industriedichtung *f* <Verpack> industrial packing
Industrieelektronik *f* <Elektrotech, Fertig> industrial electronics
Industrieelektronikröhre *f* <Elektronik> industrial electronic tube
Industriefernsehen *n* <Fernseh> business television, industrial television
Industriefotografie *f* <Foto> commercial photography
industrielle Bestrahlungsanlage *f* <Kerntech> industrial irradiator
industrielle Einleitung *f* <Wasserversorg> industrial discharge
industrielle Elektronik *f* <Elektrotech, Fertig, Telekom> industrial electronics
industrielle Kernenergie *f* <Kerntech> industrial nuclear power
industrielle Umweltverschmutzung *f* <Umweltschmutz> industrial pollution
industrielle Klärschlamm *m* <Abfall> industrial sewage sludge
industrieller Schadstoff *m* <Umweltschmutz> industrial pollutant
industrielles Fernsehen *n* *(CCTV)* <Fernseh> business television, industrial television; closed-circuit television *(CCTV)*
industrielles Nutzwasser *n* <Wasserversorg> industrial water
industrielles Schüttgutcontainersystem *n* <Verpack> industrial bulk container system
Industriemagnetron *n* <Elektronik> industrial magnetron
Industriemüll *m* <Abfall> industrial waste
Industriemülldeponie *f* <Abfall> industrial landfill
Industrienorm *f* <Comp & DV, Qual> industrial standard
Industrieofen *m* 1. <Fertig> furnace; 2. <Maschinen> furnace, industrial furnace, industrial oven
Industrieroboter *m* <Künstl Int, Sicherheit> IR, industrial robot
Industrieschadstoff *m* <Umweltschmutz> industrial pollutant
Industrieschornstein *m* <Bau> stack
Industrieschutzhelm *m* <Sicherheit> industrial safety helmet
Industriestandard *m* <Comp & DV, Qual> industrial standard, industry standard
Industrieverpackung *f* <Kerntech> industrial packaging
induzieren *v* <Elektriz, Phys> induce
induzierender Durchfluss *m* <Bau> inducing flow *(Brunnen)*
induziert *adj* 1. <Elektriz> induced *(Spannung)*; 2. <Phys> induced
induzierte Aktivierung *f*/**durch Elektronen** <Strahlphys> electron-induced activation

induzierte elektromotorische Kraft *f* <Elektriz, Elektrotech> induced electromotive force
induzierte Emission *f* <Elektronik, Strahlphys> induced emission, stimulated emission
induzierte EMK *f* <Phys> induced EMF
induzierte Kernreaktion *f* <Kerntech> artificial nuclear reaction, induced nuclear reaction
induzierte Ladung *f* <Elektriz, Elektrotech> induced charge
induzierte Radioaktivität *f* <Kerntech> induced radioactivity
induzierte Spannung *f* <Elektriz> induced voltage
induzierte Störung *f* <Comp & DV> induced interference
induzierter Anstellwinkel *m* <Lufttrans> induced attack angle
induzierter Fehler *m* <Comp & DV> induced failure
induzierter Luftwiderstand *m* <Lufttrans> induced drag
induzierter Strom *m* <Elektriz> induced current
induziertes Feld *n* <Elektriz> induced field
ineinander anordnen *v*/**satzweise** <Fertig> nest
ineinander fließen *v* <Chemie> coalesce
ineinander greifen *v* 1. <Comp & DV> interlace; 2. <Maschinen> engage, interlock, mate, mesh
ineinander greifend *adj* <Fertig> mating; interengaging *(Getriebelehre)*
ineinander zeichnen *v* <Konstzeich> draw in the mated condition
Ineinandergreifen *n* 1. <Maschinen> mating; 2. <Mechan> meshing
Ineinandergreifen *n* **der Phasen** <Trans> phase skipping
inelastische Neutronenstreuung *f* <Strahlphys, Teilphys> inelastic neutron scattering
inelastische Streuung *f* <Kerntech, Strahlphys, Teilphys> inelastic scattering
inelastischer Stoß *m* <Strahlphys, Teilphys> inelastic collision
inert *adj* <Erdöl> inert *(Petrochemie)*
inerter Abfall *m* <Abfall> inert waste
inertes Material *n* <Abfall> inert material
Inertgas *n* 1. <Fertig> inert gas; 2. <Kerntech> inert gas *(Schweißen)*; 3. <Maschinen> inert gas
Inertgasanlage *f* <Wassertrans> inert gas system *(Explosionsschutz an Bord von Öl- und Chemietankern)*
Inertgasschutzmantel *m* <Fertig> inert gas shield
Inertgasschweißen *n* <Fertig> sigma welding
Inertgasschweißen *n* **mit abschmelzender Elektrode** <Fertig> inert arc welding with a consumable electrode
Inertgasschweißen *n* **mit nicht abschmelzender Elektrode** <Fertig> inert arc welding with non-consumable electrode
inertiales Bezugssystem *n* <Phys> inertial frame
Inertialnavigationssystem *n* <Lufttrans, Raumfahrt, Wassertrans> INS, inertial navigation system
Inertialsystem *n* <Phys> inertial frame
Inertisieren *n* <Erdöl> inerting *(Sicherheit)*
inertisierter Rückstand *m* <Abfall> noncombustible residue
Inertisierung *f* <Erdöl> blanketing *(Sicherheit)*
infektiöser Abfall *m* <Abfall> anatomical waste, infectious waste, pathological waste
Inferenz *f* <Comp & DV, Künstl Int> inference
Inferenzeinheit *f* <Comp & DV> inference engine
Inferenzkomponente *f* <Künstl Int> inference engine *(eines Expertensystems)*
Inferenzmaschine *f* <Künstl Int> inference engine *(eines Expertensystems)*
Inferenzregel *f* <Künstl Int> inference rule
Inferenzstrategie *f* <Künstl Int> inference strategy, reasoning strategy

inferieren v <Künstl Int> infer
Inferieren n <Künstl Int> inference, reasoning
Infiltration f 1. <Abfall> infiltration (Schadstoffe); 2. <Wasserversorg> infiltration
Infiltrationssperre f <Kohlen> infiltration barrier
infiltrieren v <Bau, Kohlen> infiltrate; permeate
infinitesimal adj <Math> infinitesimal (unendlich klein)
Infinitesimalrechnung f <Math> calculus
Infinitum n <Math> infinity (Unendlichkeit)
Infixschreibweise f <Comp & DV> infix notation
Influenzmaschine f <Elektrotech> electrostatic generator
Infobahn f <Comp & DV, Telekom> information superhighway
Informatik f <Comp & DV> computer science, informatics
Information f <Comp & DV, Elektronik, Telekom> information
Information f **und Unterhaltung** f <Comp & DV> infotainment
Informationsabruf m <Comp & DV> information retrieval
Informationsabruf m **vom Web-Server** <Comp & DV> hit
Informationsanbieter m <Comp & DV> information provider
Informationsausgabe f <Comp & DV> information output
Informationsbelastung f **eines Datennetzes** <Comp & DV> traffic
Informationsbit n <Comp & DV, Telekom> data bit, information bit, information-carrying bit
Informationsblatt n <Comp & DV> folder
Informationscluster n <Comp & DV> cluster information
Informationseingabe f <Comp & DV> information input
Informationsempfangsstelle f <Telekom> information receiver station
Informationsentropie f <Comp & DV> information entropy
Informationsfluss m <Comp & DV> information flow
Informationsgehalt m <Comp & DV> information content
Informationsgewinnung f <Comp & DV> information gathering
Informationsgrenze f <Math> information limit (CRLB)
Informationsintegration f <Comp & DV> information integration
Informationskanal m <Telekom> information channel
Informationsquelle f <Comp & DV, Telekom> information source
Informationsrecherche f <Comp & DV> information search
Informationssammlung f <Comp & DV> information gathering
Informationsschutz m <Comp & DV> privacy of information
Informationssendestelle f <Telekom> information sending station
Informationssenke f <Telekom> information sink
Informationsservice m <Comp & DV> information service
Informationssicherheit f <Comp & DV, Telekom> information security, telecommunication security
Informationsspeicher m <Comp & DV> data storage, information storage
Informationsspeicherung f <Comp & DV> information storage
Informationsspeicherung f **und -abfrage** f <Comp & DV> ISR, information storage and retrieval
Informationsspeicherung f **und -wiederauffindung** f <Comp & DV> ISR, information storage and retrieval
Informationsstand m <Fertig> information booth (Messe)
Informationssystem n <Comp & DV, Telekom> information system, IS

Informationssysteme npl (IS) <Comp & DV> Information Systems pl (IS)
Informationstechnik f <Comp & DV, Telekom> information technology, IT (Nachrichtentechnik und Informatik)
informationstechnische Einrichtung f <Comp & DV> information technology equipment
Informationstechnologie f (IT) <Comp & DV> information technology (IT)
Informationstheorie f <Comp & DV, Elektronik, Telekom> information theory
Informationstrennzeichen n <Comp & DV> information separator
Informations-Ungleichung f <Math> information inequality (Parameterabschätzung)
Informationsverarbeitung f <Comp & DV, Elektronik> information processing
Informationsverbreitung f <Comp & DV> information dissemination
Informationsverteilungs-Account n <Comp & DV> information distribution account
Informationsverwaltungssystem n <Comp & DV> information management system
Informationszentralverbindung f <Comp & DV> information backbone
Informationszentrum n <Comp & DV> information center
Infra-Fernsprechfrequenz f <Telekom> subtelephone frequency
Infraprotein n <Chemie> infraprotein
Infrarot n (IR) <Kunststoff, Optik, Phys, Strahlphys> infrared (IR)
Infrarotabgasprüfgerät n <Kfztech> (BE) infrared exhaust gas analyser, (AE) infrared exhaust gas analyzer
Infrarotabtaster m <Gerät> infrared scanner
Infrarotbehandlung f <Strahlphys> infrared therapy
Infrarotbewegungsmelder m <Sicherheit> infrared motion alarm
Infrarotbildkonverter m <Fernseh> infrared image converter
Infrarotdetektor m <Trans> infrared detector
infrarotempfindlich adj <Strahlphys> infrared-sensitive
infrarotempfindliche Emulsion f <Foto> infrared emulsion, infrared-sensitive emulsion
Infraroterdmelder m <Raumfahrt> infrared earth sensor
Infrarotfernbedienung f <Maschinen> infrared remote control
Infrarotfilm m <Foto> infrared film
Infrarotfilter n <Strahlphys> infrared filter
Infrarotfotografie f <Foto> infrared photography
Infrarotheizung f <Heiz & Kälte, Strahlphys> infrared heating
Infrarotlaser m <Elektronik> infrared laser (IRASER)
Infrarot-LED f <Elektronik> infrared LED, infrared light-emitting diode
Infrarotlicht n <Strahlphys> infrared light
Infrarotlichtübertragung f <Telekom> infrared transmission, IR transmission, transmission by infrared light
Infrarot-Lumineszenzdiode f <Elektronik> infrared LED, infrared light-emitting diode
Infrarotphotometer n <Metrol> infrared photometer
Infrarotphotozelle f <Metrol> infrared photocell
Infrarotraumheizung f <Heiz & Kälte> infrared panel heating
Infrarotschnittstelle f <Telekom> infrared interface, optical interface
Infrarotspektralphotometer n <Labor, Metrol> infrared spectrophotometer
Infrarotspektrometer n 1. <Nichtfoss Energ> spectroradiometer; 2. <Strahlphys> infrared spectrometer

Infrarotspektroskopie f <Metrol, Strahlphys> infrared spectroscopy
Infrarotspektrum n <Strahlphys> infrared spectrum
Infrarotstrahlen mpl <Optik, Strahlphys> infrared radiation, infrared rays
Infrarotstrahlungsmesser m <Metrol> infrared detector
Infrarotübertragung f <Telekom> infrared transmission, IR transmission, transmission by infrared light
Infrarotverbindung f <Telekom> infrared link
Infraschall m <Akustik, Phys> infrasound
Infraschallbereich m <Akustik, Phys> infrasonic frequency range
Infraschallfrequenz f <Akustik, Phys> infrasonic frequency
Infrastruktur f <Comp & DV> infrastructure
Infusion f <Lebensmittel> infusion
Infusionsflasche f <Ker & Glas> infusion bottle
Ingangsetzen n <Ergon, Lufttrans, Maschinen> actuation
Ingenieur m <Mechan> engineer
Ingenieur m **für Elektrotechnik** <Elektrotech> electrical engineer
Ingenieur m **für Fernmeldetechnik** <Telekom> engineer for telecommunication, telecommunications engineer
Ingenieur m **für Hochfrequenztechnik** <Funktech> radio engineer
Ingenieur m **für Informationstechnik** <Telekom> engineer for information technology
Ingenieur m **für Nachrichtentechnik** <Telekom> communications engineer, engineer for communication technology
Ingenieurbiologie f <Fertig, Umweltschutz> engineering biology
Ingenieurbüro n <Mechan> engineering office
Ingenieurhochbau m <Bau> structural engineering
ingenieurtechnische Gestaltung f <Sicherheit> engineering design
Ingenieurwissenschaft f <Mechan> engineering
Inhaber m **eines Patents** <Patent> proprietor of a patent
Inhalt m 1. <Anstrich, Comp & DV, Math> content; 2. <Geom> volume; 3. <Kohlen, Patent> content; 4. <Wassertrans> volume
Inhalt m **der Kanalhaltung** <Nichtfoss Energ> pondage
inhaltsadressierbarer Speicher m (CAM) <Comp & DV, Künstl Int> associative memory, content-addressable memory (CAM)
Inhaltsangabe f <Verpack> contents declaration
Inhaltsstoff m <Kunststoff> ingredient
Inhaltsüberwachung f <Comp & DV> contents supervision
Inhaltsverzeichnis n <Comp & DV> contents directory
inhärent stabiler Reaktor m <Kerntech> inherently stable reactor
inhärente Adressierung f <Comp & DV> inherent addressing
inhärente Verfügbarkeit f <Lufttrans> inherent availability
Inhibitor m <Kunststoff, Wasserversorg> inhibitor
inhomogene Strahlungsquelle f <Strahlphys> nonuniform source of radiation
inhomogenes System n <Thermod> inhomogeneous system
Inhomogenität f <Ker & Glas> inhomogeneity
Initialbestätigung f <Metrol> initial verification
Initialdruck m <Maschinen> initial pressure
initialisieren v <Comp & DV> initialize
Initialisierung f <Comp & DV> initialization
Initialisierungsabsicht f <Comp & DV> scheduling intent
Initialprogrammlader m (IPL) <Comp & DV> initial program loader (IPL)

Initialzündung f <Lufttrans> priming
Initiator m 1. <Gerät> approximating pick-up; 2. <Lebensmittel> initiator
Injektion f <Chemtech, Elektronik> injection • **durch Injektion starr gekoppelt** <Telekom> injection locked
Injektionsbeton m <Bau> grouted aggregate concrete
Injektionsdüse f <Chemtech> injection nozzle
Injektionsgerät n <Bau> grouting equipment
Injektionslaser m <Elektronik> injection laser
Injektionslaserdiode f <Elektronik, Optik> ILD, injection laser diode
Injektionslogik f <Elektronik> injection logic (Mikroelektronik)
Injektionsmörtel m <Bau> grout
Injektionsschürze f <Bau> grout curtain
injektionsstabilisiert adj <Telekom> injection locked
injektionsstabilisierter Laser m <Telekom> injection-locked laser
injektionssynchronisierter Oszillator m <Elektronik> injection-locked oscillator
Injektionsverfahren n <Wassertrans> injection procedure (Bootsbau)
injektionsverriegelter Laser m <Optik> injection-locked laser
Injektionswinkel m <Kerntech> angle of injection
injektiv adv <Math> injective (eineindeutig)
injektive Abbildung <Math> injective mapping
Injektor m <Elektronik, Hydraul> injector
Injektordrossel f <Hydraul> injector throttle
Injizieren n <Elektronik> injection (Transistor)
Inklination f <Raumfahrt> inclination
Inklinationsmesser m <Kohlen> clinometer, inclinometer
Inklinationsmessgerät n <Phys> inclinometer
Inklinationswinkel m <Phys> angle of dip
inklusive ODER-Operation f <Comp & DV> inclusive OR operation
inklusive UND-Operation f <Comp & DV> inclusive AND operation
inklusives ODER-Gate n <Comp & DV> inclusive OR gate
inklusives ODER-Glied n <Comp & DV> inclusive OR circuit
inklusives UND-Gate n <Comp & DV> inclusive AND gate
inklusives UND-Glied n <Comp & DV> inclusive AND circuit
inkohärent adj <Metall, Optik, Telekom> incoherent
inkohärente Strahlung f <Phys, Telekom> incoherent radiation
inkohärenter Schall m <Akustik> incoherent sound
inkohärenter Zwilling m <Metall> incoherent twin
inkohärentes Licht n <Phys, Telekom> incoherent light
Inkohärenz f <Optik, Telekom> incoherence
inkohlen v <Kohlen> carbonize, coke
inkommensurabel adj <Math> incommensurable (nicht messbar, unvergleichbar)
inkompressibel adj <Chemie> incompressible
Inkompressibilität f <Strömphys> incompressibility
Inkompressibilität f **von Flüssigkeiten** <Thermod> incompressibility of liquids
inkompressible Flüssigkeit f <Phys> incompressible flow
inkompressible Strömung f <Strömphys> incompressible flow
Inkonstanz f <Funktech> instability (Frequenz)
Inkrafttreten n <Patent> entry into force
Inkreis m <Geom> incircle, inscribed circle
Inkreisradius m <Geom> apothem, short radius
Inkrement n <Elektronik, Math> increment (Zunahme)

Inkrementalkompilierer *m* <Comp & DV> incremental compiler
Inkrementalrechner *m* <Comp & DV> incremental computer
inkrementelle Bemaßung *f* <Gerät> incremental measure system
inkrementieren *v* <Elektronik> increment
Inland *n* <Telekom, Trans> inland
Inlands- oder Binnenemissionen *fpl* <Umweltschmutz> domestic emissions
Inlandsanruf *m* <Telekom> inland call
Inlandsflug *m* <Lufttrans> domestic flight
Inlandsflugverkehr *m* <Lufttrans> domestic service
Inlandsquelle *f* <Umweltschmutz> internal source
Inlandsverkehr *m* <Telekom> national traffic
Inline-Kopf *m* <Aufnahme> in-line head
Inline-Stereophonieband *n* <Aufnahme> in-line stereophonic tape
Innen... 1. <Fertig> female; 2. <Umweltschmutz> internal
Innenabdeckung *f* <Hydraul> inside cover
Innenabscheidungsverfahren *n* <Telekom> *(AE)* inside vapor phase oxidation, *(BE)* inside vapour phase oxidation
Innenantenne *f* <Elektrotech> indoor antenna
Innenarchitektur *f* <Bau> interior design
Innenausbau *m* <Bau> interior work
Innenauskleidung *f* <Verpack> interior lining
Innenbackenbremse *f* <Maschinen> internal expanding brake
Innenbeleuchtung *f* <Elektrotech> indoor lighting
Innenbeschichtungsprozess *m* <Telekom> *(AE)* inside vapor phase oxidation, *(BE)* inside vapour phase oxidation
Innenbeutel *m* <Verpack> bag-in-a-can
Innenboden *m* <Wassertrans> inner bottom *(Schiffbau)*
Innenbodenbeplattung *f* <Wassertrans> inner bottom plating
Innenbodenlängsverband *m* <Wassertrans> inner bottom longitudinal
Innenbordmotor *m* <Fertig> inboard engine
Innenbrenner *m* <Ker & Glas> internal burner
Innendeckel *m* <Hydraul> inside cover
innendrehen *v* <Fertig> bore
Innendrehen *n* <Maschinen> internal turning
Innendrehmeißel *m* <Maschinen> boring tool
Innendruckfestigkeit *f* <Werkprüf> inside surface strength, internal pressure strength *(Glas)*
Innendruckversuch *m* <Fertig> destructive hydrostatic test
Innendurchmesser *m* <Fertig, Maschinen, Mechan> ID, inner diameter, inside diameter
Inneneckmeißel *m* <Maschinen> internal facing tool
Inneneckrand *m* <Verpack> inside corner edge
Inneneinheit *f* <Funktech> indoor unit *(Satellitenempfänger)*
Inneneinrichtung *f* <Bau> interior fittings
Innenfläche *f* <Geom> inside face
innengekühlt *adj* <Kerntech> internally cooled
innengetrieben *adj* <Fertig> internal *(Malteserkreuz)*
Innengewinde *n* 1. <Fertig> female thread *(Kunststoffinstallationen)*; 2. <Ker & Glas> internal screw; 3. <Maschinen> female thread, inside thread, internal screw thread, internal thread • **Innengewinde schneiden** <Fertig, Maschinen> tap
Innengewindebohrer *m* <Bau> tapped valve drill
Innengewinderäumen *n* <Maschinen> internal thread broaching
Innengewindeschneiden *n* <Fertig> internal threading, tapping
Innengewindeschneider *m* <Maschinen> inside threading tool
Innenhaut *f* <Wassertrans> inner skin *(Schiffbau)*
Innenhonen *n* <Fertig> internal honing
Inneninstallation *f* <Elektrotech> internal installation
Innenisolierung *f* <Elektriz> indoor insulation
Innenkabel *n* <Fertig> indoor cable *(Faseroptik)*
Innenkern... <Kerntech> in-core
Innenlackierung *f* <Verpack> internal lacquering
innenlastig *adj* <Fertig> inboard *(Flurförderer)*
Innenleiter *m* <Telekom> inner conductor
Innenlenker *m* <Kfztech> *(BE)* saloon, *(AE)* sedan
Innenlunker *m* <Fertig> internal shrinkage
Innenmantel *m* <Telekom> inner cladding
Innenmantelung *f* <Elektriz> inner covering
Innenmessgerät *n* <Metrol> *(AE)* bore gage, *(BE)* bore gauge
Innenmessung *f (TTL-Messung)* <Foto> through-the-lens metering *(TTL metering)*
Innenmikrometer *n* <Maschinen> *(AE)* inside micrometer calipers, *(BE)* inside micrometer callipers, internal micrometer
Innennutzlast *f* <Kfztech> interior payload
Innenpoldynamo *m* <Elektriz> inner-pole dynamo, revolving-field dynamo
Innenpolgenerator *m* <Elektriz> inner-pole dynamo, inner-pole generator, revolving-field dynamo, revolving field generator
Innenputz *m* <Bau> interior plaster
Innenrad *n* <Maschinen> inside gear
Innenraster *m* <Elektronik> internal graticule *(CRT)*
Innenraum *m* <Kerntech> cavity *(Reaktordruckgefäß)*
Innenräumen *n* <Fertig, Maschinen> internal broaching
Innenräummaschine *f* <Maschinen> internal-broaching machine
Innenraumschaltanlage *f* <Elektrotech> indoor switchgear
Innenräumwerkzeug *n* <Fertig, Maschinen> internal broach
Innenreibung *f* <Maschinen> internal friction
Innenring *m* <Maschinen> inner ring
Innenriss *m* 1. <Fertig> shatter crack *(Schmiedestück)*; 2. <Metall> internal crack
Innenrückspiegel *m* <Kfztech> rear-view mirror *(Zubehör)*
Innenrundschleifen *n* <Maschinen> internal cylindrical grinding
Innenrüttler *m* <Bau> immersion vibrator, poker vibrator
Innensäule *f* <Fertig> column *(Radialbohrmaschine)*
Innenschallpegel *m* <Elektronik> interior noise level
Innenschicht *f* <Verpack> interior coating, internal lacquering
Innenschleifen *n* <Maschinen> internal grinding
Innenschleifer *m* <Maschinen> internal grinder
Innenschleifmaschine *f* <Fertig> internal-grinding machine
Innenschneidestahl *m* <Mechan> inside tool
Innenschraube *f* <Maschinen> inside screw
Innenschweißung *f* <Mechan> inside welding
Innensechskant *m* <Maschinen> hex socket, hexagon socket
Innensechskantschlüssel *m* 1. <Fertig> hexagonal recess wrench *(Kunststoffinstallationen)*; 2. <Maschinen> *(BE)* Allen key, *(AE)* Allen wrench
Innensechskantschraube *f* <Maschinen> Allen screw, hexagon socket head screw
Innenspannung *f* <Fertig> internal gripping
Innenspiegel *m* <Kfztech> driving mirror
Innenspielsteuerschieber *m* <Hydraul> inside clearance slide valve

Innenspülung f <Kohlen> internal scour
innenstehender Pfeil m <Konstzeich> inward-positioned arrowhead
Innentaster m <Maschinen> (AE) inside calipers, (BE) inside callipers, (AE) internal caliper gage, (BE) internal caliper gauge
Innenteil n 1. <Fertig> insert (Passung); 2. <Maschinen> insert, internal member; inner member (einer Passung)
Innentemperatur f <Verpack> internal temperature
Innenthermostat m <Heiz & Kälte> room thermostat
Innentrommelbelichter m <Druck> internal drum film recorder, internal drum image setter
Innenüberdeckungs-Steuerschieber m <Hydraul> inside lap slide valve
Innen- und Außentaster m <Maschinen> (AE) inside and outside calipers, (BE) inside and outside callipers
Innenverdrahtung f <Elektrotech, Telekom> indoor wiring, internal wiring
Innenverkleidung f <Wassertrans> inner lining
Innenverstärkung f <Verpack> interior strengthening bar
innenverzahntes Getriebe n <Kfztech, Maschinen> internal gear
Innenverzahnung f <Maschinen> internal toothing
Innenvolumen n <Geom> space inside
Innenvoreinström-Steuerschieber m <Hydraul> inside lead slide valve
Innenwiderstand m <Elektriz> internal resistance
Innenwinkel m <Geom> interior angle
innerbetriebliche Kommunikation f <Comp & DV> inter-office communication
innere Abdeckklappe f der Landeklappenschlitze <Lufttrans> inner shroud
innere Beschichtung f <Verpack> interior coating
innere Blockierung f <Telekom> internal blocking
innere Dämpfung f <Lufttrans> internal damping
innere Elektronen npl <Strahlphys> inner electrons
innere Elektronenschalen fpl <Strahlphys> inner orbital complex
innere Energie f 1. <Phys> internal energy; 2. <Thermod> internal energy, intrinsic energy
innere Form f <Druck> inside forme
innere Induktionsdichtung f <Verpack> induction inner seal
innere Konversion f <Kerntech, Strahlphys> internal conversion
innere Leitfähigkeit f <Elektriz> intrinsic conductivity
innere molare Energie f <Phys> molar internal energy
innere Oxidation f <Metall> internal oxidation
innere Permeabilität f <Elektriz> intrinsic permeability
innere Reibung f <Metall> internal friction
innere Rückkopplung f <Elektronik> inherent feedback (in Trioden)
innere Schirmung f <Elektrik> internal shield (bei Elektronenröhren)
innere Spannung f <Metall> internal stress
innere Störgröße f <Fertig> failure (Maschine)
innere Totalreflexion f <Optik, Wellphys> total internal reflection
innere Überlappung f <Hydraul> inside lap slide valve (Steuerschieber, Steuerkolben)
innere Umwandlung f <Kerntech, Strahlphys> internal conversion
innere Umwicklung f <Verpack> interior wrapping
innere Verpackung f <Verpack> interior packaging
innere Viskosität f <Strömphys> intrinsic viscosity
innere Wechselwinkel mpl <Geom> alternate interior angles
innerer Laufring m <Maschinen> ball inner race

innerer Photoeffekt m <Elektronik, Optik> internal photoelectric effect
innerer photoelektrischer Effekt m <Optik, Telekom> internal photoelectric effect
innerer Planet m <Raumfahrt> inner planet
innerer Türstein m <Ker & Glas> inside jamb block
innerer Weichmacher m <Kunststoff> internal plasticizer
inneres Einflugzeichen n <Lufttrans> inner marker
inneres Radgehäuse n <Lufttrans> internal wheel case
inneres Rohrende n <Maschinen> female end of a pipe
innerörtliche Verkehrsbedienung f <Trans> local traffic information
innerste Schale f mit zwei Elektronen <Kerntech> two-electron innermost shell
innewohnend adj <Anstrich> inherent
innovative Technologie f <Comp & DV, Kfztech, Nichtfoss Energ, Qual> innovative technology
Inosin n <Chemie, Lebensmittel> inosine
Inosit m <Chemie> inositol
Inositol n <Chemie, Lebensmittel> inositol
In-Pile-Kreislauf m <Kerntech> in-pile loop
Input m <Hydraul, Phys> IP, input
Input-Filter n <Raumfahrt> input filter (Weltraumfunk)
Inquadrat n <Geom> inscribed square, insquare
INS n (integriertes Navigationssystem) <Wassertrans> INS (integrated navigation system)
IN-Schnittstelle f <Telekom> intelligent network application protocol (INAP)
insektenfest adj <Verpack> insect-proof
Inselbahnhof m <Eisenbahn> island depot
Inselnetz n <Telekom> subnetwork
Inserat n <Druck> advertisement
Inseratensetzerei f <Druck> advertisement composing room
Insert-Technik f <Kunststoff> (AE) insert molding, (BE) insert moulding
insgesamt digital adj <Elektronik> all-digital
Insolation f <Nichtfoss Energ> insolation
Inspektion f 1. <Maschinen, Mechan> inspection; 2. <Wassertrans> surveying (Schiff)
Inspektionsöffnung f <Raumfahrt> inspection door (Raumschiff)
Inspektionszyklus m <Lufttrans> inspection cycle
Inspektor m <Sicherheit> factory inspector, supervisor
instabil adj <Elektriz, Elektrotech> unstable
instabile Strömungen fpl <Strömphys> unstable flows
instabile Zerspanung f <Fertig> unstable metal cutting
instabiler Abbrand m <Raumfahrt> chuffing
instabiler Kern m <Teilphys> unstable nucleus
instabiles Ausbuchten n <Kerntech> ballooning instability
instabiles Isotop n <Chemie> radioisotope
Instabilität f <Elektrotech, Verpack> instability
Instabilität f aufgrund der Oberflächenspannung <Strömphys> surface tension instabilities
Instabilität f der rotierenden Couetteströmung <Strömphys> instability of rotating Couette flow
Instabilitätserscheinungen fpl <Strömphys> instability phenomena
Installateur m <Bau> plumber
Installation f 1. <Bau> plumbing; 2. <Comp & DV> installation; setup (eines Rechners); 3. <Fertig> utility; 4. <Maschinen, Telekom> installation
Installationsobjekte npl 1. <Bau> fixtures and fittings; 2. <Metrol> fixtures
Installationsoption f <Comp & DV> set-up option
Installationsplan m <Elektriz> wiring diagram
Installationsplanung f <Comp & DV> physical planning
Installationsrohr n <Bau> conduit

Installationszeit

Installationszeit f <Comp & DV> set-up time
installieren v 1. <Bau> plumb; 2. <Comp & DV> install, mount, set up; 3. <Druck> install; 4. <Elektronik> install, mount, set up; 5. <Fertig> provide; 6. <Verpack> erect; 7. <Wassertrans> install *(Technik)*
installierte Leistung f <Elektriz> installed capacity, installed power
installierter Hubraum m <Kfztech> installed capacity
instand halten v 1. <Comp & DV> service; 2. <Maschinen> maintain, service; 3. <Mechan> maintain *(Geräte)*; service *(Werkzeuge)*
instand setzen v 1. <Bau> make good, repair; 2. <Kfztech> repair; 3. <Telekom> overhaul; 4. <Wassertrans> repair
Instandhaltung f 1. <Comp & DV> maintenance, upkeep; 2. <Eisenbahn, Fernseh, Maschinen, Mechan, Telekom> maintenance
Instandhaltungsdauer f <Qual> maintenance time
instandhaltungsgerechte Konstruktion f <Sicherheit> design for maintainability
Instandhaltungskonzept n <Qual> maintenance concept
Instandhaltungsprozessor m <Telekom> maintenance processor
Instandsetzung f 1. <Kfztech> overhaul, repair *(Motor)*; 2. <Qual> corrective maintenance; 3. <Wassertrans> repair *(eines Schiffes)*
Instandsetzungsarbeiten fpl <Bau> repairs
Instandsetzungsdauer f <Comp & DV> repair time
Instandsetzungshandbuch n <Kfztech> overhaul manual
instantiieren v <Künstl Int> instantiate
Instant-Replay n <Fernseh> instant replay
Instanz f 1. <Künstl Int> instance; 2. <Telekom> entity
Instanz f **des Systemmanagements** <Telekom> system management application entity
instationäre Strömungen fpl <Strömphys> unsteady flows
Instruktion f <Comp & DV> instruction
Instruktionsabruf m <Comp & DV> instruction fetching
Instruktionscode m <Comp & DV> instruction code
Instruktionsregister n <Comp & DV> instruction register
Instruktionsstrom m <Comp & DV> instruction stream
Instruktionszyklus m <Comp & DV> instruction cycle
Instrument n 1. <Comp & DV, Eisenbahn, Elektriz, Elektrotech, Kfztech, Lufttrans> instrument; 2. <Maschinen> apparatus, implement, instrument; 3. <Wassertrans> instrument
Instrument n **für Einzelmessungen** <Gerät> single shot instrument
Instrument n **mit Nebenwiderstand** <Gerät> shunted instrument
Instrument n **mit Nullpunkt in der Skalenmitte** <Gerät> *(AE)* center-reading instrument, *(BE)* centre-reading instrument
Instrument n **mit Shunt** <Gerät> shunted instrument
Instrument n **zur Messung von Leuchtdichte und Beleuchtungsstärke** <Optik> nitometer-luxmeter
Instrumentenanflug m <Lufttrans> instrument approach
Instrumentenanflugkarte f <Lufttrans> instrument approach chart
Instrumentenanflugpiste f <Lufttrans> instrument approach runway
Instrumentenanflugverfahren n <Lufttrans> instrument approach procedure
Instrumentenaufbau m **im Reaktorkern** <Kerntech> in-core instrument assembly
Instrumentenausrüstung f <Comp & DV> instrumentation

Instrumentenbeleuchtung f <Wassertrans> scale illumination
Instrumentenbrett n <Lufttrans> console, instrument panel
Instrumentenfehler m 1. <Gerät> instrumental error; 2. <Maschinen> instrument error; 3. <Wassertrans> index error *(Navigation)*
Instrumentenflug m <Lufttrans> instrument flying, instrument flight
Instrumentenflugberechtigung f <Lufttrans> instrument rating
Instrumentenflugberechtigung f **für Piloten** <Lufttrans> single pilot instrument rating
Instrumentenflugregeln fpl *(IFR)* <Lufttrans> instrument flight rules *(IFR)*
Instrumentenflugsimulator m **für die Grundschulung** <Lufttrans> basic instrument flight trainer
Instrumentenhöhe f <Bau> height of instrument *(Vermessung)*
Instrumentenlandesystem n *(ILS)* <Lufttrans, Raumfahrt> instrument landing system *(ILS)*
Instrumentenlandung f <Lufttrans> instrument landing
Instrumentenschalter m <Elektrotech> instrument switch
Instrumententafel f 1. <Kfztech> dash, dashboard; instrument panel *(Zubehör)*; 2. <Lufttrans, Wassertrans> instrument panel
Instrumentenverluste mpl <Elektriz> meter losses
instrumentieren v <Wassertrans> instrument
Instrumentierung f <Comp & DV> instrumentation
instrumentloses Brennelement n <Kerntech> uninstrumented fuel assembly
Intaktstabilität f <Wassertrans> intact stability *(Schiffkonstruktion)*
Integral n <Math> integral
integral versteifte Leichtmetallbeplankung f <Raumfahrt> integrally-stiffened light alloy skin *(Raumschiff)*
integrale Abtastung f <Telekom> integral sampling
Integralelement n <Labor> integral element
integrales Laufrad n <Nichtfoss Energ> integral runner *(Wasserturbine)*
Integralfunktion f <Math> integral function
Integralhelm m <Sicherheit> full-face helmet
Integralrechnung f <Math> integral calculus
Integralregler m <Labor> integral action controller
Integralschaumstoff m <Kunststoff> integral foam, integral skin foam
Integration f <Elektronik, Math, Telekom> integration
Integration f **einer Elektronikschaltung** <Elektronik> electronic circuit integration
Integrationsbereich m <Math> integral domain
Integrationsdichte f <Elektronik> integration density
Integrationsgebiet n <Math> integral domain
Integrationsschaltung f <Elektrotech> integrating circuit
Integrationsverstärkung f <Elektronik> integration gain
Integrationszeit f <Elektronik, Elektrotech> integration period, integration time
Integrator m <Elektronik, Elektrotech, Lufttrans> integrator
Integrierbeiwert m <Regelung> integral action coefficient
integrieren v <Comp & DV, Math> integrate
Integrieren n <Math> integration
integrierender Fehler m <Gerät, Qual> cumulative error
integrierender Zähler m <Gerät> totalizing counter
integrierendes Bauelement n <Elektronik> integrator
integrierendes Glied n <Regelung> integral element
integrierendes Messgerät n <Elektrotech> integrating meter
integrierendes Netz n <Lufttrans> integrating network
integrierendes Verhalten n <Regelung> integral action
Integrierkondensator m <Elektrotech> integrating capacitor

Integrierschaltung f (IS) <Comp & DV, Elektriz, Elektronik, Kontroll, Phys> integrated circuit (IC)
integriert adj 1. <Comp & DV> native; 2. <Elektronik> integrated
integrierte Brücke f <Wassertrans> integrated bridge (Navigation)
integrierte Client-Library-Schnittstelle f <Comp & DV> native client library interface
integrierte Datenbank f <Comp & DV> integrated database
integrierte Datenverarbeitung f <Comp & DV> integrated data processing
integrierte Dickschichtschaltung f <Telekom> thick layer integrated circuit
integrierte digitale Übertragung f **und Vermittlung** <Telekom> integrated digital transmission and switching
integrierte digitale Vermittlung f <Telekom> integrated digital exchange
integrierte Einheit f <Comp & DV> integrated device
integrierte Einspritzung f <Kfztech> integral injection
integrierte Funktion f <Elektronik> integrated function
integrierte Halbleiterschaltung f <Elektronik> semiconductor integrated circuit
integrierte Hausnetz-Nebenstelle f <Telekom> private integrated network exchange
integrierte Hybridkomponente f <Elektronik> integrated hybrid component
integrierte Hybridschaltung f <Telekom> hybrid integrated circuit
integrierte Injektionslogik f (IL-Logik) <Elektronik> integrated injection logic (IL)
integrierte Ladung f <Elektrotech> integrated charge
integrierte Laufzeit f <Erdöl> ITT, integrated transit time (Seismik)
integrierte Logikschaltung f <Elektronik> integrated logic circuit
integrierte optische Koppelmatrix f <Telekom> integrated optical switching matrix
integrierte optische Schaltung f (IOS) <Telekom> integrated optical circuit (IOC)
integrierte optoelektronische Schaltung f <Telekom> integrated optoelectronic circuit
integrierte Quellenfindung f <Comp & DV> IHS, integrated home system
integrierte Schaltkreise mpl **für bestimmte Funktionen** <Comp & DV> application-specific integrated circuits
integrierte Schaltkreismaske f <Elektronik> integrated circuit mask
integrierte Schaltung f (IS) <Comp & DV, Elektriz, Elektronik, Funktech, Kontroll, Phys, Telekom> integrated circuit (IC)
integrierte Schaltung f **in Metall-Gate-CMOS-Technologie** <Elektronik> metal gate CMOS integrated circuit
integrierte Siliziumschaltung f <Elektronik> silicon integrated circuit
integrierte Transistorlogik f (MTL-Logik) <Elektronik> merged transistor logic (MTL)
integrierter Bipolartransistor m <Elektronik> integrated bipolar transistor
integrierter digitaler Anschluss m <Telekom> IDA, integrated digital access
integrierter Digitalzugriff m <Comp & DV> IDA, integrated digital access
integrierter Hybridwiderstand m <Elektrotech> integrated hybrid resistor
integrierter Kindersicherheitssitz m <Kfztech> integrated child safety seat
integrierter Kondensator m <Elektrotech> integrated capacitor, on-chip capacitor
integrierter Mikrowellenschaltkreis m (MIC) <Elektronik, Funktech, Wellphys> microwave integrated circuit (MIC)
integrierter Monomode-Schaltkreis m <Telekom> single mode optical integrated circuit
integrierter MOS-Transistor m <Elektronik> integrated MOS transistor
integrierter N-Kanal-MOS-Transistor m <Elektronik> n-channel integrated MOS transistor
integrierter optischer Schalter m <Telekom> integrated optical switch
integrierter optischer Schaltkreis m (IOS) <Elektronik, Optik> integrated optical circuit, optical integrated circuit
integrierter PIN-FET-Empfänger m <Optik, Telekom> integrated PIN-FET receiver
integrierter P-Kanal-FET m <Elektronik> P-channel integrated FET
integrierter Schaltkreis m (IS) <Comp & DV, Elektriz, Elektronik, Funktech, Kontroll, Phys, Telekom> integrated circuit (IC)
integrierter Schottky-Bipolarschaltkreis m <Elektronik> Schottky bipolar-integrated circuit
integrierter Schutzmechanismus m <Comp & DV> integrated security
integrierter Sicherheitsgurt m <Kfztech> integrated safety belt
integrierter Überrollbügel m <Kfztech> integrated roll bar
integrierter Überrollkäfig m <Kfztech> integrated roll cage
integrierter verstärkter Griff m <Verpack> integral reinforced handle
integrierter Zugang m <Telekom> integrated access
integriertes Brückensystem n (IBS) <Wassertrans> integrated bridge system (IBS)
integriertes Bürokommunikationssystem n <Telekom> integrated office system
integriertes digitales Netz n <Telekom> integrated digital network (IDN)
integriertes Digitalnetz n <Telekom> integrated digital network
integriertes Filter n <Elektronik> integrated filter
integriertes Gerät n <Comp & DV> integrated device
integriertes Informationssystem n <Comp & DV> built-in help, integrated information system (Hilfe-Unterprogramm)
integriertes Navigationssystem n (INS) <Wassertrans> integrated navigation system (INS)
integriertes System n <Telekom> integrated system
integriertes Text- und Datennetz n <Telekom> integrated digital network
integriertes Verknüpfungsglied n <Elektronik> integrated logic gate
Integrierung f <Math> integration
Integrität f <Comp & DV> integrity
intelligente Chipkarte f <Comp & DV, Telekom> smart card
intelligente Datenstation f <Comp & DV> intelligent terminal, smart terminal
intelligente freiprogrammierbare Datenstation f <Comp & DV> intelligent terminal
intelligenter Roboter m <Künstl Int> intelligent robot
intelligentes Lernprogramm n (KI-Lernprogramm) <Künstl Int> intelligent tutoring system (ITS)
intelligentes Netz n <Telekom> intelligent network (IN)
intelligentes verteiltes System n <Telekom> smart distributed system

Intelligenzquotient

Intelligenzquotient *m* <Ergon> intelligence quotient
Intelligenztest *m* <Ergon> intelligence test
Intensionsgröße *f* <Phys> intensive quantity
Intensität *f (I)* 1. <Akustik, Elektriz> intensity, I; 2. <Optik> intensity, I *(proportional zur Strahlung)*
Intensitätsmodulation *f* <Telekom> intensity modulation
Intensitätspegel *m* <Elektrotech> intensity level
Intensitätsschrift *f* <Akustik> variable density recording
Intensitätsstufe *f* <Aufnahme> intensity spectrum level
Intensitätstonspur *f* <Aufnahme> variable density sound track
Intensitätsverteilung *f* <Strahlphys> intensity distribution
Intensitätswert *m* <Strahlphys> level of intensity
interagieren *v* <Comp & DV> interact
Interaktion *f* 1. <Comp & DV> *(AE)* dialog, *(BE)* dialogue, interaction; 2. <Maschinen> interaction
interaktiv *adj* 1. <Comp & DV, Kontroll> interactive; 2. <Optik> *(BE)* compact disc-interactive, *(AE)* compact disk-interactive
interaktive Bildplatte *f* <Optik> *(BE)* interactive videodisc, *(AE)* interactive videodisk
interaktive Grafikverarbeitung *f* <Comp & DV> interactive graphics
interaktive Maske *f* <Comp & DV> interactive map
interaktive Routine *f* <Comp & DV> interactive routine
interaktive Verarbeitung *f* <Comp & DV> interactive processing
interaktive Videographie *f* <Telekom> interactive videography
interaktives Fernsehen *n* <Fernseh> interactive television
interaktives Netz *n* <Telekom> interactive network
interaktives Terminal *n* <Telekom> interactive terminal
Interaktivität *f* <Comp & DV> interactivity
interatomare Kraft *f* <Strahlphys> interatomic force
interatomarer Stoß *m* <Kerntech> atom-atom collision
Interchip-Signalverzögerung *f* <Elektronik> interchip signal delay
Intercircuit-Signalverzögerung *f* <Elektronik> intercircuit signal delay
Intercity-Flugdienst *m* <Eisenbahn> intercity air service
Intercity-Flugverbindungen *fpl* <Eisenbahn> intercity air service
Intercity-Zug *m* <Eisenbahn> *(BE)* intercity train
interdigital *adj* <Telekom> interdigital
Interdigital... <Funktech, Telekom> interdigital
Interdigitalfilter *n* <Telekom> interdigital bandpass
Interdigitalleitung *f* <Phys> interdigital line *(Akusto-Elektronik)*
Interdigitalwandler *m* <Telekom> interdigital transducer
Interface *n* <Comp & DV, Telekom> interface
Interferenz *f* 1. <Comp & DV, Fernseh> interference *(Wellen)*; 2. <Funktech> interference; 3. <Optik> fringe, interference; 4. <Phys, Telekom, Wellphys> interference
Interferenz *f* **mit Datenverlust** <Elektrotech> destructive interference
Interferenz *f* **mit Informationsverlust** <Elektrotech> destructive interference
Interferenzbänder *npl* <Wellphys> interference bands
Interferenzbild *n* <Optik> interference figure
Interferenzfarben *fpl* <Foto> Newton's rings
Interferenzfilter *n* <Aufnahme, Elektronik, Optik, Phys> interference filter
Interferenzkomparator *m* <Metrol> *(AE)* gage block interferometer, *(BE)* gauge block interferometer *(Optik)*
Interferenzmaschine *f* <Comp & DV> inference machine
Interferenzmikroskop *n* <Metall, Phys> interference microscope

Interferenzmuster *n* 1. <Optik> interference fringe; 2. <Wellphys> interference pattern
Interferenzrefraktometer *n* <Optik> interference refractometer
Interferenzringe *fpl* <Phys> interference fringes
Interferenzstreifen *mpl* 1. <Fernseh> fringes; 2. <Phys> interference fringes; fringes *(gleicher Dichte)*; 3. <Wellphys> interference fringes
interferieren *v* <Phys> interfere
Interferometer *n* <Metrol, Optik, Phys, Telekom> interferometer
intergranular *adj* 1. <Fertig> intercrystalline; 2. <Metall> intergranular
Interkanalstörung *f* <Elektrotech> interchannel interference
interkontinentale Verbindung *f* <Telekom> intercontinental connection
interkristallin *adj* <Fertig> intercrystalline
interkristalline Korrosion *f* <Chemie> intercrystalline corrosion, intergranular corrosion
interkristalline Spannungsrisskorrosion *f* <Fertig> caustic embrittlement
interlaminare Festigkeit *f* <Kunststoff> interlaminar strength
interlaminare Scherfestigkeit *f* <Kunststoff> interlaminar shear strength
intermediäre Bodenart *f* <Kohlen> intermediate type of soil
intermediäres Boson *n* <Phys> intermediate boson
intermetallische Verbindung *f* <Metall> intermetallic compound
intermittierende Quelle *f* <Wasserversorg> intermittent spring
intermittierendes Luftstrahltriebwerk *n* <Lufttrans> pulsojet
intermittierendes Rauschen *n* <Akustik> intermittent noise
intermodal *adj* <Eisenbahn> intermodal
Intermodulation *f* 1. <Fernseh, Funktech> intermodulation; 2. <Raumfahrt> intermodulation *(Weltraumfunk)*; 3. <Telekom> intermodulation
Intermodulationsprodukt *n* 1. <Raumfahrt> intermodulation product *(Weltraumfunk)*; 2. <Telekom> intermodulation product
Intermodulationsrauschen *n* <Raumfahrt> intermodulation noise *(Weltraumfunk)*
Intermodulationsstörung *f* <Funktech> intermodulation interference
Intermodulationsverzerrung *f (IMD)* <Aufnahme, Funktech, Telekom> intermodulation distortion *(IMD)*
intermolekular *adj* <Chemie> intermolecular
internationale Anmeldung *f* <Patent> international application
internationale Datumsgrenze *f* <Lufttrans> international date line
internationale Einheit *f* <Elektriz> international unit
Internationale Fernmeldeunion *f (IFU)* <Telekom, Wassertrans> International Telecommunication Union, ITU *(Behörde)*
internationale Gerätekennung *f* <Telekom> IMEI, international mobile station equipment identity
internationale Hauptvermittlungsstelle *f* <Telekom> *(AE)* main international trunk-switching center, *(BE)* main international trunk-switching centre *(im Fernwahlnetz)*
internationale Identifikationsnummer *f* **für Mobilgeräte** <Mobilkom> international mobile equipment identification number *(IMEI)*
internationale Landeskennzahl *f* <Telekom> international prefix number, international telephone country code *(ITCC)*

internationale Mobil-Geräte-Identifikations-Nummer f <Mobilkom> international mobile equipment identification number (IMEI)
Internationale Organisation f für Normung (ISO) <Elektriz, Maschinen> International Standardisation Organisation (ISO)
internationale Recherchenbehörde f <Patent> international searching authority
internationale Registrierung f <Patent> international registration
Internationale Schifffahrtorganisation f (IMO) <Wassertrans> International Maritime Organization (IMO)
internationale Selbstwahl f (ISW) <Telekom> (AE) international direct dialing, (BE) international direct dialling, (AE) international direct distance dialing, (BE) international direct distance dialling (IDDD)
Internationale Strahlenschutzkommission f (ICRP) <Strahlphys> International Commission on Radiological Protection (ICRP)
internationale Teilnehmerselbstwahl f <Telekom> ISD, (AE) international subscriber dialing, (BE) international subscriber dialling
internationale Telefonauskunft f <Telekom> international directory assistance
internationale Verbindung f <Telekom> international connection
internationale Vermittlungsstelle f <Telekom> (AE) international-switching center, (BE) international-switching centre
internationaler Betriebsdienst m <Telekom> international operations service
Internationaler Code m für die Beförderung von gefährlichen Gütern mit Seeschiffen <Wassertrans> international maritime dangerous goods code, IMDG code
Internationaler Eisenbahnverband m (UIC) <Eisenbahn> International Railway Union (UIC)
internationaler Selbstwählferndienst m <Telekom> ISD, (AE) international subscriber dialing, (BE) international subscriber dialling
internationaler Standard m <Bau> international standard
internationaler Standardruß m (IRB-Ruß) <Kunststoff> industry reference black (IRB)
internationales Einheitensystem n (SI-Einheit) <Elektriz, Metrol, Phys> international system of units (SI unit)
internationales Fernamt n <Telekom> international long-distance exchange, (AE) international-switching center, (BE) international-switching centre
internationales Hauptamt n <Telekom> (AE) main international-switching center, (BE) main international-switching centre
internationales Kopfamt n <Telekom> international gateway exchange
internationales Normal n <Qual> international standard
internationales Normgewinde n <Maschinen> international standard thread
internationales paketvermitteltes Datennetz n <Telekom> international packet-switching data network
internationales paketvermitteltes Kopfamt n <Telekom> international packet-switching gateway exchange
Internationales Referenzsystem n (IRS) <Umweltschmutz> International Referral System (IRS)
Internationales Signalbuch n <Wassertrans> international code of signals, INTERCO
internationales Telegrafenalphabet n (ITA) <Lufttrans, Wassertrans> international telegraph alphabet (ITA)
Internationales Transit-Zentrum n <Raumfahrt> International Transit Centre (Weltraumfunk)

Internationales Verzeichnis n für potenziell toxische **Chemikalien** (IRPTC) <Umweltschmutz> International Register of Potentially Toxic Chemicals (IRPTC)
interne Nebenstelle f <Telekom> internal extension
interne Sortierung f <Comp & DV> internal sort
interne Uhr f <Comp & DV> clock device
interner Speicher m 1. <Comp & DV> internal memory, (AE) internal storage, (BE) internal store; 2. <Elektrotech> internal memory
interner Takt m <Telekom> internal clock
interner Wartestatus m <Comp & DV> quiesced state, quiescent state
interner Widerstand m <Elektriz> internal resistance
internes Bandlaufwerk n <Comp & DV> internal tape drive
internes Eingangssignal n <Telekom> internal input signal
internes Target n <Teilphys> internal target
Internet n <Telekom> Internet (weltweiter Verbund von Computern auf der Grundlage von Internetprotokollen)
Internet-Adresse f <Telekom> domain, Internet address
Internet-Adressen-Registrierungsstelle f <Telekom> network information center (NIC)
Internet-Adressenuntergruppe f <Telekom> Internet address subgroup
Internet-Anbieter m <Telekom> Internet provider
Internet-Arbeitsgruppe f <Telekom> Internet Engineering Task Force, WWW Consortium, W3C (Arbeitsgruppe für Internetprotokolle)
Internet-Backbone m <Comp & DV> Internet backbone
Internet-Diensteanbieter m <Telekom> Internet service provider, ISP
Internet-Dienstverwalter m <Telekom> Internet service provider (ISP)
Internet-Dokumentenadresse f <Telekom> uniform resource locator (URL)
Internet-Gateway m <Telekom> Internet gateway
Internet-Kamera f <Telekom> internet video camera, webcam
Internet-Mail-Account n <Comp & DV> Internet mail account
Internet-Mail-Client n <Comp & DV> Internet mail client
Internet-Nutzer m <Comp & DV> Internet user
Internet-PC m <Telekom> network PC
Internet-Portal n <Telekom> Internet portal
Internet-Post f <Telekom> electronic mail, e-mail
Internet-Protokoll n <Telekom> Internet protocol, IP
Internet-Seite f <Telekom> webpage
Internet-Server-Bürosoftwarepaket n <Comp & DV> suite of Internet server software
Internet-Standort m <Telekom> website
Internet-Telefonie f <Telekom> Internet telephony
Internet-Übergang m <Telekom> Internet gateway
Internet-Verschlüsselungsstandard m <Comp & DV> Internet channel encryption standard
Internet-Wörterbuch n <Telekom> dictionary online, Internet dictionary
Internet-Zugang m <Telekom> Internet access
Internet-Zugangsprotokoll n <Telekom> transmission control protocol/Internet protocol (TCP/IP)
Internet-Zugriffsprogramm n <Telekom> brower, Web-browser
Interngespräch n <Telekom> internal call
Internverbindungsleitungssatz m <Telekom> intra-office junctor circuit
Internverkehr m <Telekom> internal traffic
Interoperabilität f <Comp & DV> interoperability
Interphon n <Telekom> interphone
Interplanetar... <Raumfahrt> interplanetary

Interplanetarflug

Interplanetarflug m <Raumfahrt> interplanetary flight
Interplanetarsonde f <Raumfahrt> interplanetary probe
Interpolation f <Comp & DV, Math, Telekom> interpolation
Interpolationsfilter n <Elektronik> interpolating filter
Interpolator m <Telekom> interpolator
interpolieren v <Druck, Math> interpolate
Interpolieren n <Math> interpolation
Interpreterprogramm n <Comp & DV> interpreter
interpretierend f <Druck> punctuation
interpretierendes Programm n <Comp & DV> interpreter
Interpunktion f <Druck> punctuation
Interpunktionspunkt m <Druck> full point
Interpunktionszeichen npl <Druck> punctuation marks
Interrogator-Responsor m <Telekom> interrogator-transponder
Interrupt m <Comp & DV> interrupt
interruptgesteuertes System n <Kontroll> interrupt-driven system
Intersatellitenfunkdienst m <Raumfahrt> intersatellite service
Intersektion f <Comp & DV> intersection
intersensorische Wahrnehmung f <Ergon> intersensory perception
interstellare Materie f <Raumfahrt> interstellar matter
interstellarer Raum m <Raumfahrt> interstellar space
interstitiell adj <Bau> interstitial
Intersymbol n <Elektronik> intersymbol
Intersymbolrauschen n <Elektronik> intersymbol noise
Intersymbolstörung f 1. <Elektronik> intersymbol interference (ISI); 2. <Telekom> intersymbol interference
Intervall n <Akustik, Math> interval *(zusammenhängender Zahlbereich)*
Intervallgeschwindigkeit f <Erdöl> interval velocity *(Seismik)*
Intervall-Länge f <Metrol> interval length
Intervallrechnung f <Math> interval arithmetic
Intervallschachtelung f <Math> nest of intervals
Intervallschmierung f <Maschinen> intermittent lubrication
Intervallzeitgeber m <Comp & DV> interval timer
Interventionsstrategie f <Comp & DV> intervention strategy
Interventionsverfahren n <Comp & DV> intervention procedure
Intonation f <Akustik> intonation
intramolekular adj <Chemie, Metall> intramolecular
Intranet n <Comp & DV> Intranet
Intrinsic-Halbleiter n <Phys> intrinsic semiconductor
Intrinsic-I-Halbleiter m <Elektronik> i-type semiconductor
intrinsisch sicher adj <Sicherheit> intrinsically safe
intrinsischer Koppelverlust m <Telekom> intrinsic joint loss
Intrittfallen n <Fernseh> pulling
Intrittfallmoment n <Elektrotech> pickung-up torque, pull-in torque *(Synchronmotor)*
Intrittfallprüfung f <Elektrotech> pull-in test
Intrittkommen n <Elektrotech> paralleling pulling in synchronism, pulling into synchronism *(bei Wechselstrommaschinen, mit der Grundfrequenz)*
Intrittziehen n <Elektrotech> pulling into synchronism
Introskopie f <Kerntech> introscopy
intrusiv adj <Nichtfoss Energ> intrusive
Inulin n 1. <Chemie> alantin, dahlin, inulin; 2. <Lebensmittel> inulin
Invar n 1. <Fertig> invar *(Nickelstahl)*; 2. <Metall> invar
invariant adj <Geom> invariant

Invariante f <Geom> invariant
Inventar n <Erdöl, Verpack> inventory
invers adj <Math> inverse
Inverse f <Math> inverse
inverse Funktion f <Math> inverse function
inverse Matrix f <Math> inverse matrix
inverser Compton-Effekt m <Phys> inverse Compton effect
inverser photoelektrischer Effekt m <Elektronik> inverse photoelectric effect
inverser piezoelektrischer Effekt m <Elektrotech, Phys> inverse piezoelectric effect
inverses Verhalten n <Regelung> reverse action
Inversionskappe f <Umweltschutz> lid
Inversionsschicht f 1. <Elektronik> inversion layer *(Halbleiter)*; 2. <Funktech> inversion layer; 3. <Umweltschutz> atmospheric inversion, inversion layer
Inversionstemperatur f <Elektriz, Phys> inversion temperature
Inversprimärkriechen n <Metall> inverse primary creep
Invertase f <Lebensmittel> invertase, saccharase, sucrase
Inverter m 1. <Comp & DV> inverter, negator; 2. <Elektrotech, Telekom> inverter
Inverter/Auflade-Leistung f <Nichtfoss Energ> inverter/charger performance
Invertglas n <Ker & Glas> invert glass
invertieren v 1. <Funktech> invert; 2. <Math> invert *(Matrix)*
Invertierschaltung f <Telekom> inverter
invertierte Darstellung f <Comp & DV> reverse video
invertierte Datei f <Comp & DV> inverted file
invertierter Chip m <Elektronik> inverted chip
invertiertes Bild n <Fernseh> reverse image
invertiertes Videobild n <Comp & DV> inverse video
Invertose f <Lebensmittel> invert sugar
Invertzeitrelais n <Elektriz> inverse time relay
Invertzucker m <Lebensmittel> invert sugar
involut adj <Geom> involute
Involute f <Geom> involute
Iod n (I) <Chemie> iodine (I)
Iod... <Chemie> iodous
Iodat n <Chemie> iodate
Iodbenzol n <Chemie> iodobenzene, phenyl iodide
Iodeosin n <Chemie> iodeosin, tetraiodofluorescein
Iodhydrin n <Chemie> iodohydrin
iodieren v <Chemie> iodize
iodig adj <Chemie> iodous
Iodimetrie f <Chemie> iodometry
Iodlaser m <Elektronik> iodine laser
Iodoaurat n <Chemie> iodoaurate, tetraiodoaurate
Iodoform n <Chemie> iodoform, triiodomethane
Iodometrie f <Chemie> iodometry
iodometrisch adj <Chemie> iodometric
iodometrische Titration f <Chemie> iodometry
Iodonium n <Chemie> iodonium
Iodopsin n <Chemie> iodopsin
Iodoso... <Chemie> iodoso...
Iodosobenzol n <Chemie> iodosobenzene
Iodosolbenzyl n <Chemie> iodosobenzene
Iodwasserstoff... <Chemie> hydriodic
Iodzahl f <Lebensmittel> iodine number, iodine value
Ion n 1. <Elektriz, Elektronik> ion; 2. <Erdöl> ion *(Petrochemie)*; 3. <Funktech, Phys, Teilphys> ion
Ionen... <Strahlphys> ionic
Ionenabstand m <Chemie> interionic distance
Ionenantrieb m <Raumfahrt> ion propulsion *(Raumschiff)*
Ionenausbeute f <Strahlphys> ion yield
Ionenaustauscherharz n <Fertig> ion exchange resin *(Kunststoffinstallationen)*

Ionenaustauschglas n <Ker & Glas> ion exchange glass
Ionenaustauschisotherme f <Strahlphys> ion exchange isotherm
Ionenaustauschverfahren n <Optik, Telekom> ion exchange technique
Ionenaustausch-Wasserreiniger m <Labor> ion exchange water purifier
Ionenbeschleuniger m <Teilphys> ion accelerator
Ionenbeschuss m <Elektronik, Metall, Teilphys> ion bombardment
Ionenbeweglichkeit f <Strahlphys> ion mobility
Ionenbindung f <Strahlphys> ionic bond
Ionenbrennfleck m <Elektronik> ion burning spot
Ionencluster m <Strahlphys> ion cluster
Ionenfalle f <Elektronik, Fernseh> ion trap
Ionenfleck m <Fernseh> ion spot
Ionengetterpumpe f <Phys> ion pump
Ionenimplantation f 1. <Elektronik> ion implantation *(Halbleiter)*; 2. <Teilphys> ion implantation
Ionenlaser m <Elektronik, Maschinen> ion laser
Ionenlautsprecher m <Akustik, Aufnahme> ionic loudspeaker
Ionenleitung f <Strahlphys> ionic conductance
Ionenöffnungswinkel m <Kerntech> angle of acceptance of ions
Ionenpaar n <Strahlphys> ion pair
Ionenpolarisation f <Strahlphys> ionic polarization
Ionenpumpe f <Maschinen> ion pump
Ionenquelle f <Phys, Strahlphys> ion source
Ionenradius m <Strahlphys> ionic radius
Ionenrakete f <Raumfahrt> ion rocket
Ionenröhre f <Elektronik> gas tube
Ionenschalter m <Elektronik, Elektrotech> gas-filled relay
Ionenschubtriebwerk n <Raumfahrt> ion thruster
Ionenspaltung f <Elektriz> ionization
Ionenspektrum n <Strahlphys> ion spectrum
Ionenstärke f <Strahlphys> ionic strength
Ionenstrahl m <Elektronik, Strahlphys> ion beam
Ionenstrahl m **mit hoher Intensität** <Strahlphys> high-intensity ion beam
Ionenstrahlätzen n <Elektronik> ion beam etching
Ionenstrahlbelichtung f <Elektronik> ion exposure
Ionenstrahlbeschichtung f <Elektronik> ion beam coating, ion beam epitaxy
Ionenstrahlbildgenerator m <Elektronik> ion beam pattern generator
Ionenstrahllithographie f <Elektronik> ion beam lithography
Ionenstrahlnanolithographie f <Elektronik> ion beam nano-lithography
Ionenstrahlschattenabbildung f <Elektronik> ion beam shadow printing
Ionentriebwerk n <Raumfahrt> ion engine *(Raumschiff)*
Ionenverdampferpumpe f <Chemtech> evaporating-ion pump, evaporation-ion pump
Ionenwanderung f <Elektronik> ion migration
Ionhaushalt m <Umweltschmutz> ion budget
Ionisation f <Elektrotech, Strahlphys, Teilphys> ionization
Ionisationsargonlaser m <Elektronik> ionized argon laser
Ionisationsenergie f <Teilphys> ionization energy
Ionisationskammer f <Strahlphys, Teilphys> ionization chamber
Ionisationsmanometer n <Gerät> *(AE)* ion gage, *(BE)* ion gauge
Ionisationsmessgerät n **nach Bayard-Alpert** <Kerntech> *(AE)* Bayard-Alpert ionization gage, *(BE)* Bayard-Alpert ionization gauge
Ionisationspotenzial n <Strahlphys, Teilphys> ionization potential

Ionisationsrate f <Strahlphys, Teilphys> ionization rate
Ionisationsstrom m <Elektronik> ionization current
Ionisationsstufe f <Funktech> ledge *(Ionosphäre)*
Ionisationsvakuummeter n 1. <Gerät> *(AE)* ion gage, *(BE)* ion gauge, *(AE)* ionization vacuum gage, *(BE)* ionization vacuum gauge, *(BE)* thermionic vacuum gauge, *(AE)* thermionic vacuum gage; 2. <Heiz & Kälte> *(AE)* ionization vacuum gage, *(BE)* ionization vacuum gauge; 3. <Strahlphys, Teilphys> *(AE)* ionization gage, *(BE)* ionization gauge
Ionisationsverlust m <Strahlphys, Teilphys> ionization loss
ionischer Antrieb m <Raumfahrt> ionic propulsion *(Raumschiff)*
ionisierende Strahlung f <Elektriz, Strahlphys, Teilphys, Umweltschmutz> ionizing radiation
ionisierendes Teilchen n <Strahlphys, Teilphys> ionizing particle
ionisierter Zustand m <Strahlphys, Teilphys> ionized state
Ionisierung f <Elektriz, Elektrotech, Funktech> ionization
Ionisierungsdetektor m <Strahlphys, Teilphys> ionization detector
Ionisierungsenergie f <Strahlphys> ionization energy
Ionisierungskammer f **im Reaktorkern** <Kerntech> in-core ionization chamber
Ionisierungspotenzial n <Strahlphys, Teilphys> ionization potential
Ionisierungsschicht f <Elektriz> ionizing layer
Ionisierungsstrom m <Elektronik> ionization current
Ionisierungsverlust m <Strahlphys, Teilphys> ionization loss
Ionogramm n <Funktech> ionogram
Ionon n <Chemie> ionone
Ionophorese f <Elektriz> electrophoresis
Ionosphäre f 1. <Funktech> ionosphere; 2. <Strahlphys> ionic atmosphere, ionosphere; 3. <Teilphys> ionosphere
Ionosphärenlotung f <Funktech> sounding of the ionosphere
Ionosphärenschichten fpl <Strahlphys> ionosphere, ionospheric layers
Ionosphärensonde f <Funktech> ionosonde
ionosphärische Absorption f <Strahlphys> ionospheric absorption
ionosphärische Ausbreitung f <Funktech> ionospheric propagation
ionosphärische Streuung f <Funktech> ionospheric scatter
Ionotropie f <Chemie> ionotropy
IOS 1. <Elektronik, Optik> *(integrierter optischer Schaltkreis)* IOC *(integrated optical circuit)*; 2. <Telekom> *(integrierte optische Schaltung)* IOC *(integrated optical circuit)*
IP *(Eingabe)* <Comp & DV> IP *(input)*
Ipecin n <Chemie> emetin, emetine
Ipekakuanha f <Lebensmittel> ipecac, ipecacuanha *(Pharmakologie)*
IPL *(Initialprogrammlader)* <Comp & DV> IPL *(initial program loader)*
IR 1. <Comp & DV> *(Wiederauffinden von Informationen)* IR *(information retrieval)*; 2. <Kunststoff, Optik, Phys, Strahlphys> *(Infrarot)* IR *(infrared)*
Iraser m <Elektronik> infrared laser, iraser *(Laser)*
IRB-Ruß m *(internationaler Standardruß)* <Kunststoff> IRB *(industry reference black)*
I-Regler m *(Integralregler)* <Regelung> integral action controller
Iridium... <Chemie> iridic
Irisblende f <Phys> iris

Irisches Moos n <Lebensmittel> carrageen
Irisieren n <Meerschmutz, Optik> iridescence
Irisieren n **von Farben** <Optik> irisplay of colours
irisierendes Glas n <Ker & Glas> iridescent glass
Iron n <Chemie> irone
IRPTC *(Internationales Verzeichnis für potenziell toxische Chemikalien)* <Umweltschmutz> IRPTC *(International Register of Potentially Toxic Chemicals)*
irrational adj <Math> irrational
irrationale Zahl f 1. <Comp & DV> irrational number; 2. <Math> irrational number, surd
irreduzibles Polynom n <Comp & DV> irreducible polynomial *(nicht reduzibles Polynom)*
irreführend adj <Patent> misleading
irrelevant adj <Patent, Qual> irrelevant
irreversibel adj <Phys, Thermod> irreversible
irreversible Abschaltung f <Kerntech> irreversible shutdown
irreversibles Kolloid n <Chemie> irreversible colloid
Irrfahrt f <Comp & DV> random walk *(Statistik)*
Irrungszeichen n <Telekom> error signal
IRS *(Internationales Referenzsystem)* <Umweltschmutz> IRS *(International Referral System)*
IR-Spannungsabfall m <Elektriz> IR-drop
IS 1. <Comp & DV> *(Informationssysteme)* IS *(Information Systems)*; 2. <Comp & DV, Elektriz, Elektronik, Funktech, Kontroll, Phys, Telekom> *(Integrierschaltung, integrierte Schaltung, integrierter Schaltkreis)* IC *(integrated circuit)*; 3. <Elektronik, Funktech, Phys> *(Sättigungsstrom)* IS *(saturation current)*
ISA-Bus m <Comp & DV> ISA bus
ISA-Passungen fpl <Maschinen> ISA system of fits
Isatin n <Chemie> isatin
Isatin... <Chemie> isatic
Isatogen... <Chemie> isatogenic
Isatropic... <Chemie> isatropic
ISB *(unabhängiges Seitenband)* <Funktech> ISB *(independent sideband)*
ISB-Modulation f <Elektronik, Funktech> ISB modulation
ISDN *(diensteintegriertes digitales Netz)* <Telekom> ISDN *(integrated services digital network)*
ISDN-Adresse f <Telekom> ISDN address ISDN call number, ISDN directory number, ISDN subscriber number
ISDN-Anschluss m <Telekom> ISDN access
ISDN-Anschlusseinheit f <Telekom> ISDN line outlet, ISDN telecommunication socket *(IAE)*
ISDN-Anwenderteil n <Telekom> integrated services digital network user part, ISDN UP, ISDN user part *(ISUP)*
ISDN-Anwendungsteil n <Telekom> ISDN user part
ISDN-Basisanschluss m <Telekom> ISDN basic access, ISDN BA
ISDN-Bildfernsprecher m <Telekom> video conference terminal
ISDN-Breitbanddienst m <Telekom> B-ISDN service, Broadband Integrated Services Digital Network
ISDN-Endgerät n <Telekom> ISDN terminal
ISDN-Endgeräteanpassung f <Telekom> ISDN terminal adapter, ISDN TA
ISDN-Fernkopierer m <Telekom> ISDN-fax equipment, fax group 4 equipment
ISDN-Karte f <Telekom> ISDN controller *(PC-Karte für den ISDN-Basisanschluss)*
ISDN-Nebenstellenanlage f <Telekom> ISDN PABX, ISDN private branch exhange
ISDN-Netzabschluss m <Telekom> ISDN network termination, ISDN NT
ISDN-Netzteilnehmer-Basisanschluss m <Telekom> ISDN subscriber basic access facility *(NTBA)*
ISDN-Paketsteuerung f <Telekom> ISDN packet handler

ISDN-PC-Karte f <Telekom> ISDN controller, ISDN PC card
ISDN-Primärmultiplex-Anschluss m <Telekom> ISDN primary-rate multiplex access *(PA)*
ISDN-Primärratenanschluss m <Telekom> ISDN primary rate access
ISDN-Rufnummer f <Telekom> ISDN address, ISDN call number, ISDN directory number, ISDN subscriber number
ISDN-Stecker m <Telekom> RJ-45 plug, Western Bell plug
ISDN-Teilnehmeranschlussleitung f <Telekom> ISDN subscriber line
ISDN-Teilnehmer-Vermittlungsanlage f <Telekom> ISDN private branch exchange
ISDN-Telefon n <Telekom> ISDN telephone
ISDN-Telefon-Anwenderteil n <Telekom> telephone user part *(TUP)*
ISDN-Telefonapparat m <Telekom> digital telephone, ISDN telephone set
ISDN-Teletex-Terminal-Adapter m <Telekom> ISDN teletype terminal adapter, ISDN TTY TA
ISDN-Verbindung f <Telekom> ISDN connection
ISDN-Vermittlung f <Telekom> ISDN exchange
ISDN-Vermittlungssystem n <Telekom> ISDN switching system
ISDN-Videotelefon n <Telekom> ISDN video telephone
ISDN-Zeichengabe f <Telekom> *(AE)* ISDN signaling, *(BE)* ISDN signalling
ISDN-Zugang m <Telekom> ISDN access
isentropisch adj <Phys, Thermod> isentropic
Isethion... <Chemie> isethionic
Isethionat n <Chemie> isethionate
IS-Gehäuse n <Elektronik> integrated-circuit package
ISM *(Betriebsüberwachung)* <Telekom> ISM *(in-service monitoring)*
IS-Maschine f *(Individual-Section-Flaschenblasmaschine)* <Ker & Glas> IS machine *(individual section machine)*
ISM-Code m <Wassertrans> international safety management code, ISM Code
ISO *(Internationale Organisation für Normung)* <Elektriz, Maschinen> ISO *(International Standardisation Organisation)*
Isoalloxazin n <Chemie> isoalloxazine
Isoamyl... <Chemie> isoamyl, isopentyl
isobar adj <Thermod> isobaric
Isobare f <Kerntech, Phys, Thermod, Wassertrans> isobar
Isobarenspin m <Phys> isobaric spin
Isobathe f <Wassertrans> isobath
Isobelastung f <Lufttrans> isostress
Isoborneol n <Chemie> isoborneol
Isobutan n 1. <Chemie> isobutane, methylpropane; 2. <Erdöl> isobutane *(Kohlenwasserstoff)*
Isobuten n <Chemie> isobutene, isobutylene, methylpropene
Isobutter... <Chemie> isobutyric
Isobutyl... <Chemie> isobutyl
Isobutylalkohol m <Chemie> isopropanol, isopropylcarbinol, methylpropanol
Isochinolin n <Chemie> isoquinoline, leucoline
isochor adj <Thermod> isochore
Isochore f <Phys> isochor
Isochromatenbild n <Fertig> stress pattern
isochron adj <Comp & DV, Telekom> isochronous
isochrone Übertragung f <Comp & DV, Telekom> isochronous transmission
isochrones Glühen n <Metall> isochronal annealing
Isocinchomeron... <Chemie> isocinchomeronic
Isocroton... <Chemie> isocrotonic
Isocyan... <Chemie> isocyanic
Isocyanat n <Chemie, Kunststoff> isocyanate

Isocyanid n <Chemie> carbylamine, isocyanide
isocyclisch adj <Chemie> homocyclic, isocyclic
Isodipren n <Chemie> carene
Isodulcit n <Chemie> isodulcite
isoelektrisch adj <Elektriz> equipotential
isoelektrisch m <Chemie> isoelectric
isoelektrisches Fahrzeug n <Kfztech> isoelectric vehicle
isoelektronisch adj <Chemie> isosteric
Isofenchol n <Chemie> isofenchol
Isoflavon n <Chemie> isoflavone
Isogewichtskurve f <Lufttrans> iso-weight curve
isogonal adj <Geom> isogonal
Isokline f <Phys> isocline
Isolation f 1. <Comp & DV, Elektrotech> insulation, isolation; 2. <Heiz & Kälte> insulation, isolation *(Eigenschaft, Zustand)*; 3. <Phys, Telekom> insulation, isolation
Isolationsdurchschlag m <Elektriz> insulation breakdown
Isolationsfehler m <Elektrotech> insulation defect
Isolationsklasse f <Elektriz, Heiz & Kälte> insulation class
Isolationsmaterial n <Chemie> insulator
Isolationsmessgerät n <Elektrotech> insolation tester, megohmmeter
Isolationspapier n <Elektrotech> fish paper
Isolationsprüfer m <Elektrotech> *(BE)* earth leakage indicator, *(AE)* ground leakage indicator
Isolationsprüfgerät n <Elektriz, Gerät> insulation tester
Isolationsstrecke f <Elektrotech> insulation distance
Isolationswiderstand m <Elektrotech, Phys> insulation resistance
Isolator m 1. <Elektriz, Elektrotech, Funktech> insulator *(elektrisch)*; 2. <Fertig, Heiz & Kälte, Phys, Telekom> isolator
Isolator m mit Vergusskammer <Elektriz> pot insulator
Isolatorklemme f <Elektrotech> insulator clamp
Isoleucin n <Chemie> aminomethylpentanoic acid, isoleucine
Isolier... <Chemie, Elektrotech, Fertig, Ker & Glas, Phys, Sicherheit, Telekom> insulating
Isolierband n <Elektriz, Elektrotech> insulating tape
isolierbar adj <Chemie> isolable
Isolierbehälter m <Thermod> heat-insulated container, insulated container
Isolierblech n <Hydraul> clothing plate
Isolierdecke f <Verpack> insulating sheet
Isolierei n <Ker & Glas> porcelain insulator
isolieren v 1. <Bau, Chemie> isolate; 2. <Elektriz, Sicherheit> insulate; 3. <Fertig, Heiz & Kälte, Phys> isolate
Isolieren n <Fertig, Telekom> insulation
isolierend adj <Elektrotech, Fertig> insulating
isolierende Umhüllung f <Elektrotech> insulating covering
Isolierfähigkeit f <Verpack> insulating property
Isolierhandschuhe mpl <Sicherheit> insulating gloves
Isolierhülle f <Elektrotech> insulating covering
Isolierhülse f 1. <Elektrotech> insulating sheath; 2. <Sicherheit> protective sleeve
Isolierkörper m <Elektrotech, Telekom> insulator
Isolierlack m <Elektriz> insulating varnish
Isoliermantel m <Elektrotech> insulating sheath
Isoliermasse f <Elektrotech, Verpack> insulating compound
Isoliermaterial n 1. <Elektrotech> insulating material; 2. <Heiz & Kälte> insulant; 3. <Mechan> insulating material; 4. <Thermod> lagging
Isoliermatte f <Elektriz> insulating mat
Isoliermittel n <Bau> insulator
Isoliermuffe f <Elektrotech> insulating joint
Isolieröl n <Elektriz, Elektrotech, Erdöl> insulating oil *(Hochspannungstransformatoren)*

Isolierpapier n <Elektriz, Papier> insulating paper
Isolierplatte f 1. <Elektrotech> insulating plate; 2. <Heiz & Kälte> insulating board
Isolierporzellan n <Elektrotech> electrotechnical porcelain
Isolierrohr n 1. <Elektrotech> conduit, insulating sleeve; 2. <Maschinen> insulation pipe; 3. <Nichtfoss Energ> conduit *(elektrisch)*
Isolierscheibe f <Elektriz, Elektrotech> insulating washer
Isolierschicht f <Elektrotech, Verpack> insulating layer
Isolierschicht-Feldeffekttransistor m *(IGFET)* <Elektronik> insulated gate field-effect transistor *(IGFET)*
Isolierschienenlasche f <Eisenbahn> insulating fishplate
Isolierschlauch m 1. <Elektrotech> insulating sleeve; 2. <Kunststoff> electrical sleeving
Isolierstoff m 1. <Bau> insulating material; 2. <Fertig> insulant; 3. <Heiz & Kälte> insulant, insulator; 4. <Telekom> insulation; 5. <Verpack> insulating material
Isolierstoffklasse f <Heiz & Kälte> insulation class
Isolierstoß m <Eisenbahn> insulated rail joint
Isoliersubstrat n <Elektrotech> insulating substrate
isoliert adj <Bau, Elektrotech, Heiz & Kälte> insulated
isolierte Abschirmung f <Elektriz> insulation screen
isolierte Antennenleitung f <Elektriz, Funktech> insulated antenna cable
isolierte Durchführung f <Elektrotech> grommet
isolierte Kabelseele f <Elektriz> insulated core
isolierte Werkzeuge npl <Sicherheit> insulated tools
isolierter Draht m 1. <Elektriz> insulated conductor; 2. <Elektrotech> insulated wire
isolierter Leiter m <Elektrotech, Telekom> insulated conductor
isolierter Stromleiter m <Elektriz> insulated conductor
isoliertes Kabel n <Elektriz, Elektrotech> insulated cable
isoliertes Kabelrohr n <Elektriz> insulated conduit
isoliertes System n <Phys> isolated system
Isolierung f 1. <Bau, Elektriz> insulation; 2. <Elektrotech> insulation *(eines Leiters)*; 3. <Fertig> lagging *(Rohre)*; 4. <Heiz & Kälte> insulation *(Werkstoff)*; 5. <Hydraul> cladding, clothing; lagging *(Dampfmaschine)*; 6. <Ker & Glas, Kunststoff, Maschinen, Mechan, Nichtfoss Energ, Sicherheit, Telekom> insulation; 7. <Wassertrans> insulation *(Elektrik)*
Isolierung f zwischen Wicklungen <Elektriz> coil-to-coil insulation, interturn insulation
Isolierungsträger m <Raumfahrt> insulating substrate *(Raumschiff)*
Isolierventil n <Labor> isolating valve
Isolierverkleidung f des Nutzlastraums <Raumfahrt> payload bay insulation
Isoliervermögen n <Verpack> insulating property
Isolierwachs n <Elektrotech> insulating wax
Isolog n <Chemie> *(AE)* isolog, *(BE)* isologue
isolog adj <Chemie> isologous
Isomaltose f <Chemie> isomaltose
Isomer n <Chemie, Erdöl, Kerntech, Strahlphys> isomer
isomere Umwandlung f <Chemie> isomerization
isomerer Übergang m <Strahlphys> isomeric transition
Isomerie f <Chemie, Phys> isomerism
Isomerisation f <Chemie, Erdöl> isomerization
Isomerisierung f 1. <Chemie> isomerization; 2. <Erdöl> isomerization *(Petrochemie)*; isomerization *(Umlagerung chemischer Verbindungen in Isomere)*
Isomerismus f <Fertig> allotropy
Isometrie f <Geom> isometry
isometrisch adj <Geom> isometric
isometrische Darstellung f <Konstzeich> isometric projection
isometrische Ebene f <Geom> isometric plane
isometrische Kraft f <Ergon> isometric force

isometrische

isometrische Muskelkontraktion f <Ergon> isometric contraction
isometrische Projektion f <Geom, Konstzeich> isometric projection
Isomorphie f <Chemie> isomorphism
Isonicotin... <Chemie> isonicotinic
Isonitril n <Chemie> isocyanide, isonitrile
Isooctan n <Chemie> isooctane
Isopachenkarte f <Erdöl> isopach map (Geologie)
Isoparaffin n <Chemie> isoparaffin
Isopelletierin n <Chemie> isopelletierin
Isopentan n <Chemie> isopentane, methylbutane
Isopentyl... <Chemie> isoamyl, isopentyl
Isophorondiamin n <Kunststoff> isophorone diamine
Isoplethe f <Chemie> isopleth
Isopolysäure f <Chemie> isopoly acid
Isopren n <Chemie> isoprene, methylbutadiene
Isoprenoid n <Chemie> isoprenoid
Isopropanol n 1. <Chemie> propanol; 2. <Lebensmittel> isopropyl alcohol
Isopropenyl... <Chemie> isoallyl, isopropenyl
Isopropyl... <Chemie> isopropyl
Isopropylalkohol m <Chemie> isopropanol, propanol
Isopropylbenzol n <Chemie> cumene, isopropylbenzene
Isospin m <Strahlphys, Teilphys> isospin
Isospin-Multiplett n <Phys, Strahlphys, Teilphys> charge multiplet
Isostasie f <Nichtfoss Energ> isostasy
isoster adj <Chemie> isosteric
Isosterie f <Chemie> isosterism
isosterisch adj <Chemie> isosteric (Enzym)
isotherm adj <Mechan, Phys, Thermod> isothermal
Isothermbehälter m <Kfztech> insulated container
Isotherme f 1. <Lufttrans> isotherm; 2. <Phys> isotherm, isothermal curve, isothermal line; 3. <Thermod> isothermal curve, isothermal line; 4. <Wassertrans> isotherm
isotherme Ausdehnung f <Phys> isothermal expansion
isotherme Kompressibilität f <Phys> isothermal compressibility
isothermische Prüfung f <Metall, Qual> isothermal test
isothermische Reaktion f <Metall> isothermal reaction
isothermisches Abschrecken n <Metall> isothermal quenching
Isotone f <Phys> isotone
isotonische Muskelkontraktion f <Ergon> isotonic contraction
Isotop n <Kerntech, Phys, Strahlphys, Teilphys> isotope
Isotop n **zur industriellen Bestrahlung** <Kerntech> industrial isotope
Isotopenanalyse f <Phys> isotopic analysis
Isotopengenerator m <Raumfahrt> isotopic generator (Raumschiff)
Isotopenhäufigkeit f <Phys> isotopic abundance
Isotopenhäufigkeitsverhältnis n <Kerntech> abundance ratio
isotopenmarkiertes Material n <Kerntech> isotopically-tagged compound
Isotopenmasse f 1. <Phys> nucleon number; 2. <Phys> mass number
Isotopenmessung f <Phys> isotope measurement
Isotopenspin m <Phys> isotopic spin
Isotopentrennung f <Phys, Strahlphys, Teilphys> isotope separation
Isotopie f <Chemie> isotopy
isotrop adj 1. <Funktech> isotropic; 2. <Optik> isotropic (elektromagnetische Wellen)
Isotropantenne f <Raumfahrt> isotropic antenna (Kugelstrahler)
isotrope Abbildung f <Geom> isotropic mapping
isotrope Turbulenz f <Strömphys> isotropic turbulence
isotroper Strahler m <Funktech> isotropic radiator
isotropisch adj 1. <Raumfahrt> isotropic (Weltraumfunk); 2. <Telekom> isotropic
isotropische Verstärkung f <Raumfahrt> isotropic gain
isotropischer Gewinn m 1. <Funktech> isotropic gain (Antenne); 2. <Raumfahrt> isotropic gain (Weltraumfunk)
Isovanillin n <Chemie> isovanilline
Isoxazol n <Chemie> isoxazole
IS-Scheibe f <Elektronik> integrated-circuit wafer
Ist... <Textil> actual
Istabmaß n <Fertig> actual deviation
Istabweichung f <Gerät> actual deviation
Istdurchmesser m <Fertig> actual size (Bohrung)
Isthmus m <Wassertrans> isthmus
Istmaß n <Maschinen> actual size
IS-Trägermaterial n <Elektronik> integrated-circuit substrate
Iststrom m <Gerät> actual current
Istvorschub m <Fertig> actual feed rate
Istwert m 1. <Gerät> instantaneous value; 2. <Maschinen> actual value
Istwert m **der Teilkreisteilung** <Fertig> actual tooth spacing on pitch circle
Istwertanzeige f <Gerät> actual indication
Istwertferntasten n <Comp & DV> remote sensing
Istzustand m <Qual> actual state • **im Ist-Zustand benutzen** <Qual> use as is
ISW (internationale Selbstwahl) <Telekom> IDD (international direct dialling); IDDD (international direct distance dialling)
IT (Informationstechnologie) <Comp & DV> IT (information technology)
ITA (internationales Telegrafenalphabet) <Lufttrans, Wassertrans> ITA (international telegraph alphabet)
Itacon... <Chemie> itaconic
IT-Basis f <Comp & DV> computing foundation
IT-Berater m <Comp & DV> senior technical consultant
Iteration f 1. <Comp & DV, Umweltschmutz> iteration; 2. <Math> iteration (Wiederholung)
Iterationsmethode f <Comp & DV> iterative method
Iterationsschleife f <Comp & DV> iterative routine
iterativ adj <Comp & DV, Math> iterative (wiederholend)
iterative Operation f <Comp & DV> iterative process
iterative Suche f <Künstl Int> iterative search
iterativer Prozess m <Comp & DV> iterative process
iteratives Rechnen n <Comp & DV> iterative operation
iterieren v <Math> iterate (wiederholen)
IT-Modell n <Comp & DV> computing model
I-Träger m 1. <Bau> I-beam (gewalzt); 2. <Metall> I beam
I-Trägeraufhängung f <Kfztech> I-beam suspension
IUC (Mess- und Regeltechnik) <Elektronik> IUC (instrumentation and control)
I-Wagen m <Eisenbahn> refrigerated car, refrigerated wagon

J

j (Sprunghöhe) <Hydraul> j (height of hydraulic jump)
J 1. <Akustik, Phys> (Schallenergiefluss) J (sound-energy flux); 2. <Elektriz, Lebensmittel, Mechan, Metrol, Phys, Thermod> (Joule) J (joule); 3. <Kerntech> (Winkelmo-

mentquantenzahl) J (total angular momentum quantum number); 4. <Mechan, Thermod> *(mechanisches Wärmeäquivalent)* J *(mechanical equivalent of heat)*
Jacht f <Wassertrans> yacht
Jachthafen m <Wassertrans> marina
Jachtheck n <Wassertrans> counter stern *(Schiffbau)*
Jackson'sches Modell n <Kerntech> Jackson model
Jacquard... <Textil> jacquard
Jacquardgewebe n <Textil> jacquard fabric
Jacquardpapier n <Verpack> jacquard paper
Jacquardpappe f <Verpack> jacquard board
Jacquardwebstuhl m <Textil> jacquard loom
Jahres... <Wasserversorg> annual, yearly
Jahresabfluss m <Wasserversorg> annual runoff
Jahresdurchfluss m <Wasserversorg> annual flow
Jahresgebühr f <Patent> renewal fee
Jahresmittelwert m <Telekom> mean annual value *(Verfügbarkeit)*
jahreszeitliche Änderung f <Funktech> seasonal variation *(Wellenausbreitung)*
jährliche Last f <Wasserversorg> annual load
jährliche natürliche Hintergrundstrahlung f <Strahlphys> natural annual background radiation
jährlicher Kapazitätsfaktor m <Kohlen, Nichtfoss Energ> annual capacity factor
jährlicher Mittelwasserstand m <Wasserversorg> annual mean water level
Jalapin n <Chemie> jalapin, orizabin
Jalousie f <Bau, Ker & Glas, Kfztech> *(AE)* louver, *(BE)* louvre *(Verglasung)*
Jalousieeffekt m <Fernseh> venetian-blind effect
Jalousieöffnung f <Bau> *(AE)* louver, *(BE)* louvre
Jalousieplattenkassette f <Foto> roller-blind dark slide
Jaspégarn n <Textil> jaspe yarn
Jauche f <Abfall> slurry
Jaulen n 1. <Akustik> wow; 2. <Funktech> yoop
Jaulen n **des Plattentellers** <Aufnahme> turntable wow
Javellauge f <Fertig> javel water *(Kunststoffinstallationen)*
Jedermann-Funk m <Telekom> citizens' band radio
Jenaer Glas® n <Thermod> oven proof glass
Jerobeam f <Ker & Glas> jerobeam
Jersey m <Textil> jersey
Jervin n <Chemie> jervine
JET <Strahlphys> JET, Joint European Torus
Jetstream m <Lufttrans> jet stream
JET-Tokamak m <Kerntech, Strahlphys> JET Tokamak
Jigger m <Abfall, Textil> jigger
J-Integral-Methode f <Kerntech> J-integral method
Jitter m <Telekom> jitter, time jitter
jitterfreies Signal n <Telekom> jitter-free signal
Jitterreduktion f <Telekom> jitter reduction
j-j-Kopplung f <Kerntech, Phys> j-j coupling
Job m <Comp & DV> job
Jobabrechnung f <Comp & DV> job accounting
Jobabschnitt m <Comp & DV> job step
Jobanfang m <Comp & DV> job begin
Jobanforderung f <Comp & DV> job request
Jobanfrage f <Comp & DV> job request
Jobbefehl m <Comp & DV> job command
Jobbeschreibung f <Comp & DV> job description
Jobdatum n <Comp & DV> job date
Jobdefinition f <Comp & DV> job definition
Jobende n <Comp & DV> EOJ, end of job
Jobferneingabe f <Comp & DV> RJE, remote job entry
Jobfernverarbeitung f <Comp & DV> RJE, remote job entry
Jobfolge f <Comp & DV> job step
Jobkatalog m <Comp & DV> *(AE)* job catalog, *(BE)* job catalogue

Jobklasse f <Comp & DV> job class
joborientierte Datenstation f <Comp & DV> job-oriented terminal
joborientierte Programmiersprache f <Comp & DV> job-oriented language
Jobplanung f <Comp & DV> job scheduling
Jobschritt m <Comp & DV> job step
Jobstapel m <Comp & DV> job batch, job stack
Jobsteuerdatei f <Comp & DV> job control file
Jobsteuerprogramm n <Comp & DV> job control program, job scheduler
Jobsteuersprache f <Comp & DV> JCL, job control language
Jobsteuerung f <Comp & DV> job control
Jobstrom m <Comp & DV> job stream
Jobübergabesystem n <Comp & DV> job submission system
Jobverarbeitungssystem n <Comp & DV> job-processing system
Jobwarteschlange f <Comp & DV> job queue, job stream
Joch n 1. <Bau> frame, yoke *(Brücke)*; 2. <Elektriz> yoke; 3. <Kerntech> yoke *(Magnet)*; 4. <Kfztech> yoke *(Universalgelenk)*; 5. <Maschinen, Mechan> yoke
Jochbalken m <Bau> straining piece
Jochspule f <Elektrotech> yoke coil
Jod n *(I)* <Chemie> iodine *(I)*
Jodeldetektor m <Kerntech> Jodel detector
Jodkolben m <Labor> iodine flask
Jodsilber n <Foto> silver iodide
Jodzahl f <Kunststoff> iodine value
Johannit m <Kerntech> johannite
Johnson'sches Rauschen n <Phys> Johnson noise
Jojo-Entspinnen n <Raumfahrt> yoyo despin *(Raumschiff)*
Jokerzeichen n <Comp & DV> wildcard character
Jolle f <Wassertrans> dinghy
Josephson'sche Verbindung f <Phys> Josephson junction
Josephson'scher Effekt m <Kerntech, Phys> Josephson effect
Josephson'scher Übergang m <Elektronik, Kerntech> Josephson junction
Jost-Funktion f <Kerntech> Jost function
Joule n *(J)* <Elektriz, Lebensmittel, Mechan, Metrol, Phys, Thermod> joule *(J)*
Joule'scher Effekt m 1. <Elektriz, Elektrotech> Joule effect; 2. <Phys> Joule effect *(Wärmelehre, Magnetismus)*; 3. <Thermod> Joule effect
Joule'scher Wärmeverlust m <Elektriz> Joule's heat loss
Joule'sches Gesetz n 1. <Phys> Joule's law *(Elektrizitätslehre)*; 2. <Thermod> Joule's law
Joule-Thomson'scher Effekt m <Phys, Thermod> Joule-Kelvin expansion, Joule-Thomson effect, Joule-Thomson expansion, joule expansion
Joule-Thomson'scher Koeffizient m <Phys, Thermod> Joule-Thomson coefficient
Joule-Verlust m <Elektriz> copper loss
Joy'sche Ventilsteuerung f <Hydraul> Joy's valve-gear
Joystick m <Comp & DV, Fernseh> joystick
Joystick-Schalter m <Elektriz> joystick selector
J-Teilchen n <Phys> J particle
Juglon n <Chemie> hydroxynaphthoquinone, juglone
Jukebox-Archivierungssystem n <Optik> jukebox filing system
Jumper m <Elektrotech> jumper
Jumpstagspreize f <Wassertrans> jumper strut
Junctiondiode f <Elektriz> junction diode
Jungfernfahrt f <Wassertrans> maiden voyage

Jungfernflug m <Lufttrans> maiden flight
jungfräuliche Faser f <Ker & Glas> (AE) pristine fiber, (BE) pristine fibre
jungfräuliches Glas n <Ker & Glas> pristine glass
jungfräuliches Neutron n <Kerntech> uncollided neutron, virgin neutron
Junkers-Kalorimeter n <Thermod> continuous flow calorimeter
Jura m <Erdöl> Jurassic period (Geologie)
juristische Person f <Patent> legal person
Justier... <Gerät, Maschinen> adjusting
justierbare Blende f <Phys> adjustable aperture
justierbarer Keramikkondensator m <Elektrotech> adjustable ceramic capacitor
justierbarer Kern m <Elektrotech> adjustable core
Justiereinrichtung f <Gerät, Qual> calibration equipment
Justierelement n <Gerät> trimming element
justieren v <Maschinen> set
Justieren n 1. <Chemie> rectification; 2. <Lufttrans> setting (von Instrumenten); 3. <Metrol> adjustment
Justierfehler m <Telekom> alignment fault
Justierglied n <Gerät> adjusting element
Justierknopf m <Gerät, Mechan> adjusting knob
Justiermutter f <Fertig> (AE) leveling nut, (BE) levelling nut
Justierpotentiometer n <Elektrotech> adjustable potentiometer, trimming potentiometer
Justierring m <Gerät> adjustment ring
Justierschraube f 1. <Gerät> adjustable screw, trimming screw; 2. <Maschinen> adjusting screw, set screw; 3. <Mechan> adjusting screw
Justierspannung f <Elektriz> adjusting voltage
Justierung f 1. <Druck> adjustment; 2. <Elektronik> adjustment; calibration (von Feldgeräten); 3. <Funktech, Maschinen> adjustment
Justierwiderstand m <Gerät> trimming resistor
Just-in-Time n <Comp & DV> Just in Time, JIT
Jute f <Textil> jute
Jutedrell m <Verpack> jute sacking
Jutegarn n <Textil> jute yarn
Jutehanf m <Bau> jute fibre
Juteseil n <Maschinen> jute rope
Jutespinnerei f <Textil> jute spinning
Juteumhüllung f <Elektrotech> jute covering
juveniles Grundwasser n <Wasserversorg> juvenile water

K

k 1. <Akustik> (Wellenkonstante) k (wave constant); 2. <Chemie> (Kalium) K (potassium); 3. <Elektriz> (Kopplungskoeffizient, Phys) k (coupling coefficient); 4. <Kerntech> (Multiplikationskonstante für infinite Systeme) k (multiplication constant for an infinite system); 5. <Kerntech> (Neutronenmultiplikationskonstante) k (neutron multiplication constant); 6. <Labor> (Kilo, Kilogramm) k (kilo); 7. <Phys, Thermod> (Boltzmann'sche Konstante, Boltzmann'sche Zahl) k (Boltzmann constant)
K 1. <Akustik> (Magnetostriktionskonstante) K (magnetostriction constant); 2. <Elektriz> (Kelvin) K (kelvin); 3. <Hydraul> (Kompressionsmodul) K (bulk modulus of compression); 4. <Hydraul> (Elastizitätsmodul) K (bulk modulus of elasticity)

Ka-Band n <Funktech> Ka band (Satellite, 33–36 GHz)
Ka-Band-Radar n <Funkort> Ka-band radar, millimeter wave radar
kabbelige See f <Wassertrans> choppy sea (Seezustand)
Kabel n 1. <Aufnahme, Comp & DV, Elektriz, Elektrotech> cable, lead; 2. <Eisenbahn, Fernseh, Fertig, Funktech, Kfztech, Kunststoff, Maschinen, Mechan> cable; 3. <Optik> cable assembly; 4. <Telekom, Trans, Verpack, Wassertrans> cable • **Kabel verlegen** <Aufnahme> lay tracks
Kabel n **mit abgestufter Koppeldämpfung** <Telekom> grading coupling loss cable
Kabel n **mit losem Aufbau** <Optik> loose construction cable
Kabel n **mit metallischen Leitern** <Telekom> metal conductor cable
Kabel n **mit Nutenstruktur** <Telekom> grooved cable
Kabel n **mit separater Bleiumhüllung** <Elektriz> separately lead-sheathed cable
Kabel n **mit separater Bleiummantelung** <Elektriz> separately lead-sheathed cable
Kabel n **mit symmetrischen Adernpaaren** <Telekom> symmetrical pair cable
Kabel n **zwischen Eingabegerät und Computer** <Comp & DV> input lead
Kabelabschluss m 1. <Elektriz, Elektrotech> cable head, cable termination; 2. <Telekom> cable head
Kabelabschnitt m <Elektrotech, Telekom> cable section
Kabelabzweigdose f <Elektriz> cable junction box
Kabelabzweigkasten m <Elektriz> cable junction box
Kabelader f <Elektrotech> cable core
Kabeladernpaar n <Elektrotech, Telekom> cable pair
Kabelanschluss m 1. <Comp & DV, Elektronik, Elektrotech> cable port; 2. <Telekom> cable connection, cable port
Kabelarmatur f <Elektrotech> cable fitting
Kabelaufhängung f <Elektrotech, Kfztech> cable support
Kabelaufhängungsdraht m <Elektrotech> cable suspension wire
Kabelausgang m <Comp & DV, Elektronik, Elektrotech, Telekom> cable outlet
Kabelausschaltstrom m <Elektronik, Elektrotech> cable breaking current
Kabelausschaltvermögen n <Elektronik, Elektrotech> cable breaking capacity, cable off-load breaking capacity
Kabelausziehvorrichtung f <Elektriz> pull box
Kabelbahn f <Trans> (AE) cable railroad, (BE) cable railway
Kabelbaum m 1. <Elektrotech> cable harness; 2. <Raumfahrt> safety harness, wiring harness (Raumschiff)
Kabelbewehrung f <Fertig> (AE) cable armoring, (BE) cable armouring
Kabelbinder m 1. <Fernseh> binder; 2. <Telekom> lace
Kabelbündel n <Elektriz, Elektrotech> (BE) bunched cable, (AE) bundled cable, cable bundle
Kabeldämpfung f <Elektrotech, Telekom> cable loss
Kabeldefekt m <Elektrotech> cable defect
Kabeldose f <Elektriz> cable box
Kabeleinführung f <Heiz & Kälte, Telekom> cable entry
Kabeleinstieg m <Elektriz> cable manhole
Kabeleinziehvorrichtung f <Elektriz> cable pull box
Kabelende n <Elektrotech, Telekom> cable end
Kabelendgestell n <Elektriz, Telekom> cable support rack
Kabelendstück n <Elektriz> cable end piece
Kabelendverschluss m <Elektriz, Telekom> cable box
Kabelfehlernachweisgerät n <Elektriz> cable fault detector
Kabelfernsehanlage f <Fernseh, Telekom> cable television system

Kabelfernsehen n <Fernseh> cable TV, cable television, CATV, wired television
Kabelfernsehnetz n <Fernseh, Telekom> cable television network, multichannel video distribution system
Kabelfernsehstörung f (CATVI) <Fernseh, Funktech> cable television interference (CATVI)
Kabelfett n <Elektrotech> cable grease
Kabelformgarn n <Telekom> lacing cord
Kabelführung f 1. <Elektrotech> cable conduit, cable duct; 2. <Kfztech> cable guide (Elektrik)
Kabelgarnitur f für Deckdurchführung <Wassertrans> through deck cable fitting (Tauwerk)
kabelgefärbt adj <Textil> tow-dyed
Kabelgeschirr n <Mechan> harness
kabelgeschlagen adj <Wassertrans> cable-laid (Tauwerk)
Kabelgestell n <Elektrotech, Telekom> cable rack
Kabelgraben m <Elektriz, Elektrotech> cable trench, cable trough
Kabelhalterung f <Kfztech> cable support
Kabelisolator m <Elektrotech> cable insulator, cable isolator
Kabelisolierer m <Elektrotech> cable insulator
Kabelisolierung f <Elektriz, Elektrotech> cable covering, cable insulation
Kabeljacke f <Elektriz> cable jacket
Kabel-Kammzug-Konverter m <Textil> tow-to-top converter
Kabelkanal m 1. <Elektrotech> cable conduit, cable duct; 2. <Kerntech> raceway
Kabelkanal m des Luftschraubenblattes <Lufttrans> blade duct (Hubschrauber)
Kabelkanalzug m <Elektrotech> cable conduit
Kabelkasten m <Elektrotech, Telekom> cable box
Kabelkeller m <Telekom> underground chamber
Kabelkette f 1. <Maschinen> chain cable; 2. <Wassertrans> cable chain (Festmachen)
Kabelklemme f 1. <Elektriz, Elektrotech> (AE) armor clamp, (BE) armour clamp, cable clip, cable lug, clamp; 2. <Kfztech> cable clip (KFZ-Elektrik); 3. <Mechan> cable clamp
Kabelkompaktierung f <Optik> tight buffering
Kabelkopf m 1. <Elektrotech, Telekom> cable head; 2. <Fertig> cable plug (Kunststoffinstallationen)
Kabelkran m <Trans> cableway
Kabellänge f <Wassertrans> cable, cable length (Marinemaßeinheit)
Kabellauf m <Elektriz, Telekom> cable run
Kabelleger m <Kfztech> cable ship
Kabellegeschiff n <Telekom> cable ship
Kabellegung f <Elektrotech> cabling
Kabellehre f <Elektrotech> (AE) cable gage, (BE) cable gauge
Kabelleitungsdurchführung f <Elektrotech> grommet
Kabellogcomputerprogramm n 1. <Comp & DV, Labor> program for computer processing of wireline logs; 2. <Erdöl> program for computer processing of wireline logs (Messtechnik)
Kabellöter m <Telekom> cable jointer (Kupferkabel)
Kabelmantel m <Elektrotech, Telekom> cable sheath
Kabelmessbrücke f <Gerät> post office bridge box
Kabelmesser n 1. <Bau> hacking knife; 2. <Raumfahrt> cable cutter (Raumschiff)
Kabelmontage f <Telekom> cable assembly
Kabelmuffe f <Bau, Elektriz, Elektrotech> cable fitting, cable joint box
Kabelnetz n <Elektrotech, Fernseh, Telekom> cable network, cabled network
Kabelortungsgerät n <Elektriz> cable detector, cable locator

Kabelöse f <Elektrotech> cable lug
Kabelpaar n <Comp & DV> cable pair
Kabelprüfdraht m <Elektrotech> pilot wire
Kabelrepeater m <Elektrotech, Telekom> cable repeater
Kabelrohr n <Elektrotech> conduit
Kabelrost m <Elektrotech, Telekom> cable rack
Kabelrundfunk m <Fernseh> wired broadcasting
Kabelschacht m 1. <Elektrotech> cable shaft; 2. <Telekom> cable chamber, cable manhole, manhole
Kabelschelle f <Elektriz, Elektrotech> cable clamp, cable clip
Kabelschirm m <Elektriz> cable screen
Kabelschlitzantenne f <Funktech> radiating cable, slotted cable antenna (Funkversorgung in Tunnels)
Kabelschuh m <Elektrotech> cable lug, terminal
Kabelschutz m <Elektriz> (AE) armor, (BE) armour
Kabelschutzrohr n <Elektrotech> cable conduit
Kabelseele f <Elektrotech, Kunststoff> cable core
Kabelsendung f <Fernseh, Telekom> cable transmission
Kabelspanner m <Elektrotech> cable tensioner
Kabelspleißung f <Elektriz, Telekom> cable joint, cable splicing
Kabelstromwandler m <Elektrotech> cable current transformer
Kabelstück n <Elektrotech, Telekom> cable section
Kabelstumpf m <Elektrotech> cable end
Kabelsuchgerät n 1. <Elektriz, Elektrotech> cable detecting device, cable detector, cable locator; 2. <Telekom> cable locator
Kabeltrasse f <Telekom> cable route
Kabeltrenner m <Elektrotech> cable separator
Kabeltrommel f <Bau, Elektriz, Elektrotech, Verpack> cable drum
Kabelübertragung f <Fernseh, Telekom> cable transmission
Kabelummantelung f <Kunststoff> cable covering, cable sheathing
Kabelverbinder m <Elektriz, Elektrotech> cable connector, cable fitting
Kabelverbindung f 1. <Elektrotech> cable coupling; 2. <Fernseh> cable link; 3. <Telekom> cable communication, cable joint
Kabelverlegung f <Elektriz, Elektrotech, Telekom> cable laying
Kabelverschraubung f <Fertig> cable connection (Kunststoffinstallationen)
Kabelverstärker m <Elektrotech, Telekom> cable repeater
Kabelverteiler m <Elektrotech> cable junction
Kabelverteilpunkt m <Elektrotech> cable junction
Kabelverzweiger m 1. <Elektrotech> cable distributor, distribution cabinet, terminal box; 2. <Telekom> cable distributor
Kabelverzweigerkasten m <Telekom> cross-connect cabinet
Kabelverzweigung f <Elektrotech> cable distribution point
Kabelverzweigungspunkt m <Elektrotech> cable distribution point
Kabelweg m <Elektriz, Telekom> cable run
Kabelwinde f <Elektrotech> cable winch
Kabelzuschlag m <Telekom> slack
Kabine f 1. <Bau> cabin; 2. <Trans> car (eines Aufzugs); 3. <Wassertrans> cabin, stateroom
Kabinenbandrollsteig m <Lufttrans> (BE) cabin-type moving pavement, (AE) cabin-type moving sidewalk
Kabinendruck m <Lufttrans, Raumfahrt> cabin pressure (Raumschiff)
Kabinenförderband n <Lufttrans> cabin conveyer

Kabinenhöhe f <Lufttrans> cabin altitude
Kabinenhöhenmesser m <Lufttrans> cabin altimeter
Kabinenkreuzer m <Wassertrans> cabin cruiser *(Schifftyp)*
Kabinenschwebekorb m <Trans> cabin pulley cradle *(Bauwesen)*
Kabinenstehhöhe f <Wassertrans> cabin headroom *(Schiffbau)*
Kabinettprojektion f <Konstzeich> cabinet projection
K-Absorptionskante f <Kerntech> K-absorption edge
Kachel f <Bau> tile
kacheln v <Bau> tile
Kadmieren n <Metall> cadmium plating
Kadmium n *(Cd)* <Chemie> cadmium *(Cd)*
Kadmiumamalgam-Kryptomat n <Kerntech> kryptonate of cadmium amalgam
Kadmiumbatterie f <Elektriz> cadmium cell
Kadmium-beschichtet adj <Mechan> cadmium-plated
Kadmiumblende f <Fertig> greenockite
Kadmiumelement n <Fertig> cadmium cell
Kadmiumsulfidzelle f <Elektriz> *(AE)* cadmium sulfide cell, *(BE)* cadmium sulphide cell
Kadmiumzelle f <Elektriz> cadmium cell
Käfig m 1. <Elektriz, Elektrotech> cage, squirrel cage; 2. <Fertig> retainer *(Lager)*; 3. <Maschinen> cage
Käfiganker m <Elektrotech> cage armature *(Käfigläufer)*; squirrel cage armature
Käfigläufer m <Elektriz, Elektrotech> cage rotor, squirrel cage rotor
Käfigläufermotor m <Elektriz, Elektrotech> squirrel cage motor
Käfigrelais n <Elektrotech> cage relay
Käfigwicklung f <Elektrotech> squirrel cage winding
Kahlerit m <Kerntech> kahlerite
Kahn m <Labor> boat *(Analyse)*
Kai m <Wassertrans> dock, quay, wharf • **frei Kai** <Wassertrans> free on quay *(Seehandel)*
Kaibahn f <Eisenbahn> *(AE)* quayside railroad, *(BE)* quayside railway
Kaibandförderer m <Trans> quayside conveyor
Kainit m <Fertig> kainite
Kaistraße f <Trans> quayside roadway
Kajüte f <Wassertrans> cabin
Kajütenaufbau m <Wassertrans> cabin roof *(Schiffbau)*
Kajütsboden m <Wassertrans> cabin sole *(Schiffbau)*
Kalander m <Kunststoff, Papier, Textil> calender
Kalanderfolie f <Kunststoff> calendered film
Kalanderwalze f <Papier> bowl, calender roll
Kalanderwalzensatz m <Papier> calender stack
Kalanderwasserkasten m <Papier> calender water box
Kalandria f <Kerntech> calandria
Kalandriagefäß n <Kerntech> calandria
kalandrieren v 1. <Kunststoff> calender; 2. <Verpack> glaze *(Papier)*
Kalandrieren n <Kunststoff, Papier> calendering
Kalfaterer m <Wassertrans> caulker
kalfatern v <Wassertrans> caulk *(Schiffbau)*
Kalfaterung f <Wassertrans> caulked joint *(Schiffbau)*
Kalialaun m <Chemie> potash alum
Kaliapparat m <Labor> potash bulb
Kaliber n 1. <Erdöl> *(AE)* caliber, *(BE)* calibre; 2. <Fertig> groove; pass *(Walzen)*; 3. <Maschinen> bore, *(AE)* caliber, *(BE)* calibre; 4. <Mechan> *(AE)* gage, *(BE)* gauge; 5. <Papier> groove
Kaliberfolge f <Fertig> passes and reductions *(Walzen)*
Kaliberlehre f <Metrol> *(AE)* ring gage, *(BE)* ring gauge
Kaliberlog n <Erdöl> *(AE)* caliber log, *(BE)* calibre log *(Bohrlochmesstechnik)*
Kaliberwalze f <Fertig, Papier> grooved roll

Kalibrator m <Heiz & Kälte> calibrator
Kalibrierdruck m <Lufttrans> calibration pressure
Kalibriereinrichtung f 1. <Elektronik, Gerät> calibration equipment, calibrator; 2. <Qual> calibration equipment
kalibrieren v 1. <Bau> size; 2. <Elektriz> calibrate; 3. <Fertig> size; 4. <Labor, Maschinen, Papier, Qual> calibrate
Kalibrieren n 1. <Erdöl> calibration *(Messtechnik)*; 2. <Fertig> end sizing; 3. <Labor> calibration; 4. <Metrol> calibration *(bei Glasmessgeräten)*; 5. <Papier> calibration
Kalibriermesseinheit f <Lufttrans> calibration module
Kalibriermodul n <Lufttrans> calibration module
Kalibriernachweis m <Qual> documented verification of calibration
Kalibriernormal n <Qual> calibration standard
Kalibrierplakette f <Qual> calibration tag
kalibriert adj 1. <Erdöl> *(AE)* gaged, *(BE)* gauged; 2. <Fertig> scaled; 3. <Maschinen> calibrated; 4. <Papier> grooved
kalibrierte Bohrung f <Erdöl> *(AE)* gaged orifice, *(BE)* gauged orifice *(Messtechnik)*
kalibrierte Drossel f <Erdöl> *(AE)* gaged restriction, *(BE)* gauged restriction *(Messtechnik)*
Kalibrierung f 1. <Bau, Comp & DV, Elektriz> calibration; 2. <Elektronik> calibration *(von Messgeräten)*; 3. <Fertig> grooving, matrix *(Walzen)*; 4. <Kerntech, Lufttrans> calibration; 5. <Maschinen> calibration *(einer Feder)*; 6. <Phys, Strahlphys, Wellphys> calibration
Kalibrierung f des Signalgebers <Elektronik> signal generator calibration
Kalibrierung f eines Radars <Wellphys> radar calibration
Kaliglas n <Ker & Glas> potash glass
Kalisalpeter m <Chemie> *(AE)* potash niter, *(BE)* potash nitre, *(AE)* saltpeter, *(BE)* saltpetre *(Kaliumnitrat)*
Kalisalpeterverfahren n <Kerntech> *(AE)* saltpeter process, *(BE)* saltpetre process
Kalisalz n <Chemie> potash
Kalium n *(K)* <Chemie> potassium *(K)*
Kaliumalaun m <Chemie> potash alum
Kaliumaluminiumsulfat n <Chemie> potash alum, potassium aluminium
Kaliumcarbonat n <Chemie> potash
Kaliumchlorat n <Chemie> potassium chlorate
Kaliumchlorid n <Chemie> potassium chloride
Kaliumchromalaun m <Chemie> chrome alum
Kaliumchromsulfat n <Chemie> chrome alum
Kaliumcyanid n <Chemie> potassium cyanide
Kaliumhydrogentartrat n <Chemie> tartar
Kaliumhydroxid n <Chemie> potash, potassium hydroxide
Kaliummanganat n <Chemie> potassium manganate, potassium permanganate
Kaliumnitrat n 1. <Chemie> *(AE)* niter, *(BE)* nitre, potassium nitrate *(Kalisalpeter)*; 2. <Lebensmittel> potassium nitrate
Kaliumoxid n <Chemie> potash, potassium oxide
Kaliumpermanganat n <Chemie> potassium manganate, potassium permanganate
Kalk m <Bau, Erdöl, Fertig, Ker & Glas, Kohlen, Lebensmittel> lime *(Mineral)*
Kalkablagerung f <Lebensmittel> lime scale
Kalkanreicherung f **von Seen** <Umweltschmutz> lake liming
kalkbehandelter Bohrschlamm m <Erdöl> lime-treated mud *(Bohrtechnik)*
Kalkbeuche f <Textil> lime boil
Kalkbohrschlamm m <Erdöl> lime mud *(Bohrtechnik)*
kalken v <Bau> whitewash
Kalken n <Bau> limewashing, liming, whitewashing

kalkhaltiger Ton *m* <Bau, Kohlen, Wasserversorg> chalky clay
kalkhaltiges Wasser *n* <Wasserversorg> hard water
Kalkhydrat *n* <Bau> hydrated lime
Kalkkaseinfarbe *f* <Kunststoff> distemper
Kalkmilch *f* <Bau, Fertig> whitewash
Kalkscheidepfanne *f* <Lebensmittel> liming tank *(Zuckerfabrikation)*
Kalkscheidung *f* <Lebensmittel> lime defecation *(Zuckerfabrikation)*
Kalkschlamm *m* <Ker & Glas> lime slurry
Kalkstein *m* 1. <Bau, Fertig, Ker & Glas> limestone; 2. <Lebensmittel> calcium carbonate
Kalktünche *f* <Bau> limewash
Kalkulation *f* <Geom> calculation
Kalkung *f* 1. <Chemtech> Zucker; 2. <Lebensmittel> defecation *(zur Gewinnung des Scheidesaftes in der Zuckerherstellung)*
Kalkverfestigung *f* <Bau> lime stabilization
Kalkzuschlag *m* <Metall> limestone flux
Kalman-Filter *n* <Elektronik> Kalman filter
Kalman-Filterung *f* <Elektronik> Kalman filtering
Kalmen *fpl* <Wassertrans> doldrums *(Tiefdruckgürtel um Äquator)*
Kalomel *n* <Chemie, Elektriz> calomel
Kalomelelektrode *f* <Elektriz> calomel electrode
Kalorie *f* 1. <Ergon, Labor> cal; 2. <Lebensmittel> cal, calorie, energy
kalorienarm *adj* <Lebensmittel> low-calorie, low-energy
Kalorienverbrauch *m* <Lebensmittel> caloric expenditure
Kalorienzufuhr *f* <Ergon> caloric intake
Kalorimeter *n* <Gerät, Labor, Maschinen, Phys, Thermod> calorimeter
Kalorimeter *n* **zum Hadronnachweis** <Strahlphys, Teilphys> hadronic calorimeter
Kalorimeterbombe *f* <Labor> calorimetric bomb *(Hitzemessung)*
Kalorimetrie *f* <Phys, Thermod> calorimetry
kalorimetrisch *adj* <Chemie> calorimetric
kalorimetrische Bombe *f* <Gerät> bomb calorimeter
kalorimetrisches Thermometer *n* <Thermod> calorimetric thermometer
kalorisch *adj* <Phys, Thermod> calorific
kalorisieren *v* <Fertig> *(BE)* aluminium-impregnate, *(AE)* aluminum-impregnate
Kalorisieren *n* <Fertig> *(BE)* aluminium impregnation, *(AE)* aluminum impregnation
Kalotte *f* <Fertig> spherical surface
Kalottenfläche *f* <Fertig> impression
kalt *adj* 1. <Anstrich> cold, cool; 2. <Elektrotech> cold; 3. <Heiz & Kälte> cold, dry; 4. <Kerntech, Kohlen, Kunststoff, Metall, Thermod> cold
kalt abtrennen *v* <Maschinen> cold-shear
Kalt... <Anstrich, Maschinen, Mechan> cold
kaltabbindender Klebstoff *m* <Kunststoff> cold setting adhesive
kaltaushärten *v* <Fertig> age at room temperature
Kaltband *n* <Metall> cold-rolled strip
Kaltbearbeitung *f* <Anstrich> cold work
Kaltbiegen *n* <Mechan> cold bending
Kaltbiegeprobe *f* <Maschinen> cold-bending test
kaltbrüchiges Eisen *n* <Metall> cold-short iron
Kaltbrüchigkeit *f* <Thermod> cold brittleness
kalte Katode *f* <Elektrotech> cold cathode
kalte Lötstelle *f* 1. <Elektriz> dry joint, cold junction; 2. <Maschinen> dry joint, dry solder joint
kalte Quelle *f* <Kerntech> cold source
kalte Verbindungsstelle *f* 1. <Elektriz> cold junction; 2. <Gerät> cold junction *(beim Thermoelement)*

Kälte *f* 1. <Heiz & Kälte> cold, low temperature, refrigerating; 2. <Thermod> cold strength, coldness, cold, low temperature, refrigerating; 3. <Verpack> refrigerating
Kälteanlage *f* 1. <Heiz & Kälte, Thermod> refrigerating plant; 2. <Verpack> refrigeration machine
Kältebad *n* <Heiz & Kälte> cryogenic bath
kältebeständig *adj* 1. <Heiz & Kälte> antifreezing, nonfreezing; 2. <Lufttrans> antifreezing
Kältebeständigkeit *f* <Thermod> temperature resistance
Kältechemie *f* <Heiz & Kälte> cryochemistry
Kältedämmung *f* <Thermod> low-temperature insulation
Kältefach *n* <Heiz & Kälte> low-temperature compartment
Kältefestigkeit *f* 1. <Kunststoff> low-temperature resistance; 2. <Werkprüf> low temperature toughness
kalteingesenkt *adj* <Fertig> hobbed
kalteinsenken *v* <Fertig> broach
Kälteisolierung *f* <Thermod> low-temperature insulation
Kältekompressor *m* <Heiz & Kälte> chiller, refrigerating compressor
Kältekreislauf *m* <Heiz & Kälte> refrigeration cycle
Kältelagerraum *m* <Heiz & Kälte> chill room
Kältemaschine *f* <Heiz & Kälte, Thermod> refrigerating machine
Kältemischung *f* <Heiz & Kälte, Thermod> freezing mixture, frigorific mixture
Kältemittel *n* 1. <Heiz & Kälte> cryogen, cryogenic fluid, refrigerant, refrigerating medium; 2. <Kfztech, Maschinen, Thermod, Umweltschmutz> refrigerant
Kältemittelkreislauf *m* <Heiz & Kälte> refrigerant circuit
Kältemittelverdichter *m* <Maschinen> refrigerant compressor
Kaltendvergütung *f* <Ker & Glas> cold-end coating
Kälteofen *m* <Labor> refrigerated oven
kalter Bereich *m* <Lufttrans> cold section *(eines Strahltriebwerkes)*
kalter Biegeversuch *m* <Metall> cold bend test
kalter Fluss *m* <Thermod> cold creep
kalter Gehalt *m* <Kohlen> cold content
kalter Reaktor *m* <Kerntech> cold reactor
Kälteraum *m* <Heiz & Kälte> cold chamber, cold room
kaltes Fließen *n* <Kunststoff> cold flow
kaltes Leuchten *n* <Elektrotech> luminescence
Kälteschrank *m* <Heiz & Kälte> cold chamber, refrigeration cabinet
Kälteschrumpfung *f* <Metall, Thermod> contraction due to cold
Kältetechnik *f* 1. <Comp & DV> cryogenics; 2. <Heiz & Kälte> cryogenics, refrigeration, refrigeration engineering; 3. <Lebensmittel> refrigeration engineering; 4. <Phys> cryogenics; 5. <Raumfahrt> cryogenics *(Raumschiff)*; 6. <Thermod> cryogenics
Kältetechniker *m* 1. <Fertig> refrigeration engineer; 2. <Thermod> refrigerating engineer
kältetechnisch *adj* <Phys> cryogenic
Kälteturbine *f* <Heiz & Kälte, Lufttrans> expansion turbine
Kälteverhalten *n* 1. <Heiz & Kälte> low-temperature characteristics; 2. <Kunststoff> low-temperature performance; 3. <Maschinen> *(AE)* low-temperature behavior, *(BE)* low-temperature behaviour
Kältezentrale *f* <Heiz & Kälte> *(AE)* refrigeration control center, *(BE)* refrigeration control centre
Kaltfließen *n* <Kunststoff> cold flow, creep
Kaltfließvermögen *n* <Mechan> cold flow
Kaltfluss *m* <Mechan> cold flow
Kaltform *f* <Ker & Glas> *(AE)* cold mold, *(BE)* cold mould
Kaltformen *n* 1. <Fertig> cold working; 2. <Kunststoff> *(AE)* cold molding, *(BE)* cold moulding
Kaltfront *f* <Wassertrans> cold front *(Wetter)*
Kaltgasblasen *n* <Fertig> gas quenching

Kaltgasschubsystem

Kaltgasschubsystem *n* <Raumfahrt> cold gas thrust system *(Raumschiff)*
kaltgeformtes Holz *n* <Wassertrans> *(AE)* cold-molded wood, *(BE)* cold-moulded wood *(Schiffbau)*
Kaltgesenk *n* <Maschinen> cold die
kaltgewalzt *adj* <Thermod> cold-rolled
kaltgewalzt und ausgeglüht *adj (CRCA)* <Metall> cold-rolled and annealed *(CRCA)*
kaltgewalzter Träger *m* <Thermod> cold-rolled joist
kaltgewalztes Blech *n* <Metall> cold-rolled sheet
kaltgezogen *adj* <Thermod> cold-drawn
kaltgezogener Stahldraht *m* <Metall> cold-draw steel wire
Kaltgießen *n* <Thermod> cold casting
Kalthämmern *n* <Metall> cold hammering
Kalthärtbarkeit *f* <Fertig> strain-hardening ability
kalthärten *v* <Metall, Thermod> cold-harden
Kalthärten *n* <Kunststoff> cold setting
kalthärtender Klebstoff *m* <Kunststoff> cold setting adhesive
Kalthärtung *f* <Metall> strain hardening
Kaltkanalwerkzeug *n* <Kunststoff> *(AE)* cold runner mold, *(BE)* cold runner mould
Kaltkatodenröhre *f* <Elektronik> cold cathode tube
Kaltkautschuk *m* <Kunststoff> cold rubber
kaltkleben *v* <Thermod> cold-bond
Kaltkleben *n* <Thermod> cold bonding
Kaltklebestelle *f* <Thermod> cold start
Kaltklebstelle *f* <Thermod> cold bond
Kaltlagerung *f* <Lebensmittel> cold storage
Kaltlagerungsschaden *m* <Lebensmittel> cold storage injury
Kaltleimsystem *n* <Verpack> cold glueing system
Kaltlötstelle *f* <Elektrotech> cold junction, dry junction
Kaltluftgefrieren *n* <Lebensmittel> blast freezing
kaltlufttrocknen *v* <Thermod> dry by cold air
Kaltluftturbulenzen *fpl (CAT)* <Lufttrans> cold air turbulence *(CAT)*
Kaltluftvorhang *m* <Heiz & Kälte> air curtain
Kaltmeißel *m* <Maschinen> cold chisel
kaltnieten *v* <Fertig> clinch
kaltprägen *v* <Metall> cold-stamp
Kaltpressen *n* 1. <Kunststoff> *(AE)* cold molding, *(BE)* cold moulding; 2. <Maschinen> cold forging
Kaltpressschweißen *n* <Bau, Maschinen, Mechan> cold pressure welding
Kaltrauch *m* <Lebensmittel> cold smoke
kaltrecken *v* <Metall> cold-hammer
Kaltrecken *n* <Mechan> cold drawing, cold working
Kaltriss *m* 1. <Ker & Glas> crizzle; 2. <Werkprüf> cold crack
Kaltrissanfälligkeit *f* <Werkprüf> cold cracking risk
Kaltsäge *f* <Maschinen> cold saw
Kaltsatz *m* <Druck> cold type
Kaltschere *f* <Maschinen> cold shears
Kaltschweißstelle *f* <Fertig> shut
Kaltsprödigkeit *f* <Thermod> cold brittleness
Kaltstart *m* <Comp & DV, Elektriz, Kfztech, Maschinen, Thermod> cold start
Kaltstartlampe *f* <Elektriz> cold-start lamp
Kaltstartvorrichtung *f* <Kfztech> cold-start device *(Motor, Vergaser)*
Kaltstartzug *m* <Kfztech> choke
Kaltumformen *n* <Metall> cold reduction
Kaltverfestigung *f* 1. <Fertig> strain hardening, work hardening; 2. <Mechan> cold working; 3. <Metall> strain hardening
Kaltverfestigungskoeffizient *m* <Metall> work-hardening coefficient
kaltverformen *v* <Metall, Thermod> cold-forge

kaltverformt *adj* <Thermod> cold-forged
Kaltvulkanisation *f* <Fertig> acid cure
kaltvulkanisieren *v* <Thermod> cold-cure
Kaltvulkanisieren *n* <Thermod> cold curing
kaltvulkanisiert *adj* <Thermod> cold-cured
kaltwalzen *v* <Metall, Thermod> cold-roll
Kaltwalzen *n* <Metall, Thermod> cold rolling
Kaltwalzwerk *n* <Metall> temper mill
Kaltwassereinbruch *m* <Kerntech> intrusion of cold water
Kaltwiderstand *m* <Elektrotech> cold resistance
kaltzäher Stahl *m* <Metall> tough-at-subzero steel
kaltziehen *v* <Metall, Thermod> cold-draw
Kaltziehen *n* <Maschinen, Mechan, Metall, Thermod> cold drawing
kalzinieren *v* 1. <Fertig> calcinate; 2. <Kohlen, Metall> calcine
Kalzinieren *n* <Metall> calcination
Kalzinierofen *m* <Kohlen, Metall> calcining kiln
kalziniert *adj* <Metall> calcined
Kalzinierung *f* <Abfall> calcination
Kalziumkarbid *n* <Fertig> carbide
Kamee *f* <Ker & Glas> cameo
Kamera *f* <Elektronik, Fernseh, Foto, Funktech> camera
• **leere Kamera bedienen** <Foto> operate empty camera *(Auslöser-Test)*
Kamera *f* **mit abnehmbaren Mattscheibensucher** <Foto> camera with detachable reflex viewfinder
Kamera *f* **mit auswechselbarem Objektiv** <Foto> camera with interchangeable lens
Kamera *f* **mit Blendenverschluss** <Foto> camera with diaphragm shutter
Kamera *f* **mit CCD-Bildaufnehmer-Zeile** <Fernseh> CCD camera, charge-coupled device camera
Kamera *f* **mit CCD-Zeile** <Fernseh> CCD camera, charge-coupled device camera
Kamera *f* **mit eckigen Balgen** <Foto> square bellows camera
Kamera *f* **mit elektronischer Steuerung** <Foto> electronic-controlled camera, electronically controlled camera
Kamera *f* **mit gekoppeltem Belichtungsmesser** <Foto> camera with coupled exposure meter
Kamera *f* **mit gekoppeltem Entfernungsmesser** <Foto> camera with coupled rangefinder
Kamera *f* **mit großem Balgenauszug** <Foto> camera with large bellows extension
Kamera *f* **mit kurzem Balgenauszug** <Foto> camera with short bellows extension
Kamera *f* **mit Spiegelreflex-Fokussierung** <Foto> camera with mirror reflex focusing
Kamera *f* **mit versenkbarer Objektivfassung** <Foto> camera with collapsible mount
Kamera *f* **mit verstellbarer und schwenkbarer Standarte** <Foto> camera with rising and swinging front
Kamera *f* **mit Zeitautomatik** <Foto> aperture priority camera
Kameraanpassung *f* <Fernseh> camera matching
Kameraaufstellung *f* <Fernseh> camera line up
Kameraauszug *m* <Foto> camera extension
Kamerafassung *f* <Foto> camera mount
Kameragehäuse *n* <Foto> camera body, camera housing
Kamerahülle *f* <Foto> covering
Kamerakanal *m* <Fernseh> camera channel
Kamerakette *f* <Fernseh> camera chain
Kamerakommandosystem *n* <Fernseh> camera prompting system
Kameramann *m* <Fernseh> cameraman
Kameramonitor *m* <Fernseh> camera monitor
Kameraneiger *m* <Foto> tilt head
Kameraröhre *f* <Fernseh> camera tube

Kameraröhre f mit hochwertiger Gradation <Elektronik> high-gamma camera tube
Kameraschwenk-Potenziometer n <Aufnahme> pan pot
Kamerasignal n <Fernseh> camera signal
Kamerastativ n <Foto> camera stand
Kamerasteuerung f <Fernseh> camera control unit
Kameraumschaltung f <Fernseh> camera switching
Kamerawagen m <Fernseh> dolly
Kamin m 1. <Bau> stack; 2. <Heiz & Kälte> chimney flue; 3. <Thermod> flue
Kamineinsatzrohr n <Heiz & Kälte> flue lining
Kaminschacht m <Bau> shaft
Kaminwirkung f <Heiz & Kälte> chimney effect
Kaminzug m <Bau> (AE) draft, (BE) draught
Kamm m 1. <Bau> ridge; 2. <Nichtfoss Energ> crest *(Wellenkamm)*; 3. <Textil> comb, reed
Kammelektrode f <Raumfahrt> comb-shaped electrode *(Raumschiff)*
kämmen v 1. <Fertig> intermesh *(Zahnräder)*; 2. <Maschinen> mate, mesh, pitch; 3. <Textil> comb
Kämmen n 1. <Fertig, Maschinen> intermeshing, mating; 2. <Textil> combing
kämmend adj <Fertig> intermeshing *(Zahnräder)*
Kammer f 1. <Abfall> chamber; 2. <Bau> cavity, coffer; 3. <Chemtech> chamber; 4. <Kunststoff> cabinet; 5. <Labor> cell; 6. <Maschinen, Mechan> chamber; 7. <Wasserversorg> coffer
Kammerfilterpresse f <Abfall> chamber filter press
Kammermauer f <Wasserversorg> side wall
Kammerofen m 1. <Heiz & Kälte> box furnace; 2. <Thermod> batch furnace
Kammersäure f <Chemtech> chamber acid
Kammerschleuse f <Wasserversorg> lift lock
Kammerton m <Aufnahme> standard pitch
Kammertongenerator m <Aufnahme> standard tone generator
Kammerwand f <Wasserversorg> side wall
Kammfilter n <Elektronik, Fernseh> comb filter
Kammfilterung f <Elektronik> comb filtering
Kammgarn n <Textil> combed yarn, worsted yarn
Kammgarnanzugstoff m <Textil> worsted suiting, worsted yarn
Kammgarnnummerierung f <Textil> worsted count, worsted yarn
Kammhöhe f <Nichtfoss Energ> crest height *(Wellenkamm)*
Kammlager n <Maschinen> multicollar thrust bearing
Kämmmaschine f <Textil> combing machine
Kammprofildichtung f <Maschinen> grooved metal gasket
Kammwalze f 1. <Fertig> pinion *(Walzen)*; 2. <Textil> porcupine
Kammzug m <Textil> top *(Spinnen)*
Kammzugfärben n <Textil> top dyeing
kammzuggefärbt adj <Textil> top-dyed
Kämpfer m <Bau> impost, transom, traverse; springer *(eines Bogens)*
Kämpferholz n <Bau> impost *(Fenster)*
Kämpferlinie f <Bau> springing line
Kämpferschicht f <Bau> springing course
Kämpferstein m <Bau> springer stone
Kampfhubschrauber m <Lufttrans> combat helicopter
Kanal m 1. <Aufnahme, Bau, Comp & DV, Elektronik, Elektrotech, Fernseh> channel; 2. <Fertig> duct, flue; 3. <Funktech> channel; 4. <Heiz & Kälte> duct; 5. <Hydraul> port; 6. <Ker & Glas> canal *(des Tafelglas-Wannenofens)*; 7. <Kohlen, Kontroll> channel; 8. <Maschinen> channel, duct, passage; 9. <Mechan> channel; 10. <Nichtfoss Energ> channel, conduit; 11. <Strömphys> channel; 12. <Telekom> duct; 13. <Wassertrans> canal, channel
Kanal m im oberen Seitenband <Fernseh> channel using upper sideband
Kanal m im unteren Seitenband <Fernseh> channel using lower sideband
Kanal m mit meteorischer Streuübertragung <Funktech> meteor scatter channel
Kanalabgleich m <Aufnahme> channel phasing
Kanalabschnitt m <Wasserversorg> reach
Kanalabstand m 1. <Comp & DV, Fernseh> channel spacing; 2. <Telekom> interchannel spacing
Kanalabtastrate f <Telekom> channel sampling rate
Kanalabtastung f <Telekom> channel sampling
Kanaladapter m <Comp & DV> channel adaptor
Kanaladresswort n (KAW) <Comp & DV> channel address word (CAW)
Kanalanforderung f <Telekom> channel request
Kanalaufteilung f <Telekom> channel distribution, (AE) channeling, (AE) channelizing, (BE) channelling
Kanalausgliederung f <Telekom> channel dropping
Kanalauslastung f <Telekom> channel load
Kanalausnutzung f <Telekom> channel efficiency
Kanalausstieg m <Telekom> channel dropping
Kanalauswahl f <Telekom> channel selection
Kanalauswahlschalter m <Telekom> channel selection switch
Kanalbandbreite f <Fernseh, Funktech> channel bandwidth
Kanalbau m <Wasserversorg> canalization
Kanalbaugruppe f <Telekom> channel unit
Kanalbedarf m <Telekom> channel requirements
Kanalbefehlswort n <Comp & DV> channel command word
Kanalbelastung f <Telekom> channel loading
Kanalbelegung f <Telekom> channel occupancy
Kanalbildung f <Nichtfoss Energ> (AE) channeling, (BE) channelling
Kanalboot n <Wassertrans> canal boat *(Schifftyp)*
Kanalbündelung f <Telekom> channel multiplexing, trunking
Kanalcode m <Telekom> channel code
Kanalcodierung f <Telekom> channel coding, channel encoding
Kanaldecodierung f <Telekom> channel decoding
Kanaldefinition f <Telekom> channel definition
Kanaldotierung f <Elektronik> channel doping
Kanäle mpl <Heiz & Kälte> ducting • **in Kanälen fortleiten** <Heiz & Kälte> duct away • **mit Kanälen** <Fertig> ported
Kanaleigenschaften fpl <Telekom> channel characteristics
Kanaleinfahrt f <Wassertrans> canal entrance
Kanaleinfügung f <Telekom> channel insertion
Kanaleinrichtung f <Telekom> channel equipment
Kanaleinstieg m <Telekom> channel insertion
Kanalentzerrer m <Telekom> channel equalizer
Kanalentzerrung f <Telekom> channel equalization
Kanalfehler m <Telekom> channel error
Kanalfilter n <Elektronik> channel filter
Kanalhaltung f <Wasserversorg> reach
Kanalisation f 1. <Abfall> drain system, sewer system, sewerage system; 2. <Bau> sewer system, sewerage, sewerage system; 3. <Wasserversorg> canalization, sewerage
Kanalisationsnetz n 1. <Bau> sewerage; 2. <Wasserversorg> canalization, sewerage, sewerage system
Kanalisationssystem n <Wasserversorg> canalization
kanalisieren v <Wasserversorg> canalize
Kanalkapazität f <Comp & DV, Telekom> channel capacity

Kanalkapazität *f* eines symmetrisch gestörten Binärkanals <Telekom> capacity of a binary symmetrical disturbed channel
Kanalkennung *f* <Telekom> channel identifier
Kanalkohle *f* <Kohlen> channel coal
Kanalkompensation *f* <Aufnahme> channel balancing
Kanallandmarken *fpl* <Wassertrans> channel marks
Kanallänge *f* <Nichtfoss Energ> length of channel
Kanalrauschen *n* <Elektronik> channel noise
Kanalroute *f* <Wassertrans> channel track
Kanalruß *m* <Kunststoff> channel black
Kanalschleuse *f* <Wasserversorg> canal lock
Kanalschleusentor *n* <Wasserversorg> canal lock gate
Kanalsohle *f* <Wasserversorg> canal bottom, channel bed, channel bottom; canal bottom *(künstlicher Kanal)*
Kanalstatustabelle *f* <Comp & DV> channel status table
Kanalstatuswort *n* <Comp & DV> channel status word
Kanalsteuerung *f* <Telekom> channel control
Kanalstopper *m* <Elektrotech> channel stopper
Kanalstrahl *m* <Kerntech, Phys> canal ray
Kanalstrahlenanalyse *f* <Kerntech> canal-ray analysis
Kanalstrahlentladung *f* <Kerntech> canal-ray discharge
Kanalströmung *f* <Strömphys> channel flow, flow in channels *(Strömung im offenen Gerinne)*
Kanalsystem *n* <Heiz & Kälte> duct system, ducting
Kanaltrennfilter *n* <Elektronik> channel filter
Kanaltrennung *f* <Aufnahme> channel separation
Kanalumsetzung *f* <Elektronik> channel modulator, channel translator *(Trägerfrequenztechnik)*
Kanalverstärker *m* <Fernseh, Telekom> channel amplifier
Kanalvocoder *m* <Telekom> channel vocoder
Kanalvorwahl *f* <Fernseh> presetting of channels
Kanalwähler *m* 1. <Fernseh> channel selector, tuner; 2. <Telekom> channel selector
Kanalwählerprogrammautomatik *f* <Telekom> automatic tuner search and storage *(ATS)*
Kanalwählschalter *m* <Aufnahme> channel selector switch
Kanalweiche *f* <Telekom> channel branching filter
Kanalzuweisung *f* <Fernseh, Funktech> channel allocation
Kanalzuweisungszeit *f* <Telekom> channel allocation time
Kanalzwischenraum *m* <Comp & DV> channel spacing
Känguruh-Frachter *m* <Wassertrans> barge-carrying ship
Kannelkohle *f* <Kohlen> cannel coal
Kännelkohle *f* <Kohlen> kennel coal
Kanneneintauchkühler *m* <Lebensmittel> in-can immersion cooler
Kanne'sche Kammer *f* <Kerntech> Kanne chamber *(zur Überwachung radioaktiven Gases)*
Kannette *f* <Textil> pirn
Kanone *f* <Mechan, Phys> barrel
Kanonenbohrer *m* <Fertig> cylinder bit, half-round bit
Kanonenwirkungsgrad *m* <Fernseh> gun efficiency
kanonische Basis *f* <Math> canonical base *(Standardbasis eines Vektorraumes)*
kanonische Gesamtheit *f* <Phys> canonical ensemble
kanonische Gleichung *f* <Phys> canonical equation
kanonische Variable *f* <Phys> canonical variable
Kante *f* 1. <Bau> corner, skirt; 2. <Comp & DV, Geom> edge; 3. <Künstl Int> arc *(Graph)*; edge, link *(zwischen Knoten in Graph)*; 4. <Maschinen> edge, outline *(Graphentheorie)*; 5. <Papier, Phys, Textil, Verpack> edge • **Kanten abschlagen** <Bau> spall • **von Kante zu Kante** <Maschinen> edge-to-edge
kanten *v* <Bau> tip
Kantenabschälen *n* <Ker & Glas> edge peeling
Kantenanhebung *f* <Fernseh> edge enhancement

Kantenbeschaffenheit *f* <Metall> edge condition
Kantenbeschneiden *n* mit Schneidstift <Ker & Glas> pencil edging
Kantenbeschneidung *f* <Fertig> edge trimming
Kanteneffekt *m* <Foto> Eberhard effect, edge effect
Kantenfeinschleifen *n* <Ker & Glas> edge fine-grinding
Kantenfeuchter *m* <Papier> edge spray, edging spray
Kantenfließen *n* <Ker & Glas> edge creep
Kantengummiermaschine *f* <Verpack> edge-gumming machine
Kanteninterpretation *f* <Künstl Int> edge interpretation *(im Bildverstehen)*
Kantenkorrektur *f* <Fernseh> edge correction
Kantenkorrosion *f* <Werkprüf> edge corrosion
Kantenplatte *f* <Lufttrans> edging panel
Kantenpressung *f* <Maschinen> edge pressure
Kantenriss *m* <Ker & Glas> edge crack, edge fracture
Kantenschliff *m* <Ker & Glas> edging
Kantenschloss *n* <Bau> flush lock
Kantenschmelzen *n* <Ker & Glas> edge melting
Kantenschneider *m* <Papier> edge cutters
Kantenschneidevorgang *m* <Textil> selvedge cutting process
Kantenschutz *m* <Verpack> edge cushion, edge protection
Kantenschutzschiene *f* <Bau> nosing
Kantenstein *m* <Ker & Glas> edger block
Kantenversetzung *f* <Metall> edge dislocation
Kantenverstärkung *f* <Fernseh> edge enhancement
Kantenvorbereitung *f* <Mechan> edge preparation
Kantenwalzen *fpl* <Ker & Glas> edge bowls, edge rolls
Kantenwalzen *n* <Metall> edge rolling
Kantenwirkung *f* <Lufttrans> fringe effect
Kantholz *n* <Bau> scantling, square timber
Kanvas *m* <Textil> canvas
Kanzel *f* 1. <Lufttrans> canopy, cockpit; 2. <Wassertrans> pulpit *(Deckausrüstung)*
Kaolin *n* 1. <Druck> kaolin; 2. <Ker & Glas> china clay; 3. <Kunststoff> china clay, kaolin
Kaolinerde *f* <Druck> kaolin
Kaolinit *m* <Erdöl, Kohlen, Nichtfoss Energ> kaolinite
Kaolinschlämmen *n* <Ker & Glas> china clay washing
Kaolinsteinbruch *m* <Ker & Glas> china clay quarry
Kaon *n* 1. <Phys> kaon *(Elementarteilchen)*; 2. <Teilphys> kaon
Kap *n* <Wassertrans> cape *(Geographie)*
Kapazitanz *f* <Elektrotech, Heiz & Kälte> capacitance
Kapazität *f* 1. <Bau, Elektriz, Elektrotech, Erdöl, Funktech, Phys, Telekom> capacitance, capacity *(C)*; 2. <Foto> ampere-hour capacity *(Batterie)*; 3. <Heiz & Kälte> capacity *(C)*
Kapazität *f* einer Straße <Trans> capacity of a road
Kapazität *f* unter vorherrschenden Bedingungen <Trans> capacity under prevailing conditions
Kapazität *f* zwischen Elektroden <Funktech> interelectrode capacitance
Kapazität *f* zwischen Wicklungen <Elektriz> interturn capacitance
kapazitativer Teiler *m* <Elektrotech> capacitative voltage divider
kapazitätsarm *adj* <Gerät> anticapacitance
Kapazitätsbelastung *f* <Verpack> carrying capacity *(des Förderbands)*
Kapazitätsbrücke *f* <Elektriz, Gerät> capacitance bridge
Kapazitätsdiode *f* <Phys> capacitance diode, varactor diode
kapazitätsfrei *adj* <Elektriz, Elektrotech> noncapacitive
kapazitätsfreie Last *f* <Elektrotech> noncapacitive load

Kapazitätskasten m <Elektrotech> capacitance box
Kapazitätsklausel f <Lufttrans> capacity clause
Kapazitätskoeffizient m <Phys> capacitance coefficient
Kapazitätskontrolle f <Lufttrans> capacity control
Kapazitätsmessbrücke f 1. <Elektriz> capacity bridge; 2. <Gerät> capacitance bridge
Kapazitätsmessgerät n 1. <Elektrotech> capacitance meter; 2. <Gerät> capacitance measuring instrument
Kapazitätsmessung f <Optik> capacitance sensing
Kapazitätsrelais n <Elektriz> capacitance relay
Kapazitätsscheibe f <Optik> (BE) capacitance disc, (AE) capacitance disk (Scheibenkondensator)
Kapazitätsüberschreitung f <Comp & DV> overflow
kapazitiv adj 1. <Elektriz, Elektrotech> capacitive; 2. <Funktech, Labor, Phys, Telekom> capacitance
kapazitive Antennenanpassung f (ATC) <Funktech> aerial-tuning capacitor, antenna-tuning capacitor (ATC)
kapazitive Belastung f <Elektrotech, Telekom> capacitive load
kapazitive Komponente f <Elektriz> capacitive component
kapazitive Kopplung f 1. <Elektriz, Elektrotech> capacitance coupling, capacitive coupling; 2. <Telekom> capacitive coupling
kapazitive Last f <Elektriz, Elektrotech> capacitive load
kapazitive Reaktanz f 1. <Elektriz, Elektrotech> capacitive reactance, negative reactance; 2. <Phys> capacitive reactance
kapazitive Rückkopplung f <Elektrotech> capacitive feedback
kapazitiver Abschluss m <Elektrotech> capacitive load
kapazitiver Blindwiderstand m 1. <Elektriz, Elektrotech> capacitive reactance (XC); negative reactance; 2. <Phys> capacitive reactance
kapazitiver Dehnungsmessstreifen m <Gerät> (AE) capacitive strain gage, (BE) capacitive strain gauge
kapazitiver Druckwandler m <Phys> CPT, capacitive pressure transducer
kapazitiver Kurzschlusskolben m <Elektrotech> choke plunger
kapazitiver Ladestrom m <Elektriz> capacitive charging current
kapazitiver Spannungsteiler m <Elektrotech> capacitive voltage divider
kapazitiver Widerstand m <Elektrotech> capacitive resistance
kapazitives Dickenmessgerät n <Gerät> (AE) capacitive thickness gage, (BE) capacitive thickness gauge
kapazitives Feedback n <Elektrotech> capacitive feedback
Kapellenofen m <Heiz & Kälte> assay furnace
kapillar adj <Fertig, Maschinen> capillary
Kapillar... <Fertig, Kohlen, Strömphys> capillary
kapillaraktiv adj <Chemie> surface-active
kapillarbrechende Schicht f <Bau> anticapillary course (Straßenbau)
Kapillardruck m <Strömphys> capillary pressure
Kapillare f 1. <Erdöl, Phys> capillary tube; 2. <Strömphys> capillary
Kapillareffekt m <Erdöl, Phys> capillary action
kapillares Wasser n <Kohlen> capillary water
Kapillarfusion f <Kerntech> capillary fusion
Kapillarimeter n <Kohlen> capillarimeter
Kapillarität f <Fertig, Kohlen, Phys, Strömphys> capillarity
Kapillaritätsbruchschicht f <Kohlen> capillarity breaking layer
Kapillaritätszahl f <Strömphys> capillary number
Kapillarkraft f <Strömphys> capillary force
Kapillar-Mengenstrommesser m <Labor> capillary flowmeter

Kapillar-Rheometer n <Gerät> capillary rheometer
Kapillarrohr n <Heiz & Kälte> capillary tube
Kapillarröhrchen n <Labor, Strömphys> capillary tube
Kapillarsäule f <Gerät> capillary column (Gaschromatographie)
Kapillar-Viskosimeter n 1. <Chemtech> capillary viscometer; 2. <Gerät> capillary viscosimeter; 3. <Labor> capillary viscometer (Flüssigkeitsdurchfluss)
Kapillarwellen fpl <Strömphys> capillary waves
Kapillarwirkung f <Bau, Maschinen, Strömphys> capillarity, capillary action
Kapitälchen npl <Druck> small caps
Kapitalrückfluss m <Erdöl> return of assets (Finanzen)
Kapitänskajüte f <Wassertrans> captain's cabin
Kapitänspatent n <Wassertrans> master's certificate (Dokumente)
Kaplanschaufel f <Fertig> kaplan blade
Kappdiode f <Elektronik> limiter diode
Kappe f 1. <Elektrotech, Hydraul, Kfztech, Maschinen, Mechan> cap; 2. <Wasserversorg> cap sill (eines Schleusentors)
Kappen n 1. <Comp & DV> scissoring; 2. <Künstl Int> pruning (eines Baumes)
Kappenaufsetzer m <Verpack> capper
Kappendichtmasse f <Verpack> cap-sealing compound
Kappenisolator m <Elektrotech> cap-and-pin insulator, cap-and-rod insulator
Kappenmutter f <Mechan> cap nut
Kappenpresse f <Verpack> capping press
Kappensensor m <Verpack> missing cap detector
Kappensiegelung f <Verpack> cap sealing
Kappenversiegelungsmaschine f <Verpack> cap-sealing equipment
Kapplage f <Fertig> backing bead
Kapsel f 1. <Akustik> enclosure; 2. <Elektrotech> cap; 3. <Fertig> case; 4. <Labor> capsule; 5. <Mechan> cap; 6. <Phys> enclosure
Kapselfederdruckmessglied n <Gerät> diaphragm pressure element
Kapselfeder-Manometer n <Gerät, Metrol> (AE) pneumatic capsule gage, (BE) pneumatic capsule gauge
Kapselfederwirkdrucksensor m <Gerät> aneroid flowmeter
Kapselgehäuse n <Hydraul> enclosed casing (Turbine)
Kapselhöhenmesser m <Lufttrans> aneroid altimeter
Kapselmutter f <Mechan> cap nut
kapseln v 1. <Fertig> case, seal; 2. <Sicherheit> enclose
Kapselton m <Ker & Glas> sagger clay
Kapselung f 1. <Abfall> encapsulation; 2. <Elektrotech> encapsulation, encasing, enclosure; 3. <Fertig> containment; 4. <Sicherheit> containment, enclosure, protective enclosure
Kapselung f **der Lärmquelle** <Sicherheit> acoustic enclosure, noise enclosure
Karabinerhaken m <Maschinen> snap hook
Karat n 1. <Metall> (BE) carat, (AE) karat; 2. <Metrol> (BE) carat, carat fine, (AE) karat (Edelmetalle); carat (Edelsteine)
Karatgewicht n <Metrol> caratage
Karayagummi n <Lebensmittel> crystal gum, karaya gum
Karbid n <Fertig, Mechan, Metall> carbide
karbidbildender Zusatz m <Fertig> carbide former
Karbidentstehung f <Metall> carbide formation
Karbidrissbildung f <Metall> carbide cracking
Karbidzelle f <Fertig> carbide band
Karbolinium n <Fertig> carbolinium
Karbon n <Erdöl> carboniferous (Geologie)
Karbonatation f <Chemie> carbonatation
Karbonisation f <Kohlen> carbonization
karbonisieren v <Kohlen, Metall> carbonize, carburize

Karbonisierung

Karbonisierung f <Metall> carburization
Karbonitrieren n <Fertig> carbonitriting
Karbonitrieren n **in Gas** <Fertig> gas cyaniding
Karbonstahl m <Metall> high-carbon steel
Karborund n <Fertig, Maschinen> carborundum
Karborundschleifscheibe f <Fertig> carborundum wheel
Karborundum n <Fertig> carborundum
karburieren v 1. <Chemie> carburet; 2. <Metall> carburate, carburet
karburierend adj <Metall> (AE) carbureting, (BE) carburetting
karburiert adj <Metall> (AE) carbureted, (BE) carburetted
Karburierung f <Metall> carburation, (AE) carbureting, (BE) carburetting
Kardamomöl n <Lebensmittel> cardamom oil
Kardan... <Fertig> cardan
Kardanantrieb m <Fertig> cardan drive, universal drive
Kardandrehzapfen m <Mechan> knuckle
Kardangelenk n 1. <Fertig> cardan joint; 2. <Kfztech> universal joint (Triebstrang); 3. <Maschinen> cardan joint, gimbal joint, universal joint; 4. <Mechan> cardan joint
kardanisch gelagert adj <Raumfahrt> gimbal-mounted (Raumfahrttechnik)
kardanische Aufhängung f 1. <Maschinen> cardanic suspension, gimbal suspension; 2. <Wassertrans> gimbal
Kardanring m <Fertig, Mechan, Wassertrans> gimbal
Kardantunnel m <Kfztech> propeller shaft tunnel (Getriebe)
Kardanwelle f 1. <Fertig> cardan shaft; 2. <Kfztech> cardan shaft, propeller shaft
Karde f <Textil> card
Kardeel n <Wassertrans> strand (Tauwerk)
karden v <Textil> card
Kardenbeschlag m <Textil> card clothing
Kardengarnitur f <Textil> card clothing
Kardensaal m <Textil> card room
Karderie f <Textil> card room
kardieren v <Textil> card
Kardieren n <Textil> carding
kardierte Display-Verpackung f <Verpack> visual-carded packaging
kardierte Produkte npl <Verpack> carded packaging
Kardinal... <Fertig, Math, Wassertrans> cardinal
Kardinalpunkt m <Fertig> Gaussian point
Kardinalpunkte mpl <Wassertrans> cardinal points (Kompass)
Kardinalstriche mpl <Wassertrans> cardinal points (Kompass)
Kardinalsystem n <Wassertrans> cardinal system (Navigation)
Kardinalzahl f <Math> cardinal number (Größe einer Menge)
Kardiograph m <Ergon> cardiograph
Kardioidmikrofon n <Akustik> cardioid microphone
Kardiotachometer n <Ergon> cardiotachometer
karieren v <Fertig> (AE) checker, (BE) chequer
karierte Waren fpl <Textil> checks
Karkasse f <Textil> carcass
Karnaubawachs n <Lebensmittel> carnauba wax (Pflanzenwachs)
Karnaugh-Diagramm n <Comp & DV> Veitch diagram
Karnaugh-Tabelle f <Comp & DV> Karnaugh map
Karnies m <Bau> ogee
Karnieshobel m <Bau> ogee plane
Karnotit n <Kerntech> carnotite
Karosserie f 1. <Fertig> body; 2. <Kfztech> body, coach, coachwork; car body (Fahrzeug); 3. <Mechan> car body
Karosserie f **nach dem Baukastenprinzip** <Kfztech> unitized body

Karosseriebau m <Kfztech> coachbuilding
Karosseriehersteller m <Kfztech> body manufacturer
Karosseriewerkstatt f <Kfztech> body shop
Karosseriewerkzeuge npl <Kfztech, Maschinen> car body tooling
Karosserieziehpresse f <Fertig> body drag press
Karrageen n <Lebensmittel> carrageen
Karren m <Druck> carriage
KARS (kohärente Antistokes-Raman-Streuung) <Strahlphys, Wellphys> CARS (coherent anti-Stokes Raman scattering)
Karsthydrologie f <Wasserversorg> karst hydrology
Karstquelle f <Wasserversorg> intermittent spring, karstic spring
Karte f 1. <Bau> chart, map; 2. <Comp & DV> card; 3. <Phys> chart; 4. <Textil> card (Weben); 5. <Trans> map; 6. <Wassertrans> chart (Navigation) • **auf Karte eintragen** <Wassertrans> chart (Navigation)
Karte f **mit großem Maßstab** <Trans> large-scale map
Kartenabtastung f <Comp & DV> card sensing
Kartenbahn f <Comp & DV> card channel
Kartenbaustein/auf <Telekom> board-mounted
Kartenberichtigung f <Wassertrans> chart correction (Navigation)
Kartenbildschirm m <Wassertrans> chart display (Navigation)
Kartencode m <Comp & DV> card code
Karteneinschub m <Telekom> plug-in board
Kartenentfernung f <Funkort> ground range (Radar)
Kartenführung f <Comp & DV> card bed
Kartenkäfig m <Comp & DV> card cage
Kartenleser m <Comp & DV, Telekom> card reader
Kartenlochen n <Textil> card cutting
Kartenlocher m <Comp & DV> card punch, keypunch
Kartenmagazin n <Comp & DV> card hopper
Kartenmaßstab m 1. <Metrol> scale; 2. <Wassertrans> chart scale, map scale (Navigation)
Kartenmischer m <Comp & DV> collator
Kartennull f <Wassertrans> chart datum, map datum (Navigation, Gezeiten)
Kartensatz m <Comp & DV> card deck, card record
Kartenschlagen n <Textil> card cutting
Kartenstanzen n <Textil> card cutting
Kartenstau m <Comp & DV> card jam
Kartentelefon n <Telekom> card phone
Kartentisch m <Wassertrans> chart table, map table (Navigation)
Kartenwassertiefe f <Wassertrans> charted depth, mapped depth
Kartenwender m <Elektronik> card reverser (Computer-Peripheriegeräte)
Kartenzahlung f <Telekom> electronic cash
Kartenzuführung f <Comp & DV> card feed
kartesische Geometrie f <Geom> Cartesian geometry
kartesische Koordinaten fpl <Bau, Geom, Konstzeich, Math, Phys> Cartesian coordinates
kartesisches Koordinatengitter n <Geom> Cartesian grid
kartesisches Koordinatensystem n <Elektronik, Math> Cartesian coordinate system
kartesisches Produkt n <Comp & DV, Geom, Math> Cartesian product (Mengenverknüpfung)
Kartierer m <Bau> mapper (Vermessung)
Kartierung f <Bau> mapping (Vermessung)
Kartograph m <Bau> map maker, mapper (Vermessung)
Kartographie f <Bau> map making
Karton m 1. <Druck> cardboard, paste board, pulp board; 2. <Papier> board; 3. <Sicherheit> box; 4. <Verpack> box, cardboard, paper board, carton

Karton *m* **aus Kistenpappe** <Verpack> container board box
Kartonabfüllanlage *f* <Verpack> carton filler
Kartonabfüllmaschine *f* <Verpack> carton-filling machine
Kartonage *f* <Verpack> cardboard packaging
Kartonagen-Aufrichtmaschine *f* <Verpack> folding cardboard box erecting machine
Kartonagen-Einrichtmaschine *f* <Verpack> folding cardboard box setting machine
Kartonagenmaschine *f* <Verpack> cardboard machine
Kartonaufrichtmaschine *f* <Verpack> carton-erecting machine
Kartonaufricht- und -verschließmaschine *f* <Verpack> carton erector and closer
Kartonautomat *m* <Verpack> carton-making machine
Kartondosiermaschine *f* <Verpack> carton-dosing machine
Kartondrucker *m* **mit Einzelblatteinzug** <Verpack> sheet-fed carton printer
Kartoneinlage *f* <Verpack> cardboard backing
Kartoniereinrichtung *f* <Verpack> cartoner, cartoning equipment
Kartoniermaschine *f* <Verpack> boxing machine, cartoning machine
Kartonmaschine *f* <Papier> board machine
Kartonrecycling *n* <Verpack> paper carton recycling
Kartonröhre *f* <Verpack> cardboard tube
kartonstarkes Papier *n* <Foto> double-weight paper
Kartonverpackung *f* <Verpack> cardboard packaging
Kartonverschließautomat *m* <Verpack> package sealer
Kartonversteifung *f* <Verpack> cardboard backing
Kartonwiederverwertung *f* <Verpack> paper carton recycling
Kartusche *f* <Druck, Kfztech, Verpack> cartridge
Kartuschenfilter *n* <Kfztech> cartridge filter *(Schmierung)*
Kartuschenpapier *n* <Verpack> cartridge paper
Karusseldrehmaschine *f* 1. <Fertig> vertical turret lathe; 2. <Maschinen> vertical boring and turning mill
karweelgebaut *adj* <Wassertrans> carvel-built *(Schiffbau)*
Karyokinese *f* <Kerntech> karyokinesis
Karyolyse *f* <Kerntech> karyolisis
Karzinogen *n* 1. <Lebensmittel> carcinogen; 2. <Sicherheit> carcinogenic substance
Karzinotron *n* 1. <Elektronik> carcinotron; backward-wave oscillator *(Mikrowellentechnik)*; 2. <Phys> backward-wave oscillator, carcinotron; 3. <Telekom> carcinotron
kaschieren *v* 1. <Kunststoff> coat; 2. <Papier> paste; 3. <Verpack> coat
Kaschieren *n* 1. <Kunststoff> coating; 2. <Verpack> coating *(von Papier, Pappe)*
Kaschiermaschine *f* 1. <Papier> pasting machine; 2. <Verpack> laminating machine
Kaschierpapier *n* 1. <Papier> liner paper; 2. <Verpack> liner paper, lining paper
kaschierte Graupappe *f* <Verpack> lined chipboard
kaschierter Kunststoff *m* <Elektriz> laminated plastic
kaschiertes Gewebe *n* <Kunststoff> coated fabric
kaschiertes Papier *n* <Papier> pasted paper
Kaschierung *f* <Papier> pasting
Kascodenschaltung *f* <Funktech> cascode circuit
Käsebruch *m* <Lebensmittel> cheese curd, curd
Kasein *n* <Chemie, Lebensmittel> casein
Kaseinat *n* <Lebensmittel> caseinate
Kaseinatgummi *n* <Lebensmittel> caseinate gum
Kaseinhydrolysat *n* <Lebensmittel> casein hydrolysate
Kaseinleim *m* <Lebensmittel> casein glue
Kaserne *f* <Bau> barracks

Kaskade *f* <Elektrotech> cascade set; cascade *(Reihenschaltung)* • **in Kaskade schalten** <Elektrotech> connect in cascade
Kaskadenanordnung *f* <Elektrotech> cascade arrangement, tandem arrangement
Kaskadenfolge schweißen *v* <Fertig> cascade
Kaskadenmühle *f* <Kohlen> cascade mill
Kaskadenofen *m* <Heiz & Kälte> cascade furnace
Kaskadenprozess *f* <Strahlphys> cascade process
Kaskadensatz *m* <Elektrotech> cascaded machine set, cascaded set
Kaskadenschaltung *f* 1. <Elektriz, Elektrotech> tandem connection; cascade connection *(Reihenschaltung)*; 2. <Kerntech> tandem connection
Kaskadenteilchen *n* <Phys> Xi particle *(Elementarteilchen)*
Kaskadenübertrag *m* <Elektronik> cascaded carry *(Computertechnik)*
Kaskadenumformer *m* <Elektrotech> cascade converter, cascade motor converter
Kaskadenverbindung *f* <Kerntech> cascade connection
Kaskadenverstärker *m* <Elektronik> cascade amplifier *(mehrstufig)*
Kaskadenwandler *m* <Kerntech> cascade transformer
Kaskadenwässerung *f* <Foto> cascade washing
Kaskadieren *n* <Kohlen> cascading
Kaskoversicherung *f* <Trans> hull insurance *(Versicherung)*
Kasse *f* 1. <Comp & DV> point of sale; 2. <Verpack> counter top machine
Kassenterminal *n* <Comp & DV, Telekom, Verpack> point of sale terminal *(POS)*
Kassette *f* 1. <Abfall> landfill cell, refuse cell; subcell *(Verfüllung von Deponien)*; 2. <Bau> caisson
Kassettendecke *f* <Bau> pan ceiling
Kassettenlaufwerk *n* <Aufnahme> cartridge tape drive
Kassettenmagazin *n* <Maschinen> cartridge magazin
Kassettenrecorder *m* <Aufnahme> cassette recorder *(CR)*
Kassetten-Videorecorder *m* <Aufnahme> cartridge-type video recorder, cassette video recorder *(VCR)*
Kassiaöl *n* <Lebensmittel> cassia oil
Kassierer *m* <Trans> collector
Kasten *m* 1. <Bau> box, caisson; 2. <Druck, Fertig, Maschinen> box; 3. <Verpack> carton, case
Kasten *m* **mit Metallbeschlägen** <Verpack> metal edging case
Kastenaufricht-, -abfüll- und -verschließmaschine *f* <Verpack> case erecting, filling and closing machine
Kastendurchlass *m* <Bau> box culvert
Kastenformerei *f* <Fertig> flask moulding
kastenförmige Stabilisierung *f* <Lufttrans> box-type stiffener
kastenförmige Struktur *f* <Lufttrans> box-type structure
kastenförmige Versteifung *f* <Lufttrans> box-type stiffener
kastenförmiger Fahrbalken *m* <Trans> box-section track girder
kastengeglüht *adj* <Metall> box annealed
Kastenguss *m* <Maschinen> box casting
Kastenkaliber *n* <Fertig> box groove *(Walzen)*
Kastenkalibrierung *f* <Fertig> box pass
Kastenkipper *m* <Kohlen> tip box car
Kastenkopierrahmen *m* <Druck> printing frame
Kastenleiter *m* <Elektrotech> rectangular waveguide
Kastenmetall *n* <Metall> boxmetal
Kastenofen *m* <Heiz & Kälte> box kiln
Kastenrahmen *m* 1. <Kfztech> box-type chassis, box-type frame *(Motor)*; 2. <Maschinen> box frame, box-form frame, box-section frame

Kastenrinne

Kastenrinne f <Bau> box gutter *(Dachrinne)*; trough gutter *(Dach)*; parallel gutter *(Gebäude)*
Kastenschloss n <Bau> box lock
Kastenständer m <Fertig> box column
Kastenständerbohrmaschine f <Fertig> box-column drilling machine
Kastentisch m <Maschinen> box table
Kastenträger m <Bau> box girder
Kastenvorkalibrierung f <Fertig> box pass
Kastenzange f 1. <Fertig> square tongs; 2. <Ker & Glas> pinchers
katadioptrisch adj <Phys> catadioptric
Katalase f <Chemie, Lebensmittel> catalase
Katalog m <Comp & DV> *(AE)* catalog, *(BE)* catalogue
Katalog-Agent m <Comp & DV> catalog agent
katalogisieren v <Comp & DV> *(AE)* catalog, *(BE)* catalogue
Katalog-Server m <Comp & DV> catalog server
Katalysator m 1. <Erdöl> catalyst *(Reaktionsbeschleuniger beim Crack-Prozess)*; 2. <Kfztech> catalytic converter, *(AE)* catalytic muffler, *(BE)* catalytic silencer; catalyst *(Abgassystem, Auspuffanlage)*; 3. <Kunststoff, Textil> catalyst; 4. <Umweltschmutz> catalytic converter, *(AE)* catalytic muffler, *(BE)* catalytic silencer; 5. <Wasserversorg> catalyst
Katalysatorauspuff m <Kfztech> exhaust catalytic converter system
Katalysatorbett n <Kerntech> catalyst bed
Katalysatorgift n <Chemtech> catalytic poison
katalysatorinduzierte Deuteriumreaktion f <Kerntech> catalysed deuterium reaction
Katalyse f <Chemie, Lebensmittel> catalysis
katalysieren v <Kunststoff> catalyze
katalysierte Deuteriumreaktion f <Kerntech> catalysed deuterium reaction
katalytische Krackanlage f <Erdöl> catalytic cracking plant *(Raffinerietechnik)*
katalytische Reformierung f 1. <Chemie> catalytic reforming, catforming *(Erdöl)*; 2. <Erdöl> catalytic reforming *(Raffinerie)*
katalytische Spaltung f <Erdöl> catalytic cracking
katalytischer Inhibitor m <Chemtech> catalytic poison
katalytischer Reaktor m <Chemtech> catalytic reactor
katalytischer Umformer m <Umweltschmutz> *(BE)* catalytic converter, *(AE)* catalytic muffler, *(BE)* catalytic silencer
katalytisches Kracken n <Erdöl> catalytic cracking *(Raffinerie)*
katalytisches Reduktionsverfahren n <Umweltschmutz> catalytic process
katalytisches Reformieren n <Erdöl> catalytic reforming *(Raffinerie)*
Katamaran m 1. <Wassertrans> catamaran ship, twin-hull ship; catamaran *(Boottyp)*; 2. <Wasserversorg> catamaran dredge
Katamaran-Trägerschiff n <Trans> bacat ship, barge-aboard catamaran-ship
Katarakt m <Kohlen, Nichtfoss Energ, Wasserversorg> cataract
Kataster n <Bau> land register
Katasteramt n <Bau> land registry
Katasteraufnahme f <Bau> cadastral survey
Katastrophenausrüstung <Sicherheit> emergency equipment
Katastrophenbekämpfung f <Sicherheit> disaster control
Katastrophenschutz m <Sicherheit> disaster control, disaster prevention
Katathermometer n <Heiz & Kälte> katathermometer
Kategorie f <Patent, Qual> category

Katenoid n <Geom> catenoid
Katergol n <Raumfahrt> katergol *(Raumschiff)*
Kathedralglas n <Ker & Glas> cathedral glass
Kathepsin n <Chemie> cathepsin
Kathete f <Geom> leg
Kathetometer n 1. <Labor> cathetometer *(Längenmessung)*; 2. <Phys> cathetometer
Kation n <Elektriz, Elektrotech, Erdöl, Kohlen, Lebensmittel, Phys, Strahlphys> cation
Kationenaustauscher m <Kohlen, Umweltschmutz> cation exchanger
Kationenaustauschkapazität f 1. <Erdöl> cation exchange capacity *(Petrochemie, Elektrolyse)*; 2. <Umweltschmutz> cation exchange capacity *(Boden)*
Kationendenudationsrate f <Umweltschmutz> cation denudation rate
kationisch adj <Kohlen> cationic
kationischer Farbstoff m <Textil> basic dye
kationoid adj <Chemie> electrophilic
Katkracker m <Erdöl> cat cracker
Katode f <Elektriz, Elektronik, Elektrotech, Fernseh, Funktech, Metall, Phys> cathode
Katodendunkelraum m <Elektrotech> cathode dark space
Katodenfleck m <Elektrotech> cathode spot
Katodenfolger m <Phys> cathode follower
katodengekoppelte Gegentaktstufe f <Elektronik> long-tail pair
Katodengitter n <Fernseh> cathode screen
Katodenglimmlicht n <Phys> cathode glow
Katodenmodulation f <Elektronik> cathode modulation
Katodenschaltung f <Elektrotech> cathode circuit
Katodenstrahl m 1. <Comp & DV, Druck, Elektriz, Elektronik> cathode ray, electron ray; 2. <Fernseh, Funktech, Strahlphys> cathode ray; 3. <Telekom> cathode ray, electron ray
Katodenstrahlanzeigegerät n <Kontroll> cathode-ray display
Katodenstrahlbildschirm m <Telekom> cathode-ray screen
Katodenstrahlbündel n <Elektriz> cathode-ray pencil
Katodenstrahldisplay n <Kontroll> cathode-ray display
Katodenstrahlen mpl <Elektronik, Fernseh, Phys, Strahlphys> cathode rays
Katodenstrahloszillograph m <Elektronik> cathode-ray oscillograph *(registriert schnelle Größen)*
Katodenstrahloszilloskop n <Phys, Strahlphys> cathode-ray oscilloscope
Katodenstrahlröhre f *(KSR)* <Comp & DV, Druck, Elektriz, Elektronik, Fernseh, Funktech, Ker & Glas> cathode-ray tube *(CRT)*
Katodensumpfröhre f <Elektronik> mercury pool tube
Katodenverstärker m <Elektrotech> cathode follower
Katodenzerstäubung f <Elektrotech, Metall> cathode sputtering
katodische Korrosion f <Metall> cathodic corrosion
katodischer Schutz m <Kohlen, Metall> cathodic protection
Katodolumineszenz f <Elektrotech, Phys> cathodoluminescence
Katzenauge n 1. <Bau> cats' eyes; 2. <Kfztech> reflector *(Sicherheitszubehör, Fahrrad)*
Katzenkopf m <Maschinen> cat head, spider
Kauri-Butanolwert m <Chemie> kauri butanol number
kausales Wissen n <Künstl Int> causal knowledge
Kausalgraph m <Künstl Int> causal graph
Kausch f <Wassertrans> thimble *(Beschläge)*
kaustifizieren v <Chemie> causticize
Kaustik f <Chemie, Phys, Qual> caustic
kaustisch adj <Chemie, Qual> caustic

Kaustizität f <Chemie> causticity
Kautschuk m 1. <Chemie, Druck> caoutchouc; 2. <Elektriz> rubber; 3. <Kunststoff> raw rubber, rubber
Kavalier-Projektion f <Konstzeich> cavalier projection
Kavitation f 1. <Bau> cavitation; 2. <Erdöl> cavitation *(an Laufrädern von Wasserturbinen, Schiffsschrauben)*; 3. <Kohlen, Mechan, Metall, Nichtfoss Energ, Phys> cavitation; 4. <Raumfahrt> cavity *(Weltraumfunk)*; 5. <Strömphys> cavitation
Kavitationsversagen n <Metall> cavitation failure
Kavitationszahl f <Phys, Strömphys> cavitation number
kavitierend adj <Mechan> cavitating
KAW *(Kanaladresswort)* <Comp & DV> CAW *(channel address word)*
KB *(Kilobyte)* <Comp & DV, Telekom> KB *(kilobyte)*
K-Band n <Funktech> K band *(10,9-36 GHz)*
kcal *(Kilokalorie)* <Lebensmittel> kcal *(kilocalorie)*
keff *(effektive Neutronen-Multiplikationskonstante)* <Kerntech> keff
Kegel m 1. <Druck> body; 2. <Fernseh> cone; 3. <Fertig> taper; cotter *(Kunststoffinstallationen)*; 4. <Geom> cone; 5. <Ker & Glas> taper; 6. <Kohlen, Maschinen> cone
Kegel... <Geom> conical
Kegelanschlagschräge f <Ker & Glas> tapered stop bevel
Kegelanschnitt m <Ker & Glas> taper bevel
Kegelaufschläger m <Papier> banger
Kegelband n <Bau> T-hinge
Kegelbohren n <Fertig> taper boring
Kegelbrecher m <Kohlen> cone crusher
Kegelbremse f <Maschinen> cone brake
Kegelfeder f 1. <Fertig> volute spring; 2. <Maschinen> conical spring
kegelförmiger Abschnitt m <Elektrotech> tapered section
kegelförmiger Bestandteil m <Elektrotech> tapered section
kegelförmiger Kompressionsring m <Kfztech> tapered compression ring
kegelförmiger Trichter m <Akustik> conical horn
kegelförmiges Achsende n <Kfztech> tapered axle end
kegelförmiges Sieb n <Lebensmittel> conical sieve
Kegelfräsen n <Fertig> taper milling
Kegelgewinde n <Maschinen> taper thread
Kegelgriff m 1. <Fertig> ball handle; 2. <Maschinen> ball handle, clamping lever
Kegelhülse f <Maschinen> taper sleeve
kegelig verjüngt adj <Maschinen> tapered
Kegeligdrehen n <Maschinen> taper turning
Kegeligdrehvorrichtung f <Maschinen> taper-turning attachment
kegeliges Rohrgewinde n <Maschinen> taper pipe thread
Kegeligsenken n <Maschinen> countersinking
Kegelkopf m <Maschinen> cone head, pan head
Kegelkopfniet m 1. <Bau> cone head rivet; 2. <Maschinen> pan head rivet
Kegelkopfschraube f <Maschinen> pan head screw
Kegelkuppe f <Maschinen> blunt start, flat point
Kegelkupplung f <Kfztech, Maschinen> cone clutch
Kegellager n <Maschinen> cone bearing
Kegelleuchte f <Elektriz> cone light
Kegelnabe f <Kfztech> tapered hub *(Rad)*
Kegelniet m <Fertig> cone head rivet
Kegelpolster n <Ker & Glas> tapered pad
Kegelprober m <Kohlen> cone sampler
Kegelprojektion f <Geom> conic projection
Kegelrad n 1. <Eisenbahn, Fertig, Kfztech, Lufttrans> bevel gear; 2. <Maschinen> bevel gear, bevel wheel, conical gear, *(AE)* miter wheel, *(BE)* mitre wheel; 3. <Mechan> bevel gear

Kegelrad n **für den Radantrieb** <Kfztech> side gear *(beim Ausgleichsgetriebe)*
Kegelrad n **mit Oktoidverzahnung** <Maschinen> octoid bevel gear
Kegelrad n **und Tellerrad** n <Kfztech> crown and pinion *(Triebstrang, Differenzial)*
Kegelradantrieb m <Eisenbahn, Fertig, Maschinen> bevel gear drive
Kegelradformfräser m <Fertig> bevel-gear-formed cutter
Kegelradfräsmaschine f <Fertig> bevel gear cutting machine
Kegelradgetriebe n 1. <Fertig> obtuse-angle bevel gear; 2. <Kfztech> bevel gear set *(Hinterachse)*; 3. <Maschinen> bevel gears
Kegelradgetriebegehäuse n <Lufttrans> bevel gear housing *(Hubschrauber)*
Kegelradhobelmaschine f <Fertig> bevel gear planing machine
Kegelradpaar n <Fertig> *(AE)* miter gearing, *(BE)* mitre gearing
Kegelradumlaufgetriebe n <Fertig> bevel epicyclic train
Kegelradverzahnung f <Kfztech, Maschinen> bevel gearing
Kegelradwälzfräsmaschine f <Fertig> bevel gear generating machine
Kegelreibahle f <Maschinen> taper reamer
Kegelreibungskupplung f <Fertig> cone friction clutch
Kegelritzel n <Fertig> bevel gear pinion
Kegelrollenlager n 1. <Kfztech> taper roller bearing, tapered roller bearing; 2. <Maschinen> taper rolling bearing, tapered roller bearing, *(AE)* timken bearing
Kegelschaft m <Maschinen> taper shank
Kegelschale f <Fertig> conical shell
Kegelscheibenantrieb m <Maschinen> cone drive, cone gear
Kegelschleifen n <Fertig> taper grinding
Kegelschliff-Verbindungsstück n <Labor> cone-and-socket joint
Kegelschnecke f <Maschinen> conical worm, tapered worm
Kegelschnitt m <Geom> conic section
Kegelsenker m <Maschinen> cone countersink, countersink, rose countersink, rose-head countersink bit
Kegelseparator m <Kohlen> cone separator
Kegelsieb n <Kohlen> cone classifier
Kegelsitz m <Maschinen> conical seat
Kegelspitze f <Maschinen> truncated cone point
Kegelstift m 1. <Maschinen> taper dowel, taper pin, tapered pin; 2. <Mechan> taper pin
Kegelstumpf m <Fertig> truncated cone
Kegeltoleranz f <Maschinen> cone tolerance
Kegelventil n <Fertig, Hydraul, Maschinen> cone valve
Kegelverhältnis n <Mechan> taper
Kegelverjüngung f <Fertig> amount of taper
Kegelwinkel m 1. <Geom> cone angle; 2. <Maschinen> angle of taper
Kegelzahnrad n <Eisenbahn, Fertig, Mechan> bevel gear
Kehlbalken m <Bau> collar beam, span piece
Kehlbalkenbinder m <Bau> collar beam truss
Kehlbalkendach n <Bau> collar roof
Kehlbalkenstütze f <Bau> side post
Kehle f 1. <Akustik> throat; 2. <Bau> groove, *(BE)* plough, *(AE)* plow, valley; 3. <Fertig> throat
Kehleisen n <Fertig> vee sett; necking tool *(Schmieden)*
kehlen v <Fertig> recess
Kehlhammer m <Fertig> top fuller
Kehlhobel m <Bau> *(BE)* plough, *(AE)* plow
Kehlkopf m <Akustik> throat

Kehlkopfmikrofon

Kehlkopfmikrofon n <Akustik, Aufnahme> throat microphone
Kehlmaschine f <Fertig> (AE) molding machine, (BE) moulding machine
Kehlnaht f 1. <Bau> fillet joint (Holzbau); 2. <Fertig> fillet (Schweißen); 3. <Maschinen, Mechan> fillet weld
kehlschneiden v <Wassertrans> (AE) mold, (BE) mould (Schiffbau)
Kehrbild n <Foto> inverted image
kehren v <Bau> sweep (Kamin)
Kehrfahrzeug n 1. <Abfall> road-sweeping lorry, street cleaner, (BE) street-cleaning lorry, (AE) street-cleaning truck; 2. <Kfztech> street cleaner, (BE) street-cleaning lorry, (AE) street-cleaning truck
Kehrlage f <Telekom> reverse frequency position
Kehrmaschine f <Abfall> sweeper
Kehrmatrix f <Math> inverse matrix
Kehrwalze f <Abfall> cylinder broom (Kehrfahrzeug)
Kehrwert m 1. <Elektronik> complement; 2. <Geom> reciprocal; 3. <Math> inverse proportion, inverse ratio, multiplicative inverse, reciprocal, reciprocal value
Keil m 1. <Bau> key, wedge; 2. <Elektrotech> key; 3. <Fertig> chock, key, male spline, wedge; calm (Einsatzmesser); 4. <Maschinen> cotter, machine key, quoin, taper key, wedge; 5. <Mechan> cotter pin, key, spline, wedge; 6. <Verpack> arrowhead • **mit einem Keil spalten** <Bau> split with wedges
Keilbruch m <Metall> wedge-type fracture
Keildensitometer n <Foto> wedge densitometer
Keilfänger m <Erdöl> fishing socket (Fangwerkzeug)
Keilflankenschleifen n <Fertig> grinding of splines
keilförmig adj <Fertig> sphenoid
keilförmiger Brennraum m <Kfztech> wedge-type combustion chamber
Keillängsnut f <Fertig> keyway
Keillehre f 1. <Ker & Glas> (AE) V-gage, (BE) V-gauge; 2. <Metrol> (AE) spline gage, (BE) spline gauge
Keilloch n 1. <Maschinen> cotter slot; 2. <Mechan> key slot
Keillochhammer m <Kohlen> stone-splitting hammer
Keilmessebene f <Fertig> wedge measurement plane
Keilnabe f <Fertig> splined hub
Keilnabenprofil n <Maschinen> internal splines
Keilnut f 1. <Bau> key; 2. <Fertig> key slot, spine; 3. <Kerntech> key bed, keygroove, keyway; 4. <Maschinen> V-groove, key seating, keyway, spline; 5. <Mechan> keyway • **Keilnuten ziehen** <Fertig> keygroove
Keilnutenfräsen n <Maschinen> keyway milling, keywaying
Keilnutenfräser m <Fertig> keyway cutter
Keilnutenfräserspannfutter n <Fertig> keyway cutter chuck
Keilnutenfräsmaschine f <Maschinen> keyway-milling machine
Keilnutenräumnadel f <Maschinen> keyway broach
Keilnutenstoßen n <Maschinen> keyway slotting
Keilnutenstoßmaschine f <Maschinen> keyseater
Keilnutenziehen n <Fertig> keygrooving
Keilnutenziehwerkzeug n <Fertig> keygrooving tool
Keilnutfräser m <Maschinen> splining tool
Keilnutmaschine f <Maschinen> keywaying machine
Keilprofil n <Fertig> spline • **mit Keilprofil** <Fertig> splined
keilprofilfräsen v <Fertig> spline
Keilprofilräumnadel f <Fertig> spline broach
Keilriemen m 1. <Fertig, Heiz & Kälte> V-belt, vee belt; 2. <Kfztech> V-belt (Kühlsystem); 3. <Kunststoff> V-belt; 4. <Maschinen> vee belt; 5. <Mechan> V-belt
Keilriemenantrieb m <Maschinen> V-belt drive
Keilriemenscheibe f <Maschinen> V-belt pulley
Keilriemenspannung f <Maschinen> V-belt tension
Keilriementrieb m <Maschinen> V-belt drive
Keilrille f 1. <Maschinen> V-groove; 2. <Mechan> keyway
Keilriss m <Metall> wedge crack
Keilschieber m 1. <Maschinen> wedge-type valve; 2. <Nichtfoss Energ, Wasserversorg> sluice valve
Keilschloss n <Maschinen> gib and cotter, gib and key
Keilschweißung f <Fertig> cleft weld
Keilstein m 1. <Bau> arch stone, voussoir; 2. <Fertig> skewback
Keilstift m <Mechan> taper pin
Keilstück n <Fertig> taper parallel
Keilverzahnung f <Fertig> splining
Keilverzahnungsfräsmaschine f <Fertig> spline milling machine
Keilwelle f <Fertig, Maschinen> spline shaft, splined shaft
Keilwellenprofil n 1. <Fertig> spine; 2. <Maschinen> external splines, spline profile
Keilwellenprofil n **mit Evolventenflanken** <Maschinen> involute spline
Keilwellenverbindung f <Maschinen> spline
Keilwinkel m 1. <Fertig> wedge angle; (AE) lip angle (Spiralbohrer); 2. <Maschinen> wedge angle
Keilzapfenverbindung f <Bau> wedged mortice and tenon joint
Keilziegel m <Bau> feather-edged brick, (AE) gage brick, (BE) gauge brick
Keilzugprobe f <Metall> wedge draw test
Keim m 1. <Fertig> nucleus (Kristall); 2. <Ker & Glas> nucleus
keimfrei adj <Lebensmittel> aseptic
Keimfreitechnologie f <Sicherheit> aseptic engineering
K-Einfang m <Phys> K-capture (Kernphysik)
Kelle f <Bau> trowel • **mit Kelle abreiben** <Bau> trowel off
Kellenrückstand m <Ker & Glas> scull
Kellerbasis f <Comp & DV> stack base
Kellergeschosswand f <Bau> basement wall
Keller-Kopiereinrichtung f <Fertig> Keller attachment
Kellerspeicher m <Comp & DV> push-down stack
Kelly f <Erdöl> kelly
Kelp n <Lebensmittel> kelp
Keltapapier n <Chemie, Foto> gelatino-chloride paper
Kelvin n (K) <Elektriz, Metrol, Phys, Thermod> kelvin (K)
Kelvinbrücke f <Elektriz> Kelvin bridge
Kelvinbrückenschaltung f <Phys> Kelvin bridge
Kelvindoppelbrücke f <Elektriz> double Kelvin bridge
Kelvineffekt m <Elektriz> Kelvin effect
Kelvinformulierung f <Phys> Kelvin statement (zweiter Hauptsatz der Thermodynamik)
Kelvin'sche Skale f <Bau> Kelvin scale
Kelvin'sche Stromwaage f <Elektriz> Kelvin balance
Kelvintemperatur f 1. <Lebensmittel> absolute temperature; 2. <Phys, Raumfahrt, Thermod> Kelvin temperature
Kelvintemperaturskale f <Raumfahrt> Kelvin scale
K-Emitter m <Kerntech> K-emitter
Kennabschnitt m <Telekom> unit interval
Kennbuchstabe m <Comp & DV> code letter
Kenndämpfungswiderstand m <Phys> iterative impedance
Kenndaten npl <Telekom> characteristics (Datenblatt)
Kennelly-Heaviside-Schicht f <Phys> Kennelly-Heaviside layer
Kennfeuer n <Lufttrans> identification beacon
Kenngleichung f <Raumfahrt> characteristic equation
Kennkurve f 1. <Funktech> characteristic; 2. <Raumfahrt> characteristic curve
Kennlicht n <Lufttrans> identification light, navigation light

Kennlinie f 1. <Akustik> characteristic curve; 2. <Elektriz, Elektronik, Gerät, Phys> characteristic, characteristic curve
Kennsatz m 1. <Comp & DV> header, label, label record; 2. <Telekom> label
Kennsatzanfang m <Comp & DV> SOH, start of header
Kennung f 1. <Comp & DV> identifier, tag; 2. <Comp & DV> identification; 3. <Kontroll> identification; 4. <Telekom> answer-back, code, identification, identification code, identification signal, identifier
Kennung f der Küstenfunkstelle <Telekom> coastal station identity
Kennung f der Schiffsfunkstelle <Telekom> ship station identity
Kennungsleuchtfeuer n <Funktech> land mark beacon
Kennungsschalter m <Funkort> challenge switch *(Radar)*
Kennwert m <Geom, Telekom> characteristic, parameter, value
Kennwiderstand m <Elektrotech> image impedance
Kennwort n 1. <Comp & DV> identifier word, keyword, password; 2. <Telekom> password, keyword
Kennwortdateischutz m <Comp & DV> password protection
Kennwortschutz m <Comp & DV> password protection, password security
Kennzahl f <Telekom> code number
Kennzahl-Rufnummer-Plan m <Telekom> open-end numbering scheme, open numbering plan
Kennzahlweg m *(KZW)* <Telekom> final route *(ITG)*
Kennzeichen n 1. <Comp & DV> identifier, snowflake topology, token; 2. <Ker & Glas> badge; 3. <Kfztech> *(AE)* license plate, *(BE)* numberplate; 4. <Patent, Qual> identification; 5. <Telekom> identifier, mark, signal, token
Kennzeichengabe f <Telekom> *(AE)* signaling, *(BE)* signalling *(Telefon)*
Kennzeichenregister n <Comp & DV> flag register
kennzeichnen v 1. <Comp & DV> identify, tag; 2. <Patent, Qual, Telekom> identify; 3. <Textil> label
Kennzeichnen n 1. <Ker & Glas> badging; 2. <Textil> marking
Kennzeichnung f 1. <Comp & DV> certification; 2. <Patent, Qual> identifying marking; 3. <Telekom> identification, marking, tag, tagging; 4. <Werkprüf> marking
Kennziffer f <Telekom> prefix
kentern v <Wassertrans> capsize *(Schiff)*
Kentern bringen v/zum <Wassertrans> capsize *(Schiff)*
Kepler'sche Gesetze npl <Phys> Kepler's laws
Kepler'scher Umlauf m <Raumfahrt> Keplerian orbit
Kepler'sches Flächengesetz n <Raumfahrt> Kepler's law of areas
Kerabitumen n <Erdöl> kerogen *(Geologie)*
Kerametall n <Anstrich, Ker & Glas> ceramal, cermet
Kerametall-Beschichtung f <Anstrich> cermet coating
Keramik f <Bau, Ker & Glas, Kerntech, Mechan> ceramic
Keramikbrennofen m <Ker & Glas> ceramic kiln
Keramikfaser f <Heiz & Kälte> *(AE)* ceramic fiber, *(BE)* ceramic fibre
Keramikfliese f <Bau> glazed tile
Keramikglasur f <Ker & Glas> ceramic glaze
Keramikindustrie f <Ker & Glas> ceramic industry
Keramikkondensator m <Elektriz, Elektrotech, Funktech, Telekom> ceramic capacitor
Keramikkondensator m **mit Glasbeschichtung** <Elektrotech> glass-coated ceramic capacitor
Keramikkunst f <Ker & Glas> ceramic art
Keramikmaschine f <Ker & Glas> ceramic machine
Keramikpflasterstein m <Ker & Glas> *(BE)* ceramic pavement slab, *(AE)* ceramic sidewalk slab
Keramiktrimmer m <Elektrotech> adjustable ceramic capacitor
Keramikwandfliese f <Ker & Glas> ceramic wall tile
keramisch adj 1. <Anstrich, Elektriz, Elektrotech> ceramic; 2. <Fertig> vitrified *(Bindung)*
keramischer Brennstoff m <Kerntech> ceramic fuel
keramischer Isolator m <Elektriz> ceramic insulator
keramischer Kondensator m <Elektrotech, Phys> ceramic capacitor
keramischer Plättchenkondensator m <Elektrotech> ceramic chip capacitor
keramischer Rohrkondensator m <Elektrotech> tubular ceramic capacitor
keramisches Dual-in-line-Gehäuse n <Telekom> ceramic dual-in-line package *(CERDIP)*
keramisches Isoliermaterial n <Elektrotech> ceramic insulating material
Kerb m 1. <Fertig> rag; 2. <Kohlen> cut
Kerbbiegeversuch m <Maschinen> notch bending test
Kerbe f 1. <Bau> notch; 2. <Fertig> groove, jag, nick, notch, undercut; 3. <Kerntech> groove; 4. <Maschinen> dent, nick, notch; 5. <Mechan, Metall> notch
kerbempfindlich adj <Fertig> notch-sensitive
Kerbempfindlichkeit f <Maschinen> notch sensitivity
Kerbempfindlichkeitszahl f <Fertig> fatigue notch sensitivity
Kerbfilter n 1. <Elektronik> notch filter *(Radio)*; 2. <Funktech> notch filter *(schmalbandige Bandsperre)*; 3. <Telekom> notch filter
Kerbmessung f <Hydraul> *(AE)* notch gaging, *(BE)* notch gauging
Kerbnagel m <Maschinen> grooved pin, splined pin
Kerbschlagprobe f 1. <Kerntech> notch impact test; 2. <Mechan> notched bar impact test
Kerbschlagversuch m 1. <Anstrich> impact test; 2. <Fertig> notched bar impact test; 3. <Metrol> impact test
Kerbschlagzähigkeitswert m <Metall> impact test
Kerbstab m <Fertig> notched bar
Kerbstift m *(KS)* <Fertig, Maschinen> grooved pin, splined pin
Kerbung f 1. <Fertig> nicking; 2. <Ker & Glas> scoring
Kerbverbindung f <Bau> notch joint
kerbverzahnt adj 1. <Fertig> splined; 2. <Maschinen> serrate
Kerbverzahnung f 1. <Fertig> serration, spline; 2. <Maschinen> serration
Kerbverzahnungswälzfräser m <Fertig> serration hob
Kerbwinkel m <Metall> notch angle
Kerbwirkung f <Maschinen> notch effect
Kerbzähigkeit f 1. <Fertig> impact value; 2. <Kerntech> notch toughness; 3. <Qual> impact value
Kerma n *(K)* <Kerntech> kerma *(K)*
Kermarate f <Phys> kerma rate
Kern m 1. <Bau, Comp & DV> core; 2. <Elektriz, Elektrotech> core; slug *(Hohlleiter, Wicklung)*; 3. <Erdöl> core *(Bohrtechnik)*; 4. <Fertig> pit; limb *(Magnet)*; web *(Spiralbohrer)*; 5. <Funktech, Kerntech, Mechan> core; 6. <Optik> core *(Lichtleiter)*; 7. <Papier> core; 8. <Phys> nucleus; 9. <Raumfahrt> core; 10. <Teilphys> nucleus; 11. <Telekom> core *(elektrisch)*; 12. <Textil> core • **mit hartem Kern** <Fertig> *(AE)* hard-centered, *(BE)* hard-centred
Kernabschirmung f <Elektriz> screened core
Kernanalyse f <Erdöl> core analysis
Kernansatz m <Fertig> half dog
Kernanwendung f <Comp & DV> core application
Kernanwendungssystem n <Comp & DV> core business system
Kernbaustein m <Phys> nucleon
Kernbeton m <Bau> mass concrete
Kernbildung f <Metall> nucleation

Kernbildungsgeschwindigkeit f <Metall> nucleation rate
Kernbindungsenergie f (B) <Strahlphys, Teilphys> binding energy (B)
Kernblasmaschine f <Maschinen> core-blowing machine
Kernblech n <Elektrotech> core plate, stamping
Kernbock m <Wasserversorg> chaplet
kernbohren v <Fertig> trepan
Kernbohren n 1. <Fertig> trepanning; 2. <Maschinen> core drilling, (BE) trepanning
Kernbohrer m <Erdöl> core bit, core drill (Bohrtechnik)
Kernbohrmaschine f <Maschinen> core drill
Kernbohrmeißel m <Erdöl> core bit, annular bit (Bohrtechnik)
Kernbohrung f <Bau> core drilling
Kernbohrwerkzeug n <Erdöl> coring tool (Bohrtechnik)
Kernbrennstoff m <Kerntech> fuel
Kernbrennstofftransportbehälter m <Abfall, Sicherheit> nuclear fuel container
Kernbrennstoffwiederaufbereitungsanlage f <Abfall, Sicherheit> nuclear fuel reprocessing plant
Kernbrett n <Fertig> core board (Gießen)
Kerndeformation f <Kerntech> nuclear deformation
Kerndrehbank f <Fertig> core bar
Kerndurchmesser m 1. <Fertig> core diameter (Getriebelehre, Gewinde); 2. <Maschinen> core diameter, inside diameter; 3. <Optik> core diameter (Lichtleiter)
Kerndurchmessertoleranz f <Telekom> core diameter tolerance
Kerneisen n <Fertig> core iron (Gießen)
Kernel m <Comp & DV> kernel
Kernelement n <Elektrotech> nuclear cell
Kernenergie f <Elektriz, Kerntech, Teilphys> nuclear energy
Kernenergieanlage f <Elektrotech, Kerntech> nuclear power plant
Kernerbohrung f <Erdöl> coring
Kernerregung f <Akustik> main excitation
Kernfänger m <Kerntech> core catcher (Öltechnologie)
Kernflutsystem n <Kerntech> core-flooding train
Kernformen n <Fertig> coring (Gießen)
Kernformung f <Fertig> coremaking
Kernforschung f <Kerntech, Strahlphys> atomic research, nuclear research
Kernfusion f <Kerntech, Teilphys> nuclear fusion
Kerngewinnung f <Erdöl> coring (Bohrtechnik)
Kernguss m <Fertig> core casting
kernig adj <Textil> crisp
kerniger Griff m <Textil> crisp handle
Kernisomer n <Strahlphys> nuclear isomer
Kernisomerie f <Strahlphys> nuclear isomerism
Kernkasten m <Fertig> core box (Gießen)
Kernkopfverankerungseinheit f <Kerntech> core head plug unit
Kernkraft f <Kerntech> nuclear power
Kernkraftanteil m <Elektriz, Kerntech> nuclear tranche
Kernkraftwerk n <Elektriz, Elektrotech, Kerntech> nuclear power station
Kernkraftwerk n **mit Spitzenlast** <Kerntech> peak load nuclear power plant
Kernkraftwerk n **mit zwei Kühlkreisen** <Kerntech> two-circuit nuclear power plant
Kernkraftwerkssicherheit f <Sicherheit> nuclear plant safety
Kernladung f <Teilphys> nuclear charge
Kernladungszahl f 1. <Kerntech, Phys, Strahlphys> atomic number; 2. <Teilphys> proton number
Kernlautheit f <Akustik> main loudness
Kernleitwert m <Telekom> transfer impedance
Kernlochdurchmesser m <Maschinen> core hole
kernloser Anker m <Elektriz> coreless armature

kernloser Induktionsofen m <Heiz & Kälte> coreless induction furnace
Kernmagneton n <Teilphys> nuclear magneton
Kern-Mantel-Exzentrizität f <Telekom> core-cladding concentricity error
Kernmarke f <Fertig> core print, print (Gießen)
Kernmasse f (MN) <Kerntech> nuclear mass (MN)
Kernmitte f <Optik> (AE) core center, (BE) core centre (Lichtleiter)
Kernmittelpunkt m <Telekom> (AE) core center, (BE) core centre
Kernmodell n <Teilphys> nuclear model
Kernmodell n **mit variablem Kernträgheitsmoment** <Kerntech> variable moment of inertia model
Kernnagel m 1. <Bau> core nail; 2. <Fertig> chaplet (Getriebelehre); sprig (Gießen)
Kernobst n <Lebensmittel> pomaceous fruit
Kernpaket n <Elektrotech> core package (Transformator)
Kernphysik f 1. <Kerntech> nuclear physics; 2. <Phys> atomic physics
Kernphysik f **im mittleren Energiebereich** <Kerntech> medium-energy nuclear physics
Kernpotenzial n <Strahlphys, Teilphys> nuclear potential
Kernprobe f <Bau> core sample
Kernprüfung f <Elektriz> core test
Kernquadrupolmoment n <Phys> nuclear quadrupole moment
Kernquerschnitt m <Optik> core area
Kernradius m (r) <Kerntech> nuclear radius (r)
Kernreaktion f <Teilphys> nuclear reaction
Kernreaktionsmechanismus m <Teilphys> nuclear reaction channel
Kernreaktor m 1. <Elektriz, Elektrotech> nuclear reactor; 2. <Kerntech> atomic pile, nuclear reactor, pile; 3. <Phys> atomic pile, reactor pressure vessel; 4. <Strahlphys> atomic pile
Kernrohr n <Erdöl> core barrel
Kernsand m <Fertig> core sand (Gießen)
Kernschablone f <Fertig> core board (Gießen)
Kernschankelblech n <Metrol> core-limb lamination
Kernschatten m <Phys> úmbra
Kernspaltung f <Kerntech, Teilphys> fission, nuclear fission
Kernspaltungswirkungsquerschnitt m <Phys> fission cross section
Kernspeicher m <Comp & DV> (AE) core storage, (BE) core store
Kernspektrum n <Strahlphys> nuclear radiation spectrum
Kernspin m <Strahlphys> nuclear spin
Kernspinresonanzlog n <Kerntech> NMR log
Kernspur f <Kerntech> nuclear track
Kernstahlgerippe n <Fertig> core grid (Gießen)
Kernstrahlung f <Kerntech, Strahlphys> nuclear radiation
Kernstück n <Fertig> (AE) center, (BE) centre (Riemen)
Kernsymmetrie-Energie f <Strahlphys> nuclear symmetry energy
Kerntechnik f <Kerntech, Phys> nucleonics
kerntechnische Prüfaufsicht f 1. <Kerntech> ANIS; 2. <Qual> authorized nuclear inspector supervisor
kerntechnischer Prüfsachverständiger m <Qual> ANI, authorized nuclear inspector
Kernteilpaket n <Elektrotech> core packet (Transformator)
Kerntoleranzbereich m <Telekom> core tolerance field
Kerntoleranzfeld n <Optik> core tolerance field (Lichtleiter)
Kerntransformator m <Elektriz, Elektrotech> core-form transformer, core transformer, core-type transformer
Kerntrockenkammer f <Fertig> core stove (Gießen)
Kern- und Hülsenschliff m <Ker & Glas> graded seal

Kernverformung f <Kerntech> nuclear deformation
Kernverlust m <Elektriz> iron loss
Kernverluste mpl <Elektriz, Phys> core losses
Kernverschmelzung f <Kerntech> karyogamy
Kernwiderstand m <Telekom> mutual impedance
Kernzerschmiedung f <Fertig> hammer pipe
Kernzone f <Telekom> core area
Kernzustandsgleichung f <Strahlphys> nuclear equation of state
Kerogen n <Erdöl> kerogen
Kerosin n 1. <Erdöl> (AE) kerosene (RP-1; Destillationsprodukt); 2. <Raumfahrt> kerosene (RP-1); 3. <Thermod> (AE) kerosene, (BE) paraffin; 4. <Trans> (AE) kerosene, (BE) paraffin (RP-1)
Kerr-Effekt m <Phys> Kerr electro-optical effect (elektrooptisch); Kerr magneto-optical effect (magnetooptisch)
Kerr-Zelle f <Phys> Kerr cell
Kerze f <Kfztech> plug, spark plug (Zündung)
Kerzenfuß m <Kfztech> plug socket, spark plug socket (Zündung)
Kerzenlampe f <Elektrotech> candle lamp
Kerzenlampenhalter m <Elektrotech> candle lampholder
Kerzenschaftfassung f <Elektrotech> candle lampholder
Kessel m 1. <Eisenbahn, Kohlen, Lebensmittel, Maschinen, Mechan> boiler; 2. <Metrol> bowl (Zentrifuge); 3. <Thermod, Wassertrans> boiler (Schiffsantrieb)
Kessel m **mit Klöpperböden** <Hydraul> (BE) disc-ended boiler, (AE) disk-ended boiler
Kesselalarm m <Hydraul> boiler alarm
Kesselalarmschwimmer m <Hydraul> boiler emergency float (Sicherheitseinrichtung)
Kesselanlage f <Heiz & Kälte> boiler plant
Kesselarmaturen fpl <Hydraul> boiler fittings
Kesselausrüstung f <Hydraul> boiler fittings
Kesselbau m <Heiz & Kälte> boiler engineering
Kesselbauer m <Hydraul> boilermaker
Kesselbeschickung f <Hydraul> boiler feeding
Kesselblech n <Hydraul> boiler plate
Kesseldruck m <Maschinen> boiler pressure
Kesselersatzspeisepumpe f <Hydraul> auxiliary boiler feeder (Kesselhilfsspeisepumpe)
Kesselexplosion f <Hydraul, Sicherheit> boiler explosion
Kesselfeuerraum m <Heiz & Kälte, Hydraul> boiler furnace
Kesselfeuerung f <Hydraul> boiler furnace
Kesselflammrohr n <Hydraul> boiler flue
Kesselfront f <Hydraul> boiler front
Kesselfuchs m <Hydraul> boiler flue
Kesselführung f <Heiz & Kälte> boiler operation
Kesselhammer m <Hydraul> boiler-scaling hammer
Kesselhaus n <Heiz & Kälte, Hydraul> boiler house
Kesselherstellung f <Heiz & Kälte, Hydraul> boilermaking
Kesselhilfsspeisepumpe f <Hydraul> auxiliary boiler feeder (Kesselersatzspeisepumpe)
Kesselkohle f <Kohlen> boiler coal, steam coal
Kessellagerung f <Fertig> boiler bedding
Kesselleistung f <Heiz & Kälte> boiler capacity, boiler output
Kesselleistungsschalter m <Elektrotech> dead tank circuit-breaker
Kesselmanometer n <Gerät> (AE) boiler gage, (BE) boiler gauge
Kesselmantel m 1. <Fertig> boiler barrel; 2. <Hydraul> boiler jacket, boiler jacketing, boiler shell
Kesselnaht f <Fertig> boiler weld
Kesselniet n <Hydraul> boiler rivet
Kesselofen m <Fertig> rotary furnace
Kesselprüfblech n <Hydraul> boiler test plate
Kesselrauchgaskanal m <Hydraul> boiler flue

Kesselraum m <Hydraul> boiler room
Kesselrohr n <Maschinen> boiler tube
Kesselrohrwalze f <Hydraul> boiler tube expander
Kesselrost m <Heiz & Kälte> boiler grate
Kesselschalter m <Elektrotech> dead tank circuit-breaker
Kesselschlacke f 1. <Bau> clinker; 2. <Mechan> boiler slag
Kesselschmied m <Hydraul> boiler smith
Kesselschweißen n <Fertig> boiler welding
Kesselschweißer m <Fertig> boiler welder
Kesselschwimmer m <Hydraul> boiler float (Niveauregelung)
Kesselspeisepumpe f <Heiz & Kälte> boiler feed pump, boiler feeder
Kesselspeisewasser n <Erdöl> boiler feed water (Dampferzeugung)
Kesselspeisung f <Hydraul> boiler feeding
Kesselstein m 1. <Eisenbahn> boiler scale; 2. <Fertig> boiler scale, fur, scale; 3. <Heiz & Kälte> scale; 4. <Hydraul> boiler scale • **Kesselstein ansetzen** <Fertig> fur • **Kesselstein entfernen** <Fertig> descale
Kesselsteinablagerung f <Lebensmittel, Wasserversorg> boiler scale formation
Kesselsteinbildung f <Wasserversorg> boiler scale formation
Kesselsteinhammer m <Maschinen> boilermaker's hammer, scaling hammer
Kesselsteinlösemittel n <Heiz & Kälte> boiler-cleaning compound
Kesselsteinmittel n <Chemie> anti-incrustant
kesselsteinverhütend adj <Fertig> anti-ager scale
Kesselsteinverhütungsmittel n <Heiz & Kälte> scale inhibitor
Kesseltemperaturmessgerät n <Gerät> boiler temperature meter
Kesselüberprüfung f <Sicherheit> boiler inspection
Kesselummantelung f 1. <Heiz & Kälte> jacket; 2. <Hydraul> boiler jacketing
Kesselverkleidung f 1. <Heiz & Kälte> jacket; 2. <Hydraul> boiler jacket, boiler lagging
Kesselwagen m <Eisenbahn, Kfztech> (AE) tank car, (BE) tank wagon
Kesselwasserbehandlung f <Heiz & Kälte> boiler water purification, boiler water treatment
Kesselwasserreinigung f <Heiz & Kälte> boiler water purification, boiler water treatment
Kesselwerk n <Hydraul> boiler works
Kesselwirkungsgrad m <Heiz & Kälte> boiler efficiency
Kesselzug m <Hydraul> boiler flue
Ketazin n <Chemie> ketazine
Keten n <Chemie> ketene
Ketimin n <Chemie> ketimine
Ketobernstein... <Chemie> oxalacetic
Ketocarbonsäure f <Chemie> keto acid, oxo acid
Ketoform f <Chemie> keto form
Ketol n <Chemie> ketol
Keton n <Chemie, Kunststoff, Lebensmittel> ketone
Keton... <Chemie> ketonic
ketonartig adj <Chemie> ketonic
ketonisch adj <Chemie> ketonic
Ketosäure f <Chemie> keto acid, oxo acid
Ketsch f <Wassertrans> ketch (Segeln)
Kettatlas m <Textil> satin
Kettbaum m <Textil> beam, warp beam, weaver's beam, yarn roller • **auf den Kettbaum gewickelt** <Textil> wound onto the beam
Kettbaumfärbeapparat m <Textil> beam-dyeing machine
Kettbaumfärben n <Textil> beam dyeing
Kettbaumfärbung f <Textil> beam dyeing

Kette

Kette f 1. <Bau, Comp & DV> chain; 2. <Fertig> warp *(Drahtweben)*; line *(Fertigung)*; 3. <Kfztech, Konstzeich, Maschinen, Mechan> chain; 4. <Papier, Textil> chain, warp; 5. <Wassertrans> chain
Ketten n <Comp & DV> chaining
Kettenanschärer m <Textil> beamer
Kettenantrieb m 1. <Kfztech> chain drive *(Kraftübertragung, Motorrad)*; 2. <Maschinen> chain drive, chain transmission; 3. <Mechan> chain drive
Kettenantriebsritzel n <Kfztech> drive sprocket
Kettenbecherwerk n <Maschinen> *(AE)* chain elevator, *(BE)* chain lift
Kettenbefehl m <Comp & DV> chain
Kettenbemaßung f <Konstzeich> chain dimensioning
Kettenblock m <Mechan> chain block
Kettenbruchentwicklung f <Math> continued fraction
Kettenbrücke f <Bau> chain bridge
Kettendrucker m <Comp & DV> *(AE)* chain printer, *(BE)* train printer
Kettenfähre f <Wassertrans> chain ferry
Kettenfahrleitung f <Eisenbahn> catenary
Kettenfläche f <Geom> catenoid
Kettenflaschenzug m 1. <Bau> chain hoist; 2. <Maschinen> chain block, chain hoist, chain pulley block
Kettenförderer m <Maschinen> chain conveyor
Kettenförderung f <Kohlen> chain haulage
Kettenführung f 1. <Kfztech> chain guide *(Motor)*; 2. <Maschinen> chain guide
Kettengehäuse n <Kfztech> chain case *(Kraftübertragung, Motorrad)*
Kettengetriebe n 1. <Maschinen> chain gear, chain gearing; 2. <Mechan> sprocket wheel
kettengetrieben adj <Maschinen> chain-driven
Kettenglied n 1. <Kfztech> chain link *(Kraftübertragung, Motorrad)*; 2. <Maschinen> chain link; 3. <Mechan> link, shackle
Kettengreifer m <Wassertrans> chain grab
Kettenhebel m <Bau, Maschinen> chain lever
Kettenimpedanz f <Akustik, Elektrotech> iterative impedance
Kettenkasten m 1. <Sicherheit> chain case; 2. <Wassertrans> chain locker; cable locker *(Festmachen)*
Kettenkneifer m <Wassertrans> chain compressor *(Ankern)*
Kettenkranz m <Maschinen> sprocket wheel
Kettenkupplung f <Maschinen> chain coupling
Kettenlänge f <Kunststoff> chain length
Kettenlaufwerk n <Mechan> crawler
Kettenleiter m <Elektrotech> interactive network, lattice network
Kettenleiternetzwerk n <Elektrotech, Phys> ladder network
Kettenlinie f <Geom, Maschinen, Mechan, Phys, Wassertrans> catenary
Kettenmaßsystem n <Fertig, Gerät> incremental system
Kettenmolekül n <Chemie> chain molecule, macromolecule
Kettennuss f <Maschinen> chain sprocket, sprocket wheel
Kettenpumpe f 1. <Maschinen> chain pump; 2. <Wasserversorg> chain pump, paternoster pump
Kettenrad n 1. <Fertig> chain wheel, sprocket; 2. <Kfztech> sprocket *(Motorradgetriebe)*; 3. <Maschinen> chain sheave, chain wheel, sprocket; 4. <Mechan> sprocket wheel
Kettenradwalzfräser m <Fertig> sprocket hob
Kettenräumen n <Fertig> chain broaching
Kettenreaktion f 1. <Kerntech, Maschinen, Phys> chain reaction; 2. <Strahlphys> chain explosion

Kettenreaktion f von Neutronen bei der Kernspaltung <Kerntech, Strahlphys> chain reaction of neutrons in nuclear fission
Kettenreaktionsausmaß n <Kerntech> chain-reacting amount
Kettenregel f <Math> chain rule *(zur Differenzierung von geschachtelten Funktionen)*
Kettenrelais n <Elektrotech> link relay
Kettenrohr n <Wassertrans> chain pipe, spurling pipe *(Ankern)*
Kettenrohrzange f <Maschinen> chain wrench
Kettenrolle f <Maschinen> chain sprocket
Kettenrost m <Maschinen> chain grate
Kettensäge f <Fertig, Maschinen> chain saw
Kettenschaltung f <Elektrotech, Phys> ladder network
Kettenschärmaschine f <Textil> warper
Kettenschlaufe f <Sicherheit> chain sling
Kettenschleifer m <Papier> chain grinder
Kettenschlinge f <Maschinen> chain sling
Kettenschlüssel m <Maschinen> chain pipe wrench
Kettenschutz m 1. <Kfztech> chainguard *(Kraftübertragung, Motorrad)*; 2. <Maschinen> chain guard
Kettenspanner m 1. <Kfztech> chain tensioner *(Kraftübertragung, Motorrad)*; 2. <Papier> chain tightener
Kettenspannungsteiler m <Elektronik> ladder attenuator
Kettenstopper m <Wassertrans> chain compressor *(Ankern)*
Kettenstruktur f <Regelung> chain structure
Kettenteiler m <Elektronik> ladder attenuator
Kettenteilung f <Fertig, Maschinen> chain pitch
Kettentrieb m <Kfztech, Maschinen> chain and sprocket wheel drive
Kettentrommel f <Maschinen> chain drum
Kettenwicklung f <Elektrotech> basket winding
Kettenwiderstand m <Elektriz, Phys> iterative impedance
Kettenwirken n <Textil> warp knitting
Kettenwirkerei f <Textil> warp knitting
Kettenwirkmaschine f <Textil> warp knitting machine
Kettfaden m <Textil> end, warp thread
Kettfadenablassvorrichtung f <Textil> let-off motion
Kettfadenbruch m <Textil> breakage, warp break
Kettfadenwächter m <Textil> warp stop motion
Kettfadenwächterlamelle f <Textil> drop wire
Kettgarn n <Papier> warp yarn
Kettschären n <Textil> warping
Kettschlichten n <Textil> slasher sizing
kettschlichtgefärbt adj <Textil> slasher dyed
Kettstreifen mpl <Textil> warp streaks
Kettstuhlwirkerei f <Textil> warp knitting
Kettung f <Künstl Int> chaining
Keulenbreite f <Elektronik> beam width *(Antennentechnik)*
Keulengriff m <Maschinen> club handle
keV *(Kilo-Elektronenvolt)* <Teilphys> keV *(kilo electronvolt)*
Kevlar n <Wassertrans> kevlar *(Schiffbau)*
Keyboard n <Comp & DV> keyboard
Keyes-Verfahren n <Lebensmittel> Keyes process *(Destillation)*
KF *(Konfidenzfaktor)* <Künstl Int> CF *(confidence factor)*
kfG *(kontextfreie Grammatik)* <Künstl Int> CFG *(context-free grammar)*
K-förmiger Zahn m <Maschinen> hook tooth, hooked tooth
Kf-Wert m <Abfall, Qual> coefficient of permeability
Kfz *(Kraftfahrzeug)* <Kfztech> MC *(motorcar)*
Kfz-Sicherheitseinrichtung f <Sicherheit> road safety device
Kfz-Teile npl <Kfztech> motorcar parts

kg 1. <Labor> *(Kilo, Kilogramm)* kg *(kilogram)*; 2. <Phys> *(Kilogramm)* kg *(kilogramme)*
kgN *(kleinster gemeinsamer Nenner)* <Math> LCD *(least common denominator)*
kgT *(kleinster gemeinsamer Teiler)* <Geom> LCD *(least common denominator)*
kgV *(kleinstes gemeinsames Vielfaches)* 1. <Comp & DV> LCM *(least common multiple)*; 2. <Geom> LCM *(lowest common multiple)*
khz *(Kilohertz)* <Elektriz, Funktech> kHz *(kilohertz)*
KI *(künstliche Intelligenz)* <Künstl Int> AI *(artificial intelligence)*
Kick *m* <Erdöl> kick *(Bohrtechnik)*
Kickdown *m* <Kfztech> kickdown *(bei Automatikgetrieben)*
Kickdown-Schalter *m* <Kfztech> kickdown switch
Kickstarter *m* <Kfztech> kick starter *(Motorradmotor)*
Kiel *m* <Wassertrans> keel
Kiel *m* **zum Auflaufen auf Land** <Wassertrans> beaching keel *(Schiffbau)*
Kielbank *f* <Wassertrans> careening grid *(Instandhaltung)*
Kielboot *n* <Wassertrans> keel boat
kielbrüchig *adj* <Wassertrans> broken-backed *(Schiff)*
Kielbucht *f* <Wassertrans> sagging *(Schiffbau)*
Kielgang *m* <Wassertrans> keel strake; garboard strake *(Schiffbau)*
kielholen *v* <Wassertrans> careen *(Instandhaltung)*
Kiellegung *f* <Wassertrans> keel laying
Kielplanke *f* <Wassertrans> garboard plank
Kielplatte *f* <Wassertrans> keel plate
Kielschwein *n* <Wassertrans> keelson
Kielstapelung *f* <Wassertrans> cribbing *(Schiffbau)*
Kielträger *m* <Raumfahrt> keel
Kielwasser *n* <Meerschmutz, Strömphys, Wassertrans> wake *(Schiff)*
Kiemennetz *n* <Wassertrans> gillnet *(Fischerei)*
Kies *m* <Bau, Kohlen> gravel, pebbles
Kiesel... <Chemie> siliceous
Kieselerde *f* <Bau, Elektronik, Phys, Sicherheit> silica
Kieselerdestaub *m* <Sicherheit> silica dust
Kieselfilterschicht *f* <Abfall> gravel filter layer *(Deponie)*
Kieselgel *n* 1. <Lebensmittel> silica gel *(Trockenmittel)*; 2. <Verpack> silica gel
Kieselgur *f* <Lebensmittel> diatomaceous earth
Kieselhydrogel *n* <Kunststoff> precipitated silica
Kieselsäure *f* <Chemie> silicic acid
Kieselstein *m* <Bau> pebble, shingle
Kieselwolfram... <Chemie> silicotungstic *(Säure)*
Kiesgrube *f* <Bau, Wasserversorg> gravel pit
Kiespressdach *n* <Bau> felt and gravel roof
Kiessand *m* <Bau> grit
Kiesschotter *m* <Bau> ballast
Kiesstrand *m* <Wassertrans> shingle *(Meer)*
Kikuchi-Linie *f* <Kerntech> Kikuchi line
KI-Lernprogramm *n* *(intelligentes Lernprogramm)* <Künstl Int> ITS *(intelligent tutoring system)*
Kiln *m* <Kohlen> kiln
Kilo *n* *(Kilogramm)* <Labor> kilo, k
Kilobyte *n* *(KB)* <Comp & DV, Telekom> kilobyte, KB
Kilo-Elektronenvolt *m* *(keV)* <Teilphys> kilo electronvolt, keV
Kilogramm *n* 1. <Labor> *(Kilo, k, kg)* kilogram, kilogramme, kg; 2. <Phys> kilogram, kilogramme, kg *(SI-Einheit der Masse)*
Kilohertz *n* *(kHz)* <Elektriz, Funktech> kilocycle per second, kilohertz, kHz *(SI-Einheit der Frequenz)*
Kilokalorie *f* *(kcal)* <Lebensmittel> kilocalorie, kilogram calorie, kilogramme calorie, kcal
Kilometer *m* *(km)* <Metrol> *(AE)* kilometer, *(BE)* kilometre, km

Kilometerpunkt *m* <Eisenbahn> mileage point
Kilometerstein *m* <Trans> milestone
Kilonem *n* <Chemie> kilonem
Kilopondmeter *m* <Metrol> kilogram force meter
Kilostream *m* <Telekom> kilostream circuit *(Punkt-zu-Punkt -Datenverbindung)*
Kilovolt *n* *(kV)* <Elektrotech> kilovolt, kV
Kilovoltampere *n* <Elektriz, Elektrotech> kilovoltampere *(Einheit der Scheinleistung)*
Kilowatt *n* *(kW)* <Elektriz> kilowatt, kW *(SI-Einheit der Leistung)*
Kilowattstunde *f* *(kWh)* <Elektriz, Elektrotech> kilowatt hour, kWh *(Energieeinheit)*
Kimm *f* <Wassertrans> chine *(Teil des Schiffsrumpfes)*; visible horizon *(astronomische Navigation)*
Kimmbeplattung *f* <Wassertrans> bilge plating
Kimme *f* <Lufttrans> chine *(Wasserflugzeug)*
kimmerische Gebirgsbildung *f* <Erdöl> Cimmerian orogeny
kimmerische Orogenese *f* <Erdöl> Cimmerian orogeny *(Geologie)*
kimmerische Schichtenkontinuitätsstörung *f* <Erdöl> Cimmerian unconformity *(Geologie)*
Kimmgang *m* <Wassertrans> bilge strake
Kimmlinie *f* <Wassertrans> sea line
Kimmplatte *f* <Wassertrans> bilge plate
Kimmstringer *m* <Wassertrans> bilge stringer
Kimmstütze *f* <Wassertrans> bilge shore
Kimmtiefe *f* <Wassertrans> dip of horizon *(Navigation)*
kindersichere Verpackung *f* <Verpack> child-resistant packaging
kindersicherer Verschluss *m* <Ker & Glas> childproof finish
Kindersicherungsverschluss *m* <Verpack> CRC, child-resistant closure
Kinematik *f* <Fertig, Maschinen, Mechan> kinematics
kinematisch *adj* <Eisenbahn, Ergon, Fertig, Maschinen, Mechan, Nichtfoss Energ, Phys> kinematic
kinematische Begrenzungslinie *f* <Eisenbahn> *(AE)* kinematic gage, *(BE)* kinematic gauge
kinematische Fahrzeugbegrenzungslinie *f* <Eisenbahn> *(AE)* kinematic vehicle gage, *(BE)* kinematic vehicle gauge
kinematische Kette *f* <Maschinen> kinematic chain
kinematische Viskosität *f* <Heiz & Kälte, Maschinen, Mechan, Nichtfoss Energ, Phys> kinematic viscosity
kinematische Wirbelzähigkeit *f* *($\mathcal{E}$)* <Hydraul> kinematic eddy viscosity *($\mathcal{E}$)*
kinematische Zähigkeit *f* <Mechan, Nichtfoss Energ, Phys> kinematic viscosity
kinematischer Zwang *m* <Fertig> constraint; geometrical constraint *(Getriebelehre)*
Kinesiologie *f* <Ergon> kinesiology
Kineskop *n* <Elektronik> cinescope *(Fernsehen)*
Kinetik *f* <Maschinen, Mechan, Metall, Phys, Thermod> kinetics
Kinetik *f* **der Gase** <Thermod> gas kinetics
kinetisch *adj* <Ergon, Fertig, Kerntech, Maschinen, Mechan, Phys, Raumfahrt, Strahlphys> kinetic
kinetisch erzeugter Auftrieb *m* <Umweltschmutz> kinetically-induced buoyancy
kinetische Aufheizung *f* <Raumfahrt> kinetic heating *(Raumschiff)*
kinetische Energie *f* <Ergon, Maschinen, Mechan, Phys, Raumfahrt> kinetic energy
kinetische Gastheorie *f* <Phys, Thermod> kinetic theory of gases
kinetische Spektrophotometrie *f* <Strahlphys> kinetic spectrophotometry

kinetische

kinetische Trennung f <Kerntech> kinetic separation
kinetische Wärme f <Mechan, Phys> kinetic heat
kinetischer Isotopeneffekt m <Kerntech> kinetic isotope effect
Kingston-Ventil n <Wassertrans> Kingston valve *(Schiff)*
Kink f <Wassertrans> kink
Kipp m <Fernseh> sweep
Kipp... <Bau, Kfztech> tilting
Kippachse f <Bau> horizontal axis
kippbarer Turm m <Nichtfoss Energ> tiltable tower
Kippbecherwerk n <Trans> tipping bucket conveyor
Kippbehälter m <Kerntech> tilting basket
Kippbühne f <Trans> tipping platform
Kippe f <Kohlen> tip
kippen v 1. <Bau> tilt, tip; 2. <Kohlen> dump
Kippen n 1. <Bau> tipping; 2. <Fertig> tilting; 3. <Lufttrans> pitching
Kipper m 1. <Abfall> skip, tipper truck; 2. <Bau> tipper; 3. <Kfztech> dump car, dump truck, dump wagon; 4. <Kohlen> tipper
Kipperaufbau m <Kfztech> tilting body
kippfähiger Wagen m <Eisenbahn> tip-up car, tip-up wagon
Kippfahrzeug n <Bau> tipper
Kippfenster n <Bau> pivot-hung window
Kippfensterflügel m <Bau> pivot-hung sash
Kippflügelflugzeug n <Lufttrans> gyroplane, tilt wing plane
Kippform f <Ker & Glas> tilting mould
Kippfrequenz f <Elektronik, Fernseh, Labor, Telekom> sweep frequency
Kippglied n <Elektronik> bistable
Kipphebel m 1. <Elektrotech> tumbler; 2. <Fertig> tilting lever *(Kunststoffinstallationen)*; 3. <Kfztech> rocker, rocker arm; 4. <Maschinen> rocking arm, tumbler lever; 5. <Mechan> rocker
Kipphebelabdeckung f <Kfztech> rocker cover
Kipphebelbock m <Kfztech> rocker arm support
Kipphebeleinheit f <Kfztech> rocker arm assembly
Kipphebelgehäuse n <Kfztech> rocker box
Kipphebelschalter m <Elektrotech> tumbler switch
Kipphebelwelle f <Kfztech> rocker arm shaft
Kippkreis m <Fernseh> sweep circuit
Kippkübel m 1. <Bau> dumping bucket, tipping bucket; 2. <Meerschmutz> skip; 3. <Trans> dumping bucket, tilting skip
Kipplager n <Maschinen> rocker bearing
Kipp-LKW m <Kfztech> dump truck
Kipplore f <Trans> tilting wagon
Kippmoment n 1. <Bau> overturning moment; 2. <Elektrotech> *(AE)* breakdown torque, *(BE)* pull-out torque; 3. <Maschinen> tilting moment; 4. <Nichtfoss Energ> pitching moment
Kipppanel f <Kfztech> rocker panel
Kippregel f <Bau> *(AE)* leveling alidade, *(BE)* levelling alidade *(Vermessung)*
Kippschalter m 1. <Elektriz, Elektrotech> rocker switch, toggle switch, tumbler switch; 2. <Kontroll> toggle switch
Kippschaltung f <Elektronik> flip-flop, multivibrator
Kipp'scher Apparat m <Labor> Kipp's apparatus *(Generator)*
Kippschwingung f <Elektriz, Elektronik, Phys> relaxation oscillation
Kippstromgatter n <Fernseh> scanning gate
Kippstufenrost m <Abfall> rocking grate
Kippstuhl m <Kerntech> upender
Kipptisch m <Fertig, Maschinen> tilting table
Kippverstärker m <Elektronik, Gerät> bistable amplifier, sweep deflection amplifier

Kippversuch m <Elektrotech> *(AE)* breakdown test, *(BE)* pull-out test
Kippvorrichtung f 1. <Bau> tipping device; 2. <Kerntech> tilting device; 3. <Kfztech> dump
Kippwagen m <Kfztech> dump wagon, tipper
KI-Programmiersprache f <Künstl Int> *(AE)* AI programing language, *(BE)* AI programming language
Kirchhoff'sche Gesetze npl <Elektriz, Elektrotech> Kirchhoff's laws *(bei Stromnetzen)*
Kirchhoff'sches Strahlungsgesetz n <Strahlphys> Kirchhoff's law of emission of radiation
Kirnen n <Lebensmittel> churning *(Margarineherstellung)*
Kirschrotglut f <Metall> cherry-red heat
kissenförmige Verzeichnung f <Foto, Optik, Phys> *(AE)* negative distortion, *(BE)* pincushion distortion
Kiste f 1. <Fertig> box *(Wärmebehandlung)*; 2. <Trans> box; 3. <Verpack> carton
Kistenabfüllung f <Verpack> case packing
kistengeglüht adj <Metall> box annealed
Kistenglühen n <Fertig, Metall> box annealing
Kistenglühofen n <Metall> box annealing furnace
Kistenheber m <Maschinen> box hook
Kistenofen m <Thermod> pot furnace
Kistenöffner m <Verpack> nail puller
Kistenpalette f <Trans> box pallet
Kistenpalette f mit Gittergeflecht <Trans> box pallet with mesh
KI-System n <Künstl Int> AI system
Kitt m <Bau> putty
kittartige Formmasse f *(DMC)* <Kunststoff> *(AE)* dough-molding compound, *(BE)* dough-moulding compound *(DMC)*
kitten v 1. <Fertig> putty; 2. <Ker & Glas> cement
Kittentfernungsmesser n <Bau> hacking knife
Kittfalz m <Bau> fillister *(Fenster)*
Kittmesser n <Bau> putty knife, stopping knife
Kjeldahl-Apparat m <Labor> Kjeldahl digestion apparatus *(Stickstoffbestimmung)*
Kjeldahlmethode f <Kerntech> Kjeldahl method *(Stickstoffbestimmung)*
K-Jetronic-Einspritzanlage f <Kfztech> K-Jetronic fuel injection, continuous injection system
K-Kanten-Gamma-Densitometrie f <Kerntech> K-edge gamma densitometry
Klammer f 1. <Bau> staple; 2. <Druck> parenthesis; 3. <Fertig> staple; 4. <Maschinen> clamp, clamping band, clasp, cramp; 5. <Mechan> clip; 6. <Raumfahrt> brace *(Raumschiff)* • **in Klammern setzen** <Math> bracket together
Klammerflansch m <Maschinen> clamped tube flange
klammerfreie Schreibweise f <Comp & DV> Polish notation, parenthesis-free notation
klammerfreier Ausdruck m <Comp & DV> prefix notation
Klammerhaken m <Maschinen> dog hook
Klammern fpl 1. <Druck> brackets *(eckige)*; 2. <Fernseh> cramping; 3. <Math> brackets
Klammerpulsgenerator m <Fernseh> clamp pulse generator
Klammerungsschaltung f <Fernseh> clamping circuit
Klammerverbindungen fpl **für Rohre** <Maschinen> clamped pipe connections
Klammerzeichen n <Math> quantifier *(Logik)*
Klang m <Akustik, Aufnahme> combination sound, sound, tone; patch *(für Klangsynthese)*
Klanganalysator m <Akustik> *(BE)* sound analyser, *(AE)* sound analyzer
Klangcode m <Akustik, Aufnahme> sound code
Klangdiffusor m <Aufnahme> sound diffuser

Klangfarbe f <Akustik> quality of sound, quality of tone, timbre, timbre of sound, tonality, tone, tone colour, tone quality
Klangfarbenregler m <Aufnahme> tone control
Klangspektrograph m <Akustik> sound spectrograph
Klangverzerrung f <Aufnahme> sound distortion
Klapp... <Fertig, Foto, Lufttrans> folding
Klappachse f <Lufttrans> folding axis *(Hubschrauber)*
klappbar adj 1. <Lufttrans> folding; 2. <Mechan, Verpack> hinged
klappbarer Außenlastträger m <Lufttrans> folding pylon *(Hubschrauber)*
klappbarer Steckverschluss m <Verpack> hinged plug orifice closure
klappbares Luftschraubenblatt n <Lufttrans> folding blade *(Hubschrauber)*
Klappbrücke f <Bau> balance bridge, bascule bridge, counterpoise bridge
Klappdeckel m 1. <Fertig> hinged cover; 2. <Verpack> hinged lid
Klappe f 1. <Akustik> key; 2. <Bau> valve; 3. <Fertig> leaf; hinged cover, valve *(Kunststoffinstallationen)*; 4. <Heiz & Kälte> damper, trap; 5. <Hydraul> shutter; valve *(Absperrorgan von Rohrleitungen)*; 6. <Kerntech> valve; 7. <Labor> lid; 8. <Maschinen> bascule, cap, clack, flap; 9. <Mechan> flap, valve; 10. <Papier> flap; 11. <Raumfahrt> flap *(Raumschiff)*; 12. <Textil, Verpack> flap
Klappe f mit Schnappverschluss <Verpack> flap snap
Klappendurchflussmesser m <Gerät> airfoil flow meter, flap flow meter
Klappenflügel m <Heiz & Kälte> damper blade
Klappenführungsrippe f <Lufttrans> flap track rib
Klappenheber m <Lufttrans> flap jack
Klappenkolben m <Hydraul> bucket
Klappenkolbenpumpe f <Hydraul> bucket pump
Klappenscharnier n <Bau> flap hinge
Klappenschrank m <Telekom> switchboard
Klappenschrank m für Induktoranruf <Telekom> magneto switchboard
Klappenströmungsmesser m <Gerät> airfoil flow meter
Klappenteller m <Fertig> *(BE)* valve disc, *(AE)* valve disk *(Kunststoffinstallationen)*
Klappentext m <Druck> blurb
Klappenträger m <Fertig> clapper box
Klappenträgerklemmschraube f <Fertig> apron-clamping bolt
Klappenventil n <Hydraul> butterfly valve, clapper valve, flap valve, leaf valve
Klappenwehr n <Wasserversorg> lever weir
Klappfahrrad n <Kfztech> folding bicycle
Klappflügelflugzeug n <Lufttrans> folding-wing aircraft
Klappflügelpropeller m <Wassertrans> folding propeller *(Schiffbau)*
Klappgabel f <Lufttrans> hinge fork
Klappgelenk n des Luftschraubenblattes <Lufttrans> blade-folding hinge *(Hubschrauber)*
Klappkamera f <Foto> folding camera
Klappladen m <Bau> folding shutter
Klappschütz n <Nichtfoss Energ> tilting gate
Klappsitz m <Kfztech> tip-up seat
Klappsucher m <Foto> collapsible viewfinder
Klappsucher m mit Gegenlichtblende <Foto> folding viewfinder with hood
Klapptisch m <Maschinen> folding table
Klapptor n <Nichtfoss Energ> flap gate
Klapptür f <Bau> flap door
klar adj 1. <Eisenbahn, Optik> clear; 2. <Textil> bright; 3. <Wassertrans> clear
klar passieren v <Eisenbahn, Wassertrans> clear

Kläranlage f 1. <Abfall> clarification plant, purification plant, sewage treatment plant, sewage works, wastewater purification plant, wastewater treatment plant; 2. <Bau> clarification plant, sewage treatment plant, sewage works
Kläranlagebehälter m <Wasserversorg> filtration vat
Kläranlagenabfluss m <Wasserversorg> sewage effluent
Klärapparat m <Chemie> clarifier
Klärbecken n 1. <Abfall> clarification basin, clarification tank; 2. <Bau> settling basin; 3. <Chemtech> sedimentation basin, settling basin; 4. <Erdöl> clarifier basin *(Bohrtechnik)*; 5. <Kohlen> clarification basin, settling basin; 6. <Wasserversorg> clarification basin, clarification tank
Klärbehälter m 1. <Chemtech> sedimentation tank; 2. <Wasserversorg> filtering basin, filtering tank
klare Abbildung f <Optik> clear image
klare Konfiguration f <Lufttrans> clean configuration
klären v <Bau, Chemtech, Lebensmittel, Patent, Qual> clarify
Klären n <Chemtech> defecation
klarer Brunnen m <Bau> clear well
Klarfritte f <Ker & Glas> clear frit
Klärgas n <Thermod> digester gas
Klarglas n <Ker & Glas> clear glass
Klärgrube f 1. <Bau> dry well, settling pit; 2. <Erdöl> decanting pit *(Abscheidetechnik)*; 3. <Kohlen> settling pit; 4. <Wasserversorg> cess pit, cess pool
Klarheit f 1. <Optik> clearness *(einer Abbildung)*; 2. <Wasserversorg> clearness *(des Wassers)*
klarieren v <Eisenbahn, Wassertrans> clear
Klarifikator m <Chemie> clarifier
Klarlack m <Kunststoff> varnish
Klärleitung f <Erdöl> decanting trunk *(Abscheidetechnik)*
klarmachen v zum <Wassertrans> make ready, rig for
Klärmittel n <Lebensmittel> clarifier, fining agent
Klärschlamm m 1. <Abfall> sewage sludge, wastewater sludge; 2. <Bau> digested sludge; 3. <Wasserversorg> sewage sludge
Klarschriftleser m <Comp & DV> character reader
Klarsichtfolie f <Verpack> cellophane, cling film, film wrap, transparent film
Klarsichtfolienverpackung f <Verpack> seethrough packaging
Klarsichtkartonage f <Verpack> skin pack
Klarsichtscheibe f <Wassertrans> clear-view screen
Klarsichtschirm m <Wassertrans> clear view screen
Klärspitze f <Kohlen> settling cone
Klärsumpf m 1. <Chemtech> settling sump; 2. <Wasserversorg> settling tank
Klärteich m 1. <Bau> settling pond; 2. <Kohlen> clear pond, settling pond
Klartext m <Comp & DV, Telekom> plain text
Klärung f 1. <Chemtech> purification; decantation *(durch Abgießen)*; 2. <Erdöl> decantation *(Abscheidetechnik)*; 3. <Lebensmittel> decantation, fining; defecation *(von Lösungen)*
Klärvorrichtung f <Lebensmittel> clarifier
Klarwasser n <Wasserversorg> clear water
Klärwerk n 1. <Abfall> clarification plant, purification plant, sewage treatment plant, sewage works, wastewater treatment plant, wastewater treatment works; 2. <Bau> clarification plant, sewage treatment plant, sewage works; 3. <Wasserversorg> clarification plant
Klasse f 1. <Comp & DV> class; 2. <Gerät> class *(Messgerät)*; 3. <Maschinen> class
Klasse-AB-Verstärker m <Elektronik> class AB-amplifier
Klassenbreite f <Patent, Qual> class interval
Klassengrenze f <Patent, Qual> class boundary, class limit

Klassenmitte

Klassenmitte f <Qual> midpoint of class, midvalue of class interval
Klassierapparat m <Chemtech> classifier
klassieren v 1. <Chemtech> classify; 2. <Papier> screen; 3. <Patent, Qual> classify
Klassieren n 1. <Fertig> separation *(Erz)*; 2. <Kohlen> grading
Klassierer m <Chemtech, Kohlen> classifier
Klassierfachmann m <Kohlen> classifier
Klassiermaschine f <Maschinen> grading machine
klassierte Kohle f <Kohlen> graded coal
Klassiertrichter m <Kohlen> cone classifier
Klassierung f 1. <Bau> screening; grading *(nach Korngrößen)*; 2. <Kohlen> screening; 3. <Telekom> ordering
Klassierungsdetektor m <Trans> selective vehicle detector
Klassifikation f <Patent, Qual, Wassertrans> classification *(Handelsmarine)*
Klassifikationsgesellschaft f <Wassertrans> classification society
Klassifizierung f <Verpack> grading
klassische Thermodynamik f <Thermod> classic thermodynamics
klassischer Elektronenradius m <Phys> classical radius of the electron
klassisches Pressen n <Ker & Glas> straight pressing
Klaubarbeit f <Kohlen> sorting by hand
klauben v <Kohlen> cob
Klauben n <Kohlen> picking
Klaue f 1. <Fertig> dog; 2. <Maschinen, Mechan> claw, dog, jaw; 3. <Papier> jaw
Klauenfutter n <Maschinen> prong chuck
Klauenkupplung f 1. <Fertig> dog clutch; 2. <Kfztech> jaw clutch; dog clutch *(Getriebe)*; 3. <Maschinen> claw clutch, dog clutch, dog coupling, jaw clutch; 4. <Papier> jaw clutch
Klauenpolmaschine f <Elektrotech> claw-pole machine
Klausel f <Künstl Int> clause
Klaviersaitendraht m <Maschinen> piano string, piano wire
Klebeband n 1. <Elektrotech, Kunststoff, Mechan> adhesive tape; 2. <Verpack> adhesive tape, self-adhesive tape
Klebebindung f 1. <Druck> perfect binding, unsewn binding; 2. <Verpack> parallel glueing
Klebeblock m <Druck> perfect-bound block
Klebeemulsion f <Verpack> emulsion adhesive
Klebefläche f <Papier> adherend
Klebefolie f <Verpack> adhesive film
Klebeheftung f <Druck> perfect binding, unsewn binding
Klebemaschine f 1. <Druck> gluer; 2. <Verpack> adhesive machine, binding machine
Klebemittel n <Chemie, Fertig> agglutinant
kleben v 1. <Bau> glue, stick; 2. <Druck> paste up; 3. <Fertig> bond *(Metall)*; 4. <Kunststoff, Papier, Verpack> adhere
Kleben n 1. <Druck> gluing up; 2. <Elektrotech> sticking; 3. <Fertig> bonding *(Metalle)*; 4. <Verpack> gluing
klebend adj 1. <Chemie, Chemtech> agglutinant; 2. <Papier> adhesive
klebender Kontakt m <Elektriz> sticking contact
Klebepaste f <Druck> adhesive paste
Kleber m 1. <Bau> cementing material; 2. <Funktech> adhesive; 3. <Kunststoff> adhesive, glue; 4. <Lebensmittel> gluten; 5. <Sicherheit> adhesive
Kleberdehnbarkeit f <Lebensmittel> gluten extensibility
Klebesandform f <Fertig> *(AE)* sand and clay mold, *(BE)* sand and clay mould *(Thermitschweißen)*
Klebeseite f <Verpack> adhesive side
Klebesiegel n <Sicherheit> tamper-proof seal

Klebestelle f <Fernseh> splice *(Band)*
Klebestellengeräusch n <Aufnahme> bloop, blooping
Klebestellenvertiefung f <Aufnahme> bloop punch
Klebestreifen m <Kunststoff> adhesive tape, self-adhesive tape
Klebeumbruch m <Druck> paste-up
Klebeverbindung f <Fertig> adhesive-bonded joint, glued joint
Klebevorrichtung f <Verpack> gluing device
Klebfestigkeit f <Kunststoff> bond strength
Klebfilm m <Verpack> adhesive film
Klebfläche f <Kunststoff> adherend
Klebfuge f 1. <Kunststoff> glue line; 2. <Verpack> glued joint
Klebfügeteil n <Kunststoff> adherend
Klebkarton m <Verpack> glued box
Klebmuffe f <Fertig> solvent cement socket *(Kunststoffinstallationen)*
klebrig adj 1. <Kunststoff> sticky, tacky; 2. <Lebensmittel> sticky; 3. <Verpack> tacky
Klebrigkeit f <Lebensmittel> ropiness
Klebrigkeitsniveau n <Verpack> tack level
Klebrigmacher m 1. <Chemie> tackiness agent; 2. <Kunststoff> tackifier, tackifying agent
Klebstoff m 1. <Funktech, Kunststoff, Mechan> adhesive; 2. <Textil> glue; 3. <Verpack> adhesive glue, bonding agent, glue
klebstoffbeständig adj <Papier> adhesive-resistant
Klebstoffschicht f <Verpack> adhesive film
Klebstutzen m <Fertig> spigot *(Kunststoffinstallationen)*
Klebung f <Fertig> cementing *(Kunststoffinstallationen)*
Klebverbindung f 1. <Kunststoff> bond; 2. <Verpack> glued joint
Klebverschluss m <Verpack> binding closure
Klebzement m <Kunststoff> cement
Kleeblattantenne f <Funktech> cloverleaf antenna
Kleeblattzapfen m <Fertig> wobbler *(Walze)*
Kleesäure f <Foto> oxalic acid
Kleiderstoff m <Textil> dress material
Kleidung f 1. <Sicherheit> clothing *(Schutzmaßnahmen)*; 2. <Textil> apparel
Kleidungsstück n <Textil> garment
Kleie f <Lebensmittel> tailings; bran, break tailings *(Müllerei)*
Kleiebürste f <Lebensmittel> bran finisher
Klein... <Foto, Verpack> small
Kleinauflage f 1. <Druck> short print run, short run; 2. <Verpack> short run
Kleinbildfilm m <Foto> miniature film *(35 mm)*
Kleinbildkamera f <Foto> miniature camera *(35 mm)*
Kleinbuchstaben mpl <Druck> lc, lower case
Kleinbus m <Kfztech> microbus, minibus
Kleindarstellung f <Konstzeich> small-scale representation
Kleindruck m 1. <Druck> fine print; 2. <Labor> low pressure
Kleindruck-Manometer n <Gerät> *(BE)* low-pressure gauge, *(AE)* low-pressure gage
kleine Fahrt f <Kfztech> slow speed *(Schiff)*
kleine Gasblase f <Fertig> pepper blister, pinhead
kleine Kohlensorte f <Ker & Glas> burgee
kleine Leistung f <Elektrotech> low power
kleine Sekunde f <Akustik> minor second
kleine Septime f <Akustik> minor seventh
kleine Sexte f <Akustik> minor sixth
kleine Stückzahl f <Fertig> batch
kleine Terz f <Akustik> minor third
kleine Tonne f <Metrol> short ton
kleine und große Wartungsarbeiten fpl <Lufttrans> minor and major servicing operation
kleine Vorwölbung f auf Glas <Ker & Glas> tit

Klein-Energieerzeuger m <Nichtfoss Energ> small power producer, SPP
kleiner Ganzton m <Akustik> minor whole tone
kleiner gleich adj <Math> equal to or less than
kleiner Halbton m <Akustik> minor semitone
kleiner Holzpfropfen m <Bau> spile
kleiner Kupolofen m <Fertig> cupolette
kleiner Server m <Comp & DV> entry-level server
kleiner vernachlässigbarer Durchmesserfehler m <Metrol> minor diameter error
kleiner Wasserlauf m <Wassertrans> (AE) creek (Geographie)
kleiner Widerstand m <Phys> low resistance
kleinerer Durchmesser m <Maschinen> minor diameter
Kleinergleichzeichen n <Math> chevron
Kleinerzeichen n <Math> chevron
kleines Beiboot n <Wassertrans> cock
kleines Gewicht n <Metrol> net weight, short weight
kleines Pleuelauge n <Kfztech> connecting rod small end, small end
kleines unbemanntes Wählamt n <Telekom> CDO, community dial office
Klein-Gordon-Gleichung f <Phys> Klein-Gordon equation
Kleinhebezeug n <Maschinen> small hoists
Kleinintegration f (SSI-Schaltung) <Comp & DV, Elektronik> small-scale integration (SSI)
Kleinkamera f <Fernseh> minicam
Kleinkonverter m <Fertig> baby Bessemer converter
Kleinlader m <Elektrotech> trickle charger
Kleinladung f <Elektrotech> trickle charge
Kleinlaster m <Kfztech> (BE) light lorry, (AE) light truck
Kleinlastwagen m <Kfztech> pick-up, pick-up truck
Kleinlieferwagen m <Kfztech> minivan
Kleinlokomotive f 1. <Eisenbahn> light rail motor tractor; 2. <Elektrotech> small-power locomotive
Kleinmagnetron n <Elektronik> miniature magnetron
Kleinmotor m <Elektriz> fractional horsepower motor, integral horsepower motor, small-power motor, small-type motor
Kleinoffset n <Druck> small offset print
Kleinoffsetdruck m <Druck> small offset print
Kleinpflaster n <Bau> (BE) pebble pavement, (AE) pebble sidewalk
kleinporig adj <Fertig> close (Gefüge)
Kleinpresse f <Fertig> subpress
Kleinschalter m <Elektriz> installation switch
Klein'sche Flasche f <Geom> Klein bottle
Kleinserien mit Laschen versehen v <Verpack> batch tabbing
Kleinserienproduktion f <Maschinen> short run
Kleinsignal n <Elektronik> low-level signal, small signal
Kleinsignal-Parameter m <Elektronik> small signal parameter
Kleinsignal-Transistor m <Elektronik> small signal transistor
Kleinsignal-Verstärker m <Elektronik> small signal amplifier
Kleinsignal-Verstärkung f <Elektronik> small signal amplification
Kleinst... <Maschinen> minimum
Kleinstauflagendruck m <Druck> very short run printing
Kleinstbildkamera f <Foto> subminiature camera
Kleinstbodenstationsystem n <Telekom> VSAT system (zur Satelliten-Datenübertragung)
kleinste Leerlaufdrehzahl f <Elektriz> minimum idling speed
kleinste nachweisbare Spur f <Chemie> ultratrace (Analyse)

kleinste Strukturgröße f <Elektronik> minimum feature size
kleinste Umtastung f (MSK) <Elektronik> minimum-shift keying (MSK)
kleinster gemeinsamer Nenner m (kgN) <Math> lowest common denominator (LCD)
kleinster gemeinsamer Teiler m (kgT) <Math> least common denominator (LCD)
kleinster wahrnehmbarer Unterschied m <Ergon> just noticeable difference
Kleinsterdefunkstelle f <Funktech, Telekom> very-small aperture terminal (VSAT)
kleinstes Fehlerquadrat n <Telekom> least square error
kleinstes gemeinsames Vielfaches n (kgV) <Comp & DV, Math> least common multiple, lowest common multiple (LCM)
Kleinsthörgerät n <Akustik> insert earphone (mit Ohranpassung)
Kleinstintegration f (SSI-Schaltung) <Elektronik> small-scale integration (SSI)
Kleinstluftfahrzeugindex m <Lufttrans> miniature aircraft index
Kleinstmaß n <Maschinen> minimum size
Kleinstrelais n <Elektrotech> subminiature relay
Kleinströhre f <Elektronik> acorn tube
Kleinstsignal n <Elektronik> minimum signal
Kleinstspiel n <Maschinen> minimum clearance
Kleinstübermaß n <Maschinen> minimum interference
Kleinstzone f <Mobilkom> microcell (zellulares Netz)
Kleinwagen m <Kfztech> light vehicle
Kleinwählerzentrale f <Telekom> UAX, unit automatic exchange
Kleinware f auf Kartonunterlagen <Verpack> carded packaging
Kleinwerkzeuge npl <Maschinen> small tools
Kleinwinkelprisma n <Phys> small-angle prism
Kleinzelle f <Mobilkom> minicell (zellulares Netz)
Kleinzellensystem n <Mobilkom> small cell system
Kleister m 1. <Fertig> slipping; 2. <Kunststoff> glue
Klemm... <Maschinen> gripping
Klemmbacke f 1. <Fertig> gripping die (Stauchmaschine); 2. <Kerntech> jaw; 3. <Maschinen> gripping jaw
Klemmblock m <Ker & Glas> chuck block
Klemmbügel m <Maschinen> clamp
Klemmdiode f <Elektronik, Phys> clamping diode
Klemmdose f <Elektrotech> connection box
Klemme f 1. <Bau> clamp, cleat; 2. <Elektriz> terminal; 3. <Fertig> connector, terminal; 4. <Foto, Kfztech, Maschinen> clamp; 5. <Mechan> clamp, clip; 6. <Telekom> hub (Leitung)
Klemmeffekt m <Akustik> pinch effect
klemmen v 1. <Maschinen> jam; 2. <Mechan> clamp
Klemmen n 1. <Aufnahme, Fernseh, Telekom> clamping; 2. <Kfztech> jamming (Bremsen); 3. <Maschinen> jamming
Klemmenblock m <Elektriz, Elektrotech, Kontroll> terminal block
Klemmenbrett n <Elektrotech> terminal strip
Klemmenimpedanz f <Kontroll> terminal impedance
Klemmenkasten m <Elektriz> terminal box
Klemmenleiste f 1. <Elektriz, Elektrotech> terminal block, terminal strip; 2. <Fertig> terminal strip (Kunststoffinstallationen); 3. <Kontroll> terminal strip; 4. <Telekom> connection strip, terminal strip
Klemmenstreifen m <Elektrotech> terminal strip
Klemmgesperre n <Maschinen> silent ratchet
Klemmgriff m <Fertig> clamping lever
Klemmhebel m <Fertig> binder lever, clamping lever, lock lever

Klemmhülse

Klemmhülse f 1. <Fertig> collet; 2. <Maschinen> clamping sleeve
Klemmimpulse mpl <Fernseh> clamping pulses
Klemmkasten m <Elektrotech> conduit box
Klemmklampe f <Wassertrans> jam cleat *(Deckbeschläge)*
Klemmkopf m <Maschinen> clamping handle
Klemmkörper m <Elektrotech> clamping part
Klemmlasche f <Elektrotech> clamping lug
Klemmmechanismus m <Maschinen> clamping mechanism
Klemmmutter f <Fertig> clamping nut, locking nut
Klemmnabe f <Fertig> clamping collar *(Kunststoffinstallationen)*
Klemmplan m <Fertig> clamping plan *(für das Aufspannen eines Werkstücks)*
Klemmplatte f <Eisenbahn> rail clip
Klemmreflektor m <Foto> clamping reflector
Klemmring m 1. <Fertig> adjusting collar; 2. <Maschinen> clamp ring, clamping ring, lock ring
Klemmschaltung f <Elektronik> clamping circuit *(Schaltkreistechnik)*
Klemmschraube f 1. <Elektrotech> binding post, terminal; 2. <Fertig> clamp, clamping bolt, set screw; clamping screw *(Kunststoffinstallationen)*; 3. <Maschinen> binding screw, clamping screw, lock screw, locking screw, set bolt
Klemmsitz m <Mechan> force fit
Klemmstelle f <Elektrotech> clamping point
Klemmstück n <Fertig> block *(Stößel)*
Klemmung f 1. <Bau> clamp *(Nivellierinstrument)*; 2. <Elektronik> clamping *(Fernsehtechnik)*
Klemmvorrichtung f <Labor> clamp
Klempner m <Bau> plumber
Klempnerarbeiten fpl <Bau> plumbing
Kletterfilmverdampfer m <Lebensmittel> climbing film evaporator
Kletterschalung f <Bau> climbing forms
Klickspur f <Aufnahme> click track
Klima n <Bau, Heiz & Kälte, Nichtfoss Energ, Sicherheit> climate
Klimaaggregat n <Heiz & Kälte> air conditioner
Klimaanlage f 1. <Fertig> air-conditioning plant; 2. <Heiz & Kälte> air-conditioning plant, air-conditioning system; 3. <Kfztech> air conditioner *(Innenraum)*; 4. <Sicherheit> air-conditioning plant • **mit Klimaanlage** <Fertig> air-conditioned
Klimadecke f <Heiz & Kälte> air-handling ceiling, ventilated ceiling
Klimadetektor m <Trans> climatic detector
Klimafestigkeit f <Bau, Maschinen> weathering resistance
Klimagerät n <Heiz & Kälte, Maschinen> air conditioner
Klimakammer f <Ergon, Heiz & Kälte, Verpack, Werkprüf> climatic chamber, environmental chamber
Klimalabor n <Labor> climate testing laboratory
Klimaleuchte f <Heiz & Kälte> air-handling luminaire
Klimaprüffeld n <Werkprüf> climate testing laboratory
Klimaprüfung f <Werkprüf> climate investigation, climate test, climate testing, climatic test, climatic testing
Klimaraum m <Verpack> climatic chamber
Klimaregelung f <Heiz & Kälte> air conditioning
Klimaschutz m <Sicherheit> climatic protection
Klimatauglichkeitsprüfung f <Werkprüf> climatic robustness test
Klimatechnik f <Heiz & Kälte> air conditioning
klimatische Bedingungen fpl <Verpack> climatic conditions
klimatisieren v <Bau, Heiz & Kälte> air-condition
klimatisierte Raumluft f <Heiz & Kälte> conditioned air
klimatisierter Inspektionsraum m <Maschinen> temperature-controlled inspection room
Klimatisierung f 1. <Heiz & Kälte> air conditioning, climatization; 2. <Papier> air conditioning
Klimaversuch m <Verpack> climatic test
Klimazone f <Kohlen, Nichtfoss Energ> climate zone
Klinge f 1. <Fertig> blade *(Schraubenzieher)*; 2. <Maschinen> blade
Klingel f 1. <Comp & DV, Elektrotech> bell; 2. <Telekom> ringer
Klingeldraht m <Elektrotech> bell wire
klingeln v <Telekom> ring
Klingeln n 1. <Elektrotech> ringing *(Telefonglocke)*; 2. <Erdöl> knock; 3. <Kfztech> pinking; pinging *(Motor)*; 4. <Mechan, Telekom> ringing
Klingeln n **des Motors** <Maschinen> ringing engine
Klingeltrafo m <Elektrotech> bell transformer
Klingeltransformator m <Elektrotech> bell transformer
Klingen n <Aufnahme> ringing
Klinke f 1. <Bau> catch; 2. <Elektriz> jack; 3. <Fertig> pawl; ratchet *(Kunststoffinstallationen)*; 4. <Maschinen, Mechan> catch, dog, pawl, ratchet; 5. <Telekom> jack *(Telefon)*; jack *(Schalttafel)*
Klinkenfeder f <Fertig> pawl spring
Klinkenhülse f <Elektrotech> jack bush
Klinkenkupplung f <Maschinen> pawl coupling
Klinkenrad n 1. <Fertig> ratchet; 2. <Maschinen> dog wheel, ratchet, ratchet wheel
Klinkenradvorschub m <Maschinen> ratchet feed
Klinkenschaltwerk n <Maschinen> ratchet mechanism
Klinkenstecker m 1. <Elektriz, Elektrotech> jack; jack plug *(Schalttafel)*; 2. <Telekom> jack plug, switchboard plug • **mit Klinkenstecker angeschlossen** <Elektrotech> jacked
Klinkenstreifen m <Elektrotech> jack strip
Klinkenumschalter m <Elektriz> jack switch
Klinkenumschaltertafel f <Elektriz> jack switchboard
Klinker m <Bau, Ker & Glas> clinker, clinker brick
Klinkerkitt m <Bau, Ker & Glas> clinker cement
Klinsch m <Wassertrans> clinch *(Knoten)*
Klippe f <Wassertrans> cliff, rock *(Geographie)*
Klippegel m <Fernseh> clipping level
Klipper m <Fernseh> clipper
Klipperverstärker m <Fernseh> clipper amplifier
Klirrdämpfung f <Akustik> harmonic distortion
Klirren n <Elektronik> harmonic distortion
Klirrfaktor m 1. <Akustik, Elektronik> distortion coefficient, distortion factor, harmonic distortion factor, percentage harmonic content, ripple factor; total harmonic distortion, THD *(mit allen Harmonischen)*; 2. <Aufnahme> distortion; 3. <Telekom> harmonic distortion
Klirrfaktormessbrücke f <Aufnahme, Metrol> distortion bridge, distortion-measuring bridge, harmonic detector
Klirrfaktormesser m <Aufnahme> distortion meter
Klirrfaktormessgerät n <Aufnahme, Metrol> distortion analyzer, distortion factor meter, distortion meter
Klirrfaktormessung f <Metrol> distortion factor measurement
Klischee n <Druck> block, printing block
Klischeehersteller m <Druck> process engraver
Klischeeherstellung f <Druck> blockmaking
Klischeeprüfer m <Druck> *(AE)* type height gage, *(BE)* type height gauge
Klon m <Comp & DV> clone
Klopfbrett n <Druck> planer
klopfen v 1. <Bau> beat; 2. <Kohlen> tap; 3. <Maschinen> hammer, knock; 4. <Metall> beat; 5. <Phys> tap
Klopfen n 1. <Erdöl> knock *(Ottomotor: Kraftstoffselbstentzündung mit Verbrennung)*; 2. <Kfztech> beat, pinking; knocking *(Vorzündung des Motors)*; 3. <Maschinen> knocking; 4. <Phys> tapping

klopffeste Mischung f <Kfztech> antiknock mixture
Klopffestigkeit f 1. <Erdöl> knock rating *(Kraftstoff)*; 2. <Kfztech> antiknock resistance
Klopfschall m <Sicherheit> percussive sound
Klöppel m <Fertig> clapper
Klöppelverbindung f <Kerntech> ball coupling
Klotz m <Kohlen> block
Klotzbremse f <Eisenbahn, Maschinen> block brake, shoe brake
Klotzdruck m <Textil> block printing
Klotzen n <Textil> padding *(Färben)*
Klotzfärbung f <Textil> pad dyeing
Klotzmaschine f <Textil> pad mangle
klumpen v <Lebensmittel> cake
Klumpen m 1. <Ker & Glas> chunk, stick; 2. <Lebensmittel> caking; 3. <Papier> lump • **Klumpen bilden** <Papier> clot
Klumpenglas n <Ker & Glas> chunk glass
Kluppe f <Maschinen> screw plate
Kluppenrahmen m <Textil> clip frame
Kluppenspannrahmen m <Textil> clip stenter
Klüse f <Wassertrans> hawse *(Tauwerk)*
Klüver m <Wassertrans> jib *(Segeln)*
Klystron n <Elektronik, Funktech, Phys, Raumfahrt, Strahlphys, Telekom> klystron
Klystron n **mit drei Resonanzkammern** <Elektronik> three-cavity klystron
Klystronoszillator m <Elektronik> klystron oscillator
Klystronverstärker m <Elektronik> klystron amplifier, klystron repeater
km m *(Kilometer)* <Labor> km *(kilometer)*
Knack m <Akustik> click
Knacken n <Akustik> crackle
Knackfilter n <Telekom> click filter
Knackgeräusch n <Akustik> click
Knall m <Sicherheit> noise
Knall... <Chemie> fulminic
Knallgas n <Chemie> oxyhydrogen
Knallsignal n <Eisenbahn> torpedo
Knalltrauma n <Akustik> acoustic trauma
Knallvorrichtung f <Eisenbahn> detonator
Knallvorrichtungsexplosion f <Eisenbahn> exploding of detonator
Knarre f <Maschinen> ratchet, ratchet lever
Knäuel n <Textil> hank
Knauf m <Mechan> knob
Knautschzone f <Kfztech> crumple zone, deformable zone
Knebel m <Kohlen, Wassertrans> toggle
Knebelgriff m <Bau, Fertig, Mechan> T-handle, locking handle
Knebelkerbstift m <Fertig> grooved pin *(Kunststoffinstallationen)*
Knebelschraube f 1. <Bau> capstan-headed screw; 2. <Maschinen> T-screw, tommy screw
Kneifzange f 1. <Bau> nipper pliers, pincers, nippers; 2. <Fertig, Maschinen> nipper pliers, nippers, pincers
kneten v <Fertig> masticate *(Kunststoffe)*; temper *(Ton)*
Kneten n <Lebensmittel> kneading
Kneter m 1. <Chemtech> mixer; 2. <Kunststoff> internal mixer, kneader; 3. <Papier> kneader
Kneter m **mit gegenläufigen Schaufeln** <Kunststoff> dough mixer
Knetlegierung f <Fertig> plastic alloy
Knetmaschine f <Fertig> masticator *(Kunststoffe)*
Knick m **einer Kurve** <Maschinen> knee of a curve
Knickband n <Metall> kink band
Knickbelastung f <Trans> buckling load
knicken v <Maschinen> buckle
Knicken n 1. <Maschinen> buckling; 2. <Textil> breaking
Knickfestigkeit f <Maschinen, Metall> buckling resistance

Knickinstabilität f <Kerntech> kink instability
Knickkraft f <Fertig> critical compressive force
Knicklast f <Maschinen, Qual> buckling load
Knickpunkt m <Bau> knee
Knickspannung f <Mechan, Qual> breaking stress
Knickstab m <Fertig> column
Knickstelle f <Metall> kink
Knickung f <Maschinen> buckling
Knickversuch m <Mechan, Qual> breaking test
Knie n <Wassertrans> knee *(Schiffbau)*
Knieblech n <Wassertrans> bracket *(Schiffbau)*
Kniegelenkbolzen m <Maschinen> fulcrum pin
Kniehebel m 1. <Ker & Glas, Kunststoff> toggle; 2. <Lufttrans> bell crank *(Hubschrauber)*; 3. <Maschinen> knee joint, knuckle joint, toggle, toggle joint
Kniehebelpresse f <Ker & Glas, Maschinen> toggle press
Kniekicker m <Bau> knee kicker *(Teppichlegen)*
Knierohr n <Bau, Mechan> bend, knee bend, pipe bend
Knieschoner m <Sicherheit> knee pad
Kniestück n 1. <Bau> knee; 2. <Labor> elbow; 3. <Maschinen> knee; 4. <Wassertrans> angle *(Schiffbau)*
Kniff m <Papier> crease
Knight'sche Verschiebung f <Kerntech> Knight shift
Knistern n <Akustik, Aufnahme> crackle
Knitter m 1. <Papier> crumpling; 2. <Textil> crease
Knittererholung f <Textil> crease recovery
Knitterfestausrüstung f <Textil> crease-resist finish
Knitterfestigkeit f 1. <Textil> crease resistance, crush resistance; 2. <Qual> crush resistance
knitterfrei adj <Textil> crush-resistant
knittern v <Textil> crease, crush
Knittern n <Textil> crushing
KNN *(künstliches neuronales Netzwerk)* <Künstl Int> ANN *(artificial neural network)*
Knochen m <Akustik, Chemie, Ker & Glas, Kohlen, Textil> bone
Knochenasche f <Ker & Glas> bone ash
Knochengallerte f <Chemie> ossein
Knochenkohle f <Kohlen> animal charcoal, bone coal
Knochenleim m <Textil> bone glue
Knochenleitung f <Akustik> bone conduction
Knochenporzellan n <Ker & Glas> bone china
Knochenvibrator m <Akustik> bone vibrator
knollenförmig adj <Metall> nodular
Knopf m <Elektrotech> button
knopfförmig adj <Elektriz, Elektrotech, Kontroll, Telekom> button-shaped
Knopflochleiste f <Textil> buttonhole facing
Knopflochmikrofon n <Akustik, Aufnahme> lapel microphone
Knopflochnähautomat m <Textil> automatic buttonhole sewing machine
Knötchen n <Fertig> fish eye *(Kunststoffe)*
Knötchenbildung f <Textil> pilling
knoten v <Textil> knot
Knoten m 1. <Comp & DV, Elektriz, Fertig> node; 2. <Ker & Glas> knot; nodule *(Blaswolle)*; 3. <Künstl Int> node *(Graph)*; 4. <Nichtfoss Energ, Papier> knot; 5. <Telekom> node, junction; 6. <Textil> knot; 7. <Wassertrans> knot *(Knoten)*; knot *(Maßeinheit)*; 8. <Wellphys> node *(einer stehenden Welle)*
Knotenbahnhof m <Eisenbahn> *(AE)* railroad center, *(BE)* railway centre
Knotenblech n 1. <Fertig> junction plate; 2. <Maschinen> gusset plate
Knotendehnbarkeit f <Textil> knot extensibility
Knotenebene f <Phys> nodal plane
Knotenfänger m <Papier> knotter

Knotenfängerstoff

Knotenfängerstoff m <Papier> knotter pulp
Knotenfestigkeit f <Textil> knot tenacity
Knotenlinie f <Akustik, Aufnahme> nodal line
knotenlos adj <Textil> knotless
knotenlose Garnlänge f <Textil> knotless yarn length
Knotenprozessor m <Comp & DV> node processor
Knotenpunkt m 1. <Akustik> node *(einer Schwingung)*; 2. <Elektriz> node; 3. <Fertig> nodal point *(Schwingung)*; 4. <Phys> nodal point; 5. <Telekom> junction *(Netzwerk)*; branch point, crosspoint
Knotenpunktentwicklungsmethode f <Kerntech> nodal expansion method *(Theorie)*
Knotenpunktverbindung f <Bau> knee bracket plate
Knotenseil n <Trans> button rope *(Kabelbahn)*
Knotenstrom m <Elektriz> nodal current
Knotenvermittlung f <Telekom> tandem switching
Knotenvermittlungsstelle f *(KVSt)* <Telekom> regional exchange, switching node, tandem central office
Knotenverteiler m <Telekom> JDF, *(BE)* junction distribution frame
Knotenvorbrecher m <Papier> knot prebreaker
knotig adj <Textil> knotty
Knowledge-Engineering n <Künstl Int> KE, knowledge engineering
Knudsen-Effekt m <Kerntech> Knudsen effect *(Molekularströmung)*
Knüppel m 1. <Fertig> billet *(Walzen)*; 2. <Mechan, Metall> billet; 3. <Sicherheit> push stick
Knüppelausstoßer m <Fertig> billet pusher
Knüppelbohrmaschine f <Fertig> billet-drilling machine
Knüppelfertigwalzen n <Fertig> billet finishing
Knüppelschaltung f <Kfztech> floor shift *(Getriebe)*
Knüppelschere f <Maschinen> billet shears
Knüppelschlepper m <Fertig> billet buggy
Knüppeltasche f <Fertig> billet cradle *(Walzen)*
Knüppelwalze f 1. <Fertig> billet roll; 2. <Maschinen> billeting roll, blooming roll
Knüppelwalzwerk n <Fertig> billet mill
KnV *(Knotenvermittlung)* <Telekom> tandem switching
Koagulans n <Kunststoff> coagulating agent
Koagulation f 1. <Chemie> clotting, coagulation; 2. <Chemtech, Kunststoff> coagulation
Koagulationsbad n <Chemie> coagulating bath
Koagulationsmittel n <Kunststoff> coagulating agent
koagulieren v <Chemie> clot
Koagulieren n 1. <Chemie> clotting; 2. <Lebensmittel> coagulating
Koaguliermittel n <Chemtech> coagulator
koaguliert adj <Chemtech> coagulated
Koagulierungsflüssigkeit f <Chemtech> coagulation liquid
Koagulum n <Kunststoff> coagulum
koaleszieren v <Chemie> coalesce
koalisieren v <Chemie> coalesce
koaxial adj <Aufnahme, Elektronik, Elektrotech, Funktech, Lufttrans> coaxial
Koaxial... <Elektronik, Elektrotech, Telekom> coaxial
Koaxialantenne f <Funktech> coaxial antenna
Koaxialbalun m <Funktech> sleeve balun
Koaxialdiode f <Elektronik> coaxial diode
koaxiale Doppelleitung f <Elektrotech> coaxial-pair cable
koaxiale Festlast f <Elektrotech> coaxial-fixed load
koaxiale Leitung f <Aufnahme> coaxial line
koaxialer Hubschraubertyp m <Lufttrans> coaxial-type helicopter
koaxialer Phasenregler m <Elektrotech> coaxial phase shifter
koaxialer Phasenschieber m <Elektrotech> coaxial phase shifter
koaxialer Phasensteller m <Elektrotech> coaxial phase shifter
koaxiales Dämpfungsglied n <Elektronik> coaxial attenuator
Koaxialfilter n <Telekom> coaxial filter
Koaxialkabel n <Aufnahme, Comp & DV, Elektriz, Elektrotech, Fernseh, Phys, Telekom> coaxial cable
Koaxiallast f <Elektrotech> coaxial load
Koaxiallautsprecher m <Akustik, Aufnahme> coaxial loudspeaker
Koaxialleitung f <Elektrotech, Phys, Telekom> coaxial line
Koaxialleitungssystem n <Telekom> coaxial-line system
Koaxialmagnetron n <Elektronik> coaxial magnetron
Koaxialstecker m <Elektrotech> coaxial connector
Koaxialtopfkreis m <Elektronik> coaxial cavity
Koaxkabel n <Elektrotech> coaxial cable
Koazervat n <Chemie> coacervate
Koazervation f <Chemie> coacervation
Koazervierung f <Chemie> coacervation
Kobalt n *(Co)* <Chemie> cobalt *(Co)*
Kobalt n **60** <Kerntech> cobalt 60, radiocobalt
Kobalt-Arsen-Kies m <Hydraul> danaide
Kobalt-60-Bestrahlungsanlage f <Kerntech> cobalt 60 irradiation plant
Kobaltblüte f <Chemie> erythrine, erythrite
Kobaltbombe f <Kerntech> cobalt bomb
Kobaltglasflasche f <Ker & Glas> cobalt bottle
Kobaltlegierung f <Anstrich> cobalt alloy
Kobold m <Comp & DV> sprite
Koch... <Chemtech, Papier> boiling
Kochbecher m <Papier> beaker
köcheln v <Lebensmittel> simmer
kochen v <Thermod> boil
Kochen n 1. <Lebensmittel> boiling; 2. <Papier> digestion; 3. <Thermod> boiling
kochend adj <Chemtech, Kerntech, Lebensmittel, Thermod> boiling
kochende Gärung f <Chemtech, Lebensmittel> boiling fermentation
kochendes Fließbett n <Kerntech> boiling bed
Kochflasche f <Chemtech> boiling flask
Kochkessel m <Lebensmittel> kettle
Kochkolben m <Chemtech> boiling flask
Kochplatte f <Chemtech> boiling plate
Kochpunkt m <Lebensmittel> boiling point
Kochsalzersatz n <Lebensmittel> salt substitute
Kochsäureanlage f <Papier> acid plant
Kodachrome-Verfahren® n <Foto> dye-transfer process, kodachrome process®
Koeffizient m 1. <Elektriz> coefficient; 2. <Hydraul> coefficient *(Kontraktion)*; coefficient *(Leistung, Beaufschlagung)*; 3. <Math, Mechan, Phys, Qual> coefficient
Koeffizient m **der linearen Ausdehnung** <Phys, Qual> coefficient of linear expansion
Koeffizient m **der linearen Wärmedehnzahl** <Phys, Qual> coefficient of linear expansion
Koeffizient m **der Thermospaltung** <Phys> thermal fission factor
Koerzitivfeldstärke f <Elektriz, Elektrotech> coercive force, coercive field strength
Koerzitivkraft f 1. <Aufnahme> coercivity; 2. <Elektriz, Elektrotech, Metall> coercive force; 3. <Phys> coercive force, coercivity
Kofferdamm m <Wasserversorg> cofferdam
Kofferraum m <Kfztech> *(BE)* boot, *(BE)* rear boot, *(AE)* rear trunk, *(AE)* trunk
Kognition f <Künstl Int> cognition
Kognitionswissenschaft f <Künstl Int> cognitive science
kognitiv adj <Ergon> cognitive

kognitive Wissenschaft f <Künstl Int> cognitive science
kognitives Abbild n <Ergon> cognitive map
kohärent adj <Aufnahme, Elektriz, Elektronik, Metall, Optik, Strahlphys, Telekom, Wellphys> coherent
kohärente Ableitung f <Elektronik> coherent deduction
kohärente Antistokes-Raman-Streuung f (KARS) <Strahlphys, Wellphys> coherent anti-Stokes Raman scattering (CARS)
kohärente Differenz-PCM f <Telekom> differential coherent pulse code modulation (DCPC)
kohärente Differenz-Pulscodemodulation f <Telekom> differential coherent pulse code modulation (DCPC)
kohärente Grenze f <Metall> coherent boundary
kohärente Grenzfläche f <Metall> coherent interface
kohärente optische Übertragung f <Telekom> coherent optical transmission
kohärente Phasenumtastung f <Telekom> CPSK, coherent phase shift keying
kohärente Signalverarbeitung f <Elektronik, Telekom> coherent signal processing
kohärente Strahlung f <Optik, Telekom> coherent radiation
kohärente Übertragung f <Telekom> coherent transmission
kohärente Wellen fpl <Strahlphys, Wellphys> coherent waves
kohärenter Bereich m <Telekom> coherent area
kohärenter monochromatischer Lichtstrahl m <Strahlphys> coherent monochromatic beam (lasererzeugt)
kohärenter Schall m <Akustik> coherent sound
kohärenter Strahl m <Elektronik> coherent beam
kohärentes Impulsradar n <Funkort> coherent pulse radar
kohärentes Licht n 1. <Elektronik> coherent light (aus dem Laser); 2. <Strahlphys, Telekom, Wellphys> coherent light
kohärentes Rauschen n <Aufnahme> coherent noise
kohärentes Teilchen n <Metall> coherent particle
Kohärenz f 1. <Elektronik> coherence (Nachrichtentechnik); coherence (systematischer logischer Zusammenhang); 2. <Optik, Phys, Telekom> coherence; 3. <Wellphys> coherency
Kohärenz f **eines Laserstrahls** <Wellphys> coherency of a laser beam (über große Entfernungen hinweg)
Kohärenzbandbreite f <Telekom> coherence bandwidth
Kohärenzbereich m <Optik> coherence area
Kohärenzgrad m <Optik, Telekom> degree of coherence
Kohärenzlänge f <Optik, Phys, Telekom> coherence length
Kohärenzoszillator m <Elektronik> coherent oscillator (Radartechnik)
Kohärenzzeit f <Optik, Phys, Telekom> coherence time
Kohäsion f <Bau, Metall, Phys> cohesion
Kohäsionsenergie f <Metall> cohesive energy
Kohäsionsfestigkeit f <Kunststoff> adhesive shear strength
Kohäsionskraft f <Phys, Qual> cohesive force
kohäsionslos adj <Bau> cohesionless
kohäsionsloser Boden m <Bau> cohesionless soil
Kohle f <Kohlen> coal
Kohle f **mit hohem Schwefelgehalt** <Kohlen> high-sulfur coal
kohleartig adj <Chemie> carbonaceous
Kohleaufdampfverfahren n <Kerntech> carbon replica method
Kohlebecken n <Kohlen> coal basin
kohlebefeuert adj <Kohlen> coal-fired
Kohlebürste f <Elektriz, Elektrotech, Kfztech, Maschinen> carbon brush, graphite brush

Kohleelektrode f <Chemie, Elektrotech> carbon electrode
Kohlefadenlampe f <Elektrotech> carbon filament lamp
Kohlefaser f <Kunststoff> (AE) carbon fiber, (BE) carbon fibre
Kohleförderung f <Kohlen> coal extraction
kohlegeheizter Kessel m <Heiz & Kälte> coal-fired boiler
Kohlehalterlampe f <Elektrotech> carbon holder lamp
kohlehaltig adj <Kohlen> coal-bearing
Kohlehydrat n <Lebensmittel> carbohydrate
Kohlekontakt m <Elektrotech> carbon contact
Kohlekraftwerk n <Thermod> fossil fuel power station
Kohlelichtbogen m <Elektriz, Elektrotech, Fertig, Maschinen> carbon arc
Kohlelichtbogenlampe f <Elektriz> carbon arc lamp
Kohlelichtbogenschneiden n <Fertig, Maschinen> carbon arc cutting
Kohlelichtbogenschweißen n <Fertig> carbon arc welding
Kohlemassewiderstand m <Elektrotech> carbon composition resistor
Kohlemikrofon n <Akustik> carbon microphone
Kohlenabbau m <Kohlen> coal mining
Kohlenachbrechen n <Kohlen> breaking down coal
Kohlenauffüllung f <Kohlen> coal backing
Kohlenbecken n <Kohlen> coalfield
Kohlenbergbau m <Kohlen> coal mining
Kohlenbergwerk n <Kohlen> coal mine, coal pit, coal works
Kohlenbergwerksprengstoff m <Kohlen> coal-mining explosive
Kohlenbohrer m <Kohlen> coal drill
Kohlenbunker m <Kohlen> coal bunker
Kohlendioxid n <Chemie, Elektronik, Kohlen, Maschinen, Umweltschmutz> carbon dioxide
Kohlendioxidlaser m (CO_2-Laser) <Elektronik> carbon dioxide laser (CO_2 laser)
Kohlendioxidtreibhauseffekt m <Umweltschmutz> carbon dioxide greenhouse effect
Kohlendisulfid n <Chemie> (AE) carbon disulfide, (BE) carbon disulphide
Kohlenelektrode f <Elektriz> carbon electrode
Kohlenfaserfilz m <Raumfahrt> (AE) carbon fiber felt, (BE) carbon fibre felt (Raumschiff)
Kohlenfaserverbundstoff m <Raumfahrt> (AE) carbon fiber, (BE) carbon fibre (Raumschiff)
Kohlenfeld n <Kohlen> coalfield
Kohlenfließband n <Kohlen> coal belt
Kohlenflöz n <Kohlen> coal seam
Kohlenflözsohle f <Kohlen> coal-seam floor
Kohlenförderband n <Kohlen> coal belt
Kohlenformation f <Kohlen> coal formation
Kohlengrieß m <Kohlen> small coal
Kohlengrube f <Kohlen> coal mine, coal pit
Kohlengrus m 1. <Ker & Glas> culm; 2. <Kohlen> coal beans, duff, slack coal
Kohlenhacke f <Kohlen> coal pick
Kohlenhalde f <Kohlen> coal yard
Kohlenhauer m <Kohlen> coal breaker
Kohlenhydrat n <Chemie, Lebensmittel> carbohydrate
Kohlenklein n <Kohlen> breeze, charcoal duff, small coal
Kohlenkorb m <Kohlen> coal basket
Kohlenkübel m <Kohlen> coal basket
Kohlenladeplatz m <Eisenbahn, Wassertrans> coal wharf
Kohlenlagerplatz m <Kohlen> coal yard
Kohlenmonoxid n 1. <Chemie, Elektronik> carbon monoxide; 2. <Kfztech> carbon monoxide (Abgase); 3. <Kohlen, Maschinen, Sicherheit, Umweltschmutz> carbon monoxide

Kohlenmonoxidfilter

Kohlenmonoxidfilter n <Sicherheit> carbon monoxide filter
Kohlenmonoxidlaser m <Elektronik> carbon monoxide laser
Kohlenöl n <Kohlen> coal oil
Kohlenoxid n <Chemie> carbon monoxide
Kohlenoxidchlorid n <Chemie> phosgene
Kohlenpulver n <Kohlen> coal powder
Kohlenrutsche f <Kohlen> coal chute
Kohlensack m <Fertig> belly *(Gießen)*
Kohlensäure f <Fertig> carbon dioxide, carbonic acid *(Kunststoffinstallationen)* • **Kohlensäure entziehen** <Kohlen> decarbonate • **mit Kohlensäure versetzen** <Fertig> aerate
Kohlensäurediamid n <Chemie> carbamide, urea
Kohlensäuredichlorid n <Chemie> phosgene
Kohlensäurefeuerlöscher m <Sicherheit> carbon dioxide fire extinguisher
kohlensäurehaltig adj <Lebensmittel> carbonated
Kohlensäurelöscher m <Sicherheit> carbon dioxide fire extinguisher
kohlensaures Calcium n <Lebensmittel> calcium carbonate
Kohlensäuresättigung f <Chemie> carbonatation
Kohlenschacht m <Kohlen> coal mine
Kohlenschicht f <Kohlen> coal seam
Kohlenschichtboden m <Kohlen> bottom of a coal seam
Kohlenschiff n 1. <Kohlen> coal ship; 2. <Wassertrans> coal ship, collier
Kohlenschlacke f <Kohlen> slag
Kohlenschlamm m <Kohlen> coal sludge, coal slurry
Kohlenschlamm-Rohrleitung f <Kohlen> coal slurry pipeline
Kohlenschrämmaschine f <Kohlen> coal cracker, coal cutter
Kohlenschurre f <Kohlen> coal chute
Kohlensilo m <Kohlen> coal bunker
Kohlenstaub m 1. <Kohlen> coal powder, coaldust, pulverized coal; 2. <Maschinen> pulverized coal; 3. <Umweltschmutz> coal dust
Kohlenstaubbrenner m <Heiz & Kälte> pulverized coal burner
Kohlenstaubfeuerung f <Heiz & Kälte> pulverized coal firing
Kohlenstaubmikrofon n <Aufnahme> carbon microphone
Kohlenstoff m (C) <Chemie> carbon (C) • **auf Kohlenstoff basiert** <Anstrich> organic
kohlenstoffarm adj <Fertig> mild *(Stahl)*
kohlenstoffarmer Stahl m <Metall> low-carbon steel
Kohlenstoffdioxid n <Chemie> carbon dioxide
Kohlenstoffdisulfid n <Chemie> (AE) carbon disulfide, (BE) carbon disulphide
Kohlenstoffentziehung f <Kohlen> decarbonization
Kohlenstofffaser f <Kunststoff> (AE) carbon fiber, (BE) carbon fibre
Kohlenstofffeuerlöscher m <Sicherheit> carbon dioxide fire extinguisher
kohlenstofffrei adj <Metall> carbon-free
kohlenstoffhaltig adj <Chemie> carbonaceous
Kohlenstoffkreislauf m <Lebensmittel> carbon cycle
Kohlenstoffmassentransport m <Kerntech> carbon mass transfer
kohlenstoffreicher Stahl m <Metall> high-carbon steel
Kohlenstoffstahl m 1. <Anstrich> carbon steel; 2. <Fertig> ordinary steel; 3. <Kerntech, Kohlen, Mechan> carbon steel; 4. <Metall> carbon steel, ordinary steel
Kohlenstoffstahlstaub m <Kohlen> carbon steel dust
Kohlenstofftetrachlorid n <Chemie> carbon tetrachloride, tetrachloromethane
Kohlenstoß m <Kohlen> coalface

Kohlenstoßpflock m <Kohlen> coalface cleat
Kohlenstraße f <Kohlen> coal road
Kohlenumwandlung f <Kohlen> coal conversion
Kohlenwagen m 1. <Eisenbahn> tender; 2. <Kfztech, Kohlen> mine car
Kohlenwand f <Kohlen> coal wall
Kohlenwaschanlage f <Kohlen> coal washer
Kohlenwaschen n <Kohlen> coal washing
Kohlenwasserstoff m 1. <Chemie> hydrocarbon *(KW-Stoff)*; 2. <Erdöl, Meerschmutz, Umweltschmutz> hydrocarbon
Kohlenwasserstoff-Aerosoltreibgas n <Erdöl> hydrocarbon aerosol propellant
Kohlenwasserstoff-Einsatzprodukt n <Erdöl> hydrocarbon feedstocks *(Raffinerie)*
Kohlenwasserstofffalle f <Erdöl> hydrocarbon trap *(Geologie)*
Kohlenzeche f <Kohlen> coal mine, coal pit
Kohleschicht f <Elektrotech> carbon film
Kohleschichtwiderstand m <Elektrotech> carbon film resistor
Kohleverbrennung f <Kerntech> carbon burning
Kohleverflüssigung f <Erdöl> coal liquefaction *(Kohlehydrierung)*
Kohlevergasung f 1. <Erdöl> coal gasification *(Gaserzeugung)*; 2. <Thermod> coal gasification
Kohlevorkommen n <Kohlen> coal deposit
Kohlewiderstand m <Elektriz, Elektrotech, Phys> carbon resistor
Kohle-Zink-Zelle f <Labor> Leclanché cell
kohobieren v <Chemie> cohobate *(Destillation)*
Koinzidenz f <Aufnahme, Elektronik, Phys, Strahlphys> coincidence
Koinzidenzeffekt m <Aufnahme> coincidence effect
Koinzidenzschaltung f 1. <Elektronik> coincidence circuit *(Datenverarbeitung)*; 2. <Phys, Strahlphys> coincidence circuit
Koje f <Wassertrans> berth
Kokille f 1. <Fertig> metallic die, (AE) permanent mold, (BE) permanent mould; chill *(Gießen)*; 2. <Mechan> (AE) ingot mold, (BE) ingot mould • **in Kokillen gießen** <Fertig> chill
Kokillenausbruch m <Fertig> dressing
Kokillenguss m 1. <Fertig> chill casting, chilling, die casting, (AE) permanent-mold casting, (BE) permanent-mould casting; 2. <Maschinen> gravity die-casting; 3. <Metall> chill casting, gravity die-casting
Kokillenkleber m <Verpack> chill permanent adhesive
Kokosbast m <Wassertrans> coir *(Tauwerk)*
Koks m <Ker & Glas, Kohlen, Maschinen, Metall> coke
Koksbrechanlage f <Kohlen> coke breaker
Koksbrechen n <Kohlen> coking cracking
Koksgabel f <Kohlen> coke fork
Koksgrus m <Kohlen> coking duff
Kokskohle f <Kohlen> coking coal
Kokskorb m <Kohlen> coke basket
Koksmühle f <Kohlen> coke mill
Koksofen m <Heiz & Kälte, Ker & Glas> coke oven
Koksofengas n <Kohlen> coke oven gas
Koksroheisen n <Metall> coke iron, coke pig
Kokssäule f <Kohlen> coke column
Koksschmiedeeisen n <Metall> coke iron
Koksstaub m <Kohlen> coke dust
Kolben m 1. <Elektriz> bulb; 2. <Elektronik> cone *(einer Katodenstrahlröhre)*; 3. <Elektrotech> bulb, envelope; bulb *(Thermometer)*; 4. <Fertig> flask; 5. <Hydraul, Kfztech> piston *(Motor)*; 6. <Kunststoff> ram; 7. <Labor> flask; 8. <Maschinen> bulb; 9. <Mechan, Wassertrans> piston *(Motor)*

Kolbenabschwächer m <Elektronik, Funktech> piston attenuator *(Mikrowellen)*
Kolbenachsendichtung f <Raumfahrt> gland *(Raumschiff)*
Kolbenblitz m <Foto> flash bulb
Kolbenboden m 1. <Kfztech> piston head, piston top; piston crown *(Motor)*; 2. <Maschinen> piston head
Kolbenbohrung f <Fertig> cylinder *(Kolbenpumpe)*
Kolbenbolzen m 1. <Kfztech> *(AE)* wrist pin; *(BE)* gudgeon pin *(Motor, Kolben)*; 2. <Maschinen, Mechan> *(BE)* gudgeon pin, *(AE)* wrist pin
Kolbenbolzenauge n <Kfztech> piston pin boss
Kolbenbolzenbuchse f <Kfztech> piston-pin bushing, small end bushing
Kolbenbolzennabe f <Kfztech> piston pin boss
kolbenbolzenseitiger Pleuelkopf m <Maschinen> connecting rod small end
kolbenbolzenseitiges Ende n <Maschinen> small end
Kolbenbolzensicherung f <Kfztech> *(BE)* gudgeon pin lock, *(AE)* piston-pin lock, *(AE)* wrist pin lock
Kolbenbuchse f <Kfztech> piston boss bushing
Kolbendichtung f <Maschinen> piston packing, piston seal
Kolbendosierpumpe f <Gerät> piston-type metering pump
Kolbenentlastungskanal m <Bau> piston relief duct
Kolbenfläche f <Maschinen> piston area
Kolbenfresser m <Kfztech> piston seizing
Kolbengebläse n <Maschinen> piston blower
Kolbengeschwindigkeit f <Maschinen> piston speed
Kolbenhub m <Bau, Maschinen> piston stroke
Kolbenhublänge f <Kfztech> length of piston stroke
Kolbenklopfen n <Kfztech> piston knock
Kolbenkompressor m <Maschinen> piston compressor, positive-displacement compressor
Kolbenkopf m <Kfztech> piston top
Kolbenmanometer n <Gerät> *(AE)* piston-type pressure gage, *(BE)* piston-type pressure gauge
Kolbenmantel m 1. <Kfztech> piston body, piston skirt; 2. <Maschinen> piston surface
Kolbenmaschine f <Kfztech, Lufttrans, Maschinen, Wassertrans> reciprocating engine
Kolbenmotor m <Kfztech, Lufttrans, Wassertrans> piston engine, reciprocating engine
Kolbenplatte f <Mechan> butt plate
Kolbenpleuelstange f <Kfztech> piston connecting rod
Kolbenpumpe f <Maschinen, Wasserversorg> piston pump, reciprocating pump
Kolbenring m <Kfztech, Maschinen, Mechan> piston ring
Kolbenringfressen n <Maschinen> piston ring sticking
Kolbenringnut f <Kfztech> piston ring groove
Kolbenringnute f <Kfztech> piston ring groove
Kolbenringspanner m <Kfztech> piston ring clamp *(Werkzeug)*
Kolbenringstoßfuge f <Kfztech> piston ring gap
Kolbenringteilfuge f <Kfztech> piston ring joint
Kolbenschaft m 1. <Kfztech> piston body, piston skirt; 2. <Maschinen> piston body
Kolbenschieber m <Maschinen> piston valve
Kolbenschlag m <Kfztech> piston slap *(Motor)*
kolbenseitiger Pleuelstangenkopf m <Kfztech> connecting rod small end
kolbenseitiges Pleuelstangenende n <Kfztech> small end
Kolbenspiel n <Kfztech, Maschinen> piston clearance
Kolbenstange f 1. <Eisenbahn> piston rod; 2. <Fertig> piston rod, stem; 3. <Kfztech, Maschinen, Wassertrans> piston rod
Kolbenstange f der Pumpe <Wasserversorg> pump rod

Kolbensteg m <Kfztech> piston land *(Motor)*
Kolbensteuerventil n <Fertig> plunger valve *(Kunststoffinstallationen)*
Kolbenverdichter m 1. <Heiz & Kälte> reciprocating compressor; 2. <Maschinen> piston compressor, positive-displacement compressor, reciprocating compressor, reciprocating piston compressor
Kolbenverklemmung f <Fertig> mug lock
Kolbenweg m <Maschinen> piston travel
Kolbenzähler m <Gerät> piston-type flowmeter
Kolk m <Fertig, Maschinen> crater
Kolkbildung f <Fertig> cratering *(Spanung)*; erosion *(Spanfläche)*
Kolkung f <Fertig> crater *(Spanfläche)*
Kolkverschleiß m <Maschinen> crater wear
Kollagen n <Lebensmittel> collagen
Kollargol n <Chemie> collargol
kollationieren v <Druck> collate
Kollationiermarken fpl <Druck> collating marks
Kollationiermaschine f <Druck> collating machine
Kollationiertisch m <Druck> collating table
kollektive Anregung f bei Teilchenwechselwirkung <Strahlphys> collective excitation in particle interaction
kollektive Schutzausrüstung f <Sicherheit> group safety equipment
Kollektivkniehebelsystem n <Lufttrans> collective bellcrank *(Hubschrauber)*
Kollektivmarke f <Patent> collective mark
Kollektivmodell n <Kerntech> collective model
Kollektivsteigung f <Lufttrans> collective pitch *(Hubschrauber)*
Kollektivsteigungsanzeiger m <Lufttrans> collective pitch indicator
Kollektivsteigungssteuerung f <Lufttrans> collective pitch control *(Hubschrauber)*
Kollektivsteigungssynchronisierung f <Lufttrans> collective pitch synchronizer *(Hubschrauber)*
Kollektor m 1. <Elektriz, Elektronik, Elektrotech> commutator, drain; collector *(bei dynamoelektrischer Maschine)*; 2. <Lufttrans> manifold; 3. <Nichtfoss Energ> collector; 4. <Phys> collector *(elektrisch)*
Kollektor m des Transistors <Elektronik> transistor collector
Kollektorabdeckplatte f <Nichtfoss Energ> collector cover plate
Kollektoranreicherung f <Elektronik> collector doping
Kollektoranschluss m <Elektrotech> common collector connection
Kollektorbasisstufe f <Elektrotech> emitter follower
Kollektorbürste f <Elektriz> commutator brush
Kollektorelektrode f <Elektrotech> collector electrode
Kollektorgebiet n <Elektrotech> collector region
Kollektorkontakt m <Elektrotech> collector contact
Kollektorleistungsvermögen n <Nichtfoss Energ, Qual> collector efficiency
Kollektormotor m <Elektrotech> collector motor
Kollektorring m <Elektrotech, Phys> collector ring, slip ring
Kollektorschaltung f <Elektrotech> *(BE)* earthed collector connection, *(AE)* grounded collector connection
Kollektorstab m <Elektrotech> commutator bar
Kollektortiltwinkel m <Nichtfoss Energ> collector tilt angle
Kollektorverstärker m <Elektrotech> emitter follower
Kollektorzone f <Elektrotech> collector region
Kollergang m 1. <Fertig> pug mill, sand mill; 2. <Hydraul> pug mill; 3. <Papier> edge runner
Kollermühle f 1. <Ker & Glas> edge runner mill; 2. <Kunststoff> edge mill

kollidieren v 1. <Trans> collide; 2. <Wassertrans> collide *(mit Schiff)*
kollidieren v **mit** <Trans> collide with
kolligativ adj <Chemie> colligative
Kolliko n <Trans> collico
Kollimation f 1. <Elektronik> collimation *(Antennentechnik)*; 2. <Fertig, Nichtfoss Energ, Optik> collimation
Kollimationslicht n <Nichtfoss Energ> collimated light
Kollimator m <Fertig, Foto, Phys, Telekom> collimator
Kollimatorfehler m <Foto> collimating fault
Kollimatorlinse f <Foto> collimated lens
kollimierte Punktquelle f <Kerntech> collimated point source
kollinear adj <Geom> collinear
kollineare Laserspektroskopie f <Strahlphys> collinear laser spectroscopy
Kollision f 1. <Comp & DV, Eisenbahn> collision; 2. <Raumfahrt> collision *(Raumschiff)*; 3. <Teilphys, Telekom, Wassertrans> collision
Kollisionsenergie f <Teilphys> collision energy
Kollisionserkennung f *(CD)* <Comp & DV, Telekom> collision detection *(CD)*
Kollisionskurs m 1. <Raumfahrt> collision course *(Raumschiff)*; 2. <Wassertrans> collision course *(Navigation)*
Kollisionsschutzradar n <Funkort> anticollision radar
Kollisionsschutzschalter m <Sicherheit> collision avoidance controller
Kollisionsverhütungshilfen f <Wassertrans> collision avoidance aids *(Radar)*
Kollisionswarner m <Wassertrans> anticollision marker
Kollisionswarnsystem n <Raumfahrt> collision warning system *(Raumschiff)*
Kollodium n <Druck, Foto> collodion
Kollodiumplatte f <Foto> collodion plate
Kolloid n 1. <Ker & Glas, Kohlen, Kunststoff> colloid; 2. <Lebensmittel> colloid *(gallertartiger Stoff)*
Kolloid... <Chemie> colloidal
kolloidal adj <Chemie> colloidal
kolloidaler Bohrschlamm m <Erdöl> colloidal mud *(Bohrtechnik)*
Kolloidantrieb m <Raumfahrt> colloid propulsion *(Raumschiff)*
kolloiddisperses System n <Umweltschmutz> colloide disperse system
Kolloidmühle f <Lebensmittel> colloid mill
Kolonne f 1. <Chemie, Druck> column; 2. <Trans> platoon
Kolonnenauflösung f <Trans> platoon dispersion
Kolophonium n <Fertig, Kunststoff> rosin
kolorieren v <Druck> *(AE)* color, *(BE)* colour
Kolorimeter n <Ergon, Kunststoff, Labor, Papier, Phys, Strahlphys> colorimeter
Kolorimetrie f <Ergon, Phys, Strahlphys, Wasserversorg> colorimetry
Kolormetrie f <Wasserversorg> colorimetry
Kolumne f <Druck> column
Koma n 1. <Elektronik> coma *(kometenschweifartiger Abbildungsfehler)*; 2. <Optik> coma *(Asymmetriefehler)*
Kombi m <Kfztech> combined
Kombiflugschrauber m <Lufttrans> gyrodyne
Kombimassengutfrachter m <Wassertrans> combination bulk carrier
Kombination f 1. <Akustik, Aufnahme, Bau, Comp & DV> combination; 2. <Lufttrans> combination *(Hubschrauber)*; 3. <Maschinen> combination
Kombinationsfutter n <Maschinen> combination chuck
Kombinationsmikrofon n <Aufnahme> combination microphone
Kombinationsschloss n <Bau> combination lock, puzzle lock

Kombinationston m <Akustik> combination tone, complex tone
Kombinator m <Telekom> combiner
Kombinatorik f <Math> combinatorics
kombinatorisch adj <Elektronik, Geom, Künstl Int> combinatorial
kombinatorische Explosion f <Künstl Int> combinatorial explosion
kombinatorische Logik f <Elektronik, Künstl Int> combinatorial logic
kombinieren v <Comp & DV> combine
kombinierte Abfrage f <Comp & DV> relational query
kombinierte Axial-Radialturbine f <Hydraul> combined flow turbine
kombinierte Verpackung f <Verpack> combined packaging
kombinierter Anzeiger m **für Gleitweg und Landkurs** <Funkort, Lufttrans> glide path localizer
kombinierter Deckel-Böden-Flaschenkarton m <Verpack> coupled lid-base bottle tray
kombinierter Querneigungs- und Steigungsanzeiger m <Lufttrans> *(BE)* bank-and-pitch indicator, *(AE)* turn-and-bank indicator
kombinierter Repulsions-Induktionsmotor m <Elektrotech> repulsion-induction motor
kombinierter Verkehr m <Eisenbahn> intermodal traffic
kombinierter Verteiler m *(KV)* <Telekom> combined distribution frame *(CDF)*
kombiniertes Druck- und Kopiergerät n <Comp & DV> copier-printer, printer copier
kombiniertes Förder- und Tragseil n <Wassertrans> combined hauling and carrying rope
kombiniertes Höhen- und Querruder n <Lufttrans> elevon
kombiniertes Kabel n <Elektrotech> composite cable
kombiniertes Straßen-See-Transportsystem n <Trans> road-sea combined transport system
Kombischaltung f <Comp & DV> *(AE)* combinational circuit, *(BE)* combinatorial circuit
Kombischiff n <Wassertrans> combined vessel
Kombiwagen m <Kfztech> *(BE)* estate car, *(BE)* shooting brake, *(AE)* shooting break, *(AE)* station wagon
Kombiwerk n <Elektriz> combined heat and power station
Kombiwerkzeug n <Maschinen> combination tool
Kombizange f <Maschinen> all-purpose wrench, combination pliers
Kombüse f <Wassertrans> galley *(Schiff)*
Kombüsenausstattung f <Lufttrans, Wassertrans> galley furnishings
Kometenkern m <Raumfahrt> comet core *(Raumschiff)*
Komfortfernsprechapparat m <Telekom> added-feature telephone, comfort telephone set
Komfortklimaanlage f <Heiz & Kälte> comfort air-conditioning plant
Komforttelefon n <Telekom> added feature telephone
Komma n 1. <Akustik, Druck> comma; 2. <Math> comma, decimal point
Kommandant m <Raumfahrt, Wassertrans> commander *(Marine)*
Kommando n <Comp & DV, Raumfahrt> command
Kommandobrücke f <Wassertrans> navigation bridge *(Handelsmarine)*
Kommandoempfänger m <Raumfahrt> command receiver
Kommandokanal m <Raumfahrt> command channel
Kommandokapsel f <Raumfahrt> command module
Kommandoleitung f <Fernseh, Funktech> talkback circuit
Kommandomodul n <Raumfahrt> command module

Kommando- und Servicemodul n (CSM) <Raumfahrt> command and service module, CSM (Raumschiff)
Kommandoverbindung f <Raumfahrt> command link
Kommandozentrale f <Wassertrans> (AE) command-and-control center, (BE) command-and-control centre (Schiff)
kommen v <Telekom> over (Umschalten auf Empfang im Semi-Duplex-Funkverkehr)
kommensurabel adj <Math> commensurable (verträglich)
Kommentar m <Comp & DV> comment • **auf Kommentar setzen** <Comp & DV> comment
Kommentarspur f <Aufnahme> commentary track, narration audio track
kommentieren v <Comp & DV> comment
kommerzielle Datenverarbeitung f <Comp & DV> commercial computing
Kommerzieller und Amateurfunkdienst m (ICAS) <Funktech> Intermittent Commercial and Amateur Services (ICAS)
kommerzieller Verkehr m <Trans> revenue-earning traffic
Kommodore m <Wassertrans> commodore (Marine)
kommunale Entsorgung f <Abfall> municipal waste disposal
kommunale Kläranlage f <Abfall> municipal sewage works
kommunaler Abfall m <Abfall> municipal waste, urban solid waste, urban waste
kommunaler Güterverkehr m <Trans> public hauling
kommunales Abwasser n <Abfall> municipal sewage, municipal waste water
kommunales Wasser n <Wasserversorg> town water
Kommunikation f <Comp & DV, Kontroll, Telekom> communication
Kommunikation f **offener Systeme** (OSI) <Comp & DV, Telekom> open systems interconnection (OSI)
Kommunikation f **vor Ort** <Telekom> on-scene communications
Kommunikation f **zwischen Computern** <Comp & DV> computer-to-computer communications
Kommunikationsendgerät n <Telekom> communication terminal
Kommunikationskanal m 1. <Comp & DV> pipe; 2. <Mobilkom> traffic channel (GSM)
Kommunikationskontrollprotokoll n <Comp & DV> transmission control protocol (im Internet)
Kommunikationsmedium n <Comp & DV> communication medium
Kommunikationsnetz n <Kontroll> communication network
Kommunikationsprotokoll n <Telekom> communication protocol (Regeln für Verbindungen im Rechnernetz)
Kommunikationsschicht f <Telekom> session layer (in Netzen)
Kommunikationsschnittstelle f <Künstl Int, Telekom> communication interface
Kommunikationsserver m <Telekom> communication server
Kommunikationssicherheit f <Telekom> communication security
Kommunikationssoftware f 1. <Comp & DV> communication software, messaging software; 2. <Telekom> communication software
Kommunikationssteuerungsprotokoll n <Telekom> OSI-layer 5 protocol, session layer protocol
Kommunikationssteuerungsschicht f <Comp & DV, Telekom> OSI-layer No5, session layer
Kommunikationssystem n <Comp & DV, Telekom> communication system, messaging system
Kommunikationstheorie f <Comp & DV, Telekom> communication theory

Kommunikationsverbindung f <Comp & DV> communication link, data link
Kommutation f 1. <Elektriz, Elektrotech> commutation; 2. <Math> commutation (Vertauschung)
kommutativ adj <Math> commutative (vertauschbar)
Kommutativgesetz n <Math> commutative law (Vertauschungsgesetz)
Kommutator m <Elektriz, Elektrotech, Kfztech> commutator
Kommutatorbürste f <Kfztech> commutator brush (KFZ-Elektrik, Kollektorbürste)
Kommutator-Drehstromerregermaschine f <Elektrotech> phase advancer
Kommutator-Druckring m <Elektrotech> commutator V-ring
Kommutator-Druckringisolation f <Elektrotech> commutator V-ring insulation
Kommutatorfahne f <Elektrotech> commutator lug, commutator riser
Kommutator-Frequenzwandler m <Elektrotech> commutator-type frequency converter
Kommutatorgleichstrommotor m <Elektrotech> commutator dc motor
Kommutatorhebelschalter m <Elektriz> lever commutator switch
Kommutatorlamelle f <Elektriz, Elektrotech> commutator bar, commutator segment
Kommutatorlamelleisolation f <Elektrotech> commutator segment insulation, commutator separator
Kommutatormotor m <Elektriz, Elektrotech> commutator motor
Kommutatorrichtungsschalter m <Elektriz> commutator switch
Kommutatorring m <Elektriz> commutator ring
Kommutatorschritt m <Elektrotech> commutator pitch
Kommutatorsprühen n <Elektriz> commutator sparking
Kommutatorstab m <Elektriz> commutator bar
Kommutatorsteg m <Elektrotech> commutator segment
Kommutator-Stegisolation f <Elektrotech> commutator segment insulation
kommutierbarer Umformer m/**unter Last** <Elektriz> load-commutated converter
Kommutierpol m <Elektriz> commutating pole
Kommutierung f <Elektronik, Elektrotech> commutation
Kommutierungsachse f <Elektrotech> axis of commutation
Kommutierungsdrossel f <Elektronik> commutation inductance
Kommutierungsfehler m <Elektronik> commutation failure (beim Stromrichter)
Kommutierungs-Grenzkurven f <Elektrotech> black band
Kommutierungskapazität f <Elektrotech> commutation capacitor
Kommutierungskurve f <Elektrotech> commutation curve (Hysteresisschleife)
Kommutierungsstromübergang m <Elektronik> commutation repetitive transient
Kommutierungsversager f <Elektrotech> commutation failure
Kommutierungszeit f <Elektronik> commutation interval
kompakt adj 1. <Anstrich, Aufnahme, Elektriz, Kfztech> compact, solid; 2. <Maschinen> compact; 3. <Optik> tight
Kompaktader f <Optik> tight buffer (Lichtwellenleiter)
kompakte Hochtemperatur-Elektrolysezelle f <Kfztech> high-temperature solid electrolyte cell
kompakter Leistungsschalter m <Elektrotech> moulded-case circuit-breaker
kompaktes Mantelkabel n <Optik> tight-jacketed cable

Kompaktierung

Kompaktierung f <Abfall> compaction
Kompaktkabel n <Optik> tight construction cable
Kompaktkassette f <Aufnahme> compact cassette
Kompaktleiter m <Elektriz> compacted conductor
Kompaktor m <Abfall> landfill compactor, packer unit
Kompakt-Quarzoszillator m <Elektronik> simple-packaged crystal oscillator
Kompander m 1. <Comp & DV> compander; 2. <Elektronik> compander *(Kompressor und Expander)*; 3. <Raumfahrt> compander; 4. <Telekom> compressor-expander
Kompandersignal n <Elektronik> companded signal
Kompandierer m <Telekom> compressor-expander
kompandierte Deltamodulation f *(CDM)* <Telekom> companded delta modulation *(CDM)*
Kompandierung f <Elektronik, Funktech, Telekom> companding
Komparator m <Comp & DV, Elektronik, Kontroll, Metrol, Telekom> comparator
Komparatorschaltung f <Elektronik> comparator circuit
Kompass m 1. <Bau> *(AE)* leveling compass, *(BE)* levelling compass; 2. <Lufttrans> compass
Kompassberichtigung f <Lufttrans, Wassertrans> compass compensating
Kompassbüchse f <Phys> compass bowl
Kompassdiopter m <Bau> sight vane
Kompasseingang m <Lufttrans, Wassertrans> compass input *(Radar)*
Kompassgefäß n <Phys> compass bowl
Kompassgehäuse n <Wassertrans> binnacle
Kompasshaube f <Wassertrans> binnacle cover
Kompasshausdeckel m <Wassertrans> binnacle cover
Kompasskompensierscheibe f <Lufttrans, Wassertrans> compass compensation base
Kompasskurs m <Lufttrans, Wassertrans> compass heading *(Navigation)*
Kompassmissweisung f <Lufttrans, Wassertrans> compass variation
Kompassnadel f <Phys> compass needle
Kompasspeilung f <Lufttrans, Wassertrans> compass bearing *(Navigation)*
Kompassrose f <Lufttrans, Wassertrans> compass card, compass dial
Kompassstrich m <Wassertrans> rhumb *(Kompass)*
kompatibel adj <Comp & DV> compatible
Kompatibilität f <Comp & DV, Ergon, Fernseh, Maschinen, Telekom> compatibility
Kompendium n <Foto> matte box *(Linse)*
Kompensation f 1. <Elektriz, Elektronik, Elektrotech> compensating, compensation; 2. <Kerntech> compensating; 3. <Phys, Telekom> compensation
Kompensationsbandschreiber m <Gerät> compensating strip chart recorder, strip chart potentiometric recorder
Kompensationsentwickler m <Foto> compensating developer
Kompensationsfilter n <Foto> compensating filter
Kompensationsglied n <Aufnahme, Telekom> equalizer
Kompensationshalbleiter m <Elektrotech> compensated semiconductor
Kompensationsinstrument n <Gerät> potentiometer instrument
Kompensationskreis m 1. <Fernseh> bucking circuit; 2. <Gerät> potentiometer circuit
Kompensationsmagnet m <Wassertrans> compensating magnet *(Kompass)*
Kompensationsmessgerät n <Gerät> compensating instrument, potentiometer instrument, potentiometric meter
Kompensationsmethode f <Akustik> cancellation method

Kompensationspendel n <Phys> compensated pendulum
Kompensationsschaltung f 1. <Elektriz> compensating circuit; 2. <Gerät> compensating circuit, potentiometer circuit; 3. <Telekom> compensation circuit
Kompensationsschreiber m <Gerät> compensating recorder, null balance recorder, potentiometer recorder, self-balancing recorder
Kompensationsspannung f <Elektriz> compensating voltage
Kompensationsspule f <Aufnahme, Elektrotech> bucking coil
Kompensationsstab m <Kerntech> compensating rod
Kompensationsstrom m <Elektriz> compensating current
Kompensationstheorem n <Elektriz> compensation theorem
Kompensationsverstärker m <Elektronik> compensated amplifier
Kompensations-Voltmeter n <Gerät> compensated voltmeter
Kompensationswicklung f <Elektrotech> compensation winding
Kompensator m 1. <Elektronik, Elektrotech, Gerät> compensator; 2. <Hydraul> expansion joint; 3. <Phys> compensator
Kompensatormaschine f <Elektriz> compensator machine
Kompensatorschaltung f <Gerät> potentiometer circuit
kompensieren v <Kohlen, Phys, Wassertrans> compensate *(Kompass)*
kompensierendes Registriergerät n <Gerät> compensating recorder
Kompetenz f <Ergon> proficiency
Kompilierbefehl m <Comp & DV> compiler directive
kompilieren v <Comp & DV> compile
kompilierendes Programm n <Comp & DV> compiler generator
Kompilierer m <Comp & DV, Telekom> compiler
Kompilierfehler m <Comp & DV> compilation error
Kompilierung f <Comp & DV> compilation
Kompilierzeit f <Comp & DV> compilation time
komplan adj <Fertig> coplanar
komplanar adv <Geom> coplanar *(synonym für koplanar)*
Komplement n 1. <Comp & DV> complement; 2. <Elektronik> complement *(zur Darstellung von Negativwerten)*
Komplementär... <Foto, Phys> complementary
Komplementärausgaben fpl <Elektronik> complementary outputs
komplementäres Paar n <Elektronik> complementary pair *(Transistoren)*
Komplementärfarbe f <Foto, Phys> *(AE)* complementary color, *(BE)* complementary colour
Komplementarität f <Phys, Teilphys> complementarity
Komplementaritätsprinzip n <Phys> principle of complementarity
Komplementär-Metalloxid-Halbleiter m *(CMOS)* <Comp & DV, Elektronik> complementary metal oxide semiconductor *(CMOS)*
Komplementär-Symmetrischer Metalloxid-Halbleiter m *(COSMOS)* <Elektronik> complementary-symmetrical metal oxide semiconductor *(COSMOS)*
Komplementärtransistoren mpl <Elektronik> complementary transistors
Komplementärwinkel m <Geom> complementary angles
Komplementmenge f <Math> complementary set
Komplettdrucker m <Comp & DV> alphanumeric printer, character printer, complete character printer
komplette Bremse f <Kfztech> brake assembly
Komplettierung f <Erdöl> completion *(Bohrtechnik)*

Komplettschnitt *m* <Maschinen> combination die, compound die
komplex *adj* 1. <Comp & DV, Elektriz, Elektrotech, Geom, Kontroll, Phys> complex; 2. <Telekom> sophisticated *(Gerät)*
Komplexbildner *m* 1. <Chemie> chelating agent, complexing agent; 2. <Kerntech> complexing agent
Komplexbildung *f* <Chemie> chelation, complexing
komplexe Admittanz *f* <Elektriz> complex admittance
komplexe Größe *f* <Math> complex quantity, complex value
komplexe Impedanz *f* <Elektriz> complex impedance
komplexe Kreisverstärkung *f* <Regelung> complex loop chain
komplexe Leistung *f* <Elektriz> complex power *(AC-Leistungsberechnung)*
komplexe Permeabilität *f* <Elektriz> complex permeability
komplexe Variable *f* <Comp & DV> complex variable
komplexe Wellenform *f* <Elektriz> complex waveform
komplexe Zahl *f* <Comp & DV, Math> complex number
komplexer Brechungsindex *m* <Phys> complex refractive index
komplexer Regelfaktor *m* <Regelung> complex control factor
komplexer Scheinwiderstand *m* <Elektrotech> complex impedance
komplexer Widerstand *m* <Phys> complex impedance
komplexer Wirkleitwert *m* <Phys> complex admittance
Komplexerz *n* <Kohlen> complex ore
komplexes Netz *n* <Comp & DV> network complexity
komplexes Signal *n* <Elektronik> complex signal
Komplexität *f* <Chemie, Comp & DV> complexity
Komponente *f* 1. <Elektriz, Elektronik> component *(Funktionselement)*; 2. <Kerntech, Kohlen, Maschinen> component; 3. <Math> component *(eines Vektors)*; 4. <Mechan, Patent> component; 5. <Telekom> component *(physikalisch)*
Komponentenzusammenbau *m* <Kfztech> component assembly
Kompost *m* <Abfall> compost
Kompostaufbereitungsanlage *f* <Abfall> composting plant
Kompostaufbetrieb *m* <Abfall> composting plant
Kompostbelüftung *f* <Abfall> compost aeration
Kompostbereitung *f* <Abfall> composting
kompostierbarer Abfall *m* <Abfall> compostable waste
Kompostierung *f* <Abfall> composting *(von Hausmüll)*
Kompostierungsanlage *f* <Abfall> composting plant
Kompostierungsrückstand *m* <Abfall> composting residue
Kompostierungstechnik *f* <Abfall> composting technique
Kompostreifung *f* <Abfall> compost maturing, compost ripening
Kompostwerk *n* <Abfall> composting plant
Kompoundierung *f* <Elektrotech> compounding
Kompoundmotor *m* <Elektriz, Elektrotech> compound motor
Kompressibilität *f* <Lufttrans, Metall, Papier, Phys, Strömphys, Thermod> compressibility *(Gas)*
Kompressibilitätseffekte *mpl* <Lufttrans, Qual> compressibility effects
Kompressibilitätsfaktor *m* (z) <Thermod> compressibility factor *(z)*
Kompressibilitätswiderstand *m* <Lufttrans> compressibility drag
kompressible Strömung *f* <Strömphys> compressible flow

Kompression *f* 1. <Abfall> compaction; 2. <Aufnahme> compression; 3. <Comp & DV> compaction *(von Daten)*; 4. <Elektronik, Erdöl, Hydraul, Kfztech, Kunststoff, Papier, Phys, Thermod, Wellphys> compression
Kompressionsarbeit *f* <Maschinen> compression work
Kompressionsbeiwert *m* <Kohlen, Qual> compression index
Kompressionsfaktor *m* <Telekom> compression factor, compression ratio *(bei Datenkompression)*
Kompressionsfeder *f* <Hydraul> compression spring
Kompressionsfilter *n* <Elektronik> compression filter
Kompressionshahn *m* <Hydraul> compression cock
Kompressionshub *m* <Kfztech> compression stroke *(Motor)*
Kompressionskältemaschine *f* <Thermod> compression plant, compression refrigerator
Kompressionskammer *f* 1. <Erdöl> compression chamber *(Tieftauchtechnik)*; 2. <Hydraul, Kohlen> compression chamber
Kompressionskoeffizient *m* <Phys, Qual> coefficient of compressibility
Kompressionskühlung *f* <Heiz & Kälte> compression refrigeration
Kompressionsmesser *m* <Papier> compressometer
Kompressionsmodul *n* (K) <Hydraul> bulk modulus of compression *(K)*
Kompressionsperiode *f* <Hydraul, Qual> compression period
Kompressionspumpe *f* <Hydraul> compression pump
Kompressionspunkt *m* <Hydraul> compression point
Kompressionsraum *m* <Kfztech> compression chamber
Kompressionsring *m* <Kfztech> compression ring *(Motor)*
Kompressionsstufe *f* <Maschinen> compression stage
Kompressionstreiber *m* <Aufnahme> compression driver
Kompressions- und Dekompressionsmanager *m* <Comp & DV> audio compression manager, ACM *(für Audiodateien)*
Kompressionsverhältnis *n* 1. <Aufnahme> compression ratio; 2. <Elektronik> compression ratio *(Mikrowellen)*; 3. <Heiz & Kälte> compression ratio; 4. <Kfztech> compression ratio *(Motor)*; 5. <Maschinen, Qual, Thermod> compression ratio
Kompressionswärme *f* <Thermod> heat of compression
Kompressionszeitraum *m* <Hydraul, Qual> compression period
Kompressor *m* 1. <Erdöl> compressor *(Maschine)*; 2. <Fertig> compressor; 3. <Kfztech> compressor *(Motor)*; 4. <Labor, Maschinen, Mechan, Papier, Telekom> compressor
Kompressoranlage *f* <Maschinen> compressor plant
Kompressorenschalldämpfer *m* <Sicherheit> noise protection for compressors
Kompressschrift *f* <Druck> condensed face
komprimierbarer Abfall *m* <Abfall> compressible waste
komprimieren *v* <Comp & DV> compress, pack *(Daten)*
komprimiert *adj* <Elektronik, Fernseh, Kerntech, Telekom, Verpack> compressed
komprimierte Digitalübertragung *f* <Telekom> compressed digital transmission
komprimierte Klebstoffzufuhr *f* <Verpack> pressurized glue feed
komprimierte Sprachinformationen *fpl* <Elektronik> compressed speech
komprimierter Videopegel *m* <Fernseh> compressed video level
komprimiertes Internet-Protokoll *n* **für serielle Leitungen** <Telekom> compressed serial line Internet protocol, CSLIP

komprimiertes

komprimiertes Kernmaterial n <Kerntech> compressed nuclear matter
komprimiertes Signal n <Elektronik> compressed signal
Komprimierung f <Comp & DV, Elektronik> compression (Hochfrequenztechnik)
Kondensat n 1. <Chemtech> condensate (Produkt); 2. <Erdöl, Heiz & Kälte, Maschinen, Thermod> condensate
Kondensatablassventil n <Hydraul> pet valve
Kondensatabscheider m 1. <Chemie> condenser; 2. <Chemtech> condenser cooler; 3. <Erdöl> steam trap
Kondensatdurchdringung f <Fertig> condensate permeation (Kunststoffinstallationen)
Kondensation f 1. <Bau> condensation; 2. <Chemtech> condensation (durch Oberflächenkühlung); 3. <Fertig> condensation (Kunststoffinstallationen); 4. <Heiz & Kälte> condensation; 5. <Ker & Glas> condensation (von Doppelverglasung); 6. <Labor, Maschinen, Phys, Thermod, Umweltschmutz, Verpack> condensation
Kondensationsanlage f <Chemtech> condensing plant
Kondensationsauffanggefäß n <Labor> condensing trap
Kondensationskern m <Thermod, Umweltschmutz> condensation nucleus
Kondensationskernzähler m <Umweltschmutz> condensation nucleus counter
Kondensationspolymer n <Kunststoff> condensation polymer
Kondensationspolymerisation f <Kunststoff> condensation polymerization
Kondensationsprodukt n <Chemtech> condensate
Kondensationssatz m <Heiz & Kälte> condensing set
Kondensationssatz m **mit Zwischenüberhitzung** <Heiz & Kälte> condensing set with reheat
Kondensationssäule f <Chemtech> condensation column
Kondensationsturbine f <Heiz & Kälte> condensing turbine
Kondensationsturm m <Chemtech> cooling tower
Kondensationswärme f <Thermod> condensation heat
Kondensator m 1. <Akustik, Aufnahme, Chemtech> condenser; 2. <Comp & DV> capacitor; 3. <Elektriz, Elektrotech> capacitor, condenser; 4. <Erdöl> condenser (Kühltechnik); 5. <Fernseh, Funktech> capacitor; 6. <Kfztech> condenser; capacitor (Zündung); 7. <Kohlen> condenser; 8. <Labor, Maschinen, Papier> capacitor, condenser; 9. <Phys, Telekom> capacitor; 10. <Wassertrans> condenser
Kondensator m **für punktförmig verteilte Kapazität** <Elektrotech> lumped capacitor
Kondensator m **mit Festelektrolyt** <Elektrotech> solid electrolyte capacitor
Kondensatorblindwiderstand m <Elektrotech> reactance capacitance, resistance capacitance
Kondensatorenblock m <Elektrotech> capacitor bank
Kondensatorenreihe f <Elektrotech> capacitor bank
Kondensatorentladung f <Elektrotech> capacitor discharge
Kondensatorgruppe f <Elektrotech> bank of capacitors
Kondensatorkühler m <Chemtech> condenser cooler
Kondensatorkühlwasserpumpe f <Heiz & Kälte> condenser circulating pump
Kondensatorlautsprecher m <Aufnahme> electrostatic loudspeaker
Kondensatormikrofon n 1. <Akustik> condenser microphone; 2. <Aufnahme> capacitor microphone, condenser microphone, electrostatic microphone
Kondensatormotor m <Elektriz, Elektrotech> capacitive starting motor, capacitor motor, capacitor-run motor, capacitor split-phase motor, (AE) permanent split capacitor motor

Kondensatormotor m **mit Anlauf- und Betriebskondensator** <Elektrotech> capacitor start-and-run motor, capacitor motor, (AE) permanent split capacitor motor, two-value capacitor motor
Kondensatorplatte f <Elektriz, Phys> capacitor plate
Kondensatorreihe f <Elektrotech> bank of capacitors
Kondensatorreststrom m <Elektrotech> capacitor leakage current
Kondensatorrohr n <Maschinen> condenser tube
Kondensatorsäule f <Elektrotech> capacitor stack
Kondensatorschalter m <Elektrotech> capacitor circuit-breaker, capacitor contactor
Kondensatorspeicher m <Comp & DV> capacitor store
Kondensatorstapel m <Elektrotech> capacitor stack
Kondensatorzündung f <Kfztech> capacitor ignition
Kondensatpumpe f <Chemtech> condensate pump
Kondensatsammelgefäß n <Chemtech> condensation trap
Kondensattopf m <Hydraul> steam trap
Kondenser m <Chemie> condenser
Kondensfahne f <Lufttrans> condensation trail
kondensierbar adj <Phys> condensable
kondensieren v 1. <Chemtech> condense (Lebensmittel); 2. <Thermod> condense (Flüssigkeit); 3. <Verpack> condense
kondensiert adj <Chemtech, Thermod> condensed
kondensiertes System n <Chemtech> condensed system
Kondensierung f <Maschinen, Verpack> condensation
Kondensierungswärme f <Thermod> condensation heat
Kondensmilch f <Lebensmittel> condensed milk (gezuckert)
Kondensor m 1. <Foto> condenser; 2. <Labor> condenser (Linsen-System); 3. <Optik> condenser
Kondensorlampe f <Foto> condenser lamp
Kondensorlinse f <Foto> condensing lens
Kondensorsystem n <Foto> condenser system (Vergrößerer)
Kondensstreifen m <Lufttrans> condensation trail
Kondenstopf m <Heiz & Kälte, Papier> steam trap
Kondenswasser n <Chemtech, Heiz & Kälte> condensate
Kondenswasserablauf m <Heiz & Kälte> condensate drain
Kondenswasserabscheider m <Heiz & Kälte> steam trap
Kondenswasserheizung f <Heiz & Kälte> anticondensation heater
Kondenswasserpumpe f <Chemtech> condensate pump
Konditionieren n <Kunststoff, Papier> conditioning
konditioniert adj <Patent, Textil> conditioned
Konditionierung f <Ergon, Kunststoff> conditioning
Konduktanz f <Elektriz, Elektrotech, Phys> conductance
Konduktanzbrücke f <Elektriz> conductance bridge
Konduktionsstrom m <Elektriz> conduction current
Konfektion f <Textil> making-up
Konfektionierung f <Telekom> packaging (Bauelement)
Konfektionsklebrigkeit f <Kunststoff> green tack, tack
Konferenz f <Fernseh, Telekom> conference
Konferenz f **mit mehreren Teilnehmern** <Telekom> add on conference call, CONF (Dienstmerkmal im Euro-ISDN)
Konferenz f **mit Voranmeldung und Einwahlmöglichkeit** <Telekom> meet me conference, MMC (Dienstmerkmal im Euro-ISDN)
Konferenzbridge f <Telekom> conference bridge
Konferenzbrücke f <Telekom> conference bridge
Konferenzeinrichtung f <Telekom> bridging (ISDN)

Konferenzgespräch *n* **im Internet** <Telekom> internet relay chat
Konferenzschaltung *f* <Fernseh> conference network
Konferenzverbindung *f* <Telekom> conference call
Konfidenzfaktor *m (KF)* <Künstl Int> certainty factor, confidence factor *(CF)*; degree of truth
Konfidenzintervall *n* <Math> confidence interval
Konfidenzmenge *f* <Math> confidence set
Konfidenzniveau *n* <Math> confidence level *(Überdeckungs-Wahrscheinlichkeit)*
Konfidenzwert *m* <Künstl Int> certainty factor, confidence factor, degree of truth
Konfiguration *f* <Comp & DV, Elektronik, Metall, Phys, Telekom> configuration
Konfigurationsabhängigkeiten *fpl* <Comp & DV> configuration dependencies
Konfigurationseinstellung *f* <Comp & DV> configuration setting
Konfigurationsentropie *f* <Metall> configurational entropy
Konfigurationsmanagement *n* <Comp & DV> configuration management
Konfigurationsraum *m* <Phys> configuration space
konfigurieren *v* <Comp & DV, Kontroll> configure
konfiguriert *adj* <Comp & DV> configured-in
Konfigurierung *f* **für reduzierte Last** <Raumfahrt> reduced load configuration *(Raumschiff)*
Konflikt *m* <Telekom> contention
Konfliktauflösung *f* 1. <Künstl Int> conflict resolution; 2. <Telekom> contention control
konfliktfreie Verkehrsströme *fpl* <Trans> nonconflicting traffic flows
Konfliktlösung *f* <Künstl Int> conflict resolution
Konfliktstelle *f* <Trans> conflict point
Konfliktverkehrsströme *mpl* <Trans> conflicting traffic flows
Konformität *f* <Patent, Qual> compliance, conformity
Konformitätsbescheinigung *f* 1. <Maschinen> certificate; 2. <Patent, Qual> COC, certificate of compliance, certificate of conformance, conformity certificate
Konformitätszeichen *n* <Qual> mark of conformity
Konformitätszertifikat *n* <Patent, Qual> certificate of conformity
Konglomerat *n* <Nichtfoss Energ> conglomerate
kongruent *adj* <Geom, Math> congruent *(Zahlen)*
kongruente Dreiecke *npl* <Geom> congruent triangles
kongruente Zahlen *fpl* <Math> congruent numbers *(kongruent bezüglich eines Moduls)*
Kongruenz *f* <Geom> congruence
Koniferosid *n* <Chemie> coniferin
Königswelle *f* <Mechan> mainshaft
Königszapfen *m* <Maschinen, Mechan> *(AE)* kingbolt, *(BE)* kingpin
Koniin *n* <Chemie> conicine, coniine, propylpiperidine
konisch *adj* 1. <Akustik> conical; 2. <Fertig> taper; 3. <Geom, Ker & Glas, Labor> conical; 4. <Maschinen> tapered; 5. <Textil> conic
konisch zulaufen *v* <Fertig> taper
Konischdrehen *n* <Maschinen> conical turning, taper turning
Konischdrehvorrichtung *f* <Maschinen> taper-turning attachment
konische Hohlform *f* <Fertig> cup
konische Hülle *f* <Raumfahrt> conical shell *(Raumschiff)*
konische Klemmverbindung *f* <Maschinen> conical clamping connection
konische Kreuzspule *f* <Textil> cone
konische Nadel *f* <Maschinen> tapered needle
konische Schliffverbindung *f* <Labor> conical ground glass point

konische Unterlegscheibe *f* <Maschinen> taper washer
konische Verzahnung *f* <Maschinen> tapered teeth
konischer Becher *m* <Labor> conical beaker
konischer verstärkter Rand *m* <Ker & Glas> conical reinforced rim
konisches Horn *n* <Akustik> conical horn
konisches Rohrstück *n* <Maschinen> taper pipe
konisches Wellenende *n* <Maschinen> conical shaft end
Konizität *f* 1. <Fertig> amount of taper; 2. <Mechan> taper
konjugiert *adj* <Geom, Metall, Phys> conjugate
konjugierte Ebene *f* <Metall> conjugate plane
konjugierte Halbmesser *mpl* <Geom> conjugate radii
konjugierte Punkte *mpl* <Phys> conjugate points
konjugierte Zweige *fpl* <Elektrotech> conjugate branches
konjugierter Durchmesser *m* <Metall> conjugate diameter
konjugiertes Gleiten *n* <Metall> conjugate slip
konjugiert-komplex *adj* <Math> conjugate, conjugate-complex
konjugiert-komplexe Scheinwiderstände *mpl* <Akustik> conjugate impedances
konjugiert-komplexe Zahl *f* <Math> conjugate, conjugate-complex number
Konkatenation *f* <Elektrotech> concatenation
konkav *adj* <Geom, Ker & Glas, Optik> concave
konkave Fläche *f* <Geom> concave surface
konkave Funktion *f* <Math> concave function
konkave Krümmung *f* 1. <Ker & Glas> concave bow; 2. <Math> concave curvature
konkave Menge *f* <Math> concave set
konkaves Optikwerkzeug *n* <Ker & Glas> concave optical tool
Konkavgitter *n* **von Rowland** <Phys> concave grating of Rowland
Konkavität *f* <Optik> concavity
konkavkonvex *adj* <Optik> concavo-convex
Konkavlinse *f* <Foto> concave lens
Konkavspiegel *m* <Phys> concave mirror
Konkordanz *f* <Erdöl> conformity *(Geologie)*
konkretisieren *v* <Künstl Int> instantiate
Konkurrenz *f* <Comp & DV, Metall> competition
Konkurrenzbetrieb *m* <Comp & DV, Telekom> contention mode
Konkurrenzsituation *f* <Comp & DV> contention
Konkurrenzwachstum *n* <Metall> competition growth
konkurrierende Anforderung *f* <Comp & DV> contention
Konnektionismus *m* <Künstl Int> connectionism
Konnektivität *f* <Comp & DV> connectivity
Konnektor *m* <Elektrotech> connector
Konode *f* 1. <Fertig> conode; 2. <Metall> tie line
kononische Verteilung *f* <Phys> canonical distribution
Konservation *f* <Strömphys> conservation
konservative Kraft *f* <Phys> conservative force
Konserven *fpl* <Verpack> *(AE)* canned food, *(BE)* tinned food
Konservenbüchse *f* <Abfall> *(AE)* can, *(BE)* tin
Konservenglas *n* <Ker & Glas> *(AE)* canning jar, *(BE)* preserving jar
Konservenherstellung *f* <Lebensmittel> canning
konservieren *v* <Lebensmittel> preserve
konservierter Latex *m* <Kunststoff> preserved latex
Konservierung *f* <Erdöl, Lebensmittel, Phys, Thermod, Verpack> conservation
Konservierungsmittel *n* <Verpack> preservative
Konservierungsstoff *m* <Lebensmittel> preservative
Konsignant *m* <Trans> consignor *(Überseehandel)*
konsistent *adj* <Bau, Chemie, Comp & DV, Math> consistent

konsistente

konsistente Kopie *f* <Comp & DV> point-in-time copy
konsistente Schätzfunktion *f* <Math> consistent estimator *(Konvergenz in Wahrscheinlichkeit)*
konsistentes Back-up *n* <Comp & DV> point-in-time back-up
Konsistenz *f* 1. <Comp & DV, Kohlen, Künstl Int, Lebensmittel> consistency; 2. <Math> consistency *(einer Schätzfunktion)*
Konsistenzerhaltung *f* <Künstl Int> consistency maintenance, truth maintenance
Konsistenzgrenze *f* <Kohlen> limit of consistency
Konsistenzmesser *m* <Bau> consistometer
Konsistenzprüfung *f* <Künstl Int> consistency check
Konsistenzzahl *f* <Kohlen> consistency index
Konsole *f* 1. <Bau> bracket, corbel, support; 2. <Comp & DV> console, panel; 3. <Fertig> knee; 4. <Gerät> console *(Ein-Ausgabe-Einheit)*; 5. <Kontroll> console; 6. <Maschinen> bracket; 7. <Mechan> pad; 8. <Wassertrans> console *(Schiffelektronik)*
Konsolfräsmaschine *f* 1. <Fertig> knee-and-column milling machine; 2. <Maschinen> knee-and-column milling machine, knee-type milling machine
Konsolführung *f* <Fertig> key slide
konsolidieren *v* <Comp & DV> consolidate
Konsolidierung *f* <Kohlen> consolidation
Konsolidierungsprüfung *f* <Kohlen> consolidation test
Konsoltisch *m* <Fertig> knee table
Konsonant *m* <Akustik> consonant
konsonanter Akkord *m* <Akustik> concord
konsonanter Grundakkord *m* <Akustik> fundamental concord
konstant *adj* <Druck, Elektriz, Foto, Funktech, Lufttrans, Math, Phys, Telekom, Trans> constant
konstant geringe Batterieentladung *f* <Raumfahrt> battery drain *(Raumschiff)*
Konstant... <Elektriz, Elektronik, Maschinen, Telekom> constant
Konstantan *n* <Elektriz> constantan
Konstantausgangsspannung *f* <Elektrotech> constant output voltage
Konstantausgangsstrom *m* <Elektrotech> constant output current
Konstante *f* <Comp & DV, Funktech, Gerät, Math, Phys> constant
konstante Brennweite *f* <Foto> fixed focus
konstante Drehzahl *f* <Optik> constant angular velocity
konstante Gruppenlaufzeit *f* <Telekom> constant group delay *(Signalübertragung)*
konstante Lineargeschwindigkeit *f* <Comp & DV, Optik> constant linear velocity
konstante Spationierung *f* <Druck> monospace
konstante Verzögerung *f* <Elektriz> constant deceleration, fixed delay
konstante Winkelgeschwindigkeit *f* <Comp & DV, Gerät, Optik> constant angular velocity
konstanter Dauerlärm *m* <Sicherheit> continuous steady-state noise
konstanter Faktor *m* <Math> common ratio
konstanter Lärm *m* <Sicherheit> steady-state noise
konstanter Steigungswinkel *m* <Lufttrans> collective pitch angle
konstanter Verkehrsfluss *m* <Trans> stable flow
konstantes Feld *n* <Elektriz> constant field
Konstantfett *n* <Maschinen> set grease
Konstanthalter *m* <Telekom> regulator
Konstantlast *f* <Elektriz> constant load
Konstantlithographie *f* <Elektronik> constant lithography
Konstantspannung *f* <Elektriz, Elektronik, Labor> constant voltage, regulated voltage

Konstantspannungsquelle *f* <Elektriz, Elektrotech, Gerät> constant-voltage source
Konstantstrom *m* <Elektriz, Elektrotech, Labor> constant current
Konstantstromdynamo *m* <Elektriz> constant-current dynamo
Konstantstrom-Modulation *f* <Elektronik> constant-current modulation
Konstantstromoszillator *m* <Elektronik> constant-current oscillator
Konstantstromquelle *f* <Gerät> constant-current source
Konstantstromtransformator *m* <Elektrotech> constant-current transformer
Konstantstromversorgung *f* <Elektrotech> regulated power supply
Konstantzugwinde *f* <Wassertrans> self-tensioning winch *(Deckbeschläge)*
Konstanz *f* <Telekom> stability *(Frequenz)*
Konstitutionswasser *n* <Wasserversorg> combined water
Konstringenz *f* <Phys> constringence
konstruieren *v* 1. <Bau> construct, design; 2. <Geom> construct
konstruiert *adj* **/auf sanften Ausfall** <Kontroll> fail-soft
Konstrukteur *m* <Bau, Ergon, Maschinen> design engineer, designer
Konstruktion *f* 1. <Bau> structure; 2. <Comp & DV> design; 3. <Fertig, Geom> construction; 4. <Maschinen> design
Konstruktionsabteilung *f* <Mechan> engineering department
Konstruktionsarbeitsplatz *m* <Ergon> design workplace
konstruktionsbedingter Defekt *m* <Kerntech> design-related defect
Konstruktionsbüro *n* <Wassertrans> design department; drawing office *(Schiffbau)*
Konstruktionselement *n* <Bau> structural element
Konstruktionsendprüfung *f* <Raumfahrt> FDR, final design review
Konstruktionshöhe *f* <Bau> overall height
Konstruktionsmerkmal *n* <Maschinen, Wassertrans> constructional feature, design feature *(Schiffbau)*
Konstruktionsplan *m* <Wassertrans> construction plan *(Schiffbau)*
Konstruktionsprogramm *n* <Comp & DV> design aid
Konstruktionsriss *m* <Wassertrans> construction plan *(Schiffbau)*
Konstruktionsschaden *m* <Bau> structural damage
Konstruktionsstückliste *f* <Konstzeich> design parts list
Konstruktionssystem *n* <Comp & DV> design system
Konstruktionsüberprüfung *f* <Qual> design review
Konstruktionsvorgaben *fpl* <Qual> design input
Konstruktionswasserlinie *f* <Wassertrans> design waterline *(Schiffkonstruktion)*; load waterline *(Schiffbau)*
Konstruktionswasser-Linienebene *f* <Wassertrans> design waterplane *(Schiffskonstruktion)*
Konstruktionszeichner *m* 1. <Mechan> *(AE)* designer draftsman, *(BE)* designer draughtsman; 2. <Wassertrans> *(AE)* draftsman, *(BE)* draughtsman *(Schiffkonstruktion)*
Konstruktionszeichnung *f* <Konstzeich> design drawing
konstruktiv *adj* <Bau> structural
konstruktive Interferenz *f* <Phys, Wellphys> constructive interference
Konsumgüter *npl* <Verpack> consumer goods
Kontakt *m* 1. <Comp & DV, Elektriz, Elektrotech> contact; 2. <Fertig> contact *(Kunststoffinstallationen)*; 3. <Foto, Kfztech, Maschinen, Trans, Verpack> contact • **Kontakt herstellen** <Elektronik> contact making

Kontaktabstand *m* 1. <Elektriz, Elektrotech> contact gap, break distance; 2. <Kfztech> contact gap
Kontaktabtasten *n* <Comp & DV> contact scanning
Kontaktabzug *m* <Foto> contact print
Kontaktanordnung *f* <Elektrotech> contact arrangement
Kontaktarm *m* <Elektriz> contact blade
Kontaktarmträger *m* <Elektrotech> brush rod
Kontaktbahn *f* <Telekom> bank contact
Kontaktbemessung *f* <Elektrotech> contact rating
Kontaktbildschirm *m* <Comp & DV> touch-sensitive screen
Kontaktbombe *f* <Chemtech> catalytic bomb
Kontaktbuchse *f* <Elektrotech> female contact
Kontaktcodierung *f* **im seriellen Betrieb** <Verpack> in-line contact coding
Kontaktdetektor *m* <Trans> contact detector
Kontaktelement *n* <Elektrotech> contact, contact element
Kontakt-EMK *m* <Elektrotech, Phys> contact emf
Kontaktfeder *f* 1. <Elektriz> contact blade, contact spring; 2. <Maschinen, Wasserversorg> contact spring
Kontaktfehler *m* <Elektrotech> contact fault
Kontaktfenster *n* <Elektrotech> contact window *(bei Halbleitern)*
Kontaktfläche *f* <Maschinen> contact surface
Kontaktflattern *n* <Kfztech> contact chatter
Kontaktfroster *m* <Heiz & Kälte> contact freezer
Kontaktgefrieren *n* <Heiz & Kälte> contact freezing
Kontaktgift *n* <Umweltschmutz> catalyst poison
Kontakthaftmittel *n* <Verpack> contact adhesive
Kontakthammer *m* <Elektrotech> trembler
Kontaktkleber *m* <Kunststoff, Verpack> contact adhesive
Kontaktklebstoff *m* <Kunststoff> contact adhesive
Kontaktkopie *f* <Druck, Foto> contact print
Kontaktkopieren *n* <Foto> contact printing
Kontaktkopiergerät *n* <Foto> contact printer
Kontaktkopierrahmen *m* <Foto> contact-printing frame
Kontaktlinse *f* <Ker & Glas> contact lens
Kontaktlog *n* <Erdöl> contact log *(Bohrlochmesstechnik)*
kontaktlose Aufhängung *f* <Kfztech> noncontact suspension
kontaktlose Steuerung *f* <Kfztech> breakerless triggering
kontaktlose Transistorzündung *f* <Kfztech> contactless-transistor-ized ignition
kontaktloses Relais *n* <Elektrotech> solid state relay
Kontaktmaske *f* <Elektronik> contact mask *(Grafik)*
Kontaktmaskierung *f* <Elektronik> contact masking
Kontaktmatte *f* <Sicherheit> pressure-sensitive mat, safety switchmat
Kontaktmessinstrument *n* <Gerät> contact-measuring instrument
Kontaktmikrofon *n* <Akustik, Aufnahme> contact microphone
Kontaktnegativ *n* <Foto> contact negative
Kontaktprellen *n* <Elektriz, Kfztech> contact bounce, contact chatter
Kontaktpressen *n* <Ker & Glas> *(AE)* contact molding, *(BE)* contact moulding
Kontaktpunkt *m* <Lufttrans> contact point
Kontaktsatz *m* 1. <Elektriz, Kfztech> contact set *(Zündung)*; 2. <Telekom> bank contact
Kontaktschalter *m* <Elektrotech> touch switch
Kontaktschlammverfahren *n* <Abfall> sludge contact process
Kontaktschuh *m* <Eisenbahn> collector shoe
Kontaktspalt *m* <Elektriz> contact gap
Kontaktspannung *f* <Elektriz> contact potential
Kontaktstift *m* <Elektrotech> contact pin, pin
Kontaktstück *n* <Elektrotech> contact

Kontakt-Thermometer *n* <Gerät> contact thermometer
Kontaktunterbrecher *m* <Kfztech> breaker
Kontaktverklebung *f* <Fertig> contact bonding
Kontaktverschleiß *m* <Elektriz> contact wear
Kontaktwiderstand *m* <Elektriz, Elektrotech> contact resistance
Kontaktwinkel *m* <Kohlen> contact angle
Kontaktzunge *f* <Telekom> reed
Kontaminant *m* <Abfall, Umweltschmutz> contaminating substance
Kontamination *f* <Chemie, Kohlen, Labor> contamination
Kontaminationsschutzhaube *f* <Sicherheit, Umweltschmutz> anticontamination hood
Kontaminationsüberwachung *f* <Gerät, Umweltschmutz> contamination control, contamination monitoring
kontaminieren *v* <Kerntech> contaminate
kontaminiert *adj* <Sicherheit> contaminated
kontaminierte Deponie *f* <Abfall> contaminated waste site
kontaminierter Abfall *m* <Abfall> contaminated waste
kontaminierter Standort *m* <Abfall> contaminated site, problem site
kontaminierter Stoff *m* <Abfall, Umweltschmutz> contaminated substance
Kontaminierung *f* <Kerntech> contamination
Kontermutter *f* 1. <Fertig> locking nut; 2. <Maschinen> jam nut, locknut; 3. <Mechan> locknut
Kontern *n* <Druck> retransfer
Kontext *m* <Comp & DV> context
kontextabhängig *adj* <Comp & DV> context-sensitive, contextual
kontextfrei *adj* <Comp & DV> context-free
kontextfreie Grammatik *f* *(kfG)* <Künstl Int> CF grammar, context-free grammar
kontextsensitive Grammatik *f* *(ksG)* <Künstl Int> context-sensitive grammar *(CSG)*
Kontinentalabfall *m* <Erdöl> continental slope
Kontinentalböschung *f* <Erdöl> continental slope *(Geomorphologie)*
Kontinentalschelf *n* <Erdöl> continental shelf *(Geomorphologie)*
Kontinue-Garn *n* <Textil> continuous spun yarn, continuous yarn
kontinuierlich *adj* <Akustik, Comp & DV, Elektronik, Kfztech, Kontroll, Metall, Phys, Strahlphys, Trans> continuous
kontinuierlich abstimmbares Filter *n* <Aufnahme, Funktech> continuously tunable filter
kontinuierlich arbeitende Gewichtsdosierung *f* <Verpack> continuous motion weight filling
kontinuierlich gesteuerte Geräuschsperre *f* <Telekom> continuous-controlled squelch system
kontinuierliche Abstimmung *f* <Elektronik> continuous tuning
kontinuierliche Aushärtung *f* <Metall> continuous precipitation
kontinuierliche Emissionsmesssysteme *npl* *(CEMS)* <Kohlen> continuous emission monitoring systems *(CEMS)*
kontinuierliche Kraftstoffeinspritzung *f* <Kfztech> K-Jetronic fuel injection
kontinuierliche mechanische Zwillingsnachbildung *f* <Metall> continual mechanical twinning
kontinuierliche Strahlmodulation *f* <Fernseh> continuous beam modulation
kontinuierlicher Betrieb *m* <Comp & DV> continuous operations
kontinuierlicher Kocher *m* <Papier> continuous digester

kontinuierlicher

kontinuierlicher Laser m <Elektronik> continuous laser
kontinuierlicher Laserstrahl m <Elektronik> continuous laser beam
kontinuierlicher Stollenbau m <Kohlen> continuous mining
kontinuierlicher Transport m <Trans> continuous transport
kontinuierlicher Umlaufkühlofen m <Ker & Glas> continuous recirculation lehr
kontinuierliches Be- und Entladen n <Eisenbahn> (BE) merry-go-round (MGR)
kontinuierliches Lasern n <Elektronik> continuous laser action
kontinuierliches Spektrum n <Akustik, Elektronik, Phys, Strahlphys> continuous spectrum
kontinuierliches Walzen n **von Gussglas** <Ker & Glas> continuous casting
Kontinuität f <Comp & DV, Phys, Strömphys, Telekom, Werkprüf> continuity
Kontinuitätdiskontinuität f <Werkprüf> continuity-discontinuity
Kontinuitätsgleichung f <Phys, Strömphys> continuity equation
Kontinuitätsschleife f <Fernseh> continuity log
Kontinuitätssteuerung f <Fernseh> continuity control
Kontinuum n <Math> continuum (zusammenhängendes Kompaktum)
Kontourenmessgerät n <Metrol> contour-measuring equipment
Kontourringe mpl <Phys> contour fringes
kontrahierter Abschnitt m <Hydraul> contracted section
Kontraktion f 1. <Maschinen> contraction; 2. <Metall> constriction
Kontraktionskoeffizient m <Hydraul, Qual> contraction coefficient
Kontraktionsziffer f <Hydraul, Qual> coefficient of contraction
Kontrast m <Phys> contrast
Kontrastabschwächung f <Foto> contrast reduction
Kontrastdehnung f <Telekom> expansion (Fax)
Kontrasteffekt m <Fernseh> contrast effect
Kontrast-Filterscheibe f <Foto> filter screen
Kontrastregler m <Fernseh> contrast control
kontrastreiche Wiedergabe f <Konstzeich> high-contrast reproduction
Kontrastschwelle f <Ergon> contrast threshold
Kontraststeigerung f <Foto> increase in contrast
Kontrastverhältnis n <Fernseh, Papier> contrast ratio
Kontroll... <Aufnahme, Eisenbahn, Lufttrans, Maschinen, Sicherheit> control
Kontrollanzeige f <Lufttrans> indicator (Lichtsignal)
Kontrollart f <Umweltschmutz> indicator species
Kontrolllautsprecher m <Aufnahme> monitoring loudspeaker
Kontrollbericht m <Qual> check report
Kontrollbildschirm m <Telekom> monitor (Video)
Kontrollbrunnen m <Abfall> monitoring well, observation well
Kontrolle f 1. <Comp & DV> check, control; 2. <Kfztech, Lufttrans> control; 3. <Mechan> inspection; 4. <Telekom> control; 5. <Textil> monitoring; 6. <Verpack> check
• **unter Kontrolle** <Maschinen> under control • **unter Kontrolle bringen** <Erdöl> bring under control (Bohrung nach Eruption)
Kontrolle f **der Zeitüberwachung** <Telekom> time-out supervision
Kontrolle f **während der Fertigung** <Fertig> in-process testing
Kontrolleur m <Sicherheit> factory inspector

Kontrollfeld n <Comp & DV> control field
Kontrollgerät n <Telekom> monitor, monitor unit
Kontrollgrenze f <Qual> control limit
kontrollieren v 1. <Kontroll> inspect, monitor; 2. <Telekom> control; 3. <Textil, Umweltschmutz> monitor; 4. <Patent, Qual, Verpack> check
kontrollierte Atmosphäre f 1. <Heiz & Kälte> controlled atmosphere; 2. <Verpack> CA, controlled atmosphere
kontrollierte Gärung f <Abfall> controlled fermentation
kontrollierte Müllablagerung f <Abfall> controlled dumping, controlled tipping, sanitary landfill
kontrollierter Abflusskanal m <Wasserversorg> controlled spillway
kontrollierter Flug m <Lufttrans> controlled flight
kontrollierter Leistungsrückgang m <Comp & DV> graceful degradation
kontrollierter Luftraum m <Lufttrans> controlled airspace (auf Instrumentenflug beschränkt)
kontrolliertes Abladen n **von Schutt** <Umweltschmutz> controlled dumping
kontrolliertes Trudeln n <Lufttrans> controlled spin
Kontrollkarte f **für kumulierte Werte** <Qual> consume chart
Kontrolllämpchen n <Elektrotech> pilot light
Kontrolllampe f 1. <Eisenbahn> telltale lamp; 2. <Elektrotech> pilot lamp, pilot light
Kontrollleuchte f <Kfztech> indicator (Zubehör)
Kontrolllicht n <Lufttrans> indicator light
Kontrollloch n <Maschinen> inspection hole
Kontrollmessgerät n <Gerät, Qual> checking instrument
Kontrollmonitor m <Fernseh> image and waveform monitor, off-air monitor
Kontrollprobe f <Kohlen> control assay
Kontrollprüfung f <Qual> check test
Kontrollpunkte mpl <Geom> control points
Kontrollschranke f <Bau> control barrier
Kontrollschuss m <Erdöl> checkshot (Seismik)
Kontrollspur f <Aufnahme> guide track
Kontrollsumme f <Comp & DV> checksum
Kontrollsystem n 1. <Aufnahme> pilot system; 2. <Strahlphys> monitoring system; 3. <Telekom> control system
Kontrollton m <Aufnahme> pilot tone
Kontrollturm m <Lufttrans> control tower
Kontrolluhr f <Mechan> punch clock
Kontrollventil n <Hydraul> control valve
Kontrollverstärker m <Aufnahme> bridging amplifier, monitoring amplifier
Kontrollvorrichtung f <Qual, Verpack> checking apparatus, inspection equipment
Kontrollwaage f <Verpack> checkweigher
kontrollwiegen v <Verpack> check weigh
Kontrollzettel m <Verpack> control tag (zum Aufkleben)
Kontrollzone f 1. <Kohlen> control area; 2. <Lufttrans> control sector
Kontur f 1. <Anstrich> profile; 2. <Druck, Maschinen> outline
Konturätzen n <Fertig> chemical milling
Kontureffekt m <Akustik> contour effect
Konturenausgleich m <Comp & DV> anti-aliasing
Konturensatz m <Druck> runaround
Konturenschärfe f <Foto> acutance
konturenscharfes Zeichnen n <Konstzeich> sharp-contoured impression
konturgetreuer Überzug m <Kunststoff> conformal coating
Konus m 1. <Akustik, Aufnahme, Druck> bevel, cone; 2. <Erdöl, Geom, Ker & Glas, Kfztech, Maschinen> cone
Konusbohrer m <Erdöl> cone bit (Bohrtechnik)

Konusbohrmeißel m <Erdöl> cone bit
Konusbremse f <Maschinen> wedge brake
Konuskupplung f <Fertig, Kfztech, Maschinen> cone clutch, conical clutch
Konuslager n <Maschinen> cone-type bearing
Konuslautsprecher m <Aufnahme> cone loudspeaker
Konusmembran f <Akustik> conical diaphragm
Konusnabe f <Kfztech> tapered hub *(Rad)*
Konusresonanz f <Aufnahme> cone resonance
Konustreiber m <Fertig> drift
Konustrichter m <Akustik> conical horn
Konuswinkel m <Lufttrans> flapping angle
Konvektion f <Elektrotech, Erdöl, Heiz & Kälte, Ker & Glas, Phys, Raumfahrt, Strömphys, Thermod> convection
Konvektion f **im rotierenden Ringspalt** <Strömphys> rotating annulus convection
Konvektionsabriss m <Raumfahrt> absence of convection *(Raumschiff)*
Konvektionskühler m <Heiz & Kälte, Phys, Thermod> convection cooler
Konvektionskühlung f 1. <Heiz & Kälte> convection cooling, jet cooling; 2. <Phys, Thermod> convection cooling
Konvektionsmangel m <Raumfahrt> absence of convection *(Raumschiff)*
Konvektionsofen m <Heiz & Kälte, Phys, Thermod> convection oven
Konvektionsstrom m <Elektrotech, Ker & Glas, Phys> convection current
Konvektionsströmung f <Heiz & Kälte, Thermod> convection current
Konvektionstrocknen n <Heiz & Kälte, Phys, Thermod> convection drying
Konvektionstrockner m <Heiz & Kälte, Phys, Thermod> convection dryer
Konvektionsüberhitzer m <Heiz & Kälte, Phys, Thermod> convection superheater
Konvektionswärme f <Heiz & Kälte, Phys, Thermod> convection heat
Konvektionszahl f <Heiz & Kälte, Phys, Thermod> convection coefficient
konvektiv adj <Heiz & Kälte, Lufttrans, Phys, Strahlphys, Thermod> convective
konvektive Strömungen fpl <Strömphys> convective flows
konvektive Turbulenz f <Lufttrans> convective turbulence
konvektive Verwirbelung f <Lufttrans> convective turbulence
Konvektor m <Thermod> convector, convector heater
konventionell startendes und landendes Flugzeug n *(CTOL-Flugzeug)* <Lufttrans> conventional takeoff and landing aircraft *(CTOL aircraft)*
konventionelle Leitung f <Optik> conventional cable
konventionelle Transportpalette f <Verpack> conventional transportable pallet
konventioneller Rechner m *(CISC)* <Comp & DV> complex instruction set computer *(CISC)*
konventioneller Satz m <Druck> hot-metal typesetting
konvergent adj 1. <Geom, Math> convergent *(einem Grenzelement zustrebend)*; 2. <Optik> convergent
konvergente Geraden fpl <Geom> convergent lines
konvergente Reihe f <Math> convergent series
konvergenter Strahl m <Optik> convergent beam
Konvergenz f <Fernseh, Geom, Math> convergence
Konvergenz f **Festnetz-Mobilfunknetz** <Mobilkom, Telekom> fixed mobile convergence
Konvergenzbaugruppe f <Fernseh> convergence assembly
Konvergenzfehler m <Fernseh> convergence errors
Konvergenzkreise mpl <Fernseh> convergence circuits
konvergieren v <Math> converge
Konversion f <Elektronik, Elektrotech, Erdöl, Kerntech, Trans> conversion
Konversionsanlage f <Kerntech> conversion plant
Konversionselektronen npl <Strahlphys> conversion electrons
Konversionsfaktor m <Kerntech> conversion factor
Konversionsfrequenz f <Elektronik, Fernseh, Funktech> conversion frequency
Konversionsgewinn m <Elektronik> conversion gain *(Atomphysik)*
Konversionsgrad m <Trans> conversion degree
Konversionsöl n <Erdöl> conversion oil
Konversionsspannungsverstärkung f <Elektrotech> conversion voltage gain
Konversionswandler m <Elektrotech> conversion transducer
Konverter m 1. <Elektronik, Elektrotech, Fertig, Funktech, Heiz & Kälte> converter; 2. <Kerntech> converter reactor; 3. <Maschinen, Metall, Papier, Telekom> converter
Konvertereinsatz m <Fertig> converter charge
Konverterfutter n <Fertig> converter lining
Konverterkammzug m <Textil> converted top
Konverterkippung f <Fertig> converter tilting
Konverterreaktor m <Kerntech> thermal converter reactor
Konvertersatz m <Elektrotech> converter set
Konverterstahl m <Fertig> converter steel
konvertieren v <Comp & DV> convert
Konvertierung f <Comp & DV> conversion
Konvertierungsprogramm n <Comp & DV> conversion program
Konvertkupfer n <Chemie> black copper
konvex adj <Bau, Geom, Ker & Glas, Optik, Phys> convex
konvexe Fläche f 1. <Geom> convex surface; 2. <Ker & Glas> convex surface *(der Plankonvexlinse)*
konvexe Funktion f <Math> convex function *(Graph hat konvexe Krümmung)*
konvexe Krümmung f 1. <Ker & Glas> convex bow; 2. <Math> convex curvature
konvexe Menge f <Math> convex set
konvexer Spiegel m <Phys> convex mirror
konvexer Stab m <Bau> ovolo
konvexes Optikwerkzeug n <Ker & Glas> convex optical tool
Konvexität f <Optik> convexity
konvexkonkav adj <Optik> convexo-concave
konvexkonkave Linse f <Bau> meniscus
Konvexlinse f <Foto> convex lens
konvolutionelles Filter n <Elektronik> convolutional filter
konvolutionelles Filtern n <Elektronik> convolutional filtering
Konvolutionscode m <Telekom> convolution code, convolutional code
Konzentrat n <Kohlen, Lebensmittel> concentrate
Konzentration f 1. <Elektriz, Elektronik, Kohlen, Kunststoff, Telekom, Umweltschmutz> concentration *(c)*; 2. <Textil> strength
Konzentration f **der Umweltschadstoffe** <Umweltschmutz> ambient pollutant concentration
Konzentration f **in Bodenhöhe** <Umweltschmutz> GLC, ground level concentration *(von Schadstoffen)*
Konzentration f **in der Luft** <Umweltschmutz> atmospheric concentration
Konzentrationselektrode f <Elektrotech> focusing electrode
Konzentrationselement n <Chemtech, Elektriz> concentration cell

Konzentrationskoppelfeld

Konzentrationskoppelfeld *n* <Telekom> concentration network
Konzentrationsstufe *f* <Telekom> concentration stage
Konzentrationstabelle *f* <Kohlen> concentrating table
Konzentrationsüberspannung *f* <Raumfahrt> concentration overvoltage *(Raumschiff)*
Konzentrationsverhältnis *n* <Kohlen, Nichtfoss Energ> concentration ratio
Konzentrationszelle *f* <Elektriz> concentration cell
Konzentrator *m* <Elektronik, Heiz & Kälte, Nichtfoss Energ, Telekom> concentrator
Konzentratorkoppelfeld *n* <Telekom> concentrator switching array
Konzentratorzentrale *f* <Telekom> RSU, remote switching unit, remote switching stage
konzentrieren *v* <Kohlen> concentrate
konzentriert *adj* <Lebensmittel> concentrated
konzentrierte Salpetersäure *f* <Chemie> aqua fortis
konzentriertes Schaltelement *n* <Elektrotech, Phys> lumped circuit element
Konzentrierung *f* <Kohlen> concentration
konzentrisch *adj* <Elektriz, Geom, Maschinen, Mechan, Optik> concentric
konzentrisch verdrillter Leiter *m* <Elektriz> concentrically-stranded circular conductor
konzentrische Kreise *mpl* <Geom> concentric circles
konzentrische Wicklungen *fpl* <Elektriz> concentric windings
konzentrischer Leiter *m* <Elektriz> concentric conductor
konzentrischer Mantellichtleiter *m* <Optik> concentric optical cable
konzentrisches Dreibackenfutter *n* <Maschinen> triple jaw concentric chuck, triple jaw concentric gripping chuck
Konzentrizität *f* <Mechan> concentricity
Konzentrizitätsfehler *m* <Fertig> concentricity error *(Faseroptik)*
Konzentrizitätsfehler *m* **Kern-Bezugsfläche** <Optik, Telekom> core-reference surface concentricity error
Konzentrizitätsfehler *m* **zwischen Kern und Mantel** <Telekom> core-cladding concentricity error
Konzeptbildung *f* <Ergon> concept formation
konzeptionell *adj* <Comp & DV> conceptual
Konzeptqualität *f* <Druck> draft quality
konzeptuelle Dependenz *f* <Künstl Int> CD, conceptual dependency, CD
Konzession *f* <Erdöl> concession *(Handel, Recht)*; *(BE)* licence, *(AE)* license *(Recht)*
kooperative Emission *f* <Metall> cooperative emission
Koordinate *f* <Bau, Comp & DV, Geom, Konstzeich, Kontroll, Labor, Maschinen, Math, Phys> coordinate
Koordinaten *fpl* <Bau, Math> coordinates
Koordinatenachsen *fpl* <Math> coordinate axes
Koordinatenbemaßung *f* <Konstzeich> coordinate dimensioning
Koordinatenbohrmaschine *f* <Maschinen> coordinate boring and drilling machine
Koordinatendrucker *m* <Raumfahrt> X-Y recorder
Koordinatenfräsen *n* <Maschinen> jig milling
Koordinatenfräsmaschine *f* <Maschinen> coordinate-milling machine
Koordinatengeometrie *f* <Geom> coordinate geometry
Koordinatenmessgerät *n* <Metrol> coordinate-measuring machine
Koordinatenpaar *n* <Math> pair of coordinates
Koordinatenplotter *m* <Raumfahrt> X-Y plotter
Koordinatenpunkte *mpl* <Geom> coordinate points *pl (in einem kartesischem System)*
Koordinatenschaltersystem *n* <Telekom> crossbar system

Koordinatenschaltervermittlungsstelle *f* <Telekom> crossbar exchange
Koordinatenschleifen *n* <Maschinen> jig grinding
Koordinatenschreiber *m* <Gerät> graph plotter, two-axis plotter
Koordinatensystem *n* <Geom, Math, Phys> coordinate system
Koordinatentransformation *f* <Kontroll, Math> coordinate transformation
Koordinatenverschiebung *f* <Kontroll> coordinate displacement
Koordinatenwähler *m* <Elektrotech> cross coupling, crossbar selector
Koordinatenwandlung *f* <Elektronik> coordinate transformation
Koordinatenzeichentisch *m* <Gerät> X-Y plotting table
Koordination *f* <Ergon, Kunststoff, Metall, Qual> coordination
Koordinationsblatt *n* <Fertig, Qual> coordination sheet *(Kunststoffinstallationen)*
Koordinationszahl *f* <Metall, Qual> coordination number
koordinieren *v* <Bau> coordinate
koordinierte Verbindung *f* <Metall> coordinate linkage
Köper *m* <Textil> twill *(Weben)*
Kopf *m* 1. <Comp & DV> head, header; 2. <Maschinen> head; 3. <Mechan> crown; 4. <Telekom> head, header
• **über Kopf** <Maschinen> overhead
Kopf *m* **einer Schraube** <Maschinen> screw head
Kopf *m* **eines Hammers** <Maschinen> hammer head
Kopfabnutzung *f* <Aufnahme, Fernseh> head wear
Kopfanfangszeichen *n* <Comp & DV> SOH, start of header
Kopfanschlusskappe *f* <Elektronik> top cap *(Elektronenröhre)*
Kopfanstauchwerkzeug *n* <Lufttrans> heading tool
Kopfausrichtung *f* <Aufnahme, Comp & DV, Fernseh> head alignment
Kopfbahn *f* <Maschinen> crest track
Kopfbahnhof *m* <Eisenbahn> dead-end station, rail head, *(AE)* railroad terminus, *(BE)* railway terminus
Kopfband *n* <Bau> angle brace, angle tie, raker; knee brace, strut *(Holzbau)*
Kopf-Bandberührung *f* <Aufnahme, Fernseh> head-to-tape contact
Kopf-Bandgeschwindigkeit *f* <Aufnahme, Fernseh> head-to-tape velocity
Kopfbaugruppe *f* <Aufnahme, Fernseh> head assembly
Kopfbrett *n* <Wassertrans> headboard *(Segeln)*
Kopfdrehmaschine *f* <Maschinen> face lathe
Kopfempfindlichkeit *f* <Aufnahme, Fernseh> head response
Kopfentmagnetisierer *m* <Aufnahme> head demagnetizer
Kopfflanke *f* <Fertig> face *(Zahnrad)*
Kopfgeschirr *n* <Aufnahme> headset *(Mikrofon, Kopfhörer)*
kopfgesteuerter Motor *m* <Kfztech> overhead valve engine
kopfgesteuertes Flugzeug *n* <Lufttrans> canard wing aircraft
Kopfhaube *f* <Sicherheit> hair protector
Kopfhöhe *f* 1. <Bau> headroom; 2. <Fertig, Maschinen, Mechan> addendum
Kopfhöhe *f* **des Großrades** <Fertig> *(AE)* gear addendum
Kopfhörer *m* 1. <Akustik> circumaural earphone, earphone; 2. <Aufnahme> headphone, headset, headphones; 3. <Funktech> headphone; 4. <Telekom> headphone, headphones, headset
Kopfhörer-Anschlussbuchse *f* <Aufnahme, Funktech, Telekom> headphone jack

Kopfhörerentzerrer m <Akustik> equalizer for earphone
Kopfjustierung f <Aufnahme, Fernseh> head adjustment
Kopfkanal m <Fernseh> head channel
Kopfkreis m <Fertig, Maschinen> addendum circle, addendum line
Kopfkreisdurchmesser m <Maschinen> addendum circle
Kopfkreiszylinder m <Fertig> addendum cylinder
Kopfkürzung f <Fertig> addendum reduction
kopflastig untergehen v <Wassertrans> go down by the bows
Kopflastigkeit f <Lufttrans> nose heaviness
Kopflebensdauer f <Aufnahme, Fernseh> head life
Kopfleine f <Wassertrans> head line *(Festmachen)*
Kopfleiste f 1. <Druck> headband; 2. <Konstzeich> column heading panel
Kopfleitwerk n <Nichtfoss Energ> headworks
kopflose Schraube f <Maschinen> headless screw
Kopfnachführung f <Fernseh> head tracking
Kopfplatte f 1. <Bau> cap; 2. <Maschinen> top plate
Kopfplattform f <Eisenbahn> bay platform
Kopfpositionierungimpuls m <Fernseh> head position pulse
Kopfrad n 1. <Fernseh> head wheel; 2. <Hydraul> breast wheel
Kopfraum m <Lebensmittel> head space
Kopfrechnen n <Math> mental arithmetic
Kopfregelung f <Aufnahme, Fernseh> head banding
Kopfrohr n <Hydraul> head pipe
Kopfschraube f <Maschinen> screw with head
Kopfservoeinstellung f <Fernseh> head servo lock
Kopfspiel n 1. <Fertig> bottom clearance, clearance, crest clearance *(Zahnrad)*; 2. <Maschinen> clearance
Kopfstation f <Telekom> head end
Kopfsteg m <Druck> head
Kopfstein m <Bau> header *(Mauerwerk)*
Kopfsteinschicht f <Bau> header course
Kopfstelle f <Comp & DV, Telekom> head end
Kopfstück n 1. <Kerntech> head piece; 2. <Maschinen> head end
Kopfstütze f <Kfztech> headrest
Kopfteil m 1. <Kerntech> head piece; 2. <Telekom> overhead
Kopftrommel f <Fernseh> head drum
Kopf- und Bodenfreiheit f <Maschinen> top and bottom clearance
Kopf- und Gesichtsschutz m <Sicherheit> full face mask
Kopfverband m <Bau> header bond
Kopfverschmutzung f <Aufnahme, Fernseh> head clogging
Kopfverstärker m <Raumfahrt> head amplifier *(Weltraumfunk)*
Kopfwelle f <Kerntech> head wave
Kopfwicklung f <Aufnahme, Fernseh> head winding
Kopfzeile f <Comp & DV> header
Kopfzentrale f gateway center *(KZ)*
Kopie f 1. <Akustik> rerecording; 2. <Comp & DV, Druck, Fernseh> copy
Kopier... <Aufnahme, Fertig, Foto, Maschinen> copying
Kopierbezugsstück n <Fertig> copying master
Kopierdauer f <Foto> printing time
Kopierdrehen n <Maschinen> copy turning
Kopierdrehmaschine f <Maschinen> copying lathe
Kopierdrehmeißel m <Maschinen> copying lathe tool
Kopierecho n <Fernseh> printing echo
Kopiereffekt m <Aufnahme> print-through
kopieren v 1. <Comp & DV> copy, duplicate; 2. <Druck> copy; 3. <Fernseh> duplicate; 4. <Fertig> form; 5. <Foto> duplicate; 6. <Maschinen> copy

Kopieren n 1. <Aufnahme> dubbing; 2. <Fernseh> print-through; 3. <Fertig> forming; 4. <Foto> printing; 5. <Maschinen> copying
Kopieren n **mit Fotopapier** <Foto> printing through photo paper
kopierfähige Zeichnung f <Konstzeich> copyable drawing
Kopierfräsen n 1. <Fertig> copy milling; 2. <Maschinen> copy milling, copy turning, tracer milling
Kopierfräsmaschine f <Maschinen> copy-milling machine
Kopiergenehmigung f <Comp & DV> ATC, authorization to copy
Kopiermaschine f 1. <Foto> printer; 2. <Maschinen> copying machine, duplicator
Kopiermaske f <Foto> printing mask
Kopierschablone f <Fertig> template, templet
Kopierschutz m <Comp & DV> copy prohibit, copy protection; serial copy management system *(gegen Mehrfachkopieren)*
Kopierstift m <Maschinen> follower
kopierte Daten npl <Comp & DV> snapped data
Kopiertechnik f <Foto> printing technique
Kopiertisch m <Foto> printing stage
Kopiervorlage f <Elektronik> master pattern *(Werkzeugmaschinen)*
Kopiervorrichtung f <Fertig, Maschinen> copying attachment
koplanar adj <Geom, Phys> coplanar *(in einer Ebene liegend)*
koplanarer Lichtleiter m <Phys> coplanar waveguide
koplanarer Wellenleiter m <Phys> coplanar waveguide
kopolare Dämpfung f <Raumfahrt> copolar attenuation *(Weltraumfunk)*
Koppel f <Maschinen> coupler
Koppelanordnung f <Telekom> switching array
koppelbar adj <Eisenbahn> coupleable
Koppeldämpfung f <Telekom> coupler loss, coupling loss
Koppelebene f <Fertig> coupler plane *(Getriebelehre)*
Koppelelement n <Optik> coupler *(Lichtleiter)*
Koppelfeld n <Telekom> switching array, switching network, switching network complex, switching stage
Koppelkondensator m 1. <Elektrotech> blocking capacitor; 2. <Phys> blocking capacitor *(für Wechselspannungen)*
Koppelkurve f <Fertig> coupler curve *(Getriebelehre)*
Koppelleitung f <Phys> strapping *(eines Magnetrons)*
Koppellenkerachse f <Kfztech> dead beam axle
Koppelmatrix f <Telekom> connecting matrix *(Vermitteln)*; switching matrix
Koppelmittellinie f <Fertig> coupler link *(Getriebelehre)*
Koppeln n <Lufttrans, Wassertrans> dead reckoning *(Navigation)*
Koppelnetz n <Telekom> switching network
Koppelnetzwerk n <Gerät> coupling network
Koppelort m <Wassertrans> dead-reckoning position *(Navigation)*
Koppelproduktverwertung f <Abfall> by-product recovery
Koppelpunkt m <Telekom> switching point, crosspoint
Koppelschaltung f 1. <Gerät> coupling network; 2. <Telekom> switching circuit
Koppelspule f <Elektrotech> coupling coil
Koppelstelle f <Telekom> switching point
Koppelung f **des Funkempfängers und -senders mit Telefonnetzen** <Funktech> autopatch *(Sprechfunk)*
Koppelungsverfahren n <Lufttrans> docking procedure
Koppelverlust m <Optik, Telekom> coupling loss
Koppelvielfach n <Telekom> connecting matrix *(Vermitteln)*; switching matrix

Koppelwirkungsgrad

Koppelwirkungsgrad m <Optik, Telekom> coupling efficiency
Koppelzuordnung f <Fertig> coupler coordination *(Getriebelehre)*
Koppler m 1. <Akustik, Comp & DV, Elektrotech, Funktech, Optik> coupler; 2. <Telekom> coupler, switching unit
Koppler m **für Lichtleitfasern** <Optik> *(AE)* optical fiber coupler, *(BE)* optical fibre coupler
Kopplerverlust m <Optik> coupler loss
Kopplung f 1. <Comp & DV> coupling; 2. <Elektriz> coupling, interconnection; 3. <Elektrotech> coupler, coupling, interconnection; 4. <Fertig, Kfztech, Labor, Lufttrans> coupling; 5. <Maschinen> linkage, linking; 6. <Mechan, Phys> coupling; 7. <Telekom> coupling, switching
Kopplung f **mit nächstem Nachbaratom** <Kerntech> *(AE)* nearest neighbor coupling, *(BE)* nearest neighbour coupling *(in Kristallen)*
Kopplung f **zwischen Stufen** <Elektronik, Funktech> interstage coupling
Kopplungsfaktor m <Elektriz, Qual> coefficient of coupling
Kopplungsimpedanz f <Elektriz> coupling impedance
Kopplungskoeffizient m (k) <Elektriz, Phys> coupling coefficient *(k)*
Kopplungskondensator m <Elektriz, Phys> coupling capacitor
Kopplungskonstante f <Kerntech, Mechan, Phys> coupling constant
Kopplungsnetz n <Gerät> coupling network
Kopplungsschleife f <Elektrotech> coupling loop
Kopplungsspule f 1. <Akustik> coupler; 2. <Elektriz, Elektrotech> coupler, coupling coil
Kopplungsstecker m <Funktech, Telekom> coupler
Kopplungstrafo m <Elektrotech> coupling transformer
Kopplungstransformator m <Elektriz, Elektrotech> coupling transformer
Kopplungsübertrager m <Elektrotech> coupling transformer
Kopplungswiderstand m 1. <Elektriz> coupling resistance; 2. <Telekom> mutual impedance
Kopräzipitation f <Kerntech> coprecipitation
Kops m <Ker & Glas> cop
Korb m <Bau, Elektrotech, Lebensmittel> basket
Korbbodenspule f <Elektrotech> basket coil
Korbbodenwicklung f <Elektrotech> basket winding
Korbbogen m <Bau> basket handle arch, *(AE)* three-centered arch, *(BE)* three-centred arch
Korbgeflechtpackung f <Ker & Glas> basketweave packing
Korbornamentausbildung f <Bau> pannier
Korbspule f <Elektrotech> basket coil
Kordel f <Textil> cord
Kordeln n 1. <Fertig> diamond knurling; 2. <Maschinen> cross knurling, diamond knurling
Kork m <Bau, Ker & Glas, Labor, Maschinen> cork
Korkbohrer m <Labor> cork borer
Korkenzieherregel f <Elektriz, Elektrotech, Phys> corkscrew rule
Korkradpolitur f <Ker & Glas> cork polishing
Korkscheibe f <Maschinen> cork washer
Korkstoff m <Chemie> suberin
Korkstopfen m <Verpack> corking plug
Korkverschluss m 1. <Ker & Glas> cork finish; 2. <Labor> cork
Korn n 1. <Chemtech> particle; 2. <Kohlen> grain; 3. <Textil> particle
Kornabstumpfung f <Fertig> glazing *(Schleifscheibe)*
Kornanalyse f <Bau> grading analysis

Kornbranntwein m <Lebensmittel> grain alcohol
Körnchen n <Chemtech, Verpack> granule
Korndurchmesser m <Chemtech> particle size
körnen v 1. <Fertig> grain; 2. <Maschinen, Mechan> *(AE)* center-punch, *(BE)* centre-punch; 3. <Verpack> granulate
Körner m 1. <Fertig> puncher, punch; 2. <Maschinen> *(AE)* center punch, *(BE)* centre punch, prick punch, punch; 3. <Mechan> *(AE)* center punch, *(BE)* centre punch
Körnermarke f 1. <Fertig> *(AE)* center punch mark, *(BE)* centre punch mark; 2. <Maschinen> *(AE)* center mark, *(BE)* centre mark
Körnerspitze f <Maschinen> *(AE)* center, *(BE)* centre
Kornett n <Metall> cornet
Kornform f <Kohlen> grain shape
Kornfraktion f <Kohlen> grain fraction
Korngefüge n <Metall> grain structure
Korngrenze f <Kerntech, Metall> grain boundary
Korngrenzendiffusion f <Metall> grain boundary diffusion
Korngrenzenkorrosion f <Chemie> intergranular corrosion
Korngrenzenverschiebung f <Metall> grain boundary migration
Korngröße f 1. <Chemtech> particle size; 2. <Kohlen> grain size; 3. <Kunststoff> particle size; 4. <Maschinen, Metall> grain size • **nach Korngrößen trennen** <Bau> size
Korngrößenabstufung f <Kohlen> granulometry
Korngrößenanalyse f <Chemtech> *(BE)* particle size analyser, particle size analysis, *(AE)* particle size analyzer
Korngrößenanteil m <Kohlen> size fraction
Korngrößenbestimmung f 1. <Kohlen> size grading; 2. <Kunststoff> particle size measurement
Korngrößenkurve f <Ker & Glas> particle size curve
Korngrößenverteilung f <Bau> particle size distribution
körnig adj <Bau, Ker & Glas, Metall> granular
körniger Bruch m <Kerntech, Metall> granular fracture
körniger Korund m <Ker & Glas> emery
körniger Schliff m <Ker & Glas> sugary cut
körniges Material n <Bau> granular material
Körnigkeit f <Foto> granularity
Kornklassierung f <Kohlen> size grading
kornlos adj <Foto> grainless
Kornoberflächenentwicklung f <Foto> grain surface development
kornorientierter Stahl m <Metall> grain-oriented steel
Kornstruktur f <Kohlen> grain structure
Körnung f 1. <Akustik> graininess; 2. <Bau> grading; 3. <Comp & DV> granularity; 4. <Druck> graining; 5. <Fertig> grit; 6. <Kohlen> granulation; 7. <Metall> grain size; 8. <Papier> grain
Körnung f **des Gemenges** <Ker & Glas> granulation of the batch
Körnungsprüfung f <Ker & Glas> grain test
Körnungspunkt m <Maschinen> punch mark
kornverfeinerndes Glühen n <Thermod> grain-refining anneal
Kornverfeinung f <Metall> grain refinement
Kornzerfall m <Fertig> weld decay
Kornzerkleinerung f <Kohlen> particle size reduction
Korollar n <Math> corollary *(Folgesatz)*
Korona f 1. <Elektriz, Elektrotech, Funktech, Kunststoff> corona; 2. <Phys> corona *(Strahlenkranz)*; 3. <Strahlphys> corona
Koronaeffekt m <Elektriz> corona effect
Koronaentladung f 1. <Elektriz, Elektrotech> corona discharge, corona effect; 2. <Kunststoff, Phys, Raumfahrt> corona discharge

Koronafestigkeit f <Kunststoff> corona resistance
Koronallinien fpl <Strahlphys> coronal emission lines (Sonne)
Körper m 1. <Geom> body, solid; 2. <Maschinen, Mechan, Wasserversorg> body
Körper m **in Bewegung** <Mechan> body in motion
Körper m **in Ruhe** <Mechan> body at rest
Körperabstützung f <Ergon> body support
körperfester Drehkegel m <Fertig> polhode cone
Körperhaltung f <Ergon> posture
Körperkante f <Fertig, Konstzeich> edge
Körperlage f <Ergon> posture
Körpermaße fpl <Ergon> body dimensions
Körperschall m <Akustik, Bau> structure-borne noise, structure-borne sound
körperübertragene Schwingung f <Sicherheit> body-transmitted vibration
Korpuskel f <Elektronik, Teilphys> particle
Korpuskelstrahl m <Elektronik> particle beam
Korpuskularstrahlung f <Strahlphys> corpuscular radiation
Korrektionsfilter n <Foto> correction filter
Korrektur f 1. <Comp & DV> patch; 2. <Labor, Maschinen> correction
Korrekturabzug m <Druck> proof
Korrekturbefehl m <Elektrotech> patch (bei Rechnern, Programmen)
Korrekturfaktor m <Metrol, Qual> correction factor
Korrekturlesen n <Druck> proofreading
Korrekturlinse f 1. <Foto> correcting lens, correction lens; 2. <Optik> correcting optics
Korrekturmanöver n <Raumfahrt> (AE) correction maneuver, (BE) correction manoeuvre
Korrekturmaßnahmen fpl <Qual> corrective measures
Korrekturschaltung f <Fernseh> corrector circuit
Korrekturseiten fpl <Druck> page proofs
Korrekturzeichen npl <Druck> editing marks, proof correction marks, proofreader's marks
Korrelation f 1. <Comp & DV, Elektronik, Labor, Phys, Strömphys, Telekom> correlation; 2. <Math> correlation (Zwischenbeziehung)
Korrelationsdauer f <Gerät> correlation interval
Korrelationsdecodierung f <Telekom> correlation decoding (von Synchronisationswörtern)
Korrelationsfunktion f <Elektronik> correlation function
Korrelationsintervall n <Gerät> correlation interval
Korrelationskoeffizient m 1. <Comp & DV, Math> correlation coefficient (Statistik); 2. <Phys, Strömphys> correlation coefficient
Korrelationsmessverfahren n <Gerät> correlation-measuring procedure
Korrelationsphasenumtastung f <Telekom> correlative phase shift keying
Korrelator m <Elektronik, Telekom> correlator
korreliert adj <Math> correlated
korrespondenzfähiges Schriftbild n <Comp & DV, Druck> NLQ, near-letter quality
Korrespondenzprinzip n <Phys> correspondence principle
Korrespondenzqualität f (LQ) 1. <Comp & DV> letter-quality (LQ); 2. <Druck> letter quality (LQ)
korrespondierende Winkel mpl <Geom> corresponding angles
korrigieren v 1. <Bau> rectify; 2. <Comp & DV> debug, patch; 3. <Druck> patch; 4. <Qual> rectify
korrigierende Schutzeigenschaften fpl <Sicherheit> corrective protective properties
korrigierte Stelle f <Druck> patch
korrigierter Abzug m <Druck> clean proof

korrigiertes Ergebnis n <Metrol> corrected result
korrodierbar adj <Chemie> corrodible
korrodierender Stoff m <Sicherheit> corrosive substance
Korrosion f <Anstrich, Bau, Chemie, Erdöl, Funktech, Ker & Glas, Kerntech, Kfztech, Kohlen, Kunststoff, Maschinen, Mechan, Metall, Verpack> corrosion
Korrosion verursachender Stoff m <Anstrich> corrodent
korrosionsanfällig adj <Chemie> corrodible
Korrosionsanfälligkeit f <Kerntech> corrodibility
Korrosionsbeizen n <Kerntech> corrosion pickling
korrosionsbeständig adj <Chemie, Maschinen> corrosion-resistant
korrosionsbeständiger Edelstahl m <Maschinen, Metall> corrosion-resistant stainless steel
Korrosionsbeständigkeit f <Bau, Kunststoff, Metall> corrosion resistance
Korrosionselement n <Fertig> corrosion cell
Korrosionsermüdung f <Metall> corrosion fatigue
korrosionsgeschützt adj <Mechan> anticorrosive
Korrosionshemmer m <Kfztech, Raumfahrt, Metall> corrosion inhibitor
Korrosionshemmstoff m <Metall, Verpack> corrosion preventive
Korrosionsinhibitor m <Erdöl> anticorrosion additive
Korrosionsknoten m <Kerntech> corrosion nodule
Korrosionsmedium n <Chemie> corrodent
korrosionsresistent adj <Anstrich> corrosion-resistant
Korrosionsriss m <Kerntech, Metall> corrosion fatigue crack
Korrosionsschutz m <Elektriz, Verpack> corrosion prevention
Korrosionsschutzanstrich m <Kunststoff> anticorrosive coating
Korrosionsschutzbeschichtung f <Kerntech> anticorrosion coating
Korrosionsschutzmittel n 1. <Kfztech, Metall> corrosion inhibitor; 2. <Papier> anticorrosive agent
Korrosionsschutzpapier n 1. <Papier> antitarnish paper; 2. <Verpack> corrosion preventative paper
korrosionssicher adj <Fertig> corrosion-resistant (Kunststoffinstallationen)
Korrosionsspannungsriss m <Metall> stress corrosion cracking, SCC
Korrosionsverschleiß m <Maschinen, Metall> corrosive wear
korrosiver Abnutzungsverschleiß m <Mechan> abrasion-fretting corrosion
korrosives Wasser n <Wasserversorg> corrosive water
Korund m 1. <Ker & Glas> corundum, emery; 2. <Mechan> emery
Korundschlämmung f <Ker & Glas> emery washing
Kosekans m <Geom> cosecant
Kosekansfunktion f <Geom> cosecant
Kosinus m <Bau, Comp & DV, Fernseh, Geom, Telekom> cosine
Kosinusquadrat-Impuls m <Telekom> raised-cosine pulse
Kosinussatz m <Geom> cosine rule
Kosmetikfläschchen n <Ker & Glas> cosmetic jar
kosmisch adj <Raumfahrt> cosmic
kosmische Astronomie f <Raumfahrt> space astronomy
kosmische Geschwindigkeit f <Raumfahrt> cosmic velocity
kosmische Hintergrundstrahlung f <Phys> cosmic background radiation
kosmische Strahlung f <Raumfahrt> cosmic radiation
kosmischer Schauer m <Raumfahrt> cosmic shower

kosmisches

kosmisches Rauschen n <Funktech> cosmic noise
Kosmodrom m <Raumfahrt> cosmodrome
Kosmogonie f <Raumfahrt> cosmogony
Kosmographie f <Raumfahrt> cosmography
Kosmologie f <Raumfahrt> cosmology
Kosmonaut m <Raumfahrt> cosmonaut
Kosmos m <Raumfahrt> cosmic space
Kossel-Linie f <Kerntech> Kossel line *(in Röntgenspektren)*
Kosten fpl 1. <Patent, Qual> costs; 2. <Telekom> call charges, charge
Kosten fpl **durch Ausfallzeit** <Verpack> downtime cost
kostenabhängige Verkehrslenkung f <Telekom> least cost routing
Kostenanschlag m <Bau> tender
Kostenfestsetzung f <Patent> awarding of costs
Kostenfunktion f <Qual> cost function
kostenpflichtige Telefondienstleistung f <Telekom> premium rate service
Kostenplaner m <Bau> quantity surveyor
Kostenstelle f <Qual> *(AE)* cost center, *(BE)* cost centre
Kostenträger m <Qual> cost bearer
Kotangens m *(cot)* <Geom> cotangent *(cot)*
kotangentiale Umlaufbahn f <Raumfahrt> cotangential orbit
Kotflügel m 1. <Kfztech> *(AE)* fender, *(BE)* mudguard, *(BE)* wing *(Karosserie)*; 2. <Mechan> fender
kovalente Bindung f 1. <Kerntech> covalency; 2. <Metall> covalency, covalent bond
Kovalenz f <Kerntech, Metall> covalency *(Bindung)*; covalence *(Wertigkeit)*
Kovarianz f <Comp & DV, Math> covariance
KPK *(kritische Pigmentvolumenkonzentration)* <Kunststoff> cpvc *(critical pigment volume concentration)*
Krabbenkutter m <Wassertrans> shrimp trawler
Kracken n <Erdöl> cracking *(Raffinerie)*
Kraft f 1. <Elektrotech, Kfztech, Maschinen> energy, force; 2. <Metall> force *(F)*; 3. <Papier> power; 4. <Phys> force *(F)* • **außer Kraft gesetzt** <Kontroll> disabled • **außer Kraft setzen** <Elektrotech> override • **äußerste Kraft voraus** <Wassertrans> full ahead *(Motor)* • **äußerste Kraft zurück** <Wassertrans> full astern *(Motor)* • **in Kraft setzen** <Kontroll> activate, enable
Kraft f **der Masse** <Metall> body force
Kraftanlage f <Nichtfoss Energ> powerhouse
Kraftarm m <Fertig, Maschinen> moment arm
Kraftauslegepapier n <Verpack> kraft liner
kraftbetrieben adj <Maschinen, Sicherheit> energized, power-driven, powered
kraftbetriebenes Arbeitsmittel n <Sicherheit> powered equipment
kraftbetriebenes Handwerkzeug n <Ergon, Sicherheit> energized tool
Kraftdroschke f <Kfztech> cab
Kräfte fpl **in einer Ebene** <Phys> coplanar forces
Kräftediagramm n <Maschinen> force diagram
Kräftedreieck n 1. <Bau> triangle of forces; 2. <Phys> TWT, *(AE)* traveling wave tube, *(BE)* travelling wave tube, triangle of forces
kräftefreier Kreisel m <Lufttrans, Raumfahrt> free gyroscope
Kräfteparallelogramm n <Geom, Maschinen> parallelogram of forces
Kräfteplan m <Konstzeich> force diagram
Kräftepolygon n <Maschinen> polygon of forces
Kräfteumwandlung f <Bau> force transformation
Kräftevieleck n <Phys> polygon of forces
Kraftfahrzeug n 1. <Kfztech> automobile; 2. <Kfztech> motorcar, MC *(Kfz)*

Kraftfeld n <Fertig, Kerntech> field of force
Kraftflusslinie f <Elektrotech> line of flux
Kraftgewebe n <Papier> power fabric
kräftiger Farbton m <Kunststoff> *(AE)* deep color tone, *(BE)* deep colour tone
Kraftlinie f <Elektriz, Elektrotech, Maschinen, Phys> line of force
Kraftlinienfeld n <Elektrotech> field of force
Kraftmaschine f 1. <Kfztech> motor; 2. <Maschinen> engine
Kraftmaschine f **mit äußerer Verbrennung** <Kfztech, Maschinen> external combustion engine
Kraftmessdose f <Gerät> load cell
Kraftmesser m <Mechan> dynamometer
Kraftmessfühler m <Gerät> force sensor
Kraftmessplatte f <Ergon> force platform
Kraftpapier n <Verpack> kraft paper *(braunes Hartpapier)*
Kraftpappe f <Verpack> kraft board
Kraftsackpapier n <Verpack> kraft sack paper
Kraftschalter m <Lufttrans, Mechan> actuator
Kraftschlepper m <Kfztech> traction engine
Kraftschluss m <Bau> traction
Kraftschlussbeiwert m <Kfztech> adhesion coefficient
kraftschlüssig adj <Fertig, Maschinen> nonpositive
kraftschlüssiger Schnappverschluss m <Verpack> friction snap-on cap
Kraftschrauber m <Maschinen> power wrench
Kraftschreiber m <Mechan> dynamograph
Kraftsensor m <Gerät> force sensor
Kraftspannfutter n <Maschinen> power-operated lathe chuck
Kraftstoff m <Kfztech, Wassertrans> fuel
Kraftstoffanlage f 1. <Maschinen> engine fuel system; 2. <Wassertrans> fuel system *(Schiffantrieb)*
Kraftstoffanzeiger m <Kfztech> *(AE)* fuel gage, *(BE)* fuel gauge, fuel indicator
Kraftstoffbehälter m <Kfztech> fuel tank
Kraftstoffbehälterdeckel m <Kfztech> *(AE)* gas tank cap, *(AE)* gasoline tank cap, *(BE)* petrol tank cap
Kraftstoffdoppelfilter n <Kfztech> two-stage fuel filter
Kraftstoffdosierung f <Gerät> fuel metering
Kraftstoffdüse f <Maschinen> fuel nozzle
Kraftstoffdüsenhalterungabdeckung f <Lufttrans> fuel jet support cover
Kraftstoffeinlassventil n <Kfztech> fuel inlet valve
Kraftstoffeinspritzdüse f <Kfztech> fuel nozzle
Kraftstoffeinspritzpumpe f <Kfztech> fuel injection pump
Kraftstoffeinspritzung f <Kfztech, Thermod> fuel injection
Kraftstofffassungsvermögen n <Lufttrans> fuel load
Kraftstofffilter n 1. <Kfztech> fuel filter, *(AE)* gas filter, *(AE)* gasoline filter, *(BE)* petrol filter; 2. <Maschinen> fuel filter
Kraftstoffförderpumpe f <Raumfahrt> boost pump
Kraftstoffkolbenpumpe f <Kfztech> plunger fuel pump
Kraftstoffleitung f <Kfztech> fuel line; *(AE)* gas hose, *(AE)* gasoline hose, *(BE)* petrol hose *(Kraftstoff)*
Kraftstoffluftgemisch n <Thermod> fuel-air mixture
Kraftstoff-Luft-Verhältnis n <Kfztech> mixture ratio
Kraftstoffmessanzeiger m <Lufttrans> *(AE)* fuel gage indicator, *(BE)* fuel gauge indicator
Kraftstoffmessgeber m <Lufttrans> *(AE)* fuel gage transmitter, *(BE)* fuel gauge transmitter
Kraftstoffpumpe f 1. <Kfztech> fuel pump, *(AE)* gas pump, *(AE)* gasoline pump, *(BE)* petrol pump; 2. <Maschinen, Wassertrans> fuel pump
kraftstoffreiches Gemisch n <Kfztech> rich mixture
Kraftstoffschlauch m <Kunststoff> fuel hose
Kraftstoffsorte f **nach Oktanzahl** <Kfztech> fuel grade

Kraftstoffstandsprogrammsteuerung f <Lufttrans> fuel level pre-setting controls
Kraftstofftank m <Kfztech> (AE) gas tank, (AE) gasoline tank, (BE) petrol tank, tank
Kraftstoffumrechnungsfaktor m <Kohlen> conversion fuel factor
Kraftstoffverbrauch m <Kfztech> consumption, (AE) gas consumption, (AE) gasoline consumption, (BE) petrol consumption
Kraftstoffverbrauch m **pro Meile** <Kfztech, Wassertrans> mileage
Kraftstoffvorratsübermittler m <Lufttrans> fuel level transmitter
Kraftstoffzufuhr f <Kfztech, Wassertrans> engine fuel supply
Kraftstoffzusatz m **gegen Klopfen** <Kfztech> antiknock agent
Kraftstrom m <Elektriz, Elektrotech, Lufttrans> electric power
Kraftstromeinrichtungen fpl <Elektriz> power plant
Kraftstromkabel n <Elektriz> power cable
Kraftstromkreis m <Elektriz> power circuit
Kraftsystem n **mit durch resonanten Schwingungskreis geerdetem Mittel** <Elektriz> (BE) resonant-earthed neutral system, (AE) resonant-grounded neutral system
Kraftturm m <Nichtfoss Energ> power tower
Kraftübertragung f 1. <Elektrotech> power transmission; 2. <Kfztech> drive line, transmission; 3. <Maschinen> transmission of forces
Kraftübertragung f **durch Keilriemen** <Maschinen> V-belt transmission
Kraftübertragung f **durch Riementrieb** <Maschinen> power transmission by belt drive
Kraftübertragungsritzel n <Kfztech> transmission pinion
Kraftübertragungssystem n <Maschinen> power transmission system
Kraftübertragungsweg m <Kfztech, Lufttrans> power train
Kraftventilator m <Kfztech> power fan
Kraftverbindung f <Lufttrans> force link
Kraftvergleichsglied n <Gerät> force-balance element
Kraftvergleichsmesswandler m <Elektrotech> force-balance transducer
Kraftvergleichstransducer m <Elektrotech> force-balance transducer
Kraftverstärker m (PA) <Aufnahme> power amplifier (PA)
Kraftwagen m <Mechan> car
Kraftwagengetriebe n <Kfztech> motor-vehicle gear
Kraftwagenkupplung f <Maschinen> automotive clutch
Kraft-Wärme-Kopplung f <Elektriz> cogeneration
Kraftwerk n 1. <Elektrotech> generating plant, power plant, power station; 2. <Kohlen, Maschinen, Nichtfoss Energ, Phys, Telekom> power station
Kraftwerksteilanlage f <Kerntech> unit
Kraftzellstoff m <Papier> kraft pulp
Kraftzentrale f <Elektrotech> central power plant
Krag... <Bau> cantilever
Kragarm m 1. <Bau> cantilever, jib; 2. <Fertig> cantilever
Kragelement n <Bau> jetty
Kragendichtung f <Maschinen> collar joint
Kraglast f <Bau> cantilever load
Kragstein m <Bau> corbel
Kragträger m <Bau> cantilever beam
Kragtreppe f <Bau> hanging stairs
Krählarm m <Abfall> rabble arm
Kralle f <Bau, Maschinen> claw
Krampe f 1. <Bau> staple; 2. <Maschinen> cramp
Kran m <Maschinen, Mechan, Wassertrans> crane
Kranauslegearm m <Mechan> crane jib

Kranausleger m <Mechan> crane jib
Kranbein n <Bau> crane leg
Krandreharm m <Mechan> crane jib
Kranführer m <Bau, Wassertrans> crane operator
Kranführerhaus n <Bau> cabin, house
krängen v <Wassertrans> heel (vorübergehende Seitenneigung eines Schiffes)
Krangerüst n <Bau> framework
Krängung f <Wassertrans> heel
Krängungsausgleichsanlage f <Wassertrans> heel compensation system
Krängungsmoment n <Wassertrans> heeling moment (Schiffkonstruktion)
Krängungsversuch m <Wassertrans> inclining test (Schiffbau)
Krängungswinkel m <Wassertrans> angle of heel (Schiffkonstruktion)
Krankenbahre f <Sicherheit> stretcher
Krankenhausabfall m <Abfall> hospital waste
Kranlasthaken m <Bau> crane hook
Kranlaufbahn f <Bau> craneway
Kranleistung f <Fertig> cranage
Kranpfanne f <Fertig> crane ladle (Getriebelehre)
Kranportal n <Bau> gantry
Kransäule f <Maschinen> crane post
Kranschiene f <Bau> crane rail
Kranschiff n <Erdöl> crane barge (Schifffahrt)
Kranz m 1. <Bau> wreath; 2. <Maschinen> rim
Kranzleiste f <Bau> platband
Krarupisierung f <Elektrotech> continuous loading, krarup loading
Krarupkabel n <Elektrotech> krarup cable
Krater m 1. <Fertig> cup (Lichtbogenschweißen); 2. <Ker & Glas> dimple; 3. <Kunststoff> crater, pinhole • **Krater bilden** <Fertig> crater (Schweißen)
Kraterbildung f 1. <Fertig> cratering (Schweißen); 2. <Kunststoff> crawling
Kratzbandförderer m <Maschinen> scraper
Kratze f 1. <Fertig> skimmer; 2. <Textil> card (Wolle)
Krätze f <Fertig> dross
Kratzeisen m <Bau, Maschinen> raker, scraper
kratzen v <Anstrich, Textil> scratch
Kratzen n <Textil> carding
Kratzenbeschlag m <Textil> card clothing
Kratzer m <Anstrich, Kunststoff, Papier> scratch
Kratzer m **auf dem Schichtträger** <Foto> base scratch
Kratzfestigkeit f <Kunststoff, Werkprüf> scratch resistance
Kratzhobel m <Bau> scraper
Krause f <Textil> ruffle
Kräusel... <Textil> curling
Kräuselfaser f <Textil> (AE) crimped fiber, (BE) crimped fibre
Kräuselgarn n <Textil> crimped yarn
Kräusellack m <Kunststoff> wrinkle paint
kräuseln v 1. <Textil> crimp, curl, gather; 2. <Verpack> falten
kräuseln v/**sich** <Papier> curl
Kräuselung f 1. <Papier> curling; 2. <Textil> crimp, curl
Kräuselungsgrad m <Textil> degree of crimp
Krebserreger m <Lebensmittel> carcinogen
krebserzeugender Stoff m <Sicherheit> carcinogenic substance
Kreditkartenanruf m <Telekom> credit card call
Kreide f 1. <Erdöl> cretaceous period (Geologie); 2. <Ker & Glas> chalk
Kreideformation f <Wasserversorg> chalk formation
Kreidemergel m <Bau> chalk marl
Kreiden n <Kunststoff> chalking

Kreideschicht

Kreideschicht f <Bau, Kohlen> chalk stratum
Kreis m 1. <Elektrotech> circuit, cycle; 2. <Funktech> circuit; 3. <Geom> circle • **Kreis schließen** <Fertig> make a circuit *(Stromkreis)*
Kreis m **mit konzentriertem Schaltelement** <Elektrotech> lumped-element circuit
Kreis m **mit verteilten Elementen** <Elektrotech> distributed-element circuit
Kreisbahn f <Raumfahrt> circular route
Kreisbecken n <Abfall, Kohlen> clarification basin
Kreisbeschleuniger m <Teilphys> cyclic accelerator
Kreisbewegung f 1. <Fertig> gyration; 2. <Maschinen, Phys> circular motion
Kreisblattdiagramm n <Gerät> circular chart diagram
Kreisblattschreiber m <Gerät> circular chart recorder
Kreisblende f <Phys> circular aperture
Kreisbogen m <Geom> arc of a circle, circular arc
Kreisbüschel n <Fertig> family of circles
Kreisdiagramm n <Comp & DV> pie chart
Kreisel m <Maschinen, Mechan> gyroscope
Kreiselaufzeichnungsgerät n <Raumfahrt> gyrograph
Kreiselbrecher m <Lebensmittel> gyratory crusher
Kreiselgehäuse n <Lufttrans> gyrocaging
kreiselgesteuerte Datenvermittlung f <Lufttrans> gyro data-switching control
kreiselgestützt *adj* <Mechan> gyroscopic
Kreiselhorizont m <Lufttrans, Trans> artificial horizon, gyro horizon
Kreiselinstrumente npl <Lufttrans> gyro instruments
Kreiselkompass m 1. <Lufttrans> gyrocompass, gyroscopic compass; 2. <Maschinen> gyrostat; 3. <Wassertrans> gyrocompass *(Navigation)*
Kreiselkompasskoppelung f <Wassertrans> azimuth stabilization *(Radar)*
Kreiselkraft f <Nichtfoss Energ> gyroscopic force
Kreisellader m <Kfztech> centrifugal supercharger
Kreiselmoment n <Lufttrans> gyroscopic torque
Kreiselnullstellung f <Lufttrans> gyro resetting
Kreiselplattform f <Lufttrans> gyroscopic platform
Kreiselpumpe f 1. <Chemtech> centrifugal pump; 2. <Fertig> centrifugal pump *(Kunststoffinstallationen)*; 3. <Heiz & Kälte, Maschinen, Meerschmutz> centrifugal pump
Kreiselradius m <Phys> radius of gyration
Kreiselstabilisierung f <Trans> gyro stabilization
Kreiselsteuergerät n <Lufttrans> autogyro, autopilot
Kreiselströmungsdurchflussmesser m <Gerät> gyroscopic flow meter
Kreiselunwucht f <Lufttrans> gyro unbalance
Kreiselverstärkerstufe f <Lufttrans> gyro amplifier
Kreiselwendeanzeiger m <Raumfahrt> gyro turn indicator *(Raumschiff)*
Kreiselzurückstellung f <Lufttrans> gyro resetting
kreisen v <Raumfahrt> orbit
Kreisevolvente f <Geom> involute of a circle
Kreisfläche f <Geom> area of a circle
Kreisform f <Fertig> circularity
Kreisformel f <Geom, Math> circle formula
kreisförmig *adj* <Elektrotech, Geom> circular
kreisförmige Abtastung f <Elektronik> circular scan
kreisförmiger Riss m <Metall> penny-shaped crack
Kreisfrequenz f 1. <Akustik, Elektronik, Elektrotech> angular frequency, angular velocity; 2. <Fertig> pulsatance; 3. <Nichtfoss Energ> angular velocity; 4. <Phys> angular frequency, pulsatance
Kreisfunktion f <Geom> circle function
Kreisgeschwindigkeit f <Elektrotech> angular velocity
Kreisgleichung f <Geom> circle equation, equation of circle
Kreisgrad m <Elektriz> radian

Kreiskolbenmotor m <Kfztech> rotary engine, rotary piston engine
Kreislauf umpumpen v/im <Kohlen> recirculate
Kreislauf m **zur Wiederaufbereitung des Schutzgases** <Kerntech> blanket-reprocessing circuit
Kreislaufschmierung f <Maschinen> recirculating lubrication, recirculation lubrication
Kreislaufwasserführung f <Wasserversorg> recirculating water economy
Kreislaufwirtschaft f <Abfall> recycling economy
Kreismittelpunkt m <Geom> centre of circle
Kreisneigung f <Mechan> circular pitch
Kreispolarisation f <Funktech, Telekom> circular polarization
Kreisring m 1. <Fertig, Geom, Lufttrans> annulus; 2. <Maschinen> torus
kreisrunder Schleifkörper m <Fertig> grinding wheel
Kreissäge f <Maschinen> annular saw, circular saw
Kreissägeblatt n <Maschinen> circular saw blade
Kreissägenberührungsschutz m <Sicherheit> protective cover for blade
Kreisschnitt m <Elektrotech> pie section
Kreisschwingsieb n <Kohlen> circular-vibrating screen
Kreissektor m 1. <Bau> sector; 2. <Geom> sector of a circle
Kreisskale f <Gerät> dial scale
Kreisskalenanzeigegerät n <Gerät> round scale indicator
Kreissperre f <Elektronik> loop lock
kreisstabilisierte Plattform f <Lufttrans> gyroscopic platform
Kreisstellung f <Mechan> circular pitch
Kreisstruktur f <Regelung> closed-loop structure
Kreistangente f <Geom> tangent to the circle
Kreisumlaufbahn f <Raumfahrt> circular orbit
Kreisverkehr m <Trans> (AE) traffic rotary, (BE) traffic roundabout
Kreisverkehrsplatz m <Trans> (AE) rotary intersection, (BE) roundabout intersection
Kreisverschiebung f <Elektronik> circular shift *(Datenverarbeitung)*
Kreisverstärkung f <Elektronik> loop gain
Krempe f <Bau> flap *(Dachziegel)*
Krempel f <Textil> card *(Spinnen)*
Krempelbeschlag m <Textil> card clothing
Krempelei f <Textil> card room
krempeln v <Textil> card
Krempeln n <Textil> carding
Krempelsaal m <Textil> card room
Kreosot n <Chemie, Kohlen> creosote
Kreppen n <Papier> creping
Krepppapier n <Papier, Verpack> crepe paper
Kreppverband m <Textil> crepe bandage
Kreuz n 1. <Akustik> sharp; 2. <Comp & DV, Ker & Glas, Phys, Strahlphys, Teilphys> cross
Kreuzanschnitt m <Ker & Glas> cross bevel
Kreuzassemblierer m <Comp & DV> cross assembler
Kreuzbeschuss m <Phys, Strahlphys, Teilphys> cross bombardment
Kreuzbett n <Maschinen> crossbed
Kreuzbettfräsmaschine f <Maschinen> crossbed-milling machine
Kreuzblume f <Bau> finial
Kreuzbohrmeißel m 1. <Erdöl> star bit; cross bit *(Bohrtechnik)*; 2. <Fertig> star bit
kreuzen v 1. <Bau, Wassertrans> cross; 2. <Wassertrans> cruise *(Marine)*
Kreuzen n **vor dem Wind** <Wassertrans> downwind tacking *(Segeln)*

Kreuzer m <Wassertrans> cruiser *(Marinefahrzeug)*
Kreuzerheck n <Wassertrans> cruiser stern *(Schiffbau)*
Kreuzerjacht f <Wassertrans> cruiser *(Hochseejacht)*
Kreuzfahrt f <Wassertrans> cruise
Kreuzfahrtschiff n <Wassertrans> cruise liner, cruise ship, passenger cruiser
Kreuzfeld n <Elektronik, Telekom> crossed field
Kreuzfeldmikrowellenröhre f <Elektronik> M-type microwave tube
Kreuzfeldröhre f <Elektronik> crossed field tube
Kreuzfeldröhren fpl <Phys> crossed field tube *(Mikrowellenröhren)*
Kreuzfeldverstärker m <Elektronik, Telekom> crossed field amplifier
kreuzförmiger Kopf m <Maschinen> cruciform head
Kreuzgelenk n 1. <Kfztech> universal joint *(Triebstrang)*; 2. <Maschinen> cardan joint, universal joint
Kreuzgelenkgabel f <Fertig> universal joint yoke
kreuzgewickelte Spule f <Elektrotech> lattice-wound coil
Kreuzgewölbe n <Bau> groined vault
Kreuzglied-Kettenfilter n <Elektronik, Telekom> lattice filter
Kreuzgriff m <Maschinen> cross handle, spider, star handle
Kreuzhiebfeile f <Fertig, Maschinen> crosscut file
Kreuzknoten m <Wassertrans> carrick bend *(Knoten)*
Kreuzkompilierer m <Comp & DV> cross compiler
Kreuzkompilierung f <Comp & DV> cross compilation
Kreuzkopf m <Maschinen, Phys, Wassertrans> crosshead
Kreuzkopfbolzen m <Hydraul> pin of cross head
Kreuzkopfdieselmotor m <Wassertrans> crosshead engine *(Motor)*
Kreuzkopfverschiebungsgeschwindigkeit f <Phys> crosshead displacement rate
Kreuzkopfzapfen m <Maschinen> crosshead pin
Kreuzkopplung f <Elektrotech> cross coupling *(von Wellen)*
Kreuzkorrelation f <Telekom> cross correlation
Kreuzkorrelation f der Geschwindigkeit <Strömphys> double velocity correlations *(Turbulenz)*
Kreuzkorrelationsfunktion f <Elektronik> cross-correlation function
Kreuzkorrelator m <Elektronik> cross correlator
Kreuzlatte f <Bau> brace
Kreuzleistungsspektrum n <Gerät> cross power spectrum
Kreuzloch n <Maschinen> cross hole
Kreuzlochmutter f <Maschinen> capstan nut
Kreuzlochschraube f <Bau> capstan screw
Kreuzmaß n <Bau> cross
Kreuzmeißel m <Fertig> crosscut chisel
Kreuzmodulation f <Elektronik, Fernseh, Funktech> cross modulation
Kreuzprodukt n 1. <Geom> Cartesian product; 2. <Math> cross product *(von dreidimensionalen Vektoren)*
Kreuzrändeln n <Maschinen> cross knurling, diamond knurling
Kreuzriet n <Textil> leasing reed
Kreuzrohrstück n <Bau> cross
Kreuzscheibe f <Bau> cross-staff head
Kreuzschieber m <Fertig> saddle *(Fräsmaschine)*
Kreuzschiene f *(KS)* <Fernseh> crossbar
Kreuzschienenverteiler m <Telekom> crossbar distributor
Kreuzschienenwähler m <Elektrotech> cross coupling, crossbar selector
Kreuzschlaghammer m <Bau> cross-peen sledge hammer
Kreuzschlagseil n <Maschinen> regular-lay rope
Kreuzschliff m <Fertig> cross hatch

Kreuzschlitten m <Maschinen> compound rest, compound slide rest
Kreuzschlitz m <Maschinen> cross recess
Kreuzschlitzschraube f <Maschinen> Phillips screw
Kreuzschlitzschraubendreher m <Maschinen> Phillips screwdriver
Kreuzschlitzschraubenzieher m <Maschinen> Phillips screwdriver
Kreuzschlüssel m <Maschinen> *(BE)* spider spanner, *(AE)* spider wrench
Kreuzschnur f <Textil> lease band
Kreuzspring f <Wassertrans> cross spring *(Festmachen)*
Kreuzspule f <Textil> cheese
Kreuzspulinstrument n <Gerät> crossed coil instrument
Kreuzspulmesswerk n <Gerät> crossed coil movement
Kreuzstab m <Textil> lease rod
Kreuzstoß m <Fertig> double-tee joint
Kreuzstück n 1. <Bau> crosspiece; 2. <Maschinen> cross fitting, double junction, double tee, pipe cross
Kreuztisch m 1. <Foto> mechanical stage; 2. <Maschinen> compound table
Kreuztischeinrichtung f <Foto> mechanical stage
Kreuzumwandlung f <Comp & DV> cross assembly
Kreuzung f <Bau> junction
Kreuzungsbahnhof m <Eisenbahn> crossing station
Kreuzungsmast m <Elektriz> transposition tower
Kreuzungspunkt m <Comp & DV> intersection
Kreuzungsweiche f <Eisenbahn> slip
Kreuzungswinkel m <Trans> drift angle
Kreuzverband m <Bau, Eisenbahn> crossbond
Kreuzverschraubung f <Maschinen> cross union
kreuzverzahnter Fräser m <Fertig> alternate helical tooth cutter
kreuzverzahnter Walzenfräser m <Fertig> alternate gash plain mill
kreuzweis-bewehrte Platte f <Bau> two-way slab
Kreuzwickel m <Textil> cheese
Kreuzwickelhülse f <Textil> cheese tube
Kreuz-Yagi-Antenne f <Elektronik, Funktech> crossed Yagi array
Kreuzzahnscheibenfräser m <Fertig> alternate angle side and face cutter
Kriech... <Phys> creeping in
Kriechbeanspruchung f <Werkprüf> creep loading
Kriechbewegung f 1. <Bau, Metall> creeping motion; 2. <Strömphys> creeping motion *(in Verbindung mit zähen Strömungen)*
Kriechboden m 1. <Bau> raised floor; 2. <Heiz & Kälte> false floor, raised floor
Kriechbruchdehnung f <Metall> creep rupture elongation
Kriechdehnung f <Maschinen> creep strain
Kriecheigenschaften fpl <Mechan, Qual> creep properties
Kriechen n 1. <Bau, Erdöl, Kohlen> creep, creeping; 2. <Kunststoff> cissing, creep; 3. <Maschinen> creeping; 4. <Mechan, Metall> creep; 5. <Qual> creep, creeping
Kriecherholung f <Metall> recovery creep
Kriechfestigkeit f <Kunststoff, Qual> creep resistance
Kriechöl n <Kfztech> penetrating oil
Kriechspur f <Bau> climbing lane *(Straße)*
Kriechstrom m 1. <Elektrotech> leakage current; 2. <Fernseh> crawling current; 3. <Funktech> leak current; 4. <Telekom> leakage current
Kriechstrom m zur Erde <Elektriz> *(BE)* earth leakage current, *(AE)* ground leakage current
Kriechstromverlust m <Elektriz> leakage loss
Kriechverhalten n <Mechan, Qual> creep properties
Kriechversuch m <Qual, Werkprüf> creep test

Kriechweg

Kriechweg m <Elektrotech> leakage path
Kriechwegbildung f <Elektriz> tracking
Kriegsschiff n <Wassertrans> man-of-war, warship
Krimpen n <Wassertrans> backing *(des Windes)*
Krimpwerkzeug n <Elektriz> crimping tool *(Quetschverbinden)*
Kristall m 1. <Chemie, Elektronik> crystal; 2. <Fertig> grain; 3. <Ker & Glas, Strahlphys> crystal
Kristalldiode f <Elektronik> crystal diode
Kristallfilter n <Strahlphys> crystal filter
Kristallfläche f <Metall> facet
Kristallflächenwachstum n <Metall> facetted growth
Kristallgitter n <Chemie> crystal lattice
Kristallgitter-Filter n <Elektronik> crystal lattice filter
Kristallglas n <Ker & Glas> crystal glass
Kristallhalter m <Elektronik> crystal holder
Kristalline f <Ker & Glas> celluloid varnish
kristalliner Kieselerdestaub m <Sicherheit> crystalline silica dust
kristallinischer Bruch m <Metall> crystalline fracture
Kristallisation f <Chemie> crystallization
Kristallisationsbeginn m <Chemtech> crystallization point
Kristallisationsgefäß n <Chemtech> crystallizer
Kristallisationskeim m 1. <Fertig> initial nucleus, seed crystal; 2. <Thermod> initial nucleus
Kristallisationspunkt m <Strömphys> pour point
Kristallisierbecken n <Chemtech> crystallizing pond
Kristallisierschale f <Labor> crystallizing dish
kristallisierte Essigsäure f <Chemie> glacial acetic acid
Kristallkeim m <Fertig> seed crystal
Kristall-Laser m <Elektronik> crystal laser
Kristallmikrofon n <Aufnahme> crystal microphone
Kristallographie f <Metall> crystallography
kristallographisches Gleiten n <Metall> crystallographic slip
Kristalloszillator m <Strahlphys> crystal oscillator
Kristallperlwand f <Foto> beaded screen
Kristallplastizität f <Metall> crystal plasticity
Kristallseigerung f <Fertig> coring
Kristallskelett n <Fertig> dendrite *(Gefüge)*
Kristallspektrometer n <Strahlphys> crystal spectrometer
Kristallspiegelglas n <Bau> polished plate glass
Kristallstruktur f <Metall> crystal structure
Kristalltafelglas n <Ker & Glas> *(AE)* crystal sheet glass, *(BE)* thick sheet glass
Kristallwachstum n <Metall> crystal growth
kristallwasserfrei adj <Chemie> anhydrous
Kristallzüchtung f <Fertig> growing of crystals
Kriterium n <Comp & DV> criterion
Kritikalitätsabweichung f <Kerntech> deviation from criticality *(eines Reaktors)*
Kritikalitätsdurchlauf m <Kerntech> passage of criticality
kritisch adj <Kontroll> critical
kritisch gedämpftes Galvanometer n <Elektriz> dead beat galvanometer
kritisch werden v <Kerntech> go critical *(Reaktor)*
kritische Anordnung f <Kerntech> critical assembly
kritische Betriebsmittel npl <Comp & DV> critical resource
kritische Bruchspannung f <Metall> critical fracture stress
kritische Dämpfung f <Elektrotech, Phys> critical damping
kritische Dichte f <Trans> critical density
kritische Frequenz f 1. <Elektronik, Funktech> critical frequency; 2. <Telekom> critical frequency *(Mikrowellenmodus)*; cutoff frequency *(Wellenleitermodus)*

430

kritische Geschäftsabläufe mpl <Comp & DV> critical business operations
kritische Geschwindigkeit f <Mechan, Phys> critical speed
kritische Höhe f <Lufttrans> critical altitude
kritische Last f <Maschinen, Qual> critical load
kritische Masse f 1. <Kerntech> critical amount; 2. <Phys> critical mass
kritische Menge f <Kerntech> critical amount *(von Brennmaterial)*
kritische Pigmentvolumenkonzentration f *(KPK)* <Kunststoff> critical pigment volume concentration *(cpvc)*
kritische Reaktion f <Phys, Thermod> critical reaction
kritische Risslänge f <Metall, Qual> critical crack length
kritische Schubspannung f <Bau, Metall> critical shear strain
kritische Spannung f 1. <Bau> critical stress; 2. <Elektriz, Elektrotech> critical voltage; 3. <Metall> critical stress
kritische Temperatur f <Erdöl, Heiz & Kälte, Phys, Thermod> critical temperature
kritische Temperaturkurve f <Phys, Thermod> critical temperature curve
kritische Tragfläche f <Lufttrans> critical wing *(Lufttüchtigkeit)*
kritische Überhitzung f <Kerntech> departure from nuclear boiling
kritische Vergleichsdifferenz f <Qual> reproducibility critical difference
kritische Wärmestromdichte f <Kerntech> burnout, critical heat flow
kritische Wasserlinie f <Wasserversorg> critical water level
kritische Wellenlänge f 1. <Phys> critical wavelength; 2. <Telekom> cutoff wavelength *(Wellenleitermodus)*
kritische Wiederholdifferenz f <Qual> repeatability critical difference
kritischer Abschnitt m <Comp & DV> critical section
kritischer Anstellwinkel m <Lufttrans> angle of stall
kritischer Aufbau m <Kerntech> critical assembly
kritischer Bereich m <Math> critical region *(Hypothesenprüfung)*
kritischer Druck m <Heiz & Kälte, Phys, Thermod> critical pressure
kritischer Fehler m <Qual> critical defect, critical nonconformance
kritischer Kraftverstärker m <Lufttrans> critical power unit *(Lufttüchtigkeit)*
kritischer Lastpunkt m <Kerntech> hot spot
kritischer Pfad m <Comp & DV> critical path
kritischer Punkt m <Maschinen, Phys, Qual, Thermod> critical point
kritischer Reaktor m <Kerntech> critical reactor
kritischer Temperaturbereich m <Phys, Thermod> critical temperature range
kritischer Wasserstand m <Wasserversorg> critical water level
kritischer Weg m <Bau> critical path
kritischer Wert m <Math> critical value *(Konfidenzmenge)*
kritischer Widerstand m <Elektriz> critical resistance
kritischer Winkel m <Fertig, Optik, Phys, Telekom> critical angle
kritisches Ereignis n <Ergon> critical incident
kritisches Feld n <Elektrotech> critical field *(bei Magnetron)*
kritisches Frequenzband n <Aufnahme> critical band
kritisches Merkmal n <Qual> critical characteristic
kritisches Triebwerk n <Lufttrans> critical engine *(Lufttüchtigkeit)*
Kritischwerden n <Kerntech> divergence

Krokodilklemme f <Elektriz, Elektrotech, Maschinen> alligator clip, crocodile clip
Krokon... <Chemie> croconic
Kronen m <Verpack> crown
Kronenbecher m <Verpack> crown cup
Kronenbohrmeißel m <Erdöl> crown bit (Bohrtechnik)
Kronenkapsel f <Verpack> crown closure
Kronenkorken m <Verpack> crown cork
Kronenmutter f <Bau, Fertig, Maschinen> castellated nut, castle nut
Kronglas n <Ker & Glas> crown glass
Kronglaslinse f <Ker & Glas> crown glass lens
Kronglastropfen m <Ker & Glas> crown glass drop
Krookscher Kollisionsoperator m <Kerntech> Krook collision operator
Kropf m <Papier> backfall
kröpfen v 1. <Bau> joggle; 2. <Fertig, Maschinen> crank
Kröpfen n <Maschinen> cranking
Kröpfmaschine f 1. <Maschinen> joggling machine; 2. <Mechan> crimping tool
Kröpfung f <Maschinen, Mechan> offset
Kröseleisen n <Bau> grooving iron, rabbet iron
Krume f <Lebensmittel> crumb
Krumenbeschaffenheit f <Lebensmittel> bread texture, crumb texture
Krumenbildung f <Lebensmittel> crumb formation
Krumenelastizität f <Lebensmittel> crumb elasticity
Krumenfestigkeit f <Lebensmittel> crumb firmness
krümmen v <Bau> bend, camber, inflect
Krümmen n 1. <Eisenbahn> flexion; 2. <Mechan> bending
Krümmer m 1. <Bau> bend, elbow; 2. <Lufttrans> manifold (Flugzeugtriebwerk); 3. <Maschinen> bend, elbow, knee, manifold; 4. <Mechan> bend, knee, manifold, elbow
Krümmerüberwurf m <Bau> elbow union
krummlinig adj <Fertig, Geom, Phys> curvilinear
krummlinige Koordinate f <Geom, Phys> curvilinear coordinate
Krümmung f 1. <Akustik> camber (an Bauteilen); 2. <Bau> curve; 3. <Elektrotech> bend (Wellenleiter); 4. <Geom, Math> curvature; 5. <Ker & Glas> warp (von optischem Glas); 6. <Lufttrans> camber (Start- und Landebahn); 7. <Mechan> bend; 8. <Phys> curvature
Krümmung f der Interferenzstreifen <Fertig> band deflection
Krümmung f und Verdrehung f <Ker & Glas> (AE) bow and warp, (BE) warped sheet
Krümmung f von Flächen <Geom> curvature of surfaces
Krümmungskreis m <Geom> circle of curvature
Krümmungsmittelpunkt m <Geom, Phys, Strahlphys> (AE) center of curvature, (BE) centre of curvature
Krümmungsmittelpunktskurve f <Geom> evolute
Krümmungsradius m <Geom> radius of curvature
Krümmungs- und Torsionsprüfung f <Werkprüf> curvature-and-twisting test
Krümmungsverlust m <Elektrotech, Telekom> bending loss
Krümmungsversuch m <Metall> bend test (Rohren)
Krümmungszentrum n <Phys, Strahlphys> (AE) center of curvature, (BE) centre of curvature
krumpfecht adj <Textil> shrink-proof
krumpfen v <Textil> shrink
Krumpfen n <Textil> shrinkage
Kruskalgrenze f <Kerntech> Kruskal limit
Kruskalgrenzwert m <Kerntech> Kruskal limit
Kryo... <Thermod> cryogenic
Kryochemie f <Heiz & Kälte> cryochemistry
Kryoflüssigkeit f <Heiz & Kälte> cryogen, cryogenic fluid
kryogen adj <Raumfahrt> cryogenic (Raumschiff)

kryogene Stufe f <Raumfahrt> cryogenic stage (Raumschiff)
kryogener Brennstoff m <Raumfahrt> cryogenic fuel (Raumschiff)
kryogener Tank m <Raumfahrt> cryogenic tank (Raumschiff)
kryogener Treibstoff m <Raumfahrt> cryogenic propellant (Raumschiff)
Kryogenerator m <Heiz & Kälte> cryogenic generator
Kryoleiter m <Heiz & Kälte> cryoconductor
Kryolit m <Ker & Glas> cryolite
Kryomotor m <Heiz & Kälte> cryomotor
Kryophysik f <Heiz & Kälte> cryophysics
Kryopumpe f <Raumfahrt> cryopump (Raumschiff)
Kryoskopie f <Thermod> cryoscopy, freezing point method
Kryospule f <Heiz & Kälte> cryocoil, cryosolenoid
Kryostat m <Labor, Phys, Raumfahrt> cryostat
Kryotechnik f <Heiz & Kälte> cryoengineering
Kryotron n <Elektrotech> cryotron
Kryoturbogenerator m <Heiz & Kälte> cryoturbogenerator
Kryowicklung f <Heiz & Kälte> cryogenic winding
Kryptographie f 1. <Comp & DV> cryptography; 2. <Raumfahrt> cryptography (Weltraumfunk)
Kryptonabsorbtion f im flüssigen Kohlendioxid <Kerntech> KALC process, krypton absorption in liquid carbon dioxide
Kryptosterin n <Chemie> lanosterol
KS (Kerbstift) <Fertig, Maschinen> grooved pin, splined pin
K-Schale f <Kerntech, Phys> K-shell (Atomphysik)
ksG (kontextsensitive Grammatik) <Künstl Int> CSG (context-sensitive grammar)
KSR (Katodenstrahlröhre) <Comp & DV, Druck, Elektriz, Elektronik, Fernseh, Funktech> CRT (cathode-ray tube)
K-Strahler m <Kerntech> K-emitter
Ku-Band n <Funktech> Ku band (Satellit 15,3–17,2 GHz)
Kübel m 1. <Bau> bucket, pail; 2. <Fertig> skip, tub
Kübelfördergerät n <Bau> bucket conveyor
Kübelsitz m <Kfztech> bucket seat
Kubik... <Math> cubic
Kubikberechnung f im Erdbau <Bau> earthworks cubature
Kubikdezimeter m <Metrol> (AE) cubic decimeter, (BE) cubic decimetre
Kubikmaß n 1. <Bau, Kohlen> cubage; 2. <Metrol> cubic measure
Kubikmeter m <Metrol> cu.m., (AE) cubic meter, (BE) cubic metre
Kubikwindgeschwindigkeit f <Nichtfoss Energ> wind velocity cubed
Kubikwurzel f <Math> cube root
Kubik-Yard n <Metrol> cubic yard
Kubikzentimeter m <Metrol> cc, (AE) cubic centimeter, (BE) cubic centimetre
Kubikzoll m <Metrol> cubic inch
Kubikzoll m pro Minute (cim) <Metrol> cubic inches per minute (cim)
Kubikzollmotor m <Kfztech> cubic inch engine
kubisch adj <Geom, Math, Mechan, Metall, Phys> cubic
kubische Dilation f <Metall> cubic dilatation
kubische Gleichung f <Math> cubic equation
kubische Verzerrung f <Metall> cubic distortion
kubischer Ausdehnungskoeffizient m <Mechan, Phys, Qual> coefficient of cubic expansion
kubisches System n <Metall> cubic system
kubischflächenzentriertes Gitter n <Kerntech> (AE) face-centered cubic lattice, (BE) face-centred cubic lattice
Kubooktaeder n <Geom> cubic octahedron
Kubus m <Geom> cube

Kuchendiagramm

Kuchendiagramm n <Comp & DV> pie chart
Kufe f <Textil> vat
Kufenfahrgestell n <Lufttrans> landing skid
KüG *(Kurs über Grund)* <Wassertrans> course made good *(Navigation)*
Kugel f 1. <Bau> ball; 2. <Fertig> ball, pellet, sphere; ball *(Kunststoffinstallationen)*; 3. <Geom> sphere; 4. <Kohlen, Maschinen> ball; 5. <Phys> sphere • **Kugeln bilden** <Fertig> coalesce
Kugelabsperrhahn m <Maschinen> ball valve
Kugelarretierung f <Maschinen> ball stop
Kugelbehälter m <Fertig> spherical vessel
Kugelbemaßung f <Konstzeich> dimensioning of a sphere
Kugeldichtung f <Fertig> ball retainer *(Kunststoffinstallationen)*
Kugeldrehen n <Maschinen> spherical turning
Kugeldrehmaschine f <Fertig> ball-turning lathe
Kugeldrehsupport m <Fertig> ball-turning rest
Kugeldreieck n <Fertig> spherical triangle
Kugeldruck prüfen v <Metall> ball-test
Kugeldruckhärte f <Fertig> ball impression hardness
Kugeldrucklager n <Maschinen, Mechan> ball thrust bearing
Kugeldruckprüfung f <Metall> ball test
Kugeleindruckhärte f <Werkprüf> ball indentation hardness *(Kunststoffe)*
Kugelfall-Viscosimeter n <Labor> falling sphere viscometer *(Viskosität von Flüssigkeiten)*
Kugelfinne f <Maschinen> ball pane, ball peen
Kugelform f <Fertig> spheroidal form
kugelförmig adj 1. <Fertig, Maschinen> spherical; 2. <Metall> globular
Kugelfräseeinrichtung f <Maschinen> cherrying attachment
Kugelfräsen n <Fertig> cherrying
Kugelfräser m <Fertig> cherry
Kugelfunkenstrecke f <Elektrotech> measuring spark gap, sphere gap
Kugelfunkenwelle f <Elektrotech> sphere wave
Kugelgefüge n <Labor> spheroidized structure
kugelgelagert adj <Fertig> ball-bearing
Kugelgelenk n 1. <Fertig> ball joint, ball-and-socket joint, spherical joint; 2. <Foto> ball-and-socket joint; 3. <Kfztech> ball joint; 4. <Maschinen> ball joint, ball-and-socket joint, globe joint, socket joint; 5. <Mechan> ball-and-socket joint
Kugelgelenkgehäuse n <Maschinen> ball joint cage
Kugelgelenkkopf m <Foto, Maschinen> ball-and-socket head
Kugelgelenklager n <Fertig> ball-and-socket bearing
Kugelgestalt f <Fertig> sphericity
kugelgestrahlt adj <Anstrich> shot-peened
Kugelglühen n <Labor> spheroidizing
Kugelgraphit m <Fertig> spheroidal graphite *(Gefüge)* • **Kugelgraphit bilden** <Fertig> nodularize, spheroidize
Kugelgraphitguss m <Fertig> spheroidal graphite cast iron
Kugelhahn m 1. <Bau> ball cock; 2. <Fertig> ball valve *(Kunststoffinstallationen)*; 3. <Maschinen> ball valve
Kugelhaufen m <Kerntech> pebble bed
Kugelhaufenreaktor m <Kerntech> pebble bed reactor
kugelig adj <Fertig> nodular, spheroidal
kugelig gelagert adj <Fertig> spherically-seated
kugeliges Gelenklager n <Maschinen> spherical plain bearing
Kugelkäfig m <Maschinen> ball cage
Kugelkalotte f <Fertig> calotte, spherical indentation
Kugelkeil m <Fertig> spherical wedge
Kugelkoks m <Kohlen> globular coke

Kugelkoordinaten fpl <Geom, Phys> spherical coordinates
Kugelkoordinatensystem n <Geom> spherical coordinates system
Kugelkopfverbindung f <Maschinen> ball-ended linkage
Kugelkuppe f <Fertig> ball point
Kugellager n <Kfztech, Maschinen, Mechan> ball-bearing
Kugellageraußenring m <Fertig> ball-bearing outer race
Kugellagerfett n <Maschinen> ball-bearing grease
Kugellagerführung f <Maschinen> ball-bearing guideway
Kugellagerinnenring m <Fertig> ball-bearing inner race
Kugellagerkäfig m <Maschinen> ball-bearing cage
Kugellagerlaufbahn f <Fertig> ball track
Kugellagerring m <Kfztech> ball-bearing race
Kugelläppmaschine f <Fertig> ball-lapping machine
Kugellaufbahn f <Maschinen> ball track
Kugellaufrille f <Fertig> ball groove
Kugellehre f <Fertig, Metrol> *(AE)* ball gage, *(BE)* ball gauge
Kugellinse f <Phys> spherical lens
Kugelmahlen n <Kohlen> ball milling
Kugelmanipulator m <Kerntech> ball manipulator
Kugelmühle f 1. <Chemtech> ball mill; 2. <Fertig> ball crusher; 3. <Ker & Glas, Kohlen, Kunststoff, Labor, Lebensmittel, Maschinen, Papier> ball mill
Kugelmutter f <Kerntech> ball nut
Kugeloberwelle f <Raumfahrt> spherical harmonic
Kugelpfanne f <Fertig> ball cup
Kugelrückschlagventil n 1. <Fertig> ball check valve *(Kunststoffinstallationen)*; 2. <Maschinen> ball check valve
Kugelsacktest m <Ker & Glas> shot bag test
Kugelschale f 1. <Fertig> spherical shell; 2. <Maschinen> ball cup, race
Kugelschalensitz m <Maschinen> ball socket seat
Kugelschaltung f <Kfztech> floor shift *(Getriebe)*
Kugelschlaghärteprüfung f <Fertig> ball-impact hardness testing
Kugelschlagprüfung f <Metall> ball test
Kugelschraubtrieb m <Fertig> recirculating ball screw and nut
Kugelschussgerät n <Erdöl> perforating gun *(Bohrtechnik)*
Kugelsektor m <Geom> spherical sector
Kugelsperre f <Maschinen> ball lock
Kugelspiegel m <Labor, Telekom> spherical mirror
kugelstrahlen v <Fertig> peen
Kugelstrahlen n <Fertig, Ker & Glas> peening, shot peening
Kugelstrahler m 1. <Funktech> isotropic radiator, isotropic antenna; 2. <Telekom> spherical antenna
Kugeltank m 1. <Raumfahrt> spherical tank *(Raumschiff)*; 2. <Wassertrans> spherical tank
Kugeltank-Flüssiggastransportschiff n <Wassertrans> methane carrier with spherical tanks
Kugeltraglager n <Mechan> angular ball bearing
Kugeltrommel f <Kohlen> balling drum
Kugelumlaufbuchse f <Fertig> recirculating ball bushing
Kugelumlauflenkgetriebe n <Kfztech> recirculating ball steering gear
Kugelumlaufmutter f <Fertig> recirculating ball nut
Kugelumlaufspindel f 1. <Fertig> recirculating ball screw; 2. <Maschinen> ball screw, circulating ball spindle
Kugelventil n 1. <Bau> ball valve; 2. <Hydraul> globe valve; 3. <Kerntech> ball check valve; 4. <Maschinen> ball check valve, ball valve, spherical valve; 5. <Mechan, Papier> ball valve
Kugelverschluss m <Maschinen> ball lock
Kugelweiten n <Fertig> ballizing

Kugelwelle f <Akustik, Phys, Wellphys> spherical wave
Kugelzapfen m 1. <Fertig> ball journal; 2. <Maschinen> ball and socket
Kugelzweieck n <Fertig> lune
Kuhfuß m <Mechan> crowbar
kühl adj <Heiz & Kälte, Thermod> cool
kühl aufbewahren v <Verpack> keep cool
kühl lagern v <Verpack> keep cool
Kühl... <Maschinen> refrigerating
Kühlaggregat n <Thermod> refrigerator
Kühlanlage f 1. <Bau> cooling system; 2. <Heiz & Kälte> cooling plant, cooling system, refrigerating plant; 3. <Kerntech> cooler; 4. <Maschinen> cooling equipment, refrigerating plant; 5. <Thermod> refrigerating plant
Kühlart f <Heiz & Kälte> method of ventilation
Kühlbandeindrücke mpl <Ker & Glas> belt marks
Kühlbecken n <Wasserversorg> cooling pond
Kühlbereich m <Ker & Glas> annealing range
Kühlblech n <Phys, Raumfahrt, Thermod> heat sink
Kühlcontainer m <Eisenbahn, Heiz & Kälte, Wassertrans> refrigerated container
Kühleinrichtung f 1. <Heiz & Kälte> cooler, cooling system; 2. <Textil> cooling equipment
kühlen v 1. <Anstrich> cool; 2. <Heiz & Kälte> chill, cool, refrigerate; 3. <Ker & Glas> anneal; 4. <Mechan> chill; 5. <Papier> cool; 6. <Thermod> chill, cool, refrigerate
Kühlen n 1. <Ker & Glas> annealing; 2. <Kunststoff> chilling; 3. <Thermod> chilling (von Substanzen)
Kühler m 1. <Erdöl> cooler; 2. <Heiz & Kälte> chiller, cooler, heat exchanger, radiator; 3. <Kfztech, Mechan, Raumfahrt> radiator (Raumschiff); 4. <Thermod> cooler
Kühlerablasshahn m <Kfztech> radiator drain cock, radiator drain tap
Kühlerblock m <Kfztech> radiator core
Kühlerdeckel m <Kfztech> radiator cap
Kühlerdruckverschluss m <Kfztech> radiator pressure cap
Kühlereinfüllstutzen m <Kfztech> radiator filler neck
Kühlereinfüllverschluss m <Kfztech> radiator filler cap
Kühlerelement n <Heiz & Kälte> cooler element
Kühlerentlüftungsschlauch m <Kfztech> radiator vent hose
Kühlerflansch m <Kfztech> radiator flange
Kühlerfuß m <Kfztech> radiator support
Kühlergehäuse n <Kfztech> radiator frame
Kühlergrädigkeit f <Heiz & Kälte> temperature difference rating
Kühlergrill m <Kfztech> radiator grill
Kühlerjalousie f <Kfztech> radiator blind
Kühlerkern m <Kfztech> radiator core
Kühlerrippe f <Kfztech> radiator fin
Kühlerschlauch m <Kfztech> radiator hose
Kühlerstütze f <Kfztech> radiator support
Kühlerteilblock m <Kfztech> radiator element
Kühlerverschluss m <Kfztech> radiator cap
Kühlerwasserkasten m <Kfztech> radiator header
Kühlfahrzeug n 1. <Heiz & Kälte> refrigerated vehicle; 2. <Kfztech> (BE) insulated lorry, (AE) insulated truck; 3. <Maschinen> refrigerated vehicle; 4. <Thermod> (BE) heat-insulated lorry, (AE) heat-insulated truck, (BE) insulated lorry, (AE) insulated truck
Kühlfalle f <Chemtech, Heiz & Kälte> cold trap, condensation trap, cryotrap
Kühlfalte f <Ker & Glas> (BE) chill mark, (AE) chill wrinkle
Kühlfiltersystem n <Kerntech> component cooling filter
Kühlflüssigkeit f <Maschinen, Thermod> coolant, cooling liquid
Kühlgebläse n <Kfztech> cooling fan
Kühlgerät n <Heiz & Kälte> refrigerator
Kühlhalle f <Heiz & Kälte> cold store

Kühlhaus n <Heiz & Kälte> cold store, refrigerated warehouse
Kühlhochhaus n <Heiz & Kälte> high-rise cold store
Kühlkanal m <Kerntech> cooling channel
Kühlkasten m <Metall> monkey
Kühlkette f <Heiz & Kälte> cold chain
Kühlkörper m 1. <Elektriz, Elektrotech> heat sink; 2. <Funktech> dissipator, heat sink; 3. <Telekom> dissipator; 4. <Raumfahrt> heat sink
Kühlkreis m <Heiz & Kälte> cooling circuit
Kühlkreislauf m 1. <Heiz & Kälte> cooling circuit, ventilation circuit; 2. <Kerntech> heat-removal loop
Kühllast f <Heiz & Kälte> cooling load, heat gain
Kühlleistung f <Heiz & Kälte> cooling capacity, refrigerating capacity
Kühlleitung f 1. <Maschinen> cooling duct; 2. <Raumfahrt> coolant feed line (Raumschiff)
Kühlluft f <Heiz & Kälte> air coolant, cooling air
Kühlluftbedarf m <Heiz & Kälte> rate of coolant air required
Kühlluftdurchflussmenge f <Heiz & Kälte> rate of coolant air flow
Kühlluftgebläse n <Heiz & Kälte> cooling air fan
Kühlluftkanal m <Heiz & Kälte> cooling air duct
Kühlluftklappe f <Lufttrans> cooling flap
Kühlluftmenge f <Heiz & Kälte> rate of coolant air flow
Kühlluftregulierklappe f <Lufttrans> cowl flap (Motor, Triebwerk)
Kühlluftstrom m <Heiz & Kälte> rate of coolant air flow
Kühlluftweg m <Heiz & Kälte> cooling air passage, ventilating passage
Kühlmantel m <Heiz & Kälte, Kfztech> cooling jacket
Kühlmaschine f <Heiz & Kälte> refrigerating machine
Kühlmedium n <Heiz & Kälte> coolant, cooling agent, cooling medium
Kühlmittel n 1. <Heiz & Kälte> coolant, cooling agent, refrigerant; 2. <Kfztech> refrigerant; coolant (Motor); 3. <Maschinen> refrigerant; 4. <Phys> coolant; 5. <Thermod> refrigerant; 6. <Umweltschmutz> cooling medium, refrigerant
Kühlmittelbewegung f <Heiz & Kälte> circulation, coolant circulation
Kühlmitteldurchflussmenge f <Heiz & Kälte> rate of coolant flow
Kühlmittelpumpe f <Fertig> coolant pump
Kühlmittelstrom m <Heiz & Kälte> rate of coolant flow
Kühlmitteltemperatur f <Thermod> coolant temperature, temperature of cooling medium
Kühlmittelumlauf m <Heiz & Kälte> coolant circulation
Kühlmittelumwälzung f <Heiz & Kälte> coolant circulation
Kühlmittelzufuhr f <Maschinen> coolant supply
Kühlnagel m <Fertig> chill (Gießen)
Kühlnebel m <Heiz & Kälte> coolant mist
Kühlofen m 1. <Ker & Glas> annealing lehr, annealing furnace, annealing kiln, leer, lehr; 2. <Thermod> annealing furnace
Kühlofenband n <Ker & Glas> leer belt, lehr belt
Kühlofenbediener m <Ker & Glas> leer attendant, lehr attendant
Kühlofenhelfer m <Ker & Glas> leer assistant, lehr assistant
Kühlöl n 1. <Abfall> cutting oil; 2. <Maschinen> cooling oil
Kühlplan m <Ker & Glas> annealing schedule
Kühlraum m 1. <Heiz & Kälte> chill room, cold storage room; 2. <Kerntech> cooling cavity; 3. <Thermod> chill room; 4. <Wassertrans> refrigerating hold
Kühlraumladung f <Heiz & Kälte, Wassertrans> refrigerated cargo

Kühlraumlagerung

Kühlraumlagerung f <Lebensmittel> cold storage
Kührippe f 1. <Heiz & Kälte> fin; cooling fin *(außen)*; cooling rib *(innen)*; 2. <Kerntech> cooling fin, fin, rib; 3. <Lufttrans> fin; 4. <Maschinen> cooling fin, gill • **mit Kührippen versehen** <Heiz & Kälte> finned
Kühlrohr n <Maschinen> cooling tube
Kühlschiff n <Wassertrans> refrigeration ship, refrigerated cargo ship; cold storage ship, reefer ship *(Schifftyp)*
Kühlschlange f 1. <Heiz & Kälte> coil, condensing coil, cooling coil; 2. <Kerntech> cooling coil
Kühlschlitz m <Heiz & Kälte> cooling air duct
Kühlschmiermittel n <Fertig> coolant/lubricant, cutting oil, lubricant coolant
Kühlschmierstoffe mpl <Abfall> cutting oil
Kühlschrank m 1. <Heiz & Kälte> domestic refrigerator; 2. <Maschinen, Thermod> refrigerator
Kühlsole f <Fertig> cooling liquid *(Kunststoffinstallationen)*
Kühlspirale f <Maschinen> cooling spiral
Kühlstation f <Bau> fan station
Kühlsystem n 1. <Elektriz, Heiz & Kälte, Kfztech> cooling system; 2. <Maschinen> refrigerating system; 3. <Textil, Umweltschmutz> cooling system
Kühlsystem n für die Abschirmung <Kerntech> shield cooling system
Kühlteich m <Wasserversorg> cooling pond
Kühltransport m <Eisenbahn, Lebensmittel> refrigerated transport
Kühltunnel m <Lebensmittel> cooling tunnel
Kühlturm m <Chemtech, Heiz & Kälte, Maschinen, Textil, Thermod> cooling tower
Kühlung f 1. <Funktech> cooling; 2. <Heiz & Kälte> chilling; 3. <Ker & Glas, Maschinen> cooling; 4. <Thermod> refrigeration; 5. <Verpack> cooling
Kühlung f durch Kälteerzeugung <Heiz & Kälte> refrigeration
Kühlung f durch natürlichen Luftzug <Thermod> *(AE)* natural draft cooling, *(BE)* natural draught cooling
Kühlung f durch Verdampfung <Chemtech> evaporation cooling
Kühlung f mit Naturumlauf <Kerntech> natural circulation cooling
Kühlvorrichtung f <Papier> cooler
Kühlwagen m 1. <Eisenbahn> refrigerated car, refrigerated truck, refrigerated wagon; 2. <Heiz & Kälte> cold storage car, freezer, refrigerated truck
Kühlwalze f 1. <Heiz & Kälte, Kunststoff> chill roll; 2. <Papier> cooling cylinder
Kühlwaren fpl <Verpack> chilled goods
Kühlwasser n <Heiz & Kälte> chilled water, cooling water
Kühlwasserablass m <Kfztech> radiator draining
Kühlwasseranlage f <Wassertrans> cooling-water system
Kühlwasser-Durchflussmenge f <Heiz & Kälte> flow rate of cooling water
Kühlwassermantel m 1. <Fertig> water jacket; 2. <Kfztech> water galleries *(doppelwandig)*
Kühlwasserrohr n <Maschinen> cooling-water pipe
Kühlwasserstrom m <Heiz & Kälte> flow rate of cooling water
Kühlwassersystem n <Wassertrans> cooling-water system
Kühlzahl f <Heiz & Kälte> cooling coefficient
Kühlzone f 1. <Ker & Glas> cooling zone; 2. <Lebensmittel> cooling section
Kühlzonenbreite f <Heiz & Kälte> cooling range
Kühlzylinder m <Papier> sweat roll
Kuhn-Thomas-Reichsche-Summenformel f <Kerntech> Kuhn-Thomas-Reich sum rule
Küken n <Bau> plug *(Schliffverbindung)*

Külbel m <Ker & Glas> parison; blank *(Glas)*
Külbelaufnehmer m <Ker & Glas> parison gatherer
Külbelriss m <Ker & Glas> parison check
Külbelübertragung f aus der Vorform in die Fertigform <Ker & Glas> blank transfer
Kulierware f <Textil> weft-knitted fabric
Kulisse f <Fertig> crank, rocker; rocker arm *(Getriebelehre)*
Kulissenrad n 1. <Fertig> main driving gear *(Getriebelehre)*; 2. <Maschinen> bull gear
Kulissenstein m <Maschinen> slide block
Kulissensteuerung f <Maschinen> link motion
Kulmination f <Wassertrans> transit *(astronomische Navigation)*
Kulturplatte f <Labor> culture plate *(Bakteriologie)*
Kümo *(Küstenmotorschiff)* <Wassertrans> coastal motor vessel
kümpeln v <Fertig> dish
Kümpeln n 1. <Fertig> cupping; 2. <Maschinen> dishing
kumulative Verteilungsfunktion f <Math> cumulative distribution function, CDF
kumulative Wahrscheinlichkeit f <Qual> cumulative probability
kumulativer Ausfluss m <Nichtfoss Energ> cumulative discharge
kumulativer Fehler m <Gerät, Qual> cumulative error
kumulierender Zähler m <Gerät> accumulating counter
kumulierter Index m <Druck> bulk index
Kumulus m <Lufttrans, Wassertrans> cumulus *(Wolkenart)*
Kundenangaben angefertigt adj/nach <Verpack> custom-made
Kundenangaben entworfen adj/nach <Verpack> purpose-designed
Kundenanpassungsentwicklung f <Elektronik, Telekom> customization, tailored design
Kundenchip m <Comp & DV, Elektronik, Telekom> custom chip
Kundendienst m <Bau, Fernseh, Telekom> after-sales service
Kundendienstinformationen fpl <Comp & DV> customer support information
Kundenspezifikation hergestellt adj/nach <Comp & DV, Telekom> custom-made
kundenspezifisch adj 1. <Comp & DV> custom-built; 2. <Elektronik> custom-designed; 3. <Telekom> custom-built, custom-designed
kundenspezifisch ausgeführt adj <Elektronik, Telekom> custom-designed, custom-made, customized
kundenspezifische Anpassung f <Elektronik, Telekom> customization
kundenspezifische Großintegration f <Elektronik> custom LSI
kundenspezifischer Chip m <Comp & DV, Elektronik> custom chip
kundenspezifisches Engineering n <Comp & DV, Elektronik> customer application engineering *(CAE)*
kundenspezifisches Standardprodukt n <Telekom> customer specific standard product, CSSP
Kundenunterstützung f <Comp & DV> customer assistance
Kundt'sche Röhre f <Phys> Kundt's tube
Kunst... <Druck, Fertig, Foto, Papier, Verpack> artificial
Kunstantenne f <Funktech> artificial antenna, dummy antenna
Kunstchromopapier n <Verpack> imitation chromoboard
Kunstdruckkarton m <Druck> artboard
Kunstdruckpapier n <Druck, Papier> art paper, coated paper

Kunstfaser f <Verpack> (AE) man-made fiber, (BE) man-made fibre
Kunstharz n 1. <Bau, Elektriz> epoxy resin; 2. <Erdöl> synthetic resin (Petrochemie); 3. <Fertig> synthetic resin; epoxy resin (Kunststoffinstallationen); 4. <Kunststoff> epoxy resin, synthetic resin; 5. <Verpack> epoxy resin
Kunstharzbeton m <Kunststoff> polymer concrete
Kunstharzlack m <Kunststoff> enamel
kunstharzverleimtes Sperrholz n <Kunststoff> resin-bonded plywood
Kunstleder n <Kunststoff> artificial leather, mutation leather, leatherette
künstlich adj <Künstl Int, Papier> artificial
künstlich altern v <Thermod> age artificially
künstlich erzeugte Radioaktivität f <Strahlphys> artificial radioactivity
künstlich erzeugte Sprache f <Künstl Int> artificial speech, synthetic speech
künstlich gealtert adj <Thermod> artificially-aged
künstliche Abdichtung f <Abfall> synthetic lining
künstliche Alterung f <Abfall, Thermod> (BE) artificial ageing, (AE) artificial aging
künstliche Belüftung f <Sicherheit> artificial ventilation
künstliche Bewitterung f <Kunststoff> artificial weathering
künstliche Intelligenz f <Comp & DV> artificial intelligence, AI, machine intelligence
künstliche Kohle f <Chemie> charcoal
künstliche Radioaktivität f <Strahlphys> artificial radioactivity
künstliche Sprache f <Telekom> simulated speech
künstliche Stimme f <Akustik, Ergon> artificial voice
künstliche Störung f <Telekom> man-made noise
künstliche Übersäuerung f <Umweltschmutz> artificial acidification
künstlicher Flugsimulator m <Lufttrans> synthetic flight trainer
künstlicher Hafen m <Wassertrans> artificial harbour, artificial port
künstlicher Horizont m 1. <Lufttrans> artificial horizon; 2. <Raumfahrt> artificial horizon, gyrohorizon (Fluglageanzeiger); 3. <Trans> artificial horizon
künstlicher Kehlkopf m <Akustik> artificial larynx
künstlicher Luftzug m <Maschinen> (AE) forced draft, (BE) forced draught
künstlicher Mondsatellit m <Raumfahrt> lunar artificial satellite
künstlicher Mund m <Akustik> artificial mouth
künstlicher Nullpunkt m <Telekom> artificial neutral (eines Netzes)
künstlicher Sternpunkt m <Telekom> artificial neutral (eines Netzes)
künstlicher Zug m <Heiz & Kälte> (AE) forced draft, (BE) forced draught
künstliches Aufrauen n <Kerntech> artificial roughening
künstliches Geräusch n <Telekom> artificial noise
künstliches Neuron n <Künstl Int> artificial neuron, cell, neuron, unit (neurale Netze)
künstliches neuronales Netzwerk n (KNN) <Künstl Int> artificial neural net, artificial neural network (ANN)
künstliches Ohr n <Akustik, Ergon> artificial ear
künstliches Sehen n <Künstl Int> artificial vision, computational vision, CV
Kunstlichtfarbfilm m <Foto> (AE) artificial light color film, (BE) artificial light colour film
Kunstlichtfotografie f <Foto> artificial light photography
Kunstmilchprodukt n <Lebensmittel> nondairy product
Kunstporzellan n <Ker & Glas> artistic porcelain
Kunstschmiedearbeit f <Bau> wrought ironwork
Kunstseide f <Kunststoff> rayon
Kunstspeisefett n <Lebensmittel> manufactured edible fat (gehärtet)
Kunststein m <Bau> synthetic stone
Kunststoff m 1. <Abfall, Elektrotech, Erdöl> plastic; 2. <Fertig> plastomer; plastic (Kunststoffinstallationen); 3. <Foto> plastic; 4. <Kunststoff> plastic material; 5. <Telekom, Verpack> plastic
Kunststoffabfall m <Abfall> plastic waste
Kunststoffauskleidung f <Verpack> plastic liner
Kunststoffbehälter m 1. <Fertig> plastainer; 2. <Meerschmutz> dracone
Kunststoffbeschichtung f <Telekom> plastic coating
Kunststoffeinfassung f <Elektrotech> plastic cladding
Kunststoffentwicklungstank m <Foto> plastic developing tank
Kunststofffaser f 1. <Elektrotech> (AE) plastic fiber, (BE) plastic fibre; 2. <Telekom> (AE) all-plastic fiber, (BE) all-plastic fibre
Kunststofffaserkabel n <Elektrotech> (AE) plastic fiber cable, (BE) plastic fibre cable
Kunststofffassung f <Foto> plastic mounting
Kunststofffilm m <Verpack> plastic sheeting
Kunststofffolie f <Verpack> plastic sheeting
Kunststoffgebinde n <Abfall> plastic container
kunststoffisoliertes Kabel n <Elektrotech, Telekom> plastic-insulated cable
Kunststoff-Kabelführung f <Elektriz> insulated conduit
Kunststoffkondensator m <Elektrotech> plastic capacitor
Kunststoffmantel m <Fertig> plastics cladding (Faseroptik)
Kunststoffmembran f <Bau> synthetic membrane
Kunststoffpipeline f <Fertig> plastic pipeline
Kunststoffpressform f <Maschinen> (AE) mold for plastics, (BE) mould for plastics
Kunststoffpressmatrize f <Maschinen> extrusion die for plastics
Kunststoffrecycling n <Abfall> plastics recycling
Kunststoffriemen m <Maschinen> plastic belt
Kunststoffrohrleitung f <Fertig> plastic pipeline
Kunststoffschale f <Foto> plastic dish
Kunststoffschichtkondensator m <Elektrotech> plastic film capacitor
Kunststoffstöpsel m <Verpack> plastic plug
kunststoffummantelte Flüssigkristallanzeige f <Elektronik> plastic-sealing liquid crystal display
Kunststoffummantelung f <Telekom> plastic coating
Kunststoffverkleidung f <Elektrotech> plastic cladding
Kunststoffverpackung f <Abfall> plastic container, plastic packing material
Kunststoffverwertung f <Abfall> plastics recycling
Kunsttischlerei f <Bau> cabinet-making
Kunsttöpferwaren fpl <Ker & Glas> art pottery
Küpe f <Textil> vat (Färben)
Kupelle f <Chemie> cupel
Kupellieren n <Metall> cupellation
Kupellierofen m <Metall> cupellation furnace
Küpenfarbstoff m <Textil> vat dye
Kupfer n (Cu) <Chemie, Metall> copper (Cu)
Kupferacetat n <Fertig> cupric acetate
kupferartig adj <Chemie> cupreous
Kupferasbestdichtung f <Kfztech, Maschinen> copper asbestos gasket
kupferbedeckt adj <Elektriz> copper-clad
Kupferbeize f <Textil> copper mordant
Kupferbeizen n <Ker & Glas> copper staining
Kupfer-beschleunigter Salzsprühtest m <Metall> copper-accelerated salt-spray test, CASS test

Kupferblech

Kupferblech n 1. <Bau> copper sheet; 2. <Metall> sheet copper
Kupferblock m <Metall> copper ingot
Kupferdraht m <Bau, Elektriz, Metall> copper wire
Kupferdruck m <Druck> copperplate printing
Kupferdruckpapier n <Druck> copperplate-printing paper
Kupferdruckpresse f <Druck> copperplate-printing press
Kupfergarherd m <Metall> copper-refining furnace
kupferhaltig adj <Chemie> cupreous
Kupferkabel n <Telekom> copper cable
kupferkaschiert adj <Elektriz> copper-clad
Kupferknetlegierung f <Maschinen> wrought copper alloy
Kupferlackdraht m <Elektriz, Elektrotech> (AE) enameled copper wire, (BE) enamelled copper wire
Kupferlegierung f <Maschinen> copper alloy
Kupferleiter m <Elektriz> copper conductor
Kupferlitze f <Fertig> copper braid
Kupferlitzenabschirmung f <Elektriz> copper-braid shielding
Kupferniet m <Bau> copper rivet
Kupferoxidammoniak m <Textil> cuprammonium
Kupferoxydul-Gleichrichter m <Elektrotech> copper oxide rectifier
Kupferplatte f <Druck> copperplate
Kupferrohstein m <Metall> copper matte
Kupferschwärze f <Chemie> black copper
Kupfersulfat n 1. <Chemie> blue vitriol, (AE) copper tetraoxosulfate, (BE) copper tetraoxosulphate; 2. <Foto> (AE) copper sulfate, (BE) copper sulphate
Kupfertiefdruck m <Druck> copperplate printing
Kupferuranit n <Kerntech> copper uranite
Kupferverlust m <Elektrotech, Phys> copper loss
Kupfervitriol n <Chemie> blue vitriol, bluestone
Kupferzylinder m <Druck> copper-plated cylinder
Kupolofen m 1. <Eisenbahn> cupola; 2. <Fertig> cupola furnace, cupola; 3. <Ker & Glas, Kohlen> cupola, cupola furnace
Kuppe f 1. <Bau> crest, knoll, meniscus; 2. <Kohlen> knoll; 3. <Maschinen> point
Kuppel f 1. <Bau> cupola, dome; 2. <Eisenbahn, Fertig> cupola; 3. <Ker & Glas> dome; 4. <Kohlen> cupola
Kuppeldach n <Bau> dome roof
kuppeln v 1. <Fertig> couple; 2. <Maschinen> couple (zweier Wellen)
Kuppelschlauch m <Eisenbahn> coupling hose
Kuppelstange f <Eisenbahn> coupling rod, drawbar
Kuppeltisch m <Fertig> tandem table
Kuppelvorrichtung f <Maschinen> coupling
Kuppenkreis m <Fertig> nose circle (Kurve)
Kuppler m <Eisenbahn> shunter (Person)
Kupplung f 1. <Eisenbahn> coupling; 2. <Fertig> jigger; coupler (Getriebelehre); 3. <Kfztech> clutch; 4. <Lufttrans> coupling; 5. <Maschinen> attachment, clutch; 6. <Mechan> clutch, coupling; 7. <Phys, Wassertrans> coupling (Motor) • **Kupplung ausrücken** <Kfztech> declutch
Kupplung f mit Rücklaufsperre <Maschinen> backstopping clutch
Kupplungsaufnahme f <Kfztech> clutch pick-off
Kupplungsausrückgabel f <Kfztech> clutch fork
Kupplungsausrückgabelstrebe f <Kfztech> throw-out fork strut
Kupplungsausrückgabelzapfen m <Kfztech> throw-out fork pivot
Kupplungsausrückhebelbolzen m <Kfztech> release lever pin
Kupplungsausrückhebelfeder f <Kfztech> release lever spring
Kupplungsausrücklager n <Kfztech> clutch release stop, release bearing

Kupplungsausrücklagerbuchse f <Kfztech> release-bearing sleeve, throw-out bearing sleeve
Kupplungsausrücklagernabe f <Kfztech> release-bearing hub
Kupplungsausrückstange f <Kfztech> release rod
Kupplungsautomatik f <Kfztech, Maschinen, Mechan> automatic clutch control
Kupplungsbelag m <Kfztech, Maschinen, Mechan> clutch lining
Kupplungsbolzen m <Maschinen> coupling pin
Kupplungsbügel m <Eisenbahn> D-link
Kupplungsdämpfer m <Lufttrans> coupling buffer
Kupplungsdrucklager n <Kfztech> release bearing
Kupplungsdruckplatte f <Maschinen> clutch pressure plate
Kupplungsfeder f <Kfztech> clutch spring
Kupplungsflansch m <Maschinen> coupling flange
Kupplungsgehäuse n 1. <Kfztech> bell housing, clutch casing, clutch drum, clutch housing; 2. <Lufttrans> coupling cover (Hubschrauber); 3. <Maschinen> clutch housing
Kupplungsgelenk n <Maschinen> clutch coupling, coupling joint
Kupplungsgestänge n <Kfztech> clutch linkage; clutch pedal push-rod (Kupplung)
Kupplungshaken m 1. <Eisenbahn> coupling hook, draw-hook bar; 2. <Maschinen> coupling hook
Kupplungshauptzylinder m <Kfztech> master cylinder (Hauptzylinder für die Kupplungsbetätigung)
Kupplungshülse f 1. <Kfztech> clutch sleeve; 2. <Mechan, Phys> coupling sleeve
Kupplungsmuffe f 1. <Kfztech> clutch sleeve; 2. <Maschinen> clutch collar, coupling box, coupling sleeve
Kupplungsnehmerzylinder m <Kfztech> clutch output cylinder
Kupplungspedal n 1. <Kfztech> clutch lever, clutch pedal; 2. <Maschinen> clutch pedal
Kupplungspedalausrückhebel m <Kfztech> clutch pedal release lever
Kupplungspedalspiel n <Kfztech> clutch pedal clearance
Kupplungspuffer m <Lufttrans> coupling buffer
Kupplungsring m <Kfztech> clutch plate
Kupplungsscheibe f 1. <Kfztech> (AE) clutch disk, (BE) clutch disc, clutch plate, driven plate; 2. <Maschinen> driven plate, (BE) friction disc, (AE) friction disk
Kupplungsschlupf m <Kfztech> clutch slip
Kupplungsseil n <Kfztech> clutch cable
Kupplungsseilzug m <Kfztech> clutch cable
Kupplungsspiel n <Kfztech> clutch clearance
Kupplungsstange f <Kfztech> clutch rod
Kupplungsstück n <Elektrotech, Maschinen> coupling
Kupplungstreibscheibe f <Kfztech> clutch drive plate
Kupplungswelle f <Kfztech> clutch shaft
Kupplungszapfen m <Fertig> wobbler; palm end (Walze)
Kuproxgleichrichter m <Elektrotech> copper oxide rectifier
Kurbel f 1. <Bau> winch; 2. <Fertig> crank, winch; 3. <Kfztech> handle; crank (Motor); 4. <Maschinen> crank, crank handle
Kurbelbohrer m <Maschinen> bit stock drill
Kurbelgehäuse n <Fertig, Kfztech, Mechan> crankcase
Kurbelgehäusebelüftung f <Kfztech> crankcase ventilation
Kurbelgehäuseentlüfter m <Kfztech> crankcase breather, crankcase ventilator
Kurbelgehäuseoberteil n <Kfztech> crankcase top half
Kurbelgehäuseunterteil n <Kfztech> crankcase bottom half

Kurbelgehäusezwangseentlüftung f (PCV-Ventilation) <Kfztech> positive crankcase ventilation (PCV)
Kurbelgetriebe n <Maschinen> crank mechanism
Kurbelgriff m 1. <Maschinen> crank handle; 2. <Mechan> crankpin
Kurbelinduktor m <Elektrotech> Megger®, magneto
Kurbellager n <Maschinen> crank bearing
Kurbelpresse f <Maschinen> crank press
Kurbelrad n <Maschinen> crank wheel
Kurbelschwinge f <Fertig> oscillating lever (Getriebelehre)
Kurbelstange f <Kfztech, Mechan> connecting rod
Kurbeltrieb m 1. <Kfztech> crankshaft drive; 2. <Maschinen> crank drive, crank mechanism
Kurbelwange f 1. <Kfztech> crank web; 2. <Maschinen> crank arm, crank cheek, crank web
Kurbelwanne f <Fertig> crankcase
Kurbelwannensumpf m <Kfztech> crankcase bottom half
Kurbelwelle f <Fertig, Kfztech, Maschinen, Wassertrans> crankshaft
Kurbelwellenausrichtung f <Maschinen> crankshaft alignment
Kurbelwellenbohrer m <Maschinen> crankshaft drill
Kurbelwellendrehmaschine f <Maschinen> crankshaft lathe
Kurbelwellendrehzahl f <Kfztech> engine speed
Kurbelwellenlager n 1. <Kfztech> main bearing; crankshaft bearing (Motor); 2. <Maschinen> crankshaft bearing
Kurbelwellenlager n der Pleuelstangen <Kfztech> big end bearing
Kurbelwellenseite f <Maschinen> crank end
kurbelwellenseitiger Pleuelkopf m <Maschinen> big end, connecting rod big end
kurbelwellenseitiger Pleuelstangenkopf m <Maschinen> connecting rod big end
kurbelwellenseitiger Totpunkt m <Maschinen> (AE) crank-end dead-center, (BE) crank-end dead-centre
kurbelwellenseitiges Kolbenstangenende n <Kfztech> big end
kurbelwellenseitiges Pleuelende n <Kfztech> big end, connecting rod big end
Kurbelwellenzahnrad n <Textil> crankshaft wheel
Kurbelwellenzapfen m <Kfztech> crankpin
Kurbelzapfen m <Fertig, Kfztech, Maschinen, Mechan> crankpin
Kurs m 1. <Metrol> bearings; 2. <Raumfahrt> heading; 3. <Wassertrans> track; course (Navigation) • **auf Kurs** <Wassertrans> on course (Navigation) • **Kurs haben auf** <Wassertrans> stand for (Navigation) • **Kurs halten** <Wassertrans> keep course (Navigation) • **Kurs halten auf** <Wassertrans> head for • **Kurs nehmen auf** <Wassertrans> head for; steer for (Navigation) • **Kurs und Fahrt halten** <Wassertrans> maintain course and speed (Navigation) • **vom Kurs abdrehen** <Wassertrans> veer off course (Schiff) • **vom Kurs abgewichen** <Raumfahrt> off-course
Kurs m **durch Wasser** <Wassertrans> course to steer
Kurs m **über Grund** <Wassertrans> course made good (Navigation)
Kursabweichung f <Phys> yaw
Kursanalyse f <Ergon> course analysis
Kursänderung f 1. <Kohlen> change of course; 2. <Raumfahrt> alteration of course; 3. <Wassertrans> alteration of course; change of course (Navigation)
Kursanzeige f <Raumfahrt> heading indicator (Raumschiff)
kursanzeigender Kurswähler m <Lufttrans> course indicator selector

Kursanzeiger m <Lufttrans, Wassertrans> course indicator (Navigation)
Kurs-Blip-Impuls m <Funkort> course-blip pulse (Radar)
Kursbuch n <Eisenbahn> (AE) railroad guide, (BE) railway guide, (AE) schedule, (BE) timetable
Kursdatengeber m <Lufttrans> heading data generator
Kursdreieck n <Wassertrans> protractor
Kurseinstellung f <Lufttrans> course alignment (Instrumentenlandesystem)
Kursfehlerintegrator m <Lufttrans> heading error integrator
Kursfehler-Sychronismus-Anzeigerverstärker m <Lufttrans> heading error sychronizer amplifier
Kursfunksteuerkurs m <Lufttrans> localizer beam heading
Kursgeber m <Lufttrans> course selector
Kursindikator m <Lufttrans> course tracer
kursiv adj <Druck> italic
kursiv setzen v <Druck> italicize
Kursivschrift f <Druck> italic type
Kurskarte f <Wassertrans> track chart
Kurslineal n <Wassertrans> parallel ruler
Kurslinie f <Lufttrans> course line (Instrumentenlandesystem)
Kursmaterial n <Comp & DV> courseware
Kursschreiber m <Wassertrans> course recorder
Kursstabilität f <Lufttrans> directional stability
Kursstrich m <Wassertrans> lubber's line (Kompass)
Kurs- und Vertikalbezugssystem n <Raumfahrt> heading and vertical reference system (Raumschiff)
Kurswagen m <Eisenbahn> through wagon
Kurswähler m <Lufttrans> omnibearing selector (Flugwesen)
Kurswechsel m <Kohlen, Wassertrans> change of course (Navigation)
Kurswinkel m <Lufttrans> course angle
Kurtosis f 1. <Math> kurtosis (Maß für die Abweichung einer Verteilung von der Normalverteilung); 2. <Qual> kurtosis
Kurve f 1. <Bau> bend, curve; 2. <Fertig> calm; 3. <Geom, Math> curve; 4. <Trans> turning • **als Kurve zeichnen** <Geom> graph • **eine Kurve zeichnen** <Math> plot a curve • **in Kurve legen** <Raumfahrt> bank
Kurve f **für den mittleren Stichprobenumfang** <Qual> average sample number curve
Kurve f **gleicher Lautstärke** <Akustik> equal-loudness contours
Kurve f **gleicher Temperatur** <Phys> isotherm
Kurvenanflug m <Lufttrans> curved approach
Kurvenfaktor m <Nichtfoss Energ> curve factor
Kurvenfamilie f <Geom> family of curves
Kurvenform f <Lufttrans> cam contour
Kurvenfräseinrichtung f <Fertig> cam-milling attachment
Kurvenfräsmaschine f 1. <Fertig> cam-milling machine; 2. <Maschinen> cam copy miller, cam-milling machine
Kurvengetriebe n <Maschinen> cam mechanism
Kurvengleichung f 1. <Geom> curve equation; 2. <Math> equation of a curve
Kurvenhub m <Fertig> calm throw (Getriebelehre)
Kurvenlage f <Trans> banking (Flugzeug)
Kurvenlineal n <Fertig> curve
Kurvenmechanismus m <Fertig> calm mechanism (Getriebelehre)
Kurvenmessgerät n <Metrol> cam-measuring equipment
Kurvenparameter m <Regelung> curve parameter
Kurvenpunkt m <Konstzeich> plot-point on curves
Kurvenrolle f <Maschinen> cam roller, follower
Kurvenschar f <Geom> family of curves

Kurvenscheibe

Kurvenscheibe f 1. <Fertig> (BE) disc cam, (AE) disk cam (Getriebelehre); 2. <Maschinen> cam plate, cam wheel, (BE) disc cam, (AE) disk cam
Kurvenscheibengetriebe n <Fertig> (BE) disc cam mechanism, (AE) disk cam mechanism
Kurvenscheibensteuerung/mit <Fertig> (BE) disc-cam-operated, (AE) disk-cam-operated (Getriebelehre)
Kurvenschreiber m 1. <Comp & DV> graph plotter, plotter; 2. <Fertig> plotter; 3. <Gerät> curve plotter, graph plotter
Kurvensegment n <Geom> curve segment
Kurvensteuerung f <Fertig> cam control
Kurvenstößel m <Fertig> cam follower (Getriebelehre)
Kurventrommel f <Fertig> barrel cam, cylinder cam
Kurvenverbreiterung f <Bau> curve widening (Straßenbau)
Kurvenverzerrungsanalysator m <Gerät> (BE) distortion analyser, (AE) distortion analyzer
Kurvenwelle f <Fertig> undulation
kurz adj <Elektriz, Elektronik> short
Kurzbaulänge f <Fertig> minimum overall length (Kunststoffinstallationen)
Kurzbewitterungsversuch m <Kunststoff> accelerated weathering test
Kurzbezeichnung f <Konstzeich> abbreviated designation
kurzbrennweitiges Objektiv n <Foto> short-focus lens
Kurzdarstellung f <Comp & DV> abstract
kurze Trompete f <Wassertrans> cat's paw (Knoten)
kürzen v 1. <Math> reduce a fraction (eines Bruches); 2. <Papier> shorten
Kürzen n **von Fasern** <Papier> (AE) shortening of fibers, (BE) shortening of fibres
kurzer Impuls m <Wassertrans> short pulse
kurzer Spiralbohrer m <Maschinen> stub drill
kurzer Ton m <Wassertrans> short blast
kurzer Zapfen m <Bau> stub tenon
kurzes Glas n <Ker & Glas> short glass
kurzes Verbindungsstück n <Bau> faucet joint (von Rohren)
kurzes Zapfenloch n <Bau> stub mortise
Kürzester-Weg-Programm n <Trans> (AE) shortest path program, (BE) shortest path programme
Kurzfaservlies n <Kunststoff> chopped-strand mat
Kurzfassung f <Druck> abridged edition, abstract, summary
kurzgeschlossen adj <Elektriz, Elektrotech, Phys> short-circuited
kurzgeschlossene Windung f <Elektrotech> shorted winding
kurzgeschlossener Anker m <Elektriz, Elektrotech> short circuit armature
kurzgeschlossener Rotor m <Elektriz> short-circuited rotor
kurzgeschlossener Schleifringanker m <Elektriz> short-circuited slip-ring rotor
kurzgliedrige Kette f <Maschinen> short link chain
Kurzhals-Projektionsröhre f <Elektronik> short neck projection tube
Kurzhubhonen n <Maschinen> superfinishing
Kurzhubhonstein n <Maschinen> superfinishing honing stone, superfinishing stone
kurzhubig adj <Maschinen> short-stroke
Kurzhubmotor m <Kfztech> oversquare engine, short stroke engine
Kurzimpuls m <Elektronik> short pulse
Kurzkanal m <Elektronik> short channel
Kurzkanaltransistor m <Elektronik> short channel transistor
Kurzmitteilung f <Mobilkom, Telekom> short message

Kurznachrichten senden Kurzmitte <Mobilkom, Telekom> short message service-mobile originated (SMS-MO)
Kurznachrichtendienst m <Mobilkom, Telekom> short message service, SMS (Mobilfunk-Mitteilungen)
Kurznummer f <Telekom> abbreviated number
kurzöliges Alkydharz n <Kunststoff> short oil alkyd
Kurzprüfung f <Metall> short time test
Kurzrohr n <Hydraul> short pipe
Kurzrufnummer f <Telekom> abbreviated number
kurzschließen v <Telekom> short-circuit
Kurzschließen n <Elektrotech> shorting
Kurzschließer m <Elektriz> short-circuiting device
Kurzschluss m 1. <Comp & DV, Elektriz, Elektrotech> short, short circuit; 2. <Phys> closed circuit, short circuit; 3. <Telekom> short circuit
Kurzschluss m **zwischen Phasen** <Elektriz> interphase short circuit
Kurzschlussanker m <Elektriz, Elektrotech> closed-coil armature, short-circuited armature, squirrel cage rotor
Kurzschlussausschaltvermögen n <Elektriz, Elektrotech> short-circuit breaking capacity, short-circuit rupturing capacity
Kurzschlussbetrieb m <Telekom> back-to-back operation
Kurzschlussbrücke f <Elektrotech> bonding jumper, jumper
Kurzschlussdrossel f <Elektrotech> current-limiting reactor
Kurzschlusseinschaltvermögen n <Elektriz, Elektrotech> short-circuit making capacity
Kurzschlussfluss m <Akustik> short circuit flux
Kurzschlussimpedanz f <Elektriz, Elektrotech> short circuit impedance
Kurzschlusskäfig m <Elektrotech> squirrel cage
Kurzschlusskontakt m <Elektrotech> shorting contact
Kurzschlusskontaktschalter m <Elektrotech> shorting contact switch
Kurzschlussläufer m <Elektrotech> squirrel cage rotor
Kurzschlussläufermotor m <Elektrotech> squirrel cage motor
Kurzschlussleitung f <Elektrotech> adjustable short
Kurzschlussschalter m <Elektrotech> shorting switch
Kurzschlussschutz m <Elektrotech> short circuit protection
Kurzschlussstrom m <Elektrotech> short circuit flux
Kurzschlusswiderstand m <Elektrotech> short circuit impedance
Kurzspan m <Fertig> finely broken chip, short chip
Kurzstartflugzeug n (STOL-Flugzeug) <Lufttrans> short takeoff and landing aircraft (STOL aircraft)
Kurzstart-Kurzlande-Flugzeug n (STOL-Flugzeug) <Lufttrans> short takeoff and landing aircraft (STOL aircraft)
Kurzstart- und Landung f <Lufttrans> short takeoff and landing
Kurzstreckenfunkmodul n <Funktech> bluetooth-module
Kurzstreckengleiter m <Trans> short haul skidder
Kurzstreckenrichtfunkverbindung f <Funktech, Telekom> short-haul microwave link
Kurzstreckenseeverkehr m <Wassertrans> short-sea shipping
Kurzstreckentransport m <Trans> short distance transport
Kurzstreckenverkehrsflugzeug n <Trans> short haul airliner
Kurzstreckenzähler m <Kfztech> trip mileage indicator
Kurzversuch m <Qual> accelerated test
Kurzwahl f <Telekom> (AE) short code dialing, (BE) short code dialling
Kurzwahlnummer f <Telekom> speed dialing number

Kurzwelle f 1. <Elektriz> short wave; 2. <Funktech> short wave, high frequency (HF)
Kurzwellen fpl <Funktech> decametric waves
Kurzwellenband n <Funktech> high-frequency band, short-wave band
Kurzwellenfunkverbindung f <Funktech, Telekom> HF radio link, high-frequency radio link
Kurzwellenhörer m <Funktech> short- wave listener
Kurzwellenkanal m <Funktech, Telekom> HF channel, short-wave channel
Kurzwellenleistungsgenerator m <Elektriz, Funktech> high-frequency power generator
Kurzwellenleistungsverstärker m <Elektriz, Funktech> high-frequency power amplifier
Kurzwellenpeilgerät n <Funktech> high-frequency direction finder (HFDF)
Kurzwellenpeilsystem n <Funktech> HFDF system, high-frequency direction finding system
Kurzwellenrundfunk m high-frequency broadcasting
Kurzwellenübertragungsstrecke f <Funktech, Telekom> high-frequency link
Kurzwort n <Comp & DV> acronym
Kurzzange f <Maschinen> short-nosed pliers
Kurzzapfverbindung f <Bau> spur tenon joint
Kurzzeit... <Künstl Int, Kunststoff> short-term
Kurzzeitalterungsversuch m <Kunststoff> (BE) accelerated ageing test, (AE) accelerated aging test
Kurzzeitbetrieb m <Elektriz, Elektrotech, Elektronik, Telekom> short-time duty, short-time service
Kurzzeitermüdungsversuch m <Metall> rapid fatigue test
Kurzzeit-Erträglichkeitsgrenze f <Ergon> just tolerable limit
Kurzzeitgedächtnis n (KZG) <Ergon, Künstl Int> short-term memory (STM)
kurzzeitig adj 1. <Elektronik> short term; 2. <Fertig> transient
kurzzeitige Drift f <Elektronik> short-term drift
kurzzeitige Frequenzstabilität f <Elektronik, Fernseh, Funktech, Telekom> short-term frequency stability
kurzzeitiger Fehler m <Comp & DV> transient error
Kurzzeitleistung f <Elektriz> short time rating
Kurzzeitmessgerät n <Gerät> time interval measuring instrument
Kurzzeitnachleuchten n <Elektronik> CRT-lag
Kurzzeitprüfung f <Elektriz> short time test
Kurzzeitregistriergerät n <Gerät> time interval recorder
Kurzzeitschutz m <Elektrotech> short-term protection
Kurzzeitschwankungsrate f <Fernseh> flutter rate
Kurzzeitspannung f <Ker & Glas, Metall, Qual> temporary stress
Kurzzeit-Stromnennwert m <Elektriz> rated short-time current
Kurzzeitversuch m <Qual> accelerated test
Küste f <Wassertrans> coast (Geographie) • **an der Küste entlanglaufen** <Wassertrans> follow the coast • **auf die Küste zusteuern** <Wassertrans> stand inshore (Navigation) • **die Küste anlaufen** <Wassertrans> stand inshore (Navigation) • **die Küste entlangsegeln** <Wassertrans> coast (Seehandel) • **längs der Küste** <Wassertrans> alongshore
Küsten... <Wassertrans> coastal; littoral (Geographie)
küsteneinwärts adv <Wassertrans> inshore
Küstenendkabel n <Elektrotech> shore end cable
Küstenfahrt f <Wassertrans> coasting trade, coastwise trade; coastal trade (Seehandel)
Küstenfahrzeug n <Wassertrans> coaster, coasting vessel
küstenfern adj <Wassertrans> offshore
Küstenfluss m <Wasserversorg> coastal river

Küstenfunkstelle f 1. <Funktech, Telekom> coastal station; 2. <Wassertrans> coast radio station, CRS
Küstengewässer npl <Wassertrans> coastal waters
Küstenkarte f <Wassertrans> coastal chart, plan (Navigation)
Küstenkreuzer m <Wassertrans> coastal cruiser
Küstenlinie f <Meerschmutz, Wassertrans> shoreline; coastline (Geographie)
Küstenlotse m <Wassertrans> inshore pilot
Küstenlotsenwesen n <Wassertrans> inshore pilotage
Küstenmotorschiff n <Wassertrans> coastal motor vessel
küstennah adj <Wassertrans> inshore, offshore
küstennahe Ölbohrung f <Umweltschmutz> offshore well
Küstennavigation f <Wassertrans> coastal navigation (Navigation)
Küstenradarstation f <Funkort, Wassertrans> coastal radar station
Küstenringstraße f <Bau> coastal ring road
Küstenschiff n <Wassertrans> coastal vessel, coasting vessel
Küstenschifffahrt f <Wassertrans> cabotage; coastal navigation (Handelsmarine); coastal trade (Seehandel)
Küstenschlamm m <Wasserversorg> coastal deposit
Küstenschlick m <Wasserversorg> coastal deposit
Küstenschutz m <Wasserversorg> shore protection
Küstensockel m <Wassertrans> shelf
Küstenstation f <Telekom> coastal station
Küstenstreifen m <Wassertrans> shoreline
Küstenwache f <Wassertrans> coastguard
Küstenwachkutter m <Wassertrans> coastguard cutter
küstenwärts adv <Wassertrans> shoreward
Kutter m <Wassertrans> cutter (Schifftyp)
Küvette f <Labor> vessel
kV (Kilovolt) <Elektriz> kV (kilovolt)
KV (kombinierter Verteiler) <Telekom> CDF (combined distribution frame)
kW (Kilowatt) <Elektriz> kW (kilowatt)
K-Wert m <Heiz & Kälte> K-factor
kWh (Kilowattstunde) <Elektriz, Elektrotech> kWh (kilowatt-hour)
KWIC m <Comp & DV> KWIC, keyword in context
KWOC-Index m (Stichwortanalyse mit Text) <Comp & DV> KWOC, keyword out of context (keyword out of context)
KW-Stoff m (Kohlenwasserstoff) <Chemie> hydrocarbon
Kybernetik f <Comp & DV, Fertig, Raumfahrt> cybernetics
Kynch'sche Trennungstheorie f <Kerntech> Kynch separation theory
kyrillisches Zeichen n <Konstzeich> cyrillic letter
KZG (Kurzzeitgedächtnis) <Ergon, Künstl Int> STM (short-term memory)
K-Zustand m <Kerntech> K-state

L

l 1. <Comp & DV, Geom, Phys, Telekom> (Länge) l (length); 2. <Kerntech> (effektive Neutronenlebensdauer) l (effective neutron lifetime)
L 1. <Akustik> (Lautstärke) L (loudness); 2. <Aufnahme, Elektriz, Elektrotech, Funktech, Metrol, Phys, Telekom>

(Induktivität) L *(inductance)*; 3. <Kerntech> *(Diffusionslänge)* L *(diffusion length)*; 4. <Kerntech, Strahlphys> *(lineare Energieübertragung)* L *(linear energy transfer)*; 5. <Kerntech> *(Orbitalwinkelmomentzahl)* L *(total orbital angular momentum number)*; 6. <Mechan> *(Lagrange'sche Funktion)* L *(Lagrangian function)*; 7. <Optik> *(Luminanz)* L *(luminance)*; 8. <Thermod> *(Lorenz'sche Einheit)* L *(Lorenz unit)*

La *(Lanthan)* <Chemie> La *(lanthanium)*
Lab n <Lebensmittel> rennet
Labferment n <Lebensmittel> rennin
labiles Gleichgewicht n <Phys> unstable equilibrium
Labkasein n <Lebensmittel> rennet casein
Labor n <Elektriz, Labor, Phys> laboratory
Laboratorium n 1. <Kohlen> assay office; 2. <Labor> laboratory
Laborbedingungen fpl <Elektriz, Labor> laboratory conditions
Laborkittel m 1. <Labor> laboratory coat; 2. <Sicherheit> laboratory clothing
Laborkugelhahn m <Fertig> laboratory ball cock *(Kunststoffinstallationen)*
Labormessgerät n <Gerät> laboratory instrument
Labormühle f <Maschinen> triturator
Laborschemel m <Labor> laboratory stool
Laborstandard m <Kerntech> laboratory standard
Laborstruktur f <Phys> laboratory frame
Laborstuhl m <Labor> laboratory stool
Laborversuch m <Bau, Lebensmittel, Patent, Qual> bench test
Laburnin n <Chemie> cytisine, laburnine, ulexine
Labyrinthdichtung f 1. <Fertig> labyrinth packing; 2. <Kerntech, Maschinen> labyrinth seal
Labyrinthpackung f <Maschinen> labyrinth packing
Labyrinthring m <Maschinen> labyrinth seal ring
Laccain... <Chemie> laccaic
Laccol n <Chemie> laccol
Lachgas n <Chemie> laughing gas
Lack m 1. <Bau> lacquer; 2. <Druck, Elektriz> varnish; 3. <Kfztech> paint *(Karosserie)*; 4. <Kunststoff> varnish; 5. <Mechan> lacquer
Lackabfall m <Abfall> varnish waste
Lackdraht m <Elektriz, Elektrotech> *(AE)* enameled wire, *(BE)* enamelled wire
Lackfarbe f <Bau> enamel, lacquer, varnish
Lackfirnis m <Bau> shellac
Lackfolienaufnahme f <Aufnahme> lacquer recording
lackieren v 1. <Anstrich> paint; 2. <Fertig> japan, lacquer
Lackieren n <Bau> varnishing
Lackierkabine f <Anstrich> spray booth
Lackiermaschine f <Verpack> coating machine, lacquering machine
lackiert adj <Mechan> lacquered
Lackierung f <Bau> varnishing
Lackmuspapier n <Fertig> litmus paper
Lackrest m <Abfall> varnish waste
Lackschlamm m <Abfall> paint sludge
Lacksplitter m <Anstrich> paint chip
Lackversiegelung f <Verpack> lacquer sealing
Lactam n <Chemie> lactam
Lactamid n <Chemie> hydropropanamide, lactamide
Lactat n <Chemie> lactate
Lactid n <Chemie> lactide
Lacto-Butyrometer n <Lebensmittel> lactobutyrometer
Lactometer n <Lebensmittel> lactometer
Lacton n <Chemie> lactone
Lacton... <Chemie> lactonic
Lactonbildung f <Chemie> lactonization
Lactonisierung f <Chemie> lactonization

Lactonitril n <Chemie> hydropropanenitrile, lactonitrile
Lactose f <Chemie> lactose
Laddertron n <Elektronik> laddertron *(Mikrowellen-Klystron)*
Lade... <Bau, Comp & DV, Trans, Wassertrans> loading
Ladeadresse f <Comp & DV> load point
Ladeaggregat n <Elektrotech> charging set, charging unit *(z. B. für Stoßspannungsgeneratoren)*
Ladeanzeige f <Wassertrans> charge indicator *(Elektrik)*
Ladeband n <Bau, Kohlen> conveyor
Ladebaum m 1. <Mechan> boom, derrick, hoist; 2. <Wassertrans> cargo boom, cargo derrick, derrick, derrick boom *(Ladung)*
Ladebaumpfosten m <Wassertrans> Samson post *(Schiffbau)*
Ladebereich m 1. <Lufttrans> loading area *(Flughafen)*; 2. <Trans> loading area
Ladebrücke f 1. <Bau> deck; 2. <Eisenbahn> loading bridge
Ladebucht f <Raumfahrt> cargo bay *(Raumschiff)*
Ladedruck m 1. <Kfztech> boost pressure; 2. <Lufttrans> manifold pressure; boost pressure *(Überdruck)*
Ladedruckventil n <Kfztech> pressure boost valve
Lade-Entladezyklus m <Wassertrans> charge-discharge cycle
Ladefähigkeit f 1. <Lufttrans> load capacity; 2. <Wassertrans> cargo capacity *(Schiffkonstruktion)*
Ladefaktor m 1. <Mechan> load factor; 2. <Qual, Raumfahrt> loading factor *(Weltraumfunk)*
Ladefläche f <Maschinen> platform
Ladegerät n 1. <Elektriz, Foto, Funktech> charger; 2. <Trans> loader *(an Stromversorgungsnetz angeschlossen)*
Ladegerät n **für langsames Aufladen** <Elektrotech> trickle charger
Ladegeschirr n <Wassertrans> cargo gear, cargo-handling gear *(Ladung)*
Ladegestell n <Mechan> skid
Ladegleis n <Trans> loading siding
Ladehaken m <Wassertrans> cargo hook *(Ladung)*
Ladekai m <Wassertrans> loading dock
Ladeklappe f <Lufttrans> loading door *(Flugzeug)*
Ladekran m 1. <Eisenbahn> loading crane; 2. <Wassertrans> cargo crane *(Ladung)*
Ladeleistung f <Elektrotech> charging power *(Batterie, Kabel und Freileitungen)*
Ladeleitung f <Lufttrans> manifold
Ladelinie f <Wassertrans> loadline
Ladeliste f <Wassertrans> cargo manifest *(Dokumente)*
Ladeluke f 1. <Lufttrans> loading door *(Flugzeug)*; 2. <Raumfahrt> cargo hatch *(Raumschiff)*; 3. <Wassertrans> cargo hatchway; cargo hatch *(Ladung)*
Lademarke f <Lufttrans> load line
Lademaß n <Eisenbahn> *(AE)* loading gage, *(BE)* loading gauge
laden v 1. <Comp & DV> boot, load; 2. <Elektronik> input *(Programm)*; 3. <Funktech> load
Laden n 1. <Comp & DV> loading; 2. <Elektrotech> charging
Ladenanschlag m <Textil> beat-up
Ladenetz n <Wassertrans> net sling *(Ladung)*
Ladepfosten m <Wassertrans> king post *(Deckbeschläge)*; cargo post *(Ladung)*
Ladeplan m <Wassertrans> capacity plan, cargo plan *(Ladung)*
Ladeprogramm n <Comp & DV> loader, loading routine
Ladepumpe f <Meerschmutz, Wassertrans> cargo pump
Ladepunkt m <Comp & DV> load point
Lader m 1. <Comp & DV> loader; 2. <Kfztech> supercharger

Laderampe f 1. <Bau, Eisenbahn> loading platform; 2. <Lufttrans> loading ramp *(Flugzeug)*; 3. <Trans> loading platform
Laderaum m 1. <Kfztech> cargo hold *(Lkw)*; 2. <Raumfahrt> payload bay *(Raumschiff)*; 3. <Trans> cargo space *(Lkw)*; 4. <Wassertrans> hold *(Ladung)*; cargo hold, hold *(Schiff)*
Laderegler m <Raumfahrt> charging regulator
Ladeschalter m <Elektrotech> charging contactor *(einer Batterie)*
Ladeschaltung f <Elektriz> charging circuit
Ladeschaufel f <Bau> loading shovel
Ladeschlitz m <Fernseh> loading slot
Ladespannung f <Elektriz> charging voltage
Ladestation f <Elektrotech> charging station *(für Akkumulatoren/Batterien)*
Ladesteckdose f <Elektrotech> charging socket outlet
Ladestrom m <Elektriz> capacitance current, charging current
Ladestrommessgerät n <Gerät> charge rate meter
Ladestromstoß m <Elektriz> charging current impulse, charging current pulse, charging current surge
Ladetank m <Umweltschutz> cargo tank
Lade- und Löschbord m <Trans> skids
Ladeverdrängung f <Wassertrans> loaded displacement; heavy displacement *(Schiffskonstruktion)*; load displacement *(Schiffkonstruktion)*
Ladevorrichtung f <Telekom> charging equipment
Ladewert m <Trans> load value
Ladewiderstand m <Elektrotech> charging resistor *(Bauteil)*
Ladewinde f <Wassertrans> cargo winch *(Ladung)*
Ladezeit f <Foto> recharge time
Ladezeitkonstante f <Elektriz> charge time constant
Ladung f 1. <Elektriz, Elektrotech> charge; 2. <Erdöl> freight; 3. <Kohlen> load, shot; 4. <Lebensmittel> batch; 5. <Lufttrans> cargo; 6. <Phys> charge; 7. <Raumfahrt> batch, cargo *(Raumschiff)*; 8. <Strahlphys, Teilphys> charge; 9. <Telekom> charge *(elektrisch)*; 10. <Trans> batch, loading; 11. <Wassertrans> cargo, load
Ladungsaufbau m <Elektrotech> charge buildup
Ladungsdichte f <Elektriz, Phys, Strahlphys, Teilphys> charge density
Ladungsdichte f eines Teilchens <Phys, Strahlphys, Teilphys> charge density of particle
Ladungserhaltung f <Phys> connection, conservation of charge
ladungsgekoppeltes Bauelement n 1. <Elektronik, Telekom> charge-coupled device *(CCD)*; 2. <Phys> charge transfer device *(CTD)*
ladungsgekoppeltes Halbleiterbauelement n <Elektrotech> charge-coupled device
Ladungskompensierung f <Elektriz> charge neutralization
Ladungsleckstrom m <Elektriz> charge leakage
Ladungsmanifest n <Lufttrans, Wassertrans> cargo manifest *(Dokumente)*
Ladungsmassenverhältnis n <Phys, Strahlphys, Teilphys> charge-mass ratio
Ladungsmengenmesser m <Elektrotech, Metrol> coulombmeter, coulometer, voltameter
Ladungsmessgerät n <Elektriz> coulombmeter
Ladungsmultiplett n <Phys, Strahlphys, Teilphys> charge multiplet
Ladungsparitätssymmetrie f <Phys, Strahlphys, Teilphys> charge-parity symmetry
Ladungspumpe f <Elektrotech> charge pump
Ladungsspeicher m <Elektrotech> charge storage
Ladungsspeicherdiode f <Elektronik> CCD diode, charge-storage diode
Ladungsspeicherröhre f <Elektronik> charge-storage tube
Ladungssumme f <Elektrotech> net charge
Ladungsträger m 1. <Elektriz, Elektronik> carrier, charge carrier; 2. <Phys> carrier; 3. <Raumfahrt> charge carrier
Ladungsträgerbeweglichkeit f <Elektriz, Phys> carrier mobility
Ladungsträgerstrahl m <Elektronik> charged-particle beam
Ladungsträgervervielfachung f <Elektronik> carrier multiplication *(Mikroelektronik)*
Ladungstransferelement n *(CTD)* <Telekom> charge transfer device *(CTD)*
Ladungsübertragung f <Kerntech, Strahlphys, Teilphys> charge transfer
Ladungsübertragungsgerät n *(CTD)* <Raumfahrt> charge transfer device *(CTD)*
Ladungsumschlag m <Wassertrans> cargo handling *(Ladung)*
Ladungs- und Abfüllgeräte npl <Verpack> handling and filling equipment
Ladungsverluststrom m <Elektriz> charge leakage
Ladungsverschiebeband n <Kerntech, Strahlphys, Teilphys> charge transfer band
Ladungsverschiebeschaltung f *(CTD)* <Elektrotech> charge transfer device, CTD *(Halbleiter)*
Ladungsverstärker m <Elektronik> charge amplifier
Ladungsverzeichnis n <Wassertrans> manifest
Ladungswolke f <Phys, Strahlphys, Teilphys> charge cloud
Lage f 1. <Bau> lie, site; lay *(Geographie)*; 2. <Elektrotech> lay *(eines Kabels)*; 3. <Fertig> orientation *(der Schneide)*; 4. <Ker & Glas> course; 5. <Kohlen> layer; 6. <Maschinen, Raumfahrt> ply *(Raumschiff)*; 7. <Wasserversorg> layer • **in der ursprünglichen Lage** <Kohlen> in situ • **in Lage bringen** <Bau> locate
Lage f Papier <Druck> quire
Lagefehler m 1. <Maschinen> error of position, position error; 2. <Qual> error of position
Lagegenauigkeit f der Bohrung <Fertig> accuracy in hole positioning
Lageindex m <Ergon> position index
Lagemessgerät n <Gerät> position measuring instrument
Lagenenergie f <Maschinen> potential energy
Lagenhaftung f <Kunststoff> ply bond strength
Lagenisolierung f <Elektriz> interturn insulation
Lagenlösung f <Kunststoff> ply separation
Lagentrennung f <Kunststoff> ply separation
Lagenwicklung f <Elektrotech> layer winding
Lager n 1. <Bau> bearing, stockpile; 2. <Elektriz> bearing; 3. <Kohlen> deposit; 4. <Maschinen, Mechan, Papier, Trans> bearing
Lager n mit Weißmetallausguss <Maschinen> babbitted bearing
Lagerausguss m 1. <Maschinen> lining; 2. <Mechan> bearing lining
Lagerbehälter m <Heiz & Kälte> storage tank
Lagerbestand m <Erdöl> inventory, stock
Lagerbeständigkeit f 1. <Verpack> shelf impact, storage durability; 2. <Werkprüf> shelf life *(Kunststoffe)*
Lagerbestandsaufnahme f <Verpack> inventory
Lagerbock m 1. <Maschinen> bearing block, pedestal; 2. <Mechan> bracket, pedestal
Lagerbohrung f <Maschinen> bearing bore, bore of bearing
Lagerbuchse f 1. <Maschinen> brass; 2. <Mechan> bushing
Lagerbüchse f <Maschinen> bearing bush, bearing bushing, bushing

Lagerbunker

Lagerbunker m <Abfall> receiving bin, refuse bunker
Lagerdauer f <Foto> storage period
Lagerdeckel m 1. <Fertig> top bearing; 2. <Kfztech, Maschinen> bearing cap
Lageregelungsrakete f <Raumfahrt> attitude control rocket
Lagerfähigkeit f <Lebensmittel> keeping quality
Lagerfass n **für neue Brennelemente** <Kerntech> new element storage drum
Lagerfestigkeit f <Verpack> shelf stability
Lagerfläche f 1. <Bau> bed; 2. <Verpack> floor space
Lagergehäuse n <Maschinen, Papier> bearing housing
Lagergestell n <Ker & Glas> storage rack
Lagergut n <Abfall> fill mass, waste mass
Lagerhaus n <Eisenbahn> (AE) freight house, (BE) warehouse
Lagerhülse f <Maschinen> bearing bush, bearing bushing
lagerichtig adj <Konstzeich> in correct positional arrangement
lagerichtige Darstellung f <Konstzeich> topographical representation
Lagerkäfig m <Maschinen> bearing cage
Lagerkammer f <Abfall> storage chamber (für radioaktive Abfälle)
Lagerkonfiguration f <Raumfahrt> storage configuration (Raumschiff)
Lagerkontrolle f <Eisenbahn> stock control
Lagerluft f <Maschinen> bearing clearance
Lagermetall n 1. <Maschinen> bearing alloy, bearing metal; 2. <Metall> babbitt metal, bearing metal
lagern v <Bau> keep, stock
lagern lassen v <Kohlen> store
Lagern n <Comp & DV> store
Lagerplatz m <Bau> stockyard, yard
Lagerraum m 1. <Verpack> storage space; 2. <Wassertrans> storeroom (Ladung)
Lagerraum m **für Tiefkühlkost** <Heiz & Kälte> frozen-food storage room
Lagerraumkosten fpl <Verpack> cost of space
Lagerregal n <Verpack> bin, storing shelf
Lagerreibung f <Maschinen> bearing friction
Lagerring m <Fertig> roller race
Lagerrolle f <Fertig> bottom roller (Schleudergießmaschine)
Lagerschale f 1. <Fertig> pillow; 2. <Maschinen> bearing bush, bearing shell, liner, pillow
Lagerschale f **mit Bund** <Maschinen> flanged liner
Lagerschalenhälfte f <Maschinen> half-bushing
Lagerschild n <Maschinen> bearing plate
Lagerschmierung f <Fertig, Kfztech> bearing lubrication
Lagerseite f <Maschinen> bearing end
Lagerspiel n 1. <Fertig> diametral clearance; 2. <Maschinen> bearing clearance
Lagerstätte f 1. <Abfall> repository, storage facility; 2. <Erdöl> field, reservoir
Lagerstättendruck m <Erdöl> formation pressure, reservoir pressure
Lagerstättenmedium n <Erdöl> formation fluid
Lagerstättenvergasung f <Thermod> underground gasification
Lagerstättenwasser n <Erdöl> formation water
Lagerstuhl m <Mechan> cradle
Lagersystem n **in geregelter Umgebung** <Verpack> controlled environment storage system
Lagertank m <Erdöl, Raumfahrt> storage tank (Raumschiff)
Lagerung f 1. <Abfall> storage; 2. <Comp & DV> storage; 3. <Elektrotech> suspension (in Messwerken); 4. <Foto> storage (Fotomaterialien); 5. <Kohlen> stratification; 6. <Maschinen> bearing

Lagerungsbeständigkeit f <Kunststoff> shelf life
Lagerungsdichte f <Bau> compactness (Geologie)
Lagerungszapfen m <Maschinen> locating spigot
Lagerweißmetall n <Fertig> babbitt
Lagerwerkstoffe mpl <Maschinen> bearing materials
Lagerzapfen m 1. <Kerntech> journal; 2. <Kfztech> journal; trunnion (Universalgelenk); 3. <Maschinen> bearing journal; 4. <Mechan> journal, trunnion
lagestabilisierter Satellit m <Telekom> attitude-stabilized satellite (Satellitenkommunikation)
Lagestabilisierung f <Telekom> orientation control
Lagetoleranz f <Maschinen> positional tolerance, tolerance of position
Lagewinkel m <Fertig> orientation angle
Lagewinkel m **der Freiflächen-Orthogonalebene** <Fertig> tool flank orthogonal plane orientation angle
Lagezuordnung f <Fertig> position coordination
Lagrange'sche Betrachtungsweise f <Strömphys> Lagrangian viewpoint
Lagrange'sche Funktion f (L) <Mechan> Lagrangian function, angular momentum (L)
Lagrange'sche Gleichung f <Phys> Lagrange's equation
Lagrange'scher Multiplikator m <Phys> Lagrangian multiplier
Lagrange'scher Operator m <Phys> Lagrangian operator
Lagune f <Wasserversorg> lagoon
Laktam n <Chemie> lactam
Laktat n <Chemie> lactate
Lambda n <Kfztech, Phys> lambda
Lambda/2-Blättchen n <Funktech, Phys> half-wave plate
Lambda/4-Blättchen n <Phys> quarter-wave plate
Lambdapunkt m <Phys> lambda point (Kryotechnik)
Lambdasonde f <Kfztech> lambda probe
Lambdateilchen n <Phys> lambda particle
Lambda-Viertel-Dipol m <Telekom> quarter-wave dipole
Lambda-Viertel-Leitung f <Telekom> quarter-wave line
Lambda-Viertel-Symmetrieglied n quarter-wave balun
Lambert n <Optik> lambert
Lambertreflektor m <Optik> lambertian reflector
Lambert'scher Strahler m <Optik, Telekom> lambertian source, lambertian radiator
Lambert'sches Cosinusgesetz n <Optik> Lambert's cosine law
Lambert'sches Gesetz n <Phys> Lambert's law
Lambverschiebung f <Phys> Lamb shift
Lamellargraphitgusseisen n <Mechan> lamellar graphite cast iron
Lamelle f 1. <Fertig> lamination; 2. <Maschinen, Mechan> fin; 3. <Metall> lamella
Lamellenanker m <Elektriz> laminated armature
Lamellenfeder f <Maschinen> blade spring
Lamellenheizgerät n <Thermod> finned heater
Lamellenheizung f <Heiz & Kälte> strip heating
Lamellenkühler m 1. <Kfztech> finned radiator, ribbon cellular radiator; 2. <Thermod> finned cooler
Lamellenkupplung f 1. <Kfztech> (BE) multiple-disc clutch, (AE) multiple-disk clutch; 2. <Maschinen> (BE) multidisc clutch, (AE) multidisk clutch, (BE) multiple-disc clutch, (AE) multiple-disk clutch, multiple-plate clutch
Lamellenmagnet m <Phys> compound magnet
Lamellenschleifscheibe f <Maschinen> abrasive flap wheel
Lamellenspannung f <Elektriz> bar voltage
Lamellenstruktur f <Metall> lamellar structure
Lamellenverschluss m <Foto> leaf shutter
lamellierte Bürste f <Elektriz> laminated brush
lamellierter Magnet m <Elektriz> laminated magnet

laminar adj <Chemie, Strömphys> laminar
laminar fluidisierter Brennstoff m <Kerntech> paste fuel
Laminar... <Chemie> laminar
laminare Ablösung f <Strömphys> laminar separation
laminare Rohrströmung f <Strömphys> laminar pipe flow
laminare Strömung f <Heiz & Kälte, Lufttrans, Strömphys> laminar flow
Laminarin n <Chemie> laminarin
Laminartransistor m <Elektronik> laminar transistor
Laminat n 1. <Elektronik> laminate; 2. <Kunststoff> laminate, laminated plastic
Laminatprofil n <Kunststoff> laminated section
laminieren v <Bau, Druck, Fertig> laminate
Laminieren n 1. <Funktech> lamination (Transformatorkern); 2. <Kunststoff> lamination; 3. <Papier> laminating
laminiert adj <Elektriz, Fertig, Kunststoff, Wassertrans> laminated
laminierte Drehstabfeder f <Lufttrans> laminated torsion bar
laminiertes Plastik n <Wassertrans> laminated plastic
Laminierung f aus Siliziumstahl <Elektrotech> silicon steel lamination
Lampard-Thomson-Kondensator m <Phys> Lampard and Thomson capacitor
Lampe f <Kfztech> lamp (Zubehör)
Lampenbrett n <Foto> bank of lights
Lampenfassung f <Elektrotech, Foto> bulb cap, bulb holder, lamp cap, lamp holder, lamp socket
Lampengehäuse n <Foto> lamphouse (Projektor)
Lampenhalterung f <Eisenbahn> (BE) lamp iron
Lampenkolben m 1. <Elektriz> bulb; 2. <Ker & Glas> lamp bulb
Lampenschirm m <Elektrotech> hood, lampshade
Lampenwechsel m <Foto> lamp replacement
Lampenzylinder m <Ker & Glas> lamp chimney
LAN (lokales Netz) <Comp & DV, Telekom> LAN (local area network)
Land n 1. <Bau> earth, ground, soil; 2. <Wassertrans> land, shore • **an Land** <Wassertrans> ashore, on shore • **an Land absetzen** <Wassertrans> land (Ladung, Passagiere) • **an Land setzen** <Wassertrans> land (Passagiere) • **auf Land** <Erdöl> onshore
LAN-Datei-Server m <Comp & DV> LAN file server
Landaufnahme f <Bau> geodetic survey
Land-Bord-Alarmierung f <Telekom> shore-to-ship alerting
Landbrise f <Wassertrans> land breeze
Landcontainer m <Trans> land container
Lande... <Lufttrans, Wassertrans> landing
Landeanflug m <Raumfahrt> descent, descent path (Raumschiff)
Landeanfluggeschwindigkeit f <Lufttrans> landing approach speed
Landeauslaufstrecke f <Lufttrans> landing run
Landebahn f <Lufttrans> landing strip
Landebahnmarkierung f <Lufttrans> landing strip marker; (AE) hold-short line, (BE) lead-in line (Vorfeldmarkierung, -befeuerung)
Landebahnmittellinienverlängerung f <Lufttrans> (AE) extended runway centerline, (BE) extended runway centreline
Landebahnscheinwerfer m <Elektriz> landing area floodlight
Landebahnsichtweite f <Lufttrans> runway visual range
Landebake f <Funktech, Lufttrans> landing radio beacon
Landedeck n <Wassertrans> landing deck
Landefahrgestell n <Lufttrans> landing-gear leg

Landefahrzeug n <Raumfahrt> lander, landing vehicle (Raumschiff)
Landé-Faktor m <Phys> Landé factor
Landefolge f <Lufttrans> landing sequence
Landeführungsgerät n <Trans> localizer
Landefunkbake f <Funktech, Lufttrans> landing beacon
Landegebiet n <Raumfahrt> landing area (Raumschiff)
Landegeschwindigkeit f <Lufttrans> approach speed, landing speed
Landegestell n <Lufttrans> landing gear
Landegewicht n <Lufttrans> landing weight
Landekapsel f <Raumfahrt> landing capsule (Raumschiff)
Landekarte f <Lufttrans> landing chart
Landeklappe f <Lufttrans> landing flap; flap (Flugzeug)
Landekufe f <Lufttrans> landing skid, landing-gear leg
Landekursbake f <Funktech, Lufttrans> localizer beacon
Landekurssender m des Instrumentenlandesystems <Funktech, Lufttrans> ILS localizer, instrument landing system localizer
Landelänge f <Lufttrans> landing distance
Landeleitlinie f <Lufttrans> glide slope
Landelicht n <Lufttrans> landing light
landen v 1. <Raumfahrt> touch down (Raumschiff); 2. <Wassertrans> land
Landen n <Trans> landing (einer Ladung)
Landenge f <Wassertrans> isthmus
Landepfad m <Lufttrans> landing path
Landepiste f <Lufttrans> landing strip
Landepunkt m <Raumfahrt> touchdown point (Raumschiff)
Landerichtungsanzeiger m <Lufttrans> landing direction indicator
Länderkennung f <Comp & DV, Telekom> country code
Länderkennzahl f (LKZ) <Telekom> country code, national code, national telephone code, telephone country code, TCC (Ländervorwahlnummer)
Landerschließung f <Umweltschmutz> reclamation of land
Landeschleife f <Lufttrans> landing pattern turn
Landeskennzahl f <Telekom> national code
Landeskennziffer f (LKZ) <Telekom> country code, telephone country code
Landesstraße f <Trans> state road (in Deutschland)
Landestrahl m <Funktech, Lufttrans> landing beam
Landestrecke f <Lufttrans> landing distance
Landestufe f <Raumfahrt> descent stage, lander stage (Raumschiff)
Landesvorwahl f <Telekom> telephone country code
Landesystem n <Lufttrans, Wassertrans> landing system
Landeumlaufbahn f <Raumfahrt> descent orbit
Landeverfahren n <Lufttrans> landing procedure
Landezone f <Lufttrans> landing area
Landfahrzeug n <Raumfahrt> land vehicle (Raumschiff)
Landflugzeug n <Lufttrans> terraplane
Landgang m <Wassertrans> gangway
landgebundenes Luftkissenfahrzeug n <Trans> land air cushion vehicle
landgestütztes Seenavigationssystem n <Funktech, Wassertrans> terrestrial maritime navigation system
Landgewinnung f 1. <Umweltschmutz> reclamation of land; 2. <Wasserversorg> land reclamation
Landkabel n <Telekom> land cable
Landleitung f <Fernseh> landline
ländlicher Bezirk m <Elektrotech> rural district
Landluftwegtransport m <Trans> road-air combined transport
Landmarke f <Wassertrans> landmark (Seezeichen)
Landmesser m <Bau> surveyor
Landnetz n <Telekom> rural network
Landnutzungsplan m <Bau> land use plan

Landschaftsfotograf

Landschaftsfotograf *m* <Foto> landscape photographer
Landschaftsgestaltung *f* <Bau> landscaping
Landspitze *f* <Wassertrans> point *(Geographie)*
Landstraße *f* <Trans> secondary road
Landung *f* 1. <Lufttrans> landing; 2. <Raumfahrt> landing *(Raumschiff)*; 3. <Wassertrans> landing
Landung *f* **bei geringer Sicht** <Lufttrans> low visibility landing
Landung *f* **mit anschließendem Durchstarten** <Lufttrans> touch-and-go landing
Landung *f* **mit ausgefallenem Triebwerk** <Lufttrans> dead stick landing
Landung *f* **mit eingefahrenem Fahrwerk** <Lufttrans> wheels-up landing
Landungsboot *n* <Wassertrans> landing craft
Landungsbrücke *f* 1. <Bau> pier; 2. <Trans> landing stage; 3. <Wassertrans> jetty
Landungsgebühr *f* <Lufttrans, Wassertrans> *(BE)* landing charge, *(AE)* landing fee
Landungskosten *fpl* <Lufttrans> *(BE)* landing charge, *(AE)* landing fee
Landungsponton *m* <Wassertrans> landing pontoon
Landurlaub *m* <Wassertrans> shore leave
Landvermessung *f* <Bau> land survey, land surveying
Landvermittlungsstelle *f* <Telekom> rural switch
Landwind *m* <Wassertrans> land wind, offshore wind
landwirtschaftlicher Abfall *m* <Abfall> agricultural waste, farm waste
Landzentrale *f* <Telekom> rural exchange
Landzunge *f* <Wassertrans> headland *(Geographie)*
Landzuwachs *m* <Wasserversorg> land accretion
lang nachleuchtender Bildschirm *m* <Fernseh> long-persistence screen
Lang... <Funktech, Maschinen, Telekom> long
langbrennweitiges Objektiv *n* <Foto> long-focus lens
Langdrahtantenne *f* <Funktech> long-wire antenna
Langdrehautomat *m* <Maschinen> Swiss-type automatic
Langdrehen *n* <Maschinen> plain turning, straight turning
Langdrehmaschine *f* <Maschinen> plain-turning lathe, sliding lathe
lange Ringleitung *f* <Trans> long loop
lange Unterlängen *fpl* <Druck> long descenders
Länge *f* 1. <Comp & DV, Geom, Phys, Telekom> length *(l)*; 2. <Metrol> yardage
Länge *f* **der Erzeugenden** <Geom> slant height
Länge *f* **der Fahrzeugschlange** <Trans> queue length
Länge *f* **der Zeichenfolge** <Comp & DV> string length
Länge *f* **im stationären Zustand** <Optik> equilibrium length
Länge *f* **über alles** *(LüA)* 1. <Maschinen> overall length; 2. <Wassertrans> length overall *(LOA)*; ram *(Maßeinheit)*
Länge *f* **zwischen den Loten** <Wassertrans> length between perpendiculars *(Schiffkonstruktion)*
Längen *n* <Maschinen> elongation
Längenänderung *f* <Fertig> displacement *(Kunststoffinstallationen)*
Längenausdehnungskoeffizient *m* <Kunststoff> coefficient of linear expansion
Längeneinheit *f* <Fertig, Metrol> unit of length
Längengrad *m* <Wassertrans> longitude *(Geographie, Navigation)*
Längenkontraktion *f* <Phys> length contraction
Längenmaß *n* 1. <Math> linear measurement; 2. <Metrol> long measure
Längenmessgeräte *npl* <Metrol> dimensional measuring instruments
Längenmessinstrument *n* <Maschinen> length measuring instrument
Längenmessmaschine *f* <Gerät> length measuring machine

Längenmetazentrum *n* <Wassertrans> *(AE)* longitudinal metacenter, *(BE)* longitudinal metacentre *(Schiffbau)*
längenspezifische Strömungsresistanz *f* <Akustik> flow resistivity
Längentoleranz *f* <Fertig> length margin
längentreue Azimutalprojektion *f* <Funkort> azimuthal equidistant projection *(Kartographie)*
Längen- und Seitenverhältnis *n* <Comp & DV> aspect ratio
Längenverhältnis *n* <Mechan> aspect ratio
Längenverhältnis *n* **des Luftschraubenblattes** <Lufttrans> blade aspect ratio
Längenverkürzung *f* <Phys> length contraction
langer Impuls *m* <Wassertrans> long pulse *(Radar)*
langer Sturzbalken *m* <Bau> breastsummer
langer Ton *m* <Wassertrans> long blast
Langfaser *f* <Textil> long line
Langflachs *m* <Textil> line flax
Langfräsmaschine *f* <Maschinen> planomiller, planomilling machine
Langgewinde *n* <Maschinen> longscrew
Langhalsflasche *f* <Labor> long-necked flask
Langhalskolben *m* <Labor> long-necked flask
Langhobel *m* <Bau> trying plane
Langhobelfräsen *n* <Maschinen> planer milling
Langholzwagen *m* <Eisenbahn> timber wagon
langhubig *adj* <Maschinen> long-stroke
Langkorn *n* <Metall> elongated grain
langkörniger Zuschlag *m* <Bau> elongated aggregate
langlebig *adj* <Lebensmittel> long-life
langlebiges Isotop *n* <Strahlphys> long-lived isotope
Langlebigkeit *f* 1. <Anstrich> longevity; 2. <Kunststoff> durability
Langleine *f* <Wassertrans> longline *(Fischerei)*
Langleitungseffekt *m* <Elektronik> long-line effect *(Mikrowellen)*
länglich *adj* <Geom> oblong
Langlichtbogen *m* <Fertig> long arc
Langloch *n* <Maschinen, Mechan> elongated hole
Langlochfräsen *n* <Fertig> slotting
Langlochfräser *m* 1. <Fertig> router; 2. <Maschinen> slot drill, slot mill
Langlochfräsmaschine *f* 1. <Fertig> slot milling machine; 2. <Maschinen> slot drilling machine
Langölalkydharz *n* <Kunststoff> long oil alkyd
längs *adv* <Maschinen> lengthways, lengthwise
Längs... <Elektronik, Wassertrans> longitudinal
Längsabweichung *f* <Lufttrans> longitudinal divergence
Längsachse *f* <Lufttrans, Maschinen> longitudinal axis
langsam drehender Elektromotor *m* <Elektrotech> low-speed (electric) motor
langsam erstarrendes Glas *n* <Ker & Glas> slow-setting glass
langsam laufender Dieselmotor *m* 1. <Trans, Wassertrans> slow-running diesel engine; 2. <Wassertrans> slow-running diesel engine, slow-speed diesel engine
langsam laufender Kompressor *m* <Hydraul> slow speed compressor
langsam voraus *adv* <Wassertrans> slow ahead *(Motor)*
langsam voraus *adv*/**ganz** <Wassertrans> dead slow ahead
langsam zurück *adv* <Wassertrans> slow astern *(Motor)*
langsam zurück *adv*/**ganz** <Wassertrans> dead slow astern
langsame Gärung *f* <Abfall> slow fermentation
langsame Kompostierung *f* <Abfall> slow composting
langsame Nullpunktsveränderung *f* <Metall> zero creep
langsame Welle *f* <Elektrotech> slow wave
langsames Abschrecken *n* <Metall> slow quenching

langsames Anfahren n <Nichtfoss Energ> soft-start
langsames Gefrieren n <Lebensmittel> slow freezing
langsames Modem n <Elektronik> low-speed modem
langsames Neutron n <Phys> slow neutron
Langsamfahrsignal n <Eisenbahn> speed restriction board, speed-restricting signal
Langsamfilter n <Wasserversorg> slow sand filter
Langsamkeit f <Trans> slowness
Langsamladung f <Elektrotech> trickle charge
Langsamläufer m <Wassertrans> low-speed engine; low-speed diesel engine *(Dieselmotor)*
Langsamsandfiltration f <Wasserversorg> slow sand filtration
Langsamwellenstruktur f <Elektrotech> slow wave structure
Längsanker m <Elektriz> axial armature
Längsanschlag m <Fertig> length stop
Längsbalken m <Bau> stringer
Längsband n <Wassertrans> tie plate *(Schiffbau)*
Längsbeanspruchung f <Metall> longitudinal stress
Längsbespannung f <Wassertrans> longitudinal framing *(Schiffbau)*
Längsbewehrung f <Bau> longitudinal reinforcement
Langschienen fpl <Eisenbahn> continuous-welded rail, ribbon rails
Langschwelle f <Eisenbahn> stringer
Längsdrehen n 1. <Fertig> straight turning; 2. <Maschinen> plain turning
Längsdruck m <Maschinen> thrust
Längsfaltung f <Konstzeich> Leporello folding, longitudinal folding
Längsfehler m <Raumfahrt> along-track error
Längsfilter n <Elektronik> longitudinal filter
Längsglied n <Lufttrans> longitudinal member
Längsholz n <Bau> runner
Langsieb n <Papier> fourdrinier
Langsiebpapiermaschine f <Papier> fourdrinier paper machine
Längskeil m <Fertig> machine key
Längskomponente f <Phys> longitudinal component
Längskugellager n <Maschinen> thrust ball-bearing
Längslager n 1. <Fertig> thrust bearing; 2. <Maschinen> thrust bearing, thrust block
Längslagestabilisierung f <Lufttrans> pitch attitude
Längsleistungstransistor m <Elektronik> series pass power transistor
Längslenker m <Kfztech> trailing arm *(Radaufhängung)*
Längsmagnetisierung f <Akustik> longitudinal magnetization
Längsneigungsmesser m **des Luftschraubenblattes** <Lufttrans> blade pitch indicator *(Hubschrauber)*
Längsneigungswinkel m <Lufttrans> pitch angle
längsnuten v <Fertig> spline
Längsparität f <Comp & DV> horizontal parity
Langspielband n <Aufnahme> long-play tape
Langspielplatte f (LP) <Aufnahme> extended-play record, long-playing record *(EP)*
Längspleiß m <Wassertrans> long splice *(Tauwerk)*
Längsreibemaschine f <Lebensmittel> conche
Längsrichtung f <Druck> grain, grain direction
Längsriss m 1. <Fertig> shear draft; 2. <Wassertrans> sheer drawing, sheer plan *(Schiffkonstruktion)*
längsschiffs adv <Wassertrans> fore and aft
Längsschiffsplan m <Wassertrans> profile *(Schiffkonstruktion)*
Längsschleifen n 1. <Maschinen> traverse grinding; 2. <Papier> longitudinal grinding
Längsschlitten m <Maschinen> longitudinal slide
Längsschlitz m <Telekom> longitudinal slot
Längsschneidemaschine f <Druck> slitter
Längsschnitt m <Bau, Eisenbahn> longitudinal section
Längsschnittlinie f <Wassertrans> buttock *(Schiffbau)*
Längsschruppen n <Fertig> straight rough turning
Längsschwingung f <Maschinen> longitudinal oscillation
längsseits adj <Wassertrans> alongside
längsseits adv <Wassertrans> alongside *(Festmachen)*
längsseits gehen v <Wassertrans> draw alongside; go alongside *(Schiff)*
längsseits verholen v <Wassertrans> haul alongside
Längsspannung f 1. <Elektriz> direct-axis component of voltage; 2. <Metall> line tension
Längsspant m 1. <Raumfahrt> stringer *(Raumschiff)*; 2. <Wassertrans> longitudinal *(Schiffkonstruktion)*
Längsspiel n <Kerntech> end float
Längsspuraufzeichnung f <Aufnahme> longitudinal recording
Längsstabilität f <Lufttrans> longitudinal stability
Längsstrahler m <Funktech> endfire antenna, endfire array *(längsstrahlende Dipolanordnung)*
Längsstrom m <Elektriz> direct-axis component of current
Längsströmung f <Ker & Glas> longitudinal current
Längssummenkontrolle f <Comp & DV> LRC, longitudinal redundancy check
Langstatorkabel n <Elektrotech> long-stator cable *(Magnetschwebetechnik)*
Langstatorwicklung f <Elektrotech> long-stator winding *(Magnetschwebetechnik)*
Längstband m <Bau> tie plate
langstielige Zange f <Kerntech> long-handed tongs
Längsträger m <Kfztech> side member *(Chassis, Fahrgestell)*
Längsträgerkoppelung f <Lufttrans> longitudinal beam coupler
Längstransistor m <Elektronik> series pass transistor
Langstreckenlinienflugzeug n <Lufttrans> long-haul airliner
Langstreckennavigationskette f (LORAN) <Lufttrans, Wassertrans> long-range navigation *(loran)*
Langstreckennetz n (WAN) <Comp & DV, Telekom> wide area network *(WAN)*
Langstreckenradarsystem n <Funkort> long-range accuracy radar system *(LORAC)*
Langstreckenverbindung f <Telekom> long-haul communication
Langstreckenverkehr m <Lufttrans> long-haul service
Längstrimmung f <Lufttrans> pitch trim
Längstwellenausbreitung f <Funktech> very long wave propagation, waveguide propagation of very long waves
Längstwellenfrequenz f <Funktech> very low frequency, VLF *(Myriameterwellen)*
Längsversteifung f <Raumfahrt> stringer *(Raumschiff)*
langwegiges Signal n <Elektronik> long-way signal
Langwelle f (LW) <Funktech> long wave *(LW)*
Langwellenausbreitung f <Funktech> ground wave propagation, long wave propagation
Langwellenfrequenz f <Funktech> low frequency, LF *(Kilometerwellen)*
langwelliges fernes Infrarot n <Phys> far infrared *(Strahlung)*
Langzeit... <Lebensmittel> long-life
Langzeitbatterie f <Elektrotech> long-life battery
Langzeitdurchschlagspannungsprüfung f <Elektrotech, Metrol> long-time breakdown test, long-time breakdown testing
Langzeitgedächtnis n (LZG) <Ergon, Künstl Int> long-term memory *(LTM)*
Langzeitkonstante f <Elektronik> long-time constant

Langzeitmessung

Langzeitmessung f <Gerät> long-term measurement
Langzeitmission f <Raumfahrt> long mission *(Raumschiff)*
Langzeitprüfung f <Elektriz> long-term test
Langzeitquotient m <Kerntech> endurance ratio
Langzeitrelais n <Elektrotech> long-time relay
Langzeitstabilität f <Elektronik> long-term stability
Langzeit-Straßenbelagverhalten n *(LTPP)* <Bau> long-term pavement performance, LTTP
Langzeittauchen n <Erdöl> saturation diving
Langzeittaucher m <Erdöl> saturation diver
Langzeitverhalten n <Kerntech> *(AE)* long-term behavior, *(BE)* long-term behaviour
LAN-LAN-Kopplung f <Telekom> LAN-LAN linking, local area network linking
LAN-LAN-Übergang m <Telekom> LAN-LAN transition, local to wide area network transition
LAN-LAN-Verbund m <Telekom> LAN-LAN interconnection, local area network interconnection
Lanosterin n <Chemie> lanosterol
LAN-Schnittstelle f <Telekom> LAN interface
Lanthan n *(La)* <Chemie> lanthanium *(La)*
Lanthanoid n <Chemie> lanthanoide
Lanze f <Maschinen> lance
Lanzette f <Fertig> slick, slicker *(Formen)*
Laplace'sche Gleichung f <Phys> Laplace's equation
Laplace'sche Transformation f <Elektronik, Math, Phys> Laplace transformation
Laplace'scher Operator m <Kerntech, Phys> Laplacian operator
Lappaconitin n <Chemie> lappaconitine
Läppaste f <Fertig> lapping compound
Läppblasen fpl <Ker & Glas> lap blisters
Läppdorn m <Fertig> cylindrical lap
Läppeinrichtung f <Maschinen> lapping fixture
lappen v <Fertig> tongue
Lappen m 1. <Fertig> lobe; tang *(Schaft)*; 2. <Mechan> flap
läppen v <Fertig, Mechan> lap
Läppen n 1. <Fertig> lapping; 2. <Hydraul> lap *(Feinstschleifen)*; 3. <Maschinen> lapping
lappenförmig abgesetzt adj <Fertig> tanged
lappenförmig absetzen v <Fertig> tang
Läppkäfig m <Fertig> cage *(Werkstückträger)*
Läppmarkierung f <Ker & Glas> lap mark
Läppmaschine f <Fertig, Maschinen, Mechan> lapping machine
Läpprippen fpl <Ker & Glas> lapping ribs
Läppwerkzeug n 1. <Fertig> lap; 2. <Maschinen> lapping tool
Laptop m <Comp & DV> laptop, laptop computer; notebook *(PC im Buchformat)*
L-Arabinose f <Chemie> pectinose
Larizin n <Chemie> coniferin
Lärm m 1. <Akustik, Ergon> noise; 2. <Sicherheit> noise *(Fabrik, Verkehr)*; 3. <Wellphys> noise
Lärm m **am Arbeitsplatz** <Sicherheit> workplace noise
Lärmabschirmfolie f <Sicherheit> noise-attenuating curtain
lärmabsorbierende Platte f <Sicherheit> noise-absorbent panel
Lärmabstrahlung f <Sicherheit> noise emission
Lärmanlage f <Sicherheit> noisy equipment
lärmarm adj <Sicherheit> low-noise, quiet, silent
lärmarme Anlage f <Sicherheit> low-noise equipment
lärmarme Ausrüstung f <Sicherheit> low-noise equipment
lärmarme Gestaltung f <Sicherheit> low-noise design
lärmarmer Betrieb m <Sicherheit> quiet running *(Maschine)*

lärmarmes Gerät n <Sicherheit> low-noise equipment
Lärmbekämpfung f <Bau, Sicherheit> noise abatement, noise control, noise suppression
Lärmbekämpfungstechnik f <Sicherheit> noise control engineering
Lärmbekämpfungszone f <Sicherheit> noise abatement zone
Lärmbelästigung f <Akustik, Sicherheit, Umweltschutz> noise nuisance, noise pollution, sound pollution
Lärmbeurteilungskurven fpl <Ergon, Sicherheit> noise rating curves
Lärmbewertung f <Sicherheit> noise rating
Lärmdämmeinrichtung f <Sicherheit> noise attenuation guard
Lärmdämmung f <Sicherheit> noise attenuation
Lärmdosimeter n <Sicherheit> noise dosimeter
Lärmdosispegel m <Sicherheit> noise exposure level, sound exposure level
Lärmemission f 1. <Kerntech> acoustic emission; 2. <Sicherheit> noise emission; noise emission value *(von Maschinen)*
Lärmemissionswert m <Sicherheit> noise emission value *(von Maschinen)*
Lärmexposition f <Ergon> noise exposure
lärmintensive Anlage f <Sicherheit> noisy equipment
lärmintensive Maschine f noisy machine
Lärmkapselung f <Sicherheit> noise enclosure
Lärmkennwerte mpl <Sicherheit> noise characteristics
Lärmkontrolle f <Sicherheit> noise survey *(in Betrieben)*
Lärmmessausrüstung f <Sicherheit> noise measuring equipment
Lärmmesseinrichtung f <Sicherheit> noise measuring equipment
Lärmmessgerät n <Sicherheit> noise-measuring instrument, noise-survey instrument, noise-survey meter
Lärmmessgeräte npl <Sicherheit> noise measuring equipment
Lärmmessung f <Sicherheit> noise survey *(in Betrieben)*
Larmor'sche Frequenz f <Phys> Larmor frequency
Larmor'sche Präzession f <Phys> Larmor precession
Lärmpegel m <Sicherheit> level of noise, noise level • **mit niedrigem Lärmpegel** <Sicherheit> low-noise
Lärmpegelüberwachung f <Sicherheit> noise level control, noise level monitoring
Lärmpegelüberwachungsgerät n <Sicherheit> noise monitoring instrument
Lärmprüfung f <Sicherheit> noise survey
Lärmquelle f <Umweltschutz> noise source
Lärmreduzierung f <Sicherheit> noise control
lärmschluckende Auskleidung f <Sicherheit> noise-absorbent lining
Lärmschutz m <Sicherheit> noise control, noise protection
Lärmschutzbereich m <Sicherheit> noise abatement zone
Lärmschutzhaube f <Sicherheit> noise-protective hood
Lärmschutzhelm m <Sicherheit> noise helmet, noise-protection helmet
Lärmschutzkabine f <Sicherheit> noise-isolated cabin, noise-protection booth, sound-insulated cabin, sound-proof booth, soundproof cabin
Lärmschutzkapsel f <Sicherheit> soundproof enclosure
Lärmschutzkapseln fpl <Sicherheit> noise-attenuating headset
Lärmschutzschirm m <Sicherheit> noise shield
Lärmschutzvorrichtung f <Maschinen> noise absorption device
Lärmschutzwand f <Bau, Sicherheit> noise barrier
Lärmschutzzaun m <Bau> acoustic fencing
Lärmschutzzone f <Sicherheit> noise abatement zone

Lärmschwerhörigkeit f <Ergon> noise-induced hearing loss
Lärmübertragungsweg m <Sicherheit> noise transmission path
Lärmüberwachungsgerät n <Sicherheit> noise monitoring instrument
Lasche f 1. <Bau> shackle, tongue; 2. <Eisenbahn> rail splice; 3. <Fertig> cover plate, fish plate, shin; link plate, sideplate *(Kette)*; 4. <Maschinen> clip, link plate, plate, strap; 5. <Mechan> lug, shackle, tab; 6. <Verpack> tab
• **Kleinserien mit Laschen versehen** <Verpack> batch tabbing
Laschenkette f <Maschinen> pitch chain, pitched chain, plate link chain, sprocket chain
Laschennietung f <Maschinen> butt-strap riveting
Laschenschraube f <Fertig> fish bolt
Laschenverband m <Eisenbahn> fishplating
Laschenverbindung f 1. <Bau> scarf, splice joint; 2. <Fertig> butt joint *(Nieten)*
Laschung f <Bau> scarf *(von zwei Holzstücken)*
Laser m *(Lichtverstärkung durch stimulierte Strahlungsemission)* <Comp & DV, Druck, Elektronik, Phys, Strahlphys> laser *(light amplification by stimulated emission of radiation)*
Laser m **mit kurzer Wellenlänge** <Elektronik> short wavelength laser
Laserabgleich m <Elektronik> laser trimming
Laserabstrahlung f <Elektronik> laser emission
Laseralarmempfänger m <Elektronik> laser warning receiver
Laseranemometer n <Gerät> laser anemometer
Laseranregung f <Phys, Strahlphys> laser excitation
Laserantrieb m <Raumfahrt> laser propulsion *(Raumschiff)*
Laseraufnahmekopf m <Optik> laser pick-up head
Laseraufzeichnung f <Optik> laser optic recording
Laseraugenschutz m <Sicherheit> laser protective eyewear
Laserband n <Optik> laser optic tape
Laserbandbreite f <Elektronik> laser bandwidth
Laserbearbeitung f <Mechan> laser machining
Laserbeleuchtung f <Elektronik> laser illumination
Laserbezeichnung f <Elektronik> laser designation
Laserbildplatte f <Optik> *(BE)* laservision disc, *(AE)* laservision disk, *(BE)* laservision videodisc, *(AE)* laservision videodisk
Laserbohrer m <Elektronik> laser drill
Laserbrennschneiden n <Fertig> laser gas-jet cutting
Laserbündel n <Elektronik> laser burst
Laser-CD f <Optik> *(BE)* laser videodisc, *(AE)* laser videodisk
Lasercode m <Elektronik> laser code
Laserdiode f <Elektronik, Optik, Phys, Strahlphys, Telekom> LD, laser diode
Laserdiode f **kleiner Leistung** <Telekom> low-power laser diode
Laserdiode f **mit Einfach-Heteroübergang** <Elektronik> single heterojunction laser diode
Laserdruck m <Optik> laser printing
Laserdrucker m <Comp & DV, Druck, Optik> laser printer
Laserdrucker/Kopierer m <Optik> laser printer-copier
Laserdüse f <Optik> laserjet®
Laserentfernungsmesser m <Elektronik> laser rangefinder
Laserentfernungsmessung f <Telekom> laser telemetry
Laserführung f <Elektronik> laser guidance
Laserfusion f <Kerntech> laser fusion
Laserfusionshohlraum m <Elektronik> laser cavity
lasergeführt adj <Elektronik> laser-guided

lasergesteuerte Maschine f <Bau> laser-controlled machine
Laserglas n <Ker & Glas> laser glass
Laserglühen n <Elektronik> laser annealing
Laserhohlraum m <Elektronik> laser cavity
Laserimpuls m <Elektronik> laser pulse
laserinduzierte Fusion f <Kerntech> laser-driven fusion
Laserinterferometer n <Elektronik> laser interferometer
Laserjet® m <Optik> Laserjet®
Laserkalibrierung f <Elektronik> laser trimming
Laserkarte f <Optik> lasercard
Laserkommunikation f <Elektronik> laser communications
Laserkopf m <Optik> laser head
Laserkreisel m <Raumfahrt> laser gyro *(Raumschiff)*
Laserlichtschranke f <Sicherheit> laser barrier, laser-beam safety device
Lasermechanismus m <Optik> laser mechanism
Lasermedium n 1. <Elektronik> laser medium; 2. <Optik> laser medium, laser optic medium; 3. <Telekom> laser medium
Lasermessgerät n <Metrol> laser measuring instrument
Lasern n <Elektronik> laser action, lasing
Lasernachführer m <Elektronik> laser tracker, laser tracking
Laseroberflächenhärten n <Fertig> laser surface hardening
Laseroptikdiskette f <Optik> *(BE)* laser optic disc, *(AE)* laser optic disk
laseroptische Karte f <Optik> laser optic card
Laserplatte f <Comp & DV> optical disk
Laserscheibe f <Optik> *(BE)* laser disc, *(AE)* laser disk
Laserschmelzen n <Elektronik> laser melting
Laserschneiden n 1. <Bau, Elektronik> laser beam cutting; 2. <Maschinen> laser beam cutting, laser cutting
Laserschreiben n <Elektronik> laser scribing
Laserschreibstift m <Elektronik> laser scriber
Laserschutzbrille f <Sicherheit> laser eye-protection goggles
Laserschutzgerät n <Sicherheit> laser protective device
Laserschweißen n <Bau, Elektronik> laser welding
Laserschwelle f <Optik, Telekom> lasing threshold
Lasersensor m <Elektronik> laser sensor
Lasersondemassenspektographie f <Kerntech> laser probe mass spectrography
Laserspeicher m <Comp & DV> photodigital memory
Laserspektroskopie f <Phys, Strahlphys> laser spectroscopy
Laserstrahl m 1. <Elektronik, Kerntech> laser beam; 2. <Phys, Strahlphys> laser beam, laser light beam; 3. <Telekom> laser beam
Laserstrahl m **mit hoher Strahlungsdichte** <Elektronik> high-irradiance laser beam
Laserstrahlaufzeichnung f <Comp & DV, Fernseh> laser beam recording
Laserstrahlbearbeitung f <Fertig> laser-beam machining
Laserstrahl-Energie f <Elektronik> laser beam energy
Laserstrahlmodulation f <Elektronik> laser beam modulation
Laserstrahlschweißen n <Bau> laser beam welding
Laserstrahlung f <Elektronik, Phys, Strahlphys> laser radiation
Lasersublimationsschneiden n <Fertig> laser sublimation cutting
Lasertätigkeit f <Elektronik> lasing
Lasertechnik f <Fertig> laser engineering; laser technology
lasertechnische Schutzvorrichtung f <Sicherheit> laser engineering safeguard

Lasertransferdruck

Lasertransferdruck m <Druck> laser transfer printing
Lasertrennen n <Bau> laser cutting
Laserübergang m <Phys, Strahlphys> laser transition
Laserüberwachungssystem n <Phys, Strahlphys> laser monitoring system
Laserwaffe f <Elektronik> laser weapon
Laserwendeanzeiger m <Phys, Strahlphys> laser gyro
LASH-Schiff n (Leichtertransporter, Leichterträgerschiff) <Wassertrans> LASH carrier *(lighter aboard ship carrier)*
Last f 1. <Elektriz> load; 2. <Hydraul> load *(Ventil)*; 3. <Maschinen> load, weight; 4. <Mechan, Metall, Telekom> load • **unter Last** 1. <Kfztech> under load *(Motor)*; 2. <Maschinen> under load • **unter Last laufen** <Maschinen> run under load
lastabhängige Bremsung f <Trans> load-sensitive braking
Lastabschaltung f <Elektriz> load shedding
Lastabwurf m <Elektriz, Elektrotech> load loss, load shedding
Lastannahme f <Fertig> design load
Lastannahmen fpl <Qual> design loadings
Lastbegrenzer m <Sicherheit> load-arresting device
Lastbegrenzung f <Elektrotech> load limiting
Lastcharakteristik f <Elektrotech> load characteristic
Lastdauerkurve f <Elektriz> load duration curve
Lastenaufzug m <Trans> *(AE)* elevator, *(BE)* lift, hoist
Lastenbeförderung f **außerhalb des Hubschraubers** <Lufttrans> external load carrying
Lastenhebemagnet n <Trans> lift magnet
Lastenmaßstab m <Wassertrans> dead-weight scale *(Schiffkonstruktion)*
Laserstrahleinstellung f <Fertig> laser alignment
Lastfaktor m <Heiz & Kälte, Trans> load factor
lastfreie Prüfung f <Elektriz> no-load test
lastfreier Betrieb m 1. <Elektrotech> idle operation, no-load operation, off-load operation; 2. <Kfztech> off-load operation *(Motor)*
Lastfrequenzregelung f <Elektriz> load frequency control
Lastfuhrwerk n <Trans> *(AE)* cart, *(BE)* truck
Lasthaken m 1. <Bau> lifting hook; 2. <Lufttrans> load hook up *(Hubschrauber)*; 3. <Maschinen> lifting hook
Lasthebegerät n <Maschinen> lifting device
Lastigkeit f <Wassertrans> trim *(Schiff)*
Lästigkeitspegel m <Sicherheit> perceived noise level
Lastimpedanz f <Elektrotech, Phys, Telekom> load impedance
Last-in First-out adj (LIFO) <Comp & DV> last-in-first-out (LIFO)
Lastinduktanz f <Elektrotech> load inductance
Lastkahncontainer m <Trans> barge container
Lastkapazität f <Elektrotech> load capacitance; input capacitance *(bei Messsonden)*
Lastkennlinie f <Elektrotech> load characteristic
Lastkette f 1. <Maschinen> lifting chain, load chain; 2. <Sicherheit> load chain *(Hebevorrichtung)*
Lastkommutierung f <Elektronik> commutation load
Lastkraftwagen m *(Lkw)* <Kfztech> heavy goods vehicle, *(BE)* lorry, *(AE)* trailer, *(AE)* truck *(HGV)*
Lastkraftwagen-Pooling n <Kfztech> *(BE)* lorry pooling, *(AE)* truck pooling
Lastkurve f <Elektriz> load curve
Last-Leer-Verhältnis n <Kfztech> load no-load ratio
Lastminderung f <Elektrotech> load derating
Lastöse f <Maschinen> clevis
Lastösenbolzen m <Maschinen> clevis pin
Lastprüfung f <Elektriz> load test
Lastregler m <Nichtfoss Energ> load controller
Lastschalter m <Elektriz, Elektrotech> air-break switch, load switch, on-load switch, power circuit breaker, switch

Lastschwankung f <Elektriz> load fluctuation
Lastseil n 1. <Eisenbahn> fall rope; 2. <Fertig> hoisting rope
Lastspannung f <Elektriz> on-load voltage
Lastspiel n <Fertig> stress cycle
Lastspielzeit f <Fertig, Qual> endurance
Lastspitze f <Elektriz, Elektrotech> load peak, peak load
Lastspule f <Elektriz, Elektrotech> loading coil
Laststoß m <Elektriz> load impulse, load pulse
Laststromkreis m <Elektriz> load circuit
Lastteilungssystem n <Telekom> load-sharing system
lasttragende Konstruktion f <Sicherheit> load-bearing structure
Lasttransfer m <Telekom> load transfer
Lasttrennschalter m <Elektrotech> load-break switch, load-interrupter switch, switch disconnector
Lasttrennschalter m **mit Sicherungen** <Elektrotech> switch-disconnector-fuse
Lastübergabe f <Elektriz> load transfer
Lastverlauf m <Elektriz> load variation
Lastverlaufkurve f <Elektriz> load duration curve
Lastverteilung f <Bau, Lufttrans, Maschinen> load distribution
Lastverteilungsliste f <Lufttrans> load distribution manifest
Lastverteilungsmanifest n <Lufttrans> load distribution manifest
Lastverteilungsplatte f <Bau> base plate
Lastvielfaches n <Trans> load factor
Lastwagentransport m <Trans> *(AE)* trucking
Lastwechsel m <Fertig> alternation of stress, stress cycle
Lastwechselprüfung f <Metrol> endurance test
Lastwiderstand m 1. <Elektriz, Elektrotech> load resistor *(Bauteil)*; 2. <Phys> load resistance
Lastwinkel m <Elektriz> load angle
Lastzug m 1. <Eisenbahn> trailer train; 2. <Kfztech> *(BE)* articulated lorry, *(AE)* articulated truck
Lasur f <Kunststoff> scumble
Latensifikation f <Foto> latensification
latente Ausdehnungswärme f <Phys, Thermod> latent heat of expansion
latente Kristallisationswärme f <Phys, Thermod> latent heat of solidification
latente Schmelzwärme f 1. <Phys> latent heat of fusion; 2. <Thermod> effective latent heat of fusion, latent heat of fusion
latente Umwandlungswärme f <Phys, Thermod> latent heat of transformation
latente Verdampfungswärme f 1. <Phys> latent heat of evaporation, latent heat of vaporization; 2. <Thermod> latent heat of vaporization, latent heat of evaporation
latente Verdichtungswärme f <Phys, Thermod> latent heat of compression
latente Wärme f <Bau, Erdöl, Heiz & Kälte, Phys, Thermod> latent heat
latente Wärmelast f <Heiz & Kälte> latent heat load
latentes Bild n <Foto, Optik, Phys> latent image
Latentmodul n <Metall> latent modulus
Latenzzeit f 1. <Comp & DV> latency; 2. <Ergon> latency, latency period; 3. <Optik> latency
lateral adj <Ergon> lateral
laterale Dominanz f <Ergon> lateral dominance
laterale Hemmung f <Ergon> lateral inhibition
laterale Plasmaabscheidung f <Telekom> lateral plasma deposition
Lateralität f <Ergon> handedness
Lateralschwerpunkt m <Wassertrans> *(AE)* center of lateral resistance, *(BE)* centre of lateral resistance *(Schiffkonstruktion)*

Lateralsystem n <Wassertrans> lateral system
Lateraltransistor m <Elektronik> lateral transistor
Laterit m <Bau> laterite *(Geologie)*
lateritischer Boden m <Bau> lateritic soil
Laterne f <Eisenbahn> lamp
Latex m <Kunststoff, Textil> latex
Latexschaum m <Kunststoff> latex foam
Latexuntergrund m <Textil> latex backing
Latourmotor m <Metrol> repulsion motor with fixed double set of brushes
Latte f 1. <Bau> lath *(Holzbau)*; 2. <Maschinen> batten
Lattenholz n <Bau> lathwood
Lattenstift m <Bau> lath nail
Lattentasche f <Wassertrans> batten pocket *(Segeln)*
Lattenzaun m <Bau> paling
Laubbaum m <Bau> hardwood
Laubholz n <Papier> hardwood
Laubsäge f <Maschinen> scroll saw
Laudanidin n <Chemie> laudanine
Laudaninmethylether n <Chemie> laudanosine
Laudanosin n <Chemie> laudanosine
Laue'sches Diagramm n <Strahlphys> Laue pattern
Lauf m 1. <Comp & DV, Kfztech> running; 2. <Maschinen> action, operation, running, working; 3. <Textil> running
Laufbahn f <Maschinen> runway
Laufbandflachsauger m <Papier> moving-belt flat box
Laufboden m <Foto> baseboard
Laufbohrung f <Fertig> barrel bore *(Gewehr)*
Laufbrett n <Bau> duck board; toeboard *(Gerüst)*
Laufbrücke f <Wassertrans> flying bridge
Laufbuchse f <Eisenbahn> bushing
Laufbüchse f <Maschinen> bushing
Laufdecke f <Kfztech> cover *(amerikanischer Reifentyp)*
Laufeigenschaften fpl <Fertig> eccentricity
laufen v 1. <Comp & DV> run *(Programm)*; 2. <Maschinen> run
laufen lassen v <Comp & DV> run *(Programm)*
Laufen n <Kunststoff, Papier> running
laufende Bauarbeiten fpl <Bau, Kohlen> construction work in progress, CWIP, work-in-progress
laufende Kontrolle f <Fertig> in-process testing
laufende Maschine f/unter Last <Maschinen> loaded machine
laufende Nummer f <Foto> serial number
laufende Wartung f <Heiz & Kälte> routine maintenance
laufender Block m <Maschinen> running block
laufender Text m <Druck> running text
laufendes Gespräch n **in Vermittlung** <Telekom> switching call-in-progress
laufendes Gut n <Wassertrans> running gear, running rigging *(Tauwerk)*
Läufer m 1. <Elektriz, Elektrotech> rotor; 2. <Kfztech> rotor *(umlaufender Teil eines Generators)*; 3. <Maschinen> *(AE)* traveler; 4. <Mechan> runner; 5. <Textil> traveller
Läufer m **mit erhöhtem Widerstand** <Elektriz> increased resistance rotor
Läufer m **ohne Schleifring** <Elektrotech> nonwound rotor, rotor without slip-ring
Läuferanlasser m <Elektriz> rotor starter
Läuferbalken m <Ker & Glas> runner bar
Läuferbildung f <Kunststoff> curtaining, sagging
Läufereisen n <Elektrotech> armature core
Läufereisenverlust m <Elektrotech> armature core loss
Läuferfeld n <Elektriz> rotor field
Läuferkappe f <Elektrotech> rotor end-winding retaining ring
Läufernabe f <Elektrotech> rotor hub, spider
Läuferpaket n <Elektrotech> armature core
Läuferschicht f <Bau> stretching course
Läuferstein m <Bau> stretcher

Läuferverband m <Bau> stretching bond; running bond *(Mauerwerk)*
Läuferwelle f <Elektriz> rotor shaft
Läuferwicklung f <Elektrotech> rotor winding
Lauffähigkeit f **von Papier** <Papier> runnability of paper
Lauffläche f 1. <Eisenbahn> running surface; tread *(von Rädern)*; 2. <Maschinen> face
Laufgang m 1. <Eisenbahn> gangway; 2. <Kerntech> gallery, gangway; 3. <Mechan> catwalk; 4. <Wassertrans> alleyway *(Schiffbau)*
Laufgestell n <Trans> *(BE)* bogie, *(AE)* bogie truck, trailer
Laufgewicht n <Maschinen> jockey weight, movable weight, slider, sliding weight
Laufgewichtswaage f 1. <Metrol> Roman steelyard; 2. <Phys> steelyard
Laufkarte f <Qual> routing card, *(AE)* traveler, *(BE)* traveller
Laufkatze f 1. <Bau> *(AE)* traveling winch, *(BE)* travelling winch; crab *(eines Krans)*; 2. <Fertig, Kerntech> crab; 3. <Wassertrans> *(AE)* traveling crab, *(BE)* travelling crab
Laufkraftwerk n <Nichtfoss Energ> run-of-river station
Laufkran m 1. <Bau> overhead crane, *(AE)* traveling crane, *(BE)* travelling crane; 2. <Maschinen, Mechan> *(AE)* traveling crane, *(BE)* travelling crane
Laufkranz m <Eisenbahn, Lufttrans> tread *(Fahrwerk)*
Laufkranztrommel f <Kohlen> trommel
Lauflänge f <Textil> yardage
Laufplanke f <Wassertrans> gangplank
Laufrad n 1. <Fertig> impeller *(Kunststoffinstallationen)*; 2. <Heiz & Kälte> impeller; 3. <Kfztech> impeller *(Pumpe)*; rotor *(einer Turbine)*; 4. <Kohlen> impeller; 5. <Maschinen> impeller, rotor, runner, running wheel; 6. <Nichtfoss Energ> runner, rotor *(Wasserturbine)*
Laufradblatt n <Hydraul> runner blade *(Kreiselpumpe, Turbine, Wasserrad)*
Laufradmantel m <Lufttrans> *(AE)* chine tire, *(BE)* chine tyre *(Motor)*
Laufradschaufel f 1. <Hydraul> runner vane; runner blade *(Kreiselpumpe, Turbine, Wasserrad)*; 2. <Maschinen> rotor blade
Laufraum m <Elektronik> drift space *(in Laufzeitröhren)*
Laufraumelektrode f <Elektronik> drift tunnel
Laufrichtung f 1. <Druck> grain; 2. <Papier> machine direction, making direction
Laufrille f 1. <Maschinen> ball race, ball ring, groove; 2. <Mechan> race
Laufring m <Fertig> raceway *(Lager)*
Laufrinne f <Fertig> launder *(flüssiges Metall)*
Laufrolle f 1. <Fertig> guiding roller; 2. <Maschinen> ball race, ball ring, bearing race, runner; 3. <Wassertrans> runner *(Tauwerk)*
Laufschaufel f <Nichtfoss Energ> runner blade
Laufschicht f <Maschinen> antifriction layer
Laufschiene f 1. <Eisenbahn> running rail, slide rail; monorail *(der Einschienenbahn)*; 2. <Maschinen> runner, running rail
Laufsetzstock m <Maschinen> *(AE)* traveling steadyrest, *(BE)* travelling steadyrest
Laufsitz m **(LS)** <Maschinen> running fit
Laufstatus m <Comp & DV> running state
Laufsteg m 1. <Bau> catwalk; 2. <Mechan> catwalk, walkway
Lauftest m <Maschinen> running test
Lauftisch m <Maschinen> *(AE)* traveling table, *(BE)* travelling table
Laufwagenzeichenmaschine f <Konstzeich> carriage-type drafting machine
Laufweg m <Mechan> walkway
Laufwerk n 1. <Akustik> tape deck; 2. <Comp & DV> disk drive, drive; 3. <Maschinen> carriage

Laufwerk

Laufwerk n **für beschreibbare CD** <Optik> writable optical drive
Laufwerk n **für beschreibbare Scheiben** <Optik> writable optical disk drive
Laufwerk n **für löschbare CD** <Optik> erasable optical drive
Laufwerk n **für löschbare Scheiben** <Optik> erasable disk drive
Laufzapfen m 1. <Fertig> journal *(Walze)*; 2. <Mechan> neck
Laufzeit f 1. <Comp & DV, Elektronik> run duration, run time, running duration, running time; time delay *(eines Signals)*; delay time; 2. <Ker & Glas> running time; 3. <Telekom> delay *(Laufzeitkette)*; time delay *(Signal)*
Laufzeit f **eines Patents** <Patent> term of patent
Laufzeitdiode f <Elektronik> transit time diode
Laufzeitentzerrer m <Elektronik> delay equalizer
Laufzeiterzeugung f <Elektronik> time delay generation
Laufzeitfehler m <Fernseh> velocity error
Laufzeitfilter n <Elektronik> transit time filter
Laufzeitfiltern n <Elektronik> transit time filtering
Laufzeit-Frequenzgang m <Elektronik, Funktech, Telekom> frequency response of delay
Laufzeitgenerator m <Elektronik> delay generator
Laufzeitgerät n <Elektrotech> transit time device
Laufzeitkette f 1. <Gerät> delay network; 2. <Telekom> delay circuit
Laufzeitleistung f <Comp & DV> run-time output
Laufzeitleitung f <Elektronik> delay-time line
Laufzeitoszillator m <Elektronik> velocity-modulated oscillator
Laufzeitröhre f <Elektronik> drift tube, velocity-modulated tube
Laufzeitschaltung f <Elektronik> time delay circuit
Laufzeitspeicher m <Kerntech> delay line storage
Laufzeitstrahl m <Elektronik> velocity-modulated beam
Laufzeitsystem n <Comp & DV> run-time system
Laufzeitversion f <Comp & DV> run-time version
Laufzeitverzerrung f 1. <Aufnahme> delay distortion; 2. <Elektronik> delay distortion, time delay distortion; 3. <Funktech, Kontroll> delay distortion
Laufzeitverzögerung f <Comp & DV> propagation delay
Lauge f 1. <Chemie> alkali; 2. <Fertig> lye
laugen v <Bau, Chemie, Kohlen, Meerschmutz> leach
Laugenbeständigkeit f <Werkprüf> caustic solution resistance *(Glas)*
Laugenfestigkeit f <Kunststoff> alkali resistance
Laugenlösung f <Chemtech> leach liquor
Laugenmesser m <Lebensmittel> alkalimeter
Laugenmessung f <Chemie> alkalimetry
Laugenprüfer m <Gerät> alkaline tester
Laugensalz n <Chemie> alkali
Laugensprödigkeit f <Fertig> caustic embrittlement
Laugerei f <Kohlen> leaching plant
Laugfähigkeit f <Kohlen> leachability
Laugflüssigkeit f <Kohlen> barren solution
Laugung f <Ker & Glas> leaching
Laugzeit f <Kohlen> leaching time
Laurin... <Chemie> lauric
Lauryl... <Chemie> lauryl
Laut m <Akustik> tone
Lautbildung f <Akustik> phonation
läuten v <Telekom> ring
läutern v <Fertig> wash
Läuterung f 1. <Chemtech> purification *(Flüssigkeiten)*; 2. <Ker & Glas> refining
Läuterungsmittel n <Chemtech> purifying agent
Läuterungszone f <Ker & Glas> refining zone
Lautheit f <Phys> loudness *(Stärke der Lautempfindung)*

Lautheitsdrosselung f <Akustik> partial masked loudness
Lauthören n <Telekom> open listening
Läutprobe f <Elektriz> ringing test
Lautsprecher m 1. <Akustik> loudspeaker, speaker; 2. <Aufnahme> loudspeaker; infinite-baffle loudspeaker *(mit unendlicher Resonanzwand)*; 3. <Elektriz, Funktech, Phys> loudspeaker, speaker; 4. <Telekom> loudspeaker
Lautsprecher m **mit Richtwirkung** <Akustik> directional loudspeaker
Lautsprecherdämpfung f <Aufnahme> loudspeaker damping
Lautsprechergehäuse n <Aufnahme> loudspeaker enclosure, loudspeaker housing
Lautsprecherkombination f <Aufnahme> composite loudspeaker
Lautsprechersäule f <Aufnahme> column loudspeaker
Lautsprechersystem n <Akustik, Aufnahme> loudspeaker system
Lautsprechertrichter m <Aufnahme> loudspeaker cone, loudspeaker horn
Lautsprecherwand f <Akustik> loudspeaker baffle
Lautstärke f 1. <Akustik> loudness level, volume; 2. <Akustik> loudness *(L)*; 3. <Aufnahme> loudness level, loudness, volume; 4. <Comp & DV> volume; 5. <Ergon> loudness, sound intensity; 6. <Phys> loudness level
Lautstärke f **der Sprache** <Akustik> volume of speech
Lautstärkeanzeige f <Aufnahme> volume indicator
Lautstärkebereich m <Aufnahme> dynamic range
Lautstärkeempfindung f <Akustik> loudness
Lautstärkeentzerrer m <Aufnahme> volume equalizer
Lautstärkefunktion f <Aufnahme> loudness function
Lautstärkekurve f <Akustik> loudness function
Lautstärkemesser m <Akustik> loudness meter, sound level meter
Lautstärkemessgerät n <Gerät> volume meter
Lautstärkemuster n <Aufnahme> loudness pattern
Lautstärkenbereich m <Akustik> volume range
Lautstärkepegel m <Akustik, Ergon> loudness level
Lautstärkeregler m <Aufnahme> volume control
Lautstärkeumfang m <Akustik> dynamic range
Lautstärkeunterschiedsschwelle f <Akustik> differential threshold of sound pressure level
lauwarm adj <Thermod> tepid
Lavagestein n <Nichtfoss Energ> extrusive rocks
Lavalier-Mikrofon n <Aufnahme> Lavalier microphone
Lävan n <Chemie> levan
Lavendelkopie f <Foto> lavender print
Lävulin n <Chemie> levulin, synanthrose
Lävulin... <Chemie> levulinic
Lävulosan n <Chemie> fructosan, levulosan
Lävulose f 1. <Chemie> fructose, levulose; 2. <Lebensmittel> *(BE)* laevulose, *(AE)* levulose
Lawine f <Elektrotech> avalanche
Lawinendiode f <Elektronik, Phys> avalanche diode
Lawinendurchbruch m <Elektriz, Elektronik> avalanche breakdown *(Mikroelektronik)*
Lawinendurchbruchspannung f <Elektriz> avalanche voltage
Lawinenlaufzeitdiode f *(Impatt-Diode)* <Elektronik, Funktech, Phys> avalanche transit-time diode *(impatt diode)*
Lawinenphotodiode f <Elektronik, Optik> avalanche photodiode
Lawinenverstärkung f <Elektronik> avalanche gain
Lawson n <Chemie> lawsone
Layout n 1. <Comp & DV> layout *(zweidimensionale Darstellung eines Teils)*; 2. <Druck, Verpack> layout
Layoutbeschreibung f <Comp & DV> layout description

LB 1. <Comp & DV, Telekom> *(Ortsbatterie)* LB *(local battery)*; 2. <Elektrotech> *(lokale Batterie)* LB *(local battery)*
L-Band n 1. <Elektronik> L-band *(390-1550 MHz)*; 2. <Raumfahrt> L-band *(Weltraumfunk)*; 3. <Wassertrans> L-band *(Satellitenfunk)*
LC *(Flüssigkristall)* <Comp & DV, Elektriz> LC *(liquid crystal)*
LCD *(Flüssigkristallanzeige)* <Comp & DV, Elektriz, Elektronik, Fernseh, Labor, Telekom, Thermod> LCD *(liquid crystal display)*
LCD-Anzeigetafel f <Elektronik> LCD panel
LCD-Modul n <Elektronik> LCD module
LC-Filter n <Elektronik, Funktech> LC filter
LD$_{50}$ *(mittlere letale Dosis)* <Strahlphys> LD$_{50}$ *(median lethal dose)*
LD-Beiwert m *(Gleit und Widerstandsbeiwert)* <Lufttrans> LD ratio, lift and drag ratio *(Nutzeffekt des Luftfahrzeugs)*
L-Dock n <Trans> offshore dock
Leak n **in der Rohrleitung** <Wasserversorg> piping seepage
LEAR *(Antiprotonenring mit geringer Energie)* <Teilphys> LEAR *(Low-Energy Antiproton Ring)*
lebender Kolumnentitel m <Druck> running head, running title
lebendes Werk n <Wassertrans> quickwork *(Schiffkonstruktion)*
Lebendmasse f <Nichtfoss Energ> biomass
Lebensdauer f 1. <Anstrich> pot life *(Mehrkomponentenlacken)*; 2. <Elektrotech> lifetime; 3. <Fertig> fatigue life; service life *(Kunststoffinstallationen)*; 4. <Funktech> service life; 5. <Kerntech> service life *(eines Nutzgerätes)*; 6. <Kohlen> service life; 7. <Maschinen> life cycle, working life; 8. <Telekom> lifetime, service life • **mit Lebensdauer** <Heiz & Kälte> sealed for life
Lebensdauer f **bei Gebrauch** <Foto> working life *(Lösung)*
Lebensdauer f **der Batterie-Schwebespannung** <Elektrotech> float life *(Akkumulatoren)*
Lebensdauer f **des Brennstoffes im Reaktorkern** <Kerntech> in-core fuel life
Lebensdauer f **des Minoritätsträgers** <Elektronik> bulk lifetime *(versenkter Kanal, Transistortechnik)*
Lebensdauererwartung f <Telekom> lifetime expectancy
Lebensdauerprüfmenge f <Qual> life test quantity
Lebensdauerprüfung f 1. <Elektrotech> life test; 2. <Metrol> endurance test; 3. <Qual> life test
Lebensdauerschmierung f <Maschinen> for-life lubrication, lifetime lubrication
Lebensdauerschnelltest m <Werkprüf> accelerated life test
Lebenserhaltungsgerät n <Raumfahrt> environmental control system *(Raumschiff)*
Lebenserwartung f <Bau, Telekom> life expectancy
Lebensmittelbedarf m <Lebensmittel> food requirements
Lebensmittelbestrahlung f <Verpack> irradiation of food
Lebensmittelchemie f <Lebensmittel> food chemistry
Lebensmittelfarbstoff m <Lebensmittel> *(AE)* food color, *(BE)* food colour
lebensmittelgerechte Frischhaltefolienverpackung f <Verpack> food-grade packaging film
Lebensmittelgesetz n <Lebensmittel> food law
Lebensmittelkonservierung f <Lebensmittel> food preservation
Lebensmittelüberwachung f <Lebensmittel> food inspection
Lebensmittelverarbeitungsanlage f <Lebensmittel> food-processing plant
Lebensmittelvergiftung f <Lebensmittel> food poisoning
Lebensmittelverpackung f <Verpack> food packaging

Lebensmittelverpackungsmaschinen fpl <Verpack> food-wrapping machinery
Lebensmittelwissenschaft f <Lebensmittel> food science
Lebensmittelzusatzstoff m <Lebensmittel> food additive
Lebensraum m **im Wasser** <Wasserversorg> aquatic system
Lebensrettungsgerät n <Sicherheit> life-saving apparatus
Leberstärke f <Lebensmittel> animal starch, glycogen
lebhaft adj <Textil> bright *(Farbe)*
Leblanc'sche Schaltung f <Elektriz> Leblanc connection
Lecherleitung f <Funktech> Lecher line
Lecithin n <Lebensmittel> lecithin
leck adj <Wassertrans> leaky
leck werden v <Wassertrans> make water *(Schiff)*
Leck n 1. <Elektriz, Ker & Glas> leak; 2. <Nichtfoss Energ> leak, leakage; 3. <Phys, Sicherheit, Wassertrans> leak
• **ein Leck stopfen** 1. <Wassertrans> plug a leak; 2. <Wasserversorg> stop a leak
Leckage f 1. <Chemie> ullage; 2. <Lebensmittel> ullage *(bei Behältern)*; 3. <Meerschmutz, Nichtfoss Energ, Wasserversorg> leakage
Leckalarm m <Sicherheit> leakage warning
Leckauffanggefäß n <Kerntech> leakage interception vessel
Leckbestimmung f <Abfall> leak detection
leckdicht adj <Mechan> leak-tight
Leckdichtheit f <Kerntech> leak tightness *(des Druckgefäßes)*
Leckdruck m <Erdöl> leak-off pressure *(Bohrtechnik)*
lecken v <Bau, Sicherheit, Wassertrans> leak
Leckfeld n <Elektriz> stray field
Leckgeschwindigkeit f <Kerntech> leak rate
Leckluftlüfter m <Heiz & Kälte> make-up fan
Leckmesser m <Elektriz> leakage meter
Leckmessgerät n <Elektriz> leakage meter
Leckmodus m <Telekom> leaky mode, *(AE)* tunneling mode, *(BE)* tunnelling mode
Leckortungsgerät n <Heiz & Kälte> leak detector
Leckrate f <Kerntech> leak rate
Lecksicherung f <Wassertrans> damage control
Leckspürgerät n <Gerät> leak detector
Leckstrahl m <Telekom> leaky ray, *(BE)* tunnelling ray
Leckstrom m <Elektriz, Elektrotech, Telekom> leakage current
Leckstromkorrosion f <Elektrotech> leakage current corrosion
Lecksucher m <Gerät, Verpack> leak detector
Lecksuchgerät n <Heiz & Kälte> leak detector
Leckverlust m <Elektriz> leakage loss
Leckwarnsystem n <Sicherheit> leakage indicator system
Leckwasserpumpe f <Kerntech> leakage water pump
Leckweg m <Elektriz> leakage path
Leckwiderstand m <Phys> leakage resistance
Leclanché-Element n <Elektriz, Elektrotech, Labor> Leclanché cell
LED *(Leuchtdiode, Lumineszenzdiode, lichtemittierende Diode)* <Comp & DV, Elektriz, Elektronik, Fernseh, Optik, Phys, Telekom> LED *(light-emitting diode)*
Ledeburit m <Metall> ledeburite
Leder n <Foto, Textil> leather
Lederbalgen m <Foto> leather bellows
Lederdichtungsscheibe f <Fertig> leather gasket
Lederetui n <Foto> leather case
Lederfutteral n <Foto> leather case
Lederhobel m <Bau> spokeshave
Ledermanschette f <Fertig> leather packing
Ledermanschettendichtung f <Fertig> leather cup

Lederschürze

Lederschürze f <Sicherheit> leather apron
Lederstulpenhandschuh m <Sicherheit> leather gauntlet glove
Ledertreibriemen m <Fertig, Maschinen> leather belt
Lederzurichtung f <Fertig> leather finishing
Lee f <Wassertrans> lee
Leeküste f <Wassertrans> lee shore
leer adj <Comp & DV> blank
leer laufen v 1. <Fertig> loaf; 2. <Fertig> idle; 3. <Maschinen> run light, run on no load; 4. <Mechan> idle; 5. <Wasserversorg> drain
leer laufend adj 1. <Elektriz> idling; 2. <Maschinen> light-running
leer pumpen v <Wasserversorg> pump out
Leerbefehl m <Comp & DV> do-nothing instruction
leere Anweisung f <Comp & DV> dummy instruction, null instruction
leere Kamera bedienen v <Foto> operate empty camera (Auslöser-Test)
leere Menge f <Math> empty set
leere Zeichenfolge f <Comp & DV> null string
leeren v <Trans> dump
Leeren n <Metall> exhaustion
leerer Bereich m <Comp & DV> sparse array
leerer Datenträger m <Comp & DV> blank medium, empty medium
leeres Band n <Phys> empty band (Festkörperphysik)
leeres Magnetband n <Fernseh> blank magnetic tape
leeres Schriftelement n <Comp & DV> empty font
Leerfahrt/auf <Eisenbahn, Wassertrans> running light
Leergewicht n 1. <Kfztech> unladen weight; 2. <Mechan> dead weight
Leergut n <Verpack> empties
Leerhub m <Maschinen> idle stroke
Leerkarte f <Wassertrans> plotting chart, plotting sheet (Navigation)
Leerlauf m 1. <Elektrotech> idle running, idling, no-load, no-load operation, no-load running, open-circuit, open-circuit operation; 2. <Fertig> free motion, loafing; 3. <Kfztech> idle running, idle speed, idling (Motor); 4. <Maschinen> idle speed, no-load, no-load operation, no-load running, running without load; 5. <Mechan> idler • **im Leerlauf** <Maschinen> on no load • **im Leerlauf fahren** <Kfztech> coast • **im Leerlauf laufen** <Kfztech> idle (Motor)
Leerlauf... <Fertig> no-load
Leerlaufbetrieb m 1. <Elektriz> open circuit operation, idle mode; 2. <Kontroll> idle mode; 3. <Lufttrans> no-load operation
Leerlaufbetriebszustand m **beim Anflug** <Lufttrans> approach idling conditions
Leerlaufcharakteristik f <Lufttrans> no-load characteristic
Leerlaufdrehzahl f 1. <Elektrotech> idling speed, no-load speed; 2. <Mechan> idling speed
Leerlaufdrosselanschlag m <Lufttrans> idle throttle stop
Leerlaufdüse f <Kfztech> idle jet (Vergaser)
Leerlaufeigenschaften fpl <Elektriz> idle-load conditions, open circuit characteristics
Leerlaufen n <Maschinen> running light, running on no load
Leerlaufenergie f <Thermod> waste energy
Leerlaufimpedanz f 1. <Akustik> blocked electrical impedance; 2. <Elektriz, Elektrotech> open circuit impedance
Leerlaufkennlinie f <Elektriz, Elektrotech> no-load characteristic, open-circuit characteristic
Leerlaufkomponente f <Elektriz> idle component
Leerlauf-Kurzschluss-Verhältnis n <Elektriz, Elektrotech> short-circuit ratio
Leerlaufladung f <Elektrotech> floating charge

Leerlaufleistung f <Maschinen> idle power, idling power
Leerlaufmessung f <Elektriz> no-load testing, open-circuit test
Leerlaufmodus m <Kontroll> idle mode
Leerlaufprüfung f <Elektriz> idle-load test, no-load testing, open-circuit test
Leerlaufrolle f <Fertig> idler
Leerlaufschaltung f <Kohlen> open circuit
Leerlaufscheinwiderstand m <Elektrotech> open circuit impedance
Leerlaufspannung f <Elektriz, Elektrotech, Nichtfoss Energ> open circuit voltage
Leerlaufstellschraube f <Kfztech> idle adjustment screw (Vergaser)
Leerlaufstellung f <Kfztech> neutral position; neutral (Getriebe)
Leerlaufstrom m 1. <Comp & DV> idle current (Datenverarbeitung); 2. <Elektriz> idle current, open circuit current
Leerlaufversuch m <Elektrotech> open-circuit test (als Generator); no-load test (als Motor)
Leerlaufwiderstand m <Elektrotech> open circuit impedance
Leerlaufzeit f <Comp & DV> idle time
Leerlaufzustand m <Comp & DV, Elektrotech> idle condition, open circuit condition
Leerliste f <Comp & DV> empty list
Leerlokomotive f <Eisenbahn> light engine, light locomotive
Leerraumkoeffizient m <Kerntech> void coefficient (der Reaktivität)
Leerrückfahrt f <Trans> deadhead
Leerschritt m <Comp & DV> blank
Leerstelle f 1. <Comp & DV> blank, space; 2. <Phys> vacancy; 3. <Telekom> blank
Leerstellendiffusion f <Metall> vacancy diffusion
Leerstellenwanderung f <Metall> vacancy migration
Leerstring m <Comp & DV> empty string
Leertaste f <Comp & DV> blank key, space bar
Leertrum n <Fertig> nondriving free length (Kette)
Leeruder n <Wassertrans> lee helm
Leerwagensammelgleis n <Eisenbahn> empties siding
Leerzeichen n 1. <Comp & DV> blank, filler, idle character, ignore character, space, space character; 2. <Telekom> blank, idle character
Leerzeicheneinfügung f <Comp & DV, Telekom> idle insertion
Leerzeichentaste f <Comp & DV> blank key, idle character key, space key
Leerzeile f <Comp & DV> null line
leewärtig adj <Wassertrans> leeward
leewärts adv <Wassertrans> alee, leeward
Legacy-Code m <Comp & DV> legacy code
Legacy-System n <Comp & DV> legacy system
Legel m <Wassertrans> cringle (Segeln)
legen v <Telekom> lay
legen v/sich <Wassertrans> drop (Wind)
Legende f <Konstzeich> caption, inscription (Bild); key (Zeichenerklärung)
Legendre-Polynom n <Phys> Legendre polynomial
Legerwall m <Wassertrans> lee shore
legierbar adj <Fertig> alloyable (Metall)
Legierbarkeit f <Fertig> alloyability (Metall)
legieren v 1. <Fertig> alloy (Stahl); 2. <Metall> alloy
Legieren n <Metall> alloyage, alloying
legierte Diode f <Elektronik> alloy diode (Mikroelektronik)
legierter Grauguss m <Fertig> alloy cast iron
legierter Stahl m <Fertig, Metall> alloy steel

legierter Übergang m <Elektronik> alloy junction *(Halbleiter)*
legierter Zonenübergang m <Elektronik> alloyed junction *(Halbleiter)*
legiertes Karbid n <Metall> alloy carbide
legiertes Öl n <Fertig> doped oil
Legierung f 1. <Anstrich> alloy; 2. <Metall> alloyage, alloying, alloy; 3. <Papier> alloy
Legierung f für das Weichlöten <Maschinen> soft solder alloy
Legierungsbeschichtung f <Metall> alloy plating
Legierungsbestandteil m <Fertig> alloy component
Legierungselement n <Metall> alloying element
Legierungsgehalt m <Fertig> alloy content
Legierungsgrundgefüge n <Fertig> alloy matrix
Legierungsmetall n <Metall> steel alloy
Legierungsmethode f <Elektronik> alloying method
Legierungsplattierschicht f <Fertig> alloy cladding
Legierungstransistor m <Elektronik> alloy transistor, alloyed junction transistor, fused junction transistor *(Mikroelektronik)*
Legierungszusatz m <Fertig> alloy addition
Leguminose f <Lebensmittel> legume
Lehm m 1. <Bau> loam, pug; 2. <Fertig> loam; 3. <Hydraul> pug; 4. <Kohlen> clay, loam; 5. <Wasserversorg> loam
Lehmbaustein m <Bau> sun-dried brick
Lehmform f <Fertig> *(AE)* loam mold, *(BE)* loam mould
Lehmgehalt m <Bau, Kohlen> clay content
Lehmmörtel m <Ker & Glas> clay mortar
Lehmwall m <Abfall> clay barrier
Lehrdorn m <Maschinen, Metrol> *(AE)* plug gage, *(BE)* plug gauge
Lehre f 1. <Bau> strickle; 2. <Elektriz> *(AE)* gage, *(BE)* gauge; 3. <Erdöl> template, templet; 4. <Fertig, Gerät, Kfztech> *(AE)* gage, *(BE)* gauge *(Werkzeug)* • **mit Lehre messen** <Maschinen> *(AE)* gage, *(BE)* gauge
Lehre f der Weltentstehung <Raumfahrt> cosmology
Lehre f von den Kegelschnitten <Geom> conics
lehrenbohren v <Fertig> jig-bore
Lehrenbohren n <Fertig, Maschinen> jig boring
Lehrenbohrgerät n <Maschinen> jig boring tool
Lehrenbohrmaschine f 1. <Maschinen> jig borer, jig boring machine; 2. <Mechan> jig borer
Lehrenbohrwerk n <Fertig> jig borer
Lehrenform f <Fertig> template, templet
Lehrenschleifen n <Maschinen> jig grinding
Lehrenschleifmaschine f <Maschinen> jig grinder
Lehrgerüst n <Bau> *(AE)* center, *(BE)* centre *(eines Mauerwerksbogens)*
Lehrrad n <Fertig> gear master
Lehrring m 1. <Maschinen> *(AE)* female gage, *(BE)* female gauge, *(AE)* ring gage, *(BE)* ring gauge; 2. <Metrol> *(AE)* ring gage, *(BE)* ring gauge
Lehrsatz m <Math, Phys> theorem
Lehrsatz m von Pythagoras <Geom, Math> Pythagorean theorem
Leibung f <Bau> jamb, reveal *(Fenster, Tür)*
Leiche f <Druck> out *(Jargon)*
Leichengift n <Chemie> ptomaine
leicht adj <Mechan> lightweight
leicht angereicherter Halbleiter m <Elektronik> lightly-doped semiconductor
leicht entflammbar adj <Thermod> flammable, inflammable
leicht entflammbare Flüssigkeit f <Sicherheit> highly inflammable liquid
leicht entzündlicher Stoff m <Sicherheit> highly inflammable material

leicht salzig adj <Lebensmittel> brackish
leicht schmelzbar adj <Thermod> fusible
Leichtanschlusssteckverbinder m <Elektrotech> low-insertion-force connector
Leichtbauschnelltriebzug m <Eisenbahn> LRC, light-rapid-comfortable
Leichtbenzin n 1. <Erdöl> light distillates *(Destillationsprodukt)*; 2. <Kfztech> *(AE)* light gasoline, *(BE)* light petrol
Leichtbeton m <Bau> breeze concrete, lightweight concrete
leichte Bearbeitbarkeit f <Mechan> ease of machining
leichte Behandelbarkeit f <Mechan> ease of machining
leichte Bespulung f <Elektrotech> light loading *(Pupinisierung des Kabels)*
leichte Fraktionen f <Erdöl> light fractions
leichte Kohlenwasserstofffraktionen fpl <Erdöl> light hydrocarbon fractions
leichte Schnitte mpl <Erdöl> light fractions
leichte Schweißnaht f <Kerntech> concave weld, light weld
leichte Strahlung f <Kerntech> light radiation
leichte Wartung f <Mechan> ease of maintenance
leichter Beobachtungshubschrauber m <Lufttrans> light observation helicopter
leichter Mehrzweckhubschrauber m <Lufttrans> light multirole helicopter
leichter Nebel m <Umweltschmutz> mist
leichter Wind m <Wassertrans> light airs
Leichter m 1. <Meerschmutz> lighter; 2. <Wassertrans> barge *(Schifftyp)*; lighter *(Schiff)*
Leichter m ohne Eigenantrieb <Wassertrans> dumb barge
Leichtern n <Meerschmutz> lightening *(Ölabgabe von Schiff zu Schiff)*
Leichterschiff n <Trans> lighter carrier
Leichterträgerschiff n *(LASH-Schiff)* <Wassertrans> lighter aboard ship carrier, LASH carrier *(Schiff mit Schlepper an Bord)*
Leichtertransport m <Wassertrans> lighterage *(Handel)*
Leichtertransporter m *(LASH-Schiff)* <Wassertrans> lighter aboard ship carrier, LASH carrier *(Schiff)*
leichtes Ein- und Ausknipsen n <Kfztech> flickability
leichtes Erdöl n <Erdöl> light crude oil
leichtes Gewebe n <Textil> lightweight fabric
leichtes Heizöl n <Erdöl> domestic fuel oil *(Destillationsprodukt)*
leichtes Rohöl n <Erdöl> light crude oil
leichtes spurgeführtes Transportsystem n <Eisenbahn> light guideway transit system
leichtes Wasser n <Kerntech> ordinary water
Leichtfass n 1. <Kerntech> *(AE)* fiber drum, *(BE)* fibre drum *(aus faserverstärktem Kunststoff)*; 2. <Verpack> *(AE)* fiber drum, *(BE)* fibre drum, plywood drum
Leichtflintglas n <Ker & Glas> light flint
Leichtgewichtwabenkonstruktion f <Verpack> lightweight honeycomb structure
Leichtkolben m <Kfztech> full slipper piston
Leichtkraftrad n <Kfztech> light motorcycle *(mit Tretanlasser)*
Leichtkronglas n <Ker & Glas> light crown
Leichtmesonenspektrum n <Strahlphys> light meson spectrum
Leichtmetall n <Metall> light alloy
Leichtmetallguss m <Fertig> light metal casting *(Kunststoffinstallationen)*
Leichtmetalllegierung f <Mechan, Metall> light alloy
Leichtplatten fpl <Ker & Glas> light panels
leichtwassergekühlter Hybridreaktor m <Kerntech> LWHR, light water hybrid reactor

leichtwassergekühlter

leichtwassergekühlter Reaktor m <Kerntech> light water-cooled reactor
Leichtwasserlinie f <Wassertrans> light waterline
Leim m 1. <Chemie> sizing agent *(Papier)*; 2. <Druck> glue, size; 3. <Fertig> size; 4. <Kunststoff> glue; 5. <Papier> glue, size; 6. <Textil, Verpack> glue
Leimauftragmaschine f <Verpack> glue spreading machine, glue-gumming machine, gluing machine
leimen v <Bau, Papier> glue
Leimen n <Druck> gluing up
Leimfarbe f <Bau, Kunststoff> distemper
Leimfuge f <Kunststoff> glue line
Leimgewebe n <Lebensmittel> collagen
Leimmattierung f <Ker & Glas> glue-etching
Leimpresse f <Papier> size press
Leimung f <Papier> gluing, sizing
Leimungsprüfer m <Papier> sizing tester *(Gerät zur Überprüfung des Leimungsgrades von Papier)*
Leimwalze f <Papier> size roll
Leimzwinge f <Verpack> glue press
Leine f <Wassertrans> line *(Tauwerk)* • **Leine werfen** <Wassertrans> throw a line *(Tauwerk)*
Leinen n <Druck, Textil> linen
Leinenkanevas m <Textil> canvas
Leinenkleidung f <Textil> linen clothing
Leinenschießgerät n <Wassertrans> line thrower
Leinöl n <Bau, Wassertrans> linseed oil
Leinölfirnis m <Chemie> boiled linseed oil
Leinsaatöl n <Chemie> linseed oil
Leinwandbindung f <Papier, Textil> plain weave
L-Eisen n 1. <Bau> angle bar; 2. <Metall> L-iron
Leiste f 1. <Bau> batten; strip *(Holz)*; 2. <Druck> border; 3. <Fertig> tongue; 4. <Maschinen> ledge
Leistenhobelmaschine f <Bau> *(AE)* molding machine, *(BE)* moulding machine *(Holzbau)*
Leistung f 1. <Comp & DV> performance; 2. <Elektriz, Elektrotech> capacitance, capacity *(Maschine)*; power *(P)*; 3. <Erdöl> capacity *(Kapazität einer Bohrung)*; 4. <Fertig> performance *(Kunststoffinstallationen)*; 5. <Funktech> power; 6. <Heiz & Kälte> capacitance, capacity; 7. <Kerntech> output; 8. <Kfztech> power; 9. <Kontroll> performance; 10. <Maschinen> performance, power, *(BE)* throughput, *(AE)* thruput; 11. <Mechan> effect, power; 12. <Optik> power; 13. <Telekom> power *(elektrisch)*; performation
Leistung f bei Wasserkühlung <Heiz & Kälte> water-cooled rating
Leistung f des Turbogenerators <Kerntech> thermal steam generator output
Leistung f/höchste für längere Zeit entnehmbare <Lufttrans> METO power, maximum except takeoff power
Leistung f in Schwachlastzeit <Kerntech> off-peak power
Leistung f ohne Last <Kerntech> no-load force
Leistung f pro Flächeneinheit <Raumfahrt> power per unit area *(Raumschiff)*
Leistungsabfall m <Telekom> attenuation
Leistungsabgabe f 1. <Elektrotech> power output; 2. <Maschinen> output, power output; 3. <Nichtfoss Energ> power output; 4. <Qual> output
Leistungsaggregat n <Elektrotech> power pack
Leistungsanforderungen fpl <Maschinen> performance specification
Leistungsangabe f <Elektrotech> power rating
Leistungsangabe f eines radioaktiven Präparats <Kerntech> nameplate source strength
Leistungsaufnahme f 1. <Aufnahme> power-handling capacity; 2. <Elektrotech> power consumption, power input; 3. <Heiz & Kälte> power input
Leistungsbandbreite f <Elektronik> power bandwidth

Leistungsbegrenzer m <Maschinen> power limiter
Leistungsbelastung f <Metrol> power loading
Leistungsbereich m <Maschinen> performance range, power range
Leistungsbilanz f <Elektriz> power balance
Leistungsbilanz f einer Übertragungsstrecke <Telekom> link power budget
Leistungsdaten pl 1. <Fertig> performance specification; 2. <Telekom> performance data
Leistungsdiagramm n <Phys> indicator diagram
Leistungsdichte f <Kerntech, Nichtfoss Energ, Telekom> power density *(Antenne)*
Leistungsdichtespektrum n <Funktech> power spectral density
Leistungsdiode f <Elektronik> power diode
Leistungsdrahtwiderstand m <Elektrotech> power wire-wound resistor
Leistungseinbruch m <Kerntech> trip
Leistungselektronik f <Elektriz> power electronics
Leistungserhöhung f <Elektronik> power amplification
leistungsfähig adj <Phys> efficient
leistungsfähiges Werkzeug n <Comp & DV> power tool
Leistungsfähigkeit f 1. <Bau> capacitance, capacity; 2. <Comp & DV> power; 3. <Elektrotech, Ergon> capacity; 4. <Kerntech> canyon *(eines Kraftwerkes)*; 5. <Telekom> performance
Leistungsfaktor m <Elektriz, Elektrotech, Kunststoff, Phys> power factor
Leistungsfaktormesser m <Elektriz, Metrol> cos-j-meter, phase meter, power-factor indicator, power-factor meter
Leistungsfernmessgerät n <Gerät> telewattmeter
Leistungsflussdichte f <Raumfahrt> power flux density *(Raumschiff)*
Leistungsflussrichtungsrelais n <Elektriz> power directional relay
Leistungsganglinie f <Elektriz> load characteristic
leistungsgebundene Übertragung f <Telekom> line transmission, wire transmission
Leistungsgenerator m <Elektronik> power oscillator
Leistungsgrad m <Aufnahme> power efficiency *(Verstärker)*
Leistungsgröße f <Ergon> performance variable
Leistungshalbwertsbreite f <Funktech> half-power beamwidth
Leistungskabel n <Elektriz> power cable
Leistungskennlinien fpl <Heiz & Kälte> performance characteristics, performance curves
Leistungskoeffizient m <Nichtfoss Energ> power coefficient
Leistungskoeffizient m der Reaktivität <Kerntech> reactivity power coefficient
Leistungskomponente f <Elektriz> power component
Leistungskondensator m <Elektriz, Elektrotech> power capacitor
Leistungs-Kosten-Verhältnis n <Telekom> performance/cost ratio
Leistungskurve f 1. <Maschinen> performance curve; 2. <Nichtfoss Energ> power curve
Leistungsleitungen fpl <Elektrotech> load leads
Leistungsmerkmal n 1. <Fernseh> rating; 2. <Telekom> facility, user facility
Leistungsmesser m <Elektrotech> demand meter, wattmeter
Leistungsmessgerät n 1. <Elektriz> active energy meter; 2. <Gerät> power-measuring instrument; 3. <Phys> wattmeter
Leistungsmessung f <Comp & DV> performance measurement
Leistungsminderung f <Ergon> disability

Leistungsmodulation f <Telekom> intensity modulation
Leistungspegel m <Sicherheit> power level *(Lärm)*
Leistungsprüfung f <Elektriz, Maschinen> performance test
Leistungsquelle f <Elektrotech, Nichtfoss Energ> power source
Leistungsrate f <Comp & DV> yield
Leistungsreaktor m <Kerntech> power reactor
Leistungsregelung f <Mobilkom> power control *(Antennenleistung)*
Leistungsregler m 1. <Elektrotech> governor power controller; 2. <Raumfahrt> power conditioning unit *(Raumschiff)*
Leistungsrelais n <Elektrotech> power relay
Leistungsreserve f <Telekom> power margin *(Richtfunk, Satellit)*
Leistungsröhre f <Elektronik> power tube
Leistungsschalter m <Elektrotech> circuit breaker, power switch
Leistungsschalter m **für Kurzunterbrechung** <Elektrotech> auto-reclosing circuit breaker
Leistungsschalter m **mit Einschaltsperre** <Elektrotech> circuit-breaker with lock-out closing *(Hochspannungstechnik)*
Leistungsschalter m **mit integrierten Sicherungen** <Elektrotech> integrally fused circuit-breaker *(Hochspannungstechnik)*
Leistungsschalter m **mit Minimalauslösung** <Elektrotech> minimum circuit breaker
Leistungsschalter m **mit Wiedereinschaltung** <Elektrotech> automatic reclosing circuit-breaker
Leistungsschalterkombination f <Elektrotech> power switchgroup *(Hochspannungstechnik)*
Leistungsschreiber m <Gerät> power recorder
Leistungsschutzschalter m <Elektrotech> power-protection switch
Leistungsspektrum n <Elektrotech, Ergon> power spectrum
Leistungsspektrumsdichte f <Raumfahrt> power spectral density *(Weltraumfunk)*
Leistungsspitze f <Elektrotech> power surge
Leistungsstufe f <Maschinen> power stage
Leistungsteiler m <Elektriz, Elektrotech, Telekom> power divider
Leistungsthyristor m <Elektronik> power thyristor
Leistungstransformator m <Elektrotech> power transformer
Leistungstransistor m <Elektronik> power transistor
Leistungsübertragung f <Elektrotech> power transmission
Leistungsübertragungsfaktor m <Akustik> response to power
Leistungsumschalttransistor m <Elektronik> power-switching transistor
Leistungsumsetzer m <Elektrotech> power converter
Leistungsverbrauch m <Elektriz, Elektrotech> power consumption
Leistungsverhalten n 1. <Maschinen> performance properties; 2. <Telekom> performance
Leistungsverlust m 1. <Elektrotech> power loss; 2. <Maschinen> loss of power; 3. <Telekom> attenuation
Leistungsvermögen n 1. <Comp & DV> capacity; 2. <Kerntech> canyon *(eines Kraftwerkes)*; unit capacity *(einer Anlage)*
Leistungsversorgung f <Elektrotech> power supply
Leistungsversorgunseinrichtung f <Maschinen> engine lathe
Leistungsverstärker m 1. <Aufnahme, Elektronik, Elektrotech, Funktech> power amplifier, PA; 2. <Kfztech> power booster; 3. <Phys, Telekom> power amplifier, PA; 4. <Raumfahrt> power amplifier, PA *(Weltraumfunk)*; 5. <Telekom> power amplifier, PA
Leistungsverstärker m **mit hohem Verstärkungsgrad** <Elektronik> high-gain power amplifier
Leistungsverstärkerröhre f <Elektronik> power amplifier tube
Leistungsverstärkertransistor m <Elektronik> power amplifier transistor
Leistungsverstärkung f 1. <Aufnahme> power gain; 2. <Elektronik> power amplification, power gain; 3. <Phys, Qual> power gain
Leistungsverzeichnis n <Bau> specifications
Leistungswandler m <Elektrotech> power converter
Leistungswerte mpl <Maschinen> power ratings
Leit... <Comp & DV> master
Leitartikel m <Druck> lead
Leitbefehl m <Comp & DV> routing directive
Leitblatt n <Hydraul> guide blade; stationary blade *(Turbine, Kreiselpumpe)*
Leitblech n 1. <Bau> baffle; 2. <Chemtech> baffle plate; 3. <Fertig> baffle plate, deflector; 4. <Heiz & Kälte> baffle plate; 5. <Maschinen> baffle, deflector plate • **mit Leitblech versehen** <Fertig> baffle
Leitblechring m <Erdöl> baffle ring
Leitblock m <Wassertrans> lead block, leading block *(Beschläge)*
Leitdaten npl <Comp & DV> master data
Leiteinrichtung f <Maschinen> guide
leiten v 1. <Bau> carry; 2. <Comp & DV> route
leitend adj 1. <Anstrich> conducting, conductive; 2. <Elektriz> conductible, conducting, conductive
leitende Abschirmung f <Elektriz> conducting screen, conductor screen
leitender Auditor m <Qual> lead auditor
leitender Offizier m <Wassertrans> chief mate *(Handelsmarine, Besatzung)*
Leiter m 1. <Elektriz, Elektrotech> conductor, wire; core *(bei Glasfaserkabel)*; core *(eines elektrischen Kabels)*; 2. <Erdöl> conductor *(für Wärme, Kälte, Elektrizität)*; 3. <Phys> conductor; 4. <Sicherheit> safety ladder; 5. <Telekom> conductor
Leiteraddierer m <Elektronik> ladder adder
Leiterbagger m <Wassertrans> ladder dredge, ladder dredger *(Baggern)*
Leiterbahn f <Elektronik> interconnection trace
Leiterbahnabstand m <Elektronik> line separation
Leiterbündel n <Elektriz> (BE) bunched conductor, (AE) bundled conductor
Leiter-Leiter-Kapazität f <Elektriz> capacitance between conductors
Leiter-Leiter-Spannung f <Elektriz> line-to-line voltage
Leiterplatte f 1. <Comp & DV, Elektriz, Elektronik> printed circuit board, printed wiring board, PCB; 2. <Elektronik, Fernseh, Funktech, Telekom> printed circuit board • **außerhalb der Leiterplatte gelegen** <Elektronik> off board • **auf der Leiterplatte** <Elektronik> on board • **nicht auf der Leiterplatte** <Elektronik> off-board
Leiterplatten-Laminat n <Elektronik> printed circuit laminate
Leiterplattenstecker m <Elektronik> printed circuit connector
Leiterplattenträgermaterial n <Elektronik> printed circuit substrate
leitfähig adj <Elektriz> conductible, conducting, conductive
leitfähiger Gummi m <Kunststoff> conductive rubber
leitfähiges Fett n <Raumfahrt> conductive grease *(Raumschiff)*

Leitfähigkeit

Leitfähigkeit f 1. <Anstrich, Aufnahme> conductivity; 2. <Elektriz, Elektrotech> conductance, conductibility, conductivity; 3. <Kunststoff> conductivity; 4. <Phys> conductance, conductivity; 5. <Wasserversorg> conductivity
Leitfähigkeit f **im Einschaltzustand** <Elektrotech> on-state conductivity
Leitfähigkeitsmesser m <Labor> conductivity meter
Leitfähigkeitsmessgerät n <Gerät> conductivity measuring instrument, conductometric instrument
Leitfähigkeitsmesszelle f <Gerät> conductivity cell
Leitfähigkeitsmodulation f <Elektronik> conductivity modulation
Leitfähigkeitsschreiber m <Gerät> conductivity recorder
Leitfähigkeitszelle f 1. <Gerät> conductivity cell; 2. <Labor> conductance cell
Leitfeuer n <Wassertrans> leading light
Leitfläche f <Fernseh, Funktech, Telekom> land
Leitflügel m <Nichtfoss Energ> guide vane (Aerodynamik)
Leitflügel-Servomotor m <Nichtfoss Energ> guide vane servomotor
Leitflügelvibration f <Nichtfoss Energ> guide vane vibration
Leitgerade f <Geom> directrix
Leitgummi m <Kunststoff> conductive rubber
Leitintensität f <Aufnahme> standard reference intensity
Leitkabel n <Kerntech> leader
Leitkarte f <Comp & DV> master card
Leitkegel m <Bau> traffic cone (Verkehr)
Leitklampe f <Wassertrans> fairlead (Deckausrüstung)
Leitlinie f 1. <Geom> directrix; 2. <Wassertrans> leading line
Leitmarke f <Wassertrans> leading mark
Leitoszillator m <Phys> master oscillator
Leitpfad m <Comp & DV> routing path
Leitpfosten m <Bau> guide post
Leitplanke f 1. <Bau> guard rail, side rail; 2. <Sicherheit> crash barrier; 3. <Trans> (BE) crash barrier
Leitplatte f <Maschinen> deflector plate
Leitprozessor m <Telekom> administrative processor
Leitpunkte mpl <Geom> control points
Leitrad n <Kfztech> stator; reactor (Teil des Drehmomentwandlers)
Leitrechner m <Comp & DV> host, host computer, master
Leitrechnersystem n <Comp & DV> master computer system
Leitring m <Hydraul> guide ring
Leitrolle f 1. <Fertig> belt idler, idle pulley; 2. <Kfztech> belt idler; 3. <Maschinen> guide pulley, idler; 4. <Papier> jockey pulley
Leitschaufel f 1. <Heiz & Kälte> guide vane; 2. <Hydraul> guide blade, guide vane; stationary vane (Turbine, Pumpe); 3. <Lufttrans> compressor blade; stator vane (Kompressor); 4. <Maschinen, Mechan, Nichtfoss Energ> guide vane (Turbine)
Leitscheibe f <Maschinen> guide, guide pulley
Leitschicht f <Funktech> wave duct
Leitschiene f 1. <Bau> check rail; 2. <Maschinen> guide, guide bar
Leitseitenstrahlkoppler m <Lufttrans> lateral beam coupler
Leitspindel f 1. <Fertig> leadscrew (Drehmaschine); 2. <Maschinen> guide screw, lead screw, leading screw; 3. <Mechan> lead screw
Leitspindelmutter f <Maschinen> half-nut
Leitstand m <Raumfahrt> (AE) operation center, (BE) operation centre
Leitstange f <Maschinen> slide bar
Leitstation f 1. <Comp & DV> control terminal; 2. <Raumfahrt> network coordination station (in Funknetzwerken)
Leitstelle f <Kontroll> master
Leitstrahl m 1. <Raumfahrt> beacon; 2. <Wassertrans> guide beam • **einen Leitstrahl erfassen** <Lufttrans> capture a beam (Flugwesen)
Leitstrahlerzeuger m <Raumfahrt> beacon generator (Raumschiff)
leitstrahlgeführt adj <Raumfahrt> abeam
Leitstrahllenkung f <Lufttrans> beam following
Leittechnik f <Regelung> process control engineering
Leitton m 1. <Akustik> leading note; 2. <Aufnahme> neo-pilot tone
Leittuch n <Textil> leader cloth
Leit- und Zugspindeldrehmaschine f <Maschinen> engine lathe
Leitung f 1. <Aufnahme> line; 2. <Comp & DV, Elektriz, Elektrotech> line, link, cable, transmission line; lead (Stromdrahtversorgung); cable (elektrisch); conduction (von Strom); 3. <Erdöl> pipeline (zum Transport von flüssigen oder gasförmigen Medien); 4. <Fernseh> line; 5. <Fertig> main; 6. <Heiz & Kälte> conduction; 7. <Hydraul> guide (Turbine); 8. <Kerntech> line; 9. <Kfztech> cable; 10. <Maschinen, Mechan> duct; 11. <Phys> conduction; 12. <Telekom> line, circuit (Telefon) • **in Leitung gehen** <Telekom> go into circuit • **über die Leitung gehen** <Telekom> go via the circuit
Leitung f/**außer Betrieb befindliche** <Telekom> dead line
Leitung f **der Vorfelddienstleistungen** <Lufttrans> apron management service
Leitung f **mit Verstärker** <Telekom> amplified circuit
Leitung f **ohne Verstärker** <Telekom> unamplified circuit
Leitung-Nullleiter-Spannung f <Elektriz> line-to-neutral voltage
Leitungsabfertiger m <Elektriz> load dispatcher
Leitungsabschluss m <Comp & DV> line termination
Leitungsabschnitt m <Telekom> link
Leitungsanschluss m <Comp & DV> line adaptor
Leitungsanschlusseinrichtung f <Telekom> line connection unit
Leitungsanschlussfeld n <Elektrotech> line pad
Leitungsband n <Phys, Strahlphys> conduction band
Leitungsbelegungstaste f <Telekom> line seizure button
Leitungsbetrieb m <Elektrotech> line operation
leitungsbetrieben adj <Elektrotech> line-operated
Leitungsbündel n <Telekom> group, line group
Leitungscode m <Telekom> line code
Leitungsdraht m <Elektrotech> conductor wire; lead (Zuleitungsdraht)
Leitungsdurchgangsprüfer m <Gerät> circuit continuity tester
Leitungsende/am nahen <Elektrotech, Telekom> near-end
Leitungsendeinrichtung f <Telekom> line-terminating equipment
Leitungsendgerät n <Telekom> line terminal
Leitungs-Erde-Spannung f <Elektriz> (BE) line-to-earth voltage, (AE) line-to-ground voltage
Leitungsfehler m <Elektrotech> line fault
leitungsgebundene Störstrahlung f <Telekom> conducted spurious emission
leitungsgebundene Störung f <Telekom> conducted interference
Leitungsgerüst n <Elektrotech> lead frame
Leitungsgraben m <Bau> utility trench
Leitungshalterung f <Lufttrans> fairlead
Leitungsimpedanz f <Elektrotech> line impedance
Leitungsisolator m <Elektrotech> line insulator
Leitungskommutator m <Elektriz> line commutator
Leitungskompensation f <Fernseh> cable compensation circuits

Leitungskonfiguration f <Elektrotech> line configuration
Leitungskonzentrator m <Telekom> line concentrator, remote concentrator
Leitungskopplung f <Elektrotech> line coupling
Leitungskopplungstrafo m <Elektrotech> line-coupling transformer
Leitungskreis m <Heiz & Kälte> circuit
leitungsloser Chip-Träger m <Elektronik, Elektrotech> leadless chip carrier
Leitungsmast m <Elektriz> transmission tower
Leitungsmietgebühr f <Telekom> line rental
Leitungsmodul n <Telekom> line module
Leitungsnachbildung f <Telekom> artificial line
Leitungsnummer f <Comp & DV> line number *(Datenfernverarbeitung)*
Leitungspegel m <Aufnahme, Telekom> line level
Leitungsprüfer m <Gerät> circuit continuity tester; continuity tester *(elektrische Leitung auf Stromdurchgang)*
Leitungspumpe f <Elektrotech> conduction pump
Leitungsrahmen m <Elektrotech> lead frame
Leitungsrauschen n <Elektronik, Elektrotech> circuit noise, line noise
Leitungsregelung f <Elektrotech> line regulation
Leitungsrohr n 1. <Bau> conduit pipe; 2. <Nichtfoss Energ> conduit
Leitungssatz m <Telekom> junctor; line terminal *(Kreislauf)*
Leitungsschleife f <Telekom> loop
Leitungsschnittstelle f <Elektrotech> line interface
Leitungsschnur f <Telekom> cord *(elektrisch)*
Leitungsschutz m <Elektriz> line protection
Leitungsschutzdrossel f <Elektrotech> line-choking coil
Leitungsspannung f <Elektrotech> line voltage
leitungsstabilisierter Oszillator m <Elektronik> line-stabilized oscillator
Leitungssteuerung f <Elektrotech> line controller
Leitungsstörung f <Elektrotech, Telekom> line fault
Leitungsstrecke f <Wasserversorg> section
Leitungsstrom m 1. <Elektrotech> line current; 2. <Phys> conduction current
Leitungssystem n <Telekom> line system
Leitungstreiber m <Comp & DV> line driver
Leitungstrennschalter m <Elektriz> line breaker
Leitungsüberträger m 1. <Elektrotech> repeating coil; 2. <Telekom> line transformer
Leitungsübertragung f <Comp & DV> line communication
Leitungsüberwachung f <Fernseh> line monitor
Leitungsunterbrechung f <Elektriz> line break
Leitungsverbinder m <Elektrotech> cable connector, connector
Leitungsverbindung f über Schnittstellen <Elektrotech> line interfacing
Leitungsverlegung f <Bau> cut and cover
leitungsvermittelndes Amt n <Telekom> *(AE)* circuit switching center, *(BE)* circuit switching centre
leitungsvermittelte Verbindung f <Telekom> circuit-switched connection
leitungsvermittelter Dienst m <Telekom> switched service
leitungsvermittelter Trägerdienst m <Telekom> circuit-mode bearer service
leitungsvermitteltes Netz n <Comp & DV, Telekom> circuit-switched network
leitungsvermitteltes öffentliches Datennetz n *(CSPDN)* <Telekom> circuit-switched public data network *(CSPDN)*
Leitungsvermittlung f 1. <Comp & DV> circuit switching, line switching; 2. <Telekom> circuit switching, circuit switching system
Leitungsvermittlung f mit Drehmotorwählern <Telekom> rotary exchange circuit, switching with rotary selector

Leitungsverstärker m 1. <Aufnahme> line amplifier; 2. <Comp & DV> line driver; 3. <Elektronik> line amplifier; 4. <Telekom> line amplifier, line repeater
Leitungswähler m <Telekom> connector *(Schaltung)*
Leitungswasser n <Bau> tap water
Leitungszeichen n <Elektronik> line signal
Leitungszeichengabegerät n <Telekom> *(AE)* line-signaling equipment, *(BE)* line-signalling equipment
Leitvermögen n <Elektrotech> conductivity
Leitwalze f <Papier> guide roll, leading roll
Leitweg m 1. <Comp & DV> route; 2. <Telekom> route
Leitweganzeiger m <Comp & DV> routing indicator
Leitwegauswahl f <Telekom> route selection *(ITG)*
Leitwegbefehl m <Comp & DV> routing directive
Leitwegcode m <Comp & DV> routing code
Leitweginformationen fpl <Comp & DV> routing information
Leitwegkennung f <Telekom> routing identifier
Leitweglenkung f <Telekom> automatic routing, routing
Leitwegprogramm m <Comp & DV> router
Leitwegwahl f <Comp & DV> routing
Leitwerk n 1. <Lufttrans> horizontal stabilizer, tail unit; 2. <Telekom> control unit
Leitwerksansatzfläche f <Lufttrans> fin stub frame
Leitwert m <Elektriz, Elektrotech, Phys> admittance, conductance, susceptance
Leitwertbrücke f <Gerät> conductance bridge
Leitwertschreiber m <Gerät> conductivity recorder
Leitzahl f <Foto> guide number
LEM *(Mondlandefahrzeug, Mondlandefähre)* <Raumfahrt> LEM *(lunar excursion module)*
Lemma n <Math> lemma *(kleiner Lehrsatz)*
Lemniskate f <Geom> lemniscate
Lemniskoide f <Fertig> lemniscoid *(Wattkurve)*
Lemonal n <Chemie> citral
Lenkachse f 1. <Eisenbahn> leading axle; 2. <Kfztech> steering axle *(Lenkung, Räder)*
Lenkanschlag m <Kfztech> steering lock *(Lenkung)*
Lenkbarkeit f <Kerntech> *(AE)* maneuverability, *(BE)* manoeuvrability *(des Kraftniveaus)*
Lenkeinschlag m <Kfztech> steer angle *(Lenkung)*
lenken v 1. <Lufttrans> pilot, steer; 2. <Nichtfoss Energ> channel; 3. <Telekom> direct; 4. <Trans, Wassertrans> steer
Lenker m <Lufttrans> connecting rod
Lenkgehäuse n <Kfztech> steering gearbox
Lenkgeometrie f <Kfztech> steering geometry
Lenkgestänge n <Kfztech> steering linkage
Lenkgetriebe n <Kfztech> steering gear; steering gearbox *(Lenkung)*
Lenkhebel m 1. <Kfztech> steering arm; 2. <Lufttrans> drop arm
Lenkknüppelfuß m des Bugradfahrwerks <Lufttrans> nose gear steering base post
Lenkkopf m <Kfztech> steering head *(Motorradlenkung)*
Lenkkreis m <Kfztech> steering circle *(Lenkung)*
Lenkrad n <Kfztech> wheel, steering wheel *(Lenkung)*
Lenkradschloss n <Kfztech> steering column lock *(Lenkung)*
Lenkradseite f <Kfztech> offside
Lenkrolle f <Maschinen> caster, castor, guide pulley
Lenkrollradius m <Kfztech> offset radius
Lenksäule f <Kfztech> steering column
Lenkschubstange f <Kfztech> drag rod *(Lenkung)*
Lenkspurstange f <Kfztech> drag rod *(Lenkung)*
Lenkstockhebel m <Kfztech> pitman arm; drop arm *(Lenkung)*
Lenkung f <Kfztech> steering system, steering
Lenkungsausschlag m <Kfztech> lock

Lenkungsspiel

Lenkungsspiel n <Kfztech> steering play *(Lenkung)*
Lenkwelle f <Kfztech> steering shaft
Lenkzapfen m <Kfztech> steering knuckle pin
Lenkzapfensturz m <Kfztech> *(AE)* kingbolt inclination, *(BE)* kingpin inclination, steering axis inclination
Lenkzwischenhebel m <Kfztech> idler arm, relay arm
Lenkzwischenstange f <Kfztech> *(AE)* center link, *(BE)* centre link *(Lenkung)*
Lenzanlage f <Wassertrans> drainage system
lenzen v <Wassertrans> pump
Lenzklappe f <Wassertrans> transom flap
Lenzpumpe f <Wassertrans> bilge pump
Lenz'sche Regel f <Elektriz, Phys> Lenz's law
Lenz'sches Gesetz n <Elektriz, Phys> Lenz's law
Lenz- und Ballastsystem n <Wassertrans> pump and ballast system
Lenzvorrichtung f **der Plicht** <Wassertrans> cockpit drainage
Leonhardtit m <Kerntech> leonhardite
LEP *(Elektronen-Positronen-Kollideranlage)* <Teilphys> LEP *(large electron-positron collider)*
Leporellofalzung f 1. <Druck> concertina fold; 2. <Papier> accordion fold
Lepton n <Phys, Teilphys> lepton
Leptonenzahl f <Phys> lepton number
leptonisch adj <Teilphys> leptonic
Lernautomat m 1. <Elektronik> learning machine *(lernender Automat)*; 2. <Künstl Int> learning automaton
Lernen n **durch Wahrnehmung** <Ergon> perceptual learning
Lernen n **mit Verstärkern** <Künstl Int> reinforcement learning
lernende Banderol-Klebmaschine f <Verpack> *(AE)* intelligent labeling machine, *(BE)* intelligent labelling machine
lernende Etikettiermaschine f <Verpack> *(AE)* intelligent labeling machine, *(BE)* intelligent labelling machine
lernender Automat m <Künstl Int> learning automaton
lernendes System n <Künstl Int> learning system
lernfähig adj <Comp & DV> adaptive
lernfähiges System n <Künstl Int> learning system
Lernkurve f <Elektriz, Ergon> learning curve
Lernphase f <Elektronik> learning phase
Lernprogramm n <Comp & DV> tutorial
Lernregel f <Künstl Int> learning rule
lesbar adj/**nur einmal beschreibbar mehrfach** <Comp & DV, Optik> write-once read many times
lesbare und dauerhafte Kennzeichnung f <Qual> legible and durable marking
Lesbarkeit f <Ergon> legibility
Lesefehler m <Comp & DV> read error
Lesegerät n <Elektronik> reading gun
Lesegeschwindigkeit f <Comp & DV> read rate, reading rate
Leseglas n <Optik> reading glass
Lesekopf m 1. <Comp & DV> head, magnetic head, read head; 2. <Optik> read head
Leselaser m <Optik> read laser
Leselupe f <Optik> reading lens
lesen v 1. <Comp & DV> read, retrieve; 2. <Gerät> read *(Speicherdaten)*; 3. <Telekom> sense
Lesen n 1. <Comp & DV> read; 2. <Kohlen> picking; 3. <Telekom> readout
Lesen n **beim Schreiben** <Comp & DV> read while write
Lesen n **nach dem Schreiben** <Comp & DV> read after write
Lesen n **von Markierungen** <Comp & DV> mark reading
Leser m <Comp & DV, Druck> reader
Lese-/Schreib... <Comp & DV> read/write

Lese-/Schreibkanal m <Comp & DV> read/write channel
Lese-/Schreibkopf m <Comp & DV> read/write head
Lese-/Schreibspeicher m *(RAM)* <Comp & DV> random access memory *(RAM)*
Lese-/Schreibzugriff m <Comp & DV> read/write access
Lesesperre f <Comp & DV> fetch protection
Lesestift m <Comp & DV> wand
Lesestiftscanner m <Comp & DV> wand scanner
Lesestrahl m 1. <Elektronik> reading beam; 2. <Optik> read beam
Lesetransistor m <Elektronik> read transistor
Lese- und Schreibanforderungen fpl <Comp & DV> read and write requests
Lese-und-Schreibkopf m <Optik> read-write head
Leseverstärker m 1. <Comp & DV> read amplifier; 2. <Elektronik> read amplifier, sense amplifier
Lesezeichen n <Druck> bookmark
Lesezeit f <Comp & DV> read time
Lesezugriffszeit f <Comp & DV> read access time
Letaldosis f <Umweltschmutz> lethal dose
Lethargie f <Kerntech, Phys> lethargy
Letten m <Bau> clay
Lettenbohrer m <Erdöl> claying bar
Letter f <Druck> metal type
Letternmetall n <Druck> type metal
letzte Bildschirmmaske f <Elektronik> end screen *(eines Programms)*
letzte Notfallmaßnahme f <Lufttrans> last emergency action
letzte Schicht f <Anstrich> finish • **mit letzter Schicht versehen** <Anstrich> finish
Letztweg m <Telekom> final route *(ITG)*; last choice route
Letztwegbündel n <Telekom> last choice circuit group
Leuchtanzeige f <Comp & DV> electroluminescent display
Leuchtbalkenanzeige f <Gerät> bar graph display
Leuchtboje f <Wassertrans> light buoy *(Seezeichen)*
Leuchtdichte f 1. <Elektrotech, Fernseh> luminance; 2. <Optik> brightness; 3. <Phys, Telekom> luminance
Leuchtdichtefaktor m <Strahlphys> luminosity coefficient
Leuchtdichtemessung f <Gerät> luminance measurement
Leuchtdichtendifferenzschwelle f <Ergon> luminance difference threshold
Leuchtdichtenfaktor m <Ergon> luminance factor
Leuchtdichtenkontrast m <Ergon> luminance contrast
Leuchtdichtenunterschied m <Ergon> luminance difference
Leuchtdichtesignal n <Fernseh> luminance signal
Leuchtdichtetheorem n <Optik> brightness theorem
Leuchtdiode f *(LED)* 1. <Comp & DV, Elektriz, Elektronik> light-emitting diode, luminescent diode *(LED)*; 2. <Fernseh, Optik, Phys, Telekom> light-emitting diode *(LED)*
Leuchte f <Kfztech> light; lamp *(zur Warnung)*
leuchtend adj 1. <Elektrotech> luminescent; 2. <Telekom> illuminated; 3. <Textil> bright
leuchtend farbige Schutzkleidung f <Sicherheit> *(AE)* luminous and colored protective clothing, *(BE)* luminous and coloured protective clothing
Leuchtfaden m <Elektriz> filament
Leuchtfeuer n 1. <Lufttrans> beacon, beacon light; 2. <Wassertrans> light
Leuchtfeuertonne f <Wassertrans> light buoy *(Seezeichen)*
Leuchtfeuerverzeichnis n <Wassertrans> list of lights *(Navigation)*
Leuchtgerät n <Labor> illuminating apparatus *(Mikroskop)*
Leuchtglas n <Ker & Glas> luminescent glass
Leuchthorizont m <Lufttrans> crossbar *(Anflugsbefeuerungssystem)*

Leuchtkasten m <Foto> lightbox
Leuchtkörper m <Elektriz> luminaire
Leuchtkraft f <Lufttrans> lighting efficiency
Leuchtkugel f <Wassertrans> Very light, flare
Leuchtlupe f <Fertig> illuminated magnifier
Leuchtpatrone f <Wassertrans> flare
Leuchtpetroleum n <Erdöl> lamp oil *(Erdölfraktion)*
Leuchtpistole f <Wassertrans> Very pistol
Leuchtpunkt m <Fernseh> beam impact point
Leuchtrakete f <Wassertrans> flare
Leuchtschirm m <Elektriz, Elektronik> fluorescent screen
Leuchtsignal n <Lufttrans> indicator light
Leuchtskale f <Gerät> luminous dial
Leuchtstärke f <Strahlphys> luminous intensity
Leuchtstift m <Konstzeich> highlighter
Leuchtstoff m <Elektrotech, Fernseh, Strahlphys> fluorescent substance
Leuchtstofflampe f <Elektrotech> fluorescent lamp
Leuchtstoffpunkt m <Fernseh> phosphor dot
Leuchtstoffröhre f 1. <Elektriz> fluorescent tube, tubular incandescent lamp; 2. <Strahlphys> fluorescent discharge tube
Leuchttastatur f <Telekom> illuminated keypad
Leuchttisch m <Druck> light table
Leuchtturm m <Wassertrans> lighthouse
Leuchtturmwärter m lighthouse keeper
Leuchtwirksamkeit f <Phys> luminous efficiency
Leuchtzeit f <Strahlphys> luminosity life-time
Levothyroxin n <Chemie> thyroiodine
lexikalische Analyse f <Comp & DV> lexical analysis
lexikographische Anordnung f <Comp & DV> lexicographic order
L-förmig abgewinkeltes Anschlussbein n <Elektronik> gull wing lead
lg(x) <Math> Brigg's logarithm of x, lg(x)
LHC *(Hadronkollideranlage)* <Phys> LHC *(large hadron collider)*
Li *(Lithium)* <Chemie> Li *(lithium)*
Liapunovexponenten mpl <Strömphys> Liapunov exponents
Libelle f 1. <Mechan> bubble level, level; 2. <Metrol> level
Libellenblase f <Fertig> air bubble
Licht n <Foto, Kfztech, Phys> light
Lichtabdeckschirm m <Fernseh> flag
Lichtablenkung f <Optik> deflection
Lichtabsorption f <Strahlphys> absorption of light
lichtaktivierter Silizium-Gleichrichter m <Elektrotech> light-activated silicon-controlled rectifier
Lichtaufnehmer m <Elektronik> light sensor
Lichtausbeute f 1. <Ergon> luminous efficiency; 2. <Lufttrans> lighting efficiency; 3. <Phys> efficiency, luminous efficacy; 4. <Strahlphys> light yield
lichtbeständig adj <Verpack> stable to light
Lichtbeständigkeit f <Kunststoff> light resistance
Lichtbeugung f <Phys> diffraction of light
Lichtbildkompass m <Wassertrans> projector compass *(Kompass)*
Lichtbogen m <Elektriz> arc
Lichtbogenableiter m <Elektriz> arc arrester
lichtbogenbeständig adj <Fertig> arc-resistant
Lichtbogenbildung f <Elektrotech> arcing
Lichtbogenbrennschneiden n <Fertig> arc cutting
Lichtbogenbrennschneider m <Maschinen> arc cutter
Lichtbogendauer f <Elektriz> arc duration
Lichtbogenentladung f <Elektrotech> arc discharge
Lichtbogengleichrichter m <Elektriz, Elektrotech> arc rectifier
Lichtbogenhandschweißen n <Bau> manual arc welding
Lichtbogenheizgerät n <Maschinen> arc heater

Lichtbogenkennlinie f <Fertig> arc characteristic
Lichtbogenkontakte mpl <Elektriz> arcing contacts
Lichtbogenlampe f <Elektriz> arc lamp
Lichtbogenleistungsschalter m <Elektrotech> arc breaker
Lichtbogenleuchte f <Elektriz> arc light
Lichtbogenlöschkammer f <Elektrotech> arc quench chamber
Lichtbogenlöschung f <Elektriz, Elektrotech> arc extinction, arc quenching
Lichtbogenofen m 1. <Fertig, Heiz & Kälte> arc furnace; 2. <Kohlen> EAF, electric-arc furnace; 3. <Maschinen> arc furnace; 4. <Thermod> electric-arc furnace
Lichtbogenregler m <Elektriz> arc regulator
Lichtbogensauerstoffschweißen n <Bau> oxygen arc welding
Lichtbogenschneiden n 1. <Bau> arc cutting; 2. <Thermod> electric-arc cutting
Lichtbogenschutzarmatur f <Fertig> arcing shield
Lichtbogenschutzring m <Fertig> arcing ring
Lichtbogenschweißautomat m <Maschinen> automatic arc-welding machine
Lichtbogenschweißelektrode f <Elektriz> arc-welding electrode
Lichtbogenschweißen n 1. <Bau> arc welding, arc weld; 2. <Elektriz> arc welding; 3. <Fertig> arc welding, electric-arc welding; 4. <Thermod> electric-arc welding
Lichtbogenschweißen n **mit der Hand** <Mechan> manual arc welding
Lichtbogenschweißen n **mit Fülldrahtelektroden** <Bau> flux-cored arc welding
Lichtbogenschweißrauch m <Fertig> arc fume
Lichtbogenschweißstab m <Elektriz> arc-welding electrode
Lichtbogenspannung f <Elektriz> arc voltage *(einer Sicherung)*
Lichtbogenspitzenspannung f <Elektriz> peak arc voltage
Lichtbogenspur f <Elektriz> arc trace
Lichtbogenstrom m <Elektriz> arc current
Lichtbogenstromrichter m <Elektrotech> arc rectifier
Lichtbogentrennschalter m <Elektriz> arc breaker
Lichtbogenüberschlag m <Elektriz> arc over, arc striking
Lichtbogenunterdrückung f <Elektrotech> arc suppression
Lichtbogenunterdrückungsspule f <Elektriz> arc suppression coil *(Lichtbogenlöschspule)*
Lichtbogenzündung f <Elektriz, Elektrotech> arc ignition
Lichtbrechung f <Optik> refringency
Lichtbrechungsvermögen n <Foto> refractive power
Lichtbündel n <Optik> pencil of light
Lichtbüschel n <Optik> pencil of light
lichtchemisch adj <Phys> actinic
Lichtdämpfungssystem n <Lufttrans> dimmer *(Beleuchtung)*
lichtdicht adj <Verpack> lightproof
Lichtdiffuser m <Optik> diffuser
Lichtdruckplatte f <Druck> collotype plate
lichtdurchlässige Scheibe f <Optik> *(BE)* transparent disc, *(AE)* transparent disk
Lichtdurchlässigkeit f <Optik> transmittance
lichte Höhe f 1. <Bau> headroom; 2. <Elektrotech> clearance; 3. <Fertig> maximum daylight; 4. <Maschinen> clearance height, daylight, overall internal height; 5. <Wassertrans> headroom under beams *(zwischen den Decks)*
lichte Weite f 1. <Elektrotech> clearance; 2. <Fertig> internal diameter; bore size *(Rohr)*; 3. <Maschinen> inside diameter

Lichtechtheit

Lichtechtheit f <Textil> fastness to light
lichtelektrisch gravieren v <Elektronik> photoengrave
lichtelektrisch leitend adj <Elektronik> photoconductive
lichtelektrische Emission f <Elektronik> photoemission
lichtelektrische Verstärkung f <Elektronik> photoconductive gain
lichtelektrische Wirkung f <Elektronik> photoelectric effect
lichtelektrische Zelle f <Phys> photoconductive cell, photoelectric cell
lichtelektrischer Effekt m <Elektronik> photoelectric emission
lichtelektrischer Strom m <Telekom> light current
lichtelektrischer Verstärker m <Elektronik> photoelectric amplifier
lichtelektrischer Zerhacker m <Elektronik> electronic chopper
lichtelektrisches Gerät n <Elektronik> photoelectric device
lichtemittierende Diode f (LED) <Comp & DV, Elektriz, Elektronik, Fernseh, Optik, Phys, Telekom> light-emitting diode (LED)
lichtempfindlich adj 1. <Elektronik> light-sensitive; 2. <Foto> light-sensitive, photosensitive; 3. <Phys> photosensitive
lichtempfindliche Platte f <Druck> light-sensitive plate
Lichtempfindlichkeit f 1. <Elektronik> photosensitivity; 2. <Foto> film speed; 3. <Phys> actinism, photosensitivity; 4. <Strahlphys> sensitivity to light
Lichtempfindlichkeitskurve f <Elektronik> photoresponse curve
Lichten n <Trans> lifting
Lichtenberg'sche Figur f <Phys> Lichtenberg figure
Lichtenergie f <Strahlphys> light energy
lichter Abstand m <Bau> clear distance
lichtfest adj <Verpack> stable to light
Lichtfilter n <Druck> light filter
Lichtfleck m <Elektronik> light spot
Lichtgeschwindigkeit f 1. <Labor> velocity of light, c; 2. <Optik> speed of light in empty space, c (im Vakuum); 3. <Phys, Wellphys> speed of light, velocity of light
Lichtgitter n <Sicherheit> beam protection device
Lichtgitterschranke f <Sicherheit> light barrier
Lichtgriffel m <Comp & DV, Fernseh> light pen
Lichthof m <Elektronik> halo, halation
Lichthofbildung f 1. <Elektronik> halation (Katodenstrahlröhre); 2. <Foto> halation
Lichthofschutzschicht f <Foto> antihalo layer
Lichthupe f <Kfztech> headlamp flasher (Beleuchtung)
Lichtimpuls m <Elektronik> light pulse
lichtinduzierte Polymerisation f <Chemie> photopolymerization
Lichtintensität f <Strahlphys> intensity of light
Lichtjahr n <Phys> light year
Lichtkabel n <Elektrotech> light cable
Lichtkante f <Konstzeich> imaginary intersection
Lichtkegel m <Optik> cone of rays
Lichtleistung f <Foto> light output
Lichtleiter m <Phys> light guide
Lichtleitertechnik f <Comp & DV, Telekom> (AE) fiber optics, (BE) fibre optics
Lichtleitfähigkeits-Koeffizienten/mit negativem <Foto> light-negative
Lichtleitfähigkeits-Koeffizienten/mit positivem <Foto> light-positive
Lichtleitfaser f 1. <Elektrotech, Ker & Glas, Optik> (AE) optical fiber, (BE) optical fibre; 2. <Telekom> (AE) lightguide fiber, (BE) lightguide fibre, (AE) optical fiber, (BE) optical fibre • **mittels Lichtleitfaser übertragen** <Optik> (AE) fiber, (BE) fibre
Lichtleitfaser f mit abgestuftem Brechungsindex <Phys> (AE) step index fiber, (BE) step index fibre
Lichtleitfaser f mit kontinuierlich veränderlichem Brechungsindex <Phys> (AE) graded index fiber, (BE) graded index fibre
Lichtleitkabel n 1. <Optik> (AE) fiberoptic cable, (BE) fibreoptic cable, (AE) optical fiber cable, (BE) optical fibre cable; 2. <Telekom> (AE) optical fiber cable, (BE) optical fibre cable
Lichtleitstein m <Ker & Glas> light-directing block
Lichtleitung f <Elektrotech> light pipe
Lichtmarke f <Elektronik> light spot
Lichtmarken-Galvanometer n 1. <Gerät> luminous pointer galvanometer; 2. <Strahlphys> light beam galvanometer
Lichtmarkeninstrument n <Gerät> optical pointer instrument
Lichtmarkenleistungsmessgerät n <Gerät> luminous pointer power meter
Lichtmaschine f <Kfztech> generator (KFZ-Elektrik)
Lichtmenge f 1. <Optik> quantity of light (Q); 2. <Phys> quantity of light
Lichtmesser m <Foto> photometer
Lichtmikroskopie f <Metall> light microscopy
Lichtmodulation f <Elektronik, Fernseh> light modulation
Lichtmodulator m <Akustik> light modulator
Lichtmülle f <Phys> radiometer
Lichtnetz n <Elektrotech> (BE) mains, (AE) supply network
Lichtpause f <Fertig> print
Lichtpausfilm m <Konstzeich> diazotype film
Lichtpausmaterial n <Konstzeich> diazotype material
Lichtpauspapier n **auf Gewebe** <Konstzeich> mount diazo paper
Lichtpausschicht f <Konstzeich> diazo coating
Lichtpult n <Druck> light table
Lichtpunkt m <Fernseh> scanning spot
Lichtpunktabtaster m <Fernseh> flying-spot scanner
Lichtpunktkorrektur f <Fernseh> spot shape corrector
Lichtquelle f 1. <Elektronik, Elektrotech, Foto, Labor, Optik, Phys> light source; 2. <Strahlphys> luminous source
Lichtradar n <Funkort> laser radar, lidar, light detection and ranging, optical radar
Lichtreflexion f <Strahlphys> luminous reflectance
Lichtregelsystem n <Lufttrans> dimmer (Beleuchtung)
Lichtsatz m <Druck> photocomposition, phototypesetting
Lichtschacht m <Bau> funnel
Lichtschein m <Wassertrans> loom
Lichtschirm m <Optik> screen
Lichtschnittmikroskop n <Fertig> light-slit microscope
Lichtschnittverfahren n <Fertig> light-slit method
Lichtschranke f <Sicherheit> electro-sensitive safety device, light barrier, photoelectric curtain, photoelectric safeguard, photoelectric safety device
Lichtschreiber m <Phys> light pen
Lichtschutzfilter n <Foto> safelight filter
Lichtsensor m <Elektronik> optical sensor
Lichtsensorsignal n <Elektronik> optical sensor signal
Lichtsignal n 1. <Eisenbahn> (BE) colour light signal, light signal; 2. <Elektronik> light signal
Lichtsignal n **an Fußgängerüberwegen** <Trans> pedestrian crossing light (fußgängerbetätigt)
Lichtskalenschalter m <Foto> light scale switch
Lichtsonde f <Strahlphys> light detector
Lichtspalt m <Anstrich> graticule
Lichtspektrum n <Wellphys> light spectrum
Lichtstärke f 1. <Elektrotech, Ergon> luminous intensity; 2. <Foto> candle power; f-number (einer Linse); 3.

<Phys> luminous intensity; 4. <Strahlphys> intensity of light
Lichtstärkeeinheit f <Optik> light unit
Lichtstärkemessung f <Gerät> luminous intensity measurement
Lichtstift m <Comp & DV> light pen
Lichtstifterkennung f <Comp & DV> light-pen detection
Lichtstrahl m 1. <Fernseh> beam; 2. <Foto> light beam; 3. <Phys> beam, light beam, light ray; 4. <Teilphys> beam; 5. <Telekom> light beam; 6. <Wassertrans> beam *(Signal)*
Lichtstrahloszillograph m <Metrol> Duddell oscillograph, galvanometer oscillograph, loop oscillograph, moving-coil oscillograph
Lichtstrahlschweißen n <Elektronik> laser welding
lichtstreuender Körper m <Optik> diffuser
Lichtstreuung f <Elektronik> light scattering
Lichtstrom m 1. <Elektrotech> luminous flux; 2. <Optik> light current, optical flux; 3. <Phys> luminous flux
Lichtstrommessgerät n <Phys> lumenmeter
Lichttonspalt m <Aufnahme> recording slit
lichtundurchlässig adj <Verpack> lightproof
Lichtundurchlässigkeit f <Wellphys> opacity
Lichtverstärker m <Elektronik> light amplifier
Lichtverstärkung f durch stimulierte Strahlungsemission *(Laser)* <Comp & DV, Druck, Elektronik, Phys, Strahlphys> light amplification by stimulated emission of radiation *(laser)*
Lichtvorhang m <Sicherheit> light curtain, photoelectric curtain
Lichtweg m <Telekom> optical path
Lichtwelle f 1. <Elektronik, Elektrotech> light wave, optical guided wave; 2. <Optik, Wellphys> light wave
Lichtwellenleiter m 1. <Optik> *(AE)* optical fiber, *(BE)* optical fibre, optical waveguide, waveguide; 2. <Telekom> *(AE)* optical fiber, *(BE)* optical fibre
Lichtwellenleiter m mit konstanter Polarisation <Telekom> polarization holding fiber
Lichtwellenleiter m mit parabolischem Brechzahlprofil <Optik, Telekom> *(AE)* parabolic-index fiber, *(BE)* parabolic-index fibre
Lichtwellenleiterachse f <Telekom> *(AE)* fiber axis, *(BE)* fibre axis
Lichtwellenleiteranschluss m <Telekom> *(AE)* optical fiber pigtail, *(BE)* optical fibre pigtail
Lichtwellenleiterausbreitungseigenschaften fpl <Telekom> *(AE)* optical fiber propagation properties, *(BE)* optical fibre propagation properties
Lichtwellenleiterdämpfung f <Telekom> *(AE)* fiber loss, *(BE)* fibre loss
Lichtwellenleiterempfänger m <Telekom> *(AE)* fiber optic receiver, *(BE)* fibre optic receiver
Lichtwellenleiterendeinrichtung f <Telekom> *(AE)* fiberoptic terminal device, *(BE)* fibreoptic terminal device
Lichtwellenleiter-Freileitungskabel n <Telekom> aerial optical cable
Lichtwellenleiterglas n <Telekom> *(AE)* optical fiber glass, *(BE)* optical fibre glass
Lichtwellenleiterkabel n 1. <Comp & DV> *(AE)* fiberoptic cable, *(AE)* fiberoptic connection, *(BE)* fibreoptic cable, *(BE)* fibreoptic connection; 2. <Telekom> *(AE)* fiberoptic cable, *(BE)* fibreoptic cable, optical cable, *(AE)* optical-fiber cable, *(BE)* optical-fibre cable
Lichtwellenleiterkabelübertragung f <Telekom> *(AE)* optical-fiber cable transmission, *(BE)* optical-fibre cable transmission
Lichtwellenleiterkoppler m <Telekom> *(AE)* optical fiber coupler, *(BE)* optical fibre coupler
Lichtwellenleitermantel m <Telekom> *(AE)* fiberoptic cladding, *(BE)* fibreoptic cladding

Lichtwellenleitermehrfachverbindung f <Telekom> *(AE)* multifiber joint, *(BE)* multifibre joint
Lichtwellenleitermodem n <Elektronik, Telekom> *(AE)* fiberoptic modem, *(BE)* fibreoptic modem *(faseroptisches Modem)*
Lichtwellenleiternachrichtensystem n <Telekom> *(AE)* optical-fiber communication system, *(BE)* optical-fibre communication system
Lichtwellenleiternachrichtentechnik f <Telekom> *(AE)* optical-fiber communications, *(BE)* optical-fibre communications
Lichtwellenleiternetz n <Telekom> *(AE)* fiberoptic network, *(BE)* fibreoptic network
Lichtwellenleiteroptik f <Telekom> *(AE)* fiber optics, *(BE)* fibre optics
Lichtwellenleiterprofil n <Telekom> *(AE)* optical fiber profile, *(BE)* optical fibre profile
Lichtwellenleiter-Seekabel n <Telekom> submarine optical fiber cable, undersea fiber-optic cable
Lichtwellenleitersender m <Telekom> *(AE)* fiber optic transmitter, *(BE)* fibre optic transmitter
Lichtwellenleitersensor m <Metrol> *(AE)* optical fiber sensor, *(BE)* optical fibre sensor
Lichtwellenleiterspleiß m <Telekom> *(AE)* fiberoptic splice, *(BE)* fibreoptic splice, *(AE)* optical fiber splice, *(BE)* optical fibre splice
Lichtwellenleitersteckverbinder m <Telekom> *(AE)* optical fiber connector, *(BE)* optical fibre connector
Lichtwellenleitertaper m <Telekom> *(AE)* tapered fiber, *(BE)* tapered fibre
Lichtwellenleitertechnik f 1. <Comp & DV> *(AE)* fiber optics, *(BE)* fibre optics; 2. <Telekom> *(AE)* fiber optics, *(BE)* fibre optics, *(AE)* fiberoptic technology, *(BE)* fibreoptic technology
Lichtwellenleiterübertragung f <Telekom> *(AE)* fiberoptic transmission, *(BE)* fibreoptic transmission, *(AE)* optical fiber transmission, *(BE)* optical fibre transmission
Lichtwellenleiterübertragung f über Hochspannungsleitungen <Telekom> *(AE)* optical fiber link by means of high-voltage line, *(BE)* optical fibre link by means of high-voltage line
Lichtwellenleiter-Übertragungssystem n <Telekom> *(AE)* fiberoptic transmission system, *(BE)* fibreoptic transmission system
Lichtwellenleiterverbindung f <Telekom> *(AE)* optical fiber link, *(BE)* optical fibre link
Lichtwellenringnetzdatenschnittstelle f <Telekom> *(AE)* fiber-distributed data interface, *(BE)* fibre-distributed data interface
Lichtwert m <Foto> light value
Lichtwerteinstellring m <Foto> light value setting ring
Lichtwirkungsgrad m <Lufttrans> lighting efficiency *(Hubschrauber)*
Lichtzeichenmaschine f <Comp & DV> photoplotter
Lichtzeiger m <Comp & DV> light pen
Lichtzerhacker m <Elektronik> light chopper
Lichtzuleitung f <Elektrotech> light cable
Lidar n <Funkort> lidar, light detection and ranging
Lieberkühn-Reflektor m <Foto> Lieberkühn reflector
Lieferant m <Qual> supplier
Lieferdruck m <Elektrotech> as-supplied pressure
liefern v 1. <Elektriz> supply; 2. <Maschinen> deliver, discharge
Lieferposten m <Trans> consignment
Lieferung f 1. <Kerntech, Lebensmittel> delivery; 2. <Qual> consignment
Lieferwagen m 1. <Eisenbahn> van; 2. <Kfztech> delivery truck, van

Lieferwagen

Lieferwagen *m* **mit offener Pritsche** <Kfztech> pick-up, pick-up truck
Lieferzeichnung *f* <Qual> as-delivered condition
liegen *v* **in** <Comp & DV> reside
liegend *adj* <Maschinen> horizontal
liegende Welle *f* <Maschinen> lying shaft
liegender Motor *m* <Kfztech> horizontal engine
Liegeplatz *m* <Wassertrans> berth, shelter; mooring *(Festmachen)*
Liegepresse *f* <Papier> straight-through press
Liegesitz *m* <Kfztech> reclining seat *(Sitz)*
Liegewagen *m* <Eisenbahn> couchette coach
Liek *n* <Wassertrans> leech
Liektau *n* <Wassertrans> bolt rope *(Tauwerk)*
LIFO *(Last-in First-out)* <Comp & DV> LIFO *(last-in-first-out)*
LIFO-Prinzip *n* <Comp & DV> LIFO principle
LIFO-Speicher *m* <Comp & DV> LIFO stack, push-down store
Lift *m* <Hydraul, Maschinen, Trans> *(AE)* elevator, *(BE)* lift
Lift-On-Lift-Off-Schiff *n* <Wassertrans> lift-on lift-off vessel; lift-on lift-off ship *(Vertikalumschlag)*
Lift-On-Lift-Off-System *n* <Wassertrans> lift-on lift-off system
Ligand *m* <Chemie, Metall> ligand
Ligatur *f* <Druck> double letter
Lignan *n* <Chemie> lignan
Lignin *n* <Kunststoff> lignin
Lignit *m* <Kohlen> lignite
Ligroin *n* <Chemie> ligarine, naphtha, ligroin
Li-Li Schiff *n* <Wassertrans> lift-on lift-off ship, lift-on lift-off vessel *(Vertikalumschlag)*
Limit *n* <Maschinen> limit
Limma *n* <Akustik> limma
Limone *f* <Lebensmittel> lime *(Art kleine Zitrone)*
Limonit *m* <Chemie> *(BE)* brown haematite, *(AE)* brown hematite, limonite, brown iron ore
LIM-Verfahren *n* <Kunststoff> LIM, liquid injection moulding
Linalool *n* <Chemie> linalool
Lindan *n* <Chemie> gammexane
Lindenholz *n* <Bau> lime
LINEAC *(Linearbeschleuniger)* 1. <Elektrotech, Phys> LINAC *(linear accelerator)*; 2. <Teilphys> LINEAC *(linear accelerator)*
Lineal *n* <Metrol> straight edge
linear *adj* <Funktech, Geom, Math, Papier, Phys> linear
linear polarisierte Welle *f* 1. <Akustik, Elektrotech, Funktech, Telekom> linearly-polarized wave; 2. <Phys> linearly-polarized wave, plane-polarized wave
linear polarisierter Modus *f (LP-Modus)* <Optik, Telekom> linearly-polarized mode *(LP mode)*
linear polarisierter Wellentyp *m (LP-Mode)* <Optik, Telekom> linearly-polarized mode *(LP mode)*
linear unabhängig *adj* <Bau> linearly independent
linear variabler Differenzialtransformator *m* <Elektrotech> linear variable differential transformer
Linear... <Teilphys> linear
Linearabtastung *f* <Elektronik> linear scan
Linearbedingungen *fpl* <Elektrotech> linear conditions
Linearbeschleuniger *m (LINEAC)* 1. <Elektrotech, Phys> linear accelerator *(LINAC)*; 2. <Teilphys> linear accelerator *(LINEAC)*
Linearbetrieb *m* <Elektrotech> linear operation
Linearcode *m* <Telekom> linear code
Lineardruck *m* <Papier> linear pressure
lineare Aktivität *f* <Kerntech> linear activity *(einer linienförmigen Quelle)*
lineare Algebra *f* <Comp & DV, Math> linear algebra

lineare Amplitudenverzerrung *f* <Elektronik, Telekom> amplitude-frequency distortion
lineare Codierung *f* <Telekom> uniform encoding
lineare Dipolgruppe *f* <Funktech> collinear array
lineare Dispersion *f* <Strahlphys> linear dispersion
lineare Energieübertragung *f (L)* <Kerntech, Strahlphys> linear energy transfer *(L)*
lineare Geschwindigkeit *f* 1. <Fertig> velocity; 2. <Kfztech> straight-line speed
lineare Gleichrichtung *f* <Elektronik> linear detection
lineare Gleichung *f* <Math> linear equation
lineare integrierte Schaltung *f* <Elektronik> linear-integrated circuit
lineare Interpolation *f* <Telekom> linear interpolation
lineare Ionisation *f* <Phys> linear ionization
lineare Kalthärtung *f* <Metall> linear work hardening
lineare Kennlinie *f* <Elektronik> linear characteristic
lineare kinetische Energie *f* <Maschinen> linear kinetic energy
lineare Ladungsdichte *f* <Phys> linear charge density
lineare Liste *f* <Comp & DV> linear list
lineare Modulation *f* <Elektronik> linear modulation
lineare Näherung *f* <Telekom> linear approximation
lineare OEM-Stromversorgung *f* <Elektrotech> linear OEM power supply
lineare Optimierung *f* 1. <Comp & DV> linear optimization, linear programming; 2. <Math> linear optimization
lineare Platte *f* <Optik> *(BE)* linear disc, *(AE)* linear disk
lineare Polarisation *f* <Elektrotech> linear polarization
lineare Prädiktionscodierung *f (LPC)* <Elektronik, Telekom> linear predictive coding *(LPC)*
lineare Programmierung *f* 1. <Comp & DV> linear optimization, linear programming; 2. <Elektronik> *(AE)* linear programing, *(BE)* linear programming
lineare Regelung *f* <Elektrotech> linear control, linear regulation
lineare Schaltung *f* 1. <Phys> linear network; 2. <Telekom> linear circuit
lineare Schwingung *f* <Telekom> linear oscillation
lineare Skale *f* <Elektriz> linear scale
lineare Skalierungsberechnung *f* <Telekom> linear-scaling calculation
lineare Steuerung *f* <Elektriz> linear control
lineare Stromdichte *f (A)* <Elektriz> linear current density *(A)*
lineare Stromversorgung *f* <Elektrotech> linear power supply
lineare Thermodynamik *f* <Thermod> linear thermodynamics
lineare Unabhängigkeit *f* <Math> linear independence
lineare Vergrößerung *f* <Phys> linear magnification
lineare Verstärkung *f* <Elektrotech> linear amplification
lineare Verzerrung *f* <Aufnahme, Elektronik, Telekom> linear distortion
linearer Abschwächungskoeffizient *m* <Kerntech> linear attenuation coefficient
linearer Absorptionskoeffizient *m* <Phys> linear absorption coefficient
linearer Asynchronmotor *m* <Elektriz> linear induction motor
linearer Ausdehnungskoeffizient *m* <Phys> linear expansion coefficient
linearer Bereich *m (R)* <Kerntech> linear range *(R)*
linearer Dämpfungskoeffizient *m* <Phys> attenuation coefficient
linearer Defekt *m* <Metall> linear defect
linearer digitaler Sprachverwürfler *m* <Telekom> linear digital voice scrambler

linearer Elektromotor *m* <Elektrotech> linear electric motor
linearer Gleichrichter *m* <Elektronik> linear detector
linearer Hochleistungsmotor *m* <Kfztech> high-power linear motor
linearer Induktionsmotor *m* <Elektriz> linear induction motor
linearer Kanal *m* <Telekom> linear channel
linearer Konzentrator *m* <Nichtfoss Energ> linear concentrator
linearer Leistungsverstärker *m* <Elektronik, Funktech, Telekom> linear power amplifier
linearer Messwandler *m* <Elektrotech> linear transducer
linearer Prädiktionscodierer *m* **mit Codebuch-Erregung** <Telekom> CELP, codebook-excited linear predictive coder
linearer Schwächungskoeffizient *m* 1. <Kerntech> linear attenuation coefficient; 2. <Phys> linear absorption coefficient
linearer Spannungsanstieg *m* <Elektrotech> ramp voltage
linearer Stark-Effekt *m* <Phys> linear Stark effect
linearer Stromkreis *m* <Elektriz> linear circuit
linearer Transducer *m* <Elektrotech> linear transducer
linearer Verstärker *m* <Elektronik, Telekom> linear amplifier
linearer Vierpolstromkreis *m* <Elektriz> linear four-pole network
linearer Wandler *m* <Akustik> linear transducer
lineares Feedback-Steuersystem *n* <Elektrotech> linear feedback control system
lineares Filter *n* <Telekom> linear filter
lineares Filtern *n* <Telekom> linear filtering
lineares Polymer *n* <Kunststoff> linear polymer
lineares Stromkreiselement *n* <Elektronik> linear circuit element
lineares Stromnetz *n* <Elektriz> linear current network
Linearimpulsverstärker *m* <Elektronik> linear pulse amplifier
linearisieren *v* <Chemie> linearize
Linearisierungsglied *n* <Raumfahrt> linearizer *(Weltraumfahrt)*
Linearität *f* <Aufnahme, Elektriz, Elektronik, Funktech, Telekom> linearity
Linearität *f* **bei Festpunkteinstellung** <Regelung> terminal-based linearity
Linearität *f* **bei Nullpunkteinstellung** <Regelung> zero-based linearity
Linearitätsfehler *m* <Fernseh> linearity error
Linearitätssteuerung *f* <Fernseh> linearity control
Linearkreis *m* <Elektronik> linear circuit
Linearmaßstab *m* <Elektronik> linear scale
Linearmatrix *f* <Fernseh> linear matrix
Linearmodulator *m* <Elektronik> linear modulator
Linearmotor *m* <Kfztech> linear motor
Linearnetz *n* <Elektrotech> linear network
Linear-Potenziometer *n* <Elektrotech> linear potentiometer
Linearschaltung *f* <Elektronik> linear circuit *(für integrierte Schaltung)*
Linearschwingung *f* <Maschinen> linear oscillation
Linearskale *f* <Metrol> linear scale
Linearspannung *f* <Elektrotech> linear voltage
Linearstrahlröhre *f* <Elektronik> linear beam tube, linear tube
Linearstrahl-Rückwärtswellen-Oszillator *m* <Elektronik> linear beam backward wave oscillator
Linearstrahlverstärker *m* <Elektronik> linear beam amplifier

Linearturbine *f* <Wassertrans> linear turbine
Linearverhalten *n* <Elektronik> *(AE)* linear behavior, *(BE)* linear behaviour, linear performance
Linearwiderstand *m* <Elektriz, Elektrotech, Phys> linear resistor
Liner *m* <Erdöl> liner, liner pipe
Linie *f* 1. <Druck> line, rule; 2. <Eisenbahn> line; 3. <Elektronik> line *(der Schutzschicht)*; line *(des Spektrums)*; 4. <Geom, Ker & Glas, Strahlphys> line; 5. <Kerntech> peak *(in einem Spektrum)* • **die Linien abschnüren** <Wassertrans> lay down the lines *(Schiffkonstruktion)* • **eine Linie durchziehen** <Konstzeich> draw a continuous line
Linie *f* **gleicher Inklination** <Phys> isoclinal line
Linie *f* **in Betrieb** <Eisenbahn> line in service
Linie *f* **mit konstantem Abstand** <Geom> equidistant line
Linien *fpl* **der Paschenserie** <Strahlphys> Paschen series lines
Linienabstand *m* <Konstzeich> line spacing
Linienangriff *m* <Fertig> line application
Linienbeschickungsgeräte *npl* <Verpack> line-feeding equipment
Linienbreite *f* 1. <Konstzeich> line thickness; 2. <Strahlphys> line width
Liniendicke *f* <Metrol> thickness of lines
Liniendienst *m* <Lufttrans> scheduled service
Linienflug *m* <Lufttrans> scheduled flight
Linienflugzeug *n* <Lufttrans> airliner, liner
linienförmiges Netz *n* <Telekom> line network
Liniengeometrie *f* <Geom> line geometry
Liniengeschwindigkeit *f* <Verpack> line speed
Liniengrafik *f* <Comp & DV> line graphics
Liniengruppe *f* <Konstzeich> line group
Linienintegral *n* <Phys> line integral
Linienprofil *n* <Strahlphys> line profile
Linienprofil *n* **von Spektrallinien** <Strahlphys> profiles of spectral lines
Linienquelle *f* <Umweltschmutz> line source
Linienriss *m* <Wassertrans> lines drawing, lines plan *(Schiffkonstruktion)*
Linienschreiber *m* <Gerät> strip chart line recorder
Linienspannung *f* <Metall> line tension
Linienspektrum *n* 1. <Akustik, Optik, Phys> line spectrum; 2. <Raumfahrt> bright-line spectrum; 3. <Strahlphys> discontinuous spectrum, line spectrum; 4. <Telekom> line spectrum
Linienzugbeeinflussung *f* <Eisenbahn> continuous automatic train control
Linierung *f* <Papier> lining *(auf Schreibpapier)*
liniieren *v* <Druck> line, rule
Linkage-Editor *m* <Comp & DV> linkage editor
linke Seite *f* 1. <Druck> verso; 2. <Phys> left-hand side *(einer Gleichung)*
Linke-Hand-Regel *f* <Phys> left-hand rule
linker Rand *m* <Druck> left margin
linker Stereo-Kanal *m* <Aufnahme> left stereo channel
Link-Kopplung *f* <Elektrotech> link coupling
Link-Name <Comp & DV> linkname
Link-Relais *n* <Elektrotech> link relay
links abbiegender Verkehr *m* <Trans> left-turning traffic
links flatternd *adj* <Druck> ragged left
Links… <Comp & DV, Trans> left
Linksabbiegephase *f* <Trans> left turn phase
Linksausrichtung *f* <Comp & DV> left justification
linksbündig ausrichten *v* <Comp & DV> left justify
Linksbündigkeit *f* <Comp & DV> left justification
Linksdraht *m* <Textil> S-twist
Linksdrall *m* <Fertig> left-hand helix • **mit Linksdrall** <Fertig> left-helix

linksdrehend 464

linksdrehend adj 1. <Fertig> laevorotatory; 2. <Gerät> ccw, counterclockwise
linksdrehende polarisierte Welle f <Funktech> anticlockwise polarized wave
linksdrehende Zirkularpolarisation f 1. <Funktech> LHCP, left-hand circular polarization; 2. <Raumfahrt> LHCP, left-hand circular polarization *(Weltraumfunk)*
linksdrehender Propeller m <Wassertrans> left-handed propeller
Linksdrehung f <Textil> S-twist
Linksflanke f <Maschinen> left-hand tooth flank
linksgängig adj <Maschinen> left-hand, left-handed
Linksgewindeschraube f <Maschinen> left-handed screw
Linkshändigkeit f <Ergon> left handedness
Link-Sicherung f <Elektrotech> link fuse
Linkslauf m <Maschinen> reverse action
linksläufig adj <Fertig> left-hand helix
linksschief adj <Geom> positively skewed
linksschneidend adj <Maschinen> left-handed
linkssteigend adj <Fertig> left-hand helical
Linksverschiebung f <Comp & DV> left shift
Linol... <Chemie> linoleic
Linoleat n <Chemie> linoleate
Linolein n <Chemie> linoleine
Linolen... <Chemie> linolenic
Linolsäure f 1. <Kunststoff> linoleic acid; 2. <Lebensmittel> linoleic acid *(essenzielle Fettsäure)*
Linolschnitt m <Druck> linocut
Linotype® f <Druck> Linotype®
Linse f <Labor, Optik, Phys> lens
Linse f mit Begrenzungsblende <Phys> stopped lens
Linsenantenne f 1. <Raumfahrt> lens antenna *(Weltraumfunk)*; 2. <Telekom> lens antenna
Linseneffekte mpl <Foto> lens flares
Linsenfassung f <Foto> lens mount
Linsenfehler m <Optik> aberration
linsenförmig adj <Foto> lens-shaped
Linsenhalter m <Ker & Glas> lens holder
Linsenkopf m <Maschinen> raised head
Linsenkopfschraube f 1. <Bau> cheese-head screw; 2. <Maschinen> fillister-head screw, raised head screw
Linsenkuppe f <Maschinen> blunt start, oval point
Linsenlichtflecke mpl <Foto> lens flares
Linsenscheitel m <Foto> lens vertex
Linsenschirm m <Fernseh> flag
Linsenschraube f 1. <Bau> slotted fillister head screw; 2. <Maschinen> oval-head screw
Linsensenkkopf m <Maschinen> raised countersunk head
Linsensenkniet m <Maschinen> oval countersunk rivet
Linsensenkschraube f <Maschinen> raised countersunk head screw
Linsenvergütung f <Foto> lens coating
Linsenzwilling m <Metall> lenticular twin
Lipase f <Chemie> lipase
Lipid n <Chemie> lipid
lipoid adj <Chemie> lipoid
lipophil adj <Chemie> lipophile, lipophilic
Lipopolysaccharid n <Chemie> lipopolysaccharide
Lippe f 1. <Fertig> lip *(Dichtung)*; 2. <Maschinen> lip
Lippendichtung f 1. <Maschinen> lip seal, lip-type seal; 2. <Mechan> lip seal
Lippenlesen n <Akustik> lip reading
Lippenmikrofon n <Akustik> lip microphone
Lippensynchronisation f <Aufnahme> lip-sync
Liquiduslinie f <Metall> liquidus line
LISP *(Listenprogrammiersprache)* <Comp & DV> LISP *(list-programming language)*

Lissajous'sche Figur f <Funktech, Phys> Lissajous figure
Liste f <Comp & DV> list
Liste f freier Speicherplätze <Comp & DV> available list
Liste f qualifizierter Lieferanten <Qual> approved vendors list
Liste f zugelassener Erzeugnisse <Qual> qualified products list
Listenerstellung f <Comp & DV> report generation
Listenpreis m <Fertig> list price *(Kunststoffinstallationen)*
Listenprogrammgenerator m <Comp & DV> RPG, report program generator
Listenprogrammiersprache f *(LISP)* <Comp & DV> list-programming language *(LISP)*
Listenstruktur f <Comp & DV> list structure
Listenverarbeitung f <Comp & DV> list processing
Listing n <Comp & DV> program listing; listing *(eines Programms)*
Liter m <Metrol, Phys> *(AE)* liter, *(BE)* litre *(Einheit des Volumens)*
Literal n <Comp & DV> literal
Litermolarität f <Chemie> molarity
Lithargit m <Kunststoff> lead oxide, litharge
Lithergol n <Raumfahrt> lithergol
Lithium n *(Li)* <Chemie> lithium *(Li)*
Lithiumbatterie f <Elektrotech> lithium battery
Lithiumchloridakkumulator m <Kfztech> lithium-chlorine storage battery
Lithographie f <Druck> lithograph, lithographic print, lithography
Lithographieglas n <Ker & Glas> glass for lithography
lithographierte Packung f <Verpack> lithographed package
lithographische Platte f <Druck> lithoplate
lithographischer Vorgang m <Elektronik> lithographic process
Lithomaske f <Elektronik> lithographic mask
Lithosphäre f <Nichtfoss Energ> earth's crust, lithosphere
Litoral n <Wassertrans> littoral
litoral adj <Wassertrans> littoral *(Geographie)*
Litosphäre f <Umweltschutz> litosphere
Littletontemperatur f <Ker & Glas> Littleton softening point
Litze f 1. <Fertig, Maschinen, Phys> strand; 2. <Textil> braid, braiding, heald
Litzenbesatz m <Textil> braiding
Litzendraht m 1. <Elektriz, Elektrotech> braided wire, stranded conductor; 2. <Textil> heald wire
Litzenkabel n <Elektrotech> flexible cable
Litzenseil n <Maschinen> stranded rope
Live-Kamera f <Fernseh> live camera
Live-Sendung f <Fernseh> live broadcast
Live-Übertragung f <Fernseh> live coverage
Lizenz f <Patent> *(BE)* licence, *(AE)* license • **Lizenz erteilen** <Patent> grant a licence
Lizenzenentzug m <Patent> revocation of license
Lizenzgeber m <Patent> licensor
Lizenzgebühren fpl <Patent> royalties
LKF *(Luftkissenfahrzeug)* 1. <Kfztech> SEV *(surface effect vehicle)*; 2. <Trans> ACV *(air cushion vehicle)*; 3. <Wassertrans> *(BE)* hovercraft, *(AE)* hydroskimmer, surface effect ship
LKT *(luftgekühlte Triode)* <Elektronik> ACT *(air-cooled triode)*
Lkw m *(Lastkraftwagen)* <Kfztech> HGV *(heavy goods vehicle)*
Lkw m mit Hubladeklappe <Kfztech> tail lift truck
Lkw m mit offener Ladefläche <Kfztech> rack body truck
Lkw-Dienst m <Trans> cartage service

Lkw-Fahrer m <Trans> (AE) trucker, truck driver
Lkw-Zug m <Kfztech> road train
Lloyd'scher Spiegel m <Phys> Lloyd's mirror
lm (Lumen) <Fernseh, Metrol, Phys> lm (lumen)
LM (Lunar-Modul) <Raumfahrt> LM, lunar module (Raumschiff)
LM-Hangar m <Raumfahrt> LM hangar (Raumschiff)
L-Netz n <Elektrotech> L-network
LNG (Flüssigerdgas) <Erdöl, Thermod> LNG (liquefied natural gas)
LNG-Tanker m <Wassertrans> LNG tanker
ln(x) (logarithmus naturalis) ln(x), natural logarithm of x
Lobeliaalkaloid n <Chemie> lobelia alkaloid
Lobelin n <Chemie> lobeline
Loch n 1. <Bau> opening; 2. <Elektrotech> hole (Elektronenloch); 3. <Erdöl> hole (Bohrtechnik); 4. <Kerntech, Maschinen> hole; 5. <Phys> hole (fehlender Ladungsträger im Halbleiter) • **Loch graben** <Bau> hole
Lochabstand m <Foto> pitch (beim Film)
Lochband n <Meerschmutz> sifting belt
Lochbandsteuerung f <Fertig> punched-tape control
Lochbild n <Konstzeich> hole pattern
Lochblech n 1. <Bau> punched plate; 2. <Kerntech> perforated plate
Lochblende f <Fernseh> scanning aperture
Lochbohrung f <Bau, Kohlen> boreholing
Lochdorn m 1. <Fertig> drift, tapered punch; 2. <Maschinen> piercer
Lochdurchmesser m <Fertig> perforation (Kunststoffinstallationen)
Locheisen n 1. <Bau> punch; 2. <Maschinen> hollow punch; 3. <Mechan> broaching, punch
lochen v <Mechan> punch
Lochen n 1. <Bau, Kohlen> boring; 2. <Comp & DV> punching
Locher m <Comp & DV> perforator, punch
Löcherhalbleiter m <Elektronik> p-type semiconductor
Löcherleitung f <Elektrotech> hole conduction
Löchertheorie f <Phys> Dirac hole theory (Quantenmechanik)
Lochfeile f <Maschinen> riffler
Lochfotografie f <Foto> pinhole photography
Lochfräser m <Maschinen> arbor-type cutter
Lochfraß m 1. <Anstrich> pitting; 2. <Ker & Glas> pitting (Korrosion); 3. <Maschinen> pitting
Lochkamera f <Foto> pinhole camera
Lochkameraobjektiv n <Optik> pinpoint lens
Lochkarte f <Comp & DV> punch card, punched card
Lochkartendoppler m <Comp & DV> punched-card reproducer
Lochkartenleser m <Comp & DV> punched-card reader
Lochkartenmischer m <Comp & DV> collator
Lochkartenstanzer m <Comp & DV> card punch
Lochkreis m <Fertig> scribed circle (Bohren)
Lochlehre f <Bau> (AE) plug gage, (BE) plug gauge (Messtechnik)
Lochmaschine f <Druck> perforating machine
Lochmaske f 1. <Elektronik> shadow mask (Video); 2. <Fernseh> aperture mask, shadow mask
Lochmaskenröhre f <Fernseh> shadow mask tube
Lochmutter f <Maschinen> ring nut
Lochplatte f 1. <Fertig> boss; swage block (Schmieden); 2. <Mechan> swage block
Lochpresse f <Maschinen> perforating press, piercing press
Lochring m <Fertig> bolster (Schmieden)
Lochsäge f 1. <Bau> lock saw; 2. <Maschinen> keyhole saw
Lochscheibe f <Ker & Glas> bushing

Lochspalte f <Comp & DV> card column
Lochstanze f 1. <Bau, Mechan> drift; 2. <Papier> punching press
Lochstempel m <Maschinen> piercing die
Lochstreifen m 1. <Aufnahme> perforated tape; 2. <Comp & DV> perforated tape, punch tape, punched tape; 3. <Telekom> perforated tape, punched tape
Lochstreifendoppler m <Comp & DV> punched-tape reproducer
Lochstreifenleser m <Comp & DV> punched-tape reader, tape reader
Lochstreifenstanzer m <Comp & DV> tape punch
Lochstreifensteuerung f <Fertig> taping
Lochung f <Bau> core (Ziegel)
Lochwalze f <Papier> holy roll
Lochwerkzeug n <Fertig> punching tool
Lochzeile f <Comp & DV> row
Lochziegel m <Bau> perforated brick
locker gehaltener Diamant m <Ker & Glas> diamond held trailing
lockerbar adj <Fertig> strippable
lockere Schwelle f <Eisenbahn> (BE) pumping sleeper, (AE) pumping tie
lockeres Bandkabel n <Optik> loose flat cable
lockeres Schlauchkabel n <Optik> loose tube cable
lockern v 1. <Bau> ease (Erde); 2. <Fertig> strip (Schraube); 3. <Lebensmittel> aerate (Teig); 4. <Maschinen> slacken; 5. <Mechan> loosen; 6. <Qual> ease
lockern v/sich <Maschinen> work loose
Lockern n <Maschinen> slackening, slacking
lodernd adj <Thermod> blazing
Löffel m <Meerschmutz> shovel (Löffelbagger)
Löffelbagger m 1. <Bau> power shovel (Tiefbau); 2. <Wassertrans> dipper dredge, dipper dredger (Baggern); 3. <Wasserversorg> spoon dredge, spoon dredger
Löffelbohrer m <Bau> shell auger, spoon auger (Tiefbau)
Löffelbug m <Wassertrans> spoon bow (Schiffbau)
Löffelschaber m <Fertig> half-round scraper
Löffeltiefbagger m <Meerschmutz> backhoe
Log n 1. <Erdöl> log (Bohrtechnik); 2. <Lufttrans> log; 3. <Wassertrans> log (Schiffgeschwindigkeitsmesser)
Logarithmenpapier n <Math> logarithmic paper
logarithmisch adj <Math> logarithmic
logarithmische Darstellung f <Comp & DV> logarithmic graph
logarithmische Kennlinie f <Elektronik> logarithmic characteristic
logarithmische Skale f <Elektriz, Elektronik> logarithmic scale (Messtechnik)
logarithmische Spirale f <Geom> logarithmic spiral
logarithmischer Maßstab m <Elektronik> logarithmic scale
logarithmischer Verstärker m <Elektronik> logarithmic amplifier
logarithmischer Videoverstärker m <Elektronik, Fernseh> logarithmic video amplifier
logarithmisches Dekrement n <Elektronik, Phys> logarithmic decrement
logarithmisches Kriechen n <Metall> logarithmic creep
logarithmisches Potenziometer n <Elektrotech> logarithmic potentiometer
logarithmisches Wobbeln n <Elektronik> logarithmic sweep (Messtechnik)
logarithmisch-periodische Antenne f <Funktech> log periodic antenna
Logarithmus m 1. <Comp & DV> log, logarithm; 2. <Math> logarithm
Logarithmusfunktion f <Math> logarithmic function (Umkehrfunktion einer Exponentialfunktion)

Logatom

Logatom n <Akustik> logatom
Logbuch n 1. <Lufttrans, Trans> logbook; 2. <Wassertrans> ship's log; logbook *(Dokumente)*
Logbucheintrag m <Lufttrans, Wassertrans> log
Logformat n <Comp & DV> log format
Logger m <Wassertrans> drifter *(Schiffstyp)*
Logik f 1. <Comp & DV> logic; 2. <Elektronik, Math> logic
Logikanalysator m <Comp & DV, Elektronik> *(BE)* logic analyser, *(AE)* logic analyzer
Logikanalyse f <Comp & DV, Elektronik> logic analysis
Logikaufbau m 1. <Comp & DV> logic design; 2. <Elektronik> logic pattern
Logikbaustein m <Comp & DV, Elektronik> logic device
Logikeinheit f <Comp & DV> logic unit
Logikelement n <Comp & DV> logic element
Logikentwurf m <Comp & DV> logic design
Logikfamilie f <Comp & DV, Elektronik> logic family *(Schaltkreisfamilie)*
Logikgatter n <Phys> logic gate
Logikkonzeption f <Elektronik> logic design
Logikoperation f <Comp & DV> logic operation
Logikoperator m <Comp & DV, Elektronik> logic operator
Logikpegel m <Comp & DV> logic level
Logikraster n <Comp & DV> logic grid
Logikschaltbild n <Elektronik> logic diagram
Logikschaltkarte f <Comp & DV> logic card
Logikschaltung f 1. <Comp & DV> logic circuit; 2. <Phys> logic gate; 3. <Telekom> logic circuit
Logiksignal n <Elektronik> logic signal
Logiksimulation f <Elektronik> logic simulation
Logiksimulator m <Elektronik> logic simulator
Logiksymbol n <Comp & DV> logic symbol
Logiktaktanalyse f <Elektronik> logic timing analysis
Logiktaktsteuerung f <Elektronik> logic timing
Logiktest m <Elektronik> logic test
Logiktester m <Elektronik> logic tester
Logik- und Taktanalysator m <Elektronik> *(BE)* logic state and timing analyser, *(AE)* logic state and timing analyzer
Logikverknüpfung f <Elektronik> logic operation
Logikzeichen n <Comp & DV> logic symbol
Logikzeitmessung f <Elektronik> logic timing
Logikzustand m <Elektronik> logic state
Logikzustandsanalyse f <Elektronik> logic state analysis
logisch adj <Comp & DV> logical • **logisch eins** <Elektronik> logic high • **logisch null** <Elektronik> logic low
logisch integrierte Schaltung f <Elektronik> logic-integrated circuit
logisch schließen v <Math> implicate
logische Adressierung f <Comp & DV> logical addressing
logische Datei f <Comp & DV> logical file
logische Entscheidung f <Comp & DV> decision instruction
logische Fortschreibung f <Comp & DV> indexed sequence
logische Mikroschaltung f <Elektronik> logic microcircuit
logische Operation f <Comp & DV, Math> logical operation
logische Schlüsse mpl **pro Sekunde** <Künstl Int> LIPS; logical inferences per second *(Maß für Leistung von KI-Systemen)*
logische Seitenlänge f <Druck> logical page length
logische Variable f <Comp & DV> logical variable
logischer Befehl m <Comp & DV> logic instruction
logischer Block m <Comp & DV> logical block
logischer Kanal m <Comp & DV> logical channel
logischer Operator m <Comp & DV> logical operator
logischer Satz m <Comp & DV> logical record

logischer Schluss m 1. <Comp & DV> inference; 2. <Math> implication, inference
logisches Ausgangssignal n <Elektronik> logic output signal
logisches Diagramm n <Comp & DV> logical chart
logisches Eingangssignal n <Elektronik> logic input signal
logisches Schließen n 1. <Comp & DV> inference; 2. <Math> implication, inference
logisches Verknüpfungsglied n <Elektronik> logic operator
logisches Verschieben n <Comp & DV> logical shift
Logistik f <Comp & DV> logistics
logistische Unterstützung f <Trans> logistic support
Logleine f <Wassertrans> log line *(Tauwerk)*
lohgar adj <Fertig> tanned *(Gerberei)*
LOI *(Sauerstoffindex)* <Kunststoff> LOI *(limiting oxygen index)*
Lok f *(Lokomotive)* <Eisenbahn> locomotive
lokal adj <Comp & DV> local
lokal gemessene Geschwindigkeit f <Trans> spot speed
lokal oxidierter Übergang m <Elektronik> locally-oxided junction
Lokal... <Comp & DV> local
Lokalbetrieb m <Comp & DV> local mode
lokale Batterie f *(LB)* <Elektrotech> local battery *(LB)*
lokale Erklärung f <Comp & DV> local declaration
lokale Korrosion f <Maschinen> local corrosion
lokale Variable f <Comp & DV> local variable
lokaler Mehrpunkt-Verteildienst m <Telekom> local multipoint distribution service
lokaler Oszillator m <Elektronik> phase local oscillator
lokaler Punkt-zu-Mehrpunkt-Funkdienst m <Funktech, Telekom> local multipoint digital service
lokaler Rundfunk m <Fernseh> community broadcasting
lokales Busnetz n <Telekom> bus local area network
lokales Datennetz n <Telekom> local area network
lokales drahtloses Netz n <Mobilkom> radio-LAN, radio local area network, wireless LAN, wireless local area network
lokales Kreuzungssteuergerät n <Trans> local intersection controller
lokales Netz n *(LAN)* <Comp & DV, Telekom> local area network *(LAN)*
lokales ringförmiges Netz n <Telekom> ring LAN, ring-type LAN, ring-type local area network
lokales Steuergerät f <Trans> local controller
lokalisieren v <Comp & DV> localize
lokalisierte Interferenzen fpl <Phys> localized fringes *(in einer Schicht)*
Lokalisierung f 1. <Comp & DV> localization; 2. <Telekom> tracking *(Satellit)*
Lokalmodus m <Comp & DV> local mode
Lokaloszillator m <Funktech> local oscillator
Lokalspeicher m <Comp & DV> local memory
Lokfernsteuerung f <Eisenbahn> radio control of locomotive
Lokomotivbetriebswerk n <Eisenbahn> locomotive depot
Lokomotive f *(Lok)* <Eisenbahn> locomotive
Lokomotive f **mit direktem Antrieb** <Eisenbahn> direct drive locomotive
Lokomotive f **mit Reibungs- und Zahnradantrieb** <Eisenbahn> rack rail locomotive
Lokomotivführer m <Eisenbahn> train engineer
Lokomotivschuppen n <Eisenbahn> engine shed
Lokschuppen n <Eisenbahn> locomotive shed
Loktalsockel m <Elektronik> loctal base *(Elektronenröhren)*

Lokwartegleis n <Eisenbahn> locomotive holding track
Longifolen n <Chemie> longifolene
longitudinale zyklische Steuerungshilfe f <Lufttrans> fore-and-aft cyclic control support *(Hubschrauber)*
longitudinaler zyklischer Steuerknüppel m <Lufttrans> fore-and-aft cyclic stick *(Hubschrauber)*
Longitudinalprüfung f <Comp & DV> LRC, longitudinal redundancy check
Longitudinalwelle f <Phys, Wellphys> longitudinal wave *(Schallwellen)*
Long-Line-Effekt m <Telekom> long-line effect
LOOP-Anweisung f <Comp & DV> LOOP statement *(EDV-Programm)*
Lophin n <Chemie> lophine
LORAN *(Langstreckennavigationskette)* <Lufttrans, Wassertrans> loran *(long-range navigation)*
Lore f 1. <Eisenbahn> *(BE)* small tip wagon, *(AE)* spoil car, *(BE)* trolley; 2. <Kohlen> trolley
Lorentz-Fitzgerald'sche Kontraktion f <Phys> Lorentz-Fitzgerald contraction
Lorentz-Lorenz'sche Gleichung f <Phys> Lorentz-Lorenz formula
Lorentz'sche Bedingung f <Phys> *(AE)* Lorentz gage, *(BE)* Lorentz gauge
Lorentz'sche Kraft f <Elektrotech, Phys> Lorentz force
Lorentz'sche Transformation f <Phys> Lorentz transformation
Lorenz'sche Einheit f *(L)* <Thermod> Lorenz unit *(L)*
Lorenz'sche Konstante f <Phys> Lorenz constant *(Wärmeleitung)*
Lorinmaschine f <Lufttrans> athodyd
Los n 1. <Fertig> run; batch *(Kunststoffinstallationen)*; 2. <Qual> lot
lösbar adj 1. <Fertig> removable *(Verbindungen)*; 2. <Maschinen> detachable, separable
lösbare Verbindung f 1. <Fertig> fastening, non-permanent connection; 2. <Maschinen> detachable union
Losbrechmoment n <Elektrotech, Maschinen> break-away torque
Löschanlage f <Sicherheit> extinguishing equipment
löschbare CD f <Optik> erasable optical disk
löschbare Datenträgerscheibe f <Optik> erasable data disk
löschbarer optischer Speicher m <Optik> erasable optical storage
löschbarer programmierbarer Lesespeicher m *(EPROM)* <Comp & DV> erasable programmable read-only memory *(EPROM)*
löschbarer Speicher m <Comp & DV> erasable memory, erasable storage
löschbares optisches Datenträgermedium n <Optik> erasable optical medium
Löschbereich m <Comp & DV> purge area
löschbereit adj <Comp & DV> clear-to-zero
Löschdämpfung f <Akustik, Fernseh> erasure
Löschdrossel f 1. <Elektronik> commutating reactor; 2. <Fernseh> bulk eraser
löschen v 1. <Aufnahme> erase; 2. <Bau> quench *(Kalk)*; 3. <Comp & DV> clear, erase, zeroize; delete *(Zeichen)*; 4. <Druck> kill; 5. <Fertig> slake *(Kalk)*; 6. <Funktech> cancel; 7. <Mechan> quench; 8. <Meerschmutz> unload *(bei Schiffen)*; 9. <Telekom> cancel; 10. <Thermod> quench *(Feuer)*; 11. <Trans> dump; 12. <Wassertrans> discharge, unload, unship *(Ladung)*
Löschen n 1. <Akustik> erasure; 2. <Comp & DV> deletion, purging, reset; 3. <Elektriz> quench, blowing out; 4. <Fernseh> erasure; 5. <Telekom> deletion; 6. <Trans> landing *(einer Ladung)*; 7. <Wassertrans> landing

Löschen n **durch den Endanwender** <Comp & DV> end-user deletion
löschender Cursor m <Comp & DV> destructive cursor
löschendes Lesen n <Comp & DV> destructive read
Löscher m <Maschinen> extinguisher
Löschfrequenz f <Aufnahme> erase frequency
Löschgerät n 1. <Fernseh> eraser; 2. <Sicherheit> extinguishing equipment
Löschkalk m 1. <Bau> hydrated lime; 2. <Chemie> quicklime, slaked lime
Löschkanone f <Wassertrans> water gun *(Brandbekämpfung)*
Löschkondensator m <Elektrotech> commutating capacitor, quench capacitor
Löschkopf m <Aufnahme, Comp & DV, Fernseh> erase head
Loschmidt'sche Zahl f *(n0)* <Phys> Loschmidt number *(n0)*
Löschmittel n <Sicherheit> extinguishant
Löschpapierzwischenlage f <Fertig> blotting-paper washer
Löschphase f <Elektrotech> arc quenching phase *(eines Schalters)*
Löschrohrblitzableiter m <Elektrotech> expulsion-type lightning arrester
Löschrohrsicherung f <Elektrotech> expulsion-tube fuse
Löschsatz m <Comp & DV> deletion record
Löschschaum m <Thermod> foam
Löschschlauch m <Sicherheit> fire hose
Löschschlauchanschluss m <Sicherheit> fire hose coupling
Löschschlauchhaspel f <Sicherheit> fire hose reel
Löschspannung f <Elektrotech> extinction voltage
Löschspule f 1. <Aufnahme> bulk eraser; 2. <Elektrotech> blowout coil, quenching coil
Löschstrom m <Aufnahme, Fernseh> erasing current
Löschtaste f <Telekom> cancel key
Löschtechnik f <Sicherheit> extinguishing equipment, fire-fighting equipment
Löschtransformator m <Elektrotech> neutral compensator, neutralizing transformer
Löschtrog m <Fertig> bosh, water bosh *(Schmieden)*
Löschung f 1. <Akustik, Fernseh> erasure; 2. <Kontroll> clearing; 3. <Patent> *(AE)* cancelation, *(BE)* cancellation *(der Eintragung)*; 4. <Telekom> deletion
Löschungsvollzug m <Telekom> cancellation completed
Löschwasserpumpe f <Sicherheit> fire pump
Löschzeichen n <Comp & DV> delete character
Löschzeit f <Elektriz> gate turn-off time
Lose f <Maschinen> backlash, play, slack
lose adj <Maschinen, Papier> loose
lose Ablagerung f <Umweltschutz> bulk deposition
lose gewickelte Windungen fpl <Elektrotech> loosely-wound turns
lose Klemme f <Elektriz> loose terminal
lose Kupplung f <Comp & DV, Elektrotech, Maschinen, Phys> loose coupling
lose Querschwelle f <Eisenbahn> *(BE)* dancing sleeper, *(AE)* dancing tie
lose Riemenscheibe f <Maschinen> dead pulley, loose pulley
lose Rolle f <Maschinen> idle pulley
lose Schüttung f <Eisenbahn> loose ballasting
lose Stoßstelle f <Optik> loose buffer
lose Wolle f <Ker & Glas> loose wool
Loseblatt n <Druck, Papier> loose-leaf
Losefehler m <Lufttrans> backlash error *(Höhenmesser)*
Lösekeil m <Maschinen> loosening wedge

Lösemittel

Lösemittel *n* 1. <Chemie> dissolvent, menstruum; 2. <Chemtech> dissolvent
lösemittelabstoßend *adj* <Chemie> lyophobic
lösemittelanziehend *adj* <Chemie> lyophilic
lösen *v* 1. <Bau> ease; 2. <Kfztech> release *(die Bremse)*; 3. <Maschinen> disengage; 4. <Math> solve *(eine Gleichung)*; 5. <Papier> dissolve; 6. <Qual> ease; 7. <Wassertrans> loosen *(Tauwerk)*
Lösen *n* 1. <Eisenbahn> loosening *(einer Kupplung)*; 2. <Fertig> release
loser Flansch *m* <Fertig> backing flange *(Kunststoffinstallationen)*
loser Kabelaufbau *m* <Optik> loose cable structure
loser Transport *m* <Verpack> bulk transport
loser Umschlag *m* <Verpack> wraparound
loser Zement *m* <Bau> bulk cement
loses Formwerkzeug *n* <Kunststoff> *(AE)* portable mold, *(BE)* portable mould
loses Glas *n* <Ker & Glas> loose glass
loses Stoßen *n* <Optik> loose buffering
Losgröße *f* <Fertig, Qual> lot size
Loslager *n* <Maschinen> movable bearing
löslich *adj* <Erdöl, Maschinen> soluble
Löslichkeit *f* 1. <Chemie> dissolubility, solubility; 2. <Kohlen, Kunststoff> solubility
Löslichkeitsverbesserer *m* <Chemie> solutizer
loslösen *v* <Mechan> loosen
losmachen *v* <Wassertrans> unmoor
Losmachen *n* <Wassertrans> unmooring
Losrad *n* <Maschinen> loose wheel
Lössboden *m* <Bau> loess
Losscheibe *f* <Maschinen> loose pulley, loose wheel
losschlagen *v* <Fertig> rap *(Modell)*
Losschlagen *n* <Fertig> rapping *(Modell)*
losschrauben *v* <Maschinen> unscrew
Losschrauben *n* <Maschinen> unbolting, unscrewing
lossegeln *v* <Wassertrans> sail away
Lossitz *m* <Hydraul> loose seat *(Ventil)*
Losumfang *m* <Qual> lot size
Lösung *f* 1. <Chemie> dissolution; dilution *(verdünnt)*; 2. <Math> root, solution; 3. <Thermod> solution • **in Lösung schwimmen** <Anstrich> suspend
Lösung f einer Aufgabe <Patent> solution of a problem
Lösung f zur Entscheidungshilfe <Comp & DV> decision support solution
Lösungsbenzin *n* <Erdöl> petroleum spirit *(Destillationsprodukt)*
Lösungsglühen *n* <Metall> solution annealing
Lösungsgraph *m* <Künstl Int> solution graph
Lösungsmittel *n* 1. <Anstrich, Kohlen, Kunststoff, Lebensmittel, Metall> solvent; 2. <Papier> dissolver; 3. <Textil, Umweltschutz> solvent
Lösungsmitteldampf *m* <Kohlen> solvent vapour
Lösungsmittelextraktion *f* <Lebensmittel> solvent extraction
Lösungsmittelresistenz *f* <Anstrich> solvent resistance
Lösungsmittelretention *f* <Kunststoff> solvent retention
Lösungsmittelrückgewinnung *f* <Druck, Lebensmittel> solvent recovery
Lösungsmittelrückgewinnungsanlage *f* <Abfall> solvent recovery plant, solvent recovery unit
Lösungspolymerisation *f* <Kunststoff> solution polymerization
Lösungsschweißen *n* <Kunststoff> solvent welding
Lösungsstrategie *f* <Künstl Int> solution strategy
Lösungs- und Verdünnungs-Kältemaschine *f* <Phys> dilution refrigerator
Lösungsvermittler *m* <Chemie> solutizer
Lösungswärme *f* <Thermod> heat of solution

losweise Prüfung *f* <Qual> lot-by-lot inspection
loswerfen *v* <Wassertrans> cast off
Lot *n* 1. <Bau> bob, plummet, solder; 2. <Elektronik> lead *(vertikales Referenzmaß)*; 3. <Elektrotech> solder; 4. <Fertig> bob, plumb, solder; 5. <Funktech> solder; 6. <Geom> perpendicular; 7. <Maschinen> solder • **außer Lot** <Bau> out-of-plumb • **im Lot** <Bau> plumb • **ins Lot bringen** <Bau> plumb • **Lot einführen** <Elektriz> run solder
Lötanschluss *m* <Maschinen> soldered fitting
Lötauge *n* <Fernseh, Funktech, Telekom> land
Lötbad *n* 1. <Fertig> molten solder, solder bath; 2. <Maschinen> soldering bath
Lotband *n* <Maschinen> strip solder
lötbar *adj* 1. <Elektriz> solderable; 2. <Fertig> brazeable, solderable
lotbarer Grund *m* <Wassertrans> soundings *(Meer)*
Lötbarkeit *f* <Elektriz> solderability
Lotblei *n* <Wassertrans> sounding lead
Lötbrüchigkeit *f* <Fertig> solder embrittlement
Lötdraht *m* 1. <Fertig> solder wire; 2. <Maschinen> filler wire
loten *v* <Funktech> sound *(Navigation)*
Loten *n* <Bau> plumbing
löten *v* 1. <Fertig> solder, sweat; 2. <Funktech, Maschinen> solder
Löten *n* <Elektriz, Fertig, Maschinen> soldering
Lötfahne *f* <Fertig> soldering ear, soldering tag
Lötfläche *f* <Fernseh, Funktech, Telekom> land
Lötfluss *m* <Elektriz> soldering flux
Lötflussmittel *n* <Fertig> flux, solder flux
lötfrei *adj* <Funktech> solderless
Lötfuge *f* <Fertig> joint clearance, soldering joint gap
Lötkolben *m* 1. <Bau> soldering iron; 2. <Fertig> bit, copper bit, iron, soldering bit, soldering iron, soldering copper; 3. <Funktech, Maschinen> soldering iron
Lotkreisel *m* <Raumfahrt> gyroscopic verticant
Lötlampe *f* 1. <Bau> *(BE)* blowlamp, *(AE)* blowtorch; 2. <Fertig> *(BE)* blowlamp; 3. <Maschinen> blowlamp, blowtorch
Lotleine *f* <Wassertrans> lead line, sounding line
Lötmittel *n* 1. <Elektrotech> flux, soldering flux, solder; 2. <Fertig> flux
Lötnaht *f* 1. <Fertig> solder; 2. <Lebensmittel> soldered seam *(Konservendosen)*
Lötofen *m* <Fertig> brazier
Lötöse *f* 1. <Fertig> lug, soldering ear; 2. <Funktech> solder lug
Lötpistole *f* <Elektriz> soldering gun
lotrecht *adj* 1. <Bau> plumb; 2. <Fertig> perpendicular
lotrecht einfallend *adj* <Fertig> incident normally
Lötrohr *n* <Bau> soldering blowpipe
Lotschnur *f* <Bau> plumb line
Lötschweißung *f* <Bau> braze welding
Lotse *m* <Wassertrans> pilot
Lötseite *f* <Telekom> soldered side
lotsen *v* <Lufttrans, Wassertrans> pilot
Lotsenboot *n* <Wassertrans> pilot boat
Lotsenfunk *f* <Wassertrans> pilot radio service
Lotsenkutter *m* <Wassertrans> pilot cutter
Lotsenrevier *n* <Wassertrans> pilot waters
Lötstelle *f* 1. <Bau> wiped joint; 2. <Maschinen> soldering joint
Lotung *f* <Wassertrans> sounding
Lötung *f* <Maschinen> soldering
Lötverbindung *f* <Maschinen> soldered joint
Lötzinn *n* 1. <Bau> plumber's solder; 2. <Funktech> solder; 3. <Maschinen> tin-lead solder
Low-Level-Injektion *f* <Elektronik> low-level injection

LOX *(Flüssigsauerstoff)* <Raumfahrt, Thermod> lox *(liquid oxygen)*
Loxodrome *f* 1. <Raumfahrt> loxodromic line; 2. <Wassertrans> rhumb line *(Navigation)*
loxodromische Navigation *f* <Wassertrans> rhumb line navigation *(Navigation)*
LP *(Langspielplatte)* <Aufnahme> EP *(extended-play record)*; LP *(long-playing record)*
LPC *(lineare Prädiktionscodierung)* <Elektronik, Telekom> LPC *(linear predictive coding)*
LPC-Codierung *f* <Elektronik, Telekom> LPC-coding
LPC-Vocoder *m* <Telekom> linear predictive coding vocoder
LPG *(Flüssiggas)* <Erdöl, Heiz & Kälte, Kfztech, Thermod, Wassertrans> LPG *(liquefied petroleum gas)*
LPG-Bus *m* <Kfztech> LPG bus
LPG-Motor *m* <Kfztech> LPG engine
LPG-Tanker *m* <Wassertrans> LPG tanker
LPG-Transporter *m* <Thermod> LPG carrier
LPM *(Zeilen pro Minute)* <Comp & DV> LPM *(lines per minute)*
LP-Modus *m* *(linear polarisierter Modus, linear polarisierter Wellentyp)* <Optik, Telekom> LP mode *(linearly-polarized mode)*
L-Profil *n* <Metall> L-section
LQ *(Korrespondenzqualität)* <Comp & DV, Druck> LQ *(letter quality)*
LS *(Laufsitz)* <Maschinen> running fit
LSB *(niedrigstwertiges Bit)* <Comp & DV, Telekom> LSB *(least significant bit)*
L-Schale *f* <Phys> L-shell *(Atomphysik)*
LSI *(Großintegration, hoher Integrationsgrad)* <Comp & DV, Elektronik, Phys, Telekom> LSI *(large-scale integration)*
L-Signal *n* <Elektronik> L-signal, low-level signal
LSI-Schaltkreis *f* *(hochintegrierter Schaltkreis)* <Elektronik, Phys, Telekom> LSI circuit *(large-scale integrated circuit)*
L-S-Kopplung *f* *(Russell-Saunders-Kopplung)* <Kerntech> l-s coupling *(Russell-Saunders coupling)*
LSL *(untere Spezifikationsgrenze)* <Qual> LSL *(lower specification limit)*
L-Stein *m* <Ker & Glas> L-block
L-Träger *m* <Bau> L-beam
LTTP *(Langzeit-Straßenbelagverhalten)* <Bau> LTPP *(long-term pavement performance)*
Lu *(Lutetium)* <Chemie> Lu *(lutetium)*
LüA *(Länge über alles)* <Wassertrans> LOA *(length overall)*
Lucit *n* <Kunststoff> Lucite
Lücke *f* 1. <Bau> interstice; 2. <Comp & DV> gap; 3. <Elektrotech> gap *(Magnetkreis)*; 4. <Maschinen> gap
Lückenbildung *f* <Metall> void formation
Lückeneffekt *m* <Fernseh> gap effect
Lückenfräser *m* 1. <Maschinen> gap cutter; 2. <Mechan> angular milling cutter
Lückengrad *m* <Metall> voids fraction
Lückengrund *m* <Fertig> space bottom *(Zahnrad)*
Lückenschalter *m* <Fernseh> gapping switch
Lückenwachstum *n* <Metall> void growth
Lüder'sche Linie *f* <Fertig> line of yielding
Luft *f* 1. <Elektrotech> clearance; 2. <Maschinen> backlash, bearing slackness, clearance, play • **mit Luft als Dielektrikum ausgestattet** <Elektriz> air-dielectric • **mit Luft kühlen** <Thermod> air-cool
Luftabfluss *m* <Mechan, Phys> air discharge
Luftabgleichkondensator *m* <Elektrotech> air trimmer capacitor
Luftabscheider *m* 1. <Abfall> air separator; 2. <Lufttrans> deaerator; 3. <Maschinen> air separator

Luftabscheidung *f* <Chemtech> air separation
Luftabschluss anlassen v/unter <Thermod> pot-anneal
Luftabsorption *f* <Sicherheit> atmospheric attenuation *(Lärm)*
Luftabzug *m* 1. <Fertig> air exhaust; 2. <Wassertrans> air vent
Luftabzugsöffnung *f* <Heiz & Kälte> vent port
Luftabzugssystem *n* <Sicherheit> extraction fan system
Luftabzugventil *n* <Fertig> air escape valve
Luftagens *n* <Erdöl> atmospheric agent
Luftanalytik *f* <Sicherheit, Umweltschmutz> air analytics
Luftanlasser *m* <Maschinen> air starter
Luftansaugstutzen *m* <Lufttrans> inducer
Luftansaugung *f* 1. <Fertig> air intake; 2. <Maschinen> indraught of air
luftartig *adj* <Chemie> aeriform
Luftaufhängung *f* <Eisenbahn> air suspension
Luftaufklärung *f* <Meerschmutz> aerial reconnaissance
Luftaufklärungsflugzeug *n* <Meerschmutz> spotter plane *(für Ölverschmutzungen)*
Luftauftanksystem *n* <Lufttrans> *(AE)* refueling in-flight system, *(BE)* refuelling in-flight system
Luftausfällung *f* <Umweltschmutz> air shed
Luftauslass *m* 1. <Erdöl> air exhaust; 2. <Heiz & Kälte> air discharge, air outlet; 3. <Maschinen> air outlet; 4. <Mechan, Phys> air discharge; 5. <Wassertrans> air vent
Luftauslassdüse *f* <Mechan, Phys> air discharge nozzle
Luftaustritt *m* 1. <Bau, Kohlen> exfiltration; 2. <Heiz & Kälte> air discharge, air outlet
Luftaustrittsgitter *n* <Heiz & Kälte> air discharge grille
Luftbedarf *m* <Maschinen> air consumption
Luftbefeuchter *m* 1. <Heiz & Kälte, Lufttrans, Papier> humidifier; 2. <Sicherheit> air humidifier; 3. <Thermod> humidifier
Luftbehälter *m* 1. <Erdöl> air tank; 2. <Maschinen> air box, air receiver
Luftbetankungssonde *f* <Lufttrans> *(AE)* flight refueling probe, *(BE)* flight refuelling probe, *(AE)* in-flight refueling probe, *(BE)* in-flight refuelling probe
luftbetriebenes Messgerät *n* <Metrol> *(AE)* air-operated gage, *(BE)* air-operated gauge
Luftbewegung *f* 1. <Heiz & Kälte> air movement; 2. <Lufttrans> blast
Luftbild *n* aerial photograph, aerial view, air photo
Luftbildauswertung *f* <Kohlen> air photo interpretation
Luftbildfernerkundung *f* <Meerschmutz> airborne remote sensing
Luftbildkamera *f* <Foto> aerial mapping camera
Luftbildmesskamera *f* <Foto> aerial mapping camera
Luftbildvermessung *f* <Lufttrans> aerial survey
Luftblase *f* 1. <Fertig> cavity *(Guss)*; 2. <Mechan, Phys> air bubble
Luftblasenflotation *f* <Umweltschmutz> air flotation
Luftbohren *n* <Erdöl> air drilling *(Bohrtechnik)*
Luftbremse *f* <Lufttrans> air brake, speed brake
luftbremsen *v* <Raumfahrt> aerobrake *(Raumschiff)*
Luft-Brennstoff-Verhältnis *n* <Trans> air-fuel ratio
Luftbrücke *f* <Lufttrans> airlift
Luftbürste *f* 1. <Fertig> air knife; 2. <Kunststoff, Papier> air brush
Luftbürstenstreichmaschine *f* <Papier> airbrush coater
Luftchemie *f* <Umweltschmutz> atmospheric chemistry
Luftdamm *m* <Kfztech> air dam
Luftdämpfung *f* <Gerät> air damping *(am Messwerk)*
luftdicht *adj* 1. <Erdöl, Lebensmittel, Maschinen, Mechan> airtight; 2. <Papier> air-proof, airtight; 3. <Phys, Verpack, Wassertrans> airtight
luftdicht abgeschlossen *adj* <Elektrotech, Heiz & Kälte, Mechan> hermetically-sealed

Luftdichte

Luftdichte f <Heiz & Kälte> air density
luftdichte Versiegelung f <Verpack> hermetic seal
Luftdichtemessgerät n <Metrol, Papier, Phys> aerometer
luftdichter Verschluss m 1. <Telekom> hermetic sealing; 2. <Verpack> hermetic closure
Luftdiffusor m <Heiz & Kälte> air diffuser
Luftdrehkondensator m <Elektrotech> air variable capacitor
Luftdruck m 1. <Fertig> air pressure; 2. <Lufttrans> barometric pressure; 3. <Maschinen> air pressure; 4. <Phys, Wassertrans> atmospheric pressure
luftdruckbedingte Spektrallinie f <Phys> atmospheric line
Luftdruckmeißel m <Fertig> air chipper
Luftdruckmesser m 1. <Labor> barometer; 2. <Metrol> (AE) air gage, (BE) air gauge
Luftdruckmessgerät n <Erdöl> (AE) air pressure gage, (BE) air pressure gauge (Messtechnik)
Luftdruckmotor m <Lufttrans> air motor
Luftdruckschalter m <Lufttrans> barometric switch
Luftdruckschreiber m <Gerät, Labor, Wassertrans> barograph
Luftdruckwirkung f <Lufttrans> blast
Luftdurchflussmenge f <Heiz & Kälte> airflow rate, rate of air flow
Luftdurchflusszähler m <Maschinen> air meter
Luftdurchgang m <Maschinen> air passage
luftdurchlässig adj <Verpack> impermeable
Luftdurchlässigkeit f <Heiz & Kälte> air permeability
Luftdurchsatz m <Heiz & Kälte> rate of air flow
Luftdusche f <Papier> air shower
Luftdüse f 1. <Maschinen> air nozzle; 2. <Mechan> choke
Luftdüsenstreichmaschine f <Papier> air jet coater
Lufteinblasen n <Maschinen> air blowing
Lufteinblasvorrichtung f <Fertig> air injector
Lufteinlass m 1. <Fertig> air admission, air intake; 2. <Heiz & Kälte> air inlet, air intake; 3. <Ker & Glas> air inlet; 4. <Maschinen> air inlet; 5. <Wassertrans> air intake
Lufteinlass m **am Triebwerk** <Lufttrans> engine air intake
Lufteinlass m **mit variabler Geometrie** <Lufttrans> variable-geometry intake
Lufteinlassdüse f <Maschinen> air inlet nozzle
Lufteinlasshahn m 1. <Fertig, Maschinen> air inlet cock; 2. <Wasserversorg> bleeding cock
Lufteinlassventil n 1. <Lufttrans> air-charging valve; 2. <Wasserversorg> bleeding valve
Lufteinpressbohrung f <Erdöl> air input well (Förderung)
Lufteinschluss m 1. <Fertig> air cavity; 2. <Kunststoff> entrapped air; 3. <Strömphys> airlock (Röhren); 4. <Verpack> entrapped air
Lufteintritt m <Heiz & Kälte> air inlet, air intake
Lufteintrittsdruck m <Lufttrans> air intake pressure
Lufteintrittsöffnung f <Lufttrans> inducer
Lufteintrittsventil n <Lufttrans> air intake valve
luftelektrische Störung f <Elektrotech> atmospherics disturbance, static disturbance
lüften v 1. <Bau> vent, ventilate; 2. <Heiz & Kälte> ventilate; 3. <Heiz & Kälte> vent; 4. <Maschinen> vent; 5. <Thermod> ventilate; 6. <Trans> air
Luftentfeuchter m 1. <Elektrotech> drier; dehydrating breather (Transformator); 2. <Heiz & Kälte> dehydrating breather; 3. <Sicherheit> air dehumidifier
Lüfter m 1. <Eisenbahn, Elektriz> fan; 2. <Heiz & Kälte> blower, fan, ventilator; 3. <Kfztech> cooling fan, fan, ventilator; 4. <Lebensmittel> blower; 5. <Maschinen> ventilating fan; 6. <Mechan> exhauster, fan; 7. <Sicherheit> ventilator; 8. <Thermod> fan; 9. <Wassertrans> ventilator

Lüfter m **für zwei Richtungen** <Heiz & Kälte> bidirectional fan
Lüfter m **mit verschiebbarer Schlitzplatte** <Heiz & Kälte> hit-and-miss damper
Lüfterabdeckhaube f <Heiz & Kälte> fan cowl, fan shroud
Lüfteraggregat n <Heiz & Kälte> fan set, fan unit
Lüfterbaugruppe f <Heiz & Kälte> fan unit
Lüfterdrehzahl f <Heiz & Kälte> fan speed
Lüftergehäuse n <Heiz & Kälte> fan casing, fan enclosure, fan housing
Lüfterhaube f 1. <Heiz & Kälte> cowl, fan cowl, fan guard, shroud; 2. <Kfztech, Wassertrans> ventilator cowl (Deckbeschläge)
Lufterhitzer m <Fertig, Heiz & Kälte, Maschinen> air heater
Lüfterkeilriemen m <Kfztech> fan belt
Lüfterkennlinie f <Heiz & Kälte> fan characteristic
Lüfterkragen m <Heiz & Kälte> cowl, fan cowl, fan shroud
Lüfterkranz m <Heiz & Kälte> fan impeller
Lüfterleistung f <Heiz & Kälte> fan performance
Lüfterleistungskennlinie f <Heiz & Kälte> fan performance curve
Luftermessung f <Lufttrans> aerial motion picture survey (Filmaufnahmen)
Lüftermotor m <Heiz & Kälte> fan motor
Lüfterpumpe f <Mechan> exhaust pump
Lüfterrad n <Heiz & Kälte> fan impeller, impeller
Lüfterriemenscheibe f <Kfztech> fan pulley (Kühlsystem)
Lüfterschaufel f <Heiz & Kälte, Kfztech, Mechan, Thermod> fan blade
Lüfterstation f <Bau> fan station
Lüfterstufe f <Heiz & Kälte> fan stage
Lüfterstutzen m <Heiz & Kälte> fan cowl
Lüfterumfangsgeschwindigkeit f <Heiz & Kälte> tip speed
Lüfterwirkung f <Heiz & Kälte> fanning action
Luftfahrt f <Lufttrans> aeronautics
Luftfahrtfunkdienst m (AACS) <Raumfahrt> airways and air communications service (AACS)
Luftfahrtindustrie f <Trans> aeronautical industry
Luftfahrt-Navigationsausschuss m (ANC) <Lufttrans> International Civil Aviation Organization (ICAO)
Luftfahrtregeln fpl <Lufttrans> rules of the air
Luftfahrtregister n <Lufttrans> aeronautical register
Luftfahrttechnik f <Lufttrans> (AE) aerotechnics
Luftfahrzeug n <Lufttrans> aircraft
Luftfahrzeugklasse f <Lufttrans> aircraft classification
Luftfahrzeugleitwerk n <Lufttrans> aircraft tail unit
Luftfahrzeugradar n <Funkort> airborne radar
Luftfahrzeugrufzeichen n <Lufttrans> aircraft call signal
Luftfahrzeugstromversorgung f <Lufttrans> aircraft mains
Luftfahrzeugzulassungsbescheinigung f <Lufttrans> certificate of airworthiness
Luftfahrzeugverteilernetz n <Lufttrans> aircraft mains
Luftfeder f <Eisenbahn> air spring
Luftfeuchtemessgerät n <Gerät> air humidity meter
Luftfeuchteschreiber m <Gerät> air humidity recorder
Luftfeuchtigkeit f 1. <Anstrich> humidity; 2. <Fertig> humidity (Kunststoffinstallationen); 3. <Heiz & Kälte> humidity; 4. <Kohlen> air moisture; 5. <Phys, Thermod> humidity; 6. <Verpack> air humidity
Luftfeuchtigkeitsanzeige f <Verpack> humidity indicator
Luftfeuchtigkeitsmesser m 1. <Erdöl, Gerät> hygrometer; 2. <Heiz & Kälte, Labor> hygrometer, psychrometer (Feuchtigkeit); 3. <Maschinen, Phys, Thermod> hygrometer
Luftfeuchtigkeitsmessung f 1. <Heiz & Kälte> hygrometry; 2. <Wassertrans> humidity measurement

Luftfilm m <Kohlen> air film
Luftfilmsystem n <Wassertrans> air film system
Luftfilter n 1. <Chemtech, Heiz & Kälte> air filter; 2. <Kfztech> air cleaner, air filter *(Vergaser)*; 3. <Kohlen> air filter; 4. <Maschinen> air cleaner; 5. <Mechan, Phys> air filter
Luftfilterung f <Fertig> air cleaning
Luftfiltrationssystem n <Kfztech> air filtration system
luftförmig adj <Chemie> aeriform
Luftfotografie f <Foto> aerial photography
Luftfracht f 1. <Lufttrans> air freight; 2. <Verpack> air cargo, air freight
Luftfrachtbrief m <Lufttrans> air waybill
Luftführungsbahn f <Trans> aerial guideway
Luftführungsblech n <Heiz & Kälte> air baffle plate
Luftgang m <Wassertrans> air course *(Schiffbau)*
Luftgardine f <Kfztech> air curtain
Luftgefrierapparat m <Heiz & Kälte> air blast freezer
Luftgefrieren n <Heiz & Kälte> air blast freezing
luftgekühlt adj 1. <Elektriz> air-cooled; 2. <Fertig> air-cooled, air-dried; 3. <Heiz & Kälte, Maschinen, Papier, Thermod> air-cooled
luftgekühlte Röhre f <Elektronik> air-cooled tube
luftgekühlte Triode f *(LKT)* <Elektronik> air-cooled triode *(ACT)*
luftgekühlter Kondensator m <Heiz & Kälte> air-cooled condenser
luftgekühlter Motor m <Kfztech, Maschinen> air-cooled engine
luftgekühlter Transformator m <Elektriz> air-cooled transformer
luftgekühltes System n <Maschinen> air-cooled system
luftgelagert adj <Fertig> air-bearinged
luftgelöschter Kalk m <Chemie> air-slaked lime
Luftgeräusch n <Heiz & Kälte> aerodynamic noise
luftgeschmiert adj <Fertig> air-lubricated
Luftgeschwindigkeit f <Heiz & Kälte> air velocity
luftgesteuerte Setzmaschine f <Kohlen> Baum box, Baum jig
luftgestützter Abstandswarnanzeiger m <Lufttrans> airborne proximity warning indicator
luftgestütztes Kollisionswarnsystem n <Lufttrans> airborne collision avoidance system
luftgetrocknet adj 1. <Lebensmittel> air-dried; 2. <Papier> loft-dried *(auf Trockenboden)*; 3. <Verpack> air-dried
Lufthahn m <Fertig> air cock
Lufthärtestahl m 1. <Fertig> air hardening steel; 2. <Metall> air-hardening steel
Lufthärtungsstahl m 1. <Fertig> air hardening steel; 2. <Metall> air-hardening steel
Luftheizgerät n mit Gebläse <Thermod> unit heater
Luftheizung f <Heiz & Kälte> hot-air heater
Luftherd m <Kohlen> pneumatic table
lufthydraulisch adj <Fertig> air-hydraulic
Lufthygiene f <Umweltschutz> air pollution control
Luftinjektor m <Maschinen> air injector
Luftisolation f <Elektrotech> air insulation
luftisolierter Kondensator m <Elektriz> air capacitor
Luftisolierung f 1. <Elektriz, Elektrotech> air insulation; 2. <Heiz & Kälte> airspace insulation
Luftkabel n 1. <Elektrotech> overhead cable; 2. <Telekom> aerial cable, overhead cable
Luftkanal m 1. <Fertig> air conduit; 2. <Heiz & Kälte> air duct, duct; 3. <Kohlen> air duct; 4. <Lufttrans> airway; 5. <Maschinen> air conduit
Luftkasten m 1. <Erdöl> air box *(Raffinerie)*; 2. <Wassertrans> air tank *(Bootbau)*
Luftkern m <Elektrotech> air core *(Spule)*

Luftkernspule f <Elektriz> air core coil
Luftkerntransformator m <Elektriz> air core transformer, air transformer, air-cored transformer
Luftkissen n 1. <Kfztech> air bag; 2. <Lufttrans> air dashpot; 3. <Maschinen, Trans> air cushion
Luftkissen n mit eingegangenen Lufblasen <Lufttrans> sidewall air cushion
Luftkissen n mit Ringstrahl <Trans> peripheral jet air cushion
Luftkissen n mit starren Schürzen <Lufttrans> rigid sidewall air cushion
luftkissenbefördert adj <Trans> cushion-borne
Luftkissenboot n <Wassertrans> *(AE)* hydroskimmer, marine air cushion vehicle, marine hovercraft
Luftkissenfahrzeug n *(LKF)* 1. <Kfztech> surface effect vehicle *(SEV)*; 2. <Trans> *(AE)* aeromobile, air cushion vehicle; 3. <Wassertrans> *(BE)* hovercraft, *(AE)* hydroskimmer, surface effect ship *(Schifftyp)*
Luftkissenfahrzeug n mit festen Schürzen <Wassertrans> rigid sidewall hovercraft
Luftkissenfahrzeug n mit Luftschraubenantrieb <Trans> air propelled hovercraft
Luftkissenfahrzeug n mit starren Schürzen <Wassertrans> rigid skirt hovercraft
Luftkissenfahrzeug n mit Wasserschrauben <Wassertrans> water-propelled hovercraft
Luftkissenfahrzeughafen m <Wassertrans> hoverport
Luftkissenhaltesystem n <Trans> air cushion restraint system
Luftkissenplattform f <Wassertrans> hover pallet
Luftkissenschwebesystem n <Trans> air cushion levitation
Luftkissenschwebezug m <Wassertrans> hovertrain
Luftkissentransportrinne f <Trans> aeroglide
Luftkissenzug m <Eisenbahn> aerotrain
Luftklappe f 1. <Fertig> air choke, air valve; 2. <Heiz & Kälte> air damper; 3. <Kfztech> choke *(Vergaser)*; 4. <Maschinen> air valve
Luftklappensystem n <Trans> blown-flap system
Luftklassierer m <Abfall> air classifier, air separation plant
Luftkompressor m <Hydraul, Kfztech, Labor, Maschinen> air compressor
Luftkondensator m <Mechan, Phys> air capacitor
Luftkorrekturdüse f <Kfztech> air correction jet
Luftkorridor m <Lufttrans> air corridor
Luftkreislauf m <Heiz & Kälte> ventilation circuit
Luftkühlapparat m <Heiz & Kälte> air cooler
luftkühlen v <Heiz & Kälte, Mechan, Thermod> fan-cool
Luftkühler m <Heiz & Kälte, Kerntech> air cooler
Luftkühlkreislauf m <Heiz & Kälte> air refrigeration cycle
Luftkühlmittel n <Kerntech> air coolant
Luftkühlung f 1. <Elektrotech> air cooling; 2. <Heiz & Kälte> fan cooling; 3. <Kfztech, Maschinen> air cooling; 4. <Mechan, Thermod> fan cooling
Luftkühlungssystem n <Maschinen> air-cooling installation
Luftkursbuch n <Lufttrans> aerial timetable
Luftlager n 1. <Fertig> air bearing; 2. <Mechan> gas bearing
Luftleistung f <Heiz & Kälte> airflow rate, flow rate, rate of air flow
Luftleistungsschalter m <Elektrotech> air circuit breaker
Luftleitblech n 1. <Heiz & Kälte> air baffle plate, shroud; 2. <Kfztech> *(AE)* louver, *(BE)* louvre, spoiler *(Karosserie)*
Luftleitung f 1. <Akustik> air conduction; 2. <Fertig> air line; 3. <Ker & Glas> airline; 4. <Maschinen> air main, air pipeline, airline; 5. <Papier> airline
Luftleitungen fpl <Maschinen> air ducting

Luftloch

Luftloch n <Wassertrans> air course *(Schiffbau)*
luftlose Einspritzung f <Maschinen> solid injection
Luft-Luft-Wärmetauscher m <Heiz & Kälte, Maschinen> air-to-air heat exchanger
Luftmenge f <Heiz & Kälte> air volume
Luftmengenmesser m <Lufttrans> airflow sensor
Luftmesser n 1. <Fertig, Kunststoff> air knife; 2. <Papier> air blade, air knife
Luftmesserstreichmaschine f <Papier> air knife coater
Luftnavigationskarte f <Lufttrans> *(BE)* aeronautical chart, *(AE)* sectional chart
Luftpatentieren n <Fertig> air patenting
luftpatentiert adj <Fertig> air-patented
Luftpinsel m <Druck> aerograph
Luftporenbeton m <Bau> air-entrained concrete
Luftporenbildner m <Bau> air-entraining admixture *(Beton)*
Luftporenmittel n <Bau> air-entraining agent
Luftpostpapier n <Druck, Papier> airmail paper
Luftprobe f <Sicherheit, Umweltschmutz> air sample
Luftprobenahme f <Sicherheit, Umweltschmutz> air sampling
Luftprobenahmegerät n <Sicherheit, Umweltschmutz> air sampler, air-sampling device
Luftprobenahmetechnik f <Sicherheit> air-sampling technique
Luftpuffer m 1. <Fertig> air buffer; 2. <Lufttrans> air dashpot
Luftpumpe f <Maschinen, Mechan, Phys> air pump
Luftqualität f <Sicherheit> air quality
Luftqualitätsdaten npl <Umweltschmutz> air quality data
Luftrakel f 1. <Kunststoff> air knife; 2. <Papier> air doctor
Luftrakelfeuchtsystem n <Druck> air doctor dampening system
Luftraumbeschränkung f <Lufttrans> airspace restriction
Luftraumüberwachungsradar n 1. <Funkort> air search radar, surveillance radar; 2. <Wassertrans> air search radar *(Marine)*
Luftregulierung f <Maschinen> air regulator
Luftreibung f <Raumfahrt> air friction *(Raumschiff)*
Luftreibungsverluste mpl <Elektriz> windage losses
Luftreifen m <Kfztech, Trans> *(AE)* pneumatic tire, *(BE)* pneumatic tyre
Luftreifenstadtbahn f <Eisenbahn> *(AE)* pneumatic-tired metropolitan railroad, *(BE)* pneumatic-tyred metropolitan railway
Luftreinheit f <Umweltschmutz> air purity
Luftreiniger m 1. <Kohlen> air cleaner; 2. <Sicherheit> gas-cleaning equipment; 3. <Verpack> air purger
Luftreinigung f 1. <Kohlen> air cleaning; 2. <Sicherheit> clean air device *(Umweltschutz)*
Luftreinigungsgerät n <Sicherheit, Umweltschmutz> air-cleaning device
Luftrohr n <Maschinen> air pipe
Luftröhrenkühler m <Kfztech> honeycomb radiator
Luftsauerstoffbatterie f <Elektrotech> air depolarized battery
Luftsauerstoffelement n <Elektrotech> air depolarized element
Luftsauerstoffzelle f <Elektrotech> air depolarized cell
Luftsaugrohr n <Fertig> air intake
Luftsäulenlautsprecher m <Aufnahme> air column loudspeaker
Luftschacht m 1. <Heiz & Kälte> air duct; 2. <Sicherheit> ventilation shaft
Luftschadstoff m <Umweltschmutz> air pollutant, airborne pollutant, airborne hazardous substance, atmospheric pollutant
Luftschadstoffgehalt m <Umweltschmutz> level of air contaminants

Luftschall m <Heiz & Kälte> air noise
Luftschallemission f <Maschinen> airborne noise emitted
Luftschalter m <Elektrotech> air break switch
Luftschicht f <Kohlen> air film
Luftschichtkoeffizient m <Kohlen> air film coefficient
Luftschlauch m 1. <Bau> air hose *(Druckluft)*; 2. <Erdöl> air hoist; 3. <Kfztech> inner tube; 4. <Maschinen, Mechan, Phys> air hose
Luftschleiertür f <Bau> air curtain door
Luftschleuse f 1. <Heiz & Kälte, Lufttrans> airlock; 2. <Raumfahrt> airlock *(Raumschiff)*; 3. <Sicherheit, Wasserversorg> air sluice, airlock
Luftschleusensystem n <Kerntech> airlock system
Luftschlitz m 1. <Heiz & Kälte> air duct, duct; 2. <Mechan> *(AE)* louver, *(BE)* louvre
Luftschmierung f <Fertig> air lubrication
Luftschmutzstoff m <Umweltschmutz> atmospheric pollutant
Luftschraube f 1. <Lufttrans> airscrew; 2. <Trans> air propeller
Luftschraube, ffrei umlaufende vom Fahrtwind getriebene <Lufttrans> windmilling propeller
Luftschraubenansatz m <Lufttrans> blade root *(Hubschrauber)*
Luftschraubenblattachse f <Lufttrans> blade pocket *(Hubschrauber)*
Luftschraubenblattbelastung f <Lufttrans> blade loading *(Hubschrauber)*
Luftschraubenblattbreitenverhältnis n <Lufttrans> blade width ratio *(Hubschrauber)*
Luftschraubenblattfuß m <Lufttrans> blade shank
Luftschraubenblatthülle f <Lufttrans> blade sleeve *(Hubschrauber)*
Luftschraubenblattmanschette f <Lufttrans> blade cuff *(Hubschrauber)*
Luftschraubenblattoberseite f <Lufttrans> blade upper surface *(Hubschrauber)*
Luftschraubenblattpfeilung f <Lufttrans> blade sweep *(Hubschrauber)*
Luftschraubenblattprofil n <Lufttrans> blade profile *(Hubschrauber)*
Luftschraubenblattradius m <Lufttrans> blade radius *(Hubschrauber)*
Luftschraubenblattschaft m <Lufttrans> blade shank *(Hubschrauber)*
Luftschraubenblattschlagwinkel m <Lufttrans> bladder flapping angle *(Hubschrauber)*
Luftschraubenblattsehne f <Lufttrans> blade chord *(Hubschrauber)*
Luftschraubenblattspitze f <Lufttrans> blade tip *(Hubschrauber)*
Luftschraubenblattspitzenhütchen n <Lufttrans> blade tip cap *(Hubschrauber)*
Luftschraubenblattsteigungskoeffizient m <Lufttrans> blade lift coefficient *(Hubschrauber)*
Luftschraubenblattsteuersystem n <Lufttrans> blade control system *(Hubschrauber)*
Luftschraubenblatt- und Spaltflügelantrieb m <Lufttrans> blade and slot drive
Luftschraubenblattzeilung f <Lufttrans> blade spacing system *(Hubschrauber)*
Luftschraubendrehmoment n <Lufttrans> propeller torque
Luftschraubeneinstellwinkel m <Lufttrans> effective pitch *(Propeller)*
Luftschraubenhaube f <Lufttrans> spinner *(Propeller)*
Luftschraubennabe f <Lufttrans> propeller hub
Luftschraubenschlupf m <Lufttrans> slip of a propeller

Luftschraubensteigung f <Lufttrans> propeller pitch
Luftschraubensteigungseinstellung f <Lufttrans> blade pitch setting *(Hubschrauber)*
Luftschraubensteigungsregler m <Lufttrans> propeller governor
Luftschraubenstrahl m <Lufttrans> slipstream
Luftschraubenturbine f <Nichtfoss Energ> propeller turbine
Luftschraubenwelle f <Lufttrans> propeller shaft
Luftselbstkühlung f <Heiz & Kälte> natural air cooling
Luftsetzmaschine f <Kohlen> air jig, pneumatic jig
Luftsortierer m <Abfall> air classifier, air separator, air separation plant
Luftspalt m 1. <Comp & DV> gap *(eines Magnetkopfes)*; 2. <Elektriz, Elektrotech> air gap *(Magnet-Stromkreis)*; 3. <Erdöl> air gap *(Motor)*; 4. <Phys> air gap *(Elektromagnet)*; gap *(Magnet)*; 5. <Trans> air gap
Luftspalt m **zwischen Elektrode und Werkstück** <Fertig> arc gap
Luftspaltmagnetometer n <Elektrotech> flux-gate magnetometer
Luftspaltsensor m <Elektrotech> gap sensor *(Magnetschwebetechnik)*
Luftspeicherdieselmotor m <Kfztech> air cell diesel engine
Luftsperre f <Meerschmutz> air bubble boom
Luftspieß m <Fertig> pricker *(Gießen)*
luftspießen v <Fertig> vent *(Gießen)*
Luftspule f <Elektriz> air reactor, air coil
Luftstartanlassschalter m <Lufttrans> air start ignition switch
Luftstift m <Druck> air piston
Luftstoß m <Kohlen> air blast
Luftstrahl m <Fertig> air jet
Luftstrahltriebwerk n <Lufttrans> air breathing engine
Luftstrecke f <Elektriz> air gap, clearance
Luftstrom m 1. <Bau> air flow; 2. <Lufttrans> airflow
Luftstromschalter m <Elektrotech> air blast circuit breaker
Luftstromscheider m <Lufttrans> airstream separation
Luftströmung f 1. <Heiz & Kälte> air flow; 2. <Lufttrans, Strömphys> airflow
Luftströmungsmelder m <Heiz & Kälte> airflow indicator
Luftströmungsmesser m <Phys> anemometer
Luftströmungspuffer m <Lufttrans> friction damper
Luftströmungswächter m <Heiz & Kälte> airflow monitor, airflow proving switch
Luftströmungswiderstand m <Lufttrans> friction drag
Lufttank m <Erdöl> air tank
Lufttankflugzeug n <Lufttrans> *(AE)* refueling craft, *(BE)* refuelling craft
Lufttasche f <Fertig> air pocket
Lufttauglichkeitsprüfung f <Lufttrans> certification test
Lufttemperaturanzeige f **des Enteisers** <Lufttrans> de-icing air temperature indicator
Lufttransformator m <Elektriz, Elektrotech> air transformer, air-cored transformer, air core transformer
Lufttransport m <Lufttrans> air transport, airlift
lufttransportierter Eintrag m <Umweltschmutz> airborne input
Lufttrennschalter m <Elektriz> air break switch
Lufttrichter m <Kfztech> venturi *(Vergaser)*
lufttrocken adj <Papier> air-dry
lufttrockenes Papier n <Druck> air-dried paper, air-dry paper
Lufttrocknen n <Druck> air drying
Lufttrockner m <Eisenbahn, Heiz & Kälte> air dryer
lufttüchtig adj <Lufttrans> airworthy

Lufttüchtigkeit f <Lufttrans> airworthiness
• **Lufttüchtigkeit bescheinigen** <Lufttrans> certify as airworthy
Lufttüchtigkeitszeugnis n <Lufttrans> certificate of airworthiness
Luftturbulenzen fpl <Lufttrans> atmospheric turbulence
Luftüberschuss m <Maschinen> excess air
luftübertragen adj <Fertig> airborne
Luftüberwachung f <Meerschmutz> aerial surveillance
Luftüberwachungsanlage f <Sicherheit, Umweltschmutz> dust monitor
Luftumlauftrockner m <Papier> air float dryer
Luftumwälzofen m <Fertig> air-circulating furnace
Luftumwälzung f <Fertig, Heiz & Kälte> air circulation
Luft- und Raumfahrt f <Raumfahrt> aerospace
Luft- und Raumfahrtelektronik f <Lufttrans> avionics
Luft- und Raumfahrtmedizin f <Raumfahrt> aerospace medicine
Luft- und Seestreitkräfte fpl <Wassertrans> air and sea forces *(Marine)*
Lüftung f 1. <Heiz & Kälte> venting; 2. <Lufttrans> air renewal; 3. <Maschinen> airing, venting; 4. <Sicherheit> ventilation system
Lüftungsanlage f 1. <Heiz & Kälte> air-handling system; 2. <Kfztech, Wassertrans> ventilation system
Lüftungseinheit n <Heiz & Kälte> AH unit, air-handling unit
Lüftungsflügelfenster n <Bau> projected window
Lüftungskanal m <Bau> ventiduct, ventilation duct
Lüftungsleitung f <Heiz & Kälte> ventilation duct
Lüftungsöffnung f <Maschinen> air vent
Lüftungsschacht m 1. <Bau> funnel; 2. <Sicherheit> ventilation shaft
Lüftungsschlitz m <Heiz & Kälte> venting slot
Lüftungsstutzen m 1. <Kerntech> vent nozzle; 2. <Kfztech, Wassertrans> ventilator socket
Lüftungstür f <Bau> ventilating door
Lüftungszentrale f <Heiz & Kälte> *(AE)* ventilation control center, *(BE)* ventilation control centre
Luftventil n 1. <Fertig> air choke; 2. <Nichtfoss Energ> air valve
Luftverbrauch m <Maschinen> air consumption
Luftverdichter m <Hydraul> air compressor
Luftverdichtung f <Fertig> air compression
Luftvergiftung f <Umweltschmutz> air poisoning
Luftverhältnis n <Kfztech> air ratio
Luftverkehr m <Lufttrans> air traffic
Luftverkehrsgesellschaft f <Lufttrans> airline
Luftverkehrskontrolle f <Lufttrans> air traffic control service
Luftverkehrsregeln fpl <Lufttrans> rules of the air
Luftverkehrsschema n <Lufttrans> air traffic pattern
Luftverkehrszentrale f <Lufttrans> *(AE)* air traffic control center, *(BE)* air traffic control centre
Luftverschmutzung f <Umweltschmutz> air pollution, atmospheric pollution • **gegen Luftverschmutzung** <Mechan> antipollution
Luftverschmutzungsvorhersage f <Umweltschmutz> air pollution forecast
Luftversorgung f <Lufttrans> air supply
Luftverteiler m <Heiz & Kälte> air diffuser, diffuser
Luftverteilergehäuse n <Kfztech> plenum chamber
luftverunreinigend adj <Sicherheit, Umweltschmutz> air-contaminating, air-polluting
luftverunreinigender Stoff m <Umweltschmutz> air-polluting substance, atmospheric pollutant
Luftverunreinigung f <Umweltschmutz> air pollution
Luftverunreinigungsemission f <Umweltschmutz> air pollution emission

Luftverunreinigungsereignis

Luftverunreinigungsereignis n <Umweltschutz> air pollution incident
Luftverunreinigungsgefahrensituation f <Umweltschmutz> air pollution episode
Luftverunreinigungsstandard m <Umweltschmutz> level of pollution
Luftvorwärmer m 1. <Maschinen> air preheater; 2. <Papier> air heater
Luftvulkanisation f <Kunststoff> air cure
Luftwalze f <Papier> air roll
Luft-Wasser-Wärmetauscher m <Heiz & Kälte> air-to-water heat exchanger
Luftwechsel mpl 1. <Heiz & Kälte> air changes; 2. <Lufttrans> air renewal
Luftwechselgeschwindigkeit f <Heiz & Kälte> rate of air change
Luftwechsler m <Papier> air exchanger
Luftweg m <Lufttrans> airway; advisory route (Flugsicherung) • **auf dem Luftweg** <Lufttrans> airborne
Luftwegnavigationskarte f <Lufttrans> aeronautical route chart
Luftwiderstand m 1. <Heiz & Kälte> drag; 2. <Lufttrans> drag, ohmic resistance; 3. <Raumfahrt> drag; 4. <Trans> aerodynamic drag
Luftwiderstandsbeiwert m 1. <Kfztech> drag coefficient; 2. <Nichtfoss Energ> coefficient of drag (DD); 3. <Trans> aerodynamic drag factor
Luftwiderstandsdämpfer m <Lufttrans> compensating developer, drag damper
Luftwiderstandsmoment n <Lufttrans> drag moment
Luftwiderstandsstützstrebe f <Lufttrans> drag brace
Luftwiderstandsverspannung f <Lufttrans> drag brace
Luftwiderstandswinkel m <Lufttrans> drag angle
Luftzahl f <Kfztech> air ratio
Luftzerlegung f <Chemtech, Lebensmittel> air separation
Luftziegel m <Ker & Glas> adobe
Luftzufuhr f 1. <Bau> aeration; 2. <Lufttrans> air supply; 3. <Maschinen> air intake, air supply
Luftzuführung f <Maschinen> air inlet
Luftzuführungsgerät n <Sicherheit> supplied-air equipment (Atemschutz)
Luftzuführungsrohr n <Maschinen> air inlet pipe
Luftzug m <Bau, Maschinen> (AE) draft, (BE) draught
Luftzutritt m <Fertig> air admission
Luftzwischenraum m <Trans> air gap
Lugger m <Wassertrans> lugger (Segeln)
Luggersegel n <Wassertrans> lug (Segeln)
Luke f 1. <Lufttrans> hatch; 2. <Raumfahrt> access, hatch (Raumschiff); 3. <Wassertrans> hatch (Schiff)
Lukendeckel m <Wassertrans> hatch cover (Schiffteil)
lukendeckelloses Containerschiff n <Wassertrans> open-top container ship
Lukenöffnung f <Wassertrans> hatchway
Lukensüll n <Wassertrans> hatch coaming
Lumen n (lm) <Fernseh, Metrol, Phys> lumen, lm (SI-Einheit des Lichtstromes)
Luminanz f 1. <Elektronik, Elektrotech, Fernseh> luminance; 2. <Optik> luminance; 3. <Strahlphys> luminance
Luminanzdifferenz f <Fernseh> luminance difference
Luminanzsignal n <Fernseh> luminance signal
Luminanzträgerleistung f <Fernseh> luminance carrier output
Luminanzverstärker m <Elektronik> luminance amplifier
lumineszent adj <Elektrotech> luminescent
Lumineszenz f <Elektrotech, Fernseh, Phys, Strahlphys> luminescence
Lumineszenzdiode f (LED) <Comp & DV, Elektriz, Elektronik, Fernseh, Optik, Phys, Telekom> light-emitting diode (LED)

Lumineszenzschwelle f <Fernseh> luminescence threshold
Luminophor m <Chemie> luminophore
Luminosität f <Teilphys> luminosity
Lümmel m <Wassertrans> gooseneck (Beschläge)
Lummer-Brodhun-Würfel m <Phys> Lummer-Brodhun photometer (Photometer)
Lumpen m <Papier> rag
Lumpenbrecher m <Papier> rag breaker
Lumpenentstaubungstrommel f <Papier> rag duster
Lumpenshredder m <Papier> rag shredder
Lunar-Modul n (LM) <Raumfahrt> lunar module (LM)
Lünette f 1. <Fertig> back rest, boring-bar steady bracket, lathe steady, steady rest; 2. <Maschinen> back rest, back stay, (AE) center rest, (BE) centre rest
Lünettenständer m <Fertig> boring stay, end-support column, outer stay
lungengängiger Staub m <Sicherheit> respirable dust
lunisolares Potenzial n <Raumfahrt> luni-solar potential
Lunker m 1. <Bau> pipe; 2. <Fertig> pipe cavity, pocket, shrink hole; 3. <Ker & Glas, Kunststoff> void; 4. <Metall> blow hole, pipe
lunkerfrei adj <Fertig> pipeless
Lunkern n <Fertig> shrinking
Lunte gefärbt adj/als <Textil> slubbing dyed
Lupe f 1. <Ker & Glas> glass; 2. <Labor, Mechan> magnifying glass; 3. <Optik> lens (Vergrößerungsglas)
Lupenhalter m <Optik> lens holder
Lupinidin n <Chemie> lupinidine, sparteine
Luppe f 1. <Fertig> ball, bloom, puddle ball, puddled ball; 2. <Metall> balled iron, loop, puddle ball
Luppeneisen n <Fertig> ball iron
Luppenstab m <Fertig> puddle bar
Lüster m <Ker & Glas> (AE) luster, (BE) lustre
LUT (Bodenstation) <Wassertrans> LUT, local user terminal (Satellitennavigation)
Lutein n <Chemie> lutein, xanthophyll
Luteol n <Chemie> luteol
Luteolin n <Chemie> luteolin
Lutetium n (Lu) <Chemie> lutetium (Lu)
Lutidin n <Chemie> lutidine
Lutidin... <Chemie> lutidinic
Lutidon n <Chemie> lutidone
luven v <Wassertrans> luff
luvgierig adj <Wassertrans> weatherly • **luvgierig sein** <Wassertrans> carry weather helm (Segeln)
Luvliek n <Wassertrans> luff (Segeln)
Luvruder n <Wassertrans> weather helm (Pinnestellung)
Luvseite f <Wassertrans> weather side
luvwärtig adj <Wassertrans> windward
luvwärts adv <Wassertrans> windward
Lux n (lx) 1. <Foto> lux, lx; 2. <Metrol> lux, lx (SI-Einheit der Beleuchtungsstärke); 3. <Optik, Phys> lux, lx
Luxmeter n <Gerät, Metrol> illumination photometer, luxmeter, luxometer
Luxwert m <Foto> lux value
LW (Langwelle) <Funktech> LW (long wave)
LWL-Datenbus m <Telekom> optical data bus
LWL-Empfangseinrichtung f <Telekom> (AE) receive fiber optic terminal device, (BE) receive fibre optic terminal device
LWL-Faserkoppler m <Telekom> (AE) optic fiber coupler, (BE) optic fibre coupler
LWL-Kabel n <Telekom> (AE) fiberoptic cable, (BE) fibre-optic cable, (AE) optical fiber cable, (AE) optical fibre cable
LWL-Technik f <Fertig> (AE) optical fiber technology, (BE) optical fibre technology

LWL-Übertragungseinrichtung f <Telekom> (AE) transmit fiber optic terminal device, (BE) transmit fibre optic terminal device
LWL-Übertragungsleitung f <Telekom> (AE) optical fiber link, (BE) optical fibre link
LWL-Verbindung f <Fertig> (AE) optical fiber link, (BE) optical fibre link
LWL-Verstärker m <Telekom> optical repeater
LWL-Zwischenverstärker m <Telekom> optical regenerative repeater
lx (Lux) <Foto, Labor, Optik, Phys> lx (lux)
Lyman-Serie f <Phys> Lyman series
Lyogel n <Chemie> lyogel
lyophil adj <Chemie> lyophilic
Lyophilisation f 1. <Chemtech> lyophilization; 2. <Lebensmittel> freeze-drying
Lyophilisierung f <Chemtech> lyophilization
lyophob adj <Chemie> lyophobic
Lyosol n <Chemie> lyosol
lyraförmiger Flugsteuerungswinkelhebel m <Lufttrans> flight controls lyre-shaped bellcrank
Lysin n <Chemie> lysine
Lysolecithin n <Chemie> lysolecithin
Lyxon... <Chemie> lyxonic
Lyxose f <Chemie> lyxose
LZG (Langzeitgedächtnis) <Ergon, Künstl Int> LTM (long-term memory)

M

m 1. <Akustik> (Streukoeffizient, Öffnungskoeffizient) m (flare coefficient of horn); 2. <Akustik> (Scherelastizität) m (shear elasticity); 3. <Maschinen, Phys, Thermod> (Masse) m (mass); 4. <Metrol> (Meter) m (meter); 5. <Metrol> (Milli...) m (milli...); 6. <Phys> (Molekularmasse) m (molecular mass); 7. <Phys> (Gegeninduktivität) m (mutual inductance)
M 1. <Erdöl, Phys, Thermod> (Molekulargewicht) M (molecular weight); 2. <Erdöl, Lufttrans, Phys> (Machzahl) M (Mach number); 3. <Kerntech> (Reaktormultiplikation) M (multiplication of a reactor); 4. <Metrol> (Mega...) M (mega...)
m0 (Restmasse) <Kerntech> m0 (rest mass)
Ma (Atommasse) <Kerntech> Ma (atomic mass)
MA (Mittabstand) <Fertig, Kfztech, Maschinen> CD (centre distance)
Mäanderbildung f <Maschinen> meandering
M-Ablauf m <Fernseh> M-wrap
MAC (magnetisches Kalorimeter) <Teilphys> MAC (magnetic calorimeter)
machbarer Abbrand m <Kerntech> achievable burn-up
Machbarkeitsstudie f <Comp & DV> feasibility study
Machkompensator m <Lufttrans> Mach compensator
Machmeter n 1. <Gerät> machmeter (Geschwindigkeitsmessgerät für strömende Gase, meist mit Pitot-Rohr); 2. <Lufttrans> Mach meter
Mach'sches Prinzip n <Phys> Mach's principle
Macht f <Qual> power
Machtfunktion f <Qual> power function
Machzahl f (M) <Hydraul, Lufttrans, Phys> Mach number (M)
Mach-Zehender-Interferometrie f <Optik> axial slab interferometry

Maclurin n <Chemie> maclurin
Mac-Pherson Federbein n <Kfztech> MacPherson strut (Federung)
Magazin n 1. <Druck, Maschinen> magazine; 2. <Raumfahrt> pod (Raumschiff)
Magazinautomat m <Fertig> magazine automatic
Magazinrückwand f <Foto> magazine back
Magenta n <Druck> magenta
Magentaeinstellung f <Foto> magenta filter adjustment
Mager... <Bau, Kfztech, Kohlen, Lufttrans> lean
Magerbeton m <Bau> lean concrete
magere Schrift f <Druck> light face
magere Steinkohle f <Kohlen> close-burning coal
magerer Ton m <Bau> sandy clay
mageres Alkydharz n <Kunststoff> short oil alkyd
mageres Gemisch n <Kfztech> poor mixture
Magergemisch n <Kfztech, Lufttrans> lean mixture
Magerkohle f <Kohlen> free-burning coal, lean coal, non-bituminous coal, noncaking coal
Magermilch f <Lebensmittel> skimmed milk
Magermilchpulver n <Lebensmittel> skimmed-milk powder
Magerton m <Ker & Glas> meager clay
Magerungszusatz m <Ker & Glas> leaner
magische Zahlen fpl 1. <Phys> magic numbers (Kernphysik); 2. <Strahlphys> magic numbers (bei Kernen mit speziell hoher Bindungsenergie)
magisches Auge n <Elektronik, Elektrotech> electric eye, magic eye
magisches Quadrat n <Math> magic square
Magma n <Nichtfoss Energ> magma
Magnesia f <Chemie> bitter earth, magnesia
Magnesia... <Chemie> magnesian
Magnesit m <Kunststoff> magnesite
Magnesit-Chrom-Feuerfesterzeugnis n <Ker & Glas> magnesite chrome refractory
Magnesitstein m <Bau> magnesite brick
Magnesitziegel m <Bau> magnesite brick
Magnesium n (Mg) <Chemie> magnesium (Mg)
Magnesiumblitzlicht n <Foto> flashlight
magnesiumhaltig adj <Chemie> magnesian
Magnesiumoxid n 1. <Chemie> bitter earth, magnesium oxide; 2. <Kunststoff> calcined magnesia
Magnesiumsilberchloridelement n <Elektrotech> magnesium-silver chloride cell
Magneson n <Chemie> magneson
Magnet m <Aufnahme, Elektriz, Labor, Phys, Teilphys, Telekom> magnet
Magnetablenkung f <Kerntech> magnetic deflection
Magnetabscheider m <Abfall, Kerntech, Kohlen> magnetic separator
Magnetabscheidung f <Abfall> magnetic separation
Magnetabschirmung f 1. <Elektrotech> magnetic shielding; 2. <Phys> magnetic shield
Magnetabstoßung f <Elektrotech> magnetic repulsion
Magnetabstreifung f <Aufnahme> magnetic stripping
Magnetachse f <Raumfahrt> magnetic axis
Magnetanker m <Elektriz, Phys> keeper
Magnetankerlautsprecher m <Aufnahme> magnetic armature loudspeaker
Magnetanlasser m <Elektriz> magnetic starter
Magnetaufnahme f <Comp & DV> magnetic recording
Magnetaufnahmestandard m <Aufnahme> magnetic-recording standard
Magnetaufzeichnung f <Comp & DV> magnetic recording
Magnetbahn f <Elektrotech> Maglev system, magnetic train (Magnetschwebetechnik)
Magnetband n <Aufnahme, Comp & DV, Druck, Elektriz, Fernseh, Telekom> magnetic tape, tape

Magnetbandeinheit

Magnetbandeinheit f <Comp & DV> magnetic tape unit, streamer
Magnetband-Fernsehaufzeichnung f <Aufnahme, Fernseh> magnetic tape video recording
Magnetbandgerät n 1. <Comp & DV> magnetic tape recorder, magnetic tape unit, tape deck; 2. <Fernseh> tape unit
magnetbandgesteuerte Werkzeugmaschine f <Fertig> magnetic-tape-controlled machine tool
Magnetband-Heimvideosystem n <Fernseh> magnetic tape home video system
Magnetbandkassette f 1. <Comp & DV> magnetic tape cartridge, tape cartridge; 2. <Fernseh, Telekom> magnetic tape cartridge
Magnetbandkassettenlaufwerk n <Comp & DV> cartridge drive
Magnetbandrauschen n <Aufnahme> magnetic tape noise
Magnetband-Streamer m <Comp & DV> streaming tape drive, stringy floppy
Magnetblasenspeicher m <Comp & DV> magnetic bubble memory
Magnetbremse f <Eisenbahn, Kfztech> electromagnetic brake
Magnetdetektor m <Trans> magnetic detector
Magnetdickenmesser m <Labor> (AE) magnetic thickness gage, (BE) magnetic thickness gauge (bei Magnetrelais)
Magnetdiode f <Elektronik> magnetodiode
Magnetdoppelschicht f <Phys> magnetic shell
Magnetdraht m <Akustik> magnetic wire
Magnetdynamo m <Elektriz, Kfztech> magnetodynamo
Magneteisen n <Metall> magnetic iron
Magneteisenstein m <Metall> magnetic ore
magnetelektrische Maschine f <Wassertrans> magneto
Magnetfangwerkzeug n <Erdöl> magnetic fishing tool (Bohrtechnik)
Magnetfarbe f <Comp & DV> magnetic ink
Magnetfeld n <Elektriz, Elektrotech, Fernseh, Telekom> magnetic field
Magnetfeld n wechselnder Richtung <Elektrotech> alternating magnetic field
Magnetfeldkonfiguration f <Kerntech, Strahlphys> magnetic field configuration
Magnetfeldlinie f <Strahlphys> magnetic field line
Magnetfeldmesser m <Lufttrans> magnetometer
Magnetfeldmessgerät n <Elektriz> gaussmeter
Magnetfeldregler m <Elektrotech> field regulator
Magnetfeldröhre f <Elektronik> magnetron
Magnetfeldstärke f 1. <Elektriz> magnetic flux density; 2. <Phys> magnetic field strength; 3. <Strahlphys> magnetic intensity
Magnetfilter n <Maschinen> magnetic filter
Magnetfluss m <Elektriz, Elektrotech, Fernseh, Strahlphys> magnetic flux
Magnetflussdichte f <Elektriz, Phys> magnetic flux density
Magnetflüssigkeitsbad n <Fertig> wet bath
Magnetfutteraufspannung f <Fertig> magnetic chucking
Magnetgreifer m <Fertig> magnetic gripper
Magnethörkopf m <Akustik> reproducing magnetic head
Magnetinduktion f (B) <Elektriz, Elektrotech, Phys> magnetic induction (B)
magnetisch adj <Aufnahme, Elektriz, Phys> magnetic
magnetisch gehaltener Koppelpunkt m <Telekom> magnetically-latched crosspoint
magnetische Anordnung f <Metall> magnetic order
magnetische Anziehung f <Elektriz> magnetic attraction
magnetische Aufzeichnung f <Funktech, Telekom> magnetic recording

magnetische Ausrichtung f <Aufnahme> magnetic alignment
magnetische Beblasung f <Elektrotech> magnetic blowout
magnetische Beschichtung f <Aufnahme, Fernseh> magnetic coating
magnetische Breite f <Phys> magnetic latitude
magnetische Dämpfung f <Elektrotech> magnetic damping
magnetische Datenträger mpl <Comp & DV> magnetic media
magnetische Deklination f <Phys> magnetic declination
magnetische Depolarisation f <Strahlphys> magnetic depolarization (von Resonanzstrahlung)
magnetische Durchdringbarkeit f <Elektrotech> magnetic permeability
magnetische Durchlässigkeit f <Elektrotech> magnetic permeability
magnetische Energie f <Elektriz, Elektrotech, Phys> magnetic energy
magnetische Feldintensität f <Elektriz> magnetic field intensity
magnetische Feldlinie f <Kerntech> magnetic flux line
magnetische Feldstärke f (H) <Elektriz, Elektrotech> magnetic field strength (H)
magnetische Flasche f <Phys> magnetic bottle
magnetische Flussverkettung f 1. <Elektrotech> flux linkage; 2. <Nichtfoss Energ> magnetic flux linkage
magnetische Fokussierung f <Elektrotech> magnetic focusing
magnetische Halterung f <Maschinen> magnetic holding
magnetische Hysterese f <Elektriz, Elektrotech> magnetic hysteresis
magnetische Hystereseschleife f <Elektriz> magnetic hysteresis loop
magnetische Induktion f (B) <Telekom> magnetic induction (B)
magnetische Induktionsschleife f <Telekom> magnetic induction current loop
magnetische Isotopentrennung f <Kerntech> magnetic isotope separation
magnetische Karte f <Comp & DV> magnetic card
magnetische Kernresonanz f (NMR) <Erdöl, Teilphys> nuclear magnetic resonance (NMR)
magnetische Konstante f <Phys> magnetic constant
magnetische Kopplung f <Elektriz, Elektrotech> magnetic coupling
magnetische Kraft f <Elektriz> magnetic force
magnetische Kraftlinie f <Elektrotech> flux line
magnetische Kraftlinienstreuung f <Elektrotech> magnetic flux leakage
magnetische Kupplung f <Elektriz> magnetic clutch
magnetische Leitfähigkeit f <Elektriz, Elektrotech, Telekom> magnetoconductivity
magnetische Nord-Süd-Richtung f <Phys> magnetic meridian
magnetische Peilung f <Raumfahrt> magnetic bearing (Raumschiff)
magnetische Permeabilität f <Elektriz, Elektrotech> magnetic permeability
magnetische Permeabilität f des Vakuums <Elektrotech> magnetic permeability of free space
magnetische Polarisation f <Akustik, Elektrotech> magnetic polarization
magnetische Spannung f <Elektriz, Phys> magnetic voltage
magnetische Spiralstruktur f <Phys> helimagnetism
magnetische Stoffe mpl <Elektrotech> magnetic material

magnetische Streuung f <Elektrotech> magnetic leakage
magnetische Streuzahl f <Elektriz, Qual> coefficient of magnetic dispersion, coefficient of magnetic leakage
magnetische Verzögerung f <Lufttrans> magnetic drag
magnetische Vorspannung f <Aufnahme, Fernseh> magnetic bias
magnetische Wechselvorspannung f <Fernseh> AC magnetic biasing
magnetische Zelle f <Comp & DV> magnetic cell
magnetisch-epitaxiale Schicht f <Elektronik> magnetic epitaxial layer
magnetischer Ablassstopfen m <Maschinen> magnetic drain plug
magnetischer Anzeiger m <Maschinen> magnetic indicator
magnetischer Äquator m <Phys> magnetic equator
magnetischer Bezirk m <Elektrotech> magnetic domain
magnetischer Blasenspeicher m <Elektrotech> magnetic bubble memory
magnetischer Dipol m <Phys> magnetic dipole
magnetischer Dipolübergang m <Strahlphys> magnetic dipole transition
magnetischer Durchflussmessumformer m <Regelung> magnetic flow transducer
magnetischer Einschluss m <Phys> magnetic confinement (Plasmaphysik)
magnetischer Feldgradient m <Elektrotech> magnetic field gradient
magnetischer Fluss m <Elektrotech, Strahlphys> magnetic flux
magnetischer Flussmesser m <Elektriz> fluxmeter (Messinstrument)
magnetischer Induktionsfluss m <Aufnahme> magnetic flux
magnetischer Kreis m <Elektrotech> magnetic circuit
magnetischer Kurs m <Wassertrans> magnetic course
magnetischer Leitwert m <Elektriz, Elektrotech, Phys> permeance (P)
magnetischer Löschkopf m <Akustik> erasing head
magnetischer Meridian m <Phys> magnetic meridian
magnetischer Nord-Süd-Pol m <Phys> magnetic north/south pole
magnetischer Rissdetektor m <Maschinen> magnetic crack detector
magnetischer Scheinwiderstand m (B) <Aufnahme> magnetic induction (B)
magnetischer Speicher m <Elektronik> magnetic random access memory (MRAM)
magnetischer Streufaktor m <Elektriz, Qual> coefficient of magnetic dispersion, coefficient of magnetic leakage
magnetischer Strom m <Elektrotech> magnetic flux
magnetischer Tonabnehmer m <Aufnahme> magnetic pickup
magnetischer Verstärker m <Elektrotech, Phys, Raumfahrt> magamp, magnetic amplifier (Raumfahrt)
magnetischer Widerstand m 1. <Elektrotech> reluctance; 2. <Phys> magnetoresistance, reluctance, reluctivity
magnetischer Zentrierring m <Raumfahrt> (AE) magnetic-centering ring, (BE) magnetic-centring ring (Raumschiff)
magnetisches Aufzeichnungsmaterial n <Akustik> magnetic medium
magnetisches Aufzeichnungsmedium n <Fernseh> magnetic-recording medium
magnetisches Dipolmoment n <Phys> magnetic dipole moment
magnetisches Echo n <Aufnahme> magnetic echo

magnetisches Feld n <Aufnahme, Elektriz, Phys, Telekom> magnetic field
magnetisches Führen n <Elektrotech> magnetic guidance (Magnetschwebetechnik)
magnetisches Halterelais n <Elektrotech> magnetic latching relay
magnetisches Kalorimeter n (MAC) <Teilphys> magnetic calorimeter (MAC)
magnetisches Medium n <Comp & DV, Elektrotech> magnetic medium
magnetisches Moment n eines Müons <Strahlphys> muon magnetic moment
magnetisches Schwungrad n <Elektrotech> magnetic flywheel
magnetisches skalares Potenzial n <Phys> magnetic scalar potential
magnetisches Überlastrelais n <Elektriz> magnetic overload relay
magnetisches Wechselfeld n <Elektrotech> alternating magnetic field
magnetisieren v <Elektrotech, Phys> magnetize
magnetisiert adj <Aufnahme, Elektriz, Kerntech> magnetized
magnetisierter Kopf m <Aufnahme> magnetized head
magnetisiertes Plasma n <Kerntech> magnetized plasma
Magnetisierung f <Aufnahme, Elektriz, Elektrotech, Fernseh, Phys, Telekom> magnetization
Magnetisierungsfeld n <Elektrotech> magnetizing field
Magnetisierungskraft f <Elektriz> magnetizing force
Magnetisierungskurve f <Elektrotech> magnetization curve
Magnetisierungsmoment n <Elektrotech> magnetizing moment
Magnetisierungsschleife f <Elektriz> B/H loop
Magnetisierungsspule f <Elektriz, Elektrotech> magnetizing coil
Magnetismus m <Elektriz, Funktech, Phys> magnetism
Magnetit m 1. <Kerntech> magnetite (schwerer Betonzuschlagstoff); 2. <Metall> magnetic ore, magnetite
Magnetjoch n <Phys> yoke of magnet
Magnetkarte f 1. <Comp & DV> magnetic card; 2. <Elektronik> card (Datenkommunikation)
Magnetkartenlesegerät n <Comp & DV> magnetic card reader
Magnetkartenleser m <Comp & DV> magnetic card reader
Magnetkartenspeicher m <Comp & DV> magnetic card memory
Magnetkern m 1. <Comp & DV, Elektrotech> ferrite core, magnetic core; magnetic core (Teil eines Magnetkreises, bei Magnetspeicher); 2. <Phys> core
Magnetkernspeicher m <Elektrotech> magnetic core memory
Magnetkissen n <Trans> magnetic cushion
Magnetkissenaufhängung f <Trans> magnetic suspension
Magnetkissenzug m <Trans> magnetic cushion train
Magnetkompass m <Lufttrans, Wassertrans> magnetic compass
Magnetkopf m <Akustik, Aufnahme, Comp & DV, Fernseh> magnetic head
Magnetkopfkern m <Fernseh> magnetic head core
Magnetkopfspalt m 1. <Aufnahme, Comp & DV> head gap; 2. <Fernseh> gap, head gap, magnetic head gap
Magnetkraftschweißen n <Kerntech> magnetic force welding
Magnetkreis m <Elektriz, Elektrotech, Phys, Raumfahrt> magnetic circuit, magnetic flux guide (Raumschiff)

Magnetkupplung

Magnetkupplung f <Elektrotech, Maschinen> magnetic clutch
Magnetkurs m <Lufttrans> magnetic heading
Magnetlager npl 1. <Elektriz> magnetic bearing; 2. <Maschinen> magneto bearings; 3. <Mechan> magnetic bearing
Magnetlinse f <Elektrotech, Strahlphys> magnetic lens
Magnetlöschkopf m <Fernseh> magnetic-erasing head
Magnetmaterial n <Elektrotech> magnetic material
Magnetmoment n <Elektriz, Phys, Teilphys> magnetic moment
Magnetmonopol n <Phys> magnetic monopole
Magnetnadel f <Maschinen> magnetic needle
Magneto... magneto
Magnetoelektrizität f <Elektrotech> magnetoelectricity
Magnetogasdynamik f (MGD) <Kerntech> magnetogasdynamics (MGD)
Magnetograph m <Comp & DV> magnetographic printer
Magnetohydrodynamik f <Kerntech, Maschinen, Phys, Strömphys> magnetohydrodynamics
magnetohydrodynamisch adj <Elektrotech, Kerntech, Maschinen, Phys, Raumfahrt> magnetohydrodynamic
magnetohydrodynamische Pumpe f <Elektrotech> magnetohydrodynamic pump
magnetohydrodynamische Stromerzeugung f <Elektrotech> magnetohydrodynamic generation
magnetohydrodynamische Wandlung f <Elektrotech> magnetohydrodynamic conversion
magnetohydrodynamischer Generator m <Elektrotech> magnetohydrodynamic generator
magnetohydrodynamischer Konverter m (MHD-Konverter) <Raumfahrt> magnetohydrodynamic converter (MHD converter)
magnetohydrodynamischer Wandler m (MHD-Wandler) <Elektrotech, Kerntech> magnetohydrodynamic converter (MHD converter)
magnetohydrodynamisches Lager n <Maschinen> magnetohydrodynamic bearing
Magnetometer n <Elektriz, Erdöl, Lufttrans, Phys, Raumfahrt, Strömphys> magnetometer
Magnetometer-Ausleger m <Raumfahrt> magnetometer boom (Raumschiff)
Magnetometer-Messtechnik f 1. <Elektriz> magnetometry; 2. <Erdöl> magnetometry (Erdmagnetfeld); 3. <Phys> magnetometry
Magnetometer-Vermessung f <Erdöl> magnetometer survey (Erdmagnetfeld)
magnetomotorische Kraft f (MMK) <Elektriz, Phys, Strömphys> magnetomotive force (mmf)
Magneton n <Teilphys> magneton
magneto-optisch adj <Optik> m-o, magneto-optic, magneto-optical
magneto-optische Drehung f <Elektrotech> Faraday effect
magneto-optische Platte f 1. <Comp & DV> magneto-optical disk; 2. <Optik> (BE) m-o disc, (AE) m-o disk, (BE) magneto-optical disc, (AE) magneto-optical disk
magneto-optischer Effekt m <Optik> magneto-optical effect
magneto-optisches Medium n <Optik> magneto-optical medium
Magnetoplasma n <Kerntech> magnetoplasma
Magnetoskop n <Fernseh> magnetoscope
magnetostatisches Feld n <Kerntech, Phys> magnetostatic field
Magnetostriktion f <Akustik, Elektrotech, Phys> magnetostriction
Magnetostriktionskonstante f (K) <Akustik> magnetostriction constant (K)

Magnetostriktionslautsprecher m <Akustik> magnetostriction loudspeaker
Magnetostriktionsmikrofon n <Akustik> magnetostriction microphone
magnetostriktiver Wandler m 1. <Elektrotech> magnetostrictive transductor; 2. <Gerät> magnetostrictive transducer
magnetostriktives Material n <Elektrotech> magnetostrictive material
Magnetothek f <Comp & DV> tape library
Magnetpeilung f <Lufttrans> magnetic bearing
Magnetplatte f <Aufnahme, Comp & DV, Elektriz, Elektrotech> (BE) magnetic disc, (AE) magnetic disk
Magnetpol m <Elektriz, Phys> magnetic pole
Magnetpotenzial n <Elektrotech> magnetic potential
Magnetpulveraufschwämmung f <Fertig> magnetic paste
Magnetpulverkopplung f <Elektrotech> magnetic particle coupling
Magnetpulverkupplung f <Maschinen> magnetic particle clutch, magnetic powder clutch, powder clutch
Magnetpulverprüfung f 1. <Fertig> magnaflux testing; 2. <Maschinen> magnetic particle inspection; 3. <Mechan> magnetic particle examination
Magnetquantenzahl f 1. <Kerntech> total magnetic quantum number; 2. <Phys> magnetic quantum number
Magnetresonanz f <Elektrotech, Phys> magnetic resonance
Magnetresonanzspektroskopie f <Strahlphys> magnetic resonance spectroscopy
Magnetron n <Elektronik, Funktech> magnetron
Magnetrongenerator m <Elektronik> magnetron oscillator
Magnetronlichtbogenbildung f <Elektrotech> magnetron arcing
Magnetronverstärker m <Elektronik> magnetron amplifier
Magnetrührer m <Labor> magnetic stirrer
Magnetsättigung f <Elektriz, Elektrotech> magnetic saturation
Magnetscheibe f <Elektrotech> (BE) magnetic disc, (AE) magnetic disk
Magnetschirmung f <Elektriz> magnetic screening
Magnetschleifendetektor m <Trans> magnetic loop detector
Magnetschrift f <Comp & DV> magnetic ink
Magnetschrifterkennung f (MICR) <Comp & DV> magnetic ink character recognition (MICR)
Magnetschriftlesegerät n <Comp & DV> magnetic ink character reader
Magnetschriftleser m <Comp & DV> magnetic ink character reader
Magnetschriftzeichenerkennung f (MICR) <Comp & DV> magnetic ink character recognition (MICR)
Magnetschwebefahrzeug n 1. <Elektrotech> maglev, magnetic levitation vehicle, magnetically supported vehicle, magnetically suspended vehicle (Elektrotraktion); 2. <Trans> maglev, magnetic levitation vehicle, magnetically supported vehicle, magnetically suspended vehicle
Magnetschwebetechnik f <Phys, Trans> magnetic levitation
Magnetschweif m <Raumfahrt> magnetotail
Magnetsonde f <Kerntech> magnetic probe
Magnetsortierung f <Abfall> magnetic separation (von Müll)
Magnetspeicher m <Comp & DV> magnetic memory
Magnetspeichermedium n <Elektrotech> magnetic storage medium
Magnetspiegel m <Phys> magnetic mirror

Magnetspule f 1. <Elektriz, Elektrotech> magnet coil, solenoid; 2. <Heiz & Kälte> solenoid coil; 3. <Kfztech> solenoid *(Anlasser)*
Magnetspur f <Comp & DV> track
Magnetstreifen m <Aufnahme> magnetic stripe
Magnetstreifen-Streamer m <Comp & DV> streamer
Magnetstrom m <Phys> magnetron
Magnetsturm m <Raumfahrt> magnetic storm
Magnetsuszeptibilität f <Elektriz, Phys> magnetic susceptibility
Magnettinte f <Comp & DV> magnetic ink
Magnetton m <Akustik, Aufnahme> magnetic sound
Magnettonaufzeichnung f <Akustik, Aufnahme> magnetic recording
Magnettonband n <Akustik> magnetic tape
Magnettonfilm m <Aufnahme> magnetic-recording film
Magnettongerät n 1. <Akustik> digital audio tape recorder, digital audiomagnetic tape recorder; 2. <Comp & DV> magnetic recorder, recorder
Magnettonspur f <Aufnahme> magnetic sound track
Magnetträger m <Comp & DV> magnetic material, magnetic medium
Magnettrommel f <Comp & DV, Elektrotech> magnetic drum
Magnettrommelspeicher m <Comp & DV, Elektrotech> magnetic drum memory
Magnetübergang m <Metall> magnetic transition
Magnetvektorpotenzial n <Phys> magnetic vector potential
Magnetventil n 1. <Elektriz> solenoid valve; 2. <Fertig> solenoid valve *(Kunststoffinstallationen)*; 3. <Heiz & Kälte, Hydraul, Maschinen> magnet valve, magnetic valve, solenoid valve, solenoid-operated valve
Magnetverstärker m <Elektrotech> magamp, magnetic amplifier
Magnetwelle f <Elektrotech> magnetic wave
Magnetwerkstoff m <Comp & DV> magnetic material
Magnetwiderstand m <Elektrotech> magnetic resistance
Magnetwiedergabegerät n <Fernseh> magnetic reproducer
Magnetzähler m <Gerät> magnetic counter
Magnetzelle f <Comp & DV> magnetic cell
Magnetzündung f <Kfztech> magneto ignition
Magnetzunge f <Elektriz> keeper
Magnon m <Phys> magnon
Magnox-Reaktor m <Kerntech> magnox reactor
Magnum n <Ker & Glas> magnum
Magnus-Effekt m <Phys> Magnus effect
MAG-Schweißen n *(Metall-Aktivgas-Schweißen)* <Thermod> MAG welding *(metal active gas welding)*
Mahlanlage f 1. <Chemtech> grinding plant, pulverizing equipment; 2. <Papier> refining plant
Mahlbarkeit f 1. <Chemtech> grindability; 2. <Papier> beatability
mahlen v 1. <Chemtech> grind; 2. <Kohlen> mill; 3. <Lebensmittel> grind; 4. <Mechan> mill; 5. <Papier> beat
Mahlen n 1. <Chemtech> milling; 2. <Kohlen, Kunststoff> grinding; 3. <Lebensmittel> milling; 4. <Metall> grinding; 5. <Papier> beating, milling; 6. <Textil> grinding, milling
Mahlen n **im geschlossenen Kreislauf** <Kohlen> closed-circuit grinding
Mahlfeinheitsmessgerät n <Labor> *(AE)* fineness-of-grind gage, *(BE)* fineness-of-grind gauge
Mahlgerät n <Chemtech> grinding device
Mahlgrad m <Papier> freeness, freeness value
Mahlgradmessgerät n **nach Hegman** <Kunststoff> *(AE)* Hegman fineness of grind gage, *(BE)* Hegman fineness of grind gauge
Mahlgradprüfer m <Papier> freeness tester

Mahlholländer m <Papier> beater
Mahlkammer f <Chemtech> pulverizing chamber
Mahlkörper m <Kohlen> grinding medium
Mahlraum m <Chemtech> pulverizing chamber
Mahlring m <Chemtech> grinding ring
Mahlschüssel f <Fertig> pan *(Kollergang)*
Mahlstein m <Lebensmittel> millstone
Mahltrommel f <Chemtech> grinding drum
Mahl- und Extrahierprozess m <Kerntech> grind and leach process
Mahlverfahren n <Lebensmittel> milling process
Mail f <Comp & DV> mail message
Mailbox f <Telekom> mailbox
Mailbox-Betreiber m <Telekom> system operator
Mailbox-Symbol n <Telekom> mailbox icon, You have mail icon
Mail-Filter n <Telekom> mail filter *(E-mail)*
Mainframe-basiert adj <Comp & DV> mainframe based
Maisbrei m <Lebensmittel> hominy
Maischbottich m <Lebensmittel> mash tub
Maische f <Lebensmittel> pulp; pomace *(Bierbrauen und Keltern)*; mash *(Brauerei)*
Maischeapparat m <Lebensmittel> masher
Maischebereiter m <Lebensmittel> masher
Maischwasser n <Lebensmittel> mash liquor
Maisgrieß m <Lebensmittel> hominy grits
Maiskleber m <Chemie> maize gluten
Majolika f <Ker & Glas> majolica
Majolikafarben fpl <Ker & Glas> *(AE)* majolica colors, *(BE)* majolica colours
Majolikafliese f <Ker & Glas> majolica tile
Majolikamaler m <Ker & Glas> majolica painter
Majolikawaren fpl <Ker & Glas> majolica ware
Majorana-Kraft f <Kerntech> Majorana force
Majorität f <Elektronik, Phys> majority
Majoritätsgatter n <Elektronik> majority gate
Majoritätsladungsträger m <Phys> majority carrier
Majoritätslogik f <Elektronik> majority logic
Majoritätsschaltung f <Elektronik> majority gate
Majoritätsträger m <Elektronik> majority carrier
Majoritätsträgerdiode f <Elektronik> majority carrier diode
Majoritätsträger-Transistor m <Elektronik> majority carrier transistor
MAK m *(maximale Arbeitsplatzkonzentration)* <Kerntech> MAC *(maximum allowable concentration)*; TLV *(threshold limit value)*
Makadam m <Bau, Trans> macadam
Makeln n <Telekom> broker's call, call wait/call hold
Makeln n **zwischen Abfrageorganen** <Telekom> alternating between call keys
Makro n 1. <Comp & DV> *(Makrocode)* macro, macrocode; 2. <Comp & DV> *(Makrobefehl, Makroinstruktion)* macro, macroinstruction
Makroabfall m <Umweltschmutz> macrowaste
Makroassembler m <Comp & DV> macroassembler
Makroätzprüfverfahren n <Fertig> macroetch test
Makroätzung f <Fertig> macroetching
Makroaufruf m <Comp & DV> macrocall
Makrobefehl m *(Makro)* <Comp & DV> macrocommand, macroinstruction *(macro)*
Makrobiegeverlust m <Optik> macrobend loss
Makrobiegung f 1. <Fertig> macrobending *(Faseroptik)*; 2. <Optik, Telekom> macrobending
Makroblockschicht f <Telekom> macrobloc layer *(Bildcodierung)*
Makrobogen m <Elektrotech> macrobend
Makrocode m *(Makro)* <Comp & DV> macrocode *(macro)*
makrocyclisch adj <Chemie> macrocyclic
Makroerweiterung f <Comp & DV> macroexpansion

Makrogefüge

Makrogefüge n <Chemie> macrostructure
Makrogenerierung f <Comp & DV> macrogeneration
makrogeometrische Oberflächengestalt f <Fertig> macrogeometrical surface pattern
Makrohärte f <Maschinen> macrohardness
Makroinstruktion f (Makro) <Comp & DV> macrocommand, macroinstruction (macro)
Makrokrümmung f <Fertig> macrobending (Faseroptik)
makromodularer Dampfgenerator m <Kerntech> macromodular steam generator
Makromolekül n <Chemie, Kunststoff> macromolecule
makromolekular adj <Chemie> macromolecular
makromolekulare Dispersion f <Lebensmittel> macromolecular dispersion
makromolische Kohlenstoffverbindung f <Anstrich> organic polymer
Makroprogramm n <Trans> macroprogram
Makroprozessor m <Comp & DV> macroprocessor
Makroradiographie f <Kerntech> macroradiography
makroskopische Flussschwankung f <Kerntech> macroscopic flux variation
makroskopische Veränderliche f <Phys> macroscopic variable
makroskopischer Querschnitt m <Kerntech> cross-section density, macroscopic cross section
Makrostativ n <Foto> macrophoto stand
Makrosteuerung f <Trans> macrocontrol
Makro-Umwandler m <Comp & DV> macroprocessor
Makrozelle f <Mobilkom> macrocell (zellulares Netz)
makrozyklisch adj <Chemie> macrocyclic
Makulaturbogen m <Druck> waste sheet
Malakon n <Kerntech> malacon
Malat n <Chemie> malate
Malat... <Chemie> malic
Maleinimid n <Chemie> maleinimide
Maleinsäure f <Chemie> maleic acid
Malerarbeiten fpl <Anstrich> paintwork
Mall n <Wassertrans> (AE) mold, (BE) mould (Schiffbau)
Mallbreite f <Wassertrans> (AE) molding, (BE) moulding (Schiffbau)
Mallbrett n <Wassertrans> template, templet (Schiffbau)
Mallungen fpl <Wassertrans> doldrums (Tiefdruckgürtel um Äquator)
malnehmen v <Comp & DV> multiply
Malon... <Chemie> malonic
Malonamid n <Chemie> malonamide, propanediamide
Malonat n <Chemie> malonate
Malonester m <Chemie> malonate
Malonsäuredinitril n <Chemie> malononitrile
Maltase f <Lebensmittel> maltase
Malteserkreuzgetriebe n 1. <Fertig> Geneva stop; 2. <Maschinen> Geneva mechanism
Malteserkreuzscheibe f <Fertig> Geneva wheel
Malteserkreuztrieb m <Fertig> Geneva drive
Malus'sches Gesetz n <Phys> Malus's law
Malz n <Lebensmittel> malt
Malzbereitung f <Lebensmittel> malting
malzen v <Lebensmittel> malt
Mälzerei f <Lebensmittel> malt house, malting
Malzextrakt m <Lebensmittel> malt extract
Malzfabrik f <Lebensmittel> malt house
Management-Dienstleister m <Telekom> management service provider
Management-Informationsdatenbasis f <Telekom> management information base
Management-Informationssystem n (MIS) <Comp & DV, Telekom> management information system (MIS)
Mangan n (Mn) <Chemie> manganese (Mn)
Mangan... <Chemie> manganic, manganous

Manganat n <Chemie> manganate, manganite, manganate, permanganate
Manganbronze f <Chemie> manganese bronze
Manganbronzelegierung f <Mechan> manganese bronze
Mangandioxid n <Elektrotech> manganese dioxide
Manganite n <Chemie> manganite
Mangankupfer n <Chemie> cupromanganese
Manganmassel f <Metall> manganese pig
Mangano... <Chemie> manganous
Manganstahl m <Mechan, Metall> manganese steel
Mangansulfatbadverfahren n <Kerntech> (AE) manganous sulfate bath method, (BE) manganous sulphate bath method
Mangel m 1. <Kerntech> flaw (in Material); 2. <Textil> calender
Mängel mpl <Qual> defect, deficiency
Mängelbericht m <Qual> nonconformance report, nonconformity report
Mangelhalbleiter m <Elektronik> p-type semiconductor
Mangelleitfähigkeit f <Elektrotech> p-type conductivity
Mängelliste f <Bau, Qual> completion list
Mängelrüge f <Qual> notification of defects, notification of nonconformance
Manilahanf m <Wassertrans> Manila hemp (Tauwerk)
Manipulator m <Kerntech> manipulator
manipulieren v <Fertig> adjust; manipulate
manipulierfähiger Industrieroboter m <Fertig> manipulating industrial robot
Manipuliervorrichtung f <Fertig> manipulating device
Mannid n <Chemie> mannide
mannigfaltig adj <Geom> manifold
Mannigfaltigkeit f <Geom> manifold
Mannit m <Chemie> mannite, mannitol
Mannitan n <Chemie> mannitan
Mannithexanitrat n <Chemie> nitromannite
Mannitol n <Chemie> mannite, mannitol
Mannloch n 1. <Erdöl> manway (Behälter, Tank); 2. <Raumfahrt> manhole
Mannlochdeckel m <Mechan> manhole cover
Mannlochdichtung f <Mechan> manhole gasket
Mannlochverschluss m <Mechan> manhole cover
Mannonsäure f <Chemie> mannonic acid
Mannose f <Chemie> mannose
Mannschaft f 1. <Erdöl> crew (auf Bohrturm, Bohrinsel, Schiff); 2. <Wassertrans> crew (Besatzung) • **Mannschaft anheuern** <Wassertrans> take on hands (Besatzung)
Mannschaftsdeck n <Wassertrans> mess deck (Schiff)
Mannschaftsliste f <Wassertrans> crew list
Mannschaftsräume mpl <Wassertrans> crew's quarters
Mannschaftsschleuse f <Raumfahrt> crew entry tunnel
Manometer n 1. <Bau> (AE) gage, (BE) gauge, (AE) pressure gage, (BE) pressure gauge; 2. <Erdöl> manometer (Messtechnik); 3. <Fertig> manometer, (AE) pressure gage, (BE) pressure gauge; 4. <Gerät> (AE) gage, (BE) gauge; 5. <Heiz & Kälte, Kontroll> (AE) pressure gage, (BE) pressure gauge; 6. <Labor> manometer; 7. <Maschinen> (AE) gage, (BE) gauge, (AE) pressure gage, (BE) pressure gauge; 8. <Phys> manometer, (AE) pressure gage, (BE) pressure gauge
Manometer n **für Kessel** <Gerät> (AE) boiler gage, (BE) boiler gauge
Manometer n **mit Druckkompensation** 1. <Gerät> (AE) dead-weight gage, (BE) dead-weight gauge; 2. <Phys> (AE) dead-weight pressure gage, (BE) dead-weight pressure gauge
Manometer n **mit Kraftvergleich** 1. <Gerät> (AE) dead-weight gage, (BE) dead-weight gauge; 2. <Phys> (AE) dead-weight pressure gage, (BE) dead-weight pressure gauge

Manometerschalter m <Elektriz> manometric switch
Manöver n <Lufttrans, Wassertrans> (AE) maneuver, (BE) manoeuvre
Manövrierbarkeit f <Maschinen, Trans> (AE) maneuverability, (BE) manoeuvrability
manövrieren v <Wassertrans> (AE) maneuver, (BE) manoeuvre
manövrierfähig adj <Lufttrans, Wassertrans> (AE) maneuverable, (BE) manoeuvrable
Manövrierfähigkeit f <Lufttrans, Wassertrans> (AE) maneuverability, (BE) manoeuvrability
Manövriergeschwindigkeit f <Lufttrans> (AE) design-maneuvering speed, (BE) design-manoeuvring speed
Manövrierlast f <Lufttrans> (AE) maneuvering load, (BE) manoeuvring load
manövrierunfähig adj <Wassertrans> NUC, not under command (Schiff)
Mansarde f <Bau> attric, (AE) mansard
Mansardendach n <Bau> French roof
Mansardenwalmdach n <Bau> curb roof, double pitch roof
Manschette f 1. <Bau> collar; 2. <Fertig> sleeve (Kunststoffinstallationen); 3. <Maschinen, Mechan> collar
Mantel m 1. <Elektrotech> jacket; 2. <Hydraul> cladding; 3. <Kerntech> shell casing; shell (Umhüllung); jacket (zum Kühlen, Erwärmen); 4. <Kfztech> cover (amerikanischer Reifentyp); 5. <Labor> jacket (Glasartikel); 6. <Maschinen> jacket, skirt; 7. <Mechan> jacket; 8. <Optik> cladding (Lichtleiter); 9. <Raumfahrt> shielding (Kabel); 10. <Telekom> cladding
Manteldurchmesser m <Optik, Telekom> cladding diameter (Lichtleiter)
Mantelelektrode f <Bau> covered electrode
Manteletikett n <Verpack> wraparound label
Mantelfaser f <Optik> (AE) compound glass fiber, (BE) compound glass fibre (Lichtleiter)
Mantelfläche f <Geom> lateral area
Mantelgebläse n <Lufttrans, Maschinen> ducted fan
Mantelglas n <Ker & Glas> encapsulating glass
Mantelkabel n <Elektrotech> sheathed cable
Mantelkühlung f <Heiz & Kälte, Kerntech> jacket cooling
Mantelkurve f 1. <Fertig> barrel cam, cylinder cam, drum cam; 2. <Maschinen> barrel cam
Mantelmaterial n <Textil> jacketing
Mantelmitte f <Optik, Telekom> (AE) cladding center, (BE) cladding centre (Lichtleiter)
Mantelmittelpunkt m <Telekom> (AE) cladding center, (BE) cladding centre
Mantelmodenstripper m <Optik, Telekom> cladding mode stripper
Mantelmodus m <Optik, Telekom> cladding mode
Mantelofen m <Heiz & Kälte> jacket furnace
Mantelpropeller m <Lufttrans> ducted propeller
Mantelrohr n 1. <Bau> casing; 2. <Fertig> outer sleeve; 3. <Kohlen> pipe casing; 4. <Maschinen> jacket pipe
Mantelstein m <Ker & Glas> mantle block
Mantelstromstrahltriebwerk n <Lufttrans> bypass engine
Mantelstromtriebwerk n <Lufttrans> fan engine
Manteltoleranzfeld n <Optik, Telekom> cladding tolerance field
Manteltrafo m <Elektrotech> shell-type transformer
Manteltransformator m <Elektriz, Elektrotech> shell-type transformer
Mantisse f <Comp & DV, Math> mantissa
manuell adj <Funktech, Kontroll, Mechan> manual
manuell bediente Maschine f <Verpack> hand-operated machine
manuell betrieben adj <Kontroll> manually-operated

manuelle Betätigung f **des Fahrwerks** <Lufttrans> landing-gear manual release
manuelle Datenverarbeitung f <Comp & DV> manual data processing
manuelle Eingabe f <Comp & DV> manual entry, manual input
manuelle Fernbedienung f <Lufttrans> manual remote control
manuelle Geschicklichkeit f <Ergon> manual dexterity
manuelle Kühlofenbeschickung f <Ker & Glas> carry-in
manuelle Steuerung f **der Starterklappe** <Kfztech> manual choke control
manuelle Verstärkungseinstellung f <Elektronik> manual gain control
manuelle Zufuhr f <Verpack> hand feed
manueller Betrieb m <Telekom> manual working
manuelles Heben n <Sicherheit> manual lifting technique
Manufacturing Automation Protocol n (MAP) <Kontroll> manufacturing automation protocol, MAP (Standard für Fabriksautomatisierungsgeräte)
Manuskript n <Druck> copy, manuscript
MAP (Manufacturing Automation Protocol) <Kontroll> MAP, manufacturing automation protocol (Standard für Fabrikautomatisierungsgeräte)
Marantastärke f <Lebensmittel> arrowroot
Marbelmarkierung f <Ker & Glas> marver mark
Marbeln n <Ker & Glas> marvering
Marbeltisch m <Ker & Glas> marver
Marginalie f <Druck> side note (formal)
Marienglas n <Chemie> selenite
Marine f <Wassertrans> (BE) marine, (AE) navy
marine Verschmutzung f <Wasserversorg> aquatic pollutant
Marineingenieur m <Wassertrans> naval engineer
Marineingenieurwesen n <Wassertrans> naval architecture
Marinenachrichtenwesen n <Telekom> naval communication
Marineradar n <Wassertrans> marine radar
Marineschiff n <Wassertrans> naval vessel, navy ship
Marinewerft f <Wassertrans> dockyard, naval dockyard
maritim adj <Wassertrans> maritime
maritimer Satellit m <Raumfahrt> maritime satellite (Weltraumfunk)
Marke f 1. <Bau> datum level; 2. <Comp & DV> label, mark, marker, tag; sentinel (Hinweissymbol); 3. <Patent> trademark (Gemeinschaftsmarke); 4. <Telekom> mark, tag
Marke f **auf der Abschlussseite** <Comp & DV> trailer label
Markenname m <Patent> trade name
Marketerie f <Ker & Glas> marquetry
Markierbit n 1. <Comp & DV> marker; 2. <Telekom> marker bit
markieren v 1. <Bau> beacon; 2. <Comp & DV> flag, tag; 3. <Kontroll> strobe
Markieren n 1. <Comp & DV> flagging; 2. <Kerntech> (AE) labeling, (BE) labelling (durch chemischen Austausch)
Markierer m <Funktech, Ker & Glas, Telekom> marker
Markierfilz m <Papier> marking felt, ribbing felt
Markierfolge f <Telekom> marking sequence
Markierimpuls m 1. <Comp & DV, Elektronik> marker pulse; 2. <Telekom> strobe pulse (Radar)
Markiermaschine f <Fertig, Textil> marking machine
Markiernadel f <Bau> scratch awl
markierte Verbindung f <Kerntech> (AE) labeled compound, (BE) labelled compound (radioaktiv)

markierte

markierte Verbindung *f* mit Tritium <Kerntech> tritiated compound
markierter freier Kanal *m* <Telekom> marked idle channel
Markierung *f* 1. <Comp & DV> cue point, flag, label, marker, tab, tag; highlighting *(auf dem Bildschirm)*; 2. <Maschinen> mark, marker; 3. <Papier> marking; 4. <Phys> labelling *(mit Isotopen)*; 5. <Qual> tally; 6. <Telekom> labelling, marking, tag, tagging; 7. <Textil> labelling, marking; 8. <Wassertrans> mark
Markierung *f* **eines sicheren Gewässers** <Wassertrans> safe water mark
Markierung *f* **mit typisierender Information** <Künstl Int> tagging
Markierungsabtasten *n* <Comp & DV> mark scanning
Markierungsatom *n* <Kerntech> tracer atom
Markierungsbeleuchtung *f* <Eisenbahn> *(AE)* marker lights
Markierungsbit *n* <Comp & DV, Telekom> flag bit
Markierungsboje *f* <Wassertrans> marker buoy
Markierungseisen *n* <Bau> scribing iron
Markierungsende *n* <Comp & DV> end mark
Markierungsfunkfeuer *n* <Funktech, Lufttrans> fan marker beacon, radio marker beacon
Markierungsimpuls *m* <Elektronik> marker pulse
Markierungsintensität *f* <Comp & DV> mark density
Markierungslesen *n* <Comp & DV> mark reading, mark sensing, mark scanning
Markierungsleser *m* <Comp & DV> mark reader, mark sense device
Markierungspunkt *m* <Bau> monument *(Vermessung)*
Markierungsschild *n* <Verpack> marking label
Markierungssprache *f* <Druck> markup language
Markierungsstab *m* <Bau> arrow *(Vermessung)*
Markierungssystem *n* <Telekom> marker system
Markierungstechnik *f* <Kerntech> *(AE)* labeling technique, *(BE)* labelling technique
Markierungsverfahren *n* <Kerntech> *(AE)* labeling technique, *(BE)* labelling technique
Markisenstoff *m* <Textil> canvas
Markoff'sche Kette *f* <Math> Markov chain *(Statistik)*
Markröhre *f* <Chemie> pith
Marksensing *n* <Comp & DV> mark sensing
Marlspieker *m* <Wassertrans> marline spike *(Tauwerk)*
marmorieren *v* 1. <Bau> mottle; 2. <Papier> marble
Marmorieren *n* <Druck> marbling
marmoriert *adj* 1. <Bau> mottled; 2. <Papier> marbled
Marmorierung *f* <Papier> marbling
Marmorofen *m* <Ker & Glas> marble furnace
Marmorziehscheibe *f* <Ker & Glas> marble bushing
maron *adj* <Kunststoff> maroon
Marschflugkörper *m* 1. <Raumfahrt> cruise missile *(Raumfahrt)*; 2. <Wassertrans> cruise missile *(Militär)*
Marschgeschwindigkeit *f* <Wassertrans> cruising speed
Marshall-Probe *f* <Bau> Marshall test
Marshall-Prüfung *f* <Bau> Marshall test
Martensit *m* <Metall> martensite
martensitaushärtbarer Stahl *m* <Metall> maraging steel
martensitisches Härten *n* <Fertig> maraging
Martensitstahl *m* <Metall> martensitic steel
Martens-Prüfung *f* <Kunststoff> Martens test
Martens-Spiegelgerät *n* <Fertig> *(AE)* Martens strain gage, *(BE)* Martens strain gauge
Marxgenerator *m* <Kerntech> Marx generator
Marxschaltung *f* <Elektrotech> Marx circuit, Marx circuitry
Masche *f* 1. <Elektriz, Elektronik, Elektrotech> delta network, mesh; 2. <Kohlen> mesh; 3. <Lebensmittel> mesh *(eines Siebes)*; 4. <Raumfahrt> mesh *(Sieb)*; 5. <Telekom> mesh *(Netzwerk)*; 6. <Textil> mesh, stitch; 7. <Wasserversorg> mesh

Maschendraht *m* <Bau> chicken wire mesh, wire netting
Maschengröße *f* <Wassertrans> mesh size *(Fischerei)*
Maschennetz *n* <Comp & DV, Telekom> meshed network
Maschenschaltung *f* <Elektrotech> mesh; mesh connection *(Speicherröhrenanschlüsse)*
Maschenspeicherröhre *f* <Elektronik> mesh storage tube
Maschenstrom *m* <Elektrotech> mesh current
Maschenware *f* <Textil> knitted fabric
Maschenweite *f* 1. <Elektriz, Kohlen> mesh size; 2. <Maschinen> aperture
Maschenzahl *f* <Textil> mesh
Maschine *f* 1. <Comp & DV, Elektrotech, Ker & Glas, Maschinen> machine; 2. <Mechan> engine; 3. <Papier> machine; 4. <Wassertrans> engine • **Maschinen abschalten** <Ker & Glas> stop machines • **Maschinen anhalten** <Wassertrans> stop engines *(Schiffantrieb)* • **Maschinen stoppen** <Wassertrans> stop engines *(Schiffantrieb)*
Maschine *f* **für weich-elastische Verpackung** <Verpack> flexible packaging machine
Maschine *f* **mit Gegenhalter** <Maschinen> overarm machine
Maschine *f* **zum Auseinanderschachteln** <Verpack> denesting machine
Maschine *f* **zum Einsetzen von Fächern und Unterteilungen** <Verpack> division-inserting equipment
Maschine *f* **zum Herausheben, Packen und Abdecken von Trays** <Verpack> tray denesting, filling and lidding machine
Maschine *f* **zum Zahnkanten-Abrunden und -Entgraten** <Fertig> gear-tooth rounding and deburring machine
Maschine *f* **zum Zahnrad-Abrunden und -Abdachen** <Fertig> gear-tooth rounding-off and pointing machine
maschinell bearbeiten *v* <Fertig> machine
maschinell hergestellte Muttern *fpl* <Maschinen> machine-made nuts
maschinelle Ausrüstung *f* <Maschinen> machinery
maschinelle Intelligenz *f* <Künstl Int> computer intelligence, machine intelligence
maschinelle Qualitätskontrolle *f* <Verpack> machine version verification of production quality
maschineller Entwickler *m* <Foto> machine processor
maschineller Kohlenabbau *m* <Kohlen> machine coal-mining
maschinelles Bügeln *n* <Textil> pressing
maschinelles Fertigbearbeiten *n* <Maschinen> machine finishing
maschinelles Lernen *n* <Künstl Int> machine learning, ML
maschinelles Nieten *n* <Maschinen> machine riveting
maschinelles Polieren *n* <Metall> mechanical polishing
maschinelles Sehen *n* <Künstl Int> machine vision
maschinenabhängig *adj* <Comp & DV> machine-dependent
Maschinenabteilung *f* <Wassertrans> engine compartment *(Schiff)*
Maschinenachse *f* <Fertig> machine axis *(geometrische Zerlegung)*
Maschinenadresse *f* <Comp & DV> machine address
Maschinenanfahren *n* <Ker & Glas> machine start-up
Maschinenanlage *f* <Wassertrans> power plant
Maschinenarbeit *f* <Maschinen> machine work
Maschinenausfall *m* 1. <Comp & DV> machine failure; 2. <Sicherheit> machine breakdown
Maschinenausfallzeit *f* <Ergon> machine downtime
Maschinenausrüstung *f* 1. <Comp & DV> hardware; 2. <Maschinen> machinery

Maschinenbau m <Fertig, Maschinen> mechanical engineering
Maschinenbaufabriken fpl <Mechan> engineering facilities
Maschinenbauingenieur m <Fertig, Maschinen> mechanical engineer
Maschinenbedienung f <Comp & DV> machine operation
Maschinenbefehl m <Comp & DV> machine instruction
Maschinenbefehlscode m <Comp & DV> machine instruction code
Maschinenbefüllung f <Papier> machine fill
Maschinenbelastung f <Kontroll> machine load
Maschinenbelüftungsanlage f <Kfztech, Wassertrans> engine ventilation system
maschinenbestimmte Arbeitsgeschwindigkeit f <Ergon> machine-paced work
Maschinenbetrieb m 1. <Comp & DV> machine operation; 2. <Wassertrans> engine operation (Motor)
Maschinenbetriebsstundenanzeiger m <Wassertrans> engine hours indicator (Motor)
Maschinenbezugsachse f <Fertig> machine reference axis
Maschinenbreite f <Druck> machine width
Maschinenbügelsäge f <Maschinen> power hacksaw
Maschinenbütte f <Druck, Papier> machine chest
Maschinencode m <Comp & DV> machine code, machine language
Maschinendeckel m <Papier> machine deckle
Maschinendefekt m <Kfztech, Wassertrans> engine failure
Maschinendrehzahl f <Ker & Glas> machine speed
Maschinendruck m <Druck> printing by machine
maschineneigener Zeichensatz m <Comp & DV> native character set
Maschinenelement n <Maschinen> machine part
Maschinenfehler m <Comp & DV> machine error
Maschinenfertigungsanlagen fpl <Mechan> engineering facilities
Maschinenfundament n 1. <Maschinen> engine bed; 2. <Wassertrans> engine seating (Motor)
Maschinengestell n <Kfztech, Wassertrans> engine frame
maschinengestrichen adj <Verpack> machine-coated (Papier)
maschinengestütztes Lernen n <Comp & DV> machine learning
Maschinengewindebohrer m <Maschinen> machine tap, power tap
maschinengezogenes Antikglas n <Ker & Glas> antique drawn glass
maschinenglattes Papier n 1. <Druck> mill-finished paper; 2. <Verpack> MG paper, machine-glazed paper
Maschinengrundreibahle f <Maschinen> rose reamer
Maschinenguss m <Fertig> machine casting
Maschinenhammer m <Fertig, Maschinen> power hammer
Maschinenhaus n <Eisenbahn> engine shed
Maschineninstruktion f <Comp & DV> machine instruction
Maschineninstruktionscode m <Comp & DV> machine instruction code
Maschinenkapselung f <Sicherheit> machine enclosure
Maschinenklarschriftleser m <Metrol> optical reader for machine tools
Maschinenlauf m <Maschinen> machine run
Maschinenlauf m **unter Last** <Maschinen> machine running under load
Maschinenlaufzeit f <Maschinen> machine operating time

Maschinenleistung f <Fertig> machine capacity, machine efficiency
Maschinenlernen n <Künstl Int> machine learning, ML
maschinenlesbar adj <Comp & DV> machine-readable
maschinenlesbare Daten npl <Comp & DV> machine-readable data
maschinenlesbares Material n <Comp & DV> machine-readable material, MRM
Maschinennietung f <Maschinen> machine riveting
Maschinenöl n <Maschinen> machine oil, machinery oil
Maschinenoperation f <Comp & DV> machine operation
maschinenorientiert adj <Comp & DV> machine-oriented
maschinenorientierte Programmiersprache f <Comp & DV> LLL, low-level language
Maschinenpappe f <Verpack> machine-made board
Maschinenprogramm n <Comp & DV> machine program, object program
Maschinenprüfbedingung f <Comp & DV, Maschinen, Regelung, Qual> machine check
Maschinenprüfung f 1. <Comp & DV> machine check (Basissystem); 2. <Maschinen, Regelung, Qual> machine check
Maschinenraum m <Wassertrans> engine room (Motor)
Maschinenreibahle f <Maschinen> chucking reamer, machine reamer
Maschinenrichtung f <Druck> grain, grain direction
Maschinenrüstzeit f <Kontroll> machine set-up time
Maschinensäge f <Maschinen> power saw
Maschinensatz m 1. <Druck> mechanical typesetting; 2. <Elektrotech> cascade set • **im Maschinensatz gesetzt** <Druck> mechanically-set
Maschinenschaden m <Sicherheit> machine breakdown
Maschinenschalldämmung f <Sicherheit> machinery sound isolation
Maschinenschere f <Maschinen> machine shears, shearing machine
Maschinenschleifen n <Maschinen> machine grinding
Maschinenschlosser m <Kfztech, Maschinen, Mechan> fitter, locksmith, mechanic, metalworker
Maschinenschnitt m <Maschinen> machine cutting
Maschinenschraube f <Maschinen> machine bolt, machine screw
Maschinenschutz m <Sicherheit> guarding of machinery, mechanical safeguarding; mechanical equipment safety; machine safeguard, safeguard
Maschinenschutzeinrichtung f <Sicherheit> machine safeguard
Maschinenschutzkäfig m <Sicherheit> machine cage
Maschinenschutzvorrichtung f <Sicherheit> machine guard, machine safeguard, machinery guard, machinery guarding device, machinery safeguard
Maschinenschwingung f <Sicherheit> mechanical vibration
Maschinensendung f <Telekom> automatic transmission
Maschinensetzer m <Druck> machine compositor
Maschinenspannstock m <Maschinen> (BE) machine vice, (AE) machine vise
Maschinensprache f <Comp & DV> machine code, machine language, object code, object language
Maschinenstörung f 1. <Comp & DV> machine malfunction; 2. <Kfztech, Wassertrans> engine malfunction; 3. <Sicherheit> machine breakdown, machine failure, machine malfunction
Maschinenstreichen n <Papier> on-machine coating
Maschinenstundenkosten fpl <Fertig> overheads
Maschinentagebuch n <Wassertrans> engine room log
Maschinenteil n 1. <Comp & DV> environment division, machine element; 2. <Maschinen> machine part

Maschinentelegraf

Maschinentelegraf m <Wassertrans> engine room telegraph *(Motor)*
Maschinentisch m <Fertig, Maschinen> machine table
Maschinentrog m <Ker & Glas> machine tray
Maschinenüberholung f <Maschinen, Wassertrans> engine overhaul *(Motor)*
Maschinenübersetzung f *(MÜ)* <Künstl Int> machine translation *(MT)*
Maschinenumzäunung f <Sicherheit> machine fence
maschinenunabhängig adj <Comp & DV> machine-independent
Maschinenunterbau m <Mechan> engine pedestal
Maschinenwartung f <Kfztech, Wassertrans> engine maintenance
Maschinenwort n <Comp & DV> computer word
Maschinenzeit f 1. <Comp & DV> machine time; 2. <Fertig> time per cut
Maschinenzentrale f <Elektrotech> central power plant
Maschinenzuverlässigkeit f <Comp & DV> hardware reliability
Maschinenzyklus m <Comp & DV> machine cycle
Maser m *(Mikrowellenverstärkung durch stimulierte Strahlungsemission)* 1. <Elektronik, Funktech, Telekom> maser, microwave amplification by stimulated emission of radiation; 2. <Raumfahrt> maser, microwave amplification by stimulated emission of radiation *(Weltraumfunk)*
masern v <Fertig> grain
Maserpapier n <Papier> grained paper
Maserung f 1. <Bau> grain *(Holz)*; vein *(Holz, Stein)*; 2. <Fertig> grain, streak; 3. <Papier> streak *(von Holz)*
Maske f 1. <Comp & DV> mask; 2. <Fertig> shell *(Formmaskenverfahren)*; 3. <Foto> mask
Maskenauslauf m <Elektronik> mask runout
Maskenausrichtung f <Elektronik> mask alignment
Maskenausrichtungsschablone f <Elektronik> mask alignment jig *(Halbleiter)*
Maskenbit n <Comp & DV> mask bit
Maskengenerierung f <Elektronik> mask generation
maskenlose Lithographie f <Elektronik> maskless lithography
Maskenmikrofon n <Akustik> mask microphone
Maskenöffnung f <Elektronik> aperture mask
maskenprogrammierbar adj <Comp & DV, Elektronik> mask-programmable
maskenprogrammierbares Feld n <Elektronik> mask-programmable array
maskenprogrammierbares Filter n <Elektronik> mask-programmable filter
maskenprogrammierte Lithographie f <Elektronik> masked lithography
Maskenprogrammierung f <Comp & DV> mask programming
Maskenregister n <Comp & DV> mask register
Maskenröhre f <Elektronik> shadow mask tube *(Video)*
Maskensatz m <Elektronik> mask set
Maskenträger m <Elektronik> mask carrier
maskierbare Unterbrechung f <Comp & DV> maskable interrupt
maskieren v <Chemie> sequester
Maskierfolie f <Druck> masking film
Maskierung f <Comp & DV, Elektronik> masking
Maskierungsmittel n <Chemie> sequestering agent
Maß n 1. <Comp & DV> measure, measurement, metric; 2. <Maschinen> dimension, measure; 3. <Metrol> measure, measurement • **auf zu kleines Maß bearbeitet** <Fertig> overmachined
Maß n **der Glaubwürdigkeit** <Künstl Int> MB, measure of belief

Maßabweichung f 1. <Bau> margin; 2. <Maschinen> error of size, offsize; 3. <Qual> error of size
Maßanalyse f <Gerät> titrimetry
Maßband n <Metrol> measure, measuring tape, tape measure
Maßbeständigkeit f <Kunststoff, Werkprüf> dimensional stability
Maßbild n <Konstzeich> dimension illustration
Maßblatt n <Konstzeich> dimension sheet
Maßbuchstabe m <Konstzeich> dimension letter
Masse f 1. <Eisenbahn> *(BE)* earth, *(AE)* ground; 2. <Elektriz> *(BE)* earth, *(AE)* ground *(elektrisch)*; 3. <Elektrotech, Funktech, Kfztech> *(BE)* earth, *(AE)* ground; 4. <Maschinen> mass; 5. <Mechan> batch; 6. <Metrol> weight; 7. <Phys> bulk, mass; 8. <Telekom> *(BE)* earth, *(AE)* ground; 9. <Textil> bulk; 10. <Thermod> mass; 11. <Wassertrans> *(BE)* earth, *(AE)* ground • **an Masse gelegt** <Elektrotech> *(BE)* connected to earth, *(AE)* connected to ground • **an Masse legen** <Elektrotech> *(BE)* earth, *(AE)* ground
Masseanschluss m 1. <Elektriz, Elektrotech> *(BE)* connection to earth, *(AE)* connection to ground, *(BE)* earth connection, *(AE)* ground connection, *(BE)* earth terminal, *(AE)* ground terminal; 2. <Kfztech> *(BE)* earth connection, *(AE)* ground connection
Massedosierung f <Bau> weighbatching
Massedraht m <Elektrotech, Funktech, Telekom> *(BE)* earth wire, *(AE)* ground wire
Masseelektrode f <Kfztech> *(BE)* earth electrode, *(AE)* ground electrode
Masse-Energie-Äquivalenz f <Phys> mass-energy equivalence
Maßeinheit f 1. <Erdöl> unit of measurement; 2. <Konstzeich> dimension unit
Massekabel n 1. <Eisenbahn> *(BE)* earth cable, *(AE)* ground cable; 2. <Elektrotech> *(BE)* earth cable, *(BE)* earth lead, *(AE)* ground cable, *(AE)* ground lead; 3. <Kfztech, Wassertrans> *(BE)* earth cable, *(AE)* ground cable
Massekern m <Elektrotech> powdered iron core
Massel f <Fertig> blind riser, iron pig, pig
Masselbett n <Fertig> pig bed
Masseleiter m <Elektrotech, Funktech, Telekom> *(BE)* earth wire, *(AE)* ground wire
Masseleitung f <Elektrotech> *(BE)* earth line, *(AE)* ground line
Masselform f <Fertig> *(AE)* pig mold, *(BE)* pig mould
Massenabsorptionskoeffizient m <Kerntech> mass absorption coefficient
Massenauffahrunfall m <Trans> multiple-pile-up
Massenausgleich m 1. <Maschinen> mass balancing; 2. <Mechan> counterbalance
Massenausgleich m **am Luftschraubenblatt** <Lufttrans> blade balance weight *(Hubschrauber)*
Massenbelegung f <Phys> mass
Massenbeton m <Bau> mass concrete
Massenbezugslinie f <Kerntech> parent mass peak, parent peak *(in Massenspektrum)*
Massenbruchteil m <Kerntech> mass fraction
Massendefekt m 1. <Kerntech> mass defect, packing effect; 2. <Strahlphys> mass deficit
Massendichte f <Phys> mass
Massendurchsatz m <Lufttrans, Trans> mass flow
Masseneffekt m <Kerntech> mass effect
Massenentnahme f <Bau> borrow
Massenerhaltung f <Phys, Strömphys> conservation of mass
Massenfertigung f <Fertig, Kfztech, Wassertrans> mass production

Massenfluss *m* <Strömphys> mass flux *(durch Rohr, Masse pro Zeiteinheit)*
Massenform *f* <Metall> bed pig
Massenfraktion *f* <Kerntech> mass fraction
Massengleichgewicht *n* <Kerntech> mass balance
Massengramm *n* <Metrol> *(AE)* gram in mass
Massengut *n* 1. <Bau> bulk material; 2. <Wassertrans> bulk cargo *(Ladung)*
Massengutfrachter *m* <Wassertrans> bulk carrier
Massenimportschnittstelle *f* <Comp & DV> bulk import interface
massenimprägnierte Papierisolierung *f* <Elektriz> mass-impregnated paper insulation
Massenkraft *f* <Fertig> inertia force
Massenluftdurchsatz *m* <Lufttrans> mass airflow
Massenmittelpunkt *m* <Mechan, Phys> *(AE)* center of mass, *(BE)* centre of mass
Massenmittelpunktskoordinaten *fpl* <Mechan, Phys> *(AE)* center of mass coordinates, *(BE)* centre of mass coordinates
Massenparallelverarbeitung *f* <Künstl Int> MPP, massively parallel processing
Massenspeicher *m* <Comp & DV> bulk memory, mass storage, mass memory
Massenspeichereinheit *f* <Comp & DV> mass storage device
Massenspeichersystem *n* <Comp & DV> mass storage system
Massenspektrograph *m* <Phys> mass spectrograph
Massenspektrometer *n* 1. <Labor, Maschinen, Phys> mass spectrometer; 2. <Strahlphys> mass spectrograph
Massenspektrometrie *f* <Phys> mass spectrometry
Massenspektroskopanalyse *f* <Strahlphys> mass spectral analysis
Massenspektrum *n* <Phys, Strahlphys> mass spectrum
Massenspleißung *f* <Telekom> mass splicing *(Lichtleiterkabel)*
Massenstrom *m* <Lufttrans> mass flow
Massenstrommessgerät *n* <Gerät> mass flow meter
Massenstückgut *n* <Wassertrans> break bulk cargo
Massenträgheitsmoment *n* 1. <Lufttrans> angular momentum; 2. <Maschinen> mass moment of inertia
Massenüberschuss *m* <Phys> mass excess *(Kernphysik)*
Massenverlustberechnung *f* <Raumfahrt> mass budget *(Weltraumfunk)*
Massenwiderstand *m* <Fertig> inertness
Massenwirkungsgesetz *n* <Maschinen, Phys> law of mass action
Massenzahl *f* 1. <Kerntech> mass number; 2. <Phys> isotope number, nucleon number; 3. <Phys> mass number; 4. <Strahlphys, Teilphys> isotope number; 5. <Teilphys> mass number
Massenzuordnung *f* <Kerntech> mass assignment
Massepotenzial *n* <Raumfahrt> *(BE)* earth potential, *(AE)* ground potential *(Erdung)*
Masseschluss *m* <Telekom> *(AE)* ground fault, *(BE)* earth fault
Massezylinder *m* <Fertig> heating cylinder *(Kunststoffe)*
maßgenau *adj* <Maschinen> true-to-size
maßgerecht *adj* <Bau> true
maßgeschneidert *adj* <Elektronik, Telekom> custom-designed *(werbetechnisch)*
maßhaltig *adj* <Maschinen> true-to-size
Maßhaltigkeit *f* 1. <Fertig> *(AE)* accuracy to gage, *(BE)* accuracy to gauge *(Kunststoffinstallationen)*; 2. <Kunststoff, Werkprüf> dimensional stability
Maßhilfslinie *f* 1. <Konstzeich> projection line, witness line; 2. <Maschinen> extension line

massiv *adj* 1. <Bau> solid *(Holz, Stein)*; 2. <Mechan> heavy
Massivanode *f* <Elektrotech> heavy anode
Massivbeton *m* <Bau> mass concrete
massive Parallelität *f* <Künstl Int> MPP, massivelyparallel processing
massive Reaktion *f* <Metall> massive reaction
massive Treppenspindel *f* <Bau> newel post
massiver Guss *m* <Fertig> solid casting
massiver Leiter *m* <Elektrotech> solid conductor
massiv-parallel verarbeitende Plattform *f* *(MPP)* <Comp & DV> massively parallel processing platform, MPP
Massivschale *f* <Maschinen> solid liner
Massivumformung *f* <Maschinen> massive forming
Maßkennzeichen *n* <Konstzeich> identification marking of dimensions
Maßkette *f* <Konstzeich> chain dimensioning
Maßkolben *m* <Labor> volumetric flask
Maßlehre *f* <Metrol> *(AE)* limit gage, *(BE)* limit gauge
maßliche Überbestimmung *f* <Konstzeich> redundant dimensioning
maßliches Prüfen *n* <Metrol> measuring
Maßlinie *f* <Konstzeich, Maschinen> dimension line
Maßlinienbegrenzung *f* <Konstzeich> dimension line termination
Maßlücke *f* <Konstzeich> dimension gap
Maßnahme *f* 1. <Comp & DV> measure; 2. <Qual> action
Maßnahmen *fpl* **gegen absichtliche Störung** <Funkort> antijamming *(Radar)*
Maßpfeil *m* <Bau> arrowhead *(Vermessung)*
Maßprüfung *f* <Qual> dimensional check
Maßskizze *f* <Fertig> dimensional sketch
Maßstab *m* 1. <Bau> *(AE)* gage, *(BE)* gauge, scale; 2. <Comp & DV, Druck, Geom, Gerät, Maschinen> scale; 3. <Metrol> measuring rod, scale
Maßstab *m* **1:1** <Bau> natural scale
maßstabgetreue Darstellung *f* <Konstzeich> true-to-scale representation
maßstabgetreue Zeichnung *f* <Geom> scale drawing
maßstabgetreues Modell *n* <Geom, Wassertrans> scale model
maßstäblich verändern *v* <Gerät> scale *(normieren)*
maßstäbliche Darstellung *f* <Konstzeich> representation to scale
maßstäbliches Modell *n* <Wassertrans> scale model
Maßstabpapier *n* <Papier> scale paper
Maßstabsänderung *f* <Geom> scaling
Maßstabsangaben *fpl* <Konstzeich> scale particulars
Maßstabsanpassung *f* **einer Feder** <Metrol> scaling a spring
Maßstabsfaktor *m* <Comp & DV> scale factor • **Maßstabsfaktor festlegen** <Comp & DV> scale • **mit Maßstabsfaktor multiplizieren** <Gerät> scale *(normieren)*
maßstabsgerecht *adj* <Fertig> scale *(Modell)*
maßstabsgerechte Zeichnung *f* <Konstzeich> scaled drawing
Maßstelle *f* <Konstzeich> dimension joint
Maßsystem *n* <Metrol> system of units
Maßsystemanalyse *f* *(MSA)* <Qual> measurement system analysis, MSA
Maßtoleranz *f* 1. <Fertig> tolerance in size; 2. <Maschinen> dimensional tolerance, size margin, size tolerance
Maßverkörperung *f* 1. <Fertig> standard; 2. <Qual> material measure; 3. <Metrol> gauge, material measure, standard; material representation *(einer Einheit)*
Maßzeichnung *f* <Geom, Konstzeich> dimensional drawing

Mast

Mast m 1. <Bau> pole, post; 2. <Elektriz> pylon, tower *(für Hochspannungsleitungen)*; 3. <Funkort> mast *(Radarantenne)*; 4. <Funktech> mast, pedestal *(Antenne)*; 5. <Telekom> mast, pole; 6. <Wassertrans> mast • **den Mast einsetzen** <Wassertrans> step the mast *(Schiffbau)* • **den Mast herausnehmen** <Wassertrans> unstep the mast
Mastabspannungsseile npl <Funktech> tower guy wires *(Antenne)*
Mastantenne f <Funktech> mast antenna, monopole antenna *(isoliert abgespannter Rohrmast)*
Mastausleger m <Bau, Trans> derrick boom
Mastbacken fpl <Wassertrans> hounds
Mastenkran m <Bau> derrick
Master f <Optik, Telekom> master
Masterband n <Fernseh> master tape
Masterbatch m <Kunststoff> master batch
Mastern n **von Platten** <Optik> *(BE)* disc mastering, *(AE)* disk mastering
Masterplatte f <Optik> *(BE)* master disc
Master-Slave-Flipflop n <Elektronik> master-slave flip-flop
Master-Slave-Manipulator m <Kerntech> master-slave manipulator
Master-Slave-Rechnersystem n <Comp & DV> master-slave system
Mastfall m <Wassertrans> mast rake
Mastfuß m <Wassertrans> mast foot
Mastfußschiene f <Wassertrans> mast foot rail *(Bootbau, Deckbeschlag)*
Mastikation f <Kunststoff> mastication
Mastix m <Wassertrans> mastic *(Schiffbau)*
mastizieren v <Fertig> masticate *(Kunststoffe)*
Mastiziermittel n <Kunststoff> peptizer
Mastkran m <Wassertrans> mast crane *(Ladung)*
Mastschulter f <Wassertrans> hounds
Mastspitze f <Wassertrans> masthead
Mastspur f <Wassertrans> mast step
Maststrecke f <Telekom> pole route
Maststuhl m <Wassertrans> mast tabernacle
Masttopp m <Wassertrans> mast head *(Segeln)*
Masttransformator m <Elektriz> pole-mounted transformer, pole-type transformer
Mater f 1. <Druck> flong, matrix; 2. <Optik> master
Material n <Raumfahrt> fabric
Material n **mit angepasster Brechzahl** <Telekom> index-matching material
Material n **zur Anpassung des Index** <Optik> index-matching material
Materialabnahme f <Maschinen> stock removal
Materialabtragung f <Maschinen> material removal
Materialdispersion f 1. <Fertig> material dispersion *(Faseroptik)*; 2. <Optik, Telekom> material dispersion
Materialdispersionskoeffizient m <Telekom> material dispersion parameter
Materialermüdungsriss m <Raumfahrt> fatigue crack *(Raumschiff)*
Materialermüdungswiderstand m <Raumfahrt> fatigue strength *(Raumschiff)*
Materialfehler m 1. <Kerntech> flaw; 2. <Maschinen> material defect; 3. <Qual> defect of material; 4. <Textil> flaw; 5. <Werkprüf> material flaw
Materialfluss m <Verpack> material flow
Materialgröße f <Ker & Glas> stock size
Materialgrube f <Kohlen> borrow pit
Materialhandhabung f <Kerntech> materials handling
Materialplaner m <Ker & Glas> estimator
Materialplatte f <Ker & Glas> stock sheet
Materialprüfanstalt f <Qual> material-testing institute

Materialprüfreaktor m *(MTR)* <Kerntech> materials-testing reactor *(MTR)*
Materialprüfung f 1. <Maschinen> material testing; 2. <Qual> materials inspection, testing of materials
Materialrückgewinnung f <Abfall> material recovery
Materialstärke f <Wassertrans> scantling *(Schiffbau)*
Materialstreuung f <Optik, Telekom> material scattering
Materialtrichter m <Fertig> feed hopper *(Extrudieren)*
Materialwirtschaft f <Bau, Fertig, Kfztech> materials engineering; materials science
Materialzuführungstrommel f <Verpack> vibratory hopper
materielle Flussdichtewölbung f <Kerntech> material buckling
Materiewelle f <Elektrotech, Phys, Strahlphys> de Broglie wave
Maternpappe f <Druck> flong
Materplatte f <Optik> *(AE)* master disk
Mathematik f <Comp & DV, Math> mathematics
mathematisch adj <Math> mathematical
mathematische Analyse f <Math> mathematical analysis
mathematische Berechnungen fpl <Math> mathematical calculations
mathematische Grundlagen fpl <Comp & DV, Math> mathematical foundations
mathematische Hoffnung f <Comp & DV> expectation
mathematische Induktion f <Comp & DV, Math> mathematical induction
mathematische Logik f 1. <Comp & DV> symbolic logic; 2. <Math> mathematical logic
mathematische Operationen fpl <Math> mathematical operations
mathematische Physik f <Math, Phys> mathematical physics
mathematische Programmierung f <Comp & DV> mathematical programming
mathematische Wahrscheinlichkeit f <Math> mathematical probability
mathematischer Ausdruck m <Math> mathematical expression
mathematisches Modell n <Comp & DV, Elektronik> mathematical model
mathematisches Pendel n <Phys> simple pendulum
mathematisches Programmieren n <Comp & DV> mathematical programming
mathematisches Teilchen n <Kerntech> mathematical particle
Matrix f 1. <Bau> matrix *(Tiefbau)*; 2. <Comp & DV, Druck, Fernseh> matrix; 3. <Math> array, matrix; 4. <Metall> matrix
Matrix f **mit reellen Zahlen** <Math> matrix of real numbers
Matrixalgebra f <Math> matrix algebra
Matrixanzeige f <Telekom> matrix display
Matrixdisplay n <Telekom> matrix display
Matrixdrucker m 1. <Comp & DV> dot matrix printer, matrix printer; 2. <Druck> matrix printer
Matrixschaltung f <Fernseh> matrixing
Matrixsignalisation f <Trans> matrix signalization
Matrize f 1. <Comp & DV> matrix; 2. <Druck> matrix, stencil; 3. <Fertig> female die, swage; 4. <Optik> master
Matrizen fpl <Math> matrices
Matrizenalgebra f <Math> matrix algebra
Matrizenform f <Telekom> matrix configuration
Matrizenhalter m <Fertig> die holder *(Lochen)*
Matrizenmagazin n <Druck> matrix magazine
Matrizenmechanik f <Mechan, Phys> matrix mechanics
Matrizenpappe f <Druck> stereotype drymat

Matrizenschaltung f <Telekom> matrix circuit
Matrizenstahl m <Fertig> die steel
matt adj 1. <Bau> dull, flat; 2. <Druck> matt; 3. <Metall> dull, matt; 4. <Textil> (AE) lusterless, (BE) lustreless, matt
matt werden v <Bau> blind (Glas)
Mattappretur f <Textil> dull finish
Mattätzpaste f <Ker & Glas> matt-etching paste
Mattätzsalz n <Ker & Glas> matt-etching salt
Mattblech n <Fertig, Metall> terne plate
Matte f <Bau> mat
matte Satinierung f <Druck> English finish
Mattenbelag m <Bau> matting
Mattenbewehrung f <Bau> wire mesh reinforcement
mattes Papier n 1. <Druck> matt surface paper; 2. <Foto> matt paper
mattgeschliffener Fuß m <Ker & Glas> ground base
mattgeschliffenes Glas n <Ker & Glas> satin finish glass
Mattglas n <Ker & Glas> frosted glass
Mattglasur f <Ker & Glas> matt glaze
Mattheit f <Fertig> flatness
mattieren v <Textil> delustre
Mattieren n <Bau> deadening
mattierte Buchstaben mpl <Ker & Glas> dim letters
Mattierung f 1. <Ker & Glas> frosting; 2. <Kunststoff> matting
Mattierungsbad n <Ker & Glas> frosting bath
Mattierungsmittel n <Kunststoff> flatting agent
Mattkohle f <Kohlen> dull coal, kennel coal
Mattscheibe f <Foto> focusing screen, groundglass screen
Mattscheibe f **mit Fadenkreuz** <Foto> groundglass screen with reticule
Mattscheibe f **mit Fresnellinse** <Foto> groundglass with Fresnel lens
Mattscheibe f **mit Mikroprismenring** <Foto> groundglass screen with microprism collar
Mattscheibenrahmen m <Foto> focusing screen frame
Mattscheibenring m <Foto> matt collar
Mattschliff m <Ker & Glas> matt cutting
Mattschmelzfarbe f <Ker & Glas> (AE) matt vitrifiable color, (BE) matt vitrifiable colour
Matzen m <Lebensmittel> matzoth
Mauer f <Bau, Kohlen> wall
Mauerabdeckung f 1. <Bau> coping; 2. <Wasserversorg> coping (einer Schleuse)
Mauerband n <Bau> string
Mauerbohrer m <Fertig, Maschinen> masonry drill
Mauerecke f <Bau> quoin
Mauerfuß m <Ker & Glas> curb
Mauerhaken m <Bau> spike
Mauerkappe f <Bau> hood
Mauerkrone f <Bau> crown
Mauermantel m <Bau, Kohlen> mantle
mauern v <Bau> brick, lay, mason
Mauerspalt m <Kohlen> crevice
Mauerung f <Bau> walling
Mauervorsprung m <Bau> spur
Mauerwerk n 1. <Bau> masonry, walling; 2. <Ker & Glas> brickwork
Mauerwerksanker m <Bau> anchor
Mauerwerksarbeiten fpl <Bau> masonry work
Mauerziegel m <Bau> brick
Maul n 1. <Fertig> mouth (Zange); 2. <Maschinen> chaps
Maulhöhe f <Fertig> gap (Nietmaschine)
Maulpresse f <Kunststoff> C-frame press
Maulschlüssel m 1. <Ker & Glas> mouth tools; 2. <Maschinen> (BE) face spanner, face wrench, (BE) open spanner, open wrench, (BE) open-end spanner, open-end wrench; 3. <Mechan> (BE) open-end spanner, open-end wrench

Maulweite f <Maschinen> (BE) spanner opening, wrench opening
Maurer m <Bau> bricklayer, mason
Maurerkelle f <Bau> brick trowel
Maus f 1. <Comp & DV> mouse; 2. <Wassertrans> noseband
Maus f **mit zwei Tasten** <Comp & DV> two-button mouse
Mauseloch n <Erdöl> mousehole (Bohrturm)
Mausprogramm n <Comp & DV> mouse software
Maussoftware f <Comp & DV> mouse software
Mauszeiger m <Comp & DV> pointer
Mautbrücke f <Bau> toll bridge
Mautentrichtung f <Trans> toll payment
Mautstraße f <Trans> toll road
maximal glaubhafter Unfall m <Kerntech> maximum credible accident
maximal zulässige Abweichung f <Maschinen> maximum permissible deviation
maximal zulässige Arbeitsplatzkonzentration f <Umweltschmutz> occupational MAC, threshold limit value in the workplace
maximal zulässige Dosis f <Kerntech> maximum admissible dose
maximal zulässige Leistung f <Telekom> maximum admissible power
Maximal... <Akustik, Elektriz, Elektrotech, Kerntech, Metrol> maximum
Maximalausgabe f <Elektriz> maximum output
Maximalausschalter m <Elektrotech> maximum cutout
Maximalbelastung f <Elektrotech, Kohlen> maximum demand
maximale Arbeitsplatzkonzentration f (MAK) <Kerntech> maximum allowable concentration (MAC); threshold limit value (TLV)
maximale Bahnbreite f <Papier> maximum deckle
maximale beschnittene Bahnbreite f <Papier> maximum-trimmed machine width
maximale beschnittene Maschinenbreite f <Verpack> maximum-trimmed machine width
maximale Bewicklungsbreite f <Textil> maximum dressed width of warp
maximale Emissionskonzentration f (MEK) <Umweltschmutz> maximum emission concentration
maximale Fördermenge f <Erdöl> capacity
maximale Gesamtbelastung f <Bau, Eisenbahn> maximum total load
maximale Höhe f <Raumfahrt> ceiling
maximale Kapazität f <Papier, Qual> maximum capacity
maximale Last f <Lufttrans, Trans> maximum load
maximale Leistungsübertragung f <Elektrotech, Kohlen> maximum power transmission
maximale Momentleistung f <Raumfahrt> maximum instantaneous power
maximale Motordrehzahl f <Kfztech> peak engine speed
maximale Rauigkeitstoleranz f <Fertig> maximum acceptable roughness
maximale Rückfederungsbelastung f <Lufttrans> maximum spring-back load
maximale Schmelzrate f <Ker & Glas> maximum melting rate
maximale Sprechleistung f <Akustik> peak speech power
maximale Stromübertragung f <Elektrotech, Kohlen> maximum power transmission
maximale Wellengeschwindigkeit f <Nichtfoss Energ> maximum shaft speed
maximale Zugspannung f <Kerntech> ultimate tensile stress

Maximal-Eichungskennzeichen

Maximal-Eichungskennzeichen n <Metrol> approval sign
maximaler Axialdruck m <Nichtfoss Energ> maximum axial thrust
maximaler Durchsatz m <Erdöl> operational capacity
maximaler Durchschlupf m <Qual> average outgoing quality limit
maximaler Frequenzhub m <Raumfahrt, Telekom> peak frequency deviation
maximaler Laststrom m <Elektriz> maximum overload current
maximaler Schalldruck m <Akustik> peak sound pressure
maximaler vertikaler Raddruck m <Lufttrans> maximum wheel vertical load
maximaler Wärmefluss m <Kerntech> maximum flux heat, peak heat flux
maximaler Wert m <Gerät> crest value
maximales Biegemoment n <Raumfahrt> maximum bending moment
maximales bis mittleres Leistungsverhältnis n <Kerntech> peak heat flux
maximales Drehmoment n <Kfztech> maximum torque
Maximalleistung f 1. <Elektriz> maximum output, maximum power; 2. <Kerntech> maximum capacity
Maximalschalldruck m <Akustik> maximum sound pressure
Maximalsignal n <Elektronik> maximum signal
Maximalsignalamplitude f <Elektronik> maximum signal amplitude
Maximalspannung f 1. <Elektriz> maximum voltage; highest voltage *(eines Gerätes)*; 2. <Phys> peak voltage
Maximalstrom m <Elektriz> maximum current, peak current
Maximaltemperatur f **im Brennelementinneren** <Kerntech> maximum fuel central temperature
Maximaltiefgang m <Wassertrans> deepest draught *(Schiffkonstruktion)*
Maximalwert m 1. <Gerät> crest value; 2. <Phys> peak value
Maximierung f <Math, Telekom> maximization
Maximum n <Kerntech, Math> maximum
Maximum-Likelihood-Sequenzschätzung f <Telekom> maximum likelihood sequence estimation
Maximum-Minimum-Thermometer n <Heiz & Kälte, Labor> maximum and minimum thermometer
Maximumverkehrsaufkommen n **pro Stunde** <Trans> maximum hourly volume
Maximumzähler m <Elektrotech> maximum demand meter
Maxwell n *(Mx)* <Elektriz, Elektrotech> maxwell *(Mx)*
Maxwell'sche Gleichungen fpl <Phys> Maxwell's equations
Maxwell-Verteilung f <Phys> Maxwell distribution
Mayday n <Lufttrans, Wassertrans> mayday *(Notfall)*
mazerieren v <Chemie, Fertig> macerate
Mazerieren n <Chemie> maceration
Mazeriergefäß n <Chemie> macerator
Mazza f <Lebensmittel> matzoth
MB *(Mbyte, Megabyte)* <Comp & DV> MB *(megabyte)*
MBE *(Molekularstrahlepitaxie)* <Elektronik, Strahlphys> MBE *(molecular-beam epitaxy)*
M-box <Comp & DV> bb *(bei E-Mail)*
Mbyte n *(MB)* <Comp & DV> megabyte *(MB)*
McPherson-Federbein n <Kfztech> McPherson strut
McPherson-Federbein-Vorderachse f <Kfztech> McPherson strut front suspension
MCVD-Verfahren n <Fertig> modified chemical vapor deposition, MCVD *(Faseroptik)*

Md *(Mendelevium)* <Chemie> Md *(mendelevium)*
MDI *(Diphenylmethandiisocyanat)* <Kunststoff> MDI *(diphenylmethane diisocyanate)*
MDR *(Speicherdatenregister)* <Comp & DV> MDR *(memory data register)*
me *(Elektronenmasse)* <Chemie, Kerntech, Teilphys> me *(electron mass)*
MEA *(Means-End-Analyse, Mittel-Zweck-Analyse)* <Künstl Int> MEA *(means-end analysis)*
Means-End-Analyse f *(MEA)* <Künstl Int> means-end analysis, MEA *(Methode der Problemlösung)*
Mechanik f <Maschinen> mechanical system, mechanics
Mechaniker m <Kfztech, Maschinen, Mechan> fitter, mechanic, metalworker
Mechanikerdrehmaschine f <Maschinen> bench lathe
mechanisch adj <Maschinen> mechanical
mechanisch abgestimmter Oszillator m <Elektronik> mechanically-tuned oscillator
mechanisch abgestimmtes Magnetron n <Elektronik> mechanically-tuned magnetron
mechanisch betriebener Greifer m <Bau> mechanical grab
mechanisch betriebenes Schütz n <Elektriz> mechanical contactor
mechanische Abnutzung f <Bau> mechanical wear
mechanische Abschirmung f <Sicherheit> mechanical safeguarding; physical safeguard
mechanische Abwasserbehandlung f <Wasserversorg> primary clarification *(der ersten Stufe)*
mechanische Abwasserreinigung f <Abfall> primary sewage treatment
mechanische Admittanz f <Akustik> mechanical admittance
mechanische Aufnahme f <Akustik> mechanical recording
mechanische Auslenkung f *(CM)* <Akustik> mechanical compliance *(CM)*
mechanische Benzinpumpe f <Kfztech> mechanical fuel pump
mechanische Eigenschaften fpl 1. <Bau, Kunststoff, Maschinen> mechanical properties; 2. <Metall> mechanical properties *(von Nickelstahl)*; 3. <Strömphys> mechanical properties
mechanische Eingangsimpedanz f <Akustik> free mechanical impedance *(bei unbelastetem Ausgang)*
mechanische Energie f <Maschinen> mechanical energy
mechanische Feder f <Maschinen> mechanical spring
mechanische Fehler mpl <Fernseh> mechanical errors
mechanische Fördereinrichtung f <Maschinen> mechanical handling equipment
mechanische Impedanz f <Akustik, Elektrotech> mechanical impedance
mechanische Kabelverbindung f <Fernseh> mechanical splice
mechanische Kernimpedanz f <Akustik> transfer mechanical impedance
mechanische Klärung f <Abfall> sedimentation
mechanische Kraftübertragung f <Nichtfoss Energ> mechanical transmission
mechanische Modulation f <Elektronik> mechanical modulation
mechanische Nullstellung f <Elektriz> mechanical zero adjustment
mechanische Presse f <Maschinen> power press
mechanische Ramme f <Bau> power rammer
mechanische Reaktanz f <Akustik> mechanical reactance
mechanische Resistanz f <Akustik, Mechan> mechanical resistance

mechanische Resonanz f <Akustik, Mechan> mechanical resonance
mechanische Schwingung f <Akustik, Sicherheit> mechanical oscillation
mechanische Sortierung f <Abfall> mechanical separation
mechanische Spannung f <Sicherheit> stress
mechanische Spleißstelle f <Optik> mechanical splice
mechanische Stabilität f <Kunststoff> mechanical stability
mechanische Stoßprüfung f <Metrol> mechanical shock test
mechanische Teile npl <Maschinen> mechanical components
mechanische Tonaufzeichnung f <Aufnahme> mechanical recording
mechanische Trennung f <Abfall> automatic sorting, mechanical separation
mechanische Unstabilität f <Metall> mechanical instability
mechanische Verbindung f <Kerntech> mechanical bond
mechanische Wärmetheorie f <Phys> mechanical theory of heat
mechanische Wasseraufbereitung f <Wasserversorg> physical water treatment
mechanische Welle f <Elektrotech> mechanical wave
mechanische Werte mpl <Kunststoff> mechanical properties
mechanische Zurichtung f <Druck> mechanical overlay
mechanischer Abscheider m <Umweltschmutz> mechanical collector
mechanischer Antrieb m <Maschinen> mechanical drive
mechanischer Auslöser m <Elektriz> mechanical tripping device
mechanischer Blindleitwert m <Akustik> mechanical susceptance
mechanischer Drucker m <Comp & DV> impact printer
mechanischer Einschluss m <Abfall> physical stabilization (von Schadstoffen)
mechanischer Endanschlag m <Elektriz> mechanical end stop
mechanischer Gleichlauf m <Elektrotech> ganging
mechanischer Holzstoff m <Verpack> mechanical wood pulp
mechanischer Klassierer m <Kohlen> mechanical classifier
mechanischer optischer Schalter m <Telekom> mechanical optical switch
mechanischer Prober m <Kohlen> mechanical sampler
mechanischer Spleiß m <Telekom> mechanical splice
mechanischer Stampfer m <Bau> mechanical tamper
mechanischer Teilkopf m <Maschinen> mechanical-dividing head
mechanischer Wirkungsgrad m 1. <Ergon> mechanical efficiency (der Muskelarbeit); 2. <Maschinen, Nichtfoss Energ> mechanical efficiency
mechanischer Zerhacker m <Kerntech> mechanical chopper
mechanisches Abluftsystem n <Sicherheit> mechanical exhaust air installation
mechanisches Aufnahmegerät n <Akustik> mechanical recorder
mechanisches Auslesen n **mit konstanter Amplitude** <Akustik> constant-amplitude mechanical reading
mechanisches Auslesen n **mit konstanter Geschwindigkeit** <Akustik> constant-velocity mechanical reading
mechanisches Edieren n <Fernseh> mechanical editing

mechanisches Enthülsen n <Kerntech> mechanical decanning, mechanical decladding
mechanisches Filter n <Elektronik> mechanical filter
mechanisches Filter n **mit Scheibendraht** <Aufnahme> (BE) disc-wire-type mechanical filter, (AE) disk-wire-type mechanical filter
mechanisches Getriebe n <Maschinen> mechanical transmission system
mechanisches Luftfilter n <Heiz & Kälte> mechanical air filter
mechanisches System n <Akustik> mechanical system
mechanisches Testen n <Maschinen, Werkprüf> mechanical testing
mechanisches Verhalten n **von Werkstoffen** <Werkprüf> (AE) mechanical behavior of materials, (BE) mechanical behaviour of materials
mechanisches Wärmeäquivalent n 1. <Mechan> mechanical equivalent of heat (J); 2. <Phys> Joule's equivalent; 3. <Thermod> Joule's equivalent, thermal equivalent; 4. <Thermod> mechanical equivalent of heat (J)
Mechanismus m <Maschinen> mechanism
Mechanismus m **zur Besetzungsumkehr** <Strahlphys> population inversion mechanism
Mechanismus m **zur Populationsumkehr** 1. <Phys> laser population mechanism (bei Lasern); 2. <Strahlphys> laser population mechanism
Mechanochemie f <Chemtech> mechanochemistry
Meconin n <Chemie> dimethoxyphthalide, meconin, opianyl
medial adj <Ergon> medial
Median m <Qual> median
Medianebene f <Ergon> median plane
Mediante f <Akustik> mediant
Medianwert m <Comp & DV> median
Mediävalschrift f <Druck> old style
Mediävalziffern fpl <Druck> nonlining figures
Medien npl <Fernseh> media
Medium n 1. <Chemie> agent; 2. <Comp & DV, Druck> medium; 3. <Fertig> agent (Kunststoffinstallationen); 4. <Phys> medium
medizinisches Expertensystem n <Künstl Int> medical expert system
Meer n <Wassertrans> sea • **im Meer** <Erdöl> offshore
Meerbusen m <Wassertrans> gulf (Geographie)
Meerenge f <Wassertrans> sound; narrows, strait (Geographie)
Meeres… <Wassertrans> maritime
Meeresarm m <Wassertrans> inlet (Geographie)
Meeresboden m <Nichtfoss Energ, Wassertrans> seabed
Meeresbodenreinigung f <Erdöl> sea floor housekeeping (Offshore-Technik)
meeresbürtige Verschmutzung f <Meerschmutz> sea-based pollution
Meeresforschungsschiff n <Wassertrans> oceanographic research ship
Meeresgebiet n <Wassertrans> sea area
Meeresgrund m <Bau, Nichtfoss Energ, Wassertrans> seabed
Meereshöhe f <Wassertrans> sea level
Meereskunde f <Wassertrans> oceanography
Meeresspiegel m <Bau, Nichtfoss Energ, Wassertrans, Wasserversorg> sea
Meeresströmung f <Wassertrans> ocean current
meerestechnische Industrie f <Wassertrans> maritime industry
Meerestiefen fpl <Wassertrans> ocean deeps, ocean depths
Meeresumwelt f <Meerschmutz> marine environment
Meersalz n <Lebensmittel> sea salt

Meerwasser

Meerwasser n <Wassertrans> seawater
Meerwassereinbruch m <Wasserversorg> seawater intrusion
meerwassergekühlt adj <Thermod> sea water cooled
Mega... (M) <Metrol> mega... (M)
Megabit n **pro Sekunde** (Mb/s) <Telekom> megabit per second (Mbit/s)
Megabyte n (MB) <Comp & DV> megabyte (MB)
Megachip m <Elektronik> megachip
Megadoc® n <Optik> Megadoc®
Megadyn n <Metrol> megadyne
Megahertz n (MHz) <Elektriz, Elektrotech, Fernseh, Funktech> megahertz (MHz)
Megastream n <Telekom> megastream circuit (Punkt-zu-Punkt-Digitalverbindung)
Megawatt n <Elektriz> megawatt
Megayacht f <Wassertrans> mega-yacht
Megger® m <Elektrotech> Megger
Megohm n <Elektriz, Elektrotech> megohm
mehlig adj <Lebensmittel> farinaceous
mehliger Boden m <Bau> floury soil
Mehlschwitze f <Lebensmittel> roux
Mehltau m <Lebensmittel> mildew
Mehr... <Comp & DV, Elektronik, Elektrotech, Funktech, Telekom, Verpack> multi...
mehrachsige Spannung f <Kunststoff> multiaxial stress
mehrachsiges Schwerlastfahrzeug n <Kfztech> multiaxle heavy goods vehicle
Mehradressbefehl m <Comp & DV> multiaddress instruction
Mehradressinstruktion f <Comp & DV> multiaddress instruction
mehradrig adj <Maschinen> multiwire
mehradriges Kabel n 1. <Elektriz> multiconductor cable, multicore cable; 2. <Elektrotech, Fernseh, Telekom> multicore cable
Mehramplitudenmodulation f <Elektronik> multilevel modulation
Mehranodengleichrichter m <Elektrotech> multianode rectifier
mehratomig adj <Chemie> polyatomic
Mehrbahnenetikettiersystem n <Verpack> (AE) multilane labeling system, (BE) multilane labelling system
Mehrbahnmaschine f <Verpack> multilane machine
Mehrband... <Funktech> multiband
Mehrbandfilter n <Telekom> multiband filter
Mehrbenutzer... <Raumfahrt> multiple-access (Weltraumfunk)
Mehrbenutzersystem n <Comp & DV> multiuser system
Mehrbereichsboden m <Kohlen> multigraded soil
Mehrbereichsmessgerät n <Elektriz> multirange meter
Mehrbereichsöl n <Kfztech, Maschinen> multigrade oil
Mehrblattfeder f <Maschinen> multiple-blade spring
Mehrdecksystem n <Trans> multidecking system
Mehrdienstevermittlungssystem n <Telekom> multiservice switching system
mehrdimensionale Filterung f <Telekom> multidimensional filtering
Mehreinheitcontainer m <Verpack> multiunit container
Mehrelektronenröhre f <Elektronik> multigrid tube
Mehremittertransistor m <Elektronik> multi-emitter transistor
mehrfach adj <Elektriz> multiple
mehrfach ausnützen v <Telekom> multiplex
mehrfach gelitzter Leiter m <Elektriz> multiple-stranded conductor
mehrfach nutzen v <Comp & DV> multiplex
mehrfach ungesättigt adj <Chemie, Lebensmittel> polyunsaturated (Fettsäure)

Mehrfach... 1. <Akustik, Aufnahme> multi..., multiple; 2. <Comp & DV> multi..., multiple, multiport; 3. <Elektriz, Fernseh, Fertig, Foto, Mechan, Telekom> multi..., multiple
Mehrfachabstimmung f <Elektrotech> ganged tuning
Mehrfachabstimmungskreis m <Elektrotech> ganged circuit
Mehrfachabtastung f <Telekom> multiple sampling
Mehrfachanschluss m <Telekom> party line
Mehrfachanweisungszeile f <Comp & DV> multistatement line
Mehrfachausgangsstecker m <Fernseh> multiple-outlet plug
Mehrfachbandpassfilter n <Aufnahme> multiple-bandpass filter
Mehrfachbefehlsstrom-Einfachdatenstrom-Rechner m (MISD-Rechner) <Comp & DV> multiple-instruction single-data machine (MISD machine)
Mehrfachbefehlsstrom-Mehrfachdatenstrom-Rechner m (MIMD-Rechner) <Comp & DV> multiple-instruction multiple-data machine (MIMD machine)
Mehrfachbelegung f <Telekom> multiple seizure
Mehrfachbelichtung f <Foto> multiple exposure
Mehrfachbespielung f <Aufnahme> sound on sound
Mehrfachbetrieb m <Comp & DV> multiprocessing
Mehrfachbeugung f <Telekom> multiple diffraction
Mehrfachbildschirm m <Fernseh> multiscreen
Mehrfachbohren n <Fertig> multidrilling
Mehrfachbohrmaschine f <Mechan> multiple-drilling machine
Mehrfachbruch m <Math> complex fraction
Mehrfachburst m <Fernseh> multiburst
Mehrfachchipstrukturen fpl <Elektronik> multiple die pattern
Mehrfachcodierung f <Comp & DV, Telekom> concatenated code, multiple encoding
Mehrfachdrahtsystem n <Elektriz> multiple-wire system
Mehrfachdrehkondensator m 1. <Elektrotech> gang capacitor, ganged capacitor; 2. <Funktech> ganged capacitor
mehrfache Schutzerde f <Elektriz> (BE) protective multiple earthing, (AE) protective multiple grounding
Mehrfachecho n <Akustik> flutter echo, multiple echo
Mehrfacheinsatzorbiter m <Raumfahrt> recoverable orbiter
Mehrfacheinsatztreibsatz m <Raumfahrt> recoverable thruster
Mehrfachempfangsgewinn m <Telekom> diversity gain
Mehrfachempfangssystem n <Telekom> diversity system
Mehrfachexpansionsmaschine f <Maschinen> compound expansion engine, multiple-expansion engine
Mehrfachfallschirm m <Lufttrans> cluster
Mehrfachfaserverbinder m <Elektriz> multifibre joint (Lichtwellenleiter)
Mehrfachfestkondensator m <Elektrotech> capacitor bank
Mehrfachform f <Maschinen> (AE) multi-impression mold, (BE) multi-impression mould
Mehrfachfrequenz f (MF) <Elektronik, Funktech, Telekom> multiple frequency (MF)
Mehrfachgarn n <Textil> plied yarn
Mehrfachgesenk n <Maschinen> multiple die
Mehrfachglasiermaschine f <Ker & Glas> multipleglazing unit
Mehrfachkeilriemenantrieb m <Maschinen> multiple-V-belt drive
Mehrfachkeilwelle f <Maschinen> multispline shaft
Mehrfachkeule f <Funktech> multiple beam

Mehrfachkeulenantenne f <Funktech> multiple-beam aerial, multiple-beam antenna
Mehrfachknotenarchitektur f <Comp & DV> multi-node architecture
Mehrfachkoppler m <Comp & DV> multiplexer
Mehrfachleitungskabel n <Elektriz> multiple feeder
Mehrfachlösung f <Math> multiple solution *(für eine Variablen-Gleichung)*
Mehrfachmessgerät n <Metrol> multimeter, multipurpose instrument, multipurpose meter, universal measuring instrument
Mehrfachmikrofon n <Aufnahme> multiple microphone
Mehrfachmikrofonanordnung f <Akustik> multiple microphone
Mehrfachnocken m <Maschinen> multi-lobe cam
Mehrfachnullstelle f <Math> multiple root *(einer Funktion)*
Mehrfachnullstelle f einer Funktion <Math> multiple root
Mehrfachnutzung f <Elektronik> multiplex
Mehrfachprogrammierung f <Comp & DV> multiprogramming
Mehrfachpumpe f <Fertig> double pump
Mehrfachquantenmulden-Bauelement n <Elektronik> multiple quantum well device
Mehrfachräumen n <Maschinen> multiple broaching
Mehrfachreflektorantenne f <Funktech> multiple-reflector aerial, multiple-reflector antenna
Mehrfachröhrendüse f <Lufttrans> multitube nozzle
Mehrfachrufnummer f <Telekom> multiple subscriber number *(MSN)*
Mehrfachschalter m <Elektrotech> gang switch
Mehrfachschnitt m <Fertig> multiple blanking
mehrfach-selbstjustierende MOS Technologie f <Elektronik> multiple-self-aligned MOS technology *(MUSAMOST)*
Mehrfach-Server m <Telekom> multiple-server queue
Mehrfachspleißung f <Telekom> mass splicing *(Lichtwellenkabel)*
Mehrfachspule f <Elektrotech> multi-section coil
Mehrfachstahl m <Maschinen> gang tool
Mehrfachstichprobenentnahme f <Qual> multiple sampling
Mehrfachstichprobenprüfplan m <Qual> multiple-sampling plan
Mehrfachstichprobenprüfung f <Qual> multiple-sampling inspection
Mehrfachteilnehmerrufnummer f <Telekom> multiple subscriber number *(MSN)*
Mehrfachtelegrafie f <Telekom> muliple telegraphy
Mehrfachtonspur f <Aufnahme> multiple soundtrack
Mehrfachtraktion f <Elektrotech> multiple traction unit
Mehrfachverbindung f <Comp & DV> multipoint link
Mehrfachverstärker m <Elektronik> multistage amplifier
Mehrfachverteiler m <Telekom> multiple distributor
Mehrfachwahlmethode f <Ergon> multiple-choice method
Mehrfachwegsignale npl <Fernseh> multipath signals
Mehrfachwerkzeug n <Kunststoff> *(AE)* multi-impression mold, *(BE)* multi-impression mould
Mehrfachwicklung f <Elektriz, Elektrotech> multiple winding
Mehrfachzugriff m 1. <Raumfahrt> multiple access *(Weltraumfunk)*; 2. <Telekom> multiple access
Mehrfachzugriff m im Zeitmultiplex *(TDMA)* <Comp & DV, Elektronik, Raumfahrt, Telekom> time division multiple access *(TDMA)*
Mehrfachzugriff m mit Trägerkennung *(CSMA)* <Comp & DV, Telekom> carrier sense multiple access *(CSMA)*
Mehrfachzugriffscode m <Comp & DV> multiple-address code
Mehrfachzugriffsmethode f <Telekom> multiple-access principle, multiple-access method
Mehrfachzugriffssystem n <Comp & DV, Telekom> multiaccess system
mehrfädiges Garn n <Textil> plied yarn
Mehrfarbendruck m <Druck, Verpack> *(AE)* multicolor printing, *(BE)* multicolour printing
Mehrfarbenlichtpauspapier n <Konstzeich> *(AE)* multicolor diazotype paper, *(BE)* multicolour diazotype paper
Mehrfarbenrotationspresse f <Druck> *(AE)* multicolor rotary printing machine, *(BE)* multicolour rotary printing machine
mehrfarbig adj <Druck> polychrome
Mehrfaserkabel n <Elektrotech, Telekom> *(AE)* multifiber cable, *(BE)* multifibre cable
Mehrfaserverbindung f 1. <Optik> *(AE)* multifiber joint, *(BE)* multifibre joint *(Verbindung mehrerer Glasfasern)*; 2. <Telekom> *(AE)* multifiber joint, *(BE)* multifibre joint
Mehrfeldplatte f <Bau> continuous slab
Mehrfrequenz... <Elektronik, Funktech, Telekom> multifrequency
Mehrfrequenzantenne f <Elektronik, Funktech, Telekom> multifrequency aerial, multifrequency antenna
Mehrfrequenzcode m *(MFC)* <Telekom> multifrequency code *(MFC)*
Mehrfrequenzempfänger m <Elektronik, Funktech, Telekom> multifrequency receiver
Mehrfrequenz-Geber-Empfänger m <Elektronik, Funktech, Telekom> multifrequency sender-receiver
Mehrfrequenzgenerator m <Elektronik, Funktech, Telekom> multifrequency generator
Mehrfrequenz-Sender-Empfänger m <Elektronik, Funktech, Telekom> multifrequency sender-receiver
Mehrfrequenzwahl f *(MFW)* <Telekom> dual-tone multifrequency *(DTMF)*; *(AE)* multifrequency dialing, *(BE)* multifrequency dialling *(MFD)*
mehrgängig adj 1. <Fertig> multiple-screw *(Gewinde)*; multistart *(Schnecke)*; 2. <Maschinen> multistart *(Schnecke)*
mehrgängige Schnecke f <Maschinen> multistart worm
mehrgängige Schraube f <Maschinen> multiple-threaded screw
mehrgängiges Gewinde n <Maschinen> multiple thread, multistart thread
mehrgängiges Potenziometer n <Elektrotech> multiturn potentiometer
Mehrgewicht n <Verpack> excess weight
Mehrgitterröhre f 1. <Elektronik> multielectrode tube, multigrid tube; 2. <Funktech> multigrid valve
mehrgliedrig adj <Maschinen> multilink
Mehrkammerklystron n <Elektronik, Phys> multicavity klystron
Mehrkammerverbundrohrmühle f <Kohlen> compartment pebble mill
Mehrkanal... 1. <Aufnahme> multichannel; 2. <Comp & DV> multiport; 3. <Elektronik, Funktech> multichannel
Mehrkanalelementarlautsprecher m <Akustik> multichannel elementary loudspeaker
Mehrkanalfilter n <Telekom> multichannel filter
Mehrkanallautsprecher m <Aufnahme> multichannel loudspeaker
Mehrkanal-Mikrowellen-Kabelsystem n <Fernseh, Telekom> multichannel microwave distribution system *(Videoverteilung)*

Mehrkanalprotokoll

Mehrkanalprotokoll n <Comp & DV> multichannel protocol
Mehrkanalregister n <Comp & DV> multiport register
Mehrkanalträger m <Raumfahrt> multichannel carrier *(Weltraumfunk)*
Mehrkanalübertragung f <Comp & DV> multiplexing
Mehrkanalüberwachung f <Funktech, Wassertrans> multichannel monitoring *(Funk)*
Mehrkanalverstärker m <Elektronik> multichannel amplifier
Mehrkanalwähler m <Fernseh> multichannel selector
Mehrkolbenmotor m <Maschinen> multipiston engine
Mehrkollektortransistor m <Elektronik> multicollector transistor
Mehrkomponentenkleber m <Verpack> mixed adhesive
Mehrkopf... <Textil> multihead
Mehrkörperproblem n <Raumfahrt> many-body problem
Mehrkörperschiff n <Wassertrans> multihull ship, multihulled ship
Mehrkreisfilter n <Elektronik> multisection filter
Mehrkristallhalbleiter m <Elektronik> polycrystalline semiconductor
Mehrlagen... <Maschinen> multilayer
Mehrlagenabdeckung f <Elektronik> multilayer resist *(Leiterplatten)*
Mehrlagendickfilme mpl <Elektronik> multilayer thick films
Mehrlagendünnfilme mpl <Elektronik> multilayer thin films
Mehrlagenkarton m 1. <Papier> multilayer board; 2. <Verpack> multiple board
Mehrlagenleiterplatte f <Elektronik> multilayer printed circuit
Mehrlagensack m <Verpack> multiply sack, multiwall sack
mehrlagig adj <Verpack> multiply
mehrlagige gedruckte Schaltung f <Telekom> multilayer printed circuit
mehrlagige Spule f <Elektriz> multilayer coil
mehrlagige Wicklung f <Elektriz> multilayer coil
mehrlagiger Plattenheizkörper m <Heiz & Kälte> multi-level panel-type radiator
Mehrleiterkabel n <Elektrotech> multiconductor cable, multicore cable
mehrlinsiges Objektiv n <Foto> composite lens
Mehrlippenbohrer m <Fertig> subland drill
mehrlösige Bremse f <Eisenbahn> graduated brake
Mehrmeißeldrehmaschine f <Maschinen> multiple-tool lathe, multitool lathe
Mehrmeißelhalter m <Maschinen> turret
Mehrmodenfaser f 1. <Optik> *(AE)* multimode fiber, *(BE)* multimode fibre *(Lichtleitfaser)*; 2. <Telekom> *(AE)* multimode fiber, *(BE)* multimode fibre, *(AE)* multimode optical fiber, *(BE)* multimode optical fibre
Mehrmodengruppenlaufzeit f <Optik> multimode group decay *(Lichtleitfaser)*
Mehrmodenlaser m <Optik> multimode laser
Mehrmoden-Lichtwellenleiter m <Telekom> *(AE)* multimode optical fiber, *(BE)* multimode optical fibre
Mehrmodenverzerrung f <Optik> multimode distortion *(Lichtleitfaser)*
mehrmotoriger Hubschrauber m <Lufttrans> multi-engine helicopter
Mehrnormen-Fernsehempfänger m <Fernseh> multistandard TV receiver *(PAL, SECAM, NTSC)*
Mehrordnungsfilter n <Elektronik> multiple-order filter
mehrpaariges Kabel n <Elektriz> multipair cable
Mehrpfadbetrieb m <Comp & DV> multithreading
Mehrpfadprogramm n <Comp & DV> multithread program

Mehrphasen... <Elektrotech> polyphase
Mehrphasenauflösung f <Raumfahrt> phase ambiguity resolution *(Weltraumfunk)*
Mehrphasendifferenzialschutz m <Elektriz> balanced protection
Mehrphasendifferenzialschutzrelais n <Elektrotech> balanced protection relay
Mehrphasengenerator m <Elektrotech> polyphase generator
Mehrphaseninduktionsmotor m <Elektrotech> polyphase induction motor
Mehrphasenmotor m <Elektrotech> polyphase motor
Mehrphasenschrittmotor m <Elektrotech> multiphase, stepping motor
Mehrphasensteuergerät n <Kfztech> multiphase controller
Mehrphasenstromkreis m <Elektrotech> polyphase circuit
Mehrphasenstufenbohrer m <Maschinen> subland twist drill
Mehrphasensynchronmotor m <Elektrotech> polyphase synchronous motor
Mehrphasentransformator m <Elektrotech> polyphase transformer
Mehrphasenwattmeter n <Metrol> polyphase wattmeter
mehrphasig adj <Elektriz, Elektrotech> multiphase, polyphase
mehrphasige Schaltung f <Elektriz> polyphase circuit
mehrphasiger Motor m <Elektriz> polyphase motor
mehrphasiger Strom m <Elektriz> polyphase current
mehrphasiger Transformator m <Elektriz> polyphase transformer
mehrphasiges Netz n <Elektriz> polyphase network
Mehrplatinencomputer m <Comp & DV> multiboard computer
mehrplattiger Kondensator m <Elektriz> multiple-plate capacitor
Mehrplatzsystem n <Comp & DV> multiuser system
Mehrpol m <Elektrotech> multipole, n-terminal circuit
mehrpolig adj 1. <Elektriz> multipolar; 2. <Elektrotech> multipin
mehrpolige Buchse f <Elektriz> multiple socket
mehrpoliger Anker m <Elektriz> multipolar armature
mehrpoliger Stecker m <Elektriz> multiple plug
mehrpoliger Stecker m mit Verriegelung <Elektriz> multiconductor locking plug
mehrpoliges Filter n <Elektronik> multipole filter
Mehrpolschalter m <Elektriz> multiple switch
Mehrprogrammbetrieb m <Comp & DV> multiprogramming
Mehrprogrammsystem n <Comp & DV> multiprogramming system
Mehrprotokoll-Router m <Telekom> multiprotocol router
Mehrprozessor m <Comp & DV> multiprocessor
Mehrprozessorbetrieb m <Comp & DV> multiprocessing
Mehrprozessorsystem n <Comp & DV> multiprocessing system
Mehrprozessorverschachtelung f <Comp & DV> multiprocessor interleaving
Mehrpunkt... <Comp & DV> multidrop, multipoint
Mehrpunktglied n <Regelung> multiposition element
Mehrpunkt-Klebemaschine f <Verpack> multipoint glueing machine
Mehrpunkt-Regeleinrichtung f <Regelung> multiposition controller
Mehrpunktschalter m <Elektrotech> multiposition switch
Mehrpunktverbindung f <Comp & DV> multidrop line, multidrop link, multipoint connection, multipoint link
Mehrpunktverhalten n <Regelung> multiposition action *(von Gliedern)*

mehrreihiger Plattenheizkörper *m* <Heiz & Kälte> multibank panel-type radiator
Mehrrundsiebmaschine *f* <Papier> multivat board machine
mehrschäftiges Tau *n* <Wassertrans> multistrand rope *(Tauwerk)*
Mehrscheibenkupplung *f* 1. <Kfztech> *(BE)* multiple-disc clutch, *(AE)* multiple-disk clutch; 2. <Maschinen> *(BE)* multiple-disc clutch, *(AE)* multiple-disk clutch, multiple-plate clutch
Mehrschicht... <Anstrich, Maschinen, Raumfahrt> multilayer
Mehrschichtenglas *n* <Ker & Glas> laminated glass
Mehrschichtenglasfrontscheibe *f* <Kfztech> *(BE)* laminated windscreen, *(AE)* laminated windshield *(Karosserie)*
Mehrschichtenglasherstellung *f* <Ker & Glas> laminating
Mehrschichtenkarton *m* <Verpack> multiply board
Mehrschichtensicherheitsglas *n* <Ker & Glas> laminated safety glass
mehrschichtig *adj* 1. <Fertig, Kunststoff> laminated; 2. <Künstl Int> multilayered *(neurales Netz)*; 3. <Verpack> multilayered
mehrschichtige Filtration *f* <Wasserversorg> multilayer filtration
Mehrschichtleiterplatte *f* <Telekom> multilayer printed circuit
Mehrschichtprinzip *n* <Abfall> multibarrier principle *(Deponie)*
mehrschneidig *adj* <Maschinen> multiblade
Mehrschürzensystem *n* <Wassertrans> multiple-skirt system, multiskirt system
mehrskalig *adj* <Elektriz> multirange
mehrskaliges Messinstrument *n* <Elektriz> multi-range meter
Mehrspaltensatz *m* <Druck> multicolumn setting
Mehrspindelanordnung *f* <Fertig> multispindle arrangement
Mehrspindelautomat *m* <Fertig> multispindle automatic
Mehrspindelbauart *f* <Fertig> multispindle design
Mehrspindelbohrmaschine *f* 1. <Fertig> multispindle drilling machine; 2. <Maschinen> multiple drill, multiplespindle drilling machine
Mehrspindelbohrwerk *n* <Maschinen> multiple-boring machine
Mehrspindeldrehautomat *m* 1. <Fertig> multispindle chucking automatic, multispindle automatic; 2. <Maschinen> multispindle automatic machine
Mehrspindelfutterautomat *m* <Fertig> multispindle chucking automatic
Mehrspindelstangenautomat *m* 1. <Fertig> multispindle bar automatic; 2. <Mechan> automatic lathe
Mehrspur-Aufnahmesystem *n* <Aufnahme> multitrack recording system
Mehrspuraufzeichnung *f* <Akustik> multitrack recording
mehrspurige Strecke *f* <Eisenbahn> *(AE)* mixed-gage track, *(BE)* mixed-gauge track
Mehrstationenverbindung *f* <Comp & DV> multidrop link
Mehrstoffauflauf *m* <Papier> multistock headbox
Mehrstoffheizanlage *f* <Maschinen> multifuel heater
Mehrstofflager *n* <Maschinen> compound bearing
Mehrstoffmotor *m* <Kfztech, Thermod> multifuel engine
Mehrstrahlantenne *f* <Funktech, Phys> multiple-beam aerial, multiple-beam antenna
mehrstrahlig *adj* <Raumfahrt> multibeam *(Weltraumfunk)*
Mehrstrahlinterferenz *f* <Phys> multiple-beam interference
Mehrstrahlröhre *f* <Elektronik> multigun tube
Mehrstromgenerator *m* <Elektrotech> multiple-current generator
Mehrstückpackung *f* <Verpack> multipack
Mehrstufen... <Fertig> multiple-shot
Mehrstufengesenk *n* <Maschinen> progression dies, progressive dies
Mehrstufenkompressor *m* <Heiz & Kälte, Maschinen> multistage compressor
Mehrstufenrakete *f* <Raumfahrt> multistage rocket
Mehrstufenturbine *f* <Maschinen> multistage turbine
Mehrstufenverstärker *m* <Elektronik> multistage amplifier
mehrstufig *adj* <Fertig, Maschinen, Telekom> multi-stage
mehrstufige Behandlungsanlage *f* <Wasserversorg> comprehensive water treatment plant
mehrstufige Schaltung *f* <Telekom> multistage circuit
mehrstufige Stichprobenentnahme *f* <Qual> multistage sampling
mehrstufige Turbine *f* <Maschinen> multistage turbine
mehrstufiger Generator *m* **nach Marx** <Elektrotech> Marx impulse generator, Marx multistage generator
mehrstufiges Folgewerkzeug *n* <Maschinen> multistage progression tooling
mehrstufiges Netzwerk *n* <Telekom> multistage network
mehrstufiges System *n* <Telekom> multilevel system
Mehrsystemumgebung *f* **mit Geräten verschiedener Hersteller** <Comp & DV> multivendor environment
mehrteilig *adj* <Ker & Glas> split
mehrteilige Form *f* <Ker & Glas> *(AE)* split mold, *(BE)* split mould
mehrteilige Walzen *fpl* <Ker & Glas> split rollers
mehrteilige Zugeinheit *f* <Eisenbahn> multiple-train unit
mehrteiliger Brennstoffstab *m* <Kerntech> segmented fuel rod
mehrteiliger Entwicklungsbehälter *m* <Foto> multiunit developing tank
mehrteiliger Fernschnelltriebwagen *m* <Eisenbahn> multiple-unit train
mehrteiliges Werkzeug *n* <Kunststoff> *(AE)* split mold, *(BE)* split mould
Mehrträgerdienst *m* <Telekom> multibearer service
Mehrtrichter-Lautsprecher *m* <Aufnahme> multiple-cone loudspeaker
Mehrwegbetrieb *m* <Comp & DV> multithreading
Mehrwegefading *n* <Funktech, Telekom> multipath fading
Mehrwegeführung *f* <Telekom> redundant routing
Mehrwegereflexionen *fpl* <Funktech, Telekom> multipath reflections
Mehrwegflasche *f* 1. <Abfall> deposit bottle, returnable bottle; 2. <Verpack> recycled bottle
Mehrweggebinde *n* <Abfall> returnable pack, returnable container
Mehrwegkarton *m* <Verpack> reusable box
Mehrwegschieber *m* <Maschinen> multiple-way slide valve
Mehrwegventil *n* 1. <Fertig> L-port valve, multiport valve *(Kunststoffinstallationen)*; 2. <Maschinen> multiple-way valve
Mehrwegverpackung *f* <Verpack> returnable packaging
Mehrwert... <Comp & DV> value-added
Mehrwertdienste *mpl* <Telekom> value-added services
Mehrwertdienstnetz *n* *(VAN)* <Comp & DV, Telekom> value-added network *(VAN)*
mehrwertig *adj* <Chemie> polyvalent
mehrwertige Menge *f* <Comp & DV, Künstl Int> fuzzy set
Mehrwertigkeit *f* <Chemie> polyvalence, polyvalency
Mehrwertnetz *n* *(VAN)* <Comp & DV, Telekom> value-added network *(VAN)*

Mehrwicklungs-Transformator

Mehrwicklungs-Transformator *m* <Elektriz> multiwinding transformer
mehrzählig *adj* <Chemie> polydentate *(Komplexchemie)*
mehrzähnig *adj* <Chemie> polydentate
Mehrzweck... 1. <Chemie> multifunctional, polyfunctional; 2. <Kfztech> utility; 3. <Mechan> GP, general-purpose
Mehrzweckanhänger *m* <Kfztech> all-purpose trailer
Mehrzweckbehälter *m* <Erdöl, Wassertrans> multiservice vessel
Mehrzweckdrehmaschine *f* <Fertig> universal lathe
Mehrzweckelektrofahrzeug *n* <Kfztech> electric vehicle for general-purpose use
Mehrzweckfrachter *m* <Wassertrans> multipurpose carrier
Mehrzweckgerät *n* <Gerät> general-purpose instrument
Mehrzweckhubschrauber *m* <Lufttrans> multipurpose helicopter
Mehrzweckkühlraum *m* <Heiz & Kälte> multiple-purpose cold store
Mehrzweckmessgerät *n* <Gerät> general-purpose instrument
Mehrzweckreaktor *m* <Kerntech> multipurpose reactor
Mehrzweckrohrleitung *f* <Trans> multipurpose material pipeline
Mehrzweckschiff *n* 1. <Erdöl> multiservice vessel *(Schifffahrt)*; 2. <Wassertrans> multipurpose ship, multiservice vessel *(Handelsmarine)*
Mehrzwecktank *m* <Erdöl, Wassertrans> multiservice vessel *(Lagertechnik)*
Mehrzwecktanker *m* <Wassertrans> multipurpose tanker
Mehrzylindermaschine *f* <Kfztech, Maschinen> multicylinder engine
Mehrzylindertrockenpartie *f* <Papier> multicylinder dryer section
Meile *f* <Metrol, Trans> mile
Meilen *fpl* **pro Stunde** <Trans> miles per hour
Meilenfahrt *f* <Wassertrans> distance run *(Navigation)*
Meißel *m* 1. <Bau> bit, chisel, sett; 2. <Fertig> gad; chisel *(Spanung)*; 3. <Maschinen> tool
Meißel *m* **mit abgeschrägter Kante** <Maschinen> *(AE)* beveled-edge chisel, *(BE)* bevelled-edge chisel
Meißel *m* **mit gerader Schneidkante** <Fertig> square-nosed tool
Meißel *m* **mit hochgekröpftem Schneidkopf** <Fertig> raised-face tool
Meißelbohren *n* <Bau> boring with the bit
Meißelbohrer *m* <Erdöl> chisel bit *(Tiefbohrtechnik)*
Meißelhalter *m* <Maschinen> tool head
Meißelhalter *m* **mit Klappe und Klappenträger** <Fertig> apron *(Hobelmaschine, Waagerechtstoßmaschine)*
Meißelhalterschlitten *m* <Maschinen> tool carrier slide
Meißelhammer *m* <Bau> chipper
Meißelklappe *f* <Fertig, Maschinen> clapper
Meißelklappenträger *m* <Fertig> box
meißeln *v* 1. <Bau> chip, chisel; 2. <Fertig> chip
Meißeln *n* <Fertig, Maschinen> *(AE)* chiseling, *(BE)* chiselling
Meißelschaft *m* <Maschinen> tool shank
Meißelschlitten *m* <Fertig> downfeed slide; head slide *(Waagerechtstoßmaschine)*
Meißelvorschub *m* <Fertig> tool feed
Meißner'sche Schaltung *f* <Elektronik> Meissner oscillator
Meißner'scher Effekt *m* <Phys> Meissner effect
Meißner'scher Oszillator *m* <Elektronik> Meissner oscillator
Meistermodus *m* <Kontroll> master mode
Meisterrücksetzsignal *n* <Kontroll> master reset signal
MEK 1. <Kunststoff> *(Methylethylketon)* MEK *(methyl ethyl ketone)*; 2. <Umweltschutz> *(maximale Emissionskonzentration)* maximum emission concentration

Meker-Brenner *m* <Labor> Meker burner
Mel *n* <Akustik, Aufnahme, Metrol> mel *(Einheit der subjektiven Tonhöheempfindung)*
Melamin *n* <Chemie, Textil> melamine
Melamin-Formaldehydharz *n* *(MF)* 1. <Elektriz> melamine resin *(MF)*; 2. <Kunststoff> melamine formaldehyde resin *(MF)*
Melaminharz *n* *(MF)* 1. <Elektriz> melamine resin *(MF)*; 2. <Kunststoff> melamine formaldehyde resin *(MF)*
Melange *f* <Textil> blend
melangieren *v* <Textil> mix
Melanin *n* <Chemie> melanin
Melanterit *m* <Chemie> copperas
Meldeanlage *f* <Sicherheit> warning device
Meldeleitung *f* <Telekom> control circuit *(Telefon)*
melden *v* 1. <Comp & DV> return; 2. <Telekom> signal
meldepflichtig *adj* <Qual, Sicherheit> notifiable
meldepflichtige Abweichung *f* <Qual> reportable nonconformance
Melder *m* 1. <Elektronik> detector *(Signal- und Sicherungstechnik)*; 2. <Elektrotech, Gerät> annunciator
Melderelais *n* <Elektriz> pilot relay
Melde'scher Versuch *m* <Phys> Melde's experiment
Meldesignal *n* <Telekom> answer signal
Meldeverzug *m* <Telekom> answering delay
Meldung *f* 1. <Comp & DV> message; 2. <Eisenbahn> notice; 3. <Funktech> message; 4. <Sicherheit> notification; 5. <Telekom> message, signal
Meldungskopf *m* <Comp & DV> message header
Meldungsquelle *f* <Comp & DV> message source
Meldungstext *m* <Comp & DV> message text
Meldungsvermittlungssystem *n* <Telekom> message switching system
Meletin *n* <Chemie> meletin
Melezitose *f* <Chemie> melicitose, raffinose
Melibiose *f* <Chemie> melibiose
melieren *v* <Textil> blend
meliertes Gusseisen *n* <Fertig> mottled iron
Mellith... <Chemie> mellitic
Mellon *n* <Chemie> mellon
Member *n* <Comp & DV> member *(Teildatei)*
Membran *f* 1. <Aufnahme> diaphragm; 2. <Bau> membrane; 3. <Maschinen> diaphragm, membrane; 4. <Mechan> diaphragm; 5. <Raumfahrt> membrane *(Raumschiff)*
Membrandichtung *f* <Raumfahrt> bellows seal
Membrandruckdose *f* <Raumfahrt> bellows
Membranfeder *f* <Maschinen> diaphragm spring
Membranfederkupplung *f* <Kfztech> diaphragm clutch
Membranfilter *n* <Chemtech, Labor> membrane filter
Membranklappe *f* <Fertig> *(BE)* diaphragm disc valve, *(AE)* diaphragm disk valve *(Kunststoffinstallationen)*
Membrankraftstoffpumpe *f* <Kfztech> diaphragm fuel pump
Membranlautsprecher *m* <Akustik> membrane loudspeaker
Membranmesswerk *n* <Gerät> diaphragm movement
Membranpumpe *f* <Maschinen, Meerschmutz, Wasserversorg> diaphragm pump
Membranregler *m* <Kfztech> suction-type governor
Membranscheibe *f* <Maschinen> *(BE)* diaphragm disc, *(AE)* diaphragm disk
Membranschlüsselschalter *m* <Elektrotech> membrane keyswitch
Membransetzkasten *m* <Kohlen> diaphragm-type washbox
Membransiegelung *f* <Verpack> foil sealing
Membrantastatur *f* <Elektrotech> membrane keyboard

Membranventil n 1. <Fertig> diaphragm valve *(Kunststoffinstallationen)*; 2. <Maschinen> diaphragm valve
Membranverdichter m <Maschinen> diaphragm compressor
Membranverschließ- und Heißsiegelmaschine f <Verpack> film-applying lid and heat-sealing machine
Membranversiegelung f <Verpack> diaphragm sealing
Memory-Funktion f <Comp & DV> memory function
Mendelevium n *(Md)* <Chemie> mendelevium *(Md)*
Menge f 1. <Elektronik, Kontroll> quantity; 2. <Math> set; amount *(Quantität)*; 3. <Metrol> rate; 4. <Qual> batch
Mengenbestimmung f <Bau> measurement of quantities
Mengendosierung f <Gerät> volume dosage
Mengendurchfluss m 1. <Heiz & Kälte> mass flow rate; 2. <Phys> mass rate of flow
Mengendurchflussmessgerät n <Gerät> mass flow meter
Mengendurchsatz m <Heiz & Kälte> mass flow rate
Mengendurchschnitt m <Math> intersection
Mengenfluss m <Kerntech> mass flow
Mengenlehre f <Math> set theory
Mengenmessgerät n <Erdöl> flowmeter *(Messtechnik)*
Mengenregelklappe f <Heiz & Kälte> volume control damper
Mengenregelungsventil n <Fertig> flow control valve, flow controller
Mengenstrommesser m **von Gasblasen** <Labor> bubble flow meter
Mengenverpackungsanlage f <Verpack> flow wrapping machine
Mengenzählung f <Gerät> volume counting
Mengfutter n <Lebensmittel> mash *(Landwirtschaft)*
Meniskus m <Phys> meniscus *(Flüssigkeitsspiegel)*
Meniskuslinse f 1. <Foto> meniscus lens *(Linse)*; 2. <Phys> meniscus lens
Mennige f <Chemie> minium, red lead
Mensch-Maschine-Beziehung f <Comp & DV, Ergon> man-machine relationship
Mensch-Maschine-Dialog m <Comp & DV, Ergon> man-machine interaction, operator-computer dialogue
Mensch-Maschine-Interaktion f <Comp & DV, Ergon> man-machine interaction
Mensch-Maschine-Interface n *(MMI)* <Comp & DV, Kontroll, Künstl Int, Raumfahrt> human-machine interface, man-machine interface *(MMI)*
Mensch-Maschine-Kommunikation f <Comp & DV, Ergon> man-machine communication *(MMC)*
Mensch-Maschine-Schnittstelle f *(MMI)* <Comp & DV, Kontroll, Künstl Int, Raumfahrt> man-machine interface *(MMI)*
Mensch-Maschine-Simulation f <Comp & DV, Ergon> man-machine simulation
Mensch-Maschine-System n <Comp & DV, Ergon> man-machine system
Mensur f <Labor> measuring cylinder
mentale Belastung f <Ergon> mental load
Menthan n <Chemie> menthane
Menthandiamin n <Chemie> menthanediamine
Menthanol n <Chemie> menthanol
Menthanon n <Chemie> menthanone
Menthen n <Chemie> menthene
Menthenol n <Chemie> menthenol
Menthenon n <Chemie> menthenone
Menthofuran n <Chemie> menthofuran
Menü n <Comp & DV, Kontroll> menu
Menüanzeige f <Comp & DV> menu screen
Menüauswahl f <Comp & DV> menu selection
Menübildschirm m <Comp & DV> menu screen
menügeführt adj <Comp & DV> menu-driven

menügeführte Anwendung f <Comp & DV> menu-driven application
menügesteuert adj <Comp & DV> menu-driven
menügesteuerte Anwendung f <Comp & DV> menu-driven application
menügestützte Benutzeroberfläche f <Künstl Int> menu-based user interface
MEP *(mittlerer Nutzdruck)* <Lufttrans, Maschinen> mep *(mean effective pressure)*
Mepacrin n <Chemie> atebrin, mepacrine, quinacrine
Meprobamat n <Chemie> meprobamate
Mercaptal n <Chemie> mercaptal
Mercaptan n <Chemie> mercaptan, thiol
Mercapto... <Chemie> *(AE)* sulfhydryl, *(BE)* sulphhydryl
Mercaptoessig... <Chemie> mercapto-acetic, thioglycolic *(Säure)*
Mercaptol n <Chemie> mercaptol, thioacetol
Mercaptomerin n <Chemie> mercaptomerin
Mercatorkarte f <Wassertrans> Mercator chart *(Navigation)*
Mercatorprojektion f <Raumfahrt, Wassertrans> Mercator projection *(Navigation)*
Mercurochrom® n <Chemie> merbromin, Mercurochrome®
Mergel m <Wasserversorg> marl
Mergelton m <Wasserversorg> marly clay
Meridian m <Raumfahrt> meridian
Meridiandurchgang m <Raumfahrt> meridian transit
Meridiankreisel m <Raumfahrt> meridian gyro *(Raumschiff)*
meridional adj <Raumfahrt> meridional
Meridionalschnitt m <Phys> tangential focal line
Meridionalstrahl m <Funktech, Optik, Telekom> meridional ray
Merinogarn n <Textil> merino yarn
Merkaptan n 1. <Erdöl> mercaptan *(Petrochemie)*; 2. <Kunststoff> mercaptan
Merkmal n 1. <Comp & DV> attribute, characteristic, feature; 2. <Elektronik> characteristic; 3. <Künstl Int, Patent> characteristic, feature; 4. <Qual> characteristic
Merkmalausblendung f <Comp & DV> feature extraction
Merkmale npl/**zu schützende** <Patent> features to be protected
Merkmarke f <Aufnahme> cue mark
Merkpunkt m <Aufnahme> cue dot
merzerisieren v <Textil> causticize
Mesacon... <Chemie> mesaconic
Mesadiode f <Elektronik> mesa diode
Mesaprozess m <Elektronik> mesa process
MESFET *(Metallhalbleiter-Feldeffekttransistor)* <Elektronik> MESFET *(metal semiconductor field effect transistor)*
Mesidin n <Chemie> mesidine
Mesitylen n <Chemie> mesitylene
Mesitylen... <Chemie> mesitylenic
Mesomerie-Effekt m <Chemie> mesomeric effect
Meson n <Chemie, Phys, Teilphys> meson
Mesorcin n <Chemie> mesorcin
Mesothorium n <Chemie> mesothorium
Mesotron n <Chemie> meson
Mesoweinsäure f <Chemie> mesotartaric acid
Mesoxal... <Chemie> mesoxalic
Mesozoikum n <Erdöl> Mesozoic *(Geologie)*
Mess... <Bau, Gerät, Ker & Glas, Qual> measuring
Messabweichung f <Qual> error of measurement
Messader f 1. <Elektrotech> pilot wire *(bei Kabel)*; 2. <Telekom> pilot *(Kabel)*
Messanlage f <Gerät> measuring equipment, measuring system
Messanordnung f <Gerät> measuring arrangement

Messanschluss

Messanschluss *m* <Gerät, Metrol> measuring connector, measuring terminal, test connector
Messanzeige *f* <Maschinen> dial indicator
Messapparat *m* <Bau> measuring apparatus
Messapparatur *f* <Gerät> measuring equipment
Messaufgabe *f* <Gerät> measuring task
Messaufnehmer *m* <Heiz & Kälte> sensor
Messautomat *m* <Fertig> automatic testing device
Messband *n* <Bau> surveyor's tape
messbar *adj* <Math> commensurable, measurable
messbare Funktion *f* <Math> commensurable function, measurable function *(Messwesen)*
messbare Menge *f* <Metrol> measurable quantity
Messbarkeit *f* <Gerät> measurability
Messbarkeitsgrenze *f* <Gerät> limit of measurability
Messbereich *m* <Gerät> instrument range, measurement range, measuring range, range of measurement • **auf höheren Messbereich umschalten** <Gerät> uprange *(Mehrbereichsinstrument)* • **auf kleineren Messbereich umschalten** <Gerät> downrange
Messbereichsänderung *f* <Gerät> change in range
Messbereichsschalter *m* <Elektriz> range switch
Messbereichsendewert *m* <Gerät> full-scale point
Messbereichserweiterung *f* <Gerät> extension of the measuring range
Messbereichsüberschreitung *f* <Gerät> overrange
Messbereichsunterdrückung *f* <Gerät> suppression of range
Messbereichswahl *f* <Gerät> automatic ranging
Messbildverfahren *n* <Bau> photogrammetry
Messblättchen *n* <Metrol> *(AE)* gap gage, *(BE)* gap gauge
Messblende *f* 1. <Erdöl> orifice plate; 2. <Fertig> calibrated orifice; 3. <Gerät> orifice plate; measuring orifice *(Pneumatik)*; 4. <Heiz & Kälte> orifice plate; 5. <Maschinen> orifice meter; 6. <Optik> measuring orifice
Messbolzen *m* <Fertig> spindle
Messbrief *m* <Wassertrans> certificate of admeasurement, rating certificate *(Segelschiff)*
Messbrücke *f* 1. <Elektriz, Elektrotech, Gerät> bridge, measuring bridge, test bridge; 2. <Maschinen, Telekom> measuring bridge
Messbuchse *f* <Elektrotech> test jack
Messdaten *npl* <Metrol> measuring data
Messdatenabtastung *f* <Gerät> data sampling, measuring data sampling, measuring data scanning
Messdatenerfassung *f* <Gerät> acquisition of measured data, data logging, measurement data acquisition
Messdatengewinnung *f* <Gerät> measurement data acquisition
Messdatenregistrierung *f* <Gerät> data logging
Messdatensichtgerät *n* <Gerät> data display unit
Messdatenübertragungssystem *n* <Gerät> data transmission system
Messdatenumformung *f* <Gerät> data conversion
Messdatenumsetzer *m* <Gerät> data converter
Messdatenumsetzung *f* <Gerät> data conversion
Messdatenverarbeitung *f* <Gerät> measurement data processing, processing of measured data
Messdatenverdichtung *f* <Gerät> measuring data reduction
Messdatenverstärker *m* <Gerät> data amplifier
Messdorn *m* <Fertig> feeler
Messdruck *m* <Gerät> measuring pressure
Messdüse *f* 1. <Fertig> *(AE)* gaging jet cutlet, *(BE)* gauging jet cutlet; 2. <Maschinen> metering jet
Messe *f* <Wassertrans> mess
Messeinheit *f* <Metrol> unit of measurement
Messeinrichtung *f* <Gerät> measuring equipment, measuring set

Messelektrode *f* <Elektrotech> sensing electrode
messen *v* 1. <Bau> *(AE)* gage, *(BE)* gauge; 2. <Comp & DV> measure; 3. <Elektriz> *(AE)* gage, *(BE)* gauge, measure; 4. <Funktech> meter; 5. <Kontroll> *(AE)* gage, *(BE)* gauge, measure; 6. <Maschinen, Metrol> measure
Messen *n* 1. <Erdöl> *(AE)* gaging, *(BE)* gauging; 2. <Labor> measuring; 3. <Metrol> *(AE)* gaging, *(BE)* gauging
Messen *n* **der Beschichtungsdicke** <Metrol> coating-thickness measurement
messende Maschine *f* <Metrol> measuring machine
messendes Relais *n* <Gerät> detecting relay
Messer *m* 1. <Elektriz> *(AE)* gage, *(BE)* gauge; 2. <Maschinen> knife, meter; 3. <Mechan> blade; 4. <Papier> blade, knife
Messerfeile *f* <Maschinen> cant file, knife edge file, knife file
Messergebnis *n* <Gerät> measuring result, result of measurement
Messerhalter *m* 1. <Maschinen> blade holder; 2. <Papier> knife holder
Messerkopf *m* 1. <Fertig> face-milling cutter with inserted blades; 2. <Maschinen> cutter head, inserted blade cutter, inserted blade milling cutter, inserted tooth cutter
Messerschalter *m* 1. <Elektriz> knife switch; 2. <Kerntech> knife edge switch
Messerschneide *f* <Maschinen> knife edge
Messerwalze *f* <Maschinen> cutter wheel, knife drum
Messerzylinder *m* <Papier> knife cylinder
Messfehler *m* <Metrol> measuring error
Messflasche *f* <Labor> graduated flask, volumetric flask
Messfolgegeschwindigkeit *f* <Elektronik> conversion rate *(von AD-Wandler)*
Messfühler *m* 1. <Comp & DV> sensor; 2. <Gerät> *(AE)* gage, *(BE)* gauge; 3. <Kfztech> sensor; 4. <Meerschmutz> sensing device; 5. <Metrol> sensor; 6. <Optik> detector; 7. <Phys> sensor; 8. <Raumfahrt> sensor *(Weltraumfunk)*; 9. <Telekom> sensor
Messfunkenstrecke *f* <Elektrotech> measuring spark gap
Messgegenstand *m* <Gerät> measurand, object of measurement
Messgenauigkeit *f* <Metrol> accuracy of measurement
Messgerät *n* 1. <Eisenbahn> instrument; 2. <Elektriz> measuring instrument; 3. <Erdöl> instrument; 4. <Funktech> meter; 5. <Gerät> *(AE)* gage, *(BE)* gauge, measuring equipment, measuring set, measuring instrument; 6. <Kfztech, Kontroll, Labor> *(AE)* gage, *(BE)* gauge; 7. <Lufttrans> instrument; 8. <Meerschmutz> sensing device; 9. <Metrol> *(AE)* gage, *(BE)* gauge, measuring device, measuring instrument; 10. <Papier> *(AE)* gage, *(BE)* gauge; 11. <Wassertrans> instrument
Messgerät *n* **für elektrische Größen** <Gerät> electric measuring instrument, electrical measuring instrument
Messgerät *n* **mit Analogausgang** <Gerät> analog output instrument
Messgerät *n* **mit gedehnter Skale** <Gerät> expanded scale meter
Messgerät *n* **mit interner Messdatenverarbeitung** <Gerät> processing measuring instrument
Messgerät *n* **mit lebendigem Nullpunkt** <Gerät> live-zero instrument
Messgerät *n* **mit Nullpunkt in der Skalenmitte** <Gerät> central-zero instrument
Messgerät *n* **mit Projektionsskale** <Gerät> projected-scale instrument
Messgerät *n* **mit unterdrücktem Nullpunkt** <Gerät> set-up scale instrument
Messgerät *n* **mit Zeigeranzeige** <Metrol> *(AE)* dial indicating gage, *(BE)* dial indicating gauge

Messgerät *n* **zur Bestimmung des Berylliumgehaltes** <Kerntech> beryllium content meter, beryllium prospecting meter
Messgerätefehler *m* <Gerät> instrumental error
Messgeräteklemme *f* <Gerät> meter terminal
Messgeräteschutzschaltung *f* <Gerät> meter protecting circuit
Messgeräteskale *f* <Gerät> meter dial
Messgeräteverstärker *m* <Gerät> meter amplifier
Messgerätständer *m* <Metrol> *(AE)* gage stand, *(BE)* gauge stand
Messglas *n* <Ker & Glas> *(AE)* gage glass, *(BE)* gauge glass
Messglied *n* <Gerät> receiving element *(Messeinrichtung)*
Messgröße *f* 1. <Elektronik> measurand, measured quantity; 2. <Gerät, Metrol> measurand
Messgrößenwandler *m* <Gerät> converter
Messing *n* <Elektriz, Fertig, Maschinen, Metall> brass
• **mit Messing löten** <Mechan, Thermod> braze
Messingarbeiten *fpl* <Bau, Metall> brass works
Messingdraht *m* <Bau, Metall> brass wire
Messinggießerei *f* <Bau, Metall> brass foundry
Messingguss *m* <Metall> cast brass
Messinglot *n* <Bau, Maschinen> brass solder
Messingrevolverdrehmaschine *f* <Fertig> monitor lathe
Messingrundkopfschraube *f* <Bau> brass round-head wood screw
Messingschmied *m* <Bau, Metall> brass smith
Messingschraube *f* <Maschinen> brass screw
Messingstab *m* <Nichtfoss Energ> brass rod
Messingtype *f* <Druck> brass type
Messinstrument *n* 1. <Eisenbahn> instrument; 2. <Elektriz> measuring instrument; 3. <Erdöl> instrument; 4. <Funktech> meter; 5. <Gerät> measuring instrument; 6. <Kfztech, Kontroll> *(AE)* gage, *(BE)* gauge; 7. <Labor> *(AE)* gage, *(BE)* gauge, measuring equipment, measuring set; 8. <Lufttrans> instrument; 9. <Meerschmutz> sensing device; 10. <Metrol> measuring device, measuring instrument; 11. <Papier> *(AE)* gage, *(BE)* gauge; 12. <Phys> electrodynamometer; 13. <Wassertrans> instrument
Messinstrument *n* **mit Nebenwiderstand** <Gerät> shunted instrument
Messinstrument *n* **mit Shunt** <Gerät> shunted instrument
Messinstrumentfehler *m* <Gerät> instrumental error
Messkabel *n* <Elektrotech> measuring lead, test lead
Messkette *f* 1. <Bau> chain, engineer's chain, measuring chain; band chain *(Vermessung)*; 2. <Metrol> chain
Messklemme *f* <Elektrotech> test terminal
Messkolben *m* <Labor> graduated flask, volumetric flask
Messkopf *m* 1. <Gerät> measuring head, probe, sensing head; 2. <Phys> probe
Messkraft *f* <Gerät> measuring force
Messkrümmer *m* <Regelung> flow elbow
Messkugelfunkenstrecke *f* <Elektrotech, Metrol> measuring ball gap, measuring spark gap, measuring sphere gap
Messkunde *f* <Bau> surveying
Messkurslinie *f* <Lufttrans> slant course line
Messlatte *f* 1. <Bau> rod; staff *(Vermessung)*; speaking rod *(mit Ablesemarkierungen)*; 2. <Fertig> staff; 3. <Metrol> rod
Messlattenträger *m* <Bau> staff holder
Messlehre *f* 1. <Mechan> *(AE)* gage, *(BE)* gauge; 2. <Metrol> *(AE)* caliper, *(BE)* calliper
Messleitung *f* <Elektrotech> sensing lead, slotted line, test lead
Messleitungssonde *f* <Elektrotech> slotted line probe
Messmarke *f* <Fertig> pop mark

Messmaschine *f* 1. <Gerät> check station *(zur numerisch gesteuerten Fertigung)*; 2. <Metrol> measuring machine
Messmikrofon *n* <Akustik> standard microphone
Messmittelbeglaubigung *f* <Metrol> measuring instrument verification
Messmittelkalibrierung *f* <Metrol> measuring instrument calibration
Messnorm *f* <Metrol> measurement standard
Messnormale *f* <Telekom> etalon
Messobjekt *n* 1. <Gerät> object of measurement; 2. <Qual> device under test
Messort *m* <Gerät> measuring position, sensing point, sensor location
Messoszilloskop *n* <Elektriz> measuring oscilloscope
Messpfad *m* <Gerät> measuring path
Messpipette *f* <Labor> graduated pipette
Messplatte *f* <Aufnahme> test record
Messplatz *m* <Kerntech> measuring desk
Messprojektor *m* <Metrol> optical comparator
Messpunkt *m* <Telekom> test point
Messpunkt *m* **für Überfluglärm** <Lufttrans> flyover noise measurement point
Messpunktabtaster *m* <Kontroll> scanner
Messrahmen *m* <Eisenbahn> *(AE)* loading gage, *(BE)* loading gauge
Messreaktor *m* <Kerntech> measurements reactor, source reactor
Messrelais *n* 1. <Elektrotech> instrument-type relay, measuring relay, meter-type relay, sensing relay; 2. <Gerät> detecting relay, measuring relay
Messreproduzierbarkeit *f* <Metrol> reproducibility of measurements
Messschalter *m* <Elektrotech> instrument switch, sensing switch
Messschieber *m* 1. <Maschinen> *(AE)* caliper square, *(BE)* calliper square; 2. <Metrol> *(AE)* caliper, *(BE)* calliper, *(AE)* vernier caliper, *(BE)* vernier calliper
Messschrank *m* <Gerät> test board
Messschraube *f* <Metrol> micrometer
Messschreiber *m* <Elektriz, Labor> pen recorder
Messsender *m* <Funktech, Telekom> signal generator
Messshunt *n* <Elektrotech> instrument shunt
Messsignal *n* <Gerät> measurement signal
Messskale *f* 1. <Gerät> meter dial; 2. <Maschinen> measuring scale
Messskalenantrieb *m* <Raumfahrt> vernier motor
Messsonde *f* <Raumfahrt> measuring probe *(Raumschiff)*
Messspannenfehler *m* <Regelung> span error
Messspannenverschiebung *f* <Regelung> span shift
Messspion *m* <Metrol> *(AE)* feeler gage, *(BE)* feeler gauge
Messspitze *f* <Gerät> test prod
Messspule *f* <Elektrotech> search coil
Messstab *m* <Kfztech> dipstick *(Schmierung)*
Messstange *f* <Bau> *(AE)* gage bar, *(BE)* gauge bar
Messstelle *f* <Gerät> measuring position, sensing point
Messstelle *f* **mit analoger Messdatenerfassung** <Gerät> analog point
Messstelle *f* **mit digitaler Messdatenerfassung** <Gerät> digital point
Messstellenabtaster *m* <Gerät> scanner
Messstellentemperatur *f* <Gerät> measuring junction temperature
Messstellenumschalter *m* <Gerät> scanner
Messstellung *f* <Gerät> measuring position *(Bedienungselement)*
Messsteuerung *f* <Fertig> *(AE)* in-process gaging, *(BE)* in-process gauging, sizing
Messstift *m* 1. <Ker & Glas> *(AE)* roller gage, *(BE)* roller gauge; 2. <Metrol> feeler pin *(Messuhr)*

Messstrahl

Messstrahl *m* <Metrol> beam
Messstrecke *f* 1. <Fertig> conduit *(Windkanal)*; 2. <Phys> working section
Messstrom *m* <Gerät> measuring current
Messsystem *n* <Elektriz, Gerät> measuring system
Messtank *m* <Erdöl> *(AE)* gaging tank, *(BE)* gauging tank *(Messtechnik)*
messtechnische Ausrüstung *f* <Gerät> instrumentation
Messtemperatur *f* <Gerät> measured temperature
Messtisch *m* <Bau> plane table *(Vermessung)*
Messüberträger *m* <Telekom> test transformer
Messuhr *f* 1. <Maschinen> *(AE)* dial gage, *(BE)* dial gauge, dial indicator; 2. <Metrol> *(AE)* dial indicating gage, *(BE)* dial indicating gauge
Messumformer *m* 1. <Comp & DV> transducer; 2. <Elektriz> *(AE)* current transformer, *(BE)* mains transformer, instrument transformer; 3. <Elektrotech> transducer; 4. <Gerät> measuring transducer, transducer; 5. <Metrol> measuring transducer
Messumsetzer *m* <Gerät> measuring converter
Mess- und Prüfeinrichtungen *fpl* <Qual> measuring and test equipment
Mess- und Regeltechnik *f (IUC)* <Elektronik> instrumentation and control, IUC *(Leittechnik)*
Messung *f* 1. <Elektronik> measurement; 2. <Geom> mensuration; 3. <Labor> measuring; 4. <Metrol> measurement
Messung *f* **der Luftqualität** <Umweltschmutz> air quality measurement
Messung *f* **der Luftverschmutzung** 1. <Gerät> atmospheric pollution measurement; 2. <Sicherheit> measurement of air pollution
Messung *f* **des Linienprofils** <Strahlphys> line profile measurement *(von Spektrallinien)*
Messung *f* **durch Bewegungsgitter** <Optik> measurement by diffraction grating
Messungenauigkeit *f* <Metrol> inaccuracy of measurement
Messunsicherheit *f* <Gerät> measurement uncertainty, uncertainty of measurement
Messventil *n* <Gerät> measuring valve
Messverfahren *n* 1. <Bau> method of measurement; 2. <Metrol> measurement process
Messverkörperung *f* <Qual> material measure
Messverstärker *m* 1. <Elektronik> measuring amplifier; 2. <Gerät> instrumentation amplifier
Messwagen *m* <Fertig> carriage *(Steigerungsmessmaschine)*
Messwandler *m* 1. <Elektriz, Elektrotech> instrument transformer, transducer; 2. <Fertig> resolver; 3. <Gerät> instrument transformer, measuring transducer, transducer; 4. <Maschinen> transducer
Messwandler *m* **ohne Hilfsenergie** <Elektrotech> self-generating transducer
Messwarte *f* <Fertig> monitoring station
Messwehr *n* <Wasserversorg> measuring weir *(mit Einschnitt)*
Messwehr *n* **mit Einschnitt** <Bau> notched weir
Messwendekreisel *m* <Phys> rate gyro
Messwerk *n* 1. <Elektrotech> meter movement; 2. <Gerät> instrument movement, measuring system, measuring movement, meter movement
Messwerkaufhängung *f* <Gerät> movement suspension
Messwerkdämpfung *f* <Gerät> meter damping
Messwerkregler *m* <Gerät> control meter, movement controller, primary controller
Messwerkzeug *n* <Maschinen> measuring instrument
Messwert *m* <Elektronik, Gerät> measurand, measured value, measurement value, measuring value, observed value, test value

Messwert *m* **nach Auftreten eines Fehlers** <Gerät> postfault measuring value
Messwert *m* **vor Auftreten eines Fehlers** <Gerät> prefault measuring value
Messwertabtastung *f* <Gerät> data sampling, measuring data sampling, measuring data scanning, measuring data sampling
Messwerterfassung *f* <Gerät> data logging
Messwertgeber *m* 1. <Comp & DV> sensor, transducer; 2. <Lufttrans> angular velocity rate sensor *(Kreisfrequenzen, Winkelgeschwindigkeit)*
Messwertglättung *f* <Gerät> smoothing *(von Messkurven durch Mittelwertbildung)*
Messwertprüfung *f* <Gerät> data checking
Messwertregistrierung *f* <Gerät> data logging
Messwertübertragungssystem *n* <Gerät> data transmission system
Messwertumformer *m* <Comp & DV, Elektrotech> transducer
Messwertumformung *f* <Gerät> data conversion
Messwertumsetzer *m* <Gerät> data converter, measuring converter
Messwertverarbeitung *f* <Gerät> processing of measured data
Messwertverdichtung *f* <Gerät> measuring data reduction
Messwertverstärker *m* <Gerät> data amplifier
Messwertwandler *m* <Phys> transducer
Messwesen *n* <Metrol> metrology
Messwiderstand *m* <Elektrotech> instrument shunt, sensing resistor
Messzange *f* <Elektriz> clip-on instrument
Messzeit *f* <Gerät> detection time
Messzelle *f* <Gerät> detector cell, measuring cell
Messzylinder *m* 1. <Foto> measuring cylinder *(Entwicklungschemikalien)*; 2. <Labor> measuring cylinder
Metabisulfit *n* <Chemie> *(AE)* metabisulfite, *(BE)* metabisulphite
metabolische Wärmeproduktion *f* <Ergon> metabolic heat production
metabolisches Indican *n* <Chemie> indicane
Metabor... <Chemie> metaboric
Metaborat *n* <Chemie> dioxoborate, metaborate
Metadyne *f* <Elektronik> metadyne *(Elektromaschinen)*
Metakiesel... <Chemie> metasilicic
Metaldehyd *n* <Chemie> metaldehyde
Metall *n* <Ker & Glas, Metall> metal
Metall *n* **im Nachsaugesteiger** <Fertig, Metall> head metal *(Gießen)*
Metallabfall *m* <Abfall> metal waste
Metallabnahme *f* <Maschinen> metal removal
Metall-Aktivgas-Schweißen *n* *(MAG-Schweißen)* <Thermod> metal active gas welding *(MAG welding)*
Metallbanddrucker *m* <Comp & DV> band printer, belt printer
Metallbandsägeblatt *n* <Maschinen> metal-cutting bandsaw blade
Metallbandverschluss *m* <Verpack> metal strip closure
Metallbarometer *n* <Bau> surveying aneroid barometer *(Vermessung)*
Metallbauwerk *n* <Bau, Metall> metallic structure
Metallbearbeitung *f* <Fertig> machining of metals
Metallbedampfung *f* <Elektronik> *(AE)* vacuum metalization, *(BE)* vacuum metallization *(im Vakuum)*
Metallbelag *m* <Bau> *(AE)* metalization, *(BE)* metallization
Metallbeschichtung *f* 1. <Anstrich> metallic coating; 2. <Fernseh> metal coating; 3. <Maschinen> metallic coating
Metallbindung *f* <Metall> metallic bond

Metallbolzen m 1. <Bau> gate hook; 2. <Fertig> gudgeon
Metalldampf-Laser m <Elektronik> (AE) metal vapor laser, (BE) metal vapour laser
Metalldetektor m <Verpack> metal detector
Metalldicke f <Ker & Glas> metal depth
Metalldraht m <Elektrotech> metal filament
Metalldrückbank f <Fertig> spinning lathe
Metalldrücken n <Fertig> metal spinning
Metalldruckplatte f <Druck> metal plate
Metalleffektlack m <Kunststoff> metallic paint
Metallfaden m <Elektrotech> metal filament
Metallfließpressmatrize f <Maschinen> extrusion die for metal
Metallfolie f 1. <Fertig> spangle; 2. <Metall, Verpack> metal foil, (AE) metalized film, (BE) metallized film
Metallgehäuse n <Elektrotech> metal case
metallgekapselt adj <Mechan> metal-clad
metallgekapselte Niederspannungsschaltanlage f <Elektrotech> low-voltage metal-clad switchgear
metallgeschützt adj <Elektriz> metal-clad
Metallgesteinsstrecke f <Kohlen> metal drift
Metallglasur f <Elektronik> metal glaze
Metallgleichrichter m <Elektriz> metal rectifier
Metallgummifeder f <Maschinen> rubber-metal spring
Metallguss m <Fertig> nonferrous castings
Metallhalbleiter-Feldeffekttransistor m (MESFET) <Elektronik> metal semiconductor field effect transistor (MESFET)
Metallhalbleiter-Übergang m <Elektronik> metal semiconductor junction
Metallhydrid-Kraftstofftank m <Nichtfoss Energ> metal hydride fuel tank
Metallinertgasschweißen n (MIG-Schweißen) <Bau, Fertig, Thermod> metal inert gas welding (MIG welding)
metallisch adj <Anstrich> metallic
metallische Kupplung f <Maschinen> metal-to-metal clutch
metallische Leitung f <Elektrotech> metallic circuit
metallische Sperrfolie f <Aufnahme> metallic stop foil
metallischer Gleichrichter m <Elektrotech> metallic rectifier
metallischer Koppelpunkt m <Telekom> metallic crosspoint
metallischer Leiter m <Elektrotech> metallic conductor
metallischer Überzug m 1. <Mechan> cladding; 2. <Verpack> metallic coating
metallisches Gas n <Phys> metallic gas
metallisches Natururan n <Kerntech> metallic natural uranium
Metallisieren n 1. <Fertig> metal coating; 2. <Maschinen, Metall> (AE) metalization, (AE) metalizing, (BE) metallization, metallizing
metallisierter Bildschirm m <Elektronik> (AE) metalized screen, (BE) metallized screen
metallisierter Farbstoff m <Textil> (AE) premetalized dye, (BE) premetallized dye
metallisierter Glimmerkondensator m <Elektrotech> (AE) metalized mica capacitor, (BE) metallized mica capacitor
metallisierter Kondensator m <Elektrotech> (AE) metalized capacitor, (BE) metallized capacitor
metallisierter Papierkondensator m <Elektriz> (AE) metalized paper capacitor, (BE) metallized paper capacitor
Metallisierung f 1. <Elektronik> (AE) metalization, (BE) metallization (Plattierung, Galvanisierung); 2. <Ker & Glas> (AE) metalizing, (BE) metallizing; 3. <Phys, Raumfahrt> (AE) metalization, (BE) metallization (Raumschiff); 4. <Verpack> metal coating

Metallisierungsmaske f <Elektronik> (AE) metalization mask, (BE) metallization mask
Metallisierungsschicht f <Elektronik> (AE) metalization layer, (BE) metallization layer
Metallisolator-Feldeffekttransistor m (MISFET) <Elektronik> metal insulator semiconductor field effect transistor (MISFET)
Metallisolator-Halbleiter m (MIS) <Elektronik> metal insulator semiconductor (MIS)
metallkaschiert adj <Elektriz> metal-clad
Metallkassette f 1. <Foto> metal dark slide; 2. <Verpack> metal box
Metallkeramikbeschichtung f <Anstrich> metallic-ceramic coating
Metallkeramikwiderstand m <Elektrotech> metal glaze resistor
metallkiesgestrahlt adj <Anstrich> shot-peened
metallkiesstrahlen v <Fertig> shot-peen
Metallkiesstrahlen n <Fertig> shot-peening
Metallklumpen m <Metall> slug
Metallkonusröhre f <Elektronik> metal-cone tube
Metallkrystall m <Metall> metal crystal
Metallkunde f <Metall> metallography
Metall-Lichtbogenschweißen n <Thermod> metal arc welding
Metall-Luftbatterie f <Kfztech> metal air battery
Metallmantel m <Elektrotech> metal sheath
Metallograph m <Metall> metallographer
metallographisches Mikroskop n <Labor> metallographic microscope
Metalloid n <Metall> metalloid
metallorganisch adj <Chemie> organometallic
Metalloxid-Halbleiter m (MOS) <Comp & DV, Elektronik> metal oxide semiconductor (MOS)
Metalloxid-Silizium-Feldeffekttransistor m (MOSFET) <Elektronik> metal oxide silicon field effect transistor (MOSFET)
Metalloxid-Transistor m (MOS) <Elektronik> metal oxide semiconductor (MOS)
Metallpackung f <Maschinen> metallic packing
Metallpapier n (MP) <Verpack> (AE) metalized paper, metallic paper, (BE) metallized paper (MP)
Metallpigmentfarbe f <Kunststoff> metallic paint
Metallplatte f <Druck> metal plate
metallplattiertes Loch n <Elektriz> (AE) metalized hole, (BE) metallized hole
Metallpulver n <Metall> metal powder
Metallrohr n <Maschinen, Metall> metal tube
Metallröhre f <Elektronik> metal tube
Metallsägeblatt n <Maschinen> hacksaw blade, metal-cutting saw blade
Metallsägebogen m <Maschinen> hacksaw frame
Metallschicht f <Elektronik> metal film, (AE) metalization, (BE) metallization
Metallschichtwiderstand m <Elektrotech> metal film resistor
Metallschlauch m <Maschinen> flexible metal tube, flexible metallic hose
Metallschlitzsäge f <Maschinen> metal slitting saw
Metallschrott m <Abfall> metal waste
Metallseife f <Chemie> metallic soap
Metallsortieranlage f <Abfall> metal separator (magnetische Abtrennung)
Metallspritzen n 1. <Fertig> metal spraying; 2. <Maschinen> metal powder spraying
Metallspritzer m <Sicherheit> molten metal splash (Schutzkleidung)
Metallspritzverfahren n <Maschinen> metal spraying process

Metallsteuerelektrode

Metallsteuerelektrode f <Elektronik> metal gate
Metallsucher m <Maschinen> metal detector
Metallsuchgerät n <Verpack> metal detector
Metallthermometer n <Thermod> differential thermometer
Metalltröpfchen n <Fertig> droplet of metal *(Schweißen)*
metallüberzogen adj <Fertig> plated
Metallüberzug m 1. <Fertig, Maschinen, Metall> metal coating; 2. <Telekom> metallic coating; 3. <Verpack> metal coating • **mit Metallüberzug** <Fertig> metal-coated
Metallummantelung f <Elektriz> metallic sheath
Metallurg m <Metall> metallurgist
Metallurgie f <Fertig> metallurgy
metallurgisch adj 1. <Anstrich, Fertig, Kohlen> metallurgical *(Abfall)*; 2. <Metall> metallurgical
metallurgische Kohle f <Kohlen> metallurgical coal
metallverkleidet adj <Elektriz> metal-coated
Metallwaren fpl <Maschinen> hardware
Metallwerkzeug n <Fertig> metal die
Metallwiderstand m <Elektriz> metallic resistor
Metallwiedergewinnung f <Kohlen, Metall> metal recovery
Metallzierarbeiten fpl <Bau, Metall> art metal work *(Geländer, Tore)*
metamorphes Gestein n <Nichtfoss Energ> metamorphic rocks
Metamorphit m <Nichtfoss Energ> metamorphic rocks
Metaphosphat n <Chemie> metaphosphate
Metaphosphor... <Chemie> metaphosphoric
Metaregel f <Künstl Int> metarule
Metasilicat n <Chemie> metasilicate
Metasprache f <Comp & DV> metalanguage
metastabil adj <Strahlphys> metastable
metastabiler Zustand m <Metall> metastable state
metastabiles Atom n <Strahlphys> metastable atom
metastabiles Gleichgewicht n <Phys> metastable equilibrium
Metathese f <Chemie> metathesis
Metathesis f <Chemie> metathesis
Metazentrum n <Phys, Wassertrans> *(AE)* metacenter, *(BE)* metacentre *(Schiffbau)*
Metazinn... <Chemie> metastannic
Meteor m <Raumfahrt> meteor
Meteoreisen n <Metall> meteoric iron, siderial iron
Meteorenecho n <Raumfahrt> meteor echo
Meteoriteneinsturm m <Raumfahrt> meteorite influx
Meteoritenstaub m <Raumfahrt> meteor dust
Meteorschwarmfunkverbindung f <Raumfahrt> meteor burst link *(Weltraumfunk)*
Meteorstreuungs-Übertragungssystem n <Funktech, Telekom> meteor-burst communication system
Meter m 1. <Metrol> meter *(Messgerät)*; 2. <Metrol> *(AE)* meter, *(BE)* metre *(Messeinheit)*
Meterküvette f <Foto, Optik> meter cell
Meterspurweite f <Eisenbahn> *(AE)* meter gage, *(BE)* metre gauge
Meterware f <Foto> bulk film *(Film)*
Methacryl... <Chemie> methacrylic
Methacrylat n <Kunststoff> methacrylate
Methacrylatklebstoff m <Kunststoff> methacrylic adhesive
Methan n <Chemie, Erdöl> methane
Methan... <Chemie> methanoic
Methanal n <Chemie> formaldehyde, methanal
Methanamid n <Chemie> formamide, methanamide
Methangärung f <Abfall> alkaline fermentation, methane digestion, methane digestion, methane fermentation
Methangastanker m <Wassertrans> methane carrier
Methanol n <Chemie> methane alcohol, methanol

methanolbetriebener Wagen m <Kfztech, Nichtfoss Energ> methanol car
Methanolbrennstoffzelle f <Kfztech> methanol cell
methanolisch adj <Chemie> methanolic
Methenamin n <Chemie> methenamine, urotropine
Methion... <Chemie> methionic
Methode f **der gewichteten Rückstände** <Kerntech> Gelerkin method
Methode f **der kleinsten Quadrate** <Ergon, Math, Phys> least squares method
Methode f **des kritischen Weges** *(CPM)* <Comp & DV> critical path method *(CPM)*
Methode f **mit gebrochenem Strahl** <Optik> refracted ray method
Methoxybenzaldehyd n <Chemie> anisaldehyde, methoxybenzaldehyde
Methoxybenzen n <Chemie> anisole, methoxybenzene
Methoxyl... <Chemie> methoxyl
Methusalem m <Ker & Glas> methuselah
Methyl n <Chemie> methyl
Methylacetat n <Chemie> methyl acetate
Methylacrolein n <Chemie> crotonaldehyde, methylacrolein
Methylalkohol m 1. <Chemie> methane alcohol, methanol; 2. <Kunststoff> methyl alcohol
Methylamin n <Chemie> aminomethane, methylamine
Methylaminoessigsäure f <Chemie> *(N)*-methylaminoacetic acid, sarcosine
Methylanilin n <Chemie> methylaniline
Methylat n <Chemie> methylate
Methylbenzol n <Chemie> methylbenzene, toluene
Methylcyanid n <Chemie> acetonitrile, ethanenitrile
Methylcyclopentadienyl-Manganestricarbonyl n <Kfztech> methylcyclopentadienyl manganese tricarbonyl, MMT *(Kraftstoffzusatz)*
Methylenblautest m <Abfall> methylene blue test *(Verfahren zum Feststellen der Fäulnisfähigkeit von Wasser)*
Methylendioxybenzaldehyd m <Chemie> heliotropin, methylenedioxybenzaldehyde, piperonal
Methyleniodid n <Chemie> diiodomethane, methylene iodide
Methylethin n <Chemie> allylene, methylacetylene, propyne
Methylethylketon n *(MEK)* <Kunststoff> methyl ethyl ketone *(MEK)*
Methylgruppe f <Chemie> methyl group
methylieren v <Chemie> methylate
Methylierung f <Chemie> methylation *(organische Chemie)*
Methylindol n <Chemie> methylindole, skatole
Methylkautschuk m <Kunststoff> methyl rubber
Methylmorphin n <Chemie> codeine, methylmorphine
Methylnaphthalen n <Chemie> methylnaphthalene
Methylorange n <Chemie> helianthine
Methylphenyl... <Chemie> tolyl
Methylphenylether m 1. <Chemie> anisole, methoxybenzene; 2. <Lebensmittel> anisole
Methylphenylketon n <Chemie> acetophenone
Methylpiperidin n <Chemie> methylpiperidine, pipecoline
Methylpyridin n <Chemie> methylpyridine, pikoline
Methylquercetin n <Chemie> methylquercitin, rhamnetin
Methylradikal n <Chemie> methyl radical
Methyltertiärbutylether m <Erdöl> methyl tertiary-butyl ether *(Petrochemie)*
Metol n <Chemie> metol
Metrik f <Comp & DV> metric
metrisch adj 1. <Comp & DV> metric; 2. <Fertig> French; 3. <Funktech> metric; 4. <Metrol> metric, metrical
metrische Geometrie f <Geom> metrical geometry

metrische Pferdestärke f <Metrol> metric horsepower
metrische Tonne f <Metrol> metric ton
metrischer Punkt m <Druck> metric point
metrischer Schlüssel m <Kerntech> metric key
metrisches Gewinde n <Maschinen> metric thread
metrisches ISO-Gewinde n <Maschinen> ISO metric thread
metrisches Karat n <Metrol> metric carat
metrisches System n <Metrol> metric system
metrisches Trapezgewinde n <Maschinen> metric trapezoidal screw thread
Metteurtisch m <Druck> imposing table
MeV *(Million Elektronenvolt)* <Teilphys> MeV *(million electron volts)*
MF 1. <Elektriz> *(Melamin-Formaldehydharz, Melaminharz)* MF *(melamine resin)*; 2. <Elektronik, Funktech> *(Mittelfrequenz)* MF *(medium frequency)*; 3. <Elektronik, Funktech, Telekom> *(Mehrfachfrequenz)* MF *(multiple frequency)*; 4. <Kunststoff> *(Melamin-Formaldehydharz, Melaminharz)* MF *(melamine formaldehyde resin)*
MF-Band n <Funktech> MF band
MFC *(Mehrfrequenzcode)* <Telekom> MFC *(multifrequency code)*
MFC-Telefon n <Telekom> MFC telephone
MF-Generator m <Telekom> MF generator
MFM *(modifizierte Frequenzmodulation)* <Elektronik, Funktech, Telekom> MFM *(modified frequency modulation)*
MFW *(Mehrfrequenzwahl)* 1. <Telekom> DTMF *(dual-tone multifrequency)*; 2. <Telekom> MFD *(multifrequency dialling)*
Mg *(Magnesium)* <Chemie> Mg *(magnesium)*
MGD *(Magnetogasdynamik)* <Kerntech> MGD *(magnetogasdynamics)*
MHD-Konverter m *(magnetohydrodynamischer Konverter)* <Raumfahrt> MHD converter *(magnetohydrodynamic converter)*
MHD-Wandler m *(magnetohydrodynamischer Wandler)* <Elektrotech, Kerntech> MHD converter *(magnetohydrodynamic converter)*
m-Höhe f <Druck> x-height, z-height
MHz *(Megahertz)* <Elektriz, Elektrotech, Fernseh, Funktech> MHz *(megahertz)*
MIC *(integrierter Mikrowellenschaltkreis)* <Elektronik, Wellphys> MIC *(microwave integrated circuit)*
Michelson-Interferometer n <Phys> Michelson interferometer
Michelson-Morley-Experiment n <Phys> Michelson-Morley experiment
MICR *(Magnetschrifterkennung, Magnetschriftzeichenerkennung)* <Comp & DV> MICR *(magnetic ink character recognition)*
micronisiert adj <Kunststoff> micronized
mieten v <Wassertrans> charter *(Schiff)*
Mietleitung f 1. <Comp & DV> leased circuit, leased line; 2. <Telekom> leased circuit, leased line, private wire
Mietleitung f **mit Fernsprechqualität** <Telekom> speech-grade private wire
Mietleitungsnetz n <Comp & DV, Telekom> leased line network
Mietsoftware f <Telekom> rent-a-software
MIG-MAG-Schweißbrenner m <Bau> torch for MIG-MAG welding
Migration f 1. <Comp & DV> migration; 2. <Erdöl> migration *(Erdölgeologie)*; 3. <Kunststoff> migration
MIG-Schweißen n *(Metallinertgasschweißen)* <Bau, Fertig, Thermod> MIG welding *(metal inert gas welding)*
Mikro n *(Mikrofon)* <Akustik, Aufnahme, Elektriz, Funktech> mike *(microphone)*
Mikro... <Comp & DV, Metrol> micro...

Mikroampere n <Elektriz> microampere
Mikro-Amperemeter n <Elektrotech> microammeter
Mikrobaustein m <Comp & DV> microassembly
Mikrobefehl m <Comp & DV> microinstruction
Mikrobefehlscode m <Comp & DV> microcode
Mikrobiegeverlust m <Optik> microbend loss
Mikrobiegung f 1. <Fertig> microbend *(Faseroptik)*; 2. <Optik> microbending
mikrobielle Laugung f <Chemie> leaching
Mikrobild n <Metall> photomicrograph
Mikrochip m <Elektronik> microchip
Mikrocode m <Comp & DV> microcode
Mikrocomputer m <Comp & DV> microcomputer
Mikrocontroller m <Comp & DV> microcontroller
Mikrodevitrifikation f <Optik> microdevitrification
Mikrodiskette f <Comp & DV> microfloppy disk
Mikroeinkapselung f <Abfall> grain encapsulation *(von Sondermüll)*
Mikroelektronik f <Comp & DV> microelectronics
Mikroentglasung f <Optik> microdevitrification
Mikrofarad n <Elektriz, Elektronik> microfarad
Mikrofiche f 1. <Comp & DV> fiche, microfiche; 2. <Druck> microfiche
Mikrofichelesegerät n <Comp & DV, Druck> microfiche reader
Mikrofilm m <Comp & DV, Druck, Foto> microfilm • **auf Mikrofilm aufnehmen** <Comp & DV> microfilm
Mikrofilmaufnahmegerät n <Comp & DV> microfilm recorder
Mikrofilmaufzeichnungsgerät n <Comp & DV> microfilm recorder
Mikrofilmlesegerät n <Comp & DV> microfilm reader
Mikrofilmleser m <Comp & DV> microfilm reader
Mikrofilmlochkarte f <Comp & DV> aperture card
Mikrofilmrecorder m <Comp & DV> microfilm recorder
Mikrofilmtechnik f <Konstzeich> microcopying technique, microfilming technique
Mikrofon n *(Mikro)* <Akustik, Aufnahme, Elektriz, Funktech> microphone *(mike)*
Mikrofon n **mit herzförmiger Charakteristik** <Akustik, Aufnahme> cardioid microphone
Mikrofon n **mit Parabolreflektor** <Aufnahme> parabolic reflector microphone
Mikrofon n **mit Umgebungsgeräuschunterdrückung** <Mobilkom> antinoise microphone
Mikrofonabschirmung f <Aufnahme> microphone shield
Mikrofongalgen m <Aufnahme> microphone boom
Mikrofonie-Effekt m **in der Cochlea** <Akustik> cochlear microphonic effect
mikrofonisch adj <Aufnahme> microphonic
Mikrofonkabel n <Aufnahme> microphone cable
Mikrofonkappe f <Aufnahme> microphone blanket
Mikrofonmembran f <Aufnahme> microphone diaphragm
Mikrofonnetzteil n <Aufnahme> microphone power supply
Mikrofonstativ n <Aufnahme> microphone stand
Mikrofonstummschaltung f <Funktech> microphone cancellation
Mikrofontrafo m <Aufnahme> microphone transformer
mikrofonunabhängig adj <Aufnahme> off-mike
Mikrofonverstärker m <Aufnahme> microphone amplifier
Mikrographieverfahren n <Fertig> micrographic method *(Werkstoffanalyse)*
Mikrohärte f <Kunststoff, Metall> microhardness
Mikrohenry n <Elektriz> microhenry
Mikrohöhlung f <Optik> micropit
Mikroinstruktion f <Comp & DV> microinstruction
Mikrokanal m <Elektronik> microchannel
Mikrokanal-Bildverstärker m <Elektronik> microchannel image intensifier

Mikrokanalplatte

Mikrokanalplatte f <Elektronik> microchannel plate
Mikrokopie f <Comp & DV> microcopy
Mikrokopieaufzeichnungsgerät n <Comp & DV> microfilm recorder
mikrokristallin adj <Chemie> microcrystalline
Mikrokrümmung f 1. <Elektrotech> microbend; 2. <Fertig> microbend (Faseroptik); 3. <Telekom> microbending
Mikromanometer n <Gerät> micromanometer
Mikrometer n 1. <Gerät> micrometer; 2. <Labor> micrometer (Dickenmessung); 3. <Maschinen, Mechan> micrometer; 4. <Metrol> micrometer, micromil; 5. <Papier, Phys> micrometer
Mikrometer n **mit elektronischer Anzeige** <Metrol> electronic display micrometric head
Mikrometer n **mit Okular** <Maschinen> eye-piece micrometer
Mikrometermessuhr f <Metrol> dial indicating micrometer
Mikrometerschraube f 1. <Labor> (AE) calipers, (BE) callipers; 2. <Maschinen, Phys> micrometer screw
Mikrometerschraube f **mit Feingewinde** <Maschinen> finely threaded micrometer screw
Mikrometer-Schraubenkopf m <Metrol> micrometric head
Mikrometertaster m <Maschinen> (AE) micrometer calipers, (BE) micrometer callipers
Mikrometrie f <Maschinen> micrometry
Mikromillimeter m <Labor> (AE) micromillimeter, (BE) micromillimetre
mikrominiaturisierte Schaltung f <Comp & DV> microcircuit
Mikrominiaturisierung f <Elektronik> microminiaturization
Mikron n <Metrol> micron
Mikronfilter n <Maschinen> micronic filter
mikronisieren v <Chemie> micronize
Mikronschaltung f <Elektronik> micron circuit
Mikronschranke f <Elektronik> micron barrier
Mikroohm n <Elektrotech> microhm
Mikropfahl m <Bau> micropile
Mikrophotoansatz m <Foto> microattachment
Mikrophotogramm n <Metall> photomicrogram
Mikrophotometer n <Strahlphys> microphotometer
Mikropipette f <Labor> micropipette
Mikroplanfilm m <Comp & DV> microfiche
Mikroprogramm n <Comp & DV> microprogram, microroutine
Mikroprogrammieren n <Comp & DV> microprogramming
Mikroprozessor m (MP) <Comp & DV, Elektriz, Maschinen> microprocessor (MP)
Mikroprozessor-Chip m <Elektronik> microprocessor chip
Mikroprozessoreinheit f <Comp & DV> MPU, microprocessor unit
Mikroprozessorsteuerung f 1. <Comp & DV> microcontroller; 2. <Telekom> microprocessor control
Mikropumpe f <Maschinen> micropump
Mikrorakete f <Raumfahrt> microrocket (Raumschiff)
Mikrorauigkeit f <Fertig> microroughness
Mikroreaktor m <Chemie> microreactor
Mikrorechner m <Comp & DV> microcomputer
Mikrorechner-Betriebssystem n <Comp & DV> microcomputer operating system, microcomputer OS
Mikrorechner-Entwicklungssystem n <Comp & DV> microcomputer development system (MDS)
Mikrorechner-Hauptplatine f microcomputer card, PC main board, PC motherboard
Mikrorheologie f <Metall> microrheology

Mikrorille f <Akustik> microgroove
Mikrorillenaufnahme f <Aufnahme> microgroove recording
Mikrorillenplatte f <Aufnahme> microgroove record
Mikroriss m <Optik> microcrack
Mikroschalter m 1. <Elektriz, Elektrotech> micrometer; 2. <Fertig> microswitch (Kunststoffinstallationen); 3. <Kontroll> microswitch (Grenzschalter)
Mikroschaltkreis m <Fertig> microcircuit
Mikroschaltung f <Comp & DV, Elektronik> microcircuit
Mikroschaltungsaufbau m <Comp & DV, Elektronik> microcircuitry
Mikroschlitz... <Telekom> microslot
Mikroschraube f <Kontroll> microscrew
Mikroschubtriebwerk n <Raumfahrt> microthruster
Mikroschwerkraft f <Raumfahrt> microgravity
Mikroseigerung f <Metall> microsegregation
Mikrosekunde f <Comp & DV> microsecond
Mikrosieb n <Wasserversorg> microstrainer
Mikrosiebfilter n <Wasserversorg> microstrainer
Mikroskop n <Labor, Metrol> microscope
Mikroskop-Adapter m <Foto> microscope adaptor
mikroskopischer Staub m <Sicherheit> microscopic dust
mikroskopischer Zustand m <Strahlphys> microscopic state
mikroskopisches Kriechen n <Metall> microcreep
Mikroskop-Kondensator m <Labor> microscope condenser
Mikroskop-Objektträger m <Labor> microscope slide
Mikroskop-Objektträgerabdeckung f <Labor> microscope slide cover slip
Mikrosonde f <Strahlphys> microprobe
Mikrospritze f <Labor> microsyringe (Gaschromatographie)
Mikrostation f <Funktech, Telekom> very-small aperture terminal, VSAT
Mikrosteuerung f <Kfztech> microcontrol
Mikrostreifen m <Elektronik, Funktech, Phys> microstrip
Mikrostreifenleiterantenne f <Elektronik, Funktech> microstrip aerial, microstrip antenna
Mikrostreifenleitung f <Funktech> microstripline
Mikrostrip m 1. <Elektronik, Funktech, Phys> microstrip; 2. <Telekom> microribbon
Mikrostripantenne f <Elektronik, Funktech> microstrip aerial, microstrip antenna
Mikrostruktur f <Kohlen> microstructure
Mikrotechnik f <Comp & DV, Elektronik> microcircuitry
Mikrotom n <Labor> microtome (Mikroskopie)
Mikrotriebwerk n <Raumfahrt> microrocket
Mikrotron n <Teilphys> microtron
mikroverfilmen v 1. <Comp & DV, Foto> microfilm; 2. <Konstzeich> microcopy, microfilm
Mikroverformung f <Metall> microstrain
Mikrovolt n (μV) <Elektriz, Elektrotech> microvolt (μV)
Mikrowelle f <Elektriz, Elektronik, Funktech, Wellphys> microwave
Mikrowellenabsorption f <Telekom> microwave absorption
mikrowellenabstimmbares Filter n <Elektronik> microwave tunable filter
Mikrowellenantenne f 1. <Funktech> microwave aerial; 2. <Telekom> microwave aerial, microwave antenna, microwave link antenna
Mikrowellen-Bandpassfilter n <Elektronik, Funktech> microwave band-pass filter
Mikrowellenbegrenzer m <Elektronik> microwave limiter
Mikrowellendämpfung f <Elektronik, Funktech> microwave attenuation

Mikrowellen-Dämpfungsglied *n* <Elektronik, Funktech, Telekom> microwave attenuator
Mikrowellendiode *f* <Elektronik> microwave diode
Mikrowellen-Elektronenröhre *f* <Elektronik> microwave tube
mikrowellenfeste Verpackung *f* <Verpack> microwaveable packaging
Mikrowellenfilter *n* <Elektronik, Funktech> microwave filter
Mikrowellenfrequenz *f* <Elektronik, Funktech, Wellphys> extremely high frequency *(EHF)*; microwave frequency, millimeter waves; super high frequency *(SHF)*
Mikrowellengenerator *m* <Elektronik, Funktech> microwave generator
Mikrowellenherd *m* <Elektrotech, Lebensmittel> microwave oven
Mikrowellenhintergrundstrahlung *f* <Wellphys> microwave background radiation
Mikrowellenhohlraum *m* <Elektronik, Funktech> microwave cavity
Mikrowellenleistung *f* <Elektrotech> microwave power
Mikrowellenleistungstransistor *m* <Elektronik> microwave power transistor
Mikrowellenleistungsverstärker *m* <Elektronik, Funktech> microwave power amplifier
Mikrowellenleistungsverstärkung *f* <Elektronik, Funktech> microwave power amplification
Mikrowellenmischer *m* <Elektronik, Funktech> microwave mixer
Mikrowellenmodul *n* <Elektronik, Funktech> microwave module
Mikrowellenmodulator *m* <Funktech, Telekom> microwave modulator
Mikrowellenofen *m* <Elektrotech, Lebensmittel> microwave oven
Mikrowellenoszillator *m* <Elektronik, Funktech> microwave oscillator
Mikrowellenoszillatorröhre *f* <Elektronik> microwave oscillator tube
Mikrowellen-Phasenschieber *m* <Funktech> microwave phase changer
Mikrowellenresonator *m* <Elektronik, Funktech> microwave resonator
Mikrowellenröhre *f* <Maschinen> microwave tube
Mikrowellenröhre *f* **vom Typ M** <Elektronik> M-type microwave tube
Mikrowellenschaltkreis *m* <Elektronik> microwave circuit
Mikrowellenschaltung *f* <Funktech> microwave circuit
Mikrowellensignal *n* <Elektronik, Telekom> microwave signal
Mikrowellensignalgenerator *m* <Elektronik> microwave signal generator
Mikrowellensignalquelle *f* <Elektronik> microwave signal source
Mikrowellenspektroskopie *f* <Wellphys> microwave spectroscopy
Mikrowellenspektrum *n* <Wellphys> microwave spectrum
Mikrowellen-Sperrfilter *n* <Elektronik, Funktech> microwave band-stop filter
Mikrowellenstrahl *m* <Funktech> microwave beam
Mikrowellensubstrat *n* <Elektronik> microwave substrate
Mikrowellentechnik *f* <Funktech, Raumfahrt> microwave technology
Mikrowellen-Tiefpassfilter *n* <Elektronik, Funktech> microwave low-pass filter
Mikrowellen-Trägermaterial *n* <Elektronik> microwave substrate

Mikrowellentransistor *m* <Elektronik> microwave transistor
Mikrowellentransistorverstärker *m* <Elektronik> microwave transistor amplifier
Mikrowellenübertragung *f* <Elektrotech, Telekom> microwave transmission
Mikrowellenverbindung *f* <Wellphys> microwave link
Mikrowellenverstärker *m* <Funktech, Telekom, Wellphys> microwave amplifier
Mikrowellenverstärkung *f* <Funktech> microwave amplification
Mikrowellenverstärkung *f* **durch stimulierte Strahlungsemission** *(Maser)* 1. <Elektronik, Funktech, Telekom> microwave amplification by stimulated emission of radiation, maser; 2. <Raumfahrt> microwave amplification by stimulated emission of radiation, maser *(Weltraumfunk)*
Mikrowellenverzögerungsleitung *f* <Elektronik, Funktech> microwave delay line
Mikrowellenzirkulator *m* <Funktech> microwave circulator
Mikrozelle *f* <Mobilkom> microcell *(zellulares Netz)*
Mikrozwilling *m* <Metall> microtwin
Mikrozwischenstück *n* <Foto> microscope adaptor *(Kamera)*
Milch... <Chemie, Lebensmittel> lactic
Milch-Butyrometer *n* <Lebensmittel> lactobutyrometer
Milcheiweiß *n* <Lebensmittel> milk protein
Milchfett *n* <Lebensmittel> milk fat
Milchglas *n* <Ker & Glas> milk glass, translucent glass
milchige Trübung *f* <Ker & Glas> milkiness
Milchimitat *n* <Lebensmittel> nondairy product
Milchpulver *n* <Lebensmittel> milk powder
Milchsäure *f* <Lebensmittel> lactic acid
Milchsäurenitril *n* <Chemie> hydropropanenitrile, lactonitrile
milchsaures Salz *n* <Chemie> lactate
Milchschleuder *f* <Lebensmittel> cream separator
Milchtanker *m* <Kfztech> milk tanker
Milchzucker *m* <Chemie> lactose
Millefiori *n* <Ker & Glas> glass mosaic
Miller'sche Indizes *mpl* <Metall> Miller indices
Miller'sche-Brücke *f* <Elektriz> Miller bridge
Milli... *(m)* <Metrol> milli... *(m)*
Milliampere *n* <Elektriz, Elektrotech> milliampere
Milli-Amperemeter *n* <Elektrotech> milliammeter
Milliarde *f* <Math> *(AE)* billion; *(BE)* milliard *(veraltet)*; *(BE)* a thousand million
Milligramm *n* <Metrol> milligram
Millikan-Leiter *m* <Elektriz> Millikan conductor
Millikan-Versuch *m* <Phys> Millikan's experiment
Millimeter *m* <Metrol> *(AE)* millimeter, *(BE)* millimetre
Millimeterpapier *n* 1. <Bau> plotting paper; 2. <Math> graph paper
Millimeter-Wanderwellenröhre *f* <Elektronik> *(AE)* millimeter-wave traveling-wave tube, *(BE)* millimetre-wave travelling-wave tube
Millimeterwelle *f* <Elektronik, Phys> millimetric wave, *(AE)* millimeter wave, *(BE)* millimetre wave
Millimeterwellen *fpl (EHF)* <Funktech> extremely high frequency *(EHF)*
Millimeterwellen-Elektronenröhre *f* <Elektronik> *(AE)* millimeter-wave tube, *(BE)* millimetre-wave tube
Millimeterwellen-Magnetron *n* <Elektronik> *(AE)* millimeter-wave magnetron, *(BE)* millimetre-wave magnetron
Millimeterwellen-Quelle *f* <Elektronik> *(AE)* millimeter-wave source, *(BE)* millimetre-wave source
Millimeterwellenröhre *f* <Elektronik> *(AE)* millimeter-wave tube, *(BE)* millimetre-wave tube

Millimeterwellen-Ursprung

Millimeterwellen-Ursprung m <Elektronik> *(AE)* millimeter-wave source, *(BE)* millimetre-wave source
Millimeterwellenverstärker m <Elektronik, Funktech> *(AE)* millimeter-wave amplifier, *(BE)* millimetre-wave amplifier
Millimeterwellenverstärkung f <Elektronik, Funktech> *(AE)* millimeter-wave amplification, *(BE)* millimetre-wave amplification
Million f <Math> million
Million f Elektronenvolt *(MeV)* <Teilphys> million electron volts *(MeV)*
Millionen fpl **Befehle pro Sekunde** *(MIPS)* <Comp & DV> millions of instructions per second *(MIPS)*
Millionen fpl **logischer Inferenzen pro Sekunde** *(MLIPS)* <Künstl Int> millions of logical inferences per second *(MLIPS)*
Millionstel n <Math> millionth
Millisekunde f <Comp & DV> millisecond
Millivolt n <Elektriz, Elektrotech> millivolt
Millivoltbandschreiber m <Gerät> millivolt stripchart recorder
Millivoltmeter n <Elektriz> millivoltmeter
Milliwatt n <Elektriz> milliwatt
Milliwattmeter n <Elektriz> milliwattmeter
MIMD-Rechner m *(Mehrfachbefehlsstrom-Mehrfachdatenstrom-Rechner)* <Comp & DV> MIMD machine *(multiple-instruction multiple-data machine)*
Mindergewicht n <Verpack> short weight
minderwertiges Benzin n <Kfztech> *(AE)* low-test gasoline, *(BE)* low-test petrol
Mindest... <Lufttrans, Maschinen, Qual, Strahlphys, Telekom, Trans> minimum
Mindestabhebegeschwindigkeit f <Lufttrans> minimum unstick speed
Mindestabstandskontrolle f <Trans> control of minimum headway
Mindestannahmewahrscheinlichkeit f <Qual> minimum probability of acceptance
Mindestbestrahlungsdauer f einer Probe <Strahlphys> minimum specimen irradiation
Mindestbetrag m <Telekom> minimum amount
Mindestdrehzahl f <Maschinen> minimum speed
Mindestfreibord n <Wassertrans> minimum required freeboard
Mindestgebühr f <Telekom> initial period charge
Mindestgeschwindigkeit f <Maschinen> minimum speed
Mindestgewicht n <Verpack> minimum weight
Mindesthaltbarkeitsdatum n <Lebensmittel> best before date
Mindestleistungsanforderungen fpl <Qual> minimum grade requirements
Mindestschweißstrom m <Bau> minimum welding current
Mindestsinkhöhe f <Lufttrans> minimum descent altitude, minimum descent height
Mindestwert m <Qual> lower limiting value
Mindestwetterbedingungen fpl <Lufttrans> minimum weather conditions
Mine f <Kohlen, Wassertrans> mine
Minenleger m <Wassertrans> minelayer, minelaying ship *(Marine)*
Minenräumer m <Wassertrans> minesweeper *(Marine)*
Minensuchboot n <Wassertrans> minesweeper *(Marine)*
Mineral n <Kohlen> mineral
Mineralboden m <Kohlen> mineral soil
Mineralisator m <Chemie> mineralizer
mineralisch adj <Kohlen> mineral
mineralische Substanzen entfernen v <Chemtech> demineralize
mineralisoliertes Kabel n <Elektriz, Elektrotech> mineral-insulated cable
Mineralisolierung f <Elektriz> mineral insulation
Mineralmörser m <Labor> percussion mortar *(Schleifen)*
Mineralogie f <Erdöl, Kohlen> mineralogy
Mineralöl n <Kfztech, Kunststoff, Maschinen> mineral oil
Mineralölwirtschaft f <Erdöl> oil industry, petroleum industry
Mineralquelle f <Wasserversorg> mineral spring
mineralstofffrei adj <Umweltschmutz> mineral-matter-free
Minette f <Erdöl> minette *(Geologie)*
Mini... <Aufnahme, Comp & DV, Elektrotech, Optik> mini...
Miniatur f <Comp & DV, Elektronik, Ker & Glas, Maschinen> miniature
Miniaturelektronik f <Elektronik> microelectronics
Miniaturflasche f <Ker & Glas> miniature bottle
Miniaturisierung f <Comp & DV> miniaturization
Miniaturkippschalterreihe f <Elektronik> DIP switch, dual-in-line package switch
Miniaturkugellager n <Maschinen> miniature ball bearing
Miniaturleistungsschalter m <Elektrotech> miniature circuit breaker
Miniaturrelais n <Elektrotech> miniature relay
Miniaturstromschütz n <Elektriz> miniature circuit breaker
Miniatur-Wanderwellenröhre f <Elektronik> *(AE)* miniature traveling-wave tube, *(BE)* miniature travelling-wave tube
Minibündelkabel n <Optik> minibundle cable
Minicomputer m <Comp & DV> minicomputer
Minidisc-Deck n <Telekom> mini disc recorder and player
Minidisc-Diktiergerät n <Aufnahme> MD dictaphone, portable mini disc recorder with micro
Minidiskette f <Aufnahme, Comp & DV> mini disk, mini-floppy disk
Minikassette f <Aufnahme> minicassette
Minileistungsschalter m <Elektrotech> miniature circuit breaker
Minimal... <Comp & DV, Elektriz, Elektrotech, Funktech, Lufttrans, Math, Phys> minimal
minimale Abweichung f <Phys> minimum deviation
minimale Bewicklungsbreite f <Textil> minimum dressed width of warp
minimale fehlerfreie Paketierung/Depaketierung f <Telekom> MEFP, minimum error-free PAD
Minimalflächen fpl <Geom> minimal surfaces
Minimalgebühr f <Telekom> minimum charge
Minimalgleitweg m <Lufttrans> minimum glide path
Minimalleistungsrelais n <Elektriz> minimum power relay
Minimalphasenumtastung f *(MSK)* <Comp & DV, Funktech, Telekom> minimum-shift keying *(MSK)*
Minimalschalter m <Elektrotech> minimum circuit breaker
Minimalspannung f <Elektriz> minimum voltage
Minimax-Thermometer n <Strahlphys> minimum and maximum thermometer
Minimum n <Math> minimum
Minimumbreite f <Funkort> swing *(Funkpeilung)*
Minimumenttrübung f <Funkort> compensation *(Funkpeiler)*
Minimum-Maximum-Thermometer n <Strahlphys> minimum and maximum thermometer
Minimumstromrelais n <Elektriz> minimum current relay
Mini-Netzsystem n <Nichtfoss Energ> mini-grid system
Minirail f <Eisenbahn> minirail
Minirechner m <Comp & DV> minicomputer
Minirillenaufnahme f <Aufnahme> minigroove recording

Minischubboot n <Wassertrans> minipusher tug
Minkowski-Raum m <Phys> Minkowski space
Minorität f <Elektronik, Phys> minority
Minoritätsladungsträger m <Phys> minority carrier
Minoritätsträger m <Elektronik> minority carrier
Minoritätsträgerkanal m <Elektronik> bulk channel *(Transistortechnik)*
minus adj <Math> minus
Minusanschlussklemme f <Kfztech> minus terminal
Minusleiter m <Elektrotech> negative conductor
Minusplatte f <Kfztech> negative plate
Minuspol m 1. <Elektriz, Elektrotech> cathode, negative pole, negative terminal; 2. <Kfztech> minus terminal, negative terminal
Minuszeichen n <Math> minus sign
Minute f <Metrol, Phys> minute *(Einheit des ebenen Winkels)*
Minute f mit verminderter Dienstgüte <Telekom> degraded minute
Minutenzeiger m <Metrol> minute hand
MIPS *(Millionen Befehle pro Sekunde)* <Comp & DV> MIPS *(millions of instructions per second)*
Mirban... <Chemie> mirbane
MIS 1. <Comp & DV> <Management-Informationssystem> MIS *(management information system)*; 2. <Elektronik> *(Metallisolator-Halbleiter)* MIS *(metal insulator semiconductor)*
Misch... <Chemie, Erdöl, Fertig, Papier> mixing
Mischanlage f <Papier> proportioner
Mischapparat m <Chemtech> mixer
mischbar adj 1. <Chemie, Erdöl> miscible; 2. <Fertig> alloyable *(synthetische Harze)*
mischbare Substanz f <Umweltschmutz> miscible substance
Mischbarkeit f <Fertig> alloyability *(synthetische Harze)*
Mischbehälter m <Chemtech> mixing tank, mixing vessel
Mischbodenaushub m <Bau> muck
Mischbohrschlamm m <Erdöl> mixing mud *(Bohrtechnik)*
Mischbütte f <Papier> blending chest, mixing chest
Mischchargenlager n <Ker & Glas> mixed batch store
Mischdauer f <Bau> mixing time
Mischdeponie f <Abfall> co-disposal landfill
Mischdiode f <Elektronik> mixer diode
Mischdüse f 1. <Maschinen> inspirator; 2. <Metall> mixer *(Schweißarbeiten)*
Mischelement n <Chemie> mixed element
mischen v 1. <Anstrich, Aufnahme> mix; 2. <Bau> prepare; mix *(Beton)*; 3. <Comp & DV> merge; 4. <Fernseh> dub up, mix; 5. <Kunststoff, Maschinen> blend; 6. <Metall> alloy *(synthetische Harze)*; 7. <Papier> blend, mix; 8. <Textil> blend *(Fasern)* • **mischen an Ort und Stelle** <Bau> mix in place • **mischen in Stereo** <Aufnahme> mix down
Mischen n 1. <Bau> mixing; 2. <Comp & DV> merge, merging; 3. <Lebensmittel> blending
mischendes Sortieren n <Comp & DV> merge sort
Mischentwässerung f <Wasserversorg> combined sewer system
Mischer m 1. <Aufnahme, Druck, Elektronik> mixer *(Mikrowellen)*; 2. <Fernseh> adder, mixer; 3. <Funktech> mixer; 4. <Ker & Glas> blender; mixer *(Bediener der Mischanlage)*; 5. <Kunststoff> blender, mixer; 6. <Maschinen> mixer; 7. <Papier> blender
Mischer-Vorverstärker m <Elektronik> mixer preamplifier
Mischgarn n <Textil> blend *(Garn)*
Mischgerät n <Lufttrans> mixing unit *(Hubschrauber)*
Mischgestänge n <Lufttrans> mixer rod *(Hubschrauber)*

Mischgewebe n <Textil> union cloth
Mischgewebefärben n <Textil> union dyeing
Mischgitter n <Lufttrans> injection grid
Mischglied n <Gerät> comparing element
Mischheptode f <Elektrotech> pentagrid converter *(Mischröhre mit fünf Gittern)*
Mischkammer f <Kfztech> mixing chamber *(Vergaser)*
Mischkanalisation f 1. <Abfall> combined sewerage system; 2. <Wasserversorg> combined sewer system
Mischkarbid n <Metall> composite carbide
Mischkasten m <Ker & Glas, Papier> mixing box
Mischkniehebel m <Lufttrans> mixer bellcrank *(Hubschrauber)*
Mischkonzentrat-Flotation f <Kohlen> bulk flotation
Mischlicht n <Foto> mixed light
Mischlogikboard n <Elektronik> mixed-logic board
Mischlogikkarte f <Elektronik> mixed-logic board
Mischmaschine f <Bau, Maschinen> mixer
Mischmühle f <Chemtech> mixing mill
Mischpolymerisat n <Chemie, Erdöl, Kunststoff, Textil> copolymer
Mischpolymerisation f <Kunststoff> copolymerization
Mischprobe f <Kohlen> composite sample
Mischpult n 1. <Aufnahme> mixing desk; mixer *(im Studio)*; 2. <Fernseh> mixing desk
Mischraum m 1. <Aufnahme> mixing booth, mixing room; 2. <Ker & Glas> port neck
Mischrohr n <Maschinen> combining cone, combining nozzle, combining tube
Mischröhre f <Elektronik> mixer tube
Mischrührwerk n <Chemtech> agitating mixer
Mischsatz m <Druck> mixed styles
Mischschwebesystem n <Trans> mixed levitation
Mischsortieren n <Comp & DV> merge sort
Mischstrommotor m <Elektrotech> undulating current motor
Mischströmung f <Raumfahrt> mixed flow
Mischstufe f 1. <Elektronik, Elektrotech> converter, mixing stage; 2. <Fernseh, Funktech> mixer; 3. <Raumfahrt> mixer *(Weltraumfunk)*
Mischtafel f <Aufnahme> mixing sheet
Mischtechnik f <Chemtech> mixing technique
Mischtechnologie f <Elektronik> mixed technology
Mischtransistor m <Elektronik> mixing transistor
Mischtrommel f 1. <Fertig, Hydraul> pug mill; 2. <Mechan> batch
Misch- und Ausgabe-Vorratsautomat m <Verpack> mix and dispense storage system
Mischung f 1. <Akustik> mixing; 2. <Anstrich, Fertig> mix; 3. <Kunststoff> blend, mix; compound *(Kautschuk)*; 4. <Lebensmittel> mix; 5. <Maschinen> mixture; 6. <Papier> blend, mixing; 7. <Telekom> grading; 8. <Textil> blend
Mischung f aus Lehm und Sand <Ker & Glas> hogging
Mischungsauflösung f <Fernseh> mix dissolve
Mischungsentwurf m <Bau> mix design
Mischungslücke f <Metall> miscibility gap
Mischungsverhältnis n 1. <Bau> mix proportions; 2. <Kfztech> mixture ratio; 3. <Textil> blend ratio
Mischungswärme f <Thermod> heat of mixing
Mischventil n 1. <Maschinen> mixing valve; 2. <Wasserversorg> combined sewer system
Mischverstärker m <Fernseh> mixer amplifier
Mischverstärkung f 1. <Elektronik> conversion gain *(Elektronenröhren)*; 2. <Funktech> conversion gain
Mischvorgang m <Akustik> mixing
Mischware f <Ker & Glas> mixed ware
Mischwerk n <Chemtech> agitator
Mischzeit f <Bau> mixing time
Mischzyklus m <Bau> mixing cycle

MISD-Rechner

MISD-Rechner m (Mehrfachbefehlsstrom-Einfachdatenstrom-Rechner) <Comp & DV> MISD machine (multiple-instruction single-data machine)
MISFET (Metallisolator-Feldeffekttransistor) <Elektronik> MISFET (metal insulator semiconductor field effect transistor)
Missbrauch m <Patent> abuse
Missfärbung f <Ker & Glas> (AE) discoloration, (BE) discolouration
Mission f **in den tiefen Weltraum** <Raumfahrt> deep-space mission
Mission f **in Erdumlaufbahn** <Raumfahrt> earth-orbiting mission
Missklang m <Akustik> discordance, dissonance
misslingen v <Bau> fail
missweisende Peilung f <Lufttrans> magnetic bearing
missweisender Kurs m 1. <Lufttrans> magnetic heading; 2. <Wassertrans> magnetic course
Missweisung f 1. <Metrol> angle error; 2. <Wassertrans> magnetic declination, magnetic variation
Missweisungsgeber m <Lufttrans> deviation detector
Missweisungswinkel m <Phys> angle of magnetic declination
Mitbenutzer m <Telekom> joint user
miteinander verbinden v <Bau> interlock
miteinander verbunden adj <Telekom> linked together
Mitfällung f <Kerntech> coprecipitation
mitgehende Lünette f <Fertig> follow rest
mitgehender Setzstock m <Maschinen> follow rest
mitgeltende Unterlagen fpl <Qual> referenced documents
mithören v <Telekom> monitor
Mithören n 1. <Funktech> monitoring; 2. <Telekom> listening in; monitoring (Telefon)
Mithörschwelle f <Akustik> masked threshold
Mithörtelefon n <Telekom> observation telephone
Mitkoppeln n <Wassertrans> plotting (Radar)
Mitkopplung f <Aufnahme, Elektronik> positive feedback; regenerative feedback (Radio)
Mitlaufeffekt m <Elektriz> locking effect, pulling effect (Frequenz)
mitlaufen v <Maschinen> follow
mitlaufend adj 1. <Fertig> moving (Setzstock); live (Spitze); 2. <Maschinen> live, revolving, (AE) traveling, (BE) travelling; 3. <Telekom> (AE) traveling, (BE) travelling
mitlaufende Spitze f <Maschinen> (AE) live center, (BE) live centre, (AE) revolving center, (BE) revolving centre
mitlaufender Fehler m <Comp & DV> propagated error
mitlaufender Setzstock m <Maschinen> (AE) traveling stay, (AE) traveling steadyrest, (BE) travelling stay, (BE) travelling steadyrest
Mitläufer m <Textil> wrapper
Mitlaufgenerator m <Elektronik> tracking generator
Mitlaufoszillator m <Funktech> locked oscillator
Mitnahme f 1. <Funktech> locking (Frequenz); pulling-in; 2. <Telekom> pulling-in
Mitnahmebereich m <Elektriz> locking range, pulling-in range, pulling range
Mitnahmeeffekt m <Telekom> capture effect
Mitnahmegenerator m <Elektronik> locked oscillator
Mitnehmer m 1. <Bau> nosing (eines Türriegels); 2. <Elektronik> lock (Oszillator); 3. <Fertig> catch, striker; 4. <Maschinen> carrier, catch, dog, tappet; 5. <Trans> carrier
Mitnehmerbolzen m <Fertig> drive, driver
Mitnehmerkeil m <Fertig> driving key
Mitnehmernut f <Fertig> drive slot
Mitnehmerscheibe f 1. <Fertig> driver plate; 2. <Kfztech> driven plate; 3. <Maschinen> catch plate, driver chuck, driver plate, driving plate

Mitnehmerscheibeneinheit f <Kfztech> driven plate assembly
Mitnehmerspindel f <Maschinen> live spindle
Mitnehmerstange f <Erdöl> kelly (Bohrtechnik)
Mitnehmerstangenlager n <Erdöl> kelly bushing (Bohrtechnik)
Mitnehmerstift m <Maschinen> follower pin
Mitreaktanz f <Elektriz> positive phase-sequence reactance, positive-sequence inductive reactance, positive-sequence reactance
Mitreeder m <Wassertrans> part owner (Schiff)
Mitreißen n <Strömphys> entrainment
mitschwingen v <Fertig> covibrate
Mitschwingen n <Elektronik> resonance
mitsprechen v <Comp & DV> monitor
Mittabstand m (MA) <Fertig, Kfztech, Maschinen> (AE) center distance, (BE) centre distance (CD)
Mittagsbesteck n <Wassertrans> noon sight (astronomische Navigation)
Mitte f <Maschinen> (AE) center, (BE) centre • **von Mitte zu Mitte** <Maschinen> (AE) from center to center, (BE) from centre to centre
Mitte f **des Schiffes** <Wassertrans> midship
Mitte f **Schrifthöhe** <Konstzeich> midheight of the character
Mitteilung f 1. <Funktech> message (Datentransfer); 2. <Patent> communication (der Prüfungsabteilung); 3. <Telekom> message (Datentransfer)
Mitteilungsspeicherung f (MS) <Telekom> message storing (MS)
Mitteilungstransfer m <Telekom> message transfer
Mitteilungstransfersystem n <Telekom> message transfer system
Mitteilungsübermittlung f <Telekom> messaging
Mitteilungs-Übermittlungsdienst m <Telekom> MHS, message handling system
Mitteilungs-Übermittlungssystem n <Telekom> computer-based message system, CBMS
mittel adj <Comp & DV> mean
Mittel n 1. <Comp & DV> medium; 2. <Kohlen> agent; 3. <Math> mean; 4. <Meerschmutz> surface tension modifier (zur Modifizierung der Oberflächenspannung); 5. <Qual> mean
Mittel... 1. <Bau> (AE) center, (BE) centre; 2. <Comp & DV> medium; 3. <Eisenbahn, Elektriz, Fertig, Foto, Kohlen, Lebensmittel, Maschinen, Metall> (AE) center, (BE) centre
Mittelachsensatz m <Druck> (AE) ragged center setting, (BE) ragged centre setting
Mittelanschlussgleis n <Eisenbahn> centre siding
Mittelanzapfung f <Elektriz> (AE) center tap, (BE) centre tap
Mittelbalken m <Wassertrans> midship beam (Schiffbau)
mittelbituminös adj <Kohlen> semibituminous
Mittelblech n <Metall> medium plate
Mitteldeck n <Wassertrans> orlop deck (Schiffkonstruktion)
Mitteldecker m <Lufttrans> midwing plane
Mitteldestillate npl <Erdöl> medium distillates (Destillationsprodukt)
Mitteldruck-Dampfmaschine f <Hydraul> intermediate cylinder steam engine (Dampfmaschine)
Mitteldruckzylinder m <Hydraul> intermediate pressure cylinder (Dampfmaschine)
Mitteleisen n <Metall> medium iron
Mittelfalzhülse f <Verpack> (AE) center folding tubing, (BE) centre folding tubing
Mittelfeld n <Lufttrans> (AE) center panel, (BE) centre panel

Mittelflyer m <Textil> intermediate frame
Mittelfrequenz f (MF) <Elektronik> medium frequency (MF)
Mittelfrequenzband n <Elektronik> medium-frequency band
Mittelfrequenzerwärmung f <Elektrotech> medium-frequency heating
Mittelfrequenzheizung f <Elektrotech> medium-frequency heating
Mittelfrequenzofen m <Elektrotech> medium-frequency furnace
Mittelgang m <Eisenbahn> central gangway
Mittelgerüst n <Fertig> intermediate roll stand (Walzen)
Mittelgut n <Kohlen> finished middlings
Mittelhieb m <Maschinen> bastard cut
Mittelkasten m <Fertig> cheek, intermediate box (Formen)
Mittelkiel m <Wassertrans> (AE) center girder, (BE) centre girder (Schiffbau)
Mittellager n <Kfztech> (AE) center bearing, (BE) centre bearing
Mittellänge f <Druck> x-height, z-height
Mittellängsschott n <Wassertrans> (AE) center line bulkhead, (BE) centre-line bulkhead (Schiffkonstruktion)
Mittellast f <Trans> average load
Mittel-Leistungsverstärker m <Elektronik> medium-power amplifier
Mittel-Leiter m <Elektrotech> neutral conductor (Verteilung)
Mittellinie f 1. <Comp & DV, Geom> median; 2. <Maschinen, Mechan, Raumfahrt> (AE) center line, (BE) centre line
Mittellinienkreuz n <Konstzeich> (AE) center line cross, (BE) centre line cross
Mittellot n <Wassertrans> perpendicular amidships (Schiffkonstruktion)
mittelmäßig verarbeiteter Kanalruß m (MPC-Ruß) <Kunststoff> medium-processing channel carbon black (MPC carbon black)
Mittelmehl n <Lebensmittel> middlings
Mittelmotor m <Kfztech> (AE) center engine, (BE) centre engine, midengine
Mittelmotorfahrgestell n <Kfztech> mid-engine
mitteln v <Gerät> average
Mitteln n **von Signalen** <Elektronik> signal averaging
Mittelöl n <Kohlen> middle oil
Mittelpfosten m <Ker & Glas> mullion
Mittelpufferkupplung f <Eisenbahn> central buffer coupling
Mittelpunkt m 1. <Geom> midpoint; 2. <Maschinen, Telekom> (AE) center, (BE) centre
Mittelpunkterdung f <Elektriz> (BE) midpoint earthing, (AE) midpoint grounding
Mittelpunkttransformator m <Elektrotech> static balancer
Mittelrippe f <Metall> midrib
Mittelrumpflängsträger m <Raumfahrt> waist longeron (Raumschiff)
Mittelsäule f **eines Stativs** <Foto> central column of a tripod
Mittelschaltung f <Kfztech> floor shift (Getriebe)
Mittelschiene f <Eisenbahn> (AE) center rail, (BE) centre rail, middle rail
Mittelschiff n <Wassertrans> midbody (Schiffbau)
mittelschlächtiges Wasserrad n <Hydraul> breast wheel
Mittelschneider m <Fertig, Maschinen> intermediate tap, second tap, tap No2 (Gewinde)
Mittelschneidkante f <Nichtfoss Energ> central splitter edge

mittelschnell laufender Dieselmotor m <Wassertrans> medium-speed diesel engine
Mittelschnellläufer m <Wassertrans> medium-speed diesel engine, medium-speed engine (Dieselmotor)
mittelschwerer Lastwagen m <Kfztech> medium-duty truck
Mittelsenkrechte f <Geom> midperpendicular
Mittelspannung f 1. <Elektrotech> medium voltage; 2. <Metall> mean stress
Mittelspannungssystem n <Elektrotech> medium-voltage system
Mittelspur f <Aufnahme> (AE) center track, (BE) centre track
Mittelstreckenflugzeug n <Lufttrans> medium-range aircraft
Mittelstreckenlinienflugzeug n <Lufttrans> medium-range airliner
Mittelstreckenverkehrsflugzeug n <Lufttrans> medium-range airliner
Mittelstreifen m <Bau> central strip, middle strip, midstrip; medium (Straße)
Mittelteer m <Kohlen> middle tar
Mitteltonlautsprecher m <Aufnahme> midrange loudspeaker
Mittelung f <Strömphys> averaging (in turbulenter Strömung)
Mittelwalze f <Metall> middle roll
Mittelwelle f (MW) 1. <Fernseh> medium wave; 2. <Funktech> medium wave (MW)
Mittelwellen fpl (MW) <Funktech> medium frequency (MF)
Mittelwellenbereich m <Funktech> medium-wave band
Mittelwellenfrequenz f <Elektriz> medium frequency (MF)
Mittelwellenfrequenzband n <Funktech> medium-frequency band
Mittelwert m 1. <Comp & DV> mean, mean value, median; 2. <Elektriz> average value; 3. <Math> average, mean, mean value; 4. <Phys> average, average value, mean value • **Mittelwert bilden** <Gerät, Math> average
Mittelwertbildung f <Gerät> signal averaging (bei Signalen)
Mittelwertdetektor m <Elektrotech> average detector
mittelwertiger Boden m <Kohlen> medium-graded soil
Mittelzapfen m <Bau> king rod
Mittelzeitverhalten n <Kerntech> (AE) medium-term behavior, (BE) medium-term behaviour
Mittel-Zweck-Analyse f (MEA) <Künstl Int> means-end analysis, MEA (Methode der Problemlösung)
Mittenabgleich m <Fernseh> (AE) centering control, (BE) centring control
Mittenabstand m <Maschinen> (AE) distance between centers, (BE) distance between centres
Mittenbereich m <Qual> mean range
Mittenfrequenz f <Elektronik, Fernseh, Funktech> (AE) center frequency, (BE) centre frequency
mittengespeister Horizontaldraht m <Funktech> (AE) center-fed horizontal wire
Mittenkontakt m <Foto> hot-shoe flash contact
Mittenlinie f <Maschinen> (AE) line of centers, (BE) line of centres
Mittenloch n <Maschinen> (AE) center hole, (BE) centre hole
Mittenrichtigkeit f <Mechan> concentricity
mittige Last f <Bau> concentric load (Statik)
mittlere Abweichung f 1. <Elektriz, Elektronik> average deviation; 2. <Metall> standard deviation; 3. <Qual> mean deviation
mittlere aerodynamische Flügeltiefe f <Lufttrans> mean aerodynamic chord (Tragflügel)

mittlere

mittlere aerodynamische Sehne f <Lufttrans> mean aerodynamic chord *(Tragflügel)*
mittlere alkalische radioaktive Abfälle mpl <Kerntech> alkaline medium-level radioactive waste
mittlere Anomalität f <Raumfahrt> mean anomaly
mittlere Aufenthaltszeit f <Qual> mean abode time
mittlere Belastung f <Elektriz> average load
mittlere Belegungsdauer f <Telekom> mean holding time
mittlere Bewertungsnote f <Telekom> MOS, mean opinion score
mittlere Bindungsenergie f <Kerntech> mean bond energy
mittlere Bremsleistung f <Lufttrans> brake mean effective pressure
mittlere Durchgangsrate f <Telekom> average crossing rate
mittlere fehlerfreie Betriebszeit f <Comp & DV> MTBF, mean time between failures
mittlere Fehlerquote f <Qual> average error rate, average fault rate
mittlere Flügeltiefe f der Steuerfläche <Lufttrans> mean chord of the control surface
mittlere freie Diffusionsweglänge f <Kerntech> diffusion mean free path
mittlere freie Weglänge f <Akustik, Elektronik, Metall, Phys, Thermod> mean free path
mittlere Hauptverkehrsstunde f <Telekom> mean busy hour
mittlere Instandsetzungsdauer f <Telekom> mean time to repair
mittlere Instandsetzungszeit f *(MTTR)* <Mechan> mean time to repair *(MTTR)*
mittlere Integrationsdichte f *(MSI)* <Comp & DV> medium-scale integration *(MSI)*
mittlere Integrationstechnik f *(MSI)* <Comp & DV, Elektronik> medium-scale integration *(MSI)*
mittlere jährliche Änderung f <Wassertrans> mean annual variation *(Gezeiten)*
mittlere Lasche f der Höhenflosse <Lufttrans> *(AE)* horizontal stabilizer center fishplate, *(BE)* horizontal stabilizer centre fishplate
mittlere Lebensdauer f 1. <Comp & DV, Elektrotech> median life rate, median life time; 2. <Kerntech> mean life; 3. <Phys> average life, mean life
mittlere Leistung f <Telekom> average power
mittlere letale Dosis f 1. <Strahlphys> MLD, mean lethal dose; 2. <Strahlphys> median lethal dose
mittlere Letalkonzentration f <Umweltschmutz> median lethal concentration
mittlere Letalzeit f <Umweltschmutz> median lethal time
mittlere Luftspaltinduktion f <Elektriz> magnetic loading
mittlere Monatsdosis f <Umweltschmutz> average monthly dose
mittlere Ortszeit f <Trans> local mean time
mittlere Qualitätslage f <Qual> process average
mittlere Rauigkeit f <Maschinen> average roughness
mittlere Rautiefe f <Maschinen> average depth
mittlere Reparaturdauer f *(MTTR)* <Comp & DV, Qual, Raumfahrt> mean time to repair *(MTTR)*
mittlere Schallausstrahlung f <Akustik> average speech power *(Sprache)*
mittlere Schallintensität f <Akustik> sound power
mittlere Seehöhe f <Wassertrans> mean sea level
mittlere Sonnenzeit f <Raumfahrt> mean solar time
mittlere Spannweite f <Qual> mean range
mittlere Sprachleistung f <Akustik> average speech power

mittlere Steinkohle f <Kohlen> cob coal
mittlere Stichprobenumfangskurve f <Qual> average sample number curve
mittlere Strukturbreite f <Elektronik> average width of feature
mittlere Tageswassermenge f <Wasserversorg> mean daily flow
mittlere veranschlagte Zeit f zwischen zwei Ausfällen <Qual> assessed mean time between failures
mittlere Verfügbarkeit f <Regelung> average availability
mittlere Verzögerungszeit f <Elektronik> average propagation delay
mittlere Wartezeit f <Trans> average delay
mittlere Wassersäule f <Nichtfoss Energ> medium head
mittlere Windgeschwindigkeit f <Nichtfoss Energ> mean wind speed
mittlere Zeit f bis zum Ausfall <Telekom> mean time to failure
mittlere Zeit f bis zum ersten Ausfall <Qual, Telekom> mean time to first failure *(MTTFF)*
mittlere Zeit f bis zur Reparatur *(MTTR)* <Elektrotech> mean time to repair *(MTTR)*
mittlere Zeit f bis zur Überholung <Telekom> mean time to restore
mittlere Zeit f zur Wiederherstellung des betriebsfähigen Zustands <Qual> mean time to restore
mittlere Zeit f zwischen Ausfällen <Mechan> MTBF, mean time between failures
mittlere Zeit f zwischen Wartungsarbeiten <Qual> mean time between maintenance
mittlere Zeitdauer f zwischen Ausfällen <Raumfahrt> MTBF, mean time between failures
mittlere Zeitdauer f zwischen Entnahmen <Raumfahrt> MTBR, mean time between removals *(Ausbau)*
mittlerer Abweichungsbetrag m <Qual> mean deviation
mittlerer Anstellwinkel m <Lufttrans> mean pitch angle
mittlerer Ausfallabstand m <Comp & DV, Elektrotech, Telekom> MTBF, mean time between failures
mittlerer Druck m <Maschinen> mean pressure
mittlerer Durchmesser m für Bezugsoberfläche <Optik> average reference surface diameter
mittlerer Fehler m <Elektriz, Gerät> average error, mean error
mittlerer Fehleranteil m der Fertigung <Qual> process-average defective
mittlerer Gesamtfehler m <Comp & DV> mean square error
mittlerer Gleitwegfehler m <Lufttrans> mean glide path error
mittlerer Integrationsgrad m *(MSI)* <Comp & DV, Elektronik, Telekom> medium-scale integration *(MSI)*
mittlerer Kernabbrand m <Kerntech> core average burn-up
mittlerer Kerndurchmesser m <Optik, Telekom> average core diameter
mittlerer Manteldurchmesser m 1. <Optik> average cladding diameter *(Lichtleitfaser)*; 2. <Telekom> average cladding diameter
mittlerer Nutzdruck m *(MEP)* 1. <Lufttrans> mean effective pressure, mep *(Triebwerk)*; 2. <Maschinen> mean effective pressure, mep
mittlerer Prüfumfang m <Qual> average amount of inspection
mittlerer quadratischer Fehler m <Comp & DV> mean square error *(Präzisionsmaß für eine Schätzfunktion)*
mittlerer Rauschfaktor m <Elektronik> average noise factor
mittlerer Server m <Comp & DV> mid-range server

mittlerer Steigungswinkel m <Lufttrans> mean pitch angle
mittlerer Temperaturunterschied m <Heiz & Kälte> mean temperature difference
mittlerer Tidehub m <Nichtfoss Energ> mean tidal range
mittlerer Tiefgang m <Wassertrans> (AE) mean draft, (BE) mean draught (Schiffkonstruktion)
mittlerer Wartungsabstand m <Telekom> mean time between maintenance
mittleres Fehlerquadrat n <Elektriz> mean error square value
mittleres Infrarot n <Strahlphys> middle infrared
mittschiffs adj <Wassertrans> midship
mittschiffs adv <Wassertrans> amidships
Mittschiffsebene f <Wassertrans> (AE) center plane, (BE) centre plane (Schiffbau)
Mittschiffslinie f <Wassertrans> (AE) center line, (BE) centre line (Schiffbau); fore-and-aft line (Schiffkonstruktion)
Mitwiderstand m <Elektrotech> positive phase-sequence resistance
Mitzieheffekt m 1. <Elektriz> frequency pulling, locking effect, pulling effect; 2. <Funktech, Telekom> frequency pulling
Mix- und Püriergerät n <Verpack> mixing and blending equipment
Mizelle f <Kunststoff> micelle
MKSA-System n <Metrol> Giorgi system, Giorgi system of units, MKSA system, (AE) meter-kilogram-second-ampere system, (BE) metre-kilogram-second-ampere system
MLIPS (Millionen logischer Inferenzen pro Sekunde) <Künstl Int> MLIPS (millions of logical inferences per second)
MMI (Mensch-Maschine-Interface, Mensch-Maschine-Schnittstelle) <Comp & DV, Kontroll, Künstl Int, Raumfahrt> MMI (man-machine interface)
MMK (magnetomotorische Kraft) <Elektriz, Phys, Strömphys> mmf (magnetomotive force)
MMT (Methylcyclopentadienyl-Mangantricarbonyl) <Kfztech> MMT, Methylcyclopentadienyl Manganese Tricarbonyl (Kraftstoffzusatz)
mn (Neutronenmasse) <Kerntech, Strahlphys, Teilphys> mn (neutron mass)
Mn (Mangan) <Chemie> Mn (manganese)
MN (Kernmasse) <Kerntech> MN (nuclear mass)
Mnemonik f <Comp & DV, Künstl Int> mnemonics
mnemonisch adj <Comp & DV, Künstl Int> mnemonic
mnemonischer Code m <Comp & DV> mnemonic code
mnemonischer Name m <Comp & DV> mnemonic name
mnemonisches Symbol n <Comp & DV> mnemonic symbol
mnemotechnisch adj <Comp & DV, Künstl Int> mnemonic
Mo (Molybdän) <Chemie> Mo (molybdenum)
Möbeltischlerei f <Bau> cabinet-making
Möbelwagen m <Trans> (AE) removal truck, (BE) removal van
mobil adj <Mobilkom, Telekom> mobile
mobile Bodenstation f <Telekom> mobile earth station
mobile Hochgeschwindigkeits-Datenübertragung f <Mobilkom> high speed mobile data
mobile Kamera f <Fernseh> mobile camera
mobile Satellitenfernsehempfangsanlage f <Fernseh> mobile satellite TV receiving station
mobile Zahlung f <Mobilkom> mobile payment
mobiler Feuerlöscher m <Sicherheit> mobile fire extinguisher
mobiler Satellitenfunk m <Raumfahrt> mobile satellite communications (Kommunikation)
mobiler Satellitenfunkdienst m <Fernseh, Funktech> mobile communications by satellite, mobile satellite service, mobile satellite system (MSS)
mobiler Teilnehmer m <Mobilkom> mobile subscriber
mobiles Bezahlen n <Mobilkom> mobile payment
mobiles Einkaufen n <Telekom> mobile commerce
mobiles Funkmodem n <Mobilkom> portable mobile radio modem
mobiles Netzteil n <Nichtfoss Energ> mobile power pack
mobiles Satellitenfunksystem n <Fernseh, Funktech, Mobilkom> mobile satellite system (MSS)
Mobilfunkgerät n <Mobilkom> mobile, mobile transceiver
Mobilfunkkanal m <Funktech, Mobilkom> mobile radio channel
Mobilfunkkanal m **mit CDMA** <Mobilkom> code multiple access channel, traffic adapted channel (UMTS)
Mobilfunkkanal mit Schwunderscheinungen m <Mobilkom> fading mobile radio channel
Mobilfunknetz n <Mobilkom> mobile radio network
Mobilfunknetz n **C** <Mobilkom> car telephone system C, C-net, mobile telephone network C
Mobilfunknetzzelle f <Mobilkom> mobile radio cell, mobile radio network cell
Mobilfunksendeempfangsgerät n <Mobilkom> mobile, mobile transceiver
Mobilfunkstandard m **für das D- und E-Netz** <Mobilkom> GSM standard
Mobilfunkstationskontrolleinrichtung f <Mobilkom> base station controller (BSC)
Mobilfunkstelle f <Funktech, Mobilkom> mobile radio station
Mobilfunksystem n **der USA** <Mobilkom> advanced mobile phone system (AMPS)
Mobilfunktarif m <Mobilkom> mobile radio call charge rate, mobile radio tariff
Mobilfunkteilnehmer m <Mobilkom> mobile radio subscriber, mobile user, wireless subscriber
Mobilfunkteilnehmernummer f <Mobilkom> mobile subscriber number
Mobilfunk-Vermittlungsstelle f <Mobilkom> (AE) mobile services switching center, (BE) mobile services switching centre, MSC
Mobilfunkvermittlungsstelle f **mit Netzübergang** <Mobilkom> gateway mobile services switching center (GMSC)
Mobilfunkzelle f <Mobilkom> mobile radio cell, radio cell
Mobilfunkzellen-Bündel n <Mobilkom> mobile radio cellular cluster
Mobilkommunikation f **über Satelliten** <Mobilkom> mobile communications by satellite
Mobilkommunikationssystem n **der dritten Generation** <Mobilkom> Third Generation Mobile Communication System (TGMCS); universal mobile telecommunications system (UMTS)
Mobilkommunikationssystem n **zwischen Flugpassagier und Boden** <Mobilkom> terrestrial flight telecommunication system (TFTS)
Mobilometer n <Metrol> mobilometer (zur Festigkeitsbestimmung von Kunststoffen)
Mobilstation f (MS) <Mobilkom> mobile station (MS); vehicular station
Mobilstationsaufenthaltsnummer f <Mobilkom> mobile station roaming number
Mobiltelefon n <Mobilkom> mobile phone, mobile telephone
Mobiltelefondienst m <Mobilkom> mobile telephone service
Mobiltelefonregister n <Mobilkom> equipment identity register

Möbiusband n <Geom> Möbius strip
möblieren v <Bau> furnish
Modacryl n <Textil> modacrylic
Modalmix m <Wassertrans> intermodalism *(Transport und Logistik)*
Modalnoten fpl <Akustik> modal notes
Modalwert m <Geom> mode
Mode f <Optik> mode *(Schwingung)*
Modell n 1. <Comp & DV> model; 2. <Elektronik> pattern; 3. <Fertig> pattern *(Formguss)*; chuck *(Metalldrücken)*; 4. <Maschinen> copy, pattern; 5. <Telekom, Wassertrans> model *(Schiffbau)*
Modell n **unabhängiger Teilchen** <Phys> independent particle model *(Kernphysik)*
Modell... <Comp & DV> model
Modellarm m <Lufttrans> hold *(Luftfahrzeug)*
Modellaufspanntisch m <Maschinen> pattern table
Modellausschmelzgießen n <Fertig> investment casting
Modelldruck m <Textil> block printing
Modelldruckmaschine f <Druck> block printing machine
Modelleichung f <Umweltschmutz> model calibration
Modellformstoff m <Fertig> investment compound *(Modellausschmelzverfahren)*
Modellholz n <Fertig> pattern lumber *(Gießen)*
Modellieren n <Comp & DV, Elektronik, Math> *(AE)* modeling, *(BE)* modelling
Modellplatte f <Fertig> match plate *(Formen)*; pattern plate *(Gießen)*
Modellsand m <Fertig> facing sand *(Gießen)*
Modellschraube f <Fertig> lifting screw *(Gießen)*
Modellspitze f <Fertig> draw spike *(Formen)*
Modelltisch m <Maschinen> copyholder
Modelltischlerei f <Fertig> pattern shop
Modellversuch m <Wassertrans> model test *(Schiffbau)*
Modem n *(Modulator-Demodulator)* <Comp & DV, Elektronik, Funktech, Telekom> modem *(modulator-demodulator)* • **Modem betriebsbereit** <Comp & DV> modem-ready
Modem n **auf Leiterplattenebene** <Elektronik, Telekom> board-level modem
Modem-Anschluss m <Comp & DV, Elektronik> modem interface
Modembuchse f <Comp & DV> data jack
Modem-Eliminator m <Telekom> modem eliminator, null modem
Modem-Empfänger m <Comp & DV, Elektronik> modem receiver
Modem-Interface n <Comp & DV, Elektronik> modem interface
Modem-Schnittstelle f <Comp & DV, Elektronik> modem interface
Modem-Sender m <Comp & DV, Elektronik> modem transmitter
Modem-Wechsel m <Comp & DV, Elektronik> modem interchange
Modenabstreifer m 1. <Optik> mode stripper; 2. <Telekom> cladding mode stripper, mode stripper
Modendispersion f 1. <Optik> modal dispersion; 2. <Telekom> intermodal distortion, intramodal distortion, modal distortion, mode distortion, multimode distortion
Modendispersionsnull f <Telekom> zero intermodal dispersion *(Faseroptik)*
Modenfelddetektor m <Telekom> mode field detector
Modenfelddurchmesser m <Optik> mode field diameter
Modenfilter n <Optik, Telekom> mode filter
Modenkopplung f <Optik, Telekom> mode coupling
Modenmischer m <Optik, Telekom> mode mixer, mode scrambler
Modenrauschen n <Optik, Telekom> modal noise

Modenscrambler m <Optik, Telekom> mode mixer, mode scrambler
Modenspringen n <Optik, Telekom> mode hopping, mode jumping
Modensprung m <Optik, Telekom> mode hopping, mode jumping
Modenstripper m 1. <Optik> mode stripper; 2. <Telekom> cladding mode stripper, mode stripper
Modenumfang m <Optik> mode volume
Modenverteilung f **bei Ungleichgewicht** <Optik> non-equilibrium mode distribution
Modenverteilung f **im stationären Zustand** <Optik> equilibrium mode distribution
Modenverteilungslänge f **im stationären Zustand** <Optik> equilibrium mode distribution length
Modenvolumen n <Telekom> mode volume
Moderator m <Kerntech, Phys, Strahlphys> moderator
Moderator-Brennstoff-Verhältnis n <Kerntech> moderator-fuel ratio
Moderator-Spaltstoff-Verhältnis n <Kerntech> moderator-fuel ratio
Moderatortrimmung f <Kerntech> moderator control
moderieren v <Kerntech> moderate
moderne Schrift f <Druck> modern face
moderne Signalverarbeitung f <Elektronik> advanced signal processing
modernes Bauelement n <Gerät> advanced component
Modifikation f <Fertig> allotrope *(Stoff)*
modifizieren v <Comp & DV, Geom> modify
Modifizierer m <Kohlen> modifier
modifizierte Frequenzmodulation f *(MFM)* <Elektronik> modified frequency modulation *(MFM)*
modifizierte Stärke f <Lebensmittel> modified starch
modifiziertes System n <Textil> modified system
Modul m 1. <Funktech> module; 2. <Math> modulus *(absoluter Betrag für spezielle komplexwertige Funktionen)*; modulus *(Teiler bei einer Restklasse)*; 3. <Telekom> module, package
Modul n <Comp & DV, Elektrotech, Telekom> module
Modul n **der internationalen Raumstation** <Raumfahrt> international space station module, ISS module
modular adj <Comp & DV, Fertig> modular
modulare Arithmetik f <Comp & DV> modular arithmetic
modulare Einheit f <Maschinen> modular unit
modulare Erweiterungsmöglichkeit f <Telekom> scalability
modulare Installation f <Comp & DV> modular installation
modulares Führen n <Elektrotech> modular guidance *(Magnetschwebetechnik)*
modulares Messsystem n <Metrol> *(AE)* modular gaging system, *(BE)* modular gauging system
modulares Oberflächenreinigungsmittel n <Verpack> modular surface cleaner
modulares Programmieren n <Comp & DV> modular programming
modulares Rechnen n <Comp & DV> modular arithmetic
modulares Werkzeugsystem n <Fertig> modular tool system
Modularität f <Comp & DV> modularity
Modularprinzip n <Fertig> modular principle
Modulation f 1. <Akustik, Aufnahme, Comp & DV, Elektriz, Elektronik, Funktech> modulation; 2. <Raumfahrt> modulation *(Weltraumfunk)*; 3. <Telekom> modulation; 4. <Wellphys> modulation *(einer Welle)*
Modulation f **mit freiem Träger** <Elektronik> floating-carrier modulation
Modulation f **mit konstanter Amplitude** <Elektronik> constant-amplitude modulation

Modulation f **mit unabhängigen Seitenbändern** <Elektronik, Funktech> independent sideband modulation
Modulationsart f <Funktech> type of modulation
Modulationsband n <Elektronik> modulation band
Modulationsfaktor m <Phys> modulation factor
Modulationsfrequenz f <Elektronik, Funktech, Telekom> modulation frequency
Modulationsgitter n <Fernseh> modulation grid
Modulationsgrad m 1. <Elektronik> modulation depth, modulation factor; 2. <Phys> modulation depth
Modulationshüllkurve f <Aufnahme> modulation envelope
Modulationsindex m <Elektronik, Funktech> modulation index
Modulationskarte f <Comp & DV, Elektronik> modem board
Modulationsrauschen n <Akustik, Aufnahme, Elektronik, Fernseh, Telekom> modulation noise
Modulationssignal n <Aufnahme, Elektronik> modulating signal
Modulationsstufe f <Funktech> hopping period
Modulationstiefe f <Elektronik> modulation depth, modulation factor
Modulationsübertragung f <Elektronik> remodulation (von einem Träger auf einen anderen)
Modulations-Übertragungsfunktion f <Elektronik> modulation transfer function
Modulationsverstärker m <Elektronik> modulation amplifier
Modulationswelle f <Elektronik> modulating wave
Modulationswinkel m <Akustik> modulation angle
Modulator m 1. <Comp & DV> modulator; 2. <Elektronik> modulator (A/D-Wandler für Datensignale); 3. <Fernseh, Telekom> modulator
Modulator m **mit Null-Vorspannung** <Funktech> zero-bias modulator
Modulator-Demodulator m (Modem) <Comp & DV, Elektronik, Funktech, Telekom> modulator-demodulator (modem)
Modulatordiode f <Elektronik> modulator diode
Modulatortreiberstufe f <Elektronik> modulator driver
Modulbau m <Bau> segmental construction
Modulfräser m <Maschinen> module milling cutter
modulieren v <Aufnahme, Elektronik, Fernseh, Funktech, Phys, Telekom> modulate
modulierende Welle f <Funktech> modulating wave
modulierendes Signal n <Funktech> modulating wave
modulierte Rille f <Akustik> modulated groove
modulierte Schwingung f <Elektronik> modulated oscillation
modulierte Struktur f <Metall> modulated structure
modulierte Welle f <Aufnahme, Elektronik> modulated wave
modulierter Dauerstrichträger m <Funktech> MCW, modulated continuous wave
modulierter Oszillator m <Elektronik> modulated oscillator
modulierter Strahl m <Elektronik> modulated beam
modulierter Träger m <Elektronik, Fernseh, Funktech, Telekom> modulated carrier
moduliertes Signal n <Telekom> modulated signal
Modulkehrwert m <Fertig> diametral pitch
Modulon-Kontrolle f <Comp & DV> residue check
Modulsatz m <Elektronik> module set
Modulwälzfräser m <Fertig> module hob
Modum m 1. <Comp & DV, Elektriz, Elektronik, Erdöl> module (Teil einer Offshore-Plattform); 2. <Fertig> module (Kunststoffinstallationen); 3. <Kontroll> module; 4. <Maschinen> module, modulus

Modus m 1. <Comp & DV, Druck, Elektronik> mode; 2. <Math> mode, modal value (Modalwert)
Modusanzeiger m <Comp & DV> mode indicator
Modusauswahl f <Comp & DV> mode selection
Modusbeschreibung f <Comp & DV> mode description
Modusverteilung f <Comp & DV> mode distribution
Moduswechsel m <Comp & DV> mode change, mode switching
Mofette f <Nichtfoss Energ> mofette
Mohr'scher Quetschhahn m <Labor> Mohr's clip (Gummischlauch)
Mohr'scher Spannungs- und Trägheitskreis m <Fertig> Mohr's circles
Mohs'sche Härteskale f <Fertig> Mohs' scale
Moiré n <Fernseh> moiré
Moiré-Muster n 1. <Akustik> moiré pattern; 2. <Phys> moiré fringe
Moiré-Störung f <Fernseh> moiré (FS-Bild)
Mol n <Metrol> gram molecule
mol (Mole) <Chemie, Labor, Phys> mol (mole)
molal adj <Chemie> molal
molare Gaskonstante f <Phys> molar gas constant
molare Wärmekapazität f <Phys> molar heat capacity
molares Brechungsvermögen n <Thermod> molecular refractivity
Molarität f <Chemie> molarity
Molbruchzahl f <Metall> mole fraction
Molch m 1. <Erdöl> go-devil, pig, scraper; scraper (Gerät zum Reinigen von Rohrleitungen); 2. <Fertig> scraper (Kunststoffinstallationen)
Mole f 1. <Bau> breakwater, pier; 2. <Chemie, Metrol> mole, mol; 3. <Phys> mole, mol (Einheit der Stoffmenge); 4. <Wassertrans> jetty, mole, pier (Hafen)
Molekül n 1. <Chemie> molecule; 2. <Erdöl> molecule (Petrochemie); 3. <Phys> molecule • **innerhalb eines Moleküls** <Chemie> intramolecular
Molekular... <Elektronik, Kerntech, Phys, Thermod> molecular
Molekulardichte f (n) <Phys, Thermod> molecular density (n)
molekulare Depression f <Phys, Thermod> molecular depression of freezing point
molekulare Gefrierpunkterniedrigung f <Phys, Thermod> molecular depression of freezing point
molekulare Leitfähigkeit f <Phys, Thermod> molecular conductivity
molekulare Siedepunkterhöhung f <Phys, Thermod> molecular elevation of boiling point
molekulare Spektralanalyse f <Strahlphys> molecular spectroanalysis
molekulare Umlaufbahn f <Phys> molecular orbital
Molekularelektronik f <Elektronik> molecular electronics
molekulares Brechungsvermögen n <Phys> molecular refractivity
molekulares Feld n <Phys, Thermod> molecular field
molekulares Schwingungsniveau n <Strahlphys> molecular vibrational energy level
Molekulargewicht n (M) 1. <Erdöl> molecular weight, M (Petrochemie); 2. <Phys, Thermod> molecular weight, M
Molekularkräfte fpl <Metall> intermolecular forces
Molekularmasse f (m) <Phys> molecular mass (m)
Molekularpumpe f <Mechan, Phys> kinetic vacuum pump
Molekularstrahlepitaxie f (MBE) <Elektronik, Strahlphys> molecular-beam epitaxy (MBE)
Moleküldurchmesser m <Phys> diameter of molecule
Molekülgaslaser m <Elektronik> molecular gas laser
Moleküllaser m <Elektronik> molecular laser
Molekülspektrum n <Phys, Thermod> molecular spectrum

Molekülstrahl

Molekülstrahl *m* <Telekom> molecule beam
Molekülzahl *f (N)* <Phys> number of molecules *(N)*
Molenkopf *m* <Wassertrans> pierhead *(Hafen)*
Möller *m* 1. <Fertig> ore and flux; 2. <Metall> burden
möllern *v* <Metall> burden
Molvolumen *n* <Phys> molar volume
Molwärme *f* <Phys, Thermod> molecular heat
Molybdän *n (Mo)* <Chemie> molybdenum *(Mo)*
Molybdändisulfid *n* <Maschinen> *(AE)* molybdenum disulfide, *(BE)* molybdenum disulphide
Molybdänkarbid *n* <Fertig> molybdenum carbide
Molybdänstahl *m* <Metall> molybdenum steel
Molybdat *n* <Chemie> molybdate, tetraoxomolybdate
Moment *n* <Maschinen, Mechan> moment *(Kräftepaar)*
Momentan... <Maschinen> instantaneous
Momentanachse *f* <Maschinen> instantaneous axis
Momentananzeige *f* 1. <Gerät> instantaneous display; 2. <Phys> live display
momentane Dichte *f* **der kinetischen Energie** <Akustik> instantaneous kinetic energy
momentane Dichte *f* **der potenziellen Energie** <Akustik> instantaneous potential energy
momentane Dichte *f* **der Schallenergie** <Akustik> instantaneous acoustic energy
momentane Schalleistung *f* 1. <Akustik> instantaneous sound power; 2. <Phys> sound-energy flux
momentane Schallintensität *f* <Akustik> instantaneous sound power
momentane Sprachleistung *f* <Akustik> instantaneous speech power
momentane Sprechleistung *f* <Ergon> instantaneous speech power
momentaner Druckanstieg *m* <Fertig> surge
momentaner Nachführungsfehler *m* <Raumfahrt> instantaneous tracking error *(Weltraumfunk)*
momentaner Schalldruck *m* <Akustik> instantaneous sound pressure
Momentanfrequenz *f* <Funktech> instantaneous frequency
Momentanfrequenzmessung *f* <Funktech> instantaneous frequency measurement
Momentangeschwindigkeit *f* <Fertig> instantaneous velocity
Momentanpol *m* <Fertig> *(AE)* instant center, *(BE)* instant centre *(Getriebelehre)*
Momentanspannung *f* <Elektriz> instantaneous voltage
Momentanstrom *m* <Elektriz, Elektrotech> instantaneous current
Momentanwert *m* <Elektriz, Fertig, Gerät, Phys> instantaneous value
Momentaufnahme *f* <Comp & DV> snapshot
Momentenausgleich *m* <Maschinen> balancing of moments
Momentenbeiwert *m* <Maschinen> moment coefficient
Momentenkurve *f* <Maschinen> moment curve
Momentenlinie *f* <Maschinen> moment line
Momentgleichrichter *m* <Kfztech> torque rectifier *(Getriebe)*
Momentrelais *n* <Elektrotech> instantaneous relay
monadisch *adj* <Comp & DV> monadic, unary
monadische Operation *f* <Comp & DV> monadic operation
Mond *m* <Raumfahrt> moon
Möndchen *n* <Geom> meniscus
Mondflutintervall *n* <Nichtfoss Energ> lagging of the tide
Mondlandeeinheit *f* <Raumfahrt> lunar module
Mondlandefähre *f (LEM)* <Raumfahrt> lunar excursion module, LEM *(Apollo-Raumschiff)*
Mondlandefahrzeug *n (LEM)* <Raumfahrt> lunar excursion module, LEM *(Apollo-Raumschiff)*

Mondlandung *f* <Raumfahrt> lunar landing
Mondsonde *f* <Raumfahrt> lunar probe
Mondtag *m* <Raumfahrt> lunar day
Mondumlaufbahn *f* <Raumfahrt> lunar orbit
Mondversorgungsfahrzeug *n* <Raumfahrt> LLV, lunar logistics vehicle
Monitor *m* 1. <Aufnahme, Comp & DV, Elektriz> monitor; 2. <Fernseh> VDU, monitor, video display unit; 3. <Kontroll> VDU, monitor; 4. <Phys> VDU; 5. <Strahlphys> monitor
Monitoraufruf *m* <Comp & DV> monitor call
Monitorcode *m* <Comp & DV> monitor code
Monitorklasse *f* <Comp & DV> monitor class
Monitormodus *m* <Comp & DV> monitor mode
Monoacetin *n* <Chemie> monoacetin
Monoamid *n* <Chemie> monoamide
Monoamin *n* <Chemie> monoamine
Monoamino... <Chemie> monoamino
monoatomar *adj* <Chemie> monoatomic
Mono-Beschleunigungsanode *f* <Elektronik> monoaccelerator CRT
Monoblock-Betonschwelle *f* <Eisenbahn> *(BE)* monobloc concrete sleeper, *(AE)* monobloc concrete tie
Monoblockrad *n* <Eisenbahn> monobloc wheel, solid wheel
Monobrombenzol *n* <Chemie> mono-bromobenzene
Monochord *n* <Akustik> monochord
monochrom *adj* <Comp & DV, Druck, Foto> monochrome
monochromatisch *adj* <Optik, Phys> monochromatic
monochromatische Strahlung *f* <Optik, Strahlphys, Telekom> monochromatic radiation
monochromatisches Licht *n* <Wellphys> monochromatic light
Monochromator *m* <Optik, Phys, Telekom> monochromator
Monochromsignal *n* <Raumfahrt> monochrome signal *(Weltraumfunk)*
monoenergetisch *adj* <Phys> monoenergetic
Monoethylamin *n* <Chemie> aminoethane, ethylamine
Monofil *n* <Kunststoff> monofilament
Monofilament *n* <Kunststoff> monofilament
Monofilamentgarn *n* <Textil> monofilament yarn
Monofilgarn *n* <Textil> monofilament yarn
Monohydrat *n* <Chemie> monohydrate
Monohydrogen... <Chemie> monohydric
Monohydroxyketon *n* <Chemie> ketol
Monoklinsystem *n* <Metall> monoclinic system
monokrystallines Silizium *n* <Raumfahrt> monocrystalline silicon *(Raumschiff)*
monolithisch integrierte Schaltung *f* <Elektronik, Telekom> monolithic integrated circuit
monolithischer Mikrowellenschaltkreis *m* <Funktech, Phys> monolithic microwave integrated circuit
monolithischer Verstärker *m* <Elektronik> monolithic amplifier
monolithisches Feld *n* <Elektronik> monolithic array
monolithisches Filter *n* <Elektronik, Telekom> monolithic filter
Monom *n* <Math> monomial
Monomer *n* 1. <Chemie> monomer; 2. <Erdöl> monomer *(Petrochemie)*; 3. <Kunststoff> monomer
monomer *adj* <Kunststoff> monomeric
monomere Substanz *f* <Chemie> monomer
Monomethylamin *n* <Chemie> aminomethane, methylamine
Monomethyl-Aminophenolsulfat *n* <Chemie> *(AE)* methylaminophenol sulfate, *(BE)* methylaminophenol sulphate, metol

Monomodefaser f 1. <Optik> (AE) monomode fiber, (BE) monomode fibre; 2. <Telekom> (AE) single mode fiber, (BE) single mode fibre

Monomode-LWL-Übertragungssystem n <Telekom> (AE) mono-mode optical fiber transmission system, (BE) mono-mode optical fibre transmission system, (AE) single-mode fiber transmission system, (AE) single-mode fibre transmission system

monomolekular adj <Chemie> monomolecular, unimolecular

monomolekulare Reaktion f <Chemie> monomolecular reaction

Monomolekularfilm m <Chemie> monolayer, monomolecular layer

Mononatriumglutamat n <Lebensmittel> MSG, monosodium glutamate (Geschmacksstoff)

monophone Aufzeichnung f <Akustik> monophonic recording

monophoner Abtaster m <Akustik> monophonic pick-up

monophonische Aufnahme f <Aufnahme> monophonic recording

Monopol m <Funktech, Teilphys> monopole

Monopolantenne f 1. <Funktech> monopole aerial, monopole antenna; 2. <Telekom> unipole aerial, unipole antenna

Monoschicht f <Chemie> monolayer, monomolecular layer

Monosilan n <Chemie> monosilane, silicomethane

monostabil adj <Comp & DV, Elektronik> monostable

monostabile Kippschaltung f 1. <Elektronik> monostable multivibrator, one-shot multivibrator; 2. <Phys> monostable multivibrator

monostabile Schaltung f <Elektronik> one-shot circuit

monostabiler Multivibrator m <Elektronik> monostable multivibrator

Monostearat n <Chemie> monostearin

Monotaste f <Fernseh> monokey

monoton adj <Math> monotone (Eigenschaft reellwertiger Folgen, Funktionen, Reihen)

monotones Schließen n <Künstl Int> monotonic reasoning

Mono-Tonsystem n <Aufnahme> monophonic sound system

Monotreibstoff m <Chemie, Raumfahrt> monopropellant

Monotron n <Elektronik> monotron (Bildröhre für Prüfzwecke)

monotrope Reaktion f <Metall> monotropic reaction

Monotype® f <Druck> Monotype®

Monotypegießmaschine f <Druck> Monotype casting machine

monovalent adj <Chemie> monovalent, univalent

Monoverstärker m <Aufnahme> monoamplifier

Monovinylacetylen n <Chemie> vinylacetylene

Monoxid n <Chemie> monoxide

Monsun m <Wassertrans> monsoon

Montage f 1. <Aufnahme> editing; 2. <Bau> assembly; 3. <Elektronik> mounting; 4. <Fernseh> editing; 5. <Fertig> erection, mounting; 6. <Maschinen> assembly, mounting; 7. <Mechan> assembly; 8. <Papier> assembling; 9. <Trans> assembly

Montage f eines optischen Kabels <Telekom> optical cable assembly

Montage f oberhalb des Reaktorbodens <Kerntech> upper shell assembly

Montageautomat m <Maschinen> automatic assembly machine

Montageband n <Comp & DV, Fertig> flow line

Montagebauweise f <Bau> dry construction

Montagebetrieb m <Maschinen> assembly plant

Montageblock m <Aufnahme> editing block

Montagebock m <Maschinen> assembly jig

Montagefehler m <Fertig> installation error

Montagefolie f <Druck> mounting foil

Montagegerät n 1. <Fertig> assembly device; 2. <Maschinen> assembly machine, fitting device

montagegerechte Konstruktion f <Qual> design for assembly, DFA

Montagehalle f 1. <Mechan> assembly shop, erecting shop; 2. <Wassertrans> assembly hall, erecting shop (Schiffbau)

Montagekleber m <Druck> mounting glue

Montageloch n <Maschinen> mounting hole

Montagemaße npl <Fertig> assembly dimensions (Kunststoffinstallationen)

Montageniet m <Maschinen> field rivet, site rivet

Montageplatte f 1. <Fertig> mounting plate (Kunststoffinstallationen); 2. <Funktech> mounting plate; 3. <Maschinen> mounting base

Montagepunkt m <Trans> assembly point

Montageroboter m <Mechan> assembly robot

Montageroller m <Maschinen> dolly

Montageschraube f <Maschinen> assembling bolt, fitting bolt, mounting bolt

Montageschweißen n <Kerntech> field weld

Montagesteg m <Kerntech> catwalk

Montagestraße f <Mechan> assembly line

Montagewerkstatt f <Mechan> assembly shop

Montagewerkzeuge npl <Maschinen> assembly tools

Montagezeichnung f <Mechan> assembly drawing

Monte-Carlo-Methode f <Comp & DV> Monte-Carlo-method, random walk method

Monte-Carlo-Verfahren n <Comp & DV> Monte-Carlo-method

Montejus m <Chemtech> acid elevator

Monteur m 1. <Fertig> erector; 2. <Kfztech, Maschinen> assembler, fitter; 3. <Mechan> erector, fitter; 4. <Wassertrans> erector

montieren v 1. <Bau> assemble, fit, set up; 2. <Comp & DV, Fertig, Funktech> mount; 3. <Mechan> erect; 4. <Papier> assemble; 5. <Wassertrans> fit (Schiffteile)

montierte Zeichnung f <Konstzeich> assembled drawing

montierter Transformator m/am Mast <Elektriz> pole-mounted transformer

Montierung f <Fertig> mount

Montmorillonit m 1. <Erdöl> montmorillonite (Mineral); 2. <Kohlen> montmorillonite

Mooney-Anvulkanisationsdauer f <Kunststoff> Mooney scorch time

Mooney-Viskosität f <Kunststoff> Mooney viscosity

Moorboden m <Bau, Kohlen> swampy soil

Moosgummi n <Kunststoff> microcellular rubber

Moräne f <Kohlen> moraine

Moränenfilterschicht f <Abfall> morainic filter layer (Deponie)

Morin n <Chemie> morin

Morindin n <Chemie> morindin

Morpholin n <Chemie> morpholine, tetrahydrooxazine

Morphotropie f <Chemie> morphotropism

Morsecode m <Funktech> Morse code

Morsekegel m <Maschinen> Morse taper

Morsekegelstift m <Maschinen> Morse taper pin

Mörser m <Chemtech, Fertig, Labor> mortar

Mörserkeule f <Chemtech> pestle

Morsetaste f <Funktech> key

Mörtel m 1. <Bau> plaster; 2. <Fertig> mortar

Mörtelbett n <Bau> mortar

Mörtelmischmaschine f <Bau> mortar mixer

MOS

MOS 1. <Comp & DV> *(Metalloxid-Halbleiter)* MOS *(metal oxide semiconductor)*; 2. <Elektronik> *(Metalloxid-Halbleiter, Metalloxid-Transistor)* MOS *(metal oxide semiconductor)*
Mosaik *n* <Elektronik> mosaic
Mosaikdrucker *m* <Comp & DV> matrix printer
Mosaikkrankheit *f* <Lebensmittel> mosaic *(Pflanzenkrankheitslehre)*
Mosaikstein *m* <Ker & Glas> tessera
Moseley'sches Gesetz *n* <Phys> Moseley's law
MOSFET *(Metalloxid-Silizium-Feldeffekttransistor)* <Funktech> MOSFET *(metal oxide silicon field effect transistor)*
MOS-Gatter *n* <Elektronik> MOS gate
MOS-Kondensator *m* <Elektronik> MOS capacitor
MOS-Laufzeitkette *f* <Elektronik> MOS delay line
MOS-Leistungstransistor *m* <Elektronik> MOS power transistor
MOS-Logikschaltkreis *m* <Elektronik> MOS logic circuit
Mößbauer'scher Effekt *m* <Phys> Mössbauer effect
Most *m* <Lebensmittel> stum
MOS-Transistor *m* <Elektronik> MOS transistor
MOS-Treiber *m* <Elektronik> MOS driver
Motor *m* 1. <Elektriz, Elektrotech> motor; 2. <Kfztech> engine; 3. <Maschinen> engine, motor; 4. <Mechan, Wassertrans> engine • **den Motor absaufen lassen** <Kfztech> *(AE)* flood the carburetor, *(BE)* flood the carburettor • **den Motor voll ausfahren** <Maschinen> run an engine to its full capacity, work an engine to its full capacity
Motor *m* **für Flüssigtreibstoff** <Thermod> liquid fuel engine
Motor *m* **mit Doppelschlusswicklung** <Elektrotech> compound-winding motor
Motor *m* **mit Eigenkühlung** <Kfztech> ventilated motor
Motor *m* **mit einstellbarer Drehzahl** <Elektriz> adjustable varying speed motor
Motor *m* **mit Gangschaltung** <Mechan> geared motor
Motor *m* **mit gewickeltem Stator** <Elektrotech> wound-stator motor
Motor *m* **mit gleich bleibendem Schluckvolumen** <Fertig> constant-capacity motor
Motor *m* **mit hängenden Ventilen** <Kfztech> overhead valve engine
Motor *m* **mit hohem Anzugsmoment** <Mechan> high-torque motor
Motor *m* **mit Kompensationswicklung** <Elektriz> compensated motor, compound motor, compound-wound motor
Motor *m* **mit Kondensator für Anlauf und Betrieb** <Elektrotech> capacitor start-run motor
Motor *m* **mit phasengewickeltem Läufer** <Elektriz> phase-wound rotor motor
Motor *m* **mit Querstromspülung** <Kfztech> three-port two-stroke engine
Motor *m* **mit Schleifringläufer** <Elektrotech> wound-rotor motor, slip-ring motor
Motor *m* **mit Selbstzündung** <Kfztech> compression-ignition engine
Motor *m* **mit stehenden Ventilen** <Kfztech> L-head engine
Motor *m* **mit stehenden Zylindern** <Kfztech> vertical engine
Motor *m* **mit stellbarer Geschwindigkeit** <Elektriz> speed adjustable motor
Motor *m* **mit T-förmigem Verbrennungsraum** <Kfztech> T-head engine
Motor *m* **mit Turboaufladung** <Kfztech, Wassertrans> turbocharged engine
Motor *m* **mit übereinander angeordneten Ventilen** <Kfztech> F-Head engine
Motor *m* **mit Ventilen zu beiden Seiten** <Kfztech> T-head engine
Motor *m* **mit veränderlicher Drehzahl** <Elektrotech> speed variable motor
Motor *m* **mit zwei Drehzahlen** <Elektriz> double-speed motor
Motor *m* **mit Zylindern in V-Anordnung** <Maschinen> V-cylinder engine
motorangetrieben *adj* <Elektrotech> motor-driven
Motoranker *m* <Elektriz> motor armature
Motoranlasser *m* 1. <Elektrotech> motor starter; 2. <Kfztech, Lufttrans, Wassertrans> engine starter
Motorantrieb *m* <Elektrotech, Foto> motor drive
Motorantriebmechanismus *m* <Elektriz> motor drive mechanism
Motoraufhängung *f* <Kfztech> engine support, engine support lug
Motoraufzug *m* <Trans> powered lift
Motorauspuffsystem *n* <Maschinen> engine exhaust system
motorbetriebenes Abschöpfgerät *n* <Meerschmutz> self-propelled skimmer *(für Öl)*
Motorblock *m* <Kfztech, Maschinen, Mechan, Wassertrans> engine block
Motorboot *n* <Wassertrans> motorboat
Motorbremse *f* <Kfztech> engine brake; exhaust brake *(Bremsanlage)*
Motorbrennkammer *f* <Maschinen> engine combustion chamber
Motordrehmoment *n* <Kfztech, Lufttrans> engine torque
Motordrehzahl *f* <Kfztech> engine speed
Motordrehzahlaufnehmer *m* <Kfztech> engine speed pick-up
Motoreinstellung *f* <Kfztech> tuning
Motorenabgasanlage *f* <Kfztech, Wassertrans> engine exhaust system
Motorenauslaufzeit *f* <Lufttrans> engine coasting-down time
Motorenbenzin *n* 1. <Erdöl> motor spirit; 2. <Kfztech> *(AE)* gas, *(AE)* gasoline, *(BE)* petrol
Motorenfundament *n* <Mechan> engine pedestal
Motorengehäusebogen *m* <Lufttrans> engine support arch
Motorenhalterungsbogen *m* <Lufttrans> engine support arch
Motorenkühlanlage *f* <Wassertrans> engine cooling system
Motorenöl *n* <Erdöl> motor oil
Motorenprüfstand *m* <Kfztech, Lufttrans> engine test stand
Motorensteuerzentrale *f* <Kontroll> *(AE)* motor control center, *(BE)* motor control centre
Motorfähre *f* <Wassertrans> motor ferry
Motorfahrzeug *n* <Wassertrans> motor vessel
Motorgebläse *n* <Kfztech> engine fan
Motorgehäuse *n* <Kfztech> crankcase
Motorgenerator *m* <Elektrotech> motor generator
Motorgeneratorsatz *m* <Elektrotech> motor-generator set
Motorgetriebeeinheit *f* <Kfztech> power unit
motorgetriebene Pumpe *f* <Lufttrans> engine-driven pump
Motorhaube *f* <Kfztech> *(BE)* engine bonnet, *(AE)* engine hood; *(BE)* bonnet, *(AE)* hood *(Karosserie)*
Motorhaubenverriegelung *f* <Kfztech> *(BE)* bonnet catch, *(AE)* hood catch
Motorhubraum *m* 1. <Kfztech> engine capacity; 2. <Maschinen> capacity of an engine
motorische Aktivität *f* <Ergon> motor activity

motorisierter Förderwagen m <Kfztech> powered barrow
motorisierter Handkarren m <Kfztech> powered barrow
Motorkurbel f <Maschinen> engine crank
Motorlager n <Maschinen, Wassertrans> engine bearing
Motorleistung f <Heiz & Kälte> motor rating
Motorluftanlage f <Kfztech, Wassertrans> engine ventilation system
Motornachlauf m <Kfztech> engine second rating
Motoröl n 1. <Erdöl> motor oil; 2. <Kfztech, Maschinen, Wassertrans> engine oil
Motorpumpe f <Meerschmutz> motor pump
Motorradvergaser m <Kfztech> (AE) carburetors for motorcycles, (BE) carburettors for motorcycles
Motorrahmen m <Maschinen> engine frame
Motorraum m <Kfztech> engine compartment
Motorregler m <Wassertrans> engine governor
Motorrüstung f <Lufttrans> engine instruments
Motorschaden m 1. <Kfztech> engine failure, engine malfunction; engine breakdown (Motor); 2. <Wassertrans> engine failure, engine malfunction
Motorschalldämpfer m <Maschinen> (AE) engine muffler, (BE) engine silencer
Motorschiff n <Wassertrans> motor ship
Motorschwungrad n <Maschinen> engine flywheel
Motorsegler m <Wassertrans> auxiliary engine sailing ship, motor sailer
Motorspritze f <Meerschmutz> motor pump
Motorsteuerung f 1. <Kfztech> timing gear; 2. <Telekom> motor control
Motorstraßenhobel m <Bau> motor grader (Straßenbau)
Motorträger m <Lufttrans> engine mount
Motorträgerschraube f <Kfztech> engine support plug
Motortriebwagen m <Eisenbahn> rail coach
Motorüberholung f <Maschinen, Wassertrans> engine overhaul
Motorventil n <Maschinen> engine valve
Motorventilator m <Kfztech> engine fan
Motorverkleidung f <Lufttrans> cowling
Motorwagen m <Eisenbahn> rail motor car
Motorwartung f <Kfztech, Wassertrans> engine maintenance
Motorwelle f <Maschinen> engine shaft
moussierend adj <Lebensmittel> effervescent
mp (Protonenmasse) <Kerntech> mp (proton mass)
MP 1. <Comp & DV, Elektriz, Maschinen> (Mikroprozessor) MP (microprocessor); 2. <Verpack> (Metallpapier) MP (metallic paper)
MPC-Ruß m (mittelmäßig verarbeiteter Kanalruß) <Kunststoff> MPC carbon black (medium-processing channel carbon black)
MPTN-Architektur f <Comp & DV> Multiprotocol Transport Networking architecture (MPTN)
MS 1. <Mobilkom> (Mobilstation) MS (mobile station); 2. <Telekom> (Mitteilungsspeicherung) MS (message storing)
MSC (Funkvermittlungsstelle) <Mobilkom> MSC (mobile switching center, mobile switching centre)
M-Schale f <Phys> M-shell (Atomphysik)
MSI 1. <Comp & DV> (mittlere Integrationsdichte, mittlere Integrationstechnik, mittlerer Integrationsgrad) MSI (medium-scale integration); 2. <Elektronik> (mittlere Integrationsdichte, mittlerer Integrationsgrad, mittlerere Integrationsdichte) MSI (medium-scale integration); 3. <Telekom> (mittlerer Integrationsgrad, mittlerere Integrationsdichte) MSI (medium-scale integration)
MSI-Schaltkreis m <Telekom> MSI circuit
MSK 1. <Comp & DV, Funktech, Telekom> (Minimalphasenumtastung) MSK (minimum-shift keying); 2. <Elektronik> (kleinste Umtastung) MSK (minimum-shift keying)

MTL-Logik f (integrierte Transistorlogik) <Elektronik> MTL (merged transistor logic)
MTR (Materialprüfreaktor) <Kerntech> MTR (materials-testing reactor)
MT-Ruß m <Kunststoff> MT carbon black, medium thermal carbon black (mittlerer Thermalruß)
MTTR 1. <Comp & DV, Qual, Raumfahrt> (mittlere Reparaturdauer) MTTR (mean time to repair); 2. <Elektrotech> (mittlere Zeit bis zur Reparatur) MTTR (mean time to repair); 3. <Mechan> (mittlere Instandsetzungszeit) MTTR (mean time to repair)
mu (Atommassenkonstante) <Kerntech> mu (unified atomic mass constant)
Mü (Maschinenübersetzung) <Künstl Int> MT (machine translation)
Mucin n <Chemie> mucin
Mucoitinschwefel m <Chemie> (AE) mucoitinsulfur, (BE) mucoitinsulphur
Mucon... <Chemie> muconic
Mucoproteid n <Chemie> mucoprotein
Mucoprotein n <Chemie> mucoprotein
Muffe f 1. <Bau> socket; 2. <Fertig> bell, hose; 3. <Maschinen> collar, coupling sleeve, muff, sleeve, socket; 4. <Mechan> clamp, sleeve; 5. <Raumfahrt> gland (Raumschiff); 6. <Telekom> joint (Kabel)
Muffel f <Fertig, Ker & Glas, Metall> muffle
Muffelkühlofen m <Ker & Glas> muffle lehr
Muffelofen m <Fertig, Labor> muffle furnace
Muffelstütze f <Ker & Glas> muffle support
Muffengrund m <Fertig> root (Kunststoffinstallationen)
Muffenkupplung f <Maschinen> box coupling, butt coupling, muff coupling, sleeve coupling
Muffenrohr n 1. <Bau, Fertig> socket pipe; 2. <Maschinen> (AE) bell and spigot pipe, (BE) socket and spigot pipe
Muffenrohrverbindung f <Bau> spigot joint
Muffenschweißen n <Fertig> socket fusion jointing (Kunststoffinstallationen)
Muffenstück n <Fertig> flange
Muffenverbindung f 1. <Bau> spigot and socket joint, tailpiece; 2. <Maschinen> sleeve joint, socket joint
Muffenverbindungsrohre npl <Bau> spigot and socket joint pipes
Mühle f <Lebensmittel> mill; grinder, grinding mill (Mahlanlage); crusher (Quetschmühle)
Mühlenindustrie f <Lebensmittel> milling industry
Mühlenstaub m <Lebensmittel> mill dust
Mühlgerinne n <Wasserversorg> mill course, millrace
Mühlrad n <Maschinen> millwheel
Mühlstein m <Lebensmittel> millstone
Mulde f 1. <Erdöl> syncline; 2. <Fertig> basin, spherical depression; 3. <Labor, Maschinen> trough
Mulden... adj <Erdöl, Fertig> synclinal
Muldenkipper m 1. <Bau> dumper; 2. <Eisenbahn> skip wagon; 3. <Kfztech> dump truck, (BE) skip lorry, (AE) skip truck
Muldenkippwagen m <Meerschmutz> (BE) skip lorry, (AE) skip truck
Muldenofen m <Thermod> crucible furnace
Muldenpresse f <Textil> rotary press for press finishing
Müll m 1. <Bau> waste; 2. <Comp & DV, Verpack> (AE) garbage, (BE) rubbish; 3. <Wasserversorg> refuse • **Müll rein, Müll raus** <Comp & DV> garbage in, garbage out (GIGO)
Müllabfuhr f <Abfall> collection service, (AE) garbage collection, (BE) refuse collection, (AE) garbage disposal, (BE) refuse disposal, refuse collection service, waste collection
Müllabfuhrwagen m <Abfall, Kfztech> (BE) dust cart, (AE) garbage truck

Müllabladen

Müllabladen n <Umweltschmutz> dumping
Müllabladeplatz m <Umweltschmutz> dump ground
Müllablagerung f <Wasserversorg> refuse tipping
Müllabwurfschacht m <Abfall> refuse chute
Müllanfall m <Abfall> waste formation, waste production, waste stream
Müllaufbereitung f <Umweltschmutz> conditioning of waste
Müllaufgabetrichter m <Abfall> hopper-furnace feed chute
Müllballen m <Abfall> waste bale
Müllbehälter m <Abfall> (AE) garbage can, (BE) rubbish bin, waste container
Müllbeseitigung f <Umweltschmutz> waste disposal
Müllbeutel m 1. <Abfall> (AE) garbage bag, (BE) rubbish bag; 2. <Verpack> refuse sack
Müllbrennstoff m <Abfall> RDF, refuse-derived fuel
Müllbunker m <Abfall> receiving bunker
Müllcontainer m <Abfall> roll-out container, skip
Mülldeponie f <Abfall, Umweltschmutz> disposal facility, dumping site, landfill, sanitary landfill, waste disposal site, waste site
Mülldeponiegas n <Nichtfoss Energ, Umweltschmutz> landfill gas
Mülleimer m <Abfall> (AE) garbage can, (BE) rubbish bin
Mülleinfülltrichter m <Abfall> loading hopper
Müller-Brücke f <Metrol> Mueller bridge (zur Messung von Vierpolwiderständen)
Müllfahrzeug n 1. <Abfall> collection body, collection vehicle, (BE) dust cart, (AE) garbage truck, refuse body, refuse collection vehicle, waste collection vehicle; 2. <Kfztech> (BE) dust cart, (AE) garbage truck
Mullins-Effekt m <Kunststoff> Mullins effect
Müllkippe f 1. <Abfall> dump site, dumping site, waste dump, waste tip; 2. <Bau> disposal site; 3. <Umweltschmutz> refuse disposal site
Müllsack m <Abfall> (AE) garbage bag, (BE) rubbish bag, refuse sack
Müllsammelfahrzeug n <Abfall> collection vehicle
Müllsammlung f <Abfall> (AE) garbage collection, (BE) refuse collection
Müllschacht m <Bau, Umweltschmutz> waste chute
Müllschlacke f <Abfall> clinker, slag
Müllschlucker m <Abfall> garbage unit, refuse chute, waste unit
Müllsortierungsanlage f <Abfall> (AE) garbage sorting plant, refuse separation plant, refuse sorting plant, (BE) sorting plant
Mülltonne f <Abfall> (AE) garbage can, (BE) rubbish bin, waste container
Müllumladeanlage f <Abfall> transfer station
Müllumschlagstation f <Abfall> transfer station
Müllverbrennung f <Abfall> refuse incineration, trash burning, waste burning
Müllverbrennungsanlage f (MVA) 1. <Abfall> (AE) garbage incineration plant, (BE) refuse incineration plant, waste incineration plant, waste incinerator; 2. <Verpack> (AE) garbage incinerator, (BE) refuse incinerator
Müllverbrennungsofen m <Thermod> destructor
Müllverdichter m <Abfall> landfill compactor, refuse compactor, packer unit
Müllverdichtung f <Abfall> compaction, waste compaction
Müllverwertungsanlage f <Umweltschmutz> garbage recycling plant, refuse recycling plant, waste recovery plant
Müllwagen m 1. <Abfall> collection body, collection vehicle, (BE) dust cart, (AE) garbage truck, refuse body, refuse collection vehicle; 2. <Kfztech> (BE) dust cart, (AE) garbage truck, refuse collection lorry

Müllzerkleinerer m <Abfall> (AE) garbage grinder, (BE) refuse grinder
Multi... <Comp & DV, Elektrotech, Künstl Int, Optik, Telekom, Textil> multi...
Multibussystem n <Comp & DV> multibus system
multidimensionale Stellelemente npl <Ergon> multidimensional controls
Multifaserkabel n <Optik> (AE) multifiber cable, (BE) multifibre cable
Multifilamentgarn n <Textil> multifilament yarn
Multifilamentmaschine f <Textil> multifilament machine
Multifilgarn n <Textil> multifilament yarn
Multikomponentenfaser f <Optik> (AE) multicomponent glass fiber, (BE) multicomponent glass fibre (Glasfaser aus mehreren Komponenten)
Multimedia-PC m <Comp & DV, Telekom> PC media player
Multimeter n <Elektrotech, Fernseh> multimeter
Multimodefaser f <Telekom> (AE) multimode fiber, (AE) multimode optical fiber, (BE) multimode fibre, (BE) multimode optical fibre
Multimodelaser m <Telekom> multimode laser
multioktav abstimmbarer Oszillator m <Elektronik> multioctave-tunable oscillator
multioktav abstimmbares Filter n <Elektronik> multioctave-tunable filter
Multioktavabstimmung f <Elektronik> multioctave tuning
Multiplaybacktechnik f <Akustik> multiplayback
Multiplett n 1. <Phys> multiplet (Spektrallinien); 2. <Teilphys> multiplet
Multiplettaufspaltung f <Kerntech> multiplet splitting
Multiplex n 1. <Raumfahrt> multiplex (Weltraumfunk); 2. <Telekom> multiplex
Multiplexbetrieb m <Comp & DV, Elektronik> multiplex operation
multiplexen v <Elektronik, Telekom> multiplex
Multiplexen n <Comp & DV, Telekom> multiplexing
Multiplexen n **mit Zeitteilung** (TDM) <Comp & DV, Elektronik, Telekom> time division multiplex (TDM)
Multiplexen n **von Signalen** <Elektronik> signal multiplexing
Multiplexer m (MUX) <Comp & DV, Elektronik, Fernseh, Telekom> multiplexer (MUX)
Multiplexfrequenz f <Elektronik, Funktech, Telekom> multiplexing frequency
multiplexieren v <Elektronik, Telekom> multiplex
Multiplexieren n <Telekom> multiplexing
Multiplexkanal m 1. <Comp & DV> multiplexer channel, multiplexor channel; 2. <Elektronik> multiplex channel
Multiplexkarton m <Papier> multilayer board
Multiplexleitung f <Comp & DV> (BE) bus, highway, (AE) trunk
Multiplexmodus m <Comp & DV> multiplex mode
Multiplexsendung f <Fernseh> multiplex transmission
Multiplexübertragung f <Telekom> multiplex transmission
Multiplexverfahren n <Comp & DV, Elektronik> multiplex, multiplexing
Multiplikand m <Comp & DV> multiplicand
Multiplikation f <Comp & DV, Math> multiplication
Multiplikationskonstante f **für infinite Systeme** (k) <Kerntech> multiplication constant for an infinite system (k)
Multiplikationstabelle f <Math> multiplication table
Multiplikationszeichen n <Math> multiplication sign
multiplikativ adj <Math> multiplicative
Multiplikator m 1. <Comp & DV> multiplier; 2. <Math> multiplicand; 3. <Telekom> multiplicator
multiplizieren v <Comp & DV> multiply

multiplizieren v mit <Math> multiply by
Multipliziergerät n <Comp & DV> multiplier
Multiplizität f <Phys> multiplicity
Multipol m <Phys> multipole
Multiprocessing n <Comp & DV> multiprocessing
Multiprotokoll-Router m <Telekom> multiprotocol router
Multiprotokollunterstützung f <Comp & DV> multiprotocol support
Multiprozessor m <Comp & DV> multiprocessor
Multiprozessor-Server m <Comp & DV> multi-processor server
Multiprozessorsystem n <Comp & DV, Telekom> multiprocessor system
Multiscan-Monitor m <Comp & DV> multiscan monitor, multisync monitor
Multitasking n <Comp & DV> multitasking
multivariat adj <Ergon> multivariate
Multivibrator m <Funktech, Phys, Telekom> multivibrator
Multizellularlautsprecher m <Akustik> multicellular loudspeaker
Mumetall n <Phys> mumetal
Mundblasen n <Ker & Glas> mouth blowing
mundgeblasenes Glas n <Ker & Glas> hand-blown glass
mündliche Verhandlung f <Patent> oral proceedings
Mundloch n <Ker & Glas> port
Mundlochschürze f <Ker & Glas> port apron
Mundstück n 1. <Bau> nozzle; 2. <Ker & Glas> die; 3. <Telekom> mouthpiece
Mündung f 1. <Hydraul> orifice *(Unterwasser)*; 2. <Ker & Glas> orifice; 3. <Wassertrans> mouth *(Geographie)*
Mündungsgebiet n <Wassertrans> estuary *(Geographie)*
Munitionskasten m <Wassertrans> caisson *(Marine)*
Münzer m <Telekom> payphone
Münzfernsehen n <Fernseh> subscription television, subscription TV
Münzfernsprecher m <Telekom> coin box telephone, coin-operated telephone, coin payphone, coin telephone
Münzgold n <Metall> gold bullion
Münzrelais n <Elektriz> coin box relay *(bei Telefonsystem)*
Münzsilber n <Metall> silver bullion
Münztelefon n <Telekom> coin box telephone, coin-operated telephone, coin payphone, coin telephone
Münzzähler m <Gerät> prepayment meter, slot meter
Müon n <Phys> muon *(Elementarteilchen)*
Müonenzerfallsspuren fpl <Strahlphys> muon decay tracks
Müonneutrino n <Phys> muon neutrino *(Elementarteilchen)*
Murexid n <Chemie> murexide
Muring f <Wassertrans> mooring *(Festmachen)*
Muringgeschirr n <Wassertrans> mooring gear *(Festmachen)*
Muscarin n <Chemie> muscarine
Muschelantenne f <Funktech> shell antenna *(asymmetrische Spiegelantenne)*
muschelförmige Blisterpackung f <Verpack> clamshell blister
muscheliger Bruch m <Ker & Glas> conchoidal fracture
Muschelschale f <Verpack> clamshell
musikalisches Intervall n <Wellphys> musical interval *(zwischen zwei Noten)*
Musikautomat m <Optik> jukebox
Musikbelastbarkeit f <Aufnahme> music-power-handling capacity
Musik-CD f <Optik> *(BE)* CD audio disc, *(AE)* CD audio disk, *(BE)* compact audio disc
Musik-CD-Spieler m <Optik> CD audio player

Muskatblütenöl n <Lebensmittel> mace oil
Muskeleiweiß n <Chemie> myosin
Muskelfibrin n <Chemie> syntonin
Muskelstärke f <Ergon> muscular strength
Muster n 1. <Comp & DV> model, pattern, template, templet; 2. <Elektronik> pattern; 3. <Kohlen> sample, specimen; 4. <Kunststoff, Labor> specimen; 5. <Maschinen> copy, specimen; 6. <Mechan> specimen; 7. <Patent> design *(gewerbliches Muster, Geschmacksmuster)*; 8. <Qual> specimen; 9. <Telekom> model; 10. <Textil> pattern
Muster n **der Bruchfläche** <Mechan> breaking pattern
Musterbuch n <Druck> swatchbook
Mustererfassung f <Elektronik> pattern registration
Mustererkennung f 1. <Comp & DV> pattern recognition; 2. <Elektronik> pattern recognition; patterning *(Lithographie)*; 3. <Künstl Int> pattern recognition, PR
Mustererzeugung f <Elektronik> pattern generation
Musterklassifizierung f <Künstl Int> pattern classification
Musterlänge f <Textil> pattern length
Musterlos n <Qual> pilot lot
Musterprüfung f <Werkprüf> sampling
Musterrolle f <Wassertrans> muster roll *(Schiffsdokument)*
Mustersortierpumpe f <Labor> sampling pump
Musterung f <Fernseh, Textil> patterning
Mustervergleich m <Comp & DV, Künstl Int> pattern matching
Musterzeichnung f <Textil> pattern
Mutter f 1. <Akustik> mother; 2. <Fertig, Kfztech, Maschinen, Mechan> nut
Mutterboden m <Bau, Kohlen> topsoil
Mutterbodenabtrag m <Bau, Kohlen> topsoil stripping
Muttererde f <Kohlen> topsoil
Mutterfrequenz f <Elektronik> master frequency
Muttergestein n 1. <Erdöl> bedrock, source rock *(Geologie)*; 2. <Kohlen> bedrock
Mutterkorn n <Lebensmittel> ergot *(Getreidekrankheit)*
Mutterkornalkaloid n <Chemie> ergot alkaloid
Mutterkristall m <Elektronik> mother crystal
Mutterlauge f <Lebensmittel> mother liquor
Muttermodell n <Elektronik> master pattern
Mutternanziehmaschine f <Maschinen> nut tightener
Mutterngewindebohrer m <Maschinen> nut tap
Mutterngewindeschneidmaschine f <Maschinen> nut-threading machine
Mutternsicherung f <Maschinen> nut lock
Mutternuklid n <Kerntech, Strahlphys> parent nuclide
Mutternzange f <Maschinen> nut pliers
Mutterpause f <Konstzeich> master print, master tracing
Mutterplatte f <Akustik> mother
Mutterraumschiff n <Raumfahrt> mother ship
Mutterschiff n <Wassertrans> mother ship
Mutterschloss n <Maschinen> half-nut
Mutterschraube f <Bau> bolt and nut
Muttersteckverbinder m <Elektrotech> female connector
Muttersubstanz f <Fertig> parent
MUX *(Multiplexer)* <Comp & DV, Elektronik, Fernseh, Telekom> MUX *(multiplexer)*
MVA *(Müllverbrennungsanlage)* 1. <Abfall> *(AE)* garbage incineration plant, *(BE)* refuse incineration plant, waste incineration plant, waste incinerator; 2. <Verpack> *(AE)* garbage incinerator, *(BE)* refuse incinerator
MW *(Mittelwelle)* <Fernseh, Funktech> MW *(medium wave)*
MW-Bereich m <Funktech> MW band
Mx *(Maxwell)* <Elektriz, Elektrotech> Mx *(maxwell)*
Mycoprotein n <Lebensmittel> mycoprotein
Mycotoxin n <Lebensmittel> mycotoxin *(Pflanzenkrankheitslehre)*

Myon *n* <Teilphys> muon
Myonneutrino *n* <Teilphys> muon neutrino
Myosin *n* <Chemie> myosin
Myrcen *n* <Chemie> myrcene
Myria… <Metrol> myria
Myriagramm *n* <Metrol> myriagram
Myristin… <Chemie> myristic
Myristyl… <Chemie> myristyl
Myron… <Chemie> myronic
Myrosin *n* <Chemie> myrosin
Mytilotoxin *n* <Chemie> mytilotoxin

N

n 1. <Elektriz, Kerntech, Strahlphys, Teilphys> *(Neutron)* n *(neutron)*; 2. <Elektriz> *(Windungszahl pro Längeneinheit)* n *(turns per unit length)*; 3. <Kerntech, Metrol> *(Quantenzahl)* n *(principal quantum number)*; 4. <Phys, Thermod> *(Molekulardichte)* n *(molecular density)*
N 1. <Chemie> *(Stickstoff, Nitrogen)* N *(nitrogen)*; 2. <Elektriz, Hydraul, Labor, Strömphys> *(Newton)* N *(newton)*; 3. <Elektriz> *(Windungszahl)* N *(number of turns in a winding)*; 4. <Elektronik, Kerntech> *(Rauschleistung)* N *(noise power)*; 5. <Metrol, Optik> *(Strahlung)* N *(radiance)*; 6. <Phys> *(Molekülzahl)* N *(number of molecules)*
N$_A$ *(Avogadro'sche Zahl, Loschmidt'sche Zahl)* <Phys, Thermod> NA *(Avogadro's number, Loschmidt number)*
Na *(Natrium)* <Chemie> Na *(sodium)*
Nabe *f* 1. <Comp & DV> hub; 2. <Fertig> boss; 3. <Lufttrans> hub *(Hubschrauber)*; 4. <Maschinen> boss, hub
Nabel *m* <Ker & Glas, Raumfahrt> nose
Nabel *m* **der Glasmacherpfeife** <Ker & Glas> nose of blowpipe
Nabelschnur *f* <Raumfahrt> umbilical cable
Nabelschnuranschluss *m* <Raumfahrt> umbilical connector
Nabenabdeckplatte *f* <Lufttrans> hub cover plate
Nabenabzieher *m* 1. <Maschinen> hub extractor, hub puller; 2. <Mechan> hub puller
Nabendeckel *m* 1. <Fertig> collar cover *(Kunststoffinstallationen)*; 2. <Kfztech> hubcap *(Rad)*
Nabendeckplatte *f* <Lufttrans> hub cover plate *(Hubschrauber)*
Nabenkappe *f* <Kfztech> hubcap *(Rad)*
Nabenkippanschlag *m* <Lufttrans> hub tilt stop *(Sperre)*
Nabenunterlegring *m* <Lufttrans> hub spacer *(Hubschrauber)*
Nabenwulst *m* **am Steuerknüppel** <Lufttrans> control column boss
nachabgleichen *v* <Gerät> rebalance
nachahmen *v* <Math> mimic *(Bootstrapverfahren)*
Nachahmung *f* <Comp & DV> simulation
nacharbeiten *v* 1. <Fertig> remachine *(Spannung)*; 2. <Qual> refinish, rework
Nacharbeiten *n* 1. <Fertig> remachining *(Spannung)*; 2. <Ker & Glas> touching-up; 3. <Maschinen> remachining
Nachbar… 1. <Chemie> vicinal; 2. <Fernseh, Telekom> adjacent
Nachbarfunktionsinstanz *f* <Telekom> adjacent functional entity *(ISDN)*
Nachbarfunkzelle *f* <Mobilkom> adjacent cell
Nachbarkanal *m* <Fernseh, Telekom> adjacent channel

Nachbarkanalselektion *f* <Telekom> adjacent channel selectivity
Nachbarkanalstörung *f* <Telekom> adjacent channel interference *(ACI)*
Nachbarkanalunterdrückung *f* <Telekom> adjacent channel rejection
Nachbarschaft *f* 1. <Bau> neighbourhood; 2. <Chemie, Math> vicinity
Nachbarschaftslärm-Dauerschallpegel *m* <Sicherheit> community noise equivalent level
Nachbarzeichenstörung *f* <Telekom> intersymbol interference
Nachbarzelle *f* <Mobilkom> neighboring cell *(zellulares Netz)*
Nachbarzone *f* 1. <Mobilkom> neighboring cell *(zellulares Netz)*; 2. <Thermod> heat-affected zone
nachbearbeiten *v* <Fertig> rework; remachine
Nachbearbeitung *f* 1. <Elektronik> postprocessing; 2. <Fertig> dressing, remachining
Nachbearbeitung *f* **von Daten** <Comp & DV> postediting
Nachbedienung *f* <Mechan> local control
nachbehandeln *v* 1. <Bau> cure *(Beton)*; 2. <Fertig> retreat
Nachbehandlung *f* 1. <Bau> retreatment; 2. <Mechan> curing
Nachbehandlungsfilm *m* <Bau> curing membrane
Nachbehandlungstunnel *m* <Bau> curing tunnel
Nachbehandlungszeit *f* <Bau> cure period
Nachbelichtung *f* <Foto> postexposure
Nachbeschleunigungs-CRT *f* <Elektronik> postaccelerator CRT
Nachbesprechung *f* <Raumfahrt> debriefing
nachbessern *v* <Foto> retouch
Nachbild *n* <Ergon, Fernseh> afterimage
Nachbildung *f* <Telekom> balance, balancing, impedance simulating network
Nachbildwirkung *f* <Optik> persistence of vision
Nachblasen *n* <Fertig> Bessemer afterblow *(Thomasverfahren)*
nachbohren *v* 1. <Kfztech> rebore *(Motor, Zylinder)*; 2. <Maschinen> rebore
Nachbohren *n* <Maschinen> reboring
Nachbohrung *f* <Kfztech> reboring
Nachbrecher *m* <Kohlen> secondary crusher
Nachbrennen *n* <Metall, Strahlphys, Thermod> afterglow
Nachbrenner *m* 1. <Kfztech> afterburner *(Motor)*; 2. <Lufttrans, Maschinen, Thermod> afterburner
Nachbrennkammer *f (SCC)* <Abfall, Maschinen> afterburner chamber, secondary combustion chamber
nachdecken *v* <Textil> cross-dye
Nachdieseln *n* <Kfztech> dieseling
nachdrehen *v* 1. <Fertig> re-turn *(Spanung)*; 2. <Maschinen> re-turn
Nachdrehmaschine *f* <Maschinen> finishing lathe
nachdrucken *v* <Druck> reprint
nachdunkelndes Sonnenschutzglas *n* <Ker & Glas> solar control glass
nacheichen *v* <Fertig, Qual> recalibrate
Nacheichung *f* 1. <Fertig, Kerntech> recalibration; 2. <Metrol> calibration check, checking of the calibration, field calibration, recalibration, subsequent verification
nacheilen *v* <Fertig> lag *(Phase)*
Nacheilen *n* 1. <Lufttrans> lagging *(Hubschrauber)*; 2. <Mechan> lagging
nacheilend *adj* <Elektrotech> lagging
nacheilende Phasenverschiebung *f* <Phys> lag in phase
Nacheilung *f* <Elektronik> lag

Nacheilwinkel m <Elektriz, Maschinen> angle of lag, lagging angle
nachempfundenes Muster n <Elektronik> replicated pattern
Nachentzerrung f 1. <Aufnahme> de-emphasis, postequalization; 2. <Fernseh> de-emphasis, postemphasis
nachfedernd adj <Mechan> elastic
Nachfließen n <Kunststoff> cold flow
Nachflügel m <Lufttrans> flap (Flugzeug)
nachfolgendes System n <Comp & DV> downstream system
Nachfolger m <Comp & DV> descendant
Nachfolgesatz m <Comp & DV> trailer record
Nachformdrehen n <Maschinen> copy turning
Nachformdrehmaschine f <Maschinen> contour lathe, copying lathe
nachformen v <Maschinen> copy
Nachformen n 1. <Kunststoff> postforming; 2. <Maschinen> copying
Nachformfräsen n 1. <Fertig> copy milling; 2. <Maschinen> copy milling, profile milling, tracer milling
Nachformfräsmaschine f <Maschinen> profile-milling machine, profiler
Nachformfrässchablone f <Fertig> milling template
Nachformhobeln n <Fertig> copy planing
Nachformmaschine f <Maschinen> copier, copying machine
Nachformrolle f <Maschinen> contour follower, profiling roller
Nachformschleifen n <Maschinen> form grinding
Nachformsteuerung f <Fertig> contour control, forming control
Nachformwerkzeug n <Fertig> forming tool
nachfräsen v <Fertig> recut
Nachfräsen n <Fertig> recutting
Nachfrist f <Patent> period of grace
Nachführstation f <Raumfahrt> tracking station
Nachführung f 1. <Raumfahrt> tracking (Raumschiff); 2. <Telekom> tracking (Antenne)
Nachführungsantenne f <Raumfahrt> tracking antenna
Nachführungsfehler m <Metrol, Telekom> tracking error
Nachführungsgenauigkeit f <Raumfahrt> tracking accuracy (Weltraumfunk)
Nachfüllen n <Eisenbahn> refilling (von Bremsluft)
Nachfülllösung f <Foto> replenisher
Nachfüllung f <Lebensmittel, Papier> refill
nachgeben v 1. <Bau> give way; 2. <Maschinen> yield
Nachgeben n <Kerntech> yielding (von Metallen unter Druck)
nachgehen v <Qual> trace
nachgeordnet adj <Heiz & Kälte> down-stream
nachgeordnet adv <Bau> downstream
nachgeschaltet adj <Heiz & Kälte, Maschinen> downstream
nachgeschaltetes Bauglied n <Regelung> structural element next in line (des Messumformers)
nachgewiesene Lagerstätte f <Erdöl> proven field
nachgewiesene Reserven fpl <Erdöl> proven reserves
Nachgiebigkeit f <Metall> yielding
Nachglimmen n <Strahlphys, Thermod> afterglow
Nachglühen n <Raumfahrt> afterglow (Raumschiff)
Nachhall m 1. <Akustik> reverberation; 2. <Sicherheit> noise reverberation (Lärm)
Nachhallfeld n <Sicherheit> reverberant field (Lärm)
Nachhallmessgerät n <Aufnahme> reverberation unit
Nachhallplatte f <Aufnahme> reverberation plate
Nachhallraum m <Akustik> reverberation room
Nachhallspirale f <Aufnahme> spring reverberation unit

Nachhallzeit f (T) <Akustik, Aufnahme> reverberation time (T)
nachhärten v 1. <Anstrich> retemper (Metall); 2. <Fertig> age (Legierungen)
Nachhärten n 1. <Fertig> afterbake; 2. <Kunststoff> age hardening
nachhauen v <Fertig> recut (Feile)
Nachhauen n <Fertig> recutting (Feile)
nachkalibrieren v <Qual> recalibrate
Nachkalibrierung f <Kerntech, Qual> recalibration
Nachkalibrierungsbereich m <Qual> recalibration range
Nachklärbecken n <Abfall> final settling tank, secondary sedimentation basin, secondary settling tank
Nachklassierung f <Kohlen> sizing
Nachkommaziffern fpl **des Logarithmus** <Comp & DV> mantissa
Nachkühler m <Heiz & Kälte> after cooler
nachlassen v 1. <Telekom> go down; 2. <Wassertrans> abate (Wind)
Nachlauf m 1. <Aufnahme> tracking; 2. <Kfztech> caster action; 3. <Lufttrans> lagging (Hubschrauber); 4. <Strömphys> overtravel (Maschine); 5. <Strömphys> wake (eine Zylinder)
Nachlaufachse f <Kfztech> trailing axle
Nachlaufdelle f <Strömphys> wake depression
Nachlauf-Empfangsoszillator m <Elektronik> tracking local oscillator
Nachlaufen n 1. <Fernseh> lag, trailing; 2. <Kfztech> dieseling
nachlaufende Chrominanz f <Fernseh> lagging chrominance
Nachlauffilter n <Elektronik> tracking filter
Nachlaufintensität f <Strömphys> wake intensity
Nachlauf-Konfiguration f <Aufnahme> tracking configuration
Nachlaufoszillator m <Elektronik> tracking oscillator
Nachlaufschalter m <Sicherheit> overtravel switch
Nachlaufwinkel m <Kfztech> castor angle
Nachleuchtcharakteristik f 1. <Elektronik> persistence characteristic; 2. <Elektrotech> decay characteristic
Nachleuchtdauer f <Comp & DV, Elektriz, Elektronik> afterglow, persistence
Nachleuchten n 1. <Elektronik> phosphorescence; afterglow (Bildschirm); 2. <Fernseh> afterglow; 3. <Kerntech> hangover; 4. <Strahlphys, Thermod, Wassertrans> afterglow (Radar)
nachleuchtende Substanz f <Phys> phosphor
Nachleuchtung f <Elektriz> persistence
Nachleuchtzeit f <Elektronik> decay time (eines Elektronenstrahlröhrenbildschirms)
Nachmahlen n <Kohlen> secondary grinding
nachmessen v <Metrol> check the measurements made
Nachmittagsspitze f <Trans> pm peak
Nachprüfung f <Qual> check test, repeat test, retest
nachrangige Verarbeitung f <Comp & DV> background processing (Betriebsart)
Nachricht f 1. <Comp & DV> message; 2. <Eisenbahn> notice; 3. <Funktech, Telekom> message
Nachrichten fpl <Fernseh> news
Nachrichtenabfrage f <Comp & DV> message retrieval
Nachrichtenanfang m <Telekom> beginning of message
Nachrichtenauslauf m <Comp & DV> quiescing
Nachrichtenbehandlung f <Comp & DV> message handling
Nachrichteneinheit f <Comp & DV> message unit
Nachrichtenempfänger m <Comp & DV> message sink
Nachrichtenende n (EOM) <Comp & DV, Telekom> end of message (EOM)

Nachrichtenkanal

Nachrichtenkanal m 1. <Comp & DV, Elektrotech> communication channel; 2. <Mobilkom> traffic channel *(GSM)*; 3. <Telekom> communication channel
Nachrichtenkopf m 1. <Comp & DV> message header; 2. <Telekom> header, message header
Nachrichtenleitung f <Elektrotech, Telekom> communications line
Nachrichtennetz n 1. <Comp & DV> communication network; 2. <Telekom> communication network, telecommunication network
Nachrichtenquelle f 1. <Comp & DV> information source, message source; 2. <Telekom> information source
Nachrichtenredundanz f <Telekom> message redundancy
Nachrichtensatellit m *(ComSat)* <Raumfahrt, Telekom> communication satellite *(comsat)*
Nachrichtensatz m <Comp & DV> message set
Nachrichtensenderkette f <Fernseh> news network
Nachrichtensendung f <Fernseh> newscast
Nachrichtensenke f <Comp & DV> message sink
Nachrichtensicherheit f <Telekom> information security
Nachrichtensprecher m <Fernseh> newscaster
Nachrichtenstruktur m <Comp & DV> message structure
Nachrichtenstudio n <Fernseh> newsroom
Nachrichtentechnik f 1. <Elektriz> communications engineering; 2. <Telekom> telecommunications engineering
Nachrichtentext m <Comp & DV> message text
Nachrichtenübermittlung f 1. <Comp & DV> message routing; 2. <Telekom> messaging
Nachrichtenübertragung f <Comp & DV> message transfer
Nachrichten- und Rechnertechnik f <Comp & DV, Telekom> compunications, computer and communications *(rechnerintegrierte Kommunikationstechnik)*
Nachrichtenverarbeitung f <Comp & DV> message processing
Nachrichtenverarbeitungsgeräte npl <Wassertrans> message processing equipment *(Kommunikation)*
Nachrichtenverbindung f **Synchronsatelliten** <Telekom> synchronous orbit communication
Nachrichtenvermittlung f 1. <Comp & DV> message routing, message switching, messaging; 2. <Telekom> message switch
Nachrichtenvermittlung f **auf Hochspannungsleitung** <Telekom> power-line communication
Nachrichtenvermittlungsnetz n <Comp & DV, Telekom> message-switched network, message switching network *(MSN)*
Nachrichtenvermittlungsprozessor m <Telekom> message switching processor
Nachrichtenverteilung f <Comp & DV> message switching
Nachrichtenverteilungsnetz n <Comp & DV> message-switched network
Nachrichtenweiterleitung f <Comp & DV> message routing
Nachrichtenweitervermittlung f <Comp & DV> message transfer
Nachrichtenweitschweifigkeit f <Telekom> message redundancy
Nachrichtenwesen n <Telekom> telecommunications
Nachrüst... <Kontroll> add-on
nachrüsten v 1. <Maschinen> reequip; 2. <Nichtfoss Energ> retrofit
Nachrüsten n <Nichtfoss Energ> retrofit
Nachrüstsatz m <Telekom> add-on kit
Nachrüstung f <Comp & DV> retrofit
Nachsatz m <Comp & DV> tail, trailer, trailer record

nachschaben v 1. <Fertig> shave *(Ziehen)*; 2. <Maschinen> shave
Nachschaben n <Fertig> shaving *(Ziehen)*
Nachschieberfahrt f <Eisenbahn> pusher operation
Nachschlagetabelle f <Comp & DV> lookup table
nachschleifen v 1. <Fertig> reback *(Freiwinkel)*; reface *(Ventilsitz)*; 2. <Kohlen> regrind; 3. <Maschinen> regrind, resharpen
Nachschleifen n 1. <Fertig> regrinding; 2. <Maschinen> re-sharpening
Nachschliff m <Maschinen> resharpening
Nachschliffwinkel m <Fertig> sharpening angle
Nachschneider m 1. <Fertig> master tap *(Gewinde)*; 2. <Maschinen> master tap
Nachschub m <Lufttrans> reserves
Nachseifen n <Textil> soaping aftertreatment
Nachsetzlüfter m <Heiz & Kälte> make-up fan
Nachsetzmaschine f <Kohlen> cleaner jig
Nachsorge f <Abfall> monitoring after site closure *(einer Deponie)*
Nachspannvorrichtung f <Eisenbahn> wire strainer
Nachspur f <Kfztech> toe-out *(Vorderräder)*
nachstellbare Reibahle f <Maschinen> expanding reamer
Nachstellkeil m <Maschinen> tightening wedge
Nachstellleiste f <Fertig> adjustable gib
nachstemmen v <Fertig> recaulk *(Niet)*
Nachstemmen n <Fertig> recaulking *(Niet)*
Nachstrom m <Lufttrans> slipstream
Nachsynchronisieren n <Aufnahme> postsynchronization
nachsynchronisiertes Halbbildaustastintervall n <Fernseh> postsync field-blanking interval
Nachsynchronisierung f <Akustik> postscoring
Nachtbelastung f <Elektrotech> off-peak load
Nachtdienst m <Telekom> night service
nachteilig adj <Kerntech> unfavourable
Nachtglas n <Optik> night binocular, night glass
Nachtphase f <Raumfahrt> nocturnal phase
nachträglich formatiert adj <Comp & DV> postformatted
nachträgliche Hervorhebung f <Aufnahme> postemphasis *(von Tiefen)*
Nachtreichweite f <Lufttrans> night range
Nachtrockner m <Papier> afterdryer
Nachtsichtgerät n <Wassertrans> night vision device *(Navigation)*
Nachtspeicherheizgerät n <Heiz & Kälte> night storage heater
Nachtspeicherheizung f <Heiz & Kälte, Thermod> night storage heating
Nachtstromspeicherheizung f <Thermod> night storage heater
Nachttarif m <Elektriz> night tariff
Nachtwelle f <Lufttrans> night wave
Nachverarbeitung f <Elektronik> postprocessing
Nachverbrenner m <Kerntech> afterburner
Nachverbrennung f <Thermod> afterburning
Nachverbrennungskammer f <Kohlen> postcombustion chamber
Nachverfugen n <Bau> rejointing, repointing
Nachvermessung f <Bau> resurvey
Nachvertonen n **auf Videomagnetband** <Fernseh> videotape dubbing
Nachvollziehbarkeit f <Qual> traceability
Nachvollziehbarkeit f **der Kalibrierung** <Qual, Regelung> calibration traceability
Nachwahl f <Telekom> suffix dialing
Nachwahlkennstelle f <Comp & DV> suffix
Nachwärme f <Kerntech, Lufttrans> afterheat

nachwärmen v <Lebensmittel> scald *(Hartkäse)*
Nachwärmeofen m <Metall> reheating furnace
Nachweis m 1. <Patent, Qual> evidence; 2. <Strahlphys> detection *(von Radioaktivität)*
nachweisbare Grenze f <Qual> verifiable limit
Nachweisbarkeit f <Optik> detectivity
nachweisen v <Qual> verify
Nachweisgrenze f 1. <Gerät> detection limit, limit of detection; 2. <Optik> detection threshold
nachweispflichtig adj <Qual> requiring verification
Nachweispunkt m <Qual> witness point
Nachweisschwelle f <Optik> detection threshold
Nachweisvermögen n <Telekom> detectivity
Nachweiszeit f <Gerät> detection time
Nachwirkung f <Strahlphys> aftereffect
Nachwuchten n <Fertig> rebalancing
nachzeichnen v <Bau> trace
nachziehen v <Fernseh> smear
Nachziehen n 1. <Fernseh> smearing; 2. <Maschinen> tracing
Nackenschutz m <Sicherheit> neck shield
nackt adj <Elektrotech, Fertig> bare *(Elektrode)*
nackte Lichtbogen-Ionenquelle f <Kerntech> open arc ion source
nackter Brand m <Lebensmittel> loose smut *(Pflanzenkrankheitslehre)*
nackter Lichtbogen m <Elektrotech> open arc
nackter Reaktor m <Kerntech> bare reactor
nacktes Teilchen n <Kerntech> bare particle
Nadel f 1. <Aufnahme> stylus *(des Plattenspielers)*; 2. <Gerät, Kfztech> needle *(Vergaser)*; 3. <Labor> pointer; 4. <Maschinen> needle roller; 5. <Papier> needle, pin; 6. <Textil> needle
Nadelabweichung f <Phys> magnetic declination
Nadelätzen n <Ker & Glas> needle etching
Nadelbalken m <Textil> needle bar
Nadelbett n <Textil> needle bed
Nadelbrecher m <Kohlen> pick breaker
Nadeldrucker m 1. <Comp & DV> dot matrix printer; 2. <Druck> matrix printer
Nadeldüse f <Kfztech> needle jet *(Vergaser)*
Nadeleindringversuch m <Bau> static penetration test
Nadeleinstell-Skale f <Labor> needle dial
Nadelfeile f <Maschinen> needle file
Nadelfilz m <Papier> needled felt
nadelförmig adj <Fertig, Metall> acicular
nadelförmige Zone f <Metall> needle-shaped zone
nadelförmiger Kristall m <Metall> needle-shaped crystal
nadelförmiges Ferrit n <Metall> acicular ferrite
nadelförmiges Teilchen n <Metall> needle-shaped particle
Nadelgalvanometer n <Elektriz> moving-magnet galvanometer, needle galvanometer
Nadelgeräusch n <Elektronik> surface noise
Nadelhalter m <Aufnahme> needle holder
Nadelholz n <Bau> softwood
Nadelhülse f <Maschinen> needle bush
Nadelkäfig m <Maschinen> needle cage
Nadellager n <Maschinen> needle bearing, needle roller bearing
Nadelloch n <Optik> pinhole
Nadel-Nebengeräusch n <Aufnahme> stylus crosstalk
Nadelöler m <Maschinen> needle lubricator
Nadelpositionierung f <Textil> automatic needle positioner
Nadelrahmenspannmaschine f <Textil> pin stenter
Nadelstange f <Textil> needle bar
Nadelsteifheit f <Akustik> stiffness
Nadelstich m 1. <Foto> pinhole *(auf Negativ oder Foto)*; 2. <Papier> pinhole

Nadeltonquelle f <Akustik> pinpoint acoustic source
Nadelventil n 1. <Heiz & Kälte, Kerntech, Kfztech> needle valve; 2. <Labor> needle valve *(Gassteuerung)*; 3. <Maschinen, Nichtfoss Energ> needle valve
Nadelventilführung f <Kfztech> needle valve guide
Nadel-Wählscheibe f <Labor> needle dial
Nadelwehr n <Wasserversorg> needle dam, pin weir
Nadelzange f <Maschinen> needle-nose pliers
Nadelzunge f <Textil> latch
n-adriges Kabel n <Elektrotech> n-core cable
Nagatelit m <Kerntech> nagatelite
Nagel m 1. <Bau> nail; 2. <Maschinen> nail, spike; 3. <Mechan> nail
Nagel m **mit runder Kuppe** <Maschinen> clout nail
Nagelbohrer m <Bau> gimlet
Nageldachbinder m <Bau> nail roof truss
nageldurchtrittsicher adj <Sicherheit> puncture-resistant *(Sicherheitsschuhwerk)*
Nagelklaue f <Bau> nail claw
Nagelkopf m <Maschinen> nail head
Nagelmaschine f <Verpack> nailing machine
Nageltreiber m <Bau> nail punch
Nagelzieheisen n <Bau> nail claw
Nagelzieher m 1. <Bau> nail extractor; 2. <Verpack> nail puller
Nagetierschutz m <Fertig> rodent protection
nah adv <Math> nearby
Nahaufnahme f <Raumfahrt> close-up
Nahbereich m 1. <Funktech> close range, short range; near zone *(Antenne)*; 2. <Konstzeich> close-up range; 3. <Telekom> close-up area, direct service area
Nahbereichnavigation f <Funktech> short range navigation
Nahbereichsgerät n <Funktech> short range device *(Funkgerät)*
Nahbereichskabel n <Elektrotech> short haul cable
Nahbesprechungs-Empfindlichkeit f <Akustik> close-talking sensitivity
Nahbetrieb m <Telekom> local operation
nahe adv <Math> approximate to, nearby
Nähe/in der <Math> nearby
Naheinstellansatz m <Foto> close-up attachment
nähen v 1. <Fertig> sew; 2. <Textil> sew, stitch
Nähen n <Fertig, Textil> sewing
Näherei f <Textil> sewing
nähern v <Geom> approximate
nähern v/sich 1. <Math> approximate to *(einer Lösung oder einem Grenzwert)*; 2. <Wassertrans> approach *(Navigation)*
Näherung f <Math> approximation
Näherungsfehler m <Telekom> error of approximation
Näherungsfunktion f <Math> approximation function
Näherungslithographie f <Elektronik> proximity lithography
Näherungsschalter m 1. <Elektriz> proximity switch; 2. <Gerät> approximating pick-up; 3. <Kontroll> proximity switch
nahes Infrarot n <Strahlphys> near infrared
nahes Ultraviolett n <Strahlphys> near ultraviolet
Nahfeld n <Optik, Telekom> near field
Nahfeld-Abtastverfahren n <Optik, Telekom> near-field scanning technique
Nahfeldanalyse f <Telekom> near-field analysis
Nahfeldbereich m <Optik, Telekom> near-field region
Nahfeldbeugungsdiagramm n <Telekom> near-field diffraction pattern
Nahfeldbeugungsmuster n <Optik> near-field diffraction pattern

Nahfeldbrechungsmethode f <Telekom> refracted near-end method, refracted near-field method, refracted ray method
Nahfelddiagramm n <Telekom> near-field pattern
Nahfeldmaske f **aus vier konzentrischen Kreisen** <Optik> four-concentric-circle refractive index template; four-concentric-circle near-field template *(zur Bestimmung des Brechungsindex)*
Nahfeldmethode f **mit Brechung** <Optik> refracted near-field method
Nahfeldmuster n <Optik> near-field pattern *(Lichtbeugung)*
Nahfeldrasterverfahren n <Optik, Telekom> near-field scanning technique
Nahfeldstärke f <Raumfahrt> near-field intensity *(Weltraumfunk)*
Nahfeldstrahlungsdiagramm n <Telekom> near-field radiation pattern
Nähgarn n <Textil> sewing cotton
Nähmaschine f <Textil> sewing machine
Nahnebensprechen n <Elektrotech, Telekom> near-end crosstalk, near-end XT *(NEXT)*
Nahpeilung f <Funktech> short-path bearing
nahrhaft adj <Lebensmittel> nutritious, nutritive
Nährstoff m <Lebensmittel, Meerschmutz> nutrient
Nährstoffbedarf m <Lebensmittel> nutrient requirements
Nährstoffgehalt m <Lebensmittel> nutrient content
Nährstoffverlust m <Lebensmittel> nutrient loss
Nahrungsaufnahme f <Lebensmittel> ingestion
Nahrungsbedarf m <Lebensmittel> food requirements
Nährwert m <Lebensmittel> nutritive value
Nähseide f <Textil> sewing silk
Nahsprechmikrofon n <Aufnahme> close-talking microphone
Naht f 1. <Bau> seam *(Schweißen)*; 2. <Fertig> seam *(Nahtschweißen)*; 3. <Ker & Glas, Maschinen, Mechan, Papier, Textil> seam
Naht f **des Maschinensiebs** <Papier> seam of the machine wire
Nahtbasis f <Mechan> root
Nahtbildung f <Fertig> finning *(Gießen)*
Nahtdicke f <Fertig> throat thickness; throat *(Schweißen)*
Nahteil n <Ker & Glas> short vision segment
Nahthöhe f <Fertig> actual throat of fillet weld *(Kehlnaht)*
Nahtlinie f <Ker & Glas> seam line
nahtlos adj 1. <Fertig, Maschinen> weldless *(Rohr)*; 2. <Textil> seamless
nahtlos gewalzt adj <Maschinen> seamless rolled
nahtlos gezogenes Stahlrohr n <Maschinen> solid-drawn steel tube
nahtlos integrieren v <Comp & DV> seamlessly integrate
Nahtschweißen f 1. <Fertig> scarf weld *(Hammerschweißen)*; 2. <Kerntech> seam weld; 3. <Maschinen> seam welding
Nahtstellenüberwachung f <Qual> interface control
Nahtverriegeln n <Textil> backtacking of seams
Nahtvorbereitung f <Metall> joint preparation *(Schweißarbeiten)*
Nahverkehr m 1. <Eisenbahn> *(AE)* light rail transit, *(BE)* light rail transport; 2. <Trans> local traffic
Nahverkehrsinformation f <Trans> local traffic information
Nahverkehrszug m <Eisenbahn> *(AE)* stopping train
Name m <Comp & DV> name
Name m **der Zeichenfolge** <Comp & DV> string name
Namensaufruf m <Comp & DV> call by name
Namensschild n <Bau> scutcheon
Namenstaste f <Telekom> name key

Namensverzeichnis n <Telekom> name server
Namenswahl f <Telekom> name dialing
NAND-Funktion f *(UND-NICHT-Verknüpfung)* <Comp & DV> NAND operation *(NOT AND operation)*
NAND-Gate n <Comp & DV> NAND gate, NOT AND gate
NAND-Gatter n 1. <Elektronik> NAND gate; 2. <Phys> NAND gate
NAND-Gatter n **mit drei Eingängen** <Elektronik> three-input NAND gate
NAND-Glied n <Elektronik> NAND device
NAND-Tor n *(UND-Glied mit negiertem Ausgang)* <Comp & DV> NAND gate *(NOT AND gate)*
NAND-Verknüpfung f <Comp & DV> NAND operation
Nano... <Metrol> nano...
Nanoamperemeter n <Metrol> nanoammeter *(Verstärkervoltmeter)*
Nanobauelement n <Elektronik> nano-device
Nanocomputer m <Comp & DV> nano-computer
Nanodraht n <Elektronik> nano-wire
Nanoelektronik f <Elektronik> nano-electronics
Nanomaschine f <Elektronik> nano-machine
Nanomechanik f <Elektronik> nano-mechanics
Nanometer-Lithographie f <Elektronik> nano-lithography
Nanometerstruktur f <Elektronik> nano-structure
Nanometer-Struktur-Herstellung f <Elektronik> nano-structure fabrication
Nanoschaltkreis m <Elektronik> nano-circuit
Nanosekunde f <Comp & DV, Fernseh, Telekom> nano-second
Nanosekundenimpuls m <Elektronik> nanosecond impulse, nanosecond pulse *(Impuls im Nanosekundenbereich)*
Nanosekundenimpulsgeber m <Elektronik> nanosecond impulse generator, nanosecond pulse generator
Nanostrukturierung f **durch Abtragen** <Elektronik> top-down approach
Nanostrukturierung f **durch Zusammenfügen** <Elektronik> bottom-up approach
Nanotechnologie f <Elektronik> nano-technology
Napalm n <Thermod> napalm
Näpfchenziehversuch m <Fertig> cupping test
Naphtha f <Chemie, Thermod> naphtha
Naphthacen n <Chemie> naphthacene
Naphthalen n <Chemie> naphthalene
Naphthalendisulfonsäure f <Chemie> *(AE)* naphthalenedisulfonic acid, *(BE)* naphthalenedisulphonic acid
Naphthalin n <Chemie> naphthalene
Naphthalin... <Chemie> naphthalenic
Naphthan n <Chemie> decahydronaphthalene, decalin, naphthane
Naphthen n <Chemie> cyclane, cycloalkane, naphthene
Naphthenat n <Chemie> naphthenate
naphthenisch adj <Chemie> naphthenic
Naphthion... <Chemie> naphthionic
Naphthochinon n <Chemie> dihydrodiketonaphthalene, naphthoquinone
Naphthoe... <Chemie> naphthoic
Naphthol n <Chemie> naphthol
Naphtholieranlage f <Textil> naphtholation bath
Naphtholsulfon... <Chemie> *(AE)* naphtholsulfonic, *(BE)* naphtholsulphonic
Naphthophenantren n <Chemie> dibenzanthracene, naphthophenanthrene
Naphthoxazin n <Chemie> naphtoxazine, phenoxazine
Naphthoyl... <Chemie> naphthoyl
Naphthyl... <Chemie> naphthyl
Naphthylamin n <Chemie> naphthylamine
Naphthylbenzoylester m <Chemie> benzonaphthol

Naphthylen... <Chemie> naphthylene
Napier'scher Logarithmus m <Math> Napierian logarithm *(Neper'scher Logarithmus)*
narbig adj <Fertig> pitted
Narbmaschine f <Papier> graining machine
Narbung f 1. <Fertig> grain; 2. <Ker & Glas> pocking; 3. <Papier> graining
Narcein n <Chemie> narceine
Narcotin n <Chemie> narcotine, opianine
Naringenin n <Chemie> naringenin, trihydroxyflavanone
Naringin n <Chemie, Lebensmittel> naringin
NASA *(Nordamerikanische Weltraumbehörde)* <Raumfahrt> NASA *(National Aeronautics and Space Administration)*
Nase f 1. <Bau> nose; 2. <Elektrotech> aligning plug *(Röhre)*; 3. <Fertig> catch, cog, lug; 4. <Kfztech> stud; 5. <Maschinen> catch, lobe, lug; 6. <Raumfahrt> nose cone *(Raumschiff)*
nasenförmige Kreuzung f <Eisenbahn> swing nose crossing
Nasenkegel m <Lufttrans> nose cone
Nasenkeil m <Maschinen> gib, gib-head key, nose key
Nasenklappe f <Lufttrans> droop flap, leading-edge flap
Nasenkonus m <Aufnahme, Lufttrans> nose cone
Nasenring m <Kerntech, Kfztech> oil scraper ring *(Motor)*
Nasenstein m <Ker & Glas> plate block
Nasenwelle f <Lufttrans> bow wave
nass adj 1. <Fertig> hydrometallurgical *(Verfahren)*; 2. <Papier, Textil, Thermod> wet • **nass in nass** <Kunststoff> wet on wet
nass anzufeuchtendes Etikett n <Verpack> wet glue label
nass machen v <Thermod> wet
Nass... <Kohlen, Umweltschmutz, Wasserversorg> wet
Nassablagerung f <Umweltschmutz> wet deposition
Nassabscheider m 1. <Abfall> wet scrubber, wet scrubbing device; 2. <Chemtech> washer; 3. <Umweltschmutz> wet scrubber
Nassabscheidung f <Umweltschmutz> wet precipitation
Nassabsorber m <Abfall> wet scrubber, wet scrubbing device
Nassaluminiumkondensator m <Elektrotech> *(BE)* wet aluminium capacitor, *(AE)* wet aluminum capacitor
Nassausschuss m <Papier> wet broke
Nassbagger m 1. <Bau> grab dredger; 2. <Meerschmutz> dredger; 3. <Wasserversorg> river dredge
Nassbehandlung f <Kohlen, Maschinen> wet treatment
nassbeständig adj <Thermod> humidity-resistant
Nassdampf m 1. <Heiz & Kälte> saturated steam; 2. <Maschinen, Nichtfoss Energ, Thermod> wet steam
Nassdehnung f <Papier> damping stretch
Nässe schützen v/vor <Verpack> keep dry
Nasselement n <Elektrotech> wet cell
nässen v <Phys> wet
Nassentschwefelungsprozess m <Umweltschmutz> *(AE)* wet desulfurization process, *(BE)* wet desulphurization process
Nasserdgas n <Erdöl> wet natural gas
Nasseruptionskreuz n <Erdöl> wet tree
Nassfäule f <Bau> wet rot *(Holz)*
nassfestes Papier n <Papier> wet-strength paper
Nassfestigkeit f <Kunststoff> wet strength
Nassgas n <Erdöl> wet natural gas
Nassguss m <Fertig> *(AE)* green mold casting, *(BE)* green mould casting
Nassgusssand m <Fertig> green sand
nassklassieren v <Kohlen> classify
Nasskollodiumverfahren n <Druck, Foto> wet collodion process

Nasskupplung f <Maschinen> wet clutch
Nassluftfilter n <Heiz & Kälte> wet air filter
Nassmahlung f <Kohlen> wet grinding
Nassoxidation f <Abfall> WAO, wet air oxidation
Nasspartie f <Papier> wet end
Nassplattenverfahren n <Druck> wet-plate process
Nasspresse f <Papier> wet press
Nassprobe f <Chemie, Kohlen> wet assay
Nassputzen n <Fertig> liquid honing
Nassradom n <Telekom> wet radome
Nassreiniger m <Kerntech, Kohlen> scrubber
Nassreinigung f <Umweltschmutz> wet scrubbing
Nassschlamm m <Abfall> liquid sludge, slurry
Nassschlammdeponie f <Abfall, Umweltschmutz> wet sludge dumping site
Nassschleifen n 1. <Ker & Glas> wet polishing; 2. <Maschinen> wet grinding
Nassschleifmaschine f 1. <Fertig> wet grinder; 2. <Maschinen> wet-grinding machine
Nasssieberei f <Kohlen> wet screening
Nassspulen-Tantalkondensator m <Elektrotech> wet-slug tantalum capacitor
Nassthermometer n <Heiz & Kälte> wet-bulb thermometer
Nasstreibstoff m <Lufttrans> emulsified fuel
Nasstrennverfahren n <Abfall> wet sorting *(von Müll)*
Nass- und Trockenschleifen n <Ker & Glas> wet and dry polishing
Nasswaschanlage f <Sicherheit> wet washer
Nasswäscher m <Abfall> wet scrubber, wet scrubbing device
Nasszyklon m <Chemtech> hydrocyclone
Nasszylinderlaufbuchse f <Kfztech> wet cylinder liner
nationale Telefonauskunft f <Telekom> national directory assistance
nationaler britischer Standard m <Qual> National British Standard *(NBS)*
nationales Patent n <Patent> national patent
Nationalitätskennziffer f <Telekom> NID, nationality identification digit
Natrium n *(Na)* <Chemie> sodium *(Na)*
Natriumalginat n <Lebensmittel> sodium alginate
Natriumamid n <Chemie> sodamide
Natrium-Ammonium-Hydrogenphosphat-Tetrahydrat n <Chemie> microcosmic salt
natriumarmes Salz n <Lebensmittel> reduced sodium salt
Natriumbicarbonat n <Lebensmittel> baking soda, bicarbonate of soda, sodium bicarbonate
Natriumbogenlampe f <Strahlphys> sodium arc lamp
Natriumcaseinat n <Lebensmittel> sodium caseinate
Natrium-D-Linie f <Phys> sodium D-line *(Atomphysik)*
natriumgekühlter Reaktor m <Kerntech> sodium-cooled reactor
natriumgekühltes Ventil n <Kfztech> sodium-cooled valve
Natriumglutamat n <Lebensmittel> monosodium glutamate
Natriumhydrogencarbonat n <Lebensmittel> baking soda, bicarbonate of soda, sodium bicarbonate
Natriumhydrosulfit n <Chemie> *(AE)* hydrosulfite, *(BE)* hydrosulphite
Natriumlampe f <Elektriz> sodium lamp
Natriumpolyphosphat n <Lebensmittel> sodium polyphosphate
Natrium-Schwefel-Akkumulator m <Trans> *(AE)* sodium sulfur storage battery, *(BE)* sodium sulphur storage battery
Natriumsulfat n <Ker & Glas> saltcake

Natriumtetraborat

Natriumtetraborat n <Chemie> sodium borate
Natron n 1. <Ker & Glas> soda; 2. <Lebensmittel> baking soda; 3. <Papier> soda
Natronlauge f <Fertig> caustic soda *(Kunststoffinstallationen)*
Natronsalpeter m 1. <Chemie> nitratine; 2. <Ker & Glas> *(AE)* soda niter, *(BE)* soda nitre, sodium nitrate
Natronzellstoff m <Papier> soda pulp
Naturasphalt m <Bau> mineral pitch
Naturbaustein m <Bau> building stone
naturbelassen adj <Lebensmittel> unrefined
Natureckstein m <Bau> perpend stone *(beidseitig sichtbar)*
Naturfaser f <Textil> *(AE)* natural fiber, *(BE)* natural fibre
Naturgas n <Kohlen> conventional gas, natural gas
Naturgasbus m <Trans> bus with pressurized natural gas
Naturgasmotor m <Kfztech> natural gas engine
Naturgröße f <Konstzeich> full-scale representation, full-size representation
Naturharz n <Fertig> vegetable resin
Natur-Hochtransparentpapier n <Konstzeich> natural high transparency paper
Naturholzfarbe f <Bau> oleoresinous paint
naturidentisch adj <Lebensmittel> nature-identical
Naturkautschuk m *(NK)* <Kunststoff> natural rubber *(NR)*
Naturkonstante f <Phys> universal constant
Naturkraft f <Umweltschutz> physical agent
natürlich bewegtes Kühlmittel n <Heiz & Kälte> naturally-circulated coolant
natürlich gealtert adj <Thermod> naturally-aged
natürlich vorkommendes Element n <Kerntech> naturally occurring element
natürlich vorkommendes Radionuklid n <Strahlphys> natural radionuclide
natürliche Abdichtung f <Abfall> natural lining *(einer Deponie)*
natürliche Abnutzung f <Maschinen> wear and tear
natürliche Alterung f <Kunststoff, Thermod> *(BE)* natural ageing, *(AE)* natural aging
natürliche Belüftung f <Kerntech> natural ventilation
natürliche Breite f <Kerntech> natural width *(eines Energieniveaus)*
natürliche Entwässerung f <Wasserversorg> natural drainage
natürliche Farben fpl <Druck> true colours
natürliche Grundbelastung f <Umweltschutz> background level
natürliche Grundwasserregenerierung f <Wasserversorg> natural ground water recharge
natürliche Konvektion f <Strömphys> natural convection
natürliche Konvektionskühlung f <Heiz & Kälte, Kerntech> natural convection cooling
natürliche Koordinaten fpl <Geom> natural coordinates
natürliche Linienbreite f <Strahlphys> natural line width
natürliche Luftbewegung f <Heiz & Kälte> natural air circulation
natürliche Luftkühlung f <Heiz & Kälte> natural air cooling
natürliche Luftumwälzung f <Heiz & Kälte> natural air circulation
natürliche Radioaktivität f <Strahlphys> natural radioactivity
natürliche Säuerung f <Umweltschutz> natural acidification
natürliche Schwingung f <Maschinen> natural oscillation
natürliche Sprache f <Comp & DV, Künstl Int> NL, natural language
natürliche Sprachverarbeitung f <Künstl Int> NLP, natural language processing
natürliche Umwelt f <Umweltschmutz, Wasserversorg> natural environment
natürliche Zahl f <Comp & DV> natural number
natürlicher Böschungswinkel m <Bau> angle of repose
natürlicher Hafen m <Wassertrans> *(AE)* natural harbor, *(BE)* natural harbour
natürlicher Logarithmus m <Math> natural logarithm *(Logarithmus zur Basis e)*
natürlicher Luftstrom m <Heiz & Kälte> *(AE)* natural draft, *(BE)* natural draught
natürlicher Luftzug m <Heiz & Kälte> *(AE)* natural draft, *(BE)* natural draught
natürlicher Zug m <Heiz & Kälte> *(AE)* natural draft, *(BE)* natural draught
natürliches Altern n <Metall> *(BE)* natural ageing, *(AE)* natural aging
natürlichsprachliche Schnittstelle f <Künstl Int> NLI, natural language interface
natursaurer See m <Umweltschmutz> naturally acid lake
Natursteinplatte f <Bau> flag
Natururan n <Kerntech> natural uranium
Natururanblock m <Kerntech> natural uranium slug *(als Brennstoff)*
Natururanbrennstoff m <Kerntech> natural uranium fuel
Natururanreaktor m <Kerntech> uranium reactor
Nautik f <Wassertrans> navigation
Nautiker m <Wassertrans> navigator
nautisch adj <Wassertrans> nautical
nautisches Instrument n <Wassertrans> navigational instrument
nautisches Jahrbuch n <Wassertrans> nautical almanac
Navier-Stokes'sche Gleichung f <Phys, Strömphys> Navier-Stokes equation
Navigation f <Lufttrans, Telekom, Wassertrans> navigation
Navigation f **nach Lotreihen** <Wassertrans> navigation by sounding
Navigationsfunkhilfe f <Funktech> aids to navigation radio control *(anrac)*
Navigationsgerät n <Wassertrans> navigational instrument
Navigationsgerät n **für Horizontallage** <Lufttrans> horizontal situation indicator
Navigationshilfe f <Wassertrans> navigational aid
Navigationsleitsystem n <Funktech> navigational guidance system
Navigationsradar n <Funkort, Wassertrans> navigation radar, navigational radar
Navigationssatellitensystem n *(NNSS)* <Wassertrans> Navy Navigation Satellite System, NNSS *(der US-Marine)*
Navigator m <Wassertrans> navigator
navigatorisch adj <Wassertrans> navigational
navigieren v <Comp & DV, Wassertrans> navigate
Naviplan n <Wassertrans> naviplane
Nb *(Niobium)* <Chemie> Cb, Nb *(columbium)*
NC *(numerische Steuerung)* <Comp & DV, Elektriz, Kontroll, Maschinen, Mechan> NC *(numerical control)*
NC-Maschine f <Maschinen> NC machine
Nd *(Neodymium)* <Chemie> Nd *(neodymium)*
Nebelboje f <Wassertrans> fog buoy
Nebelhorn n <Wassertrans> foghorn *(Navigation)*
Nebelkammer f 1. <Teilphys> cloud chamber; 2. <Wellphys> cloud chamber *(um Strahlung nachzuweisen)*
Nebelleuchte f <Kfztech, Wassertrans> fog lamp
Nebelsignal n <Wassertrans> fog signal
Nebelwarnung f <Wassertrans> fog warning
Neben... 1. <Ergon> secondary; 2. <Geom> minor; 3. <Maschinen> minor, secondary
Nebenachse f <Geom> minor axis *(einer Ellipse)*; secondary axis

Nebenanschluss *m* <Telekom> extension
Nebenanschlussleitung *f* <Telekom> extension line
Nebenaufgabe *f* <Ergon> secondary task
Nebenausfall *m* <Qual> minor failure
Nebenaussendung *f* <Funktech> spurious emission
Nebenausstrahlung *f* <Funktech> spurious emission
Nebencomputer *m* <Comp & DV> slave
Nebendarstellung *f* <Konstzeich> secondary representation
Nebeneffekt *m* <Comp & DV> side effect
nebeneinander geordnetes Fenster *n* <Comp & DV> tiled window
nebeneinander liegende Zylinder *mpl* <Maschinen> side-by-side cylinders
Nebeneinanderschaltung *f* <Elektrotech> parallel arrangement
Nebenfehler *m* <Qual> minor defect, minor non-conformance, minor non-conformity
Nebenfluss *m* <Wassertrans> tributary of river *(Geographie)*
Nebengebäude *n* <Bau> outbuilding
Nebengeräusch *n* 1. <Aufnahme> ambient noise; 2. <Telekom> sidetone
nebengeschaltet *adj* <Elektriz, Elektrotech> parallel-connected
Nebengetriebe *n* <Fertig> power takeoff
Nebengleis *n* <Bau, Eisenbahn> spur track
Nebengrundwasserspiegel *m* <Wasserversorg> apparent water table
Nebenkanal *m* <Maschinen> bypass channel
Nebenkeule *f* 1. <Funktech> side lobe *(Antenne)*; 2. <Phys> side lobe; 3. <Raumfahrt> side lobe *(Weltraumfunk)*; 4. <Wassertrans> side lobe *(Radar)*
Nebenlinie *f* <Eisenbahn> branch line
Nebenluft *f* 1. <Heiz & Kälte> secondary air; 2. <Kfztech> air leak; 3. <Maschinen> secondary air
Nebenpleuelstange *f* <Fertig> auxiliary connecting rod
Nebenprodukt *n* <Erdöl, Kohlen, Maschinen, Umweltschmutz> by-product
Nebenquelle *f* <Elektronik> companion source
Nebensammler *m* <Wasserversorg> branch sewer
Nebenschluss *m* <Akustik, Elektrotech> shunt
Nebenschlussdynamo *m* <Elektrotech> shunt dynamo
Nebenschlusseinrichtung *f* <Eisenbahn> shunter
Nebenschlusselement *n* <Elektriz> shunting device
Nebenschlusserregung *f* <Elektrotech> shunt excitation
Nebenschlusskondensator *m* <Gerät> bypass capacitor
Nebenschlussmotor *m* <Elektriz, Elektrotech> shunt motor
Nebenschlussregler *m* <Elektrotech> shunt regulator
Nebenschlussschalter *m* <Elektriz> shunt switch, shunting switch
Nebenschlussspule *f* <Elektrotech> shunt coil
Nebenschlussstrom *m* <Elektrotech> shunt current
Nebenschlussstromkreis *m* <Elektrotech> shunt circuit
Nebenschlusswicklung *f* <Elektriz, Elektrotech> shunt winding
Nebenschlusswiderstand *m* <Elektriz, Elektrotech> shunt resistance, shunt resistor
Nebenschneide *f* <Maschinen> minor cutting edge
Nebenspannung *f* <Kerntech> secondary stress
Nebenspeicher *m* <Comp & DV> slave cache, slave store
Nebensprechabstand *m* <Elektronik> signal-to-crosstalk ratio
Nebensprechdämpfung *f* 1. <Aufnahme> crosstalk rejection; 2. <Telekom> crosstalk attenuation
Nebensprechdämpfungsmesser *m* <Aufnahme> crosstalk meter
Nebensprecheinheit *f* <Aufnahme> crosstalk unit
Nebensprechen *n* <Akustik, Aufnahme, Comp & DV, Telekom> crosstalk
Nebenstelle *f* <Telekom> telephone extension, extension
Nebenstellenanlage *f* **mit Handvermittlung** <Telekom> PMBX, private manual branch exchange
Nebenstellenanschluss *m* <Telekom> extension
Nebenstellenteilnehmer *m* <Telekom> extension subscriber, extension user
Nebenstellenvermittlung *f* <Telekom> PBX switchboard
Nebenstraße *f* 1. <Bau> byroad, byway; 2. <Trans> cross-country road
Nebenstromluft *f* <Raumfahrt> bypass air
Nebenstromölfilter *n* <Kfztech> bypass oil cleaner
Neben- und Hauptwartungsleistung *f* <Lufttrans> minor and major servicing operation
Nebenweganode *f* <Elektrotech> bypass anode
Nebenzipfel *m* 1. <Elektronik> side lobe; 2. <Funktech> side lobe *(Antenne)*; 3. <Phys> side lobe
Nebenzipfelunterdrückung *f* <Elektronik> side lobe cancellation
Néel'sche Temperatur *f* <Elektriz> Néel temperature
Néel'scher Punkt *m* <Phys> Néel point
Negation *f* <Comp & DV> NOT operation, negation
Negativ *n* <Druck, Foto> negative
negativ *adj* <Elektriz, Math> negative
negativ dotierte Zone *f* <Phys> negatively-doped region
negativ geerdete Anschlussklemme *f* <Kfztech> negative grounded terminal
negativ geerdeter Pol *m* <Kfztech> negative grounded terminal
negativ geladenes Ion *n* <Elektriz, Kohlen, Phys, Strahlphys> negative ion
Negativabtastung *f* <Fernseh> negative scanning
Negativabzug *m* <Foto> negative print
Negativbetrachter *m* <Foto> negative viewer
Negativbild *n* <Foto> negative image
negative Anschlussklemme *f* <Elektriz, Elektrotech, Kfztech> negative terminal
negative Beschleunigung *f* <Maschinen> minus acceleration
negative Bildphase *f* <Fernseh> negative picture phase
negative Bildschirmdarstellung *f* <Fernseh> inverse video, reverse video
negative Charakteristik *f* <Elektrotech> negative resistance characteristic
negative Elektrode *f* <Elektriz, Elektrotech> cathode, negative electrode
negative ganze Zahl *f* <Math> negative integer
negative Impedanz *f* <Elektrotech> negative impedance
negative Konduktanz *f* <Elektriz> negative conductance
negative Kontaktbank *f* <Kfztech> negative bank
negative Krümmung *f* <Geom> negative curvature
negative Ladung *f* <Elektriz, Elektrotech, Phys> negative charge
negative Logik *f* <Elektronik> negative logic
negative Magnetostriktion *f* <Elektrotech> negative magnetostriction
negative Mantelreibung *f* <Kohlen> negative skin friction
negative Platte *f* <Kfztech> negative plate
negative Profilverschiebung *f* <Fertig> diameter decrease *(Getriebelehre)*
negative Quittung *f* <Comp & DV, Telekom> NAK, negative acknowledgement
negative Reaktivität *f* <Kerntech> negative reactivity
negative Rückkopplung *f* 1. <Aufnahme, Elektronik> inverse feedback, negative feedback; 2. <Funktech> inverse feedback; 3. <Telekom, Wellphys> negative feedback

negative

negative Rückmeldung *f* <Comp & DV, Telekom> NAK, negative acknowledgement
negative Spannung *f* <Elektrotech> negative voltage
negative Spannungsgegenkopplung *f* <Elektrotech> negative voltage feedback
negative Spannungsversorgung *f* <Elektrotech> negative voltage supply
negative Speiseleitung *f* <Elektriz> negative feeder
negative Steigung *f* <Lufttrans> reverse pitch *(Propeller)*
negative Stromversorgung *f* <Elektrotech> negative power supply
negative Vorspannung *f* <Elektrotech> negative bias
negative Vorspur *f* <Kfztech> toe-out *(Vorderräder)*
negative Zahl *f* <Math> negative number
Negativecho *n* <Fernseh> negative echo
negativer Meniskus *m* <Foto> divergent meniscus
negativer Pol *m* <Elektriz, Elektrotech> cathode, negative pole
negativer Radsturz *m* <Kfztech> negative camber
negativer Reaktor *m* <Kerntech> negative reactor
negativer Spanwinkel *m* <Fertig, Maschinen> negative rake
negativer Sprung *m* <Wassertrans> reverse sheer *(Schiffbau)*
negativer Widerstand *m* <Elektriz, Elektrotech> negative resistance
negativer Winkel *m* <Geom> negative angle
negatives Glimmlicht *n* <Phys> negative glow
negatives Videosignal *n* <Fernseh> negative video signal
Negativfilm *m* <Druck> negative film
Negativhalter *m* <Foto> negative carrier
Negativlack *m* <Elektronik> negative resist
Negativlichtpausfilm *m* <Konstzeich> negative diazotype film
Negativlinse *f* <Foto> concave lens
Negativluftkissen *n* <Kfztech> negative air cushion
Negativmodulation *f* <Elektronik> negative modulation
Negativphotolack *m* <Elektronik> negative photoresist
Negativphotoresist *m* <Elektronik> negative photoresist
Negativstopfen *n* <Telekom> negative justification
Negativ-Stuffing *n* <Telekom> negative justification
Negativtasche *f* <Foto> negative sleeve
Negativwiderstandsdiode *f* <Elektronik> negative resistance diode
Negativwiderstandsoszillator *m* <Elektronik> negative resistance oscillator
Negativwiderstandsverstärker *m* <Elektronik> negative resistance amplifier
Negator *m* <Comp & DV> negator
Negatron *n* <Elektronik> negatron
Neger *m* <Fernseh> caption stand
Negierung *f* <Comp & DV> negation
Nehmerzylinder *m* <Kfztech> slave cylinder *(Bremsen, Kupplung)*
neigbarer Mattscheibenrahmen *m* <Foto> swinging back
Neigekopf *m* <Foto> tilting head *(Vergrößerer)*
neigen *v* <Bau> lean, sink
neigen v/sich <Bau> lean
Neigetechnik *f* <Elektrotech, Elektronik> tilting technology
Neigung *f* 1. <Bau> batter, falling gradient, inclination, incline, pitch, slant; slope *(eines Daches)*; 2. <Fertig> slant, splay, tilt; 3. <Geom> inclination, slope; 4. <Maschinen> incline; 5. <Mechan> bevel; 6. <Nichtfoss Energ> declination; 7. <Phys> inclination; 8. <Raumfahrt> inclination, pitch
Neigungsebene *f* <Bau> inclined plane
Neigungsmesser *m* 1. <Bau> clinometer, slope level; batter level *(Vermessung)*; 2. <Kohlen> clinometer, inclinometer; 3. <Metrol> clinometer

Neigungsverhältnis *n* <Eisenbahn> gradient ratio
Neigungswinkel *m* 1. <Fertig, Geom> angle of inclination; 2. <Ker & Glas> angle of pitch; 3. <Kfztech> camber *(Straße)*; 4. <Mechan> bevel
Neigungszeiger *m* <Eisenbahn> gradient post
Nein-Schaltung *f* <Elektronik> inverter gate
nematischer Flüssigkristall *m* <Elektronik> nematic liquid crystal
Nematode *f* <Lebensmittel> eelworm
Nennableitimpulsstrom *m* <Elektriz> nominal discharge current, rated discharge current
Nennabschaltstrom *m* <Elektriz> admissible interrupting current, rated interrupting current
Nenn-Anlaufzeit *f* <Elektrotech> nominal acceleration time *(IEC)*
Nennanschlussspannung *f* <Elektriz> nominal supply voltage, rated supply voltage
Nennansprechspannung *f* <Elektriz> nominal breakdown voltage, rated breakdown voltage
Nennanstieg *m* <Elektriz> nominal rate of rise, rated rate of rise; nominal steepness, rated steepness *(z. B. bei Stoßspannungsimpulsen)*
Nennarbeitsbedingungen *fpl* <Elektriz> nominal operating conditions, rated operating conditions
Nennaufnahmeleistung *f* <Elektriz> nominal input, nominal input power, nominal power input, rated input, rated input power, rated power input
Nennausgangsleistung *f* <Elektriz> nominal output, nominal output power, nominal power output, rated output, rated output power, rated power output
Nennausschaltleistung *f* <Elektriz> nominal breaking capacity, rated interrupting capacity
Nennausschaltstrom *m* <Elektriz> nominal breaking current, rated interrupting current
Nenn-„Aus"-Spannung *f* <Elektriz> nominal off-voltage, rated off-voltage
Nennbedingungen *fpl* <Elektrotech> rated conditions
Nennbelastbarkeit *f* <Elektrotech> power rating
Nennbetriebsbedingungen *fpl* <Raumfahrt> nominal operating conditions *(Raumschiff)*
Nennbohrung *f* <Kerntech> NB, nominal bore
Nenndaten *npl* <Phys> rating
Nenndeckenspannung *f* <Elektriz> nominal exciter ceiling voltage *(Erregermotor)*
Nenndicke *f* <Werkprüf> nominal thickness *(Glas)*
Nenndrehmoment *n* <Elektriz> rated load torque, rated torque
Nenndrehzahl *f* <Elektrotech> nominal speed, nominal speed of rotation, rated speed, rated speed of rotation
Nenndruck *m* 1. <Fertig> nominal pressure *(Kunststoffinstallationen)*; 2. <Maschinen> pressure rating
Nenndurchlassstrom *m* <Elektriz> nominal forward current, mean forward current, mean on-state current, rated forward current
Nenndurchlauf *m* <Comp & DV> rated throughput
Nenndurchmesser *m* <Maschinen> nominal diameter
Nenndurchsatz *m* <Heiz & Kälte> design water rate
Nenneingangsleistung *f* <Elektriz> nominal power input, rated power input
Nenneinschaltstrom *m* <Elektriz> nominal making current, rated making current
Nenneinschaltvermögen *n* <Elektriz> contact current-closing rating, nominal making capacity, rated making capacity
Nenn-„Ein"-Spannung *f* <Elektriz> nominal on-voltage, rated on voltage
Nenner *m* <Math> denominator
Nennerregergeschwindigkeit *f* <Elektriz> nominal exciter response, rated exciter response

Nennerregerspannung f <Elektriz> nominal field voltage, rated field voltage
Nennerregung f <Elektriz> nominal excitation, rated excitation
Nennfrequenz f <Elektriz> nominal frequency, rated frequency
Nenngehalt m <Verpack> nominal content
Nenngenauigkeit f <Elektrotech> nominal accuracy, rated accuracy
Nenngleichspannung f <Elektriz> nominal direct voltage, rated direct voltage
Nenngleichstrom m <Elektriz> nominal direct current, rated direct current
Nenngröße f 1. <Elektriz> nominal quantity, nominal value, rated quantity, rated value; 2. <Maschinen, Verpack> nominal size
Nennheizleistung f <Heiz & Kälte> rated heat output
Nennkapazität f <Elektrotech> rated capacity, rating
Nennkurzschlussspannung f <Elektriz> nominal impedance voltage, rated impedance voltage *(Transformator)*
Nennkurzschlussstrom m <Elektriz> nominal short-circuit current, rated short-circuit current
Nennlast f 1. <Elektriz, Elektrotech> nominal load, rated load; 2. <Verpack> effective load
Nennlebensdauer f <Elektronik> nominal life, nominal life duration, nominal lifetime, rated life, rated lifetime
Nennleistung f 1. <Aufnahme> power rating; 2. <Comp & DV> rated output; 3. <Elektriz, Elektrotech> rated power, power rating; 4. <Heiz & Kälte> rating; 5. <Ker & Glas> nominal power; 6. <Kerntech> rated power capacity *(eines Reaktors)*; 7. <Kohlen> rated capacity; 8. <Maschinen> power rating, rated capacity, rated power, rating
Nennleistungshöhe f <Lufttrans> rated altitude
Nennlichtstärke f <Foto> effective candle power
Nennmaß n 1. <Maschinen> nominal size; 2. <Qual> basic size
Nennmessbereich m <Metrol> rating
Nennmessgenauigkeit f <Metrol> rated accuracy
Nennöffnung f <Foto> effective aperture
Nennöffnung f eines Objektivs <Foto> effective aperture of a lens
Nennschließstrom m <Elektriz> rated making current
Nennschub m <Raumfahrt> nominal thrust *(im Vakuum)*
Nennschweißstrom m <Elektriz> rated welding current
Nennspannung f 1. <Elektriz, Elektrotech> rated voltage; nominal voltage *(eines Systems)*; 2. <Metall, Qual> nominal stress
Nennspannungsfestigkeit f der Isolierung <Elektriz> rated insulation level
Nennstärke f <Werkprüf> nominal thickness
Nennstehblitzstoßspannung f <Elektriz> nominal lightning impulse withstand voltage, rated lightning impulse withstand voltage
Nennstehschaltstoßspannung f <Elektriz> nominal switching impulse withstand voltage, rated switching impulse withstand voltage
Nennstrom m <Comp & DV, Elektriz, Elektrotech> rated current
Nennweite f 1. <Fertig> NB, nominal bore *(Kunststoffinstallationen)*; 2. <Heiz & Kälte> nominal size; 3. <Maschinen> nominal width
Nennwert m <Elektriz, Elektrotech> nominal value, rated value, rating
Nennwert m der Spannungsschritte <Elektriz> rated step voltage
Nennwert m des Durchflussstromes <Elektriz> rated through-current
Nennwert m des kurzzeitigen Stromes <Elektriz> rated short-time current
Nennwert m des Spannungsverhältnisses <Elektriz> rated voltage ratio
Nennwindgeschwindigkeit f <Nichtfoss Energ> rated wind speed
Neoabietin... <Chemie> neoabietic
Neodymium n *(Nd)* <Chemie> neodymium *(Nd)*
Neodym-Laser m <Elektronik> neodymium laser
Neoergosterin n <Chemie> neoergosterol
Neohexan n <Chemie> neohexane, trimethylbutane, triptane
neoklassischer Pincheffekt m <Kerntech> neoclassical pinch effect
Neon... <Elektriz, Elektrotech> neon
Neonbeleuchtung f <Elektriz> fluorescent lighting
Neongasanzeiger m <Elektrotech> neon indicator
Neonglimmlampe f <Elektriz> neon glow-lamp
Neonlampe f <Elektrotech> neon lamp
Neonlicht n <Elektrotech> fluorescent lighting, neon lamp
Neonröhre f 1. <Elektriz, Elektrotech> neon tube; 2. <Phys> neon fluorescent tube, neon tube
Neopren n <Bau, Chemie, Kunststoff, Verpack> neoprene
Neopren-Weichdichtung f <Kunststoff> *(AE)* neoprene molded seal, *(BE)* neoprene moulded seal
Neper n 1. <Akustik, Elektronik> neper; 2. <Phys> neper; 3. <Telekom> neper
Neper'scher Logarithmus m <Math> Neperian logarithm *(Neper'sche Näherung an den Logarithmus einer Zahl)*
Nephelin m <Chemie> nepheline
Nephelinsyenit m <Ker & Glas> nepheline syenite
Neptunium n *(Np)* <Chemie> neptunium *(Np)*
Nernst-Brücke f <Elektrotech> Nernst bridge
Nernst'scher Verteilungskoeffizient m <Metall> partition coefficient
Nerol n <Chemie> dimethyloctadienol, nerol
nervig adj <Textil> crisp
Netto... <Kerntech, Metrol, Umweltschmutz> net
Netto-Brutrate f <Kerntech> net breeding rate
Nettodonator m <Umweltschmutz> net donator
Nettofläche f <Nichtfoss Energ> net area *(eines Kollektors)*
Nettoflügelfläche f <Lufttrans> net wing area
Nettogewicht n <Verpack> net weight
Nettoladung f <Elektrotech> net charge
Nettoregistertonnage f <Wassertrans> net registered tonnage
Nettoregistertonne f <Metrol> register ton
Nettotonnage f 1. <Erdöl> net tonnage *(Handel, Schifffahrt)*; 2. <Wassertrans> net tonnage
Nettotonne f <Metrol> net ton
Nettotonnengehalt m <Wassertrans> net tonnage
Nettowärmeverlust m <Heiz & Kälte> net heat loss
Nettozeitintervall n <Trans> net time interval
Network-Computing-Umgebung f <Comp & DV> network computing environment
Netz n 1. <Bau> network *(Stromverteilung, Rohrleitungen)*; system *(von Rohren, Kabeln)*; 2. <Comp & DV> net, network; 3. <Elektriz, Elektrotech> grid, network; 4. <Fernseh> *(BE)* mains, *(AE)* supply network; 5. <Fertig> grid, mains; circuit *(Kunststoffinstallationen)*; 6. <Konstzeich> grid; 7. <Kontroll, Kunststoff> network; 8. <Raumfahrt> mesh *(Sieb)*; 9. <Telekom> network, system; 10. <Trans> network
Netz n mit Nachrichtenvermittlung <Comp & DV> message-switched network
Netz n mit Vermittlung <Telekom> switched network
Netz n zur gemeinsamen Nutzung von Betriebsmitteln <Comp & DV> resource-sharing network

Netzabschalter 528

Netzabschalter m <Elektriz> line power switch
Netzabschluss m **für Basisanschluss (NTBA)** <Telekom> network termination for basic access, NTBA (ISDN)
Netzabstand m <Fertig> interplanar spacing (Raumgitter)
Netzabtrennschalter m <Elektriz> line-isolating switch
Netzanalysator m <Elektriz> (BE) network analyser, (AE) network analyzer
Netzanalyse f <Comp & DV, Elektriz> network analysis
Netzanalysierer m <Comp & DV> (BE) network analyser, (AE) network analyzer
Netzanbieter m <Telekom> network carrier
Netzanlasser m <Elektriz> (AE) line starter, (BE) starter
Netzanpassung f <Telekom> network adapter
Netzanschluss m 1. <Elektriz, Elektrotech> line connection, mains supply, power supply; 2. <Nichtfoss Energ> grid connection
Netzanschlussfilter n <Elektronik> power supply filter
Netzanschlusskabel n <Elektrotech> mains lead
Netzanschlussteil n <Elektrotech> power pack
Netzarchitektur f <Comp & DV, Elektrotech> network architecture
netzartig adj <Lufttrans> network-like (Struktur)
Netzaufsicht f <Telekom> network supervisor
Netzausfall m 1. <Comp & DV> network failure; power outage; 2. <Elektrotech> line fault, power failure; 3. <Telekom> network breakdown
Netzausfallreserve f **mit Batterie** <Funktech> battery backup
Netzausläufer m <Elektrotech> network spur
Netzbelastungsanalyse f <Comp & DV> network load analysis
Netzbetreiber m <Telekom> network operator, network provider, carrier
netzbetrieben adj <Elektrotech> mains-operated
Netzbetriebssystem n <Comp & DV> network operating system
Netzbewehrung f <Bau> mat reinforcement
Netzbrummen n 1. <Akustik> hum; 2. <Elektriz> a.c. hum, mains hum, mains ripple, power-line hume, power-line noise, system hum
Netzdatenbank f <Comp & DV> network database
Netzebene f 1. <Comp & DV> network level; 2. <Elektriz> network sector; 3. <Fertig> face (Kristall); 4. <Telekom> network level
Netzeinwahl f <Telekom> (AE) direct outward dialing, (BE) direct outward dialling
Netz-Erde Spannung f <Elektriz> (BE) line-to-earth voltage, (AE) line-to-ground voltage
Netz-Erstanschluss m <Kerntech> first connection to grid
Netzfilter n <Elektronik> line filter
Netzflechtwerk n <Bau> netting (Draht)
Netzfolgeverhalten n <Kerntech> grid-following behaviour (eines Kernkraftwerkes)
Netzfrequenz f <Elektriz, Elektrotech> commercial power frequency, grid frequency, mains frequency
Netzführung f <Comp & DV, Telekom> network management
Netzführungszentrum n <Telekom> NMC, (AE) network management center, (BE) network management centre
Netzgebiet n <Konstzeich> ruled area
netzgekoppelt adj <Elektriz> (AE) connected to the electrical network, (BE) connected to the mains, mains-linked
Netzgerät n <Elektriz, Elektrotech> power pack, power supply unit
Netzgerät n **mit einem Ausgang** <Elektrotech> single output power supply
Netzgeräusch n <Aufnahme> (AE) current noise, mains noise

netzgespeist adj <Elektriz> mains-operated
Netzgleichrichter m <Elektriz> mains rectifier
Netzkabel n <Elektrotech> mains cable
Netzkarte f <Telekom> network map
Netzkennung f <Telekom> network identification code
Netzklemme f <Elektriz> line terminal
Netzknoten m 1. <Comp & DV, Elektriz, Elektronik, Elektrotech> node; 2. <Telekom> node, terminal node
Netzknoten m **für Funkkanalverwaltung** <Mobilkom> radio node controller (UMTS)
Netzknotenprozessor m <Comp & DV> node processor
Netzkontrollzentrum n 1. <Elektriz> net control station; 2. <Telekom> NCC, (AE) network control center, (BE) network control centre
Netzkonvergenz f <Telekom> network convergence (zwischen Festnetzen und mobilen Kommunikationsnetzen)
Netzkoppler m <Telekom> network gateway
Netzkoppler m **auf Anwendungsebene** <Telekom> application level gateway
Netzkopplung f <Elektriz> connection to mains
Netzkupplung f <Comp & DV, Elektrotech> network interconnection
Netzlinie f <Konstzeich> ruled line
Netzmantelelektronenschweißen n <Fertig> fusarc welding
Netzmasche f <Kohlen> screening mesh
Netzmigration f <Comp & DV> network consolidation
Netzmittel n <Foto, Kunststoff, Lebensmittel, Meerschmutz, Phys, Textil> wetting agent
Netzmodell n <Comp & DV> network model
Netznachbildung f <Elektriz, Metrol> line-impedance stabilization network, power-line stabilization network
netzparalleler Betrieb m <Elektriz> mains parallel operation
Netz-PC m <Comp & DV> network computer
Netzplan m 1. <Comp & DV> network diagram, network map; 2. <Telekom> network map
Netzplantechnik f <Textil> critical path analysis
Netzprotokoll n <Comp & DV> network protocol
Netzrechner m <Telekom> network computer, network PC
Netzregelung f <Elektrotech> line control
Netzrissbildung f <Bau> crazing
Netzrückspeiseunterwerk n <Elektrotech> receptive substation
Netzrückwirkung f <Elektriz> feedback to supply network, system perturbation
Netzschalter m <Comp & DV, Elektriz, Elektrotech> line switch, mains switch
Netzschicht f <Telekom> network layer (OSI)
Netzschnittstelle f <Comp & DV, Telekom> network interface
Netzschnittstellenkarte f <Comp & DV, Telekom> network interface card
Netzschutz m <Elektriz> mains protection, network limiter, network protection (Sicherung, Strombegrenzer)
Netzspannung f 1. <Elektriz, Elektrotech, Fernseh> line voltage, mains voltage; 2. <Kontroll> power mains
Netzspannungsabfall m <Elektriz> power- line voltage failure
Netzspannungsfrequenzregelung f <Elektrotech> terminal Volts/Hertz control
Netzspannungsregler m <Elektrotech> power-line voltage controller, power-line voltage regulator
Netzstation f <Comp & DV> network station
Netzsteckdose f <Elektrotech> mains socket
Netzstecker m <Elektriz, Elektrotech> mains plug, power outlet

Netzsteuerung f <Trans> network control
Netzsteuerungsschicht f <Telekom> network layer (OSI)
Netzstrom m <Elektrotech> mains current, supply current
Netzstromfrequenz f <Elektriz> power frequency
Netzstromstoß m <Elektriz> line transient current
Netzstromversorgung f <Elektrotech, Phys> mains supply
Netzstruktur f 1. <Optik> reticle, reticule; 2. <Telekom> network topology
Netzsynthese f <Elektriz> network synthesis
Netzteil n 1. <Fernseh> power supply unit; 2. <Foto> powerpack unit
Netzträger m <Wassertrans> common carrier (Satellitenfunk)
Netztransformator m <Elektriz, Elektrotech> (BE) mains transformer, power transformer
Netztrenner m <Elektriz> power switch
Netztrennschalter m <Elektriz> line breaker
Netzübergang m <Telekom> bridge, gateway, network gateway
netzüberschreitende Kommunikation f <Telekom> internetwork communication
Netzübertragungseinheit f <Comp & DV> gateway
Netzüberwacher m <Comp & DV> network controller
Netzüberwachung f und -führung f <Telekom> network supervision and management
Netzverbindungsrechner m <Telekom> gateway
Netzverbund m 1. <Comp & DV, Elektrotech> network interconnection; 2. <Telekom> interconnection
Netzverkehrslenkung f <Telekom> network routing
Netzversorgung f <Phys> mains supply
Netzversorgungsbereich m <Telekom> network coverage
Netzverwaltung f <Comp & DV, Telekom> network management
Netzverwaltungszentrale f <Telekom> NMC, (AE) network management center, (BE) network management centre
Netzverzögerung f <Comp & DV> network delay
netzweit adj <Comp & DV> network-wide
netzweite Erreichbarkeit f <Mobilkom> anywhere call pickup
Netzwerk n 1. <Comp & DV> net, network; 2. <Elektrotech> net, network, mesh; 3. <Kontroll> network; 4. <Raumfahrt> mesh network (Weltraumfunk); 5. <Telekom> network
Netzwerkadapter m <Telekom> network adapter
Netzwerkanalysator m <Comp & DV, Elektrotech, Funktech, Gerät> (BE) network analyser, (AE) network analyzer
Netzwerkanalyse f <Comp & DV, Elektriz, Elektrotech> network analysis
Netzwerkarchitektur f <Comp & DV, Elektrotech> network architecture
Netzwerkbandbreite f <Comp & DV> network bandwidth
Netzwerkbandbreite erhalten v <Comp & DV> conserve network bandwidth
netzwerkbasierte Anwendung f <Comp & DV> network-centric application
Netzwerkbetrieb m <Comp & DV> networking
Netzwerkbildner m <Chemie, Ker & Glas> network former
Netzwerkentwurf m <Comp & DV> network design
netzwerkfähige Produkte npl <Comp & DV> networked products
Netzwerkkapazität/bei maximaler <Comp & DV> at maximum network capacity
Netzwerkkonstante f <Elektrotech> network constant

Netzwerkmanager m <Comp & DV> network manager
Netzwerkmodell n <Comp & DV> network model
Netzwerkname m <Comp & DV> network name
Netzwerk-PC m <Telekom> network PC
Netzwerkplan m <Comp & DV> network diagram
Netzwerkprozessor m <Comp & DV> network processor
Netzwerksimulator m <Comp & DV> network simulator
Netzwerksteuerkanal m <Comp & DV> network control channel
Netzwerksteuerprogramm n <Comp & DV> network control program
Netzwerksynthese f <Elektrotech> network synthesis
Netzwerktheorie f <Elektrotech> network theory
Netzwerktopologie f <Comp & DV> network topology
Netzwerkumwandlung f <Funktech> network transformation
Netzwerkverwaltung f <Comp & DV, Telekom> network management
Netzwerkwandler m <Ker & Glas> network modifier
netzwerkweite Datenverarbeitung f <Comp & DV> Net computing
Netzzentrale f <Elektriz> net control station
Netzzug m <Wassertrans> haul (Fischerei)
Netzzugang m <Comp & DV> network access
Netzzugangsstelle f <Comp & DV> network access point, NAP
Netzzugriffssteuerung f <Comp & DV> network access control
Netzzugriffsüberwachung f <Comp & DV> network access control
Netzzusammenbruch m <Telekom> system black-out
neu abgleichen v <Gerät> rebalance
neu adressieren v <Telekom> relocate
neu aufbauen v <Elektronik> refresh (des Bildschirminhalts)
neu aufgebautes Bild n <Elektronik> refreshed image
neu aufziehen v <Kfztech> reline (Bremsanlage)
neu ausrüsten v 1. <Bau> re-equip; 2. <Wassertrans> refit (schiff)
neu bearbeiten v <Druck> revise
neu beginnen v <Telekom> restart
neu beschichten v <Anstrich> recoat
neu einschleifen v <Fertig, Kfztech> reseat (Ventil)
neu einstellen v <Mechan> reset
neu einstufen v <Qual> regrade
neu laden v <Comp & DV> reload
neu nachweisen v <Qual> re-prove, re-test
neu schreiben v <Comp & DV> rewrite
neu starten v <Comp & DV, Kontroll> restart
neu verfugen v <Bau> repoint
neu zulassen v <Qual> recertify
Neuanlauf m <Comp & DV> restart
Neuanzeige f <Comp & DV> refresh
Neuaufladung f <Elektrotech> recharging
Neuauflage f <Druck> new edition
Neuaufnahme f <Bau> (AE) releveling, (BE) relevelling (Vermessung)
Neuausstattung f <Wassertrans> refit (Schiff)
Neubearbeitung f <Druck> revised edition
Neubelag m <Fertig> relining
Neubescheinigung f <Qual> recertification
Neudruck m <Druck> reprint
neue Dienste mpl <Telekom> new communication services
neue Kerze f <Optik> new candle
neue österreichische Tunnelbauweise f <Bau> new Austrian tunnelling method
Neueichung f <Strahlphys> recalibration
Neueinspannung f <Fertig> rechucking (Werkstück)

neues 530

neues Brennelement *n* <Kerntech> new fuel assembly, new fuel element
neueste Version *f* <Comp & DV> update; current version (eines Programms)
neuester Stand *m* **der Technik** 1. <Bau> state-of-the-art; 2. <Kontroll> state of technology; 3. <Qual> state-of-the-art technique
Neuformatierung *f* <Comp & DV> reformatting
Neugrad *m* 1. <Fertig> centesimal degree; 2. <Foto> grad
Neuheit *f* <Patent> novelty
Neukalibrierung *f* <Strahlphys> recalibration
Neukonditionierung *f* <Raumfahrt> reconditioning (Raumschiff)
Neulackieren *n* <Kunststoff> refinishing
Neuneck *n* <Geom> nonagon
Neunerkomplement *n* <Comp & DV> nine's complement
Neuorganisation *f* **der automatischen Links** <Comp & DV> automatic link reorganization
Neuprofilieren *n* <Bau> reprofiling
Neuqualifizierung *f* <Qual> requalification
neurales Netz *n (NN)* <Comp & DV, Künstl Int> neural net, neural network (NN)
neurales Netzwerk *n (NN)* <Comp & DV, Künstl Int> neural net, neural network (NN)
Neuramin... <Chemie> neuraminic
Neurodin *n* <Chemie> neurodine
Neuron *n* <Comp & DV, Künstl Int> neuron
Neusilber *n* <Metall> German silver, nickel silver
Neustapelfähigkeit *f* <Verpack> restackability
Neustart *m* <Comp & DV> restart
Neustart *m* **nach Netzausfall** <Comp & DV> PFR, power fail restart
neutral *adj* <Elektriz, Phys> neutral
neutrale Achse *f* <Bau, Maschinen> neutral axis
neutrale Übertragung *f* <Comp & DV> neutral transmission
neutrale Zone *f* <Elektrotech> neutral zone
neutraler Anker *m* <Elektriz> neutral armature
neutraler Strom *m* <Phys> neutral current (Elementarteilchen)
neutrales Atom *n* <Teilphys> neutral atom
neutrales Element *n* <Math> neutral element
neutrales Gas *n* <Raumfahrt> neutral gas
neutrales Polarrelais *n* <Elektrotech> neutral polar relay
neutrales Relais *n* <Elektrotech> neutral relay, nonpolarized relay
neutrales unpolarisiertes Relais *n* <Elektrotech> nonpolarized relay
Neutralfilter *n* <Foto> neutral density filter
Neutralglycerid *n* <Chemie> triglyceride
Neutralisation *f* <Chemie, Elektrotech, Kohlen, Kunststoff> neutralization
Neutralisationsmittel *n* 1. <Abfall> neutralizer, neutralizing agent; 2. <Chemie> neutralizer
Neutralisator *m* 1. <Chemie> neutralizer; 2. <Metall> killing agent
Neutralisieren *n* <Kunststoff> neutralization
neutralisierter Verstärker *m* <Elektronik> neutralized amplifier
Neutralisierung *f* <Abfall, Elektriz, Funktech> neutralization
Neutralisierungsmittel *n* <Sicherheit> neutralizing agent
Neutralisierungswärme *f* <Thermod> heat of neutralization
Neutralsalzsprühnebeltest *m* <Anstrich> neutral salt spray test
Neutralsulfit *n* <Papier> (AE) neutral sulfite, (BE) neutral sulphite
Neutralsulfitzellstoff *m* <Papier> (AE) neutral sulfite pulp, (BE) neutral sulphite pulp

Neutralteilchen *n* <Phys, Teilphys> neutral particle
Neutralteilchenstrahlinjektion *f* <Kerntech> neutral atom beam injection
Neutrino *n* 1. <Strahlphys> neutrino (Elementarteilchen); 2. <Teilphys> neutrino
Neutron *n (n)* <Elektriz, Kerntech, Strahlphys, Teilphys> neutron (n)
Neutronenabschirmung *f* <Kerntech> neutron shield
Neutronenabsorber *m* <Kerntech> neutron absorber
neutronenabsorbierende Reaktion *f* <Kerntech> neutron-absorbing reaction
Neutronenabsorptionsquerschnitt *m* **Null** <Kerntech> zero neutron-absorption cross section
Neutronenaktivierungsaufzeichnung *f* <Kerntech> neutron activation logging
Neutronenaufzeichnung *f* <Nichtfoss Energ> neutron log
Neutronenausbeute *f* <Strahlphys, Teilphys> neutron yield
Neutronenausbeute *f* **per Spaltereignis** <Phys> neutron field per fission
Neutronenausbruch *m* <Kerntech> neutron burst
Neutroneneinfang *m* <Strahlphys, Teilphys> neutron capture, neutron radiative capture
Neutronenfluss *m* <Kerntech> neutron flux
Neutronen-Gammalog *n* <Erdöl> neutron gamma log (Bohrlochmesstechnik)
Neutronengift *n* <Kerntech> nuclear poison
Neutronenkonverter-Flussverstärker *m* <Kerntech> (AE) neutron converter donut, (BE) neutron converter doughnut
Neutronenlethargie *f* <Kerntech, Phys> lethargy
Neutronenlog *n* <Erdöl> neutron log (Bohrlochmesstechnik)
Neutronenmasse *f (mn)* <Kerntech, Strahlphys, Teilphys> neutron mass (mn)
Neutronenmessung *f* <Erdöl> neutron logging (Bohrlochmesstechnik)
Neutronenmultiplikationskonstante *f (k)* <Kerntech> neutron multiplication constant, k (Nuklearreaktor)
Neutronen-Neutronenlog *n* <Erdöl> neutron-neutron log (Bohrlochmesstechnik)
Neutronenquelle *f* <Kerntech> neutron source
Neutronenstern *m* <Raumfahrt> neutron star
Neutronenstrahl *m* <Strahlphys, Teilphys> neutron beam
Neutronenstrahlreaktor *m* <Kerntech> beam reactor
Neutronenstreuung *f* <Strahlphys, Teilphys> neutron scattering
Neutronenüberschuss *m* 1. <Phys> isotopic number; 2. <Strahlphys, Teilphys> neutron excess
Neutronenzahl *f* <Strahlphys, Teilphys> neutron number
Neutronenzählrohr *n* <Kerntech> neutron counter tube
Neuverdrahtung *f* <Elektriz> rewiring
Neuversuch *m* <Comp & DV> retry
Neuverteilung *f* <Elektrotech> redistribution
Neuzündung *f* <Raumfahrt> reignition (Raumschiff)
Newton *n (N)* 1. <Elektriz> newton, N (Einheit); 2. <Hydraul> newton, N; 3. <Metrol, Strömphys> newton, N (Einheit)
Newton'sche Flüssigkeit *f* <Strömphys> Newtonian fluid
Newton'sche Mechanik *f* <Phys> Newtonian mechanics
Newton'sche Ringe *mpl* <Foto, Strömphys> Newton's rings
Newton-Scheibe *f* <Optik> Newton's disc, Newton's disk
Newton'sches Abkühlungsgesetz *n* <Strömphys> Newton's law
Newton'sches Fernrohr *n* <Phys> Newtonian telescope
Nf (Niederfrequenz) 1. <Aufnahme, Elektronik, Fernseh, Funktech> AF (audio frequency); LF (low frequency); 2. <Telekom> LF (low frequency)

Nf-Generator m <Elektronik> AF oscillator
Nf-Signalgenerator m <Elektronik> AF signal generator
Nf-Störung f (Niederfrequenzstörung) <Elektronik, Funktech, Telekom> AFI (audio-frequency interference)
Nf-Weiche f <Telekom> audio-frequency splitter
Nf-Zwischenverstärker m <Elektronik> booster amplifier
n-Halbleiter m <Elektronik> n-type semiconductor
n-Heptylaldehyd m <Chemie> oenanthal
Ni (Nickel) <Chemie> Ni (nickel)
Nialamid n <Chemie> nialamide
Niccolit m <Chemie> niccolite, nickel arsenide (Mineralogie)
NiCd (Nickel-Cadmium) <Elektronik, Foto> NiCd (nickel-cadmium)
nicht abgebauter Stoß m <Kohlen> highwall
nicht abgeschirmte Strahlungsquelle f <Kerntech> unshielded source
nicht abgeschrägt adj <Fertig> (AE) unbeveled, (BE) unbevelled
nicht abgesetzter Konzentrator m <Telekom> co-located concentrator
nicht angereichertes Uran n <Kerntech> unenriched uranium
nicht angeschlossen adj <Telekom> off-line
nicht auf dem Chip befindlich <Elektronik> off chip
nicht aufgeblasen adj <Ker & Glas> not blown up
nicht ausgeglichenes Ruder n <Wassertrans> unbalanced rudder
nicht ausgelaufen adj <Fertig> misrun (Guss)
nicht ausgewuchtet adj <Maschinen> unbalanced
nicht backend adj <Kohlen> (AE) free-burning, nonbaking
nicht backende Kohle f <Kohlen> nonbaking coal
nicht beherrschter Prozess m <Qual> uncontrolled process, process out of control
nicht berichtigte Eigengeschwindigkeit f <Lufttrans> indicated airspeed
nicht beschichtetes Brennstoffteilchen n <Kerntech> uncoated fuel particle
nicht codiert adj <Telekom> uncoded
nicht definierte Taste f <Comp & DV> undefined key
nicht definierter Ausdruck m <Math> undefined expression
nicht drahtgewickelter Widerstand m <Elektrotech> nonwirewound resistor
nicht eingeschalteter Zustand m <Elektrotech> off-state
nicht eingetragene Marke f <Patent> unregistered mark
nicht eingetragene Rufnummer f <Telekom> non-listed number, non-published number
nicht eingezäunt adj <Bau> unfenced
nicht färbend adj <Kunststoff> nonstaining
nicht festgeschalteter Zeichengabekanal m <Telekom> (AE) nondedicated signaling channel, (BE) nondedicated signalling channel
nicht fluchtend adj <Fertig> misaligned
nicht funkenbildend adj <Elektrotech> nonarcing
nicht geerdet adj <Elektriz> earth-free
nicht geführter Modus m <Telekom> unbound mode
nicht gepoltes Relais n <Elektriz> nonpolarized relay
nicht gerichtete Schallquelle f <Akustik> simple acoustic source
nicht gerichteter Graph m <Künstl Int> nondirected graph, nonoriented graph, undirected graph
nicht gerichtetes Mikrofon n <Aufnahme> astatic microphone
nicht gespaltener Kernbrennstoff m <Kerntech> unfissioned nuclear fuel
nicht geteilte Platte f <Comp & DV> non-shared disk

nicht geteilter Speicher m <Comp & DV> non-shared memory
nicht gewebte Matte f <Ker & Glas> nonwoven scrim
nicht gewebtes Glasgarngelege n <Ker & Glas> nonwoven scrim
nicht gezündete Röhre f <Elektronik> unfired tube
nicht gleichgerichteter Wechselstrom m <Elektriz> unrectified ac
nicht gleichlaufend adj <Fernseh> out of sync
nicht gleichzeitiger Mehrfachzugriff m (TDMA) 1. <Comp & DV, Elektronik> time division multiple access, TDMA; 2. <Raumfahrt> time division multiple access, TDMA (Weltraumfunk); 3. <Telekom> time division multiple access, TDMA
nicht gut passen v <Metall> misfit
nicht identifiziertes Flugobjekt n (UFO) <Raumfahrt> unidentified flying object (UFO)
nicht imprägniertes Gewebe n <Textil> undipped fabric
nicht in einer Ebene liegend <Fertig> noncoplanar
nicht invertierend adj <Elektrotech> noninverting
nicht ionisierende Strahlung f <Strahlphys> nonionizing radiation
nicht isolierter Leiter m <Elektriz> bare conductor
nicht kompensierter Motor m <Elektrotech> noncompensated motor
nicht konfiguriert adj <Comp & DV> configured-out
nicht leitend adj 1. <Chemie> dielectric; 2. <Elektrotech> nonconductive
nicht leitender Zustand m <Elektrotech> nonconducting state
nicht lokalisierte Interferenzlinien fpl <Phys> nonlocalized fringes
nicht löschendes Lesen n <Comp & DV> nondestructive read
nicht lotgerecht adv <Bau> off plumb, out of plumb
nicht migrierender Weichmacher m <Kunststoff> nonmigratory plasticizer
nicht mischende Kaskade f <Kerntech> no-mixing cascade
nicht moderiertes Spaltneutron n <Kerntech> unmoderated fission neutron
nicht oxidiert adj <Chemie> unoxidized
nicht rastend adj <Telekom> nonlocking
nicht reservierter Parkplatz m <Trans> nonreserved space
nicht rostend adj <Anstrich> corrosion-resistant
nicht rostender Stahl m <Anstrich> stainless steel
nicht rotierender Stern m <Lufttrans> nonrotating star (Hubschrauber)
nicht schleifend adj <Fertig> nonrubbing (Dichtung)
nicht schreibende Taste f <Comp & DV> nonprinting key
nicht sperrender Konzentrator m <Elektrotech> nonblocking concentrator
nicht sperrender Schalter m <Elektrotech> nonblocking switch
nicht sperrendes Netzwerk n <Elektrotech> nonblocking network
nicht standardisierte Steuerspur f <Fernseh> nonstandard control track
nicht strukturierter Baum m <Comp & DV> unordered tree
nicht stürzen v <Verpack> keep upright
nicht taktzustandsgesteuerte Kippschaltung f <Elektronik> unclocked flip-flop
nicht tragende Mauerverkleidung f <Bau> veneer
nicht überdacht adj <Bau> open
nicht überlappende Tonhöhen fpl <Akustik> disjoined pitches

nicht 532

nicht umkehrend *adj* <Elektrotech> noninverting
nicht verbranntes Uran *n* <Kerntech> unburned uranium
nicht verdichtete Deponie *f* <Abfall> uncompacted tip
nicht verfahrensorientierte Programmiersprache *f* <Comp & DV> nonprocedural language
nicht verfärbend *adj* <Kunststoff> nonstaining
nicht vergilbend *adj* <Kunststoff> nonyellowing
nicht verriegelnd *adj* <Telekom> nonlocking
nicht verschmutzend *adj* <Umweltschmutz> non-polluting
nicht verschmutzt *adj* <Umweltschmutz> unpolluted
nicht verunreinigt *adj* <Sicherheit> clean
nicht völlig abgesättigt *adj* <Chemie> unsaturated *(Valenz)*
nicht vormagnetisiert *adj* <Comp & DV> unbiased
nicht weich gemacht *adj* <Kunststoff> unplasticized
nicht zerstörende Prüfung *f* <Maschinen, Phys> nondestructive test
nicht zugelassen *adj* 1. <Math> nonadmitted; 2. <Telekom> nonapproved
nicht zusammenhängendes Gleiten *n* <Metall> discontinuous glide
nicht zustande gekommenes Weiterreichen *n* <Mobilkom> missed handover
nicht zutreffender Fehler *m* <Comp & DV> false error
nichtabzählbar <Math> noncountable *(Mengenlehre)*
nichtaktiv *adj* <Comp & DV, Wassertrans> inactive
nichtamphibisches Luftkissenfahrzeug *n* <Wassertrans> nonamphibious hovercraft
nichtamtsberechtigt *adj* <Telekom> exchange-barred
nichtamtsberechtigt *adj* <Telekom> fully restricted
nichtamtsberechtigte Nebenstelle *f* <Telekom> completely restricted extension
nichtamtsberechtigter Teilnehmer *m* <Telekom> fully restricted subscriber
nichtausführbare Anweisung *f* <Comp & DV> nonoperable instruction
nichtautomatischer Webstuhl *m* <Textil> nonautomatic loom
nichtbehebbarer Fehler *m* <Comp & DV> irrecoverable error
nichtbenetzbar *adj* <Chemie> hydrophobic
nichtbetriebsbereit *adj* <Comp & DV> off-line
nichtbinärer Code *m* <Telekom> nonbinary code
nichtbindiger Boden *m* <Kohlen> noncohesive soil
Nicht-Bindungselektron *n* <Kerntech> nonbonding electron
nichtbrennbar *adj* 1. <Abfall, Verpack> incombustible; 2. <Werkprüf> nonflammable
nichtdarstellbares Zeichen *n* <Comp & DV> idle character
Nicht-Draht-Potenziometer *n* <Elektriz> nonwirewound potentiometer
Nichteigenzeit *f* <Phys> improper time
Nichteisen... <Elektriz, Fertig> nonferrous
nichtelastische Verformung *f* <Kunststoff> plastic yield
nichtentflammbar *adj* 1. <Chemie> nonflammable; 2. <Verpack> flameproof, noninflammable; 3. <Werkprüf> nonflammable
nichtentflammbare Kleidung *f* <Sicherheit> flameproof clothing
nichterfolgreiche Verbindung *f* <Telekom> unsuccessful call, unsuccessful connection
Nichterfüllung *f* <Qual> noncompliance
Nichterhaltung *f* **der Parität** <Phys> nonconservation of parity
nichtersetzbare Sicherung *f* <Elektriz> nonrenewable fuse

nichteuklidische Geometrie *f* <Geom> non-Euclidean geometry
nichteuklidische Geometrie *f* **der 1. Art** <Geom> hyperbolic geometry
nichtflüchtig *adj* <Comp & DV> nonvolatile
nichtflüchtige Bestandteile *mpl* <Kunststoff> nonvolatile content
nichtflüchtiger Schreib-Lese-Speicher *m* <Elektronik> non-volatile random-access memory *(NVM)*
nichtflüchtiger Speicher *m* <Comp & DV, Elektrotech> nonvolatile memory
Nichtflüchtiges *n* <Kunststoff> nonvolatile content
nichtfrostanfälliger Boden *m* <Kohlen> nonfrost susceptible soil
nichtfunkenbildend *adj* <Elektrotech> nonarcing
NICHT-Gatter *n* <Comp & DV> NOT gate
nicht-Gauß'sches Rauschen *n* <Telekom> non-Gaussian noise
nichtgeometrische Größe *f* <Konstzeich> nongeometrical quantity
nichtgitterförmig *adv* <Math> nonlatticed *(Grafik)*
NICHT-Glied *n* <Comp & DV> NOT circuit
nichthierarchisches System *n* <Telekom> nonhierarchical system
nichtig erklären *v*/**für** <Patent> revoke
Nichtigkeitsgründe *mpl* <Patent> grounds for revocation
nichtinduktiv *adj* <Elektriz> noninductive
nichtinduktive Last *f* <Elektriz> noninductive load
nichtinduktive Schaltung *f* <Elektriz> noninductive circuit
nichtinduktive Wicklung *f* <Elektriz> noninductive winding
nichtkapazitiv *adj* <Elektriz> noncapacitive
nichtkomplanar *adj* <Fertig, Geom> noncoplanar
nichtkompostierbarer Abfall *m* <Abfall> noncompostable waste
nichtkoplanar *adj* <Fertig, Geom> noncoplanar
Nichtleiter *m* <Elektriz, Elektrotech> nonconductor; insulator *(zwischen zwei Leitern)*
nichtlinear *adj* <Elektriz, Funktech, Math, Phys, Telekom> non-linear
nichtlineare Amplitudenverzerrung *f* <Aufnahme> amplitude distortion
nichtlineare Bedingungen *fpl* <Elektrotech> nonlinear conditions
nichtlineare Interpolation *f* <Telekom> nonlinear interpolation
nichtlineare Phasenverzerrung *f* <Raumfahrt> phase nonlinear distortion
nichtlineare Programmierung *f* <Comp & DV> nonlinear programming
nichtlineare Schaltung *f* <Telekom> nonlinear circuit
nichtlineare Schwingung *f* <Telekom> nonlinear oscillation
nichtlineare Skale *f* <Elektrotech, Metrol> nonlinear scale
nichtlineare Streuung *f* <Optik, Telekom> nonlinear scattering
nichtlineare Verstärkung *f* <Telekom> nonlinear amplification
nichtlineare Verzerrung *f* <Aufnahme, Elektronik, Telekom> nonlinear distortion
nichtlinearer Stark-Effekt *m* <Phys> nonlinear Stark effect
nichtlinearer Verstärker *m* <Elektronik> nonlinear amplifier
nichtlinearer Widerstand *m* 1. <Elektriz> nonlinear resistor; 2. <Telekom> nonlinear resistance
nichtlineares digitales Sprachsignal *n* <Telekom> nonlinear digital speech
nichtlineares Filtern *n* <Telekom> nonlinear filtering

nichtlineares Glied *n* <Elektrotech> nonlinear element
nichtlineares Netzwerk *n* <Elektrotech> nonlinear network
nichtlineares Potenziometer *n* <Elektrotech> nonlinear potentiometer
nichtlineares Programmieren *n* <Comp & DV> nonlinear programming
nichtlineares System *n* <Bau> non-linear system
nichtlösbar *adj* 1. <Fertig> permanent; 2. <Maschinen> permanent; 3. *(Math)* non-solvable
nichtlösbare Verbindung *f* <Fertig> permanent connection
nichtlöschbarer Speicher *m* <Comp & DV> nonerasable storage
Nichtlöser *m* <Kunststoff> diluent
nichtmagnetisch *adj* <Chemie, Elektriz> nonmagnetic
nichtmagnetischer Stahl *m* <Elektriz> nonmagnetic steel
nichtmaskierbare Unterbrechung *f* <Comp & DV> NMI, nonmaskable interrupt
nichtmaßhaltig *adj* <Maschinen> out of tolerance
nichtmaßhaltige Mess- und Prüfmittel *npl* <Qual> out-of-calibration devices
nichtmechanischer Drucker *m* <Comp & DV> nonimpact printer
nichtmessbar *adj* <Math> incommensurable
nichtmetallischer Einschluss *m* 1. <Fertig> sonim; 2. <Metall> nonmetallic inclusion
nichtmischbar *adj* <Chemie, Erdöl, Fertig> immiscible
nichtmischbare Flüssigkeiten *fpl* <Strömphys> immiscible fluids
nichtmittig *adj* <Raumfahrt> *(AE)* off-center, *(BE)* off-centre
nichtmonotones Schließen *n* <Künstl Int> non-monotonic reasoning
nicht-negative Zahlen *fpl* <Math> nonnegative numbers
nichtnormgemäß *adj* nonconforming
nichtnormkonform *adj* <Maschinen> bastard
nichtöffentlicher beweglicher Landfunk *m* <Mobilkom> private land mobile radio network
nichtöffentliches Fernsehen *n* *(CCTV)* <Fernseh> closed-circuit television *(CCTV)*
nichtoxidierbar *adj* <Chemie> inoxidizable, unoxidizable
nichtparametrisches Bootstrapping <Math> nonparametric bootstrapping *(Schätzverfahren)*
nichtperiodische Steigung *f* <Lufttrans> collective pitch
nichtperiodische Steigungsservosteuerung *f* <Lufttrans> collective pitch follow-up
nichtperiodische Steigungssteuerung *f* <Lufttrans> collective pitch control
nichtperiodische Steigungssychronisierung *f* <Lufttrans> collective pitch synchronizer
nichtperiodische Steigungsvoreinstellung *f* <Lufttrans> collective pitch anticipator
nichtperiodischer Nickwinkel *m* <Lufttrans> collective pitch angle
nichtperiodischer Steigungsanzeiger *m* <Lufttrans> collective pitch indicator
nichtperiodischer Steigungseinstellhebel *m* <Lufttrans> collective pitch lever
nichtperiodischer Steigungsschalter *m* <Lufttrans> collective pitch switch
nichtperiodischer Steigungswinkel *m* <Lufttrans> collective pitch angle
nichtpermanenter Speicher *m* <Comp & DV, Elektrotech> volatile memory
nichtplastisch *adj* nonplastic
nichtpolares Dielektrikum *n* <Phys> nonpolar dielectric
nicht-positive Zahlen *fpl* <Math> nonpositive numbers

Nicht-Primzahl *f* <Math> composite number
nichtprogrammierbare Datenstation *f* <Comp & DV> dumb terminal
nichtreaktive Last *f* <Elektriz> nonreactive load
nichtrechtwinklig *adj* <Bau> out-of-square
nichtrechtwinklig *adj* <Maschinen> out of square
nichtregisterhaltig *adj* <Druck> out of register
nichtrekursives Filter *n* <Elektronik, Telekom> finite impulse response filter, nonrecursive filter
nichtreproduzierbar *adj* <Verpack> nonreproductible
nichtreversierbarer Motor *m* <Elektriz> nonreversible motor
nichtreversierbarer Stecker *m* <Elektriz> nonreversible plug
nichtreziproke Schaltung *f* <Telekom> nonreciprocal circuit
nichtreziproker Wellenleiter *m* <Funktech> nonreciprocal wave guide
nichtrückgewinnbare Abfälle *mpl* <Abfall> nonrecoverable waste
nichtschaltbare Kupplung *f* <Maschinen> coupling
nichtschiffbar *adj* <Wassertrans> innavigable
nichtseetüchtig *adj* <Wassertrans> unseaworthy *(Schiff)*
nichtselbstständige Regelung *f* <Regelung> manual control
nichtspezifikationsgerecht *adj* <Bau> nonconforming
Nichtsprachkommunikation *f* <Telekom> nonvoice communication
nichtstationäre Modenverteilung *f* <Telekom> non-equilibrium mode distribution
nichtstationärer Lärm *m* <Sicherheit> non-steady state noise
nichtsynchronisiert *adj* <Fernseh> nonsync
nichttransparenter leitungsvermittelter Dienst *m* <Telekom> sixty-four kbps restricted service *(B-Kanal mit 64 kbit/s)*
nichttransparenter Trägerdienst *m* <Telekom> non-transparent bearer service
Nichtübereinstimmung *f* <Qual> nonconformity
nichtumkehrbar *adj* <Math, Phys, Thermod> irreversible
nichtumkehrbare Abschaltung *f* <Kerntech> irreversible shutdown
NICHT-UND-Schaltung *f* *(N-UND-Schaltung, NAND-Gatter)* <Phys> NOT AND gate *(NAND gate)*
Nichtverbundhülle *f* <Raumfahrt> unbonded skin *(Raumschiff)*
Nichtverfügbarkeit *f* <Qual> outage, unavailability
Nichtverfügbarkeitsdauer *f* <Qual> outage duration
Nichtverfügbarkeitsrate *f* <Qual> failure rate, outage rate
nichtvergleichbar *adj* <Math> incommensurable
nichtverstellbar *adj* <Fertig> constant-flow *(Pumpe)*
nichtverwertbar *adj* <Umweltschmutz> nonreclamable, nonreusable
nichtverwertbarer Abfall *m* <Abfall> non-reclaimable waste, non-reusable waste
nichtverwertbarer Abfallstoff *m* <Abfall> nonreclamable waste
nichtverwertbarer Rückstand *m* <Abfall> waste product
nichtwässrige Elektrolysebatterie *f* <Kfztech> nonacqueous electrolyte battery
nichtwiederholbare Messung *f* <Metrol> nonrepeatable measurement
nichtwiederverwendbar *adj* <Comp & DV> nonreusable
nichtwiederverwendbarer Atommüll *m* <Abfall> non-reusable radiactive waste
nichtwiederverwendbarer radioaktiver Abfall *m* <Abfall> nonreusable radiactive waste
Nickachse *f* 1. <Lufttrans> lateral axis, pitch axis; 2. <Raumfahrt> pitch axis *(Raumschiff)*

Nickdämpfer

Nickdämpfer m <Lufttrans> pitch damper
Nickdüse f <Raumfahrt> pitch jet *(Raumschiff)*
Nickel n *(Ni)* <Chemie> nickel *(Ni)*
Nickel-Cadmium n *(NiCd)* <Foto> nickel-cadmium *(NiCd)*
Nickel-Cadmium-Akku m <Foto> NiCad battery *(Blitzlicht)*
Nickel-Cadmium-Akkumulator m <Raumfahrt> nickel-cadmium battery *(Raumschiff)*
Nickel-Cadmium-Batterie f <Elektrotech, Kfztech> nickel-cadmium battery
Nickel-Cadmium-Stahlakkumulator m <Kfztech> nickel-cadmium battery
Nickel-Cadmium-Zelle f <Elektrotech, Funktech> nickel-cadmium cell
Nickel-Eisen-Akkumulator m <Kfztech> nickel-iron storage battery
Nickel-Eisen-Batterie f <Elektrotech> nickel-iron battery
nickelhaltig adj <Fertig> nickeliferous
Nickelhydroxid n <Raumfahrt> nickel-hydroxide *(Raumschiff)*
Nickelin m <Chemie> niccolite, nickel arsenide
Nickellegierung f <Anstrich> nickel alloy
nickelplattieren v <Fertig> nickel-clad
nickelplattiert adj <Fertig> nickel-clad
Nickelplattierung f <Elektriz> nickel plating
nickelreicher Stahl m <Metall> high-nickel steel
Nickelsilber n <Elektrotech> nickel silver
Nickelstahl m <Metall> high-nickel steel, nickel steel
Nickel-Zink-Akkumulator m <Kfztech> nickel-zinc storage battery
Nicken n 1. <Lufttrans> pitching; 2. <Raumfahrt> pitch *(Raumschiff)*
Nickkanal m <Lufttrans> pitch channel
Nickkreisel m <Raumfahrt> pitch gyro *(Raumschiff)*
Nicklage f <Raumfahrt> pitch attitude *(Raumschiff)*
Nickwinkel m <Lufttrans> pitch angle
Nickwinkelgeschwindigkeitskreisel m <Lufttrans> pitch rate gyro
Nicol-Prisma n <Phys> Nicol prism
Nicotinamid n <Chemie> niacinamide, nicotinamide
Nicotinsäureamid n <Chemie> niacinamide, nicotinamide
Nicotyrin n <Chemie> nicotyrine
Nidridhärten n <Chemie> nitridation *(Metallurgie)*
Niederbordwagen m <Eisenbahn> *(AE)* flatcat, *(BE)* platform wagon, gondola car, gondola wagon
niederbrennen v <Thermod> destroy by fire
Niederbringung f einer Bohrung <Erdöl> well boring, well drilling, well sinking *(Bohrtechnik)*
Niederbringung f einer Förderbohrung <Erdöl> production drilling *(Fördertechnik)*
Niederdruck m <Phys> low pressure
Niederdruckbrenner m 1. <Bau> low-pressure blowpipe; 2. <Heiz & Kälte> low-pressure burner
niederdrücken v <Bau> weigh down
Niederdruckgasbrenner m <Heiz & Kälte> low-pressure gas burner
Niederdruckheizung f <Heiz & Kälte> low-pressure heating
Niederdruckkessel m <Heiz & Kälte> low-pressure boiler
Niederdruckkolbenverdichter m <Maschinen> low-pressure piston compressor
Niederdruckkompressor m <Maschinen> low-pressure compressor
Niederdruckmanometer n <Gerät> *(BE)* low-pressure gauge, *(AE)* low-pressure gage
Niederdruckölbrenner m <Heiz & Kälte> low-pressure atomizer
Niederdruckprüfung f <Werkprüf> low-pressure test

Niederdruck-Quecksilberdampflampe f <Elektriz> low-pressure mercury lamp
Niederdruckregler m <Heiz & Kälte> low-pressure controller
Niederdruck-Schwimmerventil n <Heiz & Kälte> low-pressure float valve
Niederdruckspülen n <Meerschmutz> low-pressure flushing
Niederdrucktreibstofffilter n <Lufttrans> low-pressure fuel filter
Niederdruckverdichter m <Lufttrans> fan
Niederdruckwarmwasseranlage f <Heiz & Kälte> low-pressure hot-water system
Niederdruckwarmwasserkessel m <Heiz & Kälte> low-pressure hot-water boiler
Niederdruckzylinder m <Maschinen> low-pressure cylinder
niedere Dienstgüte f <Telekom> lower quality of service
niedere Programmiersprache f <Comp & DV> LLL, low-level language
niederenergetischer fokussierter Ionenstrahl m <Strahlphys> low-energy-focused ion beam
Niederenergie-Kernphysik f <Kerntech> low-energy nuclear physics
niederfrequente Rauschminderung f <Elektronik> audio noise reduction, denoising
niederfrequentes elektromagnetisches Feld n <Sicherheit> low-frequency electromagnetic field
Niederfrequenz f *(Nf)* 1. <Aufnahme, Elektronik> audio frequency *(AF)*; low frequency *(LF)*; 2. <Fernseh> audio frequency *(AF)*; 3. <Funktech> audio frequency *(AF)*; low frequency *(LF)*; 4. <Telekom> low frequency *(LF)*
Niederfrequenzausgleich m <Elektronik, Funktech, Telekom> low-frequency compensation
Niederfrequenzfilter n <Elektronik, Funktech, Telekom> low-frequency filter
Niederfrequenzgenerator m 1. <Elektriz> low-frequency generator; 2. <Elektronik, Funktech> audio-frequency oscillator
Niederfrequenzheizung f <Elektrotech> low-frequency heating
Niederfrequenz-Induktionserwärmung f <Elektrotech> low-frequency induction heating
Niederfrequenz-Induktionserwärmungsgerät n <Elektrotech> low-frequency induction heater
Niederfrequenzkanal m <Telekom> audio channel
Niederfrequenzofen m <Elektrotech> low-frequency furnace
Niederfrequenzoszillator m <Elektronik, Funktech, Telekom> low-frequency oscillator
Niederfrequenzsignal n 1. <Elektronik, Funktech> audio-frequency signal, low-frequency signal; 2. <Telekom> low-frequency signal
Niederfrequenzsignalgenerator m <Elektronik, Funktech> audio-frequency signal generator, low-frequency signal generator
Niederfrequenz-Spektrometer n <Labor> audio-frequency spectrometer
Niederfrequenzstörung f *(Nf-Störung)* <Elektronik, Funktech, Telekom> audio-frequency interference *(AFI)*
Niederfrequenztrichterlautsprecher m <Aufnahme> low-frequency horn loudspeaker
Niederfrequenzverstärker m 1. <Aufnahme> audio-frequency amplifier; 2. <Elektronik> audio-frequency amplifier, low-frequency amplifier; 3. <Funktech, Telekom> low-frequency amplifier
Niederfrequenzverstärkung f <Elektronik, Funktech, Telekom> low-frequency amplification

Niederfrequenzweiche f <Telekom> audio-frequency splitter
Niederfrequenz-Zwischenverstärker m <Elektronik> booster amplifier
Niedergang m 1. <Fertig> downstroke; 2. <Wassertrans> companionway (Schiffbau)
Niedergangsluke f <Wassertrans> companionway hatch (Schiffbau)
Niedergangspfosten m <Wassertrans> companionway post (Schiffbau)
Niedergangstreppe f <Wassertrans> companionway ladder (Schiffbau)
niederholen v 1. <Funktech> lower (Antenne); 2. <Wassertrans> haul down (Flagge, Segeln)
Niederholer m <Wassertrans> downhaul (Tauwerk)
Niederhub m <Kfztech> downstroke (Motor)
niederohmig adj <Funktech> low-impedance
niederohmiger Widerstand m <Phys> low resistance
Niederohmigkeit f <Funktech> low impedance
niederreißen v <Bau> wreck
Niederschlag m 1. <Bau> precipitation; 2. <Chemie> deposit; 3. <Chemtech, Erdöl> precipitate (Petrochemie); 4. <Fertig> deposit; 5. <Lebensmittel> bottoms; 6. <Metall> precipitate; 7. <Phys> sedimentation; 8. <Umweltschmutz> fallout (radioaktiv)
niederschlagen v 1. <Chemtech> precipitate; 2. <Fertig> deposit (Galvanotechnik)
niederschlagen v/sich <Thermod> condense
Niederschlagsammler m <Umweltschmutz> precipitation collector
Niederschlagselektrode f <Abfall, Umweltschmutz> collecting electrode
Niederschlagsfront f <Lufttrans> precipitation front
Niederschlagsgebiet n <Wasserversorg> catchment area, drainage area, drainage basin, precipitation area, rainfall area
Niederschlagsgefäß n <Chemtech> precipitation vessel
Niederschlagsmenge f <Nichtfoss Energ> rainfall
Niederschlagsmesser m 1. <Bau, Labor, Nichtfoss Energ> (AE) rain gage, (BE) rain gauge; 2. <Wasserversorg> (AE) precipitation gage, (BE) precipitation gauge, (AE) rain gage, (BE) rain gauge
Niederschlagsvorfall m <Umweltschmutz> precipitation event
Niederschlagswasser n <Wasserversorg> storage level regulation, storm sewage
Niederschlagung f <Phys> settling
Niederschraubabsperrventil n <Bau> screw-down stop valve
Niederschraubhahn m <Bau> screw-down cock
Niederschraubventil n <Bau> screw-down valve
Niederspannung f 1. <Elektriz, Elektrotech> low tension, low voltage; 2. <Telekom> low voltage
Niederspannungsanlage f <Elektrotech> low-voltage installation
Niederspannungskabel n <Elektriz> low-voltage cable
Niederspannungskreis m <Elektriz> low-voltage network
Niederspannungslastschalter m <Elektrotech> low-voltage circuit breaker
Niederspannungsleistungsschalter m <Elektrotech> low-voltage circuit breaker
Niederspannungsnetz n <Elektrotech> low-voltage mains, low-voltage power-installation network, low-voltage system
Niederspannungsrichtlinie f <Elektrotech> low-voltage directive
Niederspannungssteuerung f <Elektrotech> low-voltage regulation

Niederspannungswicklung f <Elektriz> low-voltage winding
Niederspannungszweig m <Elektrotech> low-voltage arm, low-voltage branch (z. B. einer Messbrücke)
Niedertemperaturprüfung f <Werkprüf> low-temperature test
Niedertor n <Wasserversorg> aft gate (Schleusentor)
Niedervolt-Lautsprecher m <Aufnahme> low-voltage electrostatic loudspeaker
niederwertig adj <Comp & DV> low-order
niederwertige Position f <Comp & DV> low-order position
niederwertiges Bit n <Comp & DV> low-order bit
niederwertigst adj <Comp & DV> least significant
niedrige Auflösung f <Comp & DV> low res, low resolution
niedrige Dämpfung f <Telekom> low loss (Übertragung)
• **mit niedriger Dämpfung** low-loss
niedrige Drehzahl f 1. <Kfztech> slow speed (Motor); 2. <Maschinen> low speed
niedrige Einfügungsdämpfung f <Telekom> low-insertion loss
niedrige Geschwindigkeit f 1. <Kfztech> slow speed (Motor); 2. <Maschinen, Telekom> low speed
niedrige Kapazität f <Elektrotech> low capacitance
niedrige Reynoldszahl f <Strömphys> low Reynolds number
niedrige Umlaufbahn f <Raumfahrt> low-altitude orbit
niedriger Druck m <Phys> low pressure
niedriger einstufen v <Bau> grade down (Gebäude)
niedriger Gang m <Maschinen> low gear
niedriger Schwefelgehalt m <Erdöl> (AE) low sulfur content, (BE) low sulphur content (Öl, Erdgas)
niedriger Signalpegel m <Elektronik> low-signal level
niedriger Widerstand m <Telekom> low resistance
niedriges Überlaufwehr n <Hydraul> flat-crested weir
niedrigfester Bohrschlamm m <Erdöl> low-solid mud (Bohrtechnik)
niedriggekohlt adj <Fertig> steely (Eisen)
Niedrig-Hoch-Übergang m <Elektronik> low-to-high transition
niedriglegiert adj <Anstrich> low-alloy (Stahl)
niedriglegierter Stahl m <Metall> low-alloy steel
Niedrigpegel m <Elektronik> low-logic level
Niedrigschubtriebwerk n <Raumfahrt> low-thrust motor (Raumschiff)
niedrigstabiler Bohrschlamm m <Erdöl> low-solid mud
niedrigste erreichbare Emissionsrate f <Umweltschmutz> lowest achievable emission rate
niedrigste nutzbare Frequenz f (NNF) <Funktech> lowest usable frequency (LUF)
niedrigster Niedrigwasserstand m <Wasserversorg> minimum low water
niedrigstwertige Ziffer f <Comp & DV, Telekom> LSD, least significant digit
niedrigstwertiges Bit n (LSB) <Comp & DV, Telekom> least significant bit (LSB)
Niedrigwasser n 1. <Nichtfoss Energ> ebb (Gezeiten); 2. <Wassertrans> low tide, low water; ebb (Gezeiten)
Niedrigwasserabfluss m <Wasserversorg> low-water discharge
Niedrigwasserlinie f <Wassertrans> low-water mark
Niedrigwassermarke f <Bau> low-water mark
Niedrigwasserstand m <Wasserversorg> low-water level
Nierencharakteristik f <Funktech> cardioid pattern
nierenförmiger Schlitz m <Maschinen> kidney-shaped slot
Niet m <Bau, Mechan> rivet
Nietbolzenkette f <Bau> pin chain

Niete

Niete f <Bau, Mechan> rivet
nieten v <Fertig> rivet, rivet up
Nieten n <Bau, Maschinen> riveting
Nietenbefestigung f <Maschinen> rivet fastening
Nietendöpper m <Maschinen> rivet header, rivet snap
Nietenkaltpresse f <Maschinen> rivet cold press
Nietensetzkopf m <Bau> set
Nietentreiber m <Maschinen> driver, rivet drift
Nieter m <Bau> riveter
Niethammer m <Bau, Maschinen> rivet hammer, riveting hammer
Nietkopf m <Bau, Maschinen> rivet head • **Nietköpfe machen** <Fertig> snap
Nietkopfsetzer m 1. <Bau> rivet set; 2. <Fertig> heading set
Nietlochreibahle f <Maschinen> bridge reamer, rivet-hole reamer
Nietlochsenker m <Maschinen> rivet countersink
Nietmaschine f <Bau, Maschinen> riveter, riveting machine
Nietmutter f <Maschinen> rivet nut
Nietnaht f <Maschinen, Verpack> riveted seam
Nietplatte f <Bau> riveted plate
Nietreihenabstand m <Fertig> back pitch
Nietrieren n <Maschinen> nitriding
Nietschaft m <Maschinen> rivet shank
Nietschaftdurchmesser m <Maschinen> rivet shan diameter
Nietstift m <Maschinen> rivet pin
Nietung f <Bau, Maschinen> riveting
Nietverbindung f <Bau, Maschinen> rivet joint, riveted joint
Nietverfahren n <Maschinen> riveting technique
Niobit m <Chemie> niobite
Niobium n (Nb) <Chemie> (AE) columbium, (BE) niobium (niobium)
Nippel m <Bau, Maschinen, Papier> nipple
Nipptide f <Nichtfoss Energ, Wassertrans> neap tide
Nirosta m <Anstrich> stainless steel
Nische f <Bau> housing, recess
Nit n <Optik> nit (Einheit)
Nitramin n <Chemie> nitramine
Nitrat n <Chemie, Umweltschmutz> nitrate
Nitratcellulose f <Kunststoff> cellulose nitrate
Nitrazin... <Chemie> nitrazine
Nitrid n <Chemie> nitride
Nitridreaktor m <Kerntech> (AE) nitride fueled reactor, (BE) nitride fuelled reactor
Nitridstahl m <Metall> nitriding steel
nitrieren v 1. <Chemie> nitrate, nitride; 2. <Chemie> nitrify
Nitrieren n <Chemie> nitration
nitrierhärten v <Chemie> nitride (Metallurgie)
Nitrierhärtung f <Mechan> nitride hardening
Nitrierstahl m <Mechan> nitrided steel
Nitrierung f <Chemie> nitration
nitrifizieren v <Chemie> nitrify (Bakterien)
Nitril n <Chemie> nitrile
Nitrilkautschuk m 1. <Fertig> nitrile rubber (Kunststoffinstallationen); 2. <Kunststoff> nitrile rubber
Nitrin n <Chemie> nitrine
Nitrit n <Chemie> nitrite
Nitro... <Chemie> nitro...
Nitroanilin n <Chemie> nitroaniline
Nitrobenzen n <Chemie> nitrobenzene
Nitrobenzol n <Chemie> nitrobenzene
Nitrochloroform n <Chemie> chloropicrin, trichloronitromethane
Nitroethan n <Chemie> nitroethane
Nitrofilmunterlage f <Foto> nitrate base
Nitroform n <Chemie> nitroforme
Nitrogen n (N) <Chemie> nitrogen (N)
Nitroglycerin n <Chemie> nitroglycerin
Nitrometer n <Chemie> azotometer, nitrometer
Nitromethan n <Chemie> nitromethane
Nitronaphthalin n <Chemie> nitronaphthalene
Nitrophenol n <Chemie> nitrophenol
nitros adj <Chemie> nitrous
Nitrosat n <Chemie> nitrosate
Nitrosierung f <Chemie> nitrosation
Nitrosit n <Chemie> nitrosite
Nitrosyl... <Chemie> nitrosyl
Nitrosylhydrogensulfat n <Chemie> (AE) nitrosulfuric acid, (BE) nitrosulphuric acid
Nitrotoluol n <Chemie> nitrotoluene
Nitroweinsäure f <Chemie> nitrotartaric acid
Nitrozelluloselack m <Bau> nitrocellulose lacquer
Nitryl... <Chemie> nitryl
Nitrylchlorid n <Chemie> nitryl chloride
Niveau n 1. <Bau> grade, level; 2. <Mechan, Metrol, Wasserversorg> level
Niveauanzeige f <Kerntech> level indicator
Niveauanzeiger m <Gerät> liquid level indicator
niveaugleich adj <Bau> at grade
Niveauhalten n <Kerntech> level holding
Niveaukonstanthalter m <Gerät> constant-level device
Niveaumessgerät n <Gerät> (AE) level gage, (BE) level gauge
Niveaumessung f <Gerät> level measurement
Niveauschalter m <Kontroll> level switch
Niveausteuerung f <Kontroll> level control system
Niveauverschiebung f <Kerntech> level displacement, level shift
Niveauwächter m <Heiz & Kälte> float switch
nivellieren v 1. <Bau> grade, level (Straßenbau); 2. <Metrol> bone in
Nivellieren n 1. <Bau> boning (Vermessung); 2. <Fertig> (AE) leveling, (BE) levelling
Nivelliergerät n <Bau> surveyor's level
Nivellierinstrument n 1. <Bau> (AE) leveling instrument, (BE) levelling instrument, level, transit; A-1 level (Vermessung); 2. <Gerät> (AE) leveling instrument, (BE) levelling instrument
Nivellierkreuz n <Bau> (AE) leveling rod, (BE) levelling rod (Vermessung)
Nivellierlatte f 1. <Bau> (AE) leveling pole, (BE) levelling pole, sighting rod; (AE) levelling staff, (BE) levelling staff, pole (Vermessung); 2. <Eisenbahn> grade stake; level indicator (Vermessung)
Nivellierlatte f mit Anzeige <Bau> (AE) target leveling staff, (BE) target levelling staff, (AE) target leveling rod, (BE) target levelling rod
Nivellierpunkt m <Bau> (AE) leveling point, (BE) levelling point
Nivellierung f <Bau> (AE) leveling, (BE) levelling
Nivellierwaage f 1. <Bau> air level, spirit level; 2. <Metrol> spirit level
Nivenit n <Kerntech> nivenite
Nixie-Röhre f <Elektronik> Nixie tube®
NK (Naturkautschuk) <Kunststoff> NR (natural rubber)
N-Kanal m <Elektronik> n-channel
N-Kanal-FET m <Elektronik> N-channel FET
N-Kanal-Filter n <Elektronik> n-channel filter
N-Kanal-Gerät n <Elektronik> n-channel device
N-Kanal-Metalloxid-Silizium n <Elektronik> N-channel metal-oxide silicon
N-Kanal-Silikon-Gate-MOS-Prozess m <Elektronik> n-channel silicon-gate MOS process
N-Kanal-Technologie f <Elektronik> n-channel technology

n-leitend adj <Elektronik> n-type
n-leitende Komponente f <Elektronik> n-type component
n-leitender Halbleiter m <Phys> n-type semiconductor
n-leitender Störstoff m <Elektronik> n-type impurity
n-leitendes Silizium n <Elektronik> n-type silicon
n-leitendes Trägermaterial n <Elektronik> n-type substrate
NLQ-Druckmodus m <Comp & DV, Druck> NLQ, near-letter quality
N-Methylglykokoll n <Chemie> (N)-methylaminoacetic acid, sarcosine
NMOS <Elektronik> NMOS, n-channel metal oxide semiconductor
NMOS-Chip m <Elektronik> NMOS chip
NMOS-Integrationsschaltung f <Elektronik> NMOS integrated circuit
NMOS-Komponente f <Elektronik> NMOS component
NMOS-Logik f <Elektronik> NMOS logic
NMOS-Transistor m <Elektronik> NMOS transistor
NMR (magnetische Kernresonanz) <Erdöl, Teilphys> NMR (nuclear magnetic resonance)
NN 1. <Comp & DV, Künstl Int> (neurales Netz, neurales Netzwerk) NN (neural network); 2. <Wassertrans> (Normalnull) msl (mean sea level)
NNF (niedrigste nutzbare Frequenz) <Funktech> LUF (lowest usable frequency)
NNSS (Navigationssatellitensystem) <Wassertrans> NNSS (Navy Navigation Satellite System)
No (Nobelium) <Chemie> No (nobelium)
Nobelium n (No) <Chemie> nobelium (No)
Nocke f 1. <Fertig> (BE) disc cam, (AE) disk cam (Getriebelehre); cam (Kunststoffinstallationen); 2. <Kfztech, Mechan> cam
Nocken m 1. <Kfztech> cam; 2. <Maschinen> cam, tappet
Nocken m **und Stößel** m <Maschinen> cam and follower
nockenbetätigt adj <Maschinen> cam-operated
Nockendrehen n <Maschinen> cam turning
Nockenerhebung f 1. <Fertig> cam lobe (Getriebelehre); 2. <Kfztech> cam lobe
Nockenform f 1. <Lufttrans> cam contour; 2. <Maschinen> cam profile; 3. <Mechan> cam shape
Nockenfräsmaschine f <Fertig> cam-milling machine
nockengesteuert adj <Fertig> (BE) disc-cam-operated, (AE) disk-cam-operated (Getriebelehre)
nockengesteuerter Automat m <Fertig> (BE) disc-cam-operated screw machine, (AE) disk-cam-operated screw machine
Nockenprofil n <Kfztech> cam profile
Nockenschalter m <Elektriz> cam switch
Nockenscheibe f 1. <Fertig> plate cam; 2. <Lufttrans> cam lobe; 3. <Maschinen> (BE) disc cam, (AE) disk cam
Nockenschleifmaschine f <Maschinen> cam grinder
Nockenschließwinkel m <Kfztech> cam angle
Nockensteuerung f <Fertig> cam control, cam gear
Nockenstößel m 1. <Fertig> cam follower (Getriebelehre); 2. <Kfztech> cam following; 3. <Lufttrans> cam follower; 4. <Maschinen> cam follower, cam roller; 5. <Mechan> cam follower
nockenverriegelt adj <Fertig> cam-lock
Nockenvorsprung m <Kfztech> cam lobe
Nockenwelle f <Kfztech, Maschinen, Mechan> camshaft
Nockenwellenantrieb m <Kfztech> camshaft drive, camshaft drive chain
Nockenwellenantriebsgehäuse n <Kfztech> timing gear housing
Nockenwellenbuchse f <Kfztech> camshaft bushing
Nockenwellendrehmaschine f <Fertig> camshaft lathe
Nockenwellengehäuse n <Maschinen> camshaft box
Nockenwellenhülse f <Kfztech> camshaft bushing
Nockenwellenräder npl <Maschinen> camshaft gears
Nockenwellenschalter m <Elektriz> camshaft controller
Nockenwellenschleifmaschine f <Maschinen> camshaft grinding machine
Nockenwellenspiel n <Kfztech> camshaft clearance
Nockenwirkung f <Maschinen> cam action
nodulare Korrosion f <Kerntech> nodular corrosion
Nomenklatur f <Patent, Qual> nomenclature
Nominalleistung f <Maschinen> rated capacity, rated power, rating
Nominalmerkmal n <Qual> nominal characteristic
nominelle Musikleistung f <Aufnahme> music power rating
nomineller Messbereich m <Labor> rated range
Nomogramm n 1. <Bau> nomogram; 2. <Phys> nomograph
Nonacosan n <Chemie> nonacosane
Nonan n <Chemie> nonane
Nonan... <Chemie> nonyl
Nonansäure f <Chemie> nonoic acid
Nonen n <Chemie> nonene, nonylene
Nonius m <Fertig, Maschinen, Mechan, Metrol> vernier
Noniuseinstellung f <Gerät> vernier adjustment
Noniuspotenziometer n <Elektriz> vernier potentiometer
Noniusskale f <Maschinen> vernier scale
Nonose f <Chemie> nonose
Nonstop Flug m <Lufttrans> nonstop flight
Nonstop-Transportsystem n <Trans> continuous transportation system
Nonwoven-Matte f <Kunststoff> nonwoven mat
Nonwoven-Scrim n <Ker & Glas> nonwoven scrim
Nonyl... <Chemie> nonyl
NO-OP (Nulloperation, keine Operation) <Comp & DV> NO-OP (no-operation)
NO-OP-Befehl m <Comp & DV> NO-OP instruction
Noppe f 1. <Ker & Glas> knop; 2. <Textil> nep, slub
noppen v <Textil> nap
Noppen n <Textil> napping, picking
Noppenwerkzeuge npl <Ker & Glas> knob tools
Noradrenalin n <Chemie> noradrenaline
Norator m <Phys> norator (elektrisches Bauelement)
Norbornadien n <Chemie> norbornadiene
Norbornan n <Chemie> norbornane
Nordhauser'sche Schwefelsäure f <Chemie> Nordhausen acid, oleum
nördliche Breite f <Wassertrans> northern latitude (Navigation)
Nordlicht n <Wassertrans> aurora borealis, northern lights (Wetterkunde)
Nordlichtabsorption f <Phys> auroral absorption
Nordlichtreflexion f <Phys> auroral reflexion
Nordlichtzone f <Funktech> auroral zone (Wellenausbreitung)
Nordpol m <Phys> north pole
nordstabilisiert adj <Wassertrans> north-up (Radar)
nordwärts adv <Wassertrans> north, northerly
nordwärts anliegen v <Wassertrans> stand to the north (Navigation)
Nordwind m <Wassertrans> north wind
Norephedrin n <Chemie> norephedrine
Norepinephrin n <Chemie> noradrenaline
NOR-Gatter n <Comp & DV> NOR gate
NOR-Glied n <Comp & DV> NOR circuit
Norm f <Comp & DV, Fernseh, Kontroll, Maschinen, Metall, Telekom> standard
normal adj <Phys> standard; normal (im rechten Winkel)
Normal n 1. <Qual> measurement standard, standard; 2. <Telekom> standard

Normalbatterie

Normalbatterie f <Elektrotech> standard cell
Normalbaulänge f <Fertig> standard overall length *(Kunststoffinstallationen)*
Normalbeanspruchung f <Maschinen> normal stress
Normalbedingungen fpl <Maschinen> normal conditions
Normalbelastung f <Elektrotech, Maschinen> standard load
Normalbenzin n <Kfztech> *(AE)* regular gasoline, *(BE)* regular petrol
Normalbereich m 1. <Comp & DV> normal range; 2. <Gerät> standard range
Normalbeschleunigung f <Maschinen> normal acceleration
Normalbraunglas n <Ker & Glas> neutral amber glass
Normalbruch m <Metall> normal rupture
Normaldatenfluss m <Comp & DV> normal flow
Normaldruck m <Heiz & Kälte> standard pressure
Normale f <Geom> normal
normale Abschaltung f <Kerntech> proper shutdown *(eines Reaktors)*
normale Arbeitsbedingungen fpl <Maschinen> normal working conditions
normale Betriebsbedingungen fpl <Kontroll> normal operating conditions
normale Bremsbetätigung f <Eisenbahn> normal brake application
normale Einsatzbedingungen fpl <Maschinen> normal working conditions
normale Hörfläche f <Akustik> normal auditory sensation area
normale Hör-Schmerzgrenze f <Akustik> normal threshold of painful hearing
normale Hörschwelle f <Akustik> normal hearing threshold
normale Leitung f <Telekom> ordinary line
normale Prüfung f <Qual> normal inspection
normale Schwingung f <Phys> normal mode *(linear polarisiert)*
normale Springzeitebbe f <Nichtfoss Energ> low-water ordinary spring tides
normale Springzeitflut f <Nichtfoss Energ> HWOST, high-water ordinary spring tide
normale Vakuumbremsbetätigung f <Eisenbahn> normal vacuum brake application
normale Verbindung f <Kerntech> normal coupling
normale wechselseitig betriebene Leitung f <Telekom> ordinary bothway line
Normaleinstellung f **der Objektivstandarte** <Foto> front standard adjustment
Normalelement n <Elektriz, Elektrotech, Phys> standard cell
normaler Fehler m <Comp & DV> soft error
normaler Sinkflugwinkel m <Lufttrans> normal descent angle
normaler Verkehr m <Trans> normal traffic
normaler Verkehrsfluss m <Trans> normal flow
normaler Zeeman-Effekt m <Phys> normal Zeeman effect
normales Energieniveau n <Kerntech> normal energy level
normales Wasser n <Kerntech> normal water, ordinary water
Normalfarben fpl <Druck> standard inks
normalfeucht adj <Textil> conditioned
Normalflamme f <Bau> neutral flame
Normalform f 1. <Comp & DV> normal form; 2. <Math> normal form, standard form
Normalformat n <Comp & DV> normal format
Normalfrequenz f <Funktech, Telekom> standard frequency *(SF)*

Normalfrequenz f **des Atomstrahls** <Kerntech> atomic beam frequency standard
Normalfrequenzgenerator m 1. <Aufnahme> synthesizer *(mit Frequenzsynthese)*; 2. <Elektronik, Funktech, Telekom> standard frequency synthesizer
Normalfrequenz-Vergleichscharakteristik f <Akustik> standard frequency compensation characteristics
Normalglas n <Ker & Glas> neutral glass
normalglühen v <Mechan> normalize
Normalhöhe f <Druck> standard height
Normalhörender m <Akustik> normal listener
normalisierte Bohrrate f <Erdöl> NDR, normalized drilling rate *(Geologie)*
normalisierte Nachweisbarkeit f <Optik> normalized detectivity
Normalisierung f <Comp & DV> normalization
Normalkondensator m <Elektrotech> standard capacitor
Normalkoordinaten fpl <Phys> normal coordinates
Normalkraft f <Fertig> pressure load *(Zahnrad)*
Normallehre f <Maschinen> *(AE)* standard gage, *(BE)* standard gauge
Normallichtquelle f 1. <Elektrotech> standard light source; 2. <Optik> standard source
Normalmaß n <Maschinen> *(AE)* standard gage, *(BE)* standard gauge
Normalmessbereich m <Gerät> standard measuring range, standard range
Normalmikrofon n <Aufnahme> standard microphone
Normalnull n *(NN)* <Wassertrans> mean sea level *(msl)*
• **über Normalnull, üNN** <Wassertrans> above sea level, asl
Normalobjektiv n <Foto> medium-angle lens, standard lens
Normalottokraftstoff m <Kfztech> *(AE)* regular gasoline, *(BE)* regular petrol
Normalprobe f <Qual> standard size specimen
Normalreflexion f <Telekom> specular reflection
Normalspannung f 1. <Elektrotech> normal voltage, standard voltage; 2. <Kohlen> normal stress; 3. <Maschinen> direct stress
Normalspur f <Bau, Eisenbahn> *(AE)* standard gage, *(BE)* standard gauge
Normalspurbahn f <Eisenbahn> *(AE)* standard gage railroad, *(BE)* standard gauge railway
Normalstand m <Kerntech> normal level
Normaltemperatur f 1. <Erdöl> normal temperature; 2. <Heiz & Kälte, Phys> standard temperature
Normalton m <Telekom> reference tone
Normalverkehrszeit f <Trans> off-peak period
Normalverteilung f 1. <Comp & DV, Math> Gaussian distribution, normal distribution; 2. <Ergon> normal distribution; 3. <Phys, Qual> normal distribution
Normalweißglas n <Ker & Glas> neutral white glass
Normalzeit f <Wassertrans> standard time *(Navigation)*
Normalzelle f <Elektrotech> standard cell
Normband n <Comp & DV> master tape
Normblende f <Hydraul> standard orifice
Normdruck m <Maschinen, Thermod> normal pressure
Normdüse f <Gerät> standard nozzle
Normenfestlegung f <Lufttrans> data convention
Normeninstitut n **der USA** <Qual> United States of America Standards Institute *(USASI)*
Normenwahlschalter m <Fernseh> standards selector
Normenwandler m <Fernseh> standards converter
Normenwandlung f <Fernseh> standards conversion
Normfarbsignal n <Fernseh> tristimulus signal
normgerecht adj <Konstzeich, Patent, Qual> conforming to standards
Normgestell n <Telekom> standard rack

Normgewicht n <Metrol> standard weight
normieren v <Comp & DV> scale
Normieren n <Metrol> (AE) gaging, (BE) gauging
normierte Frequenz f <Optik, Telekom> normalized frequency
normierte Nachweisbarkeit f <Telekom> normalized detectivity
normierte Packung f <Verpack> package for standardization
normierter britischer Stecker m <Elektriz> (BE) Home Office socket
Normierung f 1. <Comp & DV> scaling; 2. <Fernseh> standardization
Normierungsbit n <Comp & DV> noisy digit
Normierungsfaktor m <Comp & DV> scale factor
Normkegel m <Maschinen> standard taper
Normmotor m <Elektrotech> standard dimensioned motor, standard motor
Normorphin n <Chemie> normorphine
Normschrift f <Konstzeich> standard lettering
Normteil n <Maschinen> standard part
Normtemperatur f <Heiz & Kälte> standard temperature
Normtest m <Lufttrans, Patent, Qual> certification test
Normung f 1. <Comp & DV, Maschinen> standardization; 2. <Metall> normalization; 3. <Phys, Telekom> standardization
Normvorschrift f <Maschinen> standard specification
Nornarcein n <Chemie> nornarceine
Nornicotin n <Chemie> nornicotine
Noropian... <Chemie> noropianic
Norton-Getriebe n 1. <Fertig> Norton-type mechanism; 2. <Maschinen> Norton gearbox
Nortongetriebekasten m <Fertig> Norton box
Norton'scher Satz m <Phys> Norton's theorem
Nortonschubgetriebe n <Fertig> Norton-type feed box
Norvalin n <Chemie> aminopentanoic acid, norvaline
NOR-Verknüpfung f <Comp & DV> NOR operation
NOS (Notrufortungssender) <Funktech, Telekom> ELT (emergency locator transmitter)
Noscapin n <Chemie> narcotine, noscapine, opianine
Nose-In-Aufstellung f <Lufttrans> nose-in positioning
Nose-Out-Aufstellung f <Lufttrans> nose-out positioning
notablassen v <Lufttrans> jettison
Notabschalter m <Sicherheit> emergency stopping device (Maschinensicherheit)
Notabschaltknopf m <Sicherheit> panic button
Notabschaltstab n <Kerntech> emergency shutdown rod
Notabschaltung f 1. <Comp & DV, Elektrotech> emergency cut-out, emergency off-switching, emergency shutdown; 2. <Kerntech> emergency shutdown, scram
Notabsperrventil n <Sicherheit> emergency stop valve, shut-down valve
Notabstieg m <Lufttrans> emergency descent
notabwerfen v <Lufttrans> jettison
Notaggregat n <Elektrotech> stand-by set, emergency set
Notanlage f <Telekom> emergency installation
Notausgang m 1. <Lufttrans> escape lane; 2. <Raumfahrt> emergency escape (Raumschiff); 3. <Sicherheit> emergency escape, emergency exit, fire escape, fire exit; 4. <Thermod> fire exit
Notauslöser m <Gerät, Sicherheit> emergency release push, emergency-off release push
Notausrüstung f <Sicherheit> emergency equipment
Notausschalter m <Elektrotech, Sicherheit> emergency-off switch, emergency stop switch, emergency switch
Notausschaltung f <Elektrotech, Sicherheit> emergency switching, emergency-off switching
Notausstieg m 1. <Raumfahrt> escape hatch; 2. <Sicherheit> emergency escape

Notbake f <Raumfahrt> distress beacon (Sender)
Notbakensender m <Raumfahrt> emergency beacon (Weltraumfunk)
Notbatterie f <Elektriz, Elektrotech> emergency battery
Notbeleuchtung f <Elektriz> emergency lighting
Notbereitschaft f <Telekom> emergency attention
Notbetrieb m 1. <Raumfahrt> emergency mode (Raumschiff); 2. <Sicherheit> emergency operation
Notbremse f <Eisenbahn> emergency brake
Notdienst m <Sicherheit, Telekom> emergency service
Notdruckknopf m <Elektriz> emergency button, emergency release push-button
Notdusche f <Sicherheit> safety shower
Notebook n <Comp & DV> laptop; notebook (im Buchformat)
Notfall m <Sicherheit> emergency, emergency case
Notfallausrüstung f <Lufttrans, Sicherheit> emergency equipment
Notfallhilfe f <Trans> emergency aid
Notfallleitung f <Sicherheit> emergency control
Notfallräumung f <Sicherheit> emergency evacuation (von Gebäuden)
Notfallspur f <Lufttrans> escape lane
Notgerät n <Elektrotech> stand-by unit, emergency unit
Notglocke f <Elektriz> alarm bell
Nothalteeinrichtung f <Sicherheit> emergency stop device, emergency stop system
Notizblockspeicher m <Comp & DV> scratch pad memory
Notkompass m <Lufttrans> stand-by compass
Notkühlung f <Kerntech> emergency cooling; emergency core coolant (des Reaktorkerns)
Notlandung f 1. <Lufttrans> forced landing; ditching (im Wasser); 2. <Raumfahrt> emergency landing (Raumschiff)
Notlasche f <Eisenbahn> emergency fish-plating
Notlaufeigenschaften fpl <Fertig> resistance to galling (Lagermetalle)
Notluftschleuse f <Kerntech> emergency air lock
notorisch bekannte Marke f <Patent> well-known mark
Notortungsfeuer n <Lufttrans, Wassertrans> emergency location beacon
Notplatzfunkfeuer n <Lufttrans, Wassertrans> emergency location beacon
Notrakete f <Wassertrans> emergency rocket
Notruder n <Wassertrans> jury rudder
Notruf m 1. <Lufttrans> distress call; 2. <Telekom> emergency call
Notruffunkfeuer n mit Standortmeldung <Funktech, Telekom, Wassertrans> emergency position-indicating radio beacon (EPIRB)
Notrufnummer f <Telekom> emergency number (Telefon)
Notrufortungssender m (NOS) <Funktech, Telekom> emergency locator transmitter (ELT)
Notrufsäule f <Trans> emergency telephone
Notrufsystem n <Trans> emergency call system
Notruftelefon n <Telekom> emergency telephone
Notrutsche f 1. <Lufttrans> emergency slide, escape chute; 2. <Sicherheit> rescue chute
Notschalter m 1. <Lufttrans> crash switch; emergency stop (Aufzug); 2. <Sicherheit> emergency switching device
Notschwimmerfahrwerk n <Wassertrans> emergency flotation gear
Notsignal n 1. <Eisenbahn> danger signal; 2. <Lufttrans> distress signal
Notsignalfeuer n <Wassertrans> distress flare
Notspeicherauszug m <Comp & DV> rescue dump
Notsteuerung f <Lufttrans> emergency control

Notstopp

Notstopp m <Sicherheit> safety stop
Notstromaggregat n 1. <Elektriz, Telekom> emergency power supply; 2. <Mechan> emergency power generator
Notstrombatterie f <Elektrotech> stand-by battery
Notstromdieselaggregat n <Kerntech> emergency diesel generator
Notstromumschaltung f <Elektrotech> battery backup
Notstromversorgung f <Elektriz, Elektrotech, Telekom> backup power supply, emergency power supply
Notventil n <Mechan> safety valve
Notverfahren n <Lufttrans> emergency procedure
Notverschluss m <Sicherheit> panic bolt
Notversorgungstank m <Lufttrans> emergency supply tank
Notwasserung f <Lufttrans> ditching
Notzustand m <Kerntech> emergency condition (eines Reaktors)
Novain n <Chemie> carnitine
Novocain n <Chemie> novocaine
Novolak m <Chemie> novolac
Np 1. <Chemie> (Neptunium) Np (neptunium); 2. <Phys> (Neper) Np (neper)
N-Phenylurethan n <Chemie> (N)-phenyl urethane, phenylurea
npn-Transistor m <Elektronik> npn transistor
N-Ring m (Nasenring) <Kfztech> oil scraper ring (Motor)
NRZ-Aufzeichnung f <Comp & DV> NRZ recording, non-return-to-zero recording (Wechselschrift)
NRZ-Modulation f <Telekom> NRZ modulation, nonreturn-to-zero modulation
NRZ-Verfahren n <Comp & DV> NRZ recording, nonreturn-to-zero recording
n-schrittiger Anlasser m <Elektriz> n-step starter
n-stelliger Schalter m <Elektriz> n-way switch
n-tes Bündel n <Telekom> n-th choice group
NTSC (Amerikanischer Fernsehnormungsausschuss) <Fernseh> NTSC (National Television Standards Committee)
NTSC-Farbfernsehsystem n <Fernseh> NTSC color television system
n-Tupel n <Math> n-tuple
Nuancierung f 1. <Kunststoff> shading, tinting; 2. <Textil> shade
Nuclein n <Chemie> nuclein
Nuclein... <Chemie> nucleic
Nucleinsäure f <Chemie> nucleic acid
Nucleohiston n <Chemie> nucleohistone
Nucleon n <Phys, Teilphys> nucleon
nucleophil adj <Chemie> nucleophilic
nucleophile Kraft f <Chemie> nucleophilicity
Nucleophilie f <Chemie> nucleophilicity
Nü-Faktor m <Phys> nu-factor
nuklear adj <Kerntech> nuclear
nuklear betrieben adj <Kerntech, Raumfahrt> nuclear-powered (Raumschiff)
nuklear ungefährlicher Niederschlag m <Umweltschmutz> dry deposition
Nuklear... <Kerntech> nuclear
Nuklearantrieb m <Raumfahrt> nuclear propulsion (Raumschiff)
Nuklearaufklärungssatellit m <Raumfahrt> nuclear detection satellite
Nuklearbatterie f <Elektrotech> nuclear battery
nukleare Spinquantenzahl f (I) <Kerntech> nuclear spin quantum number (I)
nukleare Wiederaufbereitungsanlage <Abfall, Kerntech> nuclear reprocessing plant
nuklearer Abfall m <Abfall> nuclear waste

nuklearer elektromagnetischer Impuls m <Telekom> nuclear electromagnetic pulse
Nuklearlog n <Erdöl> nuclear log (Bohrlochmesstechnik)
nuklearmagnetischer Resonanzlog n <Erdöl> nuclear magnetic resonance log (Bohrlochmesstechnik)
Nuklearreaktor m mit fossilem Brennstoff <Kerntech> fossil nuclear reactor
Nuklearsicherheit f <Kerntech, Sicherheit> nuclear safety
Nuklearstromversorgung f <Kerntech, Raumfahrt> nuclear power supply (Raumschiff)
Nukleierungsfaktor m <Kerntech> nucleation factor
Nukleonenzahl f (A) <Teilphys> mass number (A)
Nukleonik f <Kerntech, Phys> nucleonics
Nukleus m <Comp & DV> nucleus
Nuklid n <Phys, Teilphys> nuclide
null setzen v/auf <Math> set to zero, zero
null stellen v/auf 1. <Comp & DV> reset, zeroize; 2. <Maschinen> reset to zero, set to zero
null werden v/zu <Phys> become zero
Null f <Math> naught, nil, null, zero • **mit Nullen auffüllen** <Comp & DV> zerofill
Nullabgleich m <Gerät> null balance (Brücke)
Nullachse f <Bau> neutral axis
Nulladressbefehl m <Comp & DV> zero-address instruction
Nullanweisung f <Comp & DV> null statement
Nullanzeigegerät n <Gerät> null instrument
Nullanzeigeinstrument n <Gerät> balance indicator
Nullanzeiger m <Comp & DV> null indicator
Nullator m <Phys> nullator (elektrisches Bauelement)
Nullauftriebswinkel m <Lufttrans> zero-lift angle
Nullband n <Kerntech> zero band
Nullbefehl m <Comp & DV> do-nothing instruction
Nulldefekte mpl <Qual> zero defects
Nulldispersions-Wellenlänge f <Telekom> zero dispersion wavelength (Faseroptik)
Nulldrehung f <Textil> zero twist
Nulldurchgang m <Telekom> zero crossing
Nulleinstellung f 1. <Fertig, Kerntech> zero setting; 2. <Maschinen> zero adjustment, zero setting
Nulleinsteuerung f <Comp & DV> zero insert
Nulleintrag m <Comp & DV> null entry
Nullenergiereaktor m <Kerntech> zero-energy reactor, zero-power reactor
Nullenunterdrückung f <Comp & DV> zero elimination, zero suppression
Nullenzirkel m 1. <Fertig> bow instrument, spring bow compass; 2. <Maschinen> bow compass, spring bow compass
Nullfehler mpl <Qual> zero defects
Null-Fluss-Aufhängung f <Kfztech> null flux suspension
Nullfolge f <Comp & DV> null sequence
Nullgalvanometer n <Elektriz> null galvanometer
Null-Grad-Spanwinkel m <Fertig> zero rake angle
Nullhubstellung f <Fertig> no-stroke position (Pumpe)
Nullimpedanz f <Elektriz> zero-phase-sequence impedance, zero-sequence field impedance, zero-sequence impedance
Nullinstrument n <Gerät> balance meter, (AE) center-reading instrument, (BE) centre-reading instrument, central-zero instrument, null instrument
Nulljustierung f <Kerntech> zero adjustment
Null-Last-Prüfung f <Elektriz> no-load test
Null-Last-Testen n <Elektriz> no-load test
Null-Lastwärmeverbrauch m <Ker & Glas> no-load heat consumption
Null-Leistungsfaktorprüfung f <Elektriz> zero-power-factor test

Null-Leistungsreaktor m <Kerntech> zero-energy reactor, zero-power reactor
Nullleiter m 1. <Elektrotech> neutral conductor *(Verteilung)*; 2. <Sicherheit> protective earth neutral
Nulllinie f <Maschinen> zero line
Null-Luminanz f <Fernseh> zero luminance
Nullmenge f <Math> null set, set with measure zero *(Menge vom Maße Null)*
Nullmethode f <Elektriz, Phys> null method
Nullmodem n <Telekom> modem eliminator, null modem
Nulloperation f *(NO-OP)* <Comp & DV> no-operation *(NO-OP)*
Nulloperationsbefehl m <Comp & DV> no-operation instruction
Nullpegel m <Aufnahme> zero level
Nullpunkt m 1. <Geom> zero point; origin *(von Koordinaten)*; 2. <Math> zero point; origin *(eines Koordinatensystems)*; 3. <Phys> neutral point; 4. <Thermod> zero
Nullpunkt-Amperemeter n <Gerät> *(AE)* center zero ammeter, *(BE)* centre zero ammeter
Nullpunktanhebung f <Regelung> zero elevation
Nullpunktdraht m <Elektriz> neutral wire
Nullpunktenergie f 1. <Kerntech> zero-energy level; 2. <Phys, Strahlphys> zero-point energy
Nullpunktfehler m <Comp & DV> zero error
Nullpunktklemme f <Elektriz> neutral terminal
Nullpunktrelais n <Elektriz> neutral relay
Nullpunktschwankungen fpl <Strahlphys> zero-point fluctuations
Nullpunktverlagerung f <Ker & Glas> zero displacement
Nullpunktverschiebung f <Comp & DV> zero shift
Nullpunktverschiebungs-Spannung f <Elektriz> neutral point displacement voltage
Nullreaktanz f <Elektriz> zero-phase-sequence reactance, zero-sequence inductive reactance, zero-sequence reactance
nullsetzen v <Kontroll> reset
Nullspannung f <Elektriz> null voltage, zero potential, zero voltage
Nullspannungsauslöserrelais n <Elektriz, Elektrotech> no-voltage release relay, no-zero voltage release relay
nullspannungsgesichert adj <Comp & DV> nonvolatile
Nullspant m <Wassertrans> midship section *(Schiffbau)*
Nullstelle f <Math> root, zero, zero coefficient *(einer Funktion)*
nullstellen v 1. <Aufnahme> zero; 2. <Comp & DV> zeroize
Nullsteller m <Fertig> zero adjuster, zero setter
Nullstellung f 1. <Comp & DV> reset; 2. <Maschinen> zero adjustment, zero setting, zeroizing
Nullstellung f **eines Messinstrumentes** <Metrol> zero of a measuring instrument
nullwertig adj <Chemie> avalent, zerovalent
Nullzeichen n <Comp & DV> null character, null
Nullzustand m <Comp & DV> quiescent state
numerisch adj 1. <Comp & DV, Kontroll> numeric; 2. <Math> numerical
numerisch gesteuert adj <Mechan> numerically-controlled
numerisch gesteuerte Lehrenbohrmaschine f <Maschinen> NC jig borer
numerische Analyse f <Comp & DV, Math> numerical analysis
numerische Apertur f <Elektrotech, Optik, Phys, Telekom> NA, numerical aperture
numerische Apertur f **der Einkopplung** <Telekom> launch numerical aperture
numerische Darstellung f <Comp & DV> numeric representation
numerische Einkoppelungsapertur f <Optik> launch numerical aperture
numerische gesteuerte Maschine f <Maschinen> numerical control machine
numerische Methode f <Math> numerical method
numerische Steuerung f *(NC)* <Comp & DV, Elektriz, Kontroll, Maschinen, Mechan> numerical control *(NC)*
numerische Strömungsmechanik f <Strömphys> computational fluid dynamics
numerische Tastatur f <Comp & DV> numeric pad
numerische Zuverlässigkeit f <Qual> numerical reliability
numerischer 10er-Block m <Comp & DV> numeric keypad, numeric pad
numerischer Code m <Comp & DV> numerical code
numerischer Tastenblock m <Comp & DV> numeric keypad, numeric pad
numerischer Wert m <Metrol> numerical value
numerisches Feld n <Comp & DV> numeric item *(COBOL)*
numerisches Literal n <Comp & DV> numeric literal
numerisches Zeichen n <Comp & DV> numeric character
Numerus m <Math> antilogarithm
Nummer f <Comp & DV> number
Nummer f **der Schiffsfunkstelle** <Telekom> ship station number
Nummer f **des gerufenen Teilnehmers** <Telekom> called subscriber number
Nummer f **des rufenden Teilnehmers** <Telekom> calling subscriber number
Nummerierapparat m 1. <Druck> numbering machine; 2. <Verpack> numbering apparatus
nummeriertes Exemplar n <Druck> numbered copy
Nummerierung f 1. <Comp & DV> numeration; 2. <Druck, Foto> numbering
Nummerierungsart f <Comp & DV> enumeration type
Nummerierwerk n <Druck> numbering machine
Nummerngeber m <Telekom> call sender
Nummernschalter m <Telekom> dial
Nummernschalterwahl f <Telekom> rotary dialing
Nummernscheibe f <Telekom> dial
Nummernschild n <Kfztech> *(AE)* license plate, *(BE)* numberplate *(Rechtsvorschriften)*
Nummerung f <Druck> numbering
N-UND-Schaltung f *(NICHT-UND-Schaltung)* <Phys> NAND gate *(NOT AND gate)*
Nur-Frachtcharterflug m <Lufttrans> all-cargo charter flight
Nur-Frachtdienst m <Lufttrans> all-cargo service, all-freight service
Nur-Frachtfluglinie f <Lufttrans> all-cargo carrier
Nur-Frachtflugzeug n <Lufttrans> all-cargo aircraft
Nur-Frachtlastfaktor m <Lufttrans> all-cargo load factor
Nur-Lese-Bit n <Comp & DV> read-only bit
Nur-Lese-Speicher m *(ROM)* <Comp & DV, Elektriz, Elektrotech> read-only memory *(ROM)*
Nur-Postdienst m <Lufttrans> all-mail service
Nussabrieb m <Kohlen> nutty slack
Nussklassiersieb n <Kohlen> nut-sizing screen
Nüstergatt n <Wassertrans> limber hole *(Schiffbau)*
Nut f 1. <Bau> flute, groove, housing, notch, quirk, rabbet, slot; 2. <Elektrotech> slot *(Hohlleiter)*; 3. <Fertig> gash, groove, rabbet, slot; groove *(Kunststoffinstallationen)*; 4. <Maschinen> flute, groove
Nutanker m <Elektrotech> slotted armature
Nutation f <Phys, Raumfahrt> nutation
Nutationdämpfung f <Raumfahrt> nutation damper *(Raumschiff)*
Nute f <Kerntech> groove
Nuteisen n <Mechan> groove-cutting chisel

nuten v 1. <Bau> channel, groove, rabbet; 2. <Fertig> keyway, match
Nuten n 1. <Bau> fluting; 2. <Maschinen> grooving
Nutenanker m <Elektriz> slotted armature
Nutenauslauf m <Maschinen> flute run-out
Nutendrehen n <Fertig> grooving
Nuteneinschleifen n <Fertig> nicking
Nutenfräsen n <Maschinen> keyway milling, keywaying, slot milling
Nutenfräser m <Maschinen> T-slot cutter, keyway cutter
Nutenfräsmaschine f <Maschinen> keyway-milling machine, slot milling machine
Nutenkeil m <Maschinen> feather key, sunk key
Nutenmeißel m 1. <Fertig> cape chisel, half-round chisel; 2. <Maschinen> grooving tool, slotting tool
Nutenschaftfräser m <Maschinen> slot mill
Nutenscheibe f <Fertig> (BE) disc cam, (AE) disk cam (Getriebelehre)
Nutenscheibenfräser m <Maschinen> slotting side and face cutter
Nutenstoßen n <Maschinen> keyway slotting
Nutenstoßmeißel m <Fertig> slotting tool
Nutenwalze f <Maschinen> grooved roll
Nuthobel m <Bau> grooving plane, (BE) plough, (AE) plow, (BE) plough plane, (AE) plow plane, rabbet plane, rebate plane
Nutkeil m <Maschinen> key
Nutmutter f <Bau, Maschinen> slotted nut
NU-Ton m (kein Anschluss unter dieser Nummer) <Telekom> NUT (number-unobtainable tone)
Nutringmanschette f <Maschinen> chevron-type seal
Nutrolle f <Maschinen> grooved pulley
Nutscheibe f <Maschinen> grooved wheel
Nut- und Federholz n <Bau> matchboard
Nut- und Federverspundung f <Bau> grooved and tongued joint
Nutzarbeit f <Maschinen> effective work
Nutzaussendung f <Telekom> wanted emission
Nutzbandbreite f <Aufnahme, Funktech> effective bandwidth
nutzbar adj 1. <Comp & DV> usable; 2. <Mechan> effective
nutzbar machen v <Wasserversorg> harness (Wasser)
nutzbare Bodenfläche f <Bau> usable floor area
nutzbare Leistung f <Elektriz> available power
nutzbarer Bereich m <Gerät> overrange (unter einschränkenden Bedingungen)
nutzbarer Speicherraum m <Wasserversorg> live storage
nutzbares Akustikzentrum n <Aufnahme> (AE) effective acoustic center, (BE) effective acoustic centre
nutzbares Gefälle n <Nichtfoss Energ> effective head
Nutzbarmachung f 1. <Metall> activation; 2. <Wasserversorg> harnessing
Nutzbremsung f <Eisenbahn> regenerative braking
Nutzdampf m <Kerntech> service steam
Nutzeffekt m <Heiz & Kälte> useful effect
nutzen v <Bau> occupy
Nutzer m <Telekom> user, subscriber
Nutzerdienstklasse f <Telekom> user class of service
nutzerfreundlich adj <Ergon, Sicherheit> user-friendly
nutzerfreundliche Anlage f <Sicherheit> user-friendly system
Nutzfahrzeug n <Kfztech> utility vehicle; commercial vehicle (Fahrzeugart)
Nutzfahrzeug... <Kfztech> utility
Nutzfeldstärke f <Funktech> signal strength
Nutzfläche f 1. <Bau> floor space; 2. <Maschinen> useful surface

Nutzförderhöhe f <Fertig> operating head
Nutzholz n <Bau> timber
Nutzkanal m **im ISDN** <Telekom> B-channel
Nutzladefaktor m <Trans> load factor (Transportflugzeuge)
Nutzlast f 1. <Bau> imposed load, live load; 2. <Elektriz> active load; 3. <Kfztech> live weight, payload; 4. <Lufttrans, Meerschmutz> payload; 5. <Raumfahrt> cargo, payload (Raumschiff); 6. <Telekom> payload; 7. <Wassertrans> safe working load, SWL (Ladegeschirr); 8. <Wasserversorg> service load
Nutzlastbedienungsgerät n <Raumfahrt> payload manipulator arm (Raumschiff)
Nutzleistung f <Elektriz> active power
Nutzleistungsturbine f <Lufttrans> free turbine
Nutzmasse f <Meerschmutz> payload
Nutzpferdestärke f <Maschinen> EHP, effective horsepower
Nutzschicht f <Textil> top layer
Nutzsignal n 1. <Elektronik> wanted signal; 2. <Telekom> desired signal
Nutzträger-Störträger-Abstand m <Telekom> wanted-to-unwanted carrier power ratio
Nutzung f **von Unternehmensdaten** <Comp & DV> mining corporate data
Nutzungsdauer f <Funktech, Kerntech, Kohlen, Telekom> service life
nutzungsgemäße Zahlungsgrundlage f <Telekom> pay-by-use basis
Nutzwärme f <Thermod> available heat
Nutzzeit f <Comp & DV> uptime (des Systems)
Nutzzyklus m <Kontroll> duty cycle
Nydrazid n <Chemie> nydrazid
Nylon n <Druck, Kunststoff> nylon
Nylonbuchse f <Kunststoff> nylon bush
Nylonseil n <Maschinen> nylon rope
nylonverstärkt adj <Verpack> nylon-reinforced
Nyquist-Demodulator m <Fernseh> Nyquist demodulator
Nyquist-Frequenz f <Telekom> Nyquist rate
Nyquist-Ortskurve f <Regelung> Nyquist plot
Nyquist-Rate f <Telekom> Nyquist rate
Nystatin n <Chemie> nystatin

#

OA (Operationsverstärker, Rechenverstärker) <Comp & DV, Elektronik, Phys> op amp (operational amplifier)
Obelisk m <Geom> obelisk
oben gehen v/**nach** <Wassertrans> go above (Schiff)
oben liegende Nockenwelle f <Kfztech, Mechan> OHC, overhead camshaft
oben liegende Nockenwelle f/**indirekt wirkende** <Kfztech> indirect overhead camshaft
obengesteuert adj <Kfztech> overhead
obengesteuerter Motor m <Kfztech> overhead valve engine, OHV engine
obengesteuertes Ventil n <Kfztech> OHV, overhead valve
Oben-Top Container m <Trans> til-top container
o-Benzoesäuresulfimid n <Chemie> saccharin
Oberantrieb m <Fertig> overcrank action

Oberband n <Fernseh, Funktech> high band
Oberbaumaterial n <Bau, Eisenbahn> permanent-way equipment *(Gleisbau)*
Oberbegriff m <Patent> preamble
Oberbekleidung f <Textil> outerwear
Oberdeck n <Wassertrans> upper deck
Oberdominante f <Akustik> submediant
Oberdruckpresse f <Kunststoff> downstroke press
obere Abdeckplatte f <Kerntech> roof shielding plate
obere Entscheidungsgrenze f <Qual> upper control limit
obere Grenzabweichung f <Qual> upper limiting deviation
obere Grenze f <Qual> upper limit
obere Grenzwellenlänge f <Elektronik> threshold wavelength
obere Ionosphäre f <Telekom> upper ionosphere
obere Kühltemperatur f <Ker & Glas> upper annealing temperature
obere Schleusenhaltung f <Wasserversorg> head bay, head crown
obere Spaltzone f <Kerntech> upper core
obere Spezifikationsgrenze f <Qual> upper specification limit
obere Tragrolle f <Maschinen> carrying idler
obere und untere Heftung f <Verpack> top and bottom stapling
obere Verbindungsplatte f <Kerntech> upper tie plate *(eines Rasterelementes)*
obere Walze f <Maschinen> upper roll
obere Zündgrenze f <Sicherheit> upper-flammable limit *(Explosionsschutz)*
obere Zwischenstufe f <Metall> upper bainite
oberer Anschluss m <Kerntech> upper end fitting; top fitting *(eines Brennelementes)*
oberer Druckraum m <Kerntech> top plenum, upper plenum
oberer Grenzwert m <Sicherheit> upper limit
oberer Kompressionsring m <Kfztech> top compression ring *(Kolben)*
oberer Mühlstein m <Lebensmittel> upper millstone
oberer Rand m <Druck> head margin
oberer Totpunkt m *(OT)* <Kfztech> *(AE)* top dead center, *(BE)* top dead centre *(top dead center)*
oberer Türriegel m <Bau> top rail *(Türrahmen)*
oberer Verdichtungsring m <Kfztech> top compression ring *(Kolben)*
oberer Verschluss m **eines Brennelements** <Kerntech> upper end plug
oberes Auffangbecken n <Kerntech> upper containment pool
oberes Hubende n <Maschinen> top of stroke of piston
oberes Pleuelauge n <Kfztech> connecting rod small end, small end
oberes Raster n <Kerntech> upper grid *(in einem Bündelelement)*
oberes Schleusentor n 1. <Nichtfoss Energ> head gate, sluicegate; 2. <Wasserversorg> up-stream water gate
oberes Seitenband n *(OSB)* <Elektronik, Funktech, Telekom> upper sideband *(USB)*
oberes Speicherbecken n <Nichtfoss Energ> upper storage basin
Oberfeuer n <Wassertrans> rear range light, upper range light
Oberfläche f 1. <Anstrich> finish, surface; 2. <Bau> face; 3. <Ker & Glas> finish; 4. <Lufttrans> skin *(Luftfahrzeug)*; 5. <Math, Papier> surface; 6. <Patent> surface *(eines Zeichnungsblattes)* • **auf der Oberfläche schwimmend** <Chemie> supernatant • **die Oberfläche abtragen** <Fertig> de-surface • **die Oberfläche durchbrechen** <Wassertrans> broach • **eben mit der Oberfläche eingebaut** <Fertig> flush-mounted
Oberflächenabfluss m <Bau> runoff
Oberflächenabtragung f <Maschinen> surface removal
oberflächenaktiv adj <Chemie> surface-active
oberflächenaktive Substanz f <Chemie> surfactant
oberflächenaktiver Bohrschlamm m <Erdöl> surfactant mud *(Bohrtechnik)*
oberflächenaktiver Stoff m 1. <Lebensmittel> surfactant; 2. <Meerschmutz> surface active agent, surfactant
oberflächenaktives Mittel n <Kohlen> surfactant
Oberflächenbehandlung f 1. <Bau> surface dressing *(Straße)*; 2. <Fertig, Ker & Glas> surface treatment; 3. <Papier> surface application
Oberflächenbereich m <Umweltschmutz> surface area
Oberflächenberieselung f <Heiz & Kälte> spray cooling
Oberflächenbeschaffenheit f 1. <Fertig, Maschinen> surface finish; 2. <Papier> appearance, surface finish
oberflächenbeschriebene CD f <Optik> *(BE)* surface-written videodisc, *(AE)* surface-written videodisk
Oberflächenbewässerung f <Wasserversorg> surface irrigation
Oberflächenbrand m <Sicherheit> surface fire
Oberflächenbrecher m <Meerschmutz> breaker board
Oberflächendruckkarte f <Wassertrans> surface pressure chart *(Wetterkunde)*
Oberflächeneffektfahrzeug n <Kfztech> surface effect vehicle
oberflächenemittierende Leuchtdiode f <Optik> surface-emitting light-emitting diode
oberflächenemittierende Lumineszenzdiode f <Optik> surface-emitting light-emitting diode
Oberflächenenergie f <Phys> surface energy
Oberflächenentlastungsanlage f <Wasserversorg> flood spillway
Oberflächenentwässerung f <Bau, Wassertrans> surface drainage
Oberflächenerdungskontakt m <Elektriz> *(BE)* surface earthing connection, *(AE)* surface grounding connection
Oberflächenfärbung f <Papier> *(AE)* surface coloring, *(BE)* surface colouring
Oberflächenfehler m <Fertig> surface imperfection
Oberflächenfestigkeit f <Papier> surface bonding strength
Oberflächenform f <Maschinen> surface profile
oberflächengehärtete Schiene f <Eisenbahn> surface-hardened rail
oberflächengehärteter Stahl m <Mechan> case-hardened steel
oberflächengetrocknet adj <Fertig> skin-dried
Oberflächengüte f 1. <Fertig, Maschinen> finish, surface finish, surface quality; 2. <Mechan> finish; 3. <Papier> surface finish
Oberflächenhärte f <Maschinen> surface hardness
Oberflächenhärtung f 1. <Fertig> surface hardening; 2. <Mechan> case hardening; 3. <Metall> surface hardening
Oberflächenhöchsttemperatur f <Sicherheit> surface temperature limit *(für bestimmte Gerätetypen)*
Oberflächeninduktion f <Elektriz> surface induction
Oberflächenintegral n <Phys> surface integral
Oberflächenionisierung f **durch Laserstrahlung** <Kerntech> laser impact surface ionization *(Massenspektroskopie)*
Oberflächenkanal m <Elektronik> surface channel
Oberflächenkontamination f <Abfall, Umweltschmutz> ground surface contamination, surface contamination
Oberflächenkühlung f <Heiz & Kälte> surface cooling
Oberflächenladung f <Elektriz, Phys> surface charge

Oberflächenladungsdichte

Oberflächenladungsdichte f <Elektriz, Phys> surface charge density
Oberflächenlängsriss m <Fertig> roke
Oberflächenleimung f <Papier> surface sizing
Oberflächenmatte f <Kunststoff> surfacing mat
Oberflächenmessgerät n <Metrol> surface measuring instrument
Oberflächenmessung f <Fertig> surface measurement
Oberflächenmethode f <Abfall> surface method *(Ablagerungstechnik)*
Oberflächenmontage f <Comp & DV, Elektriz> surface mounting
Oberflächenmontagetechnik f <Telekom> surface mounting technology
oberflächenmontierte Steckdose f <Elektriz> surface socket, surface-mounted socket
oberflächenmontiertes Bauelement n <Telekom> surface mounting device
oberflächenmontiertes Element n *(SMD)* <Elektriz> surface mounting device *(SMD)*
Oberflächenphysik f <Phys> surface physics
Oberflächenporen fpl <Ker & Glas> apparent porosity
Oberflächenprüfung f <Fertig> surface analysis
Oberflächenrauheit f <Bau, Maschinen> surface roughness
Oberflächenreibung f <Mechan> skin friction
Oberflächenrekonstruktion f <Phys> surface reconstruction *(Festkörperphysik)*
Oberflächenriss m <Labor> vent
Oberflächenrisse mpl <Bau> map cracking
Oberflächenrost m <Bau> surface rust
Oberflächenschadstoff m <Umweltschmutz> surface contaminant
Oberflächenschicht f <Meerschmutz> surface layer
Oberflächenschutzfilm m <Verpack> surface protection film
Oberflächensensor m <Gerät> surface sensor
Oberflächenspannung f <Bau, Kohlen, Kunststoff, Maschinen, Phys, Strömphys> surface tension
Oberflächenspannungsmesser m <Labor> surface tension meter
Oberflächenspannungstank m <Raumfahrt> surface tension tank *(Raumschiff)*
Oberflächenstrom m <Telekom> surface current
Oberflächenströmung f <Wassertrans> surface current *(Meer)*
Oberflächenstruktur f <Maschinen> surface texture
Oberflächentechnik f <Fertig> surface coating; surface finishing; surface machining; surface technique
Oberflächenthermometer n <Gerät> contact thermometer, surface temperature sensor
oberflächentrockener Zuschlagstoff m <Bau> saturated surface-dry aggregate
Oberflächenverbindung f <Telekom> surface connection
Oberflächenvergütung f 1. <Anstrich> finish, surface finish; 2. <Ker & Glas> blooming
Oberflächenverluste mpl <Elektriz> can loss *(durch Magnetfelder oder Wirbelstrom)*
Oberflächenvermarkung f <Bau> surface demarcation *(Vermessung)*
Oberflächenverschleiß m <Maschinen> surface wear
Oberflächenversiegelung f <Abfall> final cover *(einer Deponie)*
Oberflächenvorbehandlung f <Bau> surface preparation
Oberflächenvorwärmer m <Elektrotech> feed water heater
Oberflächenwasser n <Bau, Kohlen, Umweltschmutz, Wasserversorg> surface water

Oberflächenwasserbewirtschaftung f <Wasserversorg> surface water management
Oberflächenwassererosion f <Kohlen> surface water erosion
Oberflächenwelle f <Elektriz, Optik> surface wave
Oberflächenwellenbauelement n <Telekom> surface acoustic wave device
Oberflächenwellenleiter m <Elektriz> Goubau line, surface waveguide
Oberflächenwiderstand m <Werkprüf> bar-to-bar resistance, surface resistance
Oberflächenzeichen n <Konstzeich> surface symbol, systematic symbol
Oberflammofen m <Ker & Glas> top flame furnace
Obergautsche f <Papier> lumpbreaker
Obergerinne n <Hydraul> headrace
Obergesenk n 1. <Fertig> top die; 2. <Maschinen> upper die
Obergrenze f 1. <Math> upper bound; 2. <Telekom> upper limit
Obergurt m <Bau> top flange
Obergurtplatte f <Wassertrans> rider plate *(Schiffbau)*
oberhalb adv <Bau> upstream
Oberhaupt n **einer Schleuse** <Wasserversorg> forebay *(Hangkanal oder Oberkanal)*
Oberhieb m <Maschinen> second cut
Oberholm m <Bau> head beam
oberirdisch adj <Kohlen> overground
oberirdisches Wasser n <Wasserversorg> surface water
Oberkasten m <Fertig> cope *(Getriebelehre)*
Oberlage f <Textil> top ply
Oberleitung f <Bau, Eisenbahn, Elektriz, Elektrotech> aerial contact line, overhead contact system, overhead line, trolley wire
Oberleitungsbus m <Kfztech> electric trolley, trolley bus
Oberleitungsgelenkverbindung f <Eisenbahn> overhead-line knuckle
Oberleitungssystem n *(O-System)* 1. <Eisenbahn> overhead line system; 2. <Trans> trolley system
Oberlicht n <Wassertrans> skylight *(Deckausrüstung)*
Oberlichtaufbau m <Eisenbahn> clerestory
Oberlichtfenster n <Bau> clerestory
Oberlinie f <Druck> cap line
Oberrinne f <Wasserversorg> headrace *(eines Wasserrads)*
oberschlächtiges Wasserrad n <Wasserversorg> overshot wheel
Oberschleusendrempel m <Wasserversorg> *(AE)* head miter sill, *(BE)* head mitre sill
Oberschleusenschwelle f <Wasserversorg> *(AE)* head miter sill, *(BE)* head mitre sill
Oberschlitten m <Maschinen> top slide
Oberschwingung f 1. <Akustik, Funktech, Mechan> harmonic; 2. <Telekom> harmonic oscillation
Oberschwingung f **der Welligkeit** <Elektronik> ripple harmonic *(eines Gleichrichters)*
Oberschwingungsgehalt m <Elektronik> relative harmonic content
Oberschwingungsgenerator m <Phys> harmonic generator
Oberseite f 1. <Druck> felt side; 2. <Lufttrans> upper surface *(eines Flügels)*; 3. <Papier> top side
Oberseitenanschluss m <Elektrotech> face-up
oberste Rippe f <Lufttrans> top rib *(des Seitenflossenansatzes)*
oberstes Stockwerk n <Ker & Glas> top floor
Oberstrichwert m <Aufnahme> peak power output
Obersupport m <Maschinen> top slide rest
Oberteil n <Textil> top

Oberteil n des Spindelkastens <Maschinen> barrel
Oberteilkomplett n <Fertig> bonnet assembly *(Kunststoffinstallationen)*
Oberton m <Elektronik> overtone
Obertor n 1. <Nichtfoss Energ> head gate; 2. <Wasserversorg> sluicegate, upstream water gate; crown gate *(einer Kanalschleuse)*; head gate *(einer Schleuse)*
Obertransport m <Textil> top feed
Oberwalze f 1. <Ker & Glas> pressure roller; 2. <Maschinen, Papier> top roll
Oberwasser n 1. <Hydraul> headwater; 2. <Nichtfoss Energ> upstream head; 3. <Wasserversorg> head bay, head crown, headwater
Oberwelle f <Akustik, Elektronik> harmonic
Oberwellendämpfung f <Elektronik, Funktech> harmonic attenuation
Oberwellenerzeugung f <Telekom> harmonic generation
Oberwellenfilter n 1. <Elektronik, Telekom> harmonic filter; 2. <Raumfahrt> harmonic filter *(Weltraumfunk)*
Oberwellengehalt n <Elektronik, Funktech> harmonic content
Oberwellengenerator m <Phys, Telekom> harmonic generator
Oberwellenmessgerät n <Gerät> harmonic detector
Oberwellenquarz m <Elektronik> harmonic mode crystal
Oberwellensperrung f <Elektronik> harmonic rejection
Oberwellenstrom f <Elektrotech> current ripple
Oberwellenunterdrückung f <Elektronik, Funktech> harmonic rejection
Oberwerk n <Wassertrans> upper works *(Schiffkonstruktion)*
Oberwerksbau m <Kohlen> rise workings
Objekt n <Comp & DV> object, object variable
objektartige Anwendung f <Comp & DV> object-like application
Objektcode m <Comp & DV> object code
Objekterzeugung f <Comp & DV> object creation
Objektiv n 1. <Foto> lens *(mit Blendenvorwahl)*; 2. <Optik> objective; lens *(Linsenkombination)*; 3. <Phys> lens
Objektiv n mit großer Blende <Foto> large-aperture lens
Objektiv n mit Vorsatz-Anamorphot <Foto> anamorphotic lens
Objektivanschluss m <Foto> lens mount
Objektivanschlussplatte f <Foto> lens mounting plate
Objektivbrett n <Foto> lens panel
Objektivdeckel m <Foto> lens-cap
Objektivdetektor m <Trans> objective detector
objektivgekoppelter Belichtungsmesser m <Foto> lens-coupled exposure meter
Objektivköcher m <Foto> lens case
Objektivlinse f 1. <Labor> objective lens *(Mikroskop)*; 2. <Metall> objective lens
Objektivöffnung f <Foto> lens aperture
Objektivrohr n <Optik> body tube
Objektivsatz m <Foto> set of lenses
Objektivträger m 1. <Foto> camera front; 2. <Ker & Glas> microscope slide
Objektivtubus m <Foto> lens barrel, lens flange
Objektiv-Verschluss m <Foto> lens shutter
Objektlautstärkepegel m <Akustik> loudness level of test sound
Objektmodul n <Comp & DV> object module
objektorientiert adj <Künstl Int> object-oriented *(OO)*
objektorientierte Architektur f <Comp & DV> object-oriented architecture
objektorientierte Programmierung f *(OOP)* <Künstl Int> object-oriented programming *(OOP)*

objektorientierter Aufbau m <Comp & DV> object-oriented architecture
objektorientierter Entwurf m <Comp & DV> object-oriented design
objektorientiertes Design n <Comp & DV> object-oriented design
objektorientiertes Programmiersystem n <Comp & DV> object-oriented programming system
Objektprogramm n <Comp & DV> object program
Objektschutzanlage f <Sicherheit> object protection system *(Brandschutz)*
objektseitige Brennebene f <Foto> front focal plane
Objektsprache f <Comp & DV> object language, target language
Objekttechnologie f <Comp & DV> object technology
Objektträger m <Labor> slide *(Mikroskop)*
Objektvariable f <Comp & DV> object, object variable
obligater Aerobier m <Lebensmittel> obligate aerobe
Oblimak m <Kerntech> oblate spheromak, oblimak
OBM-Verfahren n <Metall> Q-BOP process
OBO *(Erz-Schüttgut-Öl, Flüssigkeitsmassengut)* <Wassertrans> OBO *(ore-bulk oil)*
OBO-Frachter m <Wassertrans> OBO carrier *(Schiff)*
Obsidian m <Ker & Glas> obsidian
Obus m <Elektrotech, Trans> P trolley bus
OC *(Operationscharakteristik)* 1. <Math> OC, operating characteristic *(Testtheorie)*; 2. <Qual> OC, operating characteristic
Ochsengalle f <Druck> ox gall
Ockerton m <Ker & Glas> ochrey clay
OC-Kurve f 1. <Math> OC curve, operating characteristic curve *(Testtheorie)*; 2. <Qual> operating characteristic curve
OCO *(Erz-Kohle-Öl)* <Trans> OCO *(ore-coal-oil)*
OCO-Frachter m <Trans> OCO carrier
OCR *(optische Zeichenerkennung)* <Comp & DV> OCR *(optical character recognition)*
OCR-Schriftart f <Comp & DV> OCR font
OCR-Zeichensatz m <Comp & DV> OCR font
Octacosan n <Chemie> octacosane
Octadecadienat n <Chemie> linoleate
Octadecan n <Chemie> octadecane
Octadecyl... <Chemie> octadecyl
Octagon n <Geom> octagon
Octan n <Chemie> octane
Octan... <Chemie> octane
Octanal n <Chemie> octanal
Octanoyl... <Chemie> octanoyl
octavalent adj <Chemie> octavalent
Octen n <Chemie> octene, octylene
Octin n <Chemie> caprylidene, octyne
Octose f <Chemie> octose
Octylaldehyd m <Chemie> octanal
Octylen n <Chemie> octene, octylene
ODER-Gatter n <Phys> OR circuit, OR gate
ODER-Glied n <Comp & DV, Elektronik> OR circuit, OR gate
ODER-Schaltung f <Phys> OR circuit, OR gate
ODER-Tor n <Comp & DV> OR circuit, OR gate
ODER-Verknüpfung f <Comp & DV> OR operation
ODER-Zeichen n <Elektronik> OR operator
Odometer n <Bau> odometer
Odoriermittel n <Erdöl> odorant *(Petrochemie)*
OEM-Hersteller m <Maschinen> OEM, original equipment manufacturer
OEM-Modem n <Elektronik> OEM modem
Oersted n <Elektrotech> oersted
Ofen m 1. <Anstrich> furnace; 2. <Elektrotech> heater; 3. <Heiz & Kälte, Labor, Maschinen> furnace; 4. <Metall>

Ofen

kiln; 5. <Papier> kiln, oven; 6. <Thermod> oven • **im Ofen trocknen** <Thermod> kiln-dry
Ofen *m* **mit erzwungener Konvektion** <Labor> oven with forced convection
Ofen *m* **mit natürlicher Konvektion** <Labor> oven with natural convection
Ofenalterung *f* <Kunststoff> *(BE)* oven ageing, *(AE)* oven aging
Ofenauskleidung *f* <Maschinen> furnace lining
Ofenbär *m* <Ker & Glas> bear
Ofenbediener *m* <Ker & Glas> teaser, top man
Ofenbefüllung *f* <Ker & Glas> furnace fill
Ofenbruch *m* <Metall> cadmia
ofenfestes Glas *n* <Thermod> oven proof glass
ofengetrocknet *adj* 1. <Bau> kiln-dried, KD, oven-dried *(Bauholz)*; 2. <Thermod> kiln-dried, KD, oven-dried
Ofenherdgewebe *n* <Ker & Glas> bench cloth
Ofenkoks *m* <Kohlen> oven coke
Ofenkopf *m* <Ker & Glas> port endwall
Ofenkranz *m* <Ker & Glas> port crown
Ofenkuppel *f* <Ker & Glas> crown
Ofenleistung *f* <Ker & Glas> furnace performance
Ofenloch *n* <Ker & Glas> notch, porthole
Ofenmantel *m* <Ker & Glas> shell
Ofenrast *f* <Ker & Glas> bosh
Ofenrost *m* <Maschinen> furnace grate
Ofenrückwand *f* <Ker & Glas> port back wall
Ofenruß *m* <Kunststoff> furnace black
Ofensau *f* <Fertig> sow
Ofensohle *f* <Ker & Glas> port sill
ofentrocken *adj* <Papier> bone-dry, oven-dry
ofentrocknender Lack *m* <Kunststoff> stoving finish, stoving varnish
Ofentrocknung *f* <Bau> kiln drying
Ofenwand *f* 1. <Ker & Glas> wicket wall; 2. <Maschinen> furnace wall
Ofenzug *m* <Ker & Glas> port uptake
offen *adj* 1. <Comp & DV> open; 2. <Telekom> uncoded; 3. <Textil> open-ended
offen gewickelter Anker *m* <Elektriz> open coil armature
Offenbarung *f* **der Erfindung** <Patent> disclosure of the invention
Offenblendmessung *f* <Foto> full-aperture metering *(Messen)*
offene Bauweise *f* <Bau> open cut
offene Blende *f* <Foto> full aperture
offene Deltaschaltung *f* <Elektriz> open delta connection
offene Dreieckschaltung *f* <Elektriz> open delta connection
offene Haube *f* <Papier> open hood
offene Lage *f* <Elektriz> open position
offene Menge *f* <Math> open set
offene Mischwalze *f* <Kunststoff> open mill
offene Schleife *f* <Elektriz, Lufttrans> open loop
offene See *f* <Wassertrans> offing • **auf offener See** <Wassertrans> in the offing
offene Seite *f* <Fertig> mouth
offene Sicherung *f* <Elektriz> open fuse
offene Spindeltreppe *f* <Bau> open newel stairs
offene Sprache *f* <Telekom> clear speech *(Mobiltelefon)*
offene Stirnseite *f* <Hydraul> open front
offene Stirnwand *f* <Hydraul> open front
offene Systemlösung *f* <Comp & DV> open industry solution
offene Tunnelbauweise *f* <Bau, Kohlen> cut-and-cover method
offene Unterlegscheibe *f* <Maschinen> C-spacer
offene Walzenstraße *f* <Fertig> looping mill
offene Wartezeit *f* <Kunststoff> open assembly time
offene Zelle *f* <Kunststoff> open cell
offener Abzugsgraben *m* 1. <Abfall> open drain, open sewer; 2. <Wasserversorg> open drain
offener Brennstoffkreislauf *m* <Kerntech> once-through fuel cycle, open fuel cycle
offener Güterwagen *m* 1. <Eisenbahn> *(AE)* freight truck, *(BE)* goods lorry, gondola car, gondola wagon; 2. <Trans> truck
offener Kanal *m* <Hydraul> open channel
offener Kern *m* <Elektriz> open core
offener Kopflunker *m* <Fertig> primary pipe
offener Kreis *m* <Regelung> open loop *(in binären Schaltsystemen)*
offener Kreislauf *m* <Lufttrans> open loop
offener Kühlkreis *m* <Heiz & Kälte, Kerntech> open circuit cooling
offener Leiter *m* <Elektrotech> open conductor
offener Lichtbogen *m* <Elektrotech> open arc
offener Propeller *m* <Trans> open propeller
offener Resonator *m* <Elektronik> open resonator
offener Riemen *m* <Fertig, Papier> open belt
offener Seekanal *m* <Wasserversorg> lock and inland-lake canal
offener Selbstentlade-Drehgestellgüterwagen *m* <Kfztech> bogie open self-discharge wagon
offener Standard *m* <Comp & DV> open standard
offener Stoffauflauf *m* <Papier> open headbox
offener Stromkreis *m* <Elektriz, Elektrotech> open circuit
offener Zugriff *m* <Comp & DV> open access
offenes Gesenk *n* <Maschinen> open die
offenes Gitter *n* <Elektronik> free grid *(Elektronenröhre)*
offenes Intervall *n* <Math> open interval
offenes Meer *n* <Wassertrans> high seas, open sea
offenes Polygon *n* <Geom> open polygon
offenes Regelsystem *n* <Regelung> open control system
offenes Schneideisen *n* <Maschinen> split die
offenes Standrohr *n* <Bau> open standpipe
offenes System *n* <Comp & DV, Thermod> open system
offenes Wasser *n* <Wassertrans> open water
offenkettiger Kohlenwasserstoff *m* <Kunststoff> aliphatic hydrocarbon
öffentliche Gesundheit *f* <Wasserversorg> public health
öffentliche Straße *f* <Bau> public road
öffentliche Versorgungsunternehmen *npl* <Wasserversorg> public utilities
öffentliche Wasserversorgung *f* <Wasserversorg> public water supply
öffentlicher beweglicher Landfunk *m* <Mobilkom> public land mobile telecommunications system *(PLMTS)*
öffentlicher Fernsprecher *m* <Telekom> public telephone
öffentlicher Telefonanschluss *m* **im Flugzeug** <Telekom> *(AE)* airphone
öffentlicher Transport *m* <Trans> *(AE)* public transit, *(BE)* public transport
öffentlicher Wählanschluss *m* <Telekom> public dial-up port
öffentliches Bildtelefon *n* <Elektronik> CRT-equipped public phone *(zur Bedienung mit Telefonkarte)*
öffentliches Datennetz *n (PDN)* <Comp & DV> public data network *(PDN)*
öffentliches Fernsprechamt *n* <Telekom> public telephone exchange
öffentliches Fernsprechnetz *n* <Telekom> public switched telephone network, public telephone network
öffentliches Fernsprechwählnetz *n* <Telekom> general switched telephone network, public switched telephone network

öffentliches Mobilfunknetz n <Mobilkom> public land mobile network, public land mobile radio network *(PLMRN)*
öffentliches Netz n <Telekom> public network; public telephone network *(PTN)*
öffentliches Stromversorgungsnetz n <Elektriz> public supply network
öffentliches Telefon n <Telekom> public telephone
öffentliches Transportsystem n **mit ständigem Zugang** <Trans> continuous access public transport system
offenzelliger Schaumstoff m <Kunststoff> open cell cellular plastics
Offline-... <Comp & DV, Telekom> off-line
Offline-Betrieb m <Telekom> off-line working
Offline-Drucken n <Verpack> printing off line
Offline-Edieren n <Fernseh> off-line editing
Offline-Verarbeitung f <Comp & DV> off-line data processing, off-line processing
öffnen v 1. <Elektrotech> open, break *(Stromkreis)*; 2. <Foto> open *(Blende)*; 3. <Papier> open; 4. <Textil> open *(Faser)*
Öffnen n 1. <Elektrotech> opening operation, breaking *(Stromkreis)*; 2. <Maschinen> opening
öffnender Kreis m <Elektrotech> opening circuit
Öffner m 1. <Elektriz> closing contact, normally closed contact; 2. <Elektrotech> normally closed contact; 3. <Textil> opener
Öffnung f 1. <Anstrich> crack; 2. <Bau> cutout, mouth; 3. <Elektrotech> cutout; port *(Wellenleiter, Hohlleiter)*; break *(eines Kontaktes)*; 4. <Fertig> orifice; 5. <Foto> aperture *(Linse)*; 6. <Funktech> aperture *(Antenne)*; 7. <Hydraul> port; 8. <Kohlen> aperture; 9. <Maschinen> mouth, opening, orifice; 10. <Mechan> aperture, port; 11. <Papier> opening; 12. <Telekom> aperture *(Antenne)*
Öffnungsanweisungen fpl <Verpack> instructions for opening, opening instructions
Öffnungskoeffizient m (m) <Akustik> flare coefficient of horn *(m)*
Öffnungskontakt m 1. <Elektriz> normally closed contact; 2. <Elektrotech> break contact, normally closed contact
Öffnungsmechanismus m <Verpack> opening mechanism
Öffnungsring m <Ker & Glas> orifice ring
Öffnungsverhältnis n <Phys> relative aperture
Öffnungswinkel m 1. <Elektronik> beam angle *(Antennentechnik)*; acceptance angle *(Lichtwellenleiter)*; 2. <Fertig> acceptance angle, angle of acceptance *(Faseroptik)*; 3. <Funktech> acceptance angle, beamwidth; angular aperture *(Antenne)*; 4. <Kerntech> angular acceptance
Öffnungswinkel m **eines Strahls** <Funktech, Kerntech> acceptance angle of beam
Öffnungszeit f 1. <Elektriz> break time, opening time; 2. <Elektrotech> opening time *(eines Stromkreises)*
Offset-... <Druck, Papier, Raumfahrt, Verpack> offset
Offset-Antenne f 1. <Funktech> offset antenna; 2. <Raumfahrt> offset antenna *(Weltraumfunk)*
Offset-Betrieb m <Elektronik, Telekom> offset mode *(Funktechnik)*
Offset-Druck m 1. <Druck> offset, offset printing; 2. <Papier, Verpack> offset printing
Offset-Maschine f <Druck> offset printing press
Offset-Papier n <Papier> offset paper
Offset-Phasenumtastung f <Telekom> offset phase-shift keying
Offset-Presse f 1. <Druck> offset printing press; 2. <Verpack> offset press

Offset-Reflektor m <Raumfahrt> offset reflector *(Weltraumfunk)*
Offset-Rollenrotationsmaschine f <Druck> web-fed offset rotary press
Offset-Rotationsdruckmaschine f <Verpack> offset rotary press
Offset-Streichmaschine f <Papier> offset coater
Offset-Trägersystem n <Telekom> offset carrier system
Offset-Vorlage f <Konstzeich> offset master
Offset-Walze f <Druck> offset roller
Offshore-... <Erdöl, Wassertrans> offshore oil platform
Offshore-Anlage f <Erdöl, Wassertrans> offshore
Offshore-Bohrung f 1. <Erdöl> offshore drilling; 2. <Umweltschutz> offshore well; 3. <Wassertrans> offshore drilling
Offshore-Feld n <Erdöl> offshore field
Offshore-Gas n <Erdöl> offshore gas
Offshore-Lagerstätte f <Erdöl> offshore field
Offshore-Ölförderindustrie f <Erdöl> offshore oil industry
Offshore-Ölhafen m <Wassertrans> offshore port
Offshore-Plattform f <Erdöl> offshore platform
Offshore-Technik f <Erdöl> offshore engineering
OFN *(Ortsfernsprechnetz)* <Telekom> local exchange area
Ohm n <Elektriz, Elektrotech, Metrol, Phys> ohm
Ohmmesser m <Elektriz> resistance meter
Ohmmeter n <Elektriz, Elektrotech, Phys, Telekom> ohmmeter
ohmsch adj 1. <Elektriz> ohmic; 2. <Elektrotech> resistive
Ohm'sche Belastung f <Telekom> resistive load
Ohm'sche Last f <Elektrotech> resistive load
Ohm'scher Kontakt m <Elektrotech, Phys> ohmic contact
Ohm'scher Leiter m <Phys> ohmic conductor
Ohm'scher Spannungsabfall m <Elektriz> ohmic drop
Ohm'scher Verlust m <Elektriz, Elektrotech, Phys> ohmic loss
Ohm'scher Wert m <Elektrotech> ohmic value
Ohm'scher Widerstand m <Elektriz, Elektrotech> ohmic resistance, DC resistance
Ohm'sches Gesetz n <Elektriz, Elektrotech, Phys> Ohm's law
Ohr n <Akustik> ear
Öhr n 1. <Fertig> eye *(Hammer)*; eyelet; 2. <Maschinen> ear, eye
Ohrenschutz m <Sicherheit> ear protector, earmuff, hearing protector
Ohrharmonische f <Akustik> aural harmonic
Ohrhörer m <Funktech, Telekom> earphone, insert earphone
OHV-Motor m <Kfztech> OHV engine
okkludieren v <Chemie> occlude
okkludiert adj <Chemie> occluded
Okklusion f 1. <Chemie> occlusion; 2. <Wassertrans> occluded front *(Wetterkunde)*
Ökochemie f <Chemie, Umweltschutz> ecochemistry
Ökologie f <Nichtfoss Energ, Umweltschutz> ecology
ökologische Pyramide f <Nichtfoss Energ, Umweltschutz> ecological pyramid
ökologischer Zusammenbruch m <Umweltschutz> environmental collapse
ökologisches Gleichgewicht n <Nichtfoss Energ, Umweltschutz> ecological balance, ecological equilibrium
Ökonometrie f <Math> econometrics *(Statistik)*
Ökosystem n <Umweltschutz> ecosystem
Ökotoxikologie f <Umweltschutz> ecotoxicology, environmental toxicology
ökotoxikologisch adj <Umweltschutz> ecotoxicological

Ökowagen

Ökowagen m <Kfztech, Kohlen, Nichtfoss Energ> alternative fuel vehicle
o-Kresol n <Chemie> o-cresol
Oktaeder n <Geom> octahedron *(Achteck)*
oktaedrisch adj <Chemie, Geom> octahedral
oktagonal adj <Geom> octagonal
oktal adj <Comp & DV> octal
Oktalröhre f <Elektronik> octal tube
Oktalschreibweise f <Comp & DV> octal notation
Oktalsockel m <Maschinen> octal base
Oktan n <Erdöl> octane *(Petrochemie)*
Oktanzahl f 1. <Erdöl> octane number *(Raffinerie, Erdölchemie)*; 2. <Kfztech> octane index, octane number, octane rating
Oktanzahlbestimmung f <Kfztech> ONR, octane number rating
Oktanzahlwert m <Kfztech> ONR, octane number rating
Oktav n <Druck> 8vo, octavo
Oktavband n <Elektronik, Heiz & Kälte> octave band
Oktavbandfilter n <Elektronik> octave-band filter
Oktavbandoszillator n <Elektronik> octave-band oscillator
Oktavbandpassfilter n <Aufnahme> octave filter
Oktavbandpassfiltersatz m <Aufnahme> octave filter set
Oktavbandpegel m <Sicherheit> octave-band level *(Lärm)*
Oktave f <Akustik, Phys> octave
Oktavformat n <Druck> 8vo, octavo
Oktavmittenfrequenz f <Sicherheit> octave mid-frequency *(Lärm)*
Oktavschalldruckpegel m <Sicherheit> octave sound-pressure level *(Lärm)*
Oktett n 1. <Chemie, Comp & DV> octet; 2. <Elektronik> eight-bit byte
Oktett-Ausgabe f <Elektronik> eight-bit output
Oktode f <Elektronik> octode
oktogonal adj <Geom> octagonal
Oktoidverzahnung f <Maschinen> octoid gear
Oktupol m <Elektrotech> octupole
okular adj <Optik> ocular
Okular n 1. <Foto> eyepiece lens; 2. <Labor> eyepiece *(Mikroskop)*; 3. <Optik> eyeglass, eyepiece, ocular; 4. <Phys> eyepiece
Okular n mit Fadenkreuz <Optik> webbed eyepiece
Okularmikrometer n <Optik> micrometer eyepiece
OK-Zeichen n <Telekom> OK signal
Öl n <Erdöl, Maschinen, Meerschmutz, Papier> oil • **in Öl wirbeln** <Fertig> churn *(Zahnrad)* • **Öl entdecken** <Erdöl> strike oil
Ölabdichtung f <Mechan> oil seal
Ölabfall m 1. <Abfall> residual oil; 2. <Umweltschmutz> oil waste
ölabgeschreckt adj <Fertig> oil-quenched
Ölablass m <Heiz & Kälte> oil drain
Ölablasshahn m <Heiz & Kälte> oil drain valve
Ölablassöffnung f <Kfztech> oil drain hole
Ölablassschraube f <Kfztech> drain plug; oil drain plug *(Schmierung)*
Ölablassstopfen m <Mechan> oil drain plug
Ölablassventil n <Heiz & Kälte> oil drain valve
Ölablauf m <Heiz & Kälte> oil drain
Ölabscheider m 1. <Abfall> degreaser, oil separator, skimming tank; 2. <Heiz & Kälte, Maschinen, Papier> oil separator
Ölabscheidering m <Kfztech> oil scraper ring *(Motor)*
Ölabscheidung f <Abfall> de-oiling, oil removal, oil separation
Ölabschöpfgerät n <Meerschmutz> recovery device

Ölabschöpfsystem n <Meerschmutz> recovery system
Ölabschreckung f <Fertig, Thermod> oil quenching
Ölabstreifer m <Kerntech, Kfztech> oil wiper *(Motor)*
Ölabstreifring m 1. <Kerntech> oil scraper ring, oil wiper; 2. <Kfztech> oil control ring; oil control ring *(Kolben)*; oil scraper ring *(Motor)*
Ölabziehstein m <Maschinen> oilstone
Ölanregung f <Kerntech> oil stimulation
ölanziehend adj <Chemie> oleophilic
Ölaufbereitungsanlage f <Abfall> oil regeneration plant
Ölaufnahme f <Kunststoff> oil absorption
Ölaufnahmeband n <Meerschmutz> oil mop
ölaufnehmendes Band n <Meerschmutz> oleophilic belt
ölaufnehmendes Förderband-Abschöpfgerät n <Meerschmutz> oleophilic belt skimmer
Ölaufsauger m <Umweltschmutz> skimmer
Ölausbiss m <Meerschmutz> seepage
Ölausdehnungsgefäß n <Elektrotech> oil conservator
Ölausgleichsgefäß n <Elektrotech> oil conservator
Ölaustritt m <Meerschmutz> oil leakage
Ölbad n <Labor> oil bath • **im Ölbad abgeschreckt** <Thermod> oil-quenched • **im Ölbad abschrecken** <Thermod> oil-quench
Ölbad-Anlassen n <Metall> oil tempering
Ölbadluftfilter n <Kfztech> oil-bath air filter
Ölbadluftreiniger m <Kfztech> oil-bath air cleaner
Ölbadschmierung f <Maschinen> oil-bath lubrication
Ölbecken n <Erdöl> oil basin, petroleum basin
ölbefeuert adj <Erdöl> oilfired *(Kessel, Öfen)*
Ölbehälter m <Elektrotech> oil tank
Ölbekämpfung f <Meerschmutz> oil spill response
Ölbekämpfungsschiff n 1. <Meerschmutz> oil pollution combatting ship; 2. <Umweltschmutz> oil pollution fighter
ölbenetzte Luftfilterpatrone f <Kfztech> oil-moistened air filter cartridge
Ölbergbau m <Erdöl> oil mining, petroleum mining
Ölbeseitigung f durch Trennmittel <Umweltschmutz> removal of oil by separators
Ölbeseitigungsschiff n <Umweltschmutz> oil clearance vessel
ölbeständig adj <Kunststoff> oil-resistant
Ölbohrinsel f <Erdöl, Telekom> offshore oil platform
Ölbohrung f 1. <Erdöl> petroleum well; 2. <Maschinen> oil hole
Ölbremszylinder m <Fertig> oil dashpot
Ölbrenner m <Erdöl, Heiz & Kälte, Maschinen, Thermod> oil burner
Ölbunker m <Erdöl, Wassertrans> oil bunker *(Raffinerie)*
Oldham-Kupplung f <Maschinen> Oldham coupling
öldichte Sicherheitshandschuhe mpl <Sicherheit> oilproof protective gloves
Öldichtung f <Maschinen, Mechan> oil seal
Öldruckanzeige f <Kfztech> *(AE)* oil pressure gage, *(BE)* oil pressure gauge *(Schmierung)*
Öldruckkontrollleuchte f <Kfztech> oil pressure warning light
Öldruckmesser m <Kfztech> *(AE)* oil pressure gage, *(BE)* oil pressure gauge *(Schmierung)*
Öldruckpresse f <Bau> hydraulic jack
Öldruckstoßdämpfer m <Maschinen> oil dashpot
Öldruckverstärkerpumpe f <Fertig> booster pump
Öldruckwächter m <Heiz & Kälte> oil pressure switch
Öldurchflussmenge f <Heiz & Kälte> oil flow rate
Öldüse f <Kfztech> oil jet
Oleat n <Chemie> oleate
Olefin n 1. <Chemie> alkene, olefin, olefine; 2. <Erdöl> alkene; olefin *(Petrochemie)*
Olefingehalt m <Chemie> olefinic content
olefinisch adj <Chemie> olefinic

Olein n <Chemie> olein
Öleinfüllstutzen m <Kfztech> oil filler pipe; oil filler *(Motor)*
Öleinfüllverschluss m <Kfztech> oil filler cap
Öleinlassöffnung f <Kfztech> oil inlet
Olein-Palmitin-Gemisch n <Chemie> oleo oil
Ölemulsion f <Maschinen> oil emulsion
Ölen n <Textil> oiling
Ölentferner m <Anstrich> degreaser
oleophil adj <Chemie> oleophilic
Oleoresin n <Chemie> oleoresin
Öler m <Maschinen> oiler
Oleum n <Chemie> Nordhausen acid, oleum, *(AE)* fuming sulfuric acid, *(BE)* fuming sulphuric acid
Ölfang m <Wasserversorg> oil trap
Ölfangblech n <Kerntech> oil baffle
Ölfangschale f <Fertig> sump
Ölfeld n <Erdöl> oilfield *(Erdölgeologie)*
ölfest adj <Kunststoff> oil-resistant
ölfester Schlauch m <Kunststoff> oil-resisting hose
Ölfeuerung f 1. <Bau> oil firing; 2. <Heiz & Kälte> oilfired furnace
Ölfeuerungsanlage f <Maschinen> oilfired installation
Ölfilm m <Meerschmutz> oil film, oil slick
Ölfilter n 1. <Kfztech> oil filter *(Schmierung)*; 2. <Maschinen, Mechan> oil filter
Ölfilterdichtring m <Kfztech> oil filter gasket
Ölfilterersatzelement n <Kfztech> replaceable element oil filter
Ölfleck m <Meerschmutz> oil patch, slick
Ölfluss m <Erdöl> flow of oil
ölfördernd adj <Erdöl> oil-producing
ölfreier Kompressor m <Heiz & Kälte> oil-free compressor
Ölfund m <Erdöl> oil discovery
Ölgas n <Erdöl> oil gas
ölgefeuert adj <Thermod> oil-burning, oilfired
ölgefeuerter Kessel m <Heiz & Kälte> oilfired boiler
ölgefülltes Kabel n <Elektriz> oil-filled cable
Ölgehalt m <Kunststoff> oil length
ölgehärtet adj <Thermod> oil-hardened
ölgehärtetes Sieb n <Papier> oil-tempered wire
ölgekapselter Transformator m <Elektrotech> oil-immersed transformer
ölgekühlt adj <Heiz & Kälte, Thermod> oil-cooled
ölgekühlter Transformator m <Elektriz, Elektrotech> oil-cooled transformer
ölgestreckter Kautschuk m <Kunststoff> oil-extended rubber
Ölhahn m <Maschinen> oil cock
ölhaltiger Abfall m <Abfall> oily waste
ölhaltiges Abwasser n 1. <Abfall> oleiferous waste water; 2. <Umweltschmutz> oil-containing waste water
Ölhaltigkeit f <Kunststoff> oil length
Ölhärtung f <Metall> oil-hardening
Ölharz... <Chemie> oleoresinous
Ölhaut f <Textil> oilskin
Ölheizung f <Erdöl, Heiz & Kälte> oil heating
Ölheizungskessel m <Heiz & Kälte> oilfired boiler
Öliges Wasser n <Umweltschmutz> black water
Oligomer n <Chemie, Kunststoff> oligomer
oligomer adj <Chemie> oligomeric
Oligomeres n <Chemie> oligomer
Oligomerisation f <Kunststoff> oligomerization
Oligomycin n <Chemie> oligomycin
Ölindustrie f <Erdöl> oil industry, petroleum industry
Ölisolator m <Elektrotech> oil insulator
ölisolierter Kondensator m <Elektriz> oil-immersed capacitor

Olivenscharnier n <Bau> olive knuckle hinge *(Verschluss)*
Ölkabel n <Elektrotech> oil-filled cable
Ölkanal m 1. <Heiz & Kälte> oil duct; 2. <Kfztech, Maschinen> oil channel
Ölkännchen n <Fertig> squirt oiler
Ölkanne f <Fertig> hand oiler
Ölkapselung f <Sicherheit> oil enclosure
Ölkatastrophe f <Meerschmutz> oil spill disaster
Ölkessel m <Heiz & Kälte> oilfired boiler
Ölkohlebelag m <Kfztech> oil-carbon deposit
Ölkraftwerk n <Heiz & Kälte, Thermod> oilfired power station
Ölkühler m 1. <Elektriz, Kfztech> oil cooler *(Schmierung)*; 2. <Maschinen> oil cooler, oil hole; 3. <Thermod> oil cooler
Ölkühler-Wärmeaustauscher m <Lufttrans> oil-cooler heat exchanger
Ölkühlung f <Elektriz, Maschinen> oil cooling
Öllache f <Meerschmutz> spill
Öllagerstätte f <Erdöl> oilfield
Ölleck n <Meerschmutz> oil leakage
Ölleinwand f <Verpack> oiled canvas
Ölleitblech n <Heiz & Kälte> oil baffle
Ölleitung f 1. <Erdöl> oil pipeline *(Ölindustrie)*; 2. <Kfztech> oil line; 3. <Maschinen, Trans> oil pipeline
Ölleitungsrohr n <Erdöl> oil pipe
ölloses Lager n <Maschinen> oilless bearing
öllöslich adj <Kunststoff> oil-soluble
Ölmesser m 1. <Kfztech> oil meter; 2. <Maschinen> *(AE)* oil gage, *(BE)* oil gauge, oilometer
Ölmessstab m <Kfztech> oil level stick; dipstick *(Schmierung)*
Ölmotor m <Erdöl> oil engine *(Dieselmotor, Schwerölmotor)*
Öl-Naturharz-... <Chemie> oleoresinous
Ölnebelabscheider m <Umweltschmutz> oil aerosol separator
Ölnebelkühlung f <Fertig> spray cooling
Ölnebelschmierung f <Maschinen> oil mist lubrication
Ölnute f <Maschinen> oil groove
Ölpapier n <Verpack> oil packing paper, oil-drenched paper
Ölpauspapier n <Papier> oil tracing paper
Ölpest f <Umweltschmutz> black tide, oil pollution
Ölpier m <Trans> oil pier
Ölpipeline f <Erdöl, Maschinen, Trans> oil pipeline
Ölprovinz f <Erdöl> petroleum province
Ölpumpe f 1. <Heiz & Kälte> oil pump; 2. <Kfztech> oil pump *(Schmierung)*; 3. <Maschinen> oil pump
Ölpumpendichtung f <Heiz & Kälte> oil pump gasket
Ölraffinerie f <Erdöl> refinery
Ölregenerat n <Abfall> recovered oil
Ölreservoir n <Elektrotech> oil tank
Ölring m <Maschinen> oil control ring, oil ring
OLRT *(Online-Echtzeit)* <Comp & DV> OLRT *(online real time)*
Ölrückgewinnung f <Maschinen> oil reclaiming
Ölrückgewinnungsschiff n <Umweltschmutz> oil recovery vessel
Ölrückgewinnungsschute f <Umweltschmutz> oil recovery barge
Ölrückgewinnungsskimmer m <Umweltschmutz> oil recovery skimmer
Ölrückstände mpl <Abfall> oil waste, residual oil
Ölsaat f <Lebensmittel> oilseed
Ölsand m <Erdöl> tar sand
Ölsandform f <Fertig> *(AE)* oil mold, *(BE)* oil mould *(Gießen)*

Ölsäure

Ölsäure f <Chemie, Lebensmittel> oleic acid
Ölsäureester m <Chemie> oleate
Ölschalter m <Elektrotech> oil-break switch, oil circuit breaker, oil switch; dead-tank oil circuit breaker *(mit Ölzusatzbehälter)*; live-tank oil circuit *(mit Schaltstrecke im Ölgefäß)*
Ölschauglas n <Heiz & Kälte> oil sight glass
Ölschicht f <Meerschmutz> oil film, oil slick
Ölschiefer m <Erdöl> oil shale *(Erdölgeologie)*
Ölschlamm m <Meerschmutz> chocolate mousse
Ölschlamm-Öltanker m <Wassertrans> oil-slurry oil tanker
Ölschleuderring m <Maschinen> oil slinger
Ölschlitzring m <Kfztech> slotted oil control ring
Ölsieb n <Kfztech> oil strainer
Ölspeichertank m <Erdöl> oil storage tank
Ölsperre f <Meerschmutz> jib boom, oil boom
Ölsperrengebinde n <Meerschmutz> boom pack
Ölspill-Identifikationssystem n <Meerschmutz> oil spill identification system
Ölspreizring m <Kfztech> oil expander ring
Ölspülung f <Erdöl> oil-base mud
Ölspur f <Erdöl> oil show *(Erdölsuche)*
Ölspurenanalysator m <Erdöl> *(BE)* oil show analyser, *(AE)* oil show analyzer *(Erdölsuche)*
Ölstandanzeiger m 1. <Fertig> *(AE)* contents gage, *(BE)* contents gauge; 2. <Gerät> oil level indicator; 3. <Kfztech> oil level stick
Ölstandhahn m <Fertig> bleeder
Ölstandsglas n <Heiz & Kälte> oil sight glass
Ölstandsmarkierung f <Kfztech> oil level mark *(Schmierung)*
Ölstrahl m <Kfztech> oil jet
Ölstrom m <Heiz & Kälte> oil flow rate
Ölströmungsanzeiger m <Heiz & Kälte> oil flow indicator
Ölströmungsmelder m <Heiz & Kälte> oil flow indicator
Ölströmungswächter m <Heiz & Kälte> oil flow indicator
Ölsumpfschmierung f <Kfztech> sump-type lubrication *(Motorschmierung)*
Öltank m 1. <Elektrotech> oil tank; 2. <Erdöl> oil storage tank, oil tank *(Raffinerie)*
Öltanker m 1. <Erdöl> crude oil tanker, oil tanker *(Schiffahrt)*; 2. <Trans> fuel tanker; 3. <Umweltschmutz> crude carrier; 4. <Wassertrans> oil tanker
Öltankerhaverie f <Umweltschmutz> average of an oil tanker
Öltasche f <Maschinen> oil pocket
Öltemperaturanzeige f <Lufttrans> oil temperature indicator *(Flugwesen)*
Öltemperaturmessfühler m <Lufttrans> oil temperature probe
Ölteppich m <Meerschmutz> oil film, oil slick, oil spill, spill
Ölteppichbekämpfungsausrüstung f <Meerschmutz> oil spill combatting equipment
Ölteppichbeseitigung f <Meerschmutz> oil slick sinking
Ölteppichidentifikation f <Meerschmutz> oil spill identification
Öltransformator m 1. <Elektriz> oil transformer; 2. <Elektrotech> oil-immersed transformer
Öltrennschalter m <Elektriz> oil circuit breaker
Öltropfgefäß n <Maschinen> sight feed lubricator
ölverbrennendes Kraftwerk n <Elektriz> oilfired power station
Ölverdichtungsmittel n <Umweltschmutz> oil-concentrating agent
Ölvergasung f <Erdöl> oil gasification *(Gaserzeugung)*
ölverschmutztes Abwasser n <Umweltschmutz> oil-polluted waste-water
Ölverschmutzung f **des Meeres** <Meerschmutz, Umweltschmutz> oil spill
Ölverschmutzungsnotfall m <Meerschmutz, Umweltschmutz> oil pollution emergency
Ölverschüttung f <Meerschmutz> spill
ölverseuchte Gewässer npl <Umweltschmutz> oil-contaminated waters
Ölversorgung f <Kfztech> oil feed *(Schmierung)*
Ölverteiler m <Maschinen> oil distributor
Ölwaage f <Metrol> oilometer
Ölwanne f 1. <Kfztech> oil sump, oilpan; 2. <Lufttrans> oil sump; 3. <Mechan> oilpan
Ölwannendichtung f <Kfztech> oilpan gasket
Ölwannenschutz m <Kfztech> sump guard *(Motorschmierung)*
Ölwäsche f <Erdöl> oil scrubbing *(Gasaufbereitung)*
Öl-Wasser-Berührungsfläche f <Meerschmutz> oil-water interface
Ölwechsel m 1. <Kfztech> oil change *(Schmierung)*; 2. <Maschinen> oil change
Ölzahl f <Kunststoff> oil length
Öl-Zentralheizung f <Heiz & Kälte> oilfired central heating system
Ölzerstäubungsbrenner m <Heiz & Kälte> atomizing oil burner
Ölzeug n <Textil, Wassertrans> oilskin *(wetterfeste Kleidung)*
Ölzufuhr f <Kfztech> oil feed *(Schmierung)*
Ölzuführungskanal m <Kfztech> oil gallery
Omega-Minus-Teilchen n <Phys> omega minus particle *(Elementarteilchen)*
Omnibuszug m <Kfztech> passenger road train
OMR *(optische Markierungserkennung)* <Comp & DV> OMR *(optical mark recognition)*
ON *(Ortsnetz)* 1. <Elektriz> distribution network, local network; 2. <Telekom> local network
Önanth... <Chemie> oenanthic
Önanthal n <Chemie> oenanthal
ONKz *(Ortsnetzkennzahl)* <Telekom> area code
Online-... <Comp & DV> online
Online-Datenbank f <Comp & DV> online database
Online-Drucken n <Verpack> printing on line
Online-Echtzeit f *(OLRT)* <Comp & DV> online real time *(OLRT)*
Online-Informationen fpl <Comp & DV> online content
Online-Messung f <Kerntech, Metrol> online measurement
Online-Testprogramm n <Comp & DV> OLT, online test program
Online-Testsystem n <Comp & DV> OLTS, online test system
Online-Verarbeitung f <Comp & DV> online data processing, online processing
Online-Wörterbuch n <Telekom> dictionary online, internet dictionary
OOP *(objektorientierte Programmierung)* <Künstl Int> OOP *(object-oriented programming)*
OO-Umgebung f <Comp & DV> O-O environment
opake Substanz f <Strahlphys> opaque substance
Opakglas n <Ker & Glas> opaque glass
Opal m <Teilphys> Opal
Opalglas n <Ker & Glas> opal glass
Opalinglas n <Ker & Glas> opaline
Opazimeter n <Papier> opacimeter
Opazität f <Chemie, Druck, Papier, Phys, Wellphys> opacity
OPEC *(Organisation Ölexportierender Länder)* <Erdöl> OPEC *(Organization of Petroleum-Exporting Countries)*
Open-Shop-Betrieb m <Comp & DV> hands-on operation

Open-Top-Container m <Trans> open-top container
Operand m <Comp & DV, Math> operand
Operation f 1. <Comp & DV, Telekom> operation; 2. <Math> operation *(Rechenoperation)* • **Keine Operation** <Comp & DV> no-operation *(NO-OP)*
Operationenrangfolge f <Comp & DV> operator precedence
Operationscharakteristik f *(OC)* 1. <Math> operating characteristic, OC *(Testtheorie)*; 2. <Qual> operating characteristic, OC
Operationscharakteristik-Kurve f <Math> operating characteristic curve *(OC-Kurve)*
Operationscode m <Comp & DV> op code, operation code
Operationsfolge f <Comp & DV> sequence of operations
Operationsregister n <Comp & DV> operation register
Operations-Research n *(OR)* <Comp & DV> operations research *(OR)*
Operationsschlüssel m <Comp & DV> op code, operation code
Operationstabelle f <Comp & DV> operation table
Operationsverstärker m *(OA)* <Comp & DV, Elektronik, Phys> operational amplifier *(op amp)*
Operationszyklus m <Comp & DV> memory cycle *(Basissysteme)*; machine cycle *(Mikroprozessoren)*
Operator m <Comp & DV, Math> operator
operatorbedienter Betrieb m <Comp & DV> hands-on operation
Operatorenrechnung f <Math> operational calculus
Opferanode f <Erdöl> sacrificial anode *(katodischer Korrosionsschutz)*
Opian... <Chemie> opianic
Opianyl n <Chemie> meconin, opianyl
Oppositron n <Elektronik> O-type tube
Optik f <Optik> optics
Optiklithographie f <Elektronik> optical lithography
Optikmodulator m <Elektronik> optical modulator
Optikwerkzeug n <Ker & Glas> optical tool
Optimaldetektion f <Telekom> optimum detection
optimale Abtastung f <Telekom> optimal sampling
optimale Ballung f <Elektronik> optimum bunching
optimale Dämpfung f <Elektrotech> optimum damping
optimale Lösung f <Künstl Int> optimal solution
optimale Objektausleuchtung f <Strahlphys> optimum object illumination
optimale Programmierung f <Comp & DV> minimum-access programming
optimale Regelung f <Telekom> optimal control
optimale Vormagnetisierung f <Aufnahme> optimal bias
Optimalempfang m <Telekom> optimum detection
optimaler Abbrand m <Kerntech> optimum burnup
optimaler Wassergehalt m <Bau> optimum moisture content
optimaler Weg m <Telekom> optimal path
optimaler Wiedereintrittskorridor m <Raumfahrt> optimum re-entry corridor *(Landeanflug)*
optimales Regelmodell n <Ergon> optimal control model
optimales Regelsystem n <Kerntech> optimal control system
optimales Walzen n <Kohlen> optimum grind
Optimalfilter n <Elektronik, Funkort> analog matched filter *(Radar)*
optimieren v <Comp & DV> optimize
Optimierung f <Comp & DV> optimization
Option f <Comp & DV, Erdöl> option *(Verträge)*
Optionsfeld n <Comp & DV> option field
Optionstabelle f <Comp & DV> option table
optisch aktives Material n <Optik, Telekom> optically-active material

optisch eben adj <Foto> optically-flat
optisch gekoppeltes Festkörperrelais n <Elektrotech> optically-coupled solid state relay
optisch glatte Oberfläche f <Phys> optically-smooth surface
optische Aberration f <Telekom> optical aberration
optische Absorption f <Strahlphys> optical absorption
optische Abstimmung f <Telekom> optical tuning
optische Achse f 1. <Foto, Optik> optical axis; 2. <Phys> optic axis; 3. <Telekom> *(AE)* fiber axis, *(BE)* fibre axis, optical axis
optische Aktivität f <Phys> optical activity
optische Aufzeichnung f <Akustik, Aufnahme, Telekom> optical recording
optische Ausgangsleistung f <Elektrotech, Telekom> optical output power, optical power output
optische Bank f <Foto, Metrol, Phys> optical bench
optische Bildplatte f in ROM-Technik <Optik> laser optical disk
optische Bistabilität f <Telekom> optical bistability
optische Brechung f <Telekom> optical refraction
optische Datenbank f <Optik> optical database
optische Datendiskette f <Comp & DV, Optik> optical data disk
optische Datenerfassungsstation f <Comp & DV> optical image unit
optische Dichte f *(D)* <Optik> optical density *(D)*
optische Dicke f <Optik, Telekom> optical thickness
optische Dispersion f <Telekom> optical dispersion
optische Effekte mpl <Fernseh> visual effects
optische Eigenschaft f <Telekom> optical characteristic
optische Eingangsleistung f <Elektrotech> optical input power
optische Erfassung f <Elektronik> optical detection
optische Erkennung f <Elektronik> optical sensing
optische Faser f <Elektrotech, Phys, Telekom> *(AE)* optical fiber, *(BE)* optical fibre
optische Faser f aus Allglas <Elektrotech> *(AE)* all-glass optical fiber, *(BE)* all-glass optical fibre
optische Faser f in Vollplastausführung <Elektrotech> *(AE)* all-plastic optical fiber, *(BE)* all-plastic optical fibre
optische Frequenz f <Elektronik> optical frequency
optische Hybridschaltung f <Elektronik> optical hybrid circuit
optische Informationsverarbeitung f <Telekom> optical information processing
optische Interferenz f <Telekom> optical interference
optische Isolierung f <Elektrotech> optical isolation
optische Karte f <Optik> optical card
optische Kohärenz f <Telekom> optical coherence
optische Kommunikationstechnik f <Telekom> optical communications
optische Koppelmatrix f <Telekom> optical switching matrix
optische Kopplung f <Elektrotech> optical coupling
optische Leistung f <Elektrotech, Optik, Telekom> optical power
optische Leistungsquelle f <Elektrotech> optical power source
optische logische Schaltung f <Elektronik> optical logic circuit
optische Markierungserkennung f *(OMR)* <Comp & DV> optical mark recognition *(OMR)*
optische Maske f <Elektronik> optical mask
optische Modulation f <Elektronik> optical modulation
optische monomodale Übertragung f <Telekom> optical monomode transmission
optische Nachrichtentechnik f <Telekom> optical communications

optische

optische Nachrichtenübertragung f durch die Luft <Telekom> optical communication in the atmosphere
optische Nachrichtenverbindung f <Telekom> optical communication link
optische Platte f <Comp & DV> optical disk
optische Polarisation f <Telekom> optical polarization
optische Profilschleifmaschine f <Maschinen> optical profile grinder
optische Qualitätskontrolle f <Qual> optical quality control
optische Reflektrometrie f <Telekom> optical time domain reflectometry
optische Regenerierfähigkeit f <Optik> optical regenerative power
optische Resonanz f <Elektronik> optical resonance
optische Schablone f <Elektronik> optical master
optische Schnittstelle f <Telekom> infrared interface, optical interface
optische Schutzschicht f <Elektronik> optical resist
optische Schwingung f <Elektronik> optical oscillation
optische Signalverarbeitung f <Elektronik, Strahlphys> optical signal processing
optische Signalwandlung f <Elektrotech> optical signal conversion
optische Speicherung f 1. <Optik> optical storage; 2. <Telekom> optical recording
optische Spektralanalyse f <Elektronik> optical spectral analysis
optische Spleißstelle f <Optik> optical splice
optische Strahlung f <Optik, Telekom> optical radiation
optische Tonspur f <Aufnahme> optical sound track
optische Übertragung f <Comp & DV> optical transmission
optische Übertragungsleitung f <Elektrotech, Telekom> optical transmission line
optische Übertragungsstrecke f <Telekom> optical link
optische Vermittlung f 1. <Comp & DV> optical switching; 2. <Telekom> optical exchange, optical switching
optische Verstärkung f <Elektronik> optical gain
optische Verzerrung f <Ker & Glas> optical distortion
optische Videodiskette f <Optik> optical videodisk
optische Weglänge f <Optik, Telekom> optical path length
optische Weiche f <Telekom> optical combiner
optische Welle f <Elektrotech> optical wave
optische Zeichenerkennung f (OCR) <Comp & DV, Metrol> optical character recognition (OCR)
optische Zeitbereichsreflektrometrie f <Optik, Telekom> optical time domain reflectometry
optische Zielvorrichtung f <Raumfahrt> optical sight
optischer Abscheider m <Abfall> optical sorter
optischer Abtaster m <Comp & DV> bar code scanner
optischer Alarm m <Telekom> visual alarm
optischer Aufheller m <Kunststoff> optical brightener
optischer Aufnehmer m <Elektronik> optical sensor
optischer Ausgang m <Elektrotech> optical output
optischer Autofallendetektor m <Trans> optical speed trap detector
optischer Belegleser m <Comp & DV, Druck> optical character reader
optischer Bus m <Telekom> optical data bus
optischer Datenbus m <Optik, Telekom> optical data bus
optischer Datenträger m <Comp & DV> optical medium
optischer Detektor m <Elektronik, Optik, Telekom> optical detector
optischer Drehwinkel m (α) <Optik> angle of optical rotation (α)
optischer Dünnschicht-Wellenleiter m <Elektronik> thin film optical waveguide

optischer Eingang m <Elektrotech> optical input
optischer Empfänger m <Telekom> optical receiver, (AE) receive fiberoptic terminal device, (BE) receive fibreoptic terminal device
optischer Entfernungsmesser m <Foto> optical rangefinder
optischer Festwertspeicher m (OROM) 1. <Comp & DV> optical ROM, optical read-only memory (optical ROM); 2. <Optik> optical ROM, optical read-only memory (optical read-only memory)
optischer Hohlraum m <Optik> optical cavity
optischer Hohlraumresonator m <Strahlphys, Telekom> optical cavity
optischer IC m <Elektronik> optical IC
optischer Impuls m <Elektronik> optical pulse
optischer integrierter Schaltkreis m <Elektronik, Optik> integrated optical circuit, optical integrated circuit
optischer Isolator m <Telekom> optical isolator, optoisolator
optischer Kombinierer m <Optik> optical combiner
optischer Komparator m <Metrol> optical comparator
optischer Koppelpunkt m <Telekom> optical switching crosspoint
optischer Koppler m <Elektrotech, Optik, Telekom> optical coupler, (AE) optical fiber coupler, (BE) optical fibre coupler
optischer Korrelationsanalysator m <Elektronik> optical correlator
optischer Laseraufzeichner m <Elektronik> laser optical recorder
optischer Lesekopf m <Optik> optical head
optischer Leser m <Comp & DV> optical reader, optical scanner
optischer Markierungsleser m <Comp & DV> optical mark reader
optischer Maser m <Elektronik> optical maser (Laser)
optischer Melder m <Raumfahrt> optical sensor
optischer Mittelpunkt m <Phys, Telekom> (AE) optical center, (BE) optical centre
optischer Modulator m <Telekom> optical modulator
optischer Multiplexer m <Fernseh> optical multiplexer
optischer parametrischer Oszillator m <Telekom> optical parametric oscillator
optischer Repeater m <Telekom> optical repeater
optischer Resonator m 1. <Strahlphys> optical resonator; 2. <Telekom> optical cavity, optical resonator
optischer Schalter m <Elektrotech, Telekom> optical switch
optischer Seekabelverstärker m <Telekom> submerged optical repeater
optischer Sender m <Telekom> (AE) transmit fiber optic terminal device, (BE) transmit fibre optic terminal device
optischer Sensor m <Optik> optical detector
optischer Solarreflektor m <Raumfahrt> OSR, optical solar reflector (Raumschiff)
optischer Speicher m 1. <Comp & DV> optical storage; 2. <Elektrotech> optical memory; 3. <Optik> optic storage, optical memory
optischer Spektrumsanalysator m <Elektronik> (BE) optical spectral analyser, (AE) optical spectral analyzer
optischer Spleiß m <Telekom> optical splice
optischer Steckverbinder m <Telekom> optical connector, (AE) optical fiber connector, (BE) optical fibre connector
optischer Tonkopf m <Aufnahme> optical sound head
optischer Träger m <Elektronik, Telekom> optical carrier
optischer Transceiver m <Telekom> optical transceiver
optischer Übergang m <Strahlphys> optical transition
optischer Verbinder m <Optik> optical link

optischer Verstärker m <Elektronik, Telekom> optical amplifier
optischer Weg m <Optik, Phys> optical path
optischer Wellenleiter m 1. <Elektrotech, Optik> optical waveguide; 2. <Telekom> *(AE)* optical fiber, *(AE)* optical fibre
optischer Zeichenleser m <Comp & DV, Verpack> optical character reader
optischer Zirkulator m <Telekom> optical circulator
optischer Zweig m <Phys> optical branch *(Festkörpertheorie)*
optischer Zwischenregenerator m <Telekom> optical regenerative repeater
optischer Zwischenverstärker m <Telekom> optical repeater
optisches Abmessungs-Messinstrument n <Metrol> optical instrument for dimensional measurement
optisches Abtastgerät n <Fernseh> optical scanning device
optisches Anklopfen n <Telekom> optical call waiting indication
optisches Archivierungssystem n <Optik> optical filing system
optisches Band n <Optik> optical tape
optisches Bild n <Elektronik> optical image
optisches Dämpfungsglied n <Telekom> optical attenuator
optisches digitales Teilnehmeranschlussnetz n <Telekom> digital fiber-optic subscriber line network
optisches Elektron n <Strahlphys> optical electron
optisches Erholungsvermögen n <Optik> optical regenerative power
optisches Fenster n <Optik> optical window
optisches Filter n <Elektronik, Optik, Telekom> optical filter
optisches Flintglas n <Ker & Glas> optical flint
optisches Freileitungskabel n <Telekom> aerial optical cable
optisches Gitter n <Optik> diffraction grating
optisches Glas n <Foto> optical glass
optisches Hochgeschwindigkeits-LAN n <Telekom> fiber optic broadband LAN, optical fiber high-speed LAN
optisches Instrument n <Phys> optical instrument
optisches Ionenstrahlsystem n <Strahlphys> ion beam optical system
optisches Isomer n <Chemie> enantiomer, optical isomer
optisches Kabel n 1. <Fernseh, Optik> optical cable; 2. <Telekom> *(AE)* optical fiber cable, *(AE)* optical fibre cable, optical cable
optisches Koppelfeld n <Telekom> OSN, optical switching network
optisches Logikgatter n <Elektronik> optical logic gate
optisches Logikglied n <Elektronik> optical logic gate
optisches Markierungslesen n <Comp & DV> OMR, optical mark reading
optisches Medium n <Comp & DV, Optik> optical medium
optisches Messgitter n <Metrol> optical graticule
optisches Multiplexing n <Elektronik> optical multiplexing
optisches Multiplexverfahren n <Elektronik> optical multiplex
optisches Muster n <Elektronik> optical pattern
optisches Publizieren n <Optik> optical publishing
optisches Pumpen n 1. <Elektronik> optical pumping *(Optoelektrik)*; 2. <Kerntech> optical pumping; 3. <Phys> optical pumping *(Laser)*; 4. <Strahlphys> optical pumping
optisches Pyrometer n <Phys, Strahlphys> optical pyrometer

optisches Radar n <Funkort> lidar, light detecting and ranging
optisches Relais n 1. <Elektrotech> optical relay; 2. <Optik> optical repeater
optisches Schalten n <Elektrotech> optical switching
optisches Schrittschaltwerk n <Elektronik> optical stepper
optisches Servo n <Elektrotech> optical servo
optisches Signal n 1. <Comp & DV> visual/audible signal; 2. <Telekom> optical signal
optisches Speichermedium n <Optik> optical storage medium
optisches Spektrum n <Elektronik, Optik, Strahlphys> optical spectrum
optisches System n <Elektronik> optical system
optisches Teleskop n <Raumfahrt> optical telescope
optisches Tonaufzeichnungsgerät n <Aufnahme> optical sound recorder
optisches Tonwiedergabegerät n <Aufnahme> optical sound reproducer
optisches Übertragungssystem n 1. <Optik> optical transmission system; 2. <Telekom> optical line system, optical transmission system
optisches Vermittlungssystem n <Telekom> optical switching system
optisches Zeichenerkennungssystem n <Verpack> optical character reading system
optisches Zugangsnetz n <Telekom> optical access network
Optocodierer m <Raumfahrt> optical encoder *(Weltraumfunk)*
Optoelektronik f <Comp & DV, Elektrotech, Phys, Telekom> optoelectronics
optoelektronisch adj <Optik, Telekom> optoelectronic
optoelektronisch gekoppeltes Festkörperrelais n <Elektrotech> photocoupled solid-state relay
optoelektronische Koppelmatrix f <Telekom> optoelectronic switching matrix
optoelektronische Schnittstelle f <Telekom> optoelectronic interface
optoelektronischer Chip m <Elektronik> optoelectronic chip
optoelektronischer Empfänger m <Telekom> optoelectronic receiver
optoelektronischer Isolator m <Telekom> optoisolator
optoelektronischer Koppelpunkt m <Telekom> optoelectronic crosspoint
optoelektronischer Koppler m <Elektrotech> optoelectronic coupler
optoelektronischer Schalter m <Elektrotech> optoelectronic switch
optoelektronischer Verstärker m <Elektronik> optoelectronic amplifier
optoelektronischer Wandler m <Elektrotech> optoelectronic transducer
optoelektronisches Bauelement n <Optik> optoelectronic device
optoelektronisches Gerät n <Elektrotech> optoelectronic device
Optokoppler m <Elektrotech, Telekom> optocoupler
OR *(Operations-Research)* <Comp & DV> OR *(operations research)*
Orbit m <Raumfahrt> orbit
Orbital n <Strahlphys> orbital
Orbitalarbeitsstation f <Raumfahrt> orbital workshop
orbitales Labor n <Raumfahrt> orbiting laboratory
orbitales Observatorium n <Raumfahrt> orbiting astronomical observatory
Orbitalraumfahrzeug n <Raumfahrt> orbital vehicle

Orbitalsatellit

Orbitalsatellit m <Raumfahrt> orbiting satellite *(Satellit auf Erdumlaufbahn)*
Orbitalwinkelmomentzahl f (L) <Kerntech> total orbital angular momentum number (L)
Orbitänderung f <Raumfahrt> orbit modification
Orbitdauer f <Raumfahrt> orbital period
Orbiter m <Raumfahrt> orbiter
Orbiterstufe f <Raumfahrt> orbiter stage
Orbitkorrektur f <Raumfahrt> orbit correction
Orbitnachführung f <Raumfahrt> orbit tracking
Orbitsteuerung f <Raumfahrt> orbit control
Orbittrimmen n <Raumfahrt> orbit trimming
Orbitveränderung f <Raumfahrt> orbit transfer *(in andere Umlaufbahn)*
Orbitvorhersage f <Raumfahrt> orbit prediction
Orbitzähler m <Raumfahrt> orbit counter
Orcein n <Chemie> orcein
ordentliche Reflexion f <Phys> specular reflection
ordentlicher Strahl m <Phys> ordinary ray
Ordinalmerkmal n <Qual> ordinal characteristic
Ordinalzahlen fpl <Geom> ordinals
Ordinate f 1. <Comp & DV> ordinate; 2. <Math> ordinate, vertical axis, y-axis
Ordinaten fpl <Wassertrans> ordinates *(Schiffbau)*
Ordner m <Fertig> unscrambler *(Knüppel)*
Ordnung f **der Interferenz** <Phys> order
Ordnungsaxiom n <Metall> ordering axiom
ordnungsgemäßer Systemabschluss m <Comp & DV> graceful shutdown
Ordnungsgütemaß n <Comp & DV> ordering bias
Ordnungs-Unordnungsmodell n <Kerntech> order-disorder model *(eines Atomkerns)*
Ordnungszahl f 1. <Akustik> order *(einer Harmonischen)*; 2. <Kerntech> atomic number; 3. <Math> ordinal, ordinal number; 4. <Phys, Strahlphys> atomic number
Ordnungszahlen fpl 1. <Akustik, Kerntech, Phys, Strahlphys> ordinals; 2. <Math> ordinals, ordinal numbers
Ordnung-Unordnung f <Metall> order-disorder
Organ n <Chemie> organ
Organisation f **ölexportierender Länder** *(OPEC)* <Erdöl> Organization of Petroleum-Exporting Countries (OPEC)
Organisationskanal m <Mobilkom> control channel, set-up channel
organisatorische Operation f <Comp & DV> housekeeping operation
organisch adj <Anstrich, Lebensmittel> organic
organische Bestandteile npl <Erdöl> organic matter *(Erdölgeologie)*
organische Verunreinigung f <Bau> organic impurity
organischer Abfall m <Abfall> organic waste
organischer Boden m <Kohlen> organic soil
organischer Flüssigkeitslaser m <Elektronik> organic liquid laser
organischer Flüssigkeitsmotor m <Kfztech> organic fluid engine
organischer Moderator m <Kerntech> organic moderator
organischer Stoff m <Umweltschmutz> organic matter
organischer Widerstand m <Elektrotech> organic resistor
organisches Kühlmittel n <Maschinen> organic refrigerant
organisches Metall n <Metall> organic metal
organisches Polymer n <Anstrich> organic polymer
organisches Sulfid n <Chemie> (AE) organic sulfide, (BE) organic sulphide, thioether
organogen adj <Kohlen> organogenous
Organomagnesium... <Chemie> organomagnesium

Organomagnesiumverbindung f <Chemie> organomagnesium compound
organometallisch adj <Chemie> organometallic
Organosol n <Chemie> organosol
Organzin n <Textil> organzine
Orgelpfeifenanschlag m <Ker & Glas> organ stop
orientierte Kernbildung f <Metall> oriented nucleation
orientiertes Wachstum n <Metall> oriented growth
Orientierungsfaktor m <Metall> orientation factor
Orientierungspunkt m <Bau> landmark
Original n 1. <Akustik> original; 2. <Aufnahme> master; 3. <Comp & DV> first generation, magnetic master, master; 4. <Druck, Fernseh> original • **Original erstellen** <Aufnahme> master
Originalausgabe f <Druck> original edition
Originalausstattung f <Maschinen> original equipment
Originalband n <Comp & DV> master tape
Originalbearbeitung f <Fernseh> editing on original
Originalbild n <Fernseh> key frame *(Bildsignalkompression)*
Originaldokument n <Comp & DV> source document
Originalgerätehersteller m <Telekom> original equipment manufacturer
originalgetreue Darstellung f **der Druckausgabe am Bildschirm** *(WYSIWYG)* <Comp & DV> what you see is what you get (WYSIWYG)
Originalmaßstab m <Maschinen> full scale
originalsicher Verschluss m <Verpack> tamper-evident closure
Originalvorlagen fpl <Konstzeich> original documents
O-Ring m 1. <Fertig> O-ring *(Kunststoffinstallationen)*; 2. <Kfztech> *(Runddichtring)* O-ring *(Schmierung)*; 3. <Kunststoff> O-ring; 4. <Maschinen> *(Runddichtring)* O-ring; 5. <Mechan> O-ring
Orizabin n <Chemie> jalapin, orizabin
Orlean m <Lebensmittel> anatto, annatto, bixin
Orlopdeck n <Wassertrans> orlop deck *(Schiffbau)*
Ornamentglas n <Ker & Glas> patterned glass
Ornamentierung f <Ker & Glas> figuring
Ornamentwalzglas n <Ker & Glas> figured rolled glass
Ornithur... <Chemie> ornithuric
OROM *(optischer Festwertspeicher)* <Comp & DV, Optik> OROM *(optical read-only memory)*
Orotron n <Elektronik> orotron
Orsellin... <Chemie> orsellic
Ort drucken v/vor <Verpack> print on site
Ort mischen v/vor <Bau> mix in place
Ort und Stelle/an <Mechan> in situ
Ort/vor 1. <Bau> in situ, on the spot; 2. <Kohlen> in situ
Ortbeton m <Bau> in situ concrete, site concrete • **in Ortbeton hergestellt** <Bau> cast-in-place
Ortbetonpfahl m <Kohlen> cast-in-place pile
orten v <Elektrotech, Wassertrans> locate *(Schiff, Seezeichen)*
Örterbau m <Kohlen> stall working
Orthicon n <Elektronik> orthicon *(Bildaufnahmeröhre)*
Orthit m <Kerntech> orthite
Orthoameisensäure f <Chemie> orthoformic acid
Orthoborsäure f <Chemie> boric acid, orthoboric acid
orthochromatische Emulsion f <Foto> orthochromatic emulsion
Orthodrombahn f <Raumfahrt> orthodromic track
Orthodromie f <Wassertrans> orthodromy *(Navigation)*
Orthodromprojektion f <Raumfahrt> orthodromic projection
orthogonal adj <Math> orthogonal *(Vektoren, Unterräume)*
orthogonale Modulation f <Telekom> quadrature modulation

orthogonale Polarisation f <Telekom> orthogonal polarization
orthogonale Projektion f <Geom> orthogonal projection, orthographic projection
orthogonale Signale npl <Telekom> orthogonal signals
orthogonale Zerspanung f <Fertig, Math> orthogonal cutting
Orthogonalebene f <Fertig> orthogonal plane
Orthogonalfreiwinkel m <Fertig> orthogonal clearance
orthogonalisieren v <Math> orthogonalize
Orthogonalisierung f <Geom> orthogonalization
Orthogonalität f <Math> orthogonality
Orthogonalkeilwinkel m <Fertig> orthogonal wedge angle
Orthogonalschnitt m <Fertig> orthogonal cut
orthographische Projektion f <Geom> orthogonal projection, orthographic projection
Orthokiesel... <Chemie> orthosilicic
Orthokieselsäure f <Chemie> orthosilicic acid, silicic acid, tetrasilicic acid
Orthokohlensäure f <Chemie> orthocarbonic acid
orthonormal adj <Geom> orthonormal
Orthonormalsystem n <Geom> orthonormal system
Ortho-Para-Umwandlung f <Kerntech> orthopara conversion
Orthophosphat n <Chemie> orthophosphate
Orthophosphor... <Chemie> orthophosphoric
Orthosilicat n <Chemie> orthosilicate, tetraoxosilicate
orthotrophe Werkstoffe mpl <Maschinen> orthotropic materials
Orthowasserstoff m <Chemie> orthohydrogen
Orthoxyol n <Erdöl> orthoxylene (Petrochemie)
örtlich starke Rotation f <Strömphys> locally high vorticity (Wirbelstärke)
örtliche Abzugshaube f <Sicherheit> local exhaust hood
örtliche Ausrichtung f <Elektronik> local alignment
örtliche Beanspruchung f <Metall> local stress
örtliche Emissionsquelle f <Umweltschmutz> local emission, local emission source
örtliche Nachgiebigkeit f <Metall> local yielding
örtliche Querschnittsvergrößerung f <Fertig> lateral swelling
Orts... <Telekom> local
ortsabhängige Eigenschaften fpl <Telekom> location features
Ortsamt n <Telekom> local exchange
Ortsaufteilungskabel n <Telekom> local distribution cable
Ortsbatterie f (LB) <Comp & DV, Telekom> local battery (LB)
Ortsbetrieb m <Telekom> local operation
Ortschaum m <Kunststoff> foam-in-place compound
Ortsdiversity n <Telekom> site diversity
Ortsfernsprechnetz n (OFN) <Telekom> local exchange area
ortsfest adj 1. <Telekom> landline, fixed; 2. <Fertig> stationary
ortsfeste Emissionsquelle f <Umweltschmutz> stationary emission source
ortsfeste Kapselung f <Sicherheit> fixed enclosure
ortsfester Teilnehmer m <Telekom> fixed subscriber
ortsfestes Telefon n <Telekom> stationary telephone set
Ortsfrequenz f <Fernseh> spatial frequency
Ortsgebühr f <Telekom> local charge rate, local rate
ortsgebunden adj 1. <Fertig> localized (Vektor); 2. <Kohlen> sedentary
Ortsgespräch n <Telekom> local call
Ortskapazität f <Trans> local capacity
Ortskennziffer f <Telekom> local area code, telephone local network code, telephone trunk code

Ortskurve f **des Frequenzganges** <Regelung> polar plot
Ortsleitungskonzentrator m <Telekom> local line concentrator
Ortsnetz n (ON) 1. <Elektriz> distribution network, local network; 2. <Telekom> local exchange area, local network
Ortsnetzkennzahl f (ONKz) <Telekom> area code, local area code; national destination code (NDC); trunk code
Ortstarif m <Telekom> local charge rate, local rate
Ortsteilnehmerendstelle f <Telekom> LUT, local user terminal
Ortstelefonbuch n <Telekom> local telephone directory
Ortsumgehungsstraße f <Trans> belt highway
Ortsvektor m <Geom, Phys> position vector
Ortsverbindung f <Telekom> local junction
Ortsverbindungskabel n (OVk) <Telekom> local junction cable
Ortsverbindungsleitung f (OVl) <Telekom> local junction cable
Ortsverkehr m <Telekom, Trans> local traffic
Ortsvermittlungsstelle f (OVSt) <Telekom> local exchange, metropolitan switch
Ortsverteilungsnetz n <Telekom> local distribution network
Ortsvorwahl f <Telekom> local area code, telephone local network code, telephone trunk code
Ortszeit f <Telekom> local mean time
Ortungsanlage f <Wassertrans> position-fixing equipment
Ortungsboje f <Wassertrans> radio sonobuoy (Seezeichen)
Ortungsobjekt n <Funkort> target
Os (Osmium) <Chemie> Os (osmium)
Osazon n <Chemie> osazone
OSB (oberes Seitenband) <Elektronik, Funktech, Telekom> USB (upper sideband)
OSCAR (Bahnsatellit für Amateurfunkzwecke) <Funktech, Raumfahrt> OSCAR, Orbiting Satellite Carrying Amateur Radio (Weltraumfunk)
Öse f 1. <Fertig> ear, eye, ring; 2. <Maschinen> ear, eye, lug; 3. <Mechan> lug; 4. <Verpack> eyelet; 5. <Wassertrans> eye
ösen v <Wassertrans> bail
Ösenschraube f 1. <Maschinen> eye screw; 2. <Mechan> eyebolt
OSI (Kommunikation offener Systeme) <Comp & DV, Telekom> OSI (open systems interconnection)
OSI-Schichten fpl <Telekom> OSI layers
Osmat n <Chemie> osmate, tetraoxosmate
Osmium n (Os) <Chemie> osmium (Os)
Osmol n <Chemie> osmole
Osmolarität f <Chemie> osmolarity
Osmondit m <Metall> osmondite
osmophore Gruppe f <Chemie> osmophore
Osmose f <Chemtech, Erdöl, Phys> osmosis
Osmosevorgang m <Chemtech> osmosis process
osmotisch adj <Phys> osmotic
osmotischer Druck m <Erdöl, Heiz & Kälte, Phys> osmotic pressure
OSO (Erz-Schlamm-Öl) <Wassertrans> OSO (ore-slurry-oil)
Oson n <Chemie> osone
OSO-Tanker m <Wassertrans> OSO tanker
Osotriazol n <Chemie> osotriazole
Ossein n <Chemie> ossein
Östradiol n <Chemie> oestradiol
Östriol n <Chemie> oestriol
Östron n <Chemie> oestrone, theelin

Ostwald'sches

Ostwald'sches Viskosimeter n <Labor> Ostwald viscometer
O-System n (Oberleitungssystem) <Trans> trolley system
Oszillation f 1. <Elektriz, Elektronik, Funktech> oscillation; 2. <Metall> vibration; 3. <Wellphys> oscillation
Oszillator m 1. <Elektriz, Elektronik, Funktech> oscillator; 2. <Raumfahrt> oscillator (Weltraumfunk); 3. <Telekom, Wellphys> oscillator
Oszillator m **mit negativem Widerstand** <Phys> negative resistance oscillator
Oszillator m **mit offenem Schwingungskreis** <Elektronik> open loop oscillator
Oszillator m **mit Permeabilitätsabstimmung** <Funktech> permeability-tuned oscillator
Oszillator m **mit Spannungssteuerung** <Elektronik> voltage-controlled oscillator
Oszillator m **mit verlängerter Wechselwirkung** <Elektronik> extended-interaction oscillator
Oszillator m **mit Wien-Brücke** <Elektronik> Wien bridge oscillator
Oszillator m **mit YIG-abgestimmten Transistoren** <Elektronik> YIG-tuned transistor oscillator
Oszillator m **mit YIG-Abstimmung** <Elektronik> YIG-tuned oscillator
Oszillatorabstimmraum m <Elektronik> oscillator cavity
Oszillatorbatterie f <Elektronik> bank of oscillators
Oszillatordiode f <Elektronik> diode oscillator
Oszillatorhohlraum m <Funktech> oscillator cavity
Oszillatorquarz m <Elektronik> oscillator crystal
Oszillatorreihe f <Elektronik> oscillator bank
Oszillatorspule f <Elektrotech, Funktech> oscillator coil
oszillieren v 1. <Elektronik> oscillate; 2. <Fertig> rock
Oszillieren n <Fertig> rocking
oszillierend adj 1. <Elektriz, Elektronik> oscillating (Strom); 2. <Fertig> rocking; 3. <Papier> oscillating; 4. <Telekom> oscillatory
oszillierendes Spritzrohr n <Papier> oscillating shower
Oszillogramm n <Gerät> oscillogram, oscilloscope presentation
Oszillograph m <Elektronik, Phys, Wellphys> oscillograph
oszillographieren v <Elektronik> display
Oszilloskop n 1. <Comp & DV, Elektriz, Elektronik> cathode-ray oscillograph, oscilloscope; 2. <Funktech, Phys, Telekom, Wellphys> oscilloscope
Oszilloskopbild n <Gerät> oscilloscope presentation
Oszilloskopkurve f <Elektronik> oscilloscope trace
Oszilloskopröhre f <Elektronik> oscilloscope tube
OT (oberer Totpunkt) <Kfztech> TDC (top dead center)
Otologie f <Ergon> otology
Ottogasmotor m <Kfztech> liquefied petroleum gas engine
Ottokraftstoff m <Kfztech> (AE) gas, (AE) gasoline
Ottomotor m 1. <Kfztech> (AE) gas engine, (AE) gasoline engine, (BE) petrol engine, (AE) gas motor, (AE) gasoline motor, (BE) petrol motor; 2. <Maschinen> spark ignition engine; 3. <Thermod, Wassertrans> (AE) gas engine, (AE) gasoline engine, (BE) petrol engine (Verbrennungsmotor)
Ottoverfahren n <Kfztech> Otto cycle
Outsert-Technik f <Kunststoff> (AE) outsert molding, (BE) outsert moulding
Outslot-Zeichengabe f <Telekom> outslot signaling
Outsourcing n **von Netzwerken** <Comp & DV> network outsourcing
Oval n <Geom> oval
oval adj <Geom, Metall> oval
Ovaldrehen n <Fertig> oval turning
Ovaldrehmaschine f <Maschinen> oval-turning lathe
ovale Scheibe f <Maschinen> oval pulley
ovaler Boden m <Ker & Glas> oval punt
Ovalfeile f <Maschinen> oval file
Ovalflansch m <Maschinen> oval flange
ovalförmig adj <Geom> oval-shaped
Ovalisierung f <Erdöl> ovalization (Bohrproblem)
Ovalradzähler m <Gerät> oval gear meter
Ovalschleifmaschine f <Fertig> oval grinder
Ovaltürknopf m <Bau> oval knob
Ovalzirkel m <Maschinen> oval compass
Overdrive m <Kfztech> overdrive (Getriebe, Motor)
Overhead n <Telekom> overhead
Overlay n 1. <Fernseh> overlay, overlaying; 2. <Telekom> overlay
Overlay-Netz n <Telekom> overlay network
Overshot m <Erdöl> overshot (Bohrtechnik)
Ovoglobulin n <Chemie> ovoglobulin
OVSt (Ortsvermittlungsstelle) <Telekom> local exchange
Owen-Brücke f <Elektriz> Owen bridge
OWN-Codierung f <Comp & DV> own coding
Oxalamid n <Chemie> oxamide
Oxalat n <Chemie> oxalate
Oxalessig... <Chemie> oxalacetic
Oxalsäure f <Foto, Lebensmittel> oxalic acid
Oxalsäurediamid n <Chemie> oxamide
Oxalsäuredianilid n <Chemie> oxanilide
Oxalur... <Chemie> oxaluric
Oxalursäure f <Chemie> oxaluric acid
Oxalyl... <Chemie> oxalyl
Oxalylharnstoff m <Chemie> oxalylurea, parabanic acid
Oxamid n <Chemie> oxamide
Oxamid... <Chemie> oxamic
Oxanil... <Chemie> oxanilic
Oxanilid n <Chemie> oxanilide
Oxazin n <Chemie> oxazine
Oxazol n <Chemie> oxazole
Oxeton n <Chemie> oxetone
Oxid n <Aufnahme, Chemie> oxide • **in Oxid verwandeln** <Chemie> oxidize
oxidabel adj <Chemie> oxidizable
Oxidabfall m <Kohlen> oxidic waste
Oxidans n <Chemie> oxidant
Oxidant m <Raumfahrt> oxidant (Raumschiff)
Oxidation f 1. <Anstrich> corrosion, rust; 2. <Chemie, Druck, Umweltschmutz> oxidation
Oxidation f **zum Peroxid** <Chemie> peroxidation
Oxidation-Reduktion-Reaktion f <Chemie> oxidation-reduction reaction
Oxidation-Reduktionszelle f <Labor> oxidation-reduction cell
oxidationsfähig adj <Chemie> oxidable
Oxidationsfähigkeit f <Chemie> oxidizability
Oxidationsflamme f <Chemie> oxidizing flame
Oxidationsgraben m <Abfall, Wasserversorg> oxidation ditch
oxidationshemmendes Mittel n <Fertig> anti-ager oxidant
Oxidationsmittel n 1. <Anstrich> oxidizer; 2. <Chemie> oxidizer, oxidizing agent; 3. <Lufttrans> oxydizer; 4. <Thermod> overtemperature, oxidant; 5. <Umweltschmutz> oxidizing agent
Oxidationsschutzmittel n <Kunststoff> antioxidant
Oxidationsteich m 1. <Abfall> aerated lagoon, oxidation pond, sewage oxidation pond; 2. <Bau> sewage oxidation pond; 3. <Wasserversorg> maturation pond, oxidation pond
oxidativ adj <Chemie> oxidizing
Oxidator m <Chemie> oxidant
Oxidbildung f <Aufnahme> oxide buildup
Oxidchlorid n <Chemie> chloride oxide

oxidierbar *adj* <Chemie> oxidable
oxidieren *v* 1. <Anstrich> rust; 2. <Chemie> oxygenate; 3. <Chemie> oxidize
Oxidieren *n* <Chemie> oxidation
oxidierend *adj* <Chemie> oxidizing
oxidierende Flamme *f* <Chemie> oxidizing flame
oxidierender Stoff *m* <Sicherheit> oxidizing substance
oxidiert *adj* <Chemie> oxidized
oxidierte Zellulose *f* <Chemie> oxidized cellulose
oxidiertes Metall *n* <Metall> oxidized metal
Oxidierung *f* <Elektriz> oxidation
Oxidierungsmittel *n* <Thermod> overtemperature, oxidant
oxidisch *adj* <Chemie> oxidic
Oxidkatode *f* <Elektrotech> oxide-coated cathode
Oxidkeramik *f* <Fertig> oxide ceramics
oxidkeramische Drehwerkzeuge *npl* <Maschinen> oxide ceramic lathe tools
oxidkeramischer Schneidstoff *m* <Fertig> oxide ceramic cutting material
oxidkeramisches Schneidwerkzeug *n* <Fertig> oxide ceramic cutting tool
Oxidoreduktion *f* <Chemie> oxidoreduction
Oxidsalz *n* <Chemie> oxide salt
Oxidschicht *f* <Comp & DV> oxide layer • **mit Oxidschicht** <Fertig> scummy
Oxidwandisolation *f* <Elektronik> local oxidation
Oxim *n* <Chemie> oxime
Oxomanganat *n* <Chemie> malonitrile
Oxonium *n* <Chemie> oxonium
Oxopropan... <Chemie> pyruvic
Oxopyrazolin *n* <Chemie> pyrazolone
Oxosabinan *n* <Chemie> oxosabinane, thujone
Oxosalz *n* <Chemie> oxysalt
Oxosäure *f* <Chemie> oxo acid
Oxosilan *n* <Chemie> siloxane
Oxozon *n* <Chemie> oxozone
Oxyarc-Schneiden *n* <Fertig> oxyarc cutting, oxygen arc cutting
Oxycellulose *f* <Chemie> oxidized cellulose
Oxygenase *f* <Chemie> oxygenase
oxygenieren *v* <Chemie> oxygenate
Oxysäure *f* <Chemie> oxyacid
oxytozisch *adj* <Chemie> oxytocic
Ozalid® *n* <Druck> Ozalid®
Ozalidpapier® *n* <Druck> Ozalid® paper
Ozalidverfahren® *n* <Druck> Ozalid® process
ozeanisches Becken *n* <Wassertrans> oceanic basin
Ozeankabel *n* <Telekom> sea cable, submarine cable
Ozeanographie *f* <Wassertrans> oceanography
Ozon *n* <Chemie> ozone • **in Ozon verwandeln** <Chemie> ozonize • **mit Ozon anreichern** <Chemie> ozonize
Ozonabsorption *f* <Strahlphys> ozone absorption
Ozonanlage *f* <Umweltschmutz> ozone equipment
Ozonbeständigkeit *f* <Kunststoff> ozone resistance
ozonfester Kautschuk *m* <Kunststoff> ozone-resistant rubber
Ozonfestigkeit *f* <Kunststoff> ozone resistance
Ozongerät *n* <Umweltschmutz> ozone equipment
ozonhaltig *adj* <Chemie> ozonic, ozonous, ozoniferous
Ozonid *n* <Chemie> ozonide
ozonisieren *v* <Chemie> ozonize
Ozonkonzentration *f* <Umweltschmutz> ozone concentration
Ozonloch *n* <Raumfahrt, Umweltschmutz> ozone hole
Ozonolyse *f* <Chemie> ozonolysis
Ozonolyseprodukt *n* <Chemie> ozonide
Ozonoskop *n* <Chemie> ozonoscope
Ozonosphäre *f* <Raumfahrt> ozonosphere

Ozonschaden *m* <Umweltschmutz> ozone damage
ozonschädlich *adj* <Umweltschmutz> ozone-damaging
Ozonschicht *f* <Raumfahrt, Umweltschmutz> ozone layer
Ozonspaltung *f* <Chemie> ozonolysis
Ozonverunreinigung *f* <Umweltschmutz> ozone pollution

P

p 1. <Akustik, Phys> *(Schalldruck)* p *(sound pressure)*; 2. <Aufnahme, Raumfahrt> *(Schalldruck)* p *(acoustic pressure)*; 3. <Funktech, Kerntech, Phys, Teilphys> *(Proton)* p *(proton)*; 4. <Kerntech> *(Resonanzfluchtwahrscheinlichkeit)* p *(resonance escape probability)*; 5. <Metrol> *(Piko...)* p *(pico...)*
P 1. <Chemie> *(Phosphor)* P *(phosphorus)*; 2. <Elektriz> *(Leistung)* P *(power)*; 3. <Kerntech> *(Protonenzahl)* P *(proton number)*
Pa *(Pascal)* <Metrol, Phys> Pa, pascal *(Hydrostatik)*
PA 1. <Aufnahme> *(Kraftverstärker, Leistungsverstärker)* PA *(power amplifier)*; 2. <Chemie, Kunststoff, Textil> *(Polyamid)* PA *(polyamide)*; 3. <Elektronik, Raumfahrt, Telekom> *(Leistungsverstärker)* PA *(power amplifier)*; 4. <Elektrotech> *(Endstufe, Endverstärker, Leistungsverstärker)* PA *(power amplifier)*; 5. <Funktech> *(Endstufe, Leistungsverstärker)* PA *(power amplifier)*; 6. <Phys> *(Hauptverstärker, Leistungsverstärker)* PA *(power amplifier)*
PAA *(Polyacryl, Polyacrylat)* <Chemie, Kunststoff> PAA *(polyacrylate)*
Paar *n* <Elektrotech, Maschinen> couple
Paarbildung *f* <Elektriz> pairing
Paarelektronen *npl* <Strahlphys> paired electrons
paaren *v* <Elektrotech> pair *(Kabel)*
Paarerzeugung *f* <Phys> pair production
paariges Kabel *n* <Elektrotech> paired cable
Paarigkeit *f* <Fernseh> pairing *(Video)*
Paarproduktion *f* <Teilphys> pair production
Paarvernichtung *f* <Kerntech> pair annihilation
Paarvernichtungs-Peak *n* <Strahlphys> pair annihilation peak
Paarvernichtungsphoton *n* <Strahlphys> annihilation photon
Paarvernichtungsstrahlung *f* <Strahlphys> annihilation radiation
paarverseiltes Kabel *n* 1. <Elektriz, Elektrotech> paired cable, twin cable; 2. <Telekom> paired cable, twisted pair cable
Paarverseilung *f* <Telekom> pairing *(Kabel)*
paarweise verschieden *adv* <Math> pairwise distinct
Package-Kessel *m* <Heiz & Kälte> packaged boiler
Packeis *n* <Wassertrans> pack ice
packen *v* <Comp & DV> pack
Packfong *n* <Metall> German silver
Packhaus *n* <Wassertrans> warehouse
Packmaschine *f* **für Trays** <Verpack> tray packing machine
Packpapier *n* 1. <Abfall> packaging paper, wrapping paper; 2. <Papier> wrapping; 3. <Verpack> wrapping paper
Packpresse *f* <Papier> bundling press
Packstoff *m* <Abfall> packaging material

Packung

Packung *f* **mit Wiederversiegelung** <Verpack> resealable pack
Packungsanteil *m* 1. <Optik> packing fraction *(Faserbündel)*; 2. <Phys> packing fraction
Packungsattrappe *f* <Verpack> dummy
Packungsdichte *f* 1. <Elektronik> component density *(von Bauelementen)*; 2. <Kunststoff> packing density; 3. <Optik> packing fraction *(Faserbündel)*; 4. <Phys, Telekom> packing fraction
Packungstransportvorrichtung *f* <Verpack> pack-handling equipment
Packungstreiber *m* <Ker & Glas> packing stick
Paddingmaschine *f* <Textil> pad *(Färben)*
Pagina *f* <Druck> page number
Paging-Funktion *f* <Telekom> paging facility
paginieren *v* <Comp & DV, Druck> paginate
Paginierung *f* <Comp & DV, Druck> pagination • **ohne Paginierung** <Druck> blind folio • **Paginierung vornehmen** <Comp & DV> page, paginate
Paket *n* 1. <Comp & DV> packet; 2. <Elektrotech> package *(Bit bei Datenvermittlung)*; 3. <Metall, Telekom> packet; 4. <Trans> parcel
Paketanschlussstelle *f* <Telekom> packet port
Paketaufrechnung *f* <Comp & DV> packet sequencing
Paketbetriebsart *f* <Comp & DV> packet mode
Paketbildung *f* <Telekom> packet assembly, packetization
Paketbildungs-Auflösung *f* <Telekom> packet assembly/disassembly, PAD, PAD equipment
Paketdatennetz *n* <Telekom> packet data network
Paketdatenübertragungssystem *n* **für GSM-Netze** <Telekom> general packet radio system *(GPRS)*
Paketfunkverkehr *m* 1. <Funktech> packet radio *(Datenfunk)*; 2. <Telekom> packet radio
Paketgut *n* <Verpack> *(AE)* parceled goods, *(BE)* parcelled goods
Paketieranlage *f* <Verpack> bundle-tying machine
paketieren *v* <Fertig> packet, pile
Paketieren *n* <Fertig> piling
Paketierer-Depaketierer *m* <Telekom> PAD, packet assembler-disassembler, packetizer/depacketizer
Paketiermaschine *f* <Verpack> bundling machine, parcelling machine
Paketierungs-Depaketierungseinrichtung *f* <Comp & DV> PAD, packet assembler-disassembler
Paketkarte *f* <Trans> parcel registration card
Paketlaufzeit *f* <Telekom> packet delay
Paketmodus *m* <Telekom> packet mode
paketorientierte Endeinrichtung *f* <Telekom> packet-mode terminal
Paketreihung *f* <Telekom> packet sequencing
Paketrundsendung *f* <Telekom> packet broadcasting
Paketrutsche *f* <Trans> parcels chute
Paketschalter *m* 1. <Elektrotech> gang switch; 2. <Trans> parcels counter
Paketspeicherung *f* <Telekom> packet buffering, packet storing
pakettiert *adj* <Verpack> enclosed in a packet
Paketübertragung *f* <Comp & DV, Telekom> packet transmission
Paketverlust *m* <Telekom> packet loss
Paketverlust *m* **durch Vorzeitigkeit** *(EPD)* <Telekom> early packet discard, EPD *(Paketübertragung)*
paketvermittelter Übermittlungsdienst *m* <Telekom> packet-switched bearer service
paketvermitteltes Netz *n* <Telekom> PSN, packet-switched network
Paketvermittlung *f* <Comp & DV, Elektrotech, Telekom> packet switching

Paketvermittlungsamt *n* <Telekom> packet-switching exchange
Paketvermittlungseinrichtung *f* <Elektriz, Telekom> packet switch unit *(PS)*
Paketvermittlungsknoten *m* <Telekom> packet-switching node
Paketvermittlungsnetz *n* <Telekom> packet-switching network, store-and-forward switching network
Paketvermittlungsprozessor *m* <Telekom> packet-switching processor
Paketverzögerungszeit *f* <Comp & DV> packet delay
paketweiser Datenservice *m* <Comp & DV> PSN, packet-switched network
Paketweiterleitung *f* <Telekom> packet forwarding
PAL *(programmierbare logische Anordnung)* <Comp & DV> PAL *(programmable array logic)*
Paläodruck *m* <Erdöl> *(BE)* palaeopressure, *(AE)* paleopressure *(Geologie)*
paläogener Druck *m* <Erdöl> *(BE)* palaeopressure, *(AE)* paleopressure *(Geologie)*
Paläomagnetismus *m* <Phys> *(BE)* palaeomagnetism, *(AE)* paleomagnetism
Paläozoikum *n* <Erdöl> *(BE)* Palaeozoic, *(AE)* Paleozoic *(Erdgeschichte)*
Palette *f* 1. <Fertig> pallet; 2. <Ker & Glas> pallet *(für Glasbehälter)*; 3. <Trans> pallet • **auf Paletten packen** <Verpack> palletize
Palette *f* **mit abnehmbarer Seite** <Trans> pallet with loose partition
Palettenabstreifer *m* <Anstrich> palette knife
Palettenaufsetzrahmen *m* <Trans> pallet collar
Palettenbelader *m* <Verpack> pallet loader
Palettencontainer *m* <Trans> pallet container
Palettenentlader *m* <Verpack> depalletizer
Palettenentnahme *f* <Verpack> depalletization
Palettenentnehmer *m* <Verpack> depalletizer
Palettenhaube *f* <Verpack> pallet hood
Palettenschiff *n* <Wassertrans> pallet ship, palletized cargo carrier
Palettenschrumpfverpackung *f* <Verpack> pallet shrink-wrapping
Palettenschrumpfverpackungsmaschine *f* <Verpack> pallet stretch-wrapping machine
Palettenverpackungsbandmaterial *n* <Verpack> pallet strapping material
Palettenverpackungsmaschine *f* <Verpack> pallet wrapper
palettierbar *adj* <Trans> palletizable
palettieren *v* <Trans, Verpack> palletize
Palettieren *n* <Verpack> palletizing
Palettierklebstoff *m* <Verpack> palletizing adhesive
Palettiermaschine *f* <Verpack> palletizing machine
palettierter Karton *m* <Verpack> palletized board
Palettierung *f* <Trans> palletization
Palettisierung *f* <Verpack> palletization
PAL-Farbfernsehsystem *n* <Fernseh> *(AE)* PAL color-TV system, *(BE)* PAL colour-TV system, PAL system, phase-alternating line, PAL
PAL-Farbsystem *n* <Fernseh> *(AE)* PAL color system, *(BE)* PAL colour system
Palisadenstein *m* <Ker & Glas> soldier block
Palisadenzaun *m* <Bau> palisade
Palladium *n (Pd)* <Chemie, Metall> palladium *(Pd)*
Palmitat *n* <Chemie> hexadecanoate, palmitate
Palmitin *n* <Chemie> palmitin
Palmiton *n* <Chemie> hentriacontanone, palmitone
Palmkernöl *n* <Lebensmittel> palm kernel oil
Palstek *m* <Wassertrans> bowline *(Knoten)*

PAL-System n <Comp & DV, Fernseh> PAL, phase alternation line
Palygorskit n <Erdöl> attapulgite
PAM (Pulsamplitudenmodulation) <Comp & DV, Elektronik, Funktech, Telekom> PAM (pulse amplitude modulation)
PAN 1. <Umweltschmutz> (Peroxyacetylnitrat) PAN (peroxoacetylnitrate); 2. <Umweltschmutz> (Polyacrylnitril) PAN (polyacrylonitrile)
Panamax-Schiff n <Wassertrans> Panamax vessel (Schiff mit Abmessungen für den Panamakanal-Verkehr)
panchromatisch adj <Chemie, Druck, Foto> panchromatic
panchromatische Emulsion f <Foto> panchromatic emulsion
Paneel n <Bau> panel
Panfilter n <Kohlen> pan filter
Panflavin n <Chemie> panflavine
Pankreatin n <Chemie> pancreatin
Panne f 1. <Comp & DV, Fernseh> glitch; 2. <Funktech> breakdown; 3. <Mechan> flat; 4. <Telekom> breakdown; 5. <Textil> panne; 6. <Trans> breakdown (Maschinenschaden) • **eine Panne haben** <Kfztech> break down
Panning n <Comp & DV> panning
Panorama n <Fernseh, Foto, Phys> panorama
Panoramaanzeige f (PPI-Anzeige) 1. <Funkort> plan-position indication, PPI (Radar); 2. <Phys> plan-position indication, PPI
Panoramaaufnahme f <Foto> panoramic photograph
Panoramaempfänger m <Funktech> panoramic receiver
Panoramakamera f <Foto> panoramic camera
Panoramakopf m <Foto> pan-and-tilt head
Panoramaperiskop n <Kerntech> panorama periscope
panschen v <Lebensmittel> adulterate
Pantograph m <Druck, Eisenbahn> pantograph
Pantograph-Manipulator m <Kerntech> pantograph-type manipulator
Pantokarenen fpl <Wassertrans> cross-curves (Schiffkonstruktion)
pantonale Tonleiter f <Akustik> pantonal scale
Pantothenat... <Chemie> pantothenate
Panzer m <Elektrotech> (AE) armor, (BE) armour
Panzeraderleitung f <Elektrotech> metal-sheathed conductor
Panzerbatterie f <Trans> (AE) armored battery, (BE) armoured battery
Panzerblech n <Bau> (AE) armor plate, (BE) armour plate
Panzerkabel n 1. <Elektrotech> (AE) armored cable, (BE) armoured cable, metal-sheathed cable; 2. <Maschinen, Telekom> (AE) armored cable, (BE) armoured cable
Panzern <Fertig> hard-facing by welding
Panzerplatte f <Mechan> (AE) armor plate, (BE) armour plate
Panzerplattenwalzwerk n <Fertig> (AE) armor-plate mill, (BE) armour-plate mill
Panzerung f <Elektrotech> (AE) armor, (BE) armour; metal sheath (Kabeln, Adern)
Panzerventil n <Maschinen> (AE) armored valve, (BE) armoured valve
Päonin n <Chemie> peonine
Papain n <Chemie, Lebensmittel> papain
Papaverin f <Chemie> papaverine
Papayotin n <Chemie, Lebensmittel> papain
Paperback n <Druck> paperback
Papier n <Comp & DV, Druck, Papier> paper • **einfach mit Papier isoliert** <Elektriz> single-paper-covered
Papier n mit Wasserlinien 1. <Druck> laid paper; 2. <Verpack> laid paper (mit Egoutteurrippung)
Papierabrisskante f <Comp & DV> paper tear guide
Papierauskleidung f <Verpack> case-lining paper (Karton)

Papierbahn f <Papier> web
Papierband n <Elektriz> paper tape
Papierbelichtungsmesser m <Foto> enlarging meter
Papierbeutel m <Verpack> paper bag
Papierbeutel- und Sackverschluss m <Verpack> paper bag and sack closure
Papierbogen m <Papier> sheet
Papierchromatographie f <Chemie> paper chromatography
Papierchromatographiebehälter m <Labor> paper chromatography tank
Papierchromatographiegerät n <Labor> paper chromatography apparatus
Papierchromatographietank m <Labor> paper chromatography tank
Papierdichtung f <Maschinen> paper gasket
Papiereinlage f <Verpack> case-lining paper
Papiereinzug m <Comp & DV> paper feed, paper picker
Papierende-Sensor m <Comp & DV> form stop
Papiererzeugung f <Papier> papermaking
Papierfabrik f 1. <Papier> paper mill; 2. <Verpack> board mill
Papierfach n <Comp & DV> input tray, paper tray
Papierfaser f <Verpack> (AE) paper fiber, (BE) paper fibre
Papierfilter n <Maschinen> paper filter
Papierformat n <Druck, Papier> paper size
Papierfühler m <Comp & DV> paper sensor
Papiergewicht n <Papier> weight of paper
Papiergröße f <Druck> paper size
Papierhäckselmaschine f <Verpack> shredding machine
Papierhärtegrad m <Foto> paper grade
Papierherstellung f <Druck> papermaking
Papierholz n <Abfall> pulpwood
papierisoliertes Kabel n <Elektrotech, Telekom> paper-insulated cable
Papierisolierung f <Elektriz> paper insulation
Papierkabel n <Elektrotech, Telekom> paper-insulated cable
Papierkohle f <Kohlen> paper coal, papyraceous lignite
Papierkondensator m <Elektriz, Elektrotech, Funktech> paper capacitor
Papierkopie f <Comp & DV> hard copy
Papierlaminat n <Verpack> paper laminate
Papierlocher m <Papier> punch
papierlose Korrekturabzüge mpl <Druck> soft proofs
Papiermangel m <Comp & DV> paper low
Papiermaschine f <Druck, Papier> paper machine
Papiermaschinenantrieb m <Papier> paper machine drive
Papiermaschinenausschuss m <Papier> machine broke
Papiermasse f <Druck> paper pulp
Papiernegativ n <Foto> paper negative
Papierpoliermaschine f <Ker & Glas> paper polisher
Papierrolle f <Papier> jumbo roll, reel of paper
Papiersack m <Verpack> paper sack
Papierschneidemaschine f <Druck, Verpack> guillotine, paper cutter
Papierschneider m <Druck, Verpack> guillotine, paper cutter
Papierschnitzel mpl <Verpack> paper chips
Papierspan m <Papier> shaving
Papierstapel m <Papier> stack of paper
Papierstärke f <Verpack> paper grade
papierstarkes Papier n <Foto> single weight paper
Papierstoff m 1. <Druck> paper pulp; 2. <Papier> stock
Papierstreifentransport m <Gerät> chart transport
Papiertransport m <Comp & DV> paper feed

Papiertüte

Papiertüte f <Verpack> paper bag
Papierverarbeitungsindustrie f <Verpack> paper-converting industry
Papierverpackung f <Verpack> paper wrapping
Papiervorschub m <Comp & DV> form feed, paper skip
Papierwalze f <Papier> paper roll *(Superkalander)*
Papierwolle f <Verpack> paper chips
Papierzerreißmaschine f <Verpack> shredding machine
Papierzug m <Papier> paper draw
Pappdeckel m <Papier> mill board
Pappe f <Druck, Papier, Verpack> board, cardboard, paper board, paste board, pulp board
Pappenfilz m <Papier> board felt
Pappfass n <Verpack> *(AE)* fiber drum, *(BE)* fibre drum
Papptablett n <Verpack> cardboard tray
Paraban... <Chemie> parabanic
Parabansäure f <Chemie> oxalylurea, parabanic acid
Parabel f <Geom> parabola
Parabel f **dritten Grades** <Geom> cubic parabola
Parabelgeschwindigkeit f <Raumfahrt> parabolic velocity
Parabelkriechen n <Metall> parabolic creep
Parabol... 1. <Fernseh> parabolic shading; 2. <Funktech> parabolic
Parabolantenne f 1. <Funktech, Raumfahrt> parabolic antenna; 2. <Telekom> dish aerial
Parabolantenne f **mit Gitterreflektor** <Funktech, Raumfahrt> parabolic mesh antenna
parabolisch adj <Geom> parabolic
parabolischer Orbit m <Raumfahrt> parabolic orbit
Parabolprofil n <Optik, Telekom> parabolic profile
Parabolreflektor m 1. <Fernseh> dishpan; 2. <Foto> parabolic reflector
Parabolspiegel m 1. <Funktech> parabolic reflector; 2. <Phys> parabolic mirror
Parabolspiegelantenne f <Telekom> parabolic reflector antenna
Paracetaldehyd m <Chemie> paraldehyde
Parachor m <Chemie> parachor
Paracyan n <Chemie> paracyanogen
Paradiazin n <Chemie> pyrazine
Paradigma n <Comp & DV> paradigm
Paraffin n <Erdöl> wax; *(BE)* paraffin *(Petrochemie)*
Paraffinieren n <Verpack> paraffin coating
paraffiniertes Papier n <Verpack> paraffin-impregnated paper
Paraffinkohlenwasserstoff m <Chemie, Erdöl> alkane
Paraffinpapier n <Verpack> paraffin-waxed paper
Paraffinreihe f <Erdöl> paraffin series *(Petrochemie)*
Paraffinwachs n <Elektriz> paraffin wax
Paraldehyd m <Chemie> paraldehyde
Parallaxe f <Foto, Maschinen, Phys> parallax
Parallaxenausgleich m <Gerät> parallactic compensation; compensation of parallax *(Ablesefehler)*
parallaxenfrei adj <Gerät> parallax-free
parallaxenfreier Spiegel m <Gerät> antiparallax mirror
parallel adj 1. <Comp & DV, Elektriz, Elektrotech> parallel, shunt; 2. <Funktech, Geom> parallel
parallel geschaltet adj <Elektrotech, Gerät, Phys> connected in parallel, parallel-connected
parallel geschaltete Funkenstrecken fpl <Elektriz> parallel spark gaps
parallel geschaltete Widerstände mpl <Elektriz> parallel-connected resistances
parallel gespeist adj <Lufttrans> parallel-fed
parallel gespultes Garn n <Textil> parallel-wound yarn
parallel gewickelter Dynamo m <Elektrotech> shunt-wound dynamo
parallel gewickelter Motor m <Elektrotech> shunt-wound motor

parallel laufen v <Geom> run parallel
parallel richten v <Elektronik> collimate *(Strahlen)*
parallel schalten v <Elektrotech> connect in parallel
Parallel... <Comp & DV, Math> parallel
Parallelabfragespeicher m <Comp & DV> parallel-search storage
Paralleladdierer m <Comp & DV> parallel adder
Paralleladdierwerk n <Elektronik> parallel adder
Parallelalgorithmus m <Comp & DV> parallel algorithm
Parallelaufstellung f <Lufttrans> parallel positioning
Parallelband n <Strahlphys> parallel band
Parallelbetrieb m 1. <Comp & DV> concurrent operation; 2. <Druck> parallel operation
Parallelbewegung f <Maschinen> parallel motion
Paralleldrahtleitung f <Elektrotech> parallel-wire line
Parallele f <Geom> parallel line
parallele Analog-Digital-Umsetzung f <Elektronik> analog-to-digital conversion
parallele Ein-/Ausgabe f <Comp & DV, Telekom> parallel input/output
parallele Schnittstelle f <Telekom> parallel interface
parallele Sysplex-Konfiguration f <Comp & DV> parallel sysplex configuration
parallele Verschiebung f <Geom> translation
Parallelendmaß n 1. <Fertig> block, end block; 2. <Maschinen, Mechan, Metrol> *(AE)* gage block, *(BE)* gauge block, *(AE)* slip gage, *(BE)* slip gauge
Parallelepiped n <Geom> parallelepiped
paralleler Ein-/Ausgabebaustein m <Comp & DV, Elektronik> parallel input/output chip, parallel input/output unit, parallel I/O chip *(PIO)*
paralleles digitales Signal n <Elektronik> parallel digital signal
Parallelflächner m <Geom> parallelepiped
Parallelfräser m <Maschinen> parallel milling cutter
Parallelführung f <Maschinen> parallel guide
Parallelgegenkopplung f <Elektrotech> shunt feedback
Parallelgerade f <Geom> parallel line
Parallelisieren n **der Fasern** <Textil> *(AE)* parallelization of fibers, *(BE)* parallelization of fibres
Parallelität f <Geom> parallelism
Parallelkapazität f <Elektrotech> shunt capacitance
Parallelkondensator m 1. <Gerät> bypass capacitor; 2. <Phys> bypass capacitor, shunt capacitor
Parallelkreis m <Elektriz, Telekom> parallel circuit
Parallellineal n <Maschinen> parallel ruler
Parallelmanipulator m <Kerntech> master-slave manipulator
Parallelmaus f <Comp & DV> parallel mouse
Parallelmausadapter m <Comp & DV> parallel mouse adaptor
Parallelmausanschluss m <Comp & DV> parallel mouse adaptor
Parallelmultiplizierer m <Elektronik> parallel multiplier
Parallelogramm n <Geom, Mechan> parallelogram
Parallelogramm n **der Geschwindigkeiten** <Phys> parallelogram of velocities
Parallelogramm n **der Kräfte** <Maschinen, Phys> parallelogram of forces
Parallelplattenkondensator m <Elektrotech, Phys> parallel-plate capacitor
Parallelprojektion f <Geom, Konstzeich> parallel projection
Parallelprozessorkarte f <Comp & DV> parallel card
Parallelquerlenkerradaufhängung f <Kfztech> parallel-arm-type suspension
Parallelrechnen n <Comp & DV> parallel arithmetic
Parallelrechner m *(SIMD-Rechner)* <Comp & DV> single instruction multiple-data machine *(SIMD machine)*

Parallelreißer m 1. <Fertig> (AE) marking gage, (BE) marking gauge, (AE) surface gage, (BE) surface gauge, scribing block; 2. <Maschinen> (AE) marking gage, (BE) marking gauge, (AE) surface gage, (BE) surface gauge; 3. <Mechan> (AE) surface gage, (BE) surface gauge; 4. <Metrol> (AE) surface gage, (BE) surface gauge, surface geometry meter, (AE) vernier height gage, (BE) vernier height gauge
Parallelresonanz f 1. <Akustik> antiresonance; 2. <Elektronik, Phys> parallel resonance
Parallelresonanzkreis m 1. <Elektronik> parallel resonant circuit; 2. <Phys> antiresonant circuit
Parallelschallschluckwand f <Aufnahme> parallel absorbent baffle
Parallelschaltung f 1. <Elektrotech> parallel connection, shunt; 2. <Phys> parallel connection
Parallelschere f <Fertig> guillotine plate shear, guillotine shear, guillotine shearing machine
Parallelschieber m <Maschinen> parallel slide valve
Parallelschnittstelle f 1. <Comp & DV> parallel interface, parallel port; 2. <Druck, Telekom> parallel interface
Parallelschnittstellenanschluss m <Comp & DV> parallel port
Parallelschraubstock m <Maschinen> parallel vice
Parallelschwingfrequenz f <Akustik, Elektronik> antiresonant frequency (Schaltkreistechnik)
Parallelschwingkreis m 1. <Elektronik> antiresonant circuit, parallel resonant circuit, rejector; 2. <Funktech> antiresonant circuit, rejector
Parallel-Serien-Wandler m 1. <Elektrotech> parallel-to-serial converter; 2. <Telekom> dynamicizer, serializer
Parallel-Serien-Wandlung f <Elektrotech> parallel-to-serial conversion
Parallelspeicher m <Elektrotech> parallel storage
Parallelspeisung f <Elektriz, Elektrotech> parallel feeder, shunt feed
Parallelstrahl m <Phys> parallel beam
Parallelstrahlenbündel n <Elektronik> parallel beam
Parallelstreifenabschwächer m <Elektronik> parallel-vane attenuator (Mikrowellen)
Parallelstruktur f <Regelung> parallel structure
Parallelsynchronsystem n <Telekom> parallel synchronous system
Parallelton m <Akustik> relative tone
Paralleltonart f <Akustik> relative key
Parallelübergabe f <Comp & DV> parallel transmission
Parallelübertrag m <Comp & DV> carry lookahead, lookahead
Parallelübertragung f 1. <Comp & DV> parallel transfer, parallel transmission; 2. <Telekom> parallel transmission
Parallelumsetzer m <Elektronik> flash converter, parallel converter
Parallelumsetzung f <Elektronik> flash conversion, parallel conversion
Parallelverarbeitung f <Comp & DV, Elektronik, Telekom> parallel processing
Parallelverbindung f <Elektriz> parallel connection
Parallelvierkantkeile mpl <Maschinen> square parallel keys
Parallelwiderstand m <Elektrotech, Phys> parallel resistance, shunt
Parallelzahnräder npl <Maschinen> parallel gears
Parallelzange f <Kerntech> parallel-jaw tong
Parallelzugriff m <Comp & DV> parallel access
paramagnetisch adj <Elektriz, Phys> paramagnetic
paramagnetische Elektronenspinresonanz f <Phys> EPR, electron paramagnetic resonance
paramagnetische Resonanz f <Elektriz, Elektronik, Phys> paramagnetic resonance

paramagnetische Schienenbahn f <Eisenbahn> paramagnetic rail
paramagnetischer Curiepunkt m <Strahlphys> paramagnetic Curie point
paramagnetischer Verstärker m <Elektronik> paramagnetic amplifier
Paramagnetismus m <Phys> paramagnetism
Parameter m <Comp & DV, Elektronik, Funktech, Math, Phys> parameter
Parameter m **der Materialdispersion** <Optik> material dispersion parameter
Parameterdarstellung f <Math> parametric representation (einer Kurve oder Fläche)
Parameterempfindlichkeit f <Elektrotech> parameter sensitivity
Parametergleichungen fpl <Math> parametric equations (einer Geraden, einer Ebene)
Parameterintegral n <Math> parametric integral
Parameterprofil n 1. <Optik> profile parameter; 2. <Telekom> parameter profile
Parameterprofildispersion f <Optik> profile dispersion parameter
Parametersubstitution f <Comp & DV> parameter substitution
Parameterübergabe f <Comp & DV> parameter passing
parametrisch adj <Comp & DV> parametric
parametrische Analyse f <Elektronik> parametric analysis
parametrische Galliumarsenid-Verstärkerdiode f <Elektronik> gallium arsenide parametric amplifier diode
parametrische Prüfung f <Telekom> parametric test
parametrische Schwingung f <Telekom> parametric oscillation
parametrische Verstärkung f <Elektronik> parametric amplification
parametrischer Elektronenstrahlverstärker m <Elektronik> electron beam parametric amplifier
parametrischer Laser m <Elektronik> parametric laser
parametrischer Verstärker m 1. <Elektronik, Funktech, Phys> parametric amplifier, paramp; 2. <Raumfahrt> parametric amplifier, paramp (Weltraumfunk); 3. <Telekom> parametric amplifier, paramp
parametrisches Bootstrapping n <Math> parametric bootstrapping (Schätzverfahren)
parametrisieren v <Comp & DV> initialize
Parametrisierung f 1. <Comp & DV> parameterization; 2. <Math> parameterization (einer Variablen)
Paramorphin n <Chemie> paramorphine, thebaine (Pharmazie)
Parapepton n <Chemie> parapeptone, syntonin
Pararosanilin n <Chemie> pararosaniline
parasitär adj <Elektronik> parasitic
Parasitär... <Elektronik, Elektrotech, Funktech, Kerntech, Phys, Telekom> parasitic
Parasitärantenne f <Phys> parasitic aerial
Parasitärantennenelement n <Phys> parasitic element
Parasitärdiode f <Elektronik> parasitic diode
parasitäre Kopplung f <Elektrotech, Telekom> parasitic coupling
parasitäre Kraft f <Bau> redundant stress
parasitäre Schwingung f <Funktech, Phys, Telekom> parasitic oscillation
Parasitärinduktanz f <Elektrotech> parasitic inductance
Parasitärneutroneneinfang m <Kerntech> parasitic capture
paratypisch adj <Chemie> paratypical
Parawasserstoff m <Chemie> parahydrogen
paraxialer Strahl m <Phys> paraxial ray
Paraxialstrahl m <Telekom> paraxial ray

Paraxylol

Paraxylol n <Erdöl> paraxylene *(Petrochemie)*
Pardune f <Wassertrans> backstay *(Tauwerk)*
Parenthese f <Druck> parenthesis
Parität f <Comp & DV, Phys, Strahlphys, Teilphys> parity
Paritätsanzeiger m <Comp & DV> parity flag
Paritätsbit n <Comp & DV, Telekom> parity bit
Paritätserhaltung f <Phys> conservation of parity
Paritätsfehler m <Comp & DV, Telekom> parity error
Paritätsprüfung f 1. <Comp & DV> odd-even check, parity check; 2. <Telekom> parity control, parity check
Paritätsprüfungsabbruch m <Comp & DV> parity interrupt
Park & Ride System n <Trans> park and ride
Parkbremse f <Lufttrans> parking brake
parken v <Telekom> park *(Anruf)*
parkerisieren v <Fertig> parkerize
Parkerisieren n <Fertig> parkerizing
Parkern n <Metall> parkerizing
Parkettbodenbelag m <Bau> parquet flooring
Parkfläche f <Bau> parking area
Park-Gleichung f <Elektriz> Park's equation
Parkkralle f <Kfztech> clamp *(Sperre zur Blockierung der Räder)*
Parkleuchte f <Kfztech> parking light; position light *(Beleuchtung)*
Parkorbit m <Raumfahrt> parking orbit *(Raumschiff)*
Parkplatz m <Bau> parking area
Parksperre f <Kfztech> parking lock gear
Parksperrklinke f <Kfztech> parking pawl
Parkstellung f <Kfztech> parking gear
Parkuhr f <Trans> parking meter
Parkuhrladegerät n <Trans> charger at parking meter
Parkway m <Kfztech> *(AE)* parkway
Parsec n <Phys> parsec *(astronomische Einheit)*
Parser m <Comp & DV, Künstl Int> parser
Parsing n <Comp & DV, Künstl Int> parsing
Partialbruchzerlegung f <Math> partial fraction decomposition *(Integrationsverfahren)*
Partialdruck m <Thermod> partial pressure
Partialsumme f <Math> partial sum
Partialwelle f <Teilphys> partial wave
Partie f <Qual> batch
Partiekontrolle f durch Stichproben <Qual> batch inspection by samples
partiell adj <Math> partial
partielle Ableitung f <Math> partial derivative
partielle Differenzialgleichung f <Math> partial differential equation
partielle Faktorisierung f <Math> partial factorization
partielle Integration f <Math> integration by parts
partielle Kohärenz f <Optik, Telekom> partial coherence
partielles Integrieren n <Math> integration by parts
Partienstreuung f <Qual> batch variation
Partikel f 1. <Akustik, Kohlen, Teilphys> particle; 2. <Umweltschmutz> particulate material
Partikelemissionen fpl <Kfztech> particulate emissions
partikelfilterndes Atemschutzgerät n <Sicherheit> particulate-filter respirator
Partikelgröße f <Chemtech> particle size
Partition f 1. <Comp & DV> partition; 2. <Math> partition *(Teilung, Zerlegung)*
partitionieren v 1. <Comp & DV> partition; 2. <Math> partition *(teilen, zerlegen)*
partitioniert adj <Comp & DV> partitioned
Partitionierung f <Comp & DV> partitioning
Partitur f <Aufnahme> score
Partner m <Comp & DV, Telekom> peer
Partnerinstanz f <Comp & DV, Telekom> peer entity
Parton n <Teilphys> parton

Parvolin n <Chemie> parvoline
Pascal n (Pa) 1. <Metrol> pascal, Pa *(Einheit)*; 2. <Phys> pascal, Pa *(Hydrostatik)*
Pascal'sches Dreieck n <Math> Pascal's triangle
Paschen-Back'scher Effekt m <Phys> Paschen-Back effect *(Atomphysik)*
Paschen'sche Serie f <Phys> Paschen series *(Atomphysik)*
Paschen'sches Gesetz n <Phys> Paschen's law *(Gasentladung)*
PA-Sprengstoff m <Chemie> ammonia dynamite
Passagier m <Lufttrans, Wassertrans> passenger
Passagierflugzeug n <Lufttrans> airliner, passenger aircraft
Passagierflugzeug n ohne Service <Lufttrans> skybus
Passagiergroßflugzeug n <Lufttrans> airliner
Passagierlinienschiff n <Wassertrans> ocean liner, passenger liner
Passagierliste f <Lufttrans, Trans, Wassertrans> waybill
Passagierschiff n <Wassertrans> passenger ship
Passagierschnellförderer m <Kfztech> high-speed passenger conveyor
Passagiersitz m <Lufttrans> passenger seat
Passatwinde mpl <Wassertrans> trade winds
Passbohrung f <Maschinen> locating hole
passen v <Maschinen> fit
passend adj <Math> appropriate
passend schneiden v <Bau> cut to fit
Passepartout n <Foto> slip mount
Passer m <Druck> register mark, registration mark
Passermarke f <Druck> register mark, registration mark
Passfeder f 1. <Fertig> key; 2. <Kerntech> feather key; 3. <Maschinen> feather
Passfläche f <Fertig> faying surface
Passflächen fpl <Maschinen> mating surfaces
Passgenauigkeit f <Druck> register accuracy
passiv adj 1. <Anstrich> inert; 2. <Comp & DV, Elektronik> passive
passive Alarmierung f <Telekom> passive alerting
passive Betriebsart f <Elektronik> passive mode
passive Elektrodynamikstoßdämpfung f <Raumfahrt> passive electrodynamic snubber *(Raumschiff)*
passive Komponente f <Elektronik, Kerntech> passive component
passive Kraftfahrzeugsicherheit f <Kfztech> passive motor vehicle safety
passive Last f <Elektrotech> passive load
passive Schaltung f <Phys> passive circuit
passive Sonnenenergie f <Nichtfoss Energ> passive solar energy
passive Steuerung f <Raumfahrt> passive control
passive Thermosteuerung f *(PTC)* <Raumfahrt> passive thermal control *(PTC)*
passive Transporteinheit f <Trans> passive transport unit
passive Zielsuchlenkung f <Lufttrans> homing passive guidance
passiver Dipol m <Elektrotech> passive dipole
passiver Erddruck m <Bau, Kohlen> passive earth pressure
passiver Flachwagen m <Kfztech> passive flat car
passiver Infrarotdetektor m <Trans> passive infrared detector
passiver Kreis m <Telekom> passive circuit
passiver Satellit m <Raumfahrt> passive satellite
passiver Stern m <Comp & DV> passive star
passiver Strahler m <Phys> passive aerial
passiver Stromkreis m <Elektriz> passive electric circuit, passive network

passiver Vierpol m <Elektrotech> passive quadripole
passiver Wandler m <Elektrotech> passive transducer
passives Antennenelement n <Funktech> parasitic radiator
passives Bandpassfilter n <Elektronik> passive bandpass filter
passives Bandsperrfilter n <Elektronik> passive bandstop filter
passives Bauelement n <Elektrotech, Telekom> passive component
passives Element n <Elektronik> passive element
passives Filter n <Elektronik, Telekom> passive filter
passives Filtern n <Elektronik> passive filtering
passives Glied n <Regelung> passive element
passives Insassenrückhaltesystem n <Kfztech> passive occupant restraint system
passives Netzwerk n <Elektrotech> passive network
passives Sicherheitsgurtsystem n <Kfztech> passive seat belt system
passives Sonnenenergiesystem n <Nichtfoss Energ> passive solar system
passives System n <Akustik, Nichtfoss Energ> passive system
passives Trägermaterial n <Elektronik> passive substrate
passivieren v <Anstrich, Elektronik, Phys, Raumfahrt> passivate *(Raumschiff)*
passivierter Transistor m <Elektronik> passivated transistor
Passivierung f <Elektronik, Phys> passivation
Passivierungsglas n <Ker & Glas> passivation glass
Passivierungsschicht f <Elektronik> passivation layer
Passivität f <Anstrich> inertia
Passkorrektur f <Druck> register adjustment
Passlehre f <Elektriz> *(AE)* gage, *(BE)* gauge
Passscheibe f <Maschinen> *(BE)* locating disc, *(AE)* locating disk, shim
Passschraube f 1. <Fertig> precision bolt, reamed bolt; 2. <Maschinen> dowel screw
Passstift m 1. <Bau> dowel pin; 2. <Maschinen> alignment pin, dowel pin, locating pin; 3. <Mechan> alignment pin
Passstück n <Comp & DV, Elektriz, Elektrotech, Fertig, Funktech, Labor, Maschinen, Mechan, Phys, Telekom, Textil> adaptor, adjusting unit, adjuster
Passteile npl <Maschinen> mating parts
Passtoleranz f 1. <Maschinen> tolerance of fit; 2. <Mechan> fitting tolerance
Passung f 1. <Maschinen> fit; 2. <Mechan> seat
Passungsflansch m <Mechan> mating flange
Passwort n <Comp & DV, Telekom> password
Passwort-Sicherheit f <Comp & DV> password security
Passzeichen n <Druck> register mark, registration mark
Passzugabe f <Maschinen> fitting allowance
pasteurisieren v <Thermod> pasteurize
Pasteurisieren n <Thermod> pasteurization
pasteurisiert adj <Thermod> pasteurized
Pasteurisierung f <Thermod> pasteurization
Pasteur-Pipette f <Labor> Pasteur pipette
pastöser Abfall m <Abfall> pasty waste
Patent n <Patent> patent • **ein Patent benutzen** <Patent> use a patent
Patentanker m <Wassertrans> stockless anchor
Patentanmeldung f <Patent> patent application
Patentanspruch m <Patent> claim
Patentanwalt m <Patent> patent agent
Patentbeschreibung f <Patent> patent specification
patentfähige Erfindung f <Patent> patentable invention
Patentgebühren fpl <Patent> royalties

Patentierbarkeit f <Patent> patentability
patentiert adj <Maschinen, Patent> patented
Patentinhaber m <Patent> patent proprietor, proprietor of a patent
Patentlog n <Wassertrans> patent log
Patentschäkel m <Wassertrans> snap shackle *(Beschläge)*
Patentschrift f <Patent> patent specification; specification *(gedruckt)*
Patenturkunde f <Patent> patent certificate
Patentverletzer m <Patent> infringer
pathogener Abfall m <Abfall> anatomical waste, infectious waste, pathological waste
Patina f <Fertig> patina, verdigris
patinieren v <Fertig> patinate
Patio-Tür f <Ker & Glas> patio door
Patrize f <Fertig> stamp
Patrone f 1. <Elektrotech> cartridge; 2. <Foto> cartridge *(für 35 mm Film)*; 3. <Maschinen> collet
Patronenanlasser m <Kfztech, Lufttrans, Wassertrans> combustion starter
Patronenfilter n <Kfztech> cartridge filter *(Schmierung)*
Patronensicherung f <Elektriz, Elektrotech> cartridge fuse, enclosed fuse
patronieren v <Textil> draft *(Weben)*
Patrouillenboot n <Meerschmutz, Wassertrans> patrol boat
Pattern Matching n <Künstl Int> pattern matching
Pauli'sches Ausschließungsprinzip n <Phys, Teilphys> Pauli exclusion principle
Pauli'sches Prinzip n <Strahlphys> Pauli principle
Pauschalfrachtgeld n <Erdöl> lump sum freight *(Handel, Schifffahrt)*
Pauschalgebühr f <Bau, Patent> flat-rate fee
Pauschalgebührendienst m <Telekom> flat-rate service
Pauschalpreis m <Bau> lumpsum price
Pauschaltarif m 1. <Elektriz> fixed charge tarriff, flat-rate tariff; 2. <Telekom> all-in tariff, flat-rate tariff
Pause f 1. <Akustik> interval, rest; 2. <Fernseh> pause; 3. <Telekom> gap
pausen v <Maschinen> trace
Pausenknopf m <Fernseh> pause control
pausenloses Fräsen n <Fertig> index-base milling
Pausensteuerung f <Aufnahme> pause control
Pausenzeichen n <Funktech> station identification
Pauspapier n <Maschinen> tracing paper
Pay-TV f <Fernseh> pay-by-use basis, pay TV
Pb *(Blei)* <Chemie> Pb *(lead)*
PB 1. <Comp & DV, Elektronik, Fernseh, Phys, Telekom> *(Pulsbreite)* PW *(pulse width)*; 2. <Kunststoff> *(Polybuten, Polybutylen)* PB *(polybutylene)*
P-Beiwert m <Regelung> proportional coefficient
p⁻-Bereich m <Elektronik> p⁻-region
P-Bild n <Telekom> predicted frame
PBT *(Polybutylenterephthalat)* <Elektriz, Kunststoff> PBT *(polybutylene ephtalate)*
PBX *(private Selbstwählnebenstelle)* <Telekom> PBX *(private branch exchange)*
PBX-Vermittlungsschrank m <Telekom> PBX switchboard
PC 1. <Comp & DV> *(Personal Computer)* PC *(personal computer)*; 2. <Elektriz, Kunststoff> *(Polycarbonat)* PC *(polycarbonate)*
PCB 1. <Comp & DV> *(Leiterplatte, Printplatte)* PCB *(printed circuit board)*; 2. <Elektriz, Fernseh, Funktech, Telekom> *(Leiterplatte)* PCB *(printed circuit board)*; 3. <Elektronik> *(Leiterplatte, gedruckte Schaltung)* PCB *(printed circuit board)*
PC-basierter Server m <Comp & DV> PC-based server

PC-Client

PC-Client m <Comp & DV> PC client
PC-ISDN-Anwenderprogramm-Schnittstelle f <Telekom> ISDN application program interface *(CAPI)*
PC-Karte f <Elektronik> PCB, PC-board
PCM *(Pulscodemodulation)* <Aufnahme, Comp & DV, Elektronik, Funktech, Strahlphys, Telekom> PCM *(pulse code modulation)*
PCM-Codierer m <Telekom> PCM coder
PCM-Filter n <Elektronik> PCM filter
PCM/FM-Modulation f <Elektronik> PCM/FM modulation
PCM-Multiplexer m <Elektronik, Telekom> PCM multiplexer
PCM-Multiplexing n <Elektronik, Telekom> PCM multiplexing
PCM-Primärgruppe f <Telekom> primary PCM group
PCM-System n <Telekom> PCM system
PCM-Vermittlungssystem n <Telekom> PCM switching system
PCS-Faser f *(Plastic-Clad-Silica-Faser)* <Optik, Telekom> *(AE)* PCS fiber, *(BE)* PCS fibre *(plastic-clad silica fiber)*
PCU *(externe Steuereinheit, periphere Steuereinheit)* <Comp & DV> PCU *(peripheral control unit)*
PCV-Ventilation f *(Kurbelgehäusezwangsentlüftung, rückführende Kurbelgehäuseentlüftung)* <Kfztech> PCV *(positive crankcase ventilation)*
Pd *(Palladium)* <Chemie, Metall> Pd *(palladium)*
PDAP *(Polydiallyphthalat)* <Kunststoff> PDAP *(polydiallylphthalate)*
p-Diffusion f <Elektronik> p-type diffusion
PDM 1. <Comp & DV> *(Pulsdeltamodulation)* PDM *(pulse delta modulation)*; 2. <Elektronik> *(Pulsdauermodulation)* PWM *(pulse width modulation)*; 3. <Telekom> *(Pulsdauermodulation)* PDM *(pulse duration modulation)*
PDN *(öffentliches Datennetz)* <Comp & DV, Telekom> PDN *(public data network)*
p-dotierte Basis f <Elektronik> p-doped base, p-type base
PDR *(vorläufige technische Prüfung)* <Raumfahrt> PDR *(preliminary design review)*
PD-Regelung f *(Proportional-Differenzial-Regelung)* <Labor> PD control *(proportional plus derivative control)*
PD-Regler m *(Proportional-Differenzial-Regler)* <Labor> PD controller *(proportional plus derivative controller)*
PD-Verhalten n *(Proportional-Differenzial-Verhalten)* <Regelung> PD action *(proportional plus derivative action)*
Peak m <Chemie> peak
PE-C *(chloriertes Polyethylen)* <Kunststoff> CPE *(chlorinated polyethylene)*
PEC *(Photozelle)* <Druck, Elektriz, Elektronik, Fernseh, Foto, Phys, Strahlphys> PEC *(photoelectric cell)*
Pech n <Ker & Glas, Wassertrans> pitch
Pechblende f <Kerntech> uranium black
pechen v <Wassertrans> pitch
Pechkohle f <Kohlen> pitch coal
Pechpoliermaschine f <Ker & Glas> pitch polisher
Pechstein m <Erdöl> pitchstone *(Mineral)*
Pedal n 1. <Kfztech> pedal *(Bremse, Kupplung)*; 2. <Maschinen> treadle
Pedalsteller m <Kfztech> pedal adjuster *(Bremse, Kupplung)*
Peer-to-Peer-Netz n <Comp & DV> peer-to-peer network
Pegel m 1. <Akustik> level *(Schwellwert)*; 2. <Aufnahme> level; 3. <Bau> *(AE)* gage, *(BE)* gauge, level; 4. <Eisenbahn> level indicator *(Flüssigkeit)*; 5. <Elektronik> level *(Flüssigkeit)*; 6. <Maschinen> level; 7. <Mechan> *(AE)* gage, *(BE)* gauge; 8. <Metrol, Telekom> level; 9. <Wasserversorg> *(AE)* water level gage, *(BE)* water level gauge

Pegelabfall m <Maschinen> level drop
Pegelabgleichung f <Elektronik> *(AE)* leveling, *(BE)* levelling
Pegelanzeige f <Maschinen> level indicator
Pegelausgleich m <Elektronik> level adjustment
Pegeldifferenz f <Akustik> level difference
Pegeldurchgang m <Telekom> *(BE)* level crossing, *(AE)* grade crossing
Pegeldurchgangshäufigkeit f <Telekom> level crossing rate
Pegelhalt(er)ung f <Aufnahme, Fernseh, Telekom> clamping *(durch Klemmschaltung)*
Pegellatte f <Bau> *(AE)* staff gage, *(BE)* staff gauge
Pegelmarkierung f <Kfztech> oil level mark *(Schmierung)*
Pegelmessgerät n <Gerät> decibel meter, *(AE)* level gage, *(BE)* level gauge, signal level meter
Pegelmessung f <Gerät> level measurement
Pegelregler m <Verpack> level control
Pegelschreiber m <Aufnahme> level recorder
Pegelstand m <Nichtfoss Energ> water depth
Pegelverschiebung f <Elektronik> level shifting
Pegelwaage f <Mobilkom> activity discriminator *(GSM)*
Peilempfänger m <Funkort, Wassertrans> DF receiver, direction-finding receiver
peilen v <Wassertrans> sound
Peilgerät n 1. <Gerät> bearing instrument; 2. <Lufttrans> direction finder
Peilgerät n **mit kommutierenden Antennen** <Funkort> commutated antenna direction-finder
Peilkompass m <Wassertrans> azimuth compass *(Navigation)*
Peillatte f <Wassertrans> sounding pole *(Deckausrüstung)*
Peilminimum n <Funkort> bearing minimum
Peilnullstelle f <Funkort> bearing null, bearing zero
Peilscheibe f <Wassertrans> pelorus *(Navigation)*; bearing marker *(Radar)*
Peilstrahl m *(PS)* <Funkort, Wassertrans> electronic bearing cursor, electronic bearing line *(Radar)*
Peilung f 1. <Funkort> bearing, direction finding; 2. <Raumfahrt> fix, heading *(Raumschiff)*; 3. <Wassertrans> sounding; bearing *(Navigation)*
Peilwertberichtigung f <Funkort> bearing correction
Peitschenantenne f <Funktech> whip antenna
Peitschenhiebeffekt m <Trans> whiplash effect
Pektin n <Chemie, Lebensmittel> pectin
Pektingelee n <Lebensmittel> pectin jelly
Pektinose f <Chemie> arabinose, pectinose
pektinsauer adj <Chemie> pectic
Pektinsubstanz f <Chemie> pectin
Pektisation f <Chemie> pectization
pektisch adj <Chemie> pectic
pektisieren v <Chemie> pectize
Pektose f <Chemie> pectose
Pelargonat n <Chemie> pelargonate
pelargonsauer adj <Chemie> pelargonic
Pellet n <Fertig, Kerntech, Kohlen, Kunststoff, Metall> pellet
Pelletieren n <Kunststoff> pelletizing
Pelletierin n <Chemie> pelletierine, punicine
Pelletiermaschine f <Kunststoff> pelletizer
Pelletisierung f <Abfall> pelletization
Pelletizer m <Kunststoff> pelletizer *(Kautschuk)*
Pelletstapel m <Kerntech> pellet stack *(in Brennelementen)*
Pelorus m <Wassertrans> pelorus *(Navigation)*
Peltiereffekt m <Elektriz, Phys> Peltier effect
Peltier-gekühlt adj <Funktech, Telekom> Peltier-cooled
Peltierkoeffizient m <Phys> Peltier coefficient
Peltonrad n <Nichtfoss Energ> Pelton wheel

Peltonrad *n* **mit senkrechter Welle** <Nichtfoss Energ> vertical-shaft Pelton wheel
Peltonrad *n* **mit waagerechter Welle** <Nichtfoss Energ> horizontal-shaft Pelton wheel
Peltonradturbine *f* <Nichtfoss Energ> Pelton wheel turbine
Pelz *m* <Textil> lap
Pendel *n* <Mechan> pendulum
Pendelachse *f* <Kfztech> full-floating axle; floating axle *(Kraftübertragung)*
Pendelaufhängung *f* <Trans> pendulum suspension
Pendelausschlag *m* <Mechan> pendulum deflection
Pendelbecherwerk *n* <Trans> tilt bucket elevator
Pendelbewegung *f* <Maschinen, Mechan> pendulum motion
Pendeleinlegevorrichtung *f* <Ker & Glas> reciprocating charger
Pendelfehler *m* <Mechan> pendulum error
Pendelförderung *f* <Kohlen> shuttle haulage
pendelförmig *adj* <Mechan> pendular
Pendelfräsen *n* <Maschinen> pendulum milling
Pendelgewicht *n* <Mechan> pendulum bob
Pendelhalter *m* <Maschinen> floating bush
Pendelhärte *f* <Kunststoff> pendulum hardness
Pendelkugel *f* <Mechan> pendulum sphere
Pendelkugellager *n* <Maschinen> self-aligning ball bearing
Pendellager *n* 1. <Kfztech> self-aligning bearing; 2. <Maschinen> self-aligning bearing, swivel bearing, swivel plummer block; 3. <Mechan> pendulum bearing
Pendellänge *f* <Mechan> pendulum length
Pendellinse *f* 1. <Mechan> pendulum lenticle; 2. <Phys> pendulum bob
Pendelmasse *f* <Mechan> pendulum mass
Pendelmotor *m* <Elektriz> swivel bearing motor
Pendelmühle *f* <Mechan> pendulum mill
pendeln *n* <Funktech> hunt
Pendeln *n* 1. <Elektriz, Elektrotech> hunting, oscillate, swinging *(Instrumentennadel)*; 2. <Lufttrans> shuttle
pendelnahtschweißen *v* <Fertig> weave
pendelnder Werkzeughalter *m* <Maschinen> floating tool holder
Pendelplatte *f* <Hydraul> shuttle plate
Pendelreibahle *f* <Maschinen> floating reamer
Pendelrollenlager *n* <Maschinen> self-aligning roller bearing
Pendelsäge *f* <Mechan> pendulum saw
Pendelscheibe *f* <Phys> pendulum bob
Pendelschere *f* <Maschinen> pendulum shears
Pendelschwimmer *m* <Ker & Glas> pendulum floater
Pendelschwingung *f* 1. <Lufttrans> phugoid oscillation; 2. <Mechan> pendulum swing
Pendelsignal *n* <Eisenbahn> wig-wag signal
Pendeltor *n* <Bau> swing gate
Pendeltrennschleifmaschine *f* <Fertig> oscillating-type abrasive cutting machine
Pendeltür *f* <Bau> swing door, swinging door
Pendelung *f* <Lufttrans> cycling *(Hubschrauber)*
Pendelverkehr *m* <Lufttrans> shuttle service, shuttle traffic
Pendelverkehr *m* **zwischen Flughäfen** <Trans> air shuttle
Pendelversuch *m* <Mechan> pendulum test
Pendelwirkung *f* <Lufttrans> phugoid effect
Pendlerverkehr *m* <Trans> commuter traffic, office-hour traffic
penetrant *adj* <Anstrich> penetrant
Penetrationsmessgerät *n* <Kunststoff> penetration tester

Penetrometer *n* 1. <Bau> penetrometer; 2. <Kunststoff> penetration tester; 3. <Labor> penetrometer
Penizillin-Phiole *f* <Ker & Glas> *(BE)* penicillin phial, *(AE)* penicillin vial
Pentachlorid *n* <Chemie> pentachloride
Pentacyanoferrat *n* <Chemie> prussiate
Pentaeder *n* <Geom> pentahedron
Pentagon *n* <Geom> pentagon
pentagonal *adj* <Geom> pentagonal
Pentagrid-Mischröhre *f* <Elektrotech> pentagrid tube
Pentamethylen *n* <Chemie> cyclopentane, pentamethylene
Pentamethylendiamin *n* <Chemie> pentamethylenediamine
Pentamethylenimin *n* <Chemie> pentamethyleneimide, piperidine
Pentan *n* <Chemie, Erdöl> pentane *(Petrochemie)*
Pentandiamin *n* <Chemie> pentamethylenediamine
Pentanol *n* <Chemie> amyl alcohol, pentanol
Pentanon *n* <Chemie> pentanone
Penta-Prisma *n* <Foto> pentaprism
Pentaquin *n* <Chemie> pentaquine
Pentasulfid *n* <Chemie> *(AE)* pentasulfide, *(BE)* pentasulphide
Pentathionat *n* <Chemie> pentathionate
pentavalent *adj* <Chemie> pentavalent, quinquevalent
Pentavalenz *f* <Chemie> pentavalence, quinquevalence
Penten *n* <Chemie> amylene, pentene
Penthiophen *n* <Chemie> penthiophene
Penthouse *n* <Bau> appentice, levecel, penthouse; superstructure
Pentin *n* <Chemie> pentyne, valerylene
Pentit *n* <Chemie> pentite, pentitol
Pentitol *n* <Chemie> pentite, pentitol
Pentode *f* <Elektronik, Funktech, Phys> pentode
Pentosan *n* <Chemie> pentosan
Pentose *f* <Chemie> pentose
Pentosenucleosid *n* <Chemie> pentoside
Pentosid *n* <Chemie> pentoside
Pentyl... <Chemie> amyl, pentyl
Pentylalkohol *m* <Chemie> pentanol, pentyl alcohol
Pentylentetrazol *n* <Chemie> pentylenetetrazol
Pepsin *n* <Lebensmittel> pepsin
Peptisation *f* <Lebensmittel> peptization
peptisieren *v* <Chemie> peptize
Peptisierungsmittel *n* <Kunststoff> peptizer • **mit Peptisierungsmitteln abbauen** <Chemie> peptize *(Gummi)*
Peptolyse *f* <Chemie> peptolysis
Perborat *n* <Chemie> perborate
Percarbonat *n* <Chemie> percarbonate
Perchlor... <Chemie> perchloric
Perchlorat *n* <Chemie> perchlorate
Perchlorethen *n* <Chemie> perchloroethylene, tetrachloroethylene
Perchlorethylen *n* <Chemie> perchloroethylene, tetrachloroethylene
perchloriert *adj* <Chemie> perchlorinated
Perchrom... <Chemie> perchromic
Perchromat *n* <Chemie> perchromate
Pereirin *n* <Chemie> pereirine
Peressig... <Chemie> peracetic *(Säure)*
perfektes Dielektrikum *n* <Elektriz> perfect dielectric
Perforation *f* 1. <Erdöl> casing perforation *(Bohrtechnik)*; 2. <Foto> perforation, sprocket hole; 3. <Ker & Glas> perforation
Perforationskanone *f* <Erdöl> gun perforator *(Bohrtechnik)*
Perforationsloch *n* <Foto> sprocket hole
perforieren *v* <Fertig> pierce

Perforieren

Perforieren n <Fertig> piercing
Perforiermaschine f <Verpack> perforating machine
perforiert adj <Comp & DV> perforated
perforierte Beutel mpl <Verpack> perforated bags
perforiertes Spülrohr n <Lebensmittel> sparge pipe
Perforierung f <Comp & DV> perforation
Perforierwerkzeug n <Fertig> puncher
Pergament n <Druck> parchment
Pergamentpapier n 1. <Lebensmittel> parchment paper; 2. <Verpack> grease-proof paper, parchment paper
Pergaminpapier n <Lebensmittel> glassine
Perhydrol n <Chemie> perhydrol
Periastron n <Raumfahrt> periastron
peridisches Rauschen n <Akustik> cyclic noise
Perigäum n <Phys, Raumfahrt> perigee
Perigäumsschubtriebwerk n <Raumfahrt> perigee kick motor
Perihel n <Nichtfoss Energ, Phys> perihelion
Periheldrehung f <Phys> advance
Periodat n <Chemie> periodate
Periode f 1. <Akustik> period; 2. <Elektrotech, Math, Phys> cycle, period
Periode f **einer Umlaufbahn** <Raumfahrt> orbital period
Periodenbereich m <Kerntech> period range
Periodendauer f <Akustik> cycle, period
Periodenmesskanal m <Kerntech> period-measuring channel
Periodenmischer m <Lebensmittel> batch mixer
Periodensignal n <Elektronik> periodic signal
Periodenzahl f <Akustik, Aufnahme, Comp & DV, Elektronik, Funktech, Phys> frequency, number of cycles
Periodenzähler m <Gerät> cycle counter
periodisch adj 1. <Elektriz, Elektronik> periodic; 2. <Erdöl> cyclic; 3. <Math, Phys> periodic
periodisch arbeitendes Gefriergerät n <Heiz & Kälte> batch-type freezer
periodische Blattverstellung f <Lufttrans> cyclic pitch control (Hubschrauber)
periodische Dämpfung f <Elektrotech> periodic damping
periodische Dezimalzahl f <Math> recurring decimal, periodic decimal, repeating decimal
periodische Funktion f <Elektronik, Math> periodic function
periodische Größe f <Akustik, Elektronik> periodic quantity
periodische Hilfstrimmeinrichtung f <Lufttrans> cyclic pitch servo trim (Hubschrauber)
periodische Impulsgruppe f <Elektronik> periodic pulse train
periodische Inspektion f <Maschinen> periodic inspection
periodische Rückwärtsspitzensperrspannung f <Elektriz> circuit repetitive peak reverse voltage
periodische Schallwelle f <Wellphys> periodic sound wave
periodische Steigung f <Lufttrans> cyclic pitch (Hubschrauber)
periodische Steuerstufe f <Lufttrans> cyclic control step (Hubschrauber)
periodische Strahlschwenkung f <Telekom> scanning (Antenne)
periodische Vorwärtsspitzensperrspannung f <Elektriz> circuit repetitive peak off-state voltage
periodische Welle f <Elektrotech> periodic wave
periodische Winde mpl <Wassertrans> periodical winds
periodischer Betrieb m **mit Aussetzbelastung** <Elektriz> continuous-operation periodic duty
periodischer Neuaufbau m <Elektronik> periodic refresh (Bildschirm)

periodischer Schlagwinkel m <Lufttrans> cyclic flapping angle (Hubschrauber)
periodischer Ton m <Akustik> periodic tone
periodischer Wobbeldurchgang m <Elektronik> periodic sweep
periodisches Abschalten n <Kerntech> periodic shutdown (Reaktor)
periodisches Signal n <Elektronik, Telekom> periodic signal
Periodizität f <Elektronik> periodicity
peripher adj <Comp & DV, Geom, Telekom> peripheral
periphere Baugruppe f <Telekom> peripheral module
periphere Einheit f <Comp & DV> peripheral unit
periphere Steuereinheit f (PCU) <Comp & DV> peripheral control unit (PCU)
periphere Übertragung f <Comp & DV> peripheral transfer
peripherer Prozessor m <Comp & DV, Telekom> peripheral processor
peripherer Schnittstellenadapter m <Comp & DV> peripheral interface adaptor
peripherer Speicher m <Comp & DV> (AE) backing storage, (BE) backing store, peripheral memory, peripheral storage
peripheres Brennelement n <Kerntech> peripheral fuel assembly
peripheres Steuerelement n <Kerntech> peripheral control element
Peripherie f <Bau, Geom> periphery • **durch Peripherie in der Schnelligkeit eingeschränkt** <Comp & DV> peripheral-limited
Peripherieführung f <Telekom> peripheral management
Peripheriegerät n <Comp & DV, Elektrotech, Telekom> peripheral device, peripheral equipment, peripheral unit
Peripherietechnik f (PPU) <Comp & DV> peripheral processing units (PPU)
Peripheriewinkel m <Math> peripherical triangle (am Kreis)
peripherisch adj <Telekom> peripheral
Periskop n 1. <Kerntech> periscope (Manipulator); 2. <Phys> periscope; 3. <Raumfahrt> periscope (Raumschiff); 4. <Wassertrans> periscope (U-Boot)
Periskopantenne f 1. <Funktech> beam waveguide antenna, periscope aerial, periscope antenna; 2. <Phys> periscope aerial, periscope antenna
periskopisch adj <Optik> periscopic
periskopisches Objektiv n <Foto> periscopic lens
Periskopsextant m <Raumfahrt> periscopic sextant (Raumschiff)
Peristalsis-Pumpe f <Labor> peristaltic pump
Peritektikum n <Metall> peritectoid
peritektische Reaktion f <Metall> peritectic reaction
Perkolation f <Chemie, Lebensmittel> percolation
Perkussionsbohren n <Erdöl> percussion drilling
Perkussionsschall m <Sicherheit> percussive sound
perlartiger Grat m <Fertig> flash, ridge (Brennschweißen)
Perle f **im Flaschenhals** <Ker & Glas> slug in neck
Perlglanzpigment n <Kunststoff> nacreous pigment
Perlit m <Fertig, Metall> pearlite
perlitisch adj <Fertig> pearlitic
Perlitisieren n <Metall> isothermal annealing
Perlmutperle f <Ker & Glas> mother-of-pearl bead
Perlmutpigment n <Kunststoff> nacreous pigment
Perlrohrdurchflussmessgerät n <Gerät> bubble flow meter
Perlrohrfüllstandanzeigegerät n <Gerät> bubble-type level indicator
Perlrohrfüllstandmessgerät n <Gerät> bubble pipe level meter

Perlstickerei f <Textil> beading
Perm n <Erdöl> Permian period *(Geologie)*
Permafrost m 1. <Erdöl> permafrost *(Geographie)*; 2. <Kohlen> permafrost
Permalloy n <Metall> permalloy *(Nickellegierung)*
permanente Datei f <Comp & DV> permanent file
permanente Schwellwertverschiebung f <Akustik> permanent threshold shift
permanenter Fehler m <Comp & DV> hard error, permanent error
permanenter Speicher m <Comp & DV> permanent memory
permanenterregter Generator m <Elektrotech> permanent magnetic generator *(PMG)*
Permanentmagnet m <Elektriz, Elektrotech, Maschinen, Phys, Telekom, Trans> permanent magnet
permanentmagnetische Fokussierung f <Elektrotech> permanent-magnet focusing
permanentmagnetischer Generator m <Elektrotech> permanent-magnet generator
permanentmagnetischer Kondensatormotor m <Elektrotech> permanent-magnet split-capacitor motor
permanentmagnetischer Schrittmotor m <Elektrotech> permanent-magnet stepper motor
permanentmagnetischer Synchronmotor m <Elektrotech> permanent-magnet synchronous motor *(PMSM)*
permanentmagnetisches Schweben n <Trans> levitation by permanent magnets
Permanentmagnet-Lautsprecher m <Akustik> permanent-magnet loudspeaker
Permanentmuster n <Textil> permanent patterning
Permanentspeicher m 1. <Comp & DV> nonerasable storage; 2. <Elektrotech> permanent memory
Permangan... <Chemie> permanganic
Permanganat n <Chemie> permanganate
permeabel adj <Chemie> permeable
Permeabilität f 1. <Chemie, Elektriz> permeability; 2. <Erdöl> permeability *(Geologie)*; 3. <Funktech, Kohlen, Kunststoff, Nichtfoss Energ, Phys, Werkprüf> permeability
Permeabilität f **des Vakuums** <Phys> permeability of free space
Permeabilitätsmesser m <Kohlen> permeameter
Permeabilitätsmessung f <Erdöl> permeability logging *(Messtechnik)*
Permeabilitätszahl f 1. <Elektrotech> relative permeability; 2. <Kohlen> permeability coefficient; 3. <Phys> relative permeability
Permeanz f <Elektrotech> permeance
permissiver Block m <Eisenbahn> permissive block
Permittivität f <Elektriz> permittivity
Permutation f <Comp & DV, Math> permutation
Peroxid n <Chemie, Kunststoff> peroxide
peroxidieren v <Chemie> peroxidize
Peroxidierung f <Chemie> peroxidation
Peroxoborat n <Chemie> perborate
Peroxocarbonat n <Chemie> percarbonate *(Bleichmittel)*
Peroxochromat n <Chemie> perchromate
Peroxomonoschwefel m <Chemie> *(AE)* permonosulfur, *(BE)* permonosulphur
Peroxonitrat n <Chemie> pernitrate
Peroxophosphat n <Chemie> peroxophosphate
Peroxosalpeter n <Chemie> pernitrat *(Säure)*
Peroxosäure f <Chemie> peroxy acid; peracid *(anorganisch)*
Peroxosulfat n <Chemie> *(AE)* persulfate, *(BE)* persulphate
Peroxyacetylnitrat n *(PAN)* <Umweltschmutz> peroxoacetylnitrate *(PAN)*

Peroxydischwefel m <Chemie> *(AE)* peroxydisulfur, *(BE)* peroxydisulphur *(Säure)*
Peroxysäure f <Chemie> peroxy acid; peracid *(organisch)*
Perpetuum Mobile n <Thermod> perpetual motion engine
Perrhenat n <Chemie> perrhenate
Perrhenium... <Chemie> perrhenic
Persalz n <Chemie> persalt
Persäure f <Chemie> peroxy acid; peracid *(anorganisch)*
Persenning f <Wassertrans> tarpaulin
Perseulose f <Chemie> perseulose
persistentes Öl n <Abfall> persistent oil
Person f <Eisenbahn> passenger
Personalabteil n <Lufttrans> crew compartment
Personalcomputer m *(PC)* <Comp & DV> personal computer *(PC)*
Personalroboter m <Künstl Int> personal robot
Personenaufzug m <Trans> *(AE)* passenger elevator, *(BE)* passenger lift
Personenautofähre f <Wassertrans> passenger car ferry
Personenbahnhof m <Eisenbahn, Trans> passenger station
Personenbeförderung f <Lufttrans> passenger service
personenbezogene Telekommunikation f <Telekom> universal personal telecommunication
Personendosimetrie f <Strahlphys> personal dosimetry
Personenfähre f <Wassertrans> foot-passenger ferry, passenger ferry
Personenfahrzeug n <Kfztech> passenger vehicle
personengebundenes Probenahmegerät n 1. <Sicherheit> personal sampler; 2. <Umweltschmutz> personal sampler, personal sampling unit
personengebundenes Schutzsystem n <Sicherheit> personal protective system
Personenkilometer m <Trans> *(AE)* passenger kilometer, *(BE)* passenger kilometre
Personenkraftwagen m *(PKW)* <Kfztech> car; passenger car *(Fahrzeugart)*
Personenkraftwageneinheit f *(PKW-E)* <Kfztech> passenger car unit *(PCU)*
Personenlärmdosimeter n <Sicherheit> personal noise dosimeter
Personenruf m <Telekom> paging
Personenrufanlage f <Sicherheit> staff calling installation
Personenrufdienst m <Telekom> paging service
Personenrufempfänger m <Telekom> pager
Personenrufempfänger m **für Mitteilungen** <Telekom> message pager
Personenschnellverkehr m <Trans> personal rapid transport
Personenschnellverkehrssystem n <Trans> PRT, personal rapid transit, passenger rapid transit
Personenschutz m <Sicherheit> personal protection
Personenschwebebahn f <Trans> passenger ropeway
Personensicherungsanlage f <Sicherheit> personnel safeguard
personenspezifische Ansicht f <Comp & DV> personalized view
Personenverkehr m <Trans> passenger transport
Personenwagenäquivalent n <Trans> passenger car equivalent
Personenzug m <Eisenbahn> slow train
persönliche Beratung f <Comp & DV> one-to-one consultancy
persönliche Schutzausrüstung f <Sicherheit> personal protective apparatus, personal protective equipment, personal protective gear
persönlicher Augenschutz m <Sicherheit> personal eye protector

persönlicher

persönlicher Gehörschutz m <Sicherheit> personal ear protector
persönlicher Mitteilungs-Übermittlungsdienst m <Telekom> interpersonal messaging system
persönlicher Mobilfunkanrufbeantworter m <Mobilkom> mobilbox
Perspektive f <Geom> perspective, perspective view
Perspektivenwechsel m <Geom> change of perspective
perspektivische Ansicht f <Geom> perspective view
perspektivische Transformation f <Geom> perspective transformation
Perspektivlinie f <Optik> line of direction, line of perspective, line of sight
Perspex® n <Kunststoff> Perspex®
PERT-Verfahren n (Programmbewertungs- und Überprüfungsverfahren) <Comp & DV, Raumfahrt> PERT, program evaluation and review technique
Perveanz f <Elektrotech, Telekom> perveance
Perylen n <Chemie> perylene
Perzentil n <Ergon, Qual> percentile
Perzentilwert m <Qual> percentile
Perzeption f <Künstl Int> perception
PES (Polyester) <Chemie, Elektriz, Kunststoff, Textil> PES (polyester)
PET (Polyethylen) <Chemie, Elektriz, Erdöl, Kunststoff, Textil, Verpack> PET (polyethylene)
PET-Container m <Verpack> polyethylene container
Petersen-Spule f <Elektrotech> arc-suppression coil, earth-fault neutralizer, Petersen coil
PET-Film m <Verpack> PET film
PET-Flasche f <Verpack> PET bottle
PETP (Polyethylenterephthalat) <Kunststoff> PETP (polyethylene terephthalate)
PET-Palettenüberzüge mpl <Verpack> polyethylene pallet covers
PETRIFIX-Verfahren n <Abfall> PETRIFIX process (Verfestigungsverfahren für Sonderabfälle)
Petri-Netz n <Comp & DV> Petri net
Petri-Schale f <Labor> Petri dish
Petrochemikalien fpl <Erdöl> petrochemicals
petrochemisch adj <Erdöl> petrochemical
petrochemische Anlage f <Erdöl> petrochemical plant
petrochemisches Zwischenprodukt n <Erdöl> intermediate chemical (Raffinerie)
Petrol... <Chemie> petrolic
Petrolat n <Chemie> petrolatum, petroleum jelly
Petrolatum n <Chemie> petrolatum, petroleum jelly
Petrologie f <Kohlen> petrology
Pfad m <Comp & DV, Elektrotech, Phys, Telekom> path
Pfadangabe f <Comp & DV> path
Pfadname m <Comp & DV> path name
Pfahl m 1. <Bau> pile, pole, stake; peg (Vermessung); 2. <Fertig, Kohlen> pile; 3. <Wassertrans> pile, post (zum Festmachen)
Pfahlabschnitt m <Kohlen> pile segment
Pfahlabschnitthöhe f <Bau, Kohlen> pile cut-off level
Pfahlanschluss m <Kohlen> pile joint
Pfahlart f <Kohlen> type of pile
Pfahlbau m <Bau> pilework
Pfahlbuhne f <Wasserversorg> (AE) pile groin, (BE) pile groyne
Pfahlgründung f <Erdöl> piling (Tiefbautechnik)
Pfahlgruppe f <Kohlen> pile group
Pfahlkopf m <Bau, Kohlen> pile head
Pfahlkopfplatte f <Bau, Kohlen> pile cap
Pfahllageplan m <Kohlen> pile situation plan
Pfahllänge f <Kohlen> pile length
Pfahlramme f 1. <Bau> pile driver, piling frame; 2. <Kohlen> pile driver

Pfahlrammung f <Bau, Kohlen> pile driving
Pfahlring m <Bau> pile ferrule
Pfahlschuh m <Bau, Kohlen> pile shoe
Pfahlsockel m <Kohlen> pile footing
Pfahlspitze f <Kohlen> pile point, pile tip
Pfahlspleiß m <Kohlen> pile splice
Pfahltreiben n <Bau, Kohlen> piling
Pfahlwerk n 1. <Hydraul> stockade; 2. <Kohlen> row of piles
Pfahlzieher m <Bau> pile drawer, pile extractor
Pfahlzwinge f <Bau> pile ferrule
Pfänden n <Kohlen> blocking
Pfandflasche f 1. <Abfall> deposit bottle, returnable bottle, waste bottle; 2. <Verpack> deposit bottle, returnable bottle
Pfandkasten m <Verpack> reusable box
Pfanne f <Maschinen> pan
Pfannenbär m <Metall> button
Pfannenhüttenofen n <Metall> ladle metallurgy furnace, LMF
Pfeffersandstrahlen n <Ker & Glas> peppered sandblast
Pfeifboje f <Wassertrans> whistle buoy
Pfeifen n 1. <Aufnahme> whistling; 2. <Elektronik> whistle; singing (Verstärker)
Pfeifpunkt m <Elektronik> singing point
Pfeifpunktmesser m <Metrol> singing point test set
Pfeifton m <Akustik> squealing
Pfeil m <Konstzeich> arrowhead
Pfeiler m 1. <Bau> pile, pillar; post (Architektur); pier (Wände, Brücken); 2. <Kohlen> pillar; 3. <Telekom> pillar; 4. <Wassertrans> pile (Festmachen)
Pfeilerabbau m <Kohlen> pillar drawing
Pfeilerkopf m <Bau> cutwater (Brücke)
Pfeilerstaumauer f <Bau> buttress dam
Pfeilflügel m <Lufttrans> arrowhead wing
pfeilförmiger Flügel m <Lufttrans> back-swept wing, swept back wing, swept wing
Pfeillinie f <Konstzeich> arrow line
Pfeilrad n <Maschinen> double helical gear, double helical gearwheel, herringbone gearwheel
Pfeilrädergetriebe n <Maschinen> herringbone gear
Pfeilspitzenbohrer m <Maschinen> arrow-headed drill, arrowhead drill
Pfeilstellungswinkel m <Lufttrans> sweep angle (Flugwerk)
Pfeilstirnrad n <Maschinen> herringbone gear
Pfeiltaste f <Comp & DV> cursor key
pfeilverzahnen v <Maschinen> herringbone
pfeilverzahntes Getriebe n <Maschinen> herringbone gear
pfeilverzahntes Rad n 1. <Fertig> double helical gear; 2. <Maschinen> double helical gear, double helical gearwheel, herringbone gearwheel
Pfeilverzahnung f <Maschinen> herringbone teeth
Pfeilwurzelmehl n <Lebensmittel> arrowroot
Pferdestärke f (PS) <Maschinen> horsepower (hp)
Pfette f <Bau> purlin (Dach)
Pfettenstützholz n <Bau> purlin post
Pflanzeneiweiß n <Lebensmittel> vegetable protein
Pflanzengallerte f <Chemie> pectin
pflanzentötend adj <Chemie> phytocidal
pflanzlicher Abfall m <Abfall> vegetable waste
Pflaster n <Bau> paving
Pflasterbelag m <Bau> (BE) pavement, (AE) sidewalk
Pflasterer m <Bau> (AE) pavior, (BE) paviour
Pflasterglasbaustein m <Ker & Glas> (BE) pavement light, (AE) sidewalk light
Pflasterhammer m <Bau> (AE) pavior's hammer, (BE) paviour's hammer, sledge hammer

Pflasterklotz m <Ker & Glas> paving block
pflastern v <Bau> floor, pave
Pflasterstein m <Bau> paving block, paving stone, road stone, sett
Pflegekennzeichnung f durch Etikett <Textil> *(AE)* care labeling, *(BE)* care labelling
pflegen v <Bau> attend to
Pflichtbohrung f <Erdöl> obligatory well
Pflichtenheft n 1. <Elektriz> specifications; 2. <Qual> design specifications; 3. <Telekom> specifications
Pflock m 1. <Fertig> stake; 2. <Mechan> peg
pflocken v <Fertig> peg
Pflugbagger m <Kfztech> loader
Pflugschar f <Wassertrans> *(BE)* ploughshare, *(AE)* plowshare *(Festmachen)*
Pflugscharanker m <Wassertrans> *(BE)* plough anchor, *(AE)* plow anchor
PFM *(Pulsfrequenzmodulation)* <Comp & DV, Elektronik, Funktech> PFM *(pulse frequency modulation)*
Pforte f <Bau> portal
Pfosten m 1. <Bau> pillar, post, stanchion, stile; 2. <Eisenbahn> pillar; 3. <Kohlen> post; 4. <Maschinen> upright; 5. <Trans> pillar; 6. <Wassertrans> post
Pfostenramme f <Bau> post driver
Pfropfen m 1. <Lebensmittel> bung; 2. <Wassertrans> plug
Pfropfpolymer n <Kunststoff> graft polymer
Pfropfpolymerisat n <Kunststoff> graft polymer
Pfropfpolymerisation f <Kunststoff> graft polymerization
Pfund n <Metrol> lb, pound, pound avoirdupois
Pfund n **pro Kubikfuß** <Metrol> pounds per cubic foot
Pfund n **pro Quadratzoll** <Metrol> pounds per square inch, psi
Pfund-Serie f <Phys> Pfund series
Pfusch m <Fertig> slipshod work
P-Grad m <Regelung> proportional degree
pH-Abnahme f <Umweltschmutz> pH drop
p-Halbleiter m <Phys> p-type semiconductor
p⁻-Halbleiter m <Elektronik> p⁻-semiconductor
Phantomkanal-Lautsprecher m <Aufnahme> *(AE)* phantom center channel loudspeaker, *(BE)* phantom centre channel loudspeaker
Phantomkreis m <Elektrotech> phantom circuit
Phantomleitung f <Elektrotech, Telekom> phantom line
Phantomspule f <Elektrotech> phantom coil
pharmazeutische Chemie f <Chemie> pharmaceutic chemistry, pharmaceutical chemistry
Phase f <Comp & DV, Elektriz, Elektronik, Funktech, Metall, Phys, Thermod, Trans> phase • **in Phase** <Fernseh, Telekom> phased • **in Phase nacheilen** <Phys> lag in phase • **Phasen drehen** <Elektronik> phase-shift • **Phasen schieben** <Elektronik> phase shift
Phase f **einer Schallschwingung** <Akustik> phase of an acoustical vibration
Phase f **einer sinusförmigen Größe** <Akustik> phase of a sinusoidal quantity
Phased-Array-Antenne f <Funktech> phased array antenna
Phase-Erde-Schluss m <Elektriz> *(BE)* phase-to-earth fault, *(AE)* phase-to-ground fault
Phasenabgleich m 1. <Fernseh, Funktech> phase adjustment; 2. <Telekom> phase alignment, phase tuning
Phasenabgleich m **von Lautsprechern** <Aufnahme> phasing of loudspeakers
Phasenabgleichsschalter m <Aufnahme> phasing switch
Phasenabgleichsstecker m <Aufnahme> phasing plug
Phasenabweichung f <Elektronik> phase deviation
Phasenänderung f <Elektriz> phase variation

Phasenänderungsgeschwindigkeit f <Telekom> phase change velocity
Phasenanschluss m <Elektriz> phase terminal
Phasenausgleich m <Elektriz, Telekom> phase compensation
Phasenausgleicher m <Elektriz> phase equalizer
Phasenausgleichrelais n <Elektriz> phase balance relay
Phasenbelag m <Akustik> phase change coefficient
Phasendemodulation f <Elektronik, Telekom> phase demodulation
Phasendemodulator m <Elektronik> phase demodulator
Phasendetektor m <Elektronik, Fernseh, Funktech, Telekom> phase detector
Phasendiagramm n 1. <Kerntech, Metall, Thermod> phase diagram; 2. <Trans> phase diagram, phasing diagram
Phasendifferenz f <Aufnahme, Elektronik, Phys, Wellphys> phase difference
Phasendifferenzmodulation f *(DPSK)* <Elektronik, Telekom> differential phase shift keying *(DPSK)*
Phasendifferenzwinkel m <Elektriz> angle of phase difference
Phasendiskriminator m 1. <Elektronik> phase detector, phase discriminator; 2. <Funktech> phase discriminator
Phasendreher m <Phys> gyrator
Phasendrehung f <Elektronik> phase shift
Phaseneinsteller m <Elektriz> phase changer
Phaseneinstellungseinheit f <Lufttrans> phasing unit *(Hubschrauber)*
Phasenfehler m <Elektronik, Telekom> phase error
Phasenfolge f <Elektriz, Elektronik> phase sequence
Phasenfolgegleichrichter m <Elektriz> phase sequence rectifier
Phasenfrequenzgang m <Akustik> phase frequency response curve
Phasengang m <Elektronik> phase response
Phasengang m **der Amplitude** <Telekom> phase amplitude characteristic
phasengekoppelter Laser m <Elektronik> mode-locked laser
Phasengenerator m <Elektronik> phase generator
Phasengeschwindigkeit f <Phys> phase velocity
phasengespeiste Antennengruppe f <Raumfahrt> phased array antenna *(Weltraumfunk)*
phasengesteuerte Gruppenantenne f <Funktech> phased array, phased array antenna
phasengesteuerte Zündung f <Raumfahrt> phased ignition *(Raumschiff)*
phasengleich adj <Elektronik, Phys, Telekom> in-phase
phasengleiche Antenne f <Raumfahrt> broadside antenna
Phasengleichgewicht n <Thermod> phase equilibrium
Phasengleichrichter m <Elektronik> phase demodulator, phase rectifier
Phasengrenze f <Metall> phase boundary
Phasenisolierung f <Elektriz> phase insulation
Phasenkettenoszillator m <Elektronik> phase-shift oscillator
Phasenklemme f <Elektriz> phase terminal
Phasenkoeffizient m <Optik> phase coefficient
Phasenkomparator m <Fernseh> phase comparator
Phasenkonstante f 1. <Akustik, Elektriz> (β) phase constant (β); 2. <Optik, Phys> phase constant; 3. <Telekom> phase coefficient, phase constant
phasenkontinuierliche Frequenzumtastung f *(CPFSK)* <Elektronik, Funktech, Telekom> continuous phase frequency shift keying *(CPFSK)*
Phasenkontrastmikroskop n <Labor, Phys> contrasting phase microscope, phase contrast microscope

Phasenkonverter

Phasenkonverter m <Fernseh> phase converter
Phasenkopplung f <Elektronik> mode locking *(Laser)*
Phasenlaufzeit f <Telekom> phase delay
Phasenmaß n <Telekom> phase constant
Phasenmessbrücke f <Gerät> phase bridge
Phasenmodulation f *(PM)* <Aufnahme, Comp & DV, Elektronik, Funktech, Phys, Telekom> phase modulation *(PM)*
Phasenmodulator m <Elektronik, Telekom> phase modulator
Phasennacheilung f <Elektriz, Elektronik> phase lag
Phasenquadratur f <Fernseh> quadrature
Phasenrastung f <Elektronik> phase locking
Phasenraum m <Phys, Teilphys> phase space
Phasenregel f <Kerntech> phase rule
Phasenregelkreis m *(PLL)* <Elektronik, Fernseh, Funktech, Raumfahrt, Telekom> phase-locked loop *(PLL)*
Phasenregelung f <Telekom> automatic phase control, phase control, phase regulation
Phasenregler m <Fernseh> phaser
Phasenreserve f <Elektronik> phase margin
Phasenresonanz f <Phys> velocity resonance
Phasenschiebekette f <Elektriz> phase-shifting network
Phasenschiebekondensator m <Elektriz> phase-shifting capacitor
Phasenschieber m 1. <Elektriz, Elektronik, Elektrotech> phase shifter, synchronous capacitor; 2. <Kerntech> phase shifter; 3. <Telekom> phase changer, phase shifter
Phasenschieber m **der PIN-Diode** <Elektronik> PIN diode phase shifter
Phasenschieberschaltung f <Elektriz, Phys, Telekom> phase-shifting network
Phasenschnittfrequenz f <Regelung> phase crossover frequency
Phasensignal n <Fernseh> phasing signal
Phasenspalter m <Aufnahme> phase splitter
Phasenspalteroszillator m <Elektronik> phase splitter oscillator
Phasenspannung f <Elektriz> phase voltage
Phasensprung m <Fernseh> phase shift
Phasensprungmikrofon n <Aufnahme> phase shift microphone
Phasensprungtastung f *(PSK)* <Comp & DV> phase shift keying *(PSK)*
phasenstabilisierter Demodulator m <Raumfahrt> phase-locked demodulator *(Weltraumfunk)*
Phasenstabilität f <Elektriz, Telekom> phase stability
phasenstarre Kopplung f <Telekom> phase lock
Phasensteuerung f <Elektronik, Telekom> phase control
Phasenstrom m <Elektriz, Elektrotech> phase current
Phasensynchronisation f 1. <Fernseh> phase lock, phase locking; 2. <Telekom> phase locking
Phasensynchronisationskreis m <Telekom> phase-locked loop
Phasenteiler m <Elektriz> phase splitter
Phasenteilung f <Elektriz> phase splitting
Phasentrennung f 1. <Abfall> phase separation; 2. <Elektronik> phase splitting
Phasenübergang m <Thermod> phase transformation
Phasenübergangstemperatur f <Phys> transition temperature
Phase-Nullleiter-Spannung f <Elektriz> phase-to-neutral voltage
Phasenumformer m <Elektriz, Elektronik> phase changer, phase converter
phasenumgekehrte Sekundärströme mpl <Elektrotech> phase-reversed secondary currents
Phasenumkehr f <Elektriz, Fernseh> phase reversal
Phasenumkehrer m <Elektriz> phase inverter

Phasenumkehrschalter m <Elektrotech> phase reversal switch
Phasenumkehrung f <Aufnahme, Kohlen> phase inversion
Phasenumkehrverstärker m <Elektronik> paraphase amplifier
Phasenumtastung f *(PSK)* <Elektronik, Funktech, Raumfahrt, Telekom> phase-shift keying *(PSK)*
Phasenumtastungsmodulation f 1. <Elektronik, Telekom> phase shift keyed modulation; 2. <Raumfahrt> phase shift keyed modulation *(Weltraumfunk)*
Phasenunterschied m <Elektriz, Elektronik, Fernseh, Telekom> phase difference
Phasenvariation f <Elektriz> phase variation
Phasenverbesserung f <Elektrotech> power factor correction
Phasenvergleicher m <Elektronik> phase comparator
Phasenverriegelung f <Telekom> phase locking
phasenverschiebendes Element n <Elektronik> phase-shifting element
Phasenverschiebetransformator m <Elektriz> phase-shifting transformer
Phasenverschiebung f 1. <Elektriz, Elektronik> phase displacement, phase lag, phase shift; 2. <Kerntech, Phys, Raumfahrt, Telekom, Trans> phase shift; 3. <Wellphys> phase-out
Phasenverschiebung f **um neunzig Grad** <Elektriz> quadrature phase shift
Phasenverschiebungsinduktionsschleife f <Trans> phase displacement induction loop detector
Phasenverschiebungswinkel m <Elektrotech> angle of phase difference, angle of phase displacement, angle of phase shift
phasenverschoben adj 1. <Elektriz, Elektronik> dephased, out-of-phase, phase-shifted; 2. <Fernseh, Telekom, Werkprüf> out-of-phase • **um 90 Grad phasenverschoben** <Phys> in quadrature
phasenverschobenes Signal n <Elektronik> in-quadrature signal
Phasenverteilung f <Metall> phase distribution
Phasenverzerrung f <Elektronik, Fernseh, Funktech, Telekom> phase distortion
Phasenverzögerung f <Elektriz, Elektronik, Phys> phase delay, phase lag
Phasenvoreilung f 1. <Elektriz> phase advance, phase lead; 2. <Elektronik, Phys> phase lead; 3. <Telekom> leading of phase
Phasenvoreilwinkel m <Elektriz> advance of phase angle
Phasenvorschiebung f <Elektriz> phase advance
Phasenvorschub m <Elektriz> phase lead
Phasenwendungsrelais n <Elektriz> reverse phase relay
Phasenwicklung f <Elektriz> phase winding
Phasenwinkel m 1. <Elektriz, Elektronik, Lufttrans, Nichtfoss Energ, Phys> phase angle; 2. <Wellphys> phase angle *(Schwingung)*; 3. <Werkprüf> phase angle
Phasenwinkeldifferenz f <Elektriz> phase difference
Phasenzittern n 1. <Raumfahrt> phase jitter *(Weltraumfunk)*; 2. <Telekom> phase jitter
Phaseolunatin n <Chemie> phaseolunatin
Phase-Phase-Spannung f <Elektriz> line to line voltage, phase-to-phase voltage
Phellandren n <Chemie> phellandrene
Phenacetin n <Chemie> ethoxyacetanilide, phenacetin
Phenacetur... <Chemie> phenaceturic
Phenacyl n <Chemie> phenacyl
Phenanthrachinon n <Chemie> phenanthraquinone
Phenanthridin n <Chemie> phenanthridine

Phenanthridon n <Chemie> phenanthridone
Phenanthrol n <Chemie> phenanthrol
Phenanthrolin n <Chemie> phenanthroline
Phenat n <Chemie> phenate, phenolate
Phenazin n <Chemie> azophenylene, dibenzopyrazine, phenazine
Phenazon n <Chemie> phenazone
Phenetidin n <Chemie> ethoxyaniline, phenetidine
Phenetol n <Chemie> ethoxybenzene, phenetole
Pheniramin... <Chemie> pheniramine
Phenol n <Chemie> phenol
Phenolat n <Chemie> phenate, phenolate, phenoxide
Phenolharz n 1. <Elektriz> phenolic resin; 2. <Fertig> phenolic, phenolic resin; 3. <Kunststoff> phenolic resin
Phenolharzschaumstoff m <Kunststoff> phenolic foam
phenolisch adj <Chemie> phenolic
Phenolkunststoffleiste f <Fertig> phenolic lining
Phenolphthalein n <Chemie> phenolphthalein
Phenolsulfon n <Chemie> (AE) phenolsulfon, (BE) phenolsulphon
Phenoplast m <Kunststoff> phenolic plastic
Phenosafranin n <Chemie> phenosafranine, safranin, safranine
Phenothiazin n <Chemie> phenothiazine
Phenoxazin n <Chemie> naphtoxazine, phenoxazine
Phenoxid n <Chemie> phenate, phenoxide, phenolate
Phenoxybenzol n <Chemie> diphenyl ether, phenoxybenzene
Phenthiazin n <Chemie> phenothiazine
Phenyl... <Chemie> phenyl
Phenylacetaldehyd m <Chemie> phenylacetaldehyde
Phenylacetamid n <Chemie> phenylacetamide
Phenylalanin n <Chemie> phenylalanine
Phenylamin n <Chemie, Druck> phenylamine
Phenylcarbinol n <Chemie> phenylcarbinol
Phenylchromon n <Chemie> flavone, phenylchromone
Phenylendiamin n <Chemie> phenylenediamine
Phenylessig... <Chemie> phenylacetic
Phenylethylen n <Chemie> phenylethylene
Phenylglycin n <Chemie> phenylglycine
Phenylglycol... <Chemie> phenylglycolic
Phenylglykokoll n <Chemie> phenylglycine
Phenylharnstoff m <Chemie> phenylurea
Phenylhydrazin n <Chemie> phenylhydrazine
Phenylhydrazon n <Chemie> phenylhydrazone
Phenylhydroxylamin n <Chemie> phenylhydroxylamine
phenyliert adj <Chemie> phenylated
Phenyliodid n <Chemie> iodobenzene, phenyl iodide
Phenylisocyanat n <Chemie> carbanil, phenyl isocyanate
Phenylmercaptan n <Chemie> penyl mercaptan, thiophenol
Phenylpropan n <Chemie> cumene, cumol, phenylpropane
Phenylpropion... <Chemie> phenylpropiolic
Phenylsalicylat n <Chemie> salol
Phiole f <Chemie, Ker & Glas> phial, vial
Phloretin n <Chemie> phloretin
Phloretin... <Chemie> phloretic
Phloridzin n <Chemie> phloridzin, phlorizin
Phlorol n <Chemie> phlorol
Phloryhidzin n <Chemie> phloryhizin
pH-Messer m <Kohlen, Metrol> pH meter
pH-Messgerät n <Labor> pH meter
Phon n <Akustik, Phys> phon (Einheit)
Phonem n <Akustik> phoneme
phonetische Leistung f <Akustik> phonetic power
Phonon n <Phys> phonon
Phonongasmodell n <Phys> phonon gas model

Phonovision f <Optik> phonovision
Phoron n <Chemie> diisopropylidene acetone, dimethylheptadienone, phorone
phoronomisch adj <Fertig> phoronomical
Phosgen n <Chemie> phosgene
Phosphat n <Anstrich, Chemie> phosphate
Phosphatase f <Chemie> phosphatase
Phosphatieren n 1. <Chemie> phosphatization; 2. <Metall> parkerizing
Phosphat-Opalglas n <Ker & Glas> phosphate-opal glass
Phosphid n <Chemie> phosphide
Phosphit n <Chemie> phosphite
Phosphoglycerin... <Chemie> phosphoglyceric
Phospholipid n <Lebensmittel> phospholipid
Phosphomonoesterase f <Chemie> phosphatase
Phosphonium n <Chemie> phosphonium
Phosphor m 1. <Chemie> phosphorus (P); 2. <Elektronik, Phys> phosphor (Element)
Phosphorbronze f <Elektriz, Elektrotech> phosphor bronze
Phosphoreszenz f <Elektronik, Phys, Strahlphys> phosphorescence
Phosphoreszieren n <Wassertrans> phosphorescence (Meer)
phosphoreszierender Stoff m <Chemie> phosphor
phosphoreszierendes Material n <Elektronik> phosphorescent material
phosphoreszierendes Sicherheitschild n <Sicherheit> phosphorescent safety sign
phosphorigsaures Salz n <Chemie> phosphite
Phosphormolybdän... <Chemie> phosphomolybdic
Phosphorroheisen n <Chemie> phosphoric pig iron
Phosphorsalz n <Chemie> microcosmic salt
Phosphorsäure f <Chemie> phosphoric acid
Phosphorwolframat n <Chemie> phosphatododecatungstate, phosphotungstate
Phosphoryl n <Chemie> phosphoryl
Phosphorylase f <Chemie> phosphorylase
photoaktiver Transducer m <Elektrotech> photoactive transducer
photoaktiver Wandler m <Elektrotech> photoactive transducer
photoakustische Spektrometrie f <Phys> photoacoustical spectroscopy, PAS
Photoätzung f <Druck> photoetching
Photochemie f <Foto> photochemistry
Photochemigraphie f <Druck> photoengraving
photochemisch adj <Umweltschmutz> photochemical
photochemisch wirksame Strahlen mpl <Foto> actinic rays
photochemische Strahlung f <Strahlphys> actinic light
photochemische Wirkung f <Nichtfoss Energ> photochemical effect
photochemischer Smog m <Umweltschmutz> photochemical smog
Photodetektion f <Elektronik> photodetection
Photodetektor m 1. <Elektriz, Elektronik> photocell; 2. <Optik> photodetector; 3. <Telekom> light detector
Photodiode f 1. <Elektronik> photodetector diode, photodiode; 2. <Foto> photodiode; 3. <Optik> photodetector diode, photodiode; 4. <Phys> photodiode; 5. <Telekom> photodetector diode, photodiode
Photodiodengruppe f <Elektronik> photodiode array
Photoeffekt m 1. <Elektronik> photoelectric effect; 2. <Raumfahrt> photovoltaic effect (Raumschiff); 3. <Strahlphys> photoelectric effect, photoemissive effect
photoelektrisch adj <Elektriz> photoelectric
photoelektrisch betriebenes Relais n <Elektrotech> photoelectrically-operated relay

photoelektrische

photoelektrische Einsatzschwelle f <Phys> photoelectric threshold
photoelektrische Infrarotschutzvorrichtung f <Sicherheit> light photoelectric guard
photoelektrische Lichtschranken fpl **und Scanner** m <Verpack> photoelectric light barriers and scanner
photoelektrische Schutzvorrichtung f <Sicherheit> photoelectric safeguard, photoelectric safety device
photoelektrischer Detektor m <Trans> photoelectric detector
photoelektrischer Effekt m <Optik, Phys, Strahlphys, Telekom> photoelectric effect
photoelektrischer Strom m <Elektriz> photoelectric current
photoelektrischer Tonabnehmer m <Aufnahme> photoelectric pick-up
photoelektrischer Transducer m <Elektrotech> photoelectric transducer
photoelektrisches Relais n <Elektriz, Elektrotech> photoelectric relay
Photoelektron n <Elektronik> photoelectron
Photoelektronen-Spektroskopie f 1. <Phys> photoelectron spectroscopy; 2. <Phys> electron spectroscopy for chemical analysis (ESCA)
Photoelektronen-Vervielfacher m <Elektronik> photomultiplier
Photoelektronische Röhre f <Elektronik> photosensitive tube
Photoelement n <Elektrotech, Nichtfoss Energ> photovoltaic cell
Photoemission f <Fernseh> photoemission
photoemissiv adj <Foto> photoemissive
photoempfindlich adj <Elektronik> light-sensitive
photogalvanische Zelle f <Nichtfoss Energ> photogalvanic cell
Photogenerator m <Elektrotech> photogenerator
Photogrammetrie f <Bau, Raumfahrt> photogrammetry (Raumschiff)
Photogravierverfahren n <Elektronik> photoengraving
Photohalogenid n <Foto> photohalide
Photoinitiator m <Kunststoff> photoinitiator
Photoionisation f <Phys> photoionization
Photokatode f <Elektrotech, Fernseh, Phys> photocathode
Photolack m <Elektronik> photoresist, resist
Photolacküberzug m <Elektronik> photoresist coating
photoleitende Schicht f <Optik> photoconducting layer
photoleitende Trommel f <Optik> photoconducting drum
photoleitfähig adj <Foto> photoconductive
Photoleitfähigkeit f <Elektronik, Optik, Phys, Telekom> photoconductivity
Photolithographie f <Druck, Elektronik> photolithography
Photolumineszenz f <Strahlphys> photoluminescence
Photolyse f <Nichtfoss Energ> photolysis
Photomaske f <Elektronik> photomask
photomechanisch adj <Druck> photomechanical
Photometer n <Foto, Kunststoff, Phys> photometer
Photometrie f <Phys> photometry
Photomultiplier m <Strahlphys> photomultiplier
Photon n <Optik, Phys, Strahlphys, Teilphys> photon
Photonenrauschen n <Optik, Telekom> photon noise
Photonenvervielfachung f <Strahlphys> photon amplification
photonuklearer Effekt m <Kerntech> direct photonuclear effect
Photoplotter m <Comp & DV> photoplotter
Photopolymerisation f <Chemie> photopolymerization
Photoproduktion f <Teilphys> photoproduction
Photoresist n <Elektronik> photoresist

Photoröhre f <Elektronik> phototube
Photosatz m <Druck> photocomposition, phototypesetting
Photosatzanlage f <Druck> filmsetter, phototypesetter
Photosatzmaschine f <Druck> filmsetter, phototypesetter
Photoseite f <Druck> photopage
photosensibles Glas n <Ker & Glas> photosensitive glass
Photosensor m <Elektriz, Elektronik> photosensor
Photosetzer m <Druck> photocomposer
Photospaltung f <Phys> photodisintegration (Kernphysik)
Photosphäre f <Raumfahrt> photosphere
photosphärische Absorption f <Strahlphys> photospheric absorption
Photostrom m 1. <Elektriz> photoelectric current; 2. <Optik, Telekom> photocurrent
Photosynthese f <Nichtfoss Energ> photosynthesis
Photothyristor m <Elektronik, Telekom> light-activated controlled silicon rectifier
Phototransistor m <Comp & DV, Elektronik, Strahlphys> phototransistor
Photovaristor m <Elektrotech> photovaristor
Photovernetzung f <Chemie> photopolymerization
Photovervielfacher m <Elektronik, Phys, Strahlphys> photomultiplier
Photovoltaik f <Elektrotech> photovoltaics
photovoltaische Solarstromanlage f <Elektrotech> photovoltaic solar power plant
photovoltaische Zelle f <Elektriz> solar cell (PV-Zelle)
photovoltaischer Effekt m <Elektriz> photovoltaic effect
Photovoltgenerator m <Elektrotech> photovoltaic generator
Photovoltstrom m <Elektrotech> photovoltaic current
Photowiderstand m 1. <Elektronik> photoconductive cell; 2. <Strahlphys> photoresistor
Photozelle f 1. <Comp & DV> photocell, phototransistor; 2. <Druck> photocell, photoelectric cell (PEC); 3. <Elektriz, Elektronik, Elektrotech, Fernseh> photocell; photoelectric cell (PEC); photoelectric tube, phototube, photovoltaic cell; 4. <Foto> photocell, photoelectric cell; 5. <Phys> photocell, photoelectric cell (PEC); 6. <Raumfahrt> photovoltaic cell (Raumschiff); 7. <Strahlphys> photocell, photoelectric cell (PEC); 8. <Telekom> photoelectric cell
Photozellen-Registersteuerung f <Verpack> photoelectric register control
Photozellenrelais n <Elektrotech> phototube relay
Photozellenverstärker m <Elektronik> electron multiplier phototube
Photozellenzerhacker m <Elektronik> electronic chopper
pH-Regelung f <Kohlen> pH control
pH-Regler m <Kohlen> pH controller
Phthalat n <Chemie> phthalate
Phthaldiamid n <Chemie> phthalamide
Phthalein n <Chemie> phthalein
Phthalid n <Chemie> phthalide
Phthalin n <Chemie> phthaline
Phthalocyanin n <Kunststoff> phthalocyanine
phthalsauer adj <Chemie> phthalic
Phthalsäure... <Chemie> phthalic
Phthalsäureanhydrid n <Kunststoff> phthalic anhydride
Phthalsäurediamid n <Chemie> phthalamide
Phugoidbewegung f <Lufttrans> phugoid oscillation
Phugoidschwingung f <Lufttrans> phugoid oscillation
pH-Wert m <Umweltschmutz> pH-value
physikalisch geladenes Teilchen n <Umweltschmutz> charged particle
physikalisch trocknend adj <Fertig> air-drying (Lack)
physikalische Chemie f <Chemie, Phys> physical chemistry, physicochemistry

physikalische Eigenschaften *fpl* <Phys> physical properties
physikalische Optik *f* <Optik, Phys, Telekom> physical optics
physikalische Schicht *f* <Telekom> physical layer *(OSI)*
physikalische Schnittstelle *f* <Telekom> line interface, physical interface *(OSI-Ebene 1)*
physikalische Übertragung *f* <Comp & DV> physical transmission
physikalische Verbindung *f* <Telekom> physical connection
physikalische Verbindungsschicht *f* <Telekom> OSI-layer No. 1, physical connection layer
physikalische Waage *f* <Labor> physical balance
physikalischer Speicher *m* <Elektrotech> physical memory
physikalisches Pendel *n* <Phys> compound pendulum
physiologische Lautstärkeregelung *f* <Aufnahme> loudness control
physiologisches Rauschen *n* <Akustik> physiological noise
physisch *adj* <Comp & DV> physical
physische Datei *f* <Comp & DV> physical file
physische Datenbank *f* <Comp & DV> physical database
physische Steuereinheit *f* <Comp & DV> physical control unit
physischer Satz *m* <Comp & DV> physical record
Physostigmin *n* <Chemie> eserine, physostigmine
Phytase *f* <Lebensmittel> phytase
Phytin *n* <Lebensmittel> phytin
Phytinsäure *f* <Lebensmittel> phytic acid
phytotoxisch *adj* <Chemie> phytotoxic
Pi *n* <Math> pi
PI *(Polyimid)* <Elektriz, Elektronik, Kunststoff> PI *(polyimide)*
Piazin *n* <Chemie> piazine, pyrazine
PIB *(Polyisobutylen)* <Kunststoff> PIB *(polyisobutylene)*
Picaschrift *f* <Druck> pica
Pick-Up *m* <Kfztech> pick-up, pick-up truck
Pick-Up Bahnabnahme *f* <Papier> pick-up
Pick-Up Filz *m* <Papier> pick-up felt
Picosekunde *f* <Comp & DV> picosecond
PID-Regler *m* *(Proportional-Integral-Differenzial-Regler)* <Labor> PID controller *(proportional plus integral plus derivative controller)*
PID-Steuerung *f* 1. <Elektriz> PID-control *(Proportional-Integral-Differenzial-Regelung)*; 2. <Labor> *(Proportional-Integral-Differenzial-Regler)* PID controller *(proportional plus integral plus derivative controller)*
PID-Verhalten *n* *(Proportional-Integral-Differenzial-Verhalten)* <Regelung> PID action, PID-performance *(proportional plus integral plus derivative action)*
Piek *f* <Wassertrans> peak *(Segeln)*
Piepschalter *m* <Lufttrans> beep switch *(Hubschrauber)*
Pier *m* <Wassertrans> quay; pier *(Hafen)*
Pierce-Oszillator *m* <Elektronik> Pierce oscillator
Piezo... <Metrol> piezo...
Piezoeffekt *m* <Elektrotech> piezoelectric effect
piezoelektrisch *adj* <Elektrotech, Elektronik, Fertig, Phys> piezoelectric
piezoelektrisch abgestimmtes Magnetron *n* <Elektrotech> piezoelectric-tuned magnetron
piezoelektrische Eigenschaften *fpl* <Elektrotech> piezoelectric properties
piezoelektrischer Beschleunigungsaufnehmer *m* <Heiz & Kälte> piezoelectric acceleration sensor
piezoelektrischer Detektor *m* <Trans> piezoelectric detector

piezoelektrischer Effekt *m* <Phys> piezoelectric effect
piezoelektrischer Kristall *m* <Elektrotech> piezoelectric crystal
piezoelektrischer Kristalltonabnehmer *m* <Aufnahme> piezoelectric crystal pick-up
piezoelektrischer Lautsprecher *m* <Aufnahme> piezoelectric loudspeaker
piezoelektrischer Messfühler *m* <Gerät> piezoelectric sensing element
piezoelektrischer Oszillator *m* <Elektriz, Elektrotech> piezoelectric oscillator
piezoelektrischer Resonator *m* <Elektrotech> piezoelectric resonator
piezoelektrischer Stift *m* <Optik> piezoelectric stylus
piezoelektrischer Transducer *m* <Elektrotech> piezoelectric transducer
piezoelektrischer Wandler *m* <Elektrotech> piezoelectric transducer
piezoelektrisches Element *n* <Elektrotech> piezoelectric element
piezoelektrisches Mikrofon *n* <Akustik, Aufnahme> piezoelectric microphone
piezoelektrisches Substrat *n* <Elektrotech> piezoelectric substrate
Piezoelektrizität *f* <Elektrotech> piezoelectricity
Piezometer *n* <Kohlen> piezometer
piezometrische Höhe *f* <Erdöl> piezometric head
piezometrische Karte *f* <Erdöl> piezometric map
Pigment *n* <Anstrich, Kunststoff, Textil> pigment
Pigmentfarbstoff *m* <Textil> pigment
pigmentieren *v* <Textil> pigment
pigmentierte Anilinfarbe *f* <Druck> pigmented aniline ink
Pigmentierung *f* <Textil> pigmentation
Pigmentschlamm *m* <Abfall> pigment sludge
pikieren *v* <Textil> pad
Pikkolo-Brenner *m* <Ker & Glas> Piccolo burner
Piko... *(p)* <Metrol> pico... *(p)*
Pikoamperemeter *n* <Metrol> picomicroammeter
Pikolin *n* <Chemie> methylpyridine, pikoline
Pikrat *n* <Chemie> picrate
Pikrin... <Chemie> picric
Pikrotin *n* <Chemie> picrotin
Pikryl... <Chemie> picryl
Pikrylmethyl *n* <Chemie> nitramine
Piktogramm *n* 1. <Comp & DV> icon; symbol *(bei grafischen Oberflächen)*; 2. <Druck> pictograph; 3. <Math> pictogram; 4. <Sicherheit> pictorial symbol *(Gefahrstoffe)*
Pilfer-Proof-Dichtung *f* <Verpack> pilfer-proof seal
Pillbildung *f* <Textil> pilling
pillen *v* <Textil> pill
Pilocarpidin *n* <Chemie> pilocarpidine
Pilocarpin *n* <Chemie> pilocarpine
Pilot *m* <Lufttrans, Telekom> pilot
Pilotanlage *f* <Fertig, Kohlen> pilot plant
Pilotballon *m* <Lufttrans> pilot balloon
Pilotdraht *m* <Elektrotech> pilot wire *(Messtechnik)*
Pilotenhandbuch *n* <Lufttrans> flight manual
Pilotfrequenz *f* <Telekom> pilot frequency
Pilotprojekt *n* <Fertig, Qual> pilot project
Pilotsignal *n* <Fernseh, Telekom> pilot signal
Pilotton *m* <Fernseh, Funktech> pilot tone
Pilotträger *m* <Telekom> pilot carrier
Pilzanker *m* <Wassertrans> mushroom anchor *(Festmachen)*
Pilzdach *n* <Bau> umbrella roof
Pilzdeckenplatte *f* <Bau> mushroom slab
pilzförmiger Kopf *m* <Fertig> mushroom head
pilzförmiger Stopfen *m* <Ker & Glas> mushroom stopper

Pilzisolator *m* <Elektriz, Elektrotech> mushroom insulator, umbrella isolator
Pilzkopflüfter *m* <Wassertrans> mushroom ventilator *(Deckausrüstung)*
Pilzstößel *m* <Maschinen> mushroom follower
Pilzventil *n* <Kfztech> mushroom valve
Pimar... <Chemie> pimaric
Pimelin... <Chemie> pimelic
Pi-Meson *n* <Teilphys> pi meson
PIN *(Positiv-Isolierend-Negativ)* <Elektronik> PIN *(positive-isolating-negative)*
Pinakol *n* <Chemie> pinacol, pinacone, tetramethylethyleneglycol
Pinakon *n* <Chemie> pinacol, pinacone, tetramethylethyleneglycol
Pinch-Effekt *m* <Kerntech, Phys> pinch effect
PIN-Dämpfungsdiode *f* <Elektronik> PIN attenuator diode
PIN-Diode *f* <Elektronik, Phys> PIN diode
PIN-Diodenabschwächer *m* <Elektronik> PIN diode attenuator
PIN-Diodenmodulation *f* <Elektronik> PIN diode modulation
Pi-Netz *n* <Elektrotech> pi network
Pi-Netzwerk *n* <Funktech, Phys> pi network
Ping-Pong-Verfahren *n* <Telekom> burst mode, ping pong method *(Zeitgetrenntlage-Verfahren)*
Pinkingeffekt *m* <Kunststoff> pinking effect
pinkompatibel *adj* <Telekom> pin-compatible
Pinksalz *n* <Chemie> pink salt
Pinne *f* 1. <Fertig> pane, peen; 2. <Ker & Glas> pip; 3. <Maschinen> pane, peen
Pinning *n* <Metall> pinning
Pinole *f* 1. <Fertig> quill, ram; 2. <Kunststoff> mandrel, mandril; 3. <Maschinen> quill, spindle sleeve
Pinonen *n* <Chemie> pinonene
PIN-Photodiode *f* <Elektronik, Optik> PIN photodiode
Pinsel *m* 1. <Anstrich> paint brush; 2. <Fertig> swab *(Formen)*
Pinzette *f* <Ker & Glas, Labor> tweezers
Pion *n* 1. <Phys> pion *(Teilchen)*; 2. <Teilphys> pion
Pipecolin *n* <Chemie> methylpiperidine, pipecoline
Pipeline *f* <Comp & DV, Erdöl, Maschinen, Trans, Wassertrans> pipeline
Pipeline-Betrieb *m* <Telekom> pipelining
Pipelinestruktur *f* <Kontroll> pipelined architecture
Pipeline-System *n* <Telekom> pipeline system • **im Pipeline-System verarbeiten** <Comp & DV> pipeline
Pipelinetransport *m* <Trans> pipeline transportation
Pipelineverarbeitung *f* <Comp & DV> pipelining
Pipeliningmethode *f* <Comp & DV> pipelining
Piperazin *n* <Chemie> diethylenediamine, hexahydropyrazine, piperazine
Piperidin *n* <Chemie> piperidine
Piperin... <Chemie> piperic
Piperonal *n* <Chemie> heliotropin, methylenedioxybenzaldehyde, piperonal
Piperonylaldehyd *m* <Chemie> heliotropin, methylenedioxybenzaldehyde, piperonal
Piperylen *n* <Chemie> piperylene
Pipette *f* <Labor> pipette
Pipettenständer *m* <Labor> pipette stand
Pipettierkolben *m* <Labor> pipetting bulb
Piping *n* <Erdöl> piping *(Leitungen)*
Pirani-Manometer *n* <Phys> *(AE)* Pirani gage, *(BE)* Pirani gauge
Pirani-Vakuummeter *n* <Labor> *(AE)* Pirani vacuum gage, *(BE)* Pirani vacuum gauge
Pirani-Wärmeleitungsvakuummeter *n* <Labor> *(AE)* Pirani vacuum gage, *(BE)* Pirani vacuum gauge

PI-Regler *m* *(Proportional-Integral-Regler)* <Regelung> PI controller *(proportional plus integral controller)*
Piste *f* <Lufttrans> runway
Pistenrichtung *f* <Lufttrans> runway alignment
Pistenrichtungsanzeiger *m* <Lufttrans> runway alignment indicator
Pistonierkolben *m* <Erdöl> swab
Pistonierung *f* <Erdöl> swabbing
Pistophon *n* <Akustik> pistonphone
Pitotdruck *m* <Lufttrans> impact pressure
Pitot-Rohr *n* <Phys> Pitot tube
Pivalin... <Chemie> pivalic
PI-Verhalten *n* *(Proportional-Integral-Verhalten)* <Regelung> PI action *(proportional plus integral action)*
Pi-Wicklung *f* <Elektrotech> pi winding
Pixel *m* 1. <Comp & DV> pixel; 2. <Fernseh> pixel, picture element
Pixel *mpl* **pro Zoll** <Comp & DV, Fernseh> pixels per inch
Pixmap *f* <Comp & DV> pixmap
P-Kanal *m* <Elektronik> P-channel
P-Kanal-Feldeffekttransistor *m* <Elektronik> P-channel transistor, FET
P-Kanal-FET *m* <Elektronik> P-channel FET
P-Kanal-Gerät *n* <Elektronik> P-channel device
P-Kanal-Metalloxid-Silizium *n* <Elektronik> P-channel metal-oxide silicon
P-Kanal-MOS *n* <Comp & DV> PMOS, positive metal oxide semiconductor
P-Kanal-MOS-Anreicherungstransistor *m* <Elektronik> P-channel enhancement mode MOS transistor
P-Kanal-MOS-Verarmungstransistor *m* <Elektronik> P-channel depletion mode MOS transistor
p-Kollektor *m* <Elektronik> p-type collector
p-Kresol *n* <Chemie> p-cresol
PKW *(Personenkraftwagen)* <Kfztech> car, passenger car
PKW-E *(Personenkraftwageneinheit)* <Kfztech> PCU *(passenger car unit)*
PL *(Prädikatenlogik)* <Künstl Int> PL *(predicate logic)*
PLA *(programmierbare Logikanordnung)* <Comp & DV> PLA *(programmable logic array)*
Plakatschrift *f* <Druck> lettering style
Plakettendosimeter *n* <Sicherheit> badge dosimeter
Plan *m* 1. <Bau> chart, drawing; 2. <Fertig> image *(Geschwindigkeit)*; 3. <Mechan> layout; 4. <Phys> chart; 5. <Wassertrans> plan
plan *adj* <Geom> planar
Planardiode *f* <Elektronik> planar diode
planare Diffusion *f* <Elektronik> planar diffusion
planare integrierte Schaltung *f* <Elektronik> planar integrated circuit
planare Symmetrie *f* <Bau> planar symmetry
Planar-Epitaxialdiode *f* <Elektronik> planar epitaxial diode
planarer Prozess *m* <Elektronik> planar process
Planarhohlleiter *m* <Elektrotech> planar waveguide
Planarleitung *f* <Elektronik> planar line
Planarprozess *m* <Elektronik> planar process *(Transistoren)*
Planartechnik *f* <Elektronik> planar technology *(Halbleiter)*
Planartransistor *m* <Elektronik> planar transistor
Planartriode *f* <Elektronik> planar triode
Planarwellenleiter *m* <Elektrotech> planar waveguide
planbearbeiten *v* <Maschinen> face
Planbearbeitung *f* <Maschinen> facing
Planck'sche Konstante *f (h)* <Phys, Strahlphys, Teilphys> Planck's constant *(h)*
Planck'sche Quantenhypothese *f* <Phys> Planck's law
Planck'sche Strahlungsformel *f* <Phys> Planck's radiation formula

Planck'sches Strahlungsgesetz n <Phys, Strahlphys, Thermod> Planck's radiation law
Planck'sches Wirkungsquantum n (h) <Phys, Strahlphys, Teilphys> Planck's constant (h)
plandrehen v 1. <Fertig> face; 2. <Maschinen> surface
Plandrehen n <Maschinen> face turning, facing, surfacing
Plandrehfutter n <Maschinen> facing head
Plandrehmaschine f 1. <Fertig> facing lathe, surfacing lathe; 2. <Maschinen> surface lathe
Plandrehvorrichtung f <Maschinen> facing attachment
Plandrehwerkzeug n <Maschinen> facing tool
planen v 1. <Comp & DV, Eisenbahn, Fernseh> schedule; 2. <Fertig> face; 3. <Qual, Telekom> plan, schedule
Planen n <Künstl Int> planning
Planencontainer m <Trans> tilt container, tiltainer
Planer m <Bau> designer
Planeten... <Mechan> epicyclic, planetary
Planetenbewegung f <Fertig> planetary movement
Planetengetriebe n 1. <Kfztech> planet gear; 2. <Maschinen> epicyclic gear, epicycloidal gear, planetary gear train, sun-and-planet gearing; 3. <Mechan> epicyclic gear train; 4. <Wassertrans> epicyclic gear (Motor)
Planetengetriebedifferenzial n <Kfztech> planetary gear differential
Planetengetriebesatz m <Kfztech> planetary gear set
Planetengetriebesystem n <Kfztech> planetary gear system
Planeteninneres n <Raumfahrt> planetary interior
Planetenmühle f <Labor> planetary mill (Schleifen)
Planetenrad n 1. <Fertig> planetary gear; 2. <Kfztech> pinion gear (Differenzial); 3. <Maschinen> planet gear, planet wheel, planetary gear, planetary pinion
Planetenrädersatz m <Maschinen> epicyclic gear train, epicyclic train
Planetenradgetriebe n <Fertig> planetary gearing
Planetenradträger m <Maschinen> planet carrier
Planetenritzel n <Kfztech> planetary pinion
Planetensonde f <Raumfahrt> planetary probe
Planetenspindel f <Maschinen> planet spindle, planet-action spindle
Planetenträger m <Kfztech> planet carrier
Planetenzahnradgehäuse n <Lufttrans> planet pinion cage (Hubschrauber)
Planetoid m <Raumfahrt> planetoid
Planfilm m <Foto> cut film
Planfräsen n <Maschinen> plain milling
Planfräsmaschine f <Maschinen> planomilling machine
Planglasplatte f <Metrol> optical flat
Planierarbeiten fpl <Bau> grading
planieren v 1. <Bau> grade; skim (Boden); trim (Straße); 2. <Maschinen> flatten
Planieren n 1. <Bau> grading, planing; 2. <Maschinen> planishing
Planierer m <Metall> planisher
Planiergerät n <Bau, Meerschmutz> grader
Planierhammer m <Maschinen> dresser
Planiermaschine f 1. <Bau> (AE) leveling machine, (BE) levelling machine; 2. <Trans> grader
Planierraupe f <Bau, Trans> bulldozer
Planierraupe f mit hebbarem Schild <Kfztech> (BE) tiltdozer
Planierschaufel f <Trans> grader levelling blade
Planierstange f <Bau> (AE) leveling rod, (BE) levelling rod
planiert adj <Bau> (AE) leveled, (BE) levelled
Planierwerkzeug n <Maschinen> planishing tool
Planimeter n <Bau, Geom> planimeter
Planimetrie f <Geom> plane geometry, planimetry
Planke f <Bau, Wassertrans> plank (Schiffbau)

plankonkav adj <Optik> plano-concave
plankonkave Linse f <Phys> plano-concave lens
plankonvex adj <Optik> convexo-plane, plano-convex
plankonvexe Linse f <Phys> plano-convex lens
Plankurvenfräsmaschine f <Fertig> face cam milling machine
planmäßige Betriebszeit f <Telekom> scheduled operating time
planmäßige Unterbrechung f <Telekom> foreseen interruption
planmäßige Wartung f <Comp & DV, Kerntech, Qual, Telekom> scheduled maintenance
planmäßiger Flug m <Lufttrans> scheduled flight
planmäßiges Meldesignal n <Telekom> scheduled reporting signal
Planoformat n <Druck> broadside page
planometrische Projektion f <Konstzeich> planometric projection
planparallele Diode f <Elektronik> planar diode
Planparallelfräsen n <Maschinen> straddle milling, straddling
Planrad n <Maschinen> crown gear
Planscheibe f 1. <Fertig> independent four-jaw chuck; 2. <Maschinen> face chuck, face plate
Planscheibenbefestigung f <Maschinen> face plate mounting
Planschleifen n 1. <Fertig> surface grinding; 2. <Maschinen> face grinding, surface grinding
Planschleifer m <Maschinen> face grinder
Planschleifmaschine f <Maschinen> surface grinder, surface-grinding machine
Planschlitten m <Maschinen> cross slide, slide head, slide rest
Planschnitt m <Fertig> facing cut
Planschverlust m <Fertig> churn loss (Getriebelehre)
Plansenken n <Maschinen> end facing, spot facing
Planung f <Künstl Int, Telekom> planning
Planungshöhe/unter <Bau> below grade
Planungsstrategie f <Künstl Int> planning strategy
Planverzahnung f <Maschinen> crown gearing
Planvorschub m <Maschinen> cross feed
Planzug m <Fertig> cross traverse
Plasma n <Elektronik, Phys, Raumfahrt, Teilphys> plasma
plasmaaktivierte chemische Bedampfung f <Optik> (AE) plasma-activated chemical vapor deposition, (BE) plasma-activated chemical vapour deposition
plasmaaktivierte chemische Dampfabscheidung f <Optik> (AE) plasma-activated chemical vapor deposition, (BE) plasma-activated chemical vapour deposition
plasmaaktiviertes Chemical-Vapour-Deposition-Verfahren n <Elektronik> (AE) plasma-activated chemical vapor deposition process, (BE) plasma-activated chemical vapour deposition process
plasmaaktiviertes CVD-Verfahren n <Elektronik> plasma-activated CVD process
Plasmaanzeige f <Elektronik> plasma display (Elektrolumineszenz)
Plasmaätzen n <Elektronik> plasma etching
Plasmabehandlung f <Abfall> thermoplastic solidification (Sonderabfall)
Plasmabildschirm m <Comp & DV> plasma display
Plasmachemie f <Chemie> plasma chemistry
plasmaentwickelter Photolack m <Elektronik> plasma-developed resist
Plasmalichtbogenschneiden n <Bau, Maschinen> plasma arc cutting
Plasmalichtbogenstromkollektor m <Trans> plasma arc power collector

Plasmaschneideanlage

Plasmaschneideanlage f <Elektriz> plasma cutting machine
Plasmaschneiden n <Bau> plasma cutting
Plasmaschubtriebwerk n <Raumfahrt> plasma thruster *(Raumschiff)*
Plasmaschweißanlage f <Elektriz> plasma welder
Plasmaschweißbrenner m <Bau> torch for plasma welding
Plasmaspritzen n <Metall> plasma spraying
Plasmatrennen n <Bau> plasma cutting
Plasmatriebwerk n <Raumfahrt> plasma engine *(Raumschiff)*
Plasmaumgebung f <Raumfahrt> plasma environment
Plastic-Clad-Silica-Faser f *(PCS-Faser)* <Optik, Telekom> *(AE)* plastic-clad silica fiber, *(BE)* plastic-clad silica fibre *(PCS fiber)*
Plastifikator m <Fertig> plasticizer
plastifizieren v <Fertig, Kunststoff> plasticize
Plastifizieren n <Bau> fluxing
Plastifizierung f <Fertig> plasticization
Plastikbeschichtung f <Telekom> plastic coating
Plastikfaser f 1. <Elektrotech, Optik> *(AE)* plastic fiber, *(BE)* plastic fibre; 2. <Telekom> *(AE)* all-plastic fiber, *(BE)* all-plastic fibre
Plastiksprengstoff m <Sicherheit> plastic explosive
plastikummantelte Silikafaser f <Optik> *(AE)* plastic-clad silica fiber, *(BE)* plastic-clad silica fibre
Plastikummantelung f <Telekom> plastic coating
plastisch verformen v <Fertig> fail, overstrain
plastische Deformation f <Kunststoff> plastic deformation
plastische Eigenschaften fpl <Strömphys> plastic properties
plastische Fließeigenschaften fpl <Kunststoff> plastic flow properties
plastische Instabililität f <Metall> plastic instability
plastische Verformung f 1. <Bau> plastic deformation; 2. <Kerntech> plastic yield; 3. <Kunststoff, Metall, Phys> plastic deformation
plastisches Abstumpfen n <Metall> plastic blunting
plastisches Fließen n <Kunststoff, Metall> plastic flow
plastisches Nachgeben n <Kerntech> plastic yield
plastisches Schutzelement n <Sicherheit> plastic protective element
Plastisol n <Kunststoff> plastisol
Plastizität f <Kohlen, Kunststoff, Metall> plasticity
Plastizitätsgrenze f <Bau, Kohlen, Qual> plastic limit
Plastizitätsindex m <Bau, Kohlen, Qual> plasticity index
Plastomer n <Chemie, Erdöl> plastomer *(Petrochemie)*
Plastomeres n <Chemie> plastomer
Plastometer n <Kunststoff> plastimeter
Plateau n <Erdöl> plateau *(Geologie)*
Plateauhöhe f <Erdöl> plateau level *(Geologie)*
Plateauniveau n <Erdöl> plateau level *(Geologie)*
platieren v <Fertig> bond
Platierung f <Fertig> bonding
Platin n *(Pt)* 1. <Chemie> platinum *(Pt)*; 2. <Elektriz, Metall> platinum
Platin... <Chemie> platinic
Platina n <Chemie> platina
Platine f 1. <Comp & DV> board; 2. <Fernseh, Funktech, Telekom> card; 3. <Fertig> blank; 4. <Textil> sinker
Platinenbarre f <Textil> sinker bar
Platinieren n <Metall> platinization
Platinotron n <Phys> platinotron
Platintiegel m <Labor> platinum crucible
Platinwiderstandsthermometer n <Phys> platinum resistance thermometer
platonischer Körper m <Geom> Platonic object, Platonic solid

Plättchen n 1. <Chemie> lamina; 2. <Comp & DV, Elektronik, Elektrotech> wafer, chip; 3. <Ker & Glas> split
Plättchenschneidemaschine f <Fertig> dicing machine
Plättchenverarbeitung f <Elektronik> wafer processing
Plättchenverzerrung f <Elektronik> wafer distortion
Platte f 1. <Aufnahme> *(BE)* disc, *(AE)* disk; 2. <Bau> slab, table; 3. <Comp & DV> platform; *(AE)* disk *(Daten)*; 4. <Druck> plate, printing plate *(Drucken)*; plate *(Fotografie)*; 5. <Elektriz, Elektrotech> plate *(Elektroplattierung, Galvanisierung)*; 6. <Fertig> *(BE)* disc, *(AE)* disk *(Kunststoffinstallationen)*; 7. <Ker & Glas> slab; 8. <Kfztech> plate; 9. <Kohlen> panel, slab; 10. <Kunststoff> platen, sheet; 11. <Maschinen> plate, platen; 12. <Mechan> plate, sheet • **mit Platten überziehen** <Anstrich> plate
Platte f **mit achtzig Spuren** <Comp & DV> eighty-track disk
Plätteisen n <Ker & Glas> battledore
Plattenadresse f <Comp & DV> disk address
Plattenamalgamation f <Kohlen> plate amalgamation
Plattenanschluss m <Comp & DV> disk adaptor
Plattenapplikator m <Ker & Glas> apron applicator
Plattenarchiv n <Aufnahme> record library
Plattenausfall m <Comp & DV> disk failure
Plattenbalken m <Bau> T-beam *(Massivbau)*
Plattenbandförderer m 1. <Fertig> apron conveyor; 2. <Maschinen> apron conveyor, plate conveyor
Plattenbearbeitungshalle f <Wassertrans> platers' shop *(Schiffbau)*
Plattenbelag m <Bau> flagging
Plattenbereich m <Comp & DV> disk space
Plattenbeschichtungsmaschine f <Druck> plate-coating machine
Plattenbeschickung und -entnahme f <Druck> plate change
Plattendatei f <Comp & DV> disk file
Platteneinheit f <Comp & DV> disk device, disk unit
Plattenförderband n <Trans> apron conveyor
Plattenform f <Ker & Glas> flag build
Plattenformat <Druck> plate format, plate size
Plattengang n <Wassertrans> strake *(Schiffbau)*
Plattengefrieranlage f <Heiz & Kälte> plate freezer
Plattengleiskörper m <Eisenbahn> slab track
Plattengröße f <Druck> plate size
Plattengussglas n <Ker & Glas> thick rough cast plate glass
Plattenheizkörper m <Heiz & Kälte> panel-type radiator
Platteninterferometrie f <Optik> slab interferometry
Plattenkamera f <Foto> plate camera
Plattenkapazität f <Comp & DV> disk space
Plattenkassette f 1. <Comp & DV> disk cartridge; 2. <Foto> dark slide, plate holder
Plattenkondensator m <Elektriz, Elektrotech> *(BE)* disc capacitor, *(AE)* disk capacitor, plate capacitor
Plattenkopiereffekt m <Aufnahme> record crosstalk
Plattenlaufwerk n <Comp & DV> disk drive
Plattenpaar n <Elektrotech> couple of plates *(Batterie)*
Plattenplatz m <Comp & DV> disk space
Plattenpresse f <Kunststoff> platen press
Plattenrecorder m <Elektriz> *(BE)* disc recorder, *(AE)* disk recorder
plattenresident adj <Comp & DV> disk-resident
Plattensektor m <Comp & DV> disk sector
Plattenspeicher m 1. <Comp & DV> disk storage; 2. <Telekom> disk store
Plattenspieler m 1. <Aufnahme> record player; 2. <Optik> disk player *(Schallplatten, CD)*
Plattenspielwerk n <Kontroll> disk drive
Plattenspur f <Comp & DV> disk track

Plattenstapel m 1. <Comp & DV> disk pack; 2. <Kerntech> slab pile *(Brennelemente)*
Plattensteuereinheit f <Comp & DV> disk controller
Plattenstruktur f <Metall> plate structure
Plattentektonik f <Phys> plate tectonics *(Kontinentalverschiebung)*
Plattenteller m <Akustik, Aufnahme> turntable
Plattenträger m <Bau> plate girder
Plattentrockner m <Papier> slat dryer
Plattenumdrehungsverzögerung f <Comp & DV> rotational delay
Plattenverbinder m <Fertig> plate fastener *(Riemen)*
Plattenverlegen n <Bau> flagging
Plattenvorsteven m <Wassertrans> plate stem *(Schiffbau)*
Plattenwärmeaustauscher m <Heiz & Kälte> plate heat exchanger
Plattenwechsel m <Druck> plate change
Plattenwechsler m <Akustik, Aufnahme> record changer
Plattenwiedergabekopf m <Aufnahme> record playback head
Plattenzugriff m <Comp & DV> disk access
Plattenzugriffsarm m <Comp & DV> access arm
platter Reifen m <Kfztech> flat tyre
Plattform f 1. <Bau> stage; 2. <Comp & DV, Erdöl> platform *(Offshore)*; 3. <Fertig> entablature, stillage; 4. <Maschinen> platform
Plattformausrüstung f <Erdöl> platform equipment
plattformübergreifend adj <Comp & DV> cross-platform, multiplatform
plattformübergreifende Interoperabilität f <Comp & DV> interoperable across platforms
Plattformwagen m <Eisenbahn> (AE) flatcar; four-wheeled truck *(Förderung)*
Plattgatt n <Wassertrans> flat stern, square transom stern *(Schiffbau)*
Platthammer m <Fertig> flatter *(Schmieden)*
Plattheck n <Wassertrans> flat stern *(Schiffbau)*
plattieren v <Fertig> clad
Plattieren n 1. <Elektrotech> plating; 2. <Fertig> cast coating; 3. <Maschinen> cladding
Plattierschicht f <Fertig> clad plate
Plattierung f <Elektrotech, Maschinen> plating
Platting f <Wassertrans> sennet *(Knoten)*
Platz m <Telekom> operator position; site *(Internet)*
Platz m **für konzentrierte Abfrage** <Telekom> switched loop console
Platz m **im Internet** <Telekom> website
Platzbedarf m <Fernseh> footprint
Platzcode m <Verpack> site code
platzen v 1. <Bau> burst; 2. <Maschinen> burst
Platzhalter m 1. <Comp & DV> wildcard; 2. <Druck> holding lines, position marker, position-only indication *(für Abbildungen)*; 3. <Fertig> dummy part
Platzhalterzeichen n <Comp & DV> token, wildcard character
Platzlampe f <Elektrotech> pilot lamp
Platzreservierungssystem n <Trans> passenger reservation system
Platzrundenanflug m <Lufttrans> circling approach
Platzrundenführungsbefeuerung f <Lufttrans> circling guidance light
Platzsparen n <Verpack> economy of space
platzsparend adj <Verpack> space-saving
plaudern v <Telekom> chat *(im Internet)*
Plausibilitätskontrolle f <Comp & DV, Telekom> validity check
Playback n <Fernseh> audio playback, playback
Playback-Charakteristika npl <Fernseh> playback characteristics

Playback-Kopf m <Fernseh> playback head
Playback-Verlust m <Fernseh> playback loss
Playback-Videorecorder m <Fernseh> playback VTR
p-leitende Basis f <Elektronik> p-type base
p-leitende Dotierung f <Elektronik> p-type impurity
p-leitende Epitaxialschicht f <Elektronik> p-type epitaxial layer
p-leitende implantierte Schicht f <Elektronik> p-type implanted layer
p-leitendes Silizium n <Elektronik> p-type silicon
p-leitendes Silizium-Trägermaterial n <Elektronik> p-type silicon substrate
p-Leitfähigkeit f <Elektrotech> p-type conductivity
plesiochron <Telekom> plesiochronous *(Multiplexer)*
plesiochrone Übertragung f <Telekom> plesiochronous transmission
plesiochrone Übertragungseinrichtung f <Telekom> plesiochronous transmission equipment
plesiochrones Leitungsendgerät n <Telekom> plesiochronous line terminal
plesiochrones Netz n <Telekom> plesiochronous network
Pleuel n 1. <Lufttrans> connecting rod; 2. <Maschinen, Mechan> connecting rod, rod
Pleuelaugenbuchse f <Kfztech> small end bush
Pleuelbuchse f <Kfztech> piston-pin bushing, small end bushing
Pleueldeckel m <Kfztech> connecting rod cap
Pleuelende n <Maschinen> end
Pleuelfuß m <Kfztech> connecting rod big end
Pleuelfußlager n <Kfztech> big end bearing
Pleuelkopf m <Kfztech> connecting rod small end, small end
Pleuellager n <Kfztech> big end bearing, connecting rod bearing
Pleuelschaft m <Kfztech> connecting rod shank
Pleuelstange f 1. <Eisenbahn> piston rod; 2. <Fertig> conrod; 3. <Kfztech> con rod, connecting rod, piston rod; 4. <Maschinen> connecting rod, rod; 5. <Wassertrans> piston rod *(Motor)*; connecting rod *(Schiffantrieb)*
Pleuelstangenfuß m <Kfztech> big end *(Pleuel)*
Plexiglas n <Verpack> acrylic plastic
Plexiglasverkleidung f <Lufttrans> plexiglass fairing
Plicht f <Wassertrans> cockpit *(Schiff)*
Plisseefalte f <Textil> pleat
plissieren v <Textil> pleat
Plissieren n <Textil> pleating
Plissiermaschine f <Textil> pleater, pleating machine
plissiert adj <Textil> pleated
PLL *(Phasenregelkreis)* <Elektronik, Fernseh, Funktech, Raumfahrt, Telekom> PLL *(phase-locked loop)*
PLL-Demodulator m <Elektronik, Telekom> phase-locked loop demodulator, P^2L demodulator
Plombe f 1. <Bau> sealing; 2. <Mechan> seal
Plombenzange f <Verpack> lead sealing pliers
plombieren v <Fertig> seal
Plot m <Comp & DV, Wassertrans> plot *(Navigation)*
plotten v <Wassertrans> plot the position *(Navigation)*
Plotten n <Comp & DV> plotting
Plotter m 1. <Comp & DV> graph plotter, plotter; 2. <Elektriz> plotter
Plott-Tisch m <Wassertrans> plotting table *(Navigation)*
plötzliche Querschnittskontraktion f <Hydraul> sudden contraction of cross section
plötzliche Querschnittsvergrößerung f <Hydraul> sudden enlargement of cross section
plötzlicher Ausfall m <Telekom> sudden failure
plötzlicher Übergang m <Elektronik> abrupt junction *(Halbleiter)*

plötzliches

plötzliches Schlingern n **nach der Leeseite** <Wassertrans> lee lurch
Plug-and-Play n <Telekom> plug-and-play
Plumbat n <Chemie> plumbate
Plumbikon n <Elektronik> plumbicon *(Bildaufnahmeröhre)*
Plumbit n <Chemie> plumbite
Plunger m <Ker & Glas> plunger
Plungerdorn m <Ker & Glas> plunger spike
Plungerfestsitz m <Ker & Glas> plunger sticking
Plungerhilfsmechanismus m <Ker & Glas> plunger assist mechanism
Plungerkolben m 1. <Kfztech> plunger *(Bremsen, Kupplung)*; 2. <Maschinen> plunger, plunger piston, ram
Plungerpumpe f <Wasserversorg> plunger pump
plus *adj* <Math> plus
Plusplatte f <Kfztech> positive plate
Pluspol m <Kfztech> plus terminal, positive terminal
Pluszeichen n <Math> plus sign
Pluviometer n <Bau, Labor, Nichtfoss Energ, Wasserversorg> *(AE)* rain gage, *(BE)* rain gauge
Pm *(Promethium)* <Chemie> Pm *(promethium)*
PM *(Phasenmodulation)* <Aufnahme, Comp & DV, Elektronik, Fernseh, Funktech, Phys, Telekom> PM *(phase modulation)*
p⁻-Mangelhalbleiter m <Elektronik> p⁻-semiconductor
PMC-Verfahren n *(Pulverlack-Beschichtungstechnik)* <Kunststoff> PMC *(powder mould coating)*
p-Methoxypropenylbenzol n <Chemie> anethole
PMMA *(Polymethacrylat, Polymethylmethacrylat)* <Kunststoff> PMMA *(polymethyl methacrylate)*
Pneumatik f <Kfztech, Fertig, Phys> pneumatics
pneumatisch *adj* 1. <Fertig> air-actuated, air-operated; 2. <Phys> pneumatic
pneumatisch betätigter Schalter m <Elektriz> pneumatically operated switch
pneumatische Kupplung f <Maschinen> air clutch
pneumatische Messvorrichtung f <Maschinen> *(AE)* air gage, *(BE)* air gauge
pneumatische Sortieranlage f <Abfall> pneumatic sorter
pneumatischer Auslöser m <Foto> pneumatic release
pneumatischer Bohrungsmessdorn m <Fertig, Metrol> *(AE)* air gage, *(BE)* air gauge
pneumatischer Detektor m <Trans> pneumatic detector
pneumatischer Drehzahlregler m <Kfztech> suction-type governor
pneumatischer Feinzeiger m <Fertig, Metrol> *(AE)* air gage, *(BE)* air gauge
pneumatischer Förderer m <Fertig, Maschinen> pneumatic conveyor
pneumatischer Lautsprecher m <Akustik> pneumatic loudspeaker
pneumatischer Plattenhalter m <Druck> vacuum plate holder
pneumatischer Röhrenförderer m <Trans> pneumatic pipe conveyor
pneumatisches Dichtemessgerät n <Gerät> air bubble density meter
pneumatisches Gerät n <Maschinen> pneumatic equipment
pneumatisches Sortieren n <Abfall> pneumatic classification
pneumatisches Ziehkissen n <Fertig> air cushion
pneumohydraulischer Flüssigkeitsspeicher m <Fertig> hydropneumatic accumulator
Pneumonik f <Fertig> fluid technology
pn-Gleichrichter m <Elektrotech> p-n rectifier
pn-Halbleiterdiode f <Elektriz> p-n junction diode
pnp *(positiv-negativ-positiv)* <Elektronik> p-n-p *(positive-negative-positive)*

pnpn-Gerät n <Elektronik> p-n-p-n device
pnpn-Komponente f <Elektronik> p-n-p-n component
pnp-Transistor m <Elektronik, Phys> p-n-p transistor
pn-Übergang m <Elektronik, Phys> p-n junction
Po *(Polonium)* <Chemie> Po *(polonium)*
PO 1. <Kunststoff, Textil> *(Polyolefin)* PO *(polyolefin)*; 2. <Textil> *(Polynosic-Faser)* PO *(polynosic fiber)*
Pocherz n <Metall> crushed ore
Pochholz n <Mechan> lignum vitae
pochieren v <Lebensmittel> poach
Pockennarben fpl <Ker & Glas> pockmarks
Pockholz n <Wassertrans> lignum vitae
Pod-Antrieb m <Wassertrans> pod propulsion system
Podest n <Kerntech> pedestal
Podeststufe f <Bau> landing step
Podestträger m <Bau> bearer
Podestwechselbalken m <Bau> landing trimmer
Podocarpin... <Chemie> podocarpic
Podophyllin n <Chemie> podophyllin
POGO-Effekt m <Raumschiff> pogo effect *(Raumschiff)*
Point-and-Click-Schnittstelle f <Comp & DV> point-and-click interface
Pointkontakt-Gleichrichter m <Elektriz> point contact rectifier
Poiseuille'sche Strömung f <Strömphys> Poiseuille flow
Poiseuille'sches Gesetz n <Phys> Poiseuille's law
Poisson'sche Gleichung f <Phys> Poisson's equation
Poisson'sche Konstante f (σ) <Phys> Poisson's ratio
Poisson'sche Verteilung f 1. <Comp & DV> Poisson distribution; 2. <Phys> Poisson distribution *(Statistik)*; 3. <Qual> Poisson distribution
Poisson'sche Zahl f (σ) 1. <Kohlen, Kunststoff> Poisson's ratio (σ); 2. <Mechan> Poisson's ratio
Poisson'sches Gesetz n <Raumfahrt> Poisson's law *(Raumschiff)*
Poisson'sches Verhältnis n (σ) <Mechan> Poisson's ratio (σ)
Poisson-Verkehr m <Telekom> Poisson traffic
Poisson-Verteilung f <Telekom> Poisson distribution
pökeln v <Lebensmittel> cure *(Fleisch)*
Pökeln n <Lebensmittel> salting
Pol m 1. <Elektriz, Elektrotech> pole, terminal; 2. <Funktech> pole; 3. <Kfztech> terminal; 4. <Metrol, Papier, Phys> pole
Pol m/nicht mit erweitertem ausgeprägtem <Elektriz> nonsalient pole
Polarbahn/in <Raumfahrt> polar-orbiting
Polardiagramm n <Funktech, Phys> polar diagram
polares Dielektrikum n <Phys> polar dielectric
polares Molekül n <Kunststoff> polar molecule
Polarimeter n <Labor, Optik, Phys, Strahlphys> polarimeter
Polarisation f 1. <Elektriz, Elektrotech, Foto, Funktech, Phys> polarization; 2. <Raumfahrt> polarization *(Weltraumfunk)*; 3. <Strahlphys, Telekom> polarization
Polarisation f elektromagnetischer Wellen <Elektrotech> electromagnetic wave polarization
Polarisationsapparat m <Phys> saccharimeter
Polarisationsbrille f <Foto> polarizing spectacles
Polarisationsdiplexer m <Raumfahrt> polarization diplexer *(Weltraumfunk)*
Polarisationsebene f <Funktech, Optik, Phys> plane of polarization
Polarisationsfilter n <Foto> polarizing filter
Polarisationsgerät n <Labor> polarimeter
Polarisationsgitter n <Raumfahrt> polarization grid *(Weltraumfunk)*
Polarisationsisolierung f <Raumfahrt> polarization isolation *(Weltraumfunk)*

Polarisationskopplungsdämpfung f <Telekom> polarization coupling loss
Polarisationsladung f <Phys> polarization charge
Polarisationsmikroskop n 1. <Labor> polarization microscope, polarizing microscope; 2. <Phys> polarizing microscope
Polarisationsprisma n <Phys> polarizer
Polarisationsreinheit f <Raumfahrt> polarization purity *(Weltraumfunk)*
Polarisationsstrom m <Phys> polarization current
Polarisationswinkel m 1. <Funktech> angle of polarization; 2. <Optik> angle of polarization, polarizing angle
Polarisator m 1. <Metall> polarizer; 2. <Phys> polarizer, polaroid; 3. <Raumfahrt> polarizer *(Weltraumfunk)*; 4. <Telekom> polarizer
Polarisierbarkeit f <Phys> polarizability
polarisieren v <Optik> polarize
polarisierend adj <Optik> polarizing
polarisierende Substanz f <Labor> polarizer
polarisiert adj <Strahlphys> polarized
polarisierte Wellen fpl <Wellphys> polarized waves
polarisierter Stecker m <Elektriz> nonreversible plug
polarisiertes Licht n <Foto, Phys, Strahlphys> polarized light
polarisiertes Licht n/durch Reflexion <Wellphys> polarized light by reflection
polarisiertes Relais n <Elektriz, Elektrotech> polarized relay
Polariskop n <Optik, Phys> polariscope
Polarität f <Elektriz, Elektrotech, Phys> polarity
Polaritätsprüfer m <Elektriz> polarity tester
Polaritätssteuerung f <Fernseh> polarity control
Polaritätsumkehr f <Elektriz> polarity reversal
Polaritätswechsel m <Elektriz> polarity reversal
Polaritätswechsler m <Elektriz> polarity reverser
Polaritätszeichen n <Elektriz> polarity sign
Polarkoordinaten fpl 1. <Elektronik> polar coordinates *(zweidimensional)*; 2. <Math, Phys> polar coordinates
Polarkoordinaten-Anzeigeinstrument n <Gerät> polar-coordinate-indicating instrument
Polarkoordinatensystem n <Geom> polar coordinate system
Polarlicht n <Wassertrans> aurora polaris *(Wetterkunde)*
Polarographie f <Chemie> polarography
polarographische Analyse f <Chemie> polarography
Polaroid n <Phys> polaroid
Polaroidmaterial n <Strahlphys> polaroid
Polaron n <Phys> polaron
Polarorbit m <Raumfahrt> polar orbit
Polbahn f <Fertig> centrode
Polbrücke f <Eisenbahn, Wassertrans> plate strap
Polder m 1. <Abfall> landfill cell, refuse cell; 2. <Bau> polder *(Marschland)*
Polderdeich m <Bau> polder
Polderverfahren n <Abfall> cell method
polfreies Filter n <Elektronik> Chebyshev filter
Polierasche f <Bau> putty powder
polieren v 1. <Anstrich> polish; 2. <Bau> rub; 3. <Ker & Glas> burnish; 4. <Mechan> polish; 5. <Metall> burnish
Polieren n 1. <Bau> planing; 2. <Kunststoff> polishing; 3. <Maschinen> buffing, polishing; 4. <Metall> burnishing, polishing
Polierer m <Metall> burnisher
Polierkratzer m <Ker & Glas> sleek
Polierkugel f <Akustik> advance ball
Polierleinwand f <Maschinen> crocus cloth
Poliermaschine f <Maschinen> polishing head, polishing lathe
Poliermittel n <Anstrich> silica abrasive

Polycarbonat

Polierpaste f <Mechan> grinding paste
Polierrot n <Ker & Glas> rouge
Poliersand m <Anstrich> grit
Polierscheibe f 1. <Fertig> bob, polishing wheel; 2. <Maschinen> buff, polishing wheel; 3. <Mechan> buffer; 4. <Sicherheit> abrasive wheel
Polierstahl m <Ker & Glas> burnisher
polierte Steinoberfläche f <Bau> polished stone finish
Polierung f <Mechan> burnishing
Polierwalze f <Maschinen> polishing roll
Politur f 1. <Anstrich> polish; 2. <Ker & Glas> burnishing
Politurmittel n <Anstrich> polish
Polklemme f <Elektrotech> binding post, pole terminal
Poller m <Wassertrans> bollard
Polling n <Comp & DV, Telekom> polling
polnische Schreibweise f <Comp & DV> Polish notation, parenthesis-free notation
Pol-Nullstellen-Verteilung f <Telekom> pole-zero distribution
Polonium n (Po) <Chemie> polonium *(Po)*
Polprüfer m <Gerät> pole tester
Polschichtlänge f <Kohlen> effective pile length
Polschlupf m <Elektriz> pole slip
Polschlüpfung f <Elektriz> pole slip
Polschuh m 1. <Elektrotech> pole piece, pole shoe; 2. <Phys> pole shoe
Polseite f <Telekom> pole face
Polstärke f <Phys> pole strength
Polster n <Bau, Ker & Glas> pad
Polsterbruch m <Ker & Glas> pad break
Polsterstoff m <Textil> upholstery
Polsterung f 1. <Maschinen, Papier, Textil> padding; 2. <Verpack> cushioning product
Polstück n <Aufnahme, Elektrotech> pole piece
Polumkehr f <Elektrotech> polarity reversal
Polumschaltschütz n <Elektrotech> contactor-type pole-changer
Polung f <Fernseh> polarity
Polwechselschalter m <Elektrotech> polarity-reversing switch
Polwechsler m <Elektrotech> change-pole controller
Polwechsleranlasser m <Elektriz> pole-changing starter
Polwechslerschalter m <Elektriz> pole changer switch
Polwendeschalter m <Elektrotech> reversing switch
Polyacetal n 1. <Chemie> polyacetal, polyoxymethylene; 2. <Kunststoff> polyoxymethylene
Polyacryl n (PAA) 1. <Chemie, Kunststoff> polyacrylate *(PAA)*; 2. <Textil> acrylic
Polyacrylamid n <Erdöl> polyacrylamide *(Petrochemie)*
Polyacrylat n (PAA) <Chemie, Kunststoff> polyacry-late *(PAA)*
Polyacrylnitril n 1. <Chemie> polyacrylonitrile; 2. <Kunststoff> polyacrylonitrile; 3. <Textil> acrylic; 4. <Umweltschmutz> polyacrylonitrile *(PAN)*
Polyalkohol m <Chemie> polyol
Polyamid n (PA) <Chemie, Kunststoff, Textil> polyamide *(PA)*
Polyamin n <Kunststoff> polyamine
polyatomar adj <Chemie> polyatomic
Polybutadien n <Raumfahrt> polybutadiene *(Raumschiff)*
Polybuten n (PB) <Kunststoff> polybutene, polybutylene *(PB)*
Polybutylen n (PB) <Kunststoff> polybutene, polybutylene *(PB)*
Polybutylenterephthalat n (PBT) <Elektriz, Kunststoff> polybutylene ephtalate *(PBT)*
Polycarbonat n (PC) <Elektriz, Kunststoff> polycarbonate *(PC)*

Polychloropren-Latex

Polychloropren-Latex m <Kunststoff> polychloroprene latex
polychromes Glas n <Ker & Glas> polychromatic glass
Poly-Cyanethylen n <Chemie> polycyanoethylene, polyacrylonitrile
polycyclisch adj <Chemie> polycyclic
polycyclische Aromaten npl <Kunststoff> polynuclear aromatics
Polydiallyphthalat n (PDAP) <Kunststoff> polydiallylphthalate (PDAP)
Polydimethylsiloxan n <Chemie> polydimethylsiloxane
Polyeder n <Geom, Metall> polyhedron (Vielflächner)
polyedrisch adj <Geom> polyhedral
Polyen n <Chemie> polyene
Polyester m (PES) <Chemie, Elektriz, Kunststoff, Textil> polyester (PES)
Polyesterband n <Aufnahme> polyester tape
Polyesterbildung f <Chemie> polyesterification
Polyesterdruckplatte f <Druck> polyester-based plate
Polyesterfarbe f <Bau> polyester paint
Polyesterharz n <Chemie> polyester resin
Polyesterlack m <Fertig, Kunststoff> polyester varnish
Polyesterplatte f <Druck> polyester-based plate
Polyesterschaumstoff m <Abfall> polyester foam
Polyethersulfon n <Elektriz> polyether sulfon
Polyethylen n (PET) <Chemie, Elektriz, Erdöl, Kunststoff, Textil, Verpack> (AE) polyethylene, (BE) polythene (PET)
Polyethylen n hoher Dichte <Kunststoff, Verpack> HDPE, high-density polyethylene
Polyethylenschaumstoff m <Kunststoff> polyether foam
Polyethylenterephthalat n 1. <Elektriz> polyethylene terephthalate; 2. <Kunststoff> polyethylene terephthalate
Polyformaldehyd m <Chemie> polyacetal, polyoxymethylene
polyfunktionell adj <Chemie> multifunctional, polyfunctional
Polygon n 1. <Bau> traverse; 2. <Fertig> spherical triangle; 3. <Geom> polygon (Vieleck)
polygonal adj <Geom> polygonal
polygonale Laufzeitkette f <Elektronik> polygonal delay line
polygonaler Spiegel m <Optik> polygonal mirror
Polygonelektrode f <Elektriz> polygon electrode
Polygonfläche f <Geom> polygon surface
Polygonisieren n <Metall> polygonization
Polygonprofil n <Maschinen> spline profile
Polygonschaltung f <Elektriz> polygon connection, polygonal connection
Polygonversetzung f <Metall> polygonal dislocation
Polyimid n (PI) <Elektriz, Elektronik, Kunststoff> polyimide (PI)
Polyimid-Leiterplatte f <Elektronik> polyimide printed circuit
Polyisobutylen n (PIB) <Kunststoff> polyisobutylene (PIB)
Polyisocyanat n <Kunststoff> polyisocyanate
Polyisopren n <Chemie, Kunststoff> polyisoprene
Polykondensation f <Chemie> polycondensation
polykristallines Silizium n <Elektronik> polycrystalline silicon
polymer adj <Chemie> polymeric
Polymer n 1. <Chemie, Elektriz> polymer; 2. <Erdöl> polymer (Petrochemie); 3. <Fertig, Kunststoff, Textil> polymer
Polymerbeton m <Kunststoff> polymer concrete
Polymerdichtungsmasse f <Erdöl> sealant polymer
Polymerfaser f <Optik> (AE) polymer fiber, (BE) polymer fibre
Polymerie f <Chemie> polymerism

Polymerisat n <Chemie, Kunststoff, Textil> polymer
Polymerisation f 1. <Chemie> polymerization; 2. <Erdöl> polymerization (Petrochemie); 3. <Kunststoff> polymerization
Polymerisationsgrad m <Kunststoff> degree of polymerization
Polymerisationskammer f <Textil> curing oven
Polymerisiereinrichtung f <Textil> polymerizer
polymerisieren v <Chemie, Kunststoff, Papier, Textil> polymerize
Polymerisieren n <Chemie> polymerization
Polymermantel-Quarzglasfaser f <Optik, Telekom> (AE) plastic-clad silica fiber, (BE) plastic-clad silica fibre
Polymerweichmacher m <Kunststoff> polymeric plasticizer
Polymethacrylat n (PMMA) <Kunststoff> polymethacrylate, polymethyl methacrylate (PMMA)
Polymethakrylat n <Ker & Glas> organic glass
Polymethylen n <Chemie> polymethylene
Polymethylmethacrylat n (PMMA) <Kunststoff> polymethacrylate, polymethyl methacrylate (PMMA)
Polymorphie f <Metall> polymorphism
Polynom n 1. <Comp & DV> polynomial; 2. <Math> multinomial, polynomial; 3. <Phys> polynomial
Polynomcode m <Comp & DV> polynomial code
polynomisches Filter n <Elektronik> polynomial filter
Polynosic-Faser f (PO) <Textil> (AE) polynosic fiber, (BE) polynosic fibre (PO)
Polynucleotid n <Chemie> polynucleotide
Polyol n <Chemie, Kunststoff> polyol
Polyolefin n (PO) <Kunststoff, Textil> polyolefin (PO)
Polyolefincontainer m <Verpack> polyolefin container
Polyolefinsperrfolie f <Verpack> polyolefin barrier film
Polyoxyethylen n <Chemie> polyoxyethylene
Polyoxymethylen n (POM) 1. <Chemie> polyacetal, polyoxymethylene; 2. <Kunststoff> polyoxymethylene (POM)
Polypeptid n <Chemie> polypeptide
Polyphenol... <Chemie> polyphenol
Polypropylen n (PP) <Chemie, Elektriz, Erdöl, Kunststoff, Textil> polypropylene (PP)
Polysilizium n <Elektronik> polysilicon
Polysiliziumgatter n <Elektronik> polysilicon gate
Polysiliziumschicht f <Elektronik> polysilicon layer
Polysiloxan n <Kunststoff> polysiloxane
Polysolenoidmotor m <Elektriz> tubular motor
Polystyren n (PS) <Chemie, Kunststoff> polystyrene (PS)
Polystyrol n (PS) <Chemie, Elektriz, Kunststoff> polystyrene (PS)
Polystyrol-Spritzgussetikett n mit umgekehrter **Lackierung** <Verpack> polystyrene injection in-mould label
Polysulfid n <Chemie, Kunststoff> (AE) polysulfide, (BE) polysulphide
Polysulfidplast m <Chemie> thioplast
Polyterpen n <Chemie> polyterpene
Polytetrafluorethen n (PTFE) 1. <Fertig> polytetrafluoroethylene, PTFE (Kunststoffinstallationen); 2. <Kunststoff> polytetrafluoroethylene, PTFE
Polytetrafluorethylen n (PTFE) 1. <Fertig> polytetrafluoroethylene, PTFE (Kunststoffinstallationen); 2. <Kunststoff> polytetrafluoroethylene, PTFE
Polythermfrachter m <Trans> polythermal cargo ship
Polythylen n <Erdöl> polythylene (Petrochemie)
Polyurethan n (PUR) <Kunststoff, Textil> polyurethane (PUR)
Polyurethanschaum m <Kunststoff> expanded polyurethane, polyurethane foam

Polyurethanschaumstoff m <Kunststoff> expanded polyurethane, polyurethane foam
polyvalent adj <Chemie> polyvalent
Polyvalenz f <Chemie> polyvalence, polyvalency
Polyveresterung f <Chemie> polyesterification
Polyvinyl n <Chemie, Textil> polyvinyl
Polyvinylacetal n <Kunststoff> polyvinyl acetal
Polyvinylacetat n (PVAC) <Kunststoff> polyvinyl acetate (PVAC)
Polyvinylalkohol m (PVAL) <Druck, Kunststoff> polyvinyl alcohol (PVAL)
Polyvinylalkoholschlichte f <Textil> polyvinyl alcohol size
Polyvinylbutyral n (PVB) <Kunststoff> polyvinyl butyral (PVB)
Polyvinylchlorid n (PVC) <Bau, Elektrotech, Fertig> polyvinyl chloride, PVC (Kunststoffinstallationen)
Polyvinylether m (PVE) <Kunststoff> polyvinyl ether (PVE)
Polyvinylfluorid n (PVF) <Kunststoff> polyvinyl fluoride (PVF)
Polyvinylidenchlorid n (PVDC) <Kunststoff> polyvinylidene chloride (PVDC)
Polyvinylidenfluorid n (PVDF) <Kunststoff> polyvinylidene fluoride (PVFD)
polyzyklischer aromatischer Kohlenwasserstoff m <Umweltschutz> polycyclic aromatic hydrocarbon
POM (Polyoxymethylen) <Chemie, Kunststoff> POM (polyoxymethylene)
Pond n <Fertig> gram force
Ponton m <Bau, Wassertrans> pontoon
Pontonbrücke f <Wassertrans> pontoon bridge
Pontondock n <Wassertrans> pontoon dock
Pool-Wagen m <Eisenbahn> pooling car
Poopdeck n <Wassertrans> poop deck (Bootsbau)
Popcorn-Polymere npl <Kunststoff> popcorn polymers
Popelin m <Textil> poplin
Populin n <Chemie> benzoylsalicin, populin
Pore f 1. <Bau> interstice (Gewebe); 2. <Kohlen> interstice, pore, void; 3. <Metall> void
Poren fpl <Metall> pitting
Porendruck m 1. <Erdöl> pore pressure (Geologie); 2. <Kohlen> pore pressure
Porengasdruck m <Kohlen> pore gas pressure
Porengrundwasser n <Wasserversorg> interstitial water
Porenschließer m <Kunststoff> sealer
Porenvolumen n 1. <Kohlen> pore volume; 2. <Verpack> absorptive capacity
Porenwasser n <Kohlen> interstitial water, pore water
Porenwasserdruck m <Kohlen> pore water pressure
Porenwasserdruckmesser m <Bau> pore water pressure gauge
Porenwasserüberdruck m <Kohlen> pore overpressure
Porenzeile f <Bau> linear porosity (Schweißen)
Porenziffer f <Kohlen> void ratio
Porigkeit f <Metall> porosity
poromerbeschichtetes Gewebe n <Kunststoff> poromeric coated fabric
Porometer n <Chemtech> porometer
porös adj <Anstrich, Nichtfoss Energ> porous
poröse Abdeckung f <Abfall> porous cover (einer Deponie)
poröser Absorber m <Akustik> porous absorber
poröser nicht reflektierender Körper m <Akustik> porous absorber
Porosimeter n <Papier> porosimeter
Porosität f 1. <Akustik, Anstrich, Bau> porosity; 2. <Erdöl> porosity (Geologie); 3. <Kohlen, Kunststoff, Metall, Nichtfoss Energ, Papier, Wasserversorg> porosity

Porositätslog n <Erdöl> porosity log (Messtechnik)
Porositätsprüfer m <Papier> porosity tester
Porphin n <Chemie> porphin
Porphyropsin n <Chemie> porphyropsin
Port m <Comp & DV, Druck, Telekom> port
Portal n 1. <Bau> portal; 2. <Fertig> bridge (Hobelmaschine, Fräsmaschine); 3. <Maschinen> bridge; 4. <Telekom> portal (Internet); 5. <Wasserversorg> gantry
Portalkran m 1. <Bau> gantry crane, portal crane; 2. <Eisenbahn, Kerntech, Maschinen, Wassertrans> gantry crane
Portalkranroboter m <Kontroll> gantry robot
Portalmast m <Wassertrans> portal mast
Portioniervorrichtung f <Verpack> portioning machine
Portionspackmaschine f <Fertig> portioning and packaging machine
portionsweise Zugabe f <Anstrich> titration
Portlandstein m <Bau> Portland stone
Portlandzement m <Bau> Portland cement
Portlandzementbeton m <Bau> Portland cement concrete
Portraitlinse f <Foto> portrait attachment
Porzellan n <Ker & Glas> English china, china
Porzellanabdampfschale f <Ker & Glas> porcelain evaporating basin
Porzellanbohrer m <Ker & Glas> china borer, porcelain borer
Porzellanbrennofen m <Ker & Glas> porcelain calcining furnace
Porzellandekor n <Ker & Glas> china decoration
Porzellandekoration f <Ker & Glas> porcelain decoration
Porzellandreher m <Ker & Glas> china thrower, porcelain thrower
Porzellanerde f <Ker & Glas> porcelain clay
Porzellanfadenöse f <Ker & Glas> porcelain thread guide
Porzellanfarbe f <Ker & Glas> porcelain colour
Porzellanfilterplatte f <Ker & Glas> porcelain filter plate
Porzellangeräte npl <Ker & Glas> porcelain utensils
Porzellangießmaschine f <Ker & Glas> china caster, porcelain caster
Porzellanhersteller m <Ker & Glas> porcelain maker
Porzellanindustrie f <Ker & Glas> porcelain industry
Porzellanisolator m 1. <Elektriz, Elektrotech> porcelain insulator; 2. <Ker & Glas> china insulator, porcelain insulator
Porzellanisolierung f <Elektrotech> porcelain insulation
Porzellanjaspis m <Ker & Glas> porcelain jasper
Porzellanknopf m <Ker & Glas> porcelain button
Porzellanküvette f <Ker & Glas> porcelain cell
Porzellanlack m <Ker & Glas> porcelain varnish
Porzellanmalerei f <Ker & Glas> china painting, painting on porcelain
Porzellanrohr n <Ker & Glas> porcelain tube
Porzellanscherben fpl <Ker & Glas> pot sherds
Porzellanschleifer m <Ker & Glas> porcelain polisher
Porzellantasse f <Ker & Glas> porcelain cup
Porzellantiegel m <Ker & Glas, Labor> porcelain crucible
Porzellantrichter m <Ker & Glas> porcelain funnel
Porzellanvergolder m <Ker & Glas> porcelain gilder
Porzellanverteilerkasten m <Ker & Glas> porcelain conduit box
Porzellanverzierung f <Ker & Glas> china ornamentation
Porzellanwaren fpl <Ker & Glas> porcelain goods
Porzellanzahn m <Ker & Glas> porcelain tooth
POS-Anzeige f (elektronische Kassenanzeige) <Verpack> POS display (point of sale display)
Position f <Wassertrans> position (Navigation); plot (Radar)
positionieren v <Kontroll, Maschinen> position

Positionierfehler

Positionierfehler *m* <Fertig> deviation
Positioniergenauigkeit *f* 1. <Druck> positioning accuracy; 2. <Elektronik> positional accuracy
Positioniermagnet *m* <Fernseh> beam-positioning magnet
Positionierung *f* 1. <Fernseh> registration; 2. <Kontroll> positioning; 3. <Maschinen> location, positioning
Positionierungsbewegung *f* <Ergon> positioning movement
Positionierungsfehler *m* <Elektronik> placement error
Positionierungsgenauigkeit *f* <Fernseh> registration accuracy
Positionierungsgeschwindigkeit *f* <Maschinen> positioning speed
Positionierungssteuerung *f* <Fernseh> registration control
Positioniervorrichtung *f* <Maschinen> positioner
Positionierzeit *f* <Comp & DV> seek time
Positionsanzeige *f* <Fertig> position display, position indication
Positionsanzeiger *m* <Comp & DV> cursor
Positionsbestimmung *f* <Comp & DV> rotation position sensing
Positionsgeber *m* <Elektriz> position sensor
Positionslampe *f* <Lufttrans> navigation light
Positionslaterne *f* <Wassertrans> navigation light
Positionslicht *n* 1. <Raumfahrt> position light; 2. <Wassertrans> navigation light
Positionsschalter *m* <Elektrotech> position switch
Positiv *n* 1. <Druck> positive; 2. <Foto> positive, positive image (Bild, Abzug)
positiv *adj* <Elektriz, Math> positive
positiv dotierter Bereich *m* <Phys> positively doped region
positiv geerdete Anschlussklemme *f* <Kfztech> positive-grounded terminal
positiv geerdeter Pol *m* <Kfztech> positive-grounded terminal
positive Anschlussklemme *f* <Kfztech> plus terminal, positive terminal
positive Bildphase *f* <Fernseh> positive picture phase
positive Folge *f* <Elektriz> positive sequence
positive ganze Zahl *f* <Math> positive integer
positive Gauß'sche Krümmung *f* <Geom> positive Gauss curvature
positive Klemme *f* <Elektriz> positive terminal
positive Kontaktbank *f* <Kfztech> positive bank
positive Krümmung *f* <Geom> positive curvature
positive Ladung *f* <Elektriz, Elektrotech, Phys> positive charge
positive Logik *f* <Elektronik> positive logic
positive Magnetostriktion *f* <Elektrotech> positive magnetostriction
positive Phasendrehung *f* <Elektriz> positive phase sequence
positive Platte *f* <Kfztech> positive plate
positive Profilverschiebung *f* <Fertig> diameter enlargement (Getriebelehre)
positive Quadratwurzel *f* <Math> principal square root
positive Quittung *f* <Comp & DV> positive acknowledgement
positive Rückkopplung *f* <Strahlphys> positive feedback
positive Rückmeldung *f* <Comp & DV> positive acknowledgement
positive Säule *f* <Elektronik, Phys> positive column
positive Vorspannung *f* <Elektrotech> keep-alive voltage, positive bias
positive Zahl *f* <Math> positive number

positiver Anschluss *m* <Elektriz> positive terminal
positiver Netzanschluss *m* <Elektrotech> positive power supply
positiver Photolack *m* <Elektronik> positive photoresist, positive resist
positiver Pol *m* <Elektriz> positive pole
positiver Quadrant *m* <Math> positive quadrant
positiver Radsturz *m* <Kfztech> positive camber
positiver Spanwinkel *m* <Fertig, Maschinen> positive rake
positiver Sprung *m* <Wassertrans> normal sheer (Bootbau)
positiver Winkel *m* <Geom> positive angle
positives Ion *n* <Elektriz, Phys> positive ion
Positivfilm *m* <Druck> positive film
Positiv-Isolierend-Negativ (PIN) <Elektronik> positive-isolating negative (PIN)
Positivkopie *f* <Druck> positive print
Positivmodulation *f* <Elektronik> positive modulation
positiv-negativ-positiv *adj* (pnp) <Elektronik> positive-negative-positive (p-n-p)
Positivstopfen *n* <Telekom> positive justification
Positron *n* 1. <Phys> positron (Teilchen); 2. <Teilphys> positron
Postambel *f* <Comp & DV> postamble
Postemphase *f* <Fernseh> postemphasis
Posten *m* <Wassertrans> post
Postenspeisung *f* <Ker & Glas> gob feeding
POS-Terminal *n* (elektronische Kasse) <Comp & DV, Verpack> POS terminal (point of sale terminal)
Postfach *n* <Comp & DV, Telekom> mailbox
Postfixschreibweise *f* <Comp & DV> postfix notation
Postleitzahl *f* <Telekom> codemark, post code, postal code
Post-Mortem-Programm *n* <Comp & DV> postmortem program
Post-Panamax-Schiff *n* <Wassertrans> post Panamax vessel (Schiff mit zu großen Abmessungen für den Panamaverkehr)
Postprozessor *m* <Comp & DV> postprocessor
Postprozessorprogramm *n* <Comp & DV> postprocessor
Postulat *n* <Math> postulate
Post- und Frachtterminal *m* <Lufttrans> mail and cargo terminal
Postzug *m* <Eisenbahn> mail train
Potenz *f* <Math> power • **zur zweiten Potenz** <Math> raised, second power, squared, to the power of two
Potenzgesetzindexfaser *f* <Optik> (AE) power law index fiber, (BE) power law index fibre
Potenzgesetzindexprofil *n* <Optik> power law index profile
Potenzial *n* 1. <Elektriz, Elektrotech, Funktech> potential; 2. <Phys> potential function, potential
Potenzial *n* **der Schallschnelle** <Nichtfoss Energ> velocity potential
Potenzialabfall *m* <Phys> potential drop
Potenzialbarriere *f* <Elektrotech, Phys> potential barrier
Potenzialdifferenz *f* <Elektriz, Elektrotech, Phys> PD, potential difference
potenzialfreier Ausgang *m* <Elektrotech> floating output
potenzialfreier Eingang *m* <Elektrotech> floating input
Potenzialfunktion *f* <Math, Phys> harmonic function, potential function
Potenzialgefälle *n* <Elektriz> potential gradient
Potenzialgradient *m* <Elektriz> potential gradient
Potenzialkoeffizienten *mpl* <Phys, Qual> coefficients of potential
Potenzialring *m* <Phys> guard ring

Potenzialschleife f <Elektriz> potential loop
Potenzialschranke f <Elektrotech> potential barrier
Potenzialschwelle f <Elektrotech, Phys> potential barrier
Potenzialtheorie f <Math, Phys> potential theory
Potenzialtopf m <Phys> potential well
Potenzialverschiebung f <Elektronik> potential shift
Potenzialwall m <Elektrotech> potential barrier
potenziell adj <Anstrich> potential
potenzielle Energie f <Phys> potential energy
Potenzieren n <Math> involution
Potenzierung f <Math> involution
Potenziometer n 1. <Aufnahme> potentiometer; 2. <Elektriz, Elektronik, Elektrotech> adjustable voltage divider, adjustable resistor, potentiometer; 3. <Fertig> potentiometer (Kunststoffinstallationen); 4. <Funktech> potentiometer; 5. <Labor> potentiometer (Elektrochemie); 6. <Lufttrans> adjusting potentiometer; 7. <Papier> potentiometer; 8. <Phys> compensator, potentiometer
Potenziometer n **mit numerischer Anzeige** <Elektriz> read-out potentiometer
Potenziometergeber m <Gerät> retransmitting slide wire
Potenziometer-Gleitkontakt m <Elektriz> potentiometer slider
Potenziometerschleifer m <Elektriz> potentiometer slider
potenziometrische Ebene f <Erdöl> potentiometric level
potenziometrische Höhe f <Erdöl> potentiometric head
potenziometrische Karte f <Erdöl> potentiometric map
potenziometrische Titration f <Chemie> electrometric titration
potenziometrischer Rheostat m <Elektriz> potentiometer rheostat
Potenzprofil n <Telekom> power law index profile
Potenzreihe f <Math> power series
Poti n <Aufnahme> pot
Pottasche f <Ker & Glas> potash
Pourpoint m <Kfztech> pour point (Öl)
p-Oxytoluen n <Chemie> p-cresol
Poynting'scher Satz m <Phys> Poynting's theorem
Poynting'scher Vektor m (S) <Elektriz, Phys> Poynting vector (S)
Poynting-Vektor m (S) <Telekom> Poynting vector (S)
PP (Polypropylen) <Chemie, Elektriz, Erdöl, Kunststoff, Textil> PP (polypropylene)
PPI-Anzeige f (Panoramaanzeige) <Funkort, Phys> PPI (plan-position indication)
PPI-Sichtgerät n <Funkort> plan-position indicator, PPI (Radar)
PPM 1. <Comp & DV, Elektronik, Telekom> (Pulsphasenmodulation) PPM (pulse phase modulation); 2. <Elektronik, Telekom> (Pulslagenmodulation) PPM (pulse position modulation)
PP-Schnur f <Verpack> polypropylene closure, polypropylene strap
PPU (Peripherietechnik) <Comp & DV> PPU (peripheral processing units)
Präambel f <Comp & DV> preamble
Prädikat n <Comp & DV, Künstl Int> predicate
Prädikatenlogik f (PL) <Künstl Int> predicate logic (PL)
Prädiktion f <Künstl Int> prediction
Prädiktionsalgorithmus m <Telekom> prediction algorithm
Prädiktionsbild n <Telekom> predicted frame
Prädiktionscodierung f <Telekom> prediction coding, predictive coding, predictive speech coding
Prädiktionswert m <Fernseh> estimated value (picture coding)
prädiktiv codiertes Bild n <Telekom> predictive coded picture

Präfix n <Comp & DV> prefix
Präfixschreibweise f <Comp & DV> Polish notation, parenthesis-free notation, prefix notation
Präge... <Verpack, Druck> imprint
Prägedruck m <Druck> blocking
Prägedruck m **ohne Farbe** <Druck> blind blocking, blind embossing
Prägeetikett n <Verpack> embossed label
Prägeform f <Maschinen> coining die, forming die
Prägekalander m <Verpack> embossed calender
Prägemaschine f <Druck> stamping press
prägen v 1. <Druck> (AE) mold, (BE) mould; 2. <Fertig> strike (Münzen)
Prägen n 1. <Fertig> sizing; 2. <Kunststoff> embossing; 3. <Maschinen> coining
Prägepapier n <Verpack> embossed paper
Prägepresse f 1. <Druck> stamping press; 2. <Verpack> embossing machine, embossing press
Prägerillen-Aufnahme f <Aufnahme> embossed-groove recording
Prägestempel m 1. <Fertig> hob; 2. <Maschinen> coining die, forming die
Präge- und Druckwerkzeug n <Verpack> imprinter
Pragmatik f <Comp & DV> pragmatics
Prägung f <Druck> embossing
Prahm m <Wassertrans> barge (Schifftyp)
Praktikum n <Wassertrans> certificate of pratique, pratique (Dokument)
praktische Erprobung f <Maschinen> field test
praktische Leistungsfähigkeit f <Trans> practical capacity
praktische Leistungsfähigkeit f **unter ländlichen Bedingungen** <Trans> practical capacity under rural conditions
praktische Leistungsfähigkeit f **unter städtischen Bedingungen** <Trans> practical capacity under urban conditions
praktische Prüfmethode f <Optik> practical test method
praktische Prüfungsmethode f <Telekom> practical test method
Prallblech n 1. <Fertig> baffle breaker; 2. <Kohlen> impact plate; 3. <Maschinen> baffle plate • **mit Prallblechen ausstatten** <Chemtech> baffle
Prallblechring m <Erdöl> baffle ring
Prallbrecher m <Kohlen, Maschinen> impact crusher
Prallschirm m <Chemtech> baffle plate
Prallzerfaserer m <Papier> (AE) fiberizer, (BE) fibrizer
Prandtl'sche Grenzschichttheorie f <Strömphys> Prandtl's boundary layer theory
Prandtl'sche Zahl f <Strömphys> Prandtl number
Präparationsmittel n <Textil> lubricant
präparieren v <Textil> lubricate
Präpariernadel f <Labor> dissection needle
Präsentationsgrafik f <Comp & DV> presentation graphics
Präsentierkarton m <Verpack> display box
Prasselgeräusch n <Elektriz, Elektronik> crackles, crackling noise, frying noise, rattling noise
Prasseln n <Akustik> crackle
Pratzenlagermotor m <Eisenbahn> nose-suspended motor
praxisnaher Versuch m <Werkprüf> field trial
Präzession f <Phys, Raumfahrt> precession
Präzessionswert m <Raumfahrt> precession rate
Präzessionswinkel m <Nichtfoss Energ> angle of precession
Präzessionswinkelgeschwindigkeit f <Fertig> angular velocity of precession
Präzipitat n <Chemtech> precipitate

präzise

präzise *adj* <Math> accurate
Präzision *f* 1. <Comp & DV, Math, Qual> accuracy, precision; 2. <Phys> precision
Präzisionsanflug *m* <Lufttrans> precision approach
Präzisionsanflugverfahren *n* <Lufttrans> precision approach procedure
Präzisionsanzeigeinstrument *n* <Gerät> precision indicating instrument
Präzisionsbeschichtung *f* <Anstrich> precision coating
Präzisionsdrahtwiderstand *m* <Elektrotech> precision wirewound resistor
präzisionsgefertigte Unterlegscheibe *f* <Maschinen> precision shim
Präzisionsgrad *m* <Metrol> class of accuracy
Präzisionsinstrument *n* 1. <Gerät> high-accuracy instrument, precision instrument; 2. <Phys> precision instrument
Präzisionslehre *f* <Fertig> *(AE)* precision gage, *(BE)* precision gauge
Präzisionsmaß *f* <Math> accuracy measure *(für eine Schätzfunktion)*
Präzisionsmessbrücke *f* <Metrol> precision bridge
Präzisionsmessgerät *n* <Gerät, Metrol> high-accuracy instrument, precision-measuring instrument
Präzisionsmessung *f* <Fertig, Gerät, Metrol> high-accuracy measurement, precision measurement
Präzisionsradareinstufung *f* <Lufttrans> precision radar rating
Präzisionsradarstufe *f* <Lufttrans> precision radar rating
Präzisionsschallpegelmesser *m* <Sicherheit> precision sound level meter
Präzisionsschleifen *n* <Fertig, Maschinen> precision grinding
Präzisionsschleifmaschine *f* <Fertig, Maschinen> precision grinding machine
Präzisionsschlitten *m* <Maschinen> precision slide
Präzisionswaage *f* 1. <Gerät> precision balance; 2. <Labor> balance, sensitive balance
Präzisionswerkzeugmaschinen *fpl* <Maschinen> precision machine tools
Präzisionswiderstand *m* <Elektrotech> precision resistor
Präzisionswiegen *n* <Verpack> ultrahigh accuracy weighing
p⁻-Region *f* <Elektronik> p⁻-region
P-Regler *m* <Regelung> proportional controller
Pregnan *n* <Chemie> pregnane
Prehnit... <Chemie> prehnitic
Prehnitol *n* <Chemie> prehnitene
Prellblock *m* <Eisenbahn> buffer stop block
Prellbock *m* <Eisenbahn> buffer, *(BE)* bumper, *(AE)* fender
Prellplatte *f* <Bau> baffle
Pressbalken *m* <Maschinen> crosshead
Presse *f* 1. <Druck> press, printing press; 2. <Fertig> gun, press; 3. <Maschinen, Papier> press; 4. <Sicherheit> power press; 5. <Textil> press
Presse *f* **mit verstellbarem Tisch** <Fertig> adjustable bed press
pressen *v* 1. <Aufnahme> press; 2. <Fertig> *(AE)* mold, *(BE)* mould; 3. <Maschinen> press; 4. <Mechan> jam; 5. <Papier, Textil> press
Pressen *n* 1. <Aufnahme> pressing *(Schallplatten, CDs)*; 2. <Fertig> *(AE)* compression molding, *(BE)* compression moulding, *(AE)* molding, *(BE)* moulding; 3. <Ker & Glas, Kunststoff> *(AE)* compression molding, *(BE)* compression moulding; 4. <Maschinen> pressing; 5. <Mechan> extrusion; 6. <Textil> pressing; 7. <Verpack> *(AE)* compression molding, *(BE)* compression moulding
Pressenschleifer *m* <Papier> pocket grinder

584

Pressentisch *m* <Kunststoff> platen
Pressenwalze *f* <Papier> press roll
Presser *m* <Aufnahme> compressor
Presserfußautomatik *f* <Textil> automatic presser foot lifter
Presseur *m* <Kunststoff> impression roller
Pressfehler *m* 1. <Fertig> pit; 2. <Kunststoff> *(AE)* molding defect, *(BE)* moulding defect
Pressfettschmierung *f* <Fertig> pressure grease lubrication
Pressfilter *n* <Kohlen> filter press
Pressfläche *f* <Fertig> projected area *(Kunststoffe)*
Pressform *f* 1. <Ker & Glas> *(AE)* parison mold, *(BE)* parison mould; 2. <Maschinen> *(AE)* compression mold, *(BE)* compression mould
Pressformmaschine *f* <Fertig> squeezer
Pressgaskondensator *m* <Elektrotech> compressed-gas capacitor, standard capacitor *(Hochspannungsmesstechnik)*
pressgeschweißtes Schutzgitter *n* <Sicherheit> pressure-welded safety grating
Pressglas *n* <Ker & Glas> *(AE)* molded glass, *(BE)* moulded glass, pressed glass
Pressglasglimmer *m* <Elektrotech> glass-bonded mica *(Isolator)*
Pressgrat *m* 1. <Ker & Glas> fin; 2. <Kunststoff> flash
Pressiometer *n* <Kohlen> pressiometer
Pressionsmetermodul *n* <Kohlen> pressure meter modulus
Presskammer *f* <Fertig> gooseneck
Presskohle *f* <Kohlen> briquette
Presskolben *m* <Kfztech> plunger *(Brems- und Kupplungszylinder)*
Presskraft *f* <Maschinen> pressing force
Pressling *m* 1. <Akustik, Elektrotech> *(AE)* molding, *(BE)* moulding; 2. <Ker & Glas> pressing; 3. <Kunststoff> pellet
Pressluft *f* <Erdöl, Fertig, Maschinen, Phys> air
Pressluftbohren *n* <Erdöl> air flooding
Pressluftbohrer *m* <Fertig, Mechan, Phys> air drill
Presslufteinrichtung *f* <Fertig> air hydraulic unit
Pressluftfutter *n* <Fertig> air chuck, air-operated chuck
Presslufthammer *m* 1. <Fertig> air hammer; pneumatic hammer *(Schmieden)*; 2. <Maschinen> pneumatic hammer; 3. <Mechan, Phys> air hammer
Pressluftnietmaschine *f* <Fertig> pneumatic riveter
Pressluftschalter *m* <Elektriz> air breaker
Pressmaschine *f* <Ker & Glas> presser
Pressmassenfüllstoff *m* <Fertig> macerate *(Schnitzel)*
Pressmatte *f* <Ker & Glas> blanket
Pressmattenspeiser *m* <Ker & Glas> blanket charger
Pressmattenzuführung *f* <Ker & Glas> blanket feed
Pressmüllwagen *m* <Abfall> compactor vehicle, compression vehicle, packer body
Pressnietmaschine *f* <Fertig> compression riveter
Presspappe *f* <Papier, Verpack> *(AE)* molded board, *(BE)* moulded board
Presspassung *f* 1. <Fertig> press fit; 2. <Maschinen> interference fit, press fit
Pressplatte *f* <Metall> squeeze head
Presspulver *n* <Kunststoff> *(AE)* molding powder, *(BE)* moulding powder
Pressschiene *f* <Textil> presser bar
Pressschweißen *n* 1. <Bau, Fertig> pressure welding; 2. <Maschinen> plastic welding
Pressschweißung *f* <Maschinen> pressure welding
Presssintern *n* <Chemtech> sintering under pressure
Presssitz *m* <Mechan> force fit
Pressstempel *m* 1. <Fertig> heading tool; 2. <Ker & Glas> dolly; 3. <Maschinen> die, extrusion die, stamper

Pressteil n 1. <Akustik> (AE) molding, (BE) moulding; 2. <Elektrotech> (AE) molding, (BE) moulding
Pressung f <Maschinen> compression
Pressware f <Ker & Glas> pressware
Presswerkzeug n <Maschinen> press tool
Presszyklus m <Kunststoff> (AE) molding cycle, (BE) moulding cycle
Preventertau n <Wassertrans> jumper stay (Tauwerk)
Priel m <Wassertrans> tideway
Primär... <Abfall, Comp & DV, Elektriz, Elektrotech, Erdöl, Fernseh, Fertig, Heiz & Kälte, Ker & Glas, Kfztech, Metall, Optik, Papier, Strahlphys, Telekom, Wasserversorg> primary
Primärausdruck m <Comp & DV> primary expression
Primärbatterie f <Elektrotech> primary battery
Primärbeschichtung f <Fertig, Telekom> primary coating
Primärblau n <Fernseh> blue primary
Primärbremsbacke f <Kfztech> primary shoe
Primärbrennstoff m <Kohlen> primary fuel
Primärbrennstoffelement n <Kfztech> primary fuel cell
Primärdampf m <Fertig> primary steam
Primärdatei f <Comp & DV> primary file
primäre Aufhängung f <Eisenbahn> primary suspension
primäre Erdbebenwelle f (P-Welle) <Phys> P wave
primäre Induktanz f <Elektriz> primary inductance
primäre Spule f <Kfztech> primary winding
primäre zyklische Verstellung f <Lufttrans> primary cyclic variation (Hubschrauber)
Primäreinzellinearmotor m <Kfztech> single primary type linear motor
Primärelektrode f <Elektrotech> initiating electrode, primary electrode
Primärelement n <Elektriz> galvanic cell, voltaic cell
Primäremission f <Elektronik> primary emission
primärer Anker m <Elektriz> primary armature
primärer biologischer Schild m <Kerntech> primary biological shield
primärer Mantel m <Optik> primary coating (Faser)
primärer Speicher m <Comp & DV> primary memory, primary storage, primary store
primäres Kältemittel n <Heiz & Kälte> primary refrigerant
primäres Kühlmittel n <Heiz & Kälte> primary coolant
Primärfarbe f 1. <Comp & DV, Druck, Optik, Phys> (AE) primary color, (BE) primary colour; 2. <Strahlphys> (AE) prime colors, (BE) prime colours
Primärfaser f <Ker & Glas> (AE) primary fiber, (BE) primary fibre
Primärfilter n <Wasserversorg> primary filter
Primärförderung f <Erdöl> primary recovery
Primärgrün n <Fernseh> green primary
Primärgruppe f 1. <Comp & DV, Elektrotech> primary group; 2. <Telekom> group, primary group
Primärgruppenverteiler m <Telekom> group distribution frame
Primärindex m <Comp & DV> primary index
Primärionisierung f <Strahlphys> primary ionization
Primärkarbid n <Metall> primary carbide
Primärkettenkasten m <Kfztech> primary chaincase (Motorradgetriebe)
Primärkornbildung f <Fertig> ingotism
Primärkriechen n <Metall> primary creep
Primärkühlkreislauf m <Kerntech> primary coolant circuit
Primärlichtquelle f <Strahlphys> illuminating source
Primärluft f <Heiz & Kälte> primary air
Primärmanschette f <Kfztech> primary cup
Primärmantel m <Optik> primary coating (Faser)
Primärmultiplexanschluss m <Telekom> PRA, primary rate access

Primärratenanschluss m <Telekom> PRA, primary rate access (ISDN)
Primärrot n <Fernseh> red primary
Primärschlamm m <Abfall> primary sludge
Primärschlüssel m <Comp & DV> primary key
Primärspannung f <Elektriz, Elektrotech> primary voltage
Primärspeicher m <Comp & DV> primary memory
Primärspule f <Elektrotech> primary coil
Primärstation f <Comp & DV> primary station
Primärstoffauflauf m <Papier> primary headbox
Primärstrahler m <Funktech> active antenna
Primärstrahlung f <Kerntech> primary radiation
Primärstrom m <Elektriz, Elektrotech> primary current
Primärstromkreis m <Elektriz> primary circuit
Primärteilchen n <Elektrotech> primary particle
Primärwicklung f 1. <Elektriz> primary winding; 2. <Elektrotech> primary winding; primary (Transformator); 3. <Funktech> primary winding (Transformator); 4. <Kfztech, Phys> primary winding
Primärwicklung f mit Anzapfung <Elektrotech> tapped primary winding
Primärzelle f <Elektrotech> primary cell
Primärzugriff m <Telekom> primary access
Primer m <Kunststoff> primer
Primfaktor m <Math> prime factor
Primulin n <Chemie> primuline yellow
Primzahl f <Math> prime number
Primzahligkeit f <Math> primeness
Printmedien npl <Druck> print media
Printplatte f (PCB) <Comp & DV> printed circuit board (PCB)
Prinzip n <Phys> principle
Prinzip n der Addition und Teilung <Elektronik> add-and-divide principle (Frequenzsynthese)
Prinzip n der kleinsten Wirkung <Phys> principle of least action
Prinzip n der virtuellen Arbeit <Phys> principle of virtual work
Prinzip n des gefahrlosen Ausfalls <Qual> fail-safe
Prinzip n des kleinsten Zwanges <Phys> principle of least constraint
Prinzipschaltbild n <Fertig> elementary circuit diagram
Priorität f 1. <Comp & DV, Qual> precedence, priority; 2. <Patent> priority
Prioritätplaner m <Comp & DV> priority scheduler
Prioritätrecht n <Patent> priority right
Prioritätsteuerung f <Comp & DV> priority scheduling
Prioritätstufe f <Comp & DV, Qual> precedence level
Prioritätventil n <Hydraul> priority valve
Prioritätverkettung f <Elektronik> daisy chain
Prioritätverlust m <Patent> loss of priority
Prioritätwarteschlange f <Comp & DV> priority queue
Prisenkommando n <Wassertrans> boarding party
Prisma n 1. <Fertig> solid vee, vee; 2. <Geom, Ker & Glas, Labor, Phys, Telekom> prism
prismatisch adj <Geom> prismatic
Prismenbinokular n <Optik> prism binocular
Prismenblock m <Metrol> vee block
Prismenfräser m 1. <Fertig> equal-angle cutter; 2. <Maschinen> V-form cutter, double equal angle cutter
Prismenglas n <Ker & Glas> prismatic glass
Prismenspektrograph m <Strahlphys> prismatic spectrograph
Prismenspektrum n <Strahlphys> prismatic spectrum
Prismenstück n <Fertig> V-block
Pritschenwagen m <Kfztech> pick-up, (AE) platform truck
private Datenleitung f <Telekom> data private wire
private Selbstwählnebenstelle f (PBX) <Telekom> private branch exchange (PBX)

privater Nummerierungsplan m <Telekom> private numbering plan
privater Wählanschluss m <Telekom> private dial-up port
privates Fernsprechnetz n <Telekom> private telephone network
privates Netz n <Telekom> private network
Privatfernsehen n <Fernseh> commercial TV
privilegierte Operation f <Comp & DV> privileged operation
privilegierter Befehl m <Comp & DV> privileged instruction
probabilistisch adj <Künstl Int> probabilistic
Probe f 1. <Gerät> specimen; 2. <Kohlen, Kunststoff> sample, specimen; 3. <Maschinen> specimen, test specimen; 4. <Mechan> specimen; 5. <Metall> sample; 6. <Phys> sample; 7. <Qual> sample, test, specimen, trial; 8. <Raumfahrt, Telekom> sample; 9. <Textil> trial; 10. <Wasserversorg> sample; 11. <Werkprüf> sampling
• **Probe nehmen** <Phys> take a sample • **Proben nehmen** <Telekom> sample
Probe f **mit Seiteneinschnitt** <Metall> side-grooved specimen
Probeabzug m 1. <Druck> prepress proof, proof; 2. <Foto> proof, test print
Probeaufnahme f <Foto> test shot
Probebelastung f <Maschinen> test loading
Probebohrung f **mit Spülung** <Kohlen> wash boring
Probedrucke mpl <Druck> prepress proofs
Probedrucke mpl **für die Farbseparierung** <Druck> separation proofs
Probeentnahmen fpl **gleicher Menge** <Umweltschmutz> constant-volume sampling
Probekörper m <Kunststoff> sample
Probelast f <Maschinen> test loading
Probelauf m 1. <Bau> trial run; 2. <Comp & DV> dry run; 3. <Kerntech> proving run, proving trial, trial run; 4. <Maschinen> test run
Probelöffel m <Bau> spoon auger
Probenahme f <Abfall, Erdöl, Lebensmittel, Phys, Qual, Telekom> sampling
Probenahmegerät n 1. <Labor> sampling device; 2. <Sicherheit, Umweltschmutz> sampler, sampling unit
Probenbehälter m <Kerntech> sample holder; sample admission vessel (Massenspektrometer)
Probenehmer m <Comp & DV> sampler
Probenentnahme f <Elektronik, Raumfahrt, Wasserversorg, Werkprüf> sampling
Probenentnahmeröhrchen n <Labor> sampling tube
Probenentnahmestelle f <Kohlen> sampling point
Probengröße f <Metrol> sample size
Probenhalter m <Kerntech> specimen holder
Probenimpuls m <Telekom> sample pulse
Probenplan m <Metrol> sampling plan
Probensatz m <Gerät, Telekom> sampling theorem
Probenschwenkarm m <Kerntech> sample swivel arm
Probenstrom m <Gerät> sample stream (von Fluiden)
Probenverkleinerung f <Qual> sample reduction
Probenwechsler m <Kerntech> sample changer
Probepfahl m <Kohlen> test pile
Prober m <Kohlen> sampler
Proberöhrchen n <Labor> test tube
Proberöhrchengestell n <Labor> test tube rack
Proberöhrchenhalter m <Labor> test tube holder
Probeseiten fpl <Druck> page proofs
Probesilber n <Metall> standard silver
Probestab m <Maschinen> test bar
Probestreifen m <Foto> test strip
Probestück n <Gerät, Kohlen, Kunststoff, Maschinen, Mechan, Qual> specimen

Probetiegel m <Fertig> cupel
probeweise Prüfung f <Qual> sampling test
Probewürfel m <Bau> test cube
Probezugang m <Comp & DV> trial access
Probezylinder m <Bau> test cylinder
probieren v <Phys> sample
Probierhahn m <Wasserversorg> (AE) gage cock, (BE) gauge cock, test cock, try cock
Probierofen m <Heiz & Kälte> assay furnace
Probierstift m <Maschinen> touch needle
Problem n <Comp & DV, Künstl Int> problem
Problembehebung f <Comp & DV> problem recovery
Problembeschreibung f <Comp & DV> problem description
Problembestimmung f <Comp & DV> problem determination
Problemdarstellung f <Künstl Int> problem representation
Problemdefinition f <Comp & DV> problem definition
Problemdiagnose f <Comp & DV> problem determination
Problemlösen n <Künstl Int> problem solving
Problemlösungsstrategie f <Künstl Int> problem solving strategy
Problemlösungsvorgang m <Künstl Int> problem solving
Problemmüll m <Abfall> hazardous waste
problemorientierte Programmiersprache f <Comp & DV> problem-oriented language
problemorientierte Programmiersprache f **für Geschäftsbetrieb** (COBOL) <Comp & DV> common business oriented language (COBOL)
problemorientierte Software f <Comp & DV> problem-oriented software
Procain n <Chemie> procaine
Proctor-Prüfung f <Bau> Proctor test
Proctor-Versuch m <Kohlen> Proctor compaction test
Produkt n 1. <Comp & DV> product; 2. <Math> product (Ergebnis einer Multiplikation)
Produktdarstellung f <Math> factorization
Produktdetektor m <Funktech> product detector
Produktentanker m <Wassertrans> product carrier (Spezialschiff zur Beförderung von Erdölprodukten)
Produktentwicklung f <Comp & DV> product design
Produktergonomie f <Ergon> product ergonomics
Produktfehlerberichtigung f <Comp & DV> product fix
Produktion f 1. <Comp & DV> production; 2. <Kerntech> output; 3. <Kohlen> output, yield; 4. <Maschinen> production; 5. <Nichtfoss Energ> yield
Produktion f **pro Flächeneinheit** <Ker & Glas> production per unit area
Produktionsabfall m <Abfall> process waste
Produktionsausbeute f <Elektronik> fabrication yield
Produktionsband n <Kontroll> line, production line
Produktionsbericht m <Comp & DV> production statement
Produktionsbohrung f <Erdöl> production well (Fördertechnik)
Produktionsdrehmaschine f 1. <Fertig> manufacturing lathe; 2. <Maschinen> production lathe
Produktionsdruck m <Comp & DV> production printing
Produktionseinrichtungen fpl <Fernseh> production facilities
Produktionskolonne f <Erdöl> production string (Fördertechnik)
Produktionskonsole f <Fernseh> production console
Produktionsmenge f <Papier> output
Produktionsphase f <Erdöl> production phase (Fördertechnik)

Produktionsplattform f <Erdöl> production platform *(Fördertechnik)*
Produktionsregel f <Künstl Int> condition-action rule, if-then rule, production rule
Produktionsregieraum m <Fernseh> production control room
produktionsspezifischer Abfall m <Abfall> process waste
Produktionsstätte f <Mechan> fabricating shop
Produktionssystem n <Künstl Int> production system
Produktions- und Fertigungstechnik f <Ergon> industrial engineering
Produktivität f <Qual> productive efficiency, productivity
Produktivzeit f <Comp & DV> productive time
Produktpositionierung f <Comp & DV> product positioning
Produktüberwachung f <Kfztech, Qual> product tracking
produzieren v <Fernseh> produce
professioneller Mobilfunk m <Mobilkom> professional mobile radio
Profil n 1. <Anstrich, Comp & DV> profile; 2. <Bau> section; 3. <Fertig> *(AE)* form; 4. <Kfztech> tread *(Reifen)*; 5. <Maschinen> outline, profile, section; 6. <Metall> section *(Eisen, Stahl)*; 7. <Phys> profile
Profil... <Maschinen> sectional
Profilbauch m <Phys> lower surface
Profilbezugslinie f <Maschinen> pitch line
Profildichtung f 1. <Kfztech> gasket; 2. <Maschinen> gasket, profiled gasket
Profildispersion f <Optik, Telekom> profile dispersion
Profildispersionskoeffizient m <Telekom> profile dispersion parameter
Profileisenträger m <Bau> beam
Profilfräsen n <Maschinen> form milling, profile milling
Profilfräsmaschine f <Maschinen> profile-milling machine, profiler
Profilglas n <Ker & Glas> bent glass
Profilleiter m <Elektriz> sector-shaped conductor
Profilparameter m <Telekom> profile parameter
Profilriemen m <Maschinen> profiled belt
Profilrolle f **mit ausgearbeitetem Profil** <Fertig> crusher roll
Profilschleifen n <Maschinen> profile grinding, profiling
Profilschleifer m <Maschinen> profile grinder
Profilschnitt m <Konstzeich> removed section
Profilsehne f <Lufttrans> *(BE)* aerofoil chord, *(AE)* airfoil chord, chord
Profilsehnenteil n **des Fachwerks** <Lufttrans> chord member of a truss
Profilstahl m 1. <Maschinen, Metall> section steel; 2. <Wassertrans> steel section *(Schiffbau)*
Profiltiefe f <Lufttrans> chord ratio
Profiltiefenmesser m <Kfztech> tread depth gauge *(Reifen)*
Profilton m <Ker & Glas> tread clay
Profilüberdeckung f <Maschinen> transverse contact ratio
Profilverschiebung f <Fertig> addendum correction, addendum modification, addendum shift; correction *(Zahnrad)*
Profilverschiebung f **eines Diameters** <Fertig> diameter increment *(Getriebelehre)*
Profilverschiebungsfaktor m <Fertig> addendum coefficient, addendum modification coefficient
profilwalzen v <Fertig> roll-form
Profilwiderstand m <Lufttrans> profile drag
Profilziehverfahren n <Kunststoff> pultrusion *(Extrusion)*
Progesteron n <Chemie> progesterone
Progestin n <Chemie> progesterin

Prognose f 1. <Comp & DV> forecasting; 2. <Künstl Int> prediction; 3. <Wasserversorg> forecasting
Prognosesystem n <Künstl Int> prediction system
Prognostizierungsfähigkeit f <Künstl Int> what-if capability *(Expertensystem)*
Programm n 1. <Comp & DV> program, routine, software; schedule *(Zeitplan)*; 2. <Fernseh> *(AE)* program, *(BE)* programme; 3. <Telekom> program • **alle Programme laufen** <Comp & DV> all programs run • **ins Programm aufnehmen** <Fernseh> *(AE)* program, *(BE)* programme
Programm n **zur Ablaufverfolgung** <Comp & DV> trace program
Programmabbruch m <Comp & DV> program abort
Programmablaufanlage f <Comp & DV> object machine
Programmablaufrechner m <Comp & DV> object computer
Programmablaufverfolgung f <Comp & DV> trace
Programmabsturz m <Comp & DV> crash, program crash
Programmanbieter m <Comp & DV> supplier of programming
Programmänderung f <Comp & DV> program patch
Programmanforderung f <Comp & DV> program request
Programmanweisung f <Comp & DV> program statement
Programmaufruf m <Comp & DV> program request
Programmausführung f <Comp & DV> program execution, run
Programmausrüstung f <Comp & DV> software
Programmbaustein m <Comp & DV> program unit
Programmbefehl m <Comp & DV> program instruction
Programmbereich m <Comp & DV> partition
Programmbewertungs- und Überprüfungsverfahren n (PERT) <Comp & DV, Raumfahrt> program evaluation and review technique *(PERT)*
Programmbibliothek f <Comp & DV> program library
Programmdatei f <Comp & DV> program file
Programmdokumentation f <Comp & DV> program documentation
Programmdurchlauf m <Comp & DV> program run
Programmeinheit f <Comp & DV> program unit
Programmende n <Comp & DV> end of program
Programmentwicklung f <Comp & DV> program development
Programmentwurf m <Comp & DV> program design
Programmfolge f <Comp & DV> suite of programs
Programmfunktionssymbol n <Comp & DV> soft key
Programmgeber m <Elektrotech> control timer, sequencer
Programmgenerator m <Comp & DV> program generator
programmgesteuert adj <Comp & DV> software-driven
programmgesteuerte Ausspeicherung f <Comp & DV> memory dump
programmgesteuerte Selbstwählnebenstelle f <Telekom> *(AE)* stored program control PABX, *(BE)* stored programme control PABX
programmgesteuerte Vermittlungsstelle f <Telekom> *(AE)* stored program control exchange, *(BE)* stored programme control exchange
Programmgruppe f <Comp & DV> suite of programs
Programmhandbuch n <Comp & DV, Patent, Qual> documentation
programmierbar adj <Kontroll> programmable • **vom Anwender programmierbar** <Telekom> field-programmable
programmierbare Einheit f <Comp & DV> programmable device

programmierbare

programmierbare Folgesteuerung *f* <Telekom> programmable sequencer
programmierbare Kommunikationsschnittstelle *f* <Telekom> programmable communication interface *(PCI)*
programmierbare Logikanordnung *f (PLA)* <Comp & DV> programmable logic array *(PLA)*
programmierbare logische Anordnung *f (PAL)* <Comp & DV> programmable array logic *(PAL)*
programmierbare parallele Schnittstelle *f* <Telekom> programmable parallel interface
programmierbare serielle Schnittstelle *f* <Telekom> programmable serial interface
programmierbare Speichersteuerung *f* <Kontroll> programmable memory control
programmierbare Steuerung *f* <Comp & DV, Telekom> programmable control
programmierbare Taste *f* <Comp & DV> programmable key
programmierbarer Doppelbasistransistor *m* <Elektronik> programmable unijunction transistor
programmierbarer Lesespeicher *m (PROM)* <Comp & DV> programmable read-only memory *(PROM)*
programmierbarer logischer Schaltkreis *m* <Telekom> programmable logic circuit
programmierbarer Oszillator *m* <Elektronik> programmable oscillator
programmierbarer Regler *m* <Labor> programmable controller
programmierbarer Signalgenerator *m* <Elektronik> programmable signal generator
programmierbarer Speicher *m* <Comp & DV> programmable memory
programmierbares Gerät *n* <Comp & DV, Telekom> programmable device
programmierbares logisches Feld *n* <Comp & DV> programmable logic array
programmierbares Steuergerät *n* <Labor> programmable controller
programmieren *v* <Comp & DV, Telekom> program
Programmieren *n* <Comp & DV> programming
Programmierentwicklungssystem *n* <Comp & DV> software tool
Programmierer *m* <Comp & DV> programmer
Programmierereinheit *f* <Comp & DV> programmer unit
Programmiergerät *n* <Comp & DV> programmer
Programmierhilfe *f* <Comp & DV> software tool
Programmierhochsprache *f* <Comp & DV> CCITT high-level language, high-level language, high-level programming language *(CHILL)*
Programmiersprache *f* <Comp & DV> coding language, language, programming language
Programmiersprache *f* **für Dokumente im Internet** <Telekom> Hypertext markup language *(HTML)*
Programmierstandars *mpl* <Comp & DV> programming standards
programmiert *adj* <Comp & DV> programmed
programmierte Anweisung *f* <Ergon> programmed instruction
programmierte automatische Videoaufzeichnung *f* <Aufnahme, Fernseh> programmed automatic video recording
programmierte Videokassettenaufzeichnung *f* <Aufnahme, Fernseh> programmed video cassette recording, programmed video recording
programmierte Werkzeuge *npl (APT)* <Comp & DV> automatically programmed tools *(APT)*
programmierter Halt *m* <Comp & DV> program stop
programmierter Verbrennungsmotor *m* <Kfztech> programmed combustion engine

programmiertes Servosystem *n* <Raumfahrt> programmed servosystem *(Weltraumfunk)*
Programmierung *f* **mit minimaler Zugriffszeit** <Comp & DV> minimum-access programming
Programmierwerkzeug *n* <Comp & DV> software tool
Programmkompatibilität *f* <Comp & DV> program compatibility
Programmlauf *m* <Comp & DV> program run
Programmliste *f* <Comp & DV> program listing
Programmmodifikation *f* <Comp & DV> preparation
Programmpaket *n* <Comp & DV> package
Programmparameter *m* <Comp & DV> program parameter
Programmpflege *f* <Comp & DV> program maintenance
Programmprüfung *f* <Comp & DV> program testing; program check *(Basissystem)*
Programmrumpf *m* <Comp & DV> program body
Programmsatz *m* <Comp & DV> sentence
Programmschalter *m* <Comp & DV> sense switch
Programmspeichersteuerung *f* 1. <Elektriz> stored program control *(SPC)*; 2. <Telekom> *(AE)* stored program control, *(BE)* stored programme control
Programmspezifikation *f* <Comp & DV> program specification
Programmsprache *f* <Comp & DV> language
Programmstapel *m* <Comp & DV> program stack
Programmstatuswort *n* <Comp & DV> PSW, program status word
Programmsteuerung *f* <Comp & DV> program control
Programmstruktur *f* <Comp & DV> program structure
Programmsynthese *f* <Künstl Int> program synthesis
Programmteil *m* <Comp & DV> program part
Programmtest *m* <Comp & DV> bug; program testing, program check *(virtuelle Maschinen)*
Programmtesten *n* <Comp & DV> program checkout
Programmtonspur *f* <Fernseh> programme audio track
Programmumsetzer *m* <Fernseh> *(AE)* program repeater, *(BE)* programme repeater
Programmunterlagen *fpl* <Comp & DV> program documentation
Programmverbindung *f* <Comp & DV> link editing
• **Programmverbindung herstellen** <Comp & DV> link-edit
Programmverbindungssoftware *f* <Comp & DV> linkage software
Programmverifikation *f* <Comp & DV> program verification
Programmverknüpfung *f* <Comp & DV> linkage, program linking
Programmverschiebung *f* <Comp & DV> program relocation
Programmverzahnung *f* <Comp & DV> multiprogramming
Programmverzahnungssystem *n* <Comp & DV> multiprogramming system
Programmverzeichnis *n* <Comp & DV> program directory
Programmverzweigung *f* <Comp & DV> branch
Programmverzweigungspunkt *m* <Elektronik> program branch point *(Programmablauf)*
Programmvorschauseite *f* <Fernseh> index page
Programmwähler *m* <Fernseh> *(AE)* program selector, *(BE)* programme selector
Programmwiederaufnahme *f* <Comp & DV> fall-back recovery
Programmzähler *m* <Comp & DV> program counter
Programmzeituhr *f* <Labor> *(AE)* program timer, *(BE)* programme timer
Programmzuführungskabel *n* <Telekom> trunk cable *(Kabelfernsehen)*

Programmzweig m <Comp & DV, Elektronik> branch, program branch
progressive Alterung f <Kunststoff> (BE) progressive ageing, (AE) progressive aging
Progressivsystem n <Trans> progressive system
Projekt n <Comp & DV> project
Projektänderung f <Bau> variation order
Projektentwurf m <Maschinen> project design
projektieren v <Comp & DV, Qual> project
Projektion f 1. <Bau, Druck> projection; 2. <Phys> projection (eines Vektors)
Projektion f senkrecht zur Projektionsebene <Geom> orthogonal projection, orthographic projection
Projektionsaufsatz m <Optik> viewing screen (Mikroskop)
Projektionsbild n 1. <Optik> projected image; 2. <Foto> screen image
Projektionsdiode f <Elektronik> projection diode
Projektionsebene f <Geom> plane of projection, projection plane
Projektionsfernsehen n <Fernseh> projection television
Projektionsfläche f <Bau> projected area
Projektionsgerät n <Optik> light projector, projection equipment, projector
Projektionskoordinaten fpl <Geom> projection coordinate system
Projektionslampe f <Foto> projector lamp
Projektionslänge f <Bau> projection length
Projektionslinie f <Konstzeich> projection line
Projektionslinse f <Optik> projector lense
Projektionslithographie f <Elektronik> projection lithography
Projektionsmethode f <Konstzeich> first angle projection method, projection method
Projektionsmikroskop n <Optik> projection microscope
Projektionsobjektiv n <Foto> projection lens
Projektionsrichtung f <Geom> direction of projection
Projektionsskaleninstrument n <Gerät> projected-scale instrument
Projektionsskalenmessgerät n <Gerät> projected-scale instrument
Projektionsstrahl m <Geom> projection ray
Projektionswand f <Optik> projection screen
projektive Geometrie f <Geom> projective geometry
Projektor m <Foto> projector
Projektsteuerung f <Qual> project controller
Projektstudie f <Comp & DV> feasibility study
Projektüberwacher m <Qual> project controller
Projektüberwachung f <Bau, Qual> project monitoring
Projektzeichnung f <Konstzeich> project drawing
PROM (programmierbarer Lesespeicher) <Comp & DV, Funktech> PROM (programmable read-only memory)
Promenadendeck n <Wassertrans> hurricane deck, promenade deck
Promethium n (Pm) <Chemie> promethium (Pm)
PROM-Programmierer m <Comp & DV> PROM programmer
PROM-Programmiergerät n <Comp & DV> PROM burner
Prompt n <Comp & DV> prompt
prompte Gammastrahlung f <Kerntech, Strahlphys> prompt gamma radiation
promptes Neutron n <Phys> prompt neutron
promptes Neutronen npl <Kerntech, Strahlphys> prompt neutrons
prompt-kritisch adj <Kerntech> prompt-critical
Pronation f <Ergon> pronation
Prontosil n <Chemie> prontosil
Prony'scher Zaum m <Kfztech, Maschinen> Prony brake

Propadien n <Chemie> allene, propadiene
Propagator m <Phys> propagator
Propagierung f <Künstl Int> propagation
Propan n <Erdöl> propane (Petrochemie)
Propanamin n <Chemie> aminopropane, propylamine
Propandiamid n <Chemie> malonamide, propanediamide
Propanflasche f <Sicherheit> propane cylinder
Propanol n <Chemie> isopropanol, propanol
Propanon n <Chemie> acetone, propanone
Propantanker m <Erdöl> propane tanker (Schifffahrt)
Propantankschiff n <Erdöl> propane tanker
Propeller m 1. <Lufttrans> airscrew, screw; 2. <Maschinen, Meerschmutz, Papier> propeller; 3. <Wassertrans> propeller (Schiffbau); screw (Schiffantrieb)
Propeller m mit konstanter Drehzahl <Lufttrans> constant-speed propeller
Propeller m mit umkehrbarer Steigung <Wassertrans> reversible pitch propeller
Propeller m mit veränderlicher Steigung <Lufttrans, Wassertrans> variable-pitch air propeller
Propellerantrieb m <Lufttrans> propeller drive
Propellernabe f <Wassertrans> propeller hub (Schiffbau)
Propellernuss f <Wassertrans> propeller boss (Schiffbau)
Propellerrelais n <Lufttrans> propeller relay unit (Flugzeug)
Propellerschleppdrehmoment n <Lufttrans> windmill torque
Propellerschub m <Lufttrans> propeller thrust
Propellersteigung f <Lufttrans> propeller pitch
Propellerturbine f 1. <Lufttrans> turbopropeller; 2. <Maschinen> propeller turbine
Propellerturbinenluftstrahlflugzeug n <Lufttrans> propeller turbine plane
Propellerwelle f <Lufttrans, Wassertrans> propeller shaft (Schiffbau)
Propen n <Chemie, Erdöl> propene, propylene (Petrochemie)
Propenyl... n <Chemie> isopropenyl, propenyl
Propin n <Chemie> allylene, methylacetylene, propyne
Propiol... <Chemie> propynoic
Proportion f <Geom, Math> proportion
proportional wirkender Abschwächer m <Foto> proportional reducer
proportional zu adj <Math> in proportion to
Proportional... <Math> proportional
Proportionalabweichung f <Elektronik> proportional offset
Proportionalbeiwert m <Regelung> proportional coefficient
Proportional-Differenzial-Regelung f (PD-Regelung) <Labor> proportional plus derivative control (PD control)
Proportional-Differenzial-Regler m (PD-Regler) <Labor> proportional plus derivative controller (PD controller)
Proportional-Differenzial-Verhalten n (PD-Verhalten) <Regelung> proportional plus derivative action (PD action)
proportionale Kapazität f <Elektrotech> straight line capacitance
Proportionalglied n <Regelung> proportional element
Proportional-Integral-Differenzial-Regler m (PID-Regler, PID-Steuerung) <Labor> proportional plus integral plus derivative controller (PID controller)
Proportional-Integral-Differenzial-Verhalten n (PID-Verhalten) <Labor> proportional plus integral plus derivative action (PID action)
Proportional-Integral-Regler m (PI-Regler) <Labor> proportional plus integral controller (PI controller)
Proportional-Integral-Verhalten n (PI-Verhalten) <Labor> proportional plus integral action (PI action)

Proportionalkammer

Proportionalkammer *f* <Teilphys> proportional chamber
Proportionalzähler *m* <Phys, Teilphys> proportional counter
Proportionierung *f* <Fertig> proportioning
Propyl... <Chemie> propyl
Propylamin *n* <Chemie> aminopropane, propylamine
Propylen *n* <Chemie, Erdöl> propene, propylene *(Petrochemie)*
Propylpiperidin *n* <Chemie> conicine, coniine, propylpiperidine
Propynyl... <Chemie> propargyl, propynyl
Prospekt *m* <Fertig> brochure *(Kunststoffinstallationen)*
Prospektion *f* <Erdöl> prospecting *(Lagerstättensuche)*
Protease *f* <Chemie> protease
Proteinfasern *fpl* <Textil> *(AE)* protein fibers, *(BE)* protein fibres
proteinspaltendes Enzym *n* <Chemie> protease
proteolytisches Enzym *n* <Chemie> protease
Protokoll *n* 1. <Comp & DV> log, protocol, trace; 2. <Druck, Funktech> protocol; 3. <Kontroll> record; 4. <Telekom> protocol
Protokollausgabe *f* <Comp & DV> recording output
Protokollbereich *m* <Comp & DV> bucket
Protokollhierarchie *f* <Comp & DV> protocol hierarchy
protokollieren *v* 1. <Comp & DV, Telekom> log; 2. <Kontroll> record
Protokollieren *n* **und Berichten** *n* <Comp & DV> logging and reporting
Protokollkonverter *m* <Telekom> protocol converter
Protokollumsetzer *m* <Comp & DV, Telekom> protocol converter
Protolyse *f* <Chemie> protolysis
Proton *n* *(p)* 1. <Funktech, Kerntech> proton, p; 2. <Phys> proton, p *(Elementarteilchen)*; 3. <Teilphys> proton, p
Protonenabsorptionsfähigkeit *f* <Umweltschmutz> proton-absorptive capacity
Protoneneinstrahlung *f* <Raumfahrt> proton irradiation
Protonenmasse *f (mp)* <Kerntech> proton mass *(mp)*
Protonenzahl *f* 1. <Kerntech> proton number *(P)*; 2. <Teilphys> proton number
Protopektin *n* <Chemie> pectose
Prototyp *m* <Comp & DV, Elektriz, Maschinen, Wassertrans> prototype
Prototypenbau *m* <Maschinen> prototype construction
Prototypstadium *n* <Strahlphys> prototype stage
Protuberanz *f* <Raumfahrt> solar flare
provisorisch *adj* <Bau, Patent, Qual> provisional
Proximity-Effekt *n* <Phys> proximity effect
Prozedur *f* <Comp & DV> procedure
prozedurale Programmiersprache *f* <Comp & DV> procedural language
Prozedurbibliothek *f* <Comp & DV> procedure library
Prozedurname *m* <Comp & DV> procedure name
Prozedurteil *n* <Comp & DV> procedure division
Prozedurvereinbarungsanweisung *f* <Comp & DV> declarative statement
Prozedurvereinbarungssatz *m* <Comp & DV> declarative sentence
Prozedurvereinbarungsteil *m* <Comp & DV> declarative section
Prozentrelais *n* <Elektrotech> biased relay
Prozentsatz *m* <Math> percentage
Prozentsatz *m* **fehlerfreier Sekunden** <Telekom> percentage of error-free second *(Richtfunk)*
Prozentsatz *m* **fehlerhafter Einheiten** <Qual> percent defective
prozentuale Beeinträchtigung *f* **des Hörvermögens** <Akustik> percent impairment of hearing

prozentuale Zusammensetzung *f* <Ker & Glas> percentage composition
prozentualer Gehalt *m* <Metall> percentage
prozentualer Modulationsgrad *m* <Elektronik> percentage modulation
Prozess *m* <Comp & DV, Kohlen, Papier> process
prozessabhängige Ablaufsteuerung *f* <Regelung> process-oriented sequential control
Prozessaussetzung *f* <Comp & DV> process suspension
Prozessautomatisierung *f* <Comp & DV> process automation
Prozessdatenverarbeitung *f* <Comp & DV> process control
Prozesseintrag *m* <Comp & DV> process entry
Prozessfarben *fpl* <Druck> process colours
Prozessführung *f* <Kontroll> process management
Prozessgas *n* <Kohlen> process gas
prozessintegriert *adj* <Metrol> in-process
Prozessmessgröße *f* <Gerät> measured process quantity
Prozessmesswert *m* <Gerät> process value
Prozessor *m* <Comp & DV, Telekom> processor • **durch den Prozessor begrenzt** <Comp & DV> processor-limited
Prozessor *m* **für Datenübertragung** <Telekom> communications processor
Prozessor *m* **mit komplettem Befehlssatz** *(CISC)* <Comp & DV> complex instruction set computer *(CISC)*
prozessorgesteuerte Eingabe *f* <Comp & DV> processor-controlled keying
Prozessorspeicher *m* <Comp & DV> processor storage
Prozessorstatuswort *n* <Comp & DV> PSW, processor status word
Prozessorzwischenverbindung *f* <Telekom> interprocessor link
Prozessprüfung *f* <Qual> process inspection
Prozesssignalformer *m* <Regelung> receiver element
Prozesssteuerung *f* <Comp & DV, Elektriz, Elektrotech, Ergon, Kontroll, Telekom> process control
Prozessunterbrechung *f* <Comp & DV> process interrupt
Prozess-Validierung *f* <Qual> process validation
Prozessvariable *f* <Maschinen> process variable
Prozesswert *m* <Gerät> process value
Prüf... 1. <Comp & DV, Elektrotech, Gerät, Qual> inspection, test; 2. <Verpack> check; 3. <Werkprüf> inspection, test
Prüfablauf *m* <Qual> inspection and test sequence, test procedure, test run, test sequence, testing procedure
Prüfablaufplan *m* <Qual> inspection and testing plan, inspection and test schedule, inspection and test sequence plan, inspection schedule
Prüfanlage *f* <Werkprüf> test equipment, test plant, test rig, testing plant, testing set
Prüfanordnung *f* 1. <Gerät> calibration set-up; 2. <Qual> test and examination sequence plan, test set-up
Prüfanstalt *f* <Qual> testing laboratory
Prüfanweisung *f* <Qual> inspection instruction, test instruction
Prüfaufkleber *m* <Qual> inspection sticker
Prüfaufzeichnung *f* <Qual> inspection and test records, inspection record, test record, testing record
Prüfautomat *m* 1. <Fertig> automatic tester; 2. <Metrol, Qual> automatic tester, automatic test equipment, automatic testing equipment *(ATE)*
Prüfbeanspruchung *f* <Maschinen> proof stress
Prüfbecher *m* <Labor> flow cup
Prüfbedingungen *fpl* <Qual, Verpack> test conditions
Prüfbericht *m* 1. <Metrol> inspection record; 2. <Qual> inspection record, inspection report, test record, test report

Prüftechnik

Prüfbescheinigung f <Qual> inspection certificate, test certificate
Prüfbestätigung f <Qual> test certificate
Prüfbetrieb m <Qual> testing shop
Prüfbit n 1. <Comp & DV> check bit, parity bit; 2. <Elektronik> check bit
Prüfblech n <Hydraul> test plate
Prüfbrett n <Elektriz> test board
Prüfcheckliste f <Qual> inspection checklist
Prüfdaten npl <Comp & DV> test data
Prüfdatengenerator m <Comp & DV> test data generator
Prüfdraht m <Elektrotech> pilot wire
Prüfdruck m 1. <Hydraul> test pressure; 2. <Maschinen> proof pressure
Prüfeinrichtung f <Gerät, Qual> calibration equipment, test equipment
Prüfeinrichtungen fpl <Qual> inspection and test equipment
prüfen v 1. <Bau> try; 2. <Comp & DV> sense, verify; 3. <Qual> inspect; test (Messtechnik); 4. <Sicherheit> approve
Prüfen n <Comp & DV> testing
Prüfen n **mit induzierter Überspannung** <Elektriz> induced overvoltage test
Prüfer m 1. <Comp & DV> verifier; 2. <Konstzeich> checker; 3. <Patent> examiner; 4. <Qual> examiner, inspector
Prüfergebnisse npl <Qual> test results
Prüfetikett n <Verpack> control tag
Prüffeld n 1. <Metrol> proving ground, test department, test laboratory, test panel, testing department, testing laboratory, testing panel; 2. <Telekom> test room
Prüffolge f <Qual> test sequence
Prüffolgeweg m <Kontroll> audit trail
Prüffrequenz f <Werkprüf> test frequency
Prüfgegenstand m <Qual> test item
Prüfgerät n 1. <Comp & DV> test equipment; 2. <Elektrotech> inspection instrument, test apparatus, test device, test set, tester; 3. <Gerät> calibration instrument, checking instrument, test equipment, (AE) test gage, (BE) test gauge, testing instrument; 4. <Qual> checking instrument, tester, testing apparatus, testing instrument; 5. <Verpack> inspection equipment
prüfgerechte Maßeintragung f <Konstzeich> inspection-oriented dimensioning
Prüfgeschwindigkeit f <Werkprüf> test speed
Prüfglas n <Verpack> inspection window
Prüfinstrument n <Gerät, Qual> calibration instrument, checking instrument, testing instrument
Prüfintervall m <Qual> inspection interval
Prüfkabel n <Elektrotech> test lead
Prüfklemme f <Elektrotech> test terminal
Prüfklinke f <Elektrotech> test jack
Prüflaboratorium n <Qual> testing laboratory
Prüflast f 1. <Lufttrans> proof load (Lufttüchtigkeit); 2. <Maschinen> proof load, test load
Prüflauf m <Comp & DV> test run
Prüflehre f 1. <Gerät> (AE) test gage, (BE) test gauge; 2. <Maschinen> (AE) master gage, (BE) master gauge, (AE) reference gage, (BE) reference gauge
Prüfleitung f <Elektrotech> test line, test wire
Prüfling m 1. <Gerät, Kohlen, Kunststoff, Maschinen, Mechan> specimen; 2. <Qual> device under test, part under test, specimen
Prüfliste f 1. <Comp & DV> audit trail; 2. <Patent, Qual> check list
Prüflos n <Qual> inspection lot, test lot
Prüfluke f <Raumfahrt> inspection door (Raumschiff)

Prüfmanometer n <Gerät> (AE) test gage, (BE) test gauge
Prüfmaschine f <Qual> tester, testing apparatus, testing machine
Prüfmenge f <Qual> test quantity
Prüfmerkmal n <Qual> inspection characteristic
Prüfmethode f <Qual> test method
Prüfmittel n 1. <Chemie> testing agent; 2. <Fertig> testing equipment
Prüfmittelüberwachung f <Qual> control of inspection, measuring and test equipment
Prüfmuster n 1. <Qual> sample; 2. <Telekom> test pattern
Prüfnadel f <Maschinen> test needle
Prüfniveau n <Qual> inspection level
Prüfobjekt n <Qual> device under test, unit under test
Prüfort m <Qual> place of inspection
Prüfoszillator m <Elektronik> test oscillator
Prüfpfad m <Comp & DV> audit trail (Buchungskontrolle)
prüfpflichtig adj <Qual> requiring approval, requiring official approval
Prüfplakette f <Qual> inspection sticker
Prüfplan m <Qual> inspection and test schedule, inspection plan
Prüfplanung f <Qual> inspection and test planning, inspection planning
Prüfplatz m <Telekom> test position
Prüfprisma n <Maschinen> V-block
Prüfprobe f <Qual> test sample
Prüfprogramm n 1. <Comp & DV> test routine; 2. <Maschinen> test schedule
Prüfprotokoll n 1. <Comp & DV> audit trail; 2. <Qual> inspection record, test record, test report
Prüfpunkt m 1. <Comp & DV> checkpoint; 2. <Qual> inspection point; 3. <Telekom> test point
Prüfreihenfolge f <Qual> test sequence
Prüfroutine f <Qual> test routine
Prüfsachverständiger m <Qual> authorized inspector; factory-authorized inspector (im Werk)
Prüfschalter m <Elektrotech> test switch
Prüfschärfe f 1. <Elektriz> severity of testing; 2. <Patent> degree of inspection; 3. <Qual> applicability, degree of inspection, inspection level, severity of test
Prüfschild n <Ker & Glas> test plate
Prüfschnur f <Elektrotech> test patch, test cord
Prüfschrank m <Gerät> test board
Prüfsignal n <Werkprüf> test signal
Prüfsignalgeber m <Elektronik> test signal generator
Prüfsonde f <Gerät> test probe
Prüfspannung f <Elektriz> proof voltage, test voltage
Prüfspezifikation f <Qual> inspection specification
Prüfspitze f 1. <Elektronik> test prod; 2. <Gerät> probe, test prod
Prüfspule f <Elektrotech, Fertig> test coil
Prüfspur f <Comp & DV> parity track
Prüfstab m <Maschinen> test bar, test sieve
Prüfstand m 1. <Elektriz> test bed; 2. <Kerntech> test stand; 3. <Kfztech> testing bed; 4. <Maschinen> test bed, test bench; 5. <Mechan> test bed; 6. <Qual> testing bed, testing bench; 7. <Werkprüf> test rig
Prüfstelle f <Qual> testing agency
Prüfstempel m <Qual> inspection stamp
Prüfstoff m <Chemie> reagent
Prüfstück n 1. <Maschinen> test piece, test specimen; 2. <Phys> test piece; 3. <Qual> part under test, test sample, test specimen
Prüfsumme f <Comp & DV, Elektronik> checksum (Datenverarbeitung)
Prüftechnik f <Sicherheit> test engineering

Prüftest

Prüftest m <Elektriz> proof test
Prüfturnus m <Qual> inspection interval
Prüfumfang m <Qual> amount of inspection, scope of inspection
Prüf- und Kontrollpunkt m <Qual> inspection and test point
Prüf- und Messmittel npl <Qual> measuring and test equipment
Prüfung f 1. <Comp & DV> check, test, verification, sense, testing; 2. <Kohlen> assay, trial; 3. <Kontroll, Maschinen> verification; 4. <Patent> examination; 5. <Qual> inspection, test, verification; 6. <Textil> test
Prüfung f am Boden <Lufttrans> ground test
Prüfung f auf gerade Bitzahl <Comp & DV, Telekom> even parity
Prüfung f auf plötzlichen Kurzschluss <Elektriz> sudden short-circuit test
Prüfung f auf ungerade Parität <Comp & DV> odd parity
Prüfung f bei Normalbetrieb <Metall, Qual> dynamic test
Prüfung f bei Null-Leistungsfaktor <Elektriz> zero-power-factor test
Prüfung f bei Sonneneinstrahlung <Nichtfoss Energ, Werkprüf> solar radiation test
Prüfung f gehärteter Zahnräder <Fertig> hard test (Getriebelehre)
Prüfung f im Werk <Mechan> factory inspection
Prüfung f im Windkanal <Werkprüf> wind tunnel test
Prüfungsabteilung <Qual> test department
Prüfungsanforderungen fpl <Qual, Werkprüf> test requirements
Prüfungsgrad m <Patent, Qual> degree of inspection
Prüfungsklasse f <Kohlen> assay grade
Prüfungsprotokolle npl <Patent, Qual> examination records
Prüfungswert m <Kohlen> assay value
Prüfunterlagen fpl <Qual> inspection and test documents
Prüfverfahren n 1. <Metrol> audit procedure, inspection procedure; 2. <Qual> inspection and test procedure, inspection procedure, test procedure
Prüfverfahren n für LWL <Fertig> (AE) fiber optic test procedure, (BE) fibre optic test procedure (FOTP)
Prüfvermerk m <Konstzeich> check note
Prüfvorschrift f 1. <Metrol> inspection procedure; 2. <Qual> test code, test specification
Prüfweg m <Kontroll> audit trail
Prüfwert m <Gerät> test value
Prüfzeichen n 1. <Comp & DV> check character; 2. <Qual> mark of conformity; 3. <Telekom> check character
Prüfzeichnung f <Konstzeich> appraisal drawing
Prüfzeilen-Messplatz m <Fernseh> insertion signal test set
Prüfzeilenmesssignal n <Fernseh> vertical interval test signal
Prüfzertifikat n <Patent, Qual> inspection certificate
Prüfzeugnis n <Elektriz, Qual> inspection report, proof certificate, test certificate
Prüfziffer f <Comp & DV> check digit
Prüfzustand m <Qual> inspection and test status, inspection status
Prüfzuverlässigkeit f <Qual> test reliability
Prulaurasin n <Chemie> prulaurasin
Prunetol n <Chemie> genistein, prunetol
Prussiat n <Chemie> prussiate
PS 1. <Chemie, Kunststoff> (Polystyren, Polystyrol) PS (polystyrene); 2. <Elektriz> (Polystyrol) PS (polystyrene); 3. <Maschinen> (Pferdestärke) hp (horsepower); 4. <Wassertrans> (Peilstrahl) EBL (electronic bearing line)

pseudoakustisches Log n <Erdöl> pseudosonic log
pseudoakustisches Profil n <Erdöl> pseudosonic profile
Pseudobefehl m <Comp & DV> pseudoinstruction
Pseudocode m <Comp & DV> pseudocode
pseudo-elliptisches Filter n <Raumfahrt> pseudoelliptic filter (Weltraumfunk)
Pseudooperation f <Comp & DV> pseudooperation
pseudopotenziometrische Höhe f <Erdöl> pseudopotentiometric head
Pseudorauschcode m <Telekom> pseudorandom noise code
Pseudorauschen n <Telekom> PN, pseudonoise, pseudorandom noise
Pseudosphäre f <Geom> pseudosphere
Pseudosprache f <Comp & DV> pseudolanguage
Pseudostereophonie f <Aufnahme> pseudostereophony
pseudozufällig adj <Telekom> pseudorandom
Pseudozufalls... <Telekom> pseudorandom
Pseudozufallssignal n <Telekom> pseudorandom signal
Pseudozufallszahl f <Comp & DV> pseudorandom number
PSK 1. <Comp & DV> (Phasensprungtastung) PSK (phase shift keying); 2. <Elektronik, Funktech, Raumfahrt, Telekom> (Phasenumtastung) PSK (phase-shift keying)
psophometrischer Gewichtungsfaktor m <Raumfahrt> psophometric weighting factor (Weltraumfunk)
Psychoakustik f <Ergon> psychoacoustics
psychomotorisch adj <Ergon> psychomotor
Psychotrin n <Chemie> psychotrine
Psychrometer n <Labor> psychrometer
Pt (Platin) <Chemie> Pt (platinum)
PTC (passive Thermosteuerung) <Raumfahrt> PTC (passive thermal control)
Pterin n <Chemie> pterin
PTFE (Polytetrafluorethen, Polytetrafluorethylen) <Kunststoff> PTFE (polytetrafluoroethylene)
p-Toluolsulfonyl-... <Chemie> tosyl...
Ptomain n <Chemie> ptomaine
Public-key-Zertifikat n <Comp & DV> public-key certificate
puddeln v <Metall> puddle
Puddeln n <Metall> puddling
Puddelstahl m <Metall> puddled steel
Puddelstraße f <Fertig> puddle train
Puddler m <Metall> puddler
Pudermittel n <Kunststoff> powder
puffen v <Lebensmittel> puff (Mais, Reis, Hülsenfrüchte)
Puffer m 1. <Bau> pad; 2. <Comp & DV> buffer, pushdown stack; 3. <Eisenbahn> buffer, (BE) bumper, (AE) fender; 4. <Elektrotech, Fernseh> buffer; 5. <Fertig> buffer (Spanung); 6. <Funktech, Kunststoff> buffer; 7. <Maschinen> buffer, (BE) bumper, (AE) fender
Pufferanschlag m <Eisenbahn> buffer stop
Pufferarchitektur f <Comp & DV> stack architecture
Pufferbatterie f <Eisenbahn, Elektrotech> buffer battery
Pufferberührung f <Eisenbahn> buffer contact
Pufferdynamo m <Elektriz> buffer dynamo
Pufferfaser f <Optik> (AE) buffer fiber, (BE) buffer fibre
Pufferfeder f 1. <Kfztech> cushion spring (Kupplung); 2. <Maschinen> damping spring
Pufferladegerät n <Elektriz> battery booster, trickle charger
Pufferladung f <Elektrotech> booster charge, trickle charge
Pufferlösung f <Lebensmittel> buffer solution
puffern v <Comp & DV, Funktech> buffer
Puffern n <Comp & DV, Optik> buffering (Lichtleiter)
Pufferregister n <Comp & DV> buffer register

Pufferrohr n <Optik> buffer tube
Pufferschaltung f <Elektronik> buffer circuit
Pufferspeicher m 1. <Comp & DV> buffer; 2. <Elektrotech> buffer memory; 3. <Fertig> in-process storage; 4. <Telekom> buffer memory
Puffersperren n <Eisenbahn> buffer locking
Pufferstück n <Comp & DV> pad
Puffersubstanz f <Lebensmittel> buffering agent
Pufferträger m <Eisenbahn> buffer beam
Pufferverstärker m <Elektronik, Funktech> buffer amplifier
Pulegon n <Chemie> pulegone
Pulfrich-Refraktometer n <Phys> Pulfrich refractometer
Pulldown-Menü n <Comp & DV> pull-down menu
Pulpe f <Kohlen> pulp
Pulper m <Papier> pulper
Puls m 1. <Phys, Telekom> pulse; 2. <Wassertrans> pulse (Radar)
Pulsabstandsmodulation f <Telekom> pulse interval modulation
Pulsamplitudenmodulation f (PAM) <Comp & DV, Elektronik, Funktech, Telekom> pulse amplitude modulation (PAM)
Pulsbetrieb m <Elektronik> pulsed mode, pulsed operation
Pulsbreite f (PB) <Comp & DV, Elektronik, Fernseh, Phys, Telekom> pulse width (PW)
Pulsbreitenfrequenz f <Elektronik> pulse duration frequency
Pulsbreitenmodulation f 1. <Elektronik> pulse width modulation, width coding; 2. <Telekom> pulse width modulation
Pulscode m <Elektronik> pulse code
Pulscodemodulation f (PCM) <Aufnahme, Comp & DV, Elektronik, Funktech, Strahlphys, Telekom> pulse code modulation (PCM)
Pulscode-Sprachdaten npl <Elektronik> PCVD, pulse code voice data
Pulsdauer f <Comp & DV, Elektronik, Fernseh, Phys, Telekom> pulse width
Pulsdauermodulation f (PDM) <Elektronik> pulse width modulation (PWM)
Pulsdehnung f <Elektronik> expansion
Pulsdeltamodulation f (PDM) <Comp & DV> pulse delta modulation (PDM)
Pulsfrequenz f <Elektronik> pulse frequency, pulse rate
Pulsfrequenz f 1. <Comp & DV, Elektronik, Phys, Telekom> PRF, pulse repetition frequency; 2. <Fernseh> pulse rate factor
Pulsfrequenzmodulation f (PFM) 1. <Comp & DV, Elektronik> pulse frequency modulation (PFM); 2. <Funktech> pulse frequency modulation, pulse FM
Pulsgenerator m 1. <Elektronik> pulse generator, pulser; 2. <Phys> pulse generator
Pulsgenerator m **mit einstellbarer Frequenz** <Elektronik, Elektrotech> variable impulse-rate generator, variable pulse-rate generator
Pulsgenerierung f <Elektronik> pulse generation
Pulsgerät n <Phys> pulse generator
Pulshöhe f <Elektronik> pulse height
pulsieren v <Kontroll> pulse
pulsierende Strömung f <Strömphys> pulsating flow
pulsierende Verbrennung f <Thermod> resonant burning
pulsierender Fehler m <Fernseh> fluctuating error
pulsierender Staustrahlmotor m <Lufttrans> pulsating jet engine, pulse jet
pulsierender Strom m <Elektriz> pulsating current (unterbrochener Strom)

pulsierendes Neutronenlog n <Erdöl> pulsed neutron log
Pulsintervall n <Elektronik> pulse interval
Pulslage f <Elektronik> pulse position
Pulslagenmodulation f (PPM) <Elektronik, Telekom> pulse position modulation (PPM)
Pulslagenmodulator m <Telekom> pulse phase modulator
Pulslänge f <Elektronik, Wassertrans> pulse length (Radar)
Pulslängencodierung f <Elektronik> pulse width coding
Pulslaser m <Elektronik> pulsed laser
Pulsmodulation f <Elektronik> pulse modulation
Pulsmodulator m <Elektronik> pulse modulator
pulsmoduliert adj <Elektronik> pulse-modulated
Pulsmodulierung f <Elektriz> pulse modulation
Pulsneutronenlog n <Erdöl> pulsed neutron log (Messtechnik)
Pulsostrahltriebwerk n <Lufttrans> athodyd
Pulsotriebwerk n <Lufttrans> athodyd
Pulsphase f <Elektronik> pulse phase
Pulsphasenmodulation f (PPM) <Comp & DV, Elektronik, Telekom> pulse phase modulation (PPM)
Pulspolarität f <Elektronik> pulse polarity
Pulsrahmen m <Telekom> frame
Pulsrate f <Elektronik> pulse rate
Pulstransformator m <Phys> pulse transformer
Pulsübertrager m <Phys> pulse transformer
Pulsverbesserung f <Phys> pulse regeneration
Pulsweitenmodulation f (PWM) <Elektronik> pulse width modulation (PWM)
pulsweitenmodulierte Rechteckwelle f <Elektronik> PWM inverter/squarewave
Pulswiederholrate f <Phys> PRR, pulse repetition rate
Pulszeitmodulation f <Elektronik> PTM, pulse time modulation
Pult n 1. <Gerät> desk; 2. <Kontroll> console
Pultdach n <Bau> lean-to roof, pitch roof, shed roof
Pultrusion f <Kunststoff> pultrusion
Pulver n <Kunststoff, Papier> powder
Pulverabfüllanlage f <Verpack> powder filling machine
Pulverbeschichten n <Kunststoff> powder coating
Pulverbeschichtung f <Kunststoff> powder coating
Pulverbeugungskamera f <Kerntech> X-ray powder camera (Röntgenstrahlen)
Pulverbrennschneiden n <Fertig> powder cutting
Pulverdiffraktometer n <Kerntech> X-ray powder diffractometer (Röntgenstrahlen)
Pulverfeuerlöscher m <Sicherheit> powder fire extinguisher
pulverförmig adj 1. <Chemtech> pulverized; 2. <Fertig> powdery (Kunststoffinstallationen); 3. <Kohlen> powdery; 4. <Kunststoff> powdered
pulverförmiges Erz n <Kohlen> fines
pulvergefüllt adj <Elektrotech> powder-filled
pulverig adj 1. <Fertig> powdery (Kunststoffinstallationen); 2. <Kohlen> powdery
pulverisieren v 1. <Chemie> triturate; 2. <Chemtech> grind, pulverize; 3. <Fertig> levigate; 4. <Kohlen, Maschinen> pulverize
Pulverisiermühle f <Chemtech> pulverizer
pulverisiert adj 1. <Chemtech> pulverized; 2. <Kunststoff> powdered
Pulverkautschuk m <Kunststoff> powdered rubber
Pulverkörnmaschine f <Verpack> granulating machine
Pulverlack-Beschichtungstechnik f (PMC-Verfahren) <Kunststoff> (AE) powder mold coating, (BE) powder mould coating (PMC)
Pulverlöscher m <Sicherheit> dry powder extinguisher, powder fire extinguisher (Brandbekämpfung)

Pulvermetall

Pulvermetall n <Maschinen> powder metal
Pulvermetallurgie f <Metall> powder metallurgy
Pulvermühle f <Chemtech> pulverizing equipment
Pulverpressteil n <Fertig> compact
Pulverraupe f <Fertig> accumulation of particles, particle pattern
pulvrige Formmasse f <Kunststoff> (AE) molding powder, (BE) moulding powder
Pumpe f <Maschinen, Meerschmutz, Wassertrans, Wasserversorg> pump
Pumpe f **mit beweglichem Widerlager** <Fertig> rotary abutment pump
Pumpe f **mit konstanter Fördermenge** <Maschinen> constant flow pump
pumpen v <Maschinen, Wassertrans, Wasserversorg> pump
Pumpen n 1. <Raumfahrt> transfer (Raumschiff); 2. <Wasserversorg> pumping
Pumpenaggregat n <Wasserversorg> set
Pumpenanlage f <Wasserversorg> pump station, pumping plant
Pumpenanschluss m <Wasserversorg> pump connection
Pumpenarmatur f <Erdöl> pump valve
Pumpenbagger m 1. <Bau> pump dredger; 2. <Wassertrans> pump dredge, pump dredger
Pumpendruck m <Erdöl> pump pressure
Pumpendüse f <Kfztech> acceleration jet, united injector
Pumpeneinrichtung f <Wasserversorg> pumping equipment
Pumpengehäuse n 1. <Wassertrans> pump housing; 2. <Wasserversorg> pump compartment (Schacht)
Pumpengetriebe n <Wasserversorg> pump gear
Pumpenhaus n <Wasserversorg> pump house, pump room
Pumpenkolben m <Mechan> plunger
Pumpenrad n <Maschinen> impeller
Pumpenraum m <Wasserversorg> pump house, pump room
Pumpensatz m <Wasserversorg> set
Pumpensaugbecken n <Erdöl> sump
Pumpenschacht m <Wasserversorg> pump shaft, pumping shaft
Pumpensonde f <Erdöl> pumping well
Pumpenstand m <Meerschmutz> pumping unit
Pumpenstange f <Maschinen, Wasserversorg> pump rod
Pumpensteuerung f <Kontroll> pump control system
Pumpensumpf m 1. <Fertig> sump; 2. <Wasserversorg> pump sump
Pumpenturbine f <Nichtfoss Energ> pump turbine
Pumpenzylinder m <Maschinen, Wasserversorg> pump barrel, pump cylinder
Pumpfrequenz f <Raumfahrt> pump frequency (Weltraumfunk)
Pumplicht n <Elektronik> pumping light
Pumplichtlaser m <Elektronik> optically-pumped laser
Pumpmaschine f <Wasserversorg> pumping engine
Pumpphotonen npl <Elektronik> pumping photons
Pumpprüfung f <Kohlen> pumping test
Pumpspeicherverfahren n <Nichtfoss Energ> pumped storage scheme
Pumpspeicherwerk n <Wasserversorg> pumped storage electrical power station, pumped storage plant
Pumpspray n <Verpack> pump dispenser system
Pumpstange f <Maschinen, Nichtfoss Energ> sucker rod
Pumpstation f <Wassertrans> pumping station
Pumpversuch m <Wasserversorg> pumping test
Pumpwasser n <Wasserversorg> pump water
Pumpwerk n <Wasserversorg> pump station
Pumpwerksumpf m <Kohlen> pumping pit
Punicin n <Chemie> pelletierine, punicine
Punkt m 1. <Comp & DV> point; 2. <Druck> full point, point; 3. <Geom, Math, Metrol> point; 4. <Papier> dot; 5. <Phys> point
Punkt m **der Gleichzeitigkeit** <Lufttrans> equal time point (Navigation)
Punkt m **größter Annäherung** <Wassertrans> closest point of approach (Navigation)
Punktabsaugsystem n <Sicherheit> point vacuum cleaning system
Punktanschnitt m <Kunststoff> pin gate
Punktbahn f <Fertig> point path (Getriebelehre)
Punktberührung f <Elektrotech> point contact
Punktdarstellung f <Comp & DV> dot diagram, spot diagram
Punkte mpl **je Zoll** <Comp & DV> dots per inch, dpi (Auflösung von Druckern und Scannern)
punktförmig adj <Fertig> localized (Korrosion)
punktförmig angreifende Einzellast f <Fertig> concavity, concentrated load
punktförmig verteilte Kapazität f <Elektriz, Elektrotech> lumped capacitance
punktförmige Abtastung f <Comp & DV> scanning spot beam
punktförmige Lichtquelle f <Foto> point source light
punktförmige Quelle f <Foto> point source
punktförmige Schallquelle f <Sicherheit> point sound source
punktförmiges Bild n <Foto> point image
Punktgenerator m <Elektronik> dot generator
Punktgröße f <Druck> point size
Punktierung f <Konstzeich> dotting
Punktkontakt m <Elektrotech> point contact
Punktkontakt-Detektordiode f <Elektronik> point contact detector diode
Punktkontaktdiode f <Elektronik> point contact diode
Punktkontakt-Mischdiode f <Elektronik> point contact mixer diode
Punktkontakt-Siliziumdiode f <Elektronik> point contact silicon diode
Punktkontakttransistor m <Elektronik> point contact transistor
Punktkoordinaten fpl <Geom> coordinates of a point, point coordinates
Punktladung f <Elektriz, Phys> point charge
Punktlast f <Fertig> concavity, concentrated load, concentrated mass
Punktleuchte f <Elektriz> spotlight
Punktmasse f <Fertig> particle mass
Punktmatrix f <Comp & DV, Elektronik> dot matrix
Punktmatrixdrucker m <Druck> dot matrix printer
Punktmengentheorie f <Math> point set theory
Punktquelle f <Foto, Umweltschmutz> point source
Punktraster n 1. <Comp & DV> raster; 2. <Fernseh> dot grating
Punktschweißen n 1. <Bau, Elektriz> spot welding; 2. <Maschinen> resistance spot welding
Punktschweißung f <Maschinen> spot welding
Punktsprungabtastung f <Fernseh> dot interlace scanning
Punktstrahl m <Raumfahrt> spot beam (Weltraumfunk)
Punktstrahlausleuchtung f <Telekom> spot beam coverage (Satelliten)
Punktstrahlbedeckung f <Telekom> spot beam coverage (Satellit)
Punktstrahlrichtantenne f <Funktech, Raumfahrt> spot beam antenna (Weltraumfunk)
Punktsystem n <Druck> point system

punktuelle Abbildung f <Foto> point image
punktweise Aufzeichnung f <Comp & DV> plotting
punktweise Konvergenz f <Math> point by point convergence *(einer Funktion)*
Punktwert m <Ergon> score
Punktwiderstand m <Kohlen> point resistance
punktzentrierter Strahl m <Phys> homocentric beam
Punkt-zu-Mehrpunkt-Betrieb m <Telekom> point-to-multipoint operation
Punkt-zu-Punkt-... <Comp & DV> peer-to-peer, point-to-point
Punkt-zu-Punkt-Kopplung f <Comp & DV> peer-to-peer link
Punkt-zu-Punkt-Leitung f <Comp & DV, Telekom> point-to-point line
Punkt-zu-Punkt-Protokoll n <Comp & DV> point-to-point protocol
Punkt-zu-Punkt-Transport m <Trans> point-to-point transport
Punkt-zu-Punkt-Übertragung f 1. <Comp & DV> point-to-point communication; 2. <Telekom> point-to-point transmission
Punkt-zu-Punkt-Verbindung f 1. <Comp & DV> point-to-point connection; 2. <Elektrotech> point-to-point link; 3. <Telekom> point-to-point connection, point-to-point link
Punkt-zu-Punkt-Verdrahtung f <Elektrotech> point-to-point wiring
Pupinisierung f 1. <Elektrotech> coil loading, loading; 2. <Telekom> loading
Pupinkabel n 1. <Elektrotech> coil-loaded cable, loaded cable; 2. <Telekom> coil-loaded cable
Pupinspule f <Elektriz, Elektrotech, Phys> loading coil
PUR *(Polyurethan)* <Kunststoff, Textil> PUR *(polyurethane)*
Purin n <Chemie> purine
Purpur n <Druck> magenta
Purpurfiltereinstellung f <Foto> minus green filter adjustment
Purpurin n <Chemie> purpurin
Push-Pull-Verstärker m <Elektronik> push-pull amplifier
Pütting n <Wassertrans> chain plate *(Segeln)*
Putz m <Bau> plaster *(Innenputz)* • **unter Putz legen** <Bau> conceal
Putzabstandshalter m <Bau> furring
Putzarbeiten fpl <Bau> plaster work, plastering
Putzbesen m <Bau> broom drag
Pütze f <Wassertrans> bucket
putzen v <Fertig> roll, snag; trim *(Gießen)*; dress, rattle *(Guss)*
Putzen n 1. <Fertig> rattling, rolling, scouring; dressing *(Guss)*; 2. <Ker & Glas, Metall> dressing
Putzen n **von Steingut** <Ker & Glas> fettling of earthenware
Putzerei f <Fertig> dressing *(Gießen)*
Putzkelle f <Bau> plastering trowel
Putzleiste f <Bau> counterlath
Putzstern m <Fertig> star
Putzträger m <Bau> lathing
Puzzolanerde f <Bau> pozzolana
Puzzolanzement m <Bau> pozzolanic cement
PVAC *(Polyvinylacetat)* <Kunststoff> PVAC *(polyvinyl acetate)*
PVAL *(Polyvinylalkohol)* <Druck, Kunststoff> PVAL *(polyvinyl alcohol)*
PVB *(Polyvinylbutyral)* <Kunststoff> PVB *(polyvinyl butyral)*
PVC *(Polyvinylchlorid)* <Bau, Elektrotech, Kunststoff> PVC *(polyvinyl chloride)*
PVC-Band n <Aufnahme> PVC tape
PVC-C *(chloriertes Polyvinylchlorid)* <Kunststoff> CPVC *(chlorinated polyvinyl chloride)*

PVC-Einsatz m <Verpack> PVC insert fitment
PVC-Flasche f <Verpack> PVC bottle
PVC-hart n <Kunststoff> PVC rigid, U-PVC, unplasticized PVC
PVC-Isolierung f <Elektriz> PVC insulation
PVC-U <Kunststoff> U-PVC, unplasticized PVC
PVDC *(Polyvinylidenchlorid)* <Kunststoff> PVDC *(polyvinylidene chloride)*
PVDF *(Polyvinylidenfluorid)* <Kunststoff> PVFD *(polyvinylidene fluoride)*
P/V-Diagramm n *(Druck-Volumen-Diagramm)* <Thermod> pressure volume diagram
PVE *(Polyvinylether)* <Kunststoff> PVE *(polyvinyl ether)*
PVF *(Polyvinylfluorid)* <Kunststoff> PVF *(polyvinyl fluoride)*
P-Welle f *(primäre Erdbebenwelle)* <Phys> P wave
PWM *(Pulsweitenmodulation)* <Elektronik> PWM *(pulse width modulation)*
p-Xylen n <Chemie> p-xylene
Pyknometer n 1. <Erdöl> pycnometer *(Dichtemessgerät)*; 2. <Labor> density bottle; pycnometer *(Dichte)*; 3. <Phys> pycnometer
Pylon m <Bau, Elektriz, Lufttrans> pylon
Pyramide f <Geom> pyramid
Pyramidenebene f <Metall> pyramidal plane
Pyramidengleiten n <Metall> pyramidal slip
Pyramidenhorn n <Funktech> pyramidal horn *(Antenne)*
Pyran n <Chemie> pyran
Pyranometer n <Nichtfoss Energ> pyranometer
Pyranose f <Chemie> pyranose
Pyrazin n <Chemie> piazine, pyrazine
Pyrazol n <Chemie> pyrazole
Pyrazolin n <Chemie> pyrazoline
Pyrazolon n <Chemie> pyrazolone
Pyrheliometer n <Nichtfoss Energ> pyrheliometer
Pyridazin n <Chemie> pyridazine
Pyridin n <Chemie> pyridine
Pyridin-Carbonsäureamid n <Chemie> niacinamide, nicotinamide
Pyridon n <Chemie> pyridone
Pyroarsenat n <Chemie> pyroarsenate
Pyrocatechol n <Chemie> pyrocatechol, pyrocatechin
pyroelektrisch adj <Elektriz> pyroelectric
Pyroelektrizität f <Elektriz> pyroelectricity
Pyrogallol n <Chemie> pyrogallol
Pyrogallolphthalein n <Chemie> gallein
Pyrogallus... <Chemie> pyrogallic
pyrogallussauer adj <Chemie> pyrogallic
Pyrogallussäure f <Chemie> pyrogallol
pyrogen adj <Thermod> pyrogenic
pyrogene Reaktion f <Thermod> pyrogenic reaction
pyrolitische Beschichtung f <Ker & Glas> pyrolitic coating
pyrolitische Zersetzung f <Kerntech> pyrolitic decomposition
Pyrolyse f <Abfall, Chemie, Lebensmittel> pyrolysis
pyrolytisch adj <Chemie> pyrolytic
Pyromekon... <Chemie> pyromeconic
Pyromellit... <Chemie> pyromellitic
Pyromellitsäure f <Chemie> pyromellitic acid
Pyrometallurgie f <Metall> pyrometallurgy
Pyrometer n <Elektriz, Gerät, Kohlen, Labor, Strahlphys, Thermod> pyrometer
Pyrometerschutzrohr n <Sicherheit> pyrometer protection tube
Pyrometersonde f <Elektriz> pyrometer probe
Pyrometrie f <Thermod> pyrometry
pyrometrisch adj <Thermod> pyrometric
Pyron n <Chemie> pyrone
pyrophor adj <Chemie> pyrophoric, pyrophorous

Pyrophosphat

Pyrophosphat n <Chemie> pyrophosphate
Pyrophyllit n <Erdöl> pyrophyllite *(Mineral)*
pyroschleimsauer *adj* <Chemie> pyromucic
Pyroskop n <Phys> pyroscope
Pyrostat m <Phys> pyrostat
Pyrotechnik f <Raumfahrt> pyrotechnics
Pyrrol n <Chemie> pyrrole
Pyrrolidin n <Chemie> pyrrolidine, tetrahydropyrrol, tetramethyleneimine
Pyrrolin n <Chemie> pyrroline
Pyruvat n <Chemie> pyruvate
Pyruvin... <Chemie> pyruvic
pythagoreischer Lehrsatz m <Geom, Math> Pythagorean theorem
pythagoreisches Komma n <Akustik> Pythagorean comma

Q

QAM 1. <Comp & DV, Elektronik, Telekom> *(Quadratur-Amplitudenmodulation)* QAM *(quadrature amplitude modulation)*; 2. <Elektronik> *(Quadratur-Amplitudenmodulator)* QAM *(quadrature amplitude modulator)*
Q-Bit n *(Unterscheidungsbit)* <Telekom> Q bit *(qualifier bit)*
QCD *(Quantenchromodynamik)* <Teilphys> QCD *(quantum chromodynamics)*
QC-Flugzeug n *(Quick-Change-Flugzeug)* <Lufttrans> QC aircraft *(quick-change aircraft)*
Q-Demodulator m <Elektronik> Q demodulator
QED *(Quantenelektrodynamik)* <Phys> QED *(quantum electrodynamics)*
Q-Elektron n <Kerntech> Q electron, Q shell electron
Q-Faktor m *(Gütefaktor, Qualitätsfaktor)* <Funktech, Mechan, Phys, Qual, Umweltschmutz> Q factor *(quality factor)*
Q-Faktor-Messgerät n <Metrol> Q-meter
QG *(Quartärgruppe)* <Telekom> supermaster group *(Trägerfrequenzübertragung)*
Q-Kanal m <Elektronik> Q channel
Q-Kontrolle f *(Qualitätskontrolle)* <Druck, Labor, Maschinen, Mechan, Qual, Verpack> QC *(quality control)*
Q-Maschine f <Kerntech> Q device
Q-Meter n *(Gütefaktormesser)* <Phys> Q meter
Q-Perzentil-Lebensdauer f <Qual> Q percentile life
QP-Maschine f <Kerntech> QP device
QPSK *(Vierphasenumtastung)* 1. <Elektronik> QPSK *(quaternary phase shift keying)*; 2. <Telekom> QPSK *(quadriphase shift keying)*
QS *(Qualitätssicherung)* <Kerntech, Maschinen, Mechan, Qual, Verpack> QA *(quality assurance)*
QSAM *(erweiterte Zugriffsmöglichkeit für sequenzielle Dateien)* <Comp & DV> QSAM *(queued sequential access method)*
Q-Schale f <Kerntech> Q shell
QSG *(Quasistellargalaxie)* <Raumfahrt> QSG *(quasi-stellar galaxy)*
QS-Handbuch n <Qual> QA manual
Q-Signal n <Elektronik, Fernseh> Q signal
QSO *(quasi-stellares Objekt)* <Raumfahrt> QSO *(quasi-stellar object)*

QS-Programmmodul n <Qual> *(AE)* QA program module, *(BE)* QA programme module
QS-Verfahrensanweisungen *fpl* <Qual> quality procedures
QS-Verfahrenshandbuch n <Qual> QA procedures manual
Quad-Antenne f <Funktech> cubical quad
Quader m <Geom> cuboid
Quaderstein m <Bau> quarry
Quadrant m <Geom> quadrant
Quadranten-Elektrometer n <Elektriz> quadrant electrometer
Quadrat n <Geom, Math, Metrol> square
Quadratdezimeter m <Metrol> *(AE)* square decimeter, *(BE)* square decimetre
Quadratfuß m <Metrol> square foot
quadratisch *adj* 1. <Geom> square; 2. <Math> quadratic
quadratisch abhängig *adj* <Phys> square-law
quadratisch integrierbar *adj* <Math> quadratically integrable
quadratische Funktion f <Math> quadratic function
quadratische Gleichung f <Math> quadratic equation
quadratischer Gleichrichter m <Elektrotech> square law detector
quadratischer Keil m <Maschinen> square key
quadratischer Mittelwert m <Aufnahme, Elektronik, Optik, Phys> root mean square value
quadratischer Mittelwert m **der Abweichung** <Elektriz> root mean square deviation
quadratischer Mittelwert m **der Geschwindigkeit** <Phys> mean square velocity
quadratischer Wasserstandmittelwert m <Nichtfoss Energ> root line mean square water level
quadratisches Entfernungsgesetz n <Kerntech> square law
quadratisches Mittel n *(RMS)* <Aufnahme, Elektriz, Elektronik> root mean square *(rms)*
quadratisches Profil n <Optik> quadratic profile
Quadratkeil m <Maschinen> square key
Quadratmaß n 1. <Konstzeich> square dimension; 2. <Metrol> square measure
Quadratmeile f <Metrol> square mile
Quadratmeter m <Metrol> centiare, *(AE)* square meter, *(BE)* square metre
Quadratur f <Comp & DV, Elektronik, Geom, Nichtfoss Energ, Phys> quadrature
Quadratur f **des Kreises** <Geom> squaring the circle
Quadraturachse f <Elektriz> quadrature axis
Quadratur-Achsenkomponente f <Elektriz> quadrature axis component
Quadratur-Amplitudenmodulation f *(QAM)* <Comp & DV, Elektronik, Telekom> quadrature amplitude modulation *(QAM)*
Quadratur-Amplitudenmodulationskomponenten *fpl* <Comp & DV, Elektronik, Telekom> in-phase and quadrature components, QAM components
Quadratur-Amplitudenmodulator m *(QAM)* <Elektronik> quadrature amplitude modulator *(QAM)*
Quadratur-Demodulator m <Elektronik> quadrature demodulator
Quadratur-Differenzphasenumtastung f <Telekom> quadrature differential phase shift keying *(QDPSK)*
Quadratur-Einseitenbandamplitudenmodulation f <Telekom> quadrature single-sideband AM *(QSSB-AM)*
Quadraturkomponente f <Elektriz> quadrature component
Quadraturleistung f <Elektriz> quadrature power
Quadraturmodulation f <Telekom> quadrature modulation

Quadratur-Modulator m <Elektronik> quadrature amplitude modulator
Quadratur-Phasenumtastung f <Telekom> quadrature phase shift keying *(QPSK)*
Quadraturspiegelfilter n <Telekom> quadrature mirror filter
Quadratursteuerung f <Elektriz> quadrature control
Quadratwurzel f <Math> square root
Quadratyard n <Metrol> square yard
Quadratzentimeter m <Metrol> *(AE)* square centimeter, *(BE)* square centimetre
Quadratzoll m <Metrol> square inch
quadrieren v <Math> square
Quadrierschaltung f <Telekom> squaring circuit
Quadrophonie f <Akustik> quadraphony
Quadruplex... <Fernseh> quadruplex
Quadrupol m <Phys> quadrupole
Quadrupolanordnung f <Kerntech> quadrupolar configuration
Quadrupolfeld n <Kerntech> quadrupole field
Quadrupolmassenfilter n <Elektrotech> quadrupol mass filter
Quadrupolmoment n <Phys> quadrupole moment
Quadrupolpotenzial n <Kerntech> quadrupole potential
Quadrupolresonanz f <Kerntech> quadrupole resonance
Quakbox f <Aufnahme> squawkbox
Qualifikationsanforderung f <Qual> qualification requirement
Qualifikationsnachweis m <Qual> qualification records
qualifizierter Name m <Comp & DV> qualified name
Qualität f 1. <Fertig> grade; 2. <Qual> grade, quality; 3. <Verpack> paper grade
qualitative Analyse f <Wasserversorg> qualitative analysis
qualitative Autoradiographie f <Kerntech> qualitative autoradiography
qualitatives Merkmal n 1. <Patent> attribute; 2. <Qual> attribute, qualitative characteristic
Qualitätsaudit n <Qual> quality audit
Qualitätsaufzeichnungen fpl <Qual> quality records
Qualitätsbeanstandung f <Qual> nonconformance report
qualitätsbeeinflussende Tätigkeiten fpl <Qual> activities affecting quality
Qualitätsbericht m <Qual> quality report
Qualitätsbeurteilung f <Metrol> quality assessment
Qualitätsblech n <Fertig> prime
Qualitätseinbuße f <Qual> impairment of quality
Qualitätselement n <Qual> quality element
Qualitätsfähigkeit f <Qual> quality capability
Qualitätsfähigkeits-Bestätigung f <Qual> quality verification
Qualitätsfaktor m <Funktech, Mechan, Phys, Qual, Umweltschmutz> quality factor *(Q factor)*
Qualitätsförderung f <Qual> quality improvement
Qualitätsingenieur m <Qual> quality assurance engineer
Qualitätskontrolle f *(Q-Kontrolle)* <Druck, Maschinen, Mechan, Metrol, Qual, Verpack> quality control *(QC)*
Qualitätskontrolle f **in der Fertigung** <Qual> process control
Qualitätskosten pl <Qual> quality costs, quality-related costs
Qualitätskreis m <Qual> quality loop
Qualitätskriterien npl <Maschinen> quality acceptance criteria
Qualitätslage f <Qual> quality level
Qualitätsleistung f **des Lieferanten** <Qual> supplier's quality performance
Qualitätslenkung f <Qual> quality control

Qualitätslenkung f **bei mehreren Merkmalen** <Qual> multivariate quality control
Qualitätsmanagement n <Qual> quality management
Qualitätsmangel m <Qual> quality defect
Qualitätsmarke f <Verpack> quality label
Qualitätsmerkmal n <Qual> quality characteristic, quality criterion
Qualitätsminderung f 1. <Fernseh> quality degradation; 2. <Qual> degradation of quality, impairment of quality
Qualitätsnorm f <Metrol> standard of quality
Qualitätsplanung f <Qual> quality planning
Qualitätspolitik f <Qual> quality policy
Qualitätsprodukt n <Mechan> quality product
Qualitätsprüfstelle f <Qual> quality inspection and test facility
Qualitätsprüfung f <Qual> quality inspection and testing
Qualitätsregelkarte f <Qual> control chart
Qualitätsregelkarte f **für kumulierte Werte** <Qual> cusum chart
Qualitätsregelung f **in der Fertigung** <Qual> in-process quality control
Qualitätssicherung f *(QS)* <Kerntech, Maschinen, Mechan, Qual, Verpack> quality assurance *(QA)*
Qualitätssicherung f **in Entwurf und Konstruktion** <Qual> design assurance
Qualitätssicherungsabteilung f <Qual> quality assurance department
Qualitätssicherungsauflagen fpl <Qual> quality assurance requirements
Qualitätssicherungsbeauftragter m <Qual> QA representative, quality assurance representative
Qualitätssicherungsbescheinigung f <Maschinen> quality assurance certificate
Qualitätssicherungsdokumentation f <Qual> quality documentation
Qualitätssicherungshandbuch n <Qual> quality assurance manual, quality manual
Qualitätssicherungsprüfung f <Kerntech> quality assurance examination
Qualitätssicherungssystem n <Qual> quality management system, quality system
Qualitätssicherungsverfahren n <Maschinen> quality assurance procedure
Qualitätsstand m <Qual> quality status
Qualitätssteuerung f <Qual> quality control
Qualitätssystem-Bescheinigungsmaterial n <Qual> QSCM, quality system certificate material
Qualitätssystemrevision f <Qual> quality system audit
Qualitätssystemsüberprüfung f <Qual> quality system review
Qualitätssystemsvoraussetzung f <Qual> quality system requirement, QSR
Qualitätstechnik f <Qual> quality engineering
Qualitätsüberprüfung f <Kerntech> QC, quality control
Qualitätsüberwachung f <Qual> quality surveillance
Qualitätsunterelement n <Qual> quality subelement
Qualitätsverbesserung f <Qual> quality improvement
Qualitätsverlust m <Kunststoff> deterioration
Qualitätswesen n <Qual> quality management
Qualitätszahl f <Qual> quality number
Qualitätsziel n <Qual> quality objective
Quant n <Comp & DV, Optik, Phys, Teilphys> quantum
Quantelung f <Elektronik> quantization
Quantenausbeute f <Phys, Strahlphys> quantum efficiency, quantum yield
Quantenausbeute f **der Lumineszenz** <Strahlphys> quantum yield of luminescence
Quantenchemie f <Chemie> quantum chemistry

Quantenchromodynamik

Quantenchromodynamik f (QCD) <Teilphys> quantum chromodynamics (QCD)
Quantencomputer m <Comp & DV> quantum computer
Quantenelektrodynamik f (QED) 1. <Phys> quantum electrodynamics (QED); 2. <Teilphys> quantum electrodynamics
Quantenelektronik f <Elektronik> quantum electronics
Quantenfeldtheorie f <Phys> quantum field theory
Quantengas n <Phys> quantum gas
Quanten-Hall'scher Effekt m <Phys> quantum Hall effect
Quantenhydrodynamik f <Phys> quantum hydrodynamics
Quantenmechanik f <Phys, Teilphys> quantum mechanics
quantenmechanische Linienform f <Strahlphys> quantum-mechanical line shape
Quantenmulde f <Telekom> quantum well
Quantenoptik f <Phys> quantum optics
Quantenphysik f <Phys> quantum physics
quantenrauschbegrenzter Betrieb m <Telekom> quantum-noise-limited operation
Quantenrauschen n 1. <Elektronik, Optik> quantum noise; 2. <Raumfahrt> quantization noise; 3. <Telekom> quantum noise
Quantensprung m 1. <Elektronik> quantum transition; 2. <Teilphys> quantum leap
Quantenstatistik f <Phys> quantum statistics
Quantenstatus m <Raumfahrt> quantum state
Quantentheorie f <Kerntech, Raumfahrt, Teilphys> quantum theory
Quantentheorie f **der Strahlung** <Strahlphys> quantum theory of radiation
Quantentopf m <Elektronik> quantum well
Quantenverzerrung f <Elektronik> quantization distortion
Quantenwirkungsgrad m <Optik, Phys, Strahlphys, Telekom> quantum efficiency
Quantenzahl f 1. <Kerntech> quantum number (n); 2. <Labor> principal quantum number (n); 3. <Phys, Strahlphys, Teilphys> quantum number
Quantenzustand m <Teilphys> quantum state
Quantil n <Math, Qual> quantile
Quantil n **einer Wahrscheinlichkeitsverteilung** <Math, Qual> quantile of a probability distribution
quantisieren v 1. <Comp & DV> quantize; 2. <Elektronik> digitize, quantize
Quantisierer m 1. <Comp & DV> quantizer; 2. <Elektronik> quantizer (für DPCM-Methode); 3. <Telekom> quantizer
quantisiert adj <Telekom> quantized
quantisierte Form f <Telekom> quantized form
quantisierte Größe f <Elektronik> quantized quantity
quantisierte Pulslagemodulation f <Elektronik> quantized pulse modulation
quantisiertes Signal n <Regelung, Telekom> quantized signal
quantisiertes Zeichen n (QZ) <Elektronik> quantized signal (QS)
Quantisierung f <Comp & DV, Phys> quantization
Quantisierungsfehler m <Comp & DV, Elektronik, Telekom> quantization error
Quantisierungsgeräusch n <Comp & DV> quantization noise
Quantisierungsgröße f <Comp & DV> quantization size
Quantisierungsintervall n <Telekom> quantization interval
Quantisierungspegel m <Elektronik, Telekom> quantization level
Quantisierungsrauschen n <Telekom> quantization noise
Quantisierungsstufe f <Comp & DV> quantization level
Quantität f <Elektronik, Kontroll> quantity
quantitativ bestimmen v <Math> quantify
quantitative Analyse f <Wasserversorg> quantitative analysis
quantitatives Merkmal n <Qual> quantitative characteristic
Quantor m <Math> quantifier
Quantum n <Comp & DV, Phys> quantum
Quarantäneflagge f <Wassertrans> quarantine flag; yellow flag (Flagge)
Quark n 1. <Phys> quark (Teilchen); 2. <Teilphys> quark
Quark n **mit Eigenschaft Bottom** <Phys> bottom quark
Quark n **mit Eigenschaft Charm** <Phys> charmed quark
Quark n **mit Eigenschaft Down** <Phys> down quark
Quark n **mit Farbladung Blau** <Phys> blue quark
Quark n **mit Farbladung Rot** <Phys> red quark
Quark n **mit Topeigenschaft** <Phys> top quark
Quark n **mit Upeigenschaft** <Phys> up-quark
Quark-Bag m <Teilphys> quark bag
Quarkeinschluss m <Teilphys> quark confinement
Quark-Gluon-Plasma n <Teilphys> quark-gluon plasma
Quarkteilchen n <Phys> green quark (mit Farbladung grün)
Quart n 1. <Akustik> subdominant; 2. <Metrol> (AE) dry quart, liquid quart
Quartär n <Erdöl> Quaternary (Geologie)
quartär adj <Chemie> quaternary (Oniumverbindung)
quartäre Spaltung f <Kerntech> quaternary fission
Quartärgruppe f <Telekom> supermaster group (Trägerfrequenzübertragung)
Quartation f <Chemie> quartation
Quarte f <Akustik> fourth, subdominant
Quarter n <Metrol> quarter
Quarterdeck n <Wassertrans> quarter deck
Quartett n <Phys> quartet (Spektroskopie)
Quartettmodell n <Kerntech> quartet model
Quartformat n (Quarto) <Druck> 4to, quarto (quarto)
Quartiermeister m <Wassertrans> quartermaster
Quartierung f 1. <Chemie> quartation; 2. <Metall> inquartation
Quartil n <Math, Qual> quartile
Quarto n (Quartformat) <Druck> 4to, quarto (quarto)
Quartowalzwerk n <Fertig> four-high mill
Quartz m <Metall> silica glass
Quarz m 1. <Elektronik> quartz; crystal (Schaltungen); 2. <Funktech> crystal; 3. <Phys> quartz; 4. <Telekom> crystal
Quarzeichoszillator m <Funktech> crystal calibrator
Quarzfilter n <Elektronik, Funktech, Strahlphys, Telekom> crystal filter
Quarzfrequenz f <Elektronik, Fernseh, Funktech, Telekom> crystal frequency
Quarzfrequenzauswanderung f <Elektronik, Funktech, Telekom> crystal frequency drift
Quarzfrequenzquelle f <Elektronik> quartz frequency source
quarzgesteuert adj <Elektronik> crystal-controlled
Quarzglas n <Ker & Glas, Optik, Telekom> fused quartz, fused silica, quartz glass, vitreous silica
Quarzglasbeschichtung f <Telekom> silica coating
Quarzglasfaser f <Optik, Telekom> (AE) all-silica fiber, (BE) all-silica fibre
Quarzglasfenster n <Elektronik> fused silica window
Quarzglasprimärbeschichtung f <Telekom> silica coating
Quarzhalter m <Elektronik> crystal holder
Quarzit m <Bau> quartzite

Quarzkettenfilter n <Elektronik, Funktech, Telekom> crystal ladder filter
Quarzkristall m <Comp & DV, Elektriz> quartz crystal
Quarzlaufzeitkette f <Elektronik> quartz delay line
Quarzmessfühler m <Gerät> piezoelectric sensing element
Quarzoszillator m 1. <Elektriz, Elektronik> crystal oscillator, quartz oscillator; 2. <Telekom> quartz oscillator
Quarzoszillator m **mit Temperaturkompensation** <Elektronik> temperature-compensated crystal oscillator
Quarzoszillator m **mit Temperaturregelung** <Elektronik> temperature-controlled crystal oscillator
Quarzresonator m <Elektronik, Elektrotech> piezoelectric resonator, quartz resonator
Quarzsand m <Ker & Glas> glassmaking sand
Quarzschmelze f <Telekom> fused quartz
Quarzschwinger m 1. <Elektronik> quartz oscillator; 2. <Kerntech> quartz monochromator; 3. <Phys> piezoelectric oscillator; 4. <Strahlphys> crystal oscillator
Quarzschwingungsquelle f <Elektronik> quartz frequency source
Quarzsteuerung f <Aufnahme, Lufttrans> crystal control
Quarztaktsteuerung f <Elektriz> quartz crystal clock
Quarzthermostat m <Funktech> crystal oven
Quarzverzögerungsleitung f <Elektronik, Telekom> quartz delay line
Quarzzeitbasis f <Elektronik> crystal time base
Quasi... <Kerntech, Phys> quasi
quasi-adiabatisches Kalorimeter n <Kerntech> quasi-adiabatic calorimeter
Quasi-Albedo-Methode f <Kerntech> quasi-albedo approach
quasi-binär adj <Fertig> quasi-binary
Quasibrüter m <Kerntech> quasi-breeder reactor
quasi-chemische Annäherung f <Metall> quasi-chemical approximation
quasi-integrierbar adj <Math> quasi-integrable
quasi-konstante Zustandsverteilung f <Metall> quasi-steady-state distribution
quasi-konstantes Gleiten n <Kerntech> quasi-constant slip
Quasikontakt m <Elektronik> proximity contact
Quasischeitelwertspannungsmesser m <Metrol> quasi-peak voltmeter
Quasispitzen-Detektor m <Telekom> quasi-peak detector
Quasispitzenspannung f <Telekom> quasi-peak voltage
quasi-stabiler Zustand m <Phys> quasi-steady state
quasi-statisch adj <Raumfahrt> quasi-statical
quasi-statische Ladung f <Raumfahrt> quasi-statical loading
quasi-stellare Entkopplung f <Raumfahrt> quasi-stellar decoupling
quasi-stellare Radioquelle f <Raumfahrt> QSS, quasi-stellar radio source
quasi-stellares Objekt n (QSO) <Raumfahrt> quasi-stellar object (QSO)
Quasistellargalaxie f (QSG) <Raumfahrt> quasi-stellar galaxy (QSG)
Quasistreuung f <Kerntech> quasi-scattering
Quasiteilchen n <Phys> quasi-particle
quasi-unendlich langer Riss m <Kerntech> semi-infinite crack
Quasizufallsfolge f <Math> pseudo-random sequence, quasi-random sequence (Statistik)
Quassin n <Chemie> quassin
Quaste f <Textil> tuft
quaternär adj 1. <Chemie> quaternary (Legierung); 2. <Math> quaternary

Quaternion f <Math> quaternion
Quecksilber n (Hg) <Chemie> mercury (Hg)
Quecksilber... <Chemie> mercuric
Quecksilberamalgam n <Metall> quicksilver amalgam
Quecksilberbarometer n <Phys> mercury barometer
Quecksilberbatterie f <Elektrotech> mercury battery, mercury cell
Quecksilber-benetztes Relais n <Elektriz> mercury-wetted relay
Quecksilberbogenlampe f <Phys> mercury arc lamp
Quecksilberbromid-Laser m <Elektronik> mercury-bromide laser
Quecksilber-Chlorid n <Chemie> calomel
Quecksilberdampf m <Bau, Chemie> (AE) mercury vapor, (BE) mercury vapour
Quecksilberdampfgleichrichter m <Elektriz, Elektrotech> mercury-arc converter, mercury-arc rectifier, mercury rectifier, (AE) mercury-vapor rectifier, (BE) mercury-vapour rectifier
Quecksilberdampf-Gleichrichterröhre f <Elektriz> ignitron
Quecksilberdampflampe f <Elektriz, Elektrotech> mercury arc lamp, (AE) mercury vapor lamp, (BE) mercury vapour lamp
Quecksilberdampfstromrichter m <Elektrotech> mercury arc converter
Quecksilberdampfturbine f <Maschinen> (AE) mercury vapor turbine, (BE) mercury vapour turbine
Quecksilberelement n <Elektrotech> mercury cell
Quecksilber-Fulminat n <Chemie> mercury fulminate
Quecksilbergewinnung f <Chemie> mercurification
Quecksilberkatode f <Elektrotech> mercury pool cathode
Quecksilberkippschalter m <Elektriz> mercury tilt switch
Quecksilberkontakte mpl <Elektrotech> mercury-wetted contacts
Quecksilberkontaktthermometer n <Gerät> mercury contact thermometer
Quecksilberkugel f <Chemie> mercury cup
Quecksilber-Laser m <Elektronik> mercury laser
Quecksilberlichtbogen m <Elektrotech> mercury arc
Quecksilberrelais n <Elektriz> mercury relay
Quecksilberröhre f <Elektrotech> mercury pool tube
Quecksilbersäule f <Metrol> barometric column
Quecksilberschalter m <Elektriz, Elektrotech> mercury switch
Quecksilberstrahlgleichrichter m <Elektrotech> mercury rectifier
Quecksilberthermometer n <Heiz & Kälte, Phys> mercury thermometer
Quecksilberunterbrecher m <Elektrotech> mercury interrupter
Quecksilber-Verzögerungsleitung f <Elektronik> mercury delay line
Quellbereich m <Umweltschutz> source area
Quelle f 1. <Bau> spring; 2. <Comp & DV> repository, source; 3. <Elektrotech, Hydraul, Telekom> source
Quelle f **der EMK** <Phys> source of emf
Quellelektrodenkontakt m <Elektrotech> source contact
quellen v 1. <Fertig> macerate; 2. <Textil> swell
Quellen n 1. <Kunststoff> swelling; 2. <Papier> bloating; 3. <Textil> swelling
Quellenadresse f <Comp & DV, Telekom> source address
Quellenangabe f <Patent> indication of source
Quellenbereich m <Kerntech> source range (Reaktor)
Quellencode m <Comp & DV> source code
Quellencodierung f <Telekom> source encoding

Quellendatei

Quellendatei f <Comp & DV> source file
Quellendokument n <Comp & DV> source document
Quellenelektrode f <Elektrotech> source electrode
Quellenenergie f <Kohlen, Nichtfoss Energ> source energy
Quellenfassung f <Wasserversorg> tapping water
Quellenimpedanz f <Elektrotech> source impedance
Quellenprogramm n <Comp & DV> source program
Quellensprache f <Comp & DV> source language
Quellenstärke f (S) <Kerntech> source strength (S)
Quellpunkt m <Ker & Glas> hot spot
Quellschweißen n <Kunststoff> solvent welding
Quellsprachenübersetzung f <Comp & DV> source language translation
Quellton m <Erdöl> swelling clay (Mineral)
Quellung f 1. <Fertig> expansion (Beton); swelling (Kunststoffinstallationen); 2. <Kunststoff> swelling
Quellvolumen n <Bau> bulking
Quellwasser n <Lebensmittel, Wasserversorg> spring water
Quellwiderstand m <Elektrotech> source impedance
quer zur Faserrichtung adv <Fertig> (AE) across the fiber grain, (BE) across the fibre grain
Quer... 1. <Anstrich, Fertig> lateral (Bewegungsrichtung); 2. <Wassertrans> transverse (Schiffkonstruktion)
querab adv <Wassertrans> abeam
Querabweichung f <Raumfahrt> across track error
Querachse f 1. <Lufttrans> lateral axis, pitch axis; 2. <Maschinen> transverse axis
Queranteil m 1. <Comp & DV> quadrature component; 2. <Elektrotech> quadrature-axis component (der Polradspannung)
Queraufzeichnung f <Aufnahme, Fernseh> transverse recording
Querbalken m 1. <Bau> crossbeam, cross member, traverse; 2. <Fertig> intertie; 3. <Lufttrans> crossbar (Befeuerung); 4. <Maschinen> cross member
Querbelastung f <Maschinen> lateral load, transverse load
Querbeschleunigung f <Elektrotech> lateral acceleration
Querbeschleunigungsanzeiger m <Lufttrans> lateral accelerometer
Querbespantung f <Wassertrans> transverse framing (Schiffbau)
Querbewegung f 1. <Elektrotech> lateral oscillation; 2. <Maschinen> traverse motion
Querbewehrungsstahl m <Bau> distribution steel
Querbrett n <Bau> ledger (Gerüstbau)
Querbruch m <Ker & Glas> cross break
Quercetin n <Chemie> quercetin
Quercitrin n <Chemie> quercitrin
Querdehnungszahl f <Bau> Poisson's ratio
Querdraht n <Elektrotech> cross-span, span wire
Querdurchflutung f <Elektriz> quadrature-axis component of magnetomotive force
Querfaltung f <Konstzeich> cross folding
Querfeldeinfahrzeug n <Kfztech> off-highway vehicle
Querfeldkomponente f <Comp & DV> quadrature component
Querfeldverstärkermaschine f <Elektrotech> cross-field amplifier
Querfeldvorspannung f <Aufnahme> cross-field bias
Querfließpressen n <Maschinen> lateral extrusion, sideways extrusion
Querflusslinearmotor m <Trans> transverse flux linear motor
Querflussmagnet m <Elektrotech> transverse flux magnet (Magnetschwebetechnik)
Querflussmaschine f <Trans> transverse flux machine
Querflussspule f <Elektrotech> lateral-flux coil, normal-flux coil (Magnetschwebetechnik)
Querformat n 1. <Comp & DV> landscape format; 2. <Druck> oblong size
Querführung f <Kfztech> lateral guidance
Quergefälle n <Bau> camber (Straßenbau)
Quergleiten n <Metall> cross slip
Querhaupt n <Maschinen> cross girth, top rail
Querholz n <Bau> transom
Querkeil m <Fertig, Maschinen> cotter
Querkontraktionszahl f <Kunststoff> Poisson's ratio
Querkraft f <Maschinen, Trans> lateral force
Querkugellager n <Maschinen> radial ball bearing
Querlage f <Lufttrans> bank (Flugzeug)
Querlager n <Maschinen, Papier> radial bearing
Querlenker m <Kfztech> transverse control arm (Radaufhängung)
Querlenkerarm m <Kfztech> suspension arm
quermagnetisch adj <Elektrotech> transverse magnetic
Quermagnetisierungseffekt m <Elektriz> cross-magnetizing effect
Querneigung f 1. <Bau> crossfall, slope; 2. <Trans> banking (Straße)
Querparität f <Comp & DV> vertical parity
Querparitätsprüfung f <Comp & DV> VRC, vertical redundancy check
Querprofil n <Bau> cross section
Querrichtung f <Papier> cross direction • **in Querrichtung** <Fertig> laterally
Querriegel m <Bau> strap; tie beam (Holz)
Querrippen fpl <Textil> crosswise ribs
Querrippenglas n <Ker & Glas> cross reeded glass
Querriss m <Fertig> head pull (Gießen)
Querrollenlager n <Maschinen> radial roller bearing
Querruder n <Lufttrans> aileron, wing flap
Querruderausschlag m <Lufttrans> aileron deflection
Querrudernachsteuerung f <Lufttrans> aileron follow-up
Querruderstellungsanzeiger m <Lufttrans> aileron position indicator
Querrudersteuerung f <Lufttrans> aileron control
Querrudersteuerungsrad n <Lufttrans> aileron control wheel
Quersäge f <Fertig> crosscut saw
Querschiene f <Eisenbahn> crossbar
querschiffs adv <Wassertrans> aburton, athwartships
Querschlaghammer m <Bau> cross-peen hammer
Querschlitten m 1. <Fertig> saddle (Karusselldrehmaschine, Waagerechtstoßmaschine); 2. <Maschinen> cross slide, saddle, slide head, slide rest
Querschlitten des Tisches <Fertig> apron (Waagrechtstoßmaschine)
Querschlitz m <Telekom> transverse slot
Querschneide f <Fertig> chisel edge, (AE) dead center, (BE) dead centre (Spiralbohrer)
Querschneider m <Verpack> sheet-cutting machine
Querschneidewinkel m <Fertig> angle of point (Spiralbohrer)
Querschnitt m 1. <Bau> cross section, section; 2. <Fertig> crosscut (Sägen); 3. <Maschinen> cross section, sectional view; 4. <Metall, Papier> cross section; 5. <Phys> profile; 6. <Textil> cross section; 7. <Wassertrans> cross section (Schiffbau); transverse section (Schiffkonstruktion)
Querschnitt des Luftschraubenblattes <Lufttrans> blade cross section (Hubschrauber)
Querschnittsuntersuchung f <Ergon> cross-sectional study
Querschnittszeichnung f 1. <Maschinen> cross-section drawing; 2. <Wassertrans> cross-sectional drawing (Schiffkonstruktion)

Querschott *n* <Wassertrans> transverse bulkhead *(Schiffkonstruktion)*
Querschwingung *f* <Maschinen> transverse vibration
Quersee *f* <Wassertrans> beam sea
Quersieder *m* <Heiz & Kälte> cross tube boiler
Quersiederrohrkessel *m* <Heiz & Kälte> cross tube boiler
Querspannung *f* <Elektriz> quadrature-axis component of the voltage
Querspanten *npl* <Wassertrans> transverse framing *(Schiffbau)*
Querspur-Aufzeichnung *f* <Aufnahme> cross track recording
Querstabilisator *m* <Kfztech> antiroll bar
Querstabilität *f* 1. <Kfztech> lateral stability; 2. <Wassertrans> transverse stability *(Schiffkonstruktion)*
Querstrahler *m* <Telekom> broadside array
Querstrahlruder *n* <Wassertrans> lateral thruster, side thruster *(Schiffantrieb)*
Querstrebe *f* 1. <Bau> cross wall; 2. <Kfztech> antiroll bar
Querstreifen *m* <Textil> crosswise stripe
Querstrom *m* 1. <Elektriz, Elektrotech> leakage current, quadrature-axis component of current, shunt current, wattless current; 2. <Ker & Glas> transverse current; 3. <Kfztech> cross flow *(Motor)*
Querstromgebläse *n* <Elektriz> radial fan
Querstromkopfmotor *m* *(DOHC-Motor)* <Kfztech> direct-acting overhead camshaft engine *(DOHC engine)*
Querstromkühler *m* <Kfztech> cross-flow radiator
Querstromlüfter *m* <Heiz & Kälte> centrifugal fan, cross-flow fan
Querstromofen *m* <Ker & Glas> cross-fired furnace
Querströmung *f* <Kfztech> cross flow
Querstromventilator *m* <Maschinen> cross flow fan
Querstromverteiler *m* <Papier> flow spreader
Querstromwärmeaustauscher *m* <Heiz & Kälte> cross-flow heat exchanger
Quersumme *f* <Math> horizontal check sum
Quersummenkontrolle *f* <Comp & DV> parallel balance
Quersupport *m* 1. <Fertig> *(AE)* rail tool head; rail head *(Hobelmaschine)*; 2. <Maschinen> cross slide
Querträger *m* 1. <Bau> crossbar, cross girder, cross member, wind brace; 2. <Kfztech> cross member; 3. <Maschinen> cross girder, cross member, crossbeam; 4. <Phys> crosshead
Querträgermaschine *f* <Ker & Glas> x-arm machine
Quertrimmung *f* <Lufttrans> lateral trim
Querverbandsteil *n* <Wassertrans> transverse member *(Schiffbau)*
Querverbindung *f* 1. <Bau> interconnection; 2. <Kfztech> cross rail; 3. <Kunststoff> cross link; 4. <Telekom> *(BE)* interexchange, *(AE)* interoffice, interconnection
Querverbindungsleitung *f* <Telekom> tie line *(zwischen Vermittlungen)*
Querverbindungsleitung *f* **zwischen Wählnebenstellen** <Telekom> inter-PABX tie circuit
Querverbindungsschnittstelle *f* <Telekom> tie circuit interface
Quervergrößerung *f* <Phys> transverse magnification
querverlaufend *adv* <Wassertrans> transverse *(Schiffkonstruktion)*
Querversetzen *n* <Wassertrans> swaying *(Schiffsbewegung)*
Querverstrebung *f* <Wassertrans> cross brace *(Schiffbau)*
Quervorschub *m* 1. <Fertig> lateral feed; 2. <Maschinen> transverse feed
Quervorschubleitung *f* <Lufttrans> cross-feed line

Querwand *f* 1. <Bau> cross wall; 2. <Mechan> bulkhead
Querzahl *f* <Fertig> rho ratio
Querzusammenziehung *f* <Metall> lateral contraction
Quetsch... <Bau, Fertig, Maschinen, Sicherheit, Textil> crushing
Quetsche *f* 1. <Lebensmittel> masher *(Küchengerät)*; 2. <Textil> squeezer
quetschen *v* <Elektrotech, Funktech> crimp
Quetschen *n* <Papier> squeezing
Quetschfalte *f* <Textil> inverted pleat, knife pleat
Quetschfestigkeit *f* <Maschinen> ultimate crushing strength
Quetschflüssigkeit *f* <Fertig> trapping *(Hydraulik)*
Quetschgrenze *f* <Maschinen> crushing yield point
Quetschhahn *m* <Bau> pinchcock
Quetschhülse *f* <Elektrotech> ferrule *(bei Anschlussverbindungen)*
Quetschkopf *m* <Kfztech> squish combustion chamber
Quetschmühle *f* <Lebensmittel> bruiser
Quetschschutz *m* <Sicherheit> crushing safety
Quetschtube *f* <Verpack> collapsible tube
Quetschung *f* <Sicherheit> crushing *(Maschinen)*
Quetschverbindung *f* <Elektrotech> crimped connection
Quetschversuch *m* <Werkprüf> compression test, crushing test
Quetschwalze *f* <Foto> squeegee
Quetschwalzwerk *n* <Kohlen> chat roller
Quetschwerk *n* <Textil> squeezer
Quetschwiderstand *m* <Elektrotech> pinched resistor
Quetschzone *f* <Kunststoff> pinch-off area
Quick-Change-Flugzeug *n* *(QC-Flugzeug)* <Lufttrans> quick-change aircraft *(QC aircraft)*
Quicksand *m* <Kohlen> quicksand
Quickton *m* <Kohlen> quick clay
QUIL-Gehäuse *n* <Elektronik> QUIP, quad-in-line package
Quint *n* <Akustik> fifth
Quintal *n* <Metrol> *(BE)* quintal
Quinte *f* <Akustik> fifth
Quintett *n* <Phys> quintet *(Spektroskopie)*
Quirl *m* <Lebensmittel> agitator
quirlen *v* <Chemtech> agitate
QUISAM *(erweiterte indizierte Zugriffsmöglichkeit für sequenzielle Dateien)* <Comp & DV> QUISAM *(queued unique index sequential access method)*
quittieren *v* <Comp & DV> acknowledge
Quittierung *f* <Comp & DV, Kontroll> handshake
Quittung *f* <Comp & DV, Telekom> ACK, acknowledgement acknowledgement, confirmation *(positive Rückmeldung)*
Quittungsbetrieb *m* <Comp & DV, Telekom> handshake, handshaking
Quittungsbit *n* <Telekom> acknowledgement bit
Quittungsmerker *m* <Comp & DV, Telekom> acknowledgement flag
Quittungsschalter *m* <Comp & DV, Telekom> accept switch
Quittungssignal *n* <Comp & DV, Telekom> acknowledgement signal, confirmation signal
Quittungsvollzug *m* <Comp & DV, Telekom> acknowledgement delay
Quittungszeichen *n* <Telekom> acknowledgement signal
Quotient *m* <Comp & DV, Math> quotient *(Ergebnis einer Division)*
Quotientenzweig *m* <Gerät> ratio arm *(Brückenschaltung)*
Q-Wert *m* <Phys> Q value *(Kernphysik)*

QWERTY-Tastatur *f* <Comp & DV> QWERTY keyboard *(englischsprachiges Keyboard-Layout)*
QWERTZ-Tastatur *f* <Comp & DV> QWERTZ keyboard *(deutschsprachiges Keyboard-Layout)*
QZ *(quantisiertes Zeichen)* <Elektronik> QS *(quantized signal)*

R

r 1. <Akustik> *(Entfernung von der Schallquelle)* r *(distance from source)*; 2. <Kerntech> *(Kernradius)* r *(nuclear radius)*; 3. <Optik> *(Brechungswinkel, Refraktionswinkel)* r *(angle of refraction)*; 4. <Phys> *(Brechungswinkel)* r *(angle of refraction)*
R 1. <Elektriz> *(Reluktanz)* R *(magnetic reluctance)*; 2. <Elektriz> *(Widerstand)* R *(resistance)*; 3. <Kerntech> *(Rydberg-Konstante)* R *(Rydberg constant)*; 4. <Kerntech> *(Dosis)* R *(dose rate)*; 5. <Kerntech> *(linearer Bereich)* R *(linear range)*; 6. <Phys, Thermod> *(Gaskonstante)* R *(gas constant)*; 7. <Strahlphys> *(Röntgen)* R *(röntgen)*
Rα *(Rydberg-Konstante)* <Kerntech> Rα *(Rydberg constant)*
Ra *(Radium)* <Chemie> Ra *(radium)*
Racah-Kopplung *f* <Kerntech> Racah coupling
Racemat *n* <Chemie> racemate
racemisch *adj* <Chemie> racemic
racemisieren *v* <Chemie> racemize
Racemisierung *f* <Chemie> racemization
Rachen *m* 1. <Fertig> gap *(Lehre)*; 2. <Maschinen> gap, jaws, throat
Rachenlehre *f* <Maschinen, Metrol> *(AE)* caliper gage, *(BE)* calliper gauge, *(AE)* gap gage, *(BE)* gap gauge, *(AE)* snap gage, *(BE)* snap gauge
rad <Geom> radian
Rad *(Einheit der Energiedosis)* <Strahlphys> rad, radiation absorbed dose *(veraltet)*
Rad *n* <Maschinen> wheel
Rad *n* **mit Außenverzahnung** <Maschinen> external gear
Radabnutzung *f* <Eisenbahn> wheel wear
Radabstand *m* <Eisenbahn, Kfztech> wheel base
Radabweiser *m* <Bau> spur post
Radachse *f* 1. <Kfztech> axletree *(eines Pferdewagens)*; 2. <Mechan, Phys> axis of a wheel
Radar *n* <Funkort, Phys, Strahlphys, Wassertrans> radar, radio detection and ranging
Radar *n* **mit Absolutkursdarstellung** <Funkort> true motion radar
Radar *n* **mit Abtastantenne** <Funkort> scanning radar
Radar *n* **mit elektronischer Strahlauslenkung** <Funkort> electronic scanning radar
Radar *n* **mit elektronischer Strahlschwenkung** <Funkort> electronic scanning radar
Radar *n* **mit getastetem Träger** <Funkort> pulse radar
Radar *n* **mit synthetischer Strahlungscharakteristik** <Funkort> synthetic aperture radar
Radarabdeckung *f* <Funkort> radar dome, radome
Radarabtaster *m* <Lufttrans, Strahlphys, Trans, Wassertrans> radar scanner
Radarabtastschema *n* <Lufttrans, Trans, Wassertrans> radar scan pattern

Radarabtastung *f* <Funkort, Lufttrans, Trans, Wassertrans> radar scanning
Radaranflug *m* <Lufttrans> ground-controlled approach, radar approach
Radaranflugkontrollzentrum *n* <Lufttrans> *(AE)* radar approach control center, *(BE)* radar approach control centre
Radarantenne *f* 1. <Funkort> radar aerial, radar antenna, radar scanner; 2. <Lufttrans> scanner; 3. <Trans> radar aerial, radar antenna; 4. <Wassertrans> radar aerial, radar antenna, scanner
Radarantennenverkleidung *f* <Raumfahrt> radome
Radarantwort *f* <Lufttrans, Trans, Wassertrans> radar response
Radarantwortbake *f* <Funkort, Wassertrans> radar responding beacon; racon *(Seezeichen)*
Radaraufklärung *f* <Funkort> radar reconnaissance
Radarauflösung *f* <Funkort> radar resolution
Radarauflösungsvermögen *n* <Funkort> radar resolution
Radaraufzeichnung *f* <Wassertrans> radar plotting
Radarausrüstung *f* <Funkort, Lufttrans, Trans> radar equipment
Radarauswertung *f* <Wassertrans> radar plotting
Radarbake *f* 1. <Funkort> radar beacon, radar marker beacon; 2. <Strahlphys> radar beacon; 3. <Wassertrans> radar beacon, radar marker beacon *(Seezeichen)*
Radarbeobachtung *f* <Funkort> radar observation, radar picket
Radarbeobachtungsstation *f* <Trans, Wassertrans> radar picket station
Radarbild *n* 1. <Funkort> image; 2. <Wassertrans> radar image
Radarbildschirm *m* <Funkort, Lufttrans, Trans, Wassertrans> radar screen
Radarbildübertragung *f* <Funkort> radar image transmission, radar relay
Radarbildübertragungsstrecke *f* <Funkort> radar link, radar relay, radar relay link
Radarboje *f* <Wassertrans> radar marker float
Radardaten *f* <Funkort> radar data
Radardiplexer *m* <Funkort> radar diplexer
Radardrehantenne *f* <Lufttrans, Trans, Wassertrans> radar scanner
Radarecho *n* 1. <Lufttrans> radar echo; 2. <Funktech> radar response; 3. <Trans> radar echo
Radarechoanzeige *f* <Funkort, Lufttrans> radar blip
Radarecholotung *f* <Funkort> radar sounding
Radareinstufung *f* <Lufttrans, Trans, Wassertrans> radar rating
Radarempfänger *m* <Funkort> radar receiver *(FO)*
Radarerfassung *f* <Trans> radar contact
Radarfeuer *n* <Strahlphys> radar beacon
Radarflugsicherungsdienst *m* <Trans> radar air traffic control
Radarführung *f* <Lufttrans, Trans, Wassertrans> radar vectoring
Radarfunkfeuer *n* <Lufttrans, Telekom> radar beacon
Radargast *m* <Wassertrans> radar operator *(Besatzung)*
Radargeschwindigkeitsmesser *m* <Trans> radar speed meter
radargesteuert *adj* <Wassertrans> radar-controlled
Radarhaube *f* <Lufttrans, Wassertrans> radome
Radarhöhenmesser *m* <Funkort, Lufttrans> radar altimeter
Radarhorizont *m* <Funkort> radar horizon
Radaridentifikation *f* <Funkort> radar identification, radar identification by pulse repetition frequency
Radarkennung *f* <Funkort, Lufttrans, Trans> radar identification

Radarkontrolle f <Funkort, Lufttrans, Trans> radar control
Radarkuppel f 1. <Lufttrans> radome; 2. <Wassertrans> radar dome, radome
Radarküstenbild n <Wassertrans> radar coast image
Radarleitdienst m <Trans> radar surveillance
Radarlotse m <Lufttrans, Trans> radar controller
Radarmast m <Wassertrans> radar mast
Radarnase f 1. <Funkort> radome; 2. <Lufttrans> radome; blister (Hubschrauber); 3. <Wassertrans> radome
Radarnavigation f <Lufttrans> radar navigation
Radarortung f <Wassertrans> radar detection
Radarparabolreflektor m <Funkort, Lufttrans, Trans> radar dish
Radarpeilung f <Funkort, Wassertrans> radar bearing
Radarpip n <Funkort, Wassertrans> radar pip
Radarreflektor m <Funkort, Lufttrans, Trans, Wassertrans> radar reflector
Radarreflektorboje f <Wassertrans> radar reflector buoy
Radarreichweite f <Funkort, Lufttrans, Trans, Wassertrans> radar range
Radarrelaisstation f <Funktech, Trans, Wassertrans> radar relay station
Radarröhre f <Elektronik> radar tube
Radarrückstrahlvermögen n <Funkort> radar reflectivity
Radarscanner m <Strahlphys> radar scanner
Radarschirm m <Funkort, Lufttrans, Trans, Wassertrans> radar screen
Radarschirmbild n <Funkort, Lufttrans, Trans> radar display
Radarschüssel f <Funkort, Lufttrans, Trans> radar dish
Radarsendebake f <Wassertrans> ramark (ohne Empfangsteil)
Radarsichtgerät n <Funkort, Trans, Wassertrans> radar scope
Radarsonde f <Strahlphys> radar sensor
Radarstation f <Funkort, Wassertrans> radar station
Radarstelle f <Lufttrans, Trans, Wassertrans> radar unit
Radarsteuerkurs m <Lufttrans, Trans> radar heading
Radarstörung f 1. <Funktech> radar interference; radar jamming („Lametta"-Abwurf; auch durch Düppeln); 2. <Wassertrans> radar interference
Radarstrahl m <Funkort, Wassertrans> radar beam
Radartarnung f <Funkort> radar camouflage
Radarüberwachung f 1. <Funkort, Lufttrans> radar monitoring; 2. <Trans> radar monitoring, radar surveillance
Radarvorposten m <Funkort> radar picket
Radarwellen fpl <Strahlphys> radar waves
Radarzeichnung f <Lufttrans> radar plotting
Radarzielansteuerung f <Lufttrans, Trans> radar homing
Radarzielsuchkopf m <Funkort> radar homing head
Radarzielverfolgung f <Funkort, Lufttrans, Trans> RT, radar tracking
Radaufhängung f <Kfztech> suspension system
Radauswuchtung f <Kfztech> wheel balancing
Radauswuchtungsmaschine f <Eisenbahn> wheel trueing machine
Radbefestigungskeil m <Eisenbahn> wheel wedge
Radblende f <Kfztech> hub cap
Radblockierer m <Kfztech> wheel clamp
Raddurchmesser m <Fertig> gear diameter
Radeinstellung f <Kfztech> wheel alignment
Räder npl <Kfztech, Mechan> wheels
Räderblock m 1. <Fertig> wheel train; 2. <Maschinen> cluster of gearwheels
Räderfräsmaschine f <Maschinen> gear milling machine
Rädergetriebe n 1. <Fertig> gear unit, gearbox, gears; 2. <Maschinen> gear train, train of gearing
Räderkasten m <Maschinen, Mechan> gearbox
Räderkegelwinkel m <Maschinen> gear cone angle

räderlos adj <Maschinen> gearless
Rädersatz m 1. <Kfztech> set of wheels; 2. <Maschinen> nest of gearwheels
Räderschere f 1. <Fertig> gear quadrant, quadrant plate; quadrant (Spanung); 2. <Maschinen> quadrant, quadrant plate
Räderspindelkasten m <Fertig> gear head
Rädervorgelege n <Fertig> back gear
Räderwerk n 1. <Kfztech> train of gears; 2. <Maschinen> gear set, gear train, gearing, train of gears
Radfahrweg m 1. <Bau> cycle track; 2. <Trans> cycle path
Radfelge f <Kfztech> wheel rim
Radflansch m <Kfztech> wheel flange
Radflügelflugzeug n <Lufttrans> cyclogyro
radführender Lenker m <Kfztech> control arm (Federung, Aufhängung)
Radgleitenanzeige f <Eisenbahn> wheel-slide detection
radial adj <Ergon, Geom, Mechan> radial
radial ausbaubar adj <Fertig> with union ends (Kunststoffinstallationen)
Radial... <Fertig, Maschinen, Mechan, Phys> radial
Radialanteil m **der Wellenfunktion** <Strahlphys> radial part of the wave function
Radialbeaufschlagung f <Maschinen> radial admission (einer Turbine)
Radialbelastung f <Fertig> radial loading
Radialbohrmaschine f 1. <Fertig, Maschinen> radial drill, radial drilling machine; 2. <Mechan> radial drilling machine
Radialbohrmaschine f **mit Höhenverstellung** <Fertig> adjustable radial drilling machine
Radialdichtring m <Maschinen> rotary shaft seal
radiale Ablenkelektrode f <Fernseh> radial deflecting electrode
radiale elektrische Felder npl <Strahlphys> radial electrical fields
radiale Leistungsverteilung f <Kerntech> radial power distribution
radiale Symmetrie f <Geom> radial symmetry
radiale Verschiebung f <Kerntech> radial shift
radiale Verteilungsfunktion f <Strahlphys> radial distribution function
radialer Neutronenfluss m <Kerntech> radial neutron flux
radialer Teil m **der Wellenfunktion** <Strahlphys> radial part of the wave function
radialer Tischvorschub m <Fertig> table infeed
radialer Vorschub m <Maschinen> radial feed
radiales Austauschen n <Kerntech> radial shuffling (von Brennelementen)
radiales Moment n <Lufttrans> annular momentum
radiales Umsetzen n <Kerntech> radial shuffling (von Brennelementen)
Radialflügelrad n <Heiz & Kälte> radial vane wheel
Radialgeschwindigkeit f <Nichtfoss Energ> radial velocity
Radialkolbenpumpe f <Fertig> radial piston pump
Radialkomponente f <Phys> radial component
Radialkraft f <Fertig> thrust force
Radialkugellager n <Maschinen> radial ball bearing
Radiallager n 1. <Kfztech> plain bearing; 2. <Maschinen> journal bearing, radial bearing
Radiallüfter m 1. <Elektriz> radial fan; 2. <Maschinen> rotary fan
Radialreifen m 1. <Kfztech> (AE) radial tire, (BE) radial tyre; 2. <Kunststoff> (AE) radial ply tire, (BE) radial ply tyre
Radialschlag m <Fertig, Maschinen> radial run-out
Radialspiel n <Fertig, Maschinen> radial play

Radialstrahl m <Optik> direct radial *(durch die Mitte der Apertur)*
Radialturbine f <Maschinen> radial flow turbine
Radialventilator m <Heiz & Kälte> radial fan, radial flow fan
Radialverfahren n <Fertig> radial feed method *(Schneckenfräsen)*
Radialvorschub m <Maschinen> radial feed
Radialzylinderrollenlager n <Maschinen> parallel-roller journal bearing, radial cylindrical roller bearing
Radian m <Elektriz, Phys> radian *(Bogenmaß)*
Radiant m <Elektronik, Geom> radian
Radienschablone f <Fertig> *(AE)* radius gage, *(BE)* radius gauge
radieren v <Fertig> etch
Radierfestigkeit f <Konstzeich> resistance to erasure
Radikal n 1. <Kerntech, Kunststoff> radical; 2. <Math> radical *(Lösung einer reinen Gleichung)*
radikalische Polymerisation f <Kunststoff> radical polymerization
radikalische Reaktion f <Kunststoff> free radical reaction
Radikalpolymerisation f <Kunststoff> radical polymerization
Radikalreaktion f <Kunststoff> free radical reaction
Radikand m <Math> radicand *(Term unter dem Wurzelzeichen)*
Radio n <Funktech, Phys> radio
Radioactinium n <Strahlphys> radioactinium
radioaktiv adj <Erdöl, Kerntech, Phys, Strahlphys, Teilphys> radioactive
radioaktiv kontaminiertes Wasser n <Umweltschmutz> contaminated water
radioaktiv markiertes Atom n <Kerntech> tagged atom
radioaktive Altersbestimmung f <Kerntech, Strahlphys> radioactive dating
radioaktive Halbwertszeit f <Kerntech, Strahlphys> radioactive lifetime
radioaktive Kontaminierung f <Kerntech, Strahlphys> radioactive contamination
radioaktive Markierung f <Kerntech, Strahlphys> radioactive tracer
radioaktive Verschmutzung f <Umweltschmutz> radioactive pollution
radioaktive Verseuchung f 1. <Kerntech> radioactive contamination; 2. <Sicherheit> radioactive pollution; 3. <Strahlphys> radioactive contamination
radioaktive Zerfallsrate f <Kerntech, Strahlphys> radioactive decay rate
radioaktive Zerfallsreihe f 1. <Kerntech> radioactive decay series, radioactive series; 2. <Strahlphys> radioactive series, radioactive decay series
radioaktiver Abfall m 1. <Abfall> nuclear waste, radioactive waste, radwaste; 2. <Kerntech> effluent, nuclear waste, radioactive waste; 3. <Strahlphys> radioactive waste
radioaktiver Kern m <Phys> radionuclide
radioaktiver Kohlenstoff m <Phys> radiocarbon
radioaktiver Körper m <Kerntech, Strahlphys> radioactive body
radioaktiver Niederschlag m 1. <Kerntech> fallout, radioactive fallout, rainout; 2. <Strahlphys, Umweltschmutz> radioactive fallout, rainout
radioaktiver Stoff m <Sicherheit, Umweltschmutz> radioactive substance
radioaktiver Übergang m <Kerntech, Strahlphys> radioactive change
radioaktiver Zerfall m <Kerntech, Strahlphys> radioactive disintegration, radioactive decay
radioaktives Cobalt n <Chemie> radiocobalt
radioaktives Element n <Kerntech, Strahlphys> radioactive element

radioaktives Gleichgewicht n <Kerntech, Strahlphys> radioactive equilibrium
radioaktives Isotop n 1. <Chemie> radioisotope; 2. <Kerntech> radioactive isotope; 3. <Phys> radioisotope; 4. <Strahlphys> radioactive isotope; 5. <Teilphys> radioisotope
radioaktives Log n <Erdöl> radioactive log
radioaktives Markieren n <Kerntech, Strahlphys> *(AE)* radioactive labeling, *(BE)* radioactive labelling
radioaktives Material n <Elektrotech> radioactive material
radioaktives Nuklid n <Strahlphys> radionuclide
radioaktives Präparat n <Kerntech> radiation source
radioaktives Spurenelement n <Kerntech, Strahlphys> radioactive tracer
radioaktives Standardpräparat n 1. <Kerntech> radioactive standard; 2. <Strahlphys> radioactive standard, radioactivity standard
radioaktives Strontium n <Chemie> radiostrontium
radioaktives Zerfallsgesetz n <Strahlphys> law of radioactive decay
Radioaktivität f <Kerntech, Phys, Strahlphys, Teilphys> radioactivity
Radioaktivitätslog n <Erdöl> nuclear log
Radioaktivitätsmessgerät n <Strahlphys> radioactivity meter
Radioantenne f 1. <Funktech> *(BE)* radio aerial, *(AE)* radio antenna; 2. <Kfztech> *(BE)* radio aerial; *(AE)* radio antenna *(Zubehör)*; 3. <Trans> *(BE)* radio aerial, *(AE)* radio antenna
Radioastronomie f <Phys, Raumfahrt> radio astronomy
Radioastronomieantenne f <Telekom> radioastronomical antenna
Radiochemie f <Kerntech, Strahlphys> radiochemistry
radiochemischer Abzug m <Kerntech> radiochemical fume cupboard
Radiofrequenz f *(HF)* 1. <Aufnahme, Elektronik, Fernseh, Funktech> radio frequency, RF; 2. <Telekom> radio frequency, RF *(Richtfunk)*; 3. <Wassertrans> radio frequency, RF
radiofrequenzdurchlässiger Bereich m **der Atmosphäre** <Phys> radio window
radiogen adj <Phys, Strahlphys> radiogenic
Radiogoniometer n <Phys, Strahlphys> radiogoniometer
Radiogoniometrie f <Funkort, Phys, Trans> RDF, radio direction finding
Radiographie f 1. <Fertig> radiographic examination; 2. <Phys> radiography
Radioisotop n 1. <Chemie> radioisotope; 2. <Kerntech, Strahlphys> radioactive isotope
Radioisotopengenerator m <Raumfahrt> radioisotope power generator
Radiojod n <Chemie> radioiodine
Radiokobalt m <Strahlphys> radiocobalt
Radiokohlenstoff m <Strahlphys> radiocarbon
Radiokompass m <Funkort, Lufttrans> automatic direction finder
Radiolog n <Erdöl> radioactive log *(Messtechnik)*
Radiologie f <Strahlphys> radiology
Radiolumineszenz f <Strahlphys> radioluminescence
Radiolyse f <Chemie, Kerntech, Strahlphys> radiolysis
Radiolyse f **von Wasser** <Kerntech> water radiolysis
radiolytisch adj <Kerntech> radiolytic
Radiometer n <Nichtfoss Energ> radiometer
Radiometrie f <Chemie> radiometry
radiometrische Analyse f <Strahlphys> radiometric analysis
radiometrische Bohrlochvermessung f <Nichtfoss Energ> well logging

Radiomimetikum n <Chemie> radiomimetic
Radionuklid n <Strahlphys> radionuclide
Radionuklidabscheider m <Sicherheit> radionuclide trap
Radionuklidreinheit f <Kerntech, Strahlphys> radioactive purity
Radiosonde f <Telekom> radiosonde
Radiosondenbeobachtungsstation f <Funktech> meteorological sonde receiving station, radio sonde observation station
Radiospektrum n <Strahlphys> radio spectrum
Radiostern m <Raumfahrt> radio star *(Weltraumfunk)*
Radiotelefon n **mit Frequenzdekade** <Funktech> synthezised radio telephone
Radioteleskop n <Phys> radio telescope
Radiotoxizität f <Strahlphys> radiotoxicity
Radiowelle f 1. <Elektriz> radio wave; 2. <Elektronik> radiowave; 3. <Wellphys> radio wave
Radium n *(Ra)* <Chemie> radium *(Ra)*
Radium-Emanation f <Strahlphys> radium emanation
Radius m <Bau, Maschinen, Optik, Phys> radius
Radiusdrehmeißel m <Maschinen> radius tool
Radiusfräser m <Maschinen> radius form cutter
Radiuslehre f <Maschinen> *(AE)* radius gage, *(BE)* radius gauge
Radix f <Comp & DV> radix
Radixkomplement n <Comp & DV> radix complement
Radix-minus-eins-Komplement n <Comp & DV> radix-minus-one complement
Radixpunkt m <Comp & DV> radix point
Radixschreibweise f <Comp & DV> radix notation
Radixschreibweise f **mit fester Notation** <Comp & DV> fixed-base notation
Radixschreibweise f **mit gemischter Basis** <Comp & DV> mixed-base notation, mixed-radix notation
Radiziereinrichtung f <Metrol> square root extracting device
Radkappe f <Kfztech> hubcap
Radkralle f <Kfztech> wheel clamp
Radkranz m 1. <Eisenbahn> wheel flange; 2. <Maschinen> wheel rim
Radkurve f <Geom> cycloid
Radlager n <Kfztech> wheel bearing
Radlagerspiel n <Kfztech> wheel bearing clearance
Radlast f <Bau, Maschinen, Qual> wheel load
Radmagnetron n <Phys> cavity magnetron
Radmittelpunkt m <Fertig> *(AE)* gear center, *(BE)* gear centre
Radmutter f <Kfztech> wheel nut
Radnabe f <Fertig, Trans> hub, wheel hub
Radnabenflansch m <Kfztech> hub flange
Radnachlauf m <Kfztech> caster, castor
Radom n 1. <Lufttrans> blister *(Hubschrauber)*; 2. <Telekom> radome
Radon n *(Rn)* <Phys> radon *(Rn)*
Radpaar n <Lufttrans> dual wheel
Radsatz m <Eisenbahn> wheelset
Radschacht m <Lufttrans> wheel well *(Fahrwerk)*
Radschloss n <Kfztech> hublock
Radschlupf m 1. <Eisenbahn> wheel slide; 2. <Kfztech> wheel slip
Radschutz m 1. <Maschinen> wheel guard; 2. <Sicherheit> guard
Radschwingarm m <Kfztech> wheel suspension lever
Radspur f <Bau> rut
Radstand m <Eisenbahn, Kfztech> wheelbase
Radsteg m <Eisenbahn> wheel web
Radstern m <Kfztech> *(AE)* spoke wheel center, *(BE)* spoke wheel centre
Radsturzwinkel m <Kfztech> camber angle
Radtrommel f <Hydraul> drum *(Turbine)*
Radverbinder m <Eisenbahn> wheel bond
Radvorleger m <Eisenbahn> scotch block
Radzahnbahn f <Eisenbahn> *(AE)* rack railroad, *(BE)* rack railway
Radzahnbahnbeiwagen m <Eisenbahn> *(AE)* rack railroad trailer, *(BE)* rack railway trailer
Radzahnbahnschiene f <Eisenbahn> rack track
Radzapfen m <Kfztech> spindle
Radzylinder m <Kfztech> wheel cylinder
raffen v <Textil> gather
Raffhalter m <Textil> tie back
Raffination f <Fertig> refining
Raffinerie f <Erdöl, Fertig> refinery
Raffineriegas n <Erdöl> refinery gas *(Destillationsprodukt)*
Raffinerierückstände mpl <Abfall> refinery waste
raffinieren v 1. <Chemtech> clarify; 2. <Metall> refine
Raffiniergas n <Thermod> refinery gas
Raffinierofen m <Thermod> refining furnace
raffiniertes Produkt n <Meerschmutz> refined product
Raffinose f <Chemie> melicitose, raffinose
rahmen v <Foto> frame
Rahmen m 1. <Bau> carrier, frame; frame *(Tür, Fenster)*; 2. <Comp & DV> frame; 3. <Druck> chase; 4. <Eisenbahn> frame; 5. <Foto> mounting *(Foto oder Licht)*; frame *(eines Fotos)*; 6. <Funktech> loop; 7. <Kfztech, Maschinen, Mechan, Papier> frame; 8. <Raumfahrt> frame *(Antenne)*; 9. <Telekom> bay; frame, loop *(Antenne)*; frame *(digitale Übertragung)*; 10. <Verpack> frame • **im Rahmen von etwas liegen** <Patent> fall within the scope of
Rahmen m **der Ansprüche** <Patent> scope of claims
Rahmenantenne f 1. <Funkort> frame aerial, frame antenna; 2. <Funktech> frame aerial, frame antenna, loop, loop antenna
Rahmenausrichtung f <Raumfahrt> frame alignment *(Weltraumfunk)*
Rahmenbildung f <Telekom> frame generation
Rahmenblechschere f <Mechan> guillotine shears
Rahmencodierung f <Comp & DV> skeletal coding, skeleton coding
Rahmenfederung f <Fertig> arc spring
Rahmenfehler m <Comp & DV> frame error
Rahmengestell n <Telekom> frame
Rahmengleichlauf m <Telekom> frame alignment
Rahmengleichlaufverlust m <Telekom> frame alignment loss, sync loss *(PCM)*
Rahmenkennung f <Telekom> frame marking
Rahmenlänge f <Raumfahrt> frame length *(Antenne)*
Rahmenlängsträger m <Kfztech> chassis member
Rahmenprüfzeichenfolge f frame-checking sequence *(FCS)*
Rahmenquerträger m <Kfztech> cross member, cross rail
Rahmenschlupf f <Telekom> frame slip
Rahmenspant n <Wassertrans> web frame *(Schiffbau)*
Rahmenstiel m <Bau> member
Rahmensynchronisation f <Telekom> frame alignment
Rahmensynchronisierung f 1. <Raumfahrt> frame synchronization *(Weltraumfunk)*; 2. <Telekom> frame synchronization
Rahmensystem n <Künstl Int> expert system shell
Rahmenwirkungsgrad m <Raumfahrt> frame efficiency *(Weltraumfunk)*
Rähmstück n <Bau> breastsummer, summer beam
Rahnock f <Wassertrans> yardarm *(Segeln)*
Rahsegel n <Wassertrans> square sail
Rainout n <Umweltschutz> rainout

Rakel

Rakel f 1. <Druck> doctor blade, squeegee; 2. <Kunststoff, Papier> doctor; 3. <Textil> knife, squeegee; 4. <Verpack> doctor blade
Rakelmesser n <Druck> doctor blade
Rakelstreichmaschine f <Papier> blade coater, knife coater
Rakelstreichverfahren n 1. <Kunststoff> knife spreading; 2. <Papier> blade coating
Rakelwalze f <Papier> doctor roll
Rakete f <Raumfahrt> rocket
Raketenantrieb m <Raumfahrt> rocket propulsion
Raketenbasis f <Raumfahrt> launching base (Startplatz)
Raketenflugzeug n <Raumfahrt> rocketplane
Raketenraumgleiter m <Raumfahrt> orbital glider
Raketenstart m <Raumfahrt> (AE) JATO, (BE) RATO, (AE) jet-assisted takeoff, (BE) rocket-assisted takeoff
Raketenstartanlage f <Raumfahrt> rocket launching site
Raketenstarter m <Raumfahrt> rocket launcher
Raketenstufe f **mit Fluchtgeschwindigkeit** <Raumfahrt> escape rocket stage
Raketentreibstoff m <Raumfahrt> rocket fuel, rocket propellant
Raketentriebwerk n <Maschinen, Raumfahrt> rocket engine
RAM 1. <Comp & DV> (Direktzugriffsspeicher, Lese-/Schreibspeicher, Schreib-/Lesespeicher) RAM (random access memory); 2. <Elektronik> (Direktzugriffsspeicher, Schreib-/Lesespeicher) RAM (random access memory)
Raman'sche Spektrometrie f <Strahlphys> Raman spectrometry
Raman'sche Spektroskopie f <Phys> Raman spectroscopy
Raman'sche Streuung f <Phys> Raman scattering
Raman'scher Effekt m <Phys, Strahlphys> Raman effect
RAM-Bank f <Elektrotech> bank of RAMs
RAM-Laufwerk n <Optik> read-write drive
Rammarbeiten fpl <Bau> pile driving
Rammaufsatz m <Kohlen> pileblock
Rammbär m 1. <Bau> beetle head, pile driver, piling hammer; 2. <Kohlen> pile ram
Rammbug m <Trans, Wassertrans> ram bow
Ramme f 1. <Bau> rammer, ram; 2. <Fertig> hammer, monkey, rammer; punner (Gießen); 3. <Kohlen, Mechan> ram, rammer
rammen v 1. <Bau> pile, ram; 2. <Fertig> pun (Gießen); 3. <Kohlen, Wassertrans> ram; run down (Schiff)
Rammen n 1. <Bau> piling; spiling (von Pfählen); 2. <Phys> ram
Rammgerüst n <Bau> piling frame
Rammhammer m <Kohlen> pile hammer
Rammhaube f <Bau> cap, pile cap
Rammprotokoll n <Bau> penetration record
Rammtest m <Kohlen> ram penetration test
Rammungsaufzeichnung f <Kohlen> pile driving record
Rammungsformel f <Kohlen> pile driving formula
Rampe f 1. <Bau, Elektrotech, Kohlen, Papier> ramp; 2. <Raumfahrt> pad (für Raketen)
Rampenbeleuchtung f <Foto> bank of lights
rampenförmiger Lastanstieg m <Kerntech> ramp change of load
Rampgewicht n <Lufttrans> ramp weight
Rampstatus m <Lufttrans> ramp status
Ramsden-Kreis m <Optik> eye ring
Ramsden'sches Okular n <Phys> Ramsden eyepiece
RAM-Speicher m <Comp & DV> memory random access
Rand m 1. <Bau> boundary, margin, skirt; brow (eines Abhangs); 2. <Comp & DV> edge; margin (Printout); 3. <Druck> margin; 4. <Geom> boundary, edge; 5. <Labor> lip; 6. <Maschinen> rim; 7. <Phys> fringe; 8. <Textil, Verpack> edge • **am Rand** <Math> marginal • **am Rand bündig ausrichten** <Comp & DV> justify • **am Rande befindlich** <Math> marginal
Rand... <Math> marginal
Randabstand m 1. <Fertig> edge distance (Punktschweißen); 2. <Maschinen> edge distance
Randanleimmaschine f <Verpack> margin gluer
Randausgleich m <Comp & DV> justification
Randbedingung f 1. <Comp & DV> constraint; 2. <Ergon, Math> boundary condition
Randbemerkung f <Druck> side note
Randbeschnitt m <Papier> trimmings
Randdämpfung f <Aufnahme> surface damping; edge damping (Akustik)
Randdetail n <Fernseh> corner detail
Randeffekt m <Elektrotech, Lufttrans> fringe effect
Randeinstellung f <Druck> margin settings
Rändelkopf m <Maschinen> milled head, milled knob
Rändelmeißel m <Fertig> straight knurling tool
Rändelmutter f 1. <Fertig> hand nut; 2. <Maschinen> knurled nut, milled nut
rändeln v <Fertig> straight-knurl
Rändeln n 1. <Fertig> straight knurling; 2. <Ker & Glas, Maschinen> knurling
Rändelschraube f <Maschinen> knurled screw
Rändelung f <Maschinen> knurl, knurling
Rändelwerkzeug n 1. <Fertig> straight knurl; 2. <Maschinen, Mechan> knurling tool
Randentkohlung f <Fertig> edge decarburization
Rändern n <Ker & Glas> bead down
Randfaser f <Fertig> (AE) outer fiber, (BE) outer fibre
Randfeuer n <Lufttrans> boundary light
Randkapazität f <Ker & Glas> brim capacity
Randkraft f <Trans> lateral force
Randleiste f <Druck> border, box rule
Randlinie f **eines Zeichens** <Comp & DV> character outline
Randmode f <Optik> bound mode
Randnote f <Druck> side note
Randomdatei f <Comp & DV> random file
Randperforation f <Comp & DV> running perforation
Randplatte f <Eisenbahn> bearing plate
Randschicht f <Optik, Telekom> barrier layer
Randspritzer m <Papier> trim shower
Randspurbegrenzung f <Fernseh> edge of track banding
Randstein m <Bau> (AE) curb, (BE) kerb
Randstreifen m <Bau> margin
Randverteilung f <Math> marginal distribution (einer Zufallsgröße)
Randverwerfung f <Fernseh> scallop, scalloping
Randwasser n <Erdöl> edge water (Geologie)
Randwert m <Math> boundary value
Randwertprüfung f <Comp & DV> marginal check, marginal test
Randwinkel m <Fertig> angle of contact, contact angle (Kappilarrohr)
Randwirbel m <Lufttrans> wing tip vortex
Randwirbel m **am Luftschraubenblatt** <Lufttrans> blade tip vortex
Randwulst f <Fertig> bead
Randzone f 1. <Bau> fringe area; 2. <Lufttrans> fringe
Randzonenverkehr m <Trans> suburban traffic
Rangfolgemethode f <Ergon> job-ranking method
Rangier... <Eisenbahn> shunting
Rangieranzeigevorrichtung f <Eisenbahn> classification detector
Rangierbahngleis n <Eisenbahn> classification yard line

Rangierbahnhof m <Eisenbahn> classification yard, (BE) marshalling yard, shunting yard, (AE) switching station, (AE) switchyard
Rangierdraht m <Funktech, Telekom> jumper wire
Rangiereinrichtung f <Telekom> cross-connect unit
Rangieren n <Eisenbahn> (BE) classification, shunting, (AE) switching
Rangieren n **durch Umsetzen** <Eisenbahn> shunting on level tracks
Rangierer m <Eisenbahn> (BE) pointsman, shunter, (AE) switchman (Person)
Rangiergerät n **für Sattelanhänger** <Kfztech> dolly
Rangiergleis n <Eisenbahn> (BE) classification track, (AE) switching track, (AE) marshaling track, (BE) marshalling track, shunting siding, shunting track, sorting line
Rangierleiter m <Eisenbahn> foreman shunter, (AE) foreman switcher
Rangierlok f <Eisenbahn> shunting engine, switch engine
Rangierlokomotive f <Eisenbahn, Trans> shunting engine, shunting locomotive, (AE) switch engine, (AE) switcher
Rangierverteiler m <Telekom> patch panel
Rangierwinde f <Bau> shunting winch, (AE) switching winch
Rangordnung f <Comp & DV> order of precedence • **in Rangordnung bringen** <Ergon> rank
Rangreihenfolge f **für Unterbrechungen** <Regelung> daisy chain device priority
Rapidanalyse f <Chemie> proximate analysis
Rapportzahl f <Textil> number of repeats
Rapsöl n <Lebensmittel> colza oil, rapeseed oil
rasch trennender Schalter m <Elektriz> quick-break switch
Raschel f <Textil> raschel knitting machine
Raschel-Kettenwirkmaschine f <Textil> raschel knitting machine
Raschelmaschine f <Textil> raschel knitting machine
Raser-Tiefdruck m <Verpack> intaglio printing
RA-Server m <Telekom> remote access server
Raspe f <Bau> rasp
raspeln v <Bau> rasp
Rast f 1. <Fertig> dwell (Getriebelehre); bosh (Hochofen); 2. <Ker & Glas> belly (Schachtofen)
Rastantrieb m <Mechan> rack-and-pinion
Rastblende f <Foto> click stop
Rastdeckel m <Verpack> snap-on lid
Raste f 1. <Fertig, Maschinen> notch; 2. <Mechan> catch
Raster n 1. <Bau> lattice, screen; 2. <Comp & DV> raster; 3. <Druck> screen; 4. <Elektronik> graticule, pattern; grid (bei der Leiterplattenherstellung); 5. <Fernseh> graticule, raster; 6. <Funktech> pattern (Antenne); 7. <Math> lattice; 8. <Mechan> screen; 9. <Telekom> increment; spacing (Kanal, Frequenz)
rasterabgetasteter Strahl m <Elektronik> raster-scanned beam
Rasterabstand m <Fernseh> raster pitch
Rasterabtastung f <Elektronik, Fernseh> raster scanning
Rasterabtastungs-Elektronenstrahl-Lithographie f <Elektronik> raster scan electron beam lithography
Rasterabtastungs-Katodenstrahlröhre f <Elektronik> raster scan cathode ray tube
Rasteranzeige f <Comp & DV> raster display
Rasterätzung f <Druck> halftone process
Raster-Auger-Elektronenspektroskopie f <Kerntech> scanning Auger microscopy
Rasterbild n <Druck> halftone, halftone image
Rasterbildschirm m <Comp & DV> raster screen
Rastereinheit f <Comp & DV> raster unit
Rasterelektronenmikroskop n <Elektronik, Strahlphys> scanning electron microscope

Rasterelektronenstrahl m <Elektronik> scanning electron beam
Raster-Elektronenstrahl-Lithographie f <Elektronik> scanning electron beam lithography
Raster-Elektronenstrahlsystem n <Elektronik> scanning electron beam system
Rasterelement n <Comp & DV> raster element
Rasterfeld n <Comp & DV> raster
rasterförmige Abtastung f <Comp & DV> raster scanning
Rasterfrequenzteiler m <Fernseh> field divider
Rastergenerator m <Fernseh> raster generator
Rastergrafik f <Comp & DV, Druck> raster graphics
Raster-Image-Prozessor m <Druck> raster image processor, RIP
Rasterionenmikroskopie f <Strahlphys> scanning ion microscopy
Rasterklischee n <Foto> halftone block
Rastermikroskop n <Labor> scanning microscope
Rasterpunkt m 1. <Comp & DV, Fernseh> pixel; 2. <Math> lattice point
Raster-Scan-Radar n <Wassertrans> raster scan radar (Navigation)
Rasterschere f <Fertig> ratchet (Kunststoffinstallationen)
Rastersonde f <Kerntech> grid probe
Rasterstörung f <Telekom> underlap (Faksimile)
Rasterstruktur f **der Spaltzone** <Kerntech> core grid structure
Rastersystem n <Bau> bay system
Rasterverriegelung f <Fernseh> locking
Rasterwinkel m <Druck> screen angle
Rasterwinkelung f <Druck> screen angle
Rastfrequenz f <Telekom> frame frequency
Rastgetriebe n <Fertig> dwell mechanism
Rastlinie f <Fertig> mark (Dauerbruch)
Rastplatz m <Trans> lay-by
Rastpolbahn f <Fertig> body centrode (Getriebelehre)
Rastpolkegel m <Fertig> herpolhode cone (Getriebelehre)
Rastrelais n <Elektriz> latching relay
Raststift m <Bau, Fertig> latch pin
Rastzahn m <Mechan> notch
Rate f <Comp & DV> rate
Ratenüberschreitung f <Kerntech> burnout
rationale Zahl f <Comp & DV, Math> rational number
Ratsche f 1. <Bau> ratchet; 2. <Maschinen> ratchet spanner, ratchet stop, ratchet wrench
Rattenloch n <Erdöl> mousehole
Rätter m <Kohlen> cribble, riddle, screen
Rattermarke f <Fertig> chatter mark
rattern v 1. <Fertig> chatter (Spanung); 2. <Ker & Glas> chatter; 3. <Maschinen> chatter
Rattern n 1. <Fertig> chatter (Spanung); 2. <Maschinen> chatter
rau adj <Anstrich, Papier> rough
Raub m **von Transportgütern** <Lufttrans> hijack
rauben v <Lufttrans> hijack (Transportgut)
Raubkopie f 1. <Comp & DV> pirate copy; 2. <Fernseh> pirate recording
Rauch m <Ker & Glas> smoke
Rauchabzug m 1. <Heiz & Kälte> chimney; 2. <Sicherheit> smoke-venting equipment
Rauchabzugseinrichtung f <Sicherheit> fume control device
Rauchbegrenzung f <Sicherheit> smoke control
Rauchbrandwarnanlage f <Sicherheit> smoke-detection device (Brandschutz)
Rauchdetektor m <Sicherheit> smoke-detection device (Brandschutz)

Rauchdiagramm

Rauchdiagramm n <Sicherheit> smoke chart
Rauchdichtemesser m <Sicherheit> smoke density indicator
rauchdichter Helm m <Sicherheit> smoke helmet
rauchende Schwefelsäure f <Chemie> Nordhausen acid, oleum, (AE) fuming sulfuric acid, (BE) fuming sulphuric acid
Räuchern n <Lebensmittel> smoking
Rauchfahne f <Lufttrans> exhaust trail
Rauchfang m 1. <Fertig> hood (Schmieden); 2. <Heiz & Kälte> smoke flue; 3. <Mechan> funnel
rauchfreie Zone f <Sicherheit> smokeless zone
Rauchgas n 1. <Heiz & Kälte> flue gas, smoke gas; 2. <Phys, Thermod, Umweltschmutz> flue gas
Rauchgasanalyse f <Maschinen> flue gas analysis
Rauchgasentschwefelung f <Umweltschmutz> (AE) flue gas desulfurization, (BE) flue gas desulphurization
Rauchgasentschwefelungsanlage f <Sicherheit> (AE) flue gas desulfurization installation, (BE) flue gas desulphurization installation
Rauchgasentstaubung f <Abfall> particulate collection
Rauchgasmeldungsgeber m <Sicherheit> smoke-detection device (Brandschutz)
Rauchgasreiniger m <Heiz & Kälte> flue gas dust collector
Rauchgasreiniger m <Thermod> flue gas scrubber
Rauchgasreinigung f <Thermod> flue gas scrubbing
Rauchgasvorwärmer m <Heiz & Kälte> flue gas preheater
Rauchglas n <Ker & Glas> smoked glass
Rauchkammer f <Bau, Eisenbahn> smokebox
Rauchkammerrohrwand f <Eisenbahn> smokebox tube plate
Rauchkanal m <Heiz & Kälte> smoke duct
rauchlos adj <Sicherheit> smokeless
Rauchmaske f <Lufttrans> smoke mask
Rauchmelder m <Sicherheit> smoke detector
Rauchpilz m <Strömphys> plume (thermische Strömungen)
Rauchrohr n <Heiz & Kälte> smoke tube
Rauchrohrkessel m <Heiz & Kälte> fire tube boiler, smoke tube boiler
Rauchschieber m <Heiz & Kälte> slide damper
Rauchschutztür f <Sicherheit> smoke protection door
Rauchspiegelglas n <Ker & Glas> tint plate
Rauch- und Gasalarmanlage f <Sicherheit> smoke and gas alarm installation
Rauch- und Hitzeabzugsanlage f <Sicherheit> smoke and heat exhaust installation
Rauchverbrauch m <Kohlen> consumption of smoke
rauchverzehrend adj <Chemie> fumivorous
Rauchzug m <Thermod> flue
Raudecke f <Bau> friction course
raue Oberfläche f <Maschinen> rough surface
rauen v 1. <Fertig> raise; 2. <Maschinen> roughen; 3. <Textil> card (Wolle)
Rauen n 1. <Elektrotech> brushing; 2. <Fertig> raising; 3. <Ker & Glas, Kfztech, Textil> brushing
Raugriffigkeit f <Textil> harsh handle
Rauheit f <Maschinen, Mechan> roughness
Rauheitmessgerät n <Metrol> surface measuring instrument
Rauheitsnorm f <Metrol> surface roughness standard
Rauhobel m <Bau> jack plane
Rauigkeit f <Maschinen, Mechan, Papier> roughness
Rauigkeitsprüfer m <Papier> roughness tester (der Papieroberfläche)
Raum m 1. <Bau> room; 2. <Geom, Maschinen> space; 3. <Mechan> chamber; 4. <Telekom> space

Raum m **für Hilfsfallschirme** <Raumfahrt> auxiliary parachute bay
Raum m **mit geringer Schallabsorption** <Ergon> live room
Raum... <Heiz & Kälte> ambient
Raumanzug m <Raumfahrt> spacesuit
Raumaschine f <Textil> raising machine
Raumbedarf m 1. <Maschinen> space occupied, space taken up; 2. <Mechan> bulk
Raumbeleuchtung f <Fernseh> ambient light
raumbeständig adj <Bau> sound (Beton)
Raumbildbetrachter m <Foto> stereoscope
Räumbohrer m <Erdöl> reaming bit
Raumbreite f (b) <Raumfahrt> galactic latitude (b)
Raumdichte f <Bau> density
Raum-Dichte-Verhältnis n <Trans> volume-density relationship
Raumdiversity n <Funktech> space diversity
Raumeffekt m <Aufnahme> auditory perspective
raumen v <Wassertrans> veer aft (Wind)
räumen v 1. <Bau> vacate; 2. <Maschinen> broach
Räumen n <Maschinen> broaching
Räumer m 1. <Bau, Erdöl> reamer (Bohrtechnik); 2. <Kohlen> raker
Raumfähre f <Raumfahrt> space shuttle
Raumfahrt f <Raumfahrt> aerospace, astronautics, space travel
Raumfahrtfernmessung f <Metrol, Raumfahrt> aerospace telemetry
Raumfahrtflugdatennetzwerk n <Raumfahrt> spaceflight tracking and data network
Raumfahrttransportsystem n (STS) <Raumfahrt> space transportation system (STS)
Raumfahrtzentrum n <Raumfahrt> astrodrome
Raumfahrzeug n <Raumfahrt> spacecraft
Raumfahrzeugantenne f <Raumfahrt> spaceborn antenna, spacecraft antenna
Raumfahrzeugradar n <Funkort, Raumfahrt> spaceborn radar, spacecraft radar
Raumfärbung f <Aufnahme> (AE) acoustic coloring, (BE) acoustic colouring (Radio)
raumfest adj <Raumfahrt> space-bound
Raumflug m <Raumfahrt> space flight
Raumflug m **auf Umlaufbahnen** <Raumfahrt> circular flight
Raumflugumsetzer m <Raumfahrt> orbital transfer vehicle
Raumformfräsen n <Fertig> three-dimensional tracer milling
Räumfräsen n <Maschinen> broach milling
Raumfuge f <Bau> expansion joint, running joint
raumgeteilte Vermittlung f <Telekom> space division switching, space switch
raumgeteilte Vermittlungsstelle f <Telekom> space switch
raumgeteiltes Vermittlungssystem n <Telekom> space division switching system
Raumgitter n <Phys> spatial grid
Raumgleiter m <Trans> orbital glider
Rauminhalt m 1. <Bau> content; 2. <Math> volume, cubic content; 3. <Mechan, Phys, Wassertrans> cubic capacity (Ladung)
Raumisomer n <Chemie> stereoisomer
Raummittel n <Fertig> roughener
Raumkapsel f <Raumfahrt> space capsule
Raumklangsystem n <Aufnahme> binaural sound system
Raumklimagerät n <Heiz & Kälte> room air conditioner
Raumkoordinate f <Geom> space coordinate

raumkrank adj <Raumfahrt> space-sick
Raumkurve f <Geom> space curve
Raumladung f <Phys> space charge
Raumladungsimpuls m <Elektrotech> cloud pulse (bei Ladungsspeicherröhren)
Raumladungskompensation f <Phys> space charge compensation
Raumladungskonstante f <Elektrotech, Telekom> perveance
Raumlast f <Heiz & Kälte> room load
räumlich adj 1. <Chemie> steric; 2. <Phys> spatial
räumlich zentriert adj <Metall> (AE) space-centered, (BE) space-centred
räumliche Anforderungen fpl <Sicherheit> dimensional requirements
räumliche Auflösung f <Kerntech> spatial resolution
räumliche Ausdehnung f 1. <Mechan, Phys> cubic expansivity; 2. <Telekom> coverage (Netzwerk)
räumliche Dispersion f <Phys> spatial dispersion (Festkörperphysik)
räumliche Entfernung f <Phys> distance
räumliche Geometrie f <Geom> solid geometry
räumliche Kohärenz f 1. <Phys> spatial coherence; 2. <Telekom> space coherence, spatial coherence
räumliche Ladungsdichte f <Phys> volume charge density
räumliche Lage f <Konstzeich> location in space
räumliche Modulation f <Elektronik> spatial modulation
räumliche Periode f <Elektronik> spatial period
räumliche Quantisierung f <Phys> spatial quantization
räumliche Relativität f <Raumfahrt> spatial relativity
räumliche Struktur f <Umweltschmutz> spatial pattern
räumliche Veränderlichkeit f <Umweltschmutz> spatial variability
räumliche Verteilung f <Umweltschmutz> spatial distribution
räumliche Wahrnehmung f <Ergon> space perception
räumliche Zuordnung f <Konstzeich> correlation in space
räumlicher Bereich m <Elektronik> spatial domain
räumliches Bild n <Telekom> three-dimensional image
räumliches Verhalten n <Elektronik> spatial response
räumlich-zeitliche Korrelation f <Telekom> space-time correlation
Raumlöffel m <Bau> raker
Raumlufttechnik f 1. <Heiz & Kälte> ventilation and air conditioning; 2. <Sicherheit> indoor air technology
Raumluftverschmutzung f <Sicherheit, Umweltschmutz> ambient air pollution
Raumluftverunreinigung f <Sicherheit> ambient air pollution
Räummaschine f 1. <Fertig> broach (Pilgerschrittwalze); 2. <Maschinen> broaching machine
Raummasse f <Ker & Glas> bulk density
Raummultiplex n <Telekom> space-division multiplex
Raum-Multiplex-Betrieb m <Comp & DV> space division multiplex
Raumnachführung f <Raumfahrt> space tracking
Räumnadel f <Maschinen, Mechan> broach, broaching tool, internal broach
Räumnadelziehmaschine f <Mechan> broaching machine
Räumpresse f 1. <Fertig> push-type broaching machine; 2. <Maschinen> press-type vertical broaching machine
Raumschiff n <Raumfahrt> spaceship
Raumschiff n **mit Nuklearantrieb** <Raumfahrt> nuclear-powered spacecraft
Raumschlepper m <Raumfahrt> space tug
Räumschlitten m <Fertig> broach ram, broach slide

raumschots segeln v <Wassertrans> run free, sail free, sail on a broad reach; go free, sail on a close reach (Segeln)
Raumschutzanlage f <Sicherheit> intruder alarm equipment
Raumschutzmeldungsgeber m <Sicherheit, Telekom> intruder presence detector
Raumsegment n <Raumfahrt> space segment (Weltraumfunk)
Raumsimulator m <Raumfahrt> space simulation chamber
Raumsonde f <Raumfahrt> space probe, space vehicle
Raumstation f <Raumfahrt> space station
Raumstufe f <Telekom> space stage
Raumteilung f <Bau> partitioning
Raumteilungsmultiplex n <Telekom> space-division multiplex
Raumtemperatur f 1. <Heiz & Kälte, Metall> ambient temperature; 2. <Thermod> room temperature
Raumtemperaturregler m <Heiz & Kälte> thermostat
Raumthermostat m <Heiz & Kälte, Thermod> room thermostat
Raumtiefe f <Wassertrans> registered depth (Schiffkonstruktion)
Raumtoneffekt m <Akustik> binaural effect
Raumtransporter m <Raumfahrt> space shuttle
Raumüberwachungssensor m <Sicherheit> room monitoring sensor
Raumverlust m <Raumfahrt> free space loss (Weltraumfunk)
Raumvielfachsystem n <Telekom> space division system
raumvoll und auf Tiefgang adj <Wassertrans> full and down
Raumwelle f <Funktech, Phys> sky wave
Raumwerkstatt f <Raumfahrt> space workshop
Räumwerkzeug n 1. <Fertig> helical broach; 2. <Maschinen> broaching tool, broach; 3. <Mechan> broach, broaching tool
Raumwinkel m 1. <Geom> solid angle; 2. <Metall> dihedral angle; 3. <Phys> solid angle
Räumzahn m <Fertig> raked tooth
Raum-Zeit-Beziehung f <Phys> space-time relation
Raum-Zeit-Kontinuum n <Raumfahrt> space-time continuum
Raum-Zeit-Raum-Koppelnetz n <Telekom> space-time-space network
Raum-Zeit-Umkehr f <Phys> space-time reversal
Räumzug m <Maschinen> broaching pass
Raupe f <Mechan> crawler
Raupenfahrzeug n <Maschinen> crawler vehicle
Raupenkette f <Maschinen> crawler, crawler track
Raupenrad n <Kfztech> sprocket (Motorradgetriebe)
Raupenschleifmaschine f <Papier> caterpillar grinder
Raupentraktor m <Bau, Kfztech> tracked tractor
Raupenzugmaschine f <Bau, Kfztech> tracked tractor
Rauputz m <Bau> roughcast
Rauputz m **mit Kieseln** <Bau> pebble dash
Rausch... <Funktech> noise
Rauschabstand m <Akustik, Aufnahme, Comp & DV, Elektronik, Fernseh, Telekom> signal-to-noise ratio
Rauschamplitudenverteilung f <Telekom> NAD, noise amplitude distribution
rauscharme Verstärkung f <Elektronik, Telekom> low-noise amplification
rauscharmer Konverter m <Fernseh> low-noise converter (LNC)
rauscharmer Verstärker m <Elektronik, Telekom> low-noise amplifier

rauscharmer Vorverstärker m <Elektronik, Funktech, Strahlphys> low-noise preamplifier
Rauschbild n <Fernseh> noise field
Rauschdiode f <Elektronik> noise diode
Rauschen n 1. <Akustik, Aufnahme> noise; 2. <Comp & DV> noise, static; 3. <Elektronik, Fernseh, Phys> noise
Rauschen n **der Wirbelstärke** <Strömphys> background vorticity
Rauschen n **eines Störgenerators** <Elektronik> interference generator noise
Rauschen n **kurzer Wellenlänge** <Elektronik> short wavelength noise
Rauschfaktor m <Phys> noise factor
rauschfreies Signal n <Gerät> noise-free signal
Rauschgenerator m <Elektronik, Telekom> noise generator
Rauschgrenze f <Elektronik> noise margin
Rauschkurven fpl <Sicherheit> noise characteristics
rauschleifen v <Bau> rough-down
Rauschleistung f (N) <Elektronik, Kerntech> noise power (N)
Rauschliff m <Ker & Glas> (AE) gray cutting, (BE) grey cutting
Rauschmessung f <Metrol> noise measurement
Rauschminderung f <Elektronik> audio noise reduction, denoising
Rauschminderungsrand m <Elektronik> noise limit
Rauschmodulation f <Fernseh> noise modulation
Rauschpegel m <Telekom, Wellphys> noise level
Rauschprobe f <Elektronik> noise print
Rauschquelle f <Elektronik> noise generator, noise source
Rauschsignal n <Fernseh, Funktech> noise signal
Rauschsperre f 1. <Funktech> squelch; 2. <Telekom> squelch, squelch circuit
Rauschstörungsgenerator m <Elektronik> random noise generator
Rauschstörungsquelle f <Elektronik> random noise source
Rauschstörungssignal n <Elektronik> random noise signal
Rauschstreifenbildung f <Fernseh> banding on noise
Rauschtemperatur f <Elektrotech> noise temperature
Rauschunterdrücker m 1. <Aufnahme> noise reducer, noise suppressor; 2. <Elektrotech> noise suppressor; 3. <Funktech> noise blanker
Rauschunterdrückung f 1. <Aufnahme> noise reduction, noise suppression; 2. <Elektrotech> noise suppression; 3. <Funktech> noise blanking; 4. <Gerät> (AE) noise canceling, (BE) noise cancelling
Rauschzahl f 1. <Elektronik> noise factor; 2. <Elektronik, Funktech> noise figure (F); 3. <Phys> noise factor
Raute f 1. <Druck, Fertig> diamond (Walzen); 2. <Geom> lozenge, rhomb, rhombus
Rautendrahtgitter n <Bau> diamond wire lattice
Rautenschnitt m <Ker & Glas> diamond cut pattern
Rautenvorkaliber n <Fertig> diamond pass (Walzen)
Rautiefe f <Maschinen> maximum peak-to-valley height, peak-to-valley height, roughness height
Rautiefenmesser m <Metrol> (AE) peak-to-valley height gage, (BE) peak-to-valley height gauge
Rayleigh-Fading n <Funktech> Rayleigh fading
Rayleigh-Jeans'sche Gleichung f <Phys> Rayleigh-Jeans formula
Rayleigh'sche Auflösungsbedingung f <Phys> Rayleigh criterion
Rayleigh'sche Scheibe f <Akustik> (BE) Rayleigh disc, (AE) Rayleigh disk
Rayleigh'sche Streuung f <Optik, Phys, Strahlphys> Rayleigh scattering

Rayleigh'sche Verteilung f <Math> Rayleigh distribution (Spezialfall der Weibull-Verteilung)
Rayleigh'sche Welle f <Akustik> Rayleigh wave
Rayleigh'sches Interferometer n <Phys> Rayleigh interferometer
Rayleigh'sches Refraktometer n <Phys> Rayleigh refractometer
Rayleigh-Schwund m <Funktech> Rayleigh fading
Rayleigh-Streuung f <Telekom> Rayleigh scattering
Rayleigh-Verteilung f <Math> Rayleigh distribution (spezielle Weibull Verteilung)
Rb (Rubidium) <Chemie> Rb (rubidium)
RBA (relative Byteadresse) <Comp & DV> RBA (relative byte address)
RC <Elektronik> RC, resistor-capacitor
RC-Abzweigfilter n <Elektronik> RC ladder filter
RC-Filterschaltung f <Elektronik> RC filter circuit
RC-Generator m <Elektronik> RC oscillator
RC-Oszillator m <Phys> RC oscillator
RC-Schaltklickfilter n <Funktech> snubber network
RCTL-Logik f (Widerstands-Kondensator-Transistor-Logik) <Elektronik> RCTL logic (resistor-capacitor-transistor logic)
RDB (relationale Datenbank) <Telekom> RDB (relational database)
RDSS (Satellitenfunkortungssystem) <Funkort, Trans, Wassertrans> RDSS, radio determination satellite system (Satellitenfunk)
re (Elektronenradius) <Kerntech> re (electron radius)
Re 1. <Chemie> (Rhenium) Re (rhenium); 2. <Hydraul, Lufttrans, Nichtfoss Energ, Phys, Strömphys> (Reynoldszahl) Re (Reynolds number)
Reagens n <Chemie, Foto, Kohlen, Kunststoff> reagent
Reagenz n <Kunststoff> reagent
Reagenzglas n <Labor> test tube
Reagenzglasgestell n <Labor> test tube rack
Reagenzglashalter m <Labor> test tube holder
Reagenzienflasche f <Labor> reagent bottle
Reaktanz f 1. <Aufnahme> reactance; 2. <Elektriz, Elektrotech> reactance (X); 3. <Funktech> reactance
Reaktanzabfall m <Elektrotech> reactance drop
Reaktanzdämpfer m <Elektronik> reactance attenuator
Reaktanzdiagramm n <Funktech> reactance chart
Reaktanzfrequenz-Vervielfacher m <Elektronik> reactance frequency multiplier
Reaktanzschaltung f <Elektriz> reactance circuit
Reaktanzspule f <Elektrotech> reactance coil
Reaktanzstromkreis m <Elektriz> reactance circuit
Reaktanzverstärker m <Elektronik> parametric amplifier, reactance amplifier, mavar
Reaktanzverstärkerdiode f <Elektronik> parametric amplifier diode, reactance amplifier diode
Reaktion f 1. <Anstrich> reaction (chemisch); 2. <Ergon> response; 3. <Phys> reaction
Reaktionsbereich m <Kohlen> reaction zone
Reaktionsbombe f <Labor> reaction bomb
Reaktionsenergie f (Q) <Kerntech> reaction energy (Q)
Reaktionsgeschwindigkeit f <Metall> reaction rate
Reaktionsharzbeton m <Kunststoff> polymer concrete
Reaktionskleber m <Kunststoff> two-pack adhesive
Reaktionsmittel n <Abfall> solidifying agent
Reaktionspartner m <Chemie> reactant
Reaktionsprimer m <Kunststoff> wash primer
Reaktionsrad n <Wasserversorg> reaction water wheel, reaction wheel
Reaktionsreihenfolge f <Metall> order of reaction
Reaktionsschiene f <Eisenbahn> reaction rail
Reaktionsspektroskopie f <Kerntech> reaction spectroscopy

Reaktionsstrahlschub m <Lufttrans> reaction jet propulsion
Reaktionsstrahlschubkraft f <Lufttrans> reaction jet propulsion
Reaktionsturbine f <Maschinen, Mechan, Nichtfoss Energ> reaction turbine
Reaktionsverzögerer m <Abfall> retarder, retarding agent
Reaktionswärme f <Thermod> heat of reaction
Reaktionszeit f <Ergon> reaction time, response time
reaktive Bewegung f <Lufttrans> jet propulsion
reaktive Plasmaätzung f <Elektronik> reactive plasma etching
reaktiver Schaltkreis m <Elektrotech> reactive circuit
reaktives Lösemittel n <Kunststoff> reactive solvent
Reaktivfarbstoff m <Textil> reactive dye
reaktivieren v <Chemie> reactivate
Reaktivierung f <Chemie, Kohlen> reactivation
Reaktivität f <Kerntech, Phys> reactivity
Reaktivitätsabnahme f <Kerntech> decrement in reactivity
Reaktivitätsdefizit n <Kerntech> deficit reactivity
Reaktivitätselement n <Kerntech> booster element *(beim Anfahren eines Reaktors)*
Reaktivitätsrampe f <Kerntech> reactivity ramp
Reaktivitätsrückkopplung f <Kerntech> reactivity feedback
Reaktivitätssprung m <Kerntech> reactivity surge
Reaktivitätsstab m <Kerntech> booster rod *(beim Anfahren eines Reaktors)*
Reaktivitätsverlust m <Kerntech> reactivity loss
Reaktor m 1. <Elektrotech> pile; reactor *(Atomphysik)*; 2. <Kerntech> reactor
Reaktor m **im Chargenbetrieb** <Kerntech> batch reactor
Reaktor m **mit Beryllium-Reflektor** <Kerntech> beryllium-reflected reactor
Reaktor m **mit gasförmigem Reaktorkern** <Kerntech> gaseous core reactor
Reaktor m **mit Lufteinblasung** <Trans> AIR, air injection reactor
Reaktor m **mit luftisolierter Spule** <Elektriz> dry type reactor
Reaktor m **mit Naturkühlung** <Kerntech> natural nuclear reactor
Reaktor m **mit nicht gekapselter Trockenspule** <Elektriz> dry type reactor
Reaktor m **mit Plattenelementen** <Kerntech> slab reactor
Reaktorabschaltung f <Kerntech> shutdown of a reactor
Reaktorbau m <Kerntech> reactor art
Reaktorbauteil n <Kerntech> reactor component
Reaktorbehälter m <Kerntech> reactor vessel
Reaktordruckbehälter m <Kerntech> pressure vessel
Reaktordynamik f <Kerntech> reactor dynamics
Reaktorformel f <Kerntech> reactor formula
Reaktorgebäude n <Kerntech> reactor hall
Reaktorgift n <Kerntech> nuclear poison
Reaktorgitter n <Kerntech> active lattice
Reaktorkern m <Kerntech> core
Reaktorkreislauf m <Kerntech> reactor loop
Reaktorkühlmittel n <Kerntech> reactor coolant
Reaktormultiplikation f (M) <Kerntech> multiplication of a reactor (M)
Reaktorperiode f <Kerntech> reactor period
Reaktorphysik f <Phys> reactor physics
Reaktorplanung f <Kerntech> reactor design
Reaktorsicherheit f 1. <Kerntech> reactor safety; 2. <Sicherheit> power reactor safety
Reaktorsinterung f <Kerntech> reaction sintering process
Reaktortank m <Kerntech> reactor tank
Reaktortechnik f <Kerntech> reactor engineering
Reaktorunfall m <Kerntech> nuclear accident, reactor accident
Reaktorverhalten n <Kerntech> reactor behaviour
Reaktorwand f <Kerntech> reactor wall
Reaktorzelle f <Kerntech> reactor cell
real adj 1. <Comp & DV> real; 2. <Fertig> imperfect *(Kristall)*
reale Adresse f <Comp & DV> real address
realer Typ m <Comp & DV> real type
realisierbarer Abbrand m <Kerntech> achievable burn-up
realisieren v <Comp & DV> implement
Realkomponente f <Math> real component, real part
Realspeicher m <Comp & DV> real memory
Rebe f <Lebensmittel> vine
Reboiler m <Chemie, Erdöl> reboiler *(Raffinerie)*
Rechen m <Bau, Fertig> rake
Rechen... 1. <Comp & DV> arithmetic, computing; 2. <Math> arithmetic
Rechenarten fpl <Math> arithmetic operations
Rechenbefehl m <Comp & DV> arithmetic instruction
Rechenblatt n <Comp & DV> spreadsheet
Rechendrehmelder m <Gerät> computing synchro
Rechengeschwindigkeit f <Comp & DV> computing speed
Rechengröße f <Comp & DV, Math> operand
Rechengut n <Chemie> screenings *(Abwasserbehandlung)*
Rechenleistung f <Comp & DV> computing power
Rechenmaschine f <Comp & DV> calculating machine
Rechenmodell n <Comp & DV> mathematical model
Rechenoperation f <Comp & DV, Math> arithmetic operation
Rechenprüfung f <Comp & DV> arithmetic check
rechenschaftspflichtig adj <Qual> accountable
Rechenschaltung f <Elektronik> arithmetic circuit
Rechenscheibe f <Math> circular slide rule
Rechenschieber m <Comp & DV, Maschinen, Math> slide rule
Rechenstab m <Math> slide rule
Rechenstreifen m <Comp & DV> tally
Rechensystem n <Telekom> data processing system
Rechenverfahren n **nach der Methode der finiten Elemente** <Mechan> finite element calculation method
Rechenverstärker m (OA) <Comp & DV, Elektronik, Phys> operational amplifier *(op amp)*
Rechenverstärker-Chip m <Elektronik> operational amplifier chip
Rechenverstärker-Komparator m <Elektronik> operational amplifier comparator
Rechenwerk n <Comp & DV> ALU, arithmetic and logic unit, arithmetic logic unit, arithmetic unit
Rechenzeichen n <Comp & DV> arithmetic operator
Rechenzentrum n 1. <Comp & DV> DPC, *(AE)* data processing center, *(BE)* data processing centre, *(AE)* computing center, *(BE)* computing centre; 2. <Telekom> DPC, *(AE)* data processing center, *(BE)* data processing centre
Recherche f <Comp & DV, Telekom> recherche, search
Recherche f **über das Internet** <Comp & DV, Telekom> on-line recherche
Recherchenbericht m <Patent> search report
Rechnen n **mit einem Großrechner** <Comp & DV> supercomputing
Rechner m 1. <Comp & DV> computer, computing device, computing facility, machine; 2. <Elektriz, Elektronik> computer, calculator

Rechner *m* **der fünften Generation** <Comp & DV, Künstl Int> FGC, fifth generation computer
Rechner *m* **mit variabler Wortlänge** <Comp & DV> byte machine
rechnerabhängiger Speicher *m* <Comp & DV> online storage
Rechneranwendung *f* **zur Messung und Regelung** <Metrol, Regelung> computer application to measurement and control *(CAMAC)*
rechnerbasiertes Publizieren *n* <Druck> computer-based publishing
Rechnerbetriebssystem *n* <Comp & DV> operating system *(OS)*
Rechnerfamilie *f* <Comp & DV> computer family
rechnergestützte Telefonanwendungen *fpl* <Telekom> computer supported telephony applications, CSTA *(Schnittstelle)*
rechnerische Induktion *f* <Comp & DV> mathematical induction
Rechnernetz *n* <Comp & DV, Telekom> computer network
rechnerorientiert *adj* <Comp & DV> machine-oriented
Rechnerplattform *f* <Comp & DV> computing platform
Rechnerschnittstelle *f* <Telekom> computer interface
Rechnersicherheit *f* <Comp & DV> computer security
Rechnerstellwerk *n* <Eisenbahn> computer-controlled interlocking
Rechnersystem *n* <Comp & DV> computer system
Rechnerverbund *m* <Comp & DV, Telekom> computer network
Rechnerverbundbetrieb *m* <Comp & DV> multiprocessing
Rechnerverbundsystem *n* <Comp & DV> multiprocessing system
Rechnung stellen *v*/in <Telekom> charge *(Gespräch)*
Rechnung tragen *v*/der Gewichtsdifferenz <Metrol> make allowance for difference in weight
Rechnungssturzfluggeschwindigkeit *f* <Lufttrans> design-diving speed
Recht *n* <Patent> law *(objektiv)*; right *(subjektiv)*
Recht *n* **auf ein Patent** <Patent> right to a patent
rechte Seite *f* <Druck> recto
rechte Seite *f* **einer Gleichung** <Phys> right-hand side of an equation
Rechteck *n* <Geom> oblong, rectangle
Rechteckdeckleiste *f* <Bau> square staff
Rechteckferrit *m* <Elektrotech> square loop ferrite
Rechteckgenerator *m* <Fernseh> square wave generator
rechteckig *adj* <Geom> rectangular
rechteckige Hystereseschleife *f* <Elektrotech> rectangular hysteresis loop
rechteckige Verblattung *f* <Bau> square splice
rechteckiger Keil *m* <Maschinen> rectangular key
rechteckiger Querschnitt *m* <Bau> rectangular cross-section
Rechteckimpuls *m* <Elektronik> rectangular pulse
Rechteckimpulsgeber *m* <Elektronik> rectangular impulse generator, rectangular pulse generator, square-wave generator
Rechteckoberwelle *f* <Raumfahrt> tesseral harmonic
Rechteckpotenzial *n* <Phys> square potential
Rechteckspannung *f* <Fernseh> square wave voltage
Rechteckwelle *f* 1. <Aufnahme, Elektronik, Elektrotech> square wave, rectangular wave; 2. <Phys> square wave; 3. <Telekom> square waveform
Rechteckwellengeber *m* <Elektronik> square wave sensor
Rechteckwellengenerator *m* <Aufnahme> box-car generator, rectangular pulse generator, square wave generator

Rechteckwellengenerierung *f* <Elektronik> square wave generation
Rechte-Hand-Regel *f* <Elektriz, Elektrotech, Phys> corkscrew rule, right-hand rule
rechter Rand *m* <Druck> right margin • **am rechten Rand ausrichten** <Comp & DV> right justify
rechter Stereokanal *m* <Aufnahme> right stereo channel
rechter Winkel *m* <Geom, Phys> right angle • **im rechten Winkel** <Geom> at right angles *(zu einer Geraden)*
rechts ausrichten *v* <Comp & DV> right justify
rechts flatternd *adj* <Druck> ragged right
rechts vor links *adj* <Trans> priority to the right
Rechts... <Mechan> right-hand
Rechtsabbiegerverkehr *m* <Trans> right-turning traffic
Rechtsausrichtung *f* <Comp & DV> right justification
rechtsbündig ausrichten *v* <Comp & DV> right justify
rechtsbündige Ausrichtung *f* <Comp & DV> right justification
Rechtschreibüberprüfung *f* <Comp & DV> spelling checker *(in Textprogrammen)*
rechtsdrehen *v* <Wassertrans> veer *(Wind)*
rechtsdrehend *adj* 1. <Lebensmittel> clockwise-rotating; 2. <Phys> dextrorotatory
rechtsdrehend zirkular polarisiert *adj* <Elektriz> clockwise circularly polarized, right-hand circularly polarized
rechtsdrehende Polarisation *f* <Funktech> clockwise polarization *(Welle)*
rechtsdrehende Zirkularpolarisation *f* <Funktech, Raumfahrt> right-hand circular polarization *(Weltraumfunk)*
rechtsdrehender Propeller *m* <Wassertrans> right-handed propeller
Rechtsdrehung *f* <Textil> Z-twist
Rechtsflanke *f* <Maschinen> right-hand tooth flank
rechtsgängig *adj* <Maschinen> rh, right-hand, right-handed
rechtsgängige Fräsmaschine *f* <Maschinen> right-hand milling cutter
rechtsgängige Spirale *f* <Maschinen> right-handed spiral
rechtsgängiges Gewinde *n* <Maschinen> right-hand thread
rechtsgeschäftliche Übertragung *f* <Patent> assignment *(der Patentanmeldung)*
Rechtsgewinde *n* <Maschinen> right-hand thread
Rechtsgewindeschraube *f* <Maschinen> right-hand screw, right-handed screw
rechtshändiges Koordinatensystem *n* <Phys> right-handed coordinate system
Rechtshändigkeit *f* <Ergon> dextrality
Rechtsmilch *f* <Chemie> sarcolactic
Rechtsnachfolger *m* <Patent> successor in title
Rechtspersönlichkeit *f* <Patent> legal personality
rechtsschief *adj* <Geom> negatively-skewed
rechtsschneidend *adj* <Maschinen> right-hand, right-handed
Rechtsübergang *m* <Patent> transfer
Rechtsverkehr *m* <Trans> right-hand traffic
Rechtsverschiebung *f* <Comp & DV> right shift
Rechtsvorgänger *m* <Patent> predecessor in title; legal predecessor *(eines Patents)*
Rechtsweiche *f* <Trans> right-hand turnout
rechtweisend Nord *adj* <Wassertrans> true north *(Navigation)*
rechtweisende Peilung *f* <Funkort> true bearing
rechtweisender Kurs *m* <Wassertrans> true course *(Navigation)*
Rechtwinkelphase *f* <Elektronik> quadrature phase

rechtwinklig *adj* 1. <Geom> orthogonal, rectangular, right-angled; 2. <Phys> right-angled
rechtwinklig schneiden *v* <Bau> square *(Holz)*
rechtwinklig versetzte Abknickung *f* <Konstzeich> offset
rechtwinklige axonometrische Projektion *f* <Konstzeich> right-angled axonometric projection
rechtwinklige Bewegung *f* <Phys> rectilinear motion
rechtwinklige Koordinaten *fpl* <Phys> normal coordinates
rechtwinklige Koordinatenachsen *fpl* <Math> rectangular axes
rechtwinklige Parallelprojektion *f* <Konstzeich> right-angled parallel projection
rechtwinklige Verbindung *f* <Bau> square joint
rechtwinkliger Falzhobel *m* <Bau> square rabbet plane
rechtwinkliger Hohlleiter *m* <Elektrotech, Phys> rectangular waveguide
rechtwinkliges Dreieck *n* <Geom> right-angled triangle
rechtwinkliges Knie *n* <Maschinen> right-angled bend
rechtwinkliges Koordinatensystem *n* <Math> rectangular coordinate system
rechtwinkliges Prisma *n* <Optik> right-angled prism
Rechtwinkligkeit *f* 1. <Geom> rectangularity; 2. <Maschinen> squareness
Rechtwinkligschneiden *n* <Bau> squaring
Reckalterung *f* <Fertig> *(BE)* strain ageing, *(AE)* strain aging
recken *v* 1. <Fertig> forge down, rotary-swage; 2. <Textil> stretch; draw *(Fasern)*
Recken *n* 1. <Fertig> preliminary drawing; swaging *(Rundformen)*; 2. <Textil> stretch
Reckschmieden *n* <Fertig> hammering
Recorder *m* <Telekom> recorder *(für Magnetband)*
recyceln *v* <Abfall, Umweltschmutz> recycle *(aus Altmaterial zurückgewinnen)*
recyclierte Flasche *f* <Verpack> recycled bottle
Recycling *n* <Abfall, Elektriz, Ker & Glas, Maschinen, Umweltschmutz> recycling, waste recycling
Recycling *n* **von Multimaterialien** <Verpack> multimaterial recycling
Recyclinganlage *f* <Abfall> recycling plant
recyclingfähig *adj* <Abfall> recyclable
recyclingfreundlich *adj* <Abfall> recycling-friendly
Recyclingpapier *n* <Abfall> recycled paper
Recyclingprozess *m* <Abfall, Papier> recycling process
Recyclingquote *f* <Abfall> recycling rate
Redakteur *m* <Fernseh> editor
Redaktionsschluss *m* <Druck> deadline
Redestillation *f* <Chemie> redistillation, rerun
Redistribution *f* <Elektrotech> redistribution
Redoxelement *n* <Trans> redox cell
Redoxmesszelle *f* <Gerät> redox cell
Redoxpotenzial *n* <Umweltschmutz> oxidation-reduction potential
Redoxreaktion *f* <Chemie> oxidoreduction
Reduktion *f* <Kohlen> reduction
Reduktionsbleiche *f* <Fertig> reduction bleaching
Reduktionsflamme *f* <Thermod> reducing flame
Reduktionsgetriebe *n* 1. <Maschinen> reducing gear, reduction gear; 2. <Papier> speed reducer
Reduktionskost *f* <Lebensmittel> reducing diet
Reduktionsmittel *n* 1. <Chemie> reducer, reducing agent; 2. <Kohlen> reducing agent
Reduktionsprodukt *n* <Umweltschmutz> reducing product
Reduktionsröhre *f* <Labor> reduction tube
Reduktionsventil *n* <Labor> reduction valve
Reduktor *m* <Chemie> reducer, reducing agent

redundant *adj* <Comp & DV> redundant
redundante Zahl *f* <Comp & DV> redundant number
redundante Ziffern *fpl* <Comp & DV> redundant digitals
redundanter Code *m* <Comp & DV> redundant code
redundantes Zeichen *n* <Comp & DV> redundant character
Redundanz *f* 1. <Comp & DV> redudancy; 2. <Raumfahrt> redundancy *(Weltraumfunk)*
Redundanzprüfung *f* <Comp & DV, Telekom> redundancy check
reduzibles Polynom *f* <Math> reducible polynomial *(reduzierbares Polynom)*
reduzierbares Polynom *n* 1. <Comp & DV> reducible polynomial; 2. <Math> reducible polynomial *(alle Nullstellen sind Radikale)*
reduzieren *v* 1. <Kohlen> reduce; 2. <Maschinen> set down; 3. <Metall> reduce
reduzierendes Bleichbad *n* <Textil> stoving bath
reduzierendes Gas *n* <Thermod> reducing gas
Reduziergas *n* <Thermod> reducing gas
Reduziermuffe *f* <Maschinen> reduction sleeve
Reduzierofen *m* <Thermod> reduction furnace
Reduzierraum *m* <Aufnahme> reduction room
Reduzierstück *n* 1. <Bau> reducer, reducing pipe fitting; 2. <Maschinen> reducing socket
reduzierte Koordinaten *fpl* <Phys> reduced coordinates
reduzierte Masse *f* <Phys> reduced mass
reduzierte Prüfung *f* <Qual> reduced inspection
reduzierter Betrieb *m* <Comp & DV> graceful degradation
Reduzierung *f* <Kohlen> reduction
Reduzierventil *n* <Maschinen> reducing valve
Reduzierwiderstand *m* <Elektrotech> dropping resistor
Reduzierzone *f* <Metall> reducing zone
Reede *f* <Wassertrans> roads, roadstead • **auf Reede liegen** <Wassertrans> lie in the roads *(Schiff)* • **auf Reede vor Anker liegen** <Wassertrans> anchor in the roads
Reedlyte-Glas *n* <Ker & Glas> reedlyte glass
Reed-Quecksilberrelais *n* <Elektrotech> mercury-wetted reed relay
Reed-Relais *n* 1. <Elektriz, Elektrotech> reed relay; 2. <Telekom> dry reed relay, reed relay
Reed-Relais-Koppelpunkt *m* <Telekom> reed relay crosspoint
Reed-Relais-Schalter *m* <Elektrotech> reed relay switch
Reed-Relais-Schaltnetz *n* <Elektrotech> reed relay switching network *(Fernmeldewesen)*
Reed-Relais-Schaltung *f* <Elektrotech> reed relay circuit
Reed-Schalter *m* <Elektrotech> reed switch
Reedzunge *f* <Elektriz> reed
reelle Komponente *f* <Elektrotech> real component
reelle Matrix *f* <Math> matrix of real numbers
reelle Zahl *f* <Comp & DV, Math> real number
reeller Vektorraum *m* <Math> real vector space *(Vektorraum über reelle Zahlen)*
reelles Bild *n* <Phys> real image
Referenz *f* <Comp & DV> reference
Referenzachse *f* <Akustik> reference axis
Referenzaufruf *m* <Comp & DV> call by reference
Referenzband *n* <Akustik, Fernseh> reference tape
Referenzebene *f* <Comp & DV> reference level
Referenzenergie *f* <Erdöl> reference fuel *(in Preisgleitklausel)*
Referenzflächen-Toleranzbereich *m* <Optik> reference surface tolerance field
Referenzkondensator *m* <Elektriz> reference capacitor
Referenzleerband *n* <Akustik> reference tape
Referenzmodell *n* <Telekom> reference model *(OSI-sieben-Schichten-Modell)*

Referenzpegel

Referenzpegel m <Elektronik> reference level
Referenzprogramm testen v/mit <Comp & DV> benchmark
Referenzsignal n <Raumfahrt> reference burst *(Weltraumfunk)*
Referenzspannungsquelle f <Elektriz> reference-voltage source
Referenztabelle f <Comp & DV> lookup table
Referenztestmethode-Verfahren n *(RTM-Verfahren)* <Kunststoff> reference test method *(RTM)*
Reff n <Wassertrans> reef
reffen v <Wassertrans> reef *(Segeln)*
Reffkausch f <Wassertrans> reef cringle *(Segeln)*
Reffknoten m <Wassertrans> reef knot *(Knoten)*
Refiner m <Papier> refiner
Refinermahlung f <Papier> refining
reflektieren v <Optik> reflect
reflektierend adj <Optik> reflecting
reflektierende Flüssigkristallanzeige f <Elektronik> reflective LCD
reflektierende Kleidung f <Sicherheit> reflective clothing
reflektierende Rückwandbeschichtung f <Elektronik> reflective back coating
reflektierende Schallwand f <Akustik> reflex baffle
reflektierende Scheibe f <Optik> reflective disk
reflektierte Wärme f <Nichtfoss Energ, Thermod> reflected heat
reflektierte Welle f <Elektrotech, Funktech, Phys, Wellphys> reflected wave
reflektierter Strahl m 1. <Optik> reflected ray; 2. <Phys> reflected beam; reflected ray *(Licht, Röntgenstrahl)*; 3. <Wellphys> reflected ray
reflektiertes Licht n <Optik> reflected light
reflektiertes Signal n <Telekom> reflected signal
Reflektion f **am Mond** <Funkort> moonbounce
Reflektionsabschirmung f <Raumfahrt> antireflection coating
Reflektionsbeschichtung f <Raumfahrt> antireflection coating
Reflektometer n <Funktech, Kunststoff> reflectometer
Reflektor m 1. <Elektronik, Elektrotech> reflector; repeller *(bei Elektrode)*; 2. <Foto> reflector; 3. <Funktech> reflector *(Antenne)*; 4. <Kerntech> reflector; 5. <Kfztech> reflector *(Sicherheitszubehör)*; 6. <Nichtfoss Energ, Phys, Raumfahrt> reflector
Reflektor m **mit veränderlicher Brennweite** <Foto> variable focus reflector
Reflektorantenne f 1. <Funktech> reflector antenna; 2. <Telekom> reflecting antenna
Reflektorelektrode f <Elektrotech, Fernseh> reflector electrode
Reflektorprisma n <Optik> reflecting prism
Reflektorspiegel m <Optik> reflecting mirror
Reflektorstativ n <Foto> reflector stand
Reflexebene f <Optik> plane of reflection
Reflexion f 1. <Akustik, Elektronik, Geom, Optik, Phys> reflection; 2. <Raumfahrt> reflection *(Weltraumfunk)*; 3. <Telekom> reflection; 4. <Wellphys> reflection *(an der Ionosphäre)*
Reflexion f **von Röntgenstrahlen** <Strahlphys> X-ray reflection
Reflexions-Brennkurve f <Optik> caustic by reflection
Reflexionsdichte f <Optik> reflectance density
Reflexionselektronenmikroskop n <Elektronik> reflection electron microscope
Reflexionsfaktor m 1. <Elektronik> reflection factor *(Reflexionsgrad)*; 2. <Phys> reflectance; reflection factor *(Klystron)*

reflexionsfreier Raum m 1. <Akustik> anechoic room; 2. <Phys> anechoic room, dead room
Reflexionsgesetze npl <Phys> laws of reflection
Reflexionsgitter n <Phys, Wellphys> reflection grating
Reflexionsgrad m 1. <Elektronik> reflection coefficient; 2. <Phys> reflectance; reflection factor *(Klystron)*
Reflexionskammer f <Raumfahrt> reverberation chamber
Reflexionskoeffizient m 1. <Funktech> reflection coefficient; 2. <Nichtfoss Energ> reflectivity coefficient; 3. <Optik> reflectance; 4. <Phys> reflection coefficient; 5. <Raumfahrt> reflectivity coefficient
Reflexionsmeter n <Papier> reflection meter
Reflexionsraum m <Elektronik> reflector space *(Klystron)*
Reflexionsschalldämpfer m <Kfztech> *(AE)* baffle muffler, *(BE)* baffle silencer
Reflexionsspiegel m <Foto> reflex mirror
Reflexionsstrahlungsheizung f <Heiz & Kälte> reflective radiant heating
Reflexionsverfahren n <Kerntech> reflection method
Reflexionsverlust m <Aufnahme, Elektronik> reflection loss
Reflexionsvermögen n 1. <Elektrotech> reflectance; 2. <Heiz & Kälte, Nichtfoss Energ> reflectivity; 3. <Optik, Telekom> reflectance
Reflexionswand f <Aufnahme> live end *(eines Raums)*
Reflexionswinkel m <Optik, Phys, Wellphys> angle of reflection
Reflexklystron n <Elektronik, Phys> reflex klystron
Reflexkopiermethode f <Foto> reflex printing method
Reflexlicht n <Foto> reflected light
reflexmindernde Beschichtung f <Telekom> antireflective coating
Reflexmodulation f <Elektronik> reflex bunching
Reflexphasenfokussierung f <Elektronik> reflex bunching
Reflexstrahlphotodetektor m <Trans> reflected beam photo-electric detector
Reflexsucher m <Foto> reflex viewfinder
Reformieren n <Erdöl> reforming *(Raffinerie)*
reformierende Technologie f <Nichtfoss Energ> reformer technology
Reformkost f <Lebensmittel> health food
Refraktärperiode f <Ergon> refractory period
Refraktion f <Wassertrans> refraction
Refraktionsgitter n <Labor> refraction grating
Refraktionswinkel m (r) <Optik> angle of refraction *(r)*
Refraktivität f <Optik> refractiveness, refractivity
Refraktometer n <Labor, Optik, Phys, Strahlphys> refractometer
Regal n <Bau, Labor> shelf
Regalfach n <Ker & Glas> pigeonhole
Regalmarkierungen fpl <Ker & Glas> rack marks
Regalplatz m <Verpack> shelf space
Regel f <Comp & DV, Künstl Int, Metrol> rule
Regel f **des gesunden Menschenverstands** <Künstl Int> common sense rule
Regelabweichungssignal n <Kerntech> actuating signal
Regelalgorithmus m <Kontroll> control algorithm
Regelanlasser m <Elektrotech> starting rheostat
Regelation f <Phys> regelation
regelbar adj 1. <Maschinen> variable; 2. <Mechan> adjustable
regelbare Geschwindigkeit f <Maschinen> variable velocity
regelbarer Lüfter m <Lufttrans> fan, hit-or-miss governor, ventilator *(mit verschiebbarer Schlitzplatte)*
regelbarer Motor m <Mechan> adjustable speed motor
regelbasiert adj <Künstl Int> rule-based

regelbasiertes Expertensystem *n* <Künstl Int> rule-based expert system
regelbasiertes System *n* <Comp & DV, Künstl Int> rule-based system
Regelbasis *f* <Comp & DV> rule base
Regelbereich *m* <Regelung> control range
Regelcharakteristik *f* <Elektrotech> control characteristic
Regelcontainer *m* <Trans> standard container
Regeldifferenz *f* <Regelung> system deviation
Regeleinrichtung *f* 1. <Maschinen> control unit; 2. <Regelung> automatic control equipment, closed-loop controller
Regelgenauigkeit *f* <Kontroll, Regelung, Qual> accuracy, control accuracy
Regelgerät *n* 1. <Elektrotech> control unit; 2. <Regelung> automatic control unit
regelgesteuertes System *n* <Künstl Int> rule-based system
regelgestützt *adj* <Künstl Int> rule-based
Regelgröße *f* <Elektrotech, Maschinen> controlled variable
Regelkennlinie *f* <Elektrotech> control characteristic
Regelklappe *f* <Kfztech> butterfly
Regelkompass *m* <Wassertrans> standard compass
Regelkompensator *m* <Lufttrans> adjusting potentiometer
Regelkreis *m* <Elektriz, Elektronik, Elektrotech, Regelung> closed loop control system, feedback control system
Regelkreisrückführsignal *n* <Elektronik> feedback signal
regellose Verteilung *f* <Metall> random distribution
regelmäßige Instandhaltung *f* <Telekom> routine maintenance
regelmäßige Wartung *f* <Bau> scheduled service
regelmäßiges Polygon *n* <Geom> regular polygon
regeln *v* 1. <Mechan> adjust; 2. <Raumfahrt, Telekom> control
Regeln *n* <Maschinen> governing
Regeln *fpl* **für Arbeitsabläufe** <Comp & DV> workflow rules
Regelorgan *n* <Maschinen> control device
Regelpotenziometer *n* <Elektrotech> control potentiometer
Regelröhre *f* 1. <Elektronik> remote cut-off tube; variable mu tube *(mit veränderlicher Steilheit)*; variable mutual conductance tube; 2. <Elektrotech> constant-current tube *(Licht)*
Regelsatz *m* **in einem Expertensystem** <Comp & DV> inference engine
Regelschalter *m* <Elektrotech> control switch
Regelschleife *f* <Telekom> locked loop
Regelsignal *n* <Elektronik> control signal *(zur Prozessregelung)*
Regelspur *f* <Eisenbahn> *(AE)* standard gage, *(BE)* standard gauge
Regelspurbahn *f* <Eisenbahn> *(AE)* standard gage railroad, *(BE)* standard gauge railway
Regelstab *m* <Kerntech> absorber rod, control rod
Regelstange *f* <Kfztech> control rod
Regelstrecke *f* <Regelung> closed-loop controlled system
Regelsystem *n* <Regelung> automatic control system
Regeltechnik *f* <Kontroll> control technology
Regeltrafo *m* <Elektrotech> variable ratio transformer, variable transformer
Regeltransformator *m* 1. <Elektrotech> control transformer, regulating transformer, variable ratio transformer, variable transformer; 2. <Labor> regulating transformer
Regelumspanner *m* <Elektrotech> variable ratio transformer

Regelung *f* 1. <Elektriz, Elektronik, Elektrotech> automatic control, closed loop control; 2. <Kerntech> regulating; 3. <Kfztech, Maschinen> control; 4. <Regelung> automatic control; closed-loop device *(Einrichtung)*; closed-loop control *(Vorgang)*; 5. <Telekom> regulation
Regelung *f* **des Drehzahlverhältnisses** <Regelung> speed ratio control
Regelung *f* **des relativen Gleichlaufs** <Regelung> speed ratio control
Regelung *f* **mit Anzapfung** <Elektrotech> tapped control
Regelungsbereich *m* <Elektrotech> control range, regulation range
Regelungshierarchie *f* <Kontroll> control hierarchy
Regelungssystem *n* 1. <Elektronik> automatic control system, closed-loop control system; 2. <Kontroll> control system
Regelungssystem *n* **mit Rückkopplung** <Elektrotech> feedback control system
Regelungs- und Steuerungstechnik *f* <Kontroll, Regelung> automatic control science and technology
Regelventil *n* 1. <Heiz & Kälte> servo valve; 2. <Hydraul, Kontroll> control valve; 3. <Labor> regulating valve *(Flüssigkeitsregelung)*; 4. <Lufttrans> actuator control valve; 5. <Maschinen> control valve, regulating valve; 6. <Wasserversorg> flow-regulating valve
Regelverkehr *m* <Telekom> regular routing
Regelverstärker *m* <Elektronik> control amplifier, variable gain amplifier
Regelvorgang *m* <Kontroll, Qual> control action
Regelvorrichtung *f* <Elektrotech> control gear, control unit
Regelweg *m* <Telekom> normal route
Regelwehr *n* <Wasserversorg> level control weir
Regelwiderstand *m* <Elektrotech> adjustable resistor, rheostat, variable resistor, varistance
Regenbecken *n* <Wasserversorg> rainwater catchment
Regenbogenquarz *m* <Metall> rainbow quartz
Regenerat *n* <Kunststoff> reclaimed rubber, reclaim, regenerated rubber
Regeneration *f* <Abfall, Comp & DV, Metall> regeneration
Regenerationsanlage *f* <Abfall> regeneration unit
regenerativ *adj* 1. <Nichtfoss Energ> renewable; 2. <Raumfahrt> regenerative *(Weltraumfunk)*
regenerative Energie *f* <Nichtfoss Energ> renewable energy
Regenerativfeuerung *f* <Heiz & Kälte> regenerative heating
Regenerativkühlung *f* <Heiz & Kälte, Thermod> regenerative cooling
Regenerativofen *m* 1. <Ker & Glas, Maschinen> regenerative furnace; 2. <Metall> regenerating furnace
Regenerativverfahren *n* <Maschinen> regenerative system
Regenerativzelle *f* <Eisenbahn, Kfztech> regenerative cell
Regenerator *m* <Elektronik, Maschinen, Metall, Telekom> regenerator
regenerieren *v* 1. <Chemie> reactivate; 2. <Comp & DV> regenerate
regeneriert *adj* <Elektronik> regenerated
regenerierter Kautschuk *m* <Kunststoff> reclaimed rubber, reclaim, regenerated rubber
Regenerierung *f* 1. <Chemie> reactivation; 2. <Elektronik> regeneration; 3. <Kunststoff, Meerschmutz> recovery; 4. <Raumfahrt, Telekom, Wasserversorg> regeneration
Regenfallrohr *n* <Bau> downpipe, stack pipe
Regenleiste *f* <Eisenbahn> water guttering
Regenmessgerät *n* <Bau, Labor, Nichtfoss Energ, Wasserversorg> *(AE)* rain gage, *(BE)* rain gauge

Regentrübung

Regentrübung f <Wassertrans> rain clutter *(Radar)*
Regenwasser n <Bau, Kohlen> storm water
Regenwasserfallrohr n <Bau> rainwater downpipe
Regenwassernutzung f <Umweltschmutz> utilization of rainwater
Regieanweisung f für Tonmischung <Aufnahme> dubbing cue sheet
Regieeinrichtung f <Fernseh> cuer
Regieplan m <Fernseh> cue sheet
Regieraum m <Fernseh> control room, studio control room
Regiesignal n <Fernseh> cue
Regiesignalschirm m <Fernseh> cue screen
Regiespur f <Fernseh> cue track
Regiespur-Zugriffscode m <Fernseh> cue track address code
Regiezeichen n <Aufnahme> cue
Regional... <Fernseh, Funktech, Lufttrans, Nichtfoss Energ> regional
Regionalcarrier m <Lufttrans> regional carrier
regionale Verkehrsinformationen fpl <Trans> area traffic information
regionales Netz n <Telekom> metropolitan area network *(MAN)*
Regionalfernsehen n <Fernseh> local television, regional television
Regionalflughafen m <Lufttrans> regional airport
Regionalfluglinie f <Lufttrans> regional carrier
Regionalmetamorphose f <Nichtfoss Energ> regional metamorphism
Regionalpatent n <Patent> regional patent
Regionalprogramm n <Fernseh, Funktech> *(AE)* local program, *(BE)* local programme
Regionalrundfunk m <Funktech> local broadcasting
Regionalrundfunksender m <Funktech> local broadcasting station
Regionalschienenverkehr m <Eisenbahn> *(AE)* regional railroad traffic, *(BE)* regional railway traffic
Regionalwarnfunknetz n <Funktech, Trans> regional radio warning system
Register n 1. <Comp & DV> index, register; 2. <Elektrotech, Kontroll, Telekom> register • **Register zusammenstellen** <Comp & DV> index
Register n **mit Mehrfachzugriff** <Comp & DV> multiport register
Registerauszug m <Patent> extract from the register
Registerbezeichnung f <Comp & DV> register name
Registerbrief m <Wassertrans> certificate of registration; certificate of registry *(Dokumente)*
Registerdatei f <Comp & DV> register file
Registererstellung f <Comp & DV> indexing
Registerhafen m <Wassertrans> port of registration, port of registry
registerhaltig adj <Druck> in register
registerhaltig machen v <Druck> register
Registerhaltigkeit f <Druck> backing up, registration
Registerhaltung f <Druck> backing up, register, registration
Registerlänge f <Comp & DV> register length
Registersystem n <Telekom> register-controlled system
Registertonnengehalt m <Metrol> register tonnage
Registerzuordner m <Telekom> register translator
Registraturnummer f <Druck> file number
registrieren v 1. <Comp & DV> enroll, register; 2. <Mechan> index
registrierendes Messgerät n <Gerät> recording instrument
registrierendes Temperaturmessgerät n <Kerntech> thermograph

Registriergerät n <Gerät> logging device, recording instrument
Registrierinstrument n <Gerät> recording instrument
Registrierkurve f <Gerät> recorded curve, trace line
Registriermanometer n <Gerät> recording manometer
Registrieroszillograph m <Akustik, Phys, Telekom> recording oscillograph
Registrierscheibe f <Aufnahme> recording disk
Registrierstreifen m 1. <Elektriz> strip chart; 2. <Gerät> chart, continuous diagram, recording chart; strip chart *(für Streifenschreiber)*
Registrierstreifenantriebsmotor m <Gerät> chart motor
Registrierstreifentransport m <Gerät> chart transport
Registrierstreifenwalze f <Gerät> chart drum
Registrierthermometer n 1. <Phys> thermograph; 2. <Thermod> recording thermometer
Registriertrommel f <Aufnahme> recording drum
Registrierung f 1. <Comp & DV, Lufttrans> registration; 2. <Telekom> recording *(Messen)*; 3. <Trans, Wassertrans> registration
Registrierwaage f <Gerät> recording balance
Regler m 1. <Elektrotech> control unit, controller, governor, regulator; 2. <Kontroll> controller; 3. <Labor> regulator; 4. <Maschinen> control unit, governor, regulator; 5. <Mechan> governor; 6. <Raumfahrt> control; 7. <Telekom> control, regulator *(zur Einstellung)*; 8. <Wassertrans> governor *(Motor)*
Reglerdynamo m <Elektriz> control dynamo
Reglerschaltungs-Stangenanschlag m <Lufttrans> governor control stop
Reglerstange f <Maschinen> governor rod
Reglersubstanz f <Chemtech> regulator *(Polymerisation)*
Reglerventil n 1. <Hydraul> governor valve; 2. <Mechan> check valve
Reglerwiderstand m <Kfztech> rheostat
Reglette f <Druck> clump
Regressand m <Math> regressand *(abhängige Variable der Regressionsfunktion)*
Regression f <Comp & DV, Math> regression
Regressionsanalyse f <Math> regression analysis
Regressionsfunktion f <Math> regression function
Regressionsgleichung f <Math, Qual> regression equation
Regressionskurve f <Math> regression curve *(Graph der Regressionsfunktion)*
Regressor m <Math> regressor *(unabhängige Variable der Regressionsfunktion)*
Regularisation f <Wasserversorg> regularization
regulierbar adj <Maschinen> adjustable
regulierbares Grundwehr n <Umweltschmutz> adjustable submersion weir
Reguliercharakteristik f <Fertig> flow characteristics *(Kunststoffinstallationen)*
regulieren v <Bau> adjust *(Instrument)*
Regulierschraube f <Maschinen> adjusting screw, regulating screw
Regulierung f 1. <Elektrotech> adjustment, regulation; 2. <Telekom> regulation
Regulierventil n <Kfztech> check valve
Reib... 1. <Chemie> triturating; 2. <Maschinen> frictional
Reibabnutzung f <Anstrich> fretting wear
Reibahle f <Bau, Fertig, Maschinen, Mechan> reamer
Reibahle f **mit Messereinstellung** <Fertig> adjustable blade reamer
Reibahle f **mit Spiralnuten** <Maschinen> reamer with spiral flutes
Reibahle f **mit verstellbaren Messern** <Maschinen> adjustable blade reamer
Reibahlennutenfräser m <Maschinen> reamer cutter

Reibbeiwert m <Maschinen> coefficient of friction
Reibbelag m <Fertig, Kfztech> friction lining
Reibbelagwerkstoff m <Fertig> friction material
Reibboden m <Bau> frictional soil
Reibbremse f <Maschinen> friction brake
Reibechtheit f 1. <Kunststoff> rub fastness; 2. <Textil> fastness to rubbing
Reibeisen n <Bau> rasp
reiben v <Mechan> ream
Reiben n <Maschinen> galling, reaming, reaming-out
Reiber m <Chemtech> pestle
Reibermüdung f <Metall> fretting fatigue
Reibgetriebe n <Maschinen> friction drive
Reibkegelantrieb m <Maschinen> friction cone drive
Reibkorrosion f <Anstrich> fretting corrosion
Reibkorrosion f durch Abrieb <Mechan> abrasion-fretting corrosion
Reibkugel f <Maschinen> friction ball
Reibmarkierung f <Ker & Glas> scrub mark
Reibrad n <Maschinen> friction wheel
Reibradantrieb m <Maschinen> friction wheel drive
Reibrädergetriebe n <Maschinen> friction gear
Reibradgetriebe n <Maschinen> friction gear, friction gearing
Reibring m <Maschinen> friction ring
Reibsand m <Mechan> grit
Reibschale f <Chemtech, Labor> mortar
Reibscheibe f <Maschinen> (BE) friction disc, (AE) friction disk
Reibschleifen n <Fertig> lapping
Reibschluss m <Fertig> frictional grip, frictional resistance (Getriebelehre)
reibschlüssig adj <Maschinen> frictional
Reibschweißen n <Fertig> friction welding
Reibspannung f <Metall> friction stress
Reibspindel f <Maschinen> friction screw
Reibspindelpresse f <Fertig> friction-driven screw press
Reibspur f <Fertig> galling mark
Reibstelle f <Ker & Glas> rub
Reibung f <Anstrich, Fernseh, Kohlen, Kunststoff, Maschinen, Mechan, Papier, Phys, Raumfahrt, Textil> friction
Reibung f der Ruhe <Maschinen> static friction
Reibungs... <Bau> frictional
Reibungsantrieb m <Maschinen> friction drive
Reibungsbahn f <Eisenbahn> (AE) adhesion railroad, (BE) adhesion railway
reibungsbehaftete inkompressible Strömung f <Strömphys> viscous incompressible flow
Reibungsbeiwert m <Bau, Ergon> coefficient of friction
Reibungsbremse f 1. <Kfztech> friction brake (Bremsanlage); 2. <Maschinen> friction brake
Reibungsdämpfer m <Mechan> frictional damper
Reibungsdrehmoment n <Raumfahrt> frictional torque
reibungselektrischer Detektor m <Trans> triboelectric detector
Reibungselektrizität f 1. <Bau, Elektriz, Phys> frictional electricity, static electricity, triboelectricity; 2. <Textil> static electricity
Reibungsfluss m <Kerntech> frictional flow
reibungsfrei adj 1. <Maschinen, Mechan> frictionless; 2. <Strömphys> inviscid
reibungsfrei gelagert adj <Maschinen> mounted on frictionless bearings
reibungsfreie Bewegung f <Strömphys> inviscid motion
reibungsfreie Strömung f <Strömphys> inviscid motion
reibungsfreie Strömungsverteilung f <Strömphys> inviscid flow distribution

Reibungskoeffizient m <Kunststoff, Maschinen, Mechan, Metrol, Phys, Qual, Wassertrans> coefficient of friction (Schiffkonstruktion)
Reibungskondensator m <Elektrotech> snubber capacitor
Reibungskraft f 1. <Maschinen> force of friction; 2. <Metall> friction force; 3. <Phys> frictional force
Reibungskupplung f 1. <Kfztech> friction clutch; 2. <Maschinen> friction clutch, friction coupling
Reibungsleistung f <Fertig> friction horsepower
Reibungslokomotive f <Eisenbahn> adhesion locomotive
Reibungslumineszenz f <Phys> triboluminescence
Reibungsmessgerät n <Phys> tribometer
Reibungsmühle f <Papier> attrition mill
Reibungspfahl m <Bau, Kohlen> friction pile
Reibungsschaltkreis m <Elektrotech> snubber circuit
Reibungsscharnier n <Bau> friction hinge
Reibungsschweißen n <Bau> friction welding
Reibungsspannung f <Strömphys> viscous stress
Reibungsströmung f <Lebensmittel, Maschinen, Phys> viscous flow
Reibungstriebwagen m <Eisenbahn> adhesion railcar
reibungsverhindernd adj <Maschinen> antifriction
Reibungsverlust m <Maschinen, Nichtfoss Energ> friction loss
Reibungsverschleiß m <Maschinen> attrition
Reibungswärme f 1. <Fertig, Maschinen> frictional heat; 2. <Thermod> heat caused by friction
Reibungswiderstand m 1. <Elektrotech> snubber resistor; 2. <Fertig> frictional resistance; 3. <Maschinen> friction resistance; 4. <Mechan> frictional drag; 5. <Wassertrans> frictional resistance (Schiffkonstruktion)
Reibungswinkel m <Bau, Kohlen, Maschinen> angle of friction
Reibungszahl f <Maschinen, Wassertrans> coefficient of friction (Schiffkonstruktion)
Reibzahl f <Kunststoff, Qual> coefficient of friction
reichen v **von** <Telekom> range from
Reichweite f 1. <Elektronik> range (eines Emitters); 2. <Ergon> reachable space; 3. <Gerät> coverage, range; 4. <Kfztech> cruising range (Elektrofahrzeuge); 5. <Lufttrans> range (eines Luftfahrzeugs); 6. <Maschinen> reach; 7. <Phys, Raumfahrt> range; 8. <Telekom> coverage, range, working distance; 9. <Wassertrans> range (Funk)
Reichweite f pro Ladung <Trans> range of action per charge
Reichweitenhüllkurve f <Ergon> reach envelope
Reif m 1. <Lebensmittel> bloom (gefrorene Lebensmittel); 2. <Mechan> hoop
Reifegradmesser m <Lebensmittel> tenderometer
Reifen m 1. <Eisenbahn, Kfztech> (AE) tire, (BE) tyre; 2. <Mechan> hoop, ring; 3. <Trans> (AE) tire, (BE) tyre
Reifen m ohne Profil <Kfztech> (AE) bald tire, (BE) bald tyre, (AE) smooth tire, (BE) smooth tyre
Reifengarn n <Textil> (AE) tire yarn, (BE) tyre yarn
Reifengrundgewebe n <Textil> carcass
Reifenlauffläche f <Kfztech, Kunststoff> (AE) tire tread, (BE) tyre tread
Reifenmuster n <Kfztech> tread design
Reifenprofil n 1. <Eisenbahn> (AE) tire profile, (BE) tyre profile; 2. <Kfztech> tread design, tread pattern
Reifenrohling m <Kunststoff> green tyre
Reifenüberdruck m <Kfztech> over-inflation
Reifenunterdruck m <Kfztech> under-inflation
Reifung f 1. <Erdöl> maturation (Geologie); 2. <Foto> ripening (Emulsion); 3. <Lebensmittel> (BE) ageing, (AE) aging

Reifungstemperatur

Reifungstemperatur f <Ker & Glas> maturing temperature
Reihe f 1. <Comp & DV> series; 2. <Gerät> range *(von Einrichtungen)*; 3. <Math> series; 4. <Telekom> suite *(von Einrichtung)*; 5. <Textil> course, range • **in Reihe** <Comp & DV> serial • **in Reihe geschaltet** <Elektrotech, Gerät> series-connected • **in Reihe schalten** <Elektrotech> connect in series • **in Reihen** <Phys> in series
Reihen fpl **mit linken Maschen** <Textil> rows knitted in purl
Reihen fpl **mit rechten Maschen** <Textil> wales of face stitches
Reihenanlage f <Telekom> key system
Reihenanordnung f <Elektrotech> series arrangement
Reihenbohrmaschine f 1. <Fertig> gang drill, in-line multi drill; 2. <Maschinen> gang drill
Reiheneinspritzpumpe f <Kfztech> multicylinder injection pump
Reihenfolge f <Comp & DV> sequence • **in aufgeführter Reihenfolge** <Patent, Qual> in the order specified
Reihenfolgeplanung f <Kontroll> scheduling
Reihenfolgeprüfung f <Comp & DV> sequence check
Reihenfolgezugriff m <Comp & DV> sequential access
reihenförmig angeordnete Zylinder mpl <Kfztech> in-line cylinders
Reihengegenkopplung f <Elektrotech> series feedback
reihengeschaltet adj <Elektriz> in series
reihengeschalteter Kondensator m <Elektriz> series connected capacitor
reihengewickelter Dynamo m <Elektrotech> series-wound dynamo
reihengewickelter Motor m <Elektrotech> series-wound motor
Reihenklemme f <Kfztech> terminal block
Reihenkondensator m <Elektrotech> series capacitor
Reihenmotor m <Kfztech, Maschinen, Wassertrans> in-line engine
Reihenparallelschalter m <Elektrotech> series-parallel switch
Reihenparallelschaltung f <Elektrotech> series-parallel circuit
Reihenreaktanz f <Elektrotech> series reactance
Reihenresonanzfrequenz f <Elektronik, Elektrotech, Fernseh, Funktech, Telekom> series resonance frequency
Reihenresonanzkreis m <Elektrotech> series-resonant circuit
Reihenschaltung f <Elektriz, Elektrotech, Funktech, Telekom> series connection, tandem connection
Reihenschlussanlasser m <Elektriz> series starter
Reihenschlussanlassschalter m <Elektriz> series starter
Reihenschlussdynamo m <Elektriz, Elektrotech> series-wound dynamo
Reihenschlusserregung f <Elektrotech> series excitation
Reihenschlussmaschine f <Elektriz> series-excited machine, series-wound machine
Reihenschlussmotor m <Elektriz> series motor, series-wound motor
Reihenschlusswicklung f <Elektrotech> series winding
Reihenspeisung f <Elektrotech> series feed
Reihenstandmotor m <Kfztech> in-line engine
Reihenstichprobenentnahme f <Qual> sequential sampling
Reihenstichprobenprüfplan m <Qual> sequential-sampling plan
Reihenwicklung f <Elektrotech> series winding
Reihenwiderstand m <Elektrotech> series resistance

rein darstellbar adj <Chemie> isolable
rein darstellen v <Chemie> isolate
Rein... <Abfall, Druck, Fertig, Umweltschmutz> cleaned
Reindarstellung f <Fertig> isolation
reine Flüssigphase f <Thermod> liquid only phase
reine Frachtfluglinie f <Lufttrans> all-cargo carrier
reine gasförmige Phase f <Thermod> gas-only phase, gaseous phase only
reine Mathematik f <Math> pure mathematics
reine Quarte f <Akustik> perfect fourth
reine Quinte f <Akustik> perfect fifth
reiner Frachtcharterflug m <Lufttrans> all-cargo charter flight
reiner Frachtdienst m <Lufttrans> all-cargo service, all-freight service, all-freight
reiner Frachtlastfaktor m <Lufttrans> all-cargo load factor
reiner gasförmiger Zustand m <Thermod> gas-only phase, gaseous phase only
reiner Postdienst m <Lufttrans> all-mail service
reiner Zementmörtel m <Bau> neat cement
reines Auto- und LKW-Transportschiff n <Wassertrans> car and truck carrier *(Schifftyp)*
reines Frachttransportflugzeug n <Lufttrans> all-cargo aircraft
reines Konnossement n <Wassertrans> clean bill of lading *(Dokumente)*
reines Kupfer n <Metall> pure copper
reines Spektrum n <Strahlphys> pure spectrum
Reingas n <Abfall> cleaned gas, scrubbed gas
Reinhadernpapier n <Druck> all-rag paper
Reinhaltung f <Umweltschmutz> pollution control *(Luft, Gewässer)*
reinigen v 1. <Bau> clarify; 2. <Chemtech> purify; 3. <Kohlen> clarify; 4. <Metall> purify; 5. <Papier> clean; 6. <Wassertrans> grave *(Schiffinstandhaltung)*; 7. <Wasserversorg> cleanse, purify; clean up *(Schleusen)*
Reinigen n <Papier> cleaning
Reiniger m 1. <Chemtech> purifier; 2. <Fertig, Sicherheit> cleaner
Reinigerzelle f <Kohlen> cleaner cell
Reinigung f 1. <Chemtech> defecation; 2. <Erdöl> purification, swabbing; 3. <Kohlen> clarification; 4. <Meerschmutz> clean-up; 5. <Sicherheit> cleaner; 6. <Wasserversorg> cleansing
Reinigungsahle f <Kfztech> reamer *(Werkzeug)*
Reinigungsanlage f <Wasserversorg> purification plant
Reinigungsapparat m <Chemtech> purifier, purifying apparatus
Reinigungsbeginn m <Kerntech> head end treatment *(der Brennstoffaufbereitung)*
Reinigungsbehälter m <Wasserversorg> filtration vat
Reinigungsbürste f <Druck> cleaning brush
Reinigungsfällung f <Umweltschmutz> below-cloud scavenging
Reinigungsflüssigkeit f <Anstrich> decontamination fluid
Reinigungsgerät n <Fertig> cleaner
Reinigungsgestell n <Bau> cradle machine
Reinigungsgrad m <Qual, Wasserversorg> degree of purification
Reinigungshahn m <Wasserversorg> purge cock, purging cock
Reinigungskammer f <Labor> clean room
Reinigungskolben m <Erdöl> swab
Reinigungskraft f <Chemie> detergency
Reinigungsmaschine f <Fertig> cleaning device; purifier
Reinigungsmittel n 1. <Chemtech> purifying agent; 2. <Textil> detergent
Reinigungsöffnung f <Maschinen> inspection fitting

Reinigungsvermögen n <Abfall> purification capacity
Reinigungswagen m <Bau> cradle machine
Reinjektion f <Erdöl> reinjection (Fördertechnik)
Reinkohle f <Kohlen> clean coal, cleans
Reinluft f <Heiz & Kälte> filtered air
Reinluftgebiet n <Umweltschmutz> area of pure air
Reinraum m <Elektronik, Heiz & Kälte, Verpack> clean room
Reinraumtechnik f <Heiz & Kälte> clean-room technology
Reinstkohle f <Kohlen> super-clean coal
Reinvorlage f <Umweltschmutz> net receiver
Reinwasser n <Wasserversorg> clean water, pure water
Reise f <Kohlen> campaign (Ofen)
Reisedauer f <Trans> journey time
Reisefluggeschwindigkeit f <Lufttrans> cruising speed
Reiseflughöhe f <Lufttrans> cruising altitude
Reiseflugleistung f <Lufttrans> cruising power
Reisegeschwindigkeit f 1. <Eisenbahn> schedule speed; 2. <Wassertrans> cruising speed
Reisekamera f <Foto> field camera
Reisemikroskop n <Labor> (AE) traveling microscope, (BE) travelling microscope
Reisen n im Weltall <Raumfahrt> space travel
Reiseomnibus m <Kfztech> motor coach
Reiseroute f <Wassertrans> itinerary (Navigation)
Reisezug m <Eisenbahn> passenger train
Reisezugwagen m <Eisenbahn> (AE) passenger car, (BE) passenger coach, (AE) railroad car, (BE) railway carriage
Reisezugwagen m mit schwenkbarem Wagenkasten <Eisenbahn> tilting body coach
Reismühle f <Lebensmittel> rice mill
Reiß... <Bau, Lebensmittel, Textil, Verpack, Wassertrans> drawing, tearing
Reißband n <Verpack> tear tape
Reißbrett n <Bau, Textil, Wassertrans> drawing board
Reißdehnung f <Kunststoff> elongation at break
Reißdraht m <Papier> tearing wire
reißen v 1. <Anstrich> crack; 2. <Bau> crack, spring; 3. <Fertig> fissure, scribe; 4. <Ker & Glas> craze; 5. <Kunststoff> tear; 6. <Metall> crack
Reißen n 1. <Fernseh> tearing; 2. <Werkprüf> cracking (Glas)
Reißfeder f <Fertig> drawing pen
Reißfestigkeit f 1. <Kunststoff> tear resistance, tear strength; 2. <Maschinen> resistance to tearing, ultimate tensile strength; 3. <Meerschmutz> resistance to tearing; 4. <Metall> ultimate tensile strength; 5. <Phys> tensile strength; 6. <Qual> resistance to tearing; 7. <Textil> breaking strength, tear strength; 8. <Verpack> bursting strength; 9. <Wassertrans> breaking load
Reißkegelbildung f <Fertig> cupping
Reißlack m <Fertig> brittle lacquer
Reißlänge f 1. <Kunststoff> breaking length; 2. <Papier> tensile strength
Reißlängenprüfer m <Papier> tensile strength tester
Reißlängenprüfgerät n <Papier> breaking strength tester
Reißlehre f <Bau> (AE) scribing gage, (BE) scribing gauge
Reißleinengriff m <Lufttrans> parachute release handle
Reißnadel f 1. <Bau> scratch awl; 2. <Fertig> scriber
Reißschiene f <Maschinen> T-square
Reißspitze f <Bau> scribing awl
Reißverschluss m 1. <Fertig> zip, zipper; 2. <Mechan> fastener; 3. <Textil> zip, zipper • sich mit Reißverschluss öffnen lassen <Textil> zip • sich mit Reißverschluss schließen lassen <Textil> zip
Reißwerk n <Chemie> macerator
Reißwolf m <Verpack> shredding machine

Reißzwecke f <Fertig> thumbtack
Reiterchen n <Labor> rider (Waage)
Reiteretikett n <Verpack> header label
Reitergewölbe n <Ker & Glas> rider arch
Reiterlibelle f <Bau> wye level
Reitstock m 1. <Fertig> tailstock; 2. <Maschinen> back head, back puppet, deadhead, footstock, tailstock; 3. <Mechan> headstock
Reitstockkörper m <Fertig> tailstock body
Reitstockoberteil n <Fertig> tailstock barrel
Reitstockpinole f 1. <Fertig> tailstock quill; 2. <Maschinen> tail spindle, tailstock quill
Reitstockspitze f 1. <Fertig> (AE) tailstock center, (BE) tailstock centre; 2. <Maschinen> (AE) back center, (BE) back centre, (AE) dead center, (BE) dead centre, (AE) tailstock center, (BE) tailstock centre
Reitstockunterteil n <Fertig> tailstock base
Reizstoff m <Sicherheit> irritant substance
Rekombination f <Elektronik, Kerntech, Phys> recombination
Rekombinationsanlage f <Kerntech> recombination plant
Rekombinationsbasisstrom m <Elektronik> recombination base current
Rekombinationskoeffizient m <Phys> recombination coefficient
Rekombinationsrate f <Elektronik, Phys> recombination rate
Rekombinationsvorgang m <Elektronik> recombination process
rekonfigurierbar adj <Comp & DV> reconfigurable
rekonfigurieren v <Comp & DV> reconfigure
Rekonfigurierung f <Comp & DV> reconfiguration
rekonstruieren v 1. <Bau> re-equip, rehabilitate; 2. <Comp & DV> reconstruct
Rekonstruktion f <Bau> reconstruction
Rekorder m <Elektriz> recorder
Rekristallisation f <Metall> recrystallization
Rektaszension f <Raumfahrt> right ascension (gerader Aufstieg)
rektifizieren v <Elektriz> rectify
Rektifizierwalze f <Papier> rectifier roll
rekultivieren v <Abfall> revegetate
Rekultivierung f <Bau> land restoration, revegetation
Rekuperativkühlung f <Thermod> regenerative cooling
Rekuperativofen m <Ker & Glas> recuperative furnace
Rekursion f <Comp & DV, Math> recursion
Rekursionsformel f <Math> recursion formula
rekursiv adj <Comp & DV, Math> recursive
rekursive Digitalfilterung f <Elektronik> recursive filtering
rekursive Funktion f <Comp & DV, Math> recursive function
rekursive Prozedur f <Comp & DV> recursive procedure
rekursives Digitalfilter n <Elektronik> recursive filter
rekursives Filter n <Telekom> IIR filter, infinite impulse response filter, recursive filter
Relais n 1. <Comp & DV, Elektriz, Funktech> relay; 2. <Kfztech> relay (KFZ-Elektrik); 3. <Labor> relay (Elektrizität); 4. <Phys, Telekom> relay
Relais n mit mittlerer Ruhelage <Elektriz> (AE) center stable relay, (BE) centre stable relay
Relais n mit Schutzrohrkontakten <Telekom> dry reed relay
Relais n mit Zeitverriegelung <Elektriz> time-locking relay
Relaisanker m <Elektrotech> relay armature
Relaiskern m <Elektrotech> relay core
Relaiskontakt m <Elektriz, Elektrotech> relay contact

Relaismagnet

Relaismagnet m <Elektrotech> relay magnet
Relaissatz m <Telekom> relay set
Relaisschaltsystem n <Elektrotech> relay switching system
Relaisspule f <Elektrotech> relay coil
Relaisstelle f **Basisstation-Mobilteilnehmer** <Mobilkom> base-to-mobile relay
Relaisstelle f **Mobilteilnehmer-Basisstation** <Telekom> mobile-to-base relay
Relaissummen n <Elektrotech> relay hum
Relaissystem n <Telekom> relay system
Relaiswicklung f <Elektriz> relay winding
Relaiszähler m <Gerät> magnetic counter
Relation f <Comp & DV, Math> relation
relational adj <Comp & DV> relational
relationale Abfrage f <Comp & DV> relational-style query
relationale Datenbank f <Comp & DV> RDB, relational database
relationaler Graph m <Künstl Int> relation graph
relationaler Prozessor m <Comp & DV> relational processor
Relationssymbol n <Künstl Int> relation symbol, relational operator
Relativbewegung f <Phys> relative motion
relative Adressierung f <Comp & DV> relative addressing
relative Atommasse f 1. <Kerntech> atomic weight; 2. <Phys> atomic weight, relative atomic mass; 3. <Strahlphys> atomic weight
relative Bezugsdämpfung f <Aufnahme> loudness volume equivalent
relative Byteadresse f (RBA) <Comp & DV> relative byte address (RBA)
relative Dichte f 1. <Kohlen> relative density; 2. <Kunststoff> relative density, specific gravity
relative Dielektrizitätskonstante f <Elektrotech, Phys> relative dielectric constant, relative permittivity
relative Einschaltdauer f <Elektriz> relative cyclic duration factor
relative Empfindlichkeit f <Gerät> relative sensitivity
relative Feuchte f <Heiz & Kälte, Kunststoff, Textil> relative humidity
relative Feuchtigkeit f <Heiz & Kälte, Maschinen, Papier, Textil> relative humidity
relative Flughöhe f <Lufttrans> relative altitude
relative Häufigkeit f 1. <Comp & DV, Qual> relative frequency; 2. <Math> relative frequency (Statistik); 3. <Phys> relative abundance
relative Leistung f <Elektrotech> relative power
relative Luftfeuchte f <Heiz & Kälte, Phys> relative humidity
relative Luftfeuchtigkeit f <Wassertrans, Wasserversorg> relative humidity
relative Molekularmasse f <Phys> relative molecular mass
relative Öffnung f <Phys> relative aperture
relative Permeabilität f <Elektriz, Elektrotech, Phys> relative permeability
relative Permittivität f <Elektriz> relative permittivity
relative Profiltiefe f <Lufttrans> chord ratio
relative Signalamplitude f <Elektronik> relative signal amplitude
relative Spannung f <Elektrotech> negative voltage, positive voltage
relative spektrale Strahlenverteilung f <Nichtfoss Energ> relative spectral energy distribution
relative Steigung f <Lufttrans> pitch diameter ratio (Propeller)
relative Unterschiedsschwelle f <Akustik> relative difference limit
relative Wassergeschwindigkeit f <Nichtfoss Energ> relative water velocity
relative Zähigkeit f <Strömphys> relative viscosity
Relativempfindlichkeit f <Gerät> relative sensitivity
Relativempfindlichkeit f **eines Wandlers** <Akustik> relative sensitivity of a transducer
relativer Brechungsindex m <Optik, Phys> relative refractive index
relativer Fehler m 1. <Comp & DV> relative error; 2. <Math> relative error (Statistik)
relativer Trittschallpegel m <Akustik> normalized impact sound level
relativer Verdichtungsgrad m <Phys> relative pressure coefficient
relativer Winkelabweichungsgewinn m <Akustik> relative angular deviation gain
relativer Winkelabweichungsverlust m <Akustik> relative angular deviation loss
relatives lineares Bremsvermögen n <Phys> relative linear stopping power
relatives Massenbremsvermögen n <Phys> relative mass stopping power
Relativfilter n <Elektronik> constant-percentage bandwidth filter
Relativgeschwindigkeit f 1. <Fernseh> head-to-tape speed; 2. <Kohlen, Phys> relative velocity
relativistisch adj <Raumfahrt> relativistic
relativistische Mechanik f <Phys, Raumfahrt> relativistic mechanics
relativistisches Partikel n <Raumfahrt> relativistic particle
Relativität f <Raumfahrt> relativity
Relativitätseffekt m <Raumfahrt> relativity effect
Relaxation f <Kunststoff> relaxation
Relaxationsschwingung f <Elektriz> relaxation oscillation
Relaxationszeit f 1. <Akustik> relaxation time; 2. <Elektriz> relaxation time
Release-Level m <Comp & DV> release level
Release-Papier n <Druck> release paper
Reliabilität f 1. <Ergon> reliability (von psychologischen Tests); 2. <Metrol> reliability (Statistik)
Relief n <Fertig> embossing
Reliefdruck m <Druck> relief printing
Reliefgravur f <Ker & Glas> engraving in relief
Reliefwebart f <Textil> relief weave
Reling f <Wassertrans> handrail; rail (Schiffbau)
Relingslog n <Wassertrans> Lagrangian drifter (Meereskunde); Dutchman's log (Navigation)
Relingsstütze f <Wassertrans> railing stanchion
Relingsstützenfuß m <Wassertrans> stanchion deck fitting
Reluktanz f 1. <Elektriz, Elektrotech> magnetic resistance, reluctance; 2. <Phys> reluctance
Reluktanzmotor m <Elektriz, Elektrotech, Trans> reluctance motor
Reluktanzschrittmotor m <Elektrotech> reluctance switch motor, stepping motor, variable reluctance motor
Reluktivität f <Elektriz, Phys> reluctivity
remanente Feldstärke f <Elektriz> remanent flux density
remanente Ladung f <Elektrotech> remanent charge
Remanenz f 1. <Aufnahme> remanence (Restmagnetisierung); 2. <Elektriz, Elektrotech> remanence; 3. <Phys> remanence, retentivity
Remanenzinduktion f <Elektrotech> remanent induction
Remissionsgrad m <Optik> reflectance
Rendezvous n <Raumfahrt> rendezvous • **ein Rendezvous durchführen** <Raumfahrt> rendezvous

Rendezvousbahn f <Raumfahrt> rendezvous trajectory
Rendezvousmanöver n <Raumfahrt> (AE) rendezvous maneuver, (BE) rendezvous manoeuvre
Rendezvousradar m <Funkort, Raumfahrt> rendezvous radar
Rendezvousverfahren n <Raumfahrt> rendezvous procedure
Renette f <Lebensmittel> rennet (Apfelsorte)
Rennbahn-Mikrotron n <Teilphys> race track microtron
Rennherd m <Metall> bloomery hearth
Rennin n <Lebensmittel> rennin
Rennkraftstoff m <Kerntech> racing fuel
Reoxidation f <Chemie> reoxidation
Reparatur f <Kfztech, Wassertrans> repair
Reparaturanleitung f <Kfztech> repair manual
Reparaturdauer f <Comp & DV> repair time
Reparaturlack m <Kunststoff> refinishing paint
Reparaturwerkstatt f 1. <Kfztech> garage with workshop, service station; 2. <Trans> garage
reparieren v <Bau, Kfztech, Wassertrans> make good, repair
Repertoire n <Comp & DV> repertoire
repetierbar adj <Ergon, Fertig> repeatable
Repetierbarkeit f 1. <Ergon, Fertig> repeatability; 2. <Maschinen> positioning repeatability
Replikation f **auf Feldebene** <Comp & DV> field-level replication
Reportgenerierungswerkzeug n <Comp & DV> report generation tool
Repository n <Comp & DV> repository
repräsentative Probe f <Metall> representative sample
Reprise f <Textil, Werkprüf> moisture regain (Textil)
Reproduktionskamera f <Druck> process camera, reproduction camera
reproduzierbare Messung f <Metrol> repeatable measurement
Reproduzierbarkeit f <Gerät, Qual> repeatability
reproduzieren v 1. <Aufnahme> reproduce; 2. <Foto> duplicate
reprofähige Vorlage f <Druck> CRC, camera ready copy
Reprographie f <Druck> reprography
Reprographik f <Comp & DV> reprographics
Reprokamera f 1. <Druck> process camera, reproduction camera; 2. <Foto> copy camera
Reprostativ n <Foto> copying stand, copypod
Reprotechnik f <Konstzeich> reprographic technique
Repulsion f <Elektrotech> repulsion
Repulsionskraft f <Metall> repulsive power
Repulsionsmotor m <Elektriz, Elektrotech> repulsion motor
Resazurin n <Chemie> resazurin
Reserpin n <Chemie> reserpine
Reserve f 1. <Comp & DV> stand-by; 2. <Erdöl> backup; 3. <Funktech> backup (Batterie); 4. <Meerschmutz, Telekom> stand-by
Reserveablaufteil n <Telekom> backup supervisor
Reserveanknotung f <Textil> magazine creeling
Reserveausrüstung f <Comp & DV> stand-by equipment
Reservebatterie f <Elektrotech> reserve battery, stand-by battery
Reservefäden mpl <Textil> transfer tails
Reservekanister m <Kfztech> jerrican, spare can
Reservekapazität f <Trans> RC, reserve capacity
Reservekessel m <Heiz & Kälte> stand-by boiler
Reservekühleinrichtung f <Kerntech> stand-by cooling system
Reserveleitung f <Telekom> spare line
Reservenummer f <Telekom> spare number
Reservepeilung f <Raumfahrt> backup bearing

Reserveprozessor m <Telekom> stand-by processor
Reservereaktor m <Kerntech> backup reactor
Reserveschutzvorrichtung f <Sicherheit> back-up safety device
Reserveseil n <Erdöl> backup line
Reservespur f <Comp & DV> spare track
Reservesystem n <Telekom> stand-by system
Reservetreibstoffpumpe f <Raumfahrt> fuel backup pump
reserviertes Wort n <Comp & DV> reserved word
Reservierungs-Vielfachzugriffsverfahren n <Telekom> reservation multiple access
Reservoir n 1. <Bau> reservoir; 2. <Hydraul> forebay; 3. <Kfztech> reservoir (Öl, Kraftstoff); 4. <Nichtfoss Energ> reservoir; 5. <Wasserversorg> reservoir, water tank
Reset n <Comp & DV> reset
Reset-Knopf m <Fernseh> reset knob
Reset-Set-... <Elektronik> reset-set
Reset-Set-Flipflop n (RS-Flipflop) <Elektronik> reset-set flip-flop (RS flip-flop)
Reset-Set-Kippschaltung f (RS-Kippschaltung) <Elektronik> reset-set toggle (RS toggle)
Reset-Taste f <Comp & DV> reset button
resident adj <Comp & DV> resident
residenter Programmspeicher m <Comp & DV> resident program storage
residentes Programm n <Comp & DV> resident program
Residuum n 1. <Math> residue (Rest); 2. <Meerschmutz> residue
Residuumarithmetik f <Comp & DV> residue arithmetic
Resit n 1. <Fertig> resite; 2. <Kunststoff> C-stage resin, resite
Resitol n 1. <Fertig> resitol; 2. <Kunststoff> B-stage resin
Resol n <Kunststoff> A-stage resin, resol
resonant adj <Phys> resonant
resonante Energieübertragung f <Strahlphys> resonant energy transfer
Resonanz f <Aufnahme, Elektriz, Elektronik, Funktech, Phys, Teilphys, Telekom, Wellphys> resonance
Resonanzabsorption f <Telekom> resonance absorption
Resonanzanhebung f <Elektronik> peaking
Resonanzbrücke f <Elektriz> resonance bridge
Resonanzdämpfer m <Sicherheit> (AE) resonance muffler, (BE) resonance silencer
Resonanzeinfang m <Kerntech> resonance capture (von Neutronen)
Resonanzfilter n <Aufnahme, Strahlphys> resonance filter
Resonanzfluchtwahrscheinlichkeit f (p) <Kerntech> resonance escape probability (p)
Resonanzfrequenz f 1. <Akustik, Elektronik, Funktech, Telekom, Wellphys> (fR) resonant frequency (fR); 2. <Strahlphys> resonance frequency
Resonanzhohlraum m <Optik> resonant cavity
Resonanzkörper m <Elektronik> resonant cavity, resonator
Resonanzkreis m <Elektronik, Funktech, Telekom> resonant circuit
Resonanzkreis-Induktionsschleifendetektor m <Trans> resonant-circuit induction loop detector
Resonanzkurve f <Funktech> resonance curve
Resonanzleitung f <Elektronik> resonant line
Resonanzleitungsgenerator m <Elektronik> resonant-line oscillator
Resonanzlinie f <Strahlphys> resonance line
Resonanzneutronenzähler m <Strahlphys> resonance neutron detector
Resonanzpeak m <Strahlphys> resonance peak
Resonanzschaltung f **mit Eisenkernspule** <Elektrotech> ferroresonance circuit

Resonanzschärfe

Resonanzschärfe f <Phys> sharpness of resonance
Resonanzschwingkreis m <Phys> resonant circuit
Resonanzsieb n <Abfall, Kohlen> resonance screen
Resonanzspektrum n <Wellphys> resonance spectrum
Resonanzspitze f <Aufnahme> resonance peak
Resonanzstrahlung f <Strahlphys> resonance radiation
Resonanztransformator m <Elektrotech> tuned transformer
Resonanzverbreiterung f von Spektrallinien <Strahlphys> resonance broadening of spectral lines
Resonanzverstärker m <Elektronik> tuned amplifier
Resonanzwand f <Akustik> baffle
Resonanzwellenmesser m <Funktech> absorption wavemeter
Resonanzzustände mpl **optischer Hohlraumresonatoren** <Strahlphys> resonant modes of optical cavities
Resonator m <Telekom> resonator
Resonatorgitter n <Elektronik> resonator grid
Resorcin n <Kunststoff> resorcinol, resorcin
Resorcinharz n <Kunststoff> resorcinol resin, resorcinol formaldehyde resin
Resorcinol n <Chemie> resorcin, resorcinol
Resorcinphthalein n <Chemie> fluorescein
Resorcyl... <Chemie> resorcylic
Resorufin n <Chemie> resorufine
Ressourcen fpl <Comp & DV, Meerschmutz, Umweltschmutz> resources
Ressourcenposition f <Comp & DV> resource location
Ressourcenverwaltung f <Comp & DV> operations management
Rest m <Comp & DV, Math> remainder, residue
Rest... 1. <Anstrich> residual; 2. <Chemie> residuary
Restabstand m <Elektrotech> residual gap
Restart m <Comp & DV> restart, program restart
Restauftrieb m <Wassertrans> reserve buoyancy (Schiffkonstruktion)
Restaustenit m <Metall> rest austenite
Restbitfehlerrate f <Telekom> residual bit error rate
Restbogenteile mpl <Druck> oddments
Restdämpfung f <Telekom> overall loss
Restentladung f <Elektrotech> residual discharge
Restfalte f <Konstzeich> residual fold
Restfehlerhäufigkeit f <Comp & DV, Telekom> residual error rate
Restfehlerrate f <Comp & DV, Telekom> residual error rate
Restfeuchte f <Heiz & Kälte> residual moisture
Restfeuchtigkeit f <Heiz & Kälte, Verpack> residual moisture
Restflussdichte f <Elektriz> residual flux density
Restfrequenzmodulation f <Elektronik> residual frequency modulation (Rest-FM)
Restgas n <Elektronik, Umweltschmutz> residual gas
Resthärte f <Fertig> residual hardness (Kunststoffinstallationen)
Restkapazität f <Elektrotech> residual capacitance
Restklasse f <Math> residue class
Restklassenrechnung f <Math> congruence arithmetic, modular arithmetic
Restladung f <Elektriz, Elektrotech, Fernseh> residual charge
restlich adj 1. <Chemie> residuary; 2. <Comp & DV> residual
Restluftmenge f <Heiz & Kälte> residual air volume
Restmagnetisierung f 1. <Phys> remanence; 2. <Strahlphys> residual magnetization
Restmagnetismus m <Elektriz> residual magnetism
Restmasse f (m0) <Kerntech> rest mass (m0)
Restmengenpumpe f <Meerschmutz> stripping pump

Restmüll m <Abfall> tailings (nicht verwertbare Abfallstoffe)
restrukturieren v <Comp & DV> reposition
Restschrumpf m <Textil> residual shrinkage
Restseitenband n 1. <Elektronik> vestigial sideband (VSB); 2. <Fernseh> vestigial sideband; 3. <Telekom> residual sideband, vestigial sideband
Restseitenbandfilter n <Elektronik> vestigial sideband filter
Restseitenbandsignal n <Elektronik> vestigial sideband signal
Restsignal n <Comp & DV> residual noise
Restsilber n <Foto> residual silver
Restspannung f <Bau, Ker & Glas> residual stress
Reststoff m <Abfall> remainder
Reststoffdeponie f <Abfall> residue landfill
Reststrom m <Elektriz, Elektrotech> residual current
Reststromschutzvorrichtung f <Sicherheit> residual current protective device
Restunwucht f <Maschinen> remaining unbalance
Restwassergehalt m <Optik> residual water content
Restwiderstand m <Elektrotech> residual resistance
Resultante f <Maschinen> resultant
Resultierende f <Phys> resultant
Resultierende f einer Vektoraddition <Math> vector resultant
resultierende Kraft f <Bau, Mechan> resultant force
resultierender Schub m <Lufttrans> gross thrust
resynchronisieren v <Telekom> resynchronize
Retarder m <Kfztech, Kunststoff> retarder
retardieren v <Chemie> retard
retardiertes Potenzial n <Phys> retarded potential
Reten n <Chemie> methylisopropylphenanthrene, retene
Retention f <Chemie, Elektrotech> retention
Retikulum n <Optik> reticle, reticule
Retorte f <Chemtech, Erdöl, Labor> retort • **in Retorte destillieren** <Chemtech> retort
Retortenklemme f <Labor> retort clamp
Retortenkohle f <Kohlen> retort coal
Retortenschwelen n <Chemie> retorting
Retrieval-Sprache f <Comp & DV> retrieval language
Retro-Abfolge f <Raumfahrt> retrosequence
Retro-Gradation f <Lebensmittel> retrogradation
retrograde Metamorphose f <Nichtfoss Energ> retrograde metamorphism
Retro-Pack n <Raumfahrt> retropack
Retro-Rakete f <Raumfahrt> retro-rocket
Retroschub m <Lufttrans> reverse thrust
retten v 1. <Papier> save; 2. <Sicherheit> rescue
Rettung f 1. <Lufttrans, Sicherheit> recovery, rescue; 2. <Wassertrans> recovery; rescue (Notfall)
Rettungsausrüstung f <Sicherheit> life-saving equipment
Rettungsboje f <Wassertrans> lifebuoy
Rettungsboot n <Wassertrans> lifeboat
Rettungsbootsmann m <Wassertrans> lifeboatman
Rettungsbootsstation f <Funktech, Wassertrans> lifeboat station
Rettungsbootsübung f <Wassertrans> boat drill, lifeboat drill
Rettungsdienst m **und Feuerwehr** f <Lufttrans, Sicherheit, Wassertrans> rescue and firefighting service
Rettungseinrichtung f <Sicherheit> escape device, rescue apparatus
Rettungsfahrzeug n <Kfztech> rescue vehicle
Rettungsfallschirm m <Lufttrans> emergency parachute, escape parachute
Rettungsfloß n <Sicherheit, Wassertrans> life raft

Rettungsgerät n 1. <Lufttrans> rescue apparatus, rescue equipment; 2. <Sicherheit> escape device, life-saving equipment, rescue equipment; 3. <Wassertrans> life-saving apparatus, rescue apparatus, rescue device, rescue equipment
Rettungsgerät n **für Brandeinsätze** <Sicherheit> fire rescue appliance
Rettungsgerät-Funkstelle f <Funktech> survival craft station
Rettungsgurt m <Sicherheit> fall-arresting lanyard
Rettungsgürtel m 1. <Bau> (BE) life belt; 2. <Wassertrans> life belt
Rettungshubschrauber m <Lufttrans> rescue helicopter
Rettungsinsel f <Wassertrans> life raft
Rettungsleine f <Wassertrans> lifeline
Rettungsring m 1. <Sicherheit> life-saver; 2. <Wassertrans> lifebuoy
Rettungsstation f <Sicherheit> rescue station
Rettungsturm m **für Startphase** <Raumfahrt> launch phase escape tower
Rettungsweg m **bei Feuer** <Lufttrans> fire rescue path
Rettungsweste f <Wassertrans> (BE) life jacket, (AE) life preserver, (BE) life vest
Return-Taste f <Comp & DV> return key
retuschieren v <Foto> retouch
Retuschieren n <Druck, Foto> retouching
Retuschieren n **von Bildern** <Druck> image retouching
Retuschierpinsel m <Druck> retouching brush
Reuse f <Wassertrans> pot (Fischerei)
reversibel adj <Thermod> reversible
reversible Abschaltung f <Kerntech> reversible shutdown
reversibler Transducer m <Elektrotech> reversible transducer
reversibler Wandler m <Elektrotech> reversible transducer
reversieren v <Elektronik> reverse
Reversierwalzwerk n <Fertig> reversing mill
Reversion f <Kunststoff> reversion
Reversosmose f <Chemtech> reverse osmosis
revidieren v <Druck> revise
Revision f <Qual> audit
Revisionsabzug m <Druck> press proof
Revolution f <Geom> revolution
Revolver m <Maschinen> capstan
Revolveranschlag m <Maschinen> multiposition stop
Revolverbank f <Mechan> lathe
Revolverblende f <Optik> revolving diaphragm
Revolverbohrmaschine f <Maschinen> turret-type drilling machine
Revolverdrehbank f <Mechan> capstan lathe, turret lathe
Revolverdrehen n <Maschinen> capstan turning
Revolverdrehmaschine f 1. <Fertig> capstan lathe; 2. <Maschinen> turret lathe
Revolverkopf m 1. <Fertig> turret; 2. <Maschinen> capstan tool head, turret, turret head; 3. <Mechan> turret • **ohne Revolverkopf** <Fertig> nonturret
Revolverkopfbohrmaschine f <Maschinen> turret-type drilling machine
Revolverkopfdrehmaschine f <Maschinen> turret lathe
Revolverkopfschaltstellung f <Fertig> turret head indexing position
Revolverlochzange f <Maschinen> revolving head punch
Revolverschieber m <Maschinen> turret slide
Revolverschlitten m <Maschinen> ram
Reynoldsspannung f <Strömphys> Reynolds stress
Reynoldstransporttheorem n <Strömphys> Reynolds' transport theorem
Reynoldszahl f (Re) 1. <Hydraul> Reynolds number, Re; 2. <Lufttrans> Reynolds number, Re (Aerodynamik); 3. <Nichtfoss Energ, Phys, Strömphys> Reynolds number, Re
Reynoldszahlbereich m <Strömphys> Reynolds number region
Reyon n <Kunststoff> rayon
Rezension f <Druck> review
Rezeptierung f <Kunststoff> formulation
Rezeptor m <Ergon> receptor
Rezeptur f 1. <Bau> mix design; 2. <Kunststoff> formulation
Rezepturänderung f <Kunststoff> reformulation
Rezipientenglocke f <Chemie> bell jar
reziprok adj 1. <Comp & DV, Geom> reciprocal; 2. <Maschinen> converse; 3. <Math> reciprocal
Reziproke f <Math> reciprocal
reziproke Gleitzahl f <Phys> lift-drag ratio (Verhältnis von Auftrieb zu Widerstand)
reziproke Periode f <Kerntech> reciprocal period
reziproke Schaltung f <Telekom> reciprocal circuit
reziproke Steifigkeit f <Akustik> compliance
reziproker Wandler m <Akustik> reciprocal transducer
reziproker Wert m 1. <Geom> reciprocal; 2. <Math> conjugate, inverse proportion, inverse ratio, reciprocal
reziprok-quadratisch adj <Phys> inverse-square
Reziprozitätssatz m <Elektrotech, Phys> reciprocity theorem
Rf (Rutherfordium) <Chemie> Rf (rutherfordium)
RF (Radiofrequenz) <Aufnahme, Elektriz, Elektronik, Fernseh, Funktech, Telekom> HF (high frequency); RF (radio frequency)
RG (elektronischer Rauschgenerator) <Elektronik, Gerät> ENG (electronic noise generator)
RGB (rot-grün-blau) <Comp & DV, Fernseh> RGB (red-green-blue)
RGB-Bildschirm m <Comp & DV> RGB monitor
RGB-Eingabe f <Fernseh> RGB input
RGB-Monitor m <Comp & DV, Fernseh> RGB monitor
RGB-System n <Comp & DV> RGB system
R-Gespräch n <Telekom> (AE) collect call, (BE) reverse charge call, transferred charge call • **ein R-Gespräch führen** <Telekom> (AE) make a collect call, (BE) make a reverse charge call
Rh (Rhodium) <Chemie> Rh (rhodium)
RH (Hall'scher Koeffizient) <Phys> RH (Hall coefficient)
Rhamnetin n <Chemie> methylquercitin, rhamnetin
Rhamnit m <Chemie> rhamnite, rhamnitol
Rhamnose f <Chemie> rhamnose
Rhenat n <Chemie> perrhenate
Rhenium n (Re) <Chemie> rhenium (Re)
Rhenium... <Chemie> rhenic
Rheologie f <Kohlen, Kunststoff, Metall, Phys, Strömphys> rheology
rheologische Eigenschaften fpl <Strömphys> rheological properties
rheologische Variable f <Metall> rheological variable
Rheostat m <Elektriz, Elektrotech, Labor> rheostat
Rheostatgleitschieber m <Elektriz> rheostat slider
Rheostatkontaktschleifer m <Elektriz> rheostat slider
RHIC <Teilphys> RHIC, relativistic heavy ion collider
Rhodanid n <Chemie> thiocyanate
Rhodeorhetin n <Chemie> convolvulin, rhodeorhetin
Rhodium n (Rh) <Chemie> rhodium (Rh)
Rhomboid n <Geom> rhomboid
rhomboidisch adj <Geom> rhomboid, rhomboidal
Rhombus m <Geom> rhomb, rhombus
Rhombusantenne f <Funktech> (BE) rhombic aerial, (AE) rhombic antenna
Rhombuswicklung f <Elektriz> diamond winding
Rhumbatron n <Elektronik> rhumbatron (Mikrowellen)

Rhythmus

Rhythmus m <Akustik> rhythm
Richtantenne f 1. <Elektronik> beam aerial, beam antenna; 2. <Fernseh> directional antenna; 3. <Funktech> beam aerial, beam antenna, directional array, directional antenna; 4. <Raumfahrt> directional antenna, shaped beam antenna; 5. <Telekom> directional antenna
Richtantenne f **mit passiven Strahlern** <Funktech> parasitic array
Richtantennengruppe f **mit kleinen Elementabständen** <Funktech> close-spaced array
Richtantennen-Öffnungswinkel m <Raumfahrt> half-power beamwidth *(am Half-Power-Punkt)*
Richtantrieb m <Raumfahrt> director
Richtapparat m <Maschinen> *(AE)* leveling machine, *(BE)* levelling machine
Richtbake f <Wassertrans> guide beacon *(Navigation)*
Richtblei n <Bau> plummet
Richtbohren n <Erdöl> directional drilling *(Bohrtechnik)*
Richtcharakteristik f 1. <Akustik> directivity pattern; 2. <Elektronik> directional response; 3. <Funktech> beam pattern, pattern *(Antenne)*; directional diagram, directivity pattern
Richtcharakteristikwandler m <Elektronik> polar pattern converter
Richtdiagramm n 1. <Akustik> directivity pattern; 2. <Funktech> directional diagram, directivity pattern; pattern *(Antenne)*
richten v 1. <Bau> straighten; 2. <Elektronik> direct *(Signale)*; 3. <Fertig> flatten, level *(Blech)*; 4. <Maschinen> level, straighten
Richten n 1. <Bau> straightening; 2. <Elektronik> directing *(von Signalen)*; 3. <Fertig> flattening, *(AE)* leveling, *(BE)* levelling *(Blech)*; 4. <Maschinen> *(AE)* leveling, *(BE)* levelling, straightening
Richterskale f <Bau> Richter scale
Richtfaktor m <Akustik> directivity factor
Richtfehler m <Funktech> pointing error *(Antenne)*
Richtfeuer n <Wassertrans> guide light, range light *(Navigation)*
Richtfunk m <Funktech> microwave
Richtfunkantenne f <Telekom> microwave aerial, microwave antenna
Richtfunkbake f <Funktech> directive radio beacon
Richtfunkstrahl m <Funktech> microwave beam
Richtfunkstrecke f <Fernseh, Telekom> microwave link
Richtfunksystem n <Telekom> microwave system, radio relay system
Richtfunktechnik f <Funktech> radio link engineering, radio relay engineering
Richtfunktion f <Elektronik, Funktech> gain function *(Antenne)*
Richtfunkturm m <Telekom> microwave tower
Richtfunkübertragung f <Elektrotech, Telekom> microwave transmission
Richtfunkverbindung f <Fernseh, Telekom> microwave link
Richtfunkverbindung f **innerhalb der Radiosichtweite** <Telekom> line-of-sight radio relay link
Richtfunkverbindung f **mit annähernder Radiosichtweite** <Funktech, Telekom> near line-of-sight radio relay system
Richtfunkverbindung f **mit einem Funkfeld** <Funktech, Telekom> single hop radio link, single hop radio relay link
Richtgeschwindigkeit f <Trans> advisory speed, recommended speed
Richthöhe f <Mechan> elevation
Richtigkeit f 1. <Metrol> correctness; 2. <Qual> trueness

Richtkoppler m <Elektrotech, Optik, Phys, Telekom> directional coupler
Richtkreisel m <Trans> directional gyro
Richtlatte f <Bau> batten
Richtleiter m <Phys> isolator
Richtlinie f <Elektriz, Kfztech> guideline
Richtmaschine f <Maschinen> *(AE)* leveling machine, *(BE)* levelling machine, straightener, straightening machine
Richtmikrofon n <Akustik, Aufnahme> directional microphone, unidirectional microphone
Richtplatte f <Maschinen> gib
Richtrelais n <Elektrotech> directional relay
Richtseezeichen n <Wassertrans> guiding mark
Richtsender m <Fernseh, Funktech> directional-beam transmitter
Richtspannung f <Elektrotech> rectified voltage
Richtstrahl m 1. <Elektrotech> directional beam; 2. <Funktech> beam *(Richtantenne)*; 3. <Raumfahrt> beam; 4. <Telekom> directional beam
Richtstrahlanzeige f <Elektronik> directed-beam display
Richtstrahler m <Phys> reflector • **mit Richtstrahler senden** <Funktech> beam
Richtstrahlröhre f <Elektronik> shaped beam tube
Richtstrahlumschaltung f <Raumfahrt> beam switching *(Weltraumfunk)*
Richtstrom m <Elektrotech> rectified current
Richtung f 1. <Comp & DV> route, sense; 2. <Sicherheit> direction *(Flaschenzug)* • **in zwei Richtungen arbeitend** <Comp & DV> bidirectional
Richtungsabhängigkeit f <Fertig> aelotropy, anisotropy
richtungsempfindliches Relais n <Elektriz> directional relay
Richtungsfahrbahn f <Trans> lane
Richtungsfaktor m <Akustik> directivity factor
Richtungsgleis n <Eisenbahn> sorting siding
Richtungskoppler m <Optik> directional coupler
Richtungskriterium n <Comp & DV> routing criterion
Richtungsleitung f <Telekom> isolator *(Mikrowellen)*
Richtungsmaß n <Akustik> directivity index
Richtungspeilung f <Phys> direction finding
Richtungspolarisation f <Phys> oriental polarization
Richtungsschalter m <Elektriz> direction switch
Richtungsschaufel f <Lufttrans> compressor blade, stator vane *(Kompressor)*
Richtungsschild n <Bau> direction sign
Richtungsspurkapazität f <Trans> tidal capacity
Richtungsstabilität f <Raumfahrt> directional stability
Richtungstaktschrift f <Comp & DV> PE, phase encoding
Richtungsumkehr f <Maschinen> reversing the motion
Richtungswechsel m <Trans> tidal flow
Richtungswechselbetrieb m <Trans> tidal flow system
Richtungswechselspur f <Trans> reversible lane, tidal flow lane
Richtungsweiche f 1. <Elektronik> branching filter *(Richtfunkttechnik)*; directional filter *(Sende-Empfangsweiche)*; 2. <Telekom> directional filter
Richtungswinkel m <Bau> azimuth; bearing *(Kompass)*
Richtverlust m <Raumfahrt> pointing loss *(Weltraumfunk)*
Richtwaage f <Fertig, Metrol> spirit level
Richtwalze f <Maschinen, Metall> planisher, planishing roll
Richtwert m <Qual> standard value
Richtwerte mpl **für das Flankenspiel** <Fertig> recommended backlash
Richtwirkung f 1. <Fertig> directionality; 2. <Funktech, Raumfahrt> directivity
Richtwirkungsgrad m <Elektrotech> rectification efficiency

Richtwirkungsindex m (Di) <Akustik> directivity index (Di)
RIC-Verfahren n <Kunststoff> (AE) RIC molding, (BE) RIC moulding, (AE) runnerless injection compression molding, (BE) runnerless injection compression moulding
riechstoffbildend adj <Umweltschmutz> odorous
Riefe f 1. <Fertig> stria; wheel mark (von Rad); 2. <Ker & Glas> hackle mark; 3. <Maschinen> flute, ridge, score, score mark
riefen v <Fertig, Maschinen> ridge
Riefenbildung f <Fertig, Maschinen> scoring
Riefennachlauf m <Fertig> drag (Brennschneide)
Riefung f <Fertig> striation
Riegel m 1. <Bau> beam, lock bolt, shutter; bar (eines Schlosses); 2. <Kohlen> wale, waling; 3. <Mechan> tie rod; 4. <Sicherheit> lock; 5. <Wasserversorg> crosspiece (eines Schleusentors)
Riegelfachwerkwand f <Bau> framed partition
Riegelnut f <Eisenbahn> locking notch
Riegelwand f <Bau> framework wall
Riekediagramm n <Elektronik> Rieke diagram (Elektronenröhren)
Riemann'sche Geometrie f <Geom> Riemannian geometry
Riemen m 1. <Kfztech> belt; 2. <Maschinen> band, belt; 3. <Mechan, Papier> belt; 4. <Wassertrans> oar (Rudern) • **die Riemen einlegen** <Wassertrans> ship the oars (Rudern) • **einen Riemen ausrücken** <Maschinen> throw a belt off • **einen Riemen spannen** <Kfztech, Maschinen> tighten a belt
Riemenantrieb m <Heiz & Kälte, Mechan, Papier> belt drive
Riemenaufleger m <Fertig> belt mounter
Riemenbeanspruchung f <Maschinen> belt stress
Riemenbrett n <Bau> match lining
Riemendehnschlupf m <Fertig> belt creep
Riemendehnung f <Fertig> belt stretch
Riemenfett n <Maschinen> belt grease
Riemengabel f 1. <Fertig> belt fork; 2. <Maschinen> strap fork
Riemengetriebe n <Fertig> belt drive, belt transmission
riemengetrieben adj <Fertig> belt-driven
Riemenniet m <Fertig> belt rivet
Riemenreibung f <Maschinen> belt friction
Riemenscheibe f 1. <Fertig> belt pulley; 2. <Maschinen> band pulley, pulley, sheave
Riemenscheibendrehmaschine f <Maschinen> pulley lathe, pulley turning lathe
Riemenschlaufe f <Sicherheit> belt-type sling
Riemenschloss n <Papier> belt fastener
Riemenschlupf m <Maschinen> belt slip
Riemenschutz m <Maschinen> belt guard
Riemenspannrolle f <Fertig> idler
Riemenspannung f <Maschinen> belt tension
Riementrieb m 1. <Fertig> belting; 2. <Heiz & Kälte, Maschinen> belt drive
Riementrum n <Maschinen> end
Riemenverbinder m <Fertig> belt lacer, fastener
Riemenwerkstoff m <Kunststoff> belting
Riemenzug m <Fertig> belt tension
Rieselfeld n 1. <Abfall> irrigation field; 2. <Bau, Wasserversorg> sewage farm
Rieselhilfsmittel n <Lebensmittel> anticaking agent
Rieselkolonne f <Chemtech> washing tower
Rieselkühler m <Heiz & Kälte> irrigation cooler
Rieselsieb n <Ker & Glas> shower screen
Rieselturm m <Kerntech> gas washer
Rieselwände fpl <Umweltschmutz> scrubber walls
Riesenimpulslaser m <Elektronik> giant pulse laser

Riesenpulslaser m <Elektronik> Q-switch laser, Q-switched laser
Riet n <Textil> reed
Riff n <Wassertrans> reef
Riffelglas n <Bau> ribbed glass
Riffelmuster n <Textil> network design
riffeln v 1. <Bau> groove; 2. <Fertig> (AE) checker, (BE) chequer, serrate
Riffeln n 1. <Bau> fluting; 2. <Fertig> (BE) channelling
Riffelprobenteiler m <Kohlen> riffle sampler
Riffelung f 1. <Bau> flute; 2. <Fertig> corrugation; 3. <Ker & Glas> fluting; 4. <Konstzeich> (AE) checkering, (BE) chequering
Riffelwalze f 1. <Ker & Glas> crimper; 2. <Papier> fluter
Rille f 1. <Akustik> groove; 2. <Ker & Glas> flute; 3. <Maschinen> groove, ridge; 4. <Optik> groove • **mit Rille versehen** <Ker & Glas> edge with a groove
Rillen fpl <Metall> striation
rillen v <Druck> score
Rillenabstand m <Fertig> roughing width
Rillenanordnung f <Fertig> ragging
Rillenätzung f <Elektronik> V-groove etching
Rillenfehler m <Aufnahme> tracking error
Rillengeschwindigkeit f <Akustik, Aufnahme> groove speed
Rillenhohlleiter m <Funktech> corrugated waveguide (Mikrowellen)
Rillenkugellager n <Maschinen> deep-groove ball bearing, grooved ball bearing
Rillenquerschnitt m <Akustik> groove shape
Rillenscheibe f <Maschinen> grooved wheel, sheave
Rillenschiene f <Eisenbahn> grooved rail
Rillensicherung f <Akustik> groove guard
Rillenverzerrung f <Akustik> tracing distortion
Rillenwellenleiter m <Funktech> corrugated waveguide (Mikrowellen)
Rillenwinkel m <Aufnahme> groove angle
RIM-Verfahren n <Kunststoff> RIM, (AE) reaction injection molding, (BE) reaction injection moulding
Rinde f <Papier> bark
Rindenkessel m <Papier> bark boiler
Rindenpresse f <Papier> bark press
Rindenschälmaschine f <Papier> barker
Ring m 1. <Comp & DV> ring; 2. <Erdöl> annulus (Bohrloch); 3. <Fertig> bundle (Draht); 4. <Geom> torus; 5. <Hydraul> ring (Kolben); ring (Turbine); 6. <Ker & Glas> ring; 7. <Kfztech> ring (Kolben); 8. <Maschinen, Mechan> ring; 9. <Phys> fringe; 10. <Raumfahrt> annulus, ring
Ringanker m <Elektriz> ring armature
Ringanschluss m <Telekom> ring connection
Ringantenne f <Raumfahrt> halo
Ringbalg m <Fertig> liner (Kunststoffinstallationen)
ringbildender Tscherenkov-Zähler m (RICH) <Teilphys> ring-imaging Cherenkov counter (RICH)
Ringblitz m <Foto> ring flash
Ringbolzen m 1. <Maschinen> lifting eyebolt; 2. <Wassertrans> ring bolt (Deckbeschläge)
Ringbrenner m 1. <Fertig> ring burner; 2. <Maschinen> circular-type burner
Ringbuch n <Verpack> binder
Ring-Collider m <Teilphys> ring collider
Ringdehner m <Fertig> ring expander
Ringdraht m <Fertig> bundle wire
Ringdüse f <Fertig> ring nozzle
Ringfeder f 1. <Fertig> circlip; 2. <Maschinen> annular spring
ringförmig adj 1. <Fertig> annular, toroidal; 2. <Geom> ring-shaped; 3. <Mechan, Raumfahrt> annular (Raumschiff)

ringförmige

ringförmige Anlaufscheibe *f* <Maschinen> ring-type thrust washer
ringförmige Auskehlung *f* <Fertig> spool
ringförmige Schweißnaht *f* <Kerntech> ring weld
ringförmige Spaltzone *f* <Kerntech> annular core
ringförmige Ziehwanne *f* <Ker & Glas> annular bushing
ringförmiger Collider *m* <Teilphys> ring collider
ringförmiger Kohlelichtbogen *m* <Kerntech> tubular carbon arc
ringförmiger Luftspalt *m* <Kerntech> annular air gap
ringförmiger Reaktorkern *m* <Kerntech> ring core
ringförmiger Resonator *m* <Elektronik> annular resonator *(Mikrowellentechnik)*
ringförmiger Riss *m* <Ker & Glas> annular crack *(Flaschenoberfläche)*
ringförmiger Spalt *m* <Kerntech> annular gap
ringförmiges Brennelement *n* <Kerntech> annular fuel element
ringförmiges Filmsieden *n* <Kerntech> annular film boiling
Ringgehäuse *n* **der Brennkammer** <Kfztech, Lufttrans, Wassertrans> combustion chamber annular case
Ringkabelschuh *m* <Elektrotech> ring-type cable lug
Ringkanal *m* <Kerntech> annular channel
Ringkern *m* 1. <Elektriz> toroid core, torus; 2. <Elektrotech> toroidal core; 3. <Kerntech> toroidal core *(einer Spule)*
Ringkernspule *f* <Elektriz> toroidal coil
Ringkolbenzähler *m* <Gerät> oscillating piston flowmeter
Ringkopf *m* <Aufnahme, Fernseh> ring head
Ringkörper *m* <Fertig> torus
Ringläufer *m* <Textil> traveller
Ringlehre *f* <Maschinen> *(AE)* female gage, *(BE)* female gauge, *(AE)* ring gage, *(BE)* ring gauge
Ringleitung *f* 1. <Comp & DV> loop; 2. <Elektriz, Elektrotech> ring mains
Ringleitungskoppler *m* <Telekom> rat-race
Ringligkeit *f* <Textil> barriness in the weft
Ringmagnet *m* <Fernseh, Phys> annular magnet, ring magnet
Ringmaß *n* <Metrol> *(AE)* ring gage, *(BE)* ring gauge
Ringmesskammer *f* <Gerät> annular measuring chamber *(Normalblende)*
Ringmodulator *m* <Aufnahme> ring modulator
Ringmühle *f* <Chemtech> roller mill
Ringmutter *f* <Maschinen> ring nut
Ringnetz *n* 1. <Comp & DV> loop network, ring network; 2. <Telekom> ring network
Ringnetzleitungssystem *n* <Elektriz> ring main system
Ringnut *f* 1. <Fertig> annular groove; 2. <Maschinen> ring groove
Ringofen *m* <Fertig, Ker & Glas> annular kiln
Ringquetschung *f* <Kerntech> toroidal pinch effect
Ringraum *m* <Erdöl> annular space *(Bohrloch)*
Ringschieben *n* <Comp & DV> circular shift
Ringschieber *m* <Fertig, Maschinen> annular slide valve
Ringschlüssel *m* 1. <Kfztech> *(BE)* box spanner, box wrench, *(BE)* ring spanner *(Werkzeug)*; 2. <Maschinen, Mechan> *(BE)* box spanner, box wrench, *(BE)* ring spanner
ringschmieden *v* <Fertig> saddle
Ringschmieden *n* <Fertig> saddling
Ringschmierung *f* <Maschinen> ring lubrication
Ringschraube *f* 1. <Kfztech> eyebolt *(Kupplung)*; 2. <Maschinen> lifting screw, ring bolt
Ringspalt *m* 1. <Erdöl> annulus; 2. <Fertig> annular clearance, radial clearance
Ringspant *m* 1. <Lufttrans> ring frame *(Rumpf des Luftfahrzeugs)*; 2. <Raumfahrt> annular rib

Ringspeiseleitung *f* <Elektriz> ring feeder
Ringspinnerei *f* <Textil> ring spinning
Ringspinngarn *n* <Textil> ring spun yarn
Ringspinnmaschine *f* <Textil> ring spinning frame
Ringspule *f* 1. <Elektriz> ring winding, toroidal coil; 2. <Kerntech> toroidal coil
Ringsteg *m* <Kfztech> piston land *(Motor)*
Ringstruktur *f* <Telekom> ring configuration, ring topology
Ringtisch *m* <Fertig> annular table *(Werkzeugmaschine)*
Ringtopologie *f* <Comp & DV> ring topology
Ringtransformator *m* <Elektrotech> toroidal transformer
Ringventil *n* 1. <Kfztech> poppet valve; 2. <Maschinen> annular valve
Ringwaage *f* <Gerät> ring balance
Ringwade *f* <Wassertrans> ring net *(Fischerei)*
Ringwicklung *f* <Elektriz> ring winding
Ringzähler *m* <Funktech, Gerät, Telekom> ring counter
Rinne *f* 1. <Bau> channel, trench; flute *(Rille)*; 2. <Fertig> channel *(Blechumformung)*; 3. <Mechan> chute, flute • **Rinnen ziehen** <Lufttrans> gutter
Rinnenkasten *m* <Bau> rainwater head
Rinnstein *m* <Bau> gutter
Rippe *f* 1. <Bau> web; 2. <Druck> rib; 3. <Fertig> stem; 4. <Kohlen, Lufttrans> rib *(Luftfahrzeug)*; 5. <Maschinen> fin; 6. <Mechan> fin, rib, vane; 7. <Papier, Textil> rib • **mit Rippen versehen** <Raumfahrt> rib
Rippe *f* **eines Gewebes** <Textil> wale
rippen *v* 1. <Fertig> fin *(Rohr)*; 2. <Raumfahrt> rib *(Raumschiff)*
Rippenabstand *m* <Heiz & Kälte> fin spacing
Rippenblech *n* <Metall> ribbed plate
Rippenheizkörper *m* <Heiz & Kälte> ribbed radiator
Rippenhülse *f* <Kerntech> finned can
Rippenkühler *m* 1. <Heiz & Kälte> ribbed cooler, ribbed radiator; 2. <Kfztech> ribbon cellular radiator
Rippenkühlung *f* <Kerntech> fin cooling, rib cooling
Rippenmarkierung *f* <Ker & Glas> rib mark
Rippenmuster *n* <Bau> reed
Rippenrohr *n* 1. <Fertig> gilled tube; 2. <Maschinen, Mechan> finned tube
Rippenröhrenkühler *m* <Eisenbahn> tube and fin radiator
Rippenrohrkühler *m* 1. <Heiz & Kälte> finned-tube cooler; 2. <Kfztech> finned-tube radiator
Rippenstich *m* <Textil> ribbed stitch
Rippenwirkungsgrad *m* <Heiz & Kälte> fin efficiency
Rippmasche *f* <Textil> ribbed stitch
RIP-Prozessor *m* <Druck> raster image processor, RIP
Ripptide *f* <Wassertrans> rip tide *(Gezeiten)*
Riser *m* <Erdöl> riser *(Bohrung, Leitungsbau)*
Risiko *n* <Sicherheit> hazard, risk
Risikoabschätzung *f* <Meerschmutz> risk assessment
Risikobewertung *f* <Meerschmutz> risk assessment
Risikofunktion *f* <Math> risk function *(Erwartung einer Verlustfunktion)*
Riss *m* 1. <Anstrich> crack; 2. <Bau> crack, fissure, interstice; shake *(Holz)*; projection *(Vermessung)*; break *(in Holz)*; 3. <Fertig> crevice, fissure, shake; 4. <Ker & Glas> check, tearing; 5. <Kohlen> crack; 6. <Maschinen> elevation; 7. <Mechan, Metall> crack; 8. <Papier> flaw; 9. <Raumfahrt> slit; 10. <Textil> breakage, break, burst; 11. <Wassertrans> plan *(Schiffbau)*; 12. <Wasserversorg> breach • **einen Riss zukitten** <Ker & Glas> stop a crack
Rissauffangtemperatur *f* <Metall, Qual> crack-arrest temperature
Rissausbreitungsgeschwindigkeit *f* <Kerntech> crack propagation rate
Rissausdehnungskraft *f* <Metall> crack extension force

Rissauslösung f <Kerntech> initiation of fracture
Rissbeständigkeit f <Kunststoff> crack resistance
Rissbildung f 1. <Bau, Kunststoff, Mechan> cracking; 2. <Metall> crack formation, cracking
Rissbreite f <Metall> width of splitting
Rissdetektor m <Fertig> crack detector
Risserweiterung f <Kerntech> crack-opening stretch
Rissgeschwindigkeit f <Metall, Qual> crack velocity
Risshaltetemperatur f <Metall, Qual> ductile-brittle transition temperature
rissige Dämmplatte f <Aufnahme> fissured acoustic tile
rissiger Lehm m <Bau> fissured clay
Rissingangsetzung f <Metall> crack initiation
Risskernbildung f <Metall> crack nucleation
Risslehre f <Konstzeich> template, templet
Risslinie f <Fertig> scribed line
Rissöffnungsverschiebung f <Kerntech> crack-opening displacement
Rissspitze f <Metall> crack tip
Risssucher m <Kerntech, Maschinen> crack detector
Risssuchgerät n <Kerntech> crack detector
Rissverzweigung f <Metall> crack branching
RIT (Empfängerfeinabstimmung) <Funktech> RIT (receiver incremental tuning)
Ritze f 1. <Kohlen> cut; 2. <Mechan> gap
Ritzel n 1. <Fertig> pinion (Getriebelehre); 2. <Kfztech> pinion (Getriebe); 3. <Maschinen> pinion, pinion wheel
Ritzelfräsmaschine f <Maschinen> pinion-cutting machine
Ritzelkopfhöhe f <Fertig> pinion addendum (Getriebelehre)
Ritzelwälzfräsmaschine f <Fertig> pinion-hobbing machine
Ritzelwelle f <Maschinen> pinion shaft
Ritzelwellenflansch m <Kfztech> pinion shaft flange (Triebstrang)
Ritzhärte f <Kunststoff> scratch resistance
Ritzhärteprüfer m <Phys> sclerometer
Ritzhärteprüfung f <Fertig> abrasion test
Ritz'sches Kombinationsprinzip n <Phys> Ritz combination principle
Rizinusöl n <Kunststoff> castor oil
RMS (quadratisches Mittel) <Aufnahme, Elektriz, Elektronik> rms (root mean square)
RMS-Frequenzhub m <Raumfahrt> rms frequency deviation (Weltraumfunk)
Rn (Radon) <Phys> Rn (radon)
Roadster m <Kfztech> roadster
Roaming-Fähigkeit f <Mobilkom> roaming capability
Roaming-Rufnummer f <Mobilkom> mobile station roaming number (zellulares Netz)
Roaming-Teilnehmer m <Mobilkom> roaming subscriber
Roboter m <Comp & DV, Fertig, Kontroll, Künstl Int> industrial robot, robot
Robotereinzäunung f <Sicherheit> robot enclosure
Roboterschutz m <Sicherheit> robot guarding
Roboterschutzvorrichtung f <Sicherheit> robot guarding
Robotersicherheit f <Sicherheit> robot safety
Robotersteuerung f <Fertig> robot control
Robotertechnik f <Comp & DV, Fertig> robot technology
Roboterteilebeschicker m <Fertig> robotic part loader
Robotik f <Comp & DV, Künstl Int> robotics
Robustheit f <Werkprüf> robustness
Rochellesalz n 1. <Elektrotech> Rochelle salt (piezoelektrisches Material); 2. <Lebensmittel> Rochelle salt
Rock & Roll-Aufnahme f <Aufnahme> rock-and-roll recording
Rock & Roll-Ausrüstung f <Aufnahme> rock-and-roll equipment
Rock & Roll-Mischung f <Aufnahme> rock-and-roll mixing
Rockwell-Härte f <Kunststoff> Rockwell hardness
Rockwell-Härteprüfmaschine f <Maschinen> Rockwell hardness testing machine
Rockwell-Härteprüfung f <Maschinen> Rockwell hardness test
Rod n <Metrol> perch, pole, rod
roden v <Bau> clear, grub, stub
roh adj 1. <Anstrich> rough; 2. <Lebensmittel> unrefined; 3. <Maschinen> unfinished
Roh... <Abfall, Bau, Erdöl, Fernseh, Kohlen, Konstzeich, Kunststoff, Lebensmittel, Wasserversorg> crude, raw, rough
Rohabwasser n <Wasserversorg> raw sewage
Rohband n <Fernseh> raw tape
Rohbau m <Bau> carcass, preliminary building works
Rohbauarbeiten fpl <Bau> rough work
Rohbearbeitung f <Kohlen> roughing
Rohbehauen n <Bau> rough dressing (Stein)
Rohbenzin f 1. <Erdöl> light distillates; 2. <Thermod> naphtha
Rohblatt n <Konstzeich> basic sheet
Rohblei n <Fertig> pig lead
Rohblock m <Fertig> bloom
Rohbrei m <Abfall> crude pulp
Rohdichte f 1. <Bau> bulk density, bulk; 2. <Kohlen> apparent density
rohe Unterlegscheibe f <Maschinen> blank washer
Roheisen n 1. <Fertig> pig iron; 2. <Metall> iron pig, pig iron
Roheisen n **für das Windfrischverfahren** <Fertig> Bessemer pig
Roheisen-Erz-Verfahren n <Fertig> pig-and-ore process
Roherz n 1. <Fertig> green ore; 2. <Kohlen> crude ore
Rohessigsäure f <Chemie> pyroacetic acid
Rohfaser f 1. <Ker & Glas> (AE) basic fiber, (BE) basic fibre; 2. <Lebensmittel> (AE) crude fiber, (BE) crude fibre, (AE) dietary fiber, (BE) dietary fibre
Rohfestigkeit f <Kunststoff> green strength
Rohform f <Ker & Glas> (AE) body mold, (BE) body mould
Rohgarn n <Textil> feeder yarn
Rohgewicht n <Papier> apparent specific gravity
Rohgips m <Bau> plaster rock, plaster stone
Rohglas n <Ker & Glas> base glass
Rohgrießkohle f <Kohlen> rough pea coal
Rohgussglas n <Ker & Glas> rough-cast glass
Rohgussmaß n <Konstzeich> rough-casting dimension
Rohgussspiegelglas n <Ker & Glas> rough-cast plate
Rohkante f <Ker & Glas> edge as cut
Rohkarosserie f <Kfztech> body shell
Rohkautschuk m <Kunststoff> crude rubber
Rohkohle f <Kohlen> pit coal, raw coal, rough coal, unscreened coal
Rohkollagen n <Chemie> ossein
Rohling m 1. <Fertig> solid bank; sludge (Fließ- und Strangpressen); 2. <Maschinen> blank
Rohlingsrest m <Fertig> unextruded butt (Strangpressen)
Rohmaterial n <Ker & Glas, Kunststoff> raw material
Rohmaterialien npl **für Töpferwaren** <Ker & Glas> pottery raw materials
Rohmittelkohle f <Kohlen> unwashed coal
Rohmüll m <Abfall> crude refuse, raw refuse, untreated refuse
Rohöl n <Erdöl, Kohlen, Meerschmutz, Umweltschmutz> crude, crude oil
Rohölanalyse f <Erdöl> crude assay, crude oil analysis
Rohöltanker m <Wassertrans> crude carrier
Rohpapier n <Papier> base paper, raw paper

Rohplatin

Rohplatin n <Chemie> platina
Rohprotein n <Lebensmittel> crude protein
Rohr n 1. <Bau> pipe, tube; 2. <Erdöl> pipe; 3. <Fertig> conduit; 4. <Maschinen> rigid pipe, tube; 5. <Mechan> pipe; 6. <Nichtfoss Energ> tube; 7. <Strömphys> pipe
• **durch Rohre leiten** <Bau> pipe • **Rohre verlegen** <Bau> pipe
Rohr n **aus nicht rostendem Stahl** <Maschinen> stainless steel tube
Rohr n **mit Doppelbogen** <Maschinen> gooseneck pipe
Rohr n **mit Naht** <Maschinen> seamed pipe
Rohrabschneider m <Maschinen> pipe cutter, tube cutter
Rohrabzweigung f <Maschinen> branch T
Rohranheber m <Erdöl> elevator
Rohranker m <Erdöl> tubing anchor
Rohranschluss m <Bau> pipe connection
Rohranschlussstutzen m <Maschinen> pipe nipple
Rohraufhängungsteile npl <Maschinen> pipe hanger fixtures
Rohraufweiter m <Bau> tube expander
Rohrbau m <Bau> pipework
Rohrbiegemaschine f <Maschinen> pipe-bending machine
Rohrboden m <Kerntech> tube plate
Rohrbogen m 1. <Bau> pipe bend; 2. <Maschinen> pipe bend, tube bend
Rohrbruch m <Maschinen> pipe burst
Rohrbruchsicherung f <Erdöl> automatic shutoff valve
Rohrbrücke f <Erdöl> pipe rack
Rohrbündel n 1. <Bau> tube nest; 2. <Kerntech> tube bundle
Röhrchen n <Labor> tube
Röhrchenbürste f <Labor> tube brush
Röhrchenhalter m <Labor> tube holder
Röhrchenreinigungsbürste f <Labor> tube brush
Röhrchenschneider m <Labor> tube cutter
Rohrdampfboiler m <Hydraul> tubular furnace boiler
Rohrdampfkessel m <Hydraul> tubular furnace boiler
Rohrdichter m <Bau> casing expander
Rohrdichtung f <Bau> pipe gasket
Rohrdiffusion f <Metall> pipe diffusion
Rohrdüker m <Telekom> underwater pipe (Kabelführung)
Rohrdurchlass m <Bau> pipe culvert
Rohrdurchmesser m <Maschinen> pipe diameter
Röhre f 1. <Bau, Elektronik> tube; valve (Elektronenröhre); 2. <Elektrotech> conduit; bulb (Neonleuchte); 3. <Funktech> thermionic valve; 4. <Maschinen> rigid pipe, tube; 5. <Mechan> duct; 6. <Strömphys> pipe • **in Röhren** <Strömphys> in pipes
Röhre f **mit automatischer Gittervorspannungserzeugung** <Elektronik> self-biased tube
Röhre f **mit engem Bohrloch** <Labor> narrow-bore tube
Röhre f **mit kurzer Wechselwirkung** <Elektronik> short interaction tube
Röhre f **mit Loktalsockel** <Elektronik> loctal tube (Elektronenröhren)
Röhre f **mit verlängerter Wechselwirkung** <Elektronik> extended-interaction tube
Röhre f **ohne Regelkennlinie** <Elektronik> sharp cut-off tube
Rohre npl **und Armaturen** fpl <Bau> pipes and fittings
Rohrelevator m <Bau> casing elevator
Röhrendiode f <Elektronik, Kerntech> diode tube
Röhrenfassung f <Elektrotech> tube socket
Röhrenfedermanometer n <Erdöl> (AE) Bourdon gage, (BE) Bourdon gauge
Röhrenhals m <Fernseh> tube neck
Röhrenhonmaschine f <Fertig> tube honing machine
Röhrenkabel n <Telekom> conduit cable

Röhrenkapazität f <Phys> internal capacity
Röhrenkessel m <Lebensmittel> tubular boiler
Röhrenkolben m <Elektronik, Elektrotech> bulb, valve envelope
Röhrenkühler m 1. <Kfztech> tubular radiator; 2. <Lebensmittel> tubular cooler
Röhrenleitungssystem n <Maschinen> ducting
Röhrenlibelle f <Fertig> air level
Röhrenluftvorwärmer m <Heiz & Kälte> tubular air heater
Röhrenmodulator m <Elektronik> tube modulator
Röhrenoszillator m <Elektronik> tube oscillator
Röhrensender m <Elektrotech> thermionic generator
Röhrentriode f <Elektronik> triode tube (Dreielektronenröhre)
Röhrenverstärker m <Elektronik> tube amplifier, (BE) valve amplifier
Röhrenwanddickenmesser m <Kerntech> (AE) tube thickness gage, (BE) tube thickness gauge
Röhrenwärmeaustauscher m 1. <Heiz & Kälte, Kerntech> shell and tube heat exchanger; 2. <Maschinen> tube heat exchanger
Röhrenwicklung f <Elektriz> concentric winding
Rohrfedermanometer n 1. <Erdöl> (AE) Bourdon gage, (BE) Bourdon gauge (Messtechnik); 2. <Gerät> (AE) Bourdon tube gage, (BE) Bourdon tube gauge, (AE) boundary tube gage, (BE) boundary tube gauge, spring tube manometer
Rohrfitting n <Maschinen> tapped fitting
Rohrformdorn n <Hydraul> boiler pipe shaping mandrel
rohrförmig adj <Fertig> tubular
Rohrformstücke npl <Maschinen> pipe fittings
Rohrfräser m <Maschinen> burring reamer
Rohrgewinde n 1. <Bau> pipe thread; 2. <Fertig> gas thread; 3. <Maschinen> pipe thread
Rohrgewindebohrer m 1. <Bau> pipe tap; 2. <Maschinen> pipe tap, pipe-thread tap
Rohrgewindeschneidbacke f <Fertig> pipe die
Rohrglas n <Ker & Glas> tubing glass
Rohrglocke f <Fertig> pipe cover (Kunststoffinstallationen)
Rohrhaken m 1. <Bau> pipe hook; 2. <Erdöl> pipe hook (zur Montage von Rohrleitungen)
Rohrheizkörper m <Kerntech> tubular heating element
Rohrhonmaschine f <Fertig> tube honing machine
Rohrklammer f <Erdöl> casing clamp (Bohrtechnik)
Rohrklemmkeil m <Erdöl> slip
Rohrkoje f <Wassertrans> pipe cot
Rohrkonstruktion f <Erdöl> jacket (Offshore-Technik)
Rohrkopf m <Bau> casing head
Rohrkorrosionsschutz m **aus Bitumen** <Maschinen> bitumen pipe coating
Rohrkrümmer m 1. <Bau> pipe bend, pipe knee; 2. <Maschinen> pipe bend
Rohrkupplung f <Maschinen> tube coupling
Rohrleger m <Maschinen> pipe layer
Rohrleitung f 1. <Bau> conduit, pipe, pipeline, tubing; 2. <Elektrotech> conduit; 3. <Erdöl> pipeline; 4. <Kohlen, Maschinen> pipeline, piping; 5. <Nichtfoss Energ> conduit, pipeline, tubing; 6. <Telekom> conduit (Kabel); 7. <Trans, Wassertrans> pipeline; 8. <Wasserversorg> conduit
Rohrleitungsbau m <Fertig> pipeline construction (Kunststoffinstallationen)
Rohrleitungsnetz n <Bau> pipework
Rohrleitungsplan m <Wassertrans> piping plan (Schiffbau)
Rohrleitungssystem n <Maschinen> pipework system
Rohrleitungsverlegung f <Bau> laying of pipes, pipe laying

Rohrmanschette f <Bau> pipe collar
Rohrmuffe f 1. <Bau> pipe joint; 2. <Fertig> pipe bell
Rohrmühle f <Kohlen> tube mill
Rohrnetz n 1. <Bau> system of pipes; 2. <Nichtfoss Energ> tubing
Rohrniet n <Maschinen> tubular rivet
Rohrnippel m <Maschinen> barrel nipple
Rohrplatte f <Maschinen> tube plate
Rohrplattform f <Erdöl> jacket platform (Offshore-Technik)
Rohrpostanlage f <Kerntech> rabbit system
Rohrrahmen m <Kfztech> tubular frame (Motorrad)
Rohrreduzierstück n <Bau> pipe reducer
Rohrreiniger m <Maschinen> go-devil, pipe cleaner
Rohrsäge f <Maschinen> pipe saw
Rohrsatz m <Kerntech> tube nest
Rohrschelle f 1. <Bau> casing clamp, pipe strap, tube clip; 2. <Erdöl> casing clamp; 3. <Fertig> pipe bracket; 4. <Maschinen> clip, pipe clip
Rohrschiebermotor m <Kfztech> sleeve valve engine
Rohrschlange f <Labor> serpent coil (Destillieren)
Rohrschlosser m <Bau, Maschinen> pipe fitter
Rohrschlüssel m 1. <Bau, Fertig> pipe wrench; 2. <Maschinen> tube wrench
Rohrschlüssel m **mit Zähnen** <Maschinen> alligator wrench
Rohrschneider m 1. <Bau> casing cutter, pipe cutter; 2. <Maschinen> pipe cutter
Rohrschraubstock m 1. <Bau> pipe vice; 2. <Maschinen> tube vice
Rohrschweißung f <Bau> tube welding
Rohrspirale f <Labor> serpent coil (Destillieren)
Rohrstahl m <Metall> tube steel
Rohrsteckverbindung f <Bau> spigot joint
Rohrstrang m 1. <Heiz & Kälte> pipe run; 2. <Maschinen> pipe conduit
Rohrstrangpressen n <Maschinen> tube extrusion
Rohrströmung f <Strömphys> flow in pipes, pipe flow
Rohrstutzen m <Bau, Maschinen> socket
Rohrträger m <Erdöl> pipe rack (Tiefbohrtechnik)
Rohr- und Elektroleitungsraum m <Bau> duct space
Rohrverankerung f <Erdöl> tubing anchor (Bohrtechnik)
Rohrverbindung f 1. <Bau> pipe connection, pipe coupling, pipe joint; 2. <Maschinen> pipe connection, pipe joint, pipe junction, pipe union
Rohrverbindungsstück n 1. <Bau> pipe union; 2. <Maschinen> fitting, pipe coupling
Rohrverleger m <Erdöl> lay barge
Rohrverlegeschiff n <Erdöl> lay barge (Offshore-Technik); pipe-laying barge (Offshore-Leitungsbahn)
Rohrverlegung f <Bau> pipe laying
Rohrverschluss m <Maschinen> pipe plug
Rohrverschraubung f 1. <Bau> pipe screwing, tube fitting, union; 2. <Maschinen> screwed pipe coupling
Rohrverschraubung f **mit Bolzen** <Maschinen> bolted pipe joint
Rohrverschraubungsstück n <Maschinen> screwed fitting, tapped fitting, threaded fitting
Rohrverzweigung f <Fertig, Maschinen> manifold
Rohrwelle f <Maschinen> tubular shaft
Rohrwickler m <Bau> pipe twister
Rohrwiege f <Mechan> cradle
Rohrzange f 1. <Bau> pipe tongs, pipe wrench; 2. <Erdöl> pipe tongs (Bohrtechnik); 3. <Maschinen> gas pliers
Rohrziehdorn m <Maschinen> tube-drawing mandrel
Rohrzug m <Elektrotech, Telekom> cable duct
Rohschlamm m <Abfall, Wasserversorg> raw sludge
Rohschmieröl n <Fertig> black oil
Rohschraube f <Fertig> black bolt
Rohstahl m <Metall> crude steel
Rohstoff m 1. <Erdöl> raw material; 2. <Fertig> raw material, staple; 3. <Ker & Glas, Kohlen, Kunststoff> raw material; 4. <Lebensmittel> staple; 5. <Meerschmutz> resource; 6. <Papier> raw material
Rohstoffrückgewinnung f <Abfall> resource recovery
Rohsulfat n <Lebensmittel> salt cake
Rohteil n 1. <Fertig> blank, part; 2. <Mechan> blank
Rohtoluol n <Chemie> toluol
Rohvaselin n <Chemie> petrolatum
Rohvlies n <Ker & Glas> uncured mat
Rohwasser n <Wasserversorg> raw water
Rohzink n <Fertig> spelter
Roll... <Lufttrans, Raumfahrt> rolling
Rollachse f <Lufttrans, Raumfahrt> roll axis
Rollage f <Raumfahrt> roll attitude
Rollbacken m <Fertig> die plate (Gewindewalzen)
Rollbahn f 1. <Lufttrans> taxiway; 2. <Trans> moving carpet
Rollbahnfeuer n <Lufttrans> taxiway light
Rollbahnkreuzungsmarkierung f <Lufttrans> taxiway intersection marking
Rollbahnmittellinienfeuer n <Lufttrans> (AE) taxiway centerline light, (BE) taxiway centreline light
Rollbahnmittellinienmarkierung f <Lufttrans> (AE) taxiway centerline marking, (BE) taxiway centreline marking
Rollbahnrandmarkierung f <Lufttrans> taxiway edge marker
Rollbalken m <Comp & DV> scroll bar
Rollband n <Trans> moving floor, moving walkway
Rollbewegung f <Maschinen> rolling motion
Rollbord n <Fertig> curl
Rollbrücke f <Bau> roller bridge, rolling bridge
Rolle f 1. <Bau> barrel (einer Walze); 2. <Druck, Fertig> reel; 3. <Lufttrans> roll (Kunstflug); 4. <Maschinen> caster, castor, roll, roller, sheave, wheel; 5. <Mechan> roll; 6. <Papier> reel; 7. <Phys> pulley; 8. <Wassertrans> coil (Tauwerk) • **an der Rolle perforiert** <Verpack> perforated on the reel • **von der Rolle arbeitend** <Druck> reel-fed
rollen v 1. <Comp & DV> scroll; 2. <Maschinen> roll; 3. <Raumfahrt> bank
Rollen n 1. <Comp & DV> scrolling; 2. <Lufttrans> taxiing; 3. <Maschinen> rolling
Rollenaufnehmer m <Verpack> reel lifter
Rollenbahn f 1. <Maschinen> roller path, roller track; 2. <Verpack> conveyor way
Rollenbohrmeißel m <Erdöl> roller bit (Bohrtechnik)
Rollenbreite f <Papier> reel width
rollender Schnitt m <Fernseh> wipe
rollender Start m <Lufttrans> rolling takeoff
rollendes Material n 1. <Eisenbahn> (AE) railroad vehicles, (BE) railway vehicles; 2. <Kfztech> rolling stock
Rollendruckmaschine f <Druck> reel-fed press
Rolleneinschlag- und Packmaschine f <Verpack> reel wrapping and handling equipment
Rollenetikettendruck m <Verpack> roll label printing
Rollenförderer m <Verpack> conveyor way
Rollenfreilaufkupplung f <Kfztech> overrunning clutch
Rollengegenführung f <Fertig> roller box
Rollengerüst n **mit Flaschenzug** <Kerntech> gantry with hoist
Rollengesperre n <Maschinen> roller clutch
Rollenhebel m <Mechan> roller lever
Rollenhülse f <Verpack> core of spool
Rollenkette f <Maschinen> bush-roller chain, roller chain
Rollenkorb m <Maschinen> roller cage
Rollenkühlofen m <Ker & Glas> annealing lehr with rollers (Flachglas)

Rollenkupplung

Rollenkupplung f <Kfztech, Mechan> roller clutch
Rollenlager n <Kfztech, Maschinen, Mechan> roller bearing
Rollenlagergehäuse n <Maschinen> roller-bearing box
Rollenlippklampe f <Wassertrans> roller fairlead *(Beschläge)*
Rollenlünette f <Maschinen> roller steady
Rollenoffset f <Druck> web offset
Rollenoffsetdruck m <Druck> web offset printing
Rollenpackmaschine f <Verpack> reel overwrapper
Rollenprobe f <Papier> reel sample
Rollenquetscher m <Foto> squeegee
Rollenrotationsmaschine f <Druck> web-fed rotary press
Rollenrotationspresse f <Druck> web-fed rotary press
Rollenschneidmaschine f <Papier> slitter-rewinder
Rollensteuerkette f <Kfztech> roller timing chain
Rollenstößel m <Kfztech> roller shaft, roller tappet
Rollenstromabnehmer m <Trans> trolley pole
Rollentraglager n <Mechan> angular roller bearing
Rollen- und Rotationsstanzpresse f <Verpack> roller and rotary cutting press
Rollenzuführung f <Maschinen> roller feed
Rollfeld n <Lufttrans> airfield, landing field
Rollfeldringstraße f <Lufttrans> perimeter track
Rollfilm m <Foto> roll film
Rollgabelschlüssel m 1. <Fertig> monkey wrench; 2. <Maschinen> *(BE)* adjustable spanner, monkey wrench, screw wrench
Rollgangsrahmen m <Fertig> table beam *(Walzen)*
Rollgeld n <Trans> cartage
Rollglasschneider m <Ker & Glas> glass cutting wheel
Rollhalteort m <Lufttrans> taxi holding position
Rollkanal m <Lufttrans> roll channel *(Selbststeuerung)*
Rollkondensator m <Elektrotech> paper capacitor
Rollkontakt m <Maschinen> rolling contact
Rollkreis m 1. <Geom> rolling circle; 2. <Maschinen> pitch circumference
Rollkreisel m <Raumfahrt> roll rate gyro
Rollkugel f <Comp & DV> rolling ball, trackball *(Mausäquivalent beim Notebook)*; track ball *(Zeigereinheit bei grafischen Benutzeroberflächen)*
Rollkurve f <Geom> roulette
Rollkurve f **des Kreises** <Math> cycloid *(Zykloide)*
Rollladen m <Bau> roller blind, roller shutter, shutter
Rollladenstab m <Bau> reed
Rollladentor n <Bau> rolling door
Rollladentür f <Bau> rolling door
Rollleiste f <Comp & DV> scroll bar
Rollmembrane f <Fertig> cup seal *(Kunststoffinstallationen)*
Rollmenü n <Comp & DV> drop-down menu
Rollmoment n 1. <Lufttrans> rolling moment *(Luftfahrzeug)*; 2. <Maschinen> rolling moment
Rollneigung f <Verpack> curl
Rollo n <Bau> roller blind
Roll-on-Roll-off n *(Ro-Ro)* <Trans> roll-on-roll-off, ro-ro
Roll-on-Roll-off-Anlage f *(Ro-Ro-Anlage)* <Wassertrans> roll-on roll-off dock, ro-ro dock
Roll-on-Roll-off-Frachter m *(Ro-Ro-Frachter)* <Wassertrans> roll-on roll-off vessel, ro-ro vessel *(Horizontalbeladung)*
Roll-on-Roll-off-Hafen m *(Ro-Ro-Hafen)* <Wassertrans> roll-on roll-off port, ro-ro port
Roll-on-Roll-off-Schiff n *(Ro-Ro-Schiff)* <Wassertrans> roll-on roll-off ship, roll-on roll-off vessel, ro-ro vessel
Roll-on-Roll-off-System n *(Ro-Ro-System)* <Wassertrans> roll-on roll-off system, ro-ro system *(Horizontalbeladung)*

Rollpalette f <Trans> roller pallet
Rollreffbaum m <Wassertrans> roller reefing boom
Rollreibung f <Maschinen, Phys> rolling friction
Rollreibungszahl f <Mechan, Phys, Qual> coefficient of rolling friction
Rollschacht m <Kohlen> chute
Rollschere f <Fertig> slitter
Rollscheren n <Fertig> slitting
Rollschicht f <Bau> upright course *(Mauerwerk)*
Rollsichter m <Kohlen> roll screen
Rollstabilität f <Lufttrans> rolling stability *(Luftfahrzeug)*
Rollstek m <Wassertrans> rolling hitch *(Knoten)*
Rollstuhlaufzug m <Bau, Kfztech> wheelchair lift
Rolltreppe f 1. <Bau> escalator, moving staircase, moving stairway; 2. <Trans> moving staircase, moving stairway
Rolltür f <Kfztech> roll-up door
Rollwerk n <Lufttrans> landing gear
Rollwiderstand m <Maschinen> rolling resistance
Ro-Lo-Schiff n <Wassertrans> rolo ship
ROM *(Festwertspeicher, Nur-Lese-Speicher, ROM-Speicher)* <Comp & DV, Elektriz, Elektrotech> ROM *(read-only memory)*
Romandruckpapier n <Druck> antique book paper
römische Zahlen fpl <Math> Roman numerals
römische Ziffern fpl <Druck, Math> Roman numerals
römischer Bogen m <Bau> Roman arch, round arch
ROM-Medium n <Optik> read-only medium
Rommeltrommel f <Fertig> rumble
ROM-Speicher m *(ROM)* <Comp & DV, Elektriz, Elektrotech> read-only memory *(ROM)*
Ronde f <Fertig> round blank
Röntgen n *(R)* <Strahlphys> röntgen *(R)*
Röntgenabsorptionsanalyse f <Strahlphys> X-ray absorption analysis
Röntgenabsorptionsspektrum n <Kerntech, Phys> X-ray absorption spectrum
Röntgenabstandsbelichtungsanlage f <Elektronik> X-ray proximity exposure tool
Röntgenanalyse f <Gerät, Strahlphys> X-ray analysis
Röntgenaufnahme f <Strahlphys> X-ray photograph
Röntgenausbeute f <Kerntech> X-ray yield
Röntgenbelichtungsmaske f <Elektronik> X-ray exposure mask
Röntgenbestrahlung f <Kerntech> X-ray irradiation
Röntgenbeugung f <Elektronik> X-ray diffraction
Röntgenbeugungsanalyse f <Strahlphys> X-ray diffraction analysis
Röntgenbeugungskamera f <Kerntech> X-ray diffraction camera
Röntgenbild n 1. <Foto> X-ray photograph; 2. <Ker & Glas> shadowgraph
Röntgendetektor m <Gerät> X-ray detection
Röntgendiffraktometer n <Strahlphys> X-ray diffractometer
Röntgendosismesser m <Gerät> X-ray dosimeter
Röntgendosismessgerät n <Gerät> X-ray dosimeter
Röntgen-Escape-Peak m <Strahlphys> X-ray escape peak
Röntgenfarberscheinung f <Kerntech> *(AE)* X-ray coloration, *(BE)* X-ray colouration
Röntgenfluoreszenz f <Strahlphys> X-ray fluorescence
Röntgenfluoreszenzanalyse f <Strahlphys> X-ray fluorescence analysis
Röntgenfluoreszenz-Spektrometer m <Strahlphys> fluorescent X-ray spectrometer
Röntgen-Halbleiterscheiben-Stepper m <Elektronik> X-ray wafer stepper
Röntgenhintergrundstrahlung f <Strahlphys> X-ray background radiation

Röntgenkamera f <Kerntech, Strahlphys> X-ray camera
Röntgenkontaktbelichtung f <Elektronik> X-ray contact printing
Röntgenkristallographie f <Strahlphys> X-ray crystallography
Röntgenlaser m 1. <Elektronik> X-ray laser; 2. <Kerntech> X-raser, X-ray laser; 3. <Strahlphys> X-ray laser
röntgenlithographisches Gerät n <Elektronik> X-ray lithography machine
Röntgenlumineszenz f <Kerntech> roentgenoluminescence
Röntgenluminiszenz f <Kerntech> X-ray luminescence
Röntgenmaskenjustierung f <Elektronik> X-ray mask alignment
Röntgenmessgerät n <Metrol> r-meter, roentgen meter, roentgenometer
Röntgenmetallographie f 1. <Kerntech> X-ray metallography, roentgenometallography; 2. <Strahlphys> X-ray metallography
Röntgenmeter n <Strahlphys> r-meter, roentgen meter, roentgenometer
Röntgenmikroskop n <Strahlphys> X-ray microscope
Röntgenographie f <Strahlphys> radiography
Röntgenoptik f <Phys> X-ray optics
Röntgenphoton n <Kerntech> X-ray photon
Röntgenprojektionslithographie f <Elektronik> X-ray projection lithography
Röntgenprüfung f <Kerntech> X-ray testing
Röntgenquant n <Kerntech> X-ray quantum
Röntgenquelle f <Kerntech> X-ray source
Röntgenröhre f <Elektronik, Strahlphys> X-ray tube
Röntgenschutzglas n <Ker & Glas, Sicherheit> X-ray protective glass
Röntgenspektrograph m <Strahlphys> X-ray diffractometer, X-ray spectrograph
Röntgenspektrographie f <Kerntech> X-ray spectrography
Röntgenspektrometer n <Strahlphys> X-ray spectrometer
Röntgenspektrometrie f <Kerntech> X-ray spectrometry
Röntgenspektroskopie f <Kerntech> X-ray spectroscopy
Röntgenspektrum n <Strahlphys> X-ray spectrum
Röntgen-Stepper-Maske f <Elektronik> X-ray stepper mask
Röntgenstrahl m <Elektronik> X-ray beam
röntgenstrahlangeregte Photoelektronenspektroskopie f (XPS) <Phys> X-ray photoelectron spectroscopy (XPS)
röntgenstrahlbelichtet adj <Funktech> X-ray exposed
röntgenstrahlempfindlicher Photolack m <Elektronik> X-ray resist
Röntgenstrahlen mpl <Elektriz, Metall, Raumfahrt, Strahlphys, Wellphys> X-rays
Röntgenstrahlenbrechung f <Raumfahrt, Wellphys> X-ray diffraction
Röntgenstrahlen-Näherungsdrucktechnik f <Elektronik> X-ray proximity printing
Röntgenstrahlen-Resist f <Elektronik> X-ray resist
Röntgenstrahler m <Kerntech> X-emitter
Röntgenstrahlimpuls m <Elektronik> X-ray pulse
röntgenstrahlinduziertes Photoelektronenspektrum n <Kerntech> X-ray photoelectron spectrum
Röntgenstrahllithographie f <Elektronik> X-ray lithography
Röntgenstrahlmaske f <Elektronik> X-ray mask
Röntgenstrahlmikroskop n <Kerntech> X-ray microscope
Röntgenstrahlprüfung f <Mechan> X-ray examination

Röntgenstrahltest m <Mechan> X-ray examination
Röntgenstrahlung f 1. <Phys> X-rays; 2. <Strahlphys> X-radiation, X-rays
Röntgenstrukturanalyse f <Phys> X-ray structure analysis
Röntgentesten n <Kerntech> X-ray testing
Röntgentopographie f <Metall> X-ray topography
Röntgenuntersuchung f 1. <Kerntech> X-ray inspection, X-ray inspection (eines Gerätes); 2. <Raumfahrt> X-ray inspection
Röntgenwellenlänge f <Strahlphys> X-ray wavelength
Roots-Gebläse n <Maschinen> Roots blower
Roots-Gebläseflügel m <Fertig> Roots-blower member
Ropax-Schiff n <Wassertrans> ropax vessel (Roll-on-Roll-off-Fahrgastschiff)
Roringstek m <Wassertrans> fisherman's bend (Knoten)
Ro-Ro (Roll-on-Roll-off) <Trans> ro-ro, roll-on-roll-off
Ro-Ro-Anlage f (Roll-on-Roll-off-Anlage) <Wassertrans> ro-ro dock, roll-on roll-off dock
Ro-Ro-Frachter m (Roll-on-Roll-off-Frachter) <Wassertrans> ro-ro vessel, roll-on roll-off vessel (Horizontalbeladung)
Ro-Ro-Hafen m (Roll-on-Roll-off-Hafen) <Wassertrans> ro-ro port, roll-on roll-off port
Ro-Ro-Schiff n (Roll-on-Roll-off-Schiff) <Wassertrans> ro-ro ship, ro-ro vessel, roll-on roll-off vessel (Horizontalbeladung)
Ro-Ro-System n (Roll-on-Roll-off-System) <Wassertrans> ro-ro system, roll-on roll-off system (Horizontalbeladung)
rosa Glas n <Ker & Glas> pink glass
rosa Rauschen n <Akustik, Aufnahme> pink noise
rösch gemahlener Zellstoff m <Papier> free pulp
Rose f <Elektrotech> rose
Rosenkupfer n <Chemie> rose copper
Rosette f <Ker & Glas> rosette
Rossbreiten fpl <Wassertrans> horse latitudes
Rost m 1. <Anstrich, Eisenbahn> rust; 2. <Fertig> aerugo, stain; 3. <Ker & Glas> grillage; 4. <Kfztech> rust; 5. <Lebensmittel> rust (Rostkrankheit des Getreides); 6. <Lufttrans> rust; 7. <Maschinen> grate, grating (Gitter); rust (Korrosion); 8. <Mechan, Papier> rust; 9. <Raumfahrt> lattice (Raumschiff); 10. <Textil> grid; 11. <Wassertrans> rust
Rostanfressung f <Bau> honeycombing (Beton)
rostbeständig adj <Papier> rustproof
rosten v <Anstrich, Eisenbahn, Lufttrans, Wassertrans> rust
rösten v 1. <Fertig> calcine; 2. <Lebensmittel> fire; 3. <Thermod> scorch
Rösten n 1. <Fertig> calcination; 2. <Kohlen, Lebensmittel> roasting; 3. <Textil> retting
Rostentferner m <Maschinen> rust remover
Rostfeuerung f <Maschinen> grate firing
Rostfläche f <Bau> grate area
Rostflocke f <Anstrich> rust flake
rostfrei adj <Anstrich, Foto, Metall> stainless
rostfreier Stahl m <Kfztech, Metall, Papier> stainless steel
rostfreier Stahlguss m <Fertig> cast stainless steel (Kunststoffinstallationen)
rostgeschützt adj 1. <Mechan> anticorrosive, antirust; 2. <Metall> antirust
Rostinhibitor m <Verpack> rust preventive
rostlösendes Öl n <Kfztech> penetrating oil
Röstofen m <Fertig> calciner
Rostschlitten m <Fertig> grate carriage
Rostschutz m 1. <Maschinen> rust protection, rustproofing; 2. <Verpack> rust protection

Rostschutzanstrich

Rostschutzanstrich *m* <Kunststoff> antirust coating
Rostschutzmittel *n* 1. <Eisenbahn, Lufttrans> rust inhibitor; 2. <Metall> antirust agent; 3. <Papier> antirust agent, rust inhibitor; 4. <Verpack> rust preventive; 5. <Wassertrans> rust inhibitor
Rostschutzverpackung *f* <Verpack> rust preventive packaging
Roststab *m* <Heiz & Kälte> fire bar
rostverhindernd *adj* <Mechan> anticorrosive
Rot *n* <Phys> curl *(Rotation eines Vektorfeldes)*
Rotamesser *m* <Erdöl> rotameter
Rotameter *n* <Labor, Papier> rotameter
Rotarybohren *n* <Erdöl, Kohlen> rotary drilling
Rotarybohrmeißel *m* <Erdöl> rotary bit
Rotaryscheibenbohrmeißel *m* <Erdöl> *(BE)* rotary disc bit, *(AE)* rotary disk bit
Rotation *f* 1. <Comp & DV> rotation; 2. <Elektronik> curl *(eines Vektors)*; 3. <Geom> revolution, rotation; 4. <Maschinen, Phys> rotation; 5. <Strömphys> curl *(eines Vektors)*
Rotation *f* **im Gegenuhrzeigersinn** <Mechan> anticlockwise rotation, counterclockwise rotation
Rotation *f* **im Uhrzeigersinn** <Mechan> clockwise rotation
Rotations... *adj* <Maschinen> rotary
Rotationsabfüllmaschine *f* <Verpack> rotary filler
Rotationsachse *f* <Maschinen, Mechan, Phys> axis of rotation
Rotationsauslenkung *f (CR)* <Akustik> rotational compliance *(CR)*
Rotationsdruck *m* <Druck> rotary printing
Rotationsdurchflussmesser *m* <Phys> rotameter
Rotationsellipsoid *n* <Fertig, Geom> spheroid
Rotationsfläche *f* <Geom> surface of revolution
Rotationsform *f* <Maschinen> rotational mould
rotationsfrei *adj* <Phys> irrotational, nonkinking
Rotationskegel *m* <Geom> cone of revolution, rotation cone
Rotationskolben *m* <Maschinen> rotary piston
Rotationskolbenmotor *m* <Maschinen> rotary engine
Rotationskompressor *m* <Heiz & Kälte> rotary compressor
Rotationskörper *m* <Geom> body of revolution, solid of revolution
Rotationsmaschine *f* <Druck> rotary press, rotary printing machine, rotary printing press
Rotationsofen *m* <Abfall> rotary furnace, rotary kiln
Rotationsoperation *f* <Comp & DV> rotate operation
Rotationspresse *f* <Druck> rotary press, rotary printing machine, rotary printing press
Rotationspumpe *f* <Maschinen, Wasserversorg> rotary pump
Rotationsquantenzahl *f* <Phys> rotational quantum number
Rotationsrutschung *f* <Bau> rotational slide
Rotationsschnellpresse *f* **für Tabletten** <Verpack> high-speed rotary tablet compression machine
Rotationsspektrum *n* <Phys, Strahlphys> rotational spectrum
Rotationsstabilisierung *f* <Raumfahrt> stabilization of rotation
Rotationssymmetrie *f* <Geom> rotation symmetry
Rotationstiefdruck *m* 1. <Druck> rotogravure; 2. <Kunststoff> rotogravure printing
Rotationstrockner *m* <Fertig> rotary drier
Rotationsverdichter *m* <Heiz & Kälte, Maschinen> rotary compressor
Rotationsvolumen *n* <Geom> volume of rotation
Rotationszentrum *n* <Geom> centre of rotation

Rotationszerkleinerer *m* <Abfall> comminutor
Rotationszerkleinerer *m* **mit Brechrollen** <Abfall> comminutor
Rotationszylinder *m* <Geom> cylindrical solid of revolution
rote Kante *f* <Ker & Glas> red edge
Rotenon *n* <Chemie> derrin, rotenone *(Insektizid)*
roter Blutfarbstoff *m* <Chemie> *(BE)* haemoglobin, *(AE)* hemoglobin
roter Laser *m* <Elektronik> red laser
roter Stab *m* <Nichtfoss Energ> red rod
roter Strahl *m* <Elektronik> red beam
Rotfärbung *f* <Anstrich> rust
Rotgitter *n* <Fernseh> red screen grid
rotglühend *adj* <Thermod> red-hot
Rotglut *f* <Metall> red heat
rot-grün-blau *adj (RGB)* <Comp & DV, Fernseh> red-green-blue *(RGB)*
Rotguss *m* 1. <Fertig> red brass, steam bronze; 2. <Mechan> gunmetal; 3. <Metall> cannon metal, gunmetal
rotieren *v* 1. <Comp & DV> rotate; 2. <Maschinen> rotate, turn; 3. <Raumfahrt> spin
rotierend *adj* <Maschinen> rotating, rotary
rotierende Fluide *npl* <Strömphys> rotating fluids
rotierende Körper *mpl/***um die Achse** <Strömphys> spinning bodies *(Untersuchung von Fluiden)*
rotierende obere Feststufe *f (SSUS)* <Raumfahrt> solid spinning upper stage *(SSUS)*
rotierende Pumpe *f* <Phys> rotary pump
rotierende Scheibe *f* <Maschinen> *(BE)* rotating disc, *(AE)* rotating disk
rotierende Siebbürste *f* <Papier> rotary wire brush
rotierender Ringspalt *m* <Strömphys> rotating annulus
rotierender Umformer *m* <Elektriz> dynamotor
rotierender Verdampfungsapparat *m* <Labor> rotating evaporator
rotierender Videokopf *m* <Fernseh> rotary video head
rotierender Zuführ- und Sammeltisch *m* <Verpack> rotary feeder and collecting table
rotierendes Spritzrohr *n* <Papier> rotating shower
Rotkanone *f* <Fernseh> red gun
Rotkupfer *n* <Metall> red copper
Rotkupfererz *n* <Chemie, Metall> red copper ore
rötlich verfärben *v* <Anstrich> rust
Rotmessing *n* <Metall> tombac
Rotmischer *m* <Fernseh> red madder
Rotor *m* 1. <Elektrotech> rotor, curl; 2. <Funktech> rotator *(für Antennen)*; 3. <Kfztech> trigger wheel; 4. <Kohlen, Kontroll> rotor; 5. <Lufttrans> rotor *(Hubschrauber)*; 6. <Maschinen, Nichtfoss Energ, Phys> rotor
Rotorabwind *m* <Lufttrans> rotor inflow *(Hubschrauber)*
Rotorantriebswelle *f* <Lufttrans> rotor mast *(Hubschrauber)*
Rotorblatt *n* <Lufttrans> rotor blade *(Hubschrauber)*
Rotorblätter *npl/***nach oben gestellte** <Lufttrans> coning *(Hubschrauber)*
Rotorbock *m* <Lufttrans> rotor mast *(Hubschrauber)*
Rotordrehmoment *n* <Lufttrans> rotor torque *(Hubschrauber)*
Rotordurchmesser *m* <Nichtfoss Energ> rotor diameter
Rotorfläche *f* <Lufttrans> *(BE)* disc area, *(AE)* disk area *(Hubschrauber)*
Rotorflächenbelastung *f* <Lufttrans> *(BE)* disc loading, *(AE)* disk loading
Rotorgewicht *n* <Nichtfoss Energ> rotor weight
Rotorkopf *m* <Lufttrans> rotor head *(Hubschrauber)*
Rotorkreis *m* <Lufttrans> *(BE)* rotor disc, *(AE)* rotor disk *(Hubschrauber)*
Rotorlamellierung *f* <Elektrotech> rotor lamination

Rotornabe f <Lufttrans> rotor hub *(Hubschrauber)*
Rotorplatte f <Elektrotech> rotor plate
Rotorradius m <Lufttrans> rotor radius *(Hubschrauber)*
Rotorschub m <Lufttrans> rotor thrust *(Hubschrauber)*
Rotorstrahl m <Lufttrans> rotor stream *(Hubschrauber)*
Rotorüberdrehzahl f <Lufttrans> rotor overspeed *(Hubschrauber)*
Rotorwicklung f <Elektrotech> rotor winding
Rotphase f <Trans> red phase
Rotröhre f <Elektronik> red tube
Rotrost m <Anstrich> red rust
Rot-Schwarz-Pegel m <Fernseh> red-black level
Rotsignal n <Elektronik> red signal
Rotsignaldauer f <Trans> hour of red signal indication
Rotspitzenpegel m <Fernseh> red peak level
Rotstrahl m <Elektronik, Fernseh> red beam
Rotstrahlablenkmagnet m <Fernseh> red beam magnet
Rottedeponie f <Abfall> digestion deposit
Rotten n 1. <Ker & Glas> rotting; 2. <Papier, Textil> retting *(Flachs)*
Rottezelle f <Abfall> composting drum
Rouleaudruck m <Textil> roller printing
Route f <Wassertrans> route
Routine f <Comp & DV, Kontroll, Qual> routine
Routineinspektion f <Lufttrans, Qual> routine inspection
routinemäßige Datensicherungsmaßnahme f <Comp & DV> routine backup procedure
Routineüberprüfung f <Lufttrans, Qual> routine inspection
Routing-Chart n <Wassertrans> routeing chart *(Navigation)*
Routing-System n <Wassertrans> routeing system *(Navigation)*
Roving n <Ker & Glas, Kunststoff> roving *(Glasseidenstrang zur Verstärkung von Duroplasten)*
Roving-Mikrofon n <Aufnahme> roving mike
Rovingtrommel f <Ker & Glas> roving winder
Rowland-Aufhängung f <Phys> Rowland mounting
Rowland-Kreis m <Phys> Rowland circle
Rowland-Versuch m <Phys> Rowland's experiment
RP-1 *(Kerosin)* <Erdöl, Trans> RP-1 *(kerosene)*
RSB *(Restseitenband)* <Elektronik> VSB *(vestigial sideband)*
RS-Flipflop n *(Reset-Set-Flipflop)* <Elektronik> RS flipflop *(reset-set flip-flop)*
RS-Kippschaltung f *(Reset-Set-Kippschaltung)* <Elektronik> RS toggle *(reset-set toggle)*
RTM-Verfahren n *(GFK)* 1. <Kunststoff> resin transfer moulding; 2. <Kunststoff> RTM, reference test method *(Referenztestmethode-Verfahren)*
RTOL-Flugzeug n *(verkürzt startendes und landendes Flugzeug)* <Lufttrans> RTOL aircraft *(reduced takeoff and landing aircraft)*
RTTY *(Funkfernschreiben)* <Funktech> RTTY *(radioteletype)*
Ru *(Ruthenium)* <Chemie> Ru *(ruthenium)*
Ruberythrin... <Chemie> ruberythric
Rubidium n *(Rb)* <Chemie> rubidium *(Rb)*
Rubin m <Elektronik> ruby
Rubinlaser m 1. <Elektronik> ruby laser; 2. <Strahlphys> ruby crystal laser
Rüböl n <Lebensmittel> rapeseed oil
Rubrik f <Druck> column, head
Ruck m <Trans> jolt
Ruckanzeige f <Gerät> back indication
Rückarbeitsbremsung f <Eisenbahn> regenerative braking
ruckartig ziehen v <Kfztech> hitch
ruckartige Bewegung f <Ergon> saccadic movement

Rückassemblierer m <Comp & DV> disassembler
Rückbaukammer f <Bau> dismantling chamber
Rückbewegung f <Phys> return stroke
Rückblick m <Bau> back sight *(Vermessung)*
Rückblicksystem n <Trans> rearview system
Rückbremsung f <Trans> reverse braking
Rückdampf m <Hydraul> reversing steam
Rückdämpfung f <Funktech> front-to-back ratio *(Antenne)*
Rückdiffusion f <Kerntech> back diffusion
Rückdrehen n <Wassertrans> backing *(des Windes)*
rückdrehendes Moment n <Kfztech, Lufttrans, Wassertrans> restoring moment *(Aerodynamik)*
Rückdruck m 1. <Hydraul> back pressure *(Zylinder)*; 2. <Raumfahrt> drag
Rückdruckventil n <Hydraul> back-pressure valve
Rück-EMK n <Phys> back emf
Rücken m 1. <Bau> ridge; 2. <Druck> back, spine; 3. <Maschinen> back, heel • **mit Rücken versehen** <Druck> back
Rückenappretur f <Textil> backing *(Beschichtung)*
Rückenbeschichtung f <Textil> backcoating
Rückenbeschichtungsmaterial n <Textil> backing
Rückeneinlage f <Druck> back lining
Rückenetikett n <Verpack> back label
Rückenkante f <Maschinen> heel
Rückenlehne f 1. <Ergon> backrest; 2. <Kfztech> backrest, seat back
Rückenspannung f <Metall> back stress
Rückenstütze f <Ergon> backrest
Rückentzerrung f <Elektronik> de-emphasis
rückenverstärkendes Gewebe n <Textil> backing fabric
Rückenwindanteil m <Lufttrans> downwind leg
Rückerstattung f <Patent> refund
Rückextraktionstank m <Kerntech> backwash tank
Rückfahrleuchte f <Kfztech> *(AE)* backup light, *(BE)* reversing light
Rückfahrlicht n <Kfztech> *(AE)* backup light, *(BE)* reversing light
Rückfahrscheinwerfer m <Kfztech> *(AE)* backup light, *(BE)* reversing light
Rückfaltungsfrequenz f <Elektronik> aliased frequency *(Signalverzerrung)*
rückfedern v <Maschinen> rebound
Rückfederung f 1. <Fertig> elastic recovery; 2. <Kunststoff> rebound; 3. <Maschinen> resilience, resiliency
Rückfederungsvermögen n <Kunststoff> resilience
Rückflankenverschiebung f <Elektronik> end distortion *(von Start- und Stoppsignalen bei Fernschreibern)*
Rückfluss m 1. <Chemtech> reflux *(Destillation)*; 2. <Wasserversorg> backflow
Rückflusskochen n <Chemtech> reflux boiling
Rückflusskühler m <Kerntech, Labor> reflux condenser
Rückflussleitung f <Fertig> return line
Rückflusslötung f <Raumfahrt> reflow soldering
Rückflussstrom m <Elektrotech> return current
Rückflussturbine f <Nichtfoss Energ> reverse flow turbine
Rückformung f <Kunststoff> recovery
Rückfracht f <Wassertrans> back freight, home freight, reshipment, reshipping, return cargo *(Ladung)*
Rückfrage f <Telekom> consultation call, enquiry
Rückfragehäufigkeit f <Telekom> repetition rate
rückfragen v 1. <Comp & DV> prompt; 2. <Telekom> hold for inquiry
Rückfrageplatz m <Telekom> inquiry position
Rückführen n <Fertig> recirculating
Rückführen n **der Papierbahn** <Papier> insetting *(für Mehrfarbendruck)*

rückführende

rückführende Kurbelgehäuseentlüftung f (PCV-Ventilation) <Kfztech> positive crankcase ventilation, PCV (Motor)
Rückführgröße f <Regelung> feedback signal, return signal
Rückführkreis m <Maschinen> feedback loop
Rückführsignal n <Elektronik> feedback signal (Mess- und Regelungstechnik)
Rückführtaste f <Comp & DV> return key
Rückführung f 1. <Comp & DV, Elektronik> feedback; 2. <Ker & Glas> recirculation (der Ströme in Wannenofen); 3. <Kerntech, Kontroll> feedback; 4. <Lufttrans> follow-up (Hubschrauber); 5. <Mechan, Phys> feedback; 6. <Telekom> recycling; feedback (Kontroll-Loop); 7. <Umweltschmutz> recycling; 8. <Wellphys> feedback
Rückführung f von Spaltprodukten <Kerntech> recirculation of fission products
Rückführungskreis m <Elektronik, Maschinen> feedback circuit, feedback loop
Rückführungsschleife f <Elektronik> feedback loop (bei Regelung im geschlossenen Kreis)
Rückführungssignal n <Regelung> feedback signal, return signal
Rückgabebehälter m <Verpack> returnable container
Rückgabeflasche f <Verpack> deposit bottle, returnable bottle
Rückgang m 1. <Math> decrease; 2. <Hydraul> fall (Niveau)
Rückgangslinie f <Wasserversorg> recession curve
rückgefaltetes Signal n <Elektronik> aliased signal
rückgewinnbarer Abfall m <Abfall> recoverable waste
Rückgewinnung f 1. <Abfall, Kohlen> recycling; 2. <Kunststoff, Meerschmutz> recovery
Rückgewinnungsanlage f <Abfall> reclamation plant, recovery plant
Rückgewinnungskessel m <Abfall> recovery boiler
rückgewonnene Pulpe f <Abfall> recovered pulp
rückgewonnene Wärme f <Abfall> recovered heat
Rückhalt m <Mechan> backlog
Rückhaltebecken n 1. <Bau> retention basin; 2. <Wasserversorg> detention basin, detention reservoir
Rückhaltevermögen n <Kunststoff> hold-out
Rückholen n <Telekom> call retrieval
Rückholfeder f <Kfztech, Maschinen> return spring
Rückholmechanismus m <Kerntech> return motion mechanism (für Abschaltstab)
rückhördämpfend adj <Telekom> antisidetone
Rückhördämpfungsschaltung f <Telekom> antisidetone circuit
Rückhören n <Telekom> sidetone (Telefon)
Rückkanal m 1. <Aufnahme> feedback channel; 2. <Comp & DV> reverse channel; 3. <Maschinen, Telekom> return channel
Rückkehr f <Comp & DV> return
Rückkehradresse f <Comp & DV> return address
Rückkehrbefehl m <Comp & DV> return instruction
Rückkehrcode m <Comp & DV> return code
Rückkehrkanal m <Comp & DV> return channel
Rückkehrkoeffizient m <Mechan, Phys, Qual> coefficient of restitution
Rückkontrolle f <Bau> back observation (Vermessung)
Rückkopplung f 1. <Aufnahme, Comp & DV, Elektronik, Fernseh, Funktech, Kerntech, Kontroll, Mechan, Phys, Telekom, Wellphys> feedback, response; 2. <Metall> recarburization
Rückkopplungsfaktor m <Elektronik> feedback ratio
rückkopplungsfreie Messung f <Gerät> open loop measurement
rückkopplungsfreie Relaisstation f <Elektronik> non-regenerative repeater (Radio)
rückkopplungsfreier Verstärker m <Elektronik> nonregenerative repeater (Telefon)
Rückkopplungsgrad m <Elektronik> feedback ratio
Rückkopplungskreis m 1. <Elektronik> regenerative circuit (Radio); 2. <Fernseh> feedback circuit
Rückkopplungsoszillator m <Aufnahme, Elektronik> feedback oscillator
Rückkopplungsregelung f <Elektrotech> feedback control
Rückkopplungsschaltung f <Aufnahme> feedback circuit
Rückkopplungsschleife f <Elektronik, Mechan> feedback loop
Rückkopplungsschneider m <Aufnahme> feedback cutter
Rückkopplungsschwingung f <Aufnahme> feedback oscillation
Rückkopplungssignal n <Elektronik, Regelung> feedback signal
Rückkopplungsspannung f <Elektrotech> feedback voltage
Rückkopplungsspule f <Elektrotech> feedback coil
Rückkopplungsverstärker m <Elektronik> feedback amplifier; regenerative amplifier (Radio)
Rückkopplungsverstärkung f <Elektronik> feedback amplification; regenerative amplification (Radio)
Rückkopplungswicklung f <Elektrotech> feedback winding
Rückkopplungswiderstand m <Elektrotech> feedback resistor
Rückkühler m <Wassertrans> heat exchanger (Motor)
Rücklauf m 1. <Chemtech> reflux (Destillation); 2. <Elektronik> retrace (Katodenstrahlröhren); 3. <Fernseh> back run, reverse motion; 4. <Fertig> withdrawing (Gewindebohrer); 5. <Heiz & Kälte> return; 6. <Hydraul> backflow (Kesselwasser); back stroke (Kolben); 7. <Kerntech> reverse motion; 8. <Maschinen> noncutting stroke, return movement, return stroke, return travel; 9. <Phys> return stroke; 10. <Wasserversorg> backflow
Rücklauf m des Werkzeugs <Maschinen> noncutting return of the tool
Rücklauf m von Zwischenleitungen <Telekom> re-entrant linkage
Rücklaufabtaststrahl m <Fernseh> return scanning beam
Rücklaufachse f <Kfztech> reverse idler shaft
Rücklaufaustastung f <Fernseh> flyback blanking
rücklaufende Welle f <Telekom> inward-propagating wave
Rückläufer m <Ker & Glas> runner back
Rücklaufgeschwindigkeit f <Comp & DV> rewind speed
rückläufige Skale f <Gerät> reversed scale
Rücklaufintervall n <Fernseh> return interval
Rücklaufkondensator m <Labor> reflux condenser
Rücklaufleitung fließen v/zur <Fertig> pass to exhaust
Rücklauflücke f <Fernseh> back gap
Rücklaufrad n <Kfztech> reverse idler gear (Getriebe)
Rücklaufrohr n <Heiz & Kälte> return pipe
Rücklaufschaltung f <Elektrotech> return circuit
Rücklaufschlamm m <Abfall> recycle sludge, return sludge
Rücklaufsperre f 1. <Bau> back stop; 2. <Elektriz> reverse power flow protection
Rücklaufspur f <Elektronik> retrace (Katodenstrahlröhren)
Rücklauftaste f <Comp & DV> return key
Rücklaufverhältnis n <Kerntech> reflux ratio
Rücklaufverlust m <Erdöl> loss of returns (Bohrtechnik: Bohrschlamm)

Rücklaufzeitsteuerung f <Fernseh> back timing
Rückleiter m <Elektriz> return conductor, return wire
Rückleitung f 1. <Elektrotech> return wire; 2. <Fertig> shunt line *(Hydraulik)*; 3. <Telekom> back circuit
Rückleitungsrohr n <Maschinen> return pipe
Rückleuchte f <Kfztech> tail lamp
Rückmelder m <Fertig> check-back position indicator *(Kunststoffinstallationen)*
Rückmeldesignal n <Elektronik> repeater signal
Rückmeldung f 1. <Comp & DV> ACK, acknowledgement, response; 2. <Fertig> answer-back signal, *(AE)* signaling unit, *(BE)* signalling unit *(Kunststoffinstallationen)*; 3. <Gerät> back indication; 4. <Telekom> ACK, acknowledgement
Rückmeldung f **der Signalstellung** <Eisenbahn> signal indication
Rückoxidation f <Chemie, Umweltschmutz> reoxidation
Rückprallelastizität f <Kunststoff> rebound elasticity, resilience
Rückreaktion f <Kerntech> back reaction
Rückreise f <Wassertrans> homeward passage, return voyage • **auf der Rückreise befindlich** <Wassertrans> homeward-bound
Rückruf m <Telekom> recall
Rückruf m **bei Besetzt** <Telekom> call-back on busy
Rückruf m **bei Freiwerden des besetzten Anschlusses** <Telekom> call-back when busy terminal becomes free
Rückruf m **bei Teilnehmer besetzt** <Telekom> completion of calls to busy subscriber *(ISDN)*
Rückschaltzeichen n <Comp & DV> SI character, shift-in character
Rückschlag m 1. <Elektrotech> kickback; 2. <Hydraul> back stroke *(Kolben)*; 3. <Maschinen, Mechan> recoil; 4. <Raumfahrt> backlash; 5. <Wasserversorg> kickback *(Brunnen)*
rückschlagfrei adj <Sicherheit> anti-kickback *(Werkzeug)*
rückschlagfreier Nylonhammer m <Sicherheit> recoilless nylon hammer
Rückschlagklappe f <Fertig> swing-type check valve *(Kunststoffinstallationen)*
Rückschlagschutz m <Sicherheit> flashback preventer *(Ventile und Anschlüsse)*
Rückschlagsicherung f <Bau> flame arrester
Rückschlagventil n 1. <Bau> non-return valve; 2. <Fertig> nonreturn valve; check valve *(Kunststoffinstallationen)*; 3. <Hydraul> back-pressure valve; 4. <Kfztech> check valve; check valve *(Schmierung)*; 5. <Maschinen> check valve, nonreturn valve, retention valve, return valve; 6. <Mechan> check valve; 7. <Nichtfoss Energ> nonreturn valve; 8. <Raumfahrt, Wasserversorg> check valve
Rückschlagzündung f <Kfztech> blowback
Rückschneidemethode f <Optik, Telekom> cutback technique
Rückseite f 1. <Comp & DV> back panel; 2. <Geom> back face *(eines Objekts)*; 3. <Ker & Glas> back surface *(einer Glastafel)*; 4. <Papier> back, rear
Rückseitenappretur f <Textil> back finish
rückseitig bedruckt adj <Verpack> reverse side printed
Rücksendungsetikett n <Verpack> return label
rücksetzen v <Kontroll> reset
Rücksetzen n 1. <Comp & DV, Telekom> reset; 2. <Kfztech> backing, reversing
Rücksetzknopf m <Fernseh> reset knob
Rücksetzung f <Comp & DV> reset
Rücksetzzeichen n <Comp & DV> backspace character
Rücksignal n <Elektronik> back signal
Rückspeicherung f <Comp & DV> restore

Rückspiegel m 1. <Ker & Glas> rear-view mirror; 2. <Kfztech> driving mirror, rear-view mirror
Rücksprechleitung f <Fernseh, Funktech> talkback circuit
Rücksprechmikrofon n <Aufnahme> interference microphone
Rücksprung m <Bau> recess, retreat; offset *(Mauerwerk)*
Rücksprungadresse f <Comp & DV> return address
Rücksprunghärte f <Fertig> scleroscope hardness
Rücksprunghärteprüfung f <Fertig> scleroscope method of determining hardness
Rücksprungpalette f <Trans> stevedore-type pallet
Rückspulen n <Fernseh> rewinding
Rückspulgabel f <Foto> rewind handle
Rückspulmitnehmer m <Foto> rewind cam
Rückspulspannung f <Aufnahme> rewind tension *(des Bandes)*
Rückspulvorrichtung f <Foto> rewinder
Rückspulzeit f <Comp & DV> rewind time
Rückstand m <Erdöl, Fertig, Lebensmittel, Maschinen, Meerschmutz, Umweltschmutz, Wasserversorg> residue
Rückstände mpl 1. <Bau> tailings; 2. <Erdöl> bottoms *(Absetzung von Verunreinigungen)*; bottoms *(von Destillation)*; 3. <Lebensmittel> bottoms; 4. <Papier> tailings
Rückstandssammelbehälter m <Erdöl> sump
Rückstandsdeponie f <Abfall> residue landfill
Rückstandsraffination f <Erdöl> residue refining process *(Raffinerie)*
Rückstau m 1. <Erdöl, Trans> backup; 2. <Wasserversorg> backwater
Rückstauklappe f <Bau> trap
Rückstauwasser n <Wasserversorg> backwater
Rückstauwirkung f <Nichtfoss Energ> backwater effect
Rückstellelastizität f <Comp & DV> resilience
rückstellen v 1. <Comp & DV> restore; 2. <Telekom> reset
rückstellfähig adj <Kunststoff> resilient
Rückstellfähigkeit f <Kunststoff> rebound elasticity
Rückstellfeder f 1. <Fertig> return action *(Kunststoffinstallationen)*; 2. <Kfztech> return spring; 3. <Maschinen> readjusting spring
Rückstellkraft f 1. <Aufnahme> stylus drag *(der Abspielnadel)*; 2. <Ergon, Maschinen> restoring force
Rückstellmoment n <Maschinen> righting moment
Rückstelltaste f <Comp & DV> reset button, reset key
Rückstellung f 1. <Comp & DV> restore; 2. <Fertig> restoration; 3. <Kunststoff> recovery; 4. <Telekom> reset
Rückstellzähler m <Gerät> reset counter
Rücksteuerungssystem n <Telekom> revertive control system
Rückstoß m 1. <Elektrotech> kickback, repulsion; 2. <Erdöl> kick; 3. <Hydraul> back stroke *(Kolben)*; 4. <Maschinen, Mechan> recoil
Rückstoßdüse f <Lufttrans> propelling nozzle
Rückstoßelektron n <Phys> recoil electron
Rückstoßenergie f **bei der Spaltung** <Kerntech> fission recoil
Rückstoßkern m <Phys> recoil nucleus
Rückstoßkraft f <Elektrotech> repulsion, repulsive force
Rückstoßstromversorgung f <Elektrotech> kickback power supply
rückstrahlende Markierung f <Lufttrans> retro-reflective marker *(Flughafen)*
Rückstrahler m <Kfztech> rear reflector; reflector *(Scheinwerfer)*
Rückstrahlung f <Akustik, Nichtfoss Energ> reflection
Rückstrahlungsvermögen n <Nichtfoss Energ> reflectance
Rückstreueffekt m <Kerntech> backscatter effect

rückstreuen

rückstreuen v <Optik> backscatter
rückstreuendes Material n <Kerntech> backscatterer
Rückstreufehler m <Kerntech> backscatter error
Rückstreumesser m <Kerntech> *(AE)* backscatter gage, *(BE)* backscatter gauge
Rückstreumethode f <Telekom> backscattering technique
Rückstreupeak m <Strahlphys> backscatter peak
Rückstreuung f 1. <Elektronik> backscattering; 2. <Funktech> backscatter; 3. <Optik> backscattering; 4. <Raumfahrt> backscatter; 5. <Strahlphys, Telekom> backscattering
Rückstreuverfahren n <Optik> backscattering technique
Rückstreuverlust m <Elektrotech> spill
Rückstrom m 1. <Elektrotech> return current; reverse current *(Stromerzeuger)*; 2. <Hydraul> backflow *(Kesselwasser)*
Rückstromauslösung f <Elektriz> reverse current protection
Rückstrombremsen n <Elektriz> reverse current circuit breaking
Rückstromkoeffizient m <Elektrotech> return current coefficient
Rückstromrelais n <Elektriz> reverse current relay
Rückströmung f 1. <Hydraul> backset; 2. <Ker & Glas> withdrawal current
Rücktitration f <Chemie> back titration
Rücktransformator m <Elektronik> reconverter
rücktreibende Kraft f <Mechan, Phys> restoring force
rücktreibendes Moment n <Mechan> restoring torque
Rücktrieb m <Nichtfoss Energ> drag *(Hydraulik)*
Rückübertrag m <Comp & DV> carry back
Rückverdampfer m <Erdöl> reboiler
Rückverfolgbarkeit f <Qual> traceability
rückverformen v <Fertig> recover
Rückverformung f <Kunststoff> rebound
Rückvergrößerung f <Konstzeich> re-enlargement
Rückvergrößerungsverfahren n <Konstzeich> re-enlarging process
Rückwand f 1. <Elektronik> backplate *(Gerätetechnik)*; backplane *(eines Computers)*; 2. <Foto> back *(einer Kamera)*; 3. <Verpack> back face, end panel
Rückwandanschlusssystem n <Comp & DV, Elektronik> backplane interconnect system
Rückwandbussystem n <Comp & DV> backplane bus system
Rückwandleiterplatte f <Elektronik> backplane PCB, backplane printed circuit board, motherboard
rückwärtiger Raststift m <Fertig> back pin
rückwärts bewegend adj <Raumfahrt> retrograde
rückwärts fahren v <Eisenbahn> reverse *(Zug)*
rückwärts hobeln v <Fertig> shape on the return stroke
rückwärts laufender Propeller m <Lufttrans> blade pitch reversal
Rückwärts... <Eisenbahn, Elektronik, Kfztech, Maschinen> backward
Rückwärtsabtastung f <Fernseh, Gerät> reverse scan
Rückwärtsbestätigung f <Comp & DV> reverse authentication
Rückwärtsbewegung f <Maschinen> backward motion, backward movement
Rückwärtsdiode f <Elektronik> backward diode
Rückwärtsfahren n <Eisenbahn, Kfztech> reversing
Rückwärtsfließpressen n <Maschinen> back extrusion, backward extrusion, inverted extrusion, reverse extrusion
Rückwärtsflug m <Lufttrans> backward flight
Rückwärtsgang m 1. <Kfztech> reverse gear; 2. <Mechan> reverse
Rückwärtsgangzwischenwelle f <Kfztech> reverse idler shaft

Rückwärtshub m 1. <Hydraul> back stroke *(Kolben)*; 2. <Maschinen> return stroke
Rückwärtsindikatorbit n <Telekom> backward indicator bit
Rückwärtsnähen n <Textil> reverse sewing
Rückwärtspropagierung f <Künstl Int> backpropagation, BP, backward error propagation, error back propagation *(neurale Netze)*
Rückwärtsregelung f <Telekom> feedback AGC, feedback automatic gain control
Rückwärtsschweißen n <Mechan> backhand welding
Rückwärtssortierung f <Comp & DV> backward sort
Rückwärtssperrzustand m <Elektriz> reverse blocking state
Rückwärtsspülungswasser n <Wasserversorg> backwash water
Rückwärtsstart m <Lufttrans> backward takeoff, rearward takeoff *(Hubschrauber)*
Rückwärtsstoßspitzenspannung f <Elektriz> circuit non-repetitive peak reverse voltage
Rückwärtssuche f <Künstl Int> backward search
Rückwärtssuchlauf m <Comp & DV> backward search
Rückwärtsverkettung f <Künstl Int> back chaining, backward chaining
Rückwärtswelle f 1. <Elektrotech> backward wave *(Wanderfeldröhre)*; backward wave *(Übertragungsleitung)*; 2. <Telekom> backward wave
Rückwärtswellenmagnetfeldröhre f <Elektronik> O-type tube
Rückwärtswellenoszillator m 1. <Elektronik> carcinotron *(Messtechnik, Hochfrequenztechnik)*; 2. <Elektronik, Phys> backward-wave oscillator; 3. <Telekom> carcinotron; 4. <Telekom> backward-wave oscillator
Rückwärtswellenoszillator m Typ O <Elektronik> O-type carcinotron
Rückwärtswellenoszillatorröhre f <Elektronik> BWT, backward-wave tube
Rückwärtswellenröhre f 1. <Elektronik> O-type tube; 2. <Phys> backward-wave tube
Rückwärtswellenverstärker m <Elektronik, Funktech, Telekom> BWA, backward-wave amplifier
Rückwärtszähler m 1. <Elektronik> down counter; 2. <Gerät> backward counter, count-down counter
Rückwärtszählimpuls m <Elektronik> down pulse
Rückwärtszeichen n <Telekom> backward signal
Rückwasser n 1. <Hydraul> backflow *(Kesselwasser)*; 2. <Nichtfoss Energ> tail water; 3. <Papier, Wasserversorg> backwater
Rückwasserstand m <Nichtfoss Energ> tail water level
Rückweg m <Telekom> return path
Rückweisung f <Qual> rejection
Rückweisungswahrscheinlichkeit f <Qual> probability of rejection
rückwirkende Elektrode f <Elektrotech> reflecting electrode
Rückwirkung f <Phys> reaction
Rückwirkungsfreiheit f <Gerät> absence of feedback
Rückwirkungsimpedanz f <Elektrotech> reflected impedance
Rückwirkungsleistung f <Elektrotech> reflected power
Rückwirkungsspannung f <Elektrotech> reflected voltage
Rückwirkungswiderstand m <Elektrotech> reflected resistance
Rückzahlung f <Patent> reimbursement
Rückziehungsverfahren n <Künstl Int> backtracking
Rückzugfeder f <Maschinen> drawback spring, pull-back spring, retractile spring
Rückzündung f 1. <Elektrotech> arc back *(Gleichrichter)*; 2. <Kfztech> backfire

rückzuweisende Qualitätsgrenzlage f <Qual> limiting quality level, lot tolerance percentage of defectives
rückzuweisende Qualitätslage f <Qual> rejectable quality level
Ruder n 1. <Phys> control surface; 2. <Wassertrans> oar; rudder *(Steuerruder)* • **Ruder mittschiffs legen** <Wassertrans> right the helm
Ruderanlage f <Wassertrans> helm
Ruderblatt n <Wassertrans> rudder blade
Ruderboot n <Wassertrans> rowing boat
Rudergänger m <Wassertrans> helmsman
Rudergängerin f <Wassertrans> helmswoman
Ruderhacke f <Wassertrans> skeg *(Schiffbau)*
Ruderhaken m <Wassertrans> pintle
Ruderhaus n <Wassertrans> wheel house
Ruderkette f <Wassertrans> steering chain
Ruderkoker m <Wassertrans> rudder trunk *(Schiffbau)*
Ruderlagenanzeiger m <Wassertrans> helm indicator, rudder angle indicator
Ruderleine f <Wassertrans> tiller rope *(Schiffbau)*
Rudermaschine f <Wassertrans> steering gear *(Schiffantrieb)*
Rudermoment n <Lufttrans> hinge moment
rudern v <Wassertrans> row
Ruderöse f <Wassertrans> rudder brace
Ruderpfosten m <Wassertrans> rudder post
Ruderpinne f <Wassertrans> tiller *(Bootbau)*; helm *(Schiffbau)*
Ruderquadrant m <Wassertrans> rudder quadrant
Ruderschaden m <Wassertrans> helm damage
Ruderschaft m <Wassertrans> rudder stock
Ruderservoantrieb m <Lufttrans> servo unit
Rudersorgkette f <Wassertrans> rudder chain
Ruderstand m <Wassertrans> helm *(Schiffbau)*
Ruf m <Telekom> call, ringing *(Läuten)*
Ruf m/nicht zur Verbindung führender <Telekom> lost call
Rufabwicklung f <Telekom> call handling
Rufannahmesignal n <Telekom> call acceptance signal *(Freizeichen)*
Rufannahmezeichen n <Telekom> call acceptance signal
Rufbehandlung f <Telekom> call control
Rufbetrieb m <Telekom> manual operation *(Handvermittlung)*
Rufdauer f <Telekom> ringing duration
Rufen n <Telekom> alerting
rufende Leitung f <Telekom> calling line
rufender Teilnehmer m <Telekom> caller, calling customer, calling party, calling subscriber
rufendes Telefon n <Telekom> calling telephone
Ruferkennung f <Telekom> call detection
Rufhaltung f *(RH)* <Mobilkom> call holding
Rufkanal m <Mobilkom> calling channel, paging channel
Rufkennung f <Telekom> call identification
Ruflampe f <Telekom> calling lamp
Rufmaschine f <Telekom> ringing machine
Rufmitnahme f <Telekom> follow-me *(Leistungsmerkmal)*
Rufnummer f <Telekom> directory number • **zwei separate Rufnummern** <Telekom> alternate line service *(Telefon)*
Rufnummernanzeige f <Telekom> call indicating device
Rufnummernanzeige f des Angerufenen <Telekom> called-number display
Rufnummernanzeige f des Anrufers <Telekom> calling line identification presentation, CLIP *(Dienstmerkmal des Euro-ISDN)*
Rufnummernanzeige f kommend <Telekom> connected line identification presentatiom, COLP

Rufnummernanzeigeunterdrückung f, gehend <Telekom> calling line identification repression, CLIR *(Dienstmerkmal des Euro-ISDN)*
Rufnummernanzeigeunterdrückung f, kommend <Telekom> connected line identification restriction, COLR *(Dienstmerkmal des Euro-ISDN)*
Rufnummernspeicher m <Telekom> call number memory, number finder, stored call number directory
Rufnummernverzeichnis n <Telekom> directory
Rufnummernwahl f über Spracheingabe <Telekom> voice-dialling
Rufnummernwechsel f <Telekom> number change
rufpflichtiger Punkt m <Qual> call point
Rufphase f <Telekom> ringing period
Rufportabilität f <Telekom> call portability
Rufsatz m <Telekom> ringer *(Telefon)*
Rufstrom m <Telekom> ringing current
Rufumleitung f besetzt <Telekom> call forwarding busy, CFB
Rufumleitung f fest <Telekom> call forwarding unconditional, CFU
Rufumleitung f nach Zeit <Telekom> call forwarding no reply, CFNR
Rufumlenkung f <Telekom> call rerouting
Rufunterdrückung f <Telekom> suppressed ringing *(Telefon)*
Rufverzug m <Telekom> *(AE)* postdialing delay, *(BE)* postdialling delay
Rufweite f <Wassertrans> hailing distance
Rufweiterleitung f bei Besetzt <Telekom> call diversion on busy, call forwarding on subscriber busy *(CFB)*
Rufwiederholung f <Telekom> call repetition, dialling repetition, last number recall, redialling
Rufzeichen n 1. <Funktech> call sign; 2. <Telekom> calling signal; 3. <Wassertrans> call sign *(Funk)*
Ruhe... <Fernseh> static
Ruhehörschwelle f <Akustik> threshold in quiet
Ruhekontakt m <Elektriz, Elektrotech> break contact, normally closed contact, resting contact
Ruhekufen fpl <Trans> rest skids
Ruhelage f <Elektriz, Gerät> rest position *(Zeiger)*
Ruhemasse f <Phys, Strahlphys, Teilphys> rest mass
ruhende Luftschicht f <Umweltschmutz> static air layer
ruhende Station f <Comp & DV> dormant terminal
ruhender Anker m <Elektrotech> stationary armature
ruhender Kontakt m <Elektriz> fixed contact
ruhender Verkehr m <Trans> stationary traffic
Ruheperiode f <Lufttrans> rest period
Ruhepunkt m <Mechan> *(AE)* dead center, *(BE)* dead centre
Ruhespannung f <Elektrotech> open circuit voltage
Ruhestellung f 1. <Gerät> rest position *(Zeiger)*; 2. <Maschinen> neutral position
Ruhestreifen m <Aufnahme> unmodulated track
Ruhestrom m 1. <Comp & DV> idle current *(Datenfernverarbeitung)*; 2. <Elektriz> no-load current, static current
Ruheverlust m <Elektriz> fixed loss
Ruhewert m <Elektriz> steady state value
Ruhewinkel m 1. <Bau> angle of repose; 2. <Phys> angle of friction
Ruhezustand m 1. <Comp & DV> stable state; 2. <Elektriz> steady state; 3. <Phys> state of rest; 4. <Telekom> idle state, idle condition
Ruhezustandsbetrieb m <Kontroll> stand-by mode
ruhig adj 1. <Comp & DV> quiescent; 2. <Maschinen> quiet, smooth
ruhiger Lauf m <Maschinen> quiet running, smooth running
Ruhofen m <Thermod> batch furnace

Rührapparat

Rührapparat *m* <Chemtech> agitator
Rührarm *m* <Abfall> rabble arm
rühren *v* <Papier> agitate
Rühren *n* <Ker & Glas> stirring
Rührenzange *f* <Mechan> long-nose pliers
Rührer *m* 1. <Chemtech> agitator; 2. <Ker & Glas> bubbler; thimble *(für im Hafen erschmolzenes optisches Glas)*; 3. <Kunststoff, Labor> stirrer; 4. <Lebensmittel> agitator
Rührflügel *m* <Papier> impeller
rührfrischen *v* <Fertig> puddle
Rührfrischen *n* <Fertig> puddling
Rührmaschine *f* 1. <Chemtech> agitating machine; 2. <Maschinen> agitator
Rührschaufel *f* 1. <Labor> stirrer blade; 2. <Maschinen> paddle
Rührspatel *m* <Ker & Glas> paddle
Rührwerk *n* 1. <Abfall> agitator; 2. <Erdöl> agitator *(Kohlevergasung)*; 3. <Fertig> agitator; 4. <Kunststoff> mixer; 5. <Maschinen, Papier> agitator
Rührwerkskessel *m* <Chemtech> agitating vessel
Rumpelfilter *n* <Aufnahme> rumble filter
Rumpelgeräusch *n* <Akustik> rumble
Rumpelpegel *m* <Aufnahme> rumble level
Rumpf *m* 1. <Elektrotech> carcass *(eines Elektromotors)*; 2. <Lufttrans> fuselage, hull; 3. <Wasserversorg> hull *(eines Baggers)*
Rumpf *m* **ohne Zubehör** <Lufttrans> bare fuselage, bare hull
Rumpfatom *n* <Kerntech> atomic trunk *(nach Abspaltung von Elektronen)*
Rumpfnase *f* <Funkort> radar dome, radome
Rumpfstufe *f* <Lufttrans> hull step
Rumpfverbindungsbeschlag *m* <Raumfahrt> fuselage attachment
Rund... <Bau, Elektrotech, Fertig, Maschinen, Mechan, Phys, Telekom> circular, roll, round
Rundbiegen *n* <Fertig> roll bending
Rundblickaufnahme *f* <Foto> panoramic photograph
Rundbogen *m* <Bau> Roman arch, round arch, semicircular arch
Runddichtring *m (O-Ring)* <Kfztech, Maschinen> O-ring
Runde *f* <Anstrich> lap *(beim Rennen)*
Rundeisen *n* 1. <Maschinen> round bar; 2. <Metall> round bar iron
runden *v* 1. <Comp & DV> round; 2. <Maschinen> radius; 3. <Math> round, round off
Runden *n* <Comp & DV, Math> rounding, rounding off
Runden *n* **und Binden** *n* <Druck> rounding and binding
Runden *n* **und Rückenbildung** *f* <Druck> rounding and backing
runderneuerter Reifen *m* <Kfztech> *(AE)* recapped tire, *(BE)* recapped tyre, *(AE)* remolded tire, *(BE)* remoulded tyre, *(AE)* retreaded tire, *(BE)* retreaded tyre
Runderneuerung *f* <Kfztech> recapping, *(AE)* remolding, *(BE)* remoulding, retreading *(eines Reifens)*
Rundfaxfunktion *f* <Telekom> fax broadcast
Rundfeile *f* <Maschinen> round file, round-edge file
rundförmiges Kohlenstück *n* <Kohlen> cob
Rundfräseinrichtung *f* <Maschinen> circular milling attachment
Rundfräsen *n* <Maschinen> circular milling
Rundfräsmaschine *f* <Maschinen> circular milling machine
Rundfunk *m* 1. <Fernseh> broadcasting; 2. <Funktech> broadcasting, radio; 3. <Telekom> radio broadcasting • **durch Rundfunk verbreiten** <Funktech> broadcast
Rundfunkgroßverstärker *m* <Aufnahme> PA amplifier, public address amplifier
Rundfunknetz *n* <Fernseh, Telekom> broadcasting network
Rundfunknorm *f* <Fernseh, Telekom> broadcast standard
Rundfunkqualität *f* <Fernseh, Funktech> broadcast quality
Rundfunksatellit *m* <Fernseh, Telekom> broadcasting satellite
Rundfunksatellit *m* **für Direktempfang** <Raumfahrt> direct-broadcast satellite
Rundfunksatellitendienst *m* <Raumfahrt, Telekom> broadcasting satellite service
Rundfunksender *m* 1. <Funktech, Telekom> broadcasting station, sound broadcast transmitter; 2. <Telekom> broadcast transmitter
Rundfunksendung *f* <Funktech> broadcasting • **als Rundfunksendung ausstrahlen** <Funktech> broadcast
Rundfunkstörung *f (BCI)* <Funktech> broadcast interference *(BCI)*
Rundfunkübertragung *f* <Telekom> broadcasting, transmission
Rundfunkübertragungsleitung *f* <Fernseh> program circuit
Rundfunkumsetzer *m* <Fernseh> network broadcast repeater station
Rundfunkvideografik *f* <Fernseh> broadcast videographics
Rundgesenkoberteil *n* <Maschinen> top rounding tool
Rundgewinde *n* <Maschinen> round thread
Rundgliederkette *f* <Maschinen> round link chain
Rundhaus *n* <Eisenbahn> roundhouse
Rundheck *n* <Wassertrans> elliptical stern
Rundheit *f* <Mechan> concentricity
Rundheitsmessgerät *n* <Metrol> roundness measuring instrument
Rundhobel *m* <Bau> compass plane
Rundhobeln *n* <Maschinen> circular planing
Rundhohlleiter *m* <Elektrotech, Phys, Telekom> circular waveguide
Rundhöhlung *f* <Kohlen> concave
Rundholz *n* <Fertig> spar
Rundkämmmaschine *f* <Textil> circular combing machine
Rundkammstuhl *m* <Textil> circular combing machine
Rundkeil *m* <Maschinen> round key
Rundkessel *m* <Hydraul> plain cylindrical boiler
Rundkolben *m* <Labor> round-bottomed flask
Rundkopf *m* <Maschinen> round head
Rundkopfbolzen *m* <Bau> button-head bolt
Rundkopfschraube *f* 1. <Maschinen> round-head bolt, round-head screw; 2. <Mechan> cheese-head screw
Rundkopfschraube *f* **mit Schlitz** <Bau, Maschinen> slotted round-head bolt
Rundkörper *m* <Konstzeich> round component
Rundkuppe *f* <Maschinen> rounded end
Rundlauf *m* 1. <Fertig> concentricity; 2. <Maschinen> true running
Rundlauffehler *m* 1. <Fertig> runout; 2. <Maschinen> eccentricity
rundlaufend *adj* 1. <Fertig> concentric; 2. <Maschinen> true, true-running
Rundläufer *m* 1. <Ker & Glas> rotating machine; 2. <Kunststoff> rotary table machine
Rundlaufprüfung *f* <Fertig> concentricity test
Rundlaufsenkrechtfräsmaschine *f* <Fertig> vertical-spindle rotary-table miller
Rundloch *n* <Fertig> pritchel hole *(Amboss)*
Rundlochperforation *f* <Druck> round hole perforating
Rundmagazin *n* <Foto> rotary magazine *(Diaprojektor)*

Rundmaulzange f <Fertig> hollow tongs
Rundmeißel m <Maschinen> round-nose chisel, round-nose tool
Rundmutter f <Maschinen> round nut
Rundnaht f 1. <Fertig> circumferential seam; 2. <Kerntech> girth weld *(Schweißnaht)*
Rundnahtschweißen n <Mechan> girth weld
Rundofen m <Ker & Glas> beehive kiln
Rundprobe f <Werkprüf> round specimen
Rundprofil n <Bau> round, rounds
Rundprofilinstrument n <Gerät> round edgewise pattern instrument
Rundregner m <Wasserversorg> rotating sprayer
Rundriss m <Kerntech> penny-shaped crack
Rundschaltmaschine f <Fertig> index type of machine
Rundschalttischmaschine f <Fertig> rotary
Rundschieber m <Hydraul> circular slide-valve
Rundschleifen n <Maschinen, Mechan> cylindrical grinding
Rundschleifmaschine f <Mechan> cylindrical grinder
Rundschnurring m <Maschinen> toroidal sealing ring
Rundschreiben n <Telekom> multiaddressing
Rundschuppen m <Eisenbahn> roundhouse
rundsenden v <Telekom> broadcast
Rundsenden n <Telekom> multiaddressing
Rundsichtbrille f <Sicherheit> panoramic wide-vision goggles
Rundsichtgerät n <Funkort> plan-position indicator, PPI *(Radar)*
Rundsichtpeilanzeiger m <Lufttrans> OBI, omnibearing indicator *(Flugzeug)*
Rundsichtradar n <Funkort> surveillance radar
Rundsichtvollmaske f <Sicherheit> panoramic full-face mask
Rundsieb n <Papier> vat; cylinder *(der Rundsiebpapiermaschine)*
Rundsiebpapiermaschine f <Papier> cylinder machine
Rundskale f <Kfztech> dial
Rundspruch m <Comp & DV> broadcast
Rundspruchanlage f <Lufttrans> PA system, public address system
Rundspruchmodus m <Comp & DV> broadcast mode
Rundstab m 1. <Bau> astragal, rounds; bead *(Holzbau)*; 2. <Mechan> rod
Rundstahl m 1. <Maschinen> round-nose tool; 2. <Metall> round, rounds, runner
Rundstahlkette f <Maschinen> round steel chain
Rundstrahlantenne f <Funktech> omnidirectional aerial, omnidirectional antenna
Rundstrahlkursfunkfeuer n <Funkort, Trans> omnidirectional radiorange
Rundstuhl m <Textil> circular knitting machine
Rundtaktmaschine f <Maschinen> revolving transfer machine
Rundtisch m <Maschinen> rotary table
Rundtischfräsmaschine f <Maschinen> rotary table milling machine
Rundtischschaltmaschine f <Maschinen> rotary indexing machine
Rundtörn m **mit zwei halben Schlägen** <Wassertrans> round turn and two half-hitches *(Knoten)*
Rundumetikett n <Verpack> wraparound label
Rundum-Klang m <Funktech> surround sound
Rundumlicht n <Wassertrans> all-round light *(Signal)*
Rundung f **des Rands** <Ker & Glas> rounding of the rim
Rundungsfehler m <Comp & DV, Math> rounding error
Rundzange f <Maschinen> round-nose pliers
Rundzuführdüse f <Hydraul> rounded approach orifice *(Steuerung)*

Runge f 1. <Eisenbahn> stanchion, upright; 2. <Fertig> upright; 3. <Kfztech> post *(zwischen Wagenseite und Radachse)*
Rungenpalette f <Trans> post pallet
Run-up-Bereich m <Lufttrans> run-up area
Runzelbildung f <Kunststoff> silking
runzelige Oberfläche f <Ker & Glas> cockled surface
runzelige Oberfläche f **durch ungleichmäßige Kühlung** <Ker & Glas> *(BE)* chill mark, *(AE)* chill wrinkle
Runzelkorn n <Foto> reticulation *(einer Emulsion)*
Runzellack m <Kunststoff> wrinkle paint
Rupertstropfen m <Ker & Glas> Prince Rupert drop
rupfen v <Lebensmittel> pluck *(Geflügel)*
Rupfen n 1. <Papier> picking; 2. <Textil> hessian
rupfende Kupplung f <Kfztech> grabbing clutch
Rupfestigkeit f <Papier> picking resistance
Rüsche f <Textil> ruffle
Rüschung f <Textil> gathering
Rushhour f <Trans> peak hour, rush hour
Ruß m 1. <Ker & Glas> soot; 2. <Kunststoff> carbon black
Russell-Saunders-Kopplung f *(L-S-Kopplung)* <Kerntech> Russell-Saunders coupling *(l-s coupling)*
Rußpunkt m <Maschinen> smoke point
Rußschwarz n <Erdöl> carbon black *(Petrochemie)*
Rüsteisen n <Wassertrans> chain plate *(Segeln)*
rüsten v <Maschinen> set-up
Rüsten n <Maschinen> setting up *(einer Maschine)*
Rüster m <Wassertrans> elm *(Bauholz)*
Rüstzeit f 1. <Comp & DV> set-up time; 2. <Fertig> set-up time *(Maschine)*; 3. <Druck> makeready time *(einer Druckmaschine)*; 4. <Textil> set-up time
Ruthenium n *(Ru)* <Chemie> ruthenium *(Ru)*
Ruthenium... <Chemie> ruthenic
Rutherfordium n *(Rf)* <Chemie> rutherfordium *(Rf)*
Rutherford-Streuung f <Phys> Rutherford scattering
Rutsche f 1. <Bau, Ker & Glas> chute; 2. <Maschinen, Mechan> chute, slide
Rutschen n 1. <Bau> slippage; 2. <Elektrotech> slip; 3. <Lufttrans, Trans> skidding
rutschfest adj 1. <Bau, Sicherheit> nonslip, slip-resistant; 2. <Trans> nonskidding; 3. <Wassertrans> nonslip
rutschfester Boden m <Bau, Sicherheit> non-skid floor
rutschfestes Schuhwerk n <Sicherheit> nonslip footwear
rutschfestes Sicherheitsschuhwerk n <Sicherheit> nonslip safety footwear, slip-resistant safety footwear, slip-resistant safety shoes
rutschfrei adj <Trans> nonskidding
Rutschkupplung f <Maschinen> slip clutch
rutschsicherer Bodenbelag m <Sicherheit> antislip floor covering
Rutschsicherung f *(ASD)* <Kfztech> antiskid device, antislip device, ASD *(Bremsanlage)*
Rutschung f <Kohlen> slide
Rüttelbeton m <Bau> vibrated concrete
Rüttelformmaschine f <Fertig> *(BE)* bumper
rütteln v <Kfztech> vibrate
Rütteln n <Trans> jolting
Rüttelplatte f <Abfall> tumbling station
Rüttelrost m <Fertig> shake-out
Rüttelschaffußwalze f <Bau> vibrating sheepsfoot roller
Rüttelsieb n <Chemie> vibrating screen
Rüttelsiebrost m <Kohlen> vibrating grizzly
Rütteltisch m 1. <Maschinen> concussion table; 2. <Verpack> jarring table
rüttelverdichten v <Bau> vibrate *(Beton)*
Rüttler m <Bau> vibrator
Ruzahl f <Heiz & Kälte> soot number
RWO *(Rückwärtswellenoszillator)* <Elektronik, Phys, Telekom> BWO *(backward-wave oscillator)*

R-Y-Achse f <Fernseh> R-Y axis
Rydberg-Energie f <Phys> Rydberg energy
Rydberg-Konstante f (R, Rα) <Kerntech> Rydberg constant (Rα)
R-Y-Matrix f <Fernseh> R-Y matrix
R-Y-Signal n <Fernseh> R-Y signal

S

s 1. <Elektriz> *(Schlupf)* s *(slip)*; 2. <Kerntech, Metrol, Phys> *(Spinquantenzahl)* s *(spin quantum number)*
S 1. <Elektriz, Phys> *(Poynting'scher Vektor)* S *(Poynting vector)*; 2. <Hydraul> *(Gefälle)* S *(slope)*; 3. <Kerntech> *(Quellenstärke)* S *(source strength)*; 4. <Kerntech> *(spezifische Ionisierung)* S *(specific ionization)*; 5. <Kerntech> *(Anhalteistung)* S *(stopping power)*; 6. <Kerntech> *(Spinquantengesamtzahl)* S *(total spin quantum number)*; 7. <Metrol> *(Siemens)* S *(siemens)*; 8. <Telekom> *(Zeichengabenetz)* S *(signaling network)*; *(Vertraulichkeit, Geheimhaltung)* S *(secrecy)*
Saatelement n <Kerntech> seed assembly, seed element, spike
Saatelementreaktor m <Kerntech> seed core reactor
Sabadillalkaloid n <Chemie> cevadine
Sabinan n <Chemie> sabinane, thujane
Sabine-Koeffizient m <Akustik> Sabine coefficient
Sabinen n <Chemie> sabinene, thujene
Saccharase f <Lebensmittel> invertase, sucrase; saccharase *(saccharosespaltendes Ferment)*
Saccharat n <Chemie> saccharate, sucrate
Saccharin n <Chemie> saccharin
Saccharogenamylase f <Lebensmittel> beta-amylase
Sachgebietswissen n <Künstl Int> domain knowledge
Sachverständigenabnahme f <Qual> acceptance by an authorized inspector
Sack m 1. <Kohlen, Papier> bag; 2. <Verpack> sack
Sackabfüllung f <Verpack> bag filling
Sackeinsatz m <Verpack> insertable sack
sacken v <Bau> subside *(Untergrund)*
Sackfilter n <Kohlen> bag filter
Sackförderer m <Trans> bag conveyor
Sackfüllmaschine f <Verpack> sack-filling machine
Sackkammer f <Umweltschmutz> baghouse *(Staubfilterkammer)*
Sackkarre f <Verpack> sack barrow
Sackkleid n <Textil> shift
Sackleinwand f <Textil> hessian
Sackloch n <Bau, Kohlen> blind hole *(Bohren)*
Sackmesser n <Verpack> sack knife
Sacköffnungsmaschine f <Verpack> sack-opening machine
Sackraum m <Kohlen> baghouse
Sacktank m <Lufttrans> bladder tank *(montierbar, nicht Teil des Rahmens)*
Sackung f <Abfall> settlement, settling
Sackverschließmaschine f <Verpack> sack sealer
Sackwaage f <Verpack> sack scales
Sackzunähmaschine f <Verpack> sack-closing sewing machine
Safranin n <Chemie> safranin, safranine
Safrol n <Chemie> allylmethylenedioxybenzene, safrole
Saftgehalt m <Lebensmittel> juice content

Säge f <Maschinen, Mechan> saw
Säge f mit hin- und hergehender Schnittbewegung <Fertig> reciprocating saw
Sägeband n <Fertig> saw band
Sägebank f <Bau> saw bench
Sägeblatt n 1. <Fertig> saw blade; 2. <Maschinen> blade, saw blade
Sägeblatt n mit grober Zahnteilung <Fertig> coarse-pitch blade
Sägeblattschutz m <Sicherheit> protective cover for saw blade
Sägeblock m <Bau> saw log
Sägebock m <Bau> *(AE)* buck, sawbuck, sawhorse
Sägebügel m <Maschinen> saw frame
Sägefeile f <Maschinen> saw file
Sägegrat m <Fertig> saw burr
Sägekerbe f <Maschinen> saw kerf
Sägemaschine f 1. <Bau> sawing machine; 2. <Fertig> bench saw, sawing machine; 3. <Maschinen> sawing machine
Sägemehl n <Bau> sawdust
Sägemühle f <Bau> sawmill
sägen v 1. <Bau> cut *(Holz)*; 2. <Mechan> saw
Sägen n <Bau> sawing
Sägengewinde n 1. <Fertig> buttress screwthread, buttress thread; 2. <Maschinen> buttress screwthread
Sägenut f <Fertig> saw groove, saw kerf
Sägeschärfmaschine f <Maschinen> saw-sharpening machine
Sägeschlitten m <Fertig> saw carriage
Sägeschnitt m <Bau, Fertig> saw cut
Sägespänebeton m <Bau> sawdust concrete
Sägespannkluppe f <Maschinen> saw clamp
Sägeverzahnungswälzfräser m <Fertig> sawtooth hob
Sägewelle f <Maschinen> *(AE)* saw arbor, *(BE)* saw arbour
Sägewerk n <Bau, Maschinen> sawmill
Sägezahn m <Fertig> sawtooth
Sägezahngenerator m 1. <Elektronik> relaxation oscillator *(Kippgenerator)*; 2. <Elektrotech> ramp generator; 3. <Fernseh> sawtooth generator
Sägezahngewinde n 1. <Bau> buttress thread; 2. <Maschinen> buttress screwthread
Sägezahnoszillator m <Aufnahme> sawtooth oscillator
Sägezahnriss m <Ker & Glas> serration hackle
Sägezahnschwingungen fpl <Elektriz> sawtooth oscillations
Sägezahnsignale npl <Elektrotech> sawtooth signals
Sägezahnspannung f 1. <Elektronik> linear time base; 2. <Phys> sawtooth voltage
Sägezahnstrom m <Fernseh> sawtooth current
Sägezahnteilung f <Fertig> saw pitch
Sägezahnumsetzer m <Gerät> ramp encoder
Sägezahnverschlüsseler m <Gerät> ramp encoder
Sägezahnverschlüsselung f <Gerät> sawtooth conversion
Sägezahnwellenform f <Elektrotech> ramp waveform, sawtooth waveform
saggitale Brennlinie f <Phys> sagittal focal line
saggitale Fokuslinie f <Phys> sagittal focal line
Saggitalebene f <Ergon> sagital plane
Saite f <Akustik> chord
Saitengalvanometer n <Elektriz> vibration galvanometer
Salacetol n <Chemie> salacetol
Salantol n <Chemie> salacetol, salantol
Salicin n <Chemie> salicin, saligenin
Salicyl... <Chemie> salicyl
Salicylacetol n <Chemie> salacetol, salantol
Salicylaldehyd m <Chemie> salicylaldehyde

Salicylat n <Chemie> salicylate
Salicylsäurephenylester m <Chemie> salol
Saligenin n <Chemie> salicin, saligenin
Saline f <Lebensmittel> saltern
Saling f <Wassertrans> crosstree *(Takelage)*
Salinität f <Chemie, Erdöl> salinity
Salinometer n <Kohlen> salinometer
Salmanazar f <Ker & Glas> salmanazar
Salmiak m <Fertig> ammonium chloride *(Kunststoffinstallationen)*
Salmiakgeist m <Thermod> liquid ammonia
Salol n <Chemie> salol
Salon m <Wassertrans> saloon
Salondeck n <Wassertrans> saloon deck
Salonwagen m <Eisenbahn> club car, *(AE)* parlor car, *(BE)* saloon carriage
Salpeter m <Ker & Glas, Lebensmittel> *(AE)* saltpeter, *(BE)* saltpetre
Salpeter... <Chemie> nitrous
Salpeterbildung f <Chemie> nitrification
salpetersauer adj <Chemie> nitric
Salpetersäure f <Chemie, Umweltschmutz> nitric acid
• **mit Salpetersäure behandeln** <Chemie> nitrate
Salpetrigsäureester m <Chemie> nitric ester
Salz n <Chemie> salt
Salzagens n <Kerntech> salting agent
Salzbadlöten n <Bau> salt bath brazing
salzbildend adj <Chemie> salifiable
Salzbildung f <Chemie> salification
salzbildungsfähig adj <Chemie> salifiable
Salzblase f <Ker & Glas> salt bubble
Salzdom m <Erdöl> salt dome *(Geologie)*
Salzen n <Lebensmittel> salting
Salzgehalt m <Chemie, Wassertrans, Wasserversorg> salinity
salzhaltig adj <Wasserversorg> saline
salzhaltige Luft f <Heiz & Kälte> salt-laden atmosphere
salzhaltiges Wasser n <Wasserversorg> saline water
Salzhaltigkeit f 1. <Erdöl> salinity *(Petrochemie)*; 2. <Nichtfoss Energ, Wasserversorg> salinity
Salzkissen n <Erdöl> salt pillow *(Geologie)*
Salzlösung f <Chemie> saline solution, salt liquor
Salzmutterlauge f <Chemie> bittern
Salznebel m <Werkprüf> salt mist *(Korrosionsprüfung)*
salzsauer adj <Chemie> hydrochloric
Salzsäure f <Chemie> hydrochloric acid
Salzschmelze f <Fertig> fused salt
salzschmelzengekühlt adj <Kerntech> molten-salt-cooled
Salzschmelzenreaktor m <Kerntech> molten-salt-cooled reactor
Salzsprühnebel m <Anstrich> salt spray
Salzsprühnebelprüfung f <Kunststoff> salt spray test
Salzsprühnebeltest m <Anstrich> salt spray test
Salzsprühtest m <Ker & Glas> salt spray test
Salzsprühversuch m <Kunststoff> salt spray test
Salzstock m <Erdöl> salt diapir *(Geologie)*
Salztektonik f <Erdöl> salt tectonics *(Geologie)*
Salzton m <Ker & Glas> saliferous clay
Salzwasser n 1. <Wassertrans> seawater; 2. <Wasserversorg> saline water, salt water
Salzwasserumwandlung f <Wasserversorg> saline water conversion
Samarium n *(Sm)* <Chemie> samarium *(Sm)*
Samariumeffekt m <Kerntech> samarium effect
Sammelanschluss m <Telekom> collective line, PABX line group
Sammelaufruf m <Telekom> multipolling
Sammelbatterie f <Kfztech> storage battery

Sammelbecken n <Nichtfoss Energ> storage basin, upper storage basin
Sammelbehälter m 1. <Bau> sump pan; 2. <Erdöl> receiver *(Raffinerie)*; 3. <Umweltschmutz> storage tank
Sammelelektrode f <Elektrotech> collector electrode
Sammelfahrzeug n <Abfall> waste collection vehicle
Sammelgefäß n <Labor> receiver
Sammelgleis n <Eisenbahn> main track, recessing siding
Sammelgleis n für Rangieren <Eisenbahn> advance classification track
Sammelgraben m <Wasserversorg> collecting ditch
Sammelgrube f 1. <Bau> catchpit, collecting pit; 2. <Kohlen> catchpit
Sammelgüterzug m <Eisenbahn> *(BE)* pick-up goods train, *(AE)* way freight train
Sammelgutverkehr m <Eisenbahn> groupage traffic
Sammelgutwagen m <Eisenbahn> *(AE)* groupage car, *(BE)* groupage wagon
Sammelherd m <Fertig> recipient
Sammelkanal m <Abfall> main collector, main sewer
Sammelleitung f 1. <Elektriz> bus line; 2. <Maschinen> manifold
Sammelleitungsschalter m <Elektriz> bus coupler switch
Sammellinse f 1. <Foto> convex lens; 2. <Phys, Wellphys> converging lens
sammelndes Aufschreiben n <Comp & DV> gather write
Sammelpunkt m <Trans> assembly area *(Notfall)*
Sammelreagens n <Kohlen> collecting reagent
Sammelrohr n 1. <Fertig> header; 2. <Maschinen> header, manifold; 3. <Mechan> header
Sammelruf m <Telekom> general call
Sammelschiene f 1. <Comp & DV, Elektriz> busbar; 2. <Elektrotech> bus, busbar; 3. <Maschinen> busbar
Sammelschienensystem n <Elektriz> busbar system
Sammelschienen-Trennschalter m <Elektriz> busbar-sectionalizing switch
Sammelschienenverbinder m <Elektriz> busbar coupler
Sammeltank m 1. <Abfall> collection tank; 2. <Erdöl> accumulator tank *(Raffinerie, Bohrtechnik)*
Sammelzeichnung f <Konstzeich> collection drawing
Sammler m 1. <Bau> main line; 2. <Elektrotech> secondary cell; 3. <Erdöl> manifold, pipe manifold *(Leitungsbau)*; 4. <Kohlen> collecting agent, collector
Sammlerzelle f <Elektriz, Elektrotech> accumulator cell, storage cell
Sammlung f <Elektrotech> collection *(Elektronen, Strom)*
Sammlung f von Hausmüll <Abfall> *(AE)* garbage disposal, *(BE)* refuse disposal
Sampler m <Comp & DV> sampler
Sampling n <Comp & DV> sampling
Sampling-Oszilloskop n <Gerät> sampling oscilloscope
Sampling-Theorem n <Gerät, Telekom> sampling theorem
Sampling-Verstärker m <Elektronik> sampling amplifier
Samt m <Textil> velvet
Samtdichtung f <Foto> velvet trap
Sand m <Bau, Kohlen> sand • **mit Sand abdecken** <Bau> sand • **mit Sand bestreuen** <Bau> grit
Sandäquivalent n <Bau> sand equivalent
Sandbad n <Labor> sand bath *(Erhitzen)*
Sandbelag m auf Herdbank <Ker & Glas> breezing
Sandboden m <Wassertrans> sandy bottom
Sanddecke f <Bau> blinding
Sandfalle f <Erdöl> sand trap *(Geologie)*
Sandfang m 1. <Abfall> sand filter; 2. <Bau> sand catcher; 3. <Erdöl> desander; 4. <Wasserversorg> sand trap

Sandfanganlage

Sandfanganlage f <Abfall> detritor, grit chamber, sand trap
Sandfänger m <Abfall> detritor, grit chamber, sand trap
Sandfilter n <Abfall, Kohlen> sand filter
Sandform f <Fertig> *(AE)* sand mold, *(BE)* sand mould
sandfrei adj <Erdöl> free from sand *(Bohrung)*
Sandgrund m <Wassertrans> sandy bottom
Sandguss m <Fertig> sand casting *(Zahnräder)*
sandhaltig adj <Bau> arenaceous
sandhydraulische Nassputzanlage f <Fertig> high-pressure water and sand cleaning plant, hydroblasting plant
sandhydraulisches Nassgussputzverfahren n <Fertig> high-pressure water and sand cleaning *(Gießen)*
sandig adj <Nichtfoss Energ> arenaceous
sandiger Lehm m <Bau> sandy loam
Sandkruste f <Fertig> sand skin *(Gießen)*
Sandleiste f <Fertig> lifter *(Gießen)*
Sandpapier n 1. <Druck> sandpaper; 2. <Elektriz> abrasive paper; 3. <Fertig, Mechan> sandpaper
Sandpumpe f <Erdöl> sand pump *(Seilbohren)*
Sandseil n <Erdöl> sand line *(Seilbohren)*
Sandseiltrommel f <Erdöl> sand reel *(Seilbohren)*
Sandsieber m <Fertig> sand sifter
Sandstein m <Bau, Erdöl> Geologie sandstone
Sandstrahl m <Fertig> sandblast
Sandstrahlbehandlung f <Mechan> sandblasting
Sandstrahldüse f <Ker & Glas> sandblasting nozzle
sandstrahlen v <Mechan> sandblast
Sandstrahlen n 1. <Bau> blasting, grit blasting, sandblasting; 2. <Fertig, Ker & Glas> sandblasting; 3. <Mechan> grit blasting; 4. <Meerschmutz, Wassertrans> sandblasting *(Schiffinstandhaltung)*
Sandstrahlgebläse n <Ker & Glas> sandblast apparatus
Sandstrahlmattierung f <Ker & Glas> sandblast obscuring
Sandstrahlreinigung n <Meerschmutz, Sicherheit> sandblasting
Sandstrak m <Wassertrans> garboard strake *(Schiffbau)*
Sandton m <Bau> sandy clay
Sand/Ton-Verhältnis n <Erdöl> sand-shale ratio *(Geologie)*
Sandwäsche f <Erdöl> sand washing
Sanierung f <Umweltschutz> redevelopment, rehabilitation, remedial works, renovation
sanitäre Einrichtungen fpl <Wasserversorg> sanitation
Sanitärkeramik f <Ker & Glas> sanitary ware
Sanitärtechnik f <Wasserversorg> sanitary engineering
Sanitätswache f <Wassertrans> first-aid post
Santon... <Chemie> santonic
Santonin n <Chemie> santonin
Santoninlacton n <Chemie> santonin
Saphir m <Elektronik> sapphire
Saphir-Trägermaterial n <Elektronik> sapphire substrate
Sapogenin n <Chemie> sapogenine
Saponin n <Chemie> saponin
SAR *(Such- und Rettungsdienst)* <Wassertrans> SAR, search and rescue *(Notfall)*
Sarcosin n <Chemie> *(N)*-methylaminoacetic acid, sarcosine
Sarkin n <Chemie> hypoxanthine, sarkine
SAR-Leitstelle f <Lufttrans, Wassertrans> RCC, *(AE)* rescue coordination center, *(BE)* rescue coordination centre
Satcom *(Satellitenfunk, Satellitenkommunikation)* <Funktech, Telekom, Wassertrans> satcom *(satellite communication)*
Satellit m <Phys, Telekom, Trans> satellite
Satellit m **für Navigationszwecke** <Funktech> navigational satellite

Satellit m **in Äquatorialumlaufbahn** <Raumfahrt> equatorial orbiting satellite
Satellit m **in Umlaufbahn** <Raumfahrt> orbiting satellite
Satellit m **mit erdnaher Umlaufbahn** <Funktech> LEO satellite, low-earth orbit satellite, low-earth orbiter, LEO, low-orbiting satellite
Satellit m **zur Umweltuntersuchung** <Raumfahrt> environment survey satellite
Satellit m **zur Wetterbeobachtung** <Telekom> meteorological observation satellite, meteorological satellite
Satellitenabwärtsstrecke f <Funktech> satellite downlink
Satellitenabwärtsverbindung f <Funktech, Telekom> satellite down-link
Satellitenantenne f <Telekom> *(BE)* satellite aerial, *(AE)* satellite antenna, satellite dish
Satellitenantennenempfangskonverter m <Telekom> low-noise converter *(LNC)*
Satellitenantennenempfangsumsetzer m <Telekom> low-noise block amplifier/converter *(LNB)*
Satellitenapogäumstriebwerksgruppe f <Raumfahrt> satellite apogee motor combination
Satellitenaufwärtsstrecke f <Funktech, Telekom> satellite up-link
Satellitenaufwärtsverbindung f <Funktech, Telekom> satellite up-link
Satellitenausleuchtungsbereich m <Fernseh> satellite coverage area
Satellitenbahnneigung f <Telekom> satellite orbit inclination
Satellitenbordempfänger m <Funktech, Telekom> satellite-borne receiver
Satellitenbordsender m <Funktech, Telekom> satellite-borne transmitter
Satellitendatenstation f <Comp & DV> satellite terminal
Satellitenempfangsanlage f **für DVB** <Funktech, Telekom> set top box *(STB)*
Satellitenfernmesstechnik f <Gerät> satellite telemetry
Satellitenfernsehen n <Fernseh> satellite television
Satellitenfernsprechverbindung f <Telekom> satellite telephone connection, satellite telephone link
Satellitenfunk m *(Satcom)* <Funktech, Telekom, Wassertrans> satellite communication, satellite radio *(satcom)*
Satellitenfunkortungssystem n *(RDSS)* <Funkort, Trans, Wassertrans> radio determination satellite system *(RDSS)*
Satellitenfunkverbindung f <Funktech, Telekom> satellite link, satellite radio link
satellitengestützter Mobilfunkdienst m <Mobilkom> mobile communications by satellite, mobile satellite service *(MSS)*
satellitengestützter Seenotdienst m <Funktech, Wassertrans> satellite-aided maritime search and rescue system
satellitengestützter Such- und Rettungsdienst m **auf See** <Telekom> satellite-aided maritime search and rescue system
Satelliten-Heimempfangssystem n <Telekom> satellite home-receiving system
Satellitenkanal m <Telekom> satellite channel, satellite circuit
Satellitenkanallaufzeit f <Telekom> satellite channel delay
Satellitenkanalverzögerung f <Telekom> satellite channel delay
Satellitenkommunikation f *(Satcom)* <Funktech, Telekom, Wassertrans> satellite communication, satellite radio *(satcom)*

Satellitenkommunikationssystem n <Telekom> satellite communications system
Satellitenkonstruktion f <Raumfahrt> satellite design
Satellitenlagerstätte f <Erdöl> moonpool
Satelliten-Mehrdienstesystem n <Telekom> Satellite Multiservice System
Satellitenmeteorologie f <Raumfahrt> satellite meteorology
Satellitenmobilfunk m <Mobilkom> land mobile radio satellite service, land mobile satellite service, mobile satellite communications
Satellitenmobilkommunikation f <Telekom> mobile communications by satellite
Satellitennachrichtensystem n <Telekom> satellite communications system
Satellitennavigation f *(Satnav)* <Wassertrans> satellite navigation *(satnav)*
Satellitennavigationsgerät n <Wassertrans> satellite navigator
Satellitennotfunkbake f **mit Standortangabe** <Funktech, Telekom> satellite emergency position-indicating radio beacon
Satellitennutzlast f <Raumfahrt> satellite payload
Satelliten-Querverbindung f 1. <Telekom> intersatellite link; 2. <Raumfahrt> intersatellite link *(Weltraumfunk)*
Satellitenrechner m <Comp & DV> satellite computer
Satellitenrepeater m <Telekom> satellite transponder
Satellitenrundfunk m <Funktech, Telekom> satellite broadcasting
Satellitenrundfunksendung f <Fernseh> satellite telecast
Satellitenrundfunkübertragung f <Fernseh, Funktech> broadcasting satellite transmission
Satellitenschaltung f <Raumfahrt> satellite switching
Satellitenschüssel f <Fernseh> satellite dish
Satellitenseefunksystem n <Funktech, Wassertrans> maritime satellite system
Satellitensystem-Betriebsanleitung f *(SSOG)* <Raumfahrt> satellite system operation guide, SSOG *(Weltraumfunk)*
Satellitentechnik f <Raumfahrt> satellite design
Satellitenübertragung f <Telekom> satellite transmission, transmission by satellite
Satellitenübertragungskanal m <Telekom> satellite channel, satellite circuit
Satellitenumlaufbahn f <Telekom> satellite orbit
Satellitenumlaufbahnhöhe f <Telekom> orbital altitude, orbital hight
Satellitenumlaufbahnneigung f <Telekom> inclination of satellite-orbit, orbital inclination, satellite orbit inclination
Satellitenverbindung f <Raumfahrt> satellite link
Satinage f <Papier> glazing
satinieren v <Foto> glaze
Satiniermaschine f <Verpack> glazing machine
Satnav *(Satellitennavigation)* <Wassertrans> satnav *(satellite navigation)*
satt adj 1. <Druck> saturated; 2. <Fertig> tight *(Auflage)*
Sattdampf m 1. <Heiz & Kälte, Maschinen> saturated steam; 2. <Nichtfoss Energ> wet steam; 3. <Thermod> saturated steam
sattdampfgekühlt adj <Heiz & Kälte> saturated steam-cooled
sattdampfgekühlter Reaktor m <Kerntech> saturated steam-cooled reactor
Sattel m <Maschinen> saddle
Sattel m **der Scheibenbremse** <Eisenbahn> *(BE)* disc brake calliper, *(AE)* disk brake caliper
Sattelauflieger m 1. <Kfztech> semitrailer *(Sattelschlepper, Sattelzug)*; 2. <Mechan> trailer

Sattelauflieger m **mit Plane** <Trans> tilt-type semitrailer
Sattelbahn f <Eisenbahn> monorail with straddling cars
Sattelblech n <Bau> ridge plate
Satteldach n <Bau> couple close roof, ridge roof, shed roof
Satteleinschienenbahn f <Eisenbahn> saddle bag monorail
Sattelheftung f <Druck> saddle stitching
Sattelholz n <Bau> saddle
Sattelkombination f <Trans> saddle mount combination
Sattelkraftfahrzeug n <Kfztech> *(AE)* semitrailer motor truck, *(BE)* semitrailer motor vehicle
Sattelkupplung f <Kfztech> fifth wheel *(Sattelschlepper)*
Sattelpunkt m 1. <Math> saddle point; 2. <Metall> node, saddle point
Sattelrevolverdrehmaschine f <Maschinen> capstan lathe, ram-type turret lathe
Sattelschlepper m 1. <Bau> *(BE)* bogie, *(AE)* trailer; 2. <Kfztech> articulated vehicle, *(BE)* articulated lorry, *(AE)* articulated truck
Sattelschlepperzug m <Kfztech> *(BE)* articulated lorry, *(AE)* articulated truck
Satteltrichterwagen m <Eisenbahn> hopper wagon
Sattelverschluss m <Ker & Glas> saddled finish
Sattelwagen m <Eisenbahn> saddle-bottomed car
Sattelwange f <Bau> cut stringer; open wall string *(Treppe)*
Sattelzug m 1. <Eisenbahn> articulated train; 2. <Kfztech> articulated vehicle
Sattelzugmaschine f <Kfztech> *(AE)* semitrailer towing truck, *(BE)* semitrailer towing vehicle, *(BE)* semi-trailer lorry, *(AE)* semitrailer truck
Sättigung f 1. <Comp & DV, Elektriz, Elektronik, Elektrotech, Fernseh, Funktech, Heiz & Kälte, Kerntech, Papier, Phys> saturation; 2. <Raumfahrt> saturation point *(Weltraumfunk)*
Sättigungs-Ausgangsleistung f <Elektronik> saturation output power
Sättigungs-Ausgangsstatus m <Elektronik> saturation output state
Sättigungsbedingungen fpl <Elektrotech> saturation conditions
Sättigungsbereich m <Elektronik> saturation region
Sättigungsbetriebsart f <Elektronik> saturated mode
Sättigungscharakteristik f <Strahlphys> saturation characteristic
Sättigungsdampfdruck m <Heiz & Kälte, Phys> *(AE)* saturated vapor pressure, *(BE)* saturated vapour pressure, *(AE)* saturation vapor pressure, *(BE)* saturation vapour pressure
Sättigungsdrossel f <Elektrotech> saturable core reactor, saturable reactor
Sättigungsdruck m <Thermod> saturation pressure
sättigungsfähige Drossel m <Phys> saturable reactor *(magnetischer Verstärker)*
Sättigungsgrad m 1. <Aufnahme> saturation level; 2. <Kohlen, Maschinen, Trans> degree of saturation
Sättigungshärten n <Metall> saturation hardening
Sättigungsinduktion f 1. <Elektriz> saturation magnetization; 2. <Elektrotech> saturation induction
Sättigungskonzentration f <Heiz & Kälte, Thermod> saturation concentration
Sättigungskurve f <Thermod> saturation curve
Sättigungslinie f <Thermod> saturation curve
Sättigungslogik f <Kontroll> saturation logic
Sättigungsmagnetisierung f <Elektrotech> saturation magnetization
Sättigungsmittel n <Chemie> saturant
Sättigungspunkt m <Heiz & Kälte> saturation point

Sättigungssieden

Sättigungssieden n <Kerntech> saturated boiling
Sättigungssignal n <Elektronik> saturation signal
Sättigungsspannung f <Elektrotech> saturation voltage
Sättigungsstörung f <Comp & DV> saturation noise
Sättigungsstreifenbildung f <Fernseh> banding on saturation
Sättigungsstrom m (IS) <Elektronik, Funktech, Phys> saturation current (IS)
Sättigungstemperatur f <Heiz & Kälte> saturation temperature
Sättigungstest m <Comp & DV> saturation testing
Sättigungstransformator m <Elektrotech> saturable transformer
Sättigungstransistor m <Elektronik> saturated transistor
Satz m 1. <Comp & DV> record, set; 2. <Druck> composition, matter, typesetting; 3. <Lebensmittel> batch; 4. <Maschinen> set (von Werkzeugen) • **zu einem Satz zusammengestellt** <Fertig> ganged • **zu einem Satz zusammenstellen** <Maschinen> gang
Satz m **des Pythagoras** <Math> Pythagoras' theorem
Satz m **Gewichte** <Labor> set of weights
Satz m **mit fester Länge** <Comp & DV> fixed-length record
Satz m **mit variabler Länge** <Comp & DV> variable-length record
Satz m **von Ersatzlinsen** <Foto> set of supplementary lenses
Satz m **von Patentansprüchen** <Patent> set of claims
Satz m **Wechselräder** <Maschinen> set of change wheels
Satz m **Werkzeuge** <Maschinen> tool set
Satzanordnung f <Druck> layout
Satzanweisung f <Comp & DV> record line
Satzanzahl f <Comp & DV> record count
Satzart f <Comp & DV> record type
Satzaufspannung f <Fertig> ganging
Satzbetrieb m 1. <Chemtech> batch processing, batchwise operation; 2. <Lebensmittel> batchwise operation
Satzbreite f <Druck> measure
Satzende n <Comp & DV> end of record
Satzfahnen fpl <Druck> typesetting galleys
satzfertig adj <Druck> ready for typesetting
Satzfirma f <Druck> typesetting company
Satzfläche f <Druck> live area, text area, type area
Satzformat n <Comp & DV> record format, record layout
Satzfortschreibung f <Comp & DV> record updating
Satzfräsen n <Fertig, Maschinen> gang milling
Satzfräser m <Maschinen> gang-milling cutter
Satzfräsmaschine f <Maschinen> gang-milling cutter
Satzgruppe f 1. <Comp & DV> record set; 2. <Telekom> circuit group
Satzlänge f <Comp & DV> record length
Satzmodus m <Comp & DV> record mode
satzreif adj <Druck> ready for typesetting
Satzspiegel m <Druck> live area, text area, type area
Satztrennzeichen n <Comp & DV> record separator
Satzzeichen npl <Druck> punctuation marks
Satzzeichnung f <Konstzeich> set drawing
sauber adj <Umweltschutz> non-polluting, unpolluted
saubere Technologie f <Abfall> NWT, clean technology
saubere Umgebung f <Heiz & Kälte> clean situation
sauberer Abzug m <Druck> clean proof
sauberer Schnitt m <Ker & Glas> clean cut
Sauberkeitsschicht f <Bau> blinding, subbase; base course (Tiefbau)
Säuberung f <Meerschmutz, Umweltschutz> clean-up
Säuberung f **der Küstenlinie** <Meerschmutz> shoreline clean-up
Säuberungsverfahren n <Meerschmutz> clean-up technique

sauer adj <Lebensmittel> sour
sauer werden v <Umweltschmutz> acidify
sauer zugestellt adj <Fertig> acid-lined
sauer zugestellter SM-Ofen m <Fertig> acid open-hearth furnace
Sauergas n <Erdöl> sour gas
säuerlich adj <Chemie> acidulous, sourish
säuern v 1. <Anstrich> acidify; 2. <Chemie, Lebensmittel, Papier, Textil, Umweltschmutz> acidify
Säuern n <Chemie> acidulating
säuernd adj <Papier> acidifying
Sauerstoff m <Chemie> oxygen • **mit Sauerstoff angereichert** <Chemie> oxygenated • **mit Sauerstoff anreichern** <Chemie> oxygenate
Sauerstoffalterung f **in der Druckkammer** <Kunststoff> (BE) oxygen bomb ageing, (AE) oxygen bomb aging (Kautschuk)
Sauerstoffanreicherung f 1. <Chemie> oxygenation; 2. <Ker & Glas> oxygen boosting
Sauerstoffatemgerät n 1. <Raumfahrt> oxygen respirator; 2. <Sicherheit> oxygen breathing apparatus
Sauerstoffaufblaskonverter m (BOF) <Metall> basic oxygen furnace (BOF)
Sauerstoffaufblasverfahren n <Metall> oxygen lancing
Sauerstoffblasstahl m <Metall> basic oxygen steel
Sauerstoffbrennschneiden n <Maschinen> oxycutting, oxygen cutting
Sauerstofferzeuger m <Bau> oxygen generator
Sauerstoff-Flasche f <Fertig, Raumfahrt> oxygen cylinder
sauerstofffreie Säure f <Chemie> hydracid
sauerstoffhaltig adj <Chemie> oxidic
sauerstoffhobeln v <Fertig> torch-gouge
Sauerstoffhobeln n <Fertig> gouging
Sauerstoffindex m (LOI) <Kunststoff> limiting oxygen index (LOI)
Sauerstofflanze f <Maschinen, Metall> oxygen lance
Sauerstoff-Lichtbogenschneiden n <Bau> oxyarc cutting, oxygen arc cutting
Sauerstoff-Lichtbogen-Schneidverfahren n <Fertig> oxyarc cutting, oxygen arc cutting
Sauerstoffmaske f <Raumfahrt, Sicherheit> oxygen mask
Sauerstoffofen m <Kohlen> oxygen furnace
Sauerstoffschneiden n <Mechan> oxycutting, oxygen cutting
Sauerstoffträger m 1. <Anstrich> oxidizer; 2. <Chemie> oxidant
Sauerstoffvorrat m <Raumfahrt> oxygen supply
Sauerstoffzufuhrregler m <Raumfahrt> oxygen regulator
Sauerteig m <Lebensmittel> leaven, sour dough
Säuerung f 1. <Erdöl> acidization (Bohrung); 2. <Papier> acidification
säuerungsfähig adj <Chemie, Papier> acidifiable
Säuerungsmittel n 1. <Chemie> acidifier; 2. <Textil> acidifying agent
Saug... <Bau, Ker & Glas, Meerschmutz, Wassertrans, Wasserversorg> sucking
Saugabschöpfgerät n <Meerschmutz> direct-suction skimmer (für Öl von der Wasseroberfläche)
Saugbagger m 1. <Bau> pump dredger; 2. <Wassertrans> suction dredge, suction dredger; 3. <Wasserversorg> suction dredge
Saugbagger m **mit Laderaum** <Wassertrans> suction hopper dredge, suction hopper dredger (Baggern)
Saugbaggerlöffel m <Wassertrans> dredger bucket
Saug-Blas-Verfahren n <Ker & Glas> suck-and-blow process
Saugbrenner m <Bau> low-pressure blowpipe

Saugdrosselventil *n* <Kfztech> suction throttling valve
saugen *v* <Wasserversorg> draw *(Quellwasser)*
Saugen *n* <Heiz & Kälte, Textil> suction
Sauger *m* <Mechan> exhauster
saugfähig *adj* 1. <Chemie> absorbent; 2. <Papier> absorptive
saugfähiger Beutel *m* <Verpack> moisture-absorbent bag
Saugfähigkeit *f* 1. <Papier> absorptive capacity; 2. <Verpack> absorbency
Saugfähigkeitswert *m* <Papier> absorbency value
Saugfegmaschine *f* <Abfall> vacuum street sweeper
Saugfilter *n* 1. <Abfall> suction filter; 2. <Chemtech> suction strainer; 3. <Fertig> suction filter; 4. <Kohlen> vacuum filter; 5. <Wasserversorg> suction filter
Saugform *f* <Ker & Glas> *(AE)* suction mold, *(BE)* suction mould
Sauggautsche *f* <Papier> suction couch roll
Sauggebläse *n* 1. <Heiz & Kälte> exhaust fan; 2. <Labor> aspirator
Saugheber *m* <Bau> plunger elevator
Saughöhe *f* 1. <Fertig> *(BE)* lift *(Pumpe)*; 2. <Kohlen> capillary rise; 3. <Wasserversorg> suction head, suction lift
Saughöhenprüfgerät *n* <Papier> bibliometer
Saughopperbagger *m* <Wassertrans> suction hopper dredge, suction hopper dredger
Saughub *m* <Fertig> intake stroke
Saugkammer *f* 1. <Fertig> suction port; inlet chamber *(Pumpe)*; 2. <Kfztech> suction chamber
Saugkasten *m* <Papier> suction box, vacuum box
Saugkastenbelag *m* <Papier> suction box cover
Saugkopf *m* <Meerschmutz> skimming head *(an Abschöpfgerät)*
Saugkorb *m* 1. <Fertig> suction strainer; screen assembly *(Kunststoffinstallationen)*; 2. <Meerschmutz> strainer; 3. <Wasserversorg> rose, strainer
Saugkreis *m* 1. <Elektriz> absorption circuit, acceptor, acceptor circuit; 2. <Funktech> acceptor circuit
Saugleitung *f* 1. <Fertig> suction line; 2. <Mechan> intake manifold; 3. <Sicherheit> suction hose *(Feuerwehr)*; 4. <Wasserversorg> suction pipe
Saugluft *f* 1. <Heiz & Kälte> suction air; 2. <Maschinen> *(AE)* forced draft, *(BE)* forced draught
Saugluftbremse *f* 1. <Eisenbahn> vacuum brake; 2. <Maschinen> atmospheric brake, vacuum brake
Sauglüfter *m* <Maschinen> exhaust fan, extraction fan, extractor fan, suction fan
Saugluftkühlung *f* <Heiz & Kälte> *(AE)* forced-draft cooling, *(BE)* forced-draught cooling
Sauglüftung *f* <Fertig> extract ventilation
Saugmaschine *f* <Ker & Glas> suction machine
Saugmotor *m* <Kfztech> unsupercharged engine
Saugmund *m* <Abfall> suction port *(am Kehrfahrzeug)*
Saugpapier *n* <Druck> absorbent paper
Saugpumpe *f* 1. <Hydraul> aspiration pump, aspiring pump, suction pump; 2. <Lebensmittel> suction pump; 3. <Maschinen> aspiration pump, drawing pump, double-acting pump; 4. <Papier, Textil> suction pump; 5. <Wasserversorg> lift pump, suction pump
Saugrohr *n* 1. <Bau> tail pipe; 2. <Kfztech, Maschinen> suction pipe; 3. <Nichtfoss Energ> *(AE)* draft tube, *(BE)* draught tube; 4. <Wassertrans> suction pipe; 5. <Wasserversorg> suction pipe, tail pipe
Saugschlauch *m* <Wasserversorg> suction hose
Saugseite *f* 1. <Fertig> inlet side *(Pumpe)*; 2. <Maschinen> inlet side, suction side
Saugspeisen *n* <Ker & Glas> suction feeding
Saugstrahlpumpe *f* <Labor> syphon

Saug- und Druckpumpe *f* <Wasserversorg> lift-and-force pump
Saugventil *n* 1. <Fertig> inlet valve *(Pumpe)*; 2. <Heiz & Kälte, Hydraul, Kfztech, Maschinen> suction valve
Saugventilator *m* <Mechan> exhaust fan
Saugwalzenfilz *m* <Papier> suction roll felt
Saugzug *m* 1. <Bau> *(AE)* forced draft, *(BE)* forced draught; 2. <Heiz & Kälte, Maschinen> *(AE)* induced draft, *(BE)* induced draught
Saugzugbrenner *m* <Heiz & Kälte> *(AE)* induced-draft burner, *(BE)* induced-draught burner
Saugzuglüfter *m* <Heiz & Kälte> *(AE)* induced-draft fan, *(BE)* induced-draught fan, suction fan
Saugzugventilator *m* <Maschinen> *(AE)* induced-draft fan, *(BE)* induced-draught fan
Säule *f* 1. <Bau> pillar, support, upright; head *(Wasser)*; 2. <Eisenbahn> pillar; 3. <Elektrotech> pile *(bei Batterie)*; 4. <Ker & Glas, Kohlen> pillar; 5. <Labor> cup *(eines Barometers)*; 6. <Maschinen> column, pillar; 7. <Trans> pillar
Säulenbogen *m* <Ker & Glas> pillar arch
Säulenbohrmaschine *f* <Maschinen> column-type drilling machine, pillar drill, pillar-drilling machine
Säulenführung *f* 1. <Fertig> die set; 2. <Maschinen> pillar guide
Säulengewölbe *n* <Ker & Glas> pillar arch
Säulenhals *m* <Bau> *(AE)* neck molding, *(BE)* neck moulding
Säulenisolator *m* <Elektrotech> post insulator
Säulenkran *m* <Bau> post crane
Säulenpresse *f* <Maschinen> pillar press
Säulenschaft *m* <Bau> shaft, shank
Säulen-Schnellnäher *m* <Textil> postbed high-speed seamer
Säulenwaage *f* <Labor> pillar scales
Saum *m* 1. <Bau> seam *(Stoff)*; 2. <Mechan> seam; 3. <Phys> fringe; 4. <Textil> hem, seam
säumen *v* <Textil> seam
Säumen *n* <Textil> hemming, seaming
Säure *f* <Chemie, Kfztech, Lebensmittel, Papier, Textil> acid • **durch Säure belastet** <Umweltschmutz> acid-stressed • **mit Säure behandeln** <Papier> acidize
saure Ablagerung *f* <Umweltschmutz> acid deposit
saure Ausmauerung *f* <Fertig> acid bottom and lining, acid lining
saure Erde *f* <Umweltschmutz> acid earth
saure Lösung *f* <Lebensmittel> acid solution
saure Umweltverschmutzung *f* <Umweltschmutz> acid pollution
Säureabscheidebehälter *m* <Erdöl> acid decantation drum *(Raffinerietechnik)*
Säureakzeptor *m* <Kunststoff> acid acceptor
Säureamid *n* <Lebensmittel> acid amide
Säureätzung *f* <Maschinen> acid etching
Säureauslaugung *f* <Fertig> acid leach
Säurebad *n* <Papier> acid bath
Säure-Basen-Gleichgewicht *n* <Lebensmittel> acid base balance *(in Körpersäften)*
Säurebehälter *m* <Fertig> acid reservoir
Säurebeizung *f* <Kerntech> acid pickling
Säurebelastung *f* <Umweltschmutz> acid loading, acid stress
säurebeständig *adj* 1. <Chemie> acid-proof, acid-resistant; 2. <Papier> acid-proof; 3. <Umweltschmutz> acid-proof, acid-resistant
säurebeständige Schutzhandschuhe *mpl* <Sicherheit> acid-proof protective gloves
Säurebeständigkeit *f* 1. <Kunststoff> acid resistance; 2. <Sicherheit> resistance to acids; 3. <Umweltschmutz> acid resistance

Säurebestimmung

Säurebestimmung f <Chemie, Papier> acid determination
säurebildend adj 1. <Chemie> acidic; 2. <Fertig> acid-forming
Säurebildner m <Chemie, Papier> acid former, acidifier
Säurechlorid n <Chemie, Lebensmittel> acid chloride
säuredichte Batterie f <Kfztech> nonspill battery (KFZ-Elektrik)
Säuredichtemessgerät n <Gerät> acid density meter
Säuredruckbehälter m <Erdöl> acid blow case (Raffinerie)
Säuredruckvorlage f <Chemtech> acid elevator
säureecht adj <Verpack> acid-resistant
Säureerhitzungsprobe f 1. <Erdöl> acid heat test (Raffinerie); 2. <Papier> acid heat test
Säureester m <Chemie, Lebensmittel> acid ester
Säurefällung f <Umweltschutz> acid precipitation
Säurefarbstoff m <Chemie, Textil> acid dye
säurefeste Farbe f <Bau> acid-resisting paint
säurefester Lack m <Verpack> acid-proof varnish
säurefestes Papier n <Druck> acid-proof paper
Säurefestigkeit f <Sicherheit> resistance to acids
säurefrei adj 1. <Chemie, Papier> acid-free; 2. <Verpack> acidless
säurefreier Kleber m <Foto> acid-free glue
säurefreies Öl n <Erdöl> chemically neutral oil
säurefreies Papier n <Druck, Verpack> acid-free paper
Säuregehalt m 1. <Chemie, Papier, Umweltschutz> acidity; 2. <Verpack> acid content
Säuregehaltsmessgerät n <Phys> hydrometer
Säuregehaltsmessung f <Phys> hydrometry
Säuregehaltsprüfer m <Chemtech> acidimeter
Säuregrad m 1. <Chemie> free acid; 2. <Papier> acid value; 3. <Umweltschutz> acidity level
Säuregradmesser m <Lebensmittel> acidimeter
Säuregradmessung f <Chemie, Chemtech, Lebensmittel> acidimetry
säurehaltig adj 1. <Chemie> acidiferous; 2. <Erdöl> acidic; 3. <Papier> acidiferous
Säurehärter m <Fertig> acid catalyst, acid hardening
Säurehydrazid n <Chemie> hydrazide
Säurekennzeichnen n <Ker & Glas> acid badging
Säurekitt m <Fertig> acid-proof cement
Säurekonzentration f <Umweltschutz> acid concentration
Säuremarke f <Ker & Glas> acid mark
säuremattiertes Milchglas n <Ker & Glas> acid-etched frosted glass
Säuremattierung f <Ker & Glas> acid etching
Säuremesser m <Papier> acidimeter
säuren v <Papier> sour
Säurenebel m <Umweltschutz> acid fog
Säureneutralisation f <Umweltschutz> acid neutralizing
Säurenvorstufe f <Umweltschutz> acidic precursor
Säurepolitur f <Ker & Glas> acid polishing
Säureprägen n <Ker & Glas> acid embossing
Säureprüfer m 1. <Papier> acid tester; 2. <Phys> hydrometer
Säurer m <Erdöl> acidizer
saurer Abfluss m <Umweltschutz> acid runoff
saurer Boden m <Umweltschutz> acid soil
saurer Niederschlag m <Umweltschutz> acid fallout, acid rain, acidic rain
saurer Regen m 1. <Anstrich> acid rain; 2. <Kohlen, Umweltschutz> acid rain; acid fallout (spezifisch); acidic rain (speziell)
Säureradikal n <Lebensmittel> acid radical
säureresistenter Wand- und Fußbodenbelag m <Sicherheit> acid-resisting floor and wall covering

Säurerest m <Chemie, Lebensmittel> acid radical
Säurerückgewinnung f <Fertig> acid recovery
Säurerückgewinnungsanlage f <Erdöl> acid recovery plant (Raffinerie)
saures Abgas n <Abfall> acid waste gas
saures Aerosol n <Umweltschutz> acid aerosol
saures Bad n <Foto> acid bath
saures Carbonat n <Chemie> hydrocarbonate
saures Erdgas n <Erdöl> sour gas
saures Fixierbad n <Chemtech, Foto> acid fixing bath
saures Futter n <Fertig> acid lining
saures Rohöl n <Erdöl> sour crude
saures Salz n <Lebensmittel> acid salt
saures Teilchen n <Umweltschutz> acid particle, acidic particle
saures Verfahren n <Papier> acid process
Säureschock m <Umweltschutz> acid shock
Säureschutzanzug m <Sicherheit> acid-proof suit
Säurespaltung f <Lebensmittel> acidolysis
Säurespiegel m <Kfztech> acid level
Säuretauchbad n <Chemtech> acid dipping
Säureturm m <Papier> absorption tower
Säureversprödung f <Metall> acid embrittlement
Säureverträglichkeit f <Umweltschutz> acid tolerance (des Bodens)
Säurevorbeizung f <Kerntech> acid pre-pickling
Säurewäsche f <Papier> acid washing
Säurewecker m <Lebensmittel> starter (bei Milchprodukten)
Säurewert m (AV) <Chemie> acid value (AV)
Säurewiderstandsfähigkeit f <Umweltschutz> acid tolerance
Säurezahl f 1. <Erdöl> acid number (Petrochemie); 2. <Kunststoff, Lebensmittel> acid value (SZ)
SAW-Bauelement n <Elektronik, Telekom> SAW device
SAW-Expansionsfilter n <Elektronik> SAW expansion filter
Sb (Antimon, Stibium) <Chemie> Sb (antimony)
SB (schneller Brutreaktor, schneller Brüter) <Kerntech, Phys> FBR (fast breeder reactor)
S-Bahn f (Schnellbahn) <Eisenbahn> (AE) regional express railroad, (BE) regional express railway
S-Band n <Elektronik> S-band
S-Banddiode f <Elektronik> S-band diode
SBFM (Schmalbandfrequenzmodulation) <Elektronik, Funktech, Telekom> NBFM (narrow-band frequency modulation)
S-Bogen m <Fertig> gooseneck
SBR (Styrol-Butadien-Kautschuk) <Kunststoff> SBR (styrene butadiene rubber)
SB-Tankstelle f (Selbstbedienungtankstelle) <Trans> self-service station
SBV (Sicherheitsabblasearmatur) <Erdöl> relief valve
Sc (Scandium) <Chemie> Sc (scandium)
Scandium n (Sc) <Chemie> scandium (Sc)
scannen v <Comp & DV, Druck, Wellphys> scan (Radar)
Scannen n <Comp & DV, Druck, Elektronik, Fernseh, Strahlphys> scanning
Scanner m 1. <Comp & DV> optical scanner, scanner, scanning device; 2. <Druck, Elektronik> scanner; 3. <Fernseh> optical scanning device, scanner; 4. <Kontroll, Strahlphys, Telekom> scanner
Scannerkopf m <Druck> scanning head
Scanner-Programm n <Comp & DV> scanning software
Scanner-Software f <Comp & DV> scanning software
Scatter n <Raumfahrt> scatter (Streuung)
Scatterübertragung f <Funktech> scatter propagation
Scatterverbindung f <Telekom> scatter-propagation link, scatter link

SCC 1. <Abfall> (Nachbrennkammer) SCC (secondary combustion chamber); 2. <Maschinen> (Nachbrennkammer, zweiter Brennraum) SCC (secondary combustion chamber)
Schab n <Bau> shave hook
Schabeisen n <Maschinen> scraper
schaben v 1. <Fertig> scrape (Zahnrad); 2. <Maschinen> shave; 3. <Meerschmutz> scrape
Schaben n 1. <Fertig> scraping; shaving (Zahnrad); 2. <Maschinen> shaving
Schaber m 1. <Bau> scraper; 2. <Druck> squeegee; 3. <Lebensmittel, Maschinen, Meerschmutz> scraper
Schabezahn m <Maschinen> shave tooth
Schablone f 1. <Bau> strickle, strickle board; 2. <Comp & DV> mask, template, templet; 3. <Druck> stencil; 4. <Erdöl> template; 5. <Fertig> stencil; strickle (Gießen); former (Zahnradstoßmaschine); 6. <Ker & Glas> stencil; 7. <Kfztech> template, templet (Werkzeug); 8. <Maschinen> (AE) gage, (BE) gauge, jig, pattern, template, templet; 9. <Mechan, Metrol> jig • **ohne Schablone** <Fertig> jigless
Schablonenarm m <Fertig> loam board
Schablonenbefehl m <Comp & DV> template command, templet command
schablonenformen v <Fertig> strickle
Schablonenformen n <Fertig> strickling
Schablonenformerei f <Fertig> (AE) strickle molding, (BE) strickle moulding
Schablonengrundplatte f <Fertig> loam plate (Lehmformerei)
Schablonenkern m <Fertig> backup brickwork (Gießen)
Schablonenseide f <Ker & Glas> stencil silk
Schablonenvergleich m <Comp & DV> template matching, templet matching
Schablonenwicklung f <Ker & Glas> former winding
schablonieren v <Fertig> sweep
Schablonieren n <Fertig> sweeping
Schabotte f <Maschinen> anvil bed
Schachbrettmuster n <Ker & Glas> checker pattern
Schacht m 1. <Bau> stack; manhole (Wasserbau); 2. <Erdöl> pit; 3. <Fertig> stack (Hochofen); 4. <Mechan> chute; 5. <Wasserversorg> shaft
Schachtabdeckung f <Bau> cowl, manhole cover
Schachtausmauerung f <Kohlen> stone tubbing
Schachtbrunnen m <Kohlen> sunk well
Schachtel f <Verpack> carton
Schachteleinheit f <Verpack> nesting box (aus einem Satz)
Schachtentlastungsanlage f <Wasserversorg> shaft spillway
Schachtleiter f <Wasserversorg> shaft ladder
Schachtofen m 1. <Abfall, Fertig> shaft furnace; 2. <Kohlen> shaft kiln
Schachtring m <Kohlen, Nichtfoss Energ> well casing
Schachttür f <Bau> manhole door
Schachtturbinenkammer f <Wasserversorg> flume, open flume turbine chamber
Schachtzimmerung f <Kohlen> timbering
Schaden m <Patent, Qual> damage • **Schaden erleiden** <Patent, Qual> suffer damage • **Schaden ersetzen** <Patent> compensate for damage
Schaden m **an der Ruderanlage** <Wassertrans> helm damage
Schadenlinie f <Fertig> damage curve
schadhaft adj <Elektriz, Maschinen, Qual> damaged, faulty
schadhafte Isolierung f <Elektriz> faulty insulation
schädlich adj 1. <Chemie> noxious; 2. <Kerntech> unfavourable; 3. <Sicherheit, Umweltschmutz> harmful, injurious

schädliche Substanz f <Sicherheit> harmful substance
schädlicher Widerstand m <Lufttrans> parasitic drag
Schädlichkeit f <Umweltschmutz> harmfulness
Schädling m <Lebensmittel> pest
Schadstoff m 1. <Abfall> contaminant; 2. <Meerschmutz, Sicherheit> pollutant; 3. <Umweltschmutz> harmful substance, pollutant, toxic substance
Schadstoffabfall m <Abfall> polluting waste
Schadstoffablagerung f <Umweltschmutz> pollutant deposition
schadstoffarm adj <Umweltschmutz> low-contaminant
Schadstoffausbreitung f <Umweltschmutz> propagation of pollutant
Schadstoffausleitung f <Umweltschmutz> pollutant discharge
Schadstoffaustritt m <Abfall, Umweltschmutz> spillage of harmful substances
Schadstoffbekämpfung f <Sicherheit, Umweltschmutz> pollutant control
schadstoffbelastetes Erdreich n <Umweltschmutz> pollutant-impacted ground
Schadstoffbelastung f 1. <Sicherheit> exposure to hazardous substances; 2. <Umweltschmutz> exposure to harmful substances, exposure to hazardous substances, pollution burden
Schadstoffeinleitung f <Sicherheit, Umweltschmutz> pollutant discharge
Schadstoffemission f <Umweltschmutz> emission, emission of contaminants, pollution emission
Schadstofferfassung f <Umweltschmutz> capture of pollutants
Schadstoffexposition f <Sicherheit, Umweltschmutz> exposure to hazardous substances
Schadstoffmessung f <Gerät> pollutant measurement
Schadstoffregulierung f **bei Motoren** <Maschinen> engine emission control
Schadstoffschutz m <Sicherheit, Umweltschmutz> control of harmful substances, pollutant control, protection against harmful substances
Schadstoffverdünnung f <Umweltschmutz> dilution of pollutants
Schadstoffvorstufe f <Umweltschmutz> precursor pollutant
Schadwirkung f <Umweltschmutz> harmful effect
Schaffung f **der Baufreiheit** <Bau> clearing operations
Schaffußwalze f <Bau> sheep's foot roller
Schaft m 1. <Bau, Druck> shank; 2. <Fertig> grip, (AE) rod, shank; 3. <Maschinen> shaft, shank, stem, tang; 4. <Mechan> handle, shaft
Schaftfräser m <Maschinen> end mill, end-milling cutter
Schaftfräserumfang schleifen v <Fertig> spiral
Schaftgewebe n <Textil> dobby weave fabric
Schafthülse f <Fertig> barrel
Schaftlänge f <Maschinen> shank length
Schaftmaschine f <Textil> dobby
Schaftmeißel m <Fertig> shank-type cutting tool
Schaftschraube f 1. <Bau> headless screw; 2. <Kfztech> stud; 3. <Maschinen> headless screw
Schaftschraube f **mit Schlitz** <Bau, Maschinen> slotted headless screw
Schäkel m 1. <Bau> shackle; 2. <Kfztech> shackle (Schleppvorrichtung); 3. <Maschinen, Meerschmutz> shackle; 4. <Wassertrans> sextant altitude (Beschläge); shackle (Verbindung zweier Kettenglieder)
Schäkelbolzen m <Maschinen> clevis bolt
Schäkelisolator m <Elektrotech> shackle insulator
Schakengehänge n <Eisenbahn> link suspension
Schalbretter npl <Bau> boarding
Schäldarm m <Lebensmittel> cellulose casing

Schale

Schale f 1. <Bau> layer, shell; 2. <Foto> dish; 3. <Ker & Glas> dish *(für Walze bei der Herstellung von Walzflachglas)*; 4. <Kohlen> shell *(Bau)*; 5. <Labor> basin, tray; 6. <Lebensmittel> skin; 7. <Maschinen> insert liner, liner, pan, shell; 8. <Papier, Raumfahrt> shell *(Raumschiff)*
Schäleisen n <Bau> paring chisel
Schalen n <Fertig> *(AE)* molding, *(BE)* moulding
schälen v 1. <Bau> peel; 2. <Lebensmittel> husk, peel
Schälen n <Lebensmittel> shelling; milling *(von Reis, Hafer)*
Schalenbauweise f <Bau> shell construction
Schalenfehler m <Lebensmittel> skin blemish *(Obst)*
Schalenfestigkeit f <Lebensmittel> shell strength
Schalengussform f <Mechan> chill
Schalenhartguss m 1. <Fertig> chilled cast iron; 2. <Metall> chill casting
Schalenkreuzanemometer n <Gerät> cup anemometer
Schalenkupplung f <Fertig> shell coupling *(Kunststoffinstallationen)*
Schalenmodell n <Phys> shell model *(Kernphysik)*
Schalenpackung f <Verpack> clam pack
Schalenschaukel f <Foto> dish rocker
Schalensitz m <Kfztech> bucket seat
Schalenthermometer n <Foto> dish thermometer
Schalenwärmer m <Foto> dish heater
Schälfräser m <Fertig> slab milling cutter
Schälholz n <Bau> barked timber
Schalkleiste f <Wassertrans> batten *(Segeln)*
Schalklemme f <Bau> casing clamps
Schall m <Akustik> sound • **Schall aussenden** <Wassertrans> sound
schallabsorbierend adj <Sicherheit> sound-absorbent, sound-absorptive
schallabsorbierende Platte f <Sicherheit> noise-absorbent panel
schallabsorbierende Wand f <Sicherheit> sound-absorptive panel
Schallabsorbierung f <Akustik, Aufnahme> sound absorption
Schallabsorptionsfaktor m <Sicherheit> noise reduction coefficient
Schallabsorptionsgrad m <Sicherheit> noise reduction coefficient
Schallabsorptionskoeffizient m <Sicherheit> noise reduction coefficient
Schallabsorptionswand f <Sicherheit> sound-absorption panel
Schallanalysegerät n <Akustik> *(BE)* sound analyser, *(AE)* sound analyzer
Schallaufnehmer m <Akustik> acoustic sensor
Schallaufzeichnung f <Akustik> sound recording
Schallausbreitung f <Aufnahme> sound propagation
Schallbeschleunigung f <Akustik> sound acceleration
Schallbeschleunigungspegel m <Akustik> sound acceleration level
schalldämmend adj <Sicherheit> anti-noise, sound-insulating, soundproof
schalldämmendes Material n <Aufnahme> sound-absorbing material
Schalldämmkoeffizient m <Phys> sound reduction index
Schalldämmplatte f <Sicherheit> acoustic board
Schalldämmstoff m <Sicherheit> sound-deadening material, sound-insulation material
Schalldämmung f 1. <Akustik> sound insulation, transmission loss; 2. <Aufnahme, Sicherheit> sound insulation, noise abatement
Schalldämmwand f <Sicherheit> noise-abating wall
Schalldämmzahl f <Phys> sound reduction index
schalldämpfend adj <Sicherheit> anti-noise, silencing, sound-damping

Schalldämpfer m 1. <Akustik, Bau> damper, *(AE)* muffler, *(BE)* silencer; 2. <Heiz & Kälte> attenuator; 3. <Kfztech> *(AE)* muffler, *(BE)* silencer *(Auspuffanlage)*; 4. <Lufttrans> *(AE)* aerator muffler, *(BE)* aerator silencer; 5. <Mechan> *(AE)* exhaust muffler, *(BE)* exhaust silencer
Schalldämpfer m **am Auspuff** <Mechan> *(AE)* exhaust muffler, *(BE)* exhaust silencer
Schalldämpfung f 1. <Akustik> sound absorption; 2. <Elektronik> acoustic attenuation; 3. <Heiz & Kälte> sound attenuation
Schalldämpfungskonstante f <Elektronik> acoustic attenuation constant
Schalldetektor m <Trans> sonic detector
schalldicht adj <Sicherheit> soundproof
Schalldichte f <Akustik, Phys> sound-energy density
schalldichte Fliese f <Sicherheit> soundproof tile
schalldichte Kabine f 1. <Aufnahme> sound booth, soundproofed booth; 2. <Sicherheit> soundproof booth; 3. <Telekom> soundproofed booth *(Telefon)*
schalldichte Kachel f <Sicherheit> soundproof tile
schalldichte Kammer f <Sicherheit> noise protection booth
schalldichte Tür f <Sicherheit> sound-insulated door
schalldichte Zwischenschicht f <Bau> plugging
Schalldichtepegel m <Akustik> sound-energy density level
schalldichter Helm m <Sicherheit> sound-resistant helmet
schalldichter Raum m <Aufnahme> soundproof room
Schalldosen-Adapter m <Aufnahme> phono adaptor *(für Empfänger)*
Schalldruck m 1. <Akustik> pressure; 2. <Akustik, Aufnahme, Phys, Raumfahrt> acoustic pressure, sound pressure *(Raumschiff)*
Schalldruckpegel m <Akustik, Ergon, Heiz & Kälte, Phys, Umweltschmutz> sound pressure level
Schalldruckspektrum n <Umweltschmutz> sound pressure spectrum
Schalleistung f 1. <Akustik> acoustic power; 2. <Phys> sound power
Schalleistungspegel m <Akustik, Heiz & Kälte, Phys> sound power level
Schallenergie f <Akustik, Elektrotech> sound energy
Schallenergiedichte f <Akustik, Phys> sound-energy density
Schallenergiefluss m *(J)* <Akustik, Phys> sound-energy flux *(J)*
Schallermüdung f <Metall> sonic fatigue
Schallerregung f <Akustik> sound excitation
Schallexpositionsmesser m <Sicherheit> sound exposure meter
Schallexpositionspegel m <Sicherheit> sound exposure level
Schallfeld n <Akustik> acoustical field, sound field
Schallfilter n <Akustik> acoustic filter
Schallfluss m <Akustik> volume velocity
Schallfühler m <Akustik> acoustic sensor
Schallgeber m <Akustik> sound source
Schallgenerator m <Maschinen> acoustic generator
schallgeschützt adj <Sicherheit> soundproof
schallgeschützte Kabine f <Sicherheit> soundproof booth
Schallgeschwindigkeit f 1. <Akustik> sound velocity; 2. <Phys> speed of sound; 3. <Wellphys> velocity of sound *(in Medium)*
Schallgeschwindigkeitsprofil n <Nichtfoss Energ> acoustic velocity log
Schallgewölbe n <Aufnahme> acoustic vault
Schallhärte f <Akustik> acoustic stiffness

schallharter Raum m <Aufnahme> echo chamber
Schallimpedanz f 1. <Akustik> (AI) acoustic impedance (ZA); 2. <Aufnahme, Elektrotech, Phys> acoustic impedance (ZA)
Schallimpuls m 1. <Aufnahme> sound pulse; 2. <Elektronik> acoustic pulse
Schallintensität f <Akustik, Ergon, Phys> sound intensity
Schallintensitätsdichte f <Akustik> sound intensity density
Schallintensitätspegel m <Akustik, Aufnahme> sound intensity level
Schallisolation f <Akustik> sound insulation, sound isolation
Schallisolationsmaterial n <Umweltschmutz> acoustic insulating materials
Schallisolierung f <Sicherheit> sound insulation
Schallkapsel f <Sicherheit> acoustic enclosure
Schallkapselung f <Sicherheit> acoustic enclosure
Schallknall m 1. <Lufttrans> sonic boom; 2. <Sicherheit> sonic bang
Schallkörper m <Elektronik> acoustic resonator
Schalllehre f <Akustik> acoustics
Schallloch n <Mechan> (AE) louver, (BE) louvre
Schallmachmeter n <Lufttrans> audible machmeter
Schallmesser m <Sicherheit> sound level meter
Schallmessgerät n <Sicherheit, Wellphys> sonometer
schallnah adj <Phys> transonic
schallnahe Geschwindigkeit f <Phys, Trans> transonic speed
Schallöffnung f <Aufnahme> (AE) louver, (BE) louvre (eines Lautsprechers)
Schallortung f <Wellphys> sound ranging
Schallortungsgerät n <Akustik> sound locator
Schallpegel m <Akustik, Heiz & Kälte, Sicherheit> sound level
Schallpegeldifferenz f <Akustik> sound-level difference
Schallpegelmesser m <Akustik, Ergon, Sicherheit> sound level meter
Schallpegelmessgerät n <Gerät> sound level meter, volume unit meter
Schallpegelverteilung f <Akustik> sound-level distribution
Schallplatte f <Akustik, Aufnahme> record, audio record
Schallplattenaufnahmegerät n 1. <Akustik> (BE) disc recorder; 2. <Aufnahme> (AE) disk recorder, recorder
Schallquelle f <Umweltschmutz> sound source
Schallquellenleistung f <Akustik> sound power of a source
Schallraum m <Maschinen> acoustic testing room
Schallreiz m <Akustik> acoustic stimulus
Schallrückkopplung f <Aufnahme> acoustic feedback
Schallschirm m <Umweltschmutz> baffle collector
Schallschluckdecke f <Sicherheit> sound-absorbing ceiling
Schallschluckdekor m 1. <Akustik> flat; 2. <Aufnahme> acoustic flat
schallschluckend adj <Sicherheit> noise-absorbent, sound-absorptive
schallschluckende Auskleidung f <Sicherheit> noise-absorbent lining
Schallschlucker m <Akustik> sound absorber
Schallschluckglas n <Sicherheit> noise-protective insulating glass
Schallschluckgrad m <Akustik> acoustical absorption coefficient
Schallschluckhaube f <Sicherheit> noise-protective hood
Schallschluckplatte f <Sicherheit> noise-absorbent panel
Schallschlucktür f <Sicherheit> noise abatement door
Schallschnelle f <Akustik> sound particle velocity
Schallschnellemesser m <Akustik> acoustic velocity meter
Schallschnellepegel m <Akustik> acoustic velocity level
Schallschürze f <Sicherheit> sound curtain
Schallschutz m <Sicherheit> soundproofing
Schallschutzanzug m <Sicherheit> noise protection suit
Schallschutzhelm m <Sicherheit> noise protection helmet, sound-resistant helmet
Schallschutzkabine f <Sicherheit> sound-insulated cabin, soundproof booth
Schallschutzkapsel f <Sicherheit> noise enclosure
Schallschutzkapselung f <Sicherheit> soundproof enclosure
Schallschutzvorhang m <Sicherheit> sound curtain
Schallschutzwand f 1. <Aufnahme> sound screen; 2. <Sicherheit> sound-isolating partition
Schallschwächung f <Elektronik> acoustic attenuation
Schallschwingung f <Akustik> acoustic vibration
Schallsender m <Akustik> sound source
Schallsensor m <Akustik> acoustic sensor
Schallsignal n 1. <Akustik> audio signal; 2. <Eisenbahn> sound signal; 3. <Elektronik> audio signal; 4. <Wassertrans> sound signal (Signal)
Schallspektrograph m <Akustik> sound spectrograph
Schallspektrum n <Akustik> acoustical spectrum
Schallstärke f 1. <Akustik> strength of single source; 2. <Phys> sound intensity
Schallstärkepegel m <Phys> sound intensity level
Schalltechnik f <Akustik> acoustics
schalltot adj <Akustik> acoustically inactive, acoustically inert, aphonic; anechoic, acoustically dead (reflexionsfrei)
schalltote Wand f <Aufnahme> dead end (eines Raums)
schalltote Zone f <Aufnahme> dead zone
schalltoter Raum m 1. <Akustik> anechoic room, dead room; 2. <Aufnahme> dead room; 3. <Ergon> dead room, free-field room; 4. <Phys> anechoic room, dead room
Schalltransmissionslinie f <Elektrotech> acoustic transmission line
Schalltrichter m <Akustik> foghorn, horn
Schallübertragung f <Akustik> sound transmission
Schallverzögerung f <Elektronik> acoustic delay
Schallwahrnehmung f <Akustik> auditory sensation
Schallwand f <Akustik, Aufnahme> baffle
Schallwelle f 1. <Akustik> acoustic wave, sound wave; 2. <Elektrotech> acoustic wave; 3. <Funktech, Phys, Telekom, Wellphys> sound wave
Schallwellenwiderstand m 1. <Akustik> (AI) acoustic impedance (ZA); 2. <Aufnahme, Elektrotech, Phys> acoustic impedance (ZA)
Schälmaschine f 1. <Bau> paring machine; 2. <Fertig> peeler, peeling machine; 3. <Lebensmittel> decorticator, husking machine
Schalöl n <Bau> formwork oil
Schälrad n <Fertig> skiving wheel
Schalrohr n <Bau> casing pipe
Schälschnitt m <Fertig> slabbing cut
Schalt... <Comp & DV, Elektriz, Elektrotech, Kfztech, Maschinen> switch
Schaltalgebra f <Comp & DV> logic algebra
Schaltanlage f <Elektriz, Elektrotech> switchgear
Schaltanordnung f <Elektrotech> switching arrangement
Schaltantrieb m <Maschinen> ratchet motion
Schaltautomatik f <Kfztech> automatic gear change (Getriebe)
schaltbar adj <Telekom> switchable
schaltbare Kupplung f <Maschinen> clutch
Schaltbefehl m <Comp & DV> switching command

Schaltbereichsanzeige

Schaltbereichsanzeige f <Kfztech> range indicator *(Automatikgetriebe)*
Schaltbild n 1. <Elektriz, Elektrotech> circuit diagram, circuit schematic, connection diagram, schematic, schematic circuit diagram; wiring diagram *(Verdrahtungsbild)*; 2. <Fernseh> circuit diagram; 3. <Fertig> diagrammatic circuit; 4. <Funktech> circuit diagram, schematic diagram; 5. <Telekom> circuit diagram
Schaltblock m <Elektrotech> contact block
Schaltbogen m <Elektrotech> breaking arc
Schaltbrett n 1. <Elektrotech> plugboard; 2. <Kfztech> dash, dashboard; 3. <Mechan> control panel
Schaltdauer f <Elektrotech> switching time
Schaltdifferenz f <Regelung> differential gap
Schaltdiode f <Elektronik, Telekom> switching diode
Schaltdraht m <Funktech, Telekom> jumper, jumper wire
Schalteingang m <Regelung> gate input
Schaltelement n 1. <Comp & DV> gate; 2. <Elektronik, Elektrotech> switching device, switching element; 3. <Telekom> switching device
Schaltelementmatrix f <Elektronik> cellular array
schalten v 1. <Comp & DV> switch; 2. <Elektrotech> connect *(in Reihe, parallel)*; 3. <Kfztech> put into gear; 4. <Kontroll> switch; 5. <Phys> trip; 6. <Telekom> switch, patch
Schalten n 1. <Elektrotech, Raumfahrt> switching *(Weltraumfunk)*; 2. <Maschinen> indexing, shifting
Schalter m 1. <Comp & DV, Elektriz, Elektrotech> switch, circuit breaker; switch *(Verzweigung)*; 2. <Fernseh> switch; 3. <Gerät> selector; 4. <Kontroll, Phys, Telekom> switch; 5. <Telekom> button, switch
Schalter m **auf Schalterdeck** <Elektrotech> deck switch
Schalter m **mit mehrfachen Kontakten** <Elektriz> multiple-contact switch
Schalter m **mit Sicherung** <Elektrotech> switch fuse
Schalterhalle f <Eisenbahn> passenger hall
Schaltersicherung f <Elektrotech> fuse switch
Schaltertisch m <Telekom> counter
Schaltfehler m <Elektrotech> contact fault, switch fault
Schaltfeld n <Elektrotech> switchboard
Schaltfolge f <Elektriz> switching sequence
Schaltgabel f <Kfztech> gearbox selector fork *(Getriebe)*
Schaltgenauigkeit f <Fertig> accuracy of indexing, indexability accuracy *(Revolver)*
Schaltgerät n 1. <Elektrotech> switchgear, switching device; 2. <Elektrotech> trigger box
Schaltgeschwindigkeit f <Elektrotech> switching speed
Schaltgetriebe n 1. <Fertig> intermittent gear; 2. <Kfztech> gearbox, manual transmission; manual gearbox *(Getriebe)*; 3. <Maschinen> gearbox
Schaltglied n <Maschinen> shifting link
Schaltgriff m <Elektrotech> switch handle
Schalthahn m <Bau> switch cock
Schalthäufigkeit f <Fertig> interruption frequency *(Punktschweißen)*
Schalthebel m 1. <Elektrotech> switch lever; 2. <Fertig> actuating arm, *(BE)* gear lever, *(AE)* gear shift; 3. <Kerntech> lever switch; 4. <Lufttrans> control lever; 5. <Maschinen> *(BE)* gear lever, *(AE)* gear shift
Schalthysterese f <Gerät> dead spot
Schaltjahr n <Phys> leap year
Schaltkette f <Regelung> switching chain
Schaltklinke f <Fertig, Maschinen, Mechan> pawl
Schaltknopf m <Elektrotech> contact button
Schaltknüppel m <Kfztech> *(BE)* gear lever, *(AE)* gear shift *(Getriebe)*
Schaltkontakt m <Elektrotech> make-and-break contact
Schaltkraft f <Fertig> operating force *(Kupplung)*

Schaltkreis m 1. <Elektronik> circuit *(Netzwerk mit einer oder mehreren Stromschleifen)*; 2. <Elektrotech> switching circuit; 3. <Funktech, Phys> circuit
Schaltkreis m **auf dem Chip** <Elektronik> on-chip circuit
Schaltkreis m **mit Großintegration** <Telekom> very large-scale integrated circuit
Schaltkreis m **mit mittlerem Integrationsgrad** <Telekom> medium-scale integration circuit
Schaltkreis m **zur Verhinderung von Stromspitzen** <Elektrotech> despiking circuit
Schaltkreisanalyse f <Elektrotech> circuit analysis
Schaltkreisentwurf m <Elektronik> circuit design
Schaltkreisintegration f <Elektronik> circuit integration
Schaltkreisparameter mpl <Phys> network parameters
Schaltkreisverzögerung f <Elektrotech> circuit delay
Schaltkulisse f <Kfztech> gate *(Automatikgetriebe)*
Schaltkupplung f <Maschinen, Wassertrans> clutch *(Motor)*
Schaltkurve f <Fertig> index cam
Schaltleistung f <Elektrotech> breaking capacity, making capacity
Schaltmagnet m <Fernseh> solenoid
Schaltmatrix f <Fernseh, Raumfahrt> switching matrix *(Weltraumfunk)*
Schaltmotor m <Elektriz> torque motor
Schaltmultiplexer m <Telekom> cross-connect multiplexer, switching multiplexer, switching mux
Schaltnetz n 1. <Comp & DV> *(AE)* combinational circuit, *(BE)* combinatorial circuit; 2. <Elektronik> *(AE)* combinational circuit, *(BE)* combinatorial circuit *(logische Schaltung ohne Speichervermögen)*
Schaltnetzgerät n <Elektrotech> switching power supply
Schaltnetzgerät n **mit nur einem Ausgang** <Elektrotech> single output switching power supply
Schaltnetzteil n <Telekom> switching power supply
Schaltnocke f <Fertig> contact cam *(Kunststoffinstallationen)*
Schaltplan m 1. <Elektrotech> circuit diagram, connection diagram; 2. <Kontroll> circuit diagram; 3. <Telekom> wiring diagram
Schaltplatte f <Comp & DV> *(BE)* jack panel, *(AE)* patch panel
Schaltpult n 1. <Fertig> inclined control panel; 2. <Gerät> desk; 3. <Lufttrans> control panel
Schaltrelais n <Elektriz> switch relay
Schaltröhre f <Elektrotech> switching tube
Schaltschema n 1. <Elektrotech> circuit diagram; 2. <Fertig> electrical diagram *(Kunststoffinstallationen)*
Schaltschwelle f <Kontroll> switching threshold
Schaltsignal n <Fernseh> keying signal
Schaltstange f <Fertig> actuating rod
Schaltstellung f <Fertig> setting position *(Kunststoffinstallationen)*
Schaltstromversorgung f <Elektrotech> switching power supply
Schaltstück n <Elektrotech> contact
Schaltsystem n <Fernseh> switching system
Schalttafel f 1. <Bau> control panel; 2. <Comp & DV> patch board, plugboard; 3. <Elektrotech> plugboard, switchboard; 4. <Erdöl> monkey board; 5. <Fernseh> switchboard; 6. <Funktech> panel; 7. <Kontroll> control panel; 8. <Raumfahrt> control panel; 9. <Telekom> switchboard *(Elektrik)*; patch panel; 10. <Textil> control panel; 11. <Wassertrans> switch panel *(Elektrik)*
Schalttafel f **des Bordingenieurs** <Lufttrans> flight engineer's panel
Schalttafel f **mit Steuerpult** <Kerntech> benchboard
Schalttafelbuchse f <Fernseh> hub
Schalttafelfeld n <Telekom> switchboard panel

Schalttafelinstrument n <Gerät> board-mounted instrument, panel-type instrument, switchboard panel instrument
Schalttafelmessgerät n <Gerät> panel-type instrument
Schalttafelmessinstrument n <Gerät> switchboard-type meter
Schalttaste f <Elektrotech> contact button
Schalttemperatur f <Thermod> offset temperature
Schalttisch m <Maschinen> indexing table
Schalttor n <Regelung> switching gate
Schalttransistor m <Elektronik> switching transistor
Schaltuhr f 1. <Elektriz> switch clock, timer; 2. <Foto> timer; 3. <Gerät> clock relay, time switch; 4. <Telekom> time switch
Schaltung f 1. <Comp & DV, Elektriz, Elektrotech> circuit, circuitry, switching operation; 2. <Fertig> actuation, index; 3. <Maschinen> shift; 4. <Phys> circuit; 5. <Telekom> switching
Schaltung f auf dem Chip <Elektronik> on-chip circuitry, on-chip circuit
Schaltung f auf Platine <Elektronik> on-board circuitry, on-board circuit
Schaltung f mit verzögerter Auslösung <Elektrotech> delayed-action circuit breaking
Schaltungsanalyse f <Elektrotech> circuit analysis
Schaltungsaufbau m <Elektrotech> circuit design
Schaltungsbeschreibung f <Elektrotech> circuit theory
Schaltungsentwurf m <Elektriz> circuit design
Schaltungsnetz n <Elektrotech, Telekom> network of circuit elements
Schaltunterstation f <Elektrotech> switching substation
Schaltventil n <Maschinen> on/off valve
Schaltverbindung f <Elektrotech> interconnection
Schaltverbindungsschicht f <Elektronik> interconnection layer
Schaltverlust m <Elektrotech> switching loss
Schaltverzögerung f 1. <Comp & DV> gate delay, propagation delay; 2. <Elektronik> time delay; 3. <Kontroll> switching delay; 4. <Telekom> switching delay; time delay (Relais, Schalter)
Schaltvorgang m <Elektrotech> switching
Schaltvorrichtung f <Elektrotech> switchgear
Schaltwalze f <Fertig> controller
Schaltwarte f <Heiz & Kälte, Kerntech> control room
Schaltwerk n 1. <Comp & DV> logic device; 2. <Elektriz> switching station
Schaltwerksentwurf m <Comp & DV> logic design
Schaltzeit f <Elektrotech> switching time
Schaltzelle f <Elektrotech> cell
Schaltzentrale f <Elektriz> control room
Schalung f 1. <Bau> formwork, shuttering; formwork (Betonbau); 2. <Fertig> (AE) mold, (BE) mould
Schalungsbrett n <Bau> shutter
Schalungsgerüst n <Bau> falsework
Schalwagen m <Bau> jumbo
Schalwand f <Bau> plank partition
Schamfilen n <Wassertrans> chafing (Segeln, Tauwerk)
Schamotte f <Ker & Glas> chamotte
Schamottemanschette f <Ker & Glas> collar (der Pfeife im Danner-Röhrenziehverfahren)
Schamottestein m 1. <Bau> fire brick, refractory brick; 2. <Heiz & Kälte> firebrick, refractory brick; 3. <Ker & Glas> fire brick; 4. <Labor> refractory brick; 5. <Thermod> firebrick
Schamotteton m <Ker & Glas> fireclay
Schamottetonform f <Ker & Glas> (AE) fireclay mold, (BE) fireclay mould
Schamottetontiegel m <Ker & Glas> fireclay crucible
Schamotteziegel m 1. <Bau, Heiz & Kälte> firebrick, refractory brick; 2. <Labor> fireclay brick (Ofen)

Schandeckel m <Wassertrans> covering board (Schiffbau); gunnel, gunwale (Schiffbau)
Schantungseide f <Textil> shantung
Schanzkleid n <Wassertrans> bulwark (Schiffbau)
Schanzkleidreling f <Wassertrans> bulwark rail (Schiffbau)
Schappenbohrer m <Kohlen> auger
Schar f <Fertig> group (Kurven)
scharf adj <Maschinen> sharp • **scharf am Winde** <Wassertrans> full and by (Segeln)
scharf begrenzendes Filter n <Elektronik> sharp cut-off filter
scharf begrenzte Linie f <Konstzeich> sharply bounded line
scharf begrenzter Impuls m <Elektronik> sharp pulse
scharf eingestellt adj <Foto> in focus
scharf einstellen v <Elektronik, Phys> focus
scharfe Kante f 1. <Bau> (AE) cutting edge, (BE) keen edge; 2. <Ker & Glas> arris, sharp edge
scharfe Krümmung f <Geom> sharp curvature
scharfe Kurve f <Geom> sharp curve
scharfe Oberfläche f <Ker & Glas> sharp finish
scharfe Streckgrenze f <Metall> sharp yield point
Schärfe f 1. <Akustik> sharpness; 2. <Bau> keenness (Werkzeugschneide); 3. <Ergon> acuity; 4. <Foto, Geom, Maschinen, Optik> sharpness (Bild); 5. <Qual> power
Schärfeabfall m <Foto> decrease in definition
Schärfeeinstellung f <Fernseh> sharpness control
Scharfeinstellfenster n <Foto> rangefinder window
Scharfeinstellung f <Foto> focusing
schärfen v <Maschinen> sharpen
Schärfen n 1. <Fertig> regrinding; 2. <Maschinen> sharpening
Schärfentiefe f 1. <Metall> depth of field; 2. <Optik> field depth; 3. <Phys> depth of focus
scharfer Schiffsbug m <Wassertrans> lean bow
Schärferegler m <Fernseh> focusing control
scharfkantig adj <Maschinen> sharp, sharp-edged
scharfkantige Düse f <Hydraul> sharp-edged orifice
scharfkantige Messblendenbohrung f <Hydraul> sharp-edged orifice
scharfkantige Mündung f <Hydraul> sharp-edged orifice
scharfkantiges Überlaufwehr n <Hydraul> sharp-crested weir
scharfkantiges Wehr n 1. <Hydraul> thin-edged weir; sharp-crested weir (Hydraulik); 2. <Wasserversorg> sharp-crested weir
scharfkantiges Werkzeug n <Maschinen> sharp-edged tool
Schärfmaschine f <Maschinen> sharpening machine
Scharnier n 1. <Bau> hinge, hinge joint, piano hinge; 2. <Fertig> articulation, hinge; 3. <Maschinen, Mechan> hinge • **mit Scharnier** <Maschinen> hinged • **mit Scharnier befestigen** <Bau> hinge • **mit Scharnier versehen** <Fertig> hinge
Scharnier n mit lösbaren Bolzen <Bau> loose-pin hinge
Scharnierband n <Bau> butt hinge, strap
Scharnierbandkette f <Maschinen> flat-top chain
Scharnierbeschläge mpl <Maschinen> hinge fittings
Scharniergelenk n <Bau> knuckle
Scharnierstift m <Bau> hinge pin
Scharrieren n <Bau> charring
Scharte f <Maschinen> nick
schartig adj <Mechan> jagged
Schatten m 1. <Druck, Foto> shadow; 2. <Optik, Phys> umbra
Schattenbereich m <Fernseh> shadow area
Schattenbild n <Fertig> silhouette shadow (Profilprojektor)

Schattenbildung f <Fernseh> clouding
Schattenfläche f <Druck> shadow area
Schattenkoeffizient m <Nichtfoss Energ> shading coefficient
Schattensektor m <Lufttrans> blind sector
Schattenstreifen m <Fertig> ghost, ghost line
Schattenwand f <Ker & Glas> shadow wall
schattiert adj <Ker & Glas> shaded
Schattierung f <Druck, Foto, Ker & Glas> shade
schätzen v 1. <Math, Qual> appraise, assess, estimate; 2. <Textil> assess
Schätzfunktion f <Math, Qual> estimator
Schätzung f <Math> estimation
Schätzwert m <Math, Qual> estimate
Schaubild n <Comp & DV> graph
Schaubildaufzeichner m <Labor> chart recorder
Schauer m <Teilphys> shower
Schauerteilchen n <Teilphys> shower particle
Schaufel f 1. <Bau> blade, shovel; 2. <Fertig> blade *(Turbine)*; vane; 3. <Heiz & Kälte> vane; blade *(eines Ventilators)*; 4. <Hydraul> paddle; 5. <Lufttrans> vane *(Turbomotoren)*; 6. <Maschinen> blade, bucket, paddle board, shovel; paddle *(des Ventilators, der Wasserturbine)*; 7. <Mechan> bucket, vane; 8. <Meerschmutz> shovel; 9. <Nichtfoss Energ> bucket *(Turbine)*; 10. <Wasserversorg> bucket *(eines Wasserrads)* • **eine Schaufel voll** <Bau> shovelful
Schaufelbagger m 1. <Bau> bucket dredger, shovel dredge, shovel dredger; 2. <Kfztech> shovel dredge, shovel dredger; 3. <Wasserversorg> bucket dredger
Schaufelblatt n <Hydraul> paddle
Schaufelentnahmegerät n <Abfall> scraper extractor
Schaufelflügler m <Lufttrans> cyclogyro
Schaufelgeschwindigkeit f <Nichtfoss Energ> blade speed, bucket velocity
Schaufelkranz m <Maschinen> blade ring
Schaufellader m <Meerschmutz> front-end loader
Schaufelmenge f <Nichtfoss Energ> blade quantity
Schaufeln n <Bau> shovel work
Schaufelrad n 1. <Bau> bucket wheel; 2. <Maschinen> bucket wheel, paddle wheel; 3. <Mechan> impeller; 4. <Wassertrans, Wasserversorg> paddle wheel
Schaufelradbagger m <Bau, Maschinen> bucket wheel excavator
Schaufelreihe f <Bau> line of buckets *(Förderband)*
Schaufelschrägstellung f <Bau, Nichtfoss Energ> bucket angle
Schaufelstiel m <Bau> D-handle
Schaufelteilung f <Nichtfoss Energ> blade pitch
Schaufelverjüngungsverhältnis n <Lufttrans> blade taper ratio *(Hubschrauber)*
Schaufelwerkstoffe mpl <Nichtfoss Energ> blade materials *(Turbine)*
Schaufelzeilung f <Lufttrans> blade spacing system *(Hubschrauber)*
Schauglas n 1. <Erdöl> *(AE)* gage glass, *(BE)* gauge glass; 2. <Heiz & Kälte> sight glass; 3. <Maschinen> sight feed glass
Schaukarton m <Verpack> display box
Schaukel f <Mechan> rocker
Schaukelbewegung f <Maschinen> seesaw motion, seesawing
schaukeln v <Bau> sway
Schaukeln n <Maschinen> rocking, swinging
Schaukelwelle f <Nichtfoss Energ> seiche
Schauklappe f <Mechan> inspection door
Schauklappe f **für Ausfahrverriegelung beim Landen** <Lufttrans> landing downlock optical inspection flap
Schauloch n <Ker & Glas> observation hole

Schaum m <Fertig, Ker & Glas, Kohlen, Kunststoff, Lebensmittel, Metall, Papier, Wassertrans> foam, froth, scum • **Schaum bilden** <Chemtech> foam
Schaumabscheiden n <Chemtech> foam separation
schaumartig adj <Chemtech> foamy
schaumbedeckt adj 1. <Chemtech> foamy; 2. <Fertig> scummy
Schaumbeständigkeit f <Chemtech> foam persistence
schaumbildend adj <Sicherheit, Umweltschmutz> intumescent
Schaumbildner m 1. <Chemtech> foamer, foaming agent; 2. <Sicherheit, Umweltschmutz> intumescence compound
Schaumbildung f <Chemtech, Erdöl> foaming
Schaumbildungshemmer m <Erdöl> defoaming agent
Schaumbrecher m <Chemtech> foam breaker
Schaumdecke f <Lufttrans, Sicherheit> foam blanket *(Feuerlöschung)*
schäumen v 1. <Bau> aerate *(Beton)*; 2. <Fertig> effervesce; 3. <Textil> foam
Schäumen n 1. <Fertig> frothing; 2. <Ker & Glas> foaming; 3. <Kunststoff> frothing • **zum Schäumen bringen** <Chemtech> foam
schäumend adj <Lebensmittel> effervescent
Schaumentwässerung f <Chemtech> foam drainage
Schaumfänger m <Fertig> skim bob *(Gießen)*
Schaumfeuerlöscher m <Lufttrans, Sicherheit> foam extinguisher
Schaumflotation f <Chemtech> froth flotation
schaumgekrönte Welle f <Wassertrans> whitecap *(Seezustand)*
Schaumglas n 1. <Heiz & Kälte> cellular glass; 2. <Ker & Glas> foam glass, foamed glass
Schaumgrenze f <Ker & Glas> foam line
Schaumgummi m 1. <Kunststoff> expanded rubber, foamed rubber; 2. <Mechan> expanded rubber
schaumig adj <Textil> foamy
Schaumkunstmasse f <Mechan> expanded plastic
Schaumkunststoff m <Kunststoff> cellular plastic, expanded plastic
Schaumlöffel m <Fertig> skimmer
Schaumlöschanlage f <Sicherheit> foam fire-fighting equipment
Schaumlöscher m <Sicherheit> foam fire extinguisher
Schaumlöschgerät n <Sicherheit> foam extinguishing apparatus
Schaumlöschmittel n <Sicherheit> foam extinguishing agent
Schaummittel n 1. <Lebensmittel> foaming agent; 2. <Lufttrans> foam compound *(Feuerschutzmaßnahme)*
Schaumprüfung f <Lufttrans> foaming test
Schaum-PS n *(geschäumtes Polystyrol)* <Verpack> ep *(expanded polystyrene)*
Schaumschalldämmplatte f <Sicherheit> noise abating foam panel
Schaumschicht f <Sicherheit> foam layer *(als Flammschutz)*
schaumschichtbildender Anstrich m <Kunststoff> intumescent paint
schaumschichtbildender Anstrichstoff m <Kunststoff> intumescent paint
Schaumschichttrocknung f <Lebensmittel> foam mat drying
Schaumschlacke f <Bau> foamed blast-furnace slag
Schaumstoff m 1. <Heiz & Kälte> cellular plastic; 2. <Kunststoff> foam; 3. <Mechan> expanded plastic; 4. <Textil> foam
Schaumstoffkaschierung f <Textil> foam backing

Schaumstoffschicht f <Textil> foam layer
Schaumstoffverpackung f <Verpack> plastic foam packaging
Schaumstoffverpackung f **und -auspolsterung** f <Verpack> foam packaging and cushioning
Schaumtank m <Papier> foam tank
Schaumteppich m <Lufttrans> foam carpet *(Feuerschutzmaßnahme)*
Schaumverbesserer m <Lebensmittel> lather booster
Schaumverdünnung f <Chemtech> foam dilution
Schaumverhinderungszusatz m <Maschinen> antifoaming agent
Schaumverhütungsmittel n 1. <Chemtech> foam inhibitor; 2. <Kunststoff> antifoaming agent
Schauöffnung f <Lufttrans> inspection panel *(Flugwerk)*
Schaupackung f <Verpack> dummy
Schaustück n <Verpack> window packaging
Schautropföler m <Maschinen> sight feed lubricator
Schauverpackung f <Verpack> display packaging, visual pack
Schauzeichen n <Gerät> annunciator
Scheduler m <Comp & DV> scheduler
Scheibe f 1. <Comp & DV> slice, wafer; 2. <Elektriz> dial; 3. <Elektronik> wafer *(Halbleiter)*; 4. <Ker & Glas> piece *(Tafelglas)*; 5. <Maschinen> *(BE)* disc, *(AE)* disk, flat washer, pulley, pulley wheel, round washer, sheave, washer, wheel; 6. <Textil> flange; 7. <Wassertrans> sheave *(Beschläge)*
Scheibe f **mit konstanter Drehzahl** <Optik> *(BE)* constant-angular-velocity disc, *(AE)* constant-angular-velocity disk
Scheibe f **mit konstanter Lineargeschwindigkeit** <Optik> *(BE)* constant-linear-velocity disc, *(AE)* constant-linear-velocity disk
Scheibe f **mit konstanter Winkelgeschwindigkeit** <Optik> *(BE)* constant-angular-velocity disc, *(AE)* constant-angular-velocity disk
Scheibenabschöpfer m <Umweltschutz> *(BE)* disc skimmer, *(AE)* disk skimmer
Scheibenabschöpfgerät n <Meerschmutz> *(BE)* disc skimmer, *(AE)* disk skimmer *(für Öl von der Wasseroberfläche)*
Scheibenanguss m <Kunststoff> diaphragm gate
Scheibenanker m <Elektriz, Elektrotech> *(BE)* disc armature, *(AE)* disk armature
Scheibenantenne f <Funktech> *(BE)* disc antenna, *(AE)* disk antenna
Scheibenbank f <Maschinen> bull block
Scheibenbohrmeißel m <Erdöl> *(BE)* disc bit, *(AE)* disk bit *(Bohrtechnik)*
Scheibenbremsanlage f <Bau> *(BE)* disc braking system, *(AE)* disk braking system
Scheibenbremse f <Eisenbahn, Fertig, Kfztech, Maschinen, Mechan> *(BE)* disc brake, *(AE)* disk brake
Scheibenbremsenbelag m <Kfztech> *(BE)* disc brake pad, *(AE)* disk brake pad
Scheibenfeder f <Fertig, Kfztech, Maschinen> Woodruff key
Scheibenfederkupplung f <Kfztech> diaphragm clutch
Scheibenfräser m <Maschinen> side mill, side milling cutter
Scheibengummi m <Kfztech> rubber weatherproof seal
Scheibenherstellung f <Elektronik> wafer fabrication
Scheibenklampe f <Wassertrans> cheek block *(Decksausrüstung)*
Scheibenkolben m <Hydraul> *(BE)* disc piston, *(AE)* disk piston; solid piston *(Pumpe)*
Scheibenkolbenpumpe f <Hydraul> solid piston pump
Scheibenkondensator m 1. <Elektrotech> *(BE)* disc capacitor, *(AE)* disk capacitor; 2. <Optik> *(BE)* capacitance electronic disc, *(AE)* capacitance electronic disk
Scheibenkonusantenne f <Funktech> discone antenna
Scheibenkupfer n <Chemie> rose copper
Scheibenkupplung f 1. <Fertig, Kfztech> *(BE)* disc clutch, *(AE)* disk clutch *(Kupplung)*; 2. <Maschinen> *(BE)* disc clutch, *(AE)* disk clutch, plate clutch, plate coupling; 3. <Mechan> *(BE)* disc clutch, *(AE)* disk clutch
Scheibenläufer m <Elektrotech> *(BE)* disc armature, *(AE)* disk armature
Scheibenmaske f <Elektronik> wafer mask
Scheibenmotor m <Elektrotech> pancake motor
Scheibenmühle f <Lebensmittel> *(BE)* disc mill, *(AE)* disk mill
Scheibenpolieren n <Ker & Glas> disc polishing, *(AE)* disk polishing
Scheibenrad n 1. <Fertig> *(BE)* disc, *(AE)* disk; 2. <Kfztech> *(BE)* disc centre wheel, *(AE)* disk center wheel; 3. <Maschinen> *(BE)* disc wheel, *(AE)* disk wheel
Scheibenrekorder m <Elektriz> *(BE)* disc recorder, *(AE)* disk recorder
Scheibenrelais n <Eisenbahn> vane relay
Scheibenröhre f <Elektronik> *(BE)* disc tube, *(AE)* disk tube
Scheibenschleifmaschine f <Maschinen> *(BE)* disc sanding machine, *(AE)* disk sanding machine
Scheibensignal n <Eisenbahn> *(BE)* disc signal, *(AE)* disk signal
Scheibenspeicher m <Fertig> rotary magazine
Scheibenspule f <Elektrotech> pancake coil
Scheibentriode f <Elektronik> *(BE)* disc-seal tube, *(AE)* disk-seal tube
Scheiben- und Planfräser m <Maschinen> side-and-face milling cutter
Scheibenunwucht f <Fertig> wheel imbalance
Scheibenverdampfer m <Heiz & Kälte> tray evaporator
Scheibenwaschanlage f <Kfztech> washer *(Glas)*
Scheibenwicklung f <Elektrotech> *(BE)* disc winding, *(AE)* disk winding
Scheibenwischer m <Kfztech> *(BE)* windscreen wiper, *(AE)* windshield wiper
Scheide f <Mechan> sheath
Scheidearbeit f <Kohlen> culling
Scheidebürette f <Chemtech> separating burette
Scheideglas n <Chemtech> refining glass
Scheidemittel n <Chemtech> separating agent
Scheiden n <Kohlen> sorting
Scheidepresse f <Umweltschutz> wringer
Scheideschlamm m <Chemie> scum *(Zucker)*
Scheidetrichter m <Chemtech> separating funnel
Scheidewand f <Mechan> diaphragm
Scheidung f 1. <Kohlen> sorting; 2. <Lebensmittel> defecation, liming
Schein... <Textil> mock
scheinbare Höhe f <Wassertrans> apparent altitude *(Navigation)*
scheinbarer Körpergehalt m <Kunststoff> false body
scheinbarer Wind m <Wassertrans> apparent wind *(Navigation)*
scheinbares Bild n <Foto> virtual image
Scheinenergie f <Elektriz> apparent energy
Scheinfuge f <Bau> dummy joint
Scheinleistung f <Elektriz, Elektrotech, Phys> apparent power
Scheinleistungsmessgerät n <Elektriz> apparent-power meter
Scheinleitwert m <Elektrotech, Phys> admittance
Scheinvariable f <Comp & DV> dummy variable

Scheinwerfer

Scheinwerfer m 1. <Elektrotech> searchlight; 2. <Foto> spotlight; 3. <Kfztech> headlamp *(Beleuchtung)*
Scheinwerferabschwächungsschalter m <Elektriz> dip selector switch
Scheinwerferschalter m <Elektriz, Kfztech> headlight switch
Scheinwiderstand m <Aufnahme, Elektrotech, Phys, Telekom> impedance
Scheinwiderstand m **des Lautsprechers** <Aufnahme> loudspeaker impedance
Scheinwiderstandsanpassung f <Aufnahme, Phys, Telekom> impedance matching
Scheinwiderstandsausgleicher m <Aufnahme> impedance compensator
Scheinwiderstandsbrücke f <Gerät> impedance measuring bridge
Scheinwiderstandsmessbrücke f <Gerät> impedance measuring bridge
Scheinwiderstandsverhältnis n <Aufnahme, Telekom> impedance ratio
Scheit n <Fertig> billet *(Holz)*
Scheitel m 1. <Bau> crown; soffit *(eines Bogens)*; 2. <Fertig> apex, crest, top; sagitta *(Bogen)*; 3. <Geom, Phys> vertex; 4. <Raumfahrt> apex
Scheitelkanal m <Wasserversorg> summit canal
Scheitelpunkt m <Phys> vertex
Scheitelspannungsmessgerät n <Elektriz, Gerät> peak voltmeter
Scheitelwert m 1. <Elektriz, Elektronik> peak value; 2. <Funktech> peak; 3. <Gerät> crest value; 4. <Telekom> peak value
Scheitelwertmessgerät n <Gerät> peak-reading instrument
Scheitelwertmessung f <Gerät> peak value measurement
Scheitelwinkel mpl <Geom> vertical angles
Schelf n <Wassertrans> shelf
Schellack m <Bau, Elektriz, Fertig> shellac
Schellbach-Rohr n <Ker & Glas> schelbach tubing
Schelle f <Maschinen, Mechan> clamp, clip
Schelle f **mit Schneckengewinde** <Maschinen> worm drive clamp
Schema n 1. <Bau> sketch; 2. <Comp & DV> schema; 3. <Künstl Int> frame *(in der Wissensrepräsentation)*; 4. <Maschinen> scheme
schematisches Schaltbild n <Kerntech> schematic wiring diagram
Schenkel m 1. <Bau> side, web; 2. <Fertig> grip; arm, side *(Winkel)*; handle *(Zange)*; 3. <Geom> leg, side; 4. <Maschinen, Telekom> leg *(Magnet)*
Schenkelpolläufer m <Elektrotech> salient pole rotor
Schenkelpolstator m <Elektrotech> salient pole stator
Schenkelrohr n 1. <Mechan> elbow; 2. <Metall> bent pipe
Scher... <Akustik, Fertig, Ker & Glas, Maschinen> shear
Scherbaum m <Textil> back beam
Scherbe f <Ker & Glas> cullet
Scherbeanspruchung f 1. <Bau, Maschinen, Qual> shearing strain, shearing stress; 2. <Phys> shear stress
Scherbenbrechwerk n <Ker & Glas> cullet crusher
Scherbenfänger m <Ker & Glas> cullet catcher
Scherbenrutsche f <Ker & Glas> cullet chute
Scherbenzerkleinerung f <Ker & Glas> cullet crush
scherbiges Aggregat n <Bau> flaky aggregate
Scherblatt n <Maschinen> shear blade
Scherbruch m <Bau, Metall> shear failure, shear fracture
Scherdehnung f <Bau, Metall, Phys> shear strain
Schere f 1. <Fertig> metal shears, shears; 2. <Ker & Glas> shears; 3. <Maschinen> brace, scissors, shears; 4. <Mechan> shears

Scherebene f <Bau, Fertig, Metall> shear plane
Scherelastizität f (m) <Akustik> shear elasticity (m)
scheren v 1. <Maschinen> shear; 2. <Metall> clip, shear; 3. <Textil> crop, shear, warp
Scheren n 1. <Bau> shearing; 2. <Maschinen> shear, shearing; 3. <Metall> shearing
Scherenausrichtung f <Ker & Glas> shear alignment
Scherenbolzen m <Fertig> intermediate stud *(Wechselräder)*
Scherenmarkierung f <Ker & Glas> shear mark
Scherenspray n <Ker & Glas> shear spray
Schererwärmung f <Kunststoff> shear heating
Scherfestigkeit f 1. <Bau, Fertig, Kohlen, Kunststoff> shear strength; 2. <Maschinen> resistance to shearing, shearing strength, ultimate shearing strength; 3. <Mechan, Metall> shear strength
Scherfläche f <Metall> plane of shear
Schergang m <Wassertrans> sheerstrake *(Schiffbau)*
Schergefälle n <Kunststoff> shear rate
Schergeschwindigkeit f <Kunststoff> rate of shear, shear rate
Schering-Brücke f <Elektrotech, Phys> Schering bridge
Scherkraft f 1. <Bau> shear action, shear force; 2. <Fertig> shear plane perpendicular force; 3. <Maschinen> shear force, shearing force; 4. <Mechan> shear; 5. <Phys> shearing force; 6. <Metall> shear, shear action, shear force; 7. <Qual> shear force
Schermaschine f <Kohlen, Textil> shearing machine
Schermesser n <Ker & Glas> shear blade
Schermodul n 1. <Maschinen> shear modulus (G); 2. <Phys> rigidity modulus; 3. <Phys> shear modulus (G)
Scherschichten fpl <Strömphys> shear layers *(in rotierenden Fluiden)*
Scherschichtinstabilität f <Strömphys> shear flow instability *(an Grenzfläche zweier nichtmischbarer Fluide)*
Scherschnabel m <Metall> shear lip
Scherschneidwerkzeug n <Fertig> shear-action cutting tool
Scherschnitt m <Ker & Glas> shearcut
Scherspannung f 1. <Bau> shear stress, shearing stress; 2. <Kunststoff, Metall, Phys, Strömphys> shear stress
Scherstahl m <Metall> shear steel
Scherstift m <Maschinen> shear pin
Scherströmung f <Strömphys> shear flow
Scherung f <Kunststoff, Mechan, Metall, Phys> shear
Scherversuch m <Maschinen, Qual> shear test
Scherviskosität f <Maschinen> shear viscosity
Scherwiderstand m <Fertig, Maschinen> cohesive resistance
Scherwinkel m 1. <Fertig> shear plane angle; 2. <Wassertrans> yaw angle
Scherzone f <Fertig> zone of shear
Scheuerbesen m <Wassertrans> hog
Scheuerblech n <Maschinen> chafing plate
Scheuerleiste f 1. <Bau> *(AE)* baseboard, *(AE)* mopboard, *(BE)* skirting board, washboard; 2. <Maschinen> wear strip
Scheuermittel n <Mechan> abrasive
scheuern v <Anstrich> fret *(durch Korrosion, durch Reiben)*
Scheuern n 1. <Anstrich> fretting wear; 2. <Fertig, Maschinen> galling
Scheuersand m <Anstrich> silica abrasive
Scheuerstelle f <Ker & Glas> frigger
Schicht f 1. <Anstrich> film, overlay, overlaying; 2. <Bau> course, layer; 3. <Comp & DV, Elektronik> layer; 4. <Erdöl> bed *(Geologie)*; 5. <Fertig, Ker & Glas> course *(Ziegel, Steine)*; 6. <Kohlen> bed, course, layer, stratum; 7. <Kunststoff> layer; 8. <Maschinen> ply; 9. <Metall>

layer; 10. <Papier> batch *(Papier)*; 11. <Raumfahrt> film, layer, ply; 12. <Textil> layer; 13. <Verpack> lamination; 14. <Wassertrans, Wasserversorg> layer • **in Schichten aufgebaut** <Mechan> lamellar • **Schichten bilden** <Kohlen> stratify
Schichtablagerung *f* <Elektronik, Kohlen> layer deposition
Schichtband *n* <Aufnahme> coated tape
Schichtbildaufnahme *f* <Kerntech> tomography
Schichtbildung *f* <Elektrotech> formation of layers
Schichtbürste *f* <Elektrotech> laminated brush
Schichtdicke *f* <Kunststoff> coating thickness
Schichtdickenmessung *f* <Kerntech> *(AE)* layer thickness gaging, *(BE)* layer thickness gauging
schichten *v* <Elektrotech> pile up
Schichten *n* <Ker & Glas> laying up
Schichtenanordnung *f* <Bau, Kohlen> coursing
Schichtendetektor *m* <Kohlen> level detector
Schichtenfolge *f* <Kohlen> layer sequence; bed sequence *(Geologie)*
Schichtenkarte *f* <Bau> contour map *(Vermessung)*
Schichtenspaltung *f* <Kunststoff> delamination
Schichtentrennung *f* <Fertig, Kunststoff> delamination
Schichtenwicklung *f* 1. <Elektriz> sandwich winding; 2. <Elektrotech> layer winding *(bei Bürsten)*
Schichtfestigkeit *f* <Kunststoff> interlaminar strength
Schichtgitter *n* <Phys> layer lattice *(Kristallgefüge)*
Schichtglas *n* <Ker & Glas> ply glass *(Hohlglas)*
Schichtkern *m* <Elektrotech> laminated core
Schichtkondensator *m* <Elektrotech> film capacitor
Schichtkühlung *f* <Raumfahrt> film cooling
Schichtkunststoff *m* <Wassertrans> laminated plastic
Schichtlademotor *m* <Kfztech> stratified charge engine
Schichtplatte *f* <Aufnahme> laminated record
schichtpressen *v* <Bau, Fertig> laminate
Schichtpressstoff *m* <Elektronik> laminate
Schichtprobe *f* <Umweltschmutz> whole workshift sample *(Luftprobe)*
Schichtquelle *f* <Wasserversorg> contact spring
Schichtseite *f* <Druck> emulsion side
Schichtsieden *n* <Kerntech> sheet boiling
Schichtspaltung *f* <Ker & Glas> delamination
Schichtstärke *f* <Verpack> lamination strength
Schichtstoff *m* <Kunststoff> laminate, laminated plastic
Schichtstofffolie *f* <Bau> laminated sheet
Schichtstoffherstellung *f* <Verpack> laminating
Schichtstoffprofil *n* <Kunststoff> laminated section
Schichtstofftafel *f* <Kunststoff> laminated sheet
Schichttechnik *f* <Elektronik> film deposition technique; film technology
Schichtträger *m* <Foto> film base
Schichtumwandler *m* <Raumfahrt> film transducer
Schichtung *f* 1. <Erdöl> bedding, stratification *(Geologie)*; 2. <Kohlen, Qual> stratification
Schichtung *f* **der Ladung** <Wassertrans> charge stratification
Schichtungsfestigkeit *f* <Verpack> lamination strength
Schichtverbundstoff *m* <Bau> sandwich composite
Schichtwiderstand *m* <Elektrotech> film resistor, sheet resistance
Schiebebetrieb *m* <Kfztech> overrun *(Motor)*
Schiebeboden *m* <Verpack> sliding bottom
Schiebebühne *f* <Bau> *(AE)* traveling platform, *(BE)* travelling platform
Schiebedach *n* <Kfztech> sun roof *(Karosserie)*
Schiebedeckel *m* <Verpack> sliding lid
Schiebefenster *n* <Bau> sash window, sliding sash; Yorkshire light *(Konstruktion mit Festverglasung und Hubfenster)*

Schiebefensterbeschläge *mpl* <Bau> sash hardware
Schiebefensterfeststeller *m* <Bau> sash fastener
Schiebefensterrahmen *m* <Bau> sash
Schiebefensterverschluss *m* <Bau> casement fastener
Schiebegelenk *n* <Kfztech> slip joint
Schiebegewicht *n* <Mechan> balancing weight
Schiebehülse *f* <Kfztech> sliding sleeve
Schiebeimpuls *m* <Elektronik> shift pulse
Schiebekontakt *m* <Elektrotech> sliding contact
Schiebekurve *f* <Lufttrans> flat turn *(beim Trudeln)*
Schiebelandung *f* <Lufttrans> crosswind landing
Schiebeleiter *f* <Bau> *(AE)* traveling ladder, *(BE)* travelling ladder
Schiebelok *f* <Eisenbahn> booster locomotive, *(AE)* pusher locomotive
Schiebelokomotive *f* <Eisenbahn> booster locomotive, *(AE)* pusher locomotive
schieben *v* 1. <Maschinen> slip; 2. <Mechan> push, slide
Schieben *n* <Eisenbahn> banking
Schiebenocken *m* <Maschinen> sliding cam
Schiebe-Potenziometer *n* <Elektriz> slide rheostat
Schieber *m* 1. <Bau> damper; 2. <Fertig> slide; gate valve *(Kunststoffinstallationen)*; movable head *(Messschieber)*; 3. <Heiz & Kälte> register, slide, slide damper, valve; 4. <Hydraul> valve *(Absperrorgan von Rohrleitungen)*; slide valve *(Dampfmaschine, Hydraulik)*; 5. <Maschinen> gate valve, slide valve, slider; 6. <Mechan> slide; 7. <Papier> bar; 8. <Sicherheit> push stick
Schieberad *n* 1. <Kfztech> sliding gear *(Getriebe)*; 2. <Maschinen> sliding gear
Schieberädergetriebe *n* <Maschinen> sliding gear train
Schieberadgetriebe *n* 1. <Kfztech> sliding gear transmission; 2. <Maschinen> sliding gear drive
Schieberahmen *m* <Bau> window sash
Schieberdrosselventil *n* <Hydraul> slide throttle valve *(Dampfmaschine)*
Schieberegister *n* <Comp & DV, Elektronik, Funktech, Telekom> shift register
Schieberiegel *m* <Maschinen> sliding bolt
Schiebering *m* <Maschinen> slider
Schieberkasten *m* <Hydraul> steam chest, valve chest; slide box *(Dampfmaschine)*
Schieberklappe *f* <Heiz & Kälte> slide damper
Schieberohr *n* <Optik> sliding tube *(an optischem Gerät)*
Schieberstange *f* <Hydraul> slide rod *(Dampfmaschine)*; stem *(Kolbenschieber, Steuerschieber)*; rod *(Steuerkolben, Steuerschieber)*
Schiebersteuerung *f* <Hydraul> slide valve gear
Schiebersystem *n* <Raumfahrt> shutter system
Schieberventil *n* 1. <Hydraul> gate valve; 2. <Kfztech> slide valve
Schiebeschalter *m* <Elektriz, Elektrotech> slide switch, sliding switch
Schiebesitz *m* (SS) 1. <Maschinen> close sliding fit, push fit, sliding fit; 2. <Mechan> close-sliding fit, push fit, sliding fit
Schiebetor *n* 1. <Bau> sliding gate; 2. <Wasserversorg> sash gate *(einer Schleuse)*
Schiebetür *f* <Bau> sash door, sliding door
Schiebevorrichtung *f* <Mechan> slide
Schiebewelle *f* <Maschinen> sliding shaft
Schiebewiderstand *m* <Elektriz> slide rheostat
Schiebewinkel *m* <Lufttrans> crab angle
Schieblehre *f* 1. <Maschinen> *(AE)* caliper square, *(BE)* calliper square, *(AE)* sliding calipers, *(BE)* sliding callipers, *(AE)* vernier gage, *(BE)* vernier gauge; 2. <Metrol> *(AE)* vernier caliper, *(BE)* vernier calliper
schief *adj* 1. <Bau> slanting; leaning *(Turm)*; 2. <Geom> oblique; 3. <Maschinen> out of plumb

schief

schief abgeschnittener Zylinder m <Fertig> ungula (Kegel)
Schiefbogen m <Bau> oblique arch
Schiefe f <Math> skewness (Schiefemaß einer Verteilung)
schiefe Biegung f <Fertig> bending in two planes
schiefe Brücke f <Bau> skew bridge
schiefe Ebene f <Bau, Geom> inclined plane
schiefe Umlaufbahn f <Raumfahrt> drift orbit
schiefe Verteilung f <Qual> skewed distribution
Schiefer m 1. <Bau> slate; 2. <Erdöl> shale
schiefer Kegel m <Geom> oblique cone, scalene cone
schiefer Strahl m 1. <Funktech> oblique ray; 2. <Telekom> skew ray
schiefer Winkel m <Geom> oblique angle
Schieferbedachung f <Bau> slate roof cladding
Schieferbruch m <Metall> fibrous fracture
Schieferdom m <Erdöl> shale dome (Geologie)
Schieferkohle f <Kohlen> schistous coal
Schiefernagel m <Bau> slate nail
Schieferplatte f <Bau> slate
Schieferschüttler m <Erdöl> shale shaker
Schieferstock m <Erdöl> shale diapir (Geologie)
Schieferton m <Ker & Glas> schistous clay
schiefwinkelig adj 1. <Bau> skew; 2. <Fertig> scalene
schiefwinklige axonometrische Projektion f <Konstzeich> oblique axonometric projection
schiefwinklige Brücke f <Bau> oblique bridge
schiefwinklige Parallelprojektion f <Konstzeich> oblique parallel projection
schiefwinkliges Dreieck n <Geom> oblique triangle
Schiene f 1. <Eisenbahn> rail; 2. <Fertig> bar (Schweißen); 3. <Maschinen> runner; 4. <Trans> road rail • **auf Straße und Schiene transportieren** <Trans> transport by rail and road
Schiene f **mit induktiver Rückkopplung** <Eisenbahn> inductive reaction rail
Schienen fpl <Eisenbahn> metals
schienen v <Fertig> clout
Schienenauflager n <Eisenbahn> rail bed
Schienenbeanspruchung f <Eisenbahn> rail stress
Schienenbiegepresse f <Eisenbahn> rail-bending device
Schienenbiegespannung f <Eisenbahn> rail-bending stress
Schienenbohrmaschine f <Eisenbahn> rail-boring machine, rail-drilling machine
Schienenbohrung f <Eisenbahn> rail bore
Schienenbruch m <Eisenbahn> rail break
Schienenbündel n <Eisenbahn> bundle of rails
Schienenfahrzeug n 1. <Eisenbahn> railcar, rail vehicle; 2. <Trans> railcar
Schienenfuß m <Eisenbahn> rail flange, rail foot
schienengebunden adj <Trans> rail-mounted
schienengleicher Straßenübergang m <Bau, Eisenbahn> (AE) grade crossing, (BE) level crossing
Schienengleis n 1. <Bau> tram track; 2. <Eisenbahn> track
Schienenhobelmaschine f <Eisenbahn> rail-planing machine
Schienenkabinensystem n <Eisenbahn> cabin system on rail
Schienenkontakt m <Eisenbahn> treadle
Schienenkopf m <Bau, Eisenbahn, Fertig> rail head
Schienenkopfausrundung f <Eisenbahn> rail shoulder
Schienenkran m <Eisenbahn> rail crane
Schienenlasche f <Eisenbahn> fishplate, splice bar
Schienenlegemaschine f <Eisenbahn> track-laying machine
Schienenlegen n <Eisenbahn> plate laying, rail laying, track laying

Schienennagel m 1. <Eisenbahn> dog spike; 2. <Maschinen> track spike
Schienennetz n <Eisenbahn> (AE) railroad network, (BE) railway network
Schienenomnibus m <Eisenbahn> rail motor coach, railbus
Schienenprofil n <Eisenbahn> rail profile, rail section
Schienenquerschnitt m <Eisenbahn> rail section
Schienenräumer m <Eisenbahn> cow catcher, pilot
Schienenrücker m <Eisenbahn> rail lifter, rail tongs
Schienenschleifzug m <Eisenbahn> rail-grinding train
Schienenschwebebahn f <Eisenbahn> (AE) suspended railroad, (BE) suspended railway
schienenseitiges Signal n <Eisenbahn> lineside signal
Schienensteg m <Eisenbahn> rail web
Schienenstoß m <Eisenbahn> rail joint
Schienenstoßfutterblech n <Eisenbahn> shim
Schienenstrang m <Eisenbahn> rail track
Schienenstromabnehmer m <Eisenbahn> current collector
Schienenstuhl m <Eisenbahn> chair (für Doppelkopfschienen)
Schienentankwagen m <Eisenbahn> rail tank car
Schienentragzange f <Eisenbahn> rail tongs
Schienenüberhöhung f <Eisenbahn> superelevation of rails, superelevation of track
Schienenuntersuchung f <Eisenbahn> rail inspection
Schienenverbinder m <Eisenbahn> railbond
Schienenverlegekran m <Eisenbahn> rail-laying crane
Schienenverschleißtoleranz f <Eisenbahn, Qual> rail wear tolerance
Schienenvorblock m <Fertig> rail bloom
Schiene-Straße-... <Eisenbahn> railroad
Schiene-Straße-Sattelauflieger m <Eisenbahn, Trans> railroad semitrailer
Schiene-Straße-Verkehr m <Eisenbahn, Trans> railroad-transport
schießen v <Erdöl> shoot (Seismik)
Schießhauer m <Kohlen> blaster
Schiff n 1. <Bau> bay (einer Halle, Kirche); 2. <Druck> galley; 3. <Meerschmutz, Wassertrans> boat, ship, vessel • **frei Längsseite Schiff** <Wassertrans> free alongside ship, fas (Seehandel) • **Schiff aufgeben** <Wassertrans> abandon ship (Notfall) • **Schiff in Dienst stellen** <Wassertrans> put a ship into commission
Schiff n **in Seenot** <Wassertrans> ship in distress (Notfall)
Schiff n **mit geringem Tiefgang** <Wassertrans> shallow-draught vessel
Schiff n **mit Zellcontainern** <Wassertrans> cellular container ship
schiffbar adj <Wassertrans> navigable (Gewässer)
Schiffbarkeit f <Wassertrans> navigability
Schiffbau m <Wassertrans> naval architecture, shipbuilding
Schiffbauer m <Wassertrans> shipbuilder, shipwright
Schiffbauholz n <Wassertrans> timber
Schiffbauingenieur m <Wassertrans> marine architect, naval architect
Schiffbaustahl m <Wassertrans> shipbuilding steel
Schiffbauzeichnungen fpl <Wassertrans> hull drawings
Schiffbruch m <Wassertrans> shipwreck • **Schiffbruch erleiden** <Wassertrans> suffer shipwreck
schiffbrüchig adj <Wassertrans> shipwrecked
Schifffahrt f <Wassertrans> shipping
Schifffahrtskanal m <Wassertrans> ship canal (künstliche Wasserstraße)
Schifffahrtskunde f <Wassertrans> navigation
Schifffahrtsroute f <Wassertrans> shipping lane, shipping route

Schifffahrtswarnsignal n <Telekom> navigation warning signal
Schiffsabfrage f <Telekom> ship polling
Schiffsanstrich m <Wassertrans> marine coating
Schiffsantrieb m <Erdöl> ship propulsion
Schiffsausrüstung f <Wassertrans> ship's equipment
Schiffsbeladung f <Wassertrans> ship loading
Schiffsbergung f <Wassertrans> salvaging
Schiffsboden m <Wassertrans> bottom *(Schiffbau)*
Schiffsbreite f <Wassertrans> breadth; beam *(Schiffbau)*
Schiffsdatenschreiber m <Wassertrans> voyage data recorder, VDR
Schiffsdiesel m <Wassertrans> marine diesel engine
Schiffsdieselmotor m <Wassertrans> marine diesel engine
Schiffsdieselöl n <Wassertrans> marine diesel oil
Schiffseigner m <Wassertrans> shipowner
Schiffsentwurf m <Wassertrans> ship design
Schiffsfracht f <Verpack> carriage by sea
Schiffsführung f <Wassertrans> ship handling
Schiffsgetriebe n <Wassertrans> marine gear unit
Schiffshebewerk m <Wassertrans> liftlock, ship canal lift, ship hoist
Schiffskanal m <Wassertrans> ship canal
Schiffskessel m <Heiz & Kälte> marine boiler
Schiffskompass m <Wassertrans> mariner's compass *(Navigation)*
Schiffskonstrukteur m <Wassertrans> marine architect, ship designer
Schiffskörper m <Wassertrans> hull
Schiffsküche f <Wassertrans> galley
Schiffskühlanlage f <Heiz & Kälte> marine refrigeration plant
Schiffsladung f <Wassertrans> cargo, shipload
Schiffsmakler m <Wassertrans> ship broker
Schiffsmannschaft f <Wassertrans> ship's hands *(Besatzung)*
Schiffsmaschine f <Wassertrans> marine engine
Schiffsmaschinenbau m <Wassertrans> marine engineering
Schiffsmaschineningenieur m <Wassertrans> marine engineer
Schiffsmeldesystem n <Telekom> ship reporting system
Schiffsmessbrief m <Wassertrans> certificate of measurement
Schiffsmodelversuchstank m <Wassertrans> ship model testing tank
Schiffsmotor m <Wassertrans> marine engine
Schiffsnavigation f <Wassertrans> navigation afloat
Schiffssortung f <Wassertrans> vessel location
Schiffspapiere npl <Wassertrans> ship's papers
Schiffspoller m <Wassertrans> bollard *(am Kai)*
Schiffsposition f <Wassertrans> ship's position
Schiffspropeller m <Meerschmutz> propeller
Schiffsreaktor m <Wassertrans> marine nuclear plant
Schiffsregister n <Wassertrans> ship's register
Schiffsrumpf m <Wassertrans> hull
Schiffsschleuse f <Wasserversorg> lift lock
Schiffsschraube f 1. <Meerschmutz> propeller; 2. <Trans> marine propeller; 3. <Wassertrans> water propeller, water screw; propeller *(Schiffbau)*; screw *(Schiffantrieb)*
Schiffssektion f <Wassertrans> ship section
Schiffsstandort m <Wassertrans> ship's position
Schiffstagebuch n <Wassertrans> ship's log; logbook *(Dokumente)*
Schiffstaufe f <Wassertrans> naming
Schiffstechnik f <Wassertrans> marine technology, ship technology

Schiffsträger m <Wassertrans> hull girder, ship girder *(Schiffbau)*
Schiffsturbine f <Wassertrans> marine turbine
Schiffsvermessung f <Wassertrans> tonnage measurement
Schiffsweg m <Wassertrans> track
Schiffswerft f <Wassertrans> shipyard
Schiffswinde f <Mechan> capstan
Schiffszug m <Wassertrans> bulk ship train
Schifter m <Bau> jack rafter
Schiftsparren m <Bau> jack rafter
Schild m 1. <Bau, Maschinen> shield; 2. <Mechan> plate; 3. <Trans> sign
Schilderwärmung f <Kerntech> shield heat-up
Schildkühlung f <Kerntech> shield cooling system
Schildvortrieb m <Eisenbahn> *(AE)* shield tunneling, *(BE)* shield tunnelling *(Tunnelbau)*
Schildzapfen m <Mechan> trunnion
Schillern n 1. <Ker & Glas> iridescence, iridizing; 2. <Meerschmutz> iridescence; 3. <Optik> irisation
Schimmel m <Lebensmittel> mildew
schimmelfest adj <Verpack> *(AE)* mold-resistant, *(BE)* mould-resistant
schimmernd adj <Textil> lustrous
Schindel f <Bau> shingle *(Dach)*
Schindeldach n <Bau> shingle roof
Schirm m 1. <Elektriz, Elektronik, Elektrotech> screen; shield *(für Lampe)*; 2. <Labor> shield; 3. <Mechan, Optik, Phys> screen
Schirmanguss m <Kunststoff> diaphragm gate
Schirmantenne f <Funktech> umbrella antenna
Schirmbild n 1. <Gerät> screen image; 2. <Telekom> display *(Radar)*; screen *(Monitor)*
Schirmfaktor m <Elektronik> screen factor
Schirmgitter n <Elektronik> screen grid *(Elektronenröhren)*
Schirmgitterröhre f <Elektronik> screen grid tube
Schirmhalter m <Optik> screen holder
Schirmraster n <Comp & DV> screen pattern
Schirmträger m <Elektronik> face plate *(Katodenröhre)*
Schirmung f <Elektriz> screening, shielding
Schirmungseffekt m <Elektriz> screening effect
Schlachtausbeute f <Lebensmittel> carcass dressing percentage, carcass yield *(Fleischerei)*
schlachten v <Lebensmittel> slaughter *(Schlachthaus)*; kill *(Tiere)*
Schlachten n <Lebensmittel> slaughtering *(Schlachthaus)*
Schlachthausabfall m <Abfall> abattoir waste
Schlacke f 1. <Abfall> clinker; 2. <Bau> slag; 3. <Fertig> cinder, dross, scoria, slag; 4. <Ker & Glas> slag *(Glas, das auf Hafenofenherd spritzt)*; 5. <Kohlen> scum, slag; 6. <Metall> floss, scoria, slag; 7. <Papier> slag
Schlackebad n <Fertig> molten slag
schlacken v <Fertig> scum
Schlacken n <Fertig> scumming
Schlackenabzug m <Fertig> flushing
Schlackenanhang m <Fertig> adhering slag *(Brennschneiden)*
Schlackenlauf m <Fertig> skim gate
Schlackenloch n <Fertig> pit hole
schlackenrein adj <Thermod> free from slag
Schlackenschale f <Metall> scorifier
Schlackenstein m <Metall> slag brick
Schlackenwolle f <Metall> slag wool
Schlaf m <Wassertrans> sleep
Schlafkoje f <Wassertrans> bunk
Schlafwagen m <Eisenbahn> sleeping car
Schlag m 1. <Anstrich> impact; 2. <Fertig> eccentricity; 3. <Maschinen> eccentricity, knock, lay, run-out, shock; 4.

Schlag

<Mechan> impact; 5. <Wassertrans> leg • **mit Schlag** <Textil> flared
schlagartige Ausbreitung *f* **von Flammen** <Thermod> flashover
Schlagbeanspruchung *f* <Fertig> impact stress
Schlagbelastung *f* <Verpack> impact stress
Schlagbewegung *f* <Lufttrans> flapping *(Hubschrauber)*
Schlagbiegefestigkeit *f* <Kfztech> impact resistance
Schlagbiegeprüfung *f* <Maschinen> blow bending test
Schlagbiegeversuch *m* <Bau, Maschinen, Qual> impact bending test
Schlagbohren *n* 1. <Bau> boring by percussion; 2. <Erdöl> percussion drilling *(Bohrtechnik)*; 3. <Kohlen> percussion drilling
Schlagbohrer *m* 1. <Bau> hammer drill; 2. <Maschinen> percussion drill
Schlagbohrmaschine *f* <Maschinen> hammer drill, impact drill, percussion drill
Schlagbohrschlamm *m* <Erdöl> spud mud
Schlagbrecher *m* <Kohlen> impact crusher
schlagen *v* 1. <Anstrich> agitate; 2. <Bau> beat, strike; 3. <Fertig> lash *(Gießen)*; lay *(Seil)*; wobble *(Bohrer)*; chatter *(Walzen)*; 4. <Kfztech> run out of true; 5. <Kohlen> strike; 6. <Maschinen> hammer, knock, lay, run out of true; 7. <Metall> beat
Schlagen *n* <Fertig> chatter *(Walzen)*
schlagfest *adj* 1. <Sicherheit> impact-resistant, shockproof; 2. <Verpack> shockproof
schlagfeste Pressmasse *f* <Kunststoff> *(AE)* high-impact molding compound, *(BE)* high-impact moulding compound
schlagfestes Polystyrol *n* <Kunststoff> impact polystyrene
Schlagfestigkeit *f* 1. <Kerntech, Kunststoff, Mechan, Phys> impact strength; 2. <Verpack> impact resistance; 3. <Werkprüf> impact resistance, impact strength
Schlagfestigkeit *f* **der Nabe** <Lufttrans> hub flapping stiffness *(Hubschrauber)*
Schlagfräsen *n* <Maschinen> fly cutting
schlagfrei *adj* <Fertig> balanced
Schlaggalvanisierung *f* <Elektronik> striking *(elektronischer Niederschlag)*
Schlaggelenk *n* <Lufttrans> flapping hinge *(Hubschrauber)*
Schlaggelenkbegrenzungsstift *m* <Lufttrans> flapping hinge pin *(Hubschrauber)*
Schlaghebel *m* <Mechan> ram lever
Schlagknickversuch *m* <Bau, Fertig, Qual> impact buckling test
Schlagkraft *f* <Maschinen> percussive force
Schlaglänge *f* <Fertig> lay *(Drahtseil)*
Schlaglärm *m* <Sicherheit> impact noise
Schlagloch *n* <Bau, Kfztech> pothole *(Straße)*
Schlagmeißel *m* <Erdöl> cable drilling bit *(zum Seilschlagbohren)*
Schlagmeißel *m* **für Seilschlagbohren** <Erdöl> churn drill *(Bohrtechnik)*
Schlagmikrofon *n* <Aufnahme> impact microphone
Schlagmoment *n* <Lufttrans> flapping moment *(Hubschrauber)*
Schlagniet *m* <Maschinen> percussion rivet
schlagnieten *v* <Fertig> impact-rivet
Schlagnieten *n* 1. <Fertig> impact riveting; 2. <Maschinen> percussion riveting
Schlagpressen *n* 1. <Kunststoff> forging; 2. <Maschinen> *(AE)* impact molding, *(BE)* impact moulding
Schlagprobe *f* <Verpack> impact test
Schlagprüfung *f* <Phys> impact test
Schlagschere *f* <Maschinen> guillotine shears
Schlagschmieden *n* <Maschinen> impact forging
Schlagschrauber *m* <Maschinen> *(BE)* impact spanner, *(AE)* impact wrench
Schlagseite *f* <Wassertrans> list • **Schlagseite haben** <Wassertrans> list
Schlagsieb *n* <Kohlen> impact screen
Schlagstock-Rückholfeder *f* <Textil> picking stick return spring
Schlagstrangpressen *n* <Maschinen> impact extrusion
Schlagtonhöhe *f* <Akustik> strike note
Schlagversuch *m* <Anstrich, Kunststoff, Metall, Qual, Verpack, Werkprüf> impact test
Schlagwerk *n* <Maschinen> hammer mechanism
Schlagwerkzeug *n* <Sicherheit> percussive tool
Schlagwetter *n* <Sicherheit> explosive atmosphere; flammable atmosphere *(Bergbau)*
schlagwettergeschützter Motor *m* <Elektrotech, Thermod> firedamp-proved motor
schlagwettergeschützter Schalter *m* <Thermod> firedamp-proved breaker, firedamp-proved switch
Schlagwetterschutz *m* <Sicherheit> firedamp protection *(Bergbau)*
Schlagwinkel *m* <Lufttrans> flapping angle
schlagzähes Polystyrol *n* <Kunststoff> impact polystyrene
Schlagzähigkeit *f* <Kunststoff> impact strength
Schlagzahnfräsen *n* <Maschinen> fly cutting, thread whirling
Schlagzeile *f* <Druck> catchline
Schlagzerreißversuch *m* <Metall, Qual> tensile impact test
Schlagzugversuch *m* <Metall, Qual> tensile impact test
Schlamm *m* 1. <Abfall> sludge, slurry; 2. <Bau> sediment, sludge; 3. <Ker & Glas, Kohlen> sediment, slime, slurry; 4. <Lebensmittel> sludge; 5. <Papier> slime; 6. <Wasserversorg> silt, slime, sludge
Schlammablagerungen *fpl* <Abfall> alluvial deposits
Schlammabscheider *m* <Wasserversorg> silt trap
Schlammabsetzvolumen *n* <Chemtech> settled volume *(Abwasser)*
Schlammanalyse *f* <Bau, Kohlen> sedimentation test
Schlammaufbereitung *f* <Abfall> sludge processing
Schlammbecken *n* <Kohlen> slurry pond
Schlammbehandlung *f* <Abfall> sludge processing
Schlammbelebung *f* <Abfall> activation of sludge
Schlammbelebungsverfahren *n* <Wasserversorg> activated sludge process
Schlammbeseitigung *f* 1. <Abfall> sludge removal; 2. <Wasserversorg> sludge disposal
Schlammbildungsprüfung *f* <Elektriz> sludge formation test
Schlammbohrer *m* <Erdöl> mud bit *(Bohrtechnik)*
Schlämme *f* 1. <Anstrich> slurry; 2. <Bau> whitewash
Schlammeindickung *f* 1. <Abfall> sludge thickening; 2. <Wasserversorg> sludge ripening, sludge thickening
schlämmen *v* <Fertig> levigate
Schlämmen *n* <Bau> whitewashing
Schlammentwässerung *f* <Abfall> dehydration of sludge, sludge dewatering
Schlämmeversiegelung *f* <Bau> slurry seal
Schlammfang *m* 1. <Abfall> sludge sump; 2. <Wasserversorg> silt trap
Schlammfaulbecken *n* <Abfall> stabilization pond
Schlammfaulbehälter *m* <Abfall> digestion sump, sewage digestor, sludge digestion tank, sludge digestor
Schlammfaulung *f* <Abfall> sludge digestion
Schlammfluid *n* <Erdöl> mud fluid *(Bohrtechnik)*
Schlammgehalt *m* <Wasserversorg> mud content
Schlammgewicht *n* <Erdöl> mud weight *(Tiefbohrtechnik)*

Schlammgrube f <Erdöl> mud pit *(Bohrtechnik)*
Schlammkasten m <Erdöl> mud box *(Bohrtechnik)*
Schlammkohle f <Kohlen> coal sludge, mud coal
Schlammkompostierung f <Abfall> sludge composting
Schlammkonditionierung f <Wasserversorg> sludge bulking, sludge conditioning
Schlammkonzentrat n <Kohlen> concentrated sludge
Schlämmkreide f <Fertig> whiting
Schlammkuchen m <Abfall> sludge cake
Schlammlagerraum m <Wasserversorg> silt storage space
Schlammlog n <Erdöl> mud log *(Bohrtechnik)*
Schlämmmaschine f <Chemtech> decanting machine
Schlammraum m <Kfztech> sediment chamber, sediment space *(der untere Teil des Batteriegehäuses)*
Schlammräumer m <Abfall> sludge rake
Schlammring m <Erdöl> mud ring *(Bohrtechnik)*
Schlammrückleitung f <Erdöl> mud return line *(Bohrtechnik)*
Schlammsammelbehälter m <Abfall> silt container, sludge sump
Schlammsäule f <Erdöl> mud column *(Bohrtechnik)*
Schlammschwelle f <Bau> mudsill
Schlammseil n <Erdöl> sand line
Schlammsieb n <Kohlen> slurry screen
Schlammstabilisierung f <Abfall> sludge stabilization
Schlammstapelteich m <Wasserversorg> sump
Schlammsystem n <Erdöl> mud system *(Tiefbohrtechnik)*
Schlammteich m <Wasserversorg> sump
Schlammteichverfahren n <Umweltschmutz> lagooning *(Eindämmen eines Wasserbeckens)*
Schlammtrockenbett n <Abfall> sludge drying bed
Schlammtrockner m <Abfall> sludge drier
Schlammtrocknung f <Abfall> sludge drying
Schlämmung f 1. <Chemtech> elution; 2. <Ker & Glas, Lebensmittel> elutriation
Schlammverbrennung f <Abfall, Wasserversorg> sludge incineration
Schlammverdickung f <Abfall> dehydration of sludge, sludge dewatering
Schlammverlust m <Erdöl> mud losses *(Bohrtechnik)*
Schlämmversuch m <Bau> elutriation test
Schlammverwertung f <Abfall> recycling of sludge
Schlammvulkan m 1. <Erdöl> mud volcano *(Geologie)*; 2. <Nichtfoss Energ> mud volcano
Schlammwasser n <Wasserversorg> sludge liquor
Schlammzone f <Wassertrans> mudflats *(Geographie)*
Schlange f 1. <Ker & Glas> snake; 2. <Kontroll> queue; 3. <Maschinen> coil
Schlangenbildung f <Ker & Glas> snaking
Schlangenbohrer m <Maschinen> screw auger
schlangenförmig *adj* <Bau> meandering
Schlangenkurve f <Bau> serpentine
Schlangenrohr n <Bau> coil
schlanker Balken m <Bau> slender beam
schlankgenutet *adj* <Fertig> slow-helix
Schlankheitsgrad m 1. <Bau> slenderness ratio; 2. <Lufttrans> fineness ratio *(Stromlinienaufbau)*
Schlauch m 1. <Bau> flexible hose, hose; 2. <Kfztech> inner tube; 3. <Labor> hose *(Verbindung)*; 4. <Maschinen> flexible pipe, flexible tube, hose, tube; 5. <Mechan> hose; 6. <Nichtfoss Energ> tube; 7. <Papier, Raumfahrt> hose; 8. <Umweltschmutz> bag; 9. <Wasserversorg> hose
Schlauch m **mit Textilverstärkung** <Maschinen> textile-reinforced hose
Schlauchanschluss m 1. <Bau> hose coupler; 2. <Maschinen> hose connection

Schlauchboot n <Wassertrans> inflatable boat, inflatable dinghy, rubber boat, rubber dinghy
Schlauchfilter n <Kohlen> filter bag
Schlauchfolienblasen n <Kunststoff> film blowing
Schlauchgerät n <Sicherheit> air-supplied equipment, air-supplied respirator, airline respirator *(Atemschutz)*
Schlauchhahn m <Maschinen> hose tap
Schlauchklemme f 1. <Maschinen> hose clip; 2. <Mechan> hose clamp
Schlauchkupplung f 1. <Maschinen> hose coupling; 2. <Wasserversorg> hose coupling *(für einen Gartenschlauch)*
Schlauchleitung f <Maschinen> flexible tubing, tubing
schlauchloser Reifen m <Kfztech> (AE) tubeless tire, (BE) tubeless tyre *(Reifen)*
Schlauchschelle f <Maschinen> hose clamp
Schlauchtrockenmaschine f <Textil> tubular dryer
Schlauchtülle f 1. <Fertig> connector; 2. <Maschinen> hose nozzle
Schlauchumflechtmaschine f <Kunststoff> braiding machine
Schlauchventil n 1. <Fertig> pipe compression valve *(Kunststoffinstallationen)*; 2. <Kfztech> valve
Schlaufe f <Verpack> carrying handle
schlecht werden *v* <Lebensmittel> spoil, taint
schlechte Dämmung f <Akustik> poor insulation
schlechte Isolation f <Telekom> low insulation
schlechte Isolierung f <Akustik> poor insulation
schlechte Kühlung f <Ker & Glas> bad annealing
schlechte Lötverbindung f <Elektriz> dry joint
schlechter Eingriff m <Fertig> mismating
schlechter Kontakt m <Elektrotech> bad contact
schlechter Leiter m <Elektrotech> poor conductor
schlechtes Isoliermedium n <Elektrotech> poor insulant
Schlechtgrenze f <Qual> limiting quality
Schlechtwetterausrüstung f <Wassertrans> foul weather gear
Schlechtzahl f <Qual> rejection number
Schlegel m 1. <Bau> beater; 2. <Mechan> mallet
Schleichen n <Elektrotech> crawling
schleichende Bewegung f <Metall, Strömphys> creeping motion *(zähe Strömungen)*
schleichende Verbrennung f <Kfztech> slow combustion
Schleier m 1. <Foto> fog; 2. <Ker & Glas> suspended curtain wall
Schleierkühlung f <Raumfahrt> film cooling
Schleif... <Elektrotech, Gerät, Maschinen> abrasion
Schleifband n 1. <Ker & Glas> abrasive belt *(zum Schleifen von Glas)*; 2. <Maschinen> abrasive belt
Schleifbandstandzeit f <Fertig> band life
Schleifbarkeit f <Chemtech> grindability
Schleifbock m <Maschinen> bench grinder
Schleifdrahtmessbrücke f <Elektrotech, Gerät> slide wire bridge
Schleifdruckbegrenzer m <Papier> grinding-pressure limiter
Schleife f 1. <Comp & DV, Elektriz> loop; 2. <Elektronik> loop *(geschlossener Stromweg)*; 3. <Fertig> snarl; 4. <Funktech, Ker & Glas, Kontroll, Papier> loop; 5. <Telekom> loop *(Induktion)*; 6. <Trans> turning *(Straße)*; 7. <Wassertrans> loop *(Tauwerk)*
schleifen *v* 1. <Anstrich, Bau> grind; 2. <Fertig> cut *(Edelstein)*; sandpaper *(Holz)*; grind *(Werkzeug)*; 3. <Maschinen> grind, sharpen
Schleifen n 1. <Anstrich> abrasion; 2. <Ker & Glas> cutting, grinding; 3. <Kunststoff> grinding, sanding; 4. <Maschinen> grinding, sharpening; 5. <Mechan, Papier> grinding

Schleifen

Schleifen n **mit grobkörnigem Schleifkörper** <Fertig> coarse-grain grinding
Schleifen n **und Polieren** n <Ker & Glas> grinding and polishing
schleifend adj <Papier> abrasive
Schleifengalvanometer n <Elektriz, Elektrotech> loop galvanometer
Schleifenkopplung f <Elektrotech> loop coupling
Schleifenmessung f <Elektriz> loop test
Schleifentrockner m <Papier> loop drier, loop dryer
Schleifenwicklung f <Elektrotech> lap winding, parallel winding
Schleifer m 1. <Ker & Glas> polisher *(Arbeiter)*; polisher *(Gerät zum Glasschleifen)*; 2. <Maschinen> grinder, sharpener
Schleiferei f <Ker & Glas> cutter's bay, cutting shop, polishing shop
Schleifertrog m <Papier> grinder pit
Schleiffähigkeit f <Fertig> grindability
Schleiffehler m <Ker & Glas> bloach
Schleifkohle f <Kfztech> brush
Schleifkontakt m <Elektrotech, Phys> sliding contact
Schleifkörperabrichtung f <Fertig> wheel truing
Schleiflager n <Textil> grinding-roller bearing
Schleifläufer m <Ker & Glas> polishing runner
Schleifleinen n <Maschinen> abrasive cloth
Schleifleiste f <Eisenbahn> pantograph wearing strip
Schleifmaschine f 1. <Fertig> (BE) disc grinder, (AE) disk grinder; 2. <Ker & Glas> grinding unit, polishing unit; grinder *(für Spiegelglas)*; 3. <Maschinen> grinder, grinding machine, sharpener, sharpening machine; 4. <Mechan> grinder; 5. <Papier> grinding machine
Schleifmaschinenbett n <Maschinen> grinding-machine bed
Schleifmaschinenschutz m <Sicherheit> grinder guard
Schleifmaterial n <Kohlen> abradant
Schleifmittel n 1. <Anstrich> abrasive surface; 2. <Fertig> abrading medium; 3. <Ker & Glas> grinding agent, polishing agent; 4. <Kohlen> abradant; 5. <Maschinen, Mechan> abrasive; 6. <Papier> abrasive material
Schleifmittel n **aus Aluminiumoxid** <Anstrich> alumina abrasive
Schleifpapier n 1. <Ker & Glas> polishing paper; 2. <Maschinen> abrasive sheet; 3. <Mechan> abrasive paper
Schleifpaste f <Maschinen, Mechan> grinding paste
Schleifrad n 1. <Ker & Glas> cutting wheel, grinding wheel, polishing wheel; 2. <Mechan> abrasive wheel, grinding wheel
Schleifring m 1. <Chemtech> grinding cylinder; 2. <Elektrotech> rotor slip ring; 3. <Kfztech, Phys> slip ring
Schleifringanker m <Elektrotech> wound rotor
Schleifringankermotor m <Elektrotech> slip ring induction motor
Schleifringläufer m <Elektrotech> slip-ring rotor, wound rotor
Schleifringläufermotor m <Elektrotech, Heiz & Kälte> slip ring motor
Schleifringmotor m <Elektriz, Elektrotech, Heiz & Kälte> slip ring motor
Schleifringrotor m <Elektriz, Elektrotech> slip ring rotor, wound rotor
Schleifrohstoff m <Fertig> grinding raw material
Schleifsand m <Ker & Glas> grinding sand
Schleifscheibe f 1. <Fertig> abrasive wheel, grinding wheel, wheel dresser; 2. <Maschinen> (BE) abrasive disc, (AE) abrasive disk, grinding wheel, wheel; 3. <Mechan, Sicherheit> abrasive wheel

Schleifscheibe f **mit Harzbindung** <Maschinen> resin-bonded wheel
Schleifscheibenabrichter m <Fertig> grinding wheel dresser
Schleifscheibenabrichtung f <Fertig> grinding wheel dressing
Schleifscheibenabziehwerkzeug n <Maschinen> wheel dresser
Schleifscheibenauswuchtung f <Fertig> grinding wheel balancing
Schleifscheibenfutter n <Fertig> grinding wheel chuck
Schleifscheibenverschleißvolumen n <Fertig> volume of wheel grain wear
Schleifscheibenwuchtzustand m <Fertig> grinding wheel balance
Schleifschiene f <Ker & Glas> grinding runner
Schleifsegment n 1. <Fertig> wheel segment; 2. <Maschinen> grinding segment
Schleifspäne mpl <Fertig, Maschinen> grindings
Schleifspindel f 1. <Ker & Glas> grinder spindle, beim Schleifen von Hohlglas spindle; 2. <Maschinen> grinding spindle, wheel spindle
Schleifspindelschlitten m <Fertig> wheel head slide
Schleifspindelstock m <Fertig> grinding head, wheel head
Schleifspuren fpl 1. <Fertig> wheel marks; 2. <Maschinen> grinding marks
Schleifstahl m <Lebensmittel> sharpening steel
Schleifstaub m <Fertig> grit
Schleifstein m 1. <Fertig> grinding stone, grindstone; 2. <Mechan> grinding wheel
Schleifstück n <Eisenbahn> pantograph slipper *(am Stromabnehmer)*
Schleifsupport m <Maschinen> wheel carriage
Schleifverlust m <Ker & Glas> cutting loss
Schleifvermögen n <Fertig> abrading capacity
Schleifversuch m <Fertig> abrasion test
Schleifvorrichtung f <Fertig> grinding fixture
Schleifwinkel m <Fertig> sharpening angle
Schleifzylinder m <Chemtech, Maschinen> grinding cylinder
Schleim m <Wasserversorg> slime
schleimbildend adj <Chemie> muciferous
Schleppboot n <Meerschmutz> towboat, tug boat
Schleppbremse f <Kfztech> drag brake
schleppen v 1. <Meerschmutz> tow *(Schiff)*; 2. <Trans> haul; 3. <Wassertrans> tow *(Schiff)*
Schleppen n 1. <Trans> haulage, towing; 2. <Wassertrans> dredging; towage *(Schiff)*
Schlepper m 1. <Kfztech> tractor; 2. <Lufttrans> prime mover; 3. <Meerschmutz> towboat, tug boat; 4. <Wassertrans> tug *(Schifftyp)*
Schleppflug m <Lufttrans> aerotow, areotow flight
Schleppgeschirr n <Meerschmutz> towing gear
Schleppkahn m <Wassertrans> dumb barge
Schleppkante f <Phys> trailing edge
Schleppkette f <Lufttrans> dragline
Schleppleine f <Meerschmutz, Wassertrans> towline *(Tauwerk)*
Schlepplift m <Trans> drag lift
Schlepplöffelbagger m <Bau> dragline excavator
Schleppnetz n <Wassertrans> dredge net, trawl, trawl net *(Fischerei)* • **mit Schleppnetz fischen** <Wassertrans> haul, drag *(Fischerei)*
Schleppnetzfischereifahrzeug n <Wassertrans> trawler *(Schifftyp)*
Schleppsaugbagger m <Wassertrans> trailing suction dredge, trailing suction dredger
Schleppschiff n <Wassertrans> towboat

Schleppschiffzug m <Wassertrans> towed convoy
Schleppseil n <Lufttrans> dragline
Schleppstange f <Mechan> towbar
Schlepptrosse f <Wassertrans> towrope
Schleppung f <Erdöl> drag *(von Formationsschichten)*
Schleppverband m <Wassertrans> tow train
Schleppversuchstank m <Wassertrans> ship model test tank *(Schiffbau)*
Schleppwagen m <Trans> tow car
Schleppzange f <Fertig> gripping jaws
Schleppzug m <Wassertrans> tug and tow
Schleuder f 1. <Fertig> spinner *(Schleuderguss)*; 2. <Kunststoff> centrifuge
Schleuderätzen n <Maschinen> spin etching
Schleuderbeton m <Bau> spun concrete
schleuderformgießen v <Fertig> centrifuge
Schleuderformmaschine f <Fertig> slinger *(Gießen)*
Schleuderguss m 1. <Fertig> centrispinning; 2. <Ker & Glas, Metall> centrifugal casting
Schleudergussrohr n <Fertig> spun pipe
Schleudergussstück n <Fertig> centrifugal casting
Schleudergussteil m <Fertig> spun part
Schleudergussverfahren hergestellt adj/im <Fertig> spun
Schleuderkopf m <Fertig> impeller head *(Gießen, Sandschleuderformmaschine)*
Schleuderkraft f <Raumfahrt> ejection force *(Raumschiff)*
Schleudermaschine f <Chemtech> centrifugal machine
Schleudermühle f 1. <Chemtech> centrifuge mill; 2. <Lebensmittel> disintegrator
schleudern v 1. <Fertig> jolt *(Bohrer)*; sling *(Formen)*; 2. <Kfztech> skid; 3. <Kohlen, Kunststoff> centrifuge
Schleudern n 1. <Chemtech> centrifugation; 2. <Fertig> jolt *(Bohrer)*; 3. <Kohlen> centrifuging; 4. <Raumfahrt> slingshot *(Beschleunigungsmanöver vom Orbit heraus durch Zentrifugalkraft)*; 5. <Trans> skidding
Schleuderpumpe f <Chemtech> centrifugal pump
Schleuderschutzeinrichtung f <Lufttrans> antitorque device
Schleudersitz m <Raumfahrt> ejection seat
Schleudertrennung f <Abfall> ballistic separation
Schleudertrockner m 1. <Chemtech> centrifugal hydroextractor; 2. <Textil> hydroextractor
Schleuderziehverfahren n <Ker & Glas> centrifugal drawing
Schleuse f 1. <Bau> lock; 2. <Hydraul> gate *(Dampfturbine)*; 3. <Kerntech> transfer port *(an einem Handschuhkasten)*; 4. <Nichtfoss Energ, Wasserversorg> sluice • **eine Schleuse passieren** <Wassertrans> pass through a lock
schleusen v 1. <Hydraul> gate; 2. <Wasserversorg> lock *(Schiff)*
Schleusenanlage f <Wasserversorg> lockage *(eines Schiffs)*
Schleusendeckel m <Fertig> manhole cover
Schleusendrempel m <Wasserversorg> lock sill, *(AE)* miter sill, *(BE)* mitre sill
Schleusen-Füll- und Entleersystem n <Wasserversorg> drawgate
Schleusengeld n <Wassertrans> lock dues *(Binnenwasserstraßen)*
Schleusenhaupt n <Wasserversorg> head sluices
Schleusenhaus n <Wasserversorg> lock house
Schleusenkammer f 1. <Wassertrans> lock chamber; 2. <Wasserversorg> gate chamber, lock chamber
Schleusenkanal m <Nichtfoss Energ, Wasserversorg> sluiceway
schleusenloser Kanal m <Wasserversorg> ditch canal
Schleusenplattform f <Wasserversorg> floatboard

Schleusenrahmen m <Wasserversorg> frame
Schleusenschwelle f <Wasserversorg> clap sill, lock sill, *(AE)* miter sill, *(BE)* mitre sill, sill
Schleusensohle f <Wasserversorg> floor
Schleusentor n 1. <Bau> gate; 2. <Wassertrans> lock gate, sluicegate; 3. <Wasserversorg> floodgate, gate, sluicegate
Schleusentreppe f <Wasserversorg> run of sluices
Schleusenwärter m <Wassertrans> lock keeper
Schleusung f <Nichtfoss Energ, Wasserversorg> sluicing
schlicht adj <Textil> plain
Schlichtanlage f <Textil> sizing machine
Schlichtaufnahme f <Textil> size take-up
Schlichtband n <Ker & Glas> finishing belt
Schlichtdrehmeißel m <Fertig> finish turning tool
Schlichte f 1. <Fertig> slur; wash *(Gießen)*; 2. <Ker & Glas> size; 3. <Kunststoff> size, sizing agent; 4. <Textil> sizing, sizing material; size *(Ausrüstung)*
Schlichtebad n <Textil> size bath
Schlichtemittel n 1. <Chemie> sizing agent *(Textil)*; 2. <Textil> size, sizing agent
schlichten v 1. <Fertig> dress; 2. <Maschinen> flat, plane; 3. <Textil> size
Schlichten n 1. <Fertig> finishing; 2. <Ker & Glas, Textil> sizing
Schlichten n **von Baum zu Baum** <Textil> beam-to-beam sizing
Schlichtfeile f <Fertig> smooth-cut file
Schlichtfräsen n <Maschinen> finish milling
Schlichthammer m 1. <Fertig> flatter *(Schmieden)*; 2. <Maschinen> flat-face hammer, planisher, planishing hammer
Schlichthobel m <Fertig> smooth plane
Schlichtmaschine f <Textil> sizing machine
Schlichtsorte f <Fertig> finishing grade *(Walzen)*
Schlichtwerkzeug n <Maschinen> finishing tool
Schlichtzahn m <Maschinen> finishing tooth
Schlichtzugabe f <Fertig> finish allowance
Schlick m 1. <Kohlen> slime; 2. <Wasserversorg> silt
Schlickablagerung f <Abfall> deposition of silt
Schlicker m <Fertig, Ker & Glas> slip
Schlickerguss m <Ker & Glas> slip casting
Schlickergusstiegel m <Ker & Glas> slip-cast pot
Schlickerofen m <Ker & Glas> slip kiln
Schliere f 1. <Fertig> streak, stria; 2. <Ker & Glas> ream; 3. <Optik> streak
Schlieren fpl <Ker & Glas> cords
Schlierenfotografie f <Phys> schlieren photography
schlierig adj <Ker & Glas> cordy
Schließanlage f <Sicherheit> security system
Schließbeschlag m <Bau> lock fitting
Schließblech n <Bau> lock plate, striking plate; box *(eines Schlosses)*
Schließbolzen m <Mechan> cotter pin
Schließdämpfer m <Kfztech> dashpot
Schließdeckel m <Lufttrans> blanking cover
Schließdeckel m **für Kühlluftauslass** <Lufttrans> blanking cover for air-cooling-unit outlet
Schließdruck m <Erdöl> shut-in pressure *(Lagerstättentechnik)*
Schließeffekt m <Lufttrans> blanking effect
Schließelement n <Maschinen> closing element
schließen v 1. <Bau> close, lock, shut; 2. <Druck> lock up; 3. <Elektrotech> make *(Stromkreis, Kontakt)*; 4. <Fertig> make; 5. <Künstl Int> infer; 6. <Math> implicate
Schließen n 1. <Eisenbahn> closure; 2. <Elektrotech> closure *(eines elektrischen Stromkreises)*; 3. <Künstl Int> inference, reasoning

Schließen

Schließen *n* **anhand von Standardvorgaben** <Künstl Int> default reasoning
Schließen *n* **der Form** <Druck> lock-up
Schließen *n* **und Unterbrechen** *n* <Elektrotech> make and break
Schließer *m* 1. <Elektriz> make contact, normally open contact; 2. <Elektrotech> normally open contact
Schließerrelais *n* <Elektriz> make relay
Schließer- und Öffnerelement *n* <Elektriz> make-and-break device
Schließer- und Öffnerspule *f* <Elektriz> make-and-break coil
Schließgeschwindigkeit *f* <Elektriz> closing speed
Schließkasten *m* <Bau> lock casing
Schließklappe *f* <Bau> box staple
Schließkontakt *m* 1. <Elektriz, Elektrotech> make contact, normally open contact; 2. <Fertig> maker
Schließkopf *m* <Fertig> upset point
Schließnaht *f* <Textil> closing seam
Schließnocken *m* <Maschinen> locking cam
Schließpfosten *m* <Bau> shutting post
Schließrahmen *m* <Druck> chase
Schließring *m* <Foto, Maschinen> retaining ring
Schließtisch *m* <Druck> imposing table
Schließungsbelastbarkeit *f* <Elektriz> rated making capacity
Schließungsimpuls *m* <Elektrotech> make pulse
Schließungs-Öffnungszeit *f* <Elektriz> make-and-break time
Schließvorgang *m* <Elektriz> closing operation
Schließwinkel *m* <Kfztech> dwell angle
Schließzeit *f* <Elektriz> closing time, make time
Schließzeug *n* <Druck> quoins
Schliff *m* 1. <Fertig> grinding, grind; 2. <Ker & Glas> cut; 3. <Mechan> finish
Schliffbild *n* <Fertig> grinding pattern
Schliff-Fläche *f* <Nichtfoss Energ> polished surface
Schliffstopfen *m* 1. <Labor> ground stopper; 2. <Verpack> ground-in stopper
Schliffverbindung *f* <Labor> ground glass joint
Schliffverbindungshalter *m* <Labor> ground glass joint clamp
Schlinge *f* 1. <Aufnahme> curl (des Tonbands); 2. <Fertig> snarl (Draht); 3. <Kontroll, Textil> loop; 4. <Wassertrans> loop, sling (Tauwerk)
Schlingenbildung *f* <Textil> plucking
Schlingenflor *m* <Textil> uncut pile
Schlingenflorteppich *m* <Textil> loop pile carpet
Schlingerdämpfungsanlage *f* <Wassertrans> anti-rolling device (Schiff)
Schlingerkiel *m* <Wassertrans> bilge keel
Schlingerleiste *f* <Wassertrans> fiddle
schlingern *v* <Wassertrans> lurch (Schiffbewegung)
Schlingern *n* <Wassertrans> rolling (Schiffbewegung)
Schlingern *n* **und Stampfen** *n* <Wassertrans> rolling and pitching (Schiffbewegungen)
Schlingerstabilisierung *f* <Trans> sway stabilization
Schlingerversuch *m* <Wassertrans> roll test (Schiffkonstruktion)
Schlitten *m* 1. <Eisenbahn> carriage; 2. <Maschinen> carriage, cradle, ram, slide rest, slide; 3. <Mechan> cradle; 4. <Trans> carrier
Schlittendraht *m* <Elektriz> skid wire
Schlitteneinheit *f* <Maschinen> slide unit
Schlittenkufe *f* <Mechan> runner
Schlittenrevolverdrehmaschine *f* <Maschinen> combination lathe, saddle-type turret lathe
Schlittenständer *m* <Maschinen> integral way columns

Schlitz *m* 1. <Bau> slot; 2. <Druck, Elektrotech, Fernseh> slot; 3. <Hydraul> port; 4. <Kerntech> groove; slot (in einem Rotorblatt); 5. <Maschinen> recess, slot; 6. <Papier> aperture; 7. <Phys> slit; 8. <Raumfahrt> slit (Antenne); slot; 9. <Telekom> slot; 10. <Wellphys> slit • **mit Schlitz** <Fertig> recessed (Schraube)
Schlitzabtastung *f* <Fernseh> slit scanning
Schlitzanodenmagnetron *n* <Elektronik> split anode magnetron
Schlitzantenne *f* <Funktech> slot antenna
Schlitzblende *f* <Foto> slit diaphragm
Schlitzdränung *f* <Bau> mole drainage (Tiefbau)
schlitzen *v* 1. <Bau> split, slot; 2. <Maschinen> slot; 3. <Textil> split
Schlitzen *n* 1. <Fertig> slitting (Trennschleifen); 2. <Maschinen> slitting, slotting
Schlitzfeile *f* <Maschinen> slot file, slotting file
Schlitzflügel *m* <Lufttrans> slotted wing
Schlitzfräser *m* <Maschinen> keyway cutter, slot cutter, slotting cutter
Schlitzhohlleiterabschnitt *m* <Telekom> waveguide slotted section
Schlitzkernkabel *n* <Optik> slotted core cable
Schlitzklappe *f* <Lufttrans> slot flap
Schlitzkopf *m* <Bau, Maschinen> slotted head
Schlitzleitung *f* <Phys> slot-line
Schlitzmantelkolben *m* <Kfztech> split skirt piston
Schlitzmaschine *f* 1. <Fertig> slitter; 2. <Kohlen> nicking machine
Schlitzmutter *f* <Bau, Maschinen> slotted nut
Schlitzmutterndreher *m* <Maschinen> slotted-type screwdriver
Schlitzmutternschlüssel *m* <Bau, Maschinen> slotted-type screwdriver
Schlitzniet *m* <Bau> slotted rivet
Schlitzprobe *f* <Kohlen> channel sample
Schlitzrohr *n* <Bau> slot pipe
Schlitzschraube *f* <Bau, Maschinen> slotted head screw, slotted screw
Schlitzschraube *f* **mit Zylinderkopf** <Fertig> fillister head
Schlitzsystem *n* <Raumfahrt> slit system
Schlitz- und Druckmaschine *f* <Verpack> slitting and printing machine
Schlitzverschluss *m* 1. <Fernseh> slit shutter; 2. <Foto> focal plane shutter, roller-blind shutter
Schlitzwandverfahren *n* <Abfall> slurry trenching
Schloss *n* 1. <Kfztech> lock (Karosserie); 2. <Maschinen, Sicherheit, Wassertrans> lock
Schloss *n* **ohne Klinke** <Bau> deadlock
Schlosser *m* <Kfztech, Maschinen, Mechan> fitter, locksmith, mechanic, metalworker
Schlosserhammer *m* <Maschinen> fitter's hammer, locksmith's hammer
Schlosserwerkstatt *f* <Maschinen> fitting shop
Schlossgarnitur *f* <Bau> (AE) lockset
Schlosskasten *m* 1. <Bau> box staple, case; 2. <Fertig> apron housing; 3. <Maschinen> apron
Schlossmutter *f* <Maschinen> clasp nut
Schlossriegel *m* <Bau> deadbolt
Schlossschraube *f* <Maschinen> coach bolt
Schlossschutzblech *n* <Bau> finger plate
Schlot *m* 1. <Bau> chimney stack, funnel; 2. <Erdöl> pipe (Geologie)
Schlotte *f* <Kohlen> cavity (Wasser, Kalk)
Schlotterventil *n* <Maschinen> puppet valve
Schluckbrunnen *m* <Wasserversorg> injection well
Schlucken *n* <Bau> absorption (Schall)
Schlueter-Bewegungsgleichung *f* <Kerntech> Schlueter equation of motion

Schluff m <Kohlen> silt
Schlupf m 1. <Akustik> drift; 2. <Bau> slippage; 3. <Elektriz, Elektrotech> slip (s); 4. <Mechan> backlash • **Schlupf haben** <Maschinen> slip
schlupffreier Antrieb m <Kfztech> positive drive
Schluss m 1. <Comp & DV> tail; 2. <Künstl Int> inference
Schlussanstrich m <Kunststoff> top coat
Schlussbehandlung f <Verpack> final treatment
Schlüsse ziehen v <Künstl Int> infer
Schlüssel m 1. <Bau, Comp & DV, Druck> key; 2. <Maschinen> key, screw wrench, (BE) spanner, wrench; 3. <Telekom> key
Schlüsselbart m <Bau> bit, keybit
Schlüsselbrett n <Telekom> keyshelf
Schlüsselfeile f <Maschinen> key file, warding file
Schlüsselfeld n <Comp & DV> key field
schlüsselfertig adj <Bau> turnkey
schlüsselfertiger Einbau m <Maschinen> turnkey installation
schlüsselfertiges Projekt n <Bau, Maschinen> turnkey project
schlüsselfertiges System n <Comp & DV, Telekom> turnkey system
Schlüsselformblech n <Bau> ward
Schlüsselloch n 1. <Bau> keyhole; 2. <Erdöl> key seating (Bohrtechnik)
Schlüssellochabdeckung f <Bau> key drop
Schlüssellochklappe f <Bau> scutcheon
Schlüsselschalter m 1. <Elektriz, Elektrotech> key-operated switch, keyswitch; 2. <Telekom> key-operated switch
Schlüsselschild n <Bau> key plate
Schlüsselweite f 1. <Fertig> across-flats dimensions; 2. <Maschinen> across-flats dimension, diameter across flats, width across flats
Schlüsselwort n <Comp & DV> keyword (beim Durchsuchen einer Datenbank)
Schlüsselwortkontrolle f <Comp & DV> password security
Schlüsselwortparameter m <Comp & DV> keyword parameter
Schlüsselzeichen n (SZ) <Telekom> code
Schlussflansch m <Maschinen> end flange
Schlussfolgerung f 1. <Künstl Int> inference; 2. <Math> conclusion
schlüssiger Beweis m <Patent> conclusive evidence
Schlusslaterne f <Eisenbahn> tail end marker lamp
Schlussleuchte f <Kfztech> tail light
Schlusspackwagen m <Eisenbahn> (AE) caboose, (BE) guard's van
Schlussstein m <Bau> trap; capstone, keystone (eines Gewölbes)
Schlussstrich m <Kunststoff> finishing coat
Schlussvignette f <Druck> tailpiece
Schlusszeichen n <Telekom> clearback signal
schmal gebündelter Strahl m <Strahlphys> pencil beam
Schmalband n <Comp & DV, Elektronik, Funktech, Telekom> narrow band
Schmalband... <Comp & DV, Funktech, Telekom> narrow band
Schmalband-Demodulation f <Elektronik> narrow-band demodulation
Schmalband-Einseitenband n <Funktech> narrow single sideband
Schmalband-Einseitenbandmodulation f <Funktech> narrow single sideband modulation
Schmalbandempfänger m <Telekom> narrow-band receiver
Schmalbandfernsehen n (SSTV) <Fernseh> slow scan television (SSTV)

Schmalbandfernsehen n (SSTV) <Funktech> slow scan television (SSTV)
Schmalbandfernsehsystem n <Fernseh> slow scan television system
Schmalbandfilter n <Elektronik> narrow-band filter
Schmalbandfiltern n <Elektronik> narrow-band filtering
Schmalbandfrequenzmodulation f (SBFM) <Elektronik, Funktech, Telekom> narrow-band frequency modulation (NBFM)
Schmalbandgeräusch n <Elektronik> narrow-band noise
Schmalband-ISDN n <Telekom> narrow-band ISDN
Schmalbandkoppelnetz n <Telekom> narrow-band switching network
Schmalbandkoppler m <Telekom> narrow-band switch
Schmalband-Phasenumtastung f <Telekom> narrow band phase shift keying, NBPSK
Schmalbandrauschen n <Telekom> narrow-band noise
Schmalband-Responsspektrum n <Kerntech> narrow-band response spectrum
Schmalbandröhre f <Elektronik> narrow-band tube
Schmalbandschaltung f <Elektronik> narrow-band circuit
Schmalbandsignal n <Elektronik, Telekom> narrow-band signal
Schmalband-Sperrfilter n <Elektronik> narrow-band rejection filter
Schmalband-Sprachmodulation f <Telekom> NBVM, narrow-band voice modulation
Schmalband-SSB n <Funktech> narrow ssb
Schmalband-SSB-Modulation f <Funktech> narrow ssb-modulation
Schmalbandstörung f <Elektronik> narrow-band interference
Schmalband-Tiefpassfilter n <Elektronik> narrow-band low-pass filter
Schmalband-Tiefpassfiltern n <Elektronik> narrow-band low-pass filtering
Schmalbandverstärker m <Elektronik> narrow-band amplifier
Schmalbündel n <Elektronik> pencil beam
schmale Meeresbucht f <Wassertrans> (BE) creek (Geographie)
schmale Rundfeile f <Maschinen> rat-tail file
schmaler Impuls m <Elektronik> narrow pulse
Schmalgewebe n <Textil> narrow fabric
Schmalkeilriemen m <Maschinen> narrow V belt
Schmalseite f <Fertig> end
Schmalspurbahn f <Eisenbahn> (AE) light railroad, (BE) light railway, (AE) narrow-gage railroad, (BE) narrow-gauge railway
Schmalspurbefeuerung f <Lufttrans> (AE) narrow-gage lighting system, (BE) narrow-gauge lighting system (Start- und Landebahn)
Schmalspurdiesellokomotive f <Eisenbahn> (AE) narrow-gage diesel locomotive, (BE) narrow-gauge diesel locomotive
Schmalspurschienensystem n <Eisenbahn> (AE) narrow-gage track system, (BE) narrow-gauge track system
Schmalz n <Lebensmittel> lard
Schmälzeinrichtung f <Textil> oiler
schmälzen v <Textil> lubricate
Schmälzmittel n <Textil> lubricant
Schmelz... <Ker & Glas, Maschinen, Phys, Thermod> melting
Schmelzbad n 1. <Fertig> melt; pool (Schweißen); 2. <Thermod> melting bath
schmelzbar adj <Thermod> meltable
Schmelzbereich m <Metall, Thermod> melting range
Schmelzbohren n <Kohlen> fusion drilling

Schmelzdauer f <Metall, Thermod> melting period, melting time
Schmelzdraht m <Elektriz, Elektrotech> fuse wire
Schmelzdrahtsicherung f <Elektriz> wire fuse
Schmelze f 1. <Fertig> fused metal; blow *(Gießen)*; 2. <Metall> heat melting bath
Schmelzeinsatz m <Elektrotech> fuse link
schmelzen v 1. <Anstrich> melt; 2. <Elektriz> fuse *(Sicherung)*; 3. <Lebensmittel> thaw; 4. <Metall> fuse; 5. <Papier, Textil> melt; 6. <Thermod> fuse, melt, melt down
Schmelzen n 1. <Anstrich> fusion; 2. <Fertig> founding *(Glas)*; 3. <Kohlen> smelting; 4. <Kunststoff> fusion; 5. <Papier, Textil> melting; 6. <Thermod> fusing, fusion, melting • **zum Schmelzen bringen** <Thermod> fuse
Schmelzenthalpie f <Mechan, Phys, Thermod> enthalpy of fusion
Schmelzentropie f <Mechan, Phys, Thermod> entropy of fusion
Schmelzerei f <Fertig, Metall> foundry *(Glas)*
Schmelzfarben fpl <Ker & Glas> *(AE)* vitrifiable colors, *(BE)* vitrifiable colours
Schmelzfluss m <Fertig> coalescence, melting
schmelzflüssig adj <Thermod> fusible
Schmelzformen n <Ker & Glas> fusion casting
schmelzgeschweißt adj <Maschinen> fusion-welded
Schmelzgießform f <Ker & Glas> *(AE)* font mold, *(BE)* font mould
Schmelzindex m <Kunststoff> MFI, melt flow index
Schmelzkegel m <Fertig> pyrometric cone
Schmelzkernfänger m <Kerntech> melting core catcher
Schmelzkessel m <Thermod> melting pot
Schmelzkleber m 1. <Druck> hot-melt glue; 2. <Kunststoff> hot-melt adhesive; 3. <Verpack> heat-sealing adhesive, hot-melt adhesive
Schmelzkurve f <Thermod> melting-point curve
Schmelzleiter m <Elektrotech> fuse element
Schmelzlösen n <Papier> dissolving
Schmelzlöser m <Papier> dissolver
Schmelzmittel n <Bau> flux
Schmelzofen m 1. <Kohlen> melting furnace; 2. <Kunststoff> fusing oven; 3. <Metall> melting furnace, smelter, smelting furnace; 4. <Thermod> melting furnace
Schmelzperle f <Ker & Glas> slug
Schmelzprobe f <Metall, Qual, Thermod> melting test
Schmelzpunkt m 1. <Chemtech, Kunststoff> melting point, mp; 2. <Metall> melting point, point of fusion; 3. <Papier, Textil> melting point, mp; 4. <Thermod> fusing point, melting point, mp • **den Schmelzpunkt herabsetzen** <Fertig> flux
Schmelzpunktapparat m nach Thiele <Labor> Thiele melting-point tube, Thiele tube
Schmelzpunkterniedrigung f <Thermod> lowering of the melting point
Schmelzpunktkurve f <Metall> melting-point curve
Schmelzrate f <Strömphys> melt flow rate
Schmelzschutzschild m <Raumfahrt> ablation shield *(Raumschiff)*
Schmelzschweißen n <Bau, Fertig, Maschinen, Thermod> fusion-welding
Schmelzsicherung f 1. <Elektriz> nonrenewable fuse; 2. <Elektrotech> blowout fuse; 3. <Sicherheit> safety fuse, fusible plug
Schmelzsicherung für Dampfkessel f <Hydraul> fusible plug for steam boiler *(Sicherheit)*
Schmelzspleißung f <Telekom> fusion splice
Schmelzspleißverbindung f <Optik> fusion splice
schmelztauchverzinken v <Anstrich> galvanize by hot dipping

Schmelztiegel m <Fertig, Metall, Thermod> melting crucible
Schmelzton m <Ker & Glas> fusible clay
Schmelzung f <Elektrotech> smelting
Schmelzverbindung f <Optik> fusion splice
Schmelzvorgang m 1. <Metall> smelting; 2. <Thermod> fusion, melting
Schmelzwanne f <Ker & Glas> tank
Schmelzwärme f 1. <Thermod> heat of fusion, melting heat; 2. <Metall> melting heat
Schmelzzone f <Metall> melting zone
Schmerzgrenze f <Akustik> threshold of pain
Schmidt-Zahl f <Phys> Schmidt number *(Strömungslehre)*
Schmied m <Maschinen, Metall> blacksmith
schmiedbar adj 1. <Mechan> malleable; 2. <Metall> ductile, malleable; 3. <Qual> ductile
Schmiedbarkeit f <Metall> malleability
Schmiede f 1. <Bau, Metall> smithy; 2. <Maschinen> blacksmith's shop
Schmiedearbeit f <Bau, Metall> ironwork, smithery
Schmiedebalg m <Maschinen> blacksmith's bellows
Schmiedeeisen n 1. <Bau> wrought iron; 2. <Mechan> low-carbon steel; 3. <Metall> iron
schmiedeeisern adj <Metall> wrought-iron
Schmiedegesenk n 1. <Maschinen> forging die, swage; 2. <Mechan> swage
Schmiedehammer m 1. <Fertig> forging hammer; 2. <Maschinen> blacksmith's hammer, forging hammer
Schmiedehandwerk n <Bau, Metall> smithery
Schmiedeherd m <Maschinen> blacksmith's forge
Schmiedekonus m <Fertig> draft
Schmiedemaschine f <Maschinen> forging machine
schmieden v <Anstrich, Fertig> forge
Schmieden n 1. <Bau> hammering; 2. <Fertig, Kunststoff, Maschinen> forging
Schmiedepresse f <Fertig> forging press
Schmiedeschweißen n <Bau> forge welding
Schmiedestahl m <Maschinen> forged steel
Schmiedestück n <Mechan, Metall> forging
Schmiedestück n aus Titan <Raumfahrt> titanium forging
Schmiedestückzeichnung f <Konstzeich> forging drawing
Schmiedewalzen n <Metall> roll forging
Schmiedezange f 1. <Bau> smith's pliers; 2. <Maschinen> blacksmith's tongs
Schmiedezunder m <Fertig> forge scale, forging scale
Schmiegkreis m <Geom> osculating circle
Schmierbohrung f <Maschinen> oil hole
Schmierbüchse f 1. <Fertig> grease box; 2. <Maschinen> oil cup
schmieren v 1. <Fertig> lubricate; 2. <Maschinen> grease, lubricate; 3. <Mechan> lubricate
Schmierfett n <Kfztech, Maschinen> grease
Schmierfilm m 1. <Fertig> film; 2. <Maschinen> lubricating film
Schmierkante f <Fertig> oiler
Schmierkissen n <Maschinen> oil pad, pad
Schmierlager n <Maschinen> grease bearing
Schmierloch n <Maschinen> oil hole
Schmierlötverbindung f <Fertig> wiped joint
Schmiermittel n 1. <Bau, Erdöl> lubricant; 2. <Fertig> sud; lubricant *(Kunststoffinstallationen)*; 3. <Kfztech, Maschinen, Mechan> lubricant
Schmiermittelpumpe f <Mechan> lubricating pump
Schmiermittelrückstände mpl <Abfall> waste lubricants
Schmiernippel m 1. <Kfztech> grease nipple *(Motor)*; 2. <Maschinen> grease nipple, lubricating nipple, lubrication fitting; 3. <Mechan> lubricating nipple

Schmieröl n 1. <Erdöl> motor oil; 2. <Fertig, Kfztech> lubricating oil
Schmierölfleck m <Ker & Glas> grease mark
Schmierpistole f <Bau, Kfztech> grease gun *(Werkzeug)*
Schmierplan m <Mechan> lubricating chart
Schmierstoff m 1. <Erdöl, Fertig, Maschinen> lubricant; 2. <Mechan> grease
Schmiersystem n <Kfztech, Maschinen> lubricating system
Schmierung f 1. <Fertig> lubricating; 2. <Kfztech, Kontroll, Maschinen> lubrication; 3. <Papier> greasing
Schmierung f von Hand <Maschinen> hand lubrication
Schmiervorrichtung f <Bau> lubricating unit
Schmirgel m 1. <Fertig> abrasive, emery; 2. <Ker & Glas, Mechan> emery
Schmirgelleinen n <Fertig, Mechan> emery cloth
Schmirgelpapier n 1. <Elektriz> abrasive paper; 2. <Fertig> emery paper; 3. <Mechan> abrasive paper, emery paper; 4. <Papier> rubber
Schmirgelpulver n 1. <Maschinen> emery powder; 2. <Mechan> abrasive powder
Schmirgelscheibe f <Fertig, Mechan> emery wheel
Schmitt-Trigger m <Phys> Schmitt trigger
Schmorkontakt m <Fertig> flash *(Brennschweißen)*
Schmutz m <Abfall> dirt
Schmutzbehälter m <Abfall> dust collector *(des Kehrfahrzeugs)*
Schmutzfänger m 1. <Fertig> line strainer *(Kunststoffinstallationen)*; 2. <Ker & Glas> catch pan; 3. <Kfztech> mudflap *(Zubehör)*
Schmutzfestigkeit f <Textil> resistance to soiling
Schmutzpunkt m <Papier> speck
Schmutzstoff m <Meerschmutz> pollutant
Schmutzwasser n 1. <Abfall, Bau> drain water; 2. <Umweltschmutz> wastewater
Schmutzwasserpumpe f <Kerntech> dirty-water pump
Schnabel m <Labor> lip, spout
Schnabelkipper m <Kfztech> scoop tipper
Schnabelrundkipper m <Kfztech> scoop dump car
Schnabelwagen m <Eisenbahn> *(AE)* Schnabel car, *(BE)* Schnabel wagon
Schnabelzange f <Maschinen> long-nose pliers
Schnalle f <Maschinen> buckle, clasp
Schnallenriemen m <Maschinen> buckle strap
Schnäpper m <Bau> latch bolt *(Türschloss)*
Schnäpperschloss n <Bau> spring bolt lock
Schnappfassung f <Elektrotech> snap-in socket
Schnappfeder f <Maschinen> catch spring
schnappig adj <Kunststoff> snappy
Schnappigkeit f <Kunststoff> snappiness
Schnappriegel m <Kfztech> safety catch *(Motorhaube)*
Schnappschäkel m <Wassertrans> snap shackle *(Beschläge)*
Schnappschalter m <Elektriz, Elektrotech> snap-action switch, snap-in switch
Schnappschloss n <Bau> catch, latch lock
Schnappverschluss m <Verpack> snap hinge closure, snap-on lid
Schnarchventil n <Maschinen> air valve, air-snifting valve
Schnauze f 1. <Fertig> spout; 2. <Maschinen> nozzle
Schnecke f 1. <Bau> auger; 2. <Fertig> scroll, worm; worm screw *(Schneckenpresse)*; 3. <Kunststoffinstallationen>; pressure screw worm *(Destillationsapparat)*; 5. <Maschinen> perpetual screw, screw, worm; 6. <Mechan> worm; 7. <Papier> screw
Schneckenantrieb m <Kunststoff> screw drive
Schneckenbohrer m <Bau> gimlet
Schneckenextender f <Lebensmittel> screw extruder
Schneckenfeder f <Mechan> coil spring
Schneckenfeder-Druckmesselement n <Gerät> helical pressure element
Schneckenförderer m 1. <Fertig> spiral conveyor; 2. <Maschinen> screw conveyor, spiral conveyor, worm conveyor; 3. <Papier> screw conveyor
Schneckenfräser m <Maschinen> worm milling cutter
Schneckengetriebe n 1. <Fertig> worm gearing *(Kunststoffinstallationen)*; 2. <Maschinen> worm gear pair; 3. <Mechan> worm gear
Schneckenhandbohrer m <Bau> shell gimlet
Schneckenkanal m <Kunststoff> screw channel
Schneckenlinie f <Geom, Maschinen> helix
Schnecken-Nenndurchmesser m <Kunststoff> screw diameter
Schneckenpumpe f 1. <Maschinen> spiral pump; 2. <Meerschmutz> screw pump
Schneckenrad n 1. <Maschinen> screw wheel, spiral wheel, worm gear, worm wheel; 2. <Wassertrans> worm gear
Schneckenradachsantrieb m <Kfztech> worm gear final drive
Schneckenradgetriebe n <Maschinen> worm gear pair
Schneckenradwälzfräsen n <Fertig> worm gear hobbing
Schneckensegment n <Fertig> worm segment *(Kunststoffinstallationen)*
Schneckenspalt m <Kunststoff> flight land clearance
Schneckenspiel n <Kunststoff> radial screw clearance
Schneckenspitze f <Kunststoff> screw tip
Schneckensteg m <Kunststoff> screw flight
Schneckenstrangpresse f <Kunststoff> extruder
Schneckentreppe f <Bau> spiral stairs
Schneckenverzahnung f <Maschinen> worm gearing
Schneckenwalze f <Papier> worm roll
Schneckenzahnstange f <Maschinen> worm rack
Schnee m 1. <Elektronik> grass *(Bildstörung)*; 2. <Fernseh> snow
Schneebelastung f <Bau> snow loading
Schneedetektor m <Trans> snow detector
Schneefanggitter n <Bau> snow guard
Schneeflockenkurve f <Geom> snowflake curve
Schneeflockentopologie f <Comp & DV> snowflake topology
Schneekufe f <Lufttrans> tail skid
Schneereifen m <Kfztech> *(AE)* snow tire, *(BE)* snow tyre
Schneidbacke f 1. <Fertig> screw die; 2. <Maschinen, Mechan> die
Schneidbrenner m 1. <Bau> cutting blowpipe, flame cutter; 2. <Fertig> cutting blowpipe; 3. <Maschinen> cutting torch; 4. <Mechan> flame cutter; 5. <Thermod> flame-cutting torch
Schneide f 1. <Bau> blade, cutting edge; 2. <Fertig> cutting lip *(Bohrer)*; knife edge *(Waage)*; 3. <Maschinen> bit, edge, router; 4. <Textil> cutting edge • **eine Schneide bilden** <Fertig> take an edge
Schneidebene f <Fertig> cutting plane
Schneideholz n <Bau> saw timber
Schneideisen n 1. <Fertig> die *(Gewinde)*; 2. <Maschinen> die, screwing die, stock
Schneideisenhalter m <Maschinen> die holders, die stock holder
Schneidemaschine f 1. <Bau> cutter; 2. <Druck, Foto> trimmer; 3. <Maschinen> cutting machine, shear, shears; 4. <Verpack> slit machine
Schneidemaschinenbank f <Ker & Glas> cutter's lathe
Schneidemesser n <Maschinen> blade
schneiden v 1. <Bau, Druck> cut; 2. <Geom> intersect; 3. <Maschinen, Mechan> cut; 4. <Papier> guillotine
Schneiden n <Maschinen> cutting

schneidende

schneidende Ebenen *fpl*/**sich** <Geom> intersecting planes
schneidende Kante *f* <Mechan> cutting edge
schneidende Linien *fpl*/**sich** <Geom> intersecting lines
Schneidenecke *f* <Fertig> tool corner
Schneideneingriff *m* <Fertig> tool engagement point-of-cutting action
Schneideneinsatz *m* <Fertig> insert; section *(Räumwerkzeug)*
Schneidenkopf *m* <Fertig> bit; drill tip *(Tieflochbohrer)*
Schneiden-Normalebene *f* <Fertig> tool edge normal plane
Schneidenschaft *m* <Fertig> body *(Bohrer)*
Schneidenteil *m* <Fertig> active portion *(Räumwerkzeug)*
Schneidenteil *m* **und Hals** *m* <Fertig> body *(Reibahle)*
Schneider *m* 1. <Ker & Glas> birdcage; 2. <Maschinen> cutter, shear, shears
Schneideritze *f* <Ker & Glas> scratch
Schneidestichel *m* <Akustik> cutting stylus
Schneidflüssigkeit *f* 1. <Fertig> coolant *(Spanung)*; 2. <Maschinen> cutting fluid
Schneidgewinde *n* <Maschinen> self-tapping thread
Schneidkante *f* 1. <Bau> cutting edge; 2. <Ker & Glas> cutting edge *(des Glasschneiderdiamanten)*; 3. <Maschinen> cutting edge, edge
Schneidkeramik *f* <Fertig> ceramic cutting material
Schneidkluppe *f* 1. <Fertig> die stock *(Gewinde)*; 2. <Maschinen> screw plate stock, stock and dies
Schneidkopf *m* 1. <Bau, Kohlen> cutter head *(Tunnelbau)*; 2. <Maschinen> cutter head
Schneidkopfbagger *m* <Wassertrans> cutter dredge, cutter dredger *(Baggern)*
Schneidkühlflüssigkeit *f* <Sicherheit> metalworking fluid
Schneidkühlmittel *n* <Sicherheit> metalworking fluid
Schneidlippe *f* <Maschinen> cutting lip, lip
Schneidmarke *f* <Druck> crop mark, cutting mark
Schneidmaschine *f* 1. <Lebensmittel> slicer; 2. <Maschinen> cutter, shearing machine
Schneidmetall *n* <Fertig, Metall> cutting metal
Schneidmühle *f* <Kunststoff> granulator
Schneidöl *n* 1. <Erdöl> cut oil *(Sonderöl)*; 2. <Fertig> cutting fluid; 3. <Maschinen> cutting oil
Schneidplatte *f* <Fertig> insert
Schneidrad *n* 1. <Fertig> *(BE)* disc blade, *(AE)* disk blade *(Schere)*; 2. <Maschinen> cutter, rotary shear blade
Schneidrahmen *m* <Ker & Glas> cutting frame
Schneidrücken *m* <Maschinen> land
Schneidschaftlänge *f* <Fertig> body length *(Bohrer)*
Schneidschraube *f* <Maschinen> self-cutting screw, self-tapping screw
Schneidstichel *m* <Aufnahme> recording stylus
Schneidstoff *m* <Fertig> cutting material, metal-cutting material
Schneid- und Wickelmaschine *f* <Verpack> slitting and rewinding machine
Schneidvorrichtung *f* <Ker & Glas> *(AE)* capper, *(BE)* cut-off man
Schneidwerkzeug *n* 1. <Maschinen> cutter, cutting tool, die; 2. <Mechan> cutter, cutting tool
Schneidwerkzeugwinkel *m* <Maschinen> cutting angle
Schneidzähne *mpl* <Bau> cutting teeth
Schneidzange *f* <Bau> cutting pliers
schnell ansprechende Sicherung *f* <Elektrotech> fast-acting fuse
schnell gefriergetrocknet *adj* <Lebensmittel> accelerated freeze-dried
schnell härtend *adj* <Kunststoff> fast-curing
schnell lösbare Befestigung *f* <Maschinen> quick-release fastener
schnell lösbare Rohrverbindung *f* <Maschinen> quick-release pipe coupling
schnell lösender Gurt *m* <Sicherheit> quick-release belt
schnell lyophilisiert *adj* <Lebensmittel> accelerated freeze-dried
schnell schaltender Leistungsgleichrichter *m* <Elektrotech> fast-switching power rectifier
schnell schaltender Leistungstransistor *m* <Elektronik> fast-switching power transistor
schnell spritzbarer Furnace-Ruß *m* *(FEF-Ruß)* <Kunststoff> fast extruding furnace carbon black *(FEF carbon black)*
schnell trocknen *v* <Textil> tumble
schnell trocknend *adj* <Textil, Verpack> quick-drying
schnell trocknendes Öl *n* <Kunststoff> drying oil
schnell verschließen *v* <Fertig> scuff
schnell wechselndes Signal *n* <Elektronik> fast-changing signal
Schnell... <Kerntech, Lufttrans> fast
Schnellabbinden *n* <Bau> flash set *(Zement)*
Schnellablass *m* 1. <Kerntech> dump *(des Moderators im homogenen Reaktor)*; 2. <Lufttrans> dumping *(Treibstoff)*
Schnellabrollbahn *f* <Lufttrans> exit taxiway, high-speed exit taxiway, rapid exit taxiway
Schnellabschaltstab *m* <Kerntech> scram rod
Schnellabschaltung *f* **eines Reaktors** <Kerntech> reactor trip
Schnellabschaltungskontrolle *f* <Kerntech> scram control
Schnellabstimmfilter *n* <Elektronik> fast-tuned filter
Schnellabstimmoszillator *m* <Elektronik> fast-tuned oscillator
Schnellalterungsprüfung *f* <Fertig, Kunststoff> *(BE)* accelerated ageing test, *(AE)* accelerated aging test
Schnellanschluss-Verbindungsstück *n* <Labor> quickfit connector
Schnellarbeitsstahl *m* <Fertig> high-speed steel, HSS
Schnellauslöser *m* <Maschinen> fast-acting trip
Schnellauslöseventil *n* <Kerntech> fast-acting trip valve
Schnellauslösung *f* <Elektriz> instantaneous release
Schnellbahn *f* 1. <Eisenbahn> *(AE)* high-speed railroad, *(BE)* high-speed railway; 2. <Eisenbahn> *(AE)* regional express railroad, *(BE)* regional express railway *(S-Bahn)*
Schnellbahnsystem *n* <Eisenbahn> rapid transit system
Schnellbahnwagen *m* <Eisenbahn> rapid transit car
Schnellbewitterung *f* <Kunststoff> accelerated weathering test
Schnellbohrer *m* <Mechan> high-speed drill
Schnellboot *n* <Wassertrans> patrol boat *(Marine)*
Schnellbrutreaktor *m* <Kerntech> accelerator breeder
Schnelldrehmaschine *f* <Maschinen> speed lathe
Schnelldrehmeißel *m* <Maschinen> high-speed cutting tool
Schnelldrucker *m* <Druck> high-speed printer
schnelle Diode *f* <Elektronik> fast-recovery diode
schnelle Fourier-Transformation *f* <Elektronik, Elektronik> fast Fourier transform, fast Fourier transformation
schnelle Frequenzumtastung *f* <Elektronik, Funktech, Telekom> fast frequency shift keying
schnelle Gefriertrocknung *f* <Lebensmittel> AFD, accelerated freeze-drying
schnelle Kapazitätsdiode *f* <Phys> hyperabrupt varactor diode
schnelle Leitungsvermittlung *f* <Telekom> fast circuit switch
schnelle Logik *f* <Elektronik> fast logic, high-speed logic
schnelle Luftkühlung *f* <Thermod> rapid air cooling
schnelle Lyophilisation *f* <Lebensmittel> AFD, accelerated freeze-drying

schnelle Maschenschaltung f <Elektronik> high-speed mesh
schnelle Paketvermittlung f (FPS) <Telekom> fast packet switching (FPS)
schnelle Schaltdiode f <Elektronik> high-speed switching diode
schnelle Tonhöhenschwankungen fpl <Akustik> flutter
Schnelleinfrieren n **durch Eintauchen** <Verpack> dip freezing
Schnelleinschuss m <Kerntech> fast insertion (eines Steuerstabs)
Schnellemesser m <Akustik> velocity meter
Schnellentleerung f <Lufttrans> dumping (Treibstoff)
Schnellentwickler m <Foto> fast developer
Schnellepotenzial n <Akustik> velocity potential
schneller Brüter m (SB) <Kerntech, Phys> accelerator breeder, fast breeder reactor
schneller Brutreaktor m (SB) 1. <Kerntech> accelerator breeder, fast breeder reactor; 2. <Phys> fast breeder reactor (FBR)
schneller Ionisationsstopp m <Kerntech> fast burst
schneller Kippvorgang m <Elektronik> fast sweep
schneller Papiervorschub m <Comp & DV> paper slew
schneller Pufferspeicher m <Comp & DV> cache memory
schneller Reaktor m <Kerntech> fast reactor
schneller Rücklauf m <Maschinen> fast return
schneller Schalttransistor m <Elektronik> high-speed switching transistor
Schnellerhitzung f <Lebensmittel> flash heating
schnelles Entspannen n <Ker & Glas> rapid annealing
schnelles Frequenzsprungverfahren n <Elektronik, Funktech, Telekom> fast frequency hopping
schnelles Gefrieren n <Lebensmittel> quick-freezing
schnelles Herunterregeln n <Fernseh> fast pull-down
schnelles Modem n <Elektronik> high-speed modem
schnelles Nachführen n <Lufttrans> fast slaving
schnelles Neutron n <Phys> fast neutron, prompt neutron
schnelles Peripheriegerät n <Comp & DV> fast peripheral
schnelles Playback n <Fernseh> fast playback
schnelles Richten n <Lufttrans> fast slaving
schnelles Schiff n <Wassertrans> high-speed craft, HSC
schnelles Teilchen n <Kerntech> fast particle
schnelles Zählsystem n **für Mehrfachregale** <Verpack> high-speed multirack counting system
Schnellessigbereiter m <Lebensmittel> acetifier
Schnellfähre f <Wassertrans> fast ferry
Schnellfiltration f <Lebensmittel> accelerated filtration
Schnellgang m 1. <Maschinen> fast traverse; 2. <Mechan> overdrive
Schnellgangwelle f <Maschinen> quick-motion shaft
Schnellgefriereinrichtung f <Maschinen> quick-freezing installation
Schnellgefrieren n <Heiz & Kälte> quick-freezing
Schnellheizkatode f <Fernseh> rapid heat-up cathode
Schnellkochtopf m <Lebensmittel> autoclave, pressure cooker, rapid cooking pot for electric oven, vapour-pressure cooking pot for electric oven
Schnellkompostierung f <Abfall> mechanical composting, rapid fermentation
Schnellkopie f <Fernseh> dubbing
Schnellkühlen n <Heiz & Kälte> quick-chilling, rapid chilling
Schnellkühlung f <Thermod> rapid cooling
Schnellkupplung f <Raumfahrt> quick coupler
Schnellladegerät n <Mobilkom> fast charger
Schnellladesystem n <Foto> rapid loading system

Schnellladung f 1. <Lufttrans> quick charge; 2. <Trans> boost charge
Schnellläufer m <Wassertrans> high-speed engine (Dieselmotor)
Schnellläufermotor m <Mechan> high-speed engine
Schnelllaufmotor m <Elektriz> high-speed motor
Schnelllaufzughebel m <Foto> single stroke lever
Schnellleitung f <Comp & DV> fast line
Schnelllösesystem n <Maschinen> quick-release clamping system
Schnelllösung f <Raumfahrt> quick release (Raumschiff)
Schnellmontage f <Bau> rapid assembly
Schnellnahverkehr m <Trans> (AE) rapid transit
Schnellnahverkehrshochsystem n <Eisenbahn> elevated rapid-transit system
Schnellpresse f <Druck> flat-bed cylinder press, high-speed printing press
Schnellprüfung f <Verpack> high-speed inspection
Schnellregelung f <Ergon> quickening
Schnellrelais n <Elektriz> fast-acting relay, high-speed relay
Schnellrichtrelais n <Lufttrans> fast slaving relay
Schnellrücklauf m <Maschinen> quick return
Schnellsandabscheider m <Bau> rapid sand filter
Schnellschalthebel m <Foto> rapid film advance lever
Schnellschaltrelais n <Elektriz, Elektrotech> fast-acting relay, high-speed relay
Schnellschluss m <Kerntech> scram
Schnellschlussgriffstange f <Hydraul> throttle reach-rod
Schnellschlussklappenventil n 1. <Hydraul> clack valve; 2. <Maschinen> clack valve, flap valve
Schnellschlussventil n 1. <Hydraul> throttle; 2. <Maschinen> fast-closing valve, quick-action valve; 3. <Wasserversorg> quick-closing valve
Schnellschnittstahl m <Maschinen> HSS, high-speed steel
Schnellschützen m <Textil> fly shuttle
Schnellspaltfaktor m <Kerntech> fast fission factor
Schnellspaltung f <Kerntech> fast fission
Schnellspannhebel m <Foto> rapid film advance lever, single stroke lever
Schnellspeicher m <Comp & DV> IAS, fast core, immediate access store
Schnellstahl m 1. <Maschinen> HSS, high-speed steel, super high-speed steel; 2. <Mechan> HSS, high-speed steel
Schnellstraße f <Bau> (BE) thoroughfare, (AE) thruway
Schnellstraße f **mit Mittelstreifen** <Trans> (AE) divided highway, (BE) dual carriageway
Schnelltiefkühlen n <Heiz & Kälte> quick-freezing
Schnelltrennkupplung f <Lufttrans> quick disconnect (Hydraulik)
Schnellverbinder m <Raumfahrt> quick coupler (Raumschiff)
Schnellverdampfer m <Thermod> flash boiler (in Dampfmotoren)
Schnellverkehr m <Trans> rapid traffic
Schnellverkehrsbusspur f <Trans> busway for rapid transit
Schnellverkehrssystem n **ohne Zwischenhaltestation** <Trans> nonstop rapid transit system
Schnellversuch m <Werkprüf> accelerated testing
Schnellvorschub m <Maschinen> quick feed
Schnellwaage f <Phys> steelyard
Schnellwechselbohrfutter n <Maschinen> quick-change drill chuck
Schnellwechselfutter n <Maschinen> quick-action chuck
Schnellwechselmeißel m <Maschinen> quick-change tool

Schnellwechselwerkzeughalter

Schnellwechselwerkzeughalter *m* <Maschinen> rapid-change toolholder
Schnellwelle *f* <Elektrotech> fast wave
Schnellzerreißversuch *m* <Werkprüf> high-speed tension test
Schnellzug *m* <Eisenbahn> express train, fast train; limited train *(mit Platzkartenzwang)*
Schnellzugriff *m* <Comp & DV> immediate access
Schnitt *m* 1. <Akustik> dubbing; 2. <Erdöl> fraction; 3. <Fertig> die set; sectional diagram *(Kunststoffinstallationen)*; 4. <Geom> intersection, section; 5. <Maschinen, Mechan> cut; 6. <Textil> style • **im Schnitt darstellen** <Konstzeich> represent in section
Schnittansicht *f* <Maschinen> cutaway view
Schnittbewegung *f* <Maschinen> cutting stroke
Schnittbildentfernungsmesser *m* <Foto> split image rangefinder
Schnittbrenner *m* <Labor> flat-flame burner
Schnittdarstellung *f* <Konstzeich> sectional view, sectional representation
Schnittebene *f* <Geom> section plane
Schnittteil *n* <Maschinen> blank, blanking
Schnittfestigkeit *f* <Sicherheit> resistance to cutting *(von Schutzhandschuhen)*
Schnittfläche *f* <Maschinen> cut surface
Schnittflächenschraffurlinie *f* <Fertig> section line
Schnittfuge *f* <Fertig> kerf
Schnittgang *m* <Maschinen> cutting stroke
Schnittgerade *f* <Geom> intersection line
Schnittgeschwindigkeit *f* <Maschinen> cutting speed
Schnittholz *n* <Bau> scantling
Schnittkante *f* <Druck> cutting edge
Schnittkantenzeichnung *f* <Konstzeich> section indentification
Schnittkraft *f* <Maschinen> cutting force
Schnittleistung *f* <Maschinen> cutting capacity
Schnittlinie *f* <Geom> intersection line
Schnittmarke *f* <Druck> crop mark, cutting mark
Schnittmatte *f* <Kunststoff> chopped-strand mat
Schnittmenge *f* <Comp & DV, Math> intersection
Schnitt-Modell *n* <Fertig> cutaway model *(Kunststoffinstallationen)*
Schnittplatte *f* 1. <Fertig> die shoe; 2. <Maschinen> blanking die, shearing die
Schnittpresse *f* <Fertig> blanking press
Schnittpunkt *m* 1. <Bau> intersection point; 2. <Comp & DV> intersection; 3. <Geom> point of intersection
Schnittpunkt *m* **von Linien** <Geom> line intersection
Schnittschutz *m* <Sicherheit> cutting-tool safety *(Werkzeugmaschine)*
Schnittstelle *f* 1. <Comp & DV, Elektriz, Elektronik, Kontroll, Maschinen, Phys> interface; 2. <Telekom> interface *(zwischen technischen Einheiten)*
Schnittstelle *f* **der digitalen Verbindungsleitung** <Telekom> DTI, digital trunk interface
Schnittstelle *f* **Rechner-Nebenstellenanlage** <Telekom> computer-PBX interface
Schnittstellenbereich *m* <Comp & DV> range of port
Schnittstellen-Chip *m* <Elektronik> interface chip
Schnittstelleneinheit *f* <Comp & DV> interface unit
Schnittstelleneinrichtung *f* <Telekom> interface unit *(ITG)*
Schnittstellenelement *n* <Telekom> primitive
Schnittstellenkarte *f* <Elektronik, Telekom> interface board, interface card
Schnittstellenleitung *f* <Telekom> interchange circuit, interface circuit
Schnittstellenlogik *f* <Elektronik> interface logic
Schnittstellenmodul *n* <Telekom> IM, interface module

Schnittstellenprogramm *n* <Comp & DV> interface routine
Schnittstellenschaltung *f* <Elektronik> interface circuit
Schnittstellenstandard *m* <Telekom> interface standard
Schnittstellenstandard *m* **für sicheres Homebanking** <Telekom> homebanking computer interface (HBCI)
Schnittstellensteckverbindungen *fpl* <Telekom> interface connections, interface lines
Schnittstellenstromkreis *m* <Telekom> interface circuit
Schnittstellenvoraussetzung *f* <Comp & DV> interface requirement
Schnittstempel *m* <Fertig> cutting punch
Schnitttiefe *f* 1. <Fertig> working engagement; 2. <Maschinen, Mechan> depth of cut
Schnittweiten *fpl* <Phys> conjugate points
Schnittwerkzeug *n* <Maschinen> cutter
Schnittwinkel *m* 1. <Bau> intersection angle; 2. <Fertig> cutting angle; 3. <Geom> angle of intersection; 4. <Lebensmittel> cutting angle
Schnittzeichnung *f* 1. <Bau> section drawing, sectional drawing; 2. <Maschinen> section drawing; 3. <Wassertrans> sectional drawing *(Schiffkonstruktion)*
Schnüffelventil *n* <Hydraul> blow valve
Schnur *f* 1. <Druck> string; 2. <Elektriz, Maschinen> cord
Schnürboden *m* <Wassertrans> *(AE)* mold loft, *(BE)* mould loft *(Schiffwerft)*
Schnürbodenarbeiter *m* <Wassertrans> loftsman *(Schiffbau)*
Schnürbodenverfahren *n* <Wassertrans> lofting *(Schiffbau)*
Schnurgerüst *n* <Bau> batter board *(Vermessung)*
schnurlos *adj* <Telekom> cordless
schnurlose Nebenstelle *f* <Telekom> wireless private branch exchange *(Fernsprechwesen)*
schnurlose Telekommunikationsanlage *f* <Telekom> wireless private branch exchange *(Fernsprechwesen)*
schnurloser Fernsprechschrank *m* <Telekom> cordless switchboard
schnurloser Handapparat *m* <Mobilkom> cordless telephone handset, cordless telephone set, CT set
schnurloser Klappenschrank *m* <Telekom> cordless switchboard
schnurloser Telefonapparat *m* <Mobilkom> cordless telephone handset, CT set, set cordless telephone
schnurloser Vermittlungsschrank *m* <Telekom> cordless switchboard
schnurloses Telefon *n* <Telekom> cordless telephone
Schnurlot *n* <Bau> plumb bob, plumb line
Schnurnagel *m* <Bau> line pin
Schnurstromkreis *m* <Telekom> cord circuit
Schnürverschluss *m* <Verpack> tying closure
schockgefrieren *v* <Thermod> quick-freeze
Schockgefrieren *n* <Thermod> quick-freezing
schockgefroren *adj* <Thermod> quick-frozen
Schockwelle *f* **nach Kernexplosion** <Strahlphys> nuclear shock waves
Schokoladenschaum *m* <Meerschmutz> chocolate mousse *(Öl-Wasser-Emulsion mit halbfesten Klumpen)*
Schoner *m* <Wassertrans> schooner *(Schifftyp)*
Schongang *m* <Kfztech> overdrive *(Getriebe, Motor)*
Schonganggetriebe *n* <Maschinen> overspeed gear, overspeeder
Schönheit *f* <Teilphys> beauty *(Quark-Geschmack)*
Schönschrift *f* <Comp & DV> letter-quality
Schönseite *f* <Papier> felt side *(des Papiers)*
Schön- und Wiederdruckform *f* <Druck> sheetwise form
Schönung *f* <Lebensmittel> fining

Schönungsmittel n <Lebensmittel> fining agent
Schopf m <Fertig> crop end
Schöpfbaum m <Wassertrans> brailer boom *(Fischerei)*
Schöpfbecher m <Labor> scanning electron microscope
schopfen v <Fertig> crop *(Blöcke)*
Schopfen n <Fertig> cropping
schöpfen v 1. <Ker & Glas> ladle; 2. <Papier> dip
Schöpfen n <Papier> dipping
Schöpfform f <Papier> *(AE)* mold, *(BE)* mould
Schöpfkelle f <Ker & Glas, Lebensmittel> ladle
Schopfmaschine f <Fertig> cropping shear
Schöpfrad n <Wasserversorg> scoop wheel
Schöpfrahmen m <Papier> deckle
Schöpfschaufelrad n <Trans> scoop wheel elevator, scoop wheel feeder
Schöpfwasserrad n <Wasserversorg> scoop water wheel
Schorf m <Lebensmittel> scab *(Schädlingsbekämpfung)*
Schornstein m 1. <Bau> chimney, funnel, stack; 2. <Heiz & Kälte> chimney; 3. <Mechan> funnel
Schornsteinaufsatz m <Bau, Heiz & Kälte> cowl
Schornsteinblechrinne f <Bau> fillet gutter *(kleine Rinne zwischen Dachschräge und Schornstein)*
Schornsteinkragen m <Wassertrans> funnel bonnet
Schornsteinzug m <Bau> chimney flue
Schot f <Wassertrans> sheet *(Segeln)*
Schothorn n <Wassertrans> clew *(Segeln)*
Schothornausholer m <Wassertrans> clew outhaul *(Segeln)*
Schotklemme f <Wassertrans> cam cleat *(Beschläge)*
Schotstek m <Wassertrans> becket bend, sheet bend, sheet knot *(Knoten)*
Schott n 1. <Mechan> bulkhead; 2. <Raumfahrt> bulkhead *(Raumschiff)*; 3. <Wassertrans> bulkhead *(Schiffbau)*
Schotte f <Bau> cross wall
Schottelpropeller m <Trans> Schottel propeller
Schotter m 1. <Bau> broken stone, crushed stone, gravel; metal *(Tiefbau)*; 2. <Kohlen> gravel; 3. <Trans> ballast *(Eisenbahn)*
Schotterauftragmaschine f <Bau, Trans> macadam spreader
Schotterbrecher m <Bau> stone breaker, stone crusher
Schotterdecke f <Bau, Trans> macadam
Schotterlok f <Eisenbahn> boxer engine
Schotterschüttung f <Bau> boxing
Schottersteine fpl <Bau> tailings *(übergroß)*
Schotterstraße f <Bau> broken stone road, *(AE)* metaled road, *(BE)* metalled road
Schotterunterfütterung f <Bau> hardcore
Schotterverteilungsmaschine f <Bau, Trans> road metal spreading machine
Schotterwagen m <Eisenbahn> ballast wagon
Schotterzug m <Eisenbahn> ballast train
Schottky-Barriere f <Elektronik, Phys> Schottky barrier
Schottky-Bauelement f <Elektronik> Schottky device
Schottky-Diode f <Elektronik, Phys> Schottky barrier diode, Schottky diode, hot carrier diode *(Halbleiterdiode)*
Schottky-FET m <Elektronik> Schottky barrier FET
Schottky-Gleichrichterdiode f <Elektronik> Schottky barrier rectifier diode
Schottky-Klemmdiode f <Elektronik> Schottky clamping diode
Schottky-Klemmtransistor m <Elektronik> Schottky clamped transistor
Schottky-Mischdiode f <Elektronik> Schottky barrier mixer diode
Schottky-Rauschen n <Phys> Schottky noise
Schottky-TTL f <Elektronik> Schottky TTL
Schottplatte f <Wassertrans> bulk-head plate *(Schiffbau)*

Schottversteifung f <Wassertrans> bulkhead stiffener *(Schiffbau)*
Schraffe f <Druck> serif
schraffieren v <Fertig> hachure, hatch
schraffiert adj <Druck> shaded
schraffierter Bereich m <Konstzeich> hatched area
Schraffierung f <Fertig> hachure *(technische Zeichnung)*
Schraffur f <Fertig> hatching
Schraffurmuster n <Konstzeich> hatching pattern
Schraffurwinkel m <Konstzeich> hatching angle
schräg adj 1. <Bau> *(AE)* beveled, *(BE)* bevelled, slanting; 2. <Fertig> bevel *(Kante)*; 3. <Geom> oblique
Schräg... <Bau, Fertig, Lebensmittel, Lufttrans> slant
Schrägagarkultur f <Lebensmittel> agar slant
Schrägbalken m <Bau> raker
Schrägbruch m <Metall> slant fracture
Schrägdach n <Bau> pitch roof
Schräge f 1. <Bau> batter, cant, haunch, inclination; 2. <Fertig> draft; 3. <Maschinen> incline; 4. <Metrol> bevel; 5. <Telekom> tilt
schräge Antriebswelle f <Lufttrans> inclined drive shaft *(Hubschrauber)*
schräge Bahnkreuzung f <Eisenbahn> diamond crossing, double diamond crossing
schräge Koordinaten fpl <Geom> oblique axes
schräge Normschrift f <Konstzeich> sloping-style standard lettering
schräge Serife f <Druck> oblique serif
schräge Strahlung f <Ker & Glas> skew ray
schräge Verbindung f <Bau, Maschinen> bevel joint
Schrägeingriff m 1. <Fertig> angular meshing *(Getriebelehre)*; 2. <Mechan> angular meshing
Schrägeinstechschleifmaschine f <Fertig> angle head grinding machine
schrägen v <Bau> chamfer
Schrägentfernung f <Funkort> slant range *(Radar)*
schräger Aufprallversuch m <Kfztech> oblique crash test
schräger Kanal m <Kerntech> inclined channel
schräger Strahl m <Optik> skew ray
schräges Blatt n <Bau> splayed scarf
schräges Licht n <Druck> oblique lighting
Schrägfalte f <Konstzeich> oblique fold
Schrägfuge f <Bau> bevel joint, chamfered joint
schräggeschliffen adj <Fertig> sheared *(Schnitt)*
Schrägheckfahrzeug n <Kfztech> hatchback car, hatchback model *(Fahrzeugart)*
Schrägkopfriegel m <Bau> bevel-headed bolt
Schrägkugellager n <Maschinen> angular contact ball bearing
Schrägkurslinie f <Lufttrans> slant course line
Schräglage f 1. <Druck, Geom> slant; 2. <Lufttrans> bank, banking • **in Schräglage bringen** <Wassertrans> cant
Schräglager n <Maschinen> angular contact bearing
Schräglauf m <Telekom> skew *(Faksimile)*
Schräglenker m <Kfztech> semitrailing arm *(Hinterachse)*
Schräglichtbeleuchtung f <Elektriz> oblique illumination
Schrägmotor m <Kfztech> slanter engine
Schrägpolarisation f <Elektrotech> slant polarization
Schrägrevolverkopf m <Fertig> tilted turret
Schrägrinne f <Fertig> chute
Schrägrohrmanometer n <Gerät> inclined tube manometer
Schrägschleifen n <Fertig> oblique grinding
Schrägschneidemaschine f <Fertig> angle-cutting machine
Schrägschnitt m 1. <Fertig> angle cut; 2. <Maschinen> bevel cut

Schrägsitz

Schrägsitz m <Maschinen> inclined seat; slanted seat *(eines Ventils)*
Schrägsitzmagnetventil n <Fertig> solenoid angle seat valve *(Kunststoffinstallationen)*
Schrägsitzrückschlagventil n <Fertig> angle seat check valve *(Kunststoffinstallationen)*
Schrägsitzventil n <Fertig> angle seat valve *(Kunststoffinstallationen)*
Schrägspiegler m <Funktech> offset reflector *(Antenne)*
Schrägspur... <Aufnahme, Fernseh> helical
Schrägspuraufnahme f <Aufnahme, Fernseh> helical recording
Schrägspuraufzeichnung f <Aufnahme, Fernseh> helical scan
Schrägspuredieren n <Fernseh> physical helical editing
Schrägspurverfahren n <Comp & DV> helical scan
Schrägspur-Videorecorder m <Fernseh> helical scan videotape recorder
schrägstellbare Stößelführung f <Fertig> tilting body *(Senkrechtstoßmaschine)*
schrägstellen v <Bau> tilt
Schrägstellung f 1. <Bau> skewing; 2. <Comp & DV> skew
Schrägstift m <Maschinen> angle pin
Schrägstirnrad n <Kfztech, Maschinen> helical gear
Schrägstrich m <Geom> oblique stroke, solidus
Schrägungsverhältnis n am Luftschraubenblatt <Lufttrans> blade taper ratio *(Hubschrauber)*
Schrägungswinkel m <Fertig> lead angle *(Zahnrad)*
Schrägverblattung f <Bau> splayed joint
schrägverzahnt adj 1. <Fertig> helix-toothed; 2. <Maschinen> helical
schrägverzahntes Getrieberad n <Fertig, Kfztech> helical gear
Schrägverzahnung f <Maschinen> helical teeth, spiral gearing
Schrägverzerrung f <Fernseh> skew error
Schrägwalzen n <Fertig> rotary forging
Schrägwand f <Bau> batter wall
Schrägzahnkegelrad n <Maschinen> helical bevel gear
Schrägzahnrad n 1. <Fertig> helical gear; 2. <Maschinen> helical gear, spiral gear; 3. <Mechan> helical gear
schralen v <Wassertrans> haul forward *(Wind)*
Schram m <Kohlen> kerf
Schramme f 1. <Bau> *(BE)* kerbstone; 2. <Kunststoff> scratch; 3. <Textil> scuffing
Schrammen fpl <Metall> striation
Schrammschutzplatte f <Raumfahrt> antifret plate *(Raumschiff)*
Schranke f <Fertig> barrier
Schränkeisen n <Maschinen> saw set, set
schränken v <Fertig> set *(Säge)*
Schränkmaschine f **für Sägen** <Maschinen> saw-setting machine
Schränkung f 1. <Fertig> set *(Säge)*; 2. <Maschinen> set
Schrapper m <Meerschmutz> scraper
Schrappergefäß n <Umweltschmutz> scoop
Schrappförderer m <Maschinen> scraper
Schraub... <Bau, Elektrotech, Fertig, Gerät, Maschinen, Verpack> screw
schraubbarer Linsendeckel m <Foto> screw-on lens cap
Schraubdeckel m <Verpack> continuous thread cap, screw lid, screw top
Schraube f 1. <Bau, Eisenbahn> screw; 2. <Fertig> bolt *(Kunststoffinstallationen)*; 3. <Kfztech> bolt, screw; plug *(Ablassschraube, Ablassstopfen)*; 4. <Lufttrans> screw; 5. <Maschinen> bolt, propeller, screw; 6. <Mechan, Wassertrans> screw • **Schraube anziehen** <Maschinen> tighten a screw
Schraube f **mit Bund** <Maschinen> collar screw
Schraube f **mit eingängigem Gewinde** <Maschinen> single-threaded screw
Schraube f **mit Flachgewinde** <Maschinen> square thread screw, square-threaded screw
Schraube f **mit Linksgewinde** <Maschinen> left-hand screw, left-handed screw
Schraube f **mit Rechtsgewinde** <Maschinen> right-hand screw, right-handed screw
Schraube f **mit scharfgängigem Gewinde** <Mechan> angular thread screw
Schraube f **mit Schlitz** <Maschinen> slotted screw
Schraube f **mit Spitzgewinde** <Maschinen> V-threaded screw
Schraube f **mit UN-Gewinde** <Maschinen> unified bolt
Schraube f **mit Vierkantansatz** <Maschinen> square neck bolt
Schraube f **mit zweigängigem Gewinde** <Maschinen> double-threaded screw, two-start screw
Schraube f **ohne Kopf** <Maschinen> headless screw
Schraube f **ohne Mutter** <Maschinen> screw bolt
schrauben v <Maschinen, Mechan> screw
Schrauben n <Maschinen> bolting
Schraubenautomat m <Maschinen> automatic screw machine, screw machine
Schraubenbandfeder f <Maschinen> volute spring
Schraubenbewegung f <Maschinen> screw motion
Schraubenbock m <Wassertrans> propeller bracket *(Schiffbau)*
Schraubenbolzen m 1. <Kfztech> threaded bolt; 2. <Maschinen> bolt
Schraubendrehautomat m <Maschinen> screw machine
Schraubendreher m 1. <Elektrotech, Kfztech> screwdriver *(Werkzeug)*; 2. <Maschinen> screwdriver, turnscrew
Schraubendrehereinsatz m <Maschinen> screwdriver bit, turnscrew bit
Schraubendrehmaschine f <Maschinen> screw machine, screw-cutting lathe
Schraubendruckfeder f <Maschinen> helical compression spring
Schraubenfeder f 1. <Kfztech> coil spring; 2. <Maschinen> helical spring; 3. <Mechan> coil spring; 4. <Phys> helical spring
Schraubenfederkupplung f <Kfztech> coil-spring clutch
Schraubenfedermanometer n <Gerät> helical capsule manometer, helix capsule manometer
Schraubenfläche f <Geom> helicoid
schraubenförmig adj 1. <Fertig> helical; 2. <Geom> helical, helicoid; 3. <Maschinen> helicoid, helicoidal
schraubenförmige Bewegung f <Maschinen> helicoidal motion, screw motion
schraubenförmige Nut f <Fertig> helical broaching
schraubenförmige Versetzung f <Metall> helical dislocation
Schraubenführungsklaue f <Maschinen> screw dog
Schraubenfutter n <Maschinen> screw chuck
Schraubengang m <Maschinen> convolution
Schraubengewinde n 1. <Fertig> bolt thread; 2. <Maschinen> screw thread
Schraubengewinde n **in Zoll** <Maschinen> inch screw thread
Schraubengewindelinie f <Fertig> helix
Schraubengewindeprofil n <Maschinen> screw thread profile
Schraubengewindetoleranzen fpl <Maschinen> screw thread tolerances
Schraubenkontakt m <Elektriz> screw contact
Schraubenkopf m 1. <Maschinen> bolt head, screw head; 2. <Mechan> bolt head

Schraubenkopfanstauchen n <Fertig> bolt heading
Schraubenkopffeile f <Maschinen> screw head file
Schraubenkopfschlitzen n <Fertig> screw head slotting
Schraubenkopfstauchmaschine f <Fertig> bolt-forging machine
Schraubenkörper m <Maschinen> screw body
Schraubenlehre f <Maschinen, Metrol> (AE) screw gage, (BE) screw gauge
Schraubenlinie f 1. <Fertig, Geom> helix; 2. <Maschinen> helicoid, helix • **in Schraubenlinie aufrollen** <Fertig> helix
Schraubenlinie f mit Linksdrall <Fertig> left-hand helical
Schraubenlinienabtastung f <Elektronik> helical scanning
Schraubenlochkreis m <Fertig> bolt circle (Flansch)
Schraubenlüfter m 1. <Heiz & Kälte> propeller fan; 2. <Lufttrans> propfan (Propeller)
Schraubenmutter f 1. <Bau> nut bolt; 2. <Maschinen, Mechan> nut
Schraubenpresse f <Maschinen> screw press
Schraubenpumpe f <Bau, Fertig, Maschinen, Meerschmutz> screw pump
Schraubenrad n <Maschinen> spiral wheel, worm gear
Schraubenrädergetriebe n <Maschinen> helical gear drive
Schraubenradgetriebe n <Maschinen> spiral gearing
Schraubenradpumpe f <Maschinen> mixed-flow pump
Schraubenregel f <Elektriz, Phys> cork-screw rule
Schraubenrohling m <Fertig, Maschinen> screw blank
Schraubenschaft m <Maschinen> screw body
Schraubenschaufler m <Kfztech, Meerschmutz> screw pump
Schraubenschlüssel m 1. <Fertig> (BE) spanner, wrench; 2. <Kfztech> (BE) spanner (Werkzeug); 3. <Maschinen> screw wrench, (BE) spanner, wrench; 4. <Mechan> (BE) spanner, wrench
Schraubenspindelpumpe f <Kfztech> screw pump
Schraubenstrahl m <Lufttrans> propeller wash
Schraubenumsteuerung f <Maschinen> screw reversing gear
Schraubenverbindung f 1. <Fertig> bolt joint, bolted connection, bolted union, nipple; assembly bolts (Kunststoffinstallationen); 2. <Maschinen> bolted connection, bolted joint
Schraubenverdichter m <Maschinen> screw compressor
Schraubenverschiebung f <Metall> screw dislocation
Schraubenwasser n <Meerschmutz> wake
Schraubenwelle f <Wassertrans> propeller shaft (Schiffbau)
Schraubenwinde f <Bau> screw jack
Schraubenzieher m 1. <Elektrotech, Kfztech> screwdriver (Werkzeug); 2. <Maschinen> screwdriver, turnscrew
Schraubenzwinge f <Maschinen> holdfast
Schraubfitting n <Maschinen> threaded fitting
Schraubflansch m <Maschinen> screwed flange
Schraubflasche f <Verpack> screw cap bottle
Schraubfräseinrichtung f <Maschinen> spiral milling attachment
Schraubfräsen n <Maschinen> spiral milling
Schraubgewinde n <Mechan> worm
Schraubheftzwinge f <Maschinen> G-clamp, G-cramp
Schraubhülse f <Maschinen> threaded bush
Schraubkappe f <Maschinen, Verpack> screw cap
Schraubklemme f <Elektrotech> screw-type terminal
Schraubkopf m <Elektrotech> fuse carrier (bei Stöpselsicherung)

Schraubkupplung f <Maschinen> screw coupling
Schraublehre f <Fertig, Gerät> micrometer
Schraubloch n <Bau> screw hole
Schraubmuffe f 1. <Bau> union; 2. <Maschinen> screwed fitting
Schraubölfilter n <Kfztech> screw-type oil filter
Schraubquetschung f <Kerntech> screw pinch
Schraubring m <Maschinen> screw ferrule
Schraubsicherung f <Verpack> screw locking device
Schraubsockel m <Elektrotech> screw cap (bei Glühlampen); screw base (für elektrische Lampen)
Schraubspindel f <Maschinen> jack screw, screw jack
Schraubstempel m <Bau> ratchet brace
Schraubstock m <Bau, Fertig, Maschinen, Mechan> (BE) vice, (AE) vise
Schraubstockbacke f <Maschinen> (BE) vice jaw, (AE) vise jaw
Schraubstollen m <Eisenbahn> screw spike
Schraubstück n <Maschinen> screw piece
Schraubstutzen m <Maschinen> screw socket
Schraubtiefe f <Fertig> screw penetration
Schraubtrieb m <Kfztech> inertia drive
Schraubtriebanlasser m <Kfztech> Bendix-type starter, inertia drive starting motor
Schraubventil n <Labor> screw valve (Flüssgkeitsregler)
Schraubverbindung f <Maschinen> screw joint, threaded joint
Schraubverschluss m 1. <Fertig> screw plug; 2. <Verpack> continuous thread closure, screw closure, twisting closure
Schraubvorrichtung f <Kerntech> screwing device
Schraubwerkzeug n <Maschinen> screw tool
Schraubwinde f <Maschinen> jack screw, lifting jack, screw jack, screw lifting jack
Schraubzwinge f 1. <Bau> joiner's clamp; 2. <Fertig> BWG, Birmingham Wire Gauge, C-clamp; 3. <Maschinen> C-clamp, screw clamp
Schreckschicht f <Fertig> chilling layer
Schrecksekunde f <Trans> perception-reaction time, reaction time
Schredderanlage f <Abfall> vehicle shredder (für Autos)
Schreibband n <Gerät> strip chart
Schreibbefehl m <Comp & DV> write instruction
Schreibdichte f <Comp & DV> pitch, recording density
schreiben v <Comp & DV> type, write
schreibender Drehschwingungsmesser m <Metrol> torsiograph
Schreiber m 1. <Elektriz> plotter, recorder; 2. <Gerät> recording instrument; 3. <Maschinen, Telekom> recorder
Schreiberfeder f <Gerät> recording pen
Schreiberstreifen m <Gerät> continuous diagram, recording chart
Schreibfehler m 1. <Comp & DV> write error; 2. <Druck> typo
schreibgeschützt adj <Comp & DV> read-only
schreibgeschützte Platte f <Comp & DV> read-only disk
Schreibgeschwindigkeit f 1. <Elektronik> writing speed; 2. <Gerät> tracing speed (Registriergerät)
Schreibimpuls m <Comp & DV> write pulse
Schreibkopf m <Comp & DV> head, magnetic head, record head, recording head, write head
Schreibkurve f <Gerät> trace line
Schreib-/Lese... n <Comp & DV> read/write
Schreib-/Lesekopf m <Comp & DV> read/write head
Schreib-Lesespalt m <Comp & DV> head gap (im Schreib-/Lesekopf)
Schreib-/Lesespeicher m (RAM) <Comp & DV, Elektronik> random access memory (RAM)
Schreibmarke f <Druck> cursor

Schreibmaschine

Schreibmaschine f <Comp & DV> typewriter
Schreibmaschinenpapier n <Druck> typewriter paper
Schreibmaschinenschrift f <Druck> typewriter face
Schreibmaschinenzeilenabstand m <Konstzeich> typewriter spacing
Schreibpapier n <Papier> writing paper
Schreibring m <Comp & DV> write ring, write-enable ring, write-permit ring
Schreibröhrchen n <Gerät> recording pen
Schreib-Schlitzlocher m **für Wellpappe** <Verpack> printer-slotter for corrugated board
Schreibschutz m <Comp & DV> write protect, write protection
Schreibschutzetikett n <Comp & DV> write-protect label
Schreibschutzkerbe f <Comp & DV> write-protect notch
Schreibsperre f <Comp & DV> write ring
Schreibsperreanzeiger m <Comp & DV> read-only flag
Schreibspur f <Comp & DV> recording track
Schreibstift m <Aufnahme> recording stylus
Schreibstrahlerzeuger m <Elektronik> writing gun (Oszillatoren)
Schreibtischtest m <Comp & DV> dry run
Schreibverfahren n <Comp & DV> recording mode
Schreibverstärker m <Aufnahme> record amplifier
Schreibwalze f <Comp & DV> platen
Schreibweise f <Comp & DV> notation
Schreibzeit f <Elektrotech> write time (magnetische Medien)
Schrenzpapier n <Chemie> screenings
Schrift f <Druck> face, font
Schriftart f 1. <Comp & DV> font, typeface, typestyle; 2. <Druck> font, typeface; 3. <Konstzeich> style of lettering
Schriftartänderung f <Comp & DV> font change
Schriftartladen n <Comp & DV> font downloading
Schriftartplatte f <Comp & DV> font disk
Schriftbild n 1. <Comp & DV> typeface, typestyle; 2. <Druck> face
Schriftcassette f <Comp & DV> cartridge font
Schriftfahne f <Ker & Glas> letter slip
Schriftfamilie f <Druck> type family
Schriftgarnitur f <Druck> family of weights
Schriftgießen n <Druck> type casting
Schriftgießerei f <Druck> type foundry
Schriftgrad m <Druck> type size
Schriftgröße f 1. <Druck> type size; 2. <Konstzeich> character height
Schriftgutverfilmung f <Konstzeich> filming of textual documents
Schrifthobel m <Druck> type planer
schrifthoch adj <Druck> type-high
Schrifthöhe f <Druck> height of type, height of typeface, type height, height-to-paper
Schriftkasten m <Druck> case
Schriftkegel m <Druck> body
schriftlich belegte Überwachung f <Patent, Qual> documented control
Schriftlinie f <Druck> baseline
Schriftmetall n <Druck> type metal
Schriftpräge-Auftraggerät n <Verpack> printer-applicator
Schriftschablone f <Konstzeich> lettering stencil
Schriftsetzer m <Druck> compositor, typesetter
Schriftstärke f <Druck> weight of face, weight of type, zinc plate
Schrifttype f <Druck> typeface
Schritt m 1. <Comp & DV> iteration, step; 2. <Labor> stage (Mikroskop); 3. <Regelung> step; 4. <Telekom> increment; signal component (Telegrafie); unit
Schrittabstimmung f <Elektronik> incremental tuning

Schrittfehler m <Telekom> symbol error (Übertragung)
Schrittfunktion f <Comp & DV> step function
Schrittgeschwindigkeit f <Telekom> symbol rate
schritthaltende Verarbeitung f <Comp & DV> in-line processing
schritthaltender Verbindungsaufbau m <Telekom> step-by-step operation
Schrittmacherofen m <Metall> walking beam furnace
Schrittmotor m 1. <Comp & DV> stepper motor; 2. <Elektriz, Elektrotech> stepper motor, stepping motor; 3. <Kontroll> stepper motor, stepper; 4. <Maschinen> step motor, stepper motor, stepping motor
Schrittnachführungssystem n <Raumfahrt> step track system (Weltraumfunk)
Schrittschalter m <Elektriz, Elektrotech, Telekom> step switch, stepping switch
Schrittschaltsystem n <Telekom> step-by-step system
Schrittwähler m <Telekom> stepping switch
Schrittwählersystem n <Telekom> step-by-step system (Telefon)
schrittweise adj <Kontroll> step-by-step
schrittweise adv <Kontroll> step-by-step
schrittweise Abstimmung f <Funktech> incremental tuning
schrittweise Führung f <Raumfahrt> iterative guidance
schrittweise Näherung f <Gerät> successive approximation
schrittweise Regelung f <Elektriz> step-by-step control
schrittweiser Betrieb m <Kontroll> step-by-step operation
schrittweises Positionieren n <Kontroll> stepping
Schrittweite f 1. <Elektronik, Elektrotech> increment (einer Schleife); pitch; 2. <Kontroll> step size
Schrittzähler m 1. <Comp & DV, Kontroll> step counter; 2. <Phys> pedometer
Schrittzeit f <Kontroll> step time
Schrobteil m <Fertig> sett
Schroedinger-Gleichung f <Phys> Schrodinger's equation
schroffe Klippe f <Wassertrans> bluff (Geographie)
Schröpfglas n <Ker & Glas> cupping glass
Schrotbohren n 1. <Bau> boring by shot drills; 2. <Erdöl> shot drilling
schroten v 1. <Fertig> chip; 2. <Lebensmittel> bruise, kibble
Schrotmeißel m 1. <Fertig> hot set, top chisel; 2. <Maschinen> chipping hammer
Schrotmühle f <Lebensmittel> bruiser
Schrotrauschen n 1. <Optik, Phys> shot noise; 2. <Telekom> granular noise, shot noise
Schrotrückstände mpl <Lebensmittel> break tailings
Schrotstrahlen n <Bau> grit blasting
Schrott m 1. <Mechan> junk, scrap; 2. <Qual> scrap
Schrottballen m <Abfall> scrap bundle
Schrottersatz m <Metall> scrap substitute
Schrotthändler m <Abfall, Kfztech> scrap dealer, scrap metal merchant
Schrottmeißel m <Fertig> sett
Schrottpaketierpresse f 1. <Abfall> scrap baler; 2. <Verpack> baling press
Schrottplatz m <Abfall> scrapyard
Schrottpresse f <Abfall> junk press, scrap press, scrap-baling press
Schrottsammlung f <Abfall> scrap collection
Schrottschere f <Abfall> scrap shear
Schrottschmelzverfahren n <Fertig> all-scrap process
Schrottsortierung f <Abfall> scrap sorting
Schrottverhüttung f <Abfall> scrap smelting
Schrottverwertung f <Abfall> scrap processing, scrap reuse

Schrottzusatz m <Fertig> admits of scrap
Schrotwalze f <Lebensmittel> break roller
Schrubben n <Ker & Glas> swabbing
Schrubber m <Kerntech, Kohlen> scrubber
Schrühware f <Ker & Glas> biscuit ware *(Porzellan)*
Schrumpf... <Thermod> heat-shrinkable
schrumpfbar adj <Thermod> shrinkable *(Folie)*
schrumpfecht adj <Papier> shrink-proof
schrumpfen v <Ker & Glas, Papier, Textil> shrink
Schrumpfen n 1. <Bau> shrinkage, shrinking; 2. <Maschinen> shrinkage
Schrumpffolie f 1. <Lebensmittel> shrink-film; 2. <Thermod> heat-shrinkable film, shrink-film; 3. <Verpack> shrink-film
Schrumpffolie f **mit perforierter Überlappung** <Verpack> shrink-film with perforated overlap
Schrumpffolien-Siegelmaschine f <Verpack> blister sealer
Schrumpffolienverpackung f <Verpack> blister pack
Schrumpfgrenze f <Kohlen> shrinkage limit
Schrumpfpackung f <Verpack> shrink pack
Schrumpfpalettenabdeckung f <Verpack> shrink-wrapped pallet cover
Schrumpfpassung f <Maschinen> shrink fit
Schrumpfriss m 1. <Bau> contraction crack; 2. <Fertig> contraction crack, shrinkage crack
Schrumpfrohr n <Papier> shrink sleeve
Schrumpfschachtel f <Verpack> shrink capsule
Schrumpfschlauchbeutel-Verpackungsmaschine f <Verpack> shrink sleeve wrapping machine
Schrumpfsitz m <Maschinen, Mechan> shrink fit
Schrumpfspannung f <Maschinen> contraction strain
Schrumpftoleranz f 1. <Fertig> allowance for shrinkage; 2. <Maschinen> shrinkage, shrinkage allowance
Schrumpftunnel m **für Schlauchverpackung** <Verpack> shrink tunnel for sleeving
Schrumpftunnel m **für Schlauchverschweißung** <Verpack> shrink tunnel for sleeve sealing
Schrumpfung f 1. <Fertig> consolidation; contraction *(Kunststoffinstallationen)*; 2. <Kunststoff, Maschinen, Papier, Phys> shrinkage; 3. <Telekom> shrinking; 4. <Verpack> after shrinkage
Schrumpfung f **bei Verhärtung** <Verpack> shrinkage on solidification
schrumpfverpackt adj <Verpack> shrink-wrapped *(Produkt)*
Schrumpfverpackung f 1. <Lebensmittel> shrink-wrap *(in Folie)*; 2. <Verpack> contract packaging
Schrumpfverpackungsdienst m <Verpack> contract blister packaging service
Schrumpfverpackungsmaschine f <Verpack> shrink overwrapping machine
Schrumpfzugabe f <Maschinen> shrinkage allowance
Schruppdrehen n <Fertig> rough turning
Schruppdurchgang m <Fertig> roughing pass *(Spanung)*
schruppen v 1. <Fertig> rough-cut; 2. <Maschinen> rough-cut, rough
Schruppen n <Fertig> roughing
Schruppfeile f 1. <Fertig> coarse-cut file, rough-cut file; 2. <Maschinen> rough-cut file, roughing file
Schruppfräsen n <Fertig> rough milling
Schruppfräser m <Maschinen> roughing cutter, roughing mill
Schrupphonen n <Fertig> rough honing
Schruppmeißel m 1. <Fertig> rougher; 2. <Maschinen> roughing tool
Schruppsorte f <Fertig> roughing grade *(Werkstoff)*
Schruppwerkzeug n <Maschinen> rougher

Schruppzahn m <Maschinen> roughing tooth
Schub m 1. <Bau> shear; 2. <Fertig> thrust; 3. <Lebensmittel> batch; 4. <Lufttrans> thrust; 5. <Maschinen> shear, thrust; 6. <Mechan> push; 7. <Meerschmutz, Phys> thrust; 8. <Raumfahrt> blast, thrust • **auf Schub beanspruchen** <Bau> shear
Schubabnahme f <Raumfahrt> thrust decay
Schubabschaltung f <Raumfahrt> thrust cut-off
Schubachse f <Raumfahrt> thrust axis
Schubanlage f <Lufttrans> power plant
Schubbeanspruchung f <Maschinen, Qual> shearing stress
Schubbewegung f <Maschinen> translation, translatory motion
Schubboot n <Wassertrans> push boat, push tug, pusher tug
Schubbug m <Raumfahrt> thrust cone
Schubdruck m 1. <Lufttrans> boost pressure *(Triebwerk)*; 2. <Raumfahrt> boost pressure • **Schubdruck ausüben** <Raumfahrt> boost
Schubdüse f 1. <Lufttrans> exhaust nozzle; 2. <Raumfahrt> thrust nozzle
Schuber m <Verpack> slip case
Schubfehleinstellung f <Raumfahrt> thrust misalignment
Schubfestigkeit f 1. <Bau, Kohlen, Qual> shear strength, shearing strength; 2. <Maschinen> shearing strength
Schubgelenk n <Maschinen> prismatic joint
Schubgewicht n <Lufttrans> power-weight ratio
Schubkarre f <Bau> handbarrow, wheelbarrow
Schubkarren m 1. <Bau> barrow; 2. <Fertig> wheelbarrow; 3. <Trans> barrow, wheelbarrow
Schubklauengetriebe n <Kfztech> constant-mesh gears *(Getriebe mit ständigem Eingriff)*
Schubkoeffizient m <Maschinen> reciprocal of shear modulus
Schubkonus m <Raumfahrt> thrust nozzle
Schubkraft f 1. <Bau> shear force, thrust; 2. <Fertig> longitudinal shear; 3. <Maschinen, Meerschmutz> thrust
Schubkraftverstärker m <Lufttrans> thrust augmenter
Schubkugel f <Kfztech> torque ball
Schubkurbel f <Maschinen> slider crank
Schublehre f 1. <Maschinen> *(AE)* caliper square, *(BE)* calliper square, *(AE)* sliding calipers, *(AE)* vernier gage, *(BE)* vernier gauge; 2. <Mechan, Phys> *(AE)* vernier caliper, *(BE)* vernier calliper
Schubleichter m <Wassertrans> push barge
Schubmaschine f <Eisenbahn> *(BE)* banking locomotive, *(AE)* pusher locomotive
Schubmodul n 1. <Bau> shear modulus; 2. <Kohlen> rigidity modulus; 3. <Kunststoff, Metall, Qual> shear modulus
Schubmodulation f <Raumfahrt> thrust modulation
Schubriegel m <Bau> tower bolt
Schubrohrantrieb m <Kfztech> torque tube drive *(Getriebe)*
Schubschifffahrt f <Trans> push-towing
Schubschlepper m <Wassertrans> pusher tug *(Schifftyp)*
Schubschraubtriebanlasser m <Kfztech> screw push starter
Schubschwinge f <Fertig> oscillating slider *(Getriebelehre)*
Schubspannung f 1. <Bau> shearing stress; 2. <Kunststoff> shear stress; 3. <Maschinen, Metall> shearing stress
Schubstange f 1. <Maschinen> connecting rod, rod; 2. <Mechan> connecting rod, push rod
Schubsteuerprogramm n <Raumfahrt> *(AE)* thrust program, *(BE)* thrust programme
Schubsteuerung f <Raumfahrt> boosting regulator
Schubstrebe f <Kfztech> torque arm *(Radaufhängung)*

Schubsubsystem

Schubsubsystem n <Raumfahrt> thrust subsystem
Schubtriebwerk n <Raumfahrt> booster, thruster
Schubumkehrer m <Lufttrans> thrust reverser
Schubumkehrvorrichtung f <Lufttrans> thrust reverser
Schubvektor m <Raumfahrt> thrust vector
Schubvektordüse f <Raumfahrt> thrust vectoring nozzle
Schubvektorsteuerung f <Raumfahrt> thrust vector control
Schubverband m <Wassertrans> multiple-barge convoy set, push tow
Schubverformung f <Bau, Metall> shear deformation
Schubvergrößerer m <Lufttrans> thrust augmenter
Schubwechselgetriebe n <Kfztech> sliding gear transmission, straight-toothed gearbox
Schubwelle f <Akustik> rotational wave, shear wave
Schubwert m **der Luftschraube** <Wassertrans> propeller thrust coefficient
Schubzentrum n <Raumfahrt> *(AE)* center of thrust, *(BE)* centre of thrust
Schuilingit m <Kerntech> schuilingite
Schülpe f <Fertig> scab *(Gussstück)*
Schulschiff n <Wassertrans> training ship *(Marine)*
Schulter f 1. <Fernseh> porch; 2. <Ker & Glas> shoulder; 3. <Maschinen> collar, shoulder
Schulterdecker m <Lufttrans> high-wing plane
Schulterdeckerflügel m <Lufttrans> shoulder wing *(Luftfahrzeug)*
Schulterhöhe f <Druck> shoulder height
Schulterkugellager n <Maschinen> separable ball bearing
Schulterlager n <Maschinen> separable bearing
Schulterpolster n <Textil> pad, shoulder pad
Schulterstativ n <Foto> rifle grip
Schulungszeit f **am Doppelsteuer** <Lufttrans> dual-instruction time
Schuppen m <Bau> shed
Schuppenbildung f <Lufttrans> alligatoring
Schuppenglas n <Ker & Glas> flake glass
schuppig adj <Fertig> scaly
Schur f <Textil> clip
Schürfe f <Bau> pit *(Bodenuntersuchung)*
Schürfgrube f <Kohlen> test pit
Schürfraupe f <Trans> bulldozer
Schürfschacht m <Erdöl> prospecting shaft *(Lagerstättensuche)*
Schurre f <Fertig> tip chute
Schürvorrichtung f <Heiz & Kälte> mechanical stoker
Schürze f 1. <Papier, Sicherheit> apron; 2. <Wassertrans> skirt
Schürze f **mit veränderlicher Geometrie** <Wassertrans> variable-geometry skirt
Schürzensperre f <Meerschmutz> curtain boom
Schuss m 1. <Erdöl> shot *(Erkundung von Lagerstätten)*; 2. <Fertig> roll-bent part; 3. <Papier> weft; 4. <Telekom> section *(Mastantenne)*
Schussanschlag m <Textil> beat-up
Schussbruch m <Textil> weft break
Schussdichte f <Textil> beat-up, weft density
Schussdraht m <Fertig> shute wire *(Drahtweben)*
Schüssel f 1. <Funktech> dish *(Antennenform)*; 2. <Kohlen, Metrol> bowl
Schüsselklassierer m <Kohlen> bowl classifier
Schussfaden m <Textil> pick *(Weben)*
Schussfäden mpl **je Zoll** <Textil> picks per inch
Schussfadenwächter m <Textil> weft stop motion
Schussgarn n <Papier> weft yarn
Schussgewicht n <Kunststoff> shot weight
Schussleistung f <Textil> pick rate
Schussspulmaschine f <Textil> pirn-winding machine

Schussstreifen m <Textil> bar
Schusterjunge m <Druck> orphan
Schute f <Wassertrans> dumb barge; barge *(Schifftyp)*
Schutenträger m <Wassertrans> barge carrier
Schutt m 1. <Bau> rubble, excavated material; 2. <Eisenbahn> excavated material; 3. <Fertig> debris; 4. <Kohlen> excavated material
Schuttabladeplatz m 1. <Abfall> dump site, dumping site, waste tip; 2. <Bau> dump site, waste tip
Schüttdichte f 1. <Erdöl> bulk density; 2. <Kohlen> apparent density; 3. <Kunststoff> apparent density, bulk density; 4. <Phys> bulk density
Schüttdichtemessgerät n <Gerät> bulk density meter
Schüttelapparat m <Labor> shaker
Schüttelbewegung f <Maschinen> shake, shaking motion
Schüttelherd m <Kohlen> oscillating table
schütteln v <Bau> vibrate
Schütteln n 1. <Lufttrans> buffeting; 2. <Maschinen> shaking; 3. <Raumfahrt> buffet *(Raumschiff)*
Schüttelrost m <Fertig> shaking grate
Schüttelrutsche f <Fertig> shaker conveyor
Schüttelsieb n 1. <Kohlen> griddle, vibrating screen; 2. <Verpack> vibratory sifter
Schütteltrichter m <Labor> separating funnel
Schüttelzuführer m <Mechan> vibratory feeder
Schüttgut n 1. <Bau> bulk material; 2. <Fertig> packing; 3. <Kohlen> bulk material; 4. <Verpack> bulk goods; 5. <Wassertrans> bulk cargo, dry bulk cargo *(Ladung)*
Schüttgutbehälter m **mit Gravitätsentladung** <Wassertrans> bulk container with gravity discharge
Schüttgutbehälter m **mit pneumatischer Entladung** <Wassertrans> bulk container with pressure discharge
Schüttgutcontainer m <Wassertrans> dry-bulk container
Schüttkoeffizient m *(C)* <Nichtfoss Energ> discharge coefficient *(C)*
Schüttlage f <Bau> hardcore
Schüttsteine mpl <Bau> rip-rap
Schüttwasser n <Lebensmittel> make-up water
Schüttwinkel m 1. <Bau> repose angle; 2. <Phys> angle of friction
Schutz m 1. <Anstrich, Comp & DV> protection; 2. <Elektriz, Maschinen> guard; 3. <Sicherheit> guard, protection; guard *(an einer Maschine)*; protector, safeguard
Schutz m **durch Lichtgitter** <Sicherheit> light beam protection
Schutz m **durch Lichtvorhang** <Sicherheit> light beam protection
Schutz m **mit Auslöser** <Sicherheit> trip guard
Schutz m **vor chemischen Stoffen** <Sicherheit> chemical protection
Schutz m **vor Einschaltspitzen** <Elektrotech> inrush current protection
Schütz n 1. <Elektriz> contactor; 2. <Wasserversorg> sash gate *(im Schleusentor)*
Schutzabdeckung f <Sicherheit> cover *(Getriebegehäuse)*; protective canopy
Schutzabschirmung f <Sicherheit> guard screen, guard shield
Schutzabsperrung f <Sicherheit> barrier; guarding isolation *(bei Robotern)*
Schutzabstand m 1. <Comp & DV, Funktech> guard band; 2. <Mobilkom> guard space
Schützabwehr f <Wasserversorg> floodgate
Schützanlasser m <Elektriz> contactor starter
Schutzanstrich m 1. <Ker & Glas> protective paint; 2. <Kunststoff, Papier> protective coating
Schutzanzug m <Sicherheit> protective suit, protective workwear

Schutzanzug *m* **für Rettungsdienste** <Sicherheit> protective clothing for rescue services
Schutzanzug *m* **mit Atemluftzuführung** <Sicherheit> supplied-air suit
Schutzärmel *m* <Sicherheit> protective sleeve
Schutzart *f* <Elektrotech> protective system, type of enclosure; international protection, IP *(internationaler Standard)*
Schutzaufbau *m* <Sicherheit> protective structure
Schutzausrüstung *f* <Sicherheit> protective equipment, protective material, safety equipment
Schutzband *n* <Comp & DV, Funktech> guard band
Schutzbereich *m* <Patent> extent of protection
Schutzbeschichtung *f* <Kunststoff> protective coating
Schutzbit *n* <Comp & DV> guard bit
Schutzblech *n* <Fertig, Maschinen> guard
Schutzbrett *n* <Bau> baffle board
Schutzbrille *f* <Ergon, Ker & Glas, Labor, Sicherheit> eye-protection glasses, goggles, protective goggles, protective spectacles, safety goggles
Schutzbrillenglas *n* <Sicherheit> glass for protective goggles
Schutzdach *n* 1. <Bau> shed, shelter; 2. <Sicherheit> protective canopy
Schutzdamm *m* <Bau> levee
Schutzdraht *m* <Elektriz, Elektrotech> guard wire
Schutzeigenschaft *f* <Sicherheit> protective properties
Schutzeinhausung *f* <Sicherheit> protective casing, protective enclosure, safety enclosure *(Maschinen)*
Schutzeinrichtung *f* 1. <Fertig> safety guard; 2. <Maschinen> guard, safety device, security device; 3. <Sicherheit> guard, guarding system; protector; safeguard, safety device, security device
schützen *v* 1. <Comp & DV> protect; 2. <Fertig> guard; 3. <Raumfahrt> guard; 4. <Sicherheit> guard, protect, shield
Schützenschlag *m* <Textil> picking *(Weben)*
Schützenspindel *f* <Textil> shuttle spindle
Schutzfilm *m* <Verpack> protective film
Schutzfunkenstrecke *f* <Elektriz> protective spark gap
Schutzgamaschen *fpl* <Sicherheit> protective gaiters
Schutzgas *n* 1. <Fertig> inert gas; 2. <Kerntech> blanket gas, cover gas; 3. <Maschinen> inert gas
Schutzgasabfuhrleitung *f* <Kerntech> cover gas discharge line
Schutzgasflammen *n* <Thermod> annealing under gas
Schutzgasglühen *n* <Thermod> annealing under gas
Schutzgas-Lichtbogenschweißen *n* <Fertig> inert gas arc welding
Schutzgasschweißen *n* 1. <Maschinen> inert gas-shielded welding; 2. <Mechan> gas welding
Schutzgasschweißgerät *n* <Sicherheit> protective gas welding machine
Schutzgastrennanlage *f* <Kerntech> blanket separation plant
Schutzgefäß *n* <Kerntech> guard vessel
Schutzgehäuse *n* <Sicherheit> protective casing, protective enclosure
Schutzgelände *n* <Sicherheit> safety fence
Schutzgeländer *n* <Bau> guard rail
Schutzgerüst *n* <Sicherheit> protective scaffold
Schutzgitter *n* 1. <Elektriz> guard, protecting grid; 2. <Heiz & Kälte> grille; 3. <Sicherheit> fence, fence guard, guard rail, protective screen, safety fence
Schutzgitter *n* **für Katodenstrahlröhren** <Sicherheit> protective screen for cathode ray tubes
Schutzglas *n* <Sicherheit> protective glass; filter *(Schweißarbeiten)*
Schutzgrad *m* <Sicherheit> level of protection, level of safety

Schutzhandschuhe *mpl* <Sicherheit> protective gloves
Schutzhaube *f* 1. <Heiz & Kälte> fan hood *(für Lüfter)*; 2. <Mechan> cover; 3. <Sicherheit> hood guard *(an Säge)*; hood-type protector, protective hood
Schutzhelm *m* 1. <Sicherheit> helmet, protective helmet, safety helmet; 2. <Trans> safety helmet
Schutzhülse *f* <Maschinen> protection sleeve
Schutzkäfig *m* <Sicherheit> guard, protective cage
Schutzkappe *f* 1. <Bau> boot *(am unteren Ende eines Fallrohres)*; 2. <Kfztech> boot *(am Bremszylinder)*; 3. <Maschinen> protective cap
Schutzkapsel *f* <Sicherheit> protective enclosure
Schutzkiel *m* <Wassertrans> rubbing strake
Schutzkittel *m* <Sicherheit> protective gown
Schutzklasse *f* <Fertig> safety class *(Kunststoffinstallationen)*
Schutzkleidung *f* <Sicherheit> protective clothing, safety clothing
Schutzkleidung *f* **aus Kettenringen** <Sicherheit> chain-mail garment
Schutzkleidung *f* **für chemische Arbeiten** <Sicherheit> chemical protection clothing
Schutzkleidung *f* **für Kühlräume** <Sicherheit> cold-storage protective clothing
Schutzklemme *f* <Elektriz> *(AE)* armor clamp, *(BE)* armour clamp
Schutzkontakt *m* <Elektrotech> grounding contact, sealed contact
Schutzkontaktkupplung *f* <Maschinen> shrouded coupling
Schutzkopie *f* <Comp & DV> protected master
Schutzkragen *m* <Heiz & Kälte> shroud
Schutzlack *m* <Ker & Glas> resist
Schutzleitersystem *n* <Sicherheit> protective conductor system
Schutzmanschette *f* <Lufttrans> *(BE)* boot, *(AE)* trunk
Schutzmanschettenhalter *m* <Lufttrans> boot retainer
Schutzmantel *m* 1. <Fertig> shield; 2. <Optik> protective coating
Schutzmaske *f* <Sicherheit> protective mask
Schutzmaßnahme *f* <Sicherheit> precaution, precautionary measure, preventive measure, protective measure, safeguard, safety measure, safety precaution
Schutzmittel *n* 1. <Sicherheit> guard, precautionary apparatus, safeguard, safety appliance, safety device, safety workwear; 2. *(D: Verpack)* protective agent
Schutzmittel *npl* *(D: Verpack)* means of protection, protective equipment, safety equipment
Schutznetz *n* 1. <Sicherheit> safety net; 2. <Wasserversorg> guard net
Schutzort *m* <Kohlen> blast shelter, refuge
Schutzpapier *n* <Foto> backing paper *(Rollfilm)*
Schutzplane *f* <Kfztech> *(BE)* safety bonnet, *(AE)* safety hood
Schutzrahmen *m* <Sicherheit> protective frame; overhead guard *(Gabelstapler)*
Schutzraum *m* <Bau> shelter
Schutzring *m* <Elektriz, Elektrotech, Phys> guard ring
Schutzringelektrode *f* <Elektrotech> guard ring
Schutzringkondensator *m* <Elektrotech> guard ring capacitor
Schutzrohr *n* 1. <Bau, Elektrotech> conduit; 2. <Telekom> conduit *(Kabel)*
Schutzrohrkontakt *m* <Elektrotech> reed contact, switch contact
Schutzrohrkontaktrelais *n* <Elektrotech, Telekom> reed relay
Schutzrückhaltesystem *n* <Sicherheit> protective restraint system

Schutzrücklauf

Schutzrücklauf m <Sicherheit> protective withdrawal *(Maschine)*
Schutzschaltung f <Elektriz, Elektrotech, Telekom> protection circuit, protective circuit
Schutzschicht f 1. <Anstrich> protective film; 2. <Bau> coating; 3. <Kunststoff> resist coating; 4. <Verpack> protective layer
Schutzschiene f 1. <Bau> check rail; 2. <Eisenbahn> check rail, guard rail
Schutzschild m <Fertig, Sicherheit> handscreen *(Schweißen)*; protective shield, shield
Schutzschirm m 1. <Raumfahrt> shield; 2. <Sicherheit> guard screen; protective shield, protective screen
Schutzschirmsteuerung f <Sicherheit> guard control
Schutzschlauch m 1. <Fertig> shaft casing *(biegsame Welle)*; 2. <Maschinen> casing
Schutzschranke f <Sicherheit> protective barrier
Schutzschürze f <Sicherheit> protective apron
Schutzspur f <Aufnahme> guard track
Schutzstreckensignal n <Eisenbahn> dead-section warning signal
Schutzstromstrecke f <Elektriz> protective spark gap
Schutzstufe f <Sicherheit> level of protection
Schutzsystem n <Sicherheit> guarding system
Schutzüberzug m 1. <Kunststoff> protective coating; 2. <Verpack> protective coat
Schutzüberzug m **elektrochemisch aufgebracht** <Anstrich> galvanized protective coating
Schutzumhüllung f 1. <Fertig> coating; protective enclosure; 2. <Phys> protective coating *(Lichtleiter)*; 3. <Sicherheit> coating; protective enclosure
Schutzumschlag m 1. <Druck> dust cover, jacket; 2. <Verpack> protective cover *(Buch)*
Schutzverpackung f <Verpack> protective wrapper
Schutzvorrichtung f 1. <Mechan> guard, physical safeguard, protective gear, safeguard, safety appliance, safety device, security device; 2. <Sicherheit> guard, physical safeguard, protective gear, protector, safeguard, safeguarding device, safety appliance, safety device, security device; 3. <Wasserversorg> guard net *(Überlandsystem)* • **durch Schutzvorrichtung betätigt** <Sicherheit> guard-operated
Schutzvorrichtung f **für mechanische Pressen** <Sicherheit> power press guarding
Schutzvorrichtung f **gegen Überrollen** <Sicherheit> roll-over protective structure *(Schlepper)*
Schutzvorrichtung f **mit Photozelle** <Sicherheit> photoelectric guard
Schutzvorrichtung f **mittels Laserstrahl** <Sicherheit> laser beam safety device
Schutzvorrichtungen fpl <Sicherheit> protective equipment
Schutzwagen m <Eisenbahn> match wagon, shock-absorbing wagon
Schutzwand f <Bau> screen
Schutzweiche f <Eisenbahn> catch points
Schutzwiderstand m <Elektrotech> bleeder resistor, protective resistor
Schutzwirkung f <Sicherheit> protective performance, protective power; protective properties *(Schutzkleidung)*
Schutzzone f <Kerntech> safety zone *(im Heißlabor)*
Schwabbel f <Metall> polishing wheel
Schwabbeln n <Maschinen> buffing
Schwabbelscheibe f 1. <Fertig> buff, buffing wheel, rag buffing wheel; 2. <Maschinen> buff, mop, rag wheel
schwach besetzte Matrix f <Comp & DV> sparse matrix
schwach führender Lichtwellenleiter m <Telekom> *(AE)* weakly guiding fiber, *(BE)* weakly guiding fibre

schwach leitende Faser f <Optik> *(AE)* weakly guiding fiber, *(BE)* weakly guiding fibre
schwach sieden v <Thermod> boil slowly
schwache Kopplung f <Kerntech> weak coupling
schwache Kraft f <Teilphys> weak force, weak nuclear force
schwache Quelle f <Kerntech> thin source
schwache Wechselwirkung f <Kerntech, Phys, Teilphys> weak interaction
schwacher Monofrequenzlaser m <Strahlphys> low-power single frequency laser
schwacher Positronenübergang m <Kerntech> weak positron transition
schwaches Netz n <Nichtfoss Energ> weak grid
schwaches Netzwerk n <Nichtfoss Energ> weak network
Schwachpunktdesign n <Raumfahrt> design to yield point
Schwachstellenanalyse f <Qual> weak-point analysis
Schwachstrom m <Telekom> light current
Schwachstrom-Netzschalter m <Elektriz> light switch
Schwächung f 1. <Elektrotech> attenuation, fading; absorption *(Strahleneinwirkung)*; 2. <Phys, Wellphys> attenuation
Schwächungsglied n <Aufnahme> attenuator
Schwächungskoeffizient m <Elektronik> attenuation coefficient
Schwaden m <Meerschmutz> swath
Schwalbenschwanz m <Bau, Mechan> dovetail
Schwalbenschwanzführung f <Fertig> dovetail
Schwalbenschwanznut f <Maschinen> dovetail groove
Schwalbenschwanznutenfräser m <Fertig> dovetail cutter
Schwalbenschwanzverbindung f 1. <Bau> dovetail joint, swallowtail joint; 2. <Mechan> dovetail joint
Schwammgummi m <Kunststoff> sponge rubber
Schwanenhals m <Bau> swan neck
Schwanenhalsverschluss m <Bau> S-trap *(Sanitär)*
schwanken v <Elektriz> fluctuate
Schwanken n <Gerät> jitter *(Anzeigewert, Frequenz, Impulslänge)*
schwankende Strömung f <Kerntech> unsteady flow
schwankendes Echo n <Fernseh> flutter echo
Schwankung f 1. <Elektriz> drift, fluctuation; 2. <Qual> fluctuation
Schwankung f **innerhalb der Maschine** <Ker & Glas> in-line variation
schwanzlastige Fluglage f <Lufttrans> nose-up attitude
Schwanzlastigkeit f <Lufttrans> tail heaviness
Schwanzsporn m <Lufttrans> tail skid
Schwarz... <Druck> black
Schwarzabhebung f <Fernseh> set-up
Schwarzblech n 1. <Fertig> black plate; 2. <Metall> black iron plate, black sheet
Schwärze f 1. <Druck> black; 2. <Fertig> black wash *(Gießen)*
schwärzen v 1. <Fertig> black-wash *(Gießen)*; 2. <Thermod> blacken
schwarzer Ball m <Wassertrans> black ball *(Signal)*
schwarzer Fleck m <Ker & Glas> black speck
schwarzer Körper m <Fernseh, Phys, Raumfahrt, Strahlphys, Thermod> black body
schwarzer Regelstab m <Kerntech> black absorber rod
schwarzer Strahler m <Phys, Strahlphys, Thermod> black body radiator
schwarzes Brett n <Comp & DV> bulletin board *(bei E-Mail)*
schwarzes Loch n <Phys, Raumfahrt> black hole
Schwarzfärbung f <Ker & Glas> black staining
Schwarzfichte <Wassertrans> spruce *(Holz)*

Schwarzglühen n <Metall> black annealing
Schwarzguss m <Fertig> all-black malleable cast iron
Schwarzklipper m <Fernseh> black clipper
Schwarzlicht n <Druck, Optik> black light
Schwarzlot n <Ker & Glas> black stain
Schwarzpegelregelung f <Fernseh> pedestal level control
schwarzrot adj <Metall> black red
Schwarzschild-Radius m <Phys> Schwarzschild radius
Schwarzschulter f <Fernseh> back porch
Schwarzschulterklemmen n <Fernseh> back-porch clamping
Schwarzsender m <Fernseh> bootleg
Schwarzsteuerdiode f <Fernseh> DC clamp diode
Schwarzstraßenbau m <Bau> flexible road construction
Schwärzung f 1. <Akustik> optical density; 2. <Elektriz> blackening *(des Glaskolbens einer Glühlampe)*; 3. <Ker & Glas> darkening; 4. <Phys> density
Schwärzungseinstellung f <Druck> darkness setting
Schwärzungsindex m <Ker & Glas> darkening index
Schwärzungskurve f <Foto> characteristic curve
Schwärzungsröhre f <Elektronik> skiatron
Schwarz-Weiß-... 1. <Comp & DV> monochrome *(Bildschirm)*; 2. <Druck> black and white; 3. <Foto> monochrome
Schwarz-Weiß-Bild n <Fernseh> black-and-white picture
Schwarz-Weiß-Empfänger m <Fernseh> monochrome receiver
Schwarz-Weiß-Fernsehen n <Fernseh> black and white television
Schwarz-Weiß-Überwachung f <Fernseh> black-white monitoring
Schwarzwertabhebung f <Fernseh> pedestal
Schwarzwertanhebung f <Fernseh> black lift
Schwarzwerteinstellung f <Fernseh> pedestal adjustment
Schwarzwertfrequenz f <Fernseh> black level frequency
Schwarzwertpegel m <Fernseh> black level
Schwarzwertspitze f <Fernseh> black peak
Schwarzwertunterdrückung f <Fernseh> black compression
Schwebebahn f <Trans> cableway
Schwebebus m <Lufttrans> aerobus
Schwebefähigkeit f <Hydraul> buoyancy
Schwebeflugkupplung f <Lufttrans> hover flight coupler *(Hubschrauber)*
Schwebehöhe f <Trans> hoverheight
Schwebekörper-Durchflussmesser m <Erdöl> rotameter *(Mengenmessgerät)*
Schwebekörper-Durchflussmessgerät n <Gerät> variable area flowmeter
Schwebekörper-Füllstandsmessung f <Gerät> suspended-body level measurement
Schwebeladung f <Elektrotech> floating charge
Schwebemittel n <Kunststoff> antisettling agent
Schwebemotor m <Kfztech> floating engine
Schweben n <Lufttrans> hovering *(Hubschrauber)*
Schweben n *eines Luftkissenfahrzeugs* <Trans> height hovering
schwebend adj 1. <Elektrotech> floating; 2. <Erdöl> updip *(Geologie)*; 3. <Mechan> floating
schwebender Pfahl m <Kohlen> floating pile
schwebender Schienenstoß m <Bau> suspended joint
schwebender Steueranschluss m <Elektronik> floating gate
schwebendes Grundwasser n <Wasserversorg> perched ground water
schwebendes Tröpfchen n <Umweltschmutz> suspended liquid droplet

Schwebesandglühofen m <Kerntech> fluid-bed furnace
Schwebesteuerung f <Lufttrans> hover control *(Hubschrauber)*
Schwebesystem n <Eisenbahn> suspended system
Schwebeteilchenmessung f <Gerät> measurement of suspended particulate matter *(Luftverunreinigung)*
Schwebevermögen n <Lufttrans> hovering capability *(Hubschrauber)*
Schwebezustand m <Lufttrans> hovering *(Hubschrauber)*
Schwebstoff m <Umweltschmutz> suspended particle
Schwebstofffilter n <Chemtech> air filter
Schwebstofffiltergerät n <Sicherheit> particulate-filter respirator; particulate matter respirator *(Atemschutzgerät)*
Schwebstoffteilchen n <Umweltschmutz> particulate matter
Schwebung f <Akustik, Elektronik, Phys> beat
Schwebungsfrequenz f <Elektronik, Funktech, Phys> beat frequency
Schwebungsfrequenzoszillator m *(BFO)* <Elektronik, Funktech, Phys> beat frequency oscillator *(BFO)*
Schwebungsnull f <Funktech, Telekom> zero beat
Schwebungstondetektor m <Elektronik> beat note detector
Schwefel m <Erdöl, Kunststoff> *(AE)* sulfur, *(BE)* sulphur
• **mit Schwefel behandeln** <Chemie> *(AE)* sulfurize, *(BE)* sulphurize
Schwefel... <Chemie> *(AE)* sulfurous, *(BE)* sulphurous
schwefelarme Kohle f <Kohlen> low-sulfur coal
schwefelarmes Rohöl n <Erdöl> sweet crude
Schwefelbad n <Foto> *(AE)* sulfide toning, *(BE)* sulphide toning
Schwefelchlorid n <Chemie> *(AE)* sulfur chloride, *(BE)* sulphur chloride
Schwefeldioxid n <Chemie, Lebensmittel, Umweltschmutz> *(AE)* sulfur dioxide, *(BE)* sulphur dioxide
Schwefeldioxidreduktion f <Umweltschmutz> *(AE)* sulfur dioxide reduction, *(BE)* sulphur dioxide reduction
Schwefelgehalt m <Erdöl> *(AE)* sulfur content, *(BE)* sulphur content *(Petrochemie)*
schwefelhaltig adj <Chemie> *(AE)* sulfurous, *(BE)* sulphurous
schwefelhaltiger Boden m <Kohlen> *(AE)* sulfide soil, *(BE)* sulphide soil
schwefelhaltiger Brennstoff m <Umweltschmutz> *(AE)* sulfurous combustible, *(BE)* sulphurous combustible
schwefelhaltiges Rohöl n <Erdöl> sour crude
Schwefelhaushalt m <Umweltschmutz> *(AE)* sulfur budget, *(BE)* sulphur budget
Schwefelkohlenstoff m <Chemie> *(AE)* carbon disulfide, *(BE)* carbon disulphide
Schwefelkreislauf m <Umweltschmutz> *(AE)* sulfur cycle, *(BE)* sulphur cycle
Schwefelmonochlorid n <Chemie> *(AE)* disulfur dichloride, *(BE)* disulphur dichloride
Schwefeln n <Chemie> thionation
Schwefeloxid n <Umweltschmutz> *(AE)* sulfur oxide, *(BE)* sulphur oxide
Schwefelrückgewinnungsanlage f <Abfall> *(AE)* sulfur recovery plant, *(BE)* sulphur recovery plant
Schwefelsäure f 1. <Chemie> *(AE)* sulfuric acid, *(BE)* sulphuric acid, vitriolic acid; 2. <Fertig> *(AE)* sulfuric acid, *(BE)* sulphuric acid *(Kunststoffinstallationen)*; 3. <Umweltschmutz> *(AE)* sulfuric acid, *(BE)* sulphuric acid
Schwefelsäurediamid n <Chemie> *(AE)* sulfamide, *(BE)* sulphamide
Schwefelsäureanhydrid n <Umweltschmutz> *(AE)* sulfuric anhydride, *(BE)* sulphuric anhydride
Schwefel-Sepia-Tönung f <Foto> sepia toning

Schwefeltrioxid

Schwefeltrioxid n <Chemie> (AE) sulfur trioxide, (BE) sulphur trioxide
Schwefelung f 1. <Chemie> thionation; 2. <Kohlen> (AE) sulfurization, (BE) sulphurization
Schwefelwasserstoff m 1. <Chemie> (AE) hydrogen sulfide, (BE) hydrogen sulphide; 2. <Lebensmittel> (AE) hydrogen sulfide, (BE) hydrogen sulphide, (AE) sulfurated hydrogen, (BE) sulphurated hydrogen; 3. <Umweltschmutz> (AE) hydrogen sulfide, (BE) hydrogen sulphide
schweflige Säure f <Umweltschmutz> (AE) sulfurous acid, (BE) sulphurous acid
schwefligsaures Salz n <Chemie> (AE) sulfite, (BE) sulphite
Schweif m <Raumfahrt> trail
Schweifsäge f <Maschinen> turning saw
Schweigezone f <Aufnahme> zone of silence
Schweinsleder n <Druck> pigskin
Schweiß... <Fertig, Kerntech, Maschinen, Metall, Werkprüf> welding
Schweißanzug m <Sicherheit> welder's protective clothing
Schweißauftrag m <Fertig> pad (Schweißen)
Schweißbad n <Fertig> puddle; molten pool, pool (Schweißen)
Schweißbarkeit f <Maschinen, Metall> weldability
Schweißbarkeitsversuch m <Werkprüf> weldability test
Schweißbereich m <Kerntech> weld region
Schweißbogen m <Mechan> welding arc
Schweißbrenner m 1. <Bau> acetylene blowpipe, blowpipe, welding blowpipe; 2. <Maschinen> welding torch; 3. <Mechan> welding blow-pipe, welding burner, welding torch; 4. <Thermod> welding torch
Schweißbrille f <Sicherheit> welder's goggles
Schweißbuckel m <Fertig> projection, projection weld
Schweißdraht m 1. <Bau> welding wire; 2. <Maschinen> filler wire, welding wire; 3. <Mechan> filler wire, welding rod, welding wire
Schweißdüse f <Mechan> welding nozzle
Schweißechtheit f <Textil> fastness to perspiration
Schweißelektrode f <Maschinen> welding electrode
schweißen v <Anstrich, Mechan, Papier, Thermod> weld
Schweißen n 1. <Fertig> welding; fusion (Kunststoffinstallationen); 2. <Kunststoff, Maschinen, Mechan, Thermod> welding
Schweißen n von Flicken <Fertig> patching
Schweißerhandschirm m <Bau> welding handshield
Schweißerhandschutz m <Sicherheit> welder's handshield
Schweißerhaube f <Sicherheit> welder's hood (mit automatischer Lichtdurchlässigkeitsanpassung)
Schweißerhelm m <Bau> welding helmet
Schweißerkennzeichen n <Qual> welder identification
Schweißerprüfung f <Qual> welder qualification
Schweißerschutzanzug m <Sicherheit> welder's protective clothing
Schweißerschutzausrüstung f <Sicherheit> welding protective equipment
Schweißerschutzbrille f <Sicherheit> welder's goggles, welding goggles
Schweißerschutzhandschuhe mpl <Sicherheit> welding gloves
Schweißerschutzhelm m <Sicherheit> welder's helmet
Schweißerschutzschild n <Sicherheit> face shield, welder's shield, welding shield
Schweißerschutzvorhang m <Sicherheit> welder's protective curtain
Schweißerschutzvorrichtung f <Sicherheit> welding protective equipment
Schweißervorhang m <Sicherheit> welder's protective curtain

Schweißfehler m <Mechan> welding defect
Schweißflussmittel n <Anstrich> welding flux
Schweißfolge f <Bau> welding sequence
Schweißform f <Mechan> welding die
Schweißglut f <Mechan> welding heat
Schweißgut n <Mechan> welding stock
Schweißgutausbringung f <Fertig> deposition efficiency (Schweißen)
Schweißhitze f <Metall> sweating heat, welding heat
Schweißkonstruktion f <Mechan> weldment
Schweißkrater m <Fertig> crater (Schweißen)
Schweißlage f <Mechan> pass
Schweißleistung f <Mechan> welding capacity
Schweißlinse f <Fertig> nugget (Punktschweißen)
Schweißmaschine f <Mechan> welding machine
Schweißmetall n <Metall> weld metal
Schweißmittel n 1. <Anstrich> welding flux; 2. <Mechan> welding compound
Schweißmuffe f <Fertig> fusion socket (Kunststoffinstallationen)
Schweißnabe f <Mechan> hub
Schweißnaht f 1. <Fertig> weld; 2. <Kerntech> welding seam; 3. <Maschinen> weld seam, welded body seam; 4. <Mechan> edge, seam, weld, welding seam; 5. <Metall, Papier> welded seam; 6. <Thermod> weld; 7. <Verpack> weight-filling machine
Schweißnahtfestigkeit f <Fertig> weld strength
Schweißnahtwurzel f <Bau> root of weld
Schweißofen m <Metall> balling furnace
Schweißpaste f <Maschinen> welding paste
Schweißperle f <Anstrich> weld spatter
Schweißprogramm n <Bau> (AE) welding program, (BE) welding programme
Schweißprozess m <Bau> welding process
Schweißpulver n <Fertig> unionmelt
Schweißrauchabzugsvorrichtung f <Sicherheit> welder's fume extractor
Schweißraupe f 1. <Fertig> bead, deposited metal (Schweißen); 2. <Maschinen, Mechan> bead
Schweißrichtung f <Fertig> hand of welding
Schweißschlacke f <Mechan> welding cinder
Schweißspitze f <Mechan> welding tip
Schweißspritzer m <Anstrich> weld spatter
Schweißstab m <Mechan> welding rod
Schweißstahl m 1. <Bau> wrought iron; 2. <Metall> weld steel
Schweißstelle f <Kerntech, Mechan, Metall> weld
Schweißstrom m <Mechan> welding current
Schweißstromkreis m <Bau, Mechan> welding circuit
Schweißstutzen m <Fertig> fusion spigot (Kunststoffinstallationen)
Schweißtakt m <Bau> welding cycle
Schweißtransformator m <Elektrotech> welding transformer
Schweißung f <Kerntech, Metall, Thermod> weld
Schweißung f an der Baustelle <Kerntech> site weld
Schweißung f vor Ort <Kerntech> site weld
Schweißverbindung f 1. <Fertig> fusion-welded joint (Kunststoffinstallationen); 2. <Maschinen> weld joint
Schweißverfahren n 1. <Bau> welding procedure; 2. <Maschinen> welding process
Schweißverfahrensprüfung f <Qual> welding procedure qualification
Schweißwärme f <Mechan> welding heat
Schweißwurzel f <Mechan> root
Schweißzusatz m <Mechan> filler metal
Schweißzyklus m <Bau> welding cycle
Schweizerische Normenvereinigung f (SNV) <Elektriz> Swiss Standards Association (SSA)

Schwelbrand m <Thermod> (AE) smoldering fire, (BE) smouldering fire
schwelen v <Thermod> (AE) smolder, (BE) smoulder
schwelendes Feuer n <Thermod> (AE) smoldering fire, (BE) smouldering fire
Schwelle f 1. <Akustik> threshold; 2. <Bau> sole piece, threshold; 3. <Comp & DV> threshold; 4. <Eisenbahn> (AE) cross tie, (BE) sleeper, (AE) tie; 5. <Elektronik, Ergon, Telekom> threshold
schwellen v <Wasserversorg> bulk
Schwellenbett n <Eisenbahn> (BE) sleeper-bed, (AE) tie bed
Schwellenbohrmaschine f <Eisenbahn> (BE) sleeper-drilling machine, (AE) tie-drilling machine
Schwellendechselmaschine f <Eisenbahn> (BE) sleeper-adzing machine, (AE) tie-adzing machine
Schwellendetektor m <Telekom> threshold detector
Schwelleneffekt m <Fernseh> threshold effect (Digitalfernsehen)
Schwellenenergie f <Metall, Strahlphys> threshold energy
Schwellenfeld n <Eisenbahn> (BE) distance between sleepers, (AE) distance between ties
Schwellenkennzeichnung f <Lufttrans> runway threshold marking
Schwellenoperation f <Comp & DV> threshold operation
Schwellenschraube f <Eisenbahn> (BE) sleeper screw, (AE) tie screw
Schwellenschraubenausreißgerät n <Eisenbahn> spike puller
Schwellenschraubeneindrehmaschine f <Eisenbahn> (BE) sleeper screwdriver, (AE) tie screwdriver, spike driver
Schwellenspannung f <Elektrotech, Fernseh> threshold voltage
Schwellenstrom m 1. <Elektrotech> threshold current; 2. <Optik> threshold current laser diode (Laserdiode); 3. <Strahlphys, Telekom> threshold current
Schwellenüberschreitungszahl f <Telekom> level crossing rate
Schwellenwert m 1. <Comp & DV, Telekom> threshold; 2. <Elektronik> threshold value
Schwellenwert m **für Anforderungen des Client** <Comp & DV> client's request threshold
Schwellenwertsensor m <Gerät> threshold detector
Schwellenwertsignal n <Gerät> threshold signal
Schwellholz n <Bau> sill plate
Schwellspannung f <Fertig> repeated cycle stress
Schwellwert m <Kontroll, Raumfahrt, Sicherheit> threshold
Schwellwertaudiometrie f <Sicherheit> threshold audiometry
Schwellwertdetektor m <Telekom> threshold detector
Schwellwertoperation f <Elektronik> thresholding
Schwellwertschalter m <Phys> Schmitt trigger
Schwellwertschaltung f <Kontroll> threshold circuit
Schwellwertsensor m <Gerät> threshold detector
Schwellwertsignal n <Elektronik, Gerät> threshold signal
Schwellwertsteuerung f <Kontroll> threshold control
Schwemmkegel m <Wasserversorg> alluvial cone
Schwemmland n <Bau> alluvium
Schwengelbohrloch n <Erdöl> beam well (Fördertechnik)
Schwenkachse f <Maschinen> swivel axis
Schwenkarm m <Maschinen> swinging arm, swivel arm
schwenkbar adj <Mechan> hinged
schwenkbare Rückwand f <Foto> revolving back, swinging back
schwenkbarer Reflektor m <Foto> swivel-mounted reflector

schwenkbares Brückenteil n <Trans> turntable bridge
schwenkbares Grundbrett n <Foto> tilting baseboard
schwenkbares Objektivbrett n <Foto> swing front
Schwenkbereich m <Fertig> swivel range
Schwenkbewegung f 1. <Lufttrans> flapping (Hubschrauber); hunting (Rotor); 2. <Maschinen> swinging movement
Schwenkbrücke f <Bau> swivel bridge
Schwenkdach n <Wassertrans> (AE) swiveling roof, (BE) swivelling roof
schwenken v 1. <Funktech> scan (Richtantenne); 2. <Maschinen> (BE) slew, (AE) slue, swing, swivel; 3. <Mechan> (BE) slew, (AE) slue
Schwenken n 1. <Bau> (BE) slewing, (AE) sluing; 2. <Comp & DV> panning (seitliche Verschiebung der Ansicht auf Bildschirm); 3. <Fernseh> pan and scan
Schwenken n **der Pfeife** <Ker & Glas> swinging of the pipe
Schwenken n **des Luftschraubenblattes** <Lufttrans> hunting blade (Hubschrauber)
Schwenkgelenk n <Lufttrans> drag link
Schwenkgelenk n **des Luftschraubenblattes** <Lufttrans> blade-folding hinge
Schwenkgelenkbegrenzungsstift m <Lufttrans> drag hinge pin (Hubschrauber)
Schwenkgrube f <Ker & Glas> swinging pit
Schwenkhebel m <Maschinen> (AE) swiveling lever, (BE) swivelling lever
Schwenkkörper m <Fertig> tilting yoke (Axialkolbenpumpe)
Schwenkkran m 1. <Bau> (BE) slewing crane, (AE) sluing crane, swing crane; 2. <Kerntech> (BE) slewing crane, (AE) sluing crane
Schwenkmeißelhalter m <Maschinen> swivel toolholder
Schwenkradius m <Bau> swinging round (eines Kranes)
Schwenkrinne f <Bau> swinging chute
Schwenkrolle f <Fertig> caster, castor
Schwenkrotorhubschrauber m <Lufttrans> tilting rotor helicopter
Schwenkscheibe f <Fertig> tilting box (Pumpe)
Schwenkschildplanierraupe f <Bau> angledozer
Schwenkstein m <Ker & Glas> swinging brick
Schwenktisch m <Maschinen> (AE) swiveling table, (BE) swivelling table
Schwenk- und Neigekopf m <Foto> pan-and-tilt head
Schwenkwerk n <Maschinen> slewing gear
schwer entflammbar adj <Sicherheit> flame-resistant
schwer verarbeitbarer Kanalruß m (HPC-Ruß) <Kunststoff> hard-processing channel carbon black (HPC carbon black)
schwer zugänglich adj <Bau> difficult to get at
Schwer... <Kfztech, Phys, Raumfahrt> heavy
Schwerbenzin n <Kfztech> (AE) heavy gasoline, (BE) heavy petrol
Schwerbeton m <Bau> dense concrete, heavy concrete
Schwere f <Kfztech, Phys, Raumfahrt> gravity
schwere Dünung f <Wassertrans> heavy swell (Seezustand)
schwere Fraktionen fpl <Erdöl> heavy fractions (Raffinerietechnik)
schwere Gruppe f <Kerntech> heavy group (von Spaltprodukten)
schwere Kohlenwasserstoff-Fraktionen fpl <Erdöl> heavy hydrocarbon fractions (Raffinerie)
schwere Masse f <Phys> gravitational mass
schwere Schnitte mpl <Erdöl> heavy fractions
schwere See f <Wassertrans> rough sea; heavy seas (Seezustand) • **in schwerer See laufen** <Wassertrans> fight against heavy weather, (AE) labor, (BE) labour, make heavy weather (Schiff)

Schwerefeld

Schwerefeld n <Phys> gravitational field
Schweregradientendrehmoment n <Raumfahrt> gravity gradient torque
Schweregradientenspitze f <Raumfahrt> gravity gradient boom
Schweregradientenstabilisierung f <Raumfahrt> gravity gradient stabilization
schwereloses Schweben n <Trans> air cushion levitation
Schwerelosigkeit f <Raumfahrt> weightlessness
Schwerentflammbarkeit f <Kunststoff> flame resistance
schwerer Hilfskran m <Eisenbahn> heavy breakdown crane
schwerer Holzhammer m <Bau> maul
schwerer Wasserstoff m <Chemie> deuterium
schwerer Zusammenstoß m <Trans> primary collision
schweres Heizöl n <Erdöl> residual fuel oil *(Destillationsprodukt)*
schweres Rohöl n <Erdöl> heavy crude, heavy crude oil
schweres Wasser n <Phys> heavy water
schweres Wasser n D_2O <Kerntech> heavy water *(Deuteriumoxid)*
schweres Wetter n <Wassertrans> heavy weather
Schwerewellen fpl 1. <Phys> gravitational waves, gravity waves; 2. <Strahlphys> gravitational waves
Schwerflintglas n <Ker & Glas> dense flint
Schwerflüssigkeit f <Kohlen> dense liquid
Schwerflüssigkeitsprüfung f <Kohlen> heavy-liquid test
Schwergutbaum m <Wassertrans> jumbo derrick *(Ladung)*
Schwerhörigengerät n <Akustik> hearing aid
Schwerionenfusion f <Kerntech> heavy ion fusion
Schwerionensynchrotron n *(SIS)* <Teilphys> heavy-ion synchrotron *(HIS)*
Schwerkraft f 1. <Bau, Maschinen, Papier, Phys, Raumfahrt> gravity; 2. <Umweltschmutz> gravitational force
Schwerkraftabscheider m <Abfall> gravity separator
Schwerkraftfluss m <Kerntech> gravity flow
Schwerkraftförderer m 1. <Maschinen> gravity conveyor; 2. <Verpack> conveyor way
Schwerkraftfüllmaschine f <Verpack> gravity filling machine
Schwerkraftfüllung f <Raumfahrt> gravity filling *(Raumschiff)*
Schwerkraftgießen n <Fertig> gravity casting
Schwerkraftguss m <Fertig> nonpressure casting
Schwerkraftregelstab m <Kerntech> gravity drop absorber rod
Schwerkraftsumlauf m <Heiz & Kälte> gravity circulation
Schwerkrafttrenner m <Abfall> gravity separator
Schwerkraftvakuum-Verkehrszug m <Eisenbahn> gravity vacuum transit train
Schwerkraftzusammenbruch m <Raumfahrt> gravitation collapse
Schwerkronglas n <Ker & Glas> dense crown
Schwerlast f <Kfztech, Raumfahrt> heavy-lift
Schwerlastfahrzeug n <Kfztech> *(BE)* heavy lorry, *(AE)* heavy motor truck, *(AE)* heavy truck
Schwerlastfahrzeugaufzug m <Trans> *(AE)* heavy-vehicle elevator, *(BE)* heavy-vehicle lift
Schwerlasthubschrauber m <Lufttrans> crane helicopter, heavy-lift helicopter
Schwerlastkran m 1. <Bau> goliath crane; 2. <Fertig> heavy-lift crane
Schwerlastraumträger m <Raumfahrt> heavy-lift vehicle
Schwerlastträgerrakete f *(SL-Rakete)* <Raumfahrt> heavy-lift launch vehicle, HLLV *(zum Starten)*
Schwerlastverkehr m <Trans> HGV traffic, heavy goods vehicle traffic

Schwermetall n <Kohlen, Metall, Strahlphys, Umweltschmutz> heavy metal
Schwermetall-Differenzenmethode f <Kerntech> heavy-metal difference technique
Schwerölentschwefelung f <Abfall> *(AE)* heavy-oil desulfurization, *(BE)* heavy-oil desulphurization
Schwerölmotor m <Kfztech> heavy-oil engine
Schwerölrückstand m <Abfall> heavy-oil residue
Schwerprofil n <Metall> heavy section
Schwerpunkt m 1. <Bau> *(AE)* center of gravity, *(BE)* centre of gravity; 2. <Fertig> centroid; 3. <Lufttrans, Mechan, Nichtfoss Energ, Phys, Raumfahrt> *(AE)* center of gravity, *(BE)* centre of gravity; 4. <Teilphys> *(AE)* center of mass, *(BE)* centre of mass
Schwerpunkt m **der Wasserlinienfläche** <Wassertrans> *(AE)* center of waterplane, *(BE)* centre of waterplane *(Schiffkonstruktion)*
Schwerpunkt m **des Luftschraubenblattes** <Lufttrans> blade balance
Schwerpunktenergie f <Teilphys> collision energy
Schwerpunktsystem n <Mechan, Phys> CMS, *(AE)* center-of-mass system, *(BE)* centre-of-mass system
Schwerspannstift m <Fertig> spring cotter *(Kunststoffinstallationen)*
Schwerspat m 1. <Erdöl> baryte *(Mineralogie)*; 2. <Kunststoff> baryte
Schwerstange f <Erdöl> drill collar *(Bohrtechnik)*
Schwerstoff m <Kohlen> dense medium
Schwert n <Wassertrans> *(AE)* centerboard, *(BE)* centreboard *(Schiffbau)*
Schwertfalzmaschine f <Druck> blade folder
Schwerwasser n <Kerntech> heavy water *(Deuteriumoxid)*
Schwerwasseranlage f <Kerntech> heavy-water plant
Schwerwasserdampf m <Kerntech> *(AE)* heavy-water vapor, *(BE)* heavy-water vapour
Schwerwasserentgaser m <Kerntech> heavy-water degasifier
schwerwassermoderierter Reaktor m *(SWR)* <Kerntech> heavy-water-moderated reactor *(HWR)*
schwerwassermoderierter Uranreaktor m <Kerntech> uranium heavy-water reactor
Schwerwasserreaktor m <Kerntech> heavy-water reactor
Schwerwassersprühdüse f <Kerntech> heavy-water spray nozzle
schwerwiegender Fehler m <Comp & DV> fatal error
Schwesterschiff n <Wassertrans> sister ship
Schwimmaufbereitungsanlage f <Abfall> flotation plant
Schwimmbadreaktor m <Kerntech> underwater reactor
Schwimmbagger m 1. <Bau> dredge, grab dredger; 2. <Meerschmutz> dredger; 3. <Wassertrans> dredge, dredger *(Baggern)*
Schwimmbohranlage f <Erdöl> floating rig *(Bohrtechnik)*
Schwimmbrücke f <Wassertrans> floating bridge
Schwimmdock n <Bau, Wassertrans> floating dock
Schwimmebene f <Wassertrans> waterplane *(Schiffkonstruktion)*
Schwimmen n <Wassertrans> flotation
schwimmend adj 1. <Elektrotech, Maschinen, Mechan> floating; 2. <Wassertrans> afloat
schwimmende Gründung f <Bau> floating foundation
schwimmende Landungsbrücke f <Wassertrans> landing stage
schwimmende Rückstände mpl <Abfall> floating refuse
schwimmende Umschlagsanlage f <Wassertrans> offshore floating terminal
schwimmender Gleiskörper m <Eisenbahn> floating slab track

schwimmender Kunststoffbehälter *m* <Meerschmutz> dracone
schwimmender Pfahl *m* <Bau> friction pile
schwimmender Wellenbrecher *m* <Bau> floating breakwater
schwimmendes Gebirge *n* <Kohlen> running ground
Schwimmer *m* 1. <Ker & Glas> floater; 2. <Kfztech> float; *(AE)* carburetor float, *(BE)* carburettor float *(Vergaser)*; 3. <Kohlen, Maschinen, Mechan, Wassertrans, Wasserversorg> float
Schwimmerdichtemesser *m* <Gerät> float-type densitometer
Schwimmerfahrwerk *n* <Lufttrans> floating gear *(Hubschrauber)*
Schwimmerflugzeug *n* 1. <Lufttrans> float seaplane; 2. <Trans> floatplane; 3. <Wassertrans> pontoon
Schwimmerfüllstandsmesser *m* <Gerät> float-operated level meter
Schwimmerfüllstandsmessgerät *n* <Gerät> buoyancy probe
Schwimmergehäuse *n* <Kfztech> *(AE)* carburetor float chamber, *(BE)* carburettor float chamber, float chamber *(Vergaser)*
schwimmergeregelte Alarmpfeife *f* <Hydraul> float-controlled alarm whistle
Schwimmerhahn *m* <Maschinen> ball cock
Schwimmerkammer *f* <Kfztech> *(AE)* carburetor float chamber, *(BE)* carburettor float chamber, float chamber *(Vergaser)*
Schwimmermesser *m* <Erdöl> rotameter
Schwimmernadel *f* <Kfztech> float needle
Schwimmeröffnung *f* <Ker & Glas> floater notcher
Schwimmerschalter *m* <Elektrotech, Heiz & Kälte> float switch
Schwimmerstütze *f* <Ker & Glas> floater lug
Schwimmerventil *n* 1. <Heiz & Kälte> float valve; 2. <Hydraul> float trap; 3. <Maschinen> float valve
schwimmfähig *adj* <Wassertrans> buoyant
Schwimmfähigkeit *f* <Wassertrans> buoyancy *(Schiff)*
Schwimmhaut *f* <Meerschmutz> webbing
Schwimmkammer *f* <Wassertrans> buoyancy tank *(Schiff)*
Schwimmkompass *m* <Wassertrans> liquid compass
Schwimmkragen *m* <Raumfahrt> flotation collar *(aufblasbar zur Wasserung)*
Schwimmkran *m* <Wassertrans> floating crane, pontoon crane
Schwimmring *m* <Erdöl> flotation collar *(Offshore-Technik)*
Schwimmringdichtung *f* <Maschinen> floating-ring oil seal
Schwimmsattel *m* <Kfztech> *(AE)* floating caliper, *(BE)* floating calliper
Schwimmsattelscheibenbremse *f* <Kfztech> *(AE)* floating caliper disk brake, *(BE)* floating calliper disc brake
Schwimmschicht *f* <Chemie> scum *(Wasser)*
Schwimmschirm *m* <Umweltschmutz> floating boom
Schwimmstoff *m* <Abfall> floating matter
Schwimmtank *m* <Wassertrans> flotation tank
Schwimm- und Sinkanalyse *f* <Kohlen> float-and-sink analysis
Schwimmweste *f* 1. <Bau> *(BE)* life belt, *(BE)* life jacket, *(AE)* life preserver, *(BE)* life vest; 2. <Lufttrans, Sicherheit, Wassertrans> *(BE)* life jacket, *(AE)* life preserver, *(BE)* life vest *(Seenot)*
schwinden *v* 1. <Aufnahme> fade; 2. <Bau> contract; 3. <Ker & Glas> shrink
Schwinden *n* <Bau, Metall, Telekom> shrinkage, shrinking

Schwindfuge *f* <Bau> contraction joint
Schwindmaß *n* 1. <Fertig> shrinkage; 2. <Maschinen> shrink-age, shrinkage allowance
Schwindmaßstab *m* 1. <Fertig> contraction rule, shrink rule; 2. <Qual> contraction rule
Schwindriss *m* <Bau> shrinkage crack
Schwindschutzzusatz *m* <Bau> antishrinkage admixture
Schwindung *f* 1. <Hydraul> coefficient of contraction; 2. <Ker & Glas> shrinkage; 3. <Kunststoff> *(AE)* mold shrinkage, *(BE)* mould shrinkage; 4. <Qual> coefficient of contraction
Schwindungshohlraum *m* <Fertig> shrink hole
Schwingachse *f* <Kfztech> swing axle
Schwingarm *m* 1. <Fertig> pawl arm; 2. <Kfztech> suspension arm
Schwingbalken *m* <Maschinen> walking beam
Schwingbeanspruchung *f* <Werkprüf> dynamic load
Schwingbereichsänderung *f* <Elektronik> moding *(Magnetron)*
Schwingdrossel *f* <Elektrotech> swinging choke
Schwinge *f* <Fertig> rocker; lever, oscillating link, rocker arm *(Getriebelehre)*
schwingen *v* 1. <Bau> sway; 2. <Elektriz, Elektronik> hunt, oscillate; 3. <Fertig> surge; 4. <Maschinen> swing; 5. <Wellphys> oscillate
Schwingen *n* 1. <Akustik> oscillation; 2. <Elektriz> hunting; 3. <Maschinen> swinging; 4. <Papier> swing
schwingend *adj* 1. <Aufnahme> rocking; 2. <Elektriz, Elektronik> oscillating, hunting *(Strom)*; 3. <Papier> oscillating; 4. <Telekom> oscillatory
schwingende Abtastung *f* <Fernseh> oscillatory scanning
schwingende Kugelmühle *f* <Labor> vibrating ball mill *(Schleifen)*
schwingende Saite *f* <Phys> vibrating string
schwingender Kondensator *m* <Elektriz> oscillating capacitor
schwingender Stromkreis *m* <Elektriz> oscillating circuit
schwingendes Filter *n* <Elektronik> free-bar filter
schwingendes Förderband *n* <Sicherheit> oscillating conveyor
schwingendes System *n* <Elektronik> oscillating system, oscillatory system
schwingendes Werkzeug *n* <Sicherheit> oscillatory tool
Schwinger *m* <Telekom> resonator
Schwingfestigkeit *f* <Fertig> dynamic strength
Schwingfestigkeitsprüfung *f* <Werkprüf> fatigue test
Schwingförderer *m* 1. <Lebensmittel> vibration conveyor; 2. <Maschinen> oscillating conveyor, vibrating conveyor; 3. <Mechan> push-bar conveyor
Schwinggröße *f* <Elektronik> oscillating quantity
Schwingkreis *m* 1. <Elektronik> resonating circuit; oscillating circuit *(des Oszillators)*; 2. <Funktech> oscillating circuit, oscillator circuit, resonant circuit; 3. <Telekom> resonant circuit
Schwingkristallmethode *f* <Strahlphys> oscillating crystal method
Schwingloch *n* <Gerät> dead spot
Schwingmetallfeder *f* <Maschinen> rubber-metal spring
Schwingmühle *f* <Labor> vibrating ball mill *(Schleifen)*
Schwingproben-Magnetometer *n* <Kerntech> vibrating sample magnetometer
Schwingquarz *m* 1. <Elektriz, Elektronik> crystal oscillator, quarz crystal resonator, quartz oscillator; 2. <Phys> quartz crystal oscillator
Schwingquarzfilter *n* <Elektronik> quartz crystal filter
Schwingrahmen *m* <Maschinen> swing frame
Schwingrakel *f* <Papier> oscillating doctor

Schwingschalter

Schwingschalter m <Elektrotech> rocker switch
Schwingschärfe f <Regelung> vibrational severity
Schwingschleifen n <Fertig> superfinish grinding
Schwingsieb n <Kohlen> swing sieve, vibrating screen
Schwingsitz m <Sicherheit> vibration-damping seat
Schwingspeiser mpl **und Schwingförderer** mpl <Maschinen> vibrating feeders and conveyors
Schwingspule f <Elektrotech> moving coil, oscillator coil
Schwingspulenlautsprecher m <Aufnahme> moving-coil loudspeaker
Schwingstärke f <Nichtfoss Energ> amplitude of vibration
Schwingtisch m 1. <Ker & Glas> rocking table; 2. <Sicherheit> vibration exciter
Schwingung f 1. <Akustik> cycle, oscillation; 2. <Bau, Elektriz, Elektronik> oscillation; 3. <Fertig> jar; 4. <Funktech> oscillation; 5. <Maschinen> oscillation *(eines Pendels)*; 6. <Mechan, Metall> vibration; 7. <Phys> oscillation, vibration; 8. <Telekom, Wellphys> oscillation
Schwingungsachse f <Mechan, Phys> axis of oscillation
Schwingungsamplitude f <Maschinen> oscillation amplitude
Schwingungsanalyse f <Mechan> vibration analysis
Schwingungsart f <Akustik> mode
Schwingungsausgleich m <Aufnahme, Elektronik> frequency compensation
Schwingungsbauch m 1. <Akustik> antinode, antinode of oscillation, antinode of vibration, loop; 2. <Elektriz, Elektrotech> antinode; 3. <Fertig> antinode, crest, vibration antinode
Schwingungsbauch m **einer stehenden Welle** <Wellphys> antinode of a stationary wave
Schwingungsbreite f <Akustik> total oscillation amplitude
Schwingungsbruch m <Werkprüf> fatigue fracture
schwingungsdämpfender Stoff m <Sicherheit> vibration-attenuating material
Schwingungsdämpfer m 1. <Heiz & Kälte> vibration damper; 2. <Kfztech> damper, resonance damper; vibration damper *(Motor)*; 3. <Maschinen> vibration damper
Schwingungsdauer f 1. <Elektronik> oscillation period; 2. <Maschinen> period of oscillation
Schwingungsenergie f <Kerntech> vibrational energy
Schwingungsentropie f <Metall> vibrational entropy
Schwingungserreger m <Sicherheit> vibration exciter
Schwingungserzeuger m 1. <Elektriz, Elektronik, Funktech> oscillator; 2. <Maschinen> vibrator; 3. <Phys> oscillator; 4. <Raumfahrt> oscillator *(Weltraumfunk)*; 5. <Telekom, Wellphys> oscillator
schwingungsfähiges System n <Wellphys> vibrating system
Schwingungsfestigkeit f <Maschinen> fatigue strength
Schwingungsform f <Elektronik> oscillation mode *(Wellenform)*
schwingungsfreie Befestigung f <Heiz & Kälte> antivibration mounting
schwingungsfreier Tisch m <Labor> antivibration table
Schwingungsfrequenz f <Elektronik, Telekom> oscillation frequency
Schwingungsgeschwindigkeit f <Heiz & Kälte> velocity of vibration
Schwingungsgesetze npl <Wellphys> laws of vibration *(einer eingespannten Saite)*
schwingungsisolierter Griff m <Ergon, Sicherheit> vibration-isolating handle
Schwingungsknoten m <Elektronik> node of oscillation
Schwingungslinie f <Akustik> antinodal line
Schwingungsmesseinrichtung f <Gerät> vibration-measuring equipment *(mechanisch)*
Schwingungsmessgerät n <Phys> vibrometer
Schwingungsmessung f <Heiz & Kälte> vibration measurement
Schwingungsmittelpunkt m <Mechan, Phys> *(AE)* center of oscillation, *(BE)* centre of oscillation
Schwingungsperiode f <Elektriz, Elektrotech> period of oscillation, cycle
Schwingungsquantenzahl f <Phys> vibrational quantum number
Schwingungs-Rotations-Spektrum n <Phys> vibration rotation spectrum
Schwingungsrührwerk n <Labor> vibrating stirrer
Schwingungsschutzhandschuhe mpl <Sicherheit> protective anti-vibration gloves, vibration-absorbing gloves
schwingungssicher adj <Sicherheit> antivibrating, vibration-proof
schwingungssicheres Werkzeug n <Sicherheit> antivibration tool
Schwingungsspektrum n <Phys> vibrational spectrum
Schwingungstechnik f <Sicherheit> vibration engineering
Schwingungstest m <Metrol> vibration test
Schwingungsverlauf m <Comp & DV> waveform
Schwingungsversuch m <Werkprüf> vibration test
Schwingungsweite f 1. <Maschinen> oscillation amplitude; 2. <Nichtfoss Energ> amplitude of vibration
Schwingungszahl f 1. <Akustik, Aufnahme, Comp & DV, Elektronik> frequency; 2. <Fertig> oscillation frequency; 3. <Funktech, Phys> frequency
Schwingverhalten n <Bau> sway
Schwingverschluss m <Ker & Glas> *(BE)* swing stopper finish
Schwingwelle f <Maschinen> rock shaft
Schwingzapfen m <Fertig> solenoid *(Kunststoffinstallationen)*
schwitzen v 1. <Lebensmittel> sweat *(Gießerei)*; 2. <Lebensmittel> sweat *(Ofen)*
Schwitzen n <Heiz & Kälte> sweating
Schwitzkühlung f <Heiz & Kälte> transpiration cooling
Schwitzung f <Hydraul> transpiration
Schwitzwasser n 1. <Heiz & Kälte> condensate; 2. <Wassertrans> sweat
schwojen v <Wassertrans> swing *(Schiff)*
Schwund m <Akustik, Funktech, Telekom> fade, fading
Schwund m **durch Mehrwegeausbreitung** <Funktech, Telekom> multipath fading
Schwundausfall m <Funktech> fading outage, radio fade-out
Schwundausgleich m <Funktech> automatic volume control *(AVC)*
Schwundreserve f <Telekom> link margins
Schwundtiefe f <Funktech> fade depth
Schwungkraftanlasser m <Mechan> inertial starter
Schwungkugel f <Maschinen> ball
Schwungkugelregler m <Maschinen> ball governor
Schwungmasse f 1. <Fertig> gyrating mass; 2. <Heiz & Kälte> flywheel; 3. <Kerntech> centrifugal mass, gyrating mass
Schwungrad n 1. <Heiz & Kälte> flywheel; 2. <Kfztech> flywheel *(Motor)*; 3. <Maschinen, Mechan> flywheel; 4. <Raumfahrt> momentum wheel
Schwungrad n **zur Magnetpeilung** <Raumfahrt> magnetic-bearing momentum wheel
Schwungradeffekt m <Funktech> flywheel effect
Schwungradgehäuse n <Kfztech> flywheel housing *(Motor)*
Schwungradspeicherung f <Nichtfoss Energ> flywheel storage
Schwungradzahnkranz m <Kfztech> flywheel starter ring gear *(Motor)*

Schwungring m <Heiz & Kälte> flywheel
Schwungscheibe f 1. <Heiz & Kälte> flywheel; 2. <Maschinen> inertia reel
SCN (Hinweis auf Spezifikationsänderungen) <Trans> SCN (specification change notice)
Scoop n <Labor> scanning electron microscope
Scopolamin n <Chemie> hyoscine
Scorch n <Kunststoff> scorch
Scorchneigung f <Kunststoff> scorching tendency
Scott-Schaltung f <Elektriz> Scott connection
SCPC (Ein-Kanal-pro-Träger) 1. <Raumfahrt> SCPC, single channel per carrier (Weltraumfunk); 2. <Telekom> SCPC, single channel per carrier
SCPC-System n **mit bedarfsorientierter Zuteilung** <Telekom> demand-assigned single-channel-per-carrier system
SCR (siliziumgesteuerter Gleichrichter) <Elektronik, Elektrotech> SCR (silicon-controlled rectifier)
SCR-Abstimmtransformator m <Elektrotech> SCR trimmer transformer
Scram m <Kerntech> scram
scrambeln v <Telekom> scramble
Scrambler m 1. <Raumfahrt> scrambler (Verschlüsselungsgerät); 2. <Telekom> scrambler
Scraper m <Meerschmutz> scraper
Screening n <Chemie, Elektronik> screening
SCR-geregeltes Netzgerät n <Elektrotech> SCR-regulated power supply (durch Thyristor geregelt)
Script n <Comp & DV> script
Scriptwriter m <Fernseh> scriptwriter
SCR-Konverter m <Elektrotech> SCR converter
SCR-Regelung f <Elektrotech> SCR regulation
SCR-Regler m <Elektrotech> SCR regulator
SCR-Verstärker m <Elektronik> SCR amplifier
SCR-Vorregelung f <Elektrotech> SCR preregulation
SCR-Vorregler m <Elektrotech> SCR preregulator
SCR-Wandler m <Elektrotech> SCR converter
SDLC (synchrone Datenübertragungssteuerung) <Comp & DV, Telekom> SDLC (synchronous data link control)
SDLC-Verfahren n <Comp & DV, Telekom> synchronous data link control (SDLC)
S-Draht m <Textil> S-twist
S-Drehung f <Textil> S-twist
Se (Selen, Selenium) <Chemie> Se (selenium)
Seabee-Schiff n <Wassertrans> Seabee-type vessel (Trägerschiff)
Seabeeträgerschiff n <Wassertrans> seabee carrier
SEALOSAFE-Verfahren n (Verfestigungsverfahren für Sonderabfälle) <Abfall> SEALOSAFE process
Sebacin... <Chemie> sebacic
sec 1. <Geom> (Sekans) sec, secant; 2. <Geom> (Sekante) sec, secant (von Linie)
SECAM-Farbfernsehsystem n <Fernseh> SECAM system
SECAM-System n <Fernseh> SECAM system
Sechs f <Math> six
Sechseck n <Geom> hexagon
sechseckig adj <Geom, Kerntech> hexagonal
Sechseck-Maschengitter n <Ker & Glas> hexagonal mesh (in Drahtglas)
Sechsfachrevolverkopf m <Maschinen> six-tool capstan
sechsflächig adj <Geom, Kerntech> hexahedral
Sechsflächner m <Geom> hexahedron
Sechskant m <Maschinen> hexagon • **Sechskant fräsen** <Fertig> hex
Sechskantkopf m 1. <Fertig> hexagon head; 2. <Maschinen> hex head, hexagon head, hexagonal head
Sechskantmutter f 1. <Maschinen> hex nut, hexagon nut, hexagonal nut; 2. <Mechan> hex nut

Sechskantmutterkopf m <Mechan> hexagon head
Sechskantschlüssel m <Maschinen> (BE) spanner for hexagon nuts, wrench for hexagon nuts
Sechskantschneidmutter f <Maschinen> hexagonal die nut
Sechskantschraube f 1. <Maschinen> hex bolt, hexagon bolt, hexagonal head bolt; 2. <Mechan> hexagon head screw
Sechskantschraubenschlüssel m <Mechan> hex head wrench
Sechsphasengleichrichter m <Elektrotech> six-phase rectifier
Sechsphasenspannung f <Elektriz> hexagon voltage
Sechsphasenstrom m <Elektrotech> six-phase current
Sechsspindelhalbautomat m <Fertig> six-spindle automatic screw machine
Sechstonleiter f <Akustik> hexatonic scale
sechswertig adj <Chemie> hexavalent
sechszählig adj <Chemie> hexad
Sechszylinder-V-Motor m <Kfztech> V-six engine
Sedezformat n <Druck> sixteenmo
Sediment n 1. <Bau, Erdöl> sediment (Geologie); 2. <Kohlen, Wasserversorg> sediment
Sedimentablagerung f 1. <Umweltschmutz> sedimentation; deposition (Geologie); 2. <Wasserversorg> sediment discharge
Sedimentation f 1. <Erdöl> sedimentation (Geologie); 2. <Kunststoff, Phys, Wasserversorg> sedimentation
Sedimentationsanalyse f <Kohlen> sedimentation analysis
Sedimentationsbecken n 1. <Erdöl> sedimentary basin (Geologie); 2. <Wasserversorg> sedimentation tank
Sedimentationsbehälter m <Chemtech> settling tank
Sedimentationsdauer f <Chemtech> settling time
Sedimentationsgeschwindigkeit f <Erdöl> sedimentation rate (Geologie)
Sedimentationspotenzial n <Chemtech> sedimentation potential
Sedimentbecken n <Erdöl> sedimentary basin
Sedimentboden m <Kohlen> sedimentary soil
Sedimentgestein n 1. <Bau, Erdöl> sedimentary rock (Geologie); 2. <Kohlen> sedimentary rock
sedimentieren v <Chemtech> settle
Sedimentiergefäß n <Chemtech> settling cone
Sedimentiergefäß n **nach Imhoff** <Labor> Imhoff sedimentation cone
Sedimentzuwachs m <Wasserversorg> aggradation
See f <Wassertrans> sea (Salzwasser) • **auf offener See** <Wassertrans> in the offing • **auf See** <Wassertrans> under way • **in See** <Wassertrans> under way • **von der See getragen** <Wassertrans> sea-borne • **zur See fahren** <Wassertrans> navigate • **zur See fahrend** <Wassertrans> seafaring
See m <Wassertrans> lake (Süßwasser)
See... <Wassertrans> marine, naval
Seeanker m <Wassertrans> drogue (Schleppsack); sea anchor (Schleppwiderstand)
Seeaufklärungsradar n <Wassertrans> surface search radar (Marine)
Seebeck-Effekt m 1. <Elektriz, Elektrotech> Seebeck effect (thermoelektrische Wirkung); 2. <Phys> Seebeck effect
Seebeck-Koeffizient m <Phys> Seebeck coefficient
seebeschädigt adj <Wassertrans> sea-damaged
Seecontainer m <Wassertrans> overseas container
Seedeich m 1. <Wassertrans> sea wall; 2. <Wasserversorg> sea dike
Seedrift f <Wassertrans> flotsam
See-Eigenschaften fpl <Wassertrans> seakeeping qualities

See-Erprobung

See-Erprobung f <Wassertrans> sea trials *(Schiffbau)*
Seefähigkeitszeugnis n <Wassertrans> certificate of seaworthiness *(Dokumente)*
Seefernmeldekabel n <Telekom> submarine telecommunication cable
Seefracht f <Wassertrans> shipment *(Ladung)*
Seefunkfeuer n 1. <Funktech> maritime radio beacon; 2. <Wassertrans> maritime radio beacon *(Seezeichen)*
Seefunkgespräch n 1. <Funktech> marine radio call, ship radio call; 2. <Telekom> marine call
Seefunknavigation f <Funktech> maritime radio navigation
Seefunksatellit m <Funktech, Telekom> maritime satellite
Seefunkstation f <Funktech> maritime radio station, naval radio station
Seefunktelefonat n <Telekom> ship radio call
Seefunkvermittlungsstelle f <Telekom> MSC, *(AE)* maritime switching center, *(BE)* maritime switching centre
Seegang m <Wassertrans> sea state
Seegangsecho n <Funkort> sea clutter *(Radar)*
Seegangsreflexe mpl <Wassertrans> sea clutter *(Radar)*
Seegangstrübung f <Wassertrans> sea clutter *(Radar)*
Seegerring m <Kfztech> circlip
Seegersicherung f <Fertig> circlip
Seegersicherungsring m <Maschinen> circlip ring
Seehafen m <Wassertrans> seaport, trading port; *(AE)* deep-water harbor, *(BE)* deep-water harbour *(Hafen)*
Seehandbuch n <Wassertrans> sailing directions
Seehöhe f <Wassertrans> sea level
Seehorizont m <Wassertrans> sea line
Seekabel n <Telekom> ocean cable, sea cable, submarine cable
Seekabelübertragung f <Telekom> submarine telecommunications
Seekabelverstärker m <Telekom> submerged repeater
Seekarte f 1. <Trans> sea chart; 2. <Wassertrans> hydrographic chart *(Navigation)*
Seeküste f <Wassertrans> seaboard
Seele f 1. <Bau> core; 2. <Elektriz, Elektrotech> core *(eines Drahtseils oder elektrischen Kabels)*; 3. <Fertig> core *(Kabel)*; 4. <Optik> central load-bearing element *(Lichtleiter)*; 5. <Fertig> core *(Kabel)*; 6. <Wassertrans> core *(Tauwerk)*
Seeleichter m <Wassertrans> seagoing barge
Seelotse m <Wassertrans> sea pilot
Seemannsamt n <Wassertrans> shipping office *(Behörde)*
Seemaschinist m <Wassertrans> marine engineer
seemäßig adj <Wassertrans> seaworthy
seemäßig verpackt adj <Verpack> seaworthy packaging
Seemeile f 1. <Metrol> mile; 2. <Wassertrans> nautical mile
Seenavigationsfunkdienst m **über Satelliten** <Funktech, Telekom> maritime radio navigation satellite service
Seenot f <Wassertrans> distress at sea *(Notfall)*
Seenotalarm m 1. <Telekom> distress alerting; 2. <Wassertrans> distress alert *(Notfall)*
Seenot-Funkbake f <Funkort> search and rescue beacon equipment
Seenotfunkbake f **mit Positionsmeldung** <Funktech, Telekom, Wassertrans> emergency position-indicating radio beacon *(EPIRB)*
Seenotrettungsboot n <Wassertrans> rescue boat
Seenotrettungskreuzer m <Wassertrans> *(BE)* all-weather lifeboat
Seenotrettungsleitstelle f <Lufttrans, Wassertrans> RCC, *(AE)* rescue coordination center, *(BE)* rescue coordination centre
Seenotruf m <Telekom, Wassertrans> distress signal *(Notfall)*

Seenotrufsender m <Funktech, Wassertrans> distress call transmitter, ship emergency transmitter
Seenotsender m **mit kleiner Leistung** <Funktech> low-power distress transmitter
Seenotwelle f <Funktech, Telekom> distress wavelength
Seenotwellenlänge f <Funktech, Telekom> distress wavelength
Seenotzeichen n <Telekom, Wassertrans> distress signal *(Notfall)*
Seeradar n <Wassertrans> marine radar *(Radar)*
Seeradarfrequenz f <Wassertrans> marine radar frequency *(Radar)*
Seereling f <Wassertrans> guardrail *(Deckausrüstung)*
Seeschiff n <Meerschmutz, Wassertrans> sea vessel, seagoing vessel
Seeschifffahrt f <Wassertrans> ocean navigation, sea transport
Seeschleuse f <Wassertrans> sea lock
Seestraße f <Wassertrans> sea lane
Seetransport m <Wassertrans> sea carriage, sea transport
seetüchtig adj <Wassertrans> seagoing, seaworthy; navigable *(Schiff)*
seetüchtiges Luftkissenfahrzeug n <Wassertrans> seagoing hovercraft
Seetüchtigkeit f <Meerschmutz, Wassertrans> seaworthiness
seeuntüchtig adj <Wassertrans> unseaworthy *(Schiff)*
Seeventil n <Wassertrans> Kingston valve *(Schiff)*
Seeverhältnisse npl <Wassertrans> sea conditions
Seeverkehr m <Wassertrans> sea trade
Seeverklappung f <Abfall> ocean dumping
Seeverladung f <Verpack> carriage by sea
Seevermessungsschiff n <Wassertrans> geodesic survey ship, hydrographic survey vessel *(Meereskunde)*
Seewarte f <Wassertrans> hydrographic office
seewärtige Verbindung f <Wassertrans> maritime communication
seewärts adv <Wassertrans> seawards
Seewasser n 1. <Umweltschmutz> lake water *(Süßwasser)*; 2. <Wassertrans> seawater
Seewasser-in-Rohöl-Emulsion f <Umweltschmutz> seawater-in-crude-oil emulsion
Seewassertemperatur f <Wassertrans> sea temperature
Seewassertemperatur f **an der Oberfläche** <Wassertrans> sea surface temperature
Seewasserwichte f <Wassertrans> specific gravity of seawater *(Schiffkonstruktion)*
Seeweg m <Wassertrans> sea lane, sea route
Seewind m <Wassertrans> onshore wind
Seewurf m <Wassertrans> jetsam
Seezeichen n <Wassertrans> mark, navigation mark, seamark
Seezeichentender m <Wassertrans> buoy tender *(Seezeichen)*
Seezustand m <Wassertrans> sea state
Segel n <Wassertrans> sail • **die Segel setzen** <Wassertrans> set sail
Segelboden m <Wassertrans> sail loft
Segelboot n <Wassertrans> *(AE)* sailboat, *(BE)* sailing boat
Segelfläche f <Wassertrans> sail area
Segelflugzeug n <Lufttrans> glider
Segeljolle f <Wassertrans> sailing dinghy
Segelkammer f <Wassertrans> sail locker
Segellatte f <Wassertrans> batten
Segelleinwand f <Textil> canvas
Segelmacher m <Wassertrans> sail-maker
Segelmacherwerkstatt f <Wassertrans> sail loft *(Segeln)*

segeln v <Wassertrans> sail
Segeln n <Wassertrans> sailing
Segelriss m <Wassertrans> sail plan
Segelschiff n <Wassertrans> sailing ship
Segelsport m <Wassertrans> sailing
Segelstellung bringen v/auf <Lufttrans> feather *(Luftschraube)*
Segelstellung f **der Luftschraube** <Lufttrans> feathered propeller
Segelstellung f **der Luftschraubenblätter** <Lufttrans> feathered pitch *(Luftschraube)*
Segelstellungswinkel m <Lufttrans> feathering angle
Segelstellungswirkung f <Lufttrans> feathering effect
Segeltuch n 1. <Bau, Textil> canvas; 2. <Wassertrans> sailcloth; canvas *(Segeln)*
Segeltuchplane f <Bau> tarpaulin
Segelyacht f <Wassertrans> keel boat
Segerkegel m 1. <Fertig> Seger cone, fusion cone, pyrometric cone; 2. <Ker & Glas> Seger cone
Segler m <Wassertrans> sailing ship *(Schiff)*
Segment n 1. <Comp & DV, Geom, Kerntech> segment; 2. <Maschinen> quadrant, segment
Segment n **einer Kurve** <Geom> curve segment
Segmentbogen m <Bau> segmental arch
segmentieren v 1. <Comp & DV> segment; 2. <Telekom> partition
segmentierte Abtastung f <Fernseh> segmented scanning
segmentierte Aufzeichnung f <Fernseh> segmented recording
segmentiertes Multiprozessorsystem n <Telekom> segmented multiprocessor system
Segmentierung f 1. <Comp & DV> segmentation; 2. <Künstl Int> segmentation *(von Bildern)*
Segmentkreissäge f <Maschinen> segmental circular saw
Segmentpfahl m <Kohlen> segmented pile
Segmentrad n <Maschinen> segment gear
Segmentsäge f <Maschinen> segmented saw
Segmentwehr n <Nichtfoss Energ> radial gate
Segmentwelle f <Kfztech> sector shaft
Segner'sches Wasserrad n <Wasserversorg> reaction water wheel
Sehachse f <Foto> optical axis
Sehfeld n <Ergon> visual field
Sehfeldblende f 1. <Metall> field diaphragm; 2. <Phys> field stop
Sehne f <Geom, Nichtfoss Energ> chord
Sehnenlänge f 1. <Geom> length of chord; 2. <Nichtfoss Energ> chord length
Sehnenwicklung f <Elektrotech> chord winding, fractional pitch winding
Sehrohr n <Wassertrans> periscope *(U-Boot)*
Sehschärfe f <Ergon> vision acuity, visual acuity
Sehwinkel m <Telekom> angle of sight
Seiche f <Nichtfoss Energ> seiche
seicht adj <Wassertrans> shallow *(Wasser)*
Seide f <Ker & Glas, Textil> silk
Seidenfaden m <Textil> silk filament
Seidengewebe n **mit Ripseffekt** <Textil> poult
Seidengriff m <Textil> silk-like handle
Seidenmattätzen n <Ker & Glas> satin etch
Seidenpapier n <Verpack> tissue paper
Seidenraster n <Ker & Glas> silk screen
Seidenspinnerei f <Textil> silk spinning
Seidenstoff m <Textil> silk
Seife f <Chemie> soap
Seifenerde f <Chemie> smectite
Seifenlauge f <Chemie> suds

Seifenlösung f <Fertig> soap solution *(Kunststoffinstallationen)*
Seifenschaum m <Chemie> suds
Seifenton m <Chemie> smectite
Seifenwasser n <Sicherheit> soap and water solution
seigern v <Fertig> eliquate, sweat
Seigerung f <Fertig> eliquation, liquation
Seigerungsstreifen m <Fertig> ghost, ghost line
Seignettesalz n 1. <Elektrotech> Rochelle salt *(piezoelektrisches Material)*; 2. <Lebensmittel> Rochelle salt
Seil n <Maschinen, Mechan, Papier, Textil> rope
Seilbahn f <Trans> *(AE)* cable railroad, *(BE)* cable railway, cableway
Seilbahnwagen m <Trans> carrier
Seilbremse f <Maschinen> rope brake
Seilfähre f <Wassertrans> cable ferry
Seilförderanlage f <Trans> cableway
Seilförderung f <Kohlen> rope hauling
Seilgarn n <Textil> rope yarn
Seilkinke f <Maschinen> rope kink
Seilkloben m <Maschinen> rope block
Seilöse f <Wassertrans> becket *(Tauwerk)*
Seilrettungsgerät n <Lufttrans, Sicherheit> escape rope
Seilrolle f <Maschinen> rope pulley, rope wheel
Seilrollenblock m <Erdöl> *(AE)* traveling block, *(BE)* travelling block
Seilscheibe f <Maschinen> pulley, sheave
Seilschelle f <Maschinen> rope clamp
Seilschlagbohren n <Kohlen> cable drilling, percussive rope boring
Seilschlaufe f <Sicherheit> rope-type sling
Seilschloss n <Maschinen> hook and eye
Seilschrapper m <Papier> drag scraper
Seilschwebebahn f <Eisenbahn> gondola cableway
Seilträger m <Papier> rope carrier
Seiltrieb m <Maschinen> rope drive
Seiltrommel f 1. <Fertig> rope barrel; 2. <Maschinen> drum, rope drum, winding drum; 3. <Verpack> cable reel
Seilverspannung f <Bau> guying
Seilwinde f 1. <Elektrotech> cable winch; 2. <Maschinen> cable winch, rope winch; 3. <Trans> cable winch
Seismik f <Erdöl> seismic *(Messtechnik)*
seismisch adj <Bau> seismic
seismische Erkundung f <Bau, Erdöl, Kohlen> seismic exploration
seismische Exploration f <Erdöl> seismic exploration *(Lagerstättensuche)*
seismischer Aufschluss m <Bau, Kohlen> seismic survey
seismischer Weg m <Bau, Erdöl, Kohlen> seismic path
Seismograph m <Bau, Kohlen, Phys> seismograph
Seismologie f <Phys> seismology
Seite f 1. <Bau> side; 2. <Comp & DV, Druck> page; 3. <Geom, Maschinen> side • **mit vier Seiten** <Geom> four-sided • **von der Seite beleuchten** <Foto> sidelight
Seite f **im Querformat** <Druck> oblong page
Seiten fpl **pro Minute** <Comp & DV> pages per minute
Seiten fpl **pro Stunde** <Druck> pages per hour, pph
Seitenablagerung f <Bau> spoil area *(von Bodenmaterial)*
Seitenabrufmethode f <Comp & DV> demand paging
Seitenabstand m 1. <Maschinen> side clearance; 2. <Trans> lateral clearance
Seitenabweichung f <Lufttrans> longitudinal divergence
Seitenanbau m <Ker & Glas> side pocket
Seitenaufbau m <Elektronik> lateral structure
Seitenaufprall m <Trans> side-on collision
Seitenbalken m <Comp & DV> sidebar
Seitenband n <Aufnahme, Comp & DV, Elektronik, Fernseh, Funktech, Phys, Telekom> sideband

Seitenbanddämpfung

Seitenbanddämpfung f <Elektronik, Telekom> sideband attenuation
Seitenbandfrequenz f (SF) <Elektronik, Fernseh, Funktech, Telekom> sideband frequency (SF)
Seitenbandgeräusch n <Funktech> sideband interference
Seitenbandinterferenz f <Elektronik> sideband interference
Seitenbandlegierung f <Metall> sideband alloy
Seitenbandunterdrückung f <Elektronik, Funktech> sideband suppression
Seitenbeplattung f <Wassertrans> side plating (Schiffbau)
Seitenbestimmung f <Funkort> sense determination (Funkpeilung)
Seitenblech n <Bau> side plate
Seitendeck n <Wassertrans> side deck
Seitendrehmeißel m <Maschinen> knife tool
Seitendriftlandung f <Lufttrans> lateral drift landing
Seitendrucker m <Comp & DV> page printer
Seiteneinfall m <Wassertrans> tumblehome (Schiffbau)
Seiteneinheit f <Maschinen> wing base
Seitenende n <Comp & DV> end of page
Seitenfestigkeitsverband m <Wassertrans> side construction (Schiffbau)
Seitenfläche f 1. <Bau> cheek, flank, side; 2. <Geom> face, lateral face; 3. <Maschinen> lateral face
Seitenflosse f <Lufttrans> keel; fuselage dorsal fin (in den Rumpf übergehend)
Seitenflossenansatzrahmen m <Lufttrans> fin stub frame
Seitenfrequenz f <Elektronik, Fernseh, Funktech> side frequency
seitengesteuerter Motor m 1. <Kfztech> L-head engine; 2. <Kfztech> (sv-Motor) side valve engine (sv engine)
Seitenhöhe f <Wassertrans> depth (Schiffbau)
Seitenholm m <Raumfahrt> fin post
Seiteninversion f <Fernseh> lateral inversion
Seitenkanal m <Maschinen> bypass channel
Seitenkante f <Fertig> lateral edge
Seitenkeilwinkel m <Maschinen> wedge angle
Seitenkipper m <Eisenbahn> side-dump car
Seitenklappen fpl <Verpack> folding sides; side panel (eines Kartons)
Seitenkraft f <Maschinen> lateral force
Seitenkraftkoeffizient m <Lufttrans> lateral force coefficient
Seitenlänge f 1. <Druck> length of page; 2. <Math> length of edge
Seitenleitwerksflosse f <Lufttrans> keel; tail fin
Seitenleuchte f <Kfztech> side light
Seitenmeißel m <Maschinen> side tool
Seitenneigung f <Lufttrans> bank (Flugwesen)
Seitennummerierung f <Druck> pagination
Seitenpeilung f <Wassertrans> relative bearing (Navigation)
Seitenplatte f <Fertig> end plate
Seitenpufferschraubenkupplung f <Kfztech> side buffer screw coupling
Seitenrad n <Kfztech> side gear
Seitenrahmen m 1. <Comp & DV> page frame; 2. <Verpack> side frame (eines Containers)
Seitenriss m 1. <Maschinen> side elevation; 2. <Wassertrans> sheer drawing, sheer plan (Schiffkonstruktion)
Seitenruder n 1. <Lufttrans> control surface; 2. <Phys> rudder
Seitenruderausschlag m <Lufttrans> control surface angle, rudder travel
Seitenruderbewegung f <Lufttrans> rudder travel

Seitenruderblockierung f <Lufttrans> control surface locking
Seitenruderfußhebel m <Lufttrans> rudder bar
Seitenruderhebel m <Lufttrans> rudder bar
Seitenruderholm m <Lufttrans> rudder post
Seitenruderpedal n <Lufttrans> rudder pedal
Seitenruderservoantrieb m <Lufttrans> rudder power unit
Seitenrudersteuerung f <Lufttrans> rudder control
Seitenrudertrimmklappe f <Lufttrans> rudder trim
Seitenrudertrimmlicht n <Lufttrans> rudder trim light
Seitenruderverriegelung f <Lufttrans> control surface locking
Seitenrutsch m <Phys> side slip
Seitenschneider m 1. <Maschinen> diagonal-cutting nippers, side-cutting nippers, side-cutting pliers; 2. <Raumfahrt> cable cutter (Raumschiff)
Seitenschrift f <Akustik, Aufnahme> lateral recording
Seitenschub m <Nichtfoss Energ> side thrust
Seitenschwert n <Wassertrans> leeboard
Seitensichtradar n (SLAR) <Funkort, Meerschmutz> sideways-looking airborne radar (SLAR)
Seitenspiel n <Maschinen> side clearance, side play
Seitenstabilität f <Trans> vertical stability
Seitenstahl m <Maschinen> cranked tool
Seitenstrahlpropeller m <Wassertrans> side thruster (Schiffantrieb)
Seitenstraße f <Bau> byroad, byway
Seitentabelle f <Comp & DV> page table
Seitentafel f <Fernseh> side panel (Teletext)
Seitenträger m <Wassertrans> side girder (Schiffbau)
Seitentransistor m <Elektronik> lateral transistor
Seitenumbruch m 1. <Comp & DV> page break, pagination; 2. <Druck> make-up • **Seitenumbruch erstellen** <Comp & DV> page
Seitenverbandkonstruktion f <Wassertrans> side construction (Schiffbau)
Seitenvergrößerung f <Phys> lateral magnification
Seitenverhältnis n <Mechan> aspect ratio
Seitenverkleidung f <Kfztech> side panel (Karosserie)
Seitenverriegelung f <Fernseh> side lock
Seitenversatz m <Maschinen> lateral offset
Seitenwand f **mit Lüftungsklappen** <Kfztech> sidewall with ventilation flaps
Seitenwandluftkissenfahrzeug n 1. <Lufttrans> sidewall hovercraft; 2. <Trans> sidewall-type hovercraft
Seitenwechsel m <Comp & DV> paging
Seitenwechsel m **auf Anforderung** <Comp & DV> demand paging
Seitenwind m <Lufttrans> crosswind
Seitenwinde f <Bau> side pulley
Seitenwindkomponente f <Lufttrans> longitudinal wind component
Seitenwinkel m <Nichtfoss Energ> azimuth angle
Seitenzahl f <Druck> folio, page number
Seitenzipfel m <Wassertrans> side lobe (Radar)
seitlich angebrachte Anschlussklemme f <Kfztech> side-mounted terminal
seitlich angebrachter Pol m <Kfztech> side-mounted terminal
seitlich angepresster Verschluss m <Verpack> crimp-on closure
seitlich versetzt adj <Bau> (AE) off-center, (BE) off-centre
seitlich zugeschnalltes Zusatzschubtriebwerk n <Raumfahrt> strap-on booster
seitliche Auslenkkraft f <Akustik> side thrust
seitliche Begrenzungsleuchte f <Kfztech> side marker light

seitliche Diffusion f <Elektronik> lateral diffusion
seitliche Heftung f 1. <Druck> side stitching, stab stitching; 2. <Verpack> lateral stapling *(mit Klammern)*
seitliche Hitzeübertragung f <Anstrich> lateral heat transfer
seitliche Vertiefung f <Optik> sidepit
seitlicher Abstand m <Maschinen> side clearance
seitlicher Airbag m <Kfztech> side impact air bag
seitlicher Geräuschmesspunkt m <Lufttrans> lateral noise measurement point
seitlicher Zusammenstoß m <Trans> side collision, sideswipe
seitliches Nachgeben n <Kerntech> lateral yielding
seitliches Spiel n <Maschinen> side play
Seitlichheftung f <Druck> side stitching, stab stitching
Sekans m *(sec)* <Geom> secant *(sec)*
Sekante f *(sec)* <Geom> secant, sec *(von Linie)*
Sektion f <Textil, Wassertrans> section
Sektionsschären n <Textil> section warping, sectional warping
Sektor m 1. <Comp & DV> sector; 2. <Erdöl> sector *(Geographie)*; 3. <Geom> sector
Sektorabsuchen n <Funkort> sector scanning *(Radar)*
Sektordiagramm n <Math> pie chart
Sektorenfeuer n <Wassertrans> light with sectors
Sektorschütz n <Nichtfoss Energ> sector gate
Sektorwehr n <Wasserversorg> sector weir
sekundär adj <Funktech, Strahlphys> secondary
Sekundär... <Strahlphys> secondary
Sekundärabgriff m <Elektrotech> secondary tap
Sekundärabzweigung f <Elektrotech> secondary tap
Sekundäracetat n <Textil> secondary acetate
Sekundärachse f <Geom> secondary axis
Sekundärakkumulator m <Trans> secondary storage battery
Sekundärbatterie f <Elektriz, Elektrotech> secondary battery, storage battery
Sekundärbeschichtung f <Telekom> *(AE)* fiber jacket, *(BE)* fibre jacket, secondary coating
Sekundärbremssystem n <Kfztech> secondary brake system
Sekundärbrennstoff m <Erdöl> derived fuel
Sekundärbrennstoffelement n <Kfztech> secondary fuel cell
Sekundärbuchse f <Kfztech> secondary sleeve
Sekundärdüse f <Lufttrans> secondary nozzle
sekundäre Faser f <Verpack> *(AE)* secondary fiber, *(BE)* secondary fibre
sekundäre Hülle f <Fertig> secondary coating *(Faseroptik)*
sekundäre Kernreaktion f <Kerntech> secondary nuclear reaction
sekundäre Rekristallisation f <Metall> secondary recrystallization
sekundäre Wicklung f <Elektriz> secondary winding
Sekundärelektron n <Elektronik, Strahlphys> secondary electron
Sekundärelektronenvervielfacher m <Strahlphys> secondary electron multiplier
Sekundärelement n <Elektrotech> secondary cell
Sekundäremission f <Elektronik, Phys> secondary emission
Sekundäremissionsrauschen n <Elektronik> secondary emission noise
Sekundäremissionsröhre f <Elektronik> secondary emission tube
Sekundäremissionsverhältnis n <Elektronik> secondary emission ratio
Sekundäremissionsvervielfacher m <Elektronik> photomultiplier, secondary emission multiplier

Sekundärenergie f <Erdöl> derived energy
sekundäres Kriechen n <Metall> secondary creep
sekundäres Maximum n <Optik, Phys> secondary maxima
sekundäres Schnellabschaltsystem n <Kerntech> secondary shutdown system
Sekundärfarben fpl <Druck> *(AE)* secondary colors, *(BE)* secondary colours
Sekundärförderung f <Erdöl> secondary recovery *(Fördertechnik)*
Sekundärgitteremission f <Elektronik> secondary grid emission
Sekundärgittermauerwerk n <Ker & Glas> secondary checkers
Sekundärgruppe f <Comp & DV, Telekom> super group, supergroup
Sekundärindex m <Comp & DV> secondary index
Sekundärinduktanz f <Elektrotech> secondary inductance
Sekundärionenemission f <Telekom> secondary ionic emission
Sekundärionenmassenspektrometrie f *(SIMS)* <Phys> secondary ion mass spectrometry *(SIMS)*
Sekundärionisierung f <Strahlphys> secondary ionization
Sekundärkältemittel n <Heiz & Kälte> secondary refrigerant
Sekundärkanal m <Lufttrans> secondary duct
Sekundärkern m <Erdöl> sidewall core *(Bohrtechnik)*
Sekundärklemme f <Elektrotech> secondary terminal
Sekundärkollision f <Trans> secondary collision
Sekundärkreis m <Elektrotech, Heiz & Kälte> secondary circuit
Sekundärkrümmung f <Ker & Glas> secondary curvature
Sekundärkühlmittel n <Heiz & Kälte> secondary coolant
Sekundärlichtquelle f <Strahlphys> illuminated source
Sekundärluft f <Kfztech> secondary air
Sekundärmanschette f <Kfztech> secondary cup
Sekundärnormal f <Phys> secondary standard
Sekundärprozess m <Kerntech> secondary process
Sekundärquelle f <Phys> secondary source
Sekundärreaktor m <Kerntech> secondary reactor
Sekundärrelais n <Elektriz> secondary relay
Sekundärrohstoff m 1. <Abfall> secondary material; 2. <Umweltschmutz> waste product
Sekundärseite f <Kerntech> secondary side
Sekundärsetzung f <Kohlen> secondary settlement
Sekundärspannung f <Elektriz, Elektrotech> secondary voltage
Sekundärspeicher m <Comp & DV> secondary memory, secondary storage, secondary store
Sekundärspule f 1. <Elektrotech> secondary coil, secondary winding; 2. <Kfztech> secondary winding
Sekundärstoffauflauf m <Papier> secondary headbox
Sekundärstrahler m <Funktech> parasitic radiator
Sekundärstrahlung f <Strahlphys> secondary radiation
Sekundärstrom m <Elektriz, Elektrotech> secondary current
Sekundärstromkreis m <Elektrotech> secondary circuit
Sekundärteilchen n <Kerntech> secondary particle
Sekundärteilchenradiographie f <Kerntech, Strahlphys, Teilphys> charged-particle radiography
Sekundärtrennung f <Kerntech> secondary separation
Sekundärwelle f *(S-Welle)* <Phys> secondary wave *(S-wave)*
Sekundärwicklung f <Elektrotech, Kfztech, Phys> secondary winding
Sekundärwicklung f **mit Anzapfung** <Elektrotech> tapped secondary winding

Sekundärwiderstand *m* <Elektrotech> secondary resistance
Sekundärzelle *f* <Elektriz> secondary cell
Sekunde *f* <Akustik, Metrol, Phys> second *(Zeiteinheit)*
Sekundenzeiger *m* <Maschinen> second hand
Selbstabgleich *m* <Gerät> autobalance, self-balance
selbstabgleichender Schalter *m* <Elektrotech> self-balancing switch
Selbstabnahmepapiermaschine *f* <Papier> MG machine, machine-glazing machine
Selbstabschirmung *f* <Kerntech> self-shielding
selbstabsichernde Pumpe *f* <Verpack> self-sealing pump
Selbstabsorption *f* <Kerntech> self-absorption
Selbstalterung *f* <Thermod> *(BE)* natural ageing, *(AE)* natural aging
selbstanlaufender Synchronmotor *m* <Elektrotech> self-starting synchronous motor
selbstanpassend *adj* <Comp & DV, Telekom> self-adapting
selbstanpassende Abtastung *f* <Gerät> adaptive sampling
selbstanpassendes Regelsystem *n* <Comp & DV> adaptive control system
Selbstanregung *f* <Kerntech> self-excitation
selbstansaugende Pumpe *f* <Kfztech> self-priming pump
selbstansaugende Schmutzwasserpumpe *f* <Kerntech> self-priming dirty-water pump
Selbstansteuerung *f* <Raumfahrt> homing
Selbstantrieb/mit <Trans> self-propelling
selbstaufrichtend *adj* <Wassertrans> self-righting
Selbstauslöser *m* <Foto> self-timer
Selbstbedienungstankstelle *f (SB-Tankstelle)* <Trans> self-service station
selbstdiagnostisch *adj* <Comp & DV> self-diagnostic
selbstdichtend *adj* <Raumfahrt> self-sealing
Selbstdiffusion *f* <Phys> self-diffusion
selbstdokumentierend *adj* <Comp & DV> self-documenting
selbsteinstellend *adj* <Fertig> self-adjusting *(Gerät)*
selbsteinstellende Schutzvorrichtung *f* <Sicherheit> self-adjusting guard
Selbsteinstellung *f* <Gerät> automatic adjustment
Selbstentladeeimer *m* <Bau> self-dumping bucket
Selbstentladegüterwagen *m* <Eisenbahn> *(AE)* self-discharge freight car, *(BE)* self-discharge freight wagon
Selbstentladewagen *m* <Eisenbahn> *(AE)* self-discharging car, *(BE)* self-discharging wagon
Selbstentladewaggon *m* <Eisenbahn> *(AE)* self-discharge car, *(BE)* self-discharge wagon
Selbstentladewasserkübel *m* <Wasserversorg> self-discharging water bucket
Selbstentladung *f* 1. <Elektrotech> self-discharge, self-sustained discharge; 2. <Kfztech, Raumfahrt> self-discharge
Selbstentladungs-Zeitkonstante *f* <Elektrotech> self-discharge time constant
Selbstentleerer *m* **mit Sattelboden** <Trans> saddle-bottomed self-discharging car
Selbstentzündung *f* <Raumfahrt> spontaneous ignition
Selbsterhitzung *f* <Elektrotech> self-heating
Selbsterhitzungskoeffizient *m* <Elektrotech> self-heating coefficient
selbsterregte Schwingungen *fpl* <Fertig> self-exited vibrations, self-induced vibrations
selbsterregter Leistungsoszillator *m* <Elektronik> self-excited power oscillator
selbsterregter Motor *m* <Elektriz> self-excited motor
selbsterregter Oszillator *m* <Elektronik> self-excited oscillator
Selbsterregung *f* <Elektrotech> self-excitation
selbststarrend *adj* <Fertig> air-drying *(Kern)*
selbstfahrend *adj* <Kfztech> automobile, automotive
selbstfahrender Kran *m* <Kerntech> self-propelled crane
Selbstfahrer *m* 1. <Meerschmutz> self-propelled vessel; 2. <Wassertrans> self-propelled barge *(Schifftyp)*
selbstgebacken *adj* <Lebensmittel> home-baked
selbstgefertigt *adj* <Fertig> engineered
selbstgeführtes Arbeitsteam *n* <Qual> self-managed work team
Selbstglättung *f* <Verpack> automatic decurling
Selbstgreifer *m* <Kerntech> grab *(eines Krans)*
selbsthaftender Schmelzkleber *m* <Verpack> pressure-sensitive hot-melt adhesive
selbsthärtend *adj* <Fertig> natural *(Stahl)*
Selbsthärtestahl *m* <Metall> self-hardening steel
selbstheilender Kondensator *m* <Elektrotech> self-healing capacitor
Selbstheilung *f* <Elektrotech> self-healing
selbsthemmend *adj* <Fertig> retained by friction
Selbstinduktion *f* <Elektrotech, Kfztech, Phys> self-induction
Selbstinduktionskoeffizient *m* <Elektrotech> self-inductance
Selbstinduktionsstrom *m* <Elektrotech> self-induction current
Selbstinduktivität *f* <Elektrotech, Phys> self-inductance
selbstinduziert *adj* <Phys> self-induced
Selbstionisierung *f* <Phys, Strahlphys> autoionization
Selbstjustieren *n* <Fernseh> self-adjustment
selbstjustierender Steueranschluss *m* <Elektronik> SAG, self-aligned gate
selbstjustierender Transistor *m* <Elektronik> self-aligned transistor
selbstjustierendes System *n* <Fernseh> self-controlling system
Selbstkipper *m* <Eisenbahn> self-tipping wagon
Selbstklebeband *n* <Kunststoff> self-adhesive tape
Selbstklebeetikettiermaschine *f* <Verpack> pressure-sensitive labeller
selbstklebend *adj* <Verpack> self-sealing
selbstklebende Folie *f* **für kurzfristige Schutzabdeckung** <Verpack> self-adhesive film for temporary surface protection
selbstklebender Briefumschlag *m* <Verpack> self-seal pocket envelope
selbstklebendes Abziehschildchen *n* <Verpack> easy-peel-off self-adhesive label
selbstklebendes Band *n* **aus starkem Polyethylen** <Verpack> tough polyethylene self-adhesive tape
selbstklebendes Isolierband *n* <Elektrotech> adhesive insulating tape
selbstklebendes Papier *n* <Verpack> self-adhesive paper
selbstklebendes PVC-Band *n* <Verpack> PVC pressure-sensitive tape
selbstkommutierender Umformer *m* <Elektriz> self-commutated converter
selbstkorrigierend *adj* <Comp & DV> self-correcting
selbstkorrigierender Code *m* 1. <Comp & DV, Telekom> error-correcting code, self-correcting code; 2. <Elektronik> error-correcting code
Selbstkühlung *f* 1. <Heiz & Kälte> self-cooling; 2. <Thermod> natural cooling
Selbstladenbehälter *m* <Trans> self-loading container
selbstlenzend *adj* <Wassertrans> self-draining
selbstlernender Rechner *m* <Comp & DV> self-learning machine

selbstlernendes System n <Künstl Int> learning system
Selbstleuchten n <Raumfahrt> self-luminosity
Selbstlockerung f <Kerntech> self-loosening
Selbstlüftung f <Heiz & Kälte> natural ventilation
Selbstlüftungssystem n <Verpack> self-venting system
Selbstmitlauf m <Elektronik> self-tracking *(Filter)*
selbstmitlaufendes Bandpassfilter n <Elektronik> self-tracking band-pass filter
Selbstnachführung f <Elektronik> self-tracing
selbstnachstellende Bremse f <Kfztech> self-adjusting brake
selbstnachstellende Kupplung f <Kfztech> self-adjusting clutch
selbstnachstellendes Schwimmwehr n <Meerschmutz> self-adjusting floating weir
selbstöffnender Gewindeschneidkopf m <Maschinen> self-opening diehead, self-opening screwing head
selbstorganisierende Regel f <Künstl Int> Hebbian rule
selbstorganisierendes Netz n <Künstl Int> self-organizing network
selbstorganisierendes System n <Comp & DV> self-organizing system
Selbstphasenmodulation f <Telekom> self-modulation of phase
selbstprüfender Code m <Comp & DV, Elektronik> error-detecting code
Selbstprüfung f 1. <Comp & DV> automatic check; 2. <Qual> operator inspection
selbstregelnder Schalter m <Elektrotech> self-balancing switch
selbstregelndes System n <Elektrotech> adaptive control system
selbstregelndes Wartungssystem n <Kfztech> self-regulating maintenance system
Selbstregler m <Maschinen> automatic regulator
selbstregulierende Schutzvorrichtung f <Sicherheit> self-adjusting guard
selbstreinigend adj <Maschinen> self-cleaning
selbstreinigendes Luftfilter n <Heiz & Kälte> self-cleaning air filter
Selbstreinigung f 1. <Abfall> natural purification; 2. <Wasserversorg> self-purification
Selbstreinigungskraft f <Abfall> assimilative capacity
Selbstrelativadresse f <Comp & DV> self-relative address
Selbstretter m <Sicherheit> escape self-contained breathing apparatus, pressure-demand self-contained breathing apparatus, self-contained air sampling respirator, self-contained air supply respirator *(Atemschutzgerät)*
selbstrückfallendes Relais n <Elektriz> self-resetting relay
selbstrücksetzende Schleife f <Comp & DV> self-resetting loop
selbstrückstellender Zähler m <Gerät> self-resetting counter
Selbstschalttransistor m <Elektronik> latching transistor
selbstschließend adj <Verpack> self-closing
selbstschließende Brandschutztür f <Bau> self-closing fire door
selbstschließende Tür f <Bau> self-closing door
Selbstschlussbatterie f <Bau> self-closing faucet *(Wasserhahn)*
selbstschmelzig adj <Phys> self-fluxing
selbstschmierend adj <Maschinen> self-lubricating
Selbstschmierlager n <Maschinen> self-lubricating bearing
selbstschneidendes Gewinde n <Maschinen> self-tapping thread

selbstschwingende Mischröhre f 1. <Elektronik> converter *(Elektronenröhre)*; 2. <Elektrotech> converter *(bei elektrischen Röhren)*
selbstsichernd adj <Kfztech> self-locking
selbstsichernde Mutter f <Maschinen> locknut, self-locking nut
selbstsperrend adj <Kfztech> self-locking
selbstsperrendes Ausgleichsgetriebe n <Kfztech> nonslip differential
Selbstsperrung f <Fertig> automatic interlock
Selbststabilisierungsgerät n 1. <Lufttrans> automatic stabilization equipment; 2. <Trans> automatic stabilizing equipment
selbstständig adj <Comp & DV> stand-alone
selbstständig arbeitend adj <Kontroll> autonomous
selbstständige Emission f <Elektronik> spontaneous emission
selbstständiger Docksteuerstand m <Wassertrans> off-line docking station
Selbststarter m <Lufttrans> automatic starting unit
Selbststeueranlage f <Raumfahrt, Wassertrans> autopilot
Selbststeuerungsanlage f <Lufttrans, Wassertrans> gyropilot
selbsttätig adj <Maschinen> self-acting
selbsttätig schließende Schleuse f <Wasserversorg> balance bar
selbsttätige Bremse f <Kfztech> self-acting brake
selbsttätige Feinabstimmung f <Funktech> automatic fine tuning *(ACU)*
selbsttätige Kupplung f 1. <Eisenbahn> automatic coupler, automatic coupling; 2. <Kfztech> automatic clutch, automatic coupling
selbsttätige Oxidation f <Lebensmittel> autoxidation
selbsttätige Regelung f <Regelung> automatic closed-loop control
selbsttätiger Leistungsschalter m <Elektriz> automatic circuit recloser
selbsttätiger Regler m <Maschinen> self-acting regulator
selbsttätiger Schalter m <Elektrotech> self-acting switch
Selbsttönen n 1. <Akustik> howling; 2. <Aufnahme> howling *(bei Verstärkern)*
selbsttragend adj 1. <Bau> self-contained, self-supporting; 2. <Raumfahrt> self-supporting
selbsttragende Karosserie f <Kfztech> unit construction body *(Karosserie)*
selbsttragendes Glasfaserkabel n <Telekom> self-supporting fiber-optic cable
selbsttrimmender Laderaum m <Wassertrans> self-trimming hold
Selbstüberlagerer m <Elektronik> autodyne
selbstüberwachend adj <Sicherheit> self-monitoring
selbstüberwachendes Schutzsystem n <Sicherheit> self-monitoring guard system
Selbstunterbrecher m <Elektrotech> trembler *(Anker)*
Selbstunterbrecherglocke f <Elektriz> trembler bell
selbstverriegelnd adj <Verpack> self-locking
selbstverschiebliches Programm n <Comp & DV> self-relocating program
selbstverstärkendes Polymer n <Kunststoff> self-reinforcing polymer
Selbstvulkanisation f <Kunststoff> self-vulcanization
Selbstwahl f <Telekom> *(AE)* automatic dialing, *(BE)* automatic dialling
Selbstwählfernbetrieb m <Telekom> automatic trunk working
Selbstwählferndienst m *(SWFD)* <Telekom> *(AE)* direct distance dialing *(DDD)*; *(BE)* subscriber trunk dialling *(STD)*

Selbstwähllandzentrale

Selbstwähllandzentrale f <Telekom> rural automatic exchange
Selbstwählnebenstelle f <Telekom> private branch exchange
selbstzentrierender Schraubstock m <Maschinen> (AE) self-centering vise, (BE) self-centring vice
selbstzentrierendes Spannfutter n <Maschinen> concentric chuck, scroll chuck, (AE) self-centering chuck, (BE) self-centring chuck
selbstzielsuchend adj <Raumfahrt> homing
Selbstzug m <Heiz & Kälte> (AE) natural draft, (BE) natural draught
Selbstzünder m <Kfztech> compression-ignition engine
Selbstzündung f 1. <Kfztech> self-firing, self-ignition; autoignition (Motor); compression ignition (Zündanlage bei Dieselmotor); 2. <Maschinen> self-ignition
Selbstzustellung f <Maschinen> automatic feed
SELCAL (Selektivrufsystem) <Telekom> SELCAL (selective calling system)
selektieren v <Druck> select
Selektion f <Elektronik> selectivity
selektiv galvanisierte Kontakte mpl <Elektrotech> selectively plated contacts
selektiv plattierte Kontakte mpl <Elektrotech> selectively plated contacts
selektive Beschichtung f <Nichtfoss Energ> selective coating
selektive Diffusion f <Elektronik> selective diffusion
selektive Ionenelektrode f <Labor> selective ion electrode (Elektrochemie)
selektive katalytische Reduktion f <Umweltschmutz> selective catalytic reduction
selektive Oberfläche f <Nichtfoss Energ> selective surface
selektive Prüfung f <Qual> screening inspection, screening test
selektive Reflexion f <Phys> selective reflection
selektive Rückkopplung f <Elektronik> selective feedback
selektive Stromunterbrechung f <Elektriz> discriminating circuit-breaking
selektive Verkehrsumleitung f <Trans> selective diversion of traffic
selektiver Abschwächer m <Foto> selective reducer
selektiver Bildschirmspeicher m <Comp & DV> snapshot
selektiver Schrumpf m <Textil> differential shrinkage
selektiver Schwund m <Funktech> differential fading
selektiver sequenzieller Zugriff m <Comp & DV> selective sequential access
selektiver Speicherauszug m <Comp & DV> selective dump
selektiver Zugriff m <Comp & DV> selective access
selektives Bildschirmspeichern n <Comp & DV> snapshot
selektives Fading n <Telekom> selective fading
selektives Löschen n <Comp & DV> selective erasure
selektives Protokollprogramm n <Comp & DV> snapshot
selektives Relais n <Elektriz> selective relay
selektives Schutzsystem n <Elektriz> discriminating protective system
Selektivität f <Phys> selectivity
Selektivitätsschlange f <Aufnahme> selectivity Q
Selektivruf m <Telekom> selective calling
Selektivrufsystem n (SELCAL) <Telekom> selective calling system (SELCAL)
Selektivschwund m <Telekom> selective fading
Selektivsolvens n <Erdöl> selective solvent

Selektor m <Comp & DV, Elektrotech> selector
Selektorkanal m <Comp & DV> selector channel
Selektorrelais n <Elektrotech> selector relay
Selen n (Se) <Chemie> selenium (Se)
Selen... <Chemie> selenic, selenious, selenous
Selenat n <Chemie> selenate
Selencyan... <Chemie> selenocyanic
Selengleichrichter m <Elektriz, Elektrotech, Phys> selenium rectifier
Selenid n <Chemie> selenide
Selenit n <Chemie> selenite
Selenium n (Se) <Chemie> selenium (Se)
Selen-Rubinglas n <Ker & Glas> selenium ruby glass
Selenzelle f 1. <Elektrotech> electric eye, selenium cell; 2. <Phys> selenium cell
Sellersgewinde n <Maschinen> Sellers thread, USS screw thread
Seltene-Erden-Glas n <Ker & Glas> rare-earth glass
Seltsamkeit f <Teilphys> strangeness (spezielle Quantenzahl von Hadronen)
Semantik f <Comp & DV> semantics
semantische Analyse f <Comp & DV> semantic analysis
semantischer Fehler m <Comp & DV> semantic error
semantisches Netz n <Künstl Int> associative network, semantic net, semantic network
Semaphor m <Comp & DV> semaphore
Semaphorprogramm n <Trans> (AE) semaphoric program, (BE) semaphoric programme
semiamphibisches Luftkissenfahrzeug n <Wassertrans> semiamphibious air cushion vehicle, semiamphibious hovercraft
Semicarbazid n <Chemie> semicarbazide
Semicarbazon n <Chemie> semicarbazone
Semidin n <Chemie> semidine
semihomogenes Brennelement n <Kerntech> semihomogeneous fuel element
Semikundenchip m <Elektronik> semicustom chip (halbkundenspezifischer Chip)
Semikundenschaltung f <Elektronik> semicustom circuit (halbkundenspezifische Schaltung)
semipermanente Verbindung f <Telekom> semipermanent connection
semipermeable Membran f <Phys> semipermeable membrane
Sendeantenne f 1. <Fernseh> transmitting antenna; 2. <Funktech> transmit antenna, transmitting aerial, transmitting antenna; 3. <Phys> transmitting aerial
Sendeaufforderung f 1. <Comp & DV> invitation to send; 2. <Telekom> invitation to transmit, request to send
Sendeaufforderung f „Sendeteil einschalten" <Telekom> request to send
Sendeaufruf m 1. <Comp & DV> invitation to send, polling; 2. <Funktech, Telekom> polling
Sendeaufrufintervall n <Comp & DV> polling interval
Sendeaufrufzeichen n <Comp & DV> polling character
Sendebeginnzeichen n <Funktech, Telekom> transmitter turn-on signal
Sendebeginnzeit f <Fernseh, Funktech> transmitter turn-on time
Sendebereich m 1. <Fernseh> network coverage, range, station coverage; 2. <Funktech> range, station coverage; 3. <Telekom> service area (Sender)
sendebereit adj <Comp & DV> clear-to-send, ready-to-send
Sendedatum n <Fernseh> air date
Sende-Empfangs-... <Elektronik> TR, transmit-receive, transmitting-receiving
Sende-Empfangs-Gerät n <Funktech, Telekom> transceiver

Sende-Empfangs-Weiche f 1. <Elektronik> branching filter, duplexer, transmitter-receiver circuit, transmitter-receiver filter *(Funktechnik)*; 2. <Funktech> power combiner/divider
Sendefernschreiber m <Telekom> transmit machine
Sendefolgenummer f <Telekom> send sequence number
Sendefrequenz f <Fernseh, Funktech> transmitting frequency
Sendeimpuls m <Funktech> pulse
Sendekanal m <Telekom> go channel, send channel *(Richtfunk)*
Sendekettenkennung f <Fernseh> network identification
Sendekopie f <Fernseh> transmission copy
Sendeleistung f <Fernseh, Funktech> transmitter power
Sendemikrofon n <Aufnahme> transmitting microphone
Sendemonitor m <Fernseh> air monitor
senden v 1. <Comp & DV> send, transmit; 2. <Fernseh> air, broadcast, transmit; 3. <Funktech> broadcast, transmit; 4. <Telekom> send
Senden n <Comp & DV, Telekom> sending
Senden n **von Kurznachrichten via Internet** <Mobilkom, Telekom> short message service-internet originated *(SMS-IO)*
Sendepause f <Elektrotech> off-air period
Sendeprüfung f <Fernseh> air check
Sendequalität f <Fernseh, Funktech> air quality
Sender m 1. <Fernseh, Funktech> transmitter; 2. <Telekom> sender, transmitter
Sender m **mit Solarstromversorgung** <Funktech> solar-powered transmitter, sun-powered transmitter
Sender m **mit unterdrücktem Träger** <Elektronik, Funktech> suppressed carrier transmitter
Sender m **mit Windenergieversorgung** <Telekom> wind-powered transmitter
Senderausfall m <Fernseh, Funktech> transmitter failure
Senderecht n <Fernseh, Telekom> broadcasting right
Sender-Empfänger m 1. <Comp & DV> sender-receiver, transceiver; 2. <Funktech> transceiver, transmitter-receiver
Sender-Empfänger-Zelle f <Funktech, Mobilkom> transmitter-receiver cell *(Radar)*
Sendererkennungszeichen n <Fernseh, Funktech> transmitter identification signal
Senderkennung f 1. <Fernseh> *(AE)* program identification signal, *(BE)* programme identification signal; 2. <Funktech> *(AE)* program identification, *(BE)* programme identification, *(AE)* program identification signal, *(BE)* programme identification signal, station identification
Senderkette f <Fernseh> network
Sendernetz n 1. <Elektrotech> network of transmitters *(Fernsehen)*; 2. <Fernseh> network of transmitters
Senderöhre f **mit gekühlter Anode** *(CAT)* <Elektronik> cooled-anode transmitting valve *(CAT)*
Sendersperröhre f <Elektronik> ATR-tube, antitransmit-receive tube
Sendersuchlauf m <Funktech> station finding *(Rundfunk)*
Senderzeit f <Fernseh> station time
Sendeschnittstelle f <Comp & DV> send port
Sendestation f <Comp & DV> master station *(Datenfernverarbeitung)*
Sendesystem n <Elektrotech, Telekom> transmission system
Sendetaste f <Funktech> key
Sendetermin m <Fernseh> slot
Sende-Übergabe f <Telekom> submission *(Nachrichtenübermittlung)*
Sendeüberwachungsband n <Fernseh> air check tape
Sendeweg m <Comp & DV> transmission path
Sendeweiche f <Telekom> combiner

Sendezeit f 1. <Elektrotech, Funktech> on-air period; 2. <Fernseh, Telekom> airtime, broadcasting time
Sendezentrale f <Fernseh> network control room
Sendung f 1. <Fernseh> *(AE)* program, *(BE)* programme, transmission; 2. <Telekom> send, sending; transmission • **auf Sendung** <Fernseh, Funktech> on-air • **auf Sendung gehen** <Fernseh, Funktech> switch to air • **nicht auf Sendung** <Fernseh> off-air • **vor der Sendung aufzeichnen** <Fernseh> prerecord
Sendung f **mit hoher Leistung** <Telekom> high-power transmission
Sendzimirwalzwerk n <Metall> Sendzimir mill, Z-mill
Senföl n <Lebensmittel> mustard oil, mustardseed oil
sengen v <Textil> singe
Sengen n <Textil> singeing
Sengmaschine f <Textil> singeing machine
Senkblei n 1. <Bau> bob, plumb bob; 2. <Fertig> plumb; 3. <Metrol> bob
Senke f <Elektrotech, Metall> drain, sink
senken v <Elektrotech> drain, drop, lower *(Spannung)*
Senken n 1. <Ker & Glas> sagging *(Glas)*; 2. <Maschinen> countersinking, recessing
senkendes Kissen n <Trans> height-off cushion
Senkenstrom m <Elektrotech> drain current
Senker m <Maschinen> burr, counterbore, countersink, spot face cutter
Senkgrube f <Wasserversorg> cess pit, cess pool
Senkkasten m <Maschinen, Wasserversorg> caisson
Senkkopf m 1. <Fertig> flush head; 2. <Maschinen> countersunk head, flat countersunk head
Senkkopfniet n 1. <Bau> countersunk button-head rivet; 2. <Maschinen> flush-head rivet
Senkkopfschraube f <Maschinen> countersunk-head screw, flat-head bolt, flat-head screw
Senkkopfvernietung f <Bau> countersunk riveting
Senkkörper m <Lebensmittel> sinker
Senklot n <Fertig> plummet
Senkniet m 1. <Fertig> flat countersunk head rivet, flush rivet; 2. <Maschinen> countersunk-head rivet, countersunk rivet, flat countersunk rivet
senkrecht adj 1. <Bau> upright; 2. <Geom> rectangular, vertical; 3. <Phys> normal *(im rechten Winkel)*
senkrecht adv <Bau> plumb
senkrecht arbeitende Vakuum-Siegelmaschine f <Verpack> vertical vacuum sealer
senkrecht auf adj <Geom> perpendicular to
senkrecht aufeinander adj <Math> orthogonal
senkrecht eingebauter Motor m <Kfztech> vertical engine
senkrecht startendes und landendes Flugzeug n *(VTOL-Flugzeug)* <Lufttrans> vertical takeoff and landing aircraft *(VTOL aircraft)*
senkrecht stehende Ebenen fpl <Geom> orthogonal planes, vertical planes
senkrecht wirkende Reaktionskraft f <Phys> normal reaction force
Senkrecht... <Fernseh, Maschinen> orthogonal
Senkrechtabtastung f <Fernseh> orthogonal scanning
Senkrechtaußenräummaschine f <Maschinen> vertical surface-type broaching machine
Senkrechtbohrmaschine f <Maschinen> upright drilling machine, vertical boring machine, vertical boring mill, vertical drill press, vertical drilling machine
Senkrechtdrehmaschine f <Maschinen> vertical lathe
Senkrechte f 1. <Bau> plumb; 2. <Geom> normal, perpendicular, vertical • **von der Senkrechten abweichen** <Maschinen> run out of the vertical
senkrechte Beleuchtung f <Lufttrans> overhead light
senkrechte Ebenen fpl <Geom> vertical planes

senkrechte

senkrechte Entwässerung f <Kohlen> vertical drainage
senkrechte Falten fpl <Ker & Glas> brush marks
senkrechte Komponente f <Phys> vertical component
senkrechte Rohrleitung f <Bau> stack
senkrechte Welle f <Maschinen> upright shaft
senkrechte Zugstange f <Eisenbahn> suspension rod
senkrechter Abstand f **zur stillen Wasseroberfläche** <Nichtfoss Energ> distance perpendicular to still water surface
senkrechter Ausrundungsbogen m <Eisenbahn> vertical curve
Senkrechtförderer m <Maschinen> elevator
Senkrechtförderschnecke f <Bau> screw elevator
Senkrechtfräsmaschine f 1. <Fertig> vertical-spindle milling machine; 2. <Maschinen> vertical milling machine
Senkrechtfräsvorrichtung f <Maschinen> vertical milling attachment
Senkrechtgerüst n <Fertig> edging mill (Walzen)
Senkrechtmagnetisierung f <Fernseh> perpendicular magnetization
Senkrechträummaschine f <Maschinen> vertical broaching machine
Senkrechtschlitten m <Fertig> head slide (Hobelmaschine)
Senkrechtschnitt m <Bau> vertical section
Senkrechtschnittkraft f <Fertig> vertical tool thrust
Senkrechtstab m <Bau> stile, upright
Senkrechtstoßmaschine f 1. <Fertig> slotting machine; 2. <Maschinen> vertical slotting machine
Senkrechttabulator m <Comp & DV> vertical tab
Senkrechtziehverfahren n <Ker & Glas> up-draw process
Senkring m <Ker & Glas> settle ring
Senkschacht m <Bau> caisson
Senkschlitten m <Papier> lowering cradle
Senkschraube f 1. <Bau> slotted countersunk-head screw; 2. <Fertig> flat-head screw; 3. <Maschinen> countersunk screw, countersunk-head screw, flat-head bolt, flat-head screw; 4. <Mechan> countersunk-head screw
Senkschraube f **mit Schlitz** <Maschinen> slotted countersunk-head screw
Senkschraubenmutter f <Kerntech> countersunk nut
Senkstange f <Ker & Glas> depression bar
Senkung f 1. <Bau> slump; settling (Gebäude, Gelände); 2. <Kohlen> settling; 3. <Maschinen> counterbore; 4. <Wassertrans> sinkage
Senkungsbecken n <Erdöl> subsidence basin (Stratigraphie)
Senkungskegel m <Wasserversorg> depression cone
Senkungsquelle f <Wasserversorg> depression spring
Senkwaage f <Lebensmittel> densimeter, hydrometer
Senkwasser n <Wasserversorg> percolating water
Sensibilisator m <Druck> sensitizer
sensibilisieren v <Foto> sensitize
Sensibilisierung f <Foto> sensitization
Sensibilisierungsbad n <Foto> sensitizing bath
Sensitometrie f <Akustik> sensitometry
Sensomotorik f <Ergon> sensorimotor system
sensomotorisch adj <Ergon> sensorimotor
Sensor m 1. <Comp & DV> sensor; 2. <Elektriz, Elektronik> pick-up, probe, sensor; 3. <Ergon> sensor; 4. <Gerät> probe; 5. <Kerntech, Kfztech> sensor; 6. <Maschinen> detector, probe, sensor; 7. <Phys, Telekom, Wassertrans> sensor (Messinstrumente)
Sensorbildschirm m <Comp & DV> touch-sensitive screen
sensorisch adj <Ergon> sensory
sensorische Analyse f <Lebensmittel> tasting (zur Beurteilung sensorischer Merkmale von Lebensmitteln)
sensorische Schärfe f <Ergon> sensory acuity
Sensorsignal n <Elektronik> sensor signal

Sente f <Wassertrans> ribband (Schiffbau)
Sentenriss m <Wassertrans> buttock lines (Schiffkonstruktion); plan of diagonals (Schiffbau)
Sentlatte f <Wassertrans> ribband (Schiffbau)
separate Streichmaschine f <Papier> off-machine coater
Separator m <Meerschmutz, Telekom> separator
Separatorzentrifuge f <Chemtech> centrifugal separator
Separatstreichen n <Papier> off-machine coating
Separierung f <Comp & DV> compartmentalization
Septime f <Akustik> seventh
Septum n <Elektrotech> septum
Sequencer m <Comp & DV> sequencer
Sequenz f <Akustik, Comp & DV, Kontroll, Telekom> sequence
sequenziell adj <Comp & DV, Kontroll> sequential
sequenzielle Arbeitsweise f <Comp & DV> sequential operation
sequenzielle Datei f <Comp & DV> sequential file
sequenzielle Decodierung f <Telekom> sequential decoding
sequenzielle Prüfung f <Telekom> sequential test
sequenzielle Steuerung f <Comp & DV> sequencing
sequenzielle Suche f <Comp & DV> sequential search
sequenzielle Verarbeitung f <Comp & DV> sequential processing
sequenzieller Computer m <Comp & DV> sequential computer
sequenzieller Modus m <Comp & DV> sequential mode
sequenzieller Zugriff m <Comp & DV> sequential access
sequenzielles Suchen n <Comp & DV> sequential search
Sequestiermittel n <Chemie> sequestering agent
serialisieren v <Comp & DV> serialize
Serie f 1. <Fertig> gang (Werkzeuge); 2. <Maschinen> series • **in Serie geschaltet** 1. <Gerät> series-connected; 2. <Phys> connected in series
seriell adj 1. <Comp & DV, Druck> serial; 2. <Elektriz, Kontroll, Telekom> serial
seriell erregte Maschine f <Elektriz> series-excited machine, series-wound machine
seriell erregter Dynamo m <Elektriz> series-wound dynamo
seriell erregter Motor m <Elektriz> series-wound motor
serielle Analog-Digital-Umsetzung f <Elektronik> (AE) serial analog-digital conversion
serielle Datei f <Comp & DV> serial file
serielle digitale Ausgabe f <Comp & DV> serial digital output
serielle Ein-/Ausgabe f <Comp & DV> SIO, serial input/output
serielle Form f <Elektrotech> serial form
serielle Leitung f <Elektrotech> serial line
serielle Operation f <Comp & DV> serial operation
serielle Schnittstelle f <Comp & DV, Druck, Telekom> serial interface
serielle Übertragung f <Comp & DV, Telekom> serial transfer, serial transmission
serielle Verarbeitung f <Comp & DV> serial processing
serielle Wicklung f <Elektriz> series winding
serieller A/D-Wandler m <Elektronik> serial analog-digital converter
serieller Addierer m <Comp & DV> ripple-carry adder
serieller Anschluss m <Comp & DV> serial connector
serieller Betrieb m <Comp & DV> serial operation
serieller Computer m <Comp & DV> serial computer
serieller Drucker m <Comp & DV, Druck> serial printer
serieller Ein-/Ausgabe-Baustein m <Elektronik> serial input-output chip, serial input-output unit, SIO chip
serieller Rechner m (SISD-Rechner) <Comp & DV> single instruction single-data machine (SISD machine)

serieller Speicher m <Comp & DV, Elektrotech> serial memory, serial storage
serieller Subtrahierer m <Elektronik> serial subtracter
serieller Transfer m <Comp & DV, Telekom> serial transfer
serieller Wertgeber m <Comp & DV> serial dial
serieller Zugriff m <Comp & DV, Telekom> serial access
serielles Kabel n <Elektrotech> serial line
serielles Programmieren n <Comp & DV> serial programming
seriell-parallel adj <Comp & DV> serial-parallel
seriell-parallele Umwandlung f <Elektrotech> serial-to-parallel conversion
seriell-paralleler Umwandler m <Elektrotech> serial-to-parallel converter
Serien... 1. <Phys> in series; 2. <Telekom> serial
Serienaddierer m <Comp & DV, Elektronik> serial adder
Serienarbeit f <Fertig> repetitive work
Serienbetrieb m 1. <Comp & DV> serial operation; 2. <Telekom> sequential operation
Seriencode m <Verpack> batch code
Serienfahrzeug n <Eisenbahn> production-type vehicle
Serienfertigung f <Fertig> repetitive work
Serienflugzeug n <Lufttrans> production aircraft
Seriengegenkopplung f <Elektrotech> series feedback
seriengeregelte Stromversorgung f <Elektrotech> series-regulated power supply
seriengeschaltete Spule f <Elektriz> series coil
seriengewickelter Dynamo m <Elektrotech> series-wound dynamo
seriengewickelter Motor m <Elektrotech> series-wound motor
Serienkapazität f <Elektrotech> series capacitance
Serienkollektorwiderstand m <Elektrotech> series collector resistance
Serienkondensator m <Elektrotech> series capacitor
Serienkonverter m <Elektrotech> series converter
Serienmarke f <Patent> associated mark
serienmäßig adj <Fertig> standard (Kunststoffinstallationen)
serienmäßig hergestellter Leistungsreaktor m <Kerntech> series-produced power reactor
serienmäßige Seitenruder npl <Lufttrans> serial rudders
Serienmodus m <Telekom> sequential mode
Serienmotor m <Elektrotech> series motor
Seriennummer f 1. <Comp & DV, Foto, Maschinen> serial number; 2. <Verpack> batch number
Serienparallelschalter m <Elektrotech> series-parallel switch
Serienparallelschaltung f <Elektrotech> series-parallel circuit
Serienprüfung f <Qual> batch test
Serienreaktanz f <Elektrotech> series reactance
Serienresonanz f <Phys> series resonance
Serienresonanzkreis m <Elektriz, Elektrotech> series-resonant circuit
Serienschaltung f 1. <Elektrotech> series circuit, series connection; 2. <Phys> series connection
Serienspeisung f <Elektrotech> series feed
Serienstörsignal n <Regelung> series mode signal
Serienstörsignalunterdrückung f <Regelung> series mode rejection
Serienstörsignalunterdrückungsmaß n <Regelung> series mode rejection ratio
Serientaktsignal n <Telekom> series mode signal
Serientakt-Unterdrückung f <Telekom> series mode rejection
Serientakt-Unterdrückungsmaß n <Telekom> series mode rejection ratio
Serienübertragung f <Comp & DV, Telekom> serial transmission
Serienvergleichsumsetzer m <Gerät> level-at-a-time converter
Serienwandler m <Elektrotech> series converter
Serienwicklung f <Elektrotech> series winding
Serienwiderstand m <Elektrotech> series resistance
Serienzeichen n <Patent> associated mark
Serife f <Druck> serif
serifenlose Linear-Antiqua f <Druck, Konstzeich> sans serif linear antiqua
serifenlose Schrift f <Druck, Konstzeich> sans serif
Serifenschrift f <Druck> serif
Serigraphie f <Druck> serigraphy
Serotonin n <Chemie> serotonin
Serpentin m <Bau> serpentine
Serpentinisierung f <Nichtfoss Energ> serpentinization
Server m <Comp & DV, Telekom> server
Server-basiertes HTML n <Comp & DV> server-parsed HTML
Serverfarm f <Comp & DV> server farm
Servermanager m <Comp & DV> server manager
Serversystem n <Telekom> server system
Server-zentrisch adj <Comp & DV> server-centric
Service m <Fernseh, Telekom> after-sales service
Service m für Nachrichtenintegrität <Comp & DV> message integrity service
Service m und Unterstützung <Comp & DV> service and support
Serviceaufruf m <Comp & DV> service call
Serviceklasse f (COS) <Comp & DV> class of service (COS)
Serviceknoten m <Comp & DV> service node
Serviceoszillator m <Elektronik> service oscillator
Servicestandard m <Comp & DV> service standard
Servicestufe f <Comp & DV> service level
Servierwagen m <Lufttrans> service trolley (Flugzeug)
Servo m <Kfztech> servo (Bremsen, Lenkung)
Servo m zweiter Ordnung <Elektrotech> second order servo
Servoantrieb m 1. <Elektriz> servo drive; 2. <Kfztech> booster control
Servobremse f 1. <Kfztech> power brake, power-assisted brake, servo brake; 2. <Maschinen> servo brake
Servoeinrichtung f <Maschinen> servomechanism
servogesteuert adj <Fertig> servo-acting (Kunststoffinstallationen)
Servohöhenmesser m <Lufttrans> servo altimeter
Servolenkgestängesystem n <Kfztech> linkage power-steering system
Servolenkpumpe f <Kfztech> power-steering pump
Servolenkung f <Kfztech> power steering, power-assisted steering
Servomanipulator m <Kerntech> servo manipulator
Servomechanismus m <Comp & DV, Elektrotech, Fernseh, Maschinen, Phys> servomechanism
Servomotor m <Elektrotech, Kerntech, Kontroll, Nichtfoss Energ> servomotor
Servomotorsteuerung f <Lufttrans> servo control
Servopositionierer m <Kontroll> servo positioner
Servorad n <Fernseh> servo wheel
Servoregelung f <Elektrotech, Kontroll> servo control
Servoregler m <Kontroll> servo controller
Servoschleife f <Fernseh> servo loop
Servostellglied n <Kontroll> servo positioner
Servosteuerung f 1. <Kontroll> servo controller; 2. <Lufttrans> servo control
Servosteuerungssystem n <Lufttrans> follow-up (Hubschrauber)
Servosystem n <Fernseh, Kontroll> servo system

Servosystem

Servosystem n **mit Rückführung** <Kontroll> closed-loop servo system
Servosystemabweichung f <Raumfahrt> servo system drift *(Weltraumfunk)*
Servotonrolle f <Fernseh> servo capstan
Servoventil n <Maschinen> pilot valve, servo valve
Servoverstärker m <Elektronik> servo amplifier
SES *(Bordterminal für Satellitenfunk)* <Wassertrans> SES *(ship earth station)*
Sesamöl n <Lebensmittel> sesame oil
Sesselliftbahn f <Trans> chair lift
Setzarbeit f <Kohlen> jigging
Setzbecher m <Bau> slump cone *(Ausbreitversuch)*
Setzbett n <Kohlen> jig bed
Setzbord n <Wassertrans> washboard *(Schiffbau)*
Setzeisen n <Ker & Glas> flatter
setzen v 1. <Bau> set; 2. <Bau> subside *(Untergrund)*; 3. <Druck> compose, set, typeset; 4. <Kontroll> set
Setzen n 1. <Druck> composition, typesetting; 2. <Kohlen> setting, settling; 3. <Kunststoff> permanent set, set
Setzer m <Druck> compositor, typesetter
Setzkasten m <Kohlen> jig, jig sieve, settling tank
Setzkopf m 1. <Fertig> die head *(Nieten)*; preformed head *(Niet)*; 2. <Maschinen> set head, snap head
Setzlibelle f <Maschinen> striding level
Setzmaschine f 1. <Druck> composing machine; 2. <Kohlen> jig, pan, settling tank
Setzmaschinenzeile f <Druck> slug
Setzmaß n <Bau> slump
Setzprobe f <Bau> slump test
Setzpult n <Druck> frame
Setzregal n <Druck> frame
Setzschiff n <Druck> galley
Setzstock m 1. <Fertig> end-support column, lathe steady, outer stay; 2. <Maschinen> back rest, steady-rest
Setzstufe f <Bau> riser
Setzung f 1. <Bau> subsidence; settlement *(Gebäude, Gelände)*; 2. <Erdöl> compaction; subsidence *(Geologie)*; 3. <Kunststoff> permanent set, set
Setzungsunterschied m <Bau> differential settlement
Seuche f <Lebensmittel> pest
Sext f <Akustik> sixth
Sextant m <Phys, Wassertrans> sextant *(Navigation)*
Sexte f <Akustik> submediant
Sextett n <Phys> sextet *(Spektroskopie)*
Sezierbehälter m <Labor> dissecting tray
Sezierschere f <Labor> dissecting scissors
SF 1. <Bau, Elektriz, Kerntech, Kohlen, Maschinen, Sicherheit, Trans> *(Sicherheitsfaktor)* SF *(safety factor)*; 2. <Elektronik, Fernseh, Funktech, Telekom> *(Seitenbandfrequenz)* SF *(sideband frequency)*; 3. <Elektronik, Fernseh, Funktech, Telekom> *(Signalfrequenz)* SF *(signal frequency)*; 4. <Elektronik> *(Sprachfrequenz)* VF *(voice frequency)*; SF *(speech frequency)*; 5. <Telekom> *(Sprachfrequenz)* SF *(speech frequency)*; VF *(voice frequency)*
S-förmiger Geruchverschluss m <Bau> S-trap *(Sanitär)*
S-förmiger Haken m <Maschinen> S-hook, S-shaped hook
SGML *(Standardkorrekturzeichensatz)* <Comp & DV, Druck> SGML *(Standard Generalized Markup Language)*
Shapingmaschine f 1. <Maschinen> shaping machine, shaping planer; 2. <Mechan> shaper
Shareware f <Comp & DV> shareware
Sheddach n <Bau> sawtooth roof
Shell f 1. <Comp & DV> shell; 2. <Künstl Int> shell *(eines Expertensystems)*
Shell-Prozedur f <Comp & DV> shell script
SHF *(superhohe Frequenz)* <Bau, Elektronik, Funktech, Kohlen, Kunststoff, Mechan, Phys, Thermod> SHF *(superhigh frequency)*

SHF-Signalerzeuger m <Elektronik> SHF signal generator
SHF-Signalgenerator m <Elektronik> SHF signal generator
Shippington f <Wassertrans> shipping ton
Shirting m <Textil> shirting
Shockley-Diode f <Elektronik> Shockley diode
Shockley-Versetzung f <Metall> Shockley dislocation
Shoddygewebe n <Textil> shoddy fabrics
Shore-Härte f <Kunststoff> Shore hardness
Shore-Härteprüfer m <Labor> Shore hardness tester
Shredder m 1. <Papier> shredder; 2. <Verpack> shredding machine
Shredderabfälle-Deponie f <Abfall> shredded refuse landfill
Shredding n <Abfall> shredding
Shunt m <Akustik, Elektriz, Elektrotech, Phys> shunt
Shuttle n <Eisenbahn, Lufttrans> shuttle
Shuttlehubschrauber m <Erdöl> shuttle helicopter *(Transport)*
Shuttleplatte f <Hydraul> shuttle plate
Shuttletanker m <Erdöl> shuttle tanker *(Transport)*
Si *(Silizium)* <Chemie> Si *(silicon)*
sichelförmiges Widerlager n <Fertig> crescent
sichere Belastung f <Sicherheit> safe working load
sichere Deponie f <Abfall> safe dumping ground
sichere Geschwindigkeit f <Eisenbahn, Kfztech, Lufttrans, Trans, Wassertrans> safety speed
sichere Gestaltung f <Sicherheit> safety design
sichere Höchstgeschwindigkeit f <Lufttrans> maximum threshold speed
sichere Lagerung f <Umweltschmutz> safe keeping
sichere Mindestgeschwindigkeit f <Lufttrans> minimum control speed
sichere Mindestvorführgeschwindigkeit f <Lufttrans> minimum demonstrated threshold speed
sicherer Grund m <Wassertrans> safe ground
sicherer Schutz m <Sicherheit> positive protection
sicheres Gewässer n <Wassertrans> safe water
Sicherheit f 1. <Comp & DV> security; 2. <Funktech> safety; 3. <Math> assurance; 4. <Mechan, Qual, Sicherheit, Telekom> safety
Sicherheit f **auf See** <Sicherheit, Wassertrans> marine safety
Sicherheit f **beim Betrieb von Handgeräten** <Sicherheit> safety of hand-operated machines
Sicherheit f **im Straßenverkehr** <Sicherheit> road safety
Sicherheitsabblasearmatur f *(SBV)* <Erdöl> relief valve
Sicherheitsabsperrung f <Sicherheit> safety barrier
Sicherheitsabstand m 1. <Lufttrans> clearance; 2. <Sicherheit> clearance, clearance distance, safety distance; 3. <Trans> safe headway, safety headway; 4. <Wassertrans> clearance
Sicherheitsabstand m **zwischen Frequenzbändern** <Fernseh> guard band
Sicherheitsanschlag m <Sicherheit> safety stop *(Maschine)*
Sicherheitsanschluss m <Sicherheit> safety fitting; safety fitting *(für Heißwassersysteme)*
Sicherheitsarmaturen fpl <Sicherheit> safety fittings
Sicherheitsausleger m <Erdöl> containment boom
Sicherheitsausrüstung f <Sicherheit> safety equipment, safety gear
Sicherheitsausschalter m 1. <Eisenbahn> cutout switch; 2. <Heiz & Kälte> safety cutout
Sicherheitsautomat m <Heiz & Kälte> safety cutout *(elektrisch)*
Sicherheitsbehälter m 1. <Kerntech> safety tank; 2. <Labor> safety container

Sicherheitsbenzintank m <Kfztech> (AE) safety gasoline tank, (BE) safety petrol tank
Sicherheitsberater m <Bau> safety adviser
Sicherheitsbericht m 1. <Kerntech> safety report (eines Reaktors); 2. <Qual> safety report
Sicherheitsberstscheibe f <Gerät, Sicherheit> (BE) safety disc, (AE) safety disk
Sicherheitsbolzen m <Sicherheit> security bolt
Sicherheitsbrille f <Labor, Sicherheit> safety glasses, safety goggles, safety spectacles
Sicherheitsbulletin n <Comp & DV> security bulletin
Sicherheitsdatenblatt n <Sicherheit> safety data sheet
Sicherheitsdeponie f <Abfall> safe dumping ground
Sicherheitseinrichtung f 1. <Maschinen, Qual> safety appliance; 2. <Sicherheit> safeguard; safety appliance, safety facility; safety fitting, security device (an Gebäuden)
Sicherheitserde f <Elektriz> (BE) safety earth, (AE) safety ground
Sicherheitsfachkraft f <Sicherheit> safety engineer, safety expert
Sicherheitsfahrschaltung f <Eisenbahn> deadman's handle
Sicherheitsfaktor m (SF) 1. <Bau, Elektriz, Kerntech, Kohlen> safety factor (SF); 2. <Künstl Int> certainty factor, confidence factor, degree of truth; 3. <Maschinen, Qual, Sicherheit, Trans> safety factor (SF)
Sicherheitsfarbe f <Sicherheit> (AE) safety color, (BE) safety colour
Sicherheitsfenster n <Sicherheit> security window
Sicherheitsfilm m <Druck, Foto> safety film
sicherheitsgeprüft adj <Patent, Qual, Sicherheit> approved
Sicherheitsgerät n 1. <Raumfahrt> safety unit; 2. <Qual> safety appliance; 3. <Sicherheit> safety apparatus, safety appliance
sicherheitsgerechte Gestaltung f <Ergon, Sicherheit> safety design
Sicherheitsgerüst n <Sicherheit> protective scaffold
Sicherheitsgeschirr n <Sicherheit> fall protection harness, safety harness
Sicherheitsglas n 1. <Bau, Ker & Glas> safety glass; 2. <Kfztech> multilayer glass, safety glass; 3. <Sicherheit> protective glass
Sicherheitsgriff m <Sicherheit> safety handle
Sicherheitsgurt m 1. <Kfztech> safety belt, safety harness, seat belt; 2. <Lufttrans> safety harness, seat belt; 3. <Raumfahrt> safety harness; 4. <Sicherheit> safety belt; 5. <Wassertrans> safety harness
Sicherheitsgurtspanner m <Kfztech> seat belt tensioner
Sicherheitsgurtverankerung f <Kfztech> safety belt anchorage
Sicherheitshahn m <Maschinen> safety cock
Sicherheitshaken m <Maschinen> safety hook
Sicherheitshebel m <Sicherheit> safety catch
Sicherheitshinweisschild n <Sicherheit> safety sign
Sicherheitsingenieur m <Sicherheit> safety engineer
Sicherheitskabine f <Sicherheit> safety cab (Schlepper)
Sicherheitskette f 1. <Maschinen> safety chain; 2. <Sicherheit> security chain
Sicherheitsklinke f <Sicherheit> safety pawl
Sicherheitskopie f <Comp & DV> security backup
Sicherheitskupplung f 1. <Maschinen> safety coupling; 2. <Sicherheit> safety clutch
Sicherheitslampe f <Kohlen> Davy lamp
Sicherheitslampe f des Bergmanns <Sicherheit> miner's safety lamp
Sicherheitsleitkegel m <Trans> road marker cone
Sicherheitsleitplanke f <Trans> emergency crash barrier

Sicherheitsluftventil n <Kfztech> vacuum valve
Sicherheitsmarge f <Sicherheit> safety margin
Sicherheitsmaßnahme f <Sicherheit> precaution, safety measure
Sicherheitsmindestgeschwindigkeit f beim Start <Lufttrans> minimum takeoff safety speed
Sicherheitsmindesthöhe f <Lufttrans> minimum safe altitude
Sicherheitsmotorhaube f <Kfztech> (BE) safety bonnet, (AE) safety hood
Sicherheitsmutter f <Maschinen> safety nut, self-locking nut
Sicherheitsnaht f <Textil> safety-stitch seam
Sicherheitsnetz n <Sicherheit> safety net
Sicherheitsnorm f <Telekom> safety standard
Sicherheitsplakat n <Labor> safety placard
Sicherheitsprotokoll n <Bau, Qual> safety record
Sicherheitsrahmen m <Sicherheit> protective frame, safety frame, safety structure (Schlepper)
Sicherheitsraste f <Sicherheit> safety catch
Sicherheitsregel f <Qual, Sicherheit> safety rule
Sicherheitsrichtlinie f <Bau, Eisenbahn, Kfztech, Lufttrans, Qual, Sicherheit, Trans, Wassertrans> safety recommendation
Sicherheitsrisiken npl <Comp & DV> security vulnerability
Sicherheitsrohr n <Labor> safety tube
Sicherheitsrolle f <Sicherheit> safety block (Fallschutzeinrichtung)
Sicherheitsschalter m <Elektriz, Elektrotech> safety switch
Sicherheitsschiene f <Eisenbahn> safety rail
Sicherheitsschirm m 1. <Labor> safety screen; 2. <Sicherheit> safety screen, security screen
Sicherheitsschloss n <Bau> safety lock
Sicherheitsschlüssel m <Comp & DV> security identification
Sicherheitsschranke f <Sicherheit> safety barrier
Sicherheitsschuhwerk n <Sicherheit> protective footwear, safety footwear
Sicherheitsseil n <Sicherheit> lifeline
Sicherheitsservice m <Comp & DV> security service
Sicherheitssperre f 1. <Comp & DV> interlock; 2. <Sicherheit> safety interlock
Sicherheitssprengstoff m <Sicherheit> safety explosive
Sicherheitsstandard m <Sicherheit, Telekom> safety standard
Sicherheitsstiefel m <Sicherheit> safety boot
Sicherheitsstrahlengehäuse n <Sicherheit> safety beam enclosure (Laser)
Sicherheitsstruktur f <Sicherheit> safety structure
Sicherheitssystem n 1. <Kerntech> safety system; 2. <Sicherheit> protective structure, safety equipment
Sicherheitstank m <Sicherheit> safety storage tank
Sicherheitstechnik f <Sicherheit> safety engineering
Sicherheitstechniker m <Sicherheit> safety engineer
sicherheitstechnisch gestaltet adj <Sicherheit> safety-engineered
sicherheitstechnische Anforderungen fpl <Ergon, Qual> safety requirements
sicherheitstechnische Anlage f <Qual> safety system
sicherheitstechnische Gestaltung f <Sicherheit> design for improved safety, design for safguarding, design for safety, engineering for safety, safety design
sicherheitstechnische Mittel npl <Sicherheit> safety equipment
Sicherheitstechnologie f <Comp & DV> secure technology
Sicherheitstor n 1. <Bau> safety door; 2. <Sicherheit> guard gate

Sicherheitstrichter

Sicherheitstrichter m <Labor> safety funnel
Sicherheitstür f 1. <Bau> safety door; 2. <Sicherheit> security door
Sicherheitsüberdruckventil n <Sicherheit> safety relief valve
Sicherheitsumlenkrolle f <Sicherheit> safety pulley block
Sicherheitsventil n 1. <Eisenbahn> overflow valve; 2. <Heiz & Kälte> relief valve, safety valve; 3. <Hydraul> escape valve; safety valve *(Dampfkessel)*; 4. <Kfztech> relief valve *(Schmierung)*; 5. <Maschinen> air valve; 6. <Mechan> safety valve; 7. <Nichtfoss Energ> relief valve; 8. <Sicherheit> relief safety valve, safety relief valve, safety valve
Sicherheitsverankerung f <Kfztech> seat belt anchorage
Sicherheitsverpackung f <Verpack> barrier packaging
Sicherheitsverriegelung f <Sicherheit> safety interlock, safety locking device
Sicherheitsverschluss m 1. <Sicherheit> safety clamp, safety lock; 2. <Verpack> safety closure
Sicherheitsvorhang m <Sicherheit> safety curtain
Sicherheitsvorkehrung f <Sicherheit> safety precaution
Sicherheitsvorrichtung f 1. <Maschinen, Sicherheit> protector, safety apparatus, safety appliance, safety device; 2. <Qual> safety apparatus
Sicherheitsvorschrift f 1. <Eisenbahn, Kfztech, Lufttrans> safety specification; 2. <Qual> safety regulation, safety requirement, safety specification; 3. <Sicherheit> factory safety regulation, safety code, safety instruction, safety requirement; safety regulation *(behördlich)*; 4. <Trans, Wassertrans> safety specification
Sicherheitszone f **am Ende der Start- und Landebahn** <Lufttrans> runway-end safety area
Sicherheitszündschnur f <Sicherheit> safety fuse
sichern v 1. <Bau> secure; 2. <Comp & DV> save; back up *(Daten)*; 3. <Mechan> lock; 4. <Qual> secure; 5. <Sicherheit> lock, secure
Sichern n <Comp & DV> save
Sicherung f 1. <Comp & DV> backup; 2. <Elektriz, Elektrotech, Fernseh, Funktech> fuse; 3. <Maschinen> locking device; 4. <Sicherheit> locking device, safety lock
• **durch Sicherung geschützt** <Elektrotech, Sicherheit> fuse-protected
Sicherung f **der mittleren Qualität** <Qual> average quality protection
Sicherung f **einer Qualität je Los** <Qual> lot quality protection
Sicherung f **mit Alarm und Signalgeber** <Elektriz> alarm fuse
Sicherungsanordnung f <Elektrotech> fuse array
Sicherungsautomat m <Elektriz> automatic circuit breaker
Sicherungsbereich m <Comp & DV> save area
Sicherungsblech n <Maschinen> lock plate, locking plate
Sicherungsblech n **mit Nase** <Mechan> tab washer
Sicherungsbolzen m <Wassertrans> safety pin *(Decksausrüstung)*
Sicherungsboot n <Wassertrans> *(BE)* seaward defence boat, *(AE)* seaward defense boat *(Marine)*
Sicherungsbrett n <Elektrotech> fuseboard
Sicherungsbrücke f <Elektriz> link fuse
Sicherungsbügel m <Kfztech> circlip *(Motor)*
Sicherungsdatei f <Comp & DV> save file
Sicherungsdraht m <Elektriz, Elektrotech> fuse wire
Sicherungseinsatz m <Elektriz> fuse link
Sicherungselement n <Elektriz, Elektrotech> fuse link
Sicherungsfach n <Kfztech> fuse box *(KFZ-Elektrik)*
Sicherungsfassung f <Elektrotech> fuse holder
Sicherungsfeder f <Maschinen> retaining spring
Sicherungsglied n <Elektriz> fuse, link
Sicherungsgriff m <Elektrotech> fuse carrier *(bei Rohrpatronensicherung)*
Sicherungshalter m <Elektrotech, Fernseh> fuse holder
Sicherungskasten m 1. <Elektriz, Elektrotech, Kfztech> fuse box *(KFZ-Elektrik)*; 2. <Nichtfoss Energ> consumer unit
Sicherungskette f <Sicherheit> safety chain
Sicherungskopie f <Comp & DV> backup; back-up tape, security copy *(auf Magnetband)*
Sicherungslasche f <Fernseh> record defeat tab
Sicherungsleiste f <Elektriz> fuse strip
Sicherungsmutter f 1. <Kfztech> locknut; 2. <Maschinen> locknut, pinch nut; 3. <Mechan> locknut
Sicherungsnut f <Kfztech> lock groove
Sicherungspatrone f <Elektrotech> fuse cartridge
Sicherungsposten m <Eisenbahn> flagman
Sicherungsraste f <Maschinen> safety catch
Sicherungsring m 1. <Kfztech> circlip *(Motor)*; 2. <Maschinen> circlip, retaining ring, snap ring
Sicherungsschaltung f <Eisenbahn> *(AE)* signaling wiring diagram, *(BE)* signalling wiring diagram
Sicherungsscheibe f 1. <Fertig> retaining washer; 2. <Mechan> lock washer
Sicherungsschicht f <Telekom> data link layer
Sicherungsschmelzstreifen m <Elektriz> fuse strip
Sicherungsspeicher m <Comp & DV> *(AE)* backing storage, *(BE)* backing store
Sicherungsstift m <Maschinen> locking pin
Sicherungstafel f <Elektrotech> fuseboard
Sicht f 1. <Eisenbahn, Kfztech> sight; 2. <Lufttrans> sight, visibility; 3. <Raumfahrt> optical sight; 4. <Trans, Wassertrans> sight, visibility
Sicht f **am Boden** <Lufttrans> ground visibility
Sichtanflug m <Raumfahrt> visual approach
Sichtanflugpiste f <Lufttrans> noninstrument runway
Sichtanzeige f 1. <Comp & DV> readout; 2. <Gerät> visual display; 3. <Phys> visual display unit
Sichtanzeige f **im Fahrzeug** <Kfztech> in-vehicle visual display
sichtbare Fläche f <Phys> face
sichtbare Laserlinien fpl <Strahlphys> visible laser lines
sichtbare Seite f <Geom> visible face
sichtbare Strahlung f <Optik, Telekom> visible radiation
sichtbarer Horizont m <Wassertrans> visible horizon *(astronomische Navigation)*
sichtbares Feld n <Bau, Optik> visual field
sichtbares Gebiet n <Nichtfoss Energ> visible region
sichtbares Licht n <Telekom> visible light
sichtbares Spektrum n <Phys> visible spectrum
Sichtbarkeit f <Phys> visibility
Sichtbarmachung f <Comp & DV> visualization
Sichtbedingungen fpl <Ergon, Telekom> conditions of visibility, visibility
sichten v <Chemtech> classify
Sichtentfernung f <Funktech> line-of-sight distance
Sichter m <Ker & Glas> classifier
Sichtfeld n <Optik> visual field
Sichtfenster n <Comp & DV> viewing window
Sichtfläche f <Bau> face, front
Sichtflächen fpl <Heiz & Kälte> exposed surfaces
Sichtflug m <Lufttrans> contact flight
Sichtfolie f <Konstzeich> transparent film
Sichtfunkpeiler m <Funkort> cathode-ray direction finder
Sichtgerät n 1. <Comp & DV> display, display device, display unit, video display unit, visual display unit, VDU; 2. <Funkort> display unit *(Radar)*; 3. <Gerät> cathode-ray tube display system, CRT display system, display device, indicator, oscilloscope, visual display; 4. <Raumfahrt> op-

tical sight; 5. <Telekom> terminal, visual display unit, video display unit
Sichtgitter n <Bau> screen
Sichtglas n <Verpack> inspection window, liquid level indicator
Sichtglasöler m <Maschinen> sight feed oiler
Sichtgrenze f <Lufttrans, Trans, Wassertrans> visibility limit
sichtiges Wetter n <Eisenbahn, Wassertrans> clear weather
Sichtkontrolle f <Qual> visual examination
Sichtlinie f <Optik> line of sight, line of vision
Sichtliniensignal n <Elektronik> line-of-sight signal
Sichtlinienwinkel m <Telekom> angle of sight
Sichtloch n <Maschinen> sight hole
Sichtmelder m <Gerät> annunciator
Sichtprüfmenge f <Qual> inspection test quantity
Sichtprüfung f 1. <Metrol> visual inspection; 2. <Qual> VT, visual testing, visual examination, visual inspection; 3. <Werkprüf> visual inspection
Sichtprüfung f der Fahrwerksausfahrverriegelung <Lufttrans> landing-gear downlock visual check
Sichtscheibe f 1. <Comp & DV> viewing window; 2. <Sicherheit> eyepiece *(Atemschutzgerät)*
Sichtschutz m <Abfall> screen *(einer Deponie)*
Sichttafel f <Gerät> annunciator
Sichtverhältnisse npl <Eisenbahn> sighting
Sichtvermerk m <Konstzeich> endorsement
Sichtweite f 1. <Bau, Eisenbahn> sight distance, sighting distance; 2. <Kfztech> sight distance; 3. <Lufttrans> sight distance, visibility, visibility distance; 4. <Optik> reach; 5. <Trans, Wassertrans> sight distance, visibility, visibility distance
Sichtweitenmessinstrument n <Lufttrans, Trans, Wassertrans> visibility distance measuring equipment
Sichtzeichengeber m <Raumfahrt> annunciator
Sicke f 1. <Bau> bead; 2. <Fertig> crease, dimple, reinforcing crease; 3. <Maschinen> bead
sicken v 1. <Fertig> crease; 2. <Maschinen> bead, crimp
Sicken n <Maschinen> beading, crimping
Sickenmaschine f <Fertig> flanging machine
Sickerbecken n <Abfall> infiltration basin
Sickerbrunnen m <Wasserversorg> dry well
Sickerdrainage f <Bau> rubble drain
Sickerdruck m <Bau> seepage force
Sickergrube f <Bau> soakage pit
Sickern n <Wasserversorg> seepage
Sickerschlitz m <Bau> weephole
Sickerstrahlung f <Elektrotech> leakage radiation
Sickerwasser n 1. <Abfall> leakage water, percolating water, seepage water; 2. <Bau> seepage water, soakage water; 3. <Chemie> seepage water; 4. <Kohlen> gravitational water, seepage water; 5. <Wasserversorg> leak water, percolating water
siderische Zeit f <Raumfahrt> sidereal time *(Sternzeit)*
siderischer Tag m <Phys> sidereal day
siderisches Jahr n <Phys> sidereal year
Siderit m <Metall> siderite
sideromagnetisch adj <Metall> sideromagnetic
Siderurgie f <Metall> siderurgy
Sieb n 1. <Abfall> screening equipment; 2. <Bau> screen, sieve; 3. <Elektronik> filter *(Netzwerk)*; 4. <Erdöl> strainer; 5. <Fertig> ratter, screen, sifter; 6. <Heiz & Kälte> strainer; 7. <Ker & Glas> silk screen; 8. <Kfztech> strainer; 9. <Kohlen> screen, sieve; 10. <Labor> sieve; 11. <Maschinen> strainer; 12. <Mechan> screen; 13. <Papier> wire; 14. <Textil> screen
Siebanalyse f 1. <Chemtech> sieve analysis; sieve classification *(Minerale)*; 2. <Ker & Glas> screen analysis; 3. <Kerntech> sieve analysis; 4. <Kohlen> screen analysis, sieve analysis
Siebanlage f <Kohlen> screening plant
Siebantriebswalze f <Papier> wire drive roll
Siebbereich m <Bau> grading envelope *(Diagramm)*
Siebblech n 1. <Bau> punched-plate screen; 2. <Kerntech> perforated plate; 3. <Kohlen> screen plate
Siebboden m <Chemtech> sieve bottom, sieve plate, sieve tray
Siebbohrloch n <Fertig> screen perforation *(Kunststoffinstallationen)*
Siebbüchse f <Chemtech> sieve frame
Siebdämpfung f <Elektronik> filter attenuation
Siebdrossel f <Elektrotech> filter choke, swinging choke
Siebdruck m 1. <Druck> serigraphy; 2. <Ker & Glas> screen printing; 3. <Verpack> silk screen printing
Siebdruckmaschine f <Verpack> screen printing machine
Siebdurchfall m <Kohlen> duff, underflow
Siebdurchgang m 1. <Fertig> sifting; 2. <Kohlen> screenings; 3. <Maschinen> sieving
Siebdurchlauf m <Kohlen> fines
sieben v 1. <Bau> screen *(Erde)*; 2. <Chemtech> sieve; 3. <Fertig> strain; 4. <Lebensmittel> sieve, sift; 5. <Meerschmutz> sift; 6. <Papier> screen
Sieben n 1. <Abfall> screening; 2. <Bau> grading, screening, sieving; 3. <Fertig, Meerschmutz> screening; 4. <Kohlen> screening, sieving
Sieben f <Math> seven
Siebeneck n <Fertig, Geom> heptagon
siebeneckig adj <Fertig, Geom> heptagonal
siebenflächig adj <Fertig> heptahedral
Siebenflächner m <Geom> heptahedron
Sieben-Schichten-Modell n <Telekom> seven layer model *(OSI)*
Sieben-Schicht-Referenzmodell n <Comp & DV, Telekom> seven layer reference model *(ISO/OSI)*
Siebensegmentanzeige f <Telekom> seven-bar segmented display, seven-segment display, stick display
Siebentonleiter f <Akustik> heptatonic scale
siebenwertig adj 1. <Chemie> heptavalent, septivalent; 2. <Fertig> heptavalent
Siebereifeinkohle f <Kohlen> duff
Siebfeines n <Kohlen> fines
Siebfilter n <Meerschmutz> strainer
Siebfläche f <Kohlen> screening surface
Siebfraktion f <Ker & Glas> sieve fraction
Siebgeflecht n <Kohlen> sieve mesh
Siebgeschwindigkeit f <Chemtech> sieving rate
Siebgewebe n 1. <Chemtech> sieve cloth; 2. <Kunststoff> filter screen; 3. <Papier> wire cloth
Siebglied n <Elektronik> filter element
Siebkäfig m <Fertig> screen cage *(Kunststoffinstallationen)*
Siebkette f <Funktech> ladder filter
siebklassieren v <Chemtech> sieve *(Erz)*
Siebklassierung f <Chemtech> sieve classification
Siebkohle f <Kohlen> sifted coal
Siebkondensator m <Elektrotech> filter capacitor
Siebkopf m <Labor> bolthead *(Analyse)*
Siebkopfkolben m <Labor> *(BE)* bolthead flask, *(AE)* matrass *(Analyse)*
Siebkurve f <Kohlen> grading curve
Sieblaufregler m <Papier> wire guide
Siebleder n <Papier> apron
Siebledebrett n <Papier> apron board
Siebleiderlippe f <Papier> apron lip
Siebleitwalze f <Papier> wire guide roll
Sieblinie f <Kohlen> grading curve
Siebmasche f <Fertig> mesh

Siebpartie

Siebpartie f <Papier> wire end
Siebrahmen m 1. <Chemtech> sieve frame; 2. <Papier> wire frame
Siebrest m <Abfall> screenings
Siebrohr n <Fertig> screen *(Kunststoffinstallationen)*
Siebrost m 1. <Chemtech> sieve grate; 2. <Kohlen> grizzly
Siebrückstand m <Abfall, Kohlen> screenings
Siebrückstände mpl <Lebensmittel> tailings *(Müllerei)*
Siebsatz m <Chemtech> sieve set
Siebsaugwalze f <Papier> suction roll
Siebschaltung f <Elektronik> filter circuit
Siebscheuersand m <Anstrich> mesh abrasive grit
Siebschüttler m <Labor> sieve shaker
Siebseite f <Druck, Papier> wire side
Siebspannvorrichtung f <Papier> wire stretcher
Siebtest m <Chemtech> sieve test
Siebtisch m <Kohlen> sieve table
Siebtrockner m <Chemtech> sieve drier, sieve dryer
Siebtrommel f 1. <Bau> rotary screen; 2. <Chemtech> sieve drum; 3. <Kohlen> revolving screen
Siebtrommelzentrifuge f <Kohlen> basket centrifuge
Siebtuch n 1. <Ker & Glas> bolting cloth; 2. <Papier> forming fabric; 3. <Textil> bolting fabric
Siebumlenkwalze f <Papier> wire return roll
Siebung f <Meerschmutz, Umweltschmutz> sieving
Siebwalze f <Papier> wire roll
Siebweite f <Papier> mesh
Siebwirkungsgrad m <Kohlen> screening efficiency
Siedebarometer n <Phys> hypsometer
Siedebereich m 1. <Erdöl> boiling range *(Destillation)*; 2. <Kunststoff> boiling range
Siedeblech n <Chemtech> boiling plate
sieden v <Thermod> boil
Sieden n <Lebensmittel, Thermod> boiling
Sieden n **im Behälter** <Heiz & Kälte> pool boiling
siedend adj <Thermod> boiling
Siedepunkt m <Kunststoff, Lebensmittel, Maschinen, Phys, Thermod> boiling point
Siedepunktkurve f <Thermod> *(AE)* liquid vapor equilibrium diagram, *(BE)* liquid vapour equilibrium diagram
Siedepunktserhöhung f <Maschinen> boiling point elevation
Siedereaktor m <Kerntech> boiling reactor
Siederkessel m <Hydraul> elephant boiler
Siederohrkessel m <Heiz & Kälte> water tube boiler
Siederohrwalze f <Hydraul> boiler tube expander
Siedewasserreaktor m *(SWR)* <Kerntech, Phys> boiling light water moderated reactor, boiling water reactor
Siedewasserreaktor m **mit Naturumlauf** <Kerntech> natural circulation boiling water reactor
Siedlungsabfall m <Abfall> municipal waste, urban solid waste, urban waste
Siegel n <Bau> sealing
Siegelkappe f <Verpack> sealing cap
Siegelmaschine f <Verpack> sealing machine
Siegeltemperatur f <Verpack> heat seal temperature
SI-Einheit f *(internationales Einheitensystem)* <Elektriz, Metrol, Phys> SI unit *(international system of units)*
Siemens n *(S)* <Metrol> siemens, S *(SI-Einheit des elektrischen Leitwerts)*
Siemens-Martin-Ofen m <Fertig> open hearth furnace
Siemens-Martin-Stahl m <Metall> Siemens-Martin steel, open hearth steel
Siemens-Martin-Verfahren n <Metall> Siemens-Martin process
Sievert m *(Sv)* 1. <Phys> Sievert, Sv *(Einheit)*; 2. <Strahlphys, Teilphys> Sievert, Sv
Sigma n <Math> sigma

Sigma-Delta-Modulation f <Telekom> sigma delta modulation
Sigmaschweißung f *(SMAW)* <Kerntech> shielded metal arc welding *(SMAW)*
Sigmateilchen n <Phys> sigma particle
Sigmaverstärker m <Kerntech> sigma amplifier
Signal n 1. <Akustik> signal; 2. <Bau> beacon *(Vermessung)*; 3. <Comp & DV> signal; 4. <Eisenbahn> marker, signal; 5. <Phys, Telekom, Wassertrans> signal
Signal n **in Schleife** <Telekom> looped signal
Signal n **mit harter Begrenzung** <Elektronik, Funktech, Telekom> hard-limited signal
Signal n **mit hoher Amplitude** <Elektronik> high-amplitude signal
Signal n **mit kurzer Anstiegszeit** <Elektronik> fast-rise signal
Signal n **mit niedriger Amplitude** <Elektronik> low-amplitude signal
Signalabfall m <Comp & DV> decay
Signalabfallzeit f <Comp & DV> decay time
Signalabhängigkeit f <Eisenbahn> signal interlocking
Signalagilität f <Elektronik> signal agility
Signalamplitude f <Elektronik, Telekom> signal amplitude
Signalanalysator m <Strahlphys, Telekom> *(BE)* signal analyser, *(AE)* signal analyzer
Signalanalyse f <Elektronik, Telekom> signal analysis
Signalanalyseeinrichtung f <Elektronik> *(BE)* signal analyser, *(AE)* signal analyzer
Signalanlage f <Eisenbahn> signal installation
Signalantrieb m <Eisenbahn> signal operating gear
Signalaufbereitung f <Elektronik> signal conditioning
Signalauffrischung f <Elektronik> signal regeneration
Signalaufteiler m <Fernseh> signal splitter
Signalausfall m 1. <Aufnahme, Comp & DV, Elektrotech> signal drop-out; 2. <Raumfahrt> blackout; 3. <Telekom> drop-out
Signalbandbreite f <Elektronik> signal bandwidth
Signalbegrenzung f <Elektronik> signal clipping
Signalbildung f <Elektronik> signal generation
Signalbuch n <Wassertrans> signal book
Signalbuchflagge f <Wassertrans> code flag
Signaldämpfung f **durch unkorrekte Lese-/Schreibkopfausrichtung** <Comp & DV> gap loss
Signaldehnung f <Telekom> signal extension
Signaldemodulator m <Telekom> signal detector
Signaldetektor m <Telekom> signal detector
Signal-Digitalisierer m <Elektronik> signal digitizer
Signaleinhüllende f <Telekom> signal envelope
Signaleinspeisung f <Funktech> injection, signal injection
Signaleinsteller m <Regelung> signal adjuster
Signalelektrode f <Elektronik> signal electrode
Signalerfassung f <Telekom> signal detection
Signalerzeuger m <Phys> signal generator
Signalexpandierung f <Telekom> signal expansion
Signalflagge f <Eisenbahn> signal flag
Signalflanke f <Elektronik> signal edge
Signalformung f <Comp & DV, Elektronik> signal shaping
Signalformungsfilter n <Telekom> signal-shaping filter
Signalfrequenz f *(SF)* <Elektronik, Fernseh, Funktech, Telekom> signal frequency *(SF)*
Signalgabe f <Telekom> *(AE)* signaling, *(BE)* signalling
Signalgabeentfernung f <Trans> *(AE)* signaling distance, *(BE)* signalling distance
Signalgeber m 1. <Elektronik> signal generator; 2. <Elektrotech> transducer
Signalgeber m **mit Frequenzaufbereitung** <Elektronik> synthesized signal generator

Signalgenerator m <Phys, Telekom> signal generator
Signalgruppensteuerung f <Trans> master control
Signalhorn n <Wassertrans> horn
signalisieren v <Telekom, Wassertrans> signal
Signalisieren n <Wassertrans> (BE) signalling
Signalisierung f <Comp & DV, Telekom> (AE) signaling, (BE) signalling
Signalisierungskanal m **im ISDN** <Telekom> D-channel
Signalisierungsprotokoll n <Comp & DV> (AE) signaling protocol, (BE) signalling protocol
Signalkomplex m <Fernseh> signal complex
Signalkompression f <Telekom> signal compression
Signalkomprimierungsausdehnung f <Raumfahrt> companding
Signallampe f 1. <Elektrotech> signal lamp, pilot lamp; 2. <Fernseh> tally light; 3. <Gerät> annunciator; 4. <Kontroll> signal light
Signallaterne f <Eisenbahn> signal lamp
Signallaufzeit f <Telekom> signal delay
Signalleistung f <Elektronik> signal power
Signalleitung f <Elektrotech> signal line
Signalleuchte f 1. <Fernseh> cue light; 2. <Kontroll> signal light
Signalmittelung f <Gerät> signal averaging
Signalmodellierung f <Elektronik> (AE) signal modeling, (BE) signal modelling
Signalpegel m <Elektronik, Telekom> signal level
Signalpegelmessgerät n <Gerät> signal level meter
Signalpfeife f 1. <Eisenbahn> whistle; 2. <Maschinen> alarm whistle
Signalphase f <Elektronik> signal phase
Signalplatte f <Elektronik> signal plate
Signalprozessor m <Elektronik, Fernseh, Telekom> signal processor
Signal-Rausch-Abstand m <Telekom> signal-to-noise ratio
Signal-Rausch-Verhältnis n <Gerät, Phys> signal-to-noise ratio
Signalregenerator m <Elektronik> signal regenerator
Signalschwankung f <Raumfahrt> ripple
Signalschwelle f <Telekom> signal threshold
Signalsimulierung f <Elektronik> signal simulation
Signalstation f <Eisenbahn> signal station
Signalstellung f <Eisenbahn> position of a signal
Signalstern m <Wassertrans> Very light
Signalstreuung f **an Meteoren** <Funktech> meteor scatter
Signalstruktur f <Regelung> signal structure
Signalsynthese f <Elektronik> signal synthesis
Signaltafel f <Elektrotech> annunciator, signal board
Signaltechnik f <Comp & DV> (AE) signaling system, (BE) signalling system
Signalteilung f <Eisenbahn> signal spacing
Signaltrommel f <Aufnahme> sound drum
Signalübermittlung f <Eisenbahn> signal transmission
Signalübertragung f <Telekom> signal transmission
Signalübertragung f **außerhalb des Bandes** <Raumfahrt> (AE) out-of-band signaling, (BE) out-of-band signalling
Signalumsetzer m 1. <Comp & DV, Elektronik, Fernseh, Regelung> modulator, signal converter; 2. <Telekom> modulator, signal converter
Signalumsetzerwechsel m <Comp & DV, Elektronik> modem interchange
Signalumsetzung f <Telekom> signal conversion
Signalverarbeitung f <Comp & DV, Elektronik, Künstl Int, Regelung, Telekom> signal processing
Signalverarbeitung f **auf Frequenzebene** <Elektronik, Telekom> frequency-domain signal processing
Signalverarbeitung f **in der Zeitebene** <Elektronik> time domain signal processing
Signalverarbeitungschip m <Elektronik> signal processing chip
Signalverfolgung f <Funktech> signal tracing
Signalvergleich m <Elektronik> signal comparison
Signalvergleicher m <Gerät> comparing element
Signalverstärker m <Elektronik> low-level amplifier
Signalverstärkung f <Elektronik> low-level amplification
Signalverteiler m <Telekom> signal distributor
Signalverzerrung f <Elektronik, Telekom> signal distortion
Signalverzögerung f <Elektronik, Telekom> signal delay
Signalverzögerung f **im Schaltkreis** <Elektronik> intercircuit signal delay
Signalverzögerung f **innerhalb des Chips** <Elektronik> interchip signal delay
Signalvorrichtung f <Eisenbahn> warning device
Signalwandler m <Elektrotech, Regelung, Telekom> signal converter
Signatur f <Druck> nick
Signaturnummer f <Druck> signature number
signifikantes Prüfergebnis n <Qual> significant test result
Signifikanzniveau n <Math> significance level (Irrtumswahrscheinlichkeit)
Signifikanzprüfung f <Comp & DV, Qual> significance test
Signifikanztest m 1. <Comp & DV, Math, Qual> significance test; 2. <Werkprüf> statistical test
Sikkativ n 1. <Chemie> desiccant; 2. <Chemtech> desiccative; 3. <Kunststoff> drier, dryer, drying agent, exsiccant; 4. <Thermod> desiccant, siccative
sikkativverpackt adj <Verpack> packed with siccative
Silan n 1. <Chemie> monosilane, silicomethane; 2. <Kunststoff> silane
Silbenkompandierung f <Telekom> syllabic companding
Silbentrennstrich m <Comp & DV> soft hyphen
Silbentrennungsprogramm n <Druck> hyphenation program
Silbenverständlichkeit f <Telekom> logatom articulation
Silbenverständlichkeitstest m <Telekom> syllable articulation test
Silber n (Ag) <Chemie, Metall> silver (Ag)
Silberbatterie f <Elektrotech> silver battery
Silberbeizen n <Ker & Glas> silver staining
Silberchlorid n <Elektrotech> silver chloride
Silberchloridemulsion f <Foto> silver chloride emulsion
Silberelektrode f <Labor> silver electrode (Elektrochemie)
Silberelement n <Elektrotech> silver cell
Silbergehalt m <Foto> silver content
Silbergehäuse-Tantalkondensator m <Elektrotech> silver case tantalum capacitor
Silber-Glimmer-Kondensator m <Elektrotech> silver mica capacitor
Silberhalogenid n <Foto> silver halide
Silberhalogenidpapier n <Foto> silver halide paper
Silberhalogenidschicht f <Foto> silver halide emulsion
silberhelle Stimme f <Akustik> silver voice
Silberjodid n <Foto> silver iodide
Silber-Kadmium-Batterie f <Elektrotech> silver-cadmium battery
Silber-Kadmium-Element n <Elektrotech> silver-cadmium cell
Silber-Kadmium-Zelle f <Elektrotech> silver-cadmium cell
Silberkontakt m <Elektrotech> silver contact
Silberlegierung f <Fertig, Metall> silver alloy
Silberlot n 1. <Fertig> silver solder; 2. <Maschinen> silver filler

Silberlöten

Silberlöten n <Elektriz> hard-soldering
Silberoxid n <Elektrotech> silver oxide
Silberoxidakkumulator m <Kfztech> silver oxide storage battery
Silberoxidbatterie f <Elektrotech> silver oxide battery
Silberoxidelement n <Elektrotech> silver oxide cell
Silberoxidzelle f <Elektrotech> silver oxide cell
silberplattiert adj <Fertig, Metall> silver-clad
silberplattierter Kontakt m <Elektrotech> silver-plated contact
Silberrest m <Foto> residual silver
Silberstahl m <Maschinen, Metall> silver steel
Silberzelle f <Elektrotech> silver cell
Silber-Zink-Akkuelement n <Elektrotech> silver-zinc storage cell
Silber-Zink-Akkumulator m <Elektrotech, Kfztech> silver-zinc storage battery
Silber-Zink-Akkuzelle f <Elektrotech> silver-zinc storage cell
Silber-Zink-Batterie f <Elektrotech> silver-zinc battery
Silber-Zink-Element n <Elektrotech> silver-zinc cell
Silber-Zink-Primärbatterie f <Elektrotech> silver-zinc primary battery
Silber-Zink-Primärelement n <Elektrotech> silver-zinc primary cell
Silber-Zink-Primärzelle f <Elektrotech> silver-zinc primary cell
Silber-Zink-Zelle f <Elektrotech> silver-zinc cell
Silika f <Ker & Glas> silica
Silikagel n <Verpack> blue silica gel
Silikaglas n <Ker & Glas, Optik, Telekom> vitreous silica
Silikamasse f <Chemie> silica
Silikamaterial n <Chemie> silica (Keramik)
Silikaschaum m <Ker & Glas> silica scum
Silikaschaumgrenze f <Ker & Glas> (AE) batch-melting line, (BE) silica scum line
Silikat n <Chemie> silicate
Silikatglas n <Ker & Glas, Optik, Telekom> vitreous silica
Silikatglasfaser f 1. <Optik> (AE) all-silica fiber, (BE) all-silica fibre; 2. <Telekom> (AE) all-silica fiber, (BE) all-silica fibre, (AE) silica fiber, (BE) silica fibre
Silikatglas-Lichtwellenleiter m **mit Kunststoffmantel** <Telekom> (AE) plastic-clad silica fiber, (BE) plastic-clad silica fibre
silikatisch adj <Chemie> siliceous
Silikatverbundfenster n <Raumfahrt> fused silica window (Raumschiff)
Silikid n <Chemie> silicide
Silikofluorid n <Chemie> silicofluoride
Silikon n <Elektriz, Elektrotech, Kunststoff, Wassertrans> silicone
Silikonauskleidung f <Elektrotech> silicone cladding
Silikonelastomer n <Kunststoff> silicone elastomer
Silikonflüssigkeit f <Elektriz> silicone fluid
Silikongummi n <Elektriz> silicone rubber
Silikonisieren n <Ker & Glas> siliconing
Silikonkautschuk m <Kunststoff> silicone rubber
Silikonmasse f <Wassertrans> silicone compound
Silikonverkleidung f <Elektrotech> silicone cladding
Silikonzelle f <Nichtfoss Energ> silicon cell
Silizieren n <Ker & Glas> siliconizing
Siliziumoxid n <Elektronik> silicon oxide
Silizium n 1. <Chemie> silicon (Si); 2. <Comp & DV, Elektriz, Elektronik, Elektrotech, Phys, Wassertrans> silicon
• **mit Silizium beruhigt** <Fertig> silicon-killed
Silizium n **auf Saphir** (SOS) <Elektronik> silicon-on-sapphire (SOS)
Silizium n **auf Saphir-Substrat** <Funktech> silicon on sapphire technology

Silizium n **vom Typ n** <Elektronik> n-type silicon
Silizium-Bipolartransistor m <Elektronik> silicon bipolar transistor
Siliziumbronze f <Metall> silicon bronze
Siliziumchip m <Comp & DV, Elektriz, Elektronik, Strahlphys> silicon chip
Siliziumdetektor m <Elektronik> silicon detector
Siliziumdetektordiode f <Elektronik> silicon detector diode
Siliziumdiode f <Elektronik> silicon diode
Siliziumdioxid n <Chemie> silica
siliziumdioxidhaltig adj <Chemie> siliceous
Siliziumdotierung f <Elektronik> silicon doping
Siliziumelektrostahl m <Metall> silicon electrical steel
Silizium-Epitaxialschicht f <Elektronik> silicon epitaxial layer
Silizium-FET m <Elektronik> silicon FET
Silizium-Flächendiode f <Elektronik> silicon junction diode
Silizium-Gate-Transistor m <Elektronik> silicon gate transistor
Siliziumgatter n <Elektronik> silicon gate
Siliziumgattertechnologie f <Elektronik> silicon gate technology
Siliziumgerät n <Elektronik> silicon device
siliziumgesteuerter Gleichrichter m (SCR) <Elektronik, Elektrotech> silicon-controlled rectifier (SCR)
Silizium-Gießerei f <Elektronik> silicon foundry
Siliziumgleichrichter m 1. <Elektriz, Elektrotech> silicon rectifier; 2. <Phys> silicon detector, silicon rectifier
Siliziumkarbid n 1. <Chemie> carborundum, silicon carbide; 2. <Elektrotech, Fertig, Phys> silicon carbide
Siliziumkarbidvaristor m <Elektrotech> silicon carbide varistor
Siliziumkristall m <Elektronik> silicon crystal
Siliziumkristallmischer m <Elektronik> silicon crystal mixer
Silizium-Lawinenphotodiode f <Elektronik> silicon avalanche diode, silicon avalanche photodiode
Silizium-Miniaturschaltung f <Comp & DV> silicon chip
Silizium-Mischdiode f <Elektronik> silicon mixer diode
Silizium-Nachbeschleunigungs-Fangelektrode f <Elektronik> silicon intensifier target
Siliziumnitrid n <Elektronik> silicon nitride
Siliziumoxid n 1. <Chemie> silica; 2. <Elektronik> silicon dioxide
Siliziumoxidschicht f <Elektronik> silicon dioxide layer
Silizium-Photodiode f <Elektronik> silicon photodiode
Silizium-Phototransistor m <Elektronik> silicon phototransistor
Siliziumschicht f <Elektronik> silicon layer
Siliziumsolarzelle f <Elektronik> silicon solar cell
Siliziumstahl m <Elektrotech, Metall> silicon steel
Siliziumstahlbeschichtung f <Elektrotech> silicon steel lamination
Siliziumstahlkern m <Elektrotech> silicon steel core
Silizium-Steuerelektronen-Technologie f <Elektronik> silicon gate technology
Silizium-Trägermaterial n <Elektronik> silicon substrate
Siliziumvorrichtung f <Elektrotech> silicon device
Siliziumwasserstoff m <Chemie> monosilane, silicomethane
Siliziumzähler m <Strahlphys> (AE) silicon checker, (BE) silicon counter, silicon detector
Siliziumzelle f <Elektrotech> silicon cell
Silo n 1. <Bau> bunker, hopper; 2. <Fertig> bin; 3. <Kohlen> bunker
Silodruck m <Kohlen> silo pressure
Siloxan n <Chemie> siloxane

Silt *m* <Kohlen> silt
Siltfeld *n* <Ker & Glas> silt field
Siltstein *m* <Ker & Glas> silt block
SIMD-Rechner *m* (Parallelrechner) <Comp & DV> SIMD machine *(single instruction multiple-data machine)*
SIMM *(einfaches schritthaltendes Speichermodul)* <Comp & DV> SIMM *(single in-line memory module)*
Simmerring *m* <Maschinen> shaft seal, shaft-sealing ring
Simplex *n* **auf zwei Frequenzen** <Telekom> two-frequency simplex
Simplex... <Comp & DV, Telekom> simplex
Simplex-Algorithmus *m* <Math> simplex algorithm *(lineare Optimierung)*
Simplexbetrieb *m* <Comp & DV, Telekom> simplex operation
Simplexkarton *m* <Papier> single ply board
Simplexpumpe *f* <Maschinen> simplex pump
Simplexübertragung *f* <Comp & DV> simplex transmission
SIMS *(Sekundärionenmassenspektrometrie)* <Phys> SIMS *(secondary ion mass spectrometry)*
Simsbrett *n* <Bau> fascia board
Simshobel *m* <Bau> side rabbet plane
Simulation *f* <Comp & DV, Elektriz, Elektronik, Ergon, Werkprüf> simulation
Simulationsanlage *f* <Werkprüf> simulation equipment
Simulationsprogramm *n* <Comp & DV> simulation program
Simulationssprache *f* <Comp & DV> simulation language
Simulator *m* <Comp & DV, Elektronik, Fertig, Lufttrans, Metrol, Telekom> simulator
simulieren *v* <Comp & DV> simulate
Simuliergerät *n* <Fertig, Metrol> simulator
simuliertes Ereignis *n* <Strahlphys> simulated event
simultan *adj* 1. <Comp & DV> parallel, simultaneous; 2. <Kontroll> simultaneous
simultan verwendbar *adj* <Comp & DV> re-entrant
simultan verwendbare Routine *f* <Comp & DV> re-entrant routine
simultan verwendbarer Code *m* <Comp & DV> re-entrant code
simultan verwendbares Programm *n* <Comp & DV> re-entrant program
Simultanrechner *m* <Comp & DV> parallel computer
Simultanrundfunk *m* <Fernseh> simulcast broadcasting
Simultanverarbeitung *f* <Comp & DV> multiprocessing
Simultanverarbeitungssystem *n* <Comp & DV> multiprocessing system
sin *(Sinus)* <Comp & DV, Geom> sin *(sine)*
SINAD *(Störabstand einschließlich Verzerrungen)* <Telekom> SINAD *(signal-to-noise and distortion ratio)*
SINAD-Abstand *m* <Telekom> SINAD ratio
Sinapin *n* <Chemie> sinapine
Sinapin... <Chemie> sinapic
Single-in-Line-Gehäuse *n* <Elektronik, Telekom> SIP, single in-line package *(Baustein, integrierte Schaltung)*
Single-Jersey *n* <Textil> single jersey
Single-Pair-Kabel *n* <Elektrotech> single pair cable
Singularität *f* <Phys> singularity
Singulett *n* <Phys> singlet *(Spektroskopie)*
sinken *v* 1. <Bau> sink; 2. <Wassertrans> go down, sink; founder *(Schiff)*
Sinken *n* <Hydraul> fall *(Niveau)*
Sinkflug *m* **mit Reisefluggeschwindigkeit** <Lufttrans> cruise descent
Sinkgeschwindigkeit *f* 1. <Kohlen> settling speed; 2. <Lufttrans> rate of descent
Sinkgeschwindigkeit *f* **des Wasserspiegels in der Wasserschicht** <Nichtfoss Energ> drawdown of water in aquifer
Sinkgrenzwert *m* **des Flugzeugs** <Lufttrans> limit rate of descent *(Aufsetzen)*
Sinkgut *n* 1. <Meerschmutz> sinking agent; 2. <Umweltschmutz> deposited matter
Sinkmaterial *n* <Meerschmutz> sinking agent
Sinkrate *f* <Lufttrans> rate of descent
Sinkstoff *m* <Umweltschmutz> deposited matter
Sinkstoffe *mpl* <Abfall> settleable solids
Sinnenprüfung *f* <Lebensmittel> tasting
sinnfreie Silbe *f* <Akustik> logatom
Sinnverständlichkeit *f* <Aufnahme> intelligibility
sinodisch *adj* <Fertig> sinusoidal
Sinter... <Bau, Elektrotech, Fertig, Ker & Glas> sintering
Sinteranode *f* <Elektrotech> sintered anode
Sinterbeton *m* <Bau> hooped concrete
Sintererzeugnis *n* 1. <Fertig> agglomerate; 2. <Ker & Glas> sintered refractory
Sinterglas *n* <Ker & Glas> sintered glass
Sinterkohle *f* <Kohlen> cherry coal, sintering coal
Sinterkuchen *m* <Ker & Glas> aggregate
Sintermaterial *n* <Maschinen> sintered material
Sintermetall *n* 1. <Fertig> hard carbide; 2. <Maschinen> sintered metal
Sintermetallwerkstoff *m* <Maschinen> sintered metal material
sintern *v* 1. <Chemtech> cake; 2. <Fertig> bake, coalesce, slag, vitrify; 3. <Maschinen, Metall> sinter
Sintern *n* 1. <Fertig> vitrification; 2. <Ker & Glas, Kohlen, Metall> sintering
sinternde Sandkohle *f* <Kohlen> sintering sand coal
Sinterstück *n* <Fertig> sintering
Sintertechnik *f* <Chemtech> sintering technique
Sinterteil *n* <Fertig> powder metal part
Sinterung *f* 1. <Fertig> caking, coalescence; 2. <Telekom> sintering
Sinterverfahren *n* <Chemtech> sintering technique
Sinus *m* 1. <Comp & DV> sinusoid; 2. <Comp & DV, Geom> sine *(sin)*
Sinusaufspannplatte *f* <Metrol> sine table
Sinusbedingungen *fpl* <Elektrotech> sinusoidal conditions
Sinusbussole *f* <Metrol, Phys> sine galvanometer
Sinusfeld *n* <Elektrotech> sinusoidal field
sinusförmig *adj* <Elektriz, Fertig> sinusoidal
sinusförmige Bewegung *f* <Phys> sinusoidal motion
sinusförmige Spannung *f* <Phys> sinusoidal voltage
sinusförmiger Strom *m* <Phys> sinusoidal current
sinusförmiges Signal *n* <Elektronik> sinusoidal signal
Sinusfunktion *f* 1. <Elektronik, Elektrotech> harmonic function, sinusoidal function; 2. <Math> sinus function
Sinusgröße *f* <Elektronik> sinusoidal quantity
Sinuskurve *f* 1. <Comp & DV> sine wave, sinusoid; 2. <Math> sine curve
Sinuslineal *n* <Metrol> sine bar
Sinuslinie *f* <Geom> sinusoid
Sinusmenge *f* <Elektrotech> sinusoidal quantity
Sinusoide *f* <Geom> sinusoid
Sinusoszillator *m* <Elektronik> harmonic oscillator
Sinusquantität *f* <Elektrotech> sinusoidal quantity
Sinussatz *m* <Math> sine rule
Sinusschwingung *f* <Elektronik> sinusoidal oscillation
Sinussignal *n* <Elektronik, Telekom> sinusoidal signal
Sinussignalgeber *m* <Elektronik> sinusoidal signal generator
Sinusspannung *f* <Elektrotech> sinusoidal voltage
Sinusstrom *m* <Elektrotech> sinusoidal current
Sinustabelle *f* <Math, Mechan> sine table
Sinustafel *f* <Mechan> sine table
Sinuswelle *f* <Akustik, Elektriz, Elektronik, Funktech, Geom, Phys, Wellphys> sine wave

Sinuswellenabstimmung

Sinuswellenabstimmung f <Elektronik> sine wave tuning
Sinuswellenenergie f <Nichtfoss Energ> sine wave power
Sinuswellengenerator m <Nichtfoss Energ> sine wave generator
Sinuswellenkonvergenz f <Telekom> sine wave convergence
Sinuswellenmodulation f <Elektronik> sine wave modulation
Sinuswellenoszillator m <Elektronik> sine wave oscillator
SIP-Gehäuse n <Elektronik, Telekom> SIP, single in-line package *(Baustein, integrierte Schaltung)*
Siphon m <Bau, Nichtfoss Energ> siphon, syphon
Siphon m **für Vakuumpumpe** <Labor> trap for vacuum pump
Siphonhöhe f <Nichtfoss Energ> siphon crest
Siphonüberlauf m <Nichtfoss Energ> siphon spillway
Siphonverschluss m <Bau> S-trap, syphon trap *(Sanitär)*
Sirene f 1. <Sicherheit> siren *(Alarm)*; 2. <Wassertrans> siren *(Signal)*
SIS *(Schwerionensynchrotron)* <Teilphys> HIS *(heavy-ion synchrotron)*
Sisalhanf m <Wassertrans> sisal hemp *(Tauwerk)*
Sisalseil n <Maschinen> sisal rope
S-ISDN n <Telekom> narrow-band ISDN
SISD-Rechner m *(serieller Rechner)* <Comp & DV> SISD machine *(single instruction single-data machine)*
Sitosterin n <Chemie> sitosterol
Sitosterol n <Chemie> sitosterol
Situationskalkül m <Künstl Int> situation calculus
Sitz m 1. <Hydraul> seat *(Ventil)*; 2. <Kfztech> seat; 3. <Maschinen> seat, seating; 4. <Mechan> seat
Sitzbank f <Kfztech> bench seat *(Sitze)*
Sitzbezug m <Kfztech> covering *(Autositz)*
Sitzfläche f <Maschinen> face, seat, seating, seat
Sitzkissen n <Kfztech> seat cushion
Sitzladefaktor m <Trans> load factor *(Passagierflugzeuge)*
Sitzpolster n <Kfztech> seat upholstery
Sitzung f <Comp & DV> session
Sitzungsschicht f <Comp & DV, Telekom> session layer
Sitzventil n <Fertig> globe valve *(Kunststoffinstallationen)*
skalar adj <Comp & DV, Math> scalar
Skalar m <Comp & DV, Phys> scalar
skalarer Widerstand m <Elektrotech> scalar resistor
skalares Potenzial n <Elektrotech, Phys> scalar potential
Skalarfunktion f <Comp & DV> scalar function
Skalarmessung f <Elektrotech> scalar measurement
Skalarnetzanalyse f <Elektrotech> scalar network analysis
Skalarnetzanalysegerät n <Elektrotech> *(BE)* scalar network analyser, *(AE)* scalar network analyzer
Skalarprodukt n 1. <Math> scalar product *(Vektoren)*; 2. <Phys> scalar product
Skalartyp m <Comp & DV> scalar type
Skale f 1. <Akustik, Comp & DV, Druck> scale; 2. <Elektriz> dial, scale; 3. <Geom> scale; 4. <Gerät> dial; 5. <Metrol> scale
Skale f **mit Nullpunkt links** <Gerät> left margin zero scale
Skale f **mit Nullpunkt rechts** <Gerät> right margin zero scale
Skale f **mit unterdrücktem Nullpunkt** <Gerät> set-up scale, suppressed zero scale
Skale f **mit versetztem Nullpunkt** <Gerät> displaced zero scale
Skalen fpl **der Lautstärke** <Akustik> scales of loudness
Skalen fpl **der Tonhöhe** <Akustik> scales of pitch
Skalenänderung f <Geom> scaling *(einer geometrischen Transformation)*
Skalenanzeige f <Gerät> scale indication
Skalenbereich m <Gerät> range, scale range
Skaleneinteilung f 1. <Gerät> scale; 2. <Maschinen> index, scale division; 3. <Metrol> scale spacing
Skalenendausschlag m <Gerät> full-scale deflection
Skalenendwert m <Gerät> full-scale value, maximum scale value
Skalengenauigkeit f <Gerät> scale accuracy
Skalenintervall n <Metrol> scale interval
Skalenjustierung f <Gerät> dial adjustment
Skalenlänge f <Metrol> scale length
Skalenmarke f <Gerät> hairline, memory pointer, scale mark
Skalenmarkierung f <Metrol> scale mark
Skalenmittenwert m <Gerät> midscale value
Skalennummerierung f <Metrol> scale numbering
Skalenscheibe f 1. <Gerät> dial scale; 2. <Maschinen> dial
Skalenstrich m 1. <Gerät> hairline; 2. <Maschinen> index, scale division
Skalenteilstrich m <Gerät> scale mark
Skalenteilstrichabstand m <Metrol> scale spacing
Skalenteilung f 1. <Elektriz> scale division; 2. <Gerät> scale, scale division
Skalenumschalter m <Elektriz> scale switch
Skalenwert m <Gerät> scale value
Skalenzeiger m <Gerät> dial pointer
skalierbar adj <Telekom> scalable
skalierbare Schrift f <Druck> scalable font
Skalierbarkeit f <Telekom> scalability
skalieren v <Comp & DV, Druck, Telekom> scale
Skalieren n 1. <Comp & DV> sizing; 2. <Telekom> scaling
Skalierfaktor m <Comp & DV> scale factor
Skalierung f 1. <Comp & DV> scaling; 2. <Geom> scaling *(einer geometrischen Transformation)*; 3. <Math> scaling *(Statistik)*; 4. <Ergon> rating; 5. <Gerät> scale
Skalierungsfaktor m <Elektronik, Geom> scaling factor
Skalpell n <Labor> scalpel
Skalpellklinge f <Labor> scalpel blade
Skatol n <Chemie> methylindole, skatole
Skelett n 1. <Bau> skeleton *(Gebäude)*; 2. <Maschinen> skeleton
Skelett-Träger m <Bau> skeleton girder
Skewness <Math> skewness *(Schiefemaß einer Verteilung)*
Skim-Kautschuk m <Kunststoff> skim rubber
Skimmer m 1. <Erdöl> skimmer *(Umweltschutz)*; 2. <Meerschmutz, Umweltschmutz> skimmer
Skimming n <Ker & Glas> skimming
Skineffekt m <Elektrotech, Funktech, Phys, Werkprüf> skin effect
Skinfolie f <Verpack> skin film
Skinpack n <Verpack> skin pack
Skinpackung f <Verpack> blister pack
Skizze f 1. <Lufttrans> plot *(Navigation)*; 2. <Maschinen> outline
skizzieren v <Bau> outline, sketch
Skleroprotein n <Lebensmittel> albuminoid, scleroprotein
Skleroskopapparat m <Fertig> scleroscope
Skleroskophärte f <Fertig> scleroscope hardness
Skonto m <Comp & DV> CD, cash discount
Skrubber m <Chemtech> washing tower
Skullboot m <Wassertrans> scull *(Rudern)*
skullen v <Wassertrans> scull *(Rudern)*
Skullriemen m <Wassertrans> scull *(Rudern)*
SLAR *(Seitensichtradar)* <Meerschmutz> SLAR *(sideways-looking airborne radar)*

Slatis-Siegbahn-Spektrometer *n* <Kerntech> Slatis-Siegbahn spectrometer
Slave *m* <Comp & DV> slave *(gesteuertes Gerät)*
Slave-Anwendung *f* <Comp & DV> slave application
Slave-Prozessor *f* <Comp & DV> slave processor
SLD 1. <Optik> *(Superlumineszenzdiode)* SLD *(superluminescent LED)*; 2. <Telekom> *(Superlumineszenzdiode)* SLD *(superluminescent LED)*; SRD *(superradiant diode)*
Slice-Aufbau *m* <Comp & DV> slice architecture
Slick *m* <Umweltschmutz> slick
Slipanlage *f* <Wassertrans> building slip
Slipkran *m* <Trans> boat-launching crane
slippen *v* <Wassertrans> slip *(Bootbau)*
slippen lassen *v* <Wassertrans> slip *(Tauwerk)*
Slipwagen *m* <Wassertrans> cradle
Slot *m* <Künstl Int> slot *(Schema)*
Slotted-ALOHA-Zugriffssystem *n* <Telekom> slotted ALOHA system
Slow-Scan-Television *f (SSTV)* <Fernseh> slow scan television *(SSTV)*
Slow-Scan-Videokonferenz *f* <Telekom> slow scan video conferencing
SLR *(einäugige Spiegelreflexkamera)* <Foto> SLR *(single lens reflex camera)*
SL-Rakete *f (Schwerlastträgerrakete)* <Raumfahrt> HLLV *(heavy-lift launch vehicle)*
Slurry-Tanker *m* 1. <Kfztech> slurry tanker; 2. <Wassertrans> oil-slurry oil tanker
Slushmoulding *n* <Kunststoff> *(AE)* slush molding, *(BE)* slush moulding *(Gießverfahren)*
Sm *(Samarium)* <Chemie> Sm *(samarium)*
S-Matrix *f* <Telekom> scattering matrix
SMC 1. <Elektronik> *(SMD-Bauteil)* SMC *(surface-mounted component)*; 2. <Telekom> *(Aufsetzbauelement)* SMC *(surface-mounted component)*
SMD *(Aufsetzbauelement, oberflächenmontiertes Element)* <Elektriz> SMD *(surface mounting device)*
SMD-Bauteil *n (SMC)* <Elektronik> surface-mounted component *(SMC)*
smektische Flüssigkristalle *mpl* <Elektronik> smectic liquid crystals
Smektit *m* 1. <Erdöl> smectite *(Mineral)*; 2. <Kohlen> smectite
S-Meter *n* <Funktech> S-meter
Smiley-Männchen *n* <Comp & DV> smiley *(Zeichen)*
Smith-Diagramm *n* <Funktech, Phys, Telekom> Smith chart
SM-Ofen *m* <Fertig> open hearth furnace
Smog *m* <Umweltschmutz> smog
SMPTE-Zeitcode *m (Zeitcode der Gesellschaft für Kino- und Fernsehtechniker)* <Fernseh> SMPTE time code *(Society of Motion Pictures and Television Engineers time code)*
Sn *(Seiten)* <Druck> pp *(pages)*
SNA *(Systemnetzwerkarchitektur)* <Comp & DV> SNA *(systems network architecture)*
Snaking *n* <Lufttrans> snaking *(Aerodynamik)*
SNA-Kommunikationskonzept *n* <Comp & DV> systems network architecture
Snap-In-Fassung *f* <Elektrotech> snap-in socket
Snellius'sches Gesetz *n* <Phys> Snell's law
Snubber-Kondensator *m* <Elektrotech> snubber capacitor
Snubber-Schaltkreis *m* <Elektrotech> snubber circuit
SNV *(Schweizerische Normenvereinigung)* <Elektriz> SSA *(Swiss Standards Association)*
SOAR-Bereich *m* **in Durchlassrichtung** <Elektriz> forward-bias safe operation area
Sockel *m* 1. <Bau> socket; 2. <Elektriz> female connector, socket; 3. <Elektronik> base *(Elektronikröhren)*; backplate *(Röhrenfassung)*; 4. <Elektrotech> cap *(Lampe)*; fuse base *(Sicherung)*; 5. <Fernseh> socket *(Röhre)*; 6. <Funktech> socket; 7. <Kerntech> header; 8. <Mechan> pedestal, socket; 9. <Telekom> pad, socket
Sockel *m* **von Elektronenröhren** <Elektrotech> base
Sockelleiste *f* <Ker & Glas> washboard
Sockelstift *m* <Elektrotech> base prong, base pin *(Bajonettfassung)*
Sockets basierend *adj* **/auf** <Comp & DV> sockets-based
Sofortanzeige *f* <Gerät> instantaneous display
Sofortbetrieb *m* <Telekom> no-delay operation
Sofortbildfotografie *f* <Foto> instand photography
sofortige Auslösung *f* <Maschinen> instantaneous release
Sofortkopie *f* <Comp & DV> instant copy
Sofortwahl *f* <Telekom> immediate dial
Soft-Copy *f* <Comp & DV> soft copy
Soft-Cover *m* <Druck> soft cover
Soft-Fail *m* <Comp & DV> soft fail
Soft-Fehler *m* <Comp & DV> soft error
Softproofs *fpl* <Druck> soft proofs
Softstarteinrichtung *f* <Elektriz> soft start facility, soft starter
Software *f* <Comp & DV, Elektriz, Fernseh, Phys, Telekom> software
Software *f* **zur Strukturanalyse** <Bau, Raumfahrt> structural analysis software
Softwareankündigung *f* <Comp & DV> software announcement
Softwareanpassung *f* <Comp & DV> software adaptation
Softwarebetriebsmittel *n* <Comp & DV> software resources
Software-Business-Lösung *f* <Comp & DV> software business solution
Software-Decodierung *f* <Telekom> software decoding
Softwaredesign *n* <Comp & DV> software design
Software-Engineering *n* <Comp & DV> software engineering
Software-Entwicklung *f* <Comp & DV> software development
Software-Entwurf *m* <Comp & DV> software design
Softwarekonfiguration *f* <Comp & DV> software configuration
Softwareleistungsspektrum *n* <Comp & DV> software capability
Softwaremethodik *f* <Comp & DV> software methodology
Softwarepaket *n* <Comp & DV, Phys> software package
Softwarepaket *n* **für Systemmanagement** <Comp & DV> systems management software suite
Softwareportfolio *n* <Comp & DV> portfolio of software
Softwareprodukt *n* <Telekom> software product
Softwarezustellungsbericht *m* <Comp & DV> software delivery report
Sog *m* 1. <Mechan> suction; 2. <Meerschmutz> wake
Sohlbank *f* <Bau> sill
Sohldruck *m* <Bau, Kohlen> bearing pressure
Sohle *f* 1. <Bau> floor *(Becken)*; invert *(Rohrleitung, Tunnel)*; foot wall *(Tunnelbau)*; 2. <Fertig> sole; bed *(Herd)*; 3. <Kohlen> foot wall; 4. <Maschinen, Metall> bed *(Ofen)*
Sohle *f* **des Ofenmunds** <Ker & Glas> base of neck
Sohlenmaterial *n* <Kunststoff> soling material
Sohlenplatte *f* <Erdöl> bed plate
Sohlenschuss *m* <Kohlen> blasting from the bottom
Sohlenversagen *n* <Kohlen> base failure
Sohlplatte *f* <Bau> base plate, bottom, shoe, sole
Sohlpressung *f* <Bau, Kohlen> subgrade reaction
Sohlpressungsmodul *m* <Bau, Kohlen> subgrade reaction modulus

Sohn

Sohn m <Akustik> stamper
Sohnplatte f <Akustik> stamper
Solanidin n <Chemie> solanidine
Solar... <Elektrotech, Funktech, Nichtfoss Energ> solar
Solarbatterie f <Elektrotech, Telekom> solar battery
solarbeheizter Boiler m <Fertig> solar water heater
Solardynamik f <Nichtfoss Energ> solar dynamics
solare Einstrahlung f <Nichtfoss Energ> solar irradiation
Solarelektrizität f <Elektrotech, Nichtfoss Energ> solar electricity
Solarelement n <Elektrotech, Nichtfoss Energ> solar cell
Solarenergie f <Elektrotech, Maschinen, Nichtfoss Energ, Phys> solar energy
Solarenergiekonversion f <Elektrotech, Nichtfoss Energ> solar energy conversion
Solarenergieumwandlung f <Elektrotech, Nichtfoss Energ> solar energy conversion
Solargenerator m 1. <Elektriz> solar cell generator, solar power generator; 2. <Elektrotech, Raumfahrt> solar generator; 3. <Nichtfoss Energ> solar generator, solar power generator
Solarheizungssystem n <Nichtfoss Energ> solar heating system
Solarimeter n <Nichtfoss Energ> solarimeter
Solarisation f <Foto> solarization
Solarkollektor m <Maschinen> solar collector
Solarkonstante f <Nichtfoss Energ, Phys> solar constant
Solarmobil n 1. <Kfztech> solar mobile; 2. <Nichtfoss Energ, Trans> solar-powered vehicle
Solarmodul n <Elektrotech, Nichtfoss Energ> solar module
Solarpond m <Nichtfoss Energ> solar pond
Solarstrom m <Elektrotech, Nichtfoss Energ> solar current
Solarstromtechnik f <Elektrotech> photovoltaics
Solartechnik f <Nichtfoss Energ> solar engineering, solar technology
solarthermischer Prozess m <Nichtfoss Energ> solar-thermal process
Solarwärmegewinn m <Nichtfoss Energ> solar heat gain
Solarwärmegewinnfaktor m <Nichtfoss Energ> solar heat gain factor
Solarzelle f <Elektrotech, Nichtfoss Energ, Phys, Raumfahrt, Strahlphys, Telekom, Thermod> solar cell
Solarzelle f **ohne Konzentrator** <Elektrotech> nonconcentrator solar cell
Solarzellenausleger m <Telekom> solar panel
Solarzellenflügel m <Raumfahrt> solar panel
Solarzellengruppe f <Nichtfoss Energ, Raumfahrt> solar array
Solarzellenlaken n <Nichtfoss Energ> array blanket
Solarzellenplatte f <Nichtfoss Energ> solar panel
Sole f <Kerntech> brine
Solekühlung f <Lebensmittel> brine cooling
Solenkühlsystem n <Kerntech> brine cooling system
Solenoid n <Phys> solenoid
Solenoid... n 1. <Fertig> solenoid; 2. <Elektriz, Elektrotech> solenoid *(Elektromagnet)*; 3. <Kfztech> solenoid *(Anlasser)*
Solenoidbetätigung f <Elektrotech> solenoid actuation
Solenoidfeld n <Phys> solenoidal field
Solenoidrelais n <Elektriz, Elektrotech> solenoid relay
Solenoidschrittmotor m <Elektrotech> solenoid stepper motor
Solenoidventil n <Hydraul> solenoid valve
Soletröpfchen n <Kerntech> brine droplet
Solidstate m <Elektriz> solid state
Solidstate-Relais n <Elektrotech> solid state relay
Solidstate-Signal n <Elektronik> solid state signal
Solidstate-Speicher m <Elektrotech> solid state memory device
Soliton n <Telekom> soliton *(Faseroptik)*
Sollabbrand m <Kerntech> target burn up
Sollbereich m <Gerät> desired range
Sollbruchelement n <Maschinen> breaking member
Sollbruchstelle f <Maschinen> breaking point
Sollflugbahn f <Lufttrans> required flightpath
Sollfrequenz f <Fernseh, Telekom> allocated frequency
Sollpegel m <Gerät> desired level
Sollwert m 1. <Gerät, Kerntech> desired value; 2. <Maschinen> set value
Sollwertabweichung f <Regelung> deviation from the desired set point
Sollwertanzeigeinstrument n <Gerät> rated value indicating instrument
Sollwertbereich m <Gerät> desired range
Sollwerteinsteller m **für Fernbetätigung** <Regelung> remote set-point adjuster
Sollzustand m <Gerät> desired condition
Solvens n <Chemie> dissolvent
Solventraffination f <Erdöl> solvent refining *(Raffinerie)*
Sommerdeich m <Wasserversorg> overflow dam
Sommerfeld'sche Zahl f <Nichtfoss Energ> Sommerfeld number
Sommertiefladelinie f <Wassertrans> summer load waterline *(Schiffkonstruktion)*
Sonargerät n <Strahlphys, Wassertrans> sonar *(Navigation)*
Sonde f 1. <Fertig> sound; 2. <Funktech> probe; 3. <Gerät> measuring head, probe; 4. <Kohlen, Phys, Raumfahrt> probe
Sondenmikrofon n <Akustik, Aufnahme> probe microphone
Sondenspule f <Phys> exploring coil, pick-up coil, search coil
Sonder... <Druck> special
Sonderabfall m <Abfall> hazardous waste, special waste
Sonderabfalldeponie f <Abfall> hazardous waste dump
Sonderabfallentsorgung f <Abfall> hazardous waste disposal
Sonderabfallzwischenlager n <Abfall> temporary deposit for hazardous waste
Sonderausgabe f <Druck> special edition
Sonderausstattung f <Maschinen> optional equipment
sonderberuhigter Stahl m <Fertig> abnormal steel
Sonderdienst m <Telekom> special service
Sonderfreigabe f <Qual> waiver
Sondermaschine f <Maschinen> special purpose machine
Sondermotoren mpl <Fertig> special-purpose engines
Sondermüll m <Abfall> hazardous waste, special waste
Sondermülldeponie f <Abfall> hazardous waste landfill
Sondermülleinsammlung f <Abfall> hazardous waste collection
Sondermüllverbrennungsanlage f <Abfall> hazardous waste incineration plant
Sonderschiff n <Wassertrans> special purpose ship
Sonderstahl m <Metall> special steel, specialty steel
Sonderverfahren n <Qual> special process
Sondervermerk m <Konstzeich> special note
Sonderwagen m <Eisenbahn> *(AE)* parlor car, *(BE)* saloon carriage
Sonderweiche f <Eisenbahn> special turnout
Sonderzeichen n 1. <Comp & DV> special character; 2. <Druck> pi characters, pi font, sorts, symbolic font
sondieren v <Kohlen> sound
Sondierungsbericht m <Kohlen> sounding record
Sone n <Akustik, Phys> sone

Sonnen... <Bau> solar
Sonnenabsorptionskoeffizient m <Nichtfoss Energ> solar absorption coefficient
Sonnenabsorptionsvermögen n <Nichtfoss Energ> solar absorption capacity
Sonnenaktivität f <Phys> solar activity
Sonnenazimut m <Nichtfoss Energ> solar azimuth
Sonnenbatterie f <Fernseh, Nichtfoss Energ, Telekom> solar battery
Sonnenbestrahlung f <Heiz & Kälte, Nichtfoss Energ> insolation
Sonnenblende f 1. <Foto> hood, lens hood; 2. <Kfztech> antidazzle visor *(Zubehör)*
Sonnenchemie f <Nichtfoss Energ> heliochemistry
sonnenchemisch adj <Nichtfoss Energ> heliochemical
Sonnendach n <Kfztech> sun roof *(Karosserie)*
Sonneneinstrahlung f 1. <Heiz & Kälte> insolation; 2. <Nichtfoss Energ> solar irradiation
Sonnenenergie f <Elektrotech, Maschinen, Nichtfoss Energ, Phys> solar energy • **mit Sonnenenergie betrieben** <Elektrotech, Nichtfoss Energ> solar-powered
Sonnenenergiekonversion f <Elektrotech, Nichtfoss Energ> solar energy conversion
Sonneneruption f <Raumfahrt> solar flare
Sonnenfarm f <Nichtfoss Energ> solar farm
Sonnenferne f <Phys> aphelion
Sonnenfleck m <Funktech> solar spot
Sonnenfühler m <Nichtfoss Energ, Raumfahrt> sun sensor
Sonnenhöhe f <Nichtfoss Energ> solar altitude
Sonnenkollektorplatte f <Raumfahrt> solar panel
Sonnenkraftwerk n <Elektrotech, Nichtfoss Energ> solar power plant, solar power station
Sonnenlichtabsorber m <Nichtfoss Energ, Raumfahrt> solar absorber
Sonnennähe f <Nichtfoss Energ, Phys> perihelion
Sonnenofen m <Nichtfoss Energ> solar furnace
Sonnenpaddel n <Telekom> solar panel
Sonnenprotuberanz f <Funktech> solar flare
Sonnenrad n 1. <Kfztech> sun gear; 2. <Maschinen> central gear, sun gear, sun wheel
Sonnenradsperrverzahnung f <Kfztech> sun gear lockout teeth
Sonnenradsteuerplatte f <Kfztech> sun gear control plate
Sonnenscheinmessung f <Nichtfoss Energ> sunshine monitoring
Sonnenscheinmessungsstation f <Nichtfoss Energ> sunshine monitoring station
Sonnenschutz m <Bau> screen
Sonnenschutzfarbe f <Kunststoff> shading paint
Sonnenschutzglas n <Ker & Glas> dark glass, sunglass
Sonnensegel n 1. <Nichtfoss Energ, Raumfahrt> solar sail; 2. <Wassertrans> awning
Sonnenstand m <Nichtfoss Energ> solar altitude
Sonnenstörung f <Nichtfoss Energ, Raumfahrt> sun interference
Sonnenstrahlmagnetron n <Elektronik> rising sun magnetron
Sonnenstrahlung f <Nichtfoss Energ, Strahlphys, Umweltschmutz> solar radiation
Sonnenstrahlungsdruck m <Raumfahrt> solar radiation pressure
sonnensynchrone Umlaufbahn f <Raumfahrt> sun synchronous orbit
Sonnenturm m <Nichtfoss Energ> solar tower
Sonnenwärme f <Nichtfoss Energ> solar heat
Sonnenwärmekollektor m <Nichtfoss Energ> solar collector
Sonnenwärmekonzentrator m <Nichtfoss Energ> solar concentrator
Sonnenwende f <Raumfahrt> solstice
Sonnenwendepunkt m <Raumfahrt> solstitial point
Sonnenwind m <Phys, Raumfahrt> solar wind
Sonnenzelle f <Elektrotech> solar cell
Sonometer n <Phys> sonometer
Sophorin n <Chemie> cytisine, sophorine, ulexine
Sorbens n <Meerschmutz> sorbent
Sorbinose f <Chemie> sorbose
Sorbit m <Metall> sorbite
Sorbose f <Chemie> sorbose
Sorption f <Wasserversorg> sorption
Sorptionsmittel n <Meerschmutz> sorbent
Sortier... <Comp & DV> sorting
Sortierbegriff m <Comp & DV> sort key
Sortierdatei f <Comp & DV> sort file
sortieren v 1. <Comp & DV> collate, sort; 2. <Druck> sort; 3. <Kohlen> size
Sortieren n 1. <Comp & DV> collation, sorting; 2. <Kohlen> grading
Sortierer m 1. <Comp & DV, Elektrotech> collator, sorter, sorting machine; 2. <Papier> screen
Sortierfeld n <Comp & DV> sort field
Sortierfolge f <Comp & DV> collating sequence, collation
Sortiergenerator m <Comp & DV> sort generator
Sortiermaschine f 1. <Ker & Glas> sorter; 2. <Maschinen> grading machine; 3. <Verpack> sorting machine
Sortierprogramm n <Comp & DV> sort program
Sortierprüfung f <Qual> screening inspection, screening test
Sortierschlüssel m <Comp & DV> sort key
Sortiersteige f für den Transport <Verpack> collating transit tray
Sortiertaste f <Comp & DV> sort key
Sortier- und Sammelsystem n <Verpack> collating system *(Buchbindereien)*
Sortierung f 1. <Comp & DV> collation, sort; 2. <Ker & Glas> grading
Sortierung f im Druckluftstrom <Fertig> air classification *(Pulvermetall)*
Sortierung f von Abfällen <Abfall> waste sorting
Sortierung f von Hand <Abfall> manual separation *(Müll)*
Sortiment n <Telekom, Textil> range
So-Schnittstellen-Schaltung f <Telekom> So-bus interface unit *(ISDN)*
SOS-Schaltkreis m <Elektronik> SOS logic
Sound-Karte f <Telekom> sound card
Soundtrack n <Fernseh> soundtrack
Source f <Phys> source *(Feldeffekttransistor)*
Source-Elektrode f <Elektrotech> source electrode
Source-Impedanz f <Elektrotech> source impedance
Source-Kontakt m <Elektrotech> source contact
Source-Verstärker m <Elektronik> source amplifier
Soxhlet-Apparat m <Labor> Soxhlet extractor
Soxhlet-Extraktor m <Labor> Soxhlet extractor
Soziussitz m <Kfztech> pillion *(Motorrad)*
Spachtel m <Bau> trowel
Spachtelmasse f 1. <Bau> filler, grouting compound, putty; 2. <Kerntech> filler
Spachtelmasse f für Karosseriereparaturen <Kfztech> body filler *(Karosserie)*
Spachtelmesser n <Bau> stopping knife
spachteln v 1. <Bau> putty, trowel; 2. <Fertig> prime, putty
Spachtelung f <Fertig> priming
Spachtelverbindung f <Bau> putty joint
Spalier versehen v/mit <Bau> lath
Spalling n <Ker & Glas> spalling

Spalt

Spalt *m* 1. <Bau> crack, fissure, slot; 2. <Elektrotech> gap *(Magnetkreisen)*; slot *(bei Relais)*; 3. <Fertig> die clearance; 4. <Hydraul> clearance; 5. <Ker & Glas, Kerntech, Maschinen> gap; 6. <Mechan> clearance; 7. <Metall> crack; 8. <Wellphys> slit
Spaltauskleidung *f* <Phys> slot liner
Spaltaxt *f* <Bau> *(AE)* splitting ax, *(BE)* splitting axe
Spalt-Azimut *m* <Aufnahme> gap azimuth
spaltbar *adj* 1. <Fertig> fissile; 2. <Phys> fissile, fissionable
spaltbares Isotop *n* <Phys> fertile isotope, fissile isotope
spaltbares Material *n* <Strahlphys> fissile material
Spaltblende *f* <Foto> slit diaphragm
Spaltbreite *f* 1. <Aufnahme, Elektrotech> gap length; 2. <Fernseh> gap width
Spaltbreitehalter *m* <Aufnahme> gap spacer
Spaltbruchstücke *npl* <Phys> fission fragments
Spaltdämpfung *f* <Optik> gap loss *(Interferenz)*
Spalte *f* 1. <Anstrich> crack; 2. <Comp & DV, Druck> column; 3. <Kohlen> crack; 4. <Maschinen> gap
Spalteinstellung *f* <Fernseh> gap setting
spalten *v* 1. <Fertig> fissure; 2. <Metall> cleave; 3. <Textil> split
Spalten *n* <Bau> cleaving, ripping
Spaltenabzug *m* <Druck> galley, galley proof, slip proof
Spaltensteller *m* <Comp & DV> tabulator
Spaltenwasser *n* <Kohlen> joint water
Spalter *m* <Kohlen> splitter
Spaltfestigkeit *f* 1. <Fertig> interlaminar strength; 2. <Kunststoff> interlaminar strength; adhesive shear strength *(Klebstoff)*; 3. <Metall, Qual> cleavage strength
Spaltfilter *n* <Maschinen> edge filter
Spaltflügel *m* <Lufttrans> slotted wing
Spaltfragmente *npl* <Phys> fission fragments
Spaltfuge *f* <Bau> split *(Dachsparren)*
Spaltgasplenum *n* <Kerntech> fission gas plenum
Spaltgasspeicherraum *m* <Kerntech> fission gas plenum
Spaltglas *n* <Ker & Glas> cleaved glass
Spaltgröße *f* <Ker & Glas> gap sizing
Spalthammer *m* <Bau> cleaver
Spaltkammer *f* <Kerntech> fission ionization chamber
Spaltklappe *f* <Lufttrans> slot flap
Spaltkohle *f* <Ker & Glas, Kohlen> cracking coal
Spaltlänge *f* <Akustik> gap length
Spaltmesser *n* <Sicherheit> riving knife
Spaltneutronen *npl* <Strahlphys> fission neutrons
Spaltniet *m* <Maschinen> bifurcated rivet
Spaltölfilter *n* <Kohlen> *(BE)* disc filter, *(AE)* disk filter
Spaltphasenmotor *m* <Elektrotech> split phase motor
Spaltpolmotor *m* <Elektrotech> shaded pole motor
Spaltprodukt *n* <Abfall, Kerntech> fission product
Spaltring *m* <Maschinen> split ring
Spaltriss *m* <Bau> split *(Holz)*
Spaltsäge *f* 1. <Bau> cleaving saw; 2. <Maschinen> ripping saw
Spaltstoffabbrand *m (FIFA)* <Kerntech> fissions per initial fissile atom *(FIFA)*
Spaltstoffausnutzung *f* <Kerntech> nuclear fuel utilization
Spaltstoffgitterabstand *m* <Kerntech> lattice pitch spacing
Spaltstoffstruktur *f* <Kerntech> fuel assembly
Spaltstoffverbrauch *m (FIFA)* <Kerntech> fissions per initial fissile atom *(FIFA)*
Spaltstoffverhältnis *n* <Kerntech> fissile inventory ratio
Spalttiefe *f* <Akustik, Aufnahme, Fernseh> gap depth
Spaltung *f* 1. <Bau> splitting; 2. <Chemie> breakdown; 3. <Metall> cleavage

Spaltung f mit Thermoneutronen <Kerntech> thermal neutron fission
Spaltungsfläche *f* <Metall> cleavage facet
Spaltungsionisationskammer *f* <Kerntech> fission ionization chamber
Spaltungsstörzone *f* <Kerntech> fission spike
Spaltverlust *m* <Akustik, Aufnahme, Fernseh, Optik> gap loss *(Interferenz)*
Spaltversuch *m* <Werkprüf> cleavage test
Spaltwasser *n* <Kohlen> cleft water
Spaltzone *f* <Kerntech> core *(Reaktor)*
Span *m* 1. <Bau> splinter *(Holz)*; 2. <Fertig> chip, fine, shaving; 3. <Maschinen, Mechan> chip
Spanabflussrichtung *f* <Fertig> direction of chip flow
spanabhebend bearbeiten *v* <Maschinen> machine
spanabhebende Bearbeitung *f* <Maschinen> cutting, machining
Spanabnahme *f* 1. <Fertig> linear machining; 2. <Maschinen> chip removal
Spanbrecher *m* <Maschinen> chip breaker
Spanbrecherplatte *f* <Fertig> breaker plate
Spandicke *f* <Maschinen> chip thickness
Späne *mpl* <Fertig> swarf
Späneabsaugung *f* <Sicherheit> chip exhaust *(Maschine)*
spanen *v* <Fertig> machine
Spanen *n* 1. <Fertig> cutting; 2. <Maschinen> chip removal, machining, metal cutting
Spanen n auf Gewicht <Ker & Glas> chipping to the weight
spanend bearbeitbar *adj* <Fertig> machinable
spanend bearbeiten *v* <Maschinen> machine
spanende Bearbeitbarkeit *f* <Fertig> machinability, machining property, machining quality
spanende Bearbeitung *f* <Maschinen> machining, metal cutting
Spänetrog *m* <Fertig> trough *(Spanung)*
Spanfläche *f* <Fertig> rake, tool face, top rake; face *(Werkzeug)*
Spanflächen-Orthogonalebene *f* <Fertig> tool face orthogonal plane
Spanflächen-Tangentialkraft *f* <Fertig> tool face tangential force
Spanflächen-Verschleiß *m* <Maschinen> face wear *(eines Werkzeugs)*
Spanfluss *m* <Fertig> flow of chips
Spange *f* 1. <Bau> stay bolt; 2. <Mechan> clip
Spanleistung *f* <Fertig> cutting capacity
spanlos *adj* <Fertig> nonchipping
spanlose Formgebung *f* <Maschinen> noncutting shaping
spanlose Tragflügelbefestigung *f* <Lufttrans> forged wing attachment
Spanlücke *f* <Fertig> gash
Spanmenge *f* <Fertig> quantity of metal removed
Spann... spanning
Spannarm *m* <Telekom> tension arm
Spannbacke *f* 1. <Fertig> welding jaw *(Stumpfschweißen)*; 2. <Maschinen> chuck jaw, gripping jaw, jaw
Spannbandmesswerk *n* <Gerät> taut-band movement
Spannbereich *m* <Fertig> holding capacity
Spannbeton *m* <Bau> prestressed concrete
Spannbohle *f* <Bau> strutting board
Spannbolzen *m* <Fertig> clamping bolt
Spannbuchse *f* <Fertig> holding bushing
Spannbüchse *f* <Maschinen> spring collet
Spannbügel *m* <Fertig> clip; clamp *(Kunststoffinstallationen)*

Spanndorn m <Fertig> (AE) arbor, (BE) arbour (Fräsmaschine)
Spanndraht m <Bau> stretching wire
Spanne f 1. <Maschinen> margin; 2. <Metrol> span
Spanneisen n <Fertig> clamp, crossbar
Spanneisen n **mit Stift** <Fertig> finger clamp
spannen v 1. <Bau> tighten; 2. <Fertig> hold, stress, tauten; planish (Blech); secure (Werkstück); 3. <Foto> cock; 4. <Textil> stretch
Spannen n 1. <Fertig> planishing (Blech); 2. <Maschinen> tension
Spannen n **von Hand** <Foto> manual cocking
Spanner m 1. <Fertig> clamping device; 2. <Kfztech> tensioner (Kette); 3. <Maschinen> tightener
Spannfaden m <Textil> tight pick
Spannfeder f <Maschinen> tension spring
Spannfinger m <Fertig> toe dog
Spannfutter n 1. <Fertig> chuck; 2. <Maschinen> chuck, jaw chuck; 3. <Mechan> chuck
Spanngitter n <Elektronik> frame grid (Elektronenröhre)
Spannhebel m <Foto> cocking lever
Spannhebel m **für Selbstauslöser** <Foto> self-time lever
Spannhülse f 1. <Fertig> adaptor sleeve; 2. <Maschinen> clamping sleeve
Spannhülsenlager n <Fertig> adaptor bearing
Spannhydraulik f <Maschinen> hydraulic clamping
Spannkette f <Textil> tight end
Spannkettenrad n <Fertig> idler sprocket
Spannkloben m <Maschinen> face chuck, face plate chuck, face plate dog, face plate jaw
Spannkluppe f <Maschinen> (BE) vice clamp, (AE) vise clamp
Spannkraft f <Verpack> clamping force
Spannmutter f 1. <Fertig> nut (Kunststoffinstallationen); 2. <Maschinen> coupling nut
Spannpatrone f <Maschinen> draw-in collet
Spannplatte f <Maschinen> bolster plate
Spannpratze f <Fertig> strap
Spannrahmen m <Textil> (BE) stenter frame, (AE) tenter frame
Spannring m 1. <Foto> cocking ring; 2. <Maschinen> lock ring; 3. <Mechan> expansion ring
Spannrolle f 1. <Fertig, Kfztech> belt idler; 2. <Maschinen> idler, tension roller, tightener, tightening pulley; 3. <Mechan> jockey pulley; 4. <Papier> tension roller
Spannsäule f <Kohlen> jack column
Spannscheibe f <Maschinen> tension pulley
Spannschloss n <Bau, Fertig, Maschinen> turnbuckle
Spannschlossmutter f <Maschinen> turnbuckle sleeve
Spannschraube f 1. <Fertig> clamping screw; 2. <Maschinen> clamping screw, straining screw, tension screw, tightening screw; 3. <Mechan> turnbuckle; 4. <Wassertrans> rigging screw, turnbuckle (Tauwerk)
Spannseil n 1. <Bau> stay; 2. <Maschinen> tightening cord
Spannstock m <Mechan> (BE) vice, (AE) vise
Spannstockfutter n <Maschinen> (BE) vice chuck, (AE) vise chuck
Spanntuch n <Foto> pressure cloth
Spannung f 1. <Bau> stress, tension; 2. <Elektriz, Elektrotech> potential, voltage, tension; 3. <Erdöl> stress (Werkstoff); 4. <Ergon> strain; 5. <Fertig> hold, stress; 6. <Funktech> potential; 7. <Hydraul> pressure (Dampf); 8. <Ker & Glas> strain, stress; 9. <Kfztech> voltage (Elektrik, Zündung); 10. <Kohlen> stress; 11. <Kunststoff> stress (mechanisch); 12. <Maschinen> tension; 13. <Mechan> strain; 14. <Metall> stress, tension; 15. <Papier> stress; 16. <Phys> tension, voltage; 17. <Qual> strain; 18. <Raumfahrt> potential (Weltraumfunk); 19. <Telekom> voltage; 20. <Textil> tension • **Spannung anlegen** <Elektrotech> apply a voltage • **unter Spannung setzen** <Elektrotech> apply a voltage • **unter Spannung stehend** <Elektrotech> alive, live
Spannung f **bei Belastung** <Elektrotech> on-load voltage (Klemmen)
Spannung f **gegen Erde** <Elektrotech> (BE) voltage to earth, (AE) voltage to ground
Spannung f **unter Last** <Elektrotech> closed-circuit voltage
Spannung f **zum Nullleiter** <Elektriz> voltage to neutral
Spannung f **zwischen Leiter und Erde** <Elektriz> (BE) line-to-earth voltage, (AE) line-to-ground voltage
Spannung f **zwischen Leiter und Nullpunkt** <Elektriz> line-to-neutral voltage
Spannung f **zwischen Leitern** <Elektriz> line-to-line voltage
Spannung f **zwischen Phasen** <Elektriz> phase-to-phase voltage
Spannung/Frequenz-Umwandlung f <Elektrotech> voltage-to-frequency conversion
Spannung/Frequenz-Wandler m <Elektrotech> voltage-to-frequency converter
Spannungsabbau m <Kunststoff> relaxation of stress, stress relaxation
Spannungsabfall m 1. <Elektriz, Elektrotech> drop of potential, potential drop, voltage drop; 2. <Fertig> voltage drop; 3. <Phys> potential drop, voltage drop
Spannungsabfall m **an einer Reaktanz** <Elektriz> reactance voltage drop
Spannungsabfall m **auf Impedanz** <Elektriz> impedance voltage drop
Spannungsabfall m **durch Blindwiderstand** <Elektrotech> reactance voltage drop
Spannungsabfall m **in Leitung** <Elektrotech> line voltage drop
Spannungsabfallwiderstand m <Elektrotech> voltage dropping resistor
spannungsabhängiger Widerstand m 1. <Elektrotech> voltage-dependent resistor; 2. <Phys> varistor
Spannungsanstiegsrate f <Elektrotech> rate of voltage rise
Spannungsanzeiger m 1. <Elektrotech> voltage indicator; 2. <Wassertrans> charge indicator (Elektrik)
Spannungsausgleich m <Elektriz> voltage balance
Spannungsausgleichschaltung f <Elektrotech> equipotential connection
Spannungsausgleichverbindung f <Elektrotech> equipotential connection
Spannungsbauch m 1. <Elektrotech> antinode; 2. <Kerntech> potential loop
Spannungsbegrenzer m <Elektriz, Elektrotech> voltage limiter, transient suppressor
Spannungsbegrenzung f <Elektrotech> transient suppression
Spannungsbegrenzungsstrecke f <Elektriz> voltage suppression gap
Spannungsbeobachter m <Ker & Glas> strain viewer
Spannungsbereich m <Gerät> voltage range
Spannungsdehnung f <Eisenbahn> stress expansion
Spannungsdehnungsdiagramm n <Kunststoff> stress-strain curve
Spannungsdehnungsschaubild n <Wassertrans> stress-strain diagram (Schiffbau)
Spannungsdifferenz f <Elektriz, Elektrotech, Phys> PD, potential difference, voltage difference
Spannungsdurchschlag m <Elektronik> dielectric breakdown

Spannungsentlastung *f* <Werkprüf> stress relief
Spannungserzeuger *m* <Phys> voltage generator
Spannungsfernmessgerät *n* <Gerät> televoltmeter
spannungsfrei *adj* <Elektriz> voltage-free
spannungsfrei machen *v* <Fertig> anneal *(Kunststoffe)*
Spannungsfreiglühen *n* 1. <Fertig> lonealing, temper annealing; 2. <Metall> temper annealing
spannungsführend *adj* 1. <Elektriz, Elektrotech> active, alive; 2. <Phys> live
spannungsführende Komponente *f* <Elektrotech> active component
Spannungsgefälle *n* 1. <Elektriz> potential gradient, voltage gradient; 2. <Phys> potential gradient
spannungsgeregelte Stromversorgung *f* <Elektriz> voltage-controlled power supply
spannungsgeregelter Eingang *m* <Elektrotech> voltage-controlled input
spannungsgeregelter Kondensator *m* <Elektrotech> voltage-controlled capacitor
spannungsgeregelter Oszillator *m* <Raumfahrt> VCO, voltage-controlled oscillator *(Weltraumfunk)*
spannungsgeregeltes Netzgerät *n* <Elektrotech> voltage-controlled power supply
spannungsgesteuerter Oszillator *m* <Elektronik, Telekom> VCO, voltage-controlled oscillator
Spannungsgleichhaltung *f* <Elektrotech> voltage stabilization
Spannungsgradient *m* 1. <Elektriz> voltage gradient; 2. <Phys> potential gradient
Spannungsimpuls *m* <Elektriz> voltage pulse
Spannungsknoten *m* <Elektronik> voltage node
Spannungskoeffizient *m* <Thermod> temperature pressure coefficient
Spannungskonstanthalter *m* <Elektrotech> voltage regulator
Spannungskorrosion *f* <Werkprüf> stress corrosion
Spannungskräuseln *n* <Textil> tension pucker
spannungslos *adj* <Gerät> dead
spannungslose Freigabe *f* <Elektriz, Elektrotech> no-volt release
Spannungslosigkeit *f* 1. <Elektrotech> no voltage; 2. <Gerät> absence of voltage
Spannungsmessbereich *m* <Gerät> voltage measuring range, voltage span
Spannungsmessgerät *n* 1. <Elektrotech> voltmeter; 2. <Gerät> voltage measuring instrument; 3. <Phys> voltmeter
Spannungsoptik *f* <Fertig, Optik> photoelastic optics, photoelasticity
spannungsoptisch *adj* <Fertig> photoelastic
spannungsoptischer Koeffizient *m* <Ker & Glas> stress optical coefficient
Spannungspolarität *f* <Elektrotech> voltage polarity
Spannungsquelle *f* 1. <Elektrotech> power supply, source, voltage source; 2. <Phys> voltage source
Spannungsreferenz *f* <Elektrotech> voltage reference
Spannungsreferenzdiode *f* <Elektronik> voltage reference diode
Spannungsreferenzröhre *f* <Elektrotech> voltage reference tube
Spannungsregeltransformator *m* <Elektrotech> voltage-regulating transformer
Spannungsregelung *f* <Elektriz, Elektrotech> voltage control, voltage regulation
Spannungsregler *m* 1. <Comp & DV, Elektriz, Fertig, Kfztech, Kontroll> voltage controller, voltage regulator; 2. <Textil> tension device
Spannungsreglerdynamo *m* <Elektriz> regulating dynamo

Spannungsrelais *n* <Elektrotech> voltage relay
Spannungsrelaxation *f* <Kunststoff> relaxation of stress, stress relaxation
Spannungsresonanz *f* <Elektronik> parallel resonance
Spannungsriss *m* <Werkprüf> stress crack
spannungsrissempfindlich *adj* <Fertig> prone to stress cracking *(Kunststoffinstallationen)*
Spannungsscheibe *f* <Ker & Glas> *(BE)* strain disc, *(AE)* strain disk
Spannungsschlinge *f* <Elektrotech> potential loop
Spannungsschwankung *f* <Elektriz> voltage fluctuation
Spannungsspitze *f* 1. <Elektrotech> power surge, voltage spike; 2. <Funktech> surge
Spannungssprung *m* <Elektriz> voltage jump
Spannungsstabilisator *m* <Elektrotech> voltage stabilizer
Spannungsstabilisatordiode *f* <Elektronik> voltage regulator diode
Spannungsstabilisierung *f* <Elektrotech> voltage regulation
Spannungsstoß *m* 1. <Elektrotech> surge, voltage surge; 2. <Telekom> voltage surge
Spannungsstrommessgerät *n* <Elektrotech> voltameter
Spannungsstufe *f* <Phys> voltage step
Spannungsteiler *m* 1. <Elektriz> static balancer, voltage divider; 2. <Fernseh> voltage divider; 3. <Phys> potential divider, voltage divider; 4. <Telekom> voltage divider
Spannungsteilerschaltung *f* <Gerät> voltage divider network
Spannungstensor *m* <Phys> stress tensor
Spannungstransformator *m* <Elektriz, Elektrotech> voltage transformer
Spannung-Strom Charakteristik *f* <Elektrotech> voltage current characteristic
Spannung-Strom Kennlinie *f* <Elektrotech> voltage current characteristic
Spannungsübertragungsfaktor *m* <Akustik> response to voltage
Spannungsumformer *m* <Elektriz> voltage transformer
Spannungsunterschied *m* <Elektriz> voltage difference
Spannungsverdoppler *m* <Elektriz, Phys> voltage doubler
Spannungsverdreifachung *f* <Funktech> tripling voltage
Spannungsvergleich *m* <Elektrotech> voltage comparison
Spannungsverlauf *m* <Fertig> flow of stress
Spannungsverlust *m* 1. <Elektriz, Elektrotech> voltage drop, voltage loss; 2. <Fertig> line drop, voltage drop; 3. <Phys> voltage drop
Spannungsverringerung *f* <Maschinen> stress relief
Spannungsversorgung *f* 1. <Aufnahme> power supply; 2. <Phys> power supply, supply voltage
Spannungsversorgungsuntersystem *n* <Raumfahrt> power subsystem
Spannungsverstärker *m* <Elektronik, Elektrotech> booster, voltage amplifier
Spannungsverstärkung *f* 1. <Elektronik> voltage amplification, voltage gain; 2. <Phys> voltage gain
Spannungsverstärkungsfaktor *m* <Elektrotech> voltage gain factor
Spannungsverteiler *m* <Raumfahrt> power bus
Spannungsverteilersystem *n* <Raumfahrt> power distribution network
Spannungsvervielfacher *m* <Elektriz, Elektrotech> voltage multiplier
Spannungswähler *m* <Elektrotech> voltage selector
Spannungswählschalter *m* <Elektrotech> voltage selector switch
Spannungswandler *m* 1. <Elektrotech> voltage transformer; 2. <Raumfahrt> power converter

Spannungswandler m Gleichstrom-Wechselstrom <Elektrotech> DC-AC converter
Spannungswandlung f Gleichstrom-Wechselstrom <Elektrotech> DC-AC conversion
Spannungszone f <Ker & Glas> stressed zone
Spannut f <Maschinen> flute • **ohne Spannut** <Maschinen> fluteless
Spannutlänge f <Maschinen> flute length
Spannutsteigung f <Maschinen> flute pitch
Spannvorrichtung f 1. <Fertig> workholding device, workholding fixture; 2. <Maschinen> gripping device, jig, tensioning device; 3. <Papier> stretcher
Spannwalze f <Papier> stretch roll, tightener
Spannweite f 1. <Bau> span; 2. <Gerät> range; 3. <Lufttrans> span, wing span; 4. <Phys> span (einer Tragfläche); 5. <Qual> range
Spannweitenachse f des Luftschraubenblattes <Lufttrans> blade span axis (Hubschrauber)
Spannweiten-Kontrollkarte f <Qual> range chart
Spannwirbel m <Mechan> turnbuckle
Spannzange f <Maschinen> collet chuck
Spannzeug n <Maschinen> chuck, chucking device
Spant m <Wassertrans> rib (Bootbau); frame (Schiffbau)
Spantabstand m <Wassertrans> frame spacing (Schiffbau)
Spantenriss m <Wassertrans> body plan (Schiffkonstruktion)
Spantiefe f 1. <Fertig> depth of cut, rate of cut; 2. <Maschinen> cut, depth of cut
Spantwinkel m <Wassertrans> frame angle (Schiffbau)
Spanumfangswinkel m <Fertig> angle of approach (Fräser)
Spanungsbreite f <Fertig> width of cut
Spanungsdicke f <Fertig> thickness of cut
Spanwinkel m 1. <Fertig> undercut angle (Räumwerkzeug); 2. <Maschinen> angle of rake, rake • **ohne Spanwinkel** <Fertig> neutral-rake
Sparbeize f <Fertig> pickling inhibitor, restrainer
Spardeck n <Wassertrans> spardeck
Spardiode f <Elektronik> efficiency diode
Spardüse f <Kfztech> economizer jet (Vergaser)
Spargerät n <Heiz & Kälte> economizer
Sparpackung f <Verpack> economy-size pack
Sparren m <Bau> rafter
sparsame Fahrt f <Wassertrans> economical speed
Sparschalter m <Mobilkom> power down switch
Spartein n <Chemie> lupinidine, sparteine
Spartrafo m <Elektrotech> autotransformer
Spartransformator m <Phys> autotransformer
Spat m <Erdöl> spar (Mineral)
Spätauslasssteuerschieber m <Hydraul> late release slide valve (Dampf)
Späteinlasssteuerschieber m <Hydraul> late admission slide valve (Dampf)
Spatel m <Labor> spatula
Spaten m <Bau> spade
Spatenruder n <Wassertrans> spade rudder
Spatienkeil n <Druck> spaceband
spatiieren v <Druck> space
Spatiieren n <Druck> letter spacing
spationieren v <Druck> space
Spationierung f <Druck> letter spacing
Spatium n <Druck> space
Spatprodukt n <Math> triple scalar product (dreidimensionale Vektoren)
SPDT-Schalter m 1. <Elektriz> (einpoliger Wechselschalter) SPDT switch (single pole double-throw switch); 2. <Elektrotech> (einpoliger Umschalter) SPDT switch (single pole double-throw switch)

speerförmig adj <Kerntech> javelin-shaped (Brennstab)
Speiche f <Maschinen> arm, spoke
Speichenkreuz n <Fertig> spider
Speichenrad n <Kfztech> spoke wheel
Speicher m 1. <Comp & DV> computer memory, memory, store, storage, storage device; 2. <Elektrotech> accumulator, director, memory; 3. <Fertig> rack (Spanung); 4. <Funktech> memory; 5. <Telekom> register; 6. <Wassertrans> warehouse • **im Speicher abgelegt** <Comp & DV> memory-mapped
Speicher m **mit hoher Aufzeichnungsdichte** <Comp & DV> high density storage
Speicher m **mit hoher Dichte** <Comp & DV> high-density storage
Speicher m **mit indexsequenziellem Zugriff** <Comp & DV> indexed sequential storage
Speicher m **mit seriellem Zugriff** <Comp & DV> serial access memory, serial access storage
Speicher m **mit veränderlichem Nachleuchten** <Elektronik> variable-persistence storage
Speicherabbild n <Comp & DV> memory map
Speicherabbildung f <Comp & DV> storage map
Speicherabgabe f <Comp & DV> storage out
Speicheradresse f <Comp & DV> memory location, storage location
Speicheradressregister n <Comp & DV> MAR, memory address register
Speicheraufbereitung f <Comp & DV> memory edit
Speicheraufteilungsübersicht f <Comp & DV> storage usage map
Speicherausnutzung f <Comp & DV> storage efficiency, storage utilization
Speicherauszug m <Comp & DV> dump, memory dump, storage dump • **Speicherauszug erstellen** <Comp & DV> dump
Speicherauszug m **der Änderungen** <Comp & DV> change dump
Speicherauszug m **nach Störungen** <Comp & DV> postmortem dump
Speicherauszugsdatei f <Comp & DV> dump data set
Speicherauszugsprüfung f <Comp & DV> dump check
Speicherbank f <Comp & DV> memory bank
Speicherbatterie f <Phys> storage battery
Speicherbaustein m <Comp & DV> memory chip
Speicherbauteil n <Comp & DV> memory module
Speicherbecken n <Bau, Nichtfoss Energ, Wasserversorg> reservoir basin, storage basin
Speicherbedarf m <Comp & DV> storage requirement
Speicherbelegung f <Comp & DV> storage utilization
Speicherbelegungsplan m <Comp & DV> memory map
Speicherbereich m <Comp & DV> area
Speicherbereinigung f <Comp & DV> (AE) garbage collection, (BE) rubbish collection
Speicherbereinigungsprogramm n <Comp & DV> (AE) garbage collector, (BE) rubbish collector
Speicherbetrieb m <Telekom> store-and-forward mode
Speicherbildschirm m <Elektronik> storage screen
Speicherblase f <Fertig> bladder, sac
Speicherchip m <Comp & DV> memory chip
Speicherdatenregister n (MDR) <Comp & DV> memory data register (MDR)
Speicherdatenübermittlung f <Comp & DV> store-and-forward
Speicherdichte f <Comp & DV> bit density, packing density, storage density
Speicherdirektzugriff m <Telekom> DMA, direct memory access
Speichereffekt m <Elektrotech> storage effect

Speichereinheit

Speichereinheit f 1. <Comp & DV> storage device; 2. <Telekom> SFU, SU, store-and-forward unit
Speichereintrag m <Comp & DV> storage entry
Speicherelement n 1. <Comp & DV> memory cell, storage element; 2. <Elektrotech, Telekom> storage element
Speichererweiterung f <Comp & DV> processor storage, storage expansion
Speicherfragmentierung f <Comp & DV> storage fragmentation
Speicherfunktion f <Comp & DV> memory function
Speichergerät n <Comp & DV> storage device
Speicherheizkörper m <Heiz & Kälte> storage heater
Speicherheizung f 1. <Heiz & Kälte> storage heating; 2. <Maschinen> storage heater
Speicherhierarchie f <Comp & DV> memory hierarchy, storage hierarchy
Speicherkamera f <Foto> storage camera
Speicherkapazität f 1. <Comp & DV> memory capacity, storage capacity, Diskette density; 2. <Elektrotech> memory capacity; 3. <Telekom> storage capacity
Speicherkarte f <Comp & DV> memory card
Speicherkomprimierung f <Comp & DV> memory compaction
Speicherkondensator m <Elektrotech> energy storage capacitor, reservoir capacitor
Speicherkonfiguration f <Raumfahrt> storage configuration *(Raumschiff)*
Speicherleistung f <Comp & DV> memory capacity
Speichermanagement n <Comp & DV> memory management
Speichermasche f <Elektronik> storage mesh
Speichermedium n <Comp & DV, Telekom> storage medium
Speichermodul n <Comp & DV> memory module
Speichermodum n <Comp & DV> memory module
speichern v 1. <Comp & DV> save, store; 2. <Elektrotech, Kontroll> store; 3. <Telekom> retain *(Daten)*
Speichern n <Comp & DV> save, store, storage
Speichern n **der zuletzt gewählten Rufnummer** <Telekom> storage of last dialed number
Speichern n **und Befördern** n <Kontroll> store-and-forward
Speicherorganisation f <Comp & DV> file organization
speicherorientiert adj <Comp & DV> memory-mapped
Speicherort m <Comp & DV> memory location
Speicheroszilloskop n <Gerät, Phys> storage oscilloscope
Speicherplatte f <Comp & DV> storage disk
Speicherplatz m <Comp & DV> memory location, storage location
Speicherplatzzuteilung f <Comp & DV> storage allocation
speicherprogrammierbare Rechenanlage f <Comp & DV> stored program computer
speicherprogrammierbare Steuerung f *(SPS)* <Kontroll, Regelung> programmable logic control *(PLC)*
speicherprogrammierte Rechenanlage f <Comp & DV> stored program computer
speicherprogrammierte Steuerung f <Telekom> *(AE)* stored program control, *(BE)* stored programme control
speicherprogrammierte Vermittlungsstelle f <Telekom> *(AE)* stored program control exchange, *(BE)* stored programme control exchange
speicherprogrammierter Rechner m <Kontroll> *(AE)* stored program computer, *(BE)* stored programme computer
speicherprogrammiertes Vermittlungssystem n <Telekom> *(AE)* stored program switching system, *(BE)* stored programme switching system

Speicherprüfung f <Comp & DV> storage scan
speicherresident adj <Comp & DV> memory-resident
speicherresidente Software f <Comp & DV> resident software
speicherresidentes Programm n <Comp & DV> resident software
Speicherring m <Teilphys> storage ring
Speicherröhre f <Comp & DV, Elektronik> memory tube, storage tube
Speicherröhre f **mit einem Elektronenstrahl** <Elektronik> single gun storage tube
Speicherröhre f **mit veränderlichem Nachleuchten** <Elektronik> variable-persistence storage tube
Speicherschaltdiode f <Elektronik> snap-off diode
Speicherschaltkreis m <Telekom> memory circuit
Speicherschema n <Nichtfoss Energ> storage scheme
Speicherschreibsperre f <Comp & DV> memory protection
Speicherschutz m <Comp & DV> memory protection, storage protection
Speichersee m <Bau, Nichtfoss Energ, Wasserversorg> reservoir basin
Speicherseite f <Comp & DV> memory page
Speichersicherung f <Comp & DV> storage protection
Speicherstapel m <Comp & DV> storage stack
Speicherstelle f <Comp & DV> location, storage location
Speichersteuerung f 1. <Comp & DV> memory control; 2. <Elektrotech> memory controller
Speichersubsystem n <Comp & DV> storage subsystem
Speichersystem n <Fernseh> memory system
Speichertank m **für Aktivabfälle** <Kerntech> active effluent hold-up tank
Speichertransistor m <Elektronik> memory transistor
Speichertrommel f <Fertig> magazine drum
Speichertyp m <Comp & DV> storage type
Speicher- und Adresserweiterung f <Comp & DV> storage expansion
Speicherung f 1. <Comp & DV, Meerschmutz, Teilphys> storage; 2. <Wasserversorg> damming-up
Speicherung f **und Wiederaussendung** f <Telekom> store-and-forward
Speicherungsform f <Comp & DV> file organization
Speicherungsvermögen n <Wasserversorg> storage capacity
Speicherverdichtung f <Comp & DV> memory compaction
Speichervermittlung f <Comp & DV> message switching
Speichervermittlungsnetz n <Telekom> store-and-forward switching network
Speichervermittlungsstelle f <Telekom> *(AE)* message switching center, *(BE)* message switching centre
Speicherverwaltung f <Comp & DV> memory management, memory manager
Speichervorgang m <Comp & DV> storage
Speicherzeit f <Elektronik, Elektrotech> retention time, storage time
Speicherzelle f 1. <Comp & DV> array memory, memory cell, memory location, storage cell; register *(Wortspeicher)*; 2. <Druck> storage cell; 3. <Elektrotech> cell, storage cell
Speicherzugriff m <Comp & DV, Elektrotech> memory access
Speicherzuordnung f <Comp & DV> storage allocation
Speicherzyklus m <Comp & DV> memory cycle
Speigatt n <Wassertrans> scupper
Speisebrei m <Lebensmittel> chyme *(Magen)*
Speisebrücke f <Telekom> transmission bridge
Speisekabel n 1. <Elektriz, Elektrotech> feeder, feeder cable; 2. <Fernseh> feeder cable; 3. <Funktech> feeder

Speisekopf m <Kerntech> feeder head
Speiseleitung f 1. <Elektriz, Elektrotech> feeder line, interconnecting line; 2. <Funktech> feeder line; 3. <Raumfahrt> feeder (Weltraumfunk)
speisen v <Elektrotech> feed (Strom)
Speisepumpe f <Fertig, Heiz & Kälte, Maschinen> feed pump
Speiser m <Ker & Glas> feeder
Speisereinlauf m <Ker & Glas> feeder gate
Speiserkolben m <Ker & Glas> feeder plunger
Speiseröffnung f <Ker & Glas> feeder opening
Speiserohr n <Fertig, Maschinen> feed pipe
Speiserschüssel f <Ker & Glas> feeder nose
Speiseschaltung f <Elektriz> feed circuit
Speiseschiene f <Elektriz> feeder bar
Speisespannung f <Fertig> supply voltage (Kunststoffinstallationen)
Speisestrom m <Elektrotech> energizing current, supply current
Speisesystem n <Fertig> feeding head (Gießen)
Speisetransformator m <Elektriz> feeding transformer
Speiseverfahren n <Ker & Glas> feeding process, method of feeding
Speisewalze f <Lebensmittel> feed roll (Walzenstuhl)
Speisewasser n 1. <Heiz & Kälte, Maschinen, Mechan, Papier> feedwater; 2. <Wassertrans> feedwater (Motor); 3. <Wasserversorg> feedwater
Speisewasseranschluss m <Kerntech> feedwater inlet nozzle
Speisewasseraufbereitung f <Heiz & Kälte> feedwater treatment
Speisewasserenthärtung f <Heiz & Kälte> feedwater softening
Speisewasserleitung f <Maschinen> feedwater pipe
Speisewasserpumpe f <Maschinen> feedwater pump
Speisewasserverteiler m <Kerntech> feedwater manifold
Speisewasservorwärmer m 1. <Hydraul> economizer (Dampfkessel); 2. <Wassertrans> feedwater heater (Boiler)
Speisung f 1. <Elektriz> feed; 2. <Elektrotech> energization; feed (Strom); 3. <Hydraul> feed; 4. <Telekom> power feed, power feeding, feed
Spektral... <Comp & DV, Gerät, Metall, Phys> spectral
Spektralanalysator m <Gerät> (BE) spectral analyser, (AE) spectral analyzer
Spektralanalyse f 1. <Comp & DV> spectral analysis; 2. <Metall> spectographic analysis; 3. <Phys> spectral analysis; 4. <Strahlphys> spectrum analysis; 5. <Telekom> spectral analysis
Spektralansprechgeschwindigkeit f <Optik> spectral responsivity
Spektralbänder npl <Wellphys> bands of the spectrum
Spektralbelegungsgrad m <Telekom> spectral occupancy
Spektralbeleuchtungsstärke f <Optik, Telekom> spectral irradiance
Spektralbereich m <Phys> spectral range
Spektralbestrahlungsstärke f <Optik, Telekom> spectral irradiance
Spektralbreite f <Telekom> spectral bandwidth, spectral width
Spektraldarstellung f <Akustik> spectral distribution
Spektraldichte f <Akustik, Elektronik, Phys> spectral density
Spektraldurchlassgrad m <Phys> spectral transmittance
spektrale Leistungsdichte f <Funktech> power spectral density
Spektralemissionsvermögen n <Phys> spectral emissivity
Spektralempfindlichkeit f <Telekom> spectral responsivity
Spektralfarben fpl <Strahlphys> (AE) colors of the spectrum, (BE) colours of the spectrum
Spektralfenster n <Optik, Telekom> spectral window
Spektralleuchtvermögen n <Phys> spectral luminance
Spektrallichtausbeute f <Phys> spectral luminous efficiency
Spektrallinie f 1. <Optik> line of spectrum; 2. <Phys> spectral line; 3. <Strahlphys> line, spectral line; 4. <Telekom> spectral line
Spektrallinienbreite f <Optik, Strahlphys, Telekom> spectral line width
Spektrallinienprofil n <Strahlphys> spectral line profile
Spektralphotometer n <Phys> spectrophotometer
Spektralphotometrie f <Phys> spectrophotometry
Spektralpyranometer n <Nichtfoss Energ> spectral pyranometer
Spektralreflexionsgrad m <Phys> spectral reflectance
Spektralstrahldichte f <Telekom> spectral radiance
Spektralterm m <Phys> spectral term
Spektraltonhöhe f <Akustik> spectral pitch
Spektraltransmissionsfenster n <Optik> spectral window
Spektralverteilungscharakteristik f <Elektronik> spectral characteristic
Spektralverteilungskurve f <Elektronik> spectral characterisic
Spektrograph m <Phys, Strahlphys> spectrograph
spektrographische Analyse f <Strahlphys> spectrographic analysis
Spektrometer n <Nichtfoss Energ, Phys, Telekom, Wellphys> spectrometer
Spektrometrie f <Erdöl, Telekom> spectrometry
spektrometrische Analyse f <Strahlphys> spectrometric analysis
Spektroskop n <Phys, Wellphys> spectroscope
Spektroskopie f <Phys, Strahlphys> spectroscopy
Spektrum n 1. <Aufnahme, Elektriz, Elektronik, Phys, Raumfahrt, Strahlphys> spectrum; 2. <Math> spectrum (enthält alle Eigenwerte einer Matrix)
Spektrum n **der Turbulenz** <Strömphys> spectrum of turbulence
Spektrum n **des sichtbaren Lichtes** <Wellphys> visible light spectrum
Spektrumanalysator m <Telekom> (BE) spectrum analyser, (AE) spectrum analyzer
Spender m <Labor, Lebensmittel, Verpack> dispenser
Spermidin n <Chemie> spermidine
Spermin n <Chemie> spermin, spermine
Sperr... <Elektronik, Fertig, Maschinen> barrier
Sperrausgleichsgetriebe n <Kfztech> limited slip differential
Sperrband n <Ker & Glas> stop belt
Sperrbecken n <Nichtfoss Energ, Wasserversorg> catch basin, catchment basin
Sperrbereich m 1. <Aufnahme> filter attenuation band; 2. <Elektronik> rejection band (Filter); 3. <Fernseh> guard band; 4. <Funktech> filter attenuation band
Sperrbeschichtung f <Anstrich> barrier coating
Sperrdifferenzial n <Kfztech> limited slip differential, locking differential
Sperre f 1. <Bau> barricade, stop; 2. <Comp & DV> lockout; 3. <Elektrotech> interlock, latch, locking device; choke (Wellenleiter); 4. <Funktech> rejector; 5. <Maschinen> catch, locking device, lock, stop; 6. <Meerschmutz> boom; 7. <Sicherheit> barricade, protective barrier; 8. <Telekom> lock, restriction
Sperre f **ankommender Anrufe** <Telekom> ICB, incoming-calls-barred (Benutzerservice)

Sperreingabe

Sperreingabe f <Elektronik> inhibiting input
Sperreingang m <Comp & DV, Elektronik> inhibiting input
Sperreinrichtung f 1. <Sicherheit> stop device; 2. <Telekom> barring facility
sperren v 1. <Bau> lock; insulate, stop *(Feuchtigkeit)*; 2. <Comp & DV> disable, inhibit, lock; 3. <Elektronik> inhibit *(Gatter)*; 4. <Elektrotech> interlock; 5. <Kfztech> lock up; 6. <Kontroll> block; 7. <Lufttrans> *(BE)* earth, *(AE)* ground *(Flugwesen)*; 8. <Maschinen> inhibit, lock; 9. <Qual> hold, quarantine; 10. <Telekom> lock; 11. <Wassertrans> blockade *(Schifffahrt, Hafen)*
Sperren n 1. <Eisenbahn> blocking; 2. <Elektrotech> sticking *(Schalter)*; latching *(Sperrrelais)*; 3. <Fernseh> blocking; 4. <Metall> locking; 5. <Telekom> blocking
Sperren der Verbindung <Telekom> suspension of connection
Sperren von Anrufen <Telekom> call barring
Sperrengebinde n <Meerschmutz> boom pack
Sperrenrückgewinnung f <Meerschmutz> boom retrieval *(Wasser)*
Sperrenschleppe f <Meerschmutz> boom towing
Sperrfilter n 1. <Elektrotech> absorbing filter, suppression filter; 2. <Funktech> block filter; 3. <Telekom> blocking network
Sperrgebiet n <Wassertrans> prohibited area *(Navigation)*
Sperrgitter n <Elektronik> barrier grid *(Ladung)*
Sperrgitter-Speicherröhre f <Elektronik> barrier grid storage tube
Sperrhaken m <Bau> ratchet
Sperrholz n <Bau, Kunststoff, Verpack, Wassertrans> plywood *(Bootbau)*
Sperrholzkiste f <Verpack> plywood case
Sperrholzleim m <Verpack> plywood adhesive
Sperrimpuls m <Elektronik> inhibiting pulse
Sperrklinke f 1. <Bau> catch, ratchet; 2. <Fertig> detent, dog, holding pawl, latch, locking pawl, pawl, retaining pawl, trip dog; 3. <Maschinen> detent pawl, keeper, pawl, ratchet; 4. <Mechan> catch, pawl, ratchet; 5. <Sicherheit> safety pawl
Sperrkondensator m <Elektriz, Elektrotech> blocking capacitor
Sperrkreis m <Funktech> trap *(Antennen)*
Sperrkreis m für wilde Schwingungen <Elektronik> parasitic suppressor, wave trap
Sperrlager n <Qual> hold store, quarantine store, restricted store, salvage store
Sperrmauer f 1. <Bau> barrage; 2. <Wasserversorg> dam
Sperrmüll m <Abfall> bulky waste, dump for bulky waste
Sperrmüllabfuhr f <Abfall> bulk collection
Sperrmüllsammlung f <Abfall> bulk collection
Sperrrad n 1. <Fertig> pawl wheel, ratchet; 2. <Maschinen> dog wheel
Sperrrädchen n <Maschinen> click wheel
Sperrraste f <Maschinen, Sicherheit> lock pin, safety catch
Sperrrelais n <Elektrotech> lock-up relay
Sperrrichtung f <Elektrotech> inverse direction
Sperrrichtungsbetrieb m <Elektrotech> reverse bias *(Transistoren)*
Sperrscheibe f 1. <Lufttrans> blanking plate; 2. <Mechan> lock washer
Sperrscheinwiderstand m 1. <Akustik> blocked electrical impedance; 2. <Elektrotech> blocked impedance
Sperrschicht f 1. <Bau> barrier, damp-proof course, dpc, stop, waterproofing; 2. <Elektrotech> barrier layer; 3. <Elektronik> barrier layer; depletion layer *(Transistortechnik)*; 4. <Telekom, Umweltschmutz> barrier layer
Sperrschichteffekt m <Elektrotech> photovoltaic effect
Sperrschichtelement n <Elektrotech> photovoltaic cell

Sperrschichtfeldeffekttransistor m <Elektronik> JFET, junction FET, junction field effect transistor
Sperrschichtfolie f <Verpack> barrier film
Sperrschichtgenerator m <Elektrotech> photovoltaic generator
Sperrschichtkapazität f <Elektrotech> junction capacitance
Sperrschichtkondensator m <Elektrotech> junction capacitor
Sperrschichtphotodiode f <Elektronik> depletion layer photodiode
Sperrschichtphotoeffekt m <Nichtfoss Energ, Phys> photovoltaic effect
Sperrschichtphotozelle f <Nichtfoss Energ, Phys> photovoltaic cell
Sperrschichtpolymer n <Kunststoff> barrier resin
Sperrschichtstoffe mpl <Verpack> barrier material
Sperrschichtstrom m <Elektrotech> photovoltaic current
Sperrschichttemperatur f <Elektrotech> junction temperature *(Halbleiter)*
Sperrschichtverpackung f <Verpack> barrier packaging
Sperrschichtzelle f <Elektrotech> barrier layer cell, photovoltaic cell
Sperrschrift f <Comp & DV> spacing
Sperrschwinger m <Elektronik, Phys> blocking oscillator
Sperrsignal n <Elektronik> inhibiting signal
Sperrspannung f <Phys> blocking voltage, sticking voltage
Sperrstift m <Maschinen> catch pin
Sperrstrom m <Elektrotech> off-state current, reverse current
Sperrstromverstärkung f <Elektronik> inverse gain
Sperrtor n <Sicherheit> guard gate
Sperrung f 1. <Elektrotech> interlock *(Gatter)*; 2. <Elektrotech> suspension; blocking, cutoff *(Leitfähigkeit)*; 3. <Hydraul> shutting off *(der Dampfzufuhr)*; 4. <Lufttrans> interlock; 5. <Maschinen> detent; 6. <Qual> holding, quarantining
Sperrventil n <Raumfahrt> check valve
Sperrverlust m <Raumfahrt> blocking loss *(Weltraumfunk)*
Sperrvermerk m <Qual> hold tag, stop note
Sperrverzögerungszeit f <Elektronik> reverse recovery time
Sperrvorrichtung f 1. <Elektriz, Elektrotech> blocking device, locking device; 2. <Maschinen, Sicherheit> blocking device, catch, interlocking mechanism, latch, lock, locking device, stop
Sperrwand f <Abfall> slurry wall
Sperrwandler m <Elektrotech> reverse converter
Sperrwandlung f <Elektrotech> reverse conversion
Sperrwasser n <Kerntech> seal water
Sperrwerk n <Maschinen> pawl-and-ratchet motion, ratchet-and-pawl motion
Sperrzeit f 1. <Elektronik, Elektrotech> idle period; off-period, off-time *(Gerät)*; off-period *(Schaltstück)*; 2. <Qual> quarantine period
Sperrzeit f **bei positiver Anodenspannung** <Elektrotech> blocking period
Sperrzustand m <Elektronik, Elektrotech> blocking state; off-state *(Thyristoren)*
Spezial... <Comp & DV, Fertig, Telekom, Verpack, Wassertrans> special
Spezialeffektbus m <Fernseh> special effects bus
Spezialindizierung f <Comp & DV> indexing
Spezialkleber m <Verpack> purpose-formulated adhesive
Spezialmutter f <Fertig> special nut *(Kunststoffinstallationen)*
Spezialprozessor m <Telekom> applications processor

Spezialrechner m <Comp & DV> special purpose computer
Spezialregler m <Nichtfoss Energ> dedicated controller
Spezialsoftware f <Comp & DV, Wassertrans> dedicated software
speziell angefertigt adj <Heiz & Kälte> custom-made, purpose-made
speziell entworfen adj <Verpack> purpose-designed
spezielle Relativitätstheorie f <Phys> special theory
Spezifikation f <Bau, Comp & DV, Maschinen, Qual> spec, specification
Spezifikationsblatt n <Fernseh> specification sheet
Spezifikationssprache f <Comp & DV> specification language
Spezifikations- und Beschreibungssprache f <Telekom> functional specification and description language, SDL language, specification and description language
spezifisch adj 1. <Anstrich> inherent; 2. <Phys> specific (auf Masseneinheit bezogen)
spezifische Aktivität f <Phys> specific activity
spezifische Aktivität f eines Elementes <Kerntech> element-specific activity
spezifische akustische Auslenkung f <Akustik> specific acoustic compliance
spezifische Ausstrahlung f 1. <Phys> radiant exitance; 2. <Telekom> radiance
spezifische Bruchfestigkeit f <Metall> ultimate tensile strength
spezifische Dämpfung f 1. <Raumfahrt> specific attenuation; 2. <Telekom> attenuation coefficient
spezifische Dichte f <Phys> relative density, specific gravity
spezifische Drehzahl f <Nichtfoss Energ> specific speed
spezifische elektrische Leitfähigkeit f <Elektrotech> specific conductivity
spezifische Emission f <Elektrotech> specific emission
spezifische Empfindlichkeit f <Akustik> specific sensitivity
spezifische Enthalpie f <Phys> specific enthalpy
spezifische Entropie f <Phys> specific entropy
spezifische Gibb'sche Funktion f <Phys> specific Gibbs function
spezifische Haftkraft f <Kunststoff> specific adhesion
spezifische Heizleistung f <Heiz & Kälte> specific heat output
spezifische Helmholtzfunktion f <Phys> specific Helmholtz function
spezifische innere Energie f <Phys> specific internal energy
spezifische Ionisierung f (S) <Kerntech> specific ionization (S)
spezifische Kapazität f <Elektriz> specific capacitance
spezifische Ladung f <Phys, Strahlphys, Teilphys> charge-mass ratio, specific charge
spezifische Ladung f eines Elektrons <Kerntech> electron specific charge
spezifische Lautheit f <Akustik> specific loudness
spezifische Lichtausstrahlung f 1. <Phys> luminous exitance, radiance; 2. <Strahlphys> radiance
spezifische Nachweisbarkeit f <Optik> specific detectivity
spezifische Oberfläche f <Kunststoff> specific surface area
spezifische Schallimpedanz f <Akustik, Phys> specific acoustic impedance
spezifische Schmelzwärme f <Phys> specific latent heat
spezifische spektrale Ausstrahlung f <Telekom> spectral radiance

spezifische Steife f <Raumfahrt> specific stiffness
spezifische Wärme f 1. <Heiz & Kälte, Kunststoff, Nichtfoss Energ> specific heat; 2. <Phys> specific heat capacity (c); 3. <Raumfahrt, Thermod> specific heat
spezifische Wärmekapazität f (c) <Thermod> specific heat capacity (c)
spezifischer Blindleitwert m <Akustik> specific acoustic susceptance
spezifischer Brennwert m <Heiz & Kälte> gross calorific value
spezifischer Brunnenkapazität m <Nichtfoss Energ> specific capacity of a well
spezifischer Durchgangswiderstand m <Elektrotech, Kunststoff> specific volume resistivity
spezifischer Energieverbrauch m <Elektriz> specific energy consumption
spezifischer Halbleiter m <Comp & DV> intrinsic semiconductor
spezifischer Impuls m <Raumfahrt> specific impulse
spezifischer Leitwert m <Elektriz> conductivity, specific conductance
spezifischer magnetischer Widerstand m <Phys> reluctivity
spezifischer Oberflächenwiderstand m <Elektrotech> surface resistivity
spezifischer Strömungswiderstand m <Akustik> flow resistivity
spezifischer Volumenwiderstand m 1. <Elektriz> mass resistivity; 2. <Elektrotech> bulk resistivity
spezifischer Wärmewiderstand m <Phys, Thermod> thermal resistivity
spezifischer Widerstand m 1. <Elektriz, Elektrotech> resistivity, specific resistance; 2. <Erdöl, Kohlen> resistivity; 3. <Kunststoff> electrical resistivity, resistivity
spezifisches Drehvermögen n <Phys> specific rotation
spezifisches Gewicht n 1. <Kohlen> unit weight; 2. <Phys> relative density, specific gravity
spezifisches Leistungsgewicht n <Maschinen> power-weight ratio
spezifisches Volumen n <Phys> specific volume
spezifiziertes Mittel n <Kohlen> designated agent
Spezifizierung f <Patent> spec, specification (Waren, Dienstleistungen)
Sphäre f <Geom, Phys> sphere
sphärische Aberration f <Foto, Optik, Phys, Telekom> spherical aberration
sphärische Dämmplatte f <Aufnahme> spherical baffle
sphärische Geometrie f <Geom> spherical geometry
sphärische Linse f <Phys> spherical lens
sphärischer Spiegel m <Phys> spherical mirror
sphärisches Dreieck n <Geom> spherical triangle
sphärisches Koordinatensystem n <Geom> spherical coordinates system
sphärisches Zweieck n <Fertig> spherical lune
Sphärizität f <Geom> sphericity
Sphäroguss m 1. <Fertig> nodular cast iron (Kunststoffinstallationen); 2. <Metall> nodular cast iron
Sphäroid n <Geom, Phys> spheroid
sphärolithisch adj <Fertig> spheroidal
Sphärometer n <Phys> spherometer
Sphingosin n <Chemie> sphingosine
Spickelement n <Kerntech> spike
Spicken n <Textil> oiling
Spickungspräparat n <Elektronik> implant (Radiologie)
Spiegel m <Ker & Glas, Phys> mirror
Spiegelantenne f <Funktech> reflector antenna
Spiegelbelegung f <Fertig> silvering
Spiegelbild n <Geom> mirror image

Spiegelfrequenz

Spiegelfrequenz f <Elektronik, Fernseh, Funktech> image frequency
Spiegelfrequenzstörung f <Elektronik, Fernseh> image frequency interference
Spiegelgalvanometer n 1. <Elektriz> light-spot galvanometer, mirror galvanometer, reflecting mirror galvanometer; 2. <Gerät> luminous pointer galvanometer, mirror galvanometer
Spiegelglas n 1. <Bau> plate glass; 2. <Ker & Glas> plate
Spiegelglasfehler m **durch unvollständiges Schleifen** <Ker & Glas> short finish
Spiegelheck n <Wassertrans> square transom stern, transom stern *(Schiffbau)*
Spiegelherstellung f <Ker & Glas> mirror making
spiegelige Oberfläche f <Fertig> glazing
Spiegelkasten m <Foto> reflex housing
Spiegelkerne mpl <Phys> mirror nuclei
spiegeln v <Comp & DV> shadow *(Daten)*
Spiegelnuklid n <Phys> mirror nuclide
Spiegelobjektiv n <Foto> mirror lens
Spiegelplatte f <Comp & DV> mirrored disk
Spiegelreaktor m **mit Feldumkehr** <Kerntech> field-reversed mirror reactor
Spiegelreflexion f <Phys, Telekom> specular reflection
Spiegelreflexionskoeffizient m <Telekom> specular reflection coefficient
Spiegelroheisen n <Metall> spiegel iron
Spiegelschliff m <Mechan> mirror finish
Spiegel-Server m <Comp & DV> mirrored site
Spiegelskale f <Gerät> mirror scale
Spiegelsymmetrie f <Math> axial symmetry
Spiegelteleskop n <Phys> reflecting telescope
Spiegelung f 1. <Geom> mirroring, reflection; 2. <Wellphys> reflection
Spiegelungsachse f <Geom> mirror line
spiegelverkehrte Darstellung f <Geom> mirrored representation
Spiel n 1. <Hydraul> space *(Kolben)*; 2. <Kerntech> clearance; 3. <Kfztech> free travel; backlash *(Triebstrang)*; 4. <Maschinen> allowance, backlash, clearance, play, slackness; 5. <Mechan> backlash, play • **mit Spiel** <Maschinen> loose • **Spiel geben** <Maschinen> give clearance to
spielen v <Fernseh, Mechan> play
spielfrei adj 1. <Fertig> no-play *(Kunststoffinstallationen)*; 2. <Gerät> backlash-free *(Messwerk)*
spielfreie Paarung f <Fertig> zero-backlash mating
Spielpassung f <Maschinen> clearance fit
Spielraum m 1. <Comp & DV> margin *(Zeit)*; 2. <Elektrotech, Hydraul> clearance; 3. <Maschinen> play; 4. <Mechan> clearance, play; 5. <Qual> margin
Spielsitz m <Hydraul> loose seat *(Ventil)*
Spieltheorie f <Künstl Int, Math> game theory
Spiere f <Wassertrans> spar *(Schiffbau)*; perch *(Seezeichen)*
Spierentonne f <Wassertrans> pillar buoy, spar buoy *(Seezeichen)*
Spike m <Lufttrans> spike
Spikereifen m <Kfztech> *(AE)* studded tire, *(BE)* studded tyre
Spill n 1. <Elektrotech> spill; 2. <Mechan, Wassertrans> capstan *(Schiffzubehör)*
Spilltrommel f <Wassertrans> capstan drum *(Schiffzubehör)*
Spin m <Phys, Raumfahrt, Teilphys> spin
Spinacen n <Chemie> squalene
Spinachse f <Raumfahrt> spin axis
Spinasterin n <Chemie> spinasterol
Spinaustauschkraft f <Kerntech> spin exchange force

Spin-Bahn-Kopplung f <Phys> spin orbit coupling
Spind m <Wassertrans> locker
Spindel f 1. <Bau> shaft, stem; spindle *(Drehbank)*; newel *(Treppe)*; 2. <Fertig> mandrel, mandril; spindle, stem *(Kunststoffinstallationen)*; 3. <Hydraul> rod *(Schieber)*; stem *(Ventil)*; 4. <Ker & Glas> spindle *(Glasfasern)*; 5. <Kfztech> spindle; 6. <Maschinen> mandrel, mandril, screw, spindle; 7. <Mechan> shaft, spindle
Spindel f **des Luftschraubenblattes** <Lufttrans> blade spindle *(Hubschrauber)*
Spindel f **für Drehrichtungswechsel** <Maschinen> reversing screw
Spindelaggregat n <Maschinen> spindle unit
Spindelarm m <Kfztech> spindle arm
Spindelbohrer m <Fertig> gun drill
Spindelbremse f <Eisenbahn, Kfztech, Maschinen> screw brake
Spindeldocke f <Maschinen> headstock
Spindeldrehmaschine f <Maschinen> mandrel lathe, mandril lathe
Spindelfräsmaschine f <Bau> *(AE)* spindle molding machine, *(BE)* spindle moulding machine *(Holzbau)*
Spindelkasten m 1. <Fertig> gearbox; head assembly *(Spanung)*; 2. <Maschinen> headstock
Spindelkasten m **mit Reibungskupplung** <Maschinen> friction headstock
Spindelkopf m <Maschinen> spindle head, spindle nose
Spindellagerarm m <Fertig> adjustable arm
Spindellagerplatte f <Fertig> pattern plate *(Spanung)*
Spindelnase f 1. <Fertig> nose; 2. <Maschinen> mandrel nose, mandril nose
Spindelpresse f 1. <Maschinen> fly press, screw press; 2. <Papier> screw press
Spindelstab m <Bau> spindle
Spindelstock m 1. <Fertig> headstock, head, poppet; 2. <Maschinen> head, headstock
Spindelstockspitze f 1. <Fertig> *(AE)* head center, *(BE)* head centre *(Spanung)*; 2. <Maschinen> *(AE)* headstock center, *(BE)* headstock centre
Spindeltreppe f <Bau> solid newel stair
Spindelung f <Phys> hydrometry
Spindelvorgelege n <Fertig> back gear
Spindelvorlauf m <Fertig> advance of the spindle
Spindelwange f <Bau> outside string *(Treppe)*
Spindrehimpuls m <Phys> spin angular momentum
Spinellerzeugnis n <Ker & Glas> spinel refractory
Spinnaker m <Wassertrans> spinnaker *(Segeln)*
Spinnakerbaum m <Wassertrans> spinnaker boom *(Segeln)*
Spinndüse f 1. <Ker & Glas> nozzle *(Glasfasern)*; 2. <Textil> die-spinning nozzle
Spinndüsenfilter m <Textil> spinneret filter
spinnen v <Textil> spin
Spinnen n <Textil> spinning
Spinnereiabfall m <Textil> trash
Spinnfasergarn n <Textil> spun yarn
spinngefärbt adj <Textil> spun-dyed
Spinning-Reserve f <Nichtfoss Energ> spinning reserves
Spinnkabel n <Textil> tow
Spinnkuchen m 1. <Ker & Glas> cake *(Glasfasern)*; 2. <Textil> cake
spinnkuchengefärbt adj <Textil> cake-dyed
Spinnrad n <Textil> spinning wheel
Spinnseil n <Erdöl> spinning line
Spinnverfahren n <Textil> spinning system
Spinnvlies n <Textil> spunbonded fabric
spinodale Auflösung f <Metall> spinodal decomposition
Spinquantengesamtzahl f (S) <Kerntech> total spin quantum number (S)

Spinquantenzahl *f (s)* <Kerntech, Labor, Phys> spin quantum number *(s)*
Spinschubtriebwerk *n* <Raumfahrt> spin thruster
Spinstabilisierung *f* <Raumfahrt> spin stabilization
Spintemperatur *f* <Phys> spin temperature
Spinwelle *f* <Phys> spin wave
Spion *m* <Maschinen> *(AE)* feeler gage, *(BE)* feeler gauge
Spionagesatellit *m* <Raumfahrt> spy satellite
Spiral... <Maschinen> spiral
Spiralbewegung *f* <Kerntech> helicoidal motion
Spiralbewegung *f* **von Teilchen** <Kerntech> particle spiraling
spiralbewehrte Stütze *f* <Bau> helical reinforced column
Spiralbewehrung *f* <Bau> helical reinforcement
Spiralbindung *f* <Druck> spiral binding
Spiralbohrer *m* 1. <Erdöl> spiral bit *(Bohrtechnik)*; 2. <Maschinen> jobber drill, twist bit, twist drill; 3. <Mechan> twist drill
Spiralbohrer *m* **mit Morsekegelschaft** <Maschinen> Morse taper shank twist drill
Spiralbohrer *m* **mit Zylinderschaft** <Maschinen> parallel-shank twist drill, straight shank twist drill, twist drill with parallel shank, twist drill with straight shank
Spiralbohrerschleifmaschine *f* <Maschinen> twist drill grinder
Spiraldehnung *f* <Fertig> volution
Spiraldifferenzial *f* <Kfztech> helical differential
Spirale *f* 1. <Fertig> scroll, volute; 2. <Geom, Maschinen> spiral
Spirale *f* **für mehrteiligen Entwicklungsbehälter** <Foto> multiunit tank spiral
Spiralfeder *f* 1. <Kfztech> coil spring; 2. <Maschinen> coil spring, spiral spring; 3. <Mechan> coil spring, helical spring
spiralförmig *adj* <Geom> spiral
spiralförmiger Bruch *m* <Ker & Glas> spiral fracture
Spiralfräseinrichtung *f* <Maschinen> spiral milling attachment
Spiralfräsen *n* <Maschinen> helical milling, spiral milling
Spiralfräser *m* <Maschinen> helical milling cutter, spiral milling cutter
Spiralgehäuse *n* 1. <Kfztech> volute casing; 2. <Maschinen> volute
spiralgenuteter Gewindebohrer *m* <Maschinen> spiral fluted tap
Spiralhülse *f* <Verpack> spirally-wound tube
spiralig *adj* 1. <Fertig> volute; 2. <Geom, Maschinen> spiral
spiralige Turbulenz *f* <Strömphys> spiral turbulence
Spiralinstabilität *f* <Kerntech> helical instability
Spiralkegelrad *n* <Maschinen> spiral bevel gear
Spiralkegelradgetriebe *n* <Kfztech> spiral bevel gearing
Spiralkern *m* <Elektriz> wound core
Spiralklassierer *m* <Kohlen> spiral classifier
Spirallaufschienen *fpl* <Ker & Glas> nog plate spiral runner bars
Spiralnut *f* <Maschinen> helical groove
Spiralschnecke *f* <Maschinen> spiral worm
Spiraltuner *m* <Elektronik> continuous tuner
Spiralwinkel *m* <Maschinen> spiral angle
Spiralzahnkegelrad *n* <Maschinen> spiral bevel gear
Spiran *n* <Chemie> spiran
Spiranverbindung *f* <Chemie> spiran
Spirituosen *fpl* <Lebensmittel> spirits
Spiritus *m* <Lebensmittel> spirit
Spirituslack *m* <Bau> spirit lacquer
Spirituslampe *f* <Labor> spirit lamp
spirozyklische Verbindung *f* <Chemie> spiran
spitz *adj* 1. <Maschinen> sharp; 2. <Metrol> acute *(Winkel)*

spitzbefahrene Weiche *f* <Eisenbahn> facing points
Spitzbogenkaliber *n* <Fertig> gothic pass
Spitzbohrer *m* 1. <Fertig> flat drill, spade drill; 2. <Maschinen> flat drill
Spitze *f* 1. <Elektrotech> peak, spike; 2. <Fernseh> peak; 3. <Fertig> crest *(Gewinde)*; apex *(Kegelrad)*; 4. <Funktech> peak; 5. <Geom> apex, vertex; apex *(Kegel)*; 6. <Maschinen> bit; crest *(eines Gewindes)*; nose, point; 7. <Metall> cusp; 8. <Textil> lace; 9. <Wassertrans> peak *(Festmachen)* • **mit scharfer Spitze** <Fertig> cone point
spitzen *v* <Bau> point
Spitzen *n* <Fertig> end reduction *(Schmieden)*
Spitzenamplitude *f* <Aufnahme, Elektronik> peak amplitude
Spitzenanhebungskreis *m* <Fernseh> peaking circuit
Spitzenanhebungsschaltung *f* <Fernseh> peaking network
Spitzenaufnahmepegel *m* <Aufnahme> peak recording level
Spitzenbegrenzer *m* 1. <Fernseh> peak clipper, peak limiter; 2. <Funktech> peak limiter
Spitzenbegrenzung *f* <Aufnahme, Elektronik> peak clipping, peak limitation
Spitzenbelastung *f* 1. <Elektriz> peaking capacity; 2. <Elektrotech, Kohlen> maximum demand; 3. <Telekom> peak load
Spitzenbelastungszeit *f* <Telekom> peak busy hour
Spitzende *n* <Bau> spigot
Spitzendefekt *m* <Metall> point defect
Spitzendiode *f* <Elektronik> point contact diode
Spitzendrehen *n* <Maschinen> *(AE)* turning between centers, *(BE)* turning between centres
Spitzendrehmaschine *f* <Fertig, Maschinen> *(AE)* center lathe, *(BE)* centre lathe
Spitzendrehmaschine *f* **mit Leitspindel** <Maschinen> engine lathe
Spitzendurchdringung *f* <Fernseh> tip penetration
Spitzenfaktor *m* <Elektronik> peak factor
Spitzenfrequenzhub *m* <Raumfahrt, Telekom> peak frequency deviation
Spitzengebühr *f* <Telekom> peak rate
Spitzengerät *n* <Telekom> top-of-the-line equipment
Spitzenhelligkeit *f* <Fernseh> peak brightness
Spitzenherausragen *n* <Fernseh> tip protrusion
Spitzenhöhe *f* 1. <Fernseh> tip height; 2. <Maschinen> *(AE)* center height, *(BE)* centre height, *(AE)* height of centers, *(BE)* height of centres
Spitzenimpuls-Amplitude *f* <Elektronik> peak pulse amplitude
Spitzenkegel *m* <Lufttrans> nose cone
Spitzenkontakt *m* <Elektrotech> point contact
Spitzenkonzentration *f* <Umweltschmutz> peak concentration
Spitzenlagerung *f* <Maschinen> toe bearing
Spitzenlast *f* 1. <Elektriz, Elektrotech> maximum demand, peak load; 2. <Telekom> peak load
Spitzenlastkraftwerk *n* <Elektriz> peak-load plant, peak-load power plant, peak-load power station, peaking power station
Spitzenlastpunkt *m* <Kerntech> hot spot
Spitzenleistung *f* 1. <Bau> peak capacity; 2. <Elektrotech, Fernseh> peak power; 3. <Funktech> PEP, peak envelope power; 4. <Kohlen> peak capacity; 5. <Telekom> PEP, peak envelope power
spitzenlos *adj* <Maschinen, Mechan> *(AE)* centerless, *(BE)* centreless
spitzenlose Rundschleifmaschine *f* <Mechan> *(AE)* centerless grinder, *(BE)* centreless grinder
Spitzennadelkamm *m* <Textil> bearded needle frame

Spitzenpegel *m* <Aufnahme> peak level
Spitzenprojektion *f* <Fernseh> tip projection
Spitzenregelung *f* <Fernseh> peaking control
Spitzenschallpegel *m* <Sicherheit> peak sound level
Spitzenschleifapparat *m* <Fertig> *(AE)* center grinder, *(BE)* centre grinder
Spitzenschleifen *n* <Fertig> *(AE)* on-center grinding, *(BE)* on-centre grinding
Spitzensignal *n* <Fernseh> peak signal
Spitzensignalamplitude *f* <Elektronik> peak signal amplitude
Spitzenspannung *f* <Elektrotech, Fernseh, Funktech, Phys> peak voltage
Spitzenspannung *f* **eines Senders** *(SS-Spannung)* <Funktech> peak envelope voltage *(PEV)*
Spitzenspanwinkel *m* <Fertig> top rake *(Spanung)*
Spitzensperrspannung *f* <Elektriz> peak inverse voltage
Spitzenspiel *n* <Fertig> crest clearance *(Gewinde)*
Spitzenstrom *m* <Elektriz, Elektrotech, Phys> peak current
Spitzenstundenfaktor *m* <Eisenbahn, Trans> peak hour factor
Spitzentarif *m* <Telekom> peak rate *(ISDN)*
Spitzenteilzirkel *m* <Maschinen> spring bow divider
Spitzenverkehr *m* <Trans> peak hour traffic, peak load traffic, peak period traffic
Spitzenverkehrsaufkommen *n* <Trans> peak traffic volume
Spitzenverlust *m* <Nichtfoss Energ> tip loss
Spitzenverzerrung *f* <Fernseh> peak distortion
Spitzenwasserbedarf *m* <Wasserversorg> peak water demand
Spitzenwasserdurchfluss *m* <Wasserversorg> peak water flow
Spitzenweite *f* <Maschinen> *(AE)* distance between centers, *(BE)* distance between centres
Spitzenwert *m* 1. <Elektriz, Elektronik> crest value, peak value; 2. <Funktech> peak; 3. <Gerät> crest value; 4. <Kerntech> maximum; 5. <Phys> peak value; 6. <Raumfahrt> peak factor; 7. <Telekom> peak value
Spitzenwinkel *m* <Fertig> angle of point; included angle *(Bohrer)*
Spitzenwirbel *m* **am Luftschraubenblatt** <Lufttrans> blade tip vortex *(Hubschrauber)*
Spitzenzähler *m* 1. <Elektrotech> demand meter; 2. <Gerät> excess meter *(Energieverbrauch)*
Spitzenzeit *f* 1. <Fernseh> peak time; 2. <Telekom> peak hour
Spitzenzuschaltung *f* <Fernseh> tip engagement
spitzer Maurerhammer *m* <Bau> mattock
spitzer Winkel *m* <Geom> acute angle
Spitze-Spitze-... *(SS)* 1. <Aufnahme> peak-to-peak, pp *(Hüllkurve)*; 2. <Funktech> peak-to-peak, pp
Spitze-Spitze-Amplitude *f* <Aufnahme, Elektronik> peak-to-peak amplitude
Spitze-Spitze-Signalamplitude *f* <Fernseh> peak-to-peak signal amplitude
Spitze-zu-Spitze-Wert *m* <Akustik> peak-to-peak value
Spitzformer *m* <Fertig> pegging rammer *(Gießen)*
Spitzgewinde *n* <Maschinen> V-thread, vee thread, sharp thread, sharp V thread
Spitzheit *f* <Geom> acuteness *(Winkel)*
Spitzkehre *f* <Eisenbahn> back shunt
Spitzkerbe *f* <Maschinen> V-notch
Spitzkolumne *f* <Druck> short page
Spitzkopfniet *m* <Bau> steeple head rivet
Spitzlicht *n* <Druck, Foto> highlight
Spitzlutte *f* <Kohlen> hydraulic classifier
Spitzmarke *f* <Comp & DV> heading

spitzsenken *v* <Fertig> sink
Spitzsenken *n* <Maschinen> countersinking
Spitzsenker *m* <Maschinen> countersink
Spitztonne *f* <Wassertrans> conical buoy *(Seezeichen)*
Spitzturm *m* <Bau> steeple
spitzwinkeliges V-förmiges Unterwerkzeug *n* <Fertig> acute angle die
spitzwinklig *adj* 1. <Fertig> acute-angled; 2. <Geom> acute-angled, acute-angular
spitzwinkliges Dreieck *n* <Geom> acute triangle
Spleiß *m* 1. <Elektriz> joint; 2. <Fertig> splice *(optische Fasern)*; 3. <Telekom> joint, splice; 4. <Wassertrans> splice *(Tauwerk)*
Spleißdämpfung *f* <Telekom> splice loss
Spleißdose *f* <Elektriz> splice box
spleißen *v* 1. <Fertig> splice *(optische Fasern)*; 2. <Telekom, Textil> splice *(Seil)*
Spleißen *n* <Fertig> splicing *(optische Fasern)*
Spleißer *m* <Telekom> jointer, splicer
Spleißgerät *n* <Fertig> splicer *(optische Fasern)*
Spleißkasten *m* <Elektriz> splice box
Spleißmaschine *f* <Fertig> splicer *(optische Fasern)*
Spleißstelle *f* <Optik, Telekom> splice
Spleißstelle *f* **für Lichtleitfasern** <Optik> *(AE)* optical fiber splice, *(BE)* optical fibre splice
Spleißung *f* 1. <Fernseh> splice *(Kabel)*; 2. <Textil> splicing
Spleißverbindung *f* <Fertig> splice, splicing *(optische Fasern)*
Spleißverstärkung *f* <Fertig> splice reinforcement *(optische Fasern)*
Splint *m* 1. <Bau> peg; 2. <Elektrotech> key; 3. <Fertig> cotter; 4. <Maschinen> cotter pin, forelock, key bolt, split cotter pin, split pin; 5. <Mechan> cotter pin, spline; 6. <Wassertrans> cotter pin
Splintbolzen *m* <Maschinen> cotter bolt
Splintentreiber *m* <Maschinen> driftpin, driver, key drift, rivet drift
Splintholz *n* <Bau> sap wood
Splintloch *n* <Maschinen> cotter pin hole, split pin hole
Splintverbindung *f* <Bau, Maschinen> cottered joint
Splintzieher *m* <Maschinen> cotter pin extractor, pin extractor, split pin extracting tool
Splitt *m* 1. <Anstrich> grit; 2. <Bau> chip, crushed stone, stone chipping, chippings; 3. <Kohlen> chippings
Splittdecke *f* <Bau> blinding
Splitter *m* 1. <Bau> splinter; 2. <Ker & Glas> chippings, sliver; 3. <Mechan> chip
splitterfreies Glas *n* <Kfztech> shatter-proof glass
Splitterprüfung *f* <Werkprüf> shatter test *(Glas)*
splittriger Bruch *m* <Kohlen> splintery fracture
Splittstreuer *m* <Trans> grit spreader, stone spreader
Splittstreumaschine *f* <Trans> gritter
Splittzuschlag *m* <Bau, Kohlen> crushed aggregate
Spoiler *m* 1. <Kfztech> spoiler *(Karosserie)*; 2. <Lufttrans> spoiler *(Luftfahrzeug)*
Spondeus *m* <Akustik> spondee
Spongin *n* <Chemie> spongin
Spontanaktivität *f* <Akustik> spontaneous activity
Spontanbruch *m* <Ker & Glas> spontaneous breaking
spontane Anregung *f* <Kerntech> spontaneous excitation
spontane Bremsbetätigung *f* <Eisenbahn> spontaneous brake application
spontane Emission *f* <Phys> spontaneous emission
spontane Emission *f* **von Strahlung** <Strahlphys> spontaneous emission of radiation
spontane Kernbildung *f* <Metall> spontaneous nucleation

spontane Magnetisierung f <Phys> spontaneous magnetization
spontane Selbstentzündung f <Raumfahrt> hypergolic ignition *(Treibstoff)*
spontane Spaltung f <Phys, Strahlphys> spontaneous fission
spontane Spaltungswahrscheinlichkeit f <Strahlphys> spontaneous fission probability
spontane Strahlungsemission f <Strahlphys> spontaneous emission of radiation
spontane Übergänge mpl <Strahlphys> spontaneous transitions
spontane Verbrennung f <Sicherheit> spontaneous combustion
Spontanemission f <Telekom> spontaneous emission
spontaner Zerfall m <Strahlphys> spontaneous decay
Spontanmagnetisierung f <Strahlphys> spontaneous magnetization
Sponung f <Wassertrans> rabbet
Spool-Betrieb m <Comp & DV> spooling
Spool-Einheit f <Comp & DV> spooling device
Spooler m <Comp & DV> spooler
Spooling n <Comp & DV> spooling
Spool-Programm n *(Ein-/Ausgabe parallel zu Rechenprogramm)* <Comp & DV> SPOOL *(simultaneous peripheral operations on-line)*
Sporn m <Fertig> spur
Spornrad n <Lufttrans> tail wheel
Sportsucher m <Foto> sports finder *(Kamera)*
Sportwagen m <Kfztech> sports car *(Fahrzeugart)*
Sportwagen m mit offenem Verdeck <Kfztech> cabriolet, convertible *(Fahrzeugart)*
Spotbeam-Antenne f <Funktech, Raumfahrt> spot beam antenna *(Satelliten)*
Sprach... <Akustik, Comp & DV, Telekom> speech
Sprachaktivitätsunterscheider m <Mobilkom> activity discriminator *(GSM)*
Sprachanalyse f <Telekom> speech analysis
Sprachaudiogramm n <Akustik> speech audiogram
Sprachaudiometer n <Akustik> speech audiometer
Sprachaudiometrie f <Akustik> speech audiometry
Sprachausgabe f <Comp & DV, Elektronik, Telekom> audio output, voice output, voice response
Sprachausgabe-Einheit f *(ARU)* <Comp & DV> audio response unit *(ARU)*
sprachbeschreibende Sprache f <Comp & DV> metalanguage
Sprachbetrieb m <Mobilkom> talk mode
Sprachblock m <Telekom> talkspurt *(PCM)*
Sprachbox f <Telekom> voice mailbox
Sprachbox f für Handys <Telekom> mobile voice box, voice mobilbox
Sprach-Chip m <Comp & DV> speech chip
Sprachcodierer m <Telekom> voice coder
Sprachcodierung f 1. <Raumfahrt> speech encoding; 2. <Telekom> speech coding, speech encoding; vocoding *(mit reduzierter Bitrate)*
Sprachcodierung f mit halber Bitrate <Telekom> half-rate speech coding
Sprachcodierung f mit niedriger Bitrate <Telekom> low bit-rate speech coding
Sprachcodierung f mit Standardbitrate <Mobilkom, Telekom> full-rate speech coding *(GSM: 13 kb/s)*
Sprachcodierung f mit Vorhersage <Telekom> prediction coding, predictive speech coding
Sprachdateneingabe f <Comp & DV> voice data entry
Sprach-Datennetz n <Telekom> speech data network
Sprachdemodulation f <Raumfahrt> speech detection
Sprachdienst m <Telekom> speech service

Sprache f <Comp & DV> language
Sprache f der fünften Generation <Künstl Int> fifth-generation language
Sprache f mit Blockstruktur <Comp & DV> block-structured language
Spracheingabedaten npl <Comp & DV> voice data entry
Sprachenanweisung f <Comp & DV> language statement
Sprachenprozessor m <Comp & DV> language processor
Spracherkenner m <Telekom> speech recognizer
Spracherkennung f <Comp & DV, Künstl Int> speech recognition
Spracherkennungsprogramm n <Telekom> speech recognizing program
Spracherkennungssystem n <Künstl Int> speech recognition system
Spracherzeugung f <Telekom> speech generation
Sprachfilter n <Aufnahme> speech filter
Sprachfrequenz f *(SF)* <Elektronik, Telekom> speech frequency *(SF)*; voice frequency *(VF)*
Sprachfrequenzband n <Comp & DV, Telekom> voiceband
Sprachfrequenzen fpl <Akustik> conversational frequencies
sprachgesteuerter Betrieb m <Telekom> voice-controlled operation
sprachgesteuerter Roboter m <Künstl Int> voice-operated robot
sprachgesteuerter Schalter m *(VOX)* <Funktech> voice-operated switch *(VOX)*
sprachgesteuerter Sende-Empfangsumschalter m <Telekom> voice-operated transmitter keyer
sprachgesteuertes Relais n <Telekom> voice-operated relay
sprachgesteuertes Wählen n <Telekom> *(AE)* voice dialing, *(BE)* voice dialling
Sprachinterpolation f <Raumfahrt> speech interpolation
Sprachkanal m 1. <Comp & DV> speech channel, voice channel; 2. <Mobilkom> traffic channel *(Bündelfunk)*
sprachliches Gebilde n <Comp & DV> language construct
Sprachmitteilungsdienst m <Telekom> voice messaging
Sprachmitteilungsprozessor m <Telekom> voice message processor
Sprachnetz n <Telekom> voice network
Sprachschneidegerät n <Aufnahme> speech clipper
Sprachsignal n <Telekom> speech signal
Sprachspeicher m <Telekom> speech memory
Sprachspeicherdienst m <Telekom> voice messaging
Sprachsperre f <Elektrotech> guard circuit *(Telefon)*
Sprachspitzenbegrenzung f <Funktech> speech clipping
Sprachspur f <Aufnahme> speech track, voice track
Sprachsynthese f <Comp & DV, Elektronik, Künstl Int> speech synthesis
Sprachsynthesegerät n <Telekom> speech synthesizer
Sprachsynthesizer m <Comp & DV> speech synthesizer
Sprachsynthetisator m <Telekom> speech synthesizer
Sprachübersetzer m <Comp & DV> language processor
sprach- und datenintegrierende Vermittlung f <Telekom> integrated voice-data switch
sprach- und datenintegrierende Wählnebenstelle f <Telekom> integrated voice-data PABX
Sprach- und Daten-Nebenstellenanlage f <Telekom> voice/data PBX
Sprach- und Datenpaketvermittlung f <Telekom> voice/data packet switch

Sprachverarbeitung

Sprachverarbeitung f 1. <Comp & DV, Funktech> speech processing; 2. <Künstl Int> voice processing; 3. <Telekom> speech processing
Sprachverschleierung f durch Frequenzumkehrung <Funktech, Telekom> frequency inversion privacy
Sprachverschlüsseler m <Comp & DV> vocoder
Sprachverständlichkeit f <Ergon> intelligibility
Sprachverstehen n (SU) <Künstl Int> language understanding, speech understanding
spratzen v <Metall> spit
Spratzen n <Raumfahrt> spurt (Weltraumfunk)
Spraydose f <Verpack> aerosol container
Sprayschicht f <Bau> curing membrane (Beton)
Spreader m <Trans> spreader
Spreadsheet n <Comp & DV> spreadsheet
Spread-Spectrum-Mehrfachzugriff m <Telekom> spread spectrum multiple access (SSMA)
Spread-Spectrum-Modulator m <Telekom> spread spectrum modulator
Sprechanlage f <Telekom> intercom
Sprechdatenverarbeitung f <Künstl Int> speech processing
sprechendes Straßenschild n <Trans> talking road sign
Sprecherstudio n <Aufnahme> continuity studio
sprecherunabhängiges Erkennungssystem n <Künstl Int> speaker independent recognition system
Sprechfunkgerät n 1. <Funktech> radiotelephone, walkie-talkie; 2. <Mobilkom> radiotelephone; 3. <Wassertrans> radiotelephone, walkie-talkie
Sprechgarnitur f <Telekom> headset
Sprechgebühren fpl <Mobilkom> air charges (Funkkanal)
Sprechkanal m <Telekom> working channel
sprechkanalfreier Verbindungsaufbau m <Mobilkom> OACSU, off-air call setup
Sprechkapsel f <Telekom> transmitter (Fernhörer)
Sprechkopf m <Aufnahme, Comp & DV> recording head
Sprechkreis m <Telekom> speech circuit, circuit; talking circuit (Telefon)
Sprechkreisbündel n <Telekom> circuit group, group
Sprechkreisverfügbarkeit f <Telekom> circuit availability
Sprechpegel m <Aufnahme> speech level
Sprechpegeltest m <Aufnahme> voice level test
Sprechspule f <Elektrotech> voice coil
Sprechtaste f <Aufnahme, Telekom> push-to-talk switch
Sprechweg m <Elektrotech> speech channel
Sprechwegenetzwerk n <Elektrotech> speech-path network
Sprechzeit f <Mobilkom> airtime
Sprechzeug n <Telekom> operator's telephone
Sprechzustand m <Telekom> conversation state (Telefon)
spreizbarer Gewindebohrer m <Maschinen> expanding tap, expansion tap
Spreizbuchse f <Fertig> split bushing
Spreizdorn m 1. <Fertig> (AE) expansion arbor, (BE) expansion arbour; 2. <Maschinen> (AE) expanding arbor, (BE) expanding arbour, expanding mandrel
spreizen v 1. <Fertig> straddle; 2. <Maschinen> expand; 3. <Mechan> swage
Spreizen n zwischen Querbalken <Bau> bridging
Spreizkegel m <Fertig> tapered plug
Spreizkörper m <Mechan> spreader
Spreizmodulation f <Telekom> spread spectrum modulation
Spreizmodulationstechnik f <Telekom> spread spectrum technique
Spreizreibahle f <Fertig> expansion reamer
Spreizringkupplung f <Fertig> expanding band clutch

Spreizschraube f <Maschinen> expanding screw, expansion bolt
Spreizspektrum-Phasenumtastung f <Telekom> spread spectrum phase-shift keying
Spreizspektrum-Vielfachzugriff m <Telekom> spread spectrum multiple access (SSMA)
Spreizung f 1. <Fertig> expansion (Dorn, Reibahle); 2. <Metrol> expanded scale (Messbereich)
Spreizverfahren n <Elektronik> additive method (Halbbild-Trennung)
Sprengbolzen m 1. <Mechan> explosion bolt; 2. <Raumfahrt> explosive bolt
sprengen v 1. <Bau> blast, shoot; 2. <Erdöl> shoot; 3. <Kohlen> blast; 4. <Raumfahrt> blow up
sprengen v durch Erhitzung <Kohlen> blast by heating
Sprengen n <Kohlen> blasting
Sprengfeder f <Maschinen> spring ring
Sprenggerät n <Erdöl> sparger
Sprenglochbohren n <Erdöl> shot hole drilling (Bohrtechnik)
Sprenglochstopfen m <Erdöl> shot hole plug (Bohrtechnik)
Sprengmeister m <Bau, Sicherheit> blasting foreman
Sprengniet n 1. <Maschinen> explosive rivet; 2. <Mechan> explosive-type rivet
Sprengring m 1. <Fertig> circlip; 2. <Maschinen> retainer, snap ring, spring clip; 3. <Mechan> circlip
Sprengscheibe f <Fertig> (BE) bursting disc, (AE) bursting disk
Sprengstoff m <Sicherheit> explosives
Sprengstrebe f <Bau> straining beam (Dachstuhl); strut (Holzbau)
Sprengung f 1. <Bau> blasting; 2. <Erdöl> shot
Sprengzünder m <Raumfahrt, Thermod> detonator
sprenkeln v <Textil> speckle
Spreu f <Lebensmittel> chaff
Spring f <Wassertrans> backspring (Festmachen)
springen v <Comp & DV> jump; branch (Programm)
Springen n <Werkprüf> cracking (Glas)
Springer m <Erdöl> gusher
Springflut f <Wassertrans> spring tide (Gezeiten)
Springleine f <Wassertrans> spring line (Festmachen)
Springnipptide-Zyklus m <Nichtfoss Energ> spring neap cycle
Springtide f <Nichtfoss Energ> spring tide
Springwalzwerk n <Metall> jump mill
Sprinkler m <Sicherheit, Wasserversorg> sprinkler
Sprinkleranlage f <Lufttrans, Wassertrans> fire sprinkler (Notfall)
Sprintmission f <Raumfahrt> sprint mission
Sprite m <Comp & DV> sprite
Spritz... <Bau, Foto, Kunststoff, Verpack> spray
Spritzapparat m 1. <Fertig> sprayer, spraying device; 2. <Foto> airbrush
Spritzauftrag m <Bau> spray painting (Farbe)
Spritzausblasgießmaschine f <Verpack> (AE) injection blow molding machine, (BE) injection blow moulding machine
Spritzautomat m <Fertig> automatic sprayer
Spritzbeton m <Bau> gunite, gunned concrete, sprayed concrete
Spritzblasen n <Kunststoff> (AE) injection blow molding, (BE) injection blow moulding
Spritzblasformen n <Kunststoff> (AE) injection blow molding, (BE) injection blow moulding
Spritzblasverfahren n <Kunststoff> (AE) injection blow molding, (BE) injection blow moulding
Spritzblech n <Fertig> splash baffle
Spritzdruck m <Verpack> (AE) injection molding pressure, (BE) injection moulding pressure

Spritzdüse f 1. <Maschinen> sprayer nozzle; 2. <Sicherheit> fire nozzle *(Brandschutz)*
Spritze f 1. <Fertig> gun; 2. <Labor> syringe
Spritzeinfüllung f <Verpack> injection filling
Spritzen n 1. <Bau> spatter; 2. <Ker & Glas> gunning
Spritzer m 1. <Bau> spatter; 2. <Fertig> splash guard *(Schweißen)*; 3. <Papier> squirt
Spritzerschutzanzug m <Sicherheit> splash suit
Spritzflasche f <Labor> wash bottle
Spritzform f <Maschinen> *(AE)* injection mold, *(BE)* injection mould, *(AE)* mold for thermoplastics, *(BE)* mould for thermoplastics
Spritzformen n <Kunststoff> *(AE)* injection molding, *(BE)* injection moulding
spritzgegossen adj <Fertig> die-cast *(Kunststoffe)*
Spritzgehäuse n <Labor> barrel *(Spritzgerät)*
Spritzgerät n <Fertig> sprayer, spraying device
Spritzgießen n *(IM)* 1. <Fertig> *(AE)* injection molding, *(BE)* injection moulding, pressure die-casting; 2. <Kunststoff, Verpack> *(AE)* injection molding, *(BE)* injection moulding *(IM)*
Spritzgießmaschine f *(IM)* <Kunststoff> *(AE)* injection molding machine, *(BE)* injection moulding machine
Spritzgießmasse f *(IM)* <Kunststoff> *(AE)* injection molding compound, *(BE)* injection moulding compound
Spritzgrat m <Kunststoff> flash
Spritzguss m 1. <Fertig> *(AE)* injection molding, *(BE)* injection moulding *(Kunststoffinstallationen)*; 2. <Mechan> extrusion; 3. <Verpack> *(AE)* injection molding compound, *(BE)* injection moulding compound • **aus Spritzguss** <Mechan> die-cast
Spritzgussform f <Ker & Glas, Maschinen> *(AE)* injection mold, *(BE)* injection mould
Spritzgussmaschine f 1. <Fertig> die-casting machine *(Kunststoffe)*; 2. <Sicherheit> *(AE)* injection molding machine, *(BE)* injection moulding machine
Spritzgussteile npl **mit Angussspritze** <Fertig> spray
Spritzkabine f <Anstrich> spray booth
Spritzkappe f <Wassertrans> spray hood
Spritzkühlung f <Ker & Glas> splat cooling
Spritzlack m <Kunststoff> spraying paint
Spritzlackieren n <Anstrich> spray painting
Spritzmaschine f <Fertig> sprayer, spraying machine
Spritzmetallisieren n <Fertig> metal spraying
Spritzpistole f 1. <Anstrich> spray gun; 2. <Fertig> blow gun; 3. <Verpack> airbrush
Spritzprägeverfahren n *(IC-Verfahren)* <Kunststoff> injection compression process
Spritzpressen n <Kunststoff> *(AE)* transfer molding, *(BE)* transfer moulding
Spritzpressform f <Maschinen> *(AE)* transfer mold, *(BE)* transfer mould
Spritzputz m <Bau> spatter dash
Spritzrohr n <Papier> shower
Spritzschlauch m <Wasserversorg> squirt hose
Spritzschmierung f <Eisenbahn, Kfztech, Maschinen> splash lubrication
Spritzschutz m <Maschinen> splash guard
Spritzschutzhaube f <Labor> antisplash head *(Destillation)*
Spritzschutzmaske f <Anstrich> spray mask
Spritzsee f <Wassertrans> swash *(Meer)*
Spritzung f <Fertig> injection
Spritzverdeck n <Wassertrans> spray hood
Spritzverfahren n <Druck> airbrushing
Spritzwand f <Bau> splashback
Spritzwasser n <Wassertrans> spray *(Meer)*
spritzwasserdicht adj <Fertig> splash-proof *(Kunststoffinstallationen)*

spritzwassergeschützte Entlüftungskappe f <Kfztech> splash-proof vent cap *(Autobatterie)*
Spritzwasserschutz m <Sicherheit> splash guard
Spritzwerkzeug n <Fertig> sprayer, spraying tool
Sprödbruch m 1. <Fertig> nonplastic fracture; 2. <Kerntech> brittle failure; 3. <Mechan> brittle fracture; 4. <Metall, Qual> brittle fracture, cleavage crack
Sprödbruchbeständigkeit f <Metall, Qual> brittle fracture resistance
spröde adj 1. <Kunststoff> brittle, friable; 2. <Mechan, Metall, Papier> brittle
spröder-duktiler Übergang m <Kerntech> brittle-ductile transition
Sprödigkeit f <Ker & Glas, Kunststoff, Mechan, Metall, Qual> brittleness
Sprödigkeitsbruch m <Kerntech> brittle failure
Sprödigkeitstemperatur f <Fertig> brittle point
Sprödriss m <Kerntech> brittle crack
Sprosse f 1. <Bau> tread; rail *(Fenster)*; 2. <Comp & DV> row *(Matrix)*
Sprossen fpl <Bau> rungs *(Leiter)*
Sprosseneisen n <Bau> sash bar
Sprossenschrift f <Akustik> variable density recording
Sprossenteilung f <Comp & DV> row pitch
Spruchkopf m <Raumfahrt> preamble *(Weltraumfunk)*
Sprudelbohrung f <Erdöl> gusher
Sprudeldichtemessgerät n <Gerät> bubble-type density meter
sprudelnd kochen v <Thermod> boil fast
Sprudelrohrfüllstandsmessgerät n <Gerät> bubble pipe level meter
Sprüh... <Verpack> spray
Sprühausleger m <Meerschmutz> spray boom
Sprühbehälter m <Verpack> aerosol, aerosol container
Sprühdüse f 1. <Chemtech> atomizer nozzle; 2. <Papier> pulverizer; 3. <Wasserversorg> spreader jet *(Abzweigrohr)*
Sprühen n 1. <Elektrotech> sputtering; 2. <Maschinen> spraying
Sprühentladung f 1. <Elektriz, Elektrotech> brush discharge; corona discharge *(Glimmentladung)*; 2. <Phys> brush discharge
Sprüher m <Chemtech> atomizer
Sprühkanone f <Meerschmutz> spray gun
Sprühkühlung f <Heiz & Kälte> spray cooling
Sprühöffnungen fpl <Meerschmutz> spray aperture *(für Dispergatoren)*
Sprühpfad m <Meerschmutz> spray path
Sprühstreichmaschine f <Papier> spray coater
Sprühwassersystem n <Kerntech> water spray system
Sprung m 1. <Bau> break joint *(Fassade)*; 2. <Comp & DV> jump, skip; branch *(Programm)*; 3. <Mechan> flaw; 4. <Metall> jump; 5. <Wassertrans> sheer *(Schiffbau)*; 6. <Wasserversorg> breach • **Sprung hinten** <Wassertrans> sheer aft *(Schiffkonstruktion)* • **Sprung vorn** <Wassertrans> sheer forward *(Schiffkonstruktion)*
Sprunganweisung f <Comp & DV> branch instruction
sprungartige Änderung f <Regelung> stepwise change
Sprungausfall m <Telekom> sudden failure
Sprungbefehl m <Comp & DV> jump, jump instruction
Sprungelastizität f <Kunststoff> rebound elasticity
Sprungentfernung f <Funkort> skip distance
Sprungfunktion f 1. <Comp & DV, Elektronik> jump function, step function; 2. <Telekom> step function
Sprungfunktionsgeber m <Elektronik> step function generator
Sprungfunktionsverhalten n <Elektronik> step function response
sprunghafter Vollausfall m <Qual, Sicherheit> catastrophic failure

Sprunghöhe

Sprunghöhe f 1. <Hydraul> height of hydraulic jump (j); 2. <Regelung> step height (Eingangssignal)
Sprungoperation f <Comp & DV> jump operation
Sprungrohr n <Bau> swan neck
Sprungstart m <Lufttrans> jump takeoff (Hubschrauber)
Sprungsuche f <Fernseh> shuttle search
Sprungtemperatur f <Kerntech> transition temperature
Sprungtuch n <Sicherheit> jumping sheet (Feuerwehr)
Sprungüberdeckung f <Fertig> face contact ratio (Zahnrad)
Sprungverzerrung f <Aufnahme> transient distortion
sprungweise Durchprüfung f <Comp & DV, Qual> leapfrog test
SPS 1. <Comp & DV> (Programmspeichersteuerung) SPC (stored program control); 2. <Kontroll, Labor> (speicherprogrammierbare Steuerung) PLC (programmable logic control)
SPST-Schalter m (einpoliger Ein-/Ausschalter) 1. <Elektriz, Elektrotech> SPST switch (single pole single-throw switch); 2. <Kontroll> SPST switch
Spuchstoffaustritt m <Papier> reject gate
Spuckstoff m <Chemie> screenings (Papier)
Spül... <Erdöl, Wassertrans> wash
Spülbecken n 1. <Bau> sink; 2. <Erdöl> sump
Spülbohren n 1. <Bau> jetting; 2. <Erdöl> wash boring (Bohrtechnik)
Spülbohrverfahren n <Erdöl> wash boring (Bohrtechnik)
Spülbord n <Wassertrans> washboard (Schiffbau)
Spule f 1. <Akustik> spool (für Magnetband, Papierstreifen); 2. <Aufnahme> spool; 3. <Comp & DV> reel; 4. <Elektriz, Elektrotech> bobbin, coil; 5. <Fernseh> reel; 6. <Funktech> coil; 7. <Ker & Glas> bobbin, spool; 8. <Kfztech> coil (Zündung); 9. <Mechan> coil, drum; 10. <Papier> coil; 11. <Telekom> inductance; 12. <Textil> bobbin, package • **Spulen aufstecken** <Textil> place yarn containers (Gatter)
Spule f mit Wabenwicklung <Elektrotech> lattice-wound coil
spulen v 1. <Bau> wind; 2. <Fernseh> spool; 3. <Textil> wind
Spulen n 1. <Elektrotech> reeling; 2. <Fernseh> spooling
spülen v 1. <Erdöl> purging; 2. <Maschinen> flush; 3. <Wasserversorg> blow off, scour
Spülen n 1. <Kerntech> backflushing; 2. <Maschinen> flushing
Spulenabnahme f <Textil> doffing
Spulenbelastung f <Telekom> coil loading
Spulengatter n <Textil> creel, warping creel
Spulenkern m 1. <Elektriz> coil core; 2. <Elektrotech> slug (Abstimmung), slug (Kondensatoren); core (bei Magnetrelais); 3. <Mechan> hub; 4. <Telekom> core
Spulenkörper m <Elektriz, Elektrotech> coil former, coil shell
Spulenrahmen m <Elektrotech> coil frame
Spulenwickeln n <Elektriz> coil winding
Spulfeld n <Textil> wind
Spülflüssigkeit f <Erdöl> drilling mud
Spülkanal m <Bau> water course
Spülkante f <Ker & Glas> fluxline
Spülkantenangriff m <Ker & Glas> fluxline attack
Spülkopf m <Erdöl> swivel (Bohrtechnik)
Spulkranz m <Textil> uncollapsed cake
Spülleitung f <Erdöl> mud line (Bohrtechnik)
Spülmittel n <Fertig> cleansing agent (Kunststoffinstallationen)
Spülpumpe f <Erdöl> mud pump (Bohrtechnik)
Spülpumpenarmatur f <Erdöl> mud pump valve (Bohrtechnik)
Spülrohrtour f <Erdöl> washover string (Tiefbohrtechnik)

Spülschlammanalyselog n <Erdöl> mud analysis log (Bohrtechnik)
Spülung f 1. <Erdöl> drilling mud, mud (Bohrtechnik); 2. <Ker & Glas> washing; 3. <Kerntech> scavenging; 4. <Mechan> flushing, scavenging; 5. <Meerschmutz> flushing; 6. <Nichtfoss Energ> scour; 7. <Papier> scouring; 8. <Umweltschmutz> flushing; 9. <Wassertrans> scavenging (Motor); 10. <Wasserversorg> flush, scavenging, scour, scouring; flushing (Spülwasserbohrung)
Spulvorgang m 1. <Foto> spooling; 2. <Textil> winding process
Spülvorrichtung f <Sicherheit> rinsing station
Spund m 1. <Bau> plug, spile; 2. <Lebensmittel> bung; 3. <Mechan> plug
Spundbohle f 1. <Bau> pile plank, sheet pile, sheeting pile; 2. <Kohlen> sheet pile
Spundbrett n <Bau> matchboard
Spundeisen n <Bau> tonguing iron
spunden v <Fertig> match, spung
Spundhobel m <Bau> grooving plane, matching plane, (BE) plough plane, (AE) plow plane, tongue plane, tonguing plane
Spundholzlage f <Bau> matching
Spundloch n <Lebensmittel> bunghole
Spundmaschine f <Bau> tonguing-and-grooving machine
Spundschalung f <Bau> tight sheathing
Spund- und Nutmaschine f <Bau> matching machine
Spundung f <Bau> tonguing-and-grooving
Spundverbindung f <Bau> (BE) ploughed and tongued joint, (AE) plowed and tongued joint, tongue-and-groove joint
Spundwand f 1. <Bau> sheet piling, sheeting, steel piling; 2. <Kohlen> sheet piling
Spur f 1. <Aufnahme> toe, track; 2. <Comp & DV> track; 3. <Eisenbahn> (AE) gage, (BE) gauge; 4. <Elektrotech> trace (Leiter); 5. <Fernseh> track; 6. <Fertig> channel (Lochstreifen); 7. <Kerntech, Kontroll> track; 8. <Math> trace (Spur einer quadratischen Matrix); 9. <Metrol> (AE) gage, (BE) gauge; 10. <Optik> track; 11. <Trans> (AE) gage, (BE) gauge
Spur f mit variabler Dichte <Fernseh> variable density track
Spurabstand m 1. <Aufnahme> track spacing; 2. <Fernseh> track pitch
Spuraufzeichnung f <Aufnahme> toe recording
Spurbenutzungshilfen fpl <Trans> directional aids
Spurbenutzungszählung f <Trans> directional census
Spurbreite f <Aufnahme> track width
Spurdichte f <Optik> track density
Spureinstellung f <Fernseh> track adjustment
Spuren fpl pro Inch <Comp & DV> TPI, tracks per inch
Spurenanalyse f <Wasserversorg> trace analysis
Spurenelement n <Lebensmittel, Umweltschmutz> trace element
Spurerweiterung f <Eisenbahn, Trans> (AE) gage clearance, (BE) gauge clearance
spurgebundenes Fahrzeug n <Bau, Kfztech> tracked vehicle
spurgebundenes Luftkissenfahrzeug n <Wassertrans> TACV, tracked air cushion vehicle, tracked hovercraft
spurgebundenes Massentransportsystem n <Trans> guided public mass transportation system
spurgeführt adj <Eisenbahn> railborne
spurgeführte Straße f <Trans> guided road
spurgeführtes Fahrzeug n <Trans> track-guided vehicle
spurgeführtes Luftkissenfahrzeug n <Trans> guided air cushion vehicle
spurgeführtes Transportsystem n <Trans> track-guided transport system

spurgenau laufen v <Comp & DV> track
Spurhaltesystem n <Aufnahme> sprocket hole control track system
Spurhaltung f <Fernseh> tracking, tracking control
Spurkranz m <Kfztech> flange (Rad)
Spurkranzreibung f <Eisenbahn> wheel flange friction
Spurkranzschmiermittel n <Eisenbahn> wheel flange lubricant
Spurlager n <Maschinen> step bearing, step block
Spurlinie f <Erdöl> spur line (Bohrtechnik)
Spurmechanismus m <Optik> tracking mechanism
Spurrille f <Eisenbahn> flangeway
Spurspiel n <Eisenbahn> flange-to-rail clearance
Spurspreizung f <Aufnahme> track spreading
Spurstange f 1. <Eisenbahn> tie bar, tie rod; 2. <Kfztech> tie rod; track rod (Lenkung)
Spursteigung f <Optik> track pitch
Spurstrecke f <Erdöl> spur line
Spurteilung f <Aufnahme> track pitch
Spurverfolgung f <Phys> ray tracing
Spurwähler m <Fernseh> track selector
Spurwechselbahnhof m <Eisenbahn> (AE) change-of-gage station, (BE) change-of-gauge station
Spurweite f 1. <Bau> (AE) gage, (BE) gauge; 2. <Eisenbahn> distance between rails, (AE) gage, (BE) gauge, (AE) rail gage, (BE) rail gauge, (AE) track gage, (BE) track gauge; 3. <Kfztech> track width; 4. <Mechan, Trans> (AE) gage, (BE) gauge
Spurzapfen m <Bau> pintle
SQID (supraleitfähiger Quanteninterferenzmechanismus) <Phys> SQUID (superconductive quantum interference device)
Squalan n <Chemie> squalane
Squalen n <Chemie> squalene
Square n <Metrol> square (Bauwesen)
Sr (Strontium) <Chemie> Sr (strontium)
SRAM (statischer RAM) <Comp & DV> SRAM (static RAM)
SRF-Ruß m <Kunststoff> SRF carbon black, semi-reinforcing carbon black
SR-Schaltung f (Serienresonanzkreis) <Elektrotech> series-resonant circuit
SS 1. <Fernseh, Funktech> (Spitze-Spitze-...) pp (peak-to-peak); 2. <Maschinen> (Schiebesitz) close sliding fit, push fit; 3. <Mechan> (Schiebesitz) close-sliding fit, push fit, sliding fit
SSC (supraleitfähiges Supracollider) <Teilphys> SSC (superconducting super collider)
SSI-Schaltung f (Kleinintegration) 1. <Comp & DV> SSI; 2. <Elektronik> SSI (small-scale integration, single scale integration)
SS-Spannung f (Spitzenspannung eines Senders) <Funktech> PEV (peak envelope voltage)
SSTV 1. <Fernseh> (Fernsehen mit langsamer Abtastung, Schmalbandfernsehen, Slow-Scan-Television) SSTV (slow scan television); 2. <Funktech> (Schmalbandfernsehen mit langsamer Abtastung) SSTV (slow scan television)
Staatsgespräch n <Telekom> government call
Stab m 1. <Bau> bar, member, rod; 2. <Ker & Glas> rod; 3. <Kohlen> b, bar (Holz, Metall); 4. <Maschinen> bar, rod; 5. <Mechan> rod
Stabanker m <Elektriz> bar armature
Stabantenne f <Funktech> rod antenna
Stabausdehnungs-Thermometer n 1. <Gerät> rod-and-tube thermometer; 2. <Heiz & Kälte> solid expansion thermometer
Stäbchenbakterie f <Lebensmittel> bacillus
Stabeisen n <Metall> merchant iron, rod iron
Stabelektrode f <Elektriz> rod electrode
Staberder m <Eisenbahn> (BE) earthing rod, (AE) grounding rod
Stabfeder f <Mechan> rod spring
Stabfilter n <Mechan> filter cartridge
stabförmiger Leiter m <Elektrotech> bar-type conductor
Stabhalterung f <Metall> bar (Pfahl)
stabil adj <Mechan> heavy
stabiles Gleichgewicht n <Phys> stable equilibrium
stabiles Isotop n <Phys> stable isotope
Stabilisator m 1. <Elektrotech> stabilizer, stabilizer tube; 2. <Erdöl> stabilizer (Bohrtechnik); 3. <Foto, Ker & Glas> stabilizer; 4. <Kfztech> stabilizer, stabilizer bar (Aufhängung); 5. <Kohlen, Kunststoff> stabilizer; 6. <Lufttrans> stabilizer (Luftfahrzeug); 7. <Wassertrans> stabilizer
stabilisieren v 1. <Bau> grout (Erdreich); 2. <Raumfahrt> despin (aus dem Trudeln)
Stabilisierkolonne f <Erdöl> stabilizer tower (Raffinerie)
stabilisiert adj <Raumfahrt> despun (Raumschiff)
stabilisierter Latex m <Kunststoff> stabilized latex
Stabilisierturm m <Erdöl> stabilizer tower
Stabilisierung f 1. <Elektriz, Elektrotech, Kohlen, Kunststoff> stabilization; 2. <Maschinen> steadying; 3. <Raumfahrt> stabilization (Weltraumfunk); 4. <Telekom> regulation (Spannung); 5. <Wassertrans> stabilization (Schiffkonstruktion)
Stabilisierungsanlage f <Wassertrans> (AE) stabilizing system, (BE) stabilising system
Stabilisierungseinrichtung f <Elektriz> stabilizer
Stabilisierungsfläche f <Lufttrans> stabilizer (Luftfahrzeug)
Stabilisierungsflosse f <Wassertrans> stabilizing fin
Stabilisierungsflügel m <Lufttrans> stabilizer (Luftfahrzeug)
Stabilisierungsgardine f <Wassertrans> stability curtain
Stabilisierungskreisel m <Phys> gyrostat
Stabilisierungsschiene f <Eisenbahn> stabilization rail
Stabilisierungsschürze f <Wassertrans> stability skirt
Stabilisierungssystem n 1. <Raumfahrt> despin system (Raumschiff); 2. <Wassertrans> (AE) stabilizing system, (BE) stabilising system
Stabilisierungsvorrichtung f <Elektrotech, Wassertrans> stabilization device
Stabilisierungswicklung f <Elektriz> stabilizing winding
Stabilisierungswiderstand m <Elektrotech> ballast resistor
Stabilität f <Comp & DV, Funktech, Kohlen, Lufttrans, Telekom, Wassertrans> stability
Stabilität f einer Grenzschicht <Strömphys> boundary layer stability
Stabilitätsdiagramm n <Strömphys> stability diagram
Stabilitätskurve f <Wassertrans> stability curve (Schiffkonstruktion)
Stabilitätsuntersuchung f <Strömphys> stability analysis
Stabilotron n <Phys> stabilotron
Stabkristall n <Metall> columnar crystal
Stabmagnet m <Phys> bar magnet
Stabmühle f 1. <Kohlen> rod mill; 2. <Maschinen> bar mill, rod mill
Stabprobe f <Ker & Glas> rod proof
Stab-Rohr-Methode f <Telekom> rod-in-tube technique
Stabschere f <Maschinen> bar shears
Stabschwingmühle f <Kohlen> vibrating rod mill
Stabspannung f <Fertig> stress in the bar
Stabstahl m 1. <Maschinen> bar stock; 2. <Metall> merchant bar
Stabstahlschere f <Maschinen> bar-shearing machine
Stabstahlwalzwerk n <Metall> merchant bar rolling mill

Stabstahlwerk

Stabstahlwerk n <Fertig> merchant mill
Stabstraße f <Fertig> rod mill
Stab-Temperaturregler m <Heiz & Kälte> immersion-type thermostat
Stabtransformator m <Elektriz> bar-type transformer
Stabwicklung f <Elektriz> bar winding *(Motor)*
Stachelbandführung f <Comp & DV> sprocket feed, tractor feed
Stachelrad n <Comp & DV> tractor
Stachelradwalze f <Comp & DV> sprocket, sprocket wheel
Stachydrin n <Chemie> stachydrine
Stachyose f <Chemie> stachyose
Stadiometer n <Bau> stadiometer *(Vermessung)*
Stadtbahn f <Eisenbahn> *(AE)* metropolitan railroad, *(BE)* metropolitan railway
Stadtbezirk m <Bau> ward
Stadtentsorgung f <Abfall> municipal waste disposal
Stadtentwässerung f <Abfall> municipal sewage system
städteverbindender Transport m <Trans> intercity transport
Stadtfunkrufdienst m <Funktech, Telekom> regional radio-paging service *(SfuRD)*
Stadtgas n <Erdöl> town gas
städtische Entsorgung f <Abfall> municipal waste disposal
städtische Mobilfunkteilnehmerdichte f <Mobilkom> urban mobile radio subscriber density
städtische Schnellstraße f <Trans> rapid urban artery
städtische Wasserbewirtschaftung f <Wasserversorg> urban water management
städtisches Abwasser n <Abfall> municipal waste water
städtisches Einzugsgebiet n <Wasserversorg> urban catchment
städtisches Netz n <Telekom> urban network
städtisches Straßennetz n <Trans> urban road network
städtisches Transportsystem n **ohne Zwischenhaltestation** <Trans> nonstop urban transportation
Stadt-Land-Reifen m <Kfztech> *(AE)* town-and-country tire, *(BE)* town-and-country tyre
Stadtmüll m <Wasserversorg> municipal waste
Stadtnetz n <Telekom> metropolitan network
Stadtreinigung f <Abfall> public cleansing
Stadtschnellbahn f <Eisenbahn> *(AE)* rapid transit railroad, *(BE)* rapid transit railway
Stadtstraßenbahn f <Eisenbahn> *(AE)* streetcar metro, *(BE)* tramway metro
Stadt- und Regional-S-Bahn f <Eisenbahn> *(AE)* urban and regional metropolitan railroad, *(BE)* urban and regional metropolitan railway
Stadtverkehr m 1. <Eisenbahn> *(AE)* light rail transit, *(BE)* light rail transport; 2. <Kfztech> urban cycle *(Fahrbetrieb)*; 3. <Trans> urban traffic
Stadtviertel n <Bau> block
Stadtwasser n <Wasserversorg> town water
Stadtwerke npl <Wasserversorg> public works
Staffel f <Telekom> grading, progressive grading
Staffelleitung f <Telekom> series circuit
Staffelwalze f <Metall> stepped roll
Stag n <Wassertrans> stay *(Tauwerk)* • **über Stag gehen** <Wassertrans> go-about *(Segeln)*
Staggarnat n <Wassertrans> Spanish burton *(Tauwerk)*
stagnierendes Gewässer n <Wasserversorg> stagnant water
Stagreiter m <Wassertrans> jib hank *(Segeln)*; hank *(Tauwerk)*
Stagsegel n <Wassertrans> fore staysail *(Segeln)*
Stahl m <Maschinen, Metall> steel
Stahl m **geringerer Güte** <Metall> secondary steel

Stahl... <Metall> steel
Stahlanreißlineal n <Metrol> steel straightedge
Stahlbandförderer m <Maschinen> steel band conveyor
Stahlbandkette f <Bau> steel band chain *(Vermessung)*
Stahlbandumreifung f <Verpack> steel band strapping
Stahlbau m <Bau> steel construction
Stahlbeton m 1. <Bau> reinforced concrete; 2. <Wassertrans> ferroconcrete *(Schiffbau)*
Stahlbetonrippendecke f <Bau> hollow-block floor
Stahlbildung f <Metall> acieration
Stahlblech n 1. <Fertig> iron plate; 2. <Kfztech> sheet steel *(Karosserie)*; 3. <Maschinen, Metall> sheet steel
Stahlblock m <Metall> steel ingot
Stahlbolzenkette f <Maschinen> pintle chain
Stahldraht m <Wassertrans> steel wire *(Schiffbau)*
Stahldrahtseil n <Maschinen> steel wire rope
Stahlenformung f <Elektronik> beam shaping
Stahlfeder f <Maschinen> steel spring
Stahlflechter m <Bau> steelfixer *(Stahlbeton)*
Stahlgießerei f <Metall> steel foundry
Stahlgittermast m <Bau> lattice tower
Stahlguss m 1. <Fertig> cast steel; 2. <Metall> cast steel, steel casting; 3. <Papier> cast steel
Stahlgussstück n <Wassertrans> steel casting *(Schiffbau)*
Stahlhalter m <Maschinen> tool post
stahlintensives Produkt n <Metall> steel-intensive product
Stahlkappe f <Sicherheit> steel toe cap *(Schuhwerk)*
Stahlkern m <Fertig> steel core *(Kunststoffinstallationen)*
Stahlkonstruktion f <Bau> steel construction
Stahlmantelwalze f <Bau> smooth roller
Stahlmaßband n <Metrol> steel measuring tape
Stahl-Mehrwegcontainer m **zu Demontagezwecken** <Verpack> steel re-usable CKD container
Stahlmörser m <Labor> percussion mortar *(Schleifen)*
Stahlpfahl m <Bau, Kohlen> steel pile
Stahlplattform f <Erdöl> steel platform *(Offshore-Technik)*
Stahlrad n <Eisenbahn> steel wheel
Stahlrohrblock m <Fertig> steel ingot
Stahlrohrgerüst n <Bau> tubular scaffolding
Stahlschalung f <Bau> steel forms
Stahlschienensystem n <Eisenbahn> steel rail system
Stahlschiff n <Wassertrans> steel vessel *(Schiffbau)*
Stahlschmieden n <Metall> steel forging
Stahlschrank m <Labor> steel locker
Stahlschrot m <Fertig> shot
Stahlschrott m <Abfall> steel scrap
Stahlseilfördergurt m <Maschinen> steel cord conveyor belt
Stahlstichdruck m <Mechan> die-stamping
Stahlstichprägung f <Druck> die-stamping
Stahlstreifenkolben m <Kfztech> autothermic piston
Stahlstütze f <Bau> stanchion
Stahlträger m <Bau> steel beam
Stahlverbundplatte f <Bau> bonded steel plate
Stahlvorspannglied n <Kerntech> steel tendon
Stahlwerk n <Fertig> steel works
Stahlwinkel m <Fertig> structural angle
Stahlzement m <Bau> ferrocement
Stähne f <Fertig> rap
Stall m **am Luftschraubenblatt** <Lufttrans> blade stall *(Hubschrauber)*
Stall m **am rücklaufenden Blatt** <Lufttrans> retreating blade stall *(Hubschrauber)*
Stall m **an der Luftschraubenblattspitze** <Lufttrans> blade tip stall *(Hubschrauber)*
Stamm m 1. <Bau> stem; 2. <Elektrotech> physical circuit; 3. <Telekom> physical circuit, side circuit

Stamm... <Comp & DV> master
Stammabfluss m <Umweltschmutz> stemflow
Stammband n <Aufnahme, Comp & DV> master tape
Stammblatt n <Fertig> *(BE)* disc, *(AE)* disk; body *(Säge)*
Stammbruch m <Math> unit fraction
Stammdatei f <Comp & DV> master file
Stammdateisatz m <Comp & DV> master record
Stammdaten npl 1. <Comp & DV> master data; 2. <Umweltschmutz> historical data
stammen v <Wassertrans> hail *(Hafen)*
Stammfunktion f <Math> integral function with continuous integrand, primitive, primitive function *(Integralrechnung)*
Stammholz n <Bau> standing timber
Stammkarte f <Comp & DV> master card
Stammkopie f <Comp & DV> master copy
Stammleitung f 1. <Elektrotech> physical circuit; 2. <Telekom> physical circuit, side circuit
Stammplatte f <Comp & DV> master disk
Stammsatz m <Comp & DV> master record
Stammsäure f <Chemie> parent acid
Stammzeichnung f <Konstzeich> parent drawing
Stampf... <Chemie> tamped
stampfbar adj <Fertig> rammable
Stampfbarkeit f <Fertig> rammability
Stampfbewegung f 1. <Phys> pitching; 2. <Wassertrans> pitch
Stampfdichte f <Kunststoff> tamped density
stampfen v 1. <Bau> puddle, ram, tamp; 2. <Eisenbahn> tamp; 3. <Wassertrans> pitch
Stampfen n 1. <Bau, Fertig> ramming; 2. <Wassertrans> pitching *(Schiffbewegung)*
Stampfer m 1. <Bau> rammer; 2. <Fertig> compactor, punner, rammer, tamper, tamp; pummel *(Gießen)*; 3. <Kohlen> pneumatic ram
Stampfmasse f <Ker & Glas> tamping clay
Stampfstange f <Bau> tamping rod
Stampfwinkel m <Wassertrans> angle of pitch *(Schiffbewegung)*
Stand m <Kontroll, Patent, Qual> state • **auf den neuesten Stand bringen** <Maschinen> bring up to date, update
Stand m **der Technik** 1. <Kontroll> state of the art, state of technology; 2. <Patent> prior art, state of the art; 3. <Qual> state of the art
Stand-alone-Modus m <Telekom> stand-alone mode *(Modus bei Nichtbenutzung des PC)*
Standard m 1. <Comp & DV> default, standard; 2. <Kontroll> standard
Standard-Ohm n <Elektriz> standard ohm
Standardabweichung f <Comp & DV, Ergon, Math, Phys, Qual> standard deviation
Standardabweichung f **des Mittelwerts** <Comp & DV> mean square error
Standardanwendung f <Comp & DV> off-the-shelf application
Standardband n <Fernseh> standard tape
Standardbaugruppe f <Kontroll> standard assembly
Standardbildmuster n <Fernseh> standard pattern
Standardblende f <Hydraul> standard orifice
Standarddokument m <Comp & DV> standard document
Standarddruckplatte f <Druck> standard plate
Standarddurchmesser m <Maschinen> standard diameter
Standardelement n <Phys> standard cell
Standardfehler m 1. <Math> standard error *(Wurzel aus dem mittleren quadratischen Fehler)*; 2. <Qual> standard error
Standardflottengewicht n <Lufttrans> fleet weight
Standardformat n <Comp & DV> master format

Standardfunktion f <Comp & DV> standard function
Standardhöhenmessereinstellung f <Lufttrans> standard altimeter setting
Standard-ISDN-Mehrgeräteanschluss m <Telekom> standard ISDN multipoint interface
standardisierte Hörschwelle f <Akustik> standardized threshold hearing
standardisierte Pegeldifferenz f <Akustik> standardized level difference
standardisierte Schnittstelle f <Telekom> standard inferface, standardized interface
standardisierter Trittschall m <Akustik> standardized impact sound
Standardisierung f 1. <Comp & DV> normalization, standardization; 2. <Math> standardization *(einer Zufallsgröße)*; 3. <Maschinen, Phys, Telekom> standardization
Standardkorrekturzeichensatz m *(SGML)* <Comp & DV, Druck> Standard Generalized Markup Language *(SGML)*
Standardlaufwerk n <Comp & DV> default drive
Standardlautstärkeanzeige f <Aufnahme> standard volume indicator
Standardlautstärkepegel m <Akustik> loudness level of reference sound
Standardlichtquelle f <Elektrotech> standard light source
Standardmaßeinheit f <Metrol> standard measure
standardmäßig adj <Comp & DV> standard
standardmäßige Verfügbarkeit f <Telekom> commoditization *(von Produkten)*
Standardmehrfachmesselement n <Metrol> *(AE)* standard multigaging element, *(BE)* standard multigauging element
Standardmenü n <Comp & DV> default menu
Standardmesssignal n <Fernseh> standard measuring signal
Standardmodell n <Teilphys> standard model
Standardmodus m <Comp & DV> native mode
Standardmutter f <Maschinen> standard nut
Standardnormalverteilung f <Math> standard normal distribution *(genormte Gauß'sche Verteilung)*
Standardplatte f <Druck> standard plate
Standardprotokoll n <Kontroll> standard protocol
Standardschall m <Akustik> standard sound
Standardschnittstelle f <Comp & DV> standard interface
Standardverdichtung f <Kohlen> laboratory compaction
Standardverdrängung f <Wassertrans> standard displacement
Standardweltzeit f *(UT, UTC)* <Funktech, Raumfahrt> universal time, universal time coordinated *(UTC)*
Standarte f <Foto> camera front
Standarteneinstellung f <Foto> back standard adjustment *(Balgeneinheit)*
Standartenverstellung f <Foto> lens movement
Standbeutel m <Verpack> flat-end sack
Standbild n 1. <Fernseh> freeze frame, frozen picture, still frame; 2. <Foto> still picture
Standbildübertragung f <Fernseh> static picture transmission
Standbremse f <Eisenbahn, Lufttrans> parking brake
Standbybetrieb m <Kontroll> stand-by
Standbymodus m <Kontroll, Raumfahrt> stand-by mode
Standby-Zeit f <Mobilkom> stand-by time *(Handy)*
Stander m <Wassertrans> burgee *(Flaggen)*
Ständer m 1. <Bau> timber pillar *(Holz)*; 2. <Elektriz, Elektrotech> bearing, stator; 3. <Fertig> standard, upright; 4. <Kfztech> stator; stand *(Motorrad)*; 5. <Maschinen> column, frame, holder, housing, pedestal, post, standard, support, upright
Ständerblech n <Elektrotech> stator lamination

Ständerbohrmaschine

Ständerbohrmaschine f 1. <Fertig> vertical box-column drill, vertical box-column drilling machine; 2. <Maschinen> column-type drilling machine
Ständerführung f <Maschinen> column guideway
Ständerplatte f <Elektrotech> stator plate
Ständerverbindung f <Bau> stud union
Ständerwand f <Bau> stud partition
Ständerwicklung f <Elektrotech> stator winding
Standfestigkeit f 1. <Fertig> air-dried strength; 2. <Kohlen> stability *(Sandmischung)*
Standfläche f <Maschinen> footstep
Standgerät n <Heiz & Kälte> upright unit
Standguss m <Maschinen, Metall> gravity casting
ständig mitlaufende Reserve f <Nichtfoss Energ> spinning reserves
ständig vorhanden adj <Comp & DV> resident
ständige Erreichbarkeit f <Telekom> constant availability
ständiger Belegungsdetektor m <Trans> continuous presence detector
ständiger Fehler m <Comp & DV> permanent error
ständiger Fluss m <Phys> steady flow
ständiges Echo n <Raumfahrt> permanent echo
Standkühler m <Kfztech> upright radiator
Standleitung f 1. <Comp & DV> dedicated line; 2. <Telekom> dedicated line; tie line *(zwischen Vermittlungen)*
Standmesser m <Verpack> liquid level indicator
Standmessgerät n <Gerät> *(AE)* level gage, *(BE)* level gauge
Standmessung f <Gerät> level measurement
Standort m 1. <Bau, Comp & DV, Elektriz> location; 2. <Lufttrans> radio fix; 3. <Raumfahrt> fix; 4. <Trans> radio fix; 5. <Wassertrans> radio fix; position *(Navigation)*
• **Standort festlegen** <Bau> locate
Standort m **der Deponie** <Abfall> disposal site, tip
Standortaktualisierung f <Mobilkom> location update
Standortbestimmung f <Wassertrans> positioning *(Navigation)*
Standortbestimmung f **von Fahrzeugen** <Funktech, Kfztech, Telekom> car position finding, mobile positioning, vehicle positioning
Standortdatei f <Mobilkom> LR, location register
Standortdiversity n <Telekom> site diversity
Standortkriterien npl <Kerntech, Umweltschmutz> site criteria
Standortlöschung f <Mobilkom> LCP, location cancellation procedure
Standortwahl f <Kerntech> siting
Standplatte f <Erdöl> bed plate
Standreibung f <Phys> static friction
Standrohr n 1. <Erdöl> standpipe *(Flüssigkeitsniveau)*; 2. <Raumfahrt> standpipe
Standrohrstopfen m <Fertig> standpipe adaptor, standpipe plug *(Kunststoffinstallationen)*
Standschub m <Lufttrans> static thrust
Standseilbahn f 1. <Eisenbahn> funicular, *(AE)* funicular railroad, *(BE)* funicular railway; 2. <Trans> *(AE)* cable railroad, *(BE)* cable railway, *(AE)* underground cable railroad, *(BE)* underground cable railway
standsichere Leichtmetallleiter f <Sicherheit> fall-safe light metal ladder
Standspur f <Bau, Kfztech> hard shoulder
Standverbindung f <Telekom> fixed connection
Standvermögen n <Kunststoff> nonslump properties
Standversuch m <Metall> creep test
Standzeit f 1. <Anstrich> pot life *(Lacken)*; 2. <Fertig> cutting-edge life, edge life, endurance, useful life; 3. <Maschinen> endurance, life; 4. <Qual> endurance • **mit langer Standzeit** <Fertig> long-life

Standzeitversuch m <Maschinen> tool life testing
Stange f 1. <Bau> rod; 2. <Erdöl> bar; 3. <Hydraul> rod *(Schieber)*; stem *(Ventil)*; 4. <Labor> rod *(Rührer)*; 5. <Maschinen> bar, rod, rod; 6. <Metrol> pole; 7. <Wasserversorg> rod
Stangenanguss m <Kunststoff> sprue gate
Stangenautomat m <Maschinen> bar automatic lathe
Stangenbohrer m <Maschinen> auger bit
Stangenmagazin n <Fertig> bar magazine
Stangenmaterial n <Maschinen> bar stock
Stangenriegel m <Bau> bar bolt
Stangenrissprüfgerät n <Fertig> rod crack-test instrument
Stangenrohrzug m <Metall> drawing over mandrel
Stangenrost m <Maschinen> bar screen
Stangenschwefel m <Chemie> *(AE)* roll sulfur, *(BE)* roll sulphur
Stangensystem n <Mechan> rod system
Stangenvorschub m <Maschinen> bar feed
Stangenvorschubvorrichtung f <Maschinen> bar feed mechanism
Stangenzirkel m 1. <Fertig> beam trammel; 2. <Metrol> beam compasses
Stangenzirkel m **mit Stellschraube** <Metrol> beam compasses with adjusting screw
Stangenzuführung f <Fertig> stock feeding device
Stangpressen n <Metall> extrusion
Stanniol n <Fertig> tinfoil
Stanniolverpackungsmachine f <Verpack> tinfoiling machine
Stantonzahl f <Kerntech> Stanton number
Stanzabfall m 1. <Comp & DV> chad; 2. <Fertig> punching
Stanze f 1. <Fertig> stamping press; 2. <Maschinen> blanking press; 3. <Mechan> punch, punch press; 4. <Verpack> punch
stanzen v <Bau, Maschinen, Mechan> punch
Stanzen n <Fertig> blanking, stamping
Stanzer m <Bau> punch
Stanzgewindebohrer m <Fertig> serial hand tap
Stanzmaschine f <Maschinen> punch, punching machine
Stanzpresse f 1. <Maschinen> blanking press; 2. <Mechan> punch press
Stanzteil n <Maschinen> stamping
Stanzwerkzeug n <Maschinen> blanking die, punching tool, stamping tool
Stapel m 1. <Bau> stack; 2. <Comp & DV> batch, pack, stack; 3. <Druck, Fertig> pile; 4. <Ker & Glas> stack *(Flachglas)*; stack *(Platten)*; 5. <Maschinen> stack; 6. <Mechan> file • **vom Stapel laufen lassen** <Wassertrans> launch *(Schiff)*
Stapeladresse f <Comp & DV> stack address
stapelbar adj <Verpack> stackable
Stapelbetrieb m <Telekom> batch mode
Stapelcontainer m <Trans> stackable container
Stapelfalte f <Konstzeich> stacking fold
Stapelfaser f <Ker & Glas, Textil> *(AE)* staple fiber, *(BE)* staple fibre
Stapelfasergewebe n <Ker & Glas> staple tissue
Stapelferneingabe f <Comp & DV> remote batch entry
Stapelfernverarbeitung f <Comp & DV> remote batch processing, remote batch teleprocessing
Stapelförderer m <Verpack> stacking conveyor
Stapelgarn n <Textil> stapled yarn
Stapelglasseide f <Ker & Glas> chopped strand
Stapelhöhe f <Verpack> stacking height
Stapeljobverarbeitung f <Comp & DV> stack job processing
Stapelkasten m <Verpack> stacking box

Stapellänge f <Textil> staple length
Stapellauf m <Wassertrans> launch; launching *(Schiff, Boot)*
stapeln v 1. <Bau> pile, stack, stockpile; 2. <Comp & DV> stack; 3. <Papier> stack up; 4. <Verpack> stack
Stapeln n 1. <Bau> piling; 2. <Comp & DV> stacking; 3. <Ker & Glas> stacking *(Kühlofen, Lagerraum)*; 4. <Papier> stacking
Stapelpalette f 1. <Lufttrans> pallet; 2. <Trans, Verpack> stacking pallet
Stapelsäule f <Ker & Glas> stack plume
Stapelspeicher m 1. <Comp & DV> nesting store, stack; 2. <Verpack> nesting magazine
Stapelspeicherüberlauf m <Comp & DV> stack overflow
Stapelstütze f <Wassertrans> shore
Stapelung f **von Anforderungen** <Comp & DV> request batching
Stapelverarbeitung f <Comp & DV> batch processing
Stapelverarbeitungsmodus m <Comp & DV> batch mode
Staphisagrin n <Chemie> delphinine
Stapler m 1. <Fertig> piler; 2. <Ker & Glas> stacker
Staplerarm m <Ker & Glas> stacker arm
stark adj <Optik> strong *(Linse)*
stark dotiert adj <Elektronik> heavily-doped
stark fehlerbehaftete Sekunde f <Telekom> SES, severely errored second
starke Betonmischung f <Bau> harsh mix concrete
starke Injektion f <Elektronik> high-level injection *(Halbleiter)*
starke Kraft f <Teilphys> strong force, strong nuclear force
starke Strömung f <Wassertrans> race *(Meer)*
starke Wechselwirkung f <Kerntech, Teilphys> strong interaction
Stärke f 1. <Akustik> intensity *(I)*; 2. <Druck> weight of type; 3. <Fertig> amylum; 4. <Lebensmittel> starch; 5. <Maschinen> power; 6. <Metall> strength; 7. <Papier> staple, starch; 8. <Phys> strength; 9. <Textil> starch
Stärke f **des ausfädelnden Verkehrs** <Trans> diverging volume
Stark-Effekt m <Phys> Stark effect
Stärkegranulose f <Textil> amylopectin
Stärkegummi n <Chemie, Lebensmittel> dextrin
stärkehaltig adj 1. <Lebensmittel> starchy; 2. <Papier> amylaceous
starker Seegang m <Wassertrans> heavy seas *(Seezustand)*
starker Verkehr m <Telekom> heavy traffic *(Telefonnetz)*
starker Weichmacher m <Bau> superplasticizer *(Beton)*
starkes Licht n/**äußerst** <Wellphys> intense light
starkes Negativbild n <Elektronik> strong inversion
Starkstrom... <Elektriz, Elektrotech> heavy current, power
Starkstromgleichrichter m <Elektriz> power rectifier
Starkstrominduktor m <Elektrotech> power inductor
Starkstromkabel n <Elektrotech> power cable
Starkstromleitung f <Elektrotech> power transmission line; overhead power line *(Überland)*
Starkstromleitungsarmaturen fpl <Elektrotech> overhead power line fittings *(Überland)*
Starkstromleitungsnachrichtenübertragung f <Telekom> power line communication *(PLC)*
Starkstromnetz n <Elektrotech> heavy current system, high-tension power supply, power mains, power system, power transmission network
starr adj <Maschinen> rigid
Starrachse f <Kfztech> rigid axle
starre Aussteifung f <Bau> rigid diaphragm

starre Koaxialleitung f <Elektrotech> rigid coaxial line
starre Kupplung f <Heiz & Kälte, Maschinen> rigid coupling
starre Motoraufhängung f <Lufttrans> pylon
starre Motorhalterung f <Lufttrans> pylon
starrer Flügel m <Lufttrans> fixed wing
starrer Fuß m <Foto> rigid leg *(Stativ)*
starrer Körper m <Mechan> rigid body
starrer Reflektor m <Raumfahrt> rigid reflector
starrer Rotor m <Lufttrans> rigid rotor *(Hubschrauber)*
starres freitragendes Fahrzeug n <Trans> self-supporting rigid vehicle
starres Getriebe n <Elektrotech> solid gear
Starrflügelluftfahrzeug n <Lufttrans> fixed-wing aircraft
Starrflügler m <Lufttrans> fixed-wing aircraft
Start m 1. <Comp & DV> start; 2. <Kontroll> start-up; 3. <Lufttrans> takeoff; 4. <Raumfahrt> blast, launch • **für den Start vorbereiten** <Hydraul> prime *(Dampfkessel)* • **vor dem Start** <Raumfahrt> preflight • **zum Start freigegeben** <Lufttrans> cleared for takeoff
Start m **mit Gas** <Lufttrans> power takeoff
Start m **mit Hilfsrakete** <Raumfahrt> *(AE)* JATO, *(BE)* RATO, *(AE)* jet-assisted takeoff, *(BE)* rocket-assisted takeoff
Startabbruch m <Lufttrans> aborted takeoff
Startabbruchstrecke f <Lufttrans> accelerate-stop distance
Startautomatik f <Kfztech> automatic choke
Startazimuth m <Raumfahrt> launch azimuth
Startbahn f <Lufttrans> takeoff flight path
Startbahnmarkierung f <Lufttrans> *(AE)* hold-short line, *(BE)* lead-out line *(Vorfeld)*
Startband n <Akustik> leader
Startbereich m <Lufttrans> takeoff area
Startbit n 1. <Comp & DV> start bit, start element; 2. <Telekom> start bit
starten v 1. <Bau> launch; 2. <Comp & DV> start, trigger; 3. <Kontroll> start up, start; 4. <Raumfahrt> launch
Starten n 1. <Comp & DV> triggering; 2. <Maschinen> starting
Starter m <Kfztech, Raumfahrt> starter
Starterbatterie f <Kfztech> starter battery
Starterfolgswahrscheinlichkeit f <Raumfahrt> launch success probability
Starterklappe f <Kfztech> choker plate; choke *(Vergaser)*
Starterknopf m <Kfztech> starter button *(Motor)*
Starterritzel n <Kfztech> starter motor pinion *(Motor)*
Starterzahnkranz m <Kfztech> flywheel starter ring gear *(Motor)*
Startfähigkeit f <Lufttrans> takeoff ability
Startfenster n <Raumfahrt> launch window
Startflugbahn f <Lufttrans> takeoff flight path
Startfunke m <Phys> streamer
Startfunkenkammer f <Phys> streamer chamber
Startgeschwindigkeit f <Lufttrans> liftoff speed, takeoff speed
Startgewicht n <Raumfahrt> liftoff weight
Starthilfekabel n 1. <Bau> jumper cable; 2. <Kfztech> jump lead *(Elektrik)*
Startinstruktion m <Comp & DV> initial instruction
Startkapazität f <Lufttrans> takeoff ability
Startkonfiguration f <Raumfahrt> launching configuration
Startkontrollsystem n <Lufttrans> takeoff monitoring system
Startladedruck m <Lufttrans> manifold pressure
Startlufttrichter m <Lufttrans> takeoff funnel
Startmasse f <Lufttrans> gross weight *(Flugzeug)*
Startnennleistung f <Lufttrans> takeoff power rating

Startpaket

Startpaket n <Comp & DV> starter pack
Startphase f <Lufttrans> takeoff phase
Startprogramm n 1. <Comp & DV> initial program header; 2. <Kontroll> start routine
Startrampe f <Raumfahrt> launching pad
Startrettungssystem n <Raumfahrt> launch escape system
Startrollstrecke f <Lufttrans> takeoff run
Startsignal n <Fernseh> start mark
Start-Stopp-Betrieb m <Kontroll> start-stop operation
Startturm m <Raumfahrt> launching tower
Startumgebung f <Raumfahrt> launch environment
Start- und Landebahn f <Lufttrans> runway
Start- und Landebahnbezeichnung f <Lufttrans> runway designator
Start- und Landebahnendbefeuerung f <Lufttrans> runway end light
Start- und Landebahnkreuzungsfeuer npl <Lufttrans> runway crossing lights
Start- und Landebahnmittellinie f <Lufttrans> (AE) runway centerline, (BE) runway centreline
Start- und Landebahnmittellinienbefeuerung f <Lufttrans> (AE) runway centerline light, (BE) runway centreline light
Start- und Landebahnmittellinienmarke f <Lufttrans> (AE) runway centerline marking, (BE) runway centreline marking
Start- und Landebahnneigung f <Lufttrans> runway gradient
Start- und Landebahnnummer f <Lufttrans> runway number
Start- und Landebahnränder mpl <Lufttrans> runway shoulders
Start- und Landebahnschwelle f <Lufttrans> runway threshold
Start- und Landebahnschwellenmarkierung f <Lufttrans> runway threshold marking
Start- und Landebahnstreifen mpl <Lufttrans> runway strips (Flughafen)
Start- und Landebereich m <Lufttrans> landing area (Flugplatz)
Start- und Landestreifen m <Lufttrans> strip
Start- und Zielort-Gleichung f <Trans> O-D equation, origin-destination equation
Start- und Zielortübersicht f <Trans> O-D survey, origin-destination survey
Startunfall m <Kerntech> start-up accident
Startvermögen n 1. <Comp & DV> log in, logon; 2. <Kfztech> pick-up
Startvorbereitung f <Hydraul> priming
Startwinkel m <Raumfahrt> launch azimuth
Startzeit f <Comp & DV> start time; acceleration time (Datenverarbeitung)
Statik f 1. <Bau> structural analysis; 2. <Elektrotech, Phys> static; 3. <Maschinen> statics
Statikbetrieb m <Elektrotech> static operation
Statiktransformator m <Elektrotech> static transformer
Station f <Eisenbahn, Elektrotech, Wassertrans> station (Schiff)
stationär adj <Fertig> fixed, immobile, stationary
stationäre Lichtwellen fpl <Wellphys> stationary light waves
stationäre Longitudinalwellen fpl <Wellphys> stationary longitudinal waves
stationäre Modenleistungsverteilung f <Telekom> steady state condition
stationäre Modenverteilung f <Telekom> equilibrium mode distribution, steady state condition
stationäre Phase f <Telekom> stationary phase
stationäre Schutzvorrichtung f <Sicherheit> built-in guard
stationäre Spule f <Elektriz> fixed coil
stationäre Strömung f 1. <Lufttrans> steady flow (Aerodynamik); 2. <Strömphys> steady flow
stationäre Transversalwellen fpl <Wellphys> stationary transversal waves
stationäre Umlaufbahn f <Raumfahrt> stationary orbit
stationäre Welle f <Akustik, Aufnahme, Elektrotech, Phys, Telekom, Wellphys> standing wave, stationary wave (veraltet)
stationäre Wellen fpl in der Luft <Wellphys> stationary aerial waves
stationärer Anker m <Elektriz> fixed armature
stationärer Betrieb m <Comp & DV> local mode
stationärer Flug m <Lufttrans> steady flight
stationärer Lärm m <Sicherheit> steady-state noise
stationärer Punkt m <Raumfahrt> stationary point
stationärer Rost m <Fertig> stationary grate
stationärer Satellitendienst m <Raumfahrt> fixed satellite service
stationärer Zustand m 1. <Elektrotech> steady state condition; 2. <Phys, Strahlphys> stationary state; 3. <Telekom> steady state, steady state condition; 4. <Thermod> stationary state
stationäres Aufladegerät n <Trans> stationary charger
stationäres Feld n <Elektrotech> stationary field
stationäres Kriechen n <Metall> steady state creep
stationäres Wellenmuster n <Wellphys> stationary wave pattern
Stationaritätsbedingung f <Telekom> steady state condition
Stationsaufforderung f 1. <Comp & DV> query; 2. <Telekom> enquiry (Datenübertragungssteuerung)
Stationsaufforderungszeichen n <Comp & DV, Telekom> enquiry character
Stationsnetz n <Elektrotech> network of stations
statisch adj <Comp & DV, Elektrotech> static
statisch und dynamisch auswuchten v <Heiz & Kälte> balance statically and dynamically
statische Auflagung f <Elektrotech> static charge
statische Beanspruchung f <Maschinen> static strain
statische Bedingungen fpl <Elektrotech> static conditions
statische Elektrizität f <Bau, Textil> static electricity
statische Fokussierung f <Fernseh> static focus
statische Förderhöhe f <Wasserversorg> static discharge head
statische Höhe f <Hydraul> static head (Hydraulik)
statische Kennlinie f <Elektrotech> static characteristic
statische Prüfung f <Werkprüf> static test
statische Reibung f <Maschinen, Phys> static friction
statische Saughöhe f <Wasserversorg> stem
statische Stabilität f <Lufttrans> static stability (Lufttüchtigkeit)
statische Störung f <Aufnahme> static
statische Zuordnung f <Comp & DV> static allocation
statischer Auftrieb m <Wasserversorg> static head, static lift
statischer Druck m <Akustik, Heiz & Kälte> static pressure
statischer Eintrag m <Comp & DV> static input
statischer Elutionstest m <Abfall> static leaching test
statischer Konverter m <Raumfahrt> static inverter
statischer RAM m (SRAM) <Comp & DV> static RAM (SRAM)
statischer Schub m <Lufttrans> static thrust
statischer Speicher m 1. <Comp & DV> static memory, static storage; 2. <Elektrotech> static memory

statischer Speicherauszug *m* <Comp & DV> static dump
statischer Umformer *m* <Elektriz> static converter
statischer Wechselrichter *m* <Elektrotech> static inverter
statisches Auswuchten *n* <Maschinen> static balancing
statisches Elektrizitätsfeld *n* <Elektrotech> static electric field
statisches Feld *n* <Elektrotech, Telekom> static field
statisches Gleichgewicht *n* <Maschinen> static balance
statisches Moment *n* 1. <Maschinen> static moment; 2. <Phys> moment
statisches Relais *n* <Telekom> static relay
statisches Schweben *n* <Wassertrans> static hovering
statisches Unterprogramm *n* <Comp & DV> static subroutine
Statistik *f* <Comp & DV, Math> statistics
Statistikdaten *npl* <Comp & DV> statistical data
statistische Analyse *f* <Comp & DV, Math> statistical analysis
statistische Daten *npl* <Comp & DV> statistical data
statistische Kontrolle *f* <Qual> statistical control
statistische Physik *f* <Phys> statistical physics
statistische Prozesskontrolle *f* <Qual> statistical process control
statistische Prüfung *f* 1. <Metrol> statistical check; 2. <Werkprüf> statistical test
statistische Qualitätslenkung *f* <Qual> statistical quality control
statistische Qualitätsprüfung *f* <Qual> statistical quality inspection
statistische Sicherheit *f* <Comp & DV> confidence, confidence level
statistische Streuung *f* <Kerntech> random scattering
statistische Verteilung *f* <Qual> statistical distribution
statistischer Fehler *m* <Comp & DV, Gerät> random error
statistischer Gesamtfehler *m* <Comp & DV> mean square error
statistischer Multiplexer *m* <Comp & DV, Telekom> statistical multiplexer, statmux
statistisches Gewicht *n* (*g*) <Phys> statistical weight (*g*)
statistisches Rauschen *n* <Elektronik> random noise
Stativ *n* 1. <Fertig, Foto> tripod; 2. <Labor> retort stand, stand • **mit Stativ arbeiten** <Foto> work with a tripod
Stativ *n* **mit Panoramakopf** <Foto> pan head tripod
Stativanschluss *m* <Foto> tripod bush
Stativbein *n* <Foto> tripod leg
Stativgewinde *n* <Foto> tripod bush
Stativkompass *m* <Bau> surveyor's compass
Stativkopf *m* <Foto> tripod head
Stativscheinwerfer *m* <Foto> spotlight
Stativverlängerung *f* <Foto> tripod extension
Stator *m* 1. <Elektriz, Elektrotech> stator; frame (*Elektromotor*); 2. <Kfztech> stator (*Generator*); 3. <Phys> stator
Statorgehäuse *n* <Elektrotech> stator frame
Stator-Rotor-Anlassermotor *m* <Elektriz> stator rotor starter motor
Statorspule *f* <Elektrotech> stator coil
Status *m* 1. <Comp & DV> state, status; 2. <Kontroll, Qual> status
Statusbit *n* <Comp & DV, Telekom> status bit
Statusdiagramm *n* <Comp & DV> state diagram
Statusregister *n* <Comp & DV> status register
Statuswort *n* <Comp & DV> status word
Statuszeichen *n* <Comp & DV> status character
Stau *m* 1. <Erdöl> backup (*Bohrtechnik*); 2. <Trans> (*AE*) road jam, (*BE*) tailback, (*BE*) traffic jam; 3. <Wasserversorg> backwater, damming-up

Stauanlage *f* <Wasserversorg> barrage
Staub *m* <Kohlen, Sicherheit, Umweltschmutz> dust
Staub *m* **in der Luft** <Sicherheit> airborne dust
Staubabdeckung *f* <Maschinen> dust cover
Staubabgabevorrichtung *f* <Verpack> dust-exhausting device
Staubabsauganlage *f* <Sicherheit> dust exhaust system
Staubabsaugung *f* <Sicherheit> dust extraction
Staubabsaugungsventilator *m* <Kohlen> dust exhaust fan
Staubabscheider *m* 1. <Abfall> dust separator; 2. <Kohlen> dust chamber; 3. <Kunststoff> dust collector
Staubabscheidung *f* <Abfall> particulate collection
Staubabzug *m* <Sicherheit> dust exhaust fan
Staubabzugshaube *f* <Sicherheit> dust hood
Staubabzugssystem *n* <Sicherheit> dust exhaust system
Staubalken *m* <Bau> stop log
Staubanteil *m* <Umweltschmutz> dust content
Staubaustritt *m* <Verpack> dust-exhausting device
Staubbekämpfung *f* <Chemtech> dust control
Staubbindeanlage *f* <Bau> dust suppression system
staubdicht *adj* 1. <Elektriz> dustproof; 2. <Fertig> dust-tight; 3. <Sicherheit> dustproof; 4. <Telekom> dust-proof (*Gerät*); 5. <Verpack> dust-tight, dustproof
staubdichte Brille *f* <Sicherheit> dust-tight goggles
staubdichte Konstruktion *f* <Sicherheit> dust-proof construction, dust-tight construction
Staubecken *n* <Nichtfoss Energ, Wasserversorg> catch basin, catchment basin, reservoir
Stauben *n* **von gestrichenem Papier** <Papier> powdering of coated paper
Stauberfassungseinrichtung *f* <Sicherheit> dust-collecting device
Staubexplosion *f* <Kohlen> dust explosion
Staubfangbeutel *m* <Sicherheit> dust bag (*Staubsauger, Lüfter*)
Staubfänger *m* <Kohlen> dust catcher
Staubfilter *n* 1. <Kohlen> dust collector, dust filter; 2. <Umweltschmutz> fabric filter
Staubflocke *f* <Papier> fluff
staubfreie Müllabfuhr *f* <Abfall> controlled emptying, enclosed emptying
staubfreier Raum *m* <Heiz & Kälte, Sicherheit> clean room
staubgeladene Atmosphäre *f* <Kohlen> dust-laden atmosphere
staubgeladene Umgebung *f* <Kohlen> dust-laden atmosphere
staubgeschützt *adj* <Maschinen> dustproof
staubgeschützter Motor *m* <Elektriz> dustproof motor
staubhaltig *adj* <Fertig> dust-laden (*Luft*)
staubige Luft *f* <Sicherheit> dust-laden atmosphere
Staubkappe *f* 1. <Kfztech> dust seal; 2. <Maschinen> dust cap
Staubkapsel *f* <Sicherheit> dust-collecting device
Staubkern *m* <Elektrotech> powdered iron core
Staubkohle *f* <Kohlen> dust coal, fine coal
Staubkonzentration *f* **in der Luft** <Sicherheit> airborne dust concentration
Staublech *n* <Chemtech> baffle plate
Staubmaske *f* <Fertig> dust respirator
Staubmehl *n* <Lebensmittel> mill dust
Staubmühle *f* <Chemtech> pulverizer
Staubohle *f* <Wasserversorg> flash board (*Wehr, Schleusentor*)
Staubpinsel *m* <Foto> dusting brush
Staubprobenahmegerät *n* <Sicherheit, Umweltschmutz> dust-sample collection device
Staubsauggebläse *n* <Verpack> dust aspirator

Staubschutz

Staubschutz m <Kohlen, Verpack> dust guard
Staubschutzabdeckung f <Sicherheit> dust guard
Staubschutzbalg m <Kfztech> (BE) dust boot, (AE) dust trunk (Gangschalthebel)
Staubschutzbrille f <Sicherheit> dust-tight goggles, dust-tight protective goggles
Staubschutzhaube f <Sicherheit> dust cover
Staubschutzhelm m <Sicherheit> dust protection helmet
Staubschutzmaske f <Sicherheit> dust mask
Staubschutzring m <Sicherheit> dust washer
Staubteilchen n <Umweltschmutz> dust particle
Staubtrenner m <Sicherheit> dust separator
Staub- und Sprühschutzhaube f <Sicherheit> dust and spray protective hood
Stauchautomat m <Fertig> automatic header
Stauchdraht m <Fertig> heading wire
Stauchdruckprüfung f <Werkprüf> crushing test
stauchen v 1. <Fertig> forge, gather; upset (Umformung); edge (Walzen); 2. <Papier> crush (Festigkeitsprüfung)
Stauchen n <Fertig> forging, metal gathering, shrinking
Stauchfalzmaschine f <Druck> buckle folder machine
Stauchkaliber n <Fertig> edging pass (Walzen)
Stauchkopf m <Maschinen> battered head (eines Niets)
Stauchmatrize f <Fertig> forging machine die
Stauchpresse f <Maschinen> upsetting press
Stauchschlitten m <Fertig> moving platen, (AE) traveling platen, (BE) travelling platen (Druckschweißen)
Stauchstempel m 1. <Fertig> header; 2. <Maschinen> header die, heading die
Stauchung f 1. <Fertig> linear compression, upset; 2. <Kunststoff> compression; 3. <Mechan> compressive strain; 4. <Wassertrans> strain
Stauchung f des Fahrwerkfederbeins <Lufttrans> landing-gear shock strut compression
Stauchversuch m 1. <Fertig> crushing test (für Rohre); 2. <Mechan> compression test
Stauchwiderstand m <Papier, Qual> crushing resistance
Staudamm m 1. <Nichtfoss Energ> impoundment; 2. <Wasserversorg> barrage, dam, retaining dam, river dam
Staudruck m 1. <Fertig> dynamic head; 2. <Ker & Glas, Kunststoff> back pressure; 3. <Lufttrans> dynamic pressure, impact pressure; 4. <Strömphys> stagnation pressure
Staudruckluft f <Heiz & Kälte> ram air
Staudruckmesser m <Regelung> Pitot tube
stauen v 1. <Bau> retain (Wasser); 2. <Fertig> baffle; 3. <Wasserversorg> dam
Stauen n <Wasserversorg> catching
Stauer m <Wassertrans> stevedore (Hafen)
Staufferbuchse f <Maschinen> grease cup
Staufferfett n <Fertig> grease
Staufläche f <Umweltschmutz> surface area
Stauhöhe f <Bau> head of water
Stauholz n <Wassertrans> dunnage
Stauinformationen fpl <Trans> backup service
Staukörper m <Bau> gate
Staumauer f 1. <Bau> retaining wall; 2. <Wasserversorg> retaining structure
Staumauer f aus übergroßen Schottersteinen <Bau> tailings dam
Staumenge f <Wasserversorg> catchment
Staupunkt m 1. <Erdöl> flounder point; 2. <Lufttrans> stagnation point (Aerodynamik); 3. <Strömphys> stagnation point
Stauscheibe f <Fertig> baffle plate, baffle sheet
Stauscheibendurchflussmesser m <Gerät> (BE) baffle disc flow meter, (AE) baffle disk flow meter
Stauscheiben-Durchflussmessumformer m <Regelung> target flow transducer
Stauschwelle f <Wasserversorg> drowned weir
Stausee m 1. <Nichtfoss Energ> impoundment; 2. <Wasserversorg> artificial lake, impounding reservoir
Staustrahltriebwerk n 1. <Lufttrans> aero-thermodynamic duct, ramjet; 2. <Thermod, Trans> ramjet engine
Stauung f 1. <Kohlen> congestion; 2. <Nichtfoss Energ> (AE) dike, (BE) dyke; 3. <Wassertrans> hold; stowage (Ladung)
Stauwarngerät n <Trans> queue warning sign
Stauwarnsystem n <Trans> congestion-warning system
Stauwarnzeichen n <Trans> queue warning sign
Stauwasser n 1. <Bau> perched water; 2. <Nichtfoss Energ> tail water; 3. <Wassertrans> slack water, stand of tide (Gezeiten); 4. <Wasserversorg> perched ground water; backwater (Kanal)
Stauwehr n 1. <Bau> barrage, movable dam; 2. <Wasserversorg> retaining dam
Stauwiderstand m <Trans> ram drag
Steamkracken n <Erdöl> steam cracking (Raffinerie)
Stearin n 1. <Chemie> glyceryl tristeate, stearin, tristearin; 2. <Lebensmittel> stearin
Stearyl... <Chemie> stearyl
Stechbeitel m <Mechan> chisel
Stechdrehmeißel m <Maschinen> parting tool
Stecheisen n <Mechan> broaching
Stechfase f <Bau> chamfer stop (Tischlerarbeiten)
Stechmeißel m 1. <Fertig> necking tool (Spanung); 2. <Maschinen> cutoff tool
Stechwerkzeug n <Fertig> louvring die (Blech)
Stechzirkel m 1. <Fertig> divider; 2. <Maschinen> pair of dividers; 3. <Math, Metrol, Wassertrans> dividers (Navigation)
Steckanschluss m <Elektriz> plug-in termination
steckbar adj <Elektrotech, Telekom> pluggable
steckbare Verbindungsleitung f <Comp & DV> patch cord
Steckbaugruppe f <Telekom> plug-in module
Steckblende f <Mechan> push plug
Steckbrücke f <Elektriz> jumper
Steckbuchse f 1. <Bau> lock bush; 2. <Elektriz> plug receptacle; 3. <Funktech> receptacle; 4. <Kfztech> plug socket (Elektrik)
Steckdorn m <Bau> (BE) pin spanner, pin wrench
Steckdose f 1. <Comp & DV> outlet; 2. <Elektriz, Elektrotech> connector socket, mains socket, plug box, plug connector, plug receptacle, socket; 3. <Fertig> receptacle; 4. <Kfztech> plug socket (Elektrik); 5. <Labor> electric socket; 6. <Mechan> socket; 7. <Telekom> outlet, socket outlet
Steckdosenleiste f <Elektrotech> socket board
Steckeinheit f <Telekom> plug-in unit
Steckelwalzwerk n <Fertig> Steckel mill
stecken bleiben v <Nichtfoss Energ> stall
Stecker m 1. <Comp & DV> jack; 2. <Elektriz, Elektrotech> connector, coupler, male connector, plug; 3. <Fernseh> plug; 4. <Fertig> connector; 5. <Funktech> plug; 6. <Telekom> connector, male plug, plug • **Stecker herausziehen** <Elektriz, Funktech, Telekom> unplug
Steckeradapter m <Elektrotech> plug adaptor
Steckeranschluss m <Elektrotech> plug connection
Steckerbuchse f <Elektriz> female connector
Steckerdraht m <Elektrotech> plug wire
Steckerfeld n <Fertig> pin board
Steckerkabel n <Elektrotech> patch cord
steckerkompatibel adj <Elektrotech> plug-compatible
Steckerkompatibilität f <Elektrotech, Telekom> plug compatibility
steckerkompatibler Hersteller m <Elektronik> plug-compatible manufacturer

Steckerleitung f <Elektrotech> patch cord
Steckermaß n <Metrol> (AE) plug gage, (BE) plug gauge
Steckerschalter m <Elektrotech> plug switch
Steckerschnur f <Elektrotech> patch cord
Steckerstift m 1. <Elektrotech> contact pin, pin; 2. <Fertig> ferrule (Faseroptik)
Steckertyp m **für Koaxkabel** <Funktech> connector type for coaxial cables
Steckfeld n <Fernseh> patch panel
Steckfilterfassung f <Foto> sliding filter drawer
Steckglied n <Elektriz> jumper
Steckkarte f <Telekom> plug-in board, board • **auf Steckkarte montiert** <Telekom> board-mounted
Steckkartenplatz m <Comp & DV> card slot
steckkraftloses Bauelement n <Elektronik> zero insertion force connector
Steckmodul n <Telekom> plug-in module
Steckmuffe f <Fertig> push-fit fitting (Kunststoffinstallationen)
Steckschlüssel m 1. <Fertig> (BE) socket spanner, socket wrench; 2. <Maschinen> (BE) box spanner, box wrench, (BE) socket spanner, socket wrench; 3. <Mechan> (BE) box spanner, box wrench
Steckschlüsselsatz m **im Kasten** <Kfztech> (BE) case of box spanners, case of box wrenches (Zubehör)
Steckschlüsselverriegelung f <Sicherheit> captive key interlocking
Steckschuh m <Foto> accessory shoe
Steckschwert n <Wassertrans> daggerboard, drop keel (Schiff)
Stecksicherung f <Fertig> retaining pin (Kunststoffinstallationen)
Stecksockelrelais n <Telekom> plug-in relay
Steckstift m <Bau> peg
Stecktafel f 1. <Elektrotech> plugboard; 2. <Fernseh> patch board
Steckverbinder m 1. <Elektriz, Elektrotech> connector, plug; 2. <Telekom> plug-type connector; connector (elektrisch)
Steckverbinder m **aus einem Stück** <Elektrotech> one-piece connector
Steckverbinder m **für Lichtleitfasern** <Optik> (AE) optical fiber connector, (BE) optical fibre connector
Steckverbinder m **für Lichtleitkabel** <Optik> (AE) optical fiber connector, (BE) optical fibre connector
Steckverbinder m **ohne spannungsführende Teile auf der Vorderseite** <Elektrotech> dead-front connector
Steckverbinderabschirmung f <Elektrotech> connector shield
Steckverbinderleitungen fpl <Telekom> interface circuit, interface lines
Steckverbindung f 1. <Elektriz, Elektrotech> plug-type connection; coupler (Stecker und Buchse); plug connection; pressurized connection (gasgefüllte Leitung); 2. <Mechan> socket joint; 3. <Telekom> plug and socket • **Steckverbindung unterbrechen** <Elektriz> unplug
Steckverschluss m <Ker & Glas> socket cap
Steckvorrichtung f <Elektrotech> coupler
Stefan-Boltzmann'sche Konstante f (σ) <Phys, Thermod> Stefan-Boltzmann constant (σ)
Stefan-Boltzmann'sches Gesetz n 1. <Phys> Stefan-Boltzmann law; 2. <Thermod> fourth power law
Stefan-Boltzmann'sches Strahlungsgesetz n 1. <Fernseh> fourth power law; 2. <Strahlphys> Stefan's law
Steg m 1. <Aufnahme> land; 2. <Bau> web; stem (Träger); 3. <Kerntech> catwalk; 4. <Konstzeich> ligament; 5. <Kunststoff> land; 6. <Maschinen> bridge, stud; 7. <Mechan> web (Balken)

Stegblech n <Wassertrans> tripping bracket (Schiffbau)
Stege mpl <Druck> furniture
Stegglied n <Bau> stud link
Steghohlleiter m <Bau, Funktech> ridge waveguide
Stegkette f 1. <Bau> stud link cable chain, studded link cable chain; 2. <Maschinen> stud chain, studded chain
Stegplatten fpl <Kunststoff> cellular sheet
Stehbildkamera f <Foto> still camera
Stehbolzen m 1. <Bau, Eisenbahn> stay bolt; 2. <Fertig> dowel pin; 3. <Kfztech> stud; 4. <Maschinen> stay bolt
Stehbolzengewindebohrer m 1. <Fertig> boiler tap, stay bolt tap; 2. <Maschinen> boiler stay screwing tap
stehen gebliebene Mantelfläche f <Fertig> land surface
stehen lassen v <Metall> allow to stand
Stehenbleiben n <Maschinen> stalling
stehend adj <Bau> upright
stehende Bake f <Wassertrans> fixed beacon (Seezeichen)
stehende Peilung f <Wassertrans> steady bearing (Navigation)
stehende Welle f <Akustik, Aufnahme, Elektrotech, Phys, Telekom, Wellphys> standing wave, stationary wave (veraltet)
stehender Motor m <Kfztech, Maschinen> vertical engine
stehendes Gut n <Wassertrans> standing rigging (Tauwerk)
stehendes Tauwerk n <Wassertrans> standing rigging (Tauwerk)
Stehkolben m <Labor> flat-bottomed flask
Stehlager n <Maschinen> pedestal bearing, pillow, pillow block, plummer block, plummer block bearing
Stehleiter f <Sicherheit> standing step ladder
Stehsatz m <Druck> alive matter, live matter, standing matter, standing type
Stehsetzstock m <Maschinen> steadyrest follower
Stehwelle f <Akustik, Aufnahme, Elektrotech, Phys, Telekom, Wellphys> standing wave, stationary wave (veraltet)
Stehwellenverhältnis n (SWV) 1. <Funktech> standing-wave ratio (SWR); 2. <Phys> standing-wave ratio (SWR); voltage standing wave ratio (VSWR); 3. <Telekom> standing-wave ratio (SWR); voltage standing wave ratio (VSWR)
steif adj <Textil> stiff
Steifappretur f <Textil> stiff finish
Steife f 1. <Fertig> prop; 2. <Textil> stiffness
steife Einlage f <Textil> stiffener
steife Konstruktion f <Bau> rigid construction
steifer Karton m <Verpack> rigid box
steifer Rahmen m <Bau> rigid frame
steifer Wind m <Wassertrans> moderate gale
Steifezahl m <Kohlen> compressibility coefficient
Steifheit f 1. <Akustik> stiffness; 2. <Maschinen> rigidity, rigidness, stiffness; 3. <Phys> stiffness (Feder)
steifholen v <Wassertrans> haul taut (Tauwerk)
Steifigkeit f 1. <Maschinen> rigidity, rigidness, stiffness; 2. <Papier> stiffness; 3. <Phys> rigidity; 4. <Textil> stiffness
Steifleinen n <Textil> interlining canvas
Steifmittel n <Textil> stiffening agent
steifsetzen v <Wassertrans> set taut (Tauwerk)
Steif- und Faltkartons mpl <Verpack> rigid and folding cartons
Steigaufrichter m <Verpack> tray erector
Steigeisen n <Bau> spur, step iron
steigen v 1. <Raumfahrt> climb (Raumschiff); 2. <Wassertrans> rise (Wasser)
Steigenabdichtung f <Verpack> tray sealer
steigende Bemaßung f <Konstzeich> rising dimensioning sequence

steigender

steigender Bogen *m* <Bau> rampant arch, rising arch
steigender Guss *m* <Fertig> uphill casting
Steigendgießen *n* <Fertig> bottom casting
Steiger *m* 1. <Erdöl> riser; 2. <Fertig> rising gate *(Gießen)*
steigern *v* <Elektrotech> boost
Steigerohr *n* <Elektriz> rising main
Steigersystem *n* <Fertig> risering *(Gießen)*
Steigfilmverdampfer *m* <Lebensmittel> climbing film evaporator
Steigflug *m* <Lufttrans> climb, climb cruise
Steigflug *m* mit Reisefluggeschwindigkeit <Lufttrans> cruise climb
Steigflugbeginn *m* <Lufttrans> climb out
Steigfluggeschwindigkeit *f* <Lufttrans> climb speed
Steigflugkorridor *m* <Lufttrans> climb corridor
Steigflugkurve *f* <Lufttrans> climb turn
Steigflugphase *f* <Lufttrans> climb phase
Steiggeschwindigkeit *f* <Raumfahrt> climb rate *(Raumschiff)*
Steiggradient *m* <Lufttrans> climb gradient
Steigleistung *f* <Lufttrans> climb performance
Steigleitung *f* 1. <Bau> riser, rising main; 2. <Elektriz> > riser; 3. <Erdöl> riser; marine riser *(Offshore)*; 4. <Fertig> ascending main, riser
Steignahtschweißen *n* <Mechan> uphand welding
Steigrate *f* <Lufttrans> rate of climb
Steigratenanzeige *f* <Lufttrans> indicator of the rate of climb
Steigrohr *n* 1. <Bau> riser pipe; 2. <Elektriz> rising main; 3. <Erdöl> tubing *(Bohrlochausrüstung)*; production tubing *(Fördertechnik)*
Steigstromvergaser *m* <Kfztech> *(AE)* updraft carburetor, *(BE)* updraught carburettor
Steigung *f* 1. <Bau> gradient, inclination, incline, pitch, rising gradient, slope, upgrade; 2. <Eisenbahn> gradient; 3. <Elektronik> lead *(Gewinde)*; 4. <Lufttrans> pitch; 5. <Maschinen> lead; 6. <Math> slope; 7. <Mechan> pitch *(Gewinde)*; 8. <Phys> slope; 9. <Wassertrans> pitch
Steigung *f* einer Geraden <Geom> gradient of a straight line
Steigungsänderung *f* des Luftschraubenblattes <Lufttrans> blade pitch variation *(Hubschrauber)*
Steigungsanzeiger *m* für die Luftschraube <Lufttrans> blade pitch indicator *(Hubschrauber)*
Steigungsdrehmelder *m* <Lufttrans> pitch synchro
Steigungseinstellung *f* <Lufttrans> climb setting; pitch setting *(Propeller)*
Steigungsfehler *m* 1. <Fertig> error from backlash, error of pitch; 2. <Qual> error of pitch
Steigungsgenauigkeit *f* <Fertig> pitch accuracy
Steigungsgeschwindigkeit *f* <Lufttrans> rate of climb
Steigungsgeschwindigkeitsanzeige *f* <Lufttrans> rate-of-climb indicator
Steigungshöhe *f* <Bau> rise *(Stufen)*
Steigungsinformation *f* <Lufttrans> pitch information
Steigungskompensierung *f* <Lufttrans> pitch compensation *(Hubschrauber)*
Steigungskorrekturvorrichtung *f* <Lufttrans> pitch correcting unit *(Flugkontrolle)*
Steigungsradius *m* <Lufttrans> pitch radius
Steigungsrate *f* <Lufttrans> rate of climb
Steigungssinn *m* <Fertig> hand of helix
Steigungssteuerung *f* <Lufttrans> pitch control *(Hubschrauber)*
Steigungssteuerungskompensator *m* des Luftschraubenblattes <Lufttrans> blade pitch control compensator *(Hubschrauber)*
Steigungssynchro *n* <Lufttrans> pitch synchro
Steigungsumkehr *f* <Lufttrans> pitch reversing *(Propeller)*

Steigungsverhältnis *n* <Lufttrans> advance diameter ratio *(Propeller)*
Steigungsverstellachse *f* <Lufttrans> pitch change axis *(Hubschrauber)*
Steigungsverstelldrehkreuz *n* <Lufttrans> pitch change spider *(Hubschrauber)*
Steigungsverstellholm *m* <Lufttrans> pitch change beam *(Hubschrauber)*
Steigungsverstellstange *f* <Lufttrans> pitch change rod *(Hubschrauber)*
Steigungswinkel *m* 1. <Geom> angle of elevation; 2. <Ker & Glas> angle of pitch; 3. <Lufttrans> climb angle; pitch angle *(Hubschrauber)*; 4. <Maschinen> helix angle, lead angle; 5. <Nichtfoss Energ> pitch angle
Steigungszunahme *f* <Lufttrans> pitch increase *(Hubschrauber)*
Steigzeit *f* <Elektronik> rise time
steil *adj* <Ker & Glas> steep
steil gelagerte Kohle *f* <Kohlen> edge coal
Steilanschnitt *m* <Ker & Glas> steep bevel
Steilböschung *f* <Bau> steep slope
Steildach *n* <Bau> steep roof
steile Kehrtkurve *f* <Lufttrans> steep turn
steile Straße *f* <Bau> steep road
steiler Abhang *m* <Bau> steep gradient
steiler Filter *n* <Elektronik> sharp cutting filter
steiles Einfallen *n* <Erdöl> high-angle dip *(Geologie)*
steiles Flöz *n* <Kohlen> edge seam
Steilförderer *m* <Trans> bag conveyor
steilgängig *adj* <Fertig> coarse-pitch *(Gewinde)*
Steilheit *f* 1. <Bau> steepness; 2. <Fernseh> roll-off factor *(Nyquist-Flanke)*; 3. <Phys> transconductance *(Elektronenröhre)*; 4. <Telekom> slope *(Kurve)*
Steilkantenvorbereitung *f* <Bau> square edge preparation *(Schweißen)*
Steilufer *n* <Wassertrans> bluff *(Geographie)*
Stein *m* <Bau, Kohlen> stone
Stein *m* mit runder Ecke <Ker & Glas> *(AE)* edger block, *(BE)* jamb block
Steinbearbeitung *f* <Bau> stoneworking
steinbesetzt *adj* <Mechan> *(AE)* jeweled, *(BE)* jewelled
Steinbettung *f* 1. <Bau> rock layer; 2. <Kohlen> stone bed; 3. <Wasserversorg> rock layer
Steinbohrung *f* <Kohlen> jumper boring
Steinbrecher *m* 1. <Bau> crusher, rock breaker; 2. <Kohlen> rock breaker
Steinbruch *m* <Bau> cut stone quarry, quarry, stone pit
Steindübel *m* <Bau, Kohlen> rock dowel
Steinfräse *f* <Bau> stone mill
Steinfrucht *f* <Lebensmittel> stone fruit
Steingut *n* <Ker & Glas> earthenware
Steingutdekorationsmaler *m* <Ker & Glas> earthenware decorator
Steingutglasur *f* <Ker & Glas> earthenware glazing
Steingutrohr *n* <Bau> earthenware pipe
steinig *adj* <Bau> stony
steiniger Untergrund *m* <Bau> stony ground
Steinkasten *m* <Bau> gabion
Steinkohle *f* <Kohlen> coal
Steinkohlenasche *f* <Kohlen> coal ash
Steinkohlenaufbereitung *f* <Kohlen> coal dressing
Steinkohlenflöz *n* <Kohlen> coal bed
Steinkohlengas *n* <Kohlen> coal gas
Steinkohlennaphtha *f* <Kohlen> coal naphtha
Steinkohlenschicht *f* <Kohlen> coal bed
Steinkohlenteer *m* <Kohlen> coal tar, gas tar
Steinkohlenteernaphtha *f* <Kohlen> coal-tar naphtha
Steinmeteorit *m* <Raumfahrt> aerolite *(Raumfahrt)*
Steinmetz *m* <Bau> stone dresser, stonemason

Steinmetz'scher Koeffizient m <Phys> Steinmetz coefficient
Steinmetz'sches Gesetz n <Phys> Steinmetz law *(Elektrotechnik)*
Steinobst n <Lebensmittel> stone fruit
Steinramme f <Bau> beetle
Steinsalz n <Chemie, Lebensmittel> rock salt
Steinschotter m <Trans> ballast *(Straße)*
Steinschraube f 1. <Bau> rag bolt; 2. <Maschinen> stone bolt
Steinschüttdamm m <Bau> rock-fill dam
Steinschüttung f 1. <Bau> rip-rap, rockfill, rubble; 2. <Kohlen, Wasserversorg> rockfill
Steinsetzer m <Bau> *(AE)* pavior, *(BE)* paviour
Steinsetzerhammer m <Bau> *(AE)* pavior's hammer, *(BE)* paviour's hammer
Steinsplitter m <Bau> spall
Steinwand f <Kohlen> stonewall
Steinzeug n <Ker & Glas> stoneware
Stek m <Wassertrans> knot *(Knoten)*
Stell… <Maschinen> adjusting
Stellantrieb m 1. <Comp & DV> servomechanism; 2. <Fertig> actuator unit *(Kunststoffinstallationen)*; 3. <Kontroll> actuator *(Positioniersystem)*
stellar adj <Raumfahrt> stellar
Stellaratorröhre f **in Achterform** <Kerntech> figure eight stellarator tube
Stellaufgabe f <Ergon> control task
Stellbereich m 1. <Fertig> control range, setting range *(Kunststoffinstallationen)*; 2. <Maschinen> regulating range
Stellbewegung f <Fertig> actuation, function *(Kunststoffinstallationen)*
Stellbogen m <Maschinen> quadrant
stellen v 1. <Kontroll> position; 2. <Papier> set
Stellen fpl **hinter dem Komma** <Comp & DV> fractional part
Stellenschreibimpuls m <Comp & DV> enable pulse
Stellenverschiebung f <Comp & DV> shift
Stellenwert m <Math> place value
Stellgetriebe n <Elektrotech> control gear
Stellglied n 1. <Elektrotech> actuator; 2. <Kontroll> servo positioner; actuator *(Positioniersystem)*; 3. <Maschinen> control element; 4. <Mechan> vane; 5. <Regelung> active element, final controlling element; 6. <Trans> actuator *(KFZ, Flugzeug)*
Stellgröße f <Maschinen> control variable
Stellgrößensprung m <Regelung> stepwise change of the manipulated variable
Stelling f <Wassertrans> gangway
Stellitventil n <Kfztech> stellite valve
Stellkeil m **für Schrauben** <Maschinen> screw key
Stellknopf m <Gerät> adjustment knob
Stellkraft f <Elektriz> positioning force
Stellleiste f <Maschinen> adjustable gib, gib
Stellmechanismus m 1. <Fertig> operating mechanism *(Kunststoffinstallationen)*; 2. <Gerät> adjusting mechanism
Stellmotor m 1. <Kontroll> servomotor; 2. <Mechan> actuator
Stellmutter f <Maschinen> adjusting nut, check nut, checking nut, regulating nut
Stellnocke f <Fertig> adjusting cam *(Kunststoffinstallationen)*
Stellorgan n <Fertig> regulating unit *(Kunststoffinstallationen)*
Stellort m <Regelung> control point, regulating point
Stellpotenziometer n <Gerät> adjustable potentiometer
Stellring m 1. <Gerät> adjustment ring; 2. <Maschinen> adjusting ring, set collar, setting ring; 3. <Mechan> collar

Stellschraube f 1. <Gerät> adjustable screw; 2. <Maschinen> adjusting screw, regulating screw, set bolt, set screw, temper screw
Stellstab m <Kerntech> absorber rod
Stellteil n <Ergon> control, control element
Stelltisch m <Eisenbahn> driving desk
Stellung f 1. <Ergon> posture; 2. <Kontroll> positioning
Stellungsanzeige f <Fertig> position indicator *(Kunststoffinstallationen)*
Stellungsbegrenzungsschalter m <Fertig> position limiter switch *(Kunststoffinstallationen)*
Stellungsmessgerät n <Gerät> position measuring instrument
Stellungsschalter m <Elektriz> position switch
Stellventil n <Heiz & Kälte> servo valve
Stellvertreterzeichen npl <Comp & DV> wildcard characters *(bei Angabe von Dateinamen)*
Stellwerk n <Eisenbahn> *(BE)* signal box, *(AE)* signal tower
Stellwerk n **des Rangierbahnhofs** <Eisenbahn> *(AE)* classification yard tower
Stellwerksfernsprecher m <Telekom> signal box telephone equipment
Stellwiderstand m 1. <Elektrotech> rheostat, variable resistor; 2. <Gerät> adjustable resistor
Stellwinkel m <Fertig> operating angle *(Kunststoffinstallationen)*
Stellwirkung f <Regelung> regulating action
Stellzeit f <Fertig> control time *(Kunststoffinstallationen)*
Stelzbogen m <Bau> stilted arch
Stelzenstraße f <Bau> causeway
Stemmeisen n 1. <Bau> chisel, mortise chisel; 2. <Mechan> chisel, crowbar
stemmen v <Bau> chisel
Stemmen n <Fertig> *(AE)* chiseling, *(BE)* chiselling
Stemmloch n <Bau> mortice, mortise
Stemmmaschine f 1. <Bau> mortising machine; 2. <Fertig> mortise machine; 3. <Maschinen> mortising machine
Stemmmeißel m <Bau> mortise chisel
Stemmnaht f 1. <Fertig> caulk weld; 2. <Wassertrans> caulked joint *(Schiffbau)*
Stempel m 1. <Bau> prop, punch; 2. <Kohlen> strut; 3. <Kunststoff> ram; 4. <Maschinen> die, punch, ram; 5. <Mechan> punch; 6. <Qual> stamp
Stempelätzpaste f <Ker & Glas> stamp etching paste
Stempelaufnahmeplatte f <Kohlen> punched plate
Stempelberechtigter m <Qual> stamp holder
Stempelberechtigung f <Qual> stamp authorization
Stempelfarbe f <Konstzeich> stamp pad ink
Stempelhalteplatte f <Maschinen> punch plate
Stempelhalter m <Maschinen> punch holder
Stempelüberwachung f <Qual> stamp control
Stengelglas n <Ker & Glas> stemware
Stengelkristall m <Metall> columnar crystal
Steppdecke f <Textil> comforter, continental quilt, quilt
steppen v <Textil> quilt
Step-Recovery-Diode f <Elektronik> step recovery diode
Ster n <Metrol> stere
Steradiant m <Elektronik, Phys> steradian *(Einheit)*
stereo adj <Aufnahme> stereo
Stereo n <Druck> stereotype plate
Stereoabstimmgerät n <Aufnahme> stereo tuner
Stereoaufnahme f <Aufnahme> stereo recording, stereophonic recording
Stereobandaufnahme f <Aufnahme> stereo tape recording
Stereobetrachter m <Foto> stereo viewer, stereoscope
Stereobild n <Aufnahme> stereoscopic image, three-dimensional image

Stereobildpaar

Stereobildpaar n <Foto> stereoscopic pair
Stereodecoder m <Aufnahme> stereo decoder
Stereoeffekt m <Aufnahme> stereo effect
Stereofernsehen n <Fernseh> stereovision
stereographische Projektion f <Bau, Metall> stereographic projection
Stereohilfsträger m <Aufnahme> stereo subcarrier
Stereoisomer n <Chemie> stereoisomer
Stereokamera f <Foto> stereoscopic camera
Stereokompaktanlage f <Aufnahme> three-in-one stereo component system
Stereokopfhörer m <Aufnahme> stereo headphone
Stereokopierrahmen m <Foto> transposing frame
Stereomer n <Chemie> stereoisomer
Stereomikroskop n <Labor> stereomicroscope
stereophone Tonwiedergabe f <Aufnahme> stereophonic reproduction
stereophone Übertragung f <Telekom> stereophonic transmission
Stereophonie f <Akustik, Aufnahme> stereophony
stereophonische Aufnahme f <Akustik> stereophonic recording
stereophonische Mikrorille f <Aufnahme> stereophonic microgroove
Stereoschallplatte f <Aufnahme> stereophonic record
Stereositzplatz m <Aufnahme> stereo seat
Stereoskopie f <Foto> stereoscopy
Stereoton m <Aufnahme> stereophonic sound
Stereotonabnehmer m <Akustik> stereophonic pick-up
Stereotonbandgerät n <Aufnahme> stereo tape recorder
Stereotrennung f <Aufnahme> stereo separation
Stereotypie f <Druck> stereotyping
Stereotypieplatte f <Druck> stereotype plate
Stereounterkanal m <Aufnahme> stereo subchannel
Stereovideorecorder m <Fernseh> stereo VCR
steril adj 1. <Kohlen> sterile; 2. <Lebensmittel> aseptic
sterilisiert adj <Lebensmittel> sterilized
Sterin n <Chemie> sterol
sterisch adj <Chemie> steric
Stern m <Raumfahrt> star • **Sterne betreffend** <Raumfahrt> stellar
Sternanker m <Elektriz> spider-type armature
Sternanordnung f <Fertig> radial arrangement (Zylinder)
Sternbohrmeißel m <Erdöl> star bit (Bohrtechnik)
Sternbruch m <Ker & Glas> star fracture
Sterndreieck n <Elektriz, Elektrotech, Phys> star delta, Y-delta
Sterndreieckanlasser m <Elektriz, Elektrotech> star delta starter, Y-delta starter
Sterndreieckanlasserschalter m <Elektriz> star delta starting switch
Sterndreieckanlassschalter m <Elektriz> Y-delta starting switch
Sterndreieckschalter m <Elektrotech> star delta switch
Sterndreieckschaltung f <Elektrotech> star delta connection
Sterndreiecktransformation m <Phys> star delta transformation
Sterndreieckumformung f <Elektriz> star-to-delta conversion
Sterndurchgangsdetektor m <Raumfahrt> star transit detector
sternförmiger Riss m im Porzellan <Ker & Glas> spider
sternförmiges Netz n <Telekom> radial network, star network, star-shaped network, star-type network
sternförmiges Netzwerk n <Comp & DV> active star
sternförmiges Speisesystem n <Elektriz> radial feeder system

sternförmiges System n <Elektriz> radial system
sternförmiges Verteilungssymbol n <Comp & DV> switched star
Sternfühler m <Raumfahrt> star sensor
Stern-Gerlach-Versuch m <Phys> Stern-Gerlach-experiment
sterngeschalteter Anker m <Elektriz> star-connected armature
Sterngriff m 1. <Fertig> locking handle (Kunststoffinstallationen); 2. <Maschinen> star wheel
Sternhaufen m <Raumfahrt> diffuse nebula
Sternkarte f <Wassertrans> star chart (Navigation)
Sternkeilwelle f <Maschinen> multispline shaft
Sternkompass m <Raumfahrt> astrocompass
Sternkonfiguration f <Telekom> star configuration
Sternkoppler m <Optik, Telekom> star coupler
Sternmotor m <Kfztech> radial engine
Sternnähe f <Raumfahrt> periastron
Sternnetz n <Comp & DV, Telekom> star network
Sternnetzwerk n <Comp & DV> star network
Sternpeilung f <Raumfahrt> astro fix
Sternpunkt m <Phys> neutral point (Dreiphasenstrom)
Sternpunktdraht m <Elektriz> neutral wire
Sternrad n 1. <Kfztech> star wheel; 2. <Maschinen> Geneva wheel, spoke wheel, star gear, star wheel
Sternrevolver m <Maschinen> star turret
Sternriss m 1. <Bau> star shake (Holz); 2. <Ker & Glas> star crack
Sternschaltung f 1. <Comp & DV, Elektriz, Elektrotech> star connection, Y-connection; 2. <Phys> Y-connection, star connection; 3. <Raumfahrt> star network (Weltraumfunk)
Sternspannung f <Elektriz> star voltage
Stern-Stern-Schaltung f <Elektriz> star-star connection
Sternstruktur f <Telekom> star structure
Sterntopologie f <Comp & DV> star topology
Sternverfolger m <Raumfahrt> star tracker
Sternverteilung f <Telekom> star distribution
Sternvierer m <Elektrotech> star quad
Sternviererkabel n <Elektrotech, Phys> star quad cable
Sternwalze f <Papier> starred roll
Sternzeit f <Raumfahrt> sideral time
Steroid n <Chemie> steroid
Sterol n <Chemie> sterol
stetig adj <Math> continuous (Funktion)
stetig durchstimmbar adj <Gerät> continuously tunable
stetig stellbares Getriebe n <Elektrotech> continuously variable transmission
stetig wirkende Regeleinrichtung f <Regelung> continuous controlling system
stetig wirkendes Glied n <Regelung> continuously acting element
stetige Kurve f <Geom> continuous curve
stetige Regeleinrichtung f <Regelung> continuous controlling system
stetige Zufallsgröße f <Math> continuous random variable (Wertebereich der Zufallsgröße ist ein Intervall)
stetiger Regler m <Elektronik> continuous action controller (Steuer- und Regelungstechnik)
stetiges Merkmal n 1. <Math> continuous characteristic (Wertebereich des Merkmals ist ein Intervall); 2. <Qual> continuous characteristic
Stetigförderer m <Maschinen> continuous flow conveyor
Stetigförderer m für Fußgänger <Trans> passenger conveyor
Stetigkeit f <Math> continuity
Stetigpoliermaschine f <Ker & Glas> continuous polisher
Stetigschleifer m <Papier> continuous grinder

Stetigschleifer *m* **und -polierer** *m* <Ker & Glas> continuous grinder and polisher
Steuer... <Comp & DV, Elektriz, Heiz & Kälte, Kerntech, Kohlen, Kontroll, Raumfahrt> control
Steueralgorithmus *m* <Kontroll> control algorithm
Steueranlage *f* 1. <Elektriz> control system; 2. <Wassertrans> steering system
Steueranlage *f* **eines Reaktors** <Kerntech> reactor control board
Steuerantrieb *m* <Kohlen> control driving
Steueranweisung *f* <Comp & DV> control statement
Steuerarm *m* <Fernseh> control arm
steuerbare Drossel *f* <Phys> transductor
steuerbarer Siliziumschalter *m* <Elektrotech> silicon-controlled switch
Steuerbarkeit *f* 1. <Kerntech> *(AE)* maneuverability, *(BE)* manoeuvrability *(Kraftniveau)*; 2. <Lufttrans> controllability, *(AE)* maneuverability, *(BE)* manoeuvrability
Steuerbefehl *m* 1. <Comp & DV> control command, control instruction; 2. <Kontroll> control instruction
Steuerbit *n* <Comp & DV, Telekom> control bit
Steuerblock *m* <Comp & DV> control block
Steuerbord *n* <Wassertrans> right, starboard
steuerbord *adv* <Raumfahrt> starboard
Steuerbord achteraus *adv* <Wassertrans> on the starboard quarter
Steuerbord voraus *adv* <Wassertrans> on the starboard bow
Steuerbus *m* <Comp & DV> control bus
Steuercode *m* <Comp & DV> control sequence
Steuercodes *mpl* <Comp & DV> control codes
Steuerdaten *npl* <Comp & DV> control data
Steuerdiagramm *n* <Maschinen> distribution diagram
Steuerdrossel *f* <Elektrotech> control reactor
Steuerdruck *m* 1. <Fertig> control pressure *(Kunststoffinstallationen)*; 2. <Raumfahrt> feel
Steuerdruckmechanismus *m* <Raumfahrt> feel mechanism
Steuerdruckventil *n* <Lufttrans> feel simulator valve
Steuerebene *f* <Lufttrans> control plane
Steuereinheit *f* 1. <Comp & DV> control unit, controller; 2. <Elektrotech> control unit, controller; 3. <Funktech> controller; 4. <Telekom> control unit; 5. <Trans> driving unit
Steuereinrichtung *f* 1. <Regelung> open loop control system; 2. <Telekom> control equipment
Steuerelektrode *f* 1. <Elektrotech> control electrode; 2. <Phys> gate *(Thyristor)*
Steuerfeld *n* <Comp & DV> control field
Steuerfläche *f* 1. <Lufttrans> control surface, rudder; 2. <Phys> control surface
Steuerfluss *m* <Comp & DV> control flow
Steuerfrequenz *f* <Elektronik, Fernseh, Funktech> control frequency
Steuergehäuse *n* 1. <Eisenbahn> control housing; 2. <Kfztech> timing gear housing; 3. <Maschinen> control box
Steuergehäusedeckel *m* <Kfztech> timing gear cover
Steuergerät *n* 1. <Elektriz, Elektrotech> control gear, control unit, controller; 2. <Kfztech> trigger box
Steuergestänge *n* <Hydraul> valve gear
Steuergetriebe *n* <Elektrotech> control gear
Steuergitter *n* 1. <Akustik> driving grid; 2. <Elektrotech> control grid *(Elektronenkanone)*
Steuerhebelquadrant *m* <Lufttrans> control lever quadrant
Steuerkanal *m* <Mobilkom> control channel
Steuerkennlinie *f* <Elektrotech> control characteristic
Steuerkette *f* 1. <Kfztech> timing chain; chain *(Steuerung, Spritzverstellung, Triebstrang)*; 2. <Wassertrans> steering chain

Steuerknopf *m* <Gerät> control button
Steuerknüppel *m* 1. <Comp & DV> joystick; 2. <Lufttrans> control column, control stick; 3. <Raumfahrt> joystick
Steuerknüppel *m* **für periodische Steigungssteuerung** <Lufttrans> cyclic control pitch stick
Steuerknüppel *m* **für zyklische Blattverstellung** <Lufttrans> cyclic control pitch stick, cyclic pitch stick *(Hubschrauber)*
Steuerknüppelbesatz *m* <Lufttrans> control column whipping
Steuerknüppelumwicklung *f* <Lufttrans> control column whipping
Steuerkolben *m* 1. <Fertig> actuating piston *(Hydraulik)*; 2. <Hydraul> slide valve *(Dampfmaschine)*; 3. <Maschinen> control piston
Steuerkolbenstange *f* <Hydraul> valve rod
Steuerkolbenventil *m* <Hydraul> slide valve
Steuerkompass *m* <Wassertrans> steering compass
Steuerkonsole *f* <Comp & DV, Eisenbahn> control panel
Steuerkräfteausfallanzeiger *m* <Lufttrans> artificial feel failure detector
Steuerkreis *m* <Comp & DV, Telekom> control circuit
Steuerkreisel *m* <Lufttrans> control gyro *(Hubschrauber)*
Steuerkurs *m* 1. <Lufttrans> compass heading, heading; 2. <Wassertrans> compass heading
Steuerkursfernanzeige *f* <Lufttrans> heading remote indicator
Steuerkurshalten *n* <Lufttrans> heading hold
Steuerkursinformation *f* <Lufttrans> heading information
Steuerkurs-Synchronisiereinrichtung *f* <Lufttrans> heading synchronizer
Steuerkurs-Übermittlungsumsetzer *m* <Lufttrans> heading repeater
Steuerkurswähler *m* <Lufttrans> heading selector
Steuerkurve *f* <Fertig> operating cam
Steuerleitung *f* 1. <Elektrotech> trailing cable *(Aufzug)*; control wire, pilot wire *(bei E-Antrieb)*; 2. <Telekom> control circuit
Steuerlochband *n* <Comp & DV> control tape
Steuermann *m* <Wassertrans> mate *(Handelsmarine)*
Steuermembrane *f* <Fertig> control diaphragm *(Kunststoffinstallationen)*
Steuermodus *m* <Comp & DV> control mode
steuern *v* 1. <Comp & DV> control; 2. <Kontroll> control; steer *(Richtung)*; 3. <Lufttrans, Trans> steer; 4. <Wassertrans> steer *(Navigation)*; helm *(Schiffbau)*
Steuern *n* <Regelung> open loop controlling
steuernder Basisstrom *m* <Elektronik> current base drive *(Transistoren)*
steuernder Taktgeber *m* <Telekom> master clock
steuerndes Basissignal *n* <Elektronik> base drive signal
Steueröffnung *f* <Phys> port
Steueroszillator *m* <Elektronik, Funktech> master oscillator
Steuerpodest *n* <Lufttrans> control pedestal
Steuerprogramm *n* 1. <Comp & DV> MCP, control routine, master control program; 2. <Qual> control routine; 3. <Telekom> handler *(Paketvermittlung)*; 4. <Trans> *(AE)* control program, *(BE)* control programme
Steuerpult *n* 1. <Bau> control panel; 2. <Eisenbahn> control desk; 3. <Elektriz> console, control console, control panel; 4. <Fernseh> control console; 5. <Gerät> console; 6. <Kontroll> console, control console, control panel, operator console; 7. <Lufttrans> console, control panel; 8. <Wassertrans> console *(Schiff)*
Steuerquarz *m* <Elektronik> oscillator crystal
Steuerrad *n* 1. <Wassertrans> steering wheel *(Bootbau)*; 2. <Kfztech> wheel

Steuerrakete

Steuerrakete f <Raumfahrt> control rocket
Steuerrechner m **für Vermittlungssysteme** <Telekom> switching system processor
Steuerregister n <Comp & DV> control register
Steuerrelais n <Elektriz> control relay
Steuerriemen m <Mechan> timing belt
Steuerrotor m <Lufttrans> control rotor
Steuerroutine f <Comp & DV, Qual> control routine
Steuerruder f <Lufttrans> rudder
Steuerschalter m 1. <Elektriz, Elektrotech> control switch, controller; 2. <Eisenbahn> master controller (handgesteuert)
Steuerschaltung f <Elektriz> control circuit
Steuerschaltventil n <Lufttrans> actuator control valve
Steuerschieber m <Hydraul> slide valve (Dampfmaschine, Hydraulik)
Steuerschiebergehäuse n <Hydraul> valve chest
Steuerschieberstange f <Hydraul> valve rod
Steuerschieberventil n <Hydraul> slide valve
Steuerschlitz m <Kfztech> port (Motor)
Steuerschrank m 1. <Eisenbahn> control box; 2. <Gerät> console
Steuerseil n <Maschinen> control cable
Steuersender m <Akustik> driver
Steuersignal n <Elektronik, Fernseh> control signal
Steuersprache f <Comp & DV> CL, control language
Steuerspule f <Elektrotech> drive coil
Steuerspur f <Akustik, Aufnahme, Fernseh> control track
Steuerspursignal n <Aufnahme, Fernseh> control track signal
Steuerspurzeitcode m <Aufnahme, Fernseh> control track time code
Steuerstab m <Phys> control rod (Kernreaktion)
Steuerstange f <Lufttrans> control rod
Steuerstangenresonanz f <Lufttrans> control rod resonance (Hubschrauber)
Steuerstreifen m <Akustik> control track
Steuerstrich m <Raumfahrt, Wassertrans> lubber's line (Kompass)
Steuerstufe f 1. <Elektronik> master oscillator; 2. <Lufttrans, Qual> control stage
Steuersystem n <Elektriz, Telekom> control system
Steuersystem n **mit festverdrahtetem Programm** <Telekom> (AE) wired program control system, (BE) wired programme control system
Steuertafel f <Elektriz> control board
Steuertaste f (Strg-Taste) <Comp & DV> control key (Ctrl key)
Steuerteil m <Fertig> control unit (Kunststoffinstallationen)
Steuertriebwagen m <Kfztech> driving motor car
Steuertriebwerk n <Raumfahrt> vernier motor
Steuertrommel f <Ker & Glas> timing drum
Steuer- und Anzeigegerät n <Raumfahrt> CDU, control and display unit
Steuer- und Sicherheitseinrichtung f <Heiz & Kälte> control and safety device
Steuerung f 1. <Comp & DV, Elektriz, Elektronik, Elektrotech> control (Ein/Ausstellung); control unit; 2. <Fernseh> biasing, pilot; 3. <Hydraul> valve gear (Dampflokomotive); guide (Turbine); 4. <Kfztech> control; 5. <Kontroll> controller, control; 6. <Lufttrans> control system; 7. <Maschinen, Raumfahrt> control; 8. <Regelung> open loop control; 9. <Telekom> control, steering
Steuerung f **auf hell** <Elektronik> positive modulation
Steuerung f **mittels Elektronenstrahl** <Elektronik> electronic beam steering
Steuerungsabfolge f <Kontroll, Qual> control sequence
Steuerungsart f <Lufttrans> control system
Steuerungsball m <Comp & DV> track ball

Steuerungsdämpfer m <Lufttrans> control damper
Steuerungsdiagramm n <Comp & DV> timing diagram
Steuerungseinheit f <Maschinen> control unit
Steuerungsfolge f <Kontroll, Qual> control sequence
steuerungsführende Betriebsart f <Kontroll> master mode
Steuerungsführung f <Sicherheit> jig
Steuerungshebel m <Comp & DV> joystick
Steuerungshierarchie f <Kontroll> control hierarchy
Steuerungskasten m **des Motoranlassers** <Lufttrans> engine starting control box
Steuerungspunkt m <Comp & DV> point of control
Steuerungssystem n <Kontroll> control system
Steuerungstechnik f <Regelung> control engineering
steuerungstechnisch adj <Kontroll> control-actuated; control engineering
steuerungstechnische Schutzvorrichtung f <Sicherheit> control-actuated guard
steuerungstechnische Verriegelung f <Sicherheit> control interlocking
Steuerungsübergabe f <Kontroll> control transfer, passing of control
Steuerungs- und Anzeigegeräte npl <Lufttrans> controls and indicating devices
Steuerventil n 1. <Hydraul> control valve, governor valve; 2. <Kontroll> control valve; 3. <Maschinen> control valve, distribution valve; 4. <Mechan> control valve
Steuerverhalten n 1. <Lufttrans> controllability; 2. <Raumfahrt> handling (Raumschiff)
Steuerverschlussstein m <Ker & Glas> control tweel
Steuervolumen n <Fertig> swept volume (Kunststoffinstallationen)
Steuervorgang m <Kontroll, Qual> control action
Steuerwagen m <Eisenbahn> driving trailer car
Steuerwalze f <Fertig> controller
Steuerwarte f <Elektriz, Kerntech> control room
Steuerwelle f <Kfztech> camshaft
Steuerwerk n <Elektrotech, Telekom> control unit
Steuerwinkel m <Wassertrans> steering angle (Navigation)
Steuerwort n <Comp & DV> control word
Steuerzeichen n 1. <Comp & DV> control character; 2. <Druck> functional character
Steuerzentrale f 1. <Elektriz> control room; 2. <Fernseh> (AE) control center, (BE) control centre
Stevenanlauf m <Wassertrans> forefoot (Schiffbau)
Stevenrohr n <Wassertrans> stern tube (Schiffbau)
Stibium n (Sb) <Chemie> stibium (Sb)
Stibnit m <Chemie> stibnite
Stich m 1. <Bau> camber; rise (Bogen); 2. <Textil> stitch
Stichbalken m <Bau> tail beam
Stichbogen m <Bau> flat arch, segmental arch
Sticheinreißfestigkeit f <Kunststoff> puncture resistance
Stichel m <Akustik> rake, stylus
Stichelkraft f <Akustik> stylus force
stichfest adj <Abfall> semisolid
Stichflamme f <Thermod> darting flame
Stichflammentest m <Ker & Glas> dart impact test
Stichkabel n <Bau> branch cable
Stichlängen-Einstellung f <Textil> stitch length setting
Stichleitung f 1. <Elektrotech, Funktech> stub line (Antennen); 2. <Telekom> branch line
Stichloch n <Ker & Glas> tapping hole
Stichprobe f 1. <Comp & DV> sample; 2. <Kohlen> random sampling, spot check; 3. <Maschinen> sample, spot check; 4. <Qual> random inspection, sample, spot check; 5. <Telekom> sample (Qualität)
Stichprobenahme f 1. <Ergon> sampling; 2. <Kohlen> random sample

Stichprobenauswahlsatz m <Qual> sampling fraction
Stichprobenentnahme f <Telekom> sampling *(Qualität)*
Stichprobenplan m 1. <Ergon> sampling plan; 2. <Qual> sampling scheme
Stichprobenprüfung f 1. <Qual> batch test, random sample test, sampling inspection; 2. <Werkprüf> random sample test
Stichprobenraum m <Math> sample space
Stichprobensystem n **nach einem quantitativen Merkmal** <Qual> attribute sampling system
Stichprobenumfang m <Elektronik, Qual> sample size
Stichsäge f 1. <Bau> compass saw, jig saw, keyhole saw; 2. <Maschinen> alternating saw, compass saw
Stichwort n <Elektriz> keyword • **Stichwort geben** <Aufnahme> cue
Stichwortanalyse f <Comp & DV> KWIC, keyword in context
Stichwortanalyse f **mit Text** *(KWOC-Index)* <Comp & DV> keyword out of context *(KWOC)*
Stichwortmikrofon n <Aufnahme> cue mike
Stickdioxid n <Sicherheit> nitrogen dioxide *(Luft)*
Stickoxid n 1. <Abfall> nitrogen oxide; 2. <Umweltschmutz> nitric oxide
Stickstoff m <Chemie> nitrogen
Stickstoff m **als Schutzgas** <Kerntech> nitrogen cover gas
Stickstoffbestimmungsapparat m **nach Kjeldahl** <Labor> Kjeldahl digestion apparatus
Stickstoffdioxid n 1. <Chemie> nitrogen dioxide; 2. <Umweltschmutz> nitrogen dioxide, nitrogen peroxide
stickstoffgekühlter Reaktor m <Kerntech> nitrogen-cooled reactor
stickstoffhaltig adj <Chemie> nitrogenous
Stickstoffoxid n <Chemie> nitrous oxide
Stickstoffpentoxid n <Umweltschmutz> nitrogen pentoxide
Stickstoffspülung f <Raumfahrt> nitrogen purging *(Raumschiff)*
Stickstoffwasserstoff... <Chemie> hydrazoic
Stiefel m 1. <Fertig> barrel, penstock *(Pumpe)*; 2. <Ker & Glas> boot; 3. <Maschinen> barrel
Stiel m 1. <Bau> strut *(Pfettendach)*; 2. <Ker & Glas> stem; 3. <Maschinen> handle; 4. <Mechan> handle, shaft
Stielhammer m <Fertig> helve
Stielpfanne f <Maschinen> bull ladle
Stielträger m <Ker & Glas> stem carrier
Stift m 1. <Bau> brad, dowel, pin, plug; 2. <Elektrotech> pin; 3. <Fertig> pintle; pin *(Kunststoffinstallationen)*; 4. <Kfztech> stud; 5. <Maschinen> finger, peg, pin, spike, tack; 6. <Mechan> peg, pin, stud
Stift m **mit Bund** <Maschinen> collared pin
Stiftbolzen m <Fertig> stud
Stiftbüchse f <Fertig> pin barrel
Stiftfassung f <Elektriz> bayonet socket
Stiftisolator m <Elektriz> pin insulator
Stiftkolben m <Bau> *(BE)* pin vice, *(AE)* pin vise
Stiftkontakt m <Elektriz> plug pin
Stiftlampenfassung f <Elektriz> bayonet lamp holder
Stiftloch n <Maschinen> pinhole
Stiftplotter m <Comp & DV> pen plotter
Stiftschlüssel m <Bau> *(BE)* pin spanner, pin wrench
Stiftschraube f <Maschinen> stud, tap bolt
Stiftsicherung f <Fertig> pin lock
Stiftsockel m 1. <Elektriz> bayonet cap; 2. <Fertig> pin base
Stiftstecker m <Elektrotech> pin plug
Stiftventil n <Bau> pin valve
Stigmasterin n <Chemie> stigmasterol
Stigmasterol n <Chemie> stigmasterol

stigmatische Linse f <Foto> stigmatic lens
Stilben n <Chemie> stilbene, toluylene
stille Verbrennung f <Kfztech> slow combustion
stilles Wasser n 1. <Lebensmittel> still water; 2. <Wasserversorg> quiet water
stillgelegter Abladungsplatz m abandoned site
stillgelegter Teilnehmeranschluss m <Telekom> ceased subscriber
Stilllebenfotografie f <Foto> still life photography
stilllegen v 1. <Bau> put out of service; 2. <Comp & DV> quiesce; 3. <Kohlen> deactivate
Stilllegung f 1. <Comp & DV> quiescing; 2. <Kerntech> decommissioning *(Reaktor)*
stillliegen v <Maschinen> lie idle
Stillstand m 1. <Fertig> dwell *(Getriebelehre)*; 2. <Ker & Glas> shutdown; 3. <Kfztech> stop; 4. <Lufttrans> rest period; 5. <Maschinen> standstill, stop • **zum Stillstand kommen** <Maschinen> come to a standstill
Stillstand m **der Gezeiten** <Wassertrans> stand of tide
Stillstandsgetriebe n <Fertig> dwell mechanism
Stillstandsheizung f <Heiz & Kälte> anticondensation heater
Stillstandspause f <Fertig> dwelling
Stillstandsstellung f <Fertig> dwell position
Stillstandzeit f <Erdöl> downtime
stillstehende Luft f <Nichtfoss Energ> still air
stillstehender Anker m <Elektriz> stationary armature
stillstehender Verkehr m <Trans> stationary traffic
stillstehendes Gewässer n <Wasserversorg> still water
Stillwasser n <Wassertrans> slack water *(Gezeiten)*
Stimme f <Comp & DV> voice
stimmen v <Akustik, Elektronik> tune
Stimmen n <Akustik, Elektronik> tuning
Stimmerkennung f <Raumfahrt> voice detector
Stimmgabel f <Akustik, Aufnahme, Phys, Wellphys> tuning fork
Stimmgabeloszillator m <Elektronik> fork oscillator
Stimmskale f <Aufnahme> tuning dial
Stimmton m <Akustik> pitch
Stimmung f <Akustik> pitch
stimulierte Emission f <Optik, Telekom> stimulated emission
stimulierte Strahlungsabsorption f <Strahlphys> stimulated absorption of radiation
stimulierte Strahlungsemission f <Strahlphys> stimulated emission of radiation
Stinger m <Erdöl> stinger *(Offshore)*
Stinkbrand m <Lebensmittel> bunt *(Getreidekrankheit)*
Stippe f <Kunststoff> fish eye
Stippen n <Ker & Glas> stippling
Stirlingmotor m <Maschinen> Stirling engine
Stirn f <Maschinen> end
Stirnabschreckversuch m <Fertig> end quench test
Stirnbrett n <Bau> side board
Stirndrehmeißel m 1. <Fertig> end-cut turning tool; 2. <Maschinen> facing tool
Stirnfläche f 1. <Bau> butt end; 2. <Fertig> end; 3. <Maschinen> end face • **Stirnflächen bearbeiten** <Maschinen> face
Stirnflächenbearbeitung f <Maschinen> facing
Stirnflächenbreite f <Fertig> face width
Stirnfräsen n <Maschinen> end milling, face milling
Stirnfräser m <Maschinen> face cutter, face mill, face-milling cutter, facing cutter
Stirnkantenplatte f <Ker & Glas> front lip tile
Stirnkontaktverbindung f <Telekom> butt joint
Stirnkurbel f <Maschinen> outside crank
Stirnmauer f <Bau> face wall; head wall *(Durchlass)*

Stirnrad

Stirnrad n <Maschinen> cylindrical gear, spur gear, spur wheel
Stirnrädergetriebe n <Maschinen> spur gear
Stirnradgetriebe n 1. <Kfztech> cylindrical gear pair; 2. <Maschinen> spur gear
Stirnradpaar n <Maschinen> cylindrical gear pair
Stirnradverzahnungsmaschine f <Maschinen> spur gear cutting machine
Stirnreibahle f <Fertig> bottoming reamer
Stirnschneidenfreiwinkel m <Maschinen> front clearance
Stirnseite f <Maschinen> end, face
Stirnseite f **des Kühlofens** <Ker & Glas> end of lehr
stirnseitig schneidend adj <Fertig> end-cutting
Stirnsenken n <Maschinen> end facing
Stirnsenker m <Maschinen> counterbore, end mill reamer
Stirnteilung f <Fertig> real circular pitch (Getriebelehre)
Stirnwalzenfräsen n <Maschinen> shell-end milling
Stirnwandtür f <Ker & Glas> end door
Stirnzahn m <Fertig> radial tooth (Fräser)
stochastisch adj <Comp & DV, Ergon, Math> stochastic
stochastische Abtastung f <Telekom> random sampling
stochastische Anregung f <Telekom> random excitation
stochastische Belastung f <Metall> stochastic loading
stochastische Kühlung f <Teilphys> stochastic cooling
stochastischer Prozeß m <Math, Phys> stochastic process
stochastisches Rauschen n <Telekom> stochastic noise
Stöchiometrie f <Chemie> stoichiometry
stöchiometrisch adj <Chemie> stoichiometric
stöchiometrische Zusammensetzung f <Metall> stoichiometric composition
Stock m <Fertig> body (Amboss)
Stockanker m <Wassertrans> stock anchor
Stockblender m <Kunststoff> stock blender (Kautschuk)
Stock-Car m <Kfztech> (AE) stocker
Stöckel m <Fertig> anvil stake
stocken v <Bau> point
stockloser Anker m <Wassertrans> stockless anchor
Stockpunkt m 1. <Fertig> pour point; 2. <Heiz & Kälte> solidification point; 3. <Kfztech> pour point (Öl); 4. <Kunststoff, Maschinen> solidification point
Stockschaltung f <Kfztech> floor shift (Getriebe)
Stockung f <Fertig> hang-up (Kunststoffe)
Stockwerk n <Bau> floor, storey
Stockwerkschalter m <Elektriz> landing switch
Stockzwinge f <Mechan> ferrule
Stoff m 1. <Chemie> matter, material, substance; 2. <Textil> cloth, fabric, woven fabric; 3. <Umweltschmutz> material, matter, substance
Stoffaufbereitung f <Papier> stock preparation
Stoffauflauf m <Papier> flow box, headbox
Stoffauflauflippe f <Papier> slice
Stoffbahn f <Textil> panel
Stoffballen m <Textil> roll
Stoffbespannung f <Raumfahrt> cloth (Raumschiff)
Stoffbreite f <Textil> fabric width
Stoffbütte f <Papier> stock chest, stuff
Stoffdatenblatt n <Sicherheit> material safety sheet
Stoffdichte f <Papier> consistency
Stoffdichteregler m <Papier> consistency regulator
Stoffeinlauf m <Papier> stock inlet
Stofffänger m <Papier> pulp saver, save-all
Stofffängertrog m <Papier> save-all tray
Stofffilter n <Abfall> fabric filter
stofffrei adj <Fertig> immaterial
Stoffgewicht n <Textil> fabric weight

Stoffgrund m <Textil> blotch
Stoffkasten m <Papier> drainer
Stoffleimung f <Papier> stuff sizing
Stoffmenge f <Phys> amount of substance
Stoffmengenkonzentration f <Chemie> molarity
Stoffmerkblatt n <Sicherheit> material safety sheet
Stoffmuster n <Textil> fabric sample, strike off sample, swatch
Stoffrezeptur f <Papier> furnish
stoffspezifische Haftung f <Kunststoff> specific adhesion
Stofftrennprozess m <Abfall> material separation operation
Stoffverteilung f <Papier> approach flow
Stoffwechsel m <Ergon, Lebensmittel> metabolism
Stoffwechselschlacken fpl <Abfall> metabolic waste
Stoffwechselstörung f <Lebensmittel> metabolic disorder
Stokes-Geschwindigkeit f <Ker & Glas> Stokes' velocity
Stokes'sche Theorie f <Strömphys> Stokes' theory
Stokes'sches Gesetz n <Phys> Stokes' law
STOL-Flugzeug n (Fastsenkrechtstarter, Kurzstart-Kurzlande-Flugzeug, Kurzstartflugzeug) <Lufttrans> STOL aircraft (short takeoff and landing aircraft)
Stollen m <Maschinen> cleat
Stollenholz n <Kohlen> timber
Stone n <Metrol> stone (Körpergewicht)
Stop-and-Go-Verkehr m <Trans> stop-and-go traffic
Stopf... <Eisenbahn, Telekom, Textil> stuffing
Stopfaggregat n <Eisenbahn> tamping unit
Stopfbit n <Telekom> justifying digit, stuffing digit
Stopfbüchse f 1. <Erdöl> gland (Rohrleitung); 2. <Fertig> stuffing box; 3. <Heiz & Kälte> gland, packing gland; 4. <Lebensmittel> stuffing box; 5. <Maschinen> gland, packing gland, stuffing box; 6. <Mechan> gland
Stopfbüchsenabdichtung f <Raumfahrt> gland (Raumschiff)
Stopfbüchsenbrille f 1. <Maschinen> stuffing box gland; 2. <Mechan, Wassertrans> gland
Stopfbüchsendeckel m <Kerntech> stuffing box lid
stopfbüchsenlos adj <Fertig> glandless (Kunststoffinstallationen)
Stopfeinrichtung f <Telekom> stuffing device
Stopfen m 1. <Bau> plug; 2. <Comp & DV> padding; 3. <Erdöl> plug (Rohrleitungen, Bohrloch); 4. <Fertig> bott, iron plug, stoppage, stopper; cap (Kunststoffinstallationen); 5. <Hydraul, Kfztech> plug (Ablass); 6. <Maschinen> plug, stopper; 7. <Telekom> justification, stuffing; 8. <Textil> darning, mending, plugging; 9. <Wassertrans> plug
stopfen v 1. <Bau> tamp; caulk (Fugen); pack (Schotter); 2. <Fertig> stop; 3. <Textil> darn; 4. <Textil> mend
Stopfendichtung f <Fertig> standpipe adaptor seal (Kunststoffinstallationen)
Stopfgeschwindigkeit f <Telekom> stuffing rate
Stopfmaschine f <Eisenbahn> tamping machine
Stopfnadel f <Textil> darning needle
Stopfrate f <Telekom> stuffing rate
Stopfstelle f <Textil> darn
Stopfstich m <Textil> darning stitch
Stopfwolle f <Textil> darning wool
Stopfzeichen n <Telekom> stuffing character
Stopp m 1. <Comp & DV> halt, stop; 2. <Kfztech> stop; 3. <Maschinen> check
Stoppanweisung f <Comp & DV> halt instruction
Stoppbad n <Foto> stop bath
Stoppbahn f <Lufttrans> stopway
Stoppbahnfeuer n <Lufttrans> stopway light
Stoppbit n 1. <Comp & DV> stop bit, stop element; 2. <Telekom> stop bit

Störungsmeldung

Stoppcode m <Comp & DV> stop code
Stoppelement n <Comp & DV> stop element
stoppen v <Comp & DV> stop
Stoppen n <Phys> stop
Stopperknoten m <Wassertrans> stopper knot
Stoppliste f <Comp & DV> stoplist
Stoppover m <Lufttrans> stopover
Stoppschritt m <Telekom> stop element
Stopptaste f <Fernseh> stop key
Stoppuhr f <Labor, Phys> stopwatch
Stoppzeit f <Kontroll> stop time
Stoppzylinderpresse f <Druck> stop cylinder press
Stöpsel m 1. <Ker & Glas> plug; 2. <Labor> bung; 3. <Maschinen> plug, stopper; 4. <Mechan, Telekom> plug *(Telefon)*
Stöpselfeld n <Elektrotech> plugboard
stöpseln v <Elektrotech> plug
Stöpselschalter m <Elektrotech> plug switch
Stöpselsicherung f <Elektriz> cartridge fuse
Stöpselverbindung f <Elektrotech> plug connection
Störablaufprotokollierung f <Comp & DV> postmortem review
Störabstand m <Akustik, Fernseh> signal-to-noise ratio
Störabstand m einschließlich Verzerrungen *(SINAD)* <Telekom> signal-to-noise and distortion ratio *(SINAD)*
störanfällig adj <Sicherheit> failure-prone, prone to fail, sensitive to failure, trouble-prone
Störanfälligkeit f <Sicherheit> failure-proneness
Störaustaster m <Fernseh> interference eliminator
Störaustastungsfaktor m <Raumfahrt> interference reduction factor
Störbeeinflussung f <Elektronik> interference
Störbegrenzer m <Aufnahme> noise limiter
Störecho n <Mobilkom> clutter
Störechounterdrückung f <Funkort> clutter suppression *(Radar)*
Störeinbruch m <Telekom> breakthrough
stören v 1. <Anstrich> disrupt; 2. <Bau, Qual> disturb; 3. <Raumfahrt> jam *(Weltraumfunk)*
Stören n <Funktech> jamming
störend adj 1. <Elektronik> parasitic; 2. <Funktech> spurious
störende Beeinflussung f <Fernseh> interference *(Funk)*
störende Beeinflussung f des Kabelfernsehdienstes *(CATVI)* <Fernseh, Funktech> cable television interference *(CATVI)*
Störer m <Elektronik> jammer
Störfall m 1. <Kerntech> abnormal occurrence; 2. <Qual> accident condition; 3. <Trans> incident
Störfallanalyse f <Qual> accident analysis
Störfallmelder m <Sicherheit> intrusion detector
störfest adj <Elektriz> interference-proof
Störfestigkeit f 1. <Comp & DV> interference immunity, noise immunity; 2. <Telekom> immunity to interference
Störfestigkeitsprüfung f <Elektriz> interference resistance test
Störfilter n <Raumfahrt> interference filter
Störfleck m <Elektronik> clutter *(Radar)*
störfrei adj <Sicherheit> disturbance-proof
Störfunkstelle f <Elektronik> jammer
Störgebiet n <Funktech> interference area
Störgenerator m <Elektronik> interference generator
Störgerät n <Elektronik> jammer
Störgeräusch n <Werkprüf> interference noise
Störgrößenaufschaltung f <Qual, Regelung> disturbance variable feedforward
Störhäufigkeit f <Qual> failure frequency
Störklappe f <Lufttrans> spoiler *(Luftfahrzeug)*

Störmodulation f <Funktech> incidental modulation
Störort m <Regelung> point of disturbance
Störoszillator m <Elektronik> jammer oscillator
Störpegel m 1. <Aufnahme> noise floor, noise level; 2. <Gerät, Qual, Sicherheit> disturbance level
Störpegelmessgerät n <Gerät> interference level meter
Störpegelmessplatz m <Gerät> transmission impairment measuring set
Störrauschpegel m <Umweltschmutz> background level
Störschallpegel m <Sicherheit> background noise level
Störschutzfilter n 1. <Aufnahme> noise filter; 2. <Telekom> anti-interference filter
Störschutzschirm m <Aufnahme> noise shield
Störsender m <Elektronik> jammer
Störsignal n 1. <Comp & DV> drop-in; 2. <Elektronik, Elektrotech> incident signal, interference signal, jamming signal; 3. <Funktech> jamming signal; 4. <Gerät> perturbation signal; 5. <Telekom> glitch
Störsignalpaket n <Elektrotech> drop-in package
Störspannung f <Elektrotech> noise voltage
Störstelle f <Elektronik, Elektrotech> defect, impurity
Störstellenausbreitung f <Elektronik> impurity diffusion
Störstellendichte f <Elektronik> impurity concentration, impurity density; defect density *(Halbleiter)*
Störstellenhalbleiter m <Comp & DV> extrinsic semiconductor
Störstelleninversionszone f <Elektronik, Phys> p-n junction
Störstellenleitfähigkeit f <Elektrotech> extrinsic conductivity
Störstellenleitung f <Elektrotech> defect conduction
Störstellenniveau n <Elektronik> impurity level
Störstellen-Photoleitfähigkeit f <Elektrotech> extrinsic photoconductivity
Störstellenstreuung f <Elektronik> impurity scattering
Störstrahlung f <Funktech> spurious emission
Störstrahlungspegel m <Raumfahrt> spurious emission level
Störung f 1. <Anstrich> disruption; 2. <Aufnahme> interference; 3. <Bau> breakdown; 4. <Comp & DV> failure, interference, malfunction, noise; 5. <Elektriz, Elektrotech> interference, interruption; 6. <Fernseh, Funktech> interference; 7. <Kerntech> disturbance *(Störfall)*; 8. <Maschinen> interference, malfunctioning; 9. <Metall> disorder; 10. <Phys> distortion; 11. <Qual> deficiency, disorder, disturbance, malfunction, obstacle; 12. <Raumfahrt> interference; 13. <Strömphys> disturbance, perturbation; 14. <Telekom> fault, interference, noise; disruption *(Verbindung)*
Störung f durch Industriegeräte <Elektrotech> industrial interference
Störung f durch Nachbarkanal <Aufnahme> adjacent channel interference
Störung f im Rundfunkempfänger <Funktech> broadcast receiver interference
störungsanfällig adj <Sicherheit> prone to fail
Störungsannahme f <Telekom> FRC, *(AE)* fault reception center, *(BE)* fault reception centre
Störungsanzeige f <Sicherheit> disturbance indication
störungsarme Antenne f <Funktech> anti-interference antenna
Störungsaufzeichnung f <Comp & DV> failure logging
Störungsbereich m <Fernseh> interference area
Störungsbeseitigung f <Telekom> fault clearance
Störungsdatenkarte f <Lufttrans> failure data card
Störungsdauer f 1. <Lufttrans> downtime; 2. <Qual> malfunction time; 3. <Telekom> fault time
störungsfrei adj <Sicherheit> disturbance-proof
Störungsmeldung f <Lufttrans> error signal

störungssicher *adj* <Elektrotech, Sicherheit> fail-safe
Störungssignal *n* <Lufttrans> error signal
Störungssucher *m* <Maschinen> troubleshooter
Störunterdrückung *f* 1. <Elektronik> interference rejection; 2. <Funktech, Telekom> interference suppression
Störunterdrückungssignal *n* <Elektronik> interference rejection signal
Störursache *f* <Raumfahrt> perturbation
Störwellenmethode *f* <Kerntech> distorted wave method
Stoß *m* 1. <Anstrich> impact; 2. <Bau> abutment, butt, meeting; stack *(Holz)*; 3. <Elektriz, Erdöl> surge *(Strömung)*; 4. <Fertig> joint; 5. <Lufttrans> blast; 6. <Maschinen> shock; 7. <Mechan> impulse, push; 8. <Metall> jog; 9. <Phys> collision; 10. <Raumfahrt> shock; 11. <Trans> jolt
Stoßanregung *f* <Strahlphys> impact excitation
stoßartig *adj* <Maschinen> impulsive
Stoßaufhängung *f* <Raumfahrt> shock mount
Stoßbau *m* <Kohlen> cut and fill method
Stoßbau-Kohlenschrämmaschine *f* <Kohlen> short wall coal-cutting machine
Stoßbelastung *f* 1. <Lufttrans> impact load; 2. <Verpack> impact stress
Stoßbetrieb *m* <Comp & DV> burst mode
Stoßbohren *n* <Kohlen> percussion drilling
Stoßbohrmaschine *f* <Kohlen> piston drill
Stoßbremse *f* <Kerntech> dashing vessel, dashpot
Stoßbrenner *m* <Trans> aeropulse
Stoßdämpfer *m* 1. <Bau, Fertig> shock absorber; 2. <Kerntech> dashing vessel, dashpot; 3. <Kfztech> damper; shock absorber *(Federung)*; 4. <Maschinen> dashpot, shock absorber; 5. <Mechan> shock absorber; 6. <Raumfahrt> snubber; 7. <Sicherheit> resilient isolator, shock absorber
Stoßdämpfung *f* <Akustik> transition loss
Stoßdichte *f* <Strahlphys> collision density
Stoßdruckwelle *f* <Sicherheit> shock-front pressure wave *(Explosion)*
Stößel *m* 1. <Bau> ram; 2. <Fertig> ram; slide *(Presse)*; 3. <Kfztech> tappet; 4. <Kunststoff> ram; 5. <Labor> pestle *(Schleifen)*; 6. <Maschinen> tappet, tool ram; 7. <Mechan> ram
Stößel *m* **und Mörser** *m* <Labor> pestle and mortar
Stoßelastizität *f* <Kunststoff> resilience
Stößelführung *f* <Fertig> body *(Senkrechtstoßmaschine)*
Stößelkopf *m* <Fertig> head *(Waagerechtstoßmaschine)*
Stößelschaft *m* <Kfztech> tappet stem
Stößelschutz *m* <Fertig, Sicherheit> ram guard
Stößelstange *f* <Kfztech> push rod
Stößelstangenmotor *m* <Kfztech> push rod engine
Stößelsteuerung *f* <Papier> follower trainer *(Spindelpresse)*
stoßen *v* 1. <Bau> ram; 2. <Fertig> plane, shape *(Zahnräder)*; 3. <Mechan> push
Stoßen *n* <Fertig> shaping *(Zahnräder)*
Stoßerregung *f* 1. <Strahlphys> impact excitation; 2. <Telekom> impulse excitation, shock excitation
Stoßexperiment *n* <Teilphys> collision experiment
Stoßfängerhorn *n* <Kfztech> overrider *(Karosserie)*
stoßfest *adj* <Metrol, Sicherheit> impact-resistant, resistant to impact, shock-proof
Stoßfestigkeit *f* 1. <Kunststoff> shock resistance; 2. <Maschinen, Qual> resistance to shock, resistance to impact, shock resistance; 3. <Verpack> impact resistance
Stoßfläche *f* 1. <Fertig> abutting end, abutting surface; 2. <Maschinen> abutting surface
Stoßfuge *f* <Bau> butt joint, straight joint
Stoßfunktion *f* <Elektronik> impulse function
Stoßgalvanometer *n* <Elektriz> ballistic galvanometer
Stoßgenerator *m* <Elektriz> impulse generator, pulse generator

Stoßgeräusch-Analysator *m* <Aufnahme> *(BE)* impact noise analyser, *(AE)* impact noise analyzer
stoßgeschützter Wagen *m* <Eisenbahn> *(AE)* cushion car, *(BE)* cushion wagon
stoßhaltige Schwingung *f* <Sicherheit> impact vibration
Stoßintegral *n* <Strahlphys> collision integral *(Boltzmann-Gleichung)*
Stoßionisation *f* 1. <Strahlphys> impact ionization, ionization by collision; 2. <Teilphys> ionization by collision
Stoßionisierung *f* <Strahlphys> impact ionization
Stoßkante *f* 1. <Fertig> abutting edge *(Blech)*; 2. <Mechan> edge; 3. <Textil> hem
Stoßkopf *m* <Maschinen> shaper head
Stoßkraft *f* <Maschinen> drive
Stoßlasche *f* 1. <Eisenbahn> rail splice; 2. <Maschinen> butt strap
Stoßlast *f* <Bau> dynamic loading
Stoßmaschine *f* 1. <Fertig> vertical push-cut shaper, vertical shaper; 2. <Maschinen> slotter, slotting machine
Stoßmeißel *m* 1. <Bau> slotting tool; 2. <Maschinen> shaper tool, shaping tool, slotting tool
Stoßmittelpunkt *m* <Mechan, Phys> *(AE)* center of percussion, *(BE)* centre of percussion
Stoßnaht *f* <Erdöl> abutting joint *(Schweißen)*
Stoßparameter *m* <Phys> impact parameter
Stoßplatte *f* <Mechan> butt plate
Stoßprüfung *f* <Qual, Werkprüf> impact test
Stoßratendichte *f* <Strahlphys> collision density
Stoßräumen *n* <Maschinen> push-broaching
Stoßregler *m* <Raumfahrt> bucking regulator
Stoßring *m* <Kfztech> thrust washer
Stoßrüttler *m* <Bau> jolt vibrator
Stoßschubregler *m* <Raumfahrt> buck-boost regulator
Stoßschweißen *n* <Mechan> butt welding
Stoßspannung *f* <Elektrotech> impulse voltage
Stoßspannungsgenerator *m* <Elektrotech> impulse generator
Stoßspannungsprüfung *f* <Elektrotech> impulse test
Stoßstange *f* 1. <Eisenbahn, Kfztech> *(BE)* bumper, *(AE)* fender *(Karosserie)*; 2. <Maschinen> push rod
Stoßstelle *f* <Fertig> junction
Stoßstrom *m* <Elektriz> impulse current
Stoßtest *m* <Phys> ram
Stoßtheorie *f* **der Linienverbreiterung** <Strahlphys> impact theory of line broadening
Stoßverstärker *m* <Elektronik> burst amplifier
Stoßvorrichtung *f* <Eisenbahn> buffing gear
Stoßwelle *f* <Phys, Wellphys> shock wave
Stoßwellenauslöser *m* <Raumfahrt> shock-wave initiator
Stoßwellenfront *f* <Sicherheit> shock front *(Explosion)*
straff gespannt *adj* <Fertig> taut
straffes Trumm *n* <Mechan> driving end *(Riemenantrieb)*
Strahl *m* 1. <Elektronik> ray; 2. <Fertig> half-line, jet; 3. <Kerntech> jet *(Flüssigkeit)*; 4. <Maschinen, Metall> jet; 5. <Optik> ray; 6. <Strömphys> jet; 7. <Telekom> ray
Strahlablenkung *f* 1. <Fernseh> sweep; 2. <Maschinen> deflection of beams
Strahlabschaltung *f* <Fernseh> beam cut-off
Strahlabschwächer *m* <Elektronik> beam attenuator
Strahlabschwächung *f* <Elektronik> beam attenuation
Strahlabtastung *f* <Elektronik> beam scanning
Strahlaufspaltung *f* <Elektronik> beam splitting
Strahlaufteiler *m* <Optik> beam splitter
Strahlausrichtung *f* <Fernseh> beam alignment
Strahlaustastung *f* 1. <Fernseh> beam blanking; 2. <Gerät> blanking
Strahlbohren *n* <Erdöl> jet bit drilling *(Bohrtechnik)*
Strahlbohrer *m* <Erdöl> jet bit *(Bohrtechnik)*
Strahlbreite *f* <Elektronik, Funktech, Optik> beamwidth

Strahlbündeldichte f <Elektronik> beam power density
Strahldichte f <Optik, Phys, Telekom> radiance
Strahldivergenz f <Optik, Telekom> beam divergence
Strahldurchmesser m <Optik> beam diameter
Strahldüse f 1. <Mechan> injector; 2. <Wasserversorg> jet
Strahldüsenbohren n <Kohlen> jet piercing
Strahlemittanz f <Optik> radiant emittance *(Punktgröße)*
strahlen v 1. <Anstrich> blast *(etwa mit Sand)*; 2. <Fernseh, Strahlphys> radiate
Strahlenbehandlung f <Strahlphys> radiation treatment
Strahlenbelastung f <Kerntech, Sicherheit> radiation exposure
Strahlenbrechung f 1. <Optik> refringence; 2. <Wassertrans> refraction
Strahlenbündel n 1. <Elektronik, Fertig> beam; 2. <Optik> beampencil of light; 3. <Telekom> beam
Strahlenbündelung f <Elektronik> beam focusing, beam forming
Strahlenbüschel n <Optik> beampencil of light, pencil of rays
Strahlenchemie f 1. <Kerntech> radiation chemistry; 2. <Strahlphys> radiochemistry
strahlenchemische Zersetzung f <Chemie> radiolysis
Strahlenchromatographie f <Strahlphys> radiochromatography
strahlend adj <Phys, Thermod> radiant
strahlender Kreis m <Fernseh> radiating circuit
Strahlendetektor m 1. <Labor> radiation detector; 2. <Phys> Golay cell
Strahlendiagnose f <Strahlphys> radio diagnosis
Strahlendosimeter n **mit akustischem Alarm** <Strahlphys> sound alarm radiation dosimeter
Strahlendosis f <Umweltschmutz> radiation dose
Strahlendurchlässigkeitsgrad m <Optik> transmittance
Strahlenempfindlichkeit f <Strahlphys> radio sensitivity
Strahlenexponierung f <Kerntech> radiation exposure
Strahlenexposition f <Phys> radiant exposure
Strahlengang m <Telekom> optical path
Strahlengefährdung f <Kerntech> radiation hazard
strahleninduziert adj <Strahlphys> radiation-induced
strahleninduzierte Reaktion f <Strahlphys> radiation-induced reaction
Strahlenkonzentration f <Elektronik> beam forming
Strahlenkrankheit f <Kerntech, Strahlphys> radiation sickness
Strahlenladen n <Fernseh> beam loading
Strahlenmessgerät n 1. <Phys> radiometer; 2. <Strahlphys> actinometer
Strahlenoptik f 1. <Optik> geometric optics, ray optics; 2. <Phys> geometrical optics; 3. <Telekom> ray optics
Strahlenphysik f <Kerntech, Strahlphys> radiation physics
Strahlenquelle f <Strahlphys> radiation source, radio source
Strahlenschaden m <Kerntech, Strahlphys> radiation damage
Strahlenschatten m <Funktech> shadow
Strahlenschutz m <Kerntech, Sicherheit, Strahlphys, Umweltschmutz> radiation protection
Strahlenschutzausrüstung f <Sicherheit> radiation protection equipment
Strahlenschutzbeauftragter m <Kerntech> radiation protection officer
Strahlenschutzgerät n <Sicherheit> radiation protection device
Strahlenschutzglas n <Ker & Glas> radiation shielding glass
Strahlenschutzmantel m <Sicherheit> radiation coating

Strahlenschutzraum m <Kerntech> fallout shelter
Strahlenschutztechnik f <Sicherheit> radiation protection engineering
Strahlenschutzzelle f <Kerntech> hot cell
strahlensicher adj <Kerntech, Sicherheit, Strahlphys> radiation-proof
Strahlenvernetzung f <Kunststoff> radiation cross-linking
Strahlenweg m <Optik> path of rays
Strahlenzählrohr n <Elektronik> radiation counter tube
Strahler m 1. <Elektrotech> emitter; 2. <Funktech> driven element, radiator; 3. <Heiz & Kälte, Mechan> radiator; 4. <Raumfahrt> radiator; 5. <Strahlphys> emitter; 6. <Thermod> radiator
Strahlfokussierung f <Fernseh> beam focusing
Strahlkabine f <Anstrich> blast cabinet
Strahlkegel m <Optik> cone of rays
Strahlkies m <Fertig> grit
Strahlkondensator m <Hydraul> ejector condenser
Strahlkühlung f <Teilphys> cooling
strahlläppen v <Fertig> *(AE)* vapor-blast, *(BE)* vapour-blast
Strahlläppen n <Fertig> *(AE)* vapor blasting, *(BE)* vapour blasting
Strahllärm m <Lufttrans> aerodynamic noise, airframe noise
Strahllärmdämpfer m <Lufttrans> jet noise suppressor
Strahlmischer m <Maschinen> jet mixer
Strahlmittel n <Anstrich> blast medium
Strahlmittelkorn n <Fertig> abrasive grain for blasting
Strahlöffnung f <Optik> beamwidth
Strahlpositioniersystem n <Fernseh> beam-positioning system
Strahlpositionierzeit f <Optik> radial positioning time
Strahlpulser m <Kerntech> beam pulser
Strahlpumpe f <Kerntech, Wasserversorg> jet pump
Strahlreinigen n <Bau> shot blasting
Strahlreinigung f <Fertig> impact cleaning
Strahlrichtung/in <Fernseh> downstream
Strahlrohr n 1. <Bau> tail pipe; 2. <Kerntech> beam hole *(Reaktor)*
Strahlrücklauf m 1. <Fernseh> beam return; 2. <Gerät> retrace *(Oszilloskop)*
Strahlruder n <Wassertrans> thruster
Strahlruder-Antrieb m <Wassertrans> azimuth thruster
Strahlschärfung f <Elektronik> beam sharpening
Strahlschutzgerät n <Sicherheit> beam protection device
Strahlschutzhaube n <Sicherheit> beam protection hood
Strahlsignal n <Elektronik> beam signal
Strahlspanen n <Fertig> *(AE)* vapor-blast cutting, *(BE)* vapour-blast cutting
Strahlstärke f <Elektronik> beam power
Strahlsteuerung f <Optik> radial control
Strahlstrom m <Lufttrans> jet stream
Strahlteiler m <Fernseh, Optik> beam splitter
Strahltetrode f <Elektronik> beam power tube
Strahltilt n <Fernseh> beam tilt
Strahltriebwerk n <Mechan> jet engine
Strahlturbine f <Lufttrans> jet turbine, turbojet
strahlumlenkende Düse f <Lufttrans> thrust vectoring nozzle *(Luftfahrzeug)*
Strahlumschalter m <Gerät> electron beam switch *(Elektronen)*
Strahlung f 1. <Elektriz, Elektronik, Fernseh, Funktech, Heiz & Kälte, Kerntech> radiation; 2. <Labor> radiance *(N)*; 3. <Maschinen> radiation; 4. <Optik> irradiance, radiation; 5. <Optik> radiance *(N)*; 6. <Phys, Raumfahrt, Strahlphys, Teilphys, Telekom, Thermod> radiation
• **Strahlung aussetzen** <Raumfahrt> irradiate *(Weltraumfunk)*

Strahlung

Strahlung *f* **des schwarzen Körpers** <Phys, Strahlphys, Thermod> black body radiation
Strahlung *f* **mit hoher Energie** <Strahlphys> high-energy radiation
Strahlung *f* **mit kleinem Pegel** <Strahlphys> low-level radiation
Strahlungs... 1. <Strahlphys> radiative; 2. <Thermod> radiant; radiative *(Energie)*
Strahlungsabschirmung *f* <Sicherheit> radiation shield
Strahlungsabsorption *f* <Wellphys> absorption of radiation
Strahlungsabsorptionsanalyse *f* <Kerntech> radiation absorption analysis
Strahlungsanregung *f* <Elektronik> radiation excitation
Strahlungsaufbereitung *f* <Kerntech> radiation processing
Strahlungsausbeute *f* <Phys> radiant efficiency
Strahlungsbeiwert *m* <Heiz & Kälte> radiation coefficient
Strahlungsbeständigkeit *f* <Telekom> radiation hardness
Strahlungsbündler *m* <Nichtfoss Energ> solar concentrator
Strahlungsdetektor *m* <Gerät, Strahlphys> radiation detector
Strahlungsdiagramm *n* 1. <Fernseh> beam pattern *(Richtantenne)*; 2. <Funktech> antenna pattern, beam pattern *(Antenne)*; radiation pattern; 3. <Telekom> radiation pattern
Strahlungsdichte *f* 1. <Phys> radiant energy denisty rate, radiant flux density; 2. <Strahlphys> radiant density; 3. <Telekom> radiant intensity
Strahlungsdosimetrie *f* <Strahlphys> radiation dosimetry
Strahlungsdosis *f* <Raumfahrt, Strahlphys> radiation dose
Strahlungsdruck *m* <Akustik, Phys, Strahlphys, Teilphys> radiation pressure
Strahlungseinfang *m* <Phys> radiative capture
Strahlungseinheit *f* <Strahlphys> radiation unit
Strahlungsemission *f* <Telekom> radiant emittance
Strahlungsempfänger *m* <Gerät> radiation detector
strahlungsempfindliches Papier *n* <Konstzeich> radiation-sensitive paper
Strahlungsenergie *f* 1. <Optik> radiant power; 2. <Optik> radiant energy *(U)*; 3. <Phys, Strahlphys, Telekom> radiant energy
Strahlungsenergiedichte *f* <Phys> radiant energy density
Strahlungserwärmung *f* <Strahlphys> radiation heating
strahlungserzeugt *adj* <Phys, Strahlphys> radiogenic
Strahlungsfestigkeit *f* <Kerntech> radiation resistance
Strahlungsfläche *f* <Strahlphys> radiation field *(einer Antenne)*
Strahlungsfluss *m* 1. <Optik, Phys> radiant flux; 2. <Strahlphys> flux of radiation, radiant flux; 3. <Telekom> radiant flux
Strahlungsflussdichte *f* 1. <Kerntech> radiation flux density; 2. <Optik, Strahlphys> radiant flux density; 3. <Telekom> radiant flux density; irradiance *(Stromdichte)*
Strahlungsfrequenz *f* <Kerntech> frequency of radiation
Strahlungsfunktion *f* <Nichtfoss Energ> spectral energy distribution
strahlungsgefährdeter Bereich *m* <Kerntech> radiation danger zone
strahlungsgekoppelte Antenne *f* <Funktech> parasitic aerial
Strahlungsgesetze *npl* <Strahlphys> radiation laws
Strahlungsgürtel *m* <Raumfahrt, Strahlphys> radiation belt
Strahlungshärtung *f* <Kerntech> radiation hardening

Strahlungsheizkörper *m* <Heiz & Kälte> radiant heater, radiant panel
Strahlungsheizung *f* 1. <Heiz & Kälte> radiant heater, radiant panel heating, radiant-heating system; 2. <Thermod> radiant heating
strahlungsinduziert *adj* <Strahlphys> radiation-induced
strahlungsinduzierte Aktivierung *f* <Strahlphys> radiation-induced activation
strahlungsinduzierte Mutation *f* <Strahlphys> radiation-induced mutation
Strahlungsintensität *f* 1. <Nichtfoss Energ> irradiance; 2. <Optik> radiant intensity; 3. <Phys> R, dose rate, radiant intensity; 4. <Raumfahrt> radiation intensity *(Weltraumfunk)*; 5. <Telekom> radiant intensity
Strahlungskaskade *f* <Strahlphys> radiative cascade
Strahlungskatalyse *f* <Strahlphys> radiation catalysis
Strahlungskessel *m* <Heiz & Kälte> radiant boiler
Strahlungskeule *f* <Funktech, Raumfahrt> lobe *(Weltraumfunk)*
Strahlungskollision *f* <Strahlphys> radiative collision
Strahlungskonstante *f* <Heiz & Kälte> radiation constant
Strahlungskontrolle *f* <Strahlphys> radiation monitoring
Strahlungskraft *f* <Ergon> emissivity
Strahlungskühler *m* <Thermod> radiant cooler
Strahlungskühlung *f* <Thermod> radiant cooling
Strahlungslänge *f* <Teilphys> radiation length
Strahlungsleistung *f* 1. <Fernseh> radiated power; 2. <Telekom> radiant power
Strahlungslichtbogenofen *m* <Fertig> indirect arc furnace
strahlungsloser Übergang *m* <Strahlphys> radiationless transition
Strahlungsmesser *m* <Phys> actinometer
Strahlungsmessfühler *m* <Gerät> radiation detector
Strahlungsmessgerät *n* 1. <Raumfahrt> bolometer; 2. <Strahlphys> radioactivity meter
Strahlungsmesskanal *m* <Kerntech> radiation channel
Strahlungsmessstift *m* <Kerntech> pocket dosemeter
Strahlungsmessung *f* 1. <Phys> actinometry; 2. <Strahlphys> actinometry, radiation measurements
Strahlungsmode *f* <Optik> radiation mode
Strahlungsmodus *m* <Telekom> radiation mode
Strahlungsmuster *n* 1. <Optik> radiation pattern *(Lichtleiter)*; 2. <Phys> radiation pattern; 3. <Raumfahrt> radiation pattern *(Weltraumfunk)*
Strahlungsmuster *n* **im stationären Zustand** <Optik> equilibrium radiation pattern
Strahlungsnormal *n* <Optik> standard source
Strahlungspotenzial *n* <Strahlphys> radiation potential
Strahlungspyrometer *n* <Gerät, Labor, Phys> radiation pyrometer
Strahlungspyrometer *n* **mit Thermoelement** <Gerät> thermocouple pyrometer
Strahlungsquelle *f* 1. <Elektronik, Elektrotech> radiation source; 2. <Heiz & Kälte> radiator; 3. <Kerntech, Strahlphys> radiation source
Strahlungsrekombination *f* <Elektronik> radiative recombination
strahlungssicher *adj* <Kerntech, Sicherheit> radiation-proof
Strahlungsspektrum *n* <Raumfahrt> emission spectrum
Strahlungstemperaturmesser *m* <Labor> pyrometer
Strahlungsthermometer *n* <Gerät> pyrometer
Strahlungstrockner *m* <Papier> radiant drier, radiant dryer
Strahlungsübergang *m* <Strahlphys> radiative transition
Strahlungsüberhitzer *m* <Heiz & Kälte> radiant superheater
Strahlungsübertragung *f* <Strahlphys> radiative transfer

Strahlungsüberwachung f 1. <Raumfahrt> radiation monitoring (Raumschiff); 2. <Strahlphys> radiation monitoring
Strahlungsverbrennungen fpl <Strahlphys> radiation burns
Strahlungsverlust m <Teilphys> radiation loss
Strahlungsvermögen n <Phys> radiant power
Strahlungsvernetzen n <Kunststoff> radiation cross-linking
Strahlungsvernetzung f <Kunststoff> radiation cross-linking
Strahlungswärme f <Strahlphys, Thermod> radiant heat
Strahlungswärmegewinn m <Heiz & Kälte> radiant heat gain
Strahlungswiderstand m 1. <Funktech, Kerntech> radiation resistance; 2. <Phys> radiation resistance (Antenne)
Strahlungswinkel m <Optik, Telekom> radiation angle
Strahlungswirkungsgrad m <Phys> radiant efficiency
Strahlungszahl f <Thermod> unit conductance
Strahlungszähler m <Strahlphys> radiation counter
Strahlverbreiterung f <Telekom> beam divergence
Strahlverdichter m <Maschinen> jet compressor
Strahlverdunkelung f <Elektronik> beam blanking (Katodenstrahlröhren)
Strahlverfolgung f <Fernseh> ray tracing
strahlwassergeschütz adj <Sicherheit> jetproof (Explosion)
Strahlwasserhahn m <Wasserversorg> jet cock
Strahlwiedereintrittsröhre f <Elektronik> re-entrant beam tube
Strahlwinkel m <Elektronik> beam angle
Strahlzittern n <Fernseh> beam jitter
Strahlzufuhr f <Kerntech> beam injection
Strainer m <Erdöl> strainer
Strak m <Wassertrans> strake (Schiffbau)
Stramin m <Textil> canvas (Stickereien)
Strand setzen v/auf den <Wassertrans> haul on the beach
stranden v <Wassertrans> get stranded (Schiff)
Stranden n <Wassertrans> stranding
Strandgut n <Wassertrans> jetsam
Strandlinie f <Ker & Glas> shoreline
Strandmauer f <Bau> sea wall
Strang m 1. <Fertig> looping mill; 2. <Maschinen> leg, string; 3. <Maschinen> train (eines Antriebssystem); 4. <Mechan, Papier> rope
Strangaufweitung f <Kunststoff> die swell
Strangbruchdetektor m <Ker & Glas> strand break detector
Strangeness f <Teilphys> strangeness (spezielle Quantenzahl von Hadronen)
stranggepresst adj 1. <Fertig> (AE) extrusion molded, (BE) extrusion moulded; 2. <Maschinen> extruded
stranggepresste Isolierung f <Elektriz> extruded insulation
Stranggießen n <Mechan> continuous casting
Strangpresse f 1. <Maschinen> extruder, extrusion press; 2. <Metall> extrusion press; 3. <Papier> extruder
strangpressen v <Maschinen, Metall> extrude
Strangpressen n 1. <Fertig> (AE) extrusion molding, (BE) extrusion moulding; 2. <Maschinen> extrusion
Strangpressprofil n <Mechan> extrusion
Strangpressteil n <Maschinen> extruded part
Strangpressverfahren n <Maschinen> extrusion process
Strangpresswerkzeug n 1. <Fertig> die head (Kunststoffe); 2. <Maschinen> extrusion die
Strangrohrpresse f <Verpack> extrusion machine for tubes
strapazierfähig adj <Verpack> wear-resistant
Strapazierfähigkeit f <Textil> wear resistance
Straße f 1. <Trans> road, road rail; 2. <Wassertrans> strait (Geographie) • **auf Straße und Schiene transportieren** <Trans> transport by rail and road
Straße f **in Tieflage** <Bau> sunken road
Straße f **mit geringem Verkehr** <Bau> low volume road, LVR
Straße f **mit jeweils zwei Fahrbahnen getrennt in jede Fahrtrichtung** <Trans> dual carriageway
Straßenablauf m <Bau> storm drain (Regenwasser)
Straßenabwasser n <Bau> surface water
Straßenarbeiten fpl <Trans> road works
Straßenaufreißer m <Bau, Trans> road ripper
Straßenbahn f <Eisenbahn> (AE) streetcar, (BE) tramway
Straßenbahnfahrplan m <Eisenbahn> (AE) streetcar schedule, (BE) tram schedule
Straßenbahnhaltestelle f <Eisenbahn> (AE) streetcar stop, (BE) tram stop
Straßenbahnschienen fpl <Bau, Eisenbahn> (AE) streetcar tracks, (BE) tram tracks
Straßenbahntriebwagen m <Eisenbahn> (AE) streetcar, (AE) streetcar motor coach, (BE) tramcar, (BE) tramway motor coach
Straßenbahnverkehr m **im Expressbetrieb** <Trans> (AE) express streetcar, (BE) express tramway
Straßenbahnwagen m 1. <Eisenbahn> (AE) streetcar, (BE) tram, (BE) tramcar; 2. <Trans> (AE) trolley car
Straßenbankett n <Bau> shoulder
Straßenbau m <Trans> road building, road making
Straßenbaumaschine f 1. <Bau> road making machine; 2. <Trans> road building machinery
Straßenbezeichnungsschild n <Trans> road identification sign
Straßenbrücke f <Bau, Trans> road bridge
Straßendamm m <Bau> bank
Straßendecke f <Bau> sheeting
Straßendeckenbelag m <Bau> carpet, coat
Straßendeckenbeton m <Bau> (BE) pavement-quality concrete, (AE) sidewalk-quality concrete
Straßenfahrzeugwaage f <Trans> road vehicle weighing machine
Straßenfertiger m 1. <Bau> finisher; 2. <Trans> road finishing machine
Straßengabelung f <Trans> bifurcation, road junction
straßengestützter Sender m <Trans> road-based transmitter
Straßenglätte f <Trans> skidding conditions
Straßenhaftung f <Bau> road adherence
Straßenhobel m <Bau> grader
Straßenkarte f <Trans> road map
Straßenkehricht m <Abfall> litter
Straßenkehrmaschine f <Abfall> mechanical sweeper
Straßenkreuzungspunkt m <Bau> road crossing
Straßenmeisterei f <Trans> traffic management
Straßenneigung f <Trans> road camber
Straßennetz n <Trans> road system
Straßenperle f <Ker & Glas> road bead
Straßenpflug m <Trans> road plough
Straßenplanierer m <Trans> motor grader, road grader
Straßenplatte f <Bau> pad
Straßenschienenbus m <Trans> bimodal bus, road-rail bus
Straßenschotter m <Bau> road metal
Straßentankfahrzeug n <Trans> RTC, road tank car
Straßenteermaschine f <Trans> road tarring machine
Straßenverbindung f <Trans> road communication
Straßenverkehr m <Trans> road traffic
Straßenverkehrskontrolle f <Trans> road traffic control
Straßenwalze f 1. <Bau> compactor, roller; 2. <Trans> roadroller

Straßenwalze

Straßenwalze f mit Metallrädern <Bau> iron-tired roller
Straßenzugmaschine f <Kfztech> road tractor, trailer towing vehicle
Straßenzustandsbericht m <Kfztech, Trans> road news
stratifizieren v <Kohlen> stratify
Stratigraphie f <Kohlen> stratigraphy
stratigraphische Falle f <Erdöl> stratigraphic trap *(Geologie)*
Stratocumulus m <Lufttrans> stratocumulus *(Meteorologie)*
Stratosphäre f <Nichtfoss Energ> stratosphere
Streamer m <Comp & DV> streaming tape drive; streamer *(zur Datensicherung)*
Streamerkammer f <Teilphys> streamer chamber
Streaming n <Telekom> streaming *(Daten)*; streaming video *(Internet: Echtzeit- Bewegtbildübertragung)*
Streaming-Video n <Telekom> streaming video *(Internet: Echtzeit-Bewegtbildübertragung)*
Streb m <Kohlen> bank
Strebbau m <Kohlen> longwall mining, straining work
Strebbausystem n <Kohlen> longwall system
Strebe f 1. <Bau> brace, prop, shore, spur, stay; 2. <Fertig> spur; 3. <Hydraul> brace; 4. <Kfztech> strut; 5. <Kohlen> shore; 6. <Maschinen> brace; 7. <Mechan> brace, strut, tie rod
Strebenkopf m <Bau> strutting head
strebenloser Balken m <Bau> unsupported beam
Strebepfeiler m <Bau> buttress *(Wand, Brücke)*; counterfort
Strebfront f <Kohlen> breast
Strebschrämmaschine f <Kohlen> longwall coalcutting machine
Strebstrecke f <Kohlen> gate road
Strebvorbau m <Kohlen> longwall advancing mining
Streckbalken m <Bau> binding beam, string piece
streckbar adj <Maschinen> tensile
Streckbarkeit f 1. <Kunststoff> extensibility; 2. <Maschinen> tensility
Streckbereich m <Textil> draft zone
Streckblasformen n <Kunststoff> *(AE)* stretch blow molding, *(BE)* stretch blow moulding
Streckdehnung f <Kunststoff> elongation
Strecke f 1. <Geom> line segment; 2. <Kohlen> adit; 3. <Telekom> path, range, route; 4. <Trans> route • **Strecke zurücklegen** <Trans> run
strecken v 1. <Bau> stretch; 2. <Maschinen> flatten, stretch; 3. <Textil> draw, drawing
Strecken n 1. <Fertig> drawing-out; 2. <Ker & Glas> flatting; flattening *(Zylinder)*; 3. <Textil> drawing, stretch
Strecken n **des Zylinders** <Ker & Glas> elongation of the cylinder
Streckenabschnitt m **zwischen zwei Flughäfen** <Lufttrans> mileage
Streckenbelastung f <Eisenbahn> traffic density
Streckenbeschreibung f <Lufttrans> route description
Streckenbilanz f <Funktech> link budget *(Satellitenfunk)*
Streckenblock m <Eisenbahn> block system
Streckendämpfung f <Telekom> path attenuation
Streckeneinweisungsflug m <Lufttrans> route familiarization flight
Streckenforschungsprogramm n <Trans> *(AE)* shortest route program, *(BE)* shortest route programme
Streckenhöchstgeschwindigkeit f <Eisenbahn> speed limit
Streckenisolator m <Eisenbahn> section insulator
Streckenlizenz f <Lufttrans> *(BE)* route licence, *(AE)* route license, route licensing
Streckenlokomotive f <Eisenbahn> *(AE)* road locomotive

Streckenprofil n <Telekom> path profile
Streckenprüfzug m <Eisenbahn> spot train
Streckenrelais n <Elektriz> distance relay
Streckenschalter m <Elektriz> distance switch
Streckenüberlastung f <Telekom> route congestion
Streckenverbindung f <Telekom> linking
Strecker m <Bau> binder *(Mauerwerk)*
Streckfolie f <Kunststoff, Verpack> stretch film, stretch wrapping film
streckformen v <Fertig> stretch-form
Streckformen n <Maschinen> stretch forming
Streckgesenk n <Maschinen> fuller
Streckgrenze f 1. <Anstrich, Kunststoff> yield strength; 2. <Maschinen> limit of elasticity; 3. <Mechan> elastic limit, yield point, yield strength; 4. <Metall> yield point, yield stress; 5. <Werkprüf> yield strength
Streckmaschine f <Textil> drawing frame
Streckmetall n <Mechan, Metall> expanded metal
Streckmittel n 1. <Chemtech> diluting agent; 2. <Fertig> filler *(Kunststoffe)*; 3. <Kunststoff> extender; 4. <Lebensmittel> adulterant
Streckofen m <Ker & Glas> flattening kiln
Streckspannung f <Kunststoff, Phys> yield stress
Strecktisch m <Ker & Glas> flattening table
Streckung f <Maschinen> extension
Streckverpackung f <Verpack> stretch wrapping
Streckvorrichtung f <Maschinen> strainer, stretcher
Streckwalze f <Fertig> cogging roll, cogging-down roll
Streckwalzwerk n <Fertig> elongator
Streckwerk n <Textil> drafting system, drawing frame
Streckwerkzeug n <Ker & Glas> flattening tool
streckziehen v <Fertig> stretch-form
Streckziehen n <Fertig> metal stretching
strehlen v <Maschinen> chase
Strehlen n <Maschinen> chasing
Strehler m <Maschinen> chaser
Strehlerkluppe f <Maschinen> chaser die stock
Strehlmaschine f <Maschinen> chasing lathe, chasing machine
Streichanlage f <Verpack> coating machine
Streichbalken m <Bau> head plate, trimmer, trimmer beam
Streichblech n <Fertig> sleeker *(Formen)*
Streicheisen n <Ker & Glas> *(AE)* crack-off iron, *(BE)* wetting-off iron
streichen v 1. <Anstrich> paint; 2. <Bau> coat; 3. <Comp & DV> delete; 4. <Funktech, Telekom> cancel
Streichen n 1. <Papier> coating; 2. <Textil> carding *(Spinnen)*
Streichgarnspinnerei f <Textil> woollen spinning
Streichgießverfahren n <Papier> cast coating
Streichlinien fpl <Strömphys> streaklines
Streichmaschine f 1. <Kunststoff> coating machine; 2. <Papier> coater, coating machine
Streichmaschine f mit Abquetschwalze <Papier> squeeze roll coater
Streichmaß n <Bau> *(AE)* scratch gage, *(BE)* scratch gauge
Streichmasse f 1. <Papier> *(AE)* coating color, *(BE)* coating colour, coating mixture; 2. <Verpack> coating compound
Streichmesser n <Papier> doctor blade
Streichrohpapier n <Papier> coating base paper
Streichung f <Comp & DV> deletion
Streichverfahren n <Kunststoff> spread coating
Streichwehr n <Wasserversorg> spillway
Streifen m 1. <Comp & DV> strip, stripe, tape; 2. <Gerät> chart *(für Registriergerät)*; 3. <Ker & Glas> segregation; 4. <Metall> strip; 5. <Metrol> ribbon; 6. <Papier> streak;

7. <Phys> fringe; fringes *(gleicher Neigung)*; 8. <Textil> crack, strip, stripe
Streifen *mpl* <Fertig> bands *(Herstellungsfehler)*
Streifenabreißstab *m* <Gerät> chart paper tear-off bar *(Registriergerät)*
Streifenabstand *m* <Phys> fringe separation *(Interferenz)*
Streifenantriebsmotor *m* <Gerät> chart motor
Streifenauslenkung *f* <Fertig> band deviation *(Interferenz)*
Streifenberieselung *f* <Wasserversorg> border irrigation
Streifenbildung *f* 1. <Fernseh> banding, fringing; 2. <Ker & Glas> banding
streifender Einfall *m* <Phys> grazing incidence
streifender Stoß *m* <Kerntech> glancing collision
streifender Strahl *m* <Akustik> grazing ray
Streifendetektor *m* <Raumfahrt> strip-type detector *(Raumschiff)*
Streifenformat *n* <Konstzeich> strip size
Streifenfüllung *f* <Ker & Glas> strip filling
Streifenfundament *n* <Bau> strapped footing
Streifengummierung *f* <Druck> strip gumming
Streifenleiter *f* <Elektrotech, Funktech> strip line
Streifenleitung *f* <Funktech, Phys> strip line
Streifenleitungsantenne *f* <Elektronik, Funktech> microstrip aerial, microstrip antenna
Streifenlocher *m* <Comp & DV> tape punch
Streifenmarkierung *f* <Textil> rope marking
Streifenschneider *m* <Ker & Glas> slitter
Streifenschreiber *m* <Elektriz> strip chart recorder
Streifentransport *m* <Gerät> chart feed, chart transport *(Registriergerät)*
Streifenversilberung *f* <Ker & Glas> striped silvering
Streifenvorschub *m* <Gerät> chart feed
Streifenwalze *f* <Gerät> chart drum *(Registriergerät)*
Streifenware *f* <Textil> striped fabric
Streifenwellenleiter *m* <Funktech> strip waveguide *(Mikrowellen)*
Streifenzuführung *f* <Gerät> chart feed
streifig *adj* <Papier> striped
streifiger Perlit *m* <Metall> lamellar pearlite
Streifigkeit *f* <Textil> barriness in the weft
Stretch *m* <Textil> stretch
Stretcherring *m* <Teilphys> stretcher ring
Stretchfolie *f* <Kunststoff> stretch film
Streu... <Akustik, Elektronik, Elektrotech, Kerntech, Phys> leakage, scattering, stray
Streuausbreitung *f* <Funktech> propagation by ionospheric scatter, scatter propagation
Streubaumwolle *f* <Kunststoff> cotton linters
Streubeleuchtung *f* <Elektriz> floodlighting
streuend *adj* <Phys> straying
Streufaktor *m* <Chemtech> dispersion coefficient
Streufeld *n* 1. <Fertig> leakage field *(Magnetpulververfahren)*; 2. <Telekom> leakage field
Streufeld-Methode *f* <Aufnahme> diffuse-field method
Streufluss *m* <Phys> leakage flux
Streufluss *m* **über der Fehlstelle** <Fertig> crack leakage flux
Streufolie *f* <Kerntech> scattering foil
Streuglas *n* <Ker & Glas> diffusing glass
Streugrenze *f* <Qual> dispersion limit
Streugrenzen *fpl* <Qual> limits of variation
Streukapazität *f* <Phys> stray capacitance
Streukoeffizient *m* 1. <Akustik> *(m)* flare coefficient of horn *(m)*; 2. <Telekom> scattering coefficient
Streukopplung *f* <Elektrotech> stray coupling
Streukörper *m* 1. <Optik> diffuser *(Licht)*; 2. <Telekom> diffuser
Streulicht *n* <Optik, Strahlphys> scattered light

Streumatrix *f* <Telekom> scattering matrix
Streumedium *n* <Kerntech> scattering medium
Streuneutron *n* <Kerntech> scattered neutron
Streuquerschnitt *m* <Phys> cross section, scattering cross section
Streusand *m* <Fertig> parting sand *(Gießen)*
Streuschwingung *f* <Phys> stray oscillation
Streusignalaufnahme *f* <Gerät> stray signal pick-up
Streuspektrum-Modulation *f* <Elektronik> spread spectrum modulation
Streuspektrum-Modulator *m* <Elektronik> spread spectrum modulator
Streustrahlung *f* 1. <Elektrotech> leakage radiation; 2. <Strahlphys> scattered radiation, stray radiation
Streustrom *m* <Elektrotech> stray current
Streutransformator *m* <Elektrotech> constant-current transformer
Streuung *f* 1. <Akustik> scattering; 2. <Comp & DV> scatter; 3. <Elektrotech> leakage, spill; 4. <Fertig> leak *(Strom)*; 5. <Funktech> leak, scatter *(Verluststrom)*; 6. <Kerntech, Metall, Nichtfoss Energ, Optik> scattering; 7. <Math> dispersion *(der Werte einer Zufallsgröße)*; 8. <Qual> dispersion; 9. <Strahlphys> dispersion, scattering; 10. <Telekom> dispersion, leakage, scattering
Streuung *f* **an Aerosolen** <Funktech, Phys> aerosol scattering *(Wellenausbreitung)*
Streuung *f* **der Verteilung** <Qual> variance of distribution
Streuung *f* **durch Regen** <Raumfahrt> rain scatter *(Weltraumfunk)*
Streuung *f* **im Lichtwellenleiter** <Telekom> *(AE)* fiber scattering, *(BE)* fibre scattering
Streuung *f* **mit großen Streuwinkeln** <Kerntech> wide-angle scattering
Streuungsdiagramm *n* <Math> scatter diagram *(zweidimensionale Datenerfassung)*
streuungsfreies Medium *n* <Phys> nondispersive medium
Streuungshalo *m* <Kerntech> halo of dispersion
Streuverlust *m* 1. <Akustik> transducer dissipation loss *(Wandler)*; 2. <Elektrotech> leakage loss, stray loss; 3. <Fertig> dispersion loss *(Faseroptik)*, leakage; 4. <Telekom> scattering loss
Streuvermögen *n* <Phys> dispersive power
Streuwerte *mpl* <Gerät> scattered data
Streuwinkel *m* <Phys> scattering angle
Streuwirkungsquerschnitt *m* <Phys> scattering cross section
Strg-Taste *f* *(Steuertaste)* <Comp & DV> Ctrl key *(control key)*
Strich *m* 1. <Druck> line, rule; 2. <Fertig> prime; 3. <Maschinen> stroke; 4. <Optik> groove *(Beugungsgitter)*; 5. <Textil> pile *(Gewebe)*
Strichcode *m* <Comp & DV, Telekom, Verpack> bar code
Strichcode-Lesegerät *n* <Verpack> bar code reader
Strichcode-Leser *m* <Comp & DV> bar code slot reader, bar code scanner
Strichcode-Markendrucker *m* <Verpack> bar code label printer
Strichdiagramm *n* <Comp & DV> line graph
Strichdichte *f* <Druck> stroke density
Striche *mpl* **der Kompassrose** <Wassertrans> points of the compass
Strichendmaß *n* <Fertig> *(AE)* hair gageblock, *(BE)* hair gaugeblock
Strichlehre *f* <Metrol> *(AE)* length gage, *(BE)* length gauge
Strichlinie *f* <Fertig> dashed line
Strichlistendiagramm *n* <Math> tally diagram *(Statistik)*

Strichmarkierung

Strichmarkierung *f* <Verpack> bar code
Strichmarkierungssystem *n* <Verpack> *(AE)* bar code labeling system, *(BE)* bar code labelling system
Strichplan *m* <Elektriz> single line diagram
strichpunktierter Kreis *m* <Konstzeich> chain-like circle
Strichpunktlinie *f* <Konstzeich> chain line
Strichskalenwaage *f* <Metrol> line-graduated master scales
Strichstärke *f* <Druck> weight of face, weight of type, zinc plate
Strichvorlage *f* 1. <Druck> line artwork; 2. <Konstzeich> line master
Strichzeichnung *f* 1. <Druck> line drawing; 2. <Elektriz> single line diagram; 3. <Geom> line drawing
Strick *m* <Textil> rope
Strickmaschine *f* <Fertig, Textil> knitting machine
Strickschlauch *m* <Textil> circular knitted fabric
String *m* <Druck> string
Stringer *m* 1. <Raumfahrt> stringer *(Raumschiff)*; 2. <Wassertrans> stringer *(Bootbau)*
Stringerwinkel *m* <Wassertrans> stringer angle *(Schiffbau)*
Stringkonstruktion *f* <Strahlphys> string construction
Striping *n* <Aufnahme> striping
Strippen *n* <Druck> stripping
Stroboskop *n* <Phys, Wellphys> stroboscope
Stroboskopband *n* <Aufnahme> stroboscopic tape
stroboskopischer Impuls *m* <Telekom> strobe pulse
Stroboskoplicht *n* <Raumfahrt> strobe light
Strohpappe *f* <Druck, Verpack> strawboard
Strohzellstoff *m* <Papier> straw pulp
Strohzellstoffpappe *f* <Verpack> strawboard
Strom *m* 1. <Comp & DV> stream; 2. <Elektriz, Elektrotech> current; 3. <Funktech> electricity; 4. <Phys> current; 5. <Wassertrans> current *(Navigation)* • **außer Strom setzen** <Elektrotech> de-energize • **gegen den Strom** <Wassertrans> upstream • **unter Strom** <Elektrotech> alive
Strom *m* **in Durchlassrichtung** <Elektrotech> forward current
Strom *m* **in Sperrrichtung** <Elektrotech> reverse current
Stromabfall *m* <Elektrotech> current drop
Stromabgabe *f* <Elektrotech> current output
Stromabnehmer *m* 1. <Eisenbahn> pantograph; 2. <Elektrotech> current collector
Stromabnehmerbügel *m* <Eisenbahn, Elektriz> bow *(Straßenbahn)*
Stromabnehmergestell *n* <Eisenbahn> pantograph frame
Stromabnehmersignal *n* <Eisenbahn> pantograph signal
Stromabnehmersystem *n* <Trans> power collection system
stromabwärteriger Pfeilerkopf *m* <Wasserversorg> downstream cutwater
stromabwärts *adv* <Mechan, Nichtfoss Energ, Raumfahrt, Strömphys, Wassertrans> downstream
stromabwärts gelegen *adj* <Phys, Strömphys> downstream
Stromanstiegsrate *f* <Elektrotech> rate of current rise
Stromanzeige *f* <Gerät> current reading
Stromanzeigewert *m* <Gerät> current reading
Stromaufnahme *f* <Elektrotech> current consumption
Stromaufnahme *f* **eines blockierten Läufers** <Elektriz> locked rotor current
Stromaufnahmebürste *f* <Elektrotech> current-collecting brush
stromaufwärts *adj* <Bau, Nichtfoss Energ, Strömphys> upstream

stromaufwärts *adv* <Nichtfoss Energ, Raumfahrt, Strömphys, Wassertrans> upstream
stromaufwärts gelegenes Totwasser *n* <Strömphys> upstream wake
Stromausfall *m* <Elektrotech, Kohlen> blackout, current failure
Stromausgang *m* <Elektrotech> current output, electrical output
Stromauswanderung *f* <Telekom> current drift
Strombahn *f* <Elektrotech> current path
Strombauch *m* <Elektrotech> current antinode
strombegrenzende Drosselspule *f* <Elektrotech> current-limiting reactor
strombegrenzende Schmelzsicherung *f* <Elektriz> current-limiting fuse link
strombegrenzender Leistungsschalter *m* <Elektrotech> current-limiting circuit-breaker
strombegrenzender Unterbrecher *m* <Elektriz> current-limiting circuit breaker
Strombegrenzer *m* <Elektriz, Elektrotech> current limiter
Strombegrenzung *f* <Elektrotech> current limiting
Strombegrenzungsdrossel *f* 1. <Elektriz> current-limiting inductor; 2. <Elektrotech> current-limiting reactor
Strombegrenzungsreaktanz *f* <Elektriz> current-limiting reactor
Strombegrenzungsspule *f* <Elektriz> current-limiting reactor
Strombegrenzungswiderstand *m* <Elektriz> limiting resistor
Strombereich *m* <Gerät> current range
Strombetriebslogik *f* <Elektrotech> current mode logic
Stromdichte *f* 1. <Elektriz, Elektrotech> ampere density, current density; 2. <Metall, Phys, Qual> current density
Stromdifferenzschutz *m* <Elektriz> current differential protection
Stromeingang *m* <Elektrotech> current input, electrical input
Stromelement *n* <Phys> current element
strömen *v* <Nichtfoss Energ, Strömphys, Wassertrans> flow *(Fluss, Gezeiten)*
Strömen *n* <Elektrotech> flow
Stromerhitzung *f* <Kerntech> joule heating
stromerzeugend *adj* <Elektrotech> current generating
stromerzeugende Anlage *f* <Elektrotech> current generating plant
Stromerzeuger *m* 1. <Bau> generator; 2. <Elektrotech> current generator, electric generator, generator; 3. <Wassertrans> generator *(Elektrik)*
Stromerzeugung *f* 1. <Elektrotech> electricity generation; 2. <Kerntech> electroproduction
Stromerzeugungskapazität *f* <Elektriz> generating capacity
Stromfernmessung *f* <Gerät> current telemetering
Stromfluss *m* **im Schaltkreis** <Phys> flux through a circuit
Stromflussplan *m* <Elektriz> circuit diagram
stromführend *adj* 1. <Elektrotech> active, alive, live; 2. <Phys> live
stromführende Komponente *f* <Elektrotech> active component, life component
stromführende Leitung *f* <Elektrotech> active wire, live wire
stromführende Spule *f* <Phys> current-carrying coil
stromführender Draht *m* <Elektrotech> active wire, live wire
stromführender Kreis *m* <Elektrotech> active circuit, live circuit
stromführendes Mikrofon *m* <Aufnahme> active microphone, live microphone

Stromführungskapazität f <Elektriz> current-carrying capacity
Stromfunktion f <Strömphys> stream function
Stromgebiet n <Wassertrans> water system *(Fluss)*
Stromgenerator m 1. <Kerntech> electric generator; 2. <Phys> current generator
Stromgeneratoraggregat n <Elektriz> generating set
Stromgeschwindigkeit f <Nichtfoss Energ, Wassertrans> current rate *(Meer)*
stromgesteuerte Vorrichtung f <Elektrotech> current-controlled device
stromgesteuerter Oszillator m <Elektronik> current-controlled oscillator
Stromimpuls m <Elektrotech> current pulse
Stromintensität f <Elektriz> current intensity, intensity of current
Stromkabel n <Elektriz, Elektrotech> electricity cable, power cable
Stromkabelung f <Wassertrans> rip tide *(Gezeiten)*
Stromkarte f <Wassertrans> current chart *(Navigation)*
Stromkompensator m <Gerät> current compensator
Stromkreis m <Elektriz, Elektrotech> circuit, electric circuit • **einen Stromkreis einschalten** <Elektrotech> close a circuit • **einen Stromkreis schließen** <Elektrotech> close a circuit • **im Stromkreis** <Elektriz> in circuit
Stromkreis m **für den Leerlauf und für niedrige Drehzahlen** <Kfztech> idle and low speed circuit
Stromkreisunterbrecher m <Elektriz> air interrupter *(Luftisolierung)*
Stromlauf m <Elektrotech> circuit diagram
Stromlaufplan m <Kfztech> wiring diagram
Stromleiter m <Elektrotech> conductor
Stromleiterdraht m <Elektrotech> conductor wire
Stromleiterschiene f <Elektriz> conductor rail
Stromleitung f 1. <Bau> transmission line; 2. <Elektrotech> conduction, electrical transmission line
Stromlieferung f **ans Verbrauchernetz** <Kerntech> delivery into the mains
Stromlinie f 1. <Phys> streamline; filament *(Flüssigkeit)*; 2. <Strömphys> streamline *(Flüssigkeiten, Gasen)*
Stromlinienbild n 1. <Strömphys> streamline pattern; 2. <Wasserversorg> flow pattern
Stromlinienform f <Maschinen> streamlined form
stromlinienförmig adj 1. <Kfztech> streamlined *(Karosserie)*; 2. <Lufttrans> faired; 3. <Maschinen, Phys, Raumfahrt, Strömphys> streamlined
stromlinienförmige Verkleidung f <Lufttrans> fairing
Stromlinienverkleidung f <Kfztech> streamlining
stromlos adj <Gerät> dead
stromlos geschlossen adj <Fertig> de-energized closed *(Kunststoffinstallationen)*
Stromlosigkeit f <Gerät> absence of current
Strommessbereich m <Gerät> current range
Strommesser m 1. <Elektrotech, Labor> ammeter; 2. <Nichtfoss Energ> current meter; 3. <Wassertrans> current meter *(Ozeanographie)*; 4. <Wasserversorg> current meter
Strommessgerät n <Gerät> ammeter, amperemeter, current-measuring instrument
Strommessinstrument n <Gerät> current-measuring instrument
Strommessung f <Elektrotech> current sensing
Strommesswiderstand m <Elektrotech> current-sensing resistor
Strommodulation f <Elektronik> current modulation
Stromnetz n 1. <Elektriz, Elektrotech> *(BE)* mains, mains supply, power supply circuit, power system, *(AE)* supply network; 2. <Fertig> electric mains
Strompfad m <Elektrotech> current path, path

Stromquelle f 1. <Elektrotech> current source, source; 2. <Nichtfoss Energ> current source, power source; 3. <Phys> current source
Stromrauschen n <Aufnahme> *(AE)* current noise, mains noise
Stromregelung f <Elektriz, Elektrotech> current control, current regulation
Stromregler m <Elektriz, Elektrotech> barretter, current controller, current regulator
Stromrelais n <Elektriz, Elektrotech> current relay, electric relay, electrical relay
Stromrichter m <Elektrotech> converter, current converter, rectifier, static converter
Stromrichteranlage f <Elektrotech> static converter device
Stromrichtergruppe f <Elektrotech> static converter bank
Stromrichterlokomotive f <Eisenbahn> thyristor-controlled locomotive
Stromrichtung f <Elektrotech> current direction • **in Stromrichtung** <Phys, Strömphys> downstream
Stromrückkopplung f <Elektrotech> current feedback
Stromsammlerring m <Elektriz> collector ring
Stromsammlerschuh m <Eisenbahn> collector shoe
Stromsättigung f <Elektronik> current saturation
Stromschiene f 1. <Eisenbahn> conductor rail, contact rail, live rail, third rail; 2. <Elektriz, Elektrotech> bus, busbar, conductor rail
Stromschlag m <Elektrotech, Sicherheit> electric shock
Stromschleife f <Funktech, Phys> current loop
Stromschließer m <Elektrotech> circuit closer
Stromschnelle f <Kohlen, Nichtfoss Energ> cataract
Stromschnelle-Überlauf m <Nichtfoss Energ> chute spillway
Stromschreiber m <Gerät> current recorder
Stromschutz m <Elektriz> air break switch, air interrupter *(Luftisolierung)*
Stromschwankung f <Elektriz> current fluctuation
Strom-Spannungs-Charakteristik f <Elektrotech> current-voltage characteristic
Strom-Spannungs-Kennlinie f <Elektrotech, Raumfahrt> current-voltage characteristic *(Raumschiff)*
Strom-Spannungs-Kurve f <Elektrotech> current-voltage characteristic
Stromspeicheraggregat n <Kfztech, Wassertrans> battery
Stromspitze f <Elektriz, Elektrotech> current peak
Stromstärke f <Elektriz, Elektrotech> amperage, current, current intensity
Stromstecker m <Elektriz> power outlet
Stromstoß m <Elektrotech> current pulse, current surge
Stromstoßgenerator m <Elektriz> surge generator
Stromstoßrelais n <Elektriz> latching relay
Stromstoßrelais n **mit Doppelspule** <Elektrotech> dual-coil latching relay
Stromstreckentrennung f <Elektriz> sectionalization
Stromsystem n **mit durch Impedanz geerdetem Mitteilleiter** <Elektriz> *(BE)* impedance earthed neutral, *(AE)* impedance grounded neutral
Stromsystem n **mit fest geerdetem Nullleiter** <Elektriz> *(BE)* solidly earthed neutral system, *(AE)* solidly grounded neutral system
Stromtankstelle f <Elektriz> electric power filling station
Stromteiler m <Elektriz> current divider
Stromtor-Inverter m <Elektronik> thyratron inverter
Stromtransformationsverhältnis n <Elektriz> current transformation ratio
Stromtransformator m <Elektriz, Elektrotech> *(AE)* current transformer, *(BE)* mains transformer
Stromübergang m <Nichtfoss Energ> ohmic contact

Stromübertragung

Stromübertragung f <Phys> electricity transmission
Stromübertragungsfaktor m <Akustik> response to current
Strömung f 1. <Bau, Elektrotech> flow; 2. <Erdöl> fluid flow *(Flüssigkeiten, Gasen)*; 3. <Hydraul, Nichtfoss Energ, Phys> flow; 4. <Strömphys> flow, stream; flow *(durch Blenden)*; 5. <Wassertrans> current *(Navigation)* • **in freier Strömung** <Strömphys> in free stream
Strömung f **in offenen Gerinnen** <Strömphys> flow in open channels, open channel flow
Strömung f **in offenen Kanälen** <Strömphys> flow in open channels, open channel flow
Strömungsabriss m **am Luftschraubenblatt** <Lufttrans> blade stall *(Hubschrauber)*
Strömungsabriss m **am rücklaufenden Blatt** <Lufttrans> retreating blade stall
Strömungsabriss m **an der Luftschraubenblattspitze** <Lufttrans> blade tip stall, mean aerodynamic chord *(Hubschrauber)*
Strömungsbild n <Heiz & Kälte, Wasserversorg> flow pattern
Strömungsbremse f <Kfztech> hydro-kinetic brake
Strömungsdurchsatz m <Kunststoff> flow rate, rate of flow
Strömungsdüse f 1. <Maschinen> flow nozzle; 2. <Phys> jet
Strömungsförderer m <Maschinen> flow conveyor
Strömungsgeschwindigkeit f 1. <Heiz & Kälte> flow velocity, velocity of flow; 2. <Nichtfoss Energ> current rate; 3. <Strömphys> flow speed; 4. <Wassertrans> current rate *(Meer)*; 5. <Wasserversorg> flow rate
Strömungsgeschwindigkeitsmessgerät n <Gerät> anemometer
Strömungsgetriebe n <Maschinen> fluid transmission
Strömungsinstabilität f <Strömphys> flow instability
Strömungskabine f <Sicherheit> noise downflow booth *(Lärm)*
Strömungskanal m 1. <Lufttrans, Wassertrans> wind tunnel; 2. <Wasserversorg> calibration flume, flume
Strömungskupplung f 1. <Kfztech> fluid coupling *(Kraftübertragung)*; 2. <Maschinen> hydraulic coupling
Strömungslehre f 1. <Maschinen> fluid mechanics; 2. <Strömphys> fluid dynamics
Strömungslinien fpl <Telekom> flow lines
Strömungsmechanik f <Strömphys> fluid dynamics, fluid mechanics
Strömungsmengenmessgerät n <Gerät> rate-of-flow meter
Strömungsmessgerät n <Hydraul> current meter
Strömungsmessung f <Gerät> flow measurement
Strömungsmuster n 1. <Strömphys> flow pattern; 2. <Thermod> gas flow
Strömungsregler m <Mechan> control valve
Strömungsrichtung f <Comp & DV> flow direction • **in Strömungsrichtung** <Fernseh> upstream
Strömungsrohr n <Hydraul> flow pipe
Strömungssichtbarmachung f <Strömphys> flow visualization
Strömungstechnik f <Strömphys> fluid engineering
Strömungsventil n <Lufttrans> flux valve
Strömungsvisualisierung f <Strömphys> flow visualization
Strömungswächter m <Heiz & Kälte> flow indicator
Strömungswiderstand m 1. <Akustik> flow resistance; 2. <Heiz & Kälte> drag; 3. <Hydraul> coefficient of drag *(DD)*; 4. <Maschinen> flow resistance; 5. <Phys> drag; 6. <Qual, Strömphys> resistance to flow; 7. <Trans> captation drag; 8. <Wassertrans> drag *(Schiffkonstruktion)*
Stromunterbrecher m <Elektriz, Elektrotech> circuit breaker, switch

746

Stromverband m <Bau> diagonal bond
Stromverbrauch m <Elektriz, Elektrotech> electricity consumption
Stromverdrängung f <Elektrotech> current displacement, proximity effect, skin effect
Stromverdrängungseffekt m <Elektrotech> skin effect
Stromversetzung f <Wassertrans> current set *(Navigation)*
Stromversorgung f 1. <Bau> electricity supply; 2. <Eisenbahn, Elektriz, Elektrotech> electrical power supply, power supply; 3. <Kohlen, Nichtfoss Energ, Phys, Telekom> power supply
Stromversorgung f **der Heizung** <Elektrotech> heater power supply
Stromversorgung f **mit nur einem Ausgang** <Elektrotech> single output power supply
Stromversorgungsanlage f <Telekom> power plant
Stromversorgungsmodul n <Nichtfoss Energ> containerized power pack
Stromversorgungsnetz n 1. <Elektriz, Elektrotech> electric power system, *(BE)* mains, power grid, power supply circuit, *(AE)* supply network; 2. <Kohlen> power grid
Stromversorgungssystem n **mit geerdetem Nullleiter** <Elektriz> *(BE)* earthed-neutral system, *(AE)* grounded-neutral system
Stromversorgungsunternehmen n <Elektriz> electric power supply company *(EVU)*
Stromversorgungswerk n <Elektrotech> electric power station
Stromverstärker m <Elektriz, Elektronik> current amplifier
Stromverstärkung f 1. <Aufnahme> current amplification; 2. <Elektronik> current amplification, current gain; 3. <Phys> current gain
Stromverteilung f <Elektriz> current distribution, distribution of electricity
Stromverteilungsgeräusch n <Elektronik> partition noise
Stromverteilungssystem n <Elektriz> distribution system
Stromwaage f 1. <Elektriz> ampere balance, current balance; 2. <Gerät> ampere balance; 3. <Phys> current balance
Stromwaage-Relais n <Elektriz> current balance relay
Stromwächter m <Elektrotech> current relay
Stromwandler m <Elektrotech> *(AE)* current transformer, *(BE)* mains transformer
Stromwandlung f <Elektriz> conversion
Stromwärme f <Elektriz, Elektrotech, Phys, Thermod> Joule effect
Stromwärmeverlust m <Elektriz> ohmic loss
Stromweg m <Elektrotech> current path
Stromwelle f <Elektrotech> current ripple
Stromwender m <Elektriz, Elektrotech> commutator, current reverser; collector *(dynamo-elektrische Maschine)*
Stromwendermotor m <Elektrotech, Kfztech> collector motor, commutator motor
Stromzähler m <Elektrotech> electricity meter, electricity supply meter
Stromzange f <Gerät> current probe
Stromzangeninstrument n <Elektriz> current tong test instrument
Stromzapfsäule f <Elektriz> electric power filling station
Stromzufuhr f <Bau> electricity supply
Stromzuleitung f <Elektrotech> current lead
Stromzweig m <Elektrotech> current path
Strontium n *(Sr)* <Chemie> strontium *(Sr)*
Strontium... <Chemie> strontic
Strontium-90 n <Chemie> radiostrontium

Strontiumoxid n <Chemie> strontium oxide
Strophanthin n <Chemie> strophanthin
Stropp m 1. <Meerschmutz> sling *(Anschlagmittel)*; 2. <Wassertrans> sling *(Tauwerk)*
Strouhalzahl f <Strömphys> Strouhal number
Strudel m 1. <Kohlen> vortex; 2. <Wassertrans> eddy *(Wasser)*
Struktur f 1. <Comp & DV> pattern, structure; 2. <Eisenbahn> structure; 3. <Fertig> body *(Öl)*; 4. <Kontroll> structure; 5. <Papier> formation
Struktur f der Blattunterseite <Lufttrans> bottom structure *(Hubschrauber)*
Struktur f der Flügeldruckseite <Lufttrans> bottom structure
Struktur f der Flügelunterseite <Lufttrans> bottom structure
strukturelle Falle f <Erdöl> structural tap *(Geologie)*
strukturelles Layout n <Comp & DV> taxonomy layout
strukturempfindlicher Ton m <Bau> sensitive clay
Strukturerkennung f <Comp & DV> pattern recognition
strukturiert adj <Comp & DV> patterned
strukturierte Entwicklung f <Comp & DV> structured design
strukturierte Programmierung f <Comp & DV> structured programming
strukturierte relationale Abfrage f <Comp & DV> structured relational-style query
strukturierte Suche f <Comp & DV> tree search
strukturierter Typ m <Comp & DV> structured type
strukturiertes Programmieren n <Comp & DV> structured programming
Strukturindex m <Raumfahrt> structure index
Strukturmodell n <Bau, Raumfahrt> structural model
Strukturmodell n aus finiten Elementen <Raumfahrt> finite element structural model
Strukturschaumstoff m <Kunststoff> integral foam, integral skin foam
Strukturspeicher m <Comp & DV> NVM, nonvolatile memory
Strukturteppich m <Textil> textured carpet
Strumpf m <Mechan> hose
Strumpfformen n <Textil> boarding
Strumpfwirkerei f <Textil> hose knitting
STS *(Raumfahrttransportsystem)* <Raumfahrt> STS *(space transportation system)*
Stubben m <Bau> stump
Stück/aus einem <Maschinen> in one piece
Stück färben v/in einem <Textil> dye in the piece
Stück/in einem <Maschinen> in one piece, integral
stückgefärbt adj <Textil> piece-dyed
Stuckgips m <Bau> plaster of Paris
Stückgut n 1. <Eisenbahn> less-than-carload freight, part load, part-load good; 2. <Trans> break bulk; 3. <Wassertrans> general cargo
Stückgüterladung f <Trans> package freight
Stückgutfrachter m <Wassertrans> mixed cargo ship *(Schiffstyp)*
Stückgutfrachtschiff n <Wassertrans> break bulk ship, general cargo ship
Stückgutkurswagen m <Trans> station wagon
Stückgutsendung f <Eisenbahn> less-than-carload freight shipment, part-load consignment
Stückgutverkehr m <Eisenbahn> part-load traffic
Stückkohle f <Kohlen> best coal
Stückliste f 1. <Bau> piece list; 2. <Fertig> list of parts; 3. <Maschinen> parts list
Stücknummer f <Verpack> part number
Stückware f <Textil> piece goods

Studio n mit reflexions- und schalltoter Wand <Aufnahme> live-end dead-end studio
Studioansprechsystem n <Aufnahme> studio address system
Studioauskleidung f <Aufnahme> studio lining
Studioeinrichtung f <Fernseh> studio facilities
Studiokamera f <Foto> studio camera
Studioleiter m <Fernseh> studio manager
Studiomonitor m <Fernseh> studio monitor
Studiorecorder m <Fernseh> transverse scanning recorder
Studiosendung f <Fernseh> studio broadcast
Stufe f 1. <Comp & DV> level; 2. <Ker & Glas> step; 3. <Kontroll> stage; 4. <Maschinen, Metall> step; 5. <Papier> stage
Stufe f des Umlaufgetriebes <Lufttrans> first-stage planet gear
Stufenausbeute f <Elektronik> current stage efficiency
Stufenbohrer m 1. <Fertig> stepped drill; 2. <Maschinen> multidiameter drill, step drill
Stufenbohrmeißel m <Erdöl> step bit *(Bohrtechnik)*
Stufenbreite f <Bau> going, length of step
Stufenformplotter m <Comp & DV> incremental plotter
Stufenfräsen n <Maschinen> step milling
stufenfreie Regelung f <Elektriz> stepless control
Stufengesenk n <Fertig> multistage die
Stufengitter n <Phys> echelette grating
Stufenhärten n <Fertig> marquenching, martempering
Stufenhöhe f <Bau> rise
Stufenindexfaser f <Optik, Phys> *(AE)* step index fiber, *(BE)* step index fibre
Stufenindex-Lichtwellenleiter m <Telekom> *(AE)* step index fiber, *(BE)* step index fibre
Stufenindex-LWL m <Telekom> *(AE)* step index fiber, *(BE)* step index fibre
Stufenindexprofil n <Optik> step index profile
Stufenindexprofil n im stationären Zustand <Optik> equivalent step index profile
Stufenkantenlinie f <Bau> nosing line
Stufenkeil m <Foto> step wedge
Stufenkolben m <Maschinen> differential piston, step piston
Stufenkopplung f <Elektrotech> stage coupling
stufenlos abstimmbarer Oszillator m <Elektronik, Funktech> continuously tunable oscillator
stufenlos einstellbar adj <Elektronik> continuously adjustable
stufenlos einstellbares Dämpfungsglied n <Elektronik> continuously variable attenuator
stufenlos regulierbar adj <Maschinen> infinitely variable
stufenlose Einstellung f <Elektronik> continuous adjustment
stufenlose Regelung f <Aufnahme> slide control
stufenloses Getriebe n <Maschinen> infinitely variable drive
Stufenpresse f <Maschinen> gang press
Stufenprofil n <Telekom> step index profile
Stufenrad n <Maschinen> step tooth gear, stepped gear
Stufenrädergetriebe n <Maschinen> step cone drive
Stufenregelung f <Elektriz> step-by-step control
Stufenschalter m <Elektrotech> multiple-contact switch, tap changer
Stufenscheibe f 1. <Fertig> speed cone; 2. <Maschinen> cone pulley, cone sheave, cone wheel, speed cone, step cone pulley, stepped pulley
Stufenscheibenantrieb m <Maschinen> cone pulley drive
Stufenskale f <Gerät> step scale
Stufensprung m <Fertig> progressive ratio

Stufentrenn-Retrorakete

Stufentrenn-Retrorakete f <Raumfahrt> stage separation retro-rocket
Stufenumsetzer m <Gerät> digit-at-a-time converter (AD-Umsetzung)
Stufenverschlüssler m <Gerät> digit-at-a-time converter (AD-Umsetzung)
stufenweise Reinigung f <Comp & DV> stepwise refinement
Stuffing-Bit n <Telekom> stuffing digit
stuhlfertig adj <Textil> loomstate
stuhlrohes Gewebe n <Textil> loomstate weft
Stuhlschiene f <Eisenbahn> bull-headed rail
Stulpe f <Maschinen> sleeve
stülpen v <Fertig> clinch (Ziehen)
Stülpen n <Fertig> inside-out redrawing (Ziehen)
Stulpmanschette f <Maschinen> sleeve packing
stumm geschaltetes Handy n <Mobilkom> mobile with ringing off
Stummabstimmung f <Funktech> muting device
Stummel m 1. <Kohlen> stump; 2. <Maschinen> stub
Stummelwelle f <Lufttrans> stub shaft (Hubschrauber)
Stummelwelle f/fasenringartig nach außen erweiterte <Lufttrans> bevel ring-flared stub shaft
Stummheit f <Akustik> muteness
stumpf adj 1. <Fertig, Metrol> obtuse (Winkel); 2. <Geom> obtuse; 3. <Maschinen> blunt
Stumpf m 1. <Bau> stub; 2. <Fertig> frustum; stub (Schweißen); 3. <Funktech> stub; 4. <Geom> frustum (Pyramide); 5. <Kohlen> stump; 6. <Phys> stub
stumpf beginnender Absatz m <Druck> flush paragraph
stumpf werden v <Maschinen> blunt
stumpfer Bohrmeißel m <Erdöl> dull bit (Tiefbohrtechnik)
stumpfer Stoß m <Maschinen> butt joint, straight joint
stumpfer Winkel m <Geom> obtuse angle
stumpfes Ende n <Bau> butt
Stumpffuge f <Bau> header joint (Holzbau)
Stumpfheit f <Geom> obtuseness
Stumpfnaht f 1. <Fertig> groove joint; 2. <Maschinen> butt seam
Stumpfsägenfeile f <Maschinen> blunt saw file
stumpfschweißen v <Fertig> jump-weld
Stumpfschweißen n 1. <Bau> butt welding; 2. <Fertig> butt fusion jointing (Kunststoffinstallationen); 3. <Maschinen, Mechan, Metall> butt welding
Stumpfstoß m 1. <Bau> butt joint, scarf joint; abutting joint (Schweißen); 2. <Fertig> abutting joint
stumpfstoßen v 1. <Bau> butt-joint; 2. <Fertig> bustle-joint
Stumpfstoßen n <Bau> scarf jointing
Stumpftonne f <Wassertrans> can buoy (Seezeichen)
stumpfwinklig adj <Geom> obtuse-angled, obtuse-angular
stumpfwinkliges Dreieck n <Geom> obtuse triangle
Stumpfzahn m <Fertig, Maschinen> stub tooth
Stunde f (h) <Metrol> hour (h)
Stundenkreis m <Metrol> hour circle
Stundenzeiger m <Metrol> hour hand
stürmische Überfahrt f <Wassertrans> rough crossing
Sturmklappe f <Erdöl> storm choke
Sturmlaterne f <Bau, Kohlen, Wassertrans> hurricane lamp
Sturmschutzverglasung f <Ker & Glas> antistorm glazing
Sturmsegel n <Wassertrans> storm sail
Sturz m 1. <Bau> head, header; lintel (Fenster, Tür); 2. <Phys> drop (Temperatur); 3. <Sicherheit> fall
Sturz m **eines Rades** <Kfztech> camber (Rad)

Sturzbalken m <Bau> bressumer, summer beam
stürzen v <Kohlen> dump
Sturzflug m <Lufttrans, Raumfahrt> dive • **in Sturzflug übergehen** <Raumfahrt> dive (Raumschiff)
Sturzflugbremse f <Lufttrans> diving brake
Sturzrinne f <Fertig> tip chute
Sturzsee m <Erdöl> surge
Sturzsicherung f <Sicherheit> fall-arresting device
Sturzspirale f <Lufttrans> spiral dive
Sturzwelle f <Wassertrans> breaker (Seezustand)
Sturzwinkel m <Kfztech> camber angle
Stützbalken m 1. <Bau> principal, stringer, supported beam; 2. <Kohlen> brace
Stützbock m <Bau> jack
Stützdorn m <Fertig> (AE) arbor, (BE) arbour, mandrel, mandril (Biegen)
Stütze f 1. <Bau> pillar, post, prop, shore, spur, stay, stud, strut; 2. <Fertig> prop; 3. <Hydraul> bracket; 4. <Ker & Glas> jamb; 5. <Kfztech> stand (Motorrad); 6. <Kohlen> pillar, shore; 7. <Maschinen> bracket, prop, rest, support; 8. <Mechan> brace, bracket; 9. <Wassertrans> pillar (Schiffbau)
Stützeinrichtung f <Trans> landing gear (Sattelschlepper)
Stutzen m 1. <Bau> union; 2. <Fertig> end connector (Kunststoffinstallationen); 3. <Heiz & Kälte> collar, gland; 4. <Kfztech> neck; 5. <Maschinen> connection, muff; 6. <Phys> stub
stützen v 1. <Bau> bear, buttress, carry, lean, shore, shore up, stay; 2. <Fertig, Mechan> buttress; 3. <Kohlen> shore, shore up; 4. <Wassertrans> shore up (Schiffbau)
Stützen n <Bau, Kohlen> shoring
stützend adj <Bau> supporting
Stützgleis n <Eisenbahn> supporting track
Stützholz n <Bau> propwood
Stützhülse f <Fertig> carrying bracket
Stützisolator m <Elektrotech> cap-and-pin insulator, pin insulator
Stützklappe f <Mechan> tab
Stützkraft f <Bau, Kohlen> bearing pressure
Stützlager n <Fertig> back rest
Stützlänge f <Bau> span
Stützmauer f 1. <Bau> breast wall, retaining wall, revetment, supporting wall; 2. <Ker & Glas> breast wall
Stützpfahl m <Bau> prop
Stützpfähle mpl <Bau, Kohlen> shoring
Stützpfeiler m <Bau> buttress; abutment (Holzbau)
Stützpunkt m <Geom> checkpoint
Stützrad f <Trans> stabilizing wheel
Stützrippe f <Fertig> stiffening rib
Stützrolle f <Maschinen> supporting roller
Stützsäule f <Bau> strut
Stützschiene f <Eisenbahn> supporting rail
Stützstange f 1. <Elektriz> stay pole; 2. <Lufttrans> backing bar
Stützwalze f 1. <Fertig> idle roll; 2. <Papier> antideflection roll, backing roll
Stützwinkel m 1. <Bau, Fertig> support bracket; 2. <Maschinen> angle bracket, bracket; 3. <Wassertrans> strut angle (Schiffbau)
Stützzylinder m <Fertig> plug (Sinuslineal)
STX (Textanfang) <Comp & DV> STX (start of text)
Styphninsäure f <Chemie> styphnic acid, trinitroresorcinol
Styracit m <Chemie> styracitol
Styren n <Chemie> phenylethylene, styrene, styrolene
Styrol n 1. <Chemie> phenylethylene, styrene, styrolene, vinylbenzene; 2. <Erdöl> styrene (Petrochemie)
Styrol-Butadien-Kautschuk m (SBR) <Kunststoff> styrene butadiene rubber (SBR)

Styrol-Copolymer n <Elektriz> styrol copolymer
SU (Sprachverstehen) <Künstl Int> SU (speech understanding)
Subadresse f <Telekom> subaddress (ISDN)
Subadressierung f <Telekom> subaddressing
Subchlorid n <Chemie> subchloride
Subdominante f <Akustik> subdominant
Subduktion f <Erdöl> subduction
Suberat n <Chemie> suberate
Suberin n <Chemie> suberin
Suberin... <Chemie> suberic
Suberon n <Chemie> cycloheptanone, suberone
Suberyl... <Chemie> suberyl
Subharmonische f <Akustik, Elektronik> subharmonic
subjektiv empfundene Lautheit f <Akustik> subjective loudness
subjektiver Fehler m <Gerät> personal error
subjektiver Messfehler m <Gerät> personal error
subjektiver Test m <Telekom> subjective test
subjektiver Ton m <Akustik> subjective tone
subkritisch adj <Kerntech> subcritical
subkritische Reaktion f <Strahlphys> subcritical reaction
Sublimat n <Chemie, Erdöl> sublimate
Sublimation f <Phys, Thermod> sublimation
Sublimationswärme f <Thermod> sublimation heat
Sublimatverstärkung f <Foto> mercury intensification
sublimieren v <Thermod> sublimate
Sublimieren n <Thermod> sublimation
Sublimiergefäß n <Chemie> sublimating vessel
sublimiertes Zinkoxid n <Metall> sublimated oxide of zinc
subliminal adj <Ergon> subliminal
Submillimeterwellenlänge f <Funktech> submillimeter wavelength
Subnitrat n <Chemie> subnitrate
Subnormale f <Geom> subnormal
Sub-Nyquist-Abtastung f <Telekom> sub-Nyquist-sampling (Bildcodierung)
Subreflektor m <Funktech, Raumfahrt> subreflector (Weltraumfunk)
Subroutine f <Comp & DV> subroutine
Subsatellitenpunkt m <Raumfahrt> subsatellite point
Subsegment n <Comp & DV> subsegment
Substandardschiff n <Wassertrans> substandard ship
Substantivfarbstoff m <Textil> direct dyestuff
Substanz f 1. <Ker & Glas> substance; 2. <Textil> body of the fabric • **mineralische Substanzen entfernen** <Chemtech> demineralize
Substitution f <Math> substitution (Ersetzung einer Größe)
Substitutionszeichen n <Comp & DV> SUB character, substitute character, substitution character
Substrat n 1. <Chemie> reactant (Enzyme); 2. <Comp & DV, Kunststoff, Telekom> substrate
Substrat-Transistor m <Elektronik> parasitic transistor
subsynchroner Satellit m <Raumfahrt> subsynchronous satellite
Subsystem n <Comp & DV, Raumfahrt, Telekom> subsystem
Subtangente f <Geom> subtangent
Subtask n <Comp & DV> subtask
subthermisches Neutron n <Kerntech> ultracold neutron
Subträger m <Raumfahrt> subcarrier (Weltraumfunk)
Subtrahend m <Comp & DV, Math> subtrahend
subtrahieren v <Comp & DV, Math> subtract
Subtrahierer m <Comp & DV> subtracter
Subtrahierglied n <Comp & DV> subtracter
Subtraktion f <Comp & DV, Math> subtraction

subtraktive Farbmischung f <Foto> subtractive synthesis
subtraktive Primärfarben fpl <Fernseh> subtractive primaries
Subtraktivmethode f <Elektronik> subtractive method
subtransiente Längsspannung f <Elektriz> direct-axis sub-transient voltage
subtransiente Querspannung f <Elektriz> quadrature-axis subtransient voltage
subtransienter Kurzschlusswechselstrom m <Elektriz> initial symmetrical short-circuit alternating current, subtransient short-circuit alternating current
Subtyp m <Comp & DV> subtype
Subziel n <Künstl Int> subgoal
Suchalgorithmus m <Künstl Int> search algorithm
Sucharm m <Comp & DV> seek arm
Suchbaum m <Künstl Int> search tree
Suchbegriff m <Comp & DV> search key, search pattern
Suchbereich m <Comp & DV> seek area
Suche f <Comp & DV, Künstl Int> search
Suche f in Rückwärtsrichtung <Künstl Int> backward search
Suche f in verknüpfter Liste <Comp & DV> chaining search
Suche f in Vorwärtsrichtung <Künstl Int> forward search
suchen v 1. <Comp & DV> seek; 2. <Telekom> hunt
Suchen n 1. <Comp & DV> searching; 2. <Telekom> hunting
Sucher m 1. <Foto> finder, viewfinder; 2. <Raumfahrt> viewfinder
Sucher m mit aufrechtem Bild <Foto> erect image viewfinder
Sucher m mit einem unter 45° geneigten Spiegel <Foto> right-angle finder
Sucherokular n <Foto> viewfinder eyepiece
Sucherokular n mit Korrekturlinse <Foto> viewfinder eyepiece with correcting lens
Sucherokularabdeckung f <Foto> finder hood
Suchfehler m <Comp & DV> seek error
Suchgebietsmittelpunkt m <Wassertrans> datum (Such- und Rettungsdienst)
Suchkriterium n <Comp & DV> search key
Suchlauf m 1. <Comp & DV> browsing; 2. <Funktech> station finding (Rundfunk)
Suchmaschine f <Comp & DV, Telekom> search engine (Internet)
Suchmuster n <Comp & DV> search pattern
Suchprogramm n <Comp & DV> search program
Suchscheinwerfer m <Kfztech> spotlight
Suchspule f 1. <Elektriz, Elektrotech> exploring coil, search coil; 2. <Phys> exploring coil, pick-up coil, search coil
Suchstrategie f <Künstl Int> search strategy
Suchtiefe f <Künstl Int> search depth
Such- und Rettungsdienst m (SAR) <Wassertrans> search and rescue (SAR)
Such- und Verzeichnisdienst m <Comp & DV> search and directory service
Suchwort n <Comp & DV> search word
Suchzeit f <Comp & DV, Optik> seek time
Süd zu Ost adv <Wassertrans> south by east (Kompass)
Süd zu West adv <Wassertrans> south by west (Kompass)
südliche Breite f <Wassertrans> southern latitude (Navigation)
Südlicht n <Wassertrans> aurora australis (Wetterkunde)
Südost zu Ost adv <Wassertrans> southeast by east (Kompass)

Südost

Südost zu Süd adv <Wassertrans> southeast by south (Kompass)
Sudoster m <Wassertrans> southeaster
südöstlich adj <Wassertrans> southeast
südostwärts adv <Wassertrans> southeast, southeasterly
Südostwind m <Wassertrans> southeast wind
Südpol m <Phys> South Pole
Südsüdosten m <Wassertrans> south-south-east (Kompass)
südsüdöstlich adj <Wassertrans> south-southeast
Südsüdwesten m <Wassertrans> south-southwest (Kompass)
südsüdwestlich adj <Wassertrans> south-southwest
südwärts adv <Wassertrans> south, southerly
Südwest zu Süd adv <Wassertrans> southwest by south (Kompass)
Südwest zu West adv <Wassertrans> southwest by west (Kompass)
Südwest... 1. <Wassertrans> southwest; 2. <Wassertrans> southwest (Kompass)
Südwester m <Wassertrans> southwester
südwestlich adj <Wassertrans> southwest
südwestwärts adv <Wassertrans> southwest, southwesterly
Südwestwind m <Wassertrans> southwest wind
Suffix n <Comp & DV> suffix
Suffixnotation f <Comp & DV> suffix notation
Suffixschreibweise f <Comp & DV> suffix notation
Suffizienz f <Math> sufficiency (erschöpfende Schätzfunktion)
Sugkopf m <Wasserversorg> strainer
sukzessiv adj <Kontroll> successive
sukzessive Approximation f <Gerät> successive approximation
Sulfaguanidin n <Chemie> (AE) sulfaguanidine, (BE) sulphaguanidine
Sulfamat n <Chemie> (AE) sulfamate, (BE) sulphamate
Sulfamid n <Chemie> (AE) sulfamide, (BE) sulphamide
Sulfamid... <Chemie> (AE) sulfamic, (BE) sulphamic
Sulfamidat n <Chemie> (AE) sulfamate, (BE) sulphamate
Sulfanil... <Chemie> (AE) sulfanilic, (BE) sulphanilic
Sulfanilamid n <Chemie> (AE) sulfanilamide, (BE) sulphanilamide
Sulfanilguanidin n <Chemie> (AE) sulfaguanidine, (BE) sulphaguanidine
Sulfapyridin n <Chemie> (AE) p-aminobenzenesulfamidopyridine, (BE) p-aminobenzenesulphamidopyridine, (AE) sulfapyridine, (BE) sulphapyridine
Sulfat n <Chemie, Umweltschmutz> (AE) sulfate, (BE) sulphate
Sulfatangriff m <Bau> (AE) sulfate attack, (BE) sulphate attack (Beton)
sulfatbeständiger Portlandzement m <Bau> (AE) Portland sulfate-resisting cement, (BE) Portland sulphate-resisting cement
Sulfathiazol n <Chemie> (AE) sulfathiazole, (BE) sulphathiazole
Sulfatzellstoff m <Papier> (AE) sulfate pulp, (BE) sulphate pulp
Sulfenamidbeschleuniger m <Kunststoff> (AE) sulfene amide accelerator, (BE) sulphene amide accelerator
Sulfhydryl... <Chemie> (AE) sulfhydryl, (BE) sulphhydryl
Sulfidglas n <Ker & Glas> (AE) sulfide glass, (BE) sulphide glass
Sulfin... <Chemie> (AE) sulfinic, (BE) sulphinic
Sulfinyl... <Chemie> (AE) sulfinyl, (BE) sulphinyl, thionyl
Sulfit n <Chemie> (AE) sulfite, (BE) sulphite
Sulfocyanat n <Chemie> thiocyanate

Sulfohydrat n <Chemie> (AE) sulfhydrate, (BE) sulphydrate
Sulfolan n <Chemie, Erdöl> (AE) sulfolane, (BE) sulpholane (Petrochemie)
Sulfolen n <Erdöl> (AE) sulfolene, (BE) sulpholene (Petrochemie)
Sulfon n <Chemie> (AE) sulfone, (BE) sulphone
Sulfonamid n <Chemie> (AE) sulfonamide, (BE) sulphonamide
Sulfonamidpräparat n <Chemie> (AE) sulfa drug, (BE) sulpha drug
Sulfonat n <Chemie> (AE) sulfonate, (BE) sulphonate
sulfoniert adj <Chemie> (AE) sulfonated, (BE) sulphonated
Sulfonierung f <Chemie> (AE) sulfonation, (BE) sulphonation
Sulfonsäureamid n <Chemie> (AE) sulfonamide, (BE) sulphonamide
Sulfonsäureester m <Chemie> (AE) sulfonate, (BE) sulphonate
Sulfonyl... <Chemie> (AE) sulfonyl, (AE) sulfuryl, (BE) sulphonyl, (BE) sulphuryl
Sulfurierung f <Chemie> (AE) sulfonation, (BE) sulphonation
Sulfuryl... <Chemie> (AE) sulfonyl, (AE) sulfuryl, (BE) sulphonyl, (BE) sulphuryl
Süll n <Wassertrans> coaming (Schiffbau)
Süllrahmen m <Lufttrans> coaming
Sumatrol n <Chemie> sumatrol
Summand m <Comp & DV> addend
Summe f 1. <Comp & DV> sum; 2. <Math> sum (Ergebnis einer Addition)
Summen n <Aufnahme> hum
Summen n **des Transformators** <Aufnahme> transformer hum
Summen-Differenz-Technik f <Aufnahme> sum-and-difference technique
Summenganglinie f <Nichtfoss Energ> summation hydrograph
Summenhäufigkeit f 1. <Math> cumulative frequency (Statistik); 2. <Qual> cumulative frequency
Summenhäufigkeitspolygon n <Qual> cumulative frequency polygon
Summenkurve f <Qual> cumulative curve
Summenlinie f <Qual> cumulative frequency polygon
Summenschallpegel m <Sicherheit> overall noise level
Summensignal n <Telekom> aggregate signal
Summenstanzen n <Comp & DV> summary punching
Summenstanzer m <Comp & DV> summary punch
Summenverteilung f <Qual> cumulative distribution
Summenwahrscheinlichkeit f <Qual> cumulative probability
Summer m <Elektriz, Elektrotech, Telekom, Trans> buzzer
summierender Verstärker m <Lufttrans> integrator amplifier
summierender Volumenmesser m <Lufttrans> integrating flowmeter
summierender Zähler m <Elektronik> accumulating counter
Summierer m **mit bewerteten Eingängen** <Regelung> weighted summing unit
Summierglied n <Elektronik> adder device (Schaltkreistechnik)
Summierstelle f <Regelung> summing point
Summierstufe f <Elektronik> mixer stage
Sumpf m 1. <Bau> swamp; 2. <Erdöl> sump (Bohrtechnik); 3. <Kohlen> bog, swamp; 4. <Wasserversorg> sump
Sumpfboden m <Bau, Kohlen> swampy soil
sumpfig adj <Kohlen> boggy

Sumpftorf m <Kohlen> bog peat
Sund m <Wassertrans> sound
Superaustastimpuls m <Fernseh> super-blanking pulse
Superbenzin n 1. <Erdöl> (AE) premium gasoline, (BE) premium petrol, (AE) premium grade gasoline, (BE) premium grade petrol (Destillation); 2. <Kfztech> (AE) four-star gasoline, (BE) four-star petrol, premium fuel, (AE) premium gasoline, (BE) premium petrol, (AE) premium grade gasoline, (BE) premium grade petrol
Superbrechung f <Raumfahrt> superrefraction (Weltraumfunk)
Supercomputer m <Comp & DV> number cruncher, supercomputer
Supercooling n <Phys> supercooling
Superdominante f <Akustik> submediant
Supereffizienzpunkt m <Math> superefficiency point (ungewöhnlicher Parameterwert)
Superfinieren n <Maschinen> superfinishing
Supergroßraumcontainer m <Trans> SHC, super high cube
Superhochfrequenz f <Elektriz> superhigh frequency, SHF (3000 bis 30000 MHz)
Superhochtonlautsprecher m <Aufnahme> supertweeter loudspeaker
superhohe Frequenz f (SHF) <Bau, Elektronik, Funktech, Kohlen, Kunststoff, Mechan, Phys, Thermod> superhigh frequency (SHF)
Superikonoskop n <Elektronik> image iconoscope
Superkalander m <Papier> supercalender
Superkargo m <Wassertrans> supercargo (Handelsmarine)
superkavitierender Propeller m <Wassertrans> supercavitating propeller
Superleiter m <Elektriz> superconductor
Superlumineszenz f 1. <Optik> superluminescence; 2. <Telekom> superluminescence, superradiance
Superlumineszenzdiode f (SLD) 1. <Optik> superluminescent LED (SLD); 2. <Telekom> superluminescent LED (SLD); superradiant diode (SRD)
superlumineszierende Leuchtdiode f <Optik> superluminescent LED
Superminicomputer m <Comp & DV> supermini
Supernova f <Raumfahrt> supernova
Super-Nyquist-Abtastung f <Telekom> super-Nyquist sampling
Superorthikon n <Elektronik> image orthicon
Super-Ottokraftstoff m 1. <Erdöl> (AE) premium gasoline, (BE) premium petrol; 2. <Kfztech> (AE) four-star gasoline, (BE) four-star petrol, premium fuel, (AE) premium gasoline, (BE) premium petrol
Superoxid n <Raumfahrt> peroxide
Superoxid-Ion n <Chemie> hyperoxide
Superphosphat n <Chemie> superphosphate
Superpositionsprinzip n <Phys, Wellphys> principle of superposition
Superrate f <Telekom> superrate
Superrechner m <Comp & DV> supercomputer
superschwerer Kern m <Strahlphys> superheavy nucleus
superstarker faserarmierter Wellkarton m <Verpack> (AE) heavy duty corrugated fiber board, (BE) heavy duty corrugated fibre board
superstrahlende Diode f <Optik> SRD, superradiant diode
Superstrahlung f <Optik> superradiance
Superstrings mpl <Strahlphys> superstrings
supersymmetrische Teilchen npl <Strahlphys> supersymmetrical particles
Supertanker m 1. <Erdöl> very large crude carrier; 2. <Wassertrans> mammoth tanker, supertanker

Supervisor m <Comp & DV> executive, supervisor
Supervisoraufruf m <Comp & DV> supervisor call
Superzentrifuge f <Chemie, Labor> ultracentrifuge
supin adj <Ergon> supine
Supination f <Ergon> supination
Supplementärwinkel mpl <Geom> supplementary angles
Support m 1. <Fertig> head (Hobelmaschine); 2. <Maschinen> saddle, slide rest, tool rest
Supportdrehmaschine f <Maschinen> slide lathe, slide rest lathe
Supra-Fernsprechfrequenz f <Telekom> super-telephone frequency
Suprafluidität f <Chemie, Phys> superfluidity
Supraflüssigkeit f 1. <Chemie> superfluidity; 2. <Phys> superfluid
supraleitend adj <Teilphys> superconducting
supraleitende Spule f <Trans> superconducting coil
supraleitender Magnet m <Trans> superconducting magnet
supraleitender Speicher m <Comp & DV> cryogenic memory, superconducting memory
supraleitendes Bauteil n <Elektronik> superconducting device
Supraleiter m <Elektriz, Elektronik, Phys, Telekom> superconductor
Supraleiterkabel n <Telekom> superconductor cable
supraleitfähiger Quanteninterferenzmechanismus m (SQID) <Phys> superconductive quantum interference device (SQID)
supraleitfähiges Supracollider n (SSC) <Teilphys> superconducting super collider (SSC)
Supraleitfähigkeit f <Comp & DV, Elektriz, Elektronik, Strahlphys> superconductivity
Supraleitfähigkeitsleitung f <Telekom> superconductor line
Supraleitmagnetschwebesystem n <Trans> superconducting magnet levitation
Supraleitrolle f <Trans> superconducting coil
Supraleitung f <Comp & DV, Phys, Strahlphys, Teilphys> superconductivity
supraliminal adj <Ergon> supraliminal
Surfactant m 1. <Chemie> surfactant; 2. <Phys> surface tension
Surfen n 1. <Wassertrans> surging (Bewegung in Längsrichtung, Schiffsbewegung); 2. <Telekom> surfing (Internet)
Surfen n **im Internet** <Telekom> Internet surfing
surjektiv adj <Math> surjective
suspendierter Rotor m <Nichtfoss Energ> teetered rotor
Suspension f <Kunststoff> suspension
Suspensionspolymerisation f <Kunststoff> suspension polymerization
Suspensionstechnik f <Chemie> mull technique
süßes Rohöl n <Erdöl> sweet crude
Süßholzzucker m <Chemie> glycyrrhizine
Süßstoff m <Lebensmittel> sweetener
Süßwasserbohrschlamm m <Erdöl> freshwater mud (Bohrtechnik)
Süßwasservorrat m <Wasserversorg> freshwater stock
Suszeptanz f <Elektrotech, Phys> susceptance
Suszeptibilität f <Elektriz> susceptibility
Sutton'sche Gleichung f <Kerntech> Sutton equation
Sv (Sievert) <Phys, Strahlphys, Teilphys> Sv (Sievert)
SVA (gemeinsam benutzbarer virtueller Bereich) <Comp & DV> SVA (shared virtual area)
S-Verzerrung f <Fernseh> S-distortion
sv-Motor m (seitengesteuerter Motor) <Kfztech> sv engine (side valve engine)

Sward-Härteprüfung

Sward-Härteprüfung f <Kunststoff> Sward hardness test, Sward rocker hardness test
S-Welle f *(Sekundärwelle)* <Phys> S-wave *(secondary wave)*
SWFD *(Selbstwählferndienst)* <Telekom> *(AE)* DDD, direct distance dialing; *(BE)* STD, subscriber trunk dialling
SWR 1. <Kerntech, Phys> *(Siedewasserreaktor)* BWR *(boiling water reactor)*; 2. <Kerntech> *(schwerwassermoderierter Reaktor)* HWR *(heavy-water-moderated reactor)*
SWV *(Stehwellenverhältnis)* 1. <Funktech> SWR *(standing-wave ratio)*; 2. <Phys, Telekom> SWR *(standing-wave ratio)*; VSWR *(voltage standing-wave ratio)*
Syenit m <Chemie> syenite
Sylvestren n <Chemie> sylvestrene
Symbol n <Comp & DV> symbol *(grafische Oberflächen)*; icon *(in grafischer Benutzeroberfläche)*
symbolische Adresse f <Comp & DV> symbolic address
symbolische Adressierung f <Comp & DV> symbolic addressing
symbolische Logik f <Math> mathematical logic, symbolic logic
symbolische Programmiersprache f <Comp & DV> symbolic language
symbolische Verarbeitung f <Comp & DV> symbolic processing
symbolischer Befehl m <Comp & DV> symbolic code, symbolic instruction
symbolischer Code m <Comp & DV> symbolic code
symbolischer Name m <Comp & DV> symbolic name
symbolisches Testen n <Comp & DV> symbolic debugging
Symbolrate f <Telekom> symbol rate
Symbolsatz m <Comp & DV> symbol set
Symbolschrift f <Druck> symbolic font
Symboltabelle f <Comp & DV> symbol table
Symboltaste f <Comp & DV> symbol key
Symmetrie f 1. <Math> symmetry; 2. <Telekom> balance
Symmetrie f **bezüglich Anströmung und Abströmung** <Strömphys> upstream-downstream symmetry
Symmetrieachse f 1. <Geom> axis of symmetry; 2. <Maschinen> symmetry axis
Symmetrieebene f 1. <Geom> plane of symmetry, symmetry plane; 2. <Maschinen> symmetry plane; 3. <Metall> plane of symmetry
Symmetrierglied n <Funktech> balanced-to-unbalanced transformer, balun
Symmetrierübertrager m <Funktech> balanced-to-unbalanced transformer
Symmetriezentrum n <Geom> *(AE)* center of symmetry, *(BE)* centre of symmetry
symmetrisch adj 1. <Comp & DV> symmetric, symmetrical; 2. <Math> symmetrical, symmetric • **symmetrisch gegen Erde** <Elektrotech> balanced to earth • **symmetrisch gegen Masse** <Elektrotech> balanced to earth
symmetrische Anordnung f <Elektrotech> symmetrical arrangement
symmetrische Belastung f <Elektrotech> balanced load
symmetrische digitale Teilnehmerleitung f <Telekom> symmetric digital subsriber line *(SDSL)*
symmetrische doppelte Gleisverbindung f <Eisenbahn> scissors crossing
symmetrische Funktion f <Math> symmetric function
symmetrische Leitung f 1. <Aufnahme, Elektriz, Elektrotech> balanced line; 2. <Telekom> balanced circuit
symmetrische Matrix f <Comp & DV> symmetric matrix, symmetrical matrix
symmetrische Messleitung f <Gerät> balanced measuring line

752

symmetrische Schaltung f 1. <Aufnahme> balanced circuit; 2. <Elektriz> balanced network
symmetrische Ströme mpl <Elektrotech> balanced currents
symmetrische Teilgruppe f <Telekom> balanced grading group
symmetrische Tonspur f <Akustik> symmetrical soundtrack
symmetrische Wellenfunktion f <Phys> symmetric wave function
symmetrischer Anastigmat m <Foto> symmetric anastigmat, symmetrical anastigmat
symmetrischer Eingang m <Elektrotech, Gerät> balanced input
symmetrischer Fehler m <Comp & DV> balanced error
symmetrischer Mischer m <Funktech> double-balanced mixer
symmetrischer Wandler m <Elektrotech> symmetrical transducer
symmetrisches Kabel n <Telekom> balanced cable
symmetrisches Netzwerk n <Elektrotech> balanced network
symmetrisches Schutzrelais n <Elektriz> symmetrical protective relay
symmetrisches T-Glied n <Elektrotech> H-network
Synanthrose f <Chemie> levulin, synanthrose
Synapse f <Künstl Int> neuronal interconnection, synapse
Synärese f <Kunststoff, Lebensmittel> syneresis
Synchro... <Comp & DV, Elektronik> synchronous
Synchrocompurverschluss m <Foto> synchro compur shutter
Synchrodrehmomentempfänger m <Gerät> synchro torque receiver
Synchron... <Comp & DV, Elektronik> synchronous
Synchronalternator m <Elektriz> synchronous alternator
Synchrondemodulation f <Comp & DV, Telekom> synchronous detection
synchrone Leitungsendeinrichtung f <Telekom> synchronous line terminal
synchrone Übertragung f 1. <Comp & DV> isochronous transmission; 2. <Telekom> synchronous transmission
synchrone Übertragungssteuerung f <Comp & DV, Telekom> synchronous data link control *(SDLC)*
synchroner Abzweigmultiplexer m <Telekom> synchronous add/drop multiplexer
synchroner Lauf m <Elektrotech> synchronous operation
synchroner Linearmotor m <Elektrotech> synchronous linear motor
synchroner Port m <Telekom> synchronous port
synchroner Sinuswelleninverter m <Nichtfoss Energ> synchronous sine wave inverter
synchroner Transfermodus m <Telekom> synchronous transfer mode
synchroner Wettersatellit m <Telekom> synchronous meteorological satellite
synchrones Datenübertragungsverfahren n <Comp & DV, Telekom> synchronous data link control *(SDLC)*
synchrones optisches Netz n <Telekom> synchronous optical network
Synchrongenerator m 1. <Elektriz, Elektrotech> synchronous generator; 2. <Fernseh> station sync generator
Synchrongeschwindigkeit f <Elektriz, Nichtfoss Energ> synchronous speed
Synchrongetriebe n <Kfztech> synchromesh, synchromesh transmission; constant mesh *(Getriebe)*
Synchronimpuls m <Fernseh, Telekom> synchronization pulse, sync pulse
Synchronimpulsgenerator m <Fernseh> sync pulse generator

Synchroninput m <Fernseh> sync input
Synchronisation f <Comp & DV, Fernseh, Maschinen, Telekom> synchronization, sync
Synchronisationsaustastung f <Fernseh> sync blanking
Synchronisationsbit n 1. <Comp & DV> flag bit, sync bit; 2. <Telekom> sync bit
Synchronisationsglied n <Elektrotech> synchronizer
Synchronisationsimpuls m 1. <Comp & DV, Kontroll> sync pulse; 2. <Fernseh, Telekom> synchronization pulse
Synchronisationsleerzeichen n <Comp & DV> synchronous idle
Synchronisationsmechanismus m <Comp & DV> interlock
Synchronisationsnetz n <Telekom> synchronization network
Synchronisationsstörung f <Elektronik> jitter noise; jitter *(Fernsehen)*
Synchronisationsverlust m <Telekom> out of sync
Synchronisationszeichen n <Comp & DV, Telekom> SYN, synchronous idle character
Synchronisator m <Elektrotech, Kfztech> synchronizer
Synchronisierbit n <Telekom> alignment bit
Synchronisiereinheit f <Comp & DV> timing generator
synchronisieren v <Fernseh, Telekom> sync, synchronize
Synchronisieren n <Comp & DV, Telekom> synchronization
Synchronisierer m <Comp & DV, Elektrotech> synchronizer
Synchronisierkanal m <Mobilkom> synchronization channel
Synchronisierleitung f <Kontroll> sync line, synchronizing line
Synchronisierrelais n <Elektriz> synchronizing relay
synchronisiert adj <Fernseh> in sync
synchronisierter Induktionsmotor m <Elektrotech> synchronous induction motor
synchronisiertes Getriebe n <Kfztech> synchronized transmission
Synchronisierung f 1. <Aufnahme> synchronization; 2. <Comp & DV> SYN, synchronous idle character; 3. <Kfztech, Kontroll> synchronization; 4. <Telekom> alignment, synchronization, synchronous idle character
Synchronisierungsausfall m <Telekom> loss of synchronism
Synchronisierungsbit n <Comp & DV, Telekom> sync bit
Synchronisierungsfenster n <Raumfahrt> synchronization window
Synchronisierungsimpuls m <Comp & DV, Kontroll> sync pulse
Synchronisierungskanal m <Mobilkom> synchronization channel
Synchronisierungsverlust m <Elektriz> loss of synchronism
Synchronisierungsvorrichtung f <Kfztech> synchronizer
Synchronisierungszeichen n <Comp & DV, Telekom> SYN, synchronous idle character
Synchronisiervorrichtung f <Elektriz> synchronizer
Synchronismus m <Aufnahme, Elektronik, Telekom> synchronism
Synchronität f <Maschinen> synchronism
Synchronklappe f <Aufnahme> clapper
Synchronkompensator m <Nichtfoss Energ> synchronous compensator
Synchronkondensator m <Elektrotech> synchronous capacitor
Synchronkontaktgeber m <Elektrotech> synchronizer
Synchronleichrichtung f <Elektronik> synchronous detection
Synchronmaschine f <Elektriz, Elektrotech> synchronous machine
Synchron-Modem n <Elektronik> synchronous modem
Synchronmodus m <Comp & DV, Telekom> synchronous mode
Synchronmotor m <Elektriz, Elektrotech, Phys, Trans> synchronous motor
Synchronmotor m **auf Asynchronprinzip** <Elektriz> synchronous induction motor
Synchronrechner m <Comp & DV> synchronous computer
Synchronriemen m 1. <Kfztech> timing belt *(Nockenwellenantrieb)*; 2. <Maschinen> synchronous belt, timing belt
Synchronriemenantrieb m <Maschinen> synchronous belt drive
Synchronsatellit m 1. <Raumfahrt> synchronous satellite; 2. <Telekom> geosynchronous satellite
Synchronschaltung f <Telekom> synchronous circuit
Synchronsignal n <Comp & DV, Telekom> clocked signal
Synchronsignalgemisch n <Fernseh> composite sync signal
Synchron-Sinuswelleninverter/Lader m <Nichtfoss Energ> synchronous sine wave inverter/charger
Synchronsteuerung f <Fernseh> genlock, genlocking
Synchronsystem n <Trans> simultaneous system
Synchronton m <Aufnahme> sync sound
Synchronübertragung f <Comp & DV> synchronous transmission
Synchronuhr f <Elektriz> electric synchronous clock
Synchronumformer m <Elektriz> synchronous converter
Synchronverfahren n <Comp & DV, Telekom> synchronous mode
Synchronverlust m <Fernseh> sync loss
Synchronverstärker m <Fernseh> sync amplifier
Synchronwert m <Fernseh, Telekom> sync level
Synchronwort n <Telekom> synchronization word, syncword
Synchronzähler m <Gerät> synchronous counter
Synchroton n <Phys> synchroton
Synchrotransformator m <Elektrotech> synchro transformer
Synchrotron n <Teilphys> synchrotron
Synchrotronstrahlung f 1. <Phys> synchrotron radiation; 2. <Strahlphys> synchrotron emission; 3. <Teilphys> synchrotron radiation
Synchrozentralverschluss m <Foto> synchro compur shutter
Synchrozyklotron n <Phys> synchrocyclotron
Synergismus m <Umweltschmutz> synergetic effect, synergy
Synergist m <Lebensmittel> synergist
synergistischer Effekt m <Kunststoff> synergism effect, synergistic effect
Synklinale f <Erdöl> syncline *(Geologie)*
synoptische Schalttafel f <Elektriz> synoptical switchboard
syntaktisch adj <Chemie> syntactic
syntaktisch analysieren v <Comp & DV> parse
syntaktische Analyse f <Telekom> syntactic analysis
syntaktischer Analysator m <Telekom> *(BE)* syntactic analyser, *(AE)* syntactic analyzer
Syntax f <Comp & DV> syntax
Syntaxanalyse f <Comp & DV> parsing, syntax analysis
Syntaxanalyseprogramm n <Comp & DV> *(BE)* syntax analyser, *(AE)* syntax analyzer

Syntaxfehler

Syntaxfehler *m* <Comp & DV> grammatical error, syntax error
Syntaxprüfung *f* <Comp & DV> syntax checking
Synthese *f* <Telekom> synthesis
Syntheseelastomer *n* <Erdöl> synthetic elastomer
Synthesegenerator *m* <Telekom> synthesizer
Synthesekautschuk *m* <Erdöl, Kunststoff> synthetic rubber
Syntheseöl *n* <Erdöl> synthetic crude oil
synthetisch hergestelltes Gas *n (SNG)* <Erdöl> synthetic natural gas *(SNG)*
synthetische Faser *f* <Textil> *(AE)* synthetic fiber, *(BE)* synthetic fibre
synthetische Schlichte *f* <Textil> synthetic size
synthetische Sprache *f* 1. <Künstl Int> artificial speech, synthetic speech; 2. <Telekom> synthetic speech
synthetischer Brennstoff *m* <Kohlen> synfuel *(Kohleerzeugnis)*
synthetischer Latex *m* <Kunststoff> synthetic latex
synthetisches Benzin *n* <Kfztech> *(AE)* synthetic gasoline, *(BE)* synthetic petrol
synthetisches Elastomer *n* <Erdöl> synthetic elastomer
synthetisches Gas *n* <Kohlen> syngas, synthetic gas *(Kohleerzeugnis)*
synthetisches Öl *n* 1. <Erdöl> synthetic crude; 2. <Maschinen> synthetic oil
synthetisierte Musik *f* <Elektronik> synthesized music
Synthol *n* <Chemie> synthol
Syntonin *n* <Chemie> parapeptone, syntonin
syntonisches Komma *n* <Akustik> syntonous comma
Sypersynchro-Signal *n* <Fernseh> supersync signal
Syringa... <Chemie> syringic
Sysplex-Konfiguration *f* <Comp & DV> sysplex configuration
System *n* 1. <Bau> system *(Rohren, Kabeln)*; 2. <Comp & DV, Eisenbahn> system; 3. <Telekom> system, network; 4. <Trans> system with intermediate stops • **ein System betreiben** <Comp & DV> run a system
System *n* **der dritten Generation** <Mobilkom> third generation system *(UMTS)*
System *n* **der vorlaufenden Luftschraubenblätter** <Lufttrans> advancing blade concept
System *n* **für Aktivabfälle** <Kerntech> active effluent system
System *n* **für künstliche Verzögerung** <Aufnahme> artificial delay system
System *n* **für verteilte Dateien** <Comp & DV> distributed file system
System *n* **mit Bereitschaftsbetrieb** <Telekom> hot stand-by system
System *n* **mit digitaler Modulation** <Telekom> digital modulation system
System *n* **mit Eigensicherheit** <Sicherheit> intrinsic safety system
System *n* **mit einfachem Ruf** <Telekom> combined local/toll system
System *n* **mit Entscheidungsrückmeldung** <Telekom> decision feedback system
System *n* **mit Ersatzschaltung** <Telekom> stand-by system *(Radio)*
System *n* **mit erweiterter Bandbreite** <Funktech> extended-bandwidth system
System *n* **mit festverdrahteter Logik** <Telekom> wired logic system
System *n* **mit Funktionsteilung** <Telekom> function division system
System *n* **mit heißer Reserve** <Telekom> hot stand-by system
System *n* **mit künstlicher Intelligenz** <Künstl Int> artificial intelligence system
System *n* **mit Mehrfachzugriff** <Comp & DV, Telekom> multiaccess system
System *n* **mit Motorantrieb** <Telekom> motor-driven system
System *n* **mit Registersteuerung** <Telekom> register-controlled system
System *n* **mit unterdrücktem Träger** <Telekom> suppressed carrier system
System *n* **mit versetzten Trägern** <Telekom> offset carrier system
System *n* **mit verteilter Steuerung** <Telekom> distributed control system
System *n* **mit Zeitschlitzen** <Telekom> slotted system
System *n* **mit Zentralsteuerung** <Telekom> common control system
System *n* **vorbestimmter Zeiten** <Ergon> predetermined motion time system
System *n* **zum Wiederauffinden von Informationen** <Comp & DV> information retrieval system
System *n* **zur Informationsabfrage** <Comp & DV> information retrieval system
System *n* **zur Rettung von Menschenleben bei Seenotfällen** *(GMDSS)* <Wassertrans> global marine distress and safety system *(GMDSS)*
Systemabschluss *m* <Comp & DV> closedown
Systemabsturz *m* <Comp & DV> system crash
System-Account *m* <Comp & DV> system login account
Systemanalyse *f* <Comp & DV> system analysis
Systemanalyseprogramm *n* <Comp & DV> *(BE)* system analyser program, *(AE)* system analyzer program
Systemanbieter *m* <Telekom> system provider
Systemantwort *f* <Comp & DV> response
Systemanzeige *f* <Comp & DV> prompt
Systemarchitekt *m* <Comp & DV> systems architect
Systemarchitektur *f* <Elektriz> system configuration
Systemarchitektur *f* **mit Funktionsteilung** <Telekom> function-division system architecture
systematische Ergebnisabweichung *f* <Qual> bias of result, systematic error of result
systematische Probenentnahme *f* <Comp & DV> systematic sampling
systematische Stichprobe *f* <Qual> systematic sample
systematische Stichprobenentnahme *f* <Qual> systematic sampling
systematischer Fehler *m* 1. <Elektrotech> bias; 2. <Gerät> bias error *(Messwesen)*; 3. <Phys> systematic error
systematischer Fehler *m* **des Messgeräts** <Gerät> instrumental error
systematisches Probieren *n* <Bau, Qual> trial-and-error
Systemausfall *m* <Kontroll> system failure
Systembau *m* <Bau> systems building
systembedingter Fehler *m* <Metrol> systematic error
Systembelastung *f* 1. <Comp & DV> processing load; 2. <Telekom> system load
Systembetreiber *m* <Telekom> system operator
Systembibliothek *f* <Comp & DV> system library, systems library
Systembilanz *f* <Telekom> link power budget *(Mikrowellen)*
Systemblockade *f* <Comp & DV> deadlock
Systemdämpfung *f* <Telekom> system loss, total loss *(Funkverbindung)*
Systemdämpfungsmaß *n* <Telekom> transmission loss *(TL)*
Systemdesigner *m* <Comp & DV> system designer
Systementwurfsingenieur *m* <Telekom> system designer
Systemerde *f* <Elektriz> *(BE)* system earth, *(AE)* system ground

Systemfehler *m* <Kontroll> system error
Systemfunktionsbild *n* <Elektriz> system operational diagram
Systemfunktionsplan *m* <Elektriz> system operational diagram
Systemgenerierung *f* <Comp & DV> sysgen, system generation
Systemgestaltung *f* <Ergon> system design
Systemintegrator *m* <Comp & DV> systems integrator
Systemkern *m* <Comp & DV> nucleus
Systemkonfiguration *f* <Comp & DV> system configuration
Systemkonsole *f* <Comp & DV> master console *(Basissystem)*
Systemkonstante *f (Gx)* <Akustik> system-rating constant *(Gx)*
Systemkonzeption *f* <Telekom> system design
Systemleistung *f* <Comp & DV> system performance
Systemmeldung *f* <Comp & DV> message
Systemmeldungskopf *m* <Comp & DV> message header
Systemnetzwerkarchitektur *f (SNA)* <Comp & DV> systems network architecture *(SNA)*
Systemplan *m* <Elektriz> system diagram
Systemplanung *f* <Comp & DV> system design
Systemplatte *f* <Comp & DV> system disk
Systemplattform *f* <Comp & DV> platform
Systemprogrammierung *f* <Comp & DV> systems programming
Systemprotokoll *n* <Comp & DV, Telekom> system log
Systemprüfung *f* 1. <Comp & DV> system check, system testing; 2. <Qual> system check
Systemreserve *f* <Funktech> system margin *(Richtfunk)*
Systemsicherheit *f* 1. <Comp & DV> system security; 2. <Sicherheit> systems safety
Systemsoftware *f* <Comp & DV> system software, systems software
Systemspanne *f* <Telekom> system gain
Systemsteuerkonsole *f* <Comp & DV> system control panel
Systemsteuerung *f* <Kontroll> system control
Systemsynthese *f* <Elektrotech> network synthesis
Systemtechnik *f* <Comp & DV> systems engineering
Systemteil *m* <Telekom> system component *(Netzwerk)*
Systemtheorie *f* <Telekom> system theory
Systemüberlastung *f* <Comp & DV> thrashing
Systemüberwachung *f* <Sicherheit> system monitoring
Systemurband *n* <Comp & DV> master tape
Systemverklemmung *f* <Comp & DV> deadlock
Systemverwaltung *f* <Comp & DV> housekeeping
Systemverwaltungsprozedur *f* <Comp & DV> housekeeping procedure
Systemzusammenbruch *m* <Comp & DV, Sicherheit> system crash
Systemzuverlässigkeit *f* <Sicherheit> system reliability
systolische Architektur *f* <Telekom> systolic architecture
systolische Matrix *f* <Comp & DV> systolic array
Syzygie *f* <Nichtfoss Energ> syzygy
SZ *(Säurezahl)* <Kunststoff, Lebensmittel> acid value
Szenenanalyse *f* <Künstl Int> scene analysis
Szintillation *f* <Phys, Telekom> scintillation
Szintillationskoinzidenz-Spektrometer *n* <Kerntech> scintillation coincidence spectrometer
Szintillations-Spektrometer *n* <Strahlphys> scintillation spectrometer
Szintillationszähler *m* <Gerät, Phys, Strahlphys> scintillation counter
Szintillator *m* <Strahlphys> scintillator

T

t 1. <Hydraul> *(Tiefe)* d *(depth)*; 2. <Teilphys> *(Triton)* t *(triton)*
T 1. <Akustik, Aufnahme> *(Nachhallzeit)* T *(reverberation time)*; 2. <Chemie> *(Tritium)* T *(tritium)*; 3. <Elektriz, Elektrotech, Erdöl, Kfztech, Maschinen, Mechan> *(Drehmoment)* T, torque; 4. <Fertig> *(Drehmoment)* T, torque *(Kunststoffinstallationen)*; 5. <Hydraul> *(Transpiration)* T *(transpiration)*; 6. <Labor, Phys> *(Tesla)* T *(Tesla)*; 7. <Metrol> *(absolute Temperatur)* T *(absolute temperature)*; 8. <Metrol> *(Tera...)* T *(tera...)*; 9. <Phys> *(thermodynamische Temperatur)* T *(thermodynamic temperature)*
T½ *(Halbwertszeit)* <Kerntech, Phys, Strahlphys, Teilphys> T½ *(half-life)*
Ta *(Tantal)* <Chemie> Ta *(tantalum)*
TAA *(Technische Anleitung Abfall)* <Abfall> Technical Instruction on Waste Management
Tab *m* <Comp & DV> tab
Tabakin-Potenzial *n* <Kerntech> Tabakin potential
tabellarisch *adj* <Comp & DV> tabular
tabellarische Aufstellung *f* <Comp & DV> tabulation
Tabelle *f* 1. <Bau> chart; 2. <Comp & DV> table; spreadsheet *(in der Tabellenkalkulation)*; 3. <Phys> chart
Tabellen *fpl* <Raumfahrt> ephemerides
Tabellenausgabe *f* <Comp & DV> table output
Tabellenbereich *m* <Comp & DV> table space
Tabellenform/in <Comp & DV> in tabular form
Tabellenkalkulation *f* <Comp & DV> spreadsheet
Tabellenkopf *m* <Druck> boxed head
Tabellenmaß *n* <Konstzeich> tabular dimension, tabulator dimension
Tabellensatz *m* <Druck> tabular work
Tabellensuche *f* <Comp & DV> table search
Tabellensuchoperation *f* <Comp & DV> table lookup
Tabellensuchverfahren *n* <Telekom> table look-up technique *(Sprachcodierung)*
Tabellenüberschrift *f* <Comp & DV> table header
Tabellenzeichnung *f* <Konstzeich> tabular drawing
tabellieren *v* <Comp & DV> tabulate
Tabellieren *n* <Comp & DV> tabulation
Tablett *n* 1. <Comp & DV> tablet *(Grafikverarbeitung)*; data tablet *(zur Grafikeingabe)*; 2. <Verpack> food tray
Tablette *f* <Kunststoff> pellet
Tablettenfläschchen *n* <Ker & Glas> tablet bottle
Tablettensortier- und Kontrollmaschine *f* <Verpack> tablet sorting and inspection machine
Tablettenzählvorrichtung *f* <Verpack> missing pill equipment
Tablettieren *n* 1. <Fertig> pelleting; 2. <Kunststoff> pelletizing
Tablettiermaschine *f* 1. <Fertig> pelleter; 2. <Kunststoff> pelletizer
Tabulator *m* <Comp & DV> tabulator
Tabulatorgitter *n* <Comp & DV> tab rack
Tabulatorstopp *m* <Comp & DV> tab stop
Tabulatortaste *f* <Comp & DV> tabulator key
tabulieren *v* <Comp & DV> tabulate
Tabuliertaste *f* <Comp & DV> tabulator key
T-Abzweig *m* <Bau> T-junction
Tacheometrie *f* <Bau> tacheometry *(Vermessung)*
Tachogenerator *m* <Metrol> tachodynamo, tachogenerator, tachometer generator
Tachograph *m* <Metrol> tachograph

Tachometer

Tachometer *n* 1. <Gerät> speed indicator; 2. <Kfztech> tachometer; speedometer *(Zubehör)*; 3. <Maschinen> speedometer; 4. <Metrol, Phys> tachometer
Tachometerantrieb *m* <Kfztech> speedometer drive gear
tachometrische Vermessung *f* <Bau> stadia surveying
Tachymeter *n* <Bau> tacheometer *(Vermessung)*
Tachyon *n* <Phys> tachyon
TAE-Dose *f* <Telekom> telecom socket, telecommunication socket, telephone socket
TAE-Kabel *n* <Telekom> telecom cord, telephone cord
TAE-Stecker *m* <Telekom> telecom plug, telephone plug
Tafel *f* 1. <Bau> table; plate *(einer Wand)*; 2. <Druck> plate; 3. <Elektriz, Funktech> panel; 4. <Kerntech> sheet *(dünne Platte)*; 5. <Kohlen> panel; 6. <Künstl Int> blackboard; 7. <Maschinen> plate; 8. <Phys> chart
Tafel-Abriebmaschine *f* <Kunststoff> Taber abrader
Tafelblei *n* <Bau, Metall> sheet lead
Tafelglas *n* <Ker & Glas> sheet glass
Tafelglaszuschnitte *mpl* <Ker & Glas> cut sizes
Tafelmontage *f* <Elektriz> panel mounting
täfeln *v* <Bau> pane, panel; ceil *(Decke)*
Tafelschere *f* 1. <Fertig> guillotine shearing machine, guillotine plate shear; 2. <Maschinen> plate shears, sheet shears
Täfelung *f* <Bau> panel
Taft *m* <Textil> taffeta
Tagebau *m* <Kohlen> open-pit mining, surface mine
Tagebaubergwerk *n* <Kohlen> daylight colliery
Tagebücher *npl* <Raumfahrt> ephemerides
Tagesgang *m* <Telekom> diurnal variation *(Wellenausbreitung)*
Tagesganglinie *f* <Trans> daily traffic distribution curve
Tageskilometerzähler *m* <Kfztech> odometer, trip mileage indicator; trip counter *(Instrument)*
Tagesleuchtfarbe *f* <Kunststoff> Day-Glo® paint
Tageslichtaufnahme *f* <Foto> daylight exposure
Tageslichtbeleuchtung *f* <Ergon> daylighting
Tageslichtfotografie *f* <Foto> daylight photography
Tageslichtquotient *m* <Ergon> daylight factor
Tageslichttank *m* <Foto> daylight loading tank
Tagesprobe *f* <Sicherheit, Umweltschmutz> whole-day sample
Tagesreichweite *f* <Lufttrans> day range
Tagesrückenlast *f* <Elektriz> daily base rate
Tagesverbrauch *m* <Wasserversorg> daily consumption
Tagesverkehr *m* <Trans> day traffic
Tageswanne *f* <Ker & Glas> day tank
täglich *adj* <Ergon, Raumfahrt> diurnal
Tagwasser *n* <Kohlen> surface water
Taillenabnäher *m* <Textil> waist dart
Takelage *f* <Wassertrans> rig *(Segeln)*; rigging *(Tauwerk)*
Takelplan *m* <Wassertrans> rigging drawing *(Tauwerk)*
Takelriss *m* <Wassertrans> rigging drawing *(Tauwerk)*
Takler *m* <Wassertrans> rigger
Takling *m* <Wassertrans> whipping *(Tauwerk)*
Takonit *m* <Kohlen, Metall> taconite
Takt *m* 1. <Akustik> beat, rhythm, time; 2. <Elektronik> clock; 3. <Kontroll> clock, timing; 4. <Kunststoff> cycle; 5. <Phys> stroke; 6. <Telekom> pattern, timing
Taktableitung *f* <Telekom> clock extraction, timing extraction
Taktdiagramm *n* <Comp & DV> timing diagram
takten *v* 1. <Comp & DV> clock, pace; 2. <Ergon> pace; 3. <Elektronik> pulse; 4. <Fertig, Gerät> clock
Taktfolge *f* <Elektronik> clocking sequence
Taktfrequenz *f* 1. <Elektronik, Fernseh, Funktech> clock frequency; 2. <Gerät> clock rate; 3. <Telekom> clock frequency, elementary frequency
Taktgabe *f* 1. <Kontroll> clocking; 2. <Telekom> clocking, timing
Taktgeber *m* 1. <Comp & DV> clock, clock generator, real-time clock; 2. <Elektronik> master oscillator, timing generator; 3. <Kontroll> clock, clock generator, timer; 4. <Telekom> clock generator, clock, timer
Taktgebermaß *n* <Gerät> clock rate
Taktgebung *f* <Elektronik> timing
Taktgeschwindigkeit *f* <Comp & DV> clock speed
taktgesteuertes Flipflop *n* <Elektronik> clocked flip-flop
Taktgewinnung *f* <Telekom> timing extraction
taktgleich *adj* <Elektronik> synchronous
taktil *adj* <Ergon> tactile
Taktimpuls *m* 1. <Comp & DV> clock pulse; 2. <Elektronik> clock pulse, timing pulse; 3. <Kontroll> clock pulse; 4. <Telekom> clock pulse, timing pulse
Taktintervall *n* <Elektronik> clock period
Taktizität *f* <Kunststoff> tacticity
Taktjitter *m* <Telekom> timing jitter
Taktrate *f* <Comp & DV, Gerät, Telekom> clock rate
Taktrelais *n* <Elektriz> clock relay
Taktsignal *n* 1. <Comp & DV, Telekom> clock pulse, clock signal; 2. <Elektronik> clock signal, timing signal
Taktsignalverzögerung *f* <Elektronik> clock signal skew
Taktspur *f* 1. <Aufnahme> clock track, timing track; 2. <Comp & DV> clock track
taktsteuern *v* <Gerät> clock
Taktumschaltung *f* <Kontroll> clock changeover
Taktunterbrechungseinstellung *f* <Trans> cycle split adjustment
Taktversorgung *f* <Telekom> timing distribution
Taktzyklus *m* <Comp & DV, Elektronik, Gerät, Kontroll> clock cycle, timing cycle
Talg *m* <Chemie, Fertig, Textil> tallow
talgig *adj* <Lebensmittel> tallowy
Talgöl *n* <Chemie> tallow oil
Talje *f* <Wassertrans> purchase, tackle *(Tauwerk)*
Taljereep *n* <Wassertrans> reefing pennant *(Tauwerk)*
Talkum *n* <Ker & Glas> talc
Tallöl *n* <Chemie, Kunststoff> tall oil
Talon... <Chemie> talonic
Talsohle *f* <Bau> bottom
Talsperre *f* <Wasserversorg> barrage, dam, river dam
Talstation *f* <Trans> valley station
Tambour *m* <Papier> reel spool
Tambourwalze *f* <Papier> reel drum
tan *(Tangens)* tan, tangent *(Winkelfunktion)*
Tandem *n* <Kfztech> tandem
Tandemachse *f* <Kfztech> tandem axle *(LKW)*
Tandembauart *f* <Maschinen> tandem construction
Tandembeschleuniger *m* 1. <Kerntech> tandem accelerator; 2. <Phys> tandem generator
Tandemhubschrauber *m* <Lufttrans> dual tandem helicopter, tandem rotor helicopter
Tandemmaschine *f* <Maschinen> tandem engine
Tandemmotor *m* <Elektrotech, Kontroll> tandem motor
Tandem-Potenziometer *n* <Elektriz> dual-ganged potentiometer
Tandemschaltung *f* <Elektriz> tandem connection
Tandemvibrationswalze *f* <Bau> tandem vibrating roller *(Straßenbau)*
Tandemwalzwerk *n* <Metall> tandem mill
Tangens *m (tan)* <Math> tangent, tan *(Winkelfunktion)*
Tangente *f* <Geom> tangent *(Berührgerade)*
Tangentenbussole *f* 1. <Elektrotech> tangent galvanometer; 2. <Phys> tangent compass, tangent galvanometer
Tangenten-Galvanometer *n* <Elektriz> tangent galvanometer

Tangentengleichung f <Geom> equation of the tangent
tangential adj <Geom, Mechan, Phys> tangential
Tangential... <Maschinen, Math, Phys> tangential
Tangentialbeanspruchung f <Maschinen> tangential stress
Tangentialbelastung f <Maschinen> tangential strain
Tangentialbeschleunigung f <Maschinen, Mechan> tangential acceleration
Tangentialdrehmeißel m <Fertig, Maschinen> tangential turning tool
Tangentialebene f 1. <Bau, Geom> tangent plane; 2. <Math> tangential plane
tangentialer Vorschub m <Maschinen> tangential feed
Tangentialgeschwindigkeit f <Nichtfoss Energ> tangential velocity
Tangentialkomponente f <Phys> tangential component
Tangentialnut f <Maschinen> tangential keyway
Tangentialschneidbacke f <Maschinen> tangential threading die
Tangentialschneidplatte f <Fertig> on-end insert
Tangentialschnitt m <Phys> tangential focal line
Tangentialsteuerung f <Optik> tangential control
Tangentialwälzfräsen n <Fertig> tangential hobbing
Tangentkeil m <Maschinen> tangent key, tangential key
Tangentkeilnut f <Maschinen> tangent keyway
tangierende Kreise mpl <Geom> tangent circles
Tank m 1. <Erdöl, Heiz & Kälte> tank; 2. <Kfztech> (AE) gas tank, (AE) gasoline tank, (BE) petrol tank (Kraftstoff); reservoir (Öl); 3. <Mechan, Raumfahrt, Umweltschmutz> tank; 4. <Wasserversorg> cistern
Tank m **mit Elastomermembran** <Raumfahrt> elastomer membrane tank
Tankcontainer m <Kfztech> tank container
Tankdecke f <Wassertrans> tank top (Schiffbau)
Tankdeckel m <Kfztech> filler cap, tank cap
tanken v <Kfztech, Trans> fill up
Tanken n <Kfztech, Wassertrans> (AE) refueling, (BE) refuelling
Tankentlüftung f <Raumfahrt> tank vent
Tankentwicklung f <Foto> tank development
Tanker m 1. <Erdöl> tanker (Schifffahrt); 2. <Wassertrans> liquid-bulk carrier, tanker
Tanker-Terminal m <Wassertrans> tanker terminal
Tankfahrzeug n 1. <Kfztech> (BE) tank lorry, (AE) tank truck, tanker; 2. <Meerschmutz> tanker
Tankflugzeug n <Kfztech> refueller
Tankfrachtkan m <Wassertrans> tank barge
Tankklappe f <Kfztech> filler compartment flap
Tankluke f <Wassertrans> tank hatch
Tankreaktor m <Kerntech> tank reactor
Tanksattelanhänger m <Kfztech> tank semi-trailer
Tankschiff n <Wassertrans> liquid-bulk carrier
Tankstelle f <Kfztech, Trans> filling station, (AE) gas station, (AE) gasoline station, (BE) petrol station, (BE) road petrol station
Tankventil n <Hydraul> tank valve
Tankverbindungsstück n <Raumfahrt> intertank connector (Raumschiff)
Tankwagen m 1. <Kfztech> tanker; 2. <Meerschmutz> road tanker; 3. <Wassertrans> (BE) tanker lorry, (AE) tanker truck
Tankwaggon m <Wassertrans> (BE) tanker lorry, (AE) tanker truck
Tankwall m <Erdöl> fire wall (Raffinerie)
Tankwärmer m <Foto> tank heater
Tankzwischenstück n <Raumfahrt> intertank connector (Raumschiff)
Tannat n <Chemie> tannate

Tannenbaumkristall m <Fertig> fir tree crystal, pine crystal
Tannenbaumprofil n <Fertig> fir tree profile
Tannin n <Chemie> tannin
Tanningerbstoff m <Chemie> tannin
T-Anschluss m <Bau> T-piece union
Tantal n (Ta) <Chemie> tantalum (Ta)
Tantal... <Chemie> tantalic, tantalum
Tantalanode f <Elektrotech> tantalum anode
Tantalat n <Chemie> tantalate
Tantalfestkondensator m <Elektrotech> tantalum solid capacitor
Tantalfilmkondensator m <Elektrotech> tantalum foil capacitor
Tantalkern m <Elektrotech> tantalum core, tantalum slug
Tantalkernkondensator m <Elektrotech> tantalum slug capacitor
Tantalkondensator m <Elektrotech> tantalum capacitor
Tantalnasskondensator m <Elektrotech> tantalum wet capacitor
Tantaloxid n <Elektrotech> tantalum oxide
Tantaloxidkondensator m <Elektrotech> tantalum oxide capacitor
Tantalschichtkondensator m <Elektrotech> tantalum foil capacitor
Tantiemen fpl <Druck> royalties
Tanzen n **der Leiterseile** <Elektrotech> conductor dancing, conductor galloping (Energieerzeugung)
Taperwellenleiter m <Telekom> tapered waveguide
tapezieren v <Bau> paper
Tapeziererleim m **auf Wasserbasis** <Verpack> water-based backing adhesive
Tapeziernagel m <Maschinen> tintack
Tara n 1. <Textil> tare; 2. <Trans> tare (Leergewicht); 3. <Verpack> tare (Verpackungsgewicht)
Target n <Teilphys> target
Targetbestrahlung f <Kerntech> target irradiation
Tarieren n <Lufttrans> calibration
Tarif m <Telekom> charge rate, tariff
Tarif m **für den Mobilfunkdienst** <Mobilkom> mobile radio call charge rate, mobile radio tariff
Tarif m **für den Telefondienst** <Telekom> call charge rate, telephone tariff
Tarifeinheit f <Telekom> call charge unit
Tarnverpackung f <Verpack> deceptive packaging
Tartrat n <Chemie> tartrate
Tartron... <Chemie> tartronic
Tartronoylharnstoff m <Chemie> dialuric acid, hydroxybarbituric acid, tartronoylurea
Tasche f 1. <Kohlen> pocket; 2. <Papier> pocket (des Holzschleifers)
Taschendosimeter n <Strahlphys> pocket dosemeter
Taschendosimeter n **mit Skale** <Strahlphys> graduated pocket dosimeter
Taschenentlüftungsleitung f <Papier> pocket-ventilating duct
Taschenentlüftungswalze f <Papier> pocket-ventilating roll
Taschenlampe f <Elektriz, Elektrotech> flashlight, inspection lamp, pocket lamp
Taschenrechner m 1. <Comp & DV> calculator, pocket calculator; 2. <Math, Telekom> pocket calculator
Taschenterminal n <Telekom> pocket terminal
Tasksteuerung f <Comp & DV> task management
Tast... <Fernseh, Kontroll, Telekom> touch
Tastatur f 1. <Aufnahme> keyboard (elektronische Orgel oder Synthesizer); 2. <Comp & DV, Druck, Elektriz, Fernseh, Funktech> keyboard; 3. <Telekom> keyboard, pad
Tastaturbelegung f <Comp & DV> keyboard layout

Tastaturbeschriftung

Tastaturbeschriftung f <Comp & DV> keyboard overlay
Tastaturcodierer m <Comp & DV> keyboard encoder
Tastatureingabe f <Comp & DV, Telekom> keying
Tastaturfolie f <Comp & DV> keyboard overlay
Tastaturklick m <Comp & DV, Funktech, Telekom> key click
Tastaturmaskierung f <Comp & DV> keyboard mask
Tastaturschablone f <Comp & DV> keyboard mask, keyboard overlay, keyboard template
Tastatursende-Empfangsmodus m <Comp & DV> KSR, keyboard send-receive *(Datenstation und serieller Drucker mit Bildschirmfunktion)*
Tastatursperre f <Comp & DV> keyboard lock
tastbar *adj* <Ergon> tactile
Tastbügelregler m <Gerät> chopper bar controller
Taste f 1. <Akustik> key; 2. <Aufnahme> push button; 3. <Comp & DV> button, key *(einer Tastatur)*; 4. <Druck, Elektrotech, Fernseh> key; 5. <Funktech> key; 6. <Telekom> button
Taste-Entf f *(Entfernungstaste)* <Comp & DV> delete key *(DEL key)*
tasten v 1. <Druck> keyboard; 2. <Fernseh, Funktech> key
Tastenanordnung f <Comp & DV> keyboard layout
Tastenanschlag m <Comp & DV> keystroke
Tastenblock m <Comp & DV> keypad
Tastenfeld n <Aufnahme, Comp & DV, Fernseh, Telekom> keyboard, keypad
Tastengeber m <Telekom> keyboard sender
tastengesteuert *adj* <Comp & DV> key-driven
Tastenknopf m <Kontroll> button
Tastenkürzel n <Comp & DV> keyboard shortcut
Tastennummer f <Comp & DV> key number
Tastensperre f 1. <Comp & DV> keyboard locking; 2. <Eisenbahn> lock and block
Tastenwahl f <Telekom> push-button dial
Tastenweg m <Comp & DV> key travel
Tastenwiderstand m <Comp & DV> key force
Taster m 1. <Elektriz> key switch, push button; 2. <Fertig> stylus; 3. <Labor> *(AE)* calipers, *(BE)* callipers; 4. <Maschinen> *(AE)* calipers, *(BE)* callipers, feeler; 5. <Metrol> feeler pin; *(AE)* caliper, *(BE)* calliper *(Zirkel)*
Tasterzirkel m <Metrol> *(AE)* caliper compasses, *(BE)* calliper compasses
Tastfehler m <Telekom> keying error
Tastfeld n <Comp & DV, Kontroll> touchpad
Tastfernsprecher m <Telekom> key-operated telephone
Tastflächen *fpl* <Bau> points
Tastimpuls m <Fernseh> gating pulse
Tastknopf-Befehlstafel f <Elektriz> push-button control panel
Tastknopf-Steuertafel f <Elektriz> push-button control panel
Tastkopf m 1. <Gerät> sensing head; 2. <Phys> probe
Tastlehre f <Maschinen, Mechan> *(AE)* caliper, *(BE)* calliper
Tastnase f <Fertig> follower *(Spanung)*
Tastpegel m <Fernseh> key level
Tastrelais n <Funktech> keyer relay
Tastschalterbetätigung f <Foto> push-button operation
Tastspule f 1. <Fertig> probe coil, surface probe coil; 2. <Phys> exploring coil, pick-up coil, search coil
Taststift m <Maschinen> tracer pin
Tastung f <Comp & DV> modulation
Tastverhältnis n <Phys> mark space ratio
Tastwahl f <Telekom> push-button dial
Tastwahlapparat m <Telekom> key telephone set, push-button telephone

Tastwahlfernsprechsystem n <Telekom> key telephone system
Tastwahltelefon n <Telekom> key-operated telephone, key telephone set, push-button telephone
Tastzirkel m 1. <Fertig> *(AE)* morphy caliper, *(BE)* morphy calliper; 2. <Maschinen> *(AE)* caliper compasses, *(BE)* calliper compasses, *(AE)* calipers, *(BE)* callipers, *(AE)* external and internal calipers, *(BE)* external and internal callipers
Tätigkeit f 1. <Ergon> activity, function, job; 2. <Qual> action
Tätigkeitsanforderung f <Ergon> job demand
Tätigkeitsbeschreibung f <Ergon> job description
Tätigkeitsnachweis m <Qual> proof of action
Tätigkeits- und Fehlerbericht m <Qual> status report
tatsächlich *adj* 1. <Mechan> effective; 2. <Textil> actual
tatsächliche Belastungsspitze f <Kohlen> actual peak load
tatsächliche Fahrgeschwindigkeit f <Mechan> actual running speed
tatsächliche Höchstbelastung f <Kohlen> actual peak load
tatsächliche Spaltlänge f <Fernseh> real gap length
tatsächliche Umlaufgeschwindigkeit f <Mechan> actual running speed
tatsächlicher Flugweg m <Lufttrans> actual flight path
tatsächlicher Parameter m <Comp & DV> actual parameter
tatsächlicher Zustand m <Kerntech> actual state
Tatzlagermotor m <Elektrotech> axle-hung motor, nose and axle-suspended motor, nose-suspended motor
Tatzlagerung f <Elektrotech> motor suspension bearing, nose suspension
Tau m 1. <Papier> dew; 2. <Wassertrans> rope
Taubheit f <Akustik> deafness
Taubstummheit f <Akustik> deaf-muteness
Taubucht f <Wassertrans> fake *(Tauwerk)*
Tauch... <Fertig, Kunststoff, Wassertrans> dip
Tauchbad n <Fertig> immersing bath
Tauchbeschichten n <Fertig, Kunststoff> dip coating
Tauchbeschichtung f <Kunststoff> dip coating
Taucheinfrieren n <Verpack> immersion freezing
Tauchelektrode f <Fertig> dipped electrode
tauchen v 1. <Textil> dip; 2. <Wassertrans> dive
Tauchen n 1. <Verpack> immersion coating; 2. <Wassertrans> heaving *(Schiffsbewegung)*
Taucher m <Wassertrans> diver
Taucherglocke f <Wassertrans> diving bell
Taucherkrankheit f <Wassertrans> bends
Tauchfräsen n <Maschinen> plunge cutting, plunge milling, plunge-cut milling
Tauchgerät n <Wassertrans> scuba
Tauchglocke f 1. <Erdöl> diving bell *(Tieftauchtechnik)*; 2. <Metall> bell plunger
Tauchglockendurchflussmesser m <Gerät> bell flowmeter
Tauchglocken-Manometer n 1. <Fertig> inverted-bell manometer; 2. <Gerät> *(AE)* bell pressure gage, *(BE)* bell pressure gauge, bell-type manometer
Tauchglockenwirkdruckgeber m <Gerät> bell-type difference pressure transmitter
Tauchhärtung f <Fertig> liquid hardening
Tauchheizkörper m <Elektrotech> immersion heater
Tauchkammer f <Fertig> gooseneck *(Druckguss)*
Tauchkern m <Elektrotech> plunger *(bei Relais)*
Tauchkern-Relais n <Elektrotech> plunger relay
Tauchkolben m 1. <Bau, Elektrotech, Kfztech> plunger *(Brems- und Kupplungszylinder)*; 2. <Maschinen> plunger, plunger piston, ram

Tauchkolbenmotor *m* <Kfztech, Wassertrans> trunk piston engine
Tauchkolbenpumpe *f* <Maschinen, Wasserversorg> plunger pump
Tauchlackierung *f* <Verpack> dipping process, immersion painting
Tauchlöten *n* 1. <Bau> dip brazing *(Hartlötung)*; 2. <Fertig> dip brazing
Tauchlötung *f* <Maschinen> dip soldering
Tauchmuffel *f* <Ker & Glas> immersion muffle
Tauchöl *n* <Metall> immersion oil
Tauchprobeverfahren *n* <Metall> dip test technique
Tauchpumpe *f* <Heiz & Kälte> submersible pump
Tauchschaben *n* <Maschinen> plunge shaving
Tauchschmierung *f* <Eisenbahn, Kfztech, Maschinen> splash lubrication
Tauchschwingung *f* <Wassertrans> heaving *(Schiffsbewegung)*
Tauchsicherung *f* <Sicherheit> diving security
Tauchsieder *m* <Elektriz, Elektrotech, Heiz & Kälte, Labor, Maschinen, Mechan> immersion heater
Tauchspule *f* <Elektrotech> moving coil, MC, plunger-type coil, voice coil *(bei Lautsprecher)*
Tauchspulenmikrofon *n* <Akustik, Aufnahme, Phys> moving-coil microphone
Tauchstreichverfahren *n* <Verpack> dip coating
Tauchteiler *m* <Elektronik> flap attenuator *(Mikrowellen)*
Tauchthermoelement *n* <Metrol> dip thermocouple, immersion thermocouple
Tauchüberziehen *n* <Verpack> hot-dipping
Tauchverfahren *n* <Bau> dipping method
Tauchzählrohr *n* <Heiz & Kälte> liquid flow counter tube
Tauchzelle *f* 1. <Gerät> immersion cell; 2. <Wassertrans> ballast tank *(U-Boot)*
tauen *v* <Meerschmutz> tow *(Schiff)*
Taumelkolbenzähler *m* <Gerät> nutating-piston meter
taumeln *v* 1. <Fertig> gyrate; 2. <Maschinen> wobble
Taumelscheibe *f* 1. <Fertig> Z-crank, swash plate, *(BE)* wabbling disc, *(AE)* wabbling disk, wobble plate; 2. <Maschinen> swash plate, wobble plate
Taumelscheibenmotor *m* <Maschinen> wobble plate engine
Taumelscheibenpumpe *f* <Fertig> Z-crank pump
Taumelschlag *m* <Maschinen> axial eccentricity, axial runout *(einer Wellenstirnseite)*
Tau-Neutrino *n* 1. <Phys> tauon neutrino *(Elementarteilchen)*; 2. <Teilphys> tau neutrino
Tauon *n* <Phys> tauon *(Elementarteilchen)*
Taupunkt *m* 1. <Heiz & Kälte> dew point; 2. <Lebensmittel> dew point, thawing point; 3. <Lufttrans, Maschinen, Papier, Phys> dew point
Taupunkthygrometer *n* <Metrol, Thermod> dew-point hygrometer
Taupunktmessung *f* <Ker & Glas> dew-point measurement
Taupunkttemperatur *f* <Heiz & Kälte, Kerntech, Thermod> dew-point temperature
Taurin *n* <Chemie> aminoethionic acid, taurine
Taurochol... <Chemie> taurocholic
Taurocholat *n* <Chemie> taurocholate
Tauröste *f* <Textil> dew retting
Täuschung *f* **durch Nachahmung** <Elektronik> imitative deception
Täuschungssignal *n* <Elektronik, Funktech> deception signal
Tausend *n* <Math> thousand
tausendfach *adj* <Math> thousandfold
Tausendstel *n* <Math> thousandth
Tau-Teilchen *n* <Teilphys> tau particle

Tautomer *n* <Chemie> tautomer
tautomer *adj* <Chemie> tautomeric
Tautomeres *n* <Chemie> tautomer
Tautomerisierung *f* <Chemie> tautomerization
Tautropfenglas *n* <Ker & Glas> dewdrop glass
Tauwasser *n* <Heiz & Kälte> condensate
Tauwerk *n* <Wassertrans> cordage
Taxierung *f* <Telekom> metering
Taylor'sche Entwicklung *f* <Math> Taylor expansion
Taylor'sche Reihe *f* <Math> Taylor series
Taylor'sche Zahl *f* <Strömphys> Taylor number
Taylor'scher Kegel *m* <Strahlphys> Taylor cone
Taylor'scher Satz *m* <Math> Taylor's theorem *(Analysis)*
Tb *(Terbium)* <Chemie> Tb *(terbium)*
TBKZ *(Technisches und Betriebskontrollzentrum)* <Raumfahrt> *(AE)* TOCC *(Technical and Operational Control Center)*
TBP *(Tributylphosphat)* <Kerntech> TBP *(tributylphosphate)*
TBP-Verfahren *n* <Kerntech> TBP process
Tc *(Technetium)* <Chemie> Tc *(technetium)*
TCP/IP-Zugangsknoten *m* <Comp & DV> TCP/IP access node
TD *(theoretische Dichte)* <Kerntech> TD *(theoretical density)*
TDI *(Toluendiisocyanat, Toluoldiisocyanat)* <Kunststoff> TDI *(toluyene diisocyanate)*
TDM *(Multiplexen mit Zeitteilung, Zeitmultiplexmethode, Zeitmultiplexverfahren)* <Comp & DV, Elektronik, Telekom> TDM *(time division multiplex)*
TDMA *(Mehrfachzugriff im Zeitmultiplex, nicht gleichzeitiger Mehrfachzugriff)* 1. <Comp & DV, Elektronik, Telekom> TDMA, time division multiple access; 2. <Raumfahrt> TDMA, time division multiple access *(Weltraumfunk)*
TDMA-Terminal *n* <Raumfahrt> TDMA terminal *(Weltraumfunk)*
tdw <Wassertrans> dwt *(deadweight tonnage, tons deadweight)*
Te *(Tellur)* <Chemie> Te *(tellurium)*
TE *(transversal elektrisch)* <Elektrotech, Telekom> TE *(transverse electric)*
Technetium *n* *(Tc)* <Chemie> technetium *(Tc)*
Technik *f* 1. <Elektriz, Kfztech, Kohlen> technology; 2. <Maschinen> engineering, technology; procedure, technique *(Verfahren)*; 3. <Nichtfoss Energ> technology
technische Abteilung *f* <Maschinen> engineering department
technische Anforderung *f* <Metrol, Qual> technical requirement
Technische Anleitung *f* **Abfall** *(TAA)* <Abfall> Technical Instruction on Waste Management
technische Arbeitshygiene *f* <Sicherheit> health engineering, industrial hygiene engineering
technische Beschreibung *f* 1. <Comp & DV> physical description; 2. <Eisenbahn, Qual> technical instructions
technische Betreuung *f* <Comp & DV> technical support
technische Daten *npl* <Qual, Wassertrans, Telekom> specifications
technische Daten *npl* **eines Triebwerks** <Lufttrans> engine ratings
technische Dichtmasse *f* <Kunststoff> engineering sealant
technische Einrichtung *f* <Bau> equipment
technische Einschränkung *f* <Nichtfoss Energ, Raumfahrt> technological restriction
technische Frequenz *f* <Funktech> industrial frequency
technische Gebäudeausrüstung *f* <Bau> building services

technische

technische Harze *npl* <Kunststoff> industrial resins
technische Hygiene *f* <Sicherheit> health engineering
technische Kontrolleinrichtung *f* <Sicherheit> engineering control
technische Kunststoffe *mpl* <Kunststoff> engineering plastics
technische Normen *fpl* <Maschinen> engineering standards
technische Produktdokumentation *f* <Qual> technical product documentation
technische Schutzausrüstung *f* <Sicherheit> group safety equipment
technische Störung *f* <Qual, Telekom> technical breakdown
technische Überwachung *f* <Sicherheit> engineering control
technische Zeichnung *f* <Maschinen> engineering drawing
technische Zuverlässigkeit *f* <Sicherheit> reliability engineering
technische Zwischenüberprüfung *f* <Raumfahrt> intermediate design review
technischer Arbeitsschutz *m* <Sicherheit> safety engineering
Technischer Außendienst *m* <Comp & DV> field service
technischer Bericht *m* <Fernseh, Qual, Telekom> technical report
technischer Fehler *m* <Comp & DV> malfunction
technischer Lärmschutz *m* <Sicherheit> noise control engineering
technischer Umweltschutz *m* <Umweltschmutz> environmental engineering
technisches Datenblatt *n* <Kfztech> specifications sheet
technisches Modell *n* <Raumfahrt> engineering model
technisches Natriumsulfat *n* <Lebensmittel> salt cake
Technisches und Betriebskontrollzentrum *n* (TBKZ) <Raumfahrt> (AE) Technical and Operational Control Center, TOCC (Weltraumfunk)
Technologie *f* <Kfztech, Kohlen, Nichtfoss Energ> technology
Technologie *f* des schnellen Brüters <Kerntech> fast breeder reactor technology
Technologiebewertung *f* <Qual, Sicherheit> technology assessment
Technologiesatellit *m* auf Erdumlaufbahn <Telekom> orbiting technological satellite
Technologiestand *m* <Kontroll> state of technology
Teer *m* 1. <Bau> pitch, tar; 2. <Kunststoff, Wassertrans> tar
Teeranstrich *m* <Kunststoff> tar coating
teeren *v* <Bau, Wassertrans> tar
Teerkessel *m* <Bau> tar boiler
Teerleinwand *f* <Bau> tarpaulin
Teermakadam *m* <Bau> tarmac, tarmacadam (Tiefbau)
Teerpapier *n* <Verpack> tarred board
Teerpappe *f* 1. <Bau> tarred felt; 2. <Verpack> tarred board
Teerpech *n* <Kohlen> coal-tar pitch
Teersand *m* <Erdöl> tar sand (Geologie)
Teerspritzgerät *n* <Bau> tar sprinkler
Teerspritzmaschine *f* <Bau> tar sprayer
Teer- und Bitumenkocher *m* <Bau> binder heater
Teerung *f* <Bau> tarring
Teflon® *n* <Chemie, Kunststoff> Teflon®
Teich *m* <Kohlen, Wasserversorg> pond
Teiglösung *f* <Kunststoff> skimming dough
Teigpressverfahren *n* <Ker & Glas> (AE) dough molding, (BE) dough moulding

Teil *n* 1. <Comp & DV> part; 2. <Maschinen, Mechan> component, member, part
Teil *m* der Ummantelung <Kerntech> shell section
Teilabschaltung *f* <Kerntech> partial trip (eines Reaktors)
Teilamt *n* mit Überbrückungsverkehr <Telekom> discriminating satellite exchange
teilamtsberechtigter Teilnehmer *m* <Telekom> partially restricted subscriber
Teilansicht *f* <Konstzeich> partial plan, partial view
Teilantrieb *m* <Papier> sectional drive
Teilapparat *m* 1. <Fertig> divider; 2. <Maschinen> dividing apparatus, dividing heads
Teilaufzeichnung *f* <Aufnahme> half-track recording
Teilausschnitt *m* <Konstzeich> cut
Teilbandcodierung *f* <Telekom> subband coding
teilbar *adj* <Math> divisible
teilbare Zahl *f* <Math> composite number
Teilbaum *m* <Textil> section beam
Teilbild *n* <Fernseh> field, frame, subimage
Teilbildflackern *n* <Fernseh> field rate flicker
Teilbildkonvergenz *f* <Fernseh> field convergence
Teilbildneigung *f* <Fernseh> field tilt
Teilbildrücklauf *m* <Fernseh> field flyback
Teilbildschaltung *f* <Fernseh> field gating circuit
Teilbildselektierung *f* <Comp & DV> panning
Teilbildsynchronisierung *f* <Fernseh> field sync
Teilbildzentrierungsregelung *f* <Fernseh> (AE) field-centering control, (BE) field-centring control
Teilchen *n* 1. <Akustik, Elektronik, Kohlen, Phys, Teilphys, Textil> particle; 2. <Umweltschmutz> particulate material
Teilchen *n* kleiner als Atome <Phys> subatomic particle
Teilchen *n* mit kurzer Reichweite <Kerntech> short range particle
Teilchenbeschleuniger *m* 1. <Kerntech> accelerator; 2. <Phys, Teilphys> particle accelerator
Teilchenbeschleunigung *f* <Akustik> particle acceleration
Teilchenbewegung *f* <Phys> motion of a particle
Teilchendynamik *f* <Kerntech> particle dynamics
Teilchenfamilie *f* <Teilphys> particle family
Teilchenfluss *m* <Phys> particle flux
Teilchenflussrate *f* <Phys> particle fluence rate
Teilchengeschwindigkeit *f* <Akustik> particle velocity
Teilchengröße *f* 1. <Kerntech> particle size (in Pulvern); 2. <Kunststoff, Metall> particle size
Teilchengrößenanalysator *m* <Chemtech, Kerntech> PSA, (BE) particle size analyser, (AE) particle size analyzer
Teilchengrößenmessung *f* <Kunststoff> particle size measurement
Teilchenklassierung *f* <Chemtech> particle classification
Teilchennachweis *m* <Teilphys> particle detection
Teilchenpaket *n* <Phys> bunch
Teilchenphysik *f* <Teilphys> particle physics
Teilchenschnelle *f* <Akustik> particle velocity
Teilchenschwund *m* <Kerntech> particle leakage (durch Leck)
Teilchenstoß *m* <Teilphys> particle collision
Teilchenstoßbereich *m* <Kerntech> range collision
Teilchenstrahl *m* <Teilphys> beam of particles
Teilchenstrahlung *f* <Strahlphys> corpuscular radiation
Teilchenstreuung *f* <Teilphys> particle scattering
Teilchentrennung *f* <Teilphys> particle separation
Teilchenverstärkung *f* <Metall> particle reinforcement
Teilchenzahl *f* <Metall> particle number
Teilcontainerschiff *n* <Wassertrans> semicontainer ship
Teildatei *f* <Comp & DV> member
Teildruck *m* <Phys> partial pressure
Teileherstellung *f* <Fertig> component manufacture

Teileliste *f* 1. <Fertig> list of parts; 2. <Maschinen> parts list
Teil-Ellipse *f* <Geom> partial ellipsis
teilen *v* 1. <Bau> split; 2. <Comp & DV> share; 3. <Fertig> index *(Teilkopf)*; 4. <Math> divide
Teilen *n* <Maschinen> dividing, indexing
Teilentladung *f* <Elektriz> partial discharge *(PD)*
Teilentladungseinsatz *m* <Elektriz> partial discharge inception
Teilentladungseinsatzspannung *f* <Elektriz> partial discharge inception voltage
teilentladungsfrei *adj* partial discharge-free
Teilentladungsimpulsladung *f* <Elektriz> partial discharge impulse charge, partial discharge pulse charge
Teilentladungsmessung *f* <Elektriz> corona measurement, partial discharge measurement
Teilentladungsnachweisgerät *n* <Elektriz> corona detector, partial discharge detector
Teilentladungsnormal *n* <Elektriz> partial discharge standard
Teilentladungspegel *m* <Elektriz> partial discharge level, partial inception level
Teilentladungsprüfung *f* <Elektriz> partial discharge test, partial discharge testing
Teilentladungsschwelle *f* <Elektriz> partial discharge level, partial inception level
Teilentladungszerstörung *f* <Elektriz> corona damage, corona erosion, partial discharge damage, partial discharge erosion
Teilenummer *f* <Verpack> part number
Teiler *m* 1. <Math> divisor; 2. <Metrol> submultiple *(Einheit)*; 3. <Telekom> divider
Teilerdose *f* <Elektriz> splitter box
Teilezuführrutsche *f* <Fertig> loading chute
Teilezuführung *f* <Maschinen> hopper
Teilezusammenbau *m* <Kfztech> component assembly
Teilfarb-Andrucke *mpl* <Druck> progressive proofs, progressives
Teilflächenbeschichtung *f* <Kunststoff> pattern coating, strip coating
Teilfunktion *f (Z)* <Phys> partition function *(Z)*
Teilgenauigkeit *f* <Fertig> accuracy of indexing
Teilgesamtheit *f* <Qual> subpopulation
Teilgraph *m* <Künstl Int> subgraph
Teilgruppe *f* <Telekom> subgroup *(Telefonanlage)*
teilhärten *v* <Thermod> flash-harden
Teilkammerkessel *m* <Heiz & Kälte> sectional boiler
Teilkammersystem *n* **mit elastischen Schürzen** <Wassertrans> multiple-skirted plenum chamber
Teilkanal *m* <Elektronik> subchannel
Teilkegel *m* 1. <Fertig> pitch cone *(Kegelrad)*; 2. <Maschinen> pitch cone
Teilknoten *m* <Akustik, Elektrotech> partial node
Teilkonstrukteur *m* <Fertig> detailer
Teilkopf *m* <Maschinen> dividing head, indexing head
Teilkreis *m* 1. <Maschinen> dividing circle, pitch circle; 2. <Math> limb
Teilkreisdurchmesser *m* 1. <Fertig> circle diameter; 2. <Lufttrans> pitch circle diameter, pitch diameter
Teilkurbel *f* <Fertig> index crank
Teilladung *f* <Eisenbahn> part load
Teillast *f* <Kfztech> part load
Teillinie *f* <Maschinen> pitch line
Teilmaschine *f* <Maschinen> dividing machine
Teilmenge *f* 1. <Comp & DV, Math> subset; 2. <Math> partial set
Teilmengenrelation *f* <Math> inclusion
Teilmessbereich *m* <Gerät> partial measuring range
Teilmodul *n* <Kontroll> submodule

Teilmontage *f* <Fertig> subassembling
teilmultiplexen *v* <Telekom> submultiplex
Teilnehmer *m* 1. <Comp & DV, Elektriz> subscriber; 2. <Telekom> user, subscriber *(Fernmeldnetz)*
• **Teilnehmer erreichen** <Telekom> obtain a subscriber
Teilnehmer *m* **am Mobilfunk nach GSM-Standard** <Mobilkom> digital cellular radio subscriber, GSM standard radio subscriber
Teilnehmer *m* **am öffentlichen Fernsprechdienst** <Telekom> public telephone service subscriber, telephone subscriber
Teilnehmer *m* **am zellularen digitalen Mobilfunk** <Mobilkom> digital cellular radio subscriber, GSM standard radio subscriber
Teilnehmer *m* **am zellularen Mobilfunk** <Mobilkom> cellular radio subscriber
Teilnehmer *m* **an Bord** <Telekom> on-board subscriber
Teilnehmer *m* **mit Handy** <Mobilkom> mobile subscriber
Teilnehmeradresse *f* <Telekom> party address
Teilnehmeranschluss *m* <Telekom> subscriber access, subscriber connection
Teilnehmeranschluss *m* **über Funk** <Mobilkom> radio in the loop *(RITL)*; wireless local loop
Teilnehmeranschlussbereich *m* <Telekom> customer loop
Teilnehmeranschlussleitung *f (TNL)* <Telekom> subscriber's line
Teilnehmeranschlussleitung *f* **für hohe Bitraten** <Telekom> high bit rate digital subscriber's line
Teilnehmeranschluss-Schnittstellenbaustein *m* <Telekom> subscriber line interface circuit, SLIC *(Telefon)*
Teilnehmerapparat *m* <Telekom> subscriber telephone set, subscriber set, subset
Teilnehmerbesetztsignal *n* <Telekom> subscriber-busy signal *(SSB)*
Teilnehmerbesetztton *m* <Telekom> busy tone, subscriber-busy tone
Teilnehmerbesetztzeichen *n* <Telekom> subscriber-busy signal *(SSB)*
Teilnehmerbesetztzustand *m* <Telekom> subscriber-busy condition
Teilnehmerbetrieb *m* <Elektronik, Telekom> time slicing
Teilnehmerbetriebsklasse *f* <Telekom> class of service
Teilnehmerdienst *m* <Telekom> subscriber service
Teilnehmerebene *f* <Telekom> user plane
teilnehmereigene Fernsprech-Handvermittlung *f* <Telekom> private manual exchange
Teilnehmerendeinrichtung *f* <Telekom> customer terminal, subscriber terminal, subscriber terminal equipment, terminal equipment *(TE)*
Teilnehmerendgerät *n* <Telekom> customer terminal
Teilnehmerentstörung *f* <Telekom> subscriber service
Teilnehmergerät *n* <Telekom> user equipment
Teilnehmergesprächsdichte *f* <Telekom> subscriber calling rate
Teilnehmerhauptanschluss *m* <Telekom> subscriber main connection, subscriber main station
Teilnehmeridentifizierungsmodul *n* <Telekom> subscriber identification module *(SIM)*
Teilnehmerkennung *f* <Telekom> user identification *(ISDN)*
Teilnehmerleitung *f* <Telekom> loop
Teilnehmer-Netz-Schnittstelle *f* <Telekom> user-network interface
Teilnehmersatz *m (TNS)* <Telekom> subscriber line circuit *(SLC)*
Teilnehmerschaltung *f* <Telekom> line circuit, subscriber line circuit

Teilnehmersprechstelle

Teilnehmersprechstelle f <Telekom> substation
Teilnehmer-Teilnehmer-Information f *(TNI)* <Telekom> user-to-user information *(UUI)*
Teilnehmer-Teilnehmer-Zeichengabe f <Telekom> UUS, *(AE)* user-to-user signaling, *(BE)* user-to-user signalling
Teilnehmervermittlungsstelle f *(TVSt)* <Telekom> access exchange
Teilnehmer-Zeichengabe-Übermittlungsdienst m <Telekom> *(AE)* user-signaling bearer service, *(BE)* user-signalling bearer service
Teilnehmer-zu-Teilnehmer-Protokoll n <Telekom> user-to-user protocol *(OSI Schicht 7-Protokoll)*
Teilnehmer-zu-Teilnehmer-Signalisierung f <Telekom> user-to-user signalling, UUS *(Dienstmerkmal im Euro-ISDN)*
Teilnetz n <Telekom> subnetwork
Teilpipette f <Labor> graduated pipette
Teilproblem n <Künstl Int> subproblem
Teilpunkt m <Maschinen> division point
Teilrad n <Maschinen> dividing wheel, division wheel
Teil-RAM m <Comp & DV> partial RAM
Teilreflexion f <Wellphys> partial reflection *(von Lichtwellen)*
Teilrute f <Textil> split rod
Teilsatz m <Comp & DV> subset
Teilschären n <Textil> section warping, sectional warping
Teilschärmaschine f <Textil> sectional warping machine
Teilscheibe f <Maschinen> dividing plate, division plate, index dial, index plate
Teilschere f <Maschinen> dividing shears
Teilschicht f <Telekom> sublayer
Teilschnecke f <Maschinen> dividing screw, indexing worm screw
Teilschnitt m <Konstzeich> local section
Teilschritt m <Maschinen> fractional pitch
Teilseitenanzeige f <Comp & DV> part-page display
Teilsperre f <Telekom> restricted service
Teilsteigung f <Maschinen> divided pitch
Teilstrahlungs-Pyrometer n <Gerät> spectrally selective pyrometer
Teilstrecke f <Eisenbahn> section
Teilstreckenübermittlung f <Telekom> retransmission
Teilstreckenübertragung f <Telekom> store-and-forward transmission
Teilstreckenvermittlung f <Telekom> section-by-section switching
Teilstrich m 1. <Gerät> scale division, scale mark; 2. <Labor> graduation mark
Teilstrichabstand m <Phys> spacing *(Beugungsgitter)*
Teilsystem n <Telekom> subsystem
Teilton m <Akustik> partial tone
Teilübertrag m <Comp & DV> partial carry
Teilung f 1. <Fertig> pitch *(Zahnrad)*; 2. <Maschinen> pitch; 3. <Phys> division *(der Wellenfront)*
Teilungsfehler m <Fertig, Qual> error of pitch
Teilungsfläche f <Bau> jointing plane
Teilungskorngröße f <Kohlen> partition size
Teilverbrennungsraum m <Kfztech> antechamber
Teilverfahren n <Maschinen> indexing method
Teilvermittlungsleitung f <Telekom> store-and-forward line
Teilvermittlungsstelle f <Telekom> satellite exchange
Teilversetzung f <Metall> partial dislocation
Teilvorrichtung f <Maschinen> *(AE)* index center, *(BE)* index centre
teilweise konstanter Druckveränderer m <Lufttrans> cryptosteady pressure exchanger
teilweise löschen v <Meerschmutz> lighten

teilweise verkehrsabhängiges Signal n <Trans> semi-traffic-actuated signal
teilweiser Paketverlust m <Telekom> partial packet discard
teilweises Öffnen n **der Vorform** <Ker & Glas> blank cracking
Teilwelle f <Teilphys> partial wave
Teilwinkelverhältnis n <Lufttrans> angular pitch rate
Teilzeichenfolge f <Comp & DV> substring
Teilzeichnung f <Konstzeich> detail drawing
Teilziel n <Künstl Int> subgoal
Teilzirkel m <Maschinen, Metrol> dividers
Teilzusammenbau m <Maschinen> subassembly
Teilzylinder m <Maschinen> pitch cylinder
Teilzylindermantel m <Fertig> pitch surface *(Stirnrad)*
T-Eisen n <Metall> T-iron
Tektogenese f <Erdöl> tectogenesis *(Geologie)*
Tektonik f <Erdöl> tectonics
tektonisch adj <Erdöl> tectonic *(Geologie)*
tektonischer Prozess m <Erdöl> tectonic process *(Geologie)*
Tele… <Telekom> tele
Telearbeit f <Telekom> telecommuting, tele-working
Telebox f <Telekom> mailbox
Telecine-Maschine f <Fernseh> telecine machine
Teledienst m <Telekom> teleservice
teledynamisch adj <Maschinen> teledynamic
Telefax n 1. <Comp & DV> facsimile; 2. <Funktech, Telekom> facsimile *(fax)*
Telefaxgerät n <Comp & DV, Telekom> facsimile machine
Telefon n <Telekom> telephone
Telefon n **für ISDN** <Telekom> digital telephone, ISDN telephone
Telefon n **für Mehrfrequenz- und Impuls-Wahlverfahren** <Telekom> dual-signalling telephone, MFC and pulse dialling telephone
Telefon n **für MFC- und Impulswahl** <Telekom> *(AE)* dual-signaling telephone, *(BE)* dual-signalling telephone
Telefon n **für Wählbetrieb** <Telekom> dial telephone
Telefon n **mit Anrufbeantworter** <Telekom> telephone with answering device
Telefon n **mit Display** <Telekom> telephone with display, telephone with LCD-display
Telefon n **mit Lauthörmöglichkeit** <Telekom> loudspeaking telephone, open listening telephone
Telefon n **mit LCD-Anzeige** <Telekom> telephone with display, telephone with LCD-display
Telefon n **mit zuschaltbarem Lautsprecher** <Telekom> loudspeaking telephone, open listening telephone
Telefon n **nach DECT-Standard** <Telekom> DECT-standard telephone, digital cordless telephone
Telefonanlage f <Elektrotech, Telekom> telephone facility, telephone switchgear
Telefonanrufbeantworter m <Telekom> answering machine, call answering device, telephone answerer responder
Telefonanrufbeantworter-Service m <Telekom> voice mail, voice mail box
Telefonanrufbeantworter-Service m **für Handys** <Telekom> mobile voice box, voice mobilbox
Telefonanschlusskabel n <Telekom> telecom cord, telephone set cable, telephone set cord
Telefonapparat m <Telekom> telephone subscriber set
Telefonapparat m **mit Lautsprecher** <Telekom> loudspeaker telephone set
Telefonapparat m **mit Tastenfeld** <Telekom> push-button telephone set

Telefonapparat *m* mit **Wählscheibe** <Telekom> rotary dial telephone set
Telefonbrummer *m* <Elektriz> growler *(erzeugt Störton zur Warnung)*
Telefonbuch *n* <Telekom> telephone directory
Telefondienst *m* für **Flugzeugpassagiere** <Telekom> skyphone
Telefondraht *m* <Elektrotech, Telekom> telephone wire
Telefongespräch *n* <Telekom> call, telephone call
Telefonhörkapsel *f* <Akustik, Telekom> telephone earphone
Telefon-Hotline *f* <Telekom> call center
Telefonie *f* <Telekom> radio, telephony
Telefonie-über-Internet-Protokoll *n* <Telekom> Voice-over-Internet-Protocol, VoIP
Telefoninduktionsspule *f* <Elektrotech> telephone induction coil
Telefonist *m* <Telekom> operator, *(AE)* switchboard operator, *(AE)* telephone operator, *(BE)* telephonist
Telefonkabel *n* 1. <Elektrotech> telephone cable, telephone line; 2. <Telekom> telephone cable
Telefonkabelpaar *n* <Elektrotech, Telekom> telephone cable pair
Telefonkarte *f* <Telekom> calling card, phone card
Telefonklingel *f* <Telekom> telephone bell
Telefonkonferenz *f* <Telekom> audio conference, telephone conference
Telefonleitung *f* <Elektrotech, Telekom> telephone line, telephone wire
Telefonnetz *n* <Telekom> telephone network, voice network
Telefonnummer *f* <Telekom> directory number
Telefonnummernverzeichnis *n* <Telekom> telephone number list
Telefonrelais *n* <Elektrotech> telephone relay
Telefonschalter *m* <Elektrotech> telephone switch
Telefonsonderdienst *m* <Telekom> custom calling service
Telefonsprechkapsel *f* <Akustik, Telekom> telephone transmitter
Telefonumschalter *m* <Elektrotech> telephone switch
Telefonvermittlung *f* <Telekom> telephone switching
Telefonvermittlungsschrank *m* <Elektrotech, Telekom> telephone switchgear
Telefonwählnetz *n* <Telekom> switched telephone network
Telefonzelle *f* <Telekom> telephone box
Telefonzentrale *f* 1. <Elektrotech> telephone exchange, telephon office; 2. <Telekom> switch *(Telefon)*; telephone switchgear
Telegraf *m* <Telekom> telegraph
Telegrafenamt *n* <Elektrotech> telegraph exchange, telegraph office
Telegrafenanlage *f* <Telekom> telegraph installation
Telegrafendrähte *mpl* <Foto> tram lines *(Fehlergebnisse)*
Telegrafengleichung *f* <Phys> propagation equation
Telegrafenkabel *n* <Elektrotech> telegraph cable
Telegrafie *f* <Telekom> telegraphy
Telegrafleitung *f* <Elektrotech> telegraph line
Telegramm *n* <Elektrotech, Telekom> telegram, cablegram
Telegrammwähldienst *m* *(Gentex)* <Telekom> general telegraph exchange *(gentex)*
Teleinformatik *f* <Telekom> compunications *(Datenaustausch zwischen Computern)*; teleinformatics
Telekommunikation *f* <Telekom> telecommunication *(TC)*
Telekommunikationsanbieter *m* <Telekom> telecommunication provider

Telekonferenz *f* <Comp & DV, Telekom> teleconference
Telemarketing *n* <Telekom> telemarketing
Telematik *f* <Telekom> compunications *(Datenaustausch zwischen Computern)*; telematics
Telemechanik *f* <Maschinen> telemechanics
Telemetrie *f* <Kerntech, Maschinen, Phys, Raumfahrt, Telekom> telemetry
Telemetriesteuer- und Messsystem *n (TCR)* <Raumfahrt> telemetry command and ranging subsystem, TCR *(Weltraumfunk)*
Teleobjektiv *n* <Foto> narrow-angle lens, tele-lens, telephoto lens
Telepoint® *m* <Telekom> Telepoint® *(öffentliches Funktelefonsystem von BT)*
Teleschreiben *n* <Telekom> telewriting
Teleshopping *n* <Comp & DV, Fernseh, Telekom> electronic shopping
Teleskop *n* <Raumfahrt, Wassertrans> telescope
Teleskopbrücke *f* <Lufttrans> jetway
Teleskopgabeln *fpl* <Kfztech> telescopic forks *(Motorradfederung)*
Teleskopheber *m* <Maschinen> telescope jack
Teleskoplenkrad *n* <Kfztech> telescoping steering wheel
Teleskopmontagearm *m* <Maschinen> telescopic erector arm
Teleskopschutz *m* <Sicherheit> telescopic guard
Teleskopstoßdämpfer *m* <Kfztech> telescopic shock absorber *(Federung, Aufhängung)*
Teleskopverbindung *f* <Maschinen> telescope joint
Teleskopwelle *f* <Maschinen> telescopic shaft
Teleskopzylinder *m* <Maschinen> telescopic cylinder
Telesoftware *f* <Comp & DV> telesoftware *(BTX)*
Teletext® *m* <Fernseh, Telekom> Teletext®
Teletext-Dienst *m* <Comp & DV> teletext
Televertrieb *m* <Telekom> telesales
Telex *n* <Comp & DV, Telekom> telex
Telexplatz *m* <Telekom> telex position
Telexvermittlungsstelle *f* <Telekom> telex exchange
Teller *m* <Ker & Glas> plate *(Keramik)*
Tellerbeschicker *m* <Fertig> *(BE)* disc feeder, *(AE)* disk feeder
Tellerbürste *f* <Abfall> circular broom *(an Straßenkehrmaschine)*
Tellerfeder *f* <Maschinen> *(BE)* disc spring, *(AE)* disk spring
Tellerfederkupplung *f* <Kfztech> diaphragm clutch
Tellerfederpaket *n* <Fertig> spring washer set *(Kunststoffinstallationen)*
Tellermembrane *f* <Fertig> plate diaphragm *(Kunststoffinstallationen)*
Tellermesser *n* <Papier> slitter
Tellermühle *f* <Kohlen> *(BE)* disc mill, *(AE)* disk mill
Tellerrad *n* 1. <Kfztech> crown wheel *(Triebstrang)*; ring gear *(des Ausgleichsgetriebes)*; 2. <Maschinen> crown wheel, face gear, ring gear
Tellerrad *n* und **Ritzel** *n* <Kfztech> ring and pinion
Tellerschleifer *m* <Maschinen> *(BE)* disc sanding machine, *(AE)* disk sanding machine
Tellerstößel *m* <Kfztech> flat-bottom tappet
Teller- und Kegelradgetriebe *n* <Kfztech> ring and pinion gearing *(Differenzial)*
Tellerventil *n* 1. <Fertig> poppet valve; 2. <Hydraul> *(BE)* disc valve, *(AE)* disk valve; 3. <Kfztech> mushroom valve, poppet valve; 4. <Maschinen> *(BE)* disc valve, *(AE)* disk valve
tellur *adj* <Nichtfoss Energ> telluric
Tellur *n (Te)* <Chemie> tellurium *(Te)*
Tellur... <Chemie> telluric, tellurium, tellurous
Tellurat *n* <Chemie> tellurate

Tellurid

Tellurid n <Chemie> telluride
tellurig adj <Chemie> tellurous
Tellurit m <Chemie> tellurite
Tellurmesser n <Bau> tellurometer
Tellurnitrit n 1. <Chemie> tellurium nitrate; 2. <Elektrotech> tellurium nitride
Tellurnitrit-Widerstand m <Elektrotech> tellurium nitride resistor
Tellurocker m <Chemie> tellurite *(Mineralogie)*
TEM *(transversal elektromagnetisch)* <Elektrotech, Telekom> TEM *(transverse electromagnetic)*
TEM-Modus m <Elektrotech, Telekom> TEM mode, transverse electromagnetic mode
TE-Modus m 1. <Elektrotech> H-mode, TE mode; 2. <Telekom> TE mode, transverse electric mode
Temperafarbe f <Kunststoff> distemper
Temperatur f 1. <Akustik> temperament; 2. <Erdöl, Phys, Textil, Thermod> temperature • **Temperatur halten** <Ker & Glas, Metall> soak
Temperatur f **des schwarzen Körpers** 1. <Phys> black body temperature; 2. <Strahlphys, Thermod> black body temperature
Temperatur f **unterhalb Umgebungstemperatur** <Maschinen> subambient temperature
Temperatur f **von Berührungsflächen** <Sicherheit> temperature of touchable surfaces *(von Maschinen)*
Temperaturabfall m 1. <Bau, Nichtfoss Energ> heat drop; 2. <Thermod> heat drop, temperature drop
temperaturabhängig adj <Thermod> temperature-dependent
temperaturabhängiger Widerstand m <Elektrotech> temperature-dependent resistor
Temperaturabhängigkeit f <Thermod> temperature response
Temperaturanstieg m 1. <Phys> rise in temperature; 2. <Thermod, Umweltschmutz> temperature rise
Temperaturanzeige f <Gerät> temperature indication
Temperaturausdehnungskoeffizient m <Fertig> thermal expansion coefficient
Temperaturausgleich m <Thermod> temperature balance, temperature compensation, temperature equalization, temperature equalizing
Temperaturausgleichkondensator m <Elektrotech> temperature-compensating capacitor
Temperaturausgleichnetz n <Elektrotech> temperature-compensating network
Temperaturbereich m <Thermod> temperature range
Temperaturblitz m <Kerntech> thermal flash
Temperaturcharakteristik f <Telekom> thermal characteristic
Temperaturdifferenz f <Thermod> difference in temperature, offset temperature, temperature difference
Temperaturdiffusionsverfahren n <Kerntech> temperature cycle
Temperaturerhöhung f <Phys> raising of temperature
Temperaturerniedrigung f <Phys> lowering of temperature
Temperaturfernüberwachung f <Thermod> remote temperature monitoring
temperaturfest adj <Heiz & Kälte> heat-resistant
Temperaturfühler m 1. <Gerät> temperature sensor; 2. <Thermod> temperature probe
Temperaturgefahrenkurve f <Phys, Thermod> critical temperature curve
Temperaturgefälle n <Thermod> temperature difference, temperature gradient, thermal gradient
temperaturgeregelt adj <Thermod> temperature-controlled

temperaturgeregelter Quarzoszillator m <Elektronik> temperature-controlled crystal
temperaturgeregelter Schalter m <Elektrotech> temperature-controlled switch
Temperaturgrad m <Phys, Thermod> degree of temperature
Temperaturgradient m <Thermod> thermal gradient
Temperaturkoeffizient m <Thermod> temperature coefficient
Temperaturkoeffizient m **der Kapazität** <Thermod> temperature coefficient of capacitance
Temperaturkoeffizient m **des Widerstandes** <Thermod> temperature coefficient of resistance
Temperaturkompensation f 1. <Gerät> thermal compensation; 2. <Thermod> temperature compensation
temperaturkompensierender Kondensator m <Elektrotech> temperature-compensating capacitor
temperaturkompensierendes Netzwerk n <Elektrotech> temperature-compensating network
Temperaturkurve f <Thermod> temperature curve
Temperaturleitfähigkeit f <Strömphys> diffusivity
Temperaturleitzahl f <Heiz & Kälte, Strömphys> thermal diffusivity
Temperaturlogging n <Thermod> temperature logging *(Bohrlochmesstechnik)*
Temperaturmesser m <Thermod> pyrometer
Temperaturmessgerät n <Gerät> pyrometer, temperature measuring instrument, thermal instrument
Temperaturmessgerät n **mit Thermoelement** <Gerät> thermocouple thermometer, thermoelectric thermometer
Temperaturmessinstrument n <Gerät> temperature measuring instrument
Temperaturmesskreis m **mit Thermoelement** <Gerät> thermocouple-type temperature measurement system
Temperaturmessung f <Thermod> pyrometry
Temperaturmessverfahren n <Thermod> temperature logging
Temperaturmodell n <Nichtfoss Energ, Raumfahrt> thermal model
Temperaturprofil n <Thermod> temperature profile
Temperaturregelung f <Thermod> temperature control
Temperaturregler m <Heiz & Kälte> attemperator
Temperaturschock m <Anstrich> thermal shock
Temperaturschockprüfung f <Werkprüf> temperature shock test
Temperaturschockresistenz f <Anstrich> thermal shock resistance
Temperaturschreiber m 1. <Gerät> recording pyrometer, temperature recorder; 2. <Labor> thermograph
Temperaturschwankung f <Thermod> temperature fluctuation
Temperatursicherung f <Thermod> thermal link
Temperaturskale f <Gerät> temperature scale
temperaturstabil adj <Thermod> temperature-stable
Temperaturtasche f <Erdöl, Metrol> thermowell
Temperaturumkehr f 1. <Thermod> temperature inversion; 2. <Umweltschmutz> meteorological inversion
Temperaturunterschied m <Thermod> difference in temperature
Temperaturverhältnis n <Thermod> temperature ratio
Temperaturverteilung f <Thermod> temperature distribution
Temperaturwechselbeständigkeit f <Qual, Thermod> resistance to thermal shock
Temperatur-Zeit-Test m <Thermod> thermal cycle *(in der Materialprüfung)*
Temperguss m <Fertig, Mechan, Metall> malleable cast iron

temperierte Tonleiter f <Akustik> major scale of equal temperament
Temperit m <Metall> sorbite
tempern v 1. <Anstrich> temper *(Legierungen)*; 2. <Bau> temper; 3. <Mechan> anneal; 4. <Metall> temper; 5. <Phys> anneal
Tempern n 1. <Elektronik> anneal; 2. <Fertig> malleablizing; 3. <Maschinen> tempering
Temperofen m <Mechan> annealing furnace
Tempo n <Akustik> tempo
Tempolimit n <Trans> speed limit
Tempomat m <Kfztech> cruise control device
temporär adj <Bau, Qual> temporary
temporäre Mobilteilnehmerkennung f <Mobilkom> temporary mobile subscriber identity
temporärer Kommunikationsschlüssel m <Mobilkom> temporal communication key
temporärer Staudamm m <Wasserversorg> temporary dam
TEM-Welle f <Elektrotech, Telekom> TEM wave, transverse electromagnetic wave
Tender m 1. <Eisenbahn, Meerschmutz> tender; 2. <Wassertrans> supply vessel, tender; tender boat *(Hilfsfahrzeug)*
Tenderboot n <Wassertrans> tender boat *(Hilfsboot)*
Tensid n 1. <Chemie, Kunststoff, Lebensmittel> surfactant *(Seifen und seifenartige Substanzen)*; 2. <Meerschmutz> surface active agent, surfactant
Tensor m <Math> tensor
Tensorrechnung f <Math> tensor calculus
Teppich m <Erdöl, Meerschmutz> slick
Teppichgarn n <Textil> carpet yarn
Teppichgrund m <Textil> backing
Teppichgrundgewebe n <Textil> backing for carpet
Teppichkicker m <Bau> carpet kicker
Teppichunterlage f <Textil> underlay
Tera... *(T)* <Metrol> tera... *(T)*
Terabyte n <Optik> terabyte
Teraelektronenvolt n *(TeV)* <Teilphys> tera electron volt *(TeV)*
Terbium n *(Tb)* <Chemie> terbium *(Tb)*
Terephthal... <Chemie> terephthalic
Terephthalat n <Chemie> terephthalate
Term m 1. <Comp & DV, Math> term; 2. <Metrol> level; 3. <Phys> energy level, term
Termin m <Fernseh> deadline
Terminal n 1. <Comp & DV> terminal; 2. <Erdöl> terminal *(Schifffahrt)*; 3. <Telekom> terminal
Terminal n **im Ruhezustand** <Comp & DV> dormant terminal
Terminalbereich m <Lufttrans> terminal area
Terminaleinheit f <Comp & DV> terminal device
Terminalknoten m <Künstl Int> end node, terminal node *(eines Baumes)*
Terminalserver m <Comp & DV> terminal server
Terminierung f <Kontroll> termination *(eines Prozesses)*
Terminplaner m <Comp & DV> dater
Terminprogramm n <Comp & DV> dating program
termolekular adj <Chemie> termolecular, trimolecular
ternär adj 1. <Chemie> ternary, triple; 2. <Math> ternary
ternäre Legierung f <Metall> ternary alloy
Terpentin n <Kunststoff> turpentine
Terpentinsäure f <Chemie> terebic acid
Terpinen n <Chemie> terpinene
Terpineol n <Chemie> terpineol
Terpinolen n <Chemie> terpinolene
Terpolymer n <Chemie, Kunststoff> terpolymer
Terpolymeres n <Chemie> terpolymer
Terrainaufnahme f <Bau> land surveying

Terramycin n <Chemie> oxytetracycline *(Pharmazie)*
Terrazzobeton m <Bau> granolithic concrete
Terrazzobetonauflage f <Bau> granolithic finish
terrestrische Richtfunkverbindung f <Funktech> terrestrial radio-relay link
terrestrische Verbindung f <Telekom> terrestrial link
terrestrischer digitaler Hörfunk m <Funktech> terrestrial digital audio broadcasting
terrestrischer Hörrundfunk m <Fernseh, Funktech> audio broadcasting by antenna, terrestrial audio broadcasting
terrestrischer Mobilfunkkanal m <Mobilkom> terrestrial mobile radio channel
terrestrisches digitales Fernsehen n <Fernseh> terrestrial digital video broadcasting
tert-Butylalkohol m <Chemie> trimethylcarbinol
Tertiär n <Erdöl> Tertiary era *(Geologie)*
tertiäre Förderung f <Erdöl> tertiary recovery
tertiäre Rekristallisation f <Metall> tertiary recrystallization
tertiäres Brechen n <Kohlen> tertiary crushing
Tertiärförderung f <Erdöl> EOR, enhanced oil recovery
Tertiärgruppe f *(TG)* <Telekom> mastergroup
Tertiärkraftstoff m <Kfztech> tertiary fuel
Tertiärkriechen n <Metall> tertiary creep
Tertiärwicklung f <Elektriz> tertiary winding
tervalent adj <Chemie> trivalent
Terylene® n <Chemie, Wassertrans> *(AE)* Dacron®, *(BE)* Terylene® *(Segeln)*
Terz f <Akustik> third
Tesla n *(T)* <Metrol, Phys> Tesla *(T)*
Teslaspule f <Elektrotech> Tesla coil
Tesselation f <Geom> tesselation
tesselieren v <Geom> tesselate
Tesselierung f <Geom> tesselation
Test m <Aufnahme, Comp & DV, Fernseh, Hydraul, Kerntech, Lufttrans, Qual, Raumfahrt> test
Test m **auf Dauerfestigkeit** <Lufttrans> fatigue test
Test m **bei Nulllast** <Kerntech> zero-power test
Test m **ohne Last** <Kerntech> zero-power test
Testabbrand m <Raumfahrt> test firing
Testanforderung f <Raumfahrt> test specification
Testanlage f <Raumfahrt> test rig
Testband n <Aufnahme> test tape
Testbedingung f <Qual> test condition
Testbild n <Fernseh> test pattern
Testblech n <Hydraul> test plate
Testdruck m <Hydraul> test pressure
Testeinrichtung f <Maschinen> test rig
Testelement n <Kerntech> test assembly
testen v <Qual> test
Testen n **unter Betriebsbedingungen** 1. <Mechan> environmental testing *(Gerät)*; 2. <Qual, Telekom> environmental testing
Tester m <Qual> tester, testing apparatus
Testfenster n <Comp & DV> test box
Testflug m <Lufttrans, Raumfahrt> test flight
Testgelände n <Lufttrans> test ground
Testgerüst n <Raumfahrt> test bed
Testhubschrauber m <Lufttrans> experimental helicopter
Test-Isotop n <Kerntech> isotopic tracer
Testkunde m <Qual> test customer
Testliner m <Verpack> test liner board *(für Wellpappe)*
Testmaterial n **für Acryl** <Verpack> acrylic tester
Testmuster n 1. <Comp & DV> test pattern; 2. <Elektronik> test pattern *(zur Verwendung bei Logikanalysatoren)*
Testpilot m <Lufttrans> test pilot
Testplatte f <Aufnahme> test record

Testprogramm

Testprogramm n 1. <Comp & DV> benchmark, test program; 2. <Maschinen> test schedule
Testprotokoll n <Qual> test log
Testschuss m <Raumfahrt> test firing
Testsendung f <Fernseh, Funktech> test transmission
Testsiebdurchlauf m <Fertig> test sieving
Teststrecke f <Trans> test section
Teststreifen m <Foto> test strip
Testumgebung f <Comp & DV> test bed
Testverfahren n <Maschinen> test procedure *(für Bremsanlagen)*
Testwert m <Gerät> test value
TE/TM-Schwingungsmodus m <Phys> TE/TM mode
Tetra... <Chemie> tetra...
Tetrabor... <Chemie> pyroboric *(Säure)*
Tetrabromethan n <Chemie> tetrabromoethane
Tetrabromethylen n <Chemie> tetrabromoethylene
Tetrabromid n <Chemie> tetrabromide
Tetracen n <Chemie> naphthacene, tetrazene
Tetrachlorethylen n <Chemie> perchloroethylene, tetrachloroethylene
Tetrachlorid n <Chemie> tetrachloride
Tetrachlorkohlenstoff m <Chemie> tetrachloromethane
Tetrachlormethan n <Chemie> carbon tetrachloride, tetrachloromethane
Tetrachloroplatinat n <Chemie> tetrachloroplatinate
Tetrachord m <Akustik> tetrachord
Tetradecan... <Chemie> tetradecanoic *(Säure)*
Tetraeder n <Geom> tetrahedron
Tetraethyl... <Chemie> tetraethyl
Tetraethylblei n 1. <Chemie> lead tetraethyl, tetraethyl lead, tetraethylplumbane; 2. <Kfztech> tetraethyl lead
tetragonales System n <Metall> tetragonal system
Tetrahydrid n <Chemie> pyrrolidine, tetrahydropyrrol, tetrahydride
Tetrahydrooxazin n <Chemie> morpholine, tetrahydrooxazine
Tetrahydropyrrol n <Chemie> pyrrolidine, tetrahydropyrrole
Tetraiod... <Chemie> tetraiod
Tetraiodfluorescein n <Chemie> iodeosin, tetraiodofluorescein
Tetraiodoaurat n <Chemie> iodoaurate, tetraiodoaurate
Tetraiodthyronin n <Chemie> thyroxine
Tetrakistriwolframatophosphat n <Chemie> phosphatododecatungstate, phosphotungstate
Tetralin n <Chemie> tetralin
tetramer adj <Chemie> tetrameric
Tetramethyl... <Chemie> tetramethyl
Tetramethylenimin n <Chemie> pyrrolidine, tetramethylenimine
Tetramethylethylenglycol n <Chemie> pinacol, pinacone, tetramethylethyleneglycol
Tetramin n <Chemie> tetramine
Tetranitro-N-Methylanilin n <Chemie> tetryl
Tetraoxomolybdat n <Chemie> molybdate, tetraoxomolybdate
Tetraoxygen n <Chemie> oxozone
TetraPak® m <Lebensmittel, Verpack> Tetra Pak®
Tetrasulfid n <Chemie> *(AE)* tetrasulfide, *(BE)* tetrasulphide
Tetrathion... <Chemie> tetrathionic
tetravalent adj <Chemie> quadrivalent, tetravalent
Tetravalenz f <Chemie> quadrivalence, tetravalence, tetravalency
Tetrazin n <Chemie> tetrazine
Tetrazol n <Chemie> tetrazole
Tetrode f 1. <Elektronik> tetrode tube, tetrode; 2. <Phys> tetrode

Tetrol... <Chemie> tetrolic
Tetrose f <Chemie> tetrose
Tetroxid n <Chemie> tetroxide
Tetroxo... <Chemie> silicic
Tetroxokieselsäure f <Chemie> orthosilicic acid, silicic acid, tetrasilicic acid
Tetroxoosmat n <Chemie> osmate, tetraoxoosmate
Tetroxosilicat n <Chemie> orthosilicate, tetraoxosilicate
Tetryl n <Chemie> tetralite
TEU <Wassertrans> TEU, twenty feet equivalent unit *(Ladekapazität für Standardcontainer)*
TeV *(Teraelektronenvolt)* <Teilphys> TeV *(tera electron volt)*
TE-Welle f <Elektrotech, Telekom> H-wave, TE wave, transverse electric wave
Text m 1. <Comp & DV> text *(auch E-Mail)*; 2. <Druck> text
Textanfang m *(STX)* <Comp & DV> start of text *(STX)*
Textanwendung f <Comp & DV> text management
Textaufbereitung f <Comp & DV, Druck> text editing
Textaufbereitungsprogramm n <Comp & DV, Druck> text editor
Textbaustein m <Comp & DV, Druck> text segment
Textbearbeitung f <Comp & DV, Druck> text editing, text manipulation
Textbildschirm m <Comp & DV, Druck> text screen
Textdatei f <Comp & DV, Druck> text file
Texteditor m <Comp & DV, Druck> text editor
Textende n <Comp & DV, Telekom> end of text
Text-Fax-Server m <Telekom> text/fax server *(ISDN)*
Textformatierer m <Comp & DV, Druck> text formatter
Textil... <Textil> textile
Textilabfall m <Abfall> textile waste
Textilbeschichtung f <Fertig, Textil> textile coating
Textildruck m <Fertig, Textil> textile printing
Textiletikettierung f <Textil> *(AE)* textile labeling, *(BE)* textile labelling
Textilglasfaser f <Textil> *(AE)* textile glass fiber, *(BE)* textile glass fibre
Textilgurt m <Maschinen> fabric belt
Textilgurtbandförderer m <Fertig> fabric belt conveyor
Textilriemen m 1. <Fertig> woven fabric belt; 2. <Maschinen> fabric belt
Textilstaubsammler m <Sicherheit> fabric dust collector
Textilverbundstoffe mpl <Textil> nonwovens
Textilveredlung f <Fertig, Textil> textile finishing
textilverstärktes Papier n <Papier> reinforced paper
Textkomprimierung f <Comp & DV, Druck> text compression
Text-Mailbox f <Telekom> text mailbox
Textpostfach n <Telekom> text mailbox
Textspeicher m <Comp & DV, Druck> text storage
Textspur f <Aufnahme> narration track
Textur f <Künstl Int> texture *(eines Objekts oder Bildes)*
Texturgarn n <Ker & Glas> textured yarn
texturiertes Pflanzeneiweiß n <Lebensmittel> textured vegetable protein
texturiertes Pflanzenprotein n *(TPP)* <Lebensmittel> textured vegetable protein *(TVP)*
Textverarbeitung f <Comp & DV, Druck> WP, word processing, text processing
Textverarbeitungsprogramm n <Comp & DV> word processing package
Textverarbeitungssystem n <Comp & DV, Druck> word processor
Textzeiger m 1. <Comp & DV> pointer; text pointer *(Cursor in Form von senkrechtem Strich)*; 2. <Druck> pointer, text pointer

Tf 1. <Aufnahme, Elektronik, Fernseh, Funktech> *(Tonfrequenz)* AF *(audio frequency)*; 2. <Aufnahme, Elektronik, Fernseh> *(Trägerfrequenz)* CF *(carrier frequency)*
TFEL *(Dünnschicht-Elektrolumineszenz)* <Elektronik> TFEL *(thin film electroluminescence)*
TFEL-Anzeigetechnik f <Elektronik> TFEL display technology
T-Flipflop n <Elektronik> T-flip-flop
T-förmige Nut f *(T-Nut)* <Fertig, Maschinen> tee slot *(T-slot)*
T-förmige Nutenschraube f <Maschinen> tee bolt
T-förmige Verbindung f *(T-Verbindung)* <Bau> tee piece union *(T-piece union)*
T-förmiger Koppler m *(T-Koppler)* <Optik, Telekom> tee coupler *(T-coupler)*
T-förmiges Spleiß n <Elektriz> tee joint
T-förmiges Stück n <Maschinen> T, T-piece
T-förmiges Teil n <Maschinen> tee
TF-Störung f <Elektronik, Funktech, Telekom> AFI *(audio-frequency interference)*
TG *(Tertiärgruppe)* <Telekom> mastergroup
T-Glied n <Telekom> T-junction, T-network
Tg-Wert m <Kunststoff> glass transition temperature
Th *(Thorium)* <Chemie> Th *(thorium)*
Thallium n *(Tl)* <Chemie> thallium *(Tl)*
Thallium... <Chemie> thallic, thallium
Thebain n <Chemie> dimethylmorphine, paramorphine, thebaine
Theobromin n <Chemie> dimethylxanthine, theobromine
Theodolit m <Bau> transit; theodolite *(Vermessung)*
Theophyllin n <Chemie> theophylline
Theorem n <Math, Phys> theorem
Theorembeweis m <Künstl Int> theorem proving
theorembeweisendes Programm n <Künstl Int> theorem prover
Theorembeweiser m <Künstl Int> theorem prover
theoretisch maximale numerische Apertur f <Optik, Telekom> maximum theoretical numerical aperture
theoretische Bedingungen fpl <Trans> ideal conditions
theoretische Dichte f *(TD)* <Kerntech> theoretical density *(TD)*
theoretische Grenzfrequenz f <Elektronik> theoretical cut-off frequency
theoretisches Kraftstoff-Luft-Verhältnis n <Kfztech> ideal mixture ratio
theoretisches Mischungsverhältnis n <Kfztech> ideal mixture ratio, perfect mixture ratio
Theorie f **der elektroschwachen Wechselwirkung** <Teilphys> electroweak theory
Theorie f **der großen Vereinigung aller Kräfte** <Phys> GUT, grand unified theory
Theorie f **laminarer Strömungen** <Strömphys> laminar flow theory
thermalisieren v <Chemie> thermalize
Thermalisierung f **von Neutronen** <Strahlphys, Teilphys> neutron thermalization
Thermalquelle f <Thermod> thermal spring
Thermalruß m <Kunststoff> thermal black, thermal carbon black
Thermion n <Thermod> thermion
Thermionen-Elektronenröhre f <Elektronik> hot-cathode tube
Thermionenröhre f <Elektronik> hot-cathode tube
thermionisch adj <Thermod> thermionic
thermionische Emission f <Strahlphys> thermionic emission
thermionische Strahlung f <Strahlphys> thermionic emission

thermionische Umwandlung f <Kerntech> thermionic conversion
thermisch adj <Thermod> thermal, thermic
thermisch angeregte Strömungen fpl <Strömphys> thermal flows
thermisch beschleunigte Lebensdauerprüfung f <Telekom> thermally accelerated life testing
thermisch erzeugter Auftrieb m <Umweltschmutz> thermally-induced buoyancy
thermisch gepumpter Laser m <Elektronik> thermally-pumped laser
thermisch stabil adj <Thermod> heat-stable
thermisch wirksame Masse f <Thermod> thermal mass
thermische Alterung f <Kunststoff> *(BE)* heat ageing, *(AE)* heat aging
thermische Behaglichkeitszone f <Ergon> thermal comfort zone
thermische Behandlung f <Textil> baking
thermische Diffusion f <Thermod> thermodiffusion
thermische Druckplatte f <Druck> thermal plate
thermische Eigenschaften fpl <Thermod> thermal properties
thermische Emission f <Elektrotech> thermionic emission
thermische Ersatzschaltung f <Thermod> equivalent thermal network
thermische Gleichrichtung f <Elektrotech> thermionic rectification
thermische Indifferenzzone f <Ergon> thermal indifference zone
thermische Kontraktion f <Metall, Thermod> contraction due to cold
thermische Konversion f <Elektrotech> thermionic conversion
thermische Nachverbrennung f <Kfztech> thermal post-combustion
thermische Neutralzone f <Ergon> thermal neutral zone
thermische Schneidspitze f <Thermod> thermic lance
thermische Stabilität f <Kunststoff, Verpack> heat stability
thermische Trennung f **von He und He** <Kerntech> heat flush *(in flüssigem Helium)*
thermische Wandlung f <Elektrotech> thermal conversion
thermische Wechselbeanspruchung f <Kunststoff> thermal cycling
thermische Zersetzung f 1. <Abfall> pyrolysis; 2. <Thermod> decomposition by heat
thermische Zufallsbewegung f <Strahlphys> random thermal motion *(strahlender Atome)*
thermischer Aufwind m <Nichtfoss Energ> thermal upcurrent
thermischer Ausdehnungskoeffizient m <Mechan, Phys, Qual> coefficient of thermal expansion
thermischer Dauernennstrom m <Elektriz> continuous thermal current rating
thermischer Generator m <Elektrotech> thermionic generator
thermischer Gleichrichter m <Elektrotech> thermionic rectifier
thermischer Konverter m <Elektrotech> thermionic converter
thermischer Nutzfaktor m <Phys> thermal utilization factor
thermisches Empfängerrauschen n <Telekom> thermal noise ratio
thermisches Neutron n <Phys> slow neutron
thermisches Rauschen n <Funktech, Telekom> thermal agitation noise

thermisches 768

thermisches Telefon n <Akustik> thermophone
thermisches Ungleichgewicht n <Thermod> thermal imbalance
thermisches Verbrennungsverfahren n <Abfall> incineration train
Thermistor m <Elektrotech, Phys, Telekom> thermistor
Thermistorbrücke f <Elektrotech> thermistor bridge
Thermistormesskopf m <Elektrotech> thermistor mount
Thermistorregler m <Elektrotech> thermistor control
Thermit n <Chemie, Metall> thermite
Thermitschmelzschweißen n <Fertig> aluminothermic fusion welding, nonpressure thermic welding
Thermitschweißen n <Fertig, Mechan> aluminothermic welding
Thermitschweißung f <Bau, Eisenbahn> thermit welding
Thermo... <Elektronik, Raumfahrt, Thermod> thermal
Thermoabbau m <Abfall, Kunststoff> thermal decomposition
Thermoabschalten n <Elektronik> thermal shutdown
Thermoabstimmung f <Elektronik> thermal tuning
Thermoanalyse f <Thermod> thermal analysis, thermoanalysis
Thermoausblühen n <Elektronik> thermal blooming
Thermobatterie f 1. <Elektrotech> thermal battery; 2. <Raumfahrt> thermopile
Thermobehälter m <Thermod> insulated container
Thermobild n <Thermod> thermal imaging sight
Thermobrutreaktor m <Kerntech> thermal breeding reactor
Thermochemie f <Chemie> thermochemistry
Thermodifferenzialmelder m <Thermod> rate-of-rise detector
Thermodiffusion f <Phys> thermal diffusion
Thermodiffusionsfaktor m <Phys> thermal diffusion ratio
Thermodiffusions-Koeffizient m <Phys> thermal diffusion coefficient
Thermodiffusionskonstante f (aT) <Phys> thermal diffusion constant (aT)
Thermodiffusionsverfahren n <Phys> thermal diffusion process
Thermodiffusionszahl f <Phys> thermal diffusion factor
Thermodiffusivität f <Heiz & Kälte> thermal diffusivity
Thermodiode f <Elektronik> thermal diode
Thermodissoziation f <Thermod> thermal decomposition, thermal dissociation
Thermodrucker m 1. <Comp & DV> electrothermal printer; 2. <Druck> thermal printer; 3. <Verpack> thermal transfer printer
Thermodurchbruch m <Elektronik> thermal breakdown (Halbleiter)
Thermodynamik f <Thermod> thermodynamics
thermodynamisch adj <Thermod> thermodynamic
thermodynamische Funktion f <Thermod> thermodynamic function
thermodynamische Temperatur f 1. <Hydraul> absolute temperature (θ); 2. <Phys> absolute temperature, empirical temperature, thermodynamic temperature (T); 3. <Thermod> absolute temperature, empirical temperature, thermodynamic temperature (θ)
thermodynamische Wahrscheinlichkeit f <Thermod> thermodynamic probability
thermodynamische Zustandsänderung f <Thermod> thermodynamic transformation
thermodynamische Zustandsgleichung f <Thermod> equation of thermal state, thermodynamic equation of state
thermodynamischer Vorgang m <Thermod> thermodynamic process

thermodynamisches Potenzial n <Thermod> thermodynamic potential
thermodynamisches System n <Thermod> thermodynamic system
thermoelastische Verzerrung f <Raumfahrt> thermoelastic distortion
thermoelastischer Martensit m <Metall> thermoelastic martensite
thermoelektrisch adj <Elektriz, Elektrotech, Phys, Thermod> thermoelectric, thermoelectrical
thermoelektrische Kälteerzeugung f <Heiz & Kälte> thermoelectric cooling
thermoelektrische Konversion f <Elektrotech> thermoelectric conversion
thermoelektrische Kraft f <Elektriz> thermoelectric power, thermoelectromotive force
thermoelektrische Kühlung f <Heiz & Kälte> thermoelectric cooling
thermoelektrische Säule f <Elektrotech> thermopile
thermoelektrische Sonnenenergieumwandlung f <Nichtfoss Energ> solar thermoelectric conversion
thermoelektrische Wandlung f <Elektrotech> thermoelectric conversion
thermoelektrischer Effekt m <Elektriz> thermoelectric effect
thermoelektrischer Generator m <Elektriz> thermoelectric generator
thermoelektrisches Kraftwerk n <Elektrotech> thermal-electric power plant, thermal-electric power station
thermoelektrisches Kühlelement n <Heiz & Kälte> thermoelectric cooling couple
Thermoelektrizität f <Elektriz, Phys> thermoelectricity
Thermoelement n 1. <Elektriz, Kerntech, Labor> thermocouple; 2. <Maschinen> thermoelement; 3. <Phys, Thermod> thermocouple
Thermoelementanschluss-Kompensator m <Gerät> thermocouple potentiometer
Thermoelementinstrument n <Gerät> thermocouple instrument
Thermoelement-Thermometer n <Gerät> thermocouple thermometer, thermoelectric thermometer
Thermoemission f <Phys> thermal emission
Thermofixieren n <Textil> heat setting
Thermoformen n <Kunststoff> thermoforming
Thermoform-Maschine f <Verpack> thermoform machinery
Thermoform-Verpackungssystem n <Verpack> thermoforming packaging system
Thermofühler m 1. <Gerät> temperature sensor; 2. <Thermod> temperature probe
Thermofusionsfaser f <Textil> (AE) thermobonding fiber, (BE) thermobonding fibre
Thermogeschwindigkeit f <Phys> thermal velocity
Thermogleichgewicht n <Thermod> thermal equilibrium
Thermoglühbirne f <Kfztech> thermal bulb
Thermograph m <Kerntech, Labor> thermograph
Thermographie f 1. <Phys> thermal imaging; 2. <Thermod> thermal imaging; thermograph (Bild)
Thermogravimetrie f 1. <Kunststoff> thermogravimetry; 2. <Thermod> thermal gravimetric analysis
thermogravimetrisches Analysiergerät n <Labor> (BE) thermogravimetric analyser, (AE) thermogravimetric analyzer
Thermohygrograph m <Bau> thermohygrograph
Thermoinstabilität f <Thermod> thermal instability, thermal runaway
Thermokapazitanz f <Nichtfoss Energ> thermal capacitance
Thermokette f <Elektrotech> thermopile

Thermokomponente f <Metall> thermal component
Thermokonverter m <Elektriz, Kerntech, Labor, Phys, Thermod> thermocouple converter
Thermokopierverfahren n <Konstzeich> thermocopying process
Thermokracken n <Erdöl> thermal cracking *(Raffinerie)*
thermolumineszent adj <Thermod> thermoluminescent
Thermolumineszenz f <Strahlphys, Thermod> thermoluminescence
Thermolumineszenz-Dosimeter n <Strahlphys> thermoluminescent dosimeter
Thermolyse f <Thermod> thermal decomposition, thermal dissociation, thermolysis
thermomagnetisch adj <Thermod> thermomagnetic
Thermomagnetismus m <Thermod> thermomagnetism
thermomechanischer Effekt m <Thermod> thermomechanical effect
Thermometer n 1. <Gerät> pyrometer; 2. <Heiz & Kälte, Labor> thermometer
Thermometer n **mit Fernablesung** <Heiz & Kälte> remote-reading thermometer
Thermometer n **mit Fernanzeige** <Heiz & Kälte> remote-reading thermometer
Thermometerglas n <Labor> thermometer glass
Thermometerröhre f <Phys> stem
Thermometrie f <Thermod> thermometry
thermometrisch adj <Thermod> thermometric
Thermoneutron n <Kerntech, Phys, Strahlphys> thermal neutron
thermonuklear adj <Kerntech> thermonuclear
thermonukleare Reaktion f <Kerntech> thermonuclear reaction
thermonukleare Stromerzeugung f <Kerntech> thermonuclear power generation
Thermopaar n <Phys> couple
Thermophon n <Akustik> thermophone
Thermophosphoreszenz f <Thermod> thermophosphorescence
thermoplastisch adj 1. <Fertig> heat-deformable; 2. <Kunststoff, Mechan, Wassertrans> thermoplastic *(Schiffbau)*
thermoplastischer Kautschuk m <Kunststoff> thermoplastic rubber
Thermoplastizität f <Kunststoff> thermoplasticity
Thermoplastkunststoff m <Kunststoff, Sicherheit> thermoplastic
Thermoplatte f <Druck> thermal plate
Thermoreaktor m <Kfztech> thermal exhaust manifold reactor
Thermoreformierung f <Erdöl> thermal reforming *(Raffinerie)*
Thermorelais n <Elektrotech> thermal relay
Thermosäule f 1. <Elektrotech> thermopile; 2. <Kerntech> thermal column; 3. <Phys, Raumfahrt> thermopile; 4. <Thermod> thermoelectric pile
Thermosbehälter m <Thermod> heat-insulated container
Thermoschalter m <Elektrotech> thermal switch
Thermoschild m <Kerntech> thermal shield
Thermoschutzschalter m <Thermod> thermal circuit breaker
Thermosegelflug m <Lufttrans> thermal soaring
Thermosicherung f <Thermod> thermal link
Thermosiphon m <Nichtfoss Energ> thermosiphon
Thermospaltung f <Erdöl> thermal cracking
Thermospannung f <Elektriz> contact potential
Thermospannungskompensator m <Gerät> thermocouple potentiometer
thermostabil adj <Thermod> thermostable
Thermostabilität f <Thermod> thermal stability

Thermostat n <Heiz & Kälte, Kfztech, Phys, Thermod> thermostat • **im Thermostat** <Telekom> oven-controlled
Thermostat-gesteuertes Bad n <Labor> thermostatically-controlled bath
Thermostatik f <Thermod> thermostatics
thermostatisch adj <Thermod> thermostatic
thermostatisch geregeltes Ventil n <Heiz & Kälte> thermostatically-controlled valve
thermostatisch gesteuerte Entwicklungsschale f <Foto> thermostatically-controlled developing dish
thermostatische Drossel f <Lufttrans> antisurge baffle
Thermostatregelung f <Heiz & Kälte> thermostat control
Thermostatventil n <Maschinen> thermostatic valve
Thermosteuerung f <Nichtfoss Energ, Raumfahrt> thermal control
Thermostrom m <Kerntech> thermocurrent
Thermostrommessgerät n <Elektriz> thermoamperemeter
Thermoumformer m <Elektrotech> bolometer
Thermoumformerinstrument n <Gerät> thermocouple instrument
Thermoventil n 1. <Maschinen> thermovalve; 2. <Thermod> temperature valve
Thermowaage f <Chemie> thermobalance
Thermowandler m <Elektriz, Kerntech, Labor, Phys, Thermod> thermocouple converter
Thermozersetzung f <Abfall, Kunststoff> thermal decomposition
Thévenin'scher Satz m <Phys> Thévenin's theorem
Thevetin n <Chemie> thevetin
Thialdin n <Chemie> thialdine
Thiamin n <Chemie> aneurin, thiamin
Thiazin n <Chemie> thiazine
Thiazol n <Chemie> thiazole
Thiazolin n <Chemie> dihydrothiazole, thiazoline
Thiele Rohr n <Labor> Thiele melting-point tube *(Schmelzpunkt)*
Thio... <Chemie> thio...
Thioaldehyd m <Chemie> thioaldehyde
Thioalkohol m <Chemie> *(AE)* hydrosulfide, *(BE)* hydrosulphide, thiol
Thioamid n <Chemie> thioamide
Thioarsen... <Chemie> thioarsenic *(Säure)*
Thiocarbamid n <Chemie> thiocarbamide, thiourea
Thiocarbanilid n <Chemie> (N,N)-diphenylthiourea, thiocarbanilide
Thiocarbonat n <Chemie> *(AE)* sulfocarbonate, *(BE)* sulphocarbonate, thiocarbonate
Thiocarbonyldichlorid n <Chemie> thiophosgene
Thiocyan... <Chemie> thiocyanic
Thiocyanat n <Chemie> thiocyanate
Thiodiphenylamin n <Chemie> phenothiazine, thiodiphenylamine
Thioessig... <Chemie> thioacetic *(Säure)*
Thioether m <Chemtech> *(AE)* organic sulfide, *(BE)* organic sulphide, thioether
Thioflavin n <Chemie> thioflavin
Thiofuran n <Chemie> thiophene
Thioglycol... <Chemie> mercaptoacetic, thioglycolic
Thioharnstoff m <Chemie> thiocarbamide, thiourea
Thioindigo n <Chemie> thioindigo
Thioindigorot n <Chemie> thioindigo
Thioketon n <Chemie> thioketone
Thiokohlen... <Chemie> thiocarbonic *(Säure)*
Thiol n <Chemie> *(AE)* hydrosulfide, *(BE)* hydrosulphide, thiol
Thiolat n <Chemie> mercaptide
Thion... <Chemie> thionic *(Säure)*

Thionaphthen 770

Thionaphthen n <Chemie> benzothiophene, thionaphthene
Thionin n <Chemie> thionine
Thionyl... <Chemie> thionyl
Thiopental... <Chemie> thiopental
Thiophen n <Chemie> thiophene
Thiophenol n <Chemie> penyl mercaptan, thiophenol
Thiophosgen n <Chemie> thiophosgene
Thiosäure f <Chemie> thioacid
Thioschwefel... <Chemie> *(AE)* thiosulfuric, *(BE)* thiosulphuric *(Säure)*
Thiosulfat n <Chemie> *(AE)* thiosulfate, *(BE)* thiosulphate
Thioxanthon n <Chemie> thioxanthone
Thioxen n <Chemie> thioxene
thixotrop adj <Chemie, Kohlen, Kunststoff, Phys> thixotropic
Thixotropie f <Chemie, Kohlen, Phys> thixotropy
Thomasbirne f <Fertig> basic Bessemer converter, basic converter
Thomasroheisen n <Fertig> basic Bessemer pig
Thomas-Schlacke f <Metall> Thomas's slag
Thomasstahl m 1. <Fertig> basic Bessemer steel, basic converter steel; 2. <Metall> basic Bessemer steel
Thomasverfahren n <Fertig> basic Bessemer process
Thomson'sche Brücke f <Elektriz, Elektrotech> Kelvin bridge, Thomson bridge
Thomson'sche Doppelbrücke f <Elektriz> double Thomson bridge
Thomson'sche Messbrücke f <Elektrotech> Kelvin bridge
Thomson'sche Streuung f <Phys> Thomson scattering
Thomson'scher Effekt m <Phys> Thomson effect
Thomson'scher Koeffizient m <Phys> Thomson coefficient
Thomson'scher Wirkungsquerschnitt m <Phys> Thomson cross-section
Thorianit m <Chemie> thorianite
thorierter Wolframfaden m <Elektrotech> thoriated tungsten filament
Thorit m <Kerntech> thorite
Thorium n (Th) <Chemie> thorium *(Th)*
Thorium... <Chemie> thoric
Thoriumdioxid n <Chemie> thoria
Thoriumreaktor m <Kerntech> *(AE)* thorium-fueled reactor, *(BE)* thorium-fuelled reactor
Thoriumreihe f <Strahlphys> thorium series
Thread n <Comp & DV> thread
Threose f <Chemie> threose
Thujan n <Chemie> sabinane, thujane
Thujen n <Chemie> sabinene, thujene
Thujenol n <Chemie> sabinol, thujenol, thujol
Thujon n <Chemie> oxosabinane, thujone
Thujylalkohol m <Chemie> thujyl alcohol
Thulium n (Tm) <Chemie> thulium *(Tm)*
Thymiancampher m <Chemie> thymol
Thymol n <Chemie> thymol
Thymolphthalein n <Chemie> thymolphthalein
Thyratron n <Chemie, Phys> thyratron
Thyristor m <Elektriz, Elektronik, Elektrotech, Phys, Telekom> thyristor
Thyristordiode f <Elektronik> thyristor diode
Thyristor-Koppelpunkt m <Telekom> SCR crosspoint, silicon-controlled rectifier crosspoint
Thyristor-Regelung f <Elektrotech> SCR control, SCR regulation
Thyristor-Regler m <Elektrotech> SCR controller, SCR regulator
Thyristor-Vorregelung f <Elektrotech> SCR precontrol, SCR regulation

Thyristor-Vorregler m <Elektrotech> SCR precontroller, SCR regulator
Thyristorwandler m <Elektrotech> SCR thyristor converter
Thyristor-Wechselrichter m <Elektriz> thyristor inverter
Thyronin n <Chemie> thyronine *(Biochemie)*
Thyroxin n <Chemie> thyroxine
Ti *(Titan)* <Chemie> Ti *(titanium)*
Tide f <Wassertrans> tide
Tideablauf m <Nichtfoss Energ> ebb tide
Tidenanstieg m <Wassertrans> rise of tide *(Gezeiten)*
Tidenbecken n <Wassertrans> tidal dock
Tidenfall m <Wassertrans> fall of the tide
Tidengebiet n <Meerschmutz> intertidal zone *(Zone zwischen Hoch- und Niedrigwasser)*
Tidenhub m <Wassertrans> tidal range; range of tide *(Gezeiten)*
Tidenniedrigwasser n <Wassertrans> low tide
Tidensignale npl <Wassertrans> tidal signals, tide signals
Tidenstrich m <Wassertrans> tide gate
Tidentafeln fpl <Wassertrans> tide tables
Tief n 1. <Nichtfoss Energ> fairway; 2. <Wassertrans> low *(Wetterkunde)*
Tief... <Druck, Wassertrans> deep
Tiefätzbad n <Ker & Glas> deep-etching bath
tiefätzen v <Fertig> deep-etch
Tiefätzpaste f <Ker & Glas> deep-etching paste
Tiefbagger m <Meerschmutz> backhoe
Tiefbau m <Kohlen> deep mine
Tiefbauarbeit f <Kohlen> deep mining
Tiefbauingenieur m <Bau> civil engineer
Tiefbaukohle f <Kohlen> deep-mining coal
Tiefbettfelge f <Kfztech> *(AE)* drop center rim, *(BE)* drop centre rim *(Rad)*
Tiefbettlader m <Bau> low-bed trailer, low-boy trailer
Tiefbettrahmen m <Kfztech> drop bed frame *(Fahrgestell)*
tiefbohren v <Fertig> deep-bore
Tiefbohrloch n <Kohlen> well drill hole
Tiefbrunnen m <Wasserversorg> deep well
Tiefdecker m <Lufttrans> low-wing plane
Tiefdruck m 1. <Druck> gravure, intaglio, intaglio printing; 2. <Ker & Glas> intaglio
Tiefdruckätzung f <Druck> intaglio etching
Tiefdruckgebiet n <Wassertrans> low-pressure area
Tiefdruckrinne f <Wassertrans> trough *(Wetterkunde)*
Tiefe f 1. <Hydraul> (t) depth (d); 2. <Wassertrans> depth *(Schiff, See)*
tiefe Wassersäule f <Nichtfoss Energ> low head
Tiefeinbrandelektode f <Fertig> deep-penetration electrode *(Schweißen)*
Tiefenanschlag m 1. <Fertig> depth-control stop; 2. <Maschinen> depth stop
Tiefenauslösung f <Fertig> depth trip
Tiefendosis f <Strahlphys> depth dose
Tiefengestein n <Bau> intrusive rock
Tiefenkarte f <Wassertrans> bathymetric chart *(Navigation)*
Tiefenlehre f <Maschinen, Mechan> *(AE)* depth gage, *(BE)* depth gauge
Tiefenlinie f <Wassertrans> isobath
Tiefenlot n <Wassertrans> depth sounder *(Navigation)*
Tiefenmaß n <Fertig, Metrol> *(AE)* depth gage, *(BE)* depth gauge
Tiefenmessapparat m <Mechan> *(AE)* depth gage, *(BE)* depth gauge
Tiefenmesser m 1. <Phys> bathometer; 2. <Wassertrans> depth sounder *(Navigation)*

Tiefenmesslehre f <Wassertrans> bathymetry *(Navigation)*
Tiefenmessschieber m <Maschinen, Metrol> *(AE)* vernier depth gage, *(BE)* vernier depth gauge
Tiefenmessschraube f <Fertig> *(AE)* metal depth gage, *(BE)* metal depth gauge
Tiefenmessung f 1. <Gerät> depth measurement; 2. <Metrol> *(AE)* depth gage, *(BE)* depth gauge; 3. <Nichtfoss Energ> bathymetry
Tiefenruder n <Wassertrans> hydroplane; diving rudder *(U-Boot)*
Tiefenschärfe f 1. <Fernseh> depth of focus; 2. <Foto> depth of field, depth of focus; 3. <Optik> field depth
Tiefenschärfenskale f <Foto> depth-of-field scale, depth-of-focus scale
Tiefenschrift f <Akustik> vertical recording
Tiefensehen n <Ergon> depth perception
tiefensichtig adj <Phys> orthoscopic
Tiefenskale f <Fertig> depth dial
Tiefensperre f <Aufnahme> bass cut
Tiefensperrfilter n <Aufnahme> bass-cut filter
Tiefensuche f <Künstl Int> depth-first search
Tiefentaster m <Metrol> *(AE)* depth gage, *(BE)* depth gauge
Tiefentladung f <Elektrotech> deep discharge, deep depletion
Tiefenwahrnehmung f <Ergon> depth perception
tiefer kosmischer Raum m <Raumfahrt> deep space
tiefer Polierkratzer m <Ker & Glas> deep sleek
tiefer Schwund m <Funktech> deep fading *(Wellenausbreitung)*
tiefer Spannungseinbruch m <Elektronik> deep voltage fading *(Hochfrequenztechnik)*
tiefer Weltraum m <Raumfahrt> far space
tiefes Fading n <Funktech> deep fading *(Wellenausbreitung)*
tiefes Fehlerfach n <Elektronik> deep rejection trap *(Sortiergerät)*
tiefes Probenschälchen n <Kerntech> well-type planchet
Tiefgang m <Wassertrans> *(AE)* draft, *(BE)* draught *(Schiff)*
Tiefgangsmarken fpl <Wassertrans> *(AE)* draft marks, *(BE)* draught marks *(Schiffkonstruktion)*
Tiefgefrieranlage f <Heiz & Kälte> quick-freezer
tiefgefrieren v <Thermod> deep-freeze
Tiefgefrieren n <Heiz & Kälte, Thermod> deep-freezing
Tiefgefriergerät n <Thermod> quick-freezer
Tiefgefrierschrank m <Heiz & Kälte> upright freezer
tiefgefroren adj <Thermod> deep-frozen
tiefgehen v <Wassertrans> draw *(Schiff: bestimmter Tiefgang)*
tiefgekühlt adj 1. <Heiz & Kälte> refrigerated; 2. <Lebensmittel> deep-frozen
tiefgestelltes Zeichen n <Comp & DV> subscript
tiefgezogen adj <Fertig> deep-drawn, dished
tiefgezogene Folie f <Verpack> deep-drawing foil
tiefgezogene Klarsichtfolie f <Verpack> deep-drawing film
tiefgezogener Verschluss m <Verpack> embossing closure
Tiefgründung f <Kohlen> deep foundation
tiefinelastischer Stoß m <Kerntech> deep inelastic collision
tiefkühlen v 1. <Heiz & Kälte> refrigerate; freeze *(von Lebensmitteln)*; 2. <Thermod> deep-freeze, freeze
Tiefkühlen n 1. <Heiz & Kälte> deep-freezing, freezing; 2. <Thermod> deep-freezing
Tiefkühlfahrzeug n <Heiz & Kälte> freezing trawler

Tiefkühlgerät n <Heiz & Kälte> domestic freezer, freezer, frozen-food cabinet
Tiefkühlkonzentration f <Heiz & Kälte> cryoconcentration
Tiefkühlkost f 1. <Lebensmittel> quick-frozen food; 2. <Verpack> deep-frozen food
Tiefkühlmittel n <Heiz & Kälte> freezing medium
Tiefkühlprodukt n <Verpack> frozen product
Tiefkühltechnik f <Heiz & Kälte> cryoengineering
Tiefkühltruhe f 1. <Heiz & Kälte> chest freezer; 2. <Thermod> deep freeze
Tiefkühlung f <Heiz & Kälte> superchilling
Tiefkühlverfahren n <Heiz & Kälte> freezing process
Tiefkühlzerkleinerung f <Abfall> cryogenic process
Tiefladelinie f <Wassertrans> deep-waterline *(Schiff)*
Tiefladewagen m <Eisenbahn> float, well wagon
Tieflochbohren n 1. <Fertig> deep-hole drilling; 2. <Maschinen> deep-hole boring, deep-hole drilling
Tieflochbohrer m <Fertig> deep-hole drill, drill bit, gun drill
Tieflöffelbagger m 1. <Meerschmutz> backhoe; 2. <Trans> backhoe loader
Tiefofen m 1. <Kohlen> low kiln; 2. <Metall> soaking pit
Tiefpass m <Elektronik, Lufttrans, Telekom> low-pass
Tiefpass-Abtastfilter n <Elektronik> low-pass sampled-data filter
Tiefpassfilter n <Aufnahme, Comp & DV, Elektriz, Elektronik, Phys, Telekom> low-pass filter
Tiefpassfilter n zweiter Ordnung <Elektronik> second order low-pass filter
Tiefpassfiltern n <Elektronik> low-pass filtering
Tiefpassteil m <Elektronik> low-pass section
Tiefpass-Verhalten n <Elektronik> low-pass response
Tiefpegel-Gerät n <Elektronik> low-level device
Tiefpegel-Modulation f <Elektronik> low-level modulation
Tiefpegeltransistor m <Elektronik> low-level transistor
tiefrot adj <Metall> bright red
Tiefschachtofen m <Kohlen> low-shaft furnace
Tiefschnitt m <Ker & Glas> deep cut
Tiefsee f <Wassertrans> deep-sea *(Meer)*
Tiefseedock n <Bau, Wassertrans> deep-water dock
Tiefseegraben m <Meerschmutz> trench
Tiefseekabel n <Wassertrans> deep-sea cable, submarine cable
Tiefseewellen fpl <Phys> deep-water waves
Tiefstabilisierung f <Kohlen> deep stabilization
tiefste zulässige Betriebsstellung f <Kerntech> lowest permitted operating position *(von Trimmstäben)*
tiefstehendes Zeichen n <Druck> subscript
Tieftemperatur f <Phys> low temperature
Tieftemperaturbeständigkeit f <Kunststoff> low-temperature resistance
Tieftemperaturbrechen n <Abfall> cryogenic crushing *(von Festabfällen)*
Tieftemperaturchemie f <Heiz & Kälte> cryochemistry
Tieftemperaturphysik f <Heiz & Kälte> cryophysics
Tieftemperaturtechnik f <Comp & DV, Heiz & Kälte, Phys, Raumfahrt, Thermod> cryogenics
Tieftemperatur-Thermometer n <Thermod> low-temperature thermometer
Tieftemperaturverfahren n <Thermod> low-temperature techniques
Tieftemperaturzerkleinern n <Abfall> cryogenic crushing *(von Festabfällen)*
Tieftemperaturzerkleinerung f <Abfall> cryogenic process, cryogrinding, freeze-grinding
Tieftonlautsprecher m <Aufnahme> woofer

tief-ultraviolette

tief-ultraviolette Strahlung f <Elektrotech> deep ultraviolet radiation
Tiefungswert m <Fertig> cupping ductility value
Tiefziehblech n <Fertig> deep-drawing sheet
tiefziehen v <Fertig> dish
Tiefziehen n 1. <Aufnahme> cupping; 2. <Fertig, Maschinen, Mechan> deep drawing
Tiefziehstanze f <Verpack> deep-drawing machine
Tiefziehversuch m <Fertig> cupping test
Tiefziehwerkzeug n <Maschinen> deep-drawing die
Tiegel m 1. <Fertig> crucible *(Gießen)*; 2. <Ker & Glas> pan; 3. <Labor> crucible, pan; 4. <Metall, Thermod> crucible
Tiegeldruck m <Druck> platen printing
Tiegeldruckpresse f <Druck> platen press
tiegelfreies Schmelzen n <Kerntech> floating zone melting method
Tiegelkühlofen m <Ker & Glas> pan lehr
Tiegelofen m <Fertig, Labor, Metall, Thermod> crucible furnace
Tiegelzange f <Labor, Metall> crucible tongs
tierische Abfallverwandlung f <Kohlen, Nichtfoss Energ> animal waste conversion
tierische Stärke f <Lebensmittel> animal starch
tierischer Abfall m <Abfall> animal waste
tierisches Fett n <Lebensmittel> animal fat
Tierleim m <Kunststoff, Papier> animal glue
Tiglin... <Chemie> tiglic *(Säure)*
TIG-Schweißbrenner m <Bau> torch for TIG welding
Tilgungsfähigkeit f <Fertig> absorbability *(Schwingen)*
Tilgungszeichen n <Druck> deletion mark
Timbre n <Akustik> timbre
Timesharing n <Comp & DV> time-sharing *(zeitlich geschachtelte Abarbeitung mehrerer Programme)*
Timing n <Kontroll> timing
Tinte f <Papier> ink
Tintenfestigkeit f <Papier> ink hold-out
tintenloses Strahlsystem n <Verpack> inkless ink jet system
Tintenschreiber m <Comp & DV> pen recorder
Tintenstrahldrucker m <Comp & DV, Verpack> ink jet printer
tippen v <Comp & DV> type
Tippfehler m <Comp & DV> typo *(umgangssprachlich)*
Tippfehlerquote f <Comp & DV> keying error rate
Tisch m 1. <Fertig> platen *(Hobelmaschine)*; bed *(Presse)*; 2. <Maschinen> platen, table
Tisch m **mit Teileinrichtung** <Maschinen> indexing table
Tischanschlag m <Fertig> table dog
Tischapparat m <Telekom> table set *(Telefon)*
Tischbewegung f <Fertig> working traverse
Tischbohrer m <Maschinen> bench drill
Tischbohrer m **mit Schraubstock** <Maschinen> *(BE)* bench drill with vice, *(AE)* bench drill with vise
Tischbohrmaschine f 1. <Fertig, Maschinen> bench drilling machine; 2. <Mechan> bench drill
Tischbohrmaschine f **mit Rundsäule** <Maschinen> bench pillar drilling machine
Tischcomputer m <Comp & DV> desktop computer
Tischdrehbank f <Bau, Mechan> bench lathe
Tischgerät n <Comp & DV> desktop computer
Tischgussverfahren n <Ker & Glas> table casting
Tischhobel m <Bau> bench plane
Tischlerleim m <Bau> joiner's glue
Tischmaschine f <Fertig> bench-mounted machine
Tischmikrofon n <Aufnahme> table mike
Tischrechner m <Comp & DV> desktop computer
Tisch-Reproduktionsgerät n <Foto> copy stand

Tischstativ n <Foto> table tripod
Tischvorschubbewegung f <Fertig> table feed motion
Tissue n <Verpack> tissue paper
Tissuepapier n <Papier> tissue paper
Tissuepapiermaschine f <Papier> tissue machine
Titan n *(Ti)* <Chemie> titanium *(Ti)*
Titanat n <Chemie> titanate
Titandioxid n <Chemie, Kunststoff> titanium dioxide
Titandioxidabfall m <Abfall> titanium dioxide waste *(roter Schlamm)*
Titanium... <Chemie> titanic
Titanlegierung f <Anstrich, Raumfahrt> titanium alloy
Titanoxid n <Chemie> titanium dioxide
Titansäureanhydrid n <Chemie> titanium dioxide
Titanschmieden n <Raumfahrt> titanium forging
Titanyl n <Chemie> titanyl
Titel m 1. <Comp & DV> job title; 2. <Druck> title
Titelblatt n <Druck> title page
Titelei f <Druck> front matter, prelims
Titelfeld n <Fernseh> safe title area
Titelschalter m <Fernseh> title keyer
Titelschrift f <Druck> display face, display type, titling font
Titelzeile f <Druck> header
Titer m 1. <Chemie> titre; 2. <Textil> count
Titration f <Anstrich, Chemie> titration
titrieren v <Chemie, Kohlen> titrate
Titrierung f <Chemie, Kohlen> titration
Titrimeter n <Chemie> titremeter
Titrimetrie f <Gerät> titrimetry
T-Koppler m *(T-förmiger Koppler)* <Optik, Telekom> T-coupler *(tee coupler)*
Tl *(Thallium)* <Chemie> Tl *(thallium)*
T-Leitwerk n <Lufttrans> T-tail *(Luftfahrzeug)*
TL-Triebwerk n *(Turbostrahltriebwerk)* <Lufttrans> turbojet
Tm *(Thulium)* <Chemie> Tm *(thulium)*
TM *(transversal magnetisch)* <Elektrotech, Telekom> TM *(transverse magnetic)*
TM-Modus m 1. <Elektrotech> E mode, TM mode, transverse magnetic mode; 2. <Optik> E Mode, TM mode; 3. <Telekom> E mode, TM mode, transverse magnetic mode
T-Muffe f <Bau> T-joint *(Sanitär)*
TMUX *(Transmultiplexer)* <Telekom> TMUX *(transmultiplexer)*
TM-Welle f <Elektrotech, Phys, Telekom> E wave, TM wave, transverse magnetic wave
T-Netz n <Elektrotech> T-network
TNI *(Teilnehmer-Teilnehmer-Information)* <Telekom> UUI *(user-to-user information)*
TNL *(Teilnehmeranschlussleitung)* <Telekom> subscriber's line
TNR *(gesteuerter Thermonuklearreaktor)* <Kerntech> CTR *(controlled thermonuclear reactor)*
TNS *(Teilnehmersatz)* <Telekom> SLC *(subscriber line circuit)*
TNT *(Trinitrotoluol)* <Chemie> TNT *(trinitrotoluene)*
T-Nut f *(T-förmige Nut)* <Fertig, Maschinen> T-slot *(tee slot)*
T-Nutenschraube f <Maschinen> T-bolt
Tochter... <Strahlphys> daughter
Tochtergerät n <Fernseh> slave unit
Tochterkern m <Strahlphys> daughter nucleus
Tochterkompass m <Lufttrans, Wassertrans> compass repeater, compass repeater indicator, repeater compass, repeating compass
Tochterprodukt n <Strahlphys> daughter product
Tocopherol n <Lebensmittel> tocopherol
Toilettenbecken n <Bau> *(BE)* pan

Tokamak m <Phys> tokamak
Token n <Comp & DV, Telekom> token
Token-Bus m <Comp & DV, Telekom> token bus
Token-Bus-Schnittstelle f <Telekom> token bus interface
Token-Ring m <Comp & DV, Telekom> token ring
Token-Ring-Netz n <Comp & DV, Telekom> token ring network, token-passing ring network
Token-Vielfachzugriffsverfahren n <Telekom> token-passing multiple access
Token-Zugriff m <Telekom> token access
Toleranz f 1. <Akustik, Anstrich> tolerance; 2. <Maschinen, Mechan> allowance, tolerance; 3. <Metrol, Qual> tolerance
Toleranz f **der Mittenabweichung** <Mechan> concentricity tolerance
Toleranz f **des Kerndurchmessers** <Optik> core diameter tolerance *(Lichtleiter)*
Toleranz f **des Manteldurchmessers** <Optik, Telekom> cladding diameter tolerance *(Lichtleiter)*
Toleranzbereich m <Qual> tolerance zone
Toleranzen fpl <Fertig> limits
Toleranzgrenze f 1. <Maschinen> tolerance limit; 2. <Metrol> limit of tolerance; 3. <Qual> tolerance limit
Toleranzklasse f <Maschinen> tolerance class
Toleranzprüfung f <Comp & DV> MC, marginal check, marginal test
Toleranzwert m <Sicherheit> tolerance value
toleriertes Maß n <Konstzeich> toleranced dimension
Toluchinolin n <Chemie> toluquinoline
Toluendiisocyanat n *(TDI)* <Kunststoff> toluyene diisocyanate *(TDI)*
Toluidin n <Chemie> aminotoluene, toluidine
Toluol n <Chemie, Erdöl> toluene *(Petrochemie)*
Toluoldiisocyanat n *(TDI)* <Kunststoff> toluene diisocyanate *(TDI)*
Toluonitril n <Chemie> cyanotoluene, toluonitrile
Toluyl... <Chemie> toluic
Toluyliden n <Chemie> toluylene
Tolyl... <Chemie> tolyl
Tomographie f <Kerntech, Maschinen> tomography
Ton m 1. <Akustik> sound, tone; 2. <Aufnahme> sound; 3. <Bau> argil, clay, pug; 4. <Druck, Foto> tone; 5. <Funktech> sound; 6. <Hydraul> pug; 7. <Ker & Glas> clay; adobe *(Material)*; 8. <Kohlen> clay; 9. <Telekom> tone
Ton... <Aufnahme, Comp & DV, Funktech> audio
Tonabnehmer m 1. <Akustik> reproducer; 2. <Aufnahme> pick-up
Tonabnehmerarm m <Aufnahme> pick-up arm; tone arm *(beim Plattenspieler)*
Tonabnehmerkopf m <Akustik> pick-up
Tonabnehmerstecker m <Aufnahme> phono plug
Tonabschaltung f <Raumfahrt> tone disabler
Tonabschwächer m <Elektronik> audio attenuator
Tonabstufung f <Foto> tonal gradation
tonaler Klang m <Akustik> tonal note
Tonalität f <Aufnahme> tonality
Tonarchiv n <Aufnahme> sound archive
Tonarm m <Akustik> pick-up arm
Tonart f <Akustik> key
tonartig adj <Bau> argillaceous
Tonaufnahme f <Aufnahme, Telekom> sound recording
Tonaufnahmesystem n <Aufnahme> sound recording system
Tonaufzeichnung f <Aufnahme, Telekom> audio recording, recording, sound recording; phonogram, record *(Ergebnis)*
Tonaufzeichnungsgerät n **im Cockpit** <Lufttrans> cockpit voice recorder

Tonausfall m <Aufnahme> audio dropouts
Tonbalken m <Aufnahme> sound bar
Tonband n 1. <Akustik> magnetic tape; 2. <Aufnahme> tape; 3. <Elektriz> magnetic tape
Tonband n **mit Klebestellengeräusch** <Aufnahme> blooping tape
Tonbandaufnahmegerät n <Aufnahme> recording tape deck
Tonbanddeck n <Akustik> tape deck
Tonband-Frequenzschallplatte f <Aufnahme> tone band frequency record
Tonbandgerät n 1. <Akustik> tape recorder; 2. <Aufnahme> magnetic tape recorder, tape deck, audio tape machine
Tonbandgerät n **mit zwei Rollen** <Aufnahme> reel-to-reel taperecorder
Tonbandkassette f <Aufnahme> audio tape cassette
Tonbehälter m <Ker & Glas> earthenware tank
Tondecke f <Aufnahme> sound blanket
Tondreieck n <Labor> pipeclay triangle *(Tiegelhalter)*
Toneffekt m <Aufnahme> sound boom
Toner m <Druck, Konstzeich> toner
Tonerde f <Fertig> alumina
tonerdehaltig adj <Papier> aluminiferous
Tonerdeschleifmittel n <Anstrich> alumina abrasive
Tonfilter n <Elektronik> audio filter
tonfrequent adj <Elektronik> audio-frequent
Tonfrequenz f *(Tf)* <Aufnahme, Elektronik, Fernseh, Funktech> audio frequency *(AF)*; sound frequency
Tonfrequenzband n 1. <Aufnahme> audio band; 2. <Comp & DV, Telekom> voiceband
Tonfrequenzgenerator m 1. <Elektronik> audio oscillator, audio-frequency oscillator; 2. <Funktech> audio-frequency oscillator
Tonfrequenzkanal m <Aufnahme> audio channel
Tonfrequenz-Leistungsverstärker m <Elektronik> audio power amplifier
Tonfrequenzmesser m <Aufnahme> audio-level meter
Tonfrequenz-Rückkopplung f <Aufnahme> audio feedback
Tonfrequenzsignal n <Elektronik, Funktech> audio-frequency signal
Tonfrequenz-Signalgenerator m <Elektronik, Funktech> audio-frequency signal generator
Tonfrequenz-Spektrometer n <Gerät> audio-frequency spectrometer
Tonfrequenzspektrum n <Aufnahme> audible spectrum
Tonfrequenzstörung f 1. <Elektronik> audio-frequency interference; 2. <Funktech, Telekom> audio-frequency interference
Tonfrequenzstrom m **zur Aufnahme** <Aufnahme> recording audio-frequency current
Tonfrequenzverstärker m <Aufnahme, Elektronik> audio amp, audio-frequency amplifier
Tonfrequenzverstärkung f <Elektronik, Funktech, Telekom> audio frequency amplification, audio low-frequency amplification
Tongehalt m <Bau, Kohlen> clay content
Tongemisch n <Akustik> complex sound, complex tone
Tongenerator m <Aufnahme, Telekom> tone generator
Tongewinnungsmaschine f <Ker & Glas> clay-working machine
Tongrube f <Ker & Glas> clay works
tonhaltig adj <Bau> argillaceous
tonhaltiger Bohrschlamm m <Erdöl> clay-base mud *(Bohrtechnik)*
tonhaltiger Kiessand m <Bau> hoggin
Tonheit f <Akustik> critical band rate
Tonheitsmuster n <Akustik> critical band-rate pattern

Tonheitszeitmuster

Tonheitszeitmuster *n* <Akustik> critical band-rate time pattern
Tonhobel *m* <Ker & Glas> clay cutter
Tonhöhe *f* <Akustik, Phys, Wellphys> pitch
Tonhöhenempfindung *f* <Akustik> pitch sensation
Tonhöhenschwankung *f* <Aufnahme> wow *(Jaulen)*
Tonhöhen-Unterschiedsschwelle *f* <Akustik> differential threshold of frequency
Tonhöhenverschiebung *f* <Akustik> pitch shift
tonig *adj* <Bau> argillaceous
Ton-im-Bild-Übertragung *f* <Fernseh> sound-on-vision transmission
Tonimpuls *m* <Akustik> tone burst
Tonindustrie *f* <Ker & Glas> clay industry
Toningenieur *m* <Aufnahme> audio control engineer
Tonintervall *n* <Wellphys> musical interval *(zwischen zwei Noten)*
Tonkanal *m* <Aufnahme, Fernseh> sound channel
Tonklebestelle *f* <Aufnahme> blooping patch
Tonkneter *m* <Ker & Glas> clay kneader
Tonknetmaschine *f* <Ker & Glas> clay-kneading machine
Tonkopf *m* 1. <Aufnahme> audio head; 2. <Fernseh> sound head
Tonkrug *m* <Ker & Glas> earthenware jar
Tonlage *f* <Akustik> pitch
Tonlehm *m* <Bau> clay loam
Tonlehre *f* <Aufnahme> sonics
Tonleiter *f* <Akustik> musical scale, scale
Tonlosung *f* <Wasserversorg> clay suspension
Tonlösung *f* <Foto> toning solution
Tonmaskierung *f* <Akustik> aural masking
Tonmasse *f* <Ker & Glas> clay mass
Tonmergel *m* <Wasserversorg> clay marl, marly clay
Tonmineral *n* <Bau, Kohlen> clay mineral
Tonmischapparatur *f* <Aufnahme> rerecording machine
Tonmischer *m* <Ker & Glas> clay mixer
Tonmischmaschine *f* <Ker & Glas> clay-mixing machine
Tonmischpult *n* <Aufnahme> dubbing console, sound console
Tonmischraum *m* <Aufnahme> dubbing room, sound control room
Tonmischsitzung *f* <Aufnahme> rerecording session
Tonmischstudio *n* <Aufnahme> dubbing studio
Tonmischung *f* <Akustik> dubbing
Tonmodulation *f* <Elektronik> audio modulation
Tonmodulierung *f* <Akustik, Aufnahme> sound modulation
Tonmühle *f* <Ker & Glas> clay mill
Tonnage *f* <Metrol, Wassertrans> tonnage
Tonnagegebühr *f* <Metrol> tonnage
Tonne *f* 1. <Lebensmittel, Mechan> barrel; 2. <Metrol> gross ton, long ton, metric ton, ton; 3. <Phys, Trans> barrel; 4. <Wassertrans> buoy *(Seezeichen)*
Tonnenfeder *f* <Maschinen> barrel spring
tonnenförmige Verzeichnung *f* 1. <Foto> barrel distortion; 2. <Foto, Optik> barrel-shaped distortion *(einer Linse)*; 3. <Phys> barrel distortion, barrel-shaped distortion
Tonnengewölbe *n* 1. <Bau> barrel vault, tunnel vault, wagon vault; 2. <Kohlen> tunnel vault
Tonnenlager *n* <Maschinen> barrel roller bearing, barrel-shaped roller bearing, spherical roller bearing
Tonnentragfähigkeit *f* <Wassertrans> dead-weight tons
Tonnenverzeichnung *f* <Mechan> barrel distortion
Tonperspektive *f* <Aufnahme> sound perspective
Tonplatte *f* <Ker & Glas> earthenware slab
Tonplattenpresse *f* <Ker & Glas> clay plate press
Tonqualität *f* <Telekom> sound quality
Tonquelle *f* <Elektronik> sound source

Tonregiesystem *n* <Aufnahme> audio console *(Mischpult)*
Tonrohr *n* 1. <Bau> earthenware pipe; 2. <Ker & Glas> clay pipe
Tonröhre *f* <Ker & Glas> earthenware pipe
Tonrolle *f* <Akustik, Fernseh> capstan
Tonrollenantrieb *m* <Fernseh> capstan drive
Tonsäule *f* <Akustik, Aufnahme> sound column
Tonschlamm *m* <Wasserversorg> clay silt
Tonschleife *f* <Aufnahme> sound loop
Tonschneider *m* <Akustik> cutter
Tonsieb *n* <Ker & Glas> earthenware sieve
Tonsignal *n* 1. <Akustik> audio signal; 2. <Elektronik> audio signal, tone signal; 3. <Fernseh, Funktech> audio signal; 4. <Telekom> audio signal, sound signal
Tonsignalisierung *f* <Elektronik> *(AE)* tone signaling, *(BE)* tone signalling
Tonspur *f* 1. <Akustik> soundtrack; 2. <Aufnahme> soundtrack, squeeze track; 3. <Fernseh> audio track
Tonstaub *m* <Ker & Glas> clay dust
Tonsteuerungsverstärker *m* <Aufnahme> gating amplifier
Tonstudio *n* <Aufnahme> audio control room, sound studio
Tonstufe *f* <Aufnahme> sound stage
Tonsuspension *f* <Wasserversorg> clay suspension
Tonsystem *n* <Aufnahme> sound system
Tontechnik *f* <Maschinen> acoustic engineering
Tontiegel *m* <Ker & Glas> clay crucible
Tonträger *m* *(TT)* <Aufnahme, Fernseh> sound carrier
Tonträgerdienst *m* <Telekom> audio bearer service
Tonübertragung *f* <Akustik> sound transmission
Tonumfang *m* <Akustik> scale
Tonung *f* <Foto> toning
Tonverstärker *m* <Aufnahme> audio amplifier
Tonverteilung *f* <Aufnahme> sound distribution
Tonwahl *f* *(TW)* <Telekom> *(AE)* multifrequency dialing, *(BE)* multifrequency dialling *(MFD)*
Tonwalzen *fpl* <Ker & Glas> clay rollers
Tonwarnanlage *f* <Gerät> audio alarm system
Tonwelle *f* <Akustik> capstan
Tonwert *m* <Foto> tonal value
Tonwertwiedergabe *f* <Foto> reproduction of tonal values
Tonwiedergabe *f* <Aufnahme> tone reproduction
Tonwiedergabegerät *n* <Akustik> reproducer
Tonwiedergabesystem *n* <Aufnahme> sound reproduction system
Tonziegel *m* <Ker & Glas> clay brick
Tonzusammensetzung *f* <Ker & Glas> clay composition
Tool *n* <Comp & DV> software tool, tool
Toolkit *m* <Comp & DV> tool kit
TOP *(Bürokommunikationsprotokoll)* <Telekom> TOP *(technical and office protocol)*
Top-Down-... <Comp & DV> top-down
Top-Down-Methode *f* <Comp & DV> top-down method *(Programmierung von oben nach unten)*
Top-Down-Programmierung *f* <Comp & DV> top-down programming
Top-Down-Strategie *f* <Künstl Int> top-down strategy
Top-Down-Verfahren *n* <Comp & DV> top-down method
Töpfer *m* <Ker & Glas> crockery maker
Töpfererde *f* <Ker & Glas> potter's earth
Töpfergeschirr *n* <Ker & Glas> coarse pottery
Töpferhammer *m* <Ker & Glas> potter's beetle
Töpferofen *m* <Thermod> pottery kiln
Töpferscheibe *f* <Ker & Glas> potter's wheel
Töpferton *m* <Erdöl> ball clay *(Mineral)*

Töpfertongewinnung f <Ker & Glas> potter's clay extraction
Töpferwaren fpl <Fertig> pottery
Töpferwarendekorateur m <Ker & Glas> pottery decorator
topfförmiger Feldwiderstand m <Elektriz> pot-type field rheostat
topfgeglüht adj <Metall> box annealed
Topfkern m <Funktech> pot core
Topfkreis m 1. <Elektronik> coaxial resonator; 2. <Funktech> cavity resonator; 3. <Raumfahrt> cavity *(Weltraumfunk)*
Topfofen m <Thermod> pot furnace
Topfschleifscheibe f <Fertig> cup wheel
Topfschraube f <Wassertrans> bottle screw *(Takelage, Beschläge)*
Topfzeit f <Kunststoff> pot life
topographische Bewertung f <Nichtfoss Energ> topographical assessment
Topologie f <Comp & DV, Geom> topology
topologische Eigenschaften fpl <Geom> topological properties
topotaktisch adj <Kerntech> topotactical
Topp m <Wassertrans> masthead • **vor Topp und Takel legen** <Wassertrans> lay ahull *(Schiff)*
Topplicht n <Wassertrans> masthead light *(Navigation)*
Toppnant f <Wassertrans> topping lift *(Tauwerk)*
Toppzeichen n <Wassertrans> topmark *(Seezeichen)*
Topzementierung f <Erdöl> top cementing plug *(Bohrtechnik)*
Tor n <Bau> gate
Torbernit m <Kerntech> torbernite
Torf m <Kohlen> peat
Torfkohle f <Kohlen> peat coal
Torflügelpfosten m <Wasserversorg> meeting post
torgesteuerte Diode f <Elektronik> gated diode
Torimpuls m <Elektronik> gating pulse
Tor-Katoden-Widerstand m <Elektrotech> gate-to-cathode resistor
Torkretbeton m <Bau> gunite, gunned concrete
Torkretieren n <Bau> guniting
Törn m <Wassertrans> turn *(Tauwerk)*
Törnmotor m <Elektrotech> barring motor
Törnvorrichtung f 1. <Elektrotech> barring gear *(z. B. für Turboläufer)*; 2. <Wassertrans> turning gear *(Motor)*
Toroid n <Elektriz, Kerntech> ring winding, toroid
Torpedo m 1. <Fertig> muller *(Extruder)*; 2. <Kunststoff> torpedo *(Extruder, Spritzgießmaschine)*; 3. <Wassertrans> torpedo
Torpedoboot n <Wassertrans> torpedo boat
Torpfosten m <Bau> swinging post
Torriegel m <Bau> gate latch
Torschaltung f <Phys> gate *(logischer Schaltkreis)*
Torschiff n <Wassertrans> caisson
Torschließbolzen m <Bau> barrel bolt
Torsion f <Maschinen, Metall, Phys> torsion
Torsionsbeanspruchung f <Maschinen> torsional strain, torsional stress
Torsionsfaden m <Elektrotech> torsion string
Torsionsfaden-Galvanometer n <Elektriz> torsion string galvanometer
Torsionsfeder f <Maschinen> spring subjected to torsion, torsion spring
Torsionsfestigkeit f 1. <Kfztech> torsional rigidity; 2. <Maschinen> torsion resistance
Torsionsfestigkeitsprüfer m <Papier> torsional strength tester
Torsionskabel n <Comp & DV> twisted pair cable
Torsionskonstante f <Phys> torsional constant

Torsionsmesser m <Labor, Maschinen> torsion meter
Torsionsmodul m <Phys> modulus of rigidity
Torsionsmoment n <Maschinen> torsional moment
Torsionspendel n <Phys> torsional pendulum
Torsionsschwingung f <Phys> torsional oscillation
Torsionssteifigkeit f <Maschinen> torsion resistance
Torsionsstrebe f <Kfztech> torque arm
Torsionsversuch m <Maschinen> torsion test, torsional test
Torsionswaage f <Maschinen, Metrol, Phys> torsion balance
Torsionswinkel m <Maschinen> angle of torsion
Torsteuerung f <Elektronik> gating
Torsteuerungssignal n <Elektronik> gating signal
Tortendiagramm n <Comp & DV> pie chart
Torus m 1. <Elektriz> ring winding, toroid, torus; 2. <Geom, Strahlphys> torus
Torus m **mit drei Öffnungen** <Geom> three-hole torus
Torusantenne f <Telekom> toroidal antenna
Torusfläche f <Geom> toroidal surface
Tosyl... <Chemie> tosyl...
Totalausfall m <Telekom> total failure
totale Schutzwirkung f <Sicherheit> overall protection
totale Wärmesenke f <Kerntech> ultimate heat sink
Totalreflexion f 1. <Ker & Glas> total internal reflection; 2. <Optik, Phys, Telekom> total reflection
Totalreflexionswinkel m <Optik> angle of total reflection
tote Bewegung f <Fertig> idle pass *(Walzen)*
tote Zone f 1. <Comp & DV> dead zone; 2. <Elektrotech> dead band *(Regler)*; 3. <Funktech, Mobilkom> zone of silence
Totem-Pole-Aufbau m <Elektronik> totem pole arrangement *(Verstärker)*
Totenflaute f <Wassertrans> dead calm *(Wind)*
toter Raum m <Kerntech> stagnant space
toter Winkel m <Lufttrans> blind angle
totes Werk n <Wassertrans> deadworks *(Schiffkonstruktion)*
Totgang m 1. <Kfztech> backlash *(Triebstrang)*; 2. <Mechan> backlash
totgebrannt adj <Thermod> dead burned
Totholz n <Wassertrans> deadwood *(Schiffbau)*
Totlage f <Phys> stagnation point
Totmahlen n <Kohlen> overgrinding
Totmanngriff m <Sicherheit> dead man's handle, safety control handle
Totmannschaltung f <Sicherheit> dead man's switch
Totmannsteuerung f <Trans> dead man's control
totpumpen v <Erdöl> kill *(Bohrtechnik)*
Totpunkt m 1. <Kfztech> *(AE)* dead center *(Motor, Kolben)*; 2. <Mechan> *(AE)* dead center, *(BE)* dead centre
Totrösten n <Metall> dead roasting
Totseilanker m <Erdöl> dead line anchor *(Ankertechnik)*
Totspeicher m <Comp & DV> nonerasable storage
Tötung f **durch Stromschlag** <Elektrotech> electrocution
Totwasser n <Strömphys> wake
totweicher Stahl m <Metall> dead soft steel
Totweichglühen n <Thermod> dead soft anneal
Totzeit f 1. <Comp & DV, Elektronik, Kontroll> dead time; 2. <Mechan> downtime; 3. <Phys> dead time; 4. <Telekom> idle time
Touch-Downbereich m **der Start- und Landebahn** <Lufttrans> runway touch-down zone *(Aufsetzen bei Landung)*
Touchscreen n <Comp & DV> touch screen
Touren bringen v/auf <Kfztech> rev up *(Motor)*
Tourenzähler m <Gerät> revolution counter

Townsend-Entladung

Townsend-Entladung f <Elektronik> Townsend discharge
Toxikologie f <Sicherheit, Umweltschmutz> toxicology
toxisch adj <Sicherheit, Umweltschmutz> toxic
toxische Gesamtwirkung f <Umweltschmutz> cumulative toxic effect
toxisches Abfallprodukt n <Umweltschmutz> toxic degradation product
Toxizität f <Chemie, Erdöl, Kerntech, Kunststoff, Meerschmutz> toxicity
TPP *(texturiertes Pflanzenprotein)* <Lebensmittel> TVP *(textured vegetable protein)*
TQMS *(abteilungsübergreifendes Qualitätssicherungssystem)* <Qual> TQMS *(Total Quality Management System)*
Tracer m <Kerntech> tracer
Tracer-Isotop n <Kerntech> isotopic tracer
Trackball m <Comp & DV> track ball *(Laptop)*
tracken v <Wassertrans> track *(Bewegungsdaten verfolgen; Navigation)*
Tracken n <Wassertrans> tracking *(Verfolgung von Bewegungsdaten; Navigation)*
Trafostation f <Elektriz> transformer substation
Trafostufenschalter m <Elektrotech> tap changer
Trag... <Textil> carrying
Tragant n <Lebensmittel> tragacanth
Tragantgummi n <Lebensmittel> gum tragacanth
Tragarm m <Bau> bracket, suspension arm
Tragbahre f <Sicherheit> stretcher
Tragbalken m 1. <Mechan> girder; 2. <Papier> carrying bar
tragbar adj 1. <Comp & DV> portable, transportable; 2. <Sicherheit> hand-held
tragbare Bodenstation f <Raumfahrt> transportable Earth station *(Weltraumfunk)*
tragbare Bohrmaschine f <Maschinen> portable drilling machine
tragbare Masse f <Labor> portable pulp
tragbarer Computer m <Comp & DV> laptop, laptop computer, portable
tragbarer Empfänger m <Telekom> portable receiver
tragbarer Lötofen m <Fertig> devil
tragbarer Sender m <Fernseh, Funktech> portable transmitter
tragbarer Umsetzer m <Fernseh> portable relay
tragbares Feuerlöschgerät n <Sicherheit> portable fire extinguisher
tragbares Funkmodem n <Mobilkom> portable mobile radio modem
tragbares Hausgerät n <Elektriz> portable appliance
tragbares Telefon n <Mobilkom> cell phone, *(AE)* cellular phone, portable telephone, mobile, *(BE)* mobile phone
tragbares Terminal n <Comp & DV, Wassertrans> portable terminal *(Satellitenfunk)*
Tragbeutel m <Verpack> carrier bag
Tragbild n <Maschinen> contact pattern
Tragblock m <Trans> carrying block
Trage f <Bau> handbarrow
träge adj 1. <Anstrich> inert; 2. <Raumfahrt> inertial
träge Masse f <Phys> inertial mass
träge Sicherung f <Elektrotech> delayed-action fuse, slow blow fuse
Tragebalken m 1. <Maschinen> beam; 2. <Wassertrans> girder *(Schiffbau)*
Trageeigenschaft f <Sicherheit> protective properties *(von Schutzkleidung)*
tragen v 1. <Bau> bear, carry; 2. <Comp & DV> support; 3. <Maschinen> bear; 4. <Papier> carry; 5. <Textil> carry, wear

tragend adj 1. <Bau> bearing, supporting; 2. <Maschinen> bearing
tragende Achse f <Maschinen> carrying axle
tragende Fläche f <Lufttrans> mainplane *(Luftfahrzeug)*
tragende Wand f <Bau, Kohlen> bearing wall, load-bearing wall
tragende Wände fpl <Bau> main walls *(eines Gebäudes)*
tragende Zahnflanke f <Fertig> active profile *(Getriebelehre)*
tragendes Bauteil n <Lufttrans> primary structure
Träger m 1. <Aufnahme> carrier; 2. <Bau> girder; 3. <Comp & DV> carrier; 4. <Elektronik> carrier *(Nachrichten übermittelndes System oder Organisation)*; 5. <Fernseh, Funktech> carrier; 6. <Hydraul> bracket; 7. <Kunststoff> substrate; 8. <Maschinen> arm, beam, girder, holder, support; 9. <Math> support *(einer Funktion)*; 10. <Mechan> bracket, girder; 11. <Raumfahrt> carrier *(Weltraumfunk)*; 12. <Telekom> carrier; bearer *(ISDN)*; 13. <Textil> backing for carpet; 14. <Verpack> joist; 15. <Wassertrans> girder *(Schiffbau)*
Träger m **für Frachtbehälter** <Lufttrans> cargo carrier support *(Hubschrauber)*
Trägerabfrage f <Elektronik, Telekom> CS, carrier sense
Trägerabfragesignal n <Telekom> carrier sense signal
Trägerabgleich m <Fernseh> carrier balance
Trägerabtastsignal n <Telekom> carrier sense signal
Trägeranalyse f <Kerntech> carrier analysis
Trägerauffüllung f <Elektronik> carrier replenishment
Trägerbandbreite f <Elektronik, Telekom> carrier bandwidth
Trägerdetektion f *(CD)* <Comp & DV, Elektronik, Telekom> carrier detection *(CD)*
Trägerdienst m <Telekom> bearer service
Trägerdienst m **im Paketmodus** <Telekom> packet mode bearer service
Trägerdifferenzsystem n <Fernseh> carrier difference system
Trägererfassung f <Elektronik> carrier acquisition
Trägererkennung f *(CD)* <Comp & DV, Elektronik, Telekom> carrier detection *(CD)*
Trägererzeugung f <Elektronik> carrier generation
Trägerflansch m <Bau> flange
Trägerflüssigkeit f <Fertig> carrying agent
Trägerfrequenz f *(Tf)* <Aufnahme, Elektronik, Fernseh, Funktech> carrier frequency *(CF)*
Trägerfrequenzerzeuger m <Elektronik, Fernseh, Funktech> carrier frequency oscillator
Trägerfrequenzoffset n <Elektronik, Fernseh, Funktech> carrier frequency off-set
Trägerfrequenzpegel m <Elektronik, Telekom> carrier level
Trägerfrequenzsystem n <Telekom> carrier system
Trägerfrequenzversatz m <Elektronik, Fernseh, Funktech> carrier frequency off-set
Trägerfrequenzverstärker m <Elektronik, Telekom> carrier amplifier
Trägergas n <Kerntech, Umweltschmutz> carrier gas
Träger-Geräuschabstand m <Telekom> carrier-to-noise density ratio
Trägergewebe n <Textil> backing fabric
Trägerimpuls m <Elektronik> carrier pulse
Trägerkörper m <Telekom> substrate
Trägerleistung f <Funktech> carrier power
trägerlos adj <Bau> unsupported
Trägermast m <Elektriz> supply pylon
Trägermaterial n <Anstrich, Elektronik, Elektrotech> substrate, base
Trägermaterial n **für Dickschicht-Hybridschaltung** <Elektronik> thick film hybrid circuit substrate

Trägermaterial n **für Dünnschicht-Hybridschaltung** <Elektronik> thin film hybrid circuit substrate
Trägermaterial n **vom Typ** <Elektronik> n-type substrate
Trägermodulation f <Elektronik, Telekom> carrier modulation
Trägerpaket n <Telekom> transmission burst
Trägerplatte f 1. <Fernseh> backplate; 2. <Fertig> screen base *(Lumineszenzschirm)*; 3. <Raumfahrt> support plate
Trägerrauschen n <Elektronik, Funktech, Telekom> carrier noise
Träger-Rauschprüfgerätschaft f <Elektronik> carrier noise test set
Träger-Rauschverhältnis n <Aufnahme, Funktech, Telekom> carrier-to-noise ratio
Trägerschicht f <Bau> base
Trägerschiff n <Wassertrans> barge carrier
Trägerschiffleuchter m <Trans> shipborne lighter
Trägerschwelle f <Bau> summer
Trägerschwingung f <Comp & DV> carrier
Trägersignal n <Comp & DV, Telekom> carrier signal
Trägersignal n **für mehrere Empfänger** <Raumfahrt> multidestination carrier
Trägersteuerungsmodulation f <Elektronik> controlled carrier modulation; variable carrier modulation *(Radio)*
Trägerunterdrückung f <Elektronik, Telekom> carrier suppression
Trägerunterseite f <Lufttrans> lower surface
Trägerverstärker m <Telekom> carrier amplifier
Trägerwelle f <Fernseh, Funktech, Phys, Wellphys> carrier wave, cw
Trägerwellengenerator m <Phys, Wellphys> carrier-wave generator
Trägerwellenmodulation f <Phys, Wellphys> carrier-wave modulation
Trägerzusatz m <Funktech, Telekom> carrier reinsertion
Tragetasche f <Verpack> carrier bag
tragfähig adj <Textil> wearable
tragfähiger Boden m <Kohlen> footing
Tragfähigkeit f 1. <Bau> bearing capacity, capacitance; lift *(eines Kranes)*; 2. <Erdöl> dead weight *(Gesamtzuladung)*; 3. <Kohlen> bearing capacity; 4. <Maschinen> carrying capacity, load rating, load-bearing capacity; 5. <Mechan, Nichtfoss Energ> carrying capacity; 6. <Trans> loading capacity; 7. <Wassertrans> dead weight *(Schiffkonstruktion)*; deadweight tonnage, tons deadweight, tdw
Tragfähigkeitsversuch m <Bau> bearing test
Tragfahrzeug n **mit berührungsloser Aufhängung** <Trans> support vehicle with non-contact suspension
Tragfeder f <Maschinen> suspension spring
Tragfestigkeit f <Bau, Kohlen> bearing strength
Tragfläche f 1. <Lufttrans> *(BE)* aerofoil chord, *(AE)* airfoil chord; 2. <Maschinen> bearing surface; 3. <Nichtfoss Energ> *(BE)* aerofoil, *(AE)* airfoil; 4. <Phys> wing; 5. <Wassertrans> hydrofoil
Tragflächenbelastung f <Lufttrans> wing loading
Tragflächenboot n <Wassertrans> hydrofoil
Tragflächenende n <Lufttrans> wing tip
Tragflügel m 1. <Lufttrans> *(BE)* aerofoil, *(BE)* aerofoil chord, *(AE)* airfoil, *(AE)* airfoil chord; mainplane *(Luftfahrzeug)*; 2. <Wassertrans> *(BE)* aerofoil, *(AE)* airfoil
Tragflügelansatz m <Lufttrans> wing root
Tragflügelbehälter m <Lufttrans> wing tank
Tragflügelenteisung f <Lufttrans> *(BE)* aerofoil de-icing, *(AE)* airfoil de-icing
Tragflügelenteisungsklappe f <Lufttrans> *(BE)* aerofoil de-icing valve, *(AE)* airfoil de-icing valve
Tragflügelentfrostung f <Lufttrans> *(BE)* aerofoil de-icing, *(AE)* airfoil de-icing
Tragflügelmittelstück n <Lufttrans> *(AE)* center of wing section, *(BE)* centre of wing section
Tragflügelrumpf m <Wassertrans> *(BE)* aerofoil hull, *(AE)* airfoil hull
Tragflügelwurzel f <Lufttrans> wing root
Trag-Führ-System n <Elektrotech> levitation and guiding system *(Magnetschwebetechnik)*
Trag-Führungslager n <Elektrotech> thrust and guide bearing *(bei Hydrogeneratoren)*
Traggriff m <Mechan, Verpack> carrying handle
Trägheit f <Anstrich, Heiz & Kälte, Maschinen, Mechan, Phys, Umweltschmutz> inertia
Trägheits... <Raumfahrt> inertial
Trägheitsabscheider m <Kerntech> inertial separator
Trägheitsachse f <Mechan, Phys> axis of inertia, balance axis
Trägheits-Beschleunigungsmesser m <Raumfahrt> inertial accelerometer
Trägheitseinschluss m <Phys> inertial confinement
Trägheitsellipse f <Mechan> ellipse of inertia
Trägheitshalbmesser m <Maschinen> radius of gyration
Trägheitskonstante f <Elektriz> storage-energy constant
Trägheitskraft f <Kerntech, Maschinen, Phys> inertial force
Trägheitsmelder m <Raumfahrt> inertial sensor
Trägheitsmittelpunkt m <Mechan, Phys> *(AE)* center of inertia, *(BE)* centre of inertia
Trägheitsmoment n 1. <Bau, Ergon> moment of inertia; 2. <Fertig> inertia moment; 3. <Heiz & Kälte> inertial torque; 4. <Maschinen, Wassertrans> moment of inertia *(Schiffbau)*
Trägheitsmoment n **des Blattes** <Lufttrans> blade moment of inertia *(Hubschrauber)*
Trägheitsnavigation f <Raumfahrt> inertial navigation
Trägheitsnavigationssystem n 1. <Lufttrans> INS, inertial navigation system; 2. <Raumfahrt> INS, inertial navigation system *(Raumschiff)*; 3. <Wassertrans> INS, inertial navigation system
Trägheitsplattform f <Maschinen> inertial platform
Trägheitsprodukt n <Phys> product of inertia
Trägheitsradius m <Bau> radius of gyration
Trägheitsrahmen m <Raumfahrt> inertial frame
Trägheitsschalter m <Elektrotech> inertia switch
Trägheitssteuerung f <Raumfahrt> inertial guidance
Trägheitssystem n <Raumfahrt> inertial system
Trägheitszentrum n <Mechan, Phys> *(AE)* center of inertia, *(BE)* centre of inertia
Traghülse f <Fertig> carrying sleeve
Tragkettenförderer m <Papier> arm elevator
Tragkonstruktion f <Bau> supporting structure
Tragkörper m <Sicherheit> safety frame *(Schutzbrille)*
Tragkraft f <Maschinen> load-bearing capacity
Traglager n 1. <Fertig> yoke *(Fräsdorn)*; 2. <Maschinen> journal bearing, journal box
Traglast f 1. <Bau, Kohlen> ultimate load; 2. <Fertig, Qual> collapse load
Tragleiter f <Sicherheit> portable ladder
Tragöse f <Maschinen> lifting lug
Tragplatte f <Maschinen, Papier> base plate
Tragriemen m <Foto> strap
Tragring m 1. <Kfztech> thrust collar; 2. <Maschinen> thrust block
Tragring m **für Schaltdrähte** <Telekom> jumper ring
Tragrolle f <Ker & Glas> carrying roller
Tragrumpf-Schwebekörper-Flugzeug n <Lufttrans> lifting-body aircraft
Tragschere f <Ker & Glas> shank
Tragschicht f <Bau> base course; bearing bed *(Geologie)*
Tragschichtmaterial n <Bau> base material

Tragschraube

Tragschraube f <Lufttrans> lifting rotor *(Hubschrauber)*
Tragschraubenblatt n <Lufttrans> rotor blade *(Hubschrauber)*
Tragschrauber m <Lufttrans> autogyro
Tragseil n 1. <Bau> supporting rope; 2. <Eisenbahn> track cable; 3. <Maschinen> supporting cable; 4. <Trans> carrier rope, carrying rope
Tragtiefe f <Fertig> engagement depth *(Gewinde)*
Tragtrommelroller m <Papier> pope reel
Tragvermögen n <Nichtfoss Energ, Verpack> carrying capacity
Tragwagen m <Eisenbahn> rail carrier wagon
Tragwalze f 1. <Ker & Glas> carrying roll; 2. <Papier> king roll
Tragwerk n 1. <Bau> frame, framework; 2. <Sicherheit> load-bearing structure
Tragzapfen m <Mechan> trunnion
tragzellenförmig adj <Lufttrans> cellular
Traktion f <Eisenbahn, Maschinen> traction
Traktionsnetz n <Elektriz> traction network
Traktor m 1. <Comp & DV> tractor; 2. <Kfztech> traction engine, tractor
Traktorzuführung f <Druck> tractor feed
Trampschiff n <Wassertrans> tramp, tramp ship
Trampschifffahrt f <Wassertrans> tramp shipping
Tränenblech n <Metall> bulb plate
Trangwalze f <Ker & Glas> trang roll
tränken v <Bau> saturate, temper
Tränken n 1. <Papier> soaking; 2. <Textil> impregnation, soaking
Tränklack m <Elektriz> impregnating varnish
Tränkung f 1. <Bau> impregnation *(Holz)*; 2. <Fertig> impregnation *(Sinterpressteil)*
Transaktion f <Comp & DV> transaction
Transaktionsdatei f <Comp & DV> transaction file
Transaktionssatz m <Comp & DV> transaction record
Transaktionsverarbeitung f <Comp & DV> transaction processing
Transaktionsverwaltungssoftware f <Comp & DV> transaction management software
Transatlantik-Fernmeldekabel n <Telekom> transatlantic telecommunication cable
Transcoder m <Fernseh, Telekom> transcoder
Transcontainer m <Trans> transcontainer
Transducer m <Elektriz, Elektrotech, Gerät> transducer
Transduktor m 1. <Elektrotech> magamp, magnetic amplifier, transductor; 2. <Phys> magnetic amplifier, transductor
Transfer m 1. <Ker & Glas> *(AE)* decal, *(BE)* transfer; 2. <Telekom> transfer *(Daten)*
Transfer-Gate-Schaltung f <Elektrotech> transfer gate circuit
Transferglas n <Ker & Glas> transfer glass
Transfermaschine f <Maschinen> transfer machine
Transferrate f <Telekom> transfer rate
Transferreaktion f <Kerntech> transfer reaction
Transferstraße f <Maschinen> transfer line
Transfersystemteil n <Telekom> message transfer agent
transfinite Zahl f <Math> transfinite number *(Kardinalzahl in der Mengenlehre)*
Transformation f 1. <Elektrotech> transformation; 2. <Math> transformation *(Umformung)*
Transformationsbereich m <Ker & Glas> transformation range
Transformationstemperatur f <Ker & Glas> transformation point
Transformator m <Elektriz, Elektrotech, Funktech, Phys, Telekom> transformer

Transformator m **mit Abzapfpunkten** <Elektriz> tapped transformer
Transformator m **mit Anzapfungen** <Elektrotech> tapped transformer
Transformator m **mit luftisolierter Spule** <Elektriz> dry-type transformer
Transformator m **mit nicht gekapselter Trockenspule** <Elektriz> nonencapsulated-winding dry-type transformer
Transformator m **mit offenem Kern** <Elektriz> open core transformer
Transformator m **mit radial geblechtem Kern** <Elektriz> transformer with radially laminated core
Transformator m **mit separater Wicklung** <Elektriz> separate winding transformer
Transformator m **zur Spannungserhöhung** <Elektrotech> step-up transformer
Transformatoranzapfung f <Elektrotech> transformer tap
Transformatorblech n <Elektrotech> core lamination
Transformator-EMK f <Elektriz, Phys> transformer emf
Transformatorenöl n <Elektriz> transformer oil
Transformatorenstation f <Elektriz, Elektrotech> transformer substation
transformatorgekoppelt adj <Elektriz> transformer-coupled
Transformatorisolierung f <Elektrotech> transformer isolation
Transformatorkaskade f <Elektrotech> cascaded transformers, transformer cascade
Transformatorkern m <Elektriz, Elektrotech> transformer core
Transformatorkessel m <Elektrotech> transformer tank, transformer vessel
Transformatorkopplung f 1. <Elektrotech> transformer coupling; 2. <Telekom> mutual coupling
Transformator-Leistungsfaktor m <Elektriz> transformer efficiency
Transformator-Umspannwerk n <Elektriz, Elektrotech> transformer substation
Transformatorverlust m <Elektrotech> transformer loss
transformerlose Stromversorgung f <Elektriz> transformerless power supply
transformieren v 1. <Comp & DV> transform; 2. <Math> transform *(umformen)*
Transformierung f <Elektrotech> transformation
Transfusionsflasche f <Ker & Glas> transfusion bottle
transiente Bedingungen fpl <Elektrotech> transient conditions
transiente Längsspannung f <Elektriz> direct-axis transient voltage
transiente Querspannung f <Elektriz> quadrature-axis transient voltage
Transistor m <Comp & DV, Elektriz, Phys> transistor
Transistor m **auf Chip** <Elektronik> on-chip transistor
Transistor m/**auf dem Chip befindlicher** <Elektronik> on-chip transistor
Transistor m **in Basisschaltung** <Elektronik> common-base transistor
Transistor m **in Drainschaltung** <Elektronik> common-drain transistor
Transistor m **in Emitterschaltung** <Elektronik> common-emitter transistor
Transistor m **in Gateschaltung** <Elektronik> common-gate transistor
Transistor m **in Overlaytechnik** <Elektronik> overlay transistor
Transistor m **in Source-Schaltung** <Elektronik> common source transistor

Transistorbasisschaltung f <Elektronik> transistor base circuit
Transistor-Charakteristik f <Elektronik> transistor characteristics
Transistorchip m <Elektronik> transistor chip
transistorisiert adj <Kontroll> solid state
Transistor-Leistungsverstärker m <Elektronik> transistor power amplifier
Transistor-Leistungsverstärkung f <Aufnahme> transistor power gain
Transistor-Modulator m <Elektronik> transistor modulator
Transistor-Oszillator m <Elektronik> transistor oscillator
Transistorpaar n <Elektronik> transistor pair
Transistorregler m <Kfztech> transistorized regulator
Transistorsättigung f <Elektronik> transistor saturation
Transistorschaltgerät n <Kfztech> transistor control unit
Transistor-Tetrode f <Elektronik> tetrode transistor (Triac)
Transistor-Transistor-Logik f (TTL) <Comp & DV, Elektronik> transistor-transistor logic (TTL)
Transistorverstärker m <Elektronik> transistor amplifier
Transistorverstärkung f <Elektronik> transistor amplification
Transistorvorspannung f <Elektronik> transistor bias
Transistorzündanlage f <Kfztech> transistor ignition unit, transistorized ignition system
Transitamt n <Telekom> transit exchange
Transitfernamt n <Telekom> trunk transit exchange
Transitverkehr m <Telekom> transit traffic
Translation f <Geom> translation
Translationsbewegung f <Maschinen> translation, translatory motion
Translationsfläche f <Metall> gliding plane
Transliteration f <Comp & DV> transliteration
transliterieren v <Comp & DV> transliterate
Transmission f 1. <Fertig> shafting; 2. <Maschinen, Mechan> transmission
Transmissionsdichte f <Optik> transmittance density
Transmissions-Elektronenmikroskop n <Elektronik, Kerntech, Labor, Phys> transmission electron microscope
Transmissions-Elektronenmikroskopie f <Strahlphys> transmission electron microscopy
Transmissionsfähigkeit f <Phys> transmission power
Transmissionsfenster n <Optik> transmission window
Transmissionsgitter n <Optik> transmission grating
Transmissionsgrad m <Optik> transmittance
Transmissionskette f <Maschinen> transmission chain
Transmissions-Photometer n <Gerät> haze meter
Transmissionsriemen m 1. <Maschinen> driving belt; 2. <Papier> transmission belting
Transmissionswelle f 1. <Lufttrans> connecting shaft; 2. <Maschinen> shafting
transmissive LCD-Anzeige f <Elektrotech> transmissive LCD
Transmultiplexer m (TMUX) <Telekom> transmultiplexer (TMUX)
Transomplatte f <Wassertrans> transom plate (Schiffbau)
Transparent... <Comp & DV, Telekom> transparent
transparente Scheibe f <Optik> (AE) transparent disk
Transparentemail n <Ker & Glas> transparent enamel
transparenter Trägerdienst m <Telekom> transparent bearer service
transparentes Kabinendach n <Lufttrans> canopy (Flugzeug)
Transparentetikett n <Verpack> heat transfer label (durch Wärmeübertragung aufgetragen)
Transparentglasur f <Ker & Glas> transparent glaze
Transparentpapier n <Konstzeich> transparent paper

Transparentvordruck m <Konstzeich> preprint
Transparenz f <Druck, Kunststoff, Telekom> transparency (Netzwerk)
Transpiration f (T) <Hydraul> transpiration (T)
Transpirationskühlung f <Heiz & Kälte> transpiration cooling
Transplutonium-Element n <Kerntech> transplutonium element
Transponder m 1. <Funktech> transponder; 2. <Raumfahrt> transponder (Weltraumfunk); 3. <Telekom, Wassertrans> transponder
transponieren v <Math> transpose (Matrix)
transponierendes Instrument n <Akustik> transposing instrument
Transponierte f <Math> transposed matrix (transponierte Matrix)
transponierte Matrix f <Math> transposed matrix
Transport m 1. <Comp & DV> transport; 2. <Fertig> handling; 3. <Trans> conveyance, haulage, transport; 4. <Verpack> work handling (verschiedener Werkstücke)
Transport m **in Niedrigdruckröhren** <Trans> transport in low-pressure tube
Transportanlage f <Trans> conveying plant
Transportband n 1. <Kohlen, Mechan> conveyor belt; 2. <Verpack> band conveyor, belt conveyor
Transportbehälter m <Trans, Verpack> container
Transportbeton m <Bau> ready-mixed concrete
Transportebene f <Telekom> transport layer (OSI-sieben-Schichten-Modell)
Transporteur m <Fertig> protractor
transportfähiges Windbatteriesystem n <Nichtfoss Energ> transportable wind battery system
Transporthubschrauber m <Lufttrans> transport helicopter
transportierbar adj <Comp & DV> portable, transportable
transportierbare Gasflasche f <Sicherheit> transportable gas container
transportierbare Steckdose f <Elektriz> portable socket outlet
transportieren v 1. <Textil> carry; 2. <Trans> transport; 3. <Wassertrans> ship
Transportkasten m <Fertig> tote box
Transportkette f <Maschinen> conveyor chain
Transportkiste f <Verpack> carrier box
Transportkran m <Verpack> material-handling crane
Transportleitung f <Erdöl> transmission main (Erdgas)
Transportmischer m <Bau> mixer truck
Transportmodell n <Umweltschmutz> transport model
Transportprotokoll n <Comp & DV, Telekom> transport protocol
Transportquelle f <Umweltschmutz> transportation source
Transporttraupe f <Comp & DV> tractor
Transport-Relay n <Telekom> transport relay (Übergang in OSI Schicht 4)
Transportrolle f <Maschinen> caster, castor
Transportschicht f <Telekom> transport layer (OSI-sieben-Schichten-Modell)
Transportschnecke f <Verpack> screw conveyor
Transportsicherheit f <Sicherheit> transportation safety
Transportsystem n <Bau> transportation system
Transportsystem n **mit mehreren Betriebsarten** <Trans> multiple-mode transportation system
Transport- und Installationsanweisungen fpl <Trans> handling and installation instructions
Transport- und Rettungshubschrauber m <Lufttrans> transport and rescue helicopter
Transport- und Verbindungsflugzeug n <Lufttrans> transport and communications aircraft

Transportunternehmen *n* <Trans> common carrier, haulage contractor
Transportvorrichtungen *fpl* <Verpack> handling equipment
Transportwagen *m* 1. <Bau> *(AE)* carrier car, *(BE)* carrier wagon; 2. <Kerntech> trolley; 3. <Kfztech> dolly
Transposition *f* <Akustik, Telekom> transposition
Transputer *m* 1. <Comp & DV> transputer; 2. <Kontroll> transputer
transreflektierende Flüssigkristallanzeige *f* <Elektronik> reflective LCD
Transsonikflugzeug *n* <Lufttrans> transonic aircraft
Transtainer *m* <Trans> transtainer
Transuran *n* <Kerntech> transuranic nuclide
Transuranabfall *m* <Kerntech> TRU, transuranic waste
Transurane *npl* <Strahlphys> transuranic elements
transversal elektrisch *adj (TE)* <Elektrotech, Telekom> transverse electric *(TE)*
transversal elektromagnetisch *adj (TEM)* <Elektrotech, Telekom> transverse electromagnetic *(TEM)*
transversal magnetisch *adj (TM)* <Elektrotech, Telekom> transverse magnetic *(TM)*
Transversal... <Telekom, Optik> transversal
Transversale *f* <Geom> transversal *(im Dreieck)*
transversale chromatische Aberration *f* <Phys> transverse chromatic aberration *(senkrecht zur optischen Achse)*
transversale elektrische Welle *f* <Elektrotech, Telekom> transverse electric wave
transversale elektromagnetische Welle *f* 1. <Elektrotech> transverse electromagnetic wave; 2. <Telekom> transverse electromagnetic wave, guided wave
transversale Energieverteilung *f* <Strahlphys> transverse energy distribution
transversale Interferometrie *f* <Optik, Telekom> transverse interferometry
transversale magnetische Welle *f* <Elektrotech, Telekom> transverse magnetic wave
transversale Magnetisierung *f* <Akustik> transverse magnetization
transversaler elektrischer Modus *m* <Elektrotech, Telekom> transverse electric mode
transversaler elektromagnetischer Modus *m* <Elektrotech, Telekom> transverse electromagnetic mode
transversaler magnetischer Modus *m* <Elektrotech, Telekom> transverse magnetic mode
Transversalfilter *n* <Telekom> transverse filter
Transversalglied *n* <Telekom> transversal section
Transversal-Interferometrie *f* <Optik> transverse interferometry
Transversalkomponente *f* <Phys> transverse component
Transversaltonspur *f* <Aufnahme> variable area sound track
Transversalwelle *f* <Akustik, Elektrotech, Phys, Telekom, Wellphys> transverse wave
Transverter *m* <Funktech> transverter
transzendentale Zahl *f* <Math> transcendental number
Transzendenz *f* <Math> transcendence *(von Zahlen)*
Trap *m* <Comp & DV> trap *(nicht programmierter Sprung)*
Trapatt-Diode *f* 1. <Elektronik> trapatt diode, trapped plasma avalanche time transit diode; 2. <Phys> trapatt diode, trapped plasma avalanche time transit diode
Trapez *n* 1. <Geom> trapezoid; 2. <Wassertrans> trapeze *(sailing)*
Trapezfeder *f* <Maschinen> trapezoidal spring
Trapezflügel *m* <Lufttrans> tapered wing
trapezförmig *adj* <Geom> trapezoidal

Trapezgewinde *n* <Maschinen> acme thread, trapezoidal thread
Trapezgewindeschneider *m* <Maschinen> acme thread tap
Trapezquerlenkeraufhängung *f* <Kfztech> trapezoid arm-type suspension
Trapezquerlenker-Radaufhängung *f* <Kfztech> double-wishbone suspension
Trapezverzeichnung *f* <Fernseh> keystone distortion
Trapezverzerrung *f* 1. <Elektronik> trapezoidal distortion; 2. <Fernseh> keystone distortion
trassieren *v* <Bau> locate, route
Trassierung *f* 1. <Bau> location; 2. <Telekom> routing *(von Leitungen)*
Traubenkernöl *n* <Lebensmittel> grapeseed oil
Traubenzucker *m* <Lebensmittel> grape sugar
Trauerflagge *f* <Wassertrans> mourning flag
Traufbohle *f* <Bau> eaves board
Traufbrett *n* <Bau> fascia board
Traufe *f* <Bau> eaves
Träufelwicklung *f* <Elektrotech> fed-in winding, random-coil winding
Traveller *m* <Wassertrans> *(AE)* traveler, *(BE)* traveller *(Segeln)*
Traverse *f* 1. <Bau> crossbar, crossbeam, suspension bracket; 2. <Fertig> crosshead, top beam; 3. <Maschinen> crossbeam, crosshead, top rail; 4. <Wassertrans> *(AE)* cross tie, *(BE)* sleeper *(Schiffbau)*
traversenartiges Aufnahmegestell *n* <Kerntech> bar-type pick-up base
Travertin *m* <Bau> travertine
Trawl *n* <Meerschmutz> trawl net
trawlen *v* <Wassertrans> trawl *(Fischerei)*
Trawler *m* <Wassertrans> trawler *(Schifftyp)*
Trayaufrichter *m* **und -belader** *m* <Verpack> tray erector and loader
Treber *mpl* <Lebensmittel> pomace
Treffer *m* <Comp & DV> hit
Trefferliste *f* <Comp & DV> hit list
Treffgenauigkeit *f* <Qual> accuracy of the mean
Treffpunkt *m* <Geom> point of concurrence *(von drei oder mehr Linien)*
Trehalose *f* <Chemie> trehalose
Treibanker *m* <Wassertrans> drogue, sea anchor *(Notfall)*
Treibbake *f* <Wassertrans> floating beacon *(Seezeichen)*
Treibboje *f* <Wassertrans> drifting buoy *(Wetter)*
Treibeis *n* <Wassertrans> drift ice
treiben *v* 1. <Fertig> hollow; 2. <Funktech, Maschinen> drive; 3. <Wassertrans> drift
treiben lassen *v* <Wassertrans> float
Treiben *n* <Wassertrans> flotation
treibend *adj* 1. <Elektriz, Maschinen> driving; 2. <Wassertrans> adrift *(Schiff)*
treibende Kraft *f* <Metall, Qual> driving force
treibendes Ölfeld *n* <Meerschmutz> spill
treibendes Schiff *n*/**vor Anker** <Wassertrans> dragging ship
Treiber *m* 1. <Akustik, Comp & DV> driver; 2. <Funktech> driver *(Senderstufe)*; 3. <Kontroll> drive *(Plattenspielwerk)*
Treiberimpuls *m* <Fernseh> driving pulse
Treiberprogramm *n* <Comp & DV> driver
Treiberroutine *f* <Comp & DV> peripheral driver
Treibersignale *npl* <Fernseh> driving signals
Treiberspannung *f* <Fernseh> booster voltage, drive voltage
Treiberstufe *f* <Kfztech, Raumfahrt> driver stage *(Senderverstärker)*

Treibgut *n* 1. <Abfall> floating refuse; 2. <Wassertrans> flotsam
Treibhammer *m* <Maschinen> beating hammer
Treibhauseffekt *m* <Heiz & Kälte, Kohlen, Nichtfoss Energ, Phys, Umweltschmutz> greenhouse effect
Treibholz *n* <Wassertrans> driftwood
Treibkette *f* <Maschinen> transmission chain
Treibkörper *m* <Wassertrans> surface float *(Meereskunde)*
Treibmittel *n* 1. <Kunststoff> blowing agent, foaming agent; 2. <Lebensmittel> raising agent *(Hefe, Backpulver)*; 3. <Umweltschmutz> propellant, propellent
Treibnetz *n* <Wassertrans> drift net *(Fischerei)*
Treibofen *m* <Fertig> cupel, cupellation furnace
Treibrad *n* <Maschinen> driving wheel
Treibriemen *m* <Maschinen> belt, drive belt, driving belt, transmission belt
Treibriemenabdeckung *f* <Sicherheit> belt guard
Treibsand *m* 1. <Bau> shifting sand; 2. <Kohlen> quicksand, shifting sand
Treibscheibe *f* <Kfztech> drive pulley *(Drehstromlichtmaschine)*
Treibschraube *f* <Maschinen> hammer-drive screw
Treibsitz *m (TS)* <Maschinen> drive fit
Treibstange *f* <Eisenbahn> connecting rod
Treibstoff *m* 1. <Kfztech> fuel, *(AE)* gas, *(AE)* gasoline; 2. <Raumfahrt> fuel, propellant, propellent; 3. <Wassertrans> fuel
Treibstoff *m* **für Düsentriebwerk** <Lufttrans> jet engine fuel
Treibstoffablassen *n* <Raumfahrt> fuel dumping
Treibstoffablasssystem *n* <Raumfahrt> fuel-dumping system
Treibstoffabsperrhahn *m* <Lufttrans> fuel shut-off cock
Treibstoffabsperrhahnsteuerung *f* <Lufttrans> fuel shut-off cock control link
treibstoffaufwendig *adj* <Raumfahrt> fuel-costly
Treibstoffausgleichsbehälter *m* <Lufttrans> fuel ullage box
Treibstoffdruck *m* <Raumfahrt> fuel pressure
Treibstofffilter *n* <Lufttrans> fuel filter
Treibstoffhahn *m* <Lufttrans> fuel cock
Treibstoffkühler-Wärmeaustauscher *m* <Lufttrans> fuel coolant heat exchanger
Treibstoffleitungsschacht *m* <Raumfahrt> fuel line duct
Treibstoffmangel *m* <Thermod> lack of fuel
Treibstoffmasse *f* <Raumfahrt> fuel mass, propellant mass, propellent mass
Treibstoffmassendurchsatz *m* <Lufttrans> mass fuel rate of flow
Treibstoffmesser *m* <Raumfahrt> fuel-measuring unit
Treibstoff-Oxidant-Mischungsverhältnis *n* <Raumfahrt> fuel-oxidizer mixture ratio
Treibstoffpumpe *f* <Lufttrans, Mechan, Raumfahrt> fuel pump
Treibstoffreglereinheit *f* <Lufttrans> fuel control unit
Treibstoffreserve *f* <Lufttrans> fuel reserve
Treibstoffschnellablass *m* <Lufttrans> fuel jettison
Treibstoff-Schnellablassventil *n* <Lufttrans> jettison valve
Treibstoffsorte *f* <Lufttrans> fuel grade
treibstoffsparend *adj* <Raumfahrt> fuel-efficient
Treibstoffstandgeber *m* <Lufttrans> fuel level transmitter
Treibstoffstandsensor *m* <Raumfahrt> fuel level sensor
Treibstoffsystem *n* 1. <Kfztech, Lufttrans> fuel system; 2. <Wassertrans> fuel oil system
Treibstofftank *m* <Lufttrans> fuel tank
Treibstofftankwahlschalter *m* <Lufttrans> fuel tank selector switch
Treibstofftemperaturfühler *m* <Lufttrans> fuel temperature probe
Treibstoffventil *n* <Lufttrans> fuel cock
Treibstoffverbrauch *m* <Kfztech> fuel consumption
Treibstoffversorgung *f* 1. <Kfztech> *(AE)* fueling, *(BE)* fuelling; 2. <Lufttrans> *(BE)* fuelling
Treibstoff-Vorratsprogrammsteuerung *f* <Lufttrans> fuel level pre-setting controls
Treibstoffvorrat-Wahlschalter *m* <Lufttrans> fuel level selector
Treibstoffvorschubventil *n* <Lufttrans> fuel cross-feed valve
Treibstoffzelle *f* <Raumfahrt> fuel cell
Treibstoffzufuhr *f* 1. <Kfztech> *(AE)* fueling, *(BE)* fuelling; 2. <Lufttrans> *(BE)* fuelling
Treibstoffzufuhrsteuerung *f* <Lufttrans> fuel control
Treibstoffzuladung *f* <Lufttrans> fuel load
Treibwageneinheit *f* <Eisenbahn> rail motor unit
Treibzapfen *m* <Maschinen> drive pin
Treidelroute *f* <Trans> bridletrack
Treidelweg *m* <Trans> bridleroad; bridleway *(Schiff)*
Treisegel *n* <Wassertrans> trysail *(Segeln)*
Trelliscodierung *f* <Telekom> trellis coding
Tremolo *n* <Akustik> tremolo
Trendschreiber *m* <Gerät> trend recorder
Trennautomat *m* 1. <Comp & DV> burster, decollator; 2. <Mechan> burster
Trennbarkeit *f* <Kerntech> separability
Trennbruch *m* 1. <Fertig> rupture; 2. <Metall, Qual> cleavage crack
Trenndichte *f* <Kohlen> separation density
Trenndiode *f* <Elektronik> isolation diode
Trenneinsätze *mpl* <Verpack> partitioning inserts
Trennelektrode *f* <Elektronik> splitting electrode
Trennelement *n* <Optik> isolator
trennen *v* 1. <Comp & DV> decollate; 2. <Elektriz> disconnect *(der Leitung)*; 3. <Elektrotech> isolate; disconnect *(der Leitung)*; isolate *(galvanisch)*; 4. <Kfztech> break; 5. <Maschinen> disconnect, disjoint; 6. <Telekom> disconnect *(der Verbindung)*
Trennen *n* 1. <Bau> ripping; 2. <Elektrotech> partitioning; breaking *(des Stromkreises)*; 3. <Fertig> *(BE)* disc cutting, *(AE)* disk cutting
Trennen *n* **und Ausschließen** *n* <Druck> h&j, hyphenation and justification
Trenner *m* <Elektrotech> air-break disconnector, isolating switch, isolator
Trennfestigkeit *f* <Kunststoff> bond strength
Trennfilter *n* <Elektronik, Fernseh> separation filter
Trennflüssigkeit *f* <Chemtech> separation liquid
Trennfuge *f* <Fertig> parting line
Trenngatter *n* <Elektrotech> partition gate
Trennimpuls *m* 1. <Elektriz> break impulse, break pulse *(bei Schaltgerät)*; 2. <Telekom> mark impulse, marking pulse
trenninduzierter Strom *m* <Elektrotech> break-induced current
Trennkasten *m* <Elektriz> dividing box
Trennkolonne *f* <Labor> fractionation column *(Destillieren)*
Trennkondensator *m* <Elektriz, Elektrotech> blocking capacitor
Trennkontakt *m* <Elektrotech> break contact
Trennladung *f* <Elektrotech> partitioned charge
Trennmanöver *n* <Raumfahrt> *(AE)* separation maneuver, *(BE)* separation manoeuvre
Trennmaschine *f* <Fertig> cutting-off machine
Trennmesser *n* <Maschinen> parting blade
Trennmethode *f* <Chemtech> separation process

Trennmittel

Trennmittel *n* <Kunststoff> *(AE)* mold release agent, *(BE)* mould release agent, release agent
Trennölfleck *m* <Ker & Glas> dope mark
Trennpapier *n* <Verpack> absorbents *(Fleischereierzeugnisse)*
Trennprogramm *n* <Druck> hyphenation program
Trennrakete *f* <Raumfahrt> separation rocket
Trennrelais *n* <Elektrotech> disconnect relay
Trennrohr *n* <Kerntech> calandria tube
Trennsäule *f* 1. <Chemtech> separating column; 2. <Gerät> capillary column *(Gaschromatographie)*
Trennschalter *m* 1. <Bau> circuit breaker; 2. <Elektriz, Elektrotech> circuit breaker, disconnecting switch, disconnector, isolating switch; air-break disconnector, air-break switch, air breaker, single pole switch *(mit Luftisolierung)*; 3. <Phys> isolator; 4. <Telekom> interruption key, isolator
Trennschalter *m* **mit magnetischer Löschung** <Elektriz> magnetic blowout circuit breaker
Trennschalter *m* **mit Sicherungen** <Elektrotech> disconnector-fuse
Trennschärfe *f* <Comp & DV, Elektronik, Funktech> selectivity
Trennscheibe *f* 1. <Ker & Glas> *(BE)* slitting disc, *(AE)* slitting disk; 2. <Maschinen> cutoff wheel, cutter wheel, cutting wheel, cutting-off wheel
Trennschichtpapier *n* <Papier> release-coated paper
Trennschleifen *n* 1. <Fertig> abrasive friction cutting, abrasive wheel cutting-off; 2. <Ker & Glas> *(BE)* disc grinding, *(AE)* disk grinding; 3. <Maschinen> abrasive cutting, parting-off
Trennschleifmaschine *f* 1. <Fertig> abrasive wheel cutting-off machine; 2. <Ker & Glas> *(BE)* disc grinder, *(AE)* disk grinder; 3. <Maschinen> parting-off grinder
Trennschleifscheibe *f* <Maschinen> parting-off wheel
Trennschleuder *f* 1. <Chemtech> centrifuge; 2. <Ker & Glas> spinner
Trennschneiden *n* <Fertig> splitting
Trennschritt *m* <Telekom> marking *(Doppelstrom)*
Trennschütz *n* <Elektrotech> contactor disconnector
Trennschutzschalter *m* <Elektrotech> isolating switch
Trennstelle *f* 1. <Eisenbahn> *(BE)* sleeper station, *(AE)* tie station; 2. <Elektrotech> break, test point, testing point; sectioning point *(eines Trenners)*
Trennstellung *f* 1. <Elektrotech> disconnected position *(eines Trenners)*; 2. <Telekom> splitting position
Trennstrecke *f* <Elektrotech> air break, circuit sever; isolating distance *(Hochspannung)*
Trennstufe *f* 1. <Elektrotech> buffer; 2. <Telekom> separator *(Radio)*
Trennteil *n* <Elektrotech> withdrawable part
Trenntransformator *m* <Comp & DV, Elektrotech> isolation transformer
Trenntrichter *m* <Chemtech, Labor> separating funnel
Trenntriebwerk *n* <Raumfahrt> separation motor
Trennung *f* 1. <Abfall> separation; 2. <Elektrotech> cutoff, disconnection; 3. <Erdöl> separation; 4. <Fernseh> cutoff; 5. <Kerntech> partition *(von Isotopen)*; 6. <Kohlen> separation; 7. <Maschinen> disconnection, parting; 8. <Telekom> cutoff; interruption *(Telefonist)*; disconnection *(Verbindung)*; 9. <Trans> diverging
Trennung *f* **des Luftstroms** <Lufttrans> airstream separation *(Vergaser)*
Trennung *f* **nach Korngröße** <Chemtech> particle sizing
Trennungsblech *n* <Fertig> baffle sheet
Trennungsmarke *f* <Comp & DV> group mark, group marker
Trennungsschicht *f* <Chemtech> separation layer
Trennungsstrich *m* <Druck> hyphen

Trennverfahrenprozess *m* <Erdöl> extraction process *(Raffinerie, Gasaufbereitung)*
Trennvermögen *n* <Kerntech> separating power
Trennvermögen *n* **eines Anlagenteils** <Kerntech> unit separative power
Trennverstärker *m* 1. <Aufnahme> buffer amplifier; 2. <Elektronik> isolation amplifier
Trennversuch *m* 1. <Kerntech> separative effort; 2. <Kunststoff> peel test
Trennwand *f* 1. <Abfall> slurry wall; 2. <Bau> partition wall, partition; 3. <Fertig> baffle; 4. <Maschinen> bulkhead
Trennweiche *f* 1. <Fernseh> separation filter; 2. <Funktech> separation circuit
Trennwerkzeug *n* <Maschinen> parting tool
Trennwichte *f* <Kohlen> partition density
Trennwiderstand *m* <Fertig> cohesive resistance
Trennzeichen *n* <Comp & DV> separator, slash
Trennzentrifuge *f* <Lebensmittel> separator
Treppe *f* 1. <Bau> stairs; 2. <Wassertrans> ladder *(Schiff)*
Treppenabsatz *m* <Bau> half pace
Treppenboden *m* <Bau> landing
treppenförmige Steinanordnung *f* **im Ofen** <Ker & Glas> corbel
Treppenlauf *m* <Bau> flight of stairs
Treppenlochwange *f* <Bau> face string
Treppenpfosten *m* <Bau> newel
Treppenpodest *n* <Bau> landing
Treppenschacht *m* <Bau> well
Treppenstufe *f* <Bau> step
Treppenwange *f* <Bau> string, stringer
Treppenwicklung *f* <Elektrotech> split winding, split-throw winding, stepped winding
Tresse *f* <Textil> braid
Trester *mpl* 1. <Chemie> residue *(Rückstände bei Obstsäften, Wein)*; 2. <Lebensmittel> pomace *(Rückstände beim Keltern, Bierbrauen)*
Tri *n* <Chemie> trilene
Triac *n* 1. <Elektriz> triac; 2. <Elektronik> tetrode thyristor *(Halbleitertetrode)*; 3. <Elektrotech> triac
Triacetat *n* <Textil> triacetate
Triacetin *n* <Chemie> triacetin
Triacetonamin *n* <Chemie> triacetonamin, triacetonamine
Triade *f* <Chemie> triad
Triamyl... <Chemie> triamyl
Triangulation *f* <Geom, Raumfahrt> triangulation
Triangulationspunkt *m* <Bau> triangulation point
triangulieren *v* <Geom> triangulate
Triangulierung *f* <Geom> triangulation
Triazol *n* <Chemie> triazole
Tri-Band-Handy *n* <Mobilkom> tri-band mobile phone *(GSM 900, 1800, 1900)*
Tribokorrosion *f* <Chemie> tribo-corrosion *(mechanischchemischer Verschleißprozess)*
Tribromacetaldehyd *n* <Chemie> bromal
Tribromethanal *n* <Chemie> bromal
Tribromethan *n* <Chemie> bromoform
Tributylphosphat *n* *(TBP)* <Kerntech> tributyl phosphate *(TBP)*
Tributyrin *n* <Chemie> tributyrin
Tricarballyl... <Chemie> tricarballylic *(Säure)*
Trichloressig... <Chemie> trichloroacetic *(Säure)*
Trichlorethen *n* <Chemie> trichloroethylene
Trichlorethylen *n* <Chemie> trichloroethylene
Trichlorid *n* <Chemie> trichloride
Trichlornitromethan *n* <Chemie> chloropicrin, trichloronitromethane
Trichroismus *m* <Phys> trichroism
trichroitisch *adj* <Phys> trichroic

Trichter m 1. <Bau, Chemie, Elektronik, Fertig, Ker & Glas, Labor, Lebensmittel, Mechan> funnel; 2. <Papier> hopper
Trichterantenne f <Lufttrans> horn (Funkwesen)
trichterförmige Ausweitung f <Fertig> bell
Trichterfüllgerät n <Kunststoff> hopper
Trichterlautsprecher m <Akustik, Aufnahme> horn loudspeaker
trichterloser Lautsprecher m <Akustik> direct loudspeaker, radiator loudspeaker
Trichtermühle f <Kohlen> conical mill
Trichteröffnung f <Akustik> horn mouth
Trichterrohr n <Labor> thistle funnel
Trickeinblendung f <Fernseh> electronic inlay
Trickgenerator m <Fernseh> special effects generator
Trickschieber m <Maschinen> Allan valve, trick valve
Tricosan n <Chemie> tricosane
Tricresol n <Chemie> tricresol
Tricresyl... <Chemie> tricresyl
Tricyansäuretriamid n <Chemie> melamine
tricyclisch adj <Chemie> tricyclic
Trieb m <Lebensmittel> rising power (Teiglockerung)
Triebdrehgestell n <Eisenbahn> (BE) motor bogie, (AE) motor truck
Triebdrehgestell n **mit einem Antriebsmotor** <Eisenbahn> (BE) monomotor bogie, (AE) monomotor truck
Triebfahrzeug n <Eisenbahn> tractive unit
Triebfahrzeugpark m <Eisenbahn> fleet
Triebfahrzeugpersonal n <Kfztech> driving crew
Triebkraft f 1. <Maschinen> propulsive force; 2. <Textil> agency
Triebrad n <Kfztech> pinion (Getriebe)
Triebseite f <Elektrotech> drive end
Triebspule f <Elektrotech> drive coil
Triebstange f <Maschinen> pitman
Triebstock m <Fertig> driving pin wheel, round
Triebstockgetriebe n <Maschinen> lantern gear
Triebstockverzahnung f <Maschinen> lantern gearing
Triebstockzahnrad n <Fertig> trundle wheel
Triebstrang m <Kfztech> transmission; drive train (Kraftübertragung)
Triebwagen n <Eisenbahn> power car
Triebwagenende n <Trans> rear of the railcar
Triebwasser n <Hydraul> headwater
Triebwasserkanal m <Hydraul> headrace
Triebwerk n 1. <Kfztech> power unit, train of gears; 2. <Lufttrans> engine, motor; 3. <Maschinen> engine
Triebwerk n **ohne Nachbrenner** <Lufttrans> dry engine
Triebwerkabstellen n **im Flug** <Lufttrans> engine shutdown in flight
Triebwerkdrehzahl f <Lufttrans> engine speed
Triebwerkleistung f <Lufttrans> cruising power (bei Reisegeschwindigkeit)
Triebwerksabschaltung f <Lufttrans> engine flame-out
Triebwerksanlage f <Lufttrans> power plant
Triebwerksausfall m <Lufttrans> engine flame-out
Triebwerksbefestigungen fpl <Lufttrans> engine mountings
Triebwerksblock m <Kfztech> power unit
Triebwerkschub erhöhen v <Lufttrans> advance throttle
Triebwerksdrehmoment n <Kfztech, Lufttrans> engine torque
Triebwerksdüse f <Lufttrans> engine nozzle cluster
Triebwerksgondel f <Lufttrans> engine nacelle, engine pod, nacelle
Triebwerksgondelstutzen m <Lufttrans> engine nacelle stub
Triebwerkshalterung f <Lufttrans> engine mount
Triebwerkskonus m <Raumfahrt> jet nozzle
Triebwerkskörper m <Lufttrans> engine body
Triebwerksluft-Ansaugstutzen m <Lufttrans> engine air-intake extension
Triebwerksnebenluftstrom m <Lufttrans> engine bypass air
Triebwerksprüfstand m <Lufttrans> engine test stand
Triebwerksrahmen m <Lufttrans> engine mount
Triebwerkstarter m <Lufttrans> crank switch (Anlasser)
Triebwerkstrahlsog m <Lufttrans> engine jet wash
Triebwerksunterbrechung f <Lufttrans> engine flame-out
Triebwerksuntersatz m <Lufttrans> engine stand
Triebwerkswiderlager n **und Schubgerüst** n <Lufttrans> engine mount-and-thrust structure
Triebwerkswinkelsteuerung f <Raumfahrt> engine angle command
Triebwerkverkleidung f <Lufttrans> cowling
Triebwerkwellenlager n <Lufttrans> engine shaft bearing
Trieder n <Geom> trihedron
Trifluormethan n <Chemie> fluoroform
Trifokalglas n <Ker & Glas> trifocal glass (Brille)
Triftraum m <Elektronik> drift space (Klystron)
Triftröhre f <Elektronik> drift tube
Trigger m <Comp & DV, Elektriz> trigger
Triggerdiode f <Elektronik> trigger diode
Triggerimpuls m 1. <Elektronik> trigger pulse; 2. <Fernseh> triggering pulse; 3. <Telekom> triggering lead pulse, triggering pulse
Triggerkreis m <Elektriz> trigger circuit
triggern v <Elektriz, Phys> trigger
Triggerschaltung f <Elektriz, Elektronik, Phys> trigger circuit
Triggerspannung f <Fernseh, Telekom> triggering voltage
Triggersysteme npl <Strahlphys> triggering systems
Triglycerid n <Chemie> triglyceride
Trigonometrie f <Geom> trigonometry
trigonometrisch adj <Geom> trigonometric, trigonometrical
trigonometrische Funktionen fpl <Geom, Math> trigonometrical functions
Trihydrat n <Chemie> trihydrate
Trihydroxyflavon n <Chemie> naringenin, trihydroxyflavanone
Triiodid n <Chemie> triiodide
Triiodmethan n <Chemie> iodoform, triiodomethan
triklin adj <Chemie> triclinic
triklines System n <Metall> triclinic system
Triller m <Akustik> trill
Trimaran m <Wassertrans> trimaran (Schiffskörper)
Trimellith... <Chemie> trimellitic
Trimer n <Chemie, Kunststoff> trimer
Trimerisat n <Chemie, Kunststoff> trimer
trimerisieren v <Chemie> trimerize
Trimesin... <Chemie> trimesic
Trimetallplatte f <Druck> trimetallic plate
Trimethylbenzol n <Chemie> trimethylbenzene
Trimethylen n <Chemie> cyclopropane, trimethylene
trimetrische Projektion f <Konstzeich> trimetric projection
Trimm... <Elektrotech, Kerntech> trimming
Trimm-Abschaltstab m <Kerntech> shim safety rod
Trimmauflösung f <Elektrotech> trimming resolution
Trimmelement n <Kerntech> shim assembly, shim element, shim member
trimmen v <Mechan, Raumfahrt, Wassertrans> trim (Schiff)
Trimmer m 1. <Elektriz, Elektrotech> adjustable capacitor; 2. <Funktech> padder, trimmer

Trimmerkondensator

Trimmerkondensator m 1. <Elektrotech> trimmer capacitor; 2. <Funktech> padder capacitor
Trimmlage f <Wassertrans> trim *(Schiff)*
Trimm-Nachfolge-Element n <Kerntech> automatic control assembly
Trimm-Nachfolgesteuerung f <Kerntech> automatic control assembly
Trimmpotenziometer n <Elektriz, Elektrotech> trimming potentiometer
Trimmruder n <Lufttrans> trimming tab
Trimmruder n **des Luftschraubenblattes** <Lufttrans> blade trim tab *(Hubschrauber)*
Trimmstab m <Kerntech> shim rod
Trimmstab m **mit Feinantrieb** <Kerntech> differential control rod
Trimmstabblock m <Kerntech> shim rod bank
Trimmstabilität f <Lufttrans> trim stability
Trimmsteuerung f <Lufttrans> trim control
Trimmung f 1. <Lufttrans> trimming; 2. <Raumfahrt> trim
trimolekular adj <Chemie> termolecular, trimolecular
trimorph adj <Chemie> trimorphic, trimorphous
Trimorphie f <Chemie> trimorphism *(Kristallchemie)*
Trinatrium... <Chemie> trisodium
Trinitrat n <Chemie> ternitrate, trinitrate
trinitriert adj <Chemie> trinitrated
Trinitrobenzol n <Chemie> trinitrobenzene
Trinitrotoluol n *(TNT)* <Chemie> trinitrotoluene *(TNT)*
trinkbar adj <Wasserversorg> potable
Trinkwasser n <Lebensmittel, Wasserversorg> drinking water
Trinkwasserqualität f <Wasserversorg> drinking water quality
Trinkwasserversorgung f <Wasserversorg> drinking water supply
Trinom n <Math> trinomial
Triode f 1. <Chemie> three-electrode valve, triode; 2. <Elektronik, Phys> triode
Triode-Hexode f <Elektronik> triode-hexode
Trioden-Oszillator m <Elektronik> triode oscillator
Triodenverhalten n <Elektronik> triode action
Triol n <Chemie> trihydric acid, triol
Triose f <Chemie> triose
Triowalzwerk n <Maschinen> three-high mill, three-high rolls, three-high train
Trioxid n <Chemie> teroxide, trioxide
Trioxoborsäure f <Chemie> boric acid, orthoboric acid
Trioxosilicat n <Chemie> bisilicate, metasilicate
Tripalmitin n <Chemie> tripalmitin
Tripel n <Math> triple
Tripelkarton m <Papier> three-layer board
Tripelpunkt m <Metall, Phys, Thermod> triple point
Triphenylmethan n <Chemie> triphenylmethane
Triphenylmethyl n <Chemie> trityl
Triplett n <Kerntech, Phys> triplet *(Spektroskopie)*
Triplettlinse f <Foto> triplet lens
Triptan n <Chemie> neohexane, trimethylbutane, triptane
Trisecschiff n <Trans> TRISEC ship, planing-hull-type ship
Trisilikat n <Chemie> trisilicate
Tristate-Ausgang m <Elektronik> three-state output
Tristearin n <Chemie> glyceryl tristeate, tristearin
trisubstituiert adj <Chemie> trisubstituted
Tritan n <Chemie> triphenylmethane
Trithion... <Chemie> trithionic
Tritium n *(T)* <Chemie> tritium *(T)*
Tritiumtrennung f <Kerntech> tritium extraction *(aus schwerem Wasser)*
Triton n *(t)* <Teilphys> triton *(t)*
Trittbrett n <Eisenbahn> step

Tritteisen n <Eisenbahn> foot iron
Trittplattenbremsventil n <Kfztech> treadle brake valve
Trittschall m 1. <Akustik> impact sound; 2. <Sicherheit> impact noise
Trittschalldämmung f impact sound insulation
Trittschallpegel m <Sicherheit> impact sound level
Trittschall-reduzierendes Material n <Akustik> impact sound-reducing material
Trittschallschutz m <Sicherheit> impact noise protection
Trittschall-Übertragungspegel m <Akustik, Sicherheit> impact sound transmission level
Trittstufe f <Bau> tread
trivalent adj <Chemie> ternary, triple, trivalent
Trivalenz f <Chemie> tervalence, trivalence, trivalency
Trochoide f <Geom> trochoid
Trochoiden-Massenspektrometer n <Kerntech> trochoidal mass spectrometer
trocken adj <Anstrich, Papier, Thermod> dry
trocken aufbewahren v <Verpack> keep dry
Trocken... <Anstrich, Druck, Thermod> dry
Trockenabteil n <Lebensmittel> drying section
Trockenakku m <Elektriz> dry-storage battery
Trockenakkumulator m <Elektriz> dry-storage battery
Trockenanschluss m <Elektrotech> dry connection
Trockenanschlussklemme f <Elektrotech> dry connector
Trockenapparat m <Verpack> drying machine
Trockenaufziehen n <Foto> dry mount
Trockenaufziehpresse f <Foto> dry mounting press
Trockenausschuss m <Papier> dry broke
Trockenbatterie f 1. <Elektrotech> dry battery; 2. <Phys> dry cell
Trockenbatterie f **für Glocke** <Elektriz> bell battery
Trockenbett n <Abfall> drying bed
Trockenbohrer m <Erdöl> claying bar *(Bohrtechnik)*
Trockenbohrverfahren n <Bau> dry-process boring
Trockendampf m <Heiz & Kälte, Nichtfoss Energ> dry steam
Trockendampfanteil m <Heiz & Kälte> dryness fraction of steam
Trockendock n 1. <Bau> dry dock; 2. <Wassertrans> dry dock, graving dock • **ins Trockendock gehen** <Wassertrans> drydock *(Schiff)*
trockene Abscheidung f <Abfall> dry gas cleaning
trockene Deposition f <Umweltschmutz> dry deposition
Trockenei n <Lebensmittel> powdered egg
Trockeneis n 1. <Heiz & Kälte> carbon dioxide snow, dry ice; 2. <Lebensmittel, Thermod> dry ice
Trockenelektroabscheidung f <Umweltschmutz> dry precipitation
Trockenelektrolyt n <Elektrotech> dry electrolyte
Trockenelement n <Elektrotech> dry cell
Trockenentschwefelungsprozess m <Umweltschmutz> *(AE)* dry desulfurization process, *(BE)* dry desulphurization process
trockener adiabatischer Temperaturgradient m <Umweltschmutz> dry adiabatic lapse rate *(der Atmosphäre)*
trockener Bohrschlamm m <Erdöl> dry mud *(Bohrtechnik)*
trockenes Ammoniak n <Erdöl> anhydrous ammonia
trockenes Erdgas n <Erdöl> dry natural gas, nonassociated gas
trockenes Eruptionskreuz n <Erdöl> dry tree *(Bohrlochkopfabsperrsystem)*
trockenes Präzisionsschleifen n <Maschinen> dry precision grinding
Trockenfäule f <Bau, Wassertrans> dry rot
Trockenfilter n <Maschinen> dry filter

Trockenfilz m 1. <Druck> drying felt; 2. <Papier> dry felt
Trockenfläche f <Lebensmittel> drying area
Trockenflasche f <Kfztech> receiver-dryer *(Klimaanlage)*
Trockenfrachter m <Wassertrans> dry-bulk carrier
Trockengehalt m <Papier> dry content
Trockengestell n 1. <Foto> drying frame; 2. <Labor> draining rack
Trockengewicht n 1. <Lufttrans> dry weight *(Triebwerk und Motor)*; 2. <Raumfahrt, Verpack> dry weight
Trockenhaus n <Kohlen> drying kiln
Trockenheit f <Kohlen, Textil> dryness
Trockenhitze f <Textil> dry heat
trockenhitzefixieren v <Textil> dry-heat-set
Trockenkammer f 1. <Kunststoff, Maschinen> drying cabinet; 2. <Thermod> drying chamber
Trockenklebefolie f <Foto> dry mounting tissue
Trockenkruste f <Kohlen> dry crust
Trockenkupplung f <Kfztech, Maschinen> dry clutch
Trockenlager n <Maschinen> dry bearing
Trockenlauf m <Fertig> dry run
Trockenlaufen n <Maschinen> dry running
Trockenlaufkompressor m 1. <Heiz & Kälte> oil-free compressor; 2. <Maschinen> dry-running compressor
Trockenlaufzeit f <Kunststoff> dry-cycle time
trockenlegen v <Wasserversorg> drain
Trockenlegen n 1. <Bau> dewatering *(Tiefbau)*; 2. <Wasserversorg> drainage
Trockenlöscher m <Sicherheit> dry chemical extinguisher, dry power extinguisher
Trockenmaschine f <Textil> drier, dryer
Trockenmasse f 1. <Abfall> dry matter; 2. <Raumfahrt> dry mass
Trockenmatte f <Druck> dry mat
Trockenmauer f <Bau> dry wall, dry-stone wall
Trockenmedium n <Chemie> desiccant
Trockenmilch f <Lebensmittel> dried milk, milk powder
Trockenmittel n 1. <Chemtech> dehydrating agent, dehydrator, desiccative; 2. <Kerntech> dehumidifier; 3. <Kunststoff> drier, dryer, drying agent; 4. <Lebensmittel> siccative; 5. <Thermod> desiccant, drying agent, siccative; 6. <Verpack> desiccant, drying agent
Trockenmittelbeutel m <Verpack> desiccant bag
Trockenmühle f <Kohlen> dryer mill
Trockennetztransformator m <Elektrotech> dry-type power transformer
Trockenofen m 1. <Fertig> drying stove *(Formen)*; 2. <Kohlen> drying oven; 3. <Labor> vacuum oven; 4. <Lebensmittel> drying kiln, drying oven; 5. <Maschinen> drying furnace; 6. <Papier> drying oven; 7. <Textil> baking stove, drying oven, drying stove, oven; 8. <Verpack> drying oven
Trockenoffset n <Druck> dry offset, letterset
Trockenoffsetdruck m <Verpack> dry offset printing
Trockenöl n <Kunststoff> drying oil
Trockenpartie f <Papier> dryer section
Trockenpresse f <Foto> print dryer
Trockenpulverfeuerlöscher m <Sicherheit> dry-powder fire extinguisher
Trockenraum m <Kunststoff> drying cabinet
Trockenreibung f <Maschinen> dry friction
Trockenrohdichte f <Bau, Kohlen> dry density
Trockensäule f <Labor> drying column
trocken-saurer Niederschlag m <Umweltschutz> dry acidic fallout
Trockenschale f <Fertig> core plate *(Gießen)*
Trockenschaltkreis m <Elektrotech> dry circuit
Trockenschlammdeponie f <Umweltschutz> dry-sludge disposal site
trockenschleifen v <Ker & Glas> polish till dry

Trockenschleuder f <Chemtech, Fertig, Kohlen> centrifugal drier, centrifugal dryer
Trockenschrank m 1. <Kunststoff> drying cabinet; 2. <Textil> oven
Trockenschrank m mit gleich bleibender Temperatur <Labor> constant-temperature oven
Trockenschrumpfung f <Bau> drying shrinkage
Trockenschwund m <Bau> drying shrinkage
Trockenstaub-Beseitigungsanlage f <Sicherheit> dry-dust removal installation
Trockenstoff m <Kunststoff> exsiccant
Trockensubstanz f <Abfall> dry matter
Trockensumpf m <Kfztech> dry sump *(Motor)*
Trockensumpfschmierung f <Kfztech> dry-sump lubrication
Trockentemperatur f <Heiz & Kälte> dry-bulb temperature
Trockenthermometer n <Heiz & Kälte, Thermod> dry-bulb thermometer
Trockentransformator m <Elektriz> dry-type transformer
Trockentrennung f <Abfall> dry sorting *(von Müll)*
Trockentrommel f <Bau> drying drum
Trockentunnel m <Lebensmittel, Verpack> drying tunnel
Trockenturm m <Lebensmittel> drying tower
Trocken-Überschlagspannung f <Elektriz> dry flash-over voltage
Trockenverbinder m <Elektrotech> dry connector
Trockenvermahlung f <Kohlen> dry crushing
Trockenwand f <Bau> dry-stone wall
Trockenzelle f 1. <Elektriz> dry cell *(Batterie)*; 2. <Elektrotech, Phys> dry cell
Trockenzentrifuge f <Chemtech> centrifugal drier, centrifugal dryer
Trockenziehen n <Fertig> dry drawing
Trockenzylinder m 1. <Papier> dryer cylinder; 2. <Textil> drying cylinder
Trockenzylinder m der Selbstabnahmepapiermaschine <Papier> yankee cylinder
Trockne f <Chemie> dryness • **zur Trockne eindampfen** <Chemtech> evaporate to dryness
trocknen v 1. <Anstrich> dry; 2. <Bau> bake; 3. <Chemie, Heiz & Kälte> desiccate; 4. <Kohlen> dry; 5. <Lebensmittel> desiccate; 6. <Papier> fire *(Tee)*; 7. <Thermod> desiccate, dry
Trocknen n 1. <Bau> seasoning *(Holz)*; 2. <Chemie, Erdöl> desiccation; 3. <Kohlen, Kunststoff, Papier, Textil> drying; 4. <Thermod> desiccation, drying; 5. <Verpack> desiccation
trocknend adj <Thermod> siccative
trocknendes Öl n <Lebensmittel> drying oil
Trockner m 1. <Heiz & Kälte> desiccator; 2. <Kfztech> receiver-dehydrator *(Flüssigkeitsbehälter mit Filtertrockner im Wärmekreislauf einer Klimaanlage)*; 3. <Kohlen, Maschinen, Textil, Thermod> drier, dryer; 4. <Verpack> drying machine
Trocknung f 1. <Chemtech> desiccation; 2. <Erdöl> dehydration, drying; desiccation *(Bohrtechnik)*; 3. <Kunststoff> drying; 4. <Lebensmittel> desiccation; 5. <Textil> drying; 6. <Thermod, Verpack> desiccation
Trocknungsgrad m <Bau> degree of drying
Trocknungskammer f <Thermod> drying chamber
Trocknungsleistung f <Thermod> rate of drying *(eines Trockners)*
Trocknungsmesser m <Papier> drying meter
Trocknungsmittel n <Heiz & Kälte, Lebensmittel> desiccant
Trog m 1. <Ker & Glas> tray; 2. <Labor> tray, trough; 3. <Maschinen> trough
Trogbandförderer m <Maschinen> trough conveyor
Trogbrücke f <Bau> trough bridge

Troggurtförderer

Troggurtförderer m <Maschinen> trough conveyor
Trogmischer m <Ker & Glas> trough mixer
Trogstange f <Ker & Glas> tray bar
Trommel f 1. <Fertig> roll; carrier *(Automat)*; 2. <Funktech> drum *(Fax)*; 3. <Hydraul, Kohlen> drum; 4. <Maschinen> barrel, pulley; 5. <Mechan, Meerschmutz, Papier> drum; 6. <Telekom, Verpack> drum
Trommelabschöpfgerät n <Meerschmutz> drum skimmer
Trommelabtaster m <Fernseh> drum scanner
Trommelanker m <Elektriz, Elektrotech> drum armature
Trommelanlasser m <Elektriz> drum starter
Trommelantrieb m <Maschinen> drum drive
Trommelblattdiagramm n <Gerät> drum chart diagram
Trommelbohrmaschine f <Maschinen> drum-type drilling machine
Trommelbremse f <Kfztech> drum brake
Trommeldrucker m <Comp & DV> *(BE)* barrel printer, *(AE)* drum printer
Trommelfilter n <Kohlen> drum filter
Trommelfräsmaschine f <Maschinen> drum milling machine
Trommelgeschwindigkeit f <Funktech> drum speed *(Fax)*
Trommelhöhenmesser m <Lufttrans> drum altimeter
Trommelkessel m <Hydraul> cylinder boiler
Trommelkiln m <Thermod> drum kiln
Trommelkurve f <Fertig> drum cam
Trommelmischer m 1. <Bau, Kohlen> drum mixer; 2. <Lebensmittel> barrel mixer
Trommeln n <Kunststoff> tumbling
Trommelplotter m <Comp & DV> drum plotter
Trommelrevolver m 1. <Fertig> horizontal axis turret; 2. <Maschinen> drum turret
Trommelschalter m <Elektriz> drum controller, drum switch
Trommelscheider m <Kohlen> trommel washer
Trommelschütze f <Nichtfoss Energ> drum gate
Trommelsieb n <Kohlen> drum separator, trommel
Trommelspeicher m 1. <Comp & DV> drum; 2. <Telekom> drum store
Trommeltrocknen n <Kerntech> in-drum drying
Trommeltrockner m <Maschinen, Thermod> drum drier, drum dryer
Trommelwascher m <Bau> drum washer
Trommelwelle f <Maschinen> drum shaft
Trommelwicklung f <Elektrotech> drum winding
Trommelwinde f <Maschinen> drum winch
Trompete f <Wassertrans> sheepshank *(Knoten)*
trompetenförmige Ausweitung f <Maschinen> bell mouth
Troostit m <Metall> troostite
Tropenverpackung f <Verpack> tropical packaging
Tropf m <Ker & Glas, Labor> drip
Tropfbecher m <Bau> drip cup
Tropfblech n <Maschinen> drip plate
Tröpfchen n <Umweltschutz> droplet
tropfen v <Papier> drop
Tropfen m 1. <Ker & Glas> drop, tear; 2. <Lebensmittel, Papier> drop
Tropfenabscheider m <Abfall> demister unit, mist eliminator
Tropfenbewässerung f <Fertig> drip irrigation *(Kunststoffinstallationen)*
Tropfenbildung f <Strömphys> drop formation
Tropfenwasserzeichen n <Papier> drop watermark
Tropfenzähler m 1. <Labor> dropper, dropping bottle; 2. <Phys> stalagmometer; 3. <Verpack> drop counter

Tropfenzählröhrchen n <Labor> dropper tube, dropping tube
Tropfflasche f <Labor> dropping bottle
Tropfkörper m 1. <Abfall> percolating filter, sprinkling filter, trickling filter; 2. <Umweltschmutz> trickling filter
Tropfkörperanlage f <Chemtech> biological filter *(Abwässer)*
Tropföler m 1. <Fertig> drip oiler; 2. <Maschinen> drip-feed lubricator
Tropfölschmierung f <Fertig, Maschinen> drip-feed lubrication
Tropfpunkt m <Fertig> drop point
Tropfpunktapparat m <Labor> drop-point apparatus *(Schmierfette)*
Tropfschale f <Heiz & Kälte> drip tray
Tropftrichter m <Labor> tap funnel
Tropfwasser n <Heiz & Kälte> drip water
Tropfwasserbildung f <Heiz & Kälte> dripping moisture
tropfwassergeschützt adj <Maschinen> drip-proof
Tropfzylinder m <Bau> drip cup
Tropin n <Chemie> tropine
tropischer Wirbelsturm m <Wassertrans> tropical revolving storm
Troposcatter-Richtfunksystem n <Funktech> tropospheric scatter radio relay system
Troposphäre f <Raumfahrt> troposphere
troposphärisch adj <Funktech> tropospheric
troposphärische Streuung f <Telekom> tropospheric scatter
Trosse f 1. <Maschinen> hawser; 2. <Wassertrans> cable, cluster; hawser *(schweres Tau oder Drahtseil)* • **Trosse nachschleppen** <Wassertrans> stream a warp *(Festmachen)*
Trotyl n <Chemie> trotyl
Troy-Gewicht n <Metrol> troy weight
Trub m <Lebensmittel> sludge *(bei Wein oder Bier)*
trüb adj <Textil> hazy
trübe Atmosphäre f <Nichtfoss Energ> turbid atmosphere
Trübglas n <Ker & Glas> opalescent glass
Trübheit f 1. <Nichtfoss Energ> turbidity; 2. <Textil> cloudiness
Trübstoffe mpl <Chemie> turbidity *(Wasserbehandlung)*
Trübung f 1. <Kunststoff> haze, turbidity; 2. <Nichtfoss Energ> turbidity; 3. <Textil> cloudiness; 4. <Wassertrans> clutter *(Radar)*
Trübungsanalyse f <Chemie> turbidimetry
Trübungskoeffizient m <Nichtfoss Energ> turbidity coefficient
Trübungsmesser m 1. <Gerät> haze meter; 2. <Labor> nephelometer, turbidity meter
Trübungsmittel n 1. <Ker & Glas> opacifier; 2. <Lebensmittel> cloudifier
Trübungspunkt m <Heiz & Kälte> cloud point
Truck-to-Truck-System n <Wassertrans> *(AE)* truck-to-truck system
Trudelflugerprobung f <Lufttrans> spin flight testing
Trudeln n <Lufttrans> spin
True-Motion-Radar n <Funkort> true motion radar
Truffel f <Fertig> sprue cutter; trowel *(Gießen)*
Trum n <Maschinen> end
Trümmer npl <Bau> ruins
Trümmerbeton m <Bau> broken brick concrete
Trümmergestein n <Bau, Kohlen> fragmented rocks
Truxill... <Chemie> truxillic
Truxillin n <Chemie> truxilline
tryptisch adj <Chemie> tryptic
Trysegel n <Wassertrans> trysail *(Segeln)*
TS *(Treibsitz)* <Maschinen> drive fit

T-Schaltung f <Phys> T-network
Tschebyscheff'sches Filter n <Phys> Chebyshev filter
Tschebyscheff'sches Filter n **achter Ordnung** <Elektronik> eighth-order Chebyshev filter *(polfrei)*
Tscherenkow'sche Strahlung f <Strahlphys, Teilphys> Cerenkov radiation
Tscherenkow'scher Detektor m <Strahlphys, Teilphys> Cerenkov detector
Tscherenkow'scher Effekt m <Strahlphys, Teilphys> Cerenkov effect
Tscherenkow'scher Zähler m <Strahlphys, Teilphys> Cerenkov counter
T-Spleiß n <Elektriz> T-joint
T-Stoß m <Bau> T-joint *(Sanitär)*
T-Stück n 1. <Bau> union-T; 2. <Fertig> branch tee; 3. <Maschinen> T, T-piece
TT *(Tonträger)* <Aufnahme, Fernseh> sound carrier
TTL *(Transistor-Transistor-Logik)* <Comp & DV, Elektronik> TTL *(transistor-transistor logic)*
TTL-Messung f *(Innenmessung)* <Foto> TTL metering *(through-the-lens metering)*
TT-Logik f <Comp & DV, Elektronik> transistor-transistor logic
T-Träger m <Bau> T-beam
Tubenfüll- und -ausspritzmaschine f <Verpack> tube filling and cleaning machine
Tubenfüll- und Schließmaschine f <Fertig> tube filling and closing machine
Tubenverschließmaschine f <Verpack> tube-closing machine
Tubing n <Erdöl> production tubing, tubing
Tuch n <Textil> cloth, fabric, woven fabric
Tuchfilter n <Umweltschmutz> fabric filter
Tuchpolierer m <Ker & Glas> cloth polisher
Tuchscheren n <Textil> cropping
Tuff m <Bau> tuff
Tuffgestein n <Bau> tuff
Tufting-Teppich m <Textil> tufted carpet
Tülle f <Maschinen> grommet
Tumblerschalter m <Elektrotech> tumbler switch
Tünche f <Bau> whitewash
tünchen v <Bau> whitewash
Tuner m <Fernseh, Funktech> tuner *(Frequenzabstimmvorrichtung)*
Tunnel m 1. <Bau> tunnel; 2. <Eisenbahn> *(AE)* subway, *(BE)* underground; 3. <Kohlen, Nichtfoss Energ> tunnel
Tunnelanguss m <Kunststoff> tunnel gate
Tunnelauskleidung f <Bau> tunnel lining
Tunnelbau m <Bau, Kohlen> *(AE)* tunneling, *(BE)* tunnelling
Tunnelbaumaschine f <Bau, Kohlen> *(AE)* tunneling machine, *(BE)* tunnelling machine
Tunnelbauverfahren n <Bau, Kohlen> *(AE)* tunneling technique, *(BE)* tunnelling technique
Tunnelbohrmaschine f <Bau, Kohlen> tunnel-boring machine
Tunneldiode f <Elektronik, Phys> Esaki diode, tunnel diode
Tunneldiodenverstärker m <Raumfahrt> tunnel diode amplifier *(Weltraumfunk)*
Tunneleffekt m <Elektronik, Phys> tunnel effect
Tunnelkühlofen m <Ker & Glas> tunnel lehr
Tunnelmikroskop n <Elektronik> scanning tunnel microscope, STM
Tunnelmode f <Optik> *(AE)* tunneling mode, *(BE)* tunnelling mode
Tunnelofen m <Metall> tunnel furnace
Tunnelstoß m <Bau, Kohlen> face

Tunnelstrahl m <Optik> *(AE)* tunneling ray, *(BE)* tunnelling ray
Tunnelung f <Elektronik> *(AE)* tunneling, *(BE)* tunnelling *(Halbleiter)*
Tunnelverfahren n **für weichen Boden** <Bau, Kohlen> soft ground tunneling
Tüpfelplatte f <Labor> spotting plate
Tupfen m <Textil> dot
Tür f <Bau, Kfztech> door *(Karosserie)*
Türangel f <Bau> garnet hinge, hinge
Türband n <Bau> door hinge
Turbidimetrie f <Chemie> turbidimetry
Turbine f <Elektriz, Kfztech, Maschinen, Mechan, Nichtfoss Energ, Wassertrans> turbine
Turbine f **mit innerer Beaufschlagung** <Hydraul> outward-flow turbine
Turbinenabschlussventil n <Kerntech> turbine stop valve
Turbinenantrieb m <Kfztech, Wassertrans> turbine propulsion
Turbinenblatt n <Hydraul> blade
Turbinenbohranlage f <Erdöl> turbodrill *(Bohrtechnik)*
Turbinenbohren n <Erdöl> turbine drilling *(Bohrtechnik)*
Turbinendeckband n <Lufttrans> shroud ring *(Turbinentriebwerk)*
Turbinendurchflussmesser m <Gerät> turbine flow meter
Turbinenfundament n <Kfztech, Wassertrans> turbine seating
Turbinengebäude n <Kerntech> turbine building
Turbinengehäuse n <Kfztech, Wassertrans> turbine casing
Turbinengrube f <Hydraul> turbine pit
Turbinenhaus n <Kerntech> turbine building, turbine house
Turbinenkammer f <Hydraul> turbine chamber
Turbinenkammer f **des geschlossenen Typs** <Hydraul> turbine chamber of the closed system
Turbinenkraftstoff m <Maschinen, Wassertrans> turbine fuel
Turbinenlagerung f <Kfztech, Wassertrans> turbine seating
Turbinenläufer mit Schaufeln ausrüsten v <Hydraul> provide a turbine wheel with vanes *(Montage)*
Turbinenlaufschaufel f <Fertig> turbine blade
Turbinenleistung f <Nichtfoss Energ> turbine output
Turbinenleistungsvermögen n <Nichtfoss Energ> turbine efficiency
Turbinenluftstrahltriebwerk n <Lufttrans> jet turbine engine
Turbinenmotor m <Kfztech, Wassertrans> turbine engine
Turbinenpumpe f 1. <Hydraul, Maschinen> turbine pump; 2. <Nichtfoss Energ> pump turbine; 3. <Wasserversorg> turbopump
Turbinenrad n 1. <Hydraul> turbine wheel *(Dampf-, Wasserturbine)*; 2. <Kfztech> turbine wheel; 3. <Maschinen> runner, turbine wheel; 4. <Wassertrans> turbine wheel
Turbinenschaufel f 1. <Fertig> turbine blade; 2. <Hydraul> blade; 3. <Kfztech> turbine blade; 4. <Maschinen> turbine blade, turbine vane; 5. <Nichtfoss Energ, Wassertrans> turbine blade
Turbinenstaustrahltriebwerk n <Lufttrans> turboramjet
Turbinenstufe f <Maschinen> turbine stage
Turbinentreibstoff m <Lufttrans> jet engine fuel
Turbinenzähler m <Gerät, Nichtfoss Energ> turbine flow meter, vane meter
Turbinenzug m <Eisenbahn> turbotrain
Türblatt n <Bau> leaf

Türblattquerholz

Türblattquerholz n <Bau> middle rail
Türblocken n <Eisenbahn> door blocking
Turbo... <Elektriz, Lufttrans, Phys> turbo
Turboabscheidung f <Lebensmittel> turboseparation
Turbo-Alternatorsatz m <Elektriz> turbo-alternator
Turbobohranlage f <Erdöl> turbodrill
Turbobohren n <Erdöl> turbine drilling
turboelektrischer Triebwagen m <Eisenbahn> turbo-electric motor coach
Turbogebläse n <Maschinen> turboblower
Turbogenerator m <Elektrotech, Phys> turbogenerator
Turbojet-Flugzeug n <Lufttrans> turbojet
Turbokammzug m <Textil> turbotop
Turbokompressor m <Kfztech, Wassertrans> turbocompressor
Turbokreuzer m <Wassertrans> turbocruiser
Turbolader m 1. <Kfztech> turbocharger, turbocompressor; 2. <Maschinen> turbocharger; 3. <Wassertrans> turbocharger, turbocompressor
Turboluftstrahltriebwerk n <Thermod> turbojet engine
Turbomolekularpumpe f 1. <Maschinen> turbo-molecular pump; 2. <Phys> molecular pump
Turbomotor m 1. <Kfztech> turbine engine; 2. <Thermod> turboshaft engine; 3. <Wassertrans> turbine engine
Turbopropflugzeug n <Lufttrans> turboprop
Turboproptriebwerk n <Lufttrans> propjet engine
Turbopumpe f <Maschinen, Raumfahrt, Wasserversorg> turbopump
Turbostapler m <Textil> turbostapler
Turbostrahltriebwerk n (TL-Triebwerk) <Lufttrans> turbojet
Turboverdichter m 1. <Kfztech> centrifugal supercharger; 2. <Maschinen> turbocompressor
Turbo-Wechselstromgenerator m <Elektrotech> turbo-alternator
turbulent adj <Strömphys> turbulent
turbulente Ablösung f <Strömphys> turbulent separation
turbulente Diffusion f <Kerntech> turbulent diffusion
turbulente Energie f <Strömphys> turbulent energy
turbulente Grenzschicht f <Strömphys> turbulent boundary layer
turbulente Nachströmung f <Phys> wake
turbulente Strömung f 1. <Fertig> sinuous flow; 2. <Heiz & Kälte> eddy flow; 3. <Phys, Strömphys> turbulent flow
turbulenter Fleck m <Strömphys> turbulent spot
turbulentes Wiederanlegen n <Strömphys> turbulent re-attachment
Turbulenz f <Kfztech, Nichtfoss Energ, Raumfahrt, Strömphys> turbulence
Turbulenzballen m <Strömphys> turbulent plug
turbulenzerzeugendes Gitter n <Strömphys> turbulence-generating grid
Türdrücker m <Bau> door opener
Türfalle f <Kfztech> door catch
Türflügel m <Bau> leaf, wing
Türfüllung f <Bau> door panel, panel
Türfutter n <Bau> door case, jamb lining
Türgitter n <Eisenbahn> deadlight
Turgordruck m <Lebensmittel> turgor (Pflanzenwellen: Gewebespannung)
Türgriff m <Bau, Kfztech> door handle (Tür)
Turingmaschine f <Comp & DV> Turing machine
Turingtest m <Comp & DV> Turing test
Türkantenschoner m <Bau> edge plate
Türklinke f <Bau> door handle
Türknauf m <Bau> door knob
Türkontakt m <Elektrotech> gate contact
Türlüftungsöffner m <Sicherheit> ventilation door opener

Turm m 1. <Bau> tower; 2. <Wassertrans> conning tower (U-Boot)
Turmalin n <Phys> tourmaline
Turmansatz m <Bau> stump
Turmbereich m <Ker & Glas> tower section
Turmdrehkran m 1. <Bau> tower crane; 2. <Maschinen> tower slewing crane
Turmgerüst n <Bau> tower
Turmkran m <Bau, Maschinen> tower crane
Turmpfeiler m <Bau> tower pier
Turmrohr n <Nichtfoss Energ> tower tube
Turmseilrollenblock m <Erdöl> crown block (Hebetechnik)
Turmsystem n <Abfall> tower system (Kompostierungsverfahren)
Turmtür f <Ker & Glas> tower door
Turmverbindung f <Bau> derrick girt
Turmwäscher m <Chemtech> washer
Türöffnung f <Bau> doorway
Türpfosten m <Bau> door post, jamb
Türquerriegel m <Bau> lock rail (in Schlosshöhe)
Türrahmen m <Bau> door frame
Türriegel m <Bau> bolt, latch, door bolt
Türsäule f <Kfztech> door pillar
Türscharnier n <Kfztech> door hinge
Türschließmechanismus m <Kfztech> door-locking mechanism
Türschloss n <Kfztech> door lock
Türschwelle f <Bau> doorsill, sill, threshold
Türsprechanlage f <Telekom> door interphone, gate station, interphone
Türstange f <Bau> door bar
Türstock m <Ker & Glas> goal post
Türstoßstange f <Bau> push bar
Türsturz m <Bau> browpiece, lintel
Türverkleidung f <Bau> door panel
Türverriegelung f <Kfztech> door locking
Türzarge f <Bau> (AE) buck, door case, door casing, door frame
tuschieren v <Fertig> mark
Tuschieren n <Fertig> marking
Tüten verpacken v/in <Lebensmittel, Verpack> bag
Tütennähmaschine f <Verpack> bag-stitching machine
TV (Fernsehen) <Fernseh> TV (television)
T-Verbindung f 1. <Bau> T-piece union; 2. <Elektriz> T-joint; 3. <Maschinen> T-joint, T union, tee union
T-Verbindungsstück n <Labor> T-piece, T-piece connector
T-Verschraubung f <Bau> T-piece union
T-Verzweigung f <Telekom> tee-junction, T-junction
TVI (Fernsehempfangsstörung, Fernsehstörung) <Fernseh> TVI (television interference)
TV-Kabel n <Fernseh> television cable
TVSt (Teilnehmervermittlungsstelle) <Telekom> access exchange
TW (Tonwahl) <Telekom> MFD (multifrequency dialling)
Twill m <Textil> twill
Twin-Empfänger m <Fernseh> satellite receiver, two-channel receiver
Twisted-Pair n <Elektrotech> twisted pair
Twisted-Pair Flachkabel n <Elektrotech> twisted pair flat cable
Tyndall-Effekt m <Phys> Tyndall effect
Typ n adj <Elektronik> n-type
Typanweisung f für ganzzahlige Daten <Comp & DV> integer type (Fortran)
Type f <Comp & DV> type
Typenabnahme f <Raumfahrt> type approval
Typenfreigabe f <Raumfahrt> type approval

Typenhöhe f <Druck> height of type, height of typeface
Typenkopf m <Comp & DV> print head
Typenmuster n <Qual> type sample
Typenmusteruntersuchungs- und Prüfbericht m <Qual> type sample and inspection report
Typenprüfung f 1. <Lufttrans> type test *(Turbomotor)*; 2. <Raumfahrt> type test
Typenrad n <Comp & DV, Druck> daisywheel, print wheel
Typenraddrucker m 1. <Comp & DV> *(BE)* barrel printer, daisywheel printer, *(AE)* drum printer; 2. <Druck> daisywheel printer
Typenreihe f <Maschinen> series
Typenschild n <Mechan> nameplate
Typenwalzendrucker m <Comp & DV> *(BE)* barrel printer, *(AE)* drum printer
typisierte Variable f <Künstl Int> tagged variable
Typographie f <Druck> typography
typographischer Punkt m <Druck> typographic point
Typometer m <Druck> *(AE)* line gage, *(BE)* line gauge, type scale
Typprüfbericht m <Qual> type test report
Typprüfmenge f <Qual> type test quantity
Typprüfmuster n <Qual> type test sample
Typprüfung f <Qual> type test
Typprüfungen fpl <Qual> type verifications and tests
Typprüfungsprotokoll n <Qual> type test report
Tyrosin n <Chemie> tyrosine

U

U 1. <Akustik> *(Volumengeschwindigkeit, Volumenstrom)* U *(volume current)*; 2. <Chemie> *(Uran)* U *(uranium)*; 3. <Elektriz> *(Spannung)* V *(voltage)*; 4. <Optik> *(Strahlungsenergie)* U *(radiant energy)*; 5. <Thermod> *(Wärmeübertragungskoeffizient)* U *(overall heat transfer coefficient)*
û *(wahrscheinlichste Geschwindigkeit)* <Phys> û *(most probable speed)*
U-Ablauf m <Fernseh> U-wrap
U-Bahn f <Eisenbahn> *(BE)* tube, underground, *(AE)* subway
U-Bahn-Fahrzeug n <Eisenbahn> tube vehicle
U-Bahn-Verkehr m <Eisenbahn> tube transportation
U-Bahn-Zug m <Eisenbahn> tube train
Überabtastung f 1. <Fernseh> overscan; 2. <Telekom> super-Nyquist sampling
überabzählbar <Math> noncountable *(Mengenlehre)*
überbeanspruchen v <Bau, Maschinen, Metall> overstress
Überbeanspruchung f <Bau, Maschinen, Metall> overstress
überbelasteter Ton m <Bau> overconsolidated clay
Überbelastung f 1. <Aufnahme, Elektrotech> overload; 2. <Fertig> overtensioning; 3. <Metall> overstressing; 4. <Trans> overloading *(Straße)*
Überbelegung f 1. <Telekom> congestion; 2. <Verpack> overload
überbelichten v <Foto> overexpose
überbelichteter Film m <Foto> overexposed film
überbelichtetes Bild n <Foto> overexposed picture
Überbelichtung f <Foto> overexposure
Überblattung f <Bau> overleap joint
Überblendbild n <Fernseh> cut slide

überblenden v <Comp & DV> cross-fade
Überblendregler m <Aufnahme, Elektronik> fader *(Film)*
Überblendung f <Akustik> dissolve
überbrücken v 1. <Bau, Eisenbahn> bridge; 2. <Eisenbahn> bridge over; 3. <Phys> bypass, shunt
überbrücktes H-Netzwerk n <Elektrotech> bridged-H network *(vierpolig)*
überbrücktes T-Netzwerk n <Elektrotech> bridged-T network *(vierpolig)*
Überbrückung f 1. <Bau> bridging *(von Rissen)*; 2. <Elektriz> butt contact, bypassing; 3. <Funktech, Gerät, Kfztech, Phys> bypassing
Überbrückungsdraht m <Elektrotech> jumper
Überbrückungskondensator m <Gerät, Phys> bypass capacitor, shunt capacitor
Überbrückungskontakt m <Elektrotech> bridging contact
Überbrückungsschalter m <Elektriz> bypass switch
überchlorsaures Salz n <Chemie> perchlorate
überdacht adj <Bau, Sicherheit> sheltered
überdachte Anlage f <Heiz & Kälte> sheltered installation
Überdachung f <Bau> roofing
überdecken v 1. <Anstrich> lap; 2. <Bau> overcoat; lap *(Ziegel)*
überdeckter Abzugsgraben m <Abfall> covered drain
Überdeckung f 1. <Bau> overlap, shelter; 2. <Gerät> coverage; 3. <Maschinen> overlap, profile overlap; 4. <Math> coverage *(Konfidenzmenge)*
Überdeckung f durch Geräusch <Telekom> noise masking
Überdeckungsfaktor m <Maschinen> contact ratio *(von Verzahnungen)*
Überdeckungsgrad m <Fertig> engagement factor *(Getriebelehre)*
Überdeckungswinkel m <Elektriz, Maschinen> angle of overlap, overlap angle
überdimensionieren v <Bau, Maschinen> oversize
Überdimensionierung f 1. <Bau> oversizing; 2. <Maschinen> overdimensioning, oversizing
Überdosierung f 1. <Textil> overfeed; 2. <Umweltschmutz> overdosage
Überdosis f <Umweltschmutz> OD, overdose
überdrehen v <Fertig> strip *(Gewinde)*
Überdrehungsgrad m <Fertig> overlap
Überdrehzahl f <Maschinen, Mechan> overspeed
Überdrehzahlkontrolle f <Nichtfoss Energ> overspeed control
Überdrehzahlschutz m <Maschinen> overspeed brake
Überdruck m 1. <Erdöl> overpressure *(Geologie)*; 2. <Heiz & Kälte> excess pressure; 3. <Lufttrans> overpressure; 4. <Maschinen> excess pressure
Überdrucken n <Verpack> overprinting
Überdruckkammer f <Erdöl> hyperbaric chamber *(Tauchtechnik)*
Überdruckkapselung f <Sicherheit> overpressure enclosure
Überdruck-Klimaanlage f <Heiz & Kälte> plenum system
Überdruckkühlkreislauf m <Kfztech> sealed cooling system
Überdruckmanometer n <Gerät> *(AE)* overpressure gage, *(BE)* overpressure gauge
Überdruckpumpe f für Kraftstoff <Lufttrans> booster pump
Überdruck-Schnellschlussventil n <Maschinen> pop valve
überdrucksicher adj <Sicherheit> overpressure-proof

Überdruck-Staubabzugshaube

Überdruck-Staubabzugshaube f <Sicherheit> positive-pressure-powered dust hood
Überdruckturbine f <Nichtfoss Energ> reaction turbine
Überdruckventil n 1. <Eisenbahn, Fertig, Heiz & Kälte, Hydraul> pressure relief valve, relief valve; 2. <Kfztech> pressure relief valve, relief valve *(Schmierung)*; 3. <Maschinen, Mechan> pressure relief valve, relief valve
Übereckmaß n <Maschinen> across corner dimension, width across corners
übereinander angeordnete Pressen fpl <Papier> stacked presses
Übereinstimmung f 1. <Comp & DV> conformance; 2. <Gerät> fit; 3. <Maschinen, Math> accordance
Übereinstimmungsgrad m <Trans> degree of compliance
überentwickeln v <Foto> overdevelop
Überfahrt f <Wassertrans> passage; crossing *(Navigation)*
Überfall m <Hydraul> nappe
Überfallfischgerinne n <Nichtfoss Energ> overfall-type fish pass
Überfallwehr n 1. <Nichtfoss Energ> spillway; 2. <Wasserversorg> spillway, waste weir
Überfalz m <Druck> over fold
überfalzte Fuge f <Bau> rebated joint
Überfang m <Ker & Glas> flash
Überfangen n <Ker & Glas> flashing
Überfangglas n <Ker & Glas> flashed glass
Überfangnoppe f <Ker & Glas> flashing knob
Überfangopalglas n <Ker & Glas> flashed opal
Überfangrubinglas n <Ker & Glas> flash ruby
überfärben v <Textil> cross-dye, overdye
Überfärben n <Ker & Glas> overstriking
Überfärbung f <Textil> double dyeing
Überflur... <Maschinen> floor-mounted
Überflurhydrant m 1. <Bau> pillar hydrant; 2. <Sicherheit> pillar fire hydrant
Überflussenergie f <Kohlen> spill energy
Überflussverlust m <Raumfahrt> spillover loss *(Weltraumfunk)*
überfluten v 1. <Mechan> flood; 2. <Wasserversorg> submerge; 3. <Wassertrans> flood *(Flut)*
überflutete Düse f <Lufttrans> flooded jet
Überflutung f <Wasserversorg> inundation, overflow
Überform f <Fertig> mantle; coat *(Gießen)*
überführen v <Bau> pass over *(Straße)*
Überführen n <Raumfahrt> transfer *(von Tank)*
Überführung f 1. <Bau> overbridge; 2. <Ker & Glas> carry-over
Überführungsflug m <Lufttrans> ferry flight
Überfunktion f <Metall> excess function
Übergabe f 1. <Kerntech> handover; 2. <Kontroll> transfer
Übergabeanweisung f <Kontroll> transfer instruction
Übergabe-Bestätigungsbit <Telekom> delivery confirmation bit *(ISDN)*
Übergabesignal n <Fernseh> change-over cue
Übergang m 1. <Eisenbahn> crossover; fillet radius *(von Steg zum Schienenkopf)*; 2. <Lufttrans> in-flight transition *(während des Fluges)*; 3. <Optik> joint *(Halbleiter)*
Übergang m erster Ordnung <Phys> first-order transition
Übergang m in Grundzustand <Teilphys> ground state transition
Übergang m während des Fluges <Lufttrans> flight transition
Übergang m zu neuer Serie <Fertig> job changeover
Übergang m zur Turbulenz <Strömphys> transition to turbulence
Übergang m zweiter Ordnung <Phys> second order transition

Übergangsabschnitt m <Lufttrans> transition segment *(Landung)*
Übergangsarmatur f <Fertig> transition fitting
Übergangsbahnhof m <Eisenbahn> interchange track
Übergangsbereich m <Comp & DV> transient area
Übergangsbogen m <Trans> flare-out
Übergangsbohrung f <Kfztech> bypass bore
Übergangsbrücke f <Eisenbahn> gangway
Übergangselement n <Metall> transition element
Übergangsenthalpie f <Kerntech> transition enthalpy
Übergangserscheinung f <Elektriz> transient effect, transient phenomenon
Übergangsfehler m <Comp & DV> transient error
Übergangsfläche f <Metall> interface *(zwischen Medien)*
Übergangsfunktion f <Elektronik, Elektrotech> indicial response function, step function, step response, transfer function, transient function, unit step response; transient response, transient response function *(Prozesssteuerung)*
Übergangsgleichgewicht n <Phys> transient equilibrium
Übergangsimpedanz f <Elektriz> transition impedance
Übergangskonverter m <Telekom> interworking port unit *(ISDN-Paketvermittlungsnetz)*
Übergangskriechen n <Metall> transient creep
Übergangskurve f <Trans> transition curve
Übergangsmetall n <Metall> transition metal
Übergangsmobilvermittlungsstelle f <Mobilkom> (AE) gateway mobile switching center, (BE) gateway mobile switching centre
Übergangsmuffe f <Maschinen> reducing socket
Übergangsorbit m <Raumfahrt> interim orbit, transfer orbit
Übergangspassung f <Maschinen> transition fit
Übergangsphase f <Metall> transient phase
Übergangspunkt m <Kerntech> transition point
Übergangsreibstelle f <Ker & Glas> transit rub
Übergangsrohrstück n <Maschinen> reducing pipe
Übergangsschwingung f <Elektronik> transient vibration
Übergangsstecker m <Elektrotech> plug adaptor
Übergangsstelle f 1. <Comp & DV> flowchart connector; 2. <Telekom> interface
Übergangsstück n <Bau, Maschinen> reducer
Übergangssystem n <Elektronik> transition system
Übergangsverhalten n 1. <Regelung> (AE) transient behavior, (BE) transient behaviour; 2. <Telekom> transient response
Übergangsverschraubung f <Fertig> adaptor union
Übergangsvorgang m <Kontroll> transient
Übergangswahrscheinlichkeit f <Phys> transition probability
Übergangswiderstand m <Elektriz> transition resistance
Übergangszeit f <Phys> transit time
Übergangszone f <Erdöl> transition zone *(Lagerstättentechnik)*
übergeben v 1. <Kontroll> transfer; 2. <Telekom> submit *(Nachrichten)*
übergehen v <Fertig> blend *(Form)*
übergeordnet adj <Heiz & Kälte> upstream
Übergeschwindigkeit f <Mechan> overspeed
übergesetzter Akzent m <Druck> piece accent
Übergewicht n 1. <Metrol> overweight; 2. <Verpack> excess weight
übergewichtig adj <Verpack> overweight
Überglasung f <Ker & Glas> glaze
übergroß adj <Maschinen> oversize
Übergröße f 1. <Kohlen> oversize; 2. <Maschinen> overdimension, oversize
Übergruppe f <Comp & DV, Telekom> super group, supergroup

Überhandknoten m <Wassertrans> overhand knot
Überhang m 1. <Bau> overhang; 2. <Druck> kern; 3. <Kfztech> overhang *(Karosserie)*; 4. <Maschinen> overhang
überhängende Welle f <Maschinen> overhanging shaft
überhängender Vorsteven m <Wassertrans> raking stem *(Schiffbau)*
Überhangwand f <Bau> overhanging wall
Überhangzeit f <Raumfahrt> hangover time
Überhärten n <Kunststoff> overcure
Überhärtung f <Kunststoff> overcure
Überheizen n <Elektrotech> overheating
überhitzen v 1. <Heiz & Kälte> superheat; 2. <Thermod> overheat
überhitzen v/sich <Thermod> overheat
Überhitzen n 1. <Elektrotech> overheating; 2. <Kfztech> overheating *(des Motors)*; 3. <Thermod> overheating
Überhitzer m <Heiz & Kälte, Kerntech> superheater
Überhitzerelement n <Kerntech> superheat assembly
Überhitzerschlange f <Heiz & Kälte> superheater coil
überhitzter Dampf m <Heiz & Kälte, Maschinen, Phys, Thermod> superheated steam
Überhitzung f 1. <Elektrotech, Maschinen> overheating; 2. <Metall, Phys> superheating
Überhitzungsschutz m <Sicherheit> overheat protection
Überhitzungsthermresistor m <Lufttrans> overheat thermoresistor
Überhitzungswärme abführen v <Heiz & Kälte> desuperheat
überhöhen v <Bau> bank
überhöht adj <Fertig> leptokurtic
überhöhter Druck m <Lufttrans> overpressure
Überhöhung f 1. <Akustik> camber; 2. <Bau> superelevation; 3. <Eisenbahn> cant; 4. <Fertig> leptokurtosis, step
Überhöhungsfaktor m <Kerntech> advantage factor
überholen v <Telekom> overhaul
Überholen n 1. <Trans> passing; 2. <Wassertrans> lurch *(Schiffbewegung)*
Überholklauenkupplung f <Eisenbahn> override clutch
Überholkupplung f 1. <Lufttrans> overrunning clutch *(Flugwesen)*; 2. <Maschinen> freewheel clutch, overrunning clutch
Überholsichtweite f <Trans> passing sight distance
Überholspur f 1. <Bau> passing lane; acceleration lane *(Straße)*; 2. <Trans> overtaking lane, passing lane
überholt adj <Fertig> obsolete
Überholung f 1. <Kfztech> overhaul *(eines Motors)*; 2. <Lufttrans> overhaul *(Gerätschaften)*; 3. <Wassertrans> overhaul *(Instandsetzung)*
Überhörfrequenz f <Elektronik, Strahlphys> supersonic frequency
Überhorizontausbreitung f <Strahlphys> beyond-the-horizon propagation, scatter propagation, transhorizon propagation *(von elektromagnetischen Wellen)*
Überhorizontradar n <Funkort> over-the-horizon radar
Überhub m <Raumfahrt> overdeviation *(Weltraumfunk)*
Überkapazität f 1. <Ker & Glas> wrong capacity; 2. <Lufttrans> overcapacity *(Luftransport)*
überkleben v <Konstzeich> mark
überkochen v <Thermod> boil over
Überkompoundierung f <Elektrotech> overcompounding
Überkopfförderband n <Verpack> overhead conveyor
Überkragung f <Fertig> overhang
Überkreuzungsbereich m <Elektronik> crossover area
Überkreuzverzerrung f <Aufnahme> crossover distortion
überkritische Ballung f <Elektronik> overbunching
überkritische Dämpfung f <Phys> overdamping
überkritische Drehzahl f <Elektrotech> above critical speed

überkritische Reaktion f <Strahlphys> supercritical reaction
Überladebrücke f <Wassertrans> gantry crane
überladen adj <Elektriz> overloaded
Überladung f 1. <Elektriz, Elektrotech> overcharge; 2. <Maschinen> overload
überlagern v 1. <Elektronik> superimpose, superpose; 2. <Fernseh> overlay; 3. <Telekom> heterodyne, superimpose
überlagern v/sich <Funktech> beat *(Hochfrequenztechnik)*
überlagerte Funkzone f <Mobilkom> overlaid cell
überlagerte Last f <Bau> superimposed load
Überlagerung f 1. <Anstrich> overlay, overlaying; 2. <Comp & DV, Elektronik> interference, overlay, overlaying, superimposition; blanking *(von Sendern)*; 3. <Fernseh> heterodyning; 4. <Funktech> heterodyning, interference; 5. <Phys> interference
Überlagerungsanalysator m <Aufnahme> *(BE)* heterodyne sound analyser, *(AE)* heterodyne sound analyzer
Überlagerungsempfang m <Telekom> heterodyne reception
Überlagerungsempfänger m <Funktech, Telekom> superheterodyne receiver
Überlagerungsfrequenzumsetzer m <Elektrotech> heterodyne conversion transducer
Überlagerungsnetz n <Telekom> overlay network
Überlagerungspermeabilität f <Elektronik> incremental permeability *(mit Vorpolarisierung)*
Überlagerungsprinzip n 1. <Elektronik> superposition principle *(synthetische Prüfung)*; 2. <Phys> superposition principle
Überlagerungsumsetzung f <Elektrotech, Funktech> heterodyne conversion
Überlagerungswellenmesser m <Funktech> heterodyne wavemeter
Überlagungsoszillator m <Funktech> local oscillator
Überlandleitung f 1. <Elektrotech> overhead cable, overhead power line; 2. <Telekom> overhead cable
Überlandstraße f <Trans> arterial safety road
überlappen v <Anstrich, Bau> lap
überlappen n <Metall> overlapping
überlappend anordnen v <Comp & DV> cascade
überlappend geschweißte Hülse f <Kerntech> weld overlay cladding
überlappende Klappen fpl <Verpack> overlapping flaps
überlappende Schweißnaht f <Kerntech> lap weld
überlappende Tonhöhen fpl <Akustik> conjoined pitches
überlappender Stoß m <Mechan> lap joint
Überlappnaht f <Fertig> lap weld; overlapping spot-weld *(Schweißen)*
Überlappnietung f <Fertig> lap riveting
Überlappschweißung f <Verpack> lap weld
Überlappstoß m <Fertig> shear joint
überlappte Isolierung f <Elektriz> lapped insulation
überlappte Nahtschweißung f <Fertig> lap seam-welding
überlappte Teilfuge f <Bau> lap joint
überlappte Verbindung f <Bau> lapped scarf
überlappte Wicklung f <Elektriz> lap winding
Überlappung f 1. <Anstrich> lap; 2. <Bau> lap, overlap, step joint; 3. <Comp & DV> interleaving, overlap, overlay; 4. <Druck> overlap; 5. <Eisenbahn> overlapping *(von Blockabschnitten)*; 6. <Elektronik> pipelining *(Datenverarbeitung)*; 7. <Elektrotech> overlap, overlapping; 8. <Fernseh, Telekom> overlap; 9. <Fertig> cold shut *(Schweißen)*; 10. <Hydraul> lap *(Steuerschieber und -kolben)*; 11. <Ker & Glas> lap; 12. <Maschinen> overlap;

Überlappung

13. <Phys> overlapping *(von Spektrallinien)* • **ohne Überlappung** <Anstrich> unlapped
überlappungsfreies Ventil *n* <Hydraul> lapless valve
Überlappungsnietung *f* <Verpack> lap riveting
Überlappungsschweißung *f* <Maschinen> lap welding
Überlappungsstoß *m* <Bau> lap joint, overlapping joint
Überlappungs- und Voreilsteuerschwinge *f* <Hydraul> lap and lead lever *(Dampfsteuergetriebe)*
Überlappungsventil *n* <Hydraul> lap valve
Überlast *f* 1. <Elektriz, Elektrotech> overload; 2. <Kohlen> surcharge load; 3. <Phys, Telekom> overload
Überlastabwehr *f* <Telekom> flow control *(Verkehrskontrolle)*
Überlastanzeiger *m* <Elektriz> overload indicator
Überlastbarkeit *f* 1. <Fertig> overload capacity *(Kunststoffinstallationen)*; 2. <Qual> overload capacity
überlasten *v* 1. <Bau> overstress; 2. <Fernseh, Funktech> overload; 3. <Maschinen> overstrain; 4. <Telekom> overload
überlastet *adj* <Phys, Telekom> overloaded
Überlastfaktor *m* <Elektriz> overload factor
Überlastfeder *f* <Maschinen> overload protection spring, overload spring
Überlastgrenze *f* <Regelung> overrange limit
Überlastkupplung *f* <Maschinen> overload coupling
Überlastrelais *n* <Elektriz> overload relay
Überlastschutz *m* 1. <Elektriz> overload protection; 2. <Sicherheit> load protection, overload protection; load protector
Überlastsicherung *f* <Sicherheit> overload protection, overload protective device, overload protective equipment
Überlastspannung *f* <Elektriz> overload voltage
Überlaststrom *m* <Elektriz> overload current
Überlastung *f* 1. <Comp & DV> overload, trashing; 2. <Funktech> overload; 3. <Maschinen> overload, overstrain; 4. <Raumfahrt> overload *(Weltraumfunk)*; 5. <Telekom> overload; congestion *(Netzwerke)*
Überlastungsfeder *f* <Maschinen> overload protection spring, overload spring
Überlastungskupplung *f* <Sicherheit> safety clutch
Überlastungsprüfung *f* <Elektriz> overload test
Überlastungsrelais *n* <Elektrotech> overload relay
Überlastungsschutz *m* 1. <Elektrotech> overload protection, overload protection device; 2. <Funktech> overload protection
Überlastungsschutzvorrichtung *f* <Elektrotech> overload protection device
Überlastungsstrom *m* <Elektrotech> overload current
Überlastverhältnis *n* <Elektriz> overload factor
Überlauf *m* 1. <Bau> spillway; 2. <Comp & DV> overflow; 3. <Fertig> overtravel; overflow *(Kunststoffinstallationen)*; 4. <Ker & Glas> weir; 5. <Kfztech> overflow pipe *(Kühlanlage)*; 6. <Kohlen> overflow; 7. <Maschinen> overflow, overrun; 8. <Mechan> overflow
Überlaufanzeiger *m* <Comp & DV> overflow flag
Überlaufbereich *m* <Comp & DV> overflow area
Überlaufbit *n* <Comp & DV> overflow bit
Überlaufbündel *n* <Telekom> high-usage circuit group; overflow group *(Verkehr)*
überlaufen *v* 1. <Fertig> overtravel; 2. <Maschinen> overrun; 3. <Wasserversorg> run over
Überlaufen *n* <Maschinen> overflow
Überlaufkanal *m* 1. <Hydraul> spillway canal; 2. <Nichtfoss Energ> spillway channel
Überlaufkante *f* <Bau> lip
Überlauföffnung *f* <Kfztech> overflow hole
Überlaufverfahren *n* <Ker & Glas> overflow process
Überlaufverkehr *m* <Telekom> overflow traffic

Überlaufwehr *n* 1. <Hydraul> weir; 2. <Kerntech> leaping weir
Überlaufwehrmessung *f* <Hydraul> *(AE)* notch gaging, *(BE)* notch gauging
Überlebensanzug *m* <Sicherheit, Wassertrans> survival suit *(Notfallausrüstung)*
Überlebensausrüstung *f* <Sicherheit> survival kit *(Notfallausrüstung)*
Überlebensfähigkeit *f* <Telekom> survivability *(System)*
überlegen *v*/**sich** <Wassertrans> cant over *(Schiff)*
Überleitrille *f* <Akustik, Aufnahme> lead-over groove
Übermaß *n* 1. <Maschinen> interference, overdimension, overmeasure, oversize; 2. <Mechan> oversize
Übermaß *n* **durch Losschlagen des Modells** <Fertig> rappage *(Gießen)*
übermäßige Bestrahlung *f* <Strahlphys> overexposure
übermäßiger Porenwasserdruck *m* <Bau> excessive pore water pressure
Übermaßzeichnung *f* <Konstzeich> drawing dealing with oversize parts
Übermittlung *f* <Telekom> send, transfer *(Information)*
Übermittlungsdienst *m* <Telekom> bearer service
Übermittlungsdienst *m* **über virtuelle Verbindung** <Telekom> virtual-circuit bearer service
Übermittlungsschicht *f* <Telekom> data link layer
Übermittlungssystem *n* <Comp & DV, Telekom> communication system
Übermittlungsträger *m* <Telekom> transmission bearer
Übermittlungsvorschrift *f* 1. <Comp & DV> link protocol, protocol; 2. <Druck, Funktech, Telekom> protocol
Übermodulation *f* <Aufnahme, Funktech> overmodulation
Übernahme *f* 1. <Comp & DV> inheritance; 2. <Mechan> acceptance
übernommener Fehler *m* <Comp & DV> inherited error
überoxidieren *v* <Chemie> overoxidize
überprüfen *v* 1. <Kontroll> inspect; 2. <Patent, Qual, Verpack> check
Überprüfung *f* 1. <Mechan, Qual> examination; 2. <Raumfahrt> review; 3. <Verpack> check
Überprüfung *f* **durch die Unternehmensführung** <Qual> management audit
Überprüfung *f* **durch Rundgang** <Qual, Raumfahrt> walkaround inspection *(Raumschiff)*
Überprüfung *f* **externer Links** <Comp & DV> external link checker
Überprüfung *f* **vor Inbetriebnahme** <Kerntech> precommissioning checks
Überrahmen *m* <Telekom> superframe *(PCM)*
Überreichweitenecho *n* 1. <Funkort> second-go-around, second trace echo *(Radar)*; 2. <Wassertrans> second trace echo *(Radar)*
überreif *adj* <Lebensmittel> overripe
Überrollbügel *m* <Kfztech, Sicherheit> roll bar, rollover bar
Überrollen *n* <Lufttrans> overrun
Überrollkäfig *m* <Kfztech> roll cage
übersättigen *v* 1. <Chemie> supersaturate; 2. <Thermod> oversaturate *(Dampf)*
übersättigt *adj* <Umweltschmutz> supersaturated
Übersättigung *f* <Thermod> oversaturation
übersäuern *v* <Chemie> peroxidize
Übersäuerung *f* <Lebensmittel> hyperacidity
Übersäuerung *f* **des Wassers** <Umweltschmutz> aquatic acidification
Überschall... <Wellphys> supersonic
Überschallbereich/im <Wellphys> supersonic
Überschallflugzeug *n* <Lufttrans> hypersonic aircraft, supersonic aircraft, transonic aircraft

Überschallfrequenz f 1. <Akustik, Elektronik> supersonic frequency, ultrasonic frequency; 2. <Phys> ultrasonic frequency; 3. <Strahlphys> supersonic frequency
Überschallgeschwindigkeit f 1. <Fertig> hypersonic speed; 2. <Phys> supersonic speed
Überschallknall m <Phys> sonic boom
Überschalltransport m (SST) <Lufttrans> supersonic transport (SST)
überschalten v <Fertig> override
Überschiebmuffe f 1. <Fertig> slide coupling (Kunststoffinstallationen); 2. <Maschinen> collar
Überschiebung f <Erdöl> overthrust (Geologie)
Überschlag m 1. <Elektriz, Elektrotech> arc-over, flashover, sparkover; 2. <Fertig> arcing; 3. <Phys> breakdown; 4. <Telekom> discharge
Überschlagen n 1. <Kfztech> overturning; 2. <Sicherheit, Wassertrans> roll-over
Überschlagprüfung f <Elektrotech> flash test
Überschlagspannung f <Elektriz, Elektrotech> arc overvoltage, arcing voltage, flash-over voltage, sparking voltage, spark-over voltage, withstand voltage
Überschlagspannungsprüfung f <Elektriz> withstand-voltage test, flash over test, spark over test
Überschlagsrechnung f <Bau, Qual> rough calculation
Überschlagstoßspannung f <Elektriz> impulse flashover voltage
Überschlagstrecke f <Elektrotech> flash-over distance, flash-over path
Überschlagwechselspannung f <Elektriz> a.c. flashover voltage, power-frequency flash-over voltage
Überschlagweite f <Elektriz> flash-over distance, sparking distance (eines Funkens); arcing distance (eines Lichtbogens)
überschneiden v 1. <Aufnahme> overdub; 2. <Bau> intersect
Überschneiden n <Akustik> overcutting
Überschneidungsfrequenz f <Elektronik> crossover frequency
Überschreiben n <Comp & DV> overwriting
Überschreibung f <Comp & DV> destructive addition
überschreiten v <Math> exceed
Überschreiten n <Elektrotech> overshoot
Überschrift f 1. <Comp & DV> heading; 2. <Druck> caption, head, headline, title
Überschriftzeile f <Druck> header, headline
Überschusshalbleiter m <Elektronik> n-semiconductor
überschüssig adj <Papier> odd
überschüssige Energie f <Thermod> excess energy
überschüssiger Aushubboden m <Bau> spoil (Tiefbau)
überschüssiger Boden m <Bau> spoil
Überschussreaktivität f <Kerntech> excess reactivity
Überschussstrom m <Elektriz> excess current
überschwellig adj <Ergon> supraliminal
Überschwemmung f 1. <Strömphys> drowned flow; 2. <Wassertrans> flood (Hochwasser, Fluss); 3. <Wasserversorg> overflow
Überschwemmungsgebiet n <Wasserversorg> flood plain
überschwerer Wasserstoff m <Chemie, Phys> tritium
Überschwing... <Elektrotech, Gerät> overshooting
überschwingen v <Mechan> overshoot
Überschwingen n 1. <Elektrotech> overshoot; 2. <Fernseh> slope overload (digitale Videosignalcodierung); 3. <Funktech, Telekom> overswing
Überschwingerscheinung f <Telekom> Gibbs phenomenon, overshooting phenomenon
Überschwingfaktor m <Gerät> ballistic factor (Messgerät)
Überschwingspitze f <Elektrotech> overshoot spike

Überschwingverzerrung f <Fernseh> overshoot distortion
Überschwingweite f <Regelung> transient overshoot
Übersee... <Wassertrans> overseas (Seehandel)
Überseetelefonnetz n <Telekom> intercontinental telephone network
Überseeverpackung f <Verpack> overseas packaging
Übersegler m <Wassertrans> track chart
übersetzen v <Comp & DV> compile, interpret, translate
Übersetzer m <Comp & DV> interpreter, processor
Übersetzerprogramm n <Comp & DV> interpreter, translator
Übersetzersprache f <Comp & DV> interpretative language
übersetztes Programm n <Comp & DV> object program
Übersetzung f 1. <Comp & DV> translation; 2. <Maschinen> gear ratio, multiplication; 3. <Mechan> gear ratio
Übersetzung f ins Langsame <Kfztech> transmission reduction
Übersetzungsanlage f <Comp & DV> source machine
Übersetzungsanweisung f <Comp & DV> directive
Übersetzungsgetriebe n <Maschinen> step-up gear
Übersetzungsrad n <Maschinen> translating wheel
Übersetzungsverhältnis n 1. <Elektrotech> turns ratio; 2. <Fertig> increasing ratio; 3. <Funktech> transformer ratio; 4. <Kfztech> transmission ratio; gear ratio (Getriebe); 5. <Maschinen> gear ratio; 6. <Telekom> transformer ratio
Übersetzungsverhältnis n **der Steuerung** <Lufttrans> control gearing ratio
Übersicht f 1. <Bau> survey; 2. <Kerntech> general drawing; 3. <Konstzeich> general plan
Übersichtsschalttafel f <Elektriz> synoptical switchboard
Übersichtszeichnung f 1. <Kerntech> general drawing; 2. <Mechan> general arrangement drawing
überspannen v 1. <Bau, Eisenbahn> bridge (Tal); 2. <Fertig> overtension
Überspannung f 1. <Elektriz, Elektrotech> excess voltage, overvoltage; 2. <Fertig> overtensioning (Kette); 3. <Phys> overvoltage
Überspannung f der Kondensatorbatterie <Elektriz> capacitor bank overvoltage
Überspannungsableiter m <Elektriz, Elektrotech> arrester, surge arrester, surge diverter
Überspannungsausfall m <Elektrotech> overvoltage breakdown
Überspannungsauslöser m <Gerät> overvoltage trip
Überspannungsauslösung f <Elektriz> overvoltage release
Überspannungsblitz m <Foto> photoflood bulb
Überspannungsrelais n <Elektriz, Elektrotech> overvoltage relay
Überspannungsschutz m 1. <Elektriz, Elektrotech> excess voltage protection, overvoltage protection, surge absorber, surge protection; 2. <Telekom> surge arrester
Überspannungsschutzvorrichtung f <Elektrotech> overvoltage protection device
überspielen v 1. <Aufnahme> dub; 2. <Fernseh> copy, dub
Überspielen n **von Film auf Band** <Fernseh> film-to-tape transfer
Überspielen n **von Filmen** <Fernseh> film transfer
Überspielung f <Akustik> dubbing
Übersprechdämpfung f <Akustik> crosstalk
Übersprechen n <Akustik, Aufnahme, Comp & DV, Fernseh, Phys> crosstalk
Überspringbefehl m <Comp & DV> skip instruction
überspringen v 1. <Comp & DV> skip; 2. <Elektriz> jump
Überspringen n <Comp & DV> skip

überspült

überspült *adj* <Wassertrans> awash, flooded
Überstand *m* 1. <Bau> projection; 2. <Druck> bleed
Überstaubewässerung *f* <Wasserversorg> flood irrigation
Überstauung *f* 1. <Wassertrans> overstowage; 2. <Wasserversorg> flood irrigation
überstehen *v* <Druck> bleed
übersteuern *v* <Elektrotech, Fertig> override
übersteuernd *adj* <Mechan> overriding
Übersteuerung *f* 1. <Fertig> override; 2. <Kfztech> oversteering; oversteer *(Lenkung)*; 3. <Lufttrans> override control
Übersteuerungsschalter *m* <Elektriz> override switch
Überstrahlung *f* 1. <Foto> blooming; 2. <Telekom> spillover
überstreichen *v* <Bau> top
Überstrom *m* <Elektriz, Elektrotech> overcurrent
Überström... <Elektrotech, Maschinen> overflowing
Überstromauslöser *m* <Elektrotech> overcurrent trip
Überstromausschalter *m* <Elektrotech> overcurrent circuit breaker
Überstromblockiereinrichtung *f* <Elektriz> overcurrent blocking device
Überströmen *n* <Maschinen> overflow
Überströmkanal *m* <Kfztech> transfer port *(Zweitaktmotor)*
Überstromleistungsschalter *m* <Elektrotech> overcurrent power switch
Überströmöffnung *f* <Kfztech> overflow port
Überstromrelais *n* <Elektriz> overcurrent relay
Überströmrohr *n* <Kfztech> overflow pipe *(Kühlanlage)*
Überstromschalter *m* <Elektriz> excess current switch, overcurrent switch
Überstromschutz *m* <Elektriz, Elektrotech> overcurrent protection
Überstromschutzschalter *m* <Elektrotech> overcurrent circuit breaker
Überstromunterbrecher *m* <Elektriz> overcurrent circuit breaker
Überströmventil *n* 1. <Hydraul, Kfztech, Maschinen> overflow valve; 2. <Sicherheit> relief valve
Überstruktur *f* <Fertig> superlattice, superlattice structure
überstumpfer Winkel *m* <Geom> reflex angle
übertage *adv* <Kohlen> above ground
Übertemperaturrelais *n* <Elektrotech> overtemperature-relay
Übertemperatur-Stromunterbrecher *m* <Elektriz> thermal circuit breaker
Übertönen *n* <Elektronik> capture effect *(Transmitter)*
Übertrag *m* 1. <Comp & DV, Elektronik> carry; 2. <Math> amount carried over, carry *(Rechnen)*
übertragen *v* 1. <Aufnahme, Bau> transfer; 2. <Comp & DV> carry, transfer; 3. <Funktech> transmit; 4. <Kontroll> transfer; 5. <Maschinen> convey *(Kraft)*; 6. <Math> carry over, transfer *(beim Rechnen)*; 7. <Thermod> impart *(Energie)*
übertragene Wärme *f* <Thermod> convected heat
Übertrager *m* <Funktech, Phys, Telekom> transformer
Übertragerkopplung *f* <Funktech> transformer coupling
Übertragerspule *f* <Elektrotech> repeating coil
Übertragung *f* 1. <Comp & DV, Elektriz, Kontroll, Maschinen, Mechan, Phys> transfer, translation, transmission; 2. <Telekom> transfer *(Information)*; transmission
Übertragung *f* **aus dem Speicher** <Comp & DV> copy-out
Übertragung *f* **im Klartext** <Telekom> clear transmission, plain text transmission
Übertragung *f* **in den Speicher** <Comp & DV> copy-in

Übertragung *f* **innerhalb der Sichtweite** <Telekom> line-of-sight transmission, LOS transmission
Übertragung *f* **mit Geheimhaltung** <Telekom> secure transmission
Übertragung *f* **mit Pulscodemodulation** <Telekom> PCM transmission, pulse-code modulated transmission
Übertragung *f* **mit unabhängigen Seitenbändern** <Telekom> amplitude modulation with two indepedent sidebands, independent-sideband transmission
Übertragung *f* **mit unterdrücktem Träger** <Elektronik, Funktech> suppressed carrier transmission
Übertragung *f* **mittels Lichtleitkabel** <Optik> *(AE)* fiber-optic transmission, *(BE)* fibreoptic transmission
Übertragung *f* **nur in einer Richtung** <Telekom> one-way only transmission, repeatered line transmission, simplex transmission
Übertragung *f* **über Glasfaserkabel** <Telekom> *(AE)* optical fiber cable transmission, *(BE)* optical fibre cable transmission
Übertragung *f* **über Lichtleitfasern** <Optik> *(AE)* optical fiber transmission, *(BE)* optical fibre transmission
Übertragung *f* **über Lichtwellenleiterkabel** <Optik, Telekom> fiberoptic transmission, *(AE)* optical fiber cable transmission, *(BE)* fibreoptic transmission
Übertragung *f* **unterhalb des Basisbandes** <Telekom> below baseband transmission
Übertragung *f* **von Unteradressinformationen** <Telekom> sub-addressing, SUB *(Dienstmerkmal im Euro-ISDN)*
Übertragung *f* **zwischen mehreren Stationen** <Comp & DV> multidrop transmission
Übertragungsanpassungsschicht *f* <Telekom> transmission convergence layer
Übertragungsanweisung *f* <Kontroll> transfer instruction
Übertragungsart *f* <Comp & DV> mode of transmission
Übertragungsbeiwert *m* <Regelung> transfer coefficient *(der Regelstrecke)*
Übertragungsbeiwert *m* **einer Wanderwellenröhre** <Raumfahrt> TWT transfer coefficient *(Weltraumfunk)*
Übertragungsbereich *m* 1. <Comp & DV> frequency response; 2. <Funktech, Telekom> transmitting range *(eines Senders)*
Übertragungsbereit *adj* <Telekom> ready, ready for data
Übertragungsblock *m* <Comp & DV> transmission block
Übertragungscharakteristik *f* 1. <Phys> transfer characteristic; 2. <Telekom> transmission characteristic
Übertragungscode *m* 1. <Comp & DV> transmission code; 2. <Telekom> line code, transmission code
Übertragungsdämpfung *f* 1. <Comp & DV> TL, transmission loss; 2. <Telekom> TL, transmission loss *(zwischen Sende- und Empfangsantenne)*
Übertragungseigenschaften *fpl* <Telekom> transmission performance *(des ISDN)*
Übertragungsende *n* <Comp & DV, Telekom> end of transmission
Übertragungsfaktor *m* <Akustik> response
Übertragungsfehler *m* <Comp & DV, Math, Telekom> transmission error
Übertragungsfenster *n* <Optik, Telekom> transmission window
Übertragungsfolge *f* <Comp & DV> transmission sequence
Übertragungsfrequenzgang *m* <Akustik> magnitude frequency response
Übertragungsfunktion *f* 1. <Comp & DV> transmission function; 2. <Optik, Telekom> transfer function
Übertragungsfunktion *f* **für das Grundfrequenzband** <Optik> baseband response function, baseband transfer function

Übertragungsgeschwindigkeit *f* 1. <Comp & DV> bit rate, data rate, line speed, transfer rate, transmission rate; 2. <Telekom> *(AE)* data signaling rate, *(BE)* data signalling rate, transmission rate
Übertragungsgeschwindigkeit *f* **bei Wählleitungen** <Comp & DV> switched communications speed
Übertragungsgüte *f* <Telekom> transmission quality
Übertragungs-Highway *m* <Telekom> transmission highway
Übertragungskanal *m* <Comp & DV, Telekom> communication channel, transmission channel
Übertragungskennlinie *f* <Elektronik> transfer characteristic
Übertragungskonstante *f* <Akustik> propagation coefficient
Übertragungskopf *m* <Comp & DV> transmission header
Übertragungsleitung *f* 1. <Comp & DV, Telekom> communication line, transmission line; 2. <Elektronik, Elektrotech, Phys> transmission line
Übertragungsleitung *f* **mit mehreren Anschlüssen** <Comp & DV, Telekom> multidrop circuit
Übertragungsleitungsverluste *mpl* <Elektriz> transmission-line losses
Übertragungsleitweg *m* <Telekom> transmission line
Übertragungslimit *n* <Comp & DV> transmission limit
Übertragungsmaß *n* 1. <Akustik> image transfer exponent; 2. <Telekom> image transfer exponent; propagation constant *(Kabel)*
Übertragungsmatrix *f* <Phys> transfer matrix *(Netzwerktheorie)*
Übertragungsmedium *n* <Comp & DV, Optik> transmission medium
Übertragungsmodus *m* <Comp & DV> mode of transmission
Übertragungsnetz *n* 1. <Elektriz> transmission line network; 2. <Telekom> transmission network
Übertragungsnetzknoten *m* <Telekom> transmission node
Übertragungsphase *f* **über Aufwärtsstrecke** <Telekom> uplink transmission phase
Übertragungsprotokoll *n* 1. <Comp & DV> transmission control protocol, transport protocol; 2. <Telekom> transport protocol
Übertragungsprotokoll *n* **zwischen zwei Netzen** <Telekom> border gateway protocol
Übertragungsrate *f* 1. <Comp & DV> bit rate; 2. <Elektrotech> transmission rate; 3. <Telekom> transfer rate
Übertragungsschicht *f* <Telekom> transmission layer
Übertragungsserver *m* <Comp & DV> communication server
Übertragungssicherheit *f* <Comp & DV> transmission security
Übertragungsstange *f* <Maschinen> transmission rod
Übertragungssteuerung *f* <Comp & DV> transmission control
Übertragungssteuerungsprotokoll *n* <Telekom> transmission control protocol
Übertragungssteuerungsprotokoll/Querverkehrsprotokoll *n* <Telekom> transmission control protocol/Internet protocol *(Verkehr zwischen Computern)*
Übertragungsstörung *f* <Telekom> transmission breakdown
Übertragungsstrecke *f* 1. <Funktech> transmission line; 2. <Telekom> trunk
Übertragungssystem *n* <Elektrotech, Telekom> transmission system
Übertragungssystem *n* **mit sehr hoher Kanalzahl** <Telekom> very-large-capacity transmission system

Übertragungsverfahren *n* <Elektronik> transmission mode
Übertragungsverhältnis *n* <Fernseh> transfer ratio
Übertragungsverlust *m* 1. <Optik> TL, transmission loss; 2. <Raumfahrt> TL, transmission loss *(Weltraumfunk)*; 3. <Telekom> TL, transmission loss *(Transducer)*
Übertragungswagen *m* <Fernseh> mobile control unit
Übertragungsweg *m* 1. <Comp & DV> *(BE)* bus, channel, highway, transmission path, *(AE)* trunk; 2. <Telekom> channel, transmission path
Übertragungswiderstand *m* <Elektriz> coupling impedance
Übertragungswirkungsgrad *m* <Nichtfoss Energ> transmission efficiency
Übertragungszeit *f* <Phys> transit time
Übertragungsziffer *f* <Comp & DV> carry digit
übertreffen *v* <Math> exceed
übertrocknen *v* <Papier> overdry
Über- und Unterstromrelais *n* <Elektriz> over-and-under current relay
Überverbrauchszähler *m* 1. <Elektriz> excess energy meter; 2. <Gerät> excess meter
Übervergüten *n* <Metall> *(BE)* overageing, *(AE)* overaging
überwachen *v* 1. <Aufnahme> monitor; 2. <Bau> attend to; 3. <Comp & DV, Funktech, Meerschmutz, Textil> monitor
Überwachen *n* <Fernseh> monitoring
Überwacher *m* <Kontroll, Qual> supervisor *(Person)*
überwachter Ablauf *m* <Comp & DV> attended operation
überwachter Betrieb *m* <Comp & DV> attended operation
überwachtes Abladen *n* **von Schutt** <Umweltschmutz> controlled dumping
Überwachung *f* 1. <Aufnahme> monitoring; 2. <Bau> observation; 3. <Comp & DV> monitoring; 4. <Kfztech> control; 5. <Qual, Raumfahrt> monitoring; 6. <Sicherheit> surveillance; 7. <Telekom> monitoring, supervision; 8. <Textil> monitoring
Überwachung *f* **an Ort und Stelle** <Kerntech> in situ monitoring
Überwachung *f* **der Prüfung** <Qual> monitor the review
Überwachung *f* **der Qualitätssicherung des Lieferanten** <Qual> quality assurance surveillance
Überwachung *f* **der Unterlagen** <Patent, Qual> document control
Überwachung *f* **der Wasserbeschaffenheit** <Wasserversorg> water quality monitoring
Überwachung *f* **und Instandhaltung** *f* <Qual, Telekom> monitoring and maintenance
Überwachung *f* **von Prüf- und Messmitteln** <Qual> control of inspection, measuring and test equipment
Überwachung *f* **von Qualitätsmaßnahmen** <Qual> control of quality measures
Überwachungsanlage *f* <Textil> control system
Überwachungsarmatur *f* <Fertig> process control
Überwachungsfernsehen *n* *(CCTV)* <Fernseh> closed-circuit TV *(CCTV)*
Überwachungshilfe *f* <Telekom> supervisory aid
Überwachungskontrolle *f* <Qual> supervisory control
Überwachungskopf *m* <Aufnahme> monitor head
Überwachungslampe *f* <Elektrotech> pilot lamp
Überwachungsleitung *f* <Telekom> guard circuit
Überwachungsmeldung *f* <Telekom> supervisory message
überwachungspflichtige Anlage *f* <Sicherheit> equipment subject to supervision
Überwachungsradargerät *n* <Funkort> surveillance radar unit

Überwachungsraum

Überwachungsraum m <Aufnahme> control room
Überwachungsrelais n <Elektriz> pilot relay
Überwachungssatellit m <Raumfahrt> surveillance satellite
Überwachungsschaltung f **des eigenen Amts** <Telekom> own-exchange supervisory circuit
Überwachungsstelle f <Qual> inspection agency
Überwachungssystem n 1. <Kontroll> supervising system; 2. <Qual, Strahlphys> monitoring system, supervising system, supervisory system
Überwachungston m <Telekom> supervisory tone
Überwachungs- und Leitsystem n **im Flugzeug (AWACS)** <Lufttrans> airborne warning and control system (AWACS)
Überwachungsverfahren n <Qual> monitoring procedure
Überwachungszeit f 1. <Comp & DV> monitor time; 2. <Raumfahrt> guard time (Weltraumfunk)
Überwachungszeitgeber m 1. <Comp & DV> watchdog timer; 2. <Telekom> supervisory timer
überwalzen v <Fertig> rebate
Überwasseranstrich m <Wassertrans> topside paintwork
Überwassergeschwindigkeit f <Wassertrans> surface speed (U-Boot)
Überwasserteile npl <Wassertrans> topsides (Schiffbau)
überwendlich nähen v <Textil> oversew
Überwindung f <Fertig> override
Überwucht f <Fertig> amount of overbalance
Überwuchtmasse f <Fertig> amount of overbalance
Überwurf m <Bau> lock bush
Überwurfkrümmer m <Bau> union elbow
Überwurfmutter f 1. <Fertig> clamping nut; valve nut (Kunststoffinstallationen); 2. <Maschinen> box nut, sleeve nut, union nut
überzähliger Stab m <Bau> redundant member
überziehen v 1. <Bau> surface (Material); 2. <Kunststoff, Lebensmittel> coat; 3. <Mechan> plate
Überziehen n 1. <Bau> cladding; 2. <Ker & Glas> overlay, overlaying; 3. <Kunststoff> coating; 4. <Lufttrans> stall (Strömungsabriss eines Flugzeugs); 5. <Maschinen, Phys> cladding
Überziehen n **von Stahlblech** 1. <Fertig> terne coating (Tauchverfahren); 2. <Metall> terne coating
überzogene Halbleiterscheibe f **mit Photolack** <Elektronik> resist-coated wafer
Überzug m 1. <Anstrich> overlay, overlaying, protective film; 2. <Bau> overlay, overlaying; 3. <Elektrotech> sheathing; 4. <Lebensmittel> topping (für Süßwaren, Kuchen); 5. <Mechan> film
Überzugsdicke f <Verpack> coating thickness
Überzugsmaterial n 1. <Maschinen> coating material; 2. <Verpack> coating compound
Überzwirnung f <Textil> snarl
Ubitron n <Elektronik> ubitron, undulating beam interaction electron tube
übliches Reparaturwerkzeug n <Maschinen> common repair tool
U-Bolzen m 1. <Kfztech> U-bolt (Federung, Blattfeder); 2. <Wassertrans> U-bolt (Schiffbau)
U-Boot n **(Unterseeboot)** <Wassertrans> submarine
U-Boot-Abwehr f <Wassertrans> anti-submarine warfare (ASW)
U-Boot-Bekämpfungshubschrauber m <Lufttrans> antisubmarine helicopter
U-Boot-Bunker m <Wassertrans> submarine pen (Marine)
U-Bügel m <Maschinen> stirrup, stirrup bolt
U-Eisen n 1. <Bau> channel iron; 2. <Metall> U-iron

U-Entwässerungsrohr n <Bau> U-drain
UF6 (Uranhexafluorid) <Kerntech> UF6 (uranium hexafluoride)
Ufer n 1. <Bau> bank; 2. <Wassertrans> shore, waterfront
• **über die Ufer getreten** <Wassertrans> in flood (Fluss)
• **über die Ufer treten** <Wassertrans> flood (Fluss)
Uferschutz m <Bau> bank protection (Fluss)
UFO (nicht identifiziertes Flugobjekt) <Raumfahrt> UFO (unidentified flying object)
U-förmig <Trans> U-shaped track girder
U-förmige Grundplatte f <Maschinen> U-shaped base, stirrup-shaped bed
U-förmiger Fahrbalken m <Trans> U-shaped track girder
U-förmiger Zughaken m <Maschinen> clevis
U-Formstahl m <Metall> channel section
UHF (Ultrahochfrequenz) <Elektronik, Fernseh, Funktech, Telekom, Wellphys> UHF (ultrahigh frequency)
UHF-Konverter m <Fernseh, Funktech> UHF converter
UHF-Rundfunk m <Fernseh, Funktech> UHF broadcasting
UHF-Signal n <Elektronik> UHF signal
UHF-Signalgenerator m <Elektronik> UHF signal generator
UHF-Tuner m <Fernseh> UHF tuner
Uhr f 1. <Comp & DV> clock (Symbol); 2. <Phys> chronometer
Uhrgang m <Metrol> rate of clock
Uhrglas n <Labor> watch glass
Uhrradio n <Funktech> clock radio
Uhrwerk n <Mechan> clockwork
Uhrzeigersinn drehend adj/im <Lebensmittel> clockwise-rotating
Uhrzeigersinn/gegen den <Gerät> ccw, counterclockwise
UHT (Ultrahochtemperatur) <Lebensmittel> UHT (ultrahigh temperature)
U-Klammer f <Mechan> clevis
UKW (Ultrakurzwelle) <Funktech, Wellphys> USW (ultrashort wave); very high frequency (VHF)
UKW-Bereich m <Funktech> very-high-frequency range, VHF range, VHF region
UKW-Drehfunkfeuer n **(Ultrakurzwellen-Drehfunkfeuer)** <Lufttrans> omnidirectional radio range, VOR (Funkortung); very-high frequency omnidirectional range, VHF, very-high frequency omnirage, VHFO
UKW-Drossel f <Funktech> VHF choke
UKW-Frequenzmesser m <Metrol> very-high-frequency meter
UKW-Kursfunkfeuer n <Funktech> visual and aural range
UKW-Radio n <Fernseh, Funktech> VHF radio
UKW-Rundfunk m <Funktech> FM broadcasting, VHF broadcasting
UKW-Rundstrahlkursfunkfeuer n <Trans> VHF omnidirectional radio range
UKW-Sprechfunk m <Funktech, Mobilkom> VHF radio telephone
UKW-Sprechfunkgerät n <Funktech, Mobilkom> VHF radio telephone
Ulexin n <Chemie> cytisine, ulexine
Ulme f <Wassertrans> elm (Bauholz)
Ulmin n <Chemie> ulmin
Ulmin... <Chemie> ulmic
ulnar adj <Ergon> ulnar
ULSI (Ultragroßintegration) <Elektronik> ULSI (ultralarge-scale integration)
ULSI-Schaltkreis m <Elektronik> ULSI circuit
Ultra... <Chemie, Kerntech> ultra

ultradichtes Wellenlängenmultiplex *n* <Telekom> ultradense wavelength division multiplex *(Faseroptik)*
Ultrafeinfokus *m* <Kerntech> ultrafine focus
Ultrafilter *n* 1. <Chemie> ultrafilter; 2. <Chemtech> membrane filter
Ultrafiltrat *n* <Chemie> ultrafiltrate
Ultrafiltration *f* <Chemie, Chemtech> ultra filtration
Ultrafiltrieren *n* <Chemie> ultra filtration *(Vorgang)*
Ultragroßintegration *f (ULSI)* <Elektronik> ultralarge-scale integration *(ULSI)*
Ultragroßintegration-Schaltkreis *m* <Elektronik> ultra-large-scale integration circuit
Ultrahoch… <Elektronik, Telekom, Trans> ultrahigh
Ultrahochfrequenz *f (UHF)* <Elektronik, Fernseh, Funktech, Telekom, Wellphys> ultrahigh frequency *(UHF)*
Ultrahochfrequenzwelle *f* <Wellphys> ultrahigh frequency wave
Ultrahochgeschwindigkeitsverkehr *m* <Trans> ultrahigh speed traffic
Ultrahochtemperatur *f (UHT)* <Lebensmittel> ultrahigh temperature *(UHT)*
Ultrahochvakuum *n* 1. <Heiz & Kälte, Phys> ultrahigh vacuum; 2. <Thermod> ultravacuum
Ultrakurzwelle *f (UKW)* <Funktech, Wellphys> ultrashort wave, very short wave *(USW)*
Ultrakurzwellen-Drehfunkfeuer *n* 1. <Funkort> very-high-frequency omnidirectional radio range; 2. <Lufttrans> very-high-frequency omnirange, VHFO *(UKW-Drehfunkfeuer)*
Ultrakurzwellen-Frequenz *f (UKW)* <Elektronik, Funktech> very high frequency, VHF
Ultrakurzwellen-Rundstrahlkursfunkfeuer *n* <Trans> very-high-frequency omnidirectional radio range
Ultrakurzwellen-Sprechfunkgerät *m* <Funktech> very-high-frequency radio telephone
Ultraleichtlegierung *f* <Raumfahrt> ultralight alloy
Ultramarin *n* <Chemie> ultramarine *(Farbe)*
Ultramikroanalyse *f* <Kerntech> ultramicroanalysis
Ultramikroskop *n* <Chemie> ultramicroscope
Ultramikroskopie *f* <Chemie> ultramicroscopy
Ultrapasteurisierung *f* 1. <Lebensmittel> uperization; 2. <Thermod> ultrapasteurization
Ultraschall *m* 1. <Elektriz> supersonic, ultrasonics; 2. <Phys, Strahlphys> ultrasound
Ultraschallabtragung *f* <Maschinen> ultrasonic removal
Ultraschallbad *n* <Elektriz, Labor> ultrasonic bath
Ultraschallbearbeitung *f* 1. <Kerntech> ultrasonic machining *(von Materialien)*; 2. <Maschinen> ultrasonic machining; 3. <Strahlphys> ultrasonic machining *(von Werkstücken)*
Ultraschallbohren *n* <Maschinen> ultrasonic drilling
Ultraschallbohrmaschine *f* <Fertig> ultrasonic drilling machine
Ultraschallchemie *f* <Chemie> ultrasonic chemistry
Ultraschalldetektor *m* <Trans> ultrasonic detector
Ultraschalldickenmessgerät *n* <Gerät> *(AE)* ultrasonic thickness gage, *(BE)* ultrasonic thickness gauge
Ultraschallfehlerprüfung *f* <Eisenbahn> ultrasonic flaw detection
Ultraschallfrequenz *f* <Akustik, Phys> ultrasonic frequency
Ultraschallgenerator *m* 1. <Labor> ultrasound generator; 2. <Strahlphys> ultrasonic generator
Ultraschallgeschwindigkeit *f* <Phys> hypersonic speed
Ultraschallkraftstoffzerstäuber *m* <Trans> ultrasonic fuel atomizer
Ultraschalllotung *f* <Funkort> ultrasonic sounding
Ultraschalllötung *f* <Maschinen> ultrasonic soldering

Ultraschallmaterialprüfung *f* <Maschinen> ultrasonic materials testing
Ultraschallmikroskop *n (Labor)* ultrasonic microscope
Ultraschallortung *f* <Wellphys> ultrasonic sounding
Ultraschallprüfung *f* 1. <Eisenbahn> ultrasonic probe; 2. <Kerntech> ultrasonic testing, ultrasonic examination; 3. <Maschinen, Mechan> ultrasonic examination, ultrasonic testing
Ultraschallreinigung *f* <Aufnahme, Fernseh> ultrasonic cleaning
Ultraschallreinigungsanlage *f* <Fertig> ultrasonic cleaner
Ultraschallschweißen *n* 1. <Elektriz, Kunststoff> ultrasonic welding; 2. <Thermod> ultrasonic welding, ultrasonic sealing
Ultraschallsonde *f* <Bau> ultrasonic probe
Ultraschalltechnik *f* 1. <Maschinen> ultrasonic engineering; 2. <Wellphys> ultrasonics
Ultraschalltransducer *m* <Elektriz> ultrasonic transducer
Ultraschalluntersuchung *f* 1. <Kerntech, Mechan> ultrasonic examination; 2. <Strahlphys> ultrasound scan; 3. <Wellphys> ultrasonic inspection
Ultraschallwelle *f* <Elektriz, Wellphys> ultrasonic wave
Ultraschallzerspanung *f* <Fertig> ultrasonic machining
Ultra-SCSI-RAID-Steuereinheit *f* <Comp & DV> Ultra SCSI RAID controller
ultrasonisch *adj (US)* <Elektriz> supersonic, ultrasonic *(SS)*
Ultra-Supertanker *m* <Erdöl> ULCC, ultralarge crude carrier
ultraviolett *adj (UV)* <Optik, Phys> ultraviolet *(UV)*
Ultraviolett *n* <Raumfahrt> ultraviolet
ultraviolett sichtbares Spektrophotometer *n* <Labor> ultraviolet-visible spectrophotometer
ultraviolettes Licht *n* <Kunststoff, Strahlphys> ultraviolet light
Ultraviolettfilter *n* <Foto, Sicherheit> ultraviolet filter
Ultraviolettfotografie *f* <Strahlphys> ultraviolet photography
Ultraviolettkatastrophe *f* <Phys> ultraviolet catastrophe
Ultraviolettlampe *f* <Labor> ultraviolet lamp
Ultraviolettlöschen *n* <Comp & DV> ultraviolet erasing
Ultraviolett-Mikroskop *n* <Strahlphys> ultraviolet microscope
Ultraviolettspiegel *m* <Strahlphys> ultraviolet mirror
Ultraviolettstrahlen *mpl* <Optik, Wellphys> ultraviolet rays
Ultraviolettstrahlung *f* <Nichtfoss Energ, Optik, Phys, Raumfahrt, Strahlphys, Umweltschmutz> ultraviolet radiation
Ultrazentrifuge *f* <Chemie, Kerntech, Labor, Phys> ultracentrifuge
Ultrazentrifugieren *n* 1. <Chemie> ultracentrifugation *(Vorgang)*; 2. <Kerntech> ultracentrifugation
umarbeiten *v* <Bau> redesign
Umbau *m* 1. <Bau> reconstruction; 2. <Maschinen> conversion
umbauen *v* <Bau> rebuild, reconstruct
Umbauen *n* <Bau> rebuilding
Umbausatz *m* <Maschinen> conversion kit
Umbell… <Chemie> umbellic
umbiegen *v* 1. <Mechan> crimp; 2. <Papier> angle
Umbilden *n* **eines Zuges** <Eisenbahn> reforming of a train
umbördeln *v* <Bau> bead over
umbrechen *v* <Druck> make up
umbrochene Korrekturfahnen *fpl* <Druck> page proofs
Umbruch *m* <Druck> make-up
Umbruchprogramm *n* <Druck> page make-up program

Umbuchantrag

Umbuchantrag m <Mobilkom> cell change request
Umbuchen n **bei bestehender Verbindung** <Mobilkom> in-call handover *(Zellenwechsel)*
Umbuchnachricht f <Mobilkom> roaming indication
Umcodierung f 1. <Comp & DV> transform; 2. <Telekom> transcoding, code conversion
Umdrehung f 1. <Lufttrans> *(AE)* turnaround, *(BE)* turnround *(eines Luftfahrzeugs)*; 2. <Maschinen, Papier, Phys, Raumfahrt> revolution, turn
Umdrehung f **pro Minute** <Kfztech, Phys> revolution per minute
Umdrehungsgeschwindigkeit f <Nichtfoss Energ> rotational speed
Umdrehungskegel m <Geom> cone of revolution
Umdrehungszähler m <Gerät> revolution counter
Umdrehungszählgerät n <Gerät> revolution counter
Umdruck m <Druck> transfer
umdrucken v <Druck> transfer
umfahren v <Wassertrans> round
Umfahrgleis n <Eisenbahn> loop line
Umfahrung f <Bau> bypass
Umfahrungsstrecke f <Kohlen> bypass
Umfang m 1. <Bau> perimeter, periphery; 2. <Geom> circumference, perimeter; 3. <Mechan> bulk; 4. <Patent> scope *(Umfang eines Schutzes)*; 5. <Qual> scope
• **Umfang eines Werks berechnen** <Druck> cast off
umfangreiches Büropaket n <Comp & DV> full suite
Umfangsbeanspruchung f <Metall> circumferential stress
Umfangsgeschwindigkeit f 1. <Lufttrans> circumferential speed; 2. <Nichtfoss Energ> peripheral velocity
Umfangsschleifen n <Fertig> peripheral grinding
Umfangsspannung f <Fertig> circumferential stress *(Kunststoffinstallationen)*
Umfangsteilung f <Mechan> circular pitch
umfassen v <Maschinen> encompass
umfassend adj <Comp & DV> global
umfassende Schutzwirkung f <Sicherheit> overall protection
Umfassung f <Ergon> enveloping grip
Umfassungsmauer f <Bau> enclosing wall
umflochten adj <Elektriz, Elektrotech> braided
umflochtene Leitung f <Elektriz> braided wire
umflochtener Schlauch m <Kunststoff> braided hose
umformen v 1. <Elektrotech> convert; 2. <Math> convert *(einer Gleichung)*
Umformen n 1. <Kunststoff> forming; 2. <Maschinen> forming, pressing, shaping
umformende Bearbeitung f <Maschinen> forming, shaping
Umformer m <Elektriz, Elektronik, Elektrotech, Gerät, Kontroll> converter, inverter, transducer
Umformergruppe f <Elektrotech> motor generator set
Umformersatz m <Elektrotech> motor generator set
Umformstation f <Elektriz> converter station, converting station
Umformung f 1. <Fertig> metal working; 2. <Math> convertation *(einer Gleichung)*
Umformwerk n <Elektriz> converter station, converting station
Umformwerkzeug n <Maschinen> forming tool, shaping die
Umführungsstraße f <Trans> diversion
Umführungsventil n <Labor> bypass valve
Umgang m 1. <Elektronik> convolution *(der Wicklung)*; 2. <Kerntech> gallery
Umgangsleitung f <Erdöl> bypass
umgebautes Fahrzeug n <Nichtfoss Energ> converted vehicle

umgebend adj <Ergon, Heiz & Kälte> ambient
umgebende Luft f <Umweltschmutz> ambient air
umgebendes Schutzgitter n <Sicherheit> enclosing guard
Umgebung f 1. <Anstrich> environment; 2. <Comp & DV> environment *(des Systems)*; 3. <Ergon> environment; 4. <Math> surroundings *(eines Punktes)*; 5. <Phys> surroundings
Umgebung f **für Sprachenunterstützung** <Comp & DV> language support environment
Umgebungsbedingungen fpl <Ergon, Metrol, Qual> environmental conditions
Umgebungsbelastung f <Umweltschmutz> environmental pollution
Umgebungsdrehmoment n <Raumfahrt> environmental torque
Umgebungsgas n <Kerntech> peripheral gas
Umgebungsgeräusch n <Telekom> ambient noise
Umgebungsgeräuschpegel m <Gerät, Sicherheit> ambient noise level
Umgebungslärm m <Ergon, Sicherheit> ambient noise
Umgebungsluft f <Ergon, Heiz & Kälte> ambient air
Umgebungsluftfeuchte f <Gerät, Kohlen> ambient humidity
Umgebungslufttemperatur f <Ergon, Kohlen> ambient air temperature
umgebungsluftunabhängiges Atemschutzgerät n <Sicherheit> supplied-air respirator
umgebungsluftunabhängiges Fluchtgerät n <Sicherheit> escape self-contained breathing apparatus, supplied-air respirator *(Atemschutzgerät)*
Umgebungsprüfkammer f <Qual, Raumfahrt> environmental test chamber
Umgebungsradioaktivität f <Strahlphys> environmental radioactivity
Umgebungssicherheit f <Sicherheit, Umweltschmutz> environmental safety
Umgebungstemperatur f <Fertig, Heiz & Kälte, Kontroll, Metall, Metrol, Phys, Thermod> ambient temperature
Umgebungsvariable f <Comp & DV> environment variable
umgehen v <Elektriz, Gerät, Kfztech, Kontroll> bypass
Umgehung f <Bau, Hydraul> bypass
Umgehungsgleis n <Eisenbahn> loop line
Umgehungskanal m <Wasserversorg> diversion canal
Umgehungsleitung f <Hydraul, Maschinen> bypass
Umgehungsstraße f 1. <Bau> bypass; 2. <Trans> bypass road
Umgehungsventil n 1. <Lufttrans> antisurge valve; 2. <Maschinen> bypass valve
umgekehrt adj 1. <Math> contrary, converse, inverse, opposite, reverse; 2. <Phys> reversible
umgekehrt adv <Math> conversely, inversely, oppositely, vice versa
umgekehrt proportional adj <Math> inversely proportional
umgekehrte Bildschirmdarstellung f <Fernseh> reverse video
umgekehrte Polnische Notation f *(UPN)* <Comp & DV> reverse Polish notation *(RPN)*
umgekehrte V-Antenne f <Funktech> inverted-V dipole
umgekehrter Carnot'scher Kreisprozess m <Thermod> *(AE)* vapor compression cycle, *(BE)* vapour compression cycle
umgekehrtes Steuerwerk n <Lufttrans> reversed controls
umgekehrtes Verhältnis n <Maschinen> inverse ratio, reciprocal ratio • **in umgekehrtem Verhältnis** <Maschinen> in inverse ratio

umgeschriebener Kreis m <Geom> circumcircle, circumscribed circle
umgewälzte Luft f <Heiz & Kälte> recirculated air
umgießen v <Chemtech> decant
Umgipserei f <Ker & Glas> jointing yard
umgraben v <Bau> spade
Umgreifen n <Ergon> grasping
umgrenzen v 1. <Geom> bound, delimit; 2. <Maschinen> enclose
umgrenzendes Rechteck n <Geom> bounding limits, bounding rectangle
Umhaspeln n <Textil> rewinding
umherlavieren v <Wassertrans> cast about
umhüllen v 1. <Bau> clad, coat, sheathe; 2. <Elektrotech> sheathe (Kabel); 3. <Fertig> clad; coat (Elektrode)
Umhüllen n <Bau> coating
umhüllt adj <Elektriz, Elektrotech> covered
umhüllt adj/**sehr dick** <Fertig> shielded (Elektrode)
umhüllte Stabelektrode f <Elektriz> coated rod electrode
Umhüllung f 1. <Druck> wrapping; 2. <Elektriz, Elektrotech> jacket, sheathing; 3. <Fertig> sheath (optische Faser); shielding (Elektrode); 4. <Kerntech> cladding (der Stäbe); 5. <Maschinen> shrouding, shroud; 6. <Mechan> jacket; 7. <Raumfahrt> casing, shroud (Raumschiff); 8. <Textil> wrapping
Umhüllungsmaschine f <Verpack> envelope machine
Umkehr f <Maschinen> reversal (einer Bewegung); reversing (der Drehrichtung)
Umkehranzeige f <Comp & DV> inverse video, reverse image, reverse video
Umkehrbad n <Foto> reversing bath
umkehrbar adj <Math, Thermod> reversible
umkehrbar eindeutig adj <Math> bijective
umkehrbare Abschaltung f <Kerntech> reversible shutdown
umkehrbarer Wandler m <Elektrotech> reversible transducer
umkehrbarer Zähler m <Gerät> reversible counter
Umkehrbarkeit f <Phys> reversibility
Umkehrbeschichten n <Kunststoff> reverse roll coating
Umkehrbild n <Phys> inverted image
Umkehrdampf m <Hydraul> reversing steam
Umkehrdarstellung f <Comp & DV> inverse video
Umkehreingang m <Elektronik> inverting input
Umkehremission f <Elektronik> reverse emission
umkehren v 1. <Comp & DV> invert; 2. <Elektronik, Maschinen> reverse
Umkehrentwicklung f <Foto> reversal processing
Umkehrer m <Comp & DV> inverter
Umkehrfilm m <Foto> reversal film
Umkehrfunktion f <Math> inverse function
Umkehrgetriebe n <Maschinen> reverse gear, reversing gear
Umkehrgrenzpunkt m <Lufttrans> point of no return
Umkehrkopieren n <Foto> reverse printing
Umkehrlinse f <Fernseh> beam-reversing lens
Umkehrliste f <Comp & DV> push-down list
Umkehrluftschraube f <Lufttrans> reversible pitch propeller
Umkehrmotor m <Elektriz, Elektrotech, Trans> reversible motor, reversing motor
Umkehrosmose f <Abfall, Bau> reverse osmosis
Umkehroszillator m <Elektronik> inverter oscillator
Umkehrpunkt m 1. <Fertig> (AE) dead center, (BE) dead centre; 2. <Math> reversal point; 3. <Metall> cusp (einer Kurve)
Umkehrreaktion f <Kerntech> reverse reaction
Umkehrrichtung f <Elektrotech> reverse direction

Umkehrring m <Foto> reversing ring
Umkehrrolle f <Maschinen> return pulley
Umkehrschalter m <Elektrotech> reversible switch, reversing switch
Umkehrspanne f <Gerät> dead spot
Umkehrstapel m <Comp & DV> push-down stack
Umkehrsteuerwelle f <Eisenbahn> reversing shaft
Umkehrsucher m <Foto> reversal finder
Umkehrtemperatur f <Elektriz> inversion temperature
Umkehrtransistor m <Elektronik> inverting transistor
Umkehrung f 1. <Druck> reversal; 2. <Elektronik> inversion; 3. <Telekom> reversal
Umkehrung des Luftschraubengangs <Lufttrans> reversal of the propeller pitch
Umkehrung des Steuerungsmomentes <Lufttrans> control reversal
Umkehrventil n <Hydraul> reversing valve (Dampfstrom)
Umkehrverfahren n <Druck> reversal process
Umkehrverstärker m <Elektronik> differential amplifier, inverting amplifier
Umkehrwalzenschalter m <Elektriz> reversing drum switch
umkippen v <Bau> tilt, tip, tip up
Umkippen n <Bau> tipping
umklappbares Bein n <Foto> fold-over leg (eines Stativs)
Umklappen n <Fernseh> flopover
Umklappvorgang m <Kerntech> rotating process
Umkleideraum m <Bau> changing room, locker room
Umklöppelung f <Fertig, Textil> braiding
umkonfigurierbar adj <Comp & DV> reconfigurable
umkonfigurieren v <Comp & DV> reconfigure
Umkoppelungsvorgang m <Comp & DV> routing
Umkreis m 1. <Bau> perimeter, periphery; 2. <Geom> circumcircle
umkreisen v <Raumfahrt> orbit
Umkreismittelpunkt m <Fertig, Geom> (AE) circumcenter, (BE) circumcentre
Umkreisradius m <Geom> long radius
Umkristallisierung f <Metall> recrystallization
Umladegleis n <Eisenbahn> transfer track, transshipment track
umladen v <Wassertrans> transship (Ladung)
Umladung f <Wassertrans> transshipment
Umlage f <Elektrotech> turn (Spule, Wicklung)
Umlauf m <Erdöl, Maschinen> circulation
Umlaufabwicklung f <Elektronik> deconvolution
Umlaufbahn f 1. <Funktech, Phys, Raumfahrt> orbit; 2. <Strahlphys> orbital • **in polarer Umlaufbahn** <Raumfahrt> polar-orbiting
Umlaufbahnberechnung f <Raumfahrt> orbit determination
Umlaufbahneinschuss m <Raumfahrt> orbital injection
Umlaufbahnflug m <Raumfahrt> orbital flight
Umlaufbahnhöhe f <Raumfahrt> orbital altitude, orbital height
Umlaufbahnneigung f <Raumfahrt> inclination of satellite orbit, orbit inclination, satellite orbit inclination
Umlaufbahnradiusverringerung f <Raumfahrt> orbital decay
Umlaufbahnsteuerung f <Raumfahrt> (AE) orbital maneuvering system, (BE) orbital manoeuvering system
Umlaufbahnverkleinerung f <Raumfahrt> orbit trimming
Umlaufbiegetorsionsprüfung f <Werkprüf> rotary bending and torsion fatigue test
Umlaufbiegeversuch m <Metall> rotating bending test
Umlaufdurchmesser m <Maschinen> swing
Umlaufdurchmesser m **über Bett** <Maschinen> swing of the bed, swing-over bed

Umlaufdurchmesser

Umlaufdurchmesser m **über Kröpfung** <Maschinen> swing-over gap
Umlaufdurchmesser m **über Schlitten** <Maschinen> swing of the rest, swing-over saddle
umlaufen v 1. <Funktech> orbit; 2. <Heiz & Kälte> circulate
Umlaufen n <Druck> overrun
umlaufende Kante f <Konstzeich> circumferential edge
umlaufende Nut f <Optik> groove
umlaufendes Ende n <Papier> tail end
umlaufendes Teil n <Maschinen> rotating part
Umlauffrequenz f <Phys> frequency of gyration
Umlaufgebläse n <Fertig> rotary blower
umlaufgeschmiert adj <Fertig> pressure-lubricated
Umlaufgeschwindigkeit f 1. <Aufnahme> running speed; 2. <Fertig> rate of circulation; 3. <Maschinen> rotating speed
Umlaufgetriebe n 1. <Fertig> epicyclic gear; 2. <Kfztech> angle transmission, planetary gears; 3. <Maschinen> epicyclic gear, epicycloidal gear, sun-and-planet gearing
Umlaufintegral n <Phys> circulation *(eines Vektorfeldes)*
Umlaufkessel m <Hydraul> circulating boiler
Umlaufkühlung f <Heiz & Kälte> refrigeration by circulation
Umlaufmotor m <Kfztech> rotary engine
Umlaufpumpe f <Heiz & Kälte> circulation pump
Umlaufrad n <Fertig> planetary gear
Umlaufreinigung f <Lebensmittel> CIP, cleaning in place
Umlaufschlange f <Heiz & Kälte> run-around coil
Umlaufschmierung f 1. <Fertig> circulation lubrication; 2. <Maschinen> recirculating lubrication, recirculation lubrication
Umlaufsystem n 1. <Maschinen> circulating system; 2. <Raumfahrt> circulator *(Weltraumfunk)*
Umlaufzeit f <Eisenbahn> *(AE)* turnaround time, *(BE)* turnround time
Umlegung f <Telekom> transfer *(Ruf)*
umleiten v 1. <Bau> divert; 2. <Elektriz, Gerät, Kfztech, Kontroll> bypass; 3. <Telekom> redirect; divert *(Anruf)*; 4. <Trans> divert *(Verkehr)*; 5. <Wassertrans> reroute *(Seehandel)*; 6. <Wasserversorg> deflect *(Fluss)*
Umleitung f 1. <Bau> bypass; 2. <Telekom> alternative routing, rerouting; 3. <Trans> bypass, diversion
Umleitungsempfehlung f <Trans> advisory diversion
Umleitungskanal m <Wasserversorg> diversion canal
Umlenkblech n <Bau> baffle plate
Umlenkhebel m <Kfztech> relay arm
Umlenkplatte f <Bau> baffle plate
Umlenkprisma n <Telekom> deviation prism
Umlenkrinne f <Ker & Glas> deflector chute; deflector *(Glas)*
Umlenkrolle f 1. <Fertig> tail sheave; 2. <Sicherheit> pulley block
Umlenkscheibe f <Maschinen> return pulley
Umlenkung f <Lufttrans> deflection
Umlenkwalze f <Ker & Glas> bending roller *(Libbey-Owens Verfahren)*
Umluft f <Heiz & Kälte> circulating air, recirculated air
Umluftgerät n <Heiz & Kälte> circulating air unit
umluftunabhängiges Atemschutzgerät n <Sicherheit> air-supplied respirator
ummanteln v 1. <Anstrich> encapsule; 2. <Bau> case, sheathe; encase *(mit Beton)*; 3. <Elektrotech> sheathe *(Kabel)*; 4. <Fertig> jacket
Ummanteln n <Kunststoff, Sicherheit> sheathing
ummantelt adj 1. <Fertig> coated *(Kunststoffinstallationen)*; 2. <Mechan> jacketed, lagged
ummantelte Schraube f 1. <Lufttrans> shrouded propeller; 2. <Maschinen> shrouded screw

ummanteltes Hohlglas n <Ker & Glas> cased hollow ware
ummanteltes Thermoelement n <Kerntech> sheathed thermocouple
Ummantelung f 1. <Elektriz, Elektrotech> envelope, jacket, sheath; sheathing *(eines Kabels)*; 2. <Ker & Glas> casing; 3. <Kerntech> jacketing, shell; 4. <Lufttrans> case *(Flugwesen)*; 5. <Maschinen> jacket; 6. <Mechan> casing; 7. <Meerschmutz> skirt
Ummantelungsring m <Lufttrans> shroud ring *(Turbinentriebwerk)*
ummauerter Raum m <Bau> walled enclosure
Umordnung f <Math> permutation
Umordnungsstoß m <Kerntech> rearrangement collision
U-Motor m <Kfztech> U-type engine, underfloor engine
umplanen v <Bau> redesign
Umpolarisierung f <Raumfahrt> polarity reversal
Umpolung f <Elektrotech, Telekom> polarity reversal
Umpolungsschalter m <Elektrotech> polarity-reversing switch
Umrahmung f <Bau> framing *(Trennwand)*
Umrahmungsleiste f <Druck> box rule
Umrandung f <Druck> border
umrangierbares blockierungsfreies Netz n <Telekom> rearrangeable nonblocking network
Umrechnung f <Fertig> conversion *(Kunststoffinstallationen)*
Umrechnungstabelle f <Math> conversion table
Umreifungsgeräte npl <Verpack> strapping equipment
Umreifungsmaschine f <Fertig> strapping machine
Umreifungsstahl n <Verpack> strapping steel
umreißen v <Bau> outline
Umrichter m <Elektriz, Elektrotech> converter
Umriss m 1. <Bau, Fertig> contour; 2. <Maschinen> outline • **im Umriss fräsen** <Fertig> contour
Umrissfräsen n <Maschinen> contour milling
Umrissprojektor m <Metrol> profile projector
Umrisszeichnung f <Maschinen> outline drawing
Umrollen n <Papier> rewind
Umroller m <Papier> rewinder
Umrüstanlage f <Verpack> conversion machinery
Umrüstsatz m <Maschinen> conversion kit
Umrüstung f <Maschinen> conversion, retooling
Umschalt... <Maschinen, Heiz & Kälte> reversing
umschalten v 1. <Comp & DV> shift, toggle; 2. <Kontroll> switch over; 3. <Regelung> redirect *(Signalfluss)*; 4. <Telekom> switch
Umschalter m 1. <Comp & DV> toggle; 2. <Elektrotech> change-over switch, double-throw switch, selector switch; switch *(Änderung)*; 3. <Wassertrans> change-over switch *(Elektrik)*
Umschalter m **mit beweglicher Faser** <Telekom> *(AE)* moving-fiber switch, *(BE)* moving-fibre switch
Umschalthebel m <Maschinen> reversing lever
Umschaltklappe f <Heiz & Kälte> change-over damper
Umschaltkontakt m <Elektrotech> double-throw contact
Umschaltrelais n <Elektriz, Elektrotech> change-over relay
Umschalttor n <Regelung> change-over gate
Umschaltung f 1. <Comp & DV> shift, toggle; 2. <Telekom> transfer *(Ruf)*
Umschaltung f **auf Bereitschaft** <Telekom> change-over to stand-by
Umschaltung f **auf Reserve** <Telekom> change-over to stand-by
Umschaltventil n <Hydraul> switch valve
Umschaltzeichen n <Comp & DV> shift character
Umschaltzeichen n **für Dauerumschaltung** <Comp & DV> SO character, shift-out character

Umschaltzeit f <Comp & DV> (AE) turnaround time, (BE) turnround time
Umschlag m <Trans, Wassertrans> transshipment (Schifffahrt)
Umschlaganlagen fpl <Trans> transship facilities
Umschlagen n 1. <Druck> work and turn; 2. <Verpack> overwrap
Umschlagklappe f <Verpack> tuck-in flap
Umschlagplatz m <Trans> terminal
Umschlagpumpe f <Meerschmutz> transfer pump
Umschlagpunkt m <Kerntech> transition point
Umschlagverpackung f <Verpack> overwrapping packaging
Umschlagzeit f <Elektriz, Elektrotech> transit time (Relais)
umschließen v <Maschinen, Sicherheit> enclose
umschließendes Gehäuse n <Hydraul> enclosed casing (Turbine)
Umschließungseinrichtung f <Sicherheit> containment equipment
Umschlingen n <Textil> overcasting
Umschlingung f <Fertig> contact
Umschlingungsbogen m <Fertig> arc of conduct (Getriebelehre, Riemenantrieb)
Umschlingungswinkel m <Fertig> angle of contact, contact angle
umschlossen adj <Maschinen> enclosed
umschlossenes Gerät n <Sicherheit> enclosed equipment
umschmelzen v <Fertig> recast
Umschmelzen n <Metall> refusion, remelting
Umschmelzmaschine f <Ker & Glas> (BE) burning-off and edge-melting machine, (AE) remelting machine
Umschmelzwerk n <Metall> smelting and refining works
Umschnitt m <Akustik> rerecording
umschreiben v 1. <Comp & DV> transcribe; 2. <Geom> circumscribe
umschriebener Kreis m <Geom> circumscribed circle
umsetzen v 1. <Comp & DV> convert, translate; 2. <Elektrotech> convert; 3. <Funktech, Telekom> convert (Frequenz)
Umsetzer m 1. <Comp & DV> converter; 2. <Gerät> converter; 3. <Raumfahrt> repeater (Weltraumfunk); 4. <Telekom> converter; translator (Kanal)
Umsetzerchip m <Elektronik> converter chip
Umsetzergehäuse n <Telekom> converter cabinet
Umsetzersatellit m 1. <Fernseh> relay satellite; 2. <Raumfahrt> relay satellite, repeater satellite
Umsetzersender m <Fernseh> relay transmitter
Umsetzung f 1. <Comp & DV> conversion, encoding, transform; 2. <Elektrotech> transformation; 3. <Funktech> conversion (Frequenzen); 4. <Telekom> translation
Umsetzungsoszillator m <Elektronik, Funktech> conversion oscillator
Umsetzungsprodukt n <Abfall> CFS-processed waste, CFS-treated waste, solidified waste, solidified product, solidified material
Umsetzungsprogramm n <Comp & DV> translator
umspannen v <Fertig> rechuck (Werkstück)
Umspannung f 1. <Elektrotech> transformation; 2. <Fertig> rechucking
Umspannwerk n <Elektrotech> distribution station
umspinnen v <Textil> wrap
umspringen v <Wassertrans> come-to, shift (Wind)
Umspringen n **des Wellentyps** <Elektronik> mode jump (Mikrowellen)
Umspringen n **des Windes** <Wassertrans> sudden change of wind direction
umspringende Winde mpl <Wassertrans> baffling winds

umspritzt adj <Fertig> (AE) molded, (BE) moulded (Kunststoffinstallationen)
umspulen v <Textil> rewind
Umspulen n <Textil> packaging, rewinding
umstechen v <Textil> overcast (Nähen)
Umstechen n <Textil> overcasting
umstecken v <Fertig> change (Räder)
Umstecken n <Fertig> changing (Räder)
Umsteckrad n <Fertig> pick-off change gear, pick-off gear
Umsteckwalzwerk n <Metall> looping mill
Umsteigen n <Eisenbahn> crossover (für Passagiere)
Umstellbahnhof m <Eisenbahn> shunting yard
Umstellen n **von Formeln** <Math> transposition of formulae
Umstellung f <Comp & DV> migration
Umstellungsverfahren n <Verpack> changeover procedure
umsteuerbarer Zähler m <Gerät> reversible counter
Umsteuerdampf m <Hydraul> reversing steam
Umsteuerknagge f <Fertig> reverse dog
Umsteuermöglichkeit f <Eisenbahn> reversibility
umsteuernd adj <Elektronik> reversing
Umsteuerpropeller m <Wassertrans> reversible pitch propeller; reversible pitch propeller (Bootsbau)
Umsteuerung f <Maschinen> reversing gear, reversing motion
Umsteuerung f **des Propellers** <Lufttrans> reversal of the propeller
Umsteuerungsvorrichtung f <Maschinen> reversing gear
Umsteuerungswelle f <Maschinen> reversing shaft
Umsteuerventil n 1. <Hydraul> reversing valve (Dampfstrom); 2. <Maschinen> reversing valve
Umströmung f <Wasserversorg> circulation
umstürzen v <Maschinen, Sicherheit> overturn
Umtastung f <Telekom> shift keying
Umverpackung f 1. <Abfall> overpackaging, secondary packaging; 2. <Lebensmittel> packaging
umwälzen v <Heiz & Kälte> circulate (Luft)
Umwälzheizlüfter m <Sicherheit> regenerative airheater
Umwälzpumpe f 1. <Fertig> circulating pump, circulation pump, recirculating pump; 2. <Heiz & Kälte> circulation pump; 3. <Hydraul> circulating pump; 4. <Lebensmittel> circulating pump, circulation pump, recirculating pump; 5. <Maschinen> recirculating pump
Umwälzung f 1. <Fertig> circulation; 2. <Maschinen> recirculation
Umwälzventilator m <Heiz & Kälte> ventilating fan
umwandeln v 1. <Comp & DV> convert, transform, translate; assemble (Programmiersprache); 2. <Elektrotech> convert; 3. <Math> commute
Umwandler m <Kontroll> transducer
Umwandlung f 1. <Comp & DV> transform; conversion (von Zeichencodes); 2. <Elektrotech> conversion, transformation; 3. <Kerntech> conversion, transmutation; 4. <Papier> converting; 5. <Phys> transmutation; 6. <Telekom> conversion
Umwandlung f **von Fahrbahnen in Fußgängerbereiche** <Trans> pedestrianization
Umwandlungsöl n <Erdöl> conversion oil
Umwandlungsrate f <Umweltschmutz> transformation rate
Umwandlungstemperatur f <Heiz & Kälte, Metall> transition temperature
umweben v <Textil> envelop
Umweglenkung f <Telekom> alternative routing, rerouting
Umwelt f <Umweltschmutz> environment

Umweltbedingungen

Umweltbedingungen *fpl* <Elektriz, Metrol, Nichtfoss Energ, Telekom> environmental conditions
umweltbelastend *adj* <Umweltschmutz> environment-pollutant, pollutive
umweltbelastender Stoff *m* <Umweltschmutz> environment pollutant
Umweltbelastung *f* <Nichtfoss Energ, Umweltschmutz> environmental impact, environmental pollution
Umweltfaktor *m* <Nichtfoss Energ, Umweltschmutz> ecological factor
umweltfeindlich *adj* <Umweltschmutz> pollutive
Umweltforschung *f* <Umweltschmutz> environmental investigation, environmental research
umweltfreundlich *adj* 1. <Nichtfoss Energ> environmentally benign; 2. <Umweltschmutz, Verpack> environmentally friendly, non-polluting, pollution-free, pollutive-free
umweltfreundliche Energieerzeugung *f* <Umweltschmutz> clean power generation
umweltfreundliche Kohlentechnologie *f* <Kohlen> clean coal technology
umweltfreundliche Technologie *f* 1. <Abfall> NWT, clean technology; 2. <Nichtfoss Energ> green technology; 3. <Umweltschmutz> clean technology
umweltfreundlicher Brennstoff *m* <Kfztech, Kohlen> clean fuel
umweltfreundliches Auto *n* 1. <Kfztech> clean air car, clean fuel vehicle; 2. <Nichtfoss Energ> clean air car
umweltfreundliches Fahrzeug *n* <Kfztech> clean fuel vehicle
umweltfreundliches Kraftfahrzeug *n* <Kfztech> clean fuel vehicle
umweltfreundliches Verfahren *n* <Umweltschmutz> low-pollution process
Umweltgefahr *f* <Nichtfoss Energ, Umweltschmutz> ecological menace
umweltgefährdend *adj* <Umweltschmutz> hazardous to the environment
umweltgefährdender Abfall *m* <Abfall> environmentally hazardous waste
Umweltgrenzwert *m* <Umweltschmutz> environmental threshold value
Umweltimpulse *mpl* <Ergon> exteroceptive impulses
Umweltkatastrophe *f* <Umweltschmutz> environmental disaster
Umweltkontrolle *f* <Gerät> environment monitoring
Umweltlärm *m* <Sicherheit> environmental noise
Umweltplanung *f* 1. <Nichtfoss Energ> environmental planning; 2. <Umweltschmutz> environmental planning, planned environment
umweltschädlich *adj* <Umweltschmutz> harmful to the environment
Umweltschadstoff *m* <Meerschmutz, Umweltschmutz> environment pollutant, environmental pollutant, pollutant
Umweltschmutzstoff *m* <Umweltschmutz> environmental pollutant
umweltschonende Technologie *f* <Umweltschmutz> environmental preserving technology
Umweltschutz *m* <Nichtfoss Energ, Umweltschmutz> environmental pollution control, environmental protection, environmentalism, pollution control
Umweltschutznorm *f* <Umweltschmutz> environmental protection standard
Umweltschutztechnik *f* <Umweltschmutz> environment protection technology; environmental technology; environmental engineering; pollution control equipment
Umweltsicherheit *f* <Sicherheit, Umweltschmutz> environmental safety
Umweltstress *m* <Raumfahrt, Umweltschmutz> environmental stress

Umwelttechnik *f* <Umweltschmutz> environmental engineering; environmental technology
Umwelttestverfahren *n* <Qual, Sicherheit, Umweltschmutz> environmental testing procedure
Umwelttoxikologie *f* <Umweltschmutz> environmental toxicology
Umweltüberwachung *f* <Umweltschmutz> environment monitoring, environmental supervision, environmental surveyance
Umweltüberwachungssystem *n* <Umweltschmutz> environmental monitoring system
umweltverschmutzend *adj* <Umweltschmutz> pollutive
Umweltverschmutzer *m* <Umweltschmutz> environmental pollutant
Umweltverschmutzung *f* <Umweltschmutz> environment pollution, environmental pollution
Umweltverschmutzung *f* **durch die Industrie** <Umweltschmutz> industrial pollution
Umweltverschmutzung *f* **durch Wärme** <Umweltschmutz> heat pollution
umweltverträglich *adj* <Umweltschmutz> environmentally compatible
Umweltverträglichkeit *f* <Umweltschmutz> environmental compatability, environmental compliance
Umweltverträglichkeitsprüfungsbericht *m* <Umweltschmutz> environmental impact statement
umweltverunreinigend *adj* <Umweltschmutz> pollutive
Umweltwissen *n* <Umweltschmutz> environmental science
Umweltwissenschaft *f* <Umweltschmutz> environmental science
Umwerter *m* <Telekom> translator
Umwertespeicher *m* <Telekom> translation store
Umwertung *f* <Telekom> translation
umwickeln *v* <Elektriz, Textil> rewind, wrap around
umwickelter Verschluss *m* <Verpack> taped closure
umwickeltes Garn *n* <Textil> wrapped yarn
Umwindefaser *f* <Textil> *(AE)* wrap fiber, *(BE)* wrap fibre
Umwindungsgarn *n* <Textil> covered yarn
Umzäunung *f* <Bau> boundary fence, enclosure
umzeichnen *v* <Konstzeich> redraw
Umziehgerüst *n* <Bau> *(AE)* traveling cradle, *(BE)* travelling cradle
umzwirntes Garn *n* <Textil> covered yarn
unabgeglichene Brücke *f* <Gerät> unbalanced bridge *(Messbrücke)*
unabgesättigt *adj* <Chemie> unsaturated
unabgeschirmte verdrillte Doppelleitung *f* <Telekom> unshielded twisted pair *(Telefonschnur)*
unabhängig *adj* <Heiz & Kälte> self-contained
unabhängige Energieerzeuger *mpl (IPPs)* <Kohlen> independent power producers npl, IPPs
unabhängige Erregung *f* <Elektrotech> independent excitation
unabhängige Hinterradaufhängung *f* <Kfztech> independent rear suspension
unabhängige Navigationshilfe *f* <Lufttrans> self-contained navigational aid
unabhängige Steuerung *f* <Kerntech> independent control
unabhängige Stromversorgung *f* <Eisenbahn> shore supply
unabhängige Variablen *f* <Math> independent variables
unabhängiger Patentanspruch *m* <Patent> independent claim
unabhängiges Gerät *n* <Comp & DV> self-contained equipment
unabhängiges Seitenband *n (ISB)* <Funktech> independent sideband *(ISB)*

unabhängiges System n <Kontroll> stand-alone system
unabhängiges Teilchenmodell n <Kerntech> independent particle model
unangreifbar machen v <Phys> passivate *(Korrosion)*
unannehmbar adj <Patent, Qual> unacceptable
unär adj <Comp & DV> monadic, unary
unäre Operation f <Comp & DV> monadic operation, unary operation
unaufbereitete Daten npl <Elektronik> raw data
unausgeglichener Kanal m <Aufnahme> unbalanced channel
unausgewuchtet adj <Fertig> unbalanced
unbeabsichtigtes Bremsen n <Trans> accidental braking
unbeantworteter Anruf m <Telekom> no-reply call, unanswered call
unbearbeitet adj <Maschinen> unfinished
unbearbeitete Scheibe f <Maschinen> blank washer
unbearbeitetes Teil n <Maschinen> blank
unbearbeitetes Werkstück n <Fertig> rough work, unmachined work
unbeaufsichtigter Betrieb m <Kontroll> unattended operation
Unbedenklichkeitsschwelle f <Umweltschmutz> limit of absolute safety
unbedingt adj <Comp & DV> imperative, unconditional
unbedingte Anweisung f <Comp & DV> unconditional statement
unbedingte Instruktion f <Comp & DV> imperative instruction
unbedingte Verzweigung f <Comp & DV> unconditional branch
unbedingter Sprungbefehl m <Comp & DV> unconditional jump
unbedrahteter Chip-Carrier <Elektronik, Elektrotech> leadless chip carrier
unbefugter Zugang m <Kontroll> unauthorized access
unbefugter Zugriff m <Kontroll> unauthorized access
unbegrenztes Ansprechen n **auf Impuls** *(IIR)* <Elektronik> infinite impulse response *(IIR)*
unbehauen adj <Fertig> unhewn
unbehebbarer Fehler m <Comp & DV> irrecoverable error
unbeladenes Schiff n <Wassertrans> lightship
unbeladenes Schiff n **ohne Antriebsmaschine** <Wassertrans> lightvessel
unbelastet adj <Lufttrans> no-load
unbelastete Gleichspannung f <Elektriz> no-load direct voltage
unbelegt adj <Telekom> idle; unoccupied *(Leitung)*
unbelichteter Film m <Foto> unexposed film
unbemannt adj 1. <Raumfahrt> unattended, unmanned; 2. <Raumfahrt> unattended
unbemannte Ausweichstelle f <Eisenbahn> unmanned turnout
unbemannte automatische Sendestation f <Funktech> unattended automatic transmitting station
unbemannte Maschinenräume pl <Wassertrans> unmanned machinery spaces *(UMS)*
unbemannte Raumfahrt f <Raumfahrt> unmanned space flight, unmanned space travel
unbemannter Zug m <Eisenbahn> unmanned train
unbemanntes Amt n <Telekom> unattended exchange, unmanned exchange
unbemanntes Landefahrzeug n <Raumfahrt> unmanned lander
unbemanntes Raumfahrzeug n <Raumfahrt> unattended spacecraft, unmanned spacecraft
unbemaßt adj <Konstzeich> undimensioned

unbenannter Koeffizient m <Metall> undefined coefficient
unberuhigt vergießen v <Fertig> rim
unberuhigt vergossen adj <Fertig> open, rimmed, unkilled *(Stahl)*
unberuhigt vergossener Stahl m <Fertig> open steel
unberuhigtes Vergießen n <Fertig> rimming
unbeschichtet adj <Anstrich> uncoated
unbeschränkte Mode f <Optik> unbound mode
unbeschränkter Dienst m **mit 64 kbit/s** <Telekom> sixty-four kbps unrestricted service *(ISDN)*
unbeschriftet adj <Comp & DV> blank
unbeschrifteter Datenträger m <Comp & DV> blank medium, empty medium
unbesetzt adj <Telekom> idle, unattended; unoccupied *(Leitung)*
unbespielt adj <Comp & DV> virgin
unbespielter Datenträger m <Comp & DV> virgin medium
unbespultes Kabel n <Telekom> unloaded cable
unbeständiger Verkehrsfluss m <Trans> unstable flow
unbestimmter Befehl m <Comp & DV> undefined statement
unbestimmtes Integral n <Math> indefinite integral
Unbestimmtheitsprinzip n <Phys> uncertainty principle
Unbestimmtheitsrelation f <Teilphys> uncertainty relation
unbewehrter Beton m <Bau> mass concrete, plain concrete
unbewerteter Rauschpegel m <Aufnahme> unweighted noise level
unbrennbar adj <Sicherheit> nonflammable
Unbrennbarkeitstest m <Sicherheit> noncombustibility test
UNC-Grobgewinde n <Maschinen> UNC, *(AE)* Unified National Coarse screw thread, Unified coarse thread
Undecan... <Chemie> undecanoic *(Säure)*
Undecen... <Chemie> undecylenic
Undecylen... <Chemie> undecylenic
undefinierte Anweisung f <Comp & DV> undefined statement
undefinierte Taste f <Comp & DV> undefined key
undefinierter Ausdruck m <Math> undefined expression
undefinierter Fehler m <Comp & DV> undefined error
undefinierter Satz m <Comp & DV> undefined record
undeutlich machen v <Konstzeich> detract from the clarity of *(Darstellung)*
UND-Funktion f <Comp & DV> conjunction, equivalence function, equivalence operation
UND-Gatter n <Phys> AND gate
UND-Glied n 1. <Comp & DV> AND gate; 2. <Elektronik> AND gate *(logisch)*
UND-Glied n **mit negiertem Ausgang** *(NAND-Tor)* <Comp & DV> NOT AND gate *(NAND gate)*
undicht adj <Wassertrans> leaky
undichte Stelle f <Wassertrans> leak *(Schiff)*
undichtes Brennelement n <Kerntech> leaking fuel assembly
Undichtheitserkennung f <Qual, Verpack> leakage detection
Undichtheitsprüfung f <Verpack> leakage test
Undichtigkeit f 1. <Bau> leak; 2. <Elektrotech, Nichtfoss Energ> leakage; 3. <Phys, Wassertrans> leak
undissoziierte Versetzung f <Metall> undissociated dislocation
UND-Knoten m <Künstl Int> AND node
UND-NICHT-Verknüpfung f *(NAND-Funktion)* <Comp & DV> NOT AND operation *(NAND operation)*
UND/ODER-Graph m <Künstl Int> AND/OR graph

UND-Schaltung f <Comp & DV, Elektronik, Phys> AND circuit, AND gate *(exklusive NOR-Verknüpfung)*
undurchdringlich adj <Anstrich> solid
undurchlässig adj 1. <Erdöl, Mechan, Papier> impervious; 2. <Sicherheit> impermeable; 3. <Verpack> impervious; 4. <Wasserversorg> impermeable
undurchlässiges Bohrloch n <Erdöl> tight hole
Undurchlässigkeit f <Bau, Papier, Telekom> impermeability
undurchsichtiges Medium n <Phys> opaque medium
UND-Verknüpfung f <Comp & DV> AND operation, conjunction
Und-Zeichen n <Druck> ampersand
unebener Rand m <Ker & Glas> wrinkled rim
Unebenheit f <Ker & Glas> bruise
unedel adj <Fertig> base, ignoble *(Metall)*
unedles Metall n <Fertig, Metall> base metal
unedles Thermoelement n <Gerät> base-metal thermocouple
uneffektive Sendezeit f <Telekom> ineffective airtime
uneingeschränkter Informationsübermittlungsdienst m <Telekom> unrestricted information transfer service
uneingeschränkter Übermittlungsdienst m <Telekom> unrestricted bearer service
unelastische Neutronenstreuung f <Phys> inelastic neutron scattering
unelastische Streuung f <Phys> inelastic scattering
unelastischer Stoß m <Phys> inelastic collision
Unempfindlichkeit f <Werkprüf> robustness
Unempfindlichkeitsfehler m <Regelung> dead band error
unempfindlichster Messbereich m <Gerät> least sensitive range
unendlich adj <Math> infinite • **auf unendlich einstellen** <Foto> focus for infinity
unendlich dicke Schicht f <Strahlphys> infinitely thick layer
unendlich starke Dämpfung f <Elektronik> infinite attenuation *(eines Filters)*
unendliche Impulsantwort f <Telekom> infinite impulse response *(IIR)*
unendliche Reihe f <Math> infinite series
Unendlichkeit f <Math> infinity
unentflammbar adj <Elektriz> flameproof
unentwickelt adj <Math> implicit
unerheblich adj <Patent, Qual> irrelevant
unerwünscht adj 1. <Elektronik> parasitic; 2. <Funktech> spurious
unerwünschte Amplitudenmodulation f <Elektronik> incidental amplitude modulation
unerwünschte Aussendung f <Telekom> unwanted emission
unerwünschte Frequenzmodulation f <Elektronik, Fernseh, Funktech> incidental frequency modulation
unerwünschte Modulation f <Elektronik> incidental modulation
Unfall m 1. <Mechan> emergency; 2. <Trans> accident
Unfall m **beim Heben** <Sicherheit> lifting accident
Unfallanalyse f <Sicherheit> accident analysis
unfallanfällig adj <Sicherheit> accident-prone
Unfalldatenmeldung f <Trans> accident date reporting
Unfalldetektor m <Trans> accident detector
Unfallneigung f <Ergon> accident proneness
Unfallverhütung f <Sicherheit> accident prevention
Unfallwarnsignal n <Trans> accident advisory sign, incident warning sign
Unfallwarnzeichen n <Trans> accident advisory sign
unfertig adj <Anstrich> unfinished

UNF-Feingewinde n <Maschinen> UNF, *(AE)* Unified National Fine screw thread, Unified fine thread
unformatierte Daten npl <Comp & DV> raw data
unfreiwilliger Reflex m <Ergon> involuntary reflex
ungebrannt adj <Ker & Glas, Thermod> unfired
ungebrannter Tiegel m <Ker & Glas> green pot
ungebrannter Ziegel m <Thermod> unburnt brick
ungebundene Mode f <Optik> unbound mode
ungedämpfte Schwingung f 1. <Elektronik> continuous oscillation, sustained oscillation; 2. <Phys> maintained oscillation
ungedämpfte Welle f *(CW)* <Aufnahme, Elektronik, Elektrotech, Funktech, Telekom> continuous wave *(CW)*
ungeerdet adj <Elektrotech> floating
ungefähr adj <Math> approximate
Ungefährmaß n <Konstzeich> approximate dimension
ungefärbt adj <Textil> undyed
ungefedertes Gewicht n <Kfztech> unsprung weight *(Räder, Reifen, Bremsen)*
ungeformt adj <Ker & Glas> *(AE)* unmolded, *(BE)* unmoulded
ungeglättet adj <Anstrich> rough
ungeglättete Seite f <Papier> backside
ungehärtet adj <Kunststoff> uncured
ungeklärtes häusliches Abwasser n <Abfall> raw domestic sewage
ungekürzte Länge f <Telekom> uncut length
ungeladen adj <Elektriz> uncharged
ungelagerte Schrumpfverpackung f <Verpack> unsupported shrink wrapping
ungelochter Streifen m <Telekom> unperforated tape
ungelöscht adj <Chemie> unslaked *(Kalk)*
ungelöschter Kalk m <Ker & Glas> quicklime
ungelöst adj <Fertig> quick
ungemustert adj <Textil> plain
ungenau adj <Math> inaccurate
ungenaue Logik f <Künstl Int> inexact logic
ungenaues Fluchten n <Fertig> malalignment
Ungenauigkeit f <Math> inaccuracy
Ungenauigkeitsauflösung f <Raumfahrt> ambiguity resolution
ungenehmigte Deponie f <Abfall> phantom dump
ungenießbar adj <Lebensmittel> inedible
ungenutzte Stauung f <Nichtfoss Energ> *(AE)* dead dike, *(BE)* dead dyke
ungeordnete Ablagerung f <Abfall> uncontrolled dumping, uncontrolled disposal
ungeordnete Deponie f <Abfall> fly tipping, open dump, uncontrolled tipping, wild dump
ungepaartes Elektron n <Phys> unpaired electron
ungepaartes Neutron n <Kerntech> unpaired neutron
ungepaartes Proton n <Kerntech> unpaired proton
ungepackt adj <Comp & DV> unpacked
ungepolter Elektrolytkondensator m <Elektrotech> nonpolarized electrolytic capacitor
ungerade adj <Druck, Math> odd
ungerade harmonische Schwingungen fpl <Phys> odd harmonic vibrations
ungerade Parität f <Comp & DV> odd parity
ungerade Seite f <Druck> recto
ungerade Zahl f <Math> odd number
Ungerade-Gerade-Drehung f <Kerntech> odd-even spin
Ungerade-Gerade-Kern m <Strahlphys> odd-even nucleus
Ungerade-Gerade-Spin m <Kerntech> odd-even spin
Ungerade-Ungerade-Drehung f <Kerntech> odd-odd spin
Ungerade-Ungerade-Kern m <Strahlphys> odd-odd nucleus

Ungerade-Ungerade-Spin m <Kerntech> odd-odd spin
ungeradzahlige Harmonische f <Funktech> odd harmonic
ungeradzahlige Parität f <Phys> odd parity
ungeradzahlige Zeilensprungabtastung f <Fernseh> odd-line interlaced scan
ungeregelte Kreuzung f <Trans> unsignalized junction
ungeregelte Spannung f <Elektrotech> uncontrolled voltage, unregulated voltage
ungeregeltes Bussystem n <Raumfahrt> unregulated bus system
ungerichtete Endlosfasern fpl <Kunststoff> nondirectional filaments
ungerichtete Fasern fpl <Kunststoff> (AE) randomly distributed fibers, (BE) randomly distributed fibres
ungerichteter Graph m <Künstl Int> nondirected graph, nonoriented graph, undirected graph
ungerichtetes Mikrofon n <Akustik> omnidirectional microphone
ungesättigt adj <Chemie> unsaturated
ungesättigte Kohlenstoff-Kohlenstoff-Bindung f <Erdöl> unsaturated carbon-to-carbon bond
ungesättigte Logik f <Elektronik> nonsaturated logic
ungesättigter Dampf m <Thermod> unsaturated steam
ungesättigter Kohlenwasserstoff m <Erdöl> alkene; unsaturated hydrocarbon (Petrochemie)
ungesättigter Polyester m (UP) <Kunststoff> unsaturated polyester (UP)
ungesättigter Stoff m <Chemie> unsaturate
ungesättigtes Fett n <Lebensmittel> unsaturated fat
ungeschichtet adj <Bau, Kohlen> broken
ungeschichtete Zufallsstichprobe f <Qual> simple random sample
ungeschützte Datei f <Comp & DV> scratch file, work file
ungeschützte Einzeltonnenvertäuung f (ELSBM) <Erdöl> exposed location single buoy mooring (ELSBM)
ungeschütztes Netzwerk n <Comp & DV> non-secure network
ungesiebte Kohle f <Kohlen> run of mine coal, unscreened coal
ungesinterter Pressling m <Fertig> green compact
ungespanntes Grundwasser n 1. <Kohlen> unconfined ground water; 2. <Wasserversorg> unconfined water
ungesponnene Fasern fpl <Textil> raw materials
ungestörte Probe f <Kohlen> undisturbed sample
ungestrichen adj <Anstrich> uncoated
ungetaktete Arbeit f <Ergon> self-paced work
ungeteilte Lagerbüchse f <Maschinen> unsplit bush
ungeteiltes Lager n <Maschinen> bushed bearing
ungeteiltes Schneideisen n <Maschinen> solid die
ungewaschene Kohle f <Kohlen> unwashed coal
ungiftig adj <Anstrich> nontoxic
unglasiert adj <Chemie> unglazed
ungleich adj <Math> unequal
ungleich null adj <Math> nonzero (Zahl)
ungleiche Ladungen fpl <Elektriz> opposite charges
ungleiche Pole mpl <Elektriz> unlike poles
ungleichförmige Bewegung f <Phys> nonuniform motion
Ungleichgewicht n <Maschinen> imbalance, unbalance
Ungleichheitszeichen n <Math> chevron
Ungleichlauf m <Phys> asynchronism
ungleichlaufend adj <Phys> asynchronous
ungleichmäßige Härtung f <Ker & Glas> uneven temper
ungleichmäßige Setzung f <Bau> differential settlement
ungleichmäßige Verkehrsbelastung f <Telekom> traffic load imbalance
ungleichmäßige Verkehrsverteilung f <Telekom> traffic distribution imbalance

ungleichmäßiger Boden m <Ker & Glas> slugged bottom
ungleichnamige Pole mpl <Phys> unlike poles
ungleichseitig adj <Fertig> scalene
ungleichseitiges Dreieck n <Geom> scalene triangle
Ungleichung f <Comp & DV, Math> inequality
Ungültigkeitszeichen n <Comp & DV> ignore character
ungünstig adj <Kerntech> unfavourable
ungünstiger Wind m <Wassertrans> foul wind
unidirektional adj <Telekom> unidirectional
unifarben adj <Textil> plain
Unifikation f <Künstl Int> unification (von Variablen)
Unifilaraufhängung f <Elektriz> unifilar suspension
Unifining n <Chemie> unifining
Unifizierung f <Künstl Int> unification (von Variablen)
uniform adj <Chemie> uniform (Struktur)
Unikat n <Konstzeich> unique record
Unipolar... <Comp & DV, Elektrotech> unipolar
Unipolardynamo m <Elektriz> homopolar dynamo, unipolar dynamo
unipolare Leitung f <Elektrotech> unidirectional conduction
unipolarer Transistor m <Elektronik> unipolar transistor
Unipolargenerator m <Elektrotech> homopolar generator
Unipolar-IC m <Elektronik> unipolar IC, unipolar integrated circuit
Unipolarmaschine f <Elektrotech> homopolar machine, unipolar machine
Unipolgenerator m <Elektriz> homopolar generator, unipolar generator
unisoliert adj <Fertig> unlagged (Rohr)
unisolierter Stromleiter m <Elektriz> plain conductor
Universal... <Comp & DV, Elektrotech> general-purpose, GP
Universalantenne f <Phys> multiband antenna
Universalbohrmeißel m <Fertig> universal boring tool
Universalbrücke f <Elektriz, Gerät> universal bridge
Universalchip m <Elektronik> general-purpose chip
Universalcomputer m <Comp & DV> GP computer, general-purpose computer
Universaldrehmaschine f <Maschinen> universal lathe
Universalentwicklungstank m <Foto> universal developing tank
Universalfräsmaschine f <Maschinen> universal milling machine
Universalfutter n <Maschinen> combination chuck
Universalgelenk n 1. <Kfztech> universal joint (Triebstrang); 2. <Maschinen> universal joint
Universalgerät n <Gerät> general-purpose instrument
Universalhobelmaschine f <Maschinen> universal planer
Universalinstrument n <Gerät> general-purpose instrument
Universalleiterplatte f <Elektronik> general-purpose board
Universalmanipulator m <Kerntech> universal manipulator
Universalmenge f <Comp & DV, Math> universal set
Universalmessbrücke f <Gerät> universal bridge
Universalmessgerät n <Gerät> general-purpose instrument, universal measuring instrument
Universalmessinstrument n <Gerät> universal measuring instrument
Universalmotor m <Elektriz, Elektrotech> universal motor
Universalprogramm n <Comp & DV> general-purpose program
Universalrechner m <Comp & DV> GP computer, general-purpose computer
Universalreißer m <Maschinen> scribing block

Universalrelais

Universalrelais n <Elektrotech> general-purpose relay
Universalrundschleifmaschine f <Fertig> universal cylindrical grinder
Universalschalter m <Elektriz> universal switch
Universalschleifmaschine f <Ker & Glas, Maschinen> universal grinder
Universalschlüssel m <Kfztech> (BE) adjustable spanner (Werkzeug)
Universalschneider m <Maschinen> universal cutter
Universalschnittstelle f <Telekom> general-purpose interface
Universalschraubstock m <Maschinen> (BE) universal vice, (AE) universal vise
Universalselbstentlader m <Kfztech> all-round dumping wagon
Universal-Shunt m <Elektriz> universal shunt
Universalspannfutter n <Maschinen> scroll chuck, (AE) self-centering chuck, (BE) self-centring chuck, universal chuck
Universalsprache f <Comp & DV> general-purpose language
Universalsucher m <Foto> universal viewfinder
Universalwerkzeug n <Fertig> universal tool
Universalwerkzeugschleifmaschine f <Fertig> universal cutter and tool grinding machine
Universalwindeisen n <Maschinen> universal tap wrench
Universalwinkelmesser m <Metrol> universal bevel protractor
Universalzeichensatz m <Comp & DV> universal character set
universelle persönliche Telekommunikation f <Telekom> universal personal telecommunication
universelle S-Schnittstelle f <Telekom> S-universal interface
universelle S-Schnittstellenkarte f <Telekom> S-universal interface card
universeller S-Anschluss m <Telekom> S-universal access
universelles mobiles Telekommunikationssystem n <Telekom> third generation mobile communication system (TGMCS); universal mobile telecommunications system (UMTS)
universelles Wandgesetz n <Strömphys> universal motion (in turbulenter Grenzschicht)
Univibrator m 1. <Elektronik> monostable multivibrator; 2. <Elektronik> one-shot multivibrator
UNIX-Datenstation f <Comp & DV> UNIX box
unklar kommen v <Wassertrans> foul (Festmachen, Tauwerk)
unklarer Anker m <Wassertrans> fouled anchor (Festmachen)
unklassifizierte Zuschlagstoffe mpl <Bau> all-in ballast
unkollidiertes Neutron n <Kerntech> uncollided neutron
unkontrollierte Aufzeichnung f <Aufnahme> wild recording
unkorrigierbarer Abbruchfehler m <Comp & DV> fatal error
unkorrigiertes Ergebnis n <Metrol> uncorrected result
unkristallin adj <Fertig> amorphous
unlauterer Wettbewerb m <Patent> unfair competition
unlegierter Kohlenstoffstahl m <Metall> plain carbon steel
unlegierter Stahl m 1. <Fertig> ordinary steel; 2. <Kohlen> carbon steel; 3. <Metall> carbon steel, ordinary steel
unleserlich adj <Druck> illegible
unlimitierte Laufzeitversion f <Comp & DV> unlimited run-time version
unlogarithmierter Sinus m <Geom> natural sine

unlösbar adj <Maschinen> permanent
unlösbar verbinden v <Konstzeich> connect together non-detachably
unlösbare Verbindung f <Maschinen> permanent joint
unlöslich adj <Chemie> insoluble (Petrochemie)
Unlöslichkeit f <Chemie> insolubility
unmagnetisch adj <Phys> nonmagnetic
unmaßstäblich adj <Konstzeich> not to scale
unmischbar adj <Lebensmittel> immiscible
unmittelbar adj <Telekom, Wasserversorg> direct
unmittelbare Adresse f <Comp & DV> immediate address, zero-level address
unmittelbare Adressierung f <Comp & DV> immediate addressing
unmittelbare Litzenleitung f <Optik> direct strand cable
unmittelbare manuelle Wartung f <Kerntech> direct maintenance
unmittelbare Schallaufzeichnung f <Aufnahme> direct recording
unmittelbare Schienenbefestigung f <Eisenbahn> direct rail fastening
unmittelbare Verarbeitung f <Comp & DV> demand processing, immediate processing
unmittelbarer Zugriff m <Comp & DV> immediate access
Unmittelbar-Lese-nach-Schreib-Platte f <Optik> direct read-after-write
unmodifizierte Adresse f <Comp & DV> presumptive address
unmodifizierter Befehl m <Comp & DV> presumptive instruction
unmoduliert adj <Raumfahrt> unmodulated
unmodulierte Rille f <Akustik> blank groove
üNN (über Normalnull) <Wassertrans> asl (above sea level)
Unordnung f <Metall, Qual> disorder
Unparallelität f <Fertig> out-of-parallelism
unperiodischer Impuls m <Elektronik> nonrecurrent pulse
unplastifiziert adj <Kunststoff> unplasticized
unpolares Lösemittel n <Kunststoff> nonpolar solvent
unregelmäßige Kante f <Ker & Glas> irregular edge
unregelmäßiges Garn n <Textil> irregular yarn
unregelmäßiges Polyeder n <Geom> irregular polyhedron
unregelmäßiges Polygon n <Geom> irregular polygon
unreiner Grund m <Wassertrans> foul bottom, foul ground
unrund adj <Bau> (AE) out-of-center, (BE) out-of-centre
Unrunddrehen n <Maschinen> cam turning
unrunder Verschluss m <Ker & Glas> out-of-round finish
Unrundheit f 1. <Fertig> uncircularity; 2. <Maschinen> out-of-roundness; 3. <Optik, Telekom> noncircularity (des Mantels, Kerns)
unrundkopieren v <Fertig> profile
unrundlaufen v <Kfztech, Maschinen> run out of true
unscharf adj 1. <Elektriz> fuzzy; 2. <Foto> out of focus; 3. <Ker & Glas> blurred; 4. <Künstl Int> fuzzy
unscharfe Logik f <Comp & DV, Künstl Int> fuzzy logic
unscharfe Menge f <Comp & DV, Künstl Int> fuzzy set
Unschärfe f <Fertig> flatness
unscharfer Hintergrund m <Foto> background blur
Unschärferelation f <Phys> uncertainty principle
unscharfes Bild n <Foto> fuzzy image, out of focus image
unselbstständige Entladung f <Elektrotech> nonself-sustained discharge
unsichere Umweltbedingung f <Sicherheit> unsafe environmental condition

Unsicherheit f <Comp & DV> uncertainty
unsichtbare Kante f <Konstzeich> hidden edge
unspezifischer Transistor m <Elektronik> uncommitted transistor
unspezifizierte Bitrate f <Telekom> unspecified bit rate, UBR
unstabil adj <Phys> unstable
unstabiler Bruch m <Metall> unstable fracture
Unstabilität f <Verpack> instability
unstetiger Verstärker m <Elektronik> discontinuous amplifier
unstreifiger Perlit m <Metall> nonlamellar pearlite
Unsymmetrie f 1. <Chemie, Geom> dissymmetry; 2. <Telekom> unbalance
unsymmetrisch adj 1. <Elektriz, Geom> unbalanced, asymmetrical, dissymmetrical, dissymmetric, unsymmetrical; 2. <Raumfahrt, Telekom> unbalanced
unsymmetrische Anordnung f <Elektrotech> unsymmetrical arrangement
unsymmetrische Leitung f <Elektrotech> unbalanced line
unsymmetrische Übertragung f <Telekom> unsymmetrical transmission
unsymmetrischer Ausgang m 1. <Elektrotech> unbalanced output; single-ended output (bei Vierpol); 2. <Raumfahrt> unbalanced output
unsymmetrischer Eingang m <Elektrotech> unbalanced input
unsymmetrisches Drehstromsystem n <Elektriz> unbalanced three-phase system
unsymmetrisches Filter n <Aufnahme> unbalanced filter
unsymmetrisches System n <Telekom> unbalanced system
Unterabtastung f <Telekom> sub-Nyquist-sampling, undersampling (Bildcodierung); subsampling (Datenverarbeitung)
Unteradresse f <Telekom> subaddress
Unteramboss m <Maschinen> anvil bed
Unteramt n <Telekom> dependent exchange
Unteraufgabe f <Comp & DV> subtask
Unterband n 1. <Fertig> return belt (einer Förderanlage); 2. <Telekom> lower band
Unterbandaufnahme f <Fernseh> low-band recording
Unterbandnorm f <Fernseh> low-band standard
Unterbau m 1. <Bau> base course, foundation, subbase; substructure (Straßenbau); 2. <Erdöl> substructure; 3. <Kohlen> foundation
unterbauen v <Fertig, Maschinen> shim
Unterbauen n <Fertig> shimming
Unterbaugrube f <Lufttrans> jig-pit
Unterbelastung f <Phys> unloading
unterbelichten v <Foto> underexpose
unterbelichtetes Bild n <Foto> underexposed picture
Unterbelichtung f <Foto> underexposure
Unterbeton m <Bau> blinding concrete
Unterbettungsschicht f <Bau, Kohlen> subgrade
Unterboden m <Kfztech> underbody
Unterbodenholm m <Lufttrans> floor beam
Unterbodenschutz m <Kfztech> underbody protection; undersealant (Karosserie)
unterbrechen v 1. <Anstrich> disrupt; 2. <Comp & DV, Elektrotech> interrupt, isolate; break (Stromkreis); 3. <Kontroll> interrupt
Unterbrechen n <Fernseh> cutoff
Unterbrecher m 1. <Elektriz, Elektrotech> air interrupter, break, circuit breaker, cutout, interrupter, make and break; trembler (mit Kontaktzunge); 2. <Kfztech> breaker; contact breaker (Motor); 3. <Sicherheit> cutout device (elektrische Sicherheit); 4. <Telekom> interrupter
Unterbrecherbad n <Foto> stop bath
Unterbrecherfeder f <Kfztech> breaker spring
Unterbrecherklappe f <Lufttrans> spoiler (Luftfahrzeug)
Unterbrecherkontakt m 1. <Elektrotech> break contact; 2. <Kfztech> breaker contact, point (Zündung)
Unterbrecherkontaktsteuerung f <Kfztech> breaker triggering
Unterbrecherschließwinkel m <Kfztech> dwell angle
Unterbrechung f 1. <Anstrich> disruption; 2. <Comp & DV> break, interrupt; 3. <Elektrotech> break, cutoff, cutout, disconnection, interruption; 4. <Fertig> isolation; 5. <Funktech> break-in (Sprechfunk); 6. <Kerntech> disruption (der Kettenreaktion); 7. <Maschinen> check; 8. <Phys> breakdown; 9. <Telekom> interruption
Unterbrechung f der Stromversorgung <Comp & DV> power supply interrupt
Unterbrechungsanforderung f <Comp & DV, Telekom> interrupt request
unterbrechungsfrei adj <Comp & DV> uninterruptible
unterbrechungsfreie APC-Stromversorgung f <Comp & DV> APC uninterruptible power supply
unterbrechungsfreie Stromversorgung f (USV) 1. <Kontroll> stand-by battery power supply; 2. <Telekom> uninterruptible power supply
unterbrechungsfreier Betrieb m <Comp & DV> continuous computing
unterbrechungsgesteuertes System n <Kontroll> interrupt-driven system
Unterbrechungshalt m <Comp & DV> dead halt
unterbrechungslos adj <Telekom> break-free
unterbrechungslose Stromversorgung f 1. <Elektriz> continuous power supply, uninterruptible power supply; 2. <Raumfahrt, Telekom> no-break power supply
Unterbrechungsmaske f <Comp & DV> interrupt mask
Unterbrechungspriorität f <Comp & DV> interrupt priority
Unterbrechungspunkt m <Comp & DV> breakpoint
Unterbrechungspunktbetrieb m <Kontroll> breakpoint operation
Unterbrechungsserviceprogramm n <Comp & DV> interrupt service routine
Unterbrechungssignal n <Comp & DV> interrupt signal
Unterbrechungstaste f <Comp & DV> attention key
Unterbrechungszeit f <Telekom> down-time
Unterbrechungszielsteuerung f <Comp & DV> interrupt vectoring
unterbreiten v <Patent> submit
unterbringen v 1. <Bau> house; 2. <Eisenbahn> stable
Unterbringungsplan m <Wassertrans> accommodation plan
unterbrochen adj <Telekom> broken
unterbrochen gezeichnetes Teil n <Konstzeich> interrupted view of a part, interrupted view of a compound
unterbrochene Linie f <Geom> broken line
unterbrochene Wicklung f <Elektriz> open circuit winding, open winding
unterbrochener Bogen m <Bau> broken arch
unterbrochener Stromkreis m <Elektriz> open circuit
unterbrochener Träger m <Bau> tailpiece
unterbrochener Verkehrsfluss m <Trans> interrupted traffic flow
unterbrochenes Bewegen n <Foto> intermittent agitation (Entwickler)
unterbrochenes Feuer n <Wassertrans> occulting light (Seezeichen)
unterbrochenes Gruppenfeuer n <Wassertrans> group-occulting light (Seezeichen)

unterdämpft

unterdämpft adj <Gerät> underdamped *(Messwerk)*
Unterdeck n <Wassertrans> lower deck
unterdimensionieren v <Bau> underdesign
unterdimensionierter Durchmesser m <Nichtfoss Energ> nondimensional diameter
Unterdruck m 1. <Kfztech, Maschinen> depression, negative pressure, underpressure, vacuum; 2. <Mechan> vacuum; 3. <Phys> depression, negative pressure, underpressure, vacuum
Unterdruckbremse f <Kfztech> vacuum brake
Unterdruckbremskraftverstärker m <Kfztech> vacuum-assisted power brake
unterdrücken v 1. <Comp & DV> suppress; 2. <Telekom> reject
Unterdruck-Entlastungseinrichtung f <Sicherheit> vacuum-relief device
Unterdruckförderpumpe f <Kfztech> vacuum fuel pump
Unterdruckmesser m <Maschinen> *(AE)* vacuum gage, *(BE)* vacuum gauge
Unterdruckregler m <Kfztech> suction-type governor; vacuum advance mechanism *(Zündung)*
Unterdruckregulierventil n <Kfztech> vacuum check valve
Unterdruckschlauch m <Maschinen> vacuum hose
Unterdruckseite f <Lufttrans> upper surface *(eines Flügels)*
unterdrückter Strahl m <Elektronik> blanked beam
Unterdrückung f 1. <Comp & DV, Elektrotech> suppression; 2. <Telekom> calling line identification restriction
Unterdrückung f der Nummernanzeige des gerufenen Teilnehmers <Telekom> connected-line identification restriction
Unterdrückung f von Einschaltstößen <Elektrotech> transient suppression
Unterdrückung f von Nebenempfangsstellen <Funktech> spurious response rejection ratio
Unterdrückungsanlage f <Sicherheit> suppression system *(Explosion)*
Unterdrückungsfaktor m 1. <Elektronik> rejection *(bei Differenzialverstärkern)*; 2. <Raumfahrt> suppression factor
Unterdruckventil n <Kfztech> vacuum valve
Unterdruckversteller m <Kfztech> vacuum advance mechanism *(Zündung)*
Unterdruckzündverstellung f <Kfztech> vacuum advance mechanism
untere Drehpfanne f <Eisenbahn> *(AE)* center casting, *(BE)* centre casting
untere Grenzabweichung f <Qual> lower limiting deviation
untere Grenzfrequenz f <Elektronik, Funktech, Telekom> low-frequency cutoff
untere Gurtplatte f <Bau> sole plate
untere Heizleistung f <Abfall> LCV, lower calorific value
untere Kontrollgrenze f <Qual> lower control limit
untere Kühltemperatur f <Ker & Glas> lower annealing temperature
untere Nachweisbarkeitsgrenze f <Umweltschmutz> lower limit of detectability
untere Nachweisgrenze f <Gerät> detection limit
untere Spezifikationsgrenze f <Qual> lower specification limit, LSL
untere Streckgrenze f <Metall> lower yield point
untere Treppenwange f <Bau> rough string
untere Welle f <Kfztech> lower shaft
Untereinheit f <Kontroll> subunit
unterentwickeln v <Foto> underdevelop
unterer Anschluss m <Kerntech> bottom fitting
unterer Durchlassbereich m <Telekom> low-pass band
unterer Mühlstein m <Lebensmittel> bottom millstone
unterer Querfries m <Bau> bottom rail
unterer Seitenrand m <Druck> tail
unterer Totpunkt m 1. <Fertig> pick-up point *(Hammer)*; 2. <Kfztech> *(AE)* bottom dead center, *(BE)* bottom dead centre *(Motor, Zündung)*; 3. <Mechan> BDC, *(AE)* bottom dead center, *(BE)* bottom dead centre
unteres Gelenk n <Lufttrans> lower link *(Hubschrauber)*
unteres Pleuelauge n <Kfztech> connecting rod big end
unteres Sammelbecken n <Nichtfoss Energ> lower storage basin
unteres Seitenband n *(USB)* 1. <Elektronik> lower sideband, LSB; 2. <Funktech> lower sideband, LSB *(Sprechfunk)*; 3. <Telekom> lower sideband, LSB
unteres Speicherbecken n <Nichtfoss Energ> lower storage basin
unteres Zwischenstufengefüge n <Metall> lower bainite
Unterfadenvorrat m <Textil> bobbin thread capacity
Unterfahrschutz m <Kfztech> under-run guard
Unterfahrstoßstange f <Kfztech> under-run bar, under-run bumper
Unterfangung f <Bau> underpinning, vertical shoring
Unterfangungsschalung f <Kohlen> interpit sheeting
Unterfeuer n <Wassertrans> front range light, lower range light
Unterflansch m <Bau> bottom flange
Unterflosse f <Lufttrans> lower surface *(Flugzeug)*
Unterflurbelüftung f <Heiz & Kälte> underfloor ventilation
Unterflurbewässerung f <Wasserversorg> subsurface irrigation
Unterflurkondensator m <Kerntech> underfloor condenser
Unterflurmotor m <Kfztech> underfloor engine
unterführen v <Bau> pass under
Unterführung f 1. <Bau> underbridge, underpass; 2. <Eisenbahn> *(AE)* subway, *(BE)* underground *(für Fußgänger)*
Unterfütterung f <Bau> bed
untergärige Hefe f <Lebensmittel> bottom yeast
untergehen v <Wassertrans> founder, go down, sink *(Schiff)*
untergeordnet adj 1. <Heiz & Kälte> downstream; 2. <Math> ancillary
untergeordnete Anwendung f <Comp & DV> slave application
untergeordnete Datenstation f <Comp & DV> slave station
untergeordneter Computer m <Comp & DV> slave
untergeordneter Prozessor m <Comp & DV> slave processor
untergeordneter Typ m <Comp & DV> subtype
untergeordnetes Bussystem n <Comp & DV> bus slave
Untergesenk n 1. <Fertig> bottom die, bottom swage; 2. <Maschinen> bottom die, lower die
Untergestell n 1. <Bau> *(BE)* bogie, *(AE)* trailer, carrier; 2. <Eisenbahn> frame, underframe; 3. <Fertig> base; 4. <Kerntech> underframe; 5. <Maschinen> foot, undercarriage, underframe; 6. <Raumfahrt> undercarriage; 7. <Verpack> skid base
untergliedert adj <Comp & DV> partitioned
untergliederte Datei f <Comp & DV> partitioned file
Untergraph m <Künstl Int> subgraph
Untergrenze f 1. <Math> lower bound; 2. <Telekom> lower bound, lower limit
Untergröße f <Kohlen> undersize
Untergrund m 1. <Bau> subsoil; 2. <Phys, Strahlphys, Teilphys> background

Untergrundabdichtung f <Abfall> bottom sealing
Untergrundbahn f <Eisenbahn> tube railway; tube vehicle system *(TVS)*
Untergrundbedingungen fpl <Erdöl> subsurface conditions *(Geologie)*
Untergrundbus m <Trans> underground bus
Untergrundmuster n <Druck> mechanical tint
Untergrundrauschen n <Phys> background noise
Untergruppe f <Telekom> subgroup *(telephone system)*
• **Untergruppen zusammenbauen** <Fertig> subassemble
Untergruppentrennzeichen n <Comp & DV> RS, record separator
Untergurt m 1. <Bau> bottom flange, girt, lower chord; 2. <Metall> lower chord
unterhalb adv <Bau> below, downstream
unterhalten v <Maschinen, Mechan> maintain *(Werkzeuge)*
Unterhaltungselektronik f <Elektrotech> consumer electronics, domestic electronic device, domestic electronic equipment
Unterhaupt n <Wasserversorg> tail bay *(eines Durchlasses oder Schleuse)*
Unterhitze f <Ker & Glas> bottom heat
Unterhörfrequenz f <Akustik, Elektronik> infrasonic frequency
unterirdische Ausbreitung f <Telekom> subterranean propagation
unterirdische Bewässerung f <Wasserversorg> subsurface irrigation
unterirdische Dränage f <Wasserversorg> underground drainage
unterirdische Erosion f <Wasserversorg> subsurface erosion
unterirdische Lagerung f <Thermod> underground storage
unterirdische Leitung f <Telekom> underground line
unterirdische Speicherung f <Wasserversorg> underground storage
unterirdisches Wasser n 1. <Umweltschmutz> underground water; 2. <Wasserversorg> ground water, subterranean water
Unterkanal m <Telekom> sub-channel, tributary channel
Unterkasten m 1. <Fertig> nowel; lower box *(Formen)*; drag *(Gießen)*; 2. <Wasserversorg> tail box
Unterkette f <Comp & DV> substring
Unterkolbenpresse f <Kunststoff> upstroke press
Unterkonstruktion f <Bau> substructure; furring *(für Putzauftrag)*
Unterkorn n <Kohlen> undersize
unterkritisch adj <Kerntech> subcritical
unterkritisch arbeitende Wuchtmaschine f <Elektrotech> below-resonance balancing machine
unterkritisch gedämpft adj <Gerät> underdamped *(Messwerk)*
unterkritische Menge f <Kerntech> off-critical amount
unterkritische Reaktion f <Strahlphys> subcritical reaction
unterkühlen v <Thermod> undercool *(Dampf)*; subcool *(Flüssigkeit)*
Unterkühler m <Heiz & Kälte> subcooler
unterkühlt adj 1. <Heiz & Kälte> subcooled; 2. <Phys> super-cooled
Unterkühlung f 1. <Metall, Phys> supercooling; 2. <Wassertrans> hypothermia
Unterlage f 1. <Bau> backing, ballast, base, bed, support; 2. <Fertig> base *(Kunststoffinstallationen)*; 3. <Kohlen> precoating; 4. <Lufttrans> backing bar; 5. <Maschinen> bolster; 6. <Phys> substrate; 7. <Textil, Verpack> backing

unterlagerndes Gestein n <Wasserversorg> bedrock
Unterlagsblech n <Bau> backplate
Unterlängen fpl <Druck> descenders
Unterlast f <Kerntech, Maschinen> underload
Unter-Last-Abzapfwechsel m <Elektriz> load-tap-changer
Unterlastrelais n <Elektriz> underload relay
Unterlastung f <Elektrotech, Qual> derating
Unterlastungsgrad m <Qual> derating factor
Unterlauf m <Comp & DV, Wasserversorg> underflow
Unterlegblech n <Maschinen> packing piece
Unterlegbrett n <Fertig> oddside *(für Stampfen und Stürzen)*
unterlegen v <Fertig, Maschinen> shim
Unterlegen n <Fertig> shimming
Unterlegen n **von Scheiben** <Maschinen> shimming
Unterlegscheibe f 1. <Bau, Fertig> washer; 2. <Kerntech> shim; 3. <Kfztech> plain washer; 4. <Kfztech> washer *(Schraube, Bolzen)*; 5. <Maschinen> flat washer, plain washer, shim, washer
unterlegtes Fenster n <Comp & DV> tiled window
Unterliek n <Wassertrans> foot *(Segeln)*
Unterliekstrecker m <Wassertrans> clew outhaul *(Segeln)*
Unterlizenz f <Patent> *(BE)* sublicence, *(AE)* sublicense
Untermaß n <Kohlen, Maschinen> undersize
Untermenge f <Comp & DV> subset
Untermesser n <Papier> bed knife *(Querschneider)*
untermoderiert adj <Kerntech> undermoderated *(Reaktor)*
untermoderiertes Schutzgas n <Kerntech> undermoderated blanket
Untermodulation f <Aufnahme, Elektronik> undermodulation
Unternahtriss m <Kerntech> underbead crack
Unternehmensforschung f <Comp & DV> operational research, operations research
Unternehmenskommunikation f <Comp & DV> enterprise communications
unternehmensweites System n <Comp & DV> enterprise-wide system
Unterniveau n <Phys> sublevel
Unterpflaster-Omnibus m <Trans> underground trolley bus
Unterpflasterstraßenbahn f <Trans> underground tramway
unterphosphorisch adj <Chemie> hypophosphorous
Unterproblem n <Künstl Int> subproblem
Unterprogramm n <Comp & DV> subprogram, subroutine
Unterprogrammaufruf m <Comp & DV> subroutine call
Unterprogrammbibliothek f <Comp & DV> program library, subroutine library
Unterprozess m <Comp & DV> subprocess
Unterpulverschweißen n <Bau> submerged arc-welding
Unterputz m <Bau> rendering
Unterputzschicht f <Bau> undercoat
Unterrahmen m <Telekom> subframe *(PCM)*
Unterraum m <Phys> subspace
Unterreaktivität f <Kerntech> deficit reactivity
Unterrichtsreaktor m <Kerntech> training reactor
Untersattel m <Fertig> anvil pallet, bottom pallet
Untersättigung f <Metall> undersaturation
Untersatz m <Fertig> stool *(Kokille)*
Unterschale f <Phys> subshell
Unterschall... 1. <Elektronik, Lufttrans> subsonic; 2. <Phys> infrasonic; 3. <Strahlphys> subsonic
Unterschallfrequenz f <Elektronik, Strahlphys> subsonic frequency

Unterschallgeschwindigkeit f <Phys> infrasonic speed
Unterschall-Luftfahrzeug n <Lufttrans> subsonic aircraft
unterscheidbar adj <Math> distinguishable
unterscheidend adj <Patent, Qual> distinctive
Unterscheidung f <Ergon> discrimination
Unterscheidungsbit n (Q-Bit) <Telekom> qualifier bit (Q bit)
Unterscheidungskraft f <Patent, Qual> distinctiveness
unterscheidungskräftig adj <Patent, Qual> distinctive
Unterscheidungsvermögen n <Ergon> discrimination, sensory discrimination
Unterschicht f <Textil> sublayer
Unterschied m <Comp & DV> difference
unterschiedliches Gelände n <Telekom> mixed terrain
Unterschiedsschwelle f 1. <Akustik> differential threshold, just noticeable difference; 2. <Aufnahme> difference limen
Unterschlackeschweißen n <Mechan> electroslag welding
Unterschleusendrempel m <Wasserversorg> tail mitre sill
Unterschleusenwelle f <Wasserversorg> tail mitre sill
unterschneiden v 1. <Druck> kern; 2. <Maschinen> undercut
Unterschneiden n 1. <Druck> kerning; 2. <Maschinen> undercut
Unterschneidung f <Maschinen> undercut
Unterschneidungswerte mpl <Druck> kerning values
Unterschnitt m 1. <Ker & Glas> undercut; 2. <Maschinen> interference, undercut
Unterschnittsfreiheit f <Fertig> absence of undercutting
Unterschrift f <Comp & DV> signature
unterschwellig adj <Ergon> subliminal
unterschwingen v <Elektronik> undershoot (unter Bezugswert)
Unterschwingen n <Fernseh> underswing
Untersee... <Wassertrans> submarine
Unterseeboot n (U-Boot) <Wassertrans> submarine
unterseeisch adj <Wassertrans> submarine
unterseeische Rohrleitung f <Trans> undersea pipeline
Unterseekabel n <Elektrotech, Phys> submarine cable
Unterseetanker m <Wassertrans> submarine tanker
Unterseite f 1. <Bau> soffit; sole (eines Hobels); 2. <Druck> wire side
Unterseite f des Luftschraubenblattes <Lufttrans> blade lower surface (Hubschrauber)
Unterseitenanschluss m <Elektrotech> face-down terminal
untersetzender Multivibrator m <Elektronik> dividing multivibrator
Untersetzer m 1. <Comp & DV> binary circuit; 2. <Elektronik> divider
Untersetzerschaltung f <Elektronik> dividing circuit
Untersetzung f <Maschinen> demultiplication, gear reduction, reduction ratio
Untersetzungsgetriebe n 1. <Kfztech> reduction gear (Getriebe); 2. <Maschinen> reduction gear, step-down gear
Untersetzungsverhältnis n 1. <Kfztech> gear ratio (Getriebe); 2. <Maschinen> reduction ratio
Untersicht f <Bau> intrados (eines Gewölbes); soffit (eines Balkens)
unterspannter Balken m <Bau> trussed beam
Unterspannung f <Metall> minimum stress
Unterspülung f <Bau> undermining
Unterstation f <Telekom> substation
Untersteuerung f <Kfztech> understeer (Lenkung)
unterstreichen v <Druck> underline

Unterstrom m 1. <Nichtfoss Energ> undercurrent; 2. <Wassertrans> undertow (Seezustand)
Unterstromrelais n <Elektriz> undercurrent relay
Unterströmung f 1. <Nichtfoss Energ> undercurrent; 2. <Wassertrans> undercurrent, undertow (Seezustand); 3. <Wasserversorg> undercurrent
unterstützen v 1. <Bau> back; 2. <Comp & DV> support; 3. <Mechan> boost
Unterstützung f 1. <Bau> strutting (Pfettendach); 2. <Comp & DV> support; 3. <Sicherheit> assistance
Unterstützung f von Bildkonvertierungen <Comp & DV> image conversion support
Unterstützung f von Dokumentkonvertierungen <Comp & DV> document conversion support
Unterstützungselement n <Comp & DV> support element
Unterstützungsprogramm n <Comp & DV> support program
untersuchen v 1. <Kohlen> sound; 2. <Kontroll> inspect; 3. <Math> explore
Untersuchung f 1. <Anstrich> analysis; 2. <Kohlen> assay; 3. <Werkprüf> examination
Untersuchung f des Kornaufbaus <Ker & Glas> granulometric analysis
Untersuchung f von Mikrostrukturen mit Röntgenstrahlen <Kerntech> X-ray microstructure investigation
Untersuchungsabschnitt m <Phys> test section
Untersuchungsergebnisse npl <Metrol> results of the inspection
untersynchron adj <Telekom> subsynchronous
Untersystem n <Comp & DV> subsystem
Untertagebau m <Kohlen> deep mining, drift mining
Untertagedeponie f <Abfall> subsurface repository, underground depot (UTD)
Untertage-Sicherheitsventil n <Erdöl> downhole safety valve (Bohrtechnik)
untertägig adj <Erdöl> downhole (Bohrtechnik)
untertägige Bedingungen fpl <Erdöl> downhole conditions (Bohrtechnik)
Unterteil n 1. <Elektrotech> fuse base (einer Sicherung); 2. <Fertig> body (Kunststoffinstallationen); 3. <Kontroll> module
unterteilen v <Bau> partition, split into
unterteilte Anflugbahn f <Lufttrans> segmented approach path
unterteilte Klappe f <Lufttrans> split flap (Luftfahrzeug)
unterteilter Mast m <Funktech> sectionalized tower (Antenne)
Unterteilung f <Labor> graduation
Unterteilung f eines Schiffes in wasserdichte Abteilungen <Wassertrans> compartmentation (Schiffbau)
Unterteilung in Teilbänder <Telekom> band splitting
Untertitel mpl <Fernseh> subtitles
Untertitelerzeuger m <Fernseh> subtitler
Untertitelgenerator m <Fernseh> caption generator
Untertitelscanner m <Fernseh> caption scanner
Untertitelung f <Fernseh> subtitling
Unterton m <Akustik, Elektronik> lower harmonic
Untertor n <Wasserversorg> aft gate; tail gate, tail sluice (einer Schleuse)
Unterträger m <Fernseh, Funktech> subcarrier
Unterträgerabstand m <Fernseh, Funktech> subcarrier offset
Unterträgerfrequenz f <Elektronik, Fernseh, Funktech, Telekom> subcarrier frequency
Unterträgermodulation f <Fernseh, Funktech, Telekom> subcarrier modulation
Untertransport m <Textil> drop feed
untertrocknen v <Fertig> underbake

Untertunnelung f <Bau, Kohlen> (AE) tunneling, (BE) tunnelling
Unter- und Nadeltransport m <Textil> compound feed
unterverdichtet adj <Erdöl> undercompacted (Geologie)
unterverdichtete Zone f <Erdöl> undercompacted zone (Geologie)
Unterverdichtung f <Erdöl> undercompaction (Geologie)
Unterverzeichnis n <Telekom> subdirectory
Unterwalze f 1. <Fertig> low roll (Blechbiegen); 2. <Maschinen> lower roll
Unterwanten fpl <Wassertrans> lower shrouds (Tauwerk)
Unterwaschung f <Bau> undermining
Unterwasser n 1. <Hydraul> tail water; 2. <Kohlen> underflow; 3. <Wasserversorg> tail water
Unterwasser-Atomexplosion f <Umweltschmutz> underwater atomic explosion
Unterwasserausbreitung f <Telekom> underwater propagation
Unterwasserbeton m <Bau> submerged concrete
Unterwasserbetonierrohr n <Bau> tremie pipe
Unterwasserblende f <Hydraul> submerged orifice
Unterwasserbohranlage f <Bau> marine-drilling rig
Unterwasserbohrloch n <Umweltschmutz> subsea well
Unterwasserbohrlochkopf m <Erdöl> subsea wellhead (Offshore-Technik)
Unterwasserdüse f <Hydraul> submerged orifice
Unterwasserfahrzeug n <Wassertrans> hydrospace vehicle, submarine, submarine craft, submarine ship, undersea craft, underwater ship
Unterwasserfernmeldekabel n <Telekom> submarine telecommunication cable
Unterwasserfotografie f <Foto> underwater photography
Unterwassergehäuse n <Foto> underwater housing
Unterwassergeschwindigkeit f <Wassertrans> submerged speed (U-Boot)
Unterwassergraben m <Wasserversorg> mill tail
Unterwasserhöhe f eines Wehrs <Hydraul> degree of submergence
Unterwasserkabel n <Wassertrans> submarine cable
Unterwasserkamera f <Foto> underwater camera
Unterwasserkanal m <Wasserversorg> aft bay race
Unterwasserkomplettierung f <Erdöl> subsea completion (Offshore-Technik)
Unterwasserkondensator m <Heiz & Kälte> submerged condenser
Unterwasserlautsprecher m <Aufnahme> underwater loudspeaker
Unterwassermündung f <Hydraul> submerged orifice
Unterwasserpumpe f <Wasserversorg> subaqueous pump, submerged pump
Unterwasserquelle f <Wasserversorg> submerged spring
Unterwasserschiff n <Wassertrans> underwater hull
Unterwasserschneidbrenner m <Bau> underwater cutting blowpipe
Unterwasserschweißen n <Thermod> underwater welding
Unterwassertragflügel m <Wassertrans> hydrofoil
Unterwasserturbine f <Hydraul> submerged turbine
Unterweisung f <Qual> indoctrination
Unterwerk n <Elektriz, Telekom> substation
Unterzeichenkette f <Comp & DV> substring
Unterziel n <Künstl Int> subgoal
Unterzug m 1. <Bau> beam, bearer, girder, joist, main beam, sill, trussing; 2. <Heiz & Kälte> bottom flue
Unterzugbalken m <Bau> bridging piece
Unterzustand m <Phys> sublevel
Untiefe f <Wassertrans> shoal (Geographie)
Untiefen fpl <Wassertrans> shallows
Untiefenmarkierung f <Wassertrans> isolated danger mark
unüberwachter Modus m <Comp & DV> unattended mode
unüberwachtes Drucken n <Comp & DV> unattended printing
unüberwachtes Empfangen n <Comp & DV> unattended receive
unüberwachtes Senden n <Comp & DV> unattended transmit
ununterbrochener Fluss m <Trans> uninterrupted flow
ununterbrochener Strahl m <Elektronik> continuous beam
Ununterscheidbarkeit f identischer Partikel <Phys> indistinguishability of identical particles
unverarbeitet adj <Bau, Fertig, Kohlen> raw
unverarbeiteter Brennstoff m <Kohlen> raw fuel
unverbleites Benzin n <Kfztech> (AE) unleaded gasoline, (BE) unleaded petrol
unverbrennbar adj <Fertig> incombustible
Unverbrennbarkeit f <Fertig> incombustibility
unverdaulich adj <Lebensmittel> indigestible
unverdichtete Entladung f <Elektrotech> noncondensed discharge
unverdünnt adj 1. <Chemie> undiluted; 2. <Meerschmutz> neat
unvergütet adj <Anstrich> unfinished
unverklebte Außenhaut f <Raumfahrt> unbonded skin (Raumschiff)
unverkleideter Ventilator m <Lufttrans> unducted fan (Motor, Triebwerk)
unverkokt adj <Kohlen> noncoking
unverletzt adj <Anstrich> intact
unverlierbare Mutter f <Maschinen> captive nut
unverlierbare Schraube f <Maschinen> captive screw
unverlötet adj <Funktech> solderless
unvermischt adj <Meerschmutz> neat
unvernetzt adj <Kunststoff> uncured
unverpackt adv <Kohlen> in bulk
unverschlüsselte Sprache f <Telekom> clear speech
unverschlüsselte Übertragung f <Telekom> clear transmission, plain text transmission
Unversehrtheit f <Comp & DV> integrity
unverständlich adj <Sicherheit> inaudible (z. B. Signal)
unverständliche Sprache f <Telekom> inverted speech (durch Frequenzumkehrung)
unverstellbar adj <Mechan> fixed
unverstellbare Luftschraube f <Lufttrans> fixed-pitch propeller
unverstellbarer Magnetkopf m <Comp & DV> fixed head
unverzerrt adj 1. <Comp & DV> unbiased; 2. <Elektronik> undistorted
unverzwirntes Roving n <Ker & Glas> no-twist roving
unvollkommene Elastizität f <Metall> anelasticity
unvollkommener Isolationsstoff m <Elektriz> imperfect dielectric
unvollständige Füllung f <Hydraul> cavitation
unvorhergesehene Unterbrechungen fpl <Telekom> unforeseen interruptions
unvulkanisiert adj 1. <Kunststoff> uncured; 2. <Thermod> unvulcanized
unwesentliche Abweichung f <Qual> insignificant nonconformance
Unwucht f 1. <Fertig> imbalance; 2. <Kfztech> unbalance; 3. <Maschinen> out-of-balance force, out-of-balance, runout; 4. <Mechan> unbalance

unwuchtig

unwuchtig adj 1. <Kfztech> unbalanced (Räder); 2. <Maschinen> out of balance, unbalanced
Unze f <Metrol> ounce avoirdupois, ounce, ounce troy
unzerbrechlich adj <Verpack> resistance to shattering
unzerbrechliches Glas n <Ker & Glas> unbreakable glass
unzersetzt adj <Chemie> undecomposed
Unzulänglichkeit f <Math> shortcoming
unzulässige Instruktion m <Comp & DV> illegal instruction
unzulässige Operation f <Comp & DV> illegal operation
unzulässige Verformung f <Fertig> failure
unzulässiger Befehl m <Comp & DV> illegal instruction
unzulässiges Zeichen n <Comp & DV, Telekom> illegal character
unzureichend adj <Kontroll> unsatisfactory
unzusammenhängend adj <Math> disjoint
UP (ungesättigter Polyester) <Kunststoff> UP (unsaturated polyester)
Uperisation f <Lebensmittel> uperization
UP-Harz n <Chemie> UP resin
Uplink-Frequenz f <Raumfahrt> uplink frequency (Weltraumfunk)
Uplink-Verbindung f <Raumfahrt> uplink
UPN f (umgekehrte Polnische Notation) <Comp & DV> RPN (reverse Polish notation)
U-Profil n 1. <Bau> channel; 2. <Metall> U-section
Upstream-Kanal m <Telekom> upstream channel (vom Betreiber zum Anbieter)
Uracil n <Chemie> pyrimidinedione, uracil
Uracil-D-Ribosid n <Chemie> uridine
Uran n (U) <Chemie> uranium (U)
Uran n **ohne Tochternuklide** <Kerntech> uranium free from its daughters
Uran... <Chemie> uranic, uranium, uranium, uranous
Uranabfall m <Kerntech> uranium scrap
Uranat n <Chemie> uranate
Uranbarren m <Kerntech> uranium ingot
Uranblock m <Kerntech> uranium slug
Uranbrennelement n <Kerntech> uranium fuel element
Urandikarbid n <Kerntech> uranium dicarbide
Urandioxid-Brennstoff m <Kerntech> uranium dioxide fuel
Urandioxid-Pellet n <Kerntech> uranium dioxide pellet
Uranfluorid n <Kerntech> uranic fluoride
uranführend adj <Chemie, Strahlphys> uranium-bearing
Uran-Gadolinit-Pellet n <Kerntech> urania-gadolinia pellet
Uran-Galinit n <Kerntech> uranium-galena
uranhaltiges Mineral n <Kerntech> uranium-bearing mineral
Uranhexafluorid n 1. <Kerntech> uranic fluoride; 2. <Kerntech> uranium hexafluoride (UF6)
Uraninit m <Kerntech> uraninite
Urankern m <Kerntech> uranium nucleus
Urankonversionsanlage f <Kerntech> uranium conversion plant
Urankonzentrat n <Kerntech> uranium concentrate (für homogene Reaktoren)
Uranoid n <Chemie> uranide
Uranoxid n <Chemie> uranium oxide
Uranoxid-Brennstoff m <Kerntech> uranium oxide fuel
Uranoxidgemisch n **nach Kalzinierung** <Kerntech> yellow cake (umgangssprachlich)
Uranpechblende f <Kerntech> pitchblende
Uranpecherz n <Kerntech> pitchblende
Uran-Plutonium-Kreislauf m <Kerntech> U-Pu cycle, uranium-plutonium cycle
Uranreinigung f <Kerntech> uranium refining

Uranverbindung f <Kerntech> uranium compound
Uranvorkonzentrat n <Kerntech> uranium preconcentrate
Uranzerkleinern n <Kerntech> uranium milling
Urazol n <Chemie> urazole
Urband n <Comp & DV> master tape
Ureid n <Chemie> ureide
Ureido... <Chemie> ureido...
Ureidoessigsäure f <Chemie> hydantoic acid
Ureidohydantoin n <Chemie> allantoin
ureotel adj <Chemie> ureotelic
Urethan n <Anstrich> urethane
Urform f <Ker & Glas> block
Urformen n <Kunststoff> (AE) molding, (BE) moulding
urgeformter Schlauch m <Kunststoff> (AE) molded hose, (BE) moulded hose
Uridin n <Chemie> uridine
Urinalbecken n <Bau> urinal
Urknalltheorie f <Phys> big bang theory
urladen v <Comp & DV> bootstrap
Urladen n <Comp & DV> bootstrapping
Urladeprogramm n <Comp & DV> bootstrap
Urlader m <Comp & DV> initial program loader
Urmaß n <Phys> primary standard
Urmeter m <Phys> (AE) standard meter, (BE) standard metre
Urmuster n <Phys> standard
Urnormal n <Phys> primary standard
Urobilin n <Chemie> hydrobilirubin, urobilin
Urobilinogen n <Chemie> urobilinogen
U-Rohr n <Labor> U-tube
U-Rohr-Manometer n <Heiz & Kälte> U-tube manometer
Uron... <Chemie> uronic
Urotropin n <Chemie> methenamine, urotropine
Uroxan... <Chemie> uroxanic
Urspannungsquelle f <Elektrotech> constant-voltage source
Ursprung m 1. <Comp & DV, Elektrotech> source; 2. <Geom> origin; 3. <Hydraul> source; 4. <Math> origin (eines Graphen oder Koordinatensystems); 5. <Telekom> source
Ursprung m **der EMK** <Phys> source of emf
ursprüngliches Feld n <Comp & DV> parent field
Ursprungsadresse f <Comp & DV, Telekom> source address
Ursprungsamt n <Telekom> originating exchange
Ursprungsformat n <Comp & DV> native format
Ursprungsknoten m <Telekom> originating junctor
Ursprungsmodus m <Comp & DV> native mode
Ursprungsregister n <Telekom> originating register
Ursprungsvermittlungsstelle f <Telekom> originating exchange
Urushinsäure f <Chemie> laccol
US (ultrasonisch) <Elektriz> SS, US (supersonic)
USB (unteres Seitenband) <Elektronik, Funktech, Telekom> LSB (lower sideband)
U-Schäkel m <Wassertrans> D-shackle (Beschläge)
U-Scheibe f 1. <Fertig> washer (Kunststoffinstallationen); 2. <Kfztech> washer (Unterlegscheibe Schraube, Bolzen); 3. <Maschinen> washer
US-Schweißen n <Elektriz, Kunststoff, Thermod> US welding
US-Standard-Schraubengewinde n <Maschinen> US standard thread
U-Stab m <Metall> U-bar
U-Stahl m 1. <Fertig> structural channel; 2. <Maschinen> channel section
UST-Gewinde n <Maschinen> UST, (AE) Unified Screw Thread
US-Transducer m <Elektriz> US transducer

USV *(unterbrechungsfreie Stromversorgung)* <Kontroll> stand-by battery power supply
UT *(Standardweltzeit)* <Funktech, Raumfahrt> UT *(universal time)*
UTC *(Standardweltzeit)* <Funktech, Raumfahrt> UTC *(universal time coordinated)*
UV *(ultraviolett)* <Optik, Phys> UV *(ultraviolet)*
UV-absorbierendes Glas *n* <Ker & Glas> UV-absorbing glass
UV-durchlässiges Glas *n* <Ker & Glas> UV-transmitting glass
U-Verschluss *m* <Bau> running trap
UV-Filter *n* <Foto, Sicherheit> UV filter
UV-Fotografie *f* <Strahlphys> UV photography
Uvitin... <Chemie> uvitic
UV-Katastrophe *f* <Phys> UV catastrophe
UV-Lampe *f* <Labor> UV lamp
UV-Licht *n* <Kunststoff, Strahlphys> UV light
UV-Löschen *n* <Comp & DV> UV erasing
UV-Mikroskop *n* <Strahlphys> UV microscope
UV-sichtbares Spektrophotometer *n* <Labor> UV-visible spectrophotometer
UV-Spiegel *m* <Strahlphys> UV mirror
UV-Strahlen *mpl* <Optik, Wellphys> UV rays
UV-Strahlung *f* <Nichtfoss Energ, Phys, Raumfahrt, Strahlphys, Umweltschmutz> UV radiation
Ü-Wagen *m* <Fernseh> OB van, OB vehicle
UZ *(Zenerspannung)* <Elektronik> BDV *(breakdown voltage)*

V

v 1. <Comp & DV, Phys, Textil, Thermod> *(Volumen)* v *(volume)*; 2. <Kerntech> *(Vibrationsquantenzahl)* v *(vibrational quantum number)*
V 1. <Aufnahme, Elektriz, Elektronik, Funktech, Metrol, Optik> *(Verstärkung)* A *(amplification)*; 2. <Chemie> *(Vanadium)* V *(vanadium)*; 3. <Elektriz, Elektronik, Metrol, Optik> *(Volt)* V *(volt)*
VAD *(Axialabscheideverfahren aus Dampfphase)* <Optik, Telekom> VAD, *(AE)* vapor phase axial deposition technique, *(BE)* vapour phase axial deposition technique
vage *adj* <Künstl Int> fuzzy
Vakuole *f* 1. <Chemie> void *(Zellplasma)*; 2. <Kunststoff> void
Vakuum *n* <Erdöl, Fernseh, Kfztech, Maschinen, Mechan, Papier, Phys, Raumfahrt> vacuum • **unter Vakuum** <Phys, Thermod> under vacuum • **unter Vakuum setzen** <Papier> apply vacuum to
Vakuum-Abfüllmaschine *f* <Verpack> vacuum filling machine
Vakuumabfüllung *f* <Lebensmittel> vacuum filling
Vakuumanlage *f* <Fertig> vacuum equipment
Vakuumaufdampfung *m* <Elektronik> vacuum deposition
Vakuumaufhängung *f* <Trans> vacuum suspension
vakuumbedampfen *v* <Fertig> vacuum-deposit
Vakuumbedampfung *f* <Ker & Glas> *(AE)* vapor deposition, *(BE)* vapour deposition
Vakuumbeton *m* <Bau> vacuum concrete
Vakuumblase *f* <Ker & Glas> vacuum bubble
Vakuumbremse *f* <Eisenbahn, Maschinen> vacuum brake
Vakuumdestillation *f* <Chemtech, Lebensmittel, Phys> vacuum distillation
vakuumdicht *adj* <Fertig> leakproof under vacuum *(Kunststoffinstallationen)*
Vakuumdichtung *f* <Maschinen> vacuum seal
Vakuumdiode *f* <Elektronik> vacuum diode
Vakuumentgasung *f* <Metall> vacuum degassing
Vakuum-Entladung *f* <Elektrotech> vacuum discharge
Vakuumexsikkator *m* <Labor> vacuum desiccator
Vakuumfaktor *m* 1. <Elektronik> gas ratio; 2. <Kerntech> vacuum factor
Vakuumfilter *n* <Abfall, Kohlen, Wasserversorg> vacuum filter
Vakuumfiltration *f* <Abfall, Kohlen, Labor, Lebensmittel> vacuum filtration
Vakuum-Folientransportsystem *n* <Verpack> vacuum film transport system
Vakuumform *f* <Maschinen> *(AE)* vacuum mold, *(BE)* vacuum mould
Vakuumformen *n* 1. <Ker & Glas> vacuum blowing; 2. <Kunststoff> vacuum forming
Vakuumführung *f* <Fernseh> vacuum guide
Vakuumführungssystem *n* <Fernseh> vacuum guide system
Vakuumgefäß *n* <Ker & Glas> vacuum flask
vakuumgeformt *adj* <Verpack> vacuum-formed
vakuumgeformte Verpackung *f* <Verpack> vacuum-formed package
Vakuum-Gefriertrockner *m* <Verpack> vacuum freeze-dryer
vakuumgekapselt *adj* <Elektrotech> vacuum-encapsulated
Vakuumgießen *n* <Kerntech> vacuum casting
Vakuumgitterspektrograph *m* <Wellphys> vacuum grating spectrograph
Vakuumglühanlage *f* <Metall> vacuum annealing plant
Vakuumhartlöten *n* <Bau> vacuum brazing
Vakuum-Hitzeversiegler *m* <Verpack> vacuum heat sealer
vakuumisoliert *adj* <Elektrotech> vacuum-insulated
Vakuumisolierung *f* <Elektrotech> vacuum insulation
Vakuumkolben *m* <Labor> vacuum flask
Vakuum-Kondensator *m* <Elektrotech> vacuum capacitor
Vakuum-Kontaktplattenverfahren *n* <Lebensmittel> vacuum contact plate process
Vakuumkühlung *f* <Heiz & Kälte> vacuum cooling
Vakuumleistungsschalter *m* <Elektrotech> vacuum circuit-breaker
Vakuumleitung *f* <Lebensmittel> vacuum line
Vakuum-Lichtbogen *m* <Elektrotech> vacuum arc
Vakuumluftpumpe *f* <Lebensmittel> vacuum air pump
Vakuummessgerät *n* <Metrol> *(AE)* vacuum gage, *(BE)* vacuum gauge
Vakuummessinstrument *n* <Lebensmittel> *(AE)* vacuum gage, *(BE)* vacuum gauge
Vakuummesszelle *f* <Phys> *(AE)* vacuum gage, *(BE)* vacuum gauge
Vakuummetallisierung *f* <Ker & Glas> *(AE)* vacuum metalizing, *(BE)* vacuum metallizing
Vakuummeter *n* <Lebensmittel, Maschinen> *(AE)* vacuum gage, *(BE)* vacuum gauge
Vakuumofen *m* <Maschinen> vacuum furnace
Vakuumpfanne *f* <Lebensmittel> vacuum pan *(Zuckerfabrikation)*
Vakuumplattentrockner *m* <Lebensmittel> vacuum shelf dryer

Vakuumpolarisierung

Vakuumpolarisierung f <Kerntech> vacuum polarization
Vakuumpumpe f <Labor, Lebensmittel, Maschinen, Meerschmutz, Phys> vacuum pump
Vakuumröhre f 1. <Comp & DV, Elektronik> vacuum tube; 2. <Kfztech> vacuum valve; 3. <Phys> vacuum tube
Vakuumröhrenverstärkung f <Elektronik> vacuum tube amplificaton
Vakuumsaugwagen m <Meerschmutz> vacuum truck
Vakuumschalter m <Elektriz> vacuum switch
Vakuumschaumtrocknung f <Lebensmittel> foam vacuum drying
Vakuumschmelzen n <Metall> vacuum melting
Vakuumschub m <Raumfahrt> vacuum thrust
Vakuum-Siegelmaschine f <Verpack> vacuum sealing machine
Vakuumsintern n <Metall> vacuum sintering
Vakuumsystem n <Kfztech> evacuated system
Vakuumtechnik f <Maschinen> vacuum engineering, vacuum technology
Vakuum-Thermoformungsmaschine f <Verpack> vacuum thermoforming machine
Vakuumtiefziehen n <Kunststoff> vacuum forming
Vakuumtonrolle f <Aufnahme, Fernseh> vacuum capstan
Vakuumtriode f <Elektronik> vacuum triode
Vakuumtrockenofen m <Lebensmittel> vacuum drying oven
Vakuumtrockenschrank m <Maschinen> vacuum drying cabinet
Vakuumtrockner m <Kohlen, Lebensmittel> vacuum dryer
Vakuumtrocknung f <Lebensmittel> vacuum drying
Vakuumultraviolett n <Raumfahrt> vacuum ultraviolet
Vakuumumhüllung f <Kerntech> vacuum jacket
Vakuumverdampfer m <Lebensmittel, Wasserversorg> vacuum evaporator
Vakuumverdampfung f <Heiz & Kälte> vacuum evaporation
vakuumverpackt adj <Lebensmittel> vacuum-packed
Vakuumverpackung f 1. <Maschinen> vacuum packaging; 2. <Verpack> vacuum pack, vacuum packaging
Vakuumverpackungsmaschine f <Verpack> vacuum packaging machine
Vakuumverschlussmaschine f <Verpack> vacuum closing machine
Vakuumwärmebehandlung f <Metall> vacuum heat treatment
Valenz f 1. <Chemie> valence, valency; 2. <Comp & DV> valency; 3. <Metall, Phys> valence
Valenzband n <Phys> valence band *(Halbleiterphysik)*
Valenzelektron n <Phys> valence electron *(Atomphysik)*
Valenzelektronenkonzentration f <Kerntech> valence electron concentration
Valenzzustand m <Kerntech> valence state
Valeramid n <Chemie> valeramide
Valerat n <Chemie> valerate
Valerian... <Chemie> valeric
Valeryl... <Chemie> valeryl
Valerylen n <Chemie> pentyne, valerylene
validieren v <Raumfahrt> validate
Validierung f <Comp & DV, Telekom> validation
Vanadat n <Chemie> vanadate, vanadiate
Vanadium n 1. <Chemie> vanadium *(V)*; 2. <Metall> vanadium
vanadiumhaltig adj <Chemie> vanadic, vanadiferous
Vanadiumstahl m <Metall> vanadium steel
Vanadyl... <Chemie> vanadyl
Van-Allen'scher Gürtel m <Phys, Raumfahrt> van Allen belt
Van-de-Graaff'scher Beschleuniger m <Phys> van de Graaff generator
Van-de-Graaff'scher Generator m <Elektriz, Elektrotech, Phys> van de Graaff generator
Van-der-Waals'sche Gleichung f <Phys> van der Waals equation
Van-der-Waals'scher Radius m <Phys> van der Waals radius
V-Antenne f <Funktech> V-shaped antenna
Vapotron n <Elektronik> Vapotron *(Dampfkühlungsverfahren für Leistungsröhren)*
var *(Blindleistungseinheit)* <Labor> var *(volt-amperes reactive)*
Varactor m <Elektronik> varactor *(Kapazitätsdiode)*
varactorabgestimmter Oszillator m <Elektronik> varactor-tuned oscillator
Varactorabstimmung f <Elektronik> varactor tuning
Varactorchip m <Elektronik> varactor chip
Varactordiode f <Elektronik, Phys> varactor diode
variabel adj <Kontroll, Math> variable
variabel einstellbare Einlassschieber mpl <Kerntech> variable-pitch inlet vanes
Variable f 1. <Comp & DV, Kontroll, Qual> variable; 2. <Math> variable *(Veränderliche)*
variable Dämpfung f <Elektronik> variable attenuation
variable Daten npl <Comp & DV> variable data
variable Flächenaufzeichnung f <Fernseh> variable area recording
variable Geschwindigkeit f <Maschinen> variable velocity
variable Servolenkung f <Kfztech> variable assist steering
variable Steuerzeiteinstellung f <Trans> variable valve timing *(Ventil)*
variable Ventileinstellung f <Trans> variable valve timing
variable Zeitlupe f <Fernseh> variable slow motion
Variablenbindung f <Künstl Int> variable binding
Variablenfeld n <Comp & DV, Elektrotech> variable field
Variablenformat n <Comp & DV> variable format
Variablenlänge f <Comp & DV> variable length
Variablenname m <Comp & DV> variable name
Variablenprüfung f <Qual> inspection by variables
Variablensammlung f <Comp & DV> heap
variabler Amplitudentest m <Metall> variable amplitude test
variabler Geschwindigkeitsregler m <Fernseh> variable speed control
variabler Hörfrequenzpegel m <Aufnahme> variable audio level
variabler Koaxialdämpfer m <Elektronik> variable coaxial attenuator
variabler Quarzoszillator m *(VCO)* <Funktech> variable crystal oscillator *(VXO)*
variabler Reluktanzmotor m <Kfztech> variable reluctance motor
variabler Venturi-Vergaser m *(VV-Vergaser)* <Kfztech> *(AE)* variable venturi carburetor, *(BE)* variable venturi carburettor *(VV-carburetor)*
variabler Zwischenraum m <Druck> variable space
variables Ansaugrohr n <Kfztech> variable intake manifold
variables Dämpfungsglied n 1. <Elektronik> variable attenuator, device; 2. <Phys> piston attenuator
variables Format n <Comp & DV> variable format
variables Mikrowellen-Dämpfungsglied n <Elektronik> variable microwave attenuator
Variantenzeichnung f <Konstzeich> variant drawing

Varianz f 1. <Math> variance *(Streuungsmaß einer Zufallsgröße)*; 2. <Phys, Qual> variance
Varianz f **einer Wahrscheinlichkeitsverteilung** <Qual> variance of a variate
Varianz f **einer Zufallsvariablen** <Math> variance of a random variable
Variation f <Math> variation
Variationskoeffizient m 1. <Math> coefficient of variation, variation coefficient *(Streuungsmaß einer Zufallsgröße)*; 2. <Qual> coefficient of variation
Variationsrechnung f <Math> variational calculus
variierbares Potenziometer n/**durch Magnetfeld** <Elektriz> magnetoresistor potentiometer
variiertes Uranpecherz n <Kerntech> varied pitchblende
Variokoppler m <Elektrotech> variocoupler
Variometer n <Elektrotech> inductometer, variometer
Varistor m <Elektriz, Phys> varistor
VAR-Meter n <Elektriz> apparent energy meter
Vasopressin n <Chemie> vasopressin
Vater m <Akustik, Aufnahme> master *(Original einer Plattenaufnahme)*
Vaterdatei f <Comp & DV> father file
Vaterplatte f 1. <Akustik> master; 2. <Aufnahme> metal master
Vaterstecker m <Elektrotech> male connector
Vatersteckverbinder m <Elektrotech> male connector
Vaterstück n <Fernseh> master
Vauxhall-Facette f <Ker & Glas> Vauxhall bevel
VCO *(variabler Quarzoszillator)* <Funktech> VXO *(variable crystal oscillator)*
VCR 1. <Fernseh> *(Videorecorder)* VCR *(video cassette recorder)*; 2. <Fernseh> *(Videomagnetbandgerät)* VTR *(video tape recorder)*
VD *(Viskositäts-Dichteverhältnis)* <Nichtfoss Energ> kinematic viscosity
VDE *(Verein Deutscher Elektrotechniker)* <Elektriz> Association of German Electrical Engineers
VDI *(Verein Deutscher Ingenieure)* <Elektriz> Association of German Engineers
Veitch-Diagramm n <Comp & DV> Veitch diagram
Vektor m <Comp & DV, Elektriz, Elektrotech, Geom, Math> vector • **in Vektoren umgesetzt** <Raumfahrt> vectored
Vektorabtastung f <Elektronik> vector scanning
Vektoranalysis f <Math> vector analysis *(Lineare Algebra)*
Vektorbild n <Comp & DV> vector graphics
Vektorbildschirm m <Elektronik> vector-scan cathode-ray tube
Vektor-Boson n <Phys> intermediate boson
Vektor-Elektronenstrahl-Lithographie f <Elektronik> vector-scan electron-beam lithography
Vektorenrechner m <Comp & DV> pipeline processor
Vektorfeld n <Elektriz, Elektrotech, Math> vector field
Vektorgrafik f <Comp & DV> vector graphics
Vektorgruppe f <Elektriz> vector group
vektoriert adj <Raumfahrt> vectored
Vektorkomponente f <Math, Phys> component of vector
Vektorkopplung f <Strahlphys> vector coupling
Vektormeson n <Teilphys> vector meson
Vektormodell n **des Atoms** <Phys> vector model of the atom
Vektornetzanalysator m <Elektrotech> *(BE)* vector network analyser, *(AE)* vector network analyzer
Vektornetzanalyse f <Elektrotech> vector network analysis
Vektorpotenzial n <Elektrotech> vector potential

Vektorprodukt n 1. <Geom> cross product; 2. <Math> vector product *(für drei-dimensionale Vektoren)*; 3. <Phys> vector product
Vektorprozessor m <Comp & DV> array processor, pipeline processor, vector processor
Vektorquantisierung f <Telekom> vector quantization
Vektorraum m <Math> vector space
Vektorrechner m <Comp & DV> array processor
Vektorschub m <Raumfahrt> vectored thrust
Vektorschubtriebwerk n <Raumfahrt> vectored-thrust engine
Vektorskop n <Telekom> vectorscope
Vektorstrahl m <Elektronik> vector beam
Vektorsumme f <Math> vector resultant, vector sum
Vektorunterbrechung f <Comp & DV> vectored interrupt
Vektorverarbeitung f <Comp & DV> vector processing
Velinpapier n 1. <Druck> wove paper; 2. <Papier> wove
velourieren v <Textil> raise
Velourieren n <Textil> napping
Velourisieren n <Textil> raising
Venn-Diagramm n <Comp & DV> Venn diagram
Ventil n 1. <Akustik> key; 2. <Bau> cock, valve; 3. <Fertig> valve; 4. <Hydraul> valve *(Absperrorgan von Rohrleitungen)*; 5. <Kerntech, Kfztech> valve; 6. <Maschinen> air valve, valve; 7. <Mechan, Meerschmutz, Raumfahrt> valve
Ventil n **mit konischem Sitz** <Hydraul> valve with conical seat
Ventilabdichtring m <Kfztech> valve shaft seal
Ventilator m 1. <Bau> fan, ventilating fan; 2. <Heiz & Kälte> fan, ventilator; 3. <Kfztech> cooling fan, fan; 4. <Lebensmittel> blower; 5. <Lufttrans> turbofan; 6. <Maschinen> ventilating fan; 7. <Mechan, Phys> fan; 8. <Sicherheit> ventilator; 9. <Thermod> fan
Ventilator m **mit Motor** <Sicherheit> motor-driven fan
Ventilatorflügel m <Heiz & Kälte, Kfztech, Mechan, Thermod> fan blade
Ventilatorgebläse n <Maschinen> fan blower
Ventilatorläufer m <Sicherheit> fan wheel
Ventilatorleistung f <Heiz & Kälte> fan performance
Ventilatorleistungskennlinie f <Heiz & Kälte> fan performance curve
Ventilatorlüftung f <Maschinen> fan ventilation
Ventilatorpumpe f <Papier> fan pump
Ventilatorrad n <Heiz & Kälte> fan impeller
Ventilatorriemen m 1. <Heiz & Kälte> fan belt; 2. <Kfztech> fan belt *(Kühlsystem)*; 3. <Mechan, Thermod> fan belt
Ventilatorriemenscheibe f <Kfztech> fan pulley *(Kühlsystem)*
Ventilatorverkleidung f <Kfztech> fan shroud
Ventilaufsatz m <Hydraul> cap of a valve
Ventilbewegung f <Hydraul> valve gear, valve motion
Ventildeckel m <Hydraul> cap of a valve
Ventildiagramm n <Maschinen> valve diagram
Ventildichtfläche f <Kfztech> valve mating surface
Ventildrehvorrichtung f <Kfztech> valve rotator
Ventilexzenter m <Hydraul> valve eccentric
Ventilfeder f <Kfztech, Maschinen> valve spring
Ventilführung f <Kfztech> valve guide; collet *(Motor)*
Ventilgang m <Hydraul> valve motion
Ventilgehäuse n 1. <Hydraul> valve chest; 2. <Kfztech> valve body; 3. <Maschinen> valve box
Ventilgetriebe n <Hydraul> valve gear
Ventilhahn m <Hydraul, Maschinen> valve cock
Ventilhub m 1. <Hydraul> valve motion, valve travel; 2. <Maschinen> valve lift
Ventilhubstange f <Kfztech> valve push rod
ventilieren v <Bau, Heiz & Kälte, Thermod> ventilate

ventiliert

ventiliert adj <Heiz & Kälte, Thermod> ventilated
Ventilkammer f <Maschinen> valve chamber
Ventilkappe f <Hydraul> cap of a valve
Ventilkegel m <Maschinen> valve cone
Ventilklappe f 1. <Kerntech> flap, valve flap; 2. <Kfztech> valve cap *(Reifen)*
Ventilkolben m <Hydraul> bucket
Ventilkolbenpumpe f <Hydraul> bucket pump
Ventilkörper m 1. <Hydraul> valve chest; 2. <Maschinen> valve body
Ventillauf m <Hydraul> valve gear, valve motion
ventilloser Motor m <Kfztech> valveless engine
Ventilnachschleifen n <Fertig> refacing
Ventilneueinschleifen n <Fertig> reseating
Ventilöffnung f <Maschinen> valve outlet
Ventilschaft m <Kfztech> valve shaft, valve stem
Ventilschaftabdeckung f <Kfztech> valve shaft seal
Ventilschaftführung f <Kfztech> collet *(Motor)*
Ventilschiebersitz m <Hydraul> valve seat *(Steuerkolben, Steuerschieber)*
Ventilschleifmaschine f <Fertig> valve seat grinder
Ventilschlüssel m <Bau> cock key
Ventilsitz m 1. <Hydraul> valve seat; 2. <Kerntech> seat; 3. <Kfztech> valve seat
Ventilsitzfläche f <Kfztech> valve face, valve mating surface
Ventilsitzring m 1. <Kfztech> insert; 2. <Maschinen> valve seat ring
Ventilsitzringfeder f <Kfztech> insert spring
Ventilspiel n <Kfztech, Maschinen> valve clearance
Ventilspieleinstellung f <Kfztech> valve setting
Ventilspindel f <Hydraul> valve rod, valve stem
Ventilstange f 1. <Eisenbahn> valve rod; 2. <Hydraul> valve rod, valve stem
Ventilstellungsregler m <Regelung> positioner
Ventilsteuerung f 1. <Eisenbahn> valve train; 2. <Hydraul> valve gear; 3. <Kfztech> valve control; 4. <Maschinen> trip gear
Ventilsteuerung f **beim Motor mit T-förmigem Verbrennungsraum** <Kfztech> T-head valve train
Ventilsteuerung f **beim OHV-Motor** <Kfztech> I-head valve train
Ventilsteuerzeitendiagramm n <Kfztech> valve timing diagram
Ventilstößel m 1. <Kfztech> tappet, valve lifter, valve tappet; 2. <Mechan> cam follower
Ventilstößelstange f <Kfztech> valve push rod
Ventilteller m <Kfztech> *(BE)* valve disc, *(AE)* valve disk, valve head
Ventilteller m **mit Dichtungsring** <Fertig> *(BE)* disc and seating, *(AE)* disk and seating *(Kunststoffinstallationen)*
Ventiltrieb m <Kfztech> valve gear mechanism, valve gear
Ventiltriebsritzel n <Kfztech> timing gear
Ventilüberdeckung f <Kfztech> valve lap
Ventilüberschneidung f <Kfztech> valve lap
Ventimeter n <Wassertrans> ventimeter
Venturi-Düse f <Mechan> venturi nozzle
Venturi-Messdüse f <Maschinen> venturi meter
Venturi-Messrohr n <Maschinen> venturi tube
Venturi-Rohr n 1. <Kfztech> venturi *(Vergaser)*; 2. <Maschinen> venturi tube; 3. <Phys> venturi meter
Venturi-Schlamm m <Kohlen> venturi sludge
Venturi-Schrubber m <Kohlen> venturi scrubber
Venturi-Wäscher m <Umweltschmutz> venturi scrubber
Verabredungs-Bridge f <Telekom> meet-me bridge
Verabredungskonferenz f <Telekom> meet-me conference call
Veracevin n <Chemie> cevadine

verallgemeinerte Koordinaten fpl <Phys> generalized coordinates
veraltet adj <Fertig> obsolete
Veranda f <Bau> porch, verandah
veränderbare Nachleuchtdauer f <Elektronik> variable persistence *(Katodenstrahlröhre)*
veränderlich adj 1. <Kontroll, Math> variable; 2. <Wassertrans> changeable *(Wetter)*
Veränderliche f <Math> variable
veränderliche Bitrate f <Telekom> variable bit rate
veränderliche Brennweite f <Foto> variable focal length
veränderliche Geometrie f *(VG)* <Lufttrans> variable geometry *(VG)*
veränderliche Induktanz f <Elektriz, Elektrotech> variable inductance
veränderliche Kopplungsspule f <Elektrotech> variocoupler
veränderliche Menge f <Elektrotech> variable quantity
veränderliche Mischung f <Kfztech> variable mixture
veränderliche Quantität f <Elektrotech> variable quantity
veränderliche Tragflügelgeometrie f *(VG)* <Lufttrans> variable geometry *(VG)*
veränderlicher Kondensator m <Elektriz, Elektrotech, Phys> variable capacitor
veränderlicher Widerstand m <Elektriz, Elektrotech, Phys> variable resistance, variable resistor
veränderliches Dämpfungsglied n <Elektronik> fader *(Telefontechnik)*
verändern v <Comp & DV> modify
Veränderung f <Math> variation
Veränderung f **der Metalloberfläche** <Anstrich> passivation
verankern v 1. <Bau> anchor, stay; 2. <Kohlen> brace; 3. <Meerschmutz> moor; 4. <Raumfahrt> anchor *(Raumschiff)*; 5. <Wassertrans> moor *(Festmachen)*
Verankern n 1. <Bau> anchoring; 2. <Meerschmutz> mooring
verankerte Boje f <Wassertrans> moored buoy *(Seezeichen)*
Verankerung f 1. <Bau> anchorage, bracing; 2. <Ker & Glas> bracing *(der Wannensteine)*; 3. <Kohlen> bracing; 4. <Raumfahrt> anchoring *(Raumschiff)*; 5. <Wassertrans> anchorage
Verankerungsbolzen m <Maschinen> anchor bolt
Verankerungskette f <Meerschmutz> mooring chain
Verankerungsmast m <Elektriz> anchoring tower
Verankerungsnetz n <Lufttrans> mooring harness
Verankerungsplatte f <Bau, Eisenbahn, Kohlen> anchoring plate
Verankerungsstrebe f <Maschinen> brace strut
Verankerungssystem n <Kerntech> anchorage system
veranschlagen v <Qual> assess
veranschlagte normale Nutzlast f <Lufttrans> estimated normal payload
verantwortlich adj <Qual> accountable, responsible
Verantwortlichkeit f <Qual> accountability, responsibility
Verantwortung f 1. <Qual> accountability, authority, responsibility; 2. <Sicherheit> care
Verarbeitbarkeit f <Bau> workability
verarbeiten v 1. <Comp & DV> process, serve; 2. <Funktech, Lebensmittel> process
verarbeitetes Erdöl n <Erdöl> refined petroleum *(Raffinerie)*
Verarbeitung f 1. <Bau> workmanship; 2. <Comp & DV> computation, processing; 3. <Erdöl> processing; refining *(Raffinerie)*; 4. <Kontroll> processing; 5. <Qual> workmanship; 6. <Telekom> processing

Verarbeitung *f* **auf dem Chip** <Elektronik> on-chip processing
Verarbeitung *f* **im Direktzugriff** <Comp & DV> random processing
Verarbeitung *f* **natürlicher Sprache** <Künstl Int> natural language processing
Verarbeitung *f* **von Meldungen** <Comp & DV> message handling
Verarbeitung *f* **von nur einem Glasposten** <Ker & Glas> single gob process
Verarbeitungsart *f* <Comp & DV> processing mode *(COBOL)*
Verarbeitungsbereich *m* <Ker & Glas> working range
Verarbeitungseinheit *f* <Telekom> processor
verarbeitungsfähig *adj* <Bau> workable
Verarbeitungsfolge *f* <Comp & DV> processing sequence
Verarbeitungsleistung *f* <Telekom> processing power
Verarbeitungsmodus *m* <Comp & DV> processing mode
Verarbeitungsschaltkarte *f* <Telekom> processing card
Verarbeitungsschicht *f* <Telekom> application layer
Verarbeitungsspielraum *m* <Kunststoff> pot life
Verarbeitungsstatus *m* <Comp & DV> process state
Verarbeitungsteil *m* <Comp & DV> procedure division
Verarbeitungstemperatur *f* <Ker & Glas> working temperature
Verarbeitungsvorgang *m* 1. <Comp & DV> task; 2. <Fertig> processing operation
Verarbeitungszeit *f* <Comp & DV> processing time
verarmen *v* <Kerntech> deplete
verarmter Kernbrennstoff *m* <Kerntech> depleted nuclear fuel
Verarmung *f* 1. <Elektronik> depletion *(Mikroelektronik)*; 2. <Kerntech> depletion; impoverishment *(von Erzen)*
Verarmungsschicht *f* <Phys> depletion layer
Verarmungstyp *m* <Elektronik> depletion mode
Verarmungszone *f* <Phys> depletion layer
veraschen *v* <Thermod> incinerate
Veraschungsofen *m* <Thermod> incinerator
verästeltes Wasserleitungsnetz *n* <Wasserversorg> arterial system
Veratramin *n* <Chemie> veratramine
Veratrin *n* <Chemie> cevadine, veratrine
Veratrin... <Chemie> veratric
Veratrinum *n* <Chemie> veratrine
Veratrol *n* <Chemie> veratrole
Verb *n* <Comp & DV> verb
Verband *m* <Bau> bond *(Mauerwerk)* • **im Verband legen** <Bau> bond
Verbandskasten *m* <Sicherheit> first-aid kit *(KFZ)*
Verbau *m* <Bau> lining
Verbesserer *m* <Erdöl> improver *(Wirkstoff zur Geruchs- oder Fließverbesserung)*
verbesserter Brennstoffkreislauf *m* <Kerntech> advanced fuel cycle
verbesserter Dieselmotor *m* <Kfztech> improved diesel engine
Verbesserung *f* <Lebensmittel, Patent, Qual, Sicherheit> improvement
Verbesserungsmittel *n* <Lebensmittel> improver
Verbesserungspatent *n* <Patent> improvement patent
Verbeulung *f* 1. <Kerntech> buckling *(in einem Brennelement)*; 2. <Metall> buckling
verbiegen *v* <Mechan> deform
verbinden *v* 1. <Anstrich> weld; 2. <Bau> bond, connect, joint, join, tail, tie; concrete *(zu fester Masse)*; 3. <Comp & DV> connect; bind, combine *(Programme)*; 4. <Elektrotech> patch *(mit Verbindungsleitung)*; 5. <Fernseh> patch; 6. <Fertig> agglutinate; 7. <Math> associate; 8. <Phys> connect; 9. <Telekom> connect, join; 10. <Verpack> bond to *(Verschweißen von Plastik)*
verbinden *v/sich* <Chemie> coalesce
Verbinden *n* 1. <Anstrich> fusion; 2. <Bau> joining; 3. <Telekom> jointing, linking
Verbinder *m* 1. <Elektrotech> connector; 2. <Raumfahrt> binder
Verbinder *m* **für gedruckte Schaltungen** <Elektronik> printed circuit connector
Verbinder *m* **für Lichtleitfasern** <Optik> *(AE)* optical fiber link, *(BE)* optical fibre link
verbindlich *adj* <Patent, Qual> mandatory
verbindlicher Standard *m* <Bau, Qual> mandatory standard
Verbindung *f* 1. <Anstrich> compound; 2. <Bau> connection, joint; 3. <Comp & DV> hook, interconnect, link, linkage; 4. <Eisenbahn> connection; 5. <Elektriz, Elektrotech> coupling, interconnection, joint, junction; connection *(von Stromleitern)*; 6. <Fernseh> bearding; 7. <Fertig> coupling element; 8. <Funktech> junction; 9. <Kerntech> joining, linkage; bond *(zwischen Brennmaterial und Hülle)*; 10. <Kontroll> link; 11. <Kunststoff> compound; 12. <Künstl Int> arc *(Graph)*; edge, link *(zwischen Knoten in Graph)*; 13. <Maschinen> connection, junction, linking, tie; 14. <Mechan> coupling; 15. <Optik> joint; 16. <Phys> coupling; 17. <Telekom> connection, joint, link; circuit *(Telefon)*; 18. <Wassertrans> bonding *(Schiffbau)* • **eine Verbindung halten** <Telekom> put a call on hold • **eine Verbindung herstellen** 1. <Elektrotech, Maschinen> join; 2. <Telekom> set up a call • **eine Verbindung lösen** <Maschinen> unmake a joint • **eine Verbindung trennen** <Telekom> disconnect a call
Verbindung *f* **durch Einmalklicken** <Comp & DV> single-click link
Verbindung *f* **lokaler Netze** <Telekom> LAN-LAN interconnection, local area network interconnection
Verbindung *f* **mit digitaler Modulation** <Telekom> digital modulation link
Verbindung *f* **mit einem Funkfeld** <Funktech> one-hop link
Verbindung *f* **mit optischer Sicht** <Funktech, Telekom> line-of-sight connection, line-of-sight link, LOS connection, LOS link
Verbindung *f* **ohne optische Sicht** <Funktech, Telekom> NLOS link, non-line-of-sight link
Verbindung *f* **über Glasfaserkabel** <Telekom> fiber-optic cable connection, *(AE)* fiber-optic cable link, *(BE)* fibre-optic cable connection, *(BE)* fibre-optic cable link
Verbindung *f* **über Kabel** <Telekom> cable connection, cable link
Verbindung *f* **über Lichtwellenleiter** <Telekom> *(AE)* fiber optic link, *(BE)* fibre optic link
Verbindung *f* **über Satelliten** <Telekom> satellite link
Verbindung *f* **zu Internet-Seiten** <Comp & DV> link
Verbindung *f* **zu verteilten Programmen** <Comp & DV> distributed program link *(DPL)*
Verbindungsabbau *m* <Telekom> connection tear-down, call down
Verbindungsablauf *m* <Telekom> call flow, call sequence
Verbindungsableitstrom *m* <Elektrotech> junction leakage current
Verbindungsableitung *f* <Elektrotech> junction leakage
Verbindungsanforderung *f* <Telekom> call request, connection request *(CR)*; outgoing call
Verbindungsaufbau *m* <Telekom> call set-up, connection set-up
Verbindungsaufbaubestätigung *f* <Telekom> call completion signal

Verbindungsaufbaudauer

Verbindungsaufbaudauer f <Telekom> call delay, call set-up time, connecting delay
Verbindungsaufbaugeschwindigkeit f <Telekom> setting-up time, velocity of call setting-up
Verbindungsaufbaukanal m <Telekom> control channel, set-up channel *(beim GSM)*
Verbindungsaufbauphase f <Telekom> call set-up phase
Verbindungsaufbauverzug m <Telekom> call set-up delay
Verbindungsbahn f <Eisenbahn> interchange track
Verbindungsband n <Raumfahrt> bonding strap *(Raumschiff)*
Verbindungsbearbeitung f <Telekom> call handling
Verbindungsbefehl m <Comp & DV> link
Verbindungsbogen m <Maschinen> bend coupling
Verbindungsbolzen m <Bau> bat bolt, gudgeon
Verbindungsbrücke f <Elektrotech> bonding jumper
Verbindungscomputer m <Comp & DV> gateway computer
Verbindungsdaten f <Telekom> call data
Verbindungsdauer f 1. <Comp & DV> connect time; 2. <Telekom> call duration
Verbindungsdoppelbrücke f <Lufttrans> connecting twin yoke *(Fahrwerk)*
Verbindungsdoppelgabel f <Lufttrans> connecting twin yoke
Verbindungsdraht m <Elektrotech> connecting wire, jumper
Verbindungselement n 1. <Comp & DV> link; 2. <Fertig> fastener; 3. <Maschinen> fastener, link; 4. <Mechan> link
Verbindungsfestigkeit f <Verpack> joint strength
Verbindungsflansch m <Maschinen, Raumfahrt> connecting flange
Verbindungsfuge f <Fertig> joint clearance
Verbindungsfugenbreite f <Fertig> joint clearance
Verbindungsgefäße npl <Labor> communicating vessels
Verbindungsgleis n <Eisenbahn> interchange track
Verbindungsglied n 1. <Maschinen> connecting link, coupling link, shackle; 2. <Meerschmutz> shackle
Verbindungsherstellung f <Telekom> call completion, call establishment
Verbindungsherstellzeit f <Telekom> set-up time *(Telefon)*
Verbindungskabel n 1. <Bau> interconnecting cable; 2. <Comp & DV> feeder cable, patch cable, patch cord; 3. <Elektrotech> connecting cable, interconnection cable, junction cable
Verbindungsklemme f <Elektrotech> binding post, binding clip, connecting terminal, connector, lead clamp
Verbindungskriechstrom m <Elektrotech> junction leakage current
Verbindungskristall m <Metall> compound crystal
Verbindungslasche f <Bau> backplate
Verbindungsleitung f 1. <Elektriz, Elektrotech> interconnecting feeder, interconnecting line, trunk line, patch cord; 2. <Lufttrans> interconnection; airframe bonding lead *(Flugwerk)*; 3. <Telekom> junction *(Telegrafie)*
Verbindungsleitungskabel n <Elektrotech> trunk cable
verbindungsloser Trägerdienst m <Telekom> connectionless bearer service
Verbindungsmaterial n <Kerntech> bonding material
Verbindungsmuffe f 1. <Fertig> joint sleeve; 2. <Maschinen> connector, coupling sleeve; 3. <Telekom> cable joint
Verbindungspfad m <Comp & DV> linkage path
Verbindungsplatte f 1. <Elektrotech> junction plate; 2. <Kerntech> tie plate
Verbindungsprogramm n <Comp & DV> client

Verbindungsprotokoll n <Comp & DV> link protocol *(Datenfernverarbeitung)*
Verbindungspunkt m <Elektrotech, Telekom> junction point
Verbindungsrechner m <Comp & DV> gateway computer
Verbindungsschicht f <Telekom> link layer
Verbindungsschnur f 1. <Foto> connecting cord; 2. <Telekom> cord *(Zentrale)*
Verbindungsschraube f <Maschinen> connecting screw
Verbindungsschürze f <Raumfahrt> connecting skirt
Verbindungsspeiseleitung f <Elektriz> trunk feeder
Verbindungsstange f 1. <Lufttrans> connecting rod; 2. <Maschinen> link rod, tie bar, tie rod
Verbindungsstelle f 1. <Ker & Glas> splice; 2. <Kerntech> joint; 3. <Maschinen> junction; 4. <Telekom> junction point
Verbindungssteuerung f <Comp & DV> end-to-end control
Verbindungssteuerungsprotokoll n <Telekom> call control protocol
Verbindungsstoß m <Wassertrans> butt *(Schiffbau)*
Verbindungsstrecke f <Comp & DV, Telekom> communication link
Verbindungsstück n 1. <Elektrotech> coupling, link; 2. <Labor> connector
Verbindungstopologie f <Comp & DV> interconnection topology
Verbindungstunnel m <Raumfahrt> connecting tunnel
Verbindungsumschaltung f 1. <Mobilkom> hand-off *(Mobiltelefon)*; intercell hand-off *(bei Zellenwechsel)*; 2. <Telekom> hand-off
Verbindungsumschaltung f zur nächsten Zelle <Mobilkom> intercell hand-over, intercell switching
Verbindungs- und Trennungszeichengabe f <Telekom> *(AE)* connect and disconnect signaling, *(BE)* connect and disconnect signalling
Verbindungsursprung m <Telekom> call origin
Verbindungsverlust m <Elektrotech> connection loss
Verbindungsverlustberechnung f <Raumfahrt> link budget *(Weltraumfunk)*
Verbindungsversuch m <Telekom> call attempt
Verbindungswärme f <Thermod> heat of combination
Verbindungsweg m <Telekom> communications circuit
Verbindungsweitergabe f im Diversity-Betrieb <Mobilkom> diversity handover
Verbindungswelle f <Elektrotech, Maschinen> dumb-bell shaft, *(AE)* spacer shaft
Verbindungswertigkeit f <Metall> joint efficiency *(Schweißarbeit)*
Verbindungszone f zwischen Gewölbe und Widerlager <Ker & Glas> spring zone
Verbindungszusammenstoß m <Telekom> clashing
Verbindungszustand m <Telekom> call condition
verblassen v 1. <Bau> fade; 2. <Textil> fade
verbleibende Ladung f <Elektrotech> residual charge
verbleibender Abstand m <Elektrotech> residual gap
verbleibender Widerstand m <Elektrotech> residual resistance
verbleien v 1. <Bau> plumb; 2. <Fertig> lead
verbleit adj <Fertig> lead-coated, leaded
verblenden v <Bau> face; screen *(Fenster)*; brick *(mit Ziegeln)*
Verblendstein m <Ker & Glas> facing block
Verblendung f <Bau> facing
Verblitzen n des Films <Foto> static on film *(elektrostatisch)*
verblocken v <Maschinen> interlock
Verblockung f <Maschinen> interlocking

verbolzen v 1. <Fertig> pin; 2. <Mechan> bolt
verbolzt adj <Fertig> pinned
Verbolzung f 1. <Fertig> pinning; 2. <Maschinen, Sicherheit> bolting
verborgen halten v/sich <Wassertrans> shelter *(U-Boot)*
verborgene Schicht f <Künstl Int> hidden layer *(in einem neuralen Netz)*
verborgene Strichcode-Identifikation f <Verpack> hidden bar code identification
verborgene Wärme f 1. <Bau> latent heat; 2. <Erdöl> latent heat *(Thermodynamik)*; 3. <Heiz & Kälte, Phys, Thermod> latent heat
verborgene Wärmelast f <Heiz & Kälte> latent heat load
Verbot n <Sicherheit> prohibition notice
verbotene Zerfallsart f <Strahlphys> forbidden decay mode
verbotener Übergang m <Kerntech, Strahlphys> forbidden transition
verbotenes Band n <Strahlphys> forbidden band
verbotenes Energieband n <Kerntech> forbidden energy band
Verbotsschild n <Sicherheit> prohibition sign
verbrannt adj 1. <Metall> burnt; 2. <Thermod> burned, burnt, incinerated
Verbrauch m 1. <Elektrotech> consumption; dissipation *(Energieverbrauch)*; 2. <Kfztech, Kohlen, Mechan, Nichtfoss Energ> consumption
Verbrauch m pro Kopf <Wasserversorg> per capita consumption
verbrauchen v <Maschinen> dissipate
Verbraucherstromkreis m <Telekom> load circuit
Verbrauchsdatum n <Verpack> use by date
Verbrauchsleitung f <Bau> supply pipe
Verbrauchsspitze f 1. <Elektriz> consumption peak, load peak; 2. <Kohlen, Nichtfoss Energ> consumption peak
Verbrauchszähler m <Gerät> demand meter
verbrauchte Luft f <Heiz & Kälte> vitiated air
verbreiten v <Funktech> broadcast
verbreiterbare Antenne f <Raumfahrt> deployable aerial, deployable antenna
Verbreitung f 1. <Meerschmutz> spreading; 2. <Raumfahrt> spread
Verbreitung f von Schallwellen <Elektrotech> acoustic-wave propagation
verbrennen v 1. <Kohlen> char; 2. <Metall> burn away; 3. <Thermod> burn, destroy by fire, incinerate
Verbrennen n <Chemie, Thermod> combustion
Verbrennung f 1. <Chemie> combustion; 2. <Fertig> combustion *(von Gas)*; 3. <Kunststoff, Maschinen> combustion; 4. <Metall> burning; 5. <Sicherheit> combustion; burn *(Verletzung)*; 6. <Thermod> burning, combustion
Verbrennungsanalyse f <Thermod> combustion analysis
Verbrennungsanlage f <Abfall, Umweltschmutz> incineration plant, incinerator
Verbrennungsenergie f <Thermod> combustion energy
Verbrennungsgas n 1. <Raumfahrt> combustion gas *(Raumschiff)*; 2. <Umweltschmutz> flue gas
Verbrennungsgeschwindigkeit f 1. <Maschinen> combustion speed; 2. <Thermod> rate of combustion
Verbrennungsinstabilität f <Thermod> combustion instability
Verbrennungskammer f 1. <Hydraul> combustion chamber *(Dampfkessel)*; 2. <Kfztech> combustion chamber *(Motor)*; 3. <Lufttrans, Thermod, Wassertrans> combustion chamber
Verbrennungskraftmaschine f 1. <Erdöl, Maschinen> internal combustion engine; 2. <Mechan> internal combustion machine; 3. <Trans> combustion engine
Verbrennungsluft f <Thermod> combustion air

Verbrennungsmittel n <Meerschmutz> burning agent *(für Öl)*
Verbrennungsmotor m 1. <Elektrotech, Fertig> internal combustion engine; 2. <Kfztech> explosion engine, explosion motor, internal combustion engine, motor; combustion engine *(Motor)*; 3. <Lufttrans, Maschinen, Wassertrans> combustion engine
Verbrennungsofen m 1. <Maschinen> combustion furnace; 2. <Thermod> incinerator
Verbrennungsrückstand m <Abfall> combustion residue, incineration residue, incineration ash
Verbrennungsrückstände mpl <Thermod> combustion deposits
Verbrennungsschiffchen n <Labor> combustion boat
Verbrennungsschlacke f <Abfall> incineration slag
Verbrennungssteuerung f <Wassertrans> combustion control *(Motor)*
Verbrennungstechnik f <Thermod> combustion engineering
Verbrennungswärme f 1. <Maschinen> combustion heat; 2. <Thermod> burning heat, combustion heat, heat of combustion
Verbrennungswirkungsgrad m <Raumfahrt> combustion efficiency *(Raumschiff)*
Verbrennungszahl f <Thermod> combustion index
Verbrennungszone f <Metall> combustion zone
Verbund m 1. <Bau> bond, bonding; 2. <Comp & DV> bonding, group; 3. <Raumfahrt> bonding *(Raumschiff)*
Verbund... 1. <Anstrich> composite; 2. <Raumfahrt> bonded *(Raumschiff)*
verbunden adj 1. <Bau> jointed; 2. <Comp & DV> linked *(Programme)*; 3. <Elektriz, Elektrotech> connected, coupled, interconnected; 4. <Maschinen, Phys> connected, coupled
verbundene Netze npl <Elektriz, Telekom> connected networks
verbundene Steuerorgane npl <Lufttrans> interconnected controls
verbundene Stromkreise mpl <Elektriz> linked circuits
verbundener Teilnehmer m <Telekom> connected subscriber, remote subscriber
verbundenes Mauerwerk n <Bau> bonded masonry
Verbundfestigkeit f <Kerntech> bonding strength
Verbundfolie f <Kunststoff> multilayer film
Verbundgas n <Erdöl> connection gas
Verbundglas n <Kfztech> multilayer glass
Verbundglaswindschutzscheibe f <Kfztech> *(BE)* laminated windscreen, *(AE)* laminated windshield *(Karosserie)*
Verbundgruppenbezeichnung f <Konstzeich> composite assembly drawing
Verbundhubschrauber m <Lufttrans> compound helicopter
Verbundkern m <Phys> compound nucleus
Verbundkompressor m <Kerntech> compound compressor
Verbundlager n <Maschinen> composite bearing
Verbundlenkerachse f <Kfztech> dead beam axle
Verbundmaschine f <Maschinen> compound engine
Verbundmaterial n <Fertig> composite material
Verbundmotor m 1. <Elektriz, Elektrotech> compound motor, compound-wound motor; 2. <Maschinen> composite engine
Verbundnetz n <Elektrotech> interconnection
Verbundpapier n *(BDP)* <Papier> bonded double paper *(BDP)*
Verbundpfahl m <Bau, Kohlen> composite pile
Verbundplatte f <Fertig> sandwich panel
Verbundröhre f <Elektronik> multiple-unit tube

Verbundsäule

Verbundsäule f **durch Umschnürung** <Bau> laced column
Verbundstoff m 1. <Fertig> composite *(Kunststoffe)*; 2. <Kunststoff, Verpack> composite
Verbundstoffbehälter m <Verpack> composite container
Verbundstück n <Mechan> fitting
Verbundsystem n <Comp & DV> distributed system
Verbundträger m <Bau> built-up girder, composite beam, composite girder, compound girder
Verbundverpackung f <Abfall> composite packaging
Verbundwerkstoff m 1. <Fertig> composite material; 2. <Kunststoff> composite; 3. <Raumfahrt> composite material *(Raumschiff)*
Verbundwicklung f <Elektriz, Elektrotech> compound winding
Verchromung f <Kfztech, Metall> chromium plating
verdampfen v 1. <Chemie> vaporize, volatize; 2. <Chemtech, Thermod> vaporize; 3. <Heiz & Kälte> evaporate
verdampfen lassen v <Thermod> vaporize
Verdampfen n <Chemtech> vaporization
Verdampfer m 1. <Chemtech> vaporizer *(Trennen von Stoffgemischen)*; 2. <Heiz & Kälte> evaporator; 3. <Kerntech> vaporizer; 4. <Kfztech> evaporator *(Klimaanlage)*; 5. <Labor> evaporator; 6. <Thermod> vaporizer
Verdampferschlange f <Heiz & Kälte> evaporator coil
Verdampferteil m <Heiz & Kälte> evaporator section
Verdampfschale f <Chemtech> vaporization dish
verdampft adj <Thermod> vaporized
Verdampfung f 1. <Chemie> volatilization; 2. <Druck> evaporation; 3. <Erdöl> evaporation *(Physik)*; 4. <Heiz & Kälte, Hydraul> evaporation; 5. <Kfztech> vaporization; 6. <Maschinen, Nichtfoss Energ> evaporation; 7. <Phys> vaporization; 8. <Phys, Thermod> evaporation
Verdampfung f **mit Laser** <Phys, Strahlphys> laser vaporization
Verdampfungsapparat m <Chemie> vaporizer *(Kältemaschine)*
Verdampfungsbrenner m <Thermod> vaporizing burner
Verdampfungsenthalpie f <Chemtech> evaporation enthalpy
Verdampfungsentropie f <Mechan, Phys, Thermod> entropy of vaporization
Verdampfungsfähigkeit f <Chemtech> evaporative capacity
Verdampfungsgeschwindigkeit f <Heiz & Kälte> evaporation rate
Verdampfungskessel m <Wasserversorg> evaporation pan
Verdampfungskühlung f 1. <Chemtech> evaporation cooling; 2. <Heiz & Kälte> evaporative cooling; 3. <Maschinen> evaporation cooling; 4. <Thermod> evaporative cooling
Verdampfungsleistung f <Heiz & Kälte> evaporative capacity, evaporative power
Verdampfungsmethode f <Kerntech> *(AE)* atomic vapor method, *(BE)* atomic vapour method
Verdampfungspunkt m <Chemtech, Heiz & Kälte> evaporating point
Verdampfungsratenmessgerät n <Chemtech> evaporation rate meter
Verdampfungstemperatur f <Heiz & Kälte> evaporating temperature
Verdampfungsverlust m <Chemtech, Heiz & Kälte, Wasserversorg> evaporation loss
Verdampfungsvermögen n <Heiz & Kälte> evaporative capacity
Verdampfungswärme f <Thermod> heat of vaporization
Verdaulichkeit f <Lebensmittel> digestibility

Verdauungsenzym n <Lebensmittel> digestive enzyme
Verdeck n <Kfztech> *(BE)* bonnet, *(AE)* hood *(Karosserie, Cabrio)*
verdecken v 1. <Bau> conceal; 2. <Wasserversorg> blind
Verdecken n <Aufnahme> masking *(eines Geräuschs)*
Verdecken n **von Daten** <Comp & DV> information hiding
Verdeckriegel m <Kfztech> hood catch *(Motorhaube)*
verdeckt adj <Fertig> shielded *(Lichtbogenschweißen)*
verdecktes Lichtbogenschweißen n <Bau> submerged arc-welding
Verdeckung f **durch Rauschen** <Akustik> masking by noise
Verdeckung f **durch Töne** <Akustik> masking by tones
Verdeckungseffekt m <Akustik> masking effect
Verdeckungsmaß n <Akustik> masking index
verderben v 1. <Bau> mar; 2. <Lebensmittel> spoil; taint *(Fleisch)*
verdichtbar adj <Fertig> rammable
verdichtbarer Abfall m <Abfall> compressible waste
Verdichtbarkeit f 1. <Bau> compactibility; 2. <Fertig> compactibility, rammability; 3. <Heiz & Kälte> compressibility
verdichten v 1. <Bau> compact; pack *(Straße)*; 2. <Comp & DV> pack; compress *(Daten)*; 3. <Fertig> ram; jar *(Formsand)*, pack *(Gießen)*; 4. <Metall> shingle; 5. <Papier> agglomerate; 6. <Verpack> condense
Verdichten n 1. <Fertig> ramming; jarring *(Formsand)*; 2. <Metall> shingling; 3. <Telekom> condensing
verdichtend adj <Papier> agglomerative
Verdichter m 1. <Bau> compressor, vibrator; 2. <Erdöl, Heiz & Kälte> compressor; 3. <Kfztech> compressor *(Motor)*; 4. <Maschinen> compressor; 5. <Papier> condenser
Verdichterleistung f <Heiz & Kälte> compressor rating
Verdichterleitrad n <Lufttrans> compressor stator
Verdichterrotor m <Lufttrans> compressor rotor *(Hubschrauber)*
Verdichterschaufel f <Lufttrans> compressor blade
Verdichterwirkungsgrad m <Phys> pressure coefficient
verdichtet adj <Papier> agglomerated
verdichtete Füllung f <Bau> compacted fill
verdichtete Impulsbreite f <Elektronik> compressed pulse width
Verdichtung f 1. <Abfall> compaction, compression; 2. <Bau> compaction, condensation; compression *(Boden)*; 3. <Comp & DV, Elektronik> compaction, compression; 4. <Erdöl> compaction, compression *(Geologie)*; 5. <Fertig, Heiz & Kälte, Hydraul, Kerntech, Kohlen, Kunststoff> compaction, compression; 6. <Mechan> packing; 7. <Papier> agglomeration; 8. <Phys, Thermod> compaction, compression; 9. <Wellphys> compression
Verdichtungsanlage f <Abfall> compacting machine, landfill compactor, packer unit
Verdichtungsdeponie f <Abfall> tipping with compaction
Verdichtungsdruck m <Maschinen> discharge pressure
Verdichtungsgerät n <Bau> compactor
Verdichtungsgrad m <Bau, Kohlen> degree of compaction
Verdichtungshub m <Kfztech> compression stroke *(Motor)*
Verdichtungskurve f <Thermod> compressibility curve
Verdichtungslinie f <Kohlen> compression curve
Verdichtungsraum m <Kfztech> compression chamber
Verdichtungsring m <Kfztech> compression ring *(Motor, Kolben)*
Verdichtungsstoß m <Fertig> shock
Verdichtungsstufe f <Maschinen> compression stage
Verdichtungstakt m 1. <Kfztech> compression stroke *(Motor)*; 2. <Thermod> compression stroke

Verdichtungstrend *m* <Erdöl> compaction trend *(Geologie)*
Verdichtungsverhältnis *n* 1. <Aufnahme> compression ratio; 2. <Elektronik> compression ratio *(Mikrowellen)*; 3. <Heiz & Kälte> compression ratio; 4. <Kfztech> compression ratio *(Motor)*; 5. <Maschinen, Qual, Thermod> compression ratio
Verdichtungswärme *f* <Thermod> heat of combustion, heat of compression
Verdichtungswelle *f* <Akustik> compressional wave
Verdicken *n* <Lebensmittel> thickening *(einer Flüssigkeit)*
verdickter Rand *m* <Kunststoff> fat edge
Verdickungsmittel *n* 1. <Chemtech> thickener, thickening agent *(viskositätserhöhend)*; 2. <Kunststoff, Lebensmittel> thickener, thickening agent
verdoppeln *v* 1. <Druck> double; 2. <Textil> splice *(Strümpfe)*
Verdopplung *f* <Akustik> doubling
verdrahten *v* <Fernseh, Telekom> wire
Verdrahtung *f* <Elektriz, Elektrotech, Raumfahrt> wiring
Verdrahtungsfeld *n* <Comp & DV> platter
Verdrahtungsplan *m* <Elektriz, Telekom> wiring diagram
Verdrahtungsplatte *f* <Elektronik> wiring board
verdrängen *v* <Bau> displace
Verdränger *m* 1. <Fertig> impeller *(Pumpe)*; 2. <Maschinen> displacer
Verdrängerkolben *m* <Maschinen> displacement piston
Verdrängerpumpe *f* <Hydraul> positive-displacement pump
verdrängtes Flüssigkeitsvolumen *n* <Fertig> displacement
Verdrängung *f (D)* 1. <Maschinen> displacement, D; 2. <Wassertrans> displacement, D *(Schiffkonstruktion)*; 3. <Wasserversorg> displacement, D
Verdrängung f auf Spanten <Wassertrans> *(AE)* molded displacement, *(BE)* moulded displacement *(Schiffkonstruktion)*
Verdrängungsfaktor *m* <Raumfahrt> displacement coefficient *(Raumschiff)*
Verdrängungsfehler *m* <Lufttrans> displacement error *(Instrumentenlandesystem)*
Verdrängungskühlung *f* <Heiz & Kälte> cooling by relative displacement
Verdrängungspumpe *f* <Maschinen> displacement pump, positive-displacement pump
Verdrängungsschwerpunkt *m* <Wassertrans> *(AE)* center of buoyancy, *(BE)* centre of buoyancy *(Schiffbau)*
Verdrängungsstrom *m* <Elektriz> displacement current
Verdrängungstonne *f* 1. <Metrol> displacement ton *(Schiffe)*; 2. <Wassertrans> ton of displacement *(Seezeichen)*
Verdrängungsvakuumpumpe *f* <Maschinen> positive-displacement vacuum pump
Verdrängungszähler *m* <Gerät> displacement meter, positive-displacement meter, volumetric displacement flow meter
verdrehen *v* <Bau> twist
Verdrehen *n* 1. <Bau> winding; 2. <Maschinen> twisting
Verdrehfestigkeit *f* <Maschinen, Qual> resistance to twisting
Verdrehflankenspiel *n* <Fertig> circumferential backlash *(Getriebelehre)*
Verdrehmoment *n* <Maschinen> twisting moment
Verdrehmoment *n* **des Luftschraubenblattes** <Lufttrans> blade twisting moment *(Hubschrauber)*
Verdrehrohr *n* <Kfztech> torque tube
Verdrehung *f* 1. <Maschinen> twist; 2. <Metall> torsion
Verdrehung f der Feldlinien <Kerntech> twist of the field lines

Verdrehung f durch ungebrannte Stellen <Ker & Glas> green patch distortion
Verdrehung f mit Scherung <Kerntech> twist with shear
Verdrehungsfeder *f* <Maschinen> spring subjected to torsion, torsion spring
Verdrehungsmesser *n* 1. <Fertig> troptometer; 2. <Maschinen> torsion meter
Verdrehungswinkel *m* <Maschinen> angle of twist
Verdrehverformung *f* <Metall> twisting strain
verdrillen *v* <Bau> twist
verdrillt *adj* <Comp & DV> twisted pair
verdrillte Doppelleitung *f* <Telekom> twisted pair
verdrillter Hohlleiter *m* <Elektrotech, Telekom> twisted waveguide
verdrilltes Adernpaar *n* <Comp & DV> twisted pair cable
Verdrillung *f* 1. <Elektrotech> angular twist, twisting *(einer Litzenleitung)*; transposition *(Röbelstab)*; 2. <Maschinen> twist
Verdrückungsfalle *f* <Erdöl> pinch-out trap *(Lagerstätten)*
verdübeln *v* 1. <Bau> dowel, key, peg; 2. <Fertig> peg
Verdübelung *f* <Bau> plugging
Verdunklungsvorhang *m* <Textil> blackout curtain
verdünnbar *adj* <Fertig> rarefiable
verdünnen *v* 1. <Bau> thin; 2. <Chemtech> dilute; 3. <Fertig> rarefy; 4. <Kohlen, Kunststoff, Papier> dilute
Verdünnen *n* <Chemtech> dilution *(Vorgang)*
Verdünner *m* 1. <Chemie> thinner; 2. <Kunststoff> diluent
verdünnt *adj* <Fertig> rarefied
verdünnte Legierung *f* <Metall> dilute alloy
verdünnte Lösung *f* 1. <Chemtech> dilution; 2. <Metall> dilute solution
Verdünnung *f* 1. <Chemie> thinning; dilution *(einer Flüssigkeit)*; 2. <Erdöl> dilution; 3. <Fertig> rarefaction; 4. <Ker & Glas> attenuation; 5. <Kerntech> thindown; 6. <Phys, Wellphys> rarefaction
Verdünnungslüftung *f* <Sicherheit> dilution ventilation
Verdünnungsmittel *n* 1. <Chemtech> diluting agent; 2. <Kunststoff, Papier> diluent
Verdünnungsrate *f* <Umweltschmutz> dilution rate
verdunsten *v* 1. <Chemie> volatize; 2. <Foto> evaporate *(im Vakuum)*; 3. <Heiz & Kälte> evaporate
verdunsten lassen *v* <Heiz & Kälte> evaporate
Verdunster *m* <Kfztech> evaporator *(Klimaanlage)*
Verdunstung *f* <Chemie> volatilization *(unterhalb des normalen Siedepunktes)*
Verdunstungsapparat *m* <Chemtech> evaporating apparatus
Verdunstungskälte *f* <Phys, Thermod> latent heat of evaporation
Verdunstungsmesser *m* <Chemtech> evaporimeter
veredeln *v* 1. <Metall> fine, refine; 2. <Papier> convert
Veredelung *f* 1. <Bau> *(BE)* ageing, *(AE)* aging *(Leichtmetall)*; 2. <Fertig> refining *(Stahl)*; 3. <Papier> converting
Verein m Deutscher Elektrotechniker *(VDE)* <Elektriz> Association of German Electrical Engineers
Verein m Deutscher Ingenieure *(VDI)* <Elektriz> Association of German Engineers
vereinbaren *v* <Math> agree
Vereinbarung *f* <Comp & DV> declaration, declarative statement
Vereindeutigung *f* <Künstl Int> disambiguation
vereinen *v* <Geom> join
vereinfachen *v* <Math> simplify
vereinfachte Ansicht *f* <Konstzeich> simplified view
vereinigen *v* <Geom> join
vereinigte Mengen *fpl* <Math> united sets

Vereinigung

Vereinigung f 1. <Erdöl> unitization *(Bergrecht)*; 2. <Trans> merging
Vereinigungsmenge f <Math> associated set, union of sets
vereinter Antrieb m <Raumfahrt> unified propulsion
vereisen v <Thermod> freeze
Vereisung f 1. <Lufttrans, Wassertrans> icing; 2. <Wasserversorg> ice accretion
Vereisungsanzeigerrelais n <Lufttrans, Wassertrans> ice detector relay
Vereisungsmessfühler m <Lufttrans, Wassertrans> icing probe
Vereisungswarngerät n <Lufttrans, Wassertrans> ice detector
Vereisungswiderstand m <Nichtfoss Energ> icing resistance
verengen v <Fertig> neck
verengter Flaschenhals m <Ker & Glas> choke
Verengung f 1. <Maschinen> contraction; 2. <Mechan> gooseneck, neck
Vererbung f 1. <Comp & DV> inheritance *(objekt-orientierte Programmierung)*; 2. <Künstl Int> inheritance, subtyping
Vererbungsgraph m <Künstl Int> inheritance graph
verestern v <Chemie> esterify
Veresterung f <Lebensmittel> esterification
Verfahren n 1. <Comp & DV> procedure, process; 2. <Elektriz, Kohlen> process; 3. <Maschinen> procedure; 4. <Textil> process
Verfahren n **mit Trägerabtastung** <Telekom> carrier sense system
Verfahren n **vor dem europäischen Patentamt** <Patent> proceedings before the EPO
Verfahren n **zur Hinweisanzeige** <Comp & DV> notification delivery method
Verfahren n **zur Sammlung meteorologischer und ozeanographischer Daten** <Wassertrans> oceanographic and environmental research
Verfahrenseignungsbericht m <Qual> procedure qualification record *(Schweißen)*
Verfahrensfehler m <Comp & DV> truncation error
Verfahrenshandbuch n <Qual> procedures manual
verfahrensorientierte Programmiersprache f <Comp & DV> procedural language, procedure-oriented language
Verfahrensprüfung f <Qual> process inspection and testing
Verfahrenssteuerung f <Elektrotech> process control
Verfahrenstechnik f <Erdöl, Maschinen> process engineering
Verfahrenstechniker m <Maschinen> methods engineer
Verfalldatum n <Raumfahrt> decay date
verfälschen v 1. <Comp & DV, Qual> corrupt; 2. <Lebensmittel> adulterate
Verfälschung f <Comp & DV, Qual> corruption
Verfälschungsmittel n <Lebensmittel> adulterant
verfälschungssicher adj <Verpack> tamper-proof
verfälschungssicherer Verschluss m <Verpack> tamper-proof closure
Verfalzung f <Bau> bead *(Metallbau)*
verfangen v <Sicherheit> entangle
verfangen v/sich <Sicherheit> become entangled
Verfangen n <Sicherheit> entanglement *(von Kleidung und Haaren)*
verfärben v <Kunststoff> stain
verfärben v/sich <Bau> *(AE)* color, *(BE)* colour
Verfärbung f <Kunststoff, Qual> *(AE)* discoloration, *(BE)* discolouration
Verfasserkorrektur f <Druck> author's alterations

verfaulen v <Lebensmittel> putrefy
Verfaulen bringen v/zum <Lebensmittel> putrefy
verfaulend adj <Lebensmittel> putrescent
verfault adj <Lebensmittel> putrid
verfertigen v <Mechan> fabricate
verfestigen v <Abfall> solidify
Verfestigen n <Bau> grouting
verfestigtes Kieselgeröll n <Bau> conglomerate
verfestigtes Material n <Bau> stabilized material
Verfestigung f 1. <Bau> compaction; 2. <Maschinen> hardening; 3. <Phys> solidification
Verfestigung f **von Abfällen** <Abfall> solidifying waste
Verfestigungsgrad m <Kohlen> degree of consolidation
Verfestigungsmittel n 1. <Abfall> fixative, solidifying agent; 2. <Meerschmutz> solidifier
Verfestigungsprodukt n <Abfall> CFS-processed waste, CFS-treated waste, solidified waste, solidified product, solidified material
Verfestigungsverfahren n <Abfall> fixation technique, solidification technique, stabilization technique
Verfestigungsverfahren n **für Sonderabfälle** *(PETRIFIX-Verfahren, SEALOSAFE-Verfahren, VRS-Verfahren)* <Abfall> WPC-VRS process
Verfestigungszeit f <Kunststoff> setting time
Verfilzbarkeit f <Papier> felting power
verfilzen v <Textil> mat
Verfilzen n <Papier> felting
verflanschen v <Erdöl> flange up
verflechten v <Comp & DV, Elektronik, Textil> interlace
Verflechtung f 1. <Textil> interlacing; 2. <Trans> weaving
Verflechtungsfaktor m <Trans> weaving factor
Verflechtungssteuerung f <Trans> merging control
verflüchtigen v <Chemtech, Textil> volatilize
verflüchtigen v/sich <Chemtech> evaporate *(unterhalb des normalen Siedepunktes)*
verflüssigen v <Papier> fluidify
verflüssigen v/sich <Thermod> liquefy
Verflüssigen n <Bau> fluxing
Verflüssiger m 1. <Heiz & Kälte> condenser, liquefier; 2. <Papier> fluidizer
verflüssigt adj <Thermod> liquefied
Verflüssigung f 1. <Bau> condensation *(Gase)*; 2. <Heiz & Kälte> condensation; 3. <Thermod> liquefaction
Verfolgbarkeit f <Metrol> traceability
verfolgen v <Comp & DV> monitor, trace, track
Verfolgen n <Funktech> tracing *(Signale)*
Verfolgung f 1. <Qual, Raumfahrt> tracking *(Raumschiff)*; 2. <Telekom> tracking *(Satellit)*
Verfolgungsantenne f <Raumfahrt> tracking antenna
Verfolgungsfilterdemodulator m <Raumfahrt> tracking filter demodulator *(Weltraumfunk)*
Verfolgungsradar n <Funkort, Raumfahrt> tracking radar
Verfolgungsstation f <Raumfahrt> tracking station
Verfolgungstelemetrie f **und -steuerung** f <Raumfahrt> TTC, tracking telemetry and command *(Weltraumfunk)*
verformbarer Frontbereich m <Kfztech> deformable front section
verformbarer Heckbereich m <Kfztech> deformable rear section
Verformbarkeit f 1. <Fertig> plasticity; 2. <Ker & Glas> workability
Verformbarkeit f **unter Druckbeanspruchung** <Fertig> malleability
verformen v 1. <Mechan> deform; 2. <Metall> bend
verformter Stab m <Metall> bent rod
Verformung f 1. <Bau> deformation; 2. <Kerntech, Kohlen, Kunststoff, Maschinen, Mechan, Metall, Phys, Strahlphys> deformation, distortion; 3. <Qual> distortion

Verformung f **des Blattes** <Lufttrans> blade distortion *(Hubschrauber)*
Verformung f **unter Last** <Maschinen> strain
Verformungsbruch m <Metall> ductile fracture
Verformungsenergie f <Fertig> resilience
Verformungsfestigkeit f <Kerntech> yield strength
Verformungsgrad m <Fertig> degree of deformation
Verformungsmodul m <Kohlen> compressibility modulus, deformation modulus
Verformungsriss m <Metall> ductile crack
Verfrachter m <Wassertrans> carrier; shipper *(Seehandel)*
verfügbare Benutzerzeit f <Comp & DV> available time
verfügbare Landestrecke f <Lufttrans> landing distance available
verfügbare Leistung f 1. <Aufnahme> available power *(eines Verstärkers)*; 2. <Elektriz, Maschinen, Nichtfoss Energ, Telekom> available power
verfügbare Leistungsverstärkung f <Aufnahme> available power gain
verfügbare Startabbruchstrecke f <Lufttrans> accelerate-stop distance available
verfügbare Startstrecke f <Lufttrans> takeoff distance available
Verfügbarkeit f <Comp & DV, Qual, Raumfahrt> availability
Verfügbarkeitskonzept n <Qual> availability concept
Verfügbarkeitszeit f <Qual> uptime
verfugen v <Bau> point; point *(Mauerwerk)*
Verfugen n <Bau> pointing *(Material)*
Verfugung f <Kerntech> grouting
Verfügung f <Patent, Qual> disposition
Verfüllbeton m <Bau> packing
Verfüllboden m <Umweltschutz> backfill *(Erdbau)*
verfüllte Deponie f <Abfall> complete fill
Verfüllung f 1. <Abfall> backfilling, filling; 2. <Bau> backfill
Vergabe f **gegen Höchstgebot** <Erdöl> blind auction *(Konzession)*
Vergärung f <Chemie> fermentation *(Tätigkeit)*
vergasen v <Phys> gasify
Vergaser m 1. <Ker & Glas, Maschinen, Mechan, Wassertrans> *(AE)* carburetor, *(BE)* carburettor; 2. <Kfztech> *(AE)* carburetor, *(BE)* carburettor *(Kraftstoffversorgung)*
Vergasergestänge n <Kfztech> *(AE)* carburetor linkage, *(BE)* carburettor linkage *(Vergaser)*
Vergaserglocke f <Kfztech> suction chamber
Vergaserheizmantel m <Kfztech> *(AE)* carburetor jacket, *(BE)* carburettor jacket
Vergaserknallen n <Kfztech> blowback
Vergaserkraftstoff m (VK) 1. <Erdöl> *(AE)* gas, *(AE)* gasoline, *(BE)* petrol; 2. <Kfztech> fuel, *(AE)* gas, *(AE)* gasoline, *(BE)* petrol; 3. <Thermod> *(AE)* gas, *(AE)* gasoline, *(BE)* petrol; 4. <Trans> motor spirit
Vergaserluftdüse f <Mechan> choke
Vergaserluftklappe f <Kfztech> choker plate
Vergasermischkammer f <Kfztech> *(AE)* carburetor barrel, *(BE)* carburettor barrel
Vergasermotor m 1. <Kfztech> *(AE)* carburetor engine, *(BE)* carburettor engine, normally aspirated engine; 2. <Maschinen> spark ignition engine; 3. <Thermod> *(AE)* gas engine, *(AE)* gasoline engine, *(BE)* petrol engine
Vergasernadel f <Kfztech> *(AE)* carburetor needle, *(BE)* carburettor needle
Vergaserpatschen n <Kfztech> blowback
Vergasersaugkanal m <Kfztech> *(AE)* carburetor barrel, *(BE)* carburettor barrel
Vergaserschwimmer m <Kfztech> *(AE)* carburetor float, *(BE)* carburettor float

Vergaserseilzug m <Kfztech> *(AE)* carburetor control cable, *(BE)* carburettor control cable
Vergasung f 1. <Abfall> gasification; 2. <Erdöl> gasification *(Umwandlung von Kohle oder Öl in Gas)*; 3. <Kfztech> carburation; 4. <Thermod> gasification
Vergasung f **vor Ort** <Kohlen> in-situ gasification
vergeben v 1. <Bau> place *(Auftrag)*; 2. <Fertig> place
vergießen v 1. <Bau> grout, *(AE)* mold, *(BE)* mould; 2. <Fertig, Ker & Glas> teem
Vergießen n 1. <Fertig> teeming; 2. <Kunststoff> embedding, encapsulation, potting
Vergiftung f 1. <Kerntech> poisoning *(eines Reaktors)*; 2. <Umweltschutz> poisoning
vergilben v 1. <Papier> age; 2. <Textil> yellow
Vergilben n <Papier> yellowing
vergilbungsbeständig adj <Kunststoff> nonyellowing
vergittern v <Bau> screen *(Fenster)*
verglasen v 1. <Bau> pane; 2. <Fertig> vitrify; 3. <Ker & Glas> glaze *(mit Glas versehen)*
Verglasen n <Bau, Ker & Glas> glazing *(Fenster)*
verglaste Tür f <Bau> glazed door
Verglasung f 1. <Fertig, Ker & Glas> vitrification; 2. <Kfztech> blind *(Kühler)*; 3. <Nichtfoss Energ> glazing
Verglasungsindustrie f <Ker & Glas> glazing industry
Vergleich m <Comp & DV, Gerät, Qual> comparison
vergleichbar adj <Math> commensurable *(Größen)*
Vergleichbarkeit f 1. <Math> commensurability; 2. <Qual> reproducibility
vergleichen v <Comp & DV> compare
vergleichender Versuch m <Phys> comparative test
Vergleicher m 1. <Comp & DV, Elektronik, Kontroll, Metrol> comparator *(EDV)*; 2. <Telekom> comparator
Vergleichsabstimmfrequenz f <Akustik> standard tuning frequency
Vergleichsausdruck m <Comp & DV> relational expression
Vergleichsbedingungen fpl <Qual> reproducibility conditions
Vergleichseinrichtung f <Metrol, Phys> comparator
Vergleichselektrode f <Labor> reference electrode *(Elektrochemie)*
Vergleichsfrequenz f <Akustik> SF, standard frequency
Vergleichsglied n <Gerät> comparing element
Vergleichsgrenze f <Qual> reproducibility limit
Vergleichsgröße f <Kohlen, Qual> control size
Vergleichslehre f <Maschinen> *(AE)* master gage, *(BE)* master gauge, *(AE)* reference gage, *(BE)* reference gauge
Vergleichsmesser m <Metrol> comparator *(Maschinenbau)*
Vergleichsmessung f <Gerät> comparative measurement, comparison measurement, reference measurement
Vergleichsmesswert m <Gerät> reference measurement
Vergleichsmodell n <Comp & DV> relational model
Vergleichsnormal n <Qual> reference standard
Vergleichsoberfläche f <Telekom> reference surface
Vergleichsoperator m <Comp & DV> relational operator
Vergleichsprüfmethode f <Optik> reference test method
Vergleichspunkt m <Comp & DV, Patent, Qual> benchmark
Vergleichsschall m <Akustik> reference sound
Vergleichsschaltung f <Telekom> comparison circuit
Vergleichsspannung f <Bau, Kohlen> effective stress
Vergleichsstelle f 1. <Gerät> comparison junction *(beim Thermoelement)*; 2. <Regelung> reference junction
Vergleichsstellentemperatur f <Gerät> reference junction temperature *(Thermoelement)*
Vergleichsstrecke f <Konstzeich> comparative length
Vergleichsstück n <Werkprüf> reference piece

Vergleichssystem

Vergleichssystem n <Ergon> benchmark system
Vergleichstest durchführen v <Comp & DV> benchmark
Vergleichston m <Akustik> reference tone
Vergleichswerte mpl <Akustik> comparison values
vergolden v 1. <Fertig> gild; 2. <Metall> gold-plate
Vergolden n <Fertig> gilding
vergoldeter Übergang m <Raumfahrt> gold flashing
Vergoldung f <Ker & Glas> gilding
vergorener Abfall m <Abfall> fermented waste
Vergraben n <Abfall> land burial *(von Müll)*
vergröbern v <Fertig> coarsen
vergrößern v 1. <Comp & DV> zoom in; 2. <Foto> enlarge; blow up *(Foto)*; 3. <Phys> magnify
Vergrößern n 1. <Comp & DV> zooming; 2. <Foto> enlargement process
vergrößert zeichnen v <Konstzeich> draw to a larger scale
vergrößerte Spurweite f <Eisenbahn> *(AE)* wide track gage, *(BE)* wide track gauge
vergrößertes Bild n <Phys> enlarged image
vergrößertes Sucherbild n <Foto> magnified viewfinder image
Vergrößerung f 1. <Comp & DV> zoom-in; 2. <Druck> enlargement; 3. <Foto> enlargement, enlargement print; 4. <Optik> magnification, power; 5. <Phys> magnification
Vergrößerung f **in Längsrichtung** <Phys> longitudinal magnification
Vergrößerungsabdeckrahmen m <Foto> masking frame
Vergrößerungsapparat m <Foto> enlarger
Vergrößerungsfaktor m 1. <Comp & DV> enlargement factor, zoom factor; 2. <Optik> magnification factor; 3. <Telekom> enlargement factor, zoom factor
Vergrößerungsglas n 1. <Ker & Glas> glass; 2. <Labor, Mechan, Phys> magnifying glass
Vergrößerungskamera f 1. <Druck> enlarging camera; 2. <Foto> enlarger camera, enlarging camera
Vergrößerungsmaßstab m <Konstzeich> enlargement scale
Vergrößerungspapier n <Foto> enlarging paper
Vergrößerungsrahmen m <Foto> masking frame
Vergrößerungssäule f <Foto> enlarger column
Vergrößerungsunterlage f <Foto> enlarger support
Vergrößerungsvermögen n <Phys> magnifying power
Vergussmasse f 1. <Fertig> casting compound, compound; 2. <Kerntech> grouting
vergütbar adj 1. <Fertig> heat-treatable *(Stahl)*; 2. <Thermod> heat-treatable
Vergütbarkeit f <Fertig> heat treatability *(Stahl)*
vergüten v 1. <Anstrich> finish; 2. <Bau> temper *(Stahl)*; 3. <Metall> age; 4. <Thermod> anneal *(Glas)*
Vergüten n 1. <Fertig> quenching and tempering; 2. <Thermod> anneal *(Glas)*
vergütet adj <Thermod> heat-treated; annealed *(Glas)*
vergütete Linse f <Foto, Phys> bloomed lens, coated lens
vergüteter Stahl m <Papier> annealed steel
vergütetes Objektiv n <Phys> coated lens
Vergütung f 1. <Bau> *(BE)* ageing, *(AE)* aging; 2. <Fertig> *(BE)* ageing, *(AE)* aging, *(BE)* quench ageing, *(AE)* quench aging; 3. <Foto, Phys> coating; 4. <Thermod> heat treatment
Vergütungsdiagramm n <Thermod> heat treatment diagram
Vergütungsriss m <Thermod> heat treatment crack
Vergütungsstahl m <Fertig> heat-treatable steel
verhallen v <Akustik> add reverberation; dy away *(ausklingen)*

verhallter Schall m <Sicherheit> reverberant sound
Verhalten n 1. <Bau, Ergon> *(AE)* behavior, *(BE)* behaviour; 2. <Meerschmutz> *(AE)* behavior, *(BE)* behaviour *(eines Ölteppichs)*; 3. <Math> *(AE)* behavior, *(BE)* behaviour *(einer Funktion)*; 4. <Qual> *(AE)* behavior, *(BE)* behaviour; 5. <Textil> properties
Verhalten n **eines Schiffes im Seegang** <Wassertrans> sea keeping
Verhalten n **im niederfrequenten Bereich** <Elektronik, Funktech, Telekom> low-frequency response
Verhalten n **mit fester Stellgeschwindigkeit** <Regelung> single speed floating action
Verhaltensmuster n <Ergon> *(AE)* behavior pattern, *(BE)* behaviour pattern
Verhältnis n 1. <Geom> proportion; 2. <Heiz & Kälte> ratio; 3. <Math> proportion, ratio; 4. <Phys> ratio
Verhältnis n **der spezifischen Wärmen** <Phys> ratio of specific heats
Verhältnis n **Flotte zu Ware** <Textil> liquor-to-goods ratio
Verhältnis n **Gewinn zu Rauschtemperatur** <Raumfahrt> G/T, gain-to-noise temperature ratio *(Weltraumfunk)*
Verhältnis n **Kern-Mantel** <Telekom> core-cladding ratio
Verhältnis n **Länge/Breite** <Elektrotech> aspect ratio
Verhältnis n **Reaktanz/Widerstand** *(Q)* <Akustik> ratio of reactance to resistance *(Q)*
Verhältnis n **Schub/Gewicht** <Raumfahrt> thrust-to-weight ratio
Verhältnis n **Signal zu Rauschen** <Gerät> signal-to-noise ratio *(meist Spannungsverhältnis)*
Verhältnis n **von Seitenhöhe zu Tiefgang** <Wassertrans> *(AE)* depth-to-draft ratio, *(BE)* depth-to-draught ratio *(Schiffkonstruktion)*
Verhältnis n **zwischen Förderhöhe und Widerstand** <Nichtfoss Energ> lift-to-drag ratio
Verhältnis n **zwischen Geschwindigkeit und Verkehrsdichte** <Trans> speed density relationship
Verhältnis n **zwischen Geschwindigkeit und Verkehrsfluss** <Trans> speed flow diagram, speed flow relationship
Verhältniskondensatoren mpl <Elektrotech> ratioed capacitors
Verhältnislautheit f <Akustik> relative loudness
Verhältnismessung f <Gerät> ratio measurement
Verhältnisoperator m <Comp & DV> relational operator
Verhältnisschabotte f <Fertig> anvil ratio
Verhältnistonhöhe f <Akustik> relation pitch
Verhältniswerte mpl <Akustik> ratio values
Verhältniszahl f <Math> ratio
Verhandlung f <Bau, Patent> negotiation
verhärten v 1. <Bau> set; 2. <Fertig> indurate; 3. <Metall> harden
Verhärtung f **durch Neutronenausfluss** <Kerntech> leakage hardening
verharzen v <Fertig> gum up *(Öl)*
Verharzung f <Fertig> gumming up, gummy deposit
verholen v <Wassertrans> warp *(Schiff)*
Verholen n <Wassertrans> warping *(Tauwerk)*
Verholkopf m <Wassertrans> warping head, gipsy *(Deckausrüstung)*
Verholleine f <Wassertrans> warp *(Tauwerk)*
Verholtrommel f <Wassertrans> warping drum *(Deckbeschläge)*
Verhütung f **der Wasserverschmutzung** <Umweltschmutz> prevention of water pollution
Verhütung f **von Luftverschmutzung** <Umweltschmutz> prevention of atmospheric pollution

Verifizierung f <Comp & DV, Kontroll, Maschinen, Qual> verification
Verifizierung f **mittels Grenzlehren** <Maschinen> (AE) verification by means of limit gages, (BE) verification by means of limit gauges
verjüngen v <Bau> batter, reduce
verjüngen v/**sich** <Fertig> contract, taper
verjüngt adj <Bau> (AE) beveled, (BE) bevelled; splayed (Holzbau)
verjüngter Abschnitt m <Elektrotech> tapered section
Verjüngung f 1. <Fertig> conicity, draft; 2. <Maschinen> back taper; tapering (konisch); 3. <Mechan> taper
verkabelter Haushalt m <Fernseh> cabled home
Verkabelung f <Elektriz, Elektrotech, Telekom> cabling
verkalkt adj <Metall> calcified
Verkalkung f <Metall> calcification
verkanten v <Bau> bend out of line
Verkantungsvorrichtung f <Lufttrans> (AE) leveling unit, (BE) levelling unit
verkatten v <Wassertrans> back (Anker)
Verkauf m <Erdöl> farming-out (Konzessionsbeteiligung)
Verkaufsdatum n <Verpack> sell-by date
Verkehr m <Trans> traffic
verkehrsabhängig gesteuertes Signal n <Trans> traffic-actuated signal
verkehrsabhängige Signale npl <Trans> vehicle-actuated traffic signals
verkehrsabhängige Signalgebung f <Trans> vehicle-actuated signalization
verkehrsabhängige Signalsteuerung f <Trans> vehicle-actuated control
Verkehrsablauf m <Trans> traffic flow
Verkehrsabschnitt m <Trans> traffic cut
Verkehrsabwicklung f <Telekom> traffic flow
Verkehrsampel f <Trans> traffic lights
Verkehrsampelanlage f <Elektronik> traffic signal, traffic light
Verkehrsanalysator m <Trans> (BE) traffic analyser, (AE) traffic analyzer
Verkehrsanalysedetektor m <Trans> traffic analysis detector
Verkehrsangebot n <Telekom> traffic offered
verkehrsangepasstes Steuergerät n <Trans> traffic-adjusted controller
Verkehrsaufkommen n <Trans> service volume, traffic volume
Verkehrsaufteilungsmodell n <Trans> traffic assignment model
Verkehrsaufteilungsprogramm n <Trans> (AE) traffic assignment program, (BE) traffic assignment programme
Verkehrsaufteilungssystem n <Telekom> traffic division system
Verkehrsausscheidungszahl f <Telekom> prefix (Telefon)
Verkehrsbedingungen fpl <Trans> traffic conditions
Verkehrsbelastung f <Telekom> traffic carried
Verkehrsbelastungsplan m <Trans> traffic flow diagram
verkehrsberuhigte Zone f <Trans> traffic restraint area
Verkehrsdetektor m <Trans> traffic detector
Verkehrsdichte f <Eisenbahn, Telekom, Trans> density of traffic, traffic density
Verkehrsdurchsage f <Trans> road message; traffic announcement (TA)
Verkehrserhebung f <Trans> traffic survey
Verkehrsflugzeug n <Lufttrans> passenger plane
Verkehrsfluss m <Telekom, Trans> traffic flow
Verkehrsflusssteuerung f <Trans> traffic flow control
Verkehrsfunk m <Funktech, Trans> roadside radio transmitter

Verkehrsfunkkanal m <Trans> traffic message channel (TMC)
Verkehrsfunksender m <Funktech, Trans> traffic radio transmitter
Verkehrsgüte f <Telekom> grade of service
Verkehrsinformation f <Trans> traffic information
Verkehrsinformationserkennungszeichen n <Trans> traffic information identification signal
Verkehrsinformationskanal m <Trans> traffic message channel
Verkehrsinformationssystem n <Trans> traffic message system
Verkehrsinfrastruktur f <Trans> transportation infrastructure
Verkehrskanal m 1. <Mobilkom> traffic channel (Verkehrsfunknachrichten); 2. <Telekom> traffic channel (TCH)
Verkehrskanal m **mit halber Übertragungsrate** <Telekom> half rate TCH, half rate traffic channel
Verkehrskanal m **mit voller Übertragungsrate** <Telekom> full rate TCH, full rate traffic channel
Verkehrskapazität f <Telekom> traffic-handling capability
Verkehrsklasse f <Telekom> class-of-traffic (COT)
Verkehrsklassenkennzeichen n <Telekom> class-of-traffic signal, COT signal
Verkehrsknoten m <Eisenbahn> traffic node
Verkehrsknotenpunkt m 1. <Eisenbahn> rail junction; 2. <Trans> rail junction, (AE) traffic center, (BE) traffic centre
Verkehrskontrollanlage f <Trans> traffic control installation
Verkehrskontrolle f <Trans> traffic control
Verkehrskontrollprogramm n <Trans> (AE) traffic control program, (BE) traffic control programme
Verkehrskonzentration f <Trans> traffic concentration
Verkehrskreisel m <Trans> (AE) rotary, (BE) roundabout, (AE) traffic circle
Verkehrslage f <Trans> traffic situation
Verkehrslast f 1. <Bau> rolling load, (AE) traveling load, (BE) travelling load; live load (Tiefbau); 2. <Eisenbahn, Telekom, Trans> traffic load
Verkehrsleistung f <Trans> traffic density
Verkehrsleitungsprogramm n <Trans> (AE) traffic-routing program, (BE) traffic-routing programme
Verkehrslenkung f 1. <Telekom> routing, traffic control; 2. <Trans> guidance; 3. <Wassertrans> channelization
Verkehrslenkung f **entlang dem kostengünstigsten Weg** <Telekom> least cost routing
Verkehrslenkung f **entsprechend der kürzesten Warteschlange** <Telekom> hot potato routing
Verkehrslenkung f **mit Zielsuche** <Telekom> saturation routing
Verkehrslenkungsstrategie f <Telekom> traffic routing strategy
Verkehrsleuchtnagel m <Bau> reflecting stud (auf der Straße)
Verkehrsmanagement n <Telekom> traffic management
Verkehrsmenge f <Telekom> traffic amount, traffic volume
Verkehrsmesser m <Metrol> demand meter, traffic meter
Verkehrsnachfrage f <Trans> traffic demand
Verkehrsparameter m <Trans> traffic parameter
Verkehrsplanung f <Trans> traffic planning, traffic schedule
Verkehrsprognose f <Trans> traffic forecast
Verkehrsprognoseprogramm n <Trans> (AE) traffic-forecasting program, (BE) traffic-forecasting programme
Verkehrsqualität f <Trans> level of service
Verkehrsradar n 1. <Funkort> traffic radar; 2. <Trans> road traffic radar

Verkehrsrechner

Verkehrsrechner *m* <Trans> traffic computer
Verkehrsregelung *f* <Trans> traffic regulation
Verkehrsregelung *f* **mit geschlossener Schleife** <Lufttrans> closed-loop traffic control system
Verkehrsregion *f* <Trans> traffic region
Verkehrsrichtung *f* <Eisenbahn> direction of traffic
Verkehrsschild *n* <Trans> marker
verkehrsschwache Zeit *f* 1. <Telekom> light traffic period, slack traffic period; 2. <Trans> *(AE)* light hours, lowest hourly traffic
Verkehrssicherheit *f* <Trans> road safety
Verkehrssicherheitsprogramm *n* <Trans> road safety programme
Verkehrssignal *n* <Trans> traffic signal
Verkehrssignalprogramm *n* <Trans> *(AE)* traffic signals program, *(BE)* traffic signals programme
Verkehrssignalsteuereinheit *f* <Trans> traffic signal controller
Verkehrssimulation *f* <Trans> simulation of traffic, traffic simulation
Verkehrssimulationsprogramm *n* <Trans> *(AE)* traffic simulation program, *(BE)* traffic simulation programme
Verkehrssimulator *m* <Trans> traffic simulator
Verkehrsspitze *f* <Eisenbahn, Trans> traffic peak
verkehrsstarke Zeit <Telekom> heavy hours *(Telefonnetz)*
Verkehrsstau *m* <Trans> *(BE)* traffic jam
Verkehrssteuerung *f* <Trans> control of flow
Verkehrsstrom *m* <Trans> traffic stream
Verkehrstechnik *f* <Trans> traffic engineering
Verkehrsteilnehmer *m* <Trans> road user
Verkehrstrennungsgebiet *n* <Wassertrans> traffic separation scheme *(Navigation)*
verkehrstüchtig *adj* <Kfztech> roadworthy
Verkehrsüberwachung *f* 1. <Telekom> traffic supervision; 2. <Trans> traffic surveillance
Verkehrsumlegung *f* <Trans> traffic assignment
Verkehrsumleitung *f* <Trans> diversion, diversion of traffic, traffic diversion
Verkehrsunfall *m* <Trans> traffic accident
Verkehrsverstoß *m* <Trans> traffic violation
Verkehrsvolumenzähler *m* <Trans> traffic volume meter
Verkehrsvorhersage *f* <Trans> traffic forecasting
Verkehrsvorschriften *fpl* <Trans> traffic regulations
Verkehrsweg *m* <Wassertrans> traffic lane *(Navigation)*
Verkehrswert *m* <Telekom> traffic intensity *(Einheit: Erlang)*
Verkehrszähler *m* <Trans> traffic counter
Verkehrszählung *f* <Trans> traffic census
Verkehrszählung *f* **am Querschnitt** <Trans> traffic count
Verkehrszeichen *n* <Trans> road sign, sign, traffic sign
verkehrt bombiert *adj* <Fertig> concave *(Walzen)*
verkehrt konisch *adj* <Fertig> inverted *(Gießen)*
verkeilen *v* 1. <Bau> block, key; 2. <Fertig> chock, key, wedge; 3. <Maschinen> cotter, key; 4. <Mechan> wedge
Verkeilen *n* <Fertig, Maschinen> keying
verkeilt *adj* 1. <Fertig> keyed; 2. <Mechan> splined
Verkeilung *f* 1. <Bau> keying; 2. <Maschinen> wedging
verketten *v* <Comp & DV, Kontroll> concatenate
Verketten *n* <Telekom> concatenation
verkettete Busstruktur *f* <Comp & DV> daisy chain bus
verkettete Dateien *fpl* <Comp & DV> concatenated data sets
Verkettung *f* 1. <Comp & DV, Elektronik, Elektrotech> concatenation, connection, interlinking; 2. <Künstl Int> chaining; 3. <Mechan> linkage; 4. <Telekom> concatenation
verkitten *v* 1. <Bau> cement, putty, seal, stuff; 2. <Fertig> lute

verkittete Linsen *fpl* <Ker & Glas> cemented lenses
verkittetes Glas *n* <Ker & Glas> cemented glass
verklammern *v* 1. <Bau> joggle; 2. <Fertig> cramp
Verklappung *f* <Abfall> ocean disposal of waste
Verklappung *f* **auf See** <Meerschmutz, Umweltschmutz> discharge at sea, ocean disposal
verkleben *v* <Bau> cement, stick
Verkleben *n* 1. <Bau> bonding, cementing; 2. <Kunststoff> bonding
Verklebung *f* 1. <Chemtech> agglutination; 2. <Kunststoff> bond; 3. <Mechan> bonding
Verklebung *f* **des Bootskörpers mit dem Deck** <Wassertrans> deck-hull bonding *(Schiffbau)*
verkleiden *v* <Bau> board, box up, clad, face, line; surface *(Material)*
verkleideter Träger *m* <Bau> cased beam
Verkleidung *f* 1. <Bau> cladding, facing; revetment *(Gebäude)*; 2. <Elektriz, Elektrotech> envelope, reinforcement; 3. <Erdöl> lagging; 4. <Heiz & Kälte> fairing; 5. <Hydraul> cladding, clothing; lagging *(Dampfmaschine)*; 6. <Kfztech> panel *(Karosserie)*; 7. <Lufttrans> case; fairing *(Flugzeug)*; 8. <Mechan> casing; 9. <Raumfahrt> cladding, fairing *(aerodynamisch)*
Verkleidung *f* **der Luftschraubenblattspitze** <Lufttrans> blade tip fairing *(Hubschrauber)*
Verkleidungsblech *n* <Hydraul> clothing plate
Verkleidungsmaterial *n* <Bau> sheeting
Verkleidungsübergang *m* <Lufttrans> fillet *(Flugwerk)*
verkleinern *v* <Comp & DV> zoom out
verkleinerter Maßstab *m* <Konstzeich> reduced scale
verkleinertes Modell *n* <Telekom> reduced model
Verkleinerung *f* <Foto> reduction print
Verkleinerungsgrenze *f* <Ker & Glas> diminishing stop level
Verkleinerungsmaske *f* <Foto> reduction mask
verklemmt *adj* <Raumfahrt> jammed
Verklumpung *f* <Lebensmittel> agglutination
Verknüpfbarkeit *f* <Comp & DV, Telekom> connectivity *(Teile eines Systems)*
verknüpfen *v* 1. <Comp & DV> combine; bind, concatenate *(Programme)*; 2. <Math> associate
verknüpft *adj* <Comp & DV> interlaced *(Programme)*
verknüpfte Dateien *fpl* <Comp & DV> concatenated data sets
Verknüpfung *f* 1. <Comp & DV> concatenation; 2. <Elektronik, Elektrotech> conjunction *(Schaltkreistechnik)*; connection
Verknüpfungsanweisung *f* <Comp & DV> logic instruction
Verknüpfungsbefehl *m* <Comp & DV> logic instruction
Verknüpfungsglied *n* 1. <Comp & DV> logic element, logic gate; 2. <Elektronik> gate, logic gate
Verknüpfungsschaltung *f* 1. <Comp & DV> logic circuit; 2. <Fernseh> combiner circuit
Verknüpfungstransistor *m* <Elektronik> gating transistor
verkohlen *v* 1. <Kohlen> coal; 2. <Thermod> get blackened with heat
Verkohlen *n* <Kohlen> carbonization
verkohlt *adj* <Kohlen> carbonized
verkokbar *adj* <Kohlen> coking
verkoken *v* 1. <Kohlen> carbonize, coke; 2. <Thermod> carbonize
Verkoken *n* <Kohlen> coking
Verkokung *f* <Kohlen> carbonization
Verkorkung *f* 1. <Chemie> suberification; 2. <Ker & Glas> *(BE)* bore, *(AE)* corkage
Verkorkungsmaschine *f* <Verpack> corking machine
verkratzen *v* <Anstrich> scratch

verkratzte Form f <Ker & Glas> (AE) scratched mold, (BE) scratched mould
verkupfert adj 1. <Bau> coppered; 2. <Metall> coppered, copper-plated
Verkupferung f <Ker & Glas, Metall> copper-plating, coppering
verkürzen v 1. <Phys> contract; 2. <Qual> curtail
verkürzt startendes und landendes Flugzeug n (RTOL-Flugzeug) <Lufttrans> reduced takeoff and landing aircraft (RTOL aircraft)
verkürzte Gleitwinkelbefeuerung f <Lufttrans> abbreviated visual approach slope indicator system
verkürzte Präzisionsanflugswinkelbefeuerung f <Lufttrans> abbreviated precision approach path indicator
verkürzter Bogen m <Bau> diminished arch, skeen arch, skene arch
Verkürzung f 1. <Konstzeich> shortening; foreshortening (durch ungünstige Projektion); 2. <Phys> contraction; 3. <Qual> curtailment
Verkürzungsvorsatz m <Foto> portrait attachment
Verladeband n <Trans> loading conveyor
Verladebrücke f <Eisenbahn, Trans, Wassertrans> loading bridge
verladen v 1. <Lufttrans> emplane; 2. <Trans> load; 3. <Wassertrans> ship (Ladung)
Verladeschein m <Wassertrans> shipping note
Verlag m <Druck> publishing company, publishing house
Verlagerung f <Wassertrans> shifting (Schiffbau)
Verlagswesen n <Druck> publishing industry
verlängern v <Bau> stretch
verlängert adj 1. <Geom> prolate; 2. <Mechan> elongated
verlängerte Standzeit f <Maschinen> extended tool life
verlängerter Sattelpunkt m <Metall> extended node
verlängertes Flugzeug n <Lufttrans> stretched aircraft
Verlängerung f <Telekom> extension (Zeit)
Verlängerung f **von Fristen** <Patent> extension of time limits
Verlängerungskabel n <Elektrotech> extension cable
Verlängerungsrohr n 1. <Labor> expansion tube, extension tube; 2. <Maschinen> lengthening tube
Verlängerungsspule f <Funktech> loading coil
Verlängerungsstange f <Maschinen> lengthening rod
Verlängerungsstößel m <Kohlen> follower (Pfahlgründung)
Verlängerungsstück n <Maschinen> extension piece, lengthening bar, lengthening piece
Verlängerungsstutzen m <Maschinen> extension socket
Verlängerungstubus m <Foto> extension tube
Verlangsamung f 1. <Raumfahrt> deceleration; 2. <Trans> slowing down
Verlangsamungsmesser m <Raumfahrt> decelerometer
Verlangsamungsvorrichtung f <Trans> deceleration device
verlaschen v 1. <Bau> joint; 2. <Fertig> fish, strap
Verlaschen n <Fertig> fishing, jointing
verlaschter Schienenstoß m <Eisenbahn> fishplated rail joint
verlassen v <Comp & DV> exit (Programm)
Verlässlichkeit f <Elektriz> reliability
Verlauf m 1. <Bau> course (einer Straße); 2. <Kohlen> course; 3. <Kunststoff> flow; 4. <Telekom> response
verlaufen v 1. <Bau> spread; run (Farbe); 2. <Fertig> divert (Bohrer)
Verlaufen n 1. <Fertig> deviation, drift, run-out (Bohrer); 2. <Maschinen> drift, (AE) running out of center, (BE) running out of centre, wandering
verlaufendes Weiß n <Fernseh> bleeding whites
Verläufer m <Erdöl> diverter (Bohrmeißel)
Verlauffilter n <Foto> graduated filter
Verlaufhilfsmittel n <Kunststoff> coalescing agent
Verlaufmittel n <Kunststoff> coalescing agent
Verlaufschreiber m <Elektriz> event recorder
verlegbare Steckdose f <Elektriz> portable socket outlet
Verlegelänge f <Telekom> laying length
verlegen v 1. <Bau> pave, place; lay (Leitungen); lay (Rohre); 2. <Fertig> install; 3. <Telekom> lay
Verlegen n <Bau> setting
Verlegen n **von Elektroleitungen** <Bau> electric wiring
Verleger m <Druck> publisher
Verlegung f 1. <Bau> laying; 2. <Fertig> installation (Rohre)
verleimen v <Bau, Mechan> bond
Verleimen n <Bau> bonding
Verleimmaschine f <Fertig> glueing machine
Verleimung f <Mechan> bonding
Verletzer m <Patent> infringer
Verletzung f 1. <Patent> infringement; 2. <Sicherheit> injury
Verletzung f **am Arbeitsplatz** <Sicherheit> injury in the work-place
Verletzung f **der Sicherheitsvorschriften** <Sicherheit> breach of safety rules
Verletzung f **durch Stromschlag** <Sicherheit> electrocution
verlitzen v <Fertig> strand
Verlitzung f <Fertig> stranding
verlorene Betonschalung f <Bau> permanent concrete shuttering
verlorene Schalung f <Bau> permanent shuttering
verlorener Kopf m <Fertig> shrink head; feeding head, sinkhead (Gießen)
verlöschen v <Elektriz> blow out, extinguish (Lichtbogen)
Verlust m 1. <Elektrotech> leakage, loss; 2. <Erdöl> loss (an Bohrschlamm); 3. <Maschinen, Optik, Telekom> loss
Verlust m **an äußerer Verbindungsstelle** <Optik> extrinsic joint loss, extrinsic junction loss
Verlust m **der Bildsynchronisation** <Fernseh> loss of picture lock
Verlust m **durch Antennenausrichtungsfehler** <Telekom> antenna-pointing loss
Verlust m **durch falsche Kopfausrichtung** <Aufnahme> head misalignment loss
Verlust m **durch Längsversatz** <Optik> longitudinal offset loss
Verlust m **durch Rückdiffusion** <Kerntech> back diffusion loss
Verlust m **durch seitlichen Versatz** <Optik> lateral offset loss
Verlust m **durch transversalen Versatz** <Optik> transverse offset loss
Verlust m **im Koppelelement** <Optik> coupler loss (zwischen Lichtleitern)
Verlust m **im Kühlsystem** <Heiz & Kälte> ventilating and cooling loss
Verlust m **im Ruhezustand** <Elektriz> no-load loss
Verlust m **im unbelasteten Zustand** <Elektriz> no-load loss
verlustarm adj <Elektriz> low loss
verlustarme Diode f <Elektronik> low-leakage diode, low-loss diode
verlustarme Faser f <Optik> (AE) low-loss fiber, (BE) low-loss fibre
verlustarmer Isolator m <Elektrotech> low-loss insulator
verlustarmes Dielektrikum n <Elektriz> low-loss dielectric
verlustarmes Glas n <Ker & Glas> low-loss glass
verlustarmes Kabel n <Elektriz> low-loss cable
verlustbehaftet adj <Elektrotech, Telekom> lossy

verlustbehaftete

verlustbehaftete Leitung f <Phys, Telekom> lossy line
verlustbehaftete Mode f <Optik> leaky mode *(Lichtleiter)*
verlustbehafteter Strahl m <Optik> leaky ray
Verlustbetrieb m <Telekom> loss mode working
Verlustdämpfung f <Gerät> loss attenuation
Verlustenergie-zerstreuendes Medium n <Phys> dissipative medium
Verlustfaktor m 1. <Elektriz> dissipation factor, leakage factor, loss factor, power factor; 2. <Funktech> dissipation coefficient; 3. <Lufttrans> alleviating factor
Verlustfaktormessgerät n <Metrol> loss factor meter
Verlustfaktormessung f <Metrol> *(AE)* dissipation factor test, loss factor test, *(BE)* loss tangent test
verlustfrei adj <Elektriz, Elektrotech> loss-free, lossless
Verlustfreiheit f <Funktech, Telekom> losslessness
Verlust-Funktion f <Math> loss function *(Parameterschätzung)*
Verlustleistung f 1. <Elektrotech> dissipation, loss, loss power; 2. <Funktech> dissipation; 3. <Heiz & Kälte> heat loss
Verlustleitung f <Elektrotech> lossy line
verlustlos adj <Elektrotech> lossless
verlustlose Leitung f <Telekom> zero-loss circuit
verlustloses Dielektrikum n <Elektriz> perfect dielectric
Verlustmessung f <Gerät> loss measurement
Verlustraum m <Hydraul> waste space
verlustreich adj <Elektrotech> lossy
Verlustschmierung f <Maschinen> total loss lubrication
Verluststrom m <Elektrotech> leakage current, loss current
Verlustverkehr m <Telekom> lost traffic
Verlustwärme f <Heiz & Kälte> heat loss
Verlustwiderstand m **eines Dielektrikums** <Elektriz> dielectric leakage resistance
Verlustwinkel m <Elektriz, Phys> loss angle
Verlustzeit f 1. <Comp & DV> dead time; 2. <Telekom> idle time
Vermahlung f <Kerntech> comminution
vermascht adj <Elektriz> interconnected, intermashed, mashed
vermaschtes Netzwerk n <Elektriz, Telekom> meshed network
Vermaschung f <Elektriz, Fertig> interconnection, intermashing
Vermeidung f **von Lärmbelästigung** <Umweltschmutz> prevention of noise pollution
vermengen v <Textil> blend
vermessen v 1. <Bau> conduct a survey; 2. <Meerschmutz> survey; 3. <Metrol> measure
Vermesser m <Bau> surveyor
vermessingen v <Fertig, Metall> brass
Vermessung f 1. <Bau> survey, surveying, topographical survey; 2. <Metrol, Wassertrans> survey, surveying
Vermessungsgrundlinie f <Bau> transit line
Vermessungsinstrument n <Bau> surveyor's transit; wye level, Y-level *(mit portablem Dreibockständer)*
Vermessungsstange f <Bau> stadia
Vermessungsstation f <Wassertrans> geodesic station
Vermessungstiefe f <Wassertrans> registered depth *(Schiffkonstruktion)*
Vermessungswesen n <Wassertrans> surveying
vermieten v **an** <Wassertrans> charter to *(Schiff)*
vermindern v 1. <Comp & DV> decrement; 2. <Telekom> lower
vermindertes Intervall n <Akustik> diminished interval
vermischen v 1. <Anstrich> agitate, compound; 2. <Bau, Kohlen, Textil> blend
vermischte Abwasserbehandlung f <Wasserversorg> mixed sewage treatment

vermischte Synchronsignale npl <Fernseh> mixed syncs
vermitteln v <Telekom> switch *(Telefon)*
Vermitteln n <Comp & DV> switching
vermittelndes digitales Breitband-Zugangssystem n <Telekom> switched digital broadband access system
vermittelt adj <Telekom> switched
vermittelte Leitung f <Telekom> switched circuit
Vermittlung f 1. <Comp & DV> switching; 2. <Telekom> exchange; switch, switching *(Telefon)* • **über Vermittlung hergestellt** <Telekom> switched
Vermittlung f **über virtuelle Verbindung** <Telekom> VCS, virtual-circuit switch
Vermittlungsablauf m <Telekom> call processing
Vermittlungsamt n <Telekom> *(AE)* central office switch, *(AE)* switching center, *(BE)* switching centre
Vermittlungseinheit f <Telekom> switching unit *(ISDN)*
Vermittlungseinrichtung f 1. <Comp & DV> switching equipment; 2. <Telekom> switching device *(Telefon)*; switching equipment
Vermittlungsinstanzenverbindung f <Telekom> subnetwork connection
Vermittlungsknoten m <Telekom> service switching point *(intelligentes Netzwerk)*
Vermittlungsknoten m **über virtuelle Verbindung** <Telekom> virtual-circuit switching node
Vermittlungsnetz n <Telekom> switched network
Vermittlungsplatz m <Telekom> operating position, operator position
Vermittlungsprogramm n <Telekom> switching program
Vermittlungsrechner m <Telekom> switching processor, switching system
Vermittlungsschicht f 1. <Comp & DV> network layer; 2. <Telekom> switched network layer, network layer
Vermittlungsschrank m 1. <Elektrotech> switchboard; 2. <Telekom> board, switchboard, telephone switchboard
Vermittlungsschrank-Querverbindungsleitung f <Telekom> interswitchboard tie circuit
Vermittlungsschrankreihe f <Telekom> suite of switchboards
Vermittlungsstelle f <Telekom> exchange, switch, *(AE)* switching center, *(BE)* switching centre
Vermittlungsstelle f **mit Handbetrieb** <Telekom> manual exchange
Vermittlungsstelle f **mit Zeitteilung** <Telekom> time switch *(Telefon)*
Vermittlungssystem n <Comp & DV, Telekom> switching system
Vermittlungssystem n **für mehrere Bitraten** <Telekom> multirate switching system
Vermittlungssystem n **mit Reed-Relais** <Telekom> reed relay system
Vermittlungssystem n **mit Zentralsteuerung** <Telekom> common-control switching system
Vermittlungstechnik f <Telekom> switching *(Telefon)*
Vermittlungstheorie f <Comp & DV, Telekom> switching theory
Vermittlungsverzögerung f <Regelung, Telekom> switching delay *(Netz)*
Vermittlungszentrale f **mit Zeitteilung** <Telekom> time division exchange
vermodern v <Bau> rot *(Holz)*
Vermodern n <Bau> rotting
Vermoderung f <Chemie> decomposition
vermörteln v <Bau> grout
Vermörtelung f <Bau> cement stabilization
Vernadelung f <Textil> needling
vernähen v <Fertig> sew
Vernähen n <Fertig> sewing

vernetzen v <Kunststoff> cross-link, cross-linking, cure
Vernetzen n <Kunststoff> cure
Vernetzer m <Kunststoff> cross-linking agent, curing agent
vernetztes Polyethylen n *(VPE)* <Kunststoff> cross-linked polyethylene *(XPE)*
Vernetzung f 1. <Comp & DV> network, networking; 2. <Kunststoff> cross-linking, curing
Vernetzungsgeschwindigkeit f <Kunststoff> cure rate, rate of cure
Vernetzungsmittel n <Kunststoff> cross-linking agent, curing agent
Vernetzungsstelle f <Kunststoff> cross link
Verneuil-Methode f <Optik> *(AE)* vapor phase verneuil method, *(BE)* vapour phase verneuil method *(Aufdampfung nach Verneuil)*
Vernichtung f <Teilphys> annihilation
Vernichtungsstrahlung f <Strahlphys> annihilation radiation
vernickeln v <Fertig> nickelize
vernickeltes Silberzeug n <Metall> electroplated nickel silver *(EPNS)*
vernieten v <Fertig> rivet, rivet up
Vernin n <Chemie> vernine
vernuten v <Fertig> tongue
veröffentlichen v <Druck, Patent> publish
Veröffentlichen n <Comp & DV> publishing
verpacken v <Papier> wrap • **in Tüten verpacken** <Lebensmittel, Verpack> bag
verpackt adj/als **Postversand** <Lebensmittel> mail-order-packed
verpackte Waren fpl <Verpack> *(AE)* parceled goods, *(BE)* parcelled goods
Verpackung f **aus verschäumtem Kunststoff** <Verpack> plastic foam packaging
Verpackung f/**für Mikrowellen geeignete** <Verpack> microwaveable packaging
Verpackung f **in geregelter Atmosphäre** *(CA-Verpackung)* <Verpack> controlled-atmosphere packaging *(CAP)*
Verpackungsabfall m <Abfall> packaging waste
Verpackungsband n **für Ballen** <Verpack> bale hoop
Verpackungsbandeisen n <Metall, Verpack> hoop iron
Verpackungsbereich m <Ker & Glas> packer's bay
Verpackungslinie f <Verpack> packaging line
Verpackungslinie f **für Flüssigstoffe** <Verpack> liquid packaging line
Verpackungsmaschine f 1. <Textil> packer; 2. <Verpack> boxing machine
Verpackungsmaterial n <Abfall> packaging material
Verpackungsmüll m <Abfall> packaging waste
Verpackungsprofil n <Verpack> packaging profile
Verpackungsprüfung f <Verpack> package test
Verpackungsstation f <Verpack> packaging station, packing station
Verpackungstablett n <Verpack> punnet tray *(für Obst, Kuchen)*
Verpackungs- und Entpackungsmaschine f <Verpack> case packing and unpacking *(für Ampullen)*
Verpackungszeichnung f <Konstzeich> packing drawing
Verpropfungswerkzeug n <Bau> grafting tool
verpuffen v <Sicherheit> deflagrate
Verpuffungsstrahlrohr n <Trans> aeropulse
Verputz m <Bau> coating, plaster
verputzen v <Bau> torch
Verputzen n <Bau> coating, plaster work, plastering
Verrauchung f **der Umwelt** <Umweltschutz> fumigation
verrauschte Schwarzwerte mpl <Fernseh> noisy blacks
verrauschte Synchronisierung f <Fernseh> degraded sync
verrauschtes Signal n 1. <Elektronik> noisy signal; 2. <Fernseh> degraded signal
Verrauschtsein n <Fernseh> degradation
Verreibwalze f <Druck> distributing roller mechanism
verrichten v <Telekom> perform
verriegelbar adj <Sicherheit> lockable
verriegelbare Zugangstür f <Sicherheit> lockable access door
verriegelbarer Steckverbinder m <Elektrotech> lockable connector
verriegeln v 1. <Bau> block, bolt; 2. <Comp & DV, Elektrotech> interlock, latch, lock; 3. <Fertig> interconnect; 4. <Kontroll> block; 5. <Maschinen> lock; 6. <Mechan> bolt; 7. <Sicherheit> lock
verriegelte Abschirmung f <Sicherheit> interlocked screen
verriegelte Schutzvorrichtung f <Sicherheit> interlocked guard
verriegelte Steuerung f <Sicherheit> key-locked control *(Laseranlage)*
verriegelte Weiche f <Eisenbahn> locked switch
verriegelte Zugangsstelle f <Sicherheit> interlocked access point
Verriegelung f 1. <Bau> keeper; 2. <Comp & DV> interlock, latch; 3. <Elektrotech> interlock, latching; 4. <Hydraul> shutter; 5. <Kerntech> locking; latch *(eines Steuerstabes)*; 6. <Lufttrans, Mechan> interlock; 7. <Sicherheit> access interlock, interlocking mechanism, latch, locking device; 8. <Telekom> lock
Verriegelungsbolzen m <Bau> locking bolt
Verriegelungseinrichtung f 1. <Bau> lock staple; 2. <Sicherheit> interlocking guard
Verriegelungskontakt m <Elektrotech> interlock contact
Verriegelungsrelais n <Elektriz, Elektrotech> interlock relay, interlocking relay
Verriegelungsschalter m 1. <Elektrotech> interlock switch; 2. <Sicherheit> locking switch
Verriegelungsschaltung f 1. <Elektrotech> interlock circuit; 2. <Fernseh> interlock
Verriegelungsschutz m <Sicherheit> interlocking guard
Verriegelungssignal n <Elektronik> inhibiting signal
Verriegelungssperre f <Sicherheit> interlocking gate, interlocking guard
Verriegelungssystem n <Aufnahme> interlocking system
Verriegelungsventil n <Lufttrans> lock-out valve *(Flugwesen)*
Verriegelungsvorrichtung f <Verpack> interlocking device
verringern v <Comp & DV> decrement
Verringerung f **des Jojo-Effekts** <Raumfahrt> yoyo despin *(Raumschiff)*
verrippen v <Fertig> rib
verrohren v <Erdöl> tube *(Bohrung)*
Verrohrung f <Bau> piping, tubing
Verrohrungsbühne f <Erdöl> stabbing board *(Bohrtechnik)*
Verrohrungskopf m <Erdöl> bradenhead *(Bohrung)*
Verrohrungsseil n <Bau> casing line
verrottbarer Stoff m <Abfall> putrescible matter
verrotten v <Bau> rot
verrücken v <Bau> dislodge *(Gebäude)*
verrühren v <Anstrich> agitate
versagen v 1. <Bau> break down; 2. <Mechan> fail; 3. <Phys> break down
Versagen n 1. <Elektrotech, Mechan> failure; 2. <Telekom> breakdown *(Stromversorgung)*

Versagen

Versagen *n* des Reziprozitätsgesetztes <Foto> reciprocity failure
Versagenskriterium *n* <Mechan> criterion of failure
Versalhöhe *f* <Druck> capital height, height of capital letters
Versandaktion *f* <Comp & DV> mailing
Versandfass *n* für Einzelelemente <Kerntech> single element shipping cask *(für Brennelemente)*
Versandhülle *f* <Verpack> mailing sleeve
Versandkiste *f* <Textil> packing case
Versandkontrolle *f* <Qual> shipping inspection
Versandprobe *f* <Qual> shipping sample
Versandrolle *f* <Verpack> mailing tube, postal tubes
Versandtasche *f* <Verpack> wallet-type envelope
Versandzeichnung *f* <Konstzeich> dispatch drawing
Versatz *m* 1. <Bau> break joint; 2. <Fertig> misalignment *(Faseroptik)*; 3. <Maschinen> displacement, misalignment, offset; 4. <Telekom> offset
Versatzsatz *m* <Konstzeich> parallel offset
versäubern *v* <Textil> serge
versauern *v* <Chemie> acidify *(Boden)*
Versauerung *f* <Umweltschmutz> acidification *(des Bodens)*
verschachteln *v* 1. <Comp & DV> interlace, interleave, nest; 2. <Elektronik> interleave
Verschachteln *n* 1. <Comp & DV> nesting; 2. <Telekom> interleaving
verschachtelt *adj* <Comp & DV> nested
verschachtelte Prozedur *f* <Comp & DV> nested procedure
verschachtelte Schleife *f* <Comp & DV> nested loop
verschachtelte Struktur *f* <Comp & DV> nested structure
verschachtelte Wicklung *f* <Elektrotech> banked winding, interlaced winding
verschachtelter Makroaufruf *m* <Comp & DV> nested macrocall
verschachteltes Übertragungssignal *n* <Fernseh> interleaved transmission signal
Verschachtelung *f* 1. <Comp & DV> nesting; 2. <Telekom> interleaving
Verschachtelungsebene *f* <Comp & DV> nesting level
Verschachtelungsgrad *m* <Comp & DV> nesting level
Verschachtelungsspeicher *m* <Comp & DV> nesting store
verschalen *v* <Bau> board, timber; batten *(mit Leisten)*
verschalken *v* <Wassertrans> batten down *(Luken)*
Verschalung *f* <Bau> timbering; planking *(mit Bohlen)*
verschärfte Inspektion *f* <Kerntech> tightened inspection
verschärfte Prüfung *f* <Qual> increased inspection, tightened inspection
verschärfter AQL-Wert *m* <Qual> reduced AQL value
Verschicken *n* <Comp & DV> mailing
verschiebbar *adj* 1. <Comp & DV> relocatable; 2. <Fertig> *(AE)* traveling, *(BE)* travelling
verschiebbare Adresse *f* <Comp & DV> relocatable address
verschiebbare Daten *npl* <Comp & DV> relocatable data
verschiebbarer Ausdruck *m* <Comp & DV> relocatable expression
verschiebbarer Kern *m* <Elektriz> movable core
verschiebbares Format *n* <Comp & DV> relocatable format
verschiebbares Programm *n* <Comp & DV> relocatable program
Verschiebebahnhof *m* <Eisenbahn> classification yard, *(BE)* marshalling yard, *(AE)* switchyard, shunting yard, *(AE)* switching station
Verschiebedienst *m* <Eisenbahn> shunting
Verschiebegleis *n* <Eisenbahn> classification yard line, sorting line
Verschiebelok *f* <Eisenbahn> shunter
verschieben *v* 1. <Bau> displace; 2. <Comp & DV> relocate, shift; 3. <Elektrotech> displace, shift *(Bürsten)*
verschieben *v/sich* <Wassertrans> shift *(Ladung)*
Verschieben *n* 1. <Eisenbahn> *(AE)* switching; 2. <Fertig> traversing
Verschiebewiderstand *m* <Maschinen, Qual> resistance to motion
Verschiebung *f* 1. <Bau> offset *(räumlich, zeitlich)*; 2. <Comp & DV> relocation; 3. <Comp & DV, Elektriz, Elektrotech> displacement *(D)*; shift; 4. <Funktech> drift; 5. <Kerntech> shift, travel; 6. <Kontroll> displacement *(D)*; 7. <Phys> displacement, translation; 8. <Raumfahrt> drift *(Raumschiff)*; shift *(Weltraumfunk)*; 9. <Telekom> shift; transposition *(Frequenz)*
Verschiebungsstörzone *f* <Kerntech> displacement spike
Verschiebungsstrom *m* <Elektrotech, Phys> displacement current
verschiefern *v* <Bau> split into thin sheets
verschießen *v* <Textil> fade *(Farbe)*
Verschiffung *f* <Wassertrans> shipment
Verschiffungsgewicht *n* <Wassertrans> shipping weight *(Ladung)*
verschlacken *v* <Metall> scorify
Verschlackung *f* <Kerntech> slagging *(Akkumulation von Spaltprodukten)*
verschlämmen *v* <Wasserversorg> blind
Verschlammung *f* 1. <Abfall> sludge accumulation; 2. <Chemie> silting; 3. <Wasserversorg> silting-up
verschlechtern *v/sich* <Qual, Verpack> deteriorate
verschlechterte Betriebsbedingungen *fpl* <Raumfahrt> degraded operating conditions
Verschlechterung *f* 1. <Comp & DV> degradation; 2. <Verpack> deterioration
verschleierte Sprache *f* <Telekom> inverted speech *(durch Frequenzumkehr)*
Verschleiß *m* 1. <Bau> attrition; 2. <Elektriz> wear, wear out; 3. <Fertig> cavitation *(durch Auswaschung)*; 4. <Ker & Glas> scuffing; 5. <Kerntech> galling *(durch Reibung)*; 6. <Kunststoff> abrasive wear; 7. <Maschinen> wear; 8. <Mechan> abrasion, compensation for wear, wear; 9. <Meerschmutz, Papier> wear; 10. <Textil> wear and tear • **Verschleiß ausgleichen** <Mechan> compensate for wear • **Verschleiß kompensieren** <Mechan> compensate for wear
Verschleißanzeige *f* <Maschinen> wear indicator
Verschleißausgleich *m* <Mechan> compensation for wear
verschleißbeständig *adj* <Maschinen> hardwearing, wear-resistant
Verschleißbeständigkeit *f* <Maschinen> wear resistance
Verschleißblech *n* <Maschinen> wear plate, wearing plate
verschleißen *v* <Maschinen> wear, wear down, wear off, wear out
verschleißend *adj* <Fertig> abrasive
verschleißend wirken *v* <Fertig> wear
Verschleißfaktor *m* <Mechan> abrasion factor
Verschleißfehler *m* <Kerntech> wearout defect
verschleißfest *adj* 1. <Fertig> wear-resistant; 2. <Maschinen> hardwearing, resistant to wear, wear-resistant
Verschleißfestigkeit *f* 1. <Bau> abrasion resistance *(Beton)*; 2. <Fertig> abrasion resistance, wear resistance; 3. <Kunststoff> abrasion resistance; 4. <Maschinen> re-

sistance to wear, wear resistance; 5. <Qual> resistance to wear
Verschleißfestigkeitskennzahl f <Kunststoff> abrasion resistance index
Verschleißfortschritt m <Fertig> amount of wear
verschleißfrei adj 1. <Fertig> free from wear, wearless; 2. <Maschinen> no-wear
Verschleißgrenze f <Maschinen> wear limit
Verschleißkehle f <Eisenbahn> hollow tread *(Reifen)*
Verschleißkomponente f <Kerntech> wearing element
verschleißlos adj <Maschinen> no-wear
Verschleißmarke f <Fertig> wear mark
Verschleißmarkenbreite f <Fertig> wear land value, width of wear mark
Verschleißmessung f <Fertig> wear measurement
Verschleißplatte f <Fertig> wearing plate
Verschleißprozess m <Maschinen> wear process
Verschleißprüfmaschine f <Mechan> abrasion tester
Verschleißschicht f 1. <Bau> surface dressing, veneer, wearing course; 2. <Maschinen> wearing surface
Verschleißteil n 1. <Fertig> worn part *(Kunststoffinstallationen)*; 2. <Kerntech> wearing detail, working part; 3. <Kohlen> wearing part; 4. <Maschinen> wear part, wearing part
Verschleißteilzeichnung f <Konstzeich> drawing dealing with wearing parts
Verschleißverhalten n <Fertig> *(AE)* wear behavior, *(BE)* wear behaviour
Verschleißvolumen n <Fertig> volume of metal worn away
Verschleißwert m <Bau, Qual> wear rate
Verschleißwiderstand m <Kunststoff> abrasion resistance
Verschlickung f <Chemie> silting
verschließbar adj <Sicherheit> lockable
verschließen v 1. <Anstrich> seal; 2. <Bau> cap, lock; 3. <Sicherheit> lock
Verschließen n 1. <Bau> sealing; 2. <Erdöl> plugging
Verschließmaschine f <Verpack> closing machine
verschlissener Bohrmeißel m <Erdöl> worn bit *(Tiefbohrtechnik)*
verschlossen adj <Meerschmutz> sealed
verschlossener Kanal m <Wasserversorg> locked canal
Verschluss m 1. <Bau> shutter; 2. <Erdöl> plug; 3. <Foto> shutter *(Kamera)*; 4. <Hydraul> shutter; 5. <Ker & Glas> finish *(einer Flasche)*; 6. <Lufttrans> interlocking; 7. <Mechan> blind, seal; 8. <Verpack> closure
Verschluss m **mit Außengewinde** <Ker & Glas> external-screw-thread finish
Verschluss m **mit B-Einstellung** <Foto> shutter with B setting
Verschlussauslöser m <Foto> shutter release
Verschlussauslösung f <Foto> shutter release
Verschluss-Code-Verriegelung f <Sicherheit> trapped-key interlock
Verschlussdecke f <Bau> seal coat *(Straßenbau)*
Verschlussdeckel m 1. <Fertig> lid cover; 2. <Kfztech> filler cap; 3. <Mechan> cover
Verschlüssel-Fertigungslinie f <Verpack> closure production line
verschlüsseln v 1. <Comp & DV> encode; 2. <Elektronik> encipher, encode, encrypt; 3. <Telekom> cipher, encipher, encrypt
verschlüsselt adj 1. <Raumfahrt> cryptographic *(Weltraumfunk)*; 2. <Telekom> cryptographic
verschlüsselte Nachrichtenübertragung f **ohne Schlüsselübertragung** <Telekom> encrypted message transmission without key transfer, three passage protocol
verschlüsselte Sprache f <Telekom> encrypted speech

verschlüsselte Übertragung f <Telekom> ciphered transmission, coded transmission, encrypted transmission
verschlüsseltes Videosignal n <Fernseh> coded TV-signal
Verschlüsselung f 1. <Comp & DV> encryption; 2. <Elektronik> encipherment, encoding, encryption; 3. <Funktech> encryption; 4. <Raumfahrt> cryptography, encryption; 5. <Telekom> ciphering, coding, encipherment, encryption
Verschlüsselung f **mit vollständiger Geheimhaltung** <Telekom> perfect secrecy encryption
Verschlüsselungschip m <Comp & DV, Elektronik> encoding chip, encryption chip
Verschlüsselungsmechanismus m <Comp & DV> encryption capability
Verschlüsselungsmuster n <Raumfahrt> filter template
Verschlüsselungsschema n <Elektronik> coding scheme
Verschlüsselungssystem n 1. <Comp & DV> cryptographic system; 2. <Telekom> cryptographic system, cryptosystem
Verschlüsselungsvorlage f <Raumfahrt> filter mask
Verschlusshaken m <Eisenbahn> locking hook
Verschlusshülse f <Verpack> strapping seal
Verschlusskappe f 1. <Fertig> plug; 2. <Ker & Glas> screw cap
Verschlusslamelle f <Foto> shutter blade
Verschlüssler m <Fertig> encoder
Verschlussloch n <Bau> plughole
Verschlussmutter f <Fertig> valve nut *(Kunststoffinstallationen)*
Verschlussspannknopf m <Foto> shutter-cocking knob
Verschlussstein m <Ker & Glas> tweel block
Verschlussstopfen m <Maschinen> obturating plug, sealing plug, stopper
Verschlussstück n <Maschinen> cap
Verschlusszeit f <Foto> shutter speed
Verschlusszeiteinstellknopf m <Foto> shutter speed setting knob
Verschlusszeiteinstellung f <Foto> shutter speed setting
Verschlusszeitkontrolle f <Foto> shutter speed control
verschmelzen v <Fertig> coalesce
Verschmelzung f <Ergon, Phys> fusion
verschmieden v <Fertig> cog *(Schmieden)*
verschmieren v <Fertig> lute, smear
Verschmieren n <Fertig> smearing
verschmierter Kopf m <Aufnahme, Fernseh> clogged head
verschmolzene Bifokallinsen fpl <Ker & Glas> fused bifocals
verschmolzenes Bündel n <Ker & Glas> fused bundle
verschmutzen v 1. <Kohlen, Qual> contaminate; 2. <Sicherheit> pollute
verschmutzt adj <Umweltschmutz> polluted
verschmutzter Buchstabe m <Druck> pick
verschmutztes Regenwasser n <Umweltschmutz> polluted rainwater
verschmutztes Wasser n <Wasserversorg> polluted water
Verschmutzung f <Umweltschmutz> contamination, pollution
Verschmutzung f **durch Feststoffe** <Umweltschmutz> material pollution
Verschmutzungsgrad m 1. <Heiz & Kälte> fouling factor; 2. <Qual, Umweltschmutz> degree of pollution
Verschmutzungsgrad m **des Wassers** <Umweltschmutz> pollutional index, water pollution
Verschmutzungsüberwachung f <Gerät> contamination monitoring

verschneiden

verschneiden v 1. <Bau> mix; 2. <Fertig> blend *(Flüssigkeiten)*; 3. <Lebensmittel> adulterate; 4. <Textil> blend
Verschneiden n <Lebensmittel> blending
Verschnitt m 1. <Textil> blend; 2. <Verpack> offcut
Verschnittasphalt m <Bau> cut-back asphalt
Verschnittbitumen n <Bau> cut back
Verschnittmittel n <Chemie, Kunststoff> diluent
Verschnittöl n <Erdöl> extender oil
verschobene Oberfläche f <Ker & Glas> shifted finish
verschobenes Atom n <Kerntech> displaced atom
verschränken v <Textil> interlock
verschrauben v 1. <Fertig> bolt; 2. <Maschinen> screw
Verschrauben n <Maschinen> screwing
Verschraubmaschine f <Fertig> bolting machine
Verschraubung f 1. <Bau> bolting, union; 2. <Erdöl> screwing; 3. <Fertig> union *(Kunststoffinstallationen)*; 4. <Maschinen> bolted joint, bolting, screw joint; 5. <Sicherheit> bolting
Verschraubungsteile npl <Maschinen> threaded components
verschrotten v <Qual> scrap
Verschrotten n <Kfztech> scrapping
Verschwelung f <Chemie> charring
Verschwindungspunkt m <Geom, Konstzeich> vanishing point
verschwommene Logik f <Künstl Int> fuzzy logic
verschwommenes Bild n <Phys> blurred image
Verscrambelungssteuerung f <Telekom> scrambling control
versehen adj **mit** <Telekom> fitted with
Verseifbarkeit f <Fertig> saponifiability
verseifen v <Fertig> saponify
Verseifung f <Chemie, Fertig> saponification
Verseifungsmittel n <Chemie> saponifier
Verseifungszahl f <Lebensmittel> saponification number *(Fettkennzahl)*
Verseilen n <Fertig> cabling
Verseilen n **der Fasern** <Telekom> stranding of the fibers *(optisches Kabel)*
Verseilmaschine f <Fertig> cabling machine
verseiltes Kabel n <Elektrotech> stranded cable
Verseilung f <Fertig> stranding *(Faseroptik)*
versenden v <Comp & DV> send
Versenden n <Comp & DV> mailing
Versendung f <Trans> consignment
versengen v 1. <Kohlen> char; 2. <Thermod> scorch
versengt adj <Thermod> scorched
versenkbare Antenne f <Lufttrans> flush aerial
versenkbare Linse f <Foto> flush lens
versenken v <Wassertrans> scuttle; sink *(Schiff)*
Versenken n 1. <Bau> plunging; 2. <Mechan> counterboring *(Schraubenkopf)*
versenkte Antenne f <Funktech> flush antenna, suppressed antenna
versenkte Fassung f <Foto> countersunk mount, countersunk setting; sunk mount, sunk setting *(Objektiv)*
versenkter Kanal m <Elektronik> buried channel; buried channel *(Transistortechnik)*
versenktes Einsetzen n <Foto> sunk mount, sunk setting
versenktes Schloss n <Bau> dormant lock
versetzen v <Bau> displace; stagger *(auf Lücke)*
versetzt adj <Maschinen> offset, staggered
versetzt abgestimmter Verstärker m <Elektronik> stagger-tuned amplifier
versetzt zeichnen v <Konstzeich> draw staggered
versetzte Ablage f <Comp & DV> stacker offset
versetzte Fuge f <Bau> broken joint

versetzte Köpfe mpl <Aufnahme, Fernseh> staggered heads
versetzte Schwelle f <Lufttrans> displaced threshold *(Start- und Landebahn)*
Versetzung f 1. <Akustik> transposition; 2. <Elektrotech, Phys> displacement, D; 3. <Fertig> displacement, D *(Kristall)*; 4. <Maschinen> offset; 5. <Wassertrans> drift
Versetzungsannihilation f <Metall> dislocation annihilation
Versetzungsdichte f <Metall> dislocation density
Versetzungsgeschwindigkeit f <Metall> dislocation velocity
Versetzungskern m <Metall> dislocation core
Versetzungsknickung f <Metall> dislocation kink
Versetzungsmast m <Elektriz> transposition tower
Versetzungsschutt m <Metall> dislocation debris
Versetzungsverbindung f <Metall> dislocation junction
Versetzungsverlust m <Optik> misalignment loss
verseuchen v <Kerntech, Sicherheit> contaminate
Verseuchung f 1. <Kerntech, Raumfahrt, Sicherheit> contamination; 2. <Umweltschmutz> contamination; contamination *(radioaktiv)*
Versicherungsmathematik f <Math> actuarial mathematics, actuarial science, insurance mathematics
Versickern n <Meerschmutz> seepage
Versickerung f 1. <Abfall> infiltration *(Abwasserreinigung)*; 2. <Kohlen, Wasserversorg> seepage
Versickerungsbecken n <Wasserversorg> infiltration gallery
Versickerungsbrunnen m <Wasserversorg> injection well
versiegeln v <Anstrich, Bau> seal
versiegelter Kühlkreislauf m <Kfztech> sealed cooling system
versiegeltes Präparat n <Kerntech> sealed source
Versiegelung f <Sicherheit> tamper-proof seal
Versiegelungsmittel n <Anstrich> sealant
Versiegler m <Verpack> sealing machine
Versilbern n <Fertig, Metall> silver plating *(Galvanostegie)*
versilbert adj <Fertig, Metall> silver-plated
Versilberung f <Fertig> silvering
Versilberungsförderband n <Ker & Glas> conveyor for silvering
Version f 1. <Comp & DV> release, version *(eines Programms)*; 2. <Maschinen> version
Versionenmanagement n <Comp & DV> multiple version control
versorgen v <Bau, Elektriz> supply
versorgendes Amt n <Telekom> serving exchange
Versorger m <Wassertrans> supply vessel, support vessel *(Schifftyp)*
Versorgung f 1. <Funktech> coverage; 2. <Qual> provision, supply; 3. <Telekom> feed
Versorgungsanschluss m <Raumfahrt> umbilical connector
Versorgungsbasis f <Erdöl> supply base *(Offshore-Technik)*
Versorgungsbehälter m <Kfztech> supply tank
Versorgungsbereich m 1. <Funktech> coverage area; 2. <Lufttrans> service area
Versorgungseinrichtung f <Bau, Raumfahrt> utility
Versorgungsfahrzeug n <Lufttrans> service vehicle
Versorgungsgrad m <Telekom> degree of coverage
Versorgungshauptleitung f <Elektrotech> supply main
Versorgungskanal m <Wasserversorg> delivery race
Versorgungsleitung f 1. <Bau> supply pipe, utility line; 2. <Elektrotech> feeder; 3. <Raumfahrt> umbilical cable
Versorgungsluft-Manometer n <Gerät> *(AE)* air supply gage, *(BE)* air supply gauge

Versorgungsmast m <Raumfahrt> umbilical mast
Versorgungspumpe f <Meerschmutz> supply pump
Versorgungsraumfahrzeug n <Raumfahrt> cargo vehicle
Versorgungssatellit m <Raumfahrt> utility satellite
Versorgungsschaltung f <Elektriz> feed circuit
Versorgungsschiff n <Wassertrans> replenishing ship, supply boat, support vessel, tender; supply vessel, support vessel *(Schifftyp)*
Versorgungsspannungs-Überwachung f <Regelung> power fail circuit
Versorgungstelemetrie f <Raumfahrt> housekeeping telemetry
Verspachteln n <Bau> stopping
verspannen v <Fertig> rig
Verspannung f <Eisenbahn> application *(von Laschen)*
verspannungsloser Flügel m <Lufttrans> cantilever wing
Verspannungs-Messgerät n <Labor> *(AE)* strain gage, *(BE)* strain gauge
verspätete Ausstrahlung f <Fernseh> delayed broadcast
versperren v <Bau> barricade
verspiegelte Lampe f <Foto> mirror-coated lamp
Verspiegelung f <Ker & Glas> mirror plating, silvering
verspinnbar adj <Textil> suitable for spinning
Verspinnen n <Ker & Glas, Textil> spinning
verspleißen v <Fertig> splice *(Faseroptik)*
Verspleißung f <Eisenbahn> splice joint
versplinten v <Maschinen> cotter
Versplinten n <Maschinen> cottering
Versprödung f <Kerntech, Metall> embrittlement
• **Versprödung beseitigen** <Anstrich> de-embrittle
versprühen v <Wasserversorg> spread
Versprühen n <Meerschmutz> dispersant spraying *(von Dispersionsmitteln)*
Versprühen n **eines Wasserstrahls** <Wasserversorg> spout hole
Versprüher m <Labor> atomizer
verstählen v <Metall> acierate, steel
Verstählen n 1. <Fertig> acierage; 2. <Metall> acierage, steeling
Verständigung f **durch Läuterwerk** <Eisenbahn> bell communication
Verständlichkeit f <Akustik, Funktech> intelligibility
Verständlichkeitsfaktor m 1. <Akustik> articulation index, intelligibility index; 2. <Telekom> articulation index
verstärken v 1. <Aufnahme> fade up; 2. <Bau> brace; 3. <Comp & DV> amplify; 4. <Elektronik, Elektrotech> amplify, boost; intensify *(Licht)*; 5. <Lebensmittel> enhance; 6. <Optik> amplify; 7. <Textil> splice; 8. <Wasserversorg> reinforce *(Damm, Stausee)*; 9. <Wellphys> amplify
Verstärken n <Bau> reinforcing
verstärkend adj 1. <Elektriz> amplifying; 2. <Optik> amplificatory, amplifying
verstärkender Füllstoff m <Kunststoff> reinforcing filler
verstärkendes Zweigefüge n <Raumfahrt> secondary structure
Verstärker m 1. <Aufnahme> amplifier; 2. <Comp & DV> amplifier, repeater; 3. <Elektriz, Elektronik, Elektrotech> amplifier; booster *(mit hohem Verstärkungsgrad)*; 4. <Funktech> amplifier; 5. <Phys> amplifier, repeater; 6. <Telekom> amplifier, repeater; 7. <Wellphys> amplifier
Verstärker m **auf dem Chip** <Elektronik> on-chip amplifier
Verstärker m **für zweite ZF** <Elektronik> second IF amplifier
Verstärker m **in Basisschaltung** <Elektronik> common-base amplifier
Verstärker m **in Kollektorschaltung** <Elektronik> common-collector amplifier
Verstärker m **mit geringem Rauschen** <Raumfahrt> low-noise amplifier *(Weltraumfunk)*
Verstärker m **mit hohem Verstärkungsgrad** <Elektronik> high-gain amplifier
Verstärker m **mit Lautstärkebegrenzung** <Elektronik> volume-limiting amplifier
Verstärker m **mit linearem Frequenzgang** <Elektronik, Telekom> flat amplifier
Verstärker m **mit niedriger Verstärkung** <Elektronik> low-gain amplifier
Verstärker m **mit niedriger Verstärkungsleistung** <Elektronik> small-gain amplifier
Verstärker m **mit selektiver Rückkopplung** <Elektronik> selective feedback amplifier
Verstärker m **mittlerer Leistung** <Funktech, Telekom> medium-power amplifier
Verstärkeramt n <Telekom> repeater station
Verstärkerbaugruppe f <Elektronik> amplifier module
Verstärkerbereich m <Telekom> repeater coverage area
Verstärker-Chip m <Elektronik> amplifier chip
Verstärkergestell n <Telekom> repeater deck
Verstärkergrundschaltung f <Elektronik> basic amplifier circuit
Verstärkerklasse f <Elektronik> amplifier class
Verstärkermaschine f <Elektrotech> rotary amplifier
Verstärkerrauschen n <Elektronik> amplifier noise
Verstärkerröhre f <Elektronik> amplifier tube; intensifier tube *(Bildverstärker)*
Verstärkersäule f <Chemtech> concentration column
Verstärkerschaltung f <Elektronik> amplifier circuit
Verstärkerstation f 1. <Fernseh> booster station; 2. <Telekom> repeater
Verstärkerstelle f (VS) <Telekom> repeater station
Verstärkerstufe f 1. <Aufnahme> amplifying stage; 2. <Elektronik> amplifier stage
Verstärker-Vidikon n <Elektronik> intensifier vidicon *(Bildaufnahmeröhre)*
Verstärkerzugang m <Telekom> repeater access
verstärkt adj <Fertig> heavy *(Bohrerkern)*
verstärkter Kunststoff m <Kunststoff> reinforced plastic
verstärkter Rand m <Ker & Glas> reinforced rim
verstärkter Zug m <Bau> *(AE)* forced draft, *(BE)* forced draught
verstärktes Passagierabteil n <Trans> strengthened passenger compartment
Verstärkung f 1. <Akustik> amplification, gain, transmission gain *(eines Übertragungssystems)*; 2. <Aufnahme> *(V)* amplification *(A)*; gain; 3. <Bau> haunch, reinforcement, stiffening; 4. <Elektriz, Elektronik> *(V)* amplification *(A)*; gain; 5. <Erdöl> backup; 6. <Foto> intensification; 7. <Funktech> *(V)* amplification, gain *(G)*; 8. <Funktech> amplification factor, μ *(einer Elektronenröhre)*; 9. <Hydraul> backing up *(Unterwasser)*; 10. <Ker & Glas> thickening; 11. <Maschinen> backing; 12. <Metrol> *(V)* amplification *(A)*; 13. <Optik> *(V)* amplification, gain; 14. <Papier> backing; 15. <Raumfahrt> amplification, gain *(Weltraumfunk)*; 16. <Telekom> amplification, gain; enhancement *(Absorption)*; 17. <Textil> backing, splicing; 18. <Verpack> backing
Verstärkung f **auf dem Chip** <Elektronik> on-chip amplification
Verstärkung f **der Eingangsstufe** <Elektronik> input stage gain
Verstärkung f **des geschlossenen Regelkreises** <Gerät> closed-loop gain
Verstärkung f **für Hängedisplays** <Verpack> tape hanging display reinforcement

Verstärkungsabgleich

Verstärkungsabgleich m <Elektronik> gain trimming
Verstärkungsbogen m <Bau> safety arch
Verstärkungsdrift f <Elektronik> gain drift, gain droop
Verstärkungs-Eckfrequenz f <Elektronik, Funktech, Telekom> gain crossover frequency
Verstärkungseinstellung f <Elektronik> gain adjustment, gain setting
Verstärkungselement n <Fertig> strength member
Verstärkungsfaktor m 1. <Aufnahme> amplification factor; 2. <Elektronik> amplification factor, amplifier gain; 3. <Phys> amplifier gain; 4. <Telekom> gain
Verstärkungs-Frequenzgang m <Elektronik, Funktech, Telekom> gain-frequency characteristic
Verstärkungs-Gewichtungsfaktor m <Elektronik> gain weighting factor
Verstärkungsgrad m <Comp & DV> gain
Verstärkungskurve f <Elektronik> gain curve
Verstärkungslernen n <Künstl Int> reinforcement learning
Verstärkungsmaß n <Elektronik> gain (Logarithmus des Verstärkungsfaktors)
Verstärkungsmaßänderung f <Elektronik> gain change
Verstärkungsmessung f <Gerät> gain measurement
Verstärkungsmittel n <Kunststoff> reinforcing agent
Verstärkungspfosten m <Erdöl> backup post
Verstärkungspumpen n <Aufnahme> gain pumping
Verstärkungsregelung f <Aufnahme, Comp & DV, Elektronik> gain control
Verstärkungsschaltung f <Elektronik> amplifying circuit
Verstärkungsseil n <Erdöl> backup line
Verstärkungsstütze f <Erdöl> backup post
Verstärkungstransistor m <Elektronik> amplifying transistor
Verstärkungsüberhöhung f <Elektronik> peaking
Verstärkungsveränderung f <Elektronik> gain change
Verstärkungsverhältnis n <Aufnahme> amplification ratio
Verstärkungsverschiebung f <Raumfahrt> emphasis (Weltraumfunk)
Verstärkungszug m <Eisenbahn> relief train
verstauen v <Raumfahrt, Wassertrans> stow (Ladung)
Verstehen in einer zusammenhängenden Rede <Künstl Int> discourse understanding
versteifen v 1. <Fertig> strut; 2. <Maschinen> brace
Versteifen n 1. <Bau> strutting; 2. <Raumfahrt> stiffening
versteifende Ausfütterung f <Lufttrans> backing (eines Sitzes)
Versteifung f 1. <Bau> web; lining (von Außenwänden); 2. <Fertig> web; 3. <Kerntech> boom; 4. <Maschinen> backing, stiffening
Versteifungsblech n 1. <Fertig> stiffening sheet; 2. <Maschinen> stiffening plate
Versteifungselement n 1. <Heiz & Kälte> stiffening member; 2. <Maschinen> stiffener
Versteifungsmaterial n <Verpack> reinforced packaging material
Versteifungsprofil n <Wassertrans> stiffener (Schiffbau)
Versteifungsriegel m für seitlichen Aufprall <Kfztech> side impact intrusion beam
Versteifungsrippe f <Fertig> rib
Versteifungsspreize f <Lufttrans> horizontal strut
Versteinerung f **von Schlämmen** <Abfall> sludge petrification
Verstell... <Fertig, Maschinen> adjustable
Verstellarm m <Fertig> adjustable arm (Bohrspindel)
verstellbar adj <Maschinen, Papier> adjustable
verstellbare Düse f <Maschinen> adjustable nozzle
verstellbare Reibahle f <Maschinen> adjustable reamer
verstellbare Schaufeln fpl <Maschinen> adjustable blades
verstellbare Schelle f <Gerät> adjustable strap (an Widerstand)
verstellbare Schutzvorrichtung f <Sicherheit> adjustable guard
verstellbare Skalenmarke f <Gerät> memory pointer
verstellbare Spindel f <Fertig> adjustable arm (Mehrspindelbohrmaschine)
verstellbare Spule f <Elektriz> adjustable inductance, adjustable inductance coil
verstellbare Verdrängungspumpe f <Gerät> variable displacement pump
verstellbare Vorhangwand f <Ker & Glas> adjustable curtain wall
verstellbarer Abgriff m <Gerät> adjustable strap
verstellbarer Anschlag m <Maschinen> adjustable stop
verstellbarer Antrieb m <Maschinen> variable-speed drive
verstellbarer Dreifuß m <Labor> adjustable tripod
verstellbarer Innenspiegel m <Kfztech> adjustable rearview mirror (Zubehör)
verstellbarer Messring m <Wassertrans> variable range marker (Radar)
verstellbarer Metallsägebogen m <Maschinen> adjustable hacksaw frame
verstellbarer Rückspiegel m <Kfztech> adjustable rear mirror
verstellbarer Schraubenschlüssel m 1. <Fertig> monkey wrench; 2. <Maschinen> (BE) adjustable spanner, coach wrench, monkey wrench, (BE) shifting spanner
verstellbarer Tiefenanschlag m <Maschinen> adjustable stop
verstellbarer Transformator m <Elektriz> adjustable transformer
verstellbares Mikrofon n <Aufnahme> rifle microphone
verstellbares Okular n <Foto> adjustable eyepiece
verstellbares Schleusentor n <Hydraul> shuttle
verstellbares Spanneisen n <Fertig> offset clamp
verstellbares Stativ n <Metrol> adjustable tripod (Vermessung)
verstellbares Strichmaß n <Metrol> (AE) caliper gage, (BE) calliper gauge
verstellbares Windeisen n <Maschinen> adjustable tap wrench
verstellen v <Fertig> displace (Kunststoffinstallationen)
Verstellmutter f <Maschinen> regulating nut
Verstellpropeller m 1. <Lufttrans> variable-pitch propeller, variable-pitch air propeller; 2. <Meerschmutz> variable-pitch propeller; 3. <Wassertrans> variable-pitch air propeller, variable-pitch propeller; controllable-pitch propeller (Schiffantrieb)
Verstellpumpe f <Fertig> variable delivery pump
Verstellschraube f <Wassertrans> variable-pitch propeller
Verstellstange f <Maschinen> adjusting rod
Verstellungsgestänge n <Lufttrans> blade pitch change rod (Hubschrauber)
verstemmen v 1. <Fertig> hammer-tighten; 2. <Maschinen> caulk
Verstemmen n 1. <Bau, Fertig> caulking (Niet); 2. <Maschinen> caulking
Verstemmhammer m <Maschinen> caulking hammer
Versteppung f <Umweltschutz> steppization
Verstickstahl m <Mechan> nitrided steel
verstiften v 1. <Bau> dowel, peg; 2. <Fertig> dowel pin; stud (Reparaturschweißung); 3. <Maschinen> dowel, pin
Verstiften n <Fertig> studding (Reparaturschweißung)

verstiftet *adj* <Fertig> pinned; studded *(Reparaturschweißung)*
verstimmen *v* <Elektronik> detune
verstimmte Frequenz *f* <Elektronik> off-tune frequency
Verstimmung *f* <Elektronik> detuning
verstopfen *v* 1. <Bau> block; 2. <Fertig> clog *(Schleifscheibe)*; 3. <Maschinen> clog; 4. <Mechan> pack; 5. <Mechan> jam; 6. <Papier> clog; 7. <Raumfahrt> retard *(Raumschiff)*; 8. <Wasserversorg> clog
Verstopfen *n* 1. <Erdöl> plugging *(Rohrleitungen, Bohrloch)*; 2. <Fertig> choking, tamping; 3. <Maschinen> clogging, plugging
verstopft *adj* 1. <Heiz & Kälte> choked; 2. <Maschinen> clogged
verstopfte Düse *f* <Lufttrans> choked nozzle
verstopfter Filz *m* <Ker & Glas> clogged felt
Verstopfung *f* 1. <Bau> obstruction; 2. <Kohlen, Papier> clogging; 3. <Trans> *(AE)* road jam, *(BE)* traffic jam; *(BE)* tailback *(Verkehr)*; 4. <Wasserversorg> clogging, stoppage of a water pipe
Verstopfungsgrad *m* <Trans> level of congestion *(Verkehr)*
verstöpselte Flasche *f* <Labor> stoppered bottle
verstöpselter Kolben *m* <Labor> stoppered flask
verstöpselter Messzylinder *m* <Labor> stoppered measuring cylinder
verstreben *v* 1. <Bau> brace, shore, shore up, strut; 2. <Kohlen> shore, shore up; 3. <Maschinen> brace
Verstreben *n* <Bau> strutting
Verstrebung *f* 1. <Eisenbahn> pantograph tie-bar; 2. <Hydraul, Maschinen> brace
Verstrebungsbalken *m* <Bau> straining beam, straining piece
verstrecken *v* <Textil> draw *(Spinnen)*
Verstrecken *n* <Textil> drawing
Verstreckung *f* <Textil> draft *(Spinnen)*
verstümmeln *v* <Elektrotech> clip
Verstümmelung *f* 1. <Aufnahme> clipping *(von Signalen)*; 2. <Comp & DV, Qual> corruption
Versuch *m* 1. <Gerät, Kohlen> trial; 2. <Math> attempt; 3. <Phys> test; 4. <Qual, Textil> trial
Versuch *m* **und Irrtum** *m* <Comp & DV, Qual> trial-and-error
Versuchsablaufkanal *m* <Wasserversorg> test flume
Versuchsabschnitt *m* <Bau> experimental section
Versuchsanlage *f* <Kontroll> trial equipment
Versuchsanstalt *f* <Labor> laboratory
Versuchsaufbau *m* <Comp & DV, Funktech> breadboard
Versuchsaufbau *m* **eines Schaltungsmodells** <Elektrotech> breadboard model
Versuchsaufnahme *f* <Foto> test shot
Versuchsauslegung *f* <Qual> design of experiments, DOE
Versuchsbecken *n* <Wasserversorg> experimental basin, tapping
Versuchsbedingungen *fpl* <Qual> test conditions
Versuchsbeschreibung *f* <Raumfahrt> test specification
Versuchsbohrung *f* 1. <Erdöl> trial boring *(Erdölsuche)*; 2. <Kohlen> trial boring
Versuchseinrichtung *f* 1. <Kontroll> trial equipment; 2. <Maschinen> try-out facility
Versuchsfeld *n* <Bau, Fertig, Qual> testing ground
Versuchshafen *m* <Ker & Glas> monkey pot
Versuchshubschrauber *m* <Lufttrans> experimental helicopter
Versuchslast *f* <Lufttrans> proof load *(Lufttüchtigkeit)*
Versuchslauf *m* 1. <Comp & DV> test run; 2. <Kontroll> trial run
Versuchsmodell *n* 1. <Mechan> experimental model; 2. <Werkprüf> test model
Versuchsplanung *f* <Qual> experimental design
Versuchsplatz *m* <Kerntech> test rig
Versuchspumpe *f* <Wasserversorg> test pump
Versuchsraum *m* <Lebensmittel> proving cabinet
Versuchsreihen *fpl* <Raumfahrt> experiment package
Versuchssendung *f* <Fernseh, Funktech> test transmission
Versuchsstadium *n* <Strahlphys> prototype stage
Versuchsstand *m* <Kerntech> test bay, test stand
Versuchsstrecke *f* 1. <Bau> experimental section; 2. <Kfztech> test track
Versuchswerkstatt *f* <Fertig> *(AE)* test center, *(BE)* test centre *(Kunststoffinstallationen)*
Versuchszug *m* <Eisenbahn> test train
vertäfeln *v* <Bau> wainscot
Vertäuboje *f* <Meerschmutz> mooring buoy
Vertäubung *f* <Ergon> masking
vertäuen *v* <Wassertrans> moor *(Festmachen)*
Vertäuklampe *f* <Wassertrans> mooring cleat *(Bootsbau, Deckzubehör)*
Vertäuplatz *m* <Wassertrans> mooring berth *(Hafen)*
Vertäupoller *m* <Wassertrans> mooring bitts *(Hafen)*
Vertauschen *n* **der Stecker über Kreuz** <Aufnahme> cross-plugging
Vertauschung *f* 1. <Math> commutation; 2. <Telekom> transposition *(Telegrafiezeichen)*
Vertäuung loswerfen *v* <Wassertrans> unmoor *(Festmachen)*
verteilen *v* 1. <Bau> distribute, spread; 2. <Kfztech> distribute
Verteiler *m* 1. <Comp & DV> switch; 2. <Elektriz> distributor; 3. <Kerntech> manifold; 4. <Kfztech> distributor; 5. <Lebensmittel> dispenser; 6. <Lufttrans> manifold; 7. <Maschinen> distributor; 8. <Mechan> distributor, manifold; 9. <Telekom> distribution frame, distribution board; 10. <Verpack> distributor
Verteiler *m* **für E-Mail** <Comp & DV, Telekom> e-mail distribution list
Verteilerantrieb *m* <Kfztech> distributor drive *(Zündung)*
Verteilerdose *f* <Elektriz, Elektrotech, Raumfahrt> conduit box, distribution box, junction box
Verteilereinspritzpumpe *f* <Kfztech> distributor injection pump
Verteiler-Entstörstecker *m* <Kfztech> distributor suppressor *(Zündung)*
Verteilerfinger *m* <Kfztech> distributor finger, distributor rotor; rotor arm *(Zündung)*
Verteilerkabel *n* <Elektriz> distribution cable
Verteilerkanal *m* <Nichtfoss Energ> plenum
Verteilerkappe *f* <Kfztech> distributor cap
Verteilerkasten *m* <Elektriz, Elektrotech> conduit box, distribution box, *(BE)* joint box, junction box, terminal box
Verteilerklemmschraube *f* <Kfztech> distributor clamp bolt
Verteilerläufer *m* 1. <Elektrotech> rotor; 2. <Kfztech> distributor finger, distributor rotor, distributor arm, rotor; rotor arm *(Zündung)*
Verteilerleitung *f* <Kerntech> distributed digital processing
Verteilerliste *f* <Telekom> mailing list *(elektronische Mitteilungsübermittlung)*
Verteilernetz *n* 1. <Bau> distribution network; 2. <Verpack> distribution chain
Verteilerring *m* <Kerntech> distribution ring
Verteilerrohr *n* 1. <Maschinen> header; 2. <Mechan> manifold; 3. <Wasserversorg> distributing pipe
Verteilerschaltdraht *m* <Telekom> jumper *(Telefon)*

Verteilerscheibe

Verteilerscheibe f <Kfztech> distributor cap
Verteilerschiene f <Elektriz, Elektrotech> distributing busbar, distribution bus, bus
Verteilerservice m <Verpack> distribution chain
Verteilerstück n <Fertig> spreader
Verteilerstufe f <Telekom> distribution stage
Verteilertafel f <Elektriz, Elektrotech> distributing board
Verteilertechnik f <Verpack> distribution technique
Verteilerwelle f <Kfztech> distributor shaft
Verteilerzentrale f <Telekom> (AE) distribution center, (BE) distribution centre (Telefon)
Verteilfernamt n <Telekom> (AE) distribution center, (BE) distribution centre (Telefon)
Verteilliste f <Telekom> mailing list (elektronische Mitteilungsübermittlung)
verteilt adj <Comp & DV, Math> distributed
verteilte Datenbank f <Comp & DV> distributed database
verteilte Datenverarbeitung f <Comp & DV> DDP, distributed data processing, distributed processing
verteilte Datenverarbeitungsumgebung f <Comp & DV> distributed computing environment, DCE
verteilte digitale Datenverarbeitung f <Comp & DV> distributed digital processing
verteilte Geschäftskomponenten fpl <Comp & DV> distributed business components
verteilte Induktivität f <Elektrotech> distributed inductance
verteilte Kapazität f <Elektrotech> distributed capacitance
verteilte Last f <Nichtfoss Energ> distributed load
verteilte Lastkontrolle f <Nichtfoss Energ> distributed load control
verteilte Nebenstellenanlage f <Telekom> distributed PBX
verteilte Systeme npl <Comp & DV> distributed systems
verteilte Verarbeitung f <Comp & DV> distributed processing
verteiltes Betriebssystem n <Comp & DV> distributed operating system
verteiltes Mehrfachantennensystem n <Telekom> distributed multi-antenna system
verteiltes Netz n <Comp & DV> distributed network
verteiltes System n <Telekom> dispersed system
Verteilung f 1. <Elektriz, Hydraul, Ker & Glas> distribution; 2. <Kfztech> distribution (Zündung); 3. <Maschinen> distribution; 4. <Math> distribution (der Werte einer Funktion oder Zufallsgröße); 5. <Patent> apportionment; 6. <Qual> distribution; 7. <Telekom> distribution
Verteilung f der Anwendung <Comp & DV> application partitioning
Verteilung f der Ausfallhäufigkeit <Qual> failure frequency distribution
Verteilung f mit Baumstruktur <Telekom> tree distribution
Verteilung f von festen Partikeln in der Flüssigkeit <Anstrich> dispersion
Verteilungschromatographie f <Chemie> partition chromatography
Verteilungsfehler m <Ker & Glas> defect in distribution
verteilungsfrei adj <Qual> distribution-free
Verteilungsfunktion f 1. <Math> cumulative distribution function (CDF); 2. <Phys> Q function (Quantenoptik); 3. <Qual> distribution function
Verteilungskabel n <Elektrotech> distribution cable
Verteilungskurve f <Math, Qual> distribution curve
Verteilungsnetz n 1. <Elektrotech> distribution system, distribution network; 2. <Telekom> distribution network; 3. <Wasserversorg> distribution system
Verteilungsrauschen n <Telekom> partition noise

Verteilungsrohr n <Mechan> intake manifold
Verteilungsschiefe f <Math> skewness (Schiefemaß einer Verteilung)
Verteilungssicherungskasten m <Elektriz> distribution fuse board
Verteilungssystem n <Elektrotech> distribution system
Verteilungstafel f <Telekom> distribution board (Stromversorgung)
Verteilungsunterwerk n <Elektriz> distribution substation
Verteilverstärker m <Fernseh> distribution amplifier
vertiefen v 1. <Bau> deepen, recess, hollow; 2. <Maschinen> recess
Vertiefen n <Bau> recessing
vertiefte Form f <Fertig> intaglio
Vertiefung f 1. <Bau> hollow; 2. <Fertig> dimple, recess; seating (Schraubenkopf); 3. <Ker & Glas> impression; 4. <Maschinen> cavity, depression, pit, recess
vertikal adj <Geom> vertical
Vertikal... <Elektronik, Fernseh, Phys> vertical
Vertikalablenkplatte f 1. <Elektronik> vertical deflection plate (Katodenstrahlröhre); 2. <Phys> vertical deflection plate
Vertikalablenkung f <Fernseh> vertical deflection
Vertikalablenkungsspule f <Elektrotech> vertical deflection coil
Vertikalabtastung f <Fernseh> vertical scanning
Vertikalachse f (Y-Achse) <Math> vertical axis (y-axis)
Vertikalamplitude f <Fernseh> vertical amplitude
Vertikalamplitudenregler m <Fernseh> vertical-amplitude control
Vertikalanflugsführung f <Lufttrans> approach elevation guidance
Vertikalanordnung f <Bau> vertical alignment
Vertikalaufnahme f <Aufnahme> perpendicular recording
Vertikalaustastimpuls m <Fernseh> vertical blanking pulse
Vertikalaustastlücke f <Fernseh> vertical blanking interval
Vertikalaustastung f <Elektronik, Fernseh> vertical blanking
Vertikalbezugssystemeinheit f <Raumfahrt> vertical reference unit
Vertikalbildlageregelung f <Elektronik> (AE) vertical centering, (BE) vertical centring
Vertikalbipolartransistor m <Elektronik> vertical bipolar transistor
Vertikalblättern n <Comp & DV> vertical scrolling
Vertikale f <Geom> vertical
vertikale Axe f <Phys> vertical axis
Vertikalempfindlichkeit f <Gerät> Y-sensitivity, vertical sensitivity (Oszilloskop)
Vertikalendstufe f <Elektronik> vertical output stage
vertikaler MOS-Leistungstransistor m <Elektronik> vertical power MOS transistor
vertikaler MOS-Transistor m <Elektronik> vertical MOS transistor
vertikales Richtdiagramm n <Funktech> vertical directivity diagram, vertical directivity pattern
vertikales seismisches Profil n (VSP) <Erdöl> vertical seismic profile (VSP)
Vertikal-FET m <Elektronik> vertical field-effect transistor
Vertikalflosse f <Lufttrans> keel, tail fin
Vertikalformat n <Comp & DV> vertical format
Vertikalfräsmaschine f <Maschinen> vertical milling machine
Vertikalführung f <Lufttrans> elevation guidance
Vertikalgeschwindigkeit f <Raumfahrt> vertical speed
Vertikalgeschwindigkeitsanzeige f <Raumfahrt> vertical-speed indicator

Vertikalgreifer m <Textil> vertical sewing hook
Vertikalitätstoleranz f <Ker & Glas> verticality tolerance
Vertikalkartoniermaschine f <Verpack> vertical cartoner
Vertikalkonvergenz f <Elektronik> vertical convergence
Vertikalkreisel m 1. <Lufttrans> vertical gyro; 2. <Raumfahrt> gyroscopic verticant, vertical gyro
Vertikallinearitätsregler m <Fernseh> vertical linearity control
Vertikallücke f <Fernseh> vertical interval
Vertikalmagnetisierung f <Akustik> perpendicular magnetization
Vertikalmittenabgleich m <Fernseh> (AE) vertical centering control, (BE) vertical centring control
Vertikalmodus m <Fernseh> vertical mode
Vertikalpolarisation f 1. <Elektrotech, Funktech> vertical polarization; 2. <Phys> vertical polarization (Antenne); 3. <Telekom> vertical polarization
Vertikalrichtdiagramm n <Funktech> vertical directivity pattern (Antenne)
Vertikalrohrpost f <Kerntech> vertical rabbit
Vertikalstab m <Fertig> column, strut
Vertikalstabilität f <Trans> vertical stability
Vertikalsteuerung f <Optik> vertical control
Vertikalsynchronisierimpuls m <Fernseh> vertical sync pulse
Vertikalsynchronisierung f 1. <Comp & DV> vertical tab; 2. <Fernseh> vertical lock
Vertikaltabulator m <Comp & DV> vertical tabulator
Vertikaltragsäule f <Eisenbahn> supporting column
Vertikalunterdrückung f <Comp & DV> vertical blanking
Vertikalversetzung f <Nichtfoss Energ> heaving displacement (der Boje)
Vertikalverstärker m <Elektronik> VA, vertical amplifier
Vertikalverstärkerausgang m <Elektronik> vertical-amplifier output
Vertikalverstärker-Eingang m <Elektronik> vertical-amplifier input
Vertikalverteilung f <Umweltschutz> vertical dispersion
Vertikalzylinderschleifmaschine f <Maschinen> vertical cylinder-grinding machine
Vertrag m <Bau, Patent, Qual> contract
Verträglichkeit f <Akustik, Kunststoff, Maschinen> compatibility
Verträglichkeitsprüfung f 1. <Umweltschutz> impact statement; 2. <Wasserversorg> impact study
Vertragspflichtenheft n <Bau> specifications
Vertragsstaat m <Patent> contracting state
Vertrauen n <Math, Qual> confidence
Vertrauensbereich m 1. <Math> confidence interval, confidence region, confidence set; 2. <Qual> confidence interval, confidence range, confidence region
Vertrauensgrenze f 1. <Math> confidence limit (Grenze eines Konfidenzintervalls mit eingesetzten Daten); 2. <Qual> confidence limit
Vertrauensgrenze f der Erfolgswahrscheinlichkeit <Qual> assessed reliability
Vertrauensgrenze f der mittleren Lebensdauer <Qual> assessed mean life
Vertrauensgrenze f eines Lebensdauer-Perzentils Q <Qual> assessed Q-percentile life
Vertrauensniveau n <Qual> confidence coefficient, confidence level
Vertraulichkeit f <Comp & DV> privacy
vertrimmt adj <Lufttrans> out-of-trim
Vertrimmung f <Lufttrans> out-of-trim (Luftfahrzeug)
Ver- und Entschlüsselungseinrichtung f <Telekom> encryption/decryption device (CRYPDEC)

verunreinigen v 1. <Anstrich> contaminate; 2. <Fertig> vitiate; 3. <Kohlen, Qual> contaminate; 4. <Umweltschutz> contaminate, pollute
verunreinigter Bohrschlamm m <Erdöl> contaminated mud (Bohrtechnik)
Verunreinigung f 1. <Anstrich> contaminant; 2. <Chemie> contamination; 3. <Elektronik> impurity; 4. <Ker & Glas> tramping; 5. <Kohlen> contamination; 6. <Metall> impurity
Verunreinigung f des Grundwassers <Abfall> ground water contamination, ground water pollution
Verunreinigungsquelle f <Umweltschutz> pollution source
Verunreinigungssubstanz f <Abfall> contaminant
Verursacherprinzip n <Umweltschutz> polluter-pays principle
verursachter Hörschaden m/durch Lärm <Sicherheit> noise-induced hearing impairment
verursachter Unfall m/durch Strom <Sicherheit> electrical accident
vervielfachen v <Comp & DV> multiply
Vervielfacher m 1. <Elektronik> multiplier (multipliziert Eingangssignalkurve mit Konstante); 2. <Telekom> multiplier
Vervielfacherelement n <Comp & DV> multiplexer
Vervielfachung f <Comp & DV, Math> multiplication
vervielfältigen v <Comp & DV> multiply
Vervielfältigung f <Akustik> duplicating, duplication
Vervielfältigungsgetriebe n <Maschinen> multiplying gear
Vervielfältigungsversuch m <Konstzeich> duplicating trial
verwachsene Versetzung f <Metall> grown-in dislocation
Verwachsenes n <Kohlen> true middlings
Verwählen n <Telekom> (AE) dialing error, (BE) dialling error
verwalten v <Comp & DV> host
Verwaltung f von Inhalten <Comp & DV> content management
Verwaltungsbereich m <Bau> administrative area
Verwaltungseinheit f <Telekom> management unit
Verwandlung f in Stahl <Metall> acieration
Verwandlungshubschrauber m <Lufttrans> convertiplane
Verwaschungsdüse f <Lufttrans> confusion cone
verwechslungsfrei gekennzeichnet adj <Qual> unambiguously marked
Verweil... <Comp & DV, Fertig> residence
Verweildauer f 1. <Abfall> residence time, retention time; 2. <Mechan> dwell time; 3. <Textil> dwelling time
Verweiltank m <Kerntech> delay tank
Verweilzeit f 1. <Abfall, Comp & DV> residence time; 2. <Ker & Glas> dwell time; 3. <Maschinen> dwell
Verweis m nach vorne <Comp & DV> forward reference
verwendbar adj <Comp & DV> usable
verwendbare Nebenprodukte npl <Umweltschutz> usable by-products
Verwendbarkeitsdauer f <Maschinen> working life
verwenden v <Comp & DV> use
verwerfen v 1. <Maschinen> discard; 2. <Qual> reject
verwerfen v/sich <Bau, Maschinen> warp
Verwerfen n 1. <Bau> warpage, warping; 2. <Elektronik> warping (unbrauchbarer Daten); 3. <Qual> refusal
Verwerfung f 1. <Erdöl> fault (Geologie); 2. <Patent> rejection
Verwerfung f aufwärts <Wasserversorg> uptake
Verwerfungsbecken n <Erdöl> fault basin (Geologie)
Verwerfungsfalle f <Erdöl> fault trap (Geologie)

Verwerfungsquelle f <Wasserversorg> fault spring
verwertbar adj <Abfall> recyclable
Verwertbarkeit f der Daten <Comp & DV> currency of data
verwerten v <Fertig> employ; exploit; utilize
Verwertungsquote f <Abfall> recycling rate
verwesen v <Lebensmittel> putrefy
verwesend adj <Lebensmittel> putrescent
Verwinden n <Strömphys> twisting (der Wirbelstärke)
Verwindung f <Maschinen> torsion
verwindungsfrei adj <Fertig> distortion-free
Verwindungssteifigkeit f <Maschinen> torsional strength
Verwindungsversuch m <Maschinen> torsion test, torsional test
verwirbeltes Garn n <Textil> comingle yarn
Verwirbelung f <Phys> vorticity
verwittern v <Bau> weather
verwittertes Öl n <Meerschmutz> weathered oil
Verwitterung f 1. <Bau> efflorescence, weathering; 2. <Erdöl> decomposition (Geologie); 3. <Kohlen, Meerschmutz> weathering
Verwitterungsfläche f <Bau> area of deep weathering
verwölben v <Fertig> warp
Verwölbung f <Fertig> warping
verworfen adj <Bau> warped
verworfenes Holz n <Bau> warped timber
verwunden adj <Mechan> warped
verwundenes Rad n <Trans> buckled wheel
Verwurf m <Qual> refusal
verwürfeln v <Telekom> scramble
Verwürfeln n <Telekom> scrambling
Verwürfler m <Telekom> scrambler
Verwürfler-Entwürfler m <Telekom> scrambler-descrambler
verzahnen v 1. <Bau> interlock, joggle, key; 2. <Maschinen> gear
Verzahnen n <Maschinen> gear cutting, gear-tooth generating, toothing
Verzahnen n von Schneckenrädern <Mechan> worm wheel cutting
verzahnt adj <Comp & DV> interleaved
Verzahnung f 1. <Bau> joggle, keying; toothing (Mauerwerk); 2. <Comp & DV> interleaving; 3. <Fertig> indentation; 4. <Maschinen> toothing
Verzahnung f von Schneckenrädern <Maschinen> worm cutting
Verzahnungsfehler f <Fertig> gear tooth error
Verzahnungsgeometrie f <Maschinen> geometry of gears
Verzahnungsmaschine f <Maschinen> gear-cutting machine
Verzahnungstoleranz f <Maschinen> gearing tolerance
verzapft adj <Mechan> cogged
Verzapfung f <Bau> tenon joint
verzehren v <Maschinen> absorb, dissipate
Verzehrung f <Maschinen> absorption
Verzeichnis n <Comp & DV> (AE) catalog, (BE) catalogue, dictionary, directory, record
Verzeichnisspeicher m <Telekom> directory store
Verzeichnissteuersystem n <Telekom> directory control system
Verzeichnung f <Fernseh> distortion (Bild)
verzeichnungsfrei adj <Phys> orthoscopic
verzerren v 1. <Elektronik, Telekom> distort; 2. <Fertig> strain (Material)
verzerrt adj 1. <Ker & Glas> out-of-shape; 2. <Math> biased

verzerrte Wellenmethode f <Kerntech> distorted wave method
verzerrtes Signal n <Telekom> distorted signal
Verzerrung f 1. <Akustik> distortion; 2. <Elektriz, Elektronik> distortion (Abweichung vom Ursprungssignal); 3. <Fernseh> distortion (Signal); 4. <Funktech> distortion; 5. <Math> bias (einer Schätzfunktion); 6. <Phys> distortion (Bild); 7. <Qual> bias; 8. <Telekom> distortion
Verzerrung f der zweiten Harmonischen <Elektronik> second harmonic distortion
Verzerrung f durch die dritte Harmonische <Elektronik> third harmonic distortion
Verzerrung f durch Hüllkurvenverzögerung <Raumfahrt> envelope delay distortion (Weltraumfunk)
Verzerrungsanalysator m 1. <Elektronik> (BE) harmonic analyser, (AE) harmonic analyzer; 2. <Gerät> (BE) distortion analyser, (AE) distortion analyzer
verzerrungsarme Modulation f <Elektronik> low-distortion modulation
verzerrungsbegrenzter Betrieb m <Telekom> distortion-limited operation
Verzerrungsfalle f <Aufnahme> bias trap
verzerrungsfrei adj <Elektronik> distortion-free
verzerrungsfreie Modulation f <Elektronik> distortion-free modulation, linear modulation
Verzerrungskorrektur f <Math> bias correction (Parameterschätzung)
Verzicht m <Patent, Qual> waiving
Verzichtserklärung f <Patent> waiver
Verziehen n 1. <Bau> warping, winding; 2. <Maschinen> distortion; 3. <Textil> drafting
verzierter Sturz m <Bau> platband
Verzierungen fpl <Bau> trim
Verzierungsarbeiten fpl <Wassertrans> fancywork
verzinken v 1. <Anstrich> galvanize; 2. <Bau> zinc; 3. <Fertig> galvanize; 4. <Metall> zinc
verzinktes Stahlblech n <Heiz & Kälte> galvanized sheet steel
Verzinkung f <Kunststoff> zinc coating
verzinnen v 1. <Fertig> tin, wet; 2. <Metall> tin
Verzinnen n <Bau, Ker & Glas, Metall> tinning
verzinnt adj <Chemie, Lebensmittel> (AE) canned, (BE) tinned
verzinnter Draht m <Elektrotech> tinned wire
verzinnter Leiter m <Elektriz> tinned conductor
verzinntes Blech n <Fertig, Metall> sheet tin
Verzinnung f <Elektriz> tinning
verzogen adj 1. <Bau> warped; 2. <Lufttrans> out-of-track (Hubschrauber); 3. <Maschinen> out-of-true; 4. <Mechan> warped
verzogene Antrittsstufe f <Bau> commode step
verzogene Stufe f <Bau> dancing step
Verzögerer m 1. <Elektrotech> slug (Relais); 2. <Kunststoff> retarder; 3. <Lebensmittel> inhibitor; 4. <Papier> retarder; 5. <Textil> retarding agent
verzögern v 1. <Anstrich, Bau> retard; 2. <Elektriz, Elektronik> delay (in der Ausführungsfolge); 3. <Funktech, Kontroll> delay
verzögernde Zeitablenkung f <Elektronik> delaying sweep (Oszillograph)
verzögernde Zeitbasis f <Elektronik> delaying time base
verzögert adj <Kontroll> delayed, deferred
verzögerte Abtastung f <Fernseh> delayed scanning
verzögerte Adressierung f <Comp & DV> deferred addressing
verzögerte automatische Verstärkungsregelung f <Aufnahme> delayed automatic gain control
verzögerte Belegung f <Telekom> delayed call
verzögerte Bewegung f <Maschinen> retarded motion

verzögerte Geschwindigkeit f <Maschinen> retarded velocity
verzögerte Koinzidenzspektren npl <Strahlphys> delayed coincidence spectra
verzögerte modifizierte Phasenumtastung f <Elektronik> delayed modified phase shift keying
verzögerte modulierte Phasenumtastung f <Elektronik> MPSK, modulated phase shift keying
verzögerte Regelschleife f <Telekom> delay lock loop
verzögerte selbsttätige Verstärkungsregelung f <Elektronik> delayed automatic gain control *(für Empfänger)*
verzögerte Strahlung f <Kerntech> delayed emission
verzögerter Fluss m <Kerntech> delayed flux
verzögerter Kippvorgang m <Elektronik> delaying sweep *(Oszillograph)*
verzögerter Trennschalter m <Elektriz> slow break switch
verzögertes Austastsignal n <Fernseh> DBS, delayed blanking signal
verzögertes Brechen n <Metall> delayed fracture
verzögertes Härten n <Thermod> delayed hardening
verzögertes Neutron n 1. <Kerntech> delayed neutron; 2. <Phys> delayed neutron *(Kernphysik)*; 3. <Strahlphys> delayed neutron
verzögertes Relais n <Telekom> delay mode relay
verzögertes Verkoken n <Thermod> delayed coking
verzögert-kritischer Reaktor m <Kerntech> delayed critical reactor
Verzögerung f 1. <Comp & DV> delay; 2. <Elektriz, Elektronik, Elektrotech, Fernseh> delay, lag; 3. <Fertig> lag; 4. <Funktech, Gerät, Kerntech> delay; 5. <Kfztech> deceleration; offset *(Motor)*; 6. <Kontroll> delay; 7. <Lufttrans> lag *(Hubschrauber)*; 8. <Maschinen> deceleration, retardation, retarded acceleration, retarded motion; 9. <Mechan> braking; 10. <Telekom, Trans> delay; 11. <Wasserversorg> lag
Verzögerung f **durch Konkurrenzbetrieb** <Comp & DV> contention delay
Verzögerung f **höherer Ordnung** <Elektronik> high-order delay
verzögerungsarme Regelstrecke f <Regelung> low-lag closed-loop-controlled system
Verzögerungsbad n <Foto> restraining bath *(Entwicklung)*
Verzögerungseinheit f <Regelung> lag module
Verzögerungsfreiheit f <Regelung> absence of lag *(des Reglers)*
Verzögerungsglied n <Phys> lag element
Verzögerungsglieder npl **mit Abgriffen** <Elektronik> tapped delay elements
Verzögerungskabel n <Fernseh> delay cable
Verzögerungskette f <Gerät> delay network
Verzögerungskette f **mit konstanter Laufzeit** <Elektronik> constant delay line
Verzögerungskomponente f 1. <Elektronik> delay component *(Bauteil)*; 2. <Kerntech> delay component, delay unit
Verzögerungskraft f <Phys, Trans> decelerative force, retarding force
Verzögerungsleitung f 1. <Comp & DV, Fernseh, Phys> delay line; 2. <Telekom> delay circuit, delay line
Verzögerungsleitung f **mit Abgriffen** <Elektronik> tapped delay line
Verzögerungsmittel n 1. <Bau> retarding agent *(Beton, Zement)*; 2. <Kunststoff> retarder
Verzögerungsmultivibrator m <Elektronik> delay multivibrator

Verzögerungsrelais n 1. <Elektriz> delay relay, slow-acting relay, time delay relay, time lag relay; 2. <Elektrotech> delay relay
Verzögerungsröhre f 1. <Elektronik> slow wave tube; 2. <Raumfahrt> transit time tube
Verzögerungsschalter m <Elektriz> delay switch
Verzögerungsschaltkreis m <Fernseh> delay circuit
Verzögerungsschaltung f 1. <Elektriz, Elektronik> time delay circuit; delay circuit *(Laufzeitnetzwerk)*; 2. <Gerät> delay network
Verzögerungsspur f <Bau> deceleration lane
Verzögerungssystem n <Lufttrans> lagging system
Verzögerungstank m <Kerntech> delay tank
Verzögerungsventil n <Maschinen> delay valve
Verzögerungsvorrichtung f <Trans> deceleration device
Verzögerungswicklung f <Elektriz> retardation coil
Verzögerungswinkel m 1. <Elektriz> angle of lag, current delay angle, delay angle, lag angle, retardation angle; 2. <Lufttrans> lag angle
Verzögerungszeit f 1. <Comp & DV> deceleration time; 2. <Elektriz, Elektronik> delay time, time lag; 3. <Erdöl> lagtime *(Bohrtechnik)*; 4. <Telekom> delay time
Verzug m 1. <Kunststoff> warpage; 2. <Maschinen> distortion
Verzugszeit f <Elektronik> dead time
Verzundern n <Fertig> scaling
verzweigen v 1. <Comp & DV> branch; 2. <Fertig> ramify
verzweigen v/sich <Bau> branch
verzweigtes Polymer n <Kunststoff> branched polymer
Verzweigung f 1. <Comp & DV> switch; 2. <Eisenbahn> junction; 3. <Fertig> ramification; 4. <Kerntech> manifold; 5. <Phys> branching
Verzweigungsfilter n <Elektronik> branching filter
Verzweigungskabel n (VzK) <Elektrotech, Telekom> distribution cable
Verzweigungskasten m <Elektriz> junction box
Verzweigungspunkt m 1. <Comp & DV> branch point; 2. <Elektronik, Telekom> node
Verzweigungsstück n <Optik> Y-coupler
Verzweigungsverhältnis n <Phys> branching ratio
Vestibularapparat m <Ergon> vestibular mechanism
VF 1. <Fernseh> *(Videofrequenz)* VF *(video frequency)*; 2. <Fernseh> *(Videofrequenzwandler)* VF *(video-frequency converter)*
V-Form f <Lufttrans> dihedral
V-förmig adj <Maschinen> V-shaped
V-förmige Rille f <Elektronik> V-groove
VG *(veränderliche Geometrie, veränderliche Tragflügelgeometrie)* <Lufttrans> VG *(variable geometry)*
VHF *(VHF-Bereich, Ultrakurzwellen)* <Funktech, Telekom> VHF *(very high frequency)*
VHF-Band n <Fernseh, Funktech> VHF band
VHF-Bereich m <Elektronik, Funktech> very high frequency, VHF
VHF-Drehfunkfeuer n <Funkort> VHF omnidirectional radio range
VHF-Signal n <Elektronik> VHF signal
VHF-Signalgeber m <Elektronik> VHF signal generator
VHF- und UHF-Tuner m <Fernseh> VHF and UHF tuner
VHS *(Heimvideosystem)* <Fernseh> VHS *(video home system)*
VHS-C *(Heimvideo-Aufzeichnungssystem)* <Fernseh> VHS-C *(video home system-compact)*
VI *(Viskositätsindex, Viskositätszahl)* <Kfztech, Maschinen, Strömphys, Thermod> VI *(viscosity index)*
Vibration f 1. <Elektronik> oscillation; 2. <Phys, Sicherheit> vibration
Vibrationsalarm m <Mobilkom> vibration alarm *(Handy)*

Vibrationsdämpfer 840

Vibrationsdämpfer *m* 1. <Lufttrans> vibration damper *(Luftfahrzeug)*; 2. <Sicherheit> vibration damper
Vibrationsförderer *m* 1. <Fertig> vibratory feeder, vibratory hopper; 2. <Verpack> vibratory feeder
Vibrations-Galvanometer *n* <Elektriz, Phys> vibration galvanometer
Vibrationsgenerator *m* <Raumfahrt> vibration generator
Vibrationsgeräusch *n* <Lufttrans> buzz
Vibrationsprüfung *f* <Werkprüf> vibration test
Vibrationsquantenzahl *f (v)* <Kerntech> vibrational quantum number *(v)*
Vibrationsrost *m* <Fertig> shake-out
Vibrationsrührwerk *n* <Labor> vibrating stirrer
Vibrationsschutzhandschuhe *mpl* <Sicherheit> antivibration gloves, vibration-absorbing gloves
Vibrationsschutzvorsatz *m* <Sicherheit> antivibration attachment
Vibrationssieb *n* <Chemie> vibrating screen
Vibrationsstärke *f* <Sicherheit> vibration severity
Vibrationstest *m* <Verpack> jarring test
Vibrationstisch *m* <Ker & Glas> vibrating table
Vibrationsverdichtung *f* <Kerntech> vibrocompaction
Vibrationswalze *f* <Bau> vibrating roller
Vibrato *n* <Akustik> vibrato
Vibrator *m* <Elektrotech, Maschinen> vibrator
Vibratoraufgeber *m* <Verpack> filling vibrator
vibrieren *v* 1. <Elektronik> oscillate, vibrate; 2. <Fertig> jar
Vibrieren *n* <Fertig> jarring
vibrierend *adj* 1. <Elektriz, Elektronik> oscillating, vibrating *(Strom)*; 2. <Papier> oscillating, vibrating
Vibriersieb *n* <Abfall> vibrator screen
Vibrometer *n* <Phys> vibrometer
Vicatnadel *f* <Bau> vicat needle
Vickershärte *f* 1. <Fertig> diamond-pyramid hardness, diamond-pyramid hardness number *(Walzen)*; 2. <Maschinen> Vickers hardness
Vickershärtetestgerät *n* <Maschinen> Vickers hardness testing machine
Video *n* <Fernseh> video
Video *n* **auf Abruf** <Fernseh> video-on-demand *(Kabelfernsehen)*
Videoaufzeichnung *f* <Fernseh> video recording
Videoaufzeichnung *f* **mit niedriger Geschwindigkeit** <Fernseh> long play *(LP)*
Videoaufzeichnung *f* **mit Normalgeschwindigkeit** <Fernseh> short play *(SP)*
Videoaufzeichnungsgerät *n* <Fernseh> magnetoscope
Videoausgang *m* <Fernseh> video output
Videobandabspielgerät *n* <Fernseh> videotape player
Videobandaufzeichnung *f* <Fernseh> magnetic tape video recording
Videobandbreite *f* <Ferneh> video bandwidth
Videobandrecorder *m* <Fernseh> video tape recorder
Videocassette *f* <Fernseh> video cassette
Video-Clip *n* <Fernseh> videoclip
Video-Disk *n* <Fernseh> *(BE)* video high-density disc, *(AE)* video high-density disk, *(BE)* video disc, *(AE)* video disk
Videoeingabe *f* <Fernseh> video input
Videofrequenz *f (VF)* <Fernseh> video frequency *(VF)*
Videofrequenzkanal *m* <Fernseh> video channel, video-frequency channel
Videofrequenzwandler *m (VF)* <Fernseh> video-frequency converter *(VF)*
Videogerät *n* **mit Tuner** <Fernseh> telerecorder
Videographie *f* <Fernseh> videography
Videographikadapter *m* <Fernseh> video graphic adapter *(VGA)*
Videokabel *n* <Fernseh> video cable

Videokamera *f* **mit Bandaufzeichnung** <Fernseh> camcorder
Videokarte *f* <Comp & DV> video card *(PC)*
Videokassette *f* <Fernseh> video tape cassette
Videokassettenrecordersystem *n* <Fernseh> VCR system, video cassette recorder system
Videokonferenz *f* <Telekom> video conference
Videokonferenz-Bildformat *n* <Fernseh> Video CIF, video common intermediate format
Videokonferenz-Endgerät *n* <Telekom> video conference terminal
Videokopf *m* <Fernseh> video head
Videokopfbaugruppe *f* <Fernseh> video head assembly
Videokopfjustierung *f* <Fernseh> video head alignment
Video-Langspiel *n (VLP)* <Fernseh> video long play *(VLP)*; long-playing video
Videomagnetband *n* <Fernseh> videotape
Videomagnetbandeinrichtungen *fpl* <Fernseh> videotape facilities
Videomagnetbandgerät *n (VCR)* <Fernseh> video tape recorder *(VTR)*
Videomagnetbandrecorder *m* <Fernseh> videotape recorder
Videopegel *m* <Fernseh> video level
Videoprojektor *m* <Fernseh> video projector
Videorecorder *m (VCR)* <Fernseh> telerecorder, video cassette recorder, video recorder
Videorecorderprogrammiersystem *n* <Fernseh> Show View, VCR programming system *(VPS)*
Videoschaltmatrix *f* <Fernseh> video switching matrix
Videoschaltverteiler *m* <Fernseh> video switch
Videoschaltverteilung *f* <Fernseh> video switching
Videosignal *n* 1. <Fernseh> picture signal, video signal; 2. <Phys, Telekom> video signal
Videosignalcodierung *f* <Fernseh> video coding, video signal coding
Videosignalkompression *f* <Fernseh> video compression, video signal compression
Videosignal-Kompressionsstandard *m* <Fernseh> video compression standard, video signal compression standard *(MPEG)*
Videospur *f* <Fernseh> video track
Videotelefon-Bildformat *n* <Telekom> Video QCIF, video quarter common image format
Videotex *n* <Fernseh> videotex
Videotex-Gateway *m* <Fernseh> videotex gateway
Videotex-Server *m* <Fernseh> videotex server
Videotext *n* <Fernseh, Telekom> teletext
Videothek *f* <Fernseh> video tape library
Videoübertragung *f* <Fernseh> video transmission
Videoverstärker *m* <Fernseh> video amplifier
Videoverstärkung *f* <Fernseh> video amplification
Video-Verteilverstärker *m* <Fernseh> video distribution amplifier
Videozubehör *n* <Fernseh> videoware
Vidikon *n* <Elektronik> vidicon
Vidikon-Kamera *f* <Fernseh> vidicon camera
Viehwagen *m* <Eisenbahn> cattle wagon
vielatomig *adj* <Chemie> polyatomic
vielatomige Transferreaktion *f* <Kerntech> many-nuclear transfer reaction
Vieleck *n* <Geom> polygon
vieleckig *adj* <Geom> polygonal
vielfach *adj* <Math> multiple
Vielfach *n* <Telekom> multiplex
vielfach nutzen *v* <Comp & DV> multiplex
vielfache Entwicklung *f* <Nichtfoss Energ> multiple development
Vielfachecho *n* <Aufnahme> multiple echo

Vielfachelektromessung f <Erdöl> multiple special electrical logging *(Bohrlochmessung)*
Vielfaches n <Math> multiple
Vielfaches n einer Einheit <Metrol> multiple of a unit
Vielfachheit f <Math> multiplicity *(algebraische oder geometrische)*
Vielfachleitung f 1. <Comp & DV> *(BE)* bus, highway, *(AE)* trunk; 2. <Telekom> bus
Vielfachmessinstrument n <Fernseh, Funktech> multimeter
Vielfachschrank m <Telekom> multiple switchboard
Vielfachstreuung f <Teilphys> multiple scattering
Vielfachübertrager m <Comp & DV> multiplexer
Vielfachübertragung f <Comp & DV> multiplexing, multiplex
Vielfachübertragungskanal m <Comp & DV> multiplexor channel
Vielfachzugriff m <Telekom> multiple access
Vielfachzugriff m im Zeitmultiplex <Telekom> time division multiple access
Vielfachzugriff m mit Trägerkennung *(CSMA)* <Comp & DV, Telekom> carrier sense multiple access *(CSMA)*
Vielfachzugriffsverfahren n mit Berechtigungskennzeichen <Telekom> token-passing multiple access
Vielfachzugriffsverfahren n mit Buchung <Telekom> reservation multiple access
vielflächig adj <Geom> polyhedral
Vielflächner m <Geom> polyhedron
Vielkammerklystron n <Phys> multicavity klystron
Vielkanal-Impulshöhenanalysator m <Phys> *(BE)* multichannel analyser, *(AE)* multichannel analyzer
Vielkanalprotokoll n <Comp & DV> multichannel protocol
Vielkeilverzahnung f <Fertig> multiple splining
Vielkörperzerfall m <Teilphys> multibody decay
vielphasig adj <Elektrotech> polyphase
Vielpunkt... <Comp & DV> multiplex
Vielrollenwalzwerk n **nach Sendzimir** <Metall> Sendzimir mill
Vielschlitzmagnetron n <Elektronik> multisegment magnetron
Vielschnittdrehmaschine f <Fertig> multiple-tool lathe
Vielseitigkeit f <Raumfahrt> versatility
Vielsprecher m <Telekom> high-calling-rate subscriber
Vielstoffmotor m <Kfztech, Thermod> multifuel engine
Vielstrahl-Katodenstrahlröhre f <Elektronik> multibeam CRT
Vielzweckdrehmaschine f <Fertig> general-purpose lathe
Vielzweckfaser f <Phys> *(AE)* multimode fiber, *(BE)* multimode fibre
vieradriges Kabel n <Phys> quad cable
Vieratomigkeit f <Chemie> tetratomicity
Vierbackenfutter n <Maschinen> four-jaw chuck
vierbasig adj <Chemie> quadribasic, tetrabasic
Vier-Bit-Byte n <Comp & DV> nibble, nybble
vierdimensional adj <Phys> four-dimensional
Vierdraht... <Telekom> four-wire
Vierdraht-Drehstromsystem n <Elektriz> three-phase four-wire system
Vierdrahtdurchschaltung f <Telekom> four-wire switch
Vierdrahtkoppelpunkt m <Telekom> four-wire crosspoint
Vierdrahtvermittlung f <Telekom> four-wire switch
Vierdraht-Vermittlungssystem n <Telekom> four-wire switching system
Vierdrahtverstärker m <Elektronik> four-wire repeater
Viereck n <Geom> quadrangle, quadrilateral, tetragon
viereckig adj <Geom> quadrangular, tetragonal
viereckiger Dipolrahmen m <Elektrotech> square loop

Vierende-Elträger m <Bau> open-web girder, vierendeel truss
Viereranschluss m mit Einzelanruf <Telekom> four-party line with selective ringing
Viererleitung f <Elektrotech, Telekom> phantom circuit
Vierervektor m <Phys> four-vector
Vierfach... <Foto, Kfztech> fourfold
Vierfach-Expansionsmaschine f <Maschinen> quadruple-expansion engine
Vierfachmeißelbohrer m <Erdöl> four-way bit *(Bohrtechnik)*
Vierfachmeißelhalter m <Maschinen> four-stud tool post, four-tool turret
Vierfach-Stativgestell n <Foto> fourfold tripod stand
Vierfachvergaser m <Kfztech> *(AE)* four-barrel carburetor, *(BE)* four-barrel carburettor
Vierfach-Verstärker m <Elektronik> quad-operational amplifier
vierfädige Wicklung f <Funktech> quadrifilar winding
Vier-Faktorenformel f <Kerntech> four-factor formula
Vierfarben... <Druck> *(AE)* four-color, *(BE)* four-colour
Vierfarbendruck m <Druck> *(AE)* four-color process, *(BE)* four-colour process, *(AE)* printing with four colors, *(BE)* printing with four colours
Vierfarbendruckfarbe f <Druck> *(AE)* four-color process ink, *(BE)* four-colour process ink
Vierfarbenseparation f <Druck> *(AE)* four-color separation, *(BE)* four-colour separation
Vierfarbentheorem n <Geom> *(AE)* four-color theorem, *(BE)* four-colour theorem
vierflächig adj <Geom> tetrahedral
Vierflächner m <Geom> tetrahedron
Vierflankenmethode f <Gerät> quad-slope method
Vierflügelbohrmeißel m <Erdöl> four-wing bit
Vierkanal-Verstärker m <Elektronik> four-channel amplifier
Vierkant m <Fertig> square, square end *(Kunststoffinstallationen)*
Vierkantansatz m <Maschinen> square neck
Vierkantendrehmeißel m <Fertig> square cutting tool
Vierkantfeile f <Maschinen> square file
Vierkantformmeißel m <Fertig> square forming tool
Vierkantkopf m <Maschinen> square head
Vierkantkopfschraube f <Bau> coach screw
Vierkantmutter f <Maschinen> square nut
Vierkantscheibe f <Maschinen> square washer
Vierkantschlüssel m <Maschinen> *(BE)* square spanner, *(AE)* square wrench
Vierkantschraube f 1. <Bau> square bolt; 2. <Maschinen> square head bolt
Vierkantstahl m <Metall> square
Vierkantstange f <Bau> kelly bar
Vierkantventil n <Bau> cock with square head
Vierkopf-Videorecorder m <Fernseh> quadruplex videotape recorder
Vierkurbelgetriebe n <Maschinen> four-crank mechanism
Vierleiter... <Elektrotech> four-wire
Vierleiteranlage f <Elektrotech> four-wire system
Vierleiternetz n <Elektrotech> four-wire system
Viermaster m <Wassertrans> four-master *(Segeln)*
Vierniveau-Maser m <Elektronik> four-level maser
Vierphasen... <Elektronik, Telekom> four-phase
Vierphasenmodulator m <Telekom> four-phase modulator
Vierphasenumtastung f *(QPSK)* 1. <Elektronik> quadriphase shift keying, quaternary phase shift keying *(QPSK)*; 2. <Telekom> quadrature phase-shift keying,

Vierphasenumtastung

quadriphase shift keying, quaternary phase shift keying *(QPSK)*
Vierphasenumtastung *f* **mit geglättetem Phasenverlauf** <Telekom> phase-shaped QPSK
vierphasig *adj* <Elektronik> four-phase
Vierpol *m* 1. <Elektrotech> two-port network; 2. <Phys> quadripole *(Schaltungstheorie)*; 3. <Telekom> four-terminal network
Vierpol *m* **in H-Schaltung** <Elektrotech> H-network
Vierpolanordnung *f* <Kerntech> quadrupolar configuration
Vierpoldämpfungsmaß *n* <Elektronik, Telekom> image attenuation coefficient
Vierpolfeld *n* <Kerntech> quadrupole field
Vierpolfilter *n* <Elektronik> four-pole filter
Vierpolgenerator *m* <Elektriz> four-pole generator
vierpolig *adj* <Elektriz> four-polar, four-pole
vierpolige Ausschaltkontakte *mpl* <Elektrotech> 4 PST contacts
vierpolige Umschaltkontakte *mpl* <Elektrotech> 4 PDT contacts
vierpoliger Ausschalter *m* <Elektrotech> 4 PST switch, four-pole single-throw switch
vierpoliger magnetischer Lautsprecher *m* <Aufnahme> balanced-armature loudspeaker
vierpoliger Umschalter *m* <Elektrotech> 4 PDT switch, four-pole double-throw switch
vierpoliges Ausschaltrelais *n* <Elektrotech> 4 PST relay
vierpoliges Umschaltrelais *n* <Elektrotech> 4 PDT relay
Vierpolkreuzglied *n* <Elektrotech> lattice network
Vierpolschaltung *f* <Elektriz, Elektrotech> quadripole, four-terminal network
Vierpolübertragungsmaß *n* <Akustik, Telekom> image transfer exponent
Vierpunktlager *n* <Maschinen> four-point support
Vierquadrantenbetrieb *m* <Elektriz> four-quadrant operation
Vierquadranten-Multiplizierer *m* <Elektronik> four-quadrant multiplier
Vierradantrieb *m* <Kfztech> four-wheel drive
Vierradbremsanlage *f* <Kfztech> four-wheel brake system *(Bremsanlage)*
Vierschichtdiode *f* <Elektronik> four-layer diode
Vierschichtzelle *f* <Elektronik> p-n-p-n device *(Halbleiter)*
Vierschienensystem *n* <Eisenbahn> four rail system
Vierschneidenbohrmeißel *m* <Erdöl> four-wing bit *(Bohrtechnik)*
Vierschraubenfutter *n* <Maschinen> bell chuck, cup chuck
vierseitig *adj* <Geom> four-sided, quadrilateral
vierseitige Fläche *f* <Konstzeich> four-sided area
vierseitige Skizze *f* <Verpack> four-sided sketch
vierseitiger Würfel *m* <Math> four-sided die *(für ein Zufallsexperiment)*
Vierspindelbohrmaschine *f* <Fertig> four-spindle drilling machine
Vierspur-Aufnahme *f* <Aufnahme> quarter-track recording
vierspuriger Recorder *m* <Aufnahme> four-track recorder
vierspuriges Aufzeichnen *n* <Aufnahme> four-track recording
Vierspur-Tonaufzeichnung *f* <Akustik> tetraphonic recording
Vierstoff... <Chemie> quaternary
Viertakter *m* <Maschinen> four-stroke engine
Viertakthub *m* <Kfztech> four-stroke cycle *(Motor)*
Viertaktmotor *m* 1. <Kfztech> four-stroke engine *(Motor)*; 2. <Maschinen, Mechan> four-stroke engine; 3. <Wassertrans> four-stroke engine *(Motor)*
vierte Dimension *f* <Math> fourth dimension
vierte Generation *f* <Comp & DV> fourth generation
Viertel *n* 1. <Math> fourth, quarter; 2. <Metrol> quarter
Viertelabsatz *m* <Bau> quarter space *(Treppe)*
Viertelansicht *f* <Konstzeich> quarter view
Viertelelliptikfeder *f* <Maschinen> quarter-elliptic spring
Viertelgallone *f* <Metrol> *(BE)* quart
Viertelkreis *m* <Geom, Mechan> quadrant
Viertelkreisfehler *m* <Funkort> quadrantal error *(Funkpeilung)*
Viertelkreisfräser *m* <Maschinen> corner-rounding cutters, quarter-round milling cutter
Viertelkreissims *m* <Bau> ovolo
Viertelmaske *f* <Sicherheit> quarter mask
Viertelscheffel *n* <Metrol> peck
Viertelstab *m* <Bau> astragal, quarter round
Viertelung *f* <Bau> quartering
Viertelwellenlänge *f* <Elektronik> quarter wavelength
Viertelwellenleitung *f* <Phys> quarter-wave line
Viertelwellen-Peitschenantenne *f* <Funktech> quarter-wave whip antenna
Viertelwellen-Vertikalantenne *f* **mit Gegengewicht** <Funktech> ground plane
Viertelwert *m* <Math> quartile
vierter Virialkoeffizient *m (D)* <Thermod> fourth virial coefficient *(D)*
Viertor *n* <Telekom> four-port
Vierwegehahn *m* <Wasserversorg> four-way cock
Vierwege-Palette *f* <Verpack> four-way pallet
Vierwegeventil *n* <Maschinen> four-way valve
vierwertig *adj* <Chemie> quadribasic, quadrivalent, tetravalent
Vierwertigkeit *f* <Chemie> quadrivalence, tetravalence, tetravalency
Vierzahnbohrer *m* <Erdöl> quadricone bit *(Bohrtechnik)*
Vierzylinderboxermotor *m* <Trans> flat-four engine
Vierzylinder-Motorrad *n* <Kfztech> four-cylinder motorcycle
Vierzylinderstreckwerk *n* <Textil> four-roller draw frame
Vierzylinder-V-Motor *m* <Kfztech> V-four engine
Vignettierung *f* <Foto> vignetting
Vinyl... <Chemie> vinyl
Vinylacetylen *n* <Chemie> vinylacetylene
Vinylation *f* <Chemie> vinylation
Vinylbenzen *n* <Chemie> vinylbenzene
Vinylbenzol *n* <Chemie> styrene, styrolene
Vinylethen *n* <Chemie> divinyl
Vinylethin *n* <Chemie> vinylacetylene
Vinylethylen *n* <Chemie> divinyl
vinylhomolog *adj* <Chemie> vinylogous
Vinylhomologes *n* <Chemie> vinylog
Vinyliden... <Chemie> vinylidene
Vinylierung *f* <Chemie> vinylation
Vinyllack *m* <Bau> vinyl lacquer
vinylog *adj* <Chemie> vinylogous
Vinyloges *n* <Chemie> vinylog
Vinylpyridin *n* <Chemie> vinylpyridine
Violinblock *m* <Wassertrans> fiddle block *(Deckbeschläge)*
Violur... <Chemie> violuric
Virial... <Phys> virial
Virialsatz *m* <Phys> virial theorem
Virialtheorem *n* <Phys> virial theorem
Viridin *n* <Chemie> viridine
virtuell *adj* <Comp & DV> virtual
virtuelle Adresse *f* <Comp & DV> virtual address

virtuelle Datenstation f <Comp & DV, Telekom> virtual terminal
virtuelle Dauerschaltung f <Comp & DV> permanent virtual circuit
virtuelle Heimatumgebung f <Telekom> virtual home environment
virtuelle Maschine f <Comp & DV> virtual machine
virtuelle Netzwerkdatenstation f <Comp & DV> network virtual terminal
virtuelle Platte f <Comp & DV> virtual disk
virtuelle Realität f <Künstl Int> virtual reality, VR
virtuelle Replikationsfunktion f <Comp & DV> virtual replication function
virtuelle Schallquelle f <Akustik> virtual sound source
virtuelle Speicherverwaltung f (VMS) <Comp & DV> virtual memory specification (VMS)
virtuelle Tonhöhe f <Akustik> virtual pitch
virtuelle Verbindung f <Comp & DV, Telekom> virtual circuit, virtual connection
virtuelle Verbindung f **des D-Kanals** <Telekom> D-channel virtual circuit
virtuelle Wählleitung f <Telekom> switched virtual circuit
virtueller Rufdienst m <Comp & DV> virtual call service
virtueller Speicher m <Comp & DV> virtual memory, virtual storage
virtueller Verbindungsdienst m **des B-Kanals** <Telekom> B-channel virtual circuit service
virtuelles Bild n 1. <Foto, Optik> virtual image; 2. <Wellphys> virtual image (eines Hologramms)
virtuelles Netz n **für eine geschlossene Benutzergruppe** <Telekom> virtual private network (VPN)
virtuelles Speichersystem n <Comp & DV> virtual memory system
virtuelles Teilchen n <Teilphys> virtual particle
virtuelles Terminal n <Comp & DV, Telekom> virtual terminal
Viscin n <Chemie> viscin
Viscokupplung f <Kfztech> viscous clutch (Getriebe)
Viscose f <Chemie> viscose
Visier n 1. <Optik> hole; 2. <Sicherheit> visor • **Visier schaffen** <Optik> hole sight
Visiereinrichtung f <Bau> sight
Visierfernrohr n <Bau> sighting telescope
Visiergerät n <Raumfahrt> diopter
Visiertafel f <Bau> boning rod (Vermessung)
Viskoelastizität f <Kunststoff> viscoelasticity
Viskometer n <Strömphys> viscometer
viskos adj <Chemie, Lebensmittel, Maschinen, Phys> viscous
viskose Flüssigkeit f <Lebensmittel, Maschinen, Phys> viscous fluid
viskose Strömung f <Lebensmittel, Maschinen, Phys> viscous flow
viskose Unterschicht f <Strömphys> viscous sublayer
Viskosefilament n <Textil> rayon
Viskosefilamentfaser f <Textil> rayon
viskoser Zustand m <Fertig> treacle stage (Kunststoffe)
Viskosimeter n 1. <Fertig> viscosimeter; 2. <Kunststoff> viscometer, viscosimeter; 3. <Labor, Maschinen> viscometer; 4. <Strömphys> viscosimeter; 5. <Thermod> viscosity meter
Viskosimetrie f <Fertig> viscosimetry
Viskosität f 1. <Chemie> viscosity; 2. <Erdöl> viscosity (Petrochemie); 3. <Fertig, Kohlen, Kunststoff, Maschinen, Phys, Strömphys, Thermod> viscosity
Viskositäts-Dichte-Konstante f <Thermod> viscosity gravity constant
Viskositäts-Dichteverhältnis n (VD) <Nichtfoss Energ> kinematic viscosity (Kunststoffe)

Viskositätsindex m (VI) 1. <Kfztech> viscosity index, VI (Öl); 2. <Maschinen, Strömphys, Thermod> viscosity index, VI
Viskositätskoeffizient m 1. <Phys, Qual, Strömphys> coefficient of viscosity; 2. <Thermod> viscosity coefficient
Viskositätsmessgerät n <Kunststoff> viscometer, viscosimeter
Viskositäts-Temperatur-Koeffizient m <Thermod> viscosity temperature coefficient
Viskositäts-Temperaturverhalten n <Thermod> viscosity temperature characteristics
Viskositätsverbesserer m <Erdöl> viscosity index improver (Petrochemie)
Viskositätszahl f (VI) 1. <Kfztech> viscosity index, VI (Öl); 2. <Maschinen, Strömphys, Thermod> viscosity index, VI
Visualisierung f <Comp & DV> visualization
visuell adj 1. <Ergon> visual; 2. <Optik> ocular
visuelle Standortverwaltung f <Comp & DV> visual site management
visuelle Wahrnehmung f <Ergon> visual perception
visueller Störabstand m <Aufnahme> weighted signal-to-noise ratio
visuelles Zeichenlesesystem n <Verpack> character-reading vision system
Vitellin n <Chemie> vitellin
Viterbi-Decodierung f <Telekom> Viterbi decoding
Vitrifikation f <Chemie> vitrification
Vitriolbildung f <Chemie> vitriolization
VK (Vergaserkraftstoff) <Erdöl, Kfztech, Thermod> fuel, (AE) gas, (AE) gasoline, (BE) petrol
V-Kerbe f <Kerntech> V-shaped notch
V-Kurven fpl <Elektrotech> v-curve characteristic
VLCC (Supertanker) <Erdöl> VLCC (very large crude carrier)
V-Leitwerk n <Lufttrans> V-tail
Vlies n <Papier> mat
Vliesbildung f <Ker & Glas> mat formation
Vliesseparator m <Textil> fleece separator
VLP (Video-Langspiel) <Fernseh> VLP (video long play)
VLP-Bildplatte f <Comp & DV> optical memory
VLSI (Höchstintegration) <Comp & DV, Elektronik> VLSI (very large-scale integration)
VLSI-Chip m <Comp & DV> VLSI chip
VLSI-Schaltkreis m <Elektronik, Phys> VLSI circuit
V-Meißel m <Bau> (AE) double-beveled chisel, (BE) double-bevelled chisel
VMOS-Transistor m <Elektronik> VMOS transistor
V-Motor m 1. <Kfztech> V-engine, V-type engine; 2. <Maschinen> V-cylinder engine, V-engine
V-Null-Getriebe n <Fertig> long-and-short addendum gears (Getriebelehre)
V-Nut f <Fertig> V groove
Vocoder m 1. <Aufnahme> vocoder (Sprachverschlüsselungsgerät); 2. <Comp & DV, Telekom> vocoder
Vocoder m **mit linearer Prädiktionscodierung** <Telekom> linear predictive coding vocoder
Vocoder m **mit Pitch-Anregung** <Telekom> pitch-excited vocoder
Vogelkäfigantenne f <Funktech> birdcage aerial
Vogelnest n <Ker & Glas> bird's nest
Vogelschlaggefahr f <Lufttrans> bird strike hazard
Void n <Chemie> void
Voile m <Textil> voile
volatil adj <Chemie> volatile
voll betriebsbereit adj <Maschinen> in full working order
voll durchkentern v <Wassertrans> turn turtle
voll gelaufen adj <Wassertrans> waterlogged
voll saugen v/sich <Lebensmittel> soak

voll

voll tanken v <Kfztech, Trans> fill up
Voll... <Comp & DV, Elektriz, Wassertrans> full
Volladdierer m <Comp & DV> full adder
Vollautomat m <Maschinen> fully automatic lathe
vollautomatische Blende f <Foto> fully automatic diaphragm
vollautomatische Etikettiermaschine f <Verpack> (AE) fully automatic self-adhesive labeling machine, (BE) fully automatic self-adhesive labelling machine (für selbstklebende Etiketten)
vollautomatische Landung f <Lufttrans> autoland
vollautomatische Streckverpackung f <Verpack> fully automatic stretch-wrapper pack
vollautomatisches Landesystem n <Trans> autoland system
Vollbahn f <Eisenbahn> (AE) standard gage railroad, (BE) standard gauge railway
Vollbereichslautsprecher m <Aufnahme> full-range loudspeaker
Vollbildabtastung f <Fernseh> sequential scanning
Vollbildabzug m <Foto> full-frame print
Vollbinder m <Bau> perpend stone
Volldraht m <Elektriz> solid wire
Volldruck m <Lufttrans> impact pressure
Volldruckmaschine f <Maschinen> nonexpansion engine
volle Bohrung f <Maschinen> full bore
volle Drehung f <Lufttrans> roll (Kunstflug)
volle Größe f <Maschinen> full scale
volle Rundkante f <Ker & Glas> full round edge
volle Schleife f <Lufttrans> closed loop (Kunstflug)
volle Wasserdruckhöhe f <Hydraul> full head of water
Volleinschlag m <Kfztech> steering lock (Lenkung)
vollelastischer Stoß m <Kerntech> billiard ball collision
voller Schub m <Raumfahrt> full thrust
voller Ton m <Aufnahme> round tone
volles Rohr n <Hydraul> full pipe
vollfarbig adj <Druck> (AE) full-color, (BE) full-colour
vollfette Schrift f <Druck> full-face type
vollflächig adj <Fertig> holohedral
vollfliegende Achse f <Kfztech> full-floating axle
Vollgas n <Lufttrans> full-open throttle
vollgebunden adj <Druck> full-bound, whole-bound
Vollgestängebohren n <Bau> boring by percussion with rods
völlig absorbierendes Target n <Strahlphys> total absorption target
völlig blockierungsfreies Netz n <Telekom> strictly non-blocking network
völlige Windstille f <Wassertrans> dead calm
Völligkeit f <Wassertrans> fineness (Schiffbau)
Völligkeitsgrad m <Qual, Wassertrans> coefficient of fineness (Schiffkonstruktion)
vollimprägnieren v <Bau> saturate
Vollinjektionsturbine f <Hydraul> full-injection turbine
vollisolierter Schalter m <Elektriz> all-insulated switch
Vollkammerluftkissensystem n <Trans> plenum chamber air cushion system
Vollkapselung f <Sicherheit> complete enclosure, full encapsulation (Maschine)
Vollkegel m <Fertig> external taper
Vollkeilriemen m <Fertig> solid vee-belt
Vollkernisolator m <Elektrotech> solid core-type insulator
vollkommen trocken adj <Textil> bone-dry
vollkommene Schmierung f <Maschinen> thick film lubrication
vollkommener Kristall m <Metall> perfect crystal
vollkommenes Bündel <Telekom> full availability group
Vollkreis m <Geom, Konstzeich> full circle

Vollkundenschaltung f <Elektronik> full-custom circuit (voll kundenspezifischer Schaltkreis)
Vollkunststoff-Lichtwellenleiter m <Telekom> (AE) all-plastic fiber, (BE) all-plastic fibre
Vollkurzschluss m <Elektrotech> dead short
Voll-Last f 1. <Elektriz> full load; 2. <Wassertrans> full power
Voll-Lastbündel n <Telekom> high-usage circuit group
Voll-Lastkonfiguration f <Raumfahrt> full-load configuration
Voll-Leimen n <Verpack> full gluing
Voll-Leiter m <Elektriz> solid conductor
Volllinie f <Konstzeich> continuous line
Vollmacht f 1. <Patent> authorization; 2. <Qual> authority
Vollmantelkorb m <Kohlen> bowl centrifuge (Zentrifuge)
Vollmaske f <Sicherheit> full-facepiece respirator
Vollmaterial n <Anstrich> solid
Vollniet m <Maschinen> full rivet
volloptisches Netz n <Telekom> all-optical network
Vollpappe f <Verpack> (AE) solid fiber board, (BE) solid fibre board
Vollpappenkiste f <Verpack> container board box
Vollpol m <Elektrotech> non-salient pole
Vollpolläufer m <Kfztech> smooth-core armature
Vollprüfung f <Qual> one-hundred-percent inspection
Vollrad n <Eisenbahn> solid wheel
Vollrohr n <Hydraul> full pipe
Vollschalenbauweise f <Raumfahrt> monocoque structure
Vollscheibe f 1. <Fertig> arborless wheel; 2. <Maschinen> solid pulley
Vollschmierung f <Maschinen> thick film lubrication
Vollschnitt m <Konstzeich> full section
Vollschutzanzug m <Sicherheit> full-coverage suit, fully encapsulated suit, whole-body suit
Vollschutzkleidung f <Sicherheit> full-protective clothing
vollschwarzer Pfeil m <Konstzeich> blackened arrowhead
Vollsichtbrille f <Sicherheit> panoramic wide-vision goggles
Vollsichtschutzmaske f <Sicherheit> panoramic full-face mask
Vollsperrung f <Eisenbahn> complete track load
Vollspur f 1. <Akustik, Aufnahme> full track; 2. <Eisenbahn> (AE) standard gage, (BE) standard gauge
Vollspur-Aufnahme f <Aufnahme> full-track recording
Vollspur-Aufnahmegerät n <Aufnahme> full-track recorder
vollständig adj <Math> complete
vollständig in Halbleiter-Technik adj <Elektronik> all-solid state (Ausführung)
vollständige Induktion f <Math> mathematical induction
vollständige Kapselung f <Sicherheit> complete enclosure (Maschine)
vollständige Reinigung f <Kerntech> complete purification
vollständige Verkehrsverlagerung f <Trans> complete diversion
vollständiger Schutz m <Patent> full protection
vollständiges Bildsignal n <Fernseh> composite signal
vollständiges Höhenleitwerk n <Lufttrans> empennage
vollständiges Quadrat n <Math> perfect square
Vollständigkeit f 1. <Comp & DV> integrity; 2. <Math> completeness
Vollständigkeit f <Math> completeness
Vollständigkeitsversuch m <Bau> integrity test
Vollstreckung f <Patent> enforcement
Vollstromölfilter n <Kfztech> full-flow oil filter (Schmierung)

Vollsubtrahierer m <Comp & DV> full subtractor
vollverstrecktes Garn n <Textil> fully drawn yarn
vollverteiltes Steuersystem n <Telekom> fully distributed control system
vollwandiger Träger m <Bau> I-girder
Vollwandrippe f <Raumfahrt> rib
Vollwandträger m <Bau> plate girder
Vollweg m <Elektriz, Elektrotech> full wave
Vollwegbrückenschaltung f <Elektriz> full bridge
Vollweggleichrichten n <Elektrotech> full-wave rectification
Vollweggleichrichter m <Elektriz, Elektrotech> full-wave rectifier
Vollweggleichrichtung f <Elektriz> full-wave rectification
Vollwelle f <Maschinen> solid shaft
Vollwertkost f <Lebensmittel> wholefood
vollzählige Besatzung f <Wassertrans> complement
Vollzapfen m <Bau> through tenon
Vollziegel m <Bau> solid brick
Volt n (V) 1. <Aufnahme> volt, V; 2. <Elektriz> volt, V (Einheit der Spannung); 3. <Funktech, Metrol, Optik> volt, V
Voltameter n 1. <Chemie> voltameter; 2. <Elektrotech> coulometer, voltameter, voltameter; 3. <Phys> coulometer, voltameter
Voltampere n <Elektriz> voltampere
Voltamperemeter n <Metrol> VA-meter, voltameter, voltampere meter (Scheinleistungsmesser)
Voltamperemeterstundenzähler m <Metrol> volt-ampere-hour meter
Volta'sche Säule f <Elektrotech> voltaic pile
Volta'sche Zelle f <Elektrotech> voltaic cell
Volterra-Versetzung f <Metall> Volterra dislocation
Voltmeter n <Elektrotech, Phys> voltmeter
Volt- und Frequenzabweichung f <Nichtfoss Energ> voltage and frequency deviation
Volumeinheit f <Akustik> VU, volume unit
Volumen n 1. <Comp & DV> volume; 2. <Geom> volume; 3. <Maschinen> capacity; 4. <Math> cubic content; 5. <Mechan> bulk; 6. <Metrol> cubage, cubic measure; 7. <Phys> volume; 8. <Textil> bulk; 9. <Textil, Thermod> volume
Volumen n **in Phase** (Ω) <Phys> volume in phase space (Ω)
Volumen n **von Feststoffen** <Metrol> solid measure
Volumenänderung f <Fertig> change of volume
Volumenausdehnungskoeffizient m <Thermod> expansion coefficient
Volumendiffusion f <Metall> volume diffusion
Volumendosierung f <Verpack> volume dosing
Volumendurchflussmessgerät n <Gerät> volumetric flow meter
Volumenelastizitätsmodul m (B) <Thermod> modulus of volume elasticity (B)
Volumenemissions- und Absorptionskoeffizient m <Strahlphys> volume emission and absorption coefficient
volumenerhaltende Strömungen fpl <Strömphys> isochoric flows
Volumenfraktion f <Metall> volume fraction
Volumenfraktion f **von Teilchen** <Metall> volume fraction of particles
Volumenfüllung f <Verpack> volume filling
Volumengeometrie f <Geom> CSG, constructive solid geometry
Volumengeschwindigkeit f (U) <Akustik> volume current (U)
Volumengrößenfaktor m <Metall> volume size factor
Volumenintegral n <Phys> volume integral
Volumen-Kapazitätsverhältnis n <Trans> V-C ratio, volume-capacity ratio

Volumenkraft f <Fertig> body force
Volumenmessgerät n <Fertig> capacity gauge
Volumenmessung f <Metrol> cubic measurement
Volumenmodell n <Geom> volume model
Volumenoszillator m <Elektronik, Funktech> bulk-wave oscillator (Mikrowellentechnik)
Volumenschallwelle f <Elektrotech> bulk acoustic wave
Volumenstrom m 1. <Akustik> volume current (U); 2. <Heiz & Kälte> volume flow, volume flow rate, volumetric flow
Volumenstrommessgerät n <Gerät> volumetric flow meter
Volumenstromrechner m <Gerät> volumetric flow calculator
Volumenveränderung f <Metall> volume change
Volumenverdrängung f (X) <Akustik> volume displacement (X)
Volumenwelle f <Funktech> bulk wave (microwave)
Volumenwellen-Resonator m <Elektronik, Funktech> bulk-wave resonator
Volumenzähler m <Gerät> volumetric flow meter
Volumeter n <Akustik> volumeter
volumetrische Effizienz f <Elektrotech> volumetric efficiency (Platte in Kommutatoren)
volumetrische Gleichung f <Metall> volumetric equation
volumetrischer Wirkungsgrad m <Kfztech, Maschinen, Nichtfoss Energ> volumetric efficiency
voluminös adj <Papier, Verpack> bulky
Von-Neumann-Maschine f <Comp & DV> von Neumann machine
Vorabaufnahme f <Fernseh> off-air pick-up
Vorabaufzeichnung f <Fernseh> off-air recording
Vorabdruck m <Druck> preprint
Vorabfragen n <Telekom> screening (Anrufbeantworter, Fax)
Vorabfühlschleife f <Aufnahme> presence loop
vorabspeichern v <Comp & DV> prestore
Voralterung f <Telekom> (BE) preageing, (AE) preaging
Vorankündigung f <Trans> advance information
Vorankündigungszeichen n <Trans> advance direction sign
VOR-Anlage f <Funktech> VHF omnidirectional radio range
Voranmeldungsgespräch n <Telekom> personal call
Voranode f <Elektrotech> first anode
Voranstrich m 1. <Bau> prime coat; 2. <Kunststoff> primer
Vorarbeiten n <Fertig> roughing-down
Vorarbeiter m <Ker & Glas> gaffer
Voraufbereitung f <Chemie> pretreating (Trinkwasser, Abwasser)
voraufgezeichnet adj <Comp & DV> prerecorded
Vorausanzeige f <Wassertrans> heading marker (Radar)
vorausberechnete Ausfallrate f <Qual> predicted failure rate
vorausberechnete mittlere Instandhaltungsdauer f <Qual> assessed mean active maintenance time
vorausberechnete mittlere Lebensdauer f <Qual> predicted mean life
vorausberechnetes Lebensdauer-Perzentil n **Q** <Qual> predicted Q-percentile life
Vorausexemplar n <Druck> advance copy
Vorausfahrt f <Wassertrans> headway (Schiffbewegung)
Vorauslaugen n <Kohlen> preleaching
vorausplanen v <Comp & DV, Qual> project
voraussetzen v <Math> assume
Voraussetzung f 1. <Math> assumption; 2. <Patent, Qual> requirement

voraussichtliche

voraussichtliche Abflugzeit f (ETD) <Lufttrans> estimated time of departure (ETD)
voraussichtliche Ankunftszeit f (ETA) <Lufttrans, Wassertrans> estimated time of arrival (ETA)
voraussichtliche Flugzeit f <Lufttrans> estimated flight time
voraussichtliche mittlere Instandhaltungsdauer f <Qual> predicted failure rate
voraussichtliche Offblockzeit f <Lufttrans> estimated off-block time
voraussichtliches Ausfallperzentil n <Qual> predicted Q-percentile life
vorausstabilisiert adj <Wassertrans> head-up (Radar)
Vorausströmung f <Hydraul> exhaust lead
Voraustritt m <Hydraul> exhaust lead
Vorbad n <Foto> preliminary bath
Vorband n <Aufnahme> leader tape
Vorbau m <Wassertrans> prototype (Schiffbau)
Vorbaum m <Textil> back beam
vorbearbeiten v <Comp & DV> pre-edit
vorbearbeitet adj/**auf Rohmaß** <Mechan> rough-machined
vorbearbeiteter Stahl m <Metall> preworked steel, semi-finished steel
Vorbearbeitung f <Elektronik> preprocessing
Vorbecken n <Hydraul> forebay
vorbehandeln v <Anstrich> pretreat
Vorbehandlung f 1. <Abfall, Anstrich> pretreatment; 2. <Fertig> preparatory treatment, pretreatment; 3. <Kohlen> pretreatment; 4. <Kunststoff> conditioning; 5. <Maschinen> pretreatment; 6. <Wasserversorg> preliminary treatment
vorbeifahren v <Eisenbahn> pass (am Signal)
Vorbeifahrgeräusch n <Sicherheit> vehicle passby sound
Vorbeiflug m <Raumfahrt> fly-by
Vorbeiflugeinwirkung f <Raumfahrt> fly-by effect
Vorbeipendeln n <Raumfahrt> swing-by
Vorbelastung f 1. <Elektrotech> bias; 2. <Kohlen> bias (mechanisch); 3. <Umweltschmutz> initial level of water pollution (des Wassers)
Vorbelastungsdruck m <Kohlen> preconsolidation pressure
Vorbelastungswiderstand m <Elektrotech> bleeder resistor (für Gleichrichter)
Vorbenutzung f <Patent> prior use
vorbereiten v 1. <Comp & DV> initialize; 2. <Maschinen> set-up
Vorbereiten n <Maschinen> setting, setup
vorbereitet adj <Lebensmittel> prepared
Vorbereitung f 1. <Anstrich> pretreatment; 2. <Comp & DV, Lebensmittel> preparation
Vorbereitungszeit f <Telekom> set-up time (Telefon)
Vorbeschichtung f <Fertig> primary coating
vorbespieltes Magnetband n <Aufnahme> prerecorded magnetic tape
Vorbestellung f <Telekom> advance booking (von Verbindungen)
vorbetriebliche Überprüfung f <Kerntech> precommissioning checks
vorbeugende Instandhaltung f <Bau> preventive maintenance
vorbeugende Prüfung f <Qual> preventive inspection
vorbeugende Wartung f <Comp & DV, Maschinen, Telekom> preventive maintenance
vorbildlich adj <Comp & DV> model
Vorblasen n <Ker & Glas> blow back, preblowing, puff
Vorblick m <Bau> minus sight (Vermessung)

Vorblock m 1. <Fertig> beam blank (Profil); 2. <Metall> bloom
vorblocken v <Fertig> bloom (Luppen)
Vorblock-Putzerei f <Metall> bloom yard
vorbohren v 1. <Bau> hole; 2. <Kohlen, Maschinen> predrill
Vorbohrer m <Bau> gimlet
Vorbohrloch n <Nichtfoss Energ> mousehole
Vorbohrschlamm m <Erdöl> spud mud (Flachbohrtechnik)
Vorbrecher m <Kohlen> primary crusher
Vorbrennen n **des Glashafens** <Ker & Glas> pot arching
Vordach n <Bau> canopy, porch roof
vordefiniert adj <Comp & DV> predefined
Vordehnung f <Metall> prestrain
Vorder... 1. <Maschinen> head-end, front; 2. <Wassertrans> forward (Schiff), fore... (am Schiff)
Vorderabtastung f <Fernseh> front scanning
Vorderachse f 1. <Kfztech> front axle (Räder, Kraftübertragung); 2. <Maschinen> front axle
Vorderdeck n <Wassertrans> foredeck
vordere Austastschulter f <Fernseh> front porch
vordere Einzelradaufhängung f <Kfztech> independent front suspension
vordere Hafenöffnung f <Ker & Glas> front arch
vorderer Kolben m <Kfztech> front piston, primary piston
vorderer Spalt m <Aufnahme> front gap
vorderer Verschlussstein m <Ker & Glas> front tweel
vorderer Zellenring m <Raumfahrt> forward frame section
vorderes Lot n <Wassertrans> forward perpendicular (Schiffkonstruktion)
Vorderflanke f 1. <Elektronik> leading edge (eines Impulses); 2. <Fertig> leading edge; 3. <Phys> leading edge (eines Impulses); 4. <Telekom> leading edge
Vordergabel f <Kfztech> front fork
Vorderglied n <Foto> front element (eines Objektivs)
Vordergrund m <Comp & DV> foreground • **im Vordergrund** <Comp & DV> foreground
Vordergrund... <Comp & DV> foreground
Vordergrundjob m <Comp & DV> foreground job
Vordergrundprogramm n <Comp & DV> foreground program
Vordergrundverarbeitung f <Comp & DV> foregrounding, foreground processing
Vorderkante f 1. <Druck> fore edge (eines Buches); 2. <Elektronik> leading edge (von Belegen); 3. <Fertig, Lufttrans> leading edge; 4. <Phys> leading edge (Impuls); 5. <Telekom> leading edge
Vorderkante f **des Luftschraubenblattes** <Lufttrans> blade leading edge (Hubschrauber)
Vorderkipper m <Eisenbahn> end dump wagon
Vorderplatte f <Gerät> front panel (an Gerät)
Vorderrad n <Trans> front wheel
Vorderradantrieb m <Kfztech> front-wheel drive
Vorderradaufhängung f <Kfztech> front suspension
Vorderradeinstellung f <Kfztech> front-wheel alignment
Vorderrahmen m <Foto> front frame
Vorderseite f 1. <Bau> face, front; 2. <Comp & DV> front end; 3. <Kohlen> face; 4. <Maschinen> face, front; 5. <Papier> front side
Vorderwand f <Ker & Glas> front wall
Vorderwandzelle f <Elektrotech> front-wall photovoltaic cell
Vorderzapfen m **der Kurbelwelle** <Kfztech> crankshaft front end
Vordrossel f <Kfztech> choke (Vergaser)

Vordruck m 1. <Comp & DV> form; 2. <Druck> printed form; 3. <Patent> form
Vordruckwalze f <Druck> dandy roll
Vordruckzeichnung f <Konstzeich> preprinted drawing
voreilen v 1. <Maschinen> advance; 2. <Telekom> lead *(Phase)*
Voreilen n <Fertig> leading
Voreilung f 1. <Elektriz> advance *(Phase)*; 2. <Fertig> lead *(Phase)*; 3. <Hydraul> preadmission *(Dampf)*; lead *(Steuerschieber)*; 4. <Kontroll> speed-up; 5. <Maschinen> advance, lead; 6. <Textil> overfeed
Voreilwinkel m 1. <Elektriz> advance angle, leading angle *(Phase)*; 2. <Kfztech> angle of advance; 3. <Maschinen> angle of advance, angle of lead
Voreinflugzeichen n <Lufttrans> outer marker
voreingestellte Frequenz f <Elektronik> preset frequency
voreingestelltes Potenziometer n <Elektriz> preset pot
voreinstellen v <Aufnahme, Comp & DV, Maschinen> preset
Voreinstellen n <Elektronik> presetting
Voreinstellung f <Comp & DV> presetting
Voreinstellzähler m <Gerät> preselection counter, preset counter
Voreinströmung f <Hydraul> preadmission *(Dampf)*; lead *(Steuerschieber)*
Vorentflammung f <Kfztech> advance *(Zündung)*
Vorentwurf m <Bau> preliminary design
Vorfahre m <Comp & DV> ancestor node, antecedent node *(semantische Netze)*
Vorfahrt f <Trans> right of way
Vorfahrtsrecht n 1. <Lufttrans> right of way *(Flughafen)*; 2. <Trans> priority, right of way; 3. <Wassertrans> right of way
Vorfalldatenmeldung f <Lufttrans> incident date reporting
Vorfeld n 1. <Lufttrans> ramp; 2. <Trans> apron
Vorfelddienst m <Lufttrans> apron management service, ramp services
Vorfeldeinrichtung f <Telekom> front end
Vorfeldrollbahn f <Lufttrans, Trans> apron taxiway
Vorfeldwartebereich m <Lufttrans> holding apron, holding bay *(Flughafen)*
vorfertigbearbeiten v <Fertig> semifinish
vorfertigen v <Bau> precast
Vorfilter n 1. <Elektronik> prefilter; 2. <Funktech> input filter
Vorfilter n **zweiter Ordnung** <Elektronik> second order prefilter
Vorfiltern n <Elektronik> prefiltering
Vorflügel m <Lufttrans> slat of the leading edge
Vorflutdrän m <Bau> main drain
Vorfluter m 1. <Abfall> receiving water; 2. <Wasserversorg> drainage ditch, receiving water
Vorform f 1. <Ker & Glas> blank *(der Hohlglasfertigungsmaschine)*; 2. <Optik, Telekom> preform
Vorformabdruck m <Ker & Glas> baffle mark
Vorformdeckel m <Ker & Glas> baffle *(Hohlglasfertigung)*
vorformen v <Fertig> sadden; filler *(Schweißen)*
Vorformen n 1. <Fertig> edging *(Walzen)*; 2. <Fertig> saddening *(Hämmern)*; fillering *(Schmieden)*; 3. <Kunststoff> preforming
Vorformling m <Ker & Glas, Kunststoff> preform
Vorformriss m <Ker & Glas> blank tear
Vorformspeiseöffnung f <Ker & Glas> baffle hole
Vorformtisch m <Ker & Glas> blank table
Vorfräsen n **von Zahnlücken** <Fertig> gashing
Vorfräser m <Maschinen> rougher
Vorgabe f 1. <Comp & DV> default; 2. <Qual> handicap

Vorgabe... <Comp & DV> default
Vorgabewert m <Comp & DV> default value
Vorgalvanisierbad n <Fertig> strike
Vorgang m 1. <Hydraul> event *(Drehung, Hub, Steuerschieber)*; 2. <Patent, Qual> function
Vorgänger m 1. <Comp & DV> ancestor; 2. <Künstl Int> predecessor *(in semantischen Netzen)*; ancestor node, antecedent node *(semantischen Netzen)*
Vorgängerknoten m <Künstl Int> predecessor *(in semantischen Netzen)*; ancestor node, antecedent node *(semantischen Netzen)*
Vorgarn n <Textil> rove, roving
Vorgebirge n <Wassertrans> promontory *(Geographie)*
vorgebohrter Pfahl m <Kohlen> prebored pile
vorgedruckt adj <Verpack> preprinted
vorgefertigt adj <Bau> precast
vorgeformte Faser f <Telekom> *(AE)* preformed fiber, *(BE)* preformed fibre
vorgeformtes Material n <Fertig> dummy
vorgegebene Flugbahn f <Lufttrans> assigned flight path
vorgegebener Widerstand m <Funktech> preset resistor
vorgegossen adj <Fertig> roughcast
vorgehängte Wand f <Bau> curtain wall
Vorgehensweise f <Comp & DV> approach, procedure
vorgekocht adj <Lebensmittel> precooked
vorgekühlt adj <Lebensmittel> precooled
Vorgelege n <Maschinen> countershafting, transmission
Vorgelegegetriebe n <Fertig> intermediate gearbox
Vorgelegerad n <Maschinen> countergear
Vorgelegewelle f 1. <Fertig> jack shaft; 2. <Kfztech> countershaft, countershaft gear; 3. <Maschinen> countershaft, intermediate shaft, layshaft
vorgenommene Einstellung f **beim Hersteller** <Gerät> factory setting
Vorgerüst n <Fertig> roughing stand *(Walzen)*
vorgeschaltet adj <Heiz & Kälte, Maschinen> upstream
vorgeschaltetes Filter n <Elektronik> prefilter
vorgeschliffen adj <Fertig> preground
vorgeschmiedet adj <Fertig> punched *(Bohrung)*
vorgeschmiedetes Material n <Fertig> dummy
vorgeschrieben adj <Patent, Qual> mandatory
vorgeschriebener Haltepunkt m <Qual> mandatory hold point
Vorgesenk vorformen v/im <Fertig> rough-stamp
vorgespannt adj <Phys> biased
vorgespanntes Glas n 1. <Ker & Glas> toughened glass; 2. <Trans> prestressed glass, toughened glass
vorgespanntes Lager n <Maschinen> preloaded bearing
vorgesteuertes Ventil n <Heiz & Kälte> servo-assisted valve
vorgewalzter Block m 1. <Fertig> cogged ingot; 2. <Ker & Glas> rough rolled
vorgezeichnetes Wetterverhalten n <Raumfahrt> weather pattern
vorgießen v <Fertig> rough-cast
Vorgießen n <Fertig> rough-casting
Vorgießen n **von Bohrungen** <Fertig> coring *(Spanung)*
Vorglüheinrichtung f <Kfztech> preheater *(Dieselmotor)*
Vorgriff m <Comp & DV> lookahead
Vorgruppe <Telekom> subgroup *(Telefonanlage)*
Vorhafen m <Wassertrans> offshore terminal
Vorhalle f <Bau> porch
Vorhalteinheit f <Regelung> lead module
Vorhaltverstärkung f <Regelung> derivative action gain
Vorhang m <Ker & Glas> curtain *(Abtrennung im Foucault-Verfahren zum Ziehen von Tafelglas)*
Vorhangantenne f <Funktech> curtain

Vorhangbeschichter

Vorhangbeschichter *m* <Kunststoff> curtain coater
Vorhangbeschichtung *f* <Ker & Glas> curtain coating
Vorhangbildung *f* <Kunststoff> curtaining
Vorhängeschloss *n* <Bau> padlock
vorheizen *v* <Heiz & Kälte, Thermod> preheat
vorher aufnehmen *v* <Aufnahme> prerecord
Vorherd *m* 1. <Fertig> forehearth *(Kupolofen)*; 2. <Ker & Glas> forehearth
Vorherdeingang *m* <Ker & Glas> forehearth entrance
vorhergesagte Zuverlässigkeit *f* <Qual, Raumfahrt> predicted reliability
Vorher-Nachher-Untersuchung *f* <Trans> before-and-after study
vorherrschende Winde *mpl* <Wassertrans> prevailing winds
Vorhersage *f* <Künstl Int> prediction
Vorhersagefähigkeit *f* <Qual, Umweltschutz> predictive capability
Vorhersagesystem *n* <Künstl Int> prediction system
vorhobeln *v* <Mechan> rough-plane
Vorhonen *n* <Fertig> rough honing
Vorimprägnieren *n* <Ker & Glas> prepregging
Vorkammer *f* 1. <Kfztech> antechamber; prechamber, precombustion chamber *(Dieselmotor)*; 2. <Thermod> precombustion chamber *(eines Verbrennungsmotors)*
Vorkehrung *f* <Qual, Sicherheit> precaution
Vorklärbecken *n* <Abfall> preliminary settling basin, primary settling basin, primary settlement tank
Vorklassierrost *n* <Fertig> grizzly
Vorkommnis *n* **während des Fluges** <Raumfahrt> flight occurrence
Vorkompilierer *m* <Comp & DV> preprocessor
vorkompilierter Code *m* <Comp & DV> precompiled code
vorkonfigurieren *v* <Comp & DV> preconfigure
vorkontrollieren *v* <Fernseh> preview
Vorkonzentrat *n* <Kohlen> preconcentrate
Vorkühler *m* <Heiz & Kälte> precooler
Vorlage *f* 1. <Comp & DV> master, template; 2. <Fertig, Konstzeich> master
Vorlagestück *n* <Fertig> abutting piece
Vorlast *f* <Fertig> initial load *(Werkstoffe)*
vorlastig *adj* <Wassertrans> trimmed by the head
vorlastig *adv* <Wassertrans> down by the head *(Schiff)*
Vorlauf *m* 1. <Elektrotech> pretravel; 2. <Fernseh> preroll; 3. <Fertig> lead; 4. <Heiz & Kälte> advance; 5. <Lebensmittel> first running; fore-running *(Destillation)*; 6. <Maschinen> advance, approach, forward motion
vorlaufende Welle *f* <Elektrotech> forward wave
Vorläufernetz *n* <Telekom> legacy network
Vorlauffaser *f* <Telekom> *(AE)* launching fiber, *(BE)* launching fibre
vorläufig *adj* <Bau, Patent, Qual> provisional
vorläufige Lagerung *f* <Abfall> temporary storage *(von Müll)*
vorläufige Prüfung *f* <Patent, Qual> preliminary examination
vorläufige technische Prüfung *f (PDR)* <Raumfahrt> preliminary design review *(PDR)*
vorläufiger Kostenvoranschlag *m* <Bau> preliminary cost estimate
vorläufiges Patent *n* <Patent> provisional patent
Vorlaufrohr *n* <Heiz & Kälte> flow pipe
Vorlauftemperatur *f* <Heiz & Kälte> flow temperature
Vorlaufzeit *f* 1. <Elektronik> lead time *(Zeit zwischen Produktentwurf und Fertigung)*; 2. <Fernseh> preroll time; 3. <Fertig, Maschinen, Qual> lead time
Vorlegescheibe *f* <Fertig> dummy block
Vorlegezahnrad *n* <Mechan> idler

vorlich *adv* <Wassertrans> before the beam
vorlicher als dwars *adv* <Wassertrans> forward of the beam
vorlicher als querab *adv* <Wassertrans> forward of the beam
Vorlicht *n* <Fernseh> ambient light
vormagnetisieren *v* <Funktech> bias
Vormagnetisierung *f* 1. <Aufnahme> bias, biasing, premagnetization; 2. <Comp & DV> bias; 3. <Elektrotech> magnetic bias; 4. <Fertig> bias
Vormagnetisierungsfrequenz *f* <Aufnahme> bias frequency
Vormagnetisierungsoszillator *m* <Aufnahme, Elektronik> bias oscillator *(Tonbandgeräte)*
Vormagnetisierungsquelle *f* <Elektrotech> bias source
Vormauerziegel *m* <Bau> facing brick
Vormischen *n* <Aufnahme> rough mix
Vormischerstufe *f* <Funktech> premixer
Vormischung *f* <Kunststoff> batch, master batch
Vormittagsspitze *f* <Trans> a.m. peak
Vormontage *f* 1. <Fertig, Maschinen> preassembly; 2. <Wassertrans> prefabrication *(Schiffbau)*
Vormontagestraße *f* <Fertig> preassembly line
vormontieren *v* <Fertig> preassemble *(Kunststoffinstallationen)*
vormontiert *adj* <Fertig> preassembled
vorn *adv* <Wassertrans> fore • **nach vorn** <Wassertrans> forward • **von vorn beleuchten** <Foto> front-light
Vornutenfräser *m* <Maschinen> roughing slot-mill
vorordnen *v* <Comp & DV> prestore
Vor-Ort-Einsatzleiter *m* <Meerschmutz> OSC, on-scene commander
Vorortpendlerzug *m* <Eisenbahn> commuter rail system
Vor-Ort-Prüfung *f* <Metrol> on-site test, on-site testing
Vor-Ort-Steuerung *f* <Regelung> local control
Vorortzug *m* <Eisenbahn> *(AE)* interurban train, local train
Vorpiek *f* <Wassertrans> forepeak *(Schiffbau)*
Vorplastifizieren *n* <Kunststoff> preplasticizing
vorpolieren *v* <Mechan> rough-polish
Vorpresse *f* <Papier> baby press
Vorpressen *n* <Kunststoff> preforming
Vorpressling *m* <Kunststoff> preform
Vorprodukt *n* <Textil> precursor
vorprogrammiert *adj* <Comp & DV> preprogrammed
Vorprozessor *m* <Comp & DV> preprocessor
Vorprüfung *f* <Qual> preacceptance inspection, preliminary test, preliminary testing, pretest
Vorprüfungsprotokoll *n* <Qual> preacceptance inspection report
Vorrang *m* <Comp & DV> priority
Vorrangdienst *m* <Comp & DV> priority service
vorrangige Unterbrechungsebene *f* <Comp & DV> priority interruption level
Vorrangunterbrechung *f* <Comp & DV> priority interrrupt
Vorrangventil *n* <Hydraul> priority valve
Vorrangverarbeitung *f* <Comp & DV> priority processing
Vorrat *m* <Comp & DV> repertoire
vorrätiger Durchmesser *m* <Telekom> stock diameter
Vorratsbehälter *m* 1. <Strömphys> reservoir *(für Flüssigkeit)*; 2. <Verpack> bin
Vorratsglas *n* <Ker & Glas> dispensing glass
Vorratskatode *f* <Elektrotech> dispenser cathode
Vorratslänge *f* <Telekom> slack
Vorratsschrank *m* <Labor> storage cupboard
Vorratstank *m* <Lufttrans> feeder tank
Vorratswasserheizer *m* <Heiz & Kälte> storage water heater

vorraussichtliche Anflugszeit f <Lufttrans> expected approach time
Vorreaktanz f <Elektrotech> series reactance
Vorrechner m <Telekom> front-end processor
Vorreibahle f <Maschinen> roughing reamer
Vorreiber m <Bau> casement fastener; turnbuckle *(Jalousie)*
Vorreibzahn m <Maschinen> semifinishing tooth
Vorrichtung f 1. <Maschinen> apparatus, appliance, contrivance, device, fixture, jig; 2. <Meerschmutz, Telekom> device; 3. <Textil> appliance
Vor-Rücklauf-Schalter m <Fernseh> normal-reverse switch
Vor-Rückverhältnis n <Funktech> front-to-back ratio *(Antennengewinn)*
Vor-Rückwärts-Zähler m <Elektronik> bidirectional counter, increment/decrement counter
Vorsatz m <Maschinen> attachment
Vorsatzbalgengerät n <Foto> extension bellows
Vorsatzblatt n <Druck> end sheet
Vorsatzgerät n <Maschinen> attachment
Vorsatzgerät n **für Auflicht** <Foto> incident light attachment
Vorsatzkuchen m <Ker & Glas> stopper
vorsätzliche Einleitung f <Umweltschmutz> intentional discharge *(Abwässer)*
Vorsatzlinse f <Foto> supplementary lens
Vorschalldämpfer m <Kfztech> *(AE)* premuffler, *(BE)* presilencer
Vorschaltfaser f <Telekom> *(AE)* launching fiber, *(BE)* launching fibre
Vorschaltgerät n <Elektrotech> prediction unit
Vorschaltwiderstand m <Elektrotech> ballast resistor; series resistance *(bei Gleichrichter, Anschluss)*
Vorschau f <Fernseh> preview
Vorschaumonitor m <Fernseh> preview monitor
vorschieben v 1. <Elektriz> advance *(Bürsten)*; 2. <Maschinen> feed
Vorschlaghammer m 1. <Bau> slater's hammer, sledge, sledge hammer; 2. <Maschinen> sledge hammer
vorschleifen v 1. <Fertig> pregrind; 2. <Ker & Glas> polish *(erste Phase des Schleifprozesses)*
Vorschleifen n 1. <Fertig> pregrinding; 2. <Ker & Glas> rough grinding
Vorschleuse f <Wasserversorg> head sluices
Vorschlichten n <Fertig> blocking
Vorschmelzer m <Ker & Glas> foremelter
Vorschmiedegesenk n 1. <Fertig> blanker, blocker, blocking die; 2. <Maschinen> blocking die
vorschmieden v <Fertig> edge *(Schmieden)*
Vorschmieden n <Fertig> blocking
Vorschneiden n <Ker & Glas> rough cutting
Vorschneider m 1. <Fertig> first-cut tap, tap No1, taper tap; head *(Werkzeug)*; 2. <Maschinen> first-cut tap, tap No1 *(Gewinde)*
Vorschneidzahn m <Maschinen> nicker
Vorschnell-Gießöffnung f <Verpack> flip spout closure
Vorschriften fpl <Bau, Qual, Wassertrans> specifications
Vorschub m 1. <Comp & DV> carriage; 2. <Kohlen> feed; 3. <Maschinen> feed, feeding; 4. <Mechan, Papier> feed
• **Vorschub geben** <Mechan> feed
Vorschubapparat m <Fertig> feeder
Vorschubbegrenzer f <Maschinen> feed limiter
Vorschubbereich m <Fertig> range of feeds
Vorschubgeschwindigkeit f <Maschinen> feed rate, feed speed
Vorschubgetriebe n <Maschinen> feed gear
Vorschubgetriebekasten m <Fertig> feed box
Vorschubkasten m <Fertig, Maschinen, Papier> feed box

Vorschubkomponente f <Kerntech> feed component
Vorschubkraft f <Maschinen> feed force
Vorschubkugelumlaufspindel f <Fertig> recirculating ball feed screw
Vorschublochband n <Comp & DV> control tape
Vorschubmechanismus m 1. <Fertig> advance mechanism; 2. <Maschinen> feed mechanism
Vorschubmotor m <Maschinen> feed motor
Vorschubmutter f <Fertig> feed nut
Vorschubpatrone f <Fertig> collet, feeding collet
Vorschubrad n <Fertig> feed gear *(Fräsmaschine)*
Vorschubschieber m <Fertig> pusher *(Stangenvorschub)*
Vorschubschlitten m <Maschinen> feed slide
Vorschubsperrgetriebe n <Fertig> feed-pawl mechanism
Vorschubspindel f <Fertig> feed screw
Vorschubsteuerung f <Comp & DV> carriage control
Vorschubumschaltgetriebe n <Maschinen> feed reversing gear
Vorschubwalze f <Papier> feed roll
Vorschubwechselhebel m <Fertig> feed-change lever
Vorschubwinkel m <Maschinen> angle of advance
Vorschubzahl f <Kohlen> feed rate
Vorschubzahnstange f <Fertig, Maschinen> feed rack
Vorsegel n <Wassertrans> headsail
Vorserienflugzeug n <Lufttrans> preproduction aircraft
Vorsicht f <Ergon> alertness
Vorsignal n <Eisenbahn> distant signal
Vorsignalabstand m 1. <Eisenbahn> *(AE)* presignaling distance, *(BE)* presignalling distance, warning distance; 2. <Lufttrans, Trans, Wassertrans> warning distance
Vorsignalankündigung f <Eisenbahn> outer distant signal
Vorsilbe f <Comp & DV> prefix
Vorsitzender m **der Konferenzverbindung** <Telekom> conference call chairman
Vorsorge f <Qual, Sicherheit> precaution
Vorsorgemaßnahme f <Qual, Sicherheit> precautionary measure
Vorsorgeplanung f <Meerschmutz> contingency plan
Vorspann m 1. <Comp & DV> prefix, tape header, tape leader; 2. <Fernseh> leader, tape leader
Vorspannband n <Akustik> leader
vorspannen v 1. <Funktech> bias; 2. <Maschinen> preload, pretension
Vorspannen n <Ker & Glas> tempering
Vorspannglied n <Kerntech> tendon
Vorspannung f 1. <Elektrotech, Fernseh, Fertig> bias voltage; 2. <Funktech> bias *(Halbleiter)*; 3. <Maschinen> initial tension, preload; 4. <Phys> bias voltage • **mit Vorspannung** <Fertig> compressed *(Feder)* • **ohne Vorspannung** <Funktech> unbiased, zero bias • **zu große Vorspannung** <Aufnahme> overbias
Vorspannungsbatterie f <Elektrotech> bias cell
Vorspannungskreis m <Elektrotech> bias circuit
Vorspannungsstrom m <Fernseh> biasing current
Vorspannungswicklung f <Elektrotech> bias winding
Vorspannungswiderstand m <Elektrotech> bias resistor
Vorspannverschluss m <Foto> preset shutter
vorspeichern v <Comp & DV> prestore
Vorspinnmaschine f <Textil> roving frame
Vorspring m <Wassertrans> bow spring; forward backspring *(Anlegen)*
vorspringend adj <Fertig> salient *(Teil, Winkel)*
vorspringender Teil m <Kfztech> overhang *(Karosserie)*
Vorsprung m 1. <Bau> nose; break *(Wand)*; 2. <Fertig> lobe, prominence; 3. <Maschinen> boss, lobe, nose,

Vorsprung

prominence, shoulder; 4. <Mechan> load *(eines Nockens)*
Vorsprung *m* **am Gussteil** <Ker & Glas> riser *(verursacht durch Schmelze in der Entlüftungsöffnung der Form)*
vorspulen *v* <Fernseh> fast-forward
Vorspur *f* <Kfztech> toe-in *(Vorderräder)*
Vorstag *n* <Wassertrans> forestay
Vorsteckbolzen *m* 1. <Fertig> cotter bolt; 2. <Maschinen> forelock bolt
Vorstecker *m* <Fertig> forelock
Vorsteckkeil *n* <Fertig> splint pin
vorstehen *v* <Bau> project
Vorsteuerdruckkammer *f* <Lufttrans> pilot pressure chamber
Vorsteuerventil *n* <Fertig> servo valve *(Kunststoffinstallationen)*
Vorsteven *m* <Wassertrans> stem *(Bootsbau)*
Vorstevenbeschlag *m* <Wassertrans> stem head fitting *(Bootsbau)*
Vorstreichen *n* <Papier> precoating
Vorstreichmaschine *f* <Kunststoff> bar coater
Vorstrich *m* <Fertig> cogging pass; breaking-down pass *(Walzen)*
Vorstrom *m* <Lufttrans> inflow
Vorstufe *f* 1. <Druck> prepress; 2. <Funktech> front end *(Empfänger)*
Vorstufenbetrieb *m* <Druck> prepress plant; prepress shop
Vorteiler *m* 1. <Elektronik> scaler; 2. <Funktech> prescaler
Vortreiben *n* <Bau> forcing *(Tunnelbau)*
Vortrieb *m* <Kohlen> advance
Vortriebskraft *f* <Maschinen> propelling force
Vortriebsschild *m* <Bau> shield *(Tunnelbau)*
Vortrockenzylinder *m* <Papier> baby dryer, predryer
vortrocknen *v* <Fertig> precure *(Klebeverbindung)*
Vortrocknung *f* <Fertig> predrying
vorübergehend *adj* 1. <Bau, Kontroll, Phys> transient; 2. <Qual> temporary
vorübergehend nicht erreichbar *adj* <Telekom> temporarily unavailable *(Vermittlungsplatz)*
vorübergehende Abweichung *f* <Regelung> transient deviation
vorübergehende Belastung *f* <Bau> temporary load
vorübergehende Sollwertabweichung *f* <Regelung> transient deviation from desired set point
vorübergehende Zusammenschaltung *f* <Elektrotech> patching
vorübergehender Vorgang *m* <Fertig> transient
Vor- und Rückwärtssuche *f* <Künstl Int> bidirectional search
Vorvakuum *n* <Phys> prevacuum
vorverarbeiten *v* <Comp & DV> preprocess
vorverarbeitet *adj* <Elektronik> preprocessed
vorverarbeitete Lebensmittel *npl* <Lebensmittel> convenience food
Vorverarbeitung *f* <Elektronik> preprocessing
Vorverarbeitungsprozessor *m* <Telekom> front-end processor
Vorverbrennung *f* <Thermod> precombustion
Vorverbrennungskammer *f* <Kfztech, Lufttrans, Wassertrans> combustion prechamber
Vorverdichterdruckstutzen *m* <Lufttrans> pressure inlet *(Klimatisierung)*
Vorverdichtung *f* 1. <Kfztech> supercharging; 2. <Lufttrans> supercharge; 3. <Mechan> supercharger
Vorverstärker *m* <Aufnahme, Elektronik, Funktech, Phys, Telekom> preamplifier
vorverstärkt *adj* <Elektronik> preamp

Vorverstärkung *f* 1. <Elektronik> preamplification; 2. <Raumfahrt> pre-emphasis
Vorverstärkungsfaktor *m* <Raumfahrt> pre-emphasis improvement factor
Vorversuch *m* <Kohlen, Qual> pilot test
Vorverzerrung *f* 1. <Akustik> pre-emphasis, pre-equalization, predistortion; 2. <Aufnahme, Elektronik, Phys> predistortion, pre-emphasis; 3. <Fernseh, Funktech> pre-emphasis
Vorverzerrungstechnik *f* <Telekom> predistortion technique
vorvulkanisierter Latex *m* <Kunststoff> prevulcanized latex
Vorwahl *f* <Kfztech> preselection • **mit Vorwahl** <Fertig> preoptive
Vorwähler *m* 1. <Elektronik, Heiz & Kälte> preselector; 2. <Kfztech> preselector *(Getriebe)*; 3. <Telekom> preselector
Vorwahlgangschaltung *f* <Kfztech> preselection gear change *(Getriebe)*
Vorwählgerät *n* <Heiz & Kälte> preselector
Vorwahlkennziffer *f* <Telekom> prefix number
Vorwahlmessgerät *n* <Metrol> *(AE)* presetting gage, *(BE)* presetting gauge
Vorwahlschalter *m* <Gerät> preselection switch
Vorwählschalter *m* <Heiz & Kälte> preselector
Vorwahlzähler *m* <Gerät> batching counter, predetermining counter, preselection counter, preset counter
Vorwalze *f* 1. <Fertig> bloom roll, cogging-down roll, roughing roll; 2. <Maschinen> blooming roll, breaking-down roll
vorwalzen *v* 1. <Bau> rough-down; 2. <Fertig> rough-roll, rough; 3. <Kohlen> rough
Vorwalzen *n* <Fertig> rough rolling
Vorwalzer *m* <Kohlen> rougher
Vorwalzgerüst *n* <Fertig> cogging-down stand
Vorwalzwerk *n* 1. <Fertig> cogging mill; 2. <Maschinen> blooming mill
vorwärmen *v* 1. <Fertig> recuperate; 2. <Heiz & Kälte> preheat
Vorwärmen *n* <Eisenbahn, Kunststoff, Metall> preheating
Vorwärmer *m* 1. <Erdöl> preheater *(Raffinerietechnik)*; 2. <Fertig> economizer; 3. <Heiz & Kälte> economizer, recuperator; 4. <Kerntech, Kfztech> preheater *(Dieselmotor)*
vorwärts laufendes Blatt *n* <Lufttrans> advancing blade *(Luftschraube)*
Vorwärts... <Gerät, Maschinen> forward
Vorwärtsauslösung *f* <Telekom> forward release
Vorwärtsfehlerkorrektur *f* <Comp & DV, Telekom> forward error correction
Vorwärtsfließpressen *n* <Maschinen> direct extrusion, forward extrusion
Vorwärtsführung *f* <Comp & DV> feedforward control
Vorwärtsgeschwindigkeit *f* <Lufttrans> translation speed
Vorwärtskennlinie *f* <Elektronik> forward characteristic
Vorwärtskorrektur *f* <Telekom> feed-forward correction
Vorwärtsregelung *f* <Telekom> feedforward AGC, feedforward automatic gain control
Vorwärtsrichtung *f* <Elektriz> forward-conducting direction
Vorwärts-Rückwärtsbewegung *f* <Maschinen> back-and-forth motion
Vorwärts-Rückwärtszähler *m* 1. <Elektronik> up-down counter; 2. <Gerät> bidirectional counter, forward-backward counter, up-down counter
Vorwärtssperrzeit *f* <Elektriz> circuit off-state interval, hold-off period

Waagerechtbohrmaschine

Vorwärtsstart *m* <Lufttrans> forward takeoff *(Hubschraub)*
Vorwärtsstreuung *f* 1. <Aufnahme> forward scattering; 2. <Funktech> forward scatter
Vorwärtsstrom *m* <Elektriz> forward current
Vorwärtssuche *f* <Künstl Int> forward search
Vorwärts- und Rückwärtsbewegung *f* <Maschinen> backward-and-forward motion
Vorwärtsverkettung *f* <Künstl Int> forward chaining
Vorwärtsverstärker *m* <Elektronik> forward amplifier
Vorwärtsvorspannung *f* <Elektrotech> forward bias, forward voltage
Vorwärtswellenbeschleuniger *m* <Strahlphys> progressive wave accelerator
Vorwärtswiderstand *m* <Elektriz> forward resistance
Vorwärtszähler *m* <Gerät> count-up counter, up counter
Vorwärtszeichen *n* <Telekom> forward signal
Vorwärtszweig *m* 1. <Elektrotech> forward path; 2. <Regelung> forward path *(des Regelkreises)*
vorwaschen *v* <Textil> scour
Vorwaschen *n* <Textil> scouring
Vorwiderstand *m* 1. <Elektrotech> dropping resistor, voltage multiplier; 2. <Funktech> bias resistance; 3. <Gerät> additional resistor
vorwiegend zum Lesen geeigneter Speicher *m* <Comp & DV> read-mostly memory
Vorwuchten <Maschinen> initial balancing
Vorzeichen *n* <Comp & DV, Math> sign
Vorzeichenbit *n* <Comp & DV> sign bit
Vorzeichenziffer *f* <Comp & DV> sign digit
Vorzeichnung *f* <Akustik> key signature
vorzeitig beenden *v* <Comp & DV> abend, abort
vorzeitige Beendigung *f* <Comp & DV> abnormal termination
vorzeitliche Metallurgieforschung *f* <Metall> ancient metallurgy research
Vorzerkleinerung *f* 1. <Abfall> prior crushing; 2. <Kohlen> precrushing
vorziehen *v* <Fertig> precup
Vorziehen *n* <Fertig> precupping
Vorzimmeranlage *f* <Telekom> executive-secretary system, manager/secretary station
Vorzug *m* <Comp & DV, Fertig, Lebensmittel, Qual, Raumfahrt> advantage
Vorzugs-AQL-Werte *mpl* <Qual> preferred acceptable quality levels
Vorzugsbetrieb *m* <Comp & DV> privileged operation
Vorzugsliste *f* <Raumfahrt> preferential list
Vorzugsmilch *f* <Lebensmittel> certified milk
Vorzugsreihe *f* <Fertig> preferential range *(Kunststoffinstallationen)*
Vorzugswert *m* <Funktech> preference value
Vorzündung *f* <Kfztech> advanced ignition, premature ignition; advance, preignition *(Zündung)*
Votatoranlage *f* <Lebensmittel> votator *(für kontinuierliche Margarineherstellung)*
Voute *f* <Bau> haunch
Voutenbalken *m* <Bau> haunch beam
VPE *(vernetztes Polyethylen)* <Kunststoff> XPE *(cross-linked polyethylene)*
V-Rad *n* <Maschinen> V-gear
VRS-Verfahren *n* *(Verfestigungsverfahren für Sonderabfälle)* <Abfall> WPC-VRS process
VSAT-Station *f* <Funktech, Telekom> very-small aperture terminal, VSAT
VSAT-System *n* <Telekom> VAT system, very small aperture terminal system *(zur Satelliten-Datenübertragung)*
V-Schaltung *f* <Elektriz> V-connection
V-Scheibe *f* <Maschinen> V-pulley

VSP *(vertikales seismisches Profil)* <Erdöl> VSP *(vertical seismic profile)*
V-Stellung *f* <Lufttrans> dihedral
V-Stellung *f* **des Blattes** <Lufttrans> blade tilt *(Hubschrauber)*
VSWR *(Welligkeitsfaktor)* <Funktech> VSWR *(voltage standing wave ratio)*
VTOL-Flugzeug *n* *(senkrecht startendes und landendes Flugzeug)* <Lufttrans> VTOL aircraft *(vertical takeoff and landing aircraft)*
V-Tragflächenboot *n* <Wassertrans> VEE foil craft
Vulkanfiberscheibe *f* <Maschinen> *(AE)* vulcanized fiber disk, *(BE)* vulcanized fibre disc
Vulkanisation *f* <Kunststoff, Maschinen, Thermod> vulcanization
Vulkanisationsgeschwindigkeit *f* <Kunststoff> rate of cure
Vulkanisationsverzögerer *m* <Kunststoff> antiscorching agent, retarder
vulkanisch *adj* <Erdöl> volcanic *(Geologie)*
vulkanisieren *v* <Thermod> vulcanize; heat-cure *(im Ofen)*
vulkanisiert *adj* <Thermod> vulcanized
Vulkanisierung *f* <Maschinen, Thermod> vulcanization
Vulkanit *m* 1. <Elektriz> vulcanite; 2. <Nichtfoss Energ> extrusive rocks, igneous rocks
Vulkanschlot *m* <Nichtfoss Energ> conduit *(Geologisch)*
VU-Meter *n* <Gerät> VU-meter
VV-Vergaser *m* *(variabler Venturi-Vergaser)* <Kfztech> *(AE)* VV-carburetor, *(BE)* VV-carburettor
V-X-Motor *m* <Kfztech> v-x engine
V-Zahl *f* <Optik> V-number

W

W 1. <Chemie> *(Wolfram)* W *(tungsten)*; 2. <Elektriz, Elektrotech> *(elektrische Energie)* W *(electric energy)*; 3. <Elektriz, Elektrotech, Metrol> *(Watt)* W *(watt)*; 4. <Hydraul> *(Weber'sche Zahl)* W *(Weber number)*; 5. <Kerntech> *(durchschnittliche Energie)* W *(average energy)*; 6. <Optik> *(Abstrahlung)* W *(radiant emittance)*
WA *(Wertanalyse)* <Kfztech, Qual> VA *(value analysis)*
Waage *f* 1. <Elektriz, Labor> balance; 2. <Maschinen> scales; 3. <Metrol> balance, scale; 4. <Papier> balance *(Vorrichtung)*; 5. <Phys> balance • **in Waage** <Fertig> level
Waagebalken *m* 1. <Gerät> balance beam; 2. <Lufttrans> balance arm; 3. <Maschinen> scale beam; 4. <Metrol> balance beam, beam
Waagebalkenachse *f* <Kfztech> pivot axle *(Anhänger)*
Waagebalkenaufleger *m* <Metrol> knife edge
Waagebalkenaufleger *mpl* **auf Achatplanlager** <Metrol> knife edges of balance beam resting in agate
Waagebürste *f* <Labor> balance brush
Waagelagerung *f* <Metrol> bearings
Waagengenauigkeit *f* <Metrol> accuracy of a balance
waagerecht *adj* <Bau, Fertig, Geom, Maschinen> horizontal
waagerechter Seitenschub *m* <Bau> horizontal thrust
Waagerecht... <Maschinen> horizontal
Waagerechtbohrmaschine *f* <Maschinen> horizontal boring machine, horizontal drilling machine

Waagerechtbohr- und Fräsmaschine

Waagerechtbohr- und Fräsmaschine f <Maschinen> horizontal drilling, boring and milling machine
Waagerechtbohrwerk n <Fertig> horizontal boring and milling machine
Waagerechte f <Geom> horizontal
Waagerechtfräsmaschine f <Maschinen> horizontal milling machine
Waagerechtfrässpindel f <Maschinen> horizontal milling spindle
Waagerechtjustierschraube f <Maschinen> (AE) leveling screw, (BE) levelling screw
Waagerechträummaschine f <Maschinen> horizontal broaching machine
Waagerechtschleifscheibe f <Ker & Glas> (BE) horizontal grinding disc, (AE) horizontal grinding disk
Waagerechtstoßen n <Maschinen> shaping
Waagerechtstoßmaschine f <Maschinen> shaping machine, shaping planer
Waagerechtstoßmaschine f **mit zwei Supporten** <Maschinen> double-headed shaping machine
Waagerechtziehverfahren n <Ker & Glas> horizontal drawing process
Waagrecht... <Maschinen> siehe Waagerecht...
Waagschale f 1. <Labor> scale pan, weighing boat, weighing dish; 2. <Maschinen> scale, scale pan; 3. <Metrol> scale
Wabe f <Bau, Lufttrans> honeycomb
Wabenbauweise f <Bau, Lufttrans> honeycomb construction
wabenförmige Rissbildung f <Bau> honeycombing
Wabengebilde n **für den Güterschutz** <Verpack> honeycomb protection system (Hohlraumfüller aus Papier und Pappe)
Wabengitter n <Lufttrans> honeycomb filler, honeycomb grill (aerodynamisch)
Wabenkonstruktion f <Verpack> honeycomb structure
Wabenkühler m <Kfztech> honeycomb radiator
Wabenmaterial n <Verpack> honeycomb material
Wabenstruktur f <Bau> honeycomb structure
Wabenwicklung f <Elektrotech> honeycomb winding
Wachempfänger m <Funktech> watchkeeping receiver
Wachkanal m <Comp & DV> guard channel
Wachs n 1. <Erdöl> wax (Bestandteil der Erdöle); 2. <Ker & Glas, Maschinen, Textil, Verpack> wax
Wachsamkeit f 1. <Eisenbahn> vigilance; 2. <Ergon> alertness, vigilance; 3. <Qual> vigilance
Wachsamkeitskontrolle f <Eisenbahn, Qual> vigilance control
Wachsamkeitstaste f <Eisenbahn, Qual> vigilance device
Wachsausschmelzguss m <Ker & Glas> lost wax
Wachsen n 1. <Fertig> growth (Guraguss); 2. <Lebensmittel> waxing
wachsend adv <Math> increasing
Wachsgießformen fpl <Maschinen> (AE) wax investment molds, (BE) wax investment moulds
Wachspapier n <Verpack> impregnated paper, wax paper
Wachsplatte f <Aufnahme> wax master
Wachsschutzschicht f <Ker & Glas> wax resist
Wachstuch n <Textil, Verpack> oilcloth
Wachstum n <Kerntech, Metall> growth
Wachstumsmodell n <Metall> growth pattern
Wachstumsspirale f <Metall> growth spiral
Wachstumsstufe f <Metall> growth step
Wachstumszwilling m <Metall> growth twin
wackeln v <Raumfahrt> wobble (in der Drehachse)
Wackeln n <Raumfahrt> wobble
Wade f <Wassertrans> seine net (Fischerei)
Wafer m <Comp & DV, Elektronik, Elektrotech> wafer
Wafer-Ausbeute m <Elektronik> wafer yield
Wafer-Integration f (WS-Integration) <Comp & DV, Elektronik> wafer scale integration (wsi)
Waferschalter m <Elektrotech> wafer switch
Wägefläschen n <Labor> density bottle; pycnometer (Dichtemessung)
Wägeglas n <Chemie, Labor> weighing bottle
Wägemaschine f 1. <Fertig> weighing machine; 2. <Verpack> checkweighing machine
Wagen m 1. <Druck> carriage; 2. <Eisenbahn> (AE) car, (BE) carriage, (BE) coach; 3. <Kfztech> car, passenger car; 4. <Mechan> car
Wagen m **A mit Motor** <Kfztech> carriage A containing the motor
Wagen m **mit Hecktür** <Kfztech> hatchback car, hatchback model
Wagen m **mit Pendelaufhängung** <Kfztech> car with pendulum suspension
Wagenablauf m <Eisenbahn> wagon humping
Wagenausbesserungswerkstatt f <Kfztech> car shop
Wagendach n <Eisenbahn> roof dome
Wagendeck n <Wassertrans> car deck
Wagenguss m <Fertig> bogie casting
Wagenheber m 1. <Kfztech> jack (Werkzeug); 2. <Mechan> jack
Wagenheberansatzpunkt m <Kfztech> jacking point
Wagenheberauflage f <Kfztech> jacking pad
Wagenhebewerk n <Eisenbahn> wagon hoist, wagon lift
Wagenherdofen m <Fertig> bogie furnace
Wagenkasten m 1. <Eisenbahn> body; 2. <Trans> box (Fahrzeug)
Wagenkipper m 1. <Bau> tip; 2. <Eisenbahn> dumper
Wagenladung f 1. <Eisenbahn> (AE) car load, (BE) wagon load; 2. <Trans> cart load
Wagenpark m <Kfztech> fleet
Wagenrad n <Kfztech> (BE) lorry wheel, (AE) truck wheel
Wagenrücklauf m <Telekom> carriage return
Wagenschuppen m <Eisenbahn> wagon shed
Wagenverfügung f <Eisenbahn> car distribution
Wagenzug m <Eisenbahn> train set
Waggonkippanlage f <Eisenbahn> car dumper
Wahl f <Telekom> (AE) dialing, (BE) dialling
Wahl f **aus dem PC** <Telekom> dialling from the PC
Wähl... <Comp & DV, Telekom> (AE) dialing, (BE) dialling
Wählanschluss m <Telekom> dial-up port
Wahlaufforderungszeichen n <Telekom> proceed-to-select signal
wählbar adj 1. <Comp & DV> random; 2. <Elektronik, Elektrotech> selectable
wählbare Leiterbahn f <Elektronik> chooseable wiring (Wafer)
wählbare Verarbeitung f <Comp & DV> random processing
wählbarer Impuls m <Elektronik> random pulse
Wählcode m <Telekom> (AE) dialing code, (BE) dialling code
Wahldatei f <Comp & DV> optional file
wählen v 1. <Comp & DV> poll; 2. <Telekom> dial, select
Wählen n <Telekom> (AE) dialing, (BE) dialling
Wählen n **am Handapparat** <Telekom> handset dialling (beim Schnurlostelefon)
Wählen n **bei aufliegendem Hörer** <Telekom> (AE) on-hook dialing, (BE) on-hook dialling
Wahlendezeichen n <Telekom> end-of-dialing signal
Wähler m <Comp & DV, Elektrotech, Telekom> selector
Wählfehler m <Telekom> (AE) dialing error, (BE) dialling error
wahlfrei adj <Comp & DV> optional, random
wahlfreie Verarbeitung f <Comp & DV> random processing

wahlfreier Zugriff m <Comp & DV> random access
wahlfreies Nutzerleistungsmerkmal n <Telekom> optional user facility
Wählhebel m <Kfztech> selector *(Automatikgetriebe)*
Wählimpulsfolge f <Telekom> dial impulse sequence
Wahlkomfort m <Telekom> enhanced-convenience dialling
Wählleitung f <Telekom> switched circuit
Wahlmöglichkeit f **beim Einstellen** <Comp & DV> set-up option
Wählnetz n <Comp & DV, Telekom> switched network
Wählrelais n <Elektrotech> selector relay
Wählschalter m 1. <Elektrotech> selector switch; 2. <Gerät, Telekom> selector
Wählscheibe f <Telekom> dial
Wählservice m <Comp & DV> dial service
Wählstandort m <Comp & DV> dial location
Wählsterneinrichtung f <Telekom> line concentrator; concentrator *(Fernsprechnetz)*
Wahlstufe f <Telekom> selection stage, switching stage
Wähltonverzug m <Telekom> dial tone delay
Wählvermittlung f <Telekom> automatic switching
Wählvermittlungsstelle f <Telekom> automatic switchboard
wahlweise adj <Comp & DV> optional
wahlweise anwendbare Prüfmethode f <Optik> alternative test method
wahlweiser Halt m <Comp & DV> optional stop
Wahlwiederholtaste f <Telekom> automatic call repetition key, automatic redial key, call repeat key, repetition key
Wahlwiederholung f <Telekom> last number recall, last number redial; number redial *(Merkmal)*
Wahlwiederholung f **der gespeicherten Rufnummer** <Telekom> stored number redial
Wahlwiederholung f **der letzten Rufnummer** <Telekom> last number redial
Wahlwort n <Comp & DV> optional wor
Wählzeit f <Telekom> *(AE)* dialing period, *(BE)* dialling period
wahr adj <Comp & DV, Elektriz, Kerntech, Lufttrans, Math, Metall, Raumfahrt> true
wahre Anomalie f <Raumfahrt> true anomaly
wahre Beanspruchung f <Metall> true stress
wahre Bruchspannung f <Metall> true fracture stress
wahre Dehnung f <Metall> true strain
wahre Dichte f <Kerntech> true density
wahre Eigengeschwindigkeit f <Lufttrans> TAS
wahre Fluggeschwindigkeit f <Lufttrans> TAS, true air speed
wahre Ladungen fpl <Phys> conduction charges
wahrer Längenkreis m <Bau> true meridian
wahrer Wind m <Wassertrans> true wind *(Navigation)*
Wahrheit f <Comp & DV, Math> truth
Wahrheitserhaltung f <Künstl Int> consistency maintenance, truth maintenance
Wahrheitstabelle f 1. <Comp & DV> truth table; 2. <Math> truth table *(Aussagenlogik)*
Wahrheitswert m <Comp & DV> logical value *(COBOL)*
Wahrnehmung f 1. <Ergon> cognition, perception; 2. <Künstl Int> perception
Wahrnehmungsgeschwindigkeit f <Ergon> speed of perception
wahrscheinliche Reserven fpl <Erdöl> probable reserves *(Lagerstätten)*
Wahrscheinlichkeit f 1. <Comp & DV, Ergon, Künstl Int, Phys, Qual, Telekom> probability; 2. <Math> likelihood, probability

Wahrscheinlichkeit f **der Wartezeitüberschreitung** <Telekom> probability of excess delay
Wahrscheinlichkeitsdichte f 1. <Math> probability density function, pdf; 2. <Phys> probability density
Wahrscheinlichkeitsfunktion f 1. <Math> probability function *(für diskrete Zufallsvariablen)*; 2. <Qual> probability function
Wahrscheinlichkeitsrechnung f <Künstl Int, Math> probability calculus
wahrscheinlichkeitstheoretisch adj <Künstl Int, Math> probabilistic
Wahrscheinlichkeitstheorie f <Math> probability theory
Wahrscheinlichkeitsverfahren n <Bau> probabilistic approach
Wahrscheinlichkeitsverteilung f 1. <Comp & DV> probability distribution; 2. <Math> probability distribution; 3. <Qual> probability density, probability distribution
wahrscheinlichste Geschwindigkeit f (û) <Phys> most probable speed (û)
Walfang m <Wassertrans> whaling
Walken n <Textil> fulling; milling *(Leder)*
Walm m <Bau> hip
Walmdach n <Bau> hip roof, hipped roof
Walmziegel m <Bau> hip tile
Walz... <Fertig, Maschinen> rolling
Wälz... <Fertig, Maschinen> rolling
Wälzachse f <Fertig> rolling axis
Wälzbahn f <Maschinen> pitch line
Walzbarren m <Fertig> slab
Walzbart m <Mechan> burr
Walzblock m <Metall> bloom
Walzdoppelung f <Fertig> lamination
Walze f 1. <Bau> shaft; 2. <Comp & DV> drum, platen; 3. <Druck> cylinder, roller, platen; 4. <Fertig> cylinder, drum, roll; 5. <Hydraul> drum *(Dampfdruckindikator)*; 6. <Kohlen> drum, roller; 7. <Kunststoff> cylinder, mill, roller, roll; 8. <Lebensmittel> drum, roller; 9. <Maschinen> platen, roll; 10. <Mechan> drum, roll, roller; 11. <Metall> roller; 12. <Papier> roll, roller, 13. <Textil> roller
walzen v 1. <Kohlen> mill; 2. <Maschinen> roll
Walzen n <Bau, Ker & Glas, Maschinen, Metall, Papier> rolling
Walzenanlasser m <Elektriz> drum starter
Walzenanpressdruck m <Papier> nip pressure
Walzenbeschichten n <Kunststoff> roll coating
Walzenbiegemaschine f <Fertig> bending roll
Walzenbrecher m <Kohlen> roller crusher
Walzendrehmaschine f <Maschinen> roll lathe, roll-turning lathe
Walzendruck m 1. <Druck> cylinder printing; 2. <Textil> roller printing
Walzendruckmaschine f 1. <Druck> cylinder printing machine; 2. <Verpack> rotary printing press
Walzeneinstellung f <Textil> roller setting
Walzenfräser m 1. <Fertig> cylindrical cutter; 2. <Maschinen> plain-milling cutter
Walzengusseisen n <Metall> chilled roll iron
Walzenhals m <Papier> neck
Walzenkleeblattzapfen m <Fertig> roll wobbler
Walzenkopf m <Papier> roll head
Walzenmarkierung f <Ker & Glas> roller mark
Walzenmühle f <Chemtech, Mechan> roller mill
Walzenpuffer m <Ker & Glas> roller bump
Walzenringmühle f <Kohlen> ring-roll crusher
Walzensatz m <Papier> set of rolls
Walzenschalter m <Elektriz> drum controller, drum switch
Walzenscheider m <Kohlen> drum cobber
Walzenschüsselmühle f <Kohlen> bowl mill crusher
Walzenschütze f <Nichtfoss Energ> drum gate

Walzenspalt *m* <Papier> nip
Walzenständer *m* 1. <Fertig> holster, roll housing; 2. <Maschinen> bearer, standard
Walzenstirnfräser *m* 1. <Fertig> end-face mill; 2. <Maschinen> shell-end mill
Walzenstoffauflauf *m* <Papier> roll headbox
Walzenstraße *f* <Maschinen> train of rolls
Walzenstreichverfahren *n* <Papier> roller coating
Walzenstuhl *m* <Lebensmittel> roller mill *(mit Walzen als Mahlkörper)*
Walzentasche *f* <Papier> roll pocket
Walzenträger *m* <Papier> roller beam
Walzentrockner *m* <Kohlen, Lebensmittel> drum drier, drum dryer
Walzentrog *m* <Ker & Glas> roller tray
Walzenvorschub *m* <Maschinen> roll feed
Walzenwehr *n* <Wasserversorg> roller weir
Walzfehler *m* <Fertig> cobble
Wälzfehler *m* <Fertig> overall variation, total composite error *(Getriebelehre)*
Wälzfläche *f* <Maschinen> pitch surface
Wälzfräs... <Fertig, Maschinen> generating
walzfräsen *v* <Fertig> slab-mill
Walzfräsen *n* 1. <Fertig> peripheral milling; 2. <Maschinen> roll milling
wälzfräsen *v* <Fertig, Maschinen> hob
Wälzfräsen *n* 1. <Fertig> gear hobbing, generating, hobbing; 2. <Maschinen> hobbing
Wälzfräsen *n* **im Gleichlauf** <Fertig> climb hobbing
Walzfräser *m* <Fertig> slab milling cutter
Wälzfräser *m* <Fertig, Maschinen> gear hob, generating cutter, hob, hobbing cutter
Wälzfräsmaschine *f* 1. <Fertig> generator; 2. <Maschinen> gear hobber, gear hobbing machine, hob, hobber, hobbing machine
Wälzfrässchichten *n* <Fertig> finish hob
wälzgefräst *adj* <Fertig> hobbed
wälzgelagert *adj* <Fertig> anti-ager friction-bearing
Wälzgelenk *n* <Maschinen> rolling contact joint
walzgeschmiedet *adj* <Fertig> roll-forged
Walzglas *n* <Ker & Glas> rolled glass
Walzglattglas *n* <Ker & Glas> plain-rolled glass
Walzgrat *m* <Fertig> flash; cold shut *(Walzen)*
walzhart *adj* <Fertig> as-rolled
Wälzhobelmaschine *f* <Maschinen> gear planer
Wälzkolbenzähler *m* <Gerät> oval gear meter, rod piston meter/element
Wälzkörper *m* <Maschinen> rolling meter/element
Wälzkreis *m* <Maschinen> circle of contact, pitch circle, pitch line, rolling circle
Wälzlager *n* 1. <Kfztech> roller bearing; 2. <Maschinen> antifriction bearing, rolling bearing, rolling contact bearing
Walzlegierung *f* <Mechan, Metall> rolled alloy
Walzmarkierung *f* <Ker & Glas> roll mark
Walzmaschine *f* <Maschinen> rolling machine
Walznaht *f* <Fertig> roller burr
Walzplattierdeckmetall *n* <Fertig> liner
Walzprofil *n* <Wassertrans> rolled section *(Schiffbau)*
Walzprofilieren *n* <Maschinen> profile rolling, shape rolling
Wälzpunkt *m* 1. <Fertig> pitch point *(Zahnrad)*; 2. <Maschinen> pitch point
Walzpuppe *f* <Fertig> billet *(Extruder)*
Wälzradius *m* <Maschinen> pitch radius
Wälzreibung *f* <Maschinen> sliding and rolling friction
Walzrichtung trennen *v/in* <Fertig> fishmouth
Wälzschälen *n* <Fertig> skiving
Wälzschleifen *n* <Fertig, Maschinen> grinding-generating

Wälzschleifmaschine *f* <Maschinen> gear-grinding machine
Walzschmieden *n* <Maschinen> roll forging
Walzsinter *m* <Fertig> mill scale
Walzstirnfräsen *n* <Maschinen> shell-end milling
Wälzstoßmaschine *f* <Maschinen> shaper
Walzstraße *f* <Fertig> roll line, roll train, rolling mill train
Wälztrommel *f* <Fertig> cradle *(Spanung)*
Walzverfahren *n* <Ker & Glas> rolling process
Wälzverfahren *n* <Fertig> generative process • **im Wälzverfahren herstellen** <Fertig> generate
Wälzverzahnen *n* <Maschinen> gear generating, generating
Walzwerk *n* <Fertig> rolling mill, section mill
Walzwerk *n* **für Massenfertigung** <Fertig> merchant mill
Wälzwinkel *m* <Maschinen> rolling angle
Walzzunder *m* <Fertig> mill scale, roll coating
WAN 1. <Comp & DV> WAN *(wide area network)*; 2. <Telekom> *(Weitverkehrsnetz)* WAN *(wide area network)*
Wand *f* 1. <Bau, Elektriz, Fertig, Kohlen, Labor> wall; 2. <Lufttrans> web *(eines Holms)*; 3. <Maschinen> wall; 4. <Strömphys> wall *(eines Kanals oder Rohres)*
Wandanschlussleiste *f* <Bau> fillet
Wandarm *m* <Mechan> angle bracket
wandartiger Träger *m* <Bau> deep beam
Wandbaustoffe *mpl* <Bau> walling
Wanddicke *f* <Bau, Maschinen> wall thickness
Wanddickenmessung *f* <Kerntech> *(AE)* wall thickness gaging, *(BE)* wall thickness gauging
Wanddurchführung *f* 1. <Bau> wall duct; 2. <Fertig> wall inlet fitting *(Kunststoffinstallationen)*
Wandeffekt *m* <Kohlen> wall effect
Wandeinfluss *m* <Strömphys> wall effect
Wandeinführungsisolator *m* <Elektrotech> wall bushing insulator
Wandeinsteckholz *n* <Bau> needle
Wandel *m* <Akustik> alteration
Wandelflugzeug *n* <Lufttrans> compound helicopter, convertiplane
Wander... <Bau, Telekom> *(AE)* traveling, *(BE)* travelling
Wanderfeld *n* <Elektronik> *(AE)* traveling field, *(BE)* travelling field
Wanderfeldlinearmotor *m* <Trans> *(AE)* traveling field motor, *(BE)* travelling field motor
Wanderfeldmagnetfeldröhre *f* <Elektronik> *(AE)* travelling wave magnetron, *(BE)* travelling wave magnetron
Wanderfeldmagnetron *n* 1. <Elektronik> multicavity magnetron; 2. <Funktech> travelling wave magnetron
Wanderfeldmaser *m* <Elektronik> TWM, *(AE)* traveling wave maser, *(BE)* travelling wave maser
Wanderfeldröhre *f* <Funktech, Telekom> TWT, *(AE)* traveling wave tube, *(BE)* travelling wave tube
Wanderfeldröhre *f* **für X-Band** <Elektronik> *(AE)* X-band traveling wave tube, *(BE)* X-band travelling wave tube
Wanderfeldröhrenverstärker *m* <Telekom> *(AE)* travelling wave tube amplifier, *(BE)* travelling wave tube amplifier, TWTA
Wanderfeldverstärker *m* 1. <Aufnahme> *(AE)* traveling wave acoustic amplifier, *(BE)* travelling wave acoustic amplifier; 2. <Telekom> *(AE)* traveling wave amplifier, *(BE)* travelling wave amplifier
Wanderkontrolle *f* <Qual> patrol inspection
Wanderlast *f* <Bau> moving load
Wandern *n* <Telekom> *(AE)* traveling, *(BE)* travelling; wander *(Pulscodemodulationssignal)*
wandernder Lichtpunkt *m* <Fernseh> flying spot
Wanderprüfung *f* <Qual> patrol inspection

Wanderrost m 1. <Abfall> (AE) traveling grate, (BE) travelling grate; 2. <Maschinen> travelling grate
Wanderung f <Kunststoff> migration
wanderungsbeständiger Weichmacher m <Kunststoff> nonmigratory plasticizer
Wanderwelle f <Akustik, Elektronik, Phys, Telekom> (AE) traveling wave, (BE) travelling wave
Wanderwellenantenne f 1. <Fernseh> (AE) traveling wave aerial, (BE) travelling wave aerial; 2. <Telekom> (AE) traveling wave antenna, (BE) travelling wave antenna
Wanderwellenleiter m <Telekom> (AE) traveling waveguide, (BE) travelling waveguide
Wanderwellenmotor m <Elektriz> (AE) traveling wave motor, (BE) travelling wave motor
Wanderwellenröhre f <Raumfahrt> TWT, (AE) traveling wave tube, (BE) travelling wave tube
Wanderwellenröhrenverstärker m (WWRV) <Elektronik, Raumfahrt> (AE) traveling wave tube amplifier, (BE) travelling wave tube amplifier, TWTA (Weltraumfunk)
Wandhalterung f <Bau> wall holdfast
Wandheizung f <Thermod> panel heater (Gerät); panel heating (Verfahren)
Wandinstrument n <Gerät> wall-mounted instrument
Wandkonsole f <Bau, Telekom> wall bracket
Wandkran m <Bau> wall crane
Wandlampe f <Elektriz> wall lamp
Wandler m 1. <Akustik> transducer; 2. <Elektriz, Elektronik, Elektrotech> converter, sensing element; 3. <Gerät> converter, transducer; 4. <Kfztech> converter; 5. <Kontroll> transducer; 6. <Maschinen> converter; 7. <Phys> converter, transducer; 8. <Telekom> transducer
Wandler m **zur Erzeugung negativer Widerstandskennlinien** <Elektrotech> negative impedance converter
Wandlerempfindlichkeit f <Akustik> transducer sensitivity
Wandlerverlustfaktor m <Akustik> transducer loss factor
Wandlung f **des optischen Signals** <Elektrotech> optical signal conversion
Wandnetzstecker m <Elektriz> wall outlet
Wandreibung f 1. <Maschinen> wall friction; 2. <Mechan> skin friction
Wandschale f <Bau> leaf
Wandschrank m <Labor> wall cupboard
Wandstärke f <Bau, Maschinen> wall thickness
Wandsteckdose f <Elektriz> wall outlet, wall socket
Wandsystem n <Bau> walling
Wandtafel f <Künstl Int> blackboard
Wandtäfelung f <Bau> wainscotting
Wandtemperatur f <Strömphys> wall temperature
Wandung f <Fertig, Maschinen> wall
Wandwange f <Bau> wall string
Wange f 1. <Bau> cheek, side plate; 2. <Fertig> cheek; bearer, shears (Bett); 3. <Kfztech> web (Kurbelwelle); 4. <Maschinen> cheek, shears, web
Wange f **mit eingestemmten Stufen** <Bau> housed string
Wankelmotor m <Kfztech, Maschinen, Thermod> Wankel engine
Wanne f 1. <Bau> sump pan; 2. <Ker & Glas> bath (Glaswannenofen); trough (bei der Herstellung von Behälterglas); 3. <Labor> trough; 4. <Maschinen> pan, tray
Wannenatmosphäre f <Ker & Glas> bath atmosphere
Wannenauskleidungsglas n <Ker & Glas> tank lining glass
Wannendichtung f <Bau> tanking
Wannenofen m <Ker & Glas> tank, tank furnace
Wannenofenmund m <Ker & Glas> tank neck
Wannenposition f <Fertig> downhand position (Schweißen)
Wannenstein m <Ker & Glas> tank block
Want n <Wassertrans> shroud (Takelage)
Wantspanner m <Wassertrans> rigging screw (Tauwerk)
WAP (Protokoll für drahtlose Anwendungen) <Funktech, Telekom> WAP (wireless application protocol)
WAP-Browser m <Funktech, Telekom> WAP browser
Ward-Leonard-Satz m <Elektrotech> Ward-Leonard set
Ware f <Patent, Qual> commodity
Waren fpl **unter Zollverschluss** <Wassertrans> bonded goods
Warenbank f <Textil> plaiting-down platform
Wareneingangsprüfung f <Qual> incoming inspection, receiving inspection
Warenfluss m <Verpack> flow of goods
Warengruppe f <Verpack> commodity group
Warenstrang m <Textil> rope
Warentransport m <Verpack> handling, handling of goods
Warentransportsystem n <Verpack> conveyor handling system
Warenzeichen n <Patent> trademark
Warfarin n <Chemie> warfarin
warm adj <Thermod> warm
Warm... <Kunststoff, Thermod> warm
Wärm... <Thermod> warm
warmabbindender Kleber m <Kunststoff> hot-setting adhesive
warmaushärten v <Fertig> age artificially, age with increased temperature
Warmbadhärten n <Fertig> austempering, marquenching, martempering, stepped hardening
Warmbandwalzwerk n <Metall> hot-strip mill
warme Zündkerze f <Kfztech> hot spark plug
Wärme f <Heiz & Kälte, Kerntech, Papier, Phys, Textil, Thermod> heat • **durch Wärme dehnbar** <Kohlen> dilatable • **in Wärme aushärtend** <Mechan, Kunststoff> thermosetting • **Wärme abstrahlen** <Bau> radiate
Wärme... <Thermod> caloric, thermal
Wärmeabfall m <Bau, Nichtfoss Energ, Thermod> heat drop
Wärmeabfuhr f 1. <Heiz & Kälte> heat rejection, heat removal; 2. <Kerntech> heat rejection rate (in Watt pro Stunde); 3. <Thermod> heat emission
Wärmeabführleistung f <Heiz & Kälte> heat-removal capacity
Wärmeabführvermögen n <Bau, Heiz & Kälte, Qual> heat-removal property
Wärmeabgabe f 1. <Heiz & Kälte> heat emission, heat release; 2. <Kerntech> heat release; 3. <Thermod> heat emission
Wärmeableiter m <Funktech, Telekom> dissipator
Wärmeableitung f 1. <Funktech> heat dissipation; 2. <Heiz & Kälte> heat dissipation, heat removal
Wärmeableitungsgeschwindigkeit f <Elektriz> heat dissipation
Wärmeabschirmung f <Kerntech> thermal shield
wärmeabsorbierend adj <Nichtfoss Energ, Thermod> heat-absorbing
Wärmeabsorption f <Nichtfoss Energ, Thermod> heat absorption
Wärmeabstrahlung f <Heiz & Kälte, Nichtfoss Energ> heat radiation
Wärmeaktivierung f <Metall> thermal activation
Wärmealterung f <Kunststoff> (BE) heat ageing, (AE) heat aging
Wärmeanstieg m <Thermod> heat rise
Wärmeäquivalent n <Thermod> thermal equivalent

Wärmeätzen

Wärmeätzen n <Metall> thermal etching
Wärmeaufnahme f <Heiz & Kälte> heat absorption
Wärmeausbreitung f <Heiz & Kälte> heat propagation
Wärmeausbreitungsvermögen n <Phys> thermal diffusivity
Wärmeausdehnung f 1. <Bau> heat expansion; 2. <Gerät> heat dilatation; 3. <Labor> heat expansion, thermal expansion; 4. <Maschinen, Thermod> heat dilatation, heat expansion, thermal expansion; 5. <Nichtfoss Energ> heat expansion
Wärmeausdehnungskoeffizient m <Maschinen, Thermod> thermal expansion coefficient
Wärmeausdehnungssonde f <Gerät> thermal expansion measuring element
Wärmeausgleich m <Bau, Nichtfoss Energ, Thermod> heat compensation
Wärmeausstoß m 1. <Phys> calorific output; 2. <Thermod> calorific output, heat output
Wärmeaustausch m 1. <Bau, Ergon, Fertig, Heiz & Kälte> heat exchange; 2. <Kontroll> thermal exchange; 3. <Maschinen, Nichtfoss Energ, Thermod> heat exchange
Wärmeaustauscher m 1. <Heiz & Kälte> heat exchanger, recuperator; 2. <Kerntech, Lebensmittel, Mechan, Nichtfoss Energ> heat exchanger; 3. <Thermod> heat economizer, heat exchanger
Wärmeaustauscher m mit Spiralwindungen <Kerntech> helical coil-type heat exchanger
Wärmeaustauschmedium n <Thermod> heat-exchanging medium
Wärmeaustauschrohr n <Maschinen> heat exchanger tube
Wärmebad n <Phys> heat reservoir
Wärmebeanspruchung f <Thermod> thermal stress
Wärmebedarf m <Bau, Heiz & Kälte, Nichtfoss Energ, Thermod> heat demand
wärmebehandeln v <Thermod> heat-treat
Wärmebehandlung f <Kohlen, Maschinen, Thermod> heat treatment
Wärmebelästigung f <Thermod, Umweltschmutz> thermal pollution
Wärmebelastung f 1. <Bau> heat load; 2. <Qual> heat load, thermal load; 3. <Thermod> heat load, rate of heat release, thermal pollution; 4. <Umweltschmutz> thermal load, thermal pollution
wärmebeständig adj 1. <Erdöl> thermostable; 2. <Heiz & Kälte, Phys> heat-resistant; 3. <Thermod> heat-resisting, heat-resistant, heatproof, thermostable; 4. <Verpack> heatproof
Wärmebeständigkeit f 1. <Kunststoff> heat resistance; 2. <Phys> thermal resistivity; 3. <Qual, Thermod> heat endurance, heat resistance, heat resisting, heat-resisting quality, resistance to heat, thermal resistivity, thermal stability
Wärmebewegung f <Metall> thermal agitation
Wärmebilanz f 1. <Ergon, Heiz & Kälte, Nichtfoss Energ> heat balance; 2. <Thermod> thermal balance
Wärmebild n <Thermod> heat image
Wärmebildröhre f <Elektronik> thermal-imaging tube
Wärmeblitz m <Kerntech> thermal flash
Wärmebrücke f <Heiz & Kälte, Thermod> heat bridge
Wärmedamm m <Kfztech> heat dam
wärmedämmend adj <Thermod> heat-insulating
Wärmedämmfähigkeit f 1. <Bau> heat insulation effectiveness; 2. <Thermod> heat insulation effectiveness, heat insulation power
Wärmedämmfaktor m <Phys, Qual> coefficient of thermal insulation
Wärmedämmung f 1. <Bau> thermal insulation, thermal lagging; 2. <Heiz & Kälte> thermal insulation; 3. <Thermod> heat insulation; thermal lagging (Material); thermal insulation (Verfahren); 4. <Verpack> heat insulation
Wärmedämmzahl f <Bau, Qual, Thermod> heat insulation factor
Wärmedehnung f 1. <Kohlen> dilatancy; 2. <Thermod> thermal expansion
Wärmedehnungsfuge f <Thermod> thermal expansion joint
Wärmedichte f <Bau, Kohlen, Nichtfoss Energ, Thermod> heat density
Wärmediffusion f <Thermod> thermodiffusion
Wärmedissipation f <Thermod> heat dissipation
Wärmedunst m <Thermod> heat haze
Wärmedurchbiegungstemperatur f <Werkprüf> heat deflection temperature (Kunststoffe)
Wärmedurchgang m <Heiz & Kälte, Thermod> heat transition
Wärmedurchgangszahl f <Heiz & Kälte> heat transition coefficient, thermal transmittance
Wärmedurchlass m <Thermod> heat carrying, heat conductivity
wärmedurchlässige Wand f <Phys> diathermal wall
Wärmedurchsatz m <Heiz & Kälte, Thermod> (BE) heat throughput, (AE) heat thruput
Wärmeeinflussbereich m <Thermod> heat-affected zone
Wärmeeinflusszone f <Thermod> heat-affected zone
Wärmeeinheit f <Thermod> heat unit, thermal unit
wärmeempfindlich adj <Kunststoff, Phys, Thermod> heat-sensitive
wärmeempfindliche Farbe f <Thermod> heat-sensitive paint
wärmeempfindliches Material n <Verpack> heat-sensitive material
Wärmeempfindlichkeit f <Kunststoff> heat sensitivity
Wärmeenergie f 1. <Heiz & Kälte> thermal energy; 2. <Nichtfoss Energ> heat energy; 3. <Thermod> heat energy, thermal emissivity, thermal energy
Wärmeenergiespeichersystem n <Thermod> thermal energy storage system
Wärmeentbindung f <Heiz & Kälte> heat release
Wärmeentwicklung f 1. <Bau, Nichtfoss Energ> heat build-up; 2. <Thermod> development of heat, heat build-up, heat development, heat generation
Wärmeermüdung f <Thermod> thermal fatigue
wärmeerzeugend adj <Nichtfoss Energ, Thermod> heat-generating
Wärmeerzeuger m <Heiz & Kälte, Thermod> heat generator
Wärmeerzeugung f <Thermod> heat generation
Wärmefalle f <Kerntech, Nichtfoss Energ> heat trap
wärmefest adj <Verpack> heat-resistant
Wärmefestigkeit f <Kunststoff> heat resistance
Wärmefestigkeitsgrenze f <Fertig> heat distortion point
Wärmefluss m <Heiz & Kälte, Labor, Thermod> heat flow
Wärmeflussbild n <Thermod> heat balance chart, heat balance diagram
Wärmeflussdiagramm n <Thermod> heat flow diagram
Wärmeflussmessgerät n <Gerät> heat flow meter
Wärmeformänderung f <Kunststoff> thermal deformation
Wärmeformbeständigkeit f <Kunststoff> heat distortion temperature
Wärmefortleitung f <Heiz & Kälte> thermal conduction
wärmefreisetzend adj <Heiz & Kälte> heat-released
Wärmefreisetzung f <Heiz & Kälte, Kerntech> heat release
Wärmefühler m <Thermod> heat detector, heat sensor
Wärmefunktion f <Heiz & Kälte, Kohlen, Mechan, Nichtfoss Energ, Phys, Raumfahrt, Thermod> enthalpy

wärmegedämmt adj <Heiz & Kälte> insulated against heat
Wärmegefälle n <Thermod> thermal head
Wärmegehalt m 1. <Bau, Nichtfoss Energ> heat content; 2. <Thermod> caloric content, heat content, thermal content
Wärmegerät n <Heiz & Kälte> heating appliance
Wärmegewinn m <Heiz & Kälte> heat gain
Wärmegleiche f <Lufttrans> isotherm
Wärmegleichgewicht n <Thermod> thermal balance, thermal equilibrium
Wärmegleichrichter m <Elektrotech> thermionic rectifier
Wärmegrad m <Phys, Thermod> degree of heat
Wärmegradient m **des Meeres** <Nichtfoss Energ> ocean thermal gradient, OTG
wärmehärtbar adj <Druck, Mechan, Thermod> thermosetting
wärmehärtbare Farbe f <Druck> thermosetting ink
Wärmehaushalt m 1. <Heiz & Kälte, Papier> heat balance; 2. <Phys> calorific balance; 3. <Thermod> calorific balance, heat balance, thermal balance
Wärmeinhalt m 1. <Phys> water equivalent; 2. <Thermod> thermal content
Wärmeinstabilität f <Thermod> thermal instability
Wärmeisolation f <Bau, Mechan> heat insulation
wärmeisolieren v <Fertig> lag
wärmeisolierend adj <Lufttrans, Nichtfoss Energ, Phys, Thermod> heat-insulating
wärmeisolierende Wand f <Bau, Lufttrans, Phys, Thermod> heat-insulating wall
wärmeisoliert adj <Heiz & Kälte, Mechan, Thermod> heat-insulated
Wärmeisolierung f 1. <Bau> heat-insulating jacket, thermal insulation; 2. <Heiz & Kälte> thermal insulation; 3. <Thermod> thermal insulation; lagging *(Material)*; heat insulation *(Vorgang)*; heat-insulating jacket *(angebrachte Schutzschicht)*
Wärmekapazität f 1. <Bau> heat capacity; 2. <Heiz & Kälte> heat capacity, thermal capacity; 3. <Nichtfoss Energ> heat capacity; 4. <Thermod> heat capacity, thermal bonding, thermal capacity
Wärmekapazität f **bei konstantem Druck** *(Cp)* <Labor> heat capacity at constant pressure *(Cp)*
Wärmekapazität f **bei konstantem Volumen** <Thermod> heat capacity at constant volume
Wärmekonstante f <Bau, Nichtfoss Energ, Thermod> heat constant
Wärmekontraktion f <Metall, Thermod> thermal contraction
Wärmekonvektion f <Thermod> heat convection
Wärmekraftmaschine f <Maschinen, Mechan> heat engine
Wärmekraftwerk n 1. <Elektrotech> thermal-electric power station, thermal-electric power plant; 2. <Kerntech> thermal power plant, thermal power station
Wärmekreislauf m <Heiz & Kälte> heat cycle
Wärmelagerung f <Kunststoff> *(BE)* heat ageing, *(AE)* heat aging
Wärmelast f <Heiz & Kälte> heat load
Wärmelehre f <Thermod> thermodynamics
Wärmeleistung f 1. <Heiz & Kälte> thermal output; 2. <Kerntech> thermal output *(eines Reaktors)*; 3. <Thermod> caloric power, thermal output
wärmeleitend adj <Ker & Glas, Nichtfoss Energ, Textil, Thermod> heat-conducting
wärmeleitendes Glas n <Ker & Glas> heat-conducting glass
Wärmeleitfähigkeit f 1. <Heiz & Kälte, Labor> thermal conductivity; 2. <Phys> thermal conductance; 3. <Thermod> caloric conductibility, heat conductivity, thermal conductibility, thermal diffusivity
Wärmeleitfähigkeits-Analysengerät n <Gerät> thermal conductivity gas analyzer
Wärmeleitfähigkeits-Analysenmessgerät n <Gerät> thermal conductivity gas analyzer
Wärmeleitfähigkeitsdetektor m <Gerät> thermal-conductivity detector
Wärmeleitfähigkeitskoeffizient m <Mechan, Qual> coefficient of thermal conduction
Wärmeleitung f 1. <Heiz & Kälte> thermal conduction; 2. <Thermod> caloric conductibility, heat conductivity, thermal conduction
Wärmeleitungsmesser m <Thermod> heat conductivity meter
Wärmeleitvermögen n <Heiz & Kälte, Thermod> thermal conductivity
Wärmeleitweg m <Heiz & Kälte> heat path
Wärmeleitwert m <Heiz & Kälte> thermal conductance
Wärmeleitwiderstand m <Thermod> temperature lag
Wärmeleitzahl f 1. <Ergon> heat transfer coefficient; 2. <Heiz & Kälte> thermal conductance, thermal conductivity coefficient; 3. <Kunststoff> thermal conductivity coefficient
Wärmemaschine f <Thermod> heat engine, thermal engine
Wärmemauer f <Kerntech, Thermod> heat barrier
Wärmemelder m 1. <Sicherheit> heat detector *(Brandmeldung)*; 2. <Thermod> heat detector
Wärmemeldungsgeber m <Sicherheit> heat detector *(Brandmeldung)*
Wärmemenge f 1. <Phys> quantity of heat; 2. <Thermod> amount of heat, q-gas, quantum of heat
Wärmemengenzähler m 1. <Gerät> heat meter; 2. <Heiz & Kälte> calorimetric meter
Wärmemesser m <Heiz & Kälte> calorimeter
Wärmemessung f 1. <Heiz & Kälte> calorimetry; 2. <Thermod> thermometry
Wärmemischung f <Kerntech> thermal mixing
Wärmemitführung f <Heiz & Kälte, Thermod> thermal convection
Wärmenest n <Elektriz> heat concentration
Wärmeniveau n <Thermod> thermal level
Wärmeofen m <Metall, Thermod> heating furnace
Wärmepumpe f <Heiz & Kälte, Maschinen, Thermod> heat pump
Wärmequelle f <Heiz & Kälte, Nichtfoss Energ> heat source
Wärmerauschen n 1. <Elektronik> thermal noise; 2. <Funktech> thermal agitation noise; 3. <Phys> thermal noise; 4. <Telekom> thermal agitation noise
Wärmerauschgenerator m <Elektronik> thermal noise generator
Wärmerelais n <Elektrotech> thermal relay
Wärmeriss m <Fertig> heat check, heat crack
Wärmerissbildung f <Fertig> heat checking
Wärmerückgewinnung f 1. <Abfall> heat recovery; 2. <Heiz & Kälte> heat reclamation, heat recovery; 3. <Nichtfoss Energ> heat reclamation; 4. <Thermod> heat rate, heat rate curve, heat recovery
Wärmeschaltbild n <Thermod> heat flow chart, heat flow diagram
Wärmeschalter m <Elektrotech> thermal switch
Wärmescheinwiderstand m <Elektriz> thermal impedance
Wärmeschild m <Thermod> heat shield
Wärmeschluckvermögen n <Thermod> heat-absorbing power
Wärmeschockprüfung f <Thermod> heat shock test

Wärmeschrank

Wärmeschrank m <Elektrotech> heat cabinet
wärmeschrumpfen v <Thermod> heat-shrink
Wärmeschrumpfen n <Thermod> heat shrinking, thermal contraction
Wärmeschutz m 1. <Bau> thermal lagging; 2. <Foto, Ker & Glas> heat absorbing; 3. <Thermod> thermal lagging; 4. <Verpack> heat insulation • **mit Wärmeschutz versehen** <Heiz & Kälte> insulated against heat
Wärmeschutzfilter n <Foto> heat-absorbing filter
Wärmeschutzglas n <Ker & Glas> heat-absorbing glass
Wärmeschutzschirm m <Elektrotech> heat shield
Wärmeschutzverglasung f <Ker & Glas> heat-absorbing glazing
Wärmeschutzwert m <Qual, Thermod> thermal insulation index
Wärmespannung f <Thermod> thermal stress *(mechanisch)*
Wärmespeicher m <Nichtfoss Energ, Thermod> heat accumulator
wärmespeichernd adj <Thermod> heat-retaining
Wärmespeicherung f <Ergon, Nichtfoss Energ> heat storage
Wärmespeichervermögen n <Heiz & Kälte, Thermod> thermal capacity
Wärmespektrum n <Thermod> heat spectrum, thermal spectrum
Wärmesperre f <Heiz & Kälte> thermal barrier
Wärmespiegel m <Nichtfoss Energ> heat mirror
Wärmespritzen n <Fertig> flame spraying
wärmestabil adj <Thermod> heat-stable
wärmestabilisiert adj <Thermod> heat-stabilized
Wärmestabilität f 1. <Kunststoff> heat stability; 2. <Thermod> heat stability, thermal stability; 3. <Verpack> heat stability
Wärmestau m <Nichtfoss Energ, Thermod> heat accumulation
Wärmestauung f <Nichtfoss Energ, Thermod> heat accumulation
Wärmesteigrohr n <Kfztech> heat riser tube
Wärmestoß m <Thermod> thermal shock
Wärmestrahlung f 1. <Heiz & Kälte> heat radiation; 2. <Kerntech, Labor> thermal radiation; 3. <Nichtfoss Energ> heat radiation; 4. <Phys> thermal radiation; 5. <Thermod> heat radiation, thermal radiation; 6. <Verpack> heat radiation
Wärmestrahlungdetektor m <Gerät> thermal radiation detector
Wärmestrahlungsschutzanzug m <Sicherheit> heat protective suit
Wärmestrom m <Ergon, Labor, Thermod> heat flow
Wärmestromdichte f <Kerntech> surface heat flux
Wärmestromlinie f <Thermod> heat flow line
Wärmestrommessgerät n <Gerät> heat flow meter
Wärmeströmung f 1. <Heiz & Kälte> heat flow; 2. <Thermod> heat convection
wärmesuchend adj <Thermod> heat-seeking
Wärmesystem n <Kfztech> heater system
Wärmetauscher m <Erdöl, Heiz & Kälte, Papier, Wassertrans> heat exchanger
Wärmetechnik f <Maschinen> heat engineering
Wärmetod m <Thermod> heat death
Wärmeträger m 1. <Bau> heat carrier; 2. <Fertig> coolant; 3. <Heiz & Kälte, Nichtfoss Energ> heat carrier; 4. <Thermod> heat transfer medium
Wärmeträgheit f <Heiz & Kälte, Thermod> thermal inertia
Wärmetransport m <Erdöl, Heiz & Kälte, Phys, Strömphys, Thermod> convection
Wärmetrennabziehbild n <Ker & Glas> heat-release decal

Wärmeturbine f <Kerntech> heat turbine
Wärmeübergang m 1. <Heiz & Kälte, Kunststoff> heat transfer, heat transmission; 2. <Thermod> heat transfer, heat transmission *(zwischen verschiedenen Körpern)*
Wärmeübergangsleistung f <Heiz & Kälte, Qual> heat transfer efficiency
Wärmeübergangszahl f <Ergon, Thermod> heat transfer coefficient
Wärmeübertragung f 1. <Bau> convection; 2. <Heiz & Kälte> heat transfer, heat transmission; 3. <Phys> heat transfer; 4. <Thermod> heat transfer, heat transmission
Wärmeübertragungsfläche f <Heiz & Kälte, Thermod> heat transfer surface
Wärmeübertragungskoeffizient m 1. <Phys, Qual> heat transfer coefficient; overall heat transfer coefficient (U); 2. <Thermod> overall heat transfer coefficient (U)
Wärmeübertragungsmittel n <Heiz & Kälte> heat transfer medium
Wärmeübertragungszahl f <Thermod> heat transfer coefficient
Wärmeumhüllung f <Optik> thermal wrap
Wärmeumsatz m <Nichtfoss Energ, Thermod> heat transformation
Wärmeumwandlung f **des Meeres** <Nichtfoss Energ> ocean thermal conversion
Wärmeundurchlässigkeit f <Phys> athermancy
Wärmeverbrauch m <Labor, Thermod> heat consumption
Wärmeverbrauchsmessgerät n <Gerät> heat consumption meter
Wärmeverbrauchszähler m 1. <Gerät> heat consumption meter; 2. <Heiz & Kälte> calorimetric meter
Wärmeverdampfung f <Metall> thermal evaporation
Wärmeverhalten n <Thermod> thermal properties
Wärmeverlust m <Bau, Elektriz, Maschinen, Nichtfoss Energ, Qual, Thermod> heat loss
Wärmeverlustleistung f <Elektriz> dissipated power
Wärmeverlustmode f <Optik> heat ablation mode
Wärmeverschiebung f <Nichtfoss Energ, Thermod> heat displacement
Wärmeversorgung f <Kohlen, Thermod> heat supply
Wärmewandler m <Elektrotech> thermal converter
Wärmewert m 1. <Erdöl> calorific value; 2. <Thermod> calorific value, thermal value
Wärmewiderstand m (θ) <Hydraul, Phys, Thermod> thermal resistance (θ)
Wärmewirkung f <Thermod> heat effect
Wärmewirkungsgrad m 1. <Bau, Nichtfoss Energ> heat efficiency; 2. <Phys, Thermod> heat efficiency, thermal efficiency
Wärmezähler m <Gerät> heat meter
Wärmeziffer f <Heiz & Kälte> heat transfer factor
Wärmezufuhr f <Thermod> heat input
wärmfest adj <Verpack> heatproof
warmfeste Legierung f <Mechan, Metall> high-temperature alloy
Warmformen n 1. <Kunststoff> thermoforming; 2. <Mechan> hot-forming
warmgehärtet adj <Thermod> heat-hardened
warmgeschmiedet adj <Thermod> hot-forged
warmgewalzt adj <Mechan, Thermod> hot-rolled *(Metall)*
warmgewalzter Stahl m <Metall> hot-rolled steel
warmgewalztes Stahlblech n <Fertig> latten
warmgezogen adj <Thermod> hot-drawn
warmhärten v <Thermod> heat-harden
Warmhärtung f <Thermod> heat hardening
warmkalandrieren v <Thermod> hot-roll
Warmkalandrieren n <Thermod> hot-rolling
warmkalandriert adj <Thermod> hot-rolled

Warmkautschuk m <Kunststoff> hot rubber
Wärmkurve f <Thermod> heating temperature curve
Warmlabor n <Kerntech> semihot laboratory, warm laboratory
Warmlaufbereich m <Lufttrans> run-up area *(Flughafen)*
warmlaufen v <Lufttrans> run-up *(Motor und Triebwerk)*
warmlaufen lassen v <Thermod> warm up
Warmlaufen n <Fertig> heat exchange *(Lager)*
Warmluft f <Heiz & Kälte> warm air
Warmlufterzeuger m <Heiz & Kälte> fan-assisted air heater, warm-air heater
Warmluftfront f <Wassertrans> warm front *(Wetterkunde)*
Warmluftgebläse n <Heiz & Kälte> hot-air blower
Warmluftkorridor m <Lufttrans> hot-air corridor, hot-air gallery
Warmluftleitung f <Lufttrans> hot-air duct
Warmluftregler m <Kfztech> heater control
Warmluftvorhang m <Heiz & Kälte> hot-air curtain
warmnieten v <Maschinen> hot-rivet
Warmnietung f <Maschinen> hot-riveting
Warmprägefolie f <Verpack> hot-stamping foil
Warmprägen n <Verpack> hot-stamping
Warmriss m 1. <Maschinen> heat crack, thermal crack; 2. <Metall> thermal crack
Warmschmiedegesenk n <Maschinen> hot-forging die
warmschmieden v <Thermod> hot-forge
Warmschmieden n <Thermod> hot forging
Warmstart m 1. <Comp & DV> rebooting, soft reset, warm boot, warm restart, warm start; 2. <Elektriz> hot start, hot start-up *(Kraftwerk)*; 3. <Lufttrans> hot start
Warmstauchversuch m <Werkprüf> hot-compression test
Warmumformen n <Maschinen> hot forging, warm forming
Warmverarbeitung f <Ker & Glas> hot working
warmverbinden v <Thermod> hot-bond
Warmverbindung f <Thermod> hot bonding
warmverformbar adj <Wassertrans> thermoplastic *(Schiffbau)*
warmverformen v <Thermod> heat-form
warmverformt adj <Thermod> heat-formed
Warmverformung f <Thermod> heat distortion, heat forming
warmverschweißbar adj <Thermod> heat fusible
Warmwalzband n <Metall> hot band
warmwalzen v <Thermod> hot-roll
Warmwalzen n <Thermod> hot-rolling
Warmwasser n <Heiz & Kälte> hot-water
Warmwasserboiler m **der Heizung** <Heiz & Kälte> calorifier
Warmwasserheizung f <Heiz & Kälte> hot-water heating system
Warmwasserheizungsanlage f <Heiz & Kälte> hot-water heating system
Warmwasserspeicher m <Heiz & Kälte> hot-water tank
Warmwasserzähler m <Gerät> hot-water meter *(Thermo-Hydrometer)*
Wärmzeit f <Thermod> heating time
warmziehen v <Thermod> hot-draw
Warmziehen n <Maschinen, Thermod> hot-drawing
Warmzyklon m <Kohlen> hot cyclone
Warn… <Elektrotech, Kfztech, Sicherheit> alarm, alarming, alert, warning
Warnanlage f <Sicherheit> warning device
Warnanzeige f 1. <Elektrotech> warning light; 2. <Maschinen, Sicherheit> alarm, predictor display
Warnblinkanlage f <Kfztech> hazard-warning system
Warnblinkleuchte f <Kfztech> hazard-warning lamp
Warndreieck n <Kfztech, Sicherheit> warning triangle

Warneinrichtung f <Sicherheit> warning device
Warngerät n <Bau, Kohlen, Sicherheit> warning device
Warngrenzen fpl <Qual> warning limits
Warnkleidung f <Sicherheit> warning clothing
Warnlampe f <Maschinen, Sicherheit> warning light
Warnlampe f **für niedrigen Pegelstand** <Maschinen> low-level warning light
Warnleuchte f 1. <Elektrotech> warning light; 2. <Lufttrans> indicator light; 3. <Sicherheit> warning light
Warnlicht n <Eisenbahn, Lufttrans, Trans, Wassertrans> warning light
Warnmeldung f <Eisenbahn, Lufttrans, Trans, Wassertrans> warning message
Warnrelais n <Elektriz> alarm relay
Warnschild n 1. <Sicherheit> warning sign; 2. <Verpack> caution label
Warnsignal n 1. <Eisenbahn> caution signal, warning signal; 2. <Lufttrans> warning signal; 3. <Maschinen> alarm, alarm signal; 4. <Trans, Wassertrans> warning signal
Warnspur f <Kfztech> warning track
Warnsystem n <Sicherheit> alert system
Warnthermometer n <Maschinen> alarm thermometer
Warnung f <Sicherheit> warning label
Warnvorrichtung f <Sicherheit> warning device
Warpanker m <Wassertrans> kedge anchor *(Festmachen)*
warpen v <Wassertrans> kedge *(Schiff)*
Warpen n <Wassertrans> warping *(Tauwerk)*
Warren-Motor m <Kfztech> Warren engine
Warte f <Fertig> monitoring station
Warte… <Lufttrans, Telekom> holding
Wartebereich m **für die Gepäckaufbereitung** <Lufttrans> racetrack holding pattern
Wartebetrieb m <Telekom> delay mode, suspend and resume
Warteeinrichtung f <Telekom> queueing device
Warteflugbahn f <Lufttrans> holding path
Wartefluggeschwindigkeit f <Lufttrans> holding speed
Wartegleis n <Eisenbahn> holding track
Wartehalle f <Trans> shelter
warten v 1. <Comp & DV> service; 2. <Elektriz> waite; 3. <Maschinen> service; maintain *(Werkzeuge und Geräte)*; 4. <Mechan> maintain *(Werkzeuge und Geräte)*
wartende Belegung f <Telekom> waiting call
Warteorbit m <Telekom> parking orbit
Warteordner m <Telekom> call store
Wartepunkt m <Lufttrans> holding point
Warteschaltung f **bei Teilnehmer Besetzt** <Telekom> call hold and completion to busy subscriber
Warteschlange f <Comp & DV, Kontroll, Regelung, Telekom> queue, waiting line, waiting queue
Warteschlange f **mit einer Bedieneinheit** <Telekom> single server queue
Warteschlange f **mit Priorität** <Telekom> call queuing with priority, priority queue
Warteschlangenbetrieb m 1. <Comp & DV, Regelung> queueing; 2. <Telekom> call holding, call queueing
Warteschlangenblock m <Comp & DV> queue block
Warteschlangenelement n <Comp & DV> queue element
Warteschlangengröße f <Comp & DV> queue size
Warteschlangennetz n <Telekom> queueing network
Warteschlangensteuerung f <Telekom> queue control
Warteschlangentheorie f <Comp & DV, Ergon> queueing theory
Warteschlangenverwaltung f <Comp & DV> queue management
Warteschleife f 1. <Comp & DV> wait loop; 2. <Lufttrans> flight-holding pattern, holding, holding pattern
Wartestapel m <Lufttrans> holding stack

Wartestation

Wartestation f <Comp & DV> passive station
Wartestatus m <Comp & DV> wait condition, wait state
Wartesteuerung f <Kontroll> wait control
Wartesystem n <Telekom> call queueing facility
Warteverfahren n <Lufttrans> holding procedure
Wartezeit f 1. <Comp & DV> latency, queueing time, queue time; 2. <Elektriz> delay time; 3. <Telekom> delay, queueing time
Wartezeitzähler m <Comp & DV> quiesce counter
Wartezustand m <Kontroll> wait condition
Wartezyklus m <Kontroll> wait cycle
Wartung f 1. <Bau, Comp & DV> maintenance, service; 2. <Eisenbahn, Elektriz, Fernseh, Kfztech, Lufttrans> maintenance; 3. <Maschinen> service; 4. <Mechan, Papier, Telekom> maintenance
Wartung f **am Ort** <Telekom> local maintenance
Wartungsaufzeichnung f <Lufttrans> maintenance recorder
Wartungsbereich m service area
Wartungsdauer f <Qual> maintenance time
wartungsfähig adj <Qual> maintainable
Wartungsfähigkeit f <Qual> maintainability
wartungsfrei adj 1. <Elektriz> maintenance-free; 2. <Fertig> maintenance-free (Kunststoffinstallationen); 3. <Kfztech> maintenance-free (Batterie); 4. <Telekom> maintenance free
Wartungsintervall n 1. <Comp & DV> mean time between maintenance; 2. <Maschinen> maintenance interval
Wartungslogbuch n <Lufttrans> maintenance recorder
Wartungsprogramm n <Comp & DV> service program
Wartungstunnel m <Bau> service tunnel
Wartungsvorschrift f <Lufttrans> maintenance manual
Wartungswerkstatt f <Bau> maintenance shop
Wartungswerkzeug n <Maschinen> service tool
Warze f 1. <Fertig> projection; embossment (Schweißen); 2. <Metall> button (Defekt)
Warzenblech n <Fertig> warted plate
Warzendurchmesser m <Fertig> diameter of projection (Schweißen)
Wasch... <Kohlen, Textil> washing
Waschanlage f <Fertig> washing station
Waschanleitung f <Textil> washing instructions
Waschbarkeit f <Textil> washability
Waschbeton m <Bau> exposed aggregate concrete
Waschbord n <Wassertrans> washboard (Schiffbau)
Waschbühne f <Kohlen> strake
Wäsche f 1. <Erdöl> scrubbing (Gasreinigungsverfahren); 2. <Kohlen> dressing; 3. <Papier> washup; 4. <Textil> washing; 5. <Umweltschmutz> (AE) scrubbing, (BE) stripping (Brennstoffaufbereitung)
waschecht adj <Chemie, Textil> washfast (Farbstoff)
Waschechtheit f <Textil> fastness to washing
waschen v 1. <Kohlen> clean; 2. <Papier> wash; 3. <Textil> scour; 4. <Textil> wash
Waschen n <Textil> laundering, washing
Wäscher m 1. <Chemtech> washer; 2. <Kohlen> dresser
Wäschetrockner m <Textil> drier, tumble drier
Waschflasche f 1. <Labor> wash bottle; 2. <Phys> bubbler
Waschflüssigkeit f <Chemtech> washings (verbraucht)
Waschgrieß m <Kohlen> washed smalls
Waschmaschine f <Textil> washer, washing machine
Waschmittel n <Textil> detergent
Waschprodukte npl <Kohlen> clean coal, cleans
Waschröhrchen n <Labor> washing tube
Waschtrommel f <Papier> washing drum
Waschturm m <Chemtech> washing column

Waschvollautomat m <Textil> automatic washing machine
Washout n <Kerntech> washout
Washprimer m <Kunststoff> wash primer
Wasser n <Phys, Textil> water • **auf Wasser notlanden** <Lufttrans> ditch • **in Wasser löslich** <Verpack> soluble • **Wasser absondern** <Bau> bleed (Beton absondern) • **Wasser entziehen** <Chemie> dehydrate • **Wasser machen** <Wassertrans> make water (Schiff) • **zu Wasser** <Wassertrans> by water • **zu Wasser lassen** <Wassertrans> launch (Schiff)
Wasser n/**zum Verbrauch bestimmtes** <Wasserversorg> consumption water
Wasserabdichtung f <Bau> waterproofing
Wasserabgabe f <Wasserversorg> water delivery
Wasserablasshahn m <Wasserversorg> priming cock
Wasserablenkplatte f <Kerntech> water baffle
Wasserabscheidebauwerk n <Bau> water extraction structure
Wasserabscheider m <Maschinen, Wasserversorg> water separator
Wasserabschrecken n <Metall> water quenching
Wasserabspaltung f <Chemie> dehydration
wasserabweisend adj 1. <Chemie> hydrophobic (Ionen, Atomgruppen); 2. <Textil> water-repellent
wasserabweisende Imprägnierung f <Textil> water-repellent finish
wasserabweisender Zement m <Bau> water repellent cement
wasseraktivierte Batterie f <Elektrotech> water-activated battery
Wasseranalyseausrüstung f <Labor> water analysis kit
Wasserangebot n <Wasserversorg> water supply
Wasseranlagerung f <Chemie> hydration
Wasseranschluss m <Bau> (BE) tap
Wasseraufbereitung f 1. <Abfall> water purification (bei Abwässern); 2. <Kerntech> water treatment, water-conditioning process; 3. <Kohlen> water treatment; 4. <Wasserversorg> water conditioning, water purification
Wasseraufbereitungsanlage f <Bau, Wasserversorg> water treatment plant
Wasseraufnahme f <Werkprüf> water absorption
wasseraufnehmend adj <Bau> hygroscopic
Wasseraustritt m <Heiz & Kälte> water outlet
Wasseraustrittsöffnung f <Heiz & Kälte> water outlet
Wasserbad n <Labor> water bath
Wasserbadverdampfer m <Kerntech> water bath evaporator
wasserbasiert adj <Anstrich> aqueous-based, water-based
Wasserbau m <Bau> hydraulic engineering, hydraulics, water engineering
Wasserbehälter m 1. <Bau, Eisenbahn> water tank; 2. <Textil> water container; 3. <Wassertrans> water tank; 4. <Wasserversorg> cistern, water tank
Wasserbehandlungsanlage f <Wasserversorg> water treatment plant
Wasserbeschaffenheit f <Wasserversorg> water quality
wasserbeständig adj <Textil, Verpack> water-resistant
wasserbeständiges Schleifpapier n <Maschinen> waterproof abrasive paper
Wasserbeständigkeit f <Qual, Werkprüf> hydrolytic resistance (Glas)
Wasserbewirtschaftung f <Wasserversorg> integral water management
Wasserbilanz f <Wasserversorg> water balance
Wasserbindung f <Wasserversorg> water retention
Wasserbohrschlamm m <Erdöl> water-based mud (Bohrtechnik)

Wasserbohrung f <Wasserversorg> bore, boring
Wasserbombe f <Wassertrans> depth charge *(U-Bootabwehr)*
Wasserbus m <Wassertrans> river bus *(auf Flüssen)*
Wasserchemie f <Chemie> hydrochemistry
Wasserdampf m <Chemtech, Heiz & Kälte, Hydraul, Nichtfoss Energ, Papier, Phys, Textil> steam
Wasserdampfdestillation f <Chemtech> steam distillation
Wasserdampfhemmung f <Verpack> *(AE)* water vapor barrier, *(BE)* water vapour barrier
wasserdicht adj 1. <Heiz & Kälte> waterproof, watertight; 2. <Kunststoff, Papier> waterproof; 3. <Phys> watertight; 4. <Textil> waterproof, watertight; 5. <Wassertrans> watertight
wasserdicht machen v <Bau, Heiz & Kälte> waterproof
wasserdicht versiegelte Kamera f <Foto> waterproof-sealed camera
wasserdichte Abteilung f <Wassertrans> watertight compartment *(Schiff)*
wasserdichte Bitumenisolierung f *eines Kellergeschosses* <Bau> asphalt tanking
wasserdichte Folie f <Textil> waterproof sheet
wasserdichter Stecker m <Elektriz> watertight socket outlet
wasserdichter Steckkontakt m <Elektriz> watertight socket outlet
Wasserdichtmachen n 1. <Textil> waterproofing; 2. <Wasserversorg> coffering
Wasserdruck m 1. <Bau, Heiz & Kälte> water pressure; 2. <Hydraul> head of water pressure *(resultierend aus Standhöhe oder Bewegung)*; 3. <Nichtfoss Energ> hydraulic thrust; 4. <Textil, Wasserversorg> water pressure
Wasserdruckbehälter m <Hydraul> hydraulic reservoir
Wasserdurchlässigkeit f <Kunststoff> water permeability
Wasserdurchprüfung f <Mechan> hydrotests
Wasserdüse f <Fertig> water nozzle
Wassereinbruch m <Kerntech> water ingress
Wassereinlass m 1. <Heiz & Kälte> water inlet; 2. <Wasserversorg> water inflow
Wassereinspritzung f <Erdöl> water injection
Wassereintritt m 1. <Nichtfoss Energ> water intake; 2. <Wasserversorg> water inlet
Wassereintrittsöffnung f <Heiz & Kälte> water inlet
Wasserenthärten n <Chemtech> water softening
Wasserenthärter m <Bau, Chemtech, Textil> water softener
Wasserenthärtungsanlage f <Fertig> water-softening plant
Wasserenthärtungsmittel n <Textil> water softener
Wasserentnahme f <Wasserversorg> draw-off
wasserentziehendes Mittel n <Chemtech> dehydrating agent
Wasserentziehung f <Heiz & Kälte> dehydration
Wasserentzug m <Papier> anhydration
Wassererschließung f <Umweltschmutz> water reclamation
Wasserfahrzeug n <Wassertrans> boat, vessel
Wasserfallhöhe f <Hydraul, Nichtfoss Energ> waterfall height
Wasserfarbe f <Bau> water-based paint
Wasserfass n <Bau> water butt
wasserfest adj 1. <Textil> water-resistant; 2. <Verpack> watertight
Wasserfilter n <Maschinen, Wassertrans, Wasserversorg> water filter
Wasserflugzeug n <Lufttrans, Wassertrans> hydroplane, seaplane

Wasserförderschnecke f <Meerschmutz> screw pump
Wasserfracht f <Wassertrans> waterage *(Seehandel)*
wasserfrei adj 1. <Chemie, Erdöl> anhydrous *(Petrochemie)*; 2. <Lebensmittel> anhydrous, desiccated; 3. <Papier> anhydrous; 4. <Umweltschmutz> waterless
wasserfreies Ammoniak n <Erdöl> anhydrous ammonia *(Petrochemie)*
wasserfreies Borat n <Bau> anhydrous borate
wasserfreies Natriumkarbonat n <Papier> soda ash
Wasserfreimachung f <Erdöl> dewatering *(einer Rohrleitung)*
wasserführend adj <Wasserversorg> aquiferous
wasserführende Schicht f <Wasserversorg> shaft water-bearing ground
wasserführender Boden m <Bau> water-bearing ground
Wassergang m <Wassertrans> boot topping
Wassergehalt m <Bau, Kohlen, Optik> water content
Wassergehaltsprüfung f <Fertig> moisture content test *(Form)*
wassergehärtet adj <Thermod> water-hardened
wassergekühlt adj <Elektriz, Elektronik, Heiz & Kälte, Kerntech, Kfztech, Thermod> water-cooled
wassergekühlte Röhre f <Elektronik> water-cooled tube
wassergekühlter Motor m <Kfztech> water-cooled engine
wassergekühlter Reaktor m <Kerntech> water-cooled reactor
wassergekühlter Transformator m <Elektriz> water-cooled transformer
wassergekühltes Klimagerät n <Heiz & Kälte> water-cooled air conditioning unit
wassergekühltes System n <Maschinen> water-cooled system
wassergelöst adj <Anstrich> aqueous-based, water-based
wassergelöster Feststoff m <Anstrich> water-borne slurry
wassergereinigtes Kohlengas n <Kohlen> hygas
Wassergewinnung f 1. <Umweltschmutz> water reclamation; 2. <Wasserversorg> water catchment
Wasserglas n <Ker & Glas> water glass
Wasserglätte f <Kfztech> aquaplaning
Wassergüte f <Wasserversorg> water quality
Wasserhahn m 1. <Bau> bibcock, spigot, water cock, watertap; 2. <Labor> water tap; 3. <Maschinen> bib tap, bibcock; 4. <Wasserversorg> water tap
wasserhaltig adj <Chemie, Erdöl, Papier> aqueous
wasserhaltiges Silicat n <Chemie> hydrosilicate
Wasserhaltung f <Bau> ponding *(bei Beton)*
Wasserhammer m <Strömphys> water hammer
Wasserhärte f <Wasserversorg> water hardness
Wasserhärten n <Metall> water tempering
Wasserhaushaltung f <Wasserversorg> integral water management
Wasserheizvorrichtung f <Labor> water heater
Wasserhose f <Wassertrans> waterspout
wässerig adj 1. <Chemie> aqueous; 2. <Fertig> aqueous *(Kunststoffinstallationen)*; 3. <Strömphys, Wasserversorg> aqueous
Wasserinhalt m <Phys> water content
Wasserinjektion f <Erdöl> water injection *(Sekundärfördertechnik)*
Wasserkasten m <Ker & Glas> water box
Wasserkraft f <Elektriz, Nichtfoss Energ, Thermod, Wasserversorg> hydroelectric power, hydroelectricity, water power
Wasserkraftgenerator m <Elektriz> hydroelectric generator
Wasserkraftmaschine f <Maschinen> hydraulic motor

Wasserkraftprojekt

Wasserkraftprojekt n <Bau, Nichtfoss Energ> hydroelectric project
Wasserkraftwerk n 1. <Elektriz, Elektrotech> hydroelectric generating station, hydroelectric power plant, hydroelectric power station, water power station; 2. <Nichtfoss Energ> hydroelectric generating station, hydroelectric power plant, hydroelectric power station, water power station
Wasserkühler m <Heiz & Kälte> water cooler
Wasserkühlung f 1. <Heiz & Kälte> hydrocooling; 2. <Kfztech, Maschinen> water cooling
Wasserkunde f <Bau, Kohlen> hydrology
Wasserlandung f <Lufttrans> landing on water
Wasserlauf m <Bau, Umweltschmutz> water course
Wasserleitung f 1. <Bau> aqueduct, water line; 2. <Wasserversorg> conduit
Wasserleitungsrohr n <Wasserversorg> water pipe
Wasserlinie f 1. <Druck> wire mark; 2. <Meerschmutz> water line; 3. <Wassertrans> water line; floating line *(Schiffbau)*
Wasserlinienebene f <Wassertrans> waterplane *(Schiffskonstruktion)*
Wasserlinienriss m <Wassertrans> half-breadth plan *(Schiffkonstruktion)*
Wasserlinienschwerpunkt m <Wassertrans> *(AE)* center of flotation, *(BE)* centre of flotation *(Schiffkonstruktion)*
Wasserlöscher m <Sicherheit> water fire extinguisher
wasserloser Offsetdruck m <Druck> waterless offset printing
wasserloses Offset n <Druck> waterless offset printing
wasserlöslich adj <Bau, Lebensmittel, Maschinen, Textil> water-soluble
wasserlösliches Flussmittel n <Bau> water-soluble flux
Wassermangelsicherung f <Hydraul> plug, safety plug *(Dampfkessel)*
Wassermantel m <Thermod> water jacket *(zum Kühlen oder Heizen)*
Wassermesser m <Wasserversorg> instant flowmeter
wässern v 1. <Bau, Fertig> water; 2. <Lebensmittel> macerate
Wässern n <Papier, Textil> soaking
Wassernase f <Bau> gorge
Wasseroberfläche f <Wassertrans> surface *(Meer)*
Wasseroberflächenbreite f <Hydraul> water surface width
Wasserpass m <Wassertrans> boot topping
Wasserpassanstrich m <Wassertrans> boot-topping
Wasserpassfarbe f <Wassertrans> boot-topping
Wasserprobe-Entnahmevorrichtung f <Labor> water sampler
Wasserpumpe f <Kfztech> water pump
Wasserpumpengehäuse n <Kfztech> water pump housing
Wasserpumpenzange f <Maschinen> multigrip pliers, pipe wrench
Wasserputzstrahlen n <Fertig> liquid honing
Wasserqualität f <Nichtfoss Energ, Umweltschmutz, Wasserversorg> water quality
Wasserquelle f <Hydraul> hydraulic pressure source
Wasserquerschnitt m <Wasserversorg> cross section *(eines Kanals oder Flusses)*
Wasserquotient m <Kohlen> water ratio
Wasserrad n <Maschinen, Nichtfoss Energ> water wheel
Wasserreiniger m <Textil> water purifier
Wasserreinigung f <Chemtech, Wasserversorg> water purification
Wasserreinigungsfilter n <Chemtech> water purification filter
Wasserrohr n <Heiz & Kälte, Wassertrans> water tube

Wasserröhrenkühler m <Kfztech> flanged-tube radiator
Wasserrohrkessel m 1. <Heiz & Kälte> water tube boiler; 2. <Wassertrans> water tube boiler *(Dampferzeugung)*
Wasserrückkühler m <Heiz & Kälte> water-cooled heat exchanger
Wassersand-Abschneider m <Abfall, Erdöl, Kohlen> hydrocyclone
Wassersäule f 1. <Bau, Hydraul, Nichtfoss Energ> head of water; 2. <Umweltschmutz> plume *(Unterwasser-Atomexplosion)*
Wassersäulenhöhe f <Hydraul> head *(Wasserdruck resultierend aus statischer Höhe)*
Wasserscheide f <Wasserversorg> watershed
Wasserschenkel m <Bau> throat
Wasserschicht f 1. <Nichtfoss Energ> aquifer; 2. <Wassertrans> aquifier
Wasserschlag m 1. <Fertig> water hammer *(Kunststoffinstallationen)*; 2. <Wassertrans> water hammer
Wasserschlämme f <Anstrich> water-borne slurry
Wasserschlauch m <Wasserversorg> garden hose
Wasserschleuse f <Wassertrans> sluice
Wasserschloss n <Hydraul, Wassertrans> surge tank
Wasserspeicher m 1. <Hydraul, Kerntech> water accumulator; 2. <Wasserversorg> reservoir
wassersperrende Rohrtour f <Erdöl> water string *(Bohrtechnik)*
Wasserspiegel m 1. <Bau, Erdöl> water table *(Bohrtechnik)*; 2. <Kohlen, Wassertrans> water table
Wasserstag n <Wassertrans> bob stay *(Tauwerk)*
Wasserstandsanzeiger m 1. <Maschinen> *(AE)* water gage, *(BE)* water gauge; 2. <Wasserversorg> *(AE)* water level gage, *(BE)* water level gauge, water level indicator
Wasserstandshahn m <Wasserversorg> *(AE)* gage cock, *(BE)* gauge cock, try cock
Wasserstandsmarke f 1. <Ker & Glas> watermark; 2. <Wasserversorg> *(AE)* water level gage, *(BE)* water level gauge
Wasserstandsmesser m <Wasserversorg> *(AE)* water level gage, *(BE)* water level gauge, water level indicator
Wasserstandsregler m <Wasserversorg> constant level regulator
Wasserstock m <Wasserversorg> water hydrant
Wasserstoff m (H) <Chemie> hydrogen (H)
• **Wasserstoff abspalten** <Chemie> dehydrogenate
Wasserstoffabspaltung f <Chemie> dehydrogenation
Wasserstoffelektrode f <Elektriz> hydrogen electrode, hydrogene gas electrode
Wasserstoffemissionslinie f <Strahlphys> hydrogen emission line
Wasserstoffenergietechnik f <Nichtfoss Energ> hydrogen energy technology
Wasserstoffentschwefelung f <Umweltschmutz> *(AE)* hydrodesulfurization, *(BE)* hydrodesulphurization
Wasserstoffgas n <Chemie> hydrogen gas
Wasserstoffion n <Elektronik, Phys> hydrogen ion
Wasserstofflinie f <Strahlphys> hydrogen emission line
Wasserstoffperoxid n 1. <Chemie> hydrogen peroxide; 2. <Fertig> peroxide *(Kunststoffinstallationen)*; 3. <Raumfahrt> peroxide
Wasserstoffsäure f <Chemie> hydracid
Wasserstoffsulfid n <Chemie> *(AE)* hydrogen sulfide, *(BE)* hydrogen sulphide
Wasserstoffsuperoxid n <Chemie, Raumfahrt> hydrogen peroxide
Wasserstofftank m <Raumfahrt> hydrogen tank
Wasserstoffthyratron n <Elektronik> hydrogen thyratron
wasserstoffzellengetriebenes Fahrzeug n <Kfztech, Nichtfoss Energ> hydrogen fuel cell powered car
Wasserstollen m <Wasserversorg> water adit

Wasserstrahl m <Wasserversorg> water jet; jet *(Ausströmen einer Flüssigkeit)*
Wasserstrahlantrieb m 1. <Trans> water jet propulsion; 2. <Wassertrans> hydrojet, hydrojet propulsion, water jet drive
Wasserstrahlpumpe f <Maschinen, Wassertrans> water jet pump
Wasserstrahl-Verwirbelungsverfahren n <Kunststoff> hydraulic entanglement process, hydroentanglement process
Wasserstraße f <Wassertrans> waterway
Wasserstratifikation f <Nichtfoss Energ> stratification of waters
Wasserströmungsgeschwindigkeit f <Heiz & Kälte> water flow rate
Wassertank m <Bau, Eisenbahn, Wassertrans> water tank
Wassertanker m <Bau> water tanker
Wasserthermostat m <Thermod> water thermostat
Wassertiefe f <Nichtfoss Energ> water depth
Wassertransport m <Wassertrans> water transport
Wassertrieb m <Erdöl> water drive *(Erdöl- und Erdgasförderung)*
Wasserturbine f <Maschinen, Nichtfoss Energ, Thermod> hydroturbine, water turbine
Wasserturm m <Bau> water tower
wasserundurchlässig adj 1. <Bau> impermeable, watertight; 2. <Heiz & Kälte> waterproof, watertight; 3. <Kunststoff, Papier, Textil> waterproof
Wasserundurchlässigkeit f <Bau> watertightness
Wasserung f <Lufttrans> landing on water
Wässerungstank m <Foto> wash tank
wasserunlöslich adj <Chemie, Textil> insoluble in water
Wasserventil n <Bau> water valve
Wasserverhältnis n <Kohlen> water ratio
Wasserverlust m <Wasserversorg> leakage
Wasserversorgung f <Bau, Umweltschmutz, Wasserversorg> water supply
Wasserversorgungsleitung f <Bau, Wasserversorg> water supply pipe
Wasserversorgungssystem n <Bau, Wasserversorg> water supply system
Wasserverteilung f <Bau, Wasserversorg> water distribution
wasserverunreinigender Stoff m <Umweltschmutz> water pollutant
Wasserverunreinigung f <Umweltschmutz> water pollution
Wasservorwärmer m <Heiz & Kälte> water preheater
Wasserwaage f 1. <Bau> air level, water level; 2. <Maschinen> spirit level; 3. <Mechan, Metrol> level
Wasserwagen m <Bau> water truck
Wasser-Wasser-Wärmeaustauscher m <Heiz & Kälte> water-to-water heat exchanger
Wasserzähler m 1. <Fertig> water meter *(Kunststoffinstallationen)*; 2. <Textil, Wasserversorg> water meter
Wasserzapfstelle f <Wasserversorg> water outlet, water outlet port
Wasserzeichen n <Druck, Papier> watermark
Wasserzeichenzylinder m <Papier> watermark roll
Wasser-Zement-Faktor m <Bau> water-cement ratio
Wasserzerstäuber m <Wasserversorg> water atomizer
Wasserzufluss m <Maschinen> water inlet
Wasserzulauf m 1. <Bau, Nichtfoss Energ> water intake; 2. <Wasserversorg> feed, water intake
Wasserzurückhaltung f <Wasserversorg> water retention
Wasserzwischenraum m <Kerntech> water gap
wässrige Phase f <Kohlen> aqueous phase

wässrige Suspension f <Kerntech> water suspension
wässriger Ausfluss m <Umweltschmutz> aqueous effluent
Waterless-Platte f <Druck> waterless plate
Watt n 1. <Elektriz, Elektrotech, Metrol> watt; 2. <Wassertrans> mudflats *(Geographie)*
Watte f <Textil> wad
Watte f **in Lagen** <Textil> batting
Wattenmeer n <Wassertrans> shoal *(Geographie)*
wattieren v <Textil> wad
Wattierung f 1. <Textil> padding, wadding; 2. <Verpack> air bubble cushioning, wadding
wattlos adj <Elektriz> wattless
wattlose Energie f <Elektriz> apparent energy
wattlose Komponente f <Elektriz> wattless component
wattloser Strom m <Elektriz> idle current, wattless current
Wattmesser m <Elektriz> wattmeter
Wattmessgerät n <Phys> wattmeter
Wattmeter n <Elektriz> wattmeter
Watt'sches Fissionsspektrum n <Kerntech> Watt's fission spectrum
Wattsekunde f <Elektriz> wattsecond
Wattstunde f <Elektriz, Elektrotech, Phys> watt-hour
Wattstundenmessgerät n <Phys> watt-hour meter
Wattstundenzähler m <Gerät> active energy meter
Wattzahl f <Chemie> wattage
Waugelb n <Chemie> luteolin
Wb *(Weber)* <Elektriz, Elektrotech, Metrol> Wb *(weber)*
WB *(Wissensbasis)* <Comp & DV, Künstl Int> KB *(knowledge base)*
W-Boson n 1. <Phys> W particle *(Elementarteilchen)*; 2. <Teilphys> W boson
WBS *(wissensbasiertes System)* <Comp & DV> KBS *(knowledge-based system)*
WDM *(Wellenlängenmultiplex)* <Optik, Telekom> WDM *(wavelength division multiplexing)*
Web... <Textil> weaving
Web-Anzeige f <Comp & DV> Web banner
Web-basiert adj <Comp & DV> Web-based
Web-basierte Netzverwaltung f <Telekom> web-based enterprise management *(WWW)*
Webbaum m <Textil> beam
Webbing n <Kunststoff> webbing
Webblatt n <Textil> reed
Web-Crawler m <Comp & DV> Web crawler
weben v <Textil> weave
Weben n <Textil> weaving
Weber n 1. <Elektriz> weber, Wb; 2. <Elektrotech> weber, Wb *(Einheit des magnetischen Flusses)*; 3. <Metrol> weber, Wb; 4. <Phys> weber *(Einheit des magnetischen Flusses)*
Weber'sche Zahl f *(W)* <Hydraul> Weber number *(W)*
Weberschiffchen n <Textil> shuttle
Webfach n <Textil> shed
Web-fähig adj <Comp & DV> Web-enabled
Webkante f <Bau, Papier, Textil> selvage, selvedge
Webkette f <Textil> warp
Webleinenstek m <Wassertrans> clove hitch *(Knoten)*
Webmaschine f <Textil> loom
Webschützen m <Textil> shuttle
Web-Seite f <Telekom> web page *(WWW)*
Web-Site f <Comp & DV> Web site
Webstuhl m <Textil> loom
Webstuhldrehzahl f <Textil> loom speed
Webstuhlgestell n <Textil> loom framing
Web-tauglich adj <Comp & DV> Web-enabled
Webteppich m <Textil> woven carpet
Webwarenstückanfang m <Textil> head end

Wechsel

Wechsel *m* <Bau, Elektrotech, Fernseh, Fertig, Werkprüf> alternation, change
Wechselbalken *m* <Bau> trimmer, trimmer beam
Wechselbeanspruchung *f* <Werkprüf> alternating stress, cyclic loading
Wechselburst *m* <Fernseh> alternating burst
Wechseleinheit *f* <Telekom> interchangeable plug-in unit
Wechseleinschub *m* <Telekom> interchangeable plug-in unit
Wechselfeld *n* <Elektrotech> alternating field
Wechselfestigkeit *f* <Fertig> alternate strength
Wechselfeuer *n* <Wassertrans> *(AE)* alternating colored lights, *(BE)* alternating coloured lights *(Signal)*
Wechselfluss *m* <Elektriz> alternating flux
Wechselgetriebe *n* <Maschinen> change gear, change-speed gear
Wechselinduktion *f* <Elektriz, Elektrotech> mutual induction
Wechselinduktivität *f* <Elektrotech> mutual inductance
Wechsellager *n* <Maschinen> double thrust bearing
Wechsellichtphotometer *n* <Gerät> chopped-light photometer
Wechseln *n* <Fernseh> change-over
wechselnd *adj* <Elektriz> alternating
wechselnde Ladung *f* <Bau, Metall> varying loading
Wechselobjektiv *n* <Foto> interchangeable lens
Wechselplatte *f* <Comp & DV> exchangeable disk, removable disk
wechselpolar *adj* <Elektrotech> heteropolar
wechselpolig *adj* <Elektrotech> heteropolar
Wechselrad *n* <Maschinen> change gear, change gear wheel
Wechselrädergetriebe *n* <Maschinen> change gear drive
Wechselräderkasten *m* <Maschinen> change gear box
Wechselrelais *n* <Elektriz, Elektrotech> change-over relay
Wechselrichten *n* <Elektronik> inversion
Wechselrichter *m* 1. <Elektriz> static inverter; 2. <Elektronik, Elektrotech, Phys> inverter *(WR)*
Wechselsack *m* <Foto> changing bag
Wechselschalter *m* <Elektrotech> alternate action switch, change-over switch, double-throw switch
Wechselschalter *m* **mit Unterbrechung** <Telekom> break-before-make switch
Wechselspannung *f* 1. <Elektrotech> ac voltage, AC voltage, alternating voltage; 2. <Fertig> alternate stress
Wechselstraßenzeichen *n* <Trans> variable route sign
Wechselstrom *m (AC, WS)* <Elektrotech> alternating current *(AC)*
Wechselstromdiodenschalter *m* <Elektronik> diode alternating-current switch
wechselstromgekoppelt *adj* <Kontroll> AC-coupled
Wechselstromgenerator *m* <Elektriz, Elektrotech, Phys> alternator
Wechselstrom-Gleichstrom *m (AC-GS, WS-GS)* <Elektrotech> AC-DC *(alternating current-direct current)*
Wechselstrom-Gleichstrom Umformer-Lok *f* <Eisenbahn> locomotive with AC/DC motor converter set
Wechselstromkommutatormaschine *f* <Elektrotech> a.c. commutator machine
Wechselstromkommutatormotor *m* <Elektrotech> a.c. commutator motor
Wechselstromlichtmaschine *f* <Kfztech> alternator
Wechselstrommaschine *f* <Nichtfoss Energ> alternator
Wechselstromversorgung *f* <Elektriz> alternating-current supply
Wechselumformer *m* <Elektriz> inverted rotary converter
Wechselventil *n* <Maschinen> shuttle valve
Wechselverkehrszeichen *n* <Trans> variable message sign; variable-speed message sign *(Geschwindigkeitsanzeige)*
wechselweise *adj* <Maschinen> alternating
wechselwirken *v* <Strahlphys> interact
Wechselwirkung *f* 1. <Anstrich> reaction; 2. <Elektronik, Kerntech, Labor, Maschinen, Metall, Strahlphys, Telekom> interaction
Wechselwirkung *f* **mit endlicher Reichweite** <Kerntech> finite range interaction
Wechselwirkung *f* **zwischen Strahl und Plasma** <Kerntech> beam-plasma interaction
Wechselwirkungseffekt *m* <Lufttrans> coupling
Wechselwirkungsenergie *f* <Metall> interaction energy
Wechselwirkungsfreiheit *f* <Gerät> absence of interaction
wechselwirkungslos *adv* <Strahlphys> without interacting
Wechselwirkungsraum *m* <Elektronik> interaction space *(Querfeldröhren)*
Wechselwirkungsspalt *m* <Elektronik> interaction gap *(bei Elektronenstrahlröhren)*
Wechsler *m* <Fertig> variator *(Kunststoffinstallationen)*
Wechslerschalter *m* <Elektriz> two-way switch
Weckdienst *m* <Telekom> telealerting *(ISDN)*
Wecker *m* <Telekom> ringer *(Telefon)*; prompter
Weckeruhr *f* <Labor> alarm clock
Weckglas *n* <Verpack> glass jar
Weg *m* 1. <Akustik> displacement; 2. <Elektrotech> path; 3. <Erdöl> seismic path; 4. <Funktech> path; 5. <Maschinen> daylight, deflection, stroke, travel; 6. <Phys> path; 7. <Telekom> path *(Ausbreitung)*; 8. <Trans> route
Weg *m* **in Gegenrichtung** <Telekom> return path
Wegamplitude *f* <Akustik> displacement
Wegauswahl *f* <Telekom> route selection *(ITG)*
Wegebesetzt-Zustand *m* <Telekom> congestion *(Fernsprechverkehr)*
Wegediversity *n* <Telekom> route diversity
Wegemenge *f* <Telekom> set of routes
Wegerecht *n* <Wassertrans> right of way
wegfieren *v* <Wassertrans> lower away
wegführendes Gleis *n* <Eisenbahn> down line *(von der Hauptstadt weg)*
Weglänge *f* 1. <Kerntech> path length; 2. <Lufttrans, Trans> distance covered
Wegleitstrahl *m* <Funkort, Lufttrans> glide path beam
Wegmesssystem *n* <Gerät> path-measuring system *(numerische Steuerung)*
Wegplatte *f* <Bau> pad
Wegpunkt *m* <Wassertrans> waypoint *(Navigation)*
wegräumen *v* <Bau> clear away
wegschmelzen *v* <Chemie> deliquesce
Wegschrittgröße *f* <Fertig> bit size *(Programmieren)*
Wegspeicher *m* <Telekom> path memory
Wegstreckenzähler *m* <Kfztech> odometer
Wegunterführung *f* <Bau> undergrade crossing
Wegunterschied *m* <Phys> path difference
Wegwerf... <Abfall> disposable, expendable
Wegwerfartikel *m* <Maschinen> expendable item
Wegwerfatemmaske *f* <Sicherheit> disposable respirator
Wegwerfatemschutzgerät *n* <Sicherheit> disposable respirator
Wegwerfhandschuhe *mpl* <Sicherheit> disposable gloves
Wegwerfmaske *f* <Sicherheit> disposable mask
Wegwerfölfilter *n* <Kfztech> throw-away oil filter
Wegwerfprodukt *n* <Abfall> throw-away product
Wegwerfschutzkleidung *f* <Sicherheit> disposable protective clothing

Wegwerfspritze f <Labor> disposable syringe
Wegwerfverpackung f 1. <Abfall> one-way pack; 2. <Lebensmittel> nonreturnable packaging
wehenerregend adj <Chemie> oxytocic
Wehnelt-Zylinder m <Phys> Wehnelt cylinder
Wehr n 1. <Hydraul> nappe, weir; 2. <Meerschmutz> dam, weir; 3. <Nichtfoss Energ> weir; 4. <Umweltschmutz, Wasserversorg> dam, weir
Wehrabschöpfer m <Umweltschmutz> weir skimmer
Wehrabschöpfgerät n <Meerschmutz> weir skimmer
Wehrkrone f <Wasserversorg> crest of a weir
Wehrnadel f <Wasserversorg> stop plank *(eines Wehres oder Schleuse)*
Wehrölsperre f <Meerschmutz> weir boom
Wehrverschluss m <Wasserversorg> floodgate
weich adj <Comp & DV, Elektronik, Foto, Metall, Papier, Phys, Strahlphys, Wassertrans> soft
weich machen v <Papier> soften
weich machend adj <Fertig> emollient
Weich... <Bau, Fertig, Foto, Ker & Glas, Textil> soft
weicharbeitender Entwickler m <Foto> soft contrast developer, soft effect developer
Weichdichtung f <Maschinen> flexible gasket
Weiche f 1. <Eisenbahn> *(BE)* points, *(AE)* switch, switch rail, turnout; 2. <Elektronik> band separation; 3. <Funktech> separation circuit
weiche Bremsung f <Eisenbahn> smooth braking
weiche Röhre f <Elektronik> soft tube
weiche Röntgenstrahlen mpl <Phys, Strahlphys> soft X-rays
weiche Röntgenstrahlung f <Phys, Strahlphys> soft X-rays
weiche Seite f <Papier> tender side
weiche Strahlen mpl <Strahlphys> soft radiation
weiche Strahlung f <Strahlphys> soft radiation
weiche Trennfuge f <Comp & DV> soft hyphen
Weicheisen n <Elektriz, Elektrotech, Labor, Metall, Phys> soft iron
Weicheiseninstrument n <Gerät, Metrol> electromagnetic instrument, moving-iron instrument, soft-iron instrument
Weicheisenkern m <Elektrotech> soft iron core
Weicheisenstrommessgerät n <Elektriz> moving-iron ammeter
Weicheisenvoltmeter n <Metrol> moving-iron voltmeter
weich-elastische Verpackung f <Verpack> flexible package
Weichen fpl **und Kreuzungen** fpl <Eisenbahn> switchgear
Weichenantrieb m <Elektrotech> switch drive *(Magnetschwebetechnik)*
Weichenhebel m <Eisenbahn> switch lever
Weichenheizgerät n <Eisenbahn> point heater
Weichenkreuz n <Eisenbahn> double crossover, scissors crossing
Weichenlaterne f <Eisenbahn> indicator lamp, switch-point light
Weichenseite f <Eisenbahn> turnout side
Weichenstange f <Eisenbahn> throw rod
Weichenstellstange f <Eisenbahn> point-operating stretcher
Weichenstellwerk n <Eisenbahn> *(BE)* signal box, *(AE)* signal tower
Weichenzugstange f <Eisenbahn> point rod
weicher Boden m <Bau> bad ground
weicher Fehler m <Comp & DV> soft fail
weicher Graugguss m <Metall> soft cast iron
weicher Griff m <Textil> soft handle
weicher Supraleiter m <Elektronik> soft superconductor

weiches Bromsilberpapier n <Foto> soft bromide paper
weiches Wasser n <Wasserversorg> soft water
Weichfäule f <Lebensmittel> soft rot *(bei Früchten)*
Weichfeuer n <Ker & Glas> soft fire
Weichglas n <Ker & Glas> soft glass
weichglühen v <Fertig> spheroidize
Weichglühen n 1. <Fertig> spheroidize annealing; 2. <Metall> soft annealing, softening; 3. <Thermod> dead anneal, soft anneal
Weichgummi n <Kunststoff> soft rubber
Weichharz n <Chemie> oleoresin
Weichhaut f <Fertig> bark
Weichheit f <Papier> softness
Weichholz n <Bau, Wassertrans> softwood
Weichkohle f <Kohlen> soft coal
Weichküpe f <Textil> blue vat
Weichlot n 1. <Bau, Elektriz> soft solder; 2. <Fertig> ordinary solder; 3. <Maschinen> soft solder
weichlöten v <Bau> solder
Weichlöten n <Bau, Elektriz, Maschinen> soft soldering, soldering
Weichmachen n <Metall> softening, work softening
Weichmacher m 1. <Bau, Fertig> plasticizer; 2. <Kunststoff> plasticizer, softener; 3. <Textil> softener, softening agent
weichmacherfrei adj <Kunststoff> unplasticized
weichmacherfreies PVC n <Fertig> PVC rigid, U-PVC *(Kunststoffinstallationen)*
Weichmacherlösung f <Papier> softener water
Weichmachermigration f <Kunststoff> plasticizer migration
Weichmacherwanderung f <Kunststoff> plasticizer migration
weichmagnetisches Material n <Elektrotech, Phys> soft magnetic material
Weichmetall n <Metall> soft metal
Weichpackung f <Maschinen> soft packing seal
Weichplastikradierer m <Konstzeich> soft plastic eraser
Weichporzellan n <Ker & Glas> soft porcelain
Weichroheisen n <Metall> soft pig iron
Weichrückstellung f <Comp & DV> soft reset
Weichschaum m <Kunststoff> flexible foam
weichsektoriert adj <Comp & DV> soft-sectored
weichsektorierte Platte f <Comp & DV> soft-sectored disk
Weichsektorierung f <Comp & DV> soft sectoring
Weichspülmittel n <Textil> softener *(Waschen)*
Weichstahl m <Metall> mild steel
weichstellen v <Kunststoff> plasticize
Weichstoffpackung f <Maschinen> soft packing seal
Weichüberspannung f **mit kreuzender Fahrleitung** <Elektrotech> overhead junction crossing
Weichweizen m <Lebensmittel> soft wheat
Weichzeichner m <Foto> diffuser scrim
Weichzeichnerlinse f <Foto> soft-focus lens
Weinhold-Dewar'sches Gefäß n <Chemtech, Labor> Dewar flask *(Isolierung)*
Weinhold'sches Gefäß n <Chemtech, Labor> Dewar flask *(Isolierung)*
Weinstein m 1. <Chemie> tartar; 2. <Lebensmittel> cream of tartar
Weinstock m <Lebensmittel> vine
weiß adj <Druck, Foto> white
Weiß n <Druck, Foto> white
Weißabgleich m <Fernseh> white balance
Weißanlaufen n <Kunststoff> blushing
Weißaufnahme f <Aufnahme> white recording
Weißblech n <Metall> tin plate
Weißblechabfall m <Abfall> tin plate waste

Weißblechdose

Weißblechdose f 1. <Abfall> tin-plated can; 2. <Verpack> metal can *(Konserven)*
Weißbleche npl <Fertig> menders
weiße Strahlung f <Raumfahrt> white radiation
Weißen n <Bau> whitewashing
weißer Glimmer m <Kunststoff> white mica
weißes Arsenik n <Ker & Glas> arsenic trioxide *(Arsentrioxid)*
weißes Gauß'sches Rauschen n <Telekom> uniform-spectrum random noise, white Gaussian noise
weißes Licht n <Phys> white light
weißes Phosphat-Opalglas n <Ker & Glas> white phosphate opal
weißes Rauschen n 1. <Akustik> white noise; 2. <Aufnahme> random noise, white noise; 3. <Comp & DV> Gaussian noise, white noise; 4. <Elektronik, Phys, Raumfahrt> white noise; 5. <Telekom> random noise, white noise
weißglühend adj 1. <Fertig, Strahlphys> incandescent; 2. <Metall, Thermod> white hot
weißglühender Festkörper m <Strahlphys> incandescent solid
Weißglut f <Metall, Thermod> white heat
Weißgrad m <Papier> whiteness
weißkaschierte Pappe f <Verpack> white-lined board
Weißlichtinterferenz f <Phys> white light fringe
Weißmetall n 1. <Maschinen> babbitt metal, white metal; 2. <Metall> antifriction metal, bearing metal
Weißmetallausguss m <Metall> antifrictionning
Weißmetallfutterlager n <Fertig> babbitt-lined bearing
Weißmetall-Lagerausguss m <Fertig> babbitting
Weißrauschen n 1. <Aufnahme> random noise, white noise; 2. <Elektronik> white noise
Weißrauschgenerator m 1. <Aufnahme> random noise generator; 2. <Elektronik> white noise generator
Weißrauschquelle f <Elektronik> white noise source
Weißrauschsignal n <Elektronik> white noise signal
Weiss'scher Bezirk m <Phys> Weiss domain
Weißspitze f <Fernseh> peak white, white peak
Weißwäsche f <Textil> linen
Weißwert m <Fernseh> white level
Weißwertbegrenzung f <Fernseh> white clip, white limiter
Weißwertdehnung f <Fernseh> white stretching
Weißwertkompression f <Fernseh> white compression
Weißzement m <Bau> white cement
weit abhalten v **von** <Wassertrans> give a wide berth *(Schiffführung)*
Weit... <Foto, Regelung, Telekom> broad, wide
Weite f <Kohlen> cavity
weiter Flaschenhals m <Verpack> wide-mouth neck
weiter Laufsitz m *(WL)* <Maschinen> loose fit
Weiter... <Telekom> widee
Weiterentwicklung f **der IT-Lösungen** <Comp & DV> IT advancement
weiterleiten v <Telekom> redirect, transmit
Weiterleitung f <Comp & DV> routing
Weiterreiß... <Kunststoff> tearing
Weiterreißbarkeitsprüfer m <Papier> tearing tester
Weiterreißen n <Kunststoff> tear propagation
Weiterreißprüfung f <Kunststoff> tearing test
Weiterreißversuch m <Kunststoff> tearing test
Weiterschaltbedingung f <Regelung> step-enabling condition
Weiterschaltimpuls m <Gerät> advance pulse
weitersenden v <Telekom> retransmit
Weitersendung f <Telekom> retransmission
Weiterübertragung f <Regelung> onward transmission *(des Signals)*
Weiterverarbeitung f <Regelung> processing *(des Signals)*
weiterverfolgen v <Patent> prosecute *(Anmeldung)*
Weiterverteilung f <Comp & DV> redistribution *(von E-Mail)*
Weiterverwertung f <Abfall> reclamation
Weithals... <Labor> wide mouth
Weithalsflasche f <Labor> wide-mouth bottle
Weithalskolben m <Labor> wide-necked flask
Weithalspackung f <Ker & Glas> wide-mouth container
Weithalsröhrchen n <Labor> wide-bore tube
weitreichender Lautsprecher m <Aufnahme> extended-range loudspeaker
weitreichender Transport m <Umweltschmutz> long-range transport *(von Luftschadstoffen)*
weitreichendes Präzisionsnavigationssystem n <Funktech> long-range accuracy navigation system, long-range accuracy system
Weitspannkonstruktion f <Bau> long span structure
Weitung f <Kohlen> cavity
Weitverkehrsnetz n *(WAN)* <Comp & DV, Telekom> wide area network *(WAN)*
Weitverkehrsrichtfunk m <Telekom> backbone radio relay
Weitverkehrssystem n <Telekom> wide-area system
Weitwinkelobjektiv n <Foto> panoramic lens, wide-angle lens
Weitwinkelvorsatz m <Foto> wide-angle converter
Weizen m <Lebensmittel> wheat
Weizenflugbrand m <Lebensmittel> loose smut of wheat *(Pflanzenkrankheitslehre)*
Weizenfuttermehl n <Lebensmittel> wheatmeal
Weizensteinbrand m <Lebensmittel> bunt *(Getreidekrankheit)*
welk adj <Lebensmittel> withered
Well... <Ker & Glas, Papier, Verpack> corrugated
Wellblech n <Bau, Maschinen, Metall> corrugated iron, corrugated sheet iron
Welle f 1. <Akustik> wave; 2. <Bau> axle, shaft; roller *(Rollladen)*; 3. <Comp & DV> mode; 4. <Elektriz, Elektronik, Elektrotech> shaft, wave; 5. <Eisenbahn> shaft; 6. <Fertig> shaft; 7. <Ker & Glas> buckle; 8. <Kfztech> shaft *(Motor, Triebstrang)*; 9. <Maschinen> *(AE)* arbor, *(BE)* arbour, shaft; 10. <Mechan> shaft; 11. <Papier> axle; flute *(der Wellpappe)*; 12. <Phys, Telekom> wave; 13. <Wassertrans> wave; shaft *(Schiffantrieb, Motor)*
Welle f **im Hörbereich** <Elektronik> acoustic wave
Wellen n 1. <Fertig> shafting; setting *(Zahnreihe)*; 2. <Maschinen> shafting
Wellen fpl **in Fluiden** <Strömphys> fluid waves
Wellen fpl **in Flüssigkeiten** <Strömphys> fluid waves
Wellenabschattungseffekte mpl <Telekom> wave-shadowing effects
Wellenanlage f <Wassertrans> shaft unit *(Antrieb)*
Wellenantenne f <Funktech> Beverage aerial
Wellenantrieb m 1. <Kfztech> shaft drive *(Triebstrang)*; 2. <Maschinen> shaft drive
Wellenausbreitung f <Funktech, Nichtfoss Energ, Phys, Wellphys> wave propagation
Wellenausbreitungsgeschwindigkeit f *(c)* <Hydraul> wave celerity *(c)*
Wellenausgleichskupplung f <Maschinen> resilient shaft coupling
Wellenband n <Strahlphys, Wellphys> waveband
Wellenbauch m 1. <Akustik> wave loop; 2. <Elektrotech, Wellphys> antinode
Wellenbereich m 1. <Funktech> waveband; 2. <Telekom> range
Wellenbereichsschalter m <Funktech> band switch

Wellenberg m <Nichtfoss Energ> wave crest
Wellenbeugung f <Telekom> wave diffraction
Wellenbewegung f <Wellphys> wave motion
Wellenbewegungsenergie f **pro Meter Woge** <Nichtfoss Energ> *(AE)* wave momentum per meter of crest, wave momentum per metre of crest
Wellenbezeichnung n <Maschinen> shaft designation
Wellenbrecher m 1. <Bau> breakwater; 2. <Wassertrans> breakwater, splashboard; 3. <Wasserversorg> breakwater
Wellenbrecher m **aus Bruchstein** <Bau> rubble-mound breakwater
Wellenbund m <Maschinen> shaft collar
Wellendämpfungskoeffizient m <Elektronik, Telekom> wave attenuation coefficient
Wellendichtring m <Maschinen> shaft seal, shaft-sealing ring
Wellendichtung f <Maschinen> shaft seal, shaft sealing
Wellendrahtglas n <Ker & Glas> corrugated-wired glass
Wellenende n <Maschinen> shaft end
Wellenenergie f <Wellphys> wave energy, wave power
Wellenerzeuger m <Wassertrans> wave generator *(Schiffkonstruktion)*
Wellenerzeugung f <Telekom> wave generation
Wellenfilter n <Elektriz, Wellphys> wave filter
Wellenfolge f <Akustik> wave train
Wellenform f 1. <Comp & DV, Elektriz, Elektrotech, Phys, Telekom> waveform, waveshape; 2. <Wellphys> waveform
Wellenformsynthese f <Elektronik> waveform synthesis
Wellenfortpflanzung f <Nichtfoss Energ> wave propagation
Wellenfront f <Akustik, Comp & DV, Elektrotech, Optik, Phys, Telekom, Wellphys> wavefront
Wellenfrontbereich m <Comp & DV> wavefront array
Wellenfunktion f <Phys, Teilphys, Wellphys> wave function
Wellengenerator m 1. <Wassertrans> shaft-driven generator *(Antrieb)*; 2. <Wellphys> wave generator
Wellengeschwindigkeit f <Wellphys> wave velocity
Wellenglas n <Ker & Glas> corrugated glass
Wellengleichung f <Phys, Strahlphys> wave equation
Wellengruppe f 1. <Akustik> wave train; 2. <Phys> wave group
Wellenhöhe f <Wassertrans> wave height
Welleninterferenz f <Akustik, Telekom> wave interference
Wellenkamm m <Nichtfoss Energ> wave crest
Wellenkeil m <Maschinen> shaft key
Wellenkohärenz f <Telekom> wave coherence
Wellenkonstante f (k) <Akustik> wave constant (k)
Wellenkopf m <Akustik> wavefront
Wellenkopplung f 1. <Elektrotech> shaft coupling; 2. <Telekom> wave coupling
Wellenkuppelung f <Maschinen> shaft coupling
Wellenlager n <Maschinen> shaft bearing
Wellenlänge f 1. <Akustik, Elektriz, Elektronik, Funktech, Metall, Optik, Phys, Strahlphys, Telekom> wavelength; 2. <Wassertrans> wavelength *(Seezustand)*; 3. <Wellphys> wavelength
Wellenlänge f **der maximalen Helligkeit** <Optik> peak intensity wavelength
Wellenlänge f **höchster Strahlungsintensität** <Telekom> peak intensity wavelength
Wellenlängen fpl **der Spektrallinien** <Strahlphys> wavelengths of spectral lines
Wellenlängenduplex n <Telekom> wavelength division duplex *(Faseroptik)*

Wellenlängenmultiplex n (WDM) <Optik, Telekom> wavelength division multiplexing *(WDM)*
Wellenlängen-Multiplex n **mit hoher Kanaldichte** <Telekom> high density wavelength division multiplex *(Faseroptik)*
wellenlängenselektiver Koppler m <Telekom> wavelength-selective coupler *(Faseroptik)*
Wellenlängenumschaltung f <Telekom> wavelength switching
Wellenläppmaschine f <Maschinen> shaft-lapping machine
Wellenleiter m 1. <Elektronik, Elektrotech> wave duct, waveguide; 2. <Funktech, Phys, Telekom, Wellphys> waveguide
Wellenleiterdispersion f <Optik, Telekom> waveguide dispersion
Wellenleiterfilter n <Elektronik> waveguide filter
Wellenleitermodus m <Telekom> waveguide mode
Wellenleiterphasenschieber m <Telekom> waveguide phase shifter
Wellenleitwert m <Elektrotech, Phys> admittance
Wellenmechanik f <Elektrotech, Phys, Wellphys> wave mechanics
Wellenmesser m 1. <Phys> wavemeter; 2. <Wassertrans> wavemeter *(Funk)*; 3. <Wellphys> wavemeter
Wellenoberfläche f <Elektrotech> wave surface
Wellenoptik f 1. <Optik> physical optics, wave optics; 2. <Telekom> wave optics
Wellenpaket n <Phys> wave packet
Wellenparameterfilter n <Aufnahme> composite filter
Wellenperiode f <Wassertrans> wave period *(Schiffkonstruktion)*
Wellenpferdestärke f (WPS) <Maschinen> shaft horse power
Wellenpolarisation f <Telekom> wave polarization
Wellenreiten n 1. <Wassertrans> surging *(Bewegung in Längsrichtung, Schiffsbewegung)*; 2. <Wellphys> surfing
Wellensieb n <Wellphys> wave filter
Wellenspektrum n <Wellphys> wave spectrum
Wellenstrang m <Maschinen> line, line of shafting, line shaft, shafting
Wellenstreuung f <Telekom> wave dispersion
Wellental n <Wassertrans> trough *(Meer)*
Wellentheorie f **des Lichtes** <Wellphys> wave theory of light
Wellentrennlänge f <Fertig> cutoff *(Rauheit)*
Wellentyp m 1. <Comp & DV> mode *(Datenübertragung)*; 2. <Elektronik> mode *(Mikrowellen)*
Wellentypfilter n <Elektronik> mode filter
Wellentypumwandlung f <Comp & DV> mode conversion
Wellenüberlagerung f <Akustik, Telekom> wave interference
Wellenübertragung f <Telekom> wave transmission
Wellenübertragungsmaß n 1. <Akustik> image transfer exponent; 2. <Elektronik> image transfer coefficient; 3. <Telekom> image transfer exponent, image transfer coefficient
Wellenvektor m <Elektrotech, Phys> wave vector
Wellenverbindung f <Maschinen> connection between two shafts
Wellenverstärkung f <Telekom> wave amplification
Wellenwicklung f <Elektrotech> wave winding
Wellenwiderstand m 1. <Akustik> image impedance; 2. <Elektriz, Elektrotech> characteristic admittance, image impedance, iterative impedance, surge impedance; 3. <Lufttrans> wave drag; 4. <Phys> characteristic impedance, image impedance; 5. <Wassertrans> wave resistance *(Schiffkonstruktion)*

Wellenwiderstand

Wellenwiderstand *m* **des freien Raumes** <Phys> field impedance, free-space impedance
Wellenwiderstand *m* **im Vakuum** <Phys> characteristic vacuum impedance, impedance of free space
Wellenzahl *f* <Akustik, Wellphys> wave number
Wellenzapfen *m* <Kfztech> journal of a shaft
Wellenzug *m* <Akustik, Phys, Wellphys> wave train
Welle-Teilchen-Dualismus *m* <Wellphys> wave particle duality
Welle-Teilchen-Dualität *f* <Phys> wave particle duality
Wellfaserplatte *f* <Verpack> *(AE)* corrugated fiber board, *(BE)* corrugated fibre board
Wellglas *n* <Ker & Glas> corrugated glass
wellig *adj* 1. <Fertig> undulating; 2. <Ker & Glas> wavy; 3. <Telekom> undulating
wellige Oberfläche *f* <Telekom> undulating surface
wellige Schliere *f* <Ker & Glas> wavy cord
Welligkeit *f* 1. <Akustik, Elektriz, Elektronik, Telekom> ripple; 2. <Ker & Glas> settle mark; 3. <Maschinen> waviness
Welligkeitsdämpfung *f* <Elektronik> ripple attenuation
Welligkeitsfaktor *m* <Funktech, Phys, Telekom> standing-wave ratio, voltage standing wave ratio
Welligkeitsmessgerät *n* <Gerät> ripple measuring equipment
Wellpapier *n* <Verpack> corrugated paper
Wellpappe *f* <Papier, Verpack> corrugated board, corrugated cardboard
Wellpappenkarton *m* <Verpack> corrugated board box
Wellpappenmaschine *f* <Papier> corrugator
Wellpappenprodukt *n* <Verpack> corrugated product
Wellrohr *n* 1. <Gerät> pressure bellows; 2. <Maschinen> corrugated pipe
Wellrohrmanometer *n* <Gerät> *(AE)* bellows gage, *(BE)* bellows gauge
Wellung *f* <Fertig, Ker & Glas, Papier> corrugation
Welt *f* <Raumfahrt> space
Welt *f* **mit Zugang zum Internet** <Comp & DV> Internet enabled world
Weltall *n* <Raumfahrt> outer space, space
Weltentstehungslehre *f* <Raumfahrt> cosmogony
Weltfernmeldewesen *n* <Telekom> worldwide communications
Weltmeer *n* <Wassertrans> ocean
Weltraum *m* 1. <Raumfahrt> outer space, space; 2. <Telekom> space
Weltraumbahnhof *m* <Raumfahrt> spaceport
Weltraumbeschreibung *f* <Raumfahrt> cosmography
Weltraumdruckanzug *m* <Raumfahrt> extra-vehicular pressure garment
Weltraumeignung *f* <Raumfahrt> space qualification
Weltraumfahrer *m* <Raumfahrt> cosmonaut
Weltraumfahrtbehörde *f* <Raumfahrt> space agency
Weltraumforschung *f* <Raumfahrt> space research
Weltraumfunk *m* <Funktech, Telekom> space radio communication
Weltraum-Funkverbindungsnetz *n* <Funktech, Telekom> deep space network
Weltraumhafen *m* <Raumfahrt> spaceport
Weltraumkapsel *f* <Raumfahrt> space capsule
Weltraumkommunikation *f* <Funktech, Telekom> space radio communication
Weltraumnachrichtentechnik *f* <Telekom> space communications
Weltraumobjekt *n* <Raumfahrt> object in space
Weltraumprogramm *n* <Raumfahrt> space program
Weltraumrendezvous *n* <Raumfahrt> space rendezvous
Weltraumschlepper *m* <Raumfahrt> space tug
Weltraumschuss *m* <Raumfahrt> space shot

Weltraumsonde *f* <Raumfahrt> deep-space probe, space probe
Weltraumstart *m* <Raumfahrt> space launch
Weltraumtechnik *f* <Raumfahrt> space engineering, space technology
Weltraumtelekommunikation *f* <Telekom> space communications
Weltraumtransponder *m* <Funktech, Raumfahrt, Telekom> deep-space transponder
Weltraumumgebung *f* <Raumfahrt> space environment
Weltraumzeitalter *n* <Raumfahrt> space age
Weltraumzentrum *n* <Raumfahrt> *(AE)* space center, *(BE)* space centre
weltumspannendes Funknetz *n* <Raumfahrt> worldwide network
weltumspannendes Netz *n* <Telekom> global area network *(Datennetz)*
weltweite Abdeckung *f* <Wassertrans> global coverage *(Satellitensysteme)*
weltweite Emissionen *fpl* <Umweltschmutz> global emissions
weltweite Ortssuche *f* <Telekom> global location sensing *(GLS)*
weltweite Überdeckung *f* <Funktech, Telekom> global coverage
weltweites Mobilfunksystem *n* <Mobilkom> global system for mobile communication *(GSM)*
weltweites Netz *n* <Telekom> global network
weltweites Netz *n* **der mobilen Kommunikation** <Mobilkom> global system for mobile communication
weltweites Ortungssystem *n* *(GPS)* <Funkort> global-positioning system, GPS
weltweites Positioniersystem *n* <Telekom> global positioning system *(GPS)*
Weltzeit *f* <Phys> universal time, UT
Wende *f* <Wassertrans> tack *(Segeln)*
Wendebecken *n* <Wassertrans> turning basin *(Hafen)*
Wendeformblasverfahren *n* <Ker & Glas> *(AE)* turn mold blowing, *(BE)* turn mould blowing
Wendegenauigkeit *f* <Fertig> indexability
Wendegetriebe *n* 1. <Fertig> feed-drive reverse *(Vorschub)*; 2. <Maschinen> reverse gear, reversing gear
Wendeherz *n* <Maschinen> tumbler gear
Wendekette *f* <Fertig> sling chain *(Schmieden)*
Wendeklappe *f* <Hydraul> flap valve
Wendel *f* 1. <Elektrotech> filament; 2. <Maschinen> spiral; 3. <Mechan> coil
Wendelabtastung *f* 1. <Elektronik> helical scanning; 2. <Funkort> helical scanning *(Radar)*
Wendelantenne *f* 1. <Funktech> helix, helix antenna; 2. <Telekom> corkscrew antenna, helical antenna, spiral antenna
Wendelfeder *f* <Maschinen> spiral coiled spring
wendelförmiger Wellenleiter *m* <Elektrotech, Funktech, Telekom> helix waveguide
wendelgekoppelte Wanderwellenröhre *f* <Elektronik> *(AE)* helix-traveling wave tube, *(BE)* helix-travelling wave tube
Wendelhohlleiter *m* 1. <Elektrotech, Funktech> helix waveguide; 2. <Telekom> helix waveguide, spiral waveguide
Wendeln *n* <Elektrotech> helixing
Wendelresonatorfilter *n* <Funktech> helical filter
Wendelspan bilden *v* <Fertig> helix
Wendelspur *f* <Optik> spiral track
Wendelstufe *f* <Bau> winder
Wendeltreppe *f* <Bau> corkscrew stairs, helical staircase, spindle stairs, spiral stairs
Wendemotor *m* <Elektrotech> reversible motor

wenden v 1. <Maschinen> reverse; 2. <Wassertrans> cant; turn *(Schiff)*; 3. <Wassertrans> come-to *(Schiffführung)*; turn *(Schiff)*; go-about, tack *(Segeln)*
wenden v **an/sich** <Mechan> apply for
Wenden n 1. <Fertig> rolling, rolling over; 2. <Ker & Glas> turnover; 3. <Lufttrans> *(AE)* turnaround, *(BE)* turnround *(eines Luftfahrzeugs)*; 4. <Trans> turning *(Fahrzeug)* • **im Wenden begriffen sein** <Wassertrans> be in stays *(Segeln)*
Wenden n **der Vorform** <Ker & Glas> *(AE)* blank mold turnover, *(BE)* blank mould turnover
Wendeplatz m <Bau> *(BE)* turnaround
Wendepol m <Elektriz> interpole, commutating pole
Wendepresse f <Papier> reversed press
Wendepressenfilz m <Papier> reversed press felt
Wendepunkt m 1. <Bau> turning point; 2. <Geom> point of inflection; 3. <Math> reversal point *(eines Funktionsgraphen)*
Wendeschalter m <Elektrotech> reversing switch
Wendeschiene f <Eisenbahn> reversing rail
Wendeschneidplatte f <Maschinen> indexable insert
Wendeschütz n <Elektriz> reversing contactor
Wendezug m <Eisenbahn> push-pull train
Wendung f 1. <Fertig> index *(Schneidplatte)*; 2. <Trans> turning *(Straßen)*
Wenigsprecher m <Telekom> low-calling-rate subscriber
Wenn-Dann-Regel f <Künstl Int> condition-action rule, if-then rule
Werbeblock m <Fernseh> advertising slot
Werbefotografie f <Foto> advertising photography
werben v <Druck> advertise
Werbung f <Fernseh> commercial
Werfen n <Foto> buckling *(Emulsion)*
Werft f <Wassertrans> dockyard
Werftindustrie f <Wassertrans> shipyard industry
Werftkran m <Wassertrans> shipbuilding crane
Werg n 1. <Fertig> oakum; 2. <Textil> tow; 3. <Wassertrans> oakum *(Tauwerk)*
Werk n 1. <Fertig> achievement, work; 2. <Maschinen> factory, mill, plant, works • **im Werk justiert** <Mechan> factory-adjusted
Werkbank f <Bau, Fertig, Maschinen> bench, workbench
Werkbank-Waagerechtstoßmaschine f <Maschinen> bench-type shaping machine
Werkbankzwinge f <Maschinen> holdfast
Werkblei n <Fertig> raw lead
Werkkanal m <Nichtfoss Energ, Wasserversorg> headrace canal
Werkprüfung f <Qual> manufacturer's inspection, shop test
Werksatz m <Druck> book composition
Werksbescheinigung f <Qual> certificate of compliance with order
Werkschrift f <Druck> body type, composition sizes
Werkshilfstransformator m <Kerntech> unit auxiliary transformer
Werkskontrolle f <Qual> manufacturer's quality control
Werksprüfprotokoll n <Qual> work test report
Werksprüfung f <Werkprüf> factory test
Werksprüfzeugnis n <Qual> work certificate
Werkssachverständiger m <Qual> factory-authorized inspector
Werkstatt f 1. <Fertig> shop; 2. <Ker & Glas> shop *(Arbeitsplatz)*; 3. <Maschinen> shop; 4. <Mechan> factory, workshop, shop
Werkstattabnahme f <Mechan> factory acceptance
Werkstattabnahmelehre f <Maschinen> *(AE)* factory acceptance gage, *(BE)* factory acceptance gauge
Werkstattgleis n <Eisenbahn> repair track, rip track
Werkstatthandbuch n <Kfztech> overhaul manual
Werkstattmikroskop n <Metrol> toolmaker's microscope
Werkstattniet m <Maschinen> shop rivet
Werkstattprüfgerät n <Phys> work standard
Werkstattprüfung f <Qual> shop test
Werkstattzeichnung f <Konstzeich> workshop drawing
Werkstein m <Bau> cut stone
Werkstoff m <Abfall, Maschinen, Qual, Textil, Werkprüf> materials
Werkstoffeigenschaft f <Werkprüf> materials characteristic
Werkstofffehler m <Werkprüf> materials flaw
Werkstoffkenngröße f <Werkprüf> materials characteristic
Werkstoffnachweis m <Qual> materials verification
Werkstoffnutzungszyklus m <Abfall> utilization cycle of materials
Werkstoffprüfprotokoll n <Qual> materials test certificate
Werkstoffprüfsystem n <Textil> materials-testing system
Werkstoffprüfung f 1. <Qual> materials inspection, materials testing; 2. <Werkprüf> materials testing
Werkstoffspezifikation f <Qual> materials specification
Werkstoffsteifigkeit f <Werkprüf> materials stiffness
werkstofftechnisches Qualitätsmerkmal n <Qual> materials quality feature
Werkstoffvorschub m <Fertig> bar feed
Werkstoffzugabe f <Maschinen> materials allowance
Werkstoffzylinder m <Fertig> wad
Werkstück n <Fertig, Maschinen, Mechan> workpiece
Werkstückabmessungen fpl <Maschinen> workpiece dimensions
Werkstückaufnahme f <Maschinen> workholding
Werkstückbewegung f <Fertig> workpiece motion
Werkstückdurchmesser m <Fertig> workpiece diameter
Werkstücknocken m <Fertig> workpiece cam contour
Werkstückspannvorrichtung f <Maschinen> workholding device
Werkstückstange f <Fertig> bar stock
Werkstückträger m <Fertig> pallet *(Fertigungsstraße)*; platen *(Transferstraße)* • **ohne Werkstückträger** <Fertig> nonpalletized
werksverdrahtet adj <Funktech, Telekom> shop-wired
Werktisch m <Maschinen> work table
Werkzeichnung f <Maschinen> working drawing
Werkzeug n 1. <Bau> utensil; 2. <Comp & DV> tool; 3. <Fertig> chase, tool, die; 4. <Ker & Glas> tool *(mit flacher Klinge)*; 5. <Kunststoff> *(AE)* mold, *(BE)* mould; 6. <Maschinen, Mechan> implement, tool; 7. <Textil> appliance
Werkzeug n **aus gekohltem Stahl** <Maschinen> carbon steel tool
Werkzeug n **mit Hartmetallschneide** <Mechan> carbide-tipped tool
Werkzeug n **mit Zylinderschaft** <Maschinen> parallel-shank tool
Werkzeug n **zum Drehen von der Stange** <Maschinen> bar-turning tool
Werkzeugatmung f <Kunststoff> *(AE)* mold breathing, *(BE)* mould breathing
Werkzeugauflage f <Maschinen> tool rest
Werkzeugausgabe f <Fertig> tool crib
Werkzeugbasisfreiwinkel m <Fertig> tool base clearance
Werkzeugbestimmungsgröße f <Fertig> tool element
Werkzeugbohrung f <Fertig> tool bore
Werkzeugeckenwinkel m <Fertig> tool-included angle
Werkzeugeinsatz m 1. <Kunststoff> *(AE)* mold insert, *(BE)* mould insert; 2. <Maschinen> tool bit

Werkzeug-Einstell...

Werkzeug-Einstellergänzungswinkel m <Fertig> tool approach angle
Werkzeugeinstellung f <Fertig> tooling
Werkzeugeinstellwinkel m <Fertig> tool cutting-edge angle
Werkzeugfreiwinkel m <Fertig> tool clearance
Werkzeugführung f <Fertig> die set
Werkzeugfutter n <Fertig> tool chuck
Werkzeuggeometrie f <Fertig> tool geometry
Werkzeuggriff m <Bau> shank
Werkzeughalter m <Maschinen> cutter bar, tool carrier, tool holder, tool post
Werkzeugkasten m <Maschinen> tool box
Werkzeugmacher m <Maschinen> toolmaker
Werkzeugmacherdrehmaschine f <Maschinen> toolmaker's lathe
Werkzeugmaschine f 1. <Fertig> machine tool; 2. <Kontroll> machine tool; 3. <Maschinen, Mechan> machine tool
Werkzeugmaschinenhydraulik f <Fertig> machine tool circuit
Werkzeugmaschinensteuerung f <Kontroll> machine tool control
Werkzeugneigungswinkel m <Fertig> tool cutting-edge angle, tool cutting-edge inclination
Werkzeugnis n <Qual> work test report
Werkzeugrückebene f <Fertig> tool back plane
Werkzeugrückfreiwinkel m <Fertig> tool back clearance
Werkzeugrückkeilwinkel m <Fertig> tool back wedge angle
Werkzeugrückspanwinkel m <Fertig> tool back rake
Werkzeugsatz m <Maschinen> kit, tool set
Werkzeugschaft m <Maschinen> tool shank
Werkzeugschleifen n <Maschinen> tool grinding, tool sharpening
Werkzeugschleifmaschine f <Maschinen> tool grinder, tool sharpener
Werkzeugschlitten m 1. <Maschinen> carriage, tool carriage, tool holding slide, tool slide; 2. <Mechan> carriage
Werkzeugschneide f 1. <Fertig> tool tip; 2. <Maschinen> tool edge
Werkzeugschneidenebene f <Fertig> tool cutting-edge plane
Werkzeugspannvorrichtung f <Maschinen> tool holding fixture
Werkzeugstahl m <Metall> tool steel
Werkzeugstandzeit f <Maschinen> tool life
Werkzeugteil n <Fertig> tool element
Werkzeugträger m <Maschinen> tool box
Werkzeugwechsel m <Maschinen> retooling
Werkzeugwechseleinrichtung f <Maschinen> tool changing system
Werkzeugwinkel m <Fertig, Maschinen> tool angle
Wert m <Comp & DV, Ergon, Labor, Math, Telekom> value
Wertanalyse f 1. <Ergon> value analysis, value analysis engineering; 2. <Fertig, Kfztech, Qual> value analysis, VA
Wertaufruf m <Comp & DV> call by value
Wertebereich m 1. <Comp & DV, Gerät> range; 2. <Math> domain, range (bei Variablen)
Wertebereich m einer Funktion <Math> co-domain, range, value range
Wertereihe f <Gerät> range
Werteverlauf m <Regelung> value band, value pattern
Wertigkeit f <Phys> valence, valency
wertlose Daten npl <Comp & DV> garbage
Wertstoff m <Abfall> valuable substance (verwertbarer Bestandteil des Abfalls)
Wertstoffelement n <Kerntech> valuable element
Wertstoffrückgewinnung f <Abfall> resource recovery

Wertziffer f <Mechan> quality index
wesentliche Abweichung f <Qual> significant nonconformance
wesentliches Merkmal n <Patent> essential feature
Westeuropäische Zeit f (WEZ) <Mechan> Greenwich Mean Time (GMT)
Weston-Element n <Elektriz, Elektrotech> Weston standard cell
Westwindgürtel m <Wassertrans> roaring forties
Wetter n <Nichtfoss Energ, Wassertrans> weather (Meteorologie)
Wetterbeobachtungssatellit m <Telekom> meteorological observation satellite, meteorological satellite
Wetterbeobachtungsschiff n <Wassertrans> weather ship
Wetterbericht m <Wassertrans> weather report
Wetterbericht m für Landung <Lufttrans> weather report for landing
Wetterbericht m für Start <Lufttrans> weather report for takeoff
Wetterdeck n <Wassertrans> weather deck
Wetterdrosseltür f <Kohlen> air regulator
Wetterdüse f <Kohlen> air nozzle
Wettereinzugstrecke f <Kohlen> air intake
Wetterfahne f <Bau> vane
wetterfest adj <Bau, Papier> weatherproof
wetterfeste Gummidichtung f <Kfztech> rubber weatherproof seal
Wetterfestigkeit f <Maschinen> weathering resistance, weatherproofness
wetterhart adj <Wassertrans> weather-beaten
Wetterlage f <Wassertrans> atmospheric conditions
Wettermuster n <Raumfahrt> weather pattern
Wetterradar n <Funkort> meteorological radar, weather radar
Wettersatellit m <Raumfahrt, Telekom> meteorological satellite
Wetterschacht m <Sicherheit> ventilation shaft (Bergwerk)
Wetterschutz m <Wassertrans> shelter
Wetterschutzabdeckung f <Bau> weathering
Wetterschutzdach n <Bau> hood
Wetterschutzturm m <Elektrotech> weather protecting tower (für Impulsspannungsgeneratoren)
Wettersonde f <Telekom> meteorological sonde, radio sonde
Wettersondenempfangsstation f <Telekom> meteorological sonde receiving station, radio sonde observation station
Wettersprengstoff m <Sicherheit> safety explosive
Wetterstationsschrank m <Wassertrans> weather station cabinet
Wetterwechsel m <Kohlen> circulation of the air
wetzen v <Maschinen> sharpen
Wetzstein m <Fertig> oilstone, whetstone
WEZ (Westeuropäische Zeit) <Mechan> GMT (Greenwich Mean Time)
Wheatstone'sche Brücke f <Elektriz, Elektrotech, Phys> Wheatstone bridge
Whipstock m <Erdöl> whipstock
Whisker m <Kunststoff> whisker
Whitworth-Gewinde n <Maschinen> BSW thread, British Standard Whitworth thread, Whitworth screw thread
Whole-Sale-Anbieter m <Telekom> whole-sale-carrier
Wichtekurve f <Kohlen> specific gravity curve
Wichtestufe f <Kohlen> specific gravity fraction
Wichtigkeit f <Patent, Qual> significance
wichtigste Ziffer f <Comp & DV> most significant digit

wichtigstes Zeichen n <Comp & DV> most significant character, most significant digit
Wickel... <Druck, Papier, Verpack> wraparound
Wickelfeder f <Maschinen> coil spring, coiled spring
Wickelkern m 1. <Comp & DV> hub; 2. <Elektriz> wound core
Wickelkondensator m <Elektrotech> paper capacitor
Wickelkörper m <Textil> package
wickeln v 1. <Bau> wind; 2. <Fertig> coil *(Feder)*; 3. <Maschinen> wind, wrap; 4. <Phys> wind
Wickeln n <Maschinen> winding, wrapping
Wickelpappenmaschine f <Papier> intermittent board machine, wet-board machine
Wickelplatte f <Druck> wrapround plate
Wickelrohr n <Verpack> laminated tube
Wickelwalze f <Papier> winding drum
Wickelwatte f <Textil> lap
Wickelwerk n <Textil> wind-up
Wicklung f 1. <Elektriz, Elektrotech> turn, winding; 2. <Fertig> coil *(Kunststoffinstallationen)*; 3. <Kfztech> winding *(Generator)*; 4. <Mechan> coil; 5. <Papier, Phys> winding
Wicklung f **für drei Nuten** <Elektriz> three-slot winding
Wicklung f **mit Anzapfung** <Elektrotech> tapped winding
Wicklung f **mit Schrittverlängerung** <Elektrotech> long-pitch winding
Wicklungsdrahtabstand m <Elektriz> winding pitch
Wicklungseinführung f <Elektriz> feed-in of winding
Wicklungsisolierung f <Elektriz> winding insulation
Wicklungskapazität f <Elektriz, Elektrotech> winding capacitance
Wicklungsschritt m <Elektriz> winding pitch
Wicklungsverhältnis n <Kfztech> turns ratio
Widerdruck m <Druck> backup, perfecting
Widerhaken m <Maschinen> barb
Widerhakenbolzen m <Bau> barbed bolt
Widerhall m 1. <Akustik> echo; 2. <Aufnahme> echo, reverberation; 3. <Comp & DV, Elektronik, Funktech, Phys, Wellphys> echo
Widerlager n 1. <Bau> abutment *(Architektur)*; 2. <Fertig> abutment; 3. <Ker & Glas> skewback; 4. <Mechan> dolly
Widerlagerdruck m <Fertig> abutment pressure
Widerlagerstein m <Ker & Glas> skewback block
Widerruf m <Patent> revocation
widerrufen v <Patent> revoke
widerruflich adj <Patent> revocable
Widerspruchsfreiheit f <Qual> consistency
Widerstand m 1. <Anstrich> resistance; 2. <Elektriz> resistance, resistor *(R)*; 3. <Elektrotech> resistance, resistor; 4. <Funktech> resistor *(Bauteil)*; resistance *(Wert)*; 5. <Kfztech> resistor; 6. <Kunststoff, Maschinen> resistance; 7. <Mechan> drag; 8. <Phys> resistance; 9. <Strömphys> drag *(Kraft in Strömungsrichtung)*; drag on a sphere *(bei niedrigen Reynoldszahlen)*; 10. <Telekom, Textil> resistance; 11. <Thermod> temperature resistance *(gegen Hitze oder Kälte)*; 12. <Trans> resistance *(gegen Forwärtsbewegung)* • **mit Widerstand versehen** <Elektrotech> resistive
Widerstand m **bei positiver Phasenfolge** <Elektriz> positive phase sequence resistance
Widerstand m **gegen Metallverbindung** <Anstrich> refractory metal
Widerstand m **mit Anzapfung** <Elektrotech> tapped resistor
Widerstand m **pro Einheitslänge** <Phys> resistance per unit length
Widerstand m **von Körpern** <Fertig> drag *(Strömungslehre)*

Widerstandsabfall m <Elektriz> resistance drop
Widerstandsabgleich m <Elektrotech> resistor trimming
Widerstandsachse f <Lufttrans> drag axis
Widerstandsanlasser m <Elektrotech> rheostat starter
Widerstandsanpassungs-Schaltkreis m <Phys> impedance matching network
Widerstandsaufzeichnung f <Nichtfoss Energ> resistivity log
Widerstandsbeiwert m 1. <Lufttrans> coefficient of drag, drag coefficient; 2. <Nichtfoss Energ, Qual> coefficient of drag, drag coefficient *(DD)*; 3. <Strömphys, Wassertrans> coefficient of drag, drag coefficient *(Schiffkonstruktion)*
Widerstandsbremse f 1. <Eisenbahn> dynamic brake; 2. <Trans> rheostatic brake
Widerstandsbremsung f <Eisenbahn> rheostatic braking
Widerstandsbrücke f <Gerät> resistance bridge
Widerstandsbrücke f **aus Ohm'schen Widerständen** <Gerät> resistive bridge
Widerstandsdehnungsmessstreifen m <Gerät> *(BE)* resistor gauge, *(AE)* resistor gage
Widerstandsdraht m <Elektriz, Elektrotech, Metall> resistance wire
Widerstandsdrehmoment n <Kfztech> resisting torque
Widerstandsdünnschicht f <Elektronik> resistive thin film
Widerstandelement n <Elektrotech> resistive element
widerstandsfähig adj <Kunststoff> resistant
Widerstandsfähigkeit f 1. <Anstrich> durability, resistance; 2. <Comp & DV> robustness; 3. <Kunststoff> resistance
Widerstandsfähigkeit f **gegen Temperaturschock** <Anstrich> thermal shock resistance
Widerstandsferngeber m <Gerät> retransmitting slide wire
Widerstandsgeber m <Gerät> retransmitting slide wire
Widerstandsheizung f <Elektriz, Thermod> resistance heating
Widerstandsjustierung f <Elektrotech> resistor trimming
Widerstandskapazität f <Elektronik> resistance capacity
Widerstandskapazitätskopplung f <Elektrotech> resistance capacity coupling
Widerstandskasten m <Elektrotech> resistance box
Widerstandskennlinie f <Elektrotech> load line
Widerstandskern m <Elektrotech> resistor core
Widerstandskette f <Elektrotech> resistor string
Widerstands-Kondensator-Transistor-Logik f *(RCTL-Logik)* <Elektronik> resistor-capacitor-transistor logic *(RCTL logic)*
Widerstandskopplung f <Elektrotech> resistive coupling
Widerstandskörper m <Elektrotech> resistor
Widerstandskraft f 1. <Fertig> stamina; 2. <Metall, Nichtfoss Energ> drag *(Aerodynamik)*
Widerstandslast f <Elektriz, Elektrotech> resistive load
Widerstandsleiter f <Elektrotech> resistor ladder
Widerstandslog n <Erdöl> resistivity log *(Bohrlochvermessung)*
Widerstandslötung f <Fertig> electrode soldering
Widerstandsmanometer n <Gerät> *(BE)* resistance gauge, *(AE)* resistance gage
Widerstandsmaterial n <Elektrotech> resistance material
Widerstandsmessbrücke f <Gerät> resistance bridge
Widerstandsmesser m 1. <Elektriz> resistance meter; 2. <Funktech> ohmmeter
Widerstandsmessgerät n <Elektrotech, Phys> ohmmeter
Widerstandsmoment n <Fertig> section modulus
Widerstandsnahtschweißen n <Bau> resistance seam welding

Widerstandsnetz

Widerstandsnetz *n* <Elektrotech> resistor network
Widerstandsofen *m* <Elektrotech> resistance furnace
Widerstandspunktschweißen *n* <Elektriz> resistance spot welding
Widerstandsregler *m* <Elektrotech> rheostat
Widerstandsschaltung *f* <Elektrotech, Telekom> resistive circuit
Widerstandsschleifkontakt *m* <Kfztech> rheostat-sliding contact
Widerstandsschweißen *n* <Bau, Elektriz, Thermod> resistance welding
Widerstandsschweißung *f* <Maschinen, Thermod> resistance welding
Widerstandssonde *f* <Bau> resistivity probe
Widerstandsspannungsteiler *m* <Elektriz, Elektrotech> resistive voltage divider
Widerstandsspule *f* <Elektrotech> resistance coil
Widerstandsstumpfschweißen *n* 1. <Bau> resistance butt welding; 2. <Thermod> upset welding
Widerstandstemperaturmessfühler *m* <Gerät> resistance temperature detector
Widerstandsthermometer *n* <Heiz & Kälte> resistance thermometer
Widerstandsträger *m* <Elektrotech> resistor core
Widerstandsverminderung *f* <Strömphys> drag reduction
Widerstandszelle *f* <Elektronik> photoconductive cell
Widerstandszündkerze *f* <Kfztech> resistor-type spark plug
widerstehen *v* <Anstrich> resist
Widmannstätten Platte *f* <Metall> Widmannstätten plate
Widmannstätten Struktur *f* <Metall> Widmannstätten structure
Wiedemann-Franz'sches Gesetz *n* <Phys> Wiedemann-Franz law
wieder abgleichen *v* <Gerät> rebalance
wieder andocken *v* <Raumfahrt> redock
wieder anlassen *v* 1. <Anstrich> retemper; 2. <Kfztech> restart; 3. <Lufttrans> restart *(Motor und Triebwerk)*; 4. <Wassertrans> restart
wieder anlaufen *v* <Comp & DV> restart
wieder anschließen *v* <Elektrotech> reconnect
wieder ausbrechen *v* <Sicherheit> break out again *(Feuer)*
wieder ausstrahlen *v* <Fernseh> rebroadcast
wieder einkuppeln *v* <Maschinen> reengage
wieder einrichten *v* <Maschinen> reset
wieder einschalten *v* <Comp & DV> restart
wieder einschiffen *v* <Wassertrans> re-embark *(Passagiere)*
wieder flottmachen *v* <Wassertrans> refloat *(Schiff)*
wieder laden *v* <Bau> recharge
wieder vorstellen *v* <Qual> resubmit *(Prüflos)*
wieder zusammenbauen *v* <Bau> reassemble
Wieder... <Comp & DV, Elektrotech, Kerntech, Kontroll, Maschinen, Metall> re...
Wiederanlassen *n* <Metall> retempering
Wiederanlauf *m* <Comp & DV, Kontroll> restart
Wiederanlaufbefehl *m* <Comp & DV> restart instruction
Wiederanlaufenlassen *n* <Maschinen> restart *(die Maschinen)*
Wiederanlaufroutine *f* <Comp & DV> fall-back routine
Wiederanlegen *n* **von Wirbeln** <Strömphys> reattachment of eddies
Wiederanschluss *m* <Elektrotech, Telekom> reconnection
Wiederaufarbeitung *f* <Maschinen> regeneration
wiederaufbauen *v* <Bau> rebuild, reconstruct
Wiederaufbauen *n* <Bau> rebuilding, reconstruction
wiederaufbereiten *v* <Fertig> recondition *(Formsand)*

wiederaufbereiteter Altsand *m* <Fertig> reconditioned sand
Wiederaufbereitung *f* 1. <Kerntech> reprocessing; 2. <Maschinen> reconditioning; 3. <Meerschmutz> recovery
Wiederaufbereitungsanlage *f* <Abfall> reprocessing plant
Wiederauffindbarkeit *f* <Qual> retrieval
wiederauffinden *v* <Comp & DV, Kontroll> retrieve
• **Datensicherungsbänder wiederauffinden** <Comp & DV> retrieve data storage tapes
Wiederauffinden *n* <Comp & DV> retrieval
Wiederauffinden *n* **von Informationen** *(IR)* <Comp & DV> information retrieval *(IR)*
Wiederauffinden *n* **von Nachrichten** <Comp & DV> message retrieval
Wiederauffinden *n* **von Systemmeldungen** <Comp & DV> message retrieval
Wiederauffindung *f* **von Gepäck** <Lufttrans> baggage retrieval
Wiederauffindungssystem *n* <Comp & DV> reference retrieval system
Wiederaufheizzeit *f* <Thermod> comeback
Wiederaufkohlung *f* 1. <Chemie> recarburization *(Metallurgie)*; 2. <Metall> recarburization
wiederaufladbar *adj* <Elektrotech, Foto, Funktech> rechargeable
wiederaufladbare Batterie *f* <Elektrotech> rechargeable battery
wiederaufladen *v* <Elektrotech> recharge
Wiederaufladezeit *f* <Foto> recycle time, recycling time *(Blitz)*
Wiederaufladung *f* <Elektrotech> recharging
wiederauflebend *adj* <Wasserversorg> resurgent
Wiederaufnahme *f* 1. <Aufnahme> rerecording; 2. <Comp & DV> recovery
Wiederaufnahme *f* **nach Programmstopp** <Comp & DV> checkpoint recovery
Wiederauftauchen *n* <Wassertrans> return to surface *(U-Boot)*
Wiederaufwärmen *n* **von Glas zur Weiterverarbeitung** <Ker & Glas> warming-in
Wiederausgießen *n* <Fertig> relining
Wiederausrichten *n* <Bau> throwing back into alignment *(Mauer)*
Wiederbelebungsgerät *n* <Sicherheit> resuscitation equipment
Wiederbelegungsentfernung *f* <Mobilkom> reuse distance
Wiederbelüftung *f* <Wasserversorg> reaeration
wiederbenutzbar *adj* <Comp & DV> reusable
wiederbeschreibbare CD *f* <Comp & DV> compact disk-rewritable
Wiedereinschiffung *f* <Wassertrans> re-embarkation *(Passagiere)*
Wiedereinstiegspunkt *m* <Comp & DV> restart point
Wiedereintritt *m* **in die Atmosphäre** <Raumfahrt> atmospheric re-entry
Wiedereintrittshöhe *f* <Raumfahrt> earth reentry altitude
wiedererhitzen *v* <Ker & Glas> reheat
Wiedererhitzer *m* <Wassertrans> reheat
Wiedergabe *f* 1. <Akustik> reproducing; 2. <Aufnahme> playback, replay, reproduction; 3. <Comp & DV> image; 4. <Fernseh> playback, reproducing; 5. <Foto> reproducing
Wiedergabe *f* **der Tonwerte** <Foto> reproduction of tonal values
Wiedergabecharakteristik *f* 1. <Akustik> reproducing characteristic; 2. <Aufnahme> playback characteristics, replay characteristic

Wiedergabeeigenschaft f <Fernseh> reproduction characteristic
Wiedergabegerät n <Aufnahme> sound reader
Wiedergabegeschwindigkeit f <Aufnahme> playback speed
Wiedergabegüte f <Akustik, Ergon> fidelity
Wiedergabekette f <Fernseh> reproducing chain
Wiedergabekopf m <Fernseh> reproducing head
Wiedergabemagnetkopf m <Akustik> reproducing magnetic head
Wiedergabepegel m <Aufnahme> playback level
Wiedergabesteuerung f <Aufnahme> playback control
Wiedergabesystem n <Aufnahme> playback system
Wiedergabetreue f <Ergon> fidelity • **mit hoher Wiedergabetreue** <Aufnahme> high fidelity
Wiedergabeverlust m 1. <Akustik> reproducing loss; 2. <Aufnahme> playback loss; 3. <Fernseh> reproduction loss
Wiedergabeverstärker m <Aufnahme> playback amplifier
wiedergeben v <Aufnahme> reproduce
Wiedergefrieren n <Phys> regelation
wiedergewinnen v 1. <Abfall> recover (Rohstoffe); 2. <Kohlen, Telekom> recover; 3. <Textil> retrieve
Wiedergewinnung f 1. <Abfall> recycling; recovery (von Rohstoffen); 2. <Kohlen> re-extraction, recovery; 3. <Kunststoff, Thermod> recovery; 4. <Wasserversorg> backflow
Wiedergewinnung f von Wärme <Thermod> heat rate, heat rate curve, heat recovery
Wiedergewinnungsanlage f <Abfall> reclamation plant, resource recovery plant
Wiedergewinnungssystem n <Kerntech> recovery system
wiedergewonne Energie f <Kohlen, Nichtfoss Energ> recovered energy
wiedergewonnene Ladung f <Elektrotech> recovered charge
wiedergewonnener Rohstoff m <Abfall> secondary material
wiedergewonnenes Öl n <Abfall> recovered oil
wiederherstellen v <Anstrich, Comp & DV> restore
Wiederherstellen n der Daten im Katastrophenfall <Comp & DV> disaster recovery
Wiederherstellen n der ursprünglichen Form <Telekom> regeneration
Wiederherstellung f 1. <Anstrich> restoration; 2. <Comp & DV> recovery, restore; 3. <Raumfahrt> reconditioning (Raumschiff); 4. <Telekom> restore
Wiederherstellung f des Signals <Telekom> signal restoration
Wiederherstellungsmodus m <Raumfahrt> restoration mode
Wiederherstellungsprozedur f <Comp & DV> recovery procedure
wiederholbarer Ablauf m <Kontroll> routine
Wiederholbarkeit f <Gerät> repeatability
Wiederholbedingungen fpl <Qual> repeatability conditions
wiederholen v <Comp & DV> iterate, rerun
Wiederholgenauigkeit f <Metrol> repeating accuracy
Wiederholgrenze f <Qual> repeatability limit
Wiederholpräzision f <Qual> repeatability
wiederholt überprüfen v <Metrol> recheck
wiederholter Flugplan m <Lufttrans> repetitive flight plan
wiederholter Start m mit durch Fahrtwind angetriebenem Propeller <Lufttrans> windmilling
wiederholter Verbindungsversuch m <Telekom> repeated call attempt
wiederholter Zündversuch m <Kfztech> continual relight
wiederholtes Signal n <Elektronik> repeated signal
Wiederholung f 1. <Akustik> duplication; 2. <Comp & DV> rerun, retry; 3. <Fernseh> repeat, rerun
Wiederholungsanlauf m <Comp & DV> fall-back
Wiederholungslauf m <Comp & DV> rerun
Wiederholungsprobe f <Qual> retest specimen
Wiederholungsprüfung f <Qual> repeat test, retest • **einer Wiederholungsprüfung unterziehen** <Qual> retest
Wiederholungssendung f <Telekom> retransmission
Wiederholungssignal n <Eisenbahn> repeating signal
Wiederholungstaste f <Telekom> repeat key
Wiederinbetriebnahme f <Telekom> return to service
wiederkehrende Belastung f <Bau> repeating load
wiederkehrende Impulse mpl <Elektronik> recurrent pulses
wiederkehrende Kosten fpl <Raumfahrt> recurrent cost
Wiederkehrspannung f <Elektriz> recovery voltage
Wiedernutzbarmachung f <Umweltschmutz> rehabilitation
Wiedersynthese f <Künstl Int> resynthesis
wiederurbar gemachtes Gebiet n <Umweltschmutz> reclaimed area
Wiederurbarmachung f <Umweltschmutz> recultivation (von Land oder Wasser)
Wiedervereinigungskoeffizient m <Phys> recombination coefficient
Wiederverladung f <Wassertrans> reshipment, reshipping (Ladung)
wiederverwendbar adj <Comp & DV> re-entrant, reusable
wiederverwendbare Arbeitshandschuhe mpl <Sicherheit> recycling work gloves
wiederverwendbare Datei f <Comp & DV> reusable file
wiederverwendbare Datenbestand m <Comp & DV> reusable data set
wiederverwendbarer Karton m <Verpack> reusable box
wiederverwendbares Programm n <Comp & DV> reusable routine
wiederverwendbares Verpackungsmaterial n <Verpack> reusable packaging
Wiederverwendung f 1. <Abfall> recycling, reuse (einer Pfandflasche); 2. <Kohlen> recycling
Wiederverwendung f gewerbliche Abfälle <Abfall> reuse of industrial waste
wiederverwendungsfähiges Raumschiff n <Raumfahrt> recoverable orbiter
wiederverwertbar adj <Umweltschmutz> recyclable, reusable
wiederverwertbare Verpackung f <Verpack> returnable packaging
wiederverwertbarer Treibsatz m <Raumfahrt> recoverable thruster (Raumschiff)
wiederverwertbares Abfallprodukt n <Abfall, Umweltschmutz> reusable waste product
wiederverwerten v <Abfall> recycle, reuse
wiederverwertete Flasche f <Verpack> recycled bottle
Wiederverwertung f 1. <Abfall> recycling (von Abfallstoffen); 2. <Telekom> recycling; 3. <Umweltschmutz> recovery
Wiederzündspannung f <Elektrotech> reignition voltage
Wiederzündung f <Elektrotech> reignition, restriking
Wiege f <Eisenbahn> bolster
Wiege f des Drehgestells <Eisenbahn> bogie bolster
Wiegen n <Papier> weighing
Wiegenfederung f <Eisenbahn> secondary suspension
Wien'sche Brücke f <Elektronik, Phys> Wien bridge
Wien'sches Strahlungsgesetz n <Strahlphys> Wien law

Wien'sches Verschiebungsgesetz n <Phys> Wien displacement law
Wigner-Effekt m <Phys> Wigner effect
WIG-Schweißen n <Fertig> inert arc welding with non-consumable electrode
wilde Ablagerung f <Abfall, Wasserversorg> illegal dumping
wilde Deponie f <Abfall> wild dump
wilde Müllablagerung f <Abfall> fly tipping, open dump, uncontrolled tipping
wilde Schwingung f <Funktech, Phys, Telekom> parasitic oscillation
wilde See f <Wassertrans> confused sea *(Seezustand)*
Wildzaun m <Abfall> fence *(einer Deponie)*
Willison-Kupplung f <Kfztech> Willison coupling
willkürlich gewählte Achse f <Geom> arbitrary axis
willkürlich verteilte Wicklung f <Elektriz> random winding
willkürliche Konstante f <Phys> arbitrary constant
Wilson'sche Nebelkammer f <Phys> Wilson cloud chamber
Wimpel m <Wassertrans> pennant *(Flagge)*
Wimshurstmaschine f <Elektrotech> Wimshurst machine
Winchesterplatte f <Comp & DV> Winchester disk, fixed disk, *(AE)* disk, hard disk, HD
Wind m <Lufttrans, Wassertrans> wind • **dem Wind abgekehrt** <Nichtfoss Energ> downwind • **dicht am Wind** <Wassertrans> close-hauled *(Segeln)* • **gegen den Wind** <Wassertrans> upwind • **in den Wind drehen** 1. <Lufttrans> decrab; 2. <Wassertrans> bear down • **mit dem Wind** <Nichtfoss Energ, Umweltschmutz, Wassertrans> downwind *(Segeln)* • **mit halbem Wind segeln** <Wassertrans> sail on a beam reach *(Segeln)* • **mit rauem Wind** <Wassertrans> off the wind *(Segeln)* • **scharf am Winde** <Wassertrans> full and by *(Segeln)* • **vom Wind abfallen** <Wassertrans> pay off *(Segeln)* • **vor dem Wind drehen** <Wassertrans> wear *(Segeln)* • **vor dem Wind laufen** <Wassertrans> run before the wind *(Segeln)* • **vor dem Wind segeln** <Wassertrans> run before the wind *(Segeln)*
Windbelastung f <Sicherheit> wind load *(Kran)*
windbetriebener Generator m <Elektrotech> wind-driven generator
Windbrett n <Bau> side board
Wind-Diesel-Gleichstrom-Bussystem n <Nichtfoss Energ> wind-diesel DC bus system
Wind-Diesel-System n <Nichtfoss Energ> wind-diesel system
Winddruck m <Lufttrans, Nichtfoss Energ, Wassertrans> wind pressure
Winddruckschwerpunkt m <Wassertrans> *(AE)* center of wind pressure, *(BE)* centre of wind pressure *(Schiffkonstruktion)*
Winddurchdringung f <Nichtfoss Energ> wind penetration
Winde f 1. <Bau> gin, jack, winch; 2. <Lufttrans> hoist; 3. <Maschinen> capstan, lifting jack, winch, windlass; 4. <Mechan> hoist, jack; 5. <Wassertrans> winch *(Deckbeschläge)*
Windeisen n <Maschinen> tap wrench
winden v 1. <Maschinen> wind; 2. <Textil> reel
winden v/sich <Maschinen> coil, wind
Winden n <Bau, Maschinen> winding
Windenbohrer m <Fertig> bit brace
Windenergie f <Nichtfoss Energ, Phys> wind energy
Windenkopf m <Wassertrans> gipsy head *(Deckzubehör)*
Windentrommel f <Wassertrans> winch drum *(Deckbeschläge)*

Winderhitzer m <Heiz & Kälte, Thermod> Cowper stove, blast preheater
Windflügel m 1. <Bau> vane; 2. <Heiz & Kälte, Mechan> fan blade; 3. <Nichtfoss Energ> wind vane; 4. <Thermod> fan blade
windfrischen v <Fertig> bessemerize
Windfrischen n <Fertig> converting *(Gießen)*
Windfrischstahl m 1. <Fertig> blown metal; 2. <Fertig> Bessemer steel
Windfrischverfahren n <Fertig> Bessemer process
Windgenerator m <Elektriz, Elektrotech, Nichtfoss Energ> wind generator, wind-driven generator, wind-powered generator
Windgeschwindigkeit f 1. <Lufttrans> wind speed; 2. <Nichtfoss Energ> wind velocity; 3. <Wassertrans> wind speed
Windgeschwindigkeitsmesser m 1. <Nichtfoss Energ> *(AE)* wind gage, *(BE)* wind gauge; 2. <Phys> anemometer
Windgeschwindigkeitsmessgerät n <Nichtfoss Energ> wind monitoring logger
Windgeschwindigkeitsmessung f 1. <Nichtfoss Energ> wind monitoring; 2. <Phys> anemometry
windgetriebener Generator m <Elektriz, Nichtfoss Energ> wind-powered generator
windgetriebener Stromgenerator m <Elektriz, Nichtfoss Energ> wind-powered generator
Windhose f <Wassertrans> vortex *(Wetterkunde)*
Windjammer m <Wassertrans> tall ship *(Segeln)*
Windkanal m <Bau, Lufttrans, Maschinen, Phys, Raumfahrt, Wassertrans> wind tunnel
Windkanal m mit geschlossener Messstrecke <Phys> closed-throat wind tunnel
Windkanaleinfluss m <Lufttrans> tunnel effect
Windkanaltest m <Raumfahrt> wind tunnel testing
Windkanalwaage f <Lufttrans, Wassertrans> wind tunnel balance
Windkarte f <Wassertrans> wind chart
Windkasten m 1. <Fertig> air box; 2. <Maschinen> blast box
Windkessel m 1. <Fertig> air dome, dashpot, tank; 2. <Mechan, Phys> air cylinder
Windkraft f <Nichtfoss Energ> wind power
Windkraftanlage f <Elektriz, Nichtfoss Energ> wind generator
Windkraftwerk n <Elektrotech, Nichtfoss Energ> wind-electric power station
Windmessen n **in großen Höhen** <Nichtfoss Energ> high-level wind monitoring
Windmesser m <Labor, Nichtfoss Energ, Wassertrans> anemometer
Windmotorpumpe f <Nichtfoss Energ> windmill pump
Windmühle f <Nichtfoss Energ> windmill
Windmühlenflügel m <Nichtfoss Energ> windmill vane
Windmühlenrad n 1. <Lufttrans> engine windmilling *(Hubschrauber)*; 2. <Nichtfoss Energ> windmill
Windpressung f <Fertig> blast pressure
Windpumpe f <Nichtfoss Energ> windpump
Windrad n <Nichtfoss Energ> windmill
windrecht adj <Wassertrans> wind-rode *(Festmachen)*
Windrichtungsflügel m <Nichtfoss Energ> direction vane
Windringleitung f <Fertig> bustle pipe
Windrose f <Nichtfoss Energ> wind rose
Windsack m 1. <Lufttrans> wind cone, wind sock; 2. <Wassertrans> wind cone, wind sock, wind sail
Windschatten m <Wassertrans> lee
windschiefer Flug m <Lufttrans> drifting flight
windschlüpfrig adj <Lufttrans> faired

windschnittig adj <Maschinen> streamlined
Windschutz m 1. <Aufnahme> windshield; 2. <Bau> windbreak
Windschutzscheibe f <Eisenbahn, Ker & Glas, Kfztech> (BE) windscreen, (AE) windshield
Windschutzscheibengebläse n <Kfztech> demister system (Zubehör)
Windschutzscheibe-Verbundglas n <Kfztech> laminated windshield glass
Windseite f <Wassertrans> weather side
Windsichten n <Lebensmittel> air classification, air separation
Windsichter m <Abfall> air classifier, air separator, air separation plant
Windsichtung f <Abfall> airstream sorting
Windstärke f <Lufttrans, Wassertrans> wind speed
Windstille f 1. <Nichtfoss Energ> still air; 2. <Wassertrans> calm, lull
Windstoß m 1. <Lufttrans> blast; 2. <Raumfahrt, Wassertrans> gust
Windstrebe f <Bau> wind brace
Windsurfingbrett n <Wassertrans> sailboard
Windturbine f <Maschinen, Nichtfoss Energ> wind turbine
Windturbinengenerator m <Nichtfoss Energ> wind turbine generator
Windturbineninstallation f <Nichtfoss Energ> wind turbine installation
Windturbinenpumpe f <Nichtfoss Energ> windmill pump
Windturbulenzinformation f <Nichtfoss Energ> wind turbulence information
Wind- und Seetauglichkeit f <Wassertrans> wind-and-sea state capability handling
Windung f 1. <Elektriz> turn; 2. <Fertig> coil (Feder); 3. <Maschinen> coil, spire, thread, turn, winding; 4. <Phys, Wassertrans> turn (Tauwerk)
Windungsabstand m <Elektriz> coil pitch
Windungsrichtung f <Fertig> hand of coils (Feder)
Windungsübersetzung f <Phys> turns ratio
Windungsverhältnis n <Elektriz, Phys> turns ratio
Windungszahl f (N) <Elektriz> number of turns in a winding (N)
Windungszahl f pro Längeneinheit (n) <Elektriz> turns per unit length (n)
Windungszahlverhältnis n <Elektrotech, Kfztech> turns ratio
Windverband m <Bau> wind bracing
Windversetzung f <Wassertrans> leeway
windwärtig adj <Wassertrans> windward
windwärts adv <Wassertrans> windward
Windwiderstand m <Nichtfoss Energ, Strömphys> wind resistance
Windwinkel m <Wassertrans> yaw angle
Windwirbel m <Lufttrans, Wassertrans> wind eddy
Winkel m 1. <Fertig> set square, vortex; elbow (Kunststoffinstallationen); 2. <Maschinen> elbow joint, square; 3. <Math> angle; 4. <Papier, Phys, Wassertrans> angle
• geneigt um einen Winkel von <Geom> inclined at an angle of
Winkel m der magnetischen Inklination <Phys> angle of magnetic inclination
Winkel m der Totalreflexion <Optik> angle of total reflection
Winkel m/von zwei Ebenen gebildeter <Geom> dihedral angle
Winkelabhängigkeit f <Fertig> angle dependence
Winkelabrichteinrichtung f <Fertig> angle-dressing fixture
Winkelabstand m <Telekom> angular separation
Winkelabweichung f <Elektrotech> angular deviation

Winkelantrieb m <Mechan> angle drive
Winkelband n <Bau, Fertig> angle brace, angle tie
Winkelbeschleunigung f (α) <Mechan> angular acceleration (α)
Winkelbohrung f <Fertig> ninety-degree bore (Kunststoffinstallationen)
Winkeldämpfung f <Telekom> corner loss
Winkeldiversity n <Funktech> angle diversity, angular diversity
Winkeldurchmesser m <Raumfahrt> angular diameter (Raumfahrt)
Winkeleckleiste f <Bau> nosing
Winkeleisen n 1. <Bau> angle bar, angle iron; 2. <Eisenbahn> angle bar; 3. <Maschinen> angle iron; 4. <Mechan> angle bracket
Winkeleisengelenk n <Bau> angle iron joint
Winkelendmaß n <Metrol> (AE) angle gage, (BE) angle gauge
Winkelfehler m <Gerät, Metrol> angle error
Winkelfenster n <Bau> splayed window
Winkelfräsen n <Maschinen> angle milling
Winkelfräser m 1. <Fertig> angle cutter, angular milling cutter; 2. <Maschinen> angle cutter, angular milling cutter, dovetail cutter, dovetail-milling cutter, inverse dovetail cutter, single angle cutter; 3. <Mechan> angle cutter, angular milling cutter
Winkelfrequenz f <Akustik, Elektronik, Phys> angular frequency
Winkelfunktionen fpl <Math> trigonometrical functions
Winkelgelenk n <Bau> angle joint
Winkelgeschwindigkeit f 1. <Elektrotech, Fertig> angular velocity; 2. <Lufttrans> rate of turn; 3. <Maschinen, Nichtfoss Energ, Phys, Raumfahrt> angular velocity (Raumschiff)
Winkelgeschwindigkeit f der Präzession <Nichtfoss Energ> angular velocity of precession
winkelgetreu adj <Geom> isogonal
Winkelgetriebe n 1. <Bau> (AE) miter gear, (BE) mitre gear; 2. <Kfztech> angle transmission, ring and pinion gearing; 3. <Maschinen> angle transmission
Winkelhaken m <Druck> composing stick, setting stick
Winkelhalbierende f <Geom> bisector of an angle, bisectrix
Winkelhebel m 1. <Maschinen> bell crank lever; 2. <Metall> bent lever
Winkelhebelsystem n <Mechan> bell crank system
winkelig adj <Fertig, Maschinen, Papier> angled
winkelige Biegung f <Fertig, Maschinen> angle of bend
Winkelkaliber n <Fertig> angle pass (Werkstoffe)
Winkelkonsole f <Bau, Mechan> angle bracket
Winkelkonstante f <Telekom> phase coefficient, phase constant
Winkelkopf m <Bau> cross-staff head
Winkelkreuz n <Bau> cross staff (Vermessung)
Winkelkurbel f <Maschinen> bell crank
Winkellasche f 1. <Bau> knee brace; 2. <Eisenbahn> (BE) angle fishplate, (AE) applying of angle joint bar; 3. <Fertig> angle fishplate
Winkellehre f <Fertig, Metrol> (AE) angle gage, (BE) angle gauge
Winkelmaß n 1. <Konstzeich> angular dimension; 2. <Maschinen> engineer's square; 3. <Metrol> bevel square
Winkelmesser n 1. <Fertig> protractor; 2. <Geom> goniometer, protractor; 3. <Mechan> quadrant; 4. <Metrol> angle meter, protractor
Winkelmesserskale f <Mechan> quadrant scale
Winkelmessinstrument n <Metrol> protractor

Winkelmesssystem

Winkelmesssystem n <Gerät> angular position measuring system *(Steuerungstechnik)*
Winkelmessung f <Metrol> angle measurement
Winkelmodulation f <Elektronik, Funktech> angle modulation
Winkelmomentquantenzahl f *(J)* <Kerntech> total angular momentum quantum number *(J)*
Winkelmultiplex n <Telekom> angular division multiplex
Winkelplatte f <Maschinen> angle plate
Winkelprofil n 1. <Elektronik> L-section; 2. <Mechan> angle section
Winkelreflektorantenne f <Funktech> corner reflector aerial
Winkelreibahle f <Maschinen> angled reamer
Winkelrotorgeschwindigkeit f <Lufttrans> angular rotor speed
Winkelsäule f <Foto> angled column *(Vergrößerungsgerät)*
Winkelschälversuch m <Kunststoff> T-peel test, angle-peeling test
Winkelschiene f <Fertig> set square
Winkelschleifer m <Fertig> angle grinder
Winkelschnitt m <Papier> angle cut
Winkelschraubendreher m <Maschinen> angular screwdriver, offset screwdriver
Winkelsekunde f <Phys> second of arc
Winkelstahl m <Mechan> angle steel
Winkelstellung f **der Rotorblätter/nach oben gerichtete** <Lufttrans> coning angle *(Hubschrauber)*
Winkelstirnfräser m <Maschinen> single angle cutter
Winkelstoß m <Bau> angle joint
Winkelstoß m **mit Nut und Feder** <Bau> angular grooved-and-tongued joint
Winkelstrahl m <Fertig> angle bar
Winkelstreichwalze f <Papier> angle spread roll
Winkelstück n 1. <Labor> elbow; 2. <Maschinen> bracket; 3. <Mechan> elbow
Winkelstütze f <Bau> angle bracket
Winkelstützmauer f <Bau> cantilevered wall, cantilever retaining wall
Winkelträger m <Bau> L-beam
Winkeltrieb m <Maschinen> V drive
Winkelverdrängung f *(Φ)* <Akustik> angular displacement *(Φ)*
Winkelvergrößerung f <Phys> angular magnification
Winkelversatzverlust m <Optik> angular misalignment loss
Winkelverschiebung f <Elektriz, Gerät, Lufttrans> angular displacement
Winkelverschiebungsanfälligkeit f <Lufttrans> angular displacement sensitivity
Winkelverschraubung f <Maschinen> elbow screw joint
Winkelverstärkung f <Fertig> angle bracket
Winkelvoreilung f <Fertig> angle advance
Winkelzahn m <Maschinen> straight back tooth
Winkelzahngetriebe n <Mechan> herringbone gear
Winkligbiegen n <Fertig> angled bending
Winsch f <Meerschmutz> winch *(Winde)*
Winston-Kollektor m <Nichtfoss Energ> Winston collector
Winterlagerung f <Wassertrans> winter storage *(Schiff)*
Winterreifen m <Kfztech> *(AE)* snow tire, *(BE)* snow tyre *(Reifen)*
Wippenbank f <Maschinen> pole lathe
Wippenfederung f <Eisenbahn> bow suspension
Wipper m <Kohlen> tipper
Wippkran m <Bau> luffing crane
Wippsäge f <Bau> jig saw
Wippschalter m <Elektrotech> rocker switch
Wipptisch m <Ker & Glas> tilt table
Wirbel m 1. <Eisenbahn, Elektriz, Kerntech, Kohlen, Labor, Maschinen, Mechan, Nichtfoss Energ, Phys, Raumfahrt, Strömphys> eddy, swirl, vortex; 2. <Wassertrans> swivel *(Beschläge)*; eddy, swirl, vortex *(Wasser, Wind)*; 3. <Werkprüf> eddy, swirl, vortex
Wirbel mpl **bei Strahlinstabilität** <Strömphys> vortices in jet instability
Wirbelabschöpfgerät n <Meerschmutz> vortex skimmer
wirbelbehaftet adj <Strömphys> vortical
Wirbelbettvergasung f <Chemtech> fluidized-bed gasification
Wirbeldehnung f <Strömphys> vortex stretching
Wirbeldiffusion f <Kerntech> eddy diffusion
Wirbeldurchflussmesser m <Regelung> vortex-shedding device
Wirbelerzeuger m <Wassertrans> vortex generator
wirbelfrei adj <Funktech, Geom, Phys, Strömphys> irrotational
wirbelfreie Strömung f <Phys> irrotational flow
wirbelfreies Feld n <Phys> irrotational field
Wirbelgarn n <Textil> intermingled yarn
Wirbelhaken m <Wassertrans> swivel hook *(Takelage, Beschläge)*
Wirbelkammer f <Kfztech> turbulence chamber, turbulence combustion chamber
Wirbelkern m <Strömphys> vortex core
Wirbellinie f <Strömphys> vortex line
Wirbelmeißel m <Fertig> whirling tool
wirbeln v <Nichtfoss Energ, Wassertrans> eddy
Wirbeln n <Fertig> whirling
Wirbelpaar n <Strömphys> vortex pair
Wirbelring m <Strömphys> vortex ring
Wirbelröhre f <Strömphys> vortex tube
Wirbelscheibe f <Lufttrans> *(BE)* actuator disc, *(AE)* actuator disk
Wirbelschicht f <Heiz & Kälte> fluidized bed
Wirbelschichtgefrieren n <Heiz & Kälte> fluidized-bed freezing
Wirbelschichtofen m <Chemtech> fluidized-bed furnace
Wirbelschichtröstofen m <Chemtech> fluidized-bed roasting furnace
Wirbelschichttrockner m 1. <Heiz & Kälte> *(AE)* fluidized-bed drier, *(BE)* fluidized-bed dryer; 2. <Lebensmittel> *(AE)* fluidized bed drier, *(BE)* fluidized bed dryer
Wirbelschichtverbrennung f <Umweltschutz> fluidized-bed combustion
Wirbelschichtverbrennungsanlage f <Abfall> fluidized-bed incinerator
Wirbelschichtvergasung f <Chemtech> fluidized-bed gasification
Wirbelschleppe f <Strömphys> vortex trailing
Wirbelseil n <Erdöl> spinning line
Wirbelsintern n <Chemtech> fluidized bed coating
Wirbelsinterverfahren n <Kunststoff> fluidized-bed coating
Wirbelstärke f <Strömphys> vorticity *(Rotation in Strömungsgleichungen entspricht Wirbelstärke)*
Wirbelstraße f <Strömphys> vortex street *(im Nachlauf einer ebenen Platte)*
Wirbelstreckung f <Strömphys> vortex stretching
Wirbelstrom m <Elektriz, Nichtfoss Energ, Phys, Strömphys, Werkprüf> eddy current
Wirbelstrombereich m <Kerntech> wake area
Wirbelstrombremse f 1. <Eisenbahn, Elektriz, Fernseh> eddy current brake; 2. <Fertig> *(BE)* disc brake, *(AE)* disk brake; 3. <Maschinen, Mechan> eddy current brake
Wirbelstromdämpfung f <Elektrotech> eddy-current damping, eddy-magnetic damping

Wirbelstromdurchflusszähler m <Kerntech> eddy current flowmeter
Wirbelstromgehäuse n <Maschinen> vortex chamber
Wirbelstromgleisbremse f <Eisenbahn> eddy current rail brake
Wirbelstromleistungsbremse n <Werkprüf> eddy current dynamometer
Wirbelstromraum m <Kerntech> wake space
Wirbelstromschaltung f <Elektriz> eddy current circuit
Wirbelströmung f 1. <Heiz & Kälte, Kohlen> eddy flow; 2. <Nichtfoss Energ> eddy flow, turbulence
Wirbelstromuntersuchung f <Eisenbahn> eddy current inspection
Wirbelstromverlust m <Phys, Strömphys> eddy current loss
Wirbelsturm m <Wassertrans> whirlwind
Wirbeltransportgleichung f <Strömphys> vorticity equation
Wirbelung f <Fertig> rabbling
Wirbelverteilung f <Strömphys> vortex distribution
Wirbelzerfall m <Strömphys> vortex decay
Wirbelzug m <Strömphys> vortex train
Wireline-Log n <Erdöl> wireline log *(Bohrlochmessung)*
Wire-Wrap-Verfahren n <Maschinen> wire wrap technique
Wirkanteil m <Telekom> active component
Wirkbewegung f <Maschinen> effective motion
Wirkbezugebene f <Fertig> working reference plane
Wirkdruck m <Labor, Maschinen> differential pressure
Wirkdruckdurchflussmesser m <Gerät> differential pressure flowmeter
Wirkdruckgeber m <Gerät> differential transducer
Wirkdruckmessumformer m <Gerät> differential pressure transducer
Wirkeinstellergänzungswinkel m <Fertig> working approach angle, *(AE)* working lead angle
Wirkeinstellwinkel m <Fertig> working cutting-edge angle, working minor cutting edge angle
wirken v <Textil> knit
wirkend adj <Elektrotech> active
wirkende Kraft f <Textil> agency
Wirkenergie f 1. <Elektriz> active energy; 2. <Fertig> working energy
Wirkfreiwinkel m <Fertig> working clearance
Wirkhauptschneide f <Fertig> working major cutting edge
Wirkkeilwinkel m <Fertig> working wedge angle
Wirkkomponente f <Elektrotech, Telekom> active component
Wirkkraft f <Fertig> working force
Wirklagewinkel m <Fertig> working orientation angle
Wirklänge f des Windes <Wassertrans> fetch *(Seezustand)*
Wirklast f <Elektriz, Elektrotech> active load
Wirkleistung f 1. <Elektriz, Elektrotech> active power *(Halbleiterspeicher)*; active power *(WS-Kreis)*; 2. <Fertig> working power; 3. <Labor> active power; 4. <Maschinen> effective power; 5. <Phys> active power
Wirkleistungsrelais n <Gerät> active power relay
wirklich adj <Comp & DV> actual, effective, physical, real
wirklicher Flugweg m <Lufttrans> actual flight path
Wirknebenschneide f <Fertig> working minor-cutting edge
Wirkneigungswinkel m <Fertig> working cutting-edge inclination
Wirknormalfreiwinkel m <Fertig> working normal clearance
Wirknormalkraft f <Fertig> working perpendicular force
Wirknormalspanwinkel m <Fertig> working normal rake

Wirkorthogonalebene f <Fertig> working orthogonal plane
Wirkorthogonalfreiwinkel m <Fertig> working orthogonal clearance
Wirkorthogonalkeilwinkel m <Fertig> working orthogonal wedge angle
Wirkpaar n <Fertig> working pair
Wirkrichtung f <Maschinen> effective direction
Wirkrückebene f <Fertig> working back plane
Wirkrückfreiwinkel m <Fertig> working back clearance
Wirkrückkeilwinkel m <Fertig> working back wedge angle
Wirkrückspanwinkel m <Fertig> working back rake
wirksam adj 1. <Comp & DV, Elektriz, Elektrotech, Fertig> active *(Federwindung)*; 2. <Maschinen, Mechan> effective; 3. <Phys> efficient
wirksam werden v <Maschinen> become effective
wirksame Kesselkühlfläche f <Heiz & Kälte> effective tank-cooling surface
wirksame Querschnittsfläche f <Mechan> effective cross-sectional area
wirksame Spannung f <Erdöl> effective stress *(geologie)*
wirksame Windung f <Fertig> active coil
wirksamer Druck m <Mechan> active pressure
wirksames Bildsignal n <Fernseh> effective picture signal
wirksames Mittel n <Textil> agent
Wirksamkeit f 1. <Anstrich> pot life *(Mehrkomponentenlacken)*; 2. <Lufttrans> agency; 3. <Meerschmutz, Qual> effectiveness
wirksamste Schätzfunktion f <Math> efficient estimator *(Schätzfunktion mit kleinstem mittleren quadratischen Fehler)*
Wirkschneidenebene f <Fertig> working cutting-edge plane
Wirkschneidennormalebene f <Fertig> working cutting-edge normal plane
Wirkseitenfreiwinkel m <Fertig> working side clearance
Wirkseitenkeilwinkel m <Fertig> working side wedge angle
Wirkspannung f <Elektriz, Elektrotech, Phys> active voltage
Wirkstoff m <Maschinen, Textil> agent
Wirkstrom m <Elektriz, Phys> active current
Wirktemperatur f <Qual, Thermod> effective temperature
Wirktemperaturbereich m <Qual, Thermod> effective temperature range
Wirk- und Strickmaschine f <Fertig, Textil> knitting and hosiery machine
Wirkung f 1. <Anstrich> reaction *(auf oder gegen etwas)*; 2. <Mechan> action
Wirkung f der Corioliskraft <Strömphys> Coriolis effect *(auf rotierende Fluide)*
Wirkungsablauf m <Regelung> sequence of action
Wirkungsbereich m <Raumfahrt> effective area *(Weltraumfunk)*
Wirkungsgrad m 1. <Aufnahme> efficiency *(eines Verstärkers)*; 2. <Elektriz, Funktech, Heiz & Kälte, Kerntech, Lufttrans, Maschinen> efficiency; 3. <Mechan> coefficient of efficiency, effect, efficiency; 4. <Phys> coefficient of efficiency; efficiency *(einer Wärmekraftmaschine)*; 5. <Qual> coefficient of efficiency, efficiency; 6. <Raumfahrt> efficiency *(Weltraumfunk)*; 7. <Telekom, Thermod> efficiency
Wirkungsgrad m der Quelle <Optik, Telekom> source power efficiency
Wirkungsgrad m der Verbrennung <Thermod> combustion efficiency

Wirkungsgrad

Wirkungsgrad *m* **des Blattes** <Lufttrans> blade efficiency factor *(Hubschrauber)*
Wirkungsgrad *m* **des Rotors** <Lufttrans> rotor efficiency *(Hubschrauber)*
Wirkungslinie *f* 1. <Fertig> line of action *(Kraft)*; 2. <Regelung> line of action
wirkungsmäßige Abhängigkeit *f* <Regelung> action-related dependence *(von Signalen)*
wirkungsmäßige Betrachtung *f* <Regelung> action-oriented consideration *(von Regelung, Steuerung)*
Wirkungsquantum *n* <Phys> quantum of action
Wirkungsquerschnitt *m* <Phys, Strahlphys, Teilphys> cross section
Wirkungsrichtung *f* <Regelung> direction of action
wirkungsvoll *adj* <Mechan> effective
wirkungsvolle Verpackung *f* <Verpack> efficient packaging
Wirkungsweg *m* <Regelung> path of action, signal flow path
Wirkungsweise *f* 1. <Elektrotech> operating mode; 2. <Maschinen> action, mode of operation
Wirkungszone *f* <Wasserversorg> area of influence
Wirkverbrauchsrelais *n* <Elektriz> active power relay
Wirkverbrauchszähler *m* <Elektriz> active power meter
Wirkwaren *fpl* <Textil> hosiery
Wirkwert *m* <Elektrotech> active component, real component
Wirkwinkel *m* <Fertig, Maschinen> working angle
wirtschaftlich abbaubare Lagerstätte *f* <Erdöl> commercial field *(Förderung)*
wirtschaftlich genutzte Kernenergie *f* <Kerntech> industrial nuclear power
wirtschaftliches Projekt *n* <Erdöl> economic project
Wirtschaftsinformatik *f* <Comp & DV> commercial informatics, economic informatics
Wischer *m* 1. <Fertig> squeegee; 2. <Kfztech, Maschinen> wiper
Wischerarm *m* <Kfztech> wiper arm
Wischerblatt *n* <Kfztech> wiper blade
Wischerwelle *f* <Maschinen> wiper shaft
Wischkontakt *m* <Elektriz> wiping contact
Wisch-Waschanlage *f* <Kfztech> (BE) windscreen washer, (AE) windshield washer
Wismutdraht *m* <Fertig> bismuth wire
Wismutlot *n* <Fertig> bismuth solder
Wissen *n* 1. <Comp & DV> knowledge; 2. <Ergon> cognition; 3. <Künstl Int> knowledge
Wissensakquisition *f* <Künstl Int> knowledge acquisition
wissensbasiert *adj* <Comp & DV, Künstl Int> knowledge-based
wissensbasiertes System *n* (WBS) 1. <Comp & DV> knowledge-based system *(KBS)*; 2. <Künstl Int> knowledge-based system
Wissensbasis *f* (WB) <Comp & DV, Künstl Int> knowledge base *(KB)*
wissenschaftliche Forschung <Qual> scientific research
Wissensdarstellung *f* <Comp & DV, Künstl Int> knowledge representation
Wissensdarstellungssprache *f* <Comp & DV> KRL, knowledge representation language
Wissensengineering *n* <Künstl Int> KE, knowledge engineering
Wissenserwerb *m* <Künstl Int> knowledge acquisition
Wissensrepräsentation *f* (WR) <Comp & DV, Künstl Int> knowledge representation *(KR)*
Wissenstechnik *f* <Comp & DV> KE, knowledge engineering

Wissensverarbeitung *f* <Künstl Int> knowledge processing
witterungsbedingt *adj* <Wassertrans> weather-bound *(Auslaufen eines Schiffes)*
Witterungseinflüssen aussetzen *v* <Bau> weather
Witterungsspiegel *m* <Umweltschutz> meteorological data
Witterungsverhältnisse *npl* <Umweltschutz> atmospheric conditions, meteorological conditions
WL *(weiter Laufsitz)* <Maschinen> loose fit
W-Motor *m* <Kfztech> W-type engine
Wobbelbetrieb *m* <Elektronik> sweep mode
Wobbelfrequenz *f* 1. <Elektronik, Fernseh, Labor> sweep frequency; 2. <Telekom> sweep frequency, sweep rate
wobbeln *v* <Raumfahrt> wobble
Wobbeln *n* 1. <Elektronik> sweep; 2. <Telekom> warble
Wobbelton *m* <Akustik> warble tone
Wobbler *m* <Elektronik> wobbler *(Messtechnik)*
Woge *f* 1. <Erdöl> surge *(Ozeanographie)*; 2. <Wellphys> wave *(auf See)*
wogen *v* <Wassertrans> heave
Wöhlerkurve *f* <Fertig> SIN curve, stress-number curve
Wohlklang *m* <Akustik> consonance, harmony
wohltemperierte Tonleiter *f* <Akustik> just scale, major scale of just temperament
Wohn/Bohr-Plattform *f* <Erdöl> hotel rig *(Offshore-Betrieb)*
Wohnmobil *n* <Kfztech> (AE) camper, (BE) caravan, (BE) motor caravan *(Fahrzeugart)*
Wohnplattform *f* <Erdöl> accommodation platform, flotel *(Offshore-Technik)*; hotel platform *(Offshore-Betrieb)*
Wohnraum *m* <Wassertrans> accommodation, quarter
Wohnung *f* <Bau> apartment, flat
Wohnungsbau *m* <Bau> house building
Wohnungsnetzinstallation *f* <Elektrotech> domestic electric installation
Wohnungstrennwand *f* <Bau> partition
Wohnungswesen *n* <Bau> housing
Wohnwagen *m* <Kfztech> (BE) caravan; mobile home, (AE) trailer *(für Dauerbenutzung)*
Wohnwagenanhänger *m* <Kfztech> (BE) caravan, (AE) trailer
wölben *v* 1. <Bau> vault; 2. <Fertig> camber, crown
wölben *v/sich* <Bau> camber
Wölbung *f* 1. <Bau> arch, crowning, vault; 2. <Fertig> camber, crowning *(Riemenscheibe)*; 3. <Ker & Glas> warpage; 4. <Maschinen> camber; 5. <Papier> arch; 6. <Qual> kurtosis
Wölbungsklappe *f* <Lufttrans> wing flap
Wölbungsmessgerät *n* <Phys> spherometer
Wolfram *n* (W) <Chemie> tungsten, wolfram (W)
Wolfram... <Chemie> wolframic
Wolframat *n* <Chemie> tungstate, wolframate
Wolframatosilicat *n* <Chemie> tungstosilicate
Wolframglühdraht *m* <Strahlphys> glowing tungsten filament
Wolframglühfaden *m* <Strahlphys> glowing tungsten filament
Wolframheizfaden *m* <Elektriz> tungsten filament
Wolframinertgasschweißen *n* <Mechan> tungsten inert-gas welding
Wolframinertschweißen *n* <Bau> TIG welding
Wolframit *m* <Chemie> wolframite
Wolframkarbid *n* 1. <Erdöl> tungsten carbide *(Hartmetall)*; 2. <Maschinen> tungsten carbide
Wolframspritzer *m* <Bau> tungsten spatter *(Schweißen)*
Wolframstahl *m* <Metal> tungsten steel, wolfram steel
Wolkenbruch *m* <Wassertrans> cloudburst

Wolkenhöhenmesser m <Lufttrans> ceilograph, ceilometer
Wolkenuntergrenze f <Lufttrans, Wassertrans> cloud base
wolkiger Anlauf m <Ker & Glas> paper hum
wollartig adj <Textil> woolly
Wolle f <Papier, Textil> wool
wollig adj <Textil> woolly
Wolligkeit f <Textil> woolliness
Wollkammzug m <Textil> combed top
Wollwaren fpl <Textil> woollens
Wood-Glas n <Ker & Glas> Wood's glass
Woodruff-Keil m <Kfztech, Maschinen> Woodruff key
Workflow-Management-Produkt n <Comp & DV> workflow management product
Workgroup-Anwendung f <Comp & DV> workgroup application
Workstation f <Comp & DV, Kontroll> work station
WORM (Einmalbeschreibung-Mehrfachlesen) <Optik> WORM (write once read many times)
WORM-Platte f <Comp & DV> write-once disk, write-once read many times disk
Wort n <Comp & DV> word
Wortabstand m <Comp & DV> spacing
Wortangaben fpl <Konstzeich> verbal notes
Wortbegrenzungszeichen n <Comp & DV> word delimiter
Wortebene f <Comp & DV> word plane
Worterkennung f <Künstl Int> word recognition
Worterzeugung f <Elektronik> word generation
Wortgenerator m <Elektronik> word generator (Messtechnik)
Wortlänge f <Comp & DV> word length, word size
wortorientiert adj <Comp & DV> word-oriented
Wortübertragungszeit f <Comp & DV> word time
Wortumbruch m <Comp & DV> word wrap
Wortzeichen n <Patent> word mark
Wortzeit f <Comp & DV> word time
WPS (Wellenpferdestärke) <Maschinen> shaft horse power
WR 1. <Comp & DV, Künstl Int> (Wissensrepräsentation) KR (knowledge representation); 2. <Elektronik, Elektrotech, Funktech, Phys> (Wechselrichter) inverter
Wrack n <Wassertrans> wreck
Wraparound-Banderolemaschine f <Verpack> wraparound sleeving machine
wringen v <Textil> wring
Wringen n <Papier> wringing
Wringwalze f <Papier> winger roll
WS (Wechselstrom) <Elektrotech> AC (alternating current)
WS-Adapter m <Elektrotech> AC adaptor
WS-Animeter n <Elektrotech> AC animeter
WS-Ausgang m <Elektrotech> AC output
WS-Beschichten n <Chemtech> AC bed coating
WS-Betrieb m <Elektrotech> AC operation
WS-Brücke f <Elektrotech> AC bridge
WS-Dickfilm-Elektrolumineszenzanzeige f <Elektronik> AC thick-film electroluminescent display
WS-Eingang m <Elektrotech> AC input
WS-Entladung f <Elektrotech> AC discharge
WS-Erregung f <Elektrotech> AC excitation
WS-Erzeugung f <Elektrotech> AC current generation, AC generation
WS-Feld n <Elektriz> AC field
WS-gekoppelt adj <Kontroll> AC-coupled
WS-Generator m <Elektriz, Phys> AC generator
WS-GS (Wechselstrom-Gleichstrom) <Elektrotech> AC-DC (alternating current-direct current)
WS-GS-Umsetzer m <Elektrotech> AC-DC converter

WS-GS-Umsetzung f <Elektrotech> AC-DC conversion
WS-GS-Wandler m <Elektrotech> AC-DC converter
WS-GS-Wandlung f <Elektrotech> AC-DC conversion
WS-Integration f (Wafer-Scale-Integration) <Comp & DV, Elektronik> wsi (wafer scale integration)
WS-Josephson-Effekt m <Elektronik> AC Josephson effect
WS-Kompensator m <Elektriz> AC potentiometer
WS-Komponente f <Elektriz> AC component
WS-Kondensator m <Elektriz> AC capacitor
WS-Koppler m <Telekom> AC coupler
WS-Kraft f <Elektrotech> AC electromotive force
WS-Kreis m <Elektriz> AC circuit, AC network
WS-Last f <Elektrotech> AC load
WS-Leistung f <Elektrotech> AC power
WS-Leitung f <Elektrotech> AC line
WS-Lichtbogen m <Elektrotech> AC arc
WS-Lichtbogenschweißen n <Elektrotech> AC arc welding
WS-Marker m <Wassertrans> AC marker
WS-Maschine f <Elektrotech> AC machine
WS-Messbrücke f <Elektriz> AC bridge
WS-Messinstrument n <Elektriz> AC meter
WS-Motor m <Elektriz, Elektrotech, Fertig, Phys> AC motor
WS-Netz n 1. <Elektriz> AC network; 2. <Elektrotech> AC network, AC power line
WS-Netzausfall m <Elektrotech> AC power failure
WS-Netzleitung f <Elektrotech> AC power line
WS-Quelle f <Elektrotech> AC current source
WS-Relais n 1. <Elektriz> AC relay; 2. <Elektrotech> AC armature relay, AC relay
WS-Schaltkreis m <Elektriz> AC circuit
WS-Schaltung f <Elektrotech> AC switching
WS-Schweißlichtbogen m <Fertig> AC welding arc
WS-Servomotor m <Elektrotech> AC servomotor
WS-Spannung f <Elektriz> AC voltage
WS-Stellmotor m <Elektrotech> AC servomotor
WS-Übertragungsleitung f <Elektriz> AC transmission line
WS-Versorgung f 1. <Elektriz> AC supply; 2. <Elektrotech> AC current source
WS-Versorgungssystem n <Raumfahrt> AC power system (Raumschiff)
WS-Verstärker m <Elektrotech> AC amplifier
WS-Voltmeter n <Elektriz, Elektrotech> AC voltmeter
WS-Vorspannung f <Aufnahme> AC bias
WS-Widerstand m <Elektrotech> AC resistance
W-Teilchen n <Teilphys> W particle
Wuchtbaum m <Bau> lifter
Wuchtdorn m <Fertig> balancing arbour, balancing mandrel
Wuchtdrehzahl f <Elektrotech, Maschinen> balancing speed
Wuchtebene f <Fertig> balancing plane
Wuchteinrichtung f <Fertig> balancing device
wuchten v <Fertig> balance
Wuchten n <Maschinen> balancing, counterbalancing
Wuchtfehler m <Fertig> amount of unbalance, unbalance
Wuchtgewichte npl <Maschinen> balancing weights
Wuchtgüte f <Fertig> balance quality
Wuchtmaschine f <Maschinen> balancing machine
Wuchtzustand m <Fertig> balance
Wulcherblock m <Ker & Glas> shaping block
Wulchereisen n <Ker & Glas> shaping tool
Wulst m 1. <Bau> bead, bulge, collar, flange; 2. <Fertig> ring; 3. <Ker & Glas> bead (Flasche); 4. <Kfztech> bead (Reifen); boss (der Radnabe); 5. <Maschinen> bead; 6. <Mechan> pad; 7. <Metall, Wassertrans> bulb
Wulstband n <Textil> chafer (Reifen)
Wulstblech n <Metall> bulb plate

Wulstbug

Wulstbug m <Wassertrans> bulbous bow
Wulsteisen n <Metall> bulb iron
Wulstfacette f <Ker & Glas> beaded bevel
wulstig adj <Mechan> padded
Wulstkante f <Ker & Glas> bulb edge
Wulstkern m <Kfztech> bead core
würdigen v <Math> appreciate
Würdigung f <Math> appreciation
Wurfanker m <Wassertrans> grappling hook, kedge anchor (Festmachen)
Würfel m 1. <Geom> cube; 2. <Math> dice (Statistik)
würfeln v 1. <Lebensmittel> dice; 2. <Math> dice (Statistik)
Wurfförderer m <Maschinen> throw conveyor
Wurfkette f <Erdöl> throwing chain
Wurfleine f <Wassertrans> heaving line (Tauwerk)
Wurfnetz n <Wassertrans> casting net (Fischerei)
Würze f <Lebensmittel> (AE) flavor, (AE) flavoring, (BE) flavour, (BE) flavouring
Wurzel f 1. <Comp & DV> radix, root; 2. <Math> radical, root
Wurzelfehler m <Kerntech> incomplete root penetration
Wurzel-Nyquist-Filter n <Elektronik> matched filter
Wurzelzeichen n <Math> radical sign
Wurzelziehen n <Math> evolution (aus einem beliebigen Ausdruck)
würzen v <Lebensmittel> (AE) flavor, (BE) flavour
WWRV (Wanderwellenröhrenverstärker) 1. <Elektronik> TWTA, (AE) traveling wave tube amplifier; 2. <Raumfahrt> TWTA, (BE) travelling wave tube amplifier
WYSIWYG (originalgetreue Darstellung der Druckausgabe am Bildschirm) <Comp & DV> WYSIWYG (what you see is what you get)

X

X 1. <Akustik> (Volumenverdrängung) X (volume displacement); 2. <Elektriz> (Blindwiderstand, Reaktanz) X (reactance); 3. <Kerntech> (Belastung) X (exposure)
X-Ablenkplatte f <Elektronik> horizontal deflection plate
X-Ablenkung f 1. <Elektronik> horizontal deflection (Katodenstrahlröhre); 2. <Fernseh> X-deflection
X-Ablenkverstärker m <Gerät> sweep deflection amplifier (Oszilloskop)
X-Achse f 1. <Bau, Fernseh, Geom> x-axis; 2. <Math> x-axis; 3. <Phys, Raumfahrt> x-axis (Raumschiff); 4. <Telekom> time base
Xanthat n <Chemie> xanthate
Xanthein n <Chemie> xanthein
Xanthen n <Chemie> xanthene
Xanthenol n <Chemie> xanthydrol
Xanthenon n <Chemie> xanthone
Xanthenyl n <Chemie> xanthyl
Xanthin n <Chemie> dihydroxypurine, xanthine
Xanthogen... <Chemie> xanthic, xanthogenic
Xanthogenat n <Chemie> xanthate
Xanthon n <Chemie> xanthone
Xanthophyll n <Chemie> lutein, xanthophyll
Xanthoprotein... <Chemie> xanthoproteic
Xanthosin n <Chemie> xanthosine
Xanthotoxin n <Chemie> xanthotoxin
Xanthoxylen n <Chemie> xanthoxylene
Xanthoxylin n <Chemie> xanthoxylin
Xanthydrol n <Chemie> xanthydrol

Xanthyl n <Chemie> xanthyl
X-Band n 1. <Elektronik> X-band; 2. <Telekom> X-band (Satellit; 6,2 – 10,9 GHz)
X-Band-Magnetron n <Elektronik> X-band magnetron
X-Band-Wanderfeldröhre f <Elektronik> X-band TWT, (AE) X-band traveling wave tube, (BE) X-band travelling wave tube
XC (kapazitiver Blindwiderstand) <Elektriz> XC (capacitive reactance)
Xe (Xenon) <Chemie> Xe (xenon)
X-Einheit f X-unit (Längeneinheit in der Röntgenspektroskopie)
Xenon n (Xe) <Chemie> xenon (Xe)
Xenon-Chloridlaser m <Elektronik> xenon chloride laser
Xenoneffekt m <Kerntech> xenon effect
Xenongipfel m <Kerntech> xenon peak
Xenon-Reaktivität f <Kerntech> xenon reactivity
Xenonspitze f <Kerntech> xenon peak
Xenonspitze f **nach Abschaltung** <Kerntech> xenon buildup after shutdown
Xenonvergiftung f <Kerntech> xenon poisoning effect
Xerographie f <Comp & DV, Elektriz, Elektronik> xerography
xi-Teilchen n <Phys> xi particle (Elementarteilchen)
X-Koordinate f <Phys> x-coordinate
XL (induktiver Blindwiderstand) <Elektriz> XL (inductive reactance)
XLR-Anschluss m <Fernseh> XLR connector
X-Motor m 1. <Kerntech> X-motor (eines Manipulators); 2. <Kfztech> X-type engine
XMS (Erweiterungsspeicher) <Comp & DV> XMS (extended memory specification)
XM-synchronisierter Verschluss m <Foto> XM synchronized shutter
XM-Synchroverschluss m <Foto> XM synchronized shutter
X-Naht f <Fertig> double-V groove weld
X-Nahtverbindung f <Fertig> double-V butt joint
XPS 1. <Comp & DV, Künstl Int> (Expertensystem) XPS (expert system); 2. <Phys> (röntgenstrahlangeregte Photoelektronenspektroskopie) XPS (X-ray photoelectron spectroscopy)
X-Stoß m <Fertig> double-V butt joint
X-Y Abgleich m <Fernseh> X-Y alignment
X-Y Schreiber m 1. <Elektriz> X-Y recorder; 2. <Gerät> graph plotter; 3. <Gerät> two-axis plotter
Xylen n <Kunststoff> xylene
Xylenol n <Chemie> hydroxydimethylbenzene, xylenol
Xylidin n <Chemie> xylidine
Xylit n <Chemie, Lebensmittel> xylitol (Zuckeralkohol)
Xylitol n <Chemie> xylitol
Xylol n 1. <Chemie> dimethylbenzene, xylene, xylol; 2. <Erdöl> xylene (Petrochemie); 3. <Kunststoff> xylene
Xylyl... 1. <Chemie> xylyl; 2. <Chemie> xylylene

Y

Y (Yttrium) <Chemie> Y (yttrium)
YA (akustische Admittanz) <Akustik, Elektrotech> YA (acoustic admittance)
Y-Ablenkung f 1. <Elektronik> vertical deflection; 2. <Fernseh> Y-deflection

Y-Achse f 1. <Bau, Fernseh, Geom> y-axis; 2. <Math> y-axis; 3. <Phys> y-axis
Y-Ader f <Aufnahme> Y-lead
Y-Admittanzschaltung f <Phys> Y-network
YAG *(Yttrium-Aluminium-Granat)* <Chemie> YAG *(yttrium-aluminium garnet)*
Yagein n <Chemie> harmine, yageine
Yagi-Antenne f <Telekom> Yagi antenna
YAG-Laser m <Elektronik> YAG laser *(Neodym-Laser)*
Yahoo-basierter Service m <Comp & DV> Yahoo-style service
Y-Anschluss m <Elektrotech> Y-connection
Yard n <Metrol, Textil> yard
Yawl f <Wassertrans> yawl *(Schifftyp)*
Yb *(Ytterbium)* <Chemie> Yb *(ytterbium)*
YIG *(Yttrium-Eisen-Granat)* <Chemie> YIG *(yttrium-iron garnet)*
YIG-Abstimmung f <Elektronik> YIG tuning
YIG-Bandpassfilter n <Elektronik> YIG band-pass filter
YIG-Filter n <Elektronik> YIG filter
Y-Kabel n <Fernseh> Y-cable
Y-Koordinate f <Phys> y-coordinate
Y-Koppler m <Optik, Telekom> Y-coupler
Young'scher Doppelspalt m <Phys> Young's slits
Young'scher Modul m *(E)* <Erdöl, Hydraul, Kohlen, Kunststoff, Metall, Phys> Young's modulus, E *(Elastizität)*
Yrast-Strahlung f <Kerntech> yrast radiation
Y-Schaltung f <Phys> Y-network
Y-Schnittstelle f <Telekom> Y interface, ISDN Y interface
Y-Signal n <Fernseh> Y-signal, luminance signal
Ytterbium n *(Yb)* <Chemie> ytterbium *(Yb)*
Ytterbiumoxid n <Chemie> ytterbium oxide
Yttrium n *(Y)* <Chemie> yttrium *(Y)*
Yttrium-Aluminium-Granat n *(YAG)* <Chemie> *(BE)* yttrium-aluminium garnet, *(AE)* yttrium-aluminum garnet *(YAG)*
Yttrium-Eisen-Granat n *(YIG)* <Chemie> yttrium-iron garnet *(YIG)*
Yttriumoxid n <Chemie> yttrium oxide
Yukawa-Potenzial n <Phys> Yukawa potential
Yukawa-Potenzialtopf m <Kerntech> Yukawa well
Y-Verbindung f <Elektriz, Elektrotech> forked connection, Y-connection
Y-Verbindungsstück n <Labor> Y-piece *(Zwischenstück)*
Y-Verzweigung f <Telekom> three-port coupler
Y-Vierpolparameter mpl <Elektronik> Y-parameters

Z

z *(Kompressibilitätsfaktor)* <Thermod> z *(compressibility factor)*
Z 1. <Elektriz> *(Impedanz)* Z *(impedance)*; 2. <Kerntech> *(Atomzahl)* Z *(atomic number)*; 3. <Phys> *(Teilfunktion, Zustandsfunktion)* Z *(partition function)*
ZA *(Schallimpedanz, Schallwellenwiderstand, akustische Impedanz, akustischer Scheinwiderstand)* <Aufnahme, Elektrotech, Phys> ZA *(acoustic impedance)*
Z-Achse f <Geom, Math> z-axis
Zacke f 1. <Bau> tine; 2. <Elektrotech> spike; 3. <Fertig> jag; 4. <Funkort> pip *(Radar)*

Zackenbildung f <Fernseh> serration
zackig adj <Fertig> ragged
zäh adj <Strömphys> viscid
Zähfestigkeit f <Textil> toughness
zähflüssig adj 1. <Chemie> viscous; 2. <Fertig> syrupy; 3. <Lebensmittel, Maschinen, Phys> viscous
Zähflüssigkeit f <Chemie> viscosity
Zähigkeit f 1. <Anstrich> tenacity; 2. <Erdöl, Fertig> viscosity; 3. <Kerntech> toughness; 4. <Kohlen> viscosity; 5. <Kunststoff> toughness, viscosity; 6. <Maschinen> tenacity, toughness, viscosity; 7. <Metall> toughness; 8. <Phys> toughness, viscosity; 9. <Strömphys> viscidity, viscosity *(Widerstand gegenüber Bewegung)*; 10. <Thermod> viscosity
Zähigkeitsdämpfung f <Mechan> viscous damping
Zähigkeitskoeffizient m <Phys, Qual, Strömphys> coefficient of viscosity
Zähigkeitskraft f <Lebensmittel, Maschinen, Phys> viscous force
Zähigkeitskraft f bezogen auf Volumeneinheit <Lebensmittel, Maschinen, Phys> viscous force per unit volume
Zähigkeitsmessgerät n <Strömphys> viscometer
Zähigkeitsmessung f <Strömphys> viscosity measurement
Zähigkeitsverhalten n <Fertig> *(AE)* tenacity behavior, *(BE)* tenacity behaviour
Zähigkeitswirkung f <Strömphys> action of viscosity, viscous action
Zahl f 1. <Comp & DV> number; 2. <Druck> figure; 3. <Math> number, numeral
Zählautomat m <Verpack> counting device
Zählautomat m für neun Ziffern <Verpack> nine digit counter
Zählcodierer m <Gerät> counting coder *(AD-Umsetzung)*
Zähldetektor m <Trans> passage detector
Zahlenbereich m <Comp & DV, Math> number range
Zahlendarstellung f <Comp & DV> number representation
Zahlenrollenwerk n <Gerät> counter
Zahlensystem n <Comp & DV, Math> number system
Zahlentheorie f <Math> number theory, theory of numbers
Zahlenwert m <Metrol> numerical value
Zähler m 1. <Comp & DV> counter; tally *(Summe)*; 2. <Elektriz, Elektronik, Funktech, Gerät, Kontroll> counter; 3. <Maschinen> counter, meter; 4. <Math> numerator; 5. <Telekom> counter, *(AE)* meter, *(BE)* metre *(Telefon)*
Zählergebnis n <Gerät, Math> counting result
Zählergehäuse n <Gerät> meter case
Zählerschaltuhr f <Gerät> time switch
Zählerschaltung f <Elektronik> counter circuit
Zählerstand m <Gerät> count
Zählfrequenzmessgerät n <Gerät> counter frequency meter
Zählgerät n <Gerät> counting instrument
Zählgeschwindigkeit f <Elektronik> counting rate
Zählgeschwindigkeitsmesser m <Gerät> count rate meter
Zählimpuls m <Gerät> count
Zählimpulsgeber m <Gerät> counting pulse generator
Zählmagnet m <Gerät> magnetic counter
Zählmaschine f <Fertig> counter/dispenser
Zahlmeister m <Wassertrans> purser *(Handelsmarine)*
Zählregister n <Gerät> counter
Zählrelais n <Gerät> magnetic counter
Zählrohr n 1. <Elektronik, Gerät> counter tube; 2. <Strahlphys> counting tube

Zählröhre f mit kalter Katode <Elektronik> cold-cathode counter tube
Zählrohrsonde f <Kerntech> counter tube probe
Zählschaltung f <Elektronik> counter *(Schaltkreistechnik)*
Zählschritt m <Gerät> count
Zählschritt m rückwärts <Elektronik> decrement
Zählstatistik f <Comp & DV> tabulation
Zählstelle f <Trans> counting station
Zählstrich m 1. <Comp & DV> tally; 2. <Math> tally *(auf Strichliste)*
Zählumsetzer m <Gerät> level-at-a-time converter *(AD-Umsetzer)*
Zahlung f im Einzugsverfahren <Telekom> direct-debit payment
Zählung f 1. <Gerät> count; 2. <Math> enumeration; 3. <Telekom> metering *(Telefon)*
Zählvorgang m <Gerät> count
Zählwerk n 1. <Comp & DV> register; 2. <Elektriz> counter; 3. <Elektronik> counter *(Register)*; 4. <Fertig> counter; 5. <Gerät> counter, counter mechanism; 6. <Telekom> counter, register *(Zählgerät)*
Zahlzeichen n <Druck> numeral
Zahn m 1. <Maschinen> cog, sprocket, tooth; 2. <Mechan> cog *(beim Zahnrad)*
Zahn m eines Zahnrads <Maschinen> gear tooth
Zahnanlage f <Fertig> bearing
Zahnbogen m 1. <Kfztech> sector gear; 2. <Maschinen> sector gear, sector wheel, segmental wheel
Zahnbreite f <Fertig> face width *(Kinematik)*
Zahndicke f 1. <Lufttrans> chordal thickness *(Getriebe)*; 2. <Maschinen> tooth thickness
Zahndicke f als Bogen <Fertig> arc thickness *(Getriebelehre)*
Zahndicke f im Teilkreis <Maschinen> circular thickness
Zähne mpl <Maschinen> teeth; teeth *(eines Zahnrads)* • **mit festen Zähnen** <Fertig> solid • **mit geraden Zähnen** <Maschinen> straight-tooth • **mit Zähnen versehen** <Maschinen> cog
Zähne mpl ohne Kopfkürzung <Fertig> long-addendum teeth *(Getriebelehre)*
Zahneingriff m <Maschinen> tooth contact, tooth engagement
zahnen v <Mechan> dent
Zahnflanke f <Maschinen> tooth flank
Zahnform f <Maschinen> tooth form
Zahnformfräsen n <Mechan> gear cutting
Zahnformfräser m <Maschinen> gear cutter
Zahnfuß m 1. <Fertig> dedendum *(Getriebelehre)*; 2. <Maschinen> dedendum, tooth root; 3. <Mechan> dedendum
Zahnfußhöhe f <Maschinen> dedendum, depth below pitch line
Zahngesperre n <Maschinen> ratchet-and-pawl mechanism
Zahngrund m <Fertig> gullet
Zahnhobel m <Bau> tooth plane, toothing plane
Zahnhöhe f 1. <Fertig> depth of cut *(Fräser)*; 2. <Maschinen> tooth height
Zahnhöhenkürzung f <Fertig> addendum reduction
Zahnkanten-Abrundmaschine f <Fertig> gear-tooth rounding and chamfering machine
Zahnkeilriemen m <Maschinen> toothed V-belt
Zahnkette f 1. <Fertig> gear chain; 2. <Maschinen> tooth-type chain; 3. <Textil> chain *(Reißverschluss)*
Zahnkopf m 1. <Fertig> crest *(Getriebelehre)*; 2. <Maschinen> tooth crest; 3. <Mechan> addendum
Zahnkopf m mit Stumpfverzahnung <Fertig> addendum-corrected gear

Zahnkopf m mit Zahnkopfkorrektur <Fertig> addendum-corrected gear
Zahnkopffläche f <Maschinen> crest area *(Zahnrad)*
Zahnkopfhöhe f 1. <Fertig> addendum *(Getriebelehre)*; 2. <Maschinen> addendum
Zahnkopfwinkel m <Fertig> addendum angle *(Kegelrad)*
Zahnkranz m 1. <Kfztech> ring gear *(auf Schwungrad, in das das Anlasserritzel einspurt)*; 2. <Maschinen> ring gear
Zahnkreisteilung f <Mechan> circular pitch
Zahnlücke f <Maschinen> tooth gap • **Zahnlücken vorfräsen** <Fertig> gash
Zahnlückengrund m <Maschinen> bottom land
Zahnprofil n <Maschinen> tooth profile
Zahnrad n 1. <Fertig> toothed wheel, wheel; 2. <Kfztech> pinion *(Getriebe)*; 3. <Maschinen> cogwheel, gearwheel, gear, toothed wheel; 4. <Mechan> gear ratio; 5. <Wassertrans> gear
Zahnrad n mit geraden Flanken <Maschinen> straight flank gear
Zahnrad n mit Zykloidenverzahnung <Mechan> cycloidal gear
Zahnrad n und Sperrklinke f <Maschinen> ratchet and pawl
Zahnradabzieher m <Kfztech> gear puller *(Werkzeug)*
Zahnradachsverlagerung f <Fertig> *(AE)* gear center-distance variation, *(BE)* gear centre-distance variation
Zahnradantrieb m <Maschinen> gear drive
Zahnradbahn f <Eisenbahn> *(AE)* cog railroad, *(BE)* cog railway, *(AE)* rack railroad, *(BE)* rack railway, *(AE)* rack-and-pinion railroad, *(BE)* rack-and-pinion railway
Zahnradbahnlokomotive f 1. <Eisenbahn> rack engine; 2. <Trans> rack locomotive
Zahnräder npl <Fertig> gears
Zahnradfräser m 1. <Fertig> gear cutter; 2. <Maschinen> cutter, gear cutter
Zahnradfräser Evolventenverzahnung f <Maschinen> involute gear cutter
Zahnradfräsmaschine f <Maschinen> gear milling machine, gear-cutting machine
Zahnradgebirgsbahn f <Trans> *(AE)* rack mountain railroad, *(BE)* rack mountain railway
Zahnradgetriebe n 1. <Maschinen> gearing, gears; 2. <Mechan> gear drive
Zahnradhobelmaschine f <Maschinen> gear planer
Zahnradmesszylinder m <Metrol> gear measuring cylinder
Zahnradnabe f <Fertig> gear hub
Zahnradölpumpe f <Kfztech> gear-type oil pump
Zahnradpaar n <Fertig> equal gearing
Zahnradprüfgerät n <Metrol> gear testing machine
Zahnradpumpe f 1. <Fertig> double helical pump, gear pump; 2. <Heiz & Kälte, Kfztech, Kunststoff> gear pump; 3. <Maschinen> gear pump, gear-type pump, gearwheel pump
Zahnradrohling m 1. <Fertig> wheel blank; 2. <Maschinen> gear blank
Zahnradrollmaschine f <Maschinen> gear rolling machine
Zahnradschabmaschine f <Maschinen> gear shaving machine
Zahnradschneidmaschine f <Maschinen> gear-cutting machine
Zahnradstange f <Eisenbahn> rack rail
Zahnradstoßmaschine f 1. <Fertig> shaper; 2. <Maschinen> gear shaper, gear shaping machine
Zahnradtragbild n <Maschinen> gearwheel contact pattern
Zahnradvorgelege n <Maschinen> back gear, back gearing

Zahnradwälzfräser m <Fertig> gear hob
Zahnradwechselgetriebe n <Kfztech> straight-toothed gearbox
Zahnradzahn m <Fertig> gear tooth
Zahnriemen m 1. <Kfztech> cog belt, cogged belt, notched belt *(Zündverstellung)*; 2. <Maschinen> cog belt, cogged belt
Zahnriemeneinstellung f <Kfztech> cogged belt timing, notched belt timing
Zahnring m <Maschinen> annular gear
Zahnringanker m <Elektriz> toothed ring armature
Zahnscheibe f <Maschinen> antiturn washer, tooth lock washer, toothed lock washer
Zahnsegment n 1. <Kfztech> sector gear; 2. <Maschinen> segmental wheel, toothed segment
Zahnsegmenthebel m <Fertig> quadrant lever, quadrant level
Zahnspiel n <Kfztech> backlash *(Getriebe)*
Zahnstange f 1. <Fertig> rack *(Getriebelehre)*; 2. <Kerntech> toothed rack; 3. <Kfztech> rack *(Lenkung)*; 4. <Maschinen> rack, ratch; 5. <Mechan> rack, rack-and-pinion
Zahnstangenantrieb m <Maschinen> rack-and-pinion drive
Zahnstangenfräseinrichtung f <Maschinen> rack milling attachment
Zahnstangenfräser m <Maschinen> rack cutter, rack milling cutter
Zahnstangenfräsmaschine f <Maschinen> rack milling machine
Zahnstangengetriebe n <Maschinen> rack-and-pinion drive, rack-and-pinion gear
Zahnstangenlenkung f <Kfztech> rack-and-pinion steering
Zahnstangenritzel n <Kfztech> rack pinion
Zahnstangenteilebene f <Fertig> datum plane *(Getriebelehre)*
Zahnstangentrieb m 1. <Fertig> rack-and-pinion drive; 2. <Mechan> rack-and-pinion
Zahnstangentriebrad n <Maschinen> rack wheel
Zahnstangenvorschub m <Maschinen> rack feed
Zahnstangenwinde f <Maschinen> lifting jack, rack-and-pinion jack
Zahnstein m <Bau> toothing stone *(Mauerwerk)*
Zahnteilung f <Fertig> pitch; spacing *(Reibahle)* • **mit grober Zahnteilung** <Fertig> coarse-pitch
Zahnteilung f **im Teilkreis** <Maschinen> CP, circular pitch
Zahntrieb m <Trans> rack gearing
Zahntrommel f <Maschinen> sprocket drum
Zahnweite f <Maschinen> tooth distance
zähspröde Umwandlung f <Metall> tough-brittle transition
Zange f 1. <Bau> brace *(Holz zum Zusammenhalten von Holzkonstruktionen)*; horizontal timber *(Holzbau)*; 2. <Elektriz> pliers; 3. <Fertig> clamp *(Drahtzug)*; 4. <Kfztech> pliers *(Werkzeug)*; 5. <Maschinen> pliers, tongs
Zange f **für Sicherungsringe** <Maschinen> circlip pliers
Zange f **mit gebogenem Kopf** <Maschinen> bent-nose pliers
Zangenbremse f <Eisenbahn> *(AE)* clasp brake
Zangenfutter n <Maschinen> collet chuck
Zangengreifer m <Maschinen> pince gripper
Zangenmarkierungen fpl <Ker & Glas> tong marks
Zangenseil n <Erdöl> tong line *(Bohrtechnik)*
Zangenspannfutter n <Fertig> spring collet chuck
Zangenstromwandler m <Gerät> current probe
Zangenwagen m <Fertig> gripping jaw carriage *(Rohrziehen)*

zapfen v <Lebensmittel> tap
Zapfen m 1. <Bau> gudgeon, pintle, tenon; spigot *(eines Hahns)*; 2. <Fertig> neck, spigot, trunnion; stem *(Kunststoffinstallationen)*; 3. <Kfztech> neck; 4. <Maschinen> gudgeon, journal, neck, spigot, tenon; 5. <Mechan> cog, *(AE)* faucet, journal, lug; 6. <Metall> neck *(Walzwerke)* • **mit Zapfen versehen** <Fertig> tang • **sich um einen Zapfen drehen** <Fertig> pivot
Zapfen m **der Kurbelwelle** <Kfztech> crankpin *(Motor)*
Zapfenbüchse f <Fertig> stem sleeve *(Kunststoffinstallationen)*
Zapfendichtung f <Fertig> stem seal *(Kunststoffinstallationen)*
Zapfendüse f <Kfztech> pintle-type nozzle *(Dieselmotor)*
Zapfenerweiterung f <Fertig> pin enlargement *(Getriebelehre)*
Zapfenhals m <Fertig> stem neck *(Kunststoffinstallationen)*
Zapfenkreuz n <Mechan> journal cross
Zapfenlager n 1. <Eisenbahn> *(BE)* axle box, *(AE)* journal box; 2. <Fertig> pillow; 3. <Maschinen> pin bearing
Zapfenlagerung f 1. <Fertig> pivoting; 2. <Maschinen> trunnion mounting
Zapfenlänge f <Fertig> pilot length
Zapfenloch n <Bau> mortice, mortise
Zapfenlochverbindung f <Bau> slot mortise joint
Zapfenring m <Kfztech> pivot ring
Zapfenschaltung f **unter Last** <Elektriz> on-load tap changing
Zapfenschlitz m <Bau> through mortice
Zapfenschlüssel m <Maschinen> *(BE)* pin spanner, pin wrench
Zapfenschneidemaschine f <Maschinen> tenoning machine
Zapfensenker m <Fertig> piloted counterbore
Zapfenspannglied n <Bau> shouldered tenon
Zapfenstreichmaß n <Bau> *(AE)* mortise gage, *(BE)* mortise gauge
Zapfenturbine f <Hydraul> journal turbine
Zapfenverbindung f <Bau> mortise and tenon joint
Zapfhahn m <Maschinen> *(AE)* faucet, *(BE)* tap
Zapfluftventil n <Lufttrans> air bleed valve
Zapfstelle f <Bau> *(BE)* tap
Zapfwellenschutz m <Sicherheit> power takeoff protective device *(Schlepper)*
Zaponfolie f <Kerntech> zapon foil
Zaponlack m <Kerntech> zapon lacquer
Zarge f 1. <Bau> frame; 2. <Fertig> *(AE)* can body, *(BE)* tin body
Zargenblech n <Fertig> body stock
Zargenrundung f <Fertig> body forming
Zartmacher m <Lebensmittel> tenderizer
Zäsium n <Chemie> *(BE)* caesium, *(AE)* cesium
zäsiumdotiertes Glas n <Raumfahrt> *(BE)* caesium-doped glass, *(AE)* cesium-doped glass *(Raumschiff)*
Zäsiumkatode f <Elektrotech> *(BE)* caesium cathode, *(AE)* cesium cathode
Zäsiumphotoröhre f <Elektronik> *(BE)* caesium phototube, *(AE)* cesium phototube
Zäsiumstrahlresonator m <Elektronik> *(BE)* caesium-beam resonator, *(AE)* cesium-beam resonator
Zäsiumuhr f <Raumfahrt> *(BE)* caesium clock, *(AE)* cesium clock *(Weltraumfunk)*
Zaunlatte f <Bau> pale
Zaunölsperre f <Meerschmutz> fence boom
Zaunübergang m <Bau> stile
Z-Boson n 1. <Phys> Z-particle *(Elementarteilchen)*; 2. <Teilphys> Z-boson
ZB-Vermittlungsschrank m <Telekom> common battery switchboard

Z-Diode

Z-Diode f 1. <Elektriz> Z-diode; 2. <Phys> Zener diode
Z-Diode f **mit Temperaturkompensation** <Elektronik> temperature-compensated Zener diode
Z-Draht m <Textil> Z-twist
Z-Drehung f <Textil> Z-twist
Z-Durchbruch m <Elektronik> Zener breakdown
Zeaxanthin n <Chemie> zeaxanthin
Zebrastreifen m <Trans> pedestrian crossing
Zeche f <Kohlen> mine, pit
Zechenabraum m <Abfall> colliery waste
Zeeman-Effekt m <Kerntech, Phys> Zeeman effect
Zeeman-Komponente f <Phys> Zeeman component
Zehneck n <Geom> decagon
Zehnerlogarithmus m <Comp & DV> common logarithm
Zehnertastatur f <Kontroll> ten digit keyboard
zehnflächig adj <Geom> decagonal, decahedral
Zehnflächner m <Geom> decahedron
zehnpoliges Filter n <Elektronik> ten pole filter
Zeichen n 1. <Aufnahme> signal; 2. <Comp & DV> character, mark; 3. <Kontroll> character; 4. <Telekom> signal
• **Zeichen geben** <Wassertrans> signal
Zeichen n **in Kursivdarstellung** <Comp & DV> italic character
Zeichen n **mit höchster Wertigkeit** <Comp & DV> high-order bit
Zeichen npl **pro Sekunde** (Z/sec) 1. <Comp & DV> characters per second, cps (characters per second); 2. <Druck> characters per second (cps)
Zeichen npl **pro Stunde** (Z/Std) <Druck> characters per hour, cph (characters per hour)
Zeichen npl **pro Zoll** 1. <Comp & DV> pitch; 2. <Druck> characters per inch, cpi
Zeichenabfühlung f <Comp & DV> mark scanning
Zeichenabstand m <Druck> character pitch
Zeichenabtastung f <Comp & DV> mark scanning
Zeichenanzeigeröhre f <Gerät> symbol indicator tube
Zeichenarbeit f <Konstzeich> drawing work
Zeichenart f <Comp & DV> character type
Zeichenbit n <Comp & DV> sign bit
Zeichenbreite f <Druck> character width
Zeichenbrett n <Bau, Textil, Wassertrans> drawing board (Schiffskonstruktion)
Zeichenbrett n **mit Zusatzfläche** <Konstzeich> drawing board with free margin
Zeichenbüro n <Bau> drawing office
Zeichencode m <Comp & DV> character code
Zeichendichte f 1. <Comp & DV> bit density, character density; 2. <Druck> character width
Zeichendreieck n <Maschinen> set square, triangle
Zeichendrucker m <Comp & DV> character printer
Zeichenelement n 1. <Elektronik> signal component; 2. <Telekom> signal component (Telegrafie)
Zeichenempfänger m <Telekom> signal receiver
Zeichenerkennung f 1. <Comp & DV> character recognition; pattern detection, pattern recognition (Mustererkennung); 2. <Telekom> signal recognition
Zeichenerkennungsfehler m <Telekom> character recognition error
Zeichenerkennungsfehlerrate f <Telekom> character recognition error rate
Zeichenerklärung f <Fertig> abbreviation (Kunststoffinstallationen)
Zeichenfehlerhäufigkeit f <Telekom> character error rate
Zeichenfehlerwahrscheinlichkeit f <Telekom> character error probability
Zeichenfläche f <Konstzeich> drawing area
Zeichenfolge f <Comp & DV> character string, string
Zeichenfolge f **der Länge Null** <Comp & DV> null string

Zeichenfolgenbearbeitung f <Comp & DV> string manipulation
Zeichenfolgenfunktion f <Comp & DV> string function
Zeichenfolgenoperation f <Comp & DV> string operation
Zeichenfolgenvariable f <Comp & DV> string variable
Zeichenfolgenverkettung f <Comp & DV> string concatenation
Zeichenfolie f <Konstzeich> white drawing film
Zeichengabe f <Telekom> (AE) signaling, (BE) signalling (Telefon)
Zeichengabe f **außerhalb der Zeitschlitze** <Telekom> outslot signaling
Zeichengabe f **innerhalb der Zeitlagen** <Telekom> (AE) in-slot signaling, (BE) in-slot signalling
Zeichengabe f **mit gemeinsamem Zeichenkanal** <Telekom> (AE) common channel signaling, (BE) common channel signalling
Zeichengabe f **mit getrenntem Zeichenkanal** <Telekom> (AE) out-of-band signaling, (BE) out-of-band signalling
Zeichengabe-Generator m <Elektrotech> (AE) signaling generator, (BE) signalling generator
Zeichengabeinformation f <Telekom> (AE) signaling information, (BE) signalling information
Zeichengabekanal m <Telekom> (AE) signaling channel, (BE) signalling channel
Zeichengabenetz n <Telekom> (AE) signaling network, (BE) signalling network
Zeichengabeprotokoll n <Telekom> (AE) signaling protocol, (BE) signalling protocol
Zeichengabesystem n <Telekom> (AE) signaling system, (BE) signalling system
Zeichengebung f **innerhalb des Bandes** <Raumfahrt> (AE) in-band signaling, (BE) in-band signalling (Weltraumfunk)
Zeichengenauigkeit f <Comp & DV> plotting accuracy
Zeichengenerator m <Comp & DV, Elektronik> character generator (Datenverarbeitung)
Zeichenkette f <Comp & DV, Druck> string
Zeichenkontrollgerät n <Elektronik> signal comparator (Telegrafie)
Zeichenlineal n <Metrol> ruler
Zeichenmittel n <Konstzeich> drawing instrument
zeichenorientiert adj <Comp & DV> character-oriented
Zeichenpapier n 1. <Bau> plotting paper; 2. <Geom> drawing paper
Zeichenpegel m <Telekom> signal level
Zeichenplatte f <Comp & DV> plotting board
Zeichenquantisierung f <Telekom> signal quantization
Zeichenrohr n <Konstzeich> tubular tip
Zeichensatz m 1. <Comp & DV> character set (ASCII); 2. <Druck> character set
Zeichenschiene f <Konstzeich> T-square, tee square
Zeichenstiftplotter m <Comp & DV> pen plotter
Zeichenteilmenge f <Comp & DV> character subset
Zeichentisch m <Comp & DV> plotting board
Zeichentusche f <Druck> drawing ink
Zeichentyp m <Comp & DV> character type
Zeichenumriss m <Comp & DV> character outline
Zeichenumschalter m <Telekom> character switch
Zeichenumschaltung f <Comp & DV> figures shift
Zeichenumsetzer m <Regelung, Telekom> signal converter
Zeichenumsetzung f <Telekom> signal conversion
Zeichenverzerrung f <Telekom> signal distortion (Telegrafie)
Zeichenvorrat m 1. <Comp & DV> character repertoire; 2. <Druck> character set
Zeichenwiederherstellung f <Telekom> signal restoration

Zeichenwinkel *m* <Maschinen> set square
Zeichenzuordnung *f* <Comp & DV> character assignment
zeichnen *v* <Bau, Wassertrans> draw
Zeichner *m* <Bau> *(AE)* draftsman, *(BE)* draughtsman
Zeichnung *f* 1. <Comp & DV> plot; 2. <Geom, Maschinen> drawing
Zeichnung *f* **in den Schatten** <Foto> shadow detail
Zeichnung *f* **mit vorgedruckten Darstellungen** <Konstzeich> drawing containing preprinted representations
Zeichnungsänderung *f* <Konstzeich> amendment of drawing
Zeichnungsänderungsdienst *m* <Konstzeich> drawing amendment service
Zeichnungsaustausch *m* <Konstzeich> exchange of drawings
Zeichnungsblatt *n* 1. <Bau, Konstzeich> drawing sheet; 2. <Patent> sheet of a drawing
Zeichnungsersteller *m* <Konstzeich> originator of the drawing
Zeichnungserstellung *f* <Konstzeich> preparation of drawing
Zeichnungsfeld *n* <Konstzeich> drawing area, drawing panel
zeichnungsgeprüft *adj* <Konstzeich> drawing-checked
Zeichnungsmaßstab *m* <Konstzeich> scale of the drawing
Zeichnungsnorm *f* <Konstzeich> drawing practice standard
Zeichnungsrichtlinien *fpl* <Konstzeich> principles for the preparation of drawings
Zeichnungssatz *m* <Konstzeich> set of drawing
Zeichnungsschriftfeld *n* <Konstzeich> title block
Zeichnungssystematik *f* <Konstzeich> systematic arrangement of drawings
Zeichnungsunterlage *f* <Konstzeich> preprinted drawing
Zeichnungsverfilmung *f* <Konstzeich> filming of drawings
Zeichnungsvordruck *m* <Konstzeich> preprinted drawing sheet
Zeichnungswesen *n* <Konstzeich> drawing practice
Zeiger *m* 1. <Bau> hand; 2. <Comp & DV, Eisenbahn, Elektriz> pointer; 3. <Elektrotech> pointer *(im Instrument)*; 4. <Gerät> needle; 5. <Labor> pointer; 6. <Maschinen> arm, finger, index, needle, pointer
Zeiger *m* **des Belichtungsmessers** <Foto> exposure meter needle
Zeigerarm *m* <Wassertrans> index bar *(Navigation)*
Zeigerausschlag *m* 1. <Fertig> pointer deflection; 2. <Gerät> amplitude of movement *(Messgerät)*; 3. <Metrol> pointer deflection
Zeigerdarstellung *f* <Elektrotech> phasor representation
zeigergesteuerte Unterbrechung *f* <Comp & DV> vectored interrupt
Zeigerinstrument *n* <Gerät> pointer instrument
Zeigerkette *f* <Comp & DV> pointer chain
Zeigermanometer *n* <Gerät, Heiz & Kälte> *(AE)* indicating pressure gage, *(BE)* indicating pressure gauge
Zeigerthermometer *n* <Heiz & Kälte> dial thermometer
Zeigerwaage *f* <Metall> bent-lever balance
Zeildurchschuss gesetzt *adj*/**ohne** <Druck> set solid
Zeile *f* 1. <Aufnahme> line; 2. <Comp & DV> line, row; 3. <Druck, Fernseh> line; 4. <Fertig> band *(Gießen)*
Zeilen *fpl* **pro Minute** *(LPM)* <Comp & DV> lines per minute *(LPM)*
Zeilenablenkung *f* 1. <Elektronik> horizontal deflection; 2. <Fernseh> line output
Zeilenabriss *m* <Fernseh> line tear

Zeilenabstand *m* 1. <Comp & DV> line spacing, row pitch, spacing; 2. <Druck> line spacing
Zeilenabtastdauer *f* <Fernseh> trace interval
Zeilenabtastgerät *n* <Comp & DV> raster scan device
Zeilenabtastung *f* 1. <Comp & DV> line scanning, raster scan; 2. <Elektronik> line scanning
Zeilenamplitudenregelung *f* <Fernseh> line amplitude control
Zeilenaustastpegel *m* <Fernseh> line-blanking level
Zeilenaustastung *f* <Fernseh> line blanking
Zeilencode *m* <Verpack> line code
Zeilendichte *f* <Comp & DV> line increment, scanning density
Zeilendiffusion *f* <Fernseh> line diffusion
Zeilendrucker *m* <Comp & DV> line printer
Zeilenende *n* <Comp & DV, Telekom> end of line
Zeilenfräsen *n* <Fertig> parallel-stroke milling, straight milling
Zeilenfräsmethode *f* <Fertig> line-by-line technique *(Spanung)*
Zeilenfrequenz *f* <Fernseh> horizontal frequency, line frequency
Zeilenfrequenzteiler *m* <Fernseh> line divider
Zeilengefüge *n* <Metall> banded structure
Zeilengießmaschine *f* <Druck> line caster
Zeilenkipp *m* <Fernseh> line sweep
Zeilenkippen *n* <Fernseh> line tilt
Zeilenkriechen *n* <Fernseh> line crawl
Zeilenlinearitätsregelung *f* <Fernseh> line linearity control
Zeilenmaß *n* <Druck> *(AE)* line gage, *(BE)* line gauge, type scale
Zeilenmodell *n* <Metall> band model
Zeilennummer *f* <Comp & DV> line number
zeilenorientiertes Dokument *n* <Comp & DV> line data document
Zeilenreißen *n* <Fernseh> tearing
Zeilenrücklauf *m* <Elektronik, Fernseh> line flyback
Zeilenschärfe *f* <Fernseh> line focus
Zeilenschlupf *m* <Fernseh> line slip
Zeilensprung *m* <Konstzeich> line spacing
Zeilensprungabtastung *f* 1. <Elektronik> interlaced scanning; 2. <Fernseh> interlaced scanning, line-interlaced scanning
Zeilensprungsequenz *f* <Fernseh> interlace sequence
Zeilensprungverfahren *n* 1. <Comp & DV> interlacing; 2. <Elektronik> interlaced scanning; 3. <Fernseh> interlaced scanning, interlacing, scanning interlace system
Zeilenteilung *f* <Comp & DV> scanning pitch
Zeilentrafo *m* <Fernseh> flyback transformer
Zeilentreibersignal *n* <Fernseh> line drive signal
Zeilenumbruch *m* <Comp & DV> line folding, word wrap
Zeilenvorschub *m* 1. <Comp & DV> line feed, line skipping; 2. <Telekom> line feed
Zeilenzwischenraum *m* <Druck> leading, line spacing
Zein *n* <Chemie, Kunststoff> zein
Z-Eisen *n* <Metall> Z-iron
Zeising *n* <Wassertrans> seizing
Zeit *f* 1. <Akustik, Comp & DV> time; 2. <Hydraul> duration; 3. <Phys> time • **Zeit nehmen** <Elektronik> clock
Zeit *f* **bis zum Ausfall** <Comp & DV, Maschinen> time to failure
Zeit *f* **bis zur Berührungstrockenheit** <Kunststoff> touch dry time
Zeit *f* **der Leitungsbelegung** <Telekom> occupancy
Zeit *f* **vor Entdecken eines Fehlers** <Kerntech> undetected failure time
Zeit *f* **zwischen Flugsteig zu Flugsteig** <Lufttrans> ramp-to-ramp time

Zeit

Zeit *f* **zwischen Sonnenwenden** <Raumfahrt> solstitial period
zeitabhängiges Filter *n* <Elektronik> time-varying filter
zeitabhängiges Signal *n* <Elektronik> time-varying signal
Zeitablauf *m* **beim Start** <Raumfahrt> chronology of launching
Zeitablaufdiagramm *n* <Kfztech> timing diagram
Zeitablaufplanung *f* <Comp & DV> scheduling
Zeitablaufsteuereinrichtung *f* <Kontroll> time schedule controller
Zeitableitung *f* <Phys> time derivative
Zeitablenkfrequenz *f* <Elektriz> time base frequency
Zeitablenkgenerator *m* <Elektronik> time base generator *(Katodenstrahloszillograph)*
Zeitablenkschaltung *f* 1. <Phys> time base circuit; 2. <Telekom> time base *(circuit)*
Zeitablenksignal *n* <Gerät> time base signal
Zeitablenkspannung *f* <Elektrotech> sawtooth voltage
Zeitablenkung *f* <Fernseh> sweep *(Oszilloskop)*
Zeitablenkverstärker *m* <Gerät> sweep deflection amplifier *(Oszilloskop)*
Zeitabschnitt *m* <Akustik> period
Zeitachse *f* <Elektronik> time base
Zeitansage *f* <Telekom> speaking clock
zeitaufgelöstes Spektrum *n* <Strahlphys> time-resolved spectrum
Zeitaufnahme *f* <Foto> time exposure
Zeitauslösung *f* <Comp & DV> time-out
Zeitbasis *f* <Elektronik, Phys, Telekom> time base
Zeitbasisfehler *m* <Fernseh> time base error
Zeitbasis-Korrekturschaltung *f* <Fernseh> time-base corrector
Zeitbasisspreizung *f* <Elektronik> expanded sweep *(Oszilloskop)*
Zeitbasisumsetzer *m* <Gerät> ramp encoder *(AD-Umsetzer)*
Zeitbasisverschlüsselung *f* <Gerät> sawtooth conversion *(AD-Umsetzung)*
Zeitberechnung *f* <Comp & DV> timing
Zeitbereich *m* 1. <Elektronik> time domain *(Fourieranalyse)*; 2. <Telekom> time domain
Zeitbereichsfilterung *f* <Gerät> time averaging
Zeitbereichs-Reflektometer *n* <Gerät> time domain reflectometer
Zeitbestimmung *f* <Phys> dating
Zeitblock *m* **während der Hauptsendezeit** <Fernseh> prime time slot
Zeitbruchkurve *f* <Kunststoff> time to fracture curve
Zeitcharter *f* 1. <Erdöl> time charter *(Schifffahrt)*; 2. <Wassertrans> time charter *(Seehandel)*
Zeitcode *m* <Aufnahme, Fernseh> time code
Zeitcode *m* **der Gesellschaft für Kino- und Fernsehtechniker** *(SMPTE-Zeitcode)* <Fernseh> *(AE)* Society of Motion Pictures and Television Engineers time code *(SMPTE time code)*
Zeitcode *m* **der Mittelspur** <Aufnahme> *(AE)* center track time code, *(BE)* centre track time code
Zeitdauer *f* 1. <Elektronik> time duration, time period; 2. <Hydraul> duration
Zeitdauer *f* **bis zum Abbindebeginn** <Bau> initial setting time
Zeitdehngrenze *f* <Metall, Qual> creep strain limit
Zeitdehnlinie *f* <Kunststoff> creep curve
Zeitdehnspannung *f* <Kunststoff, Qual> creep stress
Zeitdehnung *f* 1. <Gerät> time scaling; 2. <Phys> time dilation
Zeit-Demultiplextechnik *f* <Elektronik> time division demultiplexing

Zeitdiagrammessung *f* <Comp & DV> timing analysis
Zeitdifferenz *f* <Raumfahrt> hangover time
Zeitdifferenz *f* **zwischen Kanälen** <Aufnahme> inter-channel time difference
Zeit-Diversity-Empfang *m* <Telekom> time diversity reception
Zeitdrift *f* <Elektronik> time drift *(Abweichung)*
Zeitdrucker *m* <Eisenbahn> time indicator
Zeitedieren *n* <Fernseh> time code editing
Zeiteinheit *f* **je Tarifeinheit** <Telekom> metering rate
Zeiten *fpl* **niedriger Verkehrsbelastung** <Telekom> nonbusy hours
Zeitfenster *n* <Raumfahrt> time slot
Zeitfestigkeit *f* 1. <Anstrich> limited life fatigue; 2. <Fertig> endurance limit
Zeitfolge *f* 1. <Comp & DV> time series; 2. <Mechan> elapsed time
Zeitgeber *m* 1. <Comp & DV> clock register, interval timer, timing equipment; 2. <Elektronik> timing generator; 3. <Phys, Telekom> timer
zeitgesteuertes Magnetventil *n* <Kontroll> timer-controlled magnet valve
zeitgeteilte Vermittlung *f* <Telekom> time division switching, time switching
zeitgeteiltes Vermittlungssystem *n* <Telekom> time division switching system
Zeitgetrenntlage-Verfahren *n* <Telekom> burst mode
Zeitimpuls *m* <Telekom> timing pulse
Zeitintervall *m* 1. <Elektronik> time interval; period *(Telefon)*; 2. <Telekom> time slot
Zeitintervallgeber *m* <Regelung> interval timer
Zeitintervallmessgerät *n* <Gerät> time interval measuring instrument
zeitinvariantes Signal *n* <Elektronik> time-invariant signal
Zeitkanal *m* <Telekom> time slot
Zeitkohärenz *f* <Telekom> time coherence
Zeitkompression *f* <Telekom> time compression
Zeitkonstante *f* <Elektriz, Phys> time constant
Zeitkoordinate *f* <Phys> time coordinate
Zeitkorrelation *f* <Telekom> time correlation
Zeitlage *f* <Telekom> time slot
Zeitlagenwechsler *m* <Telekom> time slot interchanger
zeitlich aufgelöste Radiographie *f* <Kerntech> time-resolved radiography
zeitlich festgelegte Hauptverkehrsstunde *f* <Telekom> time-consistent busy hour
zeitliche Auflösung *f* <Umweltschmutz> temporal resolution
zeitliche Dosisverteilung *f* <Umweltschmutz> temporal dose distribution
zeitliche Kohärenz *f* 1. <Optik> temporal coherence, time coherence; 2. <Phys, Telekom> temporal coherence
zeitliche Korrelationsanalyse *f* <Kerntech> time correlation analysis
zeitliche Mittelwertbildung *f* <Gerät> time averaging
zeitliche Programmplanung *f* <Comp & DV> program scheduling
zeitliche Schwankung *f* <Umweltschmutz> temporal fluctuation, temporal variation
zeitliche Schwellwertverschiebung *f* <Akustik> temporary threshold shift
zeitlicher Lösungsablauf *m* <Comp & DV> machine time
zeitliches Zittern *n* <Elektronik> time jitter
Zeit-Licht-Leistungskurve *f* <Foto> time/light-output curve *(Blitz)*
Zeitlimit *n* <Comp & DV> time-out
Zeitlimitüberschreitung *f* <Comp & DV> watchdog time-out

Zeitlupendisk f <Fernseh> (BE) slow motion disc, (AE) slow motion disk
Zeitmarkengeber m 1. <Elektronik> time marker; 2. <Gerät> time mark generator
Zeitmarkengenerator m <Gerät> time mark generator
Zeitmarkenimpuls m <Gerät> time marker pulse
Zeitmarkierung versehen v/mit <Telekom> time-tag
Zeitmaß n 1. <Akustik> tempo; 2. <Gerät, Telekom> clock rate
Zeitmaßstab m <Gerät> timescale
Zeitmesser m <Labor> chronometer, timer
Zeitmessgerät n <Gerät> chronometer, time interval measuring instrument
Zeitmessung f <Comp & DV> timing
Zeitmittelung f <Gerät> time averaging
Zeitmodulation f <Elektronik> time modulation
Zeitmultiplex n 1. <Comp & DV, Elektronik, Phys, Raumfahrt> time division multiplex, TDM; 2. <Telekom> time division multiplex, TDM, time multiplex
Zeitmultiplexabtastregelung f <Regelung> time-shared control
Zeitmultiplexdurchschaltung f <Comp & DV> time division switching
Zeitmultiplexer m <Elektronik, Telekom> time division multiplexer
Zeitmultiplexleitung f <Telekom> highway
Zeitmultiplexmethode f (TDM) <Comp & DV, Elektronik, Telekom> time division multiplex (TDM)
Zeitmultiplexnetz n <Telekom> time division network
Zeitmultiplexsignal n <Elektronik, Telekom> time division multiplexed signal
Zeitmultiplexverfahren n (TDM) 1. <Comp & DV, Elektronik> time division multiplex (TDM); 2. <Telekom> time division multiplex, time multiplexing
Zeitmultiplexvermittlung f <Telekom> time division switching
Zeitmultiplexvermittlungssystem n <Telekom> time division switching system
Zeitmultiplex-Vielfachzugriff m **mit Vermittlung im Satelliten** <Raumfahrt> satellite-switched time-division multiple access, satellite-switched TDMA (Weltraumfunk)
Zeitnehmer m <Comp & DV> timer
zeitperiodisches Feld n <Elektrotech> time periodic field
Zeitplan m <Comp & DV, Fernseh> schedule
Zeitplanung f <Fernseh, Telekom> scheduling
Zeitplanungsprogramm n <Comp & DV> scheduler
zeitraffende Prüfung f 1. <Qual> accelerated test; 2. <Werkprüf> accelerated testing
Zeitraffer m <Fernseh> fast motion
Zeitrafferübersicht f <Trans> time lapse survey
Zeitraffung f 1. <Comp & DV> fast time scale; 2. <Gerät> time scaling
Zeitraffungsfaktor m <Qual> acceleration factor
Zeitraster m <Kerntech> time slot pattern
Zeitrasterung f <Regelung> interval timing
Zeit-Raum-Diagramm n <Trans> time-space diagram
Zeit-Raum-Zeit-Koppelnetz n <Telekom> time-space-time network
Zeitregelung f <Elektronik> timing
Zeitregulierband n <Aufnahme> timing tape
Zeitreihenanalyse f <Elektronik> time series analysis
Zeitschachtelung f <Comp & DV> time slicing
Zeitschalter m 1. <Elektriz, Elektrotech> time switch; 2. <Gerät> clock relay, time switch; 3. <Telekom> time switch
Zeitschaltrelais n <Gerät> clock relay
Zeitscheibe f <Comp & DV, Elektronik> time slice
Zeitscheibenverfahren n <Comp & DV> time slicing
Zeitschlitz m 1. <Elektronik> time slot; 2. <Telekom> slot (PCM)

Zeitschlitzeinblendung f <Telekom> blank-burst mode
Zeitschreiber m <Gerät> time recorder
Zeitschriftensatz m <Druck> magazine typesetting
Zeitsetzung f <Kontroll> timing
Zeitsignal n <Wassertrans> time signal
Zeitskale f <Gerät> timescale
Zeitspanne f <Elektronik> periodic time
Zeitsperre f <Comp & DV, Telekom> time-out
Zeitstandfestigkeit f <Kunststoff, Qual> creep strength
Zeitstandsfestigkeitslinie f <Kunststoff> time to fracture curve
Zeitstandversuch m <Kunststoff, Qual, Werkprüf> creep test
Zeitsteuertakt m <Elektronik> period pulse
Zeitsteuertakte mpl <Elektronik> periodic pulses
Zeitsteuerung f 1. <Kfztech> timing; 2. <Kontroll> time control, timing
Zeitsteuerungseinstellung f <Kfztech> timing adjustment
Zeitstufe f <Telekom> time stage
Zeitsynthese f <Elektronik> time synthesis
Zeitsynthesizer m <Elektronik> time synthesizer
Zeittakt m 1. <Fernseh> time base; 2. <Telekom> clock, metering rate, timing • **nach Zeittakt steuern** <Gerät> clock
Zeitteilung f <Elektronik> time division
Zeitüberwachung f <Telekom> time-out
Zeituhr f <Labor> timer
zeit- und codegeschachtelter Mehrfachzugriff m <Telekom> time division-code division multiple access (TD-CDMA)
zeit- und frequenzgeschachtelter Mehrfachzugriff m <Telekom> time division-frequency division multiple access (TD-FDMA)
Zeitungsdruck m <Druck> newsprint
Zeitungsdruckpapier n <Papier> newsprint
Zeitungsformat n <Druck> tabloid format (etwa 300 × 400 mm)
Zeitungsspalte f <Druck> newspaper column
Zeitverhalten n <Telekom> time characteristic, transient response
Zeitverlust m <Trans> delay
zeitverschachtelte Struktur f <Kontroll> pipelined architecture
Zeitverschiebung f <Elektronik> time shift
Zeitverschlüsseler m <Gerät> ramp encoder (AD-Umsetzer)
zeitversetzt adj <Kontroll> deferred
Zeitverzögerung f <Elektronik, Kontroll> time lag
Zeitverzögerungsrelais n <Elektrotech> time delay relay
Zeitvielfachsystem n <Telekom> time division system
Zeitvielfachzugriff m <Elektronik> time division multiple access
zeitweilig adj <Bau, Qual> temporary
zeitweilige Belastung f <Kohlen> temporary load
zeitweilige Teilnehmerkennung f <Mobilkom> temporary mobile subscriber identity
zeitweiliger Vorgang m <Fertig> transient
Zeitzählung f 1. <Elektronik> timing; 2. <Telekom> timing (Telefon)
Zeitzeichen n 1. <Elektronik> time signal; 2. <Funktech> standard time-signal; 3. <Wassertrans> time signal
Zeitzone f <Wassertrans> time zone
Zeitzyklus m <Gerät> timing cycle
Zelle f 1. <Bau, Comp & DV> cell; 2. <Elektriz> battery; 3. <Elektrotech> cell (Batterie); 4. <Kohlen, Labor> cell; 5. <Lufttrans> nacelle; 6. <Telekom> cell; 7. <Wassertrans> tank (Schiffbau)
Zellenbrenner m <Fertig> cell-type burner

Zellenbündel

Zellenbündel n <Mobilkom, Telekom> cluster *(C-Netz)*
Zellendoppelboden m <Wassertrans> cellular double bottom *(Schiffbau)*
Zellenfiltersaugtrockner m <Heiz & Kälte> filter drier, filter dryer
Zellenfundamentplatte f <Bau> cellular raft
Zellengleichrichter m <Lufttrans> honeycomb filler, honeycomb grill
Zellenkopf m <Telekom> cell header *(ATM-Adresse)*
Zellenkühler m <Kfztech> honeycomb radiator
Zellenladung f <Funktech> cell charge
Zellenpolarisation f <Elektrotech> cell polarization
Zellenschalter m <Elektrotech> multiple-contact switch
Zellglas n <Kunststoff> cellulose film
Zellgrenze f <Mobilkom> cell boundary
Zellgrenzenübergang m <Mobilkom> boundary crossing
Zellgummi n <Heiz & Kälte> cellular rubber
Zellhorn n <Verpack> celluloid
Zellophan n <Verpack> cellophane
Zellspannung f <Elektrotech> closed-circuit voltage *(Batterie)*; open circuit voltage *(offener Stromkreis)*
Zellspannung f bei Stromfluss <Elektrotech> on-load voltage *(Batterie)*
Zellstoff m 1. <Abfall> paper pulp; 2. <Druck, Papier> chemical pulp; 3. <Verpack> cellulose, woodpulp; 4. <Wasserversorg> pulp
Zellstoffentwässerungsmaschine f <Papier> wet machine
Zellstoffkocher m <Papier> digester
Zellstoffpresssystem n <Verpack> *(AE)* pulp molding system, *(BE)* pulp moulding system
Zellstoffwatte f 1. <Chemie> artificial cotton; 2. <Papier> wadding
zellular adj <Elektronik> cellular
zellulare Struktur f <Mobilkom, Telekom> cellular structure
zellularer Mobilfunk m <Mobilkom> cellular mobile radio
zellulares digitales Mobilfunksystem n <Funktech, Mobilkom> digital cellular mobile radio system, digital cellular radio system *(DCS)*
zellulares Funktelefon n <Mobilkom, Telekom> cellular radiotelephone
zellulares Mobilfunknetz n <Mobilkom> cellular mobile radio network
zellulares Mobilfunksystem n <Mobilkom> cellular mobile radio system
zellulares Netz n <Mobilkom, Telekom> cellular network
Zellularsystem n <Mobilkom, Telekom> cellular system
Zellularverfahren n <Mobilkom, Telekom> cellular technique
Zelluloid n <Verpack> celluloid
Zellulose f <Verpack> cellulose
Zellulosefasern fpl <Abfall> *(AE)* cellulose fibers, *(BE)* cellulose fibres
zellulosehaltiger Abfall m <Abfall> cellulose waste
zellulosisch adj <Textil> cellulosic
Zeltleinwand f <Textil> canvas
Zeltstoff m <Bau> canvas
Zement m <Bau, Ker & Glas, Kunststoff> cement
Zementabbindungslog n <Erdöl> cement bond log *(Bohrtechnik)*
Zementation f 1. <Erdöl> cementing; 2. <Metall> cementation
Zementationszone f <Metall> cementation zone
Zementbeton m <Bau> cement concrete
Zementbrennerei f <Ker & Glas> cement works
Zementhohldiele f <Bau> hollow core slab
zementieren v <Bau, Metall> cement
Zementieren n <Fertig> converting

zementierend adj <Bau, Metall> cementing
Zementierofen m <Metall> cementation furnace, cementing furnace
Zementierstrang m <Erdöl> cementing string
Zementiertour f <Erdöl> cementing string *(Bohrtechnik)*
Zementierung f 1. <Bau> cementation, cementing; 2. <Erdöl> cementing *(Bohrtechnik)*; 3. <Kerntech> cementation
Zementit m <Chemie> carbide of iron, cementite *(Metall)*
Zementkupfer n <Metall> cement copper
Zementmilch f <Bau> cement slurry, laitance
Zementschachtofen m <Ker & Glas> cement kiln
Zementschlämme f <Bau> cement slurry, grout, neat cement grout
Zementstahl m <Metall> blister steel, cement steel, cementation steel, cemented steel
Zementstaub m <Bau> cement dust
Zementstopfen m <Erdöl> cement plug, cementing plug *(Bohrtechnik)*
Zementverdämmung f <Erdöl> cement plug, cementing plug
Zementverfestigung f <Bau> cement stabilization
Zementverpressung f <Bau> cement injection
Zementwerk n <Bau> cement mill, cement plant, cement works
Zementwinden n <Bau> shrinkage in cement
Zener... <Phys> Zener
Zenerdiode f <Phys, Telekom> Zener diode
Zenereffekt m <Elektronik> Z-effect
Zener-Hollomon'sches Parameter m <Metall> Zener-Hollomon parameter
Zenerspannung f *(UZ)* <Elektronik> breakdown voltage *(BDV)*
Zenit m <Wassertrans> zenith *(astronomische Navigation)*
Zenitalpunkt m <Raumfahrt> zenith point
Zenitentfernung f <Raumfahrt> zenith distance
Zenitfernrohr n <Raumfahrt> zenith telescope
Zenitreduktion f <Raumfahrt> zenith reduction
Zenitvergaser m <Kfztech> *(AE)* zenith carburetor, *(BE)* zenith carburettor
Zenitwinkel m <Nichtfoss Energ> zenith angle
Zenti... *(c)* <Metrol> centi *(c)*
Zentigramm n <Metrol> centigram
Zentiliter m <Metrol> *(AE)* centiliter, *(BE)* centilitre
Zentimeter m <Metrol> *(AE)* centimeter, *(BE)* centimetre
Zentimeter-Gramm-Sekunde-System n *(CGS-System)* <Metrol> *(AE)* centimeter-gram-second system, *(BE)* centimetre-gram-second system *(CGS system)*
Zentimeterwellen fpl 1. <Funktech> super high frequency, SHF; 2. <Phys> *(AE)* centimeter waves, *(BE)* centimetre waves
Zentner m <Metrol> cwt, hundredweight, cwt, metric centner, hundredweight
zentral adj <Comp & DV> centralized
Zentral... <Comp & DV, Maschinen> central
Zentralamt n <Telekom> tertiary exchange
Zentrale f 1. <Telekom> switch *(Telefon)*; switchboard; 2. <Wassertrans> control room *(Schiff)*
zentrale Internet-Datenleitung f <Telekom> back bone
zentrale Ladevorrichtung f <Telekom> central charging equipment
zentrale Operation f <Comp & DV> centralized operation
zentrale Rechnereinheit f *(CPU)* <Comp & DV, Telekom> central processing unit *(CPU)*
zentrale Steuerung f <Telekom> central control, centralized control
zentrale Stromversorgung f <Telekom> central battery *(CB)*; central power supply

Zentraleinheit f *(CPU)* 1. <Comp & DV> central processing unit, central processor; 2. <Telekom> central processing unit, central processor *(CPU)*
zentraler Zeichengabekanal m <Telekom> *(AE)* common signaling channel, *(BE)* common signalling channel
zentrales lasttragendes Element n <Bau, Optik> central load-bearing element *(Lichtleiter)*
zentrales Steuersystem n <Telekom> centralized control system
zentrales tragendes Glied n <Optik> central strength member *(Lichtleiter)*
zentralgesteuerte Einrichtung f <Telekom> common control equipment
zentralgesteuertes System n **mit Einzelprozessor** <Telekom> single processor common-control system
zentralgesteuertes Vermittlungssystem n <Telekom> common-control switching system
Zentralheizung f <Bau, Heiz & Kälte, Thermod> central heating
Zentralheizungskessel m <Heiz & Kälte> central heating boiler
zentralisiert adj <Comp & DV> centralized
zentralisiertes System n <Telekom> centralized system
zentralisiertes Verkehrsaufteilungssystem n <Telekom> centralized traffic-division system
Zentralkanalzeichengabe f <Telekom> *(AE)* common channel signaling, *(BE)* common channel signalling
Zentralkatode f <Elektrotech> common cathode
Zentralkraft f <Maschinen, Phys> central force
Zentralprojektion f <Geom> central projection
Zentralprozessor m <Comp & DV, Telekom> central processor
Zentralrechner m <Comp & DV> mainframe computer
Zentralschmierung f 1. <Kfztech> centralized lubrication; 2. <Maschinen> central lubrication, centralized lubrication, centralized lubricating system
Zentralspeicher m <Comp & DV> CM, central memory, main memory, main store
Zentralverbindung f <Comp & DV> backbone
Zentralvermittlungsstelle f <Telekom> central switching unit, tertiary exchange
Zentralverriegelung f <Kfztech> central locking *(Türen)*
Zentralwert m <Akustik, Comp & DV, Math, Qual> median
Zentrieransatz m <Fertig> spigot
zentrierbohren v <Maschinen> *(AE)* center-drill, *(BE)* centre-drill
Zentrierbohren n <Maschinen> *(AE)* center drilling, *(BE)* centre drilling
Zentrierbohrer m 1. <Fertig> *(AE)* center drill, *(BE)* centre drill; 2. <Maschinen> *(AE)* center drill, *(BE)* centre drill, *(AE)* centering drill, *(BE)* centring drill, combination drill
Zentrierbohrung f 1. <Fertig> *(AE)* center hole, *(BE)* centre hole; 2. <Maschinen> *(AE)* centering hole, *(BE)* centring hole
Zentrierbuchse f <Maschinen> *(AE)* centering bush, *(BE)* centring bush, *(AE)* centering sleeve, *(BE)* centring sleeve
zentrieren v <Maschinen> *(AE)* center, *(BE)* centre
Zentrieren n <Maschinen> *(AE)* centering, *(BE)* centring
Zentrierer m <Erdöl> centralizer
Zentrierlinse f **mit Fadenkreuz** <Ker & Glas> *(AE)* centering lens with ruled cross, *(BE)* centring lens with ruled cross
Zentriermaschine f <Maschinen> *(AE)* centering lathe, *(BE)* centring lathe
Zentriermutter f <Maschinen> *(AE)* centering nut, *(BE)* centring nut
Zentrierring m <Fernseh> *(AE)* centering ring, *(BE)* centring ring
Zentrierschraube f <Maschinen> *(AE)* centering screw, *(BE)* centring screw
Zentrierspitze f <Maschinen> *(AE)* center point, *(BE)* centre point
Zentrierstift m <Maschinen, Mechan> *(AE)* centering pin, *(BE)* centring pin
zentriertes System n <Optik> *(AE)* centered system, *(BE)* centred system
Zentrierung f 1. <Maschinen> location; 2. <Mechan> *(AE)* centering, *(BE)* centring
Zentrierung f **der Bremsscheibe** <Kfztech> *(BE)* brake disc alignment, *(AE)* brake disk alignment *(Bremsanlage)*
Zentriervorrichtung f <Erdöl> centralizer *(Bohrtechnik)*
zentrifugal adj <Mechan, Phys> centrifugal
Zentrifugal... <Chemtech> centrifugal
Zentrifugalabschöpfgerät n <Meerschmutz> centrifugal skimmer *(für Öl)*
Zentrifugalextraktor m <Kerntech> centrifugal extractor
Zentrifugalfilter n <Kohlen> centrifugal filter
Zentrifugalhubgebläse n <Lufttrans> centrifugal flow lift fan
Zentrifugalkompressor m <Maschinen> centrifugal compressor
Zentrifugalkraft f <Chemtech, Maschinen, Phys, Raumfahrt, Strömphys, Umweltschmutz> centrifugal force
Zentrifugalmoment n <Phys> product of inertia
Zentrifugalpumpe f <Chemtech, Maschinen, Meerschmutz> centrifugal pump
Zentrifugalregler m <Nichtfoss Energ> governor
Zentrifugalreiniger m 1. <Chemtech> centrifugal cleaner; 2. <Papier> cleaner
Zentrifugalseparator m <Fertig> centrifugal separator
Zentrifugation f <Chemtech> centrifugation
Zentrifuge f 1. <Chemtech> centrifugal, centrifuge; 2. <Kunststoff, Labor, Mechan, Phys> centrifuge
Zentrifugengarn n <Textil> boxspun yarn
Zentrifugenglas n <Labor> centrifuge tube
Zentrifugentrommel f <Labor> centrifuge rotor
zentrifugieren v <Chemtech, Kohlen, Kunststoff> centrifuge
Zentrifugieren n 1. <Abfall> centrifugation, centrifuging; 2. <Kohlen> centrifuging
zentrifugierter Latex m <Kunststoff> centrifuged latex
Zentrifugierung f <Wasserversorg> centrifugation
zentrifugisch kontrollierte Schaufelteilung f <Nichtfoss Energ> centrifugally-operated blade pitch control
zentripetal adj <Phys> centripetal
Zentripetalbeschleunigung f 1. <Fertig> centripetal acceleration; 2. <Lufttrans> centrifugal acceleration, centripetal acceleration
Zentripetalkraft f <Maschinen, Phys> centripetal force
zentrisch adj <Fertig> radial *(Schubkurbelgetriebe)*
Zentrum n <Telekom> *(AE)* center, *(BE)* centre
Zentrumsbohrer m <Maschinen> *(AE)* center bit, *(BE)* centre bit
Zentrumswicklerrolle f <Papier> *(AE)* center wind reel, *(BE)* centre wind reel
Zentrumswicklung f <Papier> *(AE)* center winding, *(BE)* centre winding
zerbrechen v 1. <Bau> break, break up; 2. <Bau> spring; 3. <Kohlen> crush
Zerbrechen n <Fertig> fracture *(Kunststoffinstallationen)*
zerbrechlich adj <Qual, Verpack> fragile
Zerbrechlichkeit f <Mechan> brittleness
zerbröckeln v <Kohlen> grind
zerbröckelte Kohle f <Kohlen> coal slack
zerdrücken v <Bau, Lebensmittel, Papier> crush
Zerdrückfestigkeit f <Maschinen, Qual> resistance to crushing

Zerfall

Zerfall *m* 1. <Erdöl> decomposition; 2. <Kerntech, Phys, Strahlphys, Teilphys> decay
zerfallen *v* 1. <Chemie> slake *(Kohle)*; 2. <Maschinen> come apart; 3. <Verpack> disintegrate
Zerfallsarten *fpl* <Kerntech, Phys, Strahlphys, Teilphys> decay modes
Zerfallsenergie *f (Q)* <Kerntech> disintegration energy *(Q)*
Zerfallsgeschwindigkeit *f* <Kerntech, Phys, Strahlphys, Teilphys> decay rate
Zerfallshohlraum *m* <Kerntech> decay cavity
Zerfallskette *f* <Kerntech, Phys, Strahlphys, Teilphys> decay chain
Zerfallskonstante *f* <Kerntech, Phys, Strahlphys, Teilphys> decay constant
Zerfallskonstante *f* **bei Gammazerfall** <Strahlphys, Teilphys, Wellphys> gamma constant
Zerfallskurve *f* <Phys, Strahlphys, Teilphys> decay curve
Zerfallsrate *f* <Kerntech, Phys, Strahlphys, Teilphys> decay rate
Zerfallsteilchen *n* <Kerntech, Phys, Strahlphys, Teilphys> decay particle
Zerfallswahrscheinlichkeit *f* <Kerntech> decay constant
Zerfallszeit *f* <Elektrotech> decay time
Zerfaserer *m* <Papier> pulp machine
Zerfließbarkeit *f* <Lebensmittel> deliquescence
zerfließen *v* <Chemie> deliquesce
Zerfließen *n* <Chemie> deliquescence
zerfließend *adj* <Chemie> deliquescent
Zerhacken *n* <Elektronik> chopping *(von Gleichstrom)*
Zerhacker *m* 1. <Elektronik, Elektrotech> chopper; 2. <Funktech> alternator; 3. <Raumfahrt> DC-AC converter
Zerhackerkontrolle *f* <Eisenbahn> chopper control
Zerhackermodus *m* <Elektronik> chopped mode
Zerhackerschaltung *f* <Raumfahrt> chopper circuitry *(Raumschiff)*
Zerhackerverstärker *m* <Elektronik> chopper amplifier *(wandelt Gleichspannung in Rechteckwechselspannung um)*
zerhacktes Signal *n* <Elektronik> chopped signal
zerkeilen *v* <Bau> split with wedges
Zerkleinerer *m* 1. <Abfall> crusher; 2. <Bau> pulverizer; 3. <Maschinen> grinding machine; 4. <Kohlen> milling
zerkleinern *v* 1. <Abfall> crush; 2. <Bau> crush, pulverize; 3. <Kohlen> mill
Zerkleinern *n* 1. <Abfall> shearing; 2. <Kunststoff> crushing
zerkleinerter Abfall *m* <Abfall> crushed waste
Zerkleinerung *f* 1. <Abfall> grinding; crushing *(der Abfälle)*; 2. <Kunststoff> crushing
Zerkleinerungsanlage *f* 1. <Abfall> crusher unit; 2. <Chemtech> crushing plant
Zerkleinerungsgerät *n* <Lebensmittel> disintegrator, grinder
Zerkleinerungsgrad *m* <Kohlen> reduction ratio
Zerkleinerungsmaschine *f* 1. <Chemtech> crushing machine, grinding mill; 2. <Maschinen> crushing machine
Zerkleinerungsverfahren *n* <Abfall> crushing system
Zerkleinerungswerk *n* <Abfall> crusher unit
Zerklüftung *f* <Erdöl> fracturing *(Geologie)*
zerknittern *v* <Papier> crumple
zerknittert *adj* <Papier> crumpled
zerlegbar *adj* <Fertig> decomposable
zerlegen *v* 1. <Bau> break down; 2. <Comp & DV> parse; 3. <Kerntech> strip; 4. <Kfztech> strip down; 5. <Maschinen> decompose, disassemble, dismount, strip, take down
Zerlegen *n* <Maschinen> disassembly, taking to pieces
Zerlegung *f* 1. <Akustik> resolution; 2. <Kerntech> dismantling *(eines Brennelementes)*; 3. <Math> partition; 4. <Phys> breakdown
Zerlegung *f* **in Abschnitte** <Kerntech> sectioning technique
Zerlegung *f* **in Faktoren** <Math> decomposition in factors
Zerlegungsgleis *n* <Eisenbahn> siding for splitting up trains
zermahlen *v* 1. <Abfall> grind; 2. <Bau> mill; 3. <Lebensmittel> crush
zermalmen *v* <Kohlen> crush
Zer-Polierrot *n* <Ker & Glas> ceri-rouge
zerquetschen *v* <Lebensmittel> bruise
Zerreiben *n* 1. <Kohlen, Papier> attrition; 2. <Textil> grinding
Zerreiß... <Maschinen> tensile
zerreißen *v* 1. <Bau> break; 2. <Papier> tear
Zerreißen *n* 1. <Abfall> crushing *(von Abfällen)*; 2. <Papier> tearing
Zerreißfestigkeit *f* 1. <Kunststoff> tear resistance, tensile strength; 2. <Maschinen> tensile strength, ultimate strength; 3. <Mechan> breaking strength, tensile strength; 4. <Textil> tear strength, tensile strength
Zerreißgrenze *f* 1. <Maschinen> ultimate strength; 2. <Mechan> breaking point
Zerreißprobe *f* 1. <Maschinen> tensile specimen, tensile test piece; 2. <Mechan, Qual> breaking test; 3. <Phys> tensile test
Zerreißprüfgerät *n* <Maschinen> tensile test equipment
Zerreißprüfmaschine *f* <Maschinen> tension testing machine
Zerreißprüfstück *n* <Maschinen> tensile specimen, tensile test piece
Zerreißprüfung *f* 1. <Maschinen> tensile test; 2. <Werkprüf> breaking test
Zerreißversuch *m* <Metall> tensile test
Zerrspiegel *m* <Ker & Glas> distorting mirror
Zerschmelzen *n* <Chemie> deliquescence
Zerschneideblattsystem *n* <Konstzeich> sheet dissection system
Zerschneiden *n* <Abfall> shearing
zersetzen *v* 1. <Meerschmutz> foul; 2. <Verpack> disintegrate
Zersetzung *f* 1. <Abfall> digestion; 2. <Chemie, Comp & DV> decomposition; 3. <Kunststoff> degradation; 4. <Lebensmittel> decomposition; 5. <Verpack> deterioration
Zersetzung *f* **von Wasser durch Bestrahlung** <Kerntech> water decomposition under irradiation
Zersetzungsbereich *m* <Fertig> decomposition zone *(Kunststoffinstallationen)*
Zersetzungsmittel *n* <Chemie> decomposing agent
Zersetzungsprodukt *n* <Abfall, Umweltschmutz> residue of decomposition
Zersetzungstemperatur *f* 1. <Kunststoff> decomposition temperature; 2. <Verpack> decomposition temperature *(warme Lagerung)*
zerspalten *v* <Bau> split
zerspanbar *adj* <Fertig> machinable
Zerspanbarkeit *f* <Fertig> machinability, machinability rating
zerspanend feinstbearbeiten *v* <Fertig> superfinish
Zerspangröße *f* <Fertig> cutting variable, machining variable
Zerspankraft *f* <Maschinen> cutting force
zerspantes Volumen *n* <Fertig> volume of metal removed
Zerspanung *f* <Fertig> cutting, metal cutting, metal removal
Zerspanungsbedingung *f* <Fertig> cutting condition

Zerspanungsleistung f <Fertig> cutting efficiency
Zerspanungsvolumen n <Fertig> volume of metal removed by cutting
Zerspanvorgang m <Fertig> machining operation
Zersprühung f <Fertig> fork burst *(Funkenbild)*; explosion *(Schleiffunkenversuch)*
zerstampfen v <Lebensmittel> pound
zerstäuben v 1. <Bau> pulverize; 2. <Papier> atomize
Zerstäuber m 1. <Fertig> atomizer, nebulizer; 2. <Labor> atomizer; 3. <Papier> atomizer, sprayer
Zerstäuberdüse f 1. <Chemtech> atomizer nozzle; 2. <Fertig> diffuser jet; 3. <Lebensmittel> atomizing nozzle
Zerstäubung f 1. <Abfall> grinding *(von Abfällen)*; 2. <Elektrotech> sputtering; 3. <Erdöl> atomization *(Ölverbrennung)*
Zerstäubung f **durch Glimmentladung** <Kerntech> glow discharge sputtering
Zerstäubungsbrenner m <Heiz & Kälte, Thermod> atomizing burner
zerstörender Versuch m <Maschinen, Werkprüf> destructive test
zerstörendes Prüfverfahren n <Werkprüf> destructive testing
Zerstörer m <Wassertrans> destroyer *(Marinefahrzeug)*
Zerstörfestigkeit f <Elektriz> destruction resistance
Zerstörung f **durch Stoß** <Strahlphys> collisional destruction
zerstörungsfreie Prüfung n 1. <Maschinen, Phys> nondestructive test; 2. <Raumfahrt> NDT, nondestructive testing
zerstörungsfreie Prüfverfahren n <Maschinen> NDT, nondestructive testing
zerstörungsfreie Werkstoffprüfung f <Kerntech> nondestructive materials testing
zerstörungsfreies Prüfsystem n <Maschinen> nondestructive testing system
Zerstrahlung f <Teilphys> annihilation
Zerstrahlungsphoton n <Strahlphys> annihilation photon
zersträuben v <Fertig> reduce
zerstreuen v <Foto, Optik> diffuse *(Licht)*
Zerstreuen n **von Elektronenballungen** <Elektronik> debunching
zerstreute Reflexion f <Elektrotech> diffuse reflection
Zerstreuung f 1. <Elektrotech> dissipation; 2. <Kohlen> dispersion
Zerstreuungskreis m <Foto> circle of confusion
Zerstreuungslinse f <Phys> diverging lens
Zerstreuungsscheibchen n/am wenigsten wahrnehmbares <Foto> circle of least confusion
Zerstreuungsvermögen n <Phys> dispersive power
Zerstückeln n <Abfall> crushing
Zerstückelung f <Comp & DV> fragmentation
Zerteilung f <Chemie> deflocculation
Zertifizierung f <Comp & DV, Patent, Qual> certification
Zertifizierungsstelle f <Patent, Qual> certification body
Zertifizierungssystem n <Patent, Qual> certification system
zertrümmern v <Kohlen> fragment
Zertrümmerung f <Metall> fragmentation
zetteln v <Textil> warp
Zetteln n <Textil> beaming, warping
Zf *(Zwischenfrequenz)* <Elektronik, Funktech, Telekom> IF *(intermediate frequency)*
Zf-Durchschlag m <Funktech> IF breakthrough
Zfm *(Zwischenfrequenzmodulation)* <Elektronik, Funktech, Telekom> IFM *(intermediate frequency modulation)*
Zf-Oberwelle f <Funktech> IF harmonic
Zf-Signal n <Elektronik, Funktech> IF signal
Zf-Stufe f <Elektronik, Funktech> IF stage

Zf-Unterdrückung f <Elektronik, Funktech> IF rejection
Zf-Verstärker m <Elektronik, Funktech> IF amplifier
Zf-Verstärkung f <Elektronik, Funktech> IF amplification
Zibeton n <Chemie> cibetone, civetone
Zickzack-Flach-Nähmaschine f <Textil> zigzag flatbed sewing machine
Zickzackkerbe f <Kerntech> chevron notch
Zickzacklaufschienen fpl <Ker & Glas> nog plate chevron runner bars
Zickzackpackung f <Ker & Glas> staggered packing
Zickzackverbindung f <Elektriz> zigzag connection
Zickzackversetzung f <Metall> zigzag dislocation
Ziegel m <Ker & Glas> tile • **mit Ziegeln mauern** <Bau> brick
Ziegelbogen m <Bau> brick arch
Ziegelbrand m <Ker & Glas> batch
Ziegelbrenner m <Ker & Glas> tile burner
Ziegeldraht m <Ker & Glas> sling
Ziegelei f <Ker & Glas> brick works
Ziegelerde f <Bau> brick earth
Ziegelfabrik f <Ker & Glas> tile factory
Ziegelhersteller m <Ker & Glas> tile maker
Ziegelherstellungsmaschine f <Ker & Glas> brick and tile machine
Ziegelklammer f <Ker & Glas> tile cramp
Ziegelmauer f <Bau> brick wall
Ziegelmauerwerk n <Bau> brickwork
Ziegelofen m <Bau> brick kiln
Ziegelpflaster n <Bau> *(BE)* brick pavement, *(AE)* brick sidewalk
Ziegelpflasterung f <Bau> brick paving
Ziegelpresse f <Ker & Glas> *(AE)* brick molding machine, *(BE)* brick moulding machine, tile press
Ziegelrohling m <Ker & Glas> cob brick
Ziegelstein m <Bau, Ker & Glas> brick
Ziegelton m 1. <Bau> brick clay, loam; 2. <Ker & Glas> common clay; 3. <Kohlen> loam
Ziegelwand f <Bau> brick wall
Ziehbank f <Fertig> rack *(Ziehen)*
Ziehbrettformen n <Fertig> striking
Ziehdorn m <Ker & Glas> mandrel, mandril
Ziehdüse f 1. <Fertig> drawing die; die plate *(Draht)*; hole *(Drahtzug)*; 2. <Ker & Glas> debiteuse
Ziehdüsenblase f <Ker & Glas> debiteuse bubble
Ziehdüsenschlitz m <Ker & Glas> slot
Zieheisen n <Fertig> drawing die, drawing plate, drawplate; die plate *(Ziehen)*
ziehen v 1. <Bau> drive *(Schrauben)*; 2. <Bau> draw; 3. <Fertig> cut *(Nuten)*; 4. <Papier> draw, pull; 5. <Trans> haul
ziehen lassen v <Lebensmittel> infuse
ziehen v **und übergeben** <Comp & DV> drag and drop
Ziehen n 1. <Kunststoff> drawing; 2. <Maschinen> pulling; 3. <Papier> draw, pull; 4. <Trans> haulage
Ziehen n **eines Nagels** <Maschinen> pulling out of a nail
Ziehen n **mit Zurücklegen** <Math> drawing with replacement *(Zufallsauswahl)*
Ziehen n **ohne Zurücklegen** <Math> drawing without replacement *(Zufallsauswahl)*
Ziehen n **von Blechgefäßen** <Fertig> cupping
Ziehen n **von Schlüssen** <Künstl Int> inference, reasoning
Ziehfaden m <Textil> snag
Ziehfähigkeit f <Fertig> drawability
Ziehfett n <Fertig> drawing grease
Ziehglas n <Ker & Glas> drawn glass
Ziehgrenze f <Fertig> drawing limit
Ziehkeil m <Maschinen> draw key
Ziehkeilgetriebe n <Maschinen> draw-key transmission

Ziehkissen

Ziehkissen n <Fertig> cushion, die cushion *(Presse)*
Ziehklinge f <Bau> spokeshave
Ziehkopf m <Fertig> draw head *(Räummaschine)*
Ziehmittel n <Fertig> drawing compound
Ziehräumen n <Fertig, Maschinen> pull-broaching
Ziehräumnadel f <Fertig, Maschinen> pull broach
Ziehring m <Fertig> drawing ring; die *(Tiefziehen)*
ziehschleifen v <Fertig, Maschinen, Mechan> hone
Ziehschleifen n <Fertig, Maschinen> honing
Ziehschleifmaschine f <Mechan> honing machine
Ziehschleifwerkzeug n <Fertig> hone
Ziehschütze f <Nichtfoss Energ> sliding sluice
Ziehstab m <Ker & Glas> draw rod
Ziehstange f <Fertig> drawing clamp
Ziehstein m <Fertig> flatter *(Ziehen)*
Ziehstempel m 1. <Fertig> deep-drawing punch, drawing punch; punch *(Tiefziehen)*; 2. <Maschinen> drawing punch
Ziehstufe f <Fertig> reduction *(Ziehen)*
Ziehtrichter m <Mechan> die
Ziehtrommel f <Fertig> drawing block
Ziehturm m <Ker & Glas> drawing tower
Ziehverpackung f <Verpack> deep-drawn packaging
Ziehvorgang m <Telekom> drawing process
Ziehvorrichtung f 1. <Elektronik> active pull-up device *(Halbleitertechnik)*; 2. <Ker & Glas> drawing machine
Ziehwanne f <Ker & Glas> bushing assembly
Ziehwannengebläse n <Ker & Glas> bushing blower
Ziehwerkzeug n <Fertig> cupping tool
Ziehzange f <Fertig> gripping jaws
Ziehzange f zur Entfernung von Heftklammern <Verpack> stapling pliers
Ziehzwiebel f <Ker & Glas> onion
Ziel n <Comp & DV, Ergon, Fernseh, Funkort, Qual> target
• mit Ziel Weltraum <Raumfahrt> space-bound • über das Ziel hinausgehen <Mechan> overshoot
Ziel n der Sekundäremission <Elektronik> secondary emission target
Zielanschluss m <Telekom> terminating termination *(ISDN)*
Zielansteuerung f <Wassertrans> homing *(Navigation)*
zielbeleuchtender Laser m <Elektronik> target-illuminating laser
Zielcomputer m <Comp & DV> target computer
Zielebene f <Comp & DV> target level
Zielelektrode f <Elektrotech> target electrode
Zielen n <Bau> sighting
Zielerfassung f <Funkort> acquisition *(Radar)*
Zielerkennung f <Ergon> target detection
Zielfahrt f <Wassertrans> homing *(Navigation)*
Zielfenster n <Raumfahrt> firing window
Zielfernrohr n <Bau> sighting telescope
Zielführung f per Radio <Trans> route guidance by radio
Zielfunktion f <Math> objective function, preference function, target function *(Optimierung)*
zielgesteuertes System n <Künstl Int> goal-driven system
Zielhöhenwinkel <Funkort> angle of elevation *(Radar)*
Zielnummer f <Telekom> terminating number *(Telefon)*
Zielprogramm n <Comp & DV> object program
Zielpunktcode m <Telekom> destination point code
Zielrechner m <Comp & DV> target computer
Zielscheibe f <Teilphys> target
Zielsignal n <Elektronik> target signal
Zielsprache f <Comp & DV> object language, target language
Zielsuchkopf m <Raumfahrt> homing head
Zielsuchlenkung f <Raumfahrt> homing
Zielunterscheidung f <Funkort> target discrimination

Zielverbindung f <Telekom> terminating connection *(ISDN)*
Zielverfolgungsradar n <Funkort> tracking radar, target tracking radar; missile radar *(für Flugkörper)*
Zielverfolgungsstation f <Trans> radar tracking station
Zielverkehr m <Trans> terminating traffic
Zielvermittlungsstelle f <Telekom> destination exchange
Zielwahl f <Telekom> speed dialing *(Telefon)*
Zielweg m <Wassertrans> track
Zierbeleuchtung f <Kfztech> dummy lights
Zierbuchstabe m <Druck> swash letter
Ziergurt m <Bau> fascia
Zierleiste f 1. <Bau> batten, *(AE)* molding, *(BE)* moulding; 2. <Druck> ornamental border; 3. <Kfztech> chrome strip, trim *(Karosserie)*
Zierlinie f <Druck> ornamental rule
Zierscheibe f <Kfztech> wheel cover
Ziffer f 1. <Comp & DV> digit; 2. <Druck> figure, numeral; 3. <Math> digit, numeral
Ziffern fpl/**Linie haltende** <Druck> lining figures, modern figures
Ziffern- oder Zeichenfeld n <Comp & DV> figures case
Ziffernanzeige f <Gerät> digital display, digital readout
Ziffernblatt n <Maschinen> dial, dial plate
Ziffernradschalter m <Elektriz> thumb wheel switch
Zifferntaste f <Comp & DV> numeric key
Ziffernumschaltung f <Telekom> figure shift
Zigarrenantenne f <Raumfahrt, Telekom> cigar antenna
Zimmer n <Bau> room
Zimmerantenne f <Elektrotech> indoor antenna
Zimmerei f <Bau> carpentry
Zimmerhandwerk n <Bau> carpentry
Zimmermann m <Bau> carpenter
Zimmermannshammer m <Bau> claw hammer
Zimmermannsstek m <Wassertrans> timber hitch *(Knoten)*
Zimmertemperatur f <Phys, Thermod> room temperature
Zimtblütenöl n <Lebensmittel> cassia oil
Zimtsäurebenzylester m <Lebensmittel> benzyl cinnamate
Zimtsäureethylester m <Lebensmittel> ethyl cinnamate
Zingiberen n <Chemie> zingiberene
Zink n (Zn) <Chemie, Metall> zinc (Zn)
Zinkanode f <Wassertrans> zinc anode *(katodischer Schiffskörperschutz)*
Zinkat n <Chemie> zincate
Zinkätzung f <Druck> zinc etching
Zinkblech n <Fertig, Metall> sheet zinc
Zinkblume f <Fertig> spangle
Zinkdampf m <Kohlen> *(AE)* zinc vapor, *(BE)* zinc vapour
Zinke f 1. <Bau> tine; 2. <Maschinen> prong
Zinkeisenphosphat-Beschichtung f <Metall> anchorite
Zinkenfräser m <Fertig> dovetail cutter
Zinkkondensation f <Kohlen> zinc condensation
Zink-Luft-Akkumulator m <Kfztech> zinc-air storage battery
Zinkoxid n <Kunststoff> zinc oxide
Zinkplatte f <Druck> zinc plate
Zinkschutzüberzug m <Anstrich> galvanized protective coating
Zinküberzug m <Kunststoff> zinc coating
Zinkung f <Fertig> dovetail *(Holz)*
Zinn n (Sn) <Chemie> tin (Sn)
Zinn... <Chemie> stannic
Zinn/Chrombeschichtung f <Metall> tin/chrome plating
Zinndioxid n <Chemie> stannic oxide
Zinnfolie f <Metall> tinfoil
zinnfreier Stahl m <Metall> tin-free steel
Zinngießerei f <Fertig> pewtery

Zinnlöten n <Bau> soldering
Zinnoxid n <Chemie> stannic oxide
Zinnsäureanhydrid n <Chemie> stannic oxide
Zinnstreifen m <Ker & Glas> tin streak
Zinnwalzwerk n <Metall> tin mill
Zipfel m <Fertig> scallop *(Tiefziehen)*
Zipfelkante f <Textil> wavy selvedge
Zipfelziehen n <Fertig> earing *(beim Tiefziehen)*
Zirconat n <Chemie> zirconate
Zirconium n *(Zr)* <Chemie> zirconium *(Zr)*
Zirconium... <Chemie> zirconic
Zirconiumdioxid n <Chemie> zirconia
Zirconiumoxid n <Chemie> zirconia *(Mineralogie)*
Zirconyl... <Chemie> zirconyl
Zirkaloy n <Kerntech> zircaloy, zirconium base alloy
Zirkaloy-Hülse f <Kerntech> zircaloy cladding, zircaloy hull
Zirkel m 1. <Bau> compass; 2. <Geom> compasses, pair of compasses; 3. <Maschinen> compasses
Zirkon m <Ker & Glas> zircon
Zirkonerde f <Ker & Glas> zirconia
Zirkonerzeugnis n <Ker & Glas> zircon refractory
Zirkonoxiderzeugnis n <Ker & Glas> zirconia refractory
Zirkonschwamm m <Kerntech> zirconium sponge
zirkulär *adj* <Geom> circular
zirkular polarisierte Welle f 1. <Akustik> circularly-polarized wave; 2. <Funktech, Phys> circular-polarized wave
zirkular polarisierte Wellen *fpl* <Wellphys> circular waves
zirkularisieren v <Raumfahrt> circularize *(Umlaufbahn)*
Zirkularisierung f **der Umlaufbahn** <Raumfahrt> circularization of orbit *(Änderung in kreisförmige Umlaufbahn)*
Zirkularpolarisation f <Funktech, Phys, Telekom> circular polarization
Zirkularpolarisation f **des Lichtes** <Strahlphys> circular polarization of light
zirkularpolarisierte Hornspeisung f <Funktech> circular horn feed
Zirkulation f 1. <Erdöl> circulation *(Bohrschlamm)*; 2. <Maschinen> circulation
Zirkulationskessel m <Hydraul> circulating boiler
Zirkulationspumpe f <Hydraul> circulating pump
Zirkulationsverlust m <Erdöl> lost circulation *(Bohrschlamm)*
Zirkulator m <Telekom> circulator *(Wellenleiter)*
Zirpen n <Optik> chirping *(vogelstimmenähnliche Frequenz)*
Zischen n <Aufnahme> sibilance *(Mikrofon)*
Zischfilter n <Aufnahme> hiss filter
Zischventil n <Hydraul> pet valve
Ziselierung f <Konstzeich> chasing
Zisterne f <Wasserversorg> cistern
zitieren v <Patent> cite
Zitieren n <Patent> citation
Zitronensäure f <Lebensmittel> citric acid
Zitterbewegung f <Elektronik> dither *(Überwindung von Reibung durch periodische Erregung des zu bewegenden Teils)*
zitterfrei *adj* <Elektronik> jitter-free *(Fernsehen)*
Zittermatrix f <Fernseh> dither matrix *(digitale Bildverarbeitung)*
Zittern n <Gerät> jitter *(Zeiger)*
Zitteroszillator m <Elektronik> dither oscillator
Zivilflugvorschriften *fpl* *(CAR)* <Raumfahrt> Civil Air Regulations *(CAR)*
Zivilisationsmüll m <Abfall> waste products of civilization
Z-Koordinate f <Phys> z-coordinate
Zoll m 1. <Fertig, Metall, Metrol> in, inch; 2. <Trans> customs

Zoll *mpl* **pro Sekunde** <Aufnahme, Comp & DV> IPS, inches per second
Zollgewinde n <Maschinen> inch thread
Zollmaß n <Fertig> imperial measure
Zollschranke f <Trans> tollgate, *(AE)* turnpike
Zollschraubengewinde n <Maschinen> inch screw thread
Zollstock m 1. <Bau> *(AE)* carpenter's gage, *(BE)* carpenter's gauge; 2. <Metrol> inch rule; 3. <Textil> yardstick
Zone f 1. <Bau, Comp & DV, Funktech, Kohlen, Metall> zone; 2. <Elektrotech> belt *(einer Wicklung)*; 3. <Telekom> cell
Zone f **gleichmäßiger Beanspruchung** <Maschinen> section of uniform strength
Zone f **stärkster Beanspruchung** <Maschinen, Qual> section of maximum intensity of stress
Zonenbit n <Comp & DV> zone bit
Zonenbreite f <Elektrotech> belt spread
Zonenfaktor m <Elektrotech> *(AE)* distribution factor, spread factor; belt factor *(einer Wicklung)*
Zonenformation f <Metall> zone formation
zonengehärtetes Glas n <Ker & Glas> zone-toughened glass
Zonenhärtung f <Ker & Glas> zone toughening
Zonenreinigung f <Metall> zone refining
Zonenschmelzverfahren n <Kerntech, Metall> zone melting
Zonenteilung f <Elektrotech> belt pitch *(einer Wicklung)*
Zonenzeit f <Wassertrans> zone time; standard time *(Navigation)*
Zoom n <Comp & DV> zoom
Zoomaufnahme f <Foto> zoom picture
Zoomen n <Comp & DV> zoom, zooming *(Vergrößerung/Verkleinerung der Bildschirmdarstellung)*
Zoomhebel m <Fernseh> zoom lever
Zoomobjektiv n <Foto> zoom lens
Zoosterin n <Chemie> zoosterine, zoosterol
Zr *(Zirconium)* <Chemie> Zr *(zirconium)*
Z/Sek *(Zeichen pro Sekunde)* <Druck> cps *(characters per second)*
Z-Spannung f <Phys> Zener voltage
Z-Stab m <Metall> Z-bar
Z-Stahl m <Metall> zees
Z/Std *(Zeichen pro Stunde)* <Druck> cph *(characters per hour)*
Z-Teilchen n <Teilphys> Z-particle
Z-Transformation f <Telekom> z-transform
Zubehör n 1. <Gerät> attachment; 2. <Mechan> component; 3. <Textil> appliance
Zubehörtasche f <Foto> gadget bag
Zubehörteile *npl* <Textil> appliance parts
zubereitet *adj* <Lebensmittel> dressed *(Speise)*
Zubereitung f <Kunststoff> formulation
zubringen v <Fertig> handle *(Räumwerkzeug)*
Zubringen n <Fertig> loading
Zubringer m 1. <Eisenbahn> feeder line, feeder train; 2. <Lufttrans> feeder airline; 3. <Trans> feeder; 4. <Wassertrans> feeder ship
Zubringer m **Erde-Orbit** <Raumfahrt> earth-to-orbit shuttle
Zubringerbesen m <Abfall> feeder broom *(am Kehrfahrzeug)*
Zubringerkabel n <Comp & DV> feeder cable
Zubringerschiff n <Wassertrans> feeder ship
Zubringerstraße f <Trans> collector road
Zubringerverbindung f <Telekom> feeder link *(Satellit)*
Zubringerverkehr m <Lufttrans> shuttle service
Zubruchbauen n **des Hangenden** <Kohlen> broken working
züchten v <Fertig> grow *(Kristalle)*

Züchtung

Züchtung f <Lebensmittel> growing
Zucker m <Lebensmittel> sugar
Zuckerguss m <Lebensmittel> frosting, icing
Zuckern n <Lebensmittel> sugaring
Zuckerrohrabfallwalze f <Lebensmittel> bagasse roller
Zuckerrohrsaft m <Lebensmittel> cane juice *(Zuckerraffination)*
Zuckerzusatz/ohne <Lebensmittel> no added sugar
zudruckumformen v <Fertig> stretch-squeeze form
zuerst Abgelegtes wird als Erstes bearbeitet *(FIFO-Prinzip)* <Comp & DV> first-in-first-out *(FIFO)*
Zuerst-rein-zuerst-raus-Speicherorganisation f <Comp & DV> first in first out *(FIFO)*
Zuerst-rein-zuletzt-raus-Speicherorganisation f <Comp & DV> first in last out *(FILO)*
Zufahrt f <Bau> approach
Zufahrtdosierung f <Lufttrans> ramp metering
Zufahrtsrampe f <Bau, Trans> access ramp
Zufahrtsrinne f <Wassertrans> approach channel *(Navigation)*
Zufahrtssperrsignal n <Lufttrans> ramp closure sign
Zufahrtstraße f <Bau> access road
zufällig adj 1. <Comp & DV, Ergon, Math> random, stochastic; 2. <Metrol, Qual> random
zufällige Ergebnisabweichung f <Qual> random error of result
zufällige Phasenfehler mpl <Funktech> random phase errors
zufällige Stichprobenauswahl f <Math> random sampling
zufällige Stichprobenentnahme f <Math> random sampling
zufällige Störung f <Math> random disturbance *(Regressionsanalyse)*
zufälliger Ausfluss m <Umweltschmutz> accidental discharge
zufälliger Fehler m <Gerät, Math, Phys> random error
zufälliger Impuls m <Telekom> random pulse
Zufalls... <Comp & DV, Elektrotech, Math, Metall, Qual> random
zufallsabhängiges Signal n <Gerät> accidental signal
Zufallsabtastung f <Comp & DV> random scan
Zufallsanordnung f <Metall> random arrangement
Zufallsausfall m <Comp & DV, Elektrotech, Telekom> random failure
Zufallsbelastung f <Metall> random loading
Zufallseinflüsse mpl <Qual> chance causes
Zufallsergebnis n 1. <Elektronik> random event; 2. <Math> outcome *(Zufallsexperiment)*
Zufallsexperiment n <Math> random experiment
Zufallsfehler m 1. <Comp & DV> random error; 2. <Elektrotech> random failure; 3. <Gerät, Kohlen, Math, Metrol> random error; 4. <Qual> random error, random failure; 5. <Telekom> random error
zufallsgestreutes Stichprobenverfahren n <Chemie> random sampling
Zufallsgröße f 1. <Math> random variable; 2. <Qual> random variable
Zufallsimpuls m <Telekom> random pulse
Zufallslogik f <Elektronik> random logic
Zufallslösung f <Metall> random solution
Zufallsmodell n <Comp & DV, Math> stochastic model
Zufallsmodulation f <Akustik> accidental inflection
Zufallsnummer f <Comp & DV> random number
Zufallsprobenahme f <Qual> random sampling
Zufallssignal n <Elektronik, Telekom, Wassertrans> random signal *(Radar)*
Zufallsspannung f <Elektrotech> random voltage

Zufallsstichprobe f <Lebensmittel, Math, Qual> random sample
Zufallsstichprobenuntersuchung f <Qual> random sampling
Zufallsstreubereich m <Qual> random dispersion interval
Zufallsstreuung f 1. <Kerntech> random scattering; 2. <Qual> chance variation
Zufallsvariable f 1. <Comp & DV, Elektronik, Qual> random variable; 2. <Qual> random variable, variate
Zufallsverkehr m <Telekom> pure chance traffic
Zufallsverteilung f <Comp & DV> probability distribution
Zufallswicklung f <Elektriz, Elektrotech> random winding
Zufallszahl f <Comp & DV, Math> random number
Zufallszahlengenerator m <Comp & DV, Telekom> random number generator
Zufassungsgriff m <Ergon> seize grip
Zufluss m <Wasserversorg> inflow; influx of water *(von Wasser)*
Zufrieren n <Thermod> freeze-up
Zuführapparat m <Verpack> feeder
Zufuhrbegrenzer m <Fertig> feed limiter
Zuführeinrichtung f <Fertig> feeder
zuführen v 1. <Bau> supply; 2. <Comp & DV, Elektrotech> feed *(Strom)*
zuführen v und entlüften v <Kerntech> feed and bleed
Zuführkabel n <Elektriz> lead-in cable
Zufuhrleitung f <Kerntech> feedline
Zuführrollgang m <Fertig> feed roller table *(Walzen)*
Zuführstraße f <Bau> access road
Zufuhrtisch m <Verpack> feeding table
Zufuhrtrichter m <Verpack> feed hopper
Zuführung f 1. <Comp & DV> feed; 2. <Elektrotech> feed *(Strom)*; 3. <Hydraul, Kfztech> feed; 4. <Papier> feeding • **mit automatischer Zuführung** <Maschinen> self-feeding
Zuführung f in Längsrichtung <Kontroll> endwise feed
Zuführungsbewegung f <Fertig> feed motion *(Spanung)*
Zuführungswalze f 1. <Maschinen> feed roll, feed roller; 2. <Papier> leading-in roll
Zuführwalze f <Lebensmittel> feed roller
Zug m 1. <Bau> tension *(Festigkeitslehre)*; 2. <Fertig> run *(Schweißen)*; 3. <Ker & Glas> pull; 4. <Maschinen> *(AE)* draft, *(BE)* draught, pull, tension; 5. <Optik> draw; 6. <Raumfahrt> pull • **auf Zug wirkend** <Fertig> drawback
Zug m mit mehreren Loks <Eisenbahn> multiple unit
Zug m mit Postbeförderung <Eisenbahn> mail train
Zug m mit zwei Lokomotiven <Eisenbahn> double-heading
Zugabe f 1. <Fertig> introduction; 2. <Maschinen> allowance
Zugabstand m <Eisenbahn> train spacing
Zugabteil n <Eisenbahn> passenger compartment *(im Zug)*
Zugachse f <Metall> tensile axis
Zugang m 1. <Bau> approach; 2. <Comp & DV> access *(Datenkommunikation)*; 3. <Kohlen> access; 4. <Raumfahrt> access *(Raumschiff)*; 5. <Telekom> access, data port, port • **ankommender Zugang verhindert** <Telekom> ICB, incoming-calls-barred *(Benutzerservice)*
Zugang m über Wählleitung <Comp & DV> dial access
zugängliche Reservenquelle f <Nichtfoss Energ> accessible resource base
Zugänglichkeit f <Mechan> accessibility, ease of access
Zugangsberechtigung f <Comp & DV> password
Zugangsburstzeichen n <Telekom> access burst signal
Zugangsgebühr f <Telekom> access charge rate
Zugangsgleis n <Eisenbahn> approach track
Zugangskanal m <Telekom> access channel

Zugangskonflikt m <Telekom> concurrency
Zugangskonkurrenz f <Telekom> concurrency
Zugangskonzentrator m <Telekom> access concentrator
Zugangsleitung f <Telekom> access circuit
Zugangsnetz n <Telekom> access network
Zugangsnummer f <Telekom> access number
Zugangspunkt m <Telekom> access port, data port; port *(Netzwerk)*
Zugangsrampe f <Raumfahrt> walkway
Zugangssicherung f <Sicherheit> access interlock
Zugangssoftware f <Telekom> browser *(Internet)*
Zugangsstollen m <Bau> adit *(Tunnelbau)*
Zugangstafel f <Raumfahrt> access panel *(Raumschiff)*
Zugangstür f <Lufttrans> access door
Zugangsweg m <Telekom> access link
Zuganker m 1. <Bau> stay, tie bar, tie rod; 2. <Fertig> tie; 3. <Maschinen> tension rod; 4. <Mechan> connecting rod; 5. <Wassertrans> tie rod
Zugarm m <Comp & DV> tension arm
Zugauflösung f <Eisenbahn> splitting up trains
Zugbalken m <Bau> stretcher, tie beam
Zugband n <Fertig> tie rod
zugbeanspruchtes Element n <Maschinen> tension member
Zugbeanspruchung f <Maschinen, Mechan, Metall> tensile stress, tensile strain
zugbediente Wegübergangssicherungsanlage f <Eisenbahn> automatic level crossing safety installation
Zugbeförderungsverfahren n <Eisenbahn> method of routing
Zugbegleiter m <Eisenbahn> pilot
Zugbelastung f 1. <Maschinen> tensile strain; 2. <Metall> tensile load, tensile strain
Zugbildung f <Eisenbahn> *(AE)* marshaling, *(BE)* marshalling
Zugbinder m <Bau> through binder
Zugbrücke f <Bau> drawbridge
Zugdeichsel f <Bau> tongue
Zugdrahtseil n <Trans> haulage cable
Zug-Druck-Wechselversuch m <Werkprüf> tension/compression testing
zugehörige Verstärkung f <Elektronik> associated gain
Zugehörigkeit f <Comp & DV> membership
Zugehörigkeitskennzeichnung f <Qual> match marking
zugelassen adj 1. <Math> admitted; 2. <Patent, Qual, Trans> approved
zugeordnete Frequenz f <Comp & DV, Funktech> assigned frequency
zugerichtete Ecke f <Ker & Glas> dubbed corner
zugeschärfte Kante f <Bau> feather edge
zugesetzt adj <Maschinen> clogged
zugespitzte Faser f <Optik> *(AE)* tapered fiber, *(BE)* tapered fibre
zugewiesene Frequenz f <Fernseh, Telekom> allocated frequency
zugewiesener Kanal m <Fernseh> dedicated channel
Zugfahrer m <Eisenbahn> train driver
Zugfahrplan m <Eisenbahn> *(AE)* schedule, *(BE)* timetable, train schedule
Zugfalten fpl <Papier> ribbing
Zugfeder f 1. <Fertig> helical tension spring; 2. <Maschinen> extension spring, tension spring
Zugfestigkeit f 1. <Anstrich> tenacity; 2. <Fertig> tensile strength *(Kunststoffinstallationen)*; 3. <Kunststoff> tensile strength, ultimate tensile strength; 4. <Maschinen> resistance to tension, tensile strength, ultimate tensile strength; 5. <Mechan> tensile strength, ultimate tensile strength; 6. <Phys> tensile strength; 7. <Qual> resistance to tension; 8. <Telekom> tensile strength
Zugfestigkeit f **in ofentrockenem Zustand** <Textil> oven-dry tensile strength
Zugfestigkeitsprüfgerät n <Maschinen> tensile test equipment
Zugfestigkeitsprüfmaschine f <Maschinen> tension testing machine
Zugfestigkeitstest m <Anstrich> tensile test
Zugfolgeabstand m <Eisenbahn> headway
Zugförderung f <Eisenbahn> traction
Zuggabel f <Kfztech> drawbar *(Anhänger)*
Zuggeschirr n <Mechan> harness
Zuggewölbe n <Ker & Glas> uptake crown
Zugglied n <Maschinen, Meerschmutz> tension member
Zuggriff m <Maschinen> pull handle
Zughaken m 1. <Eisenbahn> draw hook; 2. <Kfztech> tow hook *(Anhänger)*
Zugkasten m <Hydraul> *(AE)* draft box, *(BE)* draught box
Zugkette f <Sicherheit> lifting chain
Zugklappe f 1. <Heiz & Kälte> slide damper; 2. <Raumfahrt> shutter
Zugkraft f 1. <Bau> traction; 2. <Eisenbahn, Kfztech> tractive effort; 3. <Maschinen> tensile force, traction; 4. <Metrol> tractive force; 5. <Phys> tensile force
Zugkraft/Bremskontrollhebel m <Eisenbahn> traction/brake controller
Zugkraftkontrolle f <Kfztech> traction control
Zugkraftverbindung f <Eisenbahn> traction link
Zugleiter m <Eisenbahn> traffic controller
Zugluft f <Heiz & Kälte> *(AE)* draft, *(BE)* draught
Zugmaschine f 1. <Eisenbahn> hauling engine; 2. <Kfztech> tow vehicle, traction engine, tractor, tractor unit; 3. <Lufttrans> prime mover; 4. <Maschinen> traction engine
Zugnummernmelder m <Eisenbahn> train describer
Zugorgan n <Fertig> core *(Keilriemen)*
Zugöse f <Maschinen> coupling ring
Zugpersonal n <Eisenbahn> train crew
Zugpropeller m <Lufttrans> tractor propeller
Zugprüfmaschine f <Kunststoff> tensile tester
Zugprüfung f <Qual, Werkprüf> tensile test
Zugramme f <Mechan> drop pile hammer
Zugraupe f <Fertig> string bead
Zugregister n <Heiz & Kälte> slide damper
Zugregistriereinrichtung f <Eisenbahn> train describer
Zugregler m <Heiz & Kälte> *(AE)* draft regulator, *(BE)* draught regulator
zugreifen v <Comp & DV> seek
zugreifen v **auf** <Comp & DV> access
Zugriff m <Comp & DV, Funktech, Telekom> access *(Speicher, Dokument)*
Zugriff m **nach der Warteschlangenmethode** <Comp & DV> queued access method
Zugriffsarm m <Comp & DV> actuator *(Diskette)*
Zugriffsart f <Comp & DV, Telekom> access mode
Zugriffsautorisierung f <Comp & DV> access authority
Zugriffsbefugnis f <Comp & DV> access authority
Zugriffscode m <Comp & DV> privacy
Zugriffsgeschwindigkeit f <Elektrotech> access speed *(Rechner)*
Zugriffsliste f <Comp & DV> access list
Zugriffsmethode f <Comp & DV> access method
Zugriffsöffnung f <Kerntech> access port
Zugriffspfad m <Comp & DV> access path
Zugriffsrecht n <Comp & DV> privilege
Zugriffsschlüssel m <Comp & DV> access key
Zugriffstafel f <Raumfahrt> access panel *(Raumschiff)*
Zugriffsverfahren n <Telekom> access mode

Zugriffszeit

Zugriffszeit f <Comp & DV, Druck, Elektrotech, Optik> access time
Zugrohr n <Optik> drawtube
Zugsattelzapfen m <Kfztech> fifth-wheel kingpin *(Sattelschlepper)*
Zugschaffner m <Eisenbahn> guard
Zugschalter m <Elektriz> pull switch
Zugscherversuch m <Kunststoff> lap shear test
Zugschraube f <Lufttrans> tractor propeller
Zugseil n 1. <Bau> stay; 2. <Eisenbahn> traction cable, traction rope; 3. <Maschinen> pulling rope, traction rope; 4. <Trans> hauling rope
Zugsicherung f <Eisenbahn> train protection
Zugspannung f 1. <Fertig> tensile strength *(Kunststoffinstallationen)*; 2. <Mechan, Phys> tensile stress
Zugspannungsempfindlichkeit f <Akustik> tension sensitivity
Zugspindel f <Maschinen> feed shaft
Zugspindeldrehmaschine f <Maschinen> sliding lathe
Zugstab m 1. <Bau> tie bar, tie rod; 2. <Eisenbahn> single line token *(zur Streckensicherung)*
Zugstange f 1. <Bau> anchor bar; 2. <Eisenbahn> *(AE)* draft bar, *(BE)* draught bar, drawbar; 3. <Ker & Glas> tie rod; 4. <Kfztech> drawbar, towbar *(Anhänger)*; 5. <Lufttrans> dog bone, tie bar *(Hubschrauber)*; 6. <Maschinen> drawbar, pitman, pull rod; 7. <Textil> tension bar; 8. <Wassertrans> tie rod
Zugstangenbolzen m <Kfztech> drawbar bolt *(Anhänger)*
Zugstangenfeder f <Eisenbahn> drawgear spring
Zugstangenführung f <Eisenbahn> drawbar guide
Zugstrom m <Ker & Glas> pull current
Zug-Torsionsversuch m <Werkprüf> tension/torsion testing
Zugtrum n <Fertig> driving side *(Riemen)*
Zugüberwacher m <Eisenbahn> traffic controller
Zug- und Tragseil n <Trans> hauling and carrying rope
zugverformbar adj <Fertig> ductile
Zugverformungsrest m <Kunststoff> tension set
Zugversuch m <Maschinen, Metall> tensile test
Zugvorrichtung f <Kfztech> hitch
Zugwalze f 1. <Ker & Glas> pull roll; 2. <Papier> draw roll
Zugwinkel m <Mechan> angle of traction
Zugzerlegung f **durch Abstoß** <Eisenbahn> fly shunting
Zugzulässigkeit f <Kfztech> tow rating
zuhaken v <Maschinen> clasp
Zuhaltung f <Bau> tumbler *(Türschloss)*
Zuhaltungsschloss n <Bau> tumbler lock
Zukurzkommen n <Lufttrans> undershoot
Zulage f <Heiz & Kälte> allowance
zulassen v 1. <Math> admit; 2. <Qual> approve, permit
zulässig adj <Math> admissible
zulässige Abweichung f <Anstrich> tolerance
zulässige Abweichung von der Kreisform f <Fertig> roundness tolerance
zulässige Beanspruchung f <Maschinen> safe stress
zulässige Belastung f 1. <Kohlen, Qual> permissible load; 2. <Sicherheit> permissive load; 3. <Trans> safe load
zulässige Biegebeanspruchung f <Maschinen> safe stress under bending
zulässige Bügeltemperatur f <Textil> safe ironing temperature
zulässige Fahrzeugfolgezeit f <Trans> tolerable gap between vehicles
zulässige Geschwindigkeit f <Trans> posted speed
zulässige Höchstleistung f <Telekom> maximum admissible power
zulässige Kontaktbelastung f <Elektrotech> contact rating
zulässige Landemasse f <Lufttrans> allowable landing mass
zulässige Last f 1. <Elektriz> allowable load; 2. <Fertig> design load
zulässige Maßabweichung f <Fertig> amount of variation permitted
zulässige Spannung f <Elektriz, Elektrotech> rated voltage *(Nennspannung eines Gerätes)*
zulässige Spannung f **an einer Wicklung** <Elektriz> rated voltage of a winding
zulässige Startmasse f <Lufttrans> allowable landing mass
zulässige Überlastung f <Sicherheit> permissible overload
zulässige Verzögerungszeit f <Telekom> tolerable delay time
zulässiger Patentanspruch m <Patent> admissible claim
zulässiger Störpegel m <Raumfahrt> permissible level of interference *(Weltraumfunk)*
zulässiger Strom m <Elektrotech> rated current
zulässiger Wert m <Elektriz> rated value
zulässiges Abmaß n <Maschinen> tolerance
zulässiges Gesamtgewicht n <Kfztech> permitted gross vehicle weight, total permissible laden weight; total permissible weight *(Rechtsvorschriften)*
zulässiges Spektrum n <Kerntech> allowed spectrum
zulässiges Verfahren n <Qual> qualified procedure
Zulässigkeit f <Math> admissibility
Zulassung f 1. <Elektriz> approval; 2. <Kfztech> certification, homologation *(Rechtsvorschriften)*; 3. <Lufttrans> registration; 4. <Patent, Qual> approval; 5. <Trans> approval, registration; 6. <Wassertrans> registration
Zulassungsbescheinigung f 1. <Lufttrans> certificate of registration; 2. <Patent> certificate of accreditation, certificate of registration; 3. <Qual> certificate of accreditation; 4. <Trans> approval certificate
Zulassungsprüfung f <Werkprüf> approval test
Zulassungsüberwachung f <Patent, Qual> certification review
Zulassungszeichen n <Qual> certification mark
Zulauf m 1. <Fertig> runner *(Gießen)*; 2. <Mechan> intake
Zulaufkanal m <Wasserversorg> inflow canal, intake canal
Zulaufschleuse f <Wasserversorg> intake sluice
Zulegierung f <Fertig> introduction of alloying elements *(Elektrodenumhüllung)*
zuleiten v <Mechan> feed
Zuleitung f <Elektronik, Elektrotech> lead, feeder; feed *(Strom)*; lead *(Stromdrahtversorgung)*; lead *(an Gerät befestigt)*
Zuleitungsdraht m <Elektrotech> lead wire, lead-in wire
Zuleitungskabel n <Elektrotech> feed cable
Zuleitungswellenleiter m <Elektrotech, Telekom> feed waveguide
Zuletzt-rein-zuerst-raus-Speicherorgaisation f <Comp & DV> last in first out *(LIFO)*
Zuluft f <Heiz & Kälte> fresh air, incoming air, supply air
Zuluftkanal m <Heiz & Kälte> supply duct
Zuluftrohr n <Maschinen> air inlet pipe
Zuluftventilator m <Heiz & Kälte> supply air fan
zumachen v <Kfztech> lock, lock up
zumessen v <Bau> meter
Zumessung f <Fertig> proportioning
Zumessventil n <Kfztech> proportioning valve
Zumischen n <Strömphys> entrainment
Zunahme f <Fertig, Math> increment
Zünddynamo m <Lufttrans> ignition generator
Zündeinstellmarke f <Kfztech> timing mark

Zündeinstellung f <Kfztech> ignition setting, ignition timing
Zündelektrode f <Elektrotech> starter, starter electrode
zünden v 1. <Fertig> draw, strike *(Lichtbogen)*; 2. <Kerntech> ignite *(Plasma)*; 3. <Raumfahrt> ignite
Zünden n <Fertig> drawing *(Schweißen)*
Zunder m 1. <Anstrich> rust flake; 2. <Fertig, Metall> cinder, iron scale, metal scale, scale
Zünder m 1. <Elektriz> igniter; 2. <Fertig> detonator; 3. <Maschinen, Raumfahrt> igniter
zunderverhütend adj <Fertig> anti-ager scale
Zündexperiment n <Kerntech> ignition experiment
zündfähig adj <Kfztech, Sicherheit, Thermod> explosive
zündfähiges Gemisch n <Kfztech, Sicherheit> explosive mixture
Zündflamme f <Maschinen> pilot flame
Zündfolge f <Kfztech> firing order *(Motor)*
Zündfunke m <Elektrotech, Kfztech> spark
Zündimpuls m <Elektrotech> firing pulse *(Gasröhren-Doppelwegthyristor)*
Zündkerze f 1. <Elektriz, Elektrotech> plug, spark plug; 2. <Kfztech> spark plug, sparker; sparking plug *(Motor)*; ignition plug, plug *(Zündung)*; 3. <Mechan> ignition plug
Zündkerze f mit niedrigem Wärmewert <Kfztech> hot spark plug
Zündkerzendichtung f <Kfztech> spark plug gasket
Zündkerzenelektrode f <Kfztech> spark plug electrode
Zündkerzengehäuse n <Kfztech> spark plug body, spark plug shell
Zündkerzenkabel n <Kfztech> spark plug cable, spark plug wire
Zündkerzenklemmschraube f <Kfztech> spark plug terminal
Zündkerzenloch n <Kfztech> spark plug hole
Zündkerzenspitze f <Kfztech> spark plug point
Zündkondensator m <Kfztech> ignition capacitor
Zündkontakt m <Kfztech> point *(Zündung)*
Zündkreis m <Raumfahrt> ignition circuit
Zündmagnet m 1. <Elektriz> magneto; 2. <Kfztech> ignition magneto, magneto *(Zündung)*
Zündpunkt m 1. <Erdöl> ignition point *(Physik)*; 2. <Phys> ignition point
zundrig adj <Fertig, Metall> scaly
Zündschalter m <Kfztech> ignition starter switch; ignition switch *(Zündung)*
Zündschlüssel m <Kfztech> ignition key
Zündspannung f 1. <Elektrotech> firing voltage *(Gasröhren-Doppelwegthyristor)*; 2. <Kfztech> firing voltage
Zündspule f 1. <Elektriz, Elektrotech> ignition coil; 2. <Kfztech> ignition coil; coil *(Zündung)*; 3. <Lufttrans> ignition coil
Zündspulenprimärwicklung f <Kfztech> primary winding
Zündstecker m <Kfztech> ignition plug *(Zündung)*
Zündstellwinkel m <Kfztech> timing angle *(Zündung)*
Zündstift m <Elektrotech> ignitor
Zündstiftröhre f <Elektrotech> ignitron
Zündstörung f <Funktech> ignition noise *(Funk)*
Zündtemperatur f <Maschinen> ignition temperature
Zündtransformator m <Elektrotech> ignition transformer
Zündung f 1. <Elektriz, Elektronik, Elektrotech> ignition; firing *(Magnetron)*; firing *(Gasröhren-Doppelwegthyristor)*; ignition *(Gasröhre)*; striking *(Lichtbogen)*; 2. <Fertig> striking; 3. <Kfztech, Lufttrans, Maschinen, Raumfahrt> ignition; 4. <Trans> sparking
Zündungssystem n <Kfztech> ignition system
Zündunterbrecher m <Kfztech> contact breaker *(Motor)*
Zündunterbrecherkontakt m <Kfztech> (BE) contact breaker point, points
Zündunterbrechernocken m <Kfztech> distributor cam

Zündversteller m <Kfztech> advance mechanism *(Zündung)*
Zündverstellung f <Kfztech> timing
Zündverteiler m <Kfztech, Maschinen> ignition distributor
Zündverteileranlage f <Lufttrans> ignition harness
Zündverteilung f <Kfztech> distribution *(Zündung)*
Zündvorrichtung f 1. <Fertig> primer; 2. <Heiz & Kälte> ignition device
Zündwinkel m <Elektrotech> ignition angle *(Gasentladungsröhre)*
Zündzeit f <Elektrotech> firing time *(Gasröhren-Doppelwegthyristor)*
Zündzeitpunkt m 1. <Kfztech> ignition point; point *(Zündung)*; 2. <Mechan> ignition point
Zündzeitpunkteinstellung f <Kfztech> spark timing, timing of ignition
Zündzeitpunktmarke f <Kfztech> timing mark
zunehmen v <Math> increase
zunehmend adv <Math> increasing
zunehmende Erwärmung f <Umweltschmutz> incremental heating
zunehmende Polarisation f <Elektrotech> mounting polarization
zunehmende Steigung f <Lufttrans> pitch increase *(Hubschrauber)*
Zunge f 1. <Bau> tongue; 2. <Fertig> blade *(Schieblehre)*; 3. <Labor> hand, pointer; 4. <Maschinen> tongue
Zunge f der Glasmacherpfeife <Ker & Glas> tongue
Zungenband n <Bau> T-hinge, cross-garnet hinge, cross-garnet hinge
Zungenfrequenzmesser m 1. <Gerät> vibration reed frequency meter; 2. <Labor> vibrating-reed frequency meter
Zungenfrequenzmessgerät n 1. <Gerät> reed frequency meter *(für die Netzfrequenz, Zunge unter Schutzgas)*; 2. <Labor> vibrating-reed frequency meter, vibration reed frequency meter
Zungenhorn n <Wassertrans> reed horn *(Navigation)*
Zungenkontakt m <Telekom> reed contact
Zungennadel f <Textil> latch needle
Zungenschalter m <Phys> reed switch
Zungenspitze f <Eisenbahn> tip of switch tongue *(Weiche)*
Zungenverschluss m <Eisenbahn> point lock
Zungenweiche f <Eisenbahn> point switch
zuordnen v 1. <Comp & DV> allocate, assign, translate; allocate *(Datenverarbeitung)*; 2. <Math> assign
Zuordner m 1. <Kontroll> sequencer; 2. <Telekom> translator
Zuordnung f 1. <Comp & DV, Elektrotech, Funktech> allocation; 2. <Math> assignment
Zuordnung f nach Anforderung <Raumfahrt> demand assignment *(Weltraumfunk)*
Zuordnungsanweisung f <Comp & DV> assignment statement *(Programmiersprache)*
Zuordnungsbefehl m <Comp & DV> reference instruction
Zuordnungsdatei f <Comp & DV> reference file
Zupfen n 1. <Ker & Glas> pluck; 2. <Textil> plucking
Zurichtebogen m <Druck> makeready sheet
zurichten v 1. <Bau> size; trim *(Holz)*; 2. <Druck> make ready
Zurichten n <Bau> dressing *(von Holz)*
Zurichthammer m <Bau> maul
Zurichtungsbogen m <Druck> makeready sheet
Zurring f <Wassertrans> lashing *(Ladung)*
Zurrungsplan m <Wassertrans> lashing plan *(Ladung)*
zurück adv <Wassertrans> astern *(Fahrt/Motor)*
zurückbleibend adj <Kohlen> residual
zurückbleibende Magnetisierung f 1. <Elektrotech> remanence; 2. <Phys> remanent magnetization

zurückfedern v <Maschinen> spring back
Zurückfedern n <Fertig> resilience
zurückgeben v <Comp & DV> return
zurückgelegte Fahrstrecke f <Lufttrans> mileage
zurückgelegter Kilometer m/**vom Luftfahrzeug** <Lufttrans> aircraft kilometre performed
zurückgewiesene Verbindung f <Telekom> lost call, refused call
zurückgewiesene Verbindunganforderung f <Telekom> lost call request, refused call request
zurückgewiesener Verkehr m <Telekom> lost traffic
zurückgewinnen v 1. <Abfall> recover *(Rohstoffe)*; 2. <Kohlen, Umweltschmutz> recover
Zurückgewinnung f <Kohlen> recovery
zurückgewonnenes Öl n <Umweltschmutz> recovered oil *(Altölregenerierung)*
zurückgeworfenes Signal n <Elektronik> reflected signal
zurückhalten v <Bau> retain
Zurückhaltung f <Elektrotech, Wasserversorg> retention
zurückkehren v <Comp & DV> return
Zurückklappen n <Aufnahme> foldback
Zurückklappen n **des Luftschraubenblattes** <Lufttrans> blade folding *(Hubschrauber)*
zurücklaufen lassen v <Comp & DV> rewind
Zurücknahme f <Patent> withdrawal
zurücknehmen v <Patent> abandon *(Registrierung)*
zurückrollen v <Lufttrans> backtrack
zurückschieben v <Lufttrans> push back
Zurückschieben n <Lufttrans> push-back
zurückschlagen v <Sicherheit> backfire, flash back *(Flamme)*; kick back, repel *(Werkzeug)*
Zurückschwenken n **des Luftschraubenblattes** <Lufttrans> blade folding
zurücksenden v <Phys> reflect
zurücksetzen v 1. <Bau> back, reverse; 2. <Telekom> backspace, reset
Zurücksetzen n <Comp & DV> revision rollback
Zurücksetzen n **der Konfiguration** <Comp & DV> configuration rollback
zurückspeichern v <Comp & DV> restore
zurückspringen v <Comp & DV> return
zurückspulen v <Aufnahme, Comp & DV> rewind
zurückstellen v <Comp & DV> reset
zurückverfolgen v <Comp & DV> backtrack, trace back
Zurückverfolgen n <Künstl Int> backtracking
Zurückverfolgung f <Comp & DV> backtracking
zurückweisen v <Qual> reject
Zurückweisung f 1. <Comp & DV, Qual> rejection; 2. <Patent> refusal
Zurückweisungsfehler m <Comp & DV> rejection error
Zurückweisungszahl f <Qual> rejection number
zurückwerfen v <Phys> reflect
zurückziehen v 1. <Kohlen> retreat; 2. <Patent> withdraw
zurückzuweisende Qualitätsgrenzlage f <Qual> rejectable quality level
Zusammenarbeit f <Telekom> interworking *(zwischen Netzen)*
zusammenbacken v <Lebensmittel> cake
Zusammenbacken n <Kunststoff, Lebensmittel> caking
zusammenballen v 1. <Kohlen> agglomerate; 2. <Metall> ball up *(Reifen)*
zusammenballen v/sich <Chemie> flocculate
Zusammenballen n 1. <Foto> grain clumping *(von Körnern)*; 2. <Kohlen, Metall> balling
zusammenbändeln v <Wassertrans> tie up
Zusammenbau m 1. <Bau> framing *(Balkenträger)*; 2. <Maschinen, Mechan> assembly
zusammenbauen v <Bau> assemble, mount

Zusammenbauwerkzeuge npl <Maschinen> assembly tools
Zusammenbauzeichnung f <Konstzeich, Maschinen, Mechan> assembly drawing
zusammenbinden v <Maschinen> band
zusammenblatten v <Bau> halve *(Holzbau)*
zusammenbrechen v <Bau> break down
Zusammenbruch m 1. <Bau> breakdown, failure; 2. <Elektriz, Phys> breakdown
Zusammendrücken n 1. <Fertig> flattening *(Rohr)*; 2. <Kunststoff, Maschinen> compression
Zusammenfallen n <Umweltschmutz> subsidence
zusammenfassen v <Comp & DV> abstract, consolidate
Zusammenfassung f 1. <Comp & DV> summary; 2. <Druck> abstract *(eines wissenschaftlichen Artikels)*; 3. <Patent> abstract
Zusammenfluss m <Wasserversorg> confluence
zusammenfügen v 1. <Bau> join on to; 2. <Maschinen> join
Zusammenfügen n 1. <Bau, Ker & Glas> joining; 2. <Kerntech> jointing
Zusammenfügung f <Kerntech> joining
zusammenführen v <Comp & DV> merge *(Daten)*
Zusammenführen n <Comp & DV> merging; merge *(von Daten)*
Zusammenführung f 1. <Comp & DV> junction *(von Leitungen)*; 2. <Elektrotech> fan-in
zusammengebaut gezeichnete Teile npl <Konstzeich> parts drawn in the assembled condition
zusammengedreht adj <Telekom> twisted together
zusammengefasst adj <Comp & DV> ganged
zusammengeschaltete Steuerungen fpl <Lufttrans> interconnected controls
zusammengesetzt adj <Metall> composite
zusammengesetzte Abbildung f <Math> composition
zusammengesetzte Anweisung f <Comp & DV> compound statement
zusammengesetzte Bewegung f <Mechan> compound motion
zusammengesetzte Kurbelwelle f <Maschinen> built-up crank
zusammengesetzte Mikroschaltung f <Comp & DV> microassembly
zusammengesetzte Welle f <Telekom> complex wave, composite wave
zusammengesetzte Zahl f <Math> composite number, compound number
zusammengesetzter Absorber m <Kerntech> composite absorber
zusammengesetzter Graph m <Künstl Int> composite graph
zusammengesetzter Klang m <Akustik> complex sound
zusammengesetzter Schall m <Akustik> combination sound
zusammengesetzter Ton m <Akustik> complex tone
zusammengesetztes Darstellungselement n <Geom> closed figure
zusammengesetztes logisches Element n <Comp & DV> compound logical element
zusammengesetztes Mikroskop n <Phys> compound microscope
zusammengesetztes Objektiv n <Optik> compound lens
zusammengesetztes Okular n <Optik> compound eyepiece
zusammengesetztes Zeichen n <Patent> composite mark *(aus gleichartigen Bestandteilen)*
Zusammenhang m <Comp & DV> context

zusammenhängend *adj* 1. <Comp & DV> contiguous; 2. <Math> connected, contiguous
zusammenhängende Grafiken *fpl* <Comp & DV> contiguous graphics
zusammenkitten *v* <Mechan> bond
zusammenklappbar *adj* <Bau> collapsible
zusammenkleben *v* <Chemtech, Fertig> agglutinate
Zusammenkleben *n* <Chemtech> agglutination
zusammenklumpen *v* <Chemtech, Fertig> agglutinate
zusammenkneifen *v* <Bau> punch
Zusammenlaufen *n* <Konstzeich> merging *(von Rastern)*
zusammenlegbar *adj* <Maschinen> collapsible
zusammenlegbare Faltflasche *f* <Foto> collapsible bottle *(für Entwickler)*
zusammenlegbarer Frachtcontainer *m* <Wassertrans> collapsible freight container
zusammenlegbares Teil *n* <Trans> collapsible section
Zusammennähen *n* <Textil> gathering
zusammenpassen *v* 1. <Bau> match; 2. <Maschinen> mate
zusammenpassen *v* **mit** <Maschinen> fit with
Zusammenpassen *n* <Maschinen> mating
zusammenschaltbar *adj* <Trans> joinable *(Behälter)*
zusammenschalten *v* <Funktech> interconnect
Zusammenschaltung *f* 1. <Fertig> connection; 2. <Lufttrans> interconnection
zusammenschiebbar *adj* <Maschinen> telescoping
zusammenschiebbare Lenksäule *f* <Kfztech> collapsible steering column
zusammenschiebbarer Tubus *m* <Foto> telescopic tube
zusammenschiebbares Bein *n* <Foto> sliding leg *(Stativ)*
zusammenschiebbares Stativ *n* <Foto> extension tripod, folding tripod
zusammenschiebbares Stativbein *n* <Foto> telescopic leg
zusammenschnüren *v* 1. <Maschinen> contract; 2. <Verpack> bind
Zusammenschweißen *n* <Mechan> welding
Zusammensetz-Edit *n* <Fernseh> assemble edit
zusammensetzen *v* 1. <Bau> build up; 2. <Comp & DV> assemble *(Datenverarbeitung)*
Zusammensetzen *n* <Bau> composition, matching
Zusammensetzen *n* **von Kräften** <Mechan> composition of forces
Zusammensetzung *f* 1. <Eisenbahn> structure; 2. <Kerntech, Qual> composition; 3. <Textil> analysis
zusammensintern *v* <Fertig> agglomerate
Zusammensintern *n* <Eisenbahn> sintering
Zusammenstoß *m* 1. <Eisenbahn, Kfztech, Kohlen> collision; 2. <Phys> collision, impact; 3. <Raumfahrt, Sicherheit, Wassertrans> collision
Zusammenstoß *m* **in der Luft** <Lufttrans> aerial collision
zusammenstoßen *v* **mit** 1. <Trans> collide with, run into; 2. <Wassertrans> come into collision with *(Schiff)*
Zusammentragen *n* <Druck> gathering
Zusammentragmaschine *f* <Druck> gathering machine
zusammentreffende Spitze *f* <Kohlen> coincidental peak
zusammentreffende Spitzenlast *f* <Kohlen> coincidental peak load
Zusammentreibeffekt *m* <Meerschmutz> herder effect
Zusammentreibmittel *n* <Meerschmutz> herding agent *(für Öl auf Wasseroberfläche)*
zusammenziehbar *adj* <Textil> contractile
zusammenziehbarer Boden *m* <Kohlen> contractant soil
zusammenziehen *v* <Phys> contract

zusammenziehen *v/sich* <Bau, Metall> contract
Zusammenziehen *n* <Textil> gathering *(Falten)*
Zusammenziehung *f* 1. <Kohlen> contractancy; 2. <Phys> contraction
Zusammenziehungskoeffizient *m* <Nichtfoss Energ> contraction coefficient
Zusatz *m* 1. <Druck> additive; 2. <Ker & Glas> admix; 3. <Maschinen> addition; 4. <Mechan> additive
Zusatz... 1. <Comp & DV> peripheral; 2. <Kontroll, Raumfahrt> ancillary *(Raumschiff)*
Zusatzantrieb *m* <Lufttrans> accessory drive
Zusatzaufwand *m* <Telekom> overhead
Zusatzausrüstung *f* <Mechan> ancillary equipment
Zusatzbatterie *f* <Kfztech> booster battery
Zusatzbauteil *n* <Raumfahrt> added on component *(Raumschiff)*
Zusatzbits *npl* <Telekom> overhead bits
Zusatzdämpfung *f* <Telekom> excess attenuation
Zusatzdienst *m* <Telekom> enhanced service, supplementary service
Zusatzdraht *m* <Fertig> filler rod *(Schweißen)*
Zusatzdüse *f* <Kfztech> high-speed auxiliary jet; auxiliary jet *(Vergaser)*
Zusatzdynamo *m* <Elektriz> booster dynamo
Zusatz-Edit *n* <Fernseh> add-on edit
Zusatzeinrichtung *f* <Comp & DV> feature
Zusatzfeld *n* <Comp & DV> option field
Zusatzfläche *f* <Konstzeich> free margin
Zusatzgenerator *m* <Elektriz> booster generator
Zusatzgerät *n* 1. <Kohlen, Maschinen> attachment; 2. <Telekom> accessory equipment
Zusatzgeräte *npl* <Fertig> ancillary equipment; secondary equipment *(Kunststoffinstallationen)*
Zusatzheizungssystem *n* <Nichtfoss Energ> booster heating system
Zusatzkontakt *m* <Elektrotech> auxiliary contact
Zusatzleiterplatte *f* <Elektronik> daughter board
zusätzlich *adj* 1. <Maschinen> additional; 2. <Patent> supplementary
zusätzliche Leistungen *fpl* <Bau> auxiliary work
zusätzliche Wicklung *f* <Gerät> additional winding
zusätzliches Kennzeichen *n* <Patent> additional feature
zusätzliches Merkmal *n* <Patent> additional feature
Zusatzlinse *f* <Foto> supplementary lens
Zusatzluft *f* <Kfztech> secondary air
Zusatzmaschine *f* **für Zu- und Gegenschaltung** <Elektrotech> reversible booster
Zusatzmaschine *f* **in Gegenschaltung** <Elektrotech> negative booster
Zusatzmetall *n* <Fertig> filler metal *(Schweißen)*
Zusatzmittel *n* 1. <Bau> admixture *(Beton)*; 2. <Chemie> dope
Zusatzmühle *f* <Nichtfoss Energ> booster mill
Zusatzpermeabilität *f* <Werkprüf> incremental permeability
Zusatzplatte *f* <Comp & DV> auxiliary disk
Zusatzprüfung *f* <Qual> penalty test
Zusatzrakete *f* <Raumfahrt> booster, kick rocket
Zusatzreinigung *f* <Umweltschmutz> supplementary purification
Zusatzschub *m* <Lufttrans> reheat
Zusatzschuh *m* <Foto> auxiliary shoe
Zusatzspeicher *m* <Comp & DV> add-on memory, auxiliary memory, *(AE)* auxiliary storage, *(BE)* auxiliary store, secondary memory
Zusatzspiegel *m* <Foto> auxiliary mirror
Zusatzstab *m* <Fertig> filler rod *(Schweißen)*
Zusatzstoff *m* 1. <Fertig> admixture *(Kunststoffinstallationen)*; 2. <Lebensmittel> additive; 3. <Metall> addition

zusatzstofffrei

zusatzstofffrei adj <Lebensmittel> additive-free
Zusatztank m <Raumfahrt> additional tank
Zusatztastatur f <Druck> additional keyboard
Zusatztransformator m <Elektriz, Elektrotech> auxiliary transformer, booster transformer; booster transformer *(in Gegenschaltung)*
Zusatzverstärker m <Funktech> booster amplifier
Zusatzwasser n <Heiz & Kälte, Lebensmittel> make-up water
Zusatzwasserbehälter m <Eisenbahn> auxiliary reservoir
Zusatzwecker m <Telekom> extension bell
Zusatzwicklung f <Gerät> additional winding
Zusatzwiderstand m <Gerät> additional resistor
Zusatzziffer f <Telekom> extra digit
zuschalten v 1. <Comp & DV> connect; 2. <Elektrotech> power up; switch in *(Schaltkreis schließen)*; 3. <Fertig> make
Zuschaltung f <Elektrotech> switching in
Zuschaltventil n <Maschinen> sequence valve
zuschärfen v <Fertig> scarf
Zuschauer m <Fernseh> viewer
Zuschauerbewertung f <Fernseh> rating
Zuschauerzahlen fpl <Fernseh> audience rating
Zuschlag m 1. <Bau> acceptance of tender; 2. <Heiz & Kälte> allowance; 3. <Metall> addition
Zuschlagablöseversuch m <Bau> aggregate stripping test
Zuschlagablösung f <Bau> aggregate stripping
Zuschlagabnutzungswert m <Bau> aggregate abrasion value
Zuschlagdruckfestigkeit f <Bau> aggregate crushing value
Zuschläger m <Fertig> striker *(Schmieden)*
Zuschlagstoff m 1. <Abfall> additive; 2. <Bau, Fertig> aggregate
Zuschnappklappe f <Verpack> flap snap
zuschneiden v 1. <Bau> lumber *(Holz)*; 2. <Metall> cut-to-length
Zuschnitt m <Textil> cutting
Zusetzen n <Maschinen> plugging
zuspitzen v <Bau> tip
Zuspitzung f <Mechan> taper
Zustand m 1. <Comp & DV> condition, state, status; 2. <Kontroll, Qual> status • **im galvanisierten Zustand** <Fertig> as-deposited • **in gegossenem Zustand** <Fertig> as-cast *(ohne weitere Bearbeitung)* • **in seetüchtigem Zustand** <Wassertrans> in navigable condition *(Schiff)*
Zustand m **vor der Bearbeitung** <Maschinen> premachined condition
zustande gekommener Anruf m <Telekom> completed call
zuständig adj <Patent, Qual> appropriate, responsible
Zuständigkeit f <Qual> responsibility
Zuständigkeitsgrenzen fpl <Patent, Qual> jurisdictional boundaries
Zustandsänderung f <Phys, Thermod> change of state
Zustandsanzeigelampe f <Telekom> status lamp
Zustandsbit n <Comp & DV, Telekom> status bit
Zustandsdaten npl <Telekom> status data
Zustandsdiagramm n 1. <Fertig> state diagram; 2. <Thermod> phase diagram
Zustandsdichte f <Elektronik> density of states *(Halbleiter)*
Zustandsfunktion f *(Z)* <Phys> function of state, partition function
Zustandsgleichung f 1. <Mechan, Phys> equation of state; 2. <Thermod> state equation

Zustandsgleichung f **des Atomkerns** <Strahlphys> nuclear equation of state
Zustandsgröße f <Thermod> state quantity
Zustandsprüfung f <Qual> status test
Zustandsschaubild n <Fertig> alloy diagram
Zustandsübergang m <Kontroll> state transition
Zustandsübergangsdiagramm n <Telekom> state transition diagram
Zustandswechsel m <Kontroll> state change
Zustellgetriebe n <Fertig> feeding-in mechanism *(Spanung)*
Zustellung f 1. <Maschinen> in-feed; 2. <Patent> notification
Zustimmungsblock m <Eisenbahn> permissive block
Zustöpseln n <Bau> plugging
Zustrom m 1. <Strömphys> inrush *(turbulenter Grenzschicht)*; 2. <Wasserversorg> inflow
Zuströmverhältnis n <Lufttrans> inflow ratio
Zustromwinkel m <Lufttrans> inflow angle
Zutat f <Lebensmittel> ingredient
zuteilen v <Comp & DV> allocate; allocate *(Datenverarbeitung)*
Zuteilung f <Comp & DV, Elektrotech, Funktech> allocation
zutreffend adv applicable
Zutrittsverriegelung f <Sicherheit> access interlock
zuverlässig adj 1. <Comp & DV> fault-tolerant; 2. <Telekom> reliable
zuverlässiger Transfer-Server m <Telekom> reliable transfer server
Zuverlässigkeit f 1. <Comp & DV, Elektrotech, Ergon, Maschinen, Meerschmutz, Metrol, Telekom> reliability; 2. <Qual> dependability, reliability
Zuverlässigkeit f **der Zeichengabe** <Telekom> *(AE)* signaling reliability, *(BE)* signalling reliability
Zuverlässigkeitsanalyse f <Raumfahrt> reliability analysis *(Raumschiff)*
Zuverlässigkeitsgrad m <Qual> dependability
Zuverlässigkeitsmerkmal n <Qual> reliability characteristic
Zuverlässigkeitsprüfung f <Elektrotech, Qual> reliability test, reliability testing
Zuverlässigkeitssicherung f <Qual> reliability assurance
Zuverlässigkeitstechnik f <Qual, Sicherheit> reliability engineering
Zuverlässigkeitstheorie f <Sicherheit> reliability theory
Zuwachs m 1. <Math> increment; 2. <Wasserversorg> accretion
Zuwachs... <Fertig> incremental
Zuwachsbemaßung f <Konstzeich> progressive dimensioning *(von unterbrochenen Nähten)*
Zuwachsfaktor m <Kerntech> advantage factor, build-up factor
Zuwachspermeabilität f <Werkprüf> incremental permeability
zuweisen v 1. <Comp & DV> allocate, assign; allocate *(Datenverarbeitung)*; 2. <Math> assign
Zuweisung f 1. <Comp & DV, Elektrotech> allocation; 2. <Funktech> allocation *(Frequenzband)*; 3. <Math> assignment
Zuweisung f **eines Frequenzspektrums** <Funktech> spectrum allocation
Zuweisungsanweisung f <Comp & DV> assignment statement *(Programmiersprache)*
zuwiderhandeln v <Sicherheit> contravene *(Vorschriften)*
Z-Vierpolparameter mpl <Elektronik> Z-parameters *(Widerstandsparameter)*

ZVSt *(Zentralvermittlungsstelle)* <Telekom> tertiary exchange
Zwang *m* <Heiz & Kälte, Hydraul, Kerntech> compulsion
Zwangsabschaltung *f* <Kerntech> emergency shutdown, scram
Zwangsdurchlaufkessel *m* 1. <Heiz & Kälte> forced-circulation boiler, once-through boiler; 2. <Hydraul> flash boiler, flasher
Zwangsentlüftungsanlage *f* <Sicherheit> exhaust ventilation system
Zwangskonvektion *f* <Heiz & Kälte> forced convection
Zwangskonvektionskühlofen *m* <Ker & Glas> forced-convection lehr
Zwangskraft *f* <Maschinen> constraining force
Zwangskühlung *f* <Maschinen> forced cooling
zwangsläufig *adj* <Fertig> positive
zwangsläufige Abschaltung *f* <Sicherheit> automatic shutdown
zwangsläufiger Antrieb *m* <Kfztech> positive drive
zwangsläufiger Schutz *m* <Sicherheit> positive protection
Zwangslizenz *f* <Patent> *(BE)* compulsory licence, *(AE)* compulsory license
Zwangsluftkühlung *f* <Maschinen> ducted cooling, forced-air cooling
Zwangslüftung *f* <Maschinen> forced ventilation, mechanical ventilation
Zwangsmischer *m* <Bau> pan mixer
Zwangsschmierung *f* <Maschinen> forced lubrication
Zwangsumlauf *m* <Maschinen> forced circulation, forced flow
Zwangsumlaufkessel *m* <Heiz & Kälte> forced-circulation boiler
Zwangsumlaufreaktor *m* <Kerntech> forced-circulation reactor
Zwangsumschaltung *f* <Telekom> forced handoff
Zwangsverriegelung *f* <Sicherheit> positive interlocking
Zwangswasserkühlung *f* <Maschinen> forced-water cooling
zwanzigflächig *adj* <Geom> icosahedral
Zwanzigflächner *m* <Geom> icosahedron
zweckbestimmte Anlage *f* **für Klarsichtfolie** <Verpack> dedicated food grade film plant *(zur Verpackung von Lebensmitteln)*
Zwecke *f* <Maschinen> tack
zweckgebundener Chip *m* <Elektronik> dedicated chip
zweckgestaltet *adj* <Verpack> purpose-designed
zweckmäßig *adj* <Mechan> efficient, functional, practical
zweiachsig *adj* <Geom, Metall> biaxial
zweiachsiges Drehgestell *n* <Eisenbahn> four-wheel bogie
Zweiadressbefehl *m* <Comp & DV> two-address instruction
zweiarmiger Hebel *m* <Maschinen> lever of the first kind
zweiatomig *adj* <Chemie> biatomic, diatomic
zweiatomiges Gas *n* <Phys> diatomic gas
zweiatomiges Molekül *n* <Phys, Strahlphys> diatomic molecule
zweiäugige Spiegelreflexkamera *f* <Foto> twin-lens reflex, twin-lens reflex camera
Zweibackenfutter *n* <Fertig, Maschinen> two-jaw chuck
Zweibadtonung *f* <Foto> two-bath toning
Zweibadverfahren *n* <Druck> two-bath process
Zweibadverfahren n der Dekontaminierung <Kerntech> two-bath method of decontamination
Zweiband-Handy *n* <Mobilkom> db mobile, dual band mobile
Zweibandkabel *n* <Kfztech> twin-ribbon cable
Zweibein *n* <Fertig> bipod

Zweibereichsinstrument *n* <Gerät> double-range instrument, dual-range instrument
Zweibereichsmessgerät *n* <Gerät> double-range instrument, dual-range instrument
Zweibrennstoffsystem *n* <Kfztech, Lufttrans> dual-fuel system
zweidimensional *adj* <Comp & DV, Geom, Phys> two-dimensional
zweidimensional nachformen *v* <Fertig> contour, profile
Zweidraht... <Telekom> two wire
zweidrähtiges Dreieck-Stromnetz *n* <Elektriz> two-wire delta network
zweidrähtiges Stromnetz *n* <Elektriz> two-wire network
Zweidrahtkoppelpunkt *m* <Telekom> two-wire crosspoint
Zweidrahtleitung *f* <Comp & DV, Telekom> two-wire circuit
Zweidrahtsystem *n* 1. <Elektriz> double-wire system, two-wire system; 2. <Elektrotech, Telekom> two-wire system
Zweidrahtvermittlung *f* <Telekom> two-wire switch
Zweidraht-Vermittlungssystem *n* <Telekom> two-wire switching system
Zweidraht/Vierdraht-Übergang *m* <Telekom> hybrid terminating circuit, 2wire/4wire transition
Zwei-Elektronen-Problem *n* <Kerntech> two-electron problem
Zweieranschluss *m* <Telekom> dual party line
Zweieranschlusstelefon *n* <Telekom> two-party telephone
Zweierkomplement *n* <Comp & DV> two's complement
Zweietagen-Kuppelofen *m* <Ker & Glas> double-deck crown furnace
zweifach diffundierter Transistor *m* <Elektronik> double-diffused transistor
zweifach wirkend *adj* <Maschinen> double-acting
Zweifach... <Elektronik, Ker & Glas, Kfztech> double
Zweifachdiffusion *f* <Elektronik> double diffusion
Zweifachform *f* <Ker & Glas> *(AE)* double-cavity mold, *(BE)* double-cavity mould
Zweifachkette *f* <Kfztech> duplex chain *(Kraftübertragung)*
Zweifachregler *m* <Lufttrans> dual control
Zweifachrollenkette *f* <Kfztech> double roller chain
Zweifarbe... <Druck, Elektronik, Optik> *(AE)* two-color, *(BE)* two-colour
Zweifarbendruckmaschine *f* <Druck> *(AE)* two-color press, *(BE)* two-colour press
Zweifarbenfilter *n* <Elektronik, Optik> dichroic filter
Zweifarbenspiegel *m* <Optik> dichroic mirror
zweifarbig *adj* <Druck> *(AE)* two-color, *(BE)* two-colour
zweifarbige Flüssigkristalle *npl* <Elektronik> dichroic liquid crystals
zweifarbiges Glas *n* <Ker & Glas> dichroic glass
Zweiflach *n* <Fertig> dihedron
zweiflächig *adj* <Fertig, Geom> dihedral
zweiflächige Antenne *f* <Telekom> dihedral antenna
zweiflutiger Kessel *m* <Heiz & Kälte> double-pass boiler
zweiflutiger Kühler *m* <Heiz & Kälte> double-pass heat exchanger
zweiflutiges Luftstrahltriebwerk *n* <Lufttrans> dual-flow jet engine
Zweifrequenz... <Funktech, Telekom> two-frequency
Zweifrequenzkanalbelegungsplan *m* <Telekom> *(AE)* two-frequency channeling plan, *(BE)* two-frequency channelling plan
Zweifrequenzrichtfunksystem *n* <Telekom> two-frequency radio relay system

Zweifrequenztonwahl

Zweifrequenztonwahl f <Telekom> *(AE)* two-frequency signaling, *(BE)* two-frequency signalling
Zweig m 1. <Elektriz> branch; 2. <Elektrotech> branch *(eines Netzwerkes)*; 3. <Gerät> ratio arm *(Brückenschaltung)*; 4. <Mechan, Phys> branch
Zweigangabfüllung f <Verpack> two-speed filling
Zweigangachsantrieb m <Kfztech> two-speed final drive
zweigängig adj <Maschinen> double-threaded
zweigängiges Gewinde n <Maschinen> double thread, two-start thread
Zweigleitung f 1. <Bau> branch line *(Rohrleitung)*; 2. <Elektriz, Elektrotech> branch, branch line; 3. <Wasserversorg> branch pipe
Zweigverbindung f <Elektriz> parallel connection
Zweihandschalter m <Sicherheit> two-hand control
Zweihandsicherung f <Sicherheit> two-hand control
Zweihandsteuerung f <Sicherheit> two-hand control
zweihäusige Wanne f <Ker & Glas> wasp-waisted tank
Zweihöhendotierungsprofil n <Elektronik> low-high-low doping profile
zweiholmige Seitenflosse f <Raumfahrt> twin-spar vertical fin *(Raumschiff)*
Zweihüllendesign n <Wassertrans> double-skin design *(Schiffbau)*
Zweikanalanlage f <Heiz & Kälte> dual-conduit system
Zweikanal-Einseitenband-Modulation f <Elektronik, Funktech> independent sideband modulation
Zweikanalsichtfunkpeilung f <Funkort> dual-carrier visual direction finding
Zweikomponenten... <Kunststoff, Raumfahrt> two-pack
Zweikomponentenlack m <Fertig> two-component lacquer
Zweikomponenten-Primer m <Kunststoff> two-pack primer
Zweikomponententreibstoff m <Raumfahrt> bipropellant
Zweikomponententreibstoffantrieb m <Raumfahrt> liquid bipropellant propulsion *(Raumschiff)*
Zweikontaktregler m <Kfztech> two-contact regulator
Zweikreis... <Elektronik> double-tuned
Zweikreisbremse f 1. <Kfztech> double-circuit brake, dual-circuit brake; 2. <Lufttrans> dual-circuit brake
Zweikreisbremsensystem n <Trans> separated braking circuits
Zweikreisfilter n <Elektronik> double-tuned circuit
Zweikreishohlraum m <Elektronik> double-tuned cavity
Zweikreis-TL-Triebwerk n 1. <Lufttrans> ducted-fan turbo engine; 2. <Thermod> bypass engine
Zweikreistriebwerk n 1. <Lufttrans> ducted-fan engine; 2. <Thermod> turbofan engine
Zweikreisverstärker m <Elektronik> double-tuned amplifier
Zweikreiszündanlage f <Kfztech> two-circuit ignition system
Zweikristall-Spektrometer n <Kerntech> two-circle instrument
Zweileitungsbremse f <Kfztech> twin-line brake
Zweimeißeldrehmaschine f <Maschinen> duplex lathe
zweimotorig adj <Maschinen> twin-engined
Zweinormenmonitor m <Fernseh> dual standard monitor
Zwei-Nukleonen-System n <Kerntech> two-nucleon system
zweiohrig adj <Akustik> binaural
zweiohriges Hören n <Akustik> binaural audition
Zweiphasen... <Elektrotech, Kerntech, Lufttrans, Trans> two-phase
Zweiphasen-Jet m <Lufttrans> diaphasic jet
Zweiphasenkontrollgerät n <Trans> two-phase controller
Zweiphasenkühlung f <Kerntech> two-phase cooling
Zweiphasenläufer m <Elektrotech> two-phase rotor
Zweiphasenlinearmotor m <Elektrotech> two-phase linear motor
Zweiphasenmaschine f <Elektrotech> two-phase machine
Zweiphasenmotor m <Elektriz, Elektrotech> two-phase motor
Zweiphasenreaktor m <Kerntech> two-phase reactor
Zweiphasenrotor m <Elektrotech> two-phase rotor
Zweiphasenrotorwicklung f <Elektrotech> two-phase rotor winding
Zweiphasenstator m <Elektrotech> two-phase stator
Zweiphasenstatorwicklung f <Elektrotech> two-phase stator winding
Zweiphasenstrom m <Elektrotech> two-phase current
Zweiphasenströmung f <Strömphys> two-phase flow
Zweiphasensystem n <Elektrotech> two-phase system
Zweiphasenumtastung f *(BPSK)* <Telekom> binary phase shift keying *(BPSK)*
zweiphasig adj <Elektriz, Elektrotech> biphase, two-phase
zweiphasiger Alternator m <Elektriz> two-phase alternator
zweiphasiger Strom m <Elektriz> biphase current
zweiphasiges Netz n <Elektriz> two-phase network
zweiphasiges System n <Elektriz> two-phase system
Zwei-plus-Eins-Adressbefehl m <Comp & DV> two-plus-one address instruction
Zweipol m 1. <Phys> two-terminal network; 2. <Telekom> dipole
zweipolig adj <Elektriz> bipolar
zweipolige Maschine f <Elektriz> bipolar machine
zweipolige Stromversorgung f <Elektrotech> bipolar power supply
zweipoliger Ein/Aus-Schalter m *(DPST)* <Elektrotech> double-pole single-throw *(DPST)*; double-pole single-throw switch
zweipoliger Kippschalter m <Elektriz> double-pole snap switch
zweipoliger Schalter m 1. <Elektriz> double-pole switch; 2. <Elektrotech> two-pole switch
zweipoliger Stecker m <Elektrotech> two-pin plug
zweipoliger Umschalter m *(DPDT)* <Elektrotech> double-pole double-throw *(DPDT)*
zweipoliger Verstärker m <Elektronik> bipolar amplifier
zweipoliger Wechselschalter m *(DPDT)* <Elektrotech> double-pole double-throw *(DPDT)*
zweipoliges Ein/Aus-Schaltrelais n <Elektriz> DPST relay, double-pole single-throw relay
zweipoliges Umschaltrelais n <Elektriz> double-pole double-throw relay
zweipoliges Wechselschaltrelais n <Elektriz> double-pole double-throw relay
Zweipolmotor m <Elektriz> two-pole motor
Zweipolstecker m <Elektrotech> two-pin plug
Zweipolsystem n <Elektriz> two-pole system
Zweipolwicklung f <Elektriz> bipolar winding
zweiprozessorfähig adj <Comp & DV> dual-processor enabled
Zweiprozessorfähigkeit f <Comp & DV> dual processing capability
Zweipunkt... <Kfztech, Regelung> two-step
Zweipunktglied n <Regelung> two-step action element
Zweipunktgurt m <Kfztech> safety belt *(Sicherheitszubehör)*
Zweipunktregelung f <Regelung> two-step control
Zweipunktregler m <Regelung> two-position controller
Zweipunktsignal n <Regelung> two-step signal
Zweipunktverhalten n <Regelung> two-step action

zweirädrig adj <Maschinen> two-wheeled
zweireihiges Kugellager n <Maschinen> double-row ball bearing
Zweirichtungs... <Telekom> bidirectional
Zweirichtungsanzeige f <Gerät> bidirectional read-out
Zweirichtungshobelmaschine f <Maschinen> double-cutting planing machine
Zweirichtungssignalisierung f <Eisenbahn> bi-directional signalling
Zweirichtungs-Thyristordiode f <Elektrotech> triac
Zweirichtungsverkehr m <Telekom> bidirectional traffic
Zweirichtungszähler m <Gerät> bidirectional counter, reversible counter
Zweirollen... <Aufnahme> reel-to-reel
Zweirollen-Abspielgerät n <Aufnahme> reel-to-reel player
zweisäurig adj <Chemie> diacidic *(Basen)*
Zweischeiben... <Fertig, Kfztech> *(BE)* two-disc, *(AE)* two-disk
Zweischeibenkupplung f <Kfztech> *(BE)* two-disc clutch, *(AE)* two-disk clutch, two-plate clutch
Zweischeibenschleifmaschine f 1. <Fertig> duplex grinder; 2. <Maschinen> two-wheel grinding machine
Zweischeibentrockenkupplung f <Kfztech> double-plate dry clutch
zweischenkliges Manometer n <Gerät> two-leg manometer
Zweischneidenschleifmaschine f <Ker & Glas> double-edge grinder
zweischneidig adj <Maschinen> double-edge, double-edged
zweischneidiger Bohrer m <Maschinen> fiddle drill
Zweischraubflansch m <Maschinen> two-bolted flange
Zweiseitenband n <Elektronik, Funktech> double sideband
Zweiseitenband-Modulation f <Elektronik, Funktech> double-sideband modulation
Zweiseitenband-Modulator m <Elektronik, Funktech> double-sideband modulator
Zweiseitenbandübertragung f <Funktech> double-sideband transmission
zweiseitig arbeitend adj <Mechan> double-acting
zweiseitig gerichteter Datenfluss m <Comp & DV> bidirectional flow
zweiseitig gerichtetes Mikrofon n <Akustik, Aufnahme> bidirectional microphone
zweiseitig geschliffenes Tafelglas n <Ker & Glas> twin-ground plate
zweiseitige Belüftung f <Heiz & Kälte> double-ended ventilation
zweiseitige Brennstoffbeschickung f <Kerntech> *(AE)* bidirectional refueling, *(BE)* bidirectional refuelling
zweiseitige Leiterplatte f <Elektronik> double-sided printed circuit, double-sided printed circuit board
zweiseitige Regelung f <Regelung> bilateral control
zweiseitige Steuerung f <Regelung> bilateral control
zweiseitiger Feldmagnet m <Elektrotech> double-sided field system
zweiseitiger Koppler m <Elektrotech> bidirectional coupler
zweiseitiger Pflasterhammer m <Bau> double-ended sledgehammer
zweiseitiger Spundhobel m <Bau> double-ended match plane
zweiseitiger Verstärker m <Elektronik> bilateral amplifier
zweiseitiges Getriebe n <Elektronik> bilateral gear
Zweiseitigkeit f <Papier> two sidedness
Zweispannungslokomotive f <Eisenbahn> dual voltage locomotive

Zweispindeldrehmaschine f <Maschinen> twin-screw lathe, twin-spindle lathe
Zweispitzniet m <Fertig> bifurcated rivet
Zweiständerhobelmaschine f <Maschinen> double-column planing machine, double-housing planing machine
Zweiständerpresse f <Maschinen> arch press
Zweistoff... <Chemie, Kfztech, Metall> binary
Zweistofflegierung f <Metall> binary alloy
Zweistoffmotor m <Kfztech, Lufttrans> dual-fuel engine
zweistrahlige Einspritzdüse f <Kfztech> twin-jet injection nozzle
Zweistrahlkatodenstrahlröhre f <Elektronik> dual-beam cathode-ray tube, split beam cathode-ray tube
Zweistrom... <Eisenbahn, Lufttrans> dual-current
Zweistromlokomotive f <Eisenbahn> dual-current locomotive
Zweistromtriebwerk n <Lufttrans> bypass engine, fan jet, turbofan
Zweistromtriebwerkturbine f <Lufttrans> fan jet turbine
Zweistufen... <Elektriz, Telekom> dual-level
Zweistufenplan m <Telekom> dual-level plan
Zweistufenrelais n <Elektriz> two-stage relay
Zweistufenvergaser m <Kfztech> *(AE)* two-phase carburetor, *(BE)* two-phase carburettor
zweistufige Funktion f <Elektrotech> bilevel operation
zweistufiger Betrieb m <Elektrotech> bilevel operation
zweistufiger Kompressor m <Maschinen> two-stage compressor
zweistufiger Verdichter m <Maschinen> two-stage compressor
Zweisystemkontaktunterbrecher m <Kfztech> two-system contact breaker
Zweit... <Fernseh, Kunststoff, Wassertrans> second
Zweitakt... <Kfztech> two-stroke
Zweitakter m 1. <Maschinen> twin-stroke engine; 2. <Mechan> two-stroke engine
Zweitaktgemisch n <Kfztech> *(AE)* gas-oil mixture, *(AE)* gasoline-oil mixture, *(BE)* petrol-oil mixture *(Zweitaktmotor)*
Zweitaktmotor m 1. <Kfztech> two-stroke engine; 2. <Maschinen> twin-stroke engine; 3. <Mechan, Wassertrans> two-stroke engine
Zweitaktöl n <Kfztech> two-stroke oil
Zweitanode f <Fernseh> second anode
Zweitbeschleunigereffekt m <Kunststoff> synergism effect, synergistic effect
Zweitdestillation f <Chemie> redistillation, rerun
zweite Ableitung f <Math> second differential coefficient
zweite Freifläche f <Fertig> second flank
zweite Gärung f <Lebensmittel> secondary fermentation
zweite Generation f <Comp & DV> second generation
zweite Harmonische f <Elektronik> second harmonic
zweite harmonische Einspeisung f <Elektronik> second harmonic injection
zweite Ionisationsstufe f <Phys> second ionization potential
zweite Spanfläche f <Fertig> second face
zweite Stufe f <Akustik> supertonic *(Tonleiter)*
zweite Zwischenfrequenz f <Elektronik, Fernseh, Funktech, Telekom> second intermediate frequency
zweiteilige Antriebswelle f <Kfztech> two-piece drive shaft
zweiteilige Instabilität f <Kerntech> two-stream instability
zweiteilige Kardanwelle f <Kfztech> two-piece propeller shaft
zweiteiliger Steckverbinder m <Elektrotech> two-piece connector
zweiteiliger Verbinder m <Elektrotech> two-piece connector

zweiter

zweiter Brennraum m (SCC) <Maschinen> afterburner chamber, secondary combustion chamber
zweiter Hauptsatz m **der Thermodynamik** <Phys> second law of thermodynamics
zweiter Reduktionsbrand m <Ker & Glas> second reducing firing
zweiter Überlagerungsoszillator m <Elektronik, Funktech> second local oscillator
zweiter Vergaserlufttrichter m <Kfztech> secondary barrel
zweites Deck n <Wassertrans> second deck (Schiffskonstruktion)
zweites Ionisationspotenzial n <Phys> second ionization potential
Zweitflächenspiegel m (SSM) <Raumfahrt> second surface mirror, SSM (Raumschiff)
Zweitluft f <Heiz & Kälte, Maschinen> secondary air
Zweitmantel m <Optik> secondary coating
Zweitonvorlagen fpl <Druck> bi-tones
Zweitor n <Telekom> two-port, two-port network
Zweitourenmaschine f <Druck> two-revolution press
Zweitourenpresse f <Druck> two-revolution press
Zweiträgerübertragung f <Telekom> dual-carrier transmission
Zweitumlaufecho n <Funkort, Wassertrans> second trace echo (Radar)
zweitürige Limousine f <Kfztech> (AE) coach
Zweitweg m <Telekom> second choice route
Zweiunddreißigerformat n <Druck> thirty-two-mo
Zweiwalz-Entrockner m <Lebensmittel> double-drum drier, double-drum dryer
Zweiweg... <Kfztech, Maschinen, Phys, Wasserversorg> two-way
Zweiwegdämpfungsventil n <Kfztech> two-way damper valve
Zweiweggleichrichter m <Phys> full-wave rectifier
Zweiweghahn m 1. <Labor> two-way tap; 2. <Maschinen, Wasserversorg> two-way cock
Zweiwegpalette f <Trans> two-way pallet
Zweiwegspiegel m <Ker & Glas> two-way mirror
Zweiwegventil n <Heiz & Kälte> two-way valve
zweiwertige Logik f <Künstl Int> exact logic, two-valued logic
zweiwertige Modulation f <Telekom> binary modulation
zweiwertige Phasenumtastung f <Telekom> biphase shift keying
Zweiwertigkeit f <Chemie> bivalence, divalence
Zweizonenreaktor m <Kerntech> two-zone reactor
Zweizonentransistor m (UJT) <Elektronik> programmable unijunction transistor, unijunction transistor
Zweizweckfahrzeug n <Kfztech> dual-purpose vehicle (Fahrzeugart)
Zweizylinder... <Kfztech, Maschinen> two-cylinder
Zweizylinderboxermotor m <Kfztech> flat twin engine, flat twin
Zweizylinderdruckmaschine f <Druck> two-cylinder press
Zweizylindermotor m <Maschinen> double-cylinder engine, duplex-cylinder engine
Zweizylinderspinnerei f <Textil> cotton condenser spinning
Zwergstern m <Raumfahrt> dwarf star
Zwickel m 1. <Bau> spandrel; 2. <Elektriz> filler
Zwilling m <Metall> twin
Zwillingsbildung f <Metall> twinning
Zwillingsfläche f <Metall> twinning plane
Zwillingsflugssteuerungssystem n <Lufttrans> dual flight control system
Zwillingsflugzeug n <Lufttrans> composite aircraft

Zwillingsgrenze f <Metall> twin boundary
Zwillingshahn m <Wasserversorg> twin cock
Zwillingskabel n <Elektrotech> twin cable
Zwillingslamelle f <Metall> twin lamella
Zwillingsmotor m <Maschinen> twin engines
Zwillingsparadoxon n <Phys> twin paradox
Zwillingspumpe f <Maschinen> two-throw pump
Zwillingsräder npl <Lufttrans> twin wheels (Fahrgestell)
Zwillingsräumen n <Maschinen> twin broaching
Zwillingsreaktoranlage f <Kerntech> twin-reactor station
Zwillingsscheren n <Metall> twinning shear
Zwillingssystem n <Metall> twinning system
Zwillingstriebwerk n <Lufttrans> twin engine
Zwillingstriebwerksdüsenjet m <Lufttrans> twin-engine jet aircraft
Zwillingstunnel m <Eisenbahn> twin tunnel
Zwinge f 1. <Fertig> holdfast; 2. <Labor> clamp; 3. <Maschinen> clamp, collar, collet, cramp, ferrule; 4. <Mechan> clamp, (BE) vice, (AE) vise; 5. <Optik> ferrule
Zwirn m <Textil> thread, twist
Zwirnen n 1. <Ker & Glas> twisting (von Glasfasern); 2. <Textil> throwing
Zwirnmaschine f <Textil> twister
Zwischen... <Abfall, Druck, Kfztech, Lufttrans, Phys> intermediate
Zwischenabdeckung f <Abfall> intermediate cover (einer Deponie)
Zwischenabzüge mpl <Druck> interim proofs, intermediate proofs
Zwischenanflug m <Lufttrans> intermediate approach
Zwischenanflugsposition f <Lufttrans> intermediate approach fix
Zwischenbild n <Phys> intermediate image
Zwischenbildcodierung f <Telekom> interframe coding
Zwischenbildikonoskop n <Elektronik> image iconoscope
Zwischenbildorthikon n <Elektronik> image orthicon
Zwischenboden m <Heiz & Kälte> false floor
Zwischendatei f <Comp & DV> intermediate file
Zwischendeck n <Wassertrans> between deck (Schiff); tweendeck (Schiffbau) • **im Zwischendeck** <Wassertrans> betweendecks (Schiff)
Zwischendecke f 1. <Bau> intermediate ceiling; 2. <Heiz & Kälte> false ceiling
Zwischendichtungsglas n <Ker & Glas> (BE) intermediate sealing glass, (AE) solder glass
Zwischenfalte f <Konstzeich> intermediate fold
Zwischenfassung f <Elektrotech> socket adaptor (für elektronische Röhren)
Zwischenfrequenz f (Zf) <Elektronik, Funktech, Telekom> intermediate frequency (IF)
Zwischenfrequenzdemodulator m <Funktech, Telekom> IF demodulator, IF detector, intermediate-frequency demodulator, intermediate frequency detector
Zwischenfrequenzkreis m <Funktech, Telekom> intermediate-frequency circuit
Zwischenfrequenzmodulation f (Zfm) <Elektronik, Funktech, Telekom> intermediate frequency modulation (IFM)
Zwischenfrequenzselektion f <Funktech, Telekom> intermediate-frequency selectivity
Zwischenfrequenzsignal n <Elektronik, Funktech, Telekom> intermediate frequency signal
Zwischenfrequenzverstärker m <Elektronik, Funktech, Telekom> intermediate frequency amplifier
Zwischenfrequenzverstärkerstufe f <Funktech, Telekom> IF amplifier stage, intermediate-frequency amplifier stage

Zwischenfutter n <Maschinen> cat head, spider
Zwischengehäuse n <Lufttrans> intermediate case
zwischengelagert adj <Erdöl> interlayer *(Geologie)*
zwischengeschaltet adj <Fertig> interposed
Zwischengeschoss n <Bau> intermediate storey
zwischengespeicherte Eingabe/Ausgabe f <Comp & DV> buffered input/output
Zwischengetriebe n <Maschinen> transmission gear
Zwischengitterplatz m <Phys> interstitial place
Zwischenglühen n 1. <Fertig> process annealing *(Blech)*; 2. <Maschinen> intermediate annealing, process annealing
Zwischenglühung f 1. <Fertig> intermediate softening; 2. <Metall> process annealing
Zwischengut n <Kohlen> middlings
Zwischenholm m <Lufttrans> false spar
Zwischenhülse f <Fertig> socket
Zwischenkammer f <Maschinen> receiver
Zwischenkern m <Kerntech> compound nucleus
Zwischenkopierpapier n <Konstzeich> intermediate copying paper
Zwischenkreis-Gleichstromumrichter m <Elektrotech> indirect d.c. converter
Zwischenkreisumrichter m <Elektrotech> indirect a.c. converter
Zwischenkühler m <Heiz & Kälte> intercooler
Zwischenlage f 1. <Kfztech> ply *(Reifen)*; 2. <Papier> interleaving
Zwischenlagen fpl <Verpack> cushioning product, padding
Zwischenlager n 1. <Fertig> bank *(Fließstraße)*; 2. <Maschinen> intermediate bearing
Zwischenlagerplatz m <Abfall> refuse transfer station
Zwischenlagerung f <Abfall> intermediate storage *(von Müll)*
Zwischenlagerung f **in Kühlanlagen** <Heiz & Kälte> cold storage
Zwischenlandung f **aus technischen Gründen** <Lufttrans> technical stop
Zwischenleitung f <Telekom> link
Zwischenleitungsanordnung f <Telekom> link system
Zwischenlinsenverschluss m <Foto> between the lens shutter
Zwischenmauer f <Bau> party wall
Zwischenmodenverzerrung f <Optik> intramodal distortion
Zwischenmodulation f <Aufnahme, Elektronik> intermodulation
Zwischenmodulationsprodukt n <Elektronik> intermodulation product
Zwischenmodulationsverzerrung f *(IMD)* <Elektronik> intermodulation distortion *(IMD)*
zwischenmolekular adj <Chemie, Metall> intermolecular
Zwischenplatte f 1. <Fertig> spacer *(Kunststoffinstallationen)*; 2. <Mechan> diaphragm
Zwischenpodest n <Bau> half pace
Zwischenprodukt n 1. <Fertig> in-process product; 2. <Lebensmittel> Dunst middlings *(Grieß)*
Zwischenproduktcontainer m <Verpack> intermediate bulk container
Zwischenprüfung f <Qual> in-process inspection
Zwischenpumpe f <Maschinen> booster pump
Zwischenrad n 1. <Fertig> idle gear, idler gear, idler wheel; 2. <Kfztech> idle gear *(Getriebe)*; 3. <Maschinen> idle wheel, idler wheel, idler, intermediate gear, intermediate wheel, stud wheel
Zwischenrädergetriebe n <Lufttrans> intermediate gearbox *(eines Hubschraubers)*
Zwischenraum m 1. <Comp & DV> gap, spacing; 2. <Druck> space; 3. <Fertig> interstice; 4. <Hydraul> clearance; 5. <Kerntech> gap; 6. <Kohlen> compartment; 7. <Lufttrans> clearance *(von Propeller, Flügel)*; 8. <Wassertrans> clearance • **in Zwischenräumen liegend** <Bau> interstitial • **Zwischenraum herausnehmen** <Druck> close up
Zwischenraum m **zwischen zwei Bandblöcken** <Comp & DV> IBG, interblock gap
Zwischenregenerator m <Telekom> intermediate regenerator, repeater
Zwischenregister n <Comp & DV> temporary register
zwischenschalten v <Fertig> interpose
Zwischenschaltung f <Fertig> interposition
Zwischenschicht f 1. <Lufttrans> interlining *(Luftransport)*; 2. <Telekom> interface
Zwischenschichtisolierung f <Elektriz> layer insulation
Zwischenschneide f <Fertig> drag *(Spanung)*
Zwischensetzen n <Fertig> insertion
Zwischensockel m <Elektrotech> socket adaptor
Zwischenspannungswicklung f <Elektriz> intermediate-voltage winding
Zwischensparren m <Bau> common rafter
Zwischenspeicher m 1. <Comp & DV> intermediate storage, temporary storage; 2. <Elektrotech> temporary memory; 3. <Telekom> buffer memory
Zwischenspeicherbibliothek f <Comp & DV> staging library
zwischenspeichern v <Comp & DV> buffer
Zwischenspeichern n <Comp & DV> buffering, staging
zwischenstädtischer Fluglinienverkehr m <Trans> intercity air service
Zwischenstation f <Funktech, Phys, Telekom> relay
Zwischenstecker m 1. <Elektrotech> plug adaptor; 2. <Funktech, Telekom> adapter
Zwischenstellenmodem n <Telekom> repeater modem
Zwischenstellung f <Fertig> intermediate position *(Kunststoffinstallationen)*
Zwischenstück n 1. <Bau> spacer block; 2. <Comp & DV, Elektriz, Elektrotech> adaptor; 3. <Fertig> adaptor; adaptor, transition piece *(Kunststoffinstallationen)*; 4. <Funktech, Labor, Maschinen, Mechan, Phys, Telekom, Textil> adaptor
Zwischenstück n **mit Außengewinde** <Maschinen> male adaptor
Zwischenstück n **mit Innengewinde** <Maschinen> female adaptor
Zwischenstufen-Dampfmaschine f <Hydraul> intermediate cylinder steam engine *(Dampfmaschine)*
Zwischenstufengefüge n <Fertig, Metall> bainite
Zwischenstufenzylinder m <Hydraul> intermediate pressure cylinder *(Dampfmaschine)*
Zwischensumme f <Comp & DV> batch total *(pro Stapel)*
Zwischensumme f **pro Stapel** <Comp & DV> batch total
Zwischentransformator m <Elektrotech> interstage transformer
zwischenüberhitzen v <Heiz & Kälte> reheat, superheat
Zwischenüberhitzer m <Heiz & Kälte> reheater
Zwischenüberhitzung f <Heiz & Kälte> reheating
Zwischenverbindung f <Bau> interconnection
Zwischenverdichterplattform f <Erdöl> booster platform *(Offshore-Technik; Leitungstransport)*
Zwischenverstärker m <Telekom> regenerator, repeater
Zwischenverstärker m **mit Gruppentausch** <Telekom> repeater station with frequency frogging
Zwischenverteiler m <Telekom> IDF, intermediate distribution frame
Zwischenvorgelege n <Maschinen> transmission gear
Zwischenwand f 1. <Bau> baffle, partition; 2. <Ker & Glas> midfeather

Zwischenwelle f 1. <Fertig> jack shaft; 2. <Maschinen> intermediate shaft, jack shaft
Zwischenzeilenflimmern n <Fernseh> interline flicker
Zwischenzelle f <Raumfahrt> interstage section *(Raumschiff)*
Zwitschern n 1. <Funktech> chirp *(Morsefunk)*; 2. <Raumfahrt> chirp modulation *(Weltraumfunk)*
Zwitterion n <Chemie> zwitterion
Zwitterkontakt m <Elektrotech> hermaphroditic contact
Zwittersteckverbinder m <Elektrotech> hermaphroditic connector
Zwölfeck n <Geom> dodecagon
zwölfeckig adj <Geom> dodecagonal
zwölfflächig adj <Geom> dodecahedral
Zwölfflächner m <Geom> dodecaeder, dodecahedron
Zyklenteilverfahren n <Comp & DV> cycle stealing
zyklische Abtragung f <Anstrich> cyclical erosion
zyklische Aufeinanderfolge f <Comp & DV> wraparound
zyklische Bitverschiebung f <Comp & DV> cyclic shift
zyklische Blattverstellung f <Lufttrans> cyclic pitch control *(Hubschrauber)*
zyklische Blockprüfung f *(CRC)* <Comp & DV, Elektronik, Regelung, Telekom> cyclic redundancy check *(CRC)*
zyklische Blocksicherung f *(CRC)* <Comp & DV, Elektronik, Labor, Telekom> cyclic redundancy check *(CRC)*
zyklische Erosion f <Anstrich> cyclical erosion
zyklische Hilfstrimmeinrichtung f <Lufttrans> cyclic pitch servo trim *(Hubschrauber)*
zyklische Längssteuerungsknüppelbelastung f <Lufttrans> longitudinal cyclic stick load *(Hubschrauber)*
zyklische Quersteuerungshilfe f <Lufttrans> lateral cyclic control support *(Hubschrauber)*
zyklische Stellenverschiebung f <Comp & DV> cyclic shift
zyklische Steuerstufe f <Lufttrans> cyclic control step *(Hubschrauber)*
zyklischer Blockcode m <Telekom> cyclic block code
zyklischer Code m <Telekom> cyclic code
zyklischer Einstellwinkel m <Lufttrans> cyclic pitch
zyklischer Graph m <Künstl Int> cyclic graph
zyklischer Konuswinkel m <Lufttrans> cyclic flapping angle *(Hubschrauber)*
zyklischer Schlagwinkel m <Lufttrans> cyclic flapping angle *(Hubschrauber)*
zyklischer Seitensteigungswinkel m <Lufttrans> lateral cyclic pitch *(Hubschrauber)*
zyklischer Steuerknüppel m <Lufttrans> cyclic stick *(Hubschrauber)*
Zykloide f 1. <Fertig> cycloid; 2. <Geom> cycloid *(Rollkurve des Kreises)*
Zykloidenverzahnung f <Fertig> cycloidal gear teeth, cycloidal teeth, cycloidal-profile teeth *(Getriebelehre)*
Zyklon m 1. <Erdöl> cyclone *(Abscheidetechnik)*; 2. <Kohlen, Umweltschutz> cyclone; 3. <Wassertrans> cyclone *(tropischer Wolkensturm)*
Zyklonabscheider m <Maschinen> cyclone separator
Zyklonabscheidung f <Abfall> cyclone separation, cyclone
Zyklonabschneider m <Erdöl> cyclone
Zyklone f <Wassertrans> cyclone *(Wettertief)*
Zyklonentstauber m <Umweltschutz> cyclone
Zyklonfilter n <Maschinen> cyclone filter
Zyklonofen m <Ker & Glas> cyclone furnace
Zyklopenbeton m <Bau> cyclopean concrete
zykloper Staudamm m <Wasserversorg> cyclopic barrage
Zyklotron n <Elektrotech, Phys, Strahlphys, Teilphys> cyclotron

Zyklotronfrequenz f 1. <Kerntech> electron cyclotron frequency; 2. <Strahlphys, Teilphys> cyclotron frequency
Zyklotronresonanz f <Telekom> cyclotronic resonance
Zyklotronsicherheit f <Phys, Strahlphys, Teilphys> cyclotron safety
Zyklotronstrahlung f <Phys, Strahlphys, Teilphys> cyclotron radiation
Zyklus m <Akustik, Comp & DV, Elektriz, Elektrotech, Kontroll, Kunststoff, Maschinen> cycle
Zyklus m **von Intervallen** <Akustik> cycle of intervals
Zykluszähler m <Gerät> cycle counter
Zykluszeit f <Comp & DV> cycle time
Zylinder m 1. <Comp & DV> cylinder *(Festplatte)*; 2. <Druck> cylinder; 3. <Elektrotech> solenoid *(Spule)*; 4. <Fertig> drum; barrel *(Extruder, Plastherstellung)*; 5. <Geom> cylinder; 6. <Hydraul> barrel *(Dampfmaschine)*; drum *(Dampfdruckindikator)*; 7. <Ker & Glas> cylinder *(zur Produktion von gewalztem Flachglas)*; 8. <Kfztech> cylinder *(Motor)*; 9. <Kunststoff, Maschinen, Mechan, Papier> cylinder; 10. <Verpack> drum; 11. <Wassertrans> cylinder *(Motor)*
Zylinder mpl **in V-Anordnung** <Kfztech> V-shaped cylinders
Zylinderauflager n <Fertig> barrel support *(Extruder)*
Zylinderblock m <Kfztech> cylinder block *(Motor)*
Zylinderbohrung f 1. <Kfztech> bore; 2. <Maschinen> cylinder bore
Zylinderbohrwerk n <Fertig, Maschinen> cylinder boring mill
Zylinderbuchse f <Kfztech> cylinder liner *(Motor)*
Zylinderdeckel m 1. <Eisenbahn> cylinder head; 2. <Hydraul> cylinder cover *(frontseitig)*; cylinder head *(hinten)*; cylinder cover *(rückseitig)*; 3. <Kfztech, Maschinen, Wassertrans> cylinder head *(Motor)*
Zylindereinspritzmotor m <Trans> direct injection engine
Zylinderfläche f <Geom> cylinder surface
Zylinderflansch m <Kfztech> cylinder flange *(Motor)*
Zylinderform f <Maschinen> cylindricity
Zylinderglas n <Ker & Glas> *(AE)* blown sheet, *(BE)* cylinder glass
Zylinderheizraum m <Hydraul> steam jacket
Zylinderinhalt m 1. <Kfztech> cylinder capacity; 2. <Maschinen> capacity of a cylinder
Zylinderkondensator m <Phys> cylindrical capacitor
Zylinderkoordinaten fpl <Phys> cylindrical coordinates
Zylinderkopf m 1. <Eisenbahn> cylinder head; 2. <Hydraul> cylinder head *(vorn)*; 3. <Kfztech> cylinder head, head cylinder *(Motor)*; 4. <Maschinen> cheese head, cylinder head; 5. <Wassertrans> cylinder head *(Motor)*
Zylinderkopf m **mit Innensechskant** <Maschinen> hexagon socket head
Zylinderkopfdichtung f 1. <Kfztech> cylinder head gasket, head gasket *(Motor)*; 2. <Maschinen> cylinder head gasket
Zylinderkopfniete f <Bau> cheese-head rivet
Zylinderkopfschraube f 1. <Kfztech> cylinder head bolt; 2. <Maschinen> cheese-head screw, fillister-head screw
Zylinderkörper m <Maschinen> cylinder barrel
Zylinderlaufbahn f <Kfztech> cylinder barrel *(Motor)*
Zylinderlaufbuchse f 1. <Kfztech> cylinder sleeve; cylinder liner *(Motor)*; 2. <Mechan> cylinder liner
Zylindermantel m <Fertig> cylinder barrel
Zylindermaß n <Fertig> roller
Zylinderöffnung f <Ker & Glas> opening of the cylinder
Zylinderreflektorantenne f <Telekom> cylindrical reflecting antenna
Zylinderreibahle f <Fertig> parallel reamer

Zylinderring *m* <Kfztech> cylinder ring *(Motor)*
Zylinderrollenlager *n* <Maschinen> cylindrical roller bearing
Zylinderschaft *m* <Maschinen> parallel shank, plain shank, straight shank
Zylinderschloss *n* 1. <Bau> cylinder lock, pin tumbler; 2. <Maschinen> cylinder lock
Zylinderschnecke *f* <Maschinen> cylindrical worm
Zylinderschraube *f* 1. <Fertig> cheese-head screw, fillister head; bolt, securing screw *(Kunststoffinstallationen)*; 2. <Maschinen> cheese-head screw
Zylinderschraube *m* **mit Schlitz** <Maschinen> *(BE)* slotted cheese-head screw
Zylinderstift *m* 1. <Fertig> parallel pin; 2. <Maschinen> cylindrical pin, straight pin
zylindersymmetrische Couetteströmung *f* <Strömphys> rotating Couette flow
zylindersymmetrische Couetteströmung *f* **im Ringspalt** <Strömphys> rotating Couette flow in an annulus
Zylindertrockenmaschine *f* <Textil> cylinder drying machine
Zylindertrockenschlichtmaschine *f* <Textil> cylinder-sizing machine
Zylindertrockner *m* <Textil> cylinder drying machine
Zylinderverfahren *n* <Ker & Glas> cylinder process
Zylindervolumen *n* <Kfztech> cylinder capacity
Zylinder-V-Winkel *m* <Kfztech> cylinder bank angling *(von zwei Zylinderreihen in V-Motor)*
Zylinderwand *f* 1. <Kfztech> cylinder wall *(Motor)*; 2. <Maschinen> cylinder wall
Zylinderwandung *f* <Kfztech> cylinder wall *(Motor)*
Zylinderwelle *f* <Akustik> cylindrical wave
Zylinderwicklung *f* <Elektriz, Elektrotech> concentric winding, cylindrical winding
Zylinderzange *f* <Ker & Glas> straight pincers
Zylinderziehverfahren *n* <Ker & Glas> cylinder drawing process
zylindrisch *adj* 1. <Fertig> parallel *(Schaft)*; 2. <Geom, Maschinen> cylindrical
zylindrische Bohrung *f* 1. <Fertig> parallel hole; 2. <Maschinen> cylindrical bore
zylindrische Druckform *f* <Druck> plate cylinder
zylindrische Hülle *f* <Raumfahrt> cylindrical shell *(Raumschiff)*
zylindrische Kreuzspule *f* <Textil> cheese
zylindrische Schraubenfeder *f* <Maschinen> cylindrical helical spring
zylindrische Spaltzone *f* <Kerntech> annular core
zylindrische Strahlungsquelle *f* <Kerntech> cylindrical irradiator
zylindrischer Flammrohrkessel *m* <Hydraul> cylindrical flue boiler
zylindrischer Rotationskörper *m* <Geom> cylindrical solid of revolution
zylindrisches Ausgleichsventil *n* <Nichtfoss Energ> cylindrical balanced valve
zylindrisches Mundstück *n* <Hydraul> cylindrical mouthpiece
zylindrisches Rohr-Innengewinde *n* <Fertig> BSP parallel thread *(Kunststoffinstallationen)*
zylindrisches System *n* <Abfall> drum system *(Kompostierungsverfahren)*
zylindrisches Wellenende *n* <Maschinen> cylindrical shaft end
Zymase *f* <Chemie> zymase
Zymologie *f* <Chemie> zymology
zymotisch *adj* <Chemie> zymotic

Anhänge/
Appendices

Abkürzungen / Abbreviations

α 1. <Akustik, Funktech, Phys, Strahlphys> *(Absorptionskoeffizient)* α *(absorption coefficient)*; 2. <Geom> *(Alpha)* α *(Alpha)*; 3. <Mechan> *(Winkelbeschleunigung)* α *(angular acceleration)*; 4. <Optik> *(Absorptionsfaktor)* α *(absorption factor)*; 5. <Optik> *(optischer Drehwinkel)* α *(angle of optical rotation)*

β *(Phasenkonstante)* <Akustik, Elektriz> β *(phase constant)*

ε 1. <Hydraul> *(kinematische Wirbelzähigkeit)* ε *(kinematic eddy viscosity)*; 2. <Phys> *(durchschnittliche kinetische Molekularenergie)* ε *(average molecular kinetic energy)*

μ 1. <Elektriz> *(Permeabilität)* μ *(permeability)*; 2. <Funktech> *(Verstärkung)* μ *(amplification factor)*; 3. <Thermod> μ

μ H *(Hall'sche Mobilität)* <Funktech> μ H *(Hall mobility)*

μ V *(Mikrovolt)* <Elektriz, Elektrotech> μ V *(microvolt)*

Ω *(Volumen in Phase)* <Phys> Ω *(volume in phase space)*

Φ 1. <Akustik> *(Winkelverdrängung)* Φ *(angular displacement)*; 2. <Akustik> *(Geschwindigkeitspotenzial)* Φ *(velocity potential)*

σ 1. <Bau> σ; 2. <Kohlen, Kunststoff, Mechan> σ; 3. <Phys, Thermod> σ

τ *(Relaxationszeit)* <Akustik> τ *(relaxation time)*

θ 1. <Hydraul> *(Wärmewiderstand, absolute Temperatur, thermodynamische Temperatur)* θ *(absolute temperature)*; 2. <Phys> *(Wärmewiderstand)* θ *(thermal resistance)*; 3. <Thermod> *(absolute Temperatur, thermodynamische Temperatur)* θ *(absolute temperature)*; 4. <Thermod> *(Wärmewiderstand)* θ *(thermal resistance)*

θ D *(Debye'sche Temperatur)* <Phys, Thermod> θ D *(Debye temperature)*

θ K *(Einstein'sche Temperatur)* <Thermod> θ K *(Einstein temperature)*

3-D *(dreidimensional)* <Maschinen, Phys> 3-D *(three-dimensional)*

a 1. <Akustik> *(akustische Absorption)* a *(total acoustic absorption)*; 2. <Metrol> *(Ar)* a *(are)*

A 1. <Akustik, Aufnahme, Comp & DV, Elektriz, Elektronik, Funktech, Phys, Wassertrans, Wellphys> *(Amplitude)* A *(amplitude)*; 2. <Chemie> *(Affinität)* A *(affinity)*; 3. <Elektriz, Elektrotech, Fertig, Funktech, Metrol, Phys> *(Ampere)* A *(ampere)*; 4. <Elektriz> *(lineare Stromdichte)* A *(linear current density)*; 5. <Elektrotech, Fernseh, Funktech, Phys, Wassertrans> *(Anode)* A *(anode)*; 6. <Kerntech> *(Aktivität)* A *(activity)*; 7. <Kerntech> *(Massenzahl)* A *(mass number)*; 8. <Phys> *(Aktivität, Schallstärke)* A *(activity)*; 9. <Phys> *(Isotopenmasse, Massenzahl)* A *(mass number)*; 10. <Teilphys> *(Massenzahl, Nukleonenzahl)* A *(mass number)*

Å *(Angström)* <Metrol> Å *(angstrom)*

a0 *(Bohr'scher Radius)* <Phys> a0 *(Bohr radius)*

AACS *(Luftfahrtfunkdienst)* <Raumfahrt> AACS *(airways and air communications service)*

AADT *(durchschnittliches Tagesverkehrsaufkommen pro Jahr)* <Trans> AADT *(annual average daily traffic)*

AAP *(akustischer Akzeptanzpegel)* <Akustik> ACI *(acoustic comfort index)*

AB *(akustischer Blindleitwert)* <Akustik> BA *(acoustic susceptance)*

ABC *(automatische Helligkeitsregelung)* <Fernseh> ABC *(automatic brightness control)*

AbfG *(Abfallgesetz)* <Abfall> Waste Avoidance and Management Act, Waste Disposal Act

ABS 1. <Kfztech> *(Antiblockiersystem)* ABS, antiblocking system, antilock braking system; 2. <Kunststoff> *(Acrylnitril-Butadien-Styrol)* ABS, acrylonitrile butadiene styrene *(Copolymer)*

AC 1. <Elektrotech> *(Wechselstrom)* AC *(alternating current)*; 2. <Fertig> *(Adaptivsteuerung)* AC *(adaptive control)*

ACC *(automatische Chrominanzregelung)* <Fernseh> ACC *(automatic chrominance control)*

AC-GS *(Wechselstrom-Gleichstrom)* <Elektrotech> AC-DC

ACIA *(Asynchron-Übertragungs-Schnittstellenanpasser, asynchronischer Übertragungs-Schnittstellenanpasser)* <Kontroll> ACIA *(asynchronous communications interface adaptor)*

ACN *(automatische Himmelsnavigation)* <Raumfahrt> ACN *(automatic celestial navigation)*

ACNA *(Analogrechner für Netzabgleich)* <Comp & DV> ACNA *(analog computer for net adjustment)*

ACO *(Anpassungssteuerung mit Optimierung)* <Labor> ACO *(adaptive control optimization)*

ACR *(Anflugradar)* <Raumfahrt> ACR *(approach control radar)*

ACSR *(Einseitenband mit kompandierter Amplitude)* <Funktech> ACSS *(amplitude-compandered single sideband)*

ACU *(automatische Anrufeinheit, automatisches Rufgerät)* <Raumfahrt, Telekom> ACU *(automatic calling unit)*

A/D *(Analog-Digital-...)* <Elektronik, Fernseh, Fertig> A/D *(analog-digital)*

ADF *(Radiokompass, Funkkompass, automatischer Funkpeiler)* <Funkort, Lufttrans, Telekom> ADF *(automatic direction finder)*

ADI *(duldbare tägliche Aufnahmemenge)* <Lebensmittel> ADI *(acceptable daily intake)*

ADPCM *(adaptive Differenz-Pulscodemodulation)* <Telekom> ADPCM *(adaptive differential pulse code modulation)*

ADU *(Analog-Digital-Umsetzer)* <Comp & DV, Elektronik, Fernseh, Fertig> ADC *(analog-digital converter)*
ADV *(automatische Datenverarbeitung)* <Comp & DV> ADP *(automatic data processing)*
AE *(astronomische Einheit)* <Labor> AU *(astronomical unit)*
AEC *(Amerikanischer Atomenergieverband)* <Kerntech> AEC *(Atomic Energy Commission)*
AES *(Auger'sche Elektronenspektroskopie)* <Phys, Strahlphys> AES *(Auger electron spectroscopy)*
AFGC *(automatische Frequenz- und Verstärkungsregelung)* <Elektronik, Fernseh, Funktech> AFGC *(automatic frequency and gain control)*
AFI *(automatische Fahrzeugidentifikation)* <Trans> AVI *(automatic vehicle identification)*
AFO *(automatische Fahrzeugortung)* <Trans> AVL *(automatic vehicle location)*
AFR *(automatische Frequenzregelung)* <Elektronik, Fernseh, Funktech> AFC *(automatic frequency control)*
AFS *(fester Flugfunkdienst)* <Lufttrans, Telekom> AFS *(aeronautical-fixed service)*
AFT *(automatische Scharfabstimmung)* <Funktech> AFT *(automatic fine tuning)*
AFTN *(festes Flugfunknetz)* <Lufttrans, Telekom> AFTN *(aeronautical-fixed telecommunication network)*
AG *(amerikanisches Maß)* <Maschinen> AG *(American gage)*
AGCA *(automatische Anflugsteuerung vom Boden)* <Raumfahrt> AGCA *(automatic ground-controlled approach)*
AGCL *(automatische Landesteuerung vom Boden)* <Raumfahrt> AGCL *(automatic ground-controlled landing)*
AGE *(Allylglycidether)* <Kunststoff> AGE *(allyl glycidyl ether)*
AGR *(fortgeschrittener Gas-Graphit-Reaktor)* <Kerntech> AGR *(advanced gas-cooled reactor)*
AI *(Schallimpedanz, Schallwellenwiderstand, akustische Impedanz, akustischer Scheinwiderstand)* <Akustik> ZA *(acoustic impedance)*
AIA *(Amerikanischer Luft- und Raumfahrtverband)* <Raumfahrt> AIA *(Aerospace Industries Association)*
AIS *(aeronautischer Informationsdienst)* <Raumfahrt> AIS *(aeronautical information service)*
AK *(akustische Kapazität)* <Akustik> AC *(acoustic capacitance)*
AL 1. <Akustik> *(akustischer Leitwert)* GA *(acoustic conductance)*; 2. <Telekom> *(Anschlussleitung)* subscriber's line
ALC 1. <Funktech> *(automatische Pegelregelung)* ALC *(automatic level control)*; 2. <Kfztech> *(automatischer Niveauausgleich)* ALC *(automatic level control)*
ALR *(automatische Lautstärkeregelung)* <Funktech> AVC *(automatic volume control)*
AM 1. <Akustik> *(akustische Masse)* AM *(acoustic mass)*; 2. <Aufnahme, Comp & DV, Elektriz, Elektronik, Fernseh, Funktech, Phys, Telekom, Wellphys> *(Amplitudenmodulation)* AM *(amplitude modulation)*
AME *(Atommasseneinheit)* <Kerntech> AWU *(atomic weight unit)*
AMI *(bipolare Schrittinversion)* <Telekom> AMI *(alternate mark inversion)*
AMS *(aeronautische Werkstoffnorm)* <Raumfahrt> AMS *(aeronautical material standard)*
amu *(atomare Masseneinheit)* <Kerntech> amu *(atomic mass unit)*
ANC *(Luftfahrt-Navigationsausschuss)* <Lufttrans> ICAO *(International Civil Aviation Organization)*
AOCS *(Fluglage- und Umlaufbahnkontrollsystem)* <Raumfahrt> AOCS *(attitude and orbit control system)*
AOQ *(durchschnittliche Fertigproduktqualität)* <Qual> AOQ *(average outgoing quality)*
AOQL *(durchschnittlicher Fertigproduktqualitätsgrenzwert)* <Qual> AOQL *(average outgoing quality limit)*
AOS *(automatisches Signal zur Mikrofonübergabe)* <Telekom> AOS *(automatic over signal)*
AOW *(akustische Oberflächenwelle)* <Elektronik, Telekom> SAW *(surface acoustic wave)*
APD *(Avalanchephotodiode)* <Elektronik, Optik> APD *(avalanche photodiode)*
API *(Amerikanisches Erdölinstitut)* <Erdöl> API *(American Petroleum Institute)*
APR 1. <Elektronik> *(automatische Phasenregelung)* APC *(automatic phase control)*; 2. <Fernseh> *(automatische Phasensteuerung)* APC *(automatic phase control)*
APT *(programmierte Werkzeuge)* <Comp & DV> APT *(automatically programmed tools)*
AQL *(akzeptabler Qualitätspegel, annehmbare Qualitätsgrenzlage, annehmbare Qualitätslage)* <Qual> AQL *(acceptable quality level)*
AR *(Ausgangsregister)* <Mobilkom> HLR *(home location register)*
ARGOS *(automatische Satellitenerfassung von geomagnetischen Daten)* <Wassertrans> ARGOS *(Automatic Remote Geomagnetic Observatory System)*
ARL *(akzeptabler Zuverlässigkeitspegel)* <Qual> ARL *(acceptable reliability level)*
ARPA *(automatische Radaraufnahmehilfe)* <Wassertrans> ARPA *(automatic radar plotting aid)*
ARQ *(automatische Wiederholanforderung)* <Funktech, Telekom> ARQ *(automatic repeat request)*
ARRL *(Amerikanischer Amateurdachverband)* <Funktech> ARRL *(American Radio Relay League)*
ARSR *(Flugüberwachungsradar)* <Funkort, Raumfahrt> ARSR *(air route surveillance radar)*
ARU 1. <Comp & DV> *(Sprachausgabe-Einheit)* ARU *(audio response unit)*; 2. <Funktech> *(automatische Rauschunterdrückung)* ANL *(automatic noise limiter)*
asb *(Apostilb)* <Optik> asb *(apostilb)*

ASCII *(Amerikanische Datenübertragungs-Codenorm)* <Comp & DV, Druck, Telekom> ASCII *(American Standard Code for Information Interchange)*
ASD *(Rutschsicherung)* <Kfztech> ASD *(antiskid device)*
ase *(Flugnormwirkungsgrad)* <Raumfahrt> ase *(air standard efficiency)*
ASE 1. <Künstl Int> *(automatische Spracherkennung)* ASR *(automatic speech recognition)*; 2. <Trans> *(Selbststabilisierungsgerät)* ASE *(automatic stabilizing equipment)*
ASG *(Flugnormengruppe)* <Raumfahrt> ASG *(aeronautical standards group)*
ASI 1. <Lufttrans> *(Eigengeschwindigkeitsanzeiger, Geschwindigkeitsmesser)* ASI *(airspeed indicator)*; 2. <Raumfahrt> *(Geschwindigkeitsanzeiger)* ASI *(airspeed indicator)*
ASK *(Amplitudenumtastung)* <Elektronik, Telekom> ASK *(amplitude-shift keying)*
ASL *(atomare Sicherheitslinie)* <Kerntech> ASL *(atomic safety line)*
ASLT *(fortschrittliche Festkörperlogik)* <Elektronik> ASLT *(advanced solid logic technology)*
ASME *(Amerikanische Gesellschaft der Maschinenbau-Ingenieure)* <Qual> ASME *(American Society of Mechanical Engineers)*
ASR 1. <Comp & DV> *(automatischer Sender-Empfänger)* ASR *(automatic send-receive)*; 2. <Comp & DV> *(automatisches Senden und Empfangen)* ASR *(automatic send and receive)*; 2. <Lufttrans> *(Flughafen-Überwachungsradar)* ASR *(airport surveillance radar)*
ASTM *(Amerikanische Gesellschaft für Werkstoffprüfung)* <Maschinen, Qual> ASTM *(American Society for Testing Materials)*
aT *(Thermodiffusionskonstante)* <Phys> aT *(thermal diffusion constant)*
AT *(fortschrittliche Technologie)* <Comp & DV> AT *(advanced technology)*
ATC 1. <Eisenbahn> *(automatische Zugsteuerung)* ATC, ATO *(automatic train operation)*; 2. <Funktech> *(kapazitive Antennenanpassung)* ATC *(aerial-tuning capacitor)*; 3. <Metall> *(automatischer Werkzeugwechsler)* ATC *(automatic tool changer)*
ATE *(automatische Prüfeinrichtung)* <Comp & DV> ATE *(automatic test equipment)*
ATF *(Automatikgetriebeöl)* <Kfztech> ATF *(automatic transmission fluid)*
ATI *(Antennenabstimmspule)* <Funktech> ATI *(aerial-tuning inductance)*
ATM 1. <Optik> *(Azimutal-Transversal-Mode)* ATM, azimuthal transversal mode *(optische Fasern)*; 2. <Telekom> *(asynchroner Transfermodus)* ATM, asynchronous transfer mode
ATP *(Adenosintriphosphat)* <Lebensmittel> ATP *(adenosine triphosphate)*
ATU *(Antennenanpassung)* <Funktech> ATU *(aerial-tuning unit)*
ATV *(Amateurfernsehen)* <Fernseh> ATV *(amateur television)*

AUTOPROMT *(automatisierte Maschinenwerkzeugprogrammierung)* <Fertig> AUTOPROMT *(automated programming of machine tools)*
AUTOSPOT *(automatisierte Werkzeugpositionierung)* <Fertig> AUTOSPOT *(automated system for positioning tools)*
AV *(Säurewert)* <Chemie> AV *(acid value)*
AVR *(automatische Verstärkungsregelung)* <Elektronik, Funktech, Telekom> AGC *(automatic gain control)*
AWACS *(Überwachungs- und Leitsystem im Flugzeug)* <Lufttrans> AWACS *(airborne warning and control system)*
AWG *(Amerikanische Einheit für Drahtdurchmesser)* <Metrol> AWG *(American wire gage)*
AWS *(Automatisches Warnsystem)* <Eisenbahn> AWS, Automatic Warning System *(induktive Zugbeeinflussung)*
b 1. <Kohlen, Labor> *(Bar)* b, bar *(Luftdruckeinheit)*; 2. <Raumfahrt> *(Raumbreite)* b *(galactic latitude)*
B 1. <Akustik, Elektrotech, Phys> *(Bel)* B, bel; 2. <Aufnahme> *(magnetischer Scheinwiderstand)* B *(magnetic induction)*; 3. <Chemie> *(Bor)* B *(boron)*; 4. <Elektriz, Elektrotech, Phys, Telekom> *(Magnetinduktion)* B *(magnetic induction)*; 5. <Kerntech> *(Bindungsenergie)* B *(binding energy)*; 6. <Strahlphys, Teilphys> *(Bindungsenergie, Kernbindungsenergie)* B *(binding energy)*; 7. <Thermod> *(Volumenelastizitätsmodul)* B *(modulus of volume elasticity)*
BB *(Basisband)* <Elektrotech> BB *(baseband)*
BBD 1. <Elektrotech> *(Eimerkettenspeicher)* BBD *(bucket brigade device)*; 2. <Telekom> *(Eimerkettenschaltung)* BBD *(bucket brigade device)*
BBL *(Blindlandung mit Bakenunterstützung)* <Raumfahrt> BBL *(beacons and blind landing)*
BBV *(Breitbandverstärker)* <Elektronik> wideband amplifier
BCC *(Blockprüfzeichen)* <Telekom> BCC *(block check character)*
BCD 1. <Comp & DV> *(binärcodierte Dezimalzahl, binärcodierte Drehzahl)* BCD *(binary-coded decimal)*; 2. <Funktech> *(binärcodierte Dezimalzahl)* BCD *(binary-coded decimal)*
BCI *(Rundfunkstörung)* <Telekom> BCI *(broadcast interference)*
BCS *(Bardeen-Cooper-Schrieffer)* <Comp & DV> BCS *(Bardeen-Cooper-Schrieffer)*
B/d *(Barrel pro Tag)* <Erdöl> BCD *(barrels per calendar day)*
BDP *(Verbundpapier)* <Papier> BDP *(bonded double paper)*
BE 1. <Elektriz> *(elektrischer Blindleitwert)* BE *(electric susceptance)*; 2. <Telekom> *(Basiseinheit)* BU *(base unit)*
BEG *(Bodeneffektgerät)* <Kfztech> GEM *(ground effect machine)*
BEPC *(Beijing Electron Positron Collider)* <Teilphys> BEPC *(Beijing Electron Positron Collider)*
BER *(Bitfehlerquote, Bitfehlrate)* <Comp & DV, Telekom> BER *(bit error rate)*

BfA

BfA m *(Betriebsbeauftragter für Abfall)* <Abfall> Waste Management Officer *(Berufsbezeichnung)*

BFO *(Schwebungsfrequenzoszillator)* <Elektronik, Funktech, Phys> BFO *(beat frequency oscillator)*

Bg *(geometrisches Buckling)* <Kerntech> Bg *(geometric buckling)*

BHA *(Butylhydroxyanisol)* <Lebensmittel> BHA *(butylated hydroxyanisole)*

BHT *(Butylhydroxytoluol)* <Lebensmittel> BHT *(butylated hydroxytoluene)*

BIGFET *(Feldeffekttransistor mit bipolarisoliertem Gatter)* <Elektronik> BIGFET *(bipolar-insulated gate field-effect transistor)*

BMOSFET *(Halbleiter-Feldeffekttransistor mit Rückgatter)* <Elektronik> BMOSFET *(back-gate metal-oxide semiconductor field-effect transistor)*

BOP *(Blowout-Preventer)* <Erdöl> BOP *(blowout preventer)*

BOT *(Bandanfang)* <Comp & DV> BOT *(beginning of tape)*

BPF *(Bandpassfilter)* <Aufnahme, Elektronik, Fernseh, Funktech, Phys, Telekom> BPF *(band-pass filter)*

bpi *(Bits pro Zoll)* <Comp & DV> bpi *(bits per inch)*

BPS *(Bremspferdestärke)* <Eisenbahn, Fertig, Kfztech, Mechan> BHP *(brake horsepower)*

bps *(Bits pro Sekunde)* <Comp & DV> bps *(bits per second)*

BPSK *(Zweiphasenumtastung, binäre Phasenumtastung)* <Telekom> BPSK *(binary phase shift keying)*

BRZ *(Bruttoraumzahl)* <Wassertrans> GT, gross tonnage *(Raummaß, ersetzt die Angabe in BRT)*

BS 1. <Fertig> *(Windfrischstahl)* BS *(Bessemer steel)*; 2. <Maschinen> *(Bessemerstahl)* BS *(Bessemer steel)*

BSA *(Bohrlochsohlenausrüstung)* <Erdöl> BHA *(bottom hole assembly)*

BSB *(biologischer Sauerstoffbedarf)* <Abfall, Lebensmittel, Umweltschmutz> BOD *(biological oxygen demand)*

BSS *(Britische Normenspezifikation, Britische Normvorschrift)* <Maschinen> BSS *(British Standard Specification)*

BThU *(Britische Wärmeeinheit)* <Maschinen, Metrol> BThU, British Thermal unit *(Energie)*

BTU *(Britische Wärmeeinheit)* 1. <Labor> BThU, British Thermal unit; 2. <Maschinen> *(AE)* BTU, British Thermal unit

Btx *(Bildschirmtext)* 1. <Fernseh> Videotex®; 2. <Telekom> Teletext®

BV *(Bildverstehen, Bildverständnis)* <Künstl Int> IU *(image understanding)*

BW *(Bandbreite, Bandweite)* <Aufnahme, Comp & DV, Elektronik, Fernseh, Funktech, Optik, Telekom> BW *(bandwidth)*

BZ *(Betriebszentrum)* <Telekom> *(AE)* operations center, *(BE)* operations centre

c 1. <Elektronik, Elektriz, Kohlen, Kunststoff, Telekom, Umweltschmutz> *(Konzentration)* c *(concentration)*; 2. <Hydraul> *(Wellenausbreitungsgeschwindigkeit)* c *(wave celerity)*; 3. <Metrol> *(Zenti...)* c *(centi)*; 4. <Metrol> *(Lichtgeschwindigkeit)* c *(velocity of light)*; 5. <Optik> *(Lichtgeschwindigkeit)* c *(speed of light in empty space)*; 6. <Phys> *(spezifische Wärme)* c *(specific heat capacity)*; 7. <Phys> *(Schallgeschwindigkeit)* c *(speed of sound)*; 8. <Thermod> *(spezifische Wärmekapazität)* c *(specific heat capacity)*

C 1. <Bau, Elektriz, Elektrotech, Erdöl, Heiz & Kälte> *(Kapazität)* C *(capacity)*; 2. <Chemie> *(Kohlenstoff)* C *(carbon)*; 3. <Elektriz, Elektrotech, Metrol, Phys> *(Coulomb)* C *(coulomb)*; 4. <Hydraul> *(Cauchy'sche Zahl)* C *(Cauchy coefficient)*; 5. <Hydraul> *(Chezy-Koeffizient)* C *(Chezy coefficient)*; 6. <Hydraul> *(Ausflusskoeffizient, Durchflusskoeffizient)* C *(discharge coefficient)*; 7. <Labor> *(Celsius)* C *(centigrade)*; 8. <Nichtfoss Energ> *(Schüttkoeffizient)* C *(discharge coefficient)*; 9. <Funktech, Phys, Telekom> *(Kapazität)* C *(capacitance)*

CA 1. <Heiz & Kälte> *(kontrollierte Atmosphäre)* CA *(controlled atmosphere)*; 2. <Kunststoff, Textil> *(Celluloseacetat)* CA *(cellulose acetate)*

CAD *(computergestützte Konstruktion, computergestützter Entwurf)* <Comp & DV, Elektriz, Kontroll, Mechan, Telekom, Trans> CAD *(computer-aided design)*

CADCAM *(computergestützte Konstruktion und Fertigung)* <Comp & DV> CADCAM *(computer-aided design and manufacturing)*

CAL 1. <Comp & DV> *(computergestützter Unterricht)* CAI *(computer-aided instruction)*; 2. <Comp & DV> *(computergestütztes Lernen)* CAL *(computer-aided learning)*

CAM 1. <Comp & DV, Elektriz> *(computergestützte Fertigung, computergestützte Produktion)* CAM *(computer-aided manufacturing)*; 2. <Comp & DV, Künstl Int> *(Assoziativspeicher, inhaltsadressierbarer Speicher)* CAM *(content-addressable memory)*

CAP *(computergestütztes Publizieren)* <Druck> CAP *(computer-aided publishing)*

CAPI CAPI, common application programming interface *(Software-Schnittstelle zwischen ISDN und PC)*

CAR *(Zivilflugvorschriften)* <Raumfahrt> CAR *(Civil Air Regulations)*

CASE *(computergestützte Softwareentwicklung)* <Comp & DV> CASE *(computer-aided software engineering)*

CAT 1. <Comp & DV> *(computerunterstützte Übersetzung)* CAT *(computer-assisted translation)*; 2. <Elektronik> *(Senderöhre mit gekühlter Anode)* CAT *(cooled-anode transmitting valve)*; 3. <Lufttrans> *(Kaltluftturbulenzen)* CAT *(cold air turbulence)*

CATV *(Fernsehen über Gemeinschaftsantenne)* <Fernseh> CATV *(community antenna television system)*

CATVI 1. <Fernseh> *(Kabelfernsehstörung)* CATVI *(cable television interference)*; 2. <Fernseh> *(störende Beeinflussung des Kabelfernsehdienstes) CATVI *(cable television interference)*

CAV *(computerunterstütztes Sehen)* <Künstl Int> CAV *(computer-aided vision)*

CCD *(Ladungsgekoppeltes Bauelement, CCD-Element)* <Elektronik, Fernseh, Phys, Telekom> CCD *(charge-coupled device)*

CCITT *(Internationaler Berateraussschuss für den Fernschreib- und Telefondienst)* <Telekom> CCITT *(International Telegraph and Telephone Consultative Committee)*

CCTV *(angewandtes Fernsehen, industrielles Fernsehen)* <Fernseh> CCTV *(closed-circuit television)*

cd *(Candela)* <Elektrotech, Metrol, Optik, Phys> cd *(candela)*

CD 1. <Comp & DV, Elektronik, Telekom> *(Trägerdetektion, Trägererkennung)* CD *(carrier detection)*; 2. <Comp & DV, Telekom> *(Kollisionserkennung)* CD *(collision detection)*; 3. <Comp & DV, Optik> *(Compact-Disk)* CD *(compact disk)*

CD-i *(beschreibbare CD)* <Optik> CD-I *(compact disk-interactive)*

CDM *(kompandierte Deltamodulation)* <Telekom> CDM *(companded delta modulation)*

CD-ROM 1. <Comp & DV> *(Compact-Disk ohne Schreibmöglichkeit)* CD-ROM *(compact disk read-only memory)*; 2. <Optik> *(Compact-Disk-Speicher ohne Schreibmöglichkeit)* CD-ROM *(compact disk read-only memory)*

CE *(elektrische Kapazität)* <Akustik> CE *(electric capacitance)*

CEBAF *(Gleichstromelektronenbeschleuniger)* <Teilphys> CEBAF *(continuous electron beam facility)*

CEN *(Comité Européen de Normalisation)* <Qual> CEN *(Europäischer Normungsausschuss)*

CERN *(Europäisches Kernforschungszentrum)* <Teilphys> CERN *(European Organization for Nuclear Research)*

CGA *(Farbgrafikadapter)* <Comp & DV> CGA *(colour graphics adaptor)*

Ci *(Curie)* <Phys, Strahlphys> Ci *(curie)*

CIM 1. <Comp & DV> *(CompuServe® Information Manager)* CIM *(CompuServe® Information manager)*; 2. <Comp & DV> *(computerintegrierte Fertigung)* CIM *(computer-integrated manufacture)*

cim *(Kubikzoll pro Minute)* <Labor> cim *(cubic inches per minute)*

CISC *(Prozessor mit komplettem Befehlssatz, konventioneller Rechner)* <Comp & DV> CISC *(complex instruction set computer)*

cl *(Geschwindigkeit von Längswellen)* <Akustik> cl *(velocity of longitudinal waves)*

CL 1. <Comp & DV> *(Befehlssprache, Betriebssprache)* CL *(command language)*; 2. <Hydraul, Nichtfoss Energ, Phys, Wassertrans> *(Auftriebsbeiwert, Auftriebszahl)* CL *(lift coefficient)*; 3. <Lufttrans> *(Auftriebszahl)* CL *(lift coefficient)*

CM *(mechanische Auslenkung)* <Akustik> CM *(mechanical compliance)*

CMC *(Carboxymethylcellulose)* <Kunststoff, Lebensmittel> CMC *(carboxymethylcellulose)*

CMOS *(Komplementär-Metalloxid-Halbleiter)* <Elektronik> CMOS *(complementary metal oxide semiconductor)*

CNC *(computernumerische Steuerung)* <Maschinen> CNC *(computerized numeric control)*

coax *(koaxial)* <Funktech> coax *(coaxial)*

COBOL *(problemorientierte Programmiersprache für Geschäftsbetrieb)* <Comp & DV> COBOL *(common business oriented language)*

COR *(Druckausgabeverkleinerung)* <Comp & DV> COR *(character output reduction)*

COSMOS *(Komplementär-Symmetrischer Metalloxid-Halbleiter)* <Elektronik> COSMOS *(complementary-symmetrical metal oxide semiconductor)*

cot *(Kotangens)* <Geom> cot *(cotangent)*

Cp *(Wärmekapazität bei konstantem Druck)* <Labor> Cp *(heat capacity at constant pressure)*

CPFSK *(phasenkontinuierliche Frequenzumtastung)* <Elektronik, Funktech, Telekom> CPFSK *(continuous phase frequency shift keying)*

CPK *(Chloroprenkautschuk)* <Kunststoff> CR *(chloroprene rubber)*

CPM *(Methode des kritischen Weges)* <Comp & DV> CPM *(critical path method)*

CPU *(Zentraleinheit, zentrale Rechnereinheit)* <Comp & DV, Telekom> CPU *(central processing unit)*

CR *(Rotationsauslenkung)* <Akustik> CR *(rotational compliance)*

CRC *(zyklische Blockprüfung, zyklische Blocksicherung, CRC-Prüfung)* <Comp & DV, Elektronik, Labor, Telekom> CRC *(cyclic redundancy check)*

CRCA *(kaltgewalzt und ausgeglüht)* <Metall> CRCA *(cold-rolled and annealed)*

CS *(Durchschaltevermittlung)* <Comp & DV, Telekom> CS *(circuit switching)*

CSB *(chemischer Sauerstoffbedarf)* <Umweltschmutz> COD *(chemical oxygen demand)*

CSI *(Chlorschwefelisocyanat)* <Umweltschmutz> CSI *(chlorosulfonyl isocyanate)*

CSM *(Kommando- und Servicemodul)* <Raumfahrt> CSM, command and service module *(Raumschiff)*

CSMA *(Mehrfachzugriff mit Trägerkennung)* <Comp & DV, Telekom> CSMA *(carrier sense multiple access)*

CSMA/CD *(CSMA/CD-Verfahren)* <Comp & DV, Telekom> CSMA/CD *(carrier sense multiple access with collision detection)*

CSN *(Durchschalte-Vermittlungsnetz)* <Comp & DV, Telekom> CSN *(circuit-switched network)*

CSPDN *(leitungsvermitteltes öffentliches Datennetz)* <Telekom> CSPDN *(circuit-switched public data network)*

ct *(Geschwindigkeit von Transversalwellen)* <Labor> ct *(velocity of transversal waves)*

CTA *(Cellulosetriacetat)* <Kunststoff> CTA *(cellulose triacetate)*

CTCSS *(Hilfsträgergeräuschsperre)* <Funktech> CTCSS *(continuous tone-coded squelch system)*

CTD

CTD 1. <Elektrotech> *(Ladungsverschiebeschaltung)* CTD, charge transfer device *(Halbleiter)*; 2. <Phys> *(ladungsgekoppeltes Bauelement)* CTD, charge transfer device; 3. <Raumfahrt> *(Ladungsübertragungsgerät)* CTD, charge transfer device; 4. <Telekom> *(Ladungstransferelement)* CTD, charge transfer device
CTS *(Containerschiff)* <Wassertrans> CTS *(container ship)*
CUG *(geschlossene Benutzergruppe, geschlossener Benutzerkreis)* <Comp & DV, Telekom> CUG *(closed user group)*
CVD *(Gasphasenabscheidung)* <Elektronik, Telekom> CVD *(chemical vapour deposition)*
CVS *(Teilstromentnahme nach Verdünnung)* <Umweltschmutz> CVS *(constant volume sampling)*
CW *(Dauerstrich, ungedämpfte Welle)* <Aufnahme, Elektronik, Elektrotech, Funktech, Telekom> CW *(continuous wave)*
d 1. <Chemie, Phys, Teilphys> *(Deuteron)* d *(deuteron)*; 2. <Labor> *(Dezi)* d *(deci…)*
D 1. <Akustik> *(Schwärzung)* D *(optical density)*; 2. <Chemie> *(Deuterium)* D *(deuterium)*; 3. <Elektriz> *(Verschiebung)* D *(displacement)*; 4. <Elektronik, Funktech, Phys> *(Diffusionskoeffizient)* D *(diffusion coefficient)*; 5. <Fertig, Geom, Maschinen> *(Durchmesser)* D *(diameter)*; 6. <Fertig, Phys> *(Versetzung)* D *(displacement)*; 7. <Kerntech> *(Absorptionsdosis)* D *(absorbed dose)*; 8. <Optik> *(optische Dichte)* D *(optical density)*; 9. <Strahlphys> *(absorbierte Dosis)* D *(absorbed dose)*; 10. <Thermod> *(vierter Virialkoeffizient)* D *(fourth virial coefficient)*
da *(Deka…)* <Labor> da *(deca…)*
D/A *(Digital-Analog-…)* <Aufnahme, Comp & DV, Elektronik, Labor, Telekom> D/A *(digital-analog)*
DA *(direkter Zugriff)* <Comp & DV> DA *(direct access)*
DAMA *(bedarfsgesteuerter Vielfachzugriff)* <Telekom> DAMA *(demand-assigned multiple access)*
DAT *(Digital-Audio-Tape)* <Aufnahme> DAT *(digital audio tape)*
DAU *(Digital-Analog-Umsetzer)* <Comp & DV, Elektronik, Telekom> DAC *(digital-analog converter)*
dB *(Dezibel)* <Akustik, Aufnahme, Elektronik, Funktech, Phys, Strahlphys, Umweltschmutz> dB *(decibel)*
DBA *(Datenbankadministrator, Datenbankverwalter)* <Comp & DV> DBA *(database administrator)*
dBi *(Dezibel über Isotropstrahler)* <Strahlphys> dBi *(decibels over isotropic)*
DCC *(Datenübermittlungskanal)* <Telekom> DCC *(data communication channel)*
DCD *(Datenträgerdetektor)* <Telekom> DCD *(data carrier detector)*
DCTL *(direkt gekoppelte Transistorlogik)* <Elektronik> DCTL *(direct-coupled transistor logic)*
DD 1. <Hydraul> *(Strömungswiderstand)* DD *(coefficient of drag)*; 2. <Nichtfoss Energ> *(Luftwiderstandsbeiwert, Widerstandsbeiwert)* DD *(coefficient of drag)*

DDE *(direkte Dateneingabe, direkter Dateneintrag)* <Comp & DV> DDE *(direct data entry)*
DDP *(Doppeldiodenpentode)* <Elektronik> DDP *(double diode pentode)*
DDT *(Dichlordiphenyltrichlorproäthan)* <Chemie> DDT *(dichlordiphenyltrichlorproethane)*
DEE *(Datenendeinrichtung)* <Comp & DV, Telekom> DTE *(data terminal equipment)*
DESY *(Deutsches Elektronensynchroton)* <Teilphys> DESY
DFT *(diskrete Fourier-Transformation)* <Elektronik> DFT *(discrete Fourier transform)*
DFV *(Datenfernverarbeitung)* <Comp & DV> TP *(teleprocessing)*
DGPS *(Differenzial-GPS)* <Wassertrans> DGPS, differential global positioning system *(Satellitennavigation)*
DGzRS f *(Deutsche Gesellschaft zur Rettung Schiffbrüchiger)* <Wassertrans> GLI *(German Lifeboat Institution)*
Di *(Richtwirkungsindex)* <Akustik> Di *(directivity index)*
Diac *(bidirektionale Triggerdiode)* 1. <Elektronik> diac, diode alternating-current switch *(Wechselstromdiodenschalter)*; 2. <Funktech> diac, diode alternating-current switch
DIN *(Deutsches Institut für Normung)* <Maschinen> DIN, German Standards Institution
DIVF *(digitale Vermittlungsstelle für den Fernverkehr)* <Telekom> digital trunk exchange
DIVO *(digitale Vermittlungsstelle für den Ortsverkehr)* <Telekom> digital local exchange
dl *(Deziliter)* <Labor> dl *(deciliter)*
DM *(Deltamodulation)* <Elektronik, Raumfahrt, Telekom> DM *(delta modulation)*
DMA *(Direkt-Speicherzugriff, Direktzugriffsspeicher)* <Comp & DV> DMA *(direct memory access)*
DMC *(kittartige Formmasse)* <Kunststoff> DMC *(dough-moulding compound)*
DOP *(Dioctylphthalat)* <Kunststoff> DOP *(dioctylphthalate)*
DP *(Diametral-Pitch)* <Maschinen> DP *(diametral pitch)*
DPCM *(Differenz-Pulscodemodulation)* <Elektronik, Telekom> DPCM *(differential pulse code modulation)*
DPDT *(zweipoliger Umschalter, zweipoliger Wechselschalter)* <Elektrotech> DPDT *(double-pole double-throw)*
DPSK *(Phasendifferenzmodulation, Phasendifferenzumtastung)* <Elektronik, Telekom> DPSK *(differential phase shift keying)*
DPST *(zweipoliger Ein/Aus-Schalter)* <Elektrotech> DPST *(double-pole single-throw)*
dpt *(Dioptrie)* <Optik> dpt *(dioptre)*
DRAM *(dynamischer RAM)* <Comp & DV> DRAM *(dynamic random access memory)*
DSB *(Zweiseitenband)* <Elektronik, Funktech> DSB *(double sideband)*
DSI *(digitale Sprachinterpolation)* <Raumfahrt, Telekom> DSI *(digital speech interpolation)*

DSV *(Digitalsignalverbindung)* <Telekom> digital connection
DT *(Dichtigkeitstest)* <Erdöl> LOT *(leak-off test)*
DTA *(Differenzialthermoanalyse)* <Kunststoff, Thermod, Umweltschmutz> DTA *(differential thermal analysis)*
DTL *(Dioden-Transistor-Logik)* <Elektronik> DTL *(diode transistor logic)*
DTP *(Desktop-Publishing)* <Comp & DV, Druck> DTP *(desktop publishing)*
DüE *(Datenübertragungseinrichtung)* 1. <Comp & DV> DCE *(data communication terminating equipment)*; 2. <Telekom> DCE *(data circuit terminating equipment)*
DV *(Datenverarbeitung)* <Comp & DV, Elektronik, Kontroll, Telekom> DP *(data processing)*
e *(Elektron)* <Elektriz, Elektrotech, Funktech, Phys, Teilphys> e *(electron)*
E 1. <Druck, Erdöl, Heiz & Kälte, Hydraul, Maschinen, Nichtfoss Energ, Phys, Thermod> *(Evaporation, Verdampfung)* E *(evaporation)*; 2. <Elektriz, Elektrotech, Phys> *(elektrische Feldstärke)* E *(electric field strength)*; 3. <Elektriz, Elektrotech, Kerntech, Mechan, Metrol, Phys, Thermod> *(Energie)* E *(energy)*; 4. <Elektrotech> *(elektrischer Feldvektor)* E *(electric field vector)*; 5. <Erdöl, Hydraul, Kohlen, Kunststoff, Metall, Phys> *(Young'scher Modul)* E *(Young's modulus)*; 6. <Optik> *(Energie)* power
E/A 1. <Comp & DV> *(Eingabe/Ausgabe)* I/O *(input/output)*; 2. <Elektriz> *(Eingabe/Ausgabe, Eingang/Ausgang)* I/O *(input/output)*
EAS *(äquivalente Fluggeschwindigkeit)* <Lufttrans> EAS *(equivalent airspeed)*
EC *(Ethylcellulose)* <Kunststoff> EC *(ethyl cellulose)*
ECL *(emittergekoppelte Logik)* <Comp & DV, Elektronik> ECL *(emitter-coupled logic)*
EDTV *(hochauflösendes Fernsehen)* <Fernseh> EDTV *(extended definition television)*
EDV *(elektronische Datenverarbeitung)* <Comp & DV, Elektriz, Elektronik, Kontroll> EDP *(electronic data processing)*
EEB *(elektroerosive Bearbeitung)* <Maschinen> EDM *(electro-discharge machining)*
EEPROM *(elektrisch löschbarer programmierbarer Lesespeicher)* <Elektronik> EEPROM *(electrically-erasable programmable read-only memory)*
EEROM *(elektronisch löschbarer Festwertspeicher, elektronisch löschbarer Lesespeicher)* <Comp & DV> EEROM *(electronically erasable read-only memory)*
EFS *(essenzielle Fettsäure)* <Lebensmittel> EFA *(essential fatty acid)*
EFuRD *(Europäischer Funkrufdienst)* <Telekom> European radio-paging system
EHF *(Millimeterwellen)* <Funktech> EHF *(extremely high frequency)*
E.h.t. *(Hochspannung, Höchstspannung)* <Fernseh> EHT *(extremely high tension)*

EIRP *(äquivalente isotrope Strahlungsleistung)* <Funktech, Raumfahrt> EIRP *(effective isotropically-radiated power)*
EL *(Elektrolumineszenz-Anzeige)* <Elektronik> EL *(electroluminescent display)*
ELED *(Kantenemitter-Lumineszenzdiode, kantenstrahlende Lumineszenzdiode)* <Telekom> ELED *(edge-emitting light-emitting diode)*
ELSBM *(ungeschützte Einzeltonnenvertäuung)* <Erdöl> ELSBM *(exposed location single buoy mooring)*
EMK *(elektromotorische Kraft)* <Bau, Eisenbahn, Elektriz, Elektrotech, Fernseh, Funktech, Phys> EMF *(electromotive force)*
EMS *(Expansionsspeicher-Spezifikation)* <Comp & DV> EMS *(expanded memory specification)*
EMV *(elektromagnetische Verträglichkeit)* <Elektriz, Funktech, Raumfahrt> EMC, electromagnetic compatibility *(Raumfahrt)*
EN *(Europäische Norm)* <Elektriz> European Standard
ENF *(extrem tiefe Frequenz)* <Funktech> ELF *(extremely low frequency)*
EOB *(Blockende)* <Comp & DV, Telekom> EOB *(end of block)*
EOD *(Datenende)* <Comp & DV, Telekom> EOD *(end of data)*
EOF *(Dateiende)* <Comp & DV, Telekom> EOF *(end of file)*
EOM *(Nachrichtenende)* <Comp & DV, Telekom> EOM *(end of message)*
EOT *(Bandende)* <Comp & DV> EOT *(end of tape)*
EP *(Höchstdruck)* <Maschinen> EP *(extreme pressure)*
EPIRB *(Seenot-Funkbake mit Positionsmeldung)* <Funktech, Telekom, Wassertrans> emergency position-indicating radio beacon *(Funk)*
EPM *(Äquivalent je Million)* <Umweltschmutz> EPM *(equivalent per million)*
EPNS *(versilberte Gegenstände)* <Metall> EPNS *(electroplated nickel silver)*
EPROM *(löschbarer programmierbarer Lesespeicher)* <Comp & DV> EPROM *(erasable programmable read-only memory)*
Erl *(Erlang)* <Telekom> Erl *(Erlang)*
ES 1. <Elektronik, Kerntech> *(Elektronenstrahl)* EB *(electronic beam)*; 2. <Künstl Int> *(Expertensystem)* ES *(expert system)*
ESA 1. <Raumfahrt> *(Europäische Raumfahrtbehörde)* ESA *(European Space Agency)*; 2. <Umweltschmutz> *(elektrostatischer Staubabscheider)* ESP *(electrostatic precipitator)*
ESCA *(Photoelektronenspektroskopie)* <Phys> ESCA *(electron spectroscopy for chemical analysis)*
ESR *(Elektronenspinresonanz)* <Phys, Strahlphys, Teilphys> ESR *(electron spin resonance)*
ETA *(voraussichtliche Ankunftszeit)* <Lufttrans, Wassertrans> ETA *(estimated time of arrival)*
ETD *(voraussichtliche Abflugzeit)* <Lufttrans, Wassertrans> ETD *(estimated time of departure)*

eV *(Elektronenvolt)* <Elektriz, Elektrotech, Phys, Strahlphys, Teilphys> eV *(electronvolt)*
EVA *(Ethylenvinylacetat)* <Kunststoff> EVA *(ethylene vinyl acetate)*
EVS *(Endvermittlungsstelle)* <Telekom> terminating exchange
EVSt *(Endvermittlungsstelle)* <Telekom> terminal exchange
EVU *(Elektrizitätsversorgungsunternehmen)* <Elektriz> electricity supply company
F 1. <Akustik, Aufnahme, Comp & DV, Elektronik, Funktech, Phys> *(Frequenz)* f *(frequency)*; 2. <Chemie> *(Fluor)* F *(fluorine)*; 3. <Elektriz, Elektrotech, Metrol, Phys> *(Farad)* F *(farad)*; 4. <Elektronik, Funktech> *(Rauschzahl)* F *(noise figure)*; 5. <Hydraul, Phys> *(Frouden'sche Zahl)* F *(Froude number)*; 6. <Kerntech> *(hyperfeine Quantenzahl)* F *(hyperfine quantum number)*; 7. <Metall, Phys> *(Kraft)* F *(force)*; 8. <Metall, Phys> *(freie Energie)* F *(free energy)*; 9. <Metrol> *(Fahrenheit)* F *(Fahrenheit)*; 10. <Metrol> *(Femto...)* f *(femto...)*
fA *(Antiresonanzfrequenz)* <Akustik, Elektronik> fA *(antiresonant frequency)*
FB *(Flughafenbake)* <Lufttrans, Raumfahrt> *(BE)* aerodrome beacon, *(AE)* airdrome beacon
FBAS-Signal n <Fernseh> *(AE)* composite color signal, *(BE)* composite colour signal
FCKW *(Fluorchlorokohlenwasserstoff)* <Chemie, Umweltschmutz> CFC *(chlorofluorocarbon)*
FCNE *(Flugüberwachungs- und Navigationsausrüstung)* <Lufttrans> FCNE *(flight control and navigational equipment)*
FdW *(Fahrt durchs Wasser)* <Wassertrans> speed through the water
F&E *(Forschung und Entwicklung)* <Chemtech, Comp & DV, Kfztech> R&D *(research & development)*
FEM *(Finite-Elemente-Methode)* <Maschinen> FEM *(finite elements method)*
FET *(Feldeffekttransistor)* <Comp & DV, Elektronik, Optik, Phys, Raumfahrt> FET *(field effect transistor)*
FFS *(flexibles Fertigungssystem)* <Künstl Int> FMS *(flexible manufacturing system)*
FFT *(schnelle Fourier-Transformation)* <Elektronik> FFT *(fast Fourier transform)*
FHV *(Fernleitungshauptverteiler)* <Telekom> TDF *(trunk distribution frame)*
FIFA *(Spaltstoffabbrand, Spaltstoffverbrauch)* <Kerntech> FIFA *(fissions per initial fissile atom)*
FIN *(Fahrzeug-Identifizierungsnummer)* <Kfztech> VIN *(Vehicle Identification Number)*
FIR *(finite Impulsantwort)* <Telekom> FIR *(finite impulse response)*
FKO *(Fließkommaoperation)* <Comp & DV> FLOP *(floating-point operation)*
FKP *(Fließkommaprozessor)* <Comp & DV> FPP *(floating-point processor)*
FM *(Frequenzmodulation)* <Comp & DV, Elektronik, Funktech, Phys, Telekom> FM *(frequency modulation)*

FMR *(Flüssigmetallreaktor)* <Kerntech> FSR *(flowable solids reactor)*
FPS *(schnelle Paketvermittlung)* <Telekom> FPS *(fast packet server)*
fR *(Resonanzfrequenz)* <Akustik, Elektronik, Telekom, Wellphys> fR *(resonant frequency)*
FS 1. <Comp & DV, Telekom> *(Fernschreiber)* TTY *(teletypewriter)*; 2. <Lufttrans> *(Flugsicherung)* ATC *(air traffic control)*; 3. <Maschinen> *(Festsitz)* force fit
FSK *(Frequenzumtastung)* <Comp & DV, Elektronik, Funktech, Telekom> FSK *(frequency shift keying)*
FSTV *(Breitbandfernsehen)* <Fernseh> FSTV *(fast-scan television)*
FüG *(Fahrt über Grund)* <Wassertrans> speed made good over the ground
FVSt *(Fernvermittlungsstelle)* <Telekom> *(AE)* toll exchange, *(BE)* trunk exchange, *(AE)* trunk switching center, *(BE)* trunk switching centre
g 1. <Chemie, Labor, Phys> *(Gramm)* g *(gram)*; 2. <Kerntech, Phys> *(gyromagnetisches Verhältnis)* g *(gyromagnetic ratio)*; 3. <Phys> *(statistisches Gewicht)* g *(statistical weight)*; 4. <Raumfahrt> *(Erdbeschleunigung)* g *(gravitational acceleration)*
G 1. <Aufnahme, Elektriz> *(Gauß)* G *(gauss)*; 2. <Elektronik, Ergon, Fernseh, Funktech, Raumfahrt, Telekom> *(Gewinn)* G *(gain)*; 3. <Maschinen, Phys> *(Schermodul)* G *(shear modulus)*; 4. <Metrol> *(Giga...)* G *(giga)*; 5. <Phys, Thermod> *(Gibbs'sche Funktion)* G *(Gibbs function)*
GaAs *(Galliumarsenid)* <Elektronik, Funktech, Optik, Phys> GaAs *(gallium arsenide)*
GAU *(größter anzunehmender Unfall)* <Kerntech> MCA *(maximum credible accident)*
GB *(Gigabyte)* <Comp & DV, Optik, Telekom> GB *(gigabyte)*
GFK 1. <Kunststoff> *(Glasfaserkunststoff, glasfaserverstärkter Kunststoff)* GRP, glass fibre-reinforced plastic; 2. <Verpack> *(glasfaserverstärkter Kunststoff)* GRP, glass fibre-reinforced plastic; 3. <Wassertrans> *(glasfaserverstärkter Kunststoff)* GRP, glass fibre-reinforced plastic *(Schiffbau)*
ggT *(größter gemeinsamer Teiler)* <Math> HCF *(highest common factor)*
GIGO *(Müll rein, Müll raus)* <Comp & DV> GIGO *(garbage in, garbage out)*
GII *(globale Informations-Infrastruktur)* <Comp & DV> GII *(Global Information Infrastructure)*
g/m *(Gramm pro Quadratmeter)* <Druck> gsm *(grams per square metre)*
GMDSS *(System zur Rettung von Menschenleben bei Seenotfällen)* <Wassertrans> GMDSS *(global marine distress and safety system)*
GÖV *(Gas-Öl-Verhältnis)* <Erdöl> GOR, gas-to-oil ratio *(Lagerstättentechnik)*
GP *(Glühpunkt)* <Metall> AP *(annealing point)*
GPC *(Gel-Permeations-Chromatographie)* <Kunststoff, Labor> GPC *(gel permeation chromatography)*

GPG *(Grundprimärgruppe)* <Telekom> basic group
GPS 1. <Funkort> *(weltweites Ortungssystem)* GPS, global-positioning system; 2. <Wassertrans> *(globales Positionsbestimmungssystem)* GPS, global-positioning system *(Satellitennavigation)*
GrVST *(Gruppenvermittlungsstelle)* <Telekom> GSC *(group-switching centre)*
GS *(Gleichstrom)* <Aufnahme, Comp & DV, Eisenbahn, Elektriz, Elektrotech, Fernseh, Fertig, Funktech, Phys, Telekom> DC *(direct current)*
GSZ *(Gesamtsäurezahl)* <Chemie> TAN *(total acid number)*
Gx *(Systemkonstante)* <Akustik> Gx *(system-rating constant)*
gy *(Gray)* 1. <Phys> gy, gray *(Einheit der Energiedosis)*; 2. <Teilphys> gy, gray *(Einheit der Strahlendosis)*
h 1. <Comp & DV, Funktech, Geom> *(Höhe)* h *(height)*; 2. <Metrol> *(Hekto...)* h *(hecto...)*; 3. <Metrol> *(Stunde)* h *(hour)*; 4. <Phys, Strahlphys, Teilphys> *(Planck'sche Konstante, Planck'sches Wirkungsquantum)* h *(Planck's constant)*
H 1. <Chemie> *(Wasserstoff)* H *(hydrogen)*; 2. <Elektriz, Elektrotech, Funktech, Metrol, Phys> *(Henry)* H *(henry)*; 3. <Elektriz, Elektrotech> *(magnetische Feldstärke)* H *(magnetic field strength)*; 4. <Heiz & Kälte, Kohlen, Mechan, Nichtfoss Energ, Phys, Raumfahrt, Thermod> *(Enthalpie)* H *(enthalpy)*; 5. <Hydraul> *(Hamilton'sche Funktion)* H *(Hamiltonian function)*; 6. <Optik> *(Bestrahlungsstärke)* H *(irradiance)*; 7. <Phys> *(Magnetfeldstärke)* H *(magnetic field strength)*
ha *(Hektar)* <Metrol> ha *(hectare)*
HADES *(Dielektronen-Spektrometer mit hoher Akzeptanz)* <Teilphys> HADES *(high acceptance di-electron spectrometer)*
HD 1. <Comp & DV> *(Festplatte)* HD *(hard disk)*; 2. <Comp & DV> *(halbduplex)* HDX *(half-duplex)*; 3. <Maschinen> *(Hochleistung)* HD *(heavy duty)*
HDTV *(hochauflösendes Fernsehen, hochzeiliges Fernsehverfahren)* <Fernseh> HDTV *(high-definition television)*
HERA *(Hadron-Elektron-Ring-Anlage)* <Teilphys> HERA *(hadron-electron ring collider)*
HEX 1. <Comp & DV, Geom> *(hexadezimal)* hex *(hexadecimal)*; 2. <Geom> *(Hexagon)* hex *(hexagon)*
HF 1. <Aufnahme, Elektronik, Fernseh, Funktech, Telekom> *(Funkfrequenz, Hochfrequenz)* HF *(high frequency)*; RF *(radio frequency)*; 2. <Elektriz> *(Hochfrequenz)* HF *(high frequency)*; 3. <Wassertrans> *(Hochfrequenz)* HF *(high frequency)*; RF *(radio frequency)*
hl *(Hektoliter)* <Labor> hl *(hectoliter)*
HNF *(höchste nutzbare Frequenz)* <Funktech> MUF *(maximum usable frequency)*
HPLC *(Hochdruckflüssigchromatographie)* <Labor, Lebensmittel> HPLC *(high-pressure liquid chromatography)*
HPS *(höhere Programmiersprache)* <Comp & DV, Telekom> HLL *(high-level language)*

HTR *(Hochtemperaturreaktor)* <Kerntech> HTR *(high-temperature reactor)*
HVSt *(Hauptvermittlungsstelle)* <Telekom> main exchange
HVStd *(Hauptverkehrsstunde)* <Telekom> busy hour
HVt *(Hauptverteiler)* <Telekom> MDF *(main distribution frame)*
HWZ *(Halbwertszeit)* 1. <Kerntech, Phys, Strahlphys, Teilphys> T½ *(half-life)*; 2. <Telekom> FDHM *(full duration half maximum)*
Hz *(Hertz)* <Elektriz, Elektrotech, Fernseh, Funktech, Metrol, Phys> Hz *(hertz)*
HZK *(höchstzulässige Konzentration)* <Umweltschmutz> MAC *(maximum allowable concentration)*; TLV *(threshold limit value)*
I 1. <Akustik, Elektriz> *(Intensität, Stärke)* I *(intensity)*; 2. <Chemie> *(Iod, Jod)* I *(iodine)*; 3. <Elektriz, Phys, Telekom> *(elektrischer Strom)* I *(electric current)*; 4. <Elektriz, Optik> *(Intensität)* I *(intensity)*; 5. <Kerntech> *(Energieflussdichte)* I *(energy flux density)*; 6. <Kerntech> *(nukleare Spinquantenzahl)* I *(nuclear spin quantum number)*
ICAS *(Kommerzieller und Amateurfunkdienst)* <Funktech> ICAS *(Intermittent Commercial and Amateur Services)*
ICRP *(Internationale Strahlenschutzkommission)* <Strahlphys> ICRP *(International Commission on Radiological Protection)*
ID *(Identifikation, Kennung)* <Comp & DV> identification
IFR *(Instrumentenflugregeln)* <Lufttrans> IFR *(instrument flight rules)*
IFRB *(Internationaler Ausschuss für Frequenzregistrierung)* <Raumfahrt, Telekom> IFRB *(International Frequency Registration Board)*
IFU *(Internationale Fernmeldeunion)* <Telekom, Wassertrans> ITU *(International Telecommunication Union)*
IGFET *(Isolierschicht-Feldeffekttransistor)* <Elektronik> IGFET *(insulated gate field-effect transistor)*
IIR *(unbegrenztes Ansprechen auf Impuls)* <Elektronik> IIR *(infinite impulse response)*
ILS *(Blindfluglandesystem durch Eigenpeilung, Instrumentenlandesystem)* <Funkort, Lufttrans, Raumfahrt> ILS *(instrument landing system)*
INS n *(integriertes Navigationssystem)* <Wassertrans> INS *(integrated navigation system)*
IOS 1. <Elektronik, Optik> *(integrierter optischer Schaltkreis)* IOC *(integrated optical circuit)*; 2. <Telekom> *(integrierte optische Schaltung)* IOC *(integrated optical circuit)*
IP *(Eingabe)* <Comp & DV> IP *(input)*
IPL *(Initialprogrammlader)* <Comp & DV> IPL *(initial program loader)*
IR 1. <Comp & DV> *(Wiederauffinden von Informationen)* IR *(information retrieval)*; 2. <Kunststoff, Optik, Phys, Strahlphys> *(Infrarot)* IR *(infrared)*
IRPTC *(Internationales Verzeichnis für potenziell toxische Chemikalien)* <Umweltschmutz> IRPTC

IRS

(International Register of Potentially Toxic Chemicals)
IRS *(Internationales Referenzsystem)* <Umweltschmutz> IRS *(International Referral System)*
IS 1. <Comp & DV> *(Informationssysteme)* IS *(Information Systems)*; 2. <Comp & DV, Elektriz, Elektronik, Funktech, Kontroll, Phys, Telekom> *(Integrierschaltung, integrierte Schaltung, integrierter Schaltkreis)* IC *(integrated circuit)*; 3. <Elektronik, Funktech, Phys> *(Sättigungsstrom)* IS *(saturation current)*
ISB *(unabhängiges Seitenband)* <Funktech> ISB *(independent sideband)*
ISDN *(diensteintegriertes digitales Netz)* <Telekom> ISDN *(integrated services digital network)*
ISM *(Betriebsüberwachung)* <Telekom> ISM *(in-service monitoring)*
ISO *(Internationale Organisation für Normung)* <Elektriz, Maschinen> ISO *(International Standardisation Organisation)*
ISW *(internationale Selbstwahl)* <Telekom> IDD *(international direct dialling)*; IDDD *(international direct distance dialling)*
IT *(Informationstechnologie)* <Comp & DV> IT *(information technology)*
ITA *(internationales Telegrafenalphabet)* <Lufttrans, Wassertrans> ITA *(international telegraph alphabet)*
IUC *(Mess- und Regeltechnik)* <Elektronik> IUC *(instrumentation and control)*
j *(Sprunghöhe)* <Hydraul> j *(height of hydraulic jump)*
J 1. <Akustik, Phys> *(Schallenergiefluss)* J *(sound-energy flux)*; 2. <Elektriz, Lebensmittel, Mechan, Metrol, Phys, Thermod> *(Joule)* J *(joule)*; 3. <Kerntech> *(Winkelmomentquantenzahl)* J *(total angular momentum quantum number)*; 4. <Mechan, Thermod> *(mechanisches Wärmeäquivalent)* J *(mechanical equivalent of heat)*
k 1. <Akustik> *(Wellenkonstante)* k *(wave constant)*; 2. <Chemie> *(Kalium)* K *(potassium)*; 3. <Elektriz> *(Kopplungskoeffizient, Phys)* k *(coupling coefficient)*; 4. <Kerntech> *(Multiplikationskonstante für infinite Systeme)* k *(multiplication constant for an infinite system)*; 5. <Kerntech> *(Neutronenmultiplikationskonstante)* k *(neutron multiplication constant)*; 6. <Labor> *(Kilo, Kilogramm)* k *(kilo)*; 7. <Phys, Thermod> *(Boltzmann'sche Konstante, Boltzmann'sche Zahl)* k *(Boltzmann constant)*
K 1. <Akustik> *(Magnetostriktionskonstante)* K *(magnetostriction constant)*; 2. <Elektriz> *(Kelvin)* K *(kelvin)*; 3. <Hydraul> *(Kompressionsmodul)* K *(bulk modulus of compression)*; 4. <Hydraul> *(Elastizitätsmodul)* K *(bulk modulus of elasticity)*
KARS *(kohärente Antistokes-Raman-Streuung)* <Strahlphys, Wellphys> CARS *(coherent anti-Stokes Raman scattering)*
KAW *(Kanaladresswort)* <Comp & DV> CAW *(channel address word)*
KB *(Kilobyte)* <Comp & DV, Telekom> KB *(kilobyte)*
kcal *(Kilokalorie)* <Lebensmittel> kcal *(kilocalorie)*

keff *(effektive Neutronen-Multiplikationskonstante)* <Kerntech> keff
keV *(Kilo-Elektronenvolt)* <Teilphys> keV *(kilo electronvolt)*
KF *(Konfidenzfaktor)* <Künstl Int> CF *(confidence factor)*
kfG *(kontextfreie Grammatik)* <Künstl Int> CFG *(context-free grammar)*
Kfz *(Kraftfahrzeug)* <Kfztech> MC *(motorcar)*
kg 1. <Labor> *(Kilo, Kilogramm)* kg *(kilogram)*; 2. <Phys> *(Kilogramm)* kg *(kilogramme)*
kgN *(kleinster gemeinsamer Nenner)* <Math> LCD *(least common denominator)*
kgT *(kleinster gemeinsamer Teiler)* <Geom> LCD *(least common denominator)*
kgV *(kleinstes gemeinsames Vielfaches)* 1. <Comp & DV> LCM *(least common multiple)*; 2. <Geom> LCM *(lowest common multiple)*
kHz *(Kilohertz)* <Elektriz, Funktech> kHz *(kilohertz)*
KI *(künstliche Intelligenz)* <Künstl Int> AI *(artificial intelligence)*
km m *(Kilometer)* <Labor> km *(kilometer)*
KNN *(künstliches neuronales Netzwerk)* <Künstl Int> ANN *(artificial neural network)*
KnV *(Knotenvermittlung)* <Telekom> tandem switching
KPK *(kritische Pigmentvolumenkonzentration)* <Kunststoff> cpvc *(critical pigment volume concentration)*
KS *(Kerbstift)* <Fertig, Maschinen> grooved pin, splined pin
ksG *(kontextsensitive Grammatik)* <Künstl Int> CSG *(context-sensitive grammar)*
KSR *(Katodenstrahlröhre)* <Comp & DV, Druck, Elektriz, Elektronik, Fernseh, Funktech> CRT *(cathode-ray tube)*
KüG *(Kurs über Grund)* <Wassertrans> course made good *(Navigation)*
kV *(Kilovolt)* <Elektrotech> kV *(kilovolt)*
KV *(kombinierter Verteiler)* <Telekom> CDF *(combined distribution frame)*
kW *(Kilowatt)* <Elektriz> kW *(kilowatt)*
kWh *(Kilowattstunde)* <Elektriz, Elektrotech> kWh *(kilowatt-hour)*
KZG *(Kurzzeitgedächtnis)* <Ergon, Künstl Int> STM *(short-term memory)*
l 1. <Comp & DV, Geom, Phys, Telekom> *(Länge)* l *(length)*; 2. <Kerntech> *(effektive Neutronenlebensdauer)* l *(effective neutron lifetime)*
L 1. <Akustik> *(Lautstärke)* L *(loudness)*; 2. <Aufnahme, Elektriz, Elektrotech, Funktech, Metrol, Phys, Telekom> *(Induktivität)* L *(inductance)*; 3. <Kerntech> *(Diffusionslänge)* L *(diffusion length)*; 4. <Kerntech, Strahlphys> *(lineare Energieübertragung)* L *(linear energy transfer)*; 5. <Kerntech> *(Orbitalwinkelmomentzahl)* L *(total orbital angular momentum number)*; 6. <Mechan> *(Lagrange'sche Funktion)* L *(Lagrangian function)*; 7. <Optik> *(Luminanz)* L *(luminance)*; 8. <Thermod> *(Lorenz'sche Einheit)* L *(Lorenz unit)*

LAN *(lokales Netz)* <Comp & DV, Telekom> LAN *(local area network)*
LB 1. <Comp & DV, Telekom> *(Ortsbatterie)* LB *(local battery)*; 2. <Elektrotech> *(lokale Batterie)* LB *(local battery)*
LC *(Flüssigkristall)* <Comp & DV, Elektriz> LC *(liquid crystal)*
LCD *(Flüssigkristallanzeige)* <Comp & DV, Elektriz, Elektronik, Fernseh, Labor, Telekom, Thermod> LCD *(liquid crystal display)*
LD$_{50}$ *(mittlere letale Dosis)* <Strahlphys> LD$_{50}$ *(median lethal dose)*
LEAR *(Antiprotonenring mit geringer Energie)* <Teilphys> LEAR *(Low-Energy Antiproton Ring)*
LED *(Leuchtdiode, Lumineszenzdiode, lichtemittierende Diode)* <Comp & DV, Elektriz, Elektronik, Fernseh, Optik, Phys, Telekom> LED *(light-emitting diode)*
LEM *(Mondlandefahrzeug, Mondlandefähre)* <Raumfahrt> LEM *(lunar excursion module)*
LEP *(Elektronen-Positronen-Kollideranlage)* <Teilphys> LEP *(large electron-positron collider)*
LHC *(Hadronkollideranlage)* <Phys> LHC *(large hadron collider)*
LIFO *(Last-in First-out)* <Comp & DV> LIFO *(last-in-first-out)*
LINEAC *(Linearbeschleuniger)* 1. <Elektrotech, Phys> LINAC *(linear accelerator)*; 2. <Teilphys> LINEAC *(linear accelerator)*
LISP *(Listenprogrammiersprache)* <Comp & DV> LISP *(list-programming language)*
LKF *(Luftkissenfahrzeug)* 1. <Kfztech> SEV *(surface effect vehicle)*; 2. <Trans> ACV *(air cushion vehicle)*; 3. <Wassertrans> *(BE)* hovercraft, *(AE)* hydroskimmer, surface effect ship
LKT *(luftgekühlte Triode)* <Elektronik> ACT *(air-cooled triode)*
Lkw *m (Lastkraftwagen)* <Kfztech> HGV *(heavy goods vehicle)*
lm *(Lumen)* <Fernseh, Metrol, Phys> lm *(lumen)*
LM *(Lunar-Modul)* <Raumfahrt> LM, lunar module *(Raumschiff)*
LNG *(Flüssigerdgas)* <Erdöl, Thermod> LNG *(liquefied natural gas)*
LOI *(Sauerstoffindex)* <Kunststoff> LOI *(limiting oxygen index)*
LORAN *(Langstreckennavigationskette)* <Lufttrans, Wassertrans> loran *(long-range navigation)*
LOX *(Flüssigsauerstoff)* <Raumfahrt, Thermod> lox *(liquid oxygen)*
LP *(Langspielplatte)* <Aufnahme> EP *(extended-play record)*; LP *(long-playing record)*
LPC *(lineare Prädiktionscodierung)* <Elektronik, Telekom> LPC *(linear predictive coding)*
LPG *(Flüssiggas)* <Erdöl, Heiz & Kälte, Kfztech, Thermod, Wassertrans> LPG *(liquefied petroleum gas)*
LPM *(Zeilen pro Minute)* <Comp & DV> LPM *(lines per minute)*
LQ *(Korrespondenzqualität)* <Comp & DV, Druck> LQ *(letter quality)*

LS *(Laufsitz)* <Maschinen> running fit
LSB *(niedrigstwertiges Bit)* <Comp & DV, Telekom> LSB *(least significant bit)*
LSI *(Großintegration, hoher Integrationsgrad)* <Comp & DV, Elektronik, Phys, Telekom> LSI *(large-scale integration)*
LSL *(untere Spezifikationsgrenze)* <Qual> LSL *(lower specification limit)*
LTTP *(Langzeit-Straßenbelagverhalten)* <Bau> LTPP *(long-term pavement performance)*
LüA *(Länge über alles)* <Wassertrans> LOA *(length overall)*
LUT *(Bodenstation)* <Wassertrans> LUT, local user terminal *(Satellitennavigation)*
LW *(Langwelle)* <Funktech> LW *(long wave)*
lx *(Lux)* <Foto, Labor, Optik, Phys> lx *(lux)*
LZG *(Langzeitgedächtnis)* <Ergon, Künstl Int> LTM *(long-term memory)*
m 1. <Akustik> *(Streukoeffizient, Öffnungskoeffizient)* m *(flare coefficient of horn)*; 2. <Akustik> *(Scherelastizität)* m *(shear elasticity)*; 3. <Maschinen, Phys, Thermod> *(Masse)* m *(mass)*; 4. <Metrol> *(Meter)* m *(meter)*; 5. <Metrol> *(Milli...)* m *(milli...)*; 6. <Phys> *(Molekularmasse)* m *(molecular mass)*; 7. <Phys> *(Gegeninduktivität)* m *(mutual inductance)*
M 1. <Erdöl, Phys, Thermod> *(Molekulargewicht)* M *(molecular weight)*; 2. <Erdöl, Lufttrans, Phys> *(Machzahl)* M *(Mach number)*; 3. <Kerntech> *(Reaktormultiplikation)* M *(multiplication of a reactor)*; 4. <Metrol> *(Mega...)* M *(mega...)*
m0 *(Restmasse)* <Kerntech> m0 *(rest mass)*
Ma *(Atommasse)* <Kerntech> Ma *(atomic mass)*
MA *(Mittabstand)* <Fertig, Kfztech, Maschinen> CD *(centre distance)*
MAC *(magnetisches Kalorimeter)* <Teilphys> MAC *(magnetic calorimeter)*
MAK *m (maximale Arbeitsplatzkonzentration)* <Kerntech> MAC *(maximum allowable concentration)*; TLV *(threshold limit value)*
MAP *(Manufacturing Automation Protocol)* <Kontroll> MAP, manufacturing automation protocol *(Standard für Fabrikautomatisierungsgeräte)*
MB *(Mbyte, Megabyte)* <Comp & DV> MB *(megabyte)*
MBE *(Molekularstrahlepitaxie)* <Elektronik, Strahlphys> MBE *(molecular-beam epitaxy)*
MDI *(Diphenylmethandiisocyanat)* <Kunststoff> MDI *(diphenylmethane diisocyanate)*
MDR *(Speicherdatenregister)* <Comp & DV> MDR *(memory data register)*
me *(Elektronenmasse)* <Chemie, Kerntech, Teilphys> me *(electron mass)*
MEA *(Means-End-Analyse, Mittel-Zweck-Analyse)* <Künstl Int> MEA *(means-end analysis)*
MEK 1. <Kunststoff> *(Methylethylketon)* MEK *(methyl ethyl ketone)*; 2. <Umweltschutz> *(maximale Emissionskonzentration)* maximum emission concentration
MEP *(mittlerer Nutzdruck)* <Lufttrans, Maschinen> mep *(mean effective pressure)*

MESFET

MESFET *(Metallhalbleiter-Feldeffekttransistor)* <Elektronik> MESFET *(metal semiconductor field effect transistor)*
MeV *(Million Elektronenvolt)* <Teilphys> MeV *(million electron volts)*
MF 1. <Elektriz> *(Melamin-Formaldehydharz, Melaminharz)* MF *(melamine resin)*; 2. <Elektronik, Funktech> *(Mittelfrequenz)* MF *(medium frequency)*; 3. <Elektronik, Funktech, Telekom> *(Mehrfachfrequenz)* MF *(multiple frequency)*; 4. <Kunststoff> *(Melamin-Formaldehydharz, Melaminharz)* MF *(melamine formaldehyde resin)*
MFC *(Mehrfrequenzcode)* <Telekom> MFC *(multifrequency code)*
MFM *(modifizierte Frequenzmodulation)* <Elektronik, Funktech, Telekom> MFM *(modified frequency modulation)*
MFW *(Mehrfrequenzwahl)* 1. <Telekom> DTMF *(dual-tone multifrequency)*; 2. <Telekom> MFD *(multifrequency dialling)*
MGD *(Magnetogasdynamik)* <Kerntech> MGD *(magnetogasdynamics)*
MHz *(Megahertz)* <Elektriz, Elektrotech, Fernseh, Funktech> MHz *(megahertz)*
MIC *(integrierter Mikrowellenschaltkreis)* <Elektronik, Wellphys> MIC *(microwave integrated circuit)*
MICR *(Magnetschrifterkennung, Magnetschriftzeichenerkennung)* <Comp & DV> MICR *(magnetic ink character recognition)*
MIPS *(Millionen Befehle pro Sekunde)* <Comp & DV> MIPS *(millions of instructions per second)*
MIS 1. <Comp & DV> *(Management-Informationssystem)* MIS *(management information system)*; 2. <Elektronik> *(Metallisolator-Halbleiter)* MIS *(metal insulator semiconductor)*
MISFET *(Metallisolator-Feldeffekttransistor)* <Elektronik> MISFET *(metal insulator semiconductor field effect transistor)*
MLIPS *(Millionen logischer Inferenzen pro Sekunde)* <Künstl Int> MLIPS *(millions of logical inferences per second)*
MMI *(Mensch-Maschine-Interface, Mensch-Maschine-Schnittstelle)* <Comp & DV, Kontroll, Künstl Int, Raumfahrt> MMI *(man-machine interface)*
MMK *(magnetomotorische Kraft)* <Elektriz, Phys, Strömphys> mmf *(magnetomotive force)*
MMT *(Methylcyclopentadienyl-Mangantricarbonyl)* <Kfztech> MMT, Methylcyclopentadienyl Manganese Tricarbonyl *(Kraftstoffzusatz)*
mn *(Neutronenmasse)* <Kerntech, Strahlphys, Teilphys> mn *(neutron mass)*
MN *(Kernmasse)* <Kerntech> MN *(nuclear mass)*
mol *(Mole)* <Chemie, Labor, Phys> mol *(mole)*
MOS 1. <Comp & DV> *(Metalloxid-Halbleiter)* MOS *(metal oxide semiconductor)*; 2. <Elektronik> *(Metalloxid-Halbleiter, Metalloxid-Transistor)* MOS *(metal oxide semiconductor)*
MOSFET *(Metalloxid-Silizium-Feldeffekttransistor)* <Funktech> MOSFET *(metal oxide silicon field effect transistor)*
mp *(Protonenmasse)* <Kerntech> mp *(proton mass)*

MP 1. <Comp & DV, Elektriz, Maschinen> *(Mikroprozessor)* MP *(microprocessor)*; 2. <Verpack> *(Metallpapier)* MP *(metallic paper)*
MS 1. <Mobilkom> *(Mobilstation)* MS *(mobile station)*; 2. <Telekom> *(Mitteilungsspeicherung)* MS *(message storing)*
MSC *(Funkvermittlungsstelle)* <Mobilkom> MSC *(mobile switching center, mobile switching centre)*
MSI 1. <Comp & DV> *(mittlere Integrationsdichte, mittlere Integrationstechnik, mittlerer Integrationsgrad)* MSI *(medium-scale integration)*; 2. <Elektronik> *(mittlere Integrationstechnik, mittlerer Integrationsgrad, mittlerere Integrationsdichte)* MSI *(medium-scale integration)*; 3. <Telekom> *(mittlerer Integrationsgrad, mittlerere Integrationsdichte)* MSI *(medium-scale integration)*
MSK 1. <Comp & DV, Funktech, Telekom> *(Minimalphasenumtastung)* MSK *(minimum-shift keying)*; 2. <Elektronik> *(kleinste Umtastung)* MSK *(minimum-shift keying)*
MTR *(Materialprüfreaktor)* <Kerntech> MTR *(materials-testing reactor)*
MTTR 1. <Comp & DV, Qual, Raumfahrt> *(mittlere Reparaturdauer)* MTTR *(mean time to repair)*; 2. <Elektrotech> *(mittlere Zeit bis zur Reparatur)* MTTR *(mean time to repair)*; 3. <Mechan> *(mittlere Instandsetzungszeit)* MTTR *(mean time to repair)*
mu *(Atommassenkonstante)* <Kerntech> mu *(unified atomic mass constant)*
Mü *(Maschinenübersetzung)* <Künstl Int> MT *(machine translation)*
MUX *(Multiplexer)* <Comp & DV, Elektronik, Fernseh, Telekom> MUX *(multiplexer)*
MVA *(Müllverbrennungsanlage)* 1. <Abfall> *(AE)* garbage incineration plant, *(BE)* refuse incineration plant, waste incineration plant, waste incinerator; 2. <Verpack> *(AE)* garbage incinerator, *(BE)* refuse incinerator
MW *(Mittelwelle)* <Fernseh, Funktech> MW *(medium wave)*
Mx *(Maxwell)* <Elektriz, Elektrotech> Mx *(maxwell)*
n 1. <Elektriz, Kerntech, Strahlphys, Teilphys> *(Neutron)* n *(neutron)*; 2. <Elektriz> *(Windungszahl pro Längeneinheit)* n *(turns per unit length)*; 3. <Kerntech, Metrol> *(Quantenzahl)* n *(principal quantum number)*; 4. <Phys, Thermod> *(Moleküldichte)* n *(molecular density)*
N 1. <Chemie> *(Stickstoff, Nitrogen)* N *(nitrogen)*; 2. <Elektriz, Hydraul, Labor, Strömphys> *(Newton)* N *(newton)*; 3. <Elektriz> *(Windungszahl)* N *(number of turns in a winding)*; 4. <Elektronik, Kerntech> *(Rauschleistung)* N *(noise power)*; 5. <Metrol, Optik> *(Strahlung)* N *(radiance)*; 6. <Phys> *(Molekülzahl)* N *(number of molecules)*
N_A *(Avogadro'sche Zahl, Loschmidt'sche Zahl)* <Phys, Thermod> NA *(Avogadro's number, Loschmidt number)*
NASA *(Nordamerikanische Weltraumbehörde)* <Raumfahrt> NASA *(National Aeronautics and Space Administration)*

NC *(numerische Steuerung)* <Comp & DV, Elektriz, Kontroll, Maschinen, Mechan> NC *(numerical control)*

Nf *(Niederfrequenz)* 1. <Aufnahme, Elektronik, Fernseh, Funktech> AF *(audio frequency)*; LF *(low frequency)*; 2. <Telekom> LF *(low frequency)*

NiCd *(Nickel-Cadmium)* <Elektronik, Foto> NiCd *(nickel-cadmium)*

NK *(Naturkautschuk)* <Kunststoff> NR *(natural rubber)*

NMR *(magnetische Kernresonanz)* <Erdöl, Teilphys> NMR *(nuclear magnetic resonance)*

NN 1. <Comp & DV, Künstl Int> *(neurales Netz, neurales Netzwerk)* NN *(neural network)*; 2. <Wassertrans> *(Normalnull)* msl *(mean sea level)*

NNF *(niedrigste nutzbare Frequenz)* <Funktech> LUF *(lowest usable frequency)*

NNSS *(Navigationssatellitensystem)* <Wassertrans> NNSS *(Navy Navigation Satellite System)*

NO-OP *(Nulloperation, keine Operation)* <Comp & DV> NO-OP *(no-operation)*

NOS *(Notrufortungssender)* <Funktech, Telekom> ELT *(emergency locator transmitter)*

Np 1. <Chemie> *(Neptunium)* Np *(neptunium)*; 2. <Phys> *(Neper)* Np *(neper)*

NTSC *(Amerikanischer Fernsehnormungsausschuss)* <Fernseh> NTSC *(National Television Standards Committee)*

OA *(Operationsverstärker, Rechenverstärker)* <Comp & DV, Elektronik, Phys> op amp *(operational amplifier)*

OBO *(Erz-Schüttgut-Öl, Flüssigkeitsmassengut)* <Wassertrans> OBO *(ore-bulk oil)*

OC *(Operationscharakteristik)* 1. <Math> OC, operating characteristic *(Testtheorie)*; 2. <Qual> OC, operating characteristic

OCO *(Erz-Kohle-Öl)* <Trans> OCO *(ore-coal-oil)*

OCR *(optische Zeichenerkennung)* <Comp & DV> OCR *(optical character recognition)*

OFN *(Ortsfernsprechnetz)* <Telekom> local exchange area

OLRT *(Online-Echtzeit)* <Comp & DV> OLRT *(online real time)*

OMR *(optische Markierungserkennung)* <Comp & DV> OMR *(optical mark recognition)*

ON *(Ortsnetz)* 1. <Elektriz> distribution network, local network; 2. <Telekom> local network

ONKz *(Ortsnetzkennzahl)* <Telekom> area code

OOP *(objektorientierte Programmierung)* <Künstl Int> OOP *(object-oriented programming)*

OPEC *(Organisation Ölexportierender Länder)* <Erdöl> OPEC *(Organization of Petroleum-Exporting Countries)*

OR *(Operations-Research)* <Comp & DV> OR *(operations research)*

OROM *(optischer Festwertspeicher)* <Comp & DV, Optik> OROM *(optical read-only memory)*

OSB *(oberes Seitenband)* <Elektronik, Funktech, Telekom> USB *(upper sideband)*

OSCAR *(Bahnsatellit für Amateurfunkzwecke)* <Funktech, Raumfahrt> OSCAR, Orbiting Satellite Carrying Amateur Radio *(Weltraumfunk)*

OSI *(Kommunikation offener Systeme)* <Comp & DV, Telekom> OSI *(open systems interconnection)*

OSO *(Erz-Schlamm-Öl)* <Wassertrans> OSO *(ore-slurry-oil)*

OT *(oberer Totpunkt)* <Kfztech> TDC *(top dead center)*

OVSt *(Ortsvermittlungsstelle)* <Telekom> local exchange

p 1. <Akustik, Phys> *(Schalldruck)* p *(sound pressure)*; 2. <Aufnahme, Raumfahrt> *(Schalldruck)* p *(acoustic pressure)*; 3. <Funktech, Kerntech, Phys, Teilphys> *(Proton)* p *(proton)*; 4. <Kerntech> *(Resonanzfluchtwahrscheinlichkeit)* p *(resonance escape probability)*; 5. <Metrol> *(Piko...)* p *(pico...)*

P 1. <Chemie> *(Phosphor)* P *(phosphorus)*; 2. <Elektriz> *(Leistung)* P *(power)*; 3. <Kerntech> *(Protonenzahl)* P *(proton number)*

Pa *(Pascal)* <Metrol, Phys> Pa, pascal *(Hydrostatik)*

PA 1. <Aufnahme> *(Kraftverstärker, Leistungsverstärker)* PA *(power amplifier)*; 2. <Chemie, Kunststoff, Textil> *(Polyamid)* PA *(polyamide)*; 3. <Elektronik, Raumfahrt, Telekom> *(Leistungsverstärker)* PA *(power amplifier)*; 4. <Elektrotech> *(Endstufe, Endverstärker, Leistungsverstärker)* PA *(power amplifier)*; 5. <Funktech> *(Endstufe, Leistungsverstärker)* PA *(power amplifier)*; 6. <Phys> *(Hauptverstärker, Leistungsverstärker)* PA *(power amplifier)*

PAA *(Polyacryl, Polyacrylat)* <Chemie, Kunststoff> PAA *(polyacrylate)*

PAL *(programmierbare logische Anordnung)* <Comp & DV> PAL *(programmable array logic)*

PAM *(Pulsamplitudenmodulation)* <Comp & DV, Elektronik, Funktech, Telekom> PAM *(pulse amplitude modulation)*

PAN 1. <Umweltschutz> *(Peroxyacetylnitrat)* PAN *(peroxoacetylnitrate)*; 2. <Umweltschutz> *(Polyacrylnitril)* PAN *(polyacrylonitrile)*

PB 1. <Comp & DV, Elektronik, Fernseh, Phys, Telekom> *(Pulsbreite)* PW *(pulse width)*; 2. <Kunststoff> *(Polybuten, Polybutylen)* PB *(polybutylene)*

PBT *(Polybutylenterephthalat)* <Elektriz, Kunststoff> PBT *(polybutylene ephtalate)*

PBX *(private Selbstwählnebenstelle)* <Telekom> PBX *(private branch exchange)*

PC 1. <Comp & DV> *(Personal Computer)* PC *(personal computer)*; 2. <Elektriz, Kunststoff> *(Polycarbonat)* PC *(polycarbonate)*

PCB 1. <Comp & DV> *(Leiterplatte, Printplatte)* PCB *(printed circuit board)*; 2. <Elektriz, Fernseh, Funktech, Telekom> *(Leiterplatte)* PCB *(printed circuit board)*; 3. <Elektronik> *(Leiterplatte, gedruckte Schaltung)* PCB *(printed circuit board)*

PCM *(Pulscodemodulation)* <Aufnahme, Comp & DV, Elektronik, Funktech, Strahlphys, Telekom> PCM *(pulse code modulation)*

PCU *(externe Steuereinheit, periphere Steuereinheit)* <Comp & DV> PCU *(peripheral control unit)*

PDAP *(Polydiallyphthalat)* <Kunststoff> PDAP *(polydiallylphthalate)*
PDM 1. <Comp & DV> *(Pulsdeltamodulation)* PDM *(pulse delta modulation)*; 2. <Elektronik> *(Pulsdauermodulation)* PWM *(pulse width modulation)*; 3. <Telekom> *(Pulsdauermodulation)* PDM *(pulse duration modulation)*
PDN *(öffentliches Datennetz)* <Comp & DV, Telekom> PDN *(public data network)*
PDR *(vorläufige technische Prüfung)* <Raumfahrt> PDR *(preliminary design review)*
PE-C *(chloriertes Polyethylen)* <Kunststoff> CPE *(chlorinated polyethylene)*
PEC *(Photozelle)* <Druck, Elektriz, Elektronik, Fernseh, Foto, Phys, Strahlphys> PEC *(photoelectric cell)*
PES *(Polyester)* <Chemie, Elektriz, Kunststoff, Textil> PES *(polyester)*
PET *(Polyethylen)* <Chemie, Elektriz, Erdöl, Kunststoff, Textil, Verpack> PET *(polyethylene)*
PETP *(Polyethylenterephthalat)* <Kunststoff> PETP *(polyethylene terephthalate)*
PFM *(Pulsfrequenzmodulation)* <Comp & DV, Elektronik, Funktech> PFM *(pulse frequency modulation)*
PI *(Polyimid)* <Elektriz, Elektronik, Kunststoff> PI *(polyimide)*
PIB *(Polyisobutylen)* <Kunststoff> PIB *(polyisobutylene)*
PIN *(Positiv-Isolierend-Negativ)* <Elektronik> PIN *(positive-isolating-negative)*
PKW *(Personenkraftwagen)* <Kfztech> car, passenger car
PKW-E *(Personenkraftwageneinheit)* <Kfztech> PCU *(passenger car unit)*
PL *(Prädikatenlogik)* <Künstl Int> PL *(predicate logic)*
PLA *(programmierbare Logikanordnung)* <Comp & DV> PLA *(programmable logic array)*
PLL *(Phasenregelkreis)* <Elektronik, Fernseh, Funktech, Raumfahrt, Telekom> PLL *(phase-locked loop)*
PM *(Phasenmodulation)* <Aufnahme, Comp & DV, Elektronik, Fernseh, Funktech, Phys, Telekom> PM *(phase modulation)*
PMMA *(Polymethacrylat, Polymethylmethacrylat)* <Kunststoff> PMMA *(polymethyl methacrylate)*
pnp *(positiv-negativ-positiv)* <Elektronik> p-n-p *(positive-negative-positive)*
PO 1. <Kunststoff, Textil> *(Polyolefin)* PO *(polyolefin)*; 2. <Textil> *(Polynosic-Faser)* PO *(polynosic fiber)*
POM *(Polyoxymethylen)* <Chemie, Kunststoff> POM *(polyoxymethylene)*
PP *(Polypropylen)* <Chemie, Elektriz, Erdöl, Kunststoff, Textil> PP *(polypropylene)*
PPM 1. <Comp & DV, Elektronik, Telekom> *(Pulsphasenmodulation)* PPM *(pulse phase modulation)*; 2. <Elektronik, Telekom> *(Pulslagenmodulation)* PPM *(pulse position modulation)*

PPU *(Peripherietechnik)* <Comp & DV> PPU *(peripheral processing units)*
PROM *(programmierbarer Lesespeicher)* <Comp & DV, Funktech> PROM *(programmable read-only memory)*
PS 1. <Chemie, Kunststoff> *(Polystyren, Polystyrol)* PS *(polystyrene)*; 2. <Elektriz> *(Polystyrol)* PS *(polystyrene)*; 3. <Maschinen> *(Pferdestärke)* hp *(horsepower)*; 4. <Wassertrans> *(Peilstrahl)* EBL *(electronic bearing line)*
PSK 1. <Comp & DV> *(Phasensprungtastung)* PSK *(phase shift keying)*; 2. <Elektronik, Funktech, Raumfahrt, Telekom> *(Phasenumtastung)* PSK *(phase-shift keying)*
PTC *(passive Thermosteuerung)* <Raumfahrt> PTC *(passive thermal control)*
PTFE *(Polytetrafluorethen, Polytetrafluorethylen)* <Kunststoff> PTFE *(polytetrafluoroethylene)*
PUR *(Polyurethan)* <Kunststoff, Textil> PUR *(polyurethane)*
PVAC *(Polyvinylacetat)* <Kunststoff> PVAC *(polyvinyl acetate)*
PVAL *(Polyvinylalkohol)* <Druck, Kunststoff> PVAL *(polyvinyl alcohol)*
PVB *(Polyvinylbutyral)* <Kunststoff> PVB *(polyvinyl butyral)*
PVC *(Polyvinylchlorid)* <Bau, Elektrotech, Kunststoff> PVC *(polyvinyl chloride)*
PVC-C *(chloriertes Polyvinylchlorid)* <Kunststoff> CPVC *(chlorinated polyvinyl chloride)*
PVC-U <Kunststoff> U-PVC, unplasticized PVC
PVDC *(Polyvinylidenchlorid)* <Kunststoff> PVDC *(polyvinylidene chloride)*
PVDF *(Polyvinylidenfluorid)* <Kunststoff> PVFD *(polyvinylidene fluoride)*
PVE *(Polyvinylether)* <Kunststoff> PVE *(polyvinyl ether)*
PVF *(Polyvinylfluorid)* <Kunststoff> PVF *(polyvinyl fluoride)*
PWM *(Pulsweitenmodulation)* <Elektronik> PWM *(pulse width modulation)*
QAM 1. <Comp & DV, Elektronik, Telekom> *(Quadratur-Amplitudenmodulation)* QAM *(quadrature amplitude modulation)*; 2. <Elektronik> *(Quadratur-Amplitudenmodulator)* QAM *(quadrature amplitude modulator)*
QCD *(Quantenchromodynamik)* <Teilphys> QCD *(quantum chromodynamics)*
QED *(Quantenelektrodynamik)* <Phys> QED *(quantum electrodynamics)*
QG *(Quartärgruppe)* <Telekom> supermaster group *(Trägerfrequenzübertragung)*
QPSK *(Vierphasenumtastung)* 1. <Elektronik> QPSK *(quaternary phase shift keying)*; 2. <Telekom> QPSK *(quadriphase shift keying)*
QS *(Qualitätssicherung)* <Kerntech, Maschinen, Mechan, Qual, Verpack> QA *(quality assurance)*
QSAM *(erweiterte Zugriffsmöglichkeit für sequenzielle Dateien)* <Comp & DV> QSAM *(queued sequential access method)*

QSG *(Quasistellargalaxie)* <Raumfahrt> QSG *(quasistellar galaxy)*
QSO *(quasi-stellares Objekt)* <Raumfahrt> QSO *(quasi-stellar object)*
QUISAM *(erweiterte indizierte Zugriffsmöglichkeit für sequenzielle Dateien)* <Comp & DV> QUISAM *(queued unique index sequential access method)*
QZ *(quantisiertes Zeichen)* <Elektronik> QS *(quantized signal)*
r 1. <Akustik> *(Entfernung von der Schallquelle)* r *(distance from source)*; 2. <Kerntech> *(Kernradius)* r *(nuclear radius)*; 3. <Optik> *(Brechungswinkel, Refraktionswinkel)* r *(angle of refraction)*; 4. <Phys> *(Brechungswinkel)* r *(angle of refraction)*
R 1. <Elektriz> *(Reluktanz)* R *(magnetic reluctance)*; 2. <Elektriz> *(Widerstand)* R *(resistance)*; 3. <Kerntech> *(Rydberg-Konstante)* R *(Rydberg constant)*; 4. <Kerntech> *(Dosis)* R *(dose rate)*; 5. <Kerntech> *(linearer Bereich)* R *(linear range)*; 6. <Phys, Thermod> *(Gaskonstante)* R *(gas constant)*; 7. <Strahlphys> *(Röntgen)* R *(röntgen)*
Rα *(Rydberg-Konstante)* <Kerntech> Rα *(Rydberg constant)*
Rad *(Einheit der Energiedosis)* <Strahlphys> rad, radiation absorbed dose *(veraltet)*
RAM 1. <Comp & DV> *(Direktzugriffsspeicher, Lese-/Schreibspeicher, Schreib-/Lesespeicher)* RAM *(random access memory)*; 2. <Elektronik> *(Direktzugriffsspeicher, Schreib-/Lesespeicher)* RAM *(random access memory)*
RBA *(relative Byteadresse)* <Comp & DV> RBA *(relative byte address)*
RDB *(relationale Datenbank)* <Telekom> RDB *(relational database)*
RDSS *(Satellitenfunkortungssystem)* <Funkort, Trans, Wassertrans> RDSS, radio determination satellite system *(Satellitenfunk)*
re *(Elektronenradius)* <Kerntech> re *(electron radius)*
Re 1. <Chemie> *(Rhenium)* Re *(rhenium)*; 2. <Hydraul, Lufttrans, Nichtfoss Energ, Phys, Strömphys> *(Reynoldszahl)* Re *(Reynolds number)*
RF *(Radiofrequenz)* <Aufnahme, Elektriz, Elektronik, Fernseh, Funktech, Telekom> HF *(high frequency)*; RF *(radio frequency)*
RG *(elektronischer Rauschgenerator)* <Elektronik, Gerät> ENG *(electronic noise generator)*
RGB *(rot-grün-blau)* <Comp & DV, Fernseh> RGB *(red-green-blue)*
RH *(Hall'scher Koeffizient)* <Phys> RH *(Hall coefficient)*
RIT *(Empfängerfeinabstimmung)* <Funktech> RIT *(receiver incremental tuning)*
RMS *(quadratisches Mittel)* <Aufnahme, Elektriz, Elektronik> rms *(root mean square)*
ROM *(Festwertspeicher, Nur-Lese-Speicher, ROM-Speicher)* <Comp & DV, Elektriz, Elektrotech> ROM *(read-only memory)*
Ro-Ro *(Roll-on-Roll-off)* <Trans> ro-ro, roll-on-roll-off
RP-1 *(Kerosin)* <Erdöl, Trans> RP-1 *(kerosene)*

RSB *(Restseitenband)* <Elektronik> VSB *(vestigial sideband)*
RTTY *(Funkfernschreiben)* <Funktech> RTTY *(radioteletype)*
RWO *(Rückwärtswellenoszillator)* <Elektronik, Phys, Telekom> BWO *(backward-wave oscillator)*
s 1. <Elektriz> *(Schlupf)* s *(slip)*; 2. <Kerntech, Metrol, Phys> *(Spinquantenzahl)* s *(spin quantum number)*
S 1. <Elektriz, Phys> *(Poynting'scher Vektor)* S *(Poynting vector)*; 2. <Hydraul> *(Gefälle)* S *(slope)*; 3. <Kerntech> *(Quellenstärke)* S *(source strength)*; 4. <Kerntech> *(spezifische Ionisierung)* S *(specific ionization)*; 5. <Kerntech> *(Anhalteistung)* S *(stopping power)*; 6. <Kerntech> *(Spinquantengesamtzahl)* S *(total spin quantum number)*; 7. <Metrol> *(Siemens)* S *(siemens)*; 8. <Telekom> *(Zeichengabenetz)* S *(signaling network)*; *(Vertraulichkeit, Geheimhaltung)* S *(secrecy)*
SAR *(Such- und Rettungsdienst)* <Wassertrans> SAR, search and rescue *(Notfall)*
Satcom *(Satellitenfunk, Satellitenkommunikation)* <Funktech, Telekom, Wassertrans> satcom *(satellite communication)*
Satnav *(Satellitennavigation)* <Wassertrans> satnav *(satellite navigation)*
SB *(schneller Brutreaktor, schneller Brüter)* <Kerntech, Phys> FBR *(fast breeder reactor)*
SBFM *(Schmalbandfrequenzmodulation)* <Elektronik, Funktech, Telekom> NBFM *(narrow-band frequency modulation)*
SBR *(Styrol-Butadien-Kautschuk)* <Kunststoff> SBR *(styrene butadiene rubber)*
SBV *(Sicherheitsabblasearmatur)* <Erdöl> relief valve
SCC 1. <Abfall> *(Nachbrennkammer)* SCC *(secondary combustion chamber)*; 2. <Maschinen> *(Nachbrennkammer, zweiter Brennraum)* SCC *(secondary combustion chamber)*
SCN *(Hinweis auf Spezifikationsänderungen)* <Trans> SCN *(specification change notice)*
SCPC *(Ein-Kanal-pro-Träger)* 1. <Raumfahrt> SCPC, single channel per carrier *(Weltraumfunk)*; 2. <Telekom> SCPC, single channel per carrier
SCR *(siliziumgesteuerter Gleichrichter)* <Elektronik, Elektrotech> SCR *(silicon-controlled rectifier)*
SDLC *(synchrone Datenübertragungssteuerung)* <Comp & DV, Telekom> SDLC *(synchronous data link control)*
sec 1. <Geom> *(Sekans)* sec, secant; 2. <Geom> *(Sekante)* sec, secant *(von Linie)*
SELCAL *(Selektivrufsystem)* <Telekom> SELCAL *(selective calling system)*
SES *(Bordterminal für Satellitenfunk)* <Wassertrans> SES *(ship earth station)*
SF 1. <Bau, Elektriz, Kerntech, Kohlen, Maschinen, Sicherheit, Trans> *(Sicherheitsfaktor)* SF *(safety factor)*; 2. <Elektronik, Fernseh, Funktech, Telekom> *(Seitenbandfrequenz)* SF *(sideband frequency)*; 3. <Elektronik, Fernseh, Funktech, Telekom> *(Signalfrequenz)* SF *(signal frequency)*; 4.

SGML

<Elektronik> *(Sprachfrequenz)* VF *(voice frequency)*; SF *(speech frequency)*; 5. <Telekom> *(Sprachfrequenz)* SF *(speech frequency)*; VF *(voice frequency)*

SGML *(Standardkorrekturzeichensatz)* <Comp & DV, Druck> SGML *(Standard Generalized Markup Language)*

SHF *(superhohe Frequenz)* <Bau, Elektronik, Funktech, Kohlen, Kunststoff, Mechan, Phys, Thermod> SHF *(superhigh frequency)*

SIMM *(einfaches schritthaltendes Speichermodul)* <Comp & DV> SIMM *(single in-line memory module)*

SIMS *(Sekundärionenmassenspektrometrie)* <Phys> SIMS *(secondary ion mass spectrometry)*

sin *(Sinus)* <Comp & DV, Geom> sin *(sine)*

SINAD *(Störabstand einschließlich Verzerrungen)* <Telekom> SINAD *(signal-to-noise and distortion ratio)*

SIS *(Schwerionensynchrotron)* <Teilphys> HIS *(heavy-ion synchrotron)*

SLAR *(Seitensichtradar)* <Meerschmutz> SLAR *(sideways-looking airborne radar)*

SLD 1. <Optik> *(Superlumineszenzdiode)* SLD *(superluminescent LED)*; 2. <Telekom> *(Superlumineszenzdiode)* SLD *(superluminescent LED)*; SRD *(superradiant diode)*

SLR *(einäugige Spiegelreflexkamera)* <Foto> SLR *(single lens reflex camera)*

SMC 1. <Elektronik> *(SMD-Bauteil)* SMC *(surface-mounted component)*; 2. <Telekom> *(Aufsetzbauelement)* SMC *(surface-mounted component)*

SMD *(Aufsetzbauelement, oberflächenmontiertes Element)* <Elektriz> SMD *(surface mounting device)*

Sn *(Seiten)* <Druck> pp *(pages)*

SNA *(Systemnetzwerkarchitektur)* <Comp & DV> SNA *(systems network architecture)*

SNV *(Schweizerische Normenvereinigung)* <Elektriz> SSA *(Swiss Standards Association)*

SPS 1. <Comp & DV> *(Programmspeichersteuerung)* SPC *(stored program control)*; 2. <Kontroll, Labor> *(speicherprogrammierbare Steuerung)* PLC *(programmable logic control)*

SQID *(supraleitfähiger Quanteninterferenzmechanismus)* <Phys> SQUID *(superconductive quantum interference device)*

SRAM *(statischer RAM)* <Comp & DV> SRAM *(static RAM)*

SS 1. <Fernseh, Funktech> *(Spitze-Spitze-...)* pp *(peak-to-peak)*; 2. <Maschinen> *(Schiebesitz)* close sliding fit, push fit; 3. <Mechan> *(Schiebesitz)* close-sliding fit, push fit, sliding fit

SSC *(supraleitfähiges Supracollider)* <Teilphys> SSC *(superconducting super collider)*

SSTV 1. <Fernseh> *(Fernsehen mit langsamer Abtastung, Schmalbandfernsehen, Slow-Scan-Television)* SSTV *(slow scan television)*; 2. <Funktech> *(Schmalbandfernsehen mit langsamer Abtastung)* SSTV *(slow scan television)*

STS *(Raumfahrttransportsystem)* <Raumfahrt> STS *(space transportation system)*

STX *(Textanfang)* <Comp & DV> STX *(start of text)*

SU *(Sprachverstehen)* <Künstl Int> SU *(speech understanding)*

Sv *(Sievert)* <Phys, Strahlphys, Teilphys> Sv *(Sievert)*

SVA *(gemeinsam benutzbarer virtueller Bereich)* <Comp & DV> SVA *(shared virtual area)*

SWFD *(Selbstwählferndienst)* <Telekom> *(AE)* DDD, direct distance dialing; *(BE)* STD, subscriber trunk dialling

SWR 1. <Kerntech, Phys> *(Siedewasserreaktor)* BWR *(boiling water reactor)*; 2. <Kerntech> *(schwerwassermoderierter Reaktor)* HWR *(heavy-water-moderated reactor)*

SWV *(Stehwellenverhältnis)* 1. <Funktech> SWR *(standing-wave ratio)*; 2. <Phys, Telekom> SWR *(standing-wave ratio)*; VSWR *(voltage standing-wave ratio)*

SZ *(Säurezahl)* <Kunststoff, Lebensmittel> acid value

t 1. <Hydraul> *(Tiefe)* d *(depth)*; 2. <Teilphys> *(Triton)* t *(triton)*

T 1. <Akustik, Aufnahme> *(Nachhallzeit)* T *(reverberation time)*; 2. <Chemie> *(Tritium)* T *(tritium)*; 3. <Elektriz, Elektrotech, Erdöl, Kfztech, Maschinen, Mechan> *(Drehmoment)* T, torque; 4. <Fertig> *(Drehmoment)* T, torque *(Kunststoffinstallationen)*; 5. <Hydraul> *(Transpiration)* T *(transpiration)*; 6. <Labor, Phys> *(Tesla)* T *(Tesla)*; 7. <Metrol> *(absolute Temperatur)* T *(absolute temperature)*; 8. <Metrol> *(Tera...)* T *(tera...)*; 9. <Phys> *(thermodynamische Temperatur)* T *(thermodynamic temperature)*

T½ *(Halbwertszeit)* <Kerntech, Phys, Strahlphys, Teilphys> T½ *(half-life)*

TAA *(Technische Anleitung Abfall)* <Abfall> Technical Instruction on Waste Management

TBKZ *(Technisches und Betriebskontrollzentrum)* <Raumfahrt> *(AE)* TOCC *(Technical and Operational Control Center)*

TBP *(Tributylphosphat)* <Kerntech> TBP *(tributylphosphate)*

TD *(theoretische Dichte)* <Kerntech> TD *(theoretical density)*

TDI *(Toluendiisocyanat, Toluoldiisocyanat)* <Kunststoff> TDI *(toluyene diisocyanate)*

TDM *(Multiplexen mit Zeitteilung, Zeitmultiplexmethode, Zeitmultiplexverfahren)* <Comp & DV, Elektronik, Telekom> TDM *(time division multiplex)*

TDMA *(Mehrfachzugriff im Zeitmultiplex, nicht gleichzeitiger Mehrfachzugriff)* 1. <Comp & DV, Elektronik, Telekom> TDMA, time division multiple access; 2. <Raumfahrt> TDMA, time division multiple access *(Weltraumfunk)*

TE *(transversal elektrisch)* <Elektrotech, Telekom> TE *(transverse electric)*

TEM *(transversal elektromagnetisch)* <Elektrotech, Telekom> TEM *(transverse electromagnetic)*

TeV *(Teraelektronenvolt)* <Teilphys> TeV *(tera electron volt)*
Tf 1. <Aufnahme, Elektronik, Fernseh, Funktech> *(Tonfrequenz)* AF *(audio frequency)*; 2. <Aufnahme, Elektronik, Fernseh> *(Trägerfrequenz)* CF *(carrier frequency)*
TFEL *(Dünnschicht-Elektrolumineszenz)* <Elektronik> TFEL *(thin film electroluminescence)*
TG *(Tertiärgruppe)* <Telekom> mastergroup
TM *(transversal magnetisch)* <Elektrotech, Telekom> TM *(transverse magnetic)*
TMUX *(Transmultiplexer)* <Telekom> TMUX *(transmultiplexer)*
TNI *(Teilnehmer-Teilnehmer-Information)* <Telekom> UUI *(user-to-user information)*
TNL *(Teilnehmeranschlussleitung)* <Telekom> subscriber's line
TNR *(gesteuerter Thermonuklearreaktor)* <Kerntech> CTR *(controlled thermonuclear reactor)*
TNS *(Teilnehmersatz)* <Telekom> SLC *(subscriber line circuit)*
TNT *(Trinitrotoluol)* <Chemie> TNT *(trinitrotoluene)*
TOP *(Bürokommunikationsprotokoll)* <Telekom> TOP *(technical and office protocol)*
TPP *(texturiertes Pflanzenprotein)* <Lebensmittel> TVP *(textured vegetable protein)*
TQMS *(abteilungsübergreifendes Qualitätssicherungssystem)* <Qual> TQMS *(Total Quality Management System)*
TS *(Treibsitz)* <Maschinen> drive fit
TT *(Tonträger)* <Aufnahme, Fernseh> sound carrier
TTL *(Transistor-Transistor-Logik)* <Comp & DV, Elektronik> TTL *(transistor-transistor logic)*
TV *(Fernsehen)* <Fernseh> TV *(television)*
TVI *(Fernsehempfangsstörung, Fernsehstörung)* <Fernseh> TVI *(television interference)*
TVSt *(Teilnehmervermittlungsstelle)* <Telekom> access exchange
TW *(Tonwahl)* <Telekom> MFD *(multifrequency dialling)*
U 1. <Akustik> *(Volumengeschwindigkeit, Volumenstrom)* U *(volume current)*; 2. <Chemie> *(Uran)* U *(uranium)*; 3. <Elektriz> *(Spannung)* V *(voltage)*; 4. <Optik> *(Strahlungsenergie)* U *(radiant energy)*; 5. <Thermod> *(Wärmeübertragungskoeffizient)* U *(overall heat transfer coefficient)*
û *(wahrscheinlichste Geschwindigkeit)* <Phys> û *(most probable speed)*
UF6 *(Uranhexafluorid)* <Kerntech> UF6 *(uranium hexafluoride)*
UFO *(nicht identifiziertes Flugobjekt)* <Raumfahrt> UFO *(unidentified flying object)*
UHF *(Ultrahochfrequenz)* <Elektronik, Fernseh, Funktech, Telekom, Wellphys> UHF *(ultrahigh frequency)*
UHT *(Ultrahochtemperatur)* <Lebensmittel> UHT *(ultrahigh temperature)*
UKW *(Ultrakurzwelle)* <Funktech, Wellphys> USW *(ultrashort wave)*; very high frequency *(VHF)*

ULSI *(Ultragroßintegration)* <Elektronik> ULSI *(ultralarge-scale integration)*
üNN *(über Normalnull)* <Wassertrans> asl *(above sea level)*
UP *(ungesättigter Polyester)* <Kunststoff> UP *(unsaturated polyester)*
UPN *f (umgekehrte Polnische Notation)* <Comp & DV> RPN *(reverse Polish notation)*
US *(ultrasonisch)* <Elektriz> SS, US *(supersonic)*
USB *(unteres Seitenband)* <Elektronik, Funktech, Telekom> LSB *(lower sideband)*
USV *(unterbrechungsfreie Stromversorgung)* <Kontroll> stand-by battery power supply
UT *(Standardweltzeit)* <Funktech, Raumfahrt> UT *(universal time)*
UTC *(Standardweltzeit)* <Funktech, Raumfahrt> UTC *(universal time coordinated)*
UV *(ultraviolett)* <Optik, Phys> UV *(ultraviolet)*
UZ *(Zenerspannung)* <Elektronik> BDV *(breakdown voltage)*
v 1. <Comp & DV, Phys, Textil, Thermod> *(Volumen)* v *(volume)*; 2. <Kerntech> *(Vibrationsquantenzahl)* v *(vibrational quantum number)*
V 1. <Aufnahme, Elektriz, Elektronik, Funktech, Metrol, Optik> *(Verstärkung)* A *(amplification)*; 2. <Chemie> *(Vanadium)* V *(vanadium)*; 3. <Elektriz, Elektronik, Metrol, Optik> *(Volt)* V *(volt)*
VAD *(Axialabscheideverfahren aus Dampfphase)* <Optik, Telekom> VAD, *(AE)* vapor phase axial deposition technique, *(BE)* vapour phase axial deposition technique
var *(Blindleistungseinheit)* <Labor> var *(volt-amperes reactive)*
VCO *(variabler Quarzoszillator)* <Funktech> VXO *(variable crystal oscillator)*
VCR 1. <Fernseh> *(Videorecorder)* VCR *(video cassette recorder)*; 2. <Fernseh> *(Videomagnetbandgerät)* VTR *(video tape recorder)*
VD *(Viskositäts-Dichteverhältnis)* <Nichtfoss Energ> kinematic viscosity
VDE *(Verein Deutscher Elektrotechniker)* <Elektriz> Association of German Electrical Engineers
VDI *(Verein Deutscher Ingenieure)* <Elektriz> Association of German Engineers
VF 1. <Fernseh> *(Videofrequenz)* VF *(video frequency)*; 2. <Fernseh> *(Videofrequenzwandler)* VF *(video-frequency converter)*
VG *(veränderliche Geometrie, veränderliche Tragflügelgeometrie)* <Lufttrans> VG *(variable geometry)*
VHF *(VHF-Bereich, Ultrakurzwellen)* <Funktech, Telekom> VHF *(very high frequency)*
VHS *(Heimvideosystem)* <Fernseh> VHS *(video home system)*
VHS-C *(Heimvideo-Aufzeichnungssystem)* <Fernseh> VHS-C *(video home system-compact)*
VI *(Viskositätsindex, Viskositätszahl)* <Kfztech, Maschinen, Strömphys, Thermod> VI *(viscosity index)*
VK *(Vergaserkraftstoff)* <Erdöl, Kfztech, Thermod> fuel, *(AE)* gas, *(AE)* gasoline, *(BE)* petrol

VLCC *(Supertanker)* <Erdöl> VLCC *(very large crude carrier)*

VLP *(Video-Langspiel)* <Fernseh> VLP *(video long play)*

VLSI *(Höchstintegration)* <Comp & DV, Elektronik> VLSI *(very large-scale integration)*

VPE *(vernetztes Polyethylen)* <Kunststoff> XPE *(cross-linked polyethylene)*

VSP *(vertikales seismisches Profil)* <Erdöl> VSP *(vertical seismic profile)*

VSWR *(Welligkeitsfaktor)* <Funktech> VSWR *(voltage standing wave ratio)*

W 1. <Chemie> *(Wolfram)* W *(tungsten)*; 2. <Elektriz, Elektrotech> *(elektrische Energie)* W *(electric energy)*; 3. <Elektriz, Elektrotech, Metrol> *(Watt)* W *(watt)*; 4. <Hydraul> *(Weber'sche Zahl)* W *(Weber number)*; 5. <Kerntech> *(durchschnittliche Energie)* W *(average energy)*; 6. <Optik> *(Abstrahlung)* W *(radiant emittance)*

WA *(Wertanalyse)* <Kfztech, Qual> VA *(value analysis)*

WAP *(Protokoll für drahtlose Anwendungen)* <Funktech, Telekom> WAP *(wireless application protocol)*

Wb *(Weber)* <Elektriz, Elektrotech, Metrol> Wb *(weber)*

WB *(Wissensbasis)* <Comp & DV, Künstl Int> KB *(knowledge base)*

WBS *(wissensbasiertes System)* <Comp & DV> KBS *(knowledge-based system)*

WDM *(Wellenlängenmultiplex)* <Optik, Telekom> WDM *(wavelength division multiplexing)*

WEZ *(Westeuropäische Zeit)* <Mechan> GMT *(Greenwich Mean Time)*

WL *(weiter Laufsitz)* <Maschinen> loose fit

WORM *(Einmalbeschreibung-Mehrfachlesen)* <Optik> WORM *(write once read many times)*

WPS *(Wellenpferdestärke)* <Maschinen> shaft horse power

WR 1. <Comp & DV, Künstl Int> *(Wissensrepräsentation)* KR *(knowledge representation)*; 2. <Elektronik, Elektrotech, Funktech, Phys> *(Wechselrichter)* inverter

WS *(Wechselstrom)* <Elektrotech> AC *(alternating current)*

WS-GS *(Wechselstrom-Gleichstrom)* <Elektrotech> AC-DC *(alternating current-direct current)*

WWRV *(Wanderwellenröhrenverstärker)* 1. <Elektronik> TWTA, *(AE)* traveling wave tube amplifier; 2. <Raumfahrt> TWTA, *(BE)* travelling wave tube amplifier

WYSIWYG *(originalgetreue Darstellung der Druckausgabe am Bildschirm)* <Comp & DV> WYSIWYG *(what you see is what you get)*

X 1. <Akustik> *(Volumenverdrängung)* X *(volume displacement)*; 2. <Elektriz> *(Blindwiderstand, Reaktanz)* X *(reactance)*; 3. <Kerntech> *(Belastung)* X *(exposure)*

XC *(kapazitiver Blindwiderstand)* <Elektriz> XC *(capacitive reactance)*

XL *(induktiver Blindwiderstand)* <Elektriz> XL *(inductive reactance)*

XMS *(Erweiterungsspeicher)* <Comp & DV> XMS *(extended memory specification)*

XPS 1. <Comp & DV, Künstl Int> *(Expertensystem)* XPS *(expert system)*; 2. <Phys> *(röntgenstrahlangeregte Photoelektronenspektroskopie)* XPS *(X-ray photoelectron spectroscopy)*

YA *(akustische Admittanz)* <Akustik, Elektrotech> YA *(acoustic admittance)*

YAG *(Yttrium-Aluminium-Granat)* <Chemie> YAG *(yttrium-aluminium garnet)*

YIG *(Yttrium-Eisen-Granat)* <Chemie> YIG *(yttrium-iron garnet)*

z *(Kompressibilitätsfaktor)* <Thermod> z *(compressibility factor)*

Z 1. <Elektriz> *(Impedanz)* Z *(impedance)*; 2. <Kerntech> *(Atomzahl)* Z *(atomic number)*; 3. <Phys> *(Teilfunktion, Zustandsfunktion)* Z *(partition function)*

ZA *(Schallimpedanz, Schallwellenwiderstand, akustische Impedanz, akustischer Scheinwiderstand)* <Aufnahme, Elektrotech, Phys> ZA *(acoustic impedance)*

Zf *(Zwischenfrequenz)* <Elektronik, Funktech, Telekom> IF *(intermediate frequency)*

Zfm *(Zwischenfrequenzmodulation)* <Elektronik, Funktech, Telekom> IFM *(intermediate frequency modulation)*

Z/Sek *(Zeichen pro Sekunde)* <Druck> cps *(characters per second)*

Z/Std *(Zeichen pro Stunde)* <Druck> cph *(characters per hour)*

ZVSt *(Zentralvermittlungsstelle)* <Telekom> tertiary exchange

Umrechnungstabellen / Conversion tables

1 Längenmaße / Length

		Meter metre	Zoll† inch	Fuß*† foot	Yard*† yard	Rod*† rod	Meile*† mile
1 Meter metre	=	1	39,37	3,281	1,093	0,1988	$6{,}214 \times 10^{-4}$
1 Zoll inch	=	$2{,}54 \times 10^{-2}$	1	0,083	0,02778	$5{,}050 \times 10^{-3}$	$1{,}578 \times 10^{-5}$
1 Fuß foot	=	0,3048	12	1	0,3333	0,0606	$1{,}894 \times 10^{-4}$
1 Yard yard	=	0,9144	36	3	1	0,1818	$5{,}682 \times 10^{-4}$
1 Rod rod	=	5,029	198	16,5	5,5	1	$3{,}125 \times 10^{-3}$
1 Meile mile	=	1609	63 360	5280	1760	320	1

1 Yard† (gesetzlicher Standard) = 0,914 398 41 Meter / 1 imperial standard yard = 0.914 398 41 metre
1 Yard† (wissenschaftlich) = 0,9144 Meter (genau) / 1 yard (scientific) = 0.9144 metre (exact)
1 Yard US† = 0,914 401 83 Meter / 1 US yard = 0.914 401 83 metre
1 englische Seemeile = 6080 Fuß = 1853,18 Meter / 1 English nautical mile† = 6080 ft = 1853.18 metres
1 internationale Seemeile† = 1852 Meter = 6076,12 Fuß / 1 international nautical mile = 1852 metres = 6076.12 ft

† = keine SI-Einheit / not a SI-unit
* = im deutschen Sprachraum nicht gebräuchlich / not used in German-speaking countries

2 Flächenmaße / Area

		m^2 sq. metre	$(Zoll)^2$† sq. inch	$(Fuß)^{2*}$† sq. foot	$(Yard)^{2*}$† sq. yard	Acre*† acre	$(Meile)^{2*}$† sq. mile
1 m^2 sq. metre	=	1	1550	10,76	1,196	$2{,}471 \times 10^{-4}$	$3{,}861 \times 10^{-7}$
1 $(Zoll)^2$ sq. inch	=	$6{,}452 \times 10^{-4}$	1	$6{,}944 \times 10^{-3}$	$7{,}716 \times 10^{-4}$	$1{,}594 \times 10^{-7}$	$2{,}491 \times 10^{-10}$
1 $(Fuß)^2$ sq. foot	=	0,0929	144	1	0,1111	$2{,}296 \times 10^{-5}$	$3{,}587 \times 10^{-8}$
1 $(Yard)^2$ sq. yard	=	0,8361	1296	9	1	$2{,}066 \times 10^{-4}$	$3{,}228 \times 10^{-7}$
1 Acre acre	=	$4{,}047 \times 10^3$	$6{,}273 \times 10^6$	$4{,}355 \times 10^4$	4840	1	$1{,}563 \times 10^{-3}$
1 $(Meile)^2$ sq. mile	=	$259{,}0 \times 10^4$	$4{,}015 \times 10^9$	$2{,}788 \times 10^7$	$3{,}098 \times 10^6$	640	1

1 Are† = 100 m^2 = 0,01 Hektar / 1 are = 100 sq. metres = 0.01 hectare
1 runder Querschnitt†* von $1/1000$ Zoll Durchmesser = $5{,}067 \times 10^{-10}$ m^2 = $7{,}854 \times 10^{-7}$ $(Zoll)^2$ / 1 circular mil = 5.067×10^{-10} sq. metre = 7.854×10^{-7} sq. in
1 Acre†* (gesetzlicher Standard) = 0,4047 Hektar / 1 acre (statute) = 0.4047 hectare

Umrechnungstabellen

3 Raummaße / Volume

	m^3 cubic metre	$(Zoll)^3$*† cubic inch	$(Fuß)^3$*† cubic foot	Gallone UK*† UK gallon	Gallone US*† US gallon
1 m^3 cubic metre =	1	$6{,}102 \times 10^4$	35,31	220,0	264,2
1 $(Zoll)^3$ cubic in =	$1{,}639 \times 10^{-5}$	1	$5{,}787 \times 10^{-4}$	$3{,}605 \times 10^{-3}$	$4{,}329 \times 10^{-3}$
1 $(Fuß)^3$ cubic ft =	$2{,}832 \times 10^{-2}$	1728	1	6,229	7,480
1 Gallone UK[1] UK gallon =	$4{,}546 \times 10^{-3}$	277,4	0,1605	1	1,201
1 Gallone US[2] US gallon =	$3{,}785 \times 10^{-3}$	231,0	0,1337	0,8327	1

[1] Volumen von 10 britischen Pfund H_2O bei 62 °F / volume of 10 lb of water at 62 °F
[2] Volumen von 8,328 britischen Pfund H_2O bei 60 °F / volume of 8.328 28 lb of water at 60 °F
1 m^3 = 1000 Liter / 1 cubic metre = 1000 litres
1 Acre-Fuß† = 271 328 Gallonen UK = 1233 m^3 (Kubikmeter) / 1 acre foot = 271 328 UK gallons = 1233 cubic metres
Bis 1976 war der Liter als 1000,028 cm^3 definiert (das Volumen von 1 kg H_2O bei maximaler Dichte), wurde dann aber als exakt 1000 cm^3 umdefiniert
Until 1976 the litre was equal to 1000.028 cm^3 (the volume of 1 kg of water at maximum density) but then it was revalued to be 1000 cm^3 exactly

† = keine SI-Einheit / not a SI-unit
* = im deutschen Sprachraum nicht gebräuchlich / not used in German-speaking countries

4 Winkelmaße / Angle

	Grad degree	Minute minute	Sekunde second	Radian radian	Umdrehung revolution
1 Grad degree =	1	60	3600	$1{,}745 \times 10^{-2}$	$2{,}778 \times 10^{-3}$
1 Minute minute =	$1{,}677 \times 10^{-2}$	1	60	$2{,}909 \times 10^{-4}$	$4{,}630 \times 10^{-5}$
1 Sekunde second =	$2{,}778 \times 10^{-4}$	$1{,}667 \times 10^{-2}$	1	$4{,}848 \times 10^{-6}$	$7{,}716 \times 10^{-7}$
1 Radian radian =	57,30	3438	$2{,}063 \times 10^5$	1	0,1592
1 Umdrehung revolution =	360	$2{,}16 \times 10^4$	$1{,}296 \times 10^6$	6,283	1

1 Mil†* (Artilleriemaß) = $1/64{,}000$ von 360° = 10^{-3} Radian / 1 mil = 10^{-3} radian

5 Zeit / Time

		Jahr year	mittlerer Sonnentag solar day	Stunde hour	Minute minute	Sekunde second
1 Jahr year	=	1	365,24[1]	$8,766 \times 10^3$	$5,259 \times 10^5$	$3,156 \times 10^7$
1 mittlerer Sonnentag solar day	=	$2,738 \times 10^{-3}$	1	24	1440	$8,640 \times 10^4$
1 Stunde hour	=	$1,141 \times 10^{-4}$	$4,167 \times 10^{-2}$	1	60	3600
1 Minute minute	=	$1,901 \times 10^{-6}$	$6,944 \times 10^{-4}$	$1,667 \times 10^{-2}$	1	60
1 Sekunde second	=	$3,169 \times 10^{-8}$	$1,157 \times 10^{-5}$	$2,778 \times 10^{-4}$	$1,667 \times 10^{-2}$	1

1 Jahr = 366,24 siderische Tage / 1 year = 366.24 sidereal days
1 siderischer Tag = 86 164,090 6 Sekunden / 1 sidereal day = 86 164.090 6 seconds
[1] genaue Zahl = 365,242 192 64 im Jahr 2000 AD / exact figure = 365.242 192 64 in AD 2000

† = keine SI-Einheit / not a SI-unit
* = im deutschen Sprachraum nicht gebräuchlich / not used in German-speaking countries

6 Masse / Mass

	Kilogramm kilogram	britisches Pfund*† pound	Slug*† slug	metrisches Slug*† metric slug	UK-Tonne*† UK ton	US-Tonne*† US ton	u*† u
1 Kilogramm = kilogram	1	2,205	$6,852 \times 10^{-2}$	0,1020	$9,842 \times 10^{-4}$	$11,02 \times 10^{-4}$	$6,024 \times 10^{26}$
1 britisches Pfund = pound	0,4536	1	$3,108 \times 10^{-2}$	$4,625 \times 10^{-2}$	$4,464 \times 10^{-4}$	$5,000 \times 10^{-4}$	$2,732 \times 10^{26}$
1 Slug = slug	14,59	32,17	1	1,488	$1,436 \times 10^{-2}$	$1,609 \times 10^{-2}$	$8,789 \times 10^{27}$
1 metrisches Slug = metric slug	9,806	21,62	0,6720	1	$9,652 \times 10^{-3}$	$1,081 \times 10^{-2}$	$5,907 \times 10^{27}$
1 UK-Tonne = UK ton	1016	2240	69,62	103,6	1	1,12	$6,121 \times 10^{29}$
1 US-Tonne = US ton	907,2	2000	62,16	92,51	0,8929	1	$5,465 \times 10^{29}$
1 u u	$1,660 \times 10^{-27}$	$3,660 \times 10^{-27}$	$1,137 \times 10^{-28}$	$1,693 \times 10^{-28}$	$1,634 \times 10^{-30}$	$1,829 \times 10^{-30}$	1

1 britisches Pfund† (gesetzlicher Standard) = 0,453 592 338 Kilogramm / 1 imperial standard pound = 0.453 592 338 kilogram
1 US-Pfund = 0,453 592 427 7 Kilogramm† / 1 US pound = 0.453 592 427 7 kilogram
1 internationales Pfund† = 0,453 592 37 Kilogramm / 1 international pound = 0.453 592 37 kilogram
1 Tonne† = 10^3 Kilogramm / 1 ton = 10^3 kilograms
1 Troypfund = 0,373 242 Kilogramm / 1 troy pound = 0.373 242 kilogram

Umrechnungstabellen

7 Kraft / Force

	Dyn dyne	Newton newton	Pound-Force*† pound force	Poundal*† poundal	Gram-Force*† gram force
1 Dyn dyne =	1	10^{-5}	$2{,}248 \times 10^{-6}$	$7{,}233 \times 10^{-5}$	$1{,}020 \times 10^{-3}$
1 Newton newton =	10^5	1	0,2248	7,233	102,0
1 Pound-Force pound force =	$4{,}448 \times 10^5$	4,448	1	32,17	453,6
1 Poundal poundal =	$1{,}383 \times 10^4$	0,1383	$3{,}108 \times 10^{-2}$	1	14,10
1 Gram-Force gram force =	980,7	$980{,}7 \times 10^{-5}$	$2{,}205 \times 10^{-3}$	$7{,}093 \times 10^{-2}$	1

† = keine SI-Einheit / not a SI-unit
* = im deutschen Sprachraum nicht gebräuchlich / not used in German-speaking countries

8 Leistung / Power

	Btu/h* Btu per hr	Fuß-Pfund/s* ft lb s^{-1}	Kg m/s kg metre s^{-1}	Kalorie/s* cal s^{-1}	PS*†[1] HP[2]	Watt watt
1 Btu/h Btu per hour =	1	0,2161	$2{,}987 \times 10^{-2}$	$6{,}999 \times 10^{-2}$	$3{,}929 \times 10^{-4}$	0,2931
1 Fuß-Pfund/s ft lb per second =	4,628	1	0,1383	0,3239	$1{,}818 \times 10^{-3}$	1,356
1 Kg m/s kg metre per second =	33,47	7,233	1	2,343	$1{,}315 \times 10^{-2}$	9,807
1 Kalorie/s cal per second =	14,29	3,087	$4{,}268 \times 10^{-1}$	1	$5{,}613 \times 10^{-3}$	4,187
1 PS HP =	2545	550	76,04	178,2	1	745,7
1 Watt watt =	3,413	0,7376	0,1020	0,2388	$1{,}341 \times 10^{-3}$	1

1 Watt international = 1,000 19 Watt absolut / 1 international watt = 1.000 19 absolute watt
[1] 1 PS = europäische Einheit, 1 PS = 735,498 Watt / PS (Pferdestärke) = European unit, 1 PS = 735.498 watt
[2] 1 HP = britische Einheit, 1 HP = 745,7 Watt / HP (Horsepower) = British unit, 1 HP = 745.7 watt

† = keine SI-Einheit / not a SI-unit
* = im deutschen Sprachraum nicht gebräuchlich / not used in German-speaking countries

9 Energie, Arbeit, Wärme / Energy, work, heat

		Btu*† *Btu*	Joule *joule*	Fuß-Pfund*† *ft lb*	cm^{-1} *cm^{-1}*	Kalorie *cal*	Kilowattstunde *kWh*	Elektronenvolt *electron volt*
1 Btu *Btu*	=	1	$1{,}055 \times 10^3$	778,2	$5{,}312 \times 10^{25}$	252	$2{,}930 \times 10^{-4}$	$6{,}585 \times 10^{21}$
1 Joule *joule*	=	$9{,}481 \times 10^{-4}$	1	$7{,}376 \times 10^{-1}$	$5{,}035 \times 10^{22}$	$2{,}389 \times 10^{-1}$	$2{,}778 \times 10^{-7}$	$6{,}242 \times 10^{18}$
1 Fuß-Pfund *ft lb*	=	$1{,}285 \times 10^{-3}$	1,356	1	$6{,}828 \times 10^{22}$	$3{,}239 \times 10^{-1}$	$3{,}766 \times 10^{-7}$	$8{,}464 \times 10^{18}$
1 cm^{-1} *cm^{-1}*	=	$1{,}883 \times 10^{-26}$	$1{,}986 \times 10^{-23}$	$1{,}465 \times 10^{-23}$	1	$4{,}745 \times 10^{-24}$	$5{,}517 \times 10^{-30}$	$1{,}240 \times 10^{-4}$
1 Kalorie bei 15 °C *cal 15 °C*	=	$3{,}968 \times 10^{-3}$	4,187	3,088	$2{,}108 \times 10^{23}$	1	$1{,}163 \times 10^{-6}$	$2{,}613 \times 10^{19}$
1 Kilowattstunde *kWh*	=	3412	$3{,}600 \times 10^6$	$2{,}655 \times 10^6$	$1{,}813 \times 10^{29}$	$8{,}598 \times 10^5$	1	$2{,}247 \times 10^{25}$
1 Elektronenvolt *electron volt*	=	$1{,}519 \times 10^{-22}$	$1{,}602 \times 10^{-19}$	$1{,}182 \times 10^{-19}$	$8{,}066 \times 10^3$	$3{,}827 \times 10^{-20}$	$4{,}450 \times 10^{-26}$	1

† = keine SI-Einheit / not a SI-unit
* = im deutschen Sprachraum nicht gebräuchlich / not used in German-speaking countries

10 Druck / Pressure

Umrechnungstabellen

	Normal-atmosphäre† standard atmosphere	Kg/cm⁻²† kg force cm⁻²	Dyn/cm⁻²† dyne cm⁻²	Pascal pascal	Fuß-Pfund/(Zoll)⁻²† pound force in⁻²	Fuß-Pfund/(Fuß)⁻²† pound force ft⁻²	Millibar† millibar	Torr† torr	Zoll Queck-silbersäule† barometric in Hg
1 Normalatmosphäre = standard atmosphere	1	1,033	$1,013 \times 10^6$	$1,013 \times 10^5$	14,70	2116	1013	760	29,92
1 Kg/cm⁻² = kg force cm⁻²	0,9678	1	$9,804 \times 10^5$	$9,804 \times 10^4$	14,22	2048	980,7	735,6	28,96
1 Dyn/cm⁻² = dyne cm⁻²	$9,869 \times 10^{-7}$	$10,20 \times 10^{-7}$	1	0,1	$14,50 \times 10^{-6}$	$2,089 \times 10^{-3}$	10^{-3}	$750,1 \times 10^{-6}$	$29,51 \times 10^{-6}$
1 Pascal = pascal	$9,869 \times 10^{-6}$	$10,20 \times 10^{-6}$	10	1	$14,50 \times 10^{-5}$	$3,089 \times 10^{-2}$	10^{-2}	$750,1 \times 10^{-5}$	$29,53 \times 10^{-5}$
1 Fuß-Pfund/(Zoll)⁻² = pound force in⁻²	$6,805 \times 10^{-2}$	$7,031 \times 10^{-2}$	$6,895 \times 10^4$	$6,895 \times 10^3$	1	144	68,95	51,71	2,036
1 Fuß-Pfund/(Fuß)⁻² = pound force ft⁻²	$4,725 \times 10^{-4}$	$4,882 \times 10^{-4}$	478,8	47,88	$6,944 \times 10^{-3}$	1	$47,88 \times 10^{-2}$	0,3591	$14,14 \times 10^{-3}$
1 Millibar = millibar	$0,9869 \times 10^{-3}$	$1,020 \times 10^{-3}$	10^3	10^2	$14,50 \times 10^{-3}$	2,089	1	0,7500	$29,53 \times 10^{-3}$
1 Torr = torr	$1,316 \times 10^{-3}$	$1,360 \times 10^{-3}$	$1,333 \times 10^2$	$1,333 \times 10^3$	$1,934 \times 10^{-2}$	2,784	1,333	1	$3,937 \times 10^{-2}$
Zoll Quecksilbersäule = barometric in Hg	$3,342 \times 10^{-2}$	$3,453 \times 10^{-2}$	$3,386 \times 10^4$	$3,386 \times 10^3$	$4,912 \times 10^{-1}$	70,73	33,87	25,40	1

1 Torr = 1 mm Quecksilber zu Normalbedingungen bei 13,5951 g/cm⁻³ Dichte, bei 0 °C und Schwerebeschleunigung von 980,665 cm/s⁻² / 1 torr = 1 barometric mm Hg density 13.5951 g cm⁻³ at 0 °C and acceleration due to gravity 980.665 cm/s⁻²
1 Dyn cm⁻² = 1 Barad / 1 dyne cm⁻² = 1 barad
† = keine SI-Einheit / not a SI-unit
* = im deutschen Sprachraum nicht gebräuchlich / not used in German-speaking countries

11 Magnetfluss / Magnetic flux

	Maxwell (Line)†[1] *maxwell*	Kiloline†[1] *kiloline*	Weber *weber*
1 Maxwell (1 Line) *maxwell* =	1	10^{-3}	10^{-8}
1 Kiloline *kiloline* =	10^3	1	10^{-5}
1 Weber *weber* =	10^8	10^5	1

[1] veraltet / obsolete

12 Magnetische Flussdichte / Magnetic flux density

	Gauß *gauss*	Weber/m^{-2} (Tesla) *weber m^{-2} (tesla)*	Gamma *gamma*	Maxwell/cm^{-2}† *maxwell cm^{-2}*
1 Gauß *gauss (line cm^{-2})* =	1	10^{-4}	10^5	1
1 Weber/m^{-2} (Tesla) *weber m^{-2} (tesla)* =	10^4	1	10^9	10^4
1 Gamma *gamma* =	10^{-5}	10^{-9}	1	10^{-5}
1 Maxwell/cm^{-2} *maxwell cm^{-2}* =	1	10^{-4}	10^5	1

13 Magneto-EMK / Magnetomotive force

	Ab-Amperewicklung† *abamp turn*	Amperewindung AW *amp turn*	Gilbert†[2] *gilbert*
1 Ab-Amperewicklung *abampere turn* =	1	10	12,57
1 Amperewindung AW *ampere turn* =	10^{-1}	1	1,257
1 Gilbert *gilbert* =	$7,958 \times 10^{-2}$	0,7958	1

[2] veraltet / obsolete

† = keine SI-Einheit / not a SI-unit

Umrechnungstabellen

14 Magnetische Feldstärke / Magnetic field strength

	Amperewindung/cm^{-1} amp turn cm^{-1}	Amperewindung/m^{-1} amp turn m^{-1}	Oersted oersted
1 Amperewindung/cm^{-1} amp turn cm^{-1} =	1	10^2	1,257
1 Amperewindung/m^{-1} amp turn m^{-1} =	10^{-2}	1	$1{,}257 \times 10^{-2}$
1 Oersted oersted =	0,7958	79,58	1

15 Ausleuchtung / Illumination

	Lux lux	Phot†* phot	Footcandle*† footcandle
1 Lux lm/m^2 lux (lm m^{-2}) =	1	10^{-4}	$9{,}29 \times 10^{-2}$
1 Phot lm/cm^2 phot (lm cm^{-2}) =	10^4	1	929
1 Footcandle lm/Fuß^2 foot-candle (lm ft^{-2}) =	10,76	$10{,}76 \times 10^{-4}$	1

† = keine SI-Einheit / not a SI-unit
* = im deutschen Sprachraum nicht gebräuchlich / not used in German-speaking countries

16 Leuchtdichte / Luminance

	Nit†*[1] nit	Stilb†*[1] stilb	Candela/Fuß$^{-2}$†* cd ft^{-2}	Apostilb* apostilb	Lambert lambert	Foot-Lambert†† foot-lambert
1 Nit (Candela/m^2) nit (cd m^{-2}) =	1	10^{-4}	$9{,}29 \times 10^{-2}$	π	$\pi \times 10^{-4}$	0,292
1 Stilb (Candela/cm^2) stilb (cd cm^{-2}) =	10^4	1	929	$\pi \times 10^4$	π	2920
1 Candela/Fuß^2 cd ft^{-2} =	10,76	$1{,}076 \times 10^{-3}$	1	33,8	$3{,}38 \times 10^{-3}$	π
1 Apostilb (1 m/m^2) Apostilb (1 m m^{-2}) =	$1/\pi$	$1/(\pi \times 10^4)$	$2{,}96 \times 10^{-2}$	1	10^{-4}	$9{,}29 \times 10^{-2}$
1 Lambert (1 m/cm^2) lambert (1 m cm^{-2}) =	$1/(\pi \times 10^{-4})$	$1/\pi$	296	10^4	1	929
1 Foot-Lambert foot lambert or equivalent foot candle =	3,43	$3{,}43 \times 10^{-4}$	$1/\pi$	10,76	$1{,}076 \times 10^{-3}$	1

Leuchtkraft Candela = 98,1 % der internationalen Candela / Luminous intensity of candela = 98.1 % that of international candle
1 Lumen = Lichtstrom, der von 1 Candela Leuchtkraft in den Einheitsraumwinkel abgegeben wird / 1 lumen = flux emitted by 1 candela into unit solid angle
[1] veraltet / obsolete

† = keine SI-Einheit / not a SI-unit
* = im deutschen Sprachraum nicht gebräuchlich / not used in German-speaking countries

Chemische Elemente / Chemical elements

Symbol / Symbol	Element	Element	Ordnungszahl / Atomic number
Ac	Actinium	Actinium	89
Ag	Silber	Silver	47
Al	Aluminium	Aluminium	13
Am	Americium	Americium	95
Ar	Argon	Argon	18
As	Arsen	Arsenic	33
At	Astatin	Astatine	85
Au	Gold	Gold	79
B	Bor	Boron	5
Ba	Barium	Barium	56
Be	Beryllium	Beryllium	4
Bi	Bismut	Bismuth	83
Bk	Berkelium	Berkelium	97
Br	Brom	Bromine	35
C	Kohlenstoff	Carbon	6
Ca	Calcium	Calcium	20
Cd	Cadmium	Cadmium	48
Ce	Cer	Cerium	58
Cf	Californium	Californium	98
Cl	Chlor	Chlorine	17
Cm	Curium	Curium	96
Co	Cobalt	Cobalt	27
Cr	Chrom	Chromium	24
Cs	Caesium	Caesium	55
Cu	Kupfer	Copper	29
Dy	Dysprosium	Dysprosium	66
Er	Erbium	Erbium	68
Es	Einsteinium	Einsteinium	99
Eu	Europium	Europium	63
F	Fluor	Fluorine	9
Fe	Eisen	Iron	26
Fm	Fermium	Fermium	100
Fr	Francium	Francium	87
Ga	Gallium	Gallium	31
Gd	Gadolinium	Gadolinium	64
H	Wasserstoff	Hydrogen	1
He	Helium	Helium	2
Hf	Hafnium	Hafnium	72
Hg	Quecksilber	Mercury	80
Ho	Holmium	Holmium	67
I	Iod	Iodine	53
In	Indium	Indium	49
Ir	Iridium	Iridium	77
K	Kalium	Potassium	19
Kr	Krypton	Krypton	36
La	Lanthan	Lanthanum	57
Li	Lithium	Lithium	3
Lr	Lawrentium	Lawrencium	103
Lu	Lutetium	Lutetium	71
Md	Mendelevium	Mendelevium	101
Mg	Magnesium	Magnesium	12
Mn	Mangan	Manganese	25
Mo	Molybdän	Molybdenum	42
N	Stickstoff	Nitrogen	7
Na	Natrium	Sodium	11

Umrechnungstabellen

Symbol / Symbol	Element	Element	Ordnungszahl / Atomic number
Nb	Niob	Niobium	41
Nd	Neodym	Neodymium	60
Ne	Neon	Neon	10
Ni	Nickel	Nickel	28
No	Nobelium	Nobelium	102
Np	Neptunium	Neptunium	93
O	Sauerstoff	Oxygen	8
Os	Osmium	Osmium	76
P	Phosphor	Phosphorus	15
Pa	Protaktinium	Protactinium	91
Pb	Blei	Lead	82
Pd	Palladium	Palladium	46
Pm	Promethium	Promethium	61
Po	Polonium	Polonium	84
Pr	Praseodym	Praseodymium	59
Pt	Platin	Platinum	78
Ra	Radium	Radium	88
Rb	Rubidium	Rubidium	37
Re	Rhenium	Rhenium	75
Rh	Rhodium	Rhodium	45
Rn	Radon	Radon	86
Ru	Ruthenium	Ruthenium	44
S	Schwefel	Sulphur	16
Sb	Antimon	Antimony	51
Sc	Scandium	Scandium	21
Se	Selen	Selenium	34
Si	Silicium	Silicon	14
Sm	Samarium	Samarium	62
Sn	Zinn	Tin	50
Sr	Strontium	Strontium	38
Ta	Tantal	Tantalum	73
Tb	Terbium	Terbium	65
Tc	Technetium	Technetium	43
Te	Tellur	Tellurium	52
Th	Thorium	Thorium	90
Ti	Titan	Titanium	22
Tl	Thallium	Thallium	81
Tm	Thulium	Thulium	69
U	Uran	Uranium	92
V	Vanadium	Vanadium	23
W	Wolfram	Tungsten	74
Xe	Xenon	Xenon	54
Y	Yttrium	Yttrium	39
Yb	Ytterbium	Ytterbium	70
Zn	Zink	Zinc	30
Zr	Zirkonium	Zirconium	40